AF892597

Topographic Map Symbols

Elevation markers

Contour lines	
Index contour lines	100
Depression contour lines	
Water elevation	9600
Spot elevation	x 9136

Boundaries

National	
State	
County, parish, municipal	
Township, precinct, town	
Incorporated city, village, or town	
National or state reservation	
Small park, cemetery, airport, etc.	
Land grant	

Buildings and Structures

Buildings	
School	
Church	
Cemetery	cem
Barn and warehouse	
Wells (non-water)	o oil o gas
Open-pit mine, quarry, or prospect	
Tunnel	
Benchmark	BM 8025
National Park	
Campsite	
Bridge	

Roads and Railroads

Divided highway	
Road	
Trail	
Railroad	

Geologic Map Symbols

Sedimentary Rocks

- Breccia
- Conglomerate
- Dolomite
- Limestone
- Mudstone
- Sandstone
- Siltstone
- Shale

Igneous and Metamorphic Rocks

- Extrusive
- Intrusive
- Metamorphic

Features

River	
Water well	water
Spring	
Lake	
Glacier	

HOLT

Earth Science

TEACHER EDITION

Mead A. Allison, Ph.D. • New Orleans, Louisiana
Arthur T. DeGaetano, Ph.D. • Ithaca, New York
Jay M. Pasachoff, Ph.D. • Williamstown, Massachusetts

Contents in Brief

Teacher Edition Walk-Through

Student Edition Contents in Brief

HOLT, RINEHART AND WINSTON
A Harcourt Education Company
Orlando • Austin • New York • San Diego • Toronto • London

4SCIE000015815

Discover the Earth and beyond

Students will investigate the core of the Earth and explore far-away planets with the comprehensive and up-to-date content of ***Holt Earth Science.*** Environmental science and other cross-disciplinary connections help students integrate Earth science with other studies and improve their critical-thinking skills. The engaging, easy-to-read text and vivid images encourage students to master science skills. Your teaching options will abound with engaging elements and versatile print and technology resources.

LEVELED ACTIVITIES MEET THE NEEDS OF ALL STUDENTS AND STRENGTHEN THEIR ANALYTICAL SKILLS.

- The accessible and engaging design and structure with abundant visuals ensures comprehension.
- Activities are labeled by ability level—**Basic, General,** or **Advanced**—in the **Planning Guide** and on the front cover of the ***Chapter Resource Files*** so you can easily assign exercises appropriate for each student.
- Exercises in the teacher's wrap address different **Learning Styles,** plus **Inclusion Strategies** are included so you easily provide alternative teaching techniques.

A FLEXIBLE LAB PROGRAM GIVES STUDENTS HANDS-ON EXPERIENCES TO APPLY INQUIRY SKILLS.

- The extensive laboratory program includes in-text **Chapter Labs,** two in-text **Quick Labs** for each chapter, and additional labs in the ***Chapter Resource Files*** and in the new **Lab Generator** software.
- The variety of chapter lab types—**Making Models, Inquiry,** and **Skills Practice**—meets the needs of your curriculum.
- All labs are tested by teachers for safety and efficiency, so you know they are reliable.

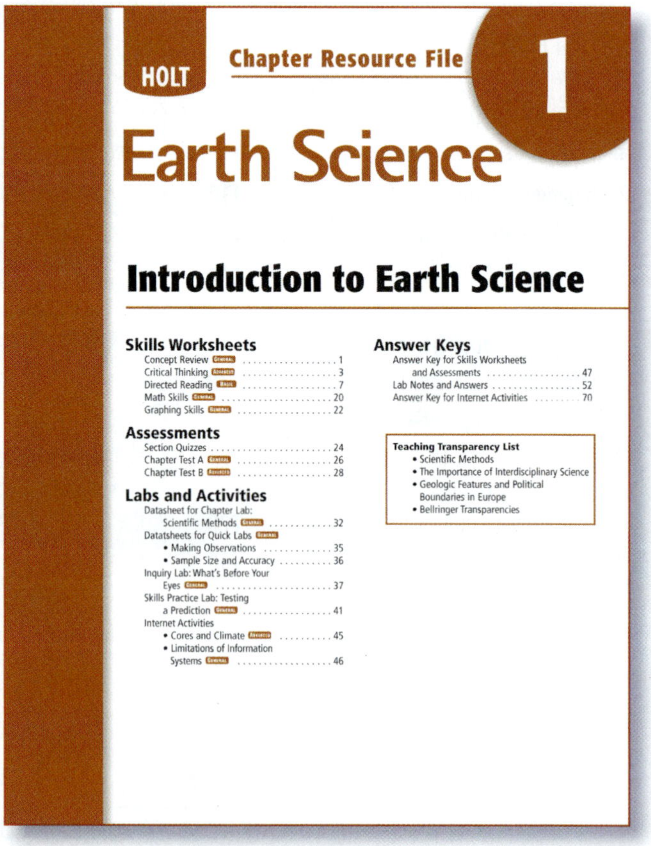

CUSTOMIZABLE, MULTI-LEVEL ASSESSMENTS LET YOU TEST YOUR WAY.

- Comprehensive section and chapter assessments allow you to check students' progress while building their critical-thinking skills.
- The program helps prepare students for testing with **Standardized Test Prep** questions in addition to reading, math, and interpreting graphics skills development throughout the program.
- Leveled assessments, such as **General** and **Advanced Chapter Tests,** allow you to tailor to students' varying abilities.
- The **ExamView® Test Generator** on the ***Teacher's One-Stop Planner*** gives you the power to customize your own assessments, and post them to **Holt Online Assessment** for automatic grading and to track student progress.

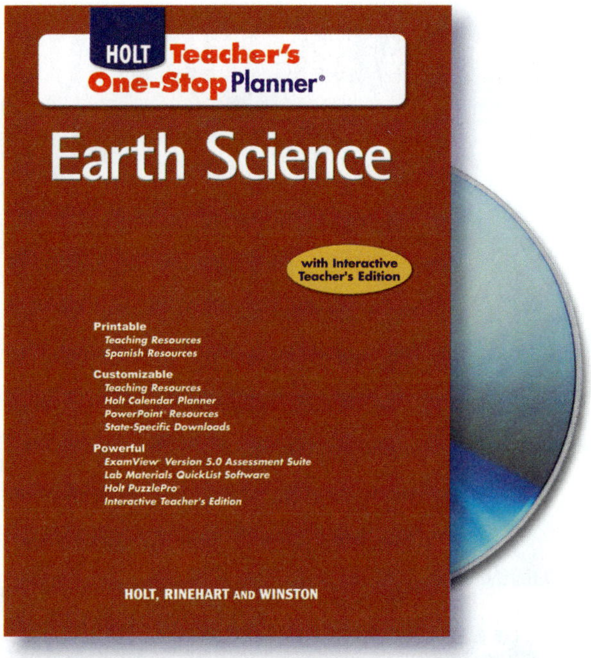

INTEGRATED TECHNOLOGY AND ONLINE RESOURCES REINFORCE STUDENTS' COMPREHENSION OF SCIENCE.

- Students can easily study at home with the ***Interactive Online Edition*** or the ***Student One Stop.***
- Video resources engage students to learn with ***Brain Food Video Quizzes,*** and ***NOVA® Videos.***
- The ***Teacher's One-Stop Planner*** makes planning a breeze with lesson plans, ***Holt Calendar Planner, PowerPoint® Resources,*** *Interactive Teacher Edition,* ***Holt PuzzlePro®,*** test generator, and more, all in one package.
- The new **Holt Lab Generator** lets you sift through our labs by keywords, standards, and other criteria. You can also modify our labs to suit you, create new labs, or add labs you get from elsewhere in the database, and get a customized materials list instantly.

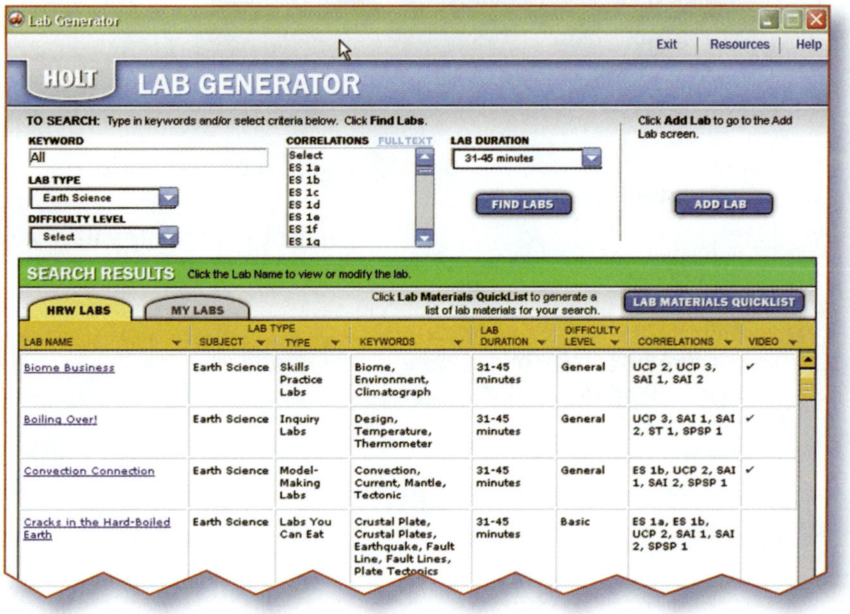

Student Edition

A textbook that motivates students to learn

ENGAGING AND WELL-ORGANIZED STUDENT PAGES MAKE SCIENCE RELEVANT AND ACCESSIBLE.

A list of the sections outlines what students will learn.

The pages are easy to navigate with outline-style headings and content grouped into chunks.

Objectives and a list of **Key Terms** draw out the key points of the text to improve reading skills.

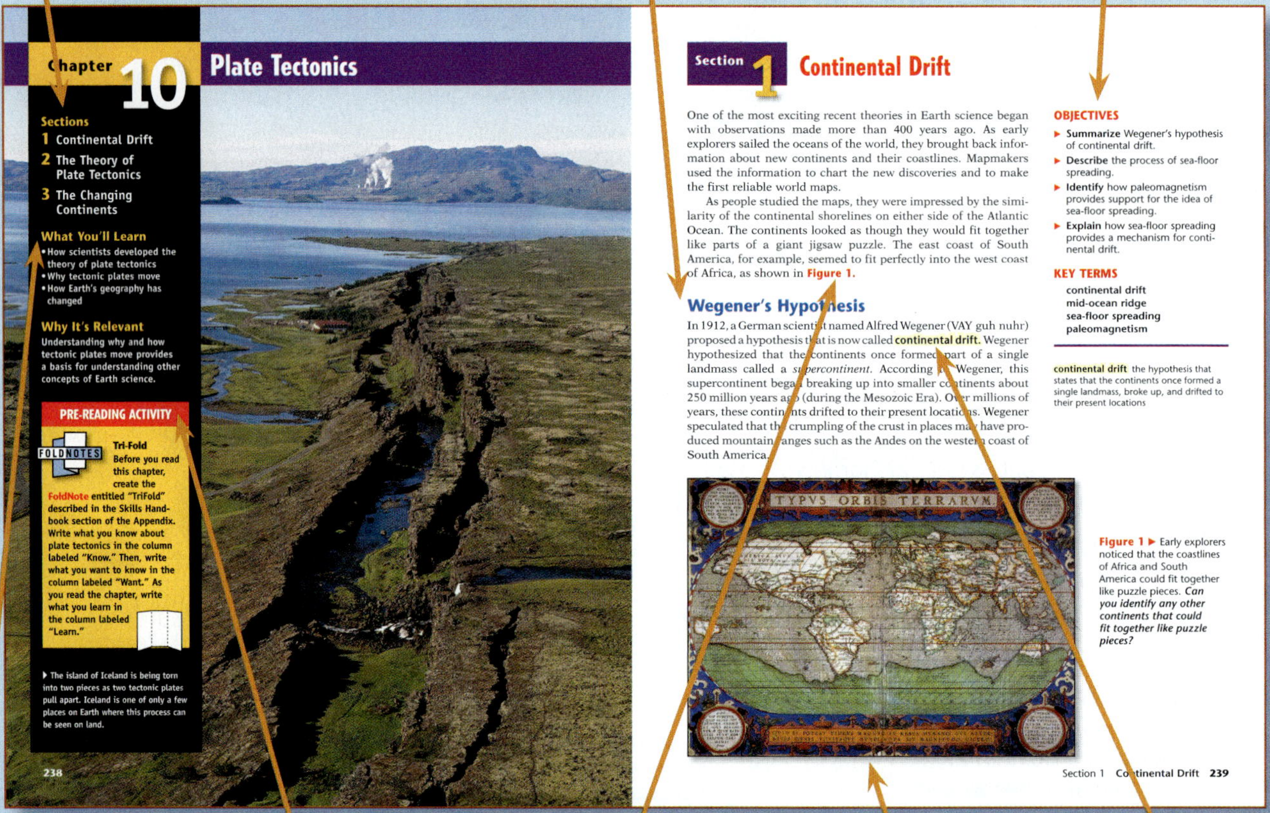

Chapter 10 Plate Tectonics

Sections

1 Continental Drift

2 The Theory of Plate Tectonics

3 The Changing Continents

What You'll Learn

- How scientists developed the theory of plate tectonics
- Why tectonic plates move
- How Earth's geography has changed

Why It's Relevant

Understanding why and how tectonic plates move provides a basis for understanding other concepts of Earth science.

PRE-READING ACTIVITY

FOLDNOTES

Tri-Fold

Before you read this chapter, create the FoldNote entitled "TriFold" described in the Skills Handbook section of the Appendix. Write what you know about plate tectonics in the column labeled "Know." Then, write what you want to know in the column labeled "Want." As you read the chapter, write what you learn in the column labeled "Learn."

▶ The island of Iceland is being torn into two pieces as two tectonic plates pull apart. Iceland is one of only a few places on Earth where this process can be seen on land.

238

Section 1 Continental Drift

One of the most exciting recent theories in Earth science began with observations made more than 400 years ago. As early explorers sailed the oceans of the world, they brought back information about new continents and their coastlines. Mapmakers used the information to chart the new discoveries and to make the first reliable world maps.

As people studied the maps, they were impressed by the similarity of the continental shorelines on either side of the Atlantic Ocean. The continents looked as though they would fit together like parts of a giant jigsaw puzzle. The east coast of South America, for example, seemed to fit perfectly into the west coast of Africa, as shown in **Figure 1.**

Wegener's Hypothesis

In 1912, a German scientist named Alfred Wegener (VAY guh nuhr) proposed a hypothesis that is now called **continental drift.** Wegener hypothesized that the continents once formed part of a single landmass called a *supercontinent.* According to Wegener, this supercontinent began breaking up into smaller continents about 250 million years ago (during the Mesozoic Era). Over millions of years, these continents drifted to their present locations. Wegener speculated that the crumpling of the crust in places may have produced mountain ranges such as the Andes on the western coast of South America.

OBJECTIVES

- **Summarize** Wegener's hypothesis of continental drift.
- **Describe** the process of sea-floor spreading.
- **Identify** how paleomagnetism provides support for the idea of sea-floor spreading.
- **Explain** how sea-floor spreading provides a mechanism for continental drift.

KEY TERMS

continental drift
mid-ocean ridge
sea-floor spreading
paleomagnetism

continental drift the hypothesis that states that the continents once formed a single landmass, broke up, and drifted to their present locations

Figure 1 ▶ Early explorers noticed that the coastlines of Africa and South America could fit together like puzzle pieces. *Can you identify any other continents that could fit together like puzzle pieces?*

Section 1 Continental Drift 239

What You'll Learn and **Why It's Relevant** engage students' interest in the chapter.

Figure references are in red to help students connect the visuals with the text.

Key Terms are highlighted in yellow and defined in the margin to develop students' vocabulary skills.

Pre-Reading Activity helps students organize the content and improve retention by creating a **FoldNote.**

Visuals are engaging and closely related to the text.

RELEVANT FEATURES ENRICH STUDENTS' UNDERSTANDING OF SCIENCE.

Maps in Action will improve students' map-reading skills, such as analyzing data, making comparisons, and inferring relationships. Corresponding transparencies and worksheets are found in ***Teaching Transparencies and Worksheets with Answer Key.***

Features—**Career Focus, Impact on Society, Eye on the Environment,** or **Science and Technology**—extend students' learning with an engaging article.

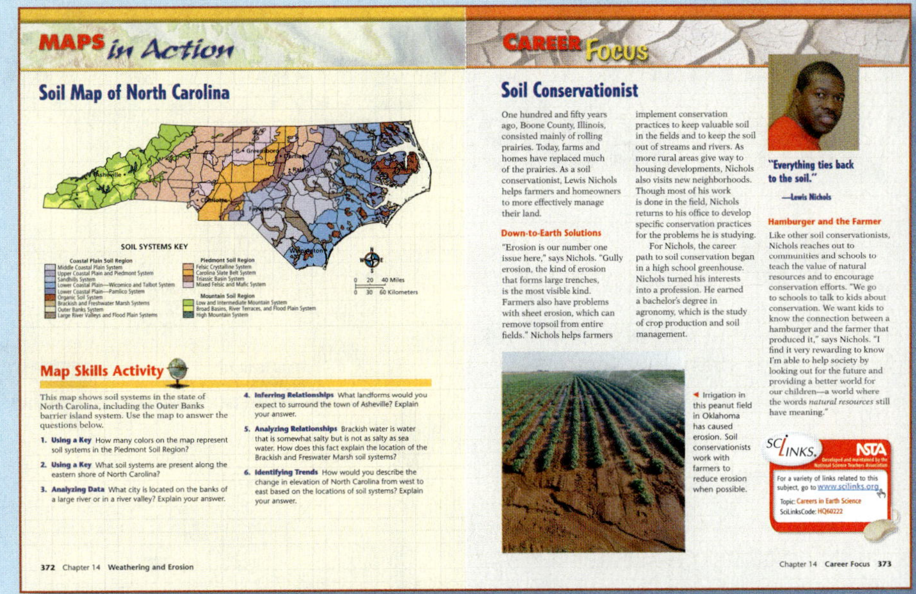

STUDENTS SHARPEN THEIR SKILLS THROUGHOUT THE PROGRAM.

Graphic Organizer guides students to outline topics and improve their comprehension.

Math Practice integrates math skills with science to help students improve their math skills.

Quick Lab gives students hands-on learning when time is limited.

In-text **Chapter Labs**—**Making Models, Inquiry,** and **Skills Practice**—provide students the opportunity to apply analytical skills while reinforcing science concepts.

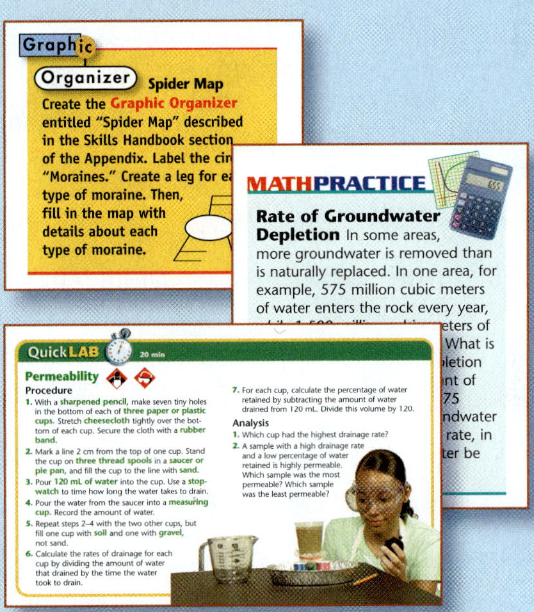

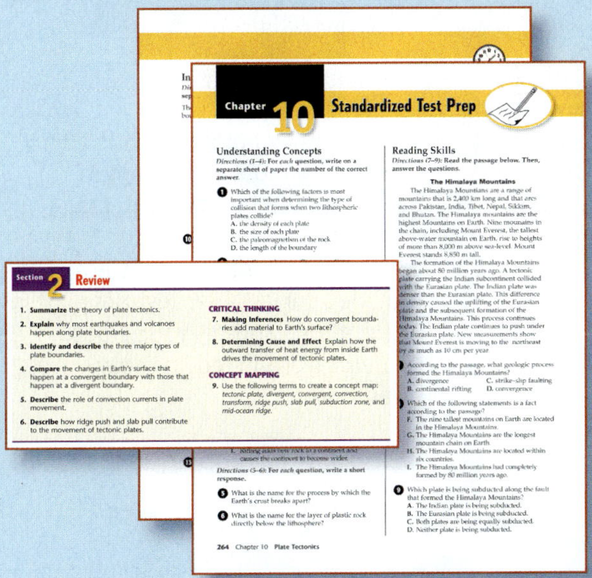

EVALUATE STUDENT PROGRESS AND PREPARE THEM FOR TESTING SUCCESS.

Section Review checks students' understanding of the section objectives.

Chapter Review is a comprehensive assessment of the chapter objectives using vocabulary, multiple-choice, short answer, critical-thinking, concept mapping, math skills, writing skills, and interpreting graphics questions.

Standardized Test Prep prepares students for standardized tests by sharpening their reading and interpreting graphics skills.

A world of possibilities with your *Teacher Edition*

PLANNING GUIDE, CHAPTER ENRICHMENT, AND LESSON CYCLE ARE THE FOUNDATION OF YOUR SUCCESS.

The concise and complete **Planning Guide** makes preparing lessons simple. The **Planning Guide**

- is an easy-to-use visual that outlines all your teaching resources.
- incorporates a **Compression Guide** to help you effectively manage your classroom time.
- integrates all review, assessments, labs, skills development, and technology resources.
- outlines section objectives to help you keep lessons focused.
- includes section correlations to the **National Science Education Standards.**

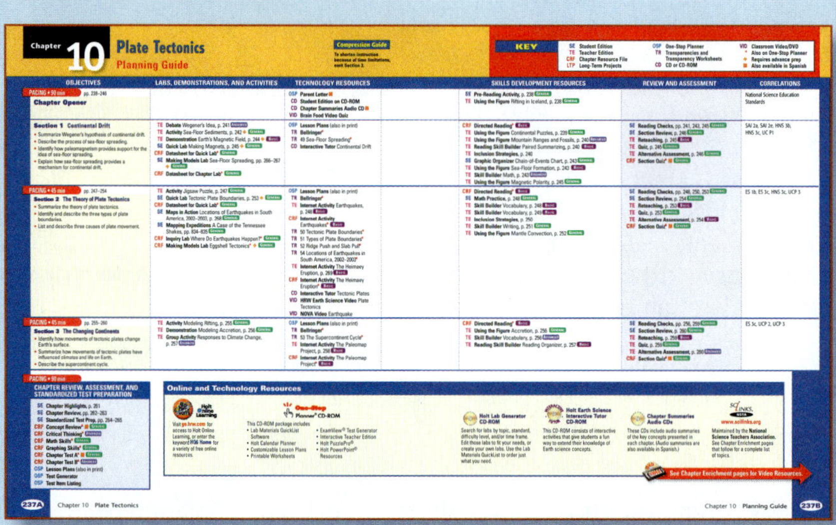
Chapter 10 Plate Tectonics
Planning Guide

In the full teacher's wrap you'll find the easy-to-follow **Lesson Cycle** including the following sections:

- **Focus** provides activities that help introduce the topic.
- **Motivate** emphasizes activities and discussions to engage students to learn.
- **Teach** provides teaching and reading strategies.
- **Close** includes reteaching strategies and a **Quiz.**

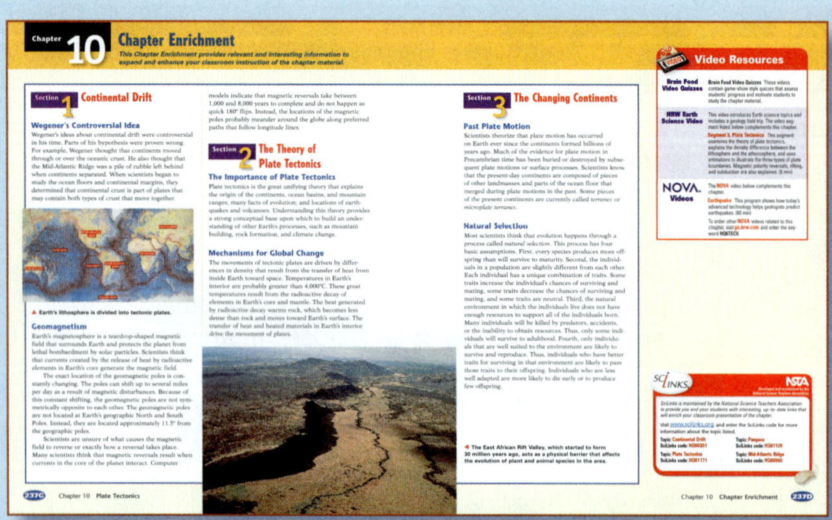
Chapter 10 Chapter Enrichment

Chapter Enrichment provides interesting facts that you can use to spark students' interest and encourage extended learning. **SciLinks®** codes are provided for another level of exploration, and all your video resources are listed for easy reference.

Teacher Edition

DIFFERENTIATED INSTRUCTION HELPS YOU TARGET ALL YOUR STUDENTS' NEEDS.

Activities in the **Planning Guide** and teacher's wrap are labeled by ability level—**Basic, General,** and **Advanced**—helping you choose appropriate activities for each student.

- **Basic** activities are designed for all students and address core skills.
- **General** activities are appropriate for most students and require more critical-thinking skills than **Basic** activities.
- **Advanced** activities are more challenging than **General** activities and require higher-thinking skills.

Learning styles—**Interpersonal, Intrapersonal, Auditory, Kinesthetic, Logical, Visual,** and **Verbal**—are highlighted throughout the lesson cycle so you can adapt material to different styles of learning. In addition, some labels identify the activities that help with **English Language Learner**s and **Co-op Learning.**

INCLUSION Strategies

- *Visually Impaired*
- *Attention Deficit Disorder*

Ask students to model the types of plate boundaries by using their hands. Have students place the index fingers of their hands together. Then, have them slide one hand under the other hand to model a convergent boundary. Have students model divergent boundaries by pulling their hands apart from the starting position. LS **Kinesthetic**

Inclusion Strategies make material accessible to all students. They are written by professionals in the field of special needs education and address many different learning exceptionalities in the classroom.

- Hearing Impaired
- Visually Impaired
- Developmentally Delayed
- Attention Deficit Disorder
- Behavior Control Issues
- Gifted and Talented

ACTIVITIES AND DEMONSTRATIONS KEEP STUDENTS INVOLVED BY CREATING RELEVANCE AND UNDERSTANDING.

Bellringer

Skill Builder

Reading Skill Builder

Cross-Disciplinary Connection

Brain Food

Misconception Alert

Internet Activity

Environmental Connection

Bellringer

Ask students where the water that flows from the faucets in their home comes from. (In most locations, the water that is used by humans comes from groundwater. In other locations, it may be purified surface water.) LS **Logical**

MISCONCEPTION ALERT

Underwater Lakes and Rivers Many students think of groundwater as an underground lake or as a river of water that flows through large open channels in rock. Explain to students that while groundwater does form underground lakes or rivers in some places, in most cases, groundwater flows slowly through tiny spaces between rock particles. The rate at which water flows depends on the permeability of the rock or sediment.

ENVIRONMENTAL CONNECTION

Harvesting Glacial Ice Alaska has more than 100,000 glaciers, covering 75,000 km². Seventy-five percent of its water is stored in these glaciers. Alaska issues permits to companies, allowing them

READING SKILL BUILDER — **BASIC**

Reading Organizer As students read this section, encourage them to take Power Notes, KWL Notes, or Two-Column Notes as described in the Skills Handbook section of the Appendix. Later, students can use these notes as a study guide for assessments. LS **Verbal** English Language Learners

Co-op Learning

Internet Activity — **ADVANCED**

Your Watershed Have students visit the Web site of the U.S. Environmental Protection Agency and go to their watershed page. Here, students can locate and learn what watershed your area is part of and find out about both surface water and groundwater resources. By analyzing the areas

Teacher Edition

Assessment options you can use

THE EXAMVIEW® VERSION 5 ASSESSMENT SUITE GIVES YOU THE POWER TO CUSTOMIZE YOUR OWN ASSESSMENTS.

With the **ExamView® Version 5 Assessment Suite** on the ***One-Stop Planner®*** you can create your own quizzes, section and chapter reviews,
and chapter tests.

- You can customize assessments by selecting from a bank of questions, organized by chapter and linked to chapter objectives.
- You can also post tests to **Holt Online Assessment,** an assessment management tool. The system automatically grades tests so you can diagnose student proficiency and track student progress.

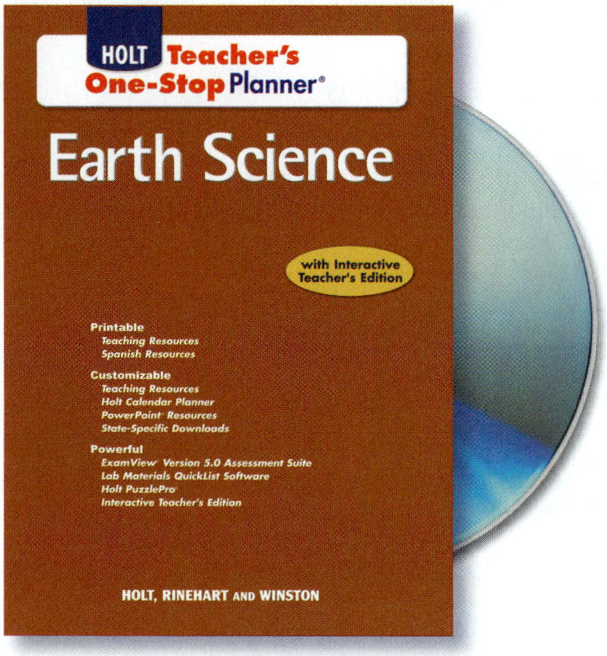

Reading Check How are sinkholes related to natural bridges? (See the Appendix for answers to Reading Checks.)

Section 3 Review

1. **Describe** glacial periods and interglacial periods.
2. **Explain** what happens to global sea level during a glacial period.
3. **Identify** the areas of Earth's surface that were covered by ice during the last glacial period.
4. **Summarize** the Milankovitch theory.
5. **Explain** how fossils of marine animals provide evidence of past ice ages.
6. **Describe** three explanations of ice ages other than the Milankovitch theory.

CRITICAL THINKING

7. **Evaluating Hypotheses** Use the information that you learned about glacial periods within an ice age to explain why volcanic eruptions and plate tectonics may not be among the causes of ice ages.
8. **Predicting Consequences** If Earth's orbit were always circular, would Earth be more likely to experience ice ages or less likely to experience ice ages? Explain your answer.

CONCEPT MAPPING

9. Use the following terms to create a concept map: *ice age, glacial period, interglacial period, Milankovitch theory, eccentricity, tilt, volcanic eruption,* and *precession.*

Quiz GENERAL

1. What is an ice age? (An ice age is a long period of climatic cooling during which the continents are glaciated repeatedly.)
2. How much of Earth's surface was covered by ice during the last glacial period? (one-third)
3. Explain the differences among Earth's eccentricity, tilt, and precession. (Eccentricity is the shape of Earth's orbit. Tilt is the angle of Earth's axis. Precession is the circular motion of Earth's axis.)

Alternative Assessment GENERAL

Poster Project Direct students to make a three-panel poster that shows Earth's ice ages and glacial and interglacial periods, the areas of Earth covered by ice during the last glacial period, and the various explanations about what causes the start of an ice age.
LS Verbal/Visual

SECTION ASSESSMENTS HELP STUDENTS FOCUS ON CHUNKS OF CONTENT.

Reading Checks are dispersed throughout the chapter to verify students' understanding of the topic. Students can check their answers in the **Appendix.**

Section Reviews help students confirm their comprehension of the text. They include critical-thinking, concept mapping, and objective questions.

Quiz in the teacher's wrap gives you a short assessment option to test students' progress.

Additional **Section Quizzes** are available in the ***Chapter Resource Files*** to assess students with matching and multiple-choice questions. The **Section Quizzes** are also available in a ***Spanish Assessment*** workbook. Easy-to-read **Answer Keys** are in the ***Chapter Resource Files.***

Alternative Assessments in the teacher's wrap give you different evaluation options, such as expository writing and concept mapping, to ensure a thorough assessment.

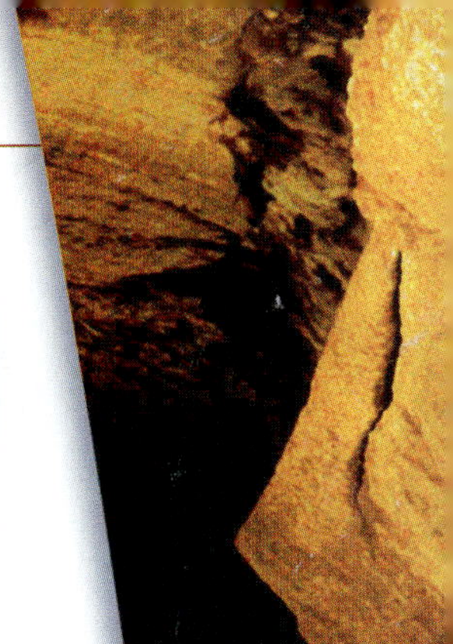

CHAPTER ASSESSMENTS TEST STUDENTS' OVERALL KNOWLEDGE OF THE SUBJECT.

Chapter Reviews help improve students' study and review habits with key terms, short answer, critical-thinking, concept mapping, math skills, writing skills, and interpreting graphics questions. Question types are similar to those found on **Chapter Tests,** making this an excellent resource for pretest practice.

Blackline masters of **Concept Review** are available in the ***Chapter Resource Files,*** the ***Study Guide,*** and the ***Spanish Study Guide*** to help students prepare for assessment.

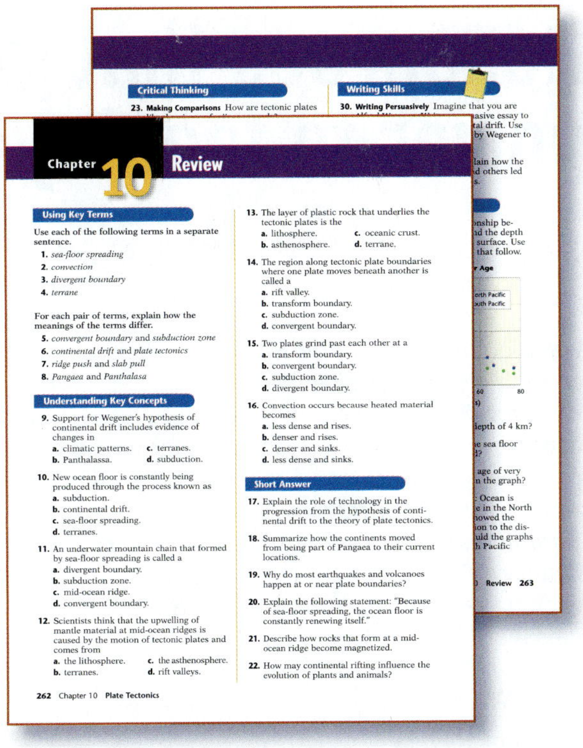

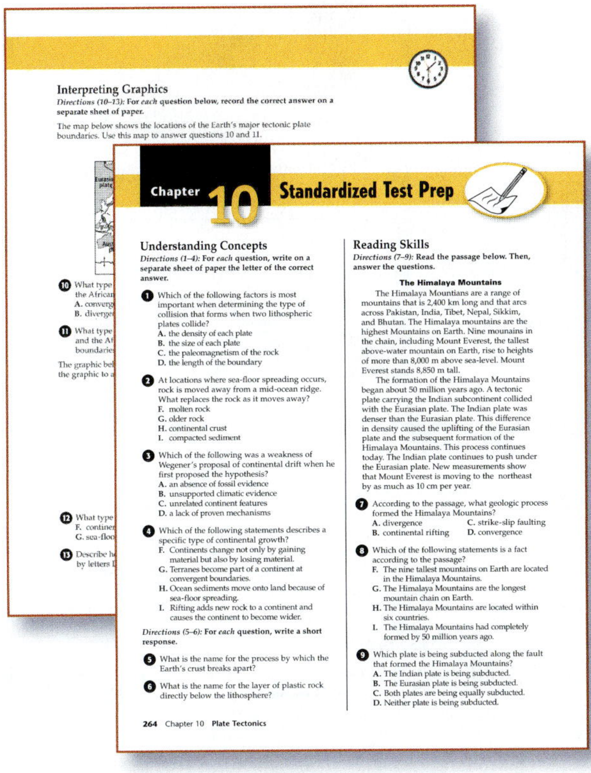

Standardized Test Preps prepare students for testing with skills practice in reading, math, and interpreting graphics. There are two full-pages of test preparation in each chapter of the *Student Edition.*

The **Standardized Test Prep Test Doctor** in the *Teacher Edition* helps you focus reteaching efforts by providing explanations about why students may have chosen incorrect multiple-choice answers and outlining clear rubrics for short-answer and extended-response questions.

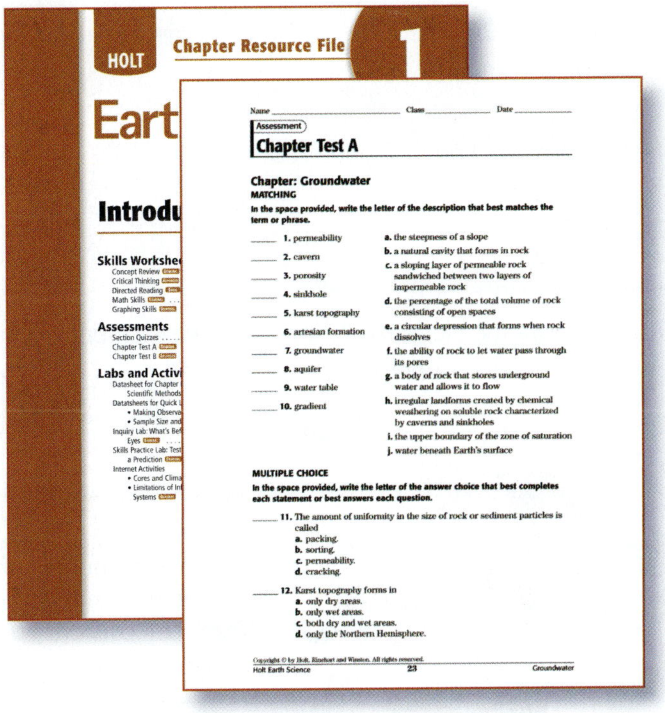

Chapter Resource Files include two levels of **Chapter Tests—General** and **Advanced**—to meet the needs of your classroom. **General Chapter Tests** are also available in the ***Spanish Assessments*** booklet. An easy-to-use answer key is in the back of the ***Chapter Resource Files.***

Assessment Resources

Resources that give you the power to teach *your way*

CHAPTER RESOURCE FILES PLACE ALL YOUR RESOURCES AT YOUR FINGERTIPS.

A well-organized ***Chapter Resource File*** is provided for each chapter of ***Holt Earth Science.*** Each ***Chapter Resource File*** has everything you need to plan and manage your lessons for each chapter in a convenient, time-saving format. You'll receive additional support from the ***Program Resource Introduction File,*** a guide to your ***Chapter Resource Files*** that contains additional teaching references. Your ***Chapter Resource Files*** include:

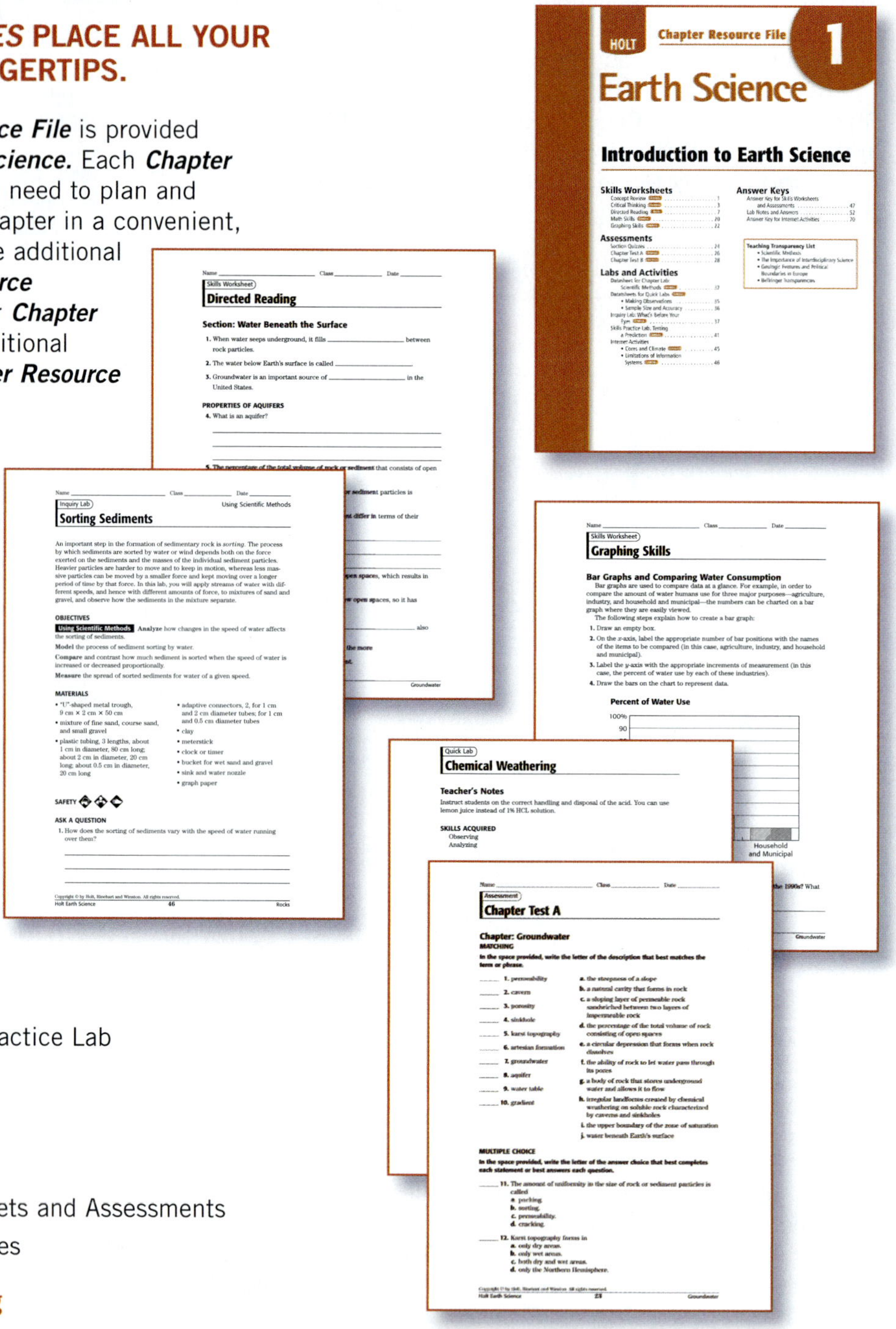

Skills Worksheets

- Concept Review
- Critical Thinking
- Directed Reading
- Math Skills
- Graphing Skills

Assessments

- Section Quizzes
- Chapter Test A: General
- Chapter Test B: Advanced

Labs and Activities

- Datasheet for Chapter Lab
- Datasheets for Quick Labs
- Inquiry Lab
- Making Models Lab or Skills Practice Lab
- Internet Activities

Answer Keys

- Lab Notes and Answers
- Answer Keys for Skills Worksheets and Assessments
- Answer Key for Internet Activities

Teaching Transparency Listing

HOLT Teacher's One-Stop Planner®

All these resources are available in a printable format on your convenient ***Teacher's One-Stop Planner®.***

ADDITIONAL RESOURCES HELP EXPAND YOUR REACH AS A TEACHER.

Study Guide helps students reinforce the skills and concepts presented in the *Student Edition* with **Concept Review** worksheets. Answers are in the corresponding ***Chapter Resource Files.***

Lesson Plans are available in print to help you prepare to teach.

Directed Reading Workbook is a guide for students to follow when reading the text to achieve better retention. Answers are in the corresponding ***Chapter Resource Files.***

Long-Term Projects with Answer Key provides long-range investigations with easy-to-follow instructions and datasheets for each exploration.

Holt Science Skills Workshop: Reading in the Content Area contains exercises to target reading skills specific to the comprehension of science concepts. Students learn to analyze text structures, recognize patterns, and organize information in ways that help them construct meaning.

Holt Science Laboratory Manager's Professional Reference is a teacher reference that explains risk management and the hazards that can occur in the classroom so you can ensure safety.

Teaching Transparencies and Worksheets with Answer Key includes full-color, engaging transparencies, complementary activities to promote students' interest, and an answer key. It also includes:

- **Bellringer Transparencies** to help you introduce each section with a thought-provoking question.
- Transparencies and worksheets that correspond with the **Maps in Action** activity in the *Student Edition* to help you lead classroom discussion.

GIVE ENGLISH-LANGUAGE LEARNERS THE TOOLS THEY NEED TO SUCCEED.

These **Spanish** resources help level the playing field for learning.

- An integrated Spanish **Glossary** in the *Student Edition*
- ***Study Guide*** in Spanish includes **Concept Review** worksheets
- ***Assessments*** in Spanish includes **Section Quizzes** and general **Chapter Tests**
- ***Chapter Summaries Audio CD Program*** in Spanish

Integrated technology takes learning beyond the classroom

THE *TEACHER'S ONE-STOP PLANNER*® IS THE ULTIMATE TEACHER PLANNING TOOL.

Planning and managing lessons has never been easier than with this convenient, all-in-one CD-ROM package that includes the following time-saving features:

Printable

- Teaching Resources
- Spanish Resources
- Transparency Masters

Customizable

- **Lesson Plans,** traditional and block-scheduling, are editable.
- **Holt Calendar Planner** is a tool that assists you in managing your time and resources by the day, week, month, or year.
- **PowerPoint® Resources** help you to design your own lectures with **PowerPoint®** presentations, an image bank, and standardized test preparation resources.

Powerful

- **ExamView® Version 5 Assessment Suite** assists you in creating quizzes and tests from a bank of hundreds of editable questions. You can also post questions to **Holt Online Assessment** for automatic grading.
- **Lab Materials QuickList Software** allows you to compile a materials list so you can easily order the items you need.
- **Holt PuzzlePro®** helps you create crossword puzzles and word searches that make learning vocabulary words fun.
- **Interactive Teacher Edition** makes your planning easy with the entire teacher text linked to related resources.

AUDIO CDs AND TUTORIAL CD-ROMs BROADEN STUDENTS' REALM OF LEARNING.

Chapter Summaries Audio CD Program provides students with audio stimulation to help them study. Also available in ***Spanish.***

Student One Stop is the entire student text, plus the Study Guide and the Directed Reading Workbook, on one disc to lighten the load of backpacks.

Earth Science Interactive Tutor CD-ROM provides a virtual experience where students can explore and investigate Earth science concepts.

Visual Concepts CD-ROM includes engaging graphics and animations that demonstrate key chapter content. Great as a student tutor or teacher presentation tool.

ATTENTION-GRABBING FOOTAGE SPARKS STUDENTS' IMAGINATION.

Brain Food Video Quizzes are game-show style quizzes that assess students' progress and motivate them to study. Available on DVD.

HRW Earth Science Videotape introduces Earth science concepts and takes students on a geology field trip.

NOVA Videos are thought-provoking videos that bring Earth science topics to life.

- **Earthquake**
- **Adrift on the Gulf Stream**
- **Hurricane!**
- **Runaway Universe**

HOLT LAB GENERATOR – NEW!

Containing our entire Holt Earth and Environmental Science laboratory bank, you can search for labs by keywords, standards, or other criteria. Print just the ones you want, add your own, or modify the ones we provide to suit your particular needs. Also includes the Lab Materials Quicklist software to give you a materials list customized for your class.

Connect to the world with current online resources

HOLT BUILDS AND MAINTAINS THE BEST ONLINE RESOURCES.

The *Interactive Online Edition* of ***Holt Earth Science*** brings the textbook to life. Students will build computer skills while learning Earth science. Your *Interactive Online Edition* includes:

- All *Student Edition* pages
- **Visual Concepts,** multimedia presentations of core concepts
- Interactive activities, such as **Concept Maps** and **Self-Check Quizzes**
- Helpful tools, such as **Science Glossary, Periodic Table,** and **Grapher**
- Web links to **go.hrw.com** and **SciLinks®**
- **Classroom Manager** and the ***One-Stop Planner®*** so you can easily create lessons and manage resources
- **Holt Online Assessment** saves you time by automatically grading tests so you can focus on improving student proficiency

Try it for yourself at **http://preview.hrw.com**

See your Holt representative for access.

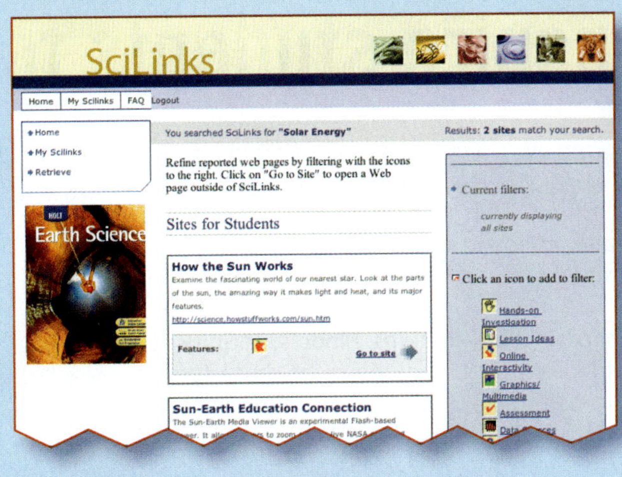

HOLT WORKS WITH NSTA TO KEEP YOU UP-TO-DATE.

SciLinks,® a web service developed and maintained by the **National Science Teachers Association (NSTA),** contains a large collection of prescreened links that include current information directly related to chapter topics.

- **SciLinks**® boxes throughout each chapter provide topics and codes so students can investigate further.
- Each topic leads to multiple links.
- Prescreening saves you valuable time searching for relevant and up-to-date Web sites.
- Sites are reviewed by science-content experts and educators.

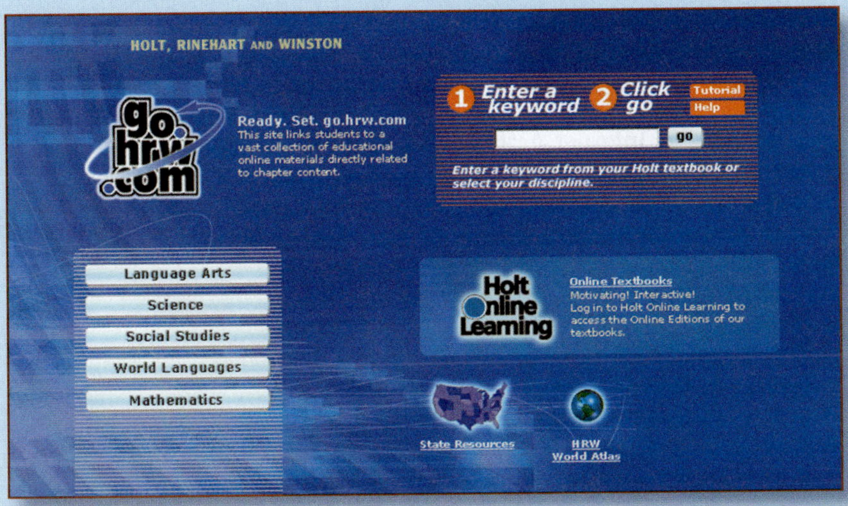

go.hrw.com broadens student learning with resources accessed by keyword. The Web site includes reference materials and resources that enrich and extend study.

Labs and Activities that inspire students to learn

IN-TEXT LABS AND ACTIVITIES IMPROVE STUDENTS' SCIENCE SKILLS.

Quick Labs are easy to do and require minimal time and materials.

Chapter Labs—Making Models, Inquiry, and **Skills Practice**—include clear procedures and help develop students' understanding of scientific methods.

Graphic Organizer guides students to outline topics to improve their comprehension.

PreReading Activity helps students organize content with a **FoldNote.**

Maps in Action features in the *Student Edition* help students develop map-reading skills. Students will become proficient in understanding maps by studying the sample map and answering related questions. You'll also find corresponding transparencies and worksheets in ***Teaching Transparencies and Worksheets with Answer Key.***

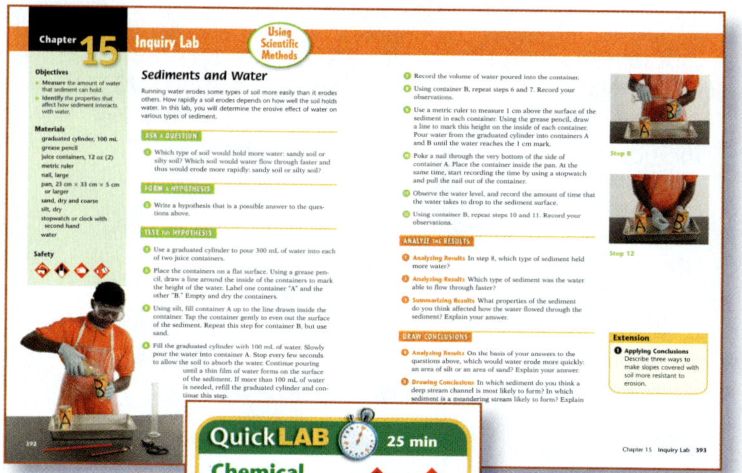

Chapter 15 Inquiry Lab — Using Scientific Methods

Sediments and Water

Objectives
- **Measure** the amount of water that sediment can hold.
- **Identify** the properties that affect how sediment interacts with water.

Materials
graduated cylinder, 100 mL
grease pencil
juice containers, 12 oz (2)
metric ruler
nail, large
pan, 23 cm × 33 cm × 5 cm or larger
sand, dry and coarse
silt, dry
stopwatch or clock with second hand
water

Safety

Running water erodes some types of soil more easily than it erodes others. How rapidly a soil erodes depends on how well the soil holds water. In this lab, you will determine the erosive effect of water on various types of sediment.

ASK A QUESTION

1. Which type of soil would hold more water: sandy soil or silty soil? Which soil would water flow through faster and thus would erode more rapidly: sandy soil or silty soil?

FORM A HYPOTHESIS

2. Write a hypothesis that is a possible answer to the questions above.

TEST THE HYPOTHESIS

3. Use a graduated cylinder to pour 300 mL of water into each of two juice containers.
4. Place the containers on a flat surface. Using a grease pencil, draw a line around the inside of the containers to mark the height of the water. Label one container "A" and the other "B." Empty and dry the containers.
5. Using silt, fill container A up to the line drawn inside the container. Tap the container gently to even out the surface of the sediment. Repeat this step for container B, but use sand.
6. Fill the graduated cylinder with 100 mL of water. Slowly pour the water into container A. Stop every few seconds to allow the soil to absorb the water. Continue pouring until a thin film of water forms on the surface of the sediment. If more than 100 mL of water is needed, refill the graduated cylinder and continue this step.
7. Record the volume of water poured into the container.
8. Using container B, repeat steps 6 and 7. Record your observations.
9. Use a metric ruler to measure 1 cm above the surface of the sediment in each container. Using the grease pencil, draw a line to mark this height on the inside of each container. Pour water from the graduated cylinder into containers A and B until the water reaches the 1 cm mark.
10. Poke a nail through the very bottom of the side of container A. Place the container inside the pan. At the same time, start recording the time by using a stopwatch and pull the nail out of the container.
11. Observe the water level, and record the amount of time that the water takes to drop to the sediment surface.
12. Using container B, repeat steps 10 and 11. Record your observations.

Step 8

Step 12

ANALYZE THE RESULTS

1. **Analyzing Results** In step 8, which type of sediment held more water?
2. **Analyzing Results** Which type of sediment was the water able to flow through faster?
3. **Summarizing Results** What properties of the sediment do you think affected how the water flowed through the sediment? Explain your answer.

DRAW CONCLUSIONS

4. **Analyzing Results** On the basis of your answers to the questions above, which would water erode more quickly: an area of silt or an area of sand? Explain your answer.
5. **Drawing Conclusions** In which sediment do you think a deep stream channel is most likely to form? In which sediment is a meandering stream likely to form? Explain

Extension

1. **Applying Conclusions** Describe three ways to make slopes covered with soil more resistant to erosion.

392 — Chapter 15 Inquiry Lab 393

QuickLAB 25 min

Chemical Weathering

Procedure

1. Place **limestone, granite, pyrite,** and **chalk chips** into separate **small beakers.**
2. Cover the rocks in **1% HCL solution.**
3. After 20 min, observe the rocks.

Analysis

1. How have the rocks changed?
2. How is this process of change like the process of chemical weathering by groundwater?

YOUR *TEACHER EDITION* INCLUDES ACTIVITIES THAT ENCOURAGE CLASS PARTICIPATION.

Bellringers help you to introduce a section with an interesting question. These are also available on transparency.

Inclusion Strategies engage students with different learning exceptionalities.

Debate gives students the opportunity to gather evidence in support of a hypothesis and identify weaknesses of a hypothesis.

Using the Figure encourages students to connect Earth science concepts to the images on the page.

Activity and **Group Activity** give you ideas to get students involved in learning Earth science.

Demonstration provides activities for you to perform to help students visualize Earth science concepts.

Discussion is a starting point for you to lead classroom dialogue.

Internet Activity sends students online to research Earth science topics. The corresponding **Internet Activity** worksheet in the ***Chapter Resource Files*** is designed to direct students' research.

Group Activity — GENERAL

Types of Glacial Lakes Tell students that there are three types of glacial lakes: kettle lakes, cirque lakes, and moraine-formed lakes. Organize the class … Have each group researc… lake and create a poster … sentation that describes … forms and what its chara… LS Verbal

Using the Figure — GENERAL

Replenishing Groundwater Have students describe the differences they see between the porous rock and the permeable rock shown in the figure at the top of this page. Ask students to speculate on how water is replenished in porous rock (by precipitation only) as compared with permeable rock. (Water may flow from one water-filled space to another.) LS Visual

INCLUSION Strategies

- *Visually Impaired*
- *Attention Deficit Disorder*

Ask students to model the types of plate boundaries using their hands. Have s… dents place the index fing… of their hands together. T… have them slide one hand under the other hand to r… a convergent boundary. H… students model divergent boundaries by pulling the… hands apart from the star… position. LS Kinesthetic

Demonstration — GENE…

Comprehension Check Demonstrate the concepts of permeability and impermeabi… Get two shallow bowls or cups. In one bo… press potters' clay around the bottom and the sides to a height of 2.5 cm; put 2.5 cm loose potting soil in the other bowl. Have … dents watch what happens as you pour ab… 60 mL of water into each bowl. The water should sit atop the clay because the clay is impermeable. The water should enter the soil because the soil is permeable. LS Visual — English Language Learners

Internet Activity — GENERAL

Sinkholes Have students research occurrences of sinkholes that have caused property damage, as in Florida in recent years. They may search area newspapers to find out what caused the sinkholes and what is being done to prevent others. A worksheet designed to direct student research on this topic can be found in the **Chapter Resource File** booklet or by visiting go.hrw.com and entering the keyword **HQ6GWEX.** LS Verbal

TEACHING SUPPORT FOR LABS AND ACTIVITIES.

The teacher's wrap provides essential information you need to lead the experiment. You'll be prepared with time requirements, lab ratings, skills acquired, and additional notes.

Useful datasheets for in-text **Chapter Labs** and **Quick Labs** are available in the ***Chapter Resource Files.*** The ***Chapter Resource Files*** also include **Lab Notes and Answers.**

Lab Manager's Professional Reference will help you ensure safety in the classroom.

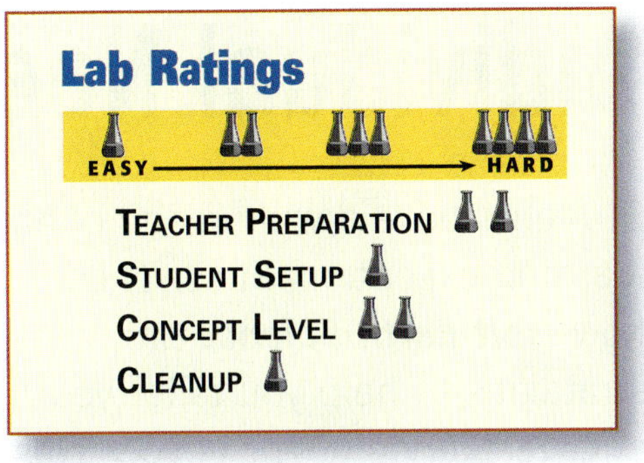

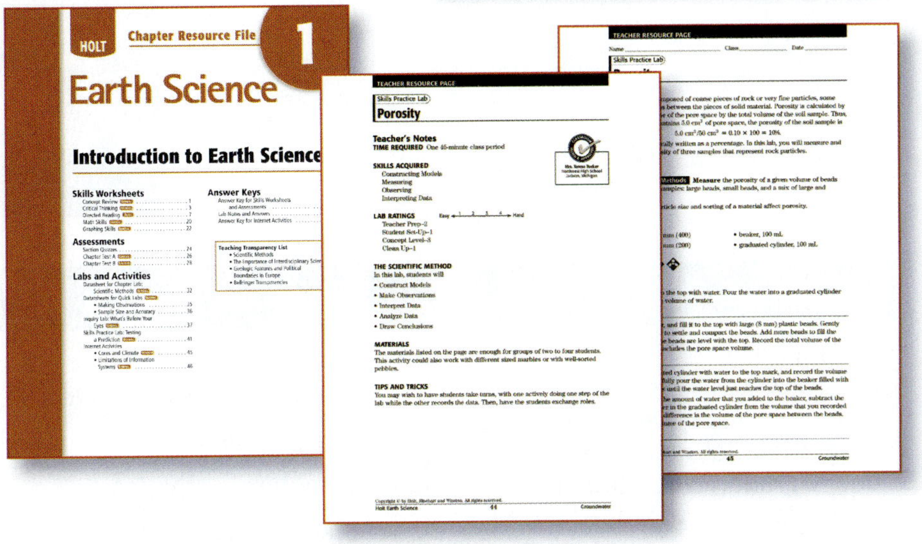

ADDITIONAL LABS EXPAND YOUR OPTIONS.

The ***Chapter Resource Files*** include more labs and activities.

- Additional **Making Models, Inquiry,** and **Skills Practice Labs** provide more hands-on experiences to help students better understand key concepts.
- **Graphing Skills** provides another opportunity for students to improve skills of making and interpreting graphs.
- **Internet Activities** encourages students to research an Earth science Web site.

Long-Term Projects with Answer Key gives students the opportunity to carry-out long-range research. The booklet includes easy-to-follow instructions, datasheets, and answer keys.

HOLT LAB GENERATOR

Every lab in the program can be browsed or searched. Use or edit the labs we provide, create new ones from our templates, or add ones from your files or from the Internet. Also includes the Lab Materials Quicklist software to give you a materials list customized for your class.

The Lab Materials Quicklist software is also available on the ***Teacher's One-Stop Planner.***

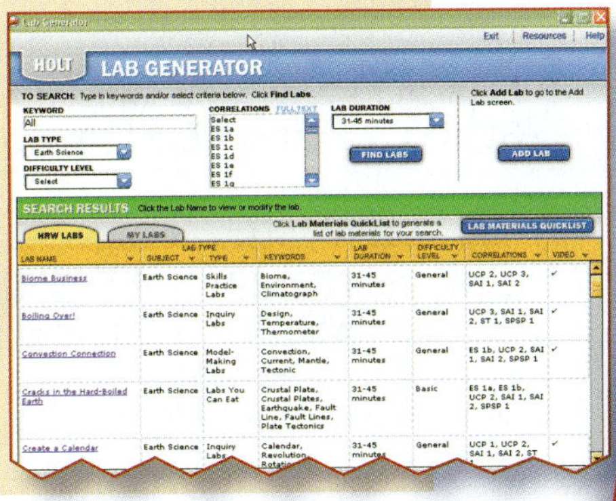

Meeting Individual Needs

Students have a wide range of abilities and learning exceptionalities. These pages show you how ***Holt Earth Science*** provides resources and strategies to help you tailor your instruction to engage every student in your classroom.

Learning exceptionality	Resources and strategies	
Learning Disabilities and Slow Learners Students who have dyslexia or dysgraphia, students reading below grade level, students having difficulty understanding abstract or complex concepts, and slow learners	• Inclusion Strategies labeled **Learning Disabled** • Activities labeled **Basic** • **Reteaching** activities • Activities labeled **Visual, Kinesthetic,** or **Auditory**	• Hands-on activities or projects • Oral presentations instead of written tests or assignments
Developmental Delays Students who are functioning far below grade level because of mental retardation, autism, or brain injury; goals are to learn or retain basic concepts	• Inclusion Strategies labeled **Developmentally Delayed** • Activities labeled **Basic**	• **Reteaching** activities • Project-based activities
Attention Deficit Disorders Students experiencing difficulty completing a task that has multiple steps, difficulty handling long assignments, or difficulty concentrating without sensory input from physical activity	• Inclusion Strategies labeled **Attention Deficit Disorder** • Activities labeled **Basic** • **Reteaching** activities • Activities labeled **Co-op Learning**	• Activities labeled **Visual, Kinesthetic,** or **Auditory** • Concepts broken into small chunks • Oral presentations instead of written tests or assignments
English as a Second Language Students learning English	• Activities labeled **English-Language Learners** • Activities labeled **Basic**	• **Reteaching** activities • Activities labeled **Visual**
Gifted and Talented Students who are performing above grade level and demonstrate aptitude in crosscurricular assignments	• Inclusion Strategies labeled **Gifted and Talented** • Activities labeled **Advanced**	• **Connection** activities • Activities that involve multiple tasks, a strong degree of independence, and student initiative

General Strategies The following strategies can help you modify instruction to help students who struggle with common classroom difficulties.

A student experiencing difficulty with...	May benefit if you...	
Beginning assignments	• Assign work in small amounts • Have the student use cooperative or paired learning • Provide varied and interesting activities	• Allow choice in assignments or projects • Reinforce participation • Seat the student closer to you
Following directions	• Gain the student's attention before giving directions • Break up the task into small steps • Give written directions rather than oral directions • Use short, simple phrases	• Stand near the student when you are giving directions • Have the student repeat directions to you • Prepare the student for changes in activity • Give visual cues by posting general routines • Reinforce improvement in or approximation of following directions
Keeping track of assignments	• Have the student use folders for assignments • Have the student use assignment notebooks	• Have the student keep a checklist of assignments and highlight assignments when they are turned in
Reading the textbook	• Provide outlines of the textbook content • Reduce the length of required reading • Allow extra time for reading • Have the students read aloud in small groups	• Have the student use peer or mentor readers • Have the student use books on tape or CD • Discuss the content of the textbook in class after reading
Staying on task	• Reduce distracting elements in the classroom • Provide a task-completion checklist • Seat the student near you	• Provide alternative ways to complete assignments, such as oral projects taped with a buddy
Behavioral or social skills	• Model the appropriate behaviors • Establish class rules, and reiterate them often • Reinforce positive behavior • Assign a mentor as a positive role model to the student • Contract with the student for expected behaviors • Reinforce the desired behaviors or any steps toward improvement	• Separate the student from any peer who stimulates the inappropriate behavior • Provide a "cooling off" period before talking with the student • Address academic/instructional problems that may contribute to disruptive behaviors • Include parents in the problem-solving process through conferences, home visits, and frequent communication
Attendance	• Recognize and reinforce attendance by giving incentives or verbal praise • Emphasize the importance of attendance by letting the student know that he or she was missed when he or she was absent	• Encourage the student's desire to be in school by planning activities that are likely to be enjoyable, giving the student a preferred responsibility to be performed in class, and involving the student in extracurricular activities • Schedule problem-solving meeting with parents, faculty, or both
Test-taking skills	• Prepare the student for testing by teaching ways to study in pairs, such as using flashcards, practice tests, and study guides, and by promoting adequate sleep, nourishment, and exercise • Decrease visual distraction by improving the visual design of the test through use of larger type, spacing, consistent layout, and shorter sentences	• During testing, allow the student to respond orally on tape or to respond using a computer; to use notes; to take breaks; to take the test in another location; to work without time constraints; or to take the test in several short sessions

Reading tools that foster comprehension

READING AIDS THROUGHOUT THE TEXT HELP STUDENTS ORGANIZE THEIR THOUGHTS.

The student text is well organized and easy to navigate to help students retain information.

- A list of upcoming sections prepares students to learn.
- A list of **Objectives** and **Key Terms** helps students to focus on essential concepts.
- **Key Terms** are highlighted and defined in the margin for quick reference.
- Every page begins with a new head so the text is easy to read.

Pre-Reading Activity prepares students for upcoming content by creating a **FoldNote.**

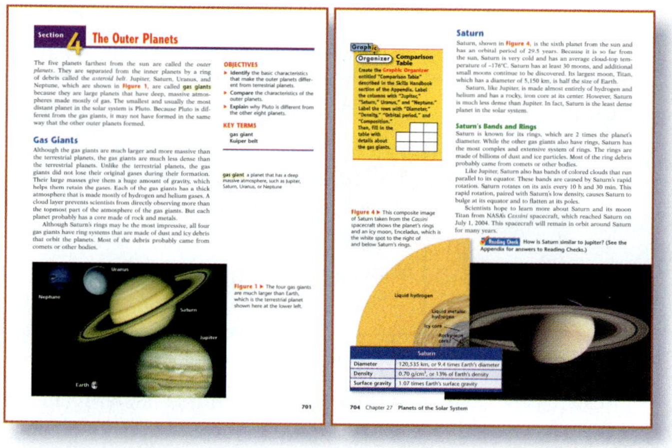

Graphic Organizer helps students understand the text by making a graphic that connects the topics.

Reading Checks give students the opportunity to verify their comprehension of the text. The answers are in the **Appendix.**

ADDITIONAL READING RESOURCES PROVIDE ANOTHER LEVEL OF SUPPORT.

Reading Skill Builder in the teacher's wrap provides a specific reading activity to reinforce students' understanding of key concepts. Additional vocabulary **Skill Builders** help you to reteach the fundamentals of a concept.

Directed Reading worksheets in the ***Chapter Resource Files*** provides a structure for students to follow when reading the text and improves retention. The worksheets are also available in a student booklet, ***Directed Reading Workbook*** and on the ***Student One Stop.*** Answers are in the ***Chapter Resource Files.***

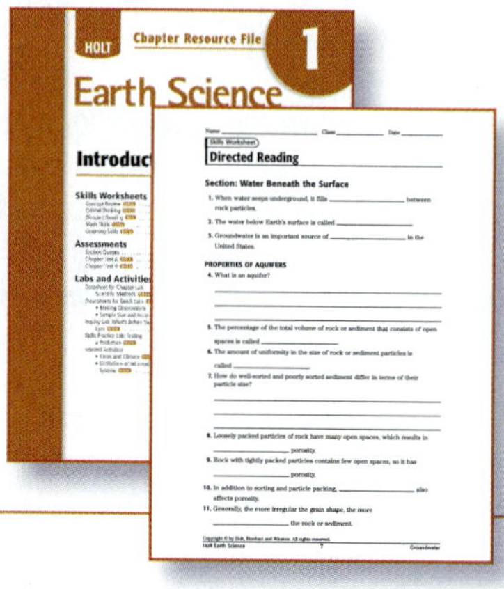

Critical Thinking worksheets in the ***Chapter Resource Files*** develop higher-thinking skills. Students make comparisons, inferences, and predictions to demonstrate their understanding of the text.

Holt Science Skills Workshop: Reading in the Content Area targets the reading skills specific to the comprehension of science texts. The ***Teacher's Edition*** comes with transparencies to lead the class.

Chapter Summaries Audio CD Program provides audio reinforcement as students review the content. Also available in ***Spanish.***

Features that incorporate math and writing skills

STUDENTS IMPROVE THEIR STUDY SKILLS BY INTEGRATING MATH WHILE LEARNING SCIENCE.

MATHPRACTICE

The Rate of Plate Movement Tectonic plates move slowly on Earth's surface. The rate of plate movement can be calculated by using the following equation:

$$rate = \frac{distance}{time}$$

In kilometers, how far would a plate that moves 4 cm per year move in 2 million years?

SKILL BUILDER — ADVANCED

Math The Atlantic Ocean is spreading at a rate of 1 to 2 cm per year, and the eastern Pacific sea floor is spreading between 3 and 8 cm per year. Have students use the average rate of spreading for the Atlantic Ocean to calculate how many years the sea floor of the Atlantic Ocean would take to spread 1 km. (1.5 cm/year; 1 km ÷ 0.000015 km/year = 66,667 years) Have students use the average rate of spreading of the Pacific Ocean to calculate how many years the sea floor of the Pacific Ocean would take to spread 1 km. (5.5 cm/year; 1 km ÷ 0.000055 km/year = 18,182 years) **LS Logical**

Math Practice links mathematics directly to the science topic being covered.

Skill Builder: Math in the teacher's wrap gives students another chance to integrate math with science.

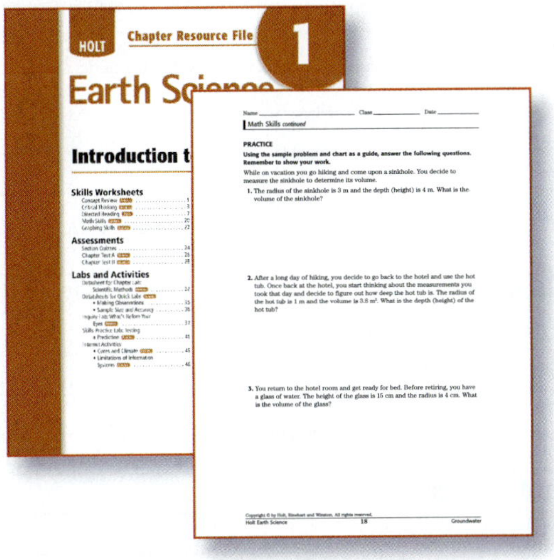

Math Skills worksheets in the ***Chapter Resource Files*** hone students' math skills by providing challenging math exercises.

Writing Skills

30. **Writing Persuasively** Imagine that you are Alfred Wegener. Write a persuasive essay to explain your idea of continental drift. Use only evidence originally used by Wegener to support his hypothesis.

31. **Communicating Main Ideas** research of Wegener, Hes to the theory of plate tect

SKILL BUILDER — ADVANCED

Writing Ask students to imagine they live on a street where a sinkhole "swallowed" several cars and damaged some homes. Have students write a letter to the editor of the local newspaper to explain how they feel about the situation and what, if anything, they think should be done to prevent future occurrences. **LS Intrapersonal**

WRITING EXERCISES REINFORCE STUDENTS' KNOWLEDGE OF SCIENCE AND FOSTER FUNDAMENTAL SKILLS.

Writing Skills questions in the **Chapter Review** allow students to practice their writing skills by developing essays that are clear and persuasive.

Skill Builder: Writing in the teacher's wrap gives students another opportunity to apply their writing skills using science topics.

Pacing Guide

Today's Earth science classroom often requires a more flexible curriculum. ***Holt Earth Science*** can help you meet a variety of needs and challenges you and your students face in the classroom. The **Pacing Guide** below shows a number of ways to adapt the program to your teaching schedule.

This **Guide** can be further adapted, allowing you to mix and match or compress the material so you can spend more time on select topics, or to allow for special projects and activities.

- **Basic** gives more time for the foundations of Earth science, especially mathematical problem-solving, with less emphasis on some advanced topics from later in the course.
- **General** provides the recommended course of study as indicated in the *Teacher Edition,* found in the individual chapter **Planning Guides** preceding each chapter.
- **Advanced** moves quickly through foundations of Earth science for students who may be comfortable with the basics, to provide additional time for advanced topics.
- **Heavy Lab/Activity** indicates ways to streamline "lecture" time to provide hands-on experience for more than a third of the blocks in the school year. (Note: even this approach does not cover all of the labs and activities that are available with ***Holt Earth Science*** and its ancillaries.)

Numbers indicate class periods recommended for the material within each chapter.	Basic	General	Advanced	Heavy Lab/ Activity
Chapter 1: Introduction to Earth Science	**6**	**5**	**4**	**0**
Section 1 What Is Earth Science?	1	1	1	0
Section 2 Science as a Process	2	2	1	0
Lab Experiments	1	0	0	0
Chapter Review and Assessment	2	2	2	0
Chapter 2: Earth as a System	**6**	**7**	**7**	**5**
Section 1 Earth: A Unique Planet	1	1	1	1
Section 2 Energy in the Earth System	2	2	2	2
Section 3 Ecology	0	1	1	0
Lab Experiments	1	1	1	0
Chapter Review and Assessment	2	2	2	2
Chapter 3: Models of the Earth	**8**	**6**	**6**	**6**
Section 1 Finding Locations on Earth	1	1	1	1
Section 2 Mapping Earth's Surface	2	1	1	0
Section 3 Types of Maps	2	1	1	1
Lab Experiments	1	1	1	2
Chapter Review and Assessment	2	2	2	2
Chapter 4: Earth Chemistry	**0**	**4**	**5**	**0**
Section 1 Matter	0	1	1	0
Section 2 Combinations of Atoms	0	1	1	0
Lab Experiments	0	0	1	0
Chapter Review and Assessment	0	2	2	0
Chapter 5: Minerals of Earth's Crust	**7**	**5**	**5**	**6**
Section 1 What Is a Mineral?	2	1	1	1
Section 2 Identifying Minerals	2	1	1	1
Lab Experiments	1	1	1	2
Chapter Review and Assessment	2	2	2	2
Chapter 6: Rocks	**9**	**7**	**7**	**8**
Section 1 Rocks and the Rock Cycle	1	1	1	1
Section 2 Igneous Rock	2	1	1	1
Section 3 Sedimentary Rock	2	1	1	1
Section 4 Metamorphic Rock	1	1	1	1
Lab Experiments	1	1	1	2
Chapter Review and Assessment	2	2	2	2

Numbers indicate class periods recommended for the material within each chapter.

	Basic	General	Advanced	Heavy Lab/ Activity
Chapter 7: Resources and Energy	**8**	**7**	**7**	**7**
Section 1 Mineral Resources	1	1	1	0
Section 2 Nonrenewable Energy	2	1	1	1
Section 3 Renewable Energy	1	1	1	1
Section 4 Resources and Conservation	1	1	1	1
Lab Experiments	1	1	1	2
Chapter Review and Assessment	2	2	2	2
Chapter 8: The Rock Record	**7**	**4**	**6**	**4**
Section 1 Determining Relative Age	2	1	1	1
Section 2 Determining Absolute Age	2	1	1	1
Section 3 The Fossil Record	0	0	1	0
Lab Experiments	1	0	1	0
Chapter Review and Assessment	2	2	2	2
Chapter 9: A View of Earth's Past	**4**	**6**	**6**	**5**
Section 1 Geologic Time	1	1	1	1
Section 2 Precambrian Time and the Paleozoic Era	0	1	1	0
Section 3 The Mesozoic and Cenozoic Eras	0	1	1	0
Lab Experiments	1	1	1	2
Chapter Review and Assessment	2	2	2	2
Chapter 10: Plate Tectonics	**7**	**8**	**8**	**8**
Section 1 Continental Drift	2	2	2	2
Section 2 The Theory of Plate Tectonics	2	2	2	2
Section 3 The Changing Continents	0	1	1	0
Lab Experiments	1	1	1	2
Chapter Review and Assessment	2	2	2	2
Chapter 11: Deformation of the Crust	**0**	**5**	**6**	**5**
Section 1 How Rock Deforms	0	2	2	0
Section 2 How Mountains Form	0	1	1	1
Lab Experiments	0	0	1	2
Chapter Review and Assessment	0	2	2	2
Chapter 12: Earthquakes	**7**	**5**	**4**	**7**
Section 1 How and Where Earthquakes Happen	2	1	1	1
Section 2 Studying Earthquakes	1	1	1	1
Section 3 Earthquakes and Society	1	1	0	1
Lab Experiments	1	0	0	2
Chapter Review and Assessment	2	2	2	2
Chapter 13: Volcanoes	**7**	**4**	**4**	**6**
Section 1 Volcanoes and Plate Tectonics	2	1	1	1
Section 2 Volcanic Eruptions	2	1	1	1
Lab Experiments	1	0	0	2
Chapter Review and Assessment	2	2	2	2
Chapter 14: Weathering and Erosion	**9**	**7**	**7**	**9**
Section 1 Weathering Processes	2	1	1	1
Section 2 Rates of Weathering	1	1	1	1
Section 3 Soil	1	1	1	1
Section 4 Erosion	2	2	2	2
Lab Experiments	1	0	0	2
Chapter Review and Assessment	2	2	2	2

Pacing Guide

Numbers indicate class periods recommended for the material within each chapter.

	Basic	General	Advanced	Heavy Lab/ Activity
Chapter 15: River Systems	**6**	**6**	**5**	**6**
Section 1 The Water Cycle	1	1	0	0
Section 2 Stream Erosion	1	1	1	1
Section 3 Stream Deposition	1	1	1	1
Lab Experiments	1	1	1	2
Chapter Review and Assessment	2	2	2	2
Chapter 16: Groundwater	**6**	**6**	**6**	**6**
Section 1 Water Beneath the Surface	2	2	2	2
Section 2 Groundwater and Chemical Weathering	1	1	1	0
Lab Experiments	1	1	1	2
Chapter Review and Assessment	2	2	2	2
Chapter 17: Glaciers	**7**	**6**	**6**	**8**
Section 1 Glaciers: Moving Ice	1	1	1	1
Section 2 Glacial Erosion and Deposition	2	2	2	2
Section 3 Ice Ages	1	1	1	1
Lab Experiments	1	0	0	2
Chapter Review and Assessment	2	2	2	2
Chapter 18: Erosion by Wind and Waves	**5**	**6**	**6**	**6**
Section 1 Wind Erosion	1	1	1	1
Section 2 Wave Erosion	1	1	1	1
Section 3 Coastal Erosion and Deposition	0	1	1	0
Lab Experiments	1	1	1	2
Chapter Review and Assessment	2	2	2	2
Chapter 19: The Ocean Basins	**6**	**6**	**5**	**4**
Section 1 The Water Planet	1	1	0	1
Section 2 Features of the Ocean Floor	1	1	1	1
Section 3 Ocean-Floor Sediments	1	1	1	0
Lab Experiments	1	1	1	0
Chapter Review and Assessment	2	2	2	2
Chapter 20: Ocean Water	**5**	**7**	**7**	**7**
Section 1 Properties of Ocean Water	1	2	2	2
Section 2 Life in the Oceans	0	1	1	0
Section 3 Ocean Resources	1	1	1	1
Lab Experiments	1	1	1	2
Chapter Review and Assessment	2	2	2	2
Chapter 21: Movements of the Ocean	**5**	**4**	**4**	**4**
Section 1 Ocean Currents	1	1	1	1
Section 2 Ocean Waves	0	0	0	0
Section 3 Tides	1	1	1	1
Lab Experiments	1	0	0	0
Chapter Review and Assessment	2	2	2	2
Chapter 22: The Atmosphere	**8**	**6**	**6**	**8**
Section 1 Characteristics of the Atmosphere	2	2	2	2
Section 2 Solar Energy and the Atmosphere	2	1	1	1
Section 3 Atmospheric Circulation	1	1	1	1
Lab Experiments	1	0	0	2
Chapter Review and Assessment	2	2	2	2

Numbers indicate class periods recommended for the material within each chapter.

	Basic	General	Advanced	Heavy Lab/ Activity
Chapter 23: Water in the Atmosphere	**0**	**5**	**5**	**7**
Section 1 Atmospheric Moisture	0	1	1	1
Section 2 Clouds and Fog	0	1	1	1
Section 3 Precipitation	0	1	1	1
Lab Experiments	0	0	0	2
Chapter Review and Assessment	0	2	2	2
Chapter 24: Weather	**7**	**7**	**7**	**6**
Section 1 Air Masses	1	1	1	0
Section 2 Fronts	1	1	1	1
Section 3 Weather Instruments	1	1	1	0
Section 4 Forecasting the Weather	1	1	1	1
Lab Experiments	1	1	1	2
Chapter Review and Assessment	2	2	2	2
Chapter 25: Climate	**7**	**5**	**5**	**7**
Section 1 Factors That Affect Climate	2	1	1	1
Section 2 Climate Zones	1	1	1	1
Section 3 Climate Change	1	1	1	1
Lab Experiments	1	0	0	2
Chapter Review and Assessment	2	2	2	2
Chapter 26: Studying Space	**7**	**6**	**6**	**8**
Section 1 Viewing the Universe	2	2	2	2
Section 2 Movements of the Earth	2	2	2	2
Lab Experiments	1	0	0	2
Chapter Review and Assessment	2	2	2	2
Chapter 27: Planets of the Solar System	**7**	**7**	**7**	**9**
Section 1 Formation of the Solar System	1	1	1	1
Section 2 Models of the Solar System	0	1	1	1
Section 3 The Inner Planets	1	1	1	1
Section 4 The Outer Planets	2	2	2	2
Lab Experiments	1	0	0	2
Chapter Review and Assessment	2	2	2	2
Chapter 28: Minor Bodies of the Solar System	**6**	**8**	**8**	**4**
Section 1 Earth's Moon	1	1	1	1
Section 2 Movements of the Moon	0	2	2	0
Section 3 Satellites of Other Planets	1	1	1	0
Section 4 Asteroids, Comets, and Meteoroids	1	1	1	1
Lab Experiments	1	1	1	0
Chapter Review and Assessment	2	2	2	2
Chapter 29: The Sun	**5**	**5**	**5**	**5**
Section 1 Structure of the Sun	1	1	1	1
Section 2 Solar Activity	1	1	1	0
Lab Experiments	1	1	1	2
Chapter Review and Assessment	2	2	2	2
Chapter 30: Stars, Galaxies, and the Universe	**6**	**8**	**8**	**7**
Section 1 Characteristics of Stars	1	1	1	0
Section 2 Stellar Evolution	0	2	2	2
Section 3 Star Groups	1	1	1	1
Section 4 The Big Bang Theory	1	1	1	1
Lab Experiments	1	1	1	1
Chapter Review and Assessment	2	2	2	2
Total	**178**	**178**	**178**	**178**

Correlation to the National Science Education Standards

The following list shows the correlation of ***Holt Earth Science*** with the **National Science Education Standards** (grades 9–12) for Earth and space science, physics, chemistry, and life science content. For further detail, see the interleaf pages before each chapter.

Standard	Code	Chapter Correlation
Unifying Concepts and Processes	**UCP**	
Systems, Order, and Organization	**UCP 1**	**1, 2, 3, 10, 16, 24, 28, 30**
Evidence, Models, and Explanation	**UCP 2**	**3, 10, 13, 18, 24, 25, 27**
Constancy, Change, and Measurement	**UCP 3**	**10, 11, 12, 14, 16, 18, 24, 27, 28, 30**
Evolution and Equilibrium	**UCP 4**	**8, 11, 13, 14, 15, 16, 17, 18, 19**
Science as Inquiry	**SAI 1**	
Abilities to Do Scientific Inquiry	**SAI 1**	
IDENTIFY QUESTIONS AND CONCEPTS THAT GUIDE SCIENTIFIC INVESTIGATIONS. Students should formulate a testable hypothesis and demonstrate the logical connections between the scientific concepts guiding a hypothesis and the design of an experiment. They should demonstrate appropriate procedures, a knowledge base, and conceptual understanding of scientific investigations.	SAI 1a	18, 24
COMMUNICATE AND DEFEND A SCIENTIFIC ARGUMENT. Students in school science programs should develop the abilities associated with accurate and effective communication. These include writing and following procedures, expressing concepts, reviewing information, summarizing data, using language appropriately, developing diagrams and charts, explaining statistical analysis, speaking clearly and logically, constructing a reasoned argument, and responding appropriately to critical comments.	SAI 1f	18, 24
Understanding About Scientific Inquiry	**SAI 2**	
Scientists usually inquire about how physical, living, or designed systems function. Conceptual principles and knowedge guide scientific inquiries. Historical and current scientific knowledge influence the design and interpretation of investigations and the evaluation of proposed explanations made by other scientists.	SAI 2a	1, 10, 24, 26
Scientists conduct investigations for a wide variety of reasons. For example, they may wish to discover new aspects of the natural world, explain recently observed phenomena, or test the conclusions of prior investigations or the predictions of current theories.	SAI 2b	1, 26

Standard	Code	Chapter Correlation
Understanding About Scientific Inquiry *(continued)*	**SAI 2**	
Scientists rely on technology to enhance the gathering and manipulation of data. New techniques and tools provide new evidence to guide inquiry and new methods to gather data, thereby contributing to the advance of science. The accuracy and precision of the data, and therefore the quality of the exploration, depends on the technology used.	SAI 2c	1, 3, 9, 19, 26, 28
Mathematics is essential in scientific inquiry. Mathematical tools and models guide and improve the posing of questions, gathering data, constructing explanations and communicating results.	SAI 2d	1, 3, 25, 27
Scientific explanations must adhere to criteria such as: a proposed explanation must be logically consistent; it must abide by the rules of evidence; it must be open to questions and possible modification; and it must be based on historical and current scientific knowledge.	SAI 2e	1, 10
Results of scientific inquiry—new knowledge and methods—emerge from different types of investigations and public communication among scientists. In communicating and defending the results of scientific inquiry, arguments must be logical and demonstrate connections between natural phenomena, investigations, and the historical body of scientific knowledge. In addition, the methods and procedures that scientists used to obtain evidence must be clearly reported to enhance opportunities for further investigation.	SAI 2f	1, 28
Earth and Space Science	**ES**	
Energy in the Earth system	**ES 1**	
The outward transfer of earth's internal heat drives convection circulation in the mantle that propels the plates comprosing earth's surface across the face of the globe.	ES 1b	10, 11
Heating of earth's surface and atmosphere by the sun drives convection within the atmosphere and oceans, producing winds and ocean currents.	ES 1c	21, 22, 23
Global climate is determined by energy transfer from the sun at and near the earth's surface. This energy transfer is influenced by dynamic processes such as cloud cover and the earth's rotation, and static conditions such as the position of mountain ranges and oceans.	ES 1d	22, 23, 25
Geochemical Cycles	**ES 2**	
Movement of matter between reservoirs is driven by the earth's internal and external sources of energy. These movements are often accompanied by a change in the physical and chemical properties of the matter. Carbon, for example, occurs in carbonate rocks such as limestone, in the atmosphere as carbon dioxide gas, in water as dissolved carbon dioxide, and in all organisms as complex molecules that control the chemistry of life.	ES 2b	6, 23

Standard	Code	Chapter Correlat
The Origin and Evolution of the Earth System	**ES 3**	
The sun, the earth, and the rest of the solar system formed from a nebular cloud of dust and gas 4.6 billion years ago. The early earth was very different from the planet we live on today.	ES 3a	27, 28
Geologic time can be estimated by observing rock sequences and using fossils to correlate the sequences at various locations. Current methods include using the known decay rates of radioactive isotopes present in rocks to measure the time since the rock was formed.	ES 3b	8, 9
Interactions among the solid earth, the oceans, the atmosphere, and organisms have resulted in the ongoing evolution of the earth system. We can observe some changes such as earthquakes and volcanic eruptions on a human time scale, but many processes such as mountain building and plate movements take place over hundreds of millions of years.	ES 3c	10, 11, 13, 14, 15, 1
Evidence for one-celled forms of life—the bacteria—extends back more than 3.5 billion years. The evolution of life caused dramatic changes in the composition of the earth's atmosphere, which did not originally contain oxygen.	ES 3d	9
The Origin and Evolution of the Universe	**ES 4**	
The origin of the universe remains one of the greatest questions in science. The "big bang" theory places the origin between 10 and 20 billion years ago, when the universe began in a hot dense state; according to this theory, the universe has been expanding ever since.	ES 4a	26, 30
Early in the history of the universe, matter, primarily the light atoms hydrogen and helium, clumped together by gravitational attraction to form countless trillions of stars. Billions of galaxies, each of which is a gravitationally bound cluster of billions of stars, now form most of the visible mass in the universe.	ES 4b	26, 29, 30
Stars produce energy from nuclear reactions, primarily the fusion of hydrogen to form helium. These and other processes in stars have led to the formation of all the other elements.	ES 4c	29, 30
Physical Science	**PS**	
Structure of Atoms	**PS 1**	**4, 7, 8**
Structure and Properties of Matter	**PS 2**	**4, 5, 7, 23, 30**
Chemical Reactions	**PS 3**	**7, 16, 22**
Motion and Forces	**PS 4**	**2, 27, 28, 29**
Conservation of Energy and the Increase in Disorder	**PS 5**	**2, 26, 29**
Interactions of Energy and Matter	**PS 6**	**12, 21, 22**

Standard	Code	Chapter Correlation
Life Science	**LS**	
Biological Evolution	LS 3	8, 9
The Interdependence of Organisms	LS 4	2, 7, 16, 20, 23
Matter, Energy, and Organization in Living Systems	LS 5	2
Science and Technology	**ST**	
Abilities of Technological Design	ST 1	12, 13, 18
Understanding About Science and Technology	ST 2	1, 28
Science in Personal and Social Perspectives	**SPSP**	
Personal Health	SPSP 1	24
Natural Hazards	SPSP 3	7, 15, 20
Risks and Benefits	SPSP 4	14, 20, 25
Natural and Human-Induced Hazards	SPSP 5	12, 13, 14, 15, 17, 18, 25
Science and Technology in Society	SPSP 6	1, 16
History and Nature of Science	**HNS**	
Nature of Science	HNS 2	3, 11, 24, 27, 28
History of Science	HNS 3	1, 3, 4, 8, 9, 10, 26, 27, 28, 29

Safety in your laboratory

Risk Asssessment

MAKING YOUR LABORATORY A SAFE PLACE TO WORK AND LEARN

Concern for safety must begin before any activity in the classroom and before students enter the lab. A careful review of the facilities should be a basic part of preparation for each school term. You should investigate the physical environment, identify any safety risks, and inspect your work areas for compliance with safety regulations.

The review of the lab should be thorough, and all safety issues must be addressed immediately. Keep a file of your review, and add to the list each year. This will allow you to continue to raise the standard of safety in your lab and classroom.

Many classroom experiments, demonstrations, and other activities are classics that have been used for years. This familiarity may lead to a comfort that can obscure inherent safety concerns. Review all experiments, demonstrations, and activities for safety concerns before presenting them to the class. Identify and eliminate potential safety hazards.

1. **Identify the Risks**
   Before introducing any activity, demonstration, or experiment to the class, analyze it and consider what could possibly go wrong. Carefully review the list of materials to make sure they are safe. Inspect the equipment in your lab or classroom to make sure it is in good working order. Read the procedures to make sure they are safe. Record any hazards or concerns you identify.

2. **Evaluate the Risks**
   Minimize the risks you identified in the last step without sacrificing learning. Remember that no activity you perform in the lab or classroom is worth risking injury. Thus, extremely hazardous activities, or those that violate your schoo policies, must be eliminated. For activitie present smaller risks, analyze each risk c to determine its likelihood. If the pedago value of the activity does not outweigh th the activity must be eliminated.

3. **Select Controls to Address Risks**
   Even low-risk activities require controls tc eliminate or minimize the risks. Make sur in devising controls you do not substitute equally or more hazardous alternative. So control methods include the following:
   - Explicit verbal and written warnings ma added or posted.
   - Equipment may be rebuilt or relocated, may be replaced, or equipment be repla entirely by safer alternatives.
   - Risky procedures may be eliminated.
   - Activities may be changed from student activities to teacher demonstrations.

4. **Implement and Review Selected Cont**
   Controls do not help if they are forgotten enforced. The implementation and review controls should be as systematic and thor as the initial analysis of safety concerns i lab and laboratory activities.

SOME SAFETY RISKS AND PREVENTATIVE CONTROLS

The following list describes several possible s hazards and controls that can be implemente resolve them. This list is not complete, but it used as a starting point to identify hazards in laboratory.

Identified risk	Preventative control
Facilities and Equipment	
Lab tables are in disrepair, room is poorly lighted and ventilated, faucets and electrical outlets do not work or are difficult to use because of their location.	Work surfaces should be level and stable. There should be adequate and ventilation. Water supplies, drains, and electrical outlets should good working order. Any equipment in a dangerous location should n used; it should be relocated or rendered inoperable.
Wiring, plumbing, and air circulation systems do not work or do not meet current specifications.	Specifications should be kept on file. Conduct a periodic review of al equipment, and document compliance. Damaged fixtures must be la as such and must be repaired as soon as possible.

Identified risk	Preventative control
Facilities and Equipment *(continued)*	
Eyewash fountains and safety showers are present, but no one knows anything about their specifications.	Ensure that eyewash fountains and safety showers meet the requirements of the ANSI standard (Z358.1).
Eyewash fountains are checked and cleaned once at the beginning of each school year. No records are kept of routine checks and maintenance on the safety showers and eyewash fountains.	Flush eyewash fountains for 5 minutes every month to remove any bacteria or other organisms from pipes. Test safety showers (measure flow in gallons per min) and eyewash fountains every 6 months and keep records of the test results.
Labs are conducted in multipurpose rooms, and equipment from other courses remains accessible.	Only the items necessary for a given activity should be available to students. All equipment should be locked away when not in use.
Students are permitted to enter or work in the lab without teacher supervision.	Lock all laboratory rooms whenever a teacher is not present. Supervising teachers must be trained in lab safety and emergency procedures.
Safety equipment and emergency procedures	
Fire and other emergency drills are infrequent, and no records or measurements are made of the results of the drills.	Always carry out critical reviews of fire or other emergency drills. Be sure that plans include alternate routes. Don't wait until an emergency to find the flaws in your plans.
Emergency evacuation plans do not include instructions for securing the lab in the event of an evacuation during a lab activity.	Plan actions in case of emergency: establish what devices should be turned off, which escape route to use, and where to meet outside the building.
Fire extinguishers are in out-of-the-way locations, not on the escape route.	Place fire extinguishers near escape routes so that they will be of use during an emergency.
Fire extinguishers are not maintained. Teachers are not trained to use them.	Document regular maintenance of fire extinguishers. Train supervisory personnel in the proper use of extinguishers. Instruct students not to use an extinguisher but to call for a teacher.
Teachers in labs and neighboring classrooms are not trained in CPR or first aid.	Teachers should receive training from the local chapter of the the American Red Cross. Certifications should be kept current with frequent refresher courses.
Teachers are not aware of their legal responsibilities in case of an injury or accident.	Review your faculty handbook for your responsibilities regarding safety in the classroom and laboratory. Contact the legal counsel for your school district to find out the extent of their support and any rules, regulations, or procedures you must follow.
Emergency procedures are not posted. Emergency numbers are kept only at the switchboard or main office. Instructions are given verbally only at the beginning of the year.	Emergency procedures should be posted at all exits and near all safety equipment. Emergency numbers should be posted at all phones, and a script should be provided for the caller to use. Emergency procedures must be reviewed periodically, and students should be reminded of them at the beginning of each activity.
Spills are handled on a case-by-case basis and are cleaned up with whatever materials happen to be on hand.	Have the appropriate equipment and materials available for cleaning up; replace them before expiration dates. Make sure students know to alert you to spilled chemicals, blood, and broken glass.
Work habits and environment	
Safety wear is only used for activities involving chemicals or hot plates.	Aprons and goggles should be worn in the lab at all times. Long hair, loose clothing, and loose jewelry should be secured.
There is no dress code established for the laboratory; students are allowed to wear sandals or open-toed shoes.	Open-toed shoes should never be worn in the laboratory. Do not allow any footwear in the lab that does not cover feet completely.
Students are required to wear safety gear but teachers and visitors are not.	Always wear safety gear in the lab. Keep extra equipment on hand for visitors.

Identified risk	Preventative control
Work habits and environment *(continued)*	
Safety is emphasized at the beginning of the term but is not mentioned later in the year.	Safety must be the first priority in all lab work. Students should be warned of risks and instructed in emergency procedures for each activity.
There is no assessment of students' knowledge and attitudes regarding safety.	Conduct frequent safety quizzes. Only students with perfect scores should be allowed to work in the lab.
You work alone during your preparation period to organize the day's labs.	Never work alone in a science laboratory or a storage area.
Safety inspections are conducted irregularly and are not documented. Teachers and administrators are unaware of what documentation will be necessary in case of a lawsuit.	Safety reviews should be frequent and regular. All reviews should be documented, and improvements must be implemented immediately. Contact legal counsel for your district to make sure your procedures will protect you in case of a lawsuit.
Purchasing, storing, and using chemicals	
The storeroom is too crowded, so you decide to keep some equipment on the lab benches.	Do not store reagents or equipment on lab benches. Keep shelves organized. Never place reactive chemicals (in bottles, beakers, flasks, wash bottles, etc.) near the edges of a lab bench.
You prepare solutions from concentrated stock to save money.	Reduce risks by ordering diluted instead of concentrated substances.
You purchase plenty of chemicals to be sure that you won't run out or to save money.	Purchase chemicals in class-size quantities. Do not purchase or have on hand more than one year's supply of each chemical.
You don't generally read labels on chemicals when preparing solutions for a lab, because you already know about a chemical.	Read each label to be sure it states the hazards and describes the precautions and first aid procedures (when appropriate) that apply to the contents in case someone else has to deal with that chemical in an emergency.
You never read the Material Safety Data Sheets (MSDSs) that come with your chemicals.	Always read the Material Safety Data Sheet (MSDS) for a chemical before using it. Follow the precautions described in the MSDS. File and organize MSDSs for all chemicals where they can be found easily in case of an emergency.
The main stockroom contains chemicals that haven't been used for years.	Do not leave bottles of chemicals unused on the shelves of the lab for more than one week or unused in the main stockroom for more than one year. Dispose of or use up any leftover chemicals.
No extra precautions are taken when flammable liquids are dispensed from their containers.	When transferring flammable liquids from bulk containers, ground the container; before transferring flammable liquids to a smaller metal container, ground both containers.
Students are told to put their broken glass and solid chemical wastes in the trash can.	Have separate containers for trash, for broken glass, and for different categories of hazardous chemical wastes.
You store chemicals alphabetically instead of by hazard class. Chemicals are stored without consideration of possible emergencies (fire, earthquake, flood, etc.), which could compound the hazard.	Use MSDSs to determine which chemicals are incompatible. Store chemicals by the hazard class indicated on the MSDS. Store chemicals that are incompatible with common fire-fighting media like water (such as alkali metals) or carbon dioxide (such as alkali and alkaline-earth metals) under conditions that eliminate the possibility of a reaction with water or carbon dioxide if it is necessary to fight a fire in the storage area.
Corrosives are kept above eye level, out of reach from any unauthorized person.	Always store corrosive chemicals on shelves below eye level. Remember, fumes from many corrosives can destroy metal cabinets and shelving.
Chemicals are kept on the stockroom floor on the days that they will be used so that they are easy to find.	Never store chemicals or other materials on floors or in the aisles of the laboratory or storeroom, even for a few minutes.

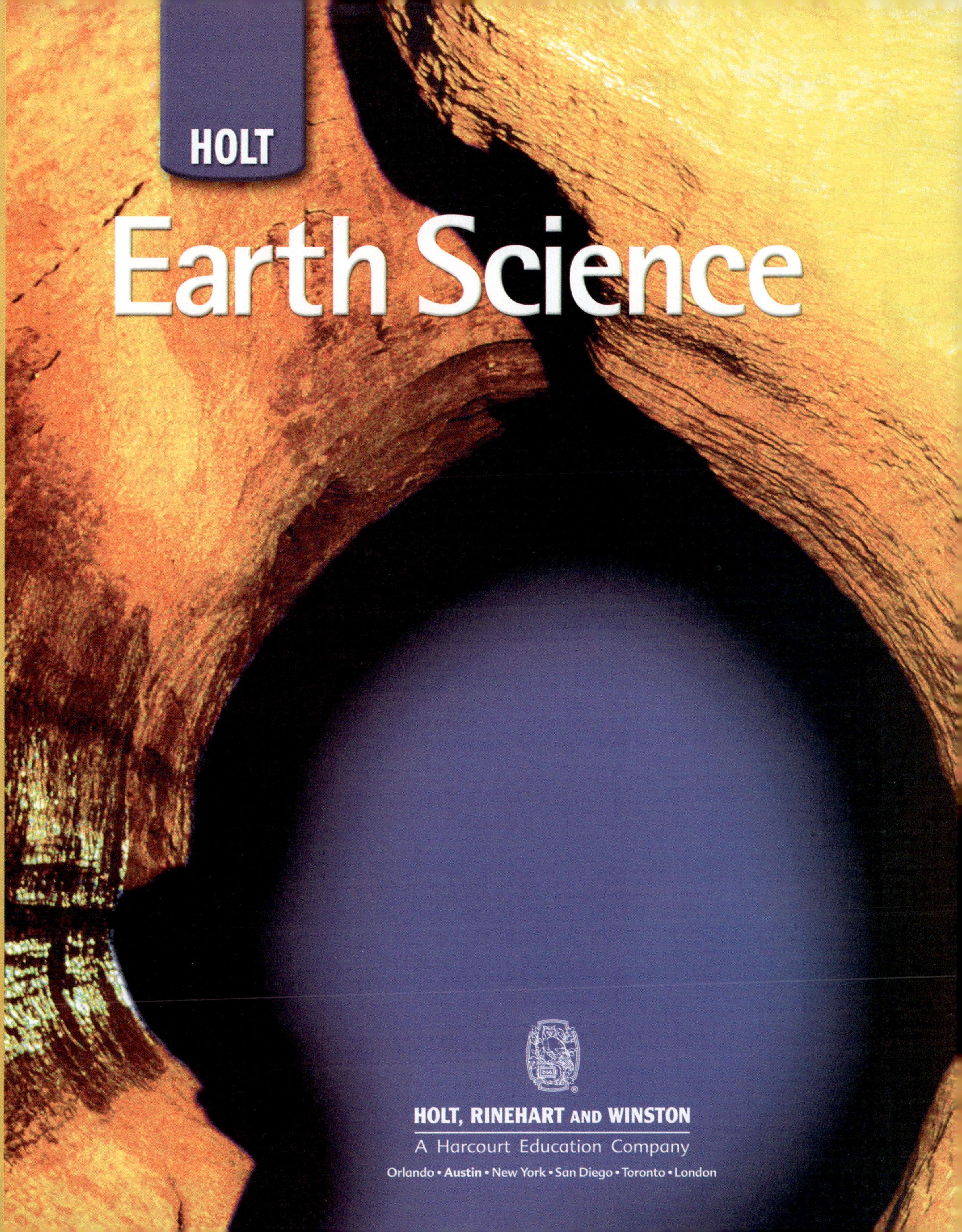

HOLT

Earth Science

HOLT, RINEHART AND WINSTON
A Harcourt Education Company
Orlando • **Austin** • New York • San Diego • Toronto • London

About the Authors

Mead A. Allison, Ph.D.
Tulane University, New Orleans, Louisiana

Mead Allison received his Ph.D. in oceanography from State University of New York. He is an associate professor of Earth and environmental sciences at Tulane University in Louisiana, where he teaches introductory geology and upper-level courses in sedimentology and marine geology.

Arthur T. DeGaetano, Ph.D.
Cornell University, Ithaca, New York

Arthur DeGaetano received his Ph.D. in meteorology from Rutgers University. He is an associate professor of Earth and atmospheric sciences at Cornell University in New York, where he teaches introductory climatology and upper-level courses in atmospheric thermodynamics and physical meteorology. Dr. DeGaetano is also the director of the Northeast Regional Climate Center.

Jay M. Pasachoff, Ph.D.
Williams College, Williamstown, Massachusetts

Jay Pasachoff received his Ph.D. in astronomy from Harvard University. He is the Field Memorial Professor of Astronomy and the director of the Hopkins Observatory at Williams College in Massachusetts, where he teaches introductory and upper-level courses in astronomy. In addition, Dr. Pasachoff has written several popular college-level astronomy textbooks.

Copyright © 2008 by Holt, Rinehart and Winston

All rights reserved. No part of this publication may be reproduced or transmitted in any form or by any means, electronic or mechanical, including photocopy, recording, or any information storage and retrieval system, without permission in writing from the publisher.

Requests for permission to make copies of any part of the work should be mailed to the following address: Permissions Department, Holt, Rinehart and Winston, 10801 N. MoPac Expressway, Building 3, Austin, Texas 78759.

ExamView is a registered trademark of FSCreations, Inc.

HOLT, ONE-STOP PLANNER and the **"Owl Design"** are trademarks licensed to Holt, Rinehart and Winston, registered in the United States of America and/or other jurisdictions.

LIVE INK is a registered trademark of Walker Reading Technologies, Inc., licensed for use by Holt, Rinehart and Winston.

NOVA® is a registered trademark of WGBH Educational Foundation.

SCILINKS is a registered trademark owned and provided by the National Science Teachers Association. All rights reserved.

Printed in the United States of America

If you have received these materials as examination copies free of charge, Holt, Rinehart and Winston retains title to the materials and they may not be resold. Resale of examination copies is strictly prohibited.

Possession of this publication in print format does not entitle users to convert this publication, or any portion of it, into electronic format.

ISBN 13: 978-0-0303-6698-7
ISBN 10: 0-03-036698-4
1 2 3 4 5 048 10 09 08 07 06

Acknowledgments

Authors

Mead A. Allison, Ph.D.
Associate Professor
Department of Earth and Environmental Sciences
Tulane University
New Orleans, Louisiana

Arthur T. DeGaetano, Ph.D.
Director, Northeast Regional Climate Center
Associate Professor
Department of Earth and Atmospheric Science
Cornell University
Ithaca, New York

Jay M. Pasachoff, Ph.D.
Director, Hopkins Observatory
Field Memorial Professor of Astronomy
Williams College
Williamstown, Massachusetts

Feature Development

Susan Feldkamp
Science Writer
Manchaca, Texas

Inclusion Specialist

Ellen McPeek Glisan
Special Needs Consultant
San Antonio, Texas

Academic Reviewers

Paul Asimow, Ph.D.
Assistant Professor of Geology and Geochemistry
Geological and Planetary Sciences
California Institute of Technology
Pasadena, California

Kenneth H. Brink, Ph.D.
Senior Scientist
Physical Oceanography Department
Woods Hole Oceanographic Institution
Woods Hole, Maine

John A. Brockhaus, Ph.D.
Professor of Geospatial Information Science
Geospatial Information Science Program
United States Military Academy
West Point, New York

Wesley N. Colley, Ph.D.
Professor of Astronomy
Department of Astronomy
University of Virginia
Portsmouth, Virginia

Terence H. Cooper, Ph.D.
Morse-Alumni Distinguished Teaching Professor
Professor of Soil and Environmental Science
Department of Soil, Water, and Climate
University of Minnesota
St. Paul, Minnesota

Roger J. Cuffey, Ph.D.
Professor of Paleontology
Department of Geosciences
Penn State University
University Park, Pennsylvania

Scott A. Darveau, Ph.D.
Associate Professor of Chemistry
Department of Chemistry
University of Nebraska at Kearney
Kearney, Nebraska

Daniela Di Iorio, Ph.D.
Assistant Professor
Department of Marine Sciences
University of Georgia
Athens, Georgia

Andrea Dobson, Ph.D.
Chair, Division of Basic Sciences and Mathematics
Associate Professor of Astronomy and General Studies
Department of Astronomy
Whitman College
Walla Walla, Washington

Turgay Ertekin, Ph.D.
Professor and Chair of Petroleum and Natural Gas Engineering
Petroleum and Natural Gas Engineering Program
Penn State University
University Park, Pennsylvania

Sandi Gardner, Ph.D.
Department Chair, Sciences
School of Arts and Sciences
Triton College
River Grove, Illinois

Tracey Graham, Ph.D.
Consultant and Instructor in Earth Science
Emporia, Kansas

Deborah Hanley, Ph.D.
Meteorologist
Florida Division of Forestry
Department of Agriculture and Consumer Services
Tallahassee, Florida

Steven Jennings, Ph.D.
Associate Professor of Geography
Geography and Environmental Studies
University of Colorado at Colorado Springs
Colorado Springs, Colorado

Christopher M. Johns-Krull, Ph.D.
Assistant Professor
Department of Physics and Astronomy
Rice University
Houston, Texas

Joel S. Leventhal, Ph.D.
Emeritus Scientist, Geochemistry
U.S. Geological Survey
Denver, Colorado

Mark Moldwin, Ph.D.
Associate Professor of Space Physics
Earth and Space Sciences
University of California, Los Angeles
Los Angeles, California

Sten Odenwald, Ph.D.
Astronomer
Astronomy and Space Physics
NASA-Goddard Space Flight Center
Greenbelt, Maryland

Henry W. Robinson, Ph.D.
Meteorologist
Office of Services
National Weather Service
Silver Spring, Maryland

Kenneth H. Rubin, Ph.D.
Associate Professor
Geology and Geophysics
University of Hawaii
Manoa, Hawaii

Dork Sahagian, Ph.D.
Professor and Director of Environmental Initiative
Lehigh University
Bethlehem, Pennsylvania

Daniel Z. Sui, Ph.D.
Professor of Geography and Holder of the Reta A. Haynes Endowed Chair in Geosciences
Department of Geography
Texas A&M University
College Station, Texas

Vatche P. Tchakerian, Ph.D.
Professor
Geosciences
Texas A&M University
College Station, Texas

Iris Moreno Totten, Ph.D.
Assistant Professor
Department of Geology
Kansas State University
Manhattan, Kansas

Dale E. Wheeler, Ph.D.
Associate Professor of Chemistry
A. R. Smith Department of Chemistry
Appalachian State University
Boone, North Carolina

Teacher Reviewers

Lowell S. Bailey
Earth Science Educator
Bedford-North Lawrence High School
Bedford, Indiana

Shawn Beightol, M.S. Ed.
Chemistry Teacher
Michael Krop Senior High School
Miami, Florida

David Blinn
Science Teacher
Wrenshall High School
Wrenshall, Minnesota

Daniel Brownstein, MAT/MA
Science Department Chair
Hastings High School
Hastings, New York

Glenn Dolphin
Earth Science Teacher
Union-Endicott High School
Endicott, New York

Alexander Dvorak
Science Teacher
Heritage School
New York, New York

Jeanne Endrikat
Science Chair
Lake Braddock Secondary School
Burke, Virginia

Anthony P. LaSalvia
Curriculum Coordinator and Science Teacher
New Lebanon Central Schools
New Lebanon, New York

Keith A. McKain
Earth Science Teacher
Colonel Richardson High School
Federalsburg, Maryland

Marie E. McKay
Earth and Environmental Science Teacher
Ashbrook High School
Gastonia, North Carolina

Mike McKee
Science Department Chair
Cypress Creek High School
Orlando, Florida

Christine V. McLelland
Distinguished Earth Science Educator in Residence
Education and Outreach
Geological Society of America
Boulder, Colorado

Tammie Niffenegger
Science Chair and Teacher
Port Washington High School
Port Washington, Wisconsin

Scott Robertson
Earth Science and Physics Teacher
North Warren Central School District
Chestertown, New York

Gabrielle Schavran
Earth Science and Chemistry Teacher
Syosset High School
Syosset, New York

Sharon M. Stroud
Earth Science Teacher and Consultant
Colorado Springs, Colorado

Teresa Tucker
Science Teacher
Northwest Community Schools
Jackson, Michigan

Lab Testing

Shawn Beightol, M.S. Ed.
Chemistry Teacher
Michael Krop Senior High School
Miami, Florida

Daniel Brownstein, MAT/MA
Science Department Chair
Hastings High School
Hastings, New York

Eric Cohen
Science Educator
Westhampton Beach High School
Westhampton Beach, New York

Alonda Droege
Science Teacher
Highline High School
Burien, Washington

Alexander Dvorak
Science Teacher
Heritage School
New York, New York

Randa Flinn, M.S.
Teacher and Science Curriculum Facilitator
Northeast High School
Broward County, Florida

Erich Landstrom
NASA SEU Educator Ambassador
Boynton Beach Community High School
Boynton Beach, Florida

Tammie Niffenegger
Science Chair and Teacher
Port Washington High School
Port Washington, Wisconsin

Scott Robertson
Earth Science and Physics Teacher
North Warren Central School District
Chestertown, New York

Teresa Tucker
Science Teacher
Northwest Community Schools
Jackson, Michigan

Standardized Test Prep Reviewers

Russell Agostaro
Program Development Specialist
Office of Funded Programs, Newburgh Enlarged City School District
Newburgh, New York

continued on page 951

CONTENTS IN BRIEF

CONTENTS

STUDYING THE EARTH — Unit 1

Chapter 1

Chapter 2

Chapter 3

Environmental Connection

COMPOSITION OF THE EARTH

Unit 2

HISTORY OF THE EARTH Unit 3

Environmental Connection

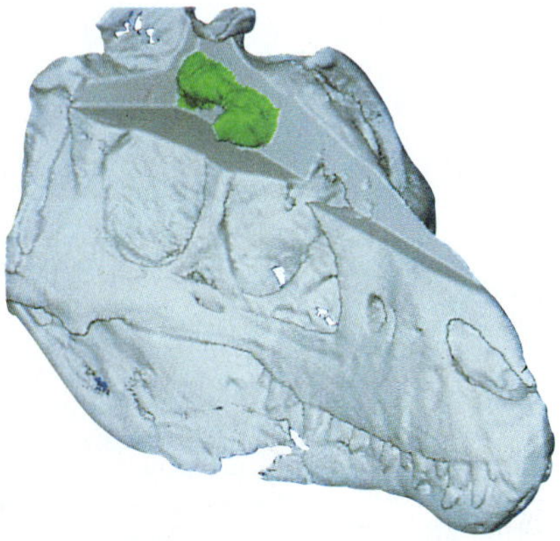

THE DYNAMIC EARTH

Unit 4

Chapter 10

Chapter 11

Chapter 12

Chapter 13

RESHAPING THE CRUST

Unit 5

Chapter 14

Chapter 15

Chapter 16

Environmental Connection

Chapter 17

Chapter 18

OCEANS Unit 6

Chapter 19

Environmental Connection

Chapter 23

Chapter 24

Chapter 25

Chapter 26

Chapter 27

Chapter 28

Environmental Connection

APPENDIX

CHAPTER LABS

Inquiry Labs

Skills Practice Labs

Making Models Labs

QUICK LABS

MAPS IN ACTION

FEATURE ARTICLES

CAREER FOCUS

SCIENCE AND TECHNOLOGY

IMPACT on Society

EYE on the Environment

Lab Safety

In the laboratory, you can engage in hands-on explorations, test your scientific hypotheses, and build practical lab skills. However, while you are working in the lab or in the field, it is your responsibility to protect yourself and your classmates by conducting yourself in a safe manner. You will avoid accidents in the lab by following directions, handling materials carefully, and taking your work seriously. Read the following safety guidelines before working in the lab. Make sure that you understand all guidelines before entering the lab.

Before You Begin

- **Read the entire activity before entering the lab.** Be familiar with the instructions before beginning an activity. Do not start an activity until you have asked your teacher to explain any parts of the activity that you do not understand.
- **Student-designed procedures or inquiry activities must be approved by your teacher before you attempt the procedures or activities.**
- **Wear the right clothing for lab work.** Before beginning work, tie back long hair, roll up loose sleeves, and put on any required personal protective equipment as directed by your teacher. Remove your wristwatch and any necklaces or jewelry that could get caught in moving parts. Avoid or confine loose clothing. Do not wear open-toed shoes, sandals, or canvas shoes.
- **Do not wear contact lenses in the lab.** Even though you will be wearing safety goggles, chemicals could get between contact lenses and your eyes and could cause irreparable eye damage. If your doctor requires that you wear contact lenses instead of glasses, then you should wear eye-cup safety goggles—similar to goggles worn for underwater swimming—in the lab. Ask your doctor or your teacher how to use eye-cup safety goggles to protect your eyes.
- **Know the location of all safety and emergency equipment used in the lab.** Know proper fire-drill procedures and the location of all fire xits. Ask your teacher where the nearest eyewash stations, safety blankets, safety shower, fire extinguisher, first-aid kit, and chemical spill kit are located. Be sure that you know how to operate the equipment safely.

While You Are Working

- **Always wear a lab apron and safety goggles.** Wear these items even if you are not working on an activity. Labs contain chemicals that can damage your clothing, skin, and eyes. If your safety goggles cloud up or are uncomfortable, ask your teacher for help.
- **NEVER work alone in the lab.** Work in the lab only when supervised by your teacher. Do not leave equipment unattended while it is in operation.
- **Perform only activities specifically assigned by your teacher.** Do not attempt any procedure without your teacher's direction. Use only materials and equipment listed in the activity or authorized by your teacher. Steps in a procedure should be performed only as described in the activity or as approved by your teacher.
- **Keep your work area neat and uncluttered.** Have only books and other materials that are needed to conduct the activity in the lab. Keep backpacks, purses, and other items in your desk, locker, or other designated storage areas.
- **Always heed safety symbols and cautions listed in activities, listed on handouts, posted in the room, provided on chemical labels, and given verbally by your teacher.** Be aware of the potential hazards of the required materials and procedures, and follow all precautions indicated.

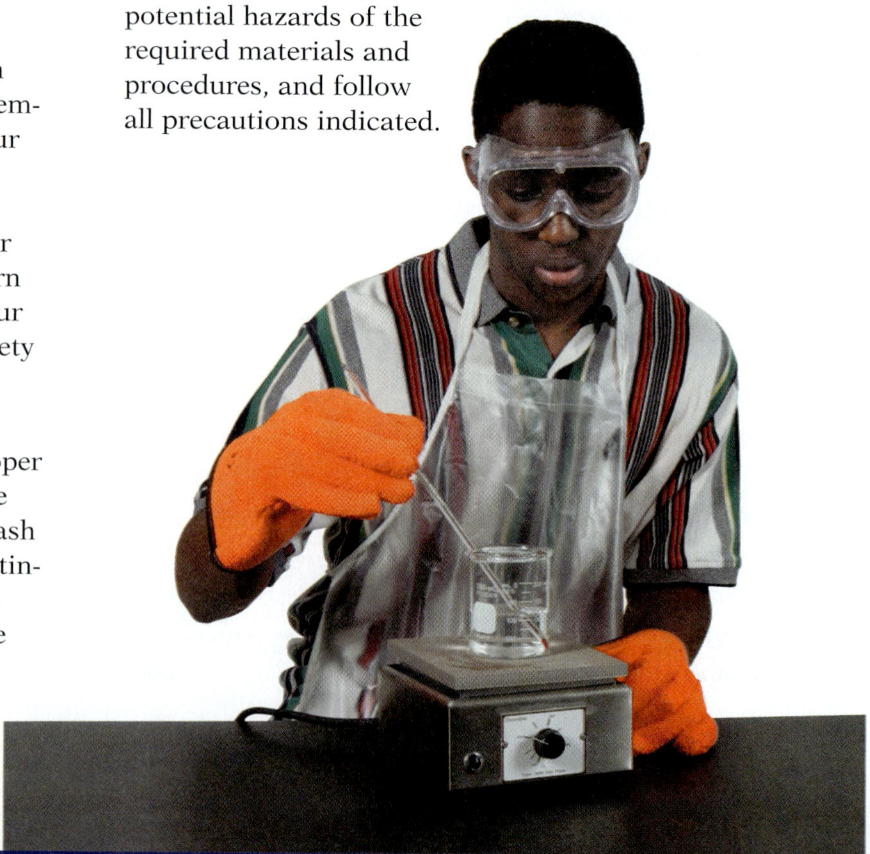

- **Be alert, and walk with care in the lab.** Be aware of others near you and your equipment.
- **Do not take food, drinks, chewing gum, or tobacco products into the lab.** Do not store or eat food in the lab.
- **NEVER taste chemicals or allow them to contact your skin.** Keep your hands away from your face and mouth, even if you are wearing gloves.
- **Exercise caution when working with electrical equipment.** Do not use electrical equipment with frayed or twisted wires. Be sure that your hands are dry before using electrical equipment. Do not let electrical cords dangle from work stations. Dangling cords can cause you to trip and can cause an electrical shock.
- **Use extreme caution when working with hot plates and other heating devices.** Keep your head, hands, hair, and clothing away from the flame or heating area. Remember that metal surfaces connected to the heated area will become hot by conduction. Gas burners should be lit only with a spark lighter, not with matches. Make sure that all heating devices and gas valves are turned off before you leave the lab. Never leave a heating device unattended when it is in use. Metal, ceramic, and glass items do not necessarily look hot when they are hot. Allow all items to cool before storing them.
- **Do not fool around in the lab.** Take your lab work seriously, and behave appropriately in the lab. Lab equipment and apparatus are not toys; never use lab time or equipment for anything other than the intended purpose. Be aware of the safety of your classmates as well as your safety at all times.

Emergency Procedures

- **Follow standard fire-safety procedures.** If your clothing catches on fire, do not run; WALK to the safety shower, stand under it, and turn it on. While doing so, call to your teacher. In case of fire, alert your teacher and leave the lab.
- **Report any accident, incident, or hazard—no matter how trivial—to your teacher immediately.** Any incident involving bleeding, burns, fainting, nausea, dizziness, chemical exposure, or ingestion should also be reported immediately to the school nurse or to a physician. If you have a close call, tell your teacher so that you and your teacher can find a way to prevent it from happening again.

- **Report all spills to your teacher immediately.** Call your teacher rather than trying to clean a spill yourself. Your teacher will tell you whether it is safe for you to clean up the spill; if it is not safe, your teacher will know how to clean up the spill.
- **If you spill a chemical on your skin, wash the chemical off in the sink and call your teacher.** If you spill a solid chemical onto your clothing, brush it off carefully without scattering it onto somebody else and call your teacher. If you get liquid on your clothing, wash it off right away by using the faucet at the sink and call your teacher.

When You Are Finished

- **Clean your work area at the conclusion of each lab period as directed by your teacher.** Broken glass, chemicals, and other waste products should be disposed of in separate, special containers. Dispose of waste materials as directed by your teacher. Put away all material and equipment according to your teacher's instructions. Report any damaged or missing equipment or materials to your teacher.
- **Wash your hands with soap and hot water after each lab period.** To avoid contamination, wash your hands at the conclusion of each lab period, and before you leaving the lab.

Safety Symbols

Before you begin working in the lab, familiarize yourself with the following safety symbols, which are used throughout your textbook, and the guidelines that you should follow when you see these symbols.

Eye Protection

- **Wear approved safety goggles as directed.** Safety goggles should be worn in the lab at all times, especially when you are working with a chemical or solution, a heat source, or a mechanical device.
- **If chemicals get into your eyes, flush your eyes immediately.** Go to an eyewash station immediately, and flush your eyes (including under the eyelids) with running water for at least 15 minutes. Use your thumb and fingers to hold your eyelids open and roll your eyeball around. While doing so, ask another student to notify your teacher.
- **Do not wear contact lenses in the lab.** Chemicals can be drawn up under a contact lens and into the eye. If you must wear contacts prescribed by a physician, tell your teacher. In this case, you must also wear approved eye-cup safety goggles to help protect your eyes.
- **Do not look directly at the sun or any light source through any optical device or lens system, and do not reflect direct sunlight to illuminate a microscope.** Such actions concentrate light rays to an intensity that can severely burn your retinas, which may cause blindness.

Clothing Protection

- **Wear an apron or lab coat at all times in the lab to prevent chemicals or chemical solutions from contacting skin or clothes.**
- **Tie back long hair, secure loose clothing, and remove loose jewelry so that they do not knock over equipment, get caught in moving parts, or come into contact with hazardous materials.**

Hand Safety

- **Do not cut an object while holding the object in your hand.** Dissect specimens in a dissecting tray.
- **Wear protective gloves when working with an open flame, chemicals, solutions, or wild or unknown plants.**
- **Use a hot mitt to handle resistors, light sources, and other equipment that may be hot.** Allow all equipment to cool before storing it.

Hygienic Care

- **Keep your hands away from your face and mouth while you are working in the lab.**
- **Wash your hands thoroughly before you leave the lab.**
- **Remove contaminated clothing immediately.** If you spill caustic substances on your skin or clothing, use the safety shower or a faucet to rinse. Remove affected clothing while you are under the shower, and call to your teacher. (It may be temporarily embarrassing to remove clothing in front of your classmates, but failure to rinse a chemical off your skin could result in permanent damage.)
- **Launder contaminated clothing separately.**

Sharp-Object Safety

- **Use extreme care when handling all sharp and pointed instruments, such as scalpels, sharp probes, and knives.**
- **Do not cut an object while holding the object in your hand.** Cut objects on a suitable work surface. Always cut in a direction away from your body.
- **Do not use double-edged razor blades in the lab.**

Glassware Safety

- **Inspect glassware before use; do not use chipped or cracked glassware.** Use heat-resistant glassware for heating materials or storing hot liquids, and use tongs or a hot mitt to handle this equipment.
- **Do not attempt to insert glass tubing into a rubber stopper without specific instructions from your teacher.**
- **Notify immediately your teacher if a piece of glassware or a light bulb breaks.** Do not attempt to clean up broken glass unless your teacher directs you to do so.

Electrical Safety

- **Do not use equipment with frayed electrical cords or loose plugs.**
- **Fasten electrical cords to work surfaces by using tape.** Doing so will prevent tripping and will ensure that equipment will not fall off the table.
- **Do not use electrical equipment near water or when your clothing or hands are wet.**
- **Hold the rubber cord when you plug in or unplug equipment.** Do not touch the metal prongs of the plug, and do not unplug equipment by pulling on the cord.
- **Wire coils on hot plates may heat up rapidly.** If heating occurs, open the switch immediately and use a hot mitt to handle the equipment.

Heating Safety

- **Be aware of any source of flames, sparks, or heat (such as open flames, electric heating coils, or hot plates) before working with flammable liquids or gases.**
- **Avoid using open flames.** If possible, work only with hot plates that have an on/off switch and an indicator light. Do not leave hot plates unattended. Do not use alcohol lamps. Turn off hot plates and open flames when they are not in use.
- **Never leave a hot plate unattended while it is turned on or while it is cooling off.**
- **Know the location of lab fire extinguishers and fire-safety blankets.**
- **Use tongs or appropriate insulated holders when handling heated objects.** Heated objects often do not appear to be hot. Do not pick up an object with your hand if it could be warm.
- **Keep flammable substances away from heat, flames, and other ignition sources.**
- **Allow all equipment to cool before storing it.**

Fire Safety

- **Know the location of lab fire extinguishers and fire-safety blankets.**
- **Know your school's fire-evacuation routes.**
- **If your clothing catches on fire, walk (do not run) to the emergency lab shower to put out the fire. If the shower is not working, STOP, DROP, and ROLL!** Smother the fire by stopping immediately, dropping to the floor, and rolling until the fire is out.

Safety with Gases

- **Do not inhale any gas or vapor unless directed to do so by your teacher.** Never inhale pure gases.
- **Handle materials that emit vapors or gases in a well-ventilated area.** This work should be done in an approved chemical fume hood.

Caustic Substances

- **If a chemical gets on your skin, on your clothing, or in your eyes, rinse it immediately and alert your teacher.**
- **If a chemical is spilled on the floor or lab bench, alert your teacher, but do not clean it up yourself unless your teacher directs you to do so.**

How to Use Your Textbook

Your Roadmap for Success with *Holt Earth Science*

Chapter 10 Plate Tectonics

Sections
1 Continental Drift
2 The Theory of Plate Tectonics
3 The Changing Continents

What You'll Learn
• How scientists developed the theory of plate tectonics
• Why tectonic plates move
• How Earth's geography has changed

Why It's Relevant
Understanding why and how tectonic plates move provides a basis for understanding other concepts of Earth science.

PRE-READING ACTIVITY
FOLDNOTES
Tri-Fold
Before you read this chapter, create the FoldNote entitled "TriFold" described in the Skills Handbook section of the Appendix. Write what you know about plate tectonics in the column labeled "Know." Then, write what you want to know in the column labeled "Want." As you read the chapter, write what you learn in the column labeled "Learn."

▶ The island of Iceland is being torn into two pieces as two tectonic plates pull apart. Iceland is one of only a few places on Earth where this process can be seen on land.

238

Section 2 The Theory of Plate Tectonics

By the 1960s, evidence supporting continental drift and sea-floor spreading led to the development of a theory called *plate tectonics*. **Plate tectonics** is the theory that explains why and how continents move and is the study of the formation of features in Earth's crust.

How Continents Move

Earth's crust and the rigid, upper part of the mantle form a layer of Earth called the **lithosphere.** The lithosphere forms the thin outer shell of Earth. It is broken into several blocks, called *tectonic plates,* that ride on a deformable layer of the mantle called the *asthenosphere* in much the same way that blocks of wood float on water. The **asthenosphere** (as THEN uh sfir) is a layer of "plastic" rock just below the lithosphere. Plastic rock is solid rock that is under great pressure and that flows very slowly, like putty does. **Figure 1** shows what tectonic plates may look like.

Earth's crust is classified into two types—*oceanic crust* and *continental crust.* Oceanic crust is dense and is made of rock that is rich in iron and magnesium. Continental crust has a low density and is made of rock that is rich in silica. Tectonic plates can include continental crust, oceanic crust, or both. The continents and oceans are carried along on the moving tectonic plates in the same way that passengers are carried by a bus.

OBJECTIVES
▶ **Summarize** the theory of plate tectonics.
▶ **Identify** and describe the three types of plate boundaries.
▶ **List** and describe three causes of plate movement.

KEY TERMS
plate tectonics
lithosphere
asthenosphere
divergent boundary
convergent boundary
transform boundary

plate tectonics the theory that explains how large pieces of the lithosphere, called *plates,* move and change shape

lithosphere the solid, outer layer of Earth that consists of the crust and the rigid upper part of the mantle

asthenosphere the solid, plastic layer of the mantle beneath the lithosphere; made of mantle rock that flows very slowly, which allows tectonic plates to move on top of it

Figure 1 ▶ Tectonic plates fit together on Earth's surface like three-dimensional puzzle pieces.

Section 2 The Theory of Plate Tectonics 247

Get Organized

Do the **Pre-Reading Activity** at the beginning of each chapter to create a **FoldNote,** which is a helpful note-taking and study aid. Use the **Graphic Organizer** activity within the chapter to organize the chapter content in a way that you understand.

STUDY TIP Go to the Skills Handbook section of the Appendix for guidance on making FoldNotes and Graphic Organizers.

Read for Meaning

Read the **Objectives** at the beginning of each section, because they will tell you what you'll need to learn. **Key Terms** are also listed for each section. Each key term is highlighted in the text and is defined in the margin. After reading each chapter, turn to the **Chapter Highlights** page and review the **Key Concepts,** which are brief summaries of the chapter's main ideas. You may want to do this even before you read the chapter.

STUDY TIP If you don't understand a definition, reread the page on which the term is introduced. The surrounding text should help make the definition easier to understand.

Be Resourceful—Use the Web

SciLinks boxes in your textbook take you to resources that you can use for science projects, reports, and research papers. Go to **scilinks.org** and type in the **SciLinks code** to find information on a topic.

Visit go.hrw.com
Find resources and reference materials that go with your textbook at **go.hrw.com.** Enter the keyword **HQ6 Home** to access the home page for your textbook.

Consider Environmental Issues

As you read each section, look for passages marked with oak leaves—one at the beginning of the passage and one at the end. These passages contain information that is directly related to important environmental issues.

Prepare for Tests

Section Reviews and **Chapter Reviews** test your knowledge of the main points of the chapter. Critical Thinking items challenge you to think about the material in different ways and in greater depth. The **standardized test prep** that is located after each Chapter Review helps you sharpen your test-taking abilities.

STUDY TIP Reread the Objectives and Chapter Highlights when studying for a test to be sure you know the material.

Use the Appendix

Your **Appendix** contains a variety of resources designed to enhance your learning experience. These resources include the **Skills Handbook,** which provides helpful study aids, **Mapping Expeditions** and **Long-Term Projects** activities, and **Reference Tables** and **Reference Maps.**

Figure 6 ▶ This stone lion sits outside Leeds Town Hall in England. It was damaged by acid precipitation, which fell regularly in Europe and North America for more than 50 years.

Acid Precipitation

Rainwater is slightly acidic because it combines with small amounts of carbon dioxide. But when fossil fuels, especially coal, are burned, nitrogen oxides and sulfur dioxides are released into the air. These compounds combine with water in the atmosphere to produce nitric acid, nitrous acid, or sulfuric acid. When these acids fall to Earth, they are called **acid precipitation.**

Acid precipitation weathers rock faster than ordinary precipitation does. In fact, many historical monuments and sculptures have been damaged by acid precipitation, as shown in **Figure 6.** Between 1940 and 1990, acid precipitation fell regularly in some cities in the United States. In 1990, the Acid Rain Control Program was added to the Clean Air Act of 1970. These regulations gave power plants 10 years to decrease sulfur dioxide emissions. The occurrence of acid precipitation has been greatly reduced since power plants have installed scrubbers that remove much of the sulfur dioxide before it can be released.

acid precipitation precipitation, such as rain, sleet, or snow, that contains a high concentration of acids, often because of the pollution of the atmosphere

Section 1 Review

1. **Identify** three agents of mechanical weathering.
2. **Describe** how ice wedging weathers rock.
3. **Explain** how two activities of plants or animals help weather rocks or soil.
4. **Compare** mechanical and chemical weathering processes.
5. **Identify** and describe three chemical processes that weather rock.
6. **Compare** hydrolysis, carbonation, and oxidation.
7. **Summarize** how acid precipitation forms.

CRITICAL THINKING

8. **Making Connections** What two agents of weathering would be rare in a desert? Explain your reasoning.
9. **Understanding Relationships** Automobile exhaust contains nitrogen oxides. How might these pollutants affect chemical weathering processes?

CONCEPT MAPPING

10. Use the following terms to create a concept map: *weathering, oxidation, mechanical weathering, ice wedging, hydrolysis, abrasion, chemical weathering, carbonation,* and *acid precipitation.*

348 Chapter 14 Weathering and Erosion

Visit Holt Online Learning

Holt Online Learning

If your teacher gives you a special password to log onto the **Holt Online Learning** site, you'll find your complete textbook on the Web. In addition, you'll find some great learning tools and practice quizzes. You'll be able to see how well you know the material from your textbook.

Unit 1 STUDYING THE EARTH

Unit 1 Outline

▶ Earth scientists can be found everywhere on Earth, braving conditions from the extreme heat of an active volcano to the frozen depths of a glacier. The Muir Glacier at Alaska's Glacier Bay National Park can be explored and studied through this ice cave.

Chapter 1 Introduction to Earth Science

Planning Guide

Compression Guide

To shorten instruction because of time limitations, omit the Chapter Lab.

OBJECTIVES	LABS, DEMONSTRATIONS, AND ACTIVITIES	TECHNOLOGY RESOURCES
PACING • 45 min pp. 4–8 **Chapter Opener**		**OSP Parent Letter** **CD Student Edition on CD-ROM** **CD Chapter Summaries Audio CD** **VID Brain Food Video Quiz**
Section 1 What Is Earth Science? • Describe two cultures that contributed to modern scientific study. • Name the four main branches of Earth science. • Discuss how Earth scientists help us understand the world around us.	TE **Discussion** Creating Calendars, p. 5 GENERAL TE **Activity** Not Your Typical Office Job, p. 6 GENERAL TE **Group Activity** Scientific Revolutions, p. 7 GENERAL TE **Group Activity** Field Geology of Your Area, p. 25 GENERAL	OSP **Lesson Plans** (also in print) TR **Bellringer*** TE **Internet Activity** Ice Cores and Climate, p. 6 ADVANCED CRF **Internet Activity** Ice Cores and Climate* ADVANCED VID **HRW Earth Science Video** Introduction to Earth Science VID **HRW Earth Science Video** Careers in Earth Science CD **Interactive Tutor** History and Future of Earth Sciences
PACING • 90 min pp. 9–16 **Section 2 Science as a Process** • Explain how science is different from other forms of human endeavor. • Identify the steps that make up scientific methods. • Analyze how scientific thought changes as new information is collected. • Explain how science affects society.	TE **Discussion** Influences on Natural Systems, p. 9 GENERAL TE **Discussion** Types of Reasoning, p. 10 GENERAL SE **Quick Lab** Making Observations, p. 11 GENERAL CRF **Datasheet for Quick Lab*** GENERAL TE **Discussion** Variables and Controls, p. 11 GENERAL SE **Quick Lab** Sample Size and Accuracy, p. 12 GENERAL CRF **Datasheet for Quick Lab*** GENERAL TE **Activity** Precision and Accuracy, p. 12 BASIC TE **Discussion** Peer-Review Pressure, p. 14 GENERAL TE **Debate** Should Science Always Be Applied?, p. 15 ADVANCED SE **Inquiry Lab** Scientific Methods, pp. 22–23 GENERAL CRF **Datasheet for Chapter Lab*** GENERAL SE **Maps in Action** Geologic Features and Political Boundaries in Europe, p. 24 GENERAL TE **Group Activity** Making Maps, p. 24 GENERAL CRF **Inquiry Lab** What's Before Your Eyes?* GENERAL CRF **Skills Practice Lab** Testing a Prediction* GENERAL	OSP **Lesson Plans** (also in print) TR **Bellringer*** TR 1 Scientific Methods* TR 2 The Importance of Interdisciplinary Science* TR 3 Geologic Features and Political Boundaries in Europe* TE **Internet Activity** Limitations of Information Systems, p. 15 GENERAL CRF **Internet Activity** Limitations of Information Systems* GENERAL VID **NOVA Video** Adrift on the Gulf Stream VID **NOVA Video** Earthquake VID **NOVA Video** Runaway Universe

PACING • 90 min

CHAPTER REVIEW, ASSESSMENT, AND STANDARDIZED TEST PREPARATION

SE **Chapter Highlights,** p. 17
SE **Chapter Review,** pp. 18–19
SE **Standardized Test Prep,** pp. 20–21
CRF **Concept Review*** GENERAL
CRF **Critical Thinking*** ADVANCED
CRF **Math Skills*** GENERAL
CRF **Graphing Skills*** GENERAL
CRF **Chapter Test A*** GENERAL
CRF **Chapter Test B*** ADVANCED
OSP **Lesson Plans** (also in print)
OSP **Test Generator**
OSP **Test Item Listing**

Online and Technology Resources

go.hrw.com Holt Online Learning

Visit **go.hrw.com** for access to Holt Online Learning, or enter the keyword **HQ6 Home** for a variety of free online resources.

One-Stop Planner® CD-ROM

This CD-ROM package includes
- Lab Materials QuickList Software
- Holt Calendar Planner
- Customizable Lesson Plans
- Printable Worksheets
- ExamView® Test Generator
- Interactive Teacher Edition
- Holt PuzzlePro®
- Holt PowerPoint® Resources

KEY

SE	Student Edition	OSP	One-Stop Planner	VID	Classroom Video/DVD
TE	Teacher Edition	TR	Transparencies and Transparency Worksheets	*	Also on One-Stop Planner
CRF	Chapter Resource File	CD	CD or CD-ROM	◆	Requires advance prep
LTP	Long-Term Projects			■	Also available in Spanish

SKILLS DEVELOPMENT RESOURCES	REVIEW AND ASSESSMENT	CORRELATIONS
SE **Pre-Reading Activity,** p. 4 GENERAL TE **Using the Figure** Racetrack Playa, p. 4 GENERAL		National Science Education Standards
CRF **Directed Reading*** BASIC TE **Using the Figure** Doing Science, p. 6 ADVANCED TE **Reading Skill Builder** Reading Organizer, p. 6 BASIC TE **Inclusion Strategies,** p. 7 TE **Skill Builder** Vocabulary, p. 7 GENERAL	SE **Reading Check,** p. 7 GENERAL SE **Section Review,** p. 8 GENERAL TE **Reteaching,** p. 8 BASIC TE **Quiz,** p. 8 GENERAL TE **Alternative Assessment,** p. 8 GENERAL CRF **Section Quiz*** ■ GENERAL	HNS 3c, SAI 2a, SAI 2b, SAI 2c, SAI 2f, SPSP 6a, SPSP 6d, ST 2a, ST 2b, ST 2d, UCP 1
CRF **Directed Reading*** BASIC TE **Using the Figure** Scientific Methods, p. 10 GENERAL TE **Inclusion Strategies,** p. 10 TE **Reading Skill Builder** Reading Organizer, p. 11 BASIC TE **Skill Builder** Math, p. 12 GENERAL SE **Math Practice,** p. 13 GENERAL TE **Using the Figure** Levels of Analysis, p. 13 BASIC SE **Graphic Organizer** Chain-of-Events Chart, p. 14 GENERAL	SE **Reading Checks,** pp. 10, 13, 14 GENERAL SE **Section Review,** p. 16 GENERAL TE **Reteaching,** p. 15 BASIC TE **Quiz,** p. 15 GENERAL TE **Alternative Assessment,** p. 16 GENERAL CRF **Section Quiz*** ■ GENERAL	SAI 2a, SAI 2b, SAI 2d, SAI 2e, SAI 2f, SPSP 6a, SPSP 6d, ST 2a, ST 2b, ST 2d, UCP 1

Holt Lab Generator CD-ROM

Search for labs by topic, standard, difficulty level, and/or time frame. Edit those labs to fit your needs, or create your own labs. Use the Lab Materials QuickList to order just what you need.

Holt Earth Science Interactive Tutor CD-ROM

This CD-ROM consists of interactive activities that give students a fun way to extend their knowledge of Earth science concepts.

Chapter Summaries Audio CDs

These CDs include audio summaries of the key concepts presented in each chapter. (Audio summaries are also available in Spanish.)

SCILINKS. NSTA

www.scilinks.org

Maintained by the **National Science Teachers Association.** See Chapter Enrichment pages that follow for a complete list of topics.

See Chapter Enrichment pages for Video Resources.

Chapter 1 Chapter Enrichment

This Chapter Enrichment provides relevant and interesting information to expand and enhance your classroom instruction of the chapter material.

Section 1 What Is Earth Science?

The Plate Tectonics Revolution

In the 1950s, scientists with the Lamont Geological Observatory (now called the Lamont-Doherty Earth Observatory) were analyzing data from the Precision Depth Recorder, a device that revealed information about the topography of the ocean floor. In 1959, these researchers published their revolutionary findings. Not only did the ocean cover great mountain chains, but at the ridges of some undersea mountains, the crust of Earth was pulling apart and new crust was forming from mantle material. The Lamont geologists had found a possible mechanism to explain Alfred Wegener's idea of continental drift.

▶ **Observations, such as those made at El Caracol, are the basis of all science.**

The question that naturally arose was, "If new crust is forming at mid-ocean ridges, why isn't the planet getting larger?" The answer was suggested by the brilliant oceanographer Harry Hess. Hess analyzed and thought deeply about the Lamont research. In 1962, Hess published *History of Ocean Basins,* which caused a revolution in the way people think about the planet they live on. In his book, Hess explained his ideas about the process of *sea-floor spreading:* that as mantle material emerged at mid-ocean rifts, it pushed older, adjacent crust away from the ridge and—very slowly—across the ocean floor. Hess hypothesized that sea-floor spreading was the agent that moved Earth's tectonic plates, which "float" on the mantle beneath them. He also thought that as Earth's plates nudged closer to the borders of continents, they collided with continental plates. In some of these collisions, the oceanic plates subducted, or sank, under the continental plates and down into the mantle. Here was crust disappearing! Hess's work suggested that Earth's crust and mantle were constantly being recycled as tectonic plates moved. In the 1960s, evidence from the study of the magnetic properties of sea-floor rocks would help confirm Hess's revolutionary ideas.

Section 2 Science as a Process

Technology and Interdisciplinary Science: Deep-Ocean Vents

In 1977, three scientists descended to the bottom of the Galapagos Rift in the Pacific Ocean—7,500 feet below the surface. These geologists were able to reach such an enormous depth because of the invention and continual improvement of submersibles, or small vehicles designed to withstand the tremendous pressures of the deep ocean. The scientists were on board *Alvin,* the U.S.'s flagship submersible. Without *Alvin,* they would never have made one of the most startling discoveries of the twentieth century.

The scientists were picking up a deep-sea rock using *Alvin's* robotic arm when they saw the water above a nearby ridge rising and shimmering. Dropping the rock to explore the unusual shimmering, the scientists came upon a scene no one on Earth had ever seen before. They saw that the rising water was shimmering because it was actually "boiling" out of a vent in ocean crust (though all other deep-ocean water is extremely cold). They were stunned by the amazing array of deep-sea animals surrounding the vent. Some of these animals—like the giant tubeworm—exist nowhere else on Earth. The scientists had discovered a completely new type of ecosystem: the deep-sea vent ecosystem. Investigation of the vent ecosystem involved scientists from numerous fields, including geology, geochemistry, geophysics, biology, and ecology. Together and separately, scientists from these fields studied the vents and the vent communities to give the world an understanding of what is likely the strangest ecosystem on Earth. Scientists from diverse fields continue to study these bizarre and fascinating vents, which have now been found in many places on the sea floor.

Using Statistics

Statistics is the science of collecting, analyzing or interpreting, and presenting data. In high school science, students will need to understand sample size and variables. They will most likely use descriptive statistics—the use of tables and graphs—to display the results of their scientific experiments.

Data may be qualitative or quantitative. Qualitative data categorize and label groups of like items. Quantitative data entail numerical information—information that measures a value. For statistics to be meaningful at the high-school level, only one variable, the factor being studied, is generally used. For example, if you want to know how much water a given plant needs to thrive, the amount of water given to the plants in the study is the only factor that should vary. All other factors, such as plant species, plant size, amount of sunlight, and room temperature, should be exactly the same. If the other factors are not the same, you cannot be sure your results measure the effects of the variable you are studying.

Sample size is also important in statistics. The larger and more random the sample is, the more accurate the results will be. For example, you cannot state with any degree of accuracy that fifty percent of students prefer rum-raisin ice cream if you question only two students. Scientists and statisticians can calculate the margin of error in any study based on the difference between the parameter in the total population and the sample statistic used to estimate it.

▶ Careful collection and examination of samples is essential to scientific inquiry.

Video Resources

Brain Food Video Quizzes

Brain Food Video Quizzes These videos contain game-show style quizzes that assess students' progress and motivate students to study the chapter material.

HRW Earth Science Video

These videos introduce Earth science topics and include a geology field trip. The video segments listed below complement this chapter.

Segment 1: Introduction to Earth Science This segment examines agents of catastrophic change and slow, steady processes that have shaped Earth since its formation. (5.5 min)

Segment 2: Careers in Earth Science This segment focuses on Earth science professionals and explores the contributions to knowledge each makes. (3 min)

NOVA Videos

The **NOVA** videos below complement this chapter.

Adrift on the Gulf Stream NOVA explores the Gulf Stream's importance to ocean life and climate from a variety of vantage points. (60 min)

Earthquake This program shows how today's advanced technology helps geologists predict earthquakes. (60 min)

Runaway Universe Join two competing teams in their quest to discover the secrets to the stars and the ultimate fate of the universe. (60 min)

To order other **NOVA** videos related to this chapter, visit **go.hrw.com** and enter the keyword **HQ6IESV**.

SciLinks

Developed and maintained by the National Science Teachers Association

SciLinks is maintained by the National Science Teachers Association to provide you and your students with interesting, up-to-date links that will enrich your classroom presentation of the chapter.

Visit www.scilinks.org and enter the SciLinks code for more information about the topic listed.

Topic: Branches of Earth Science
SciLinks code: HQ60191

Topic: Careers in Earth Science
SciLinks code: HQ60222

Topic: Scientific Methods
SciLinks code: HQ61359

Topic: Scientific Investigations
SciLinks code: HQ61358

Chapter 1

Chapter Overview

This chapter introduces students to the branches of Earth science, scientific methods, and methods of measurement and analysis that they will use in their study of Earth science. The chapter also reminds students of the relationship between science and society.

Using the Figure — GENERAL

Racetrack Playa Reinforce the idea that scientists still do not know how rocks such as the one in the photo move across the desert floor. Invite students to hypothesize how the rocks may move. Have them suggest ways they could test their hypotheses. (Answers may vary. Accept all reasonable hypotheses and feasible suggestions for testing them.) **LS Logical**

PRE-READING ACTIVITY

FOLDNOTES

Encourage students to use their FoldNote as a study guide to quiz themselves for a test on the chapter material. Students may want to create Four-Corner FoldNotes for different topics within the chapter.

Chapter 1

Introduction to Earth Science

Sections

1 What Is Earth Science?

2 Science as a Process

What You'll Learn

- What areas of study make up Earth science
- How Earth scientists study the planet
- Why the work of Earth scientists is important to society

Why It's Relevant

An understanding of Earth science is vital in determining how the environment affects human society and how human society affects the air, water, and soil of Earth.

PRE-READING ACTIVITY

FOLDNOTES

Four-Corner Fold

Before you read this chapter, create the FoldNote entitled "Four-Corner Fold" described in the Skills Handbook section of the Appendix. Label each flap of the four-corner fold with a topic from the chapter. Write what you know about each topic under the appropriate flap. As you read the chapter, add other information that you learn.

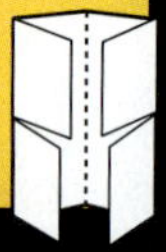

▶ The movement of rocks like this one in an area called *Racetrack Playa,* in Death Valley, California, has intrigued Earth scientists for years. Scientists have many hypotheses as to how rocks that weigh over 320 kg can move more than 880 m. However, no one knows for sure how these rocks move.

Chapter Correlations — *National Science Education Standards*

SAI 2a Scientists usually inquire about how physical, living or designed systems function. ...**(Sections 1 and 2)**

SAI 2b Scientists conduct investigations for a variety of reasons. ...**(Sections 1 and 2)**

SAI 2c Scientists rely on technology to enhance the gathering and manipulation of data. ...**(Section 1)**

SAI 2d Mathematics is essential in scientific inquiry. ... **(Section 2)**

SAI 2e Scientific explanations must adhere to criteria such as: a proposed explanation must be logically consistent; it must abide by the rules of evidence; it must be open to questions and possible modification. ...**(Section 2)**

SAI 2f Results of scientific inquiry ... emerge from different types of investigations.... **(Sections 1 and 2)**

ST 2a Scientists in different disciplines ask different questions and use different methods of investigation....**(Sections 1 and 2)**

ST 2b Science often advances with new technologies. ...**(Sections 1 and 2)**

ST 2d Science and technology are pursued for different purposes. Scientific inquiry is driven by the desire to understand the natural world, and technological design is driven by the need to meet human needs....**(Sections 1 and 2)**

SPSP 6a Science and technology are essential social enterprises.... **(Sections 1 and 2)**

SPSP 6d Individuals and society must decide on proposals involving new research and technologies....**(Sections 1 and 2)**

HNS 3C Occasionally, there are advances in science and technology that have important and long-lasting effects on science and society. Examples of such advances include plate tectonics. ...**(Section 1)**

UCP 1 Evidence consists of observations and data on which to base scientific explanations. Using evidence to understand interactions allows individuals to predict changes in natural and designed systems. . . .**(Section 2)**

Section 1 What Is Earth Science?

For thousands of years, people have looked at the world around them and wondered what forces shaped it. Throughout history, many cultures have been terrified and fascinated by seeing volcanoes erupt, feeling the ground shake during an earthquake, or watching the sky darken during an eclipse.

Some cultures developed myths or stories to explain these events. In some of these myths, angry goddesses hurled fire from volcanoes, and giants shook the ground by wrestling underneath Earth's surface. Modern science searches for natural causes and uses careful observations to explain these same events and to understand Earth and its changing landscape.

OBJECTIVES

- **Describe** two cultures that contributed to modern scientific study.
- **Name** the four main branches of Earth science.
- **Discuss** how Earth scientists help us understand the world around us.

KEY TERMS

Earth science
geology
oceanography
meteorology
astronomy

The Scientific Study of Earth

Scientific study of Earth began with careful observations. Scientists in China began keeping records of earthquakes as early as 780 BCE. The ancient Greeks compiled a catalog of rocks and minerals around 200 BCE. Other ancient peoples, including the Maya, tracked the movements of the sun, the moon, and the planets at observatories like the one shown in **Figure 1.** The Maya used these observations to create accurate calendars.

For many centuries, scientific discoveries were limited to observations of phenomena that could be seen with the unaided eye. Then, in the 16th and 17th centuries, the inventions of the microscope and the telescope made seeing previously hidden worlds possible. Eventually, the body of knowledge about Earth became known as Earth science. **Earth science** is the study of Earth and of the universe around it. Earth science, like other sciences, assumes that the causes of natural events, or phenomena, can be discovered through careful observation and experimentation.

Earth science the scientific study of Earth and the universe around it

Figure 1 ▶ El Caracol, an observatory built by the ancient Maya of Mexico, is the oldest known observatory in the Americas. Mayan calendars show the celestial movements that the Maya tracked by using observatories.

Section 1

Focus

Overview

This section defines Earth science and describes its branches of study. The section also explains why the scientific study of Earth is important.

Bellringer

Ask students why it is important to study Earth science. (Answers may vary. Accept all reasonable answers.) **LS Logical/Verbal**

Motivate

Discussion — GENERAL

Creating Calendars The Maya, Chinese, and other early peoples created calendars based on observations of the sun, moon, planets, and stars. Have students discuss why humans create calendars and how having a calendar enhances a culture and helps people. (Answers may vary. Accept all reasonable answers. Sample answer: Calendars help with agriculture and provide order to seasonal or religious rites and holidays.) **LS Logical/Verbal**

CHAPTER RESOURCES

Chapter Resource File

- Directed Reading BASIC

Technology

Transparencies
- Bellringer

Student Edition on CD-ROM

One-Stop Planner CD-ROM
- Lesson Plan

Teach

Using the Figure — ADVANCED

Doing Science Have students describe what each scientist in the figure at the bottom of the page is doing. Have them explain how technology is helping each scientist conduct his or her research. Interested students may research one type of technology shown in the figure and write about how it was developed and how it is used in Earth science. (Students should describe how each type of technology was developed and how scientists use it to gather information that may lead to new scientific discoveries.) **LS Logical/Visual**

READING SKILL BUILDER — BASIC

Reading Organizer Have students create a two-column reading organizer in which they write the branch of Earth science in one column and details about it in the adjacent column. **LS Logical** English Language Learners

Internet Activity — ADVANCED

Ice Cores and Climate Have interested students use the Internet to research how ice-core research is being used to study ancient Earth climates as a way to learn about climate change today. Students may write a report or present their findings to the class for discussion. A worksheet designed to direct student research on this topic can be found in the **Chapter Resource File** booklet or by visiting **go.hrw.com** and entering the keyword **HQ6IESX.** **LS Logical/Verbal**

CHAPTER RESOURCES

Chapter Resource File

Internet Activity
- Ice Cores and Climate ADVANCED

Branches of Earth Science

The ability to make observations improves when technology, such as new processes or equipment, is developed. Technology has allowed scientists to explore the ocean depths, Earth's unseen interior, and the vastness of space. Earth scientists have used technology and hard work to build an immense body of knowledge about Earth.

Most Earth scientists specialize in one of four major areas of study: the solid Earth, the oceans, the atmosphere, and the universe beyond Earth. Examples of Earth scientists working in these areas are shown in **Figure 2.**

geology the scientific study of the origin, history, and structure of Earth and the processes that shape Earth

oceanography the scientific study of the ocean, including the properties and movements of ocean water, the characteristics of the ocean floor, and the organisms that live in the ocean

Geology

The study of the origin, history, processes, and structure of the solid Earth is called **geology.** Geology includes many specialized areas of study. Some geologists explore Earth's crust for deposits of coal, oil, gas, and other resources. Other geologists study the forces within Earth to predict earthquakes and volcanic eruptions. Some geologists study fossils to learn more about Earth's past. Often, new knowledge forms new areas of study.

Oceanography

Oceans cover nearly three-fourths of Earth's surface. The study of Earth's oceans is called **oceanography.** Some oceanographers work on research ships that are equipped with special instruments for studying the sea. Other oceanographers study waves, tides, and ocean currents. Some oceanographers explore the ocean floor to obtain clues to Earth's history or to locate mineral deposits.

Figure 2 ▶ Fields of Study in Earth Science

Geologists who study volcanoes are called *volcanologists*. This volcanologist is measuring the magnetic and electric properties of moving lava.

This astronomer is linking a telescope with a specialized instrument called a *spectrograph*. Information gathered will help her catalog the composition of more than 100 galaxies.

This meteorologist is studying ice samples to learn about past climate. Studying past climate patterns gives scientists information about possible future changes in climate.

Activity — GENERAL

Not Your Typical Office Job Invite interested students to describe what it would be like to have a career in one of the branches of Earth science discussed in the text. Have them talk about whether it would allow them to spend time outdoors, whether it would allow them to travel, and whether it would involve adventure or danger. Have students discuss why they would or would not be interested in learning about careers in a branch of Earth science. Explain to students that most jobs in science require advanced college degrees, though some technical jobs may not. Have interested students find out about the various careers available in a particular branch of Earth science, the training needed, and what the jobs entail. **LS Intrapersonal**

Connection to HISTORY

Scientific Revolutions

Throughout history, many cultures have added to scientific knowledge. Over the last few centuries, global exploration and cultural exchanges have helped modern science change very quickly. During this time, science has aided the industrialization of countries around the world. Technology has also allowed humans to explore areas from the deep-ocean basins to the universe beyond Earth.

Advances in science usually occur through small additions to existing knowledge. The daily work of scientists and engineers normally results in step-by-step increases of understanding and improvements in the ability to meet human needs. In this way, scientists add knowledge, invent new technologies, and educate future generations of scientists.

However, some advances in science and technology occur very quickly and have effects that ripple through science and society. When a scientific advance completely changes the way that scientists think about the universe, a *scientific revolution* occurs. Scientific revolutions cause long-held ideas to be challenged and put aside for new ways of thinking and viewing the universe. Examples of recent scientific revolutions include Darwin's theory of evolution, the concept of quantum mechanics, and Einstein's general theory of relativity.

Progress in science and technology can be affected by social issues and challenges. In some cases, both scientists and nonscientists resist letting go of long-held beliefs. Many of the ideas that Einstein replaced with his theory of relativity had been in place for hundreds of years. Through many years of continued research and education, revolutionary ideas often become accepted by the scientific community and by society as a whole.

But even revolutionary ideas are continuously being tested. In 2002, experiments were performed aboard NASA's *Cassini* spacecraft to test a part of Einstein's theory of relativity. Using new technologies, scientists were able to verify Einstein's theory, which had been proposed almost 100 years earlier.

Meteorology

The study of Earth's atmosphere is called **meteorology.** Using satellites, radar, and other technologies, meteorologists study the atmospheric conditions that produce weather. Many meteorologists work as weather observers and measure factors such as wind speed, temperature, and rainfall. This weather information is then used to prepare detailed weather maps. Other meteorologists use weather maps, satellite images, and computer models to make weather forecasts. Some meteorologists study *climate,* the patterns of weather that occur over long periods of time.

meteorology the scientific study of Earth's atmosphere, especially in relation to weather and climate

astronomy the scientific study of the universe

Astronomy

The study of the universe beyond Earth is called **astronomy.** Astronomy is one of the oldest branches of Earth science. In fact, the ancient Babylonians charted the positions of planets and stars nearly 4,000 years ago. Modern astronomers use Earth-based and space-based telescopes as well as other instruments to study the sun, the moon, the planets, and the universe. Technologies such as rovers and space probes have also provided astronomers with new information about the universe.

Reading Check How has technology affected astronomy? (See the Appendix for answers to Reading Checks.)

SCILINKS. NSTA Developed and maintained by the National Science Teachers Association

For a variety of links related to this subject, go to www.scilinks.org

Topic: Branches of Earth Science
SciLinks code: HQ60191

Group Activity — GENERAL

Scientific Revolutions Divide the class into groups and have each group research one scientific revolution. These may include the work of such scientists as Copernicus, Galileo, Newton, Rutherford, Thomson, Bohr, Wegener, Hess, Einstein, Darwin, or any other scientist students choose. Have each group divide the work. Some students may describe the discoveries and how they changed how people think about themselves and the natural world. Others may research the social context and any resistance to the new ideas, while others may create the final presentation. Have students present their work to the class. **LS Interpersonal**

Co-op Learning

SKILL BUILDER — GENERAL

Vocabulary Point out that some of the Earth sciences discussed in this section end with the suffix *-ology.* Tell students that this suffix comes from the Latin *logia,* which means "science (of)." Invite students to name as many fields of science as they can that contain the suffix *-ology.* (Answers may vary.) **LS Verbal**

Answer to Reading Check

The development of telescopes, satellites, and space probes has greatly expanded astronomers' understanding of the universe.

INCLUSION Strategies

- *Learning Disabled*
- *Developmentally Delayed*

Create an outline system appropriate for this chapter and set tabs at one-half inch. For example, section heads could have a Roman numeral; the larger subheads, a capital letter; smaller subheads, Arabic numbers; and additional notes could be designated by lower case letters. Have students who have handwriting issues type chapter notes on a computer instead of writing them by hand. **LS Visual**

ENVIRONMENTAL CONNECTION

Our Unique Planet While Earth may seem to be so large that humans can have no major effect on it, human activity has been putting our environmental support system at risk. Because all of Earth's systems are related, small changes in one system may have drastic effects on other systems.

Close

Reteaching — BASIC

Picture This Provide posters or magazines that show pictures of scientists engaged in each branch of Earth science. As students read, have them correlate what is shown in the poster or magazine to the science as described in the text. English Language Learners
LS **Visual**

Quiz — GENERAL

1. What 17th century advances in technology helped advance scientific research? (the invention of the microscope and telescope)
2. What Earth processes do geologists study to try to help save lives? (earthquakes and volcanoes)

Alternative Assessment — GENERAL

Role Play Have groups of students role play being scientists in one branch of Earth science. Ask students to explain what they do and to bring in tools (or models and pictures of tools) that they would use. Encourage students to describe how their branch of Earth science helps people and expands human understanding of the world and the universe. LS **Kinesthetic**

Answers to Section Review

1. Answers may vary. Accept all reasonable answers.
2. geology, oceanography, astronomy, and meteorology
3. Answers may vary. Accept all reasonable answers.
4. Oceanographers study ocean currents, tides, or waves. Meteorologists study weather patterns and predict storms.
5. Telescopes, satellites, and space probes have expanded astronomers' view of the universe.
6. Earth scientists have helped us understand the forces that shape Earth and thus shape the environment we live in. Their understanding helps predict geologic events and helps us find Earth resources we need.
7. Both are inhospitable environments for humans and require the use of special equipment for humans to study them.
8. *Earth science* includes the fields of *geology; meteorology,* which includes the study of *climate; astronomy; oceanography;* and *environmental science.*

Figure 3 ▶ These meteorologists are risking their lives to gather information about tornadoes. If scientists can better predict when tornadoes will occur, many lives may be saved each year.

Environmental Science

Other Earth scientists study the ways in which humans interact with their environment. This relatively new field of Earth science is called *environmental science*. Environmental scientists study many issues, such as the use of natural resources, pollution, and the health of plant and animal species on Earth. Some environmental scientists study the effects of industries and technologies on the environment.

The Importance of Earth Science

Natural forces not only shape Earth but also affect life on Earth. For example, a volcanic eruption may bury a town under ash. And an earthquake may produce huge ocean waves that destroy shorelines. By understanding how natural forces shape our environment, Earth scientists, such as those in **Figure 3**, can better predict potential disasters and help save lives and property.

The work of Earth scientists also helps us understand our place in the universe. Astronomers studying distant galaxies have come up with new ideas about the origins of our universe. Geologists studying rock layers have found clues to Earth's past environments and to the evolution of life on this planet.

Earth provides the resources that make life as we know it possible. Earth also provides the materials to enrich the quality of people's lives. The fuel that powers a jet, the metal used in surgical instruments, and the paper and ink in this book all come from Earth's resources. The study of Earth science can help people gain access to Earth's resources, but Earth scientists also strive to help people use those resources wisely.

Section 1 Review

1. **Discuss** how one culture contributed to modern science.
2. **Name** the four major branches of Earth science.
3. **Describe** two specialized fields of geology.
4. **Describe** the work of oceanographers and meteorologists.
5. **Explain** how the work of astronomers has been affected by technology.

CRITICAL THINKING

6. **Analyzing Ideas** How have Earth scientists improved our understanding of the environment?
7. **Analyzing Concepts** Give two examples of how exploring space and exploring the ocean depths are similar.

CONCEPT MAPPING

8. Use the following terms to create a concept map: *Earth science, geology, meteorology, climate, environmental science, astronomy,* and *oceanography.*

CHAPTER RESOURCES

Chapter Resource File
- Section Quiz GENERAL

Workbooks
- Study Guide (also in Spanish)

Section 2 Science as a Process

OBJECTIVES

- **Explain** how science is different from other forms of human endeavor.
- **Identify** the steps that make up scientific methods.
- **Analyze** how scientific thought changes as new information is collected.
- **Explain** how science affects society.

KEY TERMS

observation
hypothesis
independent variable
dependent variable
peer review
theory

Art, architecture, philosophy, and science are all forms of human endeavor. Although artists, architects, and philosophers may use science in their work, science does not have the same goals as other human endeavors do.

The goal of science is to explain natural phenomena. Scientists ask questions about natural events and then work to answer those questions through experiments and examination. Scientific understanding moves forward through the work of many scientists, who build on the research of the generations of scientists before them.

Behavior of Natural Systems

Scientists start with the assumption that nature is understandable. Scientists also expect that similar forces in a similar situation will cause similar results. But the forces involved in natural events are complex. For example, changes in temperature and humidity can cause rain in one city, but the same changes in temperature and humidity may cause fog in another city. These different results might be due to differences in the two cities or due to complex issues, such as differences in climate.

Scientists also expect that nature is predictable, which means that the future behavior of natural forces can be anticipated. So, if scientists understand the forces and materials involved in a process, they can predict how that process will evolve. The scientists in **Figure 1,** for example, are studying ice cores in Antarctica. Ice cores can provide clues to Earth's past climate changes. Because natural systems are complex, however, a high level of understanding and predictability can be difficult to achieve. To increase their understanding, scientists follow the same basic processes of studying and describing natural events.

Figure 1 ▶ Scientists use ice cores to study past compositions of Earth's atmosphere. This information can help scientists learn about past climate changes.

Section 2

Focus

Overview

This section describes the scientific method and analyzes how scientific thought changes as new data are uncovered. The section also describes how science and society interrelate.

Bellringer

Ask students what they think the goals of the following human endeavors are: art, architecture, philosophy, sports, and politics. (Answers may vary. Accept all reasonable answers.) **LS Intrapersonal/ Interpersonal**

Motivate

Discussion — GENERAL

Influences on Natural Systems

Ask students if they have ever planted seeds. Have students list all of the reasons they can think of that different plants of the same species grow at different rates. (Answers may vary. Students may suggest that environmental variables such as soil, nutrients, water, air, and light influence how well a plant grows.) **LS Logical**

CHAPTER RESOURCES

Chapter Resource File

- **Directed Reading** BASIC
- **Inquiry Lab** What's Before Your Eyes? GENERAL
- **Skills Practice Lab** Testing a Prediction GENERAL

Technology

Transparencies
- Bellringer

Student Edition on CD-ROM

One-Stop Planner CD-ROM
- Lesson Plan

Teach

Using the Figure — GENERAL

Scientific Methods Direct students' attention to the diagram at the bottom of the page, and emphasize that the steps are a basic flowchart to be used as a guideline for scientific inquiry. Have them review the steps and identify any steps that may be substituted. (make observations and perform an experiment) Invite students to suggest different types of scientific investigation that would require one of those steps, but not both. Ask students under what conditions the final step (alter or form a new hypothesis) might be unnecessary. (if a hypothesis is supported by the observations or experiment) **LS Logical/Verbal**

Discussion — GENERAL

Types of Reasoning Explain to students that scientists use two types of reasoning to help them solve problems. Inductive reasoning, which is shown in the figure, involves starting from a specific fact or a single case and then forming a general conclusion. Deductive reasoning involves starting with a general, known principle and then drawing a conclusion about one particular case. Have students give examples of ways that they might use each type of reasoning in their everyday lives. Then, guide students in a discussion about why both types of reasoning are useful in scientific investigations. **LS Verbal**

Answer to Reading Check

Observations may lead to interesting scientific questions and may help scientists formulate reasonable and testable hypotheses.

Scientific Methods

Over time, the scientific community has developed organized and logical approaches to scientific research. These approaches are known as *scientific methods*. Scientific methods are not a set of sequential steps that scientists always follow. Rather, these methods are guidelines to scientific problem solving. **Figure 2** shows a basic flowchart of scientific methods.

observation the process of obtaining information by using the senses; the information obtained by using the senses

hypothesis an idea or explanation that is based on observations and that can be tested

Ask a Question

Scientific methods often begin with observations. **Observation** is the process of using the senses of sight, touch, taste, hearing, and smell to gather information about the world. When you see thunderclouds form in the summer sky, you are making an observation. And when you feel cool, smooth, polished marble or hear the roar of river rapids, you are making observations.

Observations can often lead to questions. What causes tornadoes to form? Why is oil discovered only in certain locations? What causes a river to change its course? Simple questions such as these have fueled years of scientific research and have been investigated through scientific methods.

SCILINKS. NSTA Developed and maintained by the National Science Teachers Association

For a variety of links related to this subject, go to www.scilinks.org

Topic: Scientific Methods

SciLinks code: HQ61359

Form a Hypothesis

Once a question has been asked and basic information has been gathered, a scientist may propose a tentative answer, which is also known as a hypothesis (hie PAHTH uh sis). A **hypothesis** (plural, *hypotheses*) is a possible explanation or solution to a problem. Hypotheses can be developed through close and careful observation. Most hypotheses are based on known facts about similar events. One example of a hypothesis is that houseplants given a large amount of sunlight will grow faster than plants given a smaller amount of sunlight. This hypothesis could be made from observing how and where other plants grow.

Reading Check Name two ways scientific methods depend on careful observations. (See the Appendix for answers to Reading Checks.)

Figure 2 ▶ Scientific Methods

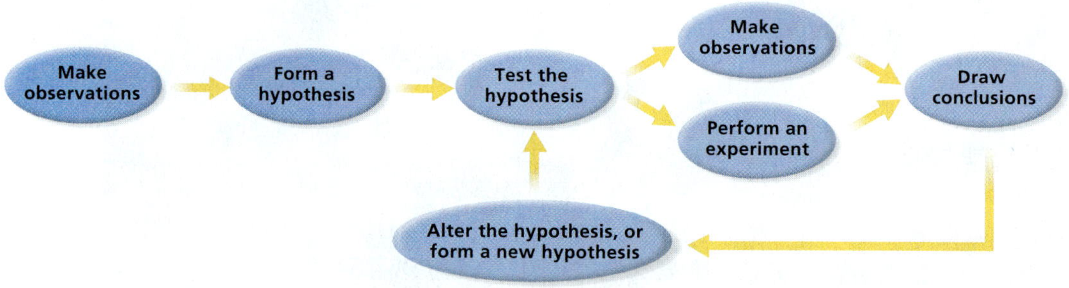

CHAPTER RESOURCES

Technology

Transparencies
- 1 Scientific Methods (with worksheet)

INCLUSION Strategies

- *Learning Disabled*
- *Behavior Control Issues*
- *Attention Deficit Disorder*

Some students can concentrate better on what they are reading if they walk while they read. Allow those who think best on the move to pace in the back of the room while reading the text. If too many students want to walk and read, or if the room is too small, ask students to read the text at home. Invite all students to try walking while they read. **LS Kinesthetic**

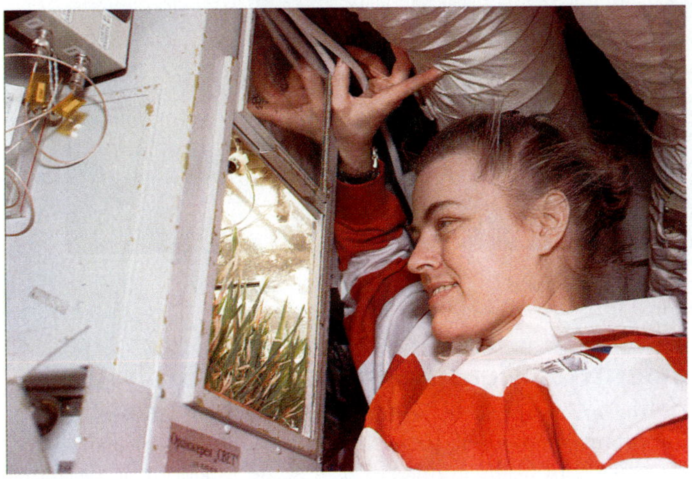

Figure 3 ▶ Astronaut Shannon Lucid observes wheat plants that are a part of an experiment aboard the *Mir* space station.

Test the Hypothesis

After a hypothesis is proposed, it is often tested by performing experiments. An *experiment* is a procedure that is carried out according to certain guidelines. Factors that can be changed in an experiment are variables. **Independent variables** are factors that can be changed by the person performing the experiment. **Dependent variables** are variables that change as a result of a change in independent variables.

In most experiments, only one independent variable is tested. For example, to test how sunlight affects plants, a scientist would grow identical plants. The plants would receive the same amount of water and fertilizer but different amounts of sunlight. Thus, sunlight would be the independent variable. How the plants respond to the different amounts of sunlight would be the dependent variable. Most experiments include a control group. A *control group* is a group that serves as a standard of comparison with another group to which the control group is identical except for one factor. In this experiment, the plants that receive a natural amount of sunlight would be the control group. An experiment that contains a control is called a *controlled experiment*. Most scientific experiments are controlled experiments. The "zero gravity" experiment shown in **Figure 3** is a controlled experiment.

Draw Conclusions

After many experiments and observations, a scientist may reach conclusions about his or her hypothesis. If the hypothesis fits the known facts, it may be accepted as true. If the experimental results differ from what was expected, the hypothesis may be changed or discarded. Expected and unexpected results lead to new questions and further study. The results of scientific inquiry may also lead to new knowledge and new methods of inquiry that further scientific aims.

independent variable in an experiment, the factor that is deliberately manipulated

dependent variable in an experiment, the factor that changes as a result of manipulation of one or more other factors (the independent variables)

Quick LAB — 5 min

Making Observations

Procedure

1. Get an ordinary **candle** of any shape and color.
2. Record all the observations you can make about the candle.
3. Light the candle with a **match**, and watch it burn for 1 min.
4. Record as many observations about the burning candle as you can. When you are finished, extinguish the flame. Record any observations.

Analysis

1. Share your results with your class. How many things that your classmates observed did you not observe? Explain this phenomenon.

Discussion — GENERAL

Variables and Controls Have students discuss why it is important to study only one independent variable at a time in an experiment. Ask how having more than one independent variable or one experimental control would affect the results of the experiment. Ask why it is important to be able to compare results involving manipulated variables with an unmanipulated control. (Students should be aware that having too many variables makes identifying the source of a change difficult. A control shows whether the independent variable resulted in any observable difference from "normal.") LS **Logical/Verbal**

Quick LAB

Skills Acquired

- Observing

Teacher's Notes You may want to explain to students that independent variables are also known as *manipulated variables* and dependent variables are also known as *responding variables*. Help students identify each during this Quick Lab.

Answers

1. Answers may vary. Accept all reasonable answers. Differences in observations may be based on wick size, slight environmental differences in different parts of the room, and how closely students are watching.

READING SKILL BUILDER — BASIC

Reading Organizer As students read about scientific methods, have them write an outline that lists each of the steps from this section and describes what each step involves. Later, students may use this outline as a study guide for assessments. LS **Verbal**

CHAPTER RESOURCES

Chapter Resource File

- Datasheet for Quick Lab GENERAL

Teach, *continued*

Activity — BASIC

Precision and Accuracy Divide students into groups to measure their desktops. Each group should use a different method of measurement. Groups should use one of the following units: inches, millimeters, or the length of the first thumb joint or a student's foot. Have groups share their measurements, then discuss which units of measure are the most and least accurate and precise. (Most accurate and precise may vary, but should be millimeters or inches. Least accurate and precise should be foot and thumb-joint length.) **LS Logical**

SKILL BUILDER — GENERAL

Math Have students measure the dimensions of objects in the classroom, such as a piece of notebook paper, the board, and a shoebox, using standard units. Then, have students use this table to convert the measurements to SI units:
1 inch = 2.54 centimeters
1 foot = 0.3048 meters
1 yard = 0.91 meters
You may have pairs of students measure the same objects and check their conversions against the measurements of other pairs of students. **LS Logical**

CHAPTER RESOURCES

Chapter Resource File

- Datasheet for Quick Lab GENERAL

Good accuracy and good precision

Poor accuracy but good precision

Good overall accuracy but poor precision

Figure 4 ▶ Accuracy and Precision

Scientific Measurements and Analysis

During an experiment, scientists must gather information. An important method of gathering information is measurement. Measurement is the comparison of some aspect of an object or event with a standard unit. Scientists around the world can compare and analyze each other's measurements because scientists use a common system of measurement called the *International System of Units,* or SI. This system includes standard measurements for length, mass, temperature, and volume. All SI units are based on intervals of 10. The Reference Tables section of the Appendix contains a chart of SI units.

Accuracy and Precision

Accuracy and precision are important in scientific measurements. *Accuracy* refers to how close a measurement is to the true value of the thing being measured. *Precision* is the exactness of the measurement. For example, a distance measured in millimeters is more precise than a distance measured in centimeters. Measurements can be precise and yet inaccurate. The relationship between accuracy and precision is shown in **Figure 4.**

QuickLAB 15 min

Sample Size and Accuracy

Procedure

1. Shuffle a deck of 52 playing cards eight times.
2. Lay out 10 cards. Record the number of red cards.
3. Reshuffle, and repeat step 2 four more times.
4. Which trials showed the highest number and lowest number of red cards? Calculate the total range of red cards by finding the difference between the highest number and lowest number.
5. Determine the mean number of red cards per trial by adding the number of red cards in the five trials and then dividing by 5.

Analysis

1. A deck of cards has 50% red cards. How close is your average to the percentage of red cards in the deck?
2. Pool the results of your classmates. How close is the new average to the percentage of red cards in the deck?
3. How does changing the sample size affect accuracy?

QuickLAB

Skills Acquired

- Observing
- Analyzing

Teacher's Notes You may have students work in pairs or in larger groups, depending on the number of card decks you have. However, the more groups that participate, the larger the sample size will be.

Answers

1. Answers may vary.
2. Answers may vary but the new average should be closer to 50%.
3. The larger the sample size is, the more accurate the results are.

Error

Error is an expression of the amount of imprecision or variation in a set of measurements. Error is commonly expressed as percentage error or as a confidence interval. Percentage error is the percentage of deviation of an experimental value from an accepted value. A *confidence interval* describes the range of values for a set percentage of measurements. For example, imagine that the average length of all of the ears of corn in a field is 23 cm, and 90% of the ears are within 3 cm of the average length. A scientist may report that the average length of all of the ears of corn in a field is 23 ± 3 cm with 90% confidence.

Observations and Models

In Earth science, using controlled experiments to test a hypothesis is often impossible. When experiments are impossible, scientists make additional observations to gather evidence. The hypothesis is then tested by examining how well the hypothesis fits or explains all of the known evidence.

Scientists also use models to simulate conditions in the natural world. A *model* is a description, representation, or imitation of an object, system, process, or concept. Scientists use several types of models, two of which are shown in **Figure 5.** Physical models are three-dimensional models that can be touched. Maps and charts are examples of graphical models.

Conceptual models are verbal or graphical models that represent how a system works or is organized. Mathematical models are mathematical equations that represent the way a system or process works. Most recently, scientists have developed computer models, which can be used to represent simple processes or complex systems. After a good computer model has been created, scientists can perform experiments by manipulating variables much as they would when performing a physical experiment.

Reading Check **Name three types of models. (See the Appendix for answers to Reading Checks.)**

MATHPRACTICE

Percentage Error Percentage error is calculated by using the following equation:

$$\text{percent error} = \left[\frac{(\text{accepted value} - \text{experimental value})}{\text{accepted value}}\right] \times 100$$

If the accepted value for the weight of a gallon of water is 3.78 kg and the measured value is 3.72 kg, what is the percentage error for the measurement? Show your work.

Figure 5 ▶ Two models of Mount Everest are shown below. The computer model on the right is used to track erosion along the Tibetan Plateau. The model on the left is a physical model.

MATHPRACTICE

Answer

% error = [(accepted value − experimental value) ÷ accepted value] × 100

[(3.78 kg − 3.72 kg) ÷ 3.78 kg] × 100 = 1.59%

SOCIAL STUDIES CONNECTION — GENERAL

Margin of Error Polls are often taken before elections to see which candidate people prefer and to attempt to forecast the outcome. As in scientific measurements, polls contain a percentage—or margin—of error. Ask students: If 47% of people polled say they like candidate A, and 44% say they like candidate B, and the margin of error is plus or minus 3%, which candidate will likely get elected? (You cannot tell because the margin of error makes them even.) Have students discuss why it is important to note the percentage of error in polls and other types of statistics that they may come across in daily life. (Because without understanding the percent error, the poll numbers may seem more conclusive than they really are.) **LS Logical**

Answer to Reading Check

Answers may vary but should include three of the following types of models: physical models, graphic models, conceptual models, computer models, and mathematical models.

TECHNOLOGY CONNECTION

Numbers Become Pictures Most computer models are visual representations of mathematical formulas. Data are entered into the formulas, and a software program solves the equations. What is seen on the computer screen is a picture of how the data act within the equation. Even a graphic such as the satellite image on this page may be created from numerical data. Satellite equipment measures the heights of objects on Earth. These measurements may then be converted into an image that shows the relative heights of objects.

Using the Figure — BASIC

Levels of Analysis Have students discuss the differences in scale they see in the two models shown in the figure. Ask students in what context each model would be most useful. (The physical model is most useful for smaller scale analysis of fine detail of the mountain itself, while the satellite image is most useful in a larger context in which the subject, Mt. Everest, is seen relative to a wide area around it.) **LS Visual/Logical**

Teach, *continued*

Graphic Organizer GENERAL

Chain-of-Events Chart

You may wish to use this activity as a pre-reading activity to assess students' prior knowledge before beginning discussion of how scientific ideas reach acceptance. You may also choose to use a similar activity as a quiz to assess knowledge of this process.

Discussion GENERAL

Peer-Review Pressure Ask students which task they would give more careful attention to—one that is to be handed in and graded or one that no one will check. Invite students to apply this concept to peer review of scientific work. Have students talk about how peer review motivates scientists to be extremely careful, accurate, and thorough in experimental design, data collection, and reporting. Invite students to discuss whether peer review is more important in science than in other fields and to give reasons for their answers. LS **Interpersonal/Intrapersonal**

Answer to Reading Check

Scientists present the results of their work at professional meetings and in scientific journals.

Figure 6 ▶ Meteorologists at a conference in California are watching the newly introduced "Science On a Sphere™" exhibit. They are wearing 3-D glasses to better see the complex and changing three-dimensional display of global temperatures.

Graphic Organizer **Chain-of-Events Chart**

Create the **Graphic Organizer** entitled "Chain-of-Events Chart" described in the Skills Handbook section of the Appendix. Then, fill in the chart with details about each step of how a hypothesis becomes an accepted scientific idea.

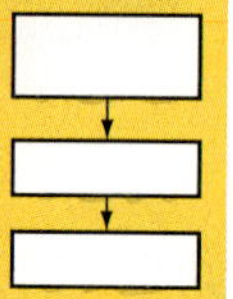

peer review the process in which experts in a given field examine the results and conclusions of a scientist's study before that study is accepted for publication

Acceptance of Scientific Ideas

When scientists reach a conclusion, they introduce their findings to the scientific community. New scientific ideas undergo review and testing by other scientists before the ideas are accepted.

Publication of Results and Conclusions

Scientists commonly present the results of their work in scientific journals or at professional meetings, such as the one shown in **Figure 6.** Results published in journals are usually written in a standard scientific format. Many journals are now being published online to allow scientists quicker access to the results of other scientists.

Peer Review

Scientists in any one research group tend to view scientific ideas similarly. Therefore, they may be biased in their experimental design or data analysis. To reduce bias, scientists submit their ideas to other scientists for peer review. **Peer review** is the process in which several experts on a given topic review another expert's work on that topic before the work gets published. These experts determine if the results and conclusions of the study merit publication. Peer reviewers commonly suggest improvements to the study, or they may determine that the results or conclusions are flawed and recommend that the study not be published. Scientists follow an ethical code that states that only valid experimental results should be published. The peer review process serves as a filter, which allows only well-supported ideas to be published.

✔ **Reading Check** **Name two places scientists present the results of their work. (See the Appendix for answers to Reading Checks.)**

MISCONCEPTION ALERT

Reproducible Truth Students may think that all scientific studies mentioned in the media are true and absolute. This is not the case. Sometimes, peer reviewers accept a scientific study because the methods were careful and correct, data were accurately analyzed, and findings were reasonable. However, this does not mean that the findings are necessarily true. A prime criterion for accepting any scientific finding is that the experiment or study be reproducible. Once a study is published, other scientists attempt to repeat the experiment. If many scientists get the same results, it is likely that the original findings were correct. But if some or many scientists do not get the same results, it is likely that the original findings were incorrect or incomplete. Many experiments must be performed with the same results before a finding is accepted. Thus, it should not be assumed that one study reveals an absolute scientific truth.

Formulating a Theory

After results are published, they usually lead to more experiments, which are designed to test and expand the original idea. This process may continue for years until the original idea is disproved, is modified, or becomes generally accepted. Sometimes, elements of different ideas are combined to form concepts that are more complete.

When an idea has undergone much testing and reaches general acceptance, that idea may help form a theory. A **theory** is an explanation that is consistent with all existing tests and observations. Theories are often based on scientific laws. A *scientific law* is a general statement that describes how the natural world behaves under certain conditions and for which no exceptions have been found. Like theories, laws are discovered through scientific research. Theories and scientific laws can be changed if conflicting information is discovered in the future.

theory the explanation for some phenomenon that is based on observation, experimentation, and reasoning; that is supported by a large quantity of evidence; and that does not conflict with any existing experimental results or observations

The Importance of Interdisciplinary Science

Scientists from many disciplines commonly contribute the information necessary to support an idea. The free exchange of ideas between fields of science allows scientists to identify explanations that fit a wide range of scientific evidence. When an explanation is supported by evidence from a variety of fields, the explanation is more likely to be accurate. New disciplines of science sometimes emerge as a result of new connections that are found between more than one branch of science. An example of the development of a widely accepted hypothesis that is based on interdisciplinary evidence is shown in **Figure 7.**

Figure 7 ▶ The hypothesis that the dinosaurs were killed by an asteroid impact was developed over many years and through the work of many scientists from different disciplines.

No dinosaur fossils exist in rock layers younger than 65 million years old.

A large impact crater about 65 million years old exists in the ocean near the Yucatan peninsula.

A layer of iridium occurs in rocks about 65 million years old all around Earth. Iridium is rare on Earth, but is common in meteoroids.

Climate models predict that a large impact would change Earth's climate and affect life on Earth.

Debate — ADVANCED

Should Science Always Be Applied? Divide students into two groups: those who think that people should always or often apply scientific discoveries to everyday life; and those who think that some consequences or side effects of knowledge gained from research should restrain people from applying findings swiftly. Have students use examples they are familiar with to support their view and counter the opposing view. LS Verbal/Logical

Close

Reteaching — BASIC

Chart of Scientific Methods Ask students to use poster-sized paper to create a chart of the different steps in the scientific method. Students should use label heads and details for each step. They may draw or find pictures to illustrate each step in the chart. LS Kinesthetic/Visual

Quiz — GENERAL

1. What is the difference between a theory and a scientific law? (A theory is an explanation that best fits all of the data; a law is a general statement to which no known exceptions exist.)
2. Why do nearly all scientific experiments use a control? (The control allows scientists to identify whether a variable had an effect on the experimental subject.)

CHAPTER RESOURCES

Chapter Resource File

Internet Activity
- Limitations of Information Systems GENERAL

Technology

Transparencies
- 2 The Importance of Interdisciplinary Science (with worksheet)

Internet Activity — GENERAL

Limitations of Information Systems Have students research common misconceptions in science, such as the following: "Earth's asthenosphere is molten," "raindrops are tear-shaped," and "clouds and steam are water vapor." Have students write a short paper that describes one misconception and then accurately explains the concept. Ask students to conclude their paper with a discussion of how the Internet can be used to disseminate both accurate and inaccurate information. Have them identify the limitations and impacts of information systems and explain how knowing the limitations of these systems allows for the effective use of such tools. A worksheet designed to direct student research on this topic can be found in the **Chapter Resource File** booklet or by visiting go.hrw.com and entering the keyword **HQ6IESX.** LS Logical

Close, *continued*

Alternative Assessment — GENERAL

Scientific Proposal Have students write a proposal in which they describe a scientific question they would like to answer by performing an experiment. Have them identify the question, explain why it is important, summarize their hypothesis, describe how they would test it, and identify the steps of the scientific method they would follow. LS **Logical**

Answers to Section Review

1. to explain what causes some natural phenomenon or to answer a question
2. Ask a question, form a hypothesis, test the hypothesis or make observations, and draw conclusions.
3. A hypothesis is an idea based on observation that can be tested; a theory is an explanation based on testing and evidence that does not conflict with existing results or observations.
4. by following a designed experiment that usually contains one independent variable and a control
5. An independent variable can be changed by the experimenter; a dependent variable changes as a result of a change in independent variables.
6. when information that conflicts with them is discovered
7. by forming a framework that all scientists follow to establish scientific truth
8. Technological development may lead to new and useful tools that help scientists explore areas they previously could not explore.
9. Sample answer: An observation that is accurate may be imprecise because the units of measurement may have been too large to identify small differences.
10. The artist is the same in that he or she observes nature carefully; the artist is different in that art is personal interpretation, not scientific measurement and reasoning.
11. Student responses may vary but should explain why they think the risk to this group is acceptable or not, based on the general value of the technology.
12. *Scientific methods* include making *observations*; forming a *hypothesis* that is tested by using an *experiment* that includes *independent variables* and *dependent variables*; and drawing *conclusions*.

Figure 8 ▶ The Alaskan pipeline has carried more than 13 billion barrels of oil since it was built in 1977. The pipeline has also sparked controversy about the potential dangers to nearby Alaskan wildlife.

Science and Society

Scientific knowledge helps us understand our world. The work of people, including scientists, is influenced by their cultural and personal beliefs. Science is a part of society, and advances in science can have important and long-lasting effects on both science and society. Examples of these far-reaching advances include the theory of plate tectonics, quantum mechanics, and the theory of evolution.

Science is also used to develop new technology, including new tools, machines, materials, and processes. Sometimes, technologies are designed to address a specific human need. In other cases, technology is an indirect result of science that was directed at another goal. For example, technology that was designed for space exploration has been used to improve computers, cars, medical equipment, and airplanes.

However, new technology may also create new problems. Scientists involved in research that leads to new technologies have an obligation to consider the possible negative effects of their work. Before making decisions about technology, people should consider the alternatives, risks, and costs and benefits to humans and to Earth. Even after such decisions are made, society often continues to debate them. For example, the Alaskan pipeline, part of which is shown in **Figure 8,** transports oil. But the transport of oil in the United States is part of an ongoing debate about how we use oil resources and how these uses affect our natural world.

Section 2 Review

1. **Describe** one reason that a scientist might conduct research.
2. **Identify** the steps that make up scientific methods.
3. **Compare** a hypothesis with a theory.
4. **Describe** how scientists test hypotheses.
5. **Describe** the difference between a dependent variable and an independent variable.
6. **Describe** the conditions under which scientific laws and theories can be changed.
7. **Summarize** how scientific methods contribute to the development of modern science.
8. **Explain** how technology can affect scientific research.

CRITICAL THINKING

9. **Analyzing Ideas** An observation can be precise but inaccurate. Do you think it is possible for an observation to be accurate but not precise? Explain.
10. **Making Comparisons** When an artist paints a picture of a natural scene, what aspects of his or her work are similar to the methods of a scientist? What aspects are different?
11. **Demonstrating Reasoned Judgment** A new technology is known to be harmful to a small group of people. How does this knowledge affect whether you would use this new technology? Explain.

CONCEPT MAPPING

12. Use the following terms to create a concept map: *independent variable, observation, experiment, dependent variable, hypothesis, scientific methods,* and *conclusion.*

CHAPTER RESOURCES

Chapter Resource File
- Section Quiz GENERAL

Workbooks
- Study Guide (also in Spanish)

Chapter 1 Highlights

Sections

1 What Is Earth Science?

Key Terms

Earth science, 5
geology, 6
oceanography, 6
meteorology, 7
astronomy, 7

Key Concepts

- Many cultures have contributed to the development of science over thousands of years.
- The four main branches of Earth science are geology, oceanography, meteorology, and astronomy.
- Earth scientists help us understand how Earth formed and the natural forces that affect human society.

2 Science as a Process

Key Terms

observation, 10
hypothesis, 10
independent variable, 11
dependent variable, 11
peer review, 14
theory, 15

Key Concepts

- Scientific research attempts to solve problems logically through scientific methods.
- Science differs from other human endeavors by following a procedure of testing to help understand natural phenomena.
- Using scientific methods, scientists develop hypotheses and theories to describe natural phenomena.
- Scientific methods include observation and experimentation.
- Science aids in the development of technology. The main aim of technology is to solve human problems.

Chapter Highlights

Alternative Assessment — GENERAL

Frontiers of Earth Science Have students think of one question they have about Earth science that could be answered by new research. Have students write down the question and the branches of Earth science (and other sciences, if appropriate) that could be involved in answering the question. Then, have students apply all appropriate steps of the scientific method to their problem. LS **Verbal/Logical**

CHAPTER RESOURCES

Chapter Resource File

- **Concept Review** GENERAL
- **Critical Thinking** ADVANCED
- **Math Skills** GENERAL
- **Graphing Skills** GENERAL
- **Chapter Test A** GENERAL
- **Chapter Test B** ADVANCED

Workbooks

- **Study Guide** (also in Spanish)
- **Assessments** (Spanish)

Technology

Classroom Videos
- Brain Food Video Quiz

HRW Earth Science Video
- Segment 1: Introduction to Earth Science
- Segment 2: Careers in Earth Science

Chapter Review

Assignment Guide

Section	Questions
1	5–6, 8–13, 20, 24–25, 34
2	1–4, 7, 14–19, 21–23, 26–29, 35–38
1 and 2	30–33

Using Key Terms

1-8. Answers may vary but should show that students understand the definitions of and differences between key terms.

Understanding Key Concepts

9. a **10.** b
11. d **12.** d
13. a **14.** b
15. d **16.** b
17. a

Short Answer

18. Accuracy is how close a measurement is to the true value; precision is the exactness of the measurement.

19. A control helps identify the response of a subject to a variable.

20. astronomy and geology

21. Answers may vary. Scientific discoveries may lead to real-world applications, such as the development of new technological tools, which in turn help scientists make more new discoveries.

22. because some explanations require data and expertise from more than one scientific discipline and because other disciplines may help support or provide evidence for findings in another field

23. Peer review ensures that reported scientific findings were achieved through the use of correct scientific methods that can be tested by other scientists.

24. by developing myths and legends

Chapter 1 Review

Using Key Terms

Use each of the following terms in a separate sentence.

1. *observation*

2. *peer review*

3. *theory*

For each pair of terms, explain how the meanings of the terms differ.

4. *hypothesis* and *theory*

5. *geology* and *astronomy*

6. *oceanography* and *meteorology*

7. *dependent variable* and *independent variable*

8. *Earth science* and *geology*

Understanding Key Concepts

9. The study of solid Earth is called
a. geology. **c.** oceanography.
b. meteorology. **d.** astronomy.

10. The Earth scientist most likely to study storms is a(n)
a. geologist. **c.** oceanographer.
b. meteorologist. **d.** astronomer.

11. The study of the origin of the solar system and the universe in general is
a. geology. **c.** meteorology.
b. ecology. **d.** astronomy.

12. How long ago were the first scientific observations about Earth made?
a. a few years ago
b. a few decades ago
c. hundreds of years ago
d. several thousand years ago

13. The Earth scientist most likely to study volcanoes is a(n)
a. geologist. **c.** oceanographer.
b. meteorologist. **d.** astronomer.

14. One possible first step in scientific problem solving is to
a. form a hypothesis.
b. ask a question.
c. test a hypothesis.
d. state a conclusion.

15. A possible explanation for a scientific problem is called a(n)
a. experiment. **c.** observation.
b. theory. **d.** hypothesis.

16. A statement that consistently and correctly explains a natural phenomenon is
a. a hypothesis. **c.** an observation.
b. a theory. **d.** a control.

17. When scientists pose questions about how nature operates and attempt to answer those questions through testing and observation, they are conducting
a. research. **c.** examinations.
b. predictions. **d.** peer reviews.

Short Answer

18. How does accuracy differ from precision in a scientific measurement?

19. Why do scientists use control groups in experiments?

20. A meteorite lands in your backyard. What two branches of Earth science would help you explain that natural event?

21. Write a short paragraph about the relationship between science and technology.

22. Give two reasons why interdisciplinary science is important to society.

23. Explain how peer review affects scientific knowledge.

24. How did some ancient cultures explain natural phenomena?

Critical Thinking

25. Making Connections How could knowing how our solar system formed affect how we see the world?

26. Evaluating Hypotheses Some scientists have hypothesized that meteorites have periodically bombarded Earth and caused mass extinctions every 26 million years. How might this hypothesis be tested?

27. Determining Cause and Effect Name some possible negative effects of a new technology that uses nuclear fuel to power cars.

28. Analyzing Ideas A scientist observes that each eruption of a volcano is preceded by a series of small earthquakes. The scientist then makes the following statement: "Earthquakes cause volcanic eruptions." Is the scientist's statement a hypothesis or a theory? Why?

29. Forming a Hypothesis You find a yellow rock and wonder if it is gold. How could you apply scientific methods to this problem?

Concept Mapping

30. Use the following terms to create a concept map: *control group, accuracy, precision, variable, technology, Earth science, experiment,* and *error*.

Math Skills

31. Making Calculations One kilogram is equal to 2.205 lb at sea level. At the same location, how many kilograms are in 100 lb?

32. Making Calculations One meter is equal to 3.281 ft. How many meters are in 5 ft?

33. Making Calculations The accepted value of the average distance between Earth and the moon is 384,467 km. If a scientist measures that the moon is 384,476 km from Earth, what is the measurement's percentage error?

Writing Skills

34. Expressing Original Ideas Imagine that you must live in a place that has all the benefits of only one of the Earth sciences. Which branch would you choose? Defend your choice in an essay.

35. Outlining Topics Explain the sequence of events that happens as a scientific hypothesis becomes a theory.

Interpreting Graphics

The graph below shows error in measuring tectonic plate movements. The blue bars represent confidence intervals. Use this graph to answer the questions that follow.

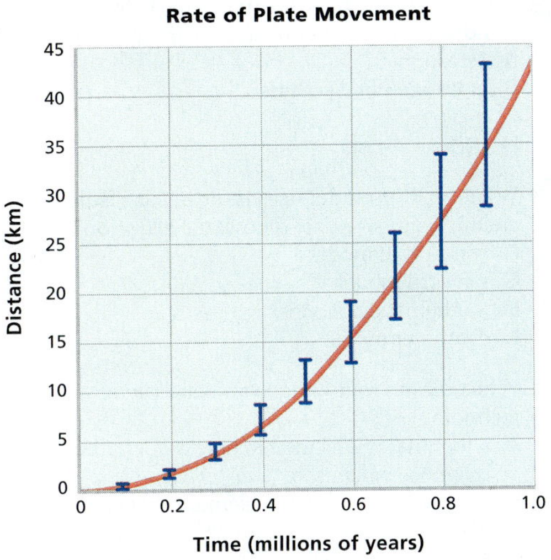

36. How much error is there in the smallest measurement of plate movement?

37. How much error is there in the largest measurement of plate movement?

38. How would you explain the difference between the error in the smallest measurement and the error in the largest measurement?

Chapter Review

Critical Thinking

25. Answers may vary. Knowing how our solar system formed helps us understand Earth's origin and how solar systems in other parts of the universe formed.

26. Answers may vary. This hypothesis might be tested by looking at ancient rocks to see if they contain evidence of meteorite material and analyzing the fossil record to see if it supports this idea.

27. Answers may vary. Possible negative effects might be leakage of radioactivity, especially in accidents, and where to store consumed, but still dangerous, nuclear waste.

28. The scientist's statement is a hypothesis because it has not yet been tested.

29. Answers may vary. Sample answer: My hypothesis may be "This rock that has a gold color is gold." Then, I would design an experiment to measure the rock's density to determine if the rock is actually gold.

Concept Mapping

30. Answers may vary but should include all of the terms listed. Sample answers appear at the end of this Teacher Edition.

Math Skills

31. 100 kg ÷ 2.205 lb/kg = 45.35 kg

32. 5 ft ÷ 3.281 ft/m = 1.52 m

33. 384,467 km − 384,476 km = −9 km; −9 km ÷ 384,467 km = −0.00002 × 100 = −0.002%

Writing Skills

34. Answers may vary. Accept all reasonable answers.

35. Answers may vary. Accept all reasonable answers.

Interpreting Graphics

36. one kilometer or less

37. about 15 kilometers

38. Answers may vary. Sample answer: The older the rock is, the harder it is to measure the rate of movement at the time the rock formed, so the higher the error value is.

Estimated Time

To give students practice under more realistic testing conditions, allow them 30 minutes to answer all of the questions in this practice test.

TEST DOCTOR

Question 2 Answer H is correct. Scientists should not force the results to match their assumptions, so answers F and I are incorrect. Greater precision will not necessarily result in a more accurate value, so answer G is incorrect. Scientists often reevaluate hypotheses and change them if the hypotheses do not fit the facts.

Question 5 Answer A is correct. Students should be clued to the fact that new technology is often designed for a specific purpose and may take the form of a tool or machine. Answer B is an advancement in science. Such an advancement may be aided by technology, but it is not an example of technology itself. Answer C demonstrates the way in which society may interact with science to help create new technologies, but a law is not an example of technology. Answer D may also be aided by technology and may represent new information, but recording observations is not an example of a new tool, process, or machine, and recording observations does not necessarily meet any human need.

Chapter 1 Standardized Test Prep

Understanding Concepts

Directions (1–5): **For *each* question, write on a separate sheet of paper the letter of the correct answer.**

1. A tested explanation of a natural phenomenon that has become widely adopted is a scientific
   A. hypothesis
   B. law
   C. theory
   D. observation

2. If experimental results do not match their predictions, scientists generally will
   F. repeat the experiment until they do match
   G. make the measurements more precise
   H. revise their working hypothesis
   I. change their experimental results

3. Scientists who study weather charts to analyze trends and to predict future weather events are
   A. astronomers
   B. environmental scientists
   C. geologists
   D. meteorologists

4. What type of model uses molded clay, soil, and chemicals to simulate a volcanic eruption?
   F. conceptual model
   G. physical model
   H. mathematical model
   I. computer model

5. Which of the following is an example of a new technology?
   A. a tool that is designed to help a doctor better diagnose patients
   B. a previously unknown element that is discovered in nature
   C. a law that is passed to fund scientists conducting new experiments
   D. scientists that record observations on the movement of a star

Directions (6–7): **For *each* question, write a short response.**

6. What is the term for the factors that change as a result of a scientific experiment?

7. Why do scientists often review one another's work before it is published?

Reading Skills

Directions (8–10): **Read the passage below. Then, answer the questions.**

Scientific Investigation

Scientists look for answers by asking questions. These questions are often answered through experimentation and observation. For example, scientists have wondered if there is some relationship between Earth's core and Earth's magnetic field.

To form their hypothesis, scientists started with what they knew: Earth has a dense, solid inner core and a molten outer core. They then created a computer model to simulate how Earth's magnetic field is generated. The model predicted that Earth's inner core spins in the same direction as the rest of Earth does but slightly faster than the surface does. If the hypothesis is correct, it might explain how Earth's magnetic field is generated. But how could the researchers test the hypothesis? Because scientists do not have the technology to drill to the core, they had to get their information indirectly. To do this, they decided to track the seismic waves that are created by earthquakes. These waves travel through Earth, and scientists can use them to infer information about the core.

8. The possibility of a connection between Earth's core and Earth's magnetic field formed the basis of the scientist's what?
   A. theory
   B. law
   C. hypothesis
   D. fact

9. To begin their investigation, the scientists first built a model. What did this model predict?
   F. Earth's outer core is molten, and the inner core is solid.
   G. Earth's inner core is molten, and the outer core is solid.
   H. Earth's inner core spins in the same direction as the rest of Earth does.
   I. Earth's outer core spins in the same direction as the rest of Earth does.

10. Why might the scientists have chosen to build a computer model of Earth, instead of a physical model of Earth?

Answers

Understanding Concepts

1. C
2. H
3. D
4. G
5. A
6. dependent variables
7. to determine the validity of the results and conclusions

Reading Skills

8. C
9. H
10. Because of the complexity of the inner workings of Earth, a computer model would be more practical and accurate than a physical model in this case.

Interpreting Graphics

11. G
12. Answers may vary. See Test Doctor for a detailed scoring rubric.

Interpreting Graphics

Directions (11–12): **For *each* question below, record the correct answer on a separate sheet of paper.**

The diagram below shows the four major areas studied by Earth scientists. Use this diagram to answer question 11.

Branches of Earth Science

11 A scientist studying the events that take place in area C would be primarily concerned with which of the following?

F. Earth's age
G. Earth's weather
H. movement of waves and tides
I. movement of the stars across the sky

Use the flowchart below to answer question 12.

Using a Scientific Method

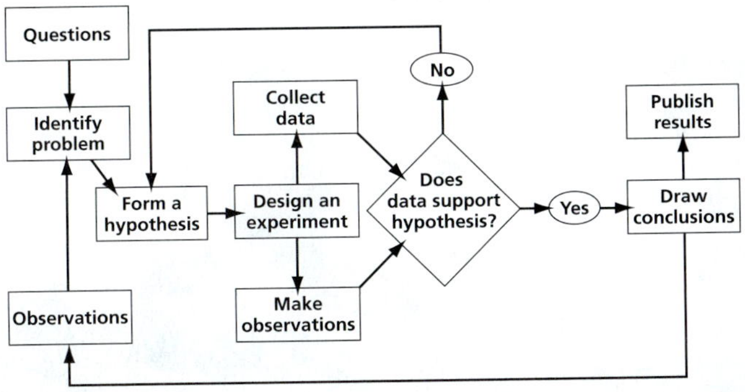

12 What are two possible outcomes of the experimental process? What would a scientist do with the information gathered during the experimental process?

Test TIP

If you are unsure of an answer, eliminate the answers that you know are wrong before choosing your response.

Standardized Test Prep

TEST DOCTOR

Question 11 Answer G is correct. Meteorologists study the weather and atmosphere in area C. Answer F is incorrect because the Earth's age is more easily determined by a geologist, who studies the history and structure of the solid Earth in area A. Answer H is incorrect because the movement of waves and tides take place in area B, not area C. Answer I is incorrect because astronomers study the movement of stars across the sky in area D.

Question 12 Full-credit answers should include the following points:

- students should demonstrate an understanding that scientific methods are logical ways of solving problems but are not sets of steps that are followed in an invariable sequence, nor that lead to an invariable outcome
- possible outomes include new observations and data collection
- a scientist analyzes data to decide whether they support or disprove the hypothesis. If they disprove the hypothesis, the hypothesis is rethought
- not only do experimental results sometimes lead to a new hypothesis, but they also often open new avenues for investigation by suggesting new questions

Test Prep Correlations *National Science Education Standards*

HNS 1b: items 2, 7
HNS 2c: items 2, 6
SAI 2a: items 3, 8, 9, 10, 11
SAI 2b: items 8, 12
SAI 2c: items 5, 10
SAI 2e: items 1, 8
UCP 2: items 4, 9, 10

CHAPTER RESOURCES

State Resources

For specific resources for your state, visit go.hrw.com and type in the keyword HSHSTR.

Scientific Methods

Teacher's Notes

Time Required

two 45-minute class periods

Lab Ratings

EASY → HARD

TEACHER PREPARATION ▲▲
STUDENT SETUP ▲
CONCEPT LEVEL ▲
CLEANUP ▲

Skills Acquired

- Observing
- Measuring
- Organizing and Analyzing Data
- Predicting
- Interpreting

The Scientific Method

In this lab, students will

- Make Observations
- Form a Hypothesis
- Test the Hypothesis
- Draw Conclusions

Materials

The materials listed on the page are enough for groups of two to four students. You may wish to have students flag or otherwise mark the puddle they are studying so they can locate it easily on return trips.

Tips and Tricks

It may help to scout some areas for puddles before students do the lab. If you cannot locate a park or similar area near your school, you may find vacant lots or cracked sidewalks that tend to hold puddles. Areas of a road very near the curb commonly develop puddles.

Using Scientific Methods

Scientific Methods

Objectives

- **Observe** natural phenomena.
- **Propose** hypotheses to explain natural phenomena.
- **Evaluate** hypotheses.

Materials

hand lens
meterstick

Not all scientists think alike, and scientists don't always agree about various concepts. However, all scientists use scientific methods, part of which are the skills of observing, inferring, and predicting. In this lab, you will apply scientific methods as you examine a place where puddles often form after rainstorms. You can study the puddle area even when the ground is dry, but it would be best to observe the area again when it is wet. Because water is one of the most effective agents of change in our environment, you should be able to make many observations.

MAKE OBSERVATIONS

1. Examine the area of the puddle and the surrounding area carefully. Make a numbered list of what can be seen, heard, smelled, or felt. Sample observations are as follows: "The ground where the puddle forms is lower than the surrounding area, and there are cracks in the soil." Remember to avoid making any suggestions of causes.

FORM A HYPOTHESIS

2. Review your observations, and write possible causes for those observations. Sample causes are as follows: "Cracks in the soil (Observation 2) may have been caused by a lack of rain (Observation 5)."

Step 1

CHAPTER RESOURCES

Chapter Resource File

- Datasheet for Chapter Lab GENERAL
- Lab Notes and Answers

3. Review your observations and possible causes, and place them into similar groups, if possible. Can one cause or set of causes explain several observations? Is each cause reasonable when compared with the others? Does any cause contradict any of the other observations?

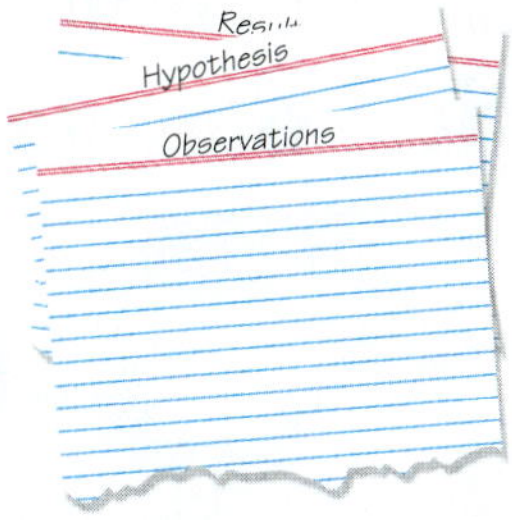

Step 2

TEST THE HYPOTHESIS

4. Based only on your hypotheses, make some predictions about what will happen at the puddle as conditions change. Describe the changes you expect and your reasoning. A sample prediction is the following: "If the puddle dries out, the crack will grow wider because of the loss of water."

5. Revisit the puddle several times to see if the changes that you observe match your predictions.

ANALYZE THE RESULTS

1. **Evaluating Results** Which of your predictions were correct, and which predictions were incorrect?

2. **Analyzing Methods** Which of your senses did you use most to make your observations? How could you improve your observations by using this sense? by using other senses?

3. **Evaluating Methods** What could you have used to measure, or quantify, many of your observations? Is quantitative observation better than qualitative observation? Explain your answer.

DRAW CONCLUSIONS

4. **Drawing Conclusions** Examine your incorrect predictions. What new knowledge have you gained from this new evidence?

5. **Analyzing Results** Reexamine your hypotheses. What can you say now about the ones that were correct?

6. **Drawing Conclusions** When knowledge is derived from observation and prediction, this process is called a *scientific method*. After reporting the results of a prediction, how might a scientist continue his or her research?

Extension

1. **Designing Experiments** Choose another small area to examine, but look for changes caused by a different factor, such as wind. Follow the steps outlined in this lab to predict changes that will occur in the area. Use scientific methods to design an experiment. Briefly describe your experiment, including how you would perform it.

Answers to Analyze the Results

1. Answers may vary. Accept all reasonable answers.
2. Answers may vary. Students may state that they used the sense of sight more than any other sense.
3. Answers may vary. Students could have measured the depth of the puddle and the degree of slope in the puddled area. They could use these measurements to calculate the volume of water in the puddle and then determine the rate at which the puddle disappeared.

Answers to Draw Conclusions

4. Answers may vary. Accept all reasonable answers.
5. Answers may vary. Accept all reasonable answers.
6. Students should recognize that after reporting the results of a prediction, a scientist would want to publish the results so other scientists can repeat the experiment to see if the results apply in all cases.

Answers to Extension

1. Answers may vary but should include an experiment that is reproducible and that follows all the steps in the scientific method.

TEACHER TESTED & APPROVED

Alonda Droege
Highline High School
Burien, WA

Geologic Features and Political Boundaries in Europe

Group Activity — GENERAL

Making Maps Have groups of students create maps of one region of the United States. Students should research the state boundaries, rivers, and topography of each state in their region. Students should use different colors or patterns to indicate rivers, boundaries, and various elevations. Have students present and explain their maps to the class. LS **Kinesthetic/Visual**

Answers to Map Skills Activity

1. rivers
2. 27
3. Italy
4. rivers
5. mountains
6. Answers may vary. These surface features represent clear physical boundaries that can be easily seen and do not change.

CHAPTER RESOURCES

Technology

Transparencies
- 3 Geologic Features and Political Boundaries in Europe (with worksheet)

MAPS in Action

Geologic Features and Political Boundaries in Europe

Map Skills Activity

This map shows the political boundaries of a part of Europe. The map also shows some surface features. Use the map to answer the questions below.

1. **Using the Key** What do the blue lines represent?
2. **Analyzing Data** How many countries are represented on the map?
3. **Examining Data** What country has two smaller countries within its borders?
4. **Applying Ideas** What type of surface features define the political boundary between Romania and Moldova and the political boundary between Switzerland and Germany?
5. **Applying Ideas** What type of surface feature defines the political boundary between Poland and the Czech Republic?
6. **Making Inferences** Why do you think that political boundaries commonly correspond with surface features?

Career Focus

Field Geologist

Imagine a summer visit to a meadow in Yellowstone National Park. Against blue skies, lush and grass-covered fields are dotted with grazing bison and elk. Field geologist Ken Pierce knows that there is more to this view than meets the eye. "Meadow areas, such as Hayden Valley," says Pierce, "are underlain by lake sediments deposited by glaciers thousands of years ago. This is why their soils hold a great deal of water and are great for plants—and for the animals that live there."

Earth's Past and Present

Pierce works in Montana for the United States Geological Survey. Pierce finds that connecting geology to biology is rewarding. "It's important to understand the geological history of the park. That's because the geology has important controls on the ecology. For example, the rock in the central part of Yellowstone is rhyolite, which forms nutrient-poor soils. This kind of rock almost always supports a desert-like ecology of lodgepole pine forests—thus, there are no bears in these areas because there's not much for bears to eat."

Geology—The Key

Like other geologists, Pierce studies the composition, structure, and history of Earth's crust. For Pierce, it was easy to see the career advantages. "You're outdoors when you're doing field geology!" says Pierce. While in the field, he observes, takes measurements, and collects samples that will later be studied in the lab. Pierce uses his observations, measurements, and samples to study the geological and biological history of the park.

Looking Ahead to the Future

Studying past events helps geologists better understand current events and sometimes predict future ones. For example, Pierce uses the data he collects to determine when the land surface was offset by faulting during past earthquakes. He uses this information to estimate the risk of future earthquakes. His research also helps predict landslides or mudflows after natural events such as wildfires. Knowing where and how often such hazards may occur helps reduce the risk to humans who visit Yellowstone National Park.

"A geologist is like a detective—he or she uses clues left in the natural world to determine what has happened in the past."

—Ken Pierce, Ph.D.

◀ Yellowstone was the first national park established in the United States.

SCiLINKS. NSTA Developed and maintained by the National Science Teachers Association

For a variety of links related to this subject, go to www.scilinks.org

Topic: Careers in Earth Science

SciLinks code: HQ60222

Career Focus

Field Geologist

Group Activity — GENERAL

Field Geology of Your Area Have students contact your state's or region's branch of the United States Geological Survey (USGS) to find out what types of rocks underlie your area and how they formed. Obtain photos of some common rocks and information on their composition and formation from the USGS, from local or school libraries, or from the Internet. Once students have obtained this information, have them work in groups to find samples of the different rocks. Students may look for the rocks in parks, playgrounds, backyards, or any other outdoor area where soil and rock are exposed. Have students create a presentation about the rocks they found, including information about the type of rock, how it formed, and how it affects—and is affected by—human activity, wildlife, and vegetation in your region. LS Kinesthetic

Extension — GENERAL

Research Have students identify four other Earth-science related careers. Have students write a brief synopsis of each career and present them to the class. LS Interpersonal

Chapter 2 Earth as a System

Planning Guide

Compression Guide

To shorten instruction because of time limitations, omit Section 3.

OBJECTIVES	LABS, DEMONSTRATIONS, AND ACTIVITIES	TECHNOLOGY RESOURCES
PACING • 45 min pp. 26–30 **Chapter Opener**		**OSP Parent Letter** **CD Student Edition on CD-ROM** **CD Chapter Summaries Audio CD** **VID Brain Food Video Quiz**
Section 1 Earth: A Unique Planet • Describe the size and shape of Earth. • Describe the compositional and structural layers of Earth's interior. • Identify the possible source of Earth's magnetic field. • Summarize Newton's law of gravitation.	TE **Discussion** Earth and its Moon, p. 27 GENERAL TE **Physics Connection** Seismic Waves, p. 28 ADVANCED	OSP **Lesson Plans** (also in print) TR **Bellringer*** TR 4 Earth's Interior* CD **Interactive Tutor** Earth's Interior CD **Interactive Tutor** Gravity, Orbits, Motion, and Forces CD **Interactive Tutor** Weight, Mass, Volume, and Density
PACING • 45 min pp. 31–38 **Section 2 Energy in the Earth System** • Compare an open system with a closed system. • List the characteristics of Earth's four major spheres. • Identify the two main sources of energy in the Earth system. • Identify four processes in which matter and energy cycle on Earth.	TE **Demonstration** Open and Closed Systems p. 31 GENERAL TE **Space Science Connection** Spheres of Other Planets and Moons, p. 33 ADVANCED TE **Physics Connection** The Laws of Thermodynamics, p. 34 ADVANCED SE **Quick Lab** Effects of Solar Energy, p. 35 GENERAL CRF **Datasheet for Quick Lab*** GENERAL TE **Group Activity** Earth's Other Cycles, p. 37 ADVANCED CRF **Inquiry Lab** Energy Transfer* GENERAL CRF **Making Models Lab** The Water Cycle* GENERAL	OSP **Lesson Plans** (also in print) TR **Bellringer*** TR 5 Open and Closed Systems* TR 6 Earth's Energy Budget* TR 7 The Nitrogen Cycle* TR 8 The Carbon Cycle* CD **Interactive Tutor** Earth Systems
PACING • 90 min pp. 39–42 **Section 3 Ecology** • Define *ecosystem.* • Identify three factors that control the balance of an ecosystem. • Summarize how energy is transferred through an ecosystem. • Describe one way that ecosystems respond to environmental change.	TE **Activity** Poster Project, p. 40 GENERAL SE **Quick Lab** Studying Ecosystems, p. 41 GENERAL CRF **Datasheet for Quick Lab*** GENERAL SE **Skills Practice Lab** Testing the Conservation of Mass, pp. 48–49 GENERAL CRF **Datasheet for Chapter Lab*** GENERAL SE **Maps in Action** Concentration of Plant Life on Earth, p. 50 GENERAL TE **Activity** Ocean-Plant Distributions, p. 50 GENERAL	OSP **Lesson Plans** (also in print) TR **Bellringer*** TR 9 A Food Web* TR 10 Concentration of Plant Life on Earth* TE **Internet Activity** Health and Biological Clocks, p. 51 ADVANCED CRF **Internet Activity** Health and Biological Clocks* ADVANCED CD **Interactive Tutor** Ice, Water, and Vapor CD **Interactive Tutor** The Water Cycle

PACING • 90 min

CHAPTER REVIEW, ASSESSMENT, AND STANDARDIZED TEST PREPARATION

SE **Chapter Highlights**, p. 43
SE **Chapter Review**, pp. 44–45
SE **Standardized Test Prep**, pp. 46–47
CRF **Concept Review*** GENERAL
CRF **Critical Thinking*** ADVANCED
CRF **Math Skills*** GENERAL
CRF **Graphing Skills*** GENERAL
CRF **Chapter Test A*** GENERAL
CRF **Chapter Test B*** ADVANCED
OSP **Lesson Plans** (also in print)
OSP **Test Generator**
OSP **Test Item Listing**

Online and Technology Resources

go.hrw.com Holt Online Learning

Visit **go.hrw.com** for access to Holt Online Learning, or enter the keyword **HQ6 Home** for a variety of free online resources.

One-Stop Planner® CD-ROM

This CD-ROM package includes
- Lab Materials QuickList Software
- Holt Calendar Planner
- Customizable Lesson Plans
- Printable Worksheets
- ExamView® Test Generator
- Interactive Teacher Edition
- Holt PuzzlePro®
- Holt PowerPoint® Resources

KEY

SE Student Edition	**OSP** One-Stop Planner	**VID** Classroom Video/DVD
TE Teacher Edition	**TR** Transparencies and Transparency Worksheets	* Also on One-Stop Planner
CRF Chapter Resource File	**CD** CD or CD-ROM	◆ Requires advance prep
LTP Long-Term Projects		■ Also available in Spanish

SKILLS DEVELOPMENT RESOURCES	REVIEW AND ASSESSMENT	CORRELATIONS
SE Pre-Reading Activity, p. 26 GENERAL **TE Using the Figure** Energy Changes, p. 26 GENERAL		National Science Education Standards
CRF Directed Reading* BASIC **SE Math Practice,** p. 28 GENERAL **TE Using the Figure** Earth's Interior Structure, p. 28 BASIC **TE Skill Builder** Writing, p. 29 ADVANCED	**SE Reading Check,** p. 28 GENERAL **SE Section Review,** p. 30 GENERAL **TE Reteaching,** p. 29 BASIC **TE Quiz,** p. 29 GENERAL **TE Alternative Assessment,** p. 29 ADVANCED **CRF Section Quiz** * ■ GENERAL	UCP 1, PS 4b
CRF Directed Reading* BASIC **TE Reading Skill Builder** Paired Summarizing, p. 32 BASIC **TE Skill Builder** Vocabulary, p. 33 GENERAL **TE Using the Figure** Overlapping of Spheres, p. 33 BASIC **TE Inclusion Strategies,** p. 33 **SE Graphic Organizer** Comparison Table, p. 34 GENERAL **TE Using the Figure** Energy Budget, p. 34 GENERAL **TE Reading Skill Builder** Reading Hint, p. 36 BASIC **TE Using the Figure** Nitrogen Cycle, p. 36 BASIC	**SE Reading Checks** pp. 32, 34, 36 GENERAL **SE Section Review,** p. 38 GENERAL **TE Homework,** p. 32 GENERAL **TE Reteaching,** p. 37 BASIC **TE Quiz,** p. 37 GENERAL **TE Alternative Assessment,** p. 37 ADVANCED **CRF Section Quiz** * ■ GENERAL	UCP 1, PS 5a, PS 5d, LS 4a, LS 5a, LS 5b, LS 5e
CRF Directed Reading* BASIC **TE Using the Figure** Members of Ecosystems, p. 39 GENERAL **TE Using the Figure** Food Webs, p. 41 BASIC **TE Inclusion Strategies,** p. 41	**SE Reading Check,** p. 40 GENERAL **SE Section Review,** p. 42 GENERAL **TE Reteaching,** p. 41 BASIC **TE Quiz,** p. 41 GENERAL **TE Alternative Assessment,** p. 42 ADVANCED **CRF Section Quiz** * ■ GENERAL	LS 4a, LS 4b, LS 4c, LS 4d, LS 4e, LS 5a, LS 5b, LS 5c, LS 5e

Holt Lab Generator CD-ROM

Search for labs by topic, standard, difficulty level, and/or time frame. Edit those labs to fit your needs, or create your own labs. Use the Lab Materials QuickList to order just what you need.

Holt Earth Science Interactive Tutor CD-ROM

This CD-ROM consists of interactive activities that give students a fun way to extend their knowledge of Earth science concepts.

Chapter Summaries Audio CDs

These CDs include audio summaries of the key concepts presented in each chapter. (Audio summaries are also available in Spanish.)

SciLinks NSTA

www.scilinks.org

Maintained by the **National Science Teachers Association.** See Chapter Enrichment pages that follow for a complete list of topics.

See Chapter Enrichment pages for Video Resources.

Chapter Enrichment

This Chapter Enrichment provides relevant and interesting information to expand and enhance your classroom instruction of the chapter material.

Section 1 Earth: A Unique Planet

Weight and Centripetal Force

The rotation of Earth affects the gravitational attraction exerted by Earth at its surface in two ways. First, the increased bulging of Earth at the equator results in the surface at the equator being farther from Earth's center than the surface at the poles is. The second effect is centripetal force, which is the force required to hold a body in a circular path. This force is familiar to anyone who has ever been in a car turning a corner quickly or whirled a ball overhead on a string. The tendency of any object to move in a straight line causes the object to fly outward from a circular path, unless some force, such as friction or tension in a string, holds it in place. It is the force of gravity that holds an object at Earth's equator in place on the surface. The net force on the object will be less at the equator than at the poles, where Earth's rotational speed is closer to zero. Because weight is equal to the net force on an object, the weight of objects at the equator is less than at the poles due to centripetal force, as well as to the greater distance from Earth's center of mass.

While the effect of centripetal force on weight at the equator is less than one percent of gravitational force due to Earth's bulging, it can be important in measurements requiring precision on the order of 0.01 m/s^2. The gravitational acceleration at the poles is 9.832 m/s^2, while the gravitational acceleration at the equator due to only Earth's bulging is 9.814 m/s^2, and is 9.780 m/s^2 when centripetal acceleration is also taken into account.

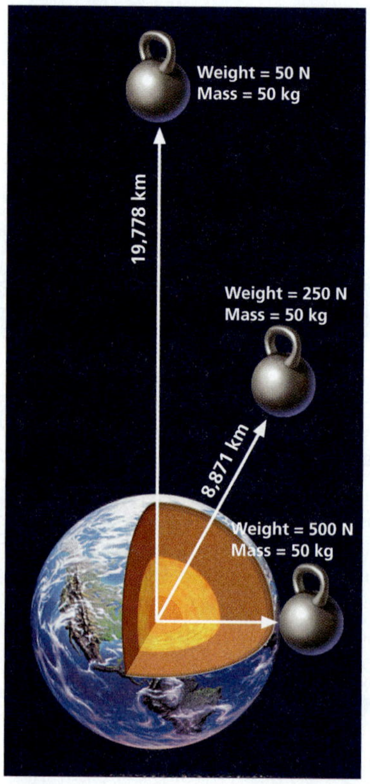

◀ **An object's mass is constant as its distance from Earth's center changes.**

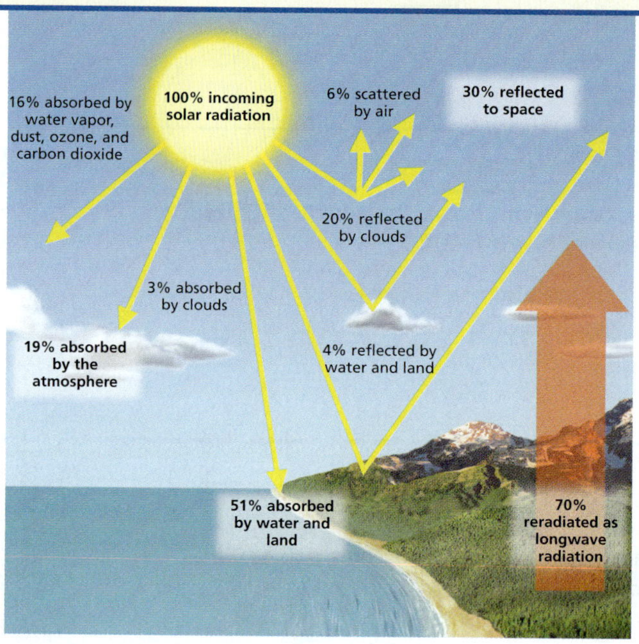

▲ **Variations in Earth's energy budget can have serious consequences on climate.**

Section 2 Energy in the Earth System

Energy Budget and Climate

While the energy provided to Earth by the sun is fairly uniform, variations in solar activity may alter the amount of energy that enters Earth's atmosphere, and thus these variations affect climate. Although the mechanism is not well understood, there appears to be a connection between the number of sunspots on the sun's surface and the amount of energy that is emitted by the sun.

Even when solar input is relatively constant, components of Earth's various "spheres" can affect how energy is distributed throughout Earth's surface. The increased concentration of carbon dioxide in the atmosphere in recent years concerns some scientists. More infrared radiation is absorbed by the atmosphere as carbon dioxide levels increase. This leads to a warming of Earth's atmosphere that resembles the warming within a greenhouse, and so is referred to as the "greenhouse effect." While carbon dioxide in the atmosphere prevents Earth from becoming too cold to support life, increases in temperature can have adverse effects, from melting ice at the poles to climatic changes, such as more drought or flooding. Damage to the ozone in the atmosphere also affects the energy budget by allowing too much ultraviolet radiation into Earth's atmosphere.

Section 3 Ecology

Organisms and their Ecosystems

The connection between an organism and its ecosystem ensures that the number of organisms stays in balance —neither too many nor too few—within that ecosystem. However, when an organism is introduced into a different ecosystem, a population imbalance may occur that can take a long time to correct. This problem has become more frequent as human activities, such as shipping and air travel, make interaction between ecosystems fast and easy. Organisms can become displaced accidentally. One example is the introduction of zebra mussels, which are native to southern European waters, to the Great Lakes area of the United States.

Before people understood the connection between organisms and ecosystems, species from one ecosystem were introduced into another to solve local problems. For instance, a desire for a rapidly growing ground cover prompted the introduction of kudzu, an Asian vine, into North America. The kudzu quickly spread. Because there were no native species to keep the kudzu under control, it now requires considerable effort and resources to prevent it's spreading throughout the southern U.S.

▼ **Energy moves through the organisms that are part of an ecosystem.**

Video Resources

Brain Food Video Quizzes

Brain Food Video Quizzes These videos contain game-show style quizzes that assess students' progress and motivate students to study the chapter material.

NOVA® Videos

To order **NOVA** videos related to this chapter, visit **go.hrw.com** and enter the keyword **HQ6EASV.**

SciLinks®

Developed and maintained by the National Science Teachers Association

SciLinks is maintained by the National Science Teachers Association to provide you and your students with interesting, up-to-date links that will enrich your classroom presentation of the chapter.

Visit www.scilinks.org and enter the SciLinks code for more information about the topic listed.

Topic: Zones of Earth
SciLinks code: HQ61684

Topic: Carbon Cycle
SciLinks code: HQ60216

Topic: Nitrogen Cycle
SciLinks code: HQ61036

Topic: Ecology
SciLinks code: HQ60462

Chapter 2

Chapter Overview

Earth is a unique planet. It is also a system in which the amount of matter is relatively constant, but through which energy flows. Earth's four spheres, which include its bodies of water, its land, its atmosphere, and its living organisms, all interact. The transfer of energy and matter on Earth takes place through various cycles. Complex relationships exist between Earth's living inhabitants and their nonliving environment.

Using the Figure — GENERAL

Energy Changes Explain that bears position themselves so that when salmon leap up the waterfall, the salmon can be easily caught. Ask students what energy changes take place when the salmon leaps upward. (Answers may vary. Sample answer: The salmon uses energy stored in its body to overcome gravity.) **LS Logical**

PRE-READING ACTIVITY

FOLDNOTES

You may want students to work in pairs to create this FoldNote. Have one student fill in the first and third columns, and have the other student fill in the second column.

Chapter 2 Earth as a System

Sections

1. Earth: A Unique Planet
2. Energy in the Earth System
3. Ecology

What You'll Learn

- How Earth is structured
- How energy and matter cycle through the Earth system
- Why living organisms are important to the Earth system

Why It's Relevant

The systems approach to studying Earth provides a way to understand the interrelated nature of the physical, chemical, and biological forces that shape the planet.

PRE-READING ACTIVITY

FOLDNOTES

Table Fold Before you read this chapter, create the FoldNote entitled "Table Fold" described in the Skills Handbook section of the Appendix. Label the columns of the table "Earth's structure," "Earth's cycles and systems," and "Ecosystems." As you read the chapter, write examples of each topic under the appropriate column.

▶ This brown bear gains energy and nutrients from eating salmon. Energy and matter move through Earth's systems in many ways, including through the eating of prey by predators.

Chapter Correlations — *National Science Education Standards*

UCP 1 The natural and designed world is complex; ... Scientists and students learn to define small portions for the convenience of investigation. The units of investigation can be referred to as "systems."... **(Sections 1 and 2)**

PS 4b Gravitation is a universal force that each mass exerts on any other mass. ...**(Section 1)**

PS 5a The total energy of the universe is constant. Energy can be transferred...However, it can never be destroyed.... **(Section 2)**

PS 5d Everything tends to become less organized and less orderly over time. Thus, in all energy transfers, the overall effect is that energy is spread out uniformly.... **(Section 2)**

LS 4a The atoms and molecules on Earth cycle among the living and nonliving components of the biosphere. **(Sections 2 and 3)**

LS 4b Energy flows through ecosystems in one direction, from photosynthetic organisms to herbivores to carnivores and decomposers. **(Section 3)**

LS 4c Organisms both cooperate and compete in ecosystems... that are stable for hundreds or thousands of years. **(Section 3)**

LS 4d Living organisms have the capacity to produce populations of infinite size, but environments ... are finite.... **(Section 3)**

LS 4e Human beings live within the world's ecosystems. Increasingly, humans modify ecosystems... . **(Section 3)**

LS 5a All matter tends toward more disorganized states. ... **(Sections 2 and 3)**

LS 5b The energy for life primarily derives from the sun. ... **(Sections 2 and 3)**

LS 5c The chemical bonds of food molecules contain energy.... **(Section 3)**

LS 5e The distribution and abundance of organisms and populations in ecosystems are limited by the availability of matter and energy .... **(Sections 2 and 3)**

Section 1 Earth: A Unique Planet

Earth is unique for several reasons. It is the only known planet in the solar system that has liquid water on its surface and an atmosphere that contains a large proportion of oxygen. Earth is also the only planet—in our solar system or in any other solar system—that is known to support life. Scientists study the characteristics of Earth that make life possible in order to know what life-supporting conditions to look for on other planets.

OBJECTIVES

- **Describe** the size and shape of Earth.
- **Describe** the compositional and structural layers of Earth's interior.
- **Identify** the possible source of Earth's magnetic field.
- **Summarize** Newton's law of gravitation.

KEY TERMS

crust
mantle
core
lithosphere
asthenosphere
mesosphere

Earth Basics

Earth is the third planet from the sun in our solar system. Earth formed about 4.6 billion years ago and is made mostly of rock. Approximately 71% of Earth's surface is covered by a relatively thin layer of water called the *global ocean*.

As viewed from space, Earth is a blue sphere covered with white clouds. Earth appears to be a perfect sphere but is actually an *oblate spheroid,* or slightly flattened sphere, as **Figure 1** shows. The spinning of Earth on its axis makes the polar regions flatten and the equatorial zone bulge. Earth's pole-to-pole circumference is 40,007 km. Its equatorial circumference is 40,074 km.

Earth's surface is relatively smooth. That is, distances between surface high points and low points are small relative to Earth's size. The difference between the height of the tallest mountain and the depth of the deepest ocean trench is about 20 km. This distance is small compared with Earth's average diameter of 12,756 km.

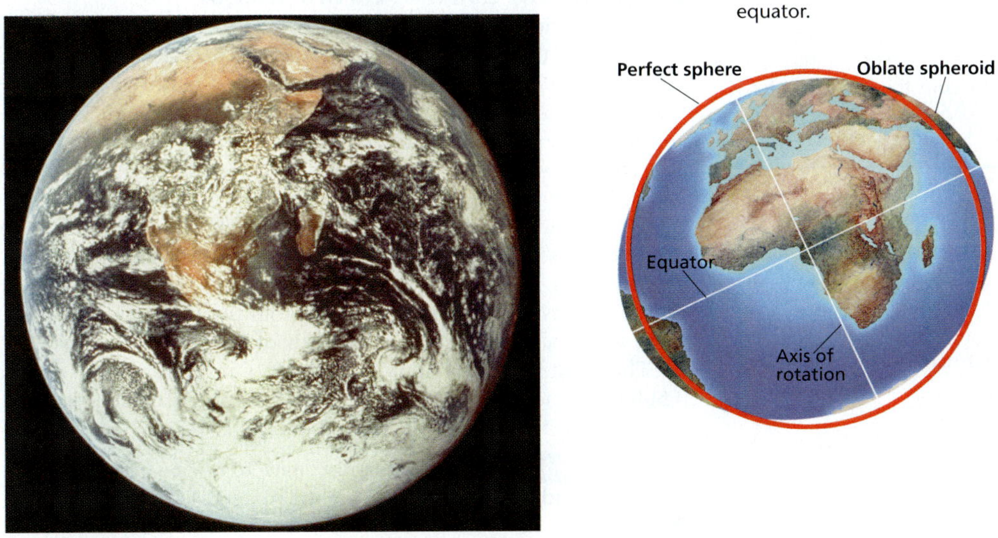

Figure 1 ▶ Although from afar Earth looks like a sphere (left), it is an oblate spheroid. In this illustration, Earth's shape has been exaggerated to show that Earth bulges at the equator.

Section 1

Focus

Overview

This section describes the structure of Earth, the basic surface features of Earth, and the nature of Earth's magnetic field.

Bellringer

Have students write down four features of Earth. These may include features not easily observed, such as the structural features of Earth's interior. (Sample answers: atmosphere, water, oceans that cover most of Earth's surface, extensive plant and animal life, ice caps, tropical rain forests, deserts, geologic activity as indicated by volcanoes and earthquakes, mountain ranges, and glaciers) LS **Verbal/Logical**

Motivate

Discussion — GENERAL

Earth and its Moon On the chalkboard make two columns labeled *Similarities* and *Differences*. Have students consider the differences between Earth and its moon and give answers for each column. (Similarities may include rocky surfaces, spherical shapes, and the sun as an energy source. Differences may include the facts that Earth has life, while the moon is lifeless; the moon has no water; and the moon has no atmosphere.) LS **Verbal/Logical**

CHAPTER RESOURCES

Chapter Resource File

- Directed Reading BASIC

Technology

Transparencies
- Bellringer

Student Edition on CD-ROM

One-Stop Planner CD-ROM
- Lesson Plan

Teach

Using the Figure — Basic

Earth's Interior Structure Have students carefully examine the information in the figure on this page. Explain to students that the specific pressures and temperatures of the interior determine whether a layer is solid or liquid. Higher temperatures tend to cause solids to melt and become liquid, but higher pressures may counter this effect and compress liquids into solids. Then, ask students which part of Earth's interior takes up the greatest volume. (the mantle) Ask which region takes up the smallest volume. (the crust) Have students use the text to create a depth scale for the layers in the figure. Ask students where the continental crust is thickest. (The crust is thickest beneath large mountain ranges.) **LS Visual**

MATH PRACTICE

Answer

time of travel for wave = thickness of layer ÷ speed of wave = (35 km ÷ 8 km/s) + (2,900 km ÷ 12 km/s) + (2,250 km ÷ 9.5 km/s) + (1,228 km ÷ 10.5 km/s)

time of travel for wave = 4 s + 242 s + 237 s + 117 s = 600 s

Answer to Reading Check

Indirect observations are the only means available for exploring Earth's interior at depths too great to be reached by drilling.

CHAPTER RESOURCES

Technology

Transparencies
- 4 Earth's Interior (with worksheet)

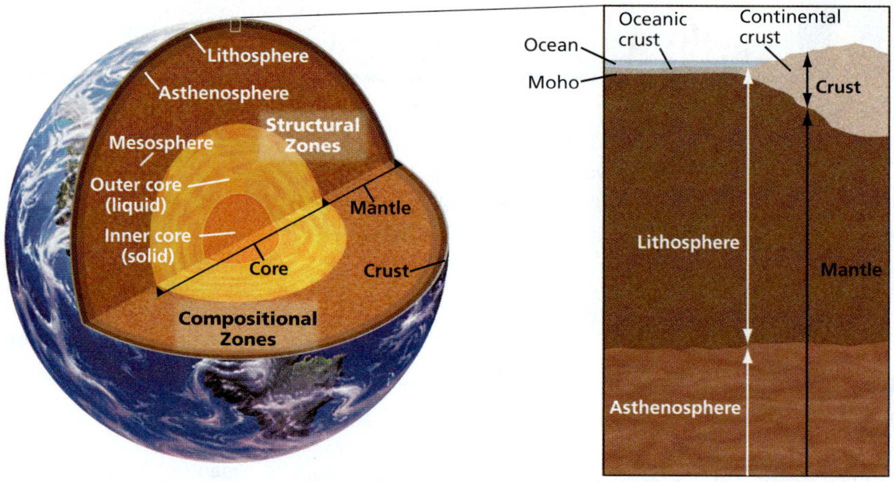

Figure 2 ▶ Changes in the speed and direction of seismic waves were used to determine the locations and properties of Earth's interior zones.

MATH PRACTICE

Speeding Waves
Earth's layers are of the following average thicknesses: crust, 35 km; mantle, 2,900 km; outer core, 2,250 km; and inner core, 1,228 km. Estimate how long a seismic wave would take to reach Earth's center if the wave's average rate of travel was 8 km/s through the crust, 12 km/s through the mantle, 9.5 km/s through the outer core, and 10.5 km/s through the inner core.

Earth's Interior

Direct observation of Earth's interior has been limited to the upper few kilometers that can be reached by drilling. So, scientists rely on indirect methods to study Earth at greater depths. For example, scientists have made important discoveries about Earth's interior through studies of seismic waves. *Seismic waves* are vibrations that travel through Earth. Earthquakes and explosions near Earth's surface produce seismic waves. By studying these waves as they travel through Earth, scientists have determined that Earth is made up of three major compositional zones and five major structural zones, as shown in **Figure 2.**

Compositional Zones of Earth's Interior

The thin, solid, outermost zone of Earth is called the **crust.** The crust makes up only 1% of Earth's mass. The crust beneath the oceans is called *oceanic crust*. Oceanic crust is only 5 to 10 km thick. The part of the crust that makes up the continents is called *continental crust*. The continental crust varies in thickness and is generally between 15 to 80 km thick. Continental crust is thickest beneath high mountain ranges.

The lower boundary of the crust, which was named for its discoverer, is called the *Mohorovičić* (MOH hoh ROH vuh CHICH) *discontinuity*, or *Moho*. The **mantle,** the layer that underlies the crust, is denser than the crust. The mantle is nearly 2,900 km thick and makes up almost two-thirds of Earth's mass.

The center of Earth is a sphere whose radius is about 3,500 km. Scientists think that this center sphere, called the **core,** is composed mainly of iron and nickel.

Reading Check **Explain why scientists have to rely on indirect observations to study Earth's interior. (See the Appendix for answers to Reading Checks.)**

crust the thin and solid outermost layer of the Earth above the mantle

mantle in Earth science, the layer of rock between Earth's crust and core

core the central part of the Earth below the mantle

PHYSICS CONNECTION — Advanced

Seismic Waves Scientists have learned about the physical structure of Earth's interior by measuring seismic waves. Seismic waves are an example of traveling waves, or mechanical waves that travel through a medium. Sound waves are the most familiar example of traveling waves. The speed of sound within a given medium depends on factors such as the density and compressibility of the medium.

Seismic waves, however, differ from sound waves in that they consist of two different types: P waves (also called *primary* or *pressure waves*) and S waves (also known as *secondary* or *shear waves*). One difference between these waves is that S waves do not travel through liquids, whereas P waves do, although more slowly than through solids.

Have students research the differences between longitudinal (or compression) waves and transverse waves, and relate these topics to P waves and S waves. Students should present their findings in a short written report or oral presentation. **LS Verbal**

Structural Zones of Earth's Interior

The three compositional zones of Earth's interior are divided into five structural zones. The uppermost part of the mantle is cool and brittle. This part of the mantle and the crust above it make up the **lithosphere,** a rigid layer 15 to 300 km thick. Below the lithosphere is a less rigid layer, known as the **asthenosphere.** The asthenosphere is about 200 to 250 km thick. Because of enormous heat and pressure, the solid rock of the asthenosphere has the ability to flow. The ability of a solid to flow is called *plasticity*. Below the asthenosphere is a layer of solid mantle rock called the **mesosphere.**

At a depth of about 2,900 km lies the boundary between the mantle and the *outer core*. Scientists think that the outer core is a dense liquid. The inner core begins at a depth of 5,150 km. The inner core is a dense, rigid solid. The inner and outer core together make up nearly one-third of Earth's mass.

lithosphere the solid, outer layer of Earth that consists of the crust and the rigid upper part of the mantle

asthenosphere the solid, plastic layer of the mantle beneath the lithosphere; made of mantle rock that flows very slowly, which allows tectonic plates to move on top of it

mesosphere literally, the "middle sphere"; the strong, lower part of the mantle between the asthenosphere and the outer core

Earth as a Magnet

Earth has two magnetic poles. The lines of force of Earth's magnetic field extend between the North geomagnetic pole and the South geomagnetic pole. Earth's magnetic field, shown in **Figure 3,** extends beyond the atmosphere and affects a region of space called the *magnetosphere*.

The source of Earth's magnetic field may be the liquid iron in Earth's outer core. Scientists hypothesize that motions within the core produce electric currents that in turn create Earth's magnetic field. However, recent research indicates that the magnetic field may have another source. Scientists have learned that the sun and moon also have magnetic fields. Because the sun contains little iron and the moon does not have a liquid outer core, discovering the sources of the magnetic fields of the sun and moon may help identify the source of Earth's magnetic field.

SCILINKS. NSTA Developed and maintained by the National Science Teachers Association

For a variety of links related to this subject, go to www.scilinks.org

Topic: Zones of Earth

SciLinks code: HQ61684

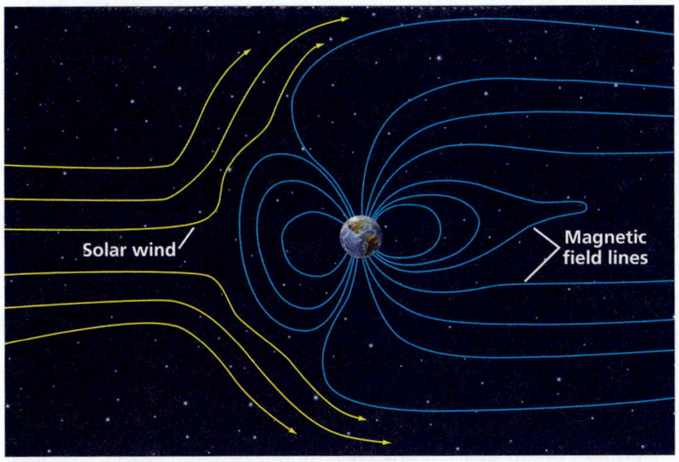

Figure 3 ▶ The magnetic field lines around Earth show the shape of Earth's magnetosphere. Earth's magnetosphere is compressed and shaped by solar wind, which is the flow of charged particles from the sun.

Close

Reteaching — BASIC

Earth's Interior Structure Draw a series of concentric circles on the board, making the two outer circles very close together and the second and third circles very far apart. To the side of the circles, write the words *crust*, *mantle*, *outer core*, and *inner core*. Have students indicate which region within the circles corresponds to each layer of Earth. LS **Visual**

Quiz — GENERAL

Determine whether each of the following statements is true or false.

1. Earth has the shape of a perfect sphere. (false)
2. The force of gravity on Earth's surface is greater at the poles than at the equator. (true)
3. Earth's core consists of two regions, one liquid and one solid. (true)

Alternative Assessment — ADVANCED

Poster Project Have students select one of the other planets in the solar system and develop a poster that compares the internal structures, shapes, surfaces, magnetic fields, and gravitational fields of the chosen planet and Earth. LS **Visual/Logical**

Teaching Tip — BASIC

Connect to Familiar Processes Students may have difficulty visualizing the properties of the asthenosphere and how solid rock can also be fluid. Have students start by thinking of thick fluids that flow slowly, like honey or syrup. Explain that some fluids are so thick that they are not observed to flow except over long periods of time. Remind them that children's plastic putty has this property.

SKILL BUILDER — ADVANCED

Writing Have students research and write an essay that compares and contrasts Earth's magnetic field with that of another planet in the solar system. Subtopics might include how each magnetic field is generated, how strong it is, and what effects it has on each planet and on the region of space that surrounds it. Encourage students to use a Venn diagram or a chart to help organize and compare the information before they begin writing. LS **Verbal/Logical**

Close, *continued*

Answers to Section Review

1. Earth is an oblate, or slightly flattened, spheroid with a polar circumference of 40,007 km and an equatorial circumference of 40,074 km.
2. Answers should include two of the following: Earth's surface is mostly covered by water; Earth has a thick atmosphere containing a large proportion of oxygen; and Earth supports life.
3. Scientists observe how seismic waves travel through Earth's interior in order to determine the physical states of Earth's deeper regions.
4. Earth's three compositional layers are the thin crust at the surface, the large mantle that lies beneath the crust, and the core, which lies at Earth's center. The five structural layers describe the interior in terms of physical properties. The crust and upper mantle thus make up the lithosphere of solid, rigid rock; beneath this lies the asthenosphere of solid but plastic rock; the mesosphere is the solid rock that makes up the rest of the mantle; the liquid outer core comes next, and the solid inner core is at Earth's center.
5. Motion within the liquid iron of Earth's outer core may produce electric currents, which in turn generate Earth's magnetic field.
6. The force of gravitational attraction between two objects increases as the masses of the objects increase and as the distance between the objects decreases.
7. A person's greater weight at the poles than at the equator suggests that the equator is farther from Earth's center than the poles are, and therefore Earth's shape is not perfectly spherical.
8. The asthenosphere is solid rock that is able to flow because of its plasticity. The mesosphere is solid rock that remains rigid.
9. On the mountain peak, I would be farther from Earth's center than at sea level, so the gravitational attraction would be slightly less. Therefore, I would weigh less.
10. Earth's interior has five compositional zones—the *lithosphere,* which consists of the crust and the rigid upper part of the *mantle;* the *asthenosphere;* the *mesosphere;* and the two regions of Earth's *core*, the liquid *outer core* and the solid *inner core*.

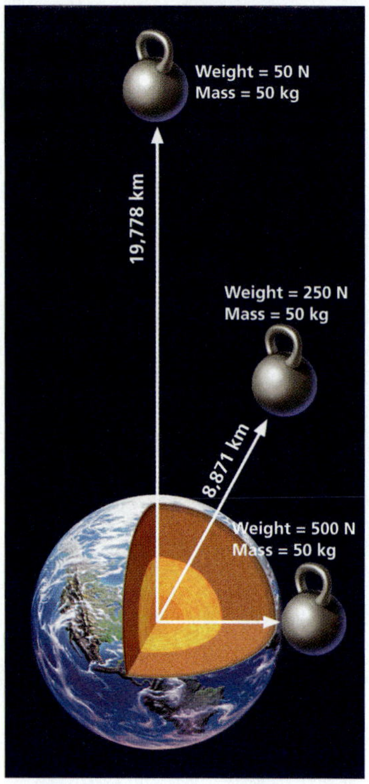

Figure 4 ▶ As the distance between an object and Earth's center increases, the weight of the object decreases. An object's mass is constant as its distance from Earth's center changes.

Earth's Gravity

Earth, like all objects in the universe, is affected by gravity. *Gravity* is the force of attraction that exists between all matter in the universe. The 17th-century scientist Isaac Newton was the first to explain the phenomenon of gravity. Newton described the effects of gravity in his *law of gravitation*. According to the law of gravitation, the force of attraction between any two objects depends on the masses of the objects and the distance between the objects. The larger the masses of two objects and the closer together the objects are, the greater the force of gravity between the objects will be.

Weight and Mass

Earth exerts a gravitational force that pulls objects toward the center of Earth. Weight is a measure of the strength of the pull of gravity on an object. The newton (N) is the SI unit used to measure weight. On Earth's surface, a kilogram of mass weighs about 10 N. The mass of an object does not change with location, but the weight of the object does. An object's weight depends on its mass and its distance from Earth's center. According to the law of gravitation, the force of gravity decreases as the distance from Earth's center increases, as shown in **Figure 4.**

Weight and Location

Weight varies according to location on Earth's surface. As you may recall, Earth spins on its axis, and this motion causes Earth to bulge near the equator. Therefore, the distance between Earth's surface and its center is greater at the equator than at the poles. This difference in distance means that your weight at the equator would be about 0.3% less than your weight at the North Pole.

Section 1 Review

1. **Describe** the size and shape of Earth.
2. **Describe** two characteristics that make Earth unique in our solar system.
3. **Summarize** how scientists learn about Earth's interior.
4. **Compare** Earth's compositional layers with its structural layers.
5. **Identify** the possible source of Earth's magnetic field.
6. **Summarize** Newton's law of gravitation.

CRITICAL THINKING

7. **Making Inferences** What does the difference between your weight at the equator and your weight at the poles suggest about the shape of Earth?
8. **Making Comparisons** How does the asthenosphere differ from the mesosphere?
9. **Analyzing Ideas** Why would you weigh less on a high mountain peak than you would at sea level?

CONCEPT MAPPING

10. Use the following terms to create a concept map: *crust, mantle, core, lithosphere, asthenosphere, mesosphere, inner core,* and *outer core.*

CHAPTER RESOURCES

Chapter Resource File

- Section Quiz GENERAL

Workbooks

- Study Guide (also in Spanish)

ENVIRONMENTAL CONNECTION

Section 2 Energy in the Earth System

Traditionally, different fields of Earth science, such as geology, oceanography, and meteorology, have been studied separately. Geologists studied Earth's rocks and interior, oceanographers studied the oceans, and meteorologists studied the atmosphere. But now, some scientists are approaching the study of Earth in a new way. They are combining knowledge of several fields of Earth science in order to study Earth as a system.

OBJECTIVES

- **Compare** an open system with a closed system.
- **List** the characteristics of Earth's four major spheres.
- **Identify** the two main sources of energy in the Earth system.
- **Identify** four processes in which matter and energy cycle on Earth.

KEY TERMS

system
atmosphere
hydrosphere
geosphere
biosphere

Earth-System Science

An organized group of related objects or components that interact to create a whole is a **system.** Systems vary in size from subatomic to the size of the universe. All systems have boundaries, and many systems have matter and energy that flow through them. Even though each system can be described separately, all systems are linked. A large and complex system, such as the Earth system, operates as a result of the combination of smaller, interrelated systems, as shown in **Figure 1.**

system a set of particles or interacting components considered to be a distinct physical entity for the purpose of study

The operation of the Earth system is a result of interaction between the two most basic components of the universe: matter and energy. *Matter* is anything that has mass and takes up space. Matter can be subatomic particles, such as protons, electrons, and neutrons. Matter can be atoms or molecules, such as oxygen atoms or water molecules, and matter can be larger objects, such as rocks, living organisms, or planets. *Energy* is defined as the ability to do work. Energy can be transferred in a variety of forms, including heat, light, vibrations, or electromagnetic waves. A system can be described by the way that matter and energy are transferred within the system or to and from other systems. Transfers of matter and energy are commonly accompanied by changes in the physical or chemical properties of the matter.

Figure 1 ▶ This threadfin butterflyfish is part of a system that includes other living organisms, such as coral. Together, the organisms are part of a larger system, a coral reef system in Micronesia.

Section 2

Focus

Overview

This section explains open and closed systems and identifies how Earth is a type of open system. Earth's four spheres are identified, and Earth's energy budget is described. Finally, the cycles in the Earth system and the effect of human activity on these cycles are examined.

Bellringer

Have students write down how energy and matter are added to and removed from the Earth system. (Answers may vary. Sample answer: Energy is added by sunlight and from Earth's interior, and is removed by radiation into space. Matter is added when meteorites collide with Earth and is removed when spacecraft are launched and by the steady loss of gases into space.) LS **Logical**

Motivate

Demonstration — GENERAL

Open and Closed Systems Boil water in pan on a hotplate. When water vapor rises out of the pan, ask students if they think the pan, hot plate, and water represent an open or closed system, and why. (The system is open because water leaves the pan in the form of water vapor.) Place a lid on the pan. Ask students if the system is now open or closed. (The system is now closed because matter no longer leaves the system.) LS **Visual/Logical**

CHAPTER RESOURCES

Chapter Resource File

- Directed Reading BASIC

Technology

Transparencies
- Bellringer

Student Edition on CD-ROM

One-Stop Planner CD-ROM
- Lesson Plan

Teach

READING SKILL BUILDER — BASIC

Paired Summarizing Group students into pairs, and have them read silently about open and closed systems. Then, have one student summarize how open and closed systems are similar and how they differ. The other student should listen to the retelling and should point out any inaccuracies or ideas that were left out. Allow students to refer to the text as needed. English Language Learners
LS Verbal

Teaching Tip — GENERAL

Connect to Prior Knowledge To help students recognize whether a system is open or closed, have them ask themselves the question "Is matter in the form of a gas entering or leaving the system?" Many systems are actually open, but because the matter that is transferred between the system and its surroundings is in the form of gas atoms or molecules, the transfer is invisible and therefore may be overlooked. If students think first about whether gas is entering or leaving the system, it will be easier for them to determine whether the system is open or closed. **LS Logical**

Answer to Reading Check

Dust and rock come to Earth from space, while hydrogen atoms from the atmosphere enter space from Earth.

Figure 2 ▶ Energy is exchanged in both the closed system (left) and the open system (right). In the open system, matter is also exchanged.

Closed Systems

A *closed system* is a system in which energy, but not matter, is exchanged with the surroundings. **Figure 2** shows a sealed jar, which is a closed system. Energy in the form of light and heat can be exchanged through the jar's sides. But because the jar is sealed, matter cannot exit or enter the system. Most aquariums are open systems because oxygen and food must be added to them, but some are closed systems. Closed-system aquariums contain a variety of organisms: plants, which produce oxygen, and aquatic animals, some of which are food for others. Some of the animals feed on the plants. Animal wastes and organic matter nourish the plants. Only sunlight enters from the surroundings.

Open Systems

An *open system* is a system in which both energy and matter are exchanged with the surroundings. The open jar in **Figure 2** is an open system. A lake is also an open system. Water molecules enter a lake through rainfall and streams. Water exits a lake through streams, evaporation, and absorption by the ground. Sunlight and air exchange heat with the lake. Wind's energy is transferred to the lake as waves.

The Earth System

Technically, all systems that make up the Earth system are open. But the Earth system is almost a closed system because matter exchange is limited. Energy enters the system in the form of sunlight and is released into space as heat. Only a small amount of dust and rock from space enters the system, and only a fraction of the hydrogen atoms in the atmosphere escape into space.

✓ Reading Check **What types of matter and energy are exchanged between Earth and space? (See the Appendix for answers to Reading Checks.)**

Homework — GENERAL

Systems at Home Have students look for various systems in and around their own homes. Have students compile a list of six to ten of these systems, noting whether they are open or closed and including a brief explanation as to why. They may list such systems as a refrigerator (a closed system, if no food is added or removed); an automobile (an open system); an electric lamp (a closed system); and a washing machine (an open system). You may wish to have students describe one item from their list for the class. **LS Verbal/Logical**

CHAPTER RESOURCES

Chapter Resource File

- **Inquiry Lab** Energy Transfer GENERAL
- **Making Models Lab** The Water Cycle GENERAL

Technology

Transparencies

- 5 Open and Closed Systems (with worksheet)

Earth's Four Spheres

Matter on Earth is in solid, liquid, and gaseous states. The Earth system is composed of four "spheres" that are storehouses of all of the planet's matter. These four spheres are shown in **Figure 3.**

The Atmosphere

The blanket of gases that surrounds Earth's surface is called the **atmosphere.** The atmosphere provides the air that you breathe and shields Earth from the sun's harmful radiation. Earth's atmosphere is made up of 78% nitrogen and 21% oxygen. The remaining 1% includes other gases, such as argon, carbon dioxide, water vapor, and helium.

The Hydrosphere

Water covers 71% of Earth's surface area, and 97% of surface water is contained in the salty oceans. The remaining 3% is fresh water. Fresh water can be found in lakes, rivers, and streams, frozen in glaciers and the polar ice sheets, and underground in soil and bedrock. All of Earth's water except the water that is in gaseous form in the atmosphere makes up the **hydrosphere.**

The Geosphere

The mostly solid part of Earth is known as the **geosphere.** This sphere includes all of the rock and soil on the surface of the continents and on the ocean floor. The geosphere also includes the solid and molten interior of Earth, which makes up the largest volume of matter on Earth. Natural processes, such as volcanism, bring matter from deep inside Earth's interior to the surface. Other processes move surface matter back into Earth's interior.

The Biosphere

Another one of the four subdivisions of the Earth system is the biosphere. The **biosphere** is composed of all of the forms of life in the geosphere, in the hydrosphere, and in the atmosphere. The biosphere also contains any organic matter that has not decomposed. Once organic matter has completely decomposed, it becomes a part of the other three spheres. The biosphere extends from the deepest parts of the ocean to the atmosphere a few kilometers above Earth's surface.

atmosphere a mixture of gases that surrounds a planet or moon

hydrosphere the portion of the Earth that is water

geosphere the mostly solid, rocky part of the Earth; extends from the center of the core to the surface of the crust

biosphere the part of Earth where life exists; includes all of the living organisms on Earth

Figure 3 ▶ The Earth system is composed of the atmosphere, hydrosphere, geosphere, and biosphere. ***Can you identify elements of the four spheres in this photo?***

SPACE SCIENCE CONNECTION — ADVANCED

Spheres of Other Planets and Moons Although Earth is unique in being the only known planet to have a biosphere, the other three spheres are found, to a greater or lesser degree, on many other planets. Some planets, like the gas giant Jupiter, are almost entirely atmosphere, and as such the study of their atmosphere constitutes most of the study of the planet. Some planets, like Mercury, may have a hydrosphere in the form of ice near the poles, but are almost exclusively geospheres. Evidence suggests that Mars once had a hydrosphere, though little remains of Martian water.

Have students research planets or large moons of planets within the solar system, with emphasis on whether the bodies possess geospheres, hydrospheres, or atmospheres. Have students note unusual features of these spheres as compared to their counterparts on Earth. Students may then present their findings in a written report, oral presentation, or poster project. LS **Verbal**

SKILL BUILDER — GENERAL

Vocabulary The names of Earth's four spheres are derived from ancient Greek. The Greek word for "ball" is *sphaira,* from which the word *sphere* is derived. The Greek roots *atmos, hydro, geo,* and *bios* mean "vapor," "water," "earth," and "life," respectively. LS **Verbal**

Using the Figure — BASIC

Overlapping of Spheres Have students look carefully at the figure on this page. Point out to them that the contents of the four spheres are not always separate, and that elements of one sphere may be found in another sphere. Thus, while water in the form of ice or liquid water is part of the hydrosphere, water in the form of water vapor is part of the atmosphere. Answer to caption question: The clouds and sky are parts of the atmosphere. The hydrosphere is represented by the lake and the snow on the mounains. The field, forest, and mountains are parts of the geosphere, and though it isn't seen in the photograph, the bottom of the lake on which the moose is standing is also part of the geosphere. The moose in the foreground, and the vegetation in the field and forest of the background are all parts of the biosphere. LS **Visual**

INCLUSION Strategies

- ***Learning Disabled***
- ***Developmentally Delayed***
- ***Hearing Impaired***

Many students struggle with larger words that are not in their day-to-day vocabularies. Give students the following word associations to help them understand and remember Earth's four spheres:
Atmosphere: Air
Hydrosphere: H_2O
Geosphere: Ground
Biosphere: Bodies and Bushes

Teach, *continued*

Graphic Organizer GENERAL

Comparison Table
You may want to use this Graphic Organizer in a game to review material before the test. Divide the class into two teams. Ask students questions about material from the Comparison Table. Give points to each team that provides correct answers.

Using the Figure — GENERAL

Energy Budget Point out to students that the energy distribution shown in the figure is for the whole Earth and that the percentages will vary for different specific locations. For instance, a desert will absorb and reflect a different amount of radiation than will a forest or an ocean. A greater quantity of carbon dioxide or water in the atmosphere will alter the absorbing properties of the atmosphere. Ask students, "How much of the total energy is reradiated into space?" (30%) "How much of this reflected energy is scattered by particles in the air?" (6%) Remind students that energy is conserved, so that all the energy that enters Earth's spheres equals the total energy that is released by them. Ask students how much energy is not absorbed by water and land. (100% − 51% = 49%)
LS Visual/Logical

Answer to Reading Check

An energy budget is the total distribution of energy to, from, and between Earth's various spheres.

CHAPTER RESOURCES

Technology

Transparencies
• 6 Earth's Energy Budget (with worksheet)

Figure 4 ▶ Incoming solar energy is balanced by solar energy reflected or reradiated by several of Earth's systems.

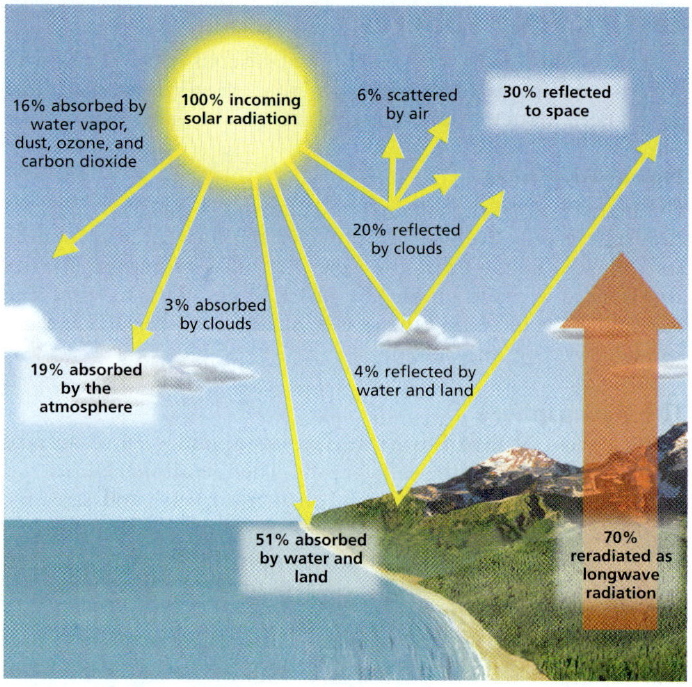

Graphic Organizer **Comparison Table**
Create the Graphic Organizer entitled "Comparison Table" described in the Skills Handbook section of the Appendix. Label the columns with "Type of Source," and "Use of Source." Label the rows with "Radiactive decay," "Convection," "Sun," and "Moon." Then, fill in the the table with details of the type and use of each energy source in Earth's system.

Earth's Energy Budget

Exchanges and flow of energy on Earth happen in predictable ways. According to the *first law of thermodynamics,* energy is transferred between systems, but it cannot be created or destroyed. The transfers of energy between Earth's spheres can be thought of as parts of an *energy budget,* in which additions in energy are balanced by subtractions. This concept is shown in **Figure 4,** which shows how solar energy is transferred through Earth's systems. Solar energy is absorbed and reflected in such a way that the solar energy input is balanced by the solar energy output. Like energy, matter can be transferred but cannot be created or destroyed.

The *second law of thermodynamics* states that when energy transfer takes place, matter becomes less organized with time. The overall effect of this natural law is that the universe's energy is spread out more and more uniformly over time.

Earth's four main spheres are open systems that can be thought of as huge storehouses of matter and energy. Matter and energy are constantly being exchanged between the spheres. This constant exchange happens through chemical reactions, radioactive decay, the radiation of energy (including light and heat), and the growth and decay of organisms.

Reading Check **Define *energy budget.* (See the Appendix for answers to Reading Checks.)**

PHYSICS CONNECTION ADVANCED

The Laws of Thermodynamics The laws of thermodynamics were discovered in the nineteenth century as the nature of heat was understood. Heat was seen as energy that could be added to or removed from a system. Because different aspects of thermodynamics were of interest at different times, the discoveries of its principles occurred in a rather haphazard order. Thus, the second law of thermodynamics, which deals with how energy becomes less useful as it is spontaneously transferred from a higher temperature object to a lower temperature object, was discovered before the first law was conclusively proven. Have students research the historical development of the four laws of thermodynamics, and have them present their findings as a bulletin board exhibit or as a Web page. **LS Verbal**

Internal Sources of Energy

When Earth formed about 4.6 billion years ago, its interior was heated by radioactive decay and gravitational contraction. Since that time, the amount of heat generated by radioactive decay has declined. But the decay of radioactive atoms still generates enough heat to keep Earth's interior hot. Earth's interior also retains much of the energy from the planet's formation.

Because Earth's interior is warmer than its surface layers, hot materials move toward the surface in a process called *convection*. As material is heated, the material's density decreases, and the hot material rises and releases heat. Cooler, denser material sinks and displaces the hot material. As a result, the heat in Earth's interior is transferred through the layers of Earth and is released at Earth's surface. On a large scale, this process drives the plate motions in the surface layers of the geosphere that create mountain ranges and ocean basins.

External Energy Sources

In order for the life-supporting processes on Earth to continue operating for billions of years, energy must be added to the Earth system. Earth's most important external energy source is the sun. Solar radiation warms Earth's atmosphere and surface. This heating causes the movement of air masses, which generates winds and ocean currents. Plants, such as the wheat shown in **Figure 5,** use solar energy to fuel their growth. Because many animals feed on plants, plants provide the energy that acts as a base for the energy flow through the biosphere. Even the chemical reactions that break down rock into soil require solar energy. Another important external source of energy is gravitational energy from the moon and sun. The pull of the sun and the moon on the oceans, combined with Earth's rotation, generates tides that cause currents and drive the mixing of ocean water.

Figure 5 ▶ Solar energy is changed into stored energy in the wheat kernels by chemical processes in the wheat plant. When the wheat is eaten, the stored energy is released from the wheat and used or stored by the consumer.

QuickLAB 10 min

Effects of Solar Energy

Procedure

1. Wrap one **small glass jar** with **black construction paper** so that no light can enter it. Get a **second glass jar.** Make sure that the second jar has a clean, transparent surface.
2. Use a **hammer** and **large nail** to punch a hole in each jar lid.
3. Place a **thermometer** through the hole in each jar lid. Place the lids tightly onto the jars.
4. Place the jars on the windowsill. Wait 5 min. Then, read the temperature from each jar's thermometer.

Analysis

1. Which jar had the higher temperature?
2. Which jar represents a system in which energy enters from outside the system?

QuickLAB

Skills Acquired

- Experimenting
- Constructing models
- Observing
- Identifying and recognizing patterns

Teacher's Notes Students should place a towel beneath the jar lids when punching holes to protect the table. The nails should have a smaller diameter than the thermometers, so that the thermometers will fit securely in the lids. Be sure that the thermometers do not touch the walls or bottom of the jar.

Answers

1. the jar that is not covered by paper
2. the jar that is not covered by paper

CHAPTER RESOURCES

Chapter Resource File

- **Datasheet for Quick Lab** GENERAL

BRAIN FOOD

Gravitational Heating Radioactive decay of material in Earth's interior is Earth's main internal source of heat. However, within Earth's core, radioactive decay is complemented by effects of gravity, which contribute to the heating of Earth's interior. Models and laboratory experiments have provided insights into these processes. Scientists think that considerable heat was produced both by gravitational pressure as material accreted to form the primitive Earth and as smaller bodies collided with Earth. However, there is uncertainty about how much of this "primitive heat" remains in Earth's interior. The separation of elements because of increased gravitational forces on denser elements may also have contributed to interior heating. Another source of gravitational heating may the slowing down of Earth's rotation because of the tidal forces exerted by the moon.

Teach, *continued*

READING SKILL BUILDER — BASIC

Reading Hint Before students read the passages under "Cycles in the Earth System," ask students to think of words they know that include the word *cycle*. (*bicycle, recycle, motorcycle*) Explain that the word *cycle* comes from the Greek word for "wheel," and in general refers to anything that returns to its starting place or repeats itself. Emphasize the word *cycle* for the passages on this page and the following page, and stress that students should be aware of how certain materials return to their initial states. English Language Learners

LS Verbal

Using the Figure — BASIC

Nitrogen Cycle Ask students in what general directions the arrows point in the figure. (In general, the arrows point from the air to the soil or water and from the soil or water to the air.) Point out that the main components in the nitrogen cycle are nitrogen in the air, which is absorbed by bacteria in the soil or water, and that chemical processes and bacteria return the nitrogen to the air. Other factors, such as animals and plants, are simply added links in this basic cycle and do not change its overall workings. Answer to caption question: Animals extract nitrogen from plants they eat and return the nitrogen to the soil through wastes and when their bodies decompose. **LS Visual/Logical**

Cycles in the Earth System

A *reservoir* is a place where matter or energy is stored. A *cycle* is a group of processes in which matter and energy repeatedly move through a series of reservoirs. Many elements on Earth cycle between reservoirs. These cycles rely on energy sources to drive them. The length of time that energy or matter spends in a reservoir can vary from a few hours to several million years.

The Nitrogen Cycle

Organisms on Earth use the element nitrogen to build proteins, which are then used to build cells. Nitrogen gas makes up 78% of the atmosphere, but most organisms cannot use the atmospheric form of nitrogen. The nitrogen must be altered, or *fixed,* before organisms can use it. Nitrogen fixing is an important step in the *nitrogen cycle,* which is shown in **Figure 6.**

In the nitrogen cycle, nitrogen moves from air to soil, from soil to plants and animals, and back to air again. Nitrogen is removed from air mainly by the action of nitrogen-fixing bacteria. These bacteria live in soil and on the roots of certain plants. The bacteria chemically change nitrogen from air into nitrogen compounds, which are vital to the growth of all plants. When animals eat plants, nitrogen compounds in the plants become part of the animals' bodies. These compounds are returned to the soil by the decay of dead animals and in animals' excretions. After nitrogen compounds enter the soil, chemical processes release nitrogen back into the atmosphere. Water-dwelling plants and animals take part in a similar nitrogen cycle.

Reading Check **Identify two nitrogen reservoirs on Earth. (See the Appendix for answers to Reading Checks.)**

Figure 6 ▶ The balance of nitrogen in the atmosphere and biosphere is maintained through the nitrogen cycle. ***What role do animals play in the nitrogen cycle?***

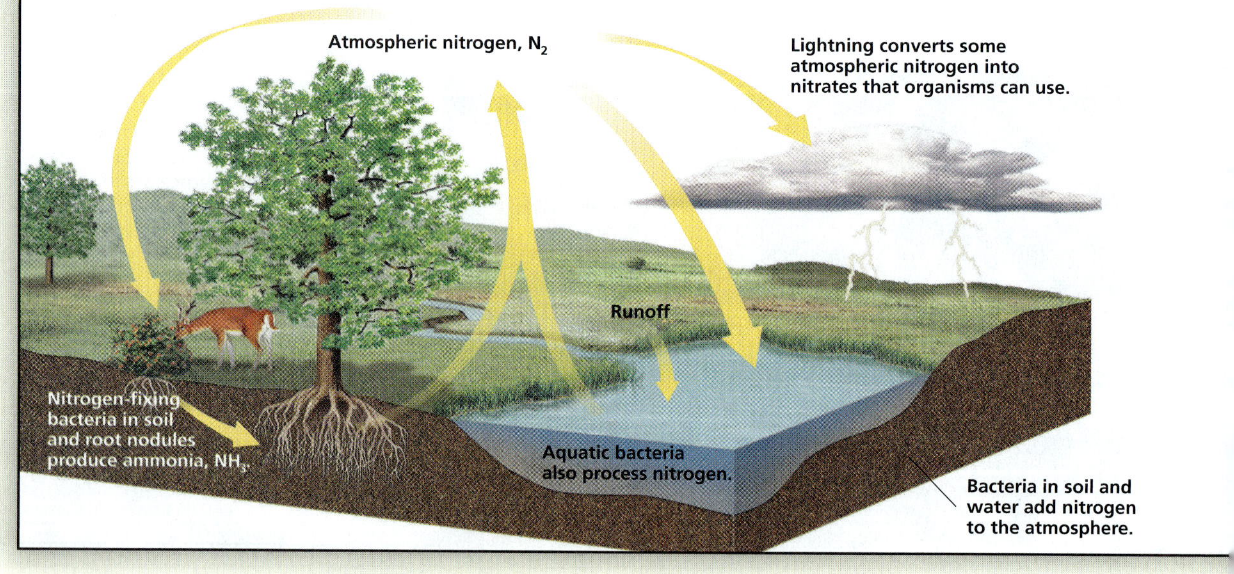

CHAPTER RESOURCES

Technology

Transparencies
- 7 The Nitrogen Cycle (with worksheet)
- 8 The Carbon Cycle (with worksheet)

Answer to Reading Check
soil and plants

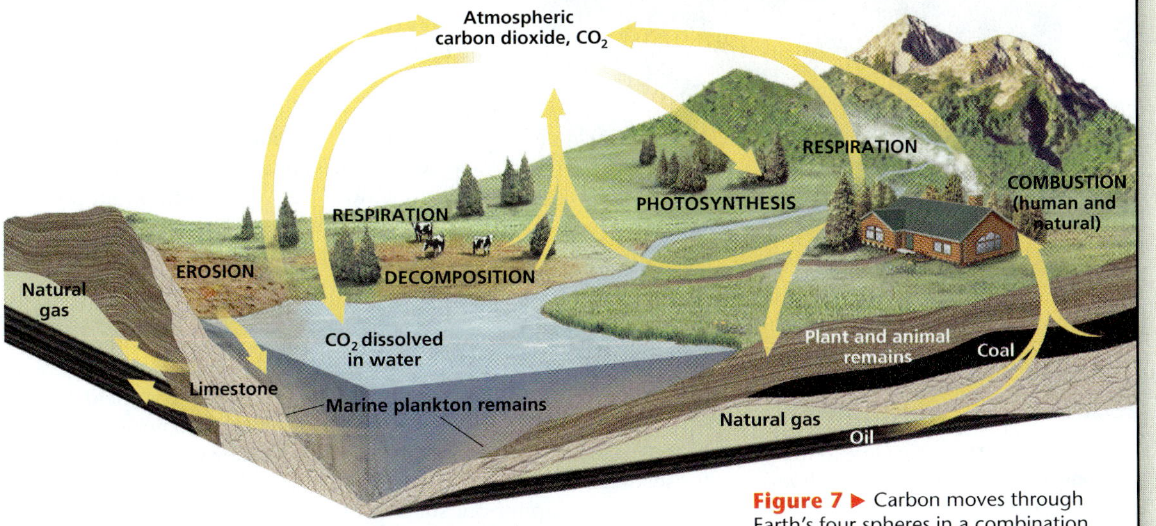

Figure 7 ▶ Carbon moves through Earth's four spheres in a combination of short-term and long-term cycles.

The Carbon Cycle

Carbon is an essential substance in the fuels used for life processes. Carbon moves through all four spheres in a process called the *carbon cycle,* as **Figure 7** shows. Part of the carbon cycle is a short-term cycle. In this short-term cycle, plants convert carbon dioxide, CO_2, from the atmosphere into carbohydrates, such as glucose, $C_6H_{12}O_6$. Then, organisms eat the plants and obtain the carbon from the carbohydrates. Next, organisms' bodies break down the carbohydrates and release some of the carbon back into the air as CO_2. Organisms also release carbon into the air through their organic wastes and by the decay of their remains, which release carbon into the air as CO_2 or as methane, CH_4.

Part of the carbon cycle is a long-term cycle in which carbon moves through Earth's four spheres over a very long time. Carbon is stored in the geosphere in buried plant or animal remains and in a type of rock called a *carbonate,* such as limestone. Carbonate forms from shells and bones of once-living organisms.

The Phosphorus Cycle

The element phosphorus is part of some molecules that organisms need to build cells. During the *phosphorus cycle,* phosphorus moves through every sphere except the atmosphere, because phosphorus is rarely a gas. Phosphorus enters soil and water when rock breaks down and when phosphorus dissolves in water. Some organisms excrete their excess phosphorus in their waste, and this phosphorus may enter soil and water. Plants absorb this phosphorus through their roots. The plants then incorporate the phosphorus into their tissues. Animals absorb the phosphorus when they eat the plants. When the animals die, the phosphorus returns to the environment through decomposition.

SCILINKS® NSTA Developed and maintained by the National Science Teachers Association

For a variety of links related to this subject, go to www.scilinks.org

Topic: Carbon Cycle
SciLinks code: HQ60216
Topic: Nitrogen Cycle
SciLinks code: HQ61036

Group Activity — ADVANCED

Earth's Other Cycles Have groups of three or four students research the oxygen, rock, and energy cycles that take place within Earth's spheres. Students should be able to describe and explain the basic steps of each cycle, and note how the cycles affect processes in various spheres (for example, how the carbon cycle affects climate through the atmospheric "greenhouse effect"). The groups that research the energy cycle should note how this is intimately connected to the water cycle, its effect on weather, and how it differs from the cycles that involve matter. Each group should then report its findings in an oral presentation. LS **Verbal/Logical**

Close

Reteaching — BASIC

Open and Closed Systems On cards or on the board, draw pictures of the following verbal descriptions, or write the descriptions themselves on the cards or board: an air conditioner that changes warm air to cool air (open system); a water wheel that does work as water flows over it (open system); a closed refrigerator that cools the air inside by removing heat to the outside of the refrigerator (closed system); a pressure cooker in which water is boiled at high pressure (closed system). Have students identify which of these systems are open and which are closed. LS **Logical**

Quiz — GENERAL

1. What is a set of interacting components in which energy, but not matter, is exchanged with the surroundings? (a closed system)
2. What part of Earth contains all liquid and solid water? (the hydrosphere)
3. What is the addition, removal, and transfer of energy between Earth's spheres an example of? (an energy budget)

Alternative Assessment — ADVANCED

Rates of Cycles Have students research the time required for the various steps to take place in the phosphorus, nitrogen, or long-term carbon cycles. Have students then describe the cycle in terms of how quickly each part of the cycle can occur, and how long it takes for a given quantity of carbon, phosphorus, or nitrogen to go through the complete cycle. Students may present their findings as a written report, oral presentation, or poster project. LS **Verbal/Logical**

Close, continued

Answers to Section Review

1. because it consists of many interacting components
2. In an open system, matter and energy are added and removed. In a closed system, only energy enters or leaves.
3. The atmosphere contains gases that sustain life and shield Earth from harmful radiation. The hydrosphere covers more than 70% of Earth's surface and is made up of all water that is not gaseous. The geosphere consists of the rock and soil of Earth's crust and the solid and molten rock within the interior. The biosphere contains all living organisms and extends from the deep ocean to the atmosphere.
4. Energy in the Earth system comes from the sun and from Earth's interior.
5. through chemical reactions, radioactive decay, radiation of heat and light, and the growth and decay of organisms
6. In the short-term cycle, CO_2 is absorbed from the atmosphere by plants, which produce carbohydrates that are consumed by organisms. Carbon is returned to the atmosphere mainly as CO_2 from animal and plant respiration and decay. In the long-term cycle, carbon also becomes buried in rock.
7. Nitrogen from the atmosphere is fixed by bacteria in the soil and water. This nitrogen is absorbed by plants, which are consumed by animals, and is returned to the soil or water when dead animals decompose. Other bacteria return nitrogen to the atmosphere.
8. Water that evaporates into the atmosphere moderates the temperature of air; water as precipitation in the geosphere causes weathering of rock; liquid water in the biosphere sustains life; water as precipitation in the hydrosphere replenishes supplies of fresh water.
9. It would add carbon dioxide to the atmosphere, because when organisms such as trees burn, CO_2 is released.
10. an open system, because matter was being added to it
11. As carbon moves through the carbon cycle, it undergoes physical and chemical changes that involve transfers of energy.
12. A *system* may be a *closed system* or an *open system* such as Earth in which both *matter* and *energy* are exchanged in the *atmosphere, hydrosphere, geosphere,* and *biosphere*.

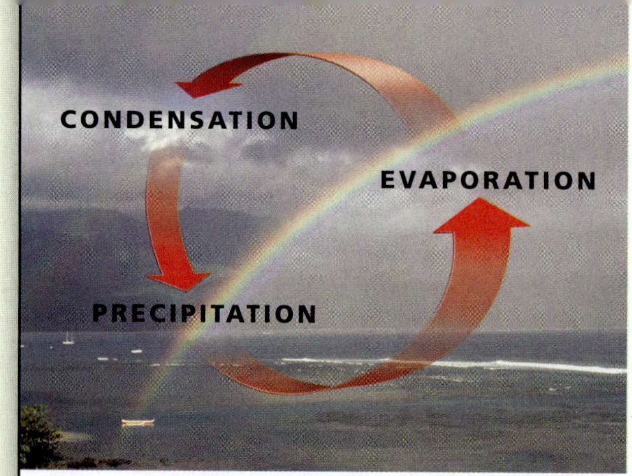

Figure 8 ▶ The water cycle is the continuous movement of water from the atmosphere to Earth's surface and back to the atmosphere.

The Water Cycle

The movement of water from the atmosphere to Earth's surface and back to the atmosphere is always taking place. This continuous movement of water is called the *water cycle*, which is shown in **Figure 8.** In the water cycle, water changes from liquid water to water vapor through the energy transfers involved in evaporation and transpiration. Evaporation occurs when energy is absorbed by liquid water and the energy changes the water into water vapor. Transpiration is the release of moisture from plant leaves. During these processes, water absorbs heat and changes state. When the water loses energy, it condenses to form water droplets, such as those that form clouds. Eventually, water falls back to Earth's surface as precipitation, such as rain, snow, or hail.

Humans and the Earth System

All natural cycles can be altered by human activities. The carbon cycle is affected when humans use fossil fuels. Fossil fuels form over millions of years. Carbon dioxide is returned to the atmospheric reservoir rapidly when humans burn these fuels. Also, both the nitrogen and phosphorus cycles are affected by agriculture. Some farming techniques can strip the soil of nitrogen and phosphorus. Many farmers replace these nutrients by using fertilizers, which can upset the balance of these elements in nature.

Section 2 Review

1. **Explain** how Earth can be considered a system.
2. **Compare** an open system with a closed system.
3. **List** two characteristics of each of Earth's four major spheres.
4. **Identify** the two main sources of energy in Earth's system.
5. **Identify** four processes in which matter and energy cycle on Earth.
6. **Explain** how carbon cycles in Earth's system.
7. **Explain** how nitrogen cycles in Earth's system.

CRITICAL THINKING

8. **Identifying Relationships** For each of Earth's four spheres, describe one way that the water cycle affects the sphere.
9. **Determining Cause and Effect** What effect, if any, would you expect a massive forest fire to have on the amount of carbon dioxide in the atmosphere? Explain your answer.
10. **Analyzing Ideas** Early Earth was constantly being bombarded by meteorites, comets, and asteroids. Was early Earth an open system or a closed system? Explain your answer.
11. **Analyzing Relationships** Explain the role of energy in the carbon cycle.

CONCEPT MAPPING

12. Use the following terms to create a concept map: *closed system, system, open system, matter, atmosphere, biosphere, energy, geosphere,* and *hydrosphere.*

CHAPTER RESOURCES

Chapter Resource File
- Section Quiz GENERAL

Workbooks
- Study Guide (also in Spanish)

ENVIRONMENTAL CONNECTION

Section 3 Ecology

One area of science in which life science and Earth science are closely linked is called *ecology*. Ecology is the study of the complex relationships between living things and their nonliving, or abiotic, environment. Some ecologists also investigate how communities of organisms change over time.

OBJECTIVES

- **Define** *ecosystem.*
- **Identify** three factors that control the balance of an ecosystem.
- **Summarize** how energy is transferred through an ecosystem.
- **Describe** one way that ecosystems respond to environmental change.

KEY TERMS

ecosystem
carrying capacity
food web

Ecosystems

Organisms on Earth inhabit many different environments. A community of organisms and the environment that the organisms inhabit is called an **ecosystem.** The terms *ecology* and *ecosystem* come from the Greek word *oikos,* which means "house." Each ecosystem on Earth is a distinct, self-supporting system. An ecosystem may be as large as an ocean or as small as a rotting log. The largest ecosystem is the entire biosphere.

Most of Earth's ecosystems contain a variety of plants and animals. Plants are important to an ecosystem because they use energy from the sun to produce their own food. Organisms that make their own food are called *producers.* Producers are a source of food for other organisms. *Consumers* are organisms that get their energy by eating other organisms. Consumers may get energy by eating producers or by eating other consumers, as the consumers shown in **Figure 1** are doing. Some consumers get energy by breaking down dead organisms. These consumers are called *decomposers*. To remain healthy, an ecosystem needs to have a balance of producers, consumers, and decomposers.

ecosystem a community of organisms and their abiotic environment

Figure 1 ▶ Vultures and a spotted hyena are feeding on an elephant carcass in Chobe National Park in Botswana. ***Name two consumers that are shown in this photo.***

Section 3

Focus

Overview

This section explains how ecosystems function, how interactions must maintain balance if the ecosystem is to thrive, and how human activities influence environmental conditions.

Bellringer

Have students write down what they think the words *ecosystem* and *ecology* mean. (Sample answers: An ecosystem is an environment in which various organisms depend on each other to exist. Ecology is the study of the function and preservation of ecosystems.) LS **Verbal**

Motivate

Using the Figure — GENERAL

Members of Ecosystems Point out that as a general rule plants are producers and animals are consumers. However, the roles of each kingdom, as well as the functions of fungi and single-celled organisms, blur the lines somewhat. Ask students to identify producers in the photograph and explain why they are producers. (The grasses, plants, and tree in the background are examples of producers, because they manufacture food for themselves.) Answer to caption question: The hyena and vultures are consumers. When alive, the elephant was a consumer. LS **Visual**

CHAPTER RESOURCES

Chapter Resource File
- Directed Reading BASIC

Technology
- Transparencies
  - Bellringer
- Student Edition on CD-ROM
- One-Stop Planner CD-ROM
  - Lesson Plan

Teach

Teaching Tip — ADVANCED

Connect to Familiar Processes Explain to students that the idea of carrying capacity is related to the economic idea of supply and demand. Draw an analogy to a store that has 100 loaves of bread on the shelf at any given time. If each consumer requires one loaf of bread, the maximum number of consumers—the carrying capacity—will be 100. Just as the store cannot supply 101 customers with the minimum number of loaves that each needs, so an ecosystem cannot supply more organisms with food than its resources allow. LS Logical

MISCONCEPTION ALERT

Catastrophic Events
Students may think that catastrophic events, such as large fires, floods, or volcanic eruptions, may so severely damage an ecosystem that it cannot recover. Point out that, while resources such as food are severely reduced for surviving populations, catastrophes may have rejuvenating effects. Floods restore nutrients to inundated areas. Many types of volcanic ash provide a soil in which vegetation thrives. Forest fires clear congested areas of vegetation and enable rapid new growth fed partly by nutritious ash from the fire.

Figure 2 ▶ The fur of this elk calf was singed in a forest fire in Yellowstone National Park. Not enough food resources remain to allow the calf to stay in this area, but the calf may return here when the area has recovered.

carrying capacity the largest population that an environment can support at any given time

Balancing Forces in Ecosystems

Organisms in an ecosystem use matter and energy. Because amounts of matter and energy in an ecosystem are limited, population growth within the ecosystem is limited, too. The largest population that an environment can support at any given time is called the **carrying capacity.** Carrying capacity depends on available resources. The carrying capacity of an ecosystem is also affected by how easily matter and energy cycle between life-forms and the environment in that ecosystem. So, a given ecosystem can support only the number of organisms that allows matter and energy to cycle efficiently through the ecosystem.

Ecological Responses to Change

Changes in any one part of an ecosystem may affect the entire system in unpredictable ways. However, in general, ecosystems react to changes in ways that maintain or restore balance in the ecosystem.

Environmental change in the form of a sudden disturbance, such as a forest fire, can greatly damage and disrupt ecosystems, as shown in **Figure 2.** But over time, organisms will migrate back into damaged areas in predictable patterns. First, grasses and fast-growing plants will start to grow. Then, shrubs and small animal species will return. Eventually, larger tree species and larger animals will return to the area. Ecosystems are resilient and tend to restore a community of organisms to its original state unless the physical environment is permanently altered.

Reading Check Explain the relationship between carrying capacity and the amount of matter and energy in an ecosystem. (See the Appendix for answers to Reading Checks.)

Connection to ENVIRONMENTAL SCIENCE

Lemmings

Lemmings are small arctic rodents. Lemmings play an important role in their ecosystem because they are the major food source for snowy owls and arctic foxes. While most animal species have a relatively constant population size, lemming populations vary greatly over a three- or four-year cycle.

When a lemming population is at its smallest, very few lemmings may be in an area. But lemmings can reproduce very quickly, and they produce large litters of up to 11 young. A female lemming can begin to produce offspring when she is only one month old. So, a very small population of lemmings can give rise to a large population in a short period of time. After a few years, the growing population of lemmings begins to use up available food resources. Eventually, lemmings begin to starve. When this happens, lemmings fight each other, and many migrate to other areas.

These mass migrations have given rise to myths that lemmings throw themselves off cliffs. Lemmings have been seen diving into the ocean, but scientists believe that lemmings do this because lemmings can swim. When lemmings swim across a stream, they can reach and populate new areas of land. When they dive into the ocean, however, they cannot reach land and often drown. When starvation, fighting, or drowning reduces a lemming population to very few members, the cycle of population growth repeats.

Activity — GENERAL

Poster Project Have students research how predator populations increase and decrease with variations in the populations of animals on which they prey. Students may use the lemming example in the text. Or, you may suggest others, such as ringed seals and polar bears, which feed on the seals. Students should find maximum and minimum numbers for given populations of predators and prey over 20 to 50 years. Posters should show pictures and explanations of steps in the cycle. Have students explain their posters to the class. LS Visual/Verbal

Answer to Reading Check

The amount of matter and energy in an ecosystem can supply a population of a given size, and no larger. This maximum population is the carrying capacity of the ecosystem.

Energy Transfer

The ultimate source of energy for almost every ecosystem is the sun. Plants capture solar energy by a chemical process called *photosynthesis*. This captured energy then flows through ecosystems from the plants, to the animals that feed on the plants, and finally to the decomposers of animal and plant remains. Matter also cycles through an ecosystem by this process.

As matter and energy cycle through an ecosystem, chemical elements are combined and recombined. Each chemical change results in either the temporary storage of energy or the loss of energy. One way to see how energy is lost as it moves through the ecosystem is to draw an energy pyramid. Producers form the base of the pyramid. Consumers that eat producers are the next level of the pyramid. Animals that eat those consumers form the upper levels of the pyramid. As you move up the pyramid, more energy is lost at each level. Therefore, the least amount of total energy is available to organisms at the top of the pyramid.

Food Chains and Food Webs

The sequence in which organisms consume other organisms can be represented by a *food chain*. However, ecosystems are complex and generally contain more organisms than are on a single food chain. In addition, many organisms eat more than just one other species. Therefore, a **food web,** such as the one shown in **Figure 3,** is used to represent the relationships between multiple food chains. Each arrow points to the organism that eats the organism at the base of the arrow.

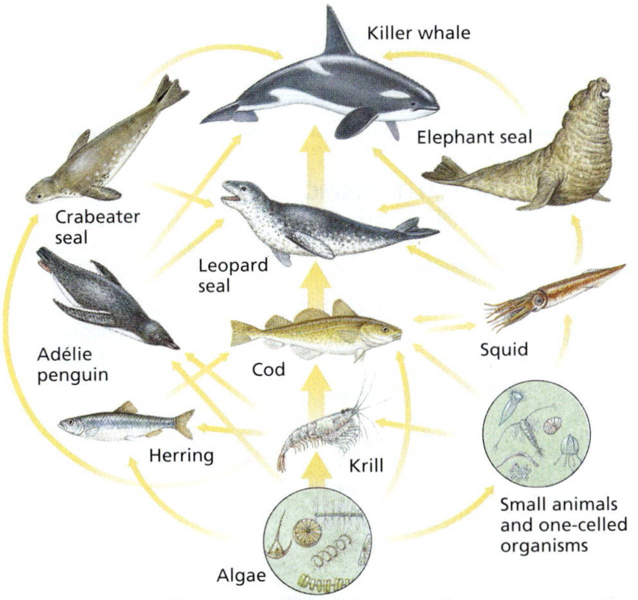

Figure 3 ▶ This food web shows how, in an ocean ecosystem, the largest organisms, such as killer whales, depend on the smallest organisms, such as algae. ***Which organisms would be near the top of a food pyramid?***

QuickLAB — 20 min

Studying Ecosystems

Procedure

1. Find a **small natural area** near your school.
2. Choose a 5 m by 5 m section of the natural area to study. This area may include the ground or vegetation such as trees or bushes.
3. Spend 10 min documenting the number and types of organisms that live in the area.

Analysis

1. How many kinds of organisms live in the area that you studied?
2. Draw a food web that describes how energy may flow through the ecosystem that you studied.

food web a diagram that shows the feeding relationships among organisms in an ecosystem

QuickLAB

Skills Acquired

- Identifying and Recognizing Patterns

Teacher's Notes Separate students into groups of two or three, and assign each group ecosystems that differ in noticeable ways.

Answers

1. Answers may vary depending on the ecosystem and time of year in which it is studied.
2. Answers may vary. For example, students should indicate that ants frequently feed on the remains of dead animals, or that birds feed on insects, and that the energy is transferred from previous consumers to the organisms that consume them.

Close

Reteaching — BASIC

Food Chains Make a series of cards on which pictures and names of the following organisms are drawn or pasted: plant, caterpillar, spider, bird, cat. Have students indicate the order in which the organisms occupy the food chain, from the lowest-level producer to the highest-level consumer. (The organisms are listed above in the correct food chain sequence.) **LS Logical**

Quiz — GENERAL

1. How does an ecosystem typically react to changes? (The ecosystem responds to restore and maintain balance.)
2. What are the three kinds of organisms in an ecosystem? (producers, consumers, and decomposers)

Using the Figure — BASIC

Food Webs Have students examine the directions in which the arrows point in the figure. Ask students which organism has all arrows pointing away from it and what this means. (algae; They are producers and the basic source of energy for the food web.) Ask students which organism has all arrows pointing toward it and what this means. (the killer whale; It is a consumer and eats more than one kind of organism.) Answer to caption question: Organisms near the top of a food pyramid would be the largest and strongest consumers. **LS Visual/Logical**

INCLUSION Strategies

- ***Behavior Control Issues***
- ***Hearing Impaired***
- ***Attention Deficit Disorder***

Students with behavior control issues are more successful learners when they have hands-on involvement. Divide the class into small groups. Scatter students with such issues among the groups. Ask them to be the "writers" for their teams. Have each team create an energy pyramid as described on this page. Have teams use old science magazines as a source of pictures for their pyramids.

CHAPTER RESOURCES

Chapter Resource File

- **Datasheet for QuickLab** GENERAL

Technology

Transparencies

- 9 A Food Web (with worksheet)

Close, continued

Alternative Assessment — Advanced

Modeling Ecosystems Have students create a poster, diorama, or computer model showing the relationships between organisms in a small ecosystem they are familiar with, such as a pond or dead log near their home. The model should include producers, consumers, and decomposers.
LS Kinesthetic/Logical

Answers to Section Review

1. An ecosystem is a community of organisms and the environment that they inhabit.
2. The biosphere sustains organisms that interact and affect each other, providing an environment for all of them.
3. how matter and energy cycle through the ecosystem, how ecosystems respond to change, and how organisms in the ecosystem interact
4. The sun's energy is stored in carbohydrates that are produced by plants by photosynthesis. This energy is absorbed by consumers that eat the plants, and is in turn absorbed by higher-level consumers. Energy from animal remains is transferred to decomposers.
5. Ecosystems will generally respond to change in a way that restores balance within the ecosystem.
6. A food chain is a sequence in which an organism is consumed by another organism. In a food web, several organisms feed upon one type of organism, which are consumed by still other organisms that compete with each other, producing a more complex pattern of consumption.
7. Good stewardship helps maintain balance within and between ecosystems, which helps to ensure their health and productivity.
8. removes vegetation that helps hold topsoil in place, otherwise leading to soil erosion; water pollution from street runoff affects the water supplies; animals may flee, disturbing neighboring ecosystems
9. Ecosystems can react to gradual changes by making small adjustments to maintain the overall balance of the system. Extreme environmental changes require dramatic adjustments by organisms and may result in permanent changes to some components of the ecosystem.
10. Energy flows from producers toward consumers that use more energy. Lower-level consumers rarely consume higher-level consumers. Producers do not consume at all.
11. *Ecology* is the study of *ecosystems*, which have a *carrying capacity* and contain a *food web* that has *producers*, *consumers*, and *decomposers*.

Figure 4 ▶ These hikers are acting responsibly by choosing to remain on marked trails in the rain forest. In this way, they are helping prevent ecological damage to the area.

Human Stewardship of the Environment

All of Earth's systems are interconnected, and changes in one system may affect the operation of other systems. Earth's ecosystems provide a wide variety of resources on which people depend. People need water and air to survive. Changes in ecosystems can affect the ability of an area to sustain a human population. For example, the quality of the atmosphere, the productivity of soils, and the availability of natural resources can affect the availability of food.

Ecological balances can be disrupted by human activity. Populations of plants and animals can be destroyed through overconsumption of resources. When humans convert large natural areas to agricultural or urban areas, natural ecosystems are often destroyed. Another serious threat to ecosystems is pollution. *Pollution* is the contamination of the environment with harmful waste products or impurities.

When people, such as those in **Figure 4,** strive to prevent ecological damage to an area, they are trying to be responsible stewards of Earth. To help ensure the ongoing health and productivity of the Earth system, many people work to use Earth's resources wisely. By using fossil fuels, land and water resources, and other natural resources wisely, many people are helping keep Earth's ecosystems in balance.

Section 3 Review

1. **Define** *ecosystem.*
2. **Explain** why the entire biosphere is an ecosystem.
3. **Identify** three factors that control the balance of an ecosystem.
4. **Summarize** how energy is transferred between the sun and consumers in an ecosystem.
5. **Describe** one way that ecosystems respond to environmental change.
6. **Compare** a food chain with a food web.
7. **Summarize** the importance of good stewardship of Earth's resources.

CRITICAL THINKING

8. **Making Inferences** Discuss two ways that the expansion of urban areas might be harmful to nearby ecosystems.
9. **Analyzing Ideas** Why would adapting to a gradual change in environment be easier for an ecosystem than adapting to a sudden disturbance would be?
10. **Making Inferences** Why does energy flow in only one direction in a given food chain of an ecosystem?

CONCEPT MAPPING

11. Use the following terms to create a concept map: *ecology, ecosystem, producer, decomposer, carrying capacity, consumer,* and *food web.*

CHAPTER RESOURCES

Chapter Resource File
- Section Quiz GENERAL

Workbooks
- Study Guide (also in Spanish)

Chapter 2 Highlights

Sections

1 Earth: A Unique Planet

Key Terms

crust, 28
mantle, 28
core, 28
lithosphere, 29
asthenosphere, 29
mesosphere, 29

Key Concepts

- Earth is an oblate spheroid that has an average diameter of 12,756 km. About 71% of Earth's surface is covered by a relatively thin layer of water.
- Seismic waves have revealed that Earth's interior is composed of a series of layers of various densities.
- Earth has a magnetic field that extends into space in a region known as the *magnetosphere*.

2 Energy in the Earth System

Key Terms

system, 31
atmosphere, 33
hydrosphere, 33
geosphere, 33
biosphere, 33

Key Concepts

- A closed system is a system in which energy enters and exits but matter remains static—neither enters nor exits. An open system is a system in which both energy and matter enter and leave.
- The Earth system can be thought of as consisting of four spheres—the geosphere, the hydrosphere, the atmosphere, and the biosphere—that influence the operation of one another.
- Matter and energy are not created or destroyed. Matter and energy cycle between Earth's systems. Energy, most of which is solar, is required to maintain this cycling.

3 Ecology

Key Terms

ecosystem, 39
carrying capacity, 40
food web, 41

Key Concepts

- An ecosystem is a community of organisms and the environment that they inhabit.
- When sudden disturbances disrupt the health of an ecosystem, the components of the ecosystem respond in ways that return the ecosystem to a balanced condition.
- One way that energy moves through ecosystems is through the eating of organisms by other organisms.
- Humans are part of the global ecosystem. Stewardship of Earth's resources is important to maintaining healthy ecosystems.

Chapter Highlights

Alternative Assessment — GENERAL

Poster Project Have students select an ecosystem and research the interactions between various organisms within the chosen ecosystem. Then, have students create a poster that shows both the energy budget for the ecosystem in general and the food web for the selected organisms. Students should include labels to identify the various interactions in each process. LS Visual/Logical

CHAPTER RESOURCES

Chapter Resource File

- Concept Review GENERAL
- Critical Thinking ADVANCED
- Math Skills GENERAL
- Graphing Skills GENERAL
- Chapter Test A GENERAL
- Chapter Test B ADVANCED

Workbooks

- **Study Guide** (also in Spanish)
- **Assessments** (Spanish)

Technology

Classroom Videos
- Brain Food Video Quiz

Chapter 2 Review

Assignment Guide

Section	Questions
1	3, 7, 9, 11, 18, 20, 32
2	1, 5–6, 10, 12–13, 15–17, 22, 24, 26, 34–37
3	2, 8, 14, 23, 25, 27, 29, 33
2 and 3	4, 19, 21, 28, 31
1–3	30

Using Key Terms

Use each of the following terms in a separate sentence.

1. *system*

2. *carrying capacity*

3. *lithosphere*

For each pair of terms, explain how the meanings of the terms differ.

4. *system* and *ecosystem*

5. *biosphere* and *geosphere*

6. *hydrosphere* and *atmosphere*

7. *mantle* and *asthenosphere*

8. *energy pyramid* and *food web*

Understanding Key Concepts

9. The diameter of Earth is greatest at the
- **a.** poles.
- **b.** equator.
- **c.** oceans.
- **d.** continents.

10. The element that makes up the largest percentage of the atmosphere is
- **a.** oxygen.
- **b.** nitrogen.
- **c.** carbon dioxide.
- **d.** ozone.

11. The gravitational attraction between two objects is determined by the mass of the two objects and the
- **a.** distance between the objects.
- **b.** weight of the objects.
- **c.** diameter of the objects.
- **d.** density of the objects.

12. Energy can enter the Earth system from internal sources through convection and from external sources through
- **a.** radioactive decay.
- **b.** wave energy.
- **c.** wind energy.
- **d.** solar energy.

13. Closed systems exchange energy but do *not* exchange
- **a.** gravity.
- **b.** matter.
- **c.** sunlight.
- **d.** heat.

14. Which of the following is *not* an ecosystem?
- **a.** a lake
- **b.** an ocean
- **c.** a tree
- **d.** an atom

15. Which of the following processes is *not* involved in the water cycle?
- **a.** evaporation
- **b.** transpiration
- **c.** nitrogen fixing
- **d.** precipitation

16. A jar with its lid on tightly is one example of a(n)
- **a.** open system.
- **b.** biosphere.
- **c.** closed system.
- **d.** ecosystem.

17. Phosphorus cycles through all spheres except the
- **a.** geosphere.
- **b.** atmosphere.
- **c.** biosphere
- **d.** hydrosphere.

Short Answer

18. What characteristic of Earth's interior is likely to be responsible for Earth's magnetic field?

19. What is the role of decomposers in the cycling of matter in the biosphere?

20. Compare the three compositional zones of Earth with the five structural zones of Earth.

21. Restate the first and second laws of thermodynamics, and explain how they relate to ecosystems on Earth.

22. Describe two ways that your daily activities affect the water cycle.

23. Explain three reasons that stewardship of Earth's resources is important.

Using Key Terms

1–8. Answers may vary but should show that students understand the definitions of and differences between key terms.

Understanding Key Concepts

9. b **10.** b
11. a **12.** d
13. b **14.** d
15. c **16.** c
17. b

Short Answer

18. the liquid iron in Earth's outer core

19. Decomposers break down organic matter and help move carbon, phosphorus, and nitrogen through their cycles.

20. The mantle, or second compositional zone, is divided into three segments that form structural zones: The lithosphere is made up of the crust (the first compositional layer) and uppermost mantle; the asthenosphere is the next layer of the mantle; and the mesosphere is the lower part of the mantle. The third compositional zone, the core, is divided into two structural zones—the outer core and the inner core.

21. The first law of thermodynamics states that energy cannot be created or destroyed. This means that all ecosystems share, transfer, and convert energy, but they do not create or destroy it. The second law of thermodynamics states that energy becomes less organized with each process. This means that as energy is transferred within or between ecosystems, it becomes less useful. An example of this is loss of energy at each step up on the food pyramid.

22. Answers may vary. Sample answer: Drinking water temporarily removes it from the supply of liquid water. Breathing and perspiring introduce water vapor into the atmosphere.

23. It keeps ecosystems in balance, keeps the Earth system healthy and productive, and protects resources, organisms, and ecosystems that are essential for human survival.

24. The atmosphere weathers rock to form sand and soil, interacts with the geosphere to recycle carbon over long periods of time, and interacts with nitrogen-fixing bacteria to return nitrogen to the geosphere.

25. Answers may vary. Sample answer: A pond contains water and has both producers and consumers to support aquatic life. Organisms may include fish, frogs, salamanders, underwater plants, and mosquitoes.

Critical Thinking

26. When fuel is burned, the products enter the atmosphere, hydrosphere, and/or geosphere. The stored energy is changed into heat or light, which may be used to do work or to provide heat.

27. Student drawings should indicate that the smaller organisms form the base of the pyramid, while progressively larger organisms form higher layers. The killer whale should be at the top of the pyramid.

24. Describe three ways in which the atmosphere interacts with the geosphere.

25. Identify two distinguishing factors of a nearby ecosystem, and name five kinds of organisms that live in that ecosystem.

Critical Thinking

26. **Analyzing Ideas** What happens to the matter and energy in fossil fuels when the fuels are burned?

27. **Making Inferences** Draw an energy pyramid that includes the organisms shown in the food web diagram in this chapter.

28. **Making Predictions** How would the removal of decomposers from Earth's biosphere affect the carbon, nitrogen, and phosphorus cycles?

29. **Analyzing Relationships** Do you think that Earth has a carrying capacity for humans? Explain your reasoning.

Concept Mapping

30. Use the following terms to create a concept map: *biosphere, magnetosphere, mantle, atmosphere, geosphere, hydrosphere, ecosystem, crust,* and *core.*

Math Skills

31. **Making Calculations** In one year, the plants in each square meter of an ecosystem obtained 1,460 kilowatt•hours (kWh) of the sun's energy by photosynthesis. In that year, each square meter of plants stored 237 kWh. What percentage of the sun's energy did the plants use for life processes in that year?

32. **Making Calculations** The average radius of Earth is 6,371 km. If the average thickness of oceanic crust is 7.5 km and the average thickness of continental crust is 35 km, what fraction of Earth's radius is each type of crust?

Writing Skills

33. **Creative Writing** If you noticed that pollution was harming a nearby lake, how would you convince your community of the need to take action to solve the problem? Describe three research tools you would use to find materials that support your opinion.

34. **Communicating Main Ideas** Explain why closed systems typically do not exist on Earth. Suggest two examples of a closed system created by humans.

Interpreting Graphics

The graphs below show the difference in energy consumption and population size in developed and developing countries. Use the graphs to answer the questions that follow.

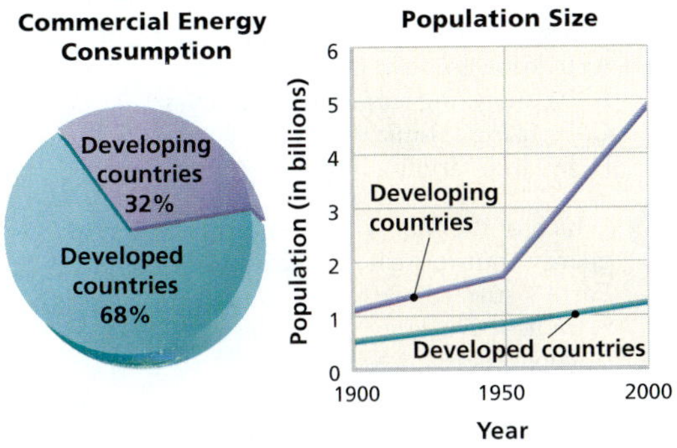

35. Describe the differences in energy consumption and population growth between developed and the developing countries.

36. Do you think that the percentage of commercial energy consumed by developing countries will increase or decrease? Explain your answer.

37. Why is information on energy consumption represented in a pie graph, while population size is shown in a line graph?

28. Because decomposers break down dead organisms, they help transfer the organisms' carbon, nitrogen, and phosphorus to soil. Without decomposers, these cycles would be slowed down or would cease.

29. yes; Earth is finite in size, and therefore has a finite amount of resources, many of which are nonrenewable. Humans require a minimum amount of certain resources to survive and remain healthy. Therefore, there is a maximum number of humans Earth can sustain.

Concept Mapping

30. Answers may vary but should include all of the terms listed. Sample answers appear at the end of this Teacher Edition.

Math Skills

31. energy stored = 1,460 kWh − 237 kWh = 1,223 kWh; percentage of energy = (1,223 kWh ÷ 1,460 kWh) × 100 = 83.8%

32. percentage of radius that is oceanic crust = (7.5 km ÷ 6,371 km) × 100 = 0.12%
percentage of radius that is continental crust = (35 km ÷ 6,371 km) × 100 = 0.55%

Writing Skills

33. Answers may vary. Accept all reasonable answers.

34. Answers may vary. Accept all reasonable answers.

Interpreting Graphics

35. Developing countries have a much higher rate of population growth than developed countries do. Developed countries use about twice as much energy as developing countries do, even though the populations of developed countries are much smaller.

36. Developing countries will probably consume a higher percentage of commercial energy over time, because population growth is increasing in developing countries.

37. The pie graph shows how separate amounts make up a total amount, while the chart of population size shows change over time.

Standardized Test Prep

Estimated Time

To give students practice under more realistic testing conditions, allow them 30 minutes to answer all of the questions in this practice test.

TEST DOCTOR

Question 3 Answer C is correct. Scientists use seismic waves to determine the composition and size of Earth's interior. Answers A and D are incorrect because scientists can drill only a few kilometers into Earth's crust, not enough to reach the interior layers nor to directly observe them. Answer B is incorrect because rock samples at the surface tell us little about the interior of Earth.

Question 4 Answer G is correct. Convection causes materials of different temperatures and densities to rise and fall in the mantle. This movement drives volcanic activity at the surface. Answer F is incorrect because radioactive decay provides only a very small amount of Earth's internal energy. Answers H and I are similar forms of energy transfer that provide little energy transfer to the surface.

Chapter 2 Standardized Test Prep

Understanding Concepts

Directions (1–5): **For *each* question, write on a separate sheet of paper the letter of the correct answer.**

1. The crust and the rigid upper part of the mantle is found in what part of the Earth?
   A. the asthenosphere
   B. the lithosphere
   C. the mesosphere
   D. the stratosphere

2. Because phosphorus rarely occurs as a gas, the phosphorus cycle mainly occurs between the
   F. biosphere, geosphere, and hydrosphere
   G. biosphere, geosphere, and atmosphere
   H. geosphere, hydrosphere, and atmosphere
   I. biosphere, hydrosphere, and atmosphere

3. How are scientists able to study the composition and size of the interior layers of Earth?
   A. by direct observation
   B. by analyzing surface rock samples
   C. by using seismic waves
   D. by deep-drilling into the interior layers

4. Which of the following methods of internal energy transfer drives volcanic activity on Earth's surface?
   F. radioactive decay
   G. convection
   H. kinetic transfer
   I. conduction

5. Earth's primary external energy source is
   A. cosmic radiation
   B. the moon
   C. distant stars
   D. the sun

Directions (6–7): **For *each* question, write a short response.**

6. What do decomposers break down to obtain energy?

7. What scientific principle states that energy can be transferred but that it cannot be created or destroyed?

Reading Skills

Directions (8–9): **Read the passage below. Then, answer the questions.**

Acid Rain

Acid rain is rain, snow, fog, dew, or sleet that has a pH that is lower than the pH of normal precipitation. Acid rain occurs primarily as a result of the combustion of fossil fuels—a process that produces, as byproducts, oxides of nitrogen and sulfur dioxide. When combined with water in the atmosphere, these compounds form nitric acid and sulfuric acid. When it falls to Earth, acid rain has profound effects. It harms forests by damaging tree leaves and bark, which leaves them vulnerable to weather, disease, and parasites. Similarly, it damages crops. And it damages aquatic ecosystems by causing the death of all but the hardiest species. Because of the extensive damage that acid rain causes, the U.S. Environmental Protection Agency limits the amount of sulfur dioxide and nitrogen oxides that can be emitted by factories, power plants, and motor vehicles.

8. According to the passage, which of the following contributes to the problem of acid rain?
   A. the use of fossil fuels in power plants and motor vehicles
   B. parasites and diseases that harm tree leaves and bark
   C. the release of nitrogen into the atmosphere by aquatic ecosystems
   D. damaged crops that release too many gases into the atmosphere

9. Which of the following statements can be inferred from the information in the passage?
   F. Acid rain is a natural problem that will correct itself if given enough time.
   G. Ecosystems damaged by acid rain adapt so that they will not be damaged in the future.
   H. Human activities are largely to blame for the problem of acid rain.
   I. Acid rain is a local phenomenon and only damages plants and animals near power plants or roadways.

Answers

Understanding Concepts

1. B
2. F
3. C
4. G
5. D
6. dead organisms
7. First Law of Thermodynamics

Reading Skills

8. A
9. H

Interpreting Graphics

10. F
11. B
12. Answers may vary. See Test Doctor for a detailed scoring rubric.

Interpreting Graphics

Directions (10–12): **For *each* question below, record the correct answer on a separate sheet of paper.**

The diagram below shows the interior layers of Earth. The layers in the diagram are representative of arrangement and are not drawn to scale. Use this diagram to answer question 10.

Structure of the Earth

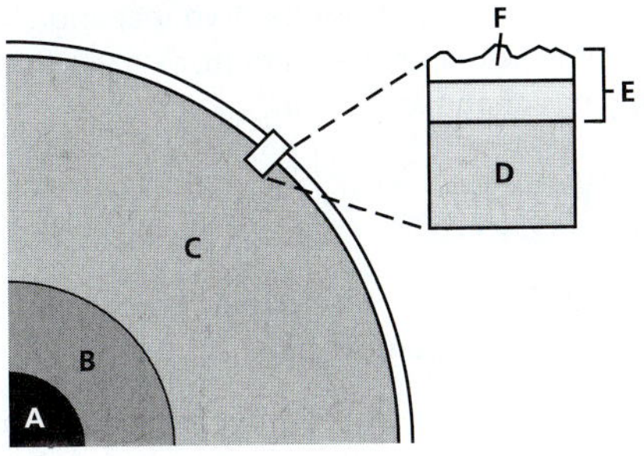

10 Which letter represents the layer of Earth known as the lithosphere?

F. layer E
G. layer D
H. layer C
I. layer A

Use the graph below, which shows predicted world-wide energy consumption by fuel type between the years 2001 and 2025, to answer questions 11 and 12.

Worldwide Energy Consumption By Source

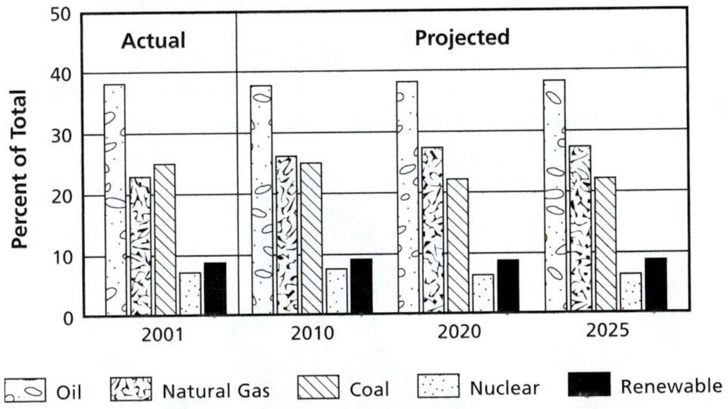

11 Which of the following sources of energy is predicted to see the greatest increase in usage between 2001 and 2025?

A. oil
B. natural gas
C. coal
D. renewables

12 What trends in energy consumption by fuel type will change over the 25 years shown on the graph above? What trends will stay the same?

Test TIP

If time permits, take short mental breaks during the test to improve your concentration.

Standardized Test Prep

TEST DOCTOR

Question 11 Answer B is correct. Answer A is incorrect because the use of oil is predicted to stay about the same over the next 25 years. Answer C is incorrect because the usage of coal is predicted to fall slightly over the next 25 years. Answer D is incorrect because the use of renewable energy sources is predicted to stay about the same.

Question 12 Full-credit answers should include the following points:
- overall, general energy trends are predicted to remain the same over the next 25 years
- oil is predicted to continue to be the leading fuel source in the future
- natural gas is expected to overtake coal by 2010 as an energy source, and the difference between the use of the two sources is expected to continue to increase steadily through 2025
- nuclear power is predicted to remain the least-used energy source and is expected to slightly decrease in usage
- fossil fuels are predicted to continue to outpace nuclear and renewable energy resources by a ratio of more than 4 to 1

Test Prep Correlations ***National Science Education Standards***

ES 1a: item 5
ES 1b: item 4
ES 2a: item 2
LS 4b: item 6
PS 5a: item 7
SAI 2c: item 3
SPSP 4b: items 8, 9
SPSP 3a: items 11, 12
SPSP 3b: items 11, 12
UCP 1: items 1, 10

CHAPTER RESOURCES

State Resources

For specific resources for your state, visit go.hrw.com and type in the keyword **HSHSTR.**

Skills Practice Lab

Testing the Conservation of Mass

Teacher's Notes

Time Required

one 45-minute class period

Lab Ratings

EASY → HARD

Teacher Preparation: 2
Student Setup: 3
Concept Level: 2
Cleanup: 2

Skills Acquired

- Experimenting
- Measuring
- Observing
- Interpreting

The Scientific Method

In this lab, students will
- Make Observations
- Test the Hypothesis
- Analyze the Results
- Draw Conclusions

Materials

The materials listed on the page are enough for groups of two students. If there are not enough beakers, graduated cylinders, or balances, have pairs or groups of students share these items.

Tips and Tricks

Have extra reactants and weighing paper available in case students make mistakes or accidentally spill the reactants. Students should be sure that the reactions have reached completion before measuring the final masses of the reaction products.

Chapter 2 **Skills Practice Lab**

Testing the Conservation of Mass

As matter cycles through the Earth system, the matter can undergo chemical changes that cause it to change state or change physical properties. However, although the matter may change form, it is not destroyed. This principle is known as the *law of conservation of mass.* In this lab, you will cause two chemicals to react to form products that differ from the two reacting chemicals. Then, you will determine whether the amount of mass in the system (the experiment) has changed.

Objectives

- **Measure** the masses of reactants and products in a chemical reaction.
- **Describe** how measuring masses of reactants and products can illustrate the law of conservation of mass.

Materials

- bag, plastic sandwich, zipper-type closure
- baking soda (sodium bicarbonate)
- balance (or scale), metric
- beaker, 400 mL
- cup, clear plastic, 150 mL (2)
- graduated cylinder, 100 mL
- paper, weighing (2 pieces)
- twist tie
- vinegar (acetic acid solution)
- water

Safety

PROCEDURE

1. On a blank sheet of paper, prepare a table like the one shown on the next page.
2. Place a piece of weighing paper on a balance. Place 4 to 5 g of baking soda on the paper. Carefully transfer the baking soda to a plastic cup.
3. Using a graduated cylinder, measure 50 mL of vinegar. Pour the vinegar into a second plastic cup.
4. Place both cups on the balance, and determine the combined mass of the cups, baking soda, and vinegar to the nearest 0.01 g. Record the combined mass in the first row of your table under "Initial mass."
5. Take the cups off the balance. Carefully and slowly pour the vinegar into the cup that contains the baking soda. To avoid splattering, add only a small amount of vinegar at a time. Gently swirl the cup to make sure that the reactants are well mixed.

Step 2

CHAPTER RESOURCES

Chapter Resource File

- Datasheet for Chapter Lab GENERAL
- Lab Notes and Answers

	Initial mass (g)	Final mass (g)	Change in mass (g)
Trial 1			
Trial 2			

DO NOT WRITE IN THIS BOOK

6. When the reaction has finished, place both cups back on the balance. Determine the combined mass to the nearest 0.01 g. Record the combined mass in the first row of your table under "Final mass."

7. Subtract final mass from initial mass, and record the difference in the first row of your table under "Change in mass."

8. Repeat step 2, but carefully transfer the baking soda to one corner of a plastic bag rather than the cup.

9. To seal the baking soda in the corner of the bag, twist the corner of the bag above the baking soda and wrap the twist tie tightly around the twisted part of the bag.

10. Add 50 mL of vinegar to the bag. Zipper-close the bag so that the vinegar cannot leak out and the bag is airtight.

11. Place the bag in the beaker, and measure the mass of the beaker, the bag, and the reactants. Record the combined mass in the second row of your table under "Initial mass."

12. Remove the twist tie from the bag, and mix the reactants.

13. When the reaction has finished, repeat steps 6 and 7 by using the beaker, bag, twist tie, and products. Record the final mass and change in mass in the table's second row.

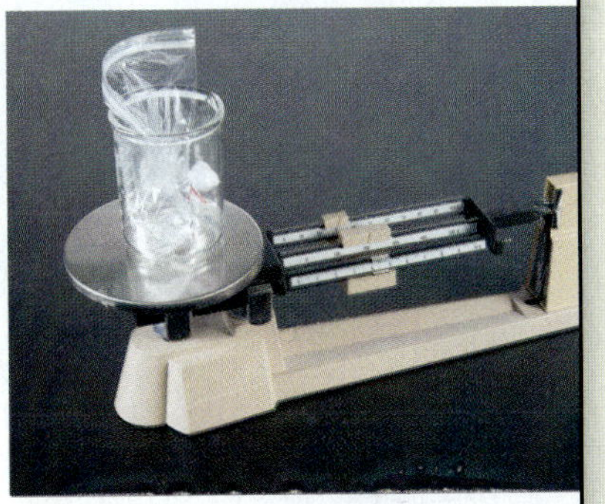

Step 11

ANALYSIS AND CONCLUSION

1. **Analyzing Data** Compare the change in mass that you calculated for the first trial with the change in mass that you calculated for the second trial. What evidence of the conservation of mass does the second trial show?

2. **Analyzing Results** Was the law of conservation of mass violated in the first trial? Explain your answer.

3. **Drawing Conclusions** Was the first trial an example of a closed system or an open system? Which type of system was the second trial? Explain your answer.

Extension

1. **Designing an Experiment** Brainstorm other ways to demonstrate the law of conservation of energy in a laboratory. Describe the materials that you would need, and describe any difficulties that you foresee.

Skills Practice Lab

Answers to Analysis and Conclusion

1. The mass of the reactants before the reaction took place equaled the mass of the products after the reaction.
2. no; The gas released during the reaction entered the atmosphere, and so was not weighed with the other reaction products.
3. The first trial was an example of an open system, because matter left the cup during the reaction, in the form of a gas. The second trial was an example of a closed system, because all matter remained inside the plastic bag before, during, and after the reaction.

Answers to Extension

1. Answers may vary. Accept all reasonable answers.

TEACHER TESTED & APPROVED

Daniel Brownstein, MA/MAT
Geology
Hastings High School
Hasting-on-Hudson, NY

Concentration of Plant Life on Earth

Activity — GENERAL

Ocean-Plant Distributions
Have students research the types of plants that are found in the oceans. Have them use this research to explain the distributions of ocean-plant life indicated in the map. Have students present their findings in a written report, an oral presentation, or a poster presentation. LS Verbal/Visual

Answers to Map Skills Activity

1. Concentrations of chlorophyll on land and in the oceans are indicated by reversed color scales. So, a high concentration in the ocean is indicated by the color red, while a high concentration on land is indicated by the color green.
2. Answers may vary. Sample answer: North Africa, central Australia, and the Arabian peninsula all have low concentrations of chlorophyll. All these areas are deserts, without much vegetation.
3. They are covered with ice.
4. near the edges of continents, especially near major river mouths; These areas are rich in nutrients that organisms need to survive.
5. Days along the equator tend to be more uniform in length than the days at higher latitudes, providing a more continuous level of sunlight, and therefore a steadier growth of plant life. Also, sunlight is more direct to these regions.

CHAPTER RESOURCES

Technology

Transparencies
- 10 Concentration of Plant Life on Earth (with worksheet)

MAPS in Action

Concentration of Plant Life on Earth

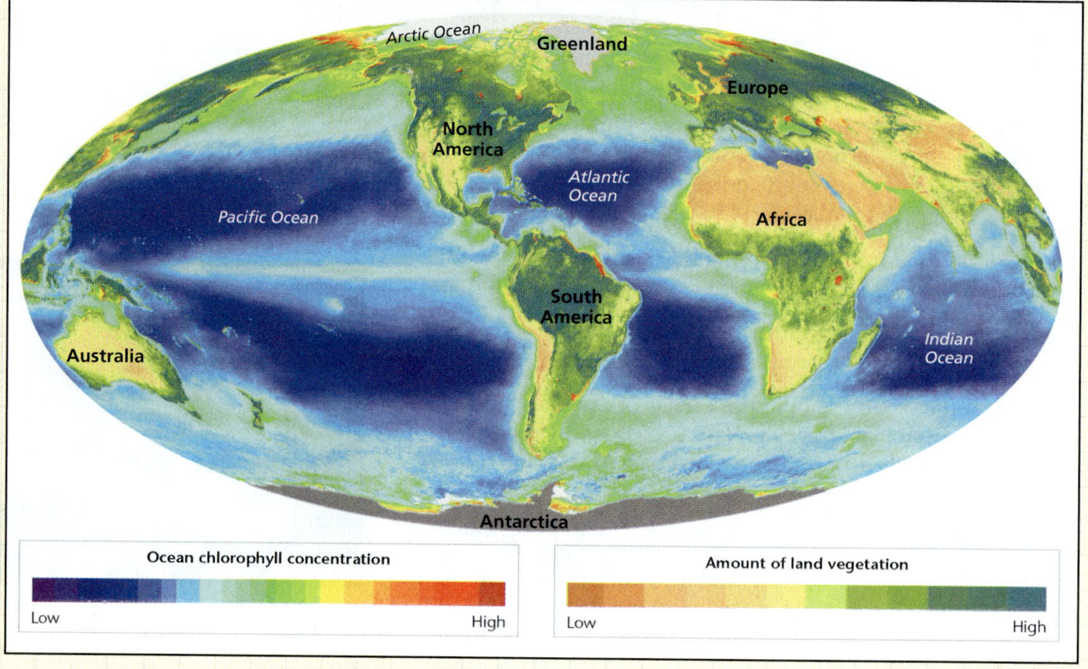

Map Skills Activity

This map shows the concentration of plant life on land and in the oceans. Each color in the key represents a concentration of plant life as indicated by the concentration of chlorophyll. The higher the concentration of chlorophyll is, the higher the concentration of plant life is. Use the map to answer the questions below.

1. **Using a Key** How can you distinguish between high chlorophyll concentration in the ocean and high chlorophyll concentration on land?
2. **Comparing Areas** List three areas that have very low chlorophyll concentration on land. What characteristics of these areas cause such low chlorophyll concentrations?
3. **Comparing Areas** Why do you think Antarctica, Greenland, and the Arctic Ocean lack chlorophyll?
4. **Identifying Trends** Where are the highest chlorophyll concentrations in the ocean located? Why do you think that these locations have high chlorophyll concentrations?
5. **Identifying Trends** Plants use sunlight and chlorophyll to produce energy. What can you infer about the amount of sunlight around the equator that could help explain why areas along the equator tend to have higher concentrations of chlorophyll than surrounding areas do?

IMPACT on Society

Biological Clocks

Humans are affected by cycles in the Earth system. Scientists think that our ancestors arranged their daily activities to correspond with daylight. They were probably awakened by sunrise and ended their workday at sunset. With the development of artificial light sources, however, humans have become less dependent on sunlight to set their daily routines. Recent research reveals surprising evidence that our bodies are still closely tied to natural, daily rhythms.

Feeling the Rhythm

Scientists have discovered that many body processes occur in 24-hour cycles called *circadian rhythms*. No one understands exactly what controls circadian rhythms, but the human body seems to have a number of internal clocks. These "biological clocks" regulate patterns of sleeping and waking, daily changes in body temperature, hormone secretions, heart rate, and blood pressure. Even moods, coordination, and memory are thought to be affected by circadian rhythms.

Broken Clocks?

Studies indicate that the cycle of darkness and light caused by Earth's rotation sets many of our biological clocks. When biological clocks get out of sync with the sun's 24-hour cycle because of long-distance travel or unusual work hours, problems may arise.

One problem is jet lag. Jet lag is the exhaustion, irritability, and insomnia that travelers suffer after a flight across several time zones. The human body may take days or weeks to reset its biological clock.

People living in areas near the poles can experience adverse effects of long periods of seasonal darkness. Extended periods of little sunlight can cause hormonal changes and mood disorders. Scientists are also beginning to understand how such mood and hormonal changes occur in people who work through the night and sleep during the day.

▲ Nocturnal animals are awake at night and sleep during the day. These animals have circadian rhythms opposite those of diurnal organisms, such as humans.

◀ Traveling over many time zones can interfere with natural patterns of sleep.

Extension

1. **Researching Information** When people travel to another time zone, their bodies stay synchronized with the cycle of the sun in the place that they left. Research ways in which travel affects circadian rhythms. Then, write a short brochure that explains how to reduce the effects of traveling across time zones.

Answer to Extension

1. Answers may vary. Student brochures should suggest ways in which travelers can adjust their circadian "clocks" to the time zone they will be traveling to before or during the flight, as well as techniques to speed adjustment once they have arrived in the different time zone.

Biological Clocks

Internet Activity —ADVANCED

Health and Biological Clocks
Have students use the Internet to research some of the health problems that can arise from extreme changes in human rhythms. These can include sleep disorders that disrupt circadian rhythms and depression that comes from living at far northern or southern latitudes. Students should investigate how these health disorders are being treated and present their findings in a written report. A worksheet designed to direct student research on this topic can be found in the **Chapter Resource File** booklet or by visiting go.hrw.com and entering the keyword **HQ6EASX.** LS **Verbal/Logical**

CHAPTER RESOURCES

Chapter Resource File

- **Internet Activity** Health and Biological Clocks ADVANCED

Models of the Earth

Planning Guide

Compression Guide

To shorten instruction because of time limitations, omit Section 2.

OBJECTIVES	LABS, DEMONSTRATIONS, AND ACTIVITIES	TECHNOLOGY RESOURCES
PACING • 45 min pp. 52–56 **Chapter Opener**	**SE Long-Term Project** Positions of Sunrise and Sunset, pp. 848–851 ADVANCED **LTP Long-Term Project** Positions of Sunrise and Sunset* ADVANCED	**OSP Parent Letter** ■ **CD Student Edition on CD-ROM** **CD Chapter Summaries Audio CD** ■ **VID Brain Food Video Quiz**
Section 1 Finding Locations on Earth • Distinguish between latitude and longitude. • Explain how latitude and longitude can be used to locate places on Earth's surface. • Explain how a magnetic compass can be used to find directions on Earth's surface.	**TE Activity** On the Grid, p. 53 ◆ GENERAL **TE History Connection** Early Cartographer, p. 54 ADVANCED	**OSP Lesson Plans** (also in print) **TR Bellringer*** **TR** 11 Parallels, Meridians, and Great Circles* **TR** 12 Magnetic Declination of the United States* **TE Internet Activity** Magnetic Declination, p. 55 GENERAL **CRF Internet Activity** Magnetic Declination* GENERAL
PACING • 45 min pp. 57–62 **Section 2 Mapping Earth's Surface** • Explain two ways that scientists get data to make maps. • Describe the characteristics and uses of three types of map projections. • Summarize how to use keys, legends, and scales to read maps.	**TE Group Activity** School Map, p. 57 ◆ GENERAL **SE Quick Lab** Making Projections, p. 58 GENERAL **CRF Datasheet for Quick Lab*** GENERAL **TE Activity** Map A-Peel, p. 58 ◆ GENERAL **TE Group Activity** Maps and Globes, p. 59 ◆ GENERAL **TE Demonstration** Sun and Shadow, p. 60 ◆ GENERAL **TE Group Activity** Chart Your Course, p. 60 ◆ ADVANCED **CRF Inquiry Lab** Scale the School* GENERAL **CRF Making Models Lab** Remote Sensing* GENERAL	**OSP Lesson Plans** (also in print) **TR Bellringer*** **TR** 13 Types of Map Projections* **CD Interactive Tutor** Mapping and Technology **CD Interactive Tutor** Satellite Measurements
PACING • 90 min pp. 63–68 **Section 3 Types of Maps** • Explain how elevation and topography are shown on a map. • Describe three types of information shown in geologic maps. • Identify two uses of soil maps.	**TE Group Activity** Terrain Models, p. 63 ◆ GENERAL **SE Quick Lab** Topographic Maps, p. 64 GENERAL **CRF Datasheet for Quick Lab*** GENERAL **TE Discussion** Value of Geologic Maps, p. 66 GENERAL **TE Group Activity** Local Soil Maps, p. 67 ◆ ADVANCED **SE Making Models Lab** Contour Maps: Island Construction, pp. 74–75 GENERAL **CRF Datasheet for Chapter Lab*** **SE Mapping Expeditions** Journey to Red River, pp. 832–833 GENERAL **SE Maps in Action** Topographic Map of the Desolation Watershed, p. 76 GENERAL **TE Activity** Local Watershed, p. 76 ◆ GENERAL **TE Group Activity** Biodiversity Maps, p. 77 ◆ GENERAL	**OSP Lesson Plans** (also in print) **TR Bellringer*** **TR** 14 Topographic Maps* **TR** 15 Topographic Map of the Desolation Watershed*

PACING • 90 min

CHAPTER REVIEW, ASSESSMENT, AND STANDARDIZED TEST PREPARATION

- **SE Chapter Highlights**, p. 69
- **SE Chapter Review**, pp. 70–71
- **SE Standardized Test Prep**, pp. 72–73
- **CRF Concept Review*** ■ GENERAL
- **CRF Critical Thinking*** ADVANCED
- **CRF Math Skills*** GENERAL
- **CRF Graphing Skills*** GENERAL
- **CRF Chapter Test A*** ■ GENERAL
- **CRF Chapter Test B*** ADVANCED
- **OSP Lesson Plans** (also in print)
- **OSP Test Generator**
- **OSP Test Item Listing**

Online and Technology Resources

go.hrw.com — Holt Online Learning

Visit go.hrw.com for access to Holt Online Learning, or enter the keyword **HQ6 Home** for a variety of free online resources.

One-Stop Planner® CD-ROM

This CD-ROM package includes
- Lab Materials QuickList Software
- Holt Calendar Planner
- Customizable Lesson Plans
- Printable Worksheets
- ExamView® Test Generator
- Interactive Teacher Edition
- Holt PuzzlePro®
- Holt PowerPoint® Resources

KEY

- SE Student Edition
- TE Teacher Edition
- CRF Chapter Resource File
- LTP Long-Term Projects
- OSP One-Stop Planner
- TR Transparencies and Transparency Worksheets
- CD CD or CD-ROM
- VID Classroom Video/DVD
- * Also on One-Stop Planner
- ◆ Requires advance prep
- ■ Also available in Spanish

SKILLS DEVELOPMENT RESOURCES	REVIEW AND ASSESSMENT	CORRELATIONS
SE **Pre-Reading Activity,** p. 52 GENERAL TE **Using the Figure** Topographic Map, p. 52 GENERAL		National Science Education Standards
CRF **Directed Reading*** BASIC TE **Skill Builder** Vocabulary, p. 54 GENERAL TE **Inclusion Strategies,** p. 54	SE **Reading Check,** p. 54 GENERAL SE **Section Review,** p. 56 GENERAL TE **Reteaching,** p. 55 BASIC TE **Quiz,** p. 55 GENERAL TE **Alternative Assessment,** p. 56 GENERAL CRF **Section Quiz*** ■ GENERAL	HNS 3c, HNS 3d, SAI 1c, SAI 1d, SAI 1f, SAI 2c, ST 1e, UCP 2
CRF **Directed Reading*** BASIC TE **Reading Skill Builder** Reading Organizer, p. 59 BASIC TE **Inclusion Strategies,** p. 60 SE **Math Practice,** p. 61 GENERAL TE **Using the Figure** Planning a Trip, p. 61 GENERAL	SE **Reading Checks,** pp. 58, 61 GENERAL SE **Section Review,** p. 62 GENERAL TE **Reteaching,** p. 61 BASIC TE **Quiz,** p. 61 GENERAL TE **Alternative Assessment,** p. 61 ADVANCED CRF **Section Quiz*** ■ GENERAL	HNS 2a, SAI 1c, SAI 1d, SAI 1f, SAI 2d, ST 1d, ST 1e, UCP 1, UCP 2
CRF **Directed Reading*** BASIC TE **Skill Builder** Math, p. 64 GENERAL TE **Reading Skill Builder** Reading Hint, p. 65 BASIC SE **Graphic Organizer** Spider Map, p. 66 GENERAL TE **Using the Figure** Map Layers, p. 66 GENERAL	SE **Reading Checks,** pp. 65, 67 GENERAL SE **Section Review,** p. 68 GENERAL TE **Reteaching,** p. 67 BASIC TE **Quiz,** p. 67 GENERAL TE **Alternative Assessment,** p. 68 GENERAL CRF **Section Quiz*** ■ GENERAL	HNS 2a, SAI 1a, SAI 1b, SAI 1c, SAI 1d, SAI 1f, SAI 2c, SAI 2d, ST 1d, ST 1e, UCP 1, UCP 2

Holt Lab Generator CD-ROM

Search for labs by topic, standard, difficulty level, and/or time frame. Edit those labs to fit your needs, or create your own labs. Use the Lab Materials QuickList to order just what you need.

Holt Earth Science Interactive Tutor CD-ROM

This CD-ROM consists of interactive activities that give students a fun way to extend their knowledge of Earth science concepts.

Chapter Summaries Audio CDs

These CDs include audio summaries of the key concepts presented in each chapter. (Audio summaries are also available in Spanish.)

SCILINKS. NSTA

www.scilinks.org

Maintained by the **National Science Teachers Association.** See Chapter Enrichment pages that follow for a complete list of topics.

See Chapter Enrichment pages for Video Resources.

Chapter 3 Chapter Enrichment

This Chapter Enrichment provides relevant and interesting information to expand and enhance your classroom instruction of the chapter material.

Section 1 Finding Locations on Earth

Invention of the Magnetic Compass

A key Chinese invention was the construction of the world's first mariner's compass. The attractive power of lodestone (magnetite) had been known in both the East and the West for over two thousand years. Chinese records describe a south-pointing spoon-shaped device as early as the Han dynasty in the second century BCE. It was originally part of fortune-telling boards on which were engraved 24 points of the compass. During the 8th century CE, Chinese scholars devised a way to magnetize iron needles by rubbing them with the mineral magnetite. By 1000 CE, the compass was refined by floating the needle in water (wet compass) or by suspending it by a silken thread (dry compass). Between 850 and 1050 CE, Chinese navigators began to use the magnetic compass to find directions at sea. As a result of this invention, during the Ming dynasty (about 1400 CE), the Chinese admiral Zheng He was able to mount major expeditions to India, Arabia, East Africa, and throughout Southeast Asia well before the Age of Exploration in the West. It wasn't until 1600 CE that the English scientist William Gilbert explained why compasses point in a north-south direction. Gilbert proposed that Earth acts as a giant magnet that has two magnetic poles. Compasses act as direction finders because they are attracted by Earth's magnetic poles.

▲ Modern magnetic compasses were invented about 1000 years ago.

◀ Ground surveys provide data for some types of maps.

Section 2 Mapping Earth's Surface

Early Maps

The world's earliest surviving map is a clay tablet made over 4,500 years ago in Babylonia. It shows the Earth as a flat disc with the city of Babylon in the center and includes the Tigris and Euphrates Rivers. Around the edge of the tablet is an ocean surrounding the world. During the second century CE, the Greek cartographer Ptolemy wrote books on the theory of mapmaking, describing the use of map projections. He created a world map complete with lines of latitude and longitude for locating positions. In the Middle Ages, simple European maps were called "T in O" maps because they showed the three continents of Asia, Africa, and Europe separated by a T-shaped body of water and ringed by the world ocean. One of the most unusual early maps was woven of sticks, twine, and seashells by the early explorers of the Pacific, the Polynesian islanders. The shells represented islands, and the sticks were the pattern of waves between them.

Remote Sensing

The techniques of remote sensing have led to the creation of maps that otherwise would have been impossible. These techniques include devices that record the radiation emitted, reflected, or absorbed by objects to build an image of the terrain. Remote sensing techniques can be used to map areas that are difficult for humans to visit, such as the depths of the ocean, remote locations on Earth, and even the surfaces of other planets and moons.

Video Resources

Brain Food Video Quizzes

Brain Food Video Quizzes These videos contain game-show style quizzes that assess students' progress and motivate students to study the chapter material.

NOVA Videos

To order **NOVA** videos related to this chapter, visit **go.hrw.com** and enter the keyword **HQ6MODV.**

Section 3 Types of Maps

The First Geologic Map

William Smith, a self-taught English geologist born in the late 1700s, created the first modern geologic map. Smith used fossils to distinguish between different rock strata. His maps used over 20 colors to depict different geologic formations.

Beyond Geography

Anything people can visualize, they can map. For example, doctors can now map the brain, blood flow, genes, and transmission patterns of disease. The availability of data from global satellite networks has made it possible to create maps that enable Earth scientists to study the interaction of Earth's many systems—the atmosphere, the land, ecosystems, oceans, weather, and human interfaces. By layering different kinds of data on a base map, it is possible to detect small changes and observe connections between the systems. Writers have created maps of imaginary landscapes from fiction such as J.R.R. Tolkein's Middle Earth, the desert planet Arrakis from Frank Herbert's Dune series, and C.S. Lewis's Narnia. Extraterrestrial maps show stars and planets. The only limit to cartography is the human imagination.

▲ The topography of a region can be shown in great detail on a topographic map.

SciLinks

Developed and maintained by the National Science Teachers Association

SciLinks is maintained by the National Science Teachers Association to provide you and your students with interesting, up-to-date links that will enrich your classroom presentation of the chapter.

Visit www.scilinks.org and enter the SciLinks code for more information about the topic listed.

Topic: Mapping
SciLinks code: HQ60910

Topic: Global Positioning System
SciLinks code: HQ60680

Topic: Cartography
SciLinks code: HQ60229

Topic: Latitude and Longitude
SciLinks code: HQ60854

Topic: Topographic Maps
SciLinks code: HQ61536

Topic: Geographic Information Systems
SciLinks code: HQ60665

Chapter 3

Chapter Overview

This chapter explains how scientists map and locate features on Earth's surface, how elevation is shown on maps, and how maps convey information about Earth.

Using the Figure — GENERAL

Topographic Map This satellite image shows the region of Southern Asia that includes India, Nepal, Tibet, and Sri Lanka. The red area north of the kite-shaped Indian subcontinent represents the Himalayas, the world's highest mountain range. The green areas represent lowlands and the yellow/orange areas represent mountains and highlands. Ask students what land feature the central area of the tear-shaped island of Sri Lanka represents. (a mountain, possibly volcanic) LS Visual

PRE-READING ACTIVITY

FOLDNOTES

You may want to use this FoldNote in a classroom discussion to review material from the chapter. On the board, write each category from the Three-Panel Flip Chart. Then, ask students to provide information for each category. Under the appropriate category on the board, write the information that students provide.

Chapter 3 Models of the Earth

Sections

1 Finding Locations on Earth

2 Mapping Earth's Surface

3 Types of Maps

What You'll Learn

- How people determine location on Earth's surface
- How people make maps
- How various types of maps are used

Why It's Relevant

Maps help people navigate and find locations on Earth's surface. Maps also help scientists study changes in Earth's surface.

PRE-READING ACTIVITY

FOLDNOTES

Three-Panel Flip Chart Before you read this chapter, create the FoldNote entitled "Three-Panel Flip Chart" described in the Skills Handbook section of the Appendix. Label the flaps of the three-panel flip chart with "Topographic maps," "Geologic maps," and "Other types of maps." As you read the chapter, write information you learn about each category under the appropriate flap.

▶ By using advanced technology, scientists can create highly accurate models of Earth. This image of India, taken from space, uses color to show the height of different land features and the depths of the ocean floor.

Chapter Correlations — *National Science Education Standards*

UCP 1 Systems, order, and organization: The natural and designed world is complex; it is too large and complicated to investigate and comprehend all at once. **(Sections 1–3)**

UCP 2 Evidence, models, and explanation: Models are tentative schemes or structures that correspond to real objects, events, or classes of events, and that have explanatory power. Models help scientists and engineers understand how things work. **(Sections 1–3)**

SAI 2c Scientists rely on technology to enhance the gathering and manipulation of data. **(Sections 1–3)**

SAI 2d Mathematics is essential in scientific inquiry. **(Sections 1–3)**

HNS 2a Science distinguishes itself from other ways of knowing and from other bodies of knowledge through the use of empirical standards, logical arguments, and skepticism... **(Sections 1–3)**

HNS 3c Occasionally, there are advances in science and technology that have important and long-lasting effects on science and society. Examples include . . . information and communication . . . . **(Sections 1–3)**

HNS 3d The historical perspective of scientific explanations demonstrates how scientific knowledge changes by evolving over time, almost always building on earlier knowledge. **(Section 1)**

Section 1 Finding Locations on Earth

Earth is very nearly a perfect sphere. A sphere has no top, bottom, or sides to use as reference points for specifying locations on its surface. However, Earth's axis of rotation can be used to establish reference points. The points at which Earth's axis of rotation intersects Earth's surface are used as reference points for defining direction. These reference points are the geographic North and South Poles. Halfway between the poles, a circle called the *equator* divides Earth into the Northern and Southern Hemispheres. A reference grid that is made up of additional circles is used to locate places on Earth's surface.

OBJECTIVES

- **Distinguish** between latitude and longitude.
- **Explain** how latitude and longitude can be used to locate places on Earth's surface.
- **Explain** how a magnetic compass can be used to find directions on Earth's surface.

KEY TERMS

parallel
latitude
meridian
longitude

Latitude

One set of circles describes positions north and south of the equator. These circles are called **parallels** because they run east and west around the world parallel to the equator. The angular distance north or south of the equator is called **latitude.**

parallel any circle that runs east and west around Earth and that is parallel to the equator; a line of latitude

latitude the angular distance north or south from the equator; expressed in degrees

Degrees of Latitude

Latitude is measured in degrees, and the equator is designated as 0° latitude. Because the distance from the equator to either of the poles is one-fourth of a circle, and a circle has 360°, the latitude of both the North Pole and the South Pole is 1/4 of 360°, or 90°, as shown in **Figure 1.** In actual distance, 1° of latitude equals 1/360 of Earth's circumference, or about 111 km.

Parallels north of the equator are labeled *N*; those south of the equator are labeled *S*. In the Northern Hemisphere, Washington, D.C., is located near a parallel that is 39° north of the equator. So, the latitude of Washington, D.C., is 39°N. Sydney, Australia is in the Southern Hemisphere and has a latitude of 34°S.

Minutes and Seconds

Each degree of latitude consists of 60 equal parts, called *minutes*. One minute (symbol: ′) of latitude equals 1.85 km. A more precise latitude for Washington, D.C., is 38°53′N. In turn, each minute is divided into 60 equal parts, called *seconds* (symbol: ″). So, the precise latitude of the center of Washington, D.C., is expressed as 38°53′23″N.

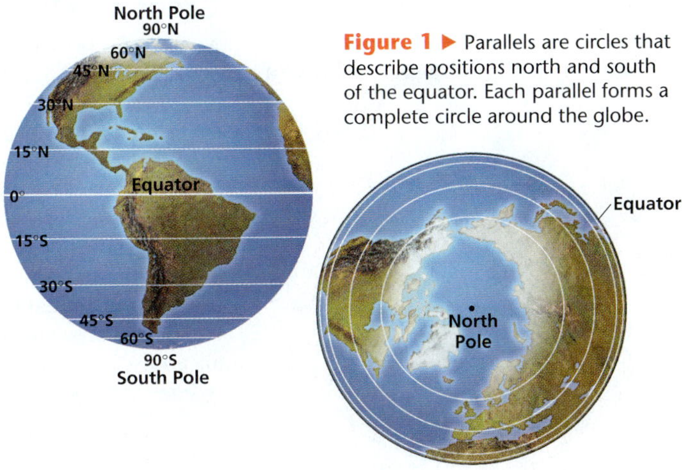

Figure 1 ▶ Parallels are circles that describe positions north and south of the equator. Each parallel forms a complete circle around the globe.

Section 1

Focus

Overview

This section explains how latitude and longitude, Earth's magnetism, and the Global Positioning System (GPS) are used to find locations on Earth.

Bellringer

Ask students to think about the route that they took to arrive at school in the morning. Have students write out a description so someone else could follow it, starting from the same location. (Answers may vary but students will probably use landmarks or cross streets to describe their route.) LS **Verbal**

Motivate

Activity — GENERAL

On the Grid Clear an area of the floor about 2 m^2. Use masking tape to lay down a series of grid lines and label them with sticky notes. Label one axis with letters and the other axis with numbers. Place objects, such as books or building blocks, within the grid. Give each student a sheet of graph paper. Have them make a map of the floor that shows the locations of all the objects. Ask students how the grid helped them map the location of objects. (The places where the grid lines cross provide reference points that can be used to describe where objects are. LS **Visual**

CHAPTER RESOURCES

Chapter Resource File
- Directed Reading BASIC

Technology

Transparencies
- Bellringer
- 11 Parallels, Meridians, and Great Circles (with worksheet)

Student Edition on CD-ROM

One-Stop Planner CD-ROM
- Lesson Plan

Teach

HISTORY CONNECTION ADVANCED

Early Cartographer Explain that the method we use for dividing the degree into subdivisions—minutes and seconds—was introduced by the Roman scholar Claudius Ptolemy in his work called *Almagest* during the second century CE. In an eight-volume work entitled *Geography*, Ptolemy compiled one of the first collections of maps. He developed the coordinate system of latitude and longitude as a way of pinpointing locations. Unfortunately, Ptolemy's maps contained errors because the estimate of Earth's size that he used was too small. Invite students to learn more about Ptolemy's contributions to Earth science. Have them share what they learn with the class in oral or written presentations. LS **Verbal**

SKILL BUILDER GENERAL

Vocabulary Lines of longitude are also known as *meridians*. The word comes from the Latin *meri-*, a variation of *medius,* which means "middle" and *dies,* meaning "day." The word once meant "noon," because all points on the same line of longitude experienced noon (and every other hour) at the same time, and therefore, were said to be located on the same meridian. Times of the day before noon were known as *ante meridian,* while times after it were *post meridian.* Today's abbreviations A.M. and P.M. come from these terms. LS **Verbal** English Language Learners

Answers to Reading Check

because the equator is the only parallel that divides Earth into halves

Figure 2 ▶ Meridians are semicircles reaching around Earth from pole to pole.

Longitude

The latitude of a particular place indicates only the place's position north or south of the equator. To determine the specific location of a place, you also need to know how far east or west that place is along its circle of latitude. East-west locations are established by using meridians. As **Figure 2** shows, a **meridian** is a semicircle (half of a circle) that runs from pole to pole.

By international agreement, one meridian was selected to be 0°. This meridian, called the *prime meridian,* passes through Greenwich, England. **Longitude** is the angular distance, measured in degrees, east or west of the prime meridian.

meridian any semicircle that runs north and south around Earth from the geographic North Pole to the geographic South Pole; a line of longitude

longitude the angular distance east or west from the prime meridian; expressed in degrees

Degrees of Longitude

Because a circle is 360°, the meridian opposite the prime meridian, halfway around the world, is labeled 180°. All locations east of the prime meridian have longitudes between 0° and 180°E. All locations west of the prime meridian have longitudes between 0° and 180°W. Washington, D.C., which lies west of the prime meridian, has a longitude of 77°W. Like latitude, longitude can be expressed in degrees, minutes, and seconds. So, a more precise location for Washington, D.C., is 38°53′23″N, 77°00′33″W.

Distance Between Meridians

The distance covered by a degree of longitude depends on where the degree is measured. At the equator, or 0° latitude, a degree of longitude equals approximately 111 km. However, all meridians meet at the poles. Because meridians meet, the distance measured by a degree of longitude decreases as you move from the equator toward the poles. At a latitude of 60°N, for example, 1° of longitude equals about 55 km. At 80°N, 1° of longitude equals only about 20 km.

Figure 3 ▶ A great-circle route from Chicago to Rome is much shorter than a route following a parallel is.

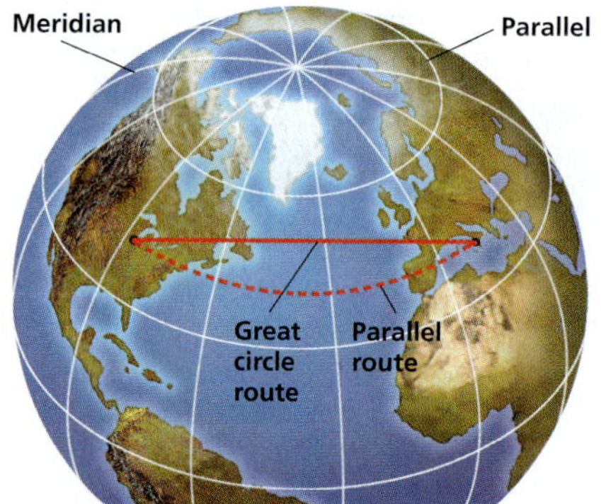

Great Circles

A great circle is often used in navigation, especially by long-distance aircraft. A *great circle* is any circle that divides the globe into halves, or marks the circumference of the globe. Any circle formed by two meridians of longitude that are directly across the globe from each other is a great circle. The equator is the only line of latitude that is a great circle. Great circles can run in any direction around the globe. Just as a straight line is the shortest distance between two points on a flat surface or plane, the route along a great circle is the shortest distance between two points on a sphere, as shown in **Figure 3.** As a result, air and sea routes often travel along great circles.

Reading Check **Why is the equator the only parallel that is a great circle? (See the Appendix for answers to Reading Checks.)**

INCLUSION Strategies

• Behavior Control Issues • Attention Deficit Disorder

One way of engaging students who have behavior control issues is to empower them by letting them make some decisions. Have students use a globe to find the latitude and longitude of major cities. Ask students to choose the cities either by thinking of them or by going to the globe and choosing them. Have students make a list of their cities and the latitude and longitude coordinates for each. LS **Logical**

BRAIN FOOD

Longitude and Time Zones Local time is a measure of the position of the sun relative to a given location. Solar noon occurs when the sun is at the highest point in the sky. As Earth rotates, at any moment, it is solar noon along one meridian. The time-zone system divides the 360° of longitude into 24 time zones—one for each hour of the day, each 15° wide—and uses the spinning Earth as a giant clock.

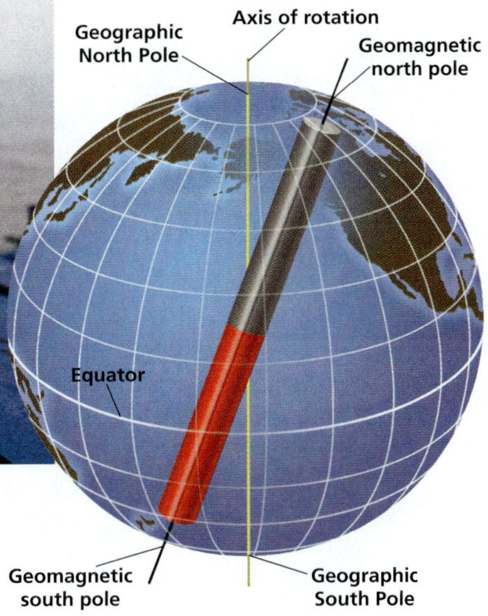

Figure 4 ▶ Earth's magnetic poles are at an angle to Earth's axis of rotation.

Finding Direction

One way to find direction on Earth is to use a magnetic compass. A magnetic compass can indicate direction because Earth has magnetic properties as if a powerful bar-shaped magnet were buried at Earth's center at an angle to Earth's axis of rotation, as shown in **Figure 4.**

The areas on Earth's surface just above where the poles of the imaginary magnet would be are called the *geomagnetic poles.* The geomagnetic poles and the geographic poles are located in different places. The needle of a compass points to the geomagnetic north pole.

Connection to GEOGRAPHY

The Prime Meridian and the International Date Line

Before 1884, most nations in the world designated an arbitrary meridian as the prime meridian for use on their land maps and nautical charts. Similarly, each country set its clocks by the meridian that it had designated as the prime meridian. Thus, each country had a different standard of time and of location. This situation made it difficult to coordinate travel and business activities.

In 1884, the International Meridian Conference met in Washington, D.C. Representatives from 25 nations came together to designate a single meridian as the prime meridian so that maps and times could be standardized around the globe. The meridian that runs through the Royal Observatory at Greenwich, England, was chosen as the prime meridian. The prime meridian marks 0° longitude.

The prime meridian runs through the Royal Observatory at Greenwich and is marked by the red line in this photo.

The designation of the Greenwich meridian as the standard prime meridian allowed an international time standard to be established. The meridian exactly opposite the prime meridian was designated as the International Date Line and marks the place where each new day officially begins at midnight. At that same moment, on the opposite side of Earth, the prime meridian marks noon of the previous day.

Close

Reteaching — Basic

Global Grid Use a ball, rubber bands, and a marking pen. Put a rubber band around the center of the ball. Draw the circle representing the ball's equator. Make a series of parallel circles above and below the equator. Ask what the circles represent. (parallels) Put a second rubber band around the middle of the ball in the opposite direction. Draw the prime meridian. Make a series of circles running from one pole to the other. Ask what these circles represent. (meridians) Point out how distance between meridians decreases as you move from the equator to the poles. **LS Visual**

Quiz — General

1. Compare what happens to longitude near the poles with what happens to latitude. (The meridians of longitude meet at the poles, so the distance between them decreases. As the circles or parallels of latitude approach the poles, they get smaller in circumference. The distances between them stay the same.)
2. What does a magnetic compass use to indicate direction? (Earth's magnetic field and the magnetic properties of metals or some minerals) **LS Verbal**

CHAPTER RESOURCES

Chapter Resource File

Internet Activity
- Magnetic Declination GENERAL

Technology

Transparencies
- 12 Magnetic Declination of the United States (with worksheet)

Internet Activity — General

Magnetic Declination Have students use the latitude and longitude of your city or town to look up the geomagnetic declination of your location by using the National Geophysical Data Center Web site. A worksheet designed to direct student research on this topic can be found in the **Chapter Resource File** booklet or by visiting go.hrw.com and entering the keyword **HQ6MODX.** **LS Logical**

Close, *continued*

Alternative Assessment — GENERAL

Cartoons Around the Globe
Have students create cartoons or comic strips that focus on one of the key concepts in this section: how the system of circles is used to locate places, how magnetic compasses indicate direction, or how the GPS satellite network makes finding places in the modern world easier. Tell students to write captions that explain the drawings. Students could use explorers such as Columbus, Lewis and Clark, or astronauts to make cartoons concrete and relevant. LS **Visual**

Answers to Section Review

1. Lines of latitude circle the globe parallel to the equator, and longitude lines run from pole to pole.
2. The intersection of latitude and longitude lines pinpoint the locations of places on the globe in degrees, minutes, and seconds. Latitude describes a place's position north or south of the equator, and longitude describes its position east or west of the prime meridian.
3. because they represent the shortest distance between two points on the globe
4. The compass needle aligns itself in a north-south direction because it points toward the geomagnetic poles. The other directions on the dial are found by reading number of degrees east or west of north.
5. Because GPS technology can give locations that are accurate to within several centimeters, GPS can be used for navigation in airplanes or ships when using compasses or landmarks is difficult or impossible.
6. Parallels are imaginary circles that are parallel with the equator; latitude is angular distance expressed in degrees north or south of the equator.
7. The circles that represent latitude are parallel and never cross one another. However, lines of longitude all pass through Earth's poles. Because Earth's circumference is greater at the equator than at the poles, the lines of latitude are farther apart at the equator than they are at the poles.
8. Locations on *Earth* are measured in *degrees, minutes,* and *seconds* by using lines of *latitude* that run *parallel* to the *equator,* and *meridians* that define lines of *longitude* and include the *prime meridian.*

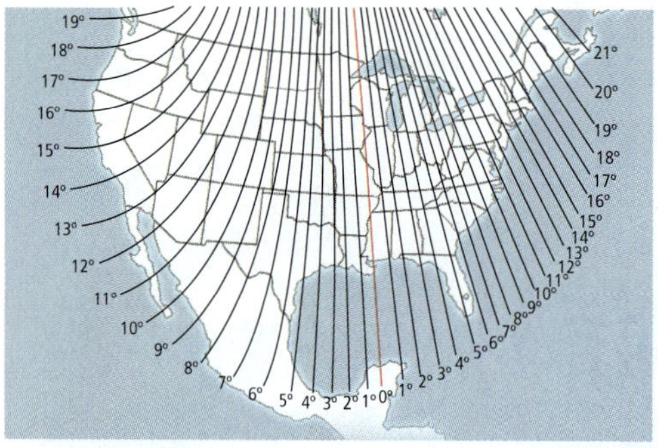

Figure 5 ▶ This map shows the magnetic declinations of the United States in 2004. The lines connect points that have the same magnetic declination.

Magnetic Declination

The angle between the direction of the geographic pole and the direction in which the compass needle points is called *magnetic declination*. In the Northern Hemisphere, magnetic declination is measured in degrees east or west of the geographic North Pole. A compass needle will align with both the geographic North Pole and the geomagnetic north pole for all locations along the line of 0° magnetic declination, which is shown as the red line in **Figure 5.**

Magnetic declination has been determined for points all over Earth. However, because Earth's magnetic field is constantly changing, the magnetic declinations of locations around the globe also change constantly. **Figure 5** shows the magnetic declinations for most of the United States in 2004. By using magnetic declination, a person can use a compass to determine geographic north for any place on Earth. Locating geographic north is important in navigation and in mapmaking.

SCILINKS. NSTA
Developed and maintained by the National Science Teachers Association

For a variety of links related to this subject, go to www.scilinks.org

Topic: Global Positioning System
SciLinks code: HQ60680

The Global Positioning System

Another way people can find their location on Earth is by using the *global positioning system,* or *GPS*. GPS is a satellite navigation system that is based on a global network of 24 satellites that transmit radio signals to Earth's surface. The first GPS satellite, known as NAVSTAR, was launched in 1978.

A GPS receiver held by a person on the ground receives signals from three satellites to calculate the latitude, longitude, and altitude of the receiver on Earth. Personal GPS receivers are accurate to within 10 to 15 m of their position, but high-tech receivers designed for military or commercial use can be accurate to within several centimeters of their location.

Section 1 Review

1. **Describe** the difference between lines of latitude and lines of longitude.
2. **Explain** how latitude and longitude are used to find specific locations on Earth.
3. **Summarize** why great-circle routes are commonly used in navigation.
4. **Explain** how a magnetic compass can be used to find directions on Earth.

CRITICAL THINKING

5. **Applying Concepts** How might GPS technology be beneficial when used in airplanes or on ships?
6. **Making Comparisons** How do parallels differ from latitude?
7. **Identifying Patterns** Explain why the distance between parallels is constant but the distance between meridians decreases as the meridians approach the poles.

CONCEPT MAPPING

8. Use the following terms to create a concept map: *equator, second, parallel, degree, Earth, minute, longitude, meridian, prime meridian,* and *latitude.*

CHAPTER RESOURCES

Chapter Resource File
- Section Quiz GENERAL

Workbooks
- Study Guide (also in Spanish)

Section 2 Mapping Earth's Surface

A globe is a familiar model of Earth. Because a globe is spherical like Earth, a globe can accurately represent the locations, relative areas, and relative shapes of Earth's surface features. A globe is especially useful in studying large surface features, such as continents and oceans. But most globes are too small to show details of Earth's surface, such as streams and highways. For that reason, a great variety of maps have been developed for studying and displaying detailed information about Earth.

OBJECTIVES

- **Explain** two ways that scientists get data to make maps.
- **Describe** the characteristics and uses of three types of map projections.
- **Summarize** how to use keys, legends, and scales to read maps.

KEY TERMS

remote sensing
map projection
legend
scale
isogram

How Scientists Make Maps

The science of making maps, called *cartography,* is a subfield of Earth science and geography. Scientists who make maps are called *cartographers.*

Cartographers use data from a variety of sources to create maps. They may collect data by conducting a field survey, shown in **Figure 1.** During a field survey, cartographers walk or drive through an area to be mapped and make measurements of that area. The information that they collect is then plotted on a map. Because surveyors cannot take measurements at every site in an area, they often use their measurements to make estimated measurements for sites between surveyed points.

By using remote sensing, cartographers can collect information about a site without being at that site. In **remote sensing,** equipment on satellites or airplanes obtain images of Earth's surface. Maps are often made by combining information from images gathered remotely with information from field surveys.

remote sensing the process of gathering and analyzing information about an object without physically being in touch with the object

Figure 1 ▶ Cartographers in the field use technology to enhance the precision of their measurements. Electronic devices can be used to measure the distance between an observer and a distant point with a high degree of accuracy.

Section 2

Focus

Overview

This section explains how cartographers obtain information to create maps, describes map projections, and explains how to use a compass rose, a legend, and scales.

Bellringer

Ask students to write a brief paragraph that summarizes different kinds of maps they have seen and what kinds of information each presents. (Students may describe floor plans of public buildings, bus or train maps, navigational charts, road maps, maps of the human body, weather maps, or maps of other planets.) LS **Verbal**

Motivate

Group Activity — GENERAL

School Map Have students work in groups to make a map of the classroom or of another area of the school. Ask them to start by making a sketch map of their area. Then, have them use tape measures, rulers, and triangles to make a more accurate map. Have them determine a scale for their map. Ask students how they think cartographers map larger areas of Earth. (Answers may vary. Students may mention field surveying, triangulating, taking aerial photographs, and using satellite images.) LS **Kinesthetic**

CHAPTER RESOURCES

Chapter Resource File

- **Directed Reading** BASIC
- **Inquiry Lab** Scale the School GENERAL
- **Making Models Lab** Remote Sensing GENERAL

Technology

Transparencies
- Bellringer

Student Edition on CD-ROM

One-Stop Planner CD-ROM
- Lesson Plan

Teach

Activity — GENERAL

Map A-Peel Give students a large orange or grapefruit. Tell students to use a marking pen to draw latitude and longitude lines on the outside of the fruit skin. Then, have them draw the rough outlines of the continents. After the ink is dry, use a knife to free the skin near the stem of each fruit and make a single cut the length of the fruit from top to bottom. By carefully slipping their fingers under the skin, students should separate the peel from the fruit in one piece. Once the peel is free, have them flatten it out. They should make additional tears in the peel to make the "map" lie flat. Have students describe any distortions produced by flattening the fruit skin. (Sample answer: Parts of the continents became separated and gaps appeared in the map.) **LS Kinesthetic**

Discussion — GENERAL

Round Earth, Flat Maps Ask students why a globe is the only completely accurate map. (Sample answer: Because the globe is three-dimensional like Earth itself, it can show all the areas in the correct shapes and relative sizes; it can also show correct directions and relative distances. Flat maps always introduce distortions by stretching or flattening Earth's curved surface.) **LS Logical**

Answer to Reading Check

Because both the parallels and the meridians are equally spaced straight lines on a cylindrical projection, the parallels and meridians form a grid.

CHAPTER RESOURCES

Chapter Resource File
- Datasheet for Quick Lab GENERAL

Technology

Transparencies
- 13 Types of Map Projections (with worksheet)

QuickLAB — 20 min

Making Projections

Procedure

1. Use a **fine-tip marker** to draw a variety of shapes on a **small glass ivy bowl** or **clear plastic hemisphere.**
2. Shine a **flashlight** through the bottom of the bowl.
3. Shape a **piece of white paper** into a cylinder around the bowl.
4. Trace the shapes projected from the bowl onto the paper.
5. Using a cone of paper, repeat steps 3 and 4.

Analysis

1. What type of projection did you create in steps 3 and 4? in step 5?
2. Compare the sizes of the shapes on the bowl with those on your papers. What areas did each projection distort?

map projection a flat map that represents a spherical surface

Map Projections

A map is a flat representation of Earth's curved surface. However, transferring a curved surface to a flat map results in a distorted image of the curved surface. An area shown on a map may be distorted in size, shape, distance, or direction. The larger the area being shown is, the greater the distortion tends to be. A map of the entire Earth would show the greatest distortion. A map of a small area, such as a city, would show only slight distortion.

Over the years, cartographers have developed several ways to transfer the curved surface of Earth onto flat maps. A flat map that represents the three-dimensional curved surface of a globe is called a **map projection.** No projection is an entirely accurate representation of Earth's surface. However, each kind of projection has certain advantages and disadvantages that must be considered when choosing a map.

Cylindrical Projections

Imagine Earth as a transparent sphere that has a light inside. If you wrapped a cylinder of paper around this lighted globe and traced the outlines of continents, oceans, parallels, and meridians, a *cylindrical projection,* shown in **Figure 2,** would result. Meridians on a cylindrical projection appear as straight, parallel lines that have an equal amount of space between them. On a globe, however, the meridians come together at the poles. A cylindrical projection is accurate near the equator but distorts distances and sizes near the poles.

Though distorted, cylindrical projections have some advantages. One advantage is that parallels and meridians form a grid, which makes locating positions easier. Also, the shapes of small areas are usually well preserved. When a cylindrical projection is used to map small areas, distortion is minimal.

Reading Check Why do meridians and parallels appear as a grid when shown on a cylindrical projection? (See the Appendix for answers to Reading Checks.)

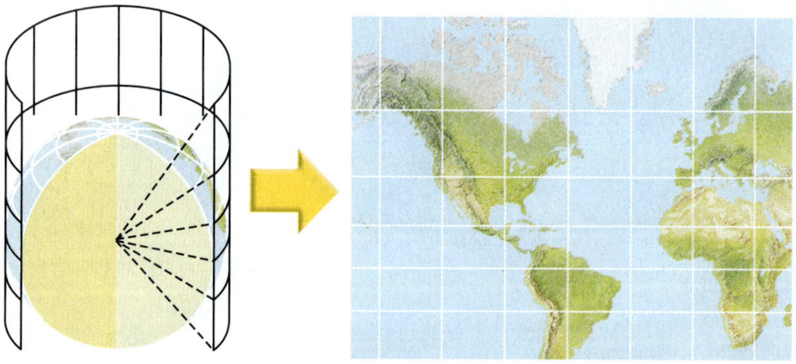

Figure 2 ▶ A light at the center of a transparent globe would project lines on a cylinder of paper (left), which would produce a cylindrical projection (right).

QuickLAB

Skills Acquired
- Recognizing patterns
- Constructing models

Teacher's Notes A spherical goldfish bowl or clear plastic hamster ball can also be used instead of the ivy bowl or plastic hemisphere. Students can add a grid to their map projections by shaping soft wire over the bowl to form the network of latitude and longitude lines.

Answers

1. Steps 3 and 4 create a cylindrical projection. Step 5 creates a conic projection.
2. On the cylindrical projection, shapes and distances closer to the poles are distorted more than those near the equator. Shapes farthest from the place where the cone came into contact with the ball are distorted the most.

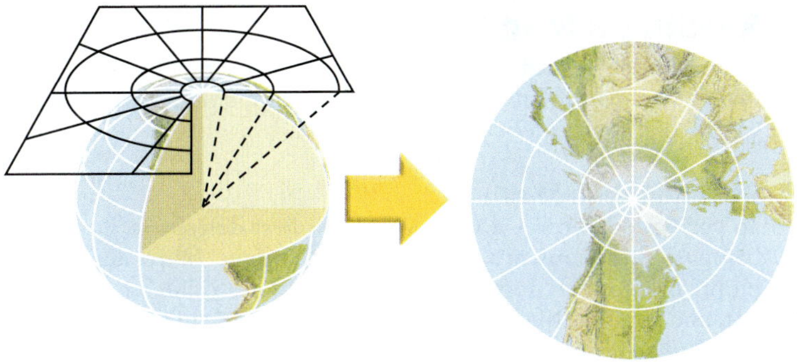

Figure 3 ▶ When a sheet of paper is placed so that it touches a lighted globe at only one point (left), the lines projected on the paper form an azimuthal projection (right).

Azimuthal Projections

A projection made by placing a sheet of paper against a transparent, lighted globe such that the paper touches the globe at only one point is called an *azimuthal* (AZ uh MYOOTH uhl) *projection,* as shown in **Figure 3.** On an azimuthal projection, little distortion occurs at the point of contact, which is commonly one of the poles. However, an azimuthal projection shows unequal spacing between parallels that causes a distortion in both direction and distance. This distortion increases as distance from the point of contact increases.

Despite distortion, an azimuthal projection is a great help to navigators in plotting routes used in air travel. As you know, a great circle is the shortest distance between any two points on the globe. When projected onto an azimuthal projection, a great circle appears as a straight line. Therefore, by drawing a straight line between any two points on an azimuthal projection, navigators can readily find a great-circle route.

Conic Projections

A projection made by placing a paper cone over a lighted globe so that the axis of the cone aligns with the axis of the globe is known as a *conic projection.* The cone touches the globe along one parallel of latitude. As shown in **Figure 4,** areas near the parallel where the cone and globe are in contact are distorted the least.

A series of conic projections may be used to increase accuracy by mapping a number of neighboring areas. Each cone touches the globe at a slightly different latitude. Fitting the adjoining areas together then produces a continuous map. Maps made in this way are called *polyconic projections.* The relative size and shape of small areas on the map are nearly the same as those on the globe.

Figure 4 ▶ A light at the center of a transparent globe would project lines on a paper cone (left), which would produce a conic projection (right).

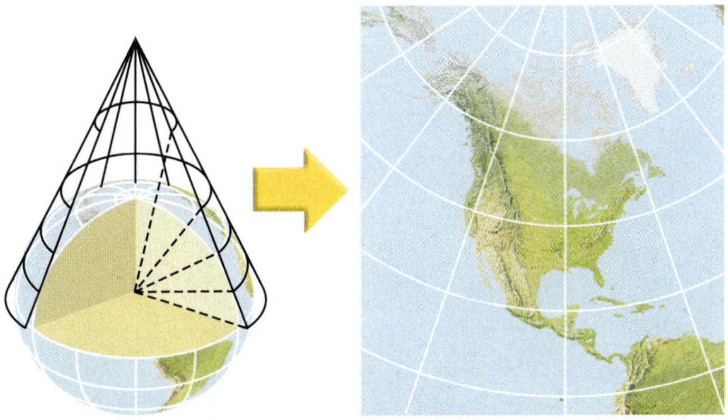

READING SKILL BUILDER — BASIC

Reading Organizer As students read this section, encourage them to take Power Notes, KWL Notes, or Two-Column Notes as described in the Skills Handbook of the Appendix. Students may want to take notes about the advantages and disadvantages of the different types of map projections. Later, students can use these notes as a study guide for assessments. LS Verbal

English Language Learners

GEOGRAPHY CONNECTION — GENERAL

Mercator Projections The most famous cylindrical map projection is the Mercator projection created by Gerhard Kramer in 1569. Born in Belgium, Kramer adopted the Latin form of his name, *Mercator,* which means "trader." Mercator designed his map for use in navigation. Because it preserves angles, compass courses (but not great circles) between two points can be read directly off his map. Its main disadvantage is that it distorts the size of many land areas. Mercator also published one of the first world atlases. Encourage students to learn more about this famous cartographer and to share what they learn with the class through an oral presentation. LS Verbal/Auditory

Group Activity — GENERAL

Maps and Globes Divide the class into small groups. Give each group a globe and several maps that use different projections. Have students compare the sizes of continents and large islands on different maps. Tell students to examine the shapes (curved or straight) and spacing of grid lines on different projections. Have them identify the kinds of distortions produced by different map projections. Bring groups together and have them discuss their conclusions. LS Visual Co-op Learning

MISCONCEPTION ALERT

Geographical Distortions Misconceptions about location and size of geographical features may be reinforced by common map projections. Cylindrical projections greatly stretch Canada and Greenland so that students may not be aware of their actual sizes. The misplacements of Europe close to the equator, of South America directly south of North America, and of most of Africa south of the equator are common geographical misconceptions.

Teach, *continued*

Demonstration — GENERAL

Sun and Shadow Cardinal directions on Earth are determined relative to the sun. To help students associate direction with the sun, use a stick and its shadow to find directions. On a sunny day, place the stick vertically in the ground and mark the tip of its shadow with a chalk mark. Wait one hour, and again mark the tip of the shadow. A line drawn from the first mark to the second gives the approximate east/west axis. (The first mark is the west end of the line and the second is the east end.) Use those directions to determine north and south. Draw an appropriate compass rose. Invite students to check the results by using a magnetic compass. **LS Kinesthetic/Visual**

Cultural Awareness — BASIC

Map Orientation Point out that it is merely by convention that north is placed at the top of most modern maps. One explanation is that most of the world's landmasses are found in the Northern Hemisphere. In medieval Europe, many maps placed Jerusalem and the east at the top. Thus, the phrase "to orient" comes from the Latin *oriens,* meaning "rising sun" and "east." The Chinese, who invented magnetic compasses, thought of the compass needle as pointing south rather than north. In the Southern Hemisphere, maps that place south at the top are very popular.

SCILINKS. NSTA Developed and maintained by the National Science Teachers Association

For a variety of links related to this subject, go to www.scilinks.org

Topic: Cartography
SciLinks code: HQ60229

Reading a Map

Maps provide information through the use of symbols. To read a map, you must understand the symbols on the map and be able to find directions and calculate distances.

Direction on a Map

To correctly interpret a map, you must first determine how the compass directions are displayed on the map. Maps are commonly drawn with north at the top, east at the right, west at the left, and south at the bottom. Parallels run from side to side, and meridians run from top to bottom. Direction should always be determined in relation to the parallels and meridians.

On maps published by the United States Geological Survey (USGS), such as the one shown in **Figure 5,** north is located at the top of the map and is marked by a parallel. The southern boundary, at the bottom of a map, is also marked by a parallel. At least two additional parallels are usually drawn in or indicated by cross hairs at 2.5′ intervals. Meridians of longitude indicate the eastern and western boundaries of USGS maps. Additional meridians may also be shown. All parallels and meridians shown on these maps are labeled in degrees, minutes, and seconds.

Many maps also include a compass rose, as shown in **Figure 5.** A *compass rose* is a symbol that indicates the cardinal directions. The *cardinal directions* are north, east, south, and west. Some maps replace the compass rose with a single arrow that points to geographic north. This arrow is generally labeled and may not always point to the top of the map.

Figure 5 ▶ Maps may show locations by marking parallels and meridians. Direction is commonly shown with a compass rose (inset).

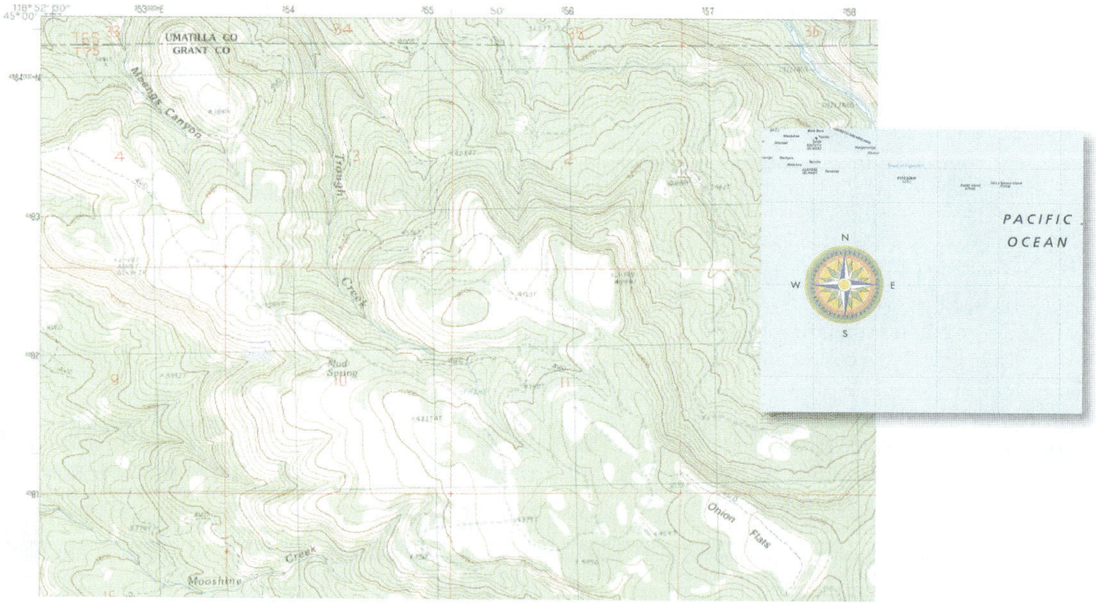

Group Activity — ADVANCED

Chart Your Course Obtain a U.S. Geological Survey map of an area near your school. Distribute photocopies of the relevant part of the map to groups of students. Visit the site with the class. Have each group select a land feature as its destination. Tell groups to draw a line from their present location to their destination, extending the line to the map border. Center a compass over the line, align the axis N-S or E-W, and read the compass bearing of the destination. Have groups locate their destination in the field. **LS Kinesthetic**

INCLUSION Strategies

- ***Gifted and Talented***

Ask students to go beyond the textbook and work together to learn more about cartographers. Tell them to create a questionnaire that has at least twelve questions they would like to ask about mapmaking. Then, have them use the Internet to find some companies that create maps and send their questionnaires to several of these firms. Ask them to compile any replies into an illustrated report for the class. **LS Verbal/Interpersonal**

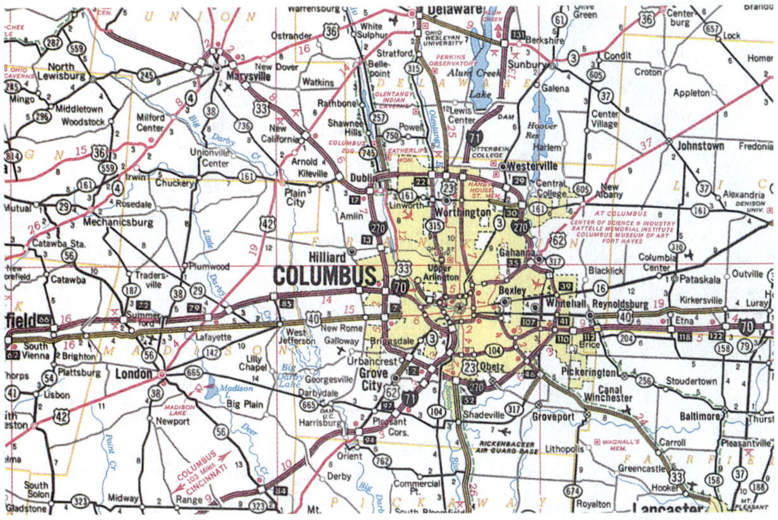

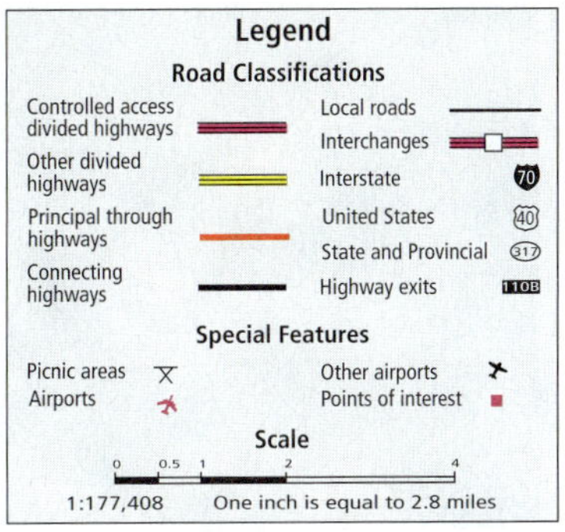

Figure 6 ▶ To be useful for estimating distance, a road map must include a map scale. Three types of map scales are shown on this legend. The map legend also explains the symbols used on this map. ***Which symbols in the map legend resemble the features that they represent?***

Symbols

Maps often have symbols for features such as cities and rivers. The symbols are explained in the map **legend,** a list of the symbols and their meanings, such as the one shown in **Figure 6.** Some symbols resemble the features that they represent. Others, such as those for towns and urban areas, are more abstract.

Map Scales

To be accurate, a map must be drawn to **scale.** The scale of a map indicates the relationship between distance shown on the map and actual distance. As **Figure 6** shows, a map scale can be expressed as a graphic scale, a fractional scale, or a verbal scale.

A *graphic scale* is a printed line that has markings on it that are similar to those on a ruler. The line represents a unit of measure, such as the kilometer or the mile. Each part of the scale represents a specific distance on Earth. To find the actual distance between two points on Earth, you first measure the distance between the points as shown on the map. Then, you compare that measurement with the map scale.

A second way of expressing scale is by using a ratio, or a *fractional scale*. For example, a fractional scale such as 1:25,000 indicates that 1 unit of distance on the map represents 25,000 of the same unit on Earth. A fractional scale remains the same with any system of measurement. In other words, the scale 1:100 could be read as 1 in. equals 100 in. or as 1 cm equals 100 cm.

A *verbal scale* expresses scale in sentence form. An example of a verbal scale is "One centimeter is equal to one kilometer." In this scale, 1 cm on the map represents 1 km on Earth.

Reading Check **Name three ways to express scale on a map.** **(See the Appendix for answers to Reading Checks.)**

legend a list of map symbols and their meanings

scale the relationship between the distance shown on a map and the actual distance

MATHPRACTICE

Determining Distance
You notice that the scale on a map of the United States says, "One centimeter equals 120 kilometers." By measuring the straight-line distance between Brooklyn, New York, and Miami, Florida, you determine that the cities are about 14.5 cm apart on the map. What is the approximate distance in kilometers between the two cities?

MATHPRACTICE

Answer
1 cm = 120 km; 14.5 cm × 120 km/cm = 1,740 km

Close

Reteaching — BASIC

Using the Legend Give students several different maps of the same geographic area—a political map, a terrain map, a road map, and a weather map. Have students use the map legends to identify the symbols each map uses to convey information. Have them compare the scales and symbols on the maps and describe how the symbols convey the purpose of each map. **LS Visual**

Quiz — GENERAL

1. What are flat maps that represent the three-dimensional spherical surface of Earth called? (map projections)
2. What part of a map explains the symbols used on that map? (the legend)
3. What is the chief advantage of a cylindrical map projection? (Parallels and meridians appear as a grid, which makes locating positions easy.)

Alternative Assessment — ADVANCED

Treasure Hunt Assign groups of students to specific areas of the school. Have them hide small prizes on school grounds and make a scale map of the area, using a graphic or verbal scale. Have them create a legend by using colors for landscape features and provide a compass rose. Have them write brief instructions to help others locate the prizes. Ask groups to exchange maps and to find the objects hidden by the other group. **LS Kinesthetic**

Using the Figure — GENERAL

Planning a Trip Invite students to use the map legend to plan a route that they would take if they were acting as navigators for a family automobile trip from Columbus, Ohio, to a nearby town shown on the map, such as London or Mechanicsburg. Answer to caption question: The highway signs are the same shape as those you would see on the roads; the picnic areas are represented by a picnic table; and airports are represented by a symbol that looks like a plane. **LS Visual**

Answer to Reading Check

by using a graphic scale, or a printed line divided into proportional parts that represent units of measure; a fractional scale, in which a ratio shows how distance on Earth relates to distance on a map; or a verbal scale, which expresses scale in sentence form

Close, *continued*

Answers to Section Review

1. Cartographers obtain information to make maps by doing surveys and by remote sensing via satellites or aerial surveys.
2. Cylindrical projections are accurate near the equator but distort distances near the poles. The parallels and meridians make a grid that makes locating positions easy. Azimuthal projections distort direction and distance because of unequal spacing of parallels, but great circles form straight lines, so plotting air travel routes is easier. When conic projections are fitted together, they produce a continuous map and preserve the relative size and shape of small areas.
3. The process of transferring a curved surface onto a flat surface causes all flat maps to distort some aspect of size, shape, distance, or direction.
4. The legend can be used to interpret the map symbols, and the scale is used to determine the relationship between distances shown on the map and actual distances.
5. Isograms connect points of equal value; isograms can be used to represent a variety of different units of measure.
6. To show a small area in useful detail, a globe would have to be huge. A flat map of a small area would be compact and have only slight distortion.
7. Maps are commonly drawn with north at the top, west to the left, east to the right, and south at the bottom. A compass rose is often included on maps to indicate the cardinal directions. If parallels and meridians are shown, they can be used to determine direction.
8. A fractional scale should be used because the scale remains the same with any system of measurement.
9. If the cone used to make the conic projection touches the globe along a parallel close to the poles, there will be little distortion and the map will be more accurate. A cylindrical projection will distort sizes and distances most near the poles.
10. *Cartography* uses many different *map projections,* including *cylindrical projections, azimuthal projections,* and *conic projections* to produce *maps,* which you can read by using a *scale* to estimate distances and the *legend* to interpret the *symbols.*

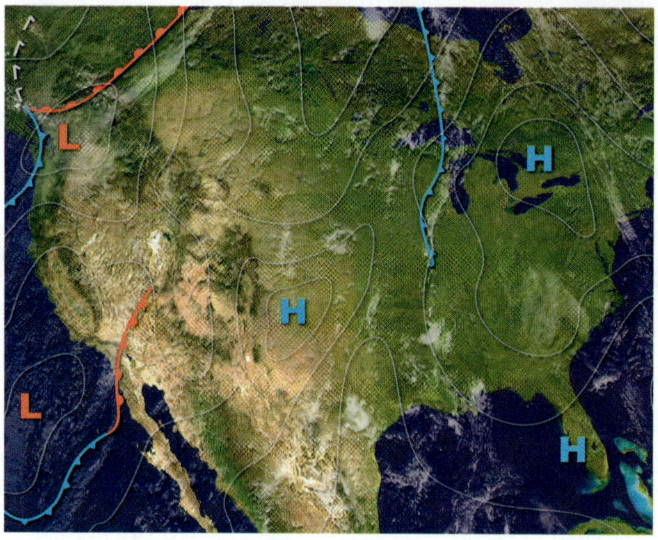

Figure 7 ▶ Areas connected by the isobars on the map share equal atmospheric pressure.

Isograms

A line on a map that represents a constant or equal value of a given quantity is an **isogram.** The prefix *iso-* is Greek for "equal." The second part of the word, *-gram,* means "drawing." This part of the word can be changed to describe the measurement being graphed. For example, when a line connects points of equal temperature, the line is called an *isotherm* because *iso-* means "equal" and *therm* means "heat." All locations along an isogram share the value that is being measured.

Isograms can be used to plot many types of data. Meteorologists use these lines to show changes in atmospheric pressure on weather maps. Isograms used in this manner on a weather map are called *isobars,* as shown in **Figure 7.** All points along an isobar share the same pressure value. Because one location cannot have two air pressures, isobars will never cross one another.

Scientists can use isograms on a map to plot data that represents almost any type of measurement. Isograms are commonly used to show areas that have similar measurements of precipitation, temperature, gravity, magnetism, density, elevation, or chemical composition.

isogram a line on a map that represents a constant or equal value of a given quantity

Section 2 Review

1. **Explain** two methods that scientists use to get the data needed to make maps.
2. **Describe** three types of map projections in terms of their different characteristics and uses.
3. **Explain** why all maps are in some way inaccurate representations.
4. **Summarize** how to use legends and scales to read maps.
5. **Describe** what isograms show.
6. **Explain** why maps are more useful than globes are for studying small areas on the surface of Earth.
7. **Summarize** how to find directions on a map.

CRITICAL THINKING

8. **Applying Concepts** If a cartographer is making a map for three countries that do not use a common unit of measurement, what type of scale should the cartographer use on the map? Explain your answer.
9. **Making Inferences** Why would a conic projection produce a better map for exploring polar regions than a cylindrical projection would?

CONCEPT MAPPING

10. Use the following terms to create a concept map: *cartography, map projection, cylindrical projection, azimuthal projection, conic projection, map, legend, scale,* and *symbol.*

CHAPTER RESOURCES

Chapter Resource File

- Section Quiz GENERAL

Workbooks

- Study Guide (also in Spanish)

Section 3 Types of Maps

Earth scientists use a wide variety of maps that show many distinct characteristics of an area. Some of these characteristics include types of rocks, differences in air pressure, and varying depths of groundwater in a region. Scientists also use maps to show locations, elevations, and surface features of Earth.

OBJECTIVES

- **Explain** how elevation and topography are shown on a map.
- **Describe** three types of information shown in geologic maps.
- **Identify** two uses of soil maps.

KEY TERMS

topography
elevation
contour line
relief

Topographic Maps

One of the most widely used maps is called a *topographic map.* Topographic maps show the surface features, or **topography,** of Earth. Most topographic maps show both natural features, such as rivers and hills, and constructed features, such as buildings and roads. Topographic maps are made by using both aerial photographs and survey points collected in the field. A topographic map shows the **elevation,** or height above sea level, of the land. Elevation is measured from *mean sea level,* the point midway between the highest and lowest tide levels of the ocean. The elevation at mean sea level is 0.

topography the size and shape of the land surface features of a region, including its relief

elevation the height of an object above sea level

Advantages of Topographic Maps

An aerial view of an island is shown in **Figure 1.** Although the drawing shows the shape of the island, it does not indicate the island's size or elevation. A typical map projection would show the island's size and shape but would not show the island's topography. A topographic map provides more detailed information about the surface of the island than either the drawing or a projection map does. The advantage of a topographic map is that it shows the island's size, shape, and elevation.

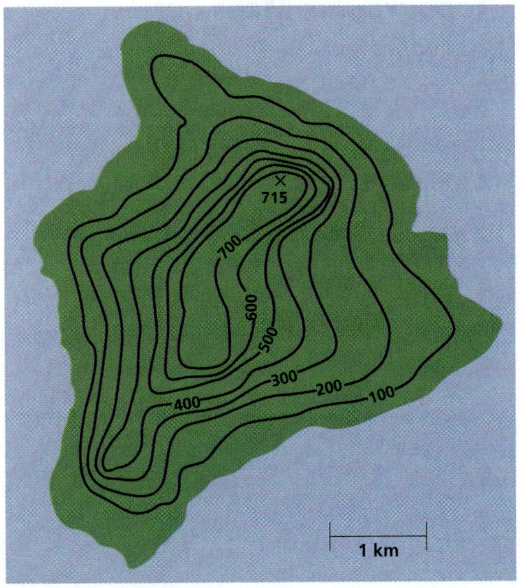

Figure 1 ▶ A drawing gives little information about the elevation of the island (left). In the topographic map (right), contour lines have been drawn to show elevation. An × marks the highest point on this map.

Section 3

Focus

Overview

This section explains how elevation and topographic features are shown on maps. The section also describes geologic maps, soil maps, and other kinds of maps used by Earth scientists.

Bellringer

Ask students to draw a picture of a hill, mountain, or valley in profile so that differences in elevation and depth are clear. Ask them how such differences might be pictured on a map. (Answers may vary. Accept all reasonable answers.) **LS Visual**

Motivate

Group Activity — GENERAL

Terrain Models Divide the class into small groups. Have them make a relief model of a familiar landform by using cardboard or foam pieces. Have them glue the layers together. Once the models are constructed, have students cover the models with papier-mâché or a flour and salt dough. Have them paint their models and add man-made features such as roads or bridges for a more realistic effect. Ask students to compare the models they created to a topographic map. (Sample answer: The model has raised features like the surface of Earth it represents, but is more cumbersome to handle.) **LS Kinesthetic**

CHAPTER RESOURCES

Chapter Resource File
- Directed Reading BASIC

Technology

Transparencies
- Bellringer
- 14 Topographic Maps (with worksheet)

Student Edition on CD-ROM

One-Stop Planner CD-ROM
- Lesson Plan

Teach

QuickLAB

Skills Acquired
- Constructing models
- Observing
- Analyzing data

Teacher's Notes Remind students to make the models no taller than the sides of the container. You could use plastic shoeboxes with clear lids for the waterproof container. To answer Analysis question 2, students can use the lids to create their topographic maps. When students have made their maps, you might wish to have them calculate the gradient of different slopes of the models.

Answers
1. 1 cm
2. Maps may vary.
3. The steeper the slope is, the closer together the contour lines are. A gentler slope will have more widely spaced contour lines.
4. The valley on my topographic map is represented by V-shaped contour lines in which the tip of the V points upstream.

CHAPTER RESOURCES

Chapter Resource File
- Datasheet for Quick Lab GENERAL

Figure 2 ▶ On a topographic map, the contour interval for this mountain would be very large because of the mountain's steep slope.

contour line a line that connects points of equal elevation on a map

relief the difference between the highest and lowest elevations in a given area

Elevation on Topographic Maps

On topographic maps, **contour lines** are used to show elevation. Each contour line is an isogram that connects points that have the same elevation. Because points at a given elevation are connected, the shape of the contour lines reflects the shape of the land.

The difference in elevation between one contour line and the next is called the *contour interval*. A cartographer chooses a contour interval suited to the scale of the map and the relief of the land. **Relief** is the difference in elevation between the highest and lowest points of the area being mapped. On maps of areas where the relief is high, such as the area shown in **Figure 2**, the contour interval may be as large as 50 or 100 m. Where the relief is low, the interval may be only 1 or 2 m.

To make reading the map easier, a cartographer makes every fifth contour line bolder than the four lines on each side of it. These bold lines, called *index contours*, are labeled by elevation. A point between two contour lines has an elevation between the elevations of the two lines. For example, if a point is halfway between the 50 and 100 m contour lines, its elevation is about 75 m. Exact elevations are marked by an × and are labeled.

QuickLAB 20 min

Topographic Maps

Procedure

1. Make a model mountain that is 6 to 8 cm high out of **modeling clay**. Work on a flat surface, and smooth out the mountain's shape. Make one side of the mountain slightly steeper than the other side.
2. Run a **paper clip** down one side of the model to form a valley that is several millimeters wide.
3. Place the model in the center of a **large waterproof container** that is at least 8 cm deep.
4. Use **tape** to hold a **ruler** upright in the container. One end of the ruler should rest on the bottom of the container. Make sure that the container is level.
5. Using the ruler as a guide, add **water** to the container to a depth of 1 cm. Use a **sharp pencil** to inscribe the clay by tracing around the model along the waterline.
6. Raise the water level 1 cm at a time until you reach the top of the model. Each time you add water to the container, inscribe another contour line in the clay along the waterline.

7. When you have finished, carefully drain the water and remove the model from the container.

Analysis

1. What is the contour interval of your model?
2. Observe your model from directly above. Try to duplicate the size and spacing of the contour lines on a sheet of paper to create a topographic map.
3. Compare the contour lines on a steep slope with those on a gentle slope. How do they differ?
4. How is a valley represented on your topographic map?

SKILL BUILDER GENERAL

Math Tell students that the contour interval for a topographic map is 25 m. Ask them what their elevation would be if they started climbing at the bottom (0 m) and crossed 3 contour lines; 5 lines; 8 lines; and 10 lines. (3 lines: 3 × 25 m = 75 m; 5 lines: 5 × 25 m = 125 m; 8 lines: 8 × 25 m = 200 m; 10 lines: 10 × 25 m = 250 m) LS **Logical**

GEOGRAPHY CONNECTION GENERAL

Quadrangles Topographic maps made by the United States Geological Survey (USGS) are called *topographic sheets* or *quadrangles*. The USGS began the project of mapping the United States in 1879. The earlier series of USGS maps represented quadrangles that covered 15' of latitude and 15' of longitude. The newer series of maps covers 7.5' of latitude and 7.5' of longitude—a smaller area, but in much greater detail. Invite interested students check out the USGS National Map on the Internet to find a map of your local area. LS **Visual**

Figure 3 ▶ The features of the area's coastal valley are represented by contour lines on the topographic map of the area.

Landforms on Topographic Maps

As shown in **Figure 3,** the spacing and the direction of contour lines indicate the shapes of the landforms represented on a topographic map. Contour lines spaced widely apart indicate that the change in elevation is gradual and that the land is relatively level. Closely spaced contour lines indicate that the change in elevation is rapid and that the slope is steep.

A contour line that bends to form a V shape indicates a valley. The bend in the V points toward the higher end of the valley. If a stream or river flows through the valley, the V in the contour line will point upstream, the direction from which the water flows. A river always flows from higher to lower elevation. The width of the V formed by the contour line shows the width of the valley.

Contour lines that form closed loops indicate a hilltop or a depression. Generally, a depression is indicated by *depression contours,* which are closed-loop contour lines that have short, straight lines perpendicular to the inside of the loop. These short lines point toward the center of the depression.

Reading Check **Why do V-shaped contour lines along a river point upstream? (See the Appendix for answers to Reading Checks.)**

Topographic Map Symbols

Symbols are used to show certain features on topographic maps. Symbol color indicates the type of feature. For example, constructed features, such as buildings, boundaries, roads, and railroads, are generally shown in black. Major highways are shown in red. Bodies of water are shown in blue, and forested areas are shown in green. Contour lines are brown or black. Often, areas whose map information has been updated based on aerial photography but not verified by field exploration are shown in purple. A key to common topographic map symbols is shown in the Reference Tables section of the Appendix.

SCILINKS. NSTA
Developed and maintained by the National Science Teachers Association

For a variety of links related to this subject, go to www.scilinks.org

Topic: Topographic Maps
SciLinks code: HQ61536

Discussion — BASIC

Comprehension Check Invite students to summarize how color is used on topographic maps. (Sample answers: Contour lines are shown in brown, green is used for forested areas, blue is used for bodies of water, roadways are red, and human built structures such as buildings or railroads are shown in black.) **LS Verbal**

HISTORY CONNECTION — ADVANCED

Public Lands The Land Ordinance System of 1785, which was signed into law by Thomas Jefferson, established the rectangular system of land survey in the Northwest Territory. Begun in Ohio and extended as the nation moved west, the survey system used latitude and longitude lines to divide public lands into townships and to subdivide townships into sections for settlement. Have students compare topographic maps of eastern states, such as Massachusetts or Virginia, that are unaffected by these practices with western states, such as Colorado or Kansas. Encourage students to look at boundaries, roads, and the grid patterns of city streets. **LS Visual**

Answer to Reading Check

Water moves from areas of higher elevation to areas of lower elevation. Because the V shape points toward higher elevation, it points upstream.

READING SKILL BUILDER — BASIC

Reading Hint Explain that the word *topography* comes from the Greek words *topos* and *graphein*. *Topos* means "place," and *graphein* means "to write." In biology, *topotype* refers to a typical specimen collected from a given locality; in literature, a *biography* describes someone's life. Similarly, a *topographer* describes the surface features of a specific place, and *topography* refers to the detailed mapping of the relief features of an area. **LS Verbal**

English Language Learners

Teach, *continued*

Using the Figure — GENERAL

Map Layers Point out that geologic maps are made by using a regular topographic map as a base and adding a layer of information about rock types. Ask students what symbols are used on geologic maps. (Answers may vary. Sample answer: Color is used to represent different rock types. The rock units are labeled with letters to indicate the geologic period when they formed. Lines show where geologic units meet. Special strike and dip symbols show angles of rock beds.) Tell students that because the geology of every area is different, geologic maps must have a legend that explains all of the colors and symbols used. In addition to the colors, the legend identifies any special geological information such as the location of fossils, faults or geologic hazards, or mineral resources. LS **Visual**

Graphic Organizer GENERAL

You may want students to work in groups to create this Spider Map. Have one student draw the map and fill in information provided by other students from the group.

Figure 4 ▶ Each color on this geologic map represents a distinct type of rock and shows where in this region that type of rock occurs.

Graphic Organizer **Spider Map**

Create the **Graphic Organizer** entitled "Spider Map" described in the Skills Handbook section of the Appendix. Label the circle "Types of maps." Create a leg for each type of map. Then, fill in the spider map with details about each type of map.

Geologic Maps

Geologic maps, such as the one shown in **Figure 4,** are designed to show the distribution of geologic features. In particular, geologic maps show the types of rocks found in a given area and the locations of faults, folds, and other structures.

Geologic maps are created on top of another map, called a *base map*. The base map provides surface features, such as topography or roads, to help identify the location of the geologic units. The base map is commonly printed in light colors or as gray lines so that the geologic information on the map is easy to read and understand.

Rock Units on Geologic Maps

A volume of rock of a given age range and rock type is a *geologic unit*. On geologic maps, geologic units are distinguished by color. Units of similar ages are generally assigned colors in the same color family, such as different shades of blue. In addition to assigning a color, geologists assign a set of letters to each rock unit. This set of letters is commonly one capital letter followed by one or more lowercase letters. The capital letter symbolizes the age of the rock, usually by geologic period. The lowercase letters represent the name of the unit or the type of rock.

Other Structures on Geologic Maps

Other markings on geologic maps are contact lines. A *contact line* indicates places at which two geologic units meet, called *contacts*. The two main types of contacts are faults and depositional contacts. Depositional contacts show where one rock layer formed above another. Faults are cracks where rocks can move past each other. Also on geologic maps are strike and dip symbols for rock beds. *Strike* indicates the direction in which the beds run, and *dip* indicates the angle at which the beds tilt.

Discussion — GENERAL

Value of Geologic Maps Explain that geologic maps are important tools used by geologists, developers and planners, and engineers. Ask students to speculate how these people might use geologic maps. (Sample answers: These maps could be used to locate mineral deposits, energy resources, and sources of groundwater. They would also be useful in identifying potential hazards such as landslides, faults, or volcanoes, or to determine which areas are best suited to different land uses, construction projects, or urban developments.) LS **Interpersonal**

Soil Maps

Another type of map that is commonly used by Earth scientists is called a *soil map*. Scientists construct soil maps to classify, map, and describe soils. Soil maps are based on soil surveys that record information about the properties of soils in a given area. Soil surveys can be performed for a variety of areas, but they are most commonly performed for a county.

The government agency that is in charge of overseeing and compiling soil data is the Natural Resources Conservation Service (NRCS). The NRCS is part of the United States Department of Agriculture (USDA). The NRCS has been mapping the distribution of soils in the United States for more than a century.

Reading Check **Why do scientists create soil maps? (See the Appendix for answers to Reading Checks.)**

Soil Surveys

A soil survey consists of three main parts: text, maps, and tables. The text of soil surveys includes general information about the geology, topography, and climate of the area being mapped. The tables describe the types and volumes of soils in the area. Soil surveys generally include two types of soil maps. The first type is a very general map that shows the approximate location of different types of soil within the area, such as the one shown in **Figure 5.** The second type shows detailed information about soils in the area.

Uses of Soil Maps

Soil maps are valuable tools for agriculture and land management. Knowing the soil properties of an area helps farmers, agricultural engineers, and government agencies identify ways to conserve and use soil and to plan sites for future development.

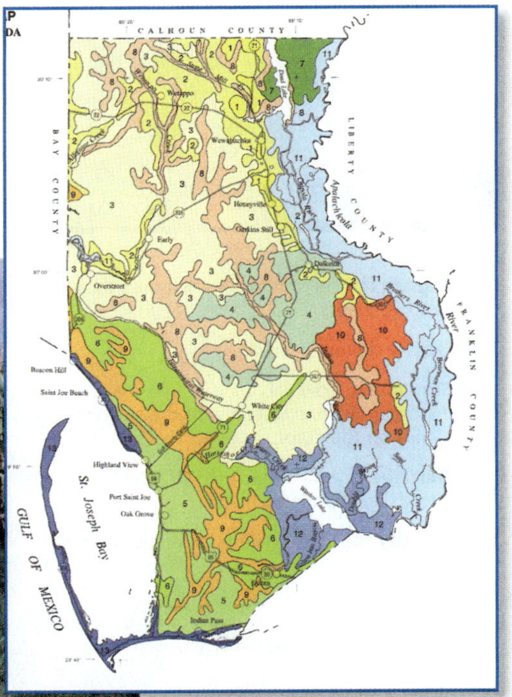

Figure 5 ▶ Scientists gather data to make a soil map by taking soil samples. Soil maps help scientists determine the potential abilities and limitations of the land to support development and agriculture.

Answer to Reading Check

Scientists create soil maps to classify, map, and describe soils.

Close

Reteaching — BASIC

Mapping Vocabulary Have students work in pairs to write definitions or to make an explanatory diagram for each of the key terms and italicized words in this section. Then, have them take turns quizzing each other to study for assessments. **LS Verbal/Visual**

Quiz — GENERAL

1. What mapping technique can be used to study changes in Earth's surface over time? (maps produced from satellite images)
2. What does the contour interval tell you about the slope of the land in a region being mapped? (The larger the contour interval the more mountainous or hilly the relief is. If a smaller interval can be used, the relief is lower.)
3. How would you plot points of equal temperature on weather maps? (by using isograms, in this case, isotherms)
4. What kind of map would you consult to determine the locations of mineral resources or natural hazards? (a geologic map)

Group Activity — ADVANCED

Local Soil Maps Divide students into groups and have each student collect soil samples from three different areas. Have them label each sample with its location. Allow the samples to dry, and have students group the soils according to particle size, color, or amount of organic matter. Next, have students create a base map by tracing the outline of a local map on paper. Invite students to pool their data and map locations of soil types, based on points where samples were collected, to see if there are any patterns. **LS Kinesthetic/Logical**

ENVIRONMENTAL CONNECTION

Soil Surveys Descriptions found in soil survey reports include the depth of each soil layer, how well water and plant roots penetrate the soil, the soil's pH, and how easily the soil can be eroded. Soil surveys describe not only the potential uses of soil but also the risks of damaging the soil or the environment. Where there is danger of erosion, a soil survey can point out how to control erosion and what plants are best for specific soils. In addition, soils are rated as to their suitability for recreational purposes, commercial uses, and habitats for wildlife.

Close, continued

Alternative Assessment — GENERAL

Song Lyrics Have students write a song about the ways in which different types of maps convey information. They could write one verse about topographic maps, another about geologic maps, and a third on soil maps. Students may find it easier to write the words first, then experiment with the rhythm and beat. After students have composed the songs, invite them to perform their songs for the class or to create an audiotape or CD recording. LS **Auditory**

Answers to Section Review

1. Contour lines connect points that have the same elevation.
2. the difference in elevation represented by the space between two adjacent contour lines
3. Contour lines that are spaced closely together indicate that the slope is steep, while those that are spaced farther apart indicate that the slope is flatter.
4. Geologic units of similar age are usually similar in color and are identified by letter symbols.
5. text: information on the geology, topography, and climate of an area; maps: the location and properties of soils found there; tables: types and volumes of soil within the area
6. to identify ways to conserve and use soil and to plan future land development
7. Sample answer: weather maps, maps of groundwater flow, and satellite maps that show changes in Earth over time
8. A V-shaped contour line indicates a valley. The bend of the V points toward higher elevation, and because water always flows from higher to lower elevation, the V points upstream.
9. Sample answer: In addition to locating rivers and roads, hikers in an unfamiliar area could easily find out where the ground is relatively level or where steep landforms are so they could plan a route.
10. A geologic map would indicate the location of any faults along which earthquakes may occur.
11. *Topographic maps* show *elevation* above the *mean sea level* and use *contour lines* and *index contours* to show elevation by using a set *contour interval.*

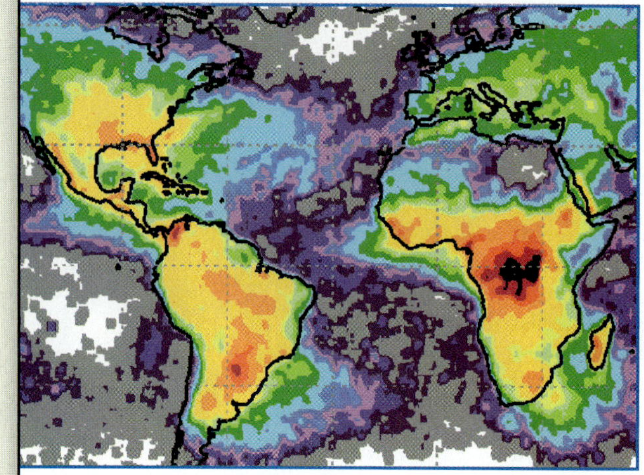

Figure 6 ▶ This map was created by using satellite data. The map shows the global distribution of lightning based on the average number of strikes per square kilometer. The highest frequency of strikes is shown in black, and the lowest frequency is shown in white.

Other Types of Maps

Earth scientists also use maps to show the location and flow of both water and air. These maps are commonly constructed by plotting data from various points around a region and then using isograms to connect the points whose data are identical.

Maps are useful to every branch of Earth science. For example, meteorologists use maps such as the one shown in **Figure 6** to record and predict weather events. Maps may be used to plot the amount of precipitation that falls in a given area. Maps are also used to show the locations of areas of high and low air pressure and the weather fronts that move across Earth's surface. These maps are updated constantly and are used by meteorologists to communicate to the public important information on daily weather conditions and emergency situations.

The location and direction of the flow of groundwater can be recorded on maps. Data from these maps can be used to determine where and when water shortages may occur. Scientists use map information to identify potential locations for power plants, waste disposal sites, and new communities.

Other types of Earth scientists use maps to study changes in Earth's surface over time. Such changes include changes in topography, changes in amounts of available resources, and changes in factors that affect climate. Maps generated by satellites are particularly useful for studying changes in Earth's surface.

Section 3 Review

1. **Explain** how elevation is shown on a topographic map.
2. **Define** *contour interval.*
3. **Summarize** how you can use information on a topographic map to compare the steepness of slopes on the map.
4. **Describe** how geologic units of similar ages are shown on a geologic map.
5. **Identify** the three main parts of a soil survey.
6. **Identify** two primary uses for soil maps.
7. **Identify** three types of maps other than topographic maps, geologic maps, or soil maps.

CRITICAL THINKING

8. **Applying Ideas** How can you use lines on a topographic map to identify the direction of river flow?
9. **Making Inferences** In what ways might topographic maps be more useful than simple map projections to someone who wants to hike in an area that he or she has never hiked in before?
10. **Identifying Patterns** What type of map would be the most useful to a scientist studying earthquake patterns: a geologic map or a topographic map?

CONCEPT MAPPING

11. Use the following terms to create a concept map: *topographic map, elevation, mean sea level, contour interval, contour line,* and *index contour.*

CHAPTER RESOURCES

Chapter Resource File
- Section Quiz GENERAL

Workbooks
- Study Guide (also in Spanish)

Chapter 3

Highlights

Sections

1 Finding Locations on Earth

Key Terms

parallel, 53
latitude, 53
meridian, 54
longitude, 54

Key Concepts

- Lines of latitude and lines of longitude form a system of intersecting circles that is used to locate places on Earth's surface.
- Parallels run east and west around Earth. Meridians run north and south from pole to pole.
- Because of Earth's magnetic field, a magnetic compass can be used to find directions on Earth's surface.

2 Mapping Earth's Surface

Key Terms

remote sensing, 57
map projection, 58
legend, 61
scale, 61
isogram, 62

Key Concepts

- Three common types of map projections are cylindrical, azimuthal, and conic projections. Each type has certain advantages and disadvantages.
- A map scale is used to find distances on a map. A legend is a list of map symbols and their meanings.
- Lines called *isograms* may be used to connect areas on a map whose properties have similar values.

3 Types of Maps

Key Terms

topography, 63
elevation, 63
contour line, 64
relief, 64

Key Concepts

- The spacing and direction of contour lines on a topographic map indicate the shapes of landforms.
- Geologic maps show the distribution of rock units and geologic structures in an area.
- Soil maps describe the types of soil located in an area.
- Earth scientists use maps to describe the movements of air and water and to study changes in Earth's surface over time.

Chapter Highlights

Alternative Assessment — ADVANCED

Mapmaker, Mapmaker Pose this scenario to students: "Suppose you want to meet a classmate at a movie theater or at a record store that is giving out free concert tickets, but he or she is unfamiliar with your neighborhood. You decide to make a map to help your friend find the way."

Have students work with a partner to make a map that shows how to reach a destination they choose in their neighborhood. Students can measure distances by pacing or by using a large ball of string. They will also need blank or graph paper, pencils, and a compass or protractor for measuring angles. Tell students to choose an appropriate scale and to include a compass rose. They should label the streets or major roads and important landmarks. Students should include a map legend for any important symbols. Have them sketch a clear route from one of the landmarks to their chosen destination. Ask them to explain why the map is a more precise method of giving directions than a verbal description is. LS Kinesthetic

CHAPTER RESOURCES

Chapter Resource File

- **Concept Review** GENERAL
- **Critical Thinking** ADVANCED
- **Math Skills** GENERAL
- **Graphing Skills** GENERAL
- **Chapter Test A** GENERAL
- **Chapter Test B** ADVANCED

Workbooks

- **Study Guide** (also in Spanish)
- **Assessments** (Spanish)

Technology

Classroom Videos
- Brain Food Video Quiz

Assignment Guide	
Section	**Questions**
1	4–5, 9–12, 17–18, 23–25, 31–33
2	1–2, 6, 13, 19–21, 30
3	3, 7–8, 14–16, 22, 26–27, 29, 34–36
1–3	28

Using Key Terms

1–8. Answers may vary but should show that students understand the definitions of and differences between key terms.

Understanding Key Concepts

9. b **10.** d
11. a **12.** d
13. c **14.** b
15. a **16.** a

Short Answer

17. One minute of latitude is equal to 1.85 km, so 1 second of latitude would be 1/60 of that distance, or 0.0308 km, or 30.8 meters.

18. Latitude is the angular distance in degrees north or south of the equator, and longitude is the angular distance in degrees east or west of the prime meridian.

19. Three main map projections are cylindrical, azimuthal, and conic. They differ in the orientation of the surface on which the map is projected: cylinder, flat sheet, or cone (respectively), and the points of contact between the globe and map. This affects the amount and type of distortion, whether latitude and longitude lines are straight or curved, and whether great circles are straight lines.

20. In cylindrical projections, latitude and longitude lines form a rectangular grid, but distances and sizes near the poles are distorted. In azimuthal projections, great circle routes are straight lines so they are useful for navigation; map distortion increases the farther one moves from the point of contact. In conic projections, maps of neighboring areas can be fitted together to increase accuracy; areas farther from the parallel where the cone and globe contact are more distorted.

21. Legends help people understand the symbols used on maps. Scales help people calculate distances in the real world from distances on the map.

22. Contour lines connect points of equal elevation on the map. Every fifth line is labeled with the elevation. The spacing of contour lines indicates the slope, whether gradual or steep. V-shaped contour lines represent valleys.

Chapter 3 Review

Using Key Terms

Use each of the following terms in a separate sentence.

1. *cartography*

2. *map projection*

3. *contour lines*

For each pair of terms, explain how the meanings of the terms differ.

4. *parallel* and *latitude*

5. *meridian* and *longitude*

6. *legend* and *scale*

7. *topography* and *relief*

8. *index contour* and *contour interval*

Understanding Key Concepts

9. The distance in degrees east or west of the prime meridian is
a. latitude.
b. longitude.
c. declination.
d. projection.

10. The distance covered by a degree of longitude
a. is 1/180 of Earth's circumference.
b. is always equal to 11 km.
c. increases as you approach the poles.
d. decreases as you approach the poles.

11. The needle of a magnetic compass points toward the
a. geomagnetic pole.
b. geographic pole.
c. parallels.
d. meridians.

12. The shortest distance between any two points on the globe is along
a. the equator.
b. a line of latitude.
c. the prime meridian.
d. a great circle.

13. If 1 cm on a map equals 1 km on Earth, the fractional scale would be written as
a. 1:1.
b. 1:100.
c. 1:100,000.
d. 1:1,000,000.

14. On a topographic map, elevation is shown by means of
a. great circles.
b. contour lines.
c. verbal scale.
d. fractional scale.

15. What type of map is commonly used to locate faults and folds in beds of rock?
a. geologic map
b. topographic map
c. soil map
d. isogram map

16. The contour interval is a measurement of
a. the change in elevation between two adjacent contour lines.
b. the distance between mean sea level and any given contour line.
c. the length of a contour line.
d. the time needed to travel between any two contour lines.

Short Answer

17. How much distance on Earth's surface does one second of latitude equal?

18. What is the difference between latitude and longitude?

19. What are the three main types of map projections? How do they differ?

20. Compare the advantages and disadvantages of the three main types of map projections.

21. How do legends and scales help people interpret maps?

22. How do contour lines on a map illustrate topography?

Critical Thinking

23. Applying Ideas What is wrong with the following location: 135°N, 185°E?

24. Identifying Trends As you move from point A to point B in the Northern Hemisphere, the length of a degree of longitude progressively decreases. In which direction are you moving?

25. Understanding Relationships Imagine that you are at a location where the magnetic declination is 0°. Describe your position relative to magnetic north and true north.

26. Making Inferences You examine a topographic map on which the contour interval is 100 m. In general, what type of terrain is probably shown on the map?

27. Applying Ideas A topographic map shows two hiking trails. Along trail A, the contour lines are widely spaced. Along contour B, the contour lines are almost touching. Which path would probably be easier and safer to follow? Why?

Concept Mapping

28. Use the following terms to create a concept map: *latitude, longitude, relief, map projection, cylindrical projection, elevation, map, azimuthal projection, contour line, conic projection, topography, legend,* and *scale.*

Math Skills

29. Making Calculations A topographic map has a contour interval of 30 m. By how many meters would your elevation change if you crossed seven contour lines?

30. Applying Quantities A map has a fractional scale of 1:24,000. How many kilometers would 3 cm on the map represent?

31. Making Calculations A city to which you are traveling is located along the same meridian as your current position but is 11° of latitude to the north of your current position. About how far away is the city?

Writing Skills

32. Writing from Research Research the navigation instrument known as the *sextant*. Make a diagram explaining how the sextant can be used to determine latitude.

33. Writing from Research Use the Internet and library resources to research global positioning systems. Write a short essay describing the different ways that GPS devices are currently being used in everyday situations. Then, make a prediction about how the technology might be used in the future.

Interpreting Graphics

The map below shows contour lines of groundwater. The lines show elevation of the water table in meters above sea level. Use the map to answer the questions that follow.

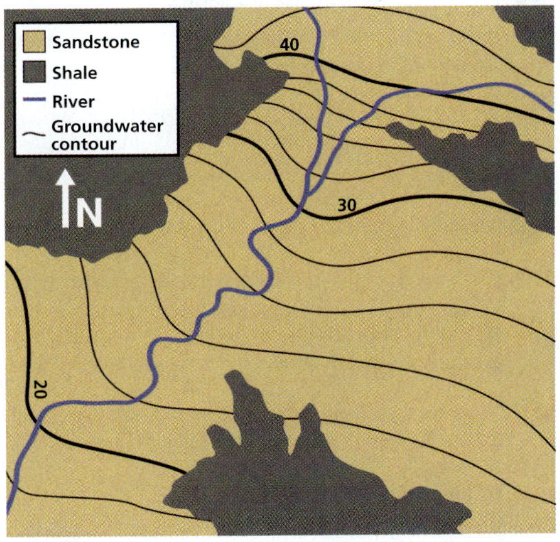

34. What is the contour interval for this map?

35. What is the highest measured level of the water table?

36. Groundwater flows from highest to lowest elevation. In which direction is the groundwater flowing?

Chapter Review

Critical Thinking

23. The latitude of the North Pole is 90°N. You cannot go any farther north than 90°. The meridian opposite the prime meridian (halfway around Earth) is 180°. At that point, the numbering starts over with W longitude. There is no such meridian as 185°E.

24. If the distance covered by a degree of longitude is decreasing from point A to point B, you are moving north toward the North Pole.

25. If you are located along a line of 0° magnetic declination, your position would be aligned with both the geomagnetic north pole and geographic north.

26. If the contour interval is as large as 100 m, the map probably represents fairly mountainous terrain.

27. Trail A, because the widely spaced contour lines indicate a gradual slope. The tightly spaced contour lines on Trail B indicate that the slope is very steep, and the path would be more dangerous.

Concept Mapping

28. Answers may vary but should include all of the terms listed. Sample answers appear at the end of this Teacher Edition.

Math Skills

29. 30 m × 7 = 210 m

30. 100 cm/m × 1000m/km = 100,000 cm/km; 24,000 cm × 3 cm = 72,000 cm ÷ 100,000 cm/km = 0.72 km

31. 11 × 111 km = 1,221 km north

Writing Skills

32. Answers may vary. Accept all reasonable answers.

33. Answers may vary. Accept all reasonable answers.

Interpreting Graphics

34. contour interval = 2 m

35. 42 m

36. The water is flowing southwest.

Estimated Time

To give students practice under more realistic testing conditions, allow them 30 minutes to answer all of the questions in this practice test.

TEST DOCTOR

Question 1 Answer B is correct. Answer A is incorrect because V-shaped contour lines are more useful in determining the direction of a the flow of a river. Answer C is incorrect because short, straight lines inside a loop indicate a depression. Answer D is icorrect because tightly spaced lines are indicative of a steep slope, not a gradual slope.

Question 8 Answer A is correct. Students should recall that great circles are the shortest distance between two points on a globe. Because great circles appear as straight lines on a gnomonic projection, such maps are often used when plotting flight plans for airplanes. Answer B is incorrect because parallels would appear as circles on the type of projection described. Answer C is incorrect because the equator, while a great circle, would appear as a parallel, a large circle. Answer D is incorrect because coastlines would appear similar to their physical shapes, with distortion in size and shape increasing as they moved away from the North Pole.

Chapter 3 Standardized Test Prep

Understanding Concepts

Directions (1–5): **For *each* question, write on a separate sheet of paper the letter of the correct answer.**

1 How can you determine whether the contours on a topographic map show a gradual slope?
A. Look for V-shaped contour lines.
B. Look for widely spaced contour lines.
C. Look for short, straight lines inside the loop.
D. Look for tightly spaced, circular contour lines.

2 How far apart would two successive index contours be on a map with a contour interval of 5 meters?
F. 5 meters
G. 10 meters
H. 20 meters
I. 25 meters.

3 What part of a road map would you use in order to measure the distance from your current location to your destination?
A. latitude lines
B. map scale
C. longitude lines
D. map legend

4 Meteorologists use isobars on a weather map in order to
F. show changes in atmospheric air pressure
G. connect points of equal temperature
H. plot local precipitation data
I. show elevation above or below sea level

5 What is the angular distance, measured in degrees, east or west of the prime meridian?
A. latitude
B. longitude
C. isogram
D. relief

Directions (6–7): **For *each* question, write a short response.**

6 At what location on Earth does each new day begin at midnight?

7 What is the latitude of the North Pole?

Reading Skills

Directions (8–10): **Read the passage below. Then, answer the questions.**

Map Projections

Earth is a sphere, and thus its surface is curved. When a curved surface is transferred to a flat map, distortions in size, shape, distance, and direction occur. To limit these distortions, cartographers have developed many ways of transferring a three-dimensional curved surface to a flat map. On Mercator projections, meridians and parallels appear as straight lines. These lines cross each other at 90° angles and form a grid. On azimuthal projections, there is little distortion at one contact point on the map, which is often one of the poles. But distortion in direction and distance increases as distance from the point of contact increases. On conic projections, the map is accurate along one parallel of latitude. Areas near this parallel are distorted the least. However, none of these maps is an entirely accurate representation of Earth's surface.

8 Which of the following appears as a straight line on a azimuthal projection, where the point of contact is the North Pole?
A. great circles
B. parallels
C. the equator
D. coastlines

9 Which of the following statements about Mercator projections is true?
F. Because latitude and longitude form a grid, plotting great circles can be done by using a straight-edged ruler.
G. Because latitude and longitude form a grid, finding specific locations is easy on a Mercator map projection.
H. Mercator maps often show the greatest distortion where the projection touched the globe.
I. Mercator maps often show polar regions as being much smaller than they actually are.

10 Why does each map described display some sort of distortion?

Answers

Understanding Concepts

1. B
2. I
3. B
4. F
5. B
6. International Date Line
7. 90ºN

Reading Skills

8. A
9. G
10. Any time a curved surface is mapped on a flat surface, distortion occurs. This distortion can take different forms and each map type produces different amounts of distortion.

Interpreting Graphics

11. C
12. Answers may vary. See Test Doctor for a detailed scoring rubric.
13. G
14. 6660 km

Interpreting Graphics

Directions (11–14): **For *each* question below, record the correct answer on a separate sheet of paper.**

Use the topographic map below to answer questions 12 and 13.

Topographic Map of the Orr River

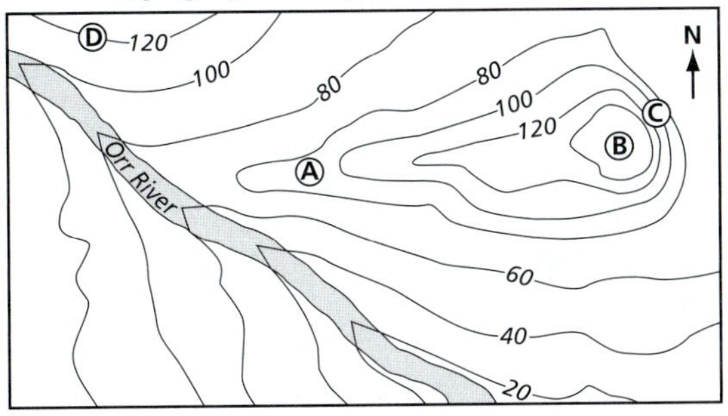

11 What location on the map has the steepest gradient?
A. location A
B. location B
C. location C
D. location D

12 In which direction is the river in the topographic map flowing? What information on the map helped you determine your answer?

The diagram below shows Earth's system of latitude and longitude lines. Lines are shown in 30° increments. Use this diagram to answer questions 14 and 15.

Latitude and Longitude

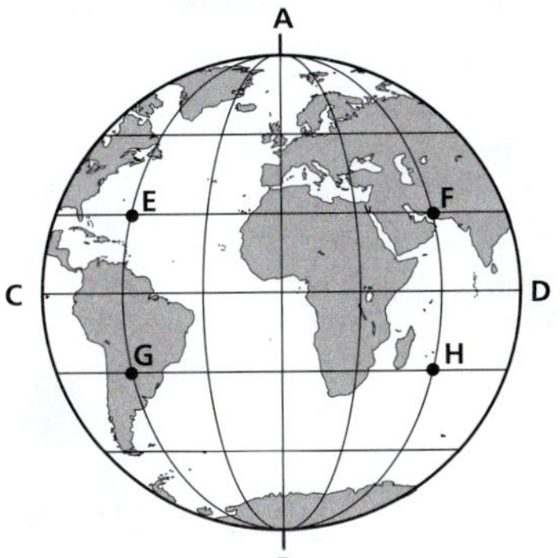

13 Which point is located at 30°N 60°E?
F. point E
G. point F
H. point G
I. point H

14 The distance between two lines of parallel that are 1° apart is about 111 km. What is the approximate distance between points G and E?

Test TIP

Choose an answer to a question based on both what you already know as well as any information presented in the question.

Standardized Test Prep

TEST DOCTOR

Question 12 Full-credit answers should include the following points:
- students should correctly determine that the river is flowing from northwest to southeast
- students should cite the fact that the topographic lines near the river form V shapes
- the tips of these V's point upstream, which allows one to determine the direction in which the river is flowing

Question 14 In order to answer this question correctly, students must correctly calculate the number of degrees between points G and E. Common mistakes include arriving at the answer 222 km. Students may make this mistake if they calculate the difference using 2°. Though there are two increments of latitude between the locations, each increment represents 30° of distance, not 1°. Another common mistake is arriving at an answer of 3,330 km. Students may make this mistake when calculating the difference in the locations to be 30°. This is the angular distance of each location compared to the equator, but the total distance between the locations is 60°. The correct answer is 6,660 km, which is calculated by multiplying 111 km by 60.

Test Prep Correlations ***National Science Education Standards***

SAI 2d: items 2, 15
SAI 2e: items 1, 9, 10, 11, 12, 13
UCP 2: items 1, 2, 3, 4, 5, 6, 7, 8, 9, 10, 11, 12, 13, 14, 15
HNS 2b: items 2, 4, 5, 6, 15

CHAPTER RESOURCES

State Resources

For specific resources for your state, visit **go.hrw.com** and type in the keyword **HSHSTR.**

Contour Maps: Island Construction

Teacher's Notes

Time Required
two 45-minute class periods

Lab Ratings
EASY → HARD

TEACHER PREPARATION (2 of 4)
STUDENT SETUP (2 of 4)
CONCEPT LEVEL (3 of 4)
CLEANUP (2 of 4)

Skills Acquired
- Constructing Models
- Identifying and Recognizing Patterns
- Interpreting
- Analyzing Data
- Measuring
- Communicating

The Scientific Method
In this lab, students will
- Make Observations
- Analyze Results
- Draw Conclusions

Materials
Materials listed are enough for groups of 2–4 students. Use waterproof clay for model construction so the model will not lose its shape when students add water.

Tips and Hints
An alternative approach is for students to sculpt the terrain from a mound of clay. If students try this, supply graph paper to serve as a grid under the map and under the plastic box so contours can be more accurately reconstructed.

Students could also build terrain models with other geologic formations such as lakes, volcanoes, or canyons and then trace the contour lines to investigate how their contours appear on topographic maps.

You may wish to have students calculate the gradient of different slopes on the model island. Provide protractors to measure the angle. Slope is measured by calculating the tangent of the surface. The tangent is calculated by dividing the vertical change in elevation by the horizontal distance. Gradient % = 100 × tan (angle).

CHAPTER RESOURCES

Chapter Resource File
- Datasheet for Chapter Lab GENERAL
- Lab Notes and Answers

Workbooks

Long-Term Projects
- Positions of Sunrise and Sunset ADVANCED

Chapter 3

Making Models Lab

Contour Maps: Island Construction

Objectives

- **Build** a scale model based on a map.
- **Identify** contour intervals and landscape features based on a map.

Materials

basin, flat (or large pan), 8 cm deep
clay, modeling (4 lb)
dowel, thick wooden (or rolling pin)
knife, plastic
paper, white
pencil
ruler, metric
scissors
topographic map from Reference Tables section of the Appendix
water

Safety

A map is a drawing that shows a simplified version of some detail of Earth's surface. There are many types of maps. Each type has its own special features and purpose. One of the most useful types of maps is the topographic map, or contour map. This type of map shows elevation and other important features of the landscape. Scientists make a contour map by using data obtained from a careful survey and photographic study of the area that the map represents.

PROCEDURE

1. Study the topographic map in the Reference Tables section of the Appendix. Record the contour interval used on the island contour map. Then, count the number of contour lines that appear on the map.
2. Use the dowel to press out as many flat pieces of clay as there were contour lines counted in Step 1. Each piece of clay should be 1 cm thick and large enough to cover the island shown on the map.
3. On a blank sheet of paper, trace the island contour map. Cut out the island from your copy of the contour map along the outermost contour line.

Step 2

4. Place this cutout on top of one of the pieces of clay. Trace the edge of the cutout in the clay. Cut the piece of clay to match the shape of the island.
5. Cut the paper tracing along the next contour line, making sure not to damage the outer ring of paper as you cut.
6. Using the new paper shape and a new layer of clay, repeat Step 4.
7. Place the paper ring from the first cut on the first clay shape that you cut out so that the outer edges of the paper ring line up with the edges of the clay. Stack the second layer of clay on the first layer so that the second layer fits inside the paper contour ring. This gives you the same contour spacing as shown on the map. Remove the paper ring.
8. Continue Steps 4–7 for each of the contour layers.
9. Use leftover clay to smooth the terraced edges into a more natural profile.
10. Make a mark inside a pan approximately 1 cm down from the rim. Put the clay model of the island into the pan, and add water to a depth of 1 cm.
11. Compare the shoreline of the model with the lines on the contour map. Continue to add water at 1 cm intervals until the water reaches the mark on the pan.

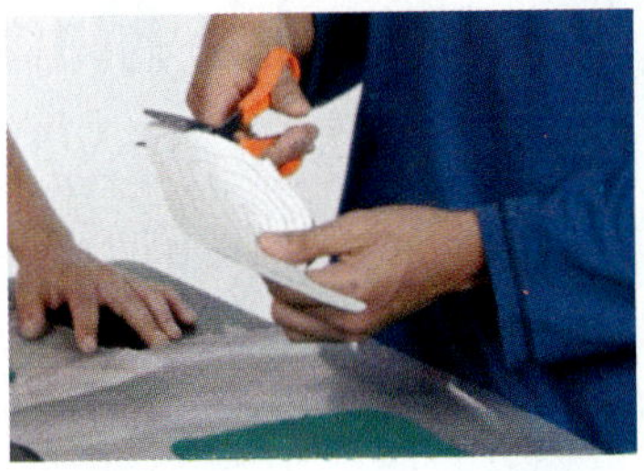

Step 5

Step 6

ANALYSIS AND CONCLUSION

1. **Making Inferences** What is the contour interval of your map?
2. **Understanding Relationships** How could you tell the steepest slope from the gentlest slope by observing the spacing of the contour lines?
3. **Analyzing Data** What is the elevation above sea level for the highest point of your model?
4. **Applying Ideas** How do you know if your model contains any areas that are below sea level? If there are any such areas, where are they and what are their elevations?
5. **Evaluating Models** What landscape feature is located at point C on your model, as indicated on the original map? What is the elevation of point B on your model?

Extension

1. **Making Predictions** From observations of your model, what conclusions can you make about where people might live on this island? Explain your answer.

Making Models Lab

Answers to Analysis and Conclusion

1. 10 meters
2. The contour lines on the steepest slope are closest together. This indicates a rapid rise in elevation over a short distance. The contour lines on the gentlest slope are spaced farther apart. This indicates a more gradual increase in elevation over a longer distance.
3. above 60 m but below 70 m
4. Near point A is an area marked by closed loops with short lines perpendicular to the inside of the loops. This indicates a depression. Because it occurs at the sea level elevation, it must extend below sea level; the elevation of the depression is at least 10 m but not greater than 20 m below mean sea level.
5. Point C represents a pass between two hills; Point B is between 20 m and 30 m above mean sea level.

Answer to Extension

1. Answers may vary. Sample answer: People would be more likely to live in the upland regions in the south-central part of the island beyond point C, where the higher elevations would be less affected by flooding from high tides or storm surges. The gentler slopes would be easier to cultivate and build upon.

CLASSROOM TESTED & APPROVED

Scott Robertson
North Warren Central School
Chestertown, NY

Topographic Map of the Desolation Watershed

Activity — GENERAL

Local Watershed Obtain a topographic map of your area that includes a local watershed. Make copies of the map and laminate them. Have students locate a stream. Have them lay a transparency over the topographic map and trace the stream in blue. Then, have them locate familiar landmarks and mark their locations on the transparency. Have them copy the contour lines onto the transparency. Next, have students locate the highest points surrounding the body of water. Explain that the watershed is the drainage basin of the stream. Have students locate the watershed boundaries. If possible, take the class on a field trip and have them locate their stream. LS **Visual**

Answers to Map Skills Activity

1. about 11,700 ft
2. southwest; The contours that cross the river valley make a V shape that points upstream.
3. the area around Bruin Creek; The contour lines near Bruin Creek are much closer together than the contour lines around Park Creek are.
4. 4520 ft
5. All three creeks flow from higher elevation to lower elevation. Also, Park Creek and Bruin Creek join Desolation Creek and then flow north-northwest.
6. 200 ft

CHAPTER RESOURCES

Workbooks

Transparencies
- 15 Topographic Map of the Desolation Watershed (with worksheet)

MAPS in Action

Topographic Map of the Desolation Watershed

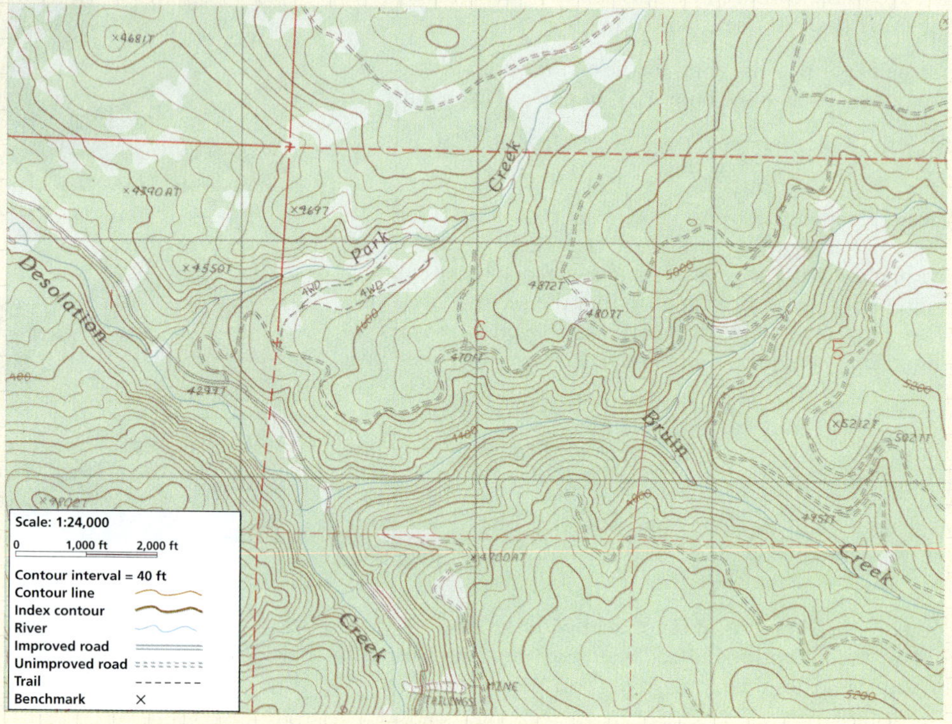

Map Skills Activity

This map, produced by the United States Geological Survey (USGS), shows the topography of the area around the Desolation watershed located in eastern Oregon. Note that the USGS always uses English units, not metric units, when measuring distance. Use the map to answer the questions below.

1. **Using a Key** What is the distance between the location on the northwest corner of the map labeled "4681T" and the location on the eastern side of the map labeled "5212T"?
2. **Analyzing Data** In what direction does Park Creek flow? How are you able to determine this information by looking at the map?
3. **Making Comparisons** Which area has the steeper slopes: the area around Park Creek or the area around Bruin Creek? How are you able to determine this information by looking at the map?
4. **Inferring Relationships** What is the elevation of the contour line that circles the point 4550T, located on the northwest portion of the map?
5. **Identifying Trends** Desolation Creek, Park Creek, and Bruin Creek enter the map from different geographic directions. Use the information on the map to determine what these creeks have in common in terms of their direction of flow.
6. **Analyzing Relationships** What is the total change in elevation between two index contours?

EYE on the Environment

Mapping Life on Earth

In 1992, researchers working in a forest in Southeast Asia made some amazing discoveries. They identified two species of deer and a species of oxen that scientists had never seen before! The researchers reported their findings and made plans for further study. Unfortunately, the forest was being cut down at an incredibly fast rate. Luckily, citizens, other scientists, and politicians got involved in trying to protect forests. Thanks to their efforts, Vietnam and Laos have begun measures to protect the remaining forests in the region.

Why All the Fuss?

The biosphere, which includes only a thin slice of Earth and its atmosphere, is the area where all life on our planet exists. For every known species, scientists estimate that there are at least eight others (mostly insects and microorganisms) that have never been identified. As forests and other natural areas are destroyed, many of these unknown species are becoming extinct.

Help from Above

Given the amount of the biosphere that has never been thoroughly studied, how do scientists identify ecosystems that are endangered? Using satellite images, scientists can identify the makeup and size of Earth's remaining unexplored natural areas. Scientists can also conduct ground surveys. By pooling their data, they can create a biodiversity map.

Biodiversity is the term used to describe the range of different organisms that live in an area. In locations that have a very high biodiversity, the variety of organisms living in a small area is large. Biodiversity maps often use color to show which parts of a region have the highest biodiversity.

▲ Before 1992, scientists had never seen the Vu Quang Ox.

Short-Term Success Story

Making and using biodiversity maps has already proven successful. In Jamaica, for example, such maps helped identify areas that may benefit from being turned into national parks. The maps also highlighted some offshore areas of high biodiversity, and officials were able to close some of these areas to fishing.

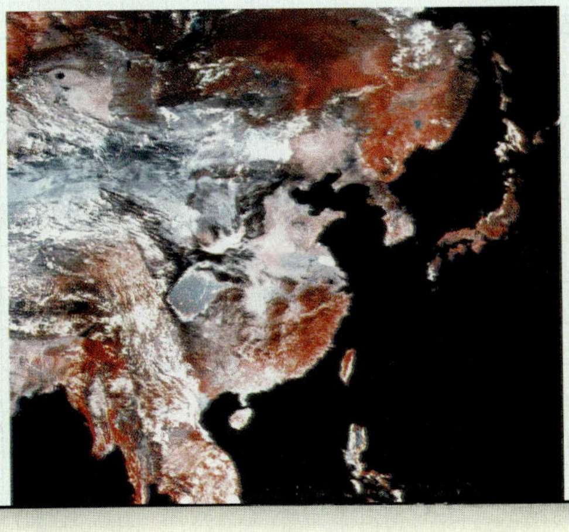

◀ This satellite image of southeast Asia shows various densities of vegetation. The redder an area is, the denser the vegetation in that area is.

Extension

1. **Making Comparisons** How does creating a biodiversity map differ from creating a topographic map? Could you create a biodiversity map that is also topographic? Explain your answer.

Eye on the Environment

Mapping Life on Earth

Group Activity — GENERAL

Biodiversity Maps Divide the class into groups of 2–4 students. Tell them to wear comfortable outdoor clothing. Have students mark off a 1 m^2 plot at the edge of a natural area near their homes or school. They can use string and stakes or use boundaries such rocks, trees or buildings to establish the borders of the observation site. Have them use string to divide the larger area into 20 cm × 20 cm squares. Tell students to count the number of plant species within each square without disturbing the organisms in the site. Have students use their data to construct a biodiversity map of their plots. They will need notebooks to record their data and graph paper to construct their maps. Make sure they can identify potentially harmful plants. Field guides to local plants may be helpful to identify their finds. You may want to combine the data collected by different groups to compare the biodiversity of plant species found in different local environments. Ask students what physical factors in the area might affect biodiversity. (Sample answers: rainfall, soil quality, temperature, and availability of nutrients.) LS **Kinesthetic/Visual**

Answer to Extension

1. In making both kinds of maps, scientists conduct ground surveys and try to identify the size and features of natural areas of the globe. They use satellite images to construct their maps, but in one case their focus is the physical features of the region and in the other it is the ecosystems. The biodiversity map could use the features of the topographic map as a base map, with information about the ecosystems added as an additional layer of colors or symbols over the contour lines.

Unit 2

COMPOSITION OF THE EARTH

Unit 2 Outline

CHAPTER 4
Earth Chemistry

CHAPTER 5
Minerals of Earth's Crust

CHAPTER 6
Rocks

CHAPTER 7
Resources and Energy

▶ The Fly Geysers of the Black Rock Desert in Nevada are surrounded by a pool of water in which many different types of minerals are dissolved. Hot, mineral-rich water periodically erupts from springs deep within Earth to create these formations.

Chapter 4 Earth Chemistry

Planning Guide

Compression Guide
To shorten instruction because of time limitations, omit the Chapter Lab.

OBJECTIVES	LABS, DEMONSTRATIONS, AND ACTIVITIES	TECHNOLOGY RESOURCES
PACING • 90 min pp. 80–86 **Chapter Opener**		**OSP** **Parent Letter** **CD** **Student Edition on CD-ROM** **CD** **Chapter Summaries Audio CD** **VID** **Brain Food Video Quiz**
Section 1 Matter • Compare chemical properties and physical properties of matter. • Describe the basic structure of an atom. • Compare atomic number, mass number, and atomic mass. • Define *isotope*. • Describe the arrangement of elements in the periodic table.	**SE** **Quick Lab** Using the Periodic Table, p. 83 GENERAL **CRF** **Datasheet for Quick Lab*** GENERAL **TE** **Activity** Poster Project, p. 83 BASIC **TE** **Astronomy Connection** Forming Elements, p. 84 ADVANCED **SE** **Inquiry Lab** Physical Properties of Elements, pp. 98–99 GENERAL **CRF** **Datasheet for Chapter Lab*** GENERAL **SE** **Maps in Action** Element Resources in the United States, p. 100 GENERAL **SE** **Mapping Expeditions** Buried Treasure, pp. 836–837 GENERAL **CRF** **Skills Practice Lab** Flame Tests* GENERAL	**OSP** **Lesson Plans** (also in print) **TR** **Bellringer*** **TR** 16 Atomic Number and Atomic Mass* **TR** 17 The Periodic Table* **TR** 20 Element Resources in the United States* **VID** **HRW Earth Science Video** Chemical Elements **CD** **Interactive Tutor** What's the Matter? (Atomic Theory and Nuclear Energy)
PACING • 45 min pp. 87–92 **Section 2 Combinations of Atoms** • Define *compound* and *molecule*. • Interpret chemical formulas. • Describe two ways that electrons form chemical bonds between atoms. • Explain the differences between compounds and mixtures.	**TE** **Activity** Composition of Compounds, p. 87 GENERAL **TE** **Physics Connection** Energy Needs, p. 89 GENERAL **SE** **Quick Lab** Compounds, p. 91 GENERAL **CRF** **Datasheet for Quick Lab*** GENERAL **CRF** **Inquiry Lab** Ionic and Covalent Conductivity* GENERAL	**OSP** **Lesson Plans** (also in print) **TR** **Bellringer*** **TR** 18 Balancing Equations* **TR** 19 Ionic Bonds and Covalent Bonds* **TE** **Internet Activity** High-Energy Physics, p. 101 ADVANCED **CRF** **Internet Activity** High-Energy Physics* ADVANCED **CD** **Interactive Tutor** Matter and Minerals

PACING • 90 min

CHAPTER REVIEW, ASSESSMENT, AND STANDARDIZED TEST PREPARATION

- **SE** **Chapter Highlights,** p. 93
- **SE** **Chapter Review,** pp. 94–95
- **SE** **Standardized Test Prep,** pp. 96–97
- **CRF** **Concept Review*** GENERAL
- **CRF** **Critical Thinking*** ADVANCED
- **CRF** **Math Skills*** GENERAL
- **CRF** **Graphing Skills*** GENERAL
- **CRF** **Chapter Test A*** GENERAL
- **CRF** **Chapter Test B*** ADVANCED
- **OSP** **Lesson Plans** (also in print)
- **OSP** **Test Generator**
- **OSP** **Test Item Listing**

Online and Technology Resources

go.hrw.com Holt Online Learning

Visit **go.hrw.com** for access to Holt Online Learning, or enter the keyword **HQ6 Home** for a variety of free online resources.

One-Stop Planner® CD-ROM

This CD-ROM package includes
- Lab Materials QuickList Software
- Holt Calendar Planner
- Customizable Lesson Plans
- Printable Worksheets
- ExamView® Test Generator
- Interactive Teacher Edition
- Holt PuzzlePro®
- Holt PowerPoint® Resources

KEY

SE	Student Edition	OSP	One-Stop Planner	VID	Classroom Video/DVD
TE	Teacher Edition	TR	Transparencies and Transparency Worksheets	*	Also on One-Stop Planner
CRF	Chapter Resource File	CD	CD or CD-ROM	◆	Requires advance prep
LTP	Long-Term Projects			■	Also available in Spanish

SKILLS DEVELOPMENT RESOURCES	REVIEW AND ASSESSMENT	CORRELATIONS
SE **Pre-Reading Activity,** p. 80 GENERAL TE **Using the Figure** Champagne Pool, p. 80 GENERAL		National Science Education Standards
CRF **Directed Reading*** BASIC TE **Using the Figure** Earth's Elements, p. 81 GENERAL SE **Graphic Organizer** Comparison Table, p. 82 GENERAL TE **Reading Skill Builder** Paired Summarizing, p. 82 BASIC TE **Inclusion Strategies,** p. 82 TE **Using the Figure** Radioactive Elements, p. 85 GENERAL	SE **Reading Check,** p. 83 GENERAL SE **Section Review,** p. 86 GENERAL TE **Homework,** p. 84 ADVANCED TE **Reteaching,** p. 85 BASIC TE **Quiz,** p. 85 GENERAL TE **Alternative Assessment,** p. 85 ADVANCED CRF **Section Quiz*** ■ GENERAL	PS 1a, PS 1b, PS 2b, HNS 3c
CRF **Directed Reading*** BASIC TE **Using the Figure** Describing Reactions, p. 88 GENERAL SE **Math Practice,** p. 88 GENERAL TE **Reading Skill Builder** Reading Hint, p. 88 BASIC TE **Inclusion Strategies,** p. 88 TE **Skill Builder** Math, p. 90 ADVANCED	SE **Reading Checks,** pp. 89, 91 GENERAL SE **Section Review,** p. 92 GENERAL TE **Homework,** p. 90 ADVANCED TE **Reteaching,** p. 91 BASIC TE **Quiz,** p. 91 GENERAL TE **Alternative Assessment,** p. 92 ADVANCED CRF **Section Quiz*** ■ GENERAL	PS 2a, PS 2e

Holt Lab Generator CD-ROM

Search for labs by topic, standard, difficulty level, and/or time frame. Edit those labs to fit your needs, or create your own labs. Use the Lab Materials QuickList to order just what you need.

Holt Earth Science Interactive Tutor CD-ROM

This CD-ROM consists of interactive activities that give students a fun way to extend their knowledge of Earth science concepts.

Chapter Summaries Audio CDs

These CDs include audio summaries of the key concepts presented in each chapter. (Audio summaries are also available in Spanish.)

SciLinks NSTA

www.scilinks.org

Maintained by the **National Science Teachers Association.** See Chapter Enrichment pages that follow for a complete list of topics.

See Chapter Enrichment pages for Video Resources.

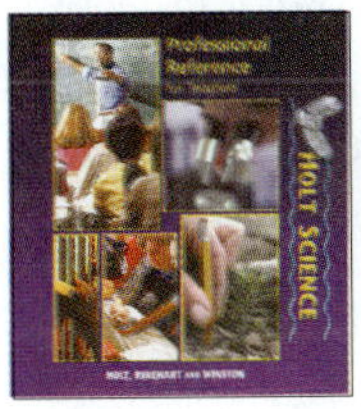

Refer to our ***Professional Reference for Teachers*** for additional teaching resources, including articles written by science education professionals about relevant and timely issues facing today's science teachers.

This Chapter Enrichment provides relevant and interesting information to expand and enhance your classroom instruction of the chapter material.

Section 1 Matter

Discovering and Naming Elements

Nine elements have been known for over 2,000 years, as evidenced by tools and artifacts made from them and references to their use. These elements—carbon, sulfur, iron, copper, silver, tin, gold, mercury, and lead—are noted in the periodic table by symbols based on their Latin names. For instance, the *C* for carbon is for the Latin *carbo,* meaning "charcoal," the form of carbon that was familiar to early cultures. *Au* stands for *aurum,* the Latin word for "gold," which may be related to a word meaning "shining dawn."

Until the year 1669, only three additional elements had been discovered: arsenic, antimony, and zinc. With the isolation of phosphorus in 1669, the process of isolating elements and identifying their properties began. This process expanded over the next three centuries. During the eighteenth century, 16 elements, including the gases hydrogen, nitrogen, and oxygen, were isolated. The nineteenth century saw the most progress in elemental chemistry: 54 elements were discovered. Of these, six were isolated by Sir Humphry Davy between 1807 and 1808. These elements, which included sodium, potassium, calcium, and barium, were isolated by means of the newly discovered process of electrolysis. In the twentieth century, 26 elements were discovered, of which almost all are radioactive, and 18 of which were created artificially.

Spectroscopy

The discovery of several elements, such as helium, rubidium, cesium, indium, and thallium were either initially made or were confirmed through spectroscopic analysis. The lines of color represent electrons jumping from one energy level to another. In doing so, they emit excess energy, which we see in the form of different light in various colors. The connection between spectroscopy and these elements is indicated by their names, most of which refer to the color of a prominent line in their spectrum. The element helium was first discovered in the sun by using spectroscopy before the element was found on Earth. Helium's name comes from *helios,* the Greek word for "sun."

Creating Elements

The first element to be synthetically produced was technetium, which was created in 1937 when scientists bombarded molybdenum with deuterons. Deuterons are particles that consist of one proton and one neutron. Although technetium has since been discovered to occur naturally, primarily in certain stars, the basic process used to create it has been used to produce a number of radioactive elements. Most of these are the *transuranium elements,* so-called because their atomic numbers are greater than that of uranium. Nuclear theory can be used to make predictions about the elements that result from bombarding nuclei with other accelerated nuclei. The process is then carried out using a particle accelerator. Many of the elements that are formed in this fashion have such short lifetimes and are

▼ **All matter can be reduced to its component elements.**

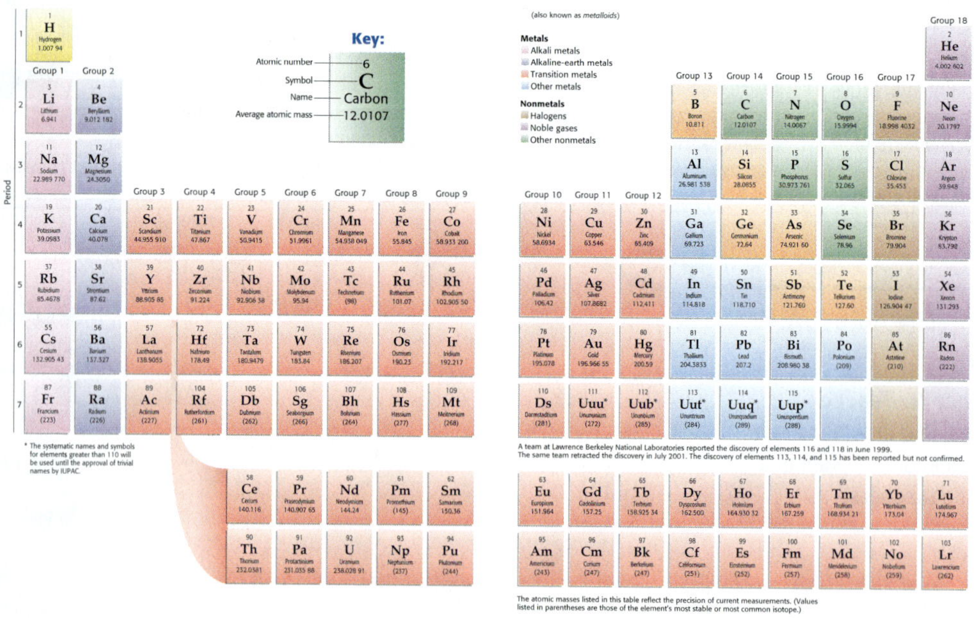

▲ **The way atoms bond determines a substance's chemical and physical properties.**

produced in such small quantities that their chemical properties are largely unknown. Nevertheless, the study of these elements is significant for confirming theories about the stability of the atomic nucleus.

Section 2 Combinations of Atoms

The Structure of Ionic Solids

Most ionic compounds are classified as salts, which are compounds whose positively-charged component, or *cation,* is any element other than H^+, and whose negatively-charged component, or anion, is any element or group other than OH^-. When solid, the cations and anions form regular structures called a *crystal lattice.* In a crystal lattice, each cation is surrounded by several anions that are electrically attracted to the cation, and around these anions are more cations attracted to the anions. To determine the regular pattern in a crystal lattice, the structure of the building block for a crystal, called the *unit cell,* must be determined. In the unit cell for sodium chloride, for instance, the sodium cation, Na^+, is at the center of a cubic unit cell, with a chloride anion, Cl^-, placed in the middle of each side of the cube. Such a crystalline structure is called *face-centered cubic.* On the other hand, cesium chloride has a unit cell in which eight chloride anions, Cl^-, are located at the corners of the cubic cell, surrounding a cesium cation, Cs^+. This arrangment is called *body-centered cubic* and is less densely packed than the face-centered crystal is. Crystal lattice arrangement in salts is frequently important for understanding the structure of minerals.

Video Resources

Brain Food Video Quizzes

Brain Food Video Quizzes These videos contain game-show style quizzes that assess students' progress and motivate students to study the chapter material.

HRW Earth Science Video

This video introduces Earth science topics and includes a geology field trip. The video segment listed below complements this chapter.

Segment 7: Chemical Elements The objective of this segment is to provide a simplified overview of the elemental composition of matter. The definition of the term *element* is given, and a brief explanation of the components of the atom is provided. Minerals are defined as being made up of elements that are chemically bonded together, and the three-dimensional structures of halite, graphite, and diamond are examined to show how the arrangement of elements and chemical bonds determines the physical properties of a mineral or a native element. (3 min)

NOVA Videos

To order **NOVA** videos related to this chapter, visit **go.hrw.com** and enter the keyword **HQ6CHMV.**

SCILINKS

NSTA Developed and maintained by the National Science Teachers Association

SciLinks is maintained by the National Science Teachers Association to provide you and your students with interesting, up-to-date links that will enrich your classroom presentation of the chapter.

Visit www.scilinks.org and enter the SciLinks code for more information about the topic listed.

Topic: The Periodic Table
SciLinks code: HQ61125

Topic: Valence Electrons
SciLinks code: HQ61591

Topic: Covalent and Ionic Bonds
SciLinks code: HQ60362

Topic: Chemical Equations
SciLinks code: HQ60269

Topic: Chemical Reactions
SciLinks code: HQ60274

Chapter 4

Chapter Overview

Matter consists of atoms, either of one type (elements), or in combination with other types (compounds). The chapter describes how electrons are arranged around an atom's nucleus, how atoms combine to form molecules, and the types of chemical bonds that hold molecules together.

Using the Figure — GENERAL

Champagne Pool The geothermal pool in the figure is heated by volcanic activity on the North Island of New Zealand. Ask students if anything in the photograph indicates this heating. (The steam rising from the pool indicates heating.) The orange color in the foreground may be caused by a form of bacteria that thrives in the pool. Although various elements are present in the water, they are not visible. **LS Visual**

PRE-READING ACTIVITY

FOLDNOTES

Encourage students to use their FoldNote as a study guide to quiz themselves for a test on the chapter material. Students may want to create Double Door FoldNotes for different topics within the chapter.

Chapter 4 Earth Chemistry

Sections

1 Matter

2 Combinations of Atoms

What You'll Learn

- How chemical structure affects physical and chemical properties of substances
- How elements combine to form compounds

Why It's Relevant

Understanding the chemical structure of substances will help you to understand the properties of the different materials that make up Earth.

PRE-READING ACTIVITY

FOLDNOTES

Double Door Before you read this chapter, create the FoldNote entitled "Double Door" described in the Skills Handbook section of the Appendix. Write "Element" on one flap of the double door and "Compound" on the other flap. As you read the chapter, compare the two topics, and write characteristics of each on the inside of the appropriate flap.

▶ New Zealand's Champagne Pool is a geothermal pool that has a temperature of 74°C! Its water contains dissolved elements such as gold, silver, mercury, sulfur, antimony, and arsenic.

Chapter Correlations ***National Science Education Standards***

PS 1a Matter is made of minute particles called atoms, and atoms are composed of even smaller components. These components have measurable properties, such as mass and electrical charge. Each atom has a positively charged nucleus surrounded by negatively charged electrons… **(Section 1)**

PS 1b The atom's nucleus is composed of protons and neutrons, which are much more massive than electrons. When an element has atoms that differ in the number of neutrons, these atoms are called different isotopes of the element. **(Section 1)**

PS 2a Atoms interact with one another by transferring or sharing electrons that are furthest from the nucleus… **(Section 2)**

PS 2b An element is composed of a single type of atom. When elements are listed in order according to the number of protons (called the atomic number), repeating patterns of physical and chemical properties identify families of elements with similar properties. This "Periodic Table" is a consequence of the repeating pattern of outermost electrons… **(Section 1)**

PS 2e Solids, liquids, and gases differ in the distances…between molecules or atoms…. **(Section 2)**

HNS 3c Occasionally, there are advances in science…that have important and long-lasting effects..., Examples of such advances include…Atomic theory. **(Section 1)**

Section 1 Matter

Every object in the universe is made of particles of some kind of substance. Scientists use the word *matter* to describe the substances of which objects are made. **Matter** is anything that takes up space and has mass. The amount of matter in any object is the *mass* of that object. All matter has observable and measurable properties. Scientists can observe the properties of a substance to identify the kind of matter that makes up that substance.

OBJECTIVES

- **Compare** chemical properties and physical properties of matter.
- **Describe** the basic structure of an atom.
- **Compare** atomic number, mass number, and atomic mass.
- **Define** *isotope.*
- **Describe** the arrangement of elements in the periodic table.

KEY TERMS

matter
element
atom
proton
electron
neutron
isotope

Properties of Matter

All matter has two types of distinguishing properties—physical properties and chemical properties. *Physical properties* are characteristics that can be observed without changing the composition of the substance. For example, physical properties include density, color, hardness, freezing point, boiling point, and the ability to conduct an electric current.

Chemical properties are characteristics that describe how a substance reacts with other substances to produce different substances. For example, a chemical property of iron is that iron reacts with oxygen to form rust. A chemical property of helium is that helium does not react with other substances to form new substances. Understanding the chemical properties of a substance requires knowing some basic information about the particles that make up all substances.

matter anything that has mass and takes up space

element a substance that cannot be separated or broken down into simpler substances by chemical means; all atoms of an element have the same atomic number

Elements

An **element** is a substance that cannot be broken down into simpler, stable substances by chemical means. Each element has a characteristic set of physical and chemical properties that can be used to identify it. **Figure 1** shows the most common elements in Earth's continental crust. More than 90 elements occur naturally on Earth. About two dozen other elements have been created in laboratories. Of the natural elements, eight make up more than 98% of Earth's crust. Every known element is represented by a symbol of one or two letters.

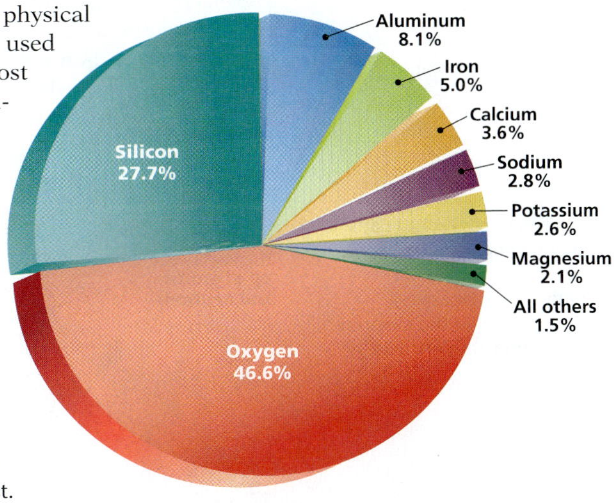

Figure 1 ▶ This graph shows the percentage of total mass that each common element makes up in Earth's continental crust.

Section 1

Focus

Overview

This section describes elements and their fundamental component, the atom. Physical and chemical properties of matter are introduced, along with the structure of atoms and their relation to each other in the periodic table.

Bellringer

Tell students that we are completely surrounded by matter. Have students give examples of matter around them. Ask them to list some characteristics of matter. Then, have them name some things that are not matter. (Answers may vary. Accept all reasonable answers.) LS **Verbal**

Motivate

Using the Figure — GENERAL

Earth's Elements Point out to students that while a number of the elements shown in the pie graph exist only as elements, most are combined with other elements to form chemical compounds, such as iron oxides, in which iron and oxygen are bonded together. Ask students to use the pie chart to identify the three most common elements in Earth's continental crust and their respective percentages. (oxygen, 46.6%; silicon, 27.7%; and aluminum, 8.1%) LS **Visual**

CHAPTER RESOURCES

Chapter Resource File

- **Directed Reading** BASIC
- **Skills Practice Lab**
  Flame Tests GENERAL

Technology

Transparencies
- Bellringer

Student Edition on CD-ROM

One-Stop Planner CD-ROM
- Lesson Plan

Teach

Graphic Organizer GENERAL

Comparison Table
You may want to use this Graphic Organizer in a game to review material before the test. Divide the class into two teams. Ask students questions about material from the Comparison Table. Give points to each team that provides correct answers.

MISCONCEPTION ALERT

Electron Cloud Explain to students that an electron cloud is an idealized representation of where an electron may be found at a particular point in time. Matter the size of tiny atoms is described as having properties of both particles and waves. Thinking of an electron as if it were spread out like a cloud within a region around the nucleus helps students understand the extended wave-like properties. Have students visualize blades of a fan moving within a region around the fan's axis. While each blade is located in only one place, pinpointing the exact location at any given moment is difficult.

Graphic Organizer **Comparison Table**

Create the Graphic Organizer entitled "Comparison Table" described in the Skills Handbook section of the Appendix. Label the columns with "Electron," "Proton," and "Neutron." Label the rows with "Charge" and "Location." Then, fill in the table with details about the charge and location of each subatomic particle.

Atoms

Elements consist of atoms. An **atom** is the smallest unit of an element that has the chemical properties of that element. Atoms cannot be broken down into smaller particles that will have the same chemical and physical properties as the atom. A single atom is so small that its size is difficult to imagine. To get an idea of how small it is, look at the thickness of this page. More than a million atoms lined up side by side would be equal to that thickness.

Atomic Structure

Even though atoms are very tiny, they are made up of smaller parts called *subatomic particles*. The three major kinds of subatomic particles are protons, electrons, and neutrons. **Protons** are subatomic particles that have a positive charge. **Electrons** are subatomic particles that have a negative charge. **Neutrons** are subatomic particles that have no charge.

atom the smallest unit of an element that maintains the chemical properties of that element

proton a subatomic particle that has a positive charge and that is located in the nucleus of an atom; the number of protons of the nucleus is the atomic number, which determines the identity of an element

electron a subatomic particle that has a negative charge

neutron a subatomic particle that has no charge and that is located in the nucleus of an atom

The Nucleus

As shown in **Figure 2**, the protons and neutrons of an atom are packed close to one another. Together they form the *nucleus*, which is a small region in the center of an atom. The nucleus has a positive charge because protons have a positive charge and neutrons have no charge.

The nucleus makes up most of an atom's mass but very little of an atom's volume. If an atom's nucleus were the size of a gumdrop, the atom itself would be as big as a football stadium. Because electrons are even smaller than protons and neutrons, the volume of an atom is mostly empty space.

The Electron Cloud

The electrons of an atom move in a certain region of space called an *electron cloud* that surrounds the nucleus. Because opposite charges attract each other, the negatively charged electrons are attracted to the positively charged nucleus. This attraction is what holds electrons in the atom.

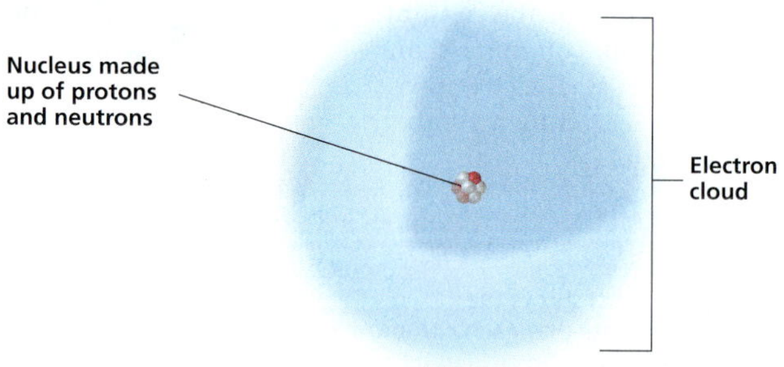

Figure 2 ▶ The nucleus of the atom is made up of protons and neutrons. The protons give the nucleus a positive charge. The negatively charged electrons are in the electron cloud that surrounds the nucleus.

INCLUSION Strategies

- Learning Disabled
- Hearing Impaired
- Developmentally Delayed

Many students better understand information when it is given visually. Help students understand the relationships among elements, atoms, electrons, protons, neutrons, and nuclei by creating a concept map.

READING SKILL BUILDER BASIC

Paired Summarizing Group students into pairs and have them read silently about atoms, atomic structure, and the atomic nucleus. Then, have students take turns summarizing the information about each subject. While one student is describing or explaining, the other student should listen to the retelling and point out any inaccuracies or ideas that were left out. Allow students to refer to the text as needed. English Language Learners
LS Verbal Co-op Learning

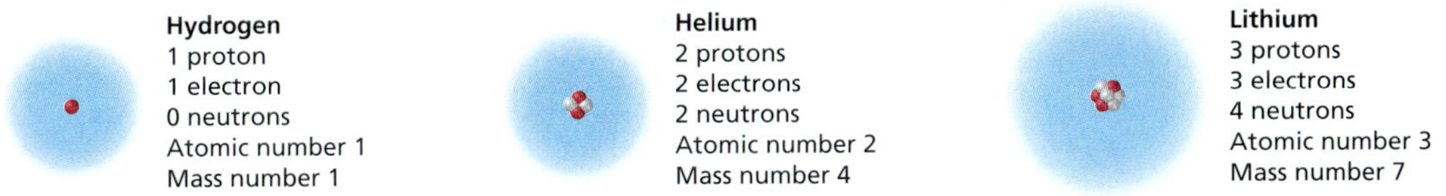

Figure 3 ▶ Hydrogen, helium, and lithium are different elements because their atoms have different atomic numbers, or different numbers of protons.

Atomic Number

The number of protons in the nucleus of an atom is called the *atomic number*. All atoms of any given element have the same atomic number. An element's atomic number sets the atoms of that element apart from the atoms of all other kinds of elements, as shown in **Figure 3.** Because an uncharged atom has an equal number of protons and electrons, the atomic number is also equal to the number of electrons in an atom of any given element.

Elements on the periodic table are ordered according to their atomic numbers. The *periodic table,* shown on the following pages, is a system for classifying elements. Elements in the same column on the periodic table have similar arrangements of electrons in their atoms. Elements that have similar arrangements of electrons have similar chemical properties.

Atomic Mass

The sum of the number of protons and neutrons in an atom is the *mass number*. The mass of a subatomic particle is too small to be expressed easily in grams. So, a special unit called the *atomic mass unit* (amu) is used. Protons and neutrons each have an atomic mass that is close to 1 amu. In contrast, electrons have much less mass than protons and neutrons do. The mass of 1 proton is equal to the combined mass of about 1,840 electrons. Because electrons add little to an atom's total mass, their mass can be ignored when calculating an atom's approximate mass.

Reading Check **What is the difference between atomic number, mass number, and atomic mass unit? (See the Appendix for answers to Reading Checks.)**

Isotopes

Although all atoms of a given element contain the same number of protons, the number of neutrons may differ. For example, while most helium atoms have two neutrons, some helium atoms have only one neutron. An atom that has the same number of protons (or the same atomic number) as other atoms of the same element do but has a different number of neutrons (and thus a different atomic mass) is called an **isotope** (IE suh TOHP).

A helium atom that has two neutrons is more massive than a helium atom that has only one neutron. Because of their different number of neutrons and their different masses, different isotopes of the same element have slightly different properties.

QuickLAB **10 min**

Using the Periodic Table

Procedure

1. Use the **periodic table** to find the atomic numbers of the following elements: carbon, iron, molybdenum, and iodine.
2. Determine the number of protons and electrons that are in each neutral atom of the elements listed in step 1.
3. Find the average atomic masses of the elements listed in step 1.

Analysis

1. Use the atomic number and average atomic mass of each element to estimate the average number of neutrons in each atom of the elements listed in step 1 of this activity.
2. Which element has the largest difference between its average number of neutrons and the number of protons? Describe any trends that you observe.

isotope an atom that has the same number of protons (or the same atomic number) as the other atoms of the same element do but that has a different number of neutrons (and thus a different atomic mass)

QuickLAB

Skills Acquired

- Collecting Data
- Classifying
- Identifying/Recognizing Patterns

Teacher's Notes Instruct students to round the average atomic masses to a whole number to simplify their calculations.

Carbon: atomic number = 6, average atomic mass = 12 amu
Iron: atomic number = 26, average atomic mass = 56 amu
Molybdenum: atomic number = 42, average atomic mass = 96 amu
Iodine: atomic number = 53, average atomic mass = 127 amu

Answers

1. Average number of neutrons:
   Carbon: 12 – 6 = 6
   Iron: 56 – 26 = 30
   Molybdenum: 96 – 42 = 54
   Iodine: 127 – 53 = 74
2. iodine; The larger the atom, the larger the difference between the number of protons and the number of neutrons.

Answer to Reading Check

The atomic number is the number of protons in an atom's nucleus. The mass number is the sum of the number of protons and the number of neutrons in an atom. The atomic mass unit is used to express the mass of subatomic particles or atoms.

Activity — BASIC

Poster Project Assign, or have each student select, an element as the subject of a poster project. Have students research the basic atomic properties of their element: atomic number, the number of electrons, mass number, number of neutrons in the most abundant isotope, and average atomic mass. Also have students list some of the element's uses and qualities. Students can give brief presentations. Alternatively, students can display posters throughout the classroom. LS **Visual**

CHAPTER RESOURCES

Chapter Resource File

- Datasheet for Quick Lab GENERAL

Technology

Transparencies

- 16 Atomic Number and Atomic Mass (with worksheet)

Teach, *continued*

ASTRONOMY CONNECTION ADVANCED

Forming Elements Hydrogen is the most abundant element in the universe. Scientists theorize that all of the hydrogen in the universe, along with some helium and lithium, was produced shortly after the Big Bang. Most other elements on Earth, however, are the products of exploding stars called *supernovas*. When stars that have masses more than 20 times that of the sun reach a certain point in their evolution, their cores collapse. Then, the outer layers of the star rapidly collapse inward as well, releasing a tremendous amount of energy in a short period of time. This energy passes outward through the star, heating and bombarding nuclear material to form the various elements in a process called *nucleosynthesis*. The star's nuclear material spreads outward from the star into interstellar space until it becomes part of new stars and their planets. Have students research the estimated amounts of certain stable elements that would be produced in an exploding Type II supernova. Compare the results to the amounts of these elements on Earth. Have students calculate how many "Earths" could be constructed from the elements exploded outward from a supernova. LS **Logical**

CHAPTER RESOURCES

Technology

Transparencies
- 17 The Periodic Table (with worksheet)

The Periodic Table of Elements

Key:

Atomic number — 6
Symbol — C
Name — Carbon
Average atomic mass — 12.0107

Period	Group 1	Group 2	Group 3	Group 4	Group 5	Group 6	Group 7	Group 8	Group 9
1	1 H Hydrogen 1.007 94								
2	3 Li Lithium 6.941	4 Be Beryllium 9.012 182							
3	11 Na Sodium 22.989 769 28	12 Mg Magnesium 24.3050							
4	19 K Potassium 39.0983	20 Ca Calcium 40.078	21 Sc Scandium 44.955 912	22 Ti Titanium 47.867	23 V Vanadium 50.9415	24 Cr Chromium 51.9961	25 Mn Manganese 54.938 045	26 Fe Iron 55.845	27 Co Cobalt 58.933 195
5	37 Rb Rubidium 85.4678	38 Sr Strontium 87.62	39 Y Yttrium 88.905 85	40 Zr Zirconium 91.224	41 Nb Niobium 92.906 38	42 Mo Molybdenum 95.94	43 Tc Technetium (98)	44 Ru Ruthenium 101.07	45 Rh Rhodium 102.905 50
6	55 Cs Cesium 132.905 4519	56 Ba Barium 137.327	57 La Lanthanum 138.905 47	72 Hf Hafnium 178.49	73 Ta Tantalum 180.947 88	74 W Tungsten 183.84	75 Re Rhenium 186.207	76 Os Osmium 190.23	77 Ir Iridium 192.217
7	87 Fr Francium (223)	88 Ra Radium (226)	89 Ac Actinium (227)	104 Rf Rutherfordium (261)	105 Db Dubnium (262)	106 Sg Seaborgium (266)	107 Bh Bohrium (264)	108 Hs Hassium (277)	109 Mt Meitnerium (268)

58 Ce Cerium 140.116	59 Pr Praseodymium 140.907 65	60 Nd Neodymium 144.242	61 Pm Promethium (145)	62 Sm Samarium 150.36
90 Th Thorium 232.038 06	91 Pa Protactinium 231.035 88	92 U Uranium 238.028 91	93 Np Neptunium (237)	94 Pu Plutonium (244)

* The systematic names and symbols for elements greater than 111 will be used until the approval of trivial names by IUPAC.

internet connect

go.hrw.com
Topic: Periodic Table
Go To: go.hrw.com
Keyword: Holt Periodic

Visit the HRW Web site for updates on the periodic table.

Homework ADVANCED

Naming the Elements Assign each student five elements from the periodic table. Have students research the origin and meaning of the names of each of their elements. Try to include in each student's list one element known since ancient or medieval times, three discovered in the nineteenth century, and one transuranium element. Have students present their findings in a short written report, an oral presentation, a poster, a skit, or a song. LS **Verbal/Auditory**

Cultural Awareness BASIC

Creating the Periodic Table The periodic table was first devised by the Russian chemist Dmitri Mendeleev in 1869. At the time, the atomic nature of matter was not widely accepted or well understood. Nevertheless, organizing elements by atomic weight and chemical properties enabled Mendeleev to discern a pattern in which elements with similar properties could be grouped together. Mendeleev even predicted the properties of the elements scandium, gallium, and germanium before they were discovered. The element mendelevium is named in his honor.

Hydrogen
Semiconductors (also known as metalloids)

Metals
- Alkali metals
- Alkaline-earth metals
- Transition metals
- Other metals

Nonmetals
- Halogens
- Noble gases
- Other nonmetals

Group 10	Group 11	Group 12	Group 13	Group 14	Group 15	Group 16	Group 17	Group 18
								2 He Helium 4.002 602
			5 B Boron 10.811	6 C Carbon 12.0107	7 N Nitrogen 14.0067	8 O Oxygen 15.9994	9 F Fluorine 18.998 4032	10 Ne Neon 20.1797
			13 Al Aluminum 26.981 5386	14 Si Silicon 28.0855	15 P Phosphorus 30.973 762	16 S Sulfur 32.065	17 Cl Chlorine 35.453	18 Ar Argon 39.948
28 Ni Nickel 58.6934	29 Cu Copper 63.546	30 Zn Zinc 65.409	31 Ga Gallium 69.723	32 Ge Germanium 72.64	33 As Arsenic 74.921 60	34 Se Selenium 78.96	35 Br Bromine 79.904	36 Kr Krypton 83.798
46 Pd Palladium 106.42	47 Ag Silver 107.8682	48 Cd Cadmium 112.411	49 In Indium 114.818	50 Sn Tin 118.710	51 Sb Antimony 121.760	52 Te Tellurium 127.60	53 I Iodine 126.904 47	54 Xe Xenon 131.293
78 Pt Platinum 195.084	79 Au Gold 196.966 569	80 Hg Mercury 200.59	81 Tl Thallium 204.3833	82 Pb Lead 207.2	83 Bi Bismuth 208.980 40	84 Po Polonium (209)	85 At Astatine (210)	86 Rn Radon (222)
110 Ds Darmstadtium (271)	111 Rg Roentgenium (272)	112 Uub* Ununbium (285)		114 Uuq* Ununquadium (289)		116 Uuh* Ununhexium (292)		

The discoveries of elements with atomic numbers 112, 114, and 116 have been reported but not fully confirmed.

63 Eu Europium 151.964	64 Gd Gadolinium 157.25	65 Tb Terbium 158.925 35	66 Dy Dysprosium 162.500	67 Ho Holmium 164.930 32	68 Er Erbium 167.259	69 Tm Thulium 168.934 21	70 Yb Ytterbium 173.04	71 Lu Lutetium 174.967
95 Am Americium (243)	96 Cm Curium (247)	97 Bk Berkelium (247)	98 Cf Californium (251)	99 Es Einsteinium (252)	100 Fm Fermium (257)	101 Md Mendelevium (258)	102 No Nobelium (259)	103 Lr Lawrencium (262)

The atomic masses listed in this table reflect the precision of current measurements. (Each value listed in parentheses is the mass number of that radioactive element's most stable or most common isotope.)

Close

Reteaching — BASIC

Atomic and Mass Numbers

Write in a column on the board the mass numbers and atomic numbers for elements widely scattered about the periodic table. Obtain the mass numbers by rounding the average atomic mass values to the nearest whole numbers. Have students use the periodic table to identify the elements. Then, have them use the masses and atomic numbers to determine the number of protons, electrons, and neutrons in each element. **LS Logical**

Quiz — GENERAL

Determine whether each of the following statements is true or false.

1. An element is made up of only one kind of atom. (true)
2. The average atomic mass is equal to the number of neutrons in an atom. (false)
3. Atoms that have the same number of protons but different numbers of neutrons are called isotopes. (true)

Alternative Assessment — ADVANCED

Quiz Show Have students create a game show in which contestants are given clues about different elements, from which they have to identify the element. The fewer clues needed to answer correctly, the more points that are awarded. The game can be designed for individual competitors or for teams. **LS Interpersonal**

Using the Figure — GENERAL

Radioactive Elements Point out to students that some elements undergo radioactive decay, becoming other elements in the process. Ask students to find technetium, the lightest radioactive element, and give its atomic number. (atomic number 43) Ask students to give the average atomic mass, rounded to a whole number, of the radioactive element polonium. (209 amu) **LS Visual/Logical**

Close, *continued*

Answers to Section Review

1. Physical properties are characteristics that can be observed without changing the composition of the substance. Chemical properties are characteristics that describe how a substance interacts with other substances to produce different kinds of matter.
2. Atoms consist of protons and neutrons in the nucleus and electrons that move in a cloud around the nucleus.
3. protons, neutrons, and electrons
4. Atomic number is the number of protons in the nucleus of an atom, as well as the number of electrons in the neutral form of the atom. Mass number is the total number of protons and neutrons in the nucleus of an atom. Average atomic mass is the weighted average of the atomic masses of the isotopes of an element.
5. Isotopes have the same number of protons as other atoms of the same element but have different numbers of neutrons.
6. The combination of atoms to form a new substance that has different properties is a result of the chemical properties of the elements.
7. Atoms of different elements have different numbers of protons in their nuclei.
8. *Matter* is made up of *atoms* that consist of *protons, neutrons,* and *electrons* and that may form *elements,* which are organized in the *periodic table* by their *atomic number.*

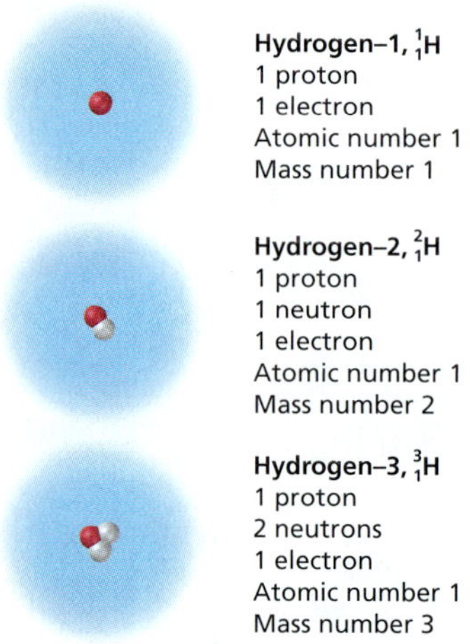

Figure 4 ▶ There are three naturally occurring isotopes of hydrogen. The main difference between these isotopes is their mass numbers.

SCiLINKS. NSTA Developed and maintained by the National Science Teachers Association

For a variety of links related to this subject, go to www.scilinks.org

Topic: Valence Electrons

SciLinks code: HQ61591

Average Atomic Mass

Because the isotopes of an element have different masses, the periodic table uses an average atomic mass for each element. The *average atomic mass* is the weighted average of the atomic masses of the naturally occurring isotopes of an element.

As shown in **Figure 4,** hydrogen has three isotopes. Each isotope has a different mass because each has a different number of neutrons. By calculating the weighted average of the atomic masses of the three naturally occurring hydrogen isotopes, you can determine the average atomic mass. As noted on the periodic table, the average atomic mass of hydrogen is 1.007 94 amu.

Valence Electrons and Periodic Properties

Based on similarities in their chemical properties, elements on the periodic table are arranged in columns, which are called *groups*. An atom's chemical properties are largely determined by the number of the outermost electrons in an atom's electron cloud. These electrons are called *valence* (VAY luhns) electrons.

Within each group, the atoms of each element generally have the same number of valence electrons. For Groups 1 and 2, the number of valence electrons in each atom is the same as that atom's group number. Atoms of elements in Groups 3–12 have 2 or more valence electrons. For Groups 13–18, the number of valence electrons in each atom is the same as that atom's group number minus 10, except for helium, He. It has only two valence electrons. Atoms in Group 18 have 8 valence electrons. When an atom has 8 valence electrons, it is considered stable, or chemically unreactive. Unreactive atoms do not easily lose or gain electrons.

Elements whose atoms have only one, two, or three valence electrons tend to lose electrons easily. These elements have metallic properties and are generally classified as *metals*. Elements whose atoms have from four to seven valence electrons are more likely to gain electrons. Many of these elements, which are in Groups 13–17, are classified as *nonmetals*.

Section 1 Review

1. **Compare** the physical properties of matter with the chemical properties of matter.
2. **Describe** the basic structure of an atom.
3. **Name** the three basic subatomic particles.
4. **Compare** atomic number, mass number, and atomic mass.
5. **Explain** how isotopes of an element differ from each other.

CRITICAL THINKING

6. **Evaluating Data** Oxygen combines with hydrogen and becomes water. Is this combination a result of the physical or chemical properties of hydrogen?
7. **Making Comparisons** What sets an atom of one element apart from atoms of all other elements?

CONCEPT MAPPING

8. Use the following terms to create a concept map: *matter, element, atom, electron, proton, neutron, atomic number,* and *periodic table.*

CHAPTER RESOURCES

Chapter Resource File
- Section Quiz GENERAL

Workbooks
- Study Guide (also in Spanish)

Section 2 Combinations of Atoms

Elements rarely occur in pure form in Earth's crust. They generally occur in combination with other elements. A substance that is made of two or more elements that are joined by chemical bonds between the atoms of those elements is called a **compound.** The properties of a compound differ from those of the elements that make up the comound, as shown in **Figure 1.**

OBJECTIVES

- **Define** *compound* and *molecule.*
- **Interpret** chemical formulas.
- **Describe** two ways that electrons form chemical bonds between atoms.
- **Explain** the differences between compounds and mixtures.

KEY TERMS

compound
molecule
ion
ionic bond
covalent bond
mixture
solution

Molecules

The smallest unit of matter that can exist by itself and retain all of a substance's chemical properties is a **molecule.** In a molecule of two or more atoms, the atoms are chemically bonded together. Some molecules consist entirely of atoms of the same element.

Some elements occur naturally as *diatomic molecules,* which are molecules that are made up of only two atoms. For example, the oxygen in the air you breathe is the diatomic molecule O_2. The *O* in this notation is the symbol for oxygen. The subscript 2 indicates the number of oxygen atoms that are bonded together.

Chemical Formulas

In any given compound, the elements that make up the compound occur in the same relative proportions. Therefore, a compound can be represented by a chemical formula. A *chemical formula* is a combination of letters and numbers that shows which elements make up a compound. It also shows the number of atoms of each element that are required to make a molecule of a compound.

The chemical formula for water is H_2O, which indicates that each water molecule consists of two atoms of hydrogen and one atom of oxygen. In a chemical formula, the subscript that appears after the symbol for an element shows the number of atoms of that element that are in a molecule. For example, in the chemical formula for water, the subscript 2 and the symbol *H* mean that two atoms of hydrogen are in each molecule of water.

compound a substance made up of atoms of two or more different elements joined by chemical bonds

molecule a group of atoms that are held together by chemical forces; a molecule is the smallest unit of matter that can exist by itself and retain all of a substance's chemical properties

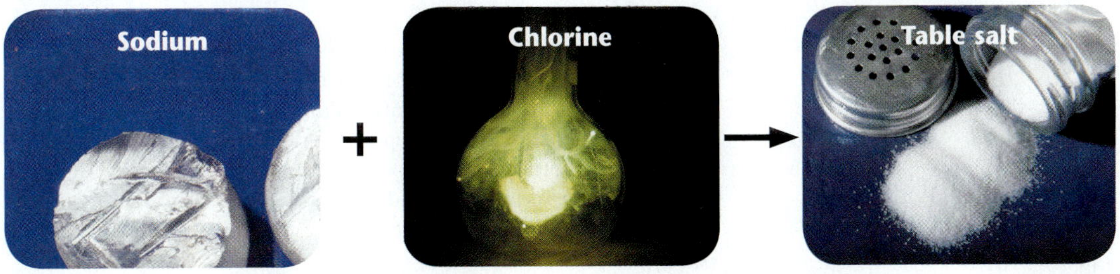

Figure 1 ▶ Sodium (Na), a silvery metal, and chlorine (Cl), a poisonous, greenish-yellow gas, are both components of table salt (sodium chloride, NaCl), which you can eat.

Section 2

Focus

Overview

This section describes molecules, chemical formulas, and how the arrangement of electrons in atoms leads to the formation of compounds. It also explains the two main types of chemical bonds, as well as mixtures and solutions.

Bellringer

Have students list three familiar compounds that each consist of two elements. Point out that there may be more than two atoms in each molecule. (Answers may vary. Three common examples are water, H_2O; table salt, NaCl; and carbon dioxide, CO_2.) **LS Logical**

Motivate

Activity — GENERAL

Composition of Compounds

Write a list of compounds on the board and have students use the periodic table to identify the elements in the compounds. Make sure the list contains at least one diatomic molecule, one compound composed of two elements, and one compound made of three or more elements. Point out that many element symbols contain two letters, and that the second letter is always lowercase. This convention is used to avoid confusion when writing a molecular formula, such as CO (carbon monoxide), or an element, such as Co (cobalt). **LS Visual/Logical**

CHAPTER RESOURCES

Chapter Resource File

- Directed Reading BASIC

Technology

Transparencies
- Bellringer

Student Edition on CD-ROM

One-Stop Planner CD-ROM
- Lesson Plan

Teach

Using the Figure — GENERAL

Describing Reactions Have students study the diagram at the top of this page. Have students count the number of atoms of each element in each compound and compare the number to the number of atoms of each element in the chemical formulas below. Ask students how the chemical formulas relate to the illustration. (The chemical formulas describe the number of each type of element in the reaction.) Answer to caption question: There are four hydrogen atoms on each side of the equation. LS **Visual**

MATHPRACTICE

Answer

unbalanced: $Mg + O_2 \rightarrow MgO$

balanced: $2Mg + O_2 \rightarrow 2MgO$

READING SKILL BUILDER — BASIC

Reading Hint As students read this page, have them balance an equation, such as

$KOH + HCl \rightarrow KCl + H_2O$

Walking through the steps of balancing an equation as they read the passage helps students understand the process. LS **Logical** English Language Learners

CHAPTER RESOURCES

Technology

Transparencies
- 18 Balancing Equations (with worksheet)

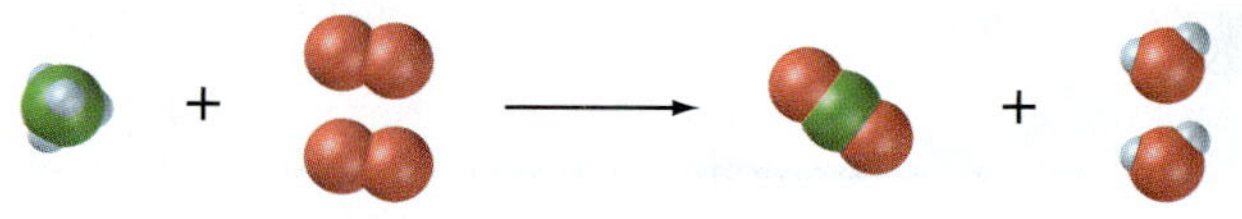

Figure 2 ▶ Methane, CH_4, and oxygen gas, O_2, react during combustion to form the products carbon dioxide, CO_2, and water, H_2O. ***How many hydrogen atoms are on each side of this reaction?***

Chemical Equations

Elements and compounds often combine through chemical reactions to form new compounds. The reaction of these elements and compounds can be described in a formula called a *chemical equation*.

Equation Structure

In a chemical equation, such as the one shown below, the *reactants*, which are on the left-hand side of the arrow, form the *products*, which are on the right-hand side of the arrow. When chemical equations are written, the arrow means "gives" or "yields."

$$CH_4 + 2O_2 \rightarrow CO_2 + 2H_2O$$

In this equation, one molecule of methane, CH_4, reacts with two molecules of oxygen, O_2, to yield one molecule of carbon dioxide, CO_2, and two molecules of water, H_2O, as shown in **Figure 2.**

MATHPRACTICE

Balancing Equations
Magnesium, Mg, reacts with oxygen gas, O_2, to form magnesium oxide, MgO. Write a balanced chemical equation for this reaction by placing the coefficients that are needed to obtain an equal number of magnesium and oxygen atoms on either side of the equation.

Balanced Equations

Chemical equations are useful for showing the types and amounts of the products that could form from a particular set of reactants. However, the equation must be balanced to show this information. A chemical equation is balanced when the number of atoms of each element on the right side of the equation is equal to the number of atoms of the same element on the left side.

To balance an equation, you cannot change chemical formulas. Changing the formulas would mean that different substances were in the reaction. To balance an equation, you must put numbers called *coefficients* in front of chemical formulas.

On the left side of the equation above, the methane molecule has four hydrogen atoms, which are indicated by the subscript 4. On the right side, each water molecule has two hydrogen atoms, which are indicated by the subscript 2. A coefficient of 2 is placed in front of the formula for water to balance the number of hydrogen atoms. A coefficient multiplies the subscript in an equation. For example, four hydrogen atoms are in the formula $2H_2O$.

A coefficient of 2 is also placed in front of the oxygen molecule on the left side of the equation so that both sides of the equation have four oxygen atoms. When the number of atoms of each element on either side of the equation is the same, the equation is balanced.

INCLUSION Strategies

- ***Attention Deficit Disorder***
- ***Behavior Control Issues***
- ***Learning Disabled***

Have students model balancing an equation. Have twelve students wear reversible basketball jerseys and pretend to be atoms of hydrogen and oxygen. Organize 3 students wearing dark jerseys (H) and 3 students wearing light jerseys (O) into the equation: $H_2 + O_2 \rightarrow H_2O$. Have students not wearing jerseys place the remaining students who are wearing jerseys into the equation to balance it. Students may have to reverse their jerseys to have the correct number of atoms. The balanced equation is $2H_2 + O_2 \rightarrow 2H_2O$. LS **Kinesthetic** Co-op Learning

Chemical Bonds

Figure 3 ▶ Scientists sometimes make physical models of molecules to better understand how chemical bonds affect the properties of compounds.

The forces that hold together the atoms in molecules are called *chemical bonds*. Chemical bonds form because of the attraction between positive and negative charges. Atoms form chemical bonds by either sharing or transferring valence electrons from one atom to another. Transferring or sharing valence electrons from one atom to another changes the properties of the substance. Variations in the forces that hold molecules together are responsible for a wide range of physical and chemical properties.

As shown in **Figure 3,** scientists can study interactions of atoms to predict which kinds of atoms will form chemical bonds together. Scientists do this by comparing the number of valence electrons that are present in a particular atom with the maximum number of valence electrons that are possible. For example, a hydrogen atom has only one valence electron. But because hydrogen can have two valence electrons, it will give up or accept another electron to reach a more chemically unreactive state.

Reading Check In what two ways do atoms form chemical bonds? (See the Appendix for answers to Reading Checks.)

Connection to CHEMISTRY

States of Matter

Most matter on Earth can be classified into three states—solid, liquid, and gas. The speed at which the particles of matter move and the distance between the particles of matter determine the state of matter.

The particles that make up solids are packed tightly together in relatively fixed positions and are not free to move much in relation to each other. So, a *solid* has a definite shape and volume. A *liquid* has a definite volume but does not have a definite shape. Instead, a liquid takes the shape of the container that holds it. The particles that make up a liquid are also tightly packed, but they are more free to move than those in a solid are. A *gas* does not have a definite volume or shape. The particles of a gas are much farther apart and move faster and more freely than those of a liquid do. Thus, a gas is a formless collection of particles that tends to expand in all directions. If a gas is not confined, the space between its particles will continue to increase.

To melt a solid, you must add heat energy. The addition of energy causes the particles to move faster. When enough energy has been added, the individual particles break away from their fixed positions. When the particles break away, the material becomes a liquid. If enough energy is added, the particles move even faster and form a gas.

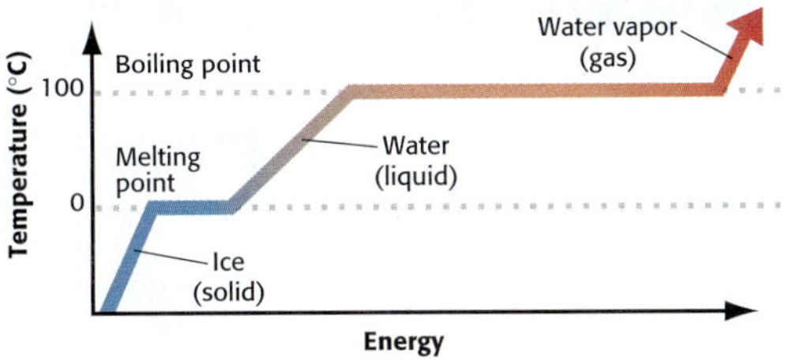

MISCONCEPTION ALERT

Orbitals Students may think of an atom as electrons orbiting the nucleus, much like planets orbiting the sun. Explain that the term *orbital* refers just to the energy level occupied by certain electrons in an atom. Point out that while electrons possess properties similar to those of objects in orbital motion, they do not actually move in a circular or elliptical path around the nucleus.

BRAIN FOOD

One of the earliest hypotheses of chemical bonding was developed by the Swedish chemist Jöns Jacob Berzelius (1779–1848). Berzelius hypothesized that all elements had either a positive or a negative charge and that only positive and negative elements would bond with each other. Berzelius's idea fell short, however, by implying that molecules that contain more than one atom of the same element could not exist because those atoms would repel each other.

Answer to Reading Check

Atoms form chemical bonds by transferring electrons or by sharing electrons.

PHYSICS CONNECTION GENERAL

Energy Needs When energy is added to a substance, either the substance's temperature increases or it changes state. During a temperature increase, the added energy contributes to the kinetic energy, or energy of motion, of the particles. The greater the average kinetic energy of the particles, the higher the temperature of the substance. At a certain temperature, it becomes easier for the forces of attraction between the particles to be overcome than for the kinetic energy of the particles to increase. At this point, the temperature of the substance remains constant and it undergoes a change of state. Have students look up the amounts of energy required for melting and boiling water and lead. Ask them to explain why more energy is needed for these substances to change from a liquid to a gas than from a solid to a liquid. (Much more energy is needed for particles to overcome all attractions between them and enter the gas phase.) LS **Logical**

Teach, *continued*

SKILL BUILDER — Advanced

Math Reinforce the fact that the charge of an ion depends both on whether electrons are added to or removed from the neutral atom and on how many electrons are involved in the transfer. Ions that give up electrons are always positively charged and are denoted by an *n*+ superscript, where *n* is the number of electrons lost. Ions that gain electrons are always negatively charged and are denoted by an *n*– superscript, where *n* is the number of electrons gained. Ask students to apply these rules, along with information from the periodic table, to identify the most stable ions for the following elements: K, Mg, Sr, Br, I, O, Mo, S. (K^{+}, Mg^{2+}, Sr^{2+}, Br^{-}, I^{-}, O^{2-}, Mo^{6+}, S^{2-})
LS Logical

MISCONCEPTION ALERT

Polyatomic Ions Students may think that ions consist of only single atoms. Explain that many ionic compounds are made up of two ions, one of which contains several atoms. Point out that in many cases, the negatively-charged ion consists of atoms that are bound together by covalent bonds, rather than by ionic bonds, but that these groups of atoms as a whole have a negative charge. Examples of polyatomic ions are OH^{-}, NO_3^{-}, CO_3^{2-}, and SO_4^{2-}. Polyatomic ions that are positively charged are less common, although the ammonium ion, NH_4^{+}, appears in some compounds.

ion an atom or molecule that has gained or lost one or more electrons and has a negative or positive charge

ionic bond the attractive force between oppositely charged ions, which form when electrons are transferred from one atom or molecule to another

SciLinks. NSTA. Developed and maintained by the National Science Teachers Association

For a variety of links related to this subject, go to www.scilinks.org

Topic: Covalent and Ionic Bonds
SciLinks code: HQ60362

Ions

When an electron is transferred from one atom to another, both atoms become charged. A particle, such as an atom or molecule, that carries a charge is called an **ion.**

Neutral sodium atoms have 11 electrons. And because a sodium atom has 1 valence electron, sodium is a Group 1 element on the periodic table. If a sodium atom loses its outermost electron, the next 8 electrons in the atom's electron cloud become the outermost electrons. Because the sodium atom now has 8 valence electrons, it is unlikely to share or transfer electrons and therefore, is stable.

However, the sodium atom is missing the 1 electron that was needed to balance the number of protons in the nucleus. When an atom no longer has a balance between positive and negative charges, it becomes an ion. The sodium atom became a positive sodium ion, Na^{+}, when the atom released its valence electron.

Suppose a chlorine atom accepts the electron that the above sodium atom lost. A chlorine atom has a total of 17 electrons, 7 of which are valence electrons. This chlorine atom now has a complete set of 8 valence electrons and is chemically stable. The extra electron, however, changes the neutral chlorine atom into a negatively charged chloride ion, Cl^{-}.

Ionic Bonds

The attractive force between oppositely charged ions that result from the transfer of electrons from one atom to another is called an **ionic bond.** A compound that forms through the transfer of electrons is called an *ionic compound*. Most ionic compounds form when electrons are transferred between the atoms of metallic and nonmetallic elements.

Sodium chloride, or common table salt, is an ionic compound. The positively charged sodium ions and negatively charged chloride ions attract one another because of their opposite charges. This attraction between the positive sodium ions and the negative chloride ions is an ionic bond. The attraction creates cube-shaped crystals, such as the table salt shown in **Figure 4.**

Figure 4 ▼

Ionic Bonds

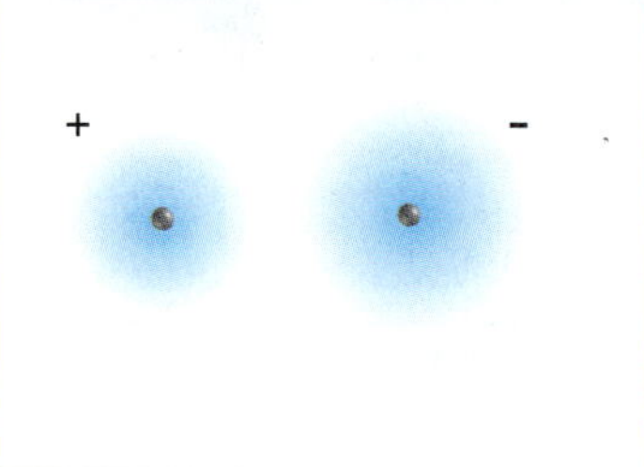

A sodium atom, Na, transfers an electron to a chlorine atom, Cl, to form table salt, NaCl. The chlorine atom with the added electron becomes a negatively charged ion, Cl^{-}, and the sodium atom, which lost the electron, becomes a positively charged ion, Na^{+}. The oppositely charged ions attract each other and form an ionic bond.

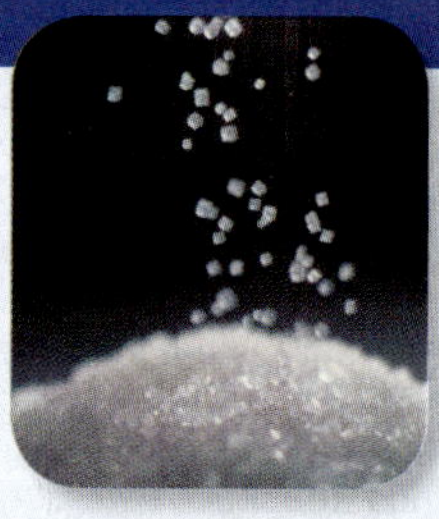

Table salt, NaCl

Homework — Advanced

Atomic and Ionic Radius The radii of atoms vary with the number of electrons in the atom's outer energy levels. For neutral atoms in a single period of the periodic table, the radius decreases as the number of electrons in the highest energy level increases. The atomic radii also increase as more high energy levels are filled.

The radius of a neutral atom is altered when that atom gives up or gains electrons to become an ion. Have students research atomic and ionic radii for a single period (row) of the periodic table. Have students then collect their findings and present them in a visual form so that comparisons between the radii of neutral and ionized atoms can be made. One way of doing this is to create a poster showing the same row of the periodic table twice, with one figure showing the radii for the neutral atoms and the other figure showing the radii for the ions. Three-dimensional or computer models can also be constructed for this purpose. **LS Kinesthetic**

Figure 5 ▼

Covalent Bonds

Covalent bonds form when atoms share one or more pairs of electrons. In a covalent compound, pairs of electrons are shared between most of the atoms that make up the compound. Chlorine gas is an example of a covalent compound in which pairs of chlorine atoms are bonded together by sharing electrons.

Chlorine gas, Cl_2

Covalent Bonds

A bond that is formed by the attraction between atoms that share electrons is a **covalent bond.** When atoms share electrons, the positive nucleus of each atom is attracted to the shared negative electrons, as shown in **Figure 5.** The pull between the positive and negative charges is the force that keeps these atoms joined.

Water is an example of a *covalent compound*—that is, a compound formed by the sharing of electrons. Two hydrogen atoms can share their single valence electrons with an oxygen atom that has six valence electrons. The sharing of electrons creates a bond and gives oxygen a stable number of eight outermost electrons. At the same time, the oxygen atom shares two of its electrons—one for each hydrogen atom—which gives each hydrogen atom a more stable number of two electrons. Thus, a water molecule consists of two atoms of hydrogen combined with one atom of oxygen.

covalent bond a bond formed when atoms share one or more pairs of electrons

Polar Covalent Bonds

In many cases, atoms that are covalently bonded do not equally share electrons. The reason for this is that the ability of atoms of some elements to attract electrons from atoms of other elements differs. A covalent bond in which the bonded atoms have an unequal attraction for the shared electrons is called a *polar covalent bond*. Water is an example of a molecule that forms as a result of polar covalent bonds. Two hydrogen atoms bond covalently with an oxygen atom and form a water molecule. Because the oxygen atom has more ability to attract electrons than the hydrogen atoms do, the electrons are not shared equally between the oxygen and hydrogen atoms. Instead, the electrons remain closer to the oxygen nucleus, which has the greater pull. As a result, the water molecule as a whole has a slightly negative charge at its oxygen end and slightly positive charges at its hydrogen ends. The slightly positive ends of a water molecule attract the slightly negative ends of other water molecules.

Reading Check Why do water molecules form from polar covalent bonds? (See the Appendix for answers to Reading Checks.)

QuickLAB — 10 min

Compounds

Procedure

1. Place **4 g of compound A** in a **clear plastic cup.**
2. Place **4 g of compound B** in a **second clear plastic cup.**
3. Observe the color and texture of each compound. Record your observations.
4. Add **5 mL of vinegar** to each cup. Record your observations.

Analysis

1. What physical and chemical differences between the two compounds did you record? How do physical properties and chemical properties differ?
2. Vinegar reacts with baking soda but not with powdered sugar. Which of these compounds is compound A, and which is compound B?

Answer to Reading Check

The oxygen atom has a larger and more positively charged nucleus than the hydrogen atoms do. As a result, the oxygen nucleus pulls the electrons from the hydrogen atoms closer to it than the hydrogen nuclei pull the shared electrons from the oxygen. This unequal attraction forms a polar-covalent bond.

Close

Reteaching — Basic

Types of Compounds Write the following compound formulas on the board, and have students identify them as being ionic or covalent: CO_2, NaCl, LiBr, H_2, MgF_2, NH_3. (covalent, ionic, ionic, covalent, ionic, covalent) **LS Visual**

Quiz — General

1. What type of compound contains elements that share electrons to form bonds? (a covalent compound)
2. What is the charge on the most stable magnesium ion? (2+)
3. What is the charge on the most stable fluorine ion? (1–)

CHAPTER RESOURCES

Chapter Resource File

- **Datasheet for Quick Lab** General
- **Inquiry Lab** Ionic and Covalent Conductivity General

Technology

Transparencies

- 19 Ionic Bonds and Covalent Bonds (with worksheet)

QuickLAB

Skills Acquired

- Experimenting
- Observing
- Inferring
- Interpreting
- Identifying/Recognizing Patterns

Teacher's Notes Compound A is powdered sugar. Compound B is baking soda. Be sure that students follow the safety precautions and clean up quickly and completely.

Answers

1. Answers may vary but physical differences should include that, when moistened, compound A is sticky, whereas compound B is not. Chemical properties include that compound B bubbles when it reacts with vinegar, whereas compound A does not. Chemical properties are properties that change as a result of chemical reactions.
2. Compound B is baking soda.

Close, *continued*

Alternative Assessment — ADVANCED

Balancing Chemical Equations
Have students select, or select for them, several unbalanced chemical reactions. Then, have students show the steps to finding the coefficients to balance the equations. Be sure all numerical steps are shown. LS Logical

Answers to Section Review

1. A compound is a substance made up of atoms of two or more different elements joined by chemical bonds. A molecule is the smallest unit of a substance that consists of two or more atoms and that keeps all of the substance's chemical properties.
2. 6 carbon (C) atoms, 12 hydrogen (H) atoms, 6 oxygen (O) atoms
3. The electrical attraction between the nucleus of an atom and the outermost electrons of the other atom causes the atoms to bond, forming a molecule.
4. Ionic bonds form between oppositely charged ions. Covalent bonds form from the attraction between atoms that share electrons.
5. The greater positive charge of the oxygen nucleus exerts a stronger attraction on the electrons of the hydrogen atom than the hydrogen nuclei exert on the electrons of the oxygen atom.
6. A mixture contains two or more substances that are not chemically combined. A compound consists of several atoms that combine in exact proportions to form a substance that has unique chemical properties.
7. sea water (NaCl in solution) and alloys such as steel
8. The substance does not chemically combine when it forms a mixture, so its chemical properties remain the same.
9. The solution would be a homogeneous mixture, and so the composition of the solution would be the same throughout. This would not necessarily be true of the other mixture.
10. Two or more atoms combine to form a *molecule*, which is the basic unit for a *compound*, which may be either an *ionic compound*, in which each element is a charged *ion*, and is held to the other ion by an *ionic bond*, or a *covalent compound*, in which atoms share electrons to form a *covalent bond*.

Figure 6 ▶ The steel that is being smelted in this steel mill in Iowa is a solution of iron, carbon, and various other metals such as nickel, chromium, and manganese.

mixture a combination of two or more substances that are not chemically combined

solution a homogeneous mixture of two or more substances that are uniformly dispersed throughout the mixture

Mixtures

On Earth, elements and compounds are generally mixed together. A **mixture** is a combination of two or more substances that are not chemically combined. The substances that make up a mixture keep their individual properties. Therefore, unlike a compound, a mixture can be separated into its parts by physical means. For example, you can use a magnet to separate a mixture of powdered sulfur, S, and iron, Fe, filings. The magnet will attract the iron, which is magnetic, and leave behind the sulfur, which is not magnetic.

Heterogeneous Mixtures

Mixtures in which two or more substances are not uniformly distributed are called *heterogeneous mixtures*. For example, the igneous rock granite is a heterogeneous mixture of crystals of the minerals quartz, feldspar, hornblende, and biotite.

Homogeneous Mixtures

In chemistry, the word *homogeneous* means "having the same composition and properties throughout." A homogeneous mixture of two or more substances that are uniformly dispersed throughout the mixture is a **solution.**

Any part of a given sample of the solution known as sea water, for example, will have the same composition. Sodium chloride, NaCl, (along with many other ionic compounds) is dissolved in sea water. The positive ends of water molecules attract negative chloride ions. And the negative end of water molecules attracts positive sodium ions. Eventually, all of the sodium and chloride ions become uniformly distributed among the water molecules.

Gases and solids can also be solutions. An *alloy* is a solution composed of two or more metals. The steel shown in **Figure 6** is an example of such a solution.

Section 2 Review

1. **Define** *compound* and *molecule.*
2. **Determine** the number of each type of atom in the following chemical formula: $C_6H_{12}O_6$.
3. **Explain** why atoms join to form molecules.
4. **Describe** the difference between ionic and covalent bonds.
5. **Explain** why a water molecule has polar covalent bonds.
6. **Compare** compounds with mixtures.
7. **Identify** two common solutions.

CRITICAL THINKING

8. **Applying Ideas** What happens to the chemical properties of a substance when it becomes part of a mixture?
9. **Evaluating Data** If you were given two mixtures and told that one is a solution, how might you determine which one is the solution?

CONCEPT MAPPING

10. Use the following terms to create a concept map: *compound, molecule, ionic compound, ionic bond, ion, covalent compound,* and *covalent bond.*

CHAPTER RESOURCES

Chapter Resource File
- Section Quiz GENERAL

Workbooks
- Study Guide (also in Spanish)

Chapter 4 Highlights

Sections

1 Matter

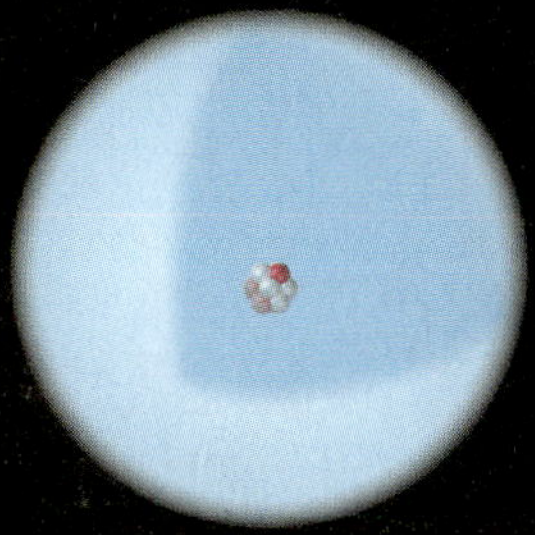

Key Terms

matter, 81
element, 81
atom, 82
proton, 82
electron, 82
neutron, 82
isotope, 83

Key Concepts

- An element is a substance that has a characteristic set of physical and chemical properties and that cannot be separated into simpler substances by chemical means.
- An atom consists of electrons surrounding a nucleus that is made up of protons and neutrons.
- The atomic number of an atom is equal to the number of protons in the atom. The mass number is equal to the sum of the protons and neutrons in the atom.
- Isotopes are atoms that have different numbers of neutrons than other atoms of the same element do.
- Elements on the periodic table are arranged in groups that are based on similarities in the chemical properties of the elements.

2 Combinations of Atoms

Key Terms

compound, 87
molecule, 87
ion, 90
ionic bond, 90
covalent bond, 91
mixture, 92
solution, 92

Key Concepts

- A compound is a substance that is made up of two or more different elements that are joined by chemical bonds between the atoms of those elements.
- A chemical formula describes which elements and how many atoms of each of those elements make up a molecule of a compound.
- Chemical bonds between atoms form when electrons are shared or transferred between the atoms.
- A mixture consists of two or more substances that are not chemically bonded. A solution is a mixture in which one substance is uniformly dispersed in another substance.

Chapter Highlights

Alternative Assessment — GENERAL

Poster Project Have students create a poster that shows the processes by which atoms bond together ionically and covalently. The poster should also depict how polar-covalent bonds form. Student illustrations should clearly show the basic nuclear structure of atoms and the correct number of electrons around the nucleus. Students should cover all relevant details in the poster. LS Logical

CHAPTER RESOURCES

Chapter Resource File

- Concept Review GENERAL
- Critical Thinking ADVANCED
- Math Skills GENERAL
- Graphing Skills GENERAL
- Chapter Test A GENERAL
- Chapter Test B ADVANCED

Workbooks

- **Study Guide** (also in Spanish)
- **Assessments** (Spanish)

Technology

Classroom Videos
- Brain Food Video Quiz

HRW Earth Science Video
- Segment 7: Chemical Elements

Chapter Review

Assignment Guide

Section	Questions
1	1, 2, 6, 9–14, 18–21, 23–24, 33–36
2	3, 7–8, 15–17, 22, 25–27, 31–32
1 and 2	4–5, 28–30

Using Key Terms

1–8. Answers may vary but should show that students understand the definitions of and differences between key terms.

Understanding Key Concepts

9. a **10.** a
11. c **12.** d
13. c **14.** d
15. c **16.** b

Short Answer

17. Chemical properties are characteristics that describe how a substance interacts with other substances; physical properties are characteristics that can be observed without changing the composition of a substance.

18. An element is a substance that cannot be broken down into a simpler substance by chemical means.

19. protons, neutrons, and electrons

20. Metallic elements tend to lose electrons easily because they have only 1, 2, or 3 valence electrons. Nonmetallic elements tend to gain electrons easily because they have 4–7 valence electrons.

21. covalent

22. NaCl contains one sodium (Na) atom and one chlorine (Cl) atom.
H_2O_2 contains two hydrogen (H) atoms and two oxygen (O) atoms.
Fe_3O_4 contains three iron (Fe) atoms and four oxygen (O) atoms.
SiO_2 contains one silicon (Si) atom and two oxygen (O) atoms.

23. The number of protons in an atom's nucleus distinguishes it from atoms of other elements.

Critical Thinking

24. If a diatomic molecule consists of atoms of one element, they will likely be bonded covalently, because both atoms are likely to have the same electrical charge.

25. Calcium chloride forms when electrons from a calcium atom transfer to two chlorine atoms, to form a calcium ion that has a +2 positive charge and two chloride ions that each have a –1 charge.

26. The chemical properties of a substance are unchanged by becoming part of a mixture, because the substance does not chemically react with the other substances in the mixture.

27. Water forms because of the chemical properties of oxygen and hydrogen.

Chapter 4 Review

Using Key Terms

Use each of the following terms in a separate sentence.

1. *matter*
2. *neutron*
3. *ion*

For each pair of terms, explain how the meanings of the terms differ.

4. *atom* and *molecule*
5. *element* and *compound*
6. *proton* and *electron*
7. *compound* and *mixture*
8. *covalent bond* and *ionic bond*

Understanding Key Concepts

9. Color and hardness are examples of a substance's
a. physical properties.
b. chemical properties.
c. atomic structure.
d. molecular properties.

10. Subatomic particles in atoms that do not carry an electric charge are called
a. neutrons. **c.** nuclei.
b. protons. **d.** ions.

11. Atoms of the same element that differ in mass are
a. ions. **c.** isotopes.
b. neutrons. **d.** molecules.

12. A combination of letters and numbers that indicates which elements make up a compound is a
a. coefficient.
b. reactant.
c. chemical bond.
d. chemical formula.

13. The type of chemical bond that forms between oppositely charged ions is a(n)
a. covalent bond.
b. mixture.
c. ionic bond.
d. solution.

14. Two or more elements whose atoms are chemically bonded form a(n)
a. mixture. **c.** nucleus.
b. ion. **d.** compound.

15. The outermost electrons in an atom's electron cloud are called
a. ions.
b. isotopes.
c. valence electrons.
d. neutrons.

16. A molecule of water, or H_2O, has one atom of
a. hydrogen. **c.** helium.
b. oxygen. **d.** osmium.

Short Answer

17. How do chemical properties differ from physical properties?

18. Define the term *element*.

19. Name three basic subatomic particles.

20. In terms of valence electrons, what is the difference between metallic elements and nonmetallic elements?

21. Which type of bonding includes the sharing of electrons?

22. Using the periodic table to help you to understand the following chemical formulas, list the name and number of atoms of each element in each compound: NaCl, H_2O_2, Fe_3O_4, and SiO_2.

23. What sets an atom of one element apart from the atoms of all other elements?

Critical Thinking

24. **Applying Ideas** Is a diatomic molecule more likely to be held together by a covalent bond or by an ionic bond? Explain your answer.

25. **Making Inferences** Calcium chloride is an ionic compound. Carbon dioxide is a covalent compound. Which of these compounds forms as a result of the transfer of electrons from one atom to another? Explain your answer.

26. **Analyzing Relationships** What happens to the chemical properties of a substance when it becomes part of a mixture? Explain your answer.

27. **Classifying Information** Oxygen combines with hydrogen to form water. Is this process due to the physical or chemical properties of oxygen and hydrogen?

Concept Mapping

28. Use the following terms to create a concept map: *chemical property, element, atom, electron, proton, neutron, atomic number, mass number, isotope, compound,* and *chemical bond*.

Math Skills

29. **Making Calculations** How many neutrons does a potassium atom have if its atomic number is 19 and its mass number is 39?

30. **Using Formulas** The covalent compound formaldehyde forms when one carbon atom, two hydrogen atoms, and one oxygen atom bond. Write a chemical formula for formaldehyde.

31. **Balancing Equations** Zinc metal, Zn, will react with hydrochloric acid, HCl, to produce hydrogen gas, H_2, and zinc chloride, $ZnCl_2$. Write and balance the chemical equation for this reaction.

Writing Skills

32. **Organizing Data** Choose one of the eight most common elements in Earth's crust, and write a brief report on the element's atomic structure, its chemical properties, and its economic importance.

33. **Communicating Main Ideas** Write a brief essay that describes how an element's chemical properties determine what other elements are likely to bond with that element.

34. **Making Comparisons** Write a brief essay that describes the differences between atoms, elements, ions, and isotopes.

Interpreting Graphics

The table below shows the average atomic masses and the atomic numbers of five elements. Use this table to answer the questions that follow.

Element	Symbol	Average atomic mass	Atomic number
Magnesium	Mg	24.3050	12
Tungsten	W	183.84	74
Copper	Cu	63.546	29
Silicon	Si	28.0855	14
Bromine	Br	79.904	35

35. If an atom of magnesium has 12 neutrons, what is its mass number?

36. Estimate the average number of neutrons in an atom of tungsten and in an atom of silicon.

37. Which element has the largest difference between the number of protons and the number of neutrons in the nucleus of one of its atoms? Explain your answer.

Chapter Review

Concept Mapping

28. Answers may vary but should include all of the terms listed. Sample answers appear at the end of this Teacher Edition.

Math Skills

29. 39 – 19 = 20 neutrons

30. CH_2O

31. $Zn + HCl \rightarrow H_2 + ZnCl_2$;

one Zn atom on each side of equation; HCl must be multiplied by 2 to give two Cl atoms and 2 H atoms on each side.

$Zn + 2HCl \rightarrow H_2 + ZnCl_2$

one Zn atom, two Cl atoms, and two H atoms on each side of equation; equation is balanced

Writing Skills

32. Answers may vary. Accept all reasonable answers.

33. Answers may vary. Accept all reasonable answers.

34. Answers may vary. Accept all reasonable answers.

Interpreting Graphics

35. Mass number = number of protons (atomic number) + number of neutrons;

for Mg: atomic number = 12;

mass number for Mg = 12 + 12 = 24

36. Number of neutrons = mass number – atomic number;

for W: 184 – 74 = 110 neutrons;

for Si: 28 – 14 = 14 neutrons

37. tungsten; The average atomic mass is close to the mass number for the element, and the difference between the average atomic mass and the atomic number, which is a measure of the number of neutrons, is greatest for tungsten (110 neutrons – 74 protons = 36). The other elements have differences between the two types of particles of less than 10.

Estimated Time

To give students practice under more realistic testing conditions, allow them 30 minutes to answer all of the questions in this practice test.

TEST DOCTOR

Question 1 Answer C is correct. Students should show an understanding that a mixture is a combination of two or more substances that are not chemically combined. Within a soil mixture, students may cite examples of elements and compounds, but these are usually isolated parts of a more comprehensive soil mixture that will be made of many types of compounds. Therefore, answers A, B, and D are incorrect.

Question 9 The correct answer is 12. Students should apply the rules explained in the passage about the placement of subscripts in order to correctly answer this question. Sulfur has two atoms so it has a subscript of 2 below and to the right of its chemical symbol. Flouride has 10 atoms so it has a subscript of 10 below and to the right of its chemical symbol. Students should add these numbers together to get the total number of atoms, 12, in one molecule of S_2F_{10}.

Chapter 4 Standardized Test Prep

Understanding Concepts

Directions (1–4): **For *each* question, write on a separate sheet of paper the number of the correct answer.**

1 Soil is an example of
- **A.** a solution
- **B.** a compound
- **C.** a mixture
- **D.** an element

2 Isotopes are atoms of the same element that has different mass numbers. This difference is caused by
- **F.** a different number of electrons in the atoms
- **G.** a different number of protons in the atoms
- **H.** a different number of neutrons in the atoms
- **I.** a different number of nuclei in the atoms

3 Which of the following statements best describes the charges of subatomic particles?
- **A.** Electrons have a negative charge, protons have a positive charge, and neutrons have no charge.
- **B.** Electrons have a positive charge, protons have a negative charge, and neutrons have a positive charge.
- **C.** Electrons have no charge, protons have a positive charge, and neutrons have a negative charge.
- **D.** In neutral atoms, protons, neutrons, and electrons have no charges.

4 An element is located on the periodic table according to
- **F.** when the element was discovered
- **G.** the letters of the element's chemical symbol
- **H.** the element's chemical name
- **I.** the element's physical and chemical properties

Directions (5–6): **For *each* question, write a short response.**

5 What is the name for an atom that has gained or lost one or more electrons and has acquired a charge?

6 Scientists use atomic numbers to help identify the atoms of different elements. How is the atomic number of an element determined?

Reading Skills

Directions (7–9): **Read the passage below. Then, answer the questions.**

Chemical Formulas

All substances can be formed by a combination of elements from a list of about 100 possible elements. Each element has a chemical symbol. A chemical formula is shorthand notation that uses chemical symbols and numbers to represent a substance. A chemical formula shows the amount of each kind of atom present in a specific molecule of a substance.

The chemical formula for water is H_2O. This formula tells you that one water molecule is composed of two atoms of hydrogen and one atom of oxygen. The 2 in the formula is a subscript. A subscript is a number written below and to the right of a chemical symbol in a formula. When a symbol, such as the O for oxygen in water's formula, has no subscript, only one atom of that element is present.

7 What does a subscript in a chemical formula represent?
- **A.** Subscripts represent the number of atoms of the chemical symbol they directly follow present in the molecule.
- **B.** Subscripts represent the number of atoms of the chemical symbol they directly precede present in the molecule.
- **C.** Subscripts represent the number of protons present in each atom's nucleus.
- **D.** Subscripts represent the total number of atoms present in a molecule.

8 Which of the following statements can be inferred from the information in the passage?
- **F.** Two atoms of hydrogen are always present in chemical formulas.
- **G.** A chemical formula indicates the elements that a molecule is made of.
- **H.** Chemical formulas can be used only to show simple molecules.
- **I.** No more than one atom of oxygen can be present in a chemical formula.

9 How many atoms would be found in a single molecule that has the chemical formula S_2F_{10}?

Answers

Understanding Concepts

1. C
2. H
3. A
4. I
5. an ion
6. by the number of protons in the atom

Reading Skills

7. A
8. G
9. 12

Interpreting Graphics

10. A
11. I
12. 2
13. B
14. Answers may vary. See Test Doctor for a detailed scoring rubric.

Interpreting Graphics

Directions (10–14): **For *each* question below, record the correct answer on a separate sheet of paper.**

The graphic below shows the upper right segment of the periodic table. Use this graphic to answer questions 10 through 12.

Segment of the Periodic Table

13	14	15	16	17	18
					2 He 4.00
5 B 10.81	6 C 12.01	7 N 14.01	8 O 16.00	9 F 19.00	10 Ne 20.18
13 Al 26.98	14 Si 28.09	15 P 30.97	16 S 32.07	17 Cl 35.45	18 Ar 39.95

10 Which pair of elements would most likely have a similar arrangement of outer electrons and have similar chemical behaviors?
- **A.** boron and aluminum
- **B.** helium and fluoride
- **C.** carbon and nitrogen
- **D.** chlorine and oxygen

11 What is the atomic mass of helium?
- **F.** 0.18
- **G.** 0.26
- **H.** 2.00
- **I.** 4.00

12 How many neutrons does the average helium atom contain?

The graphic below shows matter in three different states. Use this graphic to answer questions 13 and 14.

States of Matter

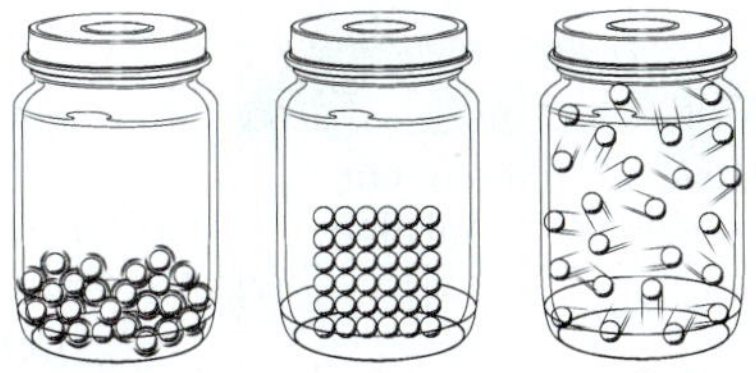

A B C

13 In what physical state is the matter in jar A?
- **A.** solid
- **B.** liquid
- **C.** gas
- **D.** plasma

14 Explain how the positions and motions of particles determine the characteristics of each state of matter.

Test TIP

Double check (with a calculator, if permitted) all mathematical computations involved in answering a question.

TEST DOCTOR

Question 12 The correct answer is 2. Students should understand that both protons and neutrons contribute equally to a neutral atom's mass, while electrons contribute little. Helium has an atomic mass of 4.00. This number is usually divided equally among protons and neutrons. Therefore, a helium atom contains two protons and two neutrons. Some students may answer 4. These students are most likely neglecting the role of neutrons in determining atomic mass.

Question 14 Full-credit answers should include the following points:
- students should demonstrate a conceptual understanding of the different properties of the three primary states of matter
- jar A has a compact, but unordered structure that fills the space in the bottom of the jar. Students should recognize this as a model of a liquid
- jar B has a definite and self-contained structure. The atoms are held in close, fixed positions. Students should recognize this as a model of solid matter
- jar C shows fast-moving atoms that have no structure and that are filling the entire space in the jar. The particles have no relationship with one another, and if the jar were not present, the distance between the particles would expand indefinitely. Students should recognize this as matter in a gaseous state

Test Prep Correlations *National Science Education Standards*

PS 1a: items 2, 3, 5, 6, 11, 12
PS 1b: items 2, 3, 6, 11, 12
PS 2a: items 4, 5, 10
PS 2b: items 4, 6, 10, 11, 12
PS 2d: item 1
PS 2e: items 13, 14
SAI 2d: items 9, 11, 12

CHAPTER RESOURCES

State Resources

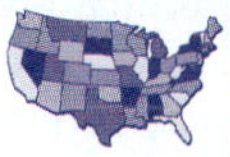

For specific resources for your state, visit go.hrw.com and type in the keyword **HSHSTR.**

Inquiry Lab

Physical Properties of Elements

Teacher's Notes

Time Required

two 45-minute class periods

Lab Ratings

EASY ——→ HARD

TEACHER PREPARATION ▲▲

STUDENT SETUP ▲▲▲

CONCEPT LEVEL ▲▲▲

CLEANUP ▲▲

Skills Acquired

- Collecting Data
- Measuring
- Organizing and Analyzing Data
- Designing Experiments
- Interpreting
- Identifying and Recognizing Patterns

The Scientific Method

In this lab, students will

- Ask a Question
- Test the Hypothesis
- Make Observations
- Analyze the Results
- Draw Conclusions

Materials

The materials listed on the page are enough for groups of two or three students. If students decide to test the heat conductivity of the metal, a hot-water bath should be set up in advance for students to use.

Chapter 4

Inquiry Lab

Using Scientific Methods

Objectives

- **Design** an experiment to test the physical properties of different metals.
- **Identify** unknown metals by comparing the data you collect and reference information.

Materials

- balance
- beakers (several)
- graduated cylinder
- hot plate
- magnet
- metal samples, unknown, similarly shaped (several)
- ruler, metric
- stopwatch
- water
- wax

Safety

Physical Properties of Elements

In this lab, you will identify samples of various metals by collecting data and comparing the data with the reference information listed in the table below. Use at least two of the physical properties listed in the table to identify each metal.

ASK A QUESTION

1. How can you use an element's physical properties to identify the element?

FORM A HYPOTHESIS

2. Use the table below to identify which physical properties you will test. Write a few sentences that describe your hypothesis and the procedure you will use to test those physical properties.

TEST THE HYPOTHESIS

3. With your lab partner(s), decide how you will use the available materials to identify each metal that you are given. Because there are many ways to measure some of the physical properties that are listed in the table below, you may need to use only some of the materials that are provided.
4. Before you start to test your hypothesis, list each step that you will need to perform.
5. After your teacher approves your plan, perform your experiment. Keep in mind that the more exact your measurements are, the easier it will be for you to identify the metals that you have been provided.
6. Record all the data that you collect and any observations that you make.

Physical Properties of Some Metals

Metal	Density (g/cm^3)	Relative hardness	Relative heat conductivity	Magnetic attraction
Aluminum, Al	2.7	28	100	No
Iron, Fe	7.9	50	34	Yes
Nickel, Ni	8.9	67	38	Yes
Tin, Sn	7.3	19	28	No
Tungsten, W	19.3	100	73	No
Zinc, Zn	7.1	28	49	No

Tips and Tricks

Students should use the same methods for measuring each sample. Students should recalibrate the balance after each weighing.

A computer component could be added to this lab by using probes to measure heat conductivity.

You may want to allow students to try brief experiments that they create, and then give students specific directions if they do not appear to have identified appropriate testing techniques. Students may encounter error in magnetism of nickel if the lab magnet is weak. Students may also encounter error in density measurements for very small volumes of metal.

CHAPTER RESOURCES

Chapter Resource File

- Datasheet for Chapter Lab GENERAL
- Lab Notes and Answers

How to Measure Physical Properties of Metals

Physical property	Description	How to measure the property
Density	mass per unit volume	If the metal is box-shaped, measure its length, height, and width, and then use these measurements to calculate the metal's volume. If the shape of the metal is irregular, add the metal to a known volume of water and determine what volume of water is displaced.
Relative hardness	how easy it is to scratch the substance	An object that has a high hardness value can scratch an object that has a lower value, but not vice versa.
Relative heat conductivity	how quickly a metal heats or cools	A metal that has a value of 100 will heat or cool twice as quickly as a metal that has a value of 50.
Magnetism	whether an object is magnetic	If a magnet placed near a metal attracts the metal, the metal is magnetic.

ANALYZE THE RESULTS

1. **Summarizing Data** Make a table that lists which physical properties you compared and what data you collected for each of the metals that you tested.
2. **Making Comparisons** Which physical properties were the easiest for you to measure and compare? Which were the most difficult to measure and compare? Explain why.
3. **Applying Ideas** What would happen if you tried to use zinc to scratch aluminum?

DRAW CONCLUSIONS

4. **Summarizing Results** Which metals were given to you? Explain how you identified each metal.
5. **Analyzing Methods** Explain why you would have difficulty distinguishing between iron and nickel unless you were to measure each metal's density.

Extension

1. **Evaluating Data** Suppose you find a metal fastener that has a density of 7 g/cm^3. What are two ways to determine whether the unknown metal is tin or zinc?

Answers to Analyze the Results

1. Answers may vary but should be consistent with the physical properties of metals.
2. Answers may vary. However, magnetic properties should be easiest to test, followed by densities, which could be directly calculated. Hardness assessment calls for subjective judgments between the various metals tested. Heat conductivity is the most difficult, given the comparative nature of the measurements and the numerous variables, such as temperature, time, and mass of water used for cooling, that need to be controlled.
3. Because aluminum and zinc have the same hardness, neither metal would be scratched when rubbed against the other.

Answers to Draw Conclusions

4. Answers may vary but should be consistent with the materials provided to each student or group.
5. Because iron and nickel are both magnetic and have nearly the same hardness, only their densities make them easily distinguishable.

Answers to Extension

1. By testing the hardness of the fastener against a sample of zinc and a sample of tin or by comparing the fastener's heat conductivity with that of the two samples, the identity of the metal could be determined.

Shawn Beightol, M.S. Ed.
Dr. Michael Krop Senior High School
Miami, FL

Answers to Map Skills Activity

1. Answers may vary but should correspond to the symbols shown within the borders of the correct state.
2. Based on the density of symbols that represent gold, students may think the gold rushes occurred in NV (Nevada) and AK (Alaska). One gold rush did occur in Alaska, but the other occurred in California. This discrepancy can be explained by considering that the symbols indicate the locations of deposits, not the value of the deposits, and that the locations of the Nevada deposits may not have been known to people in the 1800s.
3. no; The map indicates only the existence of deposits, not their size or the amount of each element that is extracted.
4. Silver recovery would be greatest in the western states that have more deposits of the other elements, including New Mexico, Arizona, California, Nevada, Washington, Idaho, Montana, Wyoming, Utah, Colorado, and Alaska.
5. Lead and zinc deposits are commonly, though not always, close together. Lead is rarely far from a zinc deposit, though zinc deposits exist in some areas, such as KY and TN, without any nearby lead deposits.

CHAPTER RESOURCES

Technology

Transparencies
- 20 Element Resources in the United States (with worksheet)

MAPS in Action

Element Resources in the United States

Map Skills Activity

This map shows the distribution of elements that are used as resources in the United States. Use the map to answer the questions below.

1. **Using a Key** Use the map to locate the state in which you live. Are any elements from the key found in your state?
2. **Analyzing Relationships** Can you identify any relationship between the locations of lead deposits and the locations of zinc deposits? Explain your reasoning.
3. **Inferring Relationships** Is it possible to use this map to find out which states have the highest production of the mineral resources that are listed on the map? Why or why not?
4. **Inferring Relationships** Silver is mainly recovered as a byproduct of the bulk mining of other metals such as copper, lead, zinc, and gold. Where in the United States do you think silver recovery would happen?
5. **Using a Key** Use the map to identify the states where uranium can be found.

SCIENCE AND TECHNOLOGY

The Smallest Particles

Matter is made of atoms. For a long time, scientists thought that atoms were the smallest particles of matter. Then, when protons, neutrons, and electrons were discovered, these particles were thought to be the smallest particles. Now, the honor goes to particles called *quarks* and *leptons*. But recent evidence suggests that quarks may be composed of smaller particles.

Energy to See

A fundamental principle of physics is that the smaller an object is, the greater the amount of energy is required to see it. Because the amount of energy needed to see subatomic particles does not exist naturally on Earth, scientists have to build up enough energy to break apart these particles so that individual parts can be isolated for study. Tevatron, a huge particle accelerator shown at right, is an underground, ring-shaped tunnel that has a 6.4 km circumference. It contains 1,000 superconducting magnets that move beams of particles at increasingly higher speeds. As the particles gain speed, they build up energy.

▶ Particle accelerators, such as this one in Illinois, are so expensive to build and operate that they require international cooperation.

When the energized particles are moving near the speed of light (299,792,458 m/s), they are directed to hit either a fixed target or particles moving in an opposite direction. On impact, the particles split. The byproducts of the particles separate and scatter. The smaller particles, including quarks and leptons, are measured by a collider detector.

Applying the Results

For now, the nature of subatomic particles may seem far removed from Earth science. However, research on the most basic particles of matter may one day influence the work of Earth scientists by changing basic theories that range from the ultimate structure of matter to the origin of the universe.

Extension

1. **Research** Find out more about quarks and leptons. Then, write a short report that describes their characteristics and where they are found in an atom.

▼ This image shows the paths of the subatomic particles created by collisions in a particle accelerator. The shape and location of each path indicates the type of particle that formed the path.

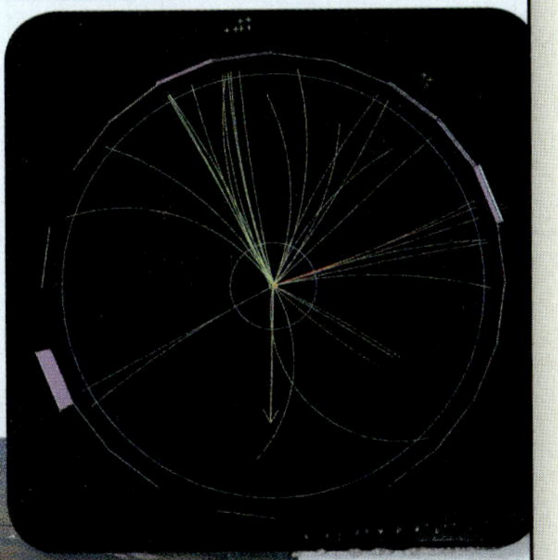

Science and Technology

The Smallest Particles

Internet Activity — ADVANCED

High-Energy Physics Have students use the Internet to research the field of high-energy physics and the various types of particle accelerators used throughout the world to study subatomic particles. Students may check the Web sites of various facilities, such as the Stanford Linear Accelerator Center (SLAC) in California, Fermilab in Illinois, and CERN in Switzerland. Have students report their findings in a written report, oral presentation, or poster presentation. A worksheet designed to direct student research on this topic can be found in the **Chapter Resource File** booklet or by visiting go.hrw.com and entering the keyword **HQ6CHMX.** LS **Verbal**

CHAPTER RESOURCES

Chapter Resource File

Internet Activity
- High-Energy Physics ADVANCED

Answers to Extension

1. Answers may vary. Accept all reasonable answers. Students should note that protons and neutrons are made up of even smaller particles, called *quarks*. Quarks have six properties, referred to as "flavors," by which they are identified, and they carry one-third of an elementary electric charge. Students should also recognize that an electron is an example of a lepton.

Chapter 5

Minerals of Earth's Crust

Planning Guide

Compression Guide

To shorten instruction because of time limitations, omit the Chapter Lab.

OBJECTIVES	LABS, DEMONSTRATIONS, AND ACTIVITIES	TECHNOLOGY RESOURCES
PACING • 45 min pp. 102–108 **Chapter Opener**		**OSP** **Parent Letter** ■ **CD** **Student Edition on CD-ROM** **CD** **Chapter Summaries Audio CD** ■ **VID** **Brain Food Video Quiz**
Section 1 What Is a Mineral? • Define *mineral.* • Compare the two main groups of minerals. • Identify the six types of silicate crystalline structures. • Describe three common nonsilicate crystalline structures.	**TE** **Group Activity** Mineral—Yes or No?, p. 103 ◆ GENERAL **TE** **Activity** Acid Test, p. 105 ◆ BASIC **SE** **Quick Lab** Modeling Tetrahedra, p. 106 GENERAL **CRF** **Datasheet for Quick Lab*** GENERAL **TE** **Group Activity** Growing Crystals, p. 106 ◆ BASIC **SE** **Maps in Action** Rock and Mineral Production in the United States, p. 122 GENERAL **CRF** **Skills Practice Lab** Copper Recovery* ◆ GENERAL	**OSP** **Lesson Plans** (also in print) **TR** **Bellringer*** **TR** 21 Characteristics of Minerals* **TR** 22 Structures of Silicate Minerals* **TR** 25 Rock and Mineral Production in the United States* **TE** **Internet Activity** Mineral Formation, p. 104 ADVANCED **CRF** **Internet Activity** Mineral Formation* ADVANCED **TE** **Internet Activity** Mining Impacts, p. 122 GENERAL **CRF** **Internet Activity** Mining Impacts* GENERAL **VID** **HRW Earth Science Video** Minerals of the Earth's Crust
PACING • 90 min pp. 109–114 **Section 2 Identifying Minerals** • Describe seven physical properties that help distinguish one mineral from another. • List five special properties that may help identify certain minerals.	**TE** **Demonstration** The Same, But Different, p. 109 ◆ GENERAL **TE** **Activity** Those Are the Breaks, p. 110 ◆ GENERAL **TE** **Activity** Guest Speaker, p. 111 ◆ GENERAL **TE** **Activity** Using Density to Identify Minerals, p. 112 ◆ GENERAL **TE** **Discussion** Crystal Powers, p. 112 GENERAL **SE** **Quick Lab** Determining Density, p. 113 GENERAL **CRF** **Datasheet for Quick Lab*** GENERAL **SE** **Skills Practice Lab** Mineral Identification, pp. 120–121 GENERAL **CRF** **Datasheet for Chapter Lab*** GENERAL **TE** **Activity** Finding Mineral Deposits, p. 123 GENERAL **SE** **Mapping Expeditions** Buried Treasure, pp. 836–837 GENERAL **CRF** **Inquiry Lab** Growing Crystals* GENERAL	**OSP** **Lesson Plans** (also in print) **TR** **Bellringer*** **TR** 23 Mohs Hardness Scale* **TR** 24 The Six Basic Crystal Systems* **CD** **Interactive Tutor** Matter and Minerals **CD** **Interactive Tutor** Weight, Mass, Volume, and Density

PACING • 90 min

CHAPTER REVIEW, ASSESSMENT, AND STANDARDIZED TEST PREPARATION

- **SE** **Chapter Highlights,** p. 115
- **SE** **Chapter Review,** pp. 116–117
- **SE** **Standardized Test Prep,** pp. 118–119
- **CRF** **Concept Review*** ■ GENERAL
- **CRF** **Critical Thinking*** ADVANCED
- **CRF** **Math Skills*** GENERAL
- **CRF** **Graphing Skills*** GENERAL
- **CRF** **Chapter Test A*** ■ GENERAL
- **CRF** **Chapter Test B*** ADVANCED
- **OSP** **Lesson Plans** (also in print)
- **OSP** **Test Generator**
- **OSP** **Test Item Listing**

Online and Technology Resources

go.hrw.com Holt Online Learning

Visit **go.hrw.com** for access to Holt Online Learning, or enter the keyword **HQ6 Home** for a variety of free online resources.

One-Stop Planner® CD-ROM

This CD-ROM package includes

- Lab Materials QuickList Software
- Holt Calendar Planner
- Customizable Lesson Plans
- Printable Worksheets
- ExamView® Test Generator
- Interactive Teacher Edition
- Holt PuzzlePro®
- Holt PowerPoint® Resources

KEY

SE Student Edition	**OSP** One-Stop Planner	**VID** Classroom Video/DVD
TE Teacher Edition	**TR** Transparencies and Transparency Worksheets	* Also on One-Stop Planner
CRF Chapter Resource File	**CD** CD or CD-ROM	◆ Requires advance prep
LTP Long-Term Projects		■ Also available in Spanish

SKILLS DEVELOPMENT RESOURCES	REVIEW AND ASSESSMENT	CORRELATIONS
SE **Pre-Reading Activity,** p. 102 GENERAL **TE** **Using the Figure** Quartz Quest, p. 102 GENERAL		National Science Education Standards
CRF **Directed Reading*** ■ BASIC **TE** **Skill Builder** Vocabulary, p. 105 GENERAL **TE** **Reading Skill Builder** Paired Summarizing, p. 106 BASIC **TE** **Inclusion Strategies,** p. 107	**SE** **Reading Checks,** pp. 105, 106 GENERAL **SE** **Section Review,** p. 108 GENERAL **TE** **Homework,** p. 105 ADVANCED **TE** **Reteaching,** p. 107 BASIC **TE** **Quiz,** p. 107 GENERAL **TE** **Alternative Assessment,** p. 107 GENERAL **CRF** **Section Quiz*** ■ GENERAL	PS 2c, PS 2f
CRF **Directed Reading*** ■ BASIC **TE** **Reading Skill Builder** Reading Organizer, p. 110 BASIC **TE** **Inclusion Strategies,** p. 110 **SE** **Graphic Organizer** Comparison Table, p. 111 GENERAL **SE** **Math Practice,** p. 112 GENERAL **TE** **Skill Builder** Writing, p. 112 ADVANCED	**SE** **Reading Checks,** pp. 111, 113 GENERAL **SE** **Section Review,** p. 114 GENERAL **TE** **Reteaching,** p. 113 BASIC **TE** **Quiz,** p. 113 GENERAL **TE** **Alternative Assessment,** p. 114 GENERAL **CRF** **Section Quiz*** ■ GENERAL	PS 2c, PS 2f

Holt Lab Generator CD-ROM

Search for labs by topic, standard, difficulty level, and/or time frame. Edit those labs to fit your needs, or create your own labs. Use the Lab Materials QuickList to order just what you need.

Holt Earth Science Interactive Tutor CD-ROM

This CD-ROM consists of interactive activities that give students a fun way to extend their knowledge of Earth science concepts.

Chapter Summaries Audio CDs

These CDs include audio summaries of the key concepts presented in each chapter. (Audio summaries are also available in Spanish.)

SciLinks NSTA

www.scilinks.org

Maintained by the **National Science Teachers Association.** See Chapter Enrichment pages that follow for a complete list of topics.

See Chapter Enrichment pages for Video Resources.

Chapter 5 Chapter Enrichment

This Chapter Enrichment provides relevant and interesting information to expand and enhance your classroom instruction of the chapter material.

Section 1 What Is a Mineral?

The Myriad Uses of Minerals

Humans use minerals from the moment we get up in the morning and brush our teeth with fluoride toothpaste, which is derived in part from the mineral fluorite, to the moment we do the same before bed. Minerals are such an integral part of our lives that we may take them for granted, but modern life could not exist without them. The walls of our homes are lined with copper wiring that carries electricity and copper pipes that carry water. The walls themselves may be made from gypsum wallboard. The concrete used to construct roads and buildings requires huge amounts of mineral-containing sand, gravel, and cement. The computers we work and play on could not exist without silicon chips. We use minerals in paint, soap, medicine, even in money. According to the U.S. Geological Survey, the average American uses about 9,000 kg of stone, sand, and gravel, 190 kg of iron, 170 kg of salt, 30 kg of aluminum, 9 kg of copper, 5 kg of lead, and 1 kg of zinc each year.

Mineral use is not new to humans. Entire eras of human history, such as the Stone Age, Bronze Age, and Iron Age, have been defined by our ability to find and process minerals. Our ancestors used minerals to make jewelry, money, decorations, pigments for painting, tools and weapons, monumental buildings, and much more.

▼ The mineral halite has been used for centuries to season food.

Biomining

To extract and process minerals from low-grade ore deposits requires a huge investment of land, equipment, and money—which may result in severe damage to the environment. As concerns about the environmental impact of mining increase and high-grade mineral ores become ever more scarce, mining companies have been turning to microbes to extract metals from low-grade ores. For example, in the presence of sulfuric acid, *Thiobacillus ferrooxidans* removes much more copper from low-grade ores than traditional methods do, with reduced economic and environmental costs. Most copper that is mined in the United States and about one-quarter of the world's copper is now processed this way. Bacteria are also being used to extract gold from low-grade ores more efficiently and at lower cost than traditional extraction methods can. Hungry microbes are even being used to clean up contaminated mine sites.

Section 2 Identifying Minerals

Animal Magnetism

It is well known that some bird species migrate thousands of miles every year, but exactly how they accurately navigate these great distances remained a mystery until recently. Scientists now think that migratory birds are able to accurately locate their destinations because they can sense Earth's magnetic field. Scientists have found tiny crystals of magnetite, a magnetic mineral, in birds' brains and skulls. It is thought that birds are able to detect the direction of the field, orient themselves, and use Earth's magnetic field, in addition to other cues, to navigate.

Birds are not the only organisms that seem to be able to sense Earth's magnetic field. Whales, sharks, rays, tuna, salmon, and honeybees all seem to have the ability. In addition, scientists have found magnetite deposits in each of them. Even certain bacteria and algae seem to be able to respond to Earth's magnetic field. Magnetotactic bacteria synthesize chains of magnetite that allow them to rotate into alignment with the north-south lines of Earth's magnetic field.

▲ Rocks like this one that contain fluorescent minerals will change color when exposed to ultraviolet light.

Synthetic Gems

While the special beauty of natural crystals may be striking, natural crystals always contain flaws. Even the most valuable and seemingly perfect gems contain microscopic imperfections, such as gas bubbles or trace amounts of other elements. Nature can make crystals no other way, but humans can. Using a variety of lab techniques, scientists can now grow certain crystals that are virtually identical to natural minerals—except that they are flawless. Sometimes the only way even an expert can tell that a gem is natural is to examine it for microscopic flaws.

The first real efforts to make artificial gems occurred in the 1800s, but the results were poor. By 1902, French mineralogist Antoine Vereuil perfected the flame-fusion method to make artificial rubies. The method is still used today to make artificial rubies, and a similar process can make sapphires, emeralds, and other gems.

In 1954, H. Tracy Hall discovered a way to make synthetic diamonds for industrial purposes. By 1970, gem-quality diamonds could be made in the lab. Today, almost 20,000 kg of synthetic diamonds are made each year and used for all kinds of tools—from tools used to quarry stone to those used to perform delicate eye surgery. Lab-grown crystals also are used in almost every electronic or optical device we use today. Synthetic rubies are used in lasers, synthetic silicon crystals are in every computer, and just about every electronic device contains a computer chip of some sort.

Video Resources

Brain Food Video Quizzes

Brain Food Video Quizzes These videos contain game-show style quizzes that assess students' progress and motivate students to study the chapter material.

HRW Earth Science Video

This video introduces Earth science topics and includes a geology field trip. The video segment listed below complements this chapter.

Segment 8: Minerals of the Earth's Crust This segment focuses on specific properties of minerals, such as cleavage, color, crystal shape, hardness, luster, and streak, which can often be used to identify mineral species. Unique properties of minerals, including luminescence, magnetism, radioactivity, refraction, and taste, which can also facilitate identification, are mentioned. (3.5 min)

NOVA® Videos

To order **NOVA** videos related to this chapter, visit **go.hrw.com** and enter the keyword **HQ6MINV.**

SciLinks®

NSTA Developed and maintained by the National Science Teachers Association

SciLinks is maintained by the National Science Teachers Association to provide you and your students with interesting, up-to-date links that will enrich your classroom presentation of the chapter.

Visit www.scilinks.org and enter the SciLinks code for more information about the topic listed.

Topic: Mineral Identification
SciLinks code: HQ60965

Topic: Minerals
SciLinks code: HQ60966

Topic: Careers in Earth Science
SciLinks code: HQ60222

Topic: Crystalline Solids
SciLinks code: HQ60369

Chapter 5

Chapter Overview

Minerals are used to make many familiar and not so familiar items, from jewelry to computers. This chapter compares the two main mineral groups—the silicates and nonsilicates—and describes their chemical and physical properties.

Using the Figure — GENERAL

Quartz Quest This photograph shows an agate (AG it), a form of the mineral chalcedony (kal SED uh nee). Chalcedony is a type of quartz. Chalcedony is microcrystalline, which means that the crystals are so small that they can be seen only by using a microscope, and commonly forms when the mineral is deposited by water. Agates, carnelian, tiger's eye, jasper, and chrysoprase are all varieties of microcrystalline quartz. Have interested students investigate the various forms of quartz and prepare posters that illustrate their findings. LS Visual

PRE-READING ACTIVITY

FOLDNOTES

You may want to assign this FoldNote activity as homework. Collect the FoldNotes to check students' understanding of the material.

Chapter 5

Minerals of Earth's Crust

Sections

1 What Is a Mineral?

2 Identifying Minerals

What You'll Learn

- What the characteristics of minerals are
- Why minerals have certain properties
- How to identify minerals

Why It's Relevant

Minerals are valued for their use in making everything from airplanes to cookware. Understanding the characteristics of minerals is also an important way to understand how Earth's processes form minerals.

PRE-READING ACTIVITY

FOLDNOTES

Double Door Before you read this chapter, create the FoldNote entitled "Double Door" described in the Skills Handbook section of the Appendix. Write "Silicate minerals" on one flap of the double door and "Nonsilicate minerals" on the other flap. As you read the chapter, compare the two topics, and write characteristics of each on the inside of the appropriate flap.

▶ Agates are a form of the mineral chalcedony, which is a type of quartz. This agate was dyed a brilliant blue to show its internal structure.

Chapter Correlations — *National Science Education Standards*

PS 2c Bonds between atoms are created when electrons are paired up by being transferred or shared. A substance composed of a single kind of atom is called an element. The atoms may bond together into molecules or crystalline solids. A compound is formed when two or more kinds of atoms bind together chemically. **(Sections 1 and 2)**

PS 2f ... atoms can bond to one another in chains, rings, and branching networks to form a variety of structures. **(Sections 1 and 2)**

Section 1 What Is a Mineral?

A ruby, a gold nugget, and a grain of salt look very different from one another, but they have one thing in common. They are minerals, the basic materials of Earth's crust. A **mineral** is a natural, usually inorganic solid that has a characteristic chemical composition, an orderly internal structure, and a characteristic set of physical properties.

OBJECTIVES

- **Define** *mineral.*
- **Compare** the two main groups of minerals.
- **Identify** the six types of silicate crystalline structures.
- **Describe** three common nonsilicate crystalline structures.

KEY TERMS

mineral
silicate mineral
nonsilicate mineral
crystal
silicon-oxygen tetrahedron

Characteristics of Minerals

To determine whether a substance is a mineral or a nonmineral, scientists ask four basic questions, as shown in **Table 1.** If the answer to all four questions is *yes,* the substance is a mineral.

First, is the substance inorganic? An inorganic substance is one that is not made up of living things or the remains of living things. Coal, for example, is organic—it is composed of the remains of ancient plants. Thus, coal is not a mineral.

Second, does the substance occur naturally? Minerals form and exist in nature. Thus, a manufactured substance, such as steel or brass, is not a mineral.

Third, is the substance a solid in crystalline form? The volcanic glass obsidian is a naturally occurring substance. However, the atoms in obsidian are not arranged in a regularly repeating crystalline structure. Thus, obsidian is not a mineral.

Finally, does the substance have a consistent chemical composition? The mineral fluorite has a consistent chemical composition of one calcium ion for every fluoride ion. Basalt, however, can have a variety of substances. The ratio of these substances commonly varies in each sample of basalt.

mineral a natural, usually inorganic solid that has a characteristic chemical composition, an orderly internal structure, and a characteristic set of physical properties

Table 1 ▶

Four Criteria for Minerals

Questions to Identify a Mineral	Coal	Brass	Obsidian	Basalt	Fluorite

Is it inorganic?	No	Yes	Yes	Yes	Yes
Does it occur naturally?		No	Yes	Yes	Yes
Is it a crystalline solid?			No	Yes	Yes
Does it have a consistent chemical composition?				No	Yes

Section 1

Focus

Overview

This section defines what a mineral is and reviews the two main groups of minerals, the silicates and nonsilicates. It describes the six common crystalline structures of silicates. It also describes the six major classes of nonsilicates.

Bellringer

Ask students to define what a mineral is and give an example. (Answers may vary. Use definitions to identify misconceptions about what a mineral is and to introduce the four characteristics that all minerals must have.) LS **Logical**

Motivate

Group Activity — GENERAL

Mineral—Yes or No? Organize students into small groups and give each group a set of samples that includes salt, sugar, ice, rock, and charcoal. Have students determine if each sample is a mineral, based on answers to the four questions scientists ask to determine if a substance is a mineral. Discuss identifications as a class. (Salt and ice are minerals. Sugar and charcoal are organic, and rock does not have a consistent chemical composition.) LS **Visual/Kinesthetic**

CHAPTER RESOURCES

Chapter Resource File

- Directed Reading BASIC

Technology

Transparencies
- Bellringer
- 21 Characteristics of Minerals (with worksheet)

Student Edition on CD-ROM

One-Stop Planner CD-ROM
- Lesson Plan

Teach

Teaching Tip — BASIC

Connect to Real Life Remind students that we use minerals regularly in everyday life. They are found in everything, including computers, watches, emery boards, dentist drills, toothpaste, and more. Have students keep a journal for a week in which they record their use of or contact with minerals or mineral-containing products. At the end of the week, discuss what types of minerals they used most often. Interested students may wish to do more research on how minerals are used in daily life. The United States Geological Survey Web site is a good place to start. LS Logical

Internet Activity — ADVANCED

Mineral Formation Invite interested students to research the three basic ways minerals form: solidification of a melt; precipitation from solution; and solid-state diffusion. Guide students to research how scientists use these processes in laboratory settings to grow flawless synthetic minerals. Students may present their findings in a written or oral report to the class. A worksheet designed to direct student research on this topic can be found in the **Chapter Resource File** booklet or by visiting go.hrw.com and entering the keyword **HQ6MINX.** LS Logical

CHAPTER RESOURCES

Chapter Resource File

Internet Activity
- Mineral Formation ADVANCED

Figure 1 ▶ Plagioclase feldspar (left), muscovite mica (center), and orthoclase feldspar (right) are 3 of the 20 common rock-forming minerals.

Kinds of Minerals

Earth scientists have identified more than 4,000 minerals, but fewer than 20 of the minerals are common. The common minerals are called *rock-forming minerals* because they form the rocks that make up Earth's crust. Three of these minerals are shown in **Figure 1.** Of the 20 rock-forming minerals, 10 are so common that they make up 90% of the mass of Earth's crust. These minerals are quartz, orthoclase, plagioclase, muscovite, biotite, calcite, dolomite, halite, gypsum, and ferromagnesian minerals. All minerals, however, can be classified into two main groups—silicate minerals and nonsilicate minerals—based on the chemical compositions of the minerals.

Silicate Minerals

A mineral that contains a combination of silicon, Si, and oxygen, O, is a **silicate mineral.** The mineral quartz has only silicon and oxygen atoms. However, other silicate minerals have one or more additional elements. Feldspars are the most common silicate minerals. The type of feldspar that forms depends on which metal combines with the silicon and oxygen atoms. Orthoclase forms when the metal is potassium, K. Plagioclase forms when the metal is sodium, Na, calcium, Ca, or both.

In addition to quartz and the feldspars, ferromagnesian minerals—which are rich in iron, Fe, and magnesium, Mg—are silicates. These minerals include olivines, pyroxenes, amphiboles, and biotite. Silicate minerals make up 96% of Earth's crust. Feldspar and quartz alone make up more than 50% of the crust.

silicate mineral a mineral that contains a combination of silicon and oxygen, and that may also contain one or more metals

Connection to CHEMISTRY

Ions and Bonds

An atom that has a positive or negative charge is called an *ion*. When an atom loses one or more electrons, the atom has more protons than it has electrons. Thus, the atom acquires a positive charge. An ion that has a positive charge is called a *cation*. All of the most abundant elements in Earth's crust except oxygen release electrons and form cations. For example, sodium, Na, atoms generally lose one electron to form a cation that has a charge of +1.

An atom that gains one or more electrons acquires a negative charge because the atom has more electrons than protons. An ion that has a negative charge is called an *anion*. Chlorine, Cl, which generally gains one electron and thus has a charge of –1, forms an anion.

Atoms and ions rarely exist alone. Most atoms and ions combine to form *compounds*. Forces called *chemical bonds* hold atoms and ions together in these compounds. When ions combine to form compounds, the ions combine in proportions that allow all of the positive charges to equal and cancel out all of the negative charges. Thus, the compound is electrically neutral. The mineral halite is a good example of this relationship. Because sodium ions have +1 charges and chlorine ions have –1 charges, one sodium ion bonds with one chlorine ion to form the compound sodium chloride, NaCl, which is the mineral halite. Thus, 1 – 1 = 0, and the compound is neutral.

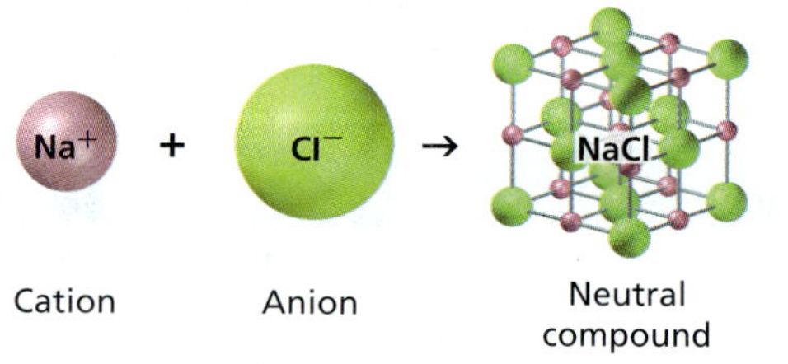

GEOLOGY CONNECTION ADVANCED

Mineraloids Some of the solid subtances we think of and call "crystals" are really glass. While glass is made from melted minerals, it does not have crystalline structure because it cools too quickly for the atoms to arrange themselves into a regular, repeating order. Because it lacks crystalline structure, glass cannot be a mineral.

Mineraloids are amorphous, inorganic substances, mineral in nature but lacking crystalline structure. An example of a mineraloid is opal, a hydrated, amorphous silicate that occurs in many beautiful colors and forms. One form, precious opal, shows flashes of different colors that depend on the size of the silica spheres in its structure. Water and mercury (which are liquid at room temperature) are also considered mineraloids.

Table 2 ▼

Major Classes of Nonsilicate Minerals		
Carbonates compounds that contain a carbonate group (CO_3)	 Dolomite, $CaMg(CO_3)_2$	 Calcite, $CaCO_3$
Halides compounds that consist of chlorine or fluorine combined with sodium, potassium, or calcium	 Halite, NaCl	 Fluorite, CaF_2
Native elements elements uncombined with other elements	 Silver, Ag	 Copper, Cu
Oxides compounds that contain oxygen and an element other than silicon	 Corundum, Al_2O_3	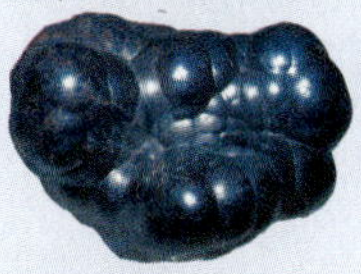 Hematite, Fe_2O_3
Sulfates compounds that contain a sulfate group (SO_4)	 Gypsum, $CaSO_4 \cdot 2H_2O$	 Anhydrite, $CaSO_4$
Sulfides compounds that consist of one or more elements combined with sulfur	 Galena, PbS	 Pyrite, FeS_2

Nonsilicate Minerals

Approximately 4% of Earth's crust is made up of minerals that do not contain compounds of silicon and oxygen, or **nonsilicate minerals.** **Table 2** organizes the six major groups of nonsilicate minerals by their chemical compositions: carbonates, halides, native elements, oxides, sulfates, and sulfides.

nonsilicate mineral a mineral that does not contain compounds of silicon and oxygen

Reading Check What compound of elements will you never find in a nonsilicate mineral? (See the Appendix for answers to Reading Checks.)

Homework — ADVANCED

Mineral Names Mineral names have many origins. Some are named after the place where they were first found or the scientist who identified them. Some names are derived from a Greek or Latin word that describes a special characteristic of the mineral. Some are named after a myth or legend associated with their origin. Have students research the origin of two or three mineral names. You can assign minerals or allow students to research on their own. Have students present their findings to the class with a brief oral report. LS Verbal

Activity — BASIC

Acid Test The group of nonsilicate minerals known as *carbonates* can be identified because they produce carbon dioxide gas when combined with an acid. Some require a strong acid such as hydrochloric acid, but many will react with a weak acid such as vinegar. Provide students with a carbonate such as calcite, a noncarbonate, and strong vinegar. Have them drop the vinegar on each sample and identify which sample is the carbonate. They should see fizzing or bubbling on the surface of the carbonate but no reaction with the noncarbonate. Note that if a sulfide is used as the noncarbonate, it also may react, but the gas is hydrogen sulfide or sulfur dioxide and has a strong smell of rotten eggs. LS Kinesthetic

Answer to Reading Check

Nonsilicates never contain compounds of silicon bonded to oxygen.

BIOLOGY CONNECTION

The Mineral Within Phosphate minerals are rare, but they are very important to animals that have bones. The non-living part of our teeth and bones is composed of microscopic crystals of calcium phosphate laid down by cells within the bone (the osteocytes). Calcium phosphate has a hexagonal crystal structure and a hardness of 5 and may make up as much as 65% of an adult bone.

SKILL BUILDER — GENERAL

Vocabulary The word *crystal* is derived from the Greek word *krystallos,* which is based on *kryos,* meaning "icy cold." The ancient Greeks thought rock crystal, colorless quartz, was ice frozen so hard that it would never melt. LS Verbal

Teach, continued

Answer to Reading Check

The building block of the silicate crystalline structure is a four-sided structure known as the *silicon-oxygen tetrahedron,* which is one silicon atom surrounded by four oxygen atoms.

QuickLAB

Skills Acquired

- Constructing Models

Teacher's Notes You can also use plastic foam balls of different sizes to represent the silicon and oxygen. Using different colored marshmallows or plastic foam balls may help students visualize the different atoms better.

Answers

1. The toothpicks represent the silicon-oxygen bonds.
2. Build a second tetrahedron and attach one toothpick to a large marshmallow of the first tetrahedron to represent the sharing of oxygens between tetrahedra.

CHAPTER RESOURCES

Chapter Resource File

- **Datasheet for Quick Lab** GENERAL
- **Skills Practice Lab** Copper Recovery GENERAL

crystal a solid whose atoms, ions, or molecules are arranged in a regular, repeating pattern

silicon-oxygen tetrahedron the basic unit of the structure of silicate minerals; a silicon ion chemically bonded to and surrounded by four oxygen ions

Quick LAB 5 min

Modeling Tetrahedra

Procedure

1. Place **four toothpicks** in a **small marshmallow.** Evenly space the toothpicks as far from each other as possible.
2. Place **four large marshmallows** on the ends of the toothpicks.

Analysis

1. In your model, what do the toothpicks represent?
2. When tetrahedra form chains or rings, the tetrahedra share electrons. If you wanted to build a chain of tetrahedra, how would you connect two tetrahedra together?

Crystalline Structure

All minerals in Earth's crust have a crystalline structure. Each type of mineral crystal is characterized by a specific geometric arrangement of atoms. A **crystal** is a solid whose atoms, ions, or molecules are arranged in a regular, repeating pattern. A large mineral crystal displays the characteristic geometry of that crystal's internal structure. The conditions under which minerals form, however, often hinder the growth of single, large crystals. As a result, minerals are commonly made up of masses of crystals that are so small that you can see them only with a microscope. But, if a crystal forms where the surrounding material is not restrictive, the mineral will develop as a single, large crystal that has one of six basic crystal shapes. Knowing the crystal shapes is helpful in identifying minerals.

One way that scientists study the structure of crystals is by using X rays. X rays that pass through a crystal and strike a photographic plate produce an image that shows the geometric arrangement of the atoms that make up the crystal.

Crystalline Structure of Silicate Minerals

Even though there are many kinds of silicate minerals, their crystalline structure is made up of the same basic building blocks. Each building block has four oxygen atoms arranged in a pyramid with one silicon atom in the center. **Figure 2** shows this four-sided structure, which is known as a **silicon-oxygen tetrahedron.**

Silicon-oxygen tetrahedra combine in different arrangements to form different silicate minerals. The various arrangements are the result of the kinds of bonds that form between the oxygen atoms of the tetrahedra and other atoms. The oxygen and silicon atoms of tetrahedra may bond with those of neighboring tetrahedra. Bonds may also form between the oxygen atoms in the tetrahedra and other elements' atoms outside of the tetrahedra.

Reading Check **What is the building block of the silicate crystalline structure? (See the Appendix for answers to Reading Checks.)**

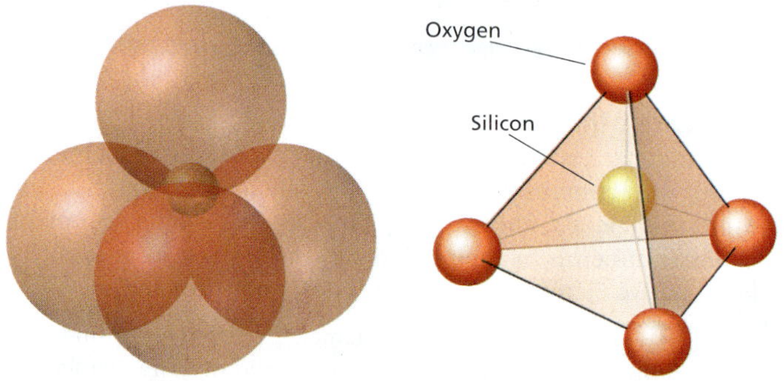

Figure 2 ▶ The structure of a silicon-oxygen tetrahedron can be shown by two different models. The model on the left represents the relative size and proximity of the atoms to one another in the molecule. The model on the right shows the tetrahedral shape of the molecule.

Group Activity — BASIC

Growing Crystals Have students work in pairs to grow crystals of sodium carbonate (washing soda) or sodium chloride (rock salt). Fill a jar with very hot water. Stir in washing soda or salt until no more will dissolve. Tie a paperclip to a piece of string. Tie the other end to a pencil. Place the pencil across the top of the jar to suspend the paperclip in the solution. Let the jar sit undisturbed for 24 hours or more. After crystals have grown, carefully remove them and observe them with a magnifying glass. LS **Visual**

READING SKILL BUILDER — BASIC

Paired Summarizing Have pairs of students read silently about the crystalline structures of silicates. Then, have one student summarize the six structures. The other student should listen to the retelling and point out any inaccuracies or ideas that were left out. Allow students to refer to the text as needed. LS **Verbal** English Language Learners

Isolated Tetrahedral Silicates and Ring Silicates

The six kinds of arrangements that tetrahedra form are shown in **Figure 3.** In minerals that have *isolated tetrahedra,* only atoms other than silicon and oxygen atoms link silicon-oxygen tetrahedra. For example, olivine is a mineral that forms when the oxygen atoms of tetrahedra bond to magnesium, Mg, and iron, Fe, atoms.

Ring silicates form when shared oxygen atoms join the tetrahedra to form three-, four-, or six-sided rings. Ionic bonds hold the rings together, and the rings align to create channels that can contain a variety of ions, molecules, and neutral atoms. Beryl and tourmaline are minerals that have ring-silicate structures.

Single-Chain Silicates and Double-Chain Silicates

In *single-chain silicates,* each tetrahedron is bonded to two others by shared oxygen atoms. In *double-chain silicates,* two single chains of tetrahedra bond to each other. Most single-chain silicate minerals are called *pyroxenes,* and those made up of double chains are called *amphiboles*.

Sheet Silicates and Framework Silicates

In the *sheet silicates,* each tetrahedron shares three oxygen atoms with other tetrahedra. The fourth oxygen atom bonds with an atom of aluminum, Al, or magnesium, Mg, which joins one sheet to another. The mica minerals, such as muscovite and biotite, are examples of sheet silicates.

In the *framework silicates,* each tetrahedron is bonded to four neighboring tetrahedra to form a three-dimensional network. Frameworks that contain only silicon-oxygen tetrahedra form the mineral quartz. The chemical formula for quartz is SiO_2. Other framework silicates, such as the feldspars, contain some tetrahedra in which atoms of aluminum or other metals substitute for some of the silicon atoms.

Figure 3 ▶ Six Kinds of Silicate-Mineral Arrangements

1 **Isolated tetrahedra** do not link with other silicon or oxygen atoms.

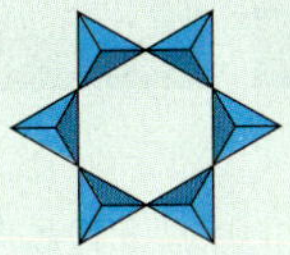

2 **Ring silicates** form rings by sharing oxygen atoms.

3 **Single-chain silicates** form a chain by sharing oxygen atoms.

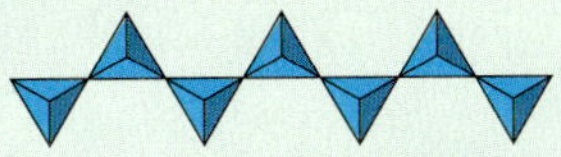

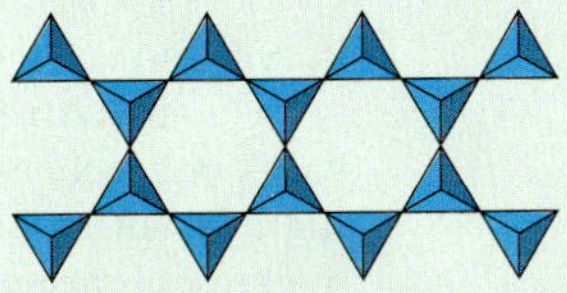

4 **Double-chain silicates** form when two single-chains of tetrahedra bond to each other.

5 **Sheet silicates** form when each tetrahedron shares three of its oxygen atoms with other tetrahedra.

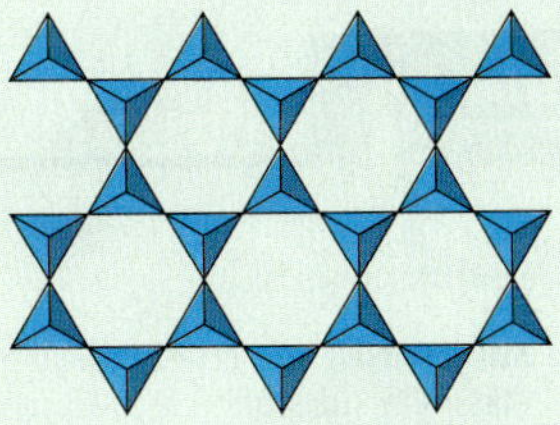

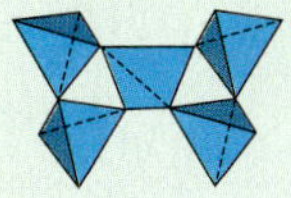

6 **Framework silicates** form when each tetrahedron is bonded to four other tetrahedra.

Close

Reteaching — Basic

Name That Mineral Have students write characteristics of silicate and nonsilicate minerals on index cards. Have them work in pairs to quiz each other about the differences between the two groups by using the cards. **LS Visual/Logical**

Quiz — General

1. What are three common crystal structures for nonsilicates? (cubes, prisms, and irregular masses)
2. What is a crystal? (a solid whose atoms, ions, or molecules are arranged in a regular, repeating pattern)
3. Name one method scientists use to study the structure of crystals? (They pass X rays through the crystals to produce an image of the geometric arrangement of the atoms in the crystal.)

Alternative Assessment — General

Flowchart Have students work in small groups to design a flowchart that classifies and describes the two main groups of minerals and their major subgroups. **LS Visual/Logical**

CHAPTER RESOURCES

Technology

Transparencies
- 22 Structures of Silicate Minerals (with worksheet)

PHYSICS CONNECTION

X-Ray Crystallography In 1912, German physicist Max von Laue passed an X-ray beam through a crystal. The beam was diffracted, creating a pattern of spots on a projection screen. Von Laue knew that diffraction occurs when X rays pass through materials that have regularly spaced obstacles. He proposed that regularly spaced atoms within crystals could explain the pattern. Today, X-ray crystallography has become an important tool in many fields and has even helped decipher the structure of DNA.

INCLUSION Strategies

- *Learning Disabled*
- *Hearing Impaired*
- *Developmentally Delayed*

Help students understand silicate-mineral arrangements by having them draw the structures. Give each student six sheets of graph paper. Have students use markers to create and label each silicate-mineral arrangement. Guide them to use the graph paper lines and intersections to make sure the triangles are even.

Close, *continued*

Answers to Section Review

1. a naturally occurring, inorganic, crystalline solid that has a consistent chemical composition and a characteristic set of physical properties
2. A mineral must be naturally occurring, be inorganic, be a crystalline solid, and have a consistent chemical composition.
3. Silicates are composed of silicon-oxygen tetrahedra, sometimes combined with other elements. Nonsilicates do not contain silicon-oxygen compounds.
4. silicon and oxygen
5. carbonates, halides, oxides, sulfates, sulfides, and native elements
6. Isolated tetrahedra do not link with other silicon or oxygen atoms. Ring silicates form rings by sharing oxygen atoms. Single-chain silicates form a chain by sharing oxygen atoms. Double-chain silicates form when two single chains of tetrahedra bond together. Sheet silicates form when each tetrahedron shares three of its oxygen atoms with other tetrahedra. Framework silicates form when each tetrahedron bonds to four other tetrahedra.
7. Nonsilicates have a wider variety of crystal structures because they have diverse chemical compositions and do not have to form tetrahedra, although some do.
8. If silicon bonded with only three oxygen atoms, it would not form tetrahedra. The Si-O compound would probably be more planar and would more likely form sheet structures when it combined, similar to how the carbons in graphite bond.
9. Gold is a mineral because it satisfies all four criteria that define a mineral. It is a native element.
10. *Minerals* are grouped into *silicate minerals,* which are composed of *silicon-oxygen tetrahedra* that can combine to form *single-chain silicates, ring silicates,* and *framework silicates;* and *nonsilicate minerals,* which form many types of *crystals*.

Figure 4 ▶ Gold (left) commonly has a dendritic shape. Halite (center) commonly has cubic crystals. Diamond (right) commonly has an octahedral crystal shape. All three of these minerals are nonsilicates.

The Crystalline Structure of Nonsilicate Minerals

Because nonsilicate minerals have diverse chemical compositions, nonsilicate minerals display a vast variety of crystalline structures. Common crystal structures for nonsilicate minerals include cubes, hexagonal prisms, and irregular masses. Some of these structures are shown in **Figure 4.**

Nonsilicates may form tetrahedra that are similar to those in silicates. However, the ions at the center of these tetrahedra are not silicon. Minerals that have the same ion at the center of the tetrahedron commonly share similar crystal structures. Thus, the classes of nonsilicate minerals can be divided into smaller groups based on the structural similarities of the minerals' crystals.

The structure of a nonsilicate crystal determines the nonsilicate's characteristics. For example, the native elements have very high densities because their crystal structures are based on the packing of atoms as close together as possible. This crystal structure is called *closest packing.* In this crystal structure, each metal atom is surrounded by 8 to 12 other metal atoms that are as close to each other as the charges of the atomic nuclei will allow.

Section 1 Review

1. **Define** *mineral.*
2. **Summarize** the characteristics that are necessary to classify a substance as a mineral.
3. **Compare** the two main groups of minerals.
4. **Identify** the two elements that are in all silicate minerals.
5. **Name** six types of nonsilicate minerals.
6. **Describe** the six main crystalline structures of silicate minerals.
7. **Explain** why nonsilicate minerals have a wider variety of crystalline structures than silicate minerals do.

CRITICAL THINKING

8. **Predicting Consequences** If silicon bonded with three oxygen atoms, how might the crystalline structures of silicate minerals be different?
9. **Applying Ideas** Gold is an inorganic substance that forms naturally in Earth's crust. Gold is also a solid and has a definite chemical composition. Is gold a mineral? Explain your answer.

CONCEPT MAPPING

10. Use the following terms to create a concept map: *mineral, crystal, silicate mineral, nonsilicate mineral, ring silicate, framework silicate, single-chain silicate,* and *silicon-oxygen tetrahedron.*

CHAPTER RESOURCES

Chapter Resource File

- Section Quiz GENERAL

Workbooks

- Study Guide (also in Spanish)

Section 2 Identifying Minerals

Earth scientists called **mineralogists** examine, analyze, and classify minerals. To identify minerals, mineralogists study the properties of the minerals. Some properties are simple to study, while special equipment may be needed to study other properties.

OBJECTIVES

- **Describe** seven physical properties that help distinguish one mineral from another.
- **List** five special properties that may help identify certain minerals.

KEY TERMS

mineralogist
streak
luster
cleavage
fracture
Mohs hardness scale
density

Physical Properties of Minerals

Each mineral has specific properties that are a result of its chemical composition and crystalline structure. These properties provide useful clues for identifying minerals. Many of these properties can be identified by simply looking at a sample of the mineral. Other properties can be identified through simple tests.

Color

One property of a mineral that is easy to observe is the mineral's color. Some minerals have very distinct colors. For example, sulfur is bright yellow, and azurite is deep blue. Color alone, however, is generally not a reliable clue for identifying a mineral sample. Many minerals are similar in color, and very small amounts of certain elements may greatly affect the color of a mineral. For example, corundum is a colorless mineral composed of aluminum and oxygen atoms. However, corundum that has traces of chromium, Cr, forms the red gem called *ruby*. Sapphire, which is a type of corundum, gets its blue color from traces of cobalt, Co, and titanium, Ti. **Figure 1** compares colorless, pure quartz with purple amethyst. Amethyst is quartz that has manganese, Mn, and iron, Fe, which cause the purple color.

Color is also an unreliable identification clue because weathered surfaces may hide the color of minerals. For example, the golden color of iron pyrite ranges from dark yellow to black when iron pyrite is weathered. When examining a mineral for color, you should inspect only the mineral's freshly exposed surfaces.

mineralogist a person who examines, analyzes, and classifies minerals

SCILINKS® NSTA Developed and maintained by the National Science Teachers Association

For a variety of links related to this subject, go to www.scilinks.org

Topic: Minerals
SciLinks code: HQ60966
Topic: Mineral Identification
SciLinks code: HQ60965

Figure 1 ▶ Pure quartz (left) is colorless. Amethyst (right) is a variety of quartz that is purple because of the presence of small amounts of manganese and iron.

Section 2

Focus

Overview

This section identifies the major properties and crystal systems of minerals. It also explains special properties that help in the identification of minerals.

Bellringer

Ask students to list properties they might use to tell one mineral from another. Use student lists to introduce the characteristics that mineralogists use to identify minerals. (Answers may vary but could include color, hardness, shape, density, and transparency.)

Motivate

Demonstration — GENERAL

The Same, But Different Show students crystals of purple fluorite, amethyst, and rose quartz. Ask students to identify which two crystals are the same mineral. Use this demonstration to explain why using only color is not a good way to identify minerals. (Fluorite and amethyst are both purple, have the same nonmetallic glassy luster, and have white streak, but they are not the same. Amethyst is purple quartz and is much harder than fluorite is; quartz does not show any cleavage, while fluorite has perfect cleavage in four directions.) LS **Visual**

CHAPTER RESOURCES

Chapter Resource File

- Directed Reading BASIC

Technology

Transparencies
- Bellringer

Student Edition on CD-ROM

One-Stop Planner CD-ROM
- Lesson Plan

Teach

Activity — GENERAL

Those Are the Breaks Cement five or six craft sticks together sandwich style with a mixture of plaster of paris (2 Tbsp), white glue (1/2 tsp), and water (2 tsp). Allow the models to dry for one hour. When dry, have students try to break apart the sticks. The stack should break apart into thinner layers, modeling the way some minerals split along one cleavage plane. For example, mica has perfect cleavage in one plane and forms thin flakes as it is cleaved. Provide a piece of mica for students to examine. Challenge students to use the craft stick model to demonstrate cleavage in more than one plane. LS **Kinesthetic**

BRAIN FOOD

Gemstones When cutting gems, such as diamonds, emeralds, and rubies, a gem cutter, or *lapidary,* may break a large stone into smaller pieces by splitting it along a cleavage plane. The lapidary determines the crystal shape, then cuts facets and grinds and polishes them to make the gem sparkle. The largest diamond ever found, the Cullinan Diamond (initial weight: 3,106 carats, or 621 g) was cut into nine large gems and many smaller ones.

READING SKILL BUILDER — BASIC

Reading Organizer As students read this section, encourage them to take Power Notes, KWL notes, or Two-Column Notes as described in the Skills Handbook section of the Appendix. Later, students can use these notes as a study guide for assessments. LS **Verbal** English Language Learners

Figure 2 ▶ All minerals have either a metallic luster, as platinum does (top), or a nonmetallic luster, as talc does (bottom).

Streak

A more reliable clue to the identity of a mineral is the color of the mineral in powdered form, which is called the mineral's **streak.** The easiest way to observe the streak of a mineral is to rub some of the mineral against a piece of unglazed ceramic tile called a *streak plate.* The streak's color may differ from the color of the solid form of the mineral. Metallic minerals generally have a dark streak. For example, the streak of gold-colored pyrite is black. For most nonmetallic minerals, however, the streak is either colorless or a very light shade of the mineral's standard color. Minerals that are harder than the ceramic plate will leave no streak.

Luster

Light that is reflected from a mineral's surface is called **luster.** A mineral is said to have a *metallic luster* if the mineral reflects light as a polished metal does, as shown in **Figure 2.** All other minerals have a *nonmetallic luster.* Mineralogists distinguish several types of nonmetallic luster. Transparent quartz and other minerals that look like glass have a glassy luster. Minerals that have the appearance of candle wax have a waxy luster. Some minerals, such as the mica minerals, have a pearly luster. Diamond is an example of a mineral that has a brilliant luster. A mineral that lacks any shiny appearance has a dull or earthy luster.

streak the color of a mineral in powdered form

luster the way in which a mineral reflects light

cleavage in geology, the tendency of a mineral to split along specific planes of weakness to form smooth, flat surfaces

fracture the manner in which a mineral breaks along either curved or irregular surfaces

Cleavage and Fracture

The tendency of a mineral to split along specific planes of weakness to form smooth, flat surfaces is called **cleavage.** When a mineral has cleavage, as shown in **Figure 3,** it breaks along flat surfaces that generally run parallel to planes of weakness in the crystal structure. For example, the mica minerals, which are sheet silicates, tend to split into parallel sheets.

Many minerals, however, do not break along cleavage planes. Instead, they **fracture,** or break unevenly, into pieces that have curved or irregular surfaces. Mineralogists describe a fracture according to the appearance of the broken surface. For example, a rough surface has an *uneven* or *irregular fracture.* A broken surface that looks like a piece of broken wood has a *splintery* or *fibrous fracture.* Curved surfaces are *conchoidal fractures* (kahng KOYD uhl FRAK chuhr), as shown in **Figure 3.**

Figure 3 ▶ Calcite is a mineral that cleaves in three directions. Quartz tends to have a conchoidal fracture.

INCLUSION Strategies

- Visually Impaired
- Hearing Impaired
- Developmentally Delayed

To help students organize the information necessary to identify minerals by their properties, guide students to associate each property with a familiar concept, object, or idea. Color: Associate sulfur's yellow with the sun or the blue of azurite with the sky. Streak: Make a streak on a piece of paper with a tube of lip balm. Luster: Feel a glass object (glassy) and a piece of talc (waxy). Cleavage and fracture: Feel a split piece of mica (cleavage) and the edges of a fractured rock (fracture). Hardness: Try to scratch a piece of wood and a rock by using a nail. Crystal shape: Feel wooden blocks that have the following shapes: square, rhombus, hexagon, and rectangle. Density: Hold a brick and a piece of plastic foam that have the same dimensions.

Table 1 ▼

Mohs Hardness Scale					
Mineral	Hardness	Common test	Mineral	Hardness	Common test
Talc	1	easily scratched by fingernail	Feldspar	6	scratches glass, but does not scratch steel
Gypsum	2	can be scratched by fingernail	Quartz	7	easily scratches both glass and steel
Calcite	3	barely can be scratched by copper penny	Topaz	8	scratches quartz
Fluorite	4	easily scratched with steel file or glass	Corundum	9	scratches topaz
Apatite	5	can be scratched by steel file or glass	Diamond	10	scratches everything

Hardness

The measure of the ability of a mineral to resist scratching is called *hardness*. Hardness does not mean "resistance to cleavage or fracture." A diamond, for example, is extremely hard but can be split along cleavage planes more easily than calcite, a softer mineral, can be split.

To determine the hardness of an unknown mineral, you can scratch the mineral against those on the **Mohs hardness scale**, which is shown in **Table 1**. This scale lists 10 minerals in order of increasing hardness. The softest mineral, talc, has a hardness of 1. The hardest mineral, diamond, has a hardness of 10. The difference in hardness between two consecutive minerals is about the same throughout the scale except for the difference between the two hardest minerals. Diamond (10) is much harder than corundum (9), which is listed on the scale before diamond.

To test an unknown mineral for hardness, you must determine the hardest mineral on the scale that the unknown mineral can scratch. For example, galena can scratch gypsum but not calcite. Thus, galena has a hardness that ranges between 2 and 3 on the Mohs hardness scale. If neither of two minerals scratches the other, the minerals have the same hardness.

The strength of the bonds between the atoms that make up a mineral's internal structure determines the hardness of that mineral. Both diamond and graphite consist only of carbon atoms. However, diamond has a hardness of 10, while the hardness of graphite is between 1 and 2. A diamond's hardness results from a strong crystalline structure in which each carbon atom is firmly bonded to four other carbon atoms. In contrast, the carbon atoms in graphite are arranged in layers that are held together by much weaker chemical bonds.

Reading Check What determines the hardness of a mineral? (See the Appendix for answers to Reading Checks.)

Mohs hardness scale the standard scale against which the hardness of minerals is rated

Graphic Organizer **Comparison Table**

Create the **Graphic Organizer** entitled "Comparison Table" described in the Skills Handbook section of the Appendix. Label the columns with "Diamond," "Corundum," "Graphite," and "Galena." Label the rows with "Hardness" and "Description." Then, fill in the table with details about the hardness and a description of each mineral.

HISTORY CONNECTION

Hardness Scale The Mohs hardness scale is named after German mineralogist and professor Friedrich Mohs. In 1822, Mohs devised the scale by arranging ten minerals according to their resistance to being scratched, from softest to hardest. He used common minerals that were readily available to him. He assigned a value of 1 to the softest, talc, and a value of 10 to the hardest, diamond. The scale is not linear; diamond is many more than 10 times harder than talc is. While the Mohs hardness scale is very useful in mineral identification, when a material's hardness must be accurately measured, a device called a *sclerometer* is used. This device measures the force needed to scratch a material with diamond.

Graphic Organizer GENERAL

Comparison Table

You may want to have students work in groups to create this Comparison Table. Have one student draw the table and fill in information provided by other students from the group.

CHAPTER RESOURCES

Technology

Transparencies
- 23 Mohs Hardness Scale (with worksheet)

Answer to Reading Check

The strength and geometric arrangement of the bonds between the atoms that make up a mineral's internal structure determine the hardness of a mineral.

Activity GENERAL

Guest Speaker Invite a rock or mineral collector to talk to your class. Rock and mineral collecting clubs exist in many cities around the country. Members are usually very willing to share their passion with others and to introduce others to the ins and outs of mineral collecting. The *Lapidary Journal* Web site has an archive page that lists local clubs by state. You could also check with a local geological survey bureau. LS **Visual/Verbal**

SKILL BUILDER — ADVANCED

Writing A mineral's habit is the way clusters of crystals grow together, creating a distinctive overall form and texture. Have students write an illustrated paper that describes some common mineral habits. LS Verbal/Visual

Activity — GENERAL

Using Density to Identify Minerals Have students weigh a mineral sample in grams. Have them fill a graduated cylinder with 200 mL of water, and then place the sample in the container. Students should read the new water level in the container and subtract that value from 200 mL. This number is the mineral's volume. To find the density, have students divide the mass (weight) by the volume (D = m/v). Check the density of the sample against density values in a field guide to minerals. LS Logical

MATHPRACTICE

Answer

$85\ g \div 34\ cm^3 = 2.5\ g/cm^3$

CHAPTER RESOURCES

Chapter Resource File

- **Inquiry Lab** Growing Crystals GENERAL

Technology

Transparencies

- 24 The Six Basic Crystal Systems (with worksheet)

Table 2 ▼

The Six Basic Crystal Systems	
Isometric or Cubic System Three axes of equal length intersect at 90° angles. 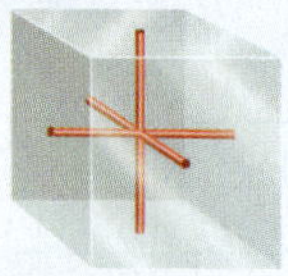	**Orthorhombic System** Three axes of unequal length intersect at 90° angles. 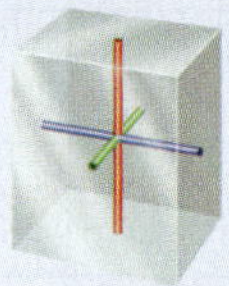
Tetragonal System Three axes intersect at 90° angles. The two horizontal axes are of equal length. The vertical axis is longer or shorter than the horizontal axes. 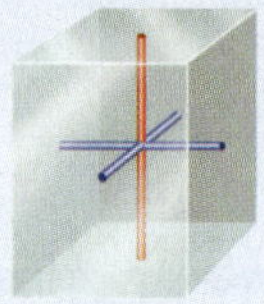	**Hexagonal System** Three horizontal axes of the same length intersect at 120° angles. The vertical axis is longer or shorter than the horizontal axes. 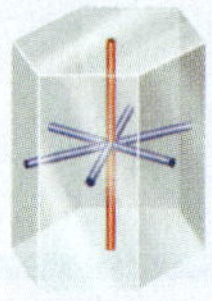
Monoclinic System Two of the three axes of unequal length intersect at 90° angles. The third axis is oblique to the others. 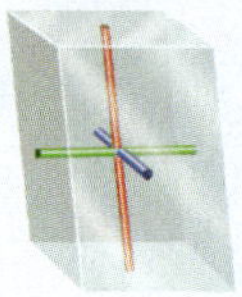	**Triclinic System** Three axes of unequal length are oblique to one another. 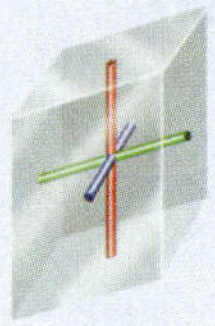

MATHPRACTICE

Calculating Density
A mineral sample has a mass (m) of 85 g and a volume (V) of 34 cm^3. Use the equation below to calculate the sample's density (D).

$$D = \frac{m}{V}$$

density the ratio of the mass of a substance to the volume of the substance; commonly expressed as grams per cubic centimeter for solids and liquids and as grams per liter for gases

Crystal Shape

A mineral crystal forms in one of six basic shapes, as shown in **Table 2.** A certain mineral always has the same general shape because the atoms that form the mineral's crystals always combine in the same geometric pattern. But the six basic shapes can become more complex as a result of environmental conditions, such as temperature and pressure, during crystal growth.

Density

When handling equal-sized specimens of various minerals, you may notice that some feel heavier than others do. For example, a piece of galena feels heavier than a piece of quartz of the same size does. However, a more precise comparison can be made by measuring the density of a sample. **Density** is the ratio of the mass of a substance to the volume of the substance.

The density of a mineral depends on the kinds of atoms that the mineral has and depends on how closely the atoms are packed. Most of the common minerals in Earth's crust have densities between 2 and 3 g/cm^3. However, the densities of minerals that contain heavy metals, such as lead, uranium, gold, and silver, range from 7 to 20 g/cm^3. Thus, density helps identify heavier minerals more readily than it helps identify lighter ones.

Discussion — GENERAL

Crystal Powers Since ancient times, crystals have been associated with magical powers. The ancient Greeks and Romans believed the future could be seen in quartz crystal balls. Emeralds were once believed to have the power to blind snakes, and rubies were thought to bring a wearer power and romance. In the 15th century, amethyst was thought to have the power to cure drunkenness. Even today, many ascribe healing powers to crystals. The Crystal Academy in Taos, New Mexico, explores the ancient art of crystal healing. Most geologists do not believe crystals have any special healing powers. Discuss crystal healing with students. What evidence would scientists need to accept that crystals have the power to heal? Have interested students research crystal healing further and report their findings to the class. LS Interpersonal

Special Properties of Minerals

All minerals exhibit the properties that were described earlier in this section. However, a few minerals have some additional, special properties that can help identify those minerals.

Fluorescence and Phosphorescence

The mineral calcite is usually white in ordinary light, but in ultraviolet light, calcite often appears red. This ability to glow under ultraviolet light is called *fluorescence*. Fluorescent minerals absorb ultraviolet light and then produce visible light of various colors, as shown in **Figure 4.**

When subjected to ultraviolet light, some minerals will continue to glow after the ultraviolet light is turned off. This property is called *phosphorescence*. It is useful in the mining of phosphorescent minerals such as eucryptite, which is an ore of lithium.

Figure 4 ▶ The fluorescent minerals calcite and willemite within this rock change colors as they are exposed to ordinary light (top) and ultraviolet light (bottom).

Chatoyancy and Asterism

In reflected light, some minerals display a silky appearance that is called *chatoyancy* (shuh TOY uhn see). This effect is also called the *cat's-eye effect*. The word *chatoyancy* comes from the French word *chat*, which means "cat," and from *oeil*, which means "eye." Chatoyancy is the result of closely packed parallel fibers within the mineral. A similar effect called *asterism* is the phenomenon in which a six-sided star shape appears when a mineral reflects light.

Reading Check **What is the difference between chatoyancy and asterism? (See the Appendix for answers to Reading Checks.)**

QuickLAB 10 min

Determining Density

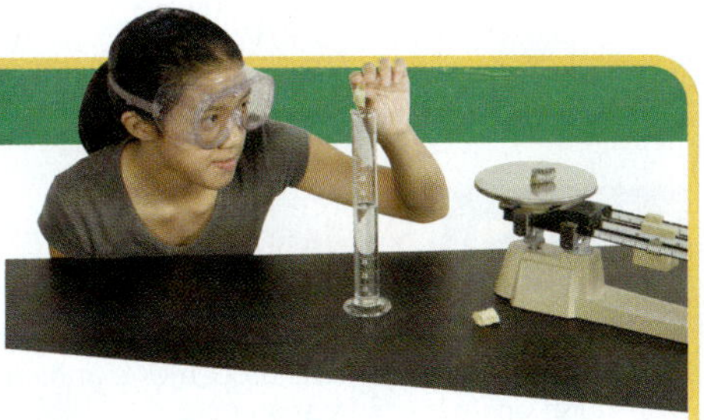

Procedure

1. Use a **triple-beam balance** to determine the mass of three similarly-sized **mineral samples** that have different masses. Record the mass of each mineral sample.
2. Fill a **graduated cylinder** with **70 mL of water.**
3. Add one mineral sample to the water in the graduated cylinder. Record the new volume after the mineral sample is added to the water.
4. Calculate the volume of the mineral sample by subtracting the 70 mL from the new volume.
5. Repeat steps 3 and 4 for the other two mineral samples.
6. Convert the volume of the mineral samples that you calculated in step 4 from milliliters to cubic centimeters by using the conversion: 1 mL = 1 cm^3.

Analysis

1. Calculate the density of each mineral sample by using the following equation:

   density = mass/volume
2. Compare the density of each mineral sample with the density of common minerals in Earth's crust. Compare the density of each mineral sample with minerals that contain a high percentage of heavy metals.
3. Do any of the mineral samples contain a high percentage of heavy metals? Explain your answer.

QuickLAB

Skills Acquired

- Collecting Data
- Classifying

Teacher's Notes: If possible, give the same specimen to more than one student and have students that have the same specimen compare observations. For your records, number the samples and keep an answer key.

Answers

1. Answers may vary depending on the samples used.
2. Answers may vary. Most common minerals have densities between 2 g/cm^3 and 3 g/cm^3. Minerals that contain heavy metals have densities between 7g/cm^3 and 20 g/cm^3.
3. Answers may vary. Mineral samples that contain a high percentage of heavy metals will have high density.

Close

Reteaching — BASIC

Peer Reviewing Have students work in pairs to write three to five questions about the material in this section. Then, have students use their questions to quiz each other. LS **Interpersonal** Co-op Learning

Quiz — GENERAL

1. What instrument can detect radioactive minerals? (Geiger counter)
2. What is the hardest substance on the Mohs scale? (diamond)
3. Amethyst is quartz with what impurities? (manganese and iron)

CHEMISTRY CONNECTION

Lighting Up Fluorescence occurs when light energy interacts with certain atoms, known as *activator elements,* in a crystal. When the atom's excited electrons fall back to their initial state, they release energy, emitting some in the form of light. Fluorescence stops once an energy source is gone. Phosphorescence continues after the light source is turned off because in some cases, excited electrons get trapped in the high-energy state. They eventually fall back to their initial energy state and give off light when they do.

Answer to Reading Check

Chatoyancy is the silky appearance of some minerals in reflected light. Asterism is the appearance of a six-sided star when a mineral reflects light.

CHAPTER RESOURCES

Chapter Resource File

- Datasheet for Quick Lab GENERAL

Close, *continued*

Alternative Assessment — GENERAL

Crystal Models Have students build models of the six crystal shapes to understand the symmetrical features of crystals. For example, they could build stick models showing the axes of symmetry, clay models of the crystal shapes, 3-dimensional paper models, or some other model of their own design. LS Visual/Kinesthetic

Answers to Section Review

1. Color, streak, luster, cleavage and fracture, hardness, and density help distinguish one mineral from another.
2. metallic and nonmetallic
3. To determine hardness, scratch a mineral with the minerals on the Mohs hardness scale or with a fingernail, a penny, or a piece of glass. The hardness of the unknown mineral is greater than the hardness of a mineral it can scratch but is less than the hardness of a mineral that scratches it.
4. Color is not a very reliable clue to the identity of a mineral because small amounts of certain elements in a crystal can greatly alter color and because weathering can change a mineral's color.
5. Answers may vary but should include five of the following: fluorescence, phosphorescence, chatoyancy, asterism, double refraction, magnetism, and radioactivity.
6. Minerals that contain iron may be attracted to magnets. Some minerals, especially magnetite, act as magnets themselves.
7. The mineral is likely to be metallic because metals generally have dark streaks and high densities.
8. One could shine UV light in mines and identify places where rocks continue to glow after the light is turned off. Phosphorescent minerals could be used as a coating to make things glow in the dark.
9. A mineral can be identified by using *luster*, which can be either *metallic luster* or *nonmetallic luster; streak*, which can be tested on a *streak plate; fracture*, which might be *conchoidal fracture;* and *hardness*, which is assessed by using the *Mohs hardness scale*.

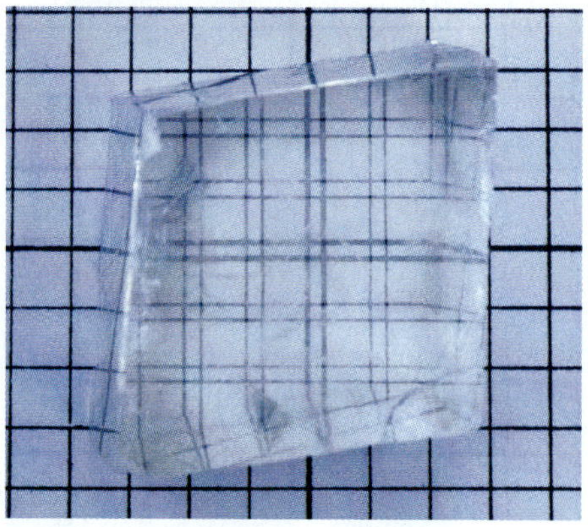

Figure 5 ▶ Some forms of the mineral calcite exhibit double refraction when light rays enter the crystal and split.

Double Refraction

Light rays bend as they pass through transparent minerals. This bending of light rays as they pass from one substance, such as air, to another, such as a mineral, is called *refraction*. Crystals of calcite and some other transparent minerals bend light in such a way that they produce a double image of any object viewed through them, as shown in **Figure 5.** This property is called *double refraction*. Double refraction takes place because light rays are split into two parts as they enter the crystal.

Magnetism

Magnets may attract small particles of some minerals that contain iron. Those minerals are also sometimes magnetic. In general, nonsilicate minerals that contain iron, such as magnetite, are more likely to be magnetic than other nonsilicate minerals are. Lodestone is a form of magnetite. Like a bar magnet, some pieces of lodestone have a north pole at one end and a south pole at the other. The needles of the first magnetic compasses were made of tiny slivers of lodestone.

Radioactivity

Some minerals have a property known as *radioactivity*. The arrangement of protons and neutrons in the nuclei of some atoms is unstable. Radioactivity results as unstable nuclei decay over time into stable nuclei by releasing particles and energy. A *Geiger counter* can be used to detect the released particles and, thus, to identify minerals that are radioactive. Uranium, U, and radium, Ra, are examples of radioactive elements. Pitchblende is the most common mineral that contains uranium. Other uranium-bearing minerals are carnotite, uraninite, and autunite.

Section 2 Review

1. **Describe** seven physical properties that help distinguish one mineral from another.
2. **Identify** the two main types of luster.
3. **Summarize** how you would determine the hardness of an unidentified mineral sample.
4. **Explain** why color is an unreliable clue to the identity of a mineral.
5. **List** five special properties that may help identify certain minerals.
6. **Explain** how magnetism can be useful for identifying minerals.

CRITICAL THINKING

7. **Evaluating Data** An unknown metal has a black streak and a density of 18 g/cm^3. Is the mineral more likely to be metallic or nonmetallic?
8. **Analyzing Methods** Explain how phosphorescence is helpful in mining eucryptite. Describe other ways in which phosphorescent minerals might be used.

CONCEPT MAPPING

9. Use the following terms to create a concept map: *luster, streak, fracture, hardness, Mohs hardness scale, streak plate, nonmetallic luster, metallic luster,* and *conchoidal fracture.*

CHAPTER RESOURCES

Chapter Resource File
- Section Quiz GENERAL

Workbooks
- Study Guide (also in Spanish)

Chapter 5 Highlights

Sections

1 What Is a Mineral?

Key Terms

mineral, 103
silicate mineral, 104
nonsilicate mineral, 105
crystal, 106
silicon-oxygen tetrahedron, 106

Key Concepts

- A mineral is a natural, usually inorganic, crystalline solid that has a characteristic chemical composition, a regularly repeating internal structure, and a characteristic set of physical properties.
- The two main groups of minerals are silicate minerals and nonsilicate minerals.
- The six major groups of nonsilicate minerals are carbonates, halides, native elements, oxides, sulfates, and sulfides.
- Silicate minerals have six types of crystalline structures based on the arrangement of the silicon-oxygen tetrahedra.

2 Identifying Minerals

Key Terms

mineralogist, 109
streak, 110
luster, 110
cleavage, 110
fracture, 110
Mohs hardness scale, 111
density, 112

Key Concepts

- Seven physical properties that help distinguish one mineral from another are color, streak, luster, cleavage and fracture, hardness, crystal shape, and density.
- Special properties such as fluorescence and phosphorescence, chatoyancy and asterism, double refraction, magnetism, and radioactivity can aid in the identification of certain minerals.

Alternative Assessment — Advanced

Mineral Displays Assign each student one of the classes of silicate or nonsilicate minerals to investigate. Have them create a poster that displays three to five of the common minerals in that class. The display should identify the chemical composition and structure of the minerals the student chose; each mineral's major characteristics and special features, if any; uses of the minerals; and anything else that might be of interest, such as origin of mineral names. Display the posters around your classroom and allow students to view each others' work. LS Visual

CHAPTER RESOURCES

Chapter Resource File

- Concept Review GENERAL
- Critical Thinking ADVANCED
- Math Skills GENERAL
- Graphing Skills GENERAL
- Chapter Test A GENERAL
- Chapter Test B ADVANCED

Workbooks

- Study Guide (also in Spanish)
- Assessments (Spanish)

Technology

Classroom Videos
- Brain Food Video Quiz

HRW Earth Science Video
- Segment 8: Minerals of the Earth's Crust

Chapter Review

Assignment Guide

Section	Questions
1	1–2, 5–6, 9–12, 17–26, 30–31, 34
2	3–4, 7–8, 13–16, 27–28, 32
1 and 2	29, 33, 35–37

Using Key Terms

1–8. Answers may vary but should show that students understand the definitions of and differences between key terms.

Understanding Key Concepts

9. a **10.** b
11. b **12.** d
13. a **14.** b
15. b **16.** c

Short Answer

17. carbonates, containing the CO_3 group; halides, consisting of chlorine or fluorine most commonly combined with sodium, potassium, or calcium; native elements, elements uncombined with other elements; oxides, compounds containing oxygen combined with elements other than silicon; sulfates containing the SO_4 group; and sulfides, compounds of elements combined with sulfur

18. quartz, orthoclase, plagioclase, muscovite, biotite, calcite, dolomite, halite, gypsum, and ferromagnesian minerals

19. Nonsilicates that have the closest-packing structure have high densities because the atoms are packed as closely together as possible, and thus, the large mass takes up the least space or volume.

20. silicates

21. none; Carbonates, halides, and sulfides are all nonsilicates.

22. Olivine is composed of isolated SiO_4 tetrahedra. The oxygen atoms of the tetrahedra bond to magnesium and iron atoms.

23. To be classified as a mineral, a substance must be an inorganic, naturally occurring crystalline solid that has a consistent chemical composition.

24. There are four oxygen atoms and one silicon atom in each silicon-oxygen tetrahedron.

Chapter 5 Review

Using Key Terms

Use each of the following terms in a separate sentence.

1. *silicon-oxygen tetrahedron*
2. *mineral*
3. *Mohs hardness scale*
4. *cleavage*

For each pair of terms, explain how the meanings of the terms differ.

5. *mineral* and *crystal*
6. *silicate mineral* and *nonsilicate mineral*
7. *luster* and *streak*
8. *fluorescence* and *phosphorescence*

Understanding Key Concepts

9. The most common silicate minerals are the
a. feldspars. **c.** carbonates.
b. halides. **d.** sulfates.

10. Ninety-six percent of Earth's crust is made up of
a. sulfur and lead.
b. silicate minerals.
c. copper and aluminum.
d. nonsilicate minerals.

11. An example of a mineral that has a basic structure consisting of isolated tetrahedra linked by atoms of other elements is
a. mica. **c.** quartz.
b. olivine. **d.** feldspar.

12. When two single chains of tetrahedra bond to each other, the result is called a
a. single-chain silicate.
b. sheet silicate.
c. framework silicate.
d. double-chain silicate.

13. The words *waxy, pearly,* and *dull* describe a mineral's
a. luster. **c.** streak.
b. hardness. **d.** fluorescence.

14. The words *uneven* and *splintery* describe a mineral's
a. cleavage. **c.** hardness.
b. fracture. **d.** luster.

15. The ratio of a mineral's mass to its volume is the mineral's
a. atomic weight. **c.** mass.
b. density. **d.** weight.

16. Double refraction is a property of some crystals of
a. mica. **c.** calcite.
b. feldspar. **d.** galena.

Short Answer

17. Describe the six major classes of nonsilicate minerals.

18. List the 10 most common rock-forming minerals.

19. Why do minerals that have the nonsilicate crystalline structure called *closest packing* have high density?

20. Which of the two main groups of minerals is more abundant in Earth's crust?

21. Which of the following mineral groups, if any, contain silicon: carbonates, halides, or sulfides?

22. Describe the tetrahedral arrangement of olivine.

23. Summarize the characteristics that a substance must have to be classified as a mineral.

24. How many oxygen ions and silicon ions are in a silicon-oxygen tetrahedron?

Critical Thinking

25. Classifying Information Natural gas is a substance that occurs naturally in Earth's crust. Is it a mineral? Explain your answer.

26. Making Comparisons Which of the following are you more likely to find in Earth's crust: the silicates feldspar and quartz or the nonsilicates copper and iron? Explain your answer.

27. Applying Ideas Iron pyrite, FeS_2, is called *fool's gold* because it looks a lot like gold. What simple test could you use to determine whether a mineral sample is gold or pyrite? Explain what the test would show.

28. Drawing Conclusions Can you determine conclusively that an unknown substance contains magnetite by using only a magnet? Explain your answer.

Concept Mapping

29. Use the following terms to create a concept map: *mineral, silicate mineral, nonsilicate mineral, silicon-oxygen tetrahedron, color, density, crystal shape, magnetism, native element, sulfate,* and *phosphorescence.*

Math Skills

30. Applying Quantities Hematite, Fe_2O_3, has three atoms of oxygen and two atoms of iron in each molecule. What percentage of the atoms in a hematite molecule are oxygen atoms?

31. Making Calculations A sample of olivine contains 3.4 billion silicon-oxygen tetrahedra. How many oxygen atoms are in the sample?

32. Applying Quantities A mineral sample has a mass of 51 g and a volume of 15 cm^3. What is the density of the mineral sample?

Writing Skills

33. Writing from Research Use the Internet or your school library to find a mineral map of the United States. Write a brief report that outlines how the minerals in your state are discovered and mined.

34. Communicating Main Ideas Write and illustrate an essay that explains how six different crystal structures form from silicon-oxygen tetrahedra.

Interpreting Graphics

This table provides information about the eight most abundant elements in Earth's crust. Use the table to answer the questions that follow.

The Eight Most Abundant Chemicals in Earth's Crust

Element	Chemical symbol	Weight (% of Earth's crust)	Volume (% of Earth's crust)*
Oxygen	O	46.60	93.8
Silicon	Si	27.72	0.9
Aluminum	Al	8.13	0.5
Iron	Fe	5.00	0.4
Calcium	Ca	3.63	1.0
Sodium	Na	2.83	1.3
Potassium	K	2.59	1.8
Magnesium	Mg	2.09	0.3
	Total	98.59	100.0

*The volume of Earth's crust comprised by all other elements is so small that it is essentially 0% when the numbers are rounded to the nearest tenth of a percent.

35. What percentage of the weight of Earth's crust is made of silicon?

36. Oxygen makes up 93.8% of Earth's crust by volume, but oxygen is only 46.60% of Earth's crust by weight. How is this possible?

37. By comparing the volume and weight percentages of aluminum and calcium, determine which element has the higher density.

Chapter Review

Critical Thinking

25. Natural gas is not a mineral because it is not a crystalline solid and because it is usually organic in origin.

26. You are more likely to find feldspar and quartz in Earth's crust because silicates make up 96% of the crust. Feldspar and quartz alone make up over 50% of the crust.

27. Streak and density would be very different for gold and pyrite. Pyrite has a green to black streak while gold has a gold streak. Gold is also much denser than pyrite is.

28. Because small particles of some minerals that contain iron may be attracted to a magnet, you cannot conclusively determine that an unknown substance contains magnetite.

Concept Mapping

29. Answers may vary but should include all of the terms listed. Sample answers appear at the end of this Teacher Edition.

Math Skills

30. 3 out of 5 atoms in a hematite molecule are oxygen. $3/5 \times 100 = 60\%$

31. Olivine is composed of isolated tetrahedra, so each tetrahedron has 4 oxygen atoms; 3.4 billion tetrahedra × 4 O atoms/tetrahedron = 13.6 billion O atoms.

32. Density = mass / volume = $51\ g \div 15\ cm^3 = 3.4\ g/cm^3$

Critical Thinking

33. Answers may vary. Accept all reasonable answers.

34. Answers may vary. Accept all reasonable answers.

Interpreting Graphics

35. 27.72%

36. Oxygen is not very dense, so it can take up a large volume without having a lot of weight.

37. Aluminum has a higher density than calcium does.

Standardized Test Prep

Estimated Time

To give students practice under more realistic testing conditions, allow them 30 minutes to answer all of the questions in this practice test.

TEST DOCTOR

Question 1 Answer C is correct. An organic substance is one that is made up of living things or the remains of living things. Coal is made up of the remains of plants, so it is organic. Minerals are inorganic. Therefore, coal is organic and not a mineral. Students who miss this question may need to review the properties of minerals.

Question 3 Each mineral has specific properties that are a result of its chemical composition and crystal structure. These properties are color, streak, luster, cleavage and fracture, hardness, crystal shape, and density.

Question 11 Full-credit answers should include the following points:
- most early metals used by humans were soft metals that could be easily worked and shaped
- copper is relatively abundant in many areas. It is easy to mine and refine. Copper is a soft metal that is easily bent and shaped
- copper resists corrosion and can be polished to a shining finish
- copper weapons and tools were superior in strength and durability to previous tools

Chapter 5 Standardized Test Prep

Understanding Concepts

Directions (1–5): **For** *each* **question, write on a separate sheet of paper the number of the correct answer.**

1. Coal is
   A. organic and a mineral
   B. inorganic and a mineral
   C. organic and not a mineral
   D. inorganic and not a mineral

2. Which of the following is one of the 10 rock-forming minerals that make up 90% of the mass of Earth's crust?
   F. quartz
   G. fluorite
   H. copper
   I. talc

3. Minerals can be identified by all of the following properties *except*
   A. specimen color
   B. specimen shreak
   C. specimen hardness
   D. specimen luster

4. All minerals in Earth's crust
   F. have a crystalline structure
   G. are classified as ring silicates
   H. are classified as pyroxenes or amphiboles
   I. have no silicon in their tetrahedral structure

5. Which mineral can be scratched by a fingernail that has a hardness of 2.5 on the Mohs scale?
   A. diamond
   B. quartz
   C. topaz
   D. talc

Directions (6–8): **For** *each* **question, write a short response.**

6. Carbonates, halides, native elements, oxides, sulfates, and sulfides are classes of what mineral group?

7. What mineral is made up of *only* the elements oxygen and silicon?

8. What property is a mineral said to have when a person is able to view double images through it?

Reading Skills

Directions (9–11): **Read the passage below. Then, answer the questions.**

Native American Copper

In North America, copper was mined at least 6,700 years ago by the Native Americans who lived on Michigan's upper peninsula. Much of this mining took place on the Isle Royale, an island located in the waters of Lake Superior.

These ancient people removed copper from the rock by using stone hammers and wedges. The rock was sometimes heated to make breaking it easier. Copper that was mined was used to make a wide variety of items for the Native Americans including jewelry, tools, weapons, fish hooks, and other objects. These objects were often marked with intricate designs. The copper mined at the Lake Superior site was traded over long distances along ancient trade routes. Copper objects from the region have been found in Ohio, Florida, the Southwest, and the Northwest.

9. According to the passage, Native Americans who mined copper
   A. used the mineral as a form of currency when buying goods from other tribes
   B. traded copper objects with other Native American tribes over a large area
   C. used the mineral to produce vastly superior weapons and armor
   D. sold it to the Native Americans living around Lake Superior

10. Which of the following statements can be inferred from the information in the passage?
   F. Copper is a very strong metal and can be forged into extremely strong items.
   G. Copper mining in the ancient world was only common in North America.
   H. Copper is a useful metal that can be forged into a wide variety of goods.
   I. Copper is a weak metal, and no items made by the ancient Native Americans remain.

11. What are some properties of copper that might have made the metal useful to Native Americans?

Answers

Understanding Concepts
1. C
2. F
3. A
4. F
5. D
6. nonsilicate minerals
7. quartz
8. double refraction

Reading Skills
9. B
10. H
11. Answers may vary. See Test Doctor for a detailed scoring rubric.

Interpreting Graphics
12. D
13. Answers may vary. See Test Doctor for a detailed scoring rubric.
14. A
15. Answers may vary. See Test Doctor for a detailed scoring rubric.

Interpreting Graphics

Directions (12–15): **For** ***each*** **question below, record the correct answer on a separate sheet of paper.**

Base your answers to questions 12 and 13 on the figure below, which shows the abundance of various elements in Earth's crust.

Elements in Earth's Crust

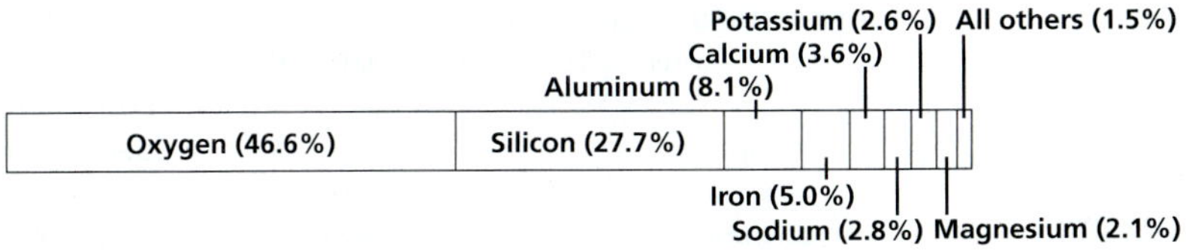

12 Hematite is composed of oxygen and what other element?

A. calcium
B. aluminum
C. sodium
D. iron

13 Silicate minerals make up about 95% of Earth's crust. However, the elements present in all minerals in this group, oxygen and silicon, make up a significantly smaller percentage of the weight of Earth's crust. How can this discrepancy be explained?

Base your answers to questions 14 and 15 on the table below, which provides information about silicate minerals.

Common Silicates

Mineral	Idealized formula	Cleavage
Olivine	$(Mg,Fe)_2SiO_4$	none
Pyroxene group	$(Mg,Fe)SiO_3$	two planes at right angles
Amphibole group	$Ca_2(Mg,Fe)_5Si_8O_{22}(OH)_2$	two planes at 60° and 120°
Micas, biotite	$K(Mg,Fe)_3AlSi_3O_{10}(OH)_2$	one plane
Micas, muscovite	$KAl_2(AlSi_3O_{10})(OH)_2$	one plane
Feldspars, orthoclase	$KAlSi_3O_8$	two planes at 90°
Feldspars, plagioclase	$(Ca,Na)AlSi_3O_8$	two planes at 90°
Quartz	SiO_2	none

14 How is the cleavage of amphibole minerals similar to that of feldspar minerals?

A. Both have two planes.
B. Both have one plane.
C. Both cleave at 60°.
D. Both cleave at 90°.

15 Which minerals are ferromagnesian? How can you identify these minerals? Predict how the chemical composition of ferromagnesian minerals affects the minerals' density and magnetic properties.

Test TIP

If a question or an answer choice contains an unfamiliar term, try to break the word into parts to determine its meaning.

Standardized Test Prep

TEST DOCTOR

Question 13 Full-credit answers should include the following points:

- by weight, oxygen and silicon make up 74.3% of Earth's crust. Minerals that contain these elements are silicate minerals, which are the most abundant minerals in Earth's crust
- however, silicate minerals may also contain additional elements. These additional elements make the total weight of silicate minerals greater than the total weight of the two base elements of oxygen and silicon
- the silicate mineral quartz is made up of only oxygen and silicon, but other silicate minerals contain one or more other elements

Question 15 Full-credit answers should include the following points:

- olivines, pyroxenes, amphiboles, and biotite are ferromagnesian minerals
- ferromagnesian minerals are rich in both iron (Fe) and magnesium (Mg)
- due to the presence of these metallic elements, the minerals formed are often dark in color
- metallic lusters and high densities are common features of ferromagnesian minerals
- often, due to the high iron content, such minerals are also magnetic

Test Prep Correlations *National Science Education Standards*

HNS 2a: items 1, 3, 4, 7, 8, 12, 14, 15
HNS 2b: items 13, 15
SPSP 3a: items 9, 10, 11
UCP 1: items 5, 6
UCP 2: items 2, 4

CHAPTER RESOURCES

State Resources

For specific resources for your state, visit go.hrw.com and type in the keyword HSHSTR.

Mineral Identification

Teacher's Notes

Time Required
one 45-minute class period

Lab Ratings
EASY → HARD

TEACHER PREPARATION (2)
STUDENT SETUP (1)
CONCEPT LEVEL (2)
CLEANUP (1)

Skills Acquired
- Observing
- Measuring
- Collecting Data
- Organizing and Analyzing Data
- Identifying and Recognizing Patterns

The Scientific Method
In this lab, students will
- Make Observations
- Analyze Results
- Draw Conclusions
- Communicate Results

Materials
The materials listed on this page are enough for groups of two to four students.

Tips and Tricks
You may want to have a field guide to minerals available for student use. Depending on the samples you provide, you could also have students perform other tests such as for magnetism, fluorescence/phosphorescence, or double refraction. Once students have tentatively identified their samples, have them compare their results with other groups and discuss any differences they find.

Objectives
- **Identify** several unknown mineral samples.
- **Evaluate** which properties of minerals are most useful in identifying mineral samples.

Materials
- file, steel
- Guide to Common Minerals (in the Reference Tables section of the Appendix)
- hand lens
- mineral samples (5)
- penny, copper
- square, glass
- streak plate

Safety
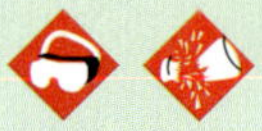

Mineral Identification

A mineral identification key can be used to compare the properties of minerals so that unknown mineral samples can be identified. Mineral properties that are often used in mineral identification keys are color, hardness, streak, luster, cleavage, and fracture. Hardness is determined by a scratch test. The Mohs hardness scale classifies minerals from 1 (soft) to 10 (hard). Streak is the color of a mineral in a finely powdered form. The streak shows less variation than the color of a sample does and thus is more useful in identification. The luster of a mineral is either metallic (having an appearance of metals) or nonmetallic. Cleavage is the tendency of a mineral to split along a plane. Planes may be in several directions. Other minerals break into irregular fragments in a process called *fracture.* In this lab, you will use these properties to classify several mineral samples.

PROCEDURE

1. Make a table with columns for sample number, color/luster, hardness, streak, cleavage/fracture, and mineral name.
2. Observe and record in your table the color of each mineral sample. Note whether the luster of each mineral is metallic or nonmetallic.

Step 4

CHAPTER RESOURCES

Chapter Resource File
- Datasheet for Chapter Lab GENERAL
- Lab Notes and Answers

Sample number	Color/ luster	Hardness	Streak	Cleavage/ fracture	Mineral name
1					
2					
3					
4					
5					

DO NOT WRITE IN THIS BOOK

3. Rub each mineral against the streak plate, and determine the color of the mineral's streak. Record your observations.

4. Using a fingernail, copper penny, glass square, and steel file, test each mineral to determine its hardness based on the Mohs hardness scale. Arrange the minerals in order of hardness. Record your observations in your table.

5. Determine whether the surface of each mineral displays cleavage or fracture. Record your observations.

6. Use the Guide to Common Minerals in the Reference Tables section of the Appendix to help you identify the mineral samples. Remember that samples of the same mineral will vary somewhat.

ANALYSIS AND CONCLUSION

1. **Analyzing Results** For each mineral, compare the streak with the color of the mineral. Which minerals have the same color as their streak? Which do not?

2. **Classifying Information** Of the mineral samples you identified, how many were silicate minerals? How many were nonsilicate minerals?

3. **Analyzing Methods** Did you find any properties that were especially useful or especially not useful in identifying each sample? Identify these properties, and explain why they were or were not useful.

4. **Evaluating Methods** If you had to write a manual to explain step by step how to identify minerals, in what order would you test different properties? Explain your answer.

Extension

1. **Understanding Relationships** Corundum, rubies, and sapphires have different colors but are considered to be the same mineral. Diamonds and graphite are made of the element carbon but are not considered to be the same mineral. Research these minerals, and explain why they are classified in this way.

Skills Practice Lab

Answers to Analysis and Conclusion

1. Answers may vary depending on the samples provided to the students. In general, nonmetals have a streak similar to their color, while metals show a dark streak.
2. Answers may vary depending on samples provided to students.
3. Answers may vary depending on samples provided to students and students' observations. Accept all reasonable answers.
4. Answers may vary. Accept all reasonable answers.

Answers to Extension

1. Graphite and diamond are both composed of carbon atoms, but have different crystalline structures. They are known as polymorphs. In graphite, each carbon atom bonds to three other carbon atoms to form 2-dimensional sheets that are weakly bonded together. The weak bonding gives graphite flexibility and lubricating properties. In diamond, each carbon atom bonds to four other carbon atoms in a strong three-dimensional network, giving diamond its incredible strength and thermal stability. Corundum, ruby, and sapphire are all considered the same mineral, even though they are different colors, because they have the same chemical composition (Al_2O_3) and the same crystalline structure (trigonal). The color differences result from impurities in the crystal. Rubies are corundum with trace amounts of chromium, while sapphires are corundum with trace amounts of cobalt and titanium.

TEACHER TESTED & APPROVED

Alexander Dvorak
Heritage School
New York City, NY

Rock and Mineral Production in the United States

Internet Activity — GENERAL

Mining Impacts Mining can have serious effects on the environment. Have interested students investigate the past and present environmental effects of mining in your state or a nearby state. They may investigate reclamation efforts or the use of technology at current mines. Have students write a persuasive essay supporting or opposing continued mining in your region based on their findings. A worksheet designed to direct student research on this topic can be found in the **Chapter Resource File** booklet or by visiting **go.hrw.com** and entering the keyword **HQ6MINX**. **LS Logical/ Interpersonal**

Answers to Map Skills Activity

1. stone
2. 3.0 + 2.7 + 1.6 + 1.0 = 8.3 ÷ 31.6 × 100 = 26.3%; CA, NV, UT, AZ, AK, NM, MT, SD, MO, TN, MN, MI, and NY
3. Answers may vary.
4. 8.2 + 5.3 = 13.5 ÷ 31.6 × 100 = 43%; CA, AZ, TX, IL, MI, OH, PA, GA, and FL
5. IL or GA, depending on where in the state the stone is mined

CHAPTER RESOURCES

Chapter Resource File

Internet Activity
- Mining Impacts GENERAL

Technology

Transparencies
- 25 Rock and Mineral Production in the United States (with worksheet)

MAPS in Action

Rock and Mineral Production in the United States

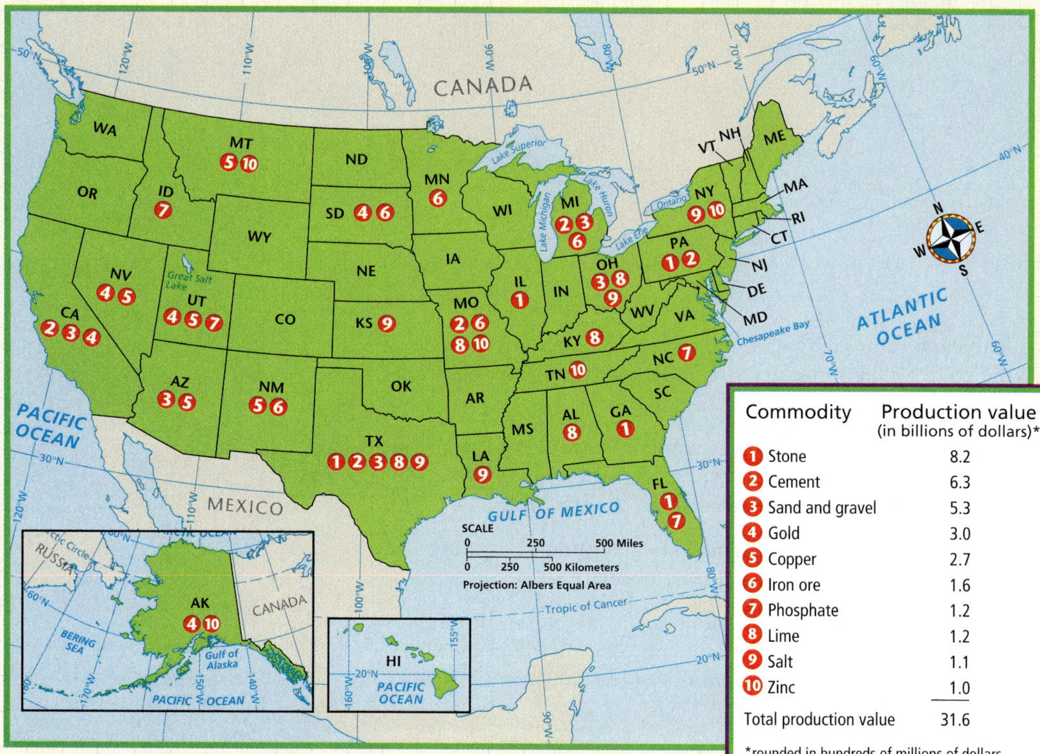

Commodity	Production value (in billions of dollars)*
1 Stone	8.2
2 Cement	6.3
3 Sand and gravel	5.3
4 Gold	3.0
5 Copper	2.7
6 Iron ore	1.6
7 Phosphate	1.2
8 Lime	1.2
9 Salt	1.1
10 Zinc	1.0
Total production value	31.6

*rounded in hundreds of millions of dollars

Map Skills Activity

This map shows the distribution of the top 10 rock and mineral commodities produced in the United States in 1999. The key provides production values for these commodities. Use the map to answer the questions below.

1. **Using a Key** Which commodity had the highest production value in 1999?
2. **Evaluating Data** Gold, copper, iron ore, and zinc are metals in the top 10 mineral commodities produced in 1999. What percentage of the total 1999 production value do these metals represent? Which states produced these metals in 1999?
3. **Using a Key** Find your state on the map. Which of the top 10 mineral commodities, if any, were produced in your state in 1999?
4. **Evaluating Data** Stone, sand, and gravel are collectively known as *aggregates*. What percentage of the total 1999 production value of the 10 commodities listed do aggregates represent? Which states were the major producers of aggregates in 1999?
5. **Analyzing Relationships** If Texas were not a producer of stone, which state would be the closest one from which people in Texas could acquire stone?

Career Focus

Mining Engineer

"At first, I wasn't interested in being a mining engineer," remembers Jami Girard-Dwyer. "I wanted to get a degree in computer science and write programs. Then, I took a course entitled 'Introduction to Mining Engineering' and got a summer job at a nearby gold mine. I became fascinated with the process of extracting ore from Earth and coming up with a finished product."

Computers and Mining

Girard-Dwyer realized that her knowledge of computers had many applications in the mining industry. "I began writing computer programs to help mining operations work more efficiently," she recalls. "I became very interested in learning how mine openings could be designed to ensure that they wouldn't collapse on workers or equipment."

Rewards

Now a mining engineer for the National Institute for Occupational Safety and Health (NIOSH), Girard-Dwyer helps keep miners safe. "The most rewarding part of my job is helping mines develop safer environments so that nobody gets hurt," she says. "I study the health and safety of the mine workers. I also do research to help prevent accidents, injuries, and fatalities." Her work frequently requires her to travel to visit mines. She says, "At a mine site, I will install monitoring equipment and computers to collect data about a particular type of problem. I will then use computers to analyze the data and make recommendations to the mining company based on the results."

Mines may be located on the surface of Earth (open-pit or strip mines) or thousands of feet underground. "Metals such as gold or silver often come to mind when people think of mining," says Girard-Dwyer. "But there are also mines that recover coal for creating energy, silica for making glass, and special industrial minerals for manufacturing everything from kitty litter to toothpaste!" Each mining operation uses unique methods to recover the minerals.

"Without mining, we would not have cars, electricity, computers, . . . or any of the items that we often take for granted in today's society."

—Jami Girard-Dwyer

◀ Miners, such as this one drilling for coal in Utah, rely on mining engineers like Girard-Dwyer to help develop safe working environments.

SCILINKS. NSTA Developed and maintained by the National Science Teachers Association

For a variety of links related to this subject, go to www.scilinks.org

Topic: Careers in Earth Science
SciLinks code: HQ60222

Career Focus

Mining Engineer

Activity — GENERAL

Finding Mineral Deposits Before minerals can be safely and economically mined, deposits must be located. Mining engineers play an important role in the prospecting and exploration of potential deposits. Have interested students research the methods used to locate and assess mineral deposits. They should prepare a poster or multimedia presentation that illustrates the process(es) involved in locating potential mineral deposits and the factors that must be considered in deciding to mine at a given location. **LS Logical/Visual**

Extension — GENERAL

Research Have students identify four other careers that are related to minerals. Have students write a brief synopsis of each career and present them to the class. (Answers may vary. Accept all reasonable answers.) **LS Verbal**

Chapter 6

Rocks

Planning Guide

Compression Guide

To shorten instruction because of time limitations, omit the Chapter Lab.

OBJECTIVES	LABS, DEMONSTRATIONS, AND ACTIVITIES	TECHNOLOGY RESOURCES
PACING • 90 min pp. 124–128 **Chapter Opener**		OSP **Parent Letter** ■ CD **Student Edition on CD-ROM** CD **Chapter Summaries Audio CD** ■ VID **Brain Food Video Quiz**
Section 1 Rocks and the Rock Cycle • Identify the three major types of rock, and explain how each type forms. • Summarize the steps in the rock cycle. • Explain Bowen's reaction series. • Summarize the factors that affect the stability of rocks.	TE **Discussion** Types of Rocks, p. 125 ◆ GENERAL SE **Skills Practice Lab** Classification of Rocks, pp. 150–151 ◆ GENERAL CRF **Datasheet for Chapter Lab*** GENERAL SE **Maps in Action** Geologic Map of Virginia, p. 152 GENERAL	OSP **Lesson Plans** (also in print) TR **Bellringer*** TR 26 The Rock Cycle* TR 27 Bowen's Reaction Series* TR 31 Geologic Map of Virginia* TE **Internet Activity** Geology of Other Planets, p. 153 ADVANCED CRF **Internet Activity** Geology of Other Planets* ADVANCED
PACING • 45 min pp. 129–134 **Section 2 Igneous Rock** • Summarize three factors that affect whether rock melts. • Describe how the cooling rate of magma and lava affects the texture of igneous rocks. • Classify igneous rocks according to their composition and texture. • Describe intrusive and extrusive igneous rock structures.	TE **Discussion** Hand Samples, p. 129 ◆ GENERAL TE **Demonstration** Partial Freezing, p. 130 ◆ GENERAL SE **Quick Lab** Crystal Formation, p. 130 ◆ GENERAL CRF **Datasheet for Quick Lab*** GENERAL TE **Geology Connection** Information from Igneous Rocks, p. 131 ADVANCED TE **Group Activity** Poster Project, p. 132 ADVANCED	OSP **Lesson Plans** (also in print) TR **Bellringer*** TR 28 Partial Melting and Fractional Crystallization* VID **CNN Video** Radon Risk
PACING • 45 min pp. 135–140 **Section 3 Sedimentary Rock** • Explain the processes of compaction and cementation. • Describe how chemical and organic sedimentary rocks form. • Describe how clastic sedimentary rock forms. • Identify seven sedimentary rock features.	TE **Demonstration** Sedimentation, p. 135 ◆ GENERAL TE **Chemistry Connection** The Carbon Cycle, p. 136 ADVANCED TE **Discussion** Comprehension Check, p. 137 GENERAL SE **Quick Lab** Graded Bedding, p. 139 ◆ GENERAL CRF **Datasheet for Quick Lab*** GENERAL CRF **Inquiry Lab** Sorting Sediments* GENERAL	OSP **Lesson Plans** (also in print) TR **Bellringer*** TR 29 Organic Limestone Formation* TE **Internet Activity** Identifying Sedimentary Rock Features, p. 138 GENERAL CRF **Internet Activity** Identifying Sedimentary Rock Features* GENERAL
PACING • 45 min pp. 141–144 **Section 4 Metamorphic Rock** • Describe the process of metamorphism. • Explain the difference between regional and contact metamorphism. • Distinguish between foliated and nonfoliated metamorphic rocks, and give an example of each.	TE **Demonstration** Metamorphism, p. 141 ◆ GENERAL TE **Activity** Poster Project, p. 142 ADVANCED CRF **Making Models Lab** Metamorphic Rock* GENERAL	OSP **Lesson Plans** (also in print) TR **Bellringer*** TR 30 Indicators of Metamorphic Conditions* CD **Interactive Tutor** Igneous, Sedimentary, and Metamorphic Rocks

PACING • 90 min

CHAPTER REVIEW, ASSESSMENT, AND STANDARDIZED TEST PREPARATION

- SE **Chapter Highlights**, p. 145
- SE **Chapter Review**, pp. 146–147
- SE **Standardized Test Prep**, pp. 148–149
- CRF **Concept Review*** ■ GENERAL
- CRF **Critical Thinking*** ADVANCED
- CRF **Math Skills*** GENERAL
- CRF **Graphing Skills*** GENERAL
- CRF **Chapter Test A*** ■ GENERAL
- CRF **Chapter Test B*** ADVANCED
- OSP **Lesson Plans** (also in print)
- OSP **Test Generator**
- OSP **Test Item Listing**

Online and Technology Resources

go.hrw.com Holt Online Learning

Visit go.hrw.com for access to Holt Online Learning, or enter the keyword **HQ6 Home** for a variety of free online resources.

One-Stop Planner® CD-ROM

This CD-ROM package includes

- Lab Materials QuickList Software
- Holt Calendar Planner
- Customizable Lesson Plans
- Printable Worksheets
- ExamView® Test Generator
- Interactive Teacher Edition
- Holt PuzzlePro®
- Holt PowerPoint® Resources

KEY

SE	Student Edition	OSP	One-Stop Planner	VID	Classroom Video/DVD
TE	Teacher Edition	TR	Transparencies and Transparency Worksheets	*	Also on One-Stop Planner
CRF	Chapter Resource File	CD	CD or CD-ROM	◆	Requires advance prep
LTP	Long-Term Projects			■	Also available in Spanish

SKILLS DEVELOPMENT RESOURCES	REVIEW AND ASSESSMENT	CORRELATIONS
SE **Pre-Reading Activity,** p. 124 GENERAL TE **Using the Figure** Eagletail Peak, p. 124 GENERAL		National Science Education Standards
CRF **Directed Reading*** BASIC TE **Using the Figure** The Rock Cycle, p. 126 GENERAL TE **Inclusion Strategies,** p. 126 SE **Graphic Organizer** Chain-of-Events Chart, p. 127 GENERAL	SE **Reading Check,** p. 127 GENERAL SE **Section Review,** p. 128 GENERAL TE **Reteaching,** p. 127 BASIC TE **Quiz,** p. 127 GENERAL TE **Alternative Assessment,** p. 128 ADVANCED CRF **Section Quiz*** ■ GENERAL	ES 2b
CRF **Directed Reading*** BASIC TE **Using the Figure** Fractional Crystallization, p. 130 BASIC TE **Reading Skill Builder** Reading Organizer, p. 131 BASIC TE **Using the Figure** Intrusive Igneous Structures, p. 133 GENERAL TE **Skill Builder** Vocabulary, p. 133 GENERAL	SE **Reading Checks,** pp. 131, 133 GENERAL SE **Section Review,** p. 134 GENERAL TE **Reteaching,** p. 133 BASIC TE **Quiz,** p. 133 GENERAL TE **Alternative Assessment,** p. 134 ADVANCED CRF **Section Quiz*** ■ GENERAL	ES 2b
CRF **Directed Reading*** BASIC TE **Using the Figure** Organic Limestone Formation, p. 136 GENERAL SE **Math Practice,** p. 137 GENERAL TE **Reading Skill Builder** Paired Summarizing, p. 137 BASIC TE **Reading Skill Builder** Reading Organizer, p. 138 BASIC	SE **Reading Checks,** pp. 137, 139 GENERAL SE **Section Review,** p. 140 GENERAL TE **Homework,** p. 136 GENERAL TE **Reteaching,** p. 139 BASIC TE **Quiz,** p. 139 GENERAL TE **Alternative Assessment,** p. 140 ADVANCED CRF **Section Quiz*** ■ GENERAL	ES 2b
CRF **Directed Reading*** BASIC TE **Reading Skill Builder** Anticipation Guide, p. 142 BASIC	SE **Reading Check,** p. 142 GENERAL SE **Section Review,** p. 144 GENERAL TE **Homework,** p. 143 ADVANCED TE **Reteaching,** p. 143 BASIC TE **Quiz,** p. 143 GENERAL TE **Alternative Assessment,** p. 144, ADVANCED CRF **Section Quiz*** ■ GENERAL	ES 2b

Holt Lab Generator CD-ROM

Search for labs by topic, standard, difficulty level, and/or time frame. Edit those labs to fit your needs, or create your own labs. Use the Lab Materials QuickList to order just what you need.

Holt Earth Science Interactive Tutor CD-ROM

This CD-ROM consists of interactive activities that give students a fun way to extend their knowledge of Earth science concepts.

Chapter Summaries Audio CDs

These CDs include audio summaries of the key concepts presented in each chapter. (Audio summaries are also available in Spanish.)

SCILINKS. NSTA

www.scilinks.org

Maintained by the **National Science Teachers Association.** See Chapter Enrichment pages that follow for a complete list of topics.

See Chapter Enrichment pages for Video Resources.

Chapter Enrichment

This Chapter Enrichment provides relevant and interesting information to expand and enhance your classroom instruction of the chapter material.

Section 1 Rocks and the Rock Cycle

Physical Stability of Igneous Rock

The zones of weakness that occur during magma crystallization have produced a number of structures that are startling for their size and remarkable regularity of shape. In some cases, these formations have occurred in intrusive structures that have been exposed over time by erosion of the material around them. A fine example of this is Devils Tower in Wyoming. This 185 m tall volcanic structure formed 40 million years ago when magma solidified beneath what was then the surface. Devils Tower is unusual not only for its height but also for the regular fluting of its steep sides. Contraction in the crystallizing phonolite caused joints to form, which created mostly pentagonal columns approximately 1.8 m across and extending the height of the tower. The columns curve near the base of the tower, an unusual feature that resulted from efficient cooling of the magma.

Similar columnar formations of basalt also appear in extrusive structures. Contraction in lava flows along the eastern side of the Sierra Nevada range in California created the Devil's Postpile, a field of hexagonal columns 60 to 90 cm across and over 18 m tall. A similar structure, the Giant's Causeway in Northern Ireland, consists of some 40,000 mostly hexagonal columns. These columns are so nearly uniform in size that they were long believed to have been artificially created.

▲ **Columnar joints occur when magma or lava cools and contracts.**

Section 2 Igneous Rock

Igneous Beaches

In most places, beach sand originates from a mix of all three types of rock and appears as various shades of brown. Some areas of the world, however, have so much igneous rock that the sands of coastal beaches consist entirely of eroded igneous rock. A number of such beaches are on the southeastern coast of the island of Hawaii. In areas where lava flowed up to and into the ocean, waves and currents pulverized the lava and transported the rock fragments to coves and inlets. The resulting beaches consist of coarse black sand. In a more unusual case, lava with a high olivine content has been deposited in a remote inlet on the southern tip of the island. Olivine (a complex silicate of magnesium and iron) is named for its olive green color, which becomes evident to anyone who sees its fine grains collected on this greenish-hued beach. This inlet is a good example of what is called a *green sand beach*.

Section 3 Sedimentary Rock

Clastic Sedimentation by Impact

The moon's 100 km thick crust lacks the plate-tectonic activity that accounts for many features on Earth. Also, the absence of air or water on the lunar surface prevents sedimentation mechanisms that are common on Earth. Yet 85% by weight of the lunar surface consists of breccia-type rocks, which are classified as sedimentary.

Lunar "sedimentation" is caused by meteorite impacts, which transport matter radially outward. The small, dusty fragments are scattered and heated to their melting point by the impact and fill in gaps between other pieces of broken rock, fusing the rock pieces together. This melting and fusing process is a different form of cementation than is found on Earth, just as the scattering from impact is not the typical means for transporting sediment on this planet. Yet, the resulting rock is recognizable as breccia. However, few planetary geologists are willing to say that most of the lunar surface is made up of sedimentary rock.

▲ The banded gneiss in this photo is an example of foliated metamorphic rock.

Section 4 Metamorphic Rock

From Limestone to Marble to Quicklime

Because the presence or addition of fluids affects the formation of metamorphic rock, the chemical composition of the resulting rock may be somewhat different from that of the source rock. Limestone and marble, though different types of rock, are chemically similar—and both have been key components in the building trades for centuries. Both of these rocks can undergo a chemical process that reduces it to its basic components: calcium oxide, or quicklime (CaO), and carbon dioxide (CO_2).

In medieval Rome, quicklime was in large demand for use in plaster to cover the surfaces of buildings. Limestone required quarrying, but there was a much handier source material available: the ruins of classical Rome. So much marble had been used in ancient Rome that medieval lime kilns (for burning the marble and creating quicklime) were provided with enough marble to produce lime for more than 500 years.

DVD VIDEO Video Resources

Brain Food Video Quizzes

Brain Food Video Quizzes These videos contain game-show style quizzes that assess students' progress and motivate students to study the chapter material.

NOVA® Videos

To order **NOVA** videos related to this chapter, visit go.hrw.com and enter the keyword **HQ6RXSV.**

SCILINKS®

NSTA Developed and maintained by the National Science Teachers Association

SciLinks is maintained by the National Science Teachers Association to provide you and your students with interesting, up-to-date links that will enrich your classroom presentation of the chapter.

Visit www.scilinks.org and enter the SciLinks code for more information about the topic listed.

Topic: The Rock Cycle
SciLinks code: HQ61319

Topic: Igneous Rock
SciLinks code: HQ60783

Topic: Sedimentary Rock
SciLinks code: HQ61365

Topic: Metamorphic Rock
SciLinks code: HQ60949

Chapter 6

Chapter Overview

Rocks are generally classified as igneous, sedimentary, or metamorphic. Rocks form in a variety of ways and have characteristic compositions and textures that help scientists classify the rocks.

Using the Figure — GENERAL

Eagletail Peak Eagletail Peak is in the Eagletail Mountains, about 100 miles west of Phoenix, Arizona. These mountains are composed mainly of sedimentary rock. Explain to students that sedimentary rock can form from rock fragments that are deposited by wind or water. Then, ask students how the rock in the photograph could have formed. (Answers may vary. Students may note that the rock that makes up the formation was deposited in flat layers a long time ago. Since then, uplift and erosion caused the rock structures that exist today to form.) LS **Visual**

PRE-READING ACTIVITY

FOLDNOTES

Have students exchange FoldNotes and check each other's notes on types of rocks. Encourage students to use a colored pen to add additional notes about igneous, sedimentary, and metamorphic rocks.

Chapter 6 Rocks

Sections

1 Rocks and the Rock Cycle

2 Igneous Rock

3 Sedimentary Rock

4 Metamorphic Rock

What You'll Learn

- How the processes that form rock determine the properties of rocks
- How scientists classify rocks

Why It's Relevant

Understanding how rocks form provides a basis for understanding the properties of different types of rocks.

PRE-READING ACTIVITY

FOLDNOTES

Pyramid Before you read this chapter, create the FoldNote entitled "Pyramid" described in the Skills Handbook section of the Appendix. Label the sides of the pyramid with "Igneous rock," "Sedimentary rock," and "Metamorphic rock." As you read the chapter, define each type of rock, and write characteristics of each type of rock on the appropriate pyramid side.

▶ Eagletail Peak stands more than 1005 m above Sonoran Desert in Arizona and is composed of sedimentary rock. Sedimentary rock is one of the three major types of rock.

Chapter Correlations	National Science Education Standards
ES 2b Movement of matter between reservoirs is driven by the earth's internal and external sources of energy. These movements are often accompanied by a change in the physical and chemical properties of the matter. **(Sections 1, 2, 3, and 4)**	

Section 1 Rocks and the Rock Cycle

The material that makes up the solid parts of Earth is known as *rock*. Rock can be a collection of one or more minerals, or rock can be made of solid organic matter. In some cases, rock is made of mineral matter that is not crystalline, such as glass. Geologists study the forces and processes that form and change the rocks of Earth's crust. Based on these studies, geologists have classified rocks into three major types by the way the rocks form.

OBJECTIVES

- **Identify** the three major types of rock, and explain how each type forms.
- **Summarize** the steps in the rock cycle.
- **Explain** Bowen's reaction series.
- **Summarize** the factors that affect the stability of rocks.

KEY TERMS

rock cycle
Bowen's reaction series

Three Major Types of Rock

Studies of volcanic activity provide information about the formation of one rock type—*igneous rock*. The word *igneous* is derived from a Latin term that means "from fire." Igneous rock forms when *magma*, or molten rock, cools and hardens. Magma is called *lava* when it is exposed at Earth's surface.

Agents of erosion, such as wind and waves, break down all types of rock into small fragments. Rocks, mineral crystals, and organic matter that have been broken into fragments are known as *sediment*. Sediment is carried away and deposited by water, ice, and wind. When sediment deposits are compressed, cemented together, and harden, *sedimentary rock* forms.

Certain forces and processes, including tremendous pressure, extreme heat, and chemical processes, also can change the form of existing rock. The rock that forms when existing rock is altered is *metamorphic rock*. The word *metamorphic* means "changed form." **Figure 1** shows an example of each major type of rock.

Figure 1 ▶ These rocks are examples of the three major rock types.

Granite (igneous)

Sandstone (sedimentary)

Gneiss (metamorphic)

Section 1

Focus

Overview

This section introduces the three basic types of rock, describes how rock changes form through the rock cycle, and describes Bowen's reaction series and the factors that affect the chemical and physical stability of rocks.

Bellringer

Have students list the names of five rocks with which they are familiar, and have them describe characteristics of the rock and identify how each type of rock is used. (Answers may vary.) **LS Verbal**

Motivate

Discussion — GENERAL

Types of Rocks Place on display for the class the igneous rocks pumice and granite, the sedimentary rocks sandstone and limestone, and the metamorphic rocks gneiss and schist. Ask students to note and discuss the common features of the two rocks of each type. (Answers may vary. Students may note that both limestone and sandstone have fine, grainy layers, that pumice is glassy and granite is made up of crystals, and that the schist and gneiss have bands of minerals or layers of crystals.) **LS Verbal/Visual**

CHAPTER RESOURCES

Chapter Resource File

- Directed Reading BASIC

Technology

Transparencies
- Bellringer

Student Edition on CD-ROM

One-Stop Planner CD-ROM
- Lesson Plan

Teach

Using the Figure — GENERAL

The Rock Cycle Emphasize to students that a set of processes causes a rock to change from one type into another or into a different form of its original type. Ask students to identify the processes that must be applied to igneous rock to change it to metamorphic rock. (chemical processes and changes in temperature and pressure) Ask students to identify what processes change metamorphic rock to sedimentary rock. (erosion, deposition, compaction, and cementation) Ask students to identify the processes that change one sedimentary rock into another sedimentary rock. (weathering, deposition, compaction, and cementation) LS **Visual**

BRAIN FOOD

Lunar Rocks According to currently accepted theory, the moon formed as a result of the collision of a large asteroid (perhaps as big as Mars) and the molten Earth about 4.6 billion years ago. This model is supported by, and accounts for, both the composition of lunar rock and its density, which is lower than Earth rock. Lunar rock came from the less dense material of Earth's surface and upper mantle. The smaller size of the moon allowed it to cool more rapidly than Earth has cooled. The absence of plate-tectonic activity and of an atmosphere has made the rock cycle of the moon much simpler and less dramatic than that of Earth.

CHAPTER RESOURCES

Technology

Transparencies
- 26 The Rock Cycle (with worksheet)
- 27 Bowen's Reaction Series (with worksheet)

Figure 2 ▶ The rock cycle illustrates the changes that igneous, sedimentary, and metamorphic rocks undergo.

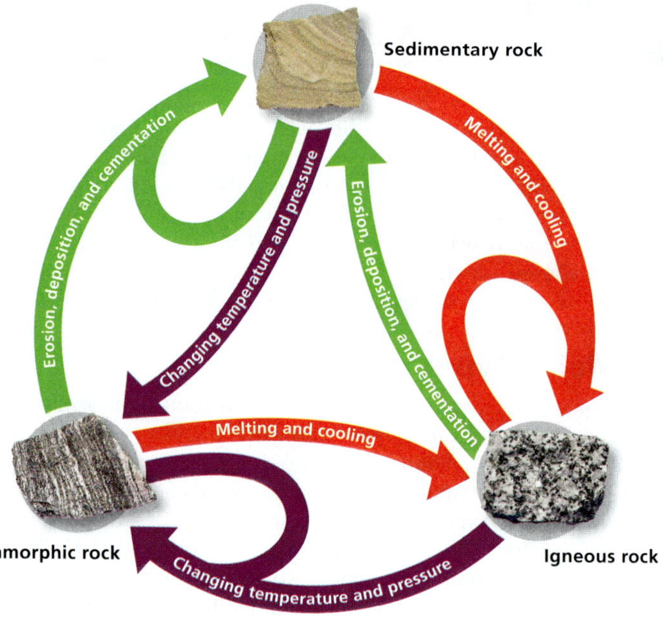

The Rock Cycle

Any of the three major types of rock can be changed into another of the three types. Geologic forces and processes cause rock to change from one type to another. This series of changes is called the **rock cycle,** which is shown in **Figure 2.**

rock cycle the series of processes in which rock forms, changes from one type to another, is destroyed, and forms again by geological processes

One starting point for examining the steps of the rock cycle is igneous rock. When a body of igneous rock is exposed at Earth's surface, a number of processes break down the igneous rock into sediment. When sediment from igneous rocks is compacted and cemented, the sediment becomes sedimentary rock. Then, if sedimentary rocks are subjected to changes in temperature and pressure, the rocks may become metamorphic rocks. Under certain temperature and pressure conditions, the metamorphic rock will melt and form magma. Then, if the magma cools, new igneous rock will form.

Much of the rock in Earth's continental crust has probably passed through the rock cycle many times during Earth's history. However, as **Figure 2** shows, a particular body of rock does not always pass through each stage of the rock cycle. For example, igneous rock may never be exposed at Earth's surface where the rock could change into sediment. Instead, the igneous rock may change directly into metamorphic rock while still beneath Earth's surface. Sedimentary rock may be broken down at Earth's surface, and the sediment may become another sedimentary rock. Metamorphic rock can be altered by heat and pressure to form a different type of metamorphic rock.

SCILINKS. NSTA Developed and maintained by the National Science Teachers Association

For a variety of links related to this subject, go to www.scilinks.org

Topic: The Rock Cycle
SciLinks code: HQ61319

INCLUSION Strategies

- Developmentally Disabled
- Hearing Impaired
- Learning Disabled

Many students can better understand an idea if they break it into individual parts. Help students to comprehend the rock cycle by asking them to write the possible changes that each type of rock can undergo. Ask them to divide a piece of paper into three parts and use each part for one type of rock. Help them get started with the sedimentary section by using the following prompts: 1: Sedimentary rock can be melted, cooled, and hardened to form igneous rock. 2: Sedimentary rock can undergo pressure and temperature changes or chemical processes to turn into ____. 3: Sedimentary rock can ____. Their responses will be repetitive, but repetitiveness will help to clarify that the same forces are acting on different types of rocks. LS **Verbal**

Properties of Rocks

All rock has physical and chemical properties that are determined by how and where the rock formed. The physical characteristics of rock reflect the chemical composition of the rock as a whole and of the individual minerals that make up the rock. The rate at which rock weathers and the way that rock breaks apart are determined by the chemical stability of the minerals in the rock. The way that minerals and rocks form is related to the stability of the rock.

Bowen's Reaction Series

In the early 1900s, a Canadian geologist named N. L. Bowen began studying how minerals crystallize from magma. He learned that as magma cools, certain minerals tend to crystallize first. As these minerals form, they remove specific elements from the magma, which changes the magma's composition. The changing composition of the magma allows different minerals that contain different elements to form. Thus, different minerals form at different times during the solidification (cooling) of magma, and they generally form in the same order.

In 1928, Bowen proposed a simplified pattern that explains the order in which minerals form as magma solidifies. This simplified flow chart is known as **Bowen's reaction series** and is shown in **Figure 3.** According to Bowen's hypothesis, minerals form in one of two ways. The first way is characterized by a gradual, continuous formation of minerals that have similar chemical compositions. The second way is characterized by sudden changes in mineral types. The pattern of mineral formation depends on the chemical composition of the magma.

Reading Check Summarize Bowen's reaction series. (See the Appendix for answers to Reading Checks.)

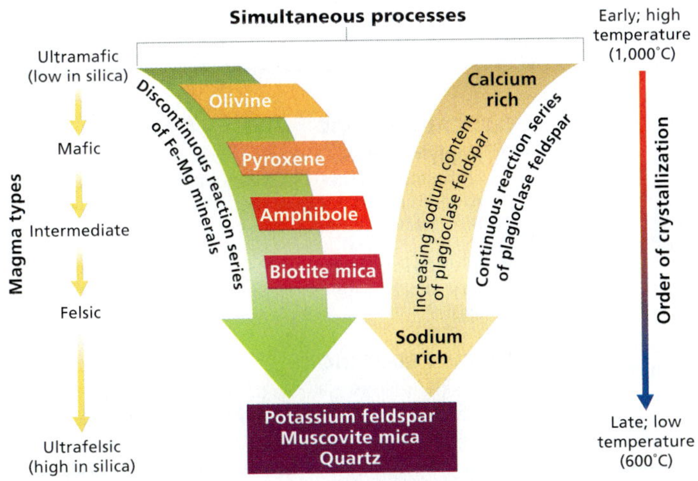

Figure 3 ▶ Different minerals crystallize at different times during the solidification of magma. Thus, as minerals crystallize from magma, the composition of the magma changes.

Graphic Organizer **Chain-of-Events Chart**

Create the Graphic Organizer entitled "Chain-of-Events Chart" in the Skills Handbook section of the Appendix. Then, fill in the chart with each step of the discontinuous reaction series of Bowen's reaction series.

Bowen's reaction series the simplified pattern that illustrates the order in which minerals crystallize from cooling magma according to their chemical composition and melting point

Graphic Organizer GENERAL

Chain-of-Events Chart

You may want to use this Graphic Organizer to assess students' prior knowledge before beginning a discussion of Bowen's reaction series. You may also choose to use a similar activity as a quiz to assess students' knowledge of Bowen's reaction series or to assess students' knowledge after students have read the description of Bowen's reaction series.

Close

Reteaching — BASIC

Rock Cycle Processes On the board, write the phrases "breaking down and compression," "changes in temperature and pressure," and "melting and cooling" on different lines. Have students indicate which processes apply to the formation of igneous, metamorphic, and sedimentary rocks.
LS **Verbal/Visual**

Quiz — GENERAL

1. What type of rock forms when small rock fragments are cemented together? (sedimentary rock)
2. What happens to igneous rock if it is melted and cooled? (The rock forms a new igneous rock.)

Teaching Tip — BASIC

Connect to Familiar Processes To help students understand how the removal of a mineral from magma determines what minerals will form from the remaining magma, have them think of a soup with many ingredients. Removing water from the mixture causes the soup to be thicker and to have a stronger taste. If, for example, potatoes and carrots are removed at the start, the soup will take on more of the flavor of meat and the remaining vegetables. Be sure that students understand that soup is a mixture, so changes to it are not chemical, as changes to solidifying magma are. LS **Logical**

Answer to Reading Check

As magma cools and solidifies, minerals crystallize out of the magma in a specific order, that depends on their melting points.

Close, *continued*

Alternative Assessment — ADVANCED

Rock Crystallization Have students create 3-dimensional models of the rock cycle. Students may use any materials they want, as long as students can explain why they chose those materials. LS Kinesthetic

Answers to Section Review

1. igneous, sedimentary, and metamorphic
2. Igneous rock forms when molten rock (magma) cools and hardens. When rock is broken down, small fragments are deposited and then compressed and cemented to form sedimentary rock. Rock that is subjected to changes in temperature or pressure or to chemical processes becomes metamorphic rock.
3. Answers may vary but should describe the formation of all three types of rock.
4. As the temperature of magma decreases, minerals crystallize out of the magma, depending on their melting points. This crystallization changes the chemical composition of the remaining magma, causing different minerals to form. This process can occur following either a continuous or a discontinuous path.
5. In general, the rocks that have the greatest chemical stability (those most resistant to weathering) are the rocks whose minerals have the greatest number of bonds between silicon and oxygen.
6. Rocks have zones of weakness determined by how and where they form. These zones may form between the layers in sedimentary rocks, or as joints in igneous rock. The rocks break more easily along these zones.
7. no; Rocks can move back and forth between any of the types of rock in the rock cycle. They may even reform as the same kind of rock for several cycles.
8. Answers may vary. Sample answer: If the sedimentary rock consists of fragments of igneous rock, this would indicate that the igneous rock broke down and changed through some process to become sedimentary rock.
9. *Rock* can be one of three main types—*igneous rock, sedimentary rock,* or *metamorphic rock*—and changes from one form to another through the *rock cycle.*

Figure 4 ▶ Devils Postpile National Monument in California is one of the world's finest examples of the igneous rock structures known as columnar joints.

Chemical Stability of Minerals

The rate at which a mineral chemically breaks down is dependent on the chemical stability of the mineral. *Chemical stability* is a measure of the tendency of a chemical compound to maintain its original chemical composition rather than break down to form a different chemical. The chemical stability of minerals is dependent on the strength of the chemical bonds between atoms in the mineral. In general, the minerals that are most resistant to weathering are the minerals that have the highest number of bonds between the elements silicon, Si, and oxygen, O.

Physical Stability of Rocks

Rocks have natural zones of weakness that are determined by how and where the rocks form. For example, sedimentary rocks may form as a series of layers of sediment. These rocks tend to break between layers. Some metamorphic rocks also tend to break in layers that form as the minerals in the rocks align during metamorphism.

Massive igneous rock structures commonly have evenly spaced zones of weakness, called *joints,* that form as the rock cools and contracts. Devils Postpile, shown in **Figure 4**, is igneous rock that has joints that cause the rock to break into columns.

Zones of weakness may also form when the rock is under intense pressure inside Earth. When rock that formed under intense pressure is uplifted to Earth's surface, decreased pressure allows the joints and fractures to open. Once these weaknesses are exposed to air and water, the processes of chemical and physical weathering begin.

Section 1 Review

1. **Identify** the three major types of rock.
2. **Explain** how each major type of rock forms.
3. **Describe** the steps in the rock cycle.
4. **Summarize** Bowen's reaction series.
5. **Explain** how the chemical stability of a mineral is affected by the bonding of the atoms in the mineral.
6. **Describe** how the conditions under which rocks form affect the physical stability of rocks.

CRITICAL THINKING

7. **Applying Ideas** Does every rock go through the complete rock cycle by changing from igneous rock to sedimentary rock, to metamorphic rock, and then back to igneous rock? Explain your answer.
8. **Identifying Relationships** How could a sedimentary rock provide evidence that the rock cycle exists?

CONCEPT MAPPING

9. Use the following terms to create a concept map: *rock, igneous rock, sedimentary rock, metamorphic rock,* and *rock cycle.*

CHAPTER RESOURCES

Chapter Resource File
- Section Quiz GENERAL

Workbooks
- Study Guide (also in Spanish)

Section 2 Igneous Rock

When magma cools and hardens, it forms **igneous rock.** Because minerals crystallize as igneous rock forms from magma, most igneous rock can be identified as *crystalline,* or made of crystals. The chemical composition of minerals in the rock and the rock's texure determine the identity of the igneous rock.

OBJECTIVES

- **Summarize** three factors that affect whether rock melts.
- **Describe** how the cooling rate of magma and lava affects the texture of igneous rocks.
- **Classify** igneous rocks according to their composition and texture.
- **Describe** intrusive and extrusive igneous rock structures.

KEY TERMS

igneous rock
intrusive igneous rock
extrusive igneous rock
felsic
mafic

The Formation of Magma

Magma forms when rock melts. The three factors that affect whether rock melts include temperature, pressure, and the presence of fluids in the rock. Rock melts when the temperature of the rock increases to above the melting point of minerals in the rock. The melting temperature is determined by the chemical composition of the minerals in the rock. Rock also melts when excess pressure is removed from rock that is hotter than its melting point. Rock may melt when fluids such as water are added to hot rock. The addition of fluids generally decreases the melting point of certain minerals in the rock, which can cause those minerals to melt.

igneous rock rock that forms when magma cools and solidifies

Partial Melting

Different minerals have different melting points, and minerals that have lower melting points are the first minerals to melt. When the first minerals melt, the magma that forms has a specific composition. As the temperature increases and as other minerals melt, the magma's composition changes. The process by which different minerals in rock melt at different temperatures is called *partial melting.* Partial melting is shown in **Figure 1.**

Figure 1 ▶ How Magma Forms by Partial Melting

This solid rock contains the minerals quartz (yellow), feldspar (gray), biotite (brown), and hornblende (green).

The first minerals that melt are quartz and some types of feldspars. The orange background represents magma.

Minerals such as biotite and hornblende generally melt last, which changes the composition of the magma.

Section 2

Focus

Overview

This section describes how igneous rocks form and what factors affect the texture and composition of igneous rocks. The section concludes with a description of igneous rock structures.

Bellringer

Have students write a short paragraph that explains where and why rock melts. (Answers may vary. Use students' answers to begin a discussion of how magma forms.) LS Logical

Motivate

Discussion — GENERAL

Hand Samples Pass around to the class samples of the following igneous rocks: basalt, pumice, granite, and obsidian. Ask students what these four rocks of very different appearance have in common. (All are made of glassy or crystalline components.) Ask students how the textures of the rocks differ. (obsidian: very smooth, no crystals; basalt: tiny crystals; granite: large crystals; pumice: no crystals, large holes.) Point out that all of the rocks formed from molten rock but the conditions under which each rock formed account for their different appearances. LS Kinesthetic/Visual

CHAPTER RESOURCES

Chapter Resource File
- Directed Reading BASIC

Technology

Transparencies
- Bellringer
- 28 Partial Melting and Fractional Crystallization (with worksheet)

Student Edition on CD-ROM

One-Stop Planner CD-ROM
- Lesson Plan

Teach

Using the Figure — BASIC

Fractional Crystallization Ask students how the process shown in the figure at the top of the page relates to the figure of partial melting at the bottom of the previous page. (The figure of fractional crystallization shows the reverse of the process of partial melting.) During fractional crystallization, minerals that have high freezing/melting termperatures crystallize first. As minerals crystallize, the composition of the magma changes, and different minerals begin to form. Ask students what a third illustration in this series would look like if it showed even cooler magma. (The amount of magma would be reduced, and more crystals, probably of different minerals, would be present.) **LS Visual/Logical**

Demonstration — GENERAL

Partial Freezing To help students visualize the process that occurs during fractional crystallization, prepare a mixture of 60 mL paraffin, 60 mL butter, and 60 mL water in a 250 mL beaker. Heat the mixture until the butter and paraffin have melted. Then, place the beaker in a 1000 mL beaker that is half full of an ice and water mixture. Have students watch the beaker and note that the wax solidifies before the butter does. Point out that if the mixture is cooled to a temperature below that of the ice bath, the water will freeze. **LS Visual**

CHAPTER RESOURCES

Chapter Resource File

- Datasheet for Quick Lab GENERAL

Figure 2 ▶ As the temperature decreases, the first minerals to crystallize from magma are minerals that have the highest freezing points, such as biotite and hornblende. As the magma changes composition and cools, minerals that have lower freezing points form.

Fractional Crystallization

When magma cools, the cooling process is the reverse of the process of partial melting. Chemicals in magma combine to form minerals, and each mineral has a different freezing point. Minerals that have the highest freezing points crystallize first. As those minerals crystallize, they remove specific chemicals from the magma and change the composition of the magma. As the composition changes, new minerals begin to form. The crystallization and removal of different minerals from the cooling magma is called *fractional crystallization* and is shown in **Figure 2.**

Minerals that form during fractional crystallization tend to settle to the bottom of the magma chamber or to stick to the ceiling and walls of the magma chamber. Crystals that form early in the process are commonly the largest because they have the longest time to grow. In some crystals, the chemical composition of the inner part of the crystal differs from the composition of the outer parts of the crystal. This difference occurs because the magma's composition changed while the crystal was growing.

QuickLAB — 20 min

Crystal Formation

Procedure

1. Add the following until **three glass jars** are 2/3 full: glass 1—water and **ice cubes;** glass 2—water at room temperature; and glass 3—hot tap water.
2. In a small **sauce pan,** mix **120 mL of Epsom salts** in **120 mL of water.** Heat the mixture on a **hot plate** over low heat. Do not let the mixture boil. Stir the mixture with a **spoon or stirring rod** until no more crystals dissolve.
3. Using a **funnel,** carefully pour equal amounts of the Epsom salts mixture into **three test tubes.** Use **tongs** to steady the test tubes as you pour. Drop a few crystals of Epsom salt into each test tube, and gently shake each one. Place one test tube into each glass jar.
4. Observe the solutions as they cool for 15 minutes. Let the glasses sit overnight, and examine the solutions again after 24 hours.

Analysis

1. In which test tube are the crystals the largest?
2. In which test tube are the crystals the smallest?
3. How does the rate of cooling affect the size of the crystals that form? Explain your answer.
4. How are the differing rates of crystal formation you observed related to igneous rock formation?
5. How would you change the procedure to obtain larger crystals of Epsom salts? Explain your answer.

QuickLAB

Skills Acquired

- Experimenting
- Predicting
- Identifying/Recognizing Patterns

Teacher's Notes Be sure that students take extra care when pouring the melted Epsom salts. Clean up spills quickly.

Answers

1. The crystals in the test tube cooled in the hot water will be largest.
2. The crystals in the test tube cooled in the ice water will be smallest.
3. The faster the liquid cools, the smaller the crystals will be.
4. Large crystals form from slowly cooling magma, and smaller crystals form from rapidly cooling magma.
5. by placing the test tube in water whose temperature is slightly lower than the temperature of the liquid Epsom salts

Textures of Igneous Rocks

Igneous rocks are classified according to where magma cools and hardens. Magma that cools deep inside the crust forms **intrusive igneous rock.** The magma that forms these rocks intrudes, or enters, into other rock masses beneath Earth's surface. The magma then slowly cools and hardens. Lava that cools at Earth's surface forms **extrusive igneous rock.**

Intrusive and extrusive igneous rocks differ from each other not only in where they form but also in the size of their crystals or grains. The texture of igneous rock is determined by the size of the crystals in the rock. The size of the crystals is determined mainly by the cooling rate of the magma. Examples of different textures of igneous rocks are shown in **Figure 3.**

intrusive igneous rock rock formed from the cooling and solidification of magma beneath Earth's surface

extrusive igneous rock rock that forms from the cooling and solidification of lava at Earth's surface

Coarse-Grained Igneous Rock

Intrusive igneous rocks commonly have large mineral crystals. The slow loss of heat allows the minerals in the cooling magma to form large, well-developed crystals. Igneous rocks that are composed of large mineral grains are described as having a *coarse-grained texture*. An example of a coarse-grained igneous rock is granite. The upper part of the continental crust is made mostly of granite.

Fine-Grained Igneous Rock

Many extrusive igneous rocks are composed of small mineral grains that cannot be seen by the unaided eye. Because these rocks form when magma cools rapidly, large crystals are unable to form. Igneous rocks that are composed of small crystals are described as having a *fine-grained texture*. Examples of common fine-grained igneous rocks are basalt and rhyolite (RIE uh LIET).

Other Igneous Rock Textures

Some igneous rock forms when magma cools slowly at first but then cools more rapidly as it nears Earth's surface. This type of cooling produces large crystals embedded within a mass of smaller ones. Igneous rock that has a mixture of large and small crystals has a *porphyritic texture* (POHR fuh RIT ik TEKS chuhr).

When a highly viscous magma cools quickly, few crystals are able to grow. If such magma contains a very small percentage of dissolved gases, a rock that has a *glassy* texture, such as obsidian forms. When this type of magma contains a large percentage of dissolved gases and cools rapidly, the gases become trapped as bubbles in the rock that forms. The rapid cooling process produces a rock full of holes called *vesicles,* such as those in pumice. This type of rock is said to have a *vesicular texture*.

Reading Check **What is the difference between fine-grained and coarse-grained igneous rock? (See the Appendix for answers to Reading Checks.)**

Figure 3 ▶ Igneous Rock Textures

READING SKILL BUILDER — BASIC

Reading Organizer Before students read this page, have them make a table that has three columns and six rows. Have them label the columns with "Texture," "Intrusive," and "Extrusive." In the "Texture" column, have students label the rows with "Coarse-grained," "Fine-grained," "Porphyritic," "Volcanic glass," and "Vesicular." As students read the page, have them insert in the correct box a brief description of the heat loss that takes place (for example, "slow heat loss" for "coarse-grained" in the "Intrusive" column), as well as whether gas is dissolved in the magma. Have students add the name of a type of igneous rock that fits each description. Later, students can use their charts as study guides for assessments. English Language Learners
LS Verbal

Answer to Reading Check

Fine-grained igneous rock forms mainly from magma that cools rapidly; coarse-grained igneous rock forms mainly from magma that cools more slowly.

GEOLOGY CONNECTION — ADVANCED

Information from Igneous Rocks Igneous rocks are affected by where they form, their rate of cooling, the presence of dissolved gases, and their chemical composition. Because of this, information about specific geological processes can be obtained by studying different types of igneous rock.

For example, the basalts that are found in the ocean crust are low in the elements rubidium and thorium, while the basalts of the continental crust have higher amounts of these elements. This difference provides information about the mechanisms by which the continental crust forms and the rate at which it forms. Similarly, a lower concentration of silica, SiO_2, which is found in differing amounts in all igneous rocks, occurs in rocks near the summits of older volcanoes. These rocks are commonly an indicator of reduced volcanic activity. Have students research what geologic processes result in the different igneous rock textures. Students should present their findings to the class. **LS Verbal/Logical**

Teach, *continued*

Group Activity — Advanced

Poster Project Divide the class into groups of three or four students, and have each group create a poster that lists the properties and provides examples of felsic, mafic, and intermediate rocks. The posters should include information about the chemical compositions, range of densities, range of melting temperatures, typical colors, and textures of each rock type, as well as common examples of each kind. Students should each do research on two or three different igneous rocks and then pool their information to create the poster. Have each group present their poster to the rest of the class. **LS Visual/Logical** Co-op Learning

MISCONCEPTION ALERT

Combination Crust
Students may be under the impression that felsic and mafic rocks never occur at the same locations. Point out that these rocks, while separate because of their chemical and physical properties, may be found in a single location, depending on the geologic history of the area. Continental crust consists of both felsic and mafic rocks.

Figure 4 ▶ Felsic rocks, such as the outcropping and hand sample shown above (left), have light coloring. Mafic rocks (right) are usually darker in color.

Composition of Igneous Rocks

The mineral composition of an igneous rock is determined by the chemical composition of the magma from which the rock formed. Each type of igneous rock has a specific mineral composition. Geologists divide igneous rock into three families—felsic, mafic (MAF ik), and intermediate. Each of the three families has a different mineral composition. Examples of rock from the felsic and mafic families are shown in **Figure 4.**

Felsic Rock

Rock in the **felsic** family forms from magma that contains a large proportion of silica. Felsic rock generally has the light coloring of its main mineral components, potassium feldspar and quartz. Felsic rock commonly also contains plagioclase feldspar, biotite mica, and muscovite mica. The felsic family includes many common rocks, such as granite, rhyolite, obsidian, and pumice.

felsic describes magma or igneous rock that is rich in feldspars and silica and that is generally light in color

mafic describes magma or igneous rock that is rich in magnesium and iron and that is generally dark in color

Mafic Rock

Rock in the **mafic** family forms from magma that contains lower proportions of silica than felsic rock does and that is rich in iron and magnesium. The main mineral components of rock in this family are plagioclase feldspar and pyroxene minerals. Mafic rock may also include dark-colored *ferromagnesian minerals*, such as hornblende. These ferromagnesian components, as well as the mineral olivine, give mafic rock a dark color. The mafic family includes the common rocks basalt and gabbro.

Intermediate Rocks

Rocks of the intermediate family are made up of the minerals plagioclase feldspar, hornblende, pyroxene, and biotite mica. Rocks in the intermediate family contain lower proportions of silica than rocks in the felsic family do but contain higher proportions of silica than rocks in the mafic family contain. Rocks in the intermediate family include diorite and andesite.

Teaching Tip

Use Technology Resources Plagioclase feldspars, which consist of aluminum silicates bound to either calcium or sodium ions, are found in both felsic and mafic rock. However, the percentage of plagioclase feldspar in magma in part determines whether the magma is felsic or mafic. Felsic magmas contain small amounts of plagioclase feldspar, and large amounts of orthoclase feldspar, or potassium aluminum silicate. Mafic magmas are made up of larger quantities of plagioclase feldspar, and contain no orthoclase feldspar, though their overall feldspar content is less than that of felsic magmas.

Encourage students to use the Internet to research the role of both kinds of feldspars in igneous rocks. A number of sites contain diagrams that show the proportions of different minerals in felsic, intermediate, and mafic magmas, as well as the olivine-rich magmas called *ultramafic*. **LS Logical**

Intrusive Igneous Rock Structures

Igneous rock masses that form underground are called *intrusions*. Intrusions form when magma intrudes, or enters, into other rock masses and then cools deep inside Earth's crust. A variety of intrusions are shown in **Figure 5.**

Batholiths and Stocks

The largest of all intrusions are called *batholiths*. Batholiths are intrusive formations that spread over at least 100 km^2 when they are exposed on Earth's surface. The word *batholith* means "deep rock." Batholiths were once thought to extend to great depths beneath Earth's surface. However, studies have determined that many batholiths extend only several thousand meters below the surface. Batholiths form the cores of many mountain ranges, such as the Sierra Nevadas in California. The largest batholith in North America forms the core of the Coast Range in British Columbia. Another type of intrusion is called a *stock*. Stocks are similar to batholiths but cover less than 100 km^2 at the surface.

Laccoliths

When magma flows between rock layers and spreads upward, it sometimes pushes the overlying rock layers into a dome. The base of the intrusion is parallel to the rock layer beneath it. This type of intrusion is called a *laccolith*. The word *laccolith* means "lake of rock." Laccoliths commonly occur in groups and can sometimes be identified by the small dome-shaped mountains they form on Earth's surface. Many laccoliths are located beneath the Black Hills of South Dakota.

Reading Check **What is the difference between stocks and batholiths? (See the Appendix for answers to Reading Checks.)**

Sills and Dikes

When magma flows between the layers of rock and hardens, a *sill* forms. A sill lies parallel to the layers of rock that surround it, even if the layers are tilted. Sills vary in thickness from a few centimeters to hundreds of meters.

Magma sometimes forces through rock layers by following existing vertical fractures or by creating new ones. When the magma solidifies, a *dike* forms. Dikes cut across rock layers rather than lying parallel to the rock layers. Dikes are common in areas of volcanic activity.

SCILINKS. NSTA Developed and maintained by the National Science Teachers Association

For a variety of links related to this subject, go to www.scilinks.org

Topic: Igneous Rock

SciLinks code: HQ60783

Figure 5 ▶ Igneous intrusions create a number of unique landforms. ***What is the difference between a dike and a sill?***

MISCONCEPTION ALERT

Exposed Intrusive Structures
Students may be confused about how intrusive structures that form below Earth's surface become exposed at the surface. Explain that parts or all of such structures are exposed when the structure is pushed up toward the surface by tectonic activity or when the material surrounding it is eroded.

SKILL BUILDER — GENERAL

Vocabulary While some of the terms for intrusive igneous structures are Greek in origin, such as *batholith* and *laccolith*, many terms can be traced back to Old English. The word *stock* comes from the Old English word *stocc*, meaning "tree trunk," which these structures vaguely resemble. *Sill* comes from the Old English *syll*, meaning "threshold." The word *dike* derives from the Old English word for "ditch" or "trench." LS **Verbal**

Using the Figure — GENERAL

Intrusive Igneous Structures
Ask students to point out the main landforms shown in the figure and describe how they are caused by intrusions. (Answers may vary. Students should point out the volcanic neck, the ridges connected to the volcanic neck and the mound in the foreground. The volcanic neck and dikes at the surface formed when overlying, less-resistant rock was eroded and the harder intrusive rock was exposed. The mound in the foreground formed when magma in the laccolith pushed the overlying layers up.) Answer to caption question: A sill is parallel to the layers of rock that it intrudes. A dike cuts across the layers it intrudes. LS **Visual/Logical**

Answer to Reading Check

A batholith is an intrusive structure that covers an area of at least 100 km^2. A stock covers an area of less than 100 km^2.

Close

Reteaching — BASIC

Igneous Rock Characteristics
Have students create a table that has two columns and four rows. Have students label the columns "Coarse-grained" and "Fine-grained." Have them label the rows "Mafic," "Felsic," "Intrusive," and "Extrusive." Then, have students fill in the table by using the names of igneous rocks or the structures that fit in each category. LS **Logical**

Quiz — GENERAL

Determine whether each of the following statements is true or false.

1. Fine-grained igneous rock results from rapid cooling. (true)
2. Mafic magmas generally produce light-colored rocks. (false)
3. During partial melting, minerals that have lower melting points melt first. (true)

Close, *continued*

Alternative Assessment — Advanced

Travel Brochures Have students create travel brochures to five locations that have prominent exposed intrusive or extrusive igneous structures. The brochures should describe the kinds of igneous rock that make up each structure and the geological processes that formed the structure. Students should include drawings or photos of the structures. LS **Visual**

Answers to Section Review

1. temperature, pressure, and the addition of fluids to the rock
2. Partial melting occurs when rock's temperature increases, causing minerals that have the lowest melting points to melt first. Fractional crystallization is the reverse process. As magma's temperature decreases, different minerals form as the composition of the magma changes.
3. When magma cools slowly, larger crystals may form, and the rock is coarse-grained. When magma cools rapidly, smaller crystals may form, and the rock is fine-grained.
4. felsic: high in potassium feldspar and quartz, with some plagioclase feldspar and mica; mafic: high in plagioclase feldspars, iron- and magnesium-rich minerals such as pyroxene and hornblende, and olivine; intermediate: plagioclase feldspar, hornblende, pyroxene, and biotite mica
5. *Batholiths* are large intrusive structures that cover an area of 100 km^2 or more. *Stocks* are intrusive structures that cover less than 100 km^2. *Laccoliths* are intrusions that push the overlying rock layers into domes and whose bases lie parallel to the rock layers beneath them. *Sills* are intrusions that are parallel to the surrounding rock layers. *Dikes* are intrusions that cut across surrounding rock layers.
6. volcanoes, volcanic necks, lava flows, lava plateaus, and tuffs
7. There can be several variables, but one that would work is to keep the rate of cooling of the magma from which crystals will form as slow as possible.
8. the felsic family, which includes potassium feldspar and quartz, and whose members are typically light in color
9. *Igneous rock* forms from *magma*, which may be *felsic*, *mafic*, or *intermediate*, and may be *coarse-grained* or *fine-grained*.

Figure 6 ▶ Shiprock, in New Mexico, is an example of a volcanic neck that was exposed by erosion.

Extrusive Igneous Rock Structures

Igneous rock masses that form on Earth's surface are called *extrusions*. A *volcano* is a vent through which magma, gases, or volcanic ash is expelled. When a volcanic eruption stops, the magma in the vent may cool to form rock. Eventually, the soft parts of the volcano are eroded by wind and water, and only the hardest rock in the vent remains. The solidified central vent is called a *volcanic neck*. Narrow dikes that sometimes radiate from the neck may also be exposed. A dramatic example of a volcanic neck called Shiprock is shown in **Figure 6.**

Extrusive igneous rock may also take other forms. Many extrusions are simply flat masses of rock called *lava flows*. A series of lava flows that cover a vast area with thick rock is known as a *lava plateau*. Volcanic ash deposits, also commonly called *tuff*, form when a volcano releases ash and other solid particles during an eruption. Tuff deposits can be several hundred meters thick and can cover areas of several hundred kilometers.

Section 2 Review

1. **Summarize** three factors that affect the melting of rock.
2. **Contrast** partial melting and fractional crystallization.
3. **Describe** how the cooling rate of magma affects the texture of igneous rock.
4. **Name** the three families of igneous rocks, and identify their specific mineral compositions.
5. **Describe** five intrusive igneous rock structures.
6. **Identify** four extrusive igneous rock structures.

CRITICAL THINKING

7. **Applying Ideas** If you wanted to create a rock that has large crystals in a laboratory, what conditions would you have to control? Explain your answer.
8. **Applying Ideas** An unidentified, light-colored igneous rock is made up of potassium feldspar and quartz. To what family of igneous rocks does the rock belong? Explain your answer.

CONCEPT MAPPING

9. Use the following terms to create a concept map: *igneous rock, magma, coarse grained, fine grained, felsic, mafic,* and *intermediate.*

CHAPTER RESOURCES

Chapter Resource File
- Section Quiz GENERAL

Workbooks
- Study Guide (also in Spanish)

Section 3

Sedimentary Rock

Loose fragments of rock, minerals, and organic material that result from natural processes, including the physical breakdown of rocks, are called *sediment*. Most sedimentary rock is made up of combinations of different types of sediment. The characteristics of sedimentary rock are determined by the source of the sediment, the way the sediment was moved, and the conditions under which the sediment was deposited.

OBJECTIVES

- **Explain** the processes of compaction and cementation.
- **Describe** how chemical and organic sedimentary rocks form.
- **Describe** how clastic sedimentary rock forms.
- **Identify** seven sedimentary rock features.

KEY TERMS

compaction
cementation
chemical sedimentary rock
organic sedimentary rock
clastic sedimentary rock

Formation of Sedimentary Rocks

After sediments form, they are generally transported by wind, water, or ice to a new location. The source of the sediment determines the sediment's composition. As the sediment moves, its characteristics change as it is physically broken down or chemically altered. Eventually, the loose sediment is deposited.

Two main processes convert loose sediment to sedimentary rock—compaction and cementation. **Compaction,** as shown in **Figure 1,** is the process in which sediment is squeezed and in which the size of the pore space between sediment grains is reduced by the weight and pressure of overlying layers. **Cementation** is the process in which sediments are glued together by minerals that are deposited by water. As water moves through the sediment, minerals precipitate from the water, surround the sediment grains, and form a cement that holds the fragments together.

Geologists classify sedimentary rocks by the processes by which the rocks form and by the composition of the rocks. There are three main classes of sedimentary rocks—chemical, organic, and clastic. These classes contain their own classifications of sedimentary rocks that are grouped based on the shape, size, and composition of the sediments that form the rocks.

compaction the process in which the volume and porosity of a sediment is decreased by the weight of overlying sediments as a result of burial beneath other sediments

cementation the process in which minerals precipitate into pore spaces between sediment grains and bind sediments together to form rock

Figure 1 ▶ **Processes That Form Sedimentary Rock**

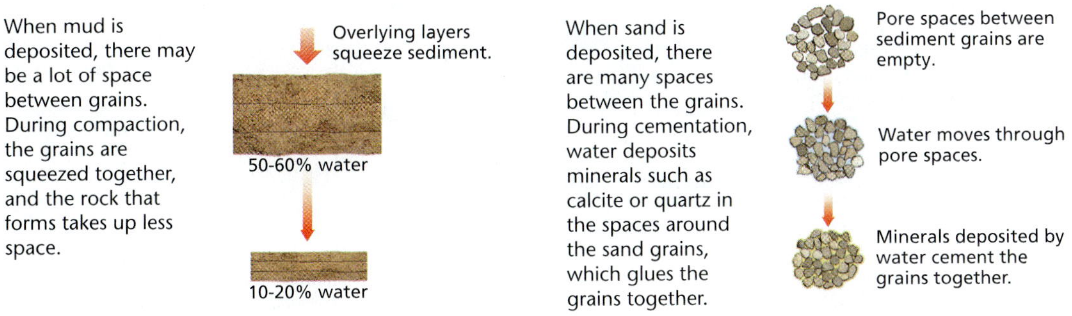

Section 3

Focus

Overview

This section explains how chemical, organic, and clastic sedimentary rocks form. It also explains how sediments change during transport and identifies seven features of sedimentary rocks.

Bellringer

Have students write down three features that they associate with sedimentary rock compared with other kinds of rock. (Answers may vary. Sample answers: Sedimentary rock may be layered, may be sandy or gritty, and may contain fossils.) LS **Verbal**

Motivate

Demonstration — GENERAL

Sedimentation Demonstrate the basic process of sedimentation by placing a few tablespoons of fine silt in a jar. Fill the jar with water, and place the lid tightly on the jar. Shake the jar, and while doing so, ask students what will happen when you set the jar down. (The silt will settle over time.) Set the jar down, and wait for two minutes. If the silt has not entirely settled, check back after 10 or 15 minutes. LS **Visual**

CHAPTER RESOURCES

Chapter Resource File

- **Directed Reading** BASIC

Technology

Transparencies
- Bellringer

Student Edition on CD-ROM

One-Stop Planner CD-ROM
- Lesson Plan

Teach

Homework — GENERAL

Chemical Sedimentary Rock Assign students to research different types of chemical sedimentary rock. Have students identify carbonate precipitates, such as calcite (chemical limestone, which is sometimes referred to as travertine). Have them also identify evaporates other than halite and gypsum (such as borax), as well as variant forms of gypsum (such as alabaster) and of halides (such as sylvite and hydrite). Students should also identify well-known locations where these various types of rock can be found. (for instance, Death Valley National Park, White Sands National Monument, and Carlsbad Caverns National Park) LS **Logical**

Using the Figure — GENERAL

Organic Limestone Formation Have students study the diagrams at the bottom of the page. Ask students if the process by which the shells become limestone is better described as compaction or cementation. (compaction) Ask students to explain their answer. (The microscopic shells are much like tiny particles of mud that can be compacted to take up less space. However, some cementation probably also occurs.) Ask students what grain size they would expect limestone to have, given this information. (Limestone should have grains the size of fine sand or grit.) LS **Logical**

CHAPTER RESOURCES

Technology

Transparencies
- 29 Organic Limestone Formation (with worksheet)

chemical sedimentary rock sedimentary rock that forms when minerals precipitate from a solution or settle from a suspension

organic sedimentary rock sedimentary rock that forms from the remains of plants or animals

Chemical Sedimentary Rock

Rock called **chemical sedimentary rock** forms from minerals that were once dissolved in water. Some chemical sedimentary rock forms when dissolved minerals precipitate out of water because of changing concentrations of chemicals.

One reason minerals precipitate is because of evaporation. When water evaporates, the minerals that were dissolved in the water are left behind. Eventually, the concentration of minerals in the remaining water becomes high enough to cause minerals to precipitate out of the water. The minerals left behind form rocks called *evaporites*. Gypsum and halite, or rock salt, are two examples of evaporites. The Bonneville Salt Flats near the Great Salt Lake in Utah are a good example of evaporite deposits.

Organic Sedimentary Rocks

The second class of sedimentary rock is **organic sedimentary rock.** Organic sedimentary rock is rock that forms from the remains of living things. Coal and some limestones are examples of organic sedimentary rocks. Coal forms from plant remains that are buried before they decay and are then compacted into matter that is composed mostly of carbon.

While chemical limestones precipitate from chemicals dissolved in water, organic limestones form when marine organisms, such as coral, clams, oysters, and plankton, remove the chemical components of the minerals calcite and aragonite from sea water. These organisms make their shells from these minerals. When they die, their shells eventually become limestone. This process of limestone formation is shown in **Figure 2.** Chalk is an example of limestone made up of the shells of tiny, one-celled marine organisms that settle to the ocean floor.

Figure 2 ▶ **Organic Limestone Formation**

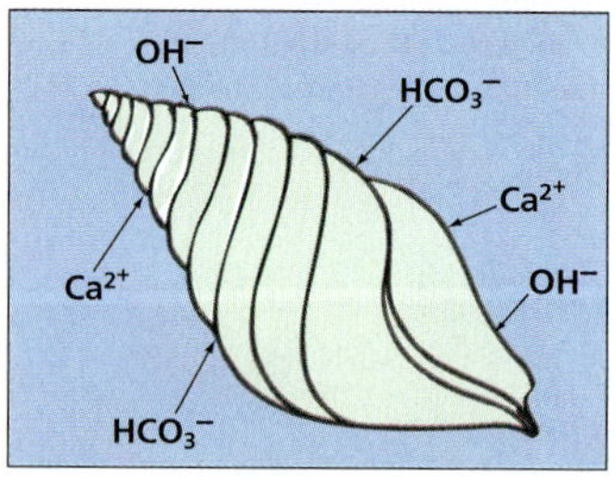

Organisms that live in lakes or oceans take chemicals from the water and produce the mineral calcium carbonate, $CaCO_3$. They use the $CaCO_3$ to build their shells or skeletons.

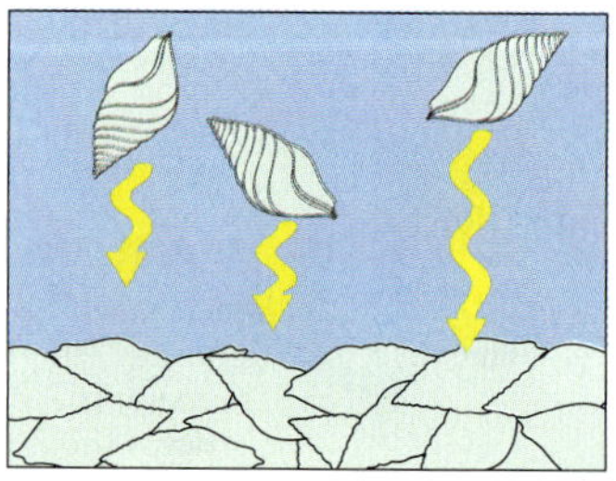

When the organisms die, the hard remains that are made of $CaCO_3$ settle to the lake or ocean floor.

The shells of the dead organisms pile up. Eventually, the layers are compacted and cemented to form limestone.

CHEMISTRY CONNECTION — ADVANCED

The Carbon Cycle The calcium carbonate that forms limestone derives in part from another rock process: volcanism. When volcanoes erupt, large amounts of carbon dioxide (CO_2) are released into the atmosphere. The carbon dioxide is then available for absorption by sea water and may form carbonic acid (H_2CO_3). The carbonic acid in turn reacts with calcium to form calcium carbonate, which effectively "stores" the absorbed carbon dioxide in the limestone that ultimately forms. When tectonic forces break down the limestone, gaseous carbon dioxide again becomes available. The carbon cycle has a slow but considerable effect on Earth by keeping atmospheric temperatures from fluctuating wildly. Have students research various steps of the carbon cycle. Students should note the length of time carbon stays in each reservoir and how each process affects the buildup of carbon dioxide in the atmosphere. Have students prepare a written report of their findings. LS **Verbal/Visual**

Figure 3 ▶ **Types of Clastic Sedimentary Rock**

Conglomerate is composed of rounded, pebble-sized fragments that are held together by a cement.

Breccia is similar to conglomerate, but breccia contains angular fragments.

Sandstone is made of small mineral grains that are cemented together.

Shale is made of flaky clay particles that compress into flat layers.

Clastic Sedimentary Rock

The third class of sedimentary rock is made of rock fragments that are carried away from their source by water, wind, or ice and left as deposits. Over time, the individual fragments may become compacted and cemented into solid rock. The rock formed from these deposits is called **clastic sedimentary rock.**

Clastic sedimentary rocks are classified by the size of the sediments they contain, as shown in **Figure 3.** One group consists of large fragments that are cemented together by finer sediments or by minerals. Rock that is composed of rounded fragments that range in size from 2 mm to boulders is called a *conglomerate*. If the fragments are angular and have sharp corners, the rock is called a *breccia* (BRECH ee uh). In conglomerates and breccias, the individual pieces of sediment can be easily seen.

Another group of clastic sedimentary rocks is made up of sand-sized grains that have been cemented together. These rocks are called *sandstone*. Because quartz is one of the hardest common minerals, quartz is the major component of most sandstones. Many sandstones have pores between the sand grains through which fluids, such as groundwater, natural gas, and crude oil, can move.

A third group of clastic sedimentary rocks, called *shale*, consists of clay-sized particles that are cemented and compacted. The flaky clay particles are usually pressed into flat layers that will easily split apart.

Reading Check **Name three groups of clastic sedimentary rock. (See the Appendix for answers to Reading Checks.)**

clastic sedimentary rock sedimentary rock that forms when fragments of preexisting rocks are compacted or cemented together

MATHPRACTICE

Sedimentation Rates The rate at which sediment accumulates is called the *sedimentation rate*. The sedimentation rate of an area is 1.5 mm per year. At this rate, how many years must pass for 10 cm of sediment to be deposited?

MATHPRACTICE

Answer

1.5 mm/year × x years
= 10 cm
1.5 mm = 0.15 cm
x years = 10 cm ÷ 0.15 cm
= 66.7 years

Answer to Reading Check

Three groups of clastic sedimentary rock are conglomerates and breccias, sandstones, and shales.

Discussion — GENERAL

Comprehension Check Ask students how texture in each of the three groups of clastic sedimentary rock indicates whether the group formed by cementation or by compaction. (The larger the clasts, or particles, in the rock, the less likely they are able to stick together without cement.) Present to students this scenario: "You find a piece of sedimentary rock in a high mountain valley. The rock has a sandy texture and no signs of internal layering. What type of rock is it? On what evidence do you base your answer?" (Answers may vary. Most likely the rock is a clastic sedimentary rock of the sandstone group. If the rock is made of tiny particles of sand or rock, the rock must be clastic. The fine grain eliminates conglomerate or breccia, and the coarseness of the grains eliminates shale.) LS **Logical**

Cultural Awareness — BASIC

Sacred Rocks The sedimentary rock forms produced by weathering and erosion commonly inspire awe. Many native peoples throughout the world consider the rock forms of their region sacred. For example, the Navajo of North America ascribe supernatural powers to the sandstone buttes and spires of Monument Valley, their tribal park in Utah and Arizona. The Aborigines of central Australia regard the vast sandstone monolith Uluru, also known as "Ayres Rock," as having an underground energy source called the "dream time," a term that also refers to a record of ancestors' activities.

READING SKILL BUILDER — BASIC

Paired Summarizing Group students into pairs and have them read silently about the three groups of clastic sedimentary rock. Then, have one student summarize the features of the rocks in each group. The other student should listen to the retelling and point out any inaccuracies or ideas that were left out. Allow students to refer to the text as needed. LS **Verbal** English Language Learners Co-op Learning

Teach, *continued*

Teaching Tip — Basic

Connect to Familiar Processes To help students understand the process of sediment sorting, draw an analogy between sorting in a moving medium (air or water) and passing sediments through a screen. If a box with a screen bottom is used to scoop up a mixture of sand, gravel, and silt, only the sand and silt will pass through the screen, while the larger gravel remains in the box. If the sand and silt mixture is put into another box with a finer screen, the silt will pass through the screen, leaving the sand behind. In the natural process of sorting, the mass of each fragment is important, in addition to its size. Also, the speed at which the wind or water moves the particles functions as the screens do. Larger particles are gradually left behind, while smaller ones are moved farther until only the finest grains remain. Sediments become better sorted the longer the sorting process continues. You may want to perform this activity as a demonstration LS **Logical**

CHAPTER RESOURCES

Chapter Resource File

- **Inquiry Lab** Sorting Sediments GENERAL

Internet Activity

- Identifying Sedimentary Rock Features GENERAL

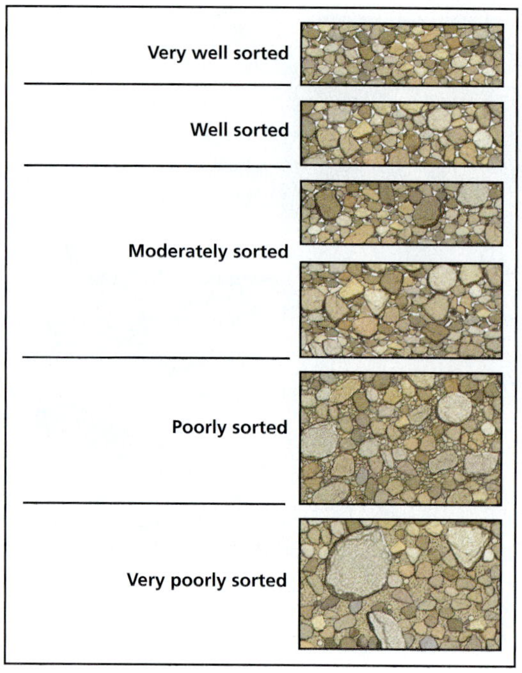

Figure 4 ▶ Sorting of Sediments

Characteristics of Clastic Sediments

The physical characteristics of sediments are determined mainly by the way sediments were transported to the place where they are deposited. Sediments are transported by four main agents: water, ice, wind, and the effects of gravity. The speed with which the agent of erosion moves affects the size of sediment particles that can be carried and the distance that the particles will move. In general, both the distance the sediment is moved and the agent that moves the sediment determine the characteristics of that sediment.

Sorting

The tendency for currents of air or water to separate sediments according to size is called *sorting*. Sediments can be well sorted, poorly sorted, or somewhere in between, as shown in **Figure 4.** In well-sorted sediments, all of the grains are roughly the same size and shape. Poorly sorted sediment consists of grains that are many different sizes. The sorting of a sediment is the result of changes in the speed of the agent that is moving the sediment. For example, when a fast-moving stream enters a lake, the speed of the water decreases sharply. Because large grains are too heavy for the current to carry, these grains are deposited first. Fine grains can stay suspended in the water for much longer than large grains can. So, fine particles are commonly deposited farther from shore or on top of coarser sediments.

Angularity

As sediment is transported from its source to where it is deposited, the particles collide with each other and with other objects in their path. These collisions can cause the particles to change size and shape. When particles first break from the source rock, they tend to be angular and uneven. Particles that have moved long distances from the source tend to be more rounded and smooth. In general, the farther sediment travels from its source, the finer and smoother the particles of sediment become.

SCILINKS. NSTA Developed and maintained by the National Science Teachers Association

For a variety of links related to this subject, go to www.scilinks.org

Topic: Sedimentary Rock
SciLinks code: HQ61365

Sedimentary Rock Features

The setting in which sediment is deposited is called a *depositional environment*. Common depositional environments include rivers, deltas, beaches, and oceans. Each depositional environment has different characteristics that create specific structures in sedimentary rock. These features allow scientists to identify the depositional environment in which the rock formed.

Internet Activity — General

Identifying Sedimentary Rock Features Have students use the Internet to locate an example of each of the sedimentary rock features described on the next page. Students may find information and photographs at Web sites for the National Park Service and the United States Geological Survey. Have students compile their findings in a written report. A worksheet designed to direct student research on this topic can be found in the **Chapter Resource File** booklet or by visiting **go.hrw.com** and entering the keyword **HQ6RXSX.** LS **Visual**

Reading Skill Builder — Basic

Reading Organizer Have students take Power Notes, KWL Notes, or Two-Column Notes, as described in the Skills Handbook section of the Appendix. Tell them to include how the speed of wind and water that transport sediments and the length of time sediments are transported determine how well-sorted and angular the sediments are. Later, students can use these notes as a study guide for assessments. LS **Verbal** English Language Learners

Stratification

Layering of sedimentary rock, as shown in **Figure 5,** is called *stratification*. Stratification occurs when the conditions of sediment deposition change. The conditions may vary when there is a change in sediment type or of depositional environment. For example, a rise in sea level may cause an area that was once a beach to become a shallow ocean, which changes the type of sediment that is deposited in the area.

Stratified layers, or *beds,* vary in thickness depending on the length of time during which sediment is deposited and how much sediment is deposited. *Massive beds*, or beds that have no internal structures, form when similar sediment is deposited for long periods of time or when a large amount of sediment is deposited at one time.

Cross-beds and Graded Bedding

Some sedimentary rocks are characterized by slanting layers called *cross-beds* that form within beds. Cross-beds, which generally form in sand dunes or river beds, are shown in **Figure 5.**

When various sizes and kinds of materials are deposited within one layer, a type of stratification called *graded bedding* may occur. Graded bedding occurs when different sizes and shapes of sediment settle to different levels. Graded beds commonly transition from largest grains on the bottom to smallest grains on the top. However, certain depositional events, such as some mudflows, may cause *reverse grading,* in which the smallest grains are on the bottom and the largest grains are on top.

Reading Check What is graded bedding? (See the Appendix for answers to Reading Checks.)

QuickLAB 10 min
Graded Bedding

Procedure

1. Place **20 mL of water** into a **small glass jar.**
2. Pour **10 mL of poorly sorted sediment** into the jar. Place a **lid** securely on the jar.
3. Shake the jar vigorously for 1 min, and then let it sit still for 5 min.
4. Observe the settled sediment.

Analysis

1. Describe any sedimentary structures you observed.
2. Name two factors responsible for the sedimentary structures you observed.

Figure 5 ▶ Examples of Sedimentary Rock Structures

QuickLAB

Skills Acquired

- Experimenting
- Identifying/Recognizing Patterns
- Predicting

Teacher's Notes You may want to use plastic jars, to prevent students from breaking glass jars. Have students clean up spills quickly and completely.

Answers

1. The sediment shows graded bedding; the largest fragments lie on the bottom of the jar, over which the next largest particles are layered, and so forth up to the top where the finest sediments are located.
2. The size and density of sediment grains cause these structures to form.

Close

Reteaching — BASIC

Sedimentary Rocks On the board, write *limestone, sandstone, breccia, evaporites,* and *shale* in one column. Write *chemical, clastic,* and *organic* in another column. Have students match the appropriate terms in the two columns. LS **Logical/Verbal**

Quiz — GENERAL

1. What reduces the angularity of clastic sediments? (The continual transport of those sediments over long periods of time causes the sediments to become rounded and smooth.)
2. What is the compound in the shells of marine animals that makes up limestone? (calcium carbonate, or $CaCO_3$)

Answer to Reading Check

Graded bedding is a type of stratification in which different sizes and types of sediments settle to different levels.

CHAPTER RESOURCES

Chapter Resource File

- Datasheet for Quick Lab GENERAL

Close, *continued*

Alternative Assessment — ADVANCED

Modeling Sedimentation Have students create a computer model or a series of detailed diagrams that show the various stages in the formation of sedimentary rock. Students should model both compaction and cementation. The models should show all of the steps that lead to the formation of sedimentary rock from deposited materials. LS **Kinesthetic/Logical**

Answers to Section Review

1. During compaction, fine grains are squeezed together and the rock that forms takes up less space. During cementation, minerals that are deposited by water glue sediments together.
2. Chemical sedimentary rocks form when chemicals in solution precipitate out of the solution and form deposits at the bottom of the body of water. Examples include gypsum and halite. Organic sedimentary rocks form when the remains of plants and animals are gradually compacted or cemented together. Examples include limestone and coal.
3. Clastic rock forms from rock fragments that were transported away from their source by wind, water, gravity, or ice, rather than by chemical processes.
4. The farther sediment is transported and the longer the transport takes, the more well-sorted and less angular the fragments become.
5. Stratification, cross-beds, graded bedding, ripple marks, mud cracks, fossils, and concretions are all features that provide information about how sediment was deposited.
6. Smooth, small rocks have probably been transported farther or for a longer period of time than angular, uneven rocks have.
7. A slow-moving stream sorts sediments more effectively because it cannot carry larger or heavier sediments as easily as it carries smaller or lighter sediments.
8. *Sedimentary rock,* is classified as *chemical sedimentary rock, organic sedimentary rock,* and *clastic sedimentary rock,* and may form by *cementation* and *compaction.*

Figure 6 ▶ This dry and mud-cracked river bed is in Nagasaki, Japan. This river bed is the site of a future dam project.

Ripple Marks

Some sedimentary rocks clearly display *ripple marks*. Ripple marks are caused by the action of wind or water on sand. When the sand becomes sandstone, the ripple marks may be preserved. When scientists find ripple marks in sedimentary rock, the scientists know that the sediment was once part of a beach or a river bed.

Mud Cracks

The ground in **Figure 6** shows mud cracks, which are another feature of sedimentary rock. Mud cracks form when muddy deposits dry and shrink. The shrinking causes the drying mud to crack. A river's flood plain or a dry lake bed is a common place to find mud cracks. Once the area is flooded again, new deposits may fill in the cracks and preserve their features when the mud hardens to solid rock.

Fossils and Concretions

The remains or traces of ancient plants and animals, called *fossils,* may be preserved in sedimentary rock. As sediments pile up, plant and animal remains are buried. Hard parts of these remains may be preserved in the rock. More often, even the hard parts dissolve and leave only impressions in the rock. Sedimentary rocks sometimes contain lumps of rock that have a composition that is different from that of the main rock body. These lumps are known as *concretions*. Concretions form when minerals precipitate from fluids and build up around a nucleus. Groundwater sometimes deposits dissolved minerals inside cavities in sedimentary rock. The minerals may crystallize inside the cavities to form a special type of rock called a *geode*.

Section 3 Review

1. **Explain** how the processes of compaction and cementation form sedimentary rock.
2. **Describe** how chemical and organic sedimentary rocks form, and give two examples of each.
3. **Describe** how clastic sedimentary rock differs from chemical and organic sedimentary rock.
4. **Explain** how the physical characteristics of sediments change during transport.
5. **Identify** seven features that you can use to identify the depositional environment in which sedimentary rocks formed.

CRITICAL THINKING

6. **Making Comparisons** Compare the histories of rounded, smooth rocks and angular, uneven rocks.
7. **Identifying Relationships** Which of the following would most effectively sort sediments: a fast-moving river or a small, slow-moving stream? Explain your answer.

CONCEPT MAPPING

8. Use the following terms to create a concept map: *cementation, clastic sedimentary rock, sedimentary rock, chemical sedimentary rock, compaction,* and *organic sedimentary rock.*

CHAPTER RESOURCES

Chapter Resource File
- Section Quiz GENERAL

Workbooks
- Study Guide (also in Spanish)

Section 4 Metamorphic Rock

The process by which heat, pressure, or chemical processes change one type of rock to another is called **metamorphism.** Most metamorphic rock, or rock that has undergone metamorphism, forms deep within Earth's crust. All metamorphic rock forms from existing igneous, sedimentary, or metamorphic rock.

OBJECTIVES

- **Describe** the process of metamorphism.
- **Explain** the difference between regional and contact metamorphism.
- **Distinguish** between foliated and nonfoliated metamorphic rocks, and give an example of each.

KEY TERMS

metamorphism
contact metamorphism
regional metamorphism
foliation
nonfoliated

Formation of Metamorphic Rocks

During metamorphism, heat, pressure, and hot fluids cause some minerals to change into other minerals. Minerals may also change in size or shape, or they may separate into parallel bands that give the rock a layered appearance. Hot fluids from magma may circulate through the rock and change the mineral composition of the rock by dissolving some materials and by adding others. All of these changes are part of metamorphism.

The type of rock that forms because of metamorphism can indicate the conditions that were in place when the original rock changed, as shown in **Figure 1.** The composition of the rock being metamorphosed, the amount and direction of heat and pressure, and the presence or absence of certain fluids cause different combinations of minerals to form.

Two types of metamorphism occur in Earth's crust. One type occurs when small volumes of rock come into contact with magma. The second type occurs when large areas of Earth's crust are affected by the heat and pressure that is caused by the movement and collisions of Earth's giant tectonic plates.

metamorphism the process in which one type of rock changes into metamorphic rock because of chemical processes or changes in temperature and pressure

Figure 1 ▶ Indicators of Metamorphic Conditions

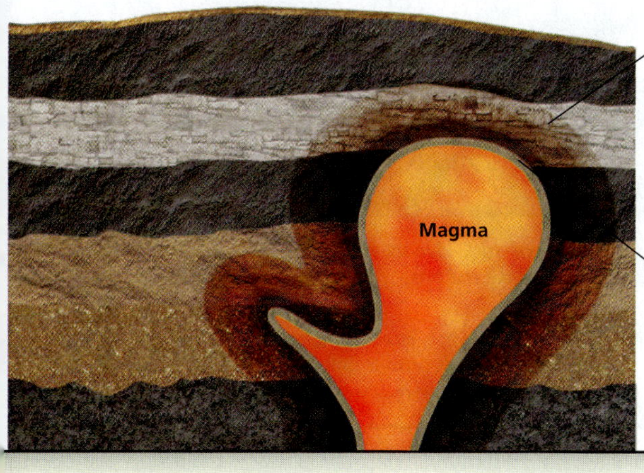

Slate is a metamorphic rock that commonly forms in the outer zone of metamorphism around a body of magma where clay-rich rock is exposed to relatively small amounts of heat.

Hornfels is a metamorphic rock that forms in the innermost zone of metamorphism, where clay-rich rock is exposed to large amounts of heat from the magma.

Section 4

Focus

Overview

This section describes how temperature and pressure changes and chemical processes change one type of rock to another. The section also explains the difference between regional and contact metamorphism and distinguishes between foliated and nonfoliated metamorphic rock.

Bellringer

Ask students to list items that change from one form to another. (Sample answers: Caterpillars become butterflies; nuts or seeds become trees; superheroes change from their secret identities; and liquid water freezes to become ice.) LS **Verbal**

Motivate

Demonstration — GENERAL

Metamorphism Sometimes, metamorphism changes a rock without changing its bulk chemical composition. Show the class samples of granite and gneiss and of limestone and marble. Have students observe similarities and differences. (Granite and gneiss may both be made of quartz and feldspars, but gneiss has bands of minerals. Marble and limestone are both composed of calcium carbonate, but marble appears to be crystalline.) LS **Visual**

CHAPTER RESOURCES

Chapter Resource File

- **Directed Reading** BASIC
- **Making Models Lab** Metamorphic Rocks GENERAL

Technology

Transparencies
- Bellringer
- 30 Indicators of Metamorphic Conditions (with worksheet)

Student Edition on CD-ROM

One-Stop Planner CD-ROM
- Lesson Plan

Teach

READING SKILL BUILDER — BASIC

Anticipation Guide Before students read this page, ask them to predict the difference between contact metamorphism and regional metamorphism. Have students write down their predictions, and then read the page to find out if their predictions were accurate. **LS Logical** English Language Learners

Answer to Reading Check

The high pressures and temperatures that result from the movement of tectonic plates may cause chemical changes in the minerals.

Activity — ADVANCED

Poster Project Have interested students research an area in which contact or regional metamorphism is known to be occurring. Students may work individually or in pairs. Have students identify the specific source of heat, pressure, or fluid introduction that causes the metamorphism, as well as the types of rock before and after the metamorphism takes place. Students should also infer from the conditions causing the metamorphism whether the stresses on the rock are applied to large areas or small areas, as indicated in the Connection to Physics feature on the Student Edition page. Students should organize their findings and present them as a poster or series of posters. **LS Logical/Visual**

contact metamorphism a change in the texture, structure, or chemical composition of a rock due to contact with magma

regional metamorphism a change in the texture, structure, or chemical composition of a rock due to changes in temperature and pressure over a large area, generally as a result of tectonic forces

Contact Metamorphism

When magma comes into contact with existing rock, heat from the magma can change the structure and mineral composition of the surrounding rock by a process called **contact metamorphism.** During contact metamorphism only a small area of rock that surrounds the hot magma is changed by the magma's heat. Hot chemical fluids moving through fractures may also cause changes in the surrounding rock during contact metamorphism.

Regional Metamorphism

Metamorphism sometimes occurs over an area of thousands of square kilometers during periods of high tectonic activity, such as when mountain ranges form. The type of metamorphism that occurs over a large area is called **regional metamorphism.**

The movement of one tectonic plate against another generates tremendous heat and pressure in the rocks at the edges of the tectonic plates. The heat and pressure cause chemical changes in the minerals of the rock. Most metamorphic rock forms as a result of regional metamorphism. However, volcanism and movement of magma often accompany tectonic activity. Thus, rocks that are formed by contact metamorphism are also commonly discovered where regional metamorphism has occurred.

Reading Check **How are minerals affected by regional metamorphism? (See the Appendix for answers to Reading Checks.)**

Connection to PHYSICS

Directed Stress

The agents that cause metamorphism are temperature, pressure, chemically active fluids, and directed stress. When one of these conditions changes in a rock's environment, the minerals in the rock move from a stable state to an unstable state. This instability causes the rock to change, or metamorphose, to reach a more stable state in the new conditions.

Stress is the amount of force per unit area.

$$\sigma = \frac{Force}{Area}$$

Stress can be caused by fluids that are trapped in the rock or by the load of overlying and surrounding rock. At about 3.3 km below the surface of Earth's crust, the pressure of overlying rock is great enough to metamorphose rock.

In general, stress affects rock equally in all directions. However, sometimes, stresses acting in particular directions exceed the mean stress on the rock. This type of stress is called *directed stress*, and it can act in three ways. *Tension* is stress that expands the rock or pulls the rock apart. *Compression* is stress that squeezes the rock. *Shear stress* is stress that pushes different parts of a body of rock in different directions.

Compression of rock during metamorphism may form tiny folds, like the ones shown here.

BRAIN FOOD

Metamorphic Gemstones Metamorphism produces some of the most beautiful gemstones. Gemstones are rare mineral crystals that can be cut and polished into jewels. The Southeast Asian country of Myanmar is the most famous source of fine rubies—red crystals of the mineral corundum—that are among the most valuable in the world. In nearby Thailand, rubies have formed during contact metamorphism when basalt metamorphoses at high temperature.

Some rubies phosphoresce with a vivid red glow when illuminated by ultraviolet light. Emeralds, a form of the mineral beryl, are usually the product of igneous and metamorphic processes. However, among the most sought-after deep green emeralds are those from Colombia, South America. These gems form by chemical sedimentary crystallization and by reactions with organic shells during contact metamorphism.

Classification of Metamorphic Rocks

Minerals in the original rock help determine the mineral composition of the metamorphosed rock. As the original rock is exposed to changes in heat and pressure, the minerals in the original rock often combine chemically to form new minerals. While metamorphic rocks are classified by chemical composition, they are first classified according to their texture. Metamorphic rocks have either a foliated texture or a nonfoliated texture.

Foliated Rocks

The metamorphic rock texture in which minerals are arranged in planes or bands is called **foliation.** Foliated rock can form in one of two ways. Extreme pressure may cause the mineral crystals in the rock to realign or regrow to form parallel bands. Foliation also occurs as minerals that have different compositions separate to produce a series of alternating dark and light bands.

Foliated metamorphic rocks include the common rocks slate, schist, and gneiss (NIES). Slate forms when pressure is exerted on the sedimentary rock shale, which contains clay minerals that are flat and thin. The fine-grained minerals in slate are compressed into thin layers, which split easily into flat sheets. Flat sheets of slate are used in building materials, such as roof tiles or walkway stones.

When large amounts of heat and pressure are exerted on slate, a coarse-grained metamorphic rock known as *schist* may form. Deep underground, intense heat and pressure may cause the minerals in schist to separate into bands as the minerals recrystallize. The metamorphosed rock that has bands of light and dark minerals is called *gneiss*. Gneiss is shown in **Figure 2.**

foliation the metamorphic rock texture in which mineral grains are arranged in planes or bands

SCILINKS. NSTA Developed and maintained by the National Science Teachers Association

For a variety of links related to this subject, go to www.scilinks.org

Topic: Metamorphic Rock
SciLinks code: HQ60949

Figure 2 ▶ Large amounts of heat and pressure may change rock into the metamorphic rock gneiss, which shows pronounced foliation.

Homework — ADVANCED

Degrees of Metamorphism Have students research metamorphic grades of rock. Students should choose a sedimentary or igneous rock and then research the temperature and pressure conditions required to transform that rock into several different forms. Have students present their findings in a short oral report. **LS Auditory**

Close

Reteaching — BASIC

Metamorphic Rocks On the board, write the words "foliated" and "nonfoliated." In a side column, write the names "slate," "quartzite," "schist," "gneiss," and "marble." Have students indicate which of the rocks fall under the first two headings. **LS Logical**

Quiz — GENERAL

1. What are the two types of metamorphism? (contact metamorphism and regional metamorphism)
2. Why do minerals in some metamorphic rocks separate into compositional bands? (Intense heat and pressure cause different minerals to separate as they recrystallize.)
3. How are marble and limestone similar? How are they different? (They have the same chemical composition. Marble is a nonfoliated metamorphic rock. Limestone, from which marble forms, is an organic sedimentary rock.)

MISCONCEPTION ALERT

Physical Separation Students may think that foliation is the result of different minerals forming within the metamorphic rock. Emphasize that foliation is a physical property of rock. Minerals may become aligned as the crystals reform in a direction perpendicular to the stress, without changing chemical composition. Separation of minerals into compositional bands may result as minerals recrystallize under conditions of great heat and pressure. In both cases, changes are because of the physical properties of rock components and not a result of a change in composition.

Close, *continued*

Alternative Assessment — ADVANCED

Metamorphic Chemistry Have students research the exact geological and chemical processes required for forming one of the following gemstones: sapphires, jade, and star sapphires. Students should present their findings in a short written report. LS **Verbal/Logical**

Answers to Section Review

1. Metamorphism is the process by which one rock changes into another kind of rock because of chemical processes or changes in temperature and pressure.
2. Contact metamorphism occurs in a small region where rock comes in contact with magma. Regional metamorphism occurs over a large area, and is generally the result of tectonic forces.
3. Foliated metamorphic rock has a texture in which mineral grains are arranged in planes or bands, whereas the texture of nonfoliated metamorphic rock does not have minerals arranged in planes or bands.
4. Slate and schist are foliated metamorphic rocks, while quartzite and marble are nonfoliated metamorphic rocks.
5. Both a butterfly and metamorphic rock undergo a change from an earlier form (a caterpillar in the case of the butterfly; a different type of rock in the case of the rock) to a new form.
6. Both rocks formed when clay-rich rock underwent contact metamorphism, but the rock that formed slate was farther from the magma than the rock that formed hornfels was.
7. Regional metamorphism is occurring where the two tectonic plates collide to uplift the Himalayas. Therefore, most metamorphism will be widespread.
8. *Metamorphic rock* forms as a result of *contact metamorphism* or *regional metamorphism,* and may be *foliated* or *nonfoliated.*

Figure 3 ▶ Marble is a nonfoliated metamorphic rock that is used as building and sculpting material.

nonfoliated the metamorphic rock texture in which mineral grains are not arranged in planes or bands

Nonfoliated Rocks

Rocks that do not have bands or aligned minerals are **nonfoliated.** Most nonfoliated metamorphic rocks share at least one of two main characteristics. First, the original rock that is metamorphosed may contain grains of only one mineral or contains very small amounts of other minerals. Thus, the rock does not form compositional bands when it is metamorphosed. Second, the original rock may contain grains that are round or square. Because the grains do not have some long and some short sides, these grains do not change position when exposed to pressure in one direction.

Quartzite is one common nonfoliated rock. Quartzite forms when quartz sandstone is metamorphosed. Because quartzite is very hard and durable, it is resistant to weathering. For this reason, quartzite remains after weaker rocks around it have eroded and may form hills or mountains.

Marble, the beautiful stone that is used for building monuments and statues, is a metamorphic rock that forms from the compression of limestone. The Parthenon, which is shown in **Figure 3,** has been standing in Greece for more than 1,400 years. However, the calcium carbonate in marble is susceptible to accelerated chemical weathering by acid rain, which is caused by air pollution. Many ancient marble structures and sculptures are being damaged by acid rain.

Section 4 Review

1. **Describe** the process of metamorphism.
2. **Explain** the difference between regional and contact metamorphism.
3. **Distinguish** between foliated and nonfoliated metamorphic rocks.
4. **Identify** two foliated metamorphic rocks and two nonfoliated metamorphic rocks.

CRITICAL THINKING

5. **Analyzing Relationships** What do a butterfly and metamorphic rock have in common?
6. **Making Comparisons** If you have samples of the two metamorphic rocks slate and hornfels, what can you say about the history of each rock?
7. **Identifying Relationships** The Himalaya Mountains are located on a boundary between two colliding tectonic plates. Would most of the metamorphic rock in that area occur in small patches or in wide regions? Explain your answer.

CONCEPT MAPPING

8. Use the following terms to create a concept map: *contact metamorphism, foliated, regional metamorphism, metamorphic rock,* and *nonfoliated.*

CHAPTER RESOURCES

Chapter Resource File

- **Section Quiz** GENERAL

Workbooks

- **Study Guide** (also in Spanish)

Chapter 6 Highlights

Sections

1 Rocks and the Rock Cycle

Key Terms

rock cycle, 126
Bowen's reaction series, 127

Key Concepts

- Rocks are classified into three major types based on how they form. These types are igneous rock, sedimentary rock, and metamorphic rock.
- In the rock cycle, rocks change from one type into another.
- The different minerals in igneous rocks form in a specific order as represented in Bowen's reaction series.

2 Igneous Rock

Key Terms

igneous rock, 129
intrusive igneous rock, 131
extrusive igneous rock, 131
felsic, 132
mafic, 132

Key Concepts

- The rate at which magma and lava cool determines the texture of igneous rock.
- Igneous rocks are divided into three families based on their mineral composition. These families are felsic, mafic, and intermediate.
- Igneous rock structures take two basic forms. They are intrusions and extrusions.

3 Sedimentary Rock

Key Terms

compaction, 135
cementation, 135
chemical sedimentary rock, 136
organic sedimentary rock, 136
clastic sedimentary rock, 137

Key Concepts

- Sedimentary rock forms in one of three ways. It may form from minerals once dissolved in water, from the remains of organisms, or from rock fragments.
- Sedimentary rocks have a number of identifiable features, including stratification, ripple marks, mud cracks, fossils, and concretions.

4 Metamorphic Rock

Key Terms

metamorphism, 141
contact metamorphism, 142
regional metamorphism, 142
foliation, 143
nonfoliated, 144

Key Concepts

- Metamorphic rock forms as a result of heat and pressure caused by hot magma or tectonic plate movement.
- Metamorphic rocks can have a foliated or nonfoliated texture.

Chapter Highlights

Alternative Assessment — GENERAL

Bulletin Board Display Have students create a display on a bulletin board that shows the processes by which rocks form and transform into other rocks. The basic form of the display should resemble the figure of the rock cycle, but students may modify the layout and the size to include other details, such as the rate of cooling and the magma type for igneous rocks, the properties of the classes of sedimentary rocks, and the textures of metamorphic rocks. **LS Logical/Visual**

CHAPTER RESOURCES

Chapter Resource File

- **Concept Review** GENERAL
- **Critical Thinking** ADVANCED
- **Math Skills** GENERAL
- **Graphing Skills** GENERAL
- **Chapter Test A** GENERAL
- **Chapter Test B** ADVANCED

Workbooks

- **Study Guide** (also in Spanish)
- **Assessments** (Spanish)

Technology

Classroom Videos
- Brain Food Video Quiz

Chapter Review

Assignment Guide

Section	Questions
1	1–3, 17, 19–20, 31
2	5, 9–12, 16, 21, 24, 29–30, 33–34
3	6, 13–14, 18, 23, 25–26
4	7–8, 15, 22, 27
2 and 4	4
1–4	28, 32

Using Key Terms

1–8. Answers may vary but should show that students understand the definitions of and differences between key terms.

Understanding Key Concepts

9. c **10.** c
11. d **12.** d
13. c **14.** a
15. d

Short Answer

16. In partial melting, as the temperature of a rock increases, the minerals that have the lowest melting point melt first. Fractional crystallization works in the opposite way. As the temperature of molten rock decreases, crystals of minerals with high freezing points form first.

17. Igneous rock forms when magma cools. Sedimentary rock forms when rock fragments are compressed and cemented together or by chemical or organic processes. Metamorphic rock forms when a rock is subjected to temperature and pressure changes or to chemical processes.

Chapter 6 Review

Using Key Terms

Use each of the following terms in a separate sentence.

1. *rock cycle*

2. *Bowen's reaction series*

3. *sediment*

For each pair of terms, explain how the meanings of the terms differ.

4. *igneous rock* and *metamorphic rock*

5. *intrusive igneous rock* and *extrusive igneous rock*

6. *chemical sedimentary rock* and *organic sedimentary rock*

7. *contact metamorphism* and *regional metamorphism*

8. *foliated* and *nonfoliated*

Understanding Key Concepts

9. Intrusive igneous rocks are characterized by a coarse-grained texture because they contain
a. heavy elements.
b. small crystals.
c. large crystals.
d. fragments of different sizes and shapes.

10. Light-colored igneous rocks are generally part of the
a. basalt family.
b. intermediate family.
c. felsic family.
d. mafic family.

11. Magma that solidifies underground forms rock masses that are known as
a. extrusions.
b. volcanic cones.
c. lava plateaus.
d. intrusions.

12. One example of an extrusion is a
a. stock. **c.** batholith.
b. dike. **d.** lava plateau.

13. Sedimentary rock formed from rock fragments is called
a. organic. **c.** clastic.
b. chemical. **d.** granite.

14. One example of chemical sedimentary rock is
a. an evaporite. **c.** sandstone.
b. coal. **d.** breccia.

15. The splitting of slate into flat layers illustrates its
a. contact metamorphism.
b. formation.
c. sedimentation.
d. foliation.

Short Answer

16. Describe partial melting and fractional crystallization.

17. Name and define the three main types of rock.

18. How do clastic sedimentary rocks differ from chemical and organic sedimentary rocks?

19. What is Bowen's reaction series?

20. What factors affect the chemical and physical stability of rock?

21. Describe three factors that affect whether rock melts.

22. Why are some metamorphic rocks foliated while others are not?

23. How does transport affect the size and shape of sediment particles?

18. Clastic sedimentary rocks form from fragments of other rocks. Chemical and organic sedimentary rock form through chemical or organic processes.

19. Bowen's reaction series is a simplified pattern that shows the order in which minerals crystallize from cooling magma according to their chemical compositions and melting points.

20. Strength of the chemical bonds between atoms in minerals affects the chemical stability of rock. Physical stability is affected by natural zones of weakness in rocks. Examples of these zones include areas between layers in sedimentary and metamorphic rock, joints formed by the cooling and contracting of igneous rock, and exfoliation layers formed by changes in pressure as rock is uplifed to Earth's surface.

21. The temperature, pressure, and presence of fluids within rock all determine the melting point of rock.

22. Foliation occurs when a rock contains two or more minerals that have grains that are neither round nor square. These conditions allow the crystals to realign under pressure or cause the minerals to separate into compositional bands. Foliation does not occur when rock is composed mainly of a single mineral.

23. As sediment particles are transported, they undergo additional weathering. The longer the transport process takes, the finer the particles become. Their shapes also become rounder and smoother as transport continues.

Critical Thinking

24. Making Inferences A certain rock is made up mostly of plagioclase feldspar and pyroxene minerals. It also includes olivine and hornblende. Will the rock have a light or dark coloring? Explain your answer.

25. Classifying Information Explain how metamorphic rock can change into either of the other two types of rock through the rock cycle.

26. Applying Ideas Imagine that you have found a piece of limestone, which is a sedimentary rock, that has strange-shaped lumps on it. Will the lumps have the same composition as the limestone? Explain your answer.

27. Analyzing Ideas Which would be easier to break, the foliated rock slate or the nonfoliated rock quartzite? Explain your answer.

Concept Mapping

28. Use the following terms to create a concept map: *rock cycle, foliated, igneous rock, intrusive, sedimentary rock, clastic sedimentary rock, metamorphic rock, chemical sedimentary rock, extrusive, organic sedimentary rock,* and *nonfoliated*.

Math Skills

29. Making Calculations The gram formula weight (weight of one mole) of the mineral quartz is 60.1 g, and the gram formula weight of magnetite is 231.5 g. If you had 4 moles of magnetite, how many moles of quartz would be equal to the weight of the magnetite?

30. Making Calculations The gram formula weight (weight of one mole) of the mineral hematite, Fe_2O_3, is 159.7 g, and the gram formula weight of magnetite, Fe_3O_4, is 231.5 g. Which of the following would weigh more: half a mole of hematite or one-third of a mole of magnetite?

Writing Skills

31. Outlining Topics Outline the essential steps in the rock cycle.

32. Writing from Research Find out what types of rock are most abundant in your state. Research the geologic processes that form those types of rock, and write a brief report that describes how the rocks in your state most likely formed.

Interpreting Graphics

The graph below is a ternary diagram that shows the classification of some igneous rocks. Refer to the Skills Handbook in the Appendix for instructions on how to read a ternary diagram. Use the diagram to answer the questions that follow.

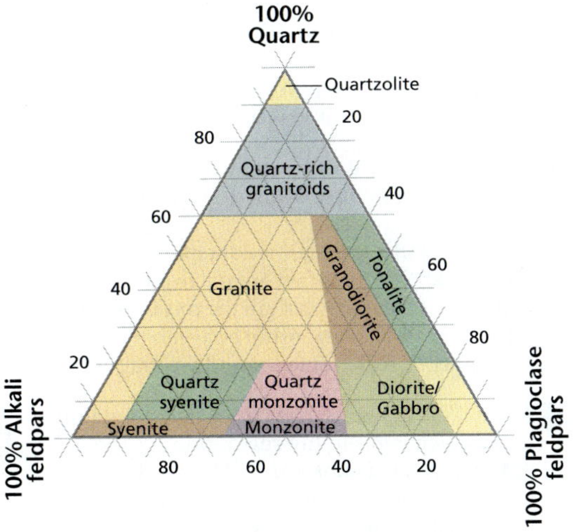

33. What is the maximum amount of quartz in a quartz syenite?

34. What would a rock that contains 30% quartz, 20% alkali feldspar, and 50% plagioclase feldspar be called?

Chapter Review

Critical Thinking

24. The minerals pyroxene, olivine, and horn-blende are ferromagnesian minerals. The rock probably formed from mafic magma and will be dark in color.

25. If metamorphic rock is uplifted to Earth's surface and is exposed to air and water, it may break down into sediment that becomes sedimentary rock. If metamorphic rock is melted and then cools, igneous rock will form.

26. The lumps are probably concretions that formed around some type of nucleus. They will not have the same composition as the limestone.

27. The slate would be more easily broken, because the grains in slate are aligned in thin layers that have weak areas between them. Nonfoliated rocks, like quartzite, do not form layers and thus do not have those zones of weakness.

Concept Mapping

28. Answers may vary but should include all of the terms listed. Sample answers appear at the end of this Teacher Edition.

Math Skills

29. (231.5 g/mol ÷ 60.1 g/mol) × 4 mol = 15.4 mol; or (231.5 g/mol × 4 mol) ÷ 60.1 g/mol = 15.4 mol

30. 159.7 g of hematite ÷ 2 = 79.9 g; 231.5 g of magnetite ÷ 3 = 77.2 g; The sample of hematite weighs more.

Writing Skills

31. Answers may vary. Accept all reasonable answers.

32. Answers may vary. Accept all reasonable answers.

Interpreting Graphics

33. 20%

34. granodiorite

Standardized Test Prep

Estimated Time

To give students practice under more realistic testing conditions, allow them 30 minutes to answer all of the questions in this practice test.

TEST DOCTOR

Question 1 Answer B is correct. Answer A is incorrect because organic remains would not withstand the high temperature conditions that produce magma, from which igneous rocks form. Answer C is incorrect because it is unlikely that organic remains would withstand the massive heat and pressure that change existing rocks into metamorphic rocks. Answer D is incorrect because felsic rock is one of the families of igneous rock. Fossils and organic remains are typically found in sedimentary rock.

Question 4 Answer H is correct. Answer F is incorrect because rocks that form last, at the lowest temperatures, tend to withstand weathering the longest. Answer G is incorrect because rocks that form first are the least stable. Answer I is incorrect because Bowen's Reaction series not only accounts for the production of different rocks from the same magma but also explains how some rocks weather faster than others. The rocks that form last are also the most chemically stable.

Chapter 6 Standardized Test Prep

Understanding Concepts

Directions (1–5): **For *each* question, write on a sheet of paper the letter of the correct answer.**

1. A rock that contains a fossil is most likely
   - **A.** igneous
   - **B.** sedimentary
   - **C.** metamorphic
   - **D.** felsic

2. The large, well-developed crystals found in some samples of granite are a sign that
   - **F.** the lava from which it formed cooled rapidly
   - **G.** the magma contained a lot of dissolved gases
   - **H.** the lava from which it formed cooled slowly
   - **I.** water deposited minerals in the rock cavities

3. How does coal differ from breccia?
   - **A.** Coal is an example of sedimentary rock, and breccia is an example of metamorphic rock.
   - **B.** Coal is an example of metamorphic rock, and breccia is an example of igneous rock.
   - **C.** Coal is an example of organic rock, and breccia is an example of clastic rock.
   - **D.** Coal is an example of clastic rock, and breccia is an example of a conglomerate.

4. How does the order in which igneous rocks form relate to their ability to resist weathering agents?
   - **F.** Rocks that form last weather faster.
   - **G.** Rocks that form first are the most resistant.
   - **H.** Rocks that form last are the most resistant.
   - **I.** There is no relationship between the order of igneous rock formation and weathering.

5. What occurs when heat from nearby magma causes changes in the surrounding rocks?
   - **A.** contact metamorphism
   - **B.** fluid metamorphism
   - **C.** intrusive metamorphism
   - **D.** regional metamorphism

Directions (6–7): **For *each* question, write a short response.**

6. What type of sedimentary rock is formed when angular clastic materials cement together?

7. What type of rock is formed when heat, pressure, and chemical processes change the physical properties of igneous rock?

Reading Skills

Directions (8–10): **Read the passage below. Then, answer the questions.**

Igneous and Sedimentary Rocks

Scientists think that Earth began as a melted mixture of many different materials. These materials underwent a physical change as they cooled and solidified. These became the first igneous rocks. Igneous rock continues to form today. Liquid rock changes from a liquid to a solid, when lava that is brought to Earth's surface by volcanoes hardens. This process can also take place far more slowly, when magma deep beneath the Earth's surface changes to a solid.

At the same time that new rocks are forming, old rocks are broken down by other processes. Weathering is the process by which wind, water, and gravity break up rock. During erosion, broken up pieces of rock are carried by water, wind, or ice and deposited as sediments elsewhere. These pieces pile up and, under heat and pressure, form sedimentary rock—rock composed of cemented fragments of older rocks.

8. Which of the following statements about the texture of sedimentary rock is most likely true?
   - **A.** Sedimentary rocks are always lumpy and made up of large pieces of older rocks.
   - **B.** Sedimentary rocks all contain alternating bands of lumpy and smooth textures.
   - **C.** Sedimentary rocks are always smooth and made up of small pieces of older rocks.
   - **D.** Sedimentary rocks have a variety of textures that depend on the size and type of pieces that make up the rock.

9. Which of the following statements can be inferred from the information in the passage?
   - **F.** Igneous rocks are the hardest form of rock.
   - **G.** Sedimentary rocks are the final stage in the life cycle of a rock.
   - **H.** Igneous rocks began forming early in Earth's history.
   - **I.** Sedimentary rocks are not affected by weathering.

10. Is igneous rock or sedimentary rock more likely to contain fossils? Explain your answer.

Answers

Understanding Concepts

1. B
2. H
3. C
4. H
5. A
6. breccias
7. metamorphic rock

Reading Skills

8. D
9. H
10. sedimentary rock; Sedimentary rocks are made up of pieces of older rocks and may contain fossils. Fossils in the original material of an igneous rock would have been destroyed when melted.

Interpreting Graphics

11. D
12. Answers may vary. See Test Doctor for a detailed scoring rubric.
13. Answers may vary. See Test Doctor for a detailed scoring rubric.

Interpreting Graphics

Directions (11–13): **For *each* question below, record the correct answer on a separate sheet of paper.**

The diagram below shows the rock cycle. Use this diagram to answer question 11.

The Rock Cycle

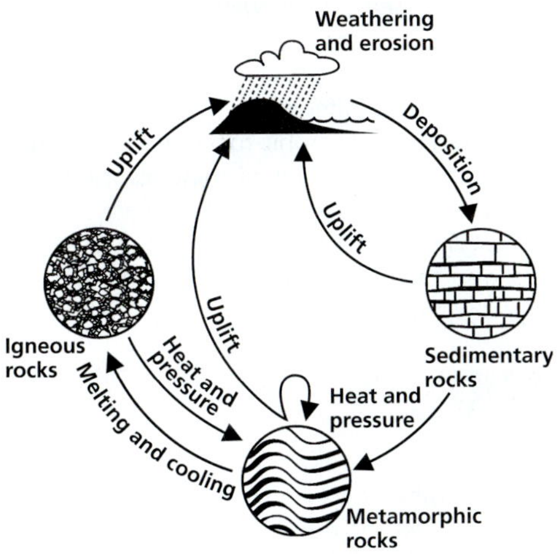

11 Which of the following processes brings rocks to Earth's surface, where they can be eroded?
A. deposition
B. weathering
C. erosion
D. uplift

Use this table to answer questions 12 and 13.

Rock Types

Rock sample	Characteristics
Rock A	multiple compacted, round, gravel-sized fragments
Rock B	coarse, well-developed, crystalline mineral grains
Rock C	small, sand-sized grains, tan coloration
Rock D	gritty texture; many small, embedded seashells

12 Is rock D igneous, sedimentary, or metamorphic? Explain the evidence that supports this classification.

13 Is rock A made up of only one mineral? Explain the evidence supporting this classification.

Test TIP

When several questions refer to the same graph, table, or diagram, or text passage, answer the questions you are sure of first.

Standardized Test Prep

TEST DOCTOR

Question 12 Full-credit answers should include the following points:
- students should be able to recognize different types of rocks by the physical characteristics of the rocks
- rock D is a sedimentary rock
- the presence of fossils or embedded seashells is a clear clue that this is a sedimentary rock
- students should demonstrate an understanding that sedimentary rocks are composed of many parts and particles that have been cemented together

Questions 13 Full-credit answers should include the following points:
- students should demonstrate an understanding that most rocks are combinations of one or more minerals
- the rock described in the table appears to be made up of more than one type of substance, so it is most likely not made of only a single mineral

Test Prep Correlations *National Science Education Standards*

ES 3c: items 5, 9
HNS 2a: items 6, 7, 8
HNS 2b: items 10, 12, 13
UCP 1: item 11
UCP 2: items 1, 2, 3, 4

CHAPTER RESOURCES

State Resources

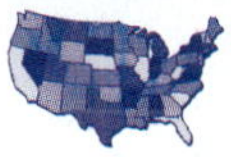

For specific resources for your state, visit go.hrw.com and type in the keyword HSHSTR.

Classification of Rocks

Teacher's Notes

Time Required

two 45-minute class periods

Lab Ratings

EASY → HARD

Teacher Preparation ▲▲
Student Setup ▲
Concept Level ▲▲
Cleanup ▲▲

Skills Acquired

- Experimenting
- Collecting Data
- Organizing and Analyzing Data
- Classifying
- Identifying and Recognizing Patterns

The Scientific Method

In this lab, students will
- Make Observations
- Analyze the Results
- Draw Conclusions

Materials

The materials listed on the page are enough for groups of two students. If two hand lenses are provided for each workstation, students could work in groups of four. You may want to use lemon juice in place of the acid.

Tips and Tricks

Students should visually identify the rocks before performing the acid test. When students test the rocks for acid reactivity, they should test all of the samples in sequence at the same time. This will reduce the number of times students will have to work with the acid, and thus reduce the chance of accidents. Make sure students wash their hands thoroughly when they are finished with the acid.

Classification of Rocks

Objectives

- USING SCIENTIFIC METHODS **Observe** the characteristics of common rocks.
- **Compare and contrast** the features of igneous, sedimentary, and metamorphic rocks.
- **Identify** igneous, sedimentary, and metamorphic rocks.

Materials

- hand lens
- hydrochloric acid, 10% dilute
- medicine dropper
- rock samples

Safety

There are many different types of igneous, sedimentary, and metamorphic rocks. Therefore, it is important to know distinguishing features of the rocks to identify the rocks. The classification of rocks is generally based on the way in which they formed their mineral composition and the size and arrangement (or texture) of their minerals.

Igneous rocks differ in the minerals they contain and the sizes of their crystals. Metamorphic rocks often look similar to igneous rocks, but they may have bands of minerals. Most sedimentary rocks are made of fragments of other rocks that are compressed and cemented together. Some common features of sedimentary rocks are parallel layers, ripple marks, cross-bedding, and the presence of fossils. In this lab, you will use these features to identify various rock samples.

PROCEDURE

1. In your notebook, make a table that has columns for sample number, description of properties, rock class, and rock name. List the numbers of the rock samples you received from your teacher.

CHAPTER RESOURCES

Chapter Resource File

- Datasheet for Chapter Lab GENERAL
- Lab Notes and Answers

2. Examine the rocks carefully. You can use a hand lens to study the fine details of the rock samples. Look for characteristics such as the shape, size, and arrangement of the mineral grains. For each sample, list in your table the distinguishing features that you observe.

Specimen	Descriptions of Properties	Rock Class	Rock Name

DO NOT WRITE IN THIS BOOK

3. Refer to the Guide to Common Rocks in the Reference Tables section of the Appendix. Compare the properties for each rock sample that you listed with the properties listed in the identification table. If you are unable to identify certain rocks, examine these rock samples again.

4. Certain rocks react with acid, which indicates that they are composed of calcite. If a rock contains calcite, the rock will bubble and release carbon dioxide. Using a medicine dropper and 10% dilute hydrochloric acid, test various samples for their reactions. **CAUTION** Wear goggles, gloves, and an apron when you work with hydrochloric acid. Wash your hands thoroughly afterward.

5. Complete your table by identifying the class of rock—igneous, sedimentary, or metamorphic—that each sample belongs to, and then name the rock.

ANALYSIS AND CONCLUSION

1. **Analyzing Methods** What properties were most useful and least useful in identifying each rock sample? Explain.

2. **Evaluating Results** Were there any samples that you found difficult to identify? Explain.

3. **Making Comparisons** Describe any characteristics common to all of the rock samples.

4. **Evaluating Ideas** How can you distinguish between a sedimentary rock and a foliated metamorphic rock if both have observable layering?

Extension

1. **Applying Conclusions** Collect a variety of rocks from your area. Use the Guide to Common Rocks to see how many you can classify. How many igneous rocks did you collect? How many sedimentary rocks did you collect? How many metamorphic rocks did you collect? After you identify the class of each rock, try to name the rock.

Answers to Analysis and Conclusion

1. Answers will vary with each student or group. Possible answers will be rock textures; the clear presence of crystals; and bands, layers, or internal structures of the rock. Least useful properties will probably include color and shape.
2. Answers will vary with each student or group. The presence of crystals, especially fine-grained ones, may make certain igneous rock difficult to identify. Nonfoliated metamorphic rock may be hard to identify. Some sedimentary rocks, though usually classifiable through their cemented texture, may not be easy to identify by name.
3. Answers will vary with each student or group. Most structural features will be common to two groups, but not to all three. Rocks of all three types may have similar chemical compositions.
4. Sedimentary rocks are made of grains that are cemented together, while foliated metamorphic rocks are made of crystals.

Answers to Extension

1. Answers will depend on the number and types of rocks collected by each student and on the ease with which they can be identified. You may want to contact the United States Geological Survey or a local geologist or bureau of geology to determine the types of rocks found in your area.

CLASSROOM TESTED & APPROVED

Scott Robertson
North Warren Central School
Chestertown, NY

Geologic Map of Virginia

MISCONCEPTION ALERT

Variations in Dates The dates given for geologic time intervals on this map are based on the ages of the rocks in the region. These dates may not match the dates in other geologic time scales because most time scales represent compiled data for several areas and include a variety of sources.

Answers to Map Skills Activity

1. Paleozoic Era, Cambrian Period
2. Charlottesville and Roanoke
3. Norfolk
4. Water in rivers and along the coastline contributed to deposition of material in the Holocene Epoch.
5. All of the Precambrian rocks formed at roughly the same time next to or on top of each other, so they are exposed near each other. The Precambrian rocks that are represented by the color brown are slightly metamorphosed rocks. The Precambrian rocks represented by the color gray are all metamorphic rocks. The Precambrian rocks represented by the patterned rock are igneous and metamorphic rocks that are older than the other two types of rock.
6. The oldest igneous rock formed between 1400 million and 980 million years ago. The oldest metamorphic rock formed between 750 million and 550 million years ago. The oldest sedimentary rock formed between 550 million and 500 million years ago.

CHAPTER RESOURCES

Technology

Transparencies
- 31 Geologic Map of Virginia (with worksheet)

MAPS in Action

Geologic Map of Virginia

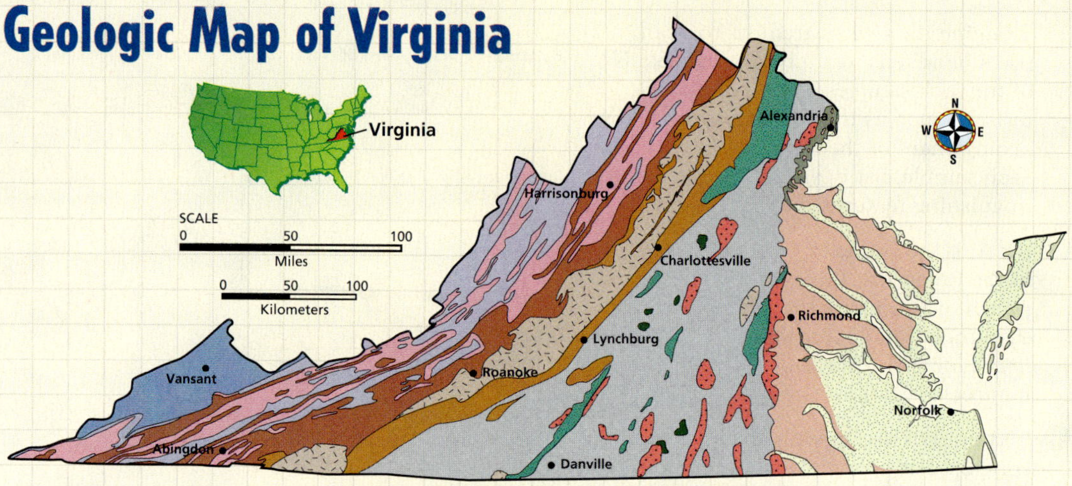

Precambrian

- **Precambrian (550–750 Ma)** Metasedimentary rocks, metarhyolite, and metabasalt
- **Precambrian (550–750 Ma)** Gneiss, schist, slate, phyllite, quartzite, and marble
- **Precambrian (980–1400 Ma)** Granite, granitic gneiss, charnockite, and layered gneiss

Paleozoic

- **Cambrian (500–550 Ma)** Dolomite, limestone, shale, and sandstone
- **Mississippian-Devonian (320–410 Ma)** Sandstone and shale with minor gypsum and coal
- **Silurian-Ordovican (410–500 Ma)** Limestone, dolomite, shale, and sandstone
- **Pennsylvanian (290–320 Ma)** Sandstone, shale, and coal
- **Paleozoic (300–500 Ma)** Granite and other felsic igneous rocks
- **Paleozoic (300–500 Ma)** Gabbro and other mafic igneous rocks

Mesozoic

- **Cretaceous (65–140 Ma)** Partly lithified sand, clay, and sandstone
- **Triassic-Jurassic (200–225 Ma)** Red and gray shale, sandstone, and conglomerate intruded by diabase and basalt

Cenozoic

- **Quaternary (20 ka–2 Ma)** Sand, mud, and gravel
- **Holocene (present–20 ka)** Sand, mud, and peat deposited in beaches, marshes, swamps, and estuaries
- **Tertiary (2–65 Ma)** Sand, mud, limy sand, and marl

Ma = millions of years
ka = thousands of years

Map Skills Activity

This map shows geologic data for the state of Virginia. The different colors indicate rocks of different ages. Use the map to answer the questions below.

1. **Using the Key** From what geologic era and period are rocks found in Roanoke, Virginia?
2. **Analyzing Data** Near which city in Virginia are the oldest rocks in the state found?
3. **Analyzing Data** Near which city in Virginia are the youngest rocks in the state found?
4. **Inferring Relationships** What feature or features helped to determine the location of the youngest rocks in Virginia?
5. **Identifying Trends** What are the differences between the three types of Precambrian rocks found in Virginia? Why do you think they are all located near each other?
6. **Analyzing Relationships** What are the age ranges of the oldest of each of the three rock types—igneous, metamorphic, and sedimentary—found in Virginia?

SCIENCE AND TECHNOLOGY

Moon Rock

NASA's Apollo missions brought 382 kg of lunar rock and soil back to Earth. In fact, moments after *Apollo 11* astronauts set foot on the moon in 1969, they began to fill two boxes with brown and gray moon rock. Later expeditions to the moon included vehicles that allowed a total of about 2,000 specimens to be collected.

Back on Earth

Geologists on Earth discovered that moon rocks are similar to Earth rocks in composition and in the way they formed. Geologists used their knowledge of Earth rock to analyze the moon rock and to learn about the geologic history of the moon.

Much of the rock material that was brought back from the moon was in a powdery form. This pulverized rock, called *regolith* (REG uh lith), covers much of the surface of the moon. Rock-dating methods show that regolith is one of the oldest materials on the moon's surface. From this information, geologists concluded that during the first billion years of the moon's existence, a shower of meteorites pulverized most of the moon's existing surface rock.

The Moon's History

The solid rocks on the moon are of two types—highland rocks and mare (MAW RAY) rocks. The highland rocks are igneous rocks that contain a large amount of plagioclase feldspar.

▶ This false-color image shows differences in the moon's composition. The blue areas have rock that contains large amounts of titanium.

Mare, which is Latin for "sea," refers to the dark areas on the lunar surface. Scientists discovered that mare rock formed after meteor showers made craters on the moon. Lava from inside the moon then poured onto the crater floors and covered large areas of the lunar surface. The lava then cooled and hardened into basalt about 4 billion years ago.

▲ *Apollo 12* astronaut Alan Bean holds a container designed to hold lunar soil (above). NASA scientist Andrea Mosie examines a volcanic moon rock (left).

Extension

1. **Making Inferences** Is mare rock classified as igneous, metamorphic, or sedimentary rock? Explain.

Science and Technology

Moon Rock

Internet Activity—ADVANCED

Geology of Other Planets Have students use the Internet to research the geology of other bodies in the solar system, most notably Mars, Venus, and Mercury. Students may check the Web sites of various facilities, such as NASA's Jet Propulsion Laboratory at the California Institute of Technology, to obtain information. Students should note geologic information that has been obtained from a variety of sources, including spacecraft that have been sent to planets, Earth-based telescopic observation, and the study of meteorites, most notably those that originated from Mars. Have students report their findings in a written report, oral presentation, or poster presentation. A worksheet designed to direct student research on this topic can be found in the **Chapter Resource File** booklet or by visiting **go.hrw.com** and entering the keyword **HQ6RXSX.** **LS Verbal/Logical**

CHAPTER RESOURCES

Chapter Resource File

Internet Activity
- Geology of Other Planets ADVANCED

Answer to Extension

1. Mare rock is classified as igneous, because the mare rock of the moon formed from lava that rose to the surface from inside the moon.

Chapter 7 Resources and Energy

Planning Guide

Compression Guide
To shorten instruction because of time limitations, omit Section 1.

OBJECTIVES	LABS, DEMONSTRATIONS, AND ACTIVITIES	TECHNOLOGY RESOURCES
PACING • 45 min pp. 154–158 **Chapter Opener**	**SE** **Long-Term Project** Air-Pollution Watch, pp. 852–855 ADVANCED **LTP** **Long-Term Project** Air-Pollution Watch* ADVANCED	**OSP** **Parent Letter** ■ **CD** **Student Edition on CD-ROM** **VID** **Brain Food Video Quiz**
Section 1 Mineral Resources • Explain what ores are and how they form. • Identify four uses for mineral resources. • Summarize two ways humans obtain mineral resources.	**TE** **Activity** Minerals at Work, p. 155 GENERAL **TE** **Group Activity** Mineral Use, p. 157 GENERAL **SE** **Mapping Expeditions** Buried Treasure, pp. 836–837 GENERAL **CRF** **Skills Practice Lab** Determination of Carbonate in Ore* ◆ GENERAL	**OSP** **Lesson Plans** (also in print) **TR** **Bellringer*** **TR** 32 The Formation of Ores and Placer Deposits* **CD** **Interactive Tutor** Renewable and Nonrenewable Resources
PACING • 45 min pp. 159–164 **Section 2 Nonrenewable Energy** • Explain why coal is a fossil fuel. • Describe how petroleum and natural gas form and how they are removed from Earth. • Summarize the processes of nuclear fission and nuclear fusion. • Explain how nuclear fission generates electricity.	**TE** **Identifying Preconceptions** Is It Renewable?, p. 159 GENERAL **TE** **Group Activity** Purifying with Pressure, p. 160 ◆ GENERAL **TE** **Demonstration** Chain Reaction, p. 162 BASIC **TE** **History Connection** Nuclear Theory, p. 162 ADVANCED **TE** **Debate** Pros and Cons, p. 163 GENERAL	**OSP** **Lesson Plans** (also in print) **TR** **Bellringer*** **TR** 33 Types of Coal* **TR** 34 Oil Traps* **TR** 35 A Nuclear Fission Reaction* **TR** 36 How a Nuclear Power Plant Generates Electricity* **TE** **Internet Activity** Resource Locations, p. 160 GENERAL **CRF** **Internet Activity** Resource Locations* GENERAL
PACING • 90 min pp. 165–168 **Section 3 Renewable Energy** • Explain how geothermal energy may be used as a substitute for fossil fuels. • Compare passive and active methods of harnessing energy from the sun. • Explain how water and wind can be harnessed to generate electricity.	**SE** **Quick Lab** Solar Collector, p. 166 GENERAL **CRF** **Datasheet for Quick Lab*** GENERAL **SE** **Inquiry Lab** Blowing in the Wind, pp. 178–179 ◆ GENERAL **CRF** **Datasheet for Chapter Lab*** GENERAL **SE** **Maps in Action** Wind Power in the United States, p. 180 GENERAL **CRF** **Inquiry Lab** The Generation of Natural Gas from Biomass* GENERAL	**OSP** **Lesson Plans** (also in print) **TR** **Bellringer*** **TR** 37 How a Hydroelectric Dam Generates Electricity* **TR** 38 Wind Power in the United States* **TE** **Internet Activity** Biomass Potential, p. 167 ADVANCED **CRF** **Internet Activity** Biomass Potential* ADVANCED
PACING • 45 min pp. 169–172 **Section 4 Resources and Conservation** • Describe two environmental impacts of mining and the use of fossil fuels. • Explain two ways the environmental impacts of mining can be reduced. • Identify three ways that you can conserve natural resources.	**TE** **Discussion** Mining, p. 169 GENERAL **SE** **Quick Lab** Reclamation, p. 170 GENERAL **CRF** **Datasheet for Quick Lab*** GENERAL **TE** **Group Activity** Savings and Recycling, p. 171 GENERAL	**OSP** **Lesson Plans** (also in print) **TR** **Bellringer*** **CD** **Interactive Tutor** Energy **TE** **Internet Activity** Alternative Energy in Your State, p. 180 GENERAL **CRF** **Internet Activity** Alternative Energy in Your State* GENERAL

PACING • 90 min

CHAPTER REVIEW, ASSESSMENT, AND STANDARDIZED TEST PREPARATION

SE **Chapter Highlights,** p. 173
SE **Chapter Review,** pp. 174–175
SE **Standardized Test Prep,** pp. 176–177
CRF **Concept Review*** ■ GENERAL
CRF **Critical Thinking*** ADVANCED
CRF **Math Skills*** GENERAL
CRF **Graphing Skills*** GENERAL
CRF **Chapter Test A*** ■ GENERAL
CRF **Chapter Test B*** ADVANCED
OSP **Lesson Plans** (also in print)
OSP **Test Generator**
OSP **Test Item Listing**

Online and Technology Resources

go.hrw.com Holt Online Learning

Visit **go.hrw.com** for access to Holt Online Learning, or enter the keyword **HQ6 Home** for a variety of free online resources.

One-Stop Planner® CD-ROM

This CD-ROM package includes
- Lab Materials QuickList Software
- Holt Calendar Planner
- Customizable Lesson Plans
- Printable Worksheets
- ExamView® Test Generator
- Interactive Teacher Edition
- Holt PuzzlePro®
- Holt PowerPoint® Resources

KEY

SE	Student Edition	OSP	One-Stop Planner	VID	Classroom Video/DVD
TE	Teacher Edition	TR	Transparencies and Transparency Worksheets	*	Also on One-Stop Planner
CRF	Chapter Resource File	CD	CD or CD-ROM	◆	Requires advance prep
LTP	Long-Term Projects			■	Also available in Spanish

SKILLS DEVELOPMENT RESOURCES	REVIEW AND ASSESSMENT	CORRELATIONS
SE **Pre-Reading Activity,** p. 154 GENERAL TE **Using the Figure** Hoover Dam, p. 154 GENERAL		National Science Education Standards
CRF **Directed Reading*** BASIC TE **Using the Figure** Slowing Water Flow, p. 156 GENERAL TE **Reading Skill Builder** Reading Organizer, p. 156 BASIC TE **Inclusion Strategies,** p. 156	SE **Reading Check,** p. 156 GENERAL SE **Section Review,** p. 158 GENERAL TE **Reteaching,** p. 158 BASIC TE **Quiz,** p. 158 GENERAL TE **Alternative Assessment,** p. 158 GENERAL CRF **Section Quiz*** ■ GENERAL	PS 3a, PS 3b, SPSP 3a
CRF **Directed Reading*** BASIC SE **Math Practice,** p. 160 GENERAL TE **Using the Figure** Rock Layers and Density, p. 161 GENERAL TE **Skill Builder** Vocabulary, p. 162 GENERAL SE **Graphic Organizer** Chain-of-Events-Chart, p. 163 GENERAL TE **Using the Figure** Electricity from Nuclear Fission, p. 163 GENERAL	SE **Reading Checks,** pp. 161, 162 GENERAL SE **Section Review,** p. 164 GENERAL TE **Homework,** p. 163 ADVANCED TE **Reteaching,** p. 164 BASIC TE **Quiz,** p. 164 GENERAL TE **Alternative Assessment,** p. 164 GENERAL CRF **Section Quiz*** ■ GENERAL	PS 1c, PS 2f, PS 3a, PS 3b, PS 3c, LS 4a, SPSP 3a
CRF **Directed Reading*** BASIC TE **Using the Figure** Natural Hot Tub, p. 165 GENERAL TE **Skill Builder** Vocabulary, p. 166 BASIC TE **Inclusion Strategies,** p. 166	SE **Reading Check,** p. 167 GENERAL SE **Section Review,** p. 168 GENERAL TE **Reteaching,** p. 167 BASIC TE **Quiz,** p. 168 GENERAL TE **Alternative Assessment,** p. 168 GENERAL CRF **Section Quiz*** ■ GENERAL	PS 3a, PS 3b, SPSP 3a
CRF **Directed Reading*** BASIC TE **Reading Skill Builder** Reading Hint, p. 170 BASIC TE **Using the Figure** Raw Recyclables, p. 171 GENERAL	SE **Reading Check,** p. 170 GENERAL SE **Section Review,** p. 172 GENERAL TE **Reteaching,** p. 171 BASIC TE **Quiz,** p. 171 GENERAL TE **Alternative Assessment,** p. 172 GENERAL CRF **Section Quiz*** ■ GENERAL	PS 3c, LS 4e, SPSP 3a

Holt Lab Generator CD-ROM

Search for labs by topic, standard, difficulty level, and/or time frame. Edit those labs to fit your needs, or create your own labs. Use the Lab Materials QuickList to order just what you need.

Holt Earth Science Interactive Tutor CD-ROM

This CD-ROM consists of interactive activities that give students a fun way to extend their knowledge of Earth science concepts.

Chapter Summaries Audio CDs

These CDs include audio summaries of the key concepts presented in each chapter. (Audio summaries are also available in Spanish.)

SCILINKS. NSTA

www.scilinks.org

Maintained by the **National Science Teachers Association.** See Chapter Enrichment pages that follow for a complete list of topics.

See Chapter Enrichment pages for Video Resources.

Chapter 7

Chapter Enrichment

This Chapter Enrichment provides relevant and interesting information to expand and enhance your classroom instruction of the chapter material.

Section 1 Mineral Resources

Metal Extraction

Before metals are extracted from ore, the ore is crushed twice. The most common means of extracting metal from ore is called *flotation,* which is used on the common metal-containing sulfide ores, such as sphalerite, ZnS, from which zinc is extracted, and cinnabar, HgS, from which mercury is extracted. The crushed ore is mixed with water. Next, chemicals that react with the ore and allow the mineral to separate out are applied. The combination of chemical reagents with water forms air bubbles, and the desired metal floats to the surface and is collected. Often, the process must be repeated several times before all of the metal is removed.

Cyanide is a common reagent used to separate metals such as gold and copper from their ores. Cyanide is part of the waste material that remains after the flotation process. Mine waste is often stored in open waste ponds or lagoons, which may pose environmental hazards. In some parts of the world—particularly Guyana, Spain, and Hungary—pond containments have failed. Tons of cyanide-laden water and silt spilled into nearby rivers or onto land, causing severe environmental damage.

◀ Copper ore that is extracted from this mine may be processed with cyanide.

▲ Nuclear wastes are stored in on-site water pools.

Section 2 Nonrenewable Energy

Nuclear Fusion

Nuclear fusion may one day be a viable alternative to consuming nonrenewable energy resources such as fossil fuels. However, several challenges remain. For example, one problem faced by fusion researchers is how to confine the rapidly moving, superhot atomic particles. A fusion power plant must be made of a material that is immune to bombardment by energetic neutrons. The plant must withstand extremely high temperatures, and it must provide a means for adding fuel to the system to sustain a continuous fusion reaction.

Yet, nuclear fusion may afford enormous benefits as an energy source. Deuterium, the fuel used in fusion, is easily extracted from ordinary water. The amounts of deuterium used are tiny, so accidental release is impossible, especially as the material would be absorbed by the container. Fusion produces absolutely no air pollution. Though radiation is produced by neutrons within the reactor, proper materials would address the problem of radioactive waste disposal. Additionally, fusion reactors, unlike fission reactors, do not produce weapons-grade nuclear materials.

Section 3 Renewable Energy

Plastics from Plants

Plastics have traditionally been made from hydrocarbons, or fossil fuels. Hydrocarbon-based plastics deplete reserves of fossil fuels, and they do not easily degrade in landfills. Increasingly, cellulose, a natural plant material, is replacing hydrocarbons in the production of plastics. Today, more than 1.5 billion pounds of pure cellulose are used to make plastics and other materials. Cellulose is a renewable resource because it is derived from the tissues of plants. Because they are made of natural plant materials, cellulose-based plastics easily degrade in landfills and are not toxic. A process called *clean fractionation* is being developed to separate cellulose from plant material efficiently, without using hazardous chemicals.

Section 4 Resources and Conservation

Methane Hydrates

The U.S. Geological Survey (USGS) estimates that methane hydrate deposits, or *clathrates,* contain almost as much carbon as deposits of all other fossil fuels combined. Although it is tempting to think of clathrates as a possible new hydrocarbon fuel source, the technological challenge of extracting frozen methane from the seabed is daunting. One major problem is the tendency for massive sea-floor slumping along continental shelves that contain clathrates. Scientists also know that the sea floor is constantly changing, which may seriously affect the stability of clathrates as they are mined.

◀ Recycling metals and finding new, renewable sources for plastics reduce the amount of waste in landfills.

Video Resources

Brain Food Video Quizzes

Brain Food Video Quizzes These videos contain game-show style quizzes that assess students' progress and motivate students to study the chapter material.

NOVA Videos

To order **NOVA** videos related to this chapter, visit **go.hrw.com** and enter the keyword **HQ6RENV.**

SciLinks

NSTA Developed and maintained by the National Science Teachers Association

SciLinks is maintained by the National Science Teachers Association to provide you and your students with interesting, up-to-date links that will enrich your classroom presentation of the chapter.

Visit www.scilinks.org and enter the SciLinks code for more information about the topic listed.

Topic: Using Mineral Resources
SciLinks code: HQ61587

Topic: Mining Minerals
SciLinks code: HQ60968

Topic: Nonrenewable Resources
SciLinks code: HQ61044

Topic: Fossil Fuels
SciLinks code: HQ60614

Topic: Nuclear Energy
SciLinks code: HQ61047

Topic: Renewable Resources
SciLinks code: HQ61291

Chapter 7

Chapter Overview

Earth's resources provide the energy required for animal, plant, and human life to exist. This chapter describes Earth's resources and the sources of renewable and nonrenewable energy. It also explores how humans obtain resources and the resulting impact on the environment.

Using the Figure —GENERAL

Hoover Dam Tell students that Hoover Dam was built across the Colorado River near Las Vegas, Nevada, in the 1930s to provide water and electricity to people moving into arid lands. Explain that, while the dam provides energy, it also affects the environment. Discuss with students what changes may occur when a dam is built. (The river's normal flow is disrupted; the reservoir behind the dam floods land that people or wildlife lived on; areas downstream from the dam may be deprived of water.) **LS Interpersonal**

PRE-READING ACTIVITY

FOLDNOTES

Encourage students to use their FoldNote as a study guide to quiz themselves for a test on the chapter material. Students may want to create Four-Corner FoldNotes for different topics within the chapter.

Chapter 7 Resources and Energy

Sections

1 Mineral Resources
2 Nonrenewable Energy
3 Renewable Energy
4 Resources and Conservation

What You'll Learn

- Which of Earth's resources humans use
- How resources are obtained
- How resource use affects the environment

Why It's Relevant

Supplies of some resources are diminishing. By understanding how these resources are used, scientists can search for sustainable, alternative resources.

PRE-READING ACTIVITY

FOLDNOTES

Four-Corner Fold

Before you read this chapter, create the FoldNote entitled "Four-Corner Fold" described in the Skills Handbook section of the Appendix. Label each flap of the four-corner fold with a topic from the chapter. Write what you know about each topic under the appropriate flap. As you read the chapter, add other information that you learn.

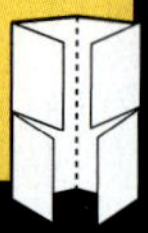

▶ Hoover Dam generates hydroelectric power. Its construction also created Lake Mead, whose water is used to irrigate more than a million acres of land in California, Arizona, and Mexico.

Chapter Correlations *National Science Education Standards*

PS 1c . . . Nuclear reactions convert a fraction of the mass of interacting particles into energy, and they can release much greater amounts of energy than atomic interactions. Fission is the splitting of a large nucleus into smaller pieces. Fusion is the joining of two nuclei . . . . **(Section 2)**

PS 2f Carbon atoms can bond to one another in chains, rings, and branching networks to form a variety of structures. . .**(Section 2)**

PS 3a Chemical reactions occur all around us. . . **(Sections 1–3)**

PS 3b Chemical reactions may release or consume energy. Some reactions such as the burning of fossil fuels release large amounts of energy by losing heat and by emitting light. Light can initiate many chemical reactions such as photosynthesis and the evolution of urban smog. **(Sections 1–3)**

PS 3c . . . Radical reactions control . . . the presence of ozone and greenhouse gases in the atmosphere, burning and processing of fossil fuels . . . . **(Sections 2 and 4)**

LS 4a The atoms and molecules on Earth cycle among living and nonliving components of the biosphere. **(Section 2)**

LS 4e Human beings live within the world's ecosystems. Increasingly, humans modify ecosystems as a result of population growth, technology, and consumption. . .**(Section 4)**

SPSP 3a Human populations use resources. . . to maintain and improve existence. . . . **(Sections 1–4)**

ENVIRONMENTAL CONNECTION

Section 1 Mineral Resources

Earth's crust contains useful mineral resources. The processes that formed many of these resources took millions of years. Scientists have identified more than 4,000 different minerals in Earth's crust. Many of these mineral resources are mined for human use.

Mineral resources can be either *metals,* such as gold, Au, silver, Ag, and aluminum, Al, or *nonmetals,* such as sulfur, S, and quartz, SiO_2. Metals can be identified by their shiny surfaces. Metals are also good conductors of heat and electricity, and they tend to bend easily when in thin sheets. Most nonmetals have a dull surface and are poor conductors of heat and electricity.

OBJECTIVES

- **Explain** what ores are and how they form.
- **Identify** four uses for mineral resources.
- **Summarize** two ways humans obtain mineral resources.

KEY TERMS

ore
lode
placer deposit
gemstone

Ores

Metallic minerals such as gold, silver, and copper, Cu, are called *native elements* and can exist in Earth's crust as nuggets of pure metal. But most other minerals in Earth's crust are *compounds* of two or more elements. Mineral deposits from which metals and nonmetals can be removed profitably are called **ores.** For example, the metal iron, Fe, can be removed from naturally occurring deposits of the minerals magnetite and hematite. Mercury, Hg, can be separated from cinnabar, and aluminum, Al, can be separated from the ore bauxite.

ore a natural material whose concentration of economically valuable minerals is high enough for the material to be mined profitably

Ores Formed by Cooling Magma

Ores form in a variety of ways, as shown in **Figure 1.** Some ores, such as chromium, Cr; nickel, Ni; and lead, Pb, ores form within cooling magma. As the magma cools, dense metallic minerals sink. As the minerals sink, layers of these minerals accumulate at the bottom of the magma chamber to form ore deposits.

Figure 1 ▶ The Formation of Ores

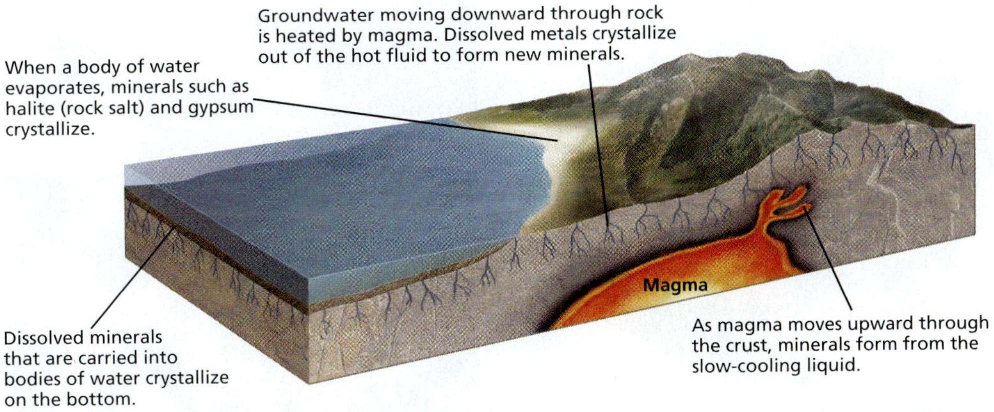

Section 1

Focus

Overview

This section summarizes Earth's mineral resources and describes how minerals are removed from ore compounds. The section also describes how humans use mineral resources.

Bellringer

Ask students to define *mineral.* (a natural, inorganic solid that has a characteristic chemical composition, an orderly internal structure, and a characteristic set of physical properties) Have them name five common minerals. (Answers may vary. Sample answers: quartz, silver, gold, salt, clay, and graphite)
LS Verbal

Motivate

Activity — GENERAL

Minerals at Work Have students list objects in the room or other common objects that are made from minerals. Ask them to identify those that are metallic or nonmetallic, those that are pure minerals, and those that are mixed with other components. (Answers may vary. Accept all reasonable answers.) **LS Visual/Logical**

CHAPTER RESOURCES

Chapter Resource File
- Directed Reading BASIC

Technology

Transparencies
- Bellringer
- 32 The Formation of Ores and Placer Deposits (with worksheet)

Student Edition on CD-ROM

One-Stop Planner CD-ROM
- Lesson Plan

Ores Formed by Contact Metamorphism

Some lead, Pb; copper, Cu; and zinc, Zn, ores form through the process of contact metamorphism. *Contact metamorphism* is a process that occurs when magma comes into contact with existing rock. Heat and chemical reactions with hot fluids from the magma can change the composition of the surrounding rock. These changes sometimes form ores.

Contact metamorphism can also form ore deposits when hot fluids called *hydrothermal solutions* move through small cracks in a large mass of rock. In this process, minerals from the surrounding rock dissolve into the hydrothermal solution. Over time, new minerals precipitate from the solution and form narrow zones of rock called *veins*. Veins commonly consist of ores of valuable heavy minerals, such as gold, Au; tin, Sn; lead, Pb; and copper, Cu. When many thick mineral veins form in a relatively small region, the ore deposit is called a **lode.** Stories of a "Mother Lode" kept people coming to California during the California gold rush in the late 1840s.

lode a mineral deposit within a rock formation

placer deposit a deposit that contains a valuable mineral that has been concentrated by mechanical action

Ores Formed by Moving Water

The movement of water helps to form ore deposits. First, tiny fragments of native elements, such as gold, Au, are released from rock as it breaks down by weathering. Then, streams carry the fragments until the currents become too weak to carry these dense metals. Finally, because of the mechanical action of the stream, the fragments become concentrated at the bottom of stream beds in **placer deposits.** A placer deposit is shown in **Figure 2.**

Reading Check Name two ways water creates ore deposits. (See the Appendix for answers to Reading Checks.)

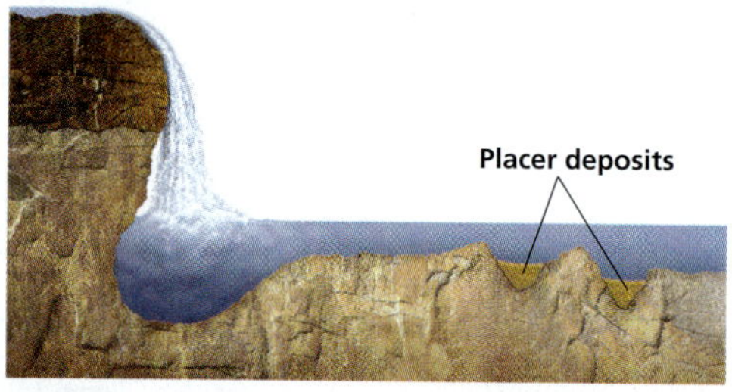

Figure 2 ▶ Placer deposits may occur at a river bend (left) or in holes downstream from a waterfall (right). Gold is a mineral that is commonly found in placer deposits. A stream carries heavy gold grains and nuggets and drops them where the current is weak.

Teach

Using the Figure — GENERAL

Slowing Water Flow Have students study the figure on this page. Ask them to describe where the ore deposits occur in each picture and then relate the locations of deposits to the stream features—the pool at the base of the waterfall and the bend. Discuss with students why these features cause the stream to deposit its load of ore fragments. (These features slow the rate of water flow. When the water slows, it can no longer carry heavy mineral particles, which sink due to gravity and collect in low areas.) **LS Visual/Logical**

READING SKILL BUILDER — BASIC

Reading Organizer Create a two-column chart to help students understand and remember the chemical abbreviations for elements discussed in this section. In one column, have students write the English word for an ore or element, and in the other column, have them write the chemical abbreviation for the element. English Language Learners **LS Verbal**

Answer to Reading Check

Water creates ore deposits by eroding rock and releasing minerals and by carrying the mineral fragments and depositing them in streambeds.

INCLUSION Strategies

• *Gifted and Talented*

Invite interested students to do further research on various uses for individual minerals. Have each student research one mineral and try to find at least five common—or not so common—uses. Students may present their findings to the class in an oral report or as a poster or other display. **LS Verbal**

CHAPTER RESOURCES

Chapter Resource File

• **Skills Practice Lab** Determination of Carbonate in Ore GENERAL

Table 1

Minerals and Their Uses	
Metallic minerals	**Uses**
Hematite and magnetite (iron)	in making steel
Galena (lead)	in car batteries; in solder
Gold, silver, and platinum	in electronics and dental work; as objects such as coins, jewelry, eating utensils, and bowls
Chalcopyrite (copper)	as wiring, in coins and jewelry, and as building ornaments
Sphalerite (zinc)	in making brass and galvanized steel
Nonmetallic minerals	**Uses**
Diamond (carbon)	in drill bits and saws (industrial grade) and in jewelry (gemstone quality)
Graphite (carbon)	in pencils, paint, lubricants, and batteries
Calcite	in cement; as building stone
Halite (salt)	in food preparation and preservation
Kaolinite (clay)	in ceramics, cement, and bricks
Quartz (sand)	as glass
Sulfur	in gunpowder, medicines, and rubber
Gypsum	in plaster and wallboard

Uses of Mineral Resources

Some metals, such as gold, Au, platinum, Pt, and silver, Ag, are prized for their beauty and rarity. Metallic ores are sources of these valuable minerals and elements. Certain rare nonmetallic minerals called **gemstones** display extraordinary brilliance and color when they are specially cut for jewelry. Other nonmetallic minerals, such as calcite and gypsum, are used as building materials. **Table 1** shows some metallic and nonmetallic minerals and their common uses.

gemstone a mineral, rock, or organic material that can be used as jewelry or an ornament when it is cut and polished

Mineral Exploration and Mining

Companies that mine and recover minerals are often looking for new areas to mine. These companies identify areas that may contain enough minerals for economic recovery through mineral exploration. In general, an area is considered for mining if it has at least 100 to 1,000 times the concentration of minerals that are found elsewhere.

During mineral exploration, people search for mineral deposits by studying local geology. Airplanes that carry special equipment are used to measure and identify patterns in magnetism, gravity, radioactivity, and rock color. Exploration teams also collect and test rock samples to determine whether the rock contains enough metal to make a mine profitable.

SciLinks. NSTA Developed and maintained by the National Science Teachers Association

For a variety of links related to this subject, go to www.scilinks.org

Topic: Using Mineral Resources
SciLinksCode: HQ61587
Topic: Mining Minerals
SciLinksCode: HQ60968

Discussion — ADVANCED

Properties of Minerals Have students review the different methods used in mineral exploration, as described in the last paragraph. Challenge students to suggest what mineral property each type of exploration is designed to detect. (Answers may vary. Examples: magnetism—minerals found in rocks that contain magnetized mineral grains, which generally consist of iron; gravity—minerals that are denser or less dense than surrounding rock; radioactivity—radioactive minerals or those found in rocks that contain radioactive materials; color—minerals that have a distinctive color or that commonly form in rocks that have a distinctive color.) LS **Logical**

ENVIRONMENTAL CONNECTION

Mountaintop Removal Discuss with the class what impact each type of mining described may have on the environment, such as the disruption of ecosystems and the pollution of water resources. Have interested students then research mountaintop-removal mining. Guide students to compare the effects of this type of mining on the environment to the effects of other mining methods. Students may present findings to the class by using pictures, drawings, or a model. LS **Verbal**

BRAIN FOOD

Alloys An alloy is a solid mixture of two or more metals, or of a metal and a nonmetal, that are fused when melted together. Common alloys include bronze, which is an alloy of copper and tin, and pewter, which is made of tin and lead or other metals. An example of an alloy of a metal and a nonmetal is steel, which is a combination of iron and up to 2% carbon. Steel may also contain other metals and nonmetals.

Group Activity — GENERAL

Mineral Use Divide the class into three groups and have each group research the specific minerals and gemstones that are used in a variety of everyday objects, such as cell phones, televisions, and soup cans. Guide students to use the library, the Internet, or both for their research. Have each group create a multimedia presentation and present their findings to the class. LS **Logical/Visual**

Close

Reteaching — Basic

Ore Minerals Have students make a table that lists commonly used metals and minerals, identifies how they are used, and predicts how they may be mined. LS **Logical**

Quiz — General

1. Why would mineral exploration teams be interested in identifying lodes? (Lodes are small regions of rich ore deposits that could potentially be mined easily for great profit because they are near the surface.)
2. What minerals are found as nodules in the ocean? (iron, manganese, and nickel)

Alternative Assessment — General

Chart Have students create a flowchart by using boxes and arrows to show how a mineral of their choice becomes part of a deposit, is mined, and then is used to make a useful object. LS **Logical/Visual**

Figure 3 ▶ With a rim diameter of 4 km and a depth of almost 1 km, the Bingham Canyon Mine in Utah is the largest copper mine in the world.

Subsurface Mining

Many mineral deposits are located below Earth's surface. These minerals are mined by miners who work underground to recover mineral deposits. These mining techniques are called *subsurface mining*.

Surface Mining

When mineral deposits are located close to Earth's surface, they may be mined by using *surface mining* methods. In these methods, the overlying rock material is stripped away to reveal the mineral deposits. A very large open-pit copper mine is shown in **Figure 3.**

Placer Mining

Minerals in placer deposits are mined by dredging. In placer mining, large buckets are attached to a floating barge. The buckets scoop up the sediments in front of the barge. Dense minerals from placer deposits are separated from the surrounding sediment. Then, the remaining sediments are released into the water.

Undersea Mining

The ocean floor also contains mineral resources. *Nodules* are lumps of minerals on the deep-ocean floor that contain iron, Fe; manganese, Mn; and nickel, Ni, and that could become economically important if they could be recovered efficiently. However, because of their location, these deposits are very difficult to mine. Mineral deposits on land can be mined less expensively than deposits on the deep-ocean floor can.

Section 1 Review

1. **Define** *ore.*
2. **Describe** three ways that ore deposits form.
3. **Identify** two uses for each of the following minerals: sulfur, copper, and diamond.
4. **Summarize** the four main types of mining.

CRITICAL THINKING

5. **Applying Ideas** Which would be a better insulator for a hot-water pipe, a metal or a nonmetal? Explain your answer.
6. **Understanding Relationships** Why are dense minerals more likely to form placer deposits than less dense minerals are?
7. **Making Inferences** Why do you think that mining on land is less costly than mining in the deep ocean is?

CONCEPT MAPPING

8. Use the following terms to create a concept map: *mineral, ore, magma, contact metamorphism, vein, lode, placer deposit,* and *mine.*

Answers to Section Review

1. An ore is a mineral deposit from which metals or nonmetals can be removed profitably.
2. Deposits form from cooling magma, contact metamorphism, or stream deposition.
3. Sulfur is used in medicine and rubber. Copper is used in wire and in coins. Diamond is used in jewelry and in saws.
4. Subsurface mining involves working underground to mine ore. Surface mining involves the removal of surface material to reach ore. Placer mining involves dredging ore from stream or lake beds. Undersea mining involves recovering ore from the ocean floor.
5. A nonmetal would be better because it does not conduct heat well.
6. Denser minerals are more likely to settle out of slowly moving water.
7. The ocean bottom can be reached only by using specialized technology and has conditions to which humans are not well adapted and for which mining procedures have not been developed.
8. A *mine* may be located over a *lode,* which is an area that is rich in *ores,* which are concentrations of *minerals* that form by stream deposition, which causes *placer deposits,* or by *contact metamorphism* or by the cooling of *magma,* both of which may form *veins.*

CHAPTER RESOURCES

Chapter Resource File
- Section Quiz GENERAL

Workbooks
- **Study Guide** (also in Spanish)

ENVIRONMENTAL CONNECTION

Section 2 Nonrenewable Energy

Many of Earth's resources are used to generate energy. Energy is used for transportation, manufacturing, and countless other things that are important to life as we know it. Energy resources that exist in limited amounts and that cannot be replaced quickly once they are used are examples of **nonrenewable resources.**

OBJECTIVES

- **Explain** why coal is a fossil fuel.
- **Describe** how petroleum and natural gas form and how they are removed from Earth.
- **Summarize** the processes of nuclear fission and nuclear fusion.
- **Explain** how nuclear fission generates electricity.

KEY TERMS

nonrenewable resource
fossil fuel
nuclear fission
nuclear fusion

Fossil Fuels

Some of the most important nonrenewable resources are buried within Earth's crust. These natural resources—coal, petroleum, and natural gas—formed from the remains of living things. Because of their organic origin, coal, petroleum, and natural gas are called **fossil fuels.** Fossil fuels consist primarily of compounds of carbon and hydrogen called *hydrocarbons*. These compounds contain stored energy originally obtained from sunlight by plants and animals that lived millions of years ago. When hydrocarbons are burned, the breaking of chemical bonds releases energy as heat and light. Much of the energy humans use every day comes from the burning of the hydrocarbons that make up fossil fuels.

Formation of Coal

The most commonly burned fossil fuel is coal. The coal deposits of today are the remains of plants that have undergone a complex process called carbonization. *Carbonization* occurs when partially decomposed plant material is buried in swamp mud and becomes peat. Bacteria consume some of the peat and release the gases methane, CH_4, and carbon dioxide, CO_2. As gases escape, the chemical content of the peat gradually changes until mainly carbon remains. The complex chemical and physical changes that produce coal happen only if oxygen in a swamp is absent. When the conditions are not right for carbonization or if the time required for coal formation has not elapsed, peat remains. Peat may be used as an energy source, as shown in **Figure 1.**

nonrenewable resource a resource that forms at a rate that is much slower than the rate at which it is consumed

fossil fuel a nonrenewable energy resource that formed from the remains of organisms that lived long ago; examples include oil, coal, and natural gas

Figure 1 ▶ Peat deposits are still forming today. Some people in Ireland and Scotland heat their houses with peat. In Ireland and Russia, peat is used to fuel some electric power plants.

Section 2

Focus

Overview

This section describes how fossil fuels—coal, petroleum, and natural gas—form and are acquired by humans. The section also explains nuclear processes and how nuclear fission can be used to generate electricity.

Bellringer

Ask students: "What would your life be like without electricity or gasoline?" (Answers may vary. Accept all reasonable answers.) LS **Intrapersonal**

Motivate

Identifying Preconceptions — GENERAL

Is It Renewable? Present students with the following list: coal, water, oil, wind, wood, sunlight, and soil. Have students identify whether each resource is renewable or nonrenewable. (Coal is nonrenewable; oil is nonrenewable; wind is renewable, wood is renewable, sunlight is renewable, and soil is nonrenewable. Students may disagree about whether water is renewable.) Use students' responses to tailor your discussion of nonrenewable resources. LS **Logical**

CHAPTER RESOURCES

Chapter Resource File
- Directed Reading BASIC

Technology

Transparencies
- Bellringer

Student Edition on CD-ROM

One-Stop Planner CD-ROM
- Lesson Plan

Teach

Group Activity — GENERAL

Purifying with Pressure This activity shows how pressure removes impurities from coal. Organize students into small groups and provide each group a basin, a towel, a measuring cup, water, and several weights, such as bricks. Have students perform the following steps:

1. Fold the towel into a square and place it in the basin.
2. Fill the measuring cup with water.
3. Slowly pour the water onto the towel to saturate it, without allowing water to accumulate in the basin. Record how much water is used.
4. Place one brick on the towel, leave for one minute, and then remove.
5. Pour the water out of the basin and measure the amount of water.
6. Repeat steps 4 and 5 twice more, once using 2 bricks and once using 3 bricks.
7. Compare the amount of water that is removed from the towel by 1, 2, and 3 bricks.

Discuss with students how increasing pressure removes impurities from the compressed substance. (The greater the pressure, the more impurities are removed.) LS Kinesthetic/Logical

MATHPRACTICE

Answer

At a rate of 4.5 billion tons/year, the coal will last about 222 years (1,000 billion tons ÷ 4.5 billion tons/year). At a rate of 10 billion tons/year, the coal will last about 100 years (1,000 billion tons ÷ 10 billion tons/year).

Figure 2 ▶ Types of Coal

Stage 1: Peat
The partial decomposition of plant remains forms a brownish-black material called *peat.*

Stage 2: Lignite
Peat is buried by other sediment. As heat and pressure increase, peat becomes lignite. Lignite is also called *brown coal.*

Stage 3: Bituminous Coal
Increased temperature and pressure turn lignite into bituminous coal, which is 80% carbon. Bituminous coal is also called *soft coal.*

Stage 4: Anthracite
Under high temperature and pressure conditions, bituminous coal eventually becomes anthracite, which is the hardest form of coal.

MATHPRACTICE

Coal Reserves
There are thought to be more than 1,000 billion tons of coal on Earth that can be mined. If 4.5 billion tons are used worldwide every year, for how many years will Earth's coal reserves last? If coal use increases to 10 billion tons per year, for how many years will Earth's coal reserves last?

Types of Coal Deposits

As peat is covered by layers of sediments, the weight of these sediments squeezes out water and gases. A denser material called *lignite* forms, as shown in the second step of **Figure 2.** The increased temperature and pressure of more sediments compacts the lignite and forms *bituminous coal.* Bituminous coal is the most abundant type of coal. Where the folding of Earth's crust produces high temperatures and pressure, bituminous coal changes into *anthracite,* the hardest form of coal. Bituminous coal is made of 80% carbon, and anthracite is made of 90% carbon. Both release a large amount of heat when they burn.

Formation of Petroleum and Natural Gas

When microorganisms and plants died in shallow prehistoric oceans and lakes, their remains accumulated on the ocean floor and lake bottoms and were buried by sediment. As more sediments accumulated, heat and pressure increased. Over millions of years, the heat and pressure caused chemical changes to convert the remains into petroleum and natural gas.

Petroleum and natural gas are mixtures of hydrocarbons. Petroleum, which is also called *oil,* is made of liquid hydrocarbons. Natural gas is made of hydrocarbons in the form of gas.

Internet Activity — GENERAL

Resource Locations Have students use the Internet to find maps that show where significant deposits of coal occur on Earth. Students may share their findings with the class. A worksheet designed to direct student research on this topic can be found in the **Chapter Resource File** booklet or by visiting go.hrw.com and entering the keyword **HQ6RENX.** LS Visual/Logical

CHAPTER RESOURCES

Chapter Resource File

Internet Activity
- Resource Locations GENERAL

Technology

Transparencies
- 33 Types of Coal (with worksheet)

Petroleum and Natural Gas Deposits

Petroleum and natural gas are very important sources of energy for transportation, farming, and many other industries. Because of their importance, petroleum and natural gas deposits are valuable and are highly sought after. Petroleum and natural gas are most often mined from permeable sedimentary rock. *Permeable rocks* have interconnected spaces through which liquids can easily flow.

As sediments accumulate and sedimentary rock forms, pressure increases. This pressure forces fluids, including water, oil, and gas, out of the pores and up through the layers of permeable rock. The fluids move upward until they reach a layer of *impermeable rock*, or rock through which liquids cannot flow, called *cap rock*. Petroleum that accumulates beneath the cap rock fills all the spaces to form an oil reservoir. Because petroleum is less dense than water, petroleum rises above any trapped water. Similarly, natural gas rises above petroleum, because natural gas is less dense than both oil and water.

Oil Traps

Geologists explore Earth's crust to discover the kinds of rock structures that may trap oil or gas. They look for oil trapped in places such as the ones shown in **Figure 3.** When a well is drilled into an oil reservoir, the petroleum and natural gas often flow to the surface. When the pressure of the overlying rock is removed, fluids rise up and out through the well.

Figure 3 ▶ Oil Traps

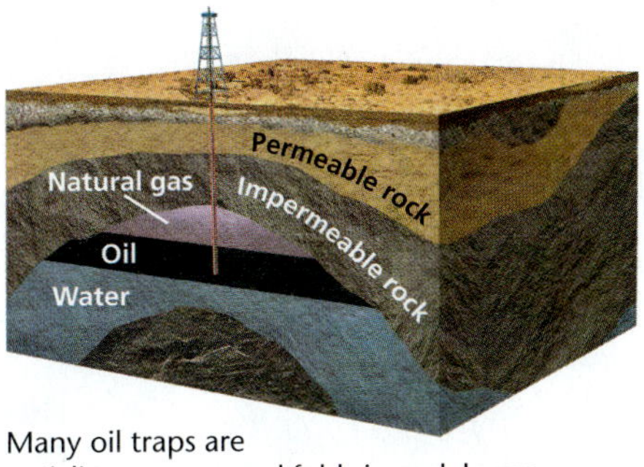

Many oil traps are anticlines, or upward folds in rock layers.

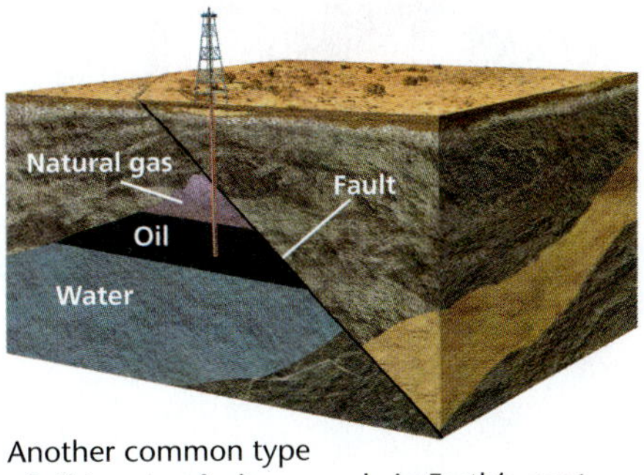

Another common type of oil trap is a fault, or crack, in Earth's crust that seals the oil- or gas-bearing formation.

Fossil-Fuel Supplies

Fossil fuels, like minerals, are nonrenewable resources. Globally, fossil fuels are one of the main sources of energy. *Crude oil*, or unrefined petroleum, is also used in the production of plastics, synthetic fabrics, medicines, waxes, synthetic rubber, insecticides, chemical fertilizers, detergents, shampoos, and many other products.

Coal is the most abundant fossil fuel in the world. Every continent has coal, but almost two-thirds of known deposits occur in three countries—the United States, Russia, and China. Scientists estimate that most of the petroleum reserves in the world have been discovered. However, scientists think that there are undiscovered natural gas reserves. There is also a relatively abundant material called *oil shale* that contains petroleum. But the cost of mining oil from shale is far greater than the present cost of recovering oil from other sedimentary rocks.

Reading Check What is cap rock? (See the Appendix for answers to Reading Checks.)

SCILINKS. NSTA — Developed and maintained by the National Science Teachers Association

For a variety of links related to this subject, go to www.scilinks.org

Topic: Nonrenewable Resources
SciLinksCode: HQ61044
Topic: Fossil Fuels
SciLinksCode: HQ60614

Using the Figure — GENERAL

Rock Layers and Density Have students study the figures that illustrate oil traps. Have students describe the characteristics of the rock layers above and below the fluids. Discuss the role that density plays in the formation of oil and natural gas deposits. Challenge students to explain how relative densities contribute to an oil "gusher." (Because the oil and natural gas cannot move through the overlying rock, as more oil and natural gas migrate up through the rock, pressure builds up. When the overlying rock is removed, the pressure causes the oil or natural gas to rush upward.) LS **Visual/Logical**

Answer to Reading Check

Cap rock is a layer of impermeable rock at the top of an oil- or natural gas-bearing formation through which fluids cannot flow.

BRAIN FOOD

Hydrocarbon Use In the U.S., the hydrocarbon most commonly used in power plants that generate electricity is bituminous coal. The purest coal, anthracite, burns more cleanly but is more expensive. In parts of Eastern Europe, lignite is the most commonly used type of coal. It has created air pollution problems in some areas. Other hydrocarbons commonly used worldwide are octane, diesel fuel, propane, oil, and butane.

MISCONCEPTION ALERT

Petroleum Deposits Students may think that petroleum and natural gas deposits are "lakes" of liquid petroleum that floats freely through empty spaces underground. Explain that petroleum and natural gas exist as liquid or gas that is held in and moves through interconnected pores and cracks in underground rocks.

CHAPTER RESOURCES

Technology

Transparencies
- 34 Oil Traps (with worksheet)

Teach, *continued*

Demonstration — BASIC

Chain Reaction To help students understand a chain reaction, have all of the students in the class stand up. Shake the hand of one student. That student should then shake hands with two other students. These two each shake the hand of two more students. The process should continue, with each student shaking the hand of two other students, until all students have joined the chain. This process should not take long. Point out that each handshake is like a neutron striking an atomic nucleus in a chain reaction, which grows rapidly unless carefully controlled. LS Kinesthetic

SKILL BUILDER — GENERAL

Vocabulary Explain that the words *nucleus* and *nuclear* derive from the Latin word forms *nuc–* and *nux–*, which mean "kernel," or "nut." Discuss with students why this is an appropriate, or inappropriate, word root for the term that describes the nucleus of an atom. LS Verbal

Answer to Reading Check

As neutrons strike neighboring nuclei, the nuclei split and release additional neutrons that strike other nuclei and cause the chain to continue.

CHAPTER RESOURCES

Technology

Transparencies
- 35 A Nuclear Fission Reaction (with worksheet)
- 36 How a Nuclear Power Plant Generates Electricity (with worksheet)

Nuclear Energy

When scientists discovered that atoms had smaller fundamental parts, scientists wondered if atoms could be split. In 1919, Ernest Rutherford first studied and explained the results of bombarding atomic nuclei with high-energy particles. In the 30 years that followed his research, nuclear (NOO klee uhr) technologies were developed that allowed atomic weapons to be made and allowed nuclear reactions to be used to generate electricity. Energy that is produced by using these technologies is called *nuclear energy*.

nuclear fission the process by which the nucleus of a heavy atom splits into two or more fragments; the process releases neutrons and energy

Nuclear Fission

One form of nuclear energy is produced by splitting the nuclei of heavy atoms. This splitting of the nucleus of a large atom into two or more smaller nuclei is called **nuclear fission.** The process of nuclear fission is shown in **Figure 4.**

The forces that hold the nucleus of an atom together are more than 1 million times stronger than the strongest chemical bonds between atoms. If a nucleus is struck by a free neutron, however, the nucleus of the atom may split. When the nucleus splits, it releases additional neutrons as well as energy. The newly released neutrons strike other nearby nuclei, which causes those nuclei to split and to release more neutrons and more energy. A chain reaction occurs as more neutrons strike neighboring atoms. If a fission reaction is allowed to continue uncontrolled, the reaction will escalate quickly and may result in an explosion. However, controlled fission produces heat that can be used to generate electricity.

Reading Check What causes a chain reaction during nuclear fission? (See the Appendix for answers to Reading Checks.)

SCILINKS. NSTA Developed and maintained by the National Science Teachers Association

For a variety of links related to this subject, go to www.scilinks.org

Topic: Nuclear Energy

SciLinksCode: HQ61047

Figure 4 ▶ An Example of a Nuclear Fission Reaction

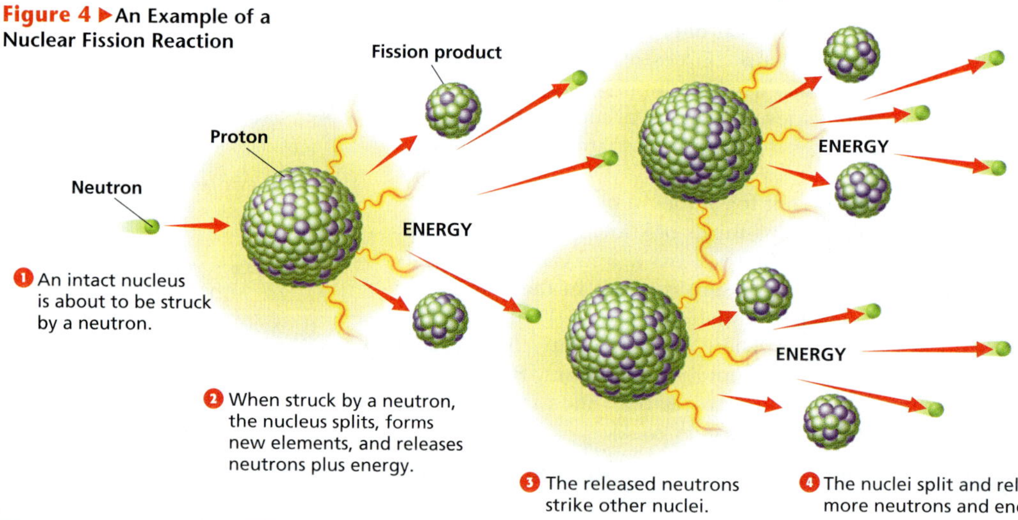

HISTORY CONNECTION — ADVANCED

Nuclear Theory The discovery of radioactivity by chemists Pierre and Marie Curie in the late nineteenth century led to greater understanding of atoms and atomic structure. Invite interested students to learn more about important scientists whose work led to the development of nuclear theory—and to Nobel Prizes for some. In addition to the Curies, you may suggest Ernest Rutherford, James Chadwick, J.J. Thomson, Albert Einstein, Enrico Fermi, and Niels Bohr.

Guide students to understand how each scientist's work advanced knowledge about nuclear theory and how consequences of the discoveries have impacted people's lives. Students may work individually or in small groups and should present their findings to the class in a brief oral report. Groups may wish to divide tasks, such as research, writing, and presentation of results. LS Verbal Co-op Learning

Figure 5 ▶ How a Nuclear Power Plant Generates Electricity

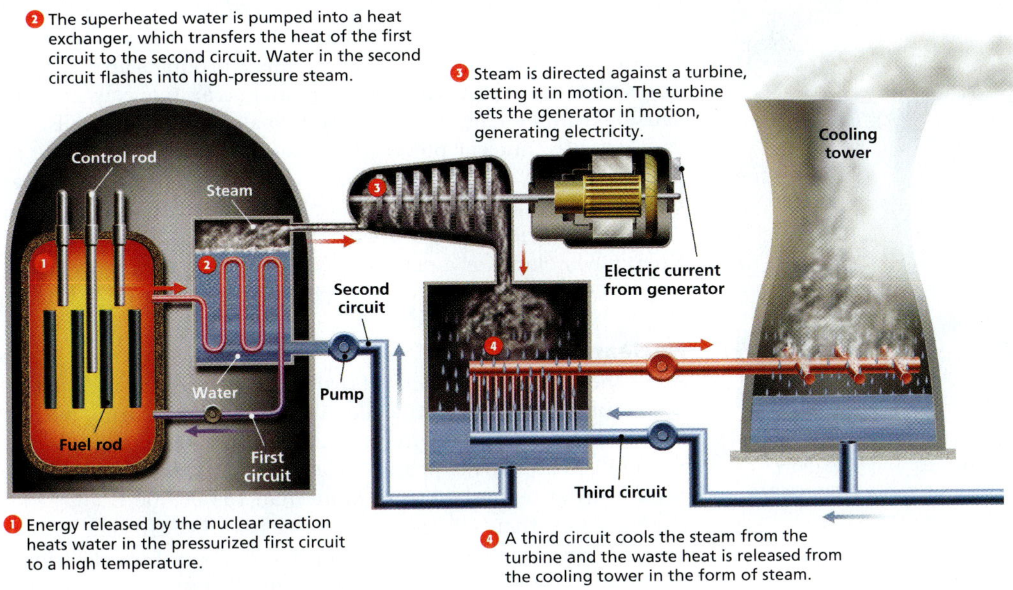

How Fission Generates Electricity

When a nuclear power plant is working correctly, the chain reaction that occurs during nuclear fission is controlled. The flow of neutrons into the fission reaction is regulated so that the reaction can be slowed down, speeded up, or stopped as needed. The specialized equipment in which controlled nuclear fission is carried out is called a *nuclear reactor*.

During fission, a tremendous amount of heat energy is released. This heat energy can in turn be used to generate electricity. **Figure 5** shows how nuclear fission inside a nuclear reactor can be used to generate electricity. Currently, only one kind of naturally occurring element is used for nuclear fission. It is a rare isotope of the element uranium called *uranium-235*, or ^{235}U. Because ^{235}U is rare, the ore that is mined is processed into fuel pellets that have a high ^{235}U content. After this process is complete, the fuel pellets are said to be uranium-enriched pellets.

These enriched fuel pellets are placed into rods to make *fuel rods*. Bundles of these fuel rods are then bombarded by neutrons. When struck by a neutron, the ^{235}U nuclei in the fuel rods split and release neutrons and energy. The resulting chain reaction causes the fuel rods to become very hot.

Water is pumped around the fuel rods to absorb and remove the heat energy. The water is then pumped into a second circuit, where the water becomes steam. The steam turns the turbines that provide power for electric generators. A third water circuit carries away excess heat and releases it into the environment.

Graphic Organizer — **Chain-of-Events Chart**

Create the **Graphic Organizer** entitled "Chain-of-Events Chart" described in the Skills Handbook section of the Appendix. Then, fill in the chart with details about each step of how electricity is generated by fission.

Debate — GENERAL

Pros and Cons Explain that radioactivity is harmful to human health and to the environment. Have students debate whether the risks of nuclear power are acceptable for the benefit of generating electricity. Encourage students to explore the health and environmental reasons for both sides of the argument. **LS Verbal/Logical**

Homework — ADVANCED

Fudging Fusion? Have interested students research the claim made by scientists in 1989 that they had achieved "cold fusion," the fusion of hydrogen atoms at room temperature. Have students report to the class how the scientific method—specifically the requirement that all experiments be reproducible—revealed the claim as a hoax. Have students explain the difference between "hot" and "cold" fusion. **LS Verbal**

Graphic Organizer — GENERAL

Chain-of-Events Chart

You may wish to use this activity to assess students' prior knowledge before beginning a discussion of the process by which electricity is generated with nuclear power. You may also choose to use a similar activity as a quiz to assess students' knowledge of nuclear fission reactions or to assess students' knowledge after students have read this page.

Using the Figure — GENERAL

Electricity from Nuclear Fission Use the figure on this page to lead a discussion of each step in the process of using nuclear fission to generate electricity. Ask them to discuss why the fuel rods are housed in a large, massively reinforced structure. Have them explain why such a structure is necessary to contain the fuel rods. (A massive structure is needed to contain radioactivity in the event of a nuclear accident or other mishap.)
LS Visual/Logical

BRAIN FOOD

$e = mc^2$ Albert Einstein developed the equation $e = mc^2$—which states that energy (e) is equal to an object's mass (m) multiplied by the speed of light (c), or 186,000 miles per second, squared—to describe the relationship of matter to energy. Although the mass of an atomic nucleus is extremely small, when you apply the equation, you can appreciate that the energy released from even a single atomic nucleus is enormous.

Close

Reteaching — Basic

Flashcards Have students make flashcards that have a main idea about one form of energy resource on one side and details on the other side. Students may work in pairs to use the flashcards to reinforce the content of this section. LS Visual

Quiz — General

1. What form of coal has the highest percentage of carbon? (anthracite)
2. At what temperatures do nuclear fusion reactions take place? (greater than 15 million°C)

Alternative Assessment — General

Letter to the Editor Have students write a letter to the editor of a local newspaper in which they advocate the use of either fossil fuels or nuclear energy. Students should argue for one type of energy by comparing it with the other. LS Logical/Verbal

Answers to Section Review

1. because they formed from the remains of organisms that died millions of years ago
2. All three form from the remains of living things that undergo chemical changes when under intense pressure and heat inside Earth. Coal forms from plant remains. Oil and natural gas commonly form from the remains of marine organisms.
3. Petroleum may be trapped in porous rocks located in anticlines, under impermeable rock layers, or by faults.
4. uranium-235
5. A nuclear chain reaction generates a great deal of heat. This heat boils water, which creates steam, which turns a turbine, which generates electricity.
6. Nuclei of hydrogen atoms combine to form the larger nuclei of helium, releasing energy.
7. because only in the last 50 years have we better understood atomic structure and developed technology to control an atomic reaction and to dispose of radioactive waste adequately
8. no, because radioactive contamination would affect the environment and kill or harm aquatic life and affect humans who eat the organisms that live in the water
9. In fission, atomic nuclei are broken apart; in fusion, the nuclei are joined together.
10. *Nonrenewable resources* include *fossil fuels* such as *petroleum, natural gas,* and *coal,* which forms by *carbonization* and includes *peat, lignite, bituminous coal,* and *anthracite coal.*

Figure 6 ▶ These water pools store radioactive wastes. The blue glow indicates that the waste products are highly radioactive.

Advantages and Disadvantages of Nuclear Fission

Nuclear power plants burn no fossil fuels and produce no air pollution. But because nuclear fission uses and produces radioactive materials that have very long half-lives, wastes must be safely stored for thousands of years. These waste products give off high doses of radiation that can destroy plant and animal cells and can cause harmful changes in the genetic material of living cells.

Currently, nuclear power plants store their nuclear wastes in dry casks or in onsite water pools, as shown in **Figure 6.** Other wastes are either stored onsite or transported to one of three disposal facilities in the United States. The U.S. Department of Energy has plans for a permanent disposal site for highly radioactive nuclear wastes.

Nuclear Fusion

All of the energy that reaches Earth from the sun is produced by a kind of nuclear reaction, called nuclear fusion. During **nuclear fusion,** the nuclei of hydrogen atoms combine to form larger nuclei of helium. This process releases energy. Fusion reactions occur only at temperatures of more than 15,000,000°C.

For more than 40 years, scientists have been trying to harness the energy released by nuclear fusion to produce electricity. More research is needed before a commercial fusion reactor can be built. If such a reactor could be built in the future, hydrogen atoms from ocean water might be used as the fuel. With ocean water as fuel, the amount of energy available from nuclear fusion would be almost limitless. Scientists also think that wastes from fusion would be much less dangerous than wastes from fission. The only byproducts of fusion are helium nuclei, which are harmless to living cells.

nuclear fusion the process by which nuclei of small atoms combine to form new, more massive nuclei; the process releases energy

Section 2 Review

1. **Explain** why coal, petroleum, and natural gas are called *fossil fuels.*
2. **Compare** how coal, petroleum, and natural gas form.
3. **Describe** the kind of rock structures in which petroleum reservoirs form.
4. **Identify** the naturally occurring element that is used for nuclear fission.
5. **Explain** how nuclear fission generates electricity.
6. **Summarize** the process of nuclear fusion.

CRITICAL THINKING

7. **Analyzing Relationships** Why have we been able to build nuclear power plants for only the last 50 years?
8. **Recognizing Relationships** Can the waste products of nuclear fission be safely disposed of in rivers or lakes? Explain your answer.
9. **Making Comparisons** How do the processes of nuclear fusion and nuclear fission differ?

CONCEPT MAPPING

10. Use the following terms to create a concept map: *nonrenewable resource, fossil fuel, coal, carbonization, peat, lignite, bituminous coal, anthracite coal, petroleum,* and *natural gas.*

CHAPTER RESOURCES

Chapter Resource File
- Section Quiz GENERAL

Workbooks
- Study Guide (also in Spanish)

ENVIRONMENTAL CONNECTION

Section 3 Renewable Energy

If current trends continue and worldwide energy needs increase, the world's supply of fossil fuels may be used up in the next 200 years. Nuclear energy does not use fossil fuels, but numerous safety concerns are associated with it. Therefore, many nations are researching alternative energy sources to ensure that safe energy resources will be available far into the future. Resources that can be replaced within a human life span or as they are used are called **renewable resources.**

OBJECTIVES

- **Explain** how geothermal energy may be used as a substitute for fossil fuels.
- **Compare** passive and active methods of harnessing energy from the sun.
- **Explain** how water and wind can be harnessed to generate electricity.

KEY TERMS

renewable resource
geothermal energy
solar energy
hydroelectric energy
biomass

Geothermal Energy

In many locations, water flows far beneath Earth's surface. This water may flow through rock that is heated by nearby magma or by hot gases that are released by magma. This water becomes heated as it flows through the rock. The hot water, or the resulting steam, is the source of a large amount of heat energy. This heat energy is called **geothermal energy,** which means "energy from the heat of Earth's interior."

Engineers and scientists have harnessed geothermal energy by drilling wells to reach the hot water. Sometimes, water is first pumped down into the hot rocks if water does not already flow through them. The resulting steam and hot water can be used as a source of heat. The steam and hot water also serve as sources of power to drive turbines, which generate electricity.

The city of San Francisco, for example, obtains some of its electricity from a geothermal power plant located in the nearby mountains. In Iceland, 85% of the homes are heated by geothermal energy. Italy and Japan have also developed power plants that use geothermal energy. A geothermal power plant is shown in **Figure 1.**

renewable resource a natural resource that can be replaced at the same rate at which the resource is consumed

geothermal energy the energy produced by heat within Earth

Figure 1 ▶ These swimmers are enjoying the hot water near a geothermal power plant in Svartsbening, Iceland.

Section 3

Focus

Overview

This section describes sources of renewable energy and how each source can substitute for fossil fuels.

Bellringer

Ask students to list things that they can renew. (Sample answers: memberships, library cards, interests, friendships) Ask them to use their list to help them write a definition of the term *renewable*. (Answers may vary.) LS **Logical**

Motivate

Using the Figure —GENERAL

Natural Hot Tub Have students study the photo at the bottom of the page. Ask students how they can tell that the water in the photo is hot. (Steam is rising from the water in the background.) Tell them that the water is a milky turquoise color because of the minerals in the water and the bacteria that live in the warm water. LS **Visual**

CHAPTER RESOURCES

Chapter Resource File

- Directed Reading BASIC

Technology

Transparencies
- Bellringer

Student Edition on CD-ROM

One-Stop Planner CD-ROM
- Lesson Plan

Teach

SKILL BUILDER — BASIC

Vocabulary Tell students that the prefix *photo-* means "light." Have students use this information to predict what the words *photosynthesis, photovoltaic,* and *photograph* mean. (the process of making something by using light, having the ability to generate electricity from light, and an image made by using light) LS **Verbal**

QuickLAB

Skills Acquired

- Experimenting
- Organizing and Analyzing Data

Teacher's Notes If your classroom has no sunny windows, you may use heat lamps. Small containers such as yogurt or cheese tubs will work well if you do not have large pans.

Answers

1. The variables are the pan lining and covering, time, and temperature. The first trial had the greatest temperature change. The last trial had the smallest temperature change.
2. The black plastic likely had the most dramatic effect on temperature change.
3. Answers may vary but should include a pan or box of water or other fluid and a lid that admits light and traps thermal energy.

Solar Energy

solar energy the energy received by Earth from the sun in the form of radiation

Another source of renewable energy is the sun. Every 15 minutes, Earth receives enough energy from the sun to meet the energy needs of the world for one year. Energy from the sun is called **solar energy.** The challenge scientists face is how to capture even a small part of the energy that travels to Earth from the sun.

Converting sunshine into heat energy can be done in two ways. A house that has windows facing the sun collects solar energy through a *passive system*. The system is passive because it does not use moving parts. Sunlight enters the house and warms the building material, which stores some heat for the evening. An *active system* includes the use of solar collectors. One type of *solar collector* is a box that has a glass top. The box is commonly placed on the roof of a building. Water circulates through tubes within the box. The sun heats the water as it moves through the tubes, which provides heat and hot water. On cloudy days, however, there may not be enough sunlight to heat the water. So, the system must use heat that was stored from previous days.

Photovoltaic cells are another active system that converts solar energy directly into electricity. Photovoltaic cells work well for small objects, such as calculators. Producing enough electricity from these cells to power cities is under investigation.

QuickLAB 30 min

Solar Collector

Procedure

1. Line the inside of a **small, shallow pan** with **black plastic.** Use **tape** to attach a **thermometer** to the inside of the pan. Fill the pan with enough **room temperature water** to cover the end of the thermometer. Fasten **plastic wrap** over the pan with a **rubber band.** Be sure you can read the thermometer.
2. Place the pan in a sunny area. Use a **stopwatch** to record the temperature every 5 min until the temperature stops rising. Discard the water.
3. Repeat steps 1 and 2, but do not cover the pan with plastic wrap.
4. Repeat steps 1 and 2, but do not line the pan with black plastic.
5. Repeat steps 1 and 2. But do not line the pan with plastic, and do not cover the pan with plastic wrap.
6. Calculate the rate of temperature change for each trial by subtracting the beginning temperature from the ending temperature. Divide this number by the time. number of minutes the temperature increased to find the rate of temperature range.

Analysis

1. What are the variables in this investigation? Which trial had the greatest rate of temperature change? the smallest rate of temperature change?
2. Which variable that you tested has the most significant effect on temperature change?
3. What materials would you use to design and build an efficient solar collector? Explain your answer.

CHAPTER RESOURCES

Chapter Resource File

- Datasheet for Quick Lab GENERAL

INCLUSION Strategies

- *Learning Disabled*
- *Visually Impaired*
- *Developmentally Delayed*

Help students create a study tool. Tell them you are going to scan each page orally. Have students who have visual impairments use audiotape to record your summary. Read the subtitles and key information under each subtitle out loud while students record. Guide students to review by listening to their recording. LS **Auditory**

ENVIRONMENTAL CONNECTION

Impact of Dams Explain that even renewable energy sources may have some negative environmental effects. For example, when rivers are blocked or altered and land behind a dam is flooded to form a reservoir, water temperature may change and land and aquatic habitats may be damaged. Guide students to weigh the advantages (electricity, recreation) and disadvantages (environmental change) of hydroelectric dams. LS **Logical**

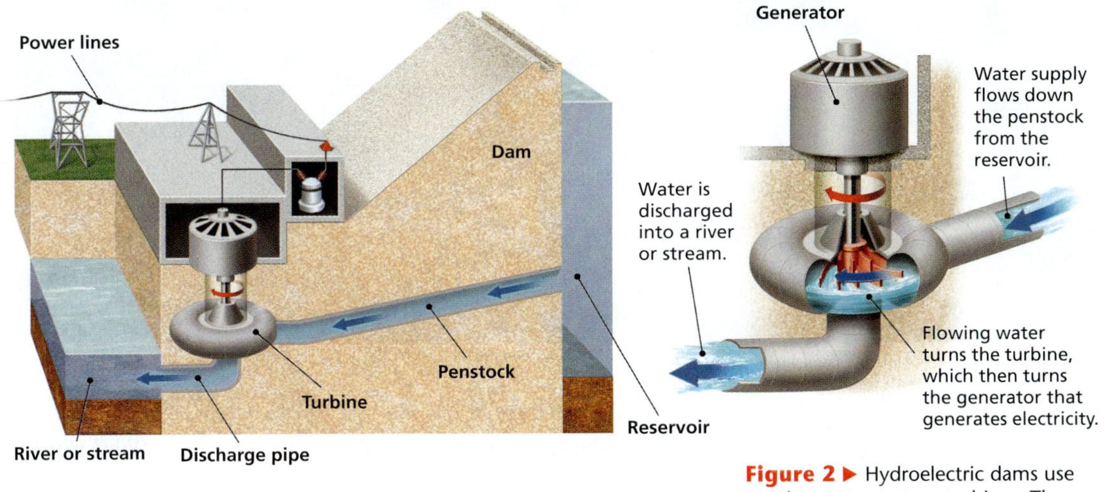

Figure 2 ▶ Hydroelectric dams use moving water to turn turbines. The movement of the turbine powers a generator that generates electricity.

Energy from Moving Water

One of the oldest sources of energy comes from moving water. Energy can be harnessed from the running water of rivers and streams or from ocean tides. In some areas of the world, energy needs can be met by **hydroelectric energy,** or the energy produced by running water. Today, 11% of the electricity in the United States comes from hydroelectric power plants. At a hydroelectric plant, massive dams hold back running water and channel the water through the plant. Inside the plant, the water spins turbines, which turn generators that produce electricity. An example of a hydroelectric plant is shown in **Figure 2.**

Another renewable source of energy for moving water is the tides. Tides are the rising and falling of sea level at certain times of the day. To make use of this tidal flow, people have built dams to trap the water at high tide and to then release it at low tide. As the water is released, it turns the turbines within the dams.

hydroelectric energy electrical energy produced by the flow of water

biomass plant material, manure, or any other organic matter that is used as an energy source

Energy from Biomass

Other renewable resources are being exploited to help supply our energy needs. Renewable energy sources that come from plant material, manure, and other organic matter, such as sawdust or paper waste, are called **biomass.** Biomass is a major source of energy in many developing countries. More than half of all trees that are cut down are used as fuel for heating or cooking. Bacteria that decompose the organic matter produce gases, such as methane, that can also be burned. Liquid fuels, such as ethanol, also form from the action of bacteria on biomass. All of these resources can be burned to generate electricity.

Reading Check Name three sources of renewable energy. (See the Appendix for answers to Reading Checks.)

SCILINKS. NSTA Developed and maintained by the National Science Teachers Association

For a variety of links related to this subject, go to www.scilinks.org

Topic: Renewable Resources
SciLinksCode: HQ61291

Internet Activity — ADVANCED

Biomass Potential Have students use the Internet to find out how much organic waste, such as wood and other plant materials, paper, and agricultural waste, Americans discard each year. Have students learn about how this waste could be used to generate power. Students should then use their findings to create a proposal for how to use America's waste to generate electricity. A worksheet designed to direct student research on this topic can be found in the **Chapter Resource File** booklet or by visiting **go.hrw.com** and entering the keyword **HQ6RENX.** LS **Verbal**

Close

Reteaching — BASIC

Energy in Action Have students demonstrate types of renewable energy by performing the following activities: solar: put a shallow bowl of water in the sun and allow the water to evaporate; wind: provide a pinwheel, and let the wind or a person's breath turn it; hydroelectric: have students hold the pinwheel upside down under a running faucet so that the flowing water turns the pinwheel. LS **Visual**

Answer to Reading Check

Answers may vary but should include three of the following: geothermal, solar, hydroelectric, and biomass.

CHAPTER RESOURCES

Chapter Resource File

- **Inquiry Lab** The Generation of Natural Gas from Biomass GENERAL

Internet Activity

- Biomass Potential ADVANCED

Technology

Transparencies

- 37 How a Hydroelectric Dam Generates Electricity (with worksheet)

BRAIN FOOD

Favorable Winds The best areas for wind energy production have winds that blow steadily at between 12.8–16 km/hr. High ridges and offshore regions are often good locations for wind farms. Tell students that for efficient electricity production, the blades on wind turbines may be more than 80 m long. You may want to use the Internet to find photos of wind turbines to illustrate their massive size.

Close, *continued*

Quiz — GENERAL

1. Why might wind energy production be unsuitable for some locations? (Some sites have weak or unreliable winds.)
2. If you were to design a house, what would you include to promote passive solar heating? (large, south-facing windows that get lots of sun)

Alternative Assessment — GENERAL

Tour Brochure Have students write and illustrate a brochure about alternative energy use in different parts of the United States. The brochure should picture and describe each renewable energy source and explain why it is appropriately located. Students may use magazines or the Internet to find information and pictures. LS **Verbal/Visual**

Answers to Section Review

1. because nonrenewable energy sources are running out
2. Heat from magma or rock heats water and produces steam that is used to boil water. The steam turns turbines that generate electricity.
3. Passive solar systems have no moving parts. Most active solar systems have pumps with moving parts and may use moving water.
4. Water released from a reservoir is channeled through turbines that turn generators to produce electricity.
5. Biomass may be burned, or it may be decomposed by bacteria to produce liquid or gas fuels that can be burned.
6. Wind and water move the blades of a turbine. The mechanical energy produced generates electricity.
7. Fossil fuels develop slowly over millions of years, while biomass can be regrown in a short time, such as a few years.
8. Answers may vary but should reflect knowledge of resources available and rationales for using them.
9. Types of *renewable resources* include *solar energy,* which may use a *passive system* or an *active system,* such as a *solar collector; geothermal energy; hydroelectric energy; biomass;* and *wind energy*.

Figure 3 ▶ The spinning blades of a windmill are connected to a generator. When winds cause the blades to spin faster, the generator produces more energy.

Energy from Wind

Wind is the movement of air over Earth's surface. Wind results from air pressure differences caused by the sun's uneven heating of Earth's surface. Wind turbines use the movement of air to convert wind energy into mechanical energy, which is used to generate electricity.

Wind energy is now being used to produce electricity in locations that have constant winds. Small, wind-driven generators are used to meet the energy needs of individual homes. *Wind farms,* such as the one shown in **Figure 3,** may have hundreds of giant wind turbines that can produce enough energy to meet the electricity needs of entire communities. However, wind generators are not practical everywhere. Even in the most favorable locations, such as in windy mountain passes, the wind does not always blow. Because the wind does not always blow, wind energy cannot be depended on as an energy source for every location.

Section 3 Review

1. **Explain** why many nations are researching alternative energy resources.
2. **Explain** how geothermal energy may be used as a substitute for fossil fuels.
3. **Describe** both passive and active methods of harnessing energy from the sun.
4. **Summarize** how electrical energy is generated from running water.
5. **Describe** how biomass can be used as fuels to generate electricity.
6. **Explain** how water and wind can be harnessed to generate electricity.

CRITICAL THINKING

7. **Making Comparisons** Both fossil fuels and biomass fuels come from plant and animal matter. Why are fossil fuels considered to be nonrenewable, and why is biomass considered to be renewable?
8. **Demonstrating Reasoned Judgement** If you were asked to construct a power plant that uses only renewable energy sources in your area, what type of energy would you use? Explain.

CONCEPT MAPPING

9. Use the following terms to create a concept map: *renewable resource, solar collector, geothermal energy, solar energy, passive system, active system, hydroelectric energy, biomass,* and *wind energy.*

CHAPTER RESOURCES

Chapter Resource File
- Section Quiz GENERAL

Workbooks
- **Study Guide** (also in Spanish)

ENVIRONMENTAL CONNECTION

Section 4 Resources and Conservation

At the present rate of use, scientists estimate that the worldwide coal reserves will last about 200 years. Many scientists also think that within the next 20 years, humans will have used half of Earth's oil supply. This limited supply of fossil fuels and other traditional energy resources has inspired research into possible new energy sources.

Scientists are also studying how the use of traditional energy sources affects Earth's ecosystems. We have learned that mining can damage or destroy fragile ecosystems. Fossil fuels and nuclear power generation may add pollution to Earth's air, water, and soil. However, people can reduce the environmental impact of their resource use. Many governments and public groups have worked to create and enforce policies that govern the use of these natural resources.

OBJECTIVES

- **Describe** two environmental impacts of mining and the use of fossil fuels.
- **Explain** two ways the environmental impacts of mining can be reduced.
- **Identify** three ways that you can conserve natural resources.

KEY TERM

conservation
recycling

Environmental Impacts of Mining

Mining for minerals can cause a variety of environmental problems. Mining may cause both air and noise pollution. Nearby water resources may also be affected by water that carries toxic substances from mining processes. Surface mining is particularly destructive to wildlife habitats. For example, surface mining often uses controlled explosions to remove layers of rock and soil, as shown in **Figure 1.** Some mining practices cause increased erosion and soil degradation. Regions above subsurface mines may sink, or subside, because of the removal of the materials below. This sinking results in the formation of sinkholes. Fires in coal mines are also very difficult to put out and are commonly left to burn out, which may take several decades or centuries.

Figure 1 ▶ The surface of this gold mine in Nevada is being blasted to remove layers of rock.

Section 4

Focus

Overview

This section describes the environmental impacts of mining and laws that help mitigate that impact. The section also covers the effects of fossil fuels on the environment and explains resource conservation, including recycling.

Bellringer

Ask students to write a paragraph that explains which renewable forms of energy they think should be developed in case a growing human population makes it likely that the world will run out of crucial resources within the next century. (Accept all reasonable answers.) **LS Logical**

Motivate

Discussion — GENERAL

Mining Some students may think of mining as an activity that occurs only underground. Explain that today, most mining, especially of coal, is done on the surface. Invite students to talk about the impact of surface mining on the people who live near the mine. Ask them: "What problems do you think these people have to deal with every day? What advantages could there be to having a mining operation in their area?" (Accept all reasonable answers.)
LS Interpersonal

CHAPTER RESOURCES

Chapter Resource File

- Directed Reading BASIC

Technology

Transparencies
- Bellringer

Student Edition on CD-ROM

One-Stop Planner CD-ROM
- Lesson Plan

Teach

QuickLAB

Skills Acquired

- Modeling
- Calculating

Teacher's Notes You may also use chocolate chip cookies in place of gelatin cups, and have students attempt to remove the chips with as little damage as possible to the rest of the cookie. Have students relate the damage done to the cookie to the damage done by surface mining.

Answers

1. reclaimed surface land
2. Answers may vary but should indicate that the uppermost layer is probably jumbled and ragged rather than smooth.
3. Answers may vary but should indicate that there will most likely be long-term damage to the area before the ecosystem can be restored.

CHAPTER RESOURCES

Chapter Resource File

- Datasheet for Quick Lab GENERAL

QuickLAB 30 min

Reclamation

Procedure

1. Use a **plastic spoon** to remove the first layer of gelatin from a **multi-layered gelatin dessert cup** into a **small bowl.**
2. Remove the next layer of gelatin, and discard it.
3. Restore the dessert cup by replacing the first layer of gelatin.

Analysis

1. What does the first layer of gelatin on the restored dessert cup represent?
2. Does the "reclaimed" dessert cup resemble the original, untouched dessert cup?
3. What factors would you address to make reclamation more successful?

Mining Regulations

In the United States, federal and state laws regulate the operation of mines. These laws were designed to prevent mining operations from contaminating local air, water, and soil resources. Some of these federal laws include the Clean Water Act; the Safe Drinking Water Act; and the Comprehensive Environmental Response, Compensation, and Liability Act. All mining operations must also comply with the federal Endangered Species Act, which protects threatened or endangered species and their habitats from being destroyed by mining practices.

Mine Reclamation

To reduce the amount of damage done to ecosystems, mining companies are required to return mined land to its original condition after mining is completed. This process, called *reclamation,* helps reduce the long-lasting environmental impact of mining. In addition to reclamation, some mining operations work hard to reduce environmental damage through frequent inspections and by using processes that reduce environmental impacts.

Fossil Fuels and the Environment

Fossil-fuel procurement affects the environment. Strip mining of coal can leave deep holes where coal was removed. Without plants and topsoil to protect it, exposed land often erodes quickly. When rocks that are exposed during mining get wet, they can weather to form acids. If runoff carries the acids into nearby rivers and streams, aquatic life may be harmed.

Fossil-fuel use also contributes to air pollution. The burning of coal that has a high sulfur content releases large amounts of sulfur dioxide, SO_2, into the atmosphere. When SO_2 combines with water in the air, acid precipitation forms. When petroleum and natural gas are burned, they also release pollutants that can damage the environment. The burning of gasoline in cars is a major contributor to air pollution. But emissions testing, which is shown in **Figure 2,** and careful maintenance help reduce the amount of pollutants released into the air. Emissions testing and maintenance includes the testing of a car's catalytic converter, a device that removes numerous pollutants from the exhaust before the exhaust leaves the car.

Figure 2 ▶ Emissions testing and maintenance of pollution-reducing devices in today's vehicles can help reduce air pollution.

Reading Check Name two ways the use of fossil fuel affects the environment. (See the Appendix for answers to Reading Checks.)

Answer to Reading Check

The use of fossil fuels affects the environment when coal is mined from the surface, which destroys the land. When fossil fuels are burned, they affect the environment by creating air pollution.

READING SKILL BUILDER BASIC

Reading Hint Write the words *conserve* and *recycle* on the board. Invite a volunteer to define each term. Ask students to relate the meaning of the word *conserve* ("to save") to the term *conservation*. Have them define the prefix *re-* ("again") and the word *cycle* ("a regular course of events that begin and end at the same point") and then apply these meanings to the meaning of the term *recycle*. LS **Verbal**

Conservation

Many people and businesses around the world have adopted practices that help reduce the negative effects of the burning of fossil fuels and the use of other natural resources. This preservation and wise use of natural resources is called **conservation.** By conserving natural resources, people can ensure that limited natural resources last longer. Conservation can also help reduce the environmental damage and amount of pollution that can result from the mining and use of natural resources.

Mineral Conservation

Earth's mineral resources are being used at a faster rate each year. Every new person added to the world's population represents a need for additional mineral resources. In developing countries, people are using more mineral resources as their countries become more industrialized. This increased demand for minerals has led many scientists to look for ways to conserve Earth's minerals.

One way to conserve minerals is to use other abundant or renewable materials in place of scarce or nonrenewable minerals. Another way to conserve minerals is by recycling them. **Recycling** is the process of using materials more than once. Some metals, such as iron, copper, and aluminum, are often recycled, as shown in **Figure 3.** Glass and many building materials can also be efficiently recycled. Recycling does require energy, but recycling uses less energy than the mining and manufacturing of new resources does.

Figure 3 ▶ These cubes are made up of metals that have been compacted and are being sent to a recycling plant. ***Can you identify the source of these metals?***

conservation the preservation and wise use of natural resources

recycling the process of recovering valuable or useful materials from waste or scrap; the process of reusing some items

Connection to ENVIRONMENTAL SCIENCE

Methane Hydrates

One potential alternative energy source that scientists are hopeful about is methane hydrates. Methane hydrates are solid, icelike crystals that have gas molecules trapped within the crystal structure of water ice. The ice crystals form a lattice that holds the methane and other gas molecules in place.

This potential new energy source is located in the frozen soil in Arctic regions and on the sea floor. Most deposits of methane hydrate are located near continental margins, where ocean life is abundant. Methane hydrates are stable in sea-floor sediments at temperatures and pressures that are common at depths of 300 m in Arctic regions and at depths of 500 m in tropical regions.

While these energy compounds may look like ice, they burn with intense flame. When methane hydrates burn, they release much less carbon dioxide than the burning of traditional fossil fuels. Scientists are researching the possibility of using methane hydrates in transportation and for generating electricity.

However, scientists are concerned about the environmental impacts of mining this resource. Also, because methane is a powerful greenhouse gas, even a modest quantity of methane released into the atmosphere could affect global temperatures.

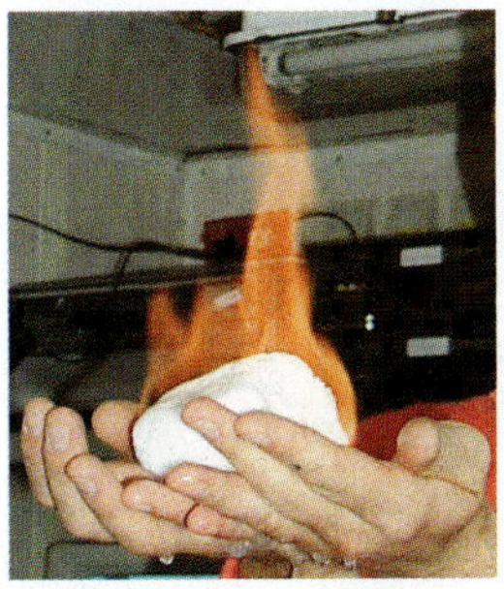

A researcher holds a chunk of burning methane hydrate.

Using the Figure — GENERAL

Raw Recyclables Have students describe the metal cubes in the figure. Invite them to discuss what processes the recycled metal must likely undergo before it is ready to be used again in products. Answer to caption question: The source of the metals is largely aluminum soda cans. **LS Visual**

Close

Reteaching — BASIC

Materials We Use Have students collect pictures from magazines that illustrate each material discussed in this section. Have students create posters that categorize the pictures as relevant to Mining, Fossil Fuels, or Recycling and Conservation. **LS Visual/ Kinesthetic**

Quiz — GENERAL

1. How can fossil fuels damage the ocean environment? (through oil spills)
2. How does an expanding human population affect resource use? (More people will need and demand more goods and services that use resources.)

ENVIRONMENTAL CONNECTION

The Cost of Energy Conservation Many people think that making a home energy efficient is expensive. Although some measures, such as buying new appliances, are expensive, many are free or inexpensive. Turning off unused lights, cleaning refrigerator coils, and air-drying clothes cost nothing. The following methods are inexpensive and will pay for themselves in less than one year: wrapping your hot water heater ($12), installing a programmable thermostat ($25), changing air filters on heating and cooling systems ($12).

Group Activity — GENERAL

Savings and Recycling Divide the class into groups. Have each group research the energy and monetary savings that accrue to the recycling of one product, such as aluminum (soda cans), paper (newspaper, boxes), plastics, or other materials. Students should compare the monetary costs and the amount of energy used to produce new material with those of processing recycled material. Have each group present its findings to the class, with illustrations and charts. **LS Logical/ Verbal** Co-op Learning

Close, *continued*

Alternative Assessment — GENERAL

Survey Tell students that recycling is only half the story and that conservation also includes buying products that contain recycled material. Have groups of students visit different types of stores and identify products that contain recycled materials. Students should make a list of the products that have the highest percentage of recycled content, then report back to the class. LS Intrapersonal

Answers to Section Review

1. destruction of wildlife habitats and pollution of air
2. by establishing and enforcing laws that reduce environmental impacts and by land reclamation
3. because fossil fuels cause environmental damage and are nonrenewable resources
4. Answers may vary but should say that reclamation is the restoration of land to its original condition.
5. Answers may vary. Sample answer: Use recycled paper and low-flow water fixtures, and bike instead of driving.
6. Recycling saves the energy that would be used to obtain resources from nature.
7. Answers may vary. Sample answer: Mining of coal may leave soil vulnerable to erosion and contamination. Spills from oil wells and pipelines can contaminate soil.
8. Energy is required both to mine minerals or cut down trees and to process these resources. Recycling uses energy mainly for processing.
9. Sample answer: establish a recycling facility; buy products made from recycled materials; establish bike lanes; promote carpooling; encourage the use of public transportation; use alternative-fuel buses; encourage the use of low-flow water fixtures; encourage the use of energy-efficient appliances; urge homeowners to install insulation; and educate people about conservation
10. Ways to reduce dependence on fossil fuels, which may cause *acid precipitation* and other harmful *environmental impacts,* include *conservation, recycling, reclamation,* using *renewable energy sources,* and finding *alternate energy sources.*

Figure 4 ▶ Fiberglass insulation is used in homes to reduce the energy required for heating and cooling.

Fossil-Fuel Conservation

Fossil fuels can be conserved by reducing the amount of energy used every day. If less energy is used, fewer fossil fuels must be burned every day to supply the smaller demand for energy. Energy can be conserved in many ways. **Figure 4** shows insulation being installed into a new house to reduce the amount of energy that will be needed for cooling and heating. Using energy-efficient appliances also reduces the amount of electricity used every day. In addition, simple actions, such as turning off lights when you leave a room and washing only full loads of laundry and dishes will reduce energy use.

Reducing the amount of driving you do also conserves fossil fuels. There is evidence that an average car produces more than 8 kg of carbon dioxide for every 3.8 L (1 gal) of gasoline burned. Even fuel-efficient and hybrid cars release some pollutants into the air. When making short trips, consider walking or riding your bicycle. You can also combine errands so that you can make fewer trips in your car.

SCILINKS. NSTA Developed and maintained by the National Science Teachers Association

For a variety of links related to this subject, go to www.scilinks.org

Topic: Conservation
SciLinks code: HQ60344

Conservation of Other Natural Resources

Conservation is important for other natural resources, such as water. Some scientists estimate that by the year 2025, the world will have a critical shortage of freshwater resources because of the increased need by a larger human population. Water can be conserved by using water-saving shower heads, faucets, and toilets. By turning off the faucet as you brush your teeth, you can conserve up to 1 gallon of water every day. You can also help conserve water by watering plants in the morning or at night and by planting native plants in your yard.

Section 4 Review

1. **Name** two environmental problems associated with the mining and use of coal.
2. **Explain** two ways the environmental impacts of mining can be reduced.
3. **Describe** two reasons why scientists are looking for alternatives to fossil fuels.
4. **Define** the term *reclamation* in your own words.
5. **Identify** three ways that you can conserve natural resources every day.
6. **State** one way that recycling can help conserve energy.

CRITICAL THINKING

7. **Analyzing Concepts** How do you think fossil-fuel use affects soil resources?
8. **Applying Ideas** Why does recycling require less energy than developing a new resource does?
9. **Drawing Conclusions** List 10 ways a small community can conserve energy and resources.

CONCEPT MAPPING

10. Use the following terms to create a concept map: *recycling, conservation, alternate energy source, renewable energy source, environmental impact, acid precipitation,* and *reclamation.*

CHAPTER RESOURCES

Chapter Resource File

- Section Quiz GENERAL

Workbooks

- Study Guide (also in Spanish)

Chapter 7 Highlights

Sections

1 Mineral Resources

Key Terms

ore, 155
lode, 156
placer deposit, 156
gemstone, 157

Key Concepts

- Ores are mineral deposits from which metallic and nonmetallic minerals can be profitably removed.
- Minerals are important sources of many useful and valuable materials.
- Humans obtain mineral resources through mining.

2 Nonrenewable Energy

nonrenewable resource, 159
fossil fuel, 159
nuclear fission, 162
nuclear fusion, 164

- Chemical and physical changes over time change the remains of plants into coal.
- Petroleum and natural gas formed from the remains of ancient microorganisms.
- Today, fossil fuels provide much of the world's energy.
- Nuclear fission can produce energy to generate electricity.

3 Renewable Energy

renewable resource, 165
geothermal energy, 165
solar energy, 166
hydroelectric energy, 167
biomass, 167

- Geothermal energy is energy from the heat of Earth's interior.
- Solar energy from the sun can be harnessed by both passive and active methods.
- Alternative sources of renewable energy include hydroelectric, tidal, solar, and wind energy.

4 Resources and Conservation

conservation, 171
recycling, 171

- Fossil fuels are nonrenewable resources. Once a nonrenewable resource is depleted, the resource may take millions of years to be replenished.
- Responsible mining operations work hard to return mined land to good condition through reclamation.
- Conservation is the preservation and wise use of natural resources.

Alternative Assessment — GENERAL

Keeping Track Have students prepare a chart that has four columns labeled, left to right, "Object/Activity," "Materials/Components," "Energy Used," and "Conservation." As students go through the day, tell them to record the following items in the appropriate columns of their chart: the objects they use and/or activities they engage in; the materials or components of the objects and how these components are obtained; the energy sources required to make and use the object; and ways in which they can conserve resources. Because their lists could be very large, consider limiting the list to a period of time or one location, such as breakfast at home or gym period at school. Have students discuss their lists with the class. **LS Logical/Interpersonal**

CHAPTER RESOURCES

Chapter Resource File

- **Concept Review** GENERAL
- **Critical Thinking** ADVANCED
- **Math Skills** GENERAL
- **Graphing Skills** GENERAL
- **Chapter Test A** GENERAL
- **Chapter Test B** ADVANCED

Workbooks

- **Study Guide** (also in Spanish)
- **Assessments** (Spanish)

Technology

Classroom Videos
- Brain Food Video Quiz

Chapter Review

Assignment Guide

Section	Questions
1	1, 5, 9, 16, 18, 23–24
2	6, 11–13, 19, 25, 27
3	2, 7–8, 14–15, 20, 22, 29, 32
4	3, 17, 21, 30–31
2 and 3	4, 26, 33–35
1–4	28

Chapter 7 Review

Using Key Terms

Use each of the following terms in a separate sentence.

1. *placer deposit*
2. *solar energy*
3. *conservation*

For each pair of terms, explain how the meanings of the terms differ.

4. *renewable resource* and *nonrenewable resource*
5. *ore* and *lode*
6. *nuclear fission* and *nuclear fusion*
7. *fossil fuel* and *biomass*
8. *geothermal energy* and *hydroelectric energy*

Understanding Key Concepts

9. Metals are known to
   a. have a dull surface.
   b. provide fuel.
   c. conduct heat and electricity well.
   d. occur only in placer deposits.

10. Energy resources that formed from the remains of once-living things are called
    a. minerals.
    b. gemstones.
    c. metals.
    d. fossil fuels.

11. Impermeable rock that occurs at the top of an oil reservoir is called
    a. coal.
    b. peat.
    c. cap rock.
    d. water.

12. Plastics, synthetic fabrics, and synthetic rubber are composed of chemicals that are derived from
    a. anthracite.
    b. petroleum.
    c. peat.
    d. minerals.

13. The splitting of the nucleus of an atom to produce energy is called
    a. geothermal energy.
    b. nuclear fission.
    c. nuclear fusion.
    d. hydroelectric power.

14. Energy experts have harnessed geothermal energy by
    a. building dams.
    b. building wind generators.
    c. drilling wells.
    d. burning coal.

15. In a hydroelectric power plant, running water produces energy by spinning a
    a. turbine.
    b. windmill.
    c. fan.
    d. reactor.

Short Answer

16. Compare the three ways that ores commonly form.
17. Name two regulations that mining operations must follow to reduce the impact they have on the environment.
18. Identify and describe the uses for three mineral resources.
19. Describe one advantage and one disadvantage of obtaining energy from nuclear fission.
20. Describe one advantage and one disadvantage to the use of solar energy.
21. Identify two ways recycling can reduce energy use.
22. Explain two ways that moving water can be used to generate electricity.
23. Compare two types of mining, and describe the possible environmental impact of each type.

Using Key Terms

1–8. Answers may vary but should show that students understand the definitions of and differences between key terms.

Understanding Key Concepts

9. c **10.** d
11. c **12.** b
13. b **14.** c
15. a

Short Answer

16. As magma cools, dense metallic minerals crystallize and sink to form deposits. When magma contacts surrounding rock, chemical reactions occur that form an ore. Water weathers and erodes rock and then carries minerals downstream, where they may be dropped where currents slow to form placer deposits.

17. Mines must follow the Clean Water Act and the Endangered Species Act.

18. Sample answer: Aluminum is used to make cans, gold is used for jewelry, and copper is used to make pots and pans.

19. advantage: no air pollutants; disadvantage: radioactive waste

20. advantage: It is a renewable resource. disadvantage: Solar energy systems may not produce as much energy when there is little sunlight.

21. Recycling a material takes less energy than mining new materials does. Also, recycling used materials takes less energy than manufacturing the same product using newly mined resources.

22. Hydroelectric dams produce electricity from moving water. Ocean tides may also be harnessed to produce energy.

23. Answers may vary. Sample answer: Subsurface mining removes ores from underground and does relatively little damage to the surface. Surface mining may disturb large areas of land and cause environmental damage by removing topsoil and vegetation. If dumped into streams, topsoil and rock may block the flow of water and upset aquatic ecosystems.

Critical Thinking

24. Answers may vary but should show that students understand the supply and demand for resources. If the rise in the price of iron were due to increased market demand, it would most likely be profitable to mine hematite. However, a price increase due to scarcity of hematite and/or increased costs to mine scarce materials may offset the potential for higher pricing due to scarcity alone. Thus, mining hematite may be less profitable.

Critical Thinking

24. Applying Ideas You learn that the price of iron is higher than it has been in 20 years. Do you think it might be profitable for a company to mine hematite? Explain your answer.

25. Understanding Relationships A certain area has extensive deposits of shale. Why might a petroleum geologist be interested in examining the area?

26. Identifying Trends Hybrid cars have efficient gasoline and electric motor combinations. They have other design elements that make them extremely fuel efficient. Do you expect that there will be more or fewer hybrid cars on the road in the future? Explain.

27. Making Inferences A certain company in your area produces ^{235}U pellets and fuel rods. With which energy source is the company involved? Explain.

Concept Mapping

28. Use the following terms to create a concept map: *resource, renewable, nonrenewable, fossil fuel, nuclear energy, geothermal energy, solar energy, hydroelectric energy,* and *conservation.*

Math Skills

29. Making Calculations In one year, the United States produced 95,000 megawatts of power from renewable energy sources. If 3% of that amount of power came from wind energy, how much energy did wind power produce that year?

30. Making Calculations A water-efficient washing machine uses 16 gallons of water per load of laundry. Older washing machines use more than 40 gallons of water per load of laundry. If you wash an average of 10 loads of laundry a month, how many gallons of water would you save in a year if you switched to the water-efficient washer?

Writing Skills

31. Researching Information A debate surrounds municipal recycling programs. Do some research, and write a paragraph explaining each side of the debate. Write another paragraph explaining your view on whether recycling programs should be continued.

32. Writing Persuasively Research the pros and cons of building dams to harness energy. Write a letter to the editor of a local newspaper to express your opinion about whether dams should be used for generating electricity.

Interpreting Graphics

The graph below shows the different contributions of various fuels to the U.S. energy supply since 1850. Use this graph to answer the questions that follow.

Energy Use in the United States

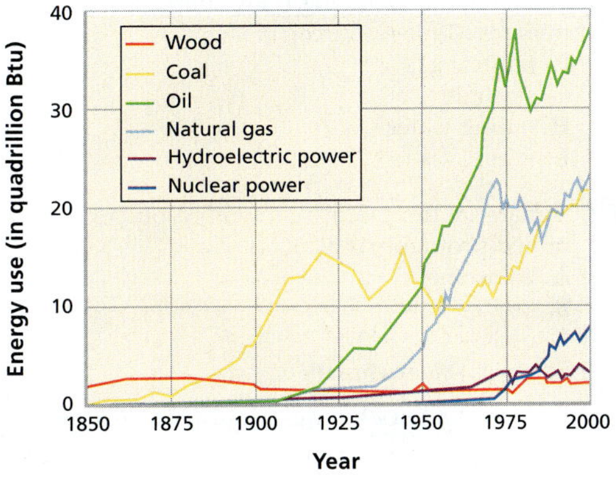

33. What were the two main energy sources used in 1875?

34. When did oil first become a more widely used energy source than coal?

35. The use of oil and natural gas rise and fall together. How do you explain this pattern?

25. The shale may be oil shale, a form of petroleum deposit, from which shale oil could be obtained by heating. Also, shale is commonly impermeable, so there may be oil trapped beneath the shale.

26. Answers may vary but should indicate that students understand that as petroleum becomes more scarce, gasoline prices may rise, causing some people to seek more fuel-efficient cars.

27. The company is involved in the nuclear energy business. Uranium is used in nuclear fission reactions.

Concept Mapping

28. Answers may vary but should include all of the terms listed. Sample answers appear at the end of this Teacher Edition.

Math Skills

29. 95,000 MW × 0.03 = 2,850 MW

30. 40 gal/load − 16 gal/load = 24 gal/load × 10 loads/month = 240 gallons of water saved per month × 12 months/year = 2,880 gallons/year

Writing Skills

31. Answers may vary. Accept all reasonable answers.

32. Answers may vary. Accept all reasonable answers.

Interpreting Graphics

33. wood and coal

34. 1950

35. Answers may vary but should indicate that oil and natural gas commonly occur together in the same deposits. Thus, they would be available from the same sources and their prices would be stable relative to each other. Therefore, it is reasonable that oil and natural gas would be used in similar quantities during the same time periods.

Estimated Time

To give students practice under more realistic testing conditions, allow them 30 minutes to answer all of the questions in this practice test.

TEST DOCTOR

Question 2 Answer I is correct. Metals are identified by their ability to conduct heat, answer F, their ability to conduct electricity, answer H, and their metallic (or shiny) surfaces, answer G. Metals may also be identified by their ability tendency to bend easily when hammered into thin sheets.

Question 3 Answer C is correct. A lode is formed from a large number of thick veins. Veins form when hot mineral solutions spread through small cracks in a large mass of rock. Answer A refers to placer deposits formed in moving water and is incorrect. Answer B is incorrect because it describes ores formed within cooling magma. Answer D refers to nodules on the ocean floor.

Question 11 Full-credit answers should include the following points:

- because the process to replenish petroleum and natural gas takes millions of years, these resources are considered nonrenewable
- nonrenewable resources are resources that form at a much slower rate than the rate at which they are consumed

Chapter 7 Standardized Test Prep

Understanding Concepts

Directions (1–5): **For** ***each*** **question, write on a separate sheet of paper the number of the correct answer.**

1 Which of the following is an example of a nonmetal mineral resource?
A. gold
B. quartz
C. aluminum
D. copper

2 Nonmetals are identified by their
F. ability to conduct heat
G. shiny surfaces
H. ability to conduct electricity
I. dull surfaces

3 A mineral deposit called a *lode* is formed by
A. metal fragments deposited in stream beds
B. layers accumulating in cooling magma
C. hot mineral solutions in cracks in rock
D. precipitation of minerals from seawater

4 Which of the following is an example of a nonrenewable resource?
F. natural gas
G. sunlight
H. falling water
I. wind

5 A material from which mineral resources can be mined profitably is a(n)
A. gemstone
B. ore
C. nodule
D. renewable resource

Directions (6–8): **For** ***each*** **question, write a short response.**

6 Federal and state laws require mining companies to return land to its original condition or better than its original condition when mining operations are completed. What is this process called?

7 What are the three forms of fossil fuels, and what form does each one take?

8 Name three common items that may be recycled to save energy and natural resources?

Reading Skills

Directions (9–11): **Read the passage below. Then, answer the questions.**

Fossil Fuels

All fossil fuels form from the buried remains of ancient organisms. But different types of fossil fuels form in different ways and from different types of organisms. Petroleum and natural gas form mainly from the remains of microscopic sea life. When these organisms die, their remains collect on the ocean floor, where they are buried by sediment. Over time, the sediment slowly becomes rock and traps the organic remains. Through physical and chemical changes over millions of years, the remains become petroleum and natural gas. Gradually, more rocks form above the rocks that contain the fossil fuels. Under the pressure of overlying rocks and sediments, the fossil fuels are able to move through permeable rocks. Permeable rocks are rocks that allow fluids, such as petroleum and natural gas, to move through them. These permeable rocks become reservoirs that hold petroleum and natural gas.

9 What process causes organic remains to turn into fossil fuels?
A. pressure caused by overlying rocks and sediments
B. the constant layering of remains from microscopic sea life
C. millions of years of physical and chemical changes
D. the movement of fluids through layers of permeable rock

10 Which of the following statements can be inferred from the information in the passage?
F. Fossil fuel formation is ongoing, and current remains may become petroleum in the future.
G. Fossil fuel formation happened millions of years ago and no longer takes place today.
H. Current petroleum and natural gas reservoirs are found only beneath the ocean floor.
I. Permeable rocks are also a good place to find other fossil fuels, such as coal.

11 Why do we consider petroleum and natural gas to be nonrenewable resources?

Answers

Understanding Concepts

1. B
2. I
3. C
4. F
5. B
6. reclamation
7. coal: solid, petroleum: liquid, and natural gas: gas
8. Answers may vary but may include glass, paper, plastic, aluminum, and rubber.

Reading Skills

9. C
10. F
11. Answers may vary. See Test Doctor for a detailed scoring rubric.

Interpreting Graphics

12. C
13. Answers may vary. See Test Doctor for a detailed scoring rubric.
14. All of the minerals listed might be used in the car manufacturing as follows: gold for computers and electronics, galena for car batteries, quartz for windows or light coverings, sulfur for tires, graphite for paint, hematite for the body and framework, and chalcopyrite for wiring.

Interpreting Graphics

Directions (12–14): **For *each* question below, record the correct answer on a separate sheet of paper.**

The graph below illustrates the sources of energy used in the United States since 1850. Future statistics are predicted based on current trends and technology development. Use this graph to answer questions 12 and 13.

U.S. Energy Use from 1850 to 2100

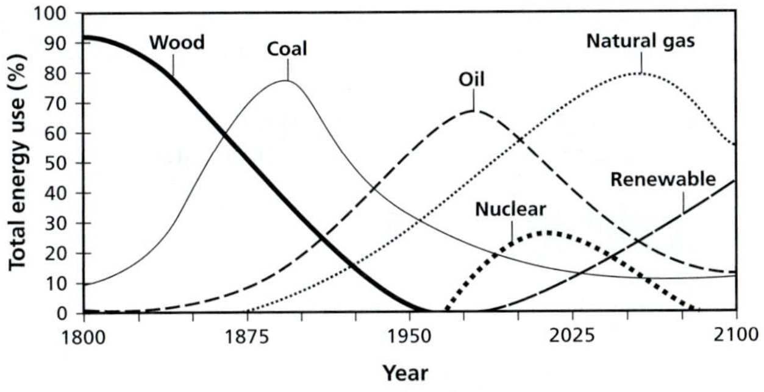

12 Which of the following is the main reason that coal became a more widely used energy source than wood in the mid-1800s?
A. Coal burns easier than wood does.
B. Coal is renewable resource, unlike wood.
C. Coal is a more efficient energy-producer than wood.
D. Coal produces fewer byproducts and waste than wood does.

13 Evaluate reasons why nuclear power is predicted to peak in usage around the year 2025, and then steadily decline in usage?

The table below shows common minerals and their uses. Use this table to answer question 14.

Minerals and Their Uses

Minerals	Uses
Gold	electronics, coins, dental work, and jewelry
Galena	solder and batteries
Quartz	glass
Sulfur	medicines, gunpowder, and rubber
Graphite	pencils, paint, and ubricants
Hematite	making steel
Chalcopyrite	coins, jewelry, and cables

14 Use your everyday knowledge of automobiles to describe the part of an automobile for which each mineral listed in the table may be used.

Test TIP

When a question refers to a graph, study the data plotted on the graph to determine any trends or anomalies before you try to answer the question.

TEST DOCTOR

Question 13 Full-credit answers should include the following points:

- students should demonstrate an understanding that technology carries both benefits and risks
- students should realize that nuclear power is a powerful, but controversial, energy source
- nuclear power has dangerous, long-lasting byproducts and the potential for serious accidents
- nuclear power produces radioactive waste, which remains hazardous for thousands of years
- nuclear power also creates the potential for a nuclear meltdown, which could release radioactivity into the atmosphere. Besides having a direct and detrimental effect on the health of local populations, radioactivity could be moved around the world by global weather patterns
- safer, cleaner, renewable resources are currently not as effective as nuclear power in many situations. As renewable resources become more efficient, they are predicted to replace nuclear power

Test Prep Correlations *National Science Education Standards*

HNS 2a: items 1, 2
HNS 2b: items 3, 9, 10
SPSP 3a: items 4, 5, 6, 7, 12, 14
SPSP 3b: items 4, 6, 7, 8, 11
SPSP 4b: items 6, 13

CHAPTER RESOURCES

State Resources

For specific resources for your state, visit go.hrw.com and type in the keyword **HSHSTR.**

Inquiry Lab

Blowing in the Wind

Teacher's Notes

Time Required

two 45-minute class periods

Lab Ratings

EASY → HARD

TEACHER PREPARATION (1)
STUDENT SETUP (1)
CONCEPT LEVEL (3)
CLEANUP (2)

Skills Acquired

- Constructing Models
- Communicating
- Designing Experiments

The Scientific Method

In this lab, students will

- Ask a Question
- Form a Hypothesis
- Test the Hypothesis
- Evaluate Results

Materials

The materials listed on the page are enough for groups of up to six students.

Tips and Tricks

If enough time or materials are not available, have student groups draw their designs. Then, have all the students in the class evaluate each design for its potential success and choose the one they think will work best. Then, have students build the chosen design, or assign the construction of several models as homework, and have the class test the models in the next class period.

Chapter 7 Inquiry Lab

Using Scientific Methods

Blowing in the Wind

Objectives

- **Prepare** a detailed sketch of your solution to the design problem.
- **Design and build** a functional windmill that lifts a specific weight as quickly as possible.

Materials

- blow-dryer, 1,500 W
- dowel or smooth rod
- foam board
- glue, white
- paper clips, large (30)
- paper cup, small (1)
- spools of thread, empty (2)
- string, 50 cm
- *optional materials for windmill blades:* foam board, paper plates, paper cups, or any other lightweight materials

Safety

MEMO
To: Division of Research and Development

Quixote Alternative Energy Systems is accepting design proposals to develop a windmill that can be used to lift window washers to the tops of buildings. As part of the design engineering team, your division has been asked to develop a working model of such a windmill. Your task is to design and build a model that can lift 30 large paper clips a vertical distance of 50 cm. The job will be given to the team whose model can lift the paper clips the fastest.

ASK A QUESTION

1. What is the best windmill design?

FORM A HYPOTHESIS

2. Brainstorm with a partner or small group of classmates to design a windmill using only the objects listed in the materials list. Sketch your design, and write a few sentences about how you think your windmill design will perform.

TEST THE HYPOTHESIS

3. Have your teacher approve your design before you begin construction. Build the base for your windmill by using glue to attach the two spools to the foam board. Make sure the spools are parallel before you glue them. Pass a dowel rod through the center of the spools. The dowel should rotate freely. Attach one end of the string securely to the dowel between the two spools.

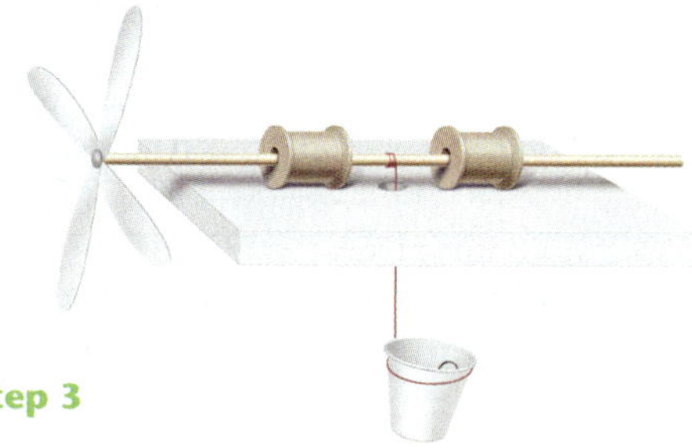

Step 3

CHAPTER RESOURCES

Chapter Resource File

- Datasheet for Chapter Lab GENERAL
- Lab Notes and Answers

Workbooks

Long-Term Projects

- Air Pollution Watch ADVANCED

4. Poke a hole through the middle of the foam board to allow the string to pass through. Place your windmill base between two lab tables or in any area that will allow the string to hang freely.
5. After you have decided on your final design, attach the windmill blades to the base.
6. If you have time, you may want to try using different material to construct your windmill blades. Test the various blades to determine whether they improve the original design. You may also want to vary the number and size of the blades on your windmill.
7. Attach the cup to the end of the string. Fill the cup with 30 paper clips. Turn on the blow-dryer, and measure the time it takes for your windmill to lift the cup.

Step 2
Sample windmill blade designs

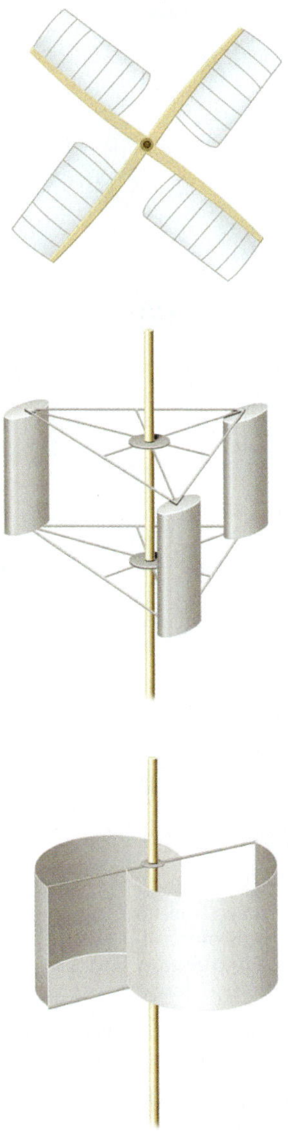

ANALYSIS AND CONCLUSION

1. **Evaluating Methods** After you test all of the designs, determine which design took the shortest amount of time to complete the test. What elements of the design do you think made it the strongest?
2. **Evaluating Models** Describe how you would change your design to make your windmill work better or faster.

Extension

1. **Research** Windmills have been used for more than 2,000 years. Research the three basic types of vertical axis machines and the applications in which they are used. Prepare a report of your findings.
2. **Making Models** Adapt your design to make a water wheel. You will find that water wheels can lift much more weight than a windmill can. Find designs on the Internet for micro-hydropower water wheels, such as the Pelton wheel, and use the designs as inspiration for your models. You can even design your own dam and reservoir.

Inquiry Lab

Answers to Analysis and Conclusion

1. Answers may vary but should include the reasons that students thought a particular design was the strongest.
2. Answers may vary. Accept all reasonable answers.

Answers to Extension

1. Answers may vary.
2. Designs may vary. Students may wish to build a model of the Pelton wheel. If students choose to draw a design for a dam and reservoir, have them explain to the class what design changes were needed to convert a wind-power design to a water-power design.

CLASSROOM TESTED & APPROVED

Mrs. Teresa Tucker
Northwest High School
Jackson, MI

Wind Power in the United States

Internet Activity — GENERAL

Alternative Energy in Your State Have students use the Internet to find national or state organizations that promote or have information about alternative energy. Have them find out what types of alternative energy are used in your state and where in the state they are used. Have them write a letter to the company, town, or county that is using the alternative energy and ask for more information about conditions at the site, how the energy is used, the technology behind the energy, and how it saves both money and resources. A worksheet designed to direct student research on this topic can be found in the **Chapter Resource File** booklet or by visiting **go.hrw.com** and entering the keyword **HQ6RENX.** **LS Verbal**

Answers to Map Skills Activity

1. WY and AK
2. CA and WA
3. TX
4. the Rocky Mountains
5. TX
6. The comparison means that these states could be a major source of wind energy, much as Saudia Arabia is a major source of oil.

CHAPTER RESOURCES

Chapter Resource File

Internet Activity
- Alternative Energy in Your State GENERAL

Technology

Transparencies
- 38 Wind Power in the United States (with worksheet)

MAPS *in Action*

Wind Power in the United States

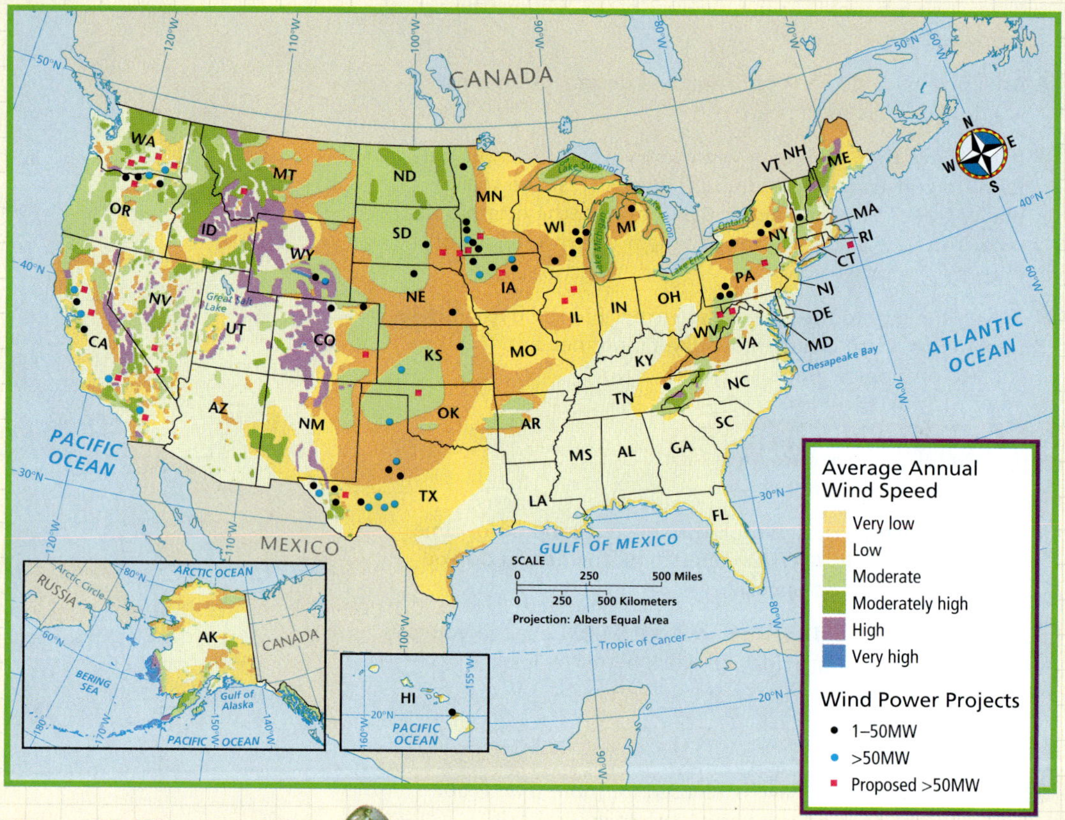

Map Skills Activity

This map shows average wind speeds and the locations of wind power projects throughout the United States. Use the map to answer the questions below:

1. **Using a Key** Name two states that have areas of very high wind speed.
2. **Using a Key** Which two states have the most proposed wind power projects?
3. **Making Comparisons** Which state has the most wind power projects currently in operation?
4. **Inferring Relationships** Examine Idaho, Wyoming, Montana, and Colorado. What landscape feature might account for the strong winds in those states?
5. **Identifying Trends** What states have more than four existing wind power projects that are larger than 50 MegaWatts (MW)?
6. **Making Comparisons** Because of their potential for wind power projects, the Great Plains states (MT, WY, CO, ND, SD, NE, KS, OK, TX, MN, and IA) have been called the "Saudi Arabia of wind energy." Why do you think this comparison has been made?

EYE on the Environment

What's Mined Is Yours

In every state of the United States, centuries of mining have left marks on the states' ecosystems. National parks alone contain more than 3,200 abandoned mining sites.

Environmental Scars

In a typical mining operation, bulldozers clear away layers of soil, which destroys vegetation and drastically changes the landscape. Streams may be diverted and cause floods in some areas.

Long-term effects can be even more serious. When discarded rock is exposed to air and rainfall, chemical reactions can occur. As a result of the chemical reactions, acids may seep into the ground and pollute nearby waterways. Even years later, plants and animals may still be unable to live in the area because of the affects of mining.

Taking Responsibility

Most mining companies take environmental issues very seriously. New techniques and a better understanding of natural processes have made reclaiming land more effective than ever. Unfortunately, reclamation is very expensive. Returning land to its original state may cost as much as $20,000 per acre. As environmental laws become tougher, mining companies must be more creative to make a profit. Because of low-grade resources and world markets, thousands of people have lost their jobs and local economies have suffered greatly when mining operations go out of business.

No Easy Answers

As the population of the United States grows, we will need more of the metal ores, fossil fuels, and other products that the mining industry provides. Fortunately, reclamation efforts are constantly improving. In addition, efforts to recycle and reuse the resources that we already have are increasing. Nonetheless, finding a compromise between our need for mineral resources and our concern for environmental damage will continue to be expensive and difficult.

▲ At the Flambeau Mine, a copper mine in Wisconsin, waster rock was stored in protected areas to minimize the impact in the enviroment.

◄ After the Flambeau Mine was closed, the waste rock and soil were replaced. A few years later, as seen in this photo, the reclaimed land was restored to its original state.

Extension

1. **Research** Find out the top three mineral resources that your state produces. In what part of the state are they located? How are the minerals extracted, and how are they used?

Eye on the Environment

What's Mined Is Yours

Group Activity — BASIC

Modern Mining Have students work in groups to research the methods used to mine a particular resource and the special requirements for reclaiming sites of those mines. If mining takes place in your state, you may suggest that some students focus on area practices. Students may research current methods of mining coal, gold, silver, copper, aluminum, or other minerals or metals. Students may be interested in the history and consequences of asbestos mining. Have students create multimedia presentations that illustrate what they have learned and then have them present their findings to the class. Invite students to share information about how some modern mining methods reduce environmental damage. LS Logical/Verbal

Answer to Extension

1. Answers may vary. Encourage students to use the Internet and the library or to send letters to your state's department of natural resources to find the information.

Unit 3 HISTORY OF THE EARTH

Unit 3 Outline

▶ Scientists learn about Earth's past by studying rocks and fossils. This fossil from the Green River formation in Wyoming is a fossilized predatory fish called *Mioplosus*. In this extraordinary fossil, was preserved a *Mioplosus* in the act of devouring a small fish.

Chapter 8 The Rock Record
Planning Guide

Compression Guide
To shorten instruction because of time limitations, omit the Chapter Lab.

OBJECTIVES	LABS, DEMONSTRATIONS, AND ACTIVITIES	TECHNOLOGY RESOURCES
PACING • 45 min pp. 184–190 **Chapter Opener**		**OSP** **Parent Letter** ■ **CD** **Student Edition on CD-ROM** **CD** **Chapter Summaries Audio CD** ■ **VID** **Brain Food Video Quiz**
Section 1 Determining Relative Age • State the principle of uniformitarianism. • Explain how the law of superposition can be used to determine the relative age of rocks. • Compare three types of unconformities. • Apply the law of crosscutting relationships to determine the relative age of rocks.	**TE** **Discussion** Rock Layers, p. 185 ◆ GENERAL **SE** **Quick Lab** What's Your Relative Age?, p. 186 GENERAL **CRF** **Datasheet for Quick Lab*** GENERAL **TE** **Discussion** Making Predictions, p. 187 GENERAL **TE** **Group Activity** Creating Layers, p. 188 ◆ GENERAL **SE** **Maps in Action** Geologic Map of Bedrock in Ohio, p. 208 GENERAL **TE** **Group Activity** Map Making, p. 208 ADVANCED **CRF** **Skills Practice Lab** Determining the Relative Age of Rock Strata* GENERAL	**OSP** **Lesson Plans** (also in print) **TR** **Bellringer*** **TR** 39 Law of Superposition* **TR** 40 Types of Unconformities* **TR** 41 Crosscutting Relationships* **TR** 43 Geologic Map of Bedrock in Ohio* **TE** **Internet Activity** Mapping Antarctica's Movement, p. 209 ADVANCED **CRF** **Internet Activity** Mapping Antarctica's Movement* ADVANCED **VID** **HRW Earth Science Video** Determining Relative Age **VID** **HRW Earth Science Video** Field Trip—Understanding Rock Layers
PACING • 45 min pp. 191–196 **Section 2 Determining Absolute Age** • Summarize the limitations of using the rates of erosion and deposition to determine the absolute age of rock formations. • Describe the formation of varves. • Explain how the process of radioactive decay can be used to determine the absolute age of rocks.	**TE** **Activity** Modeling Varves, p. 192 ◆ BASIC **TE** **Geology Connection** Glacial Lakes, p. 192 GENERAL **TE** **Discussion** Measuring Decay, p. 193 GENERAL **TE** **Chemistry Connection** Radioactive Decay of Uranium, p. 193 ADVANCED **SE** **Quick Lab** Radioactive Decay, p. 194 GENERAL **CRF** **Datasheet for Quick Lab*** GENERAL	**OSP** **Lesson Plans** (also in print) **TR** **Bellringer*** **TR** 42 Radioactive Decay and Half-Life*
PACING • 90 min pp. 197–200 **Section 3 The Fossil Record** • Describe four ways in which entire organisms can be preserved as fossils. • List five examples of fossilized traces of organisms. • Describe how index fossils can be used to determine the age of rocks.	**TE** **Demonstration** Learning Earth's History, p. 197 ◆ GENERAL **TE** **Discussion** Handprint History, p. 198 GENERAL **TE** **Group Activity** Fossils in Many Forms, p. 198 GENERAL **SE** **Making Models Lab** Types of Fossils, pp. 206–207 GENERAL **CRF** **Datasheet for Chapter Lab*** GENERAL **TE** **Activity** Spacing Footprints, p. 199 GENERAL **SE** **Mapping Expeditions** Where the Hippos Roam, pp. 840–841 GENERAL **CRF** **Inquiry Lab** Got Fossils?* GENERAL	**OSP** **Lesson Plans** (also in print) **TR** **Bellringer*** **CD** **Interactive Tutor** Evolution and Geologic Time

PACING • 90 min

CHAPTER REVIEW, ASSESSMENT, AND STANDARDIZED TEST PREPARATION

- **SE** **Chapter Highlights**, p. 201
- **SE** **Chapter Review**, pp. 202–203
- **SE** **Standardized Test Prep**, pp. 204–205
- **CRF** **Concept Review*** ■ GENERAL
- **CRF** **Critical Thinking*** ADVANCED
- **CRF** **Math Skills*** GENERAL
- **CRF** **Graphing Skills*** GENERAL
- **CRF** **Chapter Test A*** ■ GENERAL
- **CRF** **Chapter Test B*** ADVANCED
- **OSP** **Lesson Plans** (also in print)
- **OSP** **Test Generator**
- **OSP** **Test Item Listing**

Online and Technology Resources

go.hrw.com — Holt Online Learning

Visit **go.hrw.com** for access to Holt Online Learning, or enter the keyword **HQ6 Home** for a variety of free online resources.

One-Stop Planner® CD-ROM

This CD-ROM package includes
- Lab Materials QuickList Software
- Holt Calendar Planner
- Customizable Lesson Plans
- Printable Worksheets
- ExamView® Test Generator
- Interactive Teacher Edition
- Holt PuzzlePro®
- Holt PowerPoint® Resources

KEY

SE	Student Edition	OSP	One-Stop Planner	VID	Classroom Video/DVD
TE	Teacher Edition	TR	Transparencies and Transparency Worksheets	*	Also on One-Stop Planner
CRF	Chapter Resource File	CD	CD or CD-ROM	◆	Requires advance prep
LTP	Long-Term Projects			■	Also available in Spanish

SKILLS DEVELOPMENT RESOURCES	REVIEW AND ASSESSMENT	CORRELATIONS
SE **Pre-Reading Activity,** p. 184 GENERAL TE **Using the Figure** Fossil Forms, p. 184 GENERAL		National Science Education Standards
CRF **Directed Reading*** BASIC TE **Inclusion Strategies,** p. 186 TE **Using the Figure** The Law of Superposition, p. 187 GENERAL TE **Skill Builder** Vocabulary, p. 187 BASIC TE **Reading Skill Builder** Reading Hint, p. 187 BASIC SE **Graphic Organizer** Spider Map, p. 188 GENERAL TE **Using the Figure** Comparing Sedimentary Rock, p. 188 GENERAL TE **Using the Table** Analyzing Unconformities, p. 189 ADVANCED	SE **Reading Checks,** p. 186, 188 GENERAL SE **Section Review,** p. 190 GENERAL TE **Reteaching,** p. 189 BASIC TE **Quiz,** p. 189 GENERAL TE **Alternative Assessment,** p. 190 GENERAL CRF **Section Quiz*** ■ GENERAL	ES 3b, HNS 3c, UCP 4
CRF **Directed Reading*** BASIC TE **Using the Figure** Erosion Rate, p. 191 GENERAL SE **Math Practice,** p. 192 GENERAL TE **Skill Builder** Vocabulary, p. 193 ADVANCED TE **Using the Figure** Counting Decay, p. 194 BASIC TE **Using the Table** Methods for Dating Rocks, p. 195 GENERAL	SE **Reading Checks,** pp. 192, 195 GENERAL SE **Section Review,** p. 196 GENERAL TE **Reteaching,** p. 195 BASIC TE **Quiz,** pp. 196 GENERAL TE **Alternative Assessment,** p. 196 GENERAL CRF **Section Quiz*** ■ GENERAL	ES 3b, HNS 3c, UCP 4, PS 1d
CRF **Directed Reading*** BASIC TE **Reading Skill Builder** Paired Summarizing, p. 198 BASIC TE **Inclusion Strategies,** p. 198	SE **Reading Check,** p. 199 GENERAL SE **Section Review,** p. 200 GENERAL TE **Reteaching,** p. 199 BASIC TE **Quiz,** p. 200 GENERAL TE **Alternative Assessment,** p. 200 BASIC CRF **Section Quiz*** ■ GENERAL	ES 3b, HNS 3c, UCP 4, LS 3a, LS 3b

Holt Lab Generator CD-ROM

Search for labs by topic, standard, difficulty level, and/or time frame. Edit those labs to fit your needs, or create your own labs. Use the Lab Materials QuickList to order just what you need.

Holt Earth Science Interactive Tutor CD-ROM

This CD-ROM consists of interactive activities that give students a fun way to extend their knowledge of Earth science concepts.

Chapter Summaries Audio CDs

These CDs include audio summaries of the key concepts presented in each chapter. (Audio summaries are also available in Spanish.)

SCILINKS. NSTA

www.scilinks.org

Maintained by the **National Science Teachers Association.** See Chapter Enrichment pages that follow for a complete list of topics.

See Chapter Enrichment pages for Video Resources.

Chapter 8 Chapter Enrichment

This Chapter Enrichment provides relevant and interesting information to expand and enhance your classroom instruction of the chapter material.

Section 1 Determining Relative Age

The Father of Stratigraphy

William Smith (1769–1839) was an English engineer and geologist. The son of a blacksmith, Smith became a surveyor and was an avid fossil collector. Employed as a surveyor for the burgeoning mining industry, which was also building canals to transport coal, Smith was fascinated by the layers of rock uncovered by canal diggers in the southern county of Somerset, England. In 1794, he surmised that the strata, or rock layers, he saw there could be traced north across England. He headed north, and confirmed his hypothesis, finding the same strata again and again in his travels. In 1799, Smith began supervising major mine reclamation projects in Norfolk and Wales, often traveling 10,000 miles a year examining rocks. In 1815, he published his masterwork, *A Delineation of the Strata of England and Wales, with Part of Scotland*.

▲ **Stratigraphers are geologists who study the layering of sedimentary rocks.**

Smith was a keenly observant man. But he also had the ability to synthesize his observations into universal theories. He was among the first scientists to realize that different rock layers contain different fossils, and that each distinct layer contains the same fossils even when these layers are widely separated. After his death, Smith became universally acknowledged as the "Father of Stratigraphy." Stratigraphy is the science that deals with sedimentary rock layers. His rock-layer, or stratigraphic, maps of England are used by engineers and scientists to this day.

Section 2 Determining Absolute Age

The Origins of Radiometric and Carbon Dating

The origin of radioactive dating began with Ernest Rutherford's publication of *Radioactivity* (1904), in which he described the spontaneous decay of unstable isotopes of elements. A year later, John W. Stutt applied Rutherford's findings to the study of Earth materials. In America, Bertram B. Boltwood used the same data to calculate the ages of 43 different minerals. Boltwood's results indicated that some minerals were as much as 2.2 billion years old. Before long, many more Earth materials were analyzed and dated. Scientists realized that using the half-lives of short-lived isotopes generally yielded far more accurate rock ages than using the half-lives of long-lived isotopes. At first, radiometric dating used equipment to count the number of atoms present in the daughter isotopes. This cumbersome method was replaced by those using highly sensitive mass spectrometers.

Mass spectrometers also revolutionized carbon-14 dating. Willard Libby discovered the radioactive isotope carbon-14 (^{14}C) and showed that it occurs uniformly in all living organisms and decays after their death. Libby could measure the amount of ^{14}C left in dead or fossilized organic remains to date them accurately. Though scientists later found that ^{14}C does vary in organisms, they can recalibrate ^{14}C content during mass spectrometer analysis to account for these variations.

The development in 1939 of the accelerator mass spectrometer permitted scientists to analyze ^{14}C in as little as 1 mg of carbon, while eliminating interfering molecules. The accelerated mass spectrometer detects, measures, and studies the ions of isotopes.

▲ **Sedimentary rock layers that form in glacial lakes can be used to estimate the age of the deposits.**

▲ Some fossils, such as these ammonite fossils, can be used to accurately date rock layers.

The Earliest Fossils

The study of the fossil record has come a long way since William Smith noted that identical fossils were found in similar rock strata in widely divergent regions. Today, radiometric dating methods and other technologies have enabled scientists to examine the fossil remains of cyanobacteria, or blue-green algae, Earth's first inhabitants. So far, the oldest known fossil cyanobacteria have been found in Archaen greenstone from Western Australia. These single-celled organisms lived 3.5 billion years ago. Greenstones are ancient sedimentary and volcanic rocks that have been heated and deformed over time. The Australian greenstone was once on the bottom of the ocean. The fossils in this rock were found in vertical intrusions of a sedimentary rock called *chert* that had cut through the greenstone. These chert dikes were dated to nearly the same age as the greenstone into which it intruded.

Video Resources

Brain Food Video Quizzes

Brain Food Video Quizzes These videos contain game-show style quizzes that assess students' progress and motivate students to study the chapter material.

HRW Earth Science Video

This video introduces Earth science topics and includes a gology field trip. The video segments listed below complement this chapter.

Segment 12: Determining Relvative Age The history of Earth is preserved in the rock record. Geologists use their understanding of the relationships between different rock types to interpret Earth's history. This segment provides an overview of the basic principles geologists use to determine the relative ages of rocks. Included are brief descriptions of crosscutting relationships, erosional unconformities, the law of superposition, and folding and faulting. (3.5 min)

Segment 13: Field Trip—Understanding Rock Layers In this segment, Dr. Robert Roback, a former professor of geology at the University of Texas at Austin, explores the principles for determining the relative ages of rocks introduced in Segment 12. Dr. Roback takes us on a field trip to a region where igneous, metamorphic, and sedimentary rocks crop out, and he proceeds to reconstruct the area's complex geologic history. He points out important diagnostic features that have been preserved in the rocks. These features include cross-bedding in sandstone, folds in metamorphic rocks, folds in metamorphic rock cut by igneous dikes, and an erosional unconformity between sedimentary and igneous rock. (9 min)

NOVA® Videos

To order **NOVA** videos related to this chapter, visit **go.hrw.com** and enter the keyword **HQ6RECV.**

SciLinks®

NSTA Developed and maintained by the National Science Teachers Association

SciLinks is maintained by the National Science Teachers Association to provide you and your students with interesting, up-to-date links that will enrich your classroom presentation of the chapter.

Visit www.scilinks.org and enter the SciLinks code for more information about the topic listed.

Topic: Law of Superposition
SciLinks code: HQ60858

Topic: Fossil Record
SciLinks code: HQ60615

Topic: Radiometric Dating
SciLinks code: HQ61261

Topic: Relative Dating
SciLinks code: HQ61288

Chapter 8

Chapter Overview

This chapter describes the geologic record of rocks. The chapter introduces methods of dating rocks relatively and absolutely. The fossil record, processes of fossilization, and the types and importance of fossils are also covered.

Using the Figure — GENERAL

Fossil Forms Ask students what the photo shows. (fossils of organisms called *crinoids*) Point out that the detail of this fossil slab is so distinct, it looks like a sculpture, but it is, in fact, sedimentary rock. Have students speculate about how this fossil formed. (As layers of sediment rapidly accumulated over the dead bodies of these marine organisms, the sediment filled in empty spaces and preserved the details. The fossil was preserved when the sediment layers were compacted and hardened into stone.) **LS Visual**

PRE-READING ACTIVITY

FOLDNOTES

You may want students to work in pairs to create this FoldNote. Have one student fill in the first column, and have the other student fill in the last column.

Chapter 8 The Rock Record

Sections

1. Determining Relative Age
2. Determining Absolute Age
3. The Fossil Record

What You'll Learn

- How scientists determine relative age
- How scientists determine absolute age
- How fossils form

Why It's Relevant

To study Earth's 4.6 billion year history, scientists look to the information stored in rocks. Scientists must determine the age of rocks to put the events of Earth's history in order.

PRE-READING ACTIVITY

FOLDNOTES

Table Fold Before you read this chapter, create the FoldNote entitled "Table Fold" described in the Skills Handbook section of the Appendix. Label the columns of the table fold with "Absolute age" and "Relative age." As you read the chapter, write examples of each topic under the appropriate column.

▶ This slab of beautifully preserved crinoids shows each organism's 10 radial arms. Crinoids have been inhabiting aquatic environments on Earth for almost 490 million years.

Chapter Correlations — *National Science Education Standards*

ES 3b Geologic time can be estimated by observing rock sequences and using fossils to correlate the sequences at various locations. Current methods include using the known decay rates of radioactive isotopes present in rocks to measure the time since the rock was formed. **(Sections 1, 2, and 3)**

HNS 3c Occasionally, there are advances in science and technology that have important and long-lasting effects on science and society. Examples of such evidences include plate tectonics;...nuclear physics; biological evolution.... **(Sections 1, 2, and 3)**

UCP 4 Evolution is a series of changes, some gradual and some sporadic, that accounts for the present form and function of objects, organisms, and natural and designed systems. **(Sections 1, 2, and 3)**

PS 1d Structure of atoms: radioactive isotopes are unstable and undergo spontaneous nuclear reactions; a large group of identical nuclei decay in a predictable way. **(Section 2)**

LS 3a Species evolve over time.... **(Section 3)**

LS 3b The great diversity of organisms is the result of more than 3.5 billion years of evolution.... **(Section 3)**

Section 1 Determining Relative Age

Geologists estimate that Earth is about 4.6 billion years old. The idea that Earth is billions of years old originated with the work of James Hutton, an 18th-century Scottish physician and farmer. Hutton, who is shown in **Figure 1,** wrote about agriculture, weather, climate, physics, and even philosophy. Hutton was also a keen observer of the geologic changes taking place on his farm. Using scientific methods, Hutton drew conclusions based on his observations. Today, he is most famous for his ideas and writings about geology.

OBJECTIVES

- **State** the principle of uniformitarianism.
- **Explain** how the law of superposition can be used to determine the relative age of rocks.
- **Compare** three types of unconformities.
- **Apply** the law of crosscutting relationships to determine the relative age of rocks.

KEY TERMS

uniformitarianism
relative age
law of superposition
unconformity
law of crosscutting relationships

Uniformitarianism

Hutton theorized that the same forces that changed the landscape of his farm had changed Earth's surface in the past. He thought that by studying the present, people could learn about Earth's past. Hutton's principle of **uniformitarianism** is that current geologic processes, such as volcanism and erosion, are the same processes that were at work in the past. This principle is one of the basic foundations of the science of geology. Geologists later refined Hutton's ideas by pointing out that although the processes of the past and present are the same, the rates of the processes may vary over time.

uniformitarianism a principle that geologic processes that occurred in the past can be explained by current geologic processes

Figure 1 ▶ James Hutton (left) believed that studying the present is the key to understanding the past. This modern day Earth scientist (below) is looking for clues to Earth's past.

Section 1

Focus

Overview

This section describes the process for determining the relative age of rocks and some processes that resulted in the modern rock record. The section also covers uniformitarianism, the law of superposition, the principle of original horizontality, and unconformities in rock layers.

Bellringer

Ask students, "What are five visual clues that help you determine if someone is older or younger than you are?" List student responses on the board. (Answers may vary but may include color of hair, wrinkles in skin, height, general demeanor, and style of dress.) LS **Interpersonal**

Motivate

Discussion — GENERAL

Rock Layers Show students some pictures of sedimentary rock formations in which horizontal layers of rock are clearly visible. Tell them that each layer in the pictures may represent thousands of years of rock formation. Ask students which layers are the youngest and which the oldest. (The youngest layers are probably at the top, and the oldest are probably at the bottom.) LS **Visual/Logical**

CHAPTER RESOURCES

Chapter Resource File

- Directed Reading BASIC

Technology

Transparencies
- Bellringer

Student Edition on CD-ROM

One-Stop Planner CD-ROM
- Lesson Plan

Teach

QuickLAB

Skills Acquired

- Observing
- Designing Experiments
- Inferring
- Organizing Data

Teacher's Notes You may want to ensure that the groups have a mix of boys and girls who have different physical characteristics and who have different birth dates, if possible.

Answers

1. Students should use their birth dates to determine relative age.
2. The relative ages should correspond to absolute ages.
3. Unlike people, rocks may be dated according to their physical characteristics and relative positions.

Answer to Reading Check

Hutton reasoned that the extremely slow-working forces that changed the land on his farm had also slowly changed the rocks that make up Earth's crust. He concluded that large changes must happen over a period of millions of years.

CHAPTER RESOURCES

Chapter Resource File

- Datasheet for Quick Lab GENERAL

QuickLAB 20 min

What's Your Relative Age?

Procedure

1. Form a group with 5 to 10 of your classmates.
2. Work together to arrange group members in order from oldest to youngest.

Analysis

1. How did you determine your classmates' relative ages?
2. How did the relative ages compare to the absolute ages of your classmates?
3. How is this process of determining relative and absolute age different from the way scientists date rocks?

Earth's Age

Before Hutton's research was completed, many people thought that Earth was only about 6,000 years old. They also thought that all geologic features had formed at the same time. Hutton's principle of uniformitarianism raised some serious questions about Earth's age. Hutton observed that the forces that changed the land on his farm operated very slowly. He reasoned that millions of years must be needed for those same forces to create the complicated rock structures observed in Earth's crust. He concluded that Earth must be much older than previously thought. Hutton's observations and conclusions about the age of Earth encouraged other scientists to learn more about Earth's history. One way to learn about Earth's past is to determine the order in which rock layers and other rock structures formed.

Reading Check **What evidence did Hutton propose to show that Earth is very old? (See the Appendix for answers to Reading Checks.)**

Relative Age

In the same way that a history book shows an order of events, layers of rock, called *strata,* show the sequence of events that took place in the past. Using a few basic principles, scientists can determine the order in which rock layers formed. Once they know the order, a relative age can be determined for each rock layer. **Relative age** indicates that one layer is older or younger than another layer but does not indicate the rock's age in years.

Various types of rock form layers. Igneous rocks form layers when successive lava flows stack on top of each other. Some types of metamorphic rock, such as marble, also have layers. To determine the relative age of rocks, however, scientists commonly study the layers in sedimentary rocks, such as those shown in **Figure 2.**

relative age the age of an object in relation to the ages of other objects

Figure 2 ▶ The layers of sedimentary rock that make up Canyon de Chelly in Arizona were deposited over millions of years.

INCLUSION Strategies

- ***Developmentally Delayed***
- ***Hearing Impaired***
- ***Learning Disabled***

Some students need additional assistance to retain the pronunciations and meanings of the key terms presented in the section. For each of the five key terms, have students work with partners to complete the following steps:

1. Write the term at the top of a sheet of paper.
2. Write a pronunciation guide under the term.
3. Write the sentence from the textbook that first uses the term.
4. Write a new sentence that uses the term.
5. Draw a picture that shows what the term describes.
6. Share sentences and pictures with the group. LS **Verbal/Visual**

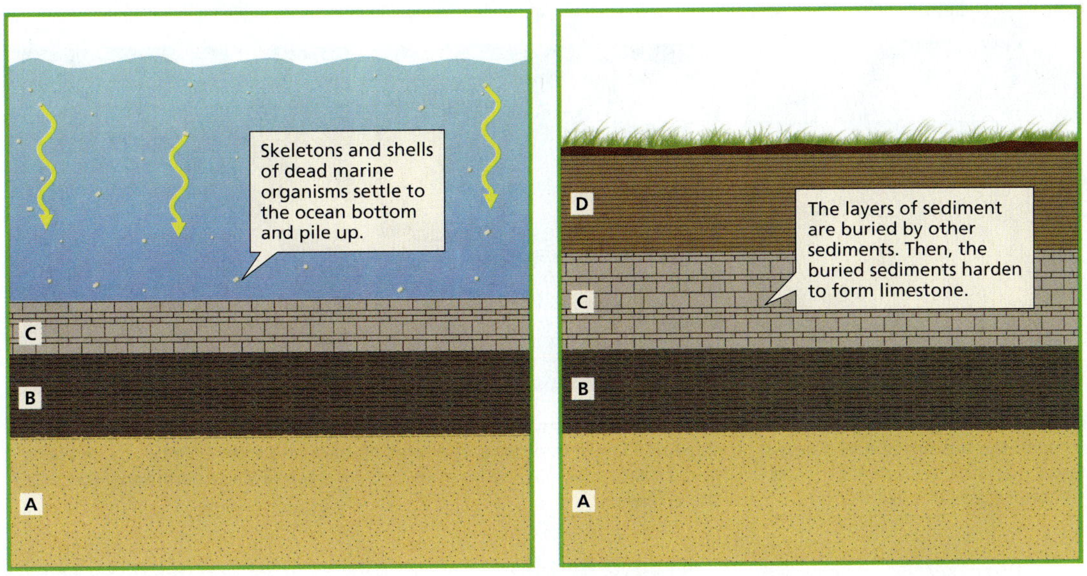

Figure 3 ▶ By applying the law of superposition, geologists are able to determine that layer D is the youngest bed in this rock section. ***According to the law of superposition, is layer B older or younger than layer C?***

Law of Superposition

Sedimentary rocks form when new sediments are deposited on top of old layers of sediment. As the sediments accumulate, they are compressed and harden into sedimentary rock layers which are called *beds*. The boundary between two beds is called a *bedding plane*.

Scientists use a basic principle called the law of superposition to determine the relative age of a layer of sedimentary rock. The **law of superposition** is that an undeformed sedimentary rock layer is older than the layers above it and younger than the layers below it. According to the law of superposition, layer A, shown in **Figure 3,** was the first layer deposited, and thus it is the oldest layer. The last layer deposited was layer D, and thus it is the youngest layer.

law of superposition the law that a sedimentary rock layer is older than the layers above it and younger than the layers below it if the layers are not disturbed

Principle of Original Horizontality

Scientists also know that sedimentary rock generally forms in horizontal layers. The *principle of original horizontality* is that sedimentary rocks left undisturbed will remain in horizontal layers. Therefore, scientists can assume that sedimentary rock layers that are not horizontal have been tilted or deformed by crustal movements that happened after the layers formed.

In some cases, tectonic forces push older layers on top of younger ones or overturn a group of rock layers. In such cases, the law of superposition cannot be easily applied. So, scientists must look for clues to the original position of layers and then apply the law of superposition.

SCILINKS® NSTA Developed and maintained by the National Science Teachers Association

For a variety of links related to this subject, go to www.scilinks.org

Topic: Law of Superposition
SciLinks code: HQ60858

Using the Figure — GENERAL

The Law of Superposition Have students look carefully at the two diagrams in the figure and describe what material overlies layer C on the left. (seawater) Point out that over millions of years, the conditions shown on the left become the conditions shown on the right, and what was once sea becomes land. Answer to caption question: Layer B is older than layer C because B is below C and was deposited earlier. LS **Logical/Visual**

Discussion — GENERAL

Making Predictions Before students read the section on original horizontality, have them discuss what might disturb the layers of rock described in the section on superposition. Then, have them read the passage and determine whether their predictions were correct. LS **Verbal**

SKILL BUILDER — BASIC

Vocabulary Tell students that the word *superposition* uses the prefix *super-*, which means "above." Ask students how knowing the meaning of this prefix can help them understand or remember the law of superposition. (The law of superposition states that rock layers on the bottom are older than those "above" or on top.) LS **Verbal** English Language Learners

READING SKILL BUILDER — BASIC

Reading Hint Tell students to look for word roots, small words, and compound words as they read this section. Finding and understanding smaller words, such as *uniform, horizontal, cross,* and *conform,* within larger words will help students understand the content. Have students work in pairs to recognize and say word roots and smaller words they find in big words. English Language Learners
LS **Verbal/Auditory**

CHAPTER RESOURCES

Technology

Transparencies
- 39 Law of Superposition (with worksheet)

Teach, *continued*

Using the Figure — GENERAL

Comparing Sedimentary Rock
Have students examine and describe each photo at the top of this page. For Graded Bedding, ask students why large particles are found at the bottom of the layer. (Gravity causes heavier particles to settle faster, so the larger particles are at the bottom of the layer.) For Cross-Beds, have students identify the strata of which the cross-beds are a part. (Each set of slanted cross-beds is part of a larger layer. Cross-bedded layers are cut off by a flat plane at the top of the layer.) For Ripple Marks, ask students to identify the direction the water moved over the sediment that became this rock. (If the peaks of ripples point in the direction of flow, the water was flowing either from the bottom right toward the top left or from the top left to the bottom right.) **LS Visual**

Graphic Organizer — GENERAL

Spider Map
You may want to have students work in groups to create this Spider Map. Have one student draw the map and fill in information provided by other students from the group.

Figure 4 ▶ Sedimentary Rock Structures

Graded Bedding Heavy particles settle to the bottom of a lake or river faster than smaller particles do and create graded beds like this one.

Cross-beds As sand slides down the slope of a large sand dune, the sand forms slanting layers like the ones shown here.

Ripple Marks When waves move back and forth on a beach, ripple marks like these commonly form.

Graphic Organizer

Spider Map
Create the Graphic Organizer entitled "Spider Map" described in the Skills Handbook section of the Appendix. Label the circle "Unconformities." Create a leg for each type of unconformity. Then, fill in the map with details about each type of unconformity.

Graded Bedding

One possible clue to the original position of rock layers is the size of the particles in the layers. In some depositional environments, the largest particles of sediment are deposited in the bottom layers, as shown in **Figure 4.** The arrangement of layers in which coarse and heavy particles are located in the bottom layers is called *graded bedding*. If larger particles are located in the top layers, the layers may have been overturned by tectonic forces.

Cross-Beds

Another clue to the original position of rock layers is in the shape of the bedding planes. When sand is deposited, sandy sediment forms curved beds at an angle to the bedding plane. These beds are called *cross-beds* and are shown in **Figure 4.** The tops of these layers commonly erode before new layers are deposited. So, the sediment appears to be curved at the bottom of the layer and to be cut off at the top. By studying the shape of the cross-beds, scientists can determine the original position of these layers.

Ripple Marks

Ripple marks are small waves that form on the surface of sand because of the action of water or wind. When the sand becomes sandstone, the ripple marks may be preserved, as shown in **Figure 4.** In undisturbed sedimentary rock layers, the crests of the ripple marks point upward. By examining the orientation of ripple marks, scientists can establish the original arrangement of the rock layers. The relative ages of the rocks can then be determined by using the law of superposition.

Reading Check **How can ripple marks indicate the original position of rock layers? (See the Appendix for answers to Reading Checks.)**

Group Activity — GENERAL

Creating Layers Give groups of students a variety of small pebbles, some sand, a plastic jar with a lid, and water. Have them shake up the sand and pebbles in the jar. Then, have them add enough water to the jar to cover the sediment, shake the jar vigorously, and let it sit for 15 minutes or more. Students should note the order in which particles settled at the bottom. Have students discuss how the experiment relates to the graded bedding of sedimentary rock. (Pebbles should form the bottom layer, with sand on top of it, and finer particles at the top.) **LS Kinesthetic**

Answer to Reading Check

Because ripple marks form at the top of a rock layer, scientists can use the orientation of the ripple marks to determine which direction was "up" when the rock layers formed.

Unconformities

Movements of Earth's crust can lift up rock layers that were buried and expose them to erosion. Then, if sediments are deposited, new rock layers form in place of the eroded layers. The missing rock layers create a break in the geologic record in the same way that pages missing from a book create a break in a story. A break in the geologic record is called an **unconformity.** An unconformity shows that deposition stopped for a period of time, and rock may have been removed by erosion before deposition resumed.

As shown in **Table 1,** there are three types of unconformities. An unconformity in which stratified rock rests upon unstratified rock is called a *nonconformity*. The boundary between a set of tilted layers and a set of horizontal layers is called an *angular unconformity*. The boundary between horizontal layers of old sedimentary rock and younger, overlying layers that are deposited on an eroded surface is called a *disconformity*. According to the law of superposition, all rocks beneath an unconformity are older than the rocks above the unconformity.

unconformity a break in the geologic record created when rock layers are eroded or when sediment is not deposited for a long period of time

Table 1 ▼

Types of Unconformities		
Type	**Example**	**Description**
Nonconformity	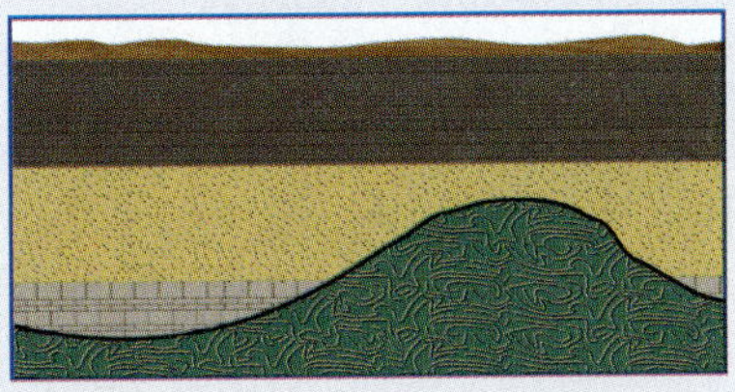	Unstratified igneous or metamorphic rock may be uplifted to Earth's surface by crustal movements. Once the rock is exposed, it erodes. Sediments may then be deposited on the eroded surface. The boundary between the new sedimentary rock and the igneous or metamorphic rock is a *nonconformity*. The boundary represents an unknown period of time during which the older rock was eroded.
Angular unconformity	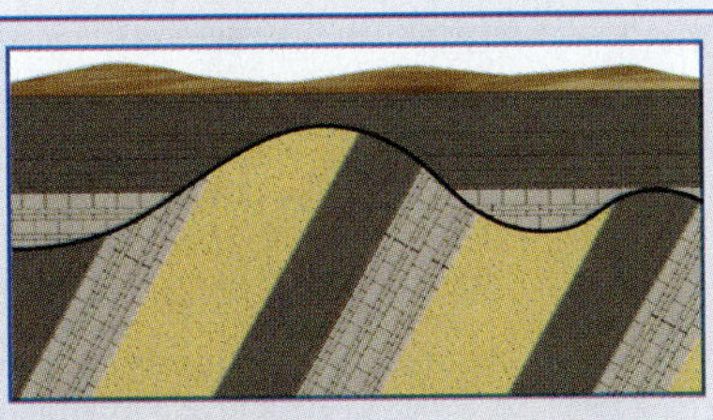	An *angular unconformity* forms when rock deposited in horizontal layers is folded or tilted and then eroded. When erosion stops, a new horizontal layer is deposited on top of a tilted layer. When the bedding planes of the older rock layers are not parallel to those of the younger rock layers deposited above them, an angular unconformity results.
Disconformity	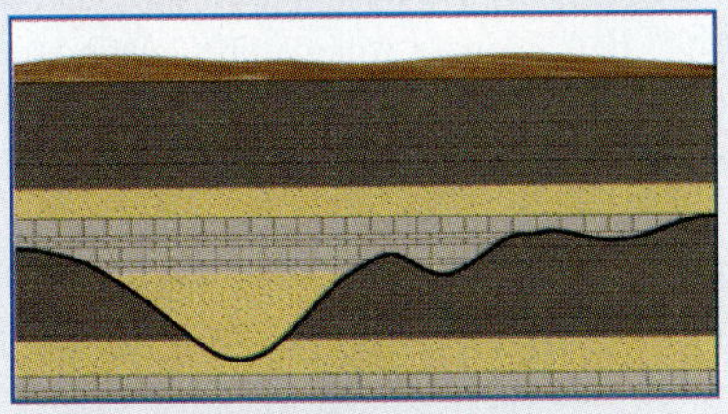	Sometimes, layers of sediments are uplifted without folding or tilting and are eroded. Eventually, the area subsides and deposition resumes. The layers on either side of the boundary are nearly horizontal. Although the rock layers look as if they were deposited continuously, a large time gap exists where the upper and lower layers meet. This gap is known as a *disconformity*.

CHAPTER RESOURCES

Chapter Resource File

- **Skills Practice Lab** Determining the Relative Age of Rock Strata GENERAL

Technology

Transparencies

- 40 Types of Unconformities (with worksheet)
- 41 Crosscutting Relationships (with worksheet)

Using the Table — ADVANCED

Analyzing Unconformities Have students study the table of unconformities. Then, have students explain which rocks are older: the rocks above or below the unconformity. (below) Ask students to summarize the differences between the types of unconformities. (The differences are based on the rocks below the unconformities. Nonconformities are underlain by metamorphic or igneous rock. Angular unconformities are underlain by tilted rock layers, while disconformities are underlain by rock layers that are parallel to the layers above the unconformity.) LS Visual

Close

Reteaching — BASIC

Changing Rocks Have students use clay of different colors to model the rock structures they learned about in this section. Students may use tools to indicate ripples and should manipulate the clay to demonstrate the forces that cause changes in rock layers. As students work with their models, ask them about the relative ages of the "rocks," or clay layers they are manipulating. LS Kinesthetic English Language Learners

Quiz — GENERAL

1. In graded bedding, what material would you find at the bottom if the rock were undisturbed? (large, heavy particles or pebbles)
2. What does the principle of original horizontality state? (Sedimentary rocks form in horizontal layers and undisturbed rock layers remain horizontal.)

Close, *continued*

Alternative Assessment — GENERAL

Illustrations Have students illustrate each type of rock formation they read about in this section. Students should label and write a brief description next to each picture. LS Visual/Verbal

Answers to Section Review

1. so they can understand the age and history of Earth
2. Answers may vary. Accept all reasonable answers.
3. According to the law, a rock layer is always younger than the rock layer below it if the layers are undisturbed.
4. An unconformity is a break in the rock record; a nonconformity is a kind of unconformity in which metamorphic or igneous rock layers are overlain by sedimentary rock layers.
5. In an angular unconformity, tilted rock layers are overlain by horizontal rock layers. In a disconformity, the rock layers above and below the unconformity are parallel.
6. The law of crosscutting relationships helps scientists understand that a body of rock or a fault is always younger than any other body of rock that it cuts through.
7. A disconformity would be harder to recognize because it has horizontal layers on both sides of the boundary. A nonconformity could be identified by the metamorphic or igneous rock below the unconformity.
8. If the fault does not cut through the unconformity, the fault is older than the layers deposited above the unconformity.
9. The *relative age* of rocks can be determined by using the *principle of original horizontality* and the *law of superposition* or by using *graded bedding*, *cross-beds*, and *ripple marks* to establish the original arrangement of layers.

Figure 5 ▶ The law of crosscutting relationships can be used to determine the relative ages of rock layers and the faults and intrusions within them.

Crosscutting Relationships

When rock layers have been disturbed by faults or intrusions, determining relative age may be difficult. A *fault* is a break or crack in Earth's crust along which rocks shift their position. An *intrusion* is a mass of igneous rock that forms when magma is injected into rock and then cools and solidifies. In such cases, scientists may apply the law of crosscutting relationships. The **law of crosscutting relationships** is that a fault or igneous intrusion is always younger than the rock layers it cuts through. If a fault or intrusion cuts through an unconformity, the fault or intrusion is younger than all the rocks it cuts through above and below the unconformity.

Figure 5 shows a series of rock layers that contains both a fault and an igneous intrusion. As you can see, an intrusion cuts across layers A, B, and C. According to the law of crosscutting relationships, the intrusion is younger than layers A, B, and C. The fault is younger than the intrusion and all four layers.

law of crosscutting relationships the principle that a fault or body of rock is younger than any other body of rock that it cuts through

Section 1 Review

1. **Explain** why it is important for scientists to be able to determine the relative age of rocks.
2. **State** the principle of uniformitarianism in your own words.
3. **Explain** how the law of superposition can be used to determine the relative age of sedimentary rock.
4. **Explain** the difference between an unconformity and a nonconformity.
5. **Compare** an angular unconformity with a disconformity.
6. **Describe** how the law of crosscutting relationships helps scientists determine the relative ages of rocks.

CRITICAL THINKING

7. **Making Comparisons** Which would be more difficult to recognize: a nonconformity or a disconformity? Explain your answer.
8. **Analyzing Relationships** Suppose that you find a series of rock layers in which a fault ends at an unconformity. Explain how you could apply the law of crosscutting relationships to determine the relative age of the fault and of the rock layers that were deposited above the unconformity.

CONCEPT MAPPING

9. Use the following terms to create a concept map: *principle of original horizontality, relative age, graded bedding, law of superposition, cross-bed,* and *ripple mark.*

CHAPTER RESOURCES

Chapter Resource File
- Section Quiz GENERAL

Workbooks
- Study Guide (also in Spanish)

Section 2 Determining Absolute Age

Relative age indicates only that one rock formation is younger or older than another rock formation. To learn more about Earth's history, scientists often need to determine the numeric age, or **absolute age,** of a rock formation.

OBJECTIVES

- **Summarize** the limitations of using the rates of erosion and deposition to determine the absolute age of rock formations.
- **Describe** the formation of varves.
- **Explain** how the process of radioactive decay can be used to determine the absolute age of rocks.

KEY TERMS

absolute age
varve
radiometric dating
half-life

absolute age the numeric age of an object or event, often stated in years before the present, as established by an absolute-dating process, such as radiometric dating

Absolute Dating Methods

Scientists use a variety of methods to measure absolute age. Some methods use geologic processes that can be observed and measured over time. Other methods measure the chemical composition of certain materials in rock.

Rates of Erosion

One way to estimate absolute age is to study rates of erosion. For example, if scientists measure the rate at which a stream erodes its bed, they can estimate the age of the stream. But determining absolute age by using the rate of erosion is practical only for geologic features that formed within the past 10,000 to 20,000 years. One example of such a feature is Niagara Falls, which is shown in **Figure 1.** For older surface features, such as the Grand Canyon, which formed over millions of years, the method is less dependable because rates of erosion may vary greatly over millions of years.

Figure 1 ▶ The rocky ledge above Niagara Falls has been eroding at a rate of about 1.3 m per year for nearly 9,900 years. *How many kilometers has the ledge been eroded in the last 9,900 years?*

Section 2

Focus

Overview

This section explains the methods used to calculate the absolute age of rocks. These methods include using erosion and deposition rates, glacial lake deposits, and radiometric dating techniques. The concepts of radioactive decay and half-life are also presented.

Bellringer

Ask students to list the ways they know to determine the exact, or absolute, age of a person or object. (Answers may vary. For a person, they can ask the person, or see a birth certificate. For an object, they may see dates stamped or printed; experts may recognize a manufacture date; or various scientific techniques may be used.) **LS Interpersonal**

Motivate

Using the Figure — GENERAL

Erosion Rate Have students use the cars and buildings in the foreground of the photo at the bottom of the page to appreciate the size of the Canadian portion of Niagara Falls. Answer to caption question: Almost 13 km of ledge have been eroded (9,900 y × 1.3 m/y = 12,870 m ÷ 1,000 = 12.87 km). **LS Logical/Visual**

CHAPTER RESOURCES

Chapter Resource File

- Directed Reading BASIC

Technology

Transparencies
- Bellringer

Student Edition on CD-ROM

One-Stop Planner CD-ROM
- Lesson Plan

Teach

MATHPRACTICE

Answer
10 m = 1,000 cm; 1,000 cm ÷ 30 cm = 33.33 × 1,000 years = 33,330 years

Activity — BASIC

Modeling Varves Have students read the text under the heading "Varve Count" on this page. Divide the class into groups. Provide each group with a glass cylinder; coarse, light-colored sand; and fine, dark-colored sand. Have students model several varves by layering sand in the cylinder. When the groups are finished, ask students to explain what the layers represent. (Each pair of layers represents a varve, or an annual deposit of sediment in a glacial lake.) Ask them why each varve has two bands. (The lower band of coarse sediment is deposited in summer as a result of the coarse sediment carried into a lake by melting snow and ice. The upper layer of finer sediment is laid down in winter when the fine sediment particles finally settle out of the water.) Have students explain what sand they chose for each band and why. (The lower band of each pair is coarse sand because heavier particles settle first. The upper band of each pair is finer mud and organic matter because the smaller particles settle more slowly.) Ask students how scientists use varves. (By counting varves in a rock deposit, scientists can estimate the age of the deposits.) LS Kinesthetic/Visual

MATHPRACTICE

Deposition If 30 cm of sediments are deposited every 1,000 years, how long would 10 m of sediments take to accumulate?

Rates of Deposition

Another way to estimate absolute age is to calculate the rate of sediment deposition. By using data collected over a long period of time, geologists can estimate the average rates of deposition for common sedimentary rocks such as limestone, shale, and sandstone. In general, about 30 cm of sedimentary rock are deposited over a period of 1,000 years. However, any given sedimentary layer that is being studied may not have been deposited at an average rate. For example, a flood can deposit many meters of sediment in just one day. In addition, the rate of deposition may change over time. Therefore, this method of determining absolute age is not always accurate; it merely provides an estimate.

Varve Count

You may know that a tree's age can be estimated by counting the growth rings in its trunk. Scientists have devised a similar method for estimating the age of certain sedimentary deposits. Some sedimentary deposits show definite annual layers, called **varves,** that consist of a light-colored band of coarse particles and a dark band of fine particles.

varve a banded layer of sand and silt that is deposited annually in a lake, especially near ice sheets or glaciers, and that can be used to determine absolute age

Varves generally form in glacial lakes. During the summer, when snow and ice melt rapidly, a rush of water can carry large amounts of sediment into a lake. Most of the coarse particles settle quickly to form a layer on the bottom of the lake. With the coming of winter, the surface of the lake begins to freeze. Fine clay particles still suspended in the water settle slowly to form a thin layer on top of the coarse sediments. A coarse summer layer and the overlying, fine winter layer make up one varve. Thus, each varve represents one year of deposition. Some varves are shown in **Figure 2.** By counting the varves, scientists can estimate the age of the sediments.

Reading Check How are varves like tree rings? (See the Appendix for answers to Reading Checks.)

Figure 2 ▶ Varves (below), which form in glacial lakes (right), can be counted to determine the absolute age of sediments.

GEOLOGY CONNECTION — GENERAL

Glacial Lakes Tell students that most glacial lakes formed after continental glaciers retreated at the end of the last glacial period. When snow and ice from the glacier melts and flows down into the lake, the meltwater carries large amounts of sediment with it. Have students do research to find out more about the formation and characteristics of glacial lakes. Ask them to contribute to a class bulletin board about these lakes. LS Visual

Answer to Reading Check
Varves are like tree rings in that varves are laid down each year. Thus, counting varves can reveal the age of sedimentary deposits.

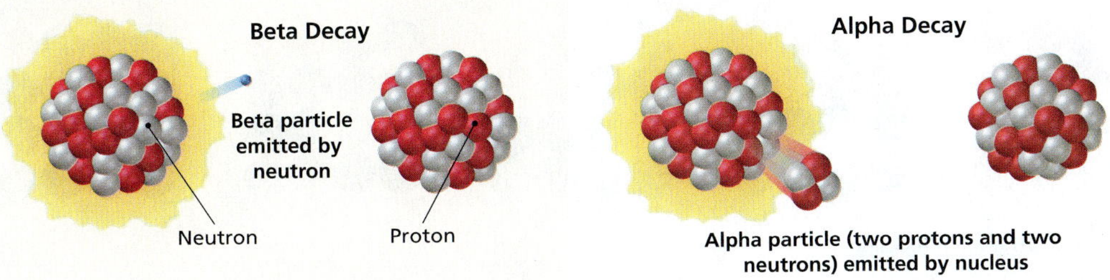

Figure 3 ▶ Beta decay and alpha decay are two forms of radioactive decay. In all forms of radioactive decay, an atom emits particles and energy.

Radiometric Dating

Rocks generally contain small amounts of radioactive material that can act as natural clocks. Atoms of the same element that have different numbers of neutrons are called *isotopes*. *Radioactive isotopes* have nuclei that emit particles and energy at a constant rate regardless of surrounding conditions. **Figure 3** shows two of the ways that radioactive isotopes decay. During the emission of the particles, large amounts of energy are released. Scientists use this natural breakdown of isotopes to accurately measure the absolute age of rocks. The method of using radioactive decay to measure absolute age is called **radiometric dating.**

As an atom emits particles and energy, the atom changes into a different isotope of the same element or an isotope of a different element. Scientists measure the concentrations of the original radioactive isotope, or *parent isotope*, and of the newly formed isotopes, or *daughter isotopes*. Using the known decay rate, the scientists compare the proportions of the parent and daughter isotopes to determine the absolute age of the rock.

radiometric dating a method of determining the absolute age of an object by comparing the relative percentages of a radioactive (parent) isotope and a stable (daughter) isotope

Connection to CHEMISTRY

Radioactive Decay of Uranium

Uranium is a radioactive element that occurs in some rocks. One form of uranium, which is represented by the symbol ^{238}U, has a mass number of 238 and an atomic number of 92. The atomic number is the number of protons in the nucleus. The mass number is the sum of the number of protons and neutrons in the nucleus. Uranium-238 is particularly useful in establishing the absolute ages of rocks.

The highly radioactive nucleus of ^{238}U spontaneously emits two protons and two neutrons in a process called *alpha decay* or alpha particle emission. The loss of protons and neutrons from the nucleus decreases both the atomic number and the mass number. After the first decay of ^{238}U, the new nucleus has a mass number of 234 and an atomic number of 90. The atom has changed into an isotope of thorium, ^{234}Th.

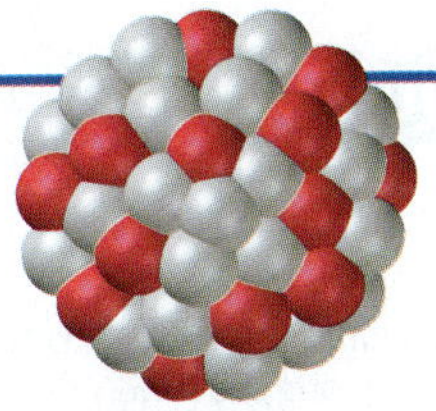

The ^{234}Th nucleus is also radioactive, and one of its neutrons changes into a proton and an electron in a process called *beta decay*. The change of the neutron to a proton increases the atomic number by one, and the thorium isotope then becomes an isotope of protactinium, ^{234}Pa, which is also radioactive.

The cycle of radioactive decay continues until a stable, or nonradioactive, form of lead is produced. A form of lead that has a mass number of 206, ^{206}Pb, is the final product of this radioactive-decay chain. In this chain of decay, ^{238}U is the parent isotope and ^{206}Pb is a daughter isotope. The time required for half of any given amount of ^{238}U to decay into ^{206}Pb is 4.5 billion years. Thus, 4.5 billion years is the half-life of ^{238}U.

Discussion — GENERAL

Measuring Decay Have students talk about why knowing the rate of radioactive decay of different elements is crucial to scientists who use radiometric dating methods to date rocks. (The age of rocks is calculated using decay rates of the radioactive elements they contain.) **LS Logical**

CHEMISTRY CONNECTION — ADVANCED

Radioactive Decay of Uranium Have students discuss what radioactive decay is and which atomic particles are involved in radioactive decay. (protons and neutrons) Ask students why ^{238}U continually transforms into other elements. (The particles that make up an element's nucleus determine the element's identity. The emission of particles changes an element's nucleus and thus transforms it into a different element.) Ask students what type of element a radioactive element eventually becomes. (a nonradioactive element) Ask students which rock is older: a rock that contains lots of ^{238}U and some ^{234}Th, or a rock that contains a little ^{238}U and a lot of ^{206}Pb. (The rock with the ^{206}Pb would be older because the ratio of parent ^{238}U to daughter nonradioactive ^{206}Pb is small. The amount of parent ^{238}U decreases with the age of the rock.) Invite interested students to research all the steps of the decay of ^{238}U into other elements, ending with ^{206}Pb. Students may use the Internet or other sources to find this information. Have students create a poster that shows how ^{238}U decays. **LS Logical/Visual**

SKILL BUILDER — ADVANCED

Vocabulary Tell students that the word *isotope* comes from the Greek *iso-* meaning "similar" or "identical," and *topos*, which means "place." Isotopes are forms of an element that have the same atomic number, but different mass numbers. The isotopes of an element have identical or very similar chemical properties. Because of this close similarity, all isotopes of an element occupy the same place on the periodic table of the elements. **LS Verbal**

CHAPTER RESOURCES

Technology

Transparencies
- 42 Radioactive Decay and Half-Life (with worksheet)

Teach, *continued*

Using the Figure — Basic

Counting Decay To reinforce the concept of half-life, have students determine the amount of the parent isotope in each square in the figure by counting out loud the number of circles shown in each square. Ask students to identify the color of the parent isotope in the figure (red) and the color of the daughter isotope (blue). Then, have them count the number of red circles in each subsequent square and tell the fraction of parent isotope and daughter isotope in each. LS Logical/Auditory

MISCONCEPTION ALERT

Radioactive Rocks
Students may think that the radioactive isotopes found in most rocks pose a significant health threat. Tell students that most radioactive materials in rocks occur in extremely minute quantities that pose little danger to living things. Radioactive decay and isotopes are all around us, but for the most part do little harm.

CHAPTER RESOURCES

Chapter Resource File

- Datasheet for Quick Lab GENERAL

Figure 4 ▶ After each half-life, one-half of the parent isotope is converted into a daughter isotope. After four half-lives, only 1/16 of the parent isotope remains.

half-life the time required for half of a sample of a radioactive isotope to break down by radioactive decay to form a daughter isotope

Half-Life

Radioactive decay happens at a relatively constant rate that is not changed by temperature, pressure, or other environmental conditions. Scientists have determined that the time required for half of any amount of a particular radioactive isotope to decay is always the same and can be determined for any isotope. Therefore, a **half-life** is the time it takes half the mass of a given amount of a radioactive isotope to decay into its daughter isotopes. If you began with 10 g of a parent isotope, you would have 5 g of that isotope after one half-life of that isotope. At the end of a second half-life, one-fourth, or 2.5 g, of the original isotope would remain. Three-fourths of the sample would now be the daughter isotope. This process is shown in **Figure 4.**

By comparing the amounts of parent and daughter isotopes in a rock sample, scientists can determine the age of the sample. The greater the percentage of daughter isotopes present in the sample, the older the rock is. But comparing parent to daughter isotopes works only when the sample has not gained or lost either parent or daughter isotopes through leaking or contamination.

QuickLAB 10 min

Radioactive Decay

Procedure

1. Use a **clock or watch that has a second hand** to record the time.
2. Wait 20 s, and then use **scissors** to carefully cut a **sheet of paper** in half. Select one piece, and set the other piece aside.
3. Repeat step 2 until nine 20 s intervals have elapsed.

Analysis

1. What does the whole piece of paper used in this investigation represent?
2. What do the pieces of paper that you set aside in each step represent?
3. What is the half-life of your paper isotope?
4. How much of your paper isotope was left after the first three intervals? after six intervals? after nine intervals? Express your answers as percentages.
5. What two factors in your model must remain constant so that your model is accurate? Explain your answer.

QuickLAB

Skills Acquired

- Constructing Models
- Observing
- Drawing Conclusions

Teacher's Notes You may want to be timekeeper and announce each 20 s interval. Or, you may have students work in pairs, with one student keeping time and the other cutting the paper.

Answers

1. The whole piece of paper represents the radioactive element, or parent isotope, before it begins to decay.
2. The pieces of paper set aside represent the fraction of the parent isotope that has decayed into daughter isotopes.
3. 20 seconds
4. 12.5%; 1.5625%; 0.1953125%
5. The length of the "half-life" interval and the fraction of the paper cut off each time (exactly 1/2) must remain constant.

Table 1 ▼

Radiometric Dating Methods				
Radiometric dating method	**Parent Isotope**	**Daughter isotope**	**Half-life**	**Effective dating range**
Radiocarbon dating	carbon-14, ^{14}C	nitrogen-14, 14N	5,730 years	less than 70,000 years
Argon-argon dating, ^{39}Ar/^{40}Ar	potassium-40, ^{40}K irradiated to form argon-39, ^{39}Ar	argon-40, ^{40}Ar	1.25 billion years	10,000 to 4.6 billion years
Potassium-argon dating, ^{40}K/^{40}Ar	potassium-40, ^{40}K	Argon-40, ^{40}Ar	1.25 billion years	50,000 to 4.6 billion years
Rubidium-strontium dating, ^{87}Rb/^{87}Sr	rubidium-87, ^{87}Rb	strontium-87, ^{87}Sr	48.8 billion years	10 million to 4.6 billion years
Uranium-lead dating, ^{235}U/^{207}Pb	uranium-235, ^{235}U	lead-207, ^{207}Pb	704 million years	10 million to 4.6 billion years
Uranium-lead dating, ^{238}U/^{206}Pb	uranium-238, ^{238}U	lead-206, ^{206}Pb	4.5 billion years	10 million to 4.6 billion years
Thorium-lead dating	thorium-232, ^{232}Th	lead-208, ^{208}Pb	14.0 billion years	greater than 200 million years

Radioactive Isotopes

The amount of time that has passed since a rock formed determines which radioactive element will give a more accurate age measurement. If too little time has passed since radioactive decay began, there may not be enough of the daughter isotope for accurate dating. If too much time has passed, there may not be enough of the parent isotope left for accurate dating.

Uranium-238, ^{238}U (which is read as "U two thirty-eight"), has an extremely long half-life of 4.5 billion years. ^{238}U is most useful for dating geologic samples that are more than 10 million years old, as long as they contain uranium. In addition to ^{238}U, several other radioactive isotopes are used to date rock samples. One such isotope is potassium-40, ^{40}K, which has a half-life of 1.25 billion years. ^{40}K occurs in mica, clay, and feldspar and is used to date rocks that are between 50,000 and 4.6 billion years old. Rubidium-87, ^{87}Rb, has a half-life of about 49 billion years. ^{87}Rb, which commonly occurs in minerals that contain ^{40}K, can be used to verify the age of rocks previously dated by using ^{40}K. **Table 1** provides a list of other radiometric dating methods.

SCILINKS. NSTA Developed and maintained by the National Science Teachers Association

For a variety of links related to this subject, go to www.scilinks.org

Topic: Radiometric Dating
SciLinks code: HQ61261

Reading Check How does the half-life of an isotope affect the accuracy of the radiometric dating method? (See the Appendix for answers to Reading Checks.)

ENVIRONMENTAL SCIENCE CONNECTION — ADVANCED

Radioactive Waste Discuss the fact that uranium used in nuclear power plants eventually becomes unusable, though still highly radioactive. This radioactive waste must be stored where it will not harm people or the environment. There have been proposals to store this waste under Yucca Mountain in Nevada, but some studies indicate that the area may be prone to earthquakes and the radioactivity may contaminate groundwater. Invite interested students to research the problem of radioactive waste storage and to debate the issue of nuclear power. **LS Logical**

Answer to Reading Check

An isotope that has an extremely long half-life will not show significant or measurable changes in a young rock. In a very old rock, an isotope that has a short half-life may have decayed to the point at which too little of the isotope is left to give an accurate age measurement. So, the estimated age of the rock must be correlated to the dating method used.

Using the Table — GENERAL

Methods for Dating Rocks Have students study the table on this page. Ask students to identify the parent isotope and its daughter isotope in each method. Have students determine which isotope should be used for dating the oldest rocks (those with half-lives of billions of years) and for dating the youngest rocks (those with shorter half-lives). Have students discuss why scientists must know the composition of a rock before choosing a method of dating. (Because different rocks contain different isotopes, scientists need to choose a dating method that matches the isotopes they expect to find in the rock.) **LS Logical**

Close

Reteaching — BASIC

Half-Life Provide groups of students with 40 pennies. Have the groups shake the pennies and then spill them in a large shallow box or tray on a table. Then, students should remove all of the coins that show "heads." Have them record this number. Have students repeat the process until all of the pennies are gone. Ask them to graph their data and to compare their data with the data from the figure that illustrates half-life on the previous page. Have them point out any inaccuracies of the model. (The data should roughly correlate with the figure. Students should recognize that heads come up statistically about half the time, so the pennies should be removed at a rate of approximately half of the pennies with each turn. In the case of the coins, each spill represents an approximate "half-life." However, the rate of radioactive decay is constant and exactly 50% of the parent material decays each half-life.) **LS Kinesthetic/Logical**

Close, *continued*

Quiz — GENERAL

1. What are radioactive isotopes? (forms of elements with nuclei that emit particles and energy at a constant rate, regardless of surrounding conditions)
2. What must be present in rock for scientists to date it using radiocarbon dating? (organic material that is less than 70,000 years old)

Alternative Assessment — GENERAL

Dating Rock Have students imagine that they have been given three rock samples to date: one sample contains ^{238}U; the second is a core from a lake bed that shows many sediment layers; the third is a young rock containing bits of bone. Have students write a description of the method they would use to date each sample and explain how the method works. **LS Logical/Verbal**

Answers to Section Review

1. Relative age is the age of an object in relation to the age of other objects. Absolute age is the numeric age of an object.
2. The rate at which sediment is deposited or eroded can vary.
3. Varves are layers of sediment that consist of one layer of sand covered by one layer of silt, that are deposited at the bottom of glacial lakes, and that represent one year of deposition.
4. By comparing the relative percentages of a radioactive (parent) isotope and a stable (daughter) isotope in a sample of rock, based on the known rate of decay of the parent, scientists can calculate the length of time since the rock formed.
5. Half-life is the time one half of a sample of a radioactive isotope takes to break down by radioactive decay to form a daughter substance.
6. Answers may vary but should reflect information from the table in this section.
7. You would use radiocarbon dating because the object is organic and because it is too young to be accurately dated by using ^{238}U.
8. Answers may vary. Accept all reasonable answers. Atomic clocks are not radioactive.
9. The *absolute age* of a rock can be determined by counting *varves* or by *radiometric dating,* in which the percentages of *parent isotopes* and *daughter isotopes* are compared, and which includes *carbon dating.*

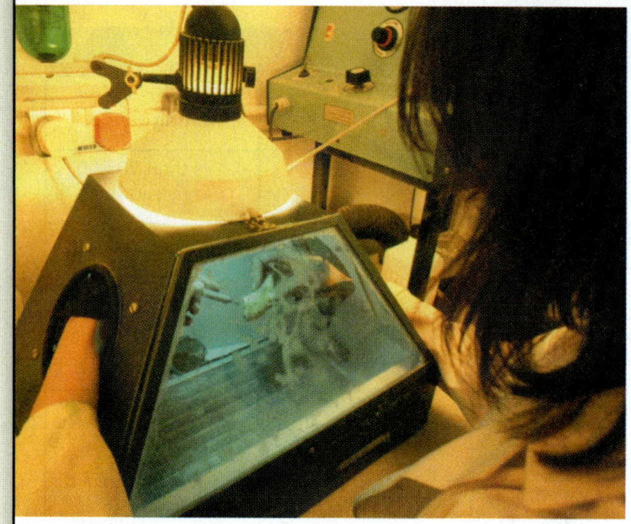

Figure 5 ▶ This scientist is preparing a mammoth's tooth for carbon-14 dating, which will determine the tooth's absolute age.

Carbon Dating

Younger rock layers may be dated indirectly by dating organic material found within the rock. The ages of wood, bones, shells, and other organic remains that are included in the layers and that are less than 70,000 years old can be determined by using a method known as *carbon-14 dating,* or *radiocarbon dating*, as shown in **Figure 5.** The isotope carbon-14, ^{14}C, combines with oxygen to form radioactive carbon dioxide, CO_2. Most CO_2 in the atmosphere contains nonradioactive carbon-12, ^{12}C. Only a small amount of CO_2 in the atmosphere contains ^{14}C.

Plants absorb CO_2, which contains either ^{12}C or ^{14}C during photosynthesis. Then, when animals eat the plants or the plant-eating animals, the ^{12}C and ^{14}C become part of the animals' body tissues. Thus, all living organisms contain both ^{12}C and ^{14}C.

To find the age of a small amount of organic material, scientists first determine the ratio of ^{14}C to ^{12}C in the sample. Then, they compare that ratio with the ratio of ^{14}C to ^{12}C known to exist in a living organism. While organisms are alive, the ratio of ^{12}C to ^{14}C remains relatively constant. When a plant or an animal dies, however, the ratio begins to change. The half-life of ^{14}C is only about 5,730 years. Because the organism is dead, it no longer absorbs ^{12}C and ^{14}C, and the amount of ^{14}C in the organism's tissues decreases steadily as the radioactive ^{14}C decays to nonradioactive nitrogen-14, ^{14}N.

Section 2 Review

1. **Differentiate** between relative and absolute age.
2. **Summarize** why calculations of absolute age based on rates of erosion and deposition can be inaccurate.
3. **Describe** varves, and describe how and where they form.
4. **Explain** how radiometric dating is used to estimate absolute age.
5. **Define** half-life, and explain how it helps determine an object's absolute age.
6. **List** three methods of radiometric dating, and explain the age range for which they are most effective.

CRITICAL THINKING

7. **Demonstrating Reasoned Judgment** Suppose you have a shark's tooth that you suspect is about 15,000 years old. Would you use ^{238}U or ^{14}C to date the tooth? Explain your answer.
8. **Making Inferences** You see an advertisement for an atomic clock. You also know that radioactive decay emits harmful radiation. Do you think the atomic clock contains decaying isotopes? Explain.

CONCEPT MAPPING

9. Use the following terms to create a concept map: *absolute age, varve, radiometric dating, parent isotope, daughter isotope,* and *carbon dating.*

CHAPTER RESOURCES

Chapter Resource File
- Section Quiz GENERAL

Workbooks
- Study Guide (also in Spanish)

Section 3 The Fossil Record

The remains of animals or plants that lived in a previous geologic time are called **fossils.** Fossils, such as the one shown in **Figure 1,** are an important source of information for finding the relative and absolute ages of rocks. Fossils also provide clues to past geologic events, climates, and the evolution of living things over time. The study of fossils is called **paleontology.**

Almost all fossils are discovered in sedimentary rock. The sediments that cover the fossils slow or stop the process of decay and protect the body of the dead organism from damage. Fossils are rarely discovered in igneous rock or metamorphic rock because intense heat, pressure, and chemical reactions that occur during the formation of these rock types destroy all organic structures.

OBJECTIVES

- **Describe** four ways in which entire organisms can be preserved as fossils.
- **List** five examples of fossilized traces of organisms.
- **Describe** how index fossils can be used to determine the age of rocks.

KEY TERMS

fossil
paleontology
trace fossil
index fossil

Interpreting the Fossil Record

The fossil record provides information about the geologic history of Earth. By revealing the ways that organisms have changed throughout the geologic past, fossils provide important clues to the environmental changes that occurred in Earth's past. For example, fossils of marine animals and plants have been discovered in areas far from any ocean. These fossils tell us that such areas were covered by an ocean in the past. Scientists can use this information to learn about how environmental changes have affected living organisms.

fossil the trace or remains of an organism that lived long ago, most commonly preserved in sedimentary rock

paleontology the scientific study of fossils

SCILINKS. NSTA Developed and maintained by the National Science Teachers Association

For a variety of links related to this subject, go to www.scilinks.org

Topic: Fossil Record
SciLinks code: HQ60615

Figure 1 ▶ Paleontologists are unearthing the remains of rhinoceroses that are 10 million years old at this site in Orchard, Nebraska.

Section 3

Focus

Overview

This section describes the fossil record and how it is interpreted. The section also covers fossil formation and the different types of fossils, and describes index fossils and their relation to the absolute age of specific rock layers.

Bellringer

Have students draw or describe the process by which they think fossils form. (Answers and drawings may vary. You may use students' answers to begin a discussion of how fossils form.) LS Logical

Motivate

Demonstration — GENERAL

Learning Earth's History Before students read this page, show them a piece of limestone with fossil shells imbedded in it or a fossil of a fish from a place that is dry land today. Ask students what the fossil reveals about the history of the area where it was found. (that the area was once underwater) Have students discuss how fossils help scientists understand the geologic history of Earth. LS Logical/Visual

CHAPTER RESOURCES

Chapter Resource File

- **Directed Reading** BASIC
- **Inquiry Lab** Got Fossils? GENERAL

Technology

Transparencies
- Bellringer

Student Edition on CD-ROM

One-Stop Planner CD-ROM
- Lesson Plan

Teach

Group Activity — GENERAL

Fossils in Many Forms Divide the class into five groups, and assign one of the five ways fossils form to each group. Have some students in each group use the Internet or library resources to collect information about how their particular type of fossil forms and under what conditions. They may also research where specimens of their specific type of fossil have been found. Ask other students to research the types of organisms most commonly preserved in that particular way. Have other students collect pictures of the fossilized organisms and use them to create a poster or to make models, if possible. (Note: Students working on mummification should concentrate on natural preservation, not artificial mummification.) Have students work together in their groups to write an oral presentation of their findings for the class.
LS Visual/Verbal Co-op Learning

READING SKILL BUILDER — BASIC

Paired Summarizing Group students into pairs, and have them read silently about each way that fossils form that is listed in the table. Then, have one student summarize one process of fossilization, while the other student listens and points out any inaccuracies or ideas that were left out. Students take turns summarizing and listening. Allow students to refer to the text as needed.
LS Verbal/Auditory English Language Learners Co-op Learning

Fossilization

Normally, dead plants and animals are eaten by other animals or decomposed by bacteria. If left unprotected, even hard parts such as bones decay and leave no trace of the organism. Only dead organisms that are buried quickly or protected from decay can become fossils. Generally, only the hard parts of organisms, such as wood, bones, shells, and teeth, become fossils. In rare cases, an entire organism may be preserved. In some types of fossils, only a replica of the original organism remains. Other fossils merely provide evidence that life once existed. **Table 1** describes different ways that fossils can form.

Table 1 ▼

How Fossils Form
Mummification Mummified remains are often found in very dry places, because most bacteria, which cause decay, cannot survive in these places. Some ancient civilizations mummified their dead by carefully extracting the body's internal organs and then wrapping the body in carefully prepared strips of cloth. 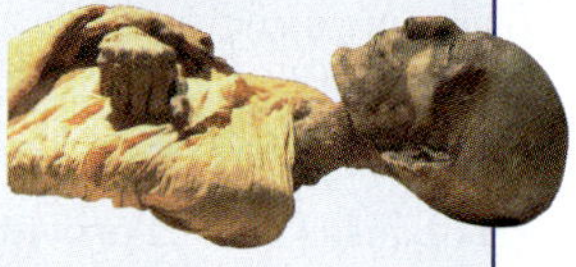
 **Amber** Hardened tree sap is called *amber*. Insects become trapped in the sticky sap and are preserved when the sap hardens. In many cases, delicate features such as legs and antennae have been preserved. In rare cases, DNA has been recovered from amber.
Tar Seeps When thick petroleum oozes to Earth's surface, the petroleum forms a tar seep. Tar seeps are commonly covered by water. Animals that come to drink the water can become trapped in the sticky tar. Other animals prey on the trapped animals and can also become trapped. The remains of the trapped animals are covered by the tar and preserved. 
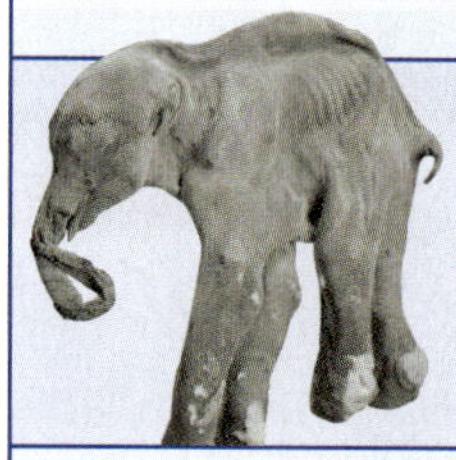 **Freezing** The low temperatures of frozen soil and ice can protect and preserve organisms. Because most bacteria cannot survive freezing temperatures, organisms that are buried in frozen soil or ice do not decay.
Petrification Mineral solutions such as groundwater replace the original organic materials that were covered by layers of sediment with new materials. Some common petrifying minerals are silica, calcite, and pyrite. The substitution of minerals for organic material often results in the formation of a nearly perfect mineral replica of the original organism. 

INCLUSION Strategies

- Hearing Impaired
- Learning Disabled
- Developmentally Delayed

Have the class read the mummification information in the table on this page as a group. Discuss the information and then ask each student to write one sentence—in his or her own words—to explain mummification and fossils. Repeat with each of the other four rows of the table. LS Auditory

Discussion — GENERAL

Handprint History Have students imagine that every year, on their birthday, they pour a layer of cement into a bucket and make a hand print in it. Ask students what the layers of cement record. (Students should recognize that the layers record how their hand changes over time.) If, in 1,000 years, someone could examine the layers, what changes would they note in each imprint? (The handprints grew for a while, then stayed the same for a long time, and finally shrank or became crooked as the maker of the imprint aged.) LS Logical

Table 2 ▼

Types of Fossils

Imprints Carbonized imprints of leaves, stems, flowers, and fish made in soft mud or clay have been found preserved in sedimentary rock. When original organic material partially decays, it leaves behind a carbon-rich film. An imprint displays the surface features of the organism.

Molds and Casts Shells often leave empty cavities called *molds* within hardened sediment. When a shell is buried, its remains eventually decay and leave an empty space. When sand or mud fills a mold and hardens, a natural cast forms. A cast is a replica of the original organism.

Coprolites Fossilized dung or waste materials from ancient animals are called *coprolites.* They can be cut into thin sections and observed through a microscope. The materials identified in these sections reveal the feeding habits of ancient animals, such as dinosaurs.

Gastroliths Some dinosaurs had stones in their digestive systems to help grind their food. In many cases, these stones, which are called *gastroliths,* survive as fossils. Gastroliths can often be recognized by their smooth, polished surfaces and by their close proximity to dinosaur remains.

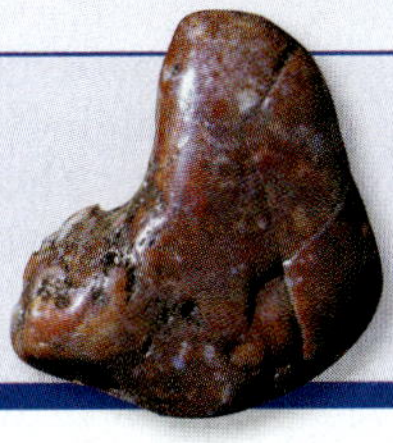

Types of Fossils

Fossils can show a remarkable amount of detail about ancient organisms. **Table 2** describes some of these fossils. In some cases, no part of the original organism survives in fossil form. But **trace fossils,** or fossilized evidence of past animal movement such as tracks, footprints, borings, and burrows, can still provide information about prehistoric life.

A trace fossil, such as the footprint of an animal, is an important clue to the animal's appearance and activities. Suppose a giant dinosaur left deep footprints in soft mud. Sand or silt may have blown or washed into the footprint so gently that the footprints remained intact. Then, more sediment may have been deposited over the prints. As time passed, the mud containing the footprints hardened into sedimentary rock and preserved the footprints. Scientists have discovered ancient footprints of reptiles, amphibians, birds, and mammals.

Reading Check **What is a trace fossil? (See the Appendix for answers to Reading Checks.)**

trace fossil a fossilized mark that formed in sedimentary rock by the movement of an animal on or within soft sediment

Activity — GENERAL

Spacing Footprints The distance between fossilized footprints can tell paleontologists how fast an ancient animal moved. Invite a few students to walk at a normal pace from one end of the classroom to the other, while other students make a chalk mark or drop a marker next to each footstep. Then, have the students run from one end of the room to the other, again marking their footsteps. Have students measure and record the distance between each set of footsteps. Ask students to draw a conclusion based on the data they've collected. (Students should conclude that the faster the movement, the farther apart the footprints are.) **LS Kinesthetic**

Close

Reteaching — BASIC

Listing Make three columns on the board, and label the columns with the following heads: How Fossils Form, Types of Fossils, and Index Fossils. Have students review the text and tell you under what heads to write the information as they read it. At the end of the chapter, have students look at the three lists and describe the differences between the categories and the usefulness of each category to scientists. **LS Verbal** English Language Learners

BIOLOGY CONNECTION — ADVANCED

Dinosaurs and Birds Ask students which type of modern animal routinely consumes small pebbles that it stores in its gizzard to aid digestion. (many types of birds) The finding of gastroliths suggests that dinosaurs did this too. In fact, paleontologists cite gastroliths as evidence of an evolutionary connection between dinosaurs and birds. Invite interested students to research other evidence for the connection between birds and dinosaurs. Have them make a poster that illustrates their findings. **LS Logical/Visual**

Answer to Reading Check

A trace fossil is fossilized evidence of past animal movement, such as tracks, footprints, borings, or burrows, that can provide information about prehistoric life.

Close, *continued*

Quiz — GENERAL

1. Why must an index fossil be clearly distinguishable from other fossils? (so that it is never confused with a fossil that may have lived in a different time period)
2. Why are index fossils useful for dating rocks? (They lived for only short spans of geologic time and exist over a large area.)
3. What is paleontology? (It is the scientific study of fossils.)

Alternative Assessment — BASIC

Captioned Pictures Ask students to create an illustrated booklet about how fossils form, types of fossils, and index fossils. Students may draw or find pictures of each category and should write an informative caption next to each picture. LS **Visual/Verbal**

Answers to Section Review

1. An entire organism may be preserved as a fossil if it is mummified, trapped in amber or in tar seeps, frozen, or petrified.
2. Four types of fossils that can be used to provide indirect evidence of organisms are imprints, molds and casts, coprolites, and gastroliths.
3. When scientists know the time interval during which a specific organism lived, they can use the presence of those fossils to date sedimentary rock layers that contain those fossils.
4. Mummification occurs in very dry regions, where bacteria cannot decay the body. Thus, much of it remains unchanged. Petrification occurs in areas where mineral solutions replace the buried organic materials with new materials, such as silica and calcite.
5. Index fossils are fossils of distinct organisms that lived during short spans of geologic time. Any rocks that contain index fossils can be given an absolute age.
6. The environment preserved the dead organisms from decay.
7. You know that the rock layers are the same age because the index fossil they contain lived only during a short span of time. The index fossil gives you the absolute age of the layers.
8. *Fossils* form when an organism is preserved through *mummification, freezing,* or *petrification* or is trapped in *amber* or in a *tar seep*.

Figure 2 ▶ If you discovered fossils such as these ammonites, you would know that the surrounding rock formed between 180 million and 206 million years ago.

Index Fossils

Paleontologists can use fossils to determine the relative ages of the rock layers in which the fossils are located. Fossils that occur only in rock layers of a particular geologic age are called **index fossils.** To be an index fossil, a fossil must meet certain requirements. First, it must be present in rocks scattered over a large region. Second, it must have features that clearly distinguish it from other fossils. Third, the organisms from which the fossil formed must have lived during a short span of geologic time. Fourth, the fossil must occur in fairly large numbers within the rock layers.

index fossil a fossil that is used to establish the age of rock layers because it is distinct, abundant, and widespread and existed for only a short span of geologic time

Index Fossils and Absolute Age

Scientists can use index fossils to estimate absolute ages of specific rock layers. Because organisms that formed index fossils lived during short spans of geologic time, the rock layer in which an index fossil was discovered can be dated accurately.

The ammonite fossils in **Figure 2** show that the rock in which the fossils were observed formed between 180 million and 206 million years ago. Scientists can also use index fossils to date rock layers in separate areas. So, an index fossil discovered in rock layers in different areas of the world indicates that the rock layers in these areas formed during the same time period.

Geologists also use index fossils to help locate rock layers that are likely to contain oil and natural gas deposits. These deposits form from plant and animal remains that change by chemical processes over millions of years.

Section 3 Review

1. **Describe** four ways in which an entire organism can be preserved as a fossil.
2. **List** four types of fossils that can be used to provide indirect evidence of organisms.
3. **Explain** how geologists use fossils to date sedimentary rock layers.
4. **Compare** the process of mummification with the process of petrification.
5. **Describe** how index fossils can be used to determine the age of rocks.

CRITICAL THINKING

6. **Applying Ideas** What two characteristics do all good sources of animal fossils have in common?
7. **Identifying Relationships** If a rock layer in Mexico and a rock layer in Australia contain the same index fossil, what do you know about the absolute ages of the layers in both places? Explain your answer.

CONCEPT MAPPING

8. Use the following terms to create a concept map: *fossil, mummification, amber, tar seep, freezing,* and *petrification.*

CHAPTER RESOURCES

Chapter Resource File

- Section Quiz GENERAL

Workbooks

- Study Guide (also in Spanish)

Chapter 8 Highlights

Sections

1 Determining Relative Age

Key Terms

uniformitarianism, 185
relative age, 186
law of superposition, 187
unconformity, 189
law of crosscutting relationships, 190

Key Concepts

- According to the principle of uniformitarianism, the forces that are changing Earth's surface today are the same forces that changed Earth's surface in the past.
- Scientists use the law of superposition to determine the relative ages of rock layers.
- Nonconformities, angular unconformities, and disconformities are interruptions in the sequence of rock layers and are collectively known as unconformities.
- Scientists use the law of crosscutting relationships to determine the relative ages of rock layers.

2 Determining Absolute Age

absolute age, 191
varve, 192
radiometric dating, 193
half-life, 194

- Absolute age is the numeric age of an object given in years.
- Because they can change over time, the rates of erosion and deposition are imprecise methods for determining absolute age of rocks.
- Varves are layers of sediment that form in glacial lakes as a result of the annual cycle of freezing and thawing of the glacier.
- Radioactive elements decay at constant and measurable rates and can be used to determine absolute age.

3 The Fossil Record

fossil, 197
paleontology, 197
trace fossil, 199
index fossil, 200

- Entire organisms may be preserved in amber, in tar seeps, or through freezing or mummification.
- Fossilized evidence of organisms includes trace fossils, imprints, molds, casts, coprolites, and gastroliths.
- Index fossils occur only in rock layers of a particular geologic age.

Chapter Highlights

Alternative Assessment — GENERAL

Modeling Rock Formations

Divide the class into groups of 5 or 6 students. Provide each group with different color clays, sufficient for them to model different types of rock formations. Provide them also with small objects, such as beans, that they can embed within layers as "fossils." Students should create their rock formations in a shallow, clear glass or plastic container. Have students work together to create the different rock formations they read about in the text, ensuring that, for instance, different color clays represent rocks of different ages. Students should be able to explain the types of rock formations they created and the relative ages of each rock stratum or formation. They should be able to explain how their "fossils" came to be preserved in the particular rock layer or formation in which the students placed them. LS Kinesthetic English Language Learners

CHAPTER RESOURCES

Chapter Resource File

- Concept Review GENERAL
- Critical Thinking ADVANCED
- Math Skills GENERAL
- Graphing Skills GENERAL
- Chapter Test A GENERAL
- Chapter Test B ADVANCED

Workbooks

- **Study Guide** (also in Spanish)
- **Assessments** (Spanish)

Technology

Classroom Videos
- Brain Food Video Quiz

HRW Earth Science Video
- Segment 12: Determining Relative Age
- Segment 13: Field Trip—Understanding Rock Layers

Chapter Review

Assignment Guide	
Section	**Questions**
1	1, 6, 9–10, 12, 15–17, 22–24, 31, 33–35
2	2–4, 8, 11, 18–19, 25, 28–30, 32
3	7, 13–14, 20–21, 26
1 and 2	5
1–3	27

Using Key Terms

1–7. Answers may vary but should show that students understand the definitions of and differences between key terms.

Understanding Key Concepts

8. b **9.** b
10. c **11.** d
12. c **13.** b
14. a

Short Answer

15. He noticed geologic changes occurring on his farm and assumed the same forces had changed Earth's surface in the past.

16. The law of superposition helps scientists determine rocks' relative ages because older rocks are at the bottom of undisturbed strata and younger rocks are at the top.

17. All are breaks in the geologic record. A nonconformity is the boundary between new sedimentary rock on top of uplifted and eroded igneous or metamorphic rock. An angular unconformity occurs when horizontal rock layers are tilted and eroded. The layers below the unconformity are not parallel to rock layers above the unconformity. A disconformity is an unconformity in which the layers below the unconformity are parallel to the layers above the unconformity.

18. Scientists know the half-life (rate of decay) of radioactive isotopes. By comparing the relative percentages of a radioactive (parent) isotope in a rock and a stable (daughter) isotope they can calculate the time at which the rock formed.

19. Three other methods of estimating absolute age are using rates of erosion, using rates of deposition, and counting varves.

20. Organisms can be preserved by mummification, or drying; by being trapped in tree sap, which hardens into amber; by being trapped and preserved in a tar seep; by freezing in ice or soil; or by petrification, a process in which minerals in groundwater replace original organic material and solidify into rock.

21. An index fossil must occur in rocks spread over a wide area; it must have features that distinguish it clearly from other fossils; it must have existed during a relatively short geologic time span; and it must occur in large numbers.

Critical Thinking

22. Answers may vary. Sample answer: He may have observed erosion and deposition over the years on his farm. He might also have noticed rock strata and concluded that they resembled layers of sediment that he saw building up through deposition.

Chapter 8 Review

Using Key Terms

Use each of the following terms in a separate sentence.

1. *uniformitarianism*
2. *varve*
3. *radiometric dating*
4. *half-life*

For each pair of terms, explain how the meanings of the terms differ.

5. *relative age* and *absolute age*
6. *law of superposition* and *law of crosscutting relationships*
7. *trace fossil* and *index fossil*

Understanding Key Concepts

8. Varves are layers of
a. limestone mixed with coarse sediments.
b. alternating coarse and fine sediments.
c. fossils.
d. sediments that have gaps that represent missing time in the rock sequence.

9. An unconformity that results when new sediments are deposited on eroded horizontal layers is a(n)
a. angular unconformity.
b. disconformity.
c. crosscut unconformity.
d. nonconformity.

10. A fault or intrusion is younger than the rock it cuts through, according to the
a. type of unconformity.
b. law of superposition.
c. law of crosscutting relationships.
d. principle of uniformitarianism.

11. The age of a rock in years is the rock's numerical age, or
a. index age. **c.** half-life age.
b. relative age. **d.** absolute age.

12. A gap in the sequence of rock layers is a(n)
a. bedding plane.
b. varve.
c. unconformity.
d. uniformity.

13. The process by which the remains of an organism are preserved by drying is called
a. petrification. **c.** erosion.
b. mummification. **d.** superposition.

14. Molds that fill with sediment sometimes produce
a. casts. **c.** coprolites.
b. gastroliths. **d.** imprints.

Short Answer

15. What prompted James Hutton to formulate the principle of uniformitarianism?

16. Describe how the law of superposition helps scientists determine relative age.

17. Compare and contrast the three types of unconformities.

18. How do scientists use radioactive decay to determine absolute age?

19. Besides radiometric dating, what are three other methods of estimating absolute age?

20. List and describe five ways that organisms can be preserved.

21. List the four characteristics that define an index fossil.

Critical Thinking

22. Making Inferences James Hutton developed the principle of uniformitarianism by observing geologic changes on his farm. What changes might he have observed?

23. Applying Ideas How might a scientist determine the original positions of the sedimentary layers beneath an angular unconformity?

24. Analyzing Relationships One intrusion cuts through all the rock layers. Another intrusion is eroded and lies beneath several layers of sedimentary rock. Which intrusion is younger? Explain your answer.

25. Analyzing Concepts A fossil that has unusual features is found in many areas on Earth. It represents a brief period of geologic time but occurs in small numbers. Would this fossil make a good index fossil? Explain.

26. Making Comparisons Compare the processes of mummification and freezing.

Concept Mapping

27. Use the following terms to create a concept map: *relative age, law of superposition, unconformity, law of crosscutting relationships, absolute age, radiometric dating, carbon dating,* and *index fossil.*

Math Skills

28. Making Calculations Scientists know that from a million grams of ^{238}U, 1/7,600 g of ^{206}Pb per year will be produced by radioactive decay. How many grams of ^{238}U would be left after 1 million years?

29. Applying Quantities A sample contains 1,000 g of an isotope that has a half-life of 500 years. How many half-lives will have to pass before the sample contains less than 10 g of the parent isotope?

30. Making Calculations The half-life of ^{238}U is 4.5 billion years. How many years would 16 g of ^{238}U take to decay into 0.5 g of ^{238}U and 15.5 g of daughter products?

Writing Skills

31. Writing Persuasively Imagine that you are James Hutton. Write a letter to a fellow scientist to convince him or her of the validity of the principle of uniformitarianism.

32. Outlining Topics Describe what happens to the amount of an isotope as it undergoes radioactive decay through three half-lives.

Interpreting Graphics

The illustration below shows crosscutting relationships in an outcrop of rock. Use this illustration to answer the questions that follow.

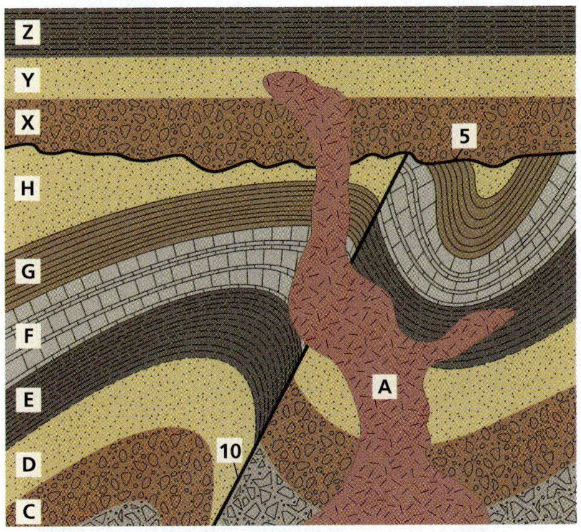

33. Is intrusion A older or younger than fault 10? Explain your answer.

34. What type of unconformity does feature 5 represent? Explain your answer.

35. Which rock formation is older: layer X or layer Y? Explain your answer.

Chapter Review

23. Answers may vary but should state that scientists know that the rocks were originally deposited in horizontal layers and that structures such as cross-beds, ripple marks, and graded bedding can help them identify the original orientation of the layers.

24. The intrusion that cuts through all the rock layers is younger according to the law of crosscutting relationships. The other intrusion is older because it has eroded and more layers have been deposited above it.

25. Answers may vary. Sample answer: If the fossil was rare and could not be found easily in rock in different areas, it would not make a good index fossil.

26. Mummification and freezing are similar in that in both the whole body is preserved and does not decay because bacteria cannot decompose the body. However, mummification tends to occur in hot, dry regions, and freezing occurs in cold, dry regions.

Concept Mapping

27. Answers may vary but should include all of the terms listed. Sample answers appear at the end of this Teacher Edition.

Math Skills

28. 1/7,600 g/y × 1,000,000 y = 1,000,000 y/7,600 y/g = 131.58 g of ^{238}U has decayed; 1,000,000 g − 131.58 g = 999,868.42 g of ^{238}U remains

29. 7 half-lives or 3,500 years

30. Five half-lives, or 5 × 4.5 billion years = 22.5 billion years

Writing Skills

31. Answers may vary. Accept all reasonable answers.

32. Answers may vary. Accept all reasonable answers.

Interpreting Graphics

33. The intrusion is younger because it cuts through the fault.

34. angular unconformity; The layers below the unconformity are folded and are tilted where they are cut by the unconformity.

35. Layer X is older, because layer Y is on top of it.

Estimated Time

To give students practice under more realistic testing conditions, allow them 30 minutes to answer all of the questions in this practice test.

TEST DOCTOR

Question 2 Answer F is correct. To answer correctly, students must recall the difference between the different types of fossils. Direct evidence of dietary habits can be found in coprolites—the fossilized dung from ancient animals.

Question 3 Answer B is correct. To answer correctly, students must recall the difference between absolute age and relative age. Answers A, C, and D are incorrect because they focus on finding the relative age of rock, not the absolute age of rock.

Question 10 Full-credit answers should include the following points:

- fossils that include the soft parts of animals are rare and may include impressions of organs or muscles
- scientists can use these animal parts to learn more about the internal structures and body systems of ancient animals
- scientists can compare the internal systems of ancient animals to the internal systems of modern animals in order to see how different animals and body systems have changed over time

Chapter 8 Standardized Test Prep

Understanding Concepts

Directions (1–5): **For *each* question, write on a separate sheet of paper the letter of the correct answer.**

1 A scientist used radiometric dating during an investigation. The scientist used this method because he or she wanted to determine the
A. relative age of rocks
B. absolute age of rocks
C. climate of a past era
D. fossil types in a rock

2 Fossils that provide direct evidence of the feeding habits of ancient animals are known as
F. coprolites
G. molds and casts
H. imprints
I. trace fossils

3 One way to estimate the absolute age of rock is
A. nonconformity
B. varve count
C. the law of superposition
D. the law of crosscutting relationships

4 To be an index fossil, a fossil must
F. be present in rocks that are scattered over a small geographic area
G. contain remains of organisms that lived for a long period of geologic time
H. occur in small numbers within the rock layers
I. have features that clearly distinguish it from other fossils

5 Which of the following statements best describes the relationship between the law of superposition and the principle of original horizontality?
A. Both describe the deposition of sediments in horizontal layers.
B. Both conclude that Earth is more than 100,000 years old.
C. Both indicate the absolute age of layers of rock.
D. Both recognize that the geologic processes in the past are the same as those at work now.

Directions (6): **For *each* question, write a short response.**

6 What is the name for a type of fossil that can be used to establish the age of rock?

Reading Skills

Directions (7–10): **Read the passage below. Then, answer the questions on a separate sheet of paper.**

Illinois Nodules

Around three hundred million years ago, the region that is now Illinois had a very different climate. Swamps and marshes covered much of the area. Scientists estimate that no fewer than 500 species lived in this ancient environment. Today, the remains of these organisms are found preserved within structures known as nodules. Nodules are round or oblong structures that are usually composed of cemented sediments. Sometimes, these nodules contain the fossilized hard parts of plants and animals. The Illinois nodules are extremely rare because many contain finely detailed impressions of the soft parts of the organisms together with the hard parts. Because they are rare, these nodules are desired for their incredible scientific value and may be found in fossil collections around the world.

7 According to the passage above, which of the following statements about nodules is correct?
A. Nodules are rarely round or oblong.
B. Nodules are usually composed of cemented sediments.
C. Nodules are rarely found outside of Illinois.
D. Nodules will always contain fossils.

8 What is the most unusual feature of the nodules found in modern-day Illinois?
F. their bright coloration
G. the fact that they come in many more unusual shapes that other nodules
H. the fact that they contain both the soft and hard parts of animals
I. their extremely heavy weight

9 Which of the following statements can be inferred from the information in the passage?
A. Illinois nodules are sought by scientists.
B. Nodules can be purchased from the state.
C. Similar nodules can be found in nearby Iowa.
D. Nodules contain dinosaur fossils.

10 What might scientists learn from nodules that contains the soft and hard parts of an animal?

Answers

Understanding Concepts

1. B
2. F
3. B
4. I
5. A
6. index fossil

Reading Skills

7. B
8. H
9. A
10. Answers may vary. See Test Doctor for a detailed scoring rubric.

Interpreting Graphics

11. B
12. D
13. Answers may vary. See Test Doctor for a detailed scoring rubric.

Interpreting Graphics

Directions (11–13): **For *each* question below, record the correct answer on a separate sheet of paper.**

The graph below shows the rate of radioactive decay. Use this graph to answer question 11.

Radioactive Decay

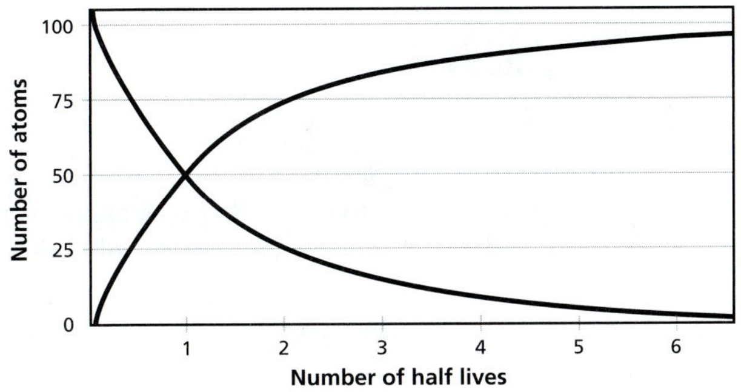

11 How many half-lives have passed when the number of daughter atoms is approximately three times the number of parent atoms?

A. one
B. two
C. three
D. four

The diagram below shows crosscutting taking place in layers of rock. Use this diagram to answer questions 12 and 13.

Layers of Rock with a Crosscutting Fault

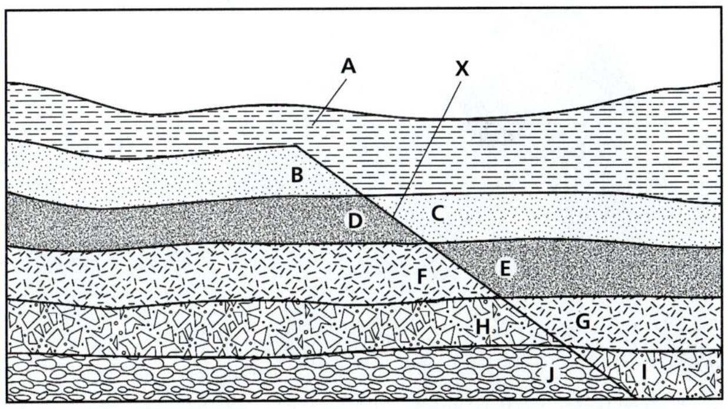

12 Which of the letter combinations below belonged to the same layer of rock before the fault disrupted the layer?

A. C and D
B. C and F
C. G and I
D. G and F

13 Which is older, structure B or structure X? Explain your answer. What structure shown on the diagram is the youngest?

Test TIP

If time permits, take short mental breaks to improve your concentration during a test.

TEST DOCTOR

Question 11 Answer B is correct. Students should be able to apply their own knowledge of how half-lives work to the chart to determine that it will take 2 half-lives for the number of daughter atoms to be 3 times the number of parent atoms. As the chart shows, after one half-life the amounts of daughter and parent atoms will be equal. After two half-lives the parent atoms atoms will have lost 75% of their original total, making up only 25% of the sample.

Question 13 Full-credit answers should include the following points:

- structure X is a simple fault—which by definition is younger than the rock it cuts through
- rock layer B must have formed before fault X occurred
- rock layer A is the youngest structure shown on the diagram. The unbroken layer on top is the youngest structure shown in the diagram. This layer must have formed after the fault. If it had formed before the fault, it would be broken in the same way that the other rock layers were broken

Test Prep Correlations — *National Science Education Standards*

ES 3b: items 1, 3, 5, 6, 11, 12, 13
HNS 2a: items 1, 2, 3, 4, 5, 6, 9, 10, 12, 13
HNS 2b: items 9, 10, 13
LS 3c: items 2, 4, 6
PS 1d: items 1, 11
SAI 2c: items 1, 11
ES 3b: items 1, 11

CHAPTER RESOURCES

State Resources

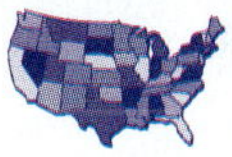

For specific resources for your state, visit **go.hrw.com** and type in the keyword **HSHSTR.**

Types of Fossils

Teacher's Notes

Time Required

one 45-minute class period

Lab Ratings

EASY ——— HARD

Teacher Preparation ▲▲
Student Setup ▲▲
Concept Level ▲▲
Cleanup ▲▲▲

Skills Acquired

- Constructing Models
- Classifying
- Interpreting Models

The Scientific Method

In this lab, students will

- Make Observations
- Analyze the Results
- Draw Conclusions

Materials

The materials listed on the page are enough for groups of two to four students. You may have your first-period students leave all fresh materials prepared for the rest of your classes.

Tips and Tricks

To save time, you may wish to divide the class into three groups, with each group making only one type of fossil and then displaying and discussing it with the whole class. Or, you may want to prepare a batch of plaster of Paris for the whole class to use while the students are working on the first part of the lab. Keep the bucket of plaster moist and covered until the students are ready to use it.

Objectives

- USING SCIENTIFIC METHODS **Model** the way different types of fossils form.
- **Demonstrate** how certain types of fossils form.

Materials

- clay, modeling
- container, plastic
- hard objects such as a shell, key, paper clip, or coin
- leaf
- newspaper
- paper, carbon, soft
- paper, white (1 sheet)
- pencil (or wood dowel)
- plaster of Paris
- spoon, plastic
- tweezers
- water
- wax paper

Safety

Types of Fossils

Paleontologists study fossils to find evidence of the kinds of life and conditions that existed on Earth in the past. Fossils are the remains of ancient plants and animals or evidence of their presence. In this lab, you will use various methods to make models of fossils.

PROCEDURE

1. Place a ball of modeling clay on a flat surface that is covered with wax paper.
2. Press the clay down to form a flat disk about 8 cm in diameter. Turn the clay over so that the smooth, flat surface is facing up.
3. Choose a small, hard object. Press the object onto the clay carefully so that you do not disturb the indentation. Is the indentation left by the object a mold or a cast? What features of the object are best shown in the indentation? Sketch the indentation.
4. On a second piece of smooth, flat clay, make a shallow imprint to represent the burrow or footprint of an animal. Sketch your fossil imprint.

Step 3

CHAPTER RESOURCES

Chapter Resource File

- Datasheet for Chapter Lab GENERAL
- Lab Notes and Answers

5. Fill a plastic container with water to a depth of 1 to 2 cm. Stir in enough plaster of Paris to make a paste that has the consistency of whipped cream.

6. Using the plastic spoon, fill both indentations with plaster. Allow excess plaster to run over the edges of the imprints. Let the plaster set for about 15 minutes until it hardens.

7. After the plaster has hardened, remove both pieces of plaster from the clay. Do the pieces of hardened plaster represent molds or casts?

8. Place the carbon paper on a flat surface with the carbon facing up. Gently place the leaf on the carbon paper, and cover it with several sheets of newspaper. Roll the pencil or wooden dowel back and forth across the surface of the newspaper several times, and press firmly to bring the leaf into full contact with the carbon paper.

9. Remove the newspaper. Lift the leaf by using the tweezers, and place it on a clean sheet of paper with the carbon-coated side facing down. Cover the leaf with clean wax paper, and roll your pencil across the surface of the wax paper.

10. Remove the wax paper and leaf. Observe and describe the carbon print left by the leaf.

Step 6

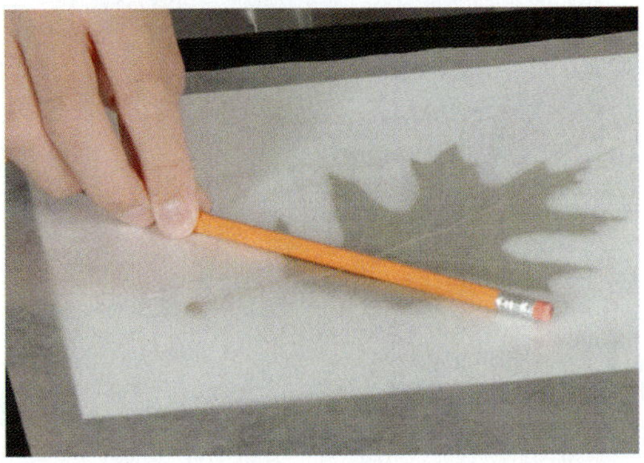

Step 9

ANALYSIS AND CONCLUSION

1. **Analyzing Results** Look at the molds and casts made by others in your class. Identify as many of the objects used to make the molds and casts as you can.

2. **Making Comparisons** How does the carbon print you made differ from an actual carbonized imprint fossil?

3. **Applying Ideas** Trace fossils are evidence of the movement of an animal on or within soft sediment. Why are imprints, molds, and casts not considered trace fossils?

Extension

1. **Making Predictions** Which organism—a rabbit, a housefly, an earthworm, or a snail—would be most likely to form fossils? Which of the organisms would leave trace fossils? Explain.

Making Models Lab

Answers to Procedure

3. The indentation left by the object is a mold.
7. The pieces of hardened plaster represent casts.

Answers to Analysis and Conclusion

1. Student responses may vary depending on the objects used, but students should be encouraged to notice details in the molds and casts that give clues to the identity of the object.
2. An actual carbonized imprint fossil is made from the carbon contained in the organism itself.
3. Imprints, molds, and casts are not trace fossils because they are permanent remains of an organism formed after the organism has died. Trace fossils are marks left while the organism was alive and moving.

Answers to Extension

1. Answers may vary, but students should recognize that an organism with a hard shell, like the snail, or one with bones, like the rabbit, would be most likely to form a fossil. The fly might leave a fossil in amber. All of the organisms may leave trace fossils in soft sediment. The earthworm is most likely to leave a trace fossil of its movement through soil.

CLASSROOM TESTED & APPROVED

Mrs. Teresa Tucker
Northwest High School
Jackson, MI

Geologic Map of Bedrock in Ohio

Group Activity —— Advanced

Map Making Have groups of students make maps of your state's bedrock geology, using the map on this page as a model. Student maps need not be highly detailed. If your state has a complex bedrock geology, tell students to map the most prominent or abundant types of bedrock, or have them concentrate on the bedrock beneath your county or region. **LS Visual**

Answers to Map Skills Activity

1. the Permian, Pennsylvanian, Mississippian, Devonian, Silurian, and Ordovician
2. The youngest bedrock is in the southeastern part of the state; the oldest bedrock is in the southwestern part of the state.
3. In traveling east to west, the bedrock would generally become older.
4. The bedrock layers are tilted upwards. The more horizontal surface of Earth cuts across the angled layers and thus exposes them. If the layers were horizontal, only the top, or youngest, layer would be exposed.
5. because Mississippian rock was formed in the Mississippian Period just after the rock formed in the Devonian Period
6. To find early reptile fossils, you'd look in Pennsylvanian rock, represented by the color blue, which occurs in a large stripe extending to the southwest from eastern Ohio.

CHAPTER RESOURCES

Technology

Transparencies
- 43 Geologic Map of Bedrock in Ohio (with worksheet)

MAPS in Action

Geologic Map of Bedrock in Ohio

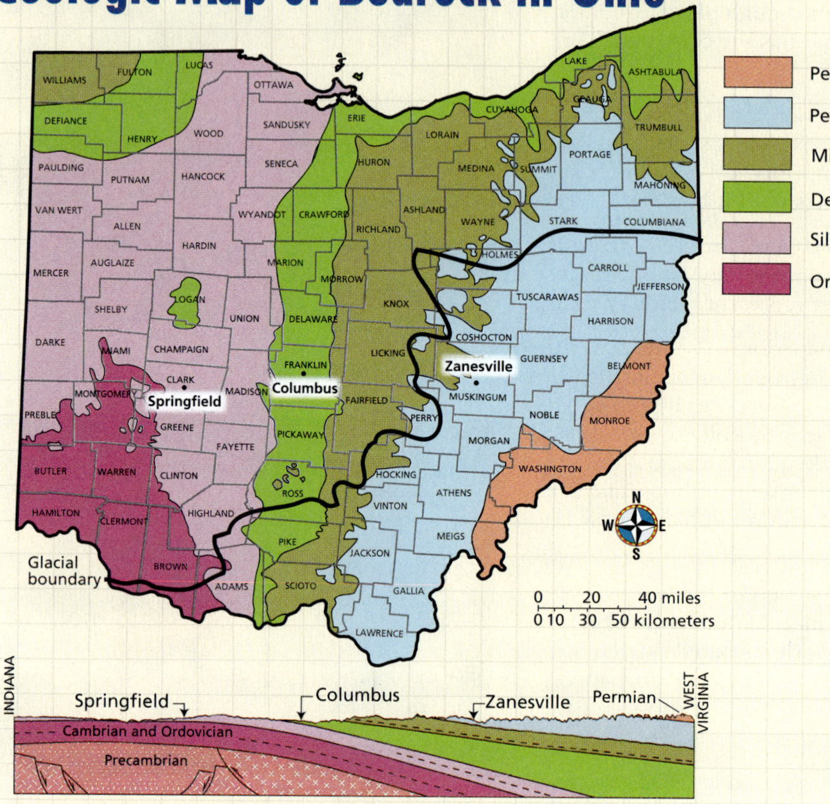

Map Skills Activity

This map shows the ages of the bedrock in Ohio. Bedrock is the solid rock that lies underneath all surface soil and other loose material. Use the map to answer the questions below.

1. **Using the Key** What geologic periods are represented in the bedrock of Ohio?
2. **Analyzing Data** Where is the youngest bedrock in Ohio? Where is the oldest bedrock in Ohio?
3. **Identifying Trends** If you traveled from east to west across Ohio, how would the ages of the rock beneath you change?
4. **Analyzing Relationships** Based on the geologic cross section, how does the shape of the rock beds cause rocks of different ages to be exposed at different places in Ohio?
5. **Identifying Relationships** Why is Mississippian rock likely to be located next to Devonian rock?
6. **Using the Key** The first reptiles appeared in the fossil record during the Pennsylvanian Period. If you wanted to look for fossils of these reptiles in Ohio, in what part of the state would you look? Explain your answer.

Clues to Climate Change

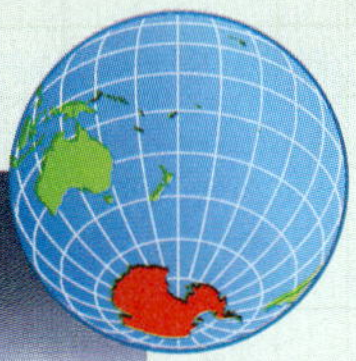

It is hard to imagine that the bitter cold climate of Antarctica was once much warmer than it is now and that the icy landscape supported thick forests. But studies of fossils found in ocean-floor sediments off the coast of Antarctica indicate climate change.

◀ Today, Antarctica is populated only by organisms that can survive extreme cold, such as these penguins.

Finding Fossil Clues

Scientists working on the Ocean Drilling Project have found samples of sediments from 60 million years ago to the present day. These samples reveal much about the complex history of climatic changes in Antarctica.

Fossil spores and pollen that are more than 39 million years old indicate that beech-tree forests once grew in Antarctica. Samples of sediment that are 37 million to 60 million years old contained soil that normally occurs in warm, humid climates. Thus, scientists learned that prior to 37 million years ago the climate in Antarctica was temperate.

Polar Beaches?

Fossils of freshwater organisms indicate that organisms were carried from Antarctic lakes to the ocean floor by rivers as recently as 20 million years ago. The fossils are evidence that the climate was once warm enough for unfrozen lakes to exist.

Scientists also discovered fossils of marine organisms that can live only in sunny coastal waters. Those found off the east coast were more than 15 million years old, and those found off the west coast were more than 4.8 million years old.

◀ About 60 million years ago, Antarctica may have looked like this Brazilian rain forest.

Extension

1. **Research** When did the ice shelves that now cover the ocean off the east and west coasts of Antarctica form? Research this topic, and write a report about your findings.

Eye on the Environment

Clues to Climate Change

Internet Activity — ADVANCED

Mapping Antarctica's Movement Have interested groups of students use the Internet to research Antarctica's movement across the planet. They should create animations or time-lapse maps that begin with Pangaea and end with Antarctica's present position, showing as many intermediate positions as possible. At each place on the map where the Antarctic continent is drawn, students should supply a time, in millions of years ago, when Antarctica occupied that position. Other continents may or may not be included in the map. Students should also add information to the map about how the Antarctic climate and environment changed as its position changed. They may include pictures or information about Antarctic flora and fauna at any or all of the continent's positions on the map. Have students present their map and a report of what they learned to the class. A worksheet designed to direct student research on this topic can be found in the **Chapter Resource File** booklet or by visiting **go.hrw.com** and entering the keyword **HQ6RECX.** LS **Visual/Logical**

CHAPTER RESOURCES

Chapter Resource File

Internet Activity
- Mapping Antarctica's Movement ADVANCED

Answer to Extension

1. Answers may vary. Students should know that the Antarctic ice sheet formed over a relatively short period (about 50,000 years) approximately 34 million years ago.

Chapter 9

A View of Earth's Past

Planning Guide

Compression Guide

To shorten instruction because of time limitations, omit Sections 2 and 3.

OBJECTIVES	LABS, DEMONSTRATIONS, AND ACTIVITIES	TECHNOLOGY RESOURCES
PACING • 90 min pp. 210–214 **Chapter Opener**		**OSP Parent Letter** ■ **CD Student Edition on CD-ROM** **CD Chapter Summaries Audio CD** ■ **VID Brain Food Video Quiz**
Section 1 Geologic Time • Summarize how scientists worked together to develop the geologic column. • List the major divisions of geologic time.	**TE Discussion** Dating Rocks, p. 212 ADVANCED **SE Quick Lab** Geologic Time Scale, p. 212 GENERAL **CRF Datasheet for Quick Lab*** GENERAL **SE Skills Practice Lab** History in the Rocks, pp. 232–233 GENERAL **CRF Datasheet for Chapter Lab*** GENERAL **CRF Inquiry Lab** Future Earth* GENERAL	**OSP Lesson Plans** (also in print) **TR Bellringer*** **TR** 44 The Geologic Time Scale* **TE Internet Activity** Extinct Organisms, p. 213 GENERAL **CRF Internet Activity** Extinct Organisms* GENERAL **TE Internet Activity** Imaging Technologies, p. 235 ADVANCED **CRF Internet Activity** Imaging Technologies* ADVANCED
PACING • 45 min pp. 215–220 **Section 2 Precambrian Time and the Paleozoic Era** • Summarize how evolution is related to geologic change. • Identify two characteristics of Precambrian rock. • Identify one major geologic and two major biological developments during the Paleozoic Era.	**SE Quick Lab** Chocolate Candy Survival, p. 217 ◆ GENERAL **CRF Datasheet for Quick Lab*** GENERAL **TE Discussion** Drawing Conclusions, p. 217 GENERAL **TE Discussion** Body of Evidence, p. 218 BASIC **TE Debate** The Current Extinction?, p. 219 ADVANCED	**OSP Lesson Plans** (also in print) **TR Bellringer*** **TR** 45 Similar Skeletal Structures of Mammals* **TR** 46 Precambrian Time and the Paleozoic Era* **TE Internet Activity** Supercontinents, p. 218 ADVANCED **CRF Internet Activity** Supercontinents* ADVANCED
PACING • 45 min pp. 221–226 **Section 3 The Mesozoic and Cenozoic Eras** • List the periods of the Mesozoic and Cenozoic Eras. • Identify two major geologic and biological developments during the Mesozoic Era. • Identify two major geologic and biological developments during the Cenozoic Era.	**TE Discussion** More Resources?, p. 221 GENERAL **TE Discussion** Flowering Plants, p. 223 GENERAL **TE Debate** Dinosaur Extinction, p. 223 ADVANCED **TE Environmental Science Connection** The Isolation of Antarctica, p. 224 ADVANCED **SE Mapping Expeditions** Where the Hippos Roam, pp. 840–841 GENERAL **TE Group Activity** Dinosaur Map, p. 234 GENERAL **SE Maps in Action** Fossil Evidence for Gondwanaland, p. 234 GENERAL **CRF Making Models Lab** Dinosaur Hunt* GENERAL	**OSP Lesson Plans** (also in print) **TR Bellringer*** **TR** 47 The Mesozoic and Cenozoic Eras* **TR** 48 Fossil Evidence for Gondwanaland* **TE Internet Activity** Online Museums, p. 223 GENERAL **CRF Internet Activity** Online Museums* GENERAL **CD Interactive Tutor** Evolution and Geologic Time

PACING • 90 min

CHAPTER REVIEW, ASSESSMENT, AND STANDARDIZED TEST PREPARATION

- **SE Chapter Highlights**, p. 227
- **SE Chapter Review**, pp. 228–229
- **SE Standardized Test Prep**, pp. 230–231
- **CRF Concept Review*** ■ GENERAL
- **CRF Critical Thinking*** ADVANCED
- **CRF Math Skills*** GENERAL
- **CRF Graphing Skills*** GENERAL
- **CRF Chapter Test A*** ■ GENERAL
- **CRF Chapter Test B*** ADVANCED
- **OSP Lesson Plans** (also in print)
- **OSP Test Generator**
- **OSP Test Item Listing**

Online and Technology Resources

go.hrw.com Holt Online Learning

Visit go.hrw.com for access to Holt Online Learning, or enter the keyword **HQ6 Home** for a variety of free online resources.

One-Stop Planner® CD-ROM

This CD-ROM package includes
- Lab Materials QuickList Software
- Holt Calendar Planner
- Customizable Lesson Plans
- Printable Worksheets
- ExamView® Test Generator
- Interactive Teacher Edition
- Holt PuzzlePro®
- Holt PowerPoint® Resources

KEY

SE	Student Edition	OSP	One-Stop Planner	VID	Classroom Video/DVD		
TE	Teacher Edition	TR	Transparencies and Transparency Worksheets	*	Also on One-Stop Planner		
CRF	Chapter Resource File	CD	CD or CD-ROM	◆	Requires advance prep		
LTP	Long-Term Projects			■	Also available in Spanish		

SKILLS DEVELOPMENT RESOURCES	REVIEW AND ASSESSMENT	CORRELATIONS
SE **Pre-Reading Activity,** p. 210 GENERAL TE **Using the Figure** Hypacrosaurus, p. 210 GENERAL		National Science Education Standards
CRF **Directed Reading*** BASIC TE **Using the Figure** Rock Layers, p. 211 GENERAL TE **Inclusion Strategies,** p. 212 TE **Skill Builder** Vocabulary, p. 213 GENERAL	SE **Reading Check,** p. 211 GENERAL SE **Section Review,** p. 214 GENERAL TE **Reteaching,** p. 213 BASIC TE **Quiz,** p. 214 GENERAL TE **Alternative Assessment,** p. 214 ADVANCED CRF **Section Quiz*** ■ GENERAL	HNS 3c, ES 3b, ES 3d, SAI 2c, LS 3a, LS 3b, LS 3c
CRF **Directed Reading*** BASIC TE **Using the Figure** Comparing Bones, p. 215 GENERAL TE **Using the Figure** Geologic Time, p. 216 GENERAL TE **Inclusion Strategies,** p. 216 TE **Using the Figure** Sea to Land, p. 219 GENERAL SE **Graphic Organizer** Spider Map, p. 219 GENERAL	SE **Reading Checks,** pp. 216, 218 GENERAL SE **Section Review,** p. 220 GENERAL TE **Homework,** p. 217 ADVANCED TE **Reteaching,** p. 219 BASIC TE **Quiz,** p. 220 GENERAL TE **Alternative Assessment,** p. 220 GENERAL CRF **Section Quiz*** ■ GENERAL	ES 3d, LS 3a, LS 3b, LS 3c
CRF **Directed Reading*** BASIC TE **Using the Figure** Comparing Characteristics, p. 222 GENERAL TE **Reading Skill Builder** Reading Organizer, p. 222 BASIC TE **Skill Builder** Vocabulary, p. 223 GENERAL TE **Using the Figure** Mammal Characteristics, p. 224 GENERAL TE **Skill Builder** Math, p. 224 BASIC	SE **Reading Checks,** pp. 222, 225 GENERAL SE **Section Review,** p. 226 GENERAL TE **Homework,** p. 225 ADVANCED TE **Reteaching,** p. 225 BASIC TE **Quiz,** p. 225 GENERAL TE **Alternative Assessment,** p. 226 GENERAL CRF **Section Quiz*** ■ GENERAL	LS 3c

Holt Lab Generator CD-ROM

Search for labs by topic, standard, difficulty level, and/or time frame. Edit those labs to fit your needs, or create your own labs. Use the Lab Materials QuickList to order just what you need.

Holt Earth Science Interactive Tutor CD-ROM

This CD-ROM consists of interactive activities that give students a fun way to extend their knowledge of Earth science concepts.

Chapter Summaries Audio CDs

These CDs include audio summaries of the key concepts presented in each chapter. (Audio summaries are also available in Spanish.)

SCILINKS. NSTA

www.scilinks.org

Maintained by the **National Science Teachers Association.** See Chapter Enrichment pages that follow for a complete list of topics.

See Chapter Enrichment pages for Video Resources.

Chapter 9 Chapter Enrichment

This Chapter Enrichment provides relevant and interesting information to expand and enhance your classroom instruction of the chapter material.

Section 1 Geologic Time

Billions of Years in the Making

Some students may think of the 4 billion years of Precambrian time as generally barren and inconsequential. On the contrary, the main life forms during this time were blue-green algae and other photosynthetic unicellular organisms. The oxygen in our atmosphere is a direct result of the billions of years of photosynthesis carried out by these "insignificant" organisms.

The atmosphere of the early Earth contained no oxygen. There is evidence that by 2.8 billion years ago oxygen was beginning to accumulate in the atmosphere from photosynthesis, or the process by which plants use carbon dioxide and the energy in sunlight to make their own food. The organisms emit oxygen as a byproduct into the atmosphere. Scientists have found microfossils of photosynthetic organisms, some of which closely resemble modern forms, from about 2.2 billion years ago. Chemical and other analyses show that by this time the atmosphere was on its way to becoming the oxygen-rich atmosphere nearly all organisms need to survive today.

▲ Scientists use clues in rock layers to study Earth's history.

Section 2 Precambrian Time and the Paleozoic Era

The Burgess Shale

The Burgess Shale is a section of Cambrian rock located in the Canadian Rockies. This formation contains the largest known collection of fossils from the Middle Cambrian Epoch. Since its discovery in 1909, the Burgess Shale has yielded more than 60,000 fossil specimens. Fossils of similar fauna have been found in Middle Cambrian rocks in Greenland and elsewhere, though in far less abundance and variety. The fossils found in the Burgess Shale are primarily marine arthropods, worms, sponges, echinoderms, mollusks, brachiopods, annelids, and coelenterates. (Coelenterates are radially symmetrical invertebrate animals, such as corals, jellyfish, and sea anemones.)

Scientists think that the Burgess Shale contains so many fossils because the area was probably covered by a mudslide that quickly buried and preserved the organisms, including their soft parts—many in exquisite and astonishing detail. Though some Burgess Shale fossils belong to established phyla and yield valuable clues about the evolution of modern forms, many are unique. Fossils of bizarre fauna such as *Hallucigenia, Wiwaxia,* and *Opabinia* are unlike anything known today. One lucky survivor of this calamity, an insignificant fossil chordate, may be the ancestor of us all.

Section 3 The Mesozoic and Cenozoic Eras

John Ostrom and "Terrible Claw"

Deinonychus antirrhopus is thought to have lived in the early Cretaceous Period and was discovered by paleontologist John Ostrom in south central Montana in 1964. Ostrom gave it this name because of the large, scythe-like claw on each of its feet. *Deinonychus* comes from two Greek words meaning "terrible claw" and *antirrhopus,* also Greek, means "upturned." Ostrom noticed that there were many similarities between *Deinonychus* and *Archaeopteryx*. His discovery not only revived the old idea that birds evolved from dinosaurs, but also suggested

▲ **Large predatory dinosaurs, such as *Tyrannosaurus,* were probably ectothermic.**

that at least some dinosaurs might also have been warm-blooded (*endothermic*), as modern birds are. Now, as a result of Ostrom's work, many scientists think this to be the case.

The depiction by Robert Bakker in Ostrom's 1969 publication about *Deinonychus* showed a fairly small but dynamic and athletic predator. Ostrom's article was groundbreaking, being the first to challenge the long-held belief that all dinosaurs were slow and lumbering.

Deinonychus ate meat and was likely a superb hunter. *Deinonychus* probably hunted in packs, enabling it to successfully bring down dinosaurs much larger than itself, such as the plant-eating *Tennontosaurus*. Some scientists think that these packs of *Deinonychus* hunted at night—behavior that surely required endothermy.

However, *Deinonychus* might not have been the only warm-blooded dinosaur. Some scientists think that the group of dinosaurs called *therapods* had a feather-like covering that may have provided insulation, a necessity for endotherms. Also, the nesting behavior of other therapods suggests that they fed and raised their young over relatively long periods of time, a behavior common to endotherms, but not to ectotherms. Ectotherms usually leave their young immediately once they're born. Also, CT scans of the fossilized remains of one theropod show that it had a four-chambered heart, just as modern endothermic animals do, as opposed to the three-chambered heart typical of modern cold-blooded animals.

Video Resources

Brain Food Video Quizzes

Brain Food Video Quizzes These videos contain game-show style quizzes that assess students' progress and motivate students to study the chapter material.

NOVA Videos

To order **NOVA** videos related to this chapter, visit **go.hrw.com** and enter the keyword **HQ6VEPV.**

SciLinks

Developed and maintained by the National Science Teachers Association

SciLinks is maintained by the National Science Teachers Association to provide you and your students with interesting, up-to-date links that will enrich your classroom presentation of the chapter.

Visit www.scilinks.org and enter the SciLinks code for more information about the topic listed.

Topic: Geologic Time Scale
SciLinks code: HQ60669

Topic: Mass Extinctions
SciLinks code: HQ60916

Topic: Geologic Periods and Epochs
SciLinks code: HQ60667

Chapter 9

Chapter Overview

Earth is billions of years old, and its changes and the evolution of its life-forms have been divided into time periods. This chapter explains geologic time and describes how Earth and its life forms have changed over geologic time.

Using the Figure —GENERAL

Hypacrosaurus This illustration shows what one type of dinosaur that lived more than 65 million years ago may have looked like. Ask students how the artist knew what colors to use for the dinosaurs' skin. (The artist did not know the skin color for sure. This type of illustration requires some artistic freedom.) LS Visual/Logical

PRE-READING ACTIVITY

FOLDNOTES You may want to use this FoldNote in a classroom discussion to review material from the chapter. On the board, write each category from the Two-Panel Flip Chart. Then, ask students to provide information from each category. Under the appropriate category on the board, write the information that students provide.

Chapter 9 A View of Earth's Past

Sections

1 Geologic Time
2 Precambrian Time and the Paleozoic Era
3 The Mesozoic and Cenozoic Eras

What You'll Learn

- How geologic time is divided
- What organisms lived during each geologic period

Why It's Relevant

The geologic time scale provides a framework for understanding the geologic processes that shape Earth. The fossil record shows that Earth is a constantly changing planet.

PRE-READING ACTIVITY

FOLDNOTES **Two-Panel Flip Chart** Before you read this chapter, create the FoldNote entitled "Two-Panel Flip Chart" described in the Skills Handbook section of the Appendix. Label the flaps of the two-panel flip chart with "Geologic Time Scale" and "Geologic History." As you read the chapter, write information you learn about each category under the appropriate flap.

▶ This illustration shows an artist's idea of how a mother *Hypacrosaurus* may have looked as she fed her hatchlings. Because most dinosaur fossils are only fossilized bone, many other characteristics, such as skin color, are left to our imagination.

Chapter Correlations — *National Science Education Standards*

HNS 3c Occasionally, there are advances in science and technology that have important and long-lasting effects on science and society.... **(Section 1)**

ES 3b Geologic time can be estimated by observing rock sequences and using fossils to correlate the sequences at various locations. Current methods include using the known decay rates of radioactive isotopes present in rocks to measure the time since the rock was formed. **(Section 1)**

ES 3d Evidence for one-celled forms of life—the bacteria—extends back more than 3.5 billion years. The evolution of life caused dramatic changes in the composition of the earth's atmosphere, which did not originally contain oxygen. **(Sections 1 and 2)**

SAI 2c Scientists rely on technology to enhance the gathering and manipulation of data. New techniques and tools provide new evidence to guide inquiry and new methods to gather data.... **(Section 1)**

LS 3a Species evolve over time. Evolution is the consequence of the interactions of (1) the potential for a species to increase its numbers, (2) the genetic variability of offspring due to mutation and recombination of genes, (3) a finite supply of the resources required for life, and (4) the ensuing selection by the environment of those offspring better able to survive and leave offspring. **(Sections 1 and 2)**

LS 3b The great diversity of organisms is the result of more than 3.5 billion years of evolution that has filled every available niche with life forms. **(Sections 1 and 2)**

LS 3c Natural selection and its evolutionary consequences provide a scientific explanation for the fossil record of ancient life forms.... **(Sections 1, 2, and 3)**

Section 1 Geologic Time

Earth's surface is constantly changing. Mountains form and erode; oceans rise and recede. As conditions on Earth's surface change, some organisms flourish and then later become extinct. Evidence of change is recorded in the rock layers of Earth's crust. To describe the sequence and length of this change, scientists have developed a *geologic time scale*. This scale outlines the development of Earth and of life on Earth.

OBJECTIVES

- **Summarize** how scientists worked together to develop the geologic column.
- **List** the major divisions of geologic time.

KEY TERMS

geologic column
era
period
epoch

The Geologic Column

By studying fossils and applying the principle that old layers of rock are below young layers, 19th-century scientists determined the relative ages of sedimentary rock in different areas around the world. No single area on Earth contained a record of all geologic time. So, scientists combined their observations to create a standard arrangement of rock layers. As shown in the example in **Figure 1,** this ordered arrangement of rock layers is called a **geologic column.** A geologic column represents a timeline of Earth's history. The oldest rocks are at the bottom of the column.

geologic column an ordered arrangement of rock layers that is based on the relative ages of the rocks and in which the oldest rocks are at the bottom

Rock layers in a geologic column are distinguished by the types of rock the layers are made of and by the kinds of fossils the layers contain. Fossils in the upper, more-recent layers resemble modern plants and animals. Most of the fossils in the lower, older layers are of plants and animals that are different from those living today. In fact, many of the fossils discovered in old layers are from species that have been extinct for millions of years.

Reading Check Where would you find fossils of extinct animals on a geologic column? (See the Appendix for answers to Reading Checks.)

Figure 1 ▶ By combining observations of rock layers in areas A, B, and C, scientists can construct a geologic column. ***Why is relative position important in determining the ages of rock layers?***

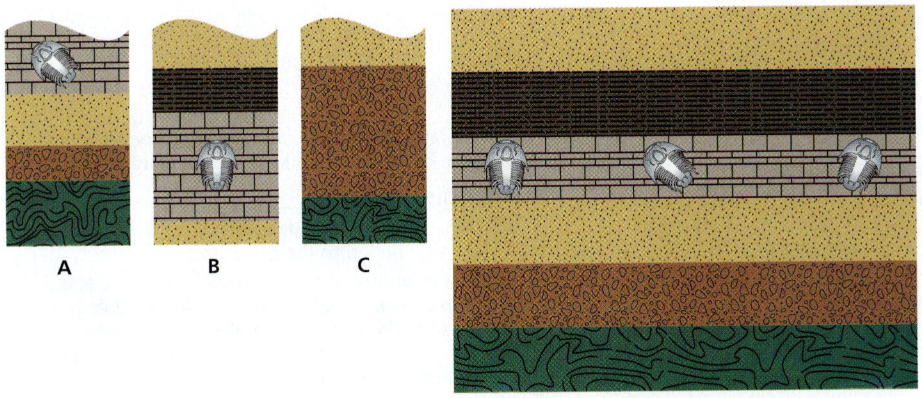

Section 1

Focus

Overview

This section describes what geologic time is and how scientists divide geologic time.

Bellringer

Ask students to think about and list or describe what types of changes they think might have taken place on Earth during its more than 4-billion-year history. (Answers may vary. Accept all reasonable answers.) **LS Verbal**

Motivate

Using the Figure — GENERAL

Rock Layers Have students examine and talk about the figure on this page. Then, use the question in the caption as the starting point for a discussion about geologic time. Answer to caption question: Relative position can tell you the relative ages of rock layers. **LS Visual**

Answer to Reading Check

You would find fossils of extinct animals in older layers of a geologic column.

CHAPTER RESOURCES

Chapter Resource File
- Directed Reading BASIC

Technology

Transparencies
- Bellringer

Student Edition on CD-ROM

One-Stop Planner CD-ROM
- Lesson Plan

Teach

Discussion — ADVANCED

Dating Rocks Tell students that, over time, Earth processes move and fold rocks. Have students discuss why radiometric dating technology is so useful to paleontologists. (It allows them to give numeric dates to rock layers.) Invite interested students to research how radiometric dating technology works and to present the information to the class. LS **Verbal/ Logical**

Quick LAB

Skills Acquired

- Measuring
- Identifying and Recognizing Patterns

Teacher's Notes You may want to make sure students understand how to use a scale in which one measure represents another, in this case, a period of time. Monitor students as they work with the adding machine tape and meter sticks.

Answers

1. Humans first appeared in the Pleistocene Epoch. The scale length is less than 2 mm.
2. The Paleozoic, Mesozoic, and Cenozoic Eras cover about 542 million years, or about 12% of geologic time. The Precambrian Era represents about 88% of geologic time.

Figure 2 ▶ This scientist is collecting rock samples that contain fossilized fungal spores that date the rock to the Triassic Period.

Using a Geologic Column

When the first geologic columns were being developed, scientists estimated the ages of rock layers by using factors such as the average rates of sediment deposition. The development of radiometric dating methods, however, allowed scientists to determine the absolute ages of rock layers with more accuracy.

Scientists can now use geologic columns to estimate the age of rock layers that cannot be dated radiometrically. To determine the layer's age, scientists compare a given rock layer with a similar layer in a geologic column that contains the same fossils or that has the same relative position. If the two layers match, they likely formed at about the same time. The scientist in **Figure 2** is investigating the ages of sedimentary rocks.

Divisions of Geologic Time

The geologic history of Earth is marked by major changes in Earth's surface, climate, and types of organisms. Geologists use these indicators to divide the geologic time scale into smaller units. Rocks grouped within each unit contain similar fossils. In fact, a unit of geologic time is generally characterized by fossils of a dominant life-form. A simplified geologic time scale is shown in **Table 1.**

Because Earth's history is so long, Earth scientists commonly use abbreviations when they discuss geologic time. For example, Ma stands for *mega-annum,* which means "one million years."

Quick LAB 30 min

Geologic Time Scale

Procedure

1. Copy the table shown at right onto a piece of **paper.**
2. Complete the table by using the scale 1 cm is equal to 10 million years.
3. Lay a **5 m strip of adding-machine paper** flat on a hard surface. Use a **meterstick,** a **metric ruler,** and a **pencil** to mark off the beginning and end of Precambrian time according to the time scale you calculated. Do the same for the three eras. Label each time division, and color each a different color with **colored pencils.**
4. Pick two periods from the geologic time scale. Using the same scale that was used in step 2, calculate the scale length for each period listed. Mark the boundaries of each period on the paper strip, and label the periods on your scale.

Era	Length of time (years)	Scale length
Precambrian	4,058,000,000	
Paleozoic	291,000,000	DO NOT WRITE IN THIS BOOK
Mesozoic	185,500,000	
Cenozoic	65,500,000 (to present)	

5. Decorate your strip by adding names or drawings of the organisms that lived in each division of time.

Analysis

1. When did humans appear? What is the scale length from that period to the present?
2. Add the lengths of the Paleozoic, Mesozoic, and Cenozoic Eras. What percentage of the geologic time scale do these eras combined represent? What percentage of the geologic time scale does Precambrian time represent?

CHAPTER RESOURCES

Chapter Resource File

- Datasheet for Quick Lab GENERAL

INCLUSION Strategies

- *Learning Disabled*
- *Attention Deficit Disorder*
- *Hearing Impaired Learners*

Have students draw and label pictures to accompany the text. Students should use all of the key terms that appear in bold type throughout the chapter as labels. They may also write descriptions on their pictures that help them remember key concepts in the chapter. LS **Visual/Verbal**

Career

Paleontologist A paleontologist is a scientist who studies prehistoric and ancient life-forms on Earth. The figure on this page shows a paleontologist at work. Paleontologists seek out rock formations to search for fossils. They analyze fossils to understand how organisms have changed over geologic time, and sometimes find fossils of ancient life-forms never before known. Nearly all paleontologists need advanced college degrees, with courses in geology, zoology, evolutionary biology, and ecology.

Table 1 ▼

Geologic Time Scale				
Era	**Period**	**Epoch**	**Beginning of interval in Ma**	**Characteristics from geologic and fossil evidence**
Cenozoic 	Quaternary	Holocene	0.0115	The last glacial period ends; complex human societies develop.
		Pleistocene	1.8	Woolly mammoths, rhinos, and humans appear.
	Tertiary	Pliocene	5.3	Large carnivores (bears, lions) appear.
		Miocene	23.0	Grazing herds are abundant; raccoons and wolves appear.
		Oligocene	33.9	Deer, pigs, camels, cats, and dogs appear.
		Eocene	55.8	Horses, flying squirrels, bats, and whales appear.
		Paleocene	65.5	Age of mammals begins; first primates appear.
Mesozoic 	Cretaceous		146	Flowering plants and modern birds appear; mass extinctions mark the end of the Mesozoic Era.
	Jurassic		200	Dinosaurs are the dominant life-form; primitive birds and flying reptiles appear.
	Triassic		251	Dinosaurs appear; ammonites are common; cycads and conifers are abundant; and mammals appear.
Paleozoic 	Permian		299	Pangaea comes together; mass extinctions mark the end of the Paleozoic Era.
	Carboniferous	Pennsylvanian Period	318	Giant cockroaches and dragonflies are common; coal deposits form; and reptiles appear.
		Mississippian Period	359	Amphibians flourish; brachiopods are common in oceans; and forests and swamps cover most land.
	Devonian		416	Age of fishes begins; amphibians appear; and giant horsetails, ferns, and seed-bearing plants develop.
	Silurian		444	Eurypterids, land plants and animals appear.
	Ordovician		488	Echinoderms appear; brachiopods increase; trilobites decline; graptolites flourish; atmosphere reaches modern O_2-rich state.
	Cambrian		542	Shelled marine invertebrates appear; trilobites and brachiopods are common. First vertebrates appear.
Precambrian time			4,600	The Earth forms; continental shields appear; fossils are rare; and stromatolites are the most common organism.

MISCONCEPTION ALERT

The Human Time Frame Some students may be under the impression that humans coexisted with dinosaurs, or with even earlier life-forms. Explain to students that hominids have been on Earth for less than 2 million years and that the dinosaurs became extinct 65 million years ago.

SKILL BUILDER — GENERAL

Vocabulary Tell students that the suffix *-zoic* is related to the word "zoo." The suffix *-zoic* refers to the life-forms that existed during a specific geologic time. Many geologic time periods are named after the organisms that were common on Earth during that time. LS Verbal/Auditory English Language Learners

Close

Reteaching — BASIC

Relative Time Have the class make a large poster or mural that shows the relative scale of each geologic eon and era they learn about in this chapter. The project will help them understand that each division is defined, not by a given time (that is, all eons or eras are not of equal length), but by the organisms that dominated during that time period. Students may label and describe each geologic time period on the poster or mural. LS Visual/Kinesthetic

CHAPTER RESOURCES

Chapter Resource File

- **Making Models Lab** Future Earth GENERAL

Internet Activity

- Extinct Organisms GENERAL

Technology

Transparencies

- 44 The Geologic Time Scale (with worksheet)

Internet Activity — GENERAL

Extinct Organisms Have students work in groups to find pictures of organisms that were common during one geologic period. Each group member may look for pictures of one organism or one class of organisms. Have students create posters that show the life-forms they researched. Invite advanced students to write about the characteristics of organisms from their time period. A worksheet designed to direct student research on this topic can be found in the **Chapter Resource File** booklet or by visiting go.hrw.com and entering the keyword **HQ6VEPX.** LS Visual/Verbal Co-op Learning

Close, *continued*

Quiz — GENERAL

1. In a geologic column, where would you expect to find the oldest fossils? (in the older rock layers)
2. On what are most geologic time divisions based? (on the organisms that dominated in that time period)
3. Why do scientists use radiometric dating for rocks? (It provides an absolute, or numerical, age.)

Alternative Assessment — ADVANCED

Descriptive Essay Have students use the time scale table as a starting point to write an essay that describes the history of life on Earth. **LS Verbal/Logical**

Answers to Section Review

1. They had to work together because different rocks occur in different parts of the world.
2. Answers may vary. Sample answer: In the Devonian period, fishes dominated Earth and amphibians first appeared.
3. A geologic column is useful because it shows the relative ages of rocks and fossils.
4. year, age, epoch, period, era, eon
5. Paleozoic Era: 291 million years; Mesozoic Era: 186 million years; Cenozoic Era: 65.5 million years
6. The geologic column follows geologic time in that it consists of the rocks that formed during each division of geologic time.
7. The scientist would want to know the age of the rock in which the fossil was found.
8. It would change because we would have to rethink the entire evolution of animals and of conditions on Earth back to the Paleozoic.
9. *Geologic time* can be divided into *Precambrian time*, the *Paleozoic Era*, the *Mesozoic Era*, and the *Cenozoic Era*, which are each divided into shorter time units called *periods*, which include even shorter time units called *epochs*.

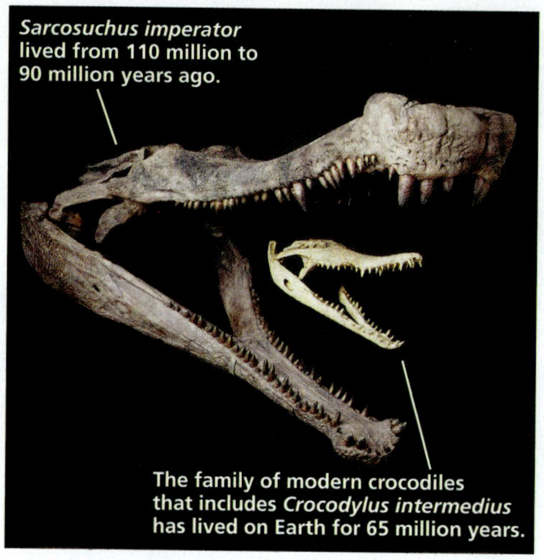

Figure 3 ▶ Crocodilians have lived on Earth for more than two geologic eras without major anatomical changes.

Eons and Eras

The largest unit of geologic time is an *eon*. Geologic time is divided into four eons—the Hadean eon, the Archean eon, the Proterozoic eon, and the Phanerozoic eon. The first three eons of Earth's history are part of a time interval commonly known as *Precambrian time*. This 4 billion year interval contains most of Earth's history. Very few fossils exist in early Precambrian rocks, so dividing Precambrian time into smaller time units is difficult.

After Precambrian time the Phanerozoic eon began. This eon, as well as most eons, is divided into smaller units of geologic time called **eras.** The first era of the Phanerozoic eon was the *Paleozoic Era* which lasted about 291 million years. Paleozoic rocks contain fossils of a wide variety of marine and terrestrial life forms. After the Paleozoic Era, the *Mesozoic Era* began and lasted about 186 million years. Mesozoic fossils include early forms of birds and of reptiles, such as the giant crocodilian shown in **Figure 3.** The present geologic era is the *Cenozoic Era,* which began about 65 million years ago. Fossils of mammals are common in Cenozoic rocks.

era a unit of geologic time that includes two or more periods

period a unit of geologic time that is longer than an epoch but shorter than an era

epoch a subdivision of geologic time that is longer than an age but shorter than a period

Periods and Epochs

Eras are divided into shorter time units called **periods.** Each period is characterized by specific fossils and is usually named for the location in which the fossils were first discovered. Where the rock record is most complete and least deformed, a detailed fossil record may allow scientists to divide periods into shorter time units called **epochs.** Epochs may be divided into smaller units of time called *ages*. Ages are defined by the occurrence of distinct fossils in the fossil record.

Section 1 Review

1. **Summarize** the reasons that many scientists had to work together to develop the geologic column.
2. **Describe** the major events in any one period of geologic time.
3. **Explain** why constructing geologic columns is useful to Earth scientists.
4. **List** the following units of time in order of length from shortest to longest: *year, period, era, eon, age,* and *epoch.*
5. **Name** the three eras of the Phanerozoic Eon, and identify how long each one lasted.
6. **Compare** geologic time with the geologic column.

CRITICAL THINKING

7. **Analyzing Relationships** When a scientist discovers a new type of fossil, what characteristic of the rock around the fossil would he or she want to learn first?
8. **Predicting Consequences** How would our understanding of Earth's past change if a scientist discovered a mammal fossil from the Paleozoic Era?

CONCEPT MAPPING

9. Use the following terms to create a concept map: *geologic time, Precambrian time, Paleozoic Era, Mesozoic Era, Cenozoic Era, period,* and *epoch.*

CHAPTER RESOURCES

Chapter Resource File
- Section Quiz GENERAL

Workbooks
- Study Guide (also in Spanish)

Section 2 Precambrian Time and the Paleozoic Era

History is a record of past events. Just as the history of civilizations is written in books, the geologic history of Earth is recorded in rock layers. The types of rock and the fossils that occur in each layer reveal information about the environment when the layer formed. For example, the presence of a limestone layer in a region indicates that the area was once covered by water.

OBJECTIVES

- **Summarize** how evolution is related to geologic change.
- **Identify** two characteristics of Precambrian rock.
- **Identify one** major geologic and two major biological developments during the Paleozoic Era.

KEY TERMS

evolution
Precambrian time
Paleozoic Era

Evolution

Fossils indicate the kinds of organisms that lived when rock formed. By examining rock layers and fossils, scientists have discovered evidence that species of living things have changed over time. Scientists call this process evolution. **Evolution** is the gradual development of new organisms from preexisting organisms. Scientists think that evolution occurs by means of natural selection. Evidence for evolution includes the similarity in skeletal structures of animals, as shown in **Figure 1.** The theory of evolution by natural selection was proposed in 1859 by Charles Darwin, an English naturalist.

evolution a heritable change in the characteristics within a population from one generation to the next; the development of new types of organisms from preexisting types of organisms over time.

Evolution and Geologic Change

Major geologic and climatic changes can affect the ability of some organisms to survive. For example, dramatic changes in sea level greatly affect organisms that live in coastal areas. By using geologic evidence, scientists try to determine how environmental changes affected organisms in the past. The fossil record shows that some organisms survived environmental changes while other organisms disappeared. Scientists use fossils to learn why some organisms survived long periods of time without changing while other organisms changed or became extinct.

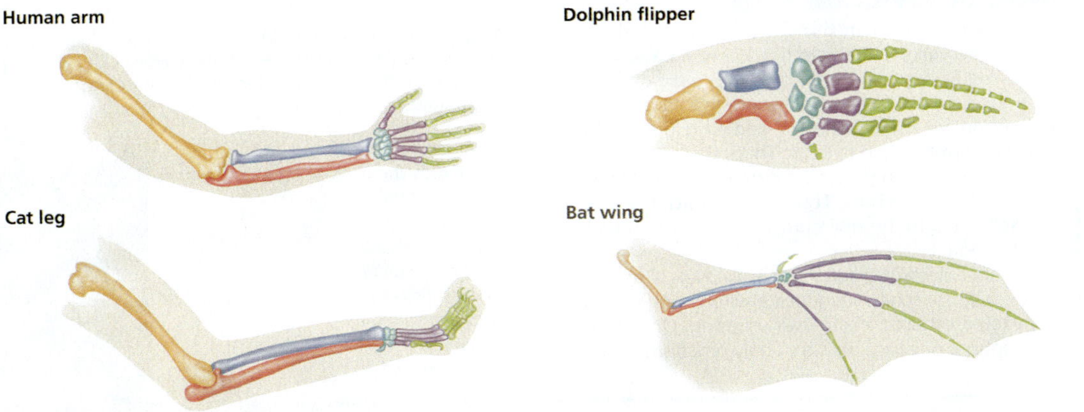

Figure 1 ▶ Bones in the front limbs of these animals are similar even though the limbs are used in different ways. Similar structures indicate a common ancestor.

Section 2

Focus

Overview

This section describes how biological evolution is related to geologic changes. It also describes the evolution of life forms during Precambrian time and the Paleozoic Era.

Bellringer

Invite students to speculate how changing conditions on Earth, such as an ice age or rising sea levels, would lead to physical changes in plants and animals. Invite students to use their imaginations in describing these changes. (Answers may vary. Accept all reasonable answers.) **LS Logical/Verbal**

Motivate

Using the Figure — GENERAL

Comparing Bones Have students examine the front limb bones shown in the figure. Ask them to describe the similarities and differences in the bones of the four animals. Ask students to describe any evidence that indicates that these animals had a common ancestor. (Students may point out differences in bone length, but should also describe similarities in the number of bones and their placement in each of the front limb skeletons.) **LS Visual/Logical**

CHAPTER RESOURCES

Chapter Resource File

- Directed Reading BASIC

Technology

Transparencies
- Bellringer
- 45 Similar Skeletal Structures of Mammals (with worksheet)

Student Edition on CD-ROM

One-Stop Planner CD-ROM
- Lesson Plan

Teach

Using the Figure — GENERAL

Geologic Time Tell students that the timescale at the top of the page shows the relative lengths of the eras to scale. To answer the caption question, students will have to calculate the distance represented by an interval of time, such as 2 mm = 65 million years.

Answer to caption question: 3,650 million years ago (3.65 billion years ago) LS **Visual/Logical**

Teaching Tip — BASIC

Connect to Prior Knowledge Students may better understand why Precambrian rocks are difficult to study if you describe for them the processes of rock deformation. Include information about metamorphism, weathering and erosion, and folding and uplifting caused by the motion of tectonic plates. Explain to students that these forces have affected Precambrian rocks for more than 3 billion years, altering them greatly since they first formed. English Language Learners

Answer to Reading Check

Earth is approximately 4.6 billion years old.

CHAPTER RESOURCES

Technology

Transparencies
- 46 Precambrian Time and the Paleozoic Era (with worksheet)

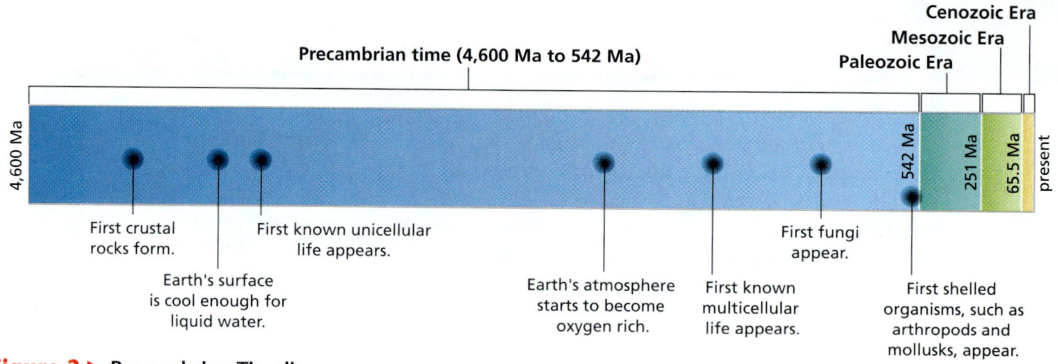

Figure 2 ▶ **Precambrian Timeline** *How many million years ago did the first unicellular life appear?*

Precambrian Time

Most scientists agree that Earth formed about 4.6 billion years ago as a large cloud, or *nebula,* spun around the newly formed sun. As material spun around the sun, particles of matter began to clump together and eventually formed Earth and the other planets of the solar system. The time interval that began with the formation of Earth and ended about 542 million years ago is known as **Precambrian time.** This division of geologic time makes up about 88% of Earth's history, as shown in **Figure 2.**

Precambrian time the interval of time in the geologic time scale from Earth's formation to the beginning of the Paleozoic era, from 4.6 billion to 542 million years ago

Even though Precambrian time makes up such a large part of Earth's history, we know relatively little about what happened during that time. We lack information partly because the Precambrian rock record is difficult to interpret. Most Precambrian rocks have been so severely deformed and altered by tectonic activity that the original order of rock layers is rarely identifiable.

Reading Check How old is Earth? (See the Appendix for answers to Reading Checks.)

Connection to BIOLOGY

Natural Selection

In part, evolution occurs through a process called *natural selection.* Natural selection has four basic principles. First, every species produces more offspring than will survive to maturity. Second, individuals in a population are slightly different, and each individual has a unique combination of traits. Third, the environment does not have enough resources to support all of the individuals that are born. Fourth, only individuals that are well suited to the environment are likely to survive and reproduce.

Natural selection ensures that individuals who have better traits for surviving in their environment are more likely to pass those traits to their offspring. One of the assumptions of evolution is that only organisms that can adapt to the environmental changes will survive. Organisms that cannot survive—in other words, those that are unfit to live and reproduce in the changing environment—become extinct.

Because their fur hides them from predators, rabbits that are adapted to survive in the arctic are white. Brown rabbits are adapted to survive in other environments.

INCLUSION Strategies

- **Visually Impaired**

Pair each visually impaired student with a non-visually impaired student. Have the sighted student tape a toothpick above the time division lines on a copy of the geologic timeline. As the visually impaired student runs a finger over the timeline, the sighte d student explains what time period the space between the toothpicks represents. Have students discuss the relative lengths of time periods based on the space between the toothpicks. LS **Kinesthetic**

ENVIRONMENTAL CONNECTION

Marine Mammal Evolution Scientists think that about 50 million years ago, the land environment changed and less food was available to land herbivores. As a result, ancient relatives of today's hoofed mammals, which likely resembled modern pigs or cows, increasingly sought food along the seashore. As more of these ancient animals came to the seaside to feed, competition may have forced some to swim farther out and to dive deeper in the water to find food. Marine mammals evolved from these ancient ungulates in less than 8 million years.

Precambrian Rocks

Large areas of exposed Precambrian rocks, called *shields,* exist on every continent. Precambrian shields are the result of several hundred million years of volcanic activity, mountain building, sedimentation, and metamorphism. After they were metamorphosed and deformed, the rocks of North America's Precambrian shield were uplifted and exposed at Earth's surface. Nearly half of the valuable mineral deposits in the world occur in the rocks of Precambrian shields. These valuable minerals include nickel, iron, gold, and copper.

Precambrian Life

Fossils are rare in Precambrian rocks, probably because Precambrian life-forms lacked bones, shells, or other hard parts that commonly form fossils. Also, Precambrian rocks are extremely old. Some date back nearly 3.9 billion years. Over this long period of time, volcanic activity, erosion, and extensive crustal movements, such as folding and faulting, probably destroyed most of the fossils that may have formed during Precambrian time.

Of the few Precambrian fossils that have been discovered, the most common are *stromatolites,* or reeflike deposits formed by blue-green algae. Stromatolites form today in warm, shallow waters, as shown in **Figure 3.** The presence of stromatolite fossils in Precambrian rocks indicates that shallow seas covered much of Earth during periods of Precambrian time. Imprints of marine worms, jellyfish, and single-celled organisms have also been discovered in rocks from late Precambrian Time.

QuickLAB 10 min

Chocolate Candy Survival

Procedure

1. Lay a piece of colorful cloth on a table.
2. Randomly sprinkle a handful of candy-coated chocolate bits on the cloth.
3. Look away for 1 min.
4. For 10 s, pick up chocolate bits one at a time. Record the colors of candy you picked up.
5. Repeat steps 1–4 with a piece of colorful cloth that has a different pattern.

Analysis

1. What colors were you more likely to pick up in the first trial? What about those candies made you pick them up?
2. When you changed the color of the cloth, did you change the color of candies you picked up?
3. How could camouflage help an organism survive?

Figure 3 ▶ Stromatolites, which are mats of blue-green algae, are the most common Precambrian fossils.

QuickLAB

Skills Acquired

- Observing
- Analyzing

Teacher's Notes You may want to make sure that the cloths closely match at least one color of candy-coated chocolate bits or that the cloth has a colorful and busy pattern. This lab can also be performed by using circles of construction paper instead of candy and extending the time in step 4 to 20 s.

Answers

1. Students are more likely to pick up the candy that did not match the cloth color because it was most easily seen.
2. Students likely picked up different color candy when it was on a different color cloth.
3. By blending into the background (the cloth), the camouflaged candy was not eaten. This applies to prey animals in the wild; camouflage helps them avoid predators.

CHAPTER RESOURCES

Chapter Resource File

- Datasheet for Quick Lab GENERAL

Discussion — GENERAL

Drawing Conclusions Have a group discussion with students about why stromatolites are the most abundant Precambrian fossils. Invite students to refer to the figure to draw conclusions about the physical characteristics that enable stromatolites to fossilize. Have students talk about how stromatolites are similar to and different from the coral reefs that occur in today's oceans. (Stromatolites are produced by thread-like algae that form sticky mats, which trap carbonate mud, which may be preserved.) LS Verbal/Visual

Homework — ADVANCED

Survivor: The Real Story Tell students that the life-forms that make up stromatolites are among the earliest on Earth. They are also among the most successful, having survived more than 3 billion years. Have students research blue-green algae, and bacteria, if possible, to find out what makes these life-forms so successful at survival. Students may prepare an illustrated report to present to the class. LS Verbal/Visual

Teach, continued

Internet Activity — ADVANCED

Supercontinents Have groups of interested students use the Internet to research supercontinents that preceded Pangaea. Invite one group to research the events that led to the creation of Rodinia (the name is Russian for "homeland"), a supercontinent that existed from 1.1 billion to 750 million years ago. Have students in this group create a map and locate today's continents as they existed in Rodinia. A second group of students should map the opening of the Panthalassic Ocean, the separation of Pannotia and the Congo craton, and the formation of Laurentia, Gondwana, and Baltica. Again, students may want to find current contents in these prehistoric landforms. Have groups present their findings to the class. A worksheet designed to direct student research on this topic can be found in the **Chapter Resource File** booklet or by visiting **go.hrw.com** and entering the keyword **HQ6VEPX**. LS **Verbal/Logical**

Answer to Reading Check

Answers may vary but should include three of the following: trilobites, brachiopods, jellyfish, worms, snails, and sponges.

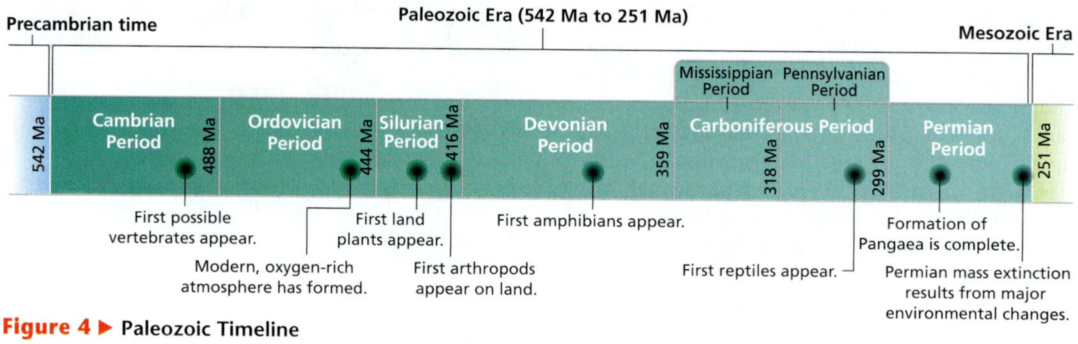

Figure 4 ▶ Paleozoic Timeline

The Paleozoic Era

As shown in **Figure 4,** the geologic era that began about 542 million years ago and ended about 251 million years ago is called the **Paleozoic Era.** At the beginning of the Paleozoic Era, Earth's landmasses were scattered around the world. By the end of the Paleozoic Era, these landmasses had collided to form the supercontinent Pangaea. This tectonic activity created new mountain ranges and lifted large areas of land above sea level.

Unlike Precambrian rocks, Paleozoic rocks hold an abundant fossil record. The number of plant and animal species on Earth increased dramatically at the beginning of the Paleozoic Era. Because of this rich fossil record, North American geologists have divided the Paleozoic Era into seven periods.

Paleozoic Era the geologic era that followed Precambrian time and that lasted from 542 million to 251 million years ago

The Cambrian Period

The Cambrian Period is the first period of the Paleozoic Era. A variety of marine life-forms appeared during this period. These Cambrian life-forms were more advanced than previous life-forms and quickly displaced the primitive organisms as the dominant life-forms. The explosion of Cambrian life may have been partly due to the warm, shallow seas that covered much of the continents during the time period. Marine *invertebrates*, or animals that do not have backbones, thrived in the warm waters. The most common of the Cambrian invertebrates were *trilobites*, such as the one shown in **Figure 5.** Scientists use many trilobites as *index fossils* to date rocks to the Cambrian Period.

The second most common animals of the Cambrian Period were the *brachiopods*, a group of shelled animals. Fossils indicate that at least 15 different families of brachiopods existed during this period. A few kinds of brachiopods exist today, but modern brachiopods are rare. Other common Cambrian invertebrates include worms, jellyfish, snails, and sponges. However, no evidence of land-dwelling plants or animals has been discovered in Cambrian rocks.

Figure 5 ▶ During the early Paleozoic Era, various types of trilobites, such as this fossilized trilobite from the genus *Moducia*, flourished in the warm, shallow seas.

Reading Check Name three common invertebrates from the Cambrian Period. (See the Appendix for answers to Reading Checks.)

Discussion — BASIC

Body of Evidence Have students discuss why the bodies of some ancient animals, such as shelled invertebrates, are well fossilized, while those of other animals, such as non-shelled invertebrates are not, or are poorly fossilized. Ask students what key body characteristic increases an organism's likelihood of being fossilized. (Animals that have hard body parts, such as shells, teeth, or bones, are far more likely to be preserved as fossils than animals whose bodies are entirely soft. Soft tissue decays far more readily and rapidly than shell or bone does.) LS **Verbal**

CHAPTER RESOURCES

Chapter Resource File

- Internet Activity
  - Supercontinents ADVANCED

Figure 6 ▶ During the Silurian Period, eurypterids lived in shallow lagoons. Eurypterids had one pair of legs for swimming and had four or five pairs for walking.

The Ordovician Period

During the Ordovician (AWR duh VISH uhn) Period populations of trilobites began to shrink. Clamlike brachiopods and cephalopod mollusks became the dominant invertebrate life-forms. Large numbers of corals appeared. Colonies of tiny invertebrates called *graptolites* also flourished in the oceans, and primitive fish appeared. By this period, *vertebrates,* or animals that have backbones, had appeared. The most primitive vertebrates were fish. Unlike modern fish, Ordovician fish did not have jaws or teeth and their bodies were covered with thick, bony plates. During the Ordovician Period, as during the Cambrian Period and Precambrian times, there was no plant life on land.

The Silurian Period

Vertebrate and invertebrate marine life continued to thrive during the Silurian Period. Echinoderms, relatives of modern sea stars, and corals became more common. Scorpion-like sea creatures called *eurypterids* (yoo RIP tuhr IDZ), such as the one shown in **Figure 6,** also existed during the Silurian Period. Fossils of giant eurypterids nearly 3 m long have been discovered in western New York. Near the end of this period, the earliest land plants as well as animals, such as scorpions, evolved on land.

The Devonian Period

The Devonian Period is called the *Age of Fishes* because fossils of many bony fishes were discovered in rocks of this period. One type of fish, called a *lungfish,* had the ability to breathe air. Other air-breathing fish, called *rhipidistians,* (RIE puh DIS tee uhnz) had strong fins that may have allowed them to crawl onto the land for short periods of time. The first amphibians, from the genus *Ichthyostega* (IK thee oh STEG uh), probably evolved from rhipidistians. *Ichthyostega,* which resembled huge salamanders, are thought to be the ancestors of modern amphibians such as frogs and toads. During the Devonian Period, land plants, such as giant horsetails, ferns, and seed-bearing plants, also began to develop. In the sea, brachiopods and mollusks continued to thrive.

Graphic Organizer **Spider Map**

Create the Graphic Organizer entitled "Spider Map" described in the Skills Handbook section of the Appendix. Label the circle "Periods of the Paleozoic Era." Create a leg for each period in the Paleozoic Era. Then, fill in the map with details about each time period.

Using the Figure — GENERAL

Sea to Land Have students describe the eurypterid shown in the figure. Then, show students a large picture of a modern scorpion (with its legs apparent). Ask students what similarities they see in the two animals. Ask them to generate a hypothesis about why legs developed on the Silurian eurypterids, and then to speculate how this adaptation likely led to the emergence of the earliest scorpion-like animals on land. (Answers may vary. Eurypterids likely developed legs for moving around on the sea bottom or in shallow water. The evolution of legs used in the water enabled the first arthropods that left the water to move around on land.) **LS Visual/Logical**

Graphic Organizer GENERAL

Spider Map

You may want to have students work in groups to create this Spider Map. Have one student draw the map and fill in information provided by other students from the group.

Close

Reteaching — BASIC

Life in the Paleozoic Tape a large, long sheet of paper to the board to represent the Paleozoic Era. Have students divide the paper into periods and then draw organisms that lived during each period. Students should label each period and each animal they draw. **LS Visual/Kinesthetic**

Debate — ADVANCED

The Current Extinction? Some scientists today think that 50% of all modern species will be extinct in a matter of a few centuries, which could make this event the worst mass extinction ever. Divide students into two groups. Have one group research and support the idea that a mass extinction is *not* occurring at this time. Have the other group research and support the idea that the worst mass extinction in Earth's history *is* occurring now. Then, have the groups debate in front of the class. **LS Verbal**

Teaching Tip — ADVANCED

Have Students Teach Others The evolution of lungs, which enabled animal life to thrive on land, is one of the major steps in the evolution of life on Earth. Have students research the evolution of lungs from gills. Invite them to give an illustrated presentation to the class to explain how this evolutionary change occurred. **LS Verbal**

Close, *continued*

Quiz — GENERAL

1. What is an index fossil? (a common fossil that defines a geologic time interval)
2. What is the origin of most of the coal we use today? (compacted plants of the Carboniferous Period)
3. When did the first primitive lungs develop in animals? (in the Devonian Period)

Alternative Assessment — GENERAL

Television Script Have students write a television script for a show about conditions and life during the Paleozoic Era. Students may read their scripts aloud to the class. LS **Verbal/Auditory**

Answers to Section Review

1. Geologic changes result in changing environmental conditions. As a result, populations either evolve or become extinct.
2. Most Precambrian rocks are highly deformed and contain few fossils.
3. Most Precambrian organisms were tiny and soft bodied. Also, the rocks are highly deformed and altered from their original state.
4. Cambrian: trilobites; Ordovician: mollusks; Silurian: echinoderms; Devonian: fishes; Carboniferous: amphibians; Permian: reptiles
5. Many types of fossil fishes have been found in rocks from this period.
6. Life-forms that became extinct during the Permian Period include many types of marine invertebrates.
7. When Pangaea formed, inland seas disappeared and organisms that lived in those seas evolved or became extinct.
8. The fossil record of Precambrian time contains few organisms.
9. The fossil record shows that life-forms from later geologic times have physical characteristics that could have resulted from the adaptation of organisms from earlier geologic times. Also, as the environment changed, new organisms appeared and others became extinct.
10. The *Paleozoic Era* is divided into seven periods, including the *Cambrian Period,* during which *invertebrates* dominated; the *Ordovician Period,* during which *vertebrates* became dominant; and the *Silurian Period.*

Figure 7 ▶ During the Carboniferous Period, crinoids, such as the one shown here, were common in the oceans. Crinoids are thought to be ancestors of modern sea lillies.

The Carboniferous Period

During the Carboniferous Period, the climate was generally warm, and the humidity was extremely high over most of the world. Forests and swamps covered much of the land. Coal deposits in Pennsylvania, Ohio, and West Virginia are the fossilized remains of these forests and swamps. During this period, the rock in which some major oil deposits occur also formed. *Carboniferous* means "carbon bearing." In North America, the Carboniferous Period is divided into the Mississippian and Pennsylvanian Periods.

Amphibians and fish continued to flourish during the Carboniferous Period. *Crinoids,* like the one shown in **Figure 7,** were common in the oceans. Insects, such as giant cockroaches and dragonflies, were common on land. Toward the end of the Pennsylvanian Period, vertebrates that were adapted to life on land appeared. These early reptiles resembled large lizards.

The Permian Period

The Permian Period marks the end of the Paleozoic Era. A *mass extinction* of a large number of Paleozoic life-forms occurred at the end of the Permian Period. The continents had joined to form the supercontinent Pangaea. The collision of tectonic plates created the Appalachian Mountains. On the northwest side of the mountains, areas of desert and dry savanna climates developed. The shallow inland seas that had covered much of Earth disappeared. As the seas retreated, many species of marine invertebrates, including trilobites and eurypterids, became extinct. However, fossils indicate that reptiles and amphibians survived the environmental changes and dominated Earth in the millions of years that followed the Paleozoic Era.

Section 2 Review

1. **Summarize** how evolution is related to geologic change.
2. **Identify** two characteristics of most Precambrian rocks.
3. **Explain** why fossils are rare in Precambrian rocks.
4. **Identify** one life-form from each of the six periods of the Paleozoic Era.
5. **Explain** why the Devonian Period is commonly called the *Age of Fishes.*
6. **Describe** the kinds of life-forms that became extinct during the mass extinction at the end of the Permian Period.

CRITICAL THINKING

7. **Drawing Conclusions** Identify one way in which the formation of Pangaea affected Paleozoic life.
8. **Identifying Relationships** Why is Precambrian time—about 88% of geologic time—not divided into smaller units based on the fossil record?
9. **Analyzing Processes** Explain two ways in which the geologic record of the Paleozoic Era supports the theory of evolution.

CONCEPT MAPPING

10. Use the following terms to create a concept map: *Paleozoic Era, invertebrate, Cambrian Period, Ordovician Period, vertebrate,* and *Silurian Period.*

CHAPTER RESOURCES

Chapter Resource File

- Section Quiz GENERAL

Workbooks

- Study Guide (also in Spanish)

Section 3 The Mesozoic and Cenozoic Eras

At the end of the Permian Period, 90% of marine organisms and more than 70% of land organisms died. This episode during which an enormous number of species died, or **mass extinction,** left many resources available for the surviving life-forms. Because resources and space were readily available, an abundance of new life-forms appeared. These new life-forms evolved, and some flourished while others eventually became extinct.

The Mesozoic Era

As shown in **Figure 1,** the geologic era that began about 251 million years ago and ended about 65 million years ago is called the **Mesozoic Era.** Earth's surface changed dramatically during the Mesozoic Era. As Pangaea broke into smaller continents, the tectonic plates drifted and collided. These collisions uplifted mountain ranges such as the Sierra Nevada in California and the Andes in South America. Shallow seas and marshes covered much of the land. In general, the climate was warm and humid.

Conditions during the Mesozoic Era favored the survival of reptiles. Lizards, turtles, crocodiles, snakes, and a variety of dinosaurs flourished during the Mesozoic Era. Thus, this era is also known as the *Age of Reptiles*. The Mesozoic Era has a rich fossil record and is divided into three periods.

OBJECTIVES

- **List** the periods of the Mesozoic and Cenozoic Eras.
- **Identify** two major geologic and biological developments during the Mesozoic Era.
- **Identify** two major geologic and biological developments during the Cenozoic Era.

KEY TERMS

mass extinction
Mesozoic Era
Cenozoic Era

mass extinction an episode during which large numbers of species become extinct

Mesozoic Era the geologic era that lasted from 251 million to 65.5 million years ago; also called the *Age of Reptiles*

Figure 1 ▶ Mesozoic Timeline

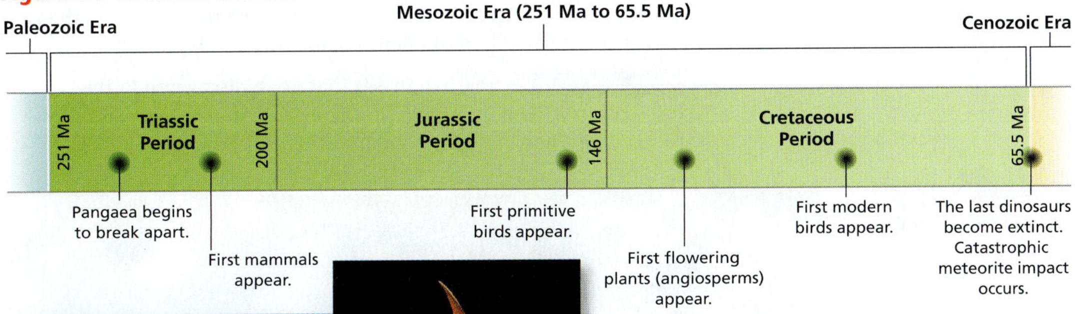

Icthyosaurus lived during the Triassic and Jurassic Periods.

The aerial *Quetzalcoatlus* lived during the Cretaceous Period.

Section 3

Focus

Overview

This section describes the conditions and the organisms of the Mesozoic Era and introduces the concept of mass extinction. The section discusses the life-forms of the Cenozoic Era and its periods and epochs, including the emergence of humans.

Bellringer

Ask students to identify movies, TV shows, or books in which dinosaurs or the Mesozoic Era have been portrayed. Have students decide whether they think the movie, TV show, or book was a realistic representation of the time period. (Answers may vary. Accept all reasonable answers.)

Motivate

Discussion — GENERAL

More Resources? Invite students to discuss why there would be more, not fewer, resources available after a mass extinction or after a period during which a large percentage of species died out. (The resources previously used by all of the extinct species now would become available to other organisms.) **LS Logical/Verbal**

CHAPTER RESOURCES

Chapter Resource File

- **Directed Reading** BASIC
- **Inquiry Lab** Dinosaur Hunt GENERAL

Technology

Transparencies
- Bellringer
- 47 The Mesozoic and Cenozoic Eras(with worksheet)

Student Edition on CD-ROM

One-Stop Planner CD-ROM
- Lesson Plan

Teach

Using the Figure — GENERAL

Comparing Characteristics Have students examine the picture of the *Archaeopteryx* on this page. Invite them to describe the physical characteristics of this Jurassic animal. Ask them to explain the similarities they observe between the *Archaeopteryx* and modern birds. (Students should note the similarities in head and beak shape, eye placement, the bones of the wings, and the preserved image of the feathers extending from the skeleton.) LS **Visual/Verbal**

READING SKILL BUILDER — BASIC

Reading Organizer Have students make a three-column chart. Have them label the columns with "Period," "Life-Forms," and "Characteristics." Have students fill in the chart as they read this page and the next page, listing the life-forms and their unique characteristics for each period of the Mesozoic Era. Students may also draw pictures of the life-forms and their characteristics in the appropriate column, next to the written description. English Language Learners LS **Verbal/Visual**

Answer to Reading Check

Answers may vary but could include *Archaeopteryx*, pterosaurs, *Apatosaurus*, and *Stegosaurus*.

Figure 2 ▶ A group of dinosaurs from the genus *Coelophysis* raced through a Triassic conifer forest in what is now New Mexico.

The Triassic Period

Dinosaurs appeared during the Triassic Period of the Mesozoic Era. Some dinosaurs were the size of squirrels. Others weighed as much as 15 tons and were nearly 30 m long. However, most of the dinosaurs of the Triassic Period were about 4 m to 5 m long and moved very quickly. As shown in **Figure 2**, these dinosaurs roamed through lush forests of cone-bearing trees and *cycads*, which are plants that resemble the palm trees of today.

Reptiles called *ichthyosaurs* lived in the Triassic oceans. New forms of marine invertebrates also evolved. The most distinctive was the ammonite, a type of shellfish that is similar to the modern nautilus. Ammonites serve as Mesozoic index fossils. The first mammals, small rodent-like forest dwellers, also appeared.

The Jurassic Period

Dinosaurs became the dominant life-form during the Jurassic Period. Fossil records indicate that two major groups of dinosaurs evolved. These groups are distinguished by their hip-bone structures. One group, called *saurischians*, or "lizard-hipped" dinosaurs, included herbivores, which are plant eaters, and carnivores, which are meat eaters. Among the largest saurischians were herbivores of the genus *Apatosaurus*, first known as *Brontosaurus*, which weighed up to 50 tons and grew to 25 m long.

The other major group of Jurassic dinosaurs, called *ornithischians*, or "bird-hipped" dinosaurs, were herbivores. Among the best known of the ornithischians were herbivores of the genus *Stegosaurus*, which were about 9 m long and about 3 m tall at the hips. In addition, flying reptiles called *pterosaurs* were common during the Jurassic Period. Like modern bats, pterosaurs flew on skin-covered wings. Fossils of the earliest birds, such as the one shown in **Figure 3**, also occur in Jurassic rocks.

Reading Check **Name two fossils that were discovered in the fossil record of the Jurassic Period. (See the Appendix for answers to Reading Checks.)**

Figure 3 ▶ The *Archaeopteryx* (AWR kee AUP tuhr IKS) was one of the first birds that appeared during the Jurassic Period.

BRAIN FOOD

The Dinosaur in the Birdcage? Though still hotly debated, there is some evidence that birds are the descendents of dinosaurs—that, in effect, dinosaurs did not die out, because some evolved into birds. The most compelling evidence of this connection is the morphological similarity between dinosaurs and birds. In addition, some dinosaur species formed huge nesting colonies, as do many of today's birds. Young hatched from eggs, and recent evidence suggests that dinosaur parents raised their young in the nest, bringing them food until they grew old enough to leave the nest. This behavior is common among birds but not among modern reptiles, whose young generally fend for themselves once they hatch. In contrast, many scientists doubt the bird-dinosaur connection and think that dinosaurs formed a third group, unlike either modern reptiles or birds.

Figure 4 ▶ This 41-foot long *Tyrannosaurus rex* was discovered near Faith, South Dakota. This specimen, named Sue, was displayed in the Field Museum in Chicago in 2000.

The Cretaceous Period

Dinosaurs continued to dominate Earth during the Cretaceous Period. Among the most spectacular dinosaurs was the carnivore *Tyrannosaurus rex,* such as the one shown in **Figure 4.** The *Tyrannosaurus rex* stood nearly 6 m tall and had huge jaws with sharp teeth that were up to 15 cm long. Also, among the common Cretaceous dinosaurs were the armored *ankylosaurs,* horned dinosaurs called *ceratopsians,* and duck-billed dinosaurs called *hadrosaurs.*

Plant life had become very sophisticated by the Cretaceous Period. The earliest flowering plants, or *angiosperms,* appeared during this period. The most common of these plants were trees such as magnolias and willows. Later, trees such as maples, oaks, and walnuts became abundant. Angiosperms became so successful that they are the dominant type of land plant today.

The Cretaceous-Tertiary Mass Extinction

The Cretaceous Period ended in another mass extinction. No dinosaur fossils have been found in rocks that formed after the Cretaceous Period. Some scientists believe that this extinction was caused by environmental changes that were the result of the movement of continents and increased volcanic activity.

However, many scientists accept the *impact hypothesis* as the explanation for the extinction of the last dinosaurs. This hypothesis is that about 65 million years ago, a giant meteorite crashed into Earth. The impact of the collision raised enough dust to block the sun's rays for many years. As Earth's climate became cooler, plant life began to die, and many animal species became extinct. As the dust settled over Earth, the dust formed a layer of iridium-laden rock. Iridium is a substance that is uncommon in rocks on Earth but that is common in meteorites.

SCILINKS. NSTA Developed and maintained by the National Science Teachers Association

For a variety of links related to this subject, go to www.scilinks.org

Topic: Mass Extinctions
SciLinks code: HQ60916
Topic: Geologic Time Scale
SciLinks code: HQ60669

Internet Activity — GENERAL

Online Museums Have students use an Internet search engine to locate the Web sites of museums that have online exhibits about dinosaurs. Students may want to check Web sites of the Smithsonian Institution, the American Museum of Natural History, or the Museum of Paleontology at the University of California at Berkeley, for example. After students view an online exhibit, have them write a report that describes the exhibit or a museum guide to the exhibit, or have them create a poster or other artwork that illustrates what they learned about dinosaurs. A worksheet designed to direct student research on this topic can be found in the **Chapter Resource File** booklet or by visiting **go.hrw.com** and entering the keyword **HQ6VEPX.** LS **Visual/Verbal**

Discussion — GENERAL

Flowering Plants Tell students that, although the Cretaceous Period is most famous for its dinosaurs, the advent of flowering plants is even more significant. Have students list flowering plants they know about and the animals that depend on them. (Students should list grasses and grains and include humans in the list of dependent animals.) Have students speculate about when these species of animals must have evolved. (after angiosperms evolved) You may also discuss the intimate connection between some angiosperms and their pollinators and how they must have evolved together. (for example, the yucca plant and yucca moth) LS **Logical/ Verbal**

Debate — ADVANCED

Dinosaur Extinction Have students form two groups to debate the causes of the mass extinction at the end of the Cretaceous Period. Ask one group of students to research and present evidence supporting the idea that climate/environmental changes caused the extinction. Have the other group research and support the impact hypothesis. After the debate, invite all students to ask questions of the debaters, to comment on the different points of view, and to offer their own viewpoints on the topic. LS **Logical/Verbal**

SKILL BUILDER — GENERAL

Vocabulary Explain to students that *dino-* comes from the Greek adjective *deinos,* which means "terrifying." The Greek word *sauros,* which means "lizard," is the source for *-saur.* Thus, *dinosaur* means "terrible lizard." Lizards are reptiles and in many ways resemble dinosaurs, which thrived in the Age of Reptiles. Ask students how this information can help them remember that the Mesozoic Era is the "Age of Reptiles." LS **Verbal/Logical**

CHAPTER RESOURCES

Chapter Resource File

Internet Activity
- Online Museums GENERAL

Teach, *continued*

BIOLOGY CONNECTION ADVANCED

Primitive Mammals Tell students that some types of primitive mammals still exist today. *Monotremes,* such as today's platypus, lay eggs and produce milk through their skin pores. *Marsupials,* including the kangaroo and the opossum, have live young that undergo most of their development in the mother's exterior pouch. Invite students to learn about the evolution of these mammal groups and about how they differ from placental mammals such as humans. Have them present their findings as oral reports to the class. LS **Verbal**

Using the Figure GENERAL

Mammal Characteristics Have students describe the mammalian characteristics of the tarsier shown on this page. (The tarsier is covered with hair.) Tell students that the characteristics that they can't see include that its body contains fat as insulation, the tarsier is warm-blooded, and its body produces milk to feed its offspring. Answer to caption question: Hair and fat help mammals survive in cool climates. Because mammals are warm-blooded, they do not need to absorb heat from their environment. LS **Logical/Verbal**

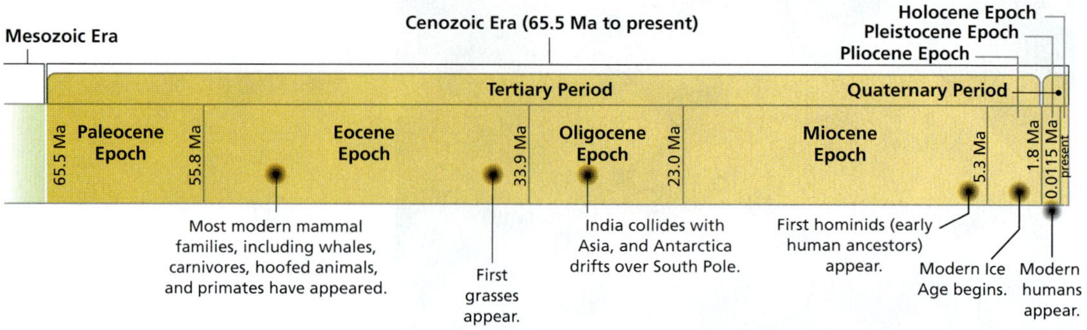

Figure 5 ▶ Cenozoic Timeline

The Cenozoic Era

Cenozoic Era the current geologic era, which began 65.5 million years ago; also called the *Age of Mammals*

As shown in **Figure 5,** the **Cenozoic Era** is the division of geologic time that began about 65 million years ago and that includes the present period. During this era, the continents moved to their present-day positions. As tectonic plates collided, huge mountain ranges, such as the Alps and the Himalayas in Eurasia, formed.

During the Cenozoic Era, dramatic changes in climate have occurred. At times, continental ice sheets covered nearly one-third of Earth's land. As temperatures decreased during the ice ages, new species that were adapted to life in cooler climates appeared. Mammals became the dominant life-form and underwent many changes. The Cenozoic Era is thus commonly called the *Age of Mammals*.

The Quaternary and Tertiary Periods

The Cenozoic Era is divided into two periods. The Tertiary Period includes the time before the last ice age. The Quaternary Period began with the last ice age and includes the present. These periods have been divided into seven epochs. The Paleocene, Eocene, Oligocene, Miocene, and Pliocene Epochs make up the *Tertiary Period*. The Pleistocene and Holocene Epochs make up the *Quaternary Period*.

Figure 6 ▶ The *tarsier* is the sole modern survivor of a group of primates common during the earlier Cenozoic Era. ***Why are mammals better suited to cool climates than reptiles are?***

The Paleocene and Eocene Epochs

The fossil record indicates that during the Paleocene Epoch many new mammals, such as small rodents, evolved. The first primates also evolved during the Paleocene Epoch. A modern survivor of an early primate group is shown in **Figure 6.**

Other mammals, including the earliest known ancestor of the horse, evolved during the Eocene Epoch. Fossil records indicate that the first whales, flying squirrels, and bats appeared during this epoch. Small reptiles continued to flourish. Worldwide, temperatures dropped by about 4°C at the end of the Eocene Epoch.

ENVIRONMENTAL SCIENCE CONNECTION ADVANCED

The Isolation of Antarctica As Pangaea broke apart, the continent that became Antarctica drifted far south and became completely surrounded by the Southern Ocean, cutting it off from other ocean currents. The climate became extremely cold, and glaciers formed. Invite students to discuss how glaciation affected Earth, the ocean, and Earth's climate. (Answers should indicate that glaciation cooled Earth's climate and lowered sea levels as Antarctic ice took up vast quantities of water.) LS **Logical/Verbal**

SKILL BUILDER BASIC

Math The earliest mammals were tiny and rodent-like and appeared in the Mesozoic Era around the time of the first dinosaurs, about 200 million years ago. The great explosion of mammal diversity and dominance on Earth did not begin until about 65 million years ago. How long did mammals exist on Earth before they began to assume a dominant role? (200 million years – 65 million years = 135 million years) LS **Logical**

The Oligocene and Miocene Epochs

During the Oligocene Epoch, the Indian subcontinent began to collide with the Eurasian continent, which caused the uplifting of the Himalayas. The worldwide climate became significantly cooler and drier. This change in climate favored grasses as well as cone-bearing and hardwood trees. Many early mammals became extinct. However, large species of deer, pigs, horses, camels, cats, and dogs flourished. Marine invertebrates, especially clams and snails, also continued to flourish.

During the Miocene Epoch, circumpolar currents formed around Antarctica, and the modern Antarctic icecap began to form. By the late Miocene Epoch, tectonic forces and dropping sea levels caused the Mediterranean Sea to dry up and refill several times. The largest known land mammals existed during this epoch. Miocene rocks also contain fossils of horses, camels, deer, rhinoceroses, pigs, raccoons, wolves, foxes, and the earliest saber-toothed cats, which are now extinct. The earliest human ancestors may date to this epoch.

SCILINKS. NSTA Developed and maintained by the National Science Teachers Association

For a variety of links related to this subject, go to www.scilinks.org

Topic: Geologic Periods and Epochs
SciLinks code: HQ60667

The Pliocene Epoch

During the Pliocene Epoch, predators—including members of the bear, dog, and cat families—evolved into modern forms. Herbivores, such as the giant ground sloth shown in **Figure 7,** flourished. The first modern horses also appeared in this epoch.

Toward the end of the Pliocene, dramatic climatic changes occurred, and the continental ice sheets began to spread. With more and more water locked in ice, sea level fell. The Bering land bridge appeared between Eurasia and North America. Changes in Earth's crust between North America and South America formed the Central American land bridge. Various species migrated between the continents across these two land bridges.

Reading Check **Why did sea level fall in the Pliocene Epoch? (See the Appendix for answers to Reading Checks.)**

Figure 7 ▶ Giant ground sloths lived during the late Pliocene in parts of North America and South America. These slow-moving leaf-eaters could grow as large as an African bull elephant and weigh as much as 5 tons.

BRAIN FOOD

Circumpolar Waters One of the reasons Antarctica is the coldest place on Earth is because it is completely encircled by the Southern Ocean, which blocks warmer ocean currents from approaching the continent. About 25 million years ago, shifting tectonic plates caused a cold, deep-ocean belt of water to form around the Antarctic continent. The belt of cold water flows continuously around Antarctica, isolating it from warmer ocean currents. Only the deepest, coldest waters from currents north of the Antarctic penetrate the Southern Ocean, and then only at great depth. Most of the water that rises to the Southern Ocean surface is not warmed; instead it is chilled by freezing Antarctic air masses, and sinks again to the cold Antarctic Bottom Water.

Homework — ADVANCED

Horse History Scientists think that today's horse is one of a small number of remnants of an evolutionary line that once included many species. Invite students to research the evolution of the horse and other members of the genus *Equus,* such as zebras, bringing in drawings of early and intermediate forms between *Eohippus* and modern horses. Students may offer their own explanations about why so few *Equus* species remain today while so many became extinct. **LS Visual/Logical**

Answer to Reading Check

During ice ages, water from the ocean was frozen as ice, so the amount of liquid water in the seas decreased and sea level fell.

Close

Reteaching — BASIC

Create a Brochure Organize students into groups. Have each group create a travel brochure for one of the geologic periods or epochs in this section. Direct students to divide up the tasks of designing, writing, and illustrating the brochure. Brochures should include information about environmental conditions and the animals and plants of the time. Post the brochures in the class where all students can see and read them. **LS Visual/Verbal** Co-op Learning

Quiz — GENERAL

1. What effect did the land bridges have on animal distribution during the ice ages? (They allowed animals to move from one continent to another.)
2. What is one possible cause of the mass extinction during the Cretaceous Era? (One possible cause is a meteorite impact.)

Close, *continued*

Alternative Assessment — GENERAL

Timeline Have pairs of students work together to produce their own timeline of the Mesozoic and Cenozoic Eras. Students should label each era and epoch, and list or draw the dominant animals and plants of the time. **LS Verbal/ Visual**

Answers to Section Review

1. Triassic: ammonites; Jurassic: dinosaurs; Cretaceous: flowering plants
2. Geologic: The breakup of Pangaea began, and a major meteorite impacted Earth. Biological: Birds appeared, and flowering plants appeared.
3. Tertiary Period: Paleocene Epoch: rodents, Eocene Epoch: horses, Oligocene Epoch: cats, Miocene Epoch: grazing herds, Pliocene Epoch: bears; Quaternary Period: Pleistocene Epoch: mammoths, Holocene Epoch: humans
4. Geologic: The continents moved to their present locations and ice sheets advanced and retreated several times. Biological: Mammals became dominant, and humans developed.
5. Ice ages led to the rise of hairy, warm-blooded mammals that could survive the cold.
6. We are living in the Cenozoic Era, the Quaternary Period, and the Holocene Epoch.
7. A combination of environmental changes and a meteorite impact caused the extinction of the dinosaurs and many other organisms, leaving many niches open for mammalian species to fill. Mammals were able to survive worldwide cooling and the following ice ages.
8. Scientists probably used evidence of the last ice age to divide the Cenozoic Era into the Tertiary and Quaternary Periods.
9. I would look for evidence of the meteorite impact that happened at that time.
10. The *Mesozoic Era* is known as the *Age of Reptiles,* included the *Triassic Period,* the *Jurassic Period,* and the *Cretaceous Period,* and was followed by the *Cenozoic Era,* which is known as the *Age of Mammals,* and included the *Tertiary Period* and the *Quaternary Period.*

Figure 8 ▶ This painting from the Stone Age was made by early humans between 15,000 and 13,000 years ago in a cave in Lascaux, France.

The Pleistocene Epoch

The Pleistocene Epoch began 1.8 million years ago. In Eurasia and North America, ice sheets advanced and retreated several times. Some animals had characteristics that allowed them to endure the cold climate, such as the thick fur that covered woolly mammoths and woolly rhinoceroses. Many other species survived by moving to warmer regions. Some species, such as giant ground sloths and dire wolves, became extinct.

Fossils of the earliest modern humans (*Homo sapiens*) were discovered in Pleistocene sediments. Evidence of modern humans, such as the cave painting shown in **Figure 8,** indicates that early humans may have been hunters.

The Holocene Epoch

The Holocene Epoch, which includes the present, began about 11,500 years ago, as the last glacial period ended. As the ice sheets melted, sea level rose about 140 m, and the coastlines took on their present shapes. The North American Great Lakes also formed as the last ice sheets retreated. During the early Holocene Epoch, modern humans developed agriculture and began to make and use tools made of bronze and iron.

Human history is extremely brief. If you think of the entire history of Earth as one year, the first multicellular organisms would have appeared in September. The dinosaurs would have disappeared at 8 P.M. on December 26. Modern humans would have would not have appeared until 11:48 P.M. on December 31.

Section 3 Review

1. **List** the periods of the Mesozoic Era, and describe one major life-form in each division.
2. **Identify** two major geologic and two major biological developments of the Mesozoic Era.
3. **List** the periods and epochs of the Cenozoic Era, and describe one major life-form in each division.
4. **Identify** two major geologic and two major biological developments of the Cenozoic Era.
5. **Explain** how the ice ages affected animal life during the Cenozoic Era.
6. **Identify** the era, period, and epoch we are in today.
7. **Describe** the worldwide environmental changes that set the stage for the Age of Mammals.

CRITICAL THINKING

8. **Drawing Conclusions** Explain the criteria scientists may have used for dividing the Cenozoic Era into the Tertiary and Quaternary Periods.
9. **Identifying Relationships** Suppose that you are a geologist who is looking for the boundary between the Cretaceous and Tertiary Periods in an outcrop. What characteristics would you look for to determine the location of the boundary? Explain your answer.

CONCEPT MAPPING

10. Use the following terms to create a concept map: *Mesozoic Era, Age of Reptiles, Jurassic Period, Triassic Period, Cretaceous Period, Cenozoic Era, Age of Mammals, Tertiary Period,* and *Quaternary Period.*

CHAPTER RESOURCES

Chapter Resource File

- Section Quiz GENERAL

Workbooks

- Study Guide (also in Spanish)

Chapter 9 Highlights

Sections

1 Geologic Time

Key Terms

geologic column, 211
era, 214
period, 214
epoch, 214

Key Concepts

- The geologic column is based on observations of the relative ages of rock layers throughout the world.
- Geologists used major changes in Earth's climate and extinctions recorded in the fossil record to divide the geologic time scale into smaller units.
- Geologic time is subdivided into eons, eras, periods, epochs, and ages.

2 Precambrian Time and the Paleozoic Era

Key Terms

evolution, 215
Precambrian time, 216
Paleozoic Era, 218

Key Concepts

- Precambrian rocks may contain valuable minerals but few fossils.
- Evolution is the gradual development of organisms from other organisms. Evidence for the theory of evolution occurs throughout the fossil record.
- The rock record reveals the evolution of marine invertebrates and vertebrates during the Paleozoic Era.

3 The Mesozoic and Cenozoic Eras

Key Terms

mass extinction, 221
Mesozoic Era, 221
Cenozoic Era, 224

Key Concepts

- The rock record of the Mesozoic Era reveals an environment that favored the development of reptiles.
- The Mesozoic Era ended with a mass extinction that included the extinction of the dinosaurs.
- The rock record of the Cenozoic Era includes the present period and reveals the rise of mammals as a predominant life-form.

Chapter Highlights

Alternative Assessment — ADVANCED

Science Consultants Have students pretend that they have been hired by a theme-park owner to produce four new theme parks. The owner wants to create a theme park for every geologic era from Precambrian time through the Cenozoic Era (but not including the Holocene Epoch). Each theme park will feature the plants and animals of the era divided into periods and epochs, with informational displays and activities for visitors. Have students write a proposal and draw a plan for each theme park, detailing its features, displays, and activities. Students may wish to create a model. Students may work in groups, with each group covering one era. Groups should present their plans and illustrations to the class. LS **Verbal/Visual/Kinesthetic**

CHAPTER RESOURCES

Chapter Resource File

- **Concept Review** GENERAL
- **Critical Thinking** ADVANCED
- **Math Skills** GENERAL
- **Graphing Skills** GENERAL
- **Chapter Test A** GENERAL
- **Chapter Test B** ADVANCED

Workbooks

- **Study Guide** (also in Spanish)
- **Assessments** (Spanish)

Technology

Classroom Videos
- Brain Food Video Quiz

Chapter Review

Assignment Guide	
Section	**Questions**
1	1–5, 7–11, 20
2	12, 13, 22, 24
3	6, 9, 14–17, 19, 23, 26
2 and 3	18, 25
1–3	21, 27–30

Using Key Terms

1–7. Answers may vary but should show that students understand the definitions of and differences between key terms.

Understanding Key Concepts

8. b
9. b
10. a
11. b
12. b
13. d
14. b
15. c
16. a
17. d

Short Answer

18. The earliest plants were probably blue-green algae that lived in the oceans during Precambrian time. During the Silurian and Devonian Periods of the Paleozoic Era, land plants such as ferns and seed-bearing plants developed. By the Carboniferous Period, forests of such plants covered the land. During the Cretaceous Period of the Mesozoic Era, the earliest flowering plants, or angiosperms, developed. They were very successful and are the dominant land plants today.
19. Answers may vary but may cite the ideas that climate change caused the extinctions or that a meteorite impact led to the Cretaceous-Tertiary extinction. Evidence for the meteorite impact theory includes iridium-rich rock layers and evidence of a crater in the ocean near the Yucatan peninsula that are the right age.
20. Scientists use rock types to distinguish the layers, as well as the kinds of fossils found in the layers.
21. Answers may vary. Sample answer: fish and reptiles; Their long-term success may be due to their ability to adapt to changing conditions.

Critical Thinking

22. Most Precambrian rocks have been so greatly changed by tectonic activity that it is difficult to identify the original order of layers. This same activity may have destroyed fossils, leaving too few to help scientists divide Precambrian time into periods. In addition, many Precambrian organisms may have lacked the hard parts that usually form fossils.
23. You would expect to find the element carbon in these deposits.

Chapter 9 Review

Using Key Terms

Use each of the following terms in a separate sentence.

1. *evolution*
2. *geologic column*
3. *period*

For each pair of terms, explain how the meanings of the terms differ.

4. *era* and *epoch*
5. *period* and *era*
6. *Mesozoic Era* and *Cenozoic Era*
7. *Precambrian time* and *Paleozoic Era*

Understanding Key Concepts

8. The geologic time scale is a
   a. scale for weighing rocks.
   b. scale that divides Earth's history into time intervals.
   c. rock record of Earth's past.
   d. collection of the same kind of rocks.

9. Scientists are able to determine the absolute ages of most rock layers in a geologic column by using
   a. the law of superposition.
   b. radiometric dating.
   c. rates of deposition.
   d. rates of erosion.

10. To determine the age of a specific rock, scientists might correlate it with a layer in a geologic column that has the same relative position and
    a. fossil content. **c.** temperature.
    b. weight. **d.** density.

11. Geologic periods can be divided into
    a. eras. **c.** days.
    b. epochs. **d.** months.

12. Precambrian time ended about
    a. 4.6 billion years ago.
    b. 542 million years ago.
    c. 65 million years ago.
    d. 25 thousand years ago.

13. The most common fossils that occur in Precambrian rocks are
    a. graptolites. **c.** eurypterids.
    b. trilobites. **d.** stromatolites.

14. The first vertebrates appeared during
    a. Precambrian time. **c.** the Mesozoic Era.
    b. the Paleozoic Era. **d.** the Cenozoic Era.

15. The *Age of Reptiles* is the name commonly given to
    a. Precambrian time. **c.** the Mesozoic Era.
    b. the Paleozoic Era. **d.** the Cenozoic Era.

16. The first flowering plants appeared during the
    a. Cretaceous Period.
    b. Triassic Period.
    c. Carboniferous Period.
    d. Ordovician Period.

17. The *Age of Mammals* is the name commonly given to
    a. Precambrian time. **c.** the Mesozoic Era.
    b. the Paleozoic Era. **d.** the Cenozoic Era.

Short Answer

18. Write a short paragraph that describes the evolution of plants that is indicated by the fossil record.
19. Describe the events that may have led to the Cretaceous-Tertiary mass extinction. What evidence have scientists discovered that supports their hypothesis?
20. Describe the criteria that scientists use to divide a geologic column into different layers.

21. Identify two organisms that are found in the fossil record of a different geologic era but that are still living on Earth today. Identify what characteristic(s) have given them their long-term success.

Critical Thinking

22. **Analyzing Ideas** Why can Precambrian time not be divided into periods by using fossils?

23. **Applying Ideas** Many coal and oil deposits formed during the Carboniferous Period. What element would you expect to find in both oil and coal?

24. **Identifying Relationships** What information in the geologic record might lead scientists to infer that shallow seas covered much of Earth during the Paleozoic Era?

25. **Making Comparisons** Compare the causes of the Permian mass extinction with those of the Cretaceous mass extinction.

Concept Mapping

26. Use the following terms to create a concept map: *geologic time, Paleozoic Era, Mesozoic Era, stromatolite, Precambrian time, eurypterid, crinoid, Cenozoic Era, trilobite, saurischian, ornithischian, dinosaur, mammal,* and *human*.

Math Skills

27. **Scientific Notation** Write the beginning and end dates of each geologic era in scientific notation.

28. **Making Calculations** The Methuselah tree in California is 4.6×10^3 years old. How many times older than this tree is Earth ?

Writing Skills

29. **Creative Writing** Write an essay about a trip back in time that includes descriptions of the organisms that lived during one of the geologic periods described in this chapter.

30. **Writing from Research** Research the discoveries made by British anthropologists Louis S. B. Leakey and Mary Leakey in Olduvai Gorge in Tanzania, Africa. Write a report about your findings.

Interpreting Graphics

The graph below shows average global temperatures since Precambrian time. Use this graph to answer the questions that follow.

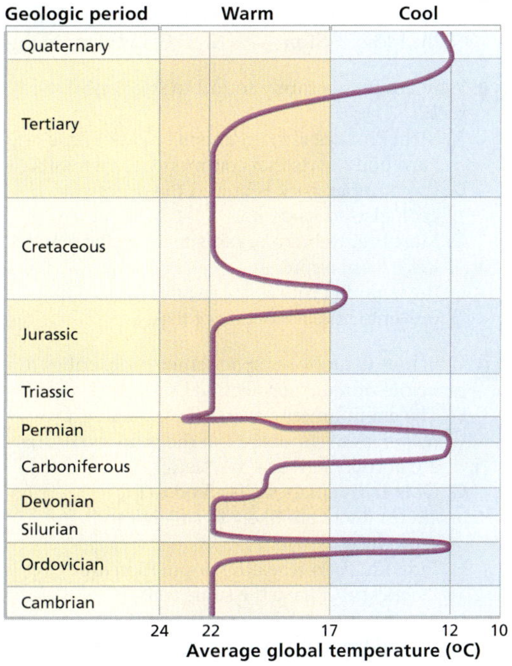

31. During which two periods were Earth's average global temperature the highest?

32. During which periods did Earth's average global temperature decrease?

33. Based on the graph, could climate change have caused the Permian mass extinction? Is climate change a likely cause of the mass extinction at the Cretaceous-Tertiary boundary? Explain your answer.

Chapter Review

24. If scientists find fossils of marine organisms in rocks from all over the world and from places where no seas exist today, they might infer that seas were once more widespread on Earth.

25. The Permian mass extinction was likely caused by environmental changes that resulted from the formation of Pangaea, whereas the Cretaceous extinction was likely caused by a meteor impact or by climate change.

Concept Mapping

26. Answers may vary but should include all of the terms listed. Sample answers appear at the end of this Teacher Edition.

Math Skills

27. Precambrian time began 4.6×10^9 years ago and ended 5.42×10^8 years ago. The Paleozoic Era began 5.42×10^8 years ago and ended 2.51×10^8 years ago. The Mesozoic Era began 2.51×10^8 years ago and ended 6.55×10^7 years ago. The Cenozoic Era began 6.55×10^7 years ago and has not ended.

28. $4.6 \times 1,000 = 4,600$ years. Earth is 4.6 billion years old, so Earth is 1,000,000 times older than the Methuselah tree.

Writing Skills

29. Answers may vary. Accept all reasonable answers.

30. Answers may vary. Accept all reasonable answers.

Interpreting Graphics

31. Permian and Triassic

32. Ordovician, Devonian, Carboniferous, Triassic, Jurassic, and Tertiary

33. Answers may vary, but should state that climate change is a possible cause of the mass extinction at the Permian-Triassic boundary because the temperature changed so significantly at the end of the Permian Period. However, temperature did not change significantly at the Cretaceous-Tertiary boundary, and therefore is not as likely a cause of that extinction event.

Estimated Time

To give students practice under more realistic testing conditions, allow them 30 minutes to answer all of the questions in this practice test.

TEST DOCTOR

Question 1 Answer B is correct. According to the scant fossil record, life on Earth during Precambrian time was primitive, so answer A is incorrect. Dinosaurs first appeared during the Triassic Period but were not the primary life-form, so answer C is incorrect. Though they continued to dominate during the Cretaceous Period, answer D, dinosaurs did not become the dominant life form until the Jurassic Period.

Question 4 Answer G is the best answer choice. Answer F is incorrect because ecosystems have limited carrying capacities. Answer H is incorrect because individuals in a population are not identical and, in fact, similarity does not foster natural selection. Answer I is incorrect because only some offspring live until maturity.

Question 9 Answer G is correct. Speed might be determined by fossilized footprints, not by teeth. The color of the dinosaur's skin would require a frozen sample, but even then, the color would be difficult to determine. Teeth would not help when determining the dinosaur's mating habits.

Chapter 9 Standardized Test Prep

Understanding Concepts

Directions (1–4): **For *each* question, write on a separate sheet of paper the letter of the correct answer.**

1. Dinosaurs first became the dominant life-forms during which geologic period?
   A. Quaternary Period **C.** Triassic Period
   B. Jurassic Period **D.** Cretaceous Period

2. Pangaea broke into separate continents during
   F. the Paleozoic Era **H.** the Cenozoic Era
   G. the Mesozoic Era **I.** Precambrian time

3. Why are fossils rarely found in Precambrian rock?
   A. Most Precambrian organisms did not have hard body parts that commonly form fossils.
   B. Precambrian rock is buried too deeply for geologists to study it.
   C. Most Precambrian organisms were too small to leave fossil remains.
   D. Precambrian rock is made of a material that prevented the formation of fossils.

4. Which of the following statements describes a principle of natural selection?
   F. The environment has more than enough resources to support all of the individuals that are born in a given ecosystem.
   G. Only individuals well-suited to the environment are likely to survive and reproduce.
   H. Individuals in a healthy population are identical and have the same traits.
   I. Most species produce plentiful offspring that will all live until maturity and reproduce.

Directions (5–7): **For *each* question, write a short response.**

5. What is the term for the largest unit of geologic time?

6. What is the term for the gradual development of organisms from other organisms by means of natural selection?

7. Why is the Cenozoic Era also known as the Age of the Mammals?

Reading Skills

Directions (8–11): **Read the passage below. Then, answer the questions below on a separate sheet of paper.**

The Discovery of a Dinosaur

In 1995, paleontologist Paul Sereno was working in a previously unexplored region of Morocco when his team made an astounding discovery—an enormous dinosaur skull. The skull was nearly 1.6 m long. Given the size of the skull, Sereno concluded that the skeleton of the animal that it came from must have been about 14 m long—about as long as a full-sized school bus. The dinosaur was even larger than the *Tyrannosaurus rex.* The newly discovered dinosaur was thought to be 90 million years old. It most likely chased other dinosaurs by running on large, powerful hind legs, and its bladelike teeth must have meant certain death for its prey.

8. Which of the following is evidence that the dinosaur described in the passage above was most likely a predator?
   A. It had sharp, bladelike teeth.
   B. It had a large skeleton and powerful hind legs used for running.
   C. It was found next to the bones of a smaller animal.
   D. It was more than 90 million years old.

9. What types of information do you think that fossilized teeth provide about an organism?
   F. the color of its skin
   G. the types of food it ate
   H. the speed at which it ran
   I. the mating habits it had

10. According to the passage, which of the following statements is true?
    A. This dinosaur was most likely a predator.
    B. This skull belonged to a large Tyrannosaurus rex.
    C. This dinosaur had powerful arms.
    D. This dinosaur ate mainly plants and berries.

11. What are some methods that scientists might have used to determine that the age of the dinosaur skull was 90 million years old?

Answers

Part A

1. B
2. G
3. A
4. G
5. an era
6. evolution
7. Mammals became the dominant life-forms and underwent many evolutionary changes during this era.

Part B

8. A
9. G
10. A
11. Answers may vary. See Test Doctor for a detailed scoring rubric.

Part C

12. D
13. F
14. the mass extinction of the dinosaurs
15. During the Little Ice Age, glaciers did not increase in size by very much, but local freezing dates came earlier and thawing dates came later.

Interpreting Graphics

Directions (12–15): **For *each* question below, record the correct answer on a separate sheet of paper.**

The timeline below shows the time divisions of the Mesozoic and Cenozoic eras. Use this timeline to answer questions 12 through 14.

The Mesozoic and Cenozoic Eras

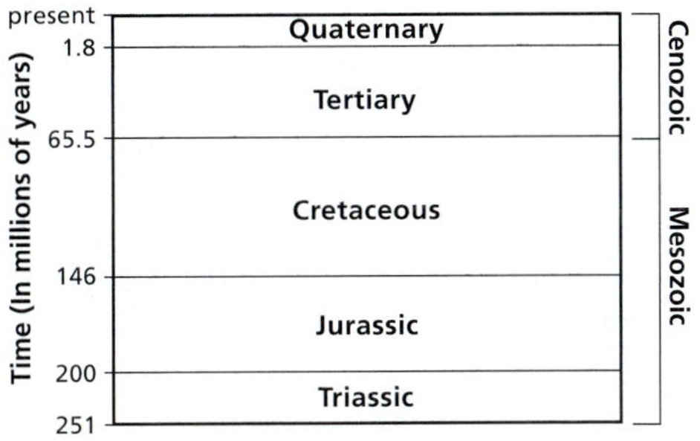

12 Human civilization developed during which of the following periods of time?
A. Triassic Period
B. Jurassic Period
C. Tertiary Period
D. Quaternary Period

13 If Earth formed about 4.6 billion years ago, what percentage of Earth's total history did the Cenozoic Era fill?
F. about 1.5%
G. about 10.5%
H. about 15%
I. about 50%

14 Which event coincides with the start of the Cenozoic Era?

The graph below shows data on global temperature changes during the last millennium. Use this graph to answer question 15.

The Medieval Warm Period and the Little Ice Age

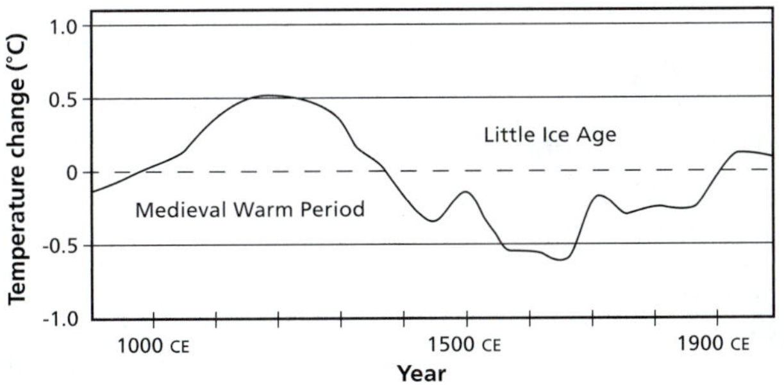

15 How do you think the temperature changes during the Little Ice Age of the Middle Ages affected the freezing and thawing of global waters? Explain your answer.

Test TIP

Simply keeping a positive attitude during any test will help you focus on the test and likely improve your score.

Standardized Test Prep

ASSESSMENT DOCTOR

Question 11 Full-credit answers should include the following points:
- an understanding that scientists not only can, but usually do, use multiple means of authentication when dating the remains of ancient plants and animals
- scientists could also use nearby index fossils and stratigraphic layers

Question 13 Answer F is correct. To answer this question correctly, a student needs to divide 65 million years (the length of the Cenozoic era) by 4.6 billion years (the age of the Earth). To save room on the calculator screen, some students may find it easier to drop the extra zeroes and simply divide 65 by 4,600. The answer is then multiplied by 100 to convert it to a percentage. The correct answer is 1.5%.

Test Prep Correlations *National Science Education Standards*

HNS 2c: items 1, 2, 3, 6
LS 3a: items 4, 6, 8
LS 3c: items 4, 6, 8
SAI 2c: item 11
SAI 2d: item 13
SAI 2e: items 11, 15
ST 2c: items 8, 9, 11, 15
UCP 4: items 4, 6

CHAPTER RESOURCES

State Resources

For specific resources for your state, visit **go.hrw.com** and type in the keyword **HSHSTR.**

Skills Practice Lab

History in the Rocks

Teacher's Notes

Time Required
one 45-minute class period

Lab Ratings
EASY → HARD

Teacher Preparation: 1
Student Setup: 1
Concept Level: 3
Cleanup: 1

Skills Acquired
- Identifying
- Analyzing
- Recognizing Patterns
- Interpreting Models

The Scientific Method
In this lab, students will
- Observe and Compare
- Analyze Data
- Determine Relationships
- Draw Conclusions

Materials
The materials listed on the page are enough for individual students.

Tips and Tricks
You may wish to have pairs of students work together to discuss their analysis of the fossil record as given in the lab. The following list describes the order of layers in each column of Figure B, from bottom to top:

Column 1: limestone, sandstone with Ordovician cephalopod, shale with Mississippian blastoid, limestone with Cretaceous echinoid, sandstone with Pennsylvanian brachiopod, shale with Triassic cephalopod.

Column 2: shale with Ordovician trilobite, shale with Silurian brachiopod, limestone with Devonian trilobite, shale with Cretaceous gastropod, shale with Cretaceous shark tooth, limestone with Cretaceous cephalopod.

Column 3: shale with Silurian brachiopod, limestone with Devonian brachiopod, shale with no fossils, sandstone with Pennsylvanian brachiopod, limestone with Devonian trilobite, shale with Ordovician trilobite.

Column 4: metamorphic rock (no fossils), limestone with no fossils, shale with Ordovician trilobite, sandstone with Mississippian cephalopod, shale with Triassic cephalopod, limestone with Cretaceous echinoid.

Chapter 9 **Skills Practice Lab**

History in the Rocks

Objectives
- **Apply** the law of superposition to sample rock columns.
- **Demonstrate** the use of index fossils in determining relative and absolute ages.
- **Evaluate** the usefulness of different methods used for determining relative and absolute age.

Materials
paper
pencil

Geologists have discovered much about the geologic history of Earth by studying the arrangement of fossils in rock layers, as well as by studying the arrangement of the rock layers themselves. Fossils provide clues about the environment in which the organism that formed the fossil existed. Scientists can determine the age of the rocks in which fossils occur because the ages of many fossils have been determined by radiometric dating of associated igneous rocks. Radiometric dating, fossil age, and rock arrangement are all used to determine changes that have occurred in the arrangement of the rock layers through geologic time. In this lab, you will discover how the geologic history of an area can be determined by examining the arrangement of fossils and rock layers.

Figure A

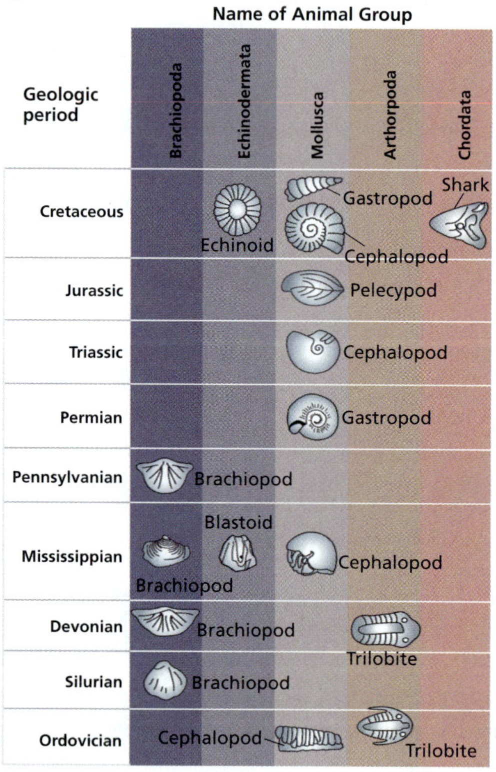

PROCEDURE

1. Study the index fossils shown in **Figure A.** Note their placement in related groups and the geologic periods in which they lived.
2. Select one of the four fossil arrangements shown in **Figure B.** This figure shows how some of these fossils may occur in a series of rock layers. Record the number of the arrangement that you are using.
3. Using **Figure A,** identify all the fossils in your arrangement and the geologic time in which the organisms that formed the fossils lived.
4. List the fossil names in order from bottom to top.
5. Do the fossils in your arrangement appear in the order of geologic time?
6. Do the fossils in your arrangement show a complete sequence of geologic periods? If not, which periods are missing?
7. Repeat steps 2–6 with each of the other three fossil arrangements.

CHAPTER RESOURCES

Chapter Resource File
- Datasheet for Chapter Lab GENERAL
- Lab Notes and Answers

Figure B

ANALYSIS AND CONCLUSION

1. **Analyzing Processes** What processes or events might explain the order in which each of the fossil arrangements was found?

2. **Evaluating Assumptions** Based on your observations in the procedure, why is it necessary that a fossil be found in a wide variety of geographic areas to be considered an index fossil?

3. **Explaining Events** Study arrangement 3 in **Figure B.** Note that there is a rock layer that contains no fossils between two rock layers that contain fossils. How might this have occurred?

Extension

1. **Examining Data** Collect fossils in your area. Identify the fossils you have collected, and describe what your area was like when the organisms existed.

2. **Research** Find out how index fossils are used to help petroleum geologists locate oil reservoirs. Then, use that information to give an oral report to your class.

Answers to Procedure

5. Column 1: no; Column 2: yes; Column 3: no; Column 4: yes
6. Column 1: no, missing periods: Silurian, Devonian, Permian, and Jurassic; Column 2: no, missing periods: Mississippian, Pennsylvanian, Permian, Triassic, and Jurassic; Column 3: no, missing periods: Mississippian, Permian, Triassic, Jurassic, and Cretaceous; Column 4: no, missing periods: Silurian, Devonian, Pennsylvanian, Permian, and Jurassic

Answers to Analysis and Conclusion

1. Deformation, including faulting or folding, of rocks in which fossils occur might explain an apparently out-of-order fossil.
2. If a fossil is found only in a small area, it cannot be used to date rocks in other areas of the world.
3. The rock may have formed in a depositional environment in which organisms did not thrive or in which the remains of organisms were not preserved.

Answers to Extension

1. Answers may vary. Accept all reasonable answers. Students may find a variety of fossils in their area.
2. Answers may vary. Accept all reasonable answers. Many of the index fossils that are used by petroleum geologists are microscopic or must be viewed with some magnification.

TEACHER TESTED & APPROVED

Alexander Dvorak
Heritage School
New York City, NY

Maps in Action

Fossil Evidence for Gondwanaland

Group Activity — GENERAL

Dinosaur Map Divide the class into three groups. Have each group of students research the species and distribution of dinosaurs in North America, Europe, or Asia. Students may divide up project tasks, and have some students do the research online, some research in the library, and others make a map with pictures of each dinosaur species at the site where that species' fossils were found. Students should indicate on their map the geologic time period during which each dinosaur species lived at that site. Ask each group to make an oral presentation to the class, using its map to illustrate major points. LS **Visual/Logical**

Co-op Learning

Answers to Map Skills Activity

1. South America, Africa, Australia, Asia (India), and Antarctica
2. Africa, Asia (India), and Antarctica
3. *Mesosaurus*
4. South America, Antarctica, Asia (India), and Australia were likely touching Africa.
5. Africa, Australia, and Asia (India) were likely touching Antarctica.
6. Answers may vary. Sample answer: The organisms did not appear in other parts of the world that were not part of the Gondwanaland part of Pangaea. Thus, it is more likely that they evolved together at the same time.
7. Answers may vary. Students may state that because the trend has been for the continents to separate, they would likely continue to move farther apart.

CHAPTER RESOURCES

Technology

Transparencies
- 48 Fossil Evidence for Gondwanaland (with worksheet)

MAPS in Action

Fossil Evidence for Gondwanaland

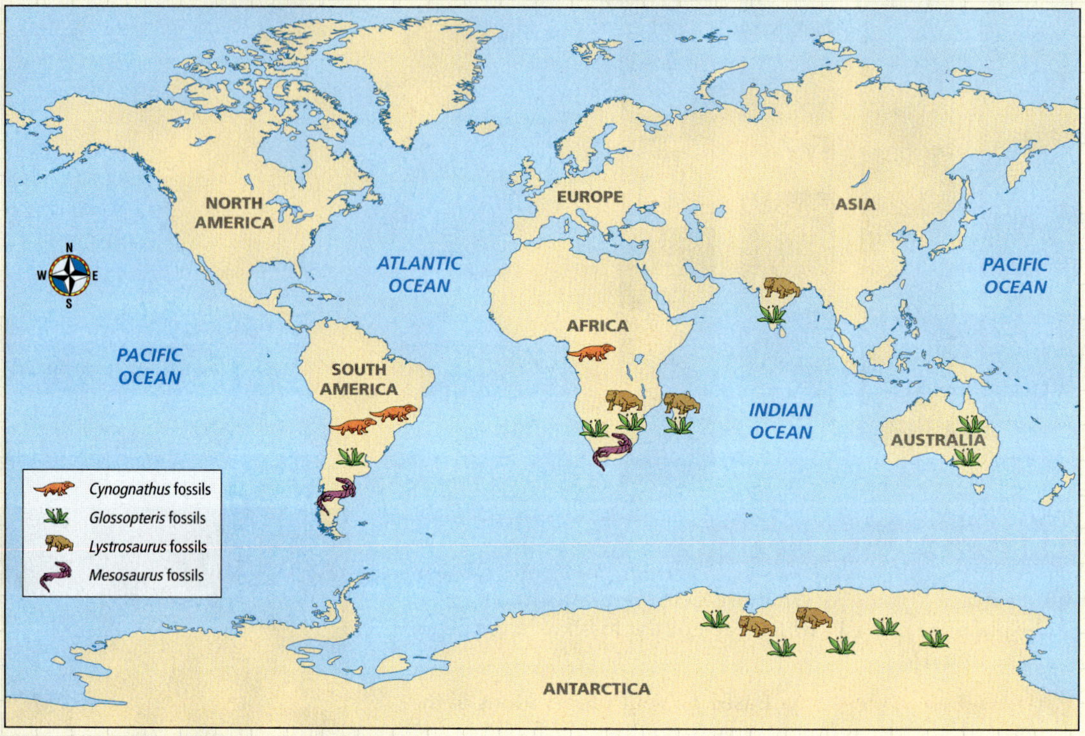

Map Skills Activity

This map shows areas where selected fossils have been found. Use the map to answer the questions below.

1. **Using the Key** On which continents have fossils from ferns of the genus *Glossopteris* been found?
2. **Using the Key** On which continents have fossils of organisms from the genus *Lystrosaurus* been found?
3. **Making Comparisons** Which fossil shown on the map was spread over the smallest area?
4. **Inferring Relationships** Based on the map, what continents were connected to Africa when the continents formed a supercontinent?
5. **Inferring Relationships** Based on the map, what continents were connected to Antarctica when the continents formed a supercontinent?
6. **Analyzing Relationships** How would you argue against a claim that ferns from the genus *Glossopteris* evolved independently on separate continents or were transported between continents that were not connected? Explain your answer.
7. **Identifying Trends** If the continents were to continue the motion they have had since the time the continents formed Gondwanaland, would you expect the east coast of South America and the west coast of Africa to be moving closer together or farther apart? Explain your answer.

SCIENCE AND TECHNOLOGY

CT Scanning Fossils

Paleontologists studying dinosaur bones want to learn as much as possible from the fossils they find. However, scientists often have to destroy a fossil to look inside it.

▼ This fossil is the skull of a *Nanotyrannus lancensis*.

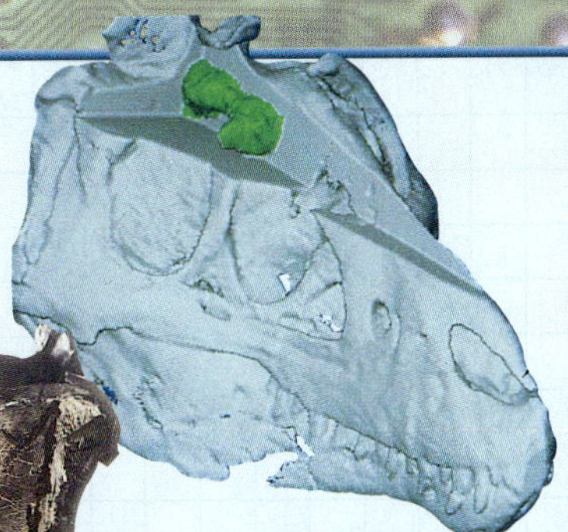

▲ This CT image shows the size and location of the dinosaur's brain.

The Price of Discovery

Usually, paleontologists examine the inside of a fossil by grinding away the specimen layer by layer. Unfortunately, by the time all of the fossil's internal structures are revealed, the specimen is completely destroyed.

Sectioning a fossil by layers also takes a lot of time. Scientists must carefully document each fresh surface because it will be destroyed later to uncover the next surface. Scientists record their observations by measuring, drawing, photographing, and making an imprint of each new surface.

New Uses for Technology

But now, some paleontologists can use *computerized axial tomography,* or CT scanning, to examine certain fossils without destroying them. CT scanning was originally developed for diagnosing illnesses in people. A CT scanner bombards an object with X rays from all angles within a single plane. The scanner measures the absorption of this radiation to create a map of the different densities within the object. Denser areas absorb more of the X rays' energy and appear white or light gray. Less dense areas appear dark gray or black. Once the object has been scanned in one plane, it is moved less than a millimeter and another scan is performed. This process is repeated until the entire object has been scanned. A computer imaging program then assembles the cross sections into a picture that can be viewed from any angle.

CT scanning allows scientists to study some extremely fragile fossils that cannot be studied by using traditional methods. It also allows scientists to study rare or unique fossils that must be preserved. Using CT scans, paleontologists can study the interior of fossils in a much less destructive way and in much less time.

◀ High-resolution CT scanners provide unparalleled views of the internal structures of fossils.

Extension

1. **Making Comparisons** Research another use of CT scans. Write a brief report about your findings.

Science and Technology

CT Scanning Fossils

Internet Activity —ADVANCED

Imaging Technologies Students may enjoy visiting sites that provide information on how CT (computed tomography) scans are used in medicine. Interested students may also research MRI (magnetic resonance imaging) and PET (positron emission tomography) technologies and their uses in medical diagnosis. Students may find how these imaging technologies work and may research how, or if, they are used by paleontologists. If they are not yet used to study fossils, students may write a proposal based on their research for an experiment that would determine whether or not these imaging technologies are useful in paleontological research. A worksheet designed to direct student research on this topic can be found in the **Chapter Resource File** booklet or by visiting **go.hrw.com** and entering the keyword **HQ6VEPX.** LS **Verbal**

CHAPTER RESOURCES

Chapter Resource File

- **Internet Activity**
  - Imaging Technologies ADVANCED

Answer to Extension

1. Student reports may vary. The most common use of this technology is for human medicine.

Unit 4 THE DYNAMIC EARTH

Unit 4 Outline

CHAPTER 10
Plate Tectonics

CHAPTER 11
Deformation of the Crust

CHAPTER 12
Earthquakes

CHAPTER 13
Volcanoes

▶ This photo was taken from the space shuttle *Endeavor* as Russia's Kliuchevskoi volcano erupted on September 30, 1994. The volcanic cloud reached 60,000 ft into the atmosphere, and wind carried the ash as far as 640 mi from the volcano.

Chapter 10 Plate Tectonics

Planning Guide

Compression Guide

To shorten instruction because of time limitations, omit Section 3.

OBJECTIVES	LABS, DEMONSTRATIONS, AND ACTIVITIES	TECHNOLOGY RESOURCES
PACING • 90 min pp. 238–246 **Chapter Opener**		**OSP Parent Letter** ■ **CD Student Edition on CD-ROM** **CD Chapter Summaries Audio CD** ■ **VID Brain Food Video Quiz**
Section 1 Continental Drift • Summarize Wegener's hypothesis of continental drift. • Describe the process of sea-floor spreading. • Identify how paleomagnetism provides support for the idea of sea-floor spreading. • Explain how sea-floor spreading provides a mechanism for continental drift.	**TE Debate** Wegener's Idea, p. 241 ADVANCED **TE Activity** Sea-Floor Sediments, p. 242 ◆ GENERAL **TE Demonstration** Earth's Magnetic Field, p. 244 ◆ BASIC **SE Quick Lab** Making Magnets, p. 245 ◆ GENERAL **CRF Datasheet for Quick Lab*** GENERAL **SE Making Models Lab** Sea-Floor Spreading, pp. 266–267 ◆ GENERAL **CRF Datasheet for Chapter Lab*** GENERAL	**OSP Lesson Plans** (also in print) **TR Bellringer*** TR 49 Sea-Floor Spreading* **CD Interactive Tutor** Continental Drift
PACING • 45 min pp. 247–254 **Section 2 The Theory of Plate Tectonics** • Summarize the theory of plate tectonics. • Identify and describe the three types of plate boundaries. • List and describe three causes of plate movement.	**TE Activity** Jigsaw Puzzle, p. 247 GENERAL **SE Quick Lab** Tectonic Plate Boundaries, p. 253 ◆ GENERAL **CRF Datasheet for Quick Lab*** GENERAL **SE Maps in Action** Locations of Earthquakes in South America, 2002–2003, p. 268 GENERAL **SE Mapping Expeditions** A Case of the Tennessee Shakes, pp. 834–835 GENERAL **CRF Inquiry Lab** Where Do Earthquakes Happen?* GENERAL **CRF Making Models Lab** Eggshell Tectonics* ◆ GENERAL	**OSP Lesson Plans** (also in print) **TR Bellringer*** **TE Internet Activity** Earthquakes, p. 248 BASIC **CRF Internet Activity** Earthquakes* BASIC TR 50 Tectonic Plate Boundaries* TR 51 Types of Plate Boundaries* TR 52 Ridge Push and Slab Pull* TR 54 Locations of Earthquakes in South America, 2002–2003* **TE Internet Activity** The Heimaey Eruption, p. 269 BASIC **CRF Internet Activity** The Heimaey Eruption* BASIC **CD Interactive Tutor** Tectonic Plates **VID HRW Earth Science Video** Plate Tectonics **VID NOVA Video** Earthquake
PACING • 45 min pp. 255–260 **Section 3 The Changing Continents** • Identify how movements of tectonic plates change Earth's surface. • Summarize how movements of tectonic plates have influenced climates and life on Earth. • Describe the supercontinent cycle.	**TE Activity** Modeling Rifting, p. 255 GENERAL **TE Demonstration** Modeling Accretion, p. 256 GENERAL **TE Group Activity** Responses to Climate Change, p. 257 ADVANCED	**OSP Lesson Plans** (also in print) **TR Bellringer*** TR 53 The Supercontinent Cycle* **TE Internet Activity** The Paleomap Project, p. 258 BASIC **CRF Internet Activity** The Paleomap Project* BASIC

PACING • 90 min

CHAPTER REVIEW, ASSESSMENT, AND STANDARDIZED TEST PREPARATION

- **SE Chapter Highlights,** p. 261
- **SE Chapter Review,** pp. 262–263
- **SE Standardized Test Prep,** pp. 264–265
- **CRF Concept Review*** ■ GENERAL
- **CRF Critical Thinking*** ADVANCED
- **CRF Math Skills*** GENERAL
- **CRF Graphing Skills*** GENERAL
- **CRF Chapter Test A*** ■ GENERAL
- **CRF Chapter Test B*** ADVANCED
- **OSP Lesson Plans** (also in print)
- **OSP Test Generator**
- **OSP Test Item Listing**

Online and Technology Resources

go.hrw.com Holt Online Learning

Visit **go.hrw.com** for access to Holt Online Learning, or enter the keyword **HQ6 Home** for a variety of free online resources.

One-Stop Planner® CD-ROM

This CD-ROM package includes

- Lab Materials QuickList Software
- Holt Calendar Planner
- Customizable Lesson Plans
- Printable Worksheets
- ExamView® Test Generator
- Interactive Teacher Edition
- Holt PuzzlePro®
- Holt PowerPoint® Resources

KEY

- **SE** Student Edition
- **TE** Teacher Edition
- **CRF** Chapter Resource File
- **LTP** Long-Term Projects
- **OSP** One-Stop Planner
- **TR** Transparencies and Transparency Worksheets
- **CD** CD or CD-ROM
- **VID** Classroom Video/DVD
- * Also on One-Stop Planner
- ◆ Requires advance prep
- ■ Also available in Spanish

SKILLS DEVELOPMENT RESOURCES	REVIEW AND ASSESSMENT	CORRELATIONS
SE Pre-Reading Activity, p. 238 GENERAL **TE Using the Figure** Rifting in Iceland, p. 238 GENERAL		National Science Education Standards
CRF Directed Reading* BASIC **TE Using the Figure** Continental Puzzles, p. 239 GENERAL **TE Using the Figure** Mountain Ranges and Fossils, p. 240 ADVANCED **TE Reading Skill Builder** Paired Summarizing, p. 240 BASIC **TE Inclusion Strategies,** p. 240 **SE Graphic Organizer** Chain-of-Events Chart, p. 243 GENERAL **TE Using the Figure** Sea-Floor Formation, p. 243 BASIC **TE Skill Builder** Math, p. 243 ADVANCED **TE Using the Figure** Magnetic Polarity, p. 245 GENERAL	**SE Reading Checks,** pp. 241, 243, 245 GENERAL **SE Section Review,** p. 246 GENERAL **TE Reteaching,** p. 245 BASIC **TE Quiz,** p. 245 GENERAL **TE Alternative Assessment,** p. 246 GENERAL **CRF Section Quiz*** ■ GENERAL	SAI 2a, SAI 2e, HNS 3b, HNS 3c, UC P1
CRF Directed Reading* BASIC **SE Math Practice,** p. 248 GENERAL **TE Skill Builder** Vocabulary, p. 248 BASIC **TE Skill Builder** Vocabulary, p. 249 BASIC **TE Inclusion Strategies,** p. 250 **TE Skill Builder** Writing, p. 251 GENERAL **TE Using the Figure** Mantle Convection, p. 252 GENERAL	**SE Reading Checks,** pp. 248, 250, 253 GENERAL **SE Section Review,** p. 254 GENERAL **TE Reteaching,** p. 253 BASIC **TE Quiz,** p. 253 GENERAL **TE Alternative Assessment,** p. 254 BASIC **CRF Section Quiz*** ■ GENERAL	ES 1b, ES 3c, HNS 3c, UCP 3
CRF Directed Reading* BASIC **TE Using the Figure** Accretion, p. 256 GENERAL **TE Skill Builder** Vocabulary, p. 256 ADVANCED **TE Reading Skill Builder** Reading Organizer, p. 257 BASIC	**SE Reading Checks,** pp. 256, 259 GENERAL **SE Section Review,** p. 260 GENERAL **TE Reteaching,** p. 259 BASIC **TE Quiz,** p. 259 GENERAL **TE Alternative Assessment,** p. 260 ADVANCED **CRF Section Quiz*** ■ GENERAL	ES 3c, UCP 2, UCP 3

Holt Lab Generator CD-ROM

Search for labs by topic, standard, difficulty level, and/or time frame. Edit those labs to fit your needs, or create your own labs. Use the Lab Materials QuickList to order just what you need.

Holt Earth Science Interactive Tutor CD-ROM

This CD-ROM consists of interactive activities that give students a fun way to extend their knowledge of Earth science concepts.

Chapter Summaries Audio CDs

These CDs include audio summaries of the key concepts presented in each chapter. (Audio summaries are also available in Spanish.)

SCILINKS. NSTA

www.scilinks.org

Maintained by the **National Science Teachers Association.** See Chapter Enrichment pages that follow for a complete list of topics.

See Chapter Enrichment pages for Video Resources.

Chapter 10 Chapter Enrichment

This Chapter Enrichment provides relevant and interesting information to expand and enhance your classroom instruction of the chapter material.

Section 1 Continental Drift

Wegener's Controversial Idea

Wegener's ideas about continental drift were controversial in his time. Parts of his hypothesis were proven wrong. For example, Wegener thought that continents moved through or over the oceanic crust. He also thought that the Mid-Atlantic Ridge was a pile of rubble left behind when continents separated. When scientists began to study the ocean floors and continental margins, they determined that continental crust is part of plates that may contain both types of crust that move together.

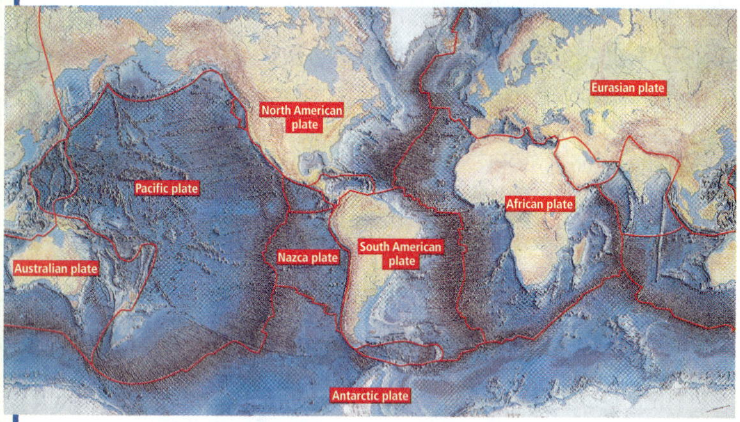

▲ **Earth's lithosphere is divided into tectonic plates.**

Geomagnetism

Earth's magnetosphere is a teardrop-shaped magnetic field that surrounds Earth and protects the planet from lethal bombardment by solar particles. Scientists think that currents created by the release of heat by radioactive elements in Earth's core generate the magnetic field.

The exact location of the geomagnetic poles is constantly changing. The poles can shift up to several miles per day as a result of magnetic disturbances. Because of this constant shifting, the geomagnetic poles are not symmetrically opposite to each other. The geomagnetic poles are not located at Earth's geographic North and South Poles. Instead, they are located approximately 11.5° from the geographic poles.

Scientists are unsure of what causes the magnetic field to reverse or exactly how a reversal takes place. Many scientists think that magnetic reversals result when currents in the core of the planet interact. Computer models indicate that magnetic reversals take between 1,000 and 8,000 years to complete and do not happen as quick 180° flips. Instead, the locations of the magnetic poles probably meander around the globe along preferred paths that follow longitude lines.

Section 2 The Theory of Plate Tectonics

The Importance of Plate Tectonics

Plate tectonics is the great unifying theory that explains the origin of the continents, ocean basins, and mountain ranges; many facts of evolution; and locations of earthquakes and volcanoes. Understanding this theory provides a strong conceptual base upon which to build an understanding of other Earth's processes, such as mountain building, rock formation, and climate change.

Mechanisms for Global Change

The movements of tectonic plates are driven by differences in density that result from the transfer of heat from inside Earth toward space. Temperatures in Earth's interior are probably greater than 4,000°C. These great temperatures result from the radioactive decay of elements in Earth's core and mantle. The heat generated by radioactive decay warms rock, which becomes less dense than rock and moves toward Earth's surface. The transfer of heat and heated materials in Earth's interior drive the movement of plates.

Section 3 The Changing Continents

Past Plate Motion

Scientists theorize that plate motion has occurred on Earth ever since the continents formed billions of years ago. Much of the evidence for plate motion in Precambrian time has been buried or destroyed by subsequent plate motions or surface processes. Scientists know that the present-day continents are composed of pieces of other landmasses and parts of the ocean floor that merged during plate motions in the past. Some pieces of the present continents are currently called *terranes* or *microplate terranes*.

Natural Selection

Most scientists think that evolution happens through a process called *natural selection*. This process has four basic assumptions. First, every species produces more offspring than will survive to maturity. Second, the individuals in a population are slightly different from each other. Each individual has a unique combination of traits. Some traits increase the individual's chances of surviving and mating, some traits decrease the chances of surviving and mating, and some traits are neutral. Third, the natural environment in which the individuals live does not have enough resources to support all of the individuals born. Many individuals will be killed by predators, accidents, or the inability to obtain resources. Thus, only some individuals will survive to adulthood. Fourth, only individuals that are well suited to the environment are likely to survive and reproduce. Thus, individuals who have better traits for surviving in that environment are likely to pass those traits to their offspring. Individuals who are less well adapted are more likely to die early or to produce few offspring.

◀ **The East African Rift Valley, which started to form 30 million years ago, acts as a physical barrier that affects the evolution of plant and animal species in the area.**

Video Resources

Brain Food Video Quizzes

Brain Food Video Quizzes These videos contain game-show style quizzes that assess students' progress and motivate students to study the chapter material.

HRW Earth Science Video

This video introduces Earth science topics and includes a geology field trip. The video segment listed below complements this chapter.

Segment 3, Plate Tectonics This segment examines the theory of plate tectonics, explains the density difference between the lithosphere and the athenosphere, and uses animations to illustrate the three types of plate boundaries. Magnetic polarity reversals, rifting, and subduction are also explained. (6 min)

NOVA® Videos

The **NOVA** video below complements this chapter.

Earthquake This program shows how today's advanced technology helps geologists predict earthquakes. (60 min)

To order other **NOVA** videos related to this chapter, visit go.hrw.com and enter the keyword **HQ6TECV.**

SciLinks®

Developed and maintained by the National Science Teachers Association

SciLinks is maintained by the National Science Teachers Association to provide you and your students with interesting, up-to-date links that will enrich your classroom presentation of the chapter.

Visit www.scilinks.org and enter the SciLinks code for more information about the topic listed.

Topic: Continental Drift
SciLinks code: HQ60351

Topic: Plate Tectonics
SciLinks code: HQ61171

Topic: Pangaea
SciLinks code: HQ61105

Topic: Mid-Atlantic Ridge
SciLinks code: HQ60960

Chapter 10

Chapter Overview

Earth is composed of layers of rock that have different compositions, structural features, and densities. Changes that occur at Earth's surface are driven by processes inside Earth that result from the transfer of heat and differences in density.

Using the Figure — GENERAL

Rifting in Iceland This photograph shows the axis of the Mid-Atlantic Ridge as it tears apart the island of Iceland. New crust forms along the spreading center of the Mid-Atlantic Ridge. Ask students to use the buildings in the background to estimate the size of the rift. The steam in the background may be a geothermal power plant. Ask students to determine why Iceland has so many geothermal power plants. (Because new crust forms here, magma transfers a lot of heat to the surrounding rock. Also, the new crust is thin.)

PRE-READING ACTIVITY

FOLDNOTES

You may want to collect students' FoldNotes to determine their prior knowledge. You may also want to modify your lesson plan to answer questions that students list in the "Want" column of their table.

Chapter 10 Plate Tectonics

Sections

1 Continental Drift

2 The Theory of Plate Tectonics

3 The Changing Continents

What You'll Learn

- How scientists developed the theory of plate tectonics
- Why tectonic plates move
- How Earth's geography has changed

Why It's Relevant

Understanding why and how tectonic plates move provides a basis for understanding other concepts of Earth science.

PRE-READING ACTIVITY

FOLDNOTES

Tri-Fold Before you read this chapter, create the FoldNote entitled "TriFold" described in the Skills Handbook section of the Appendix. Write what you know about plate tectonics in the column labeled "Know." Then, write what you want to know in the column labeled "Want." As you read the chapter, write what you learn in the column labeled "Learn."

The island of Iceland is being torn into two pieces as two tectonic plates pull apart. Iceland is one of only a few places on Earth where this process can be seen on land.

Chapter Correlations — *National Science Education Standards*

ES 1b The outward transfer of earth's internal heat drives convection circulation in the mantle that propels the plates comprising earth's surface across the face of the globe. **(Section 2)**

ES 3c Interactions among the solid earth, the oceans, the atmosphere, and organisms have resulted in the ongoing evolution of the earth system. . . . **(Section 2)**

SAI 2a . . . Historical and current scientific knowledge influence the design and interpretation of investigations and the evaluation of proposed explanations made by other scientists. **(Section 1)**

SAI 2e Scientific explanations must adhere to criteria such as: a proposed explanation must be logically consistent; it must abide by the rules of evidence; it must be open to questions and possible modification . . . **(Section 1)**

HNS 3c Occasionally, there are advances in science and technology that have important and long-lasting effects on science and society. Examples of such evidences include . . . plate tectonics . . . **(Section 1)**

UCP 1 Evidence consists of observations and data on which to base scientific explanations. Using evidence to understand interactions allows individuals to predict changes in natural and designed systems. . . . **(Sections 1 and 3)**

UCP 2 Evolution is a series of changes, some gradual and some sporadic, that accounts for the present form and function of objects, organisms, and natural and designed systems. . . . **(Section 3)**

UCP 3 . . . Interactions within and among systems result in change. Changes vary in rate, scale, and pattern, including trends and cycles. **(Section 2)**

Section 1 Continental Drift

One of the most exciting recent theories in Earth science began with observations made more than 400 years ago. As early explorers sailed the oceans of the world, they brought back information about new continents and their coastlines. Mapmakers used the information to chart the new discoveries and to make the first reliable world maps.

As people studied the maps, they were impressed by the similarity of the continental shorelines on either side of the Atlantic Ocean. The continents looked as though they would fit together like parts of a giant jigsaw puzzle. The east coast of South America, for example, seemed to fit perfectly into the west coast of Africa, as shown in **Figure 1.**

OBJECTIVES

- **Summarize** Wegener's hypothesis of continental drift.
- **Describe** the process of sea-floor spreading.
- **Identify** how paleomagnetism provides support for the idea of sea-floor spreading.
- **Explain** how sea-floor spreading provides a mechanism for continental drift.

KEY TERMS

continental drift
mid-ocean ridge
sea-floor spreading
paleomagnetism

Wegener's Hypothesis

In 1912, a German scientist named Alfred Wegener (VAY guh nuhr) proposed a hypothesis that is now called **continental drift.** Wegener hypothesized that the continents once formed part of a single landmass called a *supercontinent.* According to Wegener, this supercontinent began breaking up into smaller continents about 200 million years ago (during the Mesozoic Era). Over millions of years, these continents drifted to their present locations. Wegener speculated that the crumpling of the crust in places may have produced mountain ranges such as the Andes on the western coast of South America.

continental drift the hypothesis that states that the continents once formed a single landmass, broke up, and drifted to their present locations

Figure 1 ▶ Early explorers noticed that the coastlines of Africa and South America could fit together like puzzle pieces. ***Can you identify any other continents that could fit together like puzzle pieces?***

Section 1

Focus

Overview

This section describes the idea of continental drift. The section describes evidence for continental drift, the process of sea-floor spreading, and flaws of the hypothesis of continental drift.

Bellringer

Propose the following scenario to students: "Two islands are 1,000 km apart. The rocks on both islands have the same mineral composition, the same rare fossils, and mountains of identical age and structure." Ask students to suggest a hypothesis to explain these similarities. (Students may hypothesize that the two islands were once joined but then split apart and somehow moved to their present locations.) LS **Logical**

Motivate

Using the Figure — GENERAL

Continental Puzzles Use the question in the caption for the figure on this page as the starting point for a discussion. Answer to caption question: Other locations that may appear to fit like puzzle pieces include east Africa and the Arabian peninsula and Madagascar and east Africa. LS **Logical**

CHAPTER RESOURCES

Chapter Resource File

- Directed Reading BASIC

Technology

Transparencies
- Bellringer

Student Edition on CD-ROM

One-Stop Planner CD-ROM
- Lesson Plan

Teach

Using the Figure — Advanced

Mountain Ranges and Fossils Have students study the map on this page. Have them identify the ages of mountain ranges in the Northern Hemisphere that match. Have them identify the ranges of the organisms identified in the Southern Hemisphere. Then, have them draw a diagram that shows how the continents may have looked when they were connected. They can use the mountain ranges and distribution of organisms as a guide. (Students' drawings should show Greenland connected to Scandinavia and northern Britain, the east coast of North America connected to Europe and northwestern Africa, and South America connected to Africa, Madagascar, and India.) LS **Visual**

READING SKILL BUILDER — Basic

Paired Summarizing Group students into pairs, and have them read silently about the evidence for continental drift. Then, have one student summarize the idea of continental drift. The other student should listen to the retelling and should point out any inaccuracies or ideas that were left out. Allow students to refer to the text as needed.
LS **Verbal/Auditory** English Language Learners

Figure 2 ▶ Fossils of *Mesosaurus*, such as the one shown below, were found in both South America and western Africa. Mountain chains of similar ages also exist on different continents, as shown in the map at right.

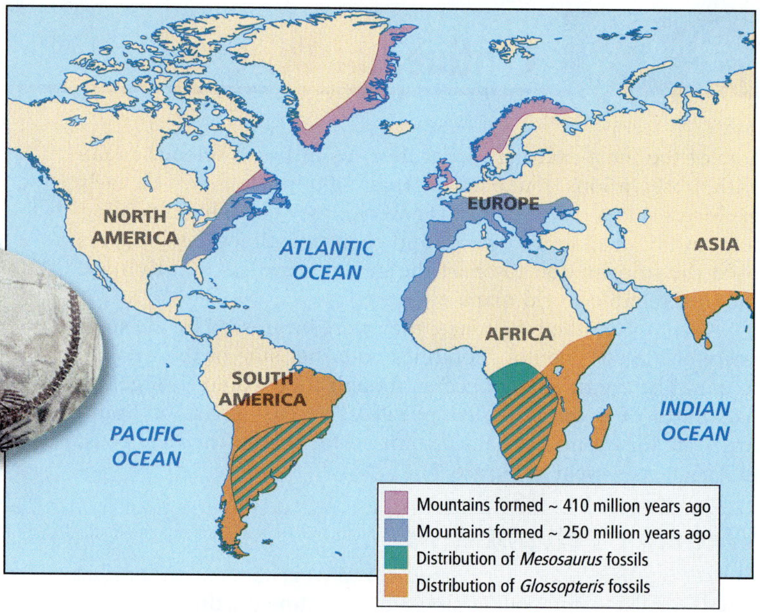

SCILINKS. NSTA Developed and maintained by the National Science Teachers Association

For a variety of links related to this subject, go to www.scilinks.org

Topic: Continental Drift
SciLinks code: HQ60351

Fossil Evidence

In addition to seeing the similarities in the coastlines of the continents, Wegener found other evidence to support his hypothesis. He reasoned that if the continents had once been joined, fossils of the same plants and animals should be found in areas that had once been connected. Wegener knew that identical fossils of *Mesosaurus,* a small, extinct land reptile, had been found in both South America and western Africa. *Mesosaurus,* a fossil of which is shown in **Figure 2,** lived 270 million years ago (during the Paleozoic Era). Wegener knew that it was unlikely that these reptiles had swum across the Atlantic Ocean. He also saw no evidence that land bridges had once connected the continents. So, he concluded that South America and Africa had been joined at one time in the past.

Evidence from Rock Formations

Geologic evidence also supported Wegener's hypothesis of continental drift. The ages and types of rocks in the coastal regions of widely separated areas, such as western Africa and eastern South America, matched closely. Mountain chains that ended at the coastline of one continent seemed to continue on other continents across the ocean, as shown in **Figure 2.** The Appalachian Mountains, for example, extend northward along the eastern coast of North America, and mountains of similar age and structure are found in Greenland, Scotland, and northern Europe. If the continents are assembled into a model supercontinent, the mountains of similar age fit together in continuous chains.

INCLUSION Strategies

• *Learning Disabled* • *Attention Deficit Disorder*

Ask students to copy all key terms that appear in bold type throughout the chapter into a Vocabulary Notebook. Students should also copy the margin definitions next to each key term. They can copy the sentence in which each key term appears to help them understand the word in the context of the chapter. The notebooks can be used for independent study. LS **Verbal**

Climatic Evidence

Changes in climatic patterns also suggest that the continents have not always been located where they are now. Geologists discovered layers of debris from ancient glaciers in southern Africa and South America. Today, those areas have climates that are too warm for glaciers to form. Other fossil evidence—such as the plant fossil shown in **Figure 3**—indicated that tropical or subtropical swamps covered areas that now have much colder climates. Wegener suggested that if the continents were once joined and positioned differently, evidence of climatic differences would be easy to explain.

Missing Mechanisms

Despite the evidence that supports the hypothesis of continental drift, Wegener's ideas were strongly opposed. Other scientists of the time rejected the mechanism by which Wegener proposed that the continents moved. Wegener suggested that the continents plowed through the rock of the ocean floor. However, this idea was easily disproved by geologic evidence. Wegener spent the rest of his life searching for a mechanism that would gain scientific consensus. Unfortunately, Wegener died in 1930 before he identified a plausible explanation.

Reading Check **Why did many scientists reject Wegener's hypothesis of continental drift? (See the Appendix for answers to Reading Checks.)**

Figure 3 ▶ The climate of Antarctica was not always as harsh and cold as it is today. When the plant that became this fossil lived, the climate of Antarctica was warm and tropical.

Debate — ADVANCED

Wegener's Idea Have students form two groups to debate the merits and flaws of Wegener's hypothesis. Have one group of students research and present specific examples of evidence that supports the idea of continental drift. Have the other students identify weaknesses of the hypothesis and present alternative explanations for the evidence suggested by the first group. **LS Logical/Verbal** Co-op Learning

Discussion — GENERAL

Making Predictions Have a discussion about mechanisms for continental drift. Have students suggest possible mechanisms that may drive the movement of continents. Try to get students to imagine the amount of force needed to move an object as massive as a continent by comparing the size and mass of a continent to the size and mass of aircraft carriers (about 100,000 metric tons) and space vehicles (launch mass = 2,040,000 kg; launch thrust = 34,677 kN). **LS Verbal**

Answer to Reading Check

Many scientists rejected Wegener's hypothesis because the mechanism that Wegener suggested was easily disproved by geologic evidence.

HISTORY CONNECTION

Alfred Lothar Wegener Alfred Wegener was born in Germany in 1880. He received a doctorate of astronomy in 1905 and served at the aeronautical observatory at Lindenberg early in his scientific career. He joined the German army at the outbreak of World War I but was injured shortly after joining. After the war, he did research in meteorology in Hamburg, Germany, for the German government. He was a professor of meteorology at the University of Graz in Austria from 1924 to 1930. He went on four polar expeditions between 1906 and 1930. On the last of these expeditions, he visited Greenland to try to determine the thickness of the Greenland ice sheet and the rate of drift of Greenland. At the end of that expedition, he died while rescuing some colleagues.

Teach, *continued*

Activity — GENERAL

Sea-Floor Sediments Have pairs of students model the depth of sea-floor sediments by following these simple steps.

1. Place a piece of paper on a flat surface.
2. Hold a shoe box of confetti so that confetti falls from one side rather than from a corner.
3. While moving the piece of paper, slowly sprinkle the confetti over the paper in a single line.
4. Measure the thickness of the confetti at different points along the line. Where is it thinnest? (nearest where you stopped sprinkling) Where is it thickest? (farther from where you stopped sprinkling) Why? (because it fell on the thickest point for the longest time) LS Kinesthetic/Visual

PHYSICS CONNECTION

Sonar Researchers used sonar to discover that the ocean floor is not flat. In the 1950s, scientists broadcast sound waves toward the sea floor and measured how long the waves took to return. The echoes revealed the existence of oceanic valleys and mountains. In short, the ocean floors turned out to be as varied as the continents! Scientists were most amazed to find a chain of undersea mountains snaking thousands of kilometers around the globe—the mid-ocean ridges.

Figure 4 ▶ Black smokers are vents on the sea floor that form as hot, mineral-rich water rushes from the hot rock at mid-ocean ridges and mixes with the surrounding cold ocean water. This photo was taken from a submersible.

Mid-Ocean Ridges

The evidence that Wegener needed to support his hypothesis was discovered nearly two decades after his death. The evidence lay on the ocean floor. In 1947, a group of scientists set out to map the Mid-Atlantic Ridge. The Mid-Atlantic Ridge is part of a system of **mid-ocean ridges,** which are undersea mountain ranges through the center of which run steep, narrow valleys. A special feature of mid-ocean ridges is shown in **Figure 4.** While studying the Mid-Atlantic Ridge, scientists noticed two surprising trends. First, they noticed that the sediment that covers the sea floor is thinner closer to a ridge than it is farther from the ridge, as shown in **Figure 5.** This evidence suggests that sediment has been settling on the sea floor farther from the ridge for a longer time than it has been settling near the ridge. Scientists then examined the remains of tiny ocean organisms found in the sediment to date the sediment. The distribution of these organisms showed that the closer the sediment is to a ridge, the younger the sediment is. This evidence indicates that rocks closer to the ridge are younger than rocks farther from the ridge.

Second, scientists learned that the ocean floor is very young. While rocks on land are as old as 3.8 billion years, none of the oceanic rocks are more than 190 million years old. Radiometric dating also showed evidence that sea-floor rocks closer to a mid-ocean ridge are younger than sea-floor rocks farther from a ridge.

mid-ocean ridge a long, undersea mountain chain that has a steep, narrow valley at its center, that forms as magma rises from the asthenosphere, and that creates new oceanic lithosphere (sea floor) as tectonic plates move apart

Figure 5 ▶ Rocks closer to a mid-ocean ridge are younger than rocks farther from the ridge. In addition, rocks closer to the ridge are covered with less sediment, which indicates that sediment has had less time to settle on them.

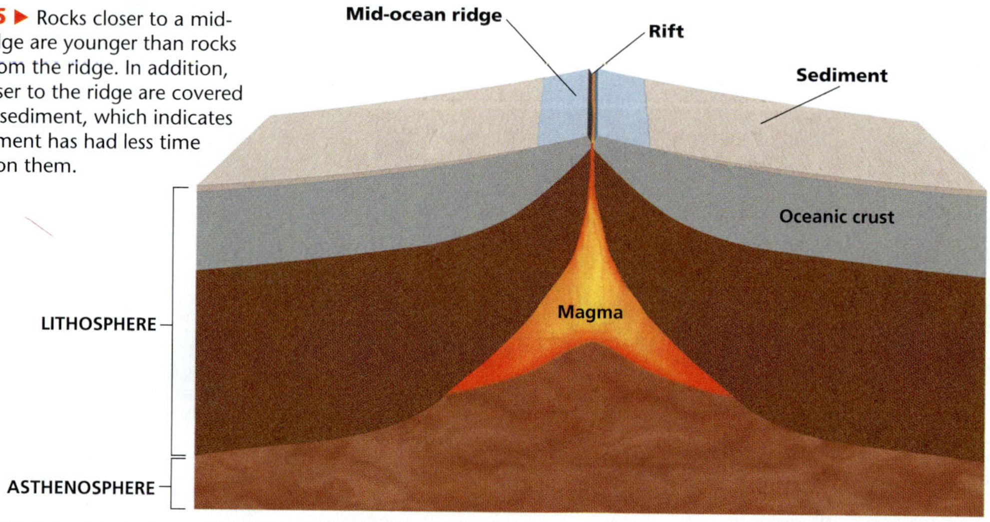

BRAIN FOOD

Ridges and Rises At the Mid-Atlantic Ridge, approximately 2.5 to 5 cm of new sea floor forms each year. At a ridge, the topography is rough because a chain of small volcanoes is linked by fissure eruptions. At the East-Pacific Rise, as much as 15 cm of new sea floor forms each year. At a rise, the topography is smoother because the crust is made of flat lava flows that erupt from large, long fissures.

Sea-Floor Spreading

In the late 1950s, a geologist named Harry Hess suggested a new hypothesis. He proposed that the valley at the center of the ridge was a crack, or *rift,* in Earth's crust. At this rift, molten rock, or *magma*, from deep inside Earth rises to fill the crack. As the ocean floor moves away from the ridge, rising magma cools and solidifies to form new rock that replaces the ocean floor. This process is shown in **Figure 6.** Robert Dietz, another geologist, named this process by which new ocean lithosphere (sea floor) forms as magma rises to Earth's surface and solidifies at a mid-ocean ridge as **sea-floor spreading.** Hess suggested that if the ocean floor is moving, the continents might be moving, too. Hess thought that sea-floor spreading was the mechanism that Wegener had failed to find.

Still, Hess's ideas were only hypotheses. More evidence for sea-floor spreading would come years later, in the mid-1960s. This evidence would be discovered through **paleomagnetism,** the study of the magnetic properties of rocks.

Reading Check How does new sea floor form? (See the Appendix for answers to Reading Checks.)

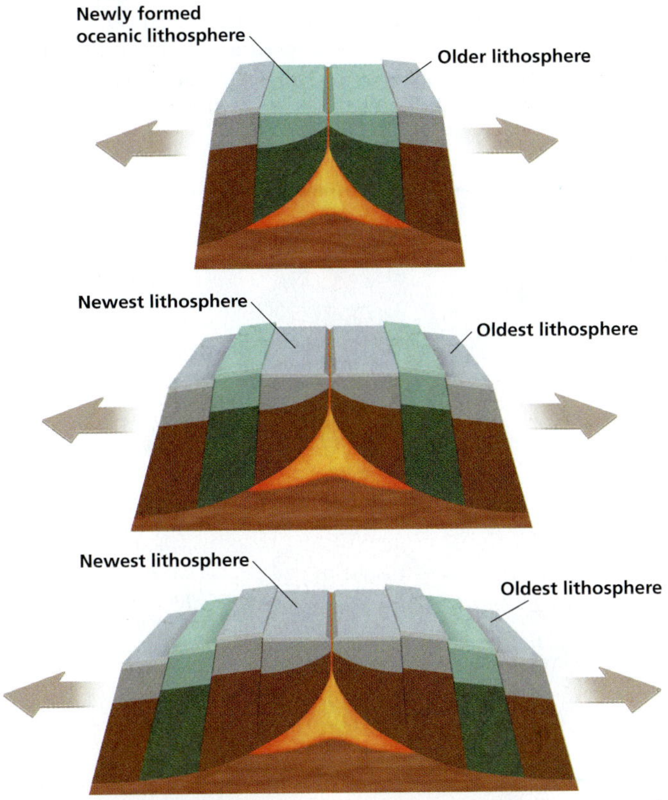

sea-floor spreading the process by which new oceanic lithosphere (sea floor) forms as magma rises to Earth's surface and solidifies at a mid-ocean ridge

paleomagnetism the study of the alignment of magnetic minerals in rock, specifically as it relates to the reversal of Earth's magnetic poles; also the magnetic properties that rock acquires during formation

Graphic Organizer **Chain-of-Events Chart**

Create the **Graphic Organizer** entitled "Chain-of-Events Chart" described in the Skills Handbook section of the Appendix. Then, fill in the chart with details about each step of sea-floor spreading.

Figure 6 ▸ As the ocean floor spreads apart at a mid-ocean ridge, magma rises to fill the rift and then cools to form new rock. As this process is repeated over millions of years, new sea floor forms.

Graphic Organizer GENERAL

Chain-of-Events Chart

You may wish to use this activity as a pre-reading activity to assess students' prior knowledge before beginning discussion of sea-floor spreading. You may also choose to use a similar activity as a quiz to assess knowledge of the hypothesis of continental drift or after students have read the description of sea-floor spreading.

Using the Figure — BASIC

Sea-Floor Formation Have students analyze the figure that illustrates sea-floor spreading. Have students identify the relative age of each block in each diagram by following the green block through the frames. Ask students to write a list of steps that describe the process of sea-floor spreading. (1. Ocean floor pulls apart. 2. New sea floor forms in the crack. 3. Ocean floor pulls apart again, and the process repeats.) **LS Visual**

Answer to Reading Check

New sea floor forms as magma rises to fill the rift that forms when two plates pull apart at a divergent boundary.

SKILL BUILDER — ADVANCED

Math The Atlantic Ocean is spreading at a rate of 1 to 2 cm per year, and the eastern Pacific sea floor is spreading between 3 and 8 cm per year. Have students use the average rate of spreading for the Atlantic Ocean to calculate how many years the sea floor of the Atlantic Ocean would take to spread 1 km. (1.5 cm/year; 1 km ÷ 0.000015 km/year = 66,667 years) Have students use the average rate of spreading of the Pacific Ocean to calculate how many years the sea floor of the Pacific Ocean would take to spread 1 km. (5.5 cm/year; 1 km ÷ 0.000055 km/year = 18,182 years) **LS Logical**

CHAPTER RESOURCES

Technology

Transparencies
- 49 Sea-Floor Spreading (with worksheet)

Teach, *continued*

Demonstration — BASIC

Earth's Magnetic Field To help students visualize how iron-bearing minerals in molten rocks align with Earth's magnetic field, place a magnet in the center of an overhead projector so that the north pole faces the top of the projector. Place a sheet of clear acetate over the magnet. Sprinkle some iron filings on the acetate. The iron filings will align with the lines of force of the magnet. Lift the acetate, and move the magnet so that the north and south poles face the sides of the projector. Explain that when Earth's magnetic field reverses, the iron-bearing minerals that form from molten rock align according to the new magnetic field. **LS Visual**

Discussion — GENERAL

Magnetic Polarity Reversals On average, over the last 10 million years, Earth's magnetic field has reversed four or five times every 1 million years. However, the last magnetic field reversal occured about 800,000 years ago. Based on measurements of Earth's magnetic field, some scientists think that Earth's magnetic field is in the early stages of a reversal, but no one knows how long a magnetic reversal takes to complete. Ask students to consider what might happen today if Earth's magnetic field reversed. **LS Logical**

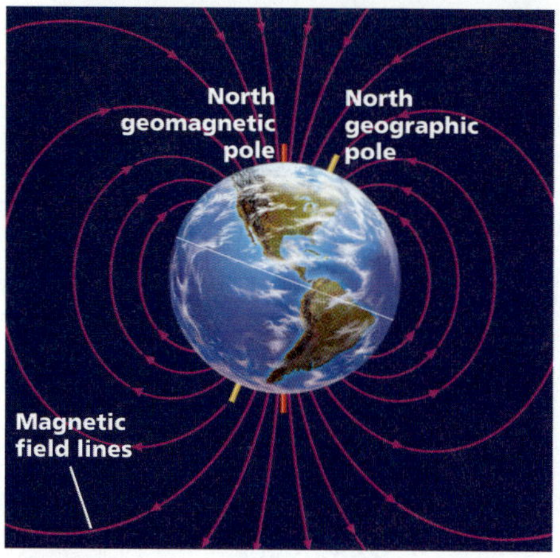

Figure 7 ▶ Earth acts as a giant magnet because of currents in Earth's core.

Paleomagnetism

If you have ever used a compass to determine direction, you know that Earth acts as a giant magnet. Earth has north and south geomagnetic poles, as shown in **Figure 7.** The compass needle aligns with the field of magnetic force that extends from one pole to the other.

As magma solidifies to form rock, iron-rich minerals in the magma align with Earth's magnetic field in the same way that a compass needle does. When the rock hardens, the magnetic orientation of the minerals becomes permanent. This residual magnetism of rock is called *paleomagnetism*.

Magnetic Reversals

Geologic evidence suggests that Earth's magnetic field has not always pointed north, as it does now. Scientists have discovered rocks whose magnetic orientations point opposite of Earth's current magnetic field. Scientists have dated rocks of different magnetic polarities. All rocks with magnetic fields that point north, or *normal polarity,* are classified in the same time periods. All rocks with magnetic fields that point south, or *reversed polarity,* also fell into specific time periods. When scientists placed these periods of normal and reverse polarity in chronological order, they discovered a pattern of alternating normal and reversed polarity in the rocks. Scientists used this pattern to create the *geomagnetic reversal time scale*.

Connection to PHYSICS

What Makes Materials Magnetic?

Some materials are magnetic, while others are not. So, what makes a material magnetic? All matter is composed of atoms. In atoms, electrons are the negatively charged particles that move around the nucleus. The motion of electrons in an atom produces magnetic fields that can give the atom a north pole and a south pole.

In most materials, the magnetic fields of individual atoms are not aligned, so the materials are not magnetic. However, in some materials, such as the iron in some rocks, the atoms group together in tiny regions called *domains*. The atoms in a domain are arranged so that the north and south poles of the atoms are aligned to create a stronger magnetic field than that of a single atom. If most of the domains in an object are also aligned, their magnetic fields combine to make the whole object magnetic.

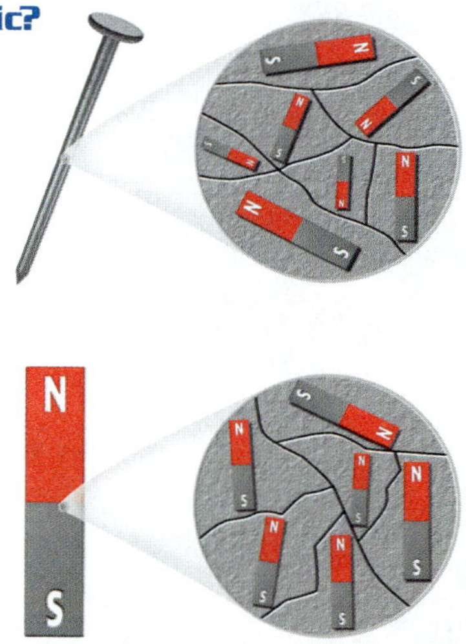

If the domains in an object are randomly arranged, the magnetic fields of individual domains cancel each other out and the object does not have magnetic properties.

If most of the domains in an object are aligned, the magnetic fields of individual domains combine to make the object magnetic.

MISCONCEPTION ALERT

Mineral Magnetism Students may think that crystals of magnetic minerals align themselves to Earth's magnetic field. Explain to students that as igneous rock forms at a mid-ocean ridge, minerals form in such a way that the magnetic fields of the minerals are aligned with Earth's magnetic field. Some students may ask why the magnetic fields of the minerals do not realign themselves when the next magnetic reversal occurs. The answer is that the rock must be heated beyond the mineral's Curie point for the mineral to lose its original magnetism. The Curie point is the temperature at which the electron orbitals in the atoms in a mineral no longer overlap because the crystal lattice has expanded too much. When the orbitals no longer overlap, the spin moments of the electrons do not align, and magnetism is lost. Unless magnetized rock is heated to a temperature above its Curie point, it will hold onto its original magnetism. The Curie temperature for the mineral magnetite is ~580°C.

Magnetic Symmetry

As scientists were learning about the age of the sea floor, they also were finding puzzling magnetic patterns on the ocean floor. The scientists used the geomagnetic reversal time scale to help them unravel the mystery of these magnetic patterns.

Scientists noticed that the striped magnetic pattern on one side of a mid-ocean ridge is a mirror image of the striped pattern on the other side of the ridge. These patterns are shown in **Figure 8.** When drawn on maps of the ocean floor, these patterns showed alternating bands of normal and reversed polarity that match the geomagnetic reversal time scale. Scientists suggested that as new sea floor forms at a mid-ocean ridge, the new sea floor records reversals in Earth's magnetic field.

By matching the magnetic patterns on each side of a mid-ocean ridge to the geomagnetic reversal time scale, scientists could assign ages to the sea-floor rocks. The scientists found that the ages of sea-floor rocks were also symmetrical. The youngest rocks were at the center, and older rocks were farther away on either side of the ridge. The only place on the sea floor that new rock forms is at the rift in a mid-ocean ridge. Thus, the patterns indicate that new rock forms at the center of a ridge and then moves away from the center in opposite directions. Thus, the symmetry of magnetic patterns—and the symmetry of ages of sea-floor rocks—supports Hess's idea of sea-floor spreading.

Reading Check **How are magnetic patterns in sea-floor rock evidence of sea-floor spreading? (See the Appendix for answers to Reading Checks.)**

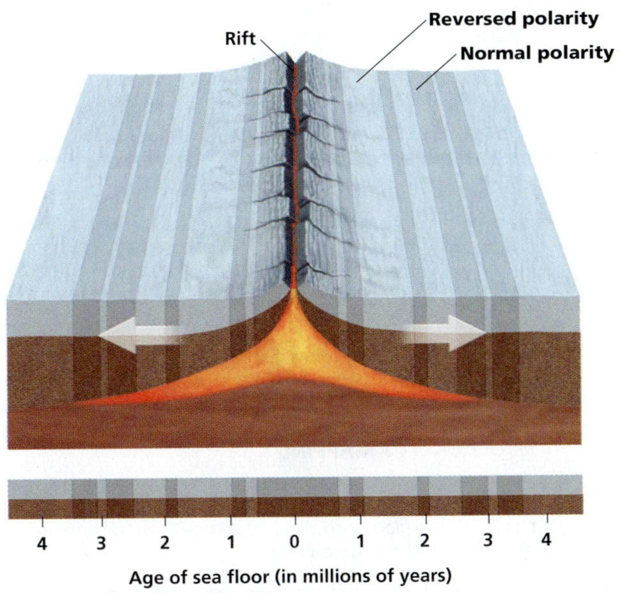

Figure 8 ▶ The stripes in the sea floor shown here illustrate Earth's alternating magnetic field. Dark stripes represent normal polarity, while the lighter stripes represent reversed polarity. ***What is the polarity of the rocks closest to the rift?***

QuickLAB 10 min

Making Magnets

Procedure

1. Slide one end of a **bar magnet** down the side of a **5 inch iron nail** 10 times. Always slide the magnet in the same direction.
2. Hold the nail over a small pile of **steel paperclips.** Record what happens.
3. Slide the bar magnet back and forth 10 times down the side of the nail. Repeat step 2.

Analysis

1. What was the effect of sliding the magnet down the nail in one direction? in different directions?
2. How does this lab demonstrate the idea of domains?

QuickLAB

Skills Acquired

- Experimenting
- Observing

Teacher's Notes You may want to check the nails before class to make sure that they are not magnetized before the lab. Tell students not to drop the nails; otherwise, the nails may lose their magnetism.

Answers

1. Sliding the magnet down the nail in one direction magnetized the nail. Sliding the magnet in different directions demagnetized the nail.
2. The domains in the nail were not aligned before the magnet was used.

Close

Reteaching — BASIC

Evidence for Continental Drift Ask students to build a small model of the ocean floor. Then, have students use paint, colored pencils, or markers to draw magnetic polarity stripes on the sea floor. On the sea floor, students should make symmetrical stripes that mirror each other on either side of the mid-ocean ridges. **LS Kinesthetic**

Quiz — GENERAL

Determine whether each of the following statements is true or false.

1. Evidence for the hypothesis of continental drift includes land bridges between continents. (false)
2. Many scientists rejected the hypothesis of continental drift because it did not explain how continents move. (true)
3. Continental drift can be partially explained by sea-floor spreading. (true)

CHAPTER RESOURCES

Chapter Resource File

- **Datasheet for Quick Lab** GENERAL

Answer to Reading Check

The symmetrical magnetic patterns in sea-floor rocks show that rock formed at one place (at a ridge) and then broke apart and moved away from the center in opposite directions.

Using the Figure — GENERAL

Magnetic Polarity Use the figure on this page to lead a discussion of magnetic symmetry. Ask students to suggest other examples of symmetry. (faces and bodies) Answer to caption question: The rocks closest to the rift have normal polarity, because they are forming now, when Earth's magnetic field has normal polarity. **LS Visual**

Close, *continued*

Alternative Assessment — GENERAL

Persuasive Essay Have students write a persuasive essay that describes their stance on sea-floor spreading as a mechanism for continental drift. Students can choose to support sea-floor spreading as a mechanism, or they may argue against it. Whichever stance each student chooses, he or she must provide supporting reasoning and evidence. LS **Logical**

Answers to Section Review

1. The coastlines of the continents seem to fit together like pieces of a jigsaw puzzle.
2. The hypothesis of continental drift is supported by evidence from fossils, rock formations, locations of mountain ranges, and climatic evidence.
3. At rifts in Earth's crust along mid-ocean ridges, magma from the mantle rises to Earth's surface and cools to form new rock. As that rock cools, it moves away from the rift, and new magma rises to fill the empty space.
4. Rocks with normal polarity formed during distinct time spans, which differ from the time spans in which rocks with reverse polarity formed. Because these same patterns are found all over the planet, scientists agree that Earth's magnetic field must have reversed at various times in Earth's history.
5. The patterns of magnetic reversals are mirrored on both sides of a mid-ocean ridge, which indicates that the magma hardened to rock at the center and then moved away from the center through sea-floor spreading.
6. Scientists use radiometric dating, sediment thickness, fossil evidence in sediments, and magnetic reversal patterns to date sea-floor rocks.
7. As new rock forms at the mid-ocean ridge, the previously existing rock moves farther from the ridge. Thus, the rock at the ridge is the youngest, and the rock farthest from the ridge is the oldest.
8. As new sea floor forms, the older sea floor moves away from the spreading center. As the sea floor moves, it carries continents along with it.
9. *Sea-floor spreading* is the mechnism by which new ocean floor forms at *mid-ocean ridges*; was identified by using *paleomagnetism*; and is a mechanism that could drive *continental drift*, which could form *supercontinents* and is supported by evidence from studying *fossils*, *climate*, and *geologic evidence*.

Figure 9 ▶ Scientists collected samples of these sedimentary rocks in California and used the magnetic properties of the samples to date the rocks by using the geomagnetic reversal time scale.

Wegener Redeemed

Another group of scientists discovered that the reversal patterns seen in rocks on the sea floor also appeared in rocks on land, such as those shown in **Figure 9.** The reversals in the land rocks matched the geomagnetic reversal time scale. Because the same pattern occurs in rocks of the same ages on both land and the sea floor, scientists became confident that magnetic patterns show changes over time. Thus, the idea of sea-floor spreading gained further favor in the scientific community.

Scientists reasoned that sea-floor spreading provides a way for the continents to move over Earth's surface. Continents are carried by the widening sea floor in much the same way that objects are moved by a conveyer belt. The molten rock from a rift cools, hardens, and then moves away in the opposite direction on both sides of the ridge. Here, at last, was the mechanism that verified Wegener's hypothesis of continental drift.

Section 1 Review

1. **Describe** the observation that first led to Wegener's hypothesis of continental drift.
2. **Summarize** the evidence that supports Wegener's hypothesis.
3. **Compare** sea-floor spreading and the formation of mid-ocean ridges.
4. **Explain** how scientists know that Earth's magnetic poles have reversed many times during Earth's history.
5. **Identify** how magnetic symmetry can be used as evidence of sea-floor spreading.
6. **Explain** how scientists date sea-floor rocks.

CRITICAL THINKING

7. **Making Inferences** How does evidence that sea-floor rocks farther from a ridge are older than rocks closer to the ridge support the idea of sea-floor spreading?
8. **Analyzing Ideas** Explain how sea-floor spreading provides an explanation for how continents may move over Earth's surface.

CONCEPT MAPPING

9. Use the following terms to create a concept map: *continental drift, paleomagnetism, fossils, climate, sea-floor spreading, geologic evidence, supercontinent,* and *mid-ocean ridge.*

CHAPTER RESOURCES

Chapter Resource File
- Section Quiz GENERAL

Workbooks
- Study Guide (also in Spanish)

Section 2 The Theory of Plate Tectonics

By the 1960s, evidence supporting continental drift and sea-floor spreading led to the development of a theory called *plate tectonics*. **Plate tectonics** is the theory that explains why and how continents move and is the study of the formation of features in Earth's crust.

How Continents Move

Earth's crust and the rigid, upper part of the mantle form a layer of Earth called the **lithosphere.** The lithosphere forms the thin outer shell of Earth. It is broken into several blocks, called *tectonic plates,* that ride on a deformable layer of the mantle called the *asthenosphere* in much the same way that blocks of wood float on water. The **asthenosphere** (as THEN uh sfir) is a layer of "plastic" rock just below the lithosphere. Plastic rock is solid rock that is under great pressure and that flows very slowly, like putty does. **Figure 1** shows what tectonic plates may look like.

Earth's crust is classified into two types—*oceanic crust* and *continental crust.* Oceanic crust is dense and is made of rock that is rich in iron and magnesium. Continental crust has a low density and is made of rock that is rich in silica. Tectonic plates can include continental crust, oceanic crust, or both. The continents and oceans are carried along on the moving tectonic plates in the same way that passengers are carried by a bus.

OBJECTIVES

- **Summarize** the theory of plate tectonics.
- **Identify** and describe the three types of plate boundaries.
- **List** and describe three causes of plate movement.

KEY TERMS

plate tectonics
lithosphere
asthenosphere
divergent boundary
convergent boundary
transform boundary

plate tectonics the theory that explains how large pieces of the lithosphere, called *plates,* move and change shape

lithosphere the solid, outer layer of Earth that consists of the crust and the rigid upper part of the mantle

asthenosphere the solid, plastic layer of the mantle beneath the lithosphere; made of mantle rock that flows very slowly, which allows tectonic plates to move on top of it

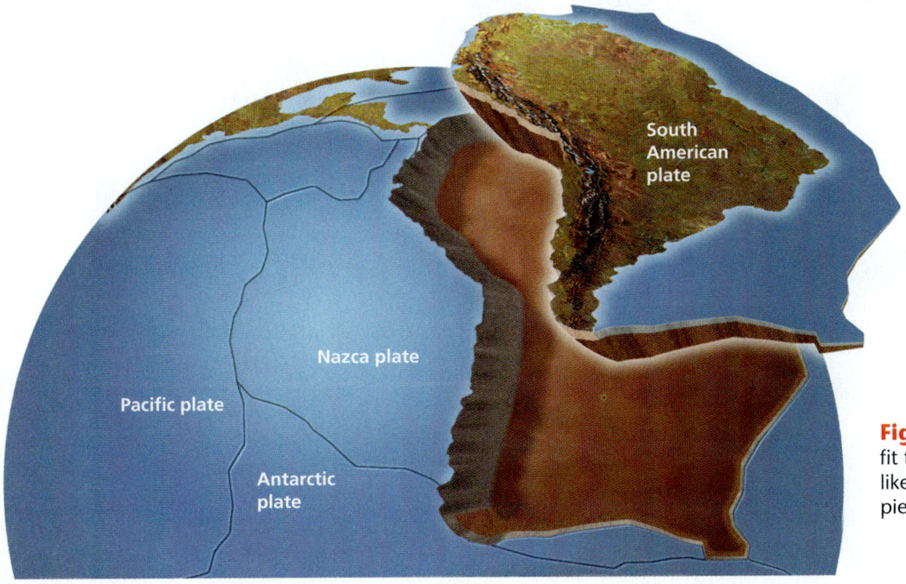

Figure 1 ▶ Tectonic plates fit together on Earth's surface like three-dimensional puzzle pieces.

Section 2

Focus

Overview

In this section, students learn about the theory of plate tectonics. The section describes the movements of tectonic plates, types of plate boundaries, and driving forces of plate tectonics.

Bellringer

Ask students how many continents exist on Earth's surface, and have them list the continents. (six: Australia, Antarctica, North America, South America, Eurasia, and Africa) Explain to students that scientists have identified at least 30 tectonic plates—10 to 15 major plates and many smaller subplates. LS **Logical**

Motivate

Activity — GENERAL

Jigsaw Puzzle Bring a 25-piece jigsaw puzzle to class, and challenge students to put the puzzle together. Explain to students that Earth's surface is like a spherical jigsaw puzzle, but instead of being small pieces of cardboard, the pieces are giant slabs of rock that are millions of square kilometers in area and billions of tons in weight. LS **Visual/Kinesthetic**

CHAPTER RESOURCES

Chapter Resource File

- **Directed Reading** BASIC
- **Inquiry Lab** Where Do Earthquakes Happen? GENERAL
- **Making Models Lab** Eggshell Tectonics GENERAL

Technology

Transparencies
- Bellringer

Student Edition on CD-ROM

One-Stop Planner CD-ROM
- Lesson Plan

MATHPRACTICE

Answer

rate = distance ÷ time;
1 cm = 0.00001 km;
0.00004 km/1 year = x km = 2,000,000 years; x km = 0.00004 km/year × 2,000,000 years = 80 km

SKILL BUILDER — BASIC

Vocabulary The word *tectonics* is from the Greek word *tektonikos,* which means "construction." Tectonics is the study of the formation of features in Earth's crust. Ask students to explain how this root word can help them understand the geologic features that result from plate tectonics. **LS Verbal**

Internet Activity — BASIC

Earthquakes Students may be interested in knowing what areas of Earth's surface have the most earthquakes each day, month, and year. The Incorporated Research Institutions for Seismicity has an online service known as the Seismic Monitor. This site constantly updates a global map of recent earthquake activity. Students can visit this site to learn where the most recent large earthquakes have occurred. A worksheet designed to direct student research on this topic can be found in the **Chapter Resource File** booklet or by visiting **go.hrw.com** and entering the keyword **HQ6TECX.** **LS Visual**

Answer to Reading Check

Scientists use the locations of earthquakes, volcanoes, trenches, and mid-ocean ridges to outline tectonic plates.

CHAPTER RESOURCES

Chapter Resource File

- **Internet Activity** Earthquakes BASIC

Technology

Transparencies
- 50 Tectonic Plate Boundaries (with worksheet)

MATHPRACTICE

The Rate of Plate Movement Tectonic plates move slowly on Earth's surface. The rate of plate movement can be calculated by using the following equation:

$$rate = \frac{distance}{time}$$

In kilometers, how far would a plate that moves 4 cm per year move in 2 million years?

Tectonic Plates

Scientists have identified about 15 major tectonic plates. While many plates are bordered by major surface features, such as mountain ranges or deep trenches in the oceans, the boundaries of the plates are not always easy to identify. As shown in **Figure 2,** the familiar outlines of the continents and oceans do not always match the outlines of plate boundaries. Some plate boundaries are located within continents far from mountain ranges.

Earthquakes

Scientists identify plate boundaries primarily by studying data from earthquakes. When tectonic plates move, sudden shifts can occur along their boundaries. These sudden movements are called *earthquakes*. Frequent earthquakes in a given zone are evidence that two or more plates may meet in that area.

Volcanoes

The locations of volcanoes can also help identify the locations of plate boundaries. Some volcanoes form when plate motions generate magma that erupts on Earth's surface. For example, the Pacific Ring of Fire is a zone of active volcanoes that encircles the Pacific Ocean. This zone is also one of Earth's major earthquake zones. The characteristics of this zone indicate that the Pacific Ocean is surrounded by plate boundaries.

Reading Check **How do scientists identify locations of plate boundaries? (See the Appendix for answers to Reading Checks.)**

Figure 2 ▶ Tectonic plates may contain both oceanic and continental crust. Notice that the boundaries of plates do not always match the outlines of continents.

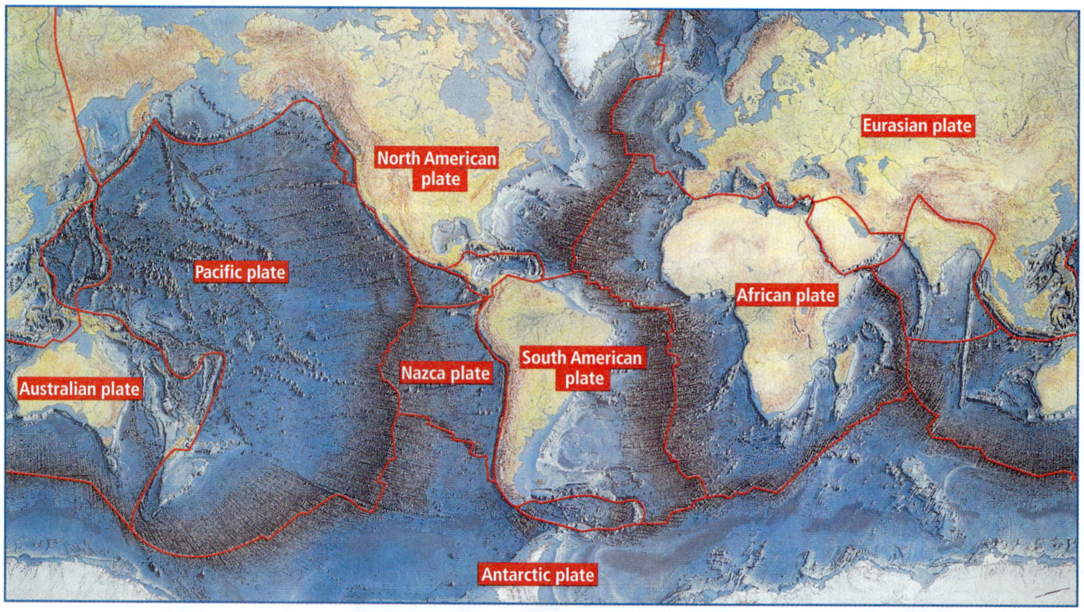

PHYSICS CONNECTION

Earthquakes An earthquake is a shaking of Earth's crust that is a result of the release of energy. Earthquakes happen when blocks of crust that were locked in place suddenly jolt as they slide past each other. The movement of the blocks creates a series of seismic waves that radiate outward in all directions from the point where the rocks move. Because these waves move through the center of Earth as well as around Earth's surface, scientists use the behavior of seismic waves to study the structure of Earth's interior. Students may better understand this process if it is compared to shaking a wrapped gift to get clues about what's inside it.

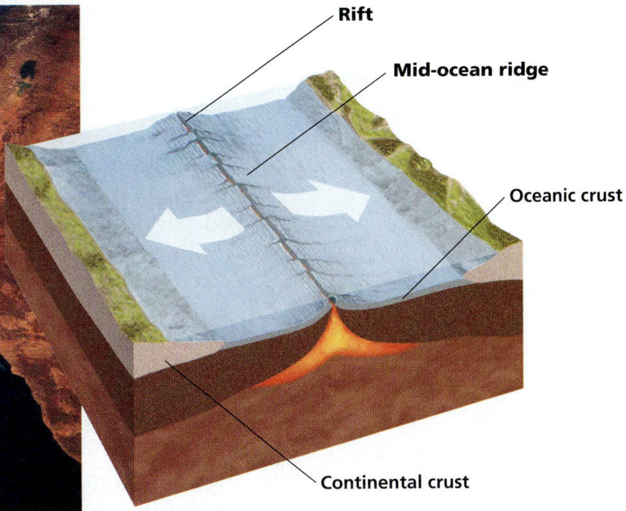

Figure 3 ▶ At divergent boundaries, plates separate. A divergent boundary exists in the Red Sea between the Arabian Peninsula and Africa.

Types of Plate Boundaries

Some of the most dramatic changes in Earth's crust, such as earthquakes and volcanic eruptions, happen along plate boundaries. Plate boundaries may be in the middle of the ocean floor, around the edges of continents, or even within continents. There are three types of plate boundaries. These plate boundaries are divergent boundaries, convergent boundaries, and transform boundaries. Each plate boundary is associated with a characteristic type of geologic activity.

Divergent Boundaries

The way that plates move relative to each other determines how the plate boundary affects Earth's surface. At a **divergent boundary,** two plates move away from each other. A divergent boundary is illustrated in **Figure 3.**

divergent boundary the boundary between tectonic plates that are moving away from each other

At divergent boundaries, magma from the asthenosphere rises to the surface as the plates move apart. The magma then cools to form new oceanic lithosphere. The newly formed rock at the ridge is warm and light. This warm, light rock sits higher than the surrounding sea floor because it's less dense. This rock forms undersea mountain ranges known as *mid-ocean ridges.* Along the center of a mid-ocean ridge is a *rift valley,* a narrow valley that forms where the plates separate.

Most divergent boundaries are located on the ocean floor. However, rift valleys may also form where continents are separated by plate movement. For example, the Red Sea occupies a huge rift valley formed by the separation of the African plate and the Arabian plate, as shown in **Figure 3.**

Discussion — Basic

Crustal Chemistry Tell students that oceanic crust is composed mainly of mafic rock, or rock that has a lot of heavy, iron- and magnesium-rich minerals. Tell students that continental crust is composed of lighter, silica-rich minerals. Ask students to think about a heavy china plate and a plastic foam plate floating on water. Ask them which plate is likely to sink beneath the other plate. (The china plate is more likely to sink, and the plastic foam plate is more likely to float.) Ask students to think about how this analogy could help them understand convergent boundaries. LS **Logical**

SKILL BUILDER — Basic

Vocabulary The word *divergent* is derived from the Latin *dis-*, which means "apart," and *vergere*, which means "to turn." Discuss with students the definitions of other words that begin with the prefix dis-, such as *distant* (the condition of being separated in space or time), *distinguish* (to separate or tell apart by differences), *distribute* (to divide and give out in shares), or *distract* (to draw attention away).
LS **Verbal** English Language Learners

MISCONCEPTION ALERT

Shape of Divergent Boundaries
Students may think of divergent boundaries as continuous, open cracks in Earth's surface. However, divergent boundaries are generally a series of smaller, offset cracks. The magma that rises at a rift flows upward through a series of small, vertical vents rather than flowing as a giant mass through a large, wide crack in the ocean floor.

CHAPTER RESOURCES

Technology

Transparencies
- 51 Types of Plate Boundaries (with worksheet)

Teach, *continued*

CHEMISTRY CONNECTION

Plate Boundary Volcanics The composition of igneous rock varies at different plate boundaries. In general, mafic and ultramafic rocks, or rocks that have very high iron and magnesium content, form at divergent plate boundaries. At convergent boundaries, igneous rocks that have intermediate compositions, or rocks with moderate amounts of both iron and silicate minerals, form. These differences are the result of the type of rock that melted to form the magma. At divergent boundaries, magma rises from the iron-rich mantle. At convergent boundaries, the crust of the overriding plate undergoes partial melting to form magma. Because continental crust and sea-floor sediments are generally high in silica, this chemical makes up more of the magma at convergent boundaries than at divergent boundaries.

INCLUSION Strategies

- *Visually Impaired*
- *Attention Deficit Disorder*

Ask students to model the types of plate boundaries by using their hands. Have students place the index fingers of their hands together. Then, have them slide one hand under the other hand to model a convergent boundary. Have students model divergent boundaries by pulling their hands apart from the starting position. LS **Kinesthetic**

Answer to Reading Check

Collisions at convergent boundaries can happen between two oceanic plates, between two continental plates, or between one oceanic plate and one continental plate.

convergent boundary the boundary between tectonic plates that are colliding

Convergent Boundaries

As plates pull apart at one boundary, they push into neighboring plates at other boundaries. **Convergent boundaries** are boundaries that form where two plates collide.

Three types of collisions can happen at convergent boundaries. One type happens when oceanic lithosphere collides with continental lithosphere, as shown in **Figure 4.** Because oceanic lithosphere is denser, it *subducts,* or sinks, under the less dense continental lithosphere. The region along a plate boundary where one plate moves under another plate is called a *subduction zone.* Deep-ocean trenches form at subduction zones. As the oceanic plate subducts, it heats up and releases fluids into the mantle above it. The addition of these fluids causes material in the overlying mantle to melt to form magma. The magma rises to the surface and forms volcanic mountains.

A second type of collision happens when two plates made of continental lithosphere collide. In this type of collision, neither plate subducts because neither plate is dense enough to subduct under the other plate. Instead, the colliding edges crumple and thicken, which causes uplift that forms large mountain ranges. The Himalaya Mountains formed in this type of collision.

The third type of collision happens between two plates that are made of oceanic lithosphere. One plate subducts under the other plate, and a deep-ocean trench forms. Fluids released from the subducted plate cause mantle rock to melt and form magma. The magma rises to the surface to form an *island arc,* which is a chain of volcanic islands. Japan is an example of an island arc.

Reading Check **Describe the three types of collisions that happen at convergent boundaries. (See the Appendix for answers to Reading Checks.)**

Figure 4 ▶ Plates collide at convergent boundaries. The islands of Japan are formed by the subduction of the Pacific plate and the Philippine plate under the Eurasian plate.

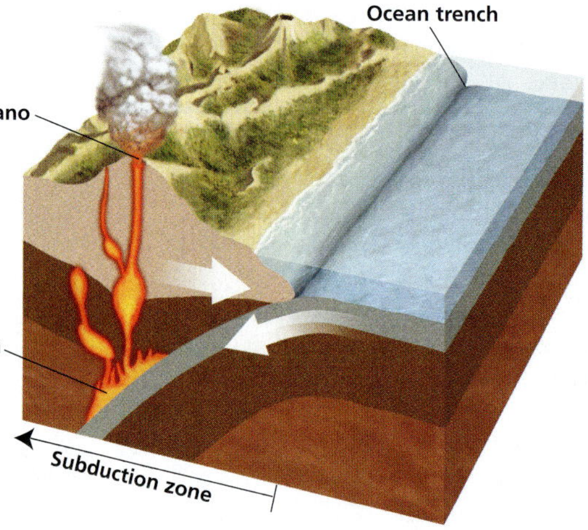

MISCONCEPTION ALERT

Magma Students may think that the magma that forms volcanoes at subduction zones forms from the melting of the subducting slab. However, studies indicate that the subducting plate maintains most of its structural integrity as it dives deep into the mantle. Many scientists think that plates remain intact until they reach the boundary between the mantle and the liquid outer core. The magma that forms at subduction zones results from the partial melting of asthenosphere and of the lithosphere of the overriding plate as fluids from the surface of the subducting plate combine with the rock material there.

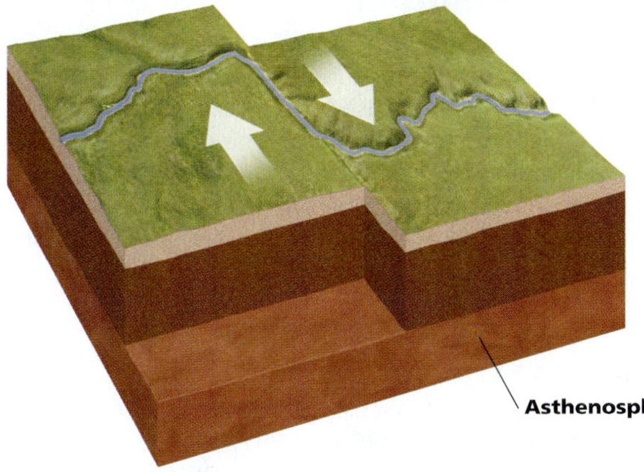

Figure 5 ▶ Plates slide past each other at transform boundaries. The course of the stream in the photo changed because the plates moved past each other at the San Andreas Fault in California.

Transform Boundaries

The boundary at which two plates slide past each other horizontally, as shown in **Figure 5,** is called a **transform boundary.** However, the plate edges usually do not slide along smoothly. Instead, they scrape against each other in a series of sudden spurts of motion that are felt as earthquakes. Unlike other types of boundaries, transform boundaries do not produce magma. The San Andreas Fault in California is a major transform boundary between the North American plate and the Pacific plate.

Transform motion also occurs along mid-ocean ridges. Short segments of a mid-ocean ridge are connected by transform boundaries called *fracture zones*.

Table 1 summarizes the three types of plate boundaries. The table also describes how each type of plate boundary changes Earth's surface and includes examples of each type of plate boundary.

transform boundary the boundary between tectonic plates that are sliding past each other horizontally

Table 1 ▼

Plate Boundary Summary

Type of boundary	Description	Example
Divergent	plates moving away from each other to form rifts and mid-ocean ridges	North American and Eurasian plates at the Mid-Atlantic Ridge
Convergent	plates moving toward each other and colliding to form ocean trenches, mountain ranges, volcanoes, and island arcs	South American and Nazca plates at the Chilean trench along the west coast of South America
Transform	plates sliding past each other while moving in opposite directions	North American and Pacific plates at the San Andreas Fault in California

MISCONCEPTION ALERT

Plate-Boundary Zones Not all plate boundaries are as simple as the main types discussed in this section. In some places, the boundaries are not well defined because the deformation that accompanies the plate movement extends over a broad belt, called a *plate-boundary zone*. Plate-boundary zones commonly involve at least two large plates and one or more *microplates,* smaller fragments of plates between the larger plates. These zones tend to have complicated geological structures and earthquake patterns. Examples of plate-boundary zones include the Mediterranean-Alpine region between the Eurasian and African Plates and the Himalayan region between the Eurasian and Indian plates.

SKILL BUILDER — GENERAL

Writing Ask students to research a specific plate boundary and to write a short essay about that plate boundary. The report must include where the boundary is located, which plates meet at that boundary, and what type of boundary it is. Students may also include what geologic features are found near the boundary, such as mountain ranges or volcanoes, and how plate boundaries affect humans. **LS Verbal**

BRAIN FOOD

Fracture Zones Oceanic fracture zones are ocean-floor valleys that horizontally offset segments of mid-ocean ridges. Some of these zones are hundreds to thousands of kilometers long. Examples of large fracture zones include the Blanco, Mendocino, Murray, and Molokai Fracture Zones off the coast of California and Mexico. These zones are presently inactive, but the offsets of the patterns of magnetic striping provide evidence of previous transform-fault activity. The San Andreas Fault, one of the few transform faults exposed on land, connects a short segment of the South Gorda–Juan de Fuca–Explorer Ridge in the Pacific Ocean to a segment of the East Pacific Rise in the Gulf of California.

Teach, *continued*

Teaching Tip — BASIC

Connect to Familiar Processes Students may comprehend the process of mantle convection better if you compare the process of mantle convection with a process with which students may be familiar—boiling water. Ask students to think about how convection currents work in a pot of water on the stove. (As water heats up from the bottom, some of the hot water rises over the center of heat. The heated water is commonly seen as rising bubbles. When some water rises over the center of heat, cooler water from other parts of the pan sink and replace the rising water. This water is heated in turn, and the process repeats.) **LS Visual** English Language Learners

Using the Figure — GENERAL

Mantle Convection Ask students to analyze the relationship between where lithosphere dives into the mantle and where colder mantle material sinks. (Lithosphere sinks into the asthenosphere near where colder mantle material sinks.) Answer to caption question: Divergent boundaries form where rising mantle material reaches the lithosphere. **LS Visual**

Causes of Plate Motion

The movement of tectonic plates is part of the mantle convection system. *Convection* is the movement of heated material due to differences in density that are caused by differences in temperatures. This process can be modeled by boiling water in a pot on the stove. As the water at the bottom of the pot is heated, the water at the bottom expands and becomes less dense than the cooler water above it. The cooler, denser water sinks, and the warmer water rises to the surface to create a cycle called a *convection cell.*

Mantle Convection

Scientists think that Earth is also a convecting system. Energy generated by Earth's core and radioactivity within the mantle heat mantle material. This heated material rises through the cooler, denser material around it. As the hot material rises, the cooler, denser material flows away from the hot material and sinks into the mantle to replace the rising material. As the mantle material moves, it drags the overlying tectonic plates along with it, as shown in **Figure 6.**

Convection currents and the resulting drag on the bottoms of tectonic plates can explain many aspects of plate movement. But scientists have identified two specific mechanisms of convection that help drive the process of plate movement.

Figure 6 ▶ Scientists think that tectonic plates are part of a convection system. ***How is the rising of hot material related to the location of divergent boundaries?***

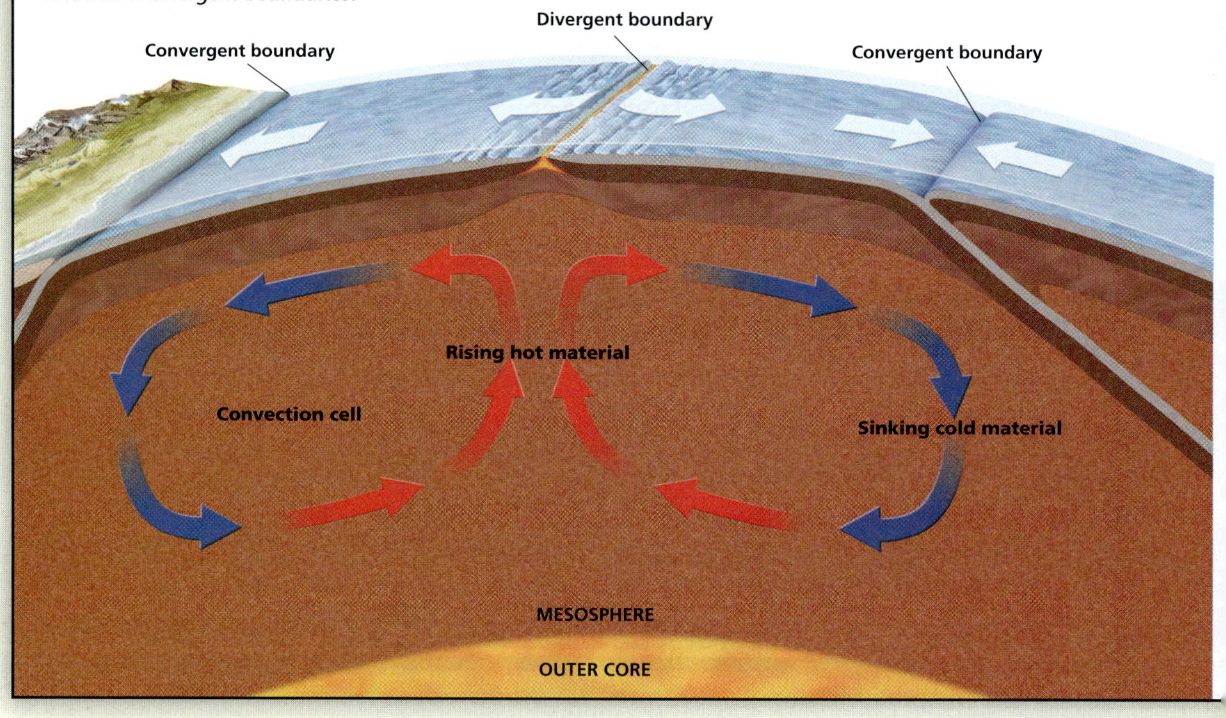

BRAIN FOOD

Speeding Plates Tectonic plates that have the largest proportion of their boundaries subducting under other plates tend to be the fastest-moving tectonic plates. The fastest-spreading center in the world is the East Pacific Rise, at which the plates move apart by as much as 15 cm per year. This ridge separates the Pacific plate from the Nazca and Cocos plates. The Pacific plate is subducting beneath the Eurasian, Philippine, and Australian plates on the western side of the Pacific Ocean. The Nazca and Cocos plates are subducting beneath the South American, Caribbean, and North American plates to the east. This evidence supports the idea that slab pull is a major factor in the movement of tectonic plates.

Ridge Push

Newly formed rock at a mid-ocean ridge is warm and less dense than older rock nearby. The warm, less dense rock is elevated above nearby rock, and older, denser rock slopes downward away from the ridge. As the newer, warmer rock cools and becomes denser, it begins to sink into the mantle and pull away from the ridge.

As the cooling rock sinks, the asthenosphere below it exerts force on the rest of the plate. This force is called *ridge push*. This force pushes the rest of the plate away from the mid-ocean ridge. Ridge push is illustrated in **Figure 7.**

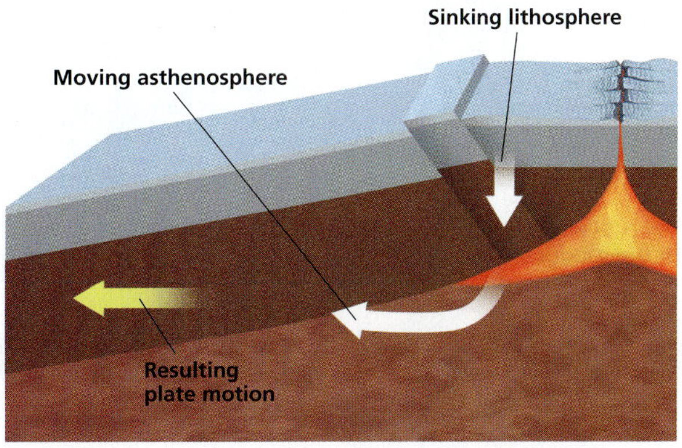

Figure 7 ▶ As the cooling lithosphere sinks, the asthenosphere moves away from the sinking lithosphere and pushes on the bottom of the plate. The plate moves in the direction that it is pushed by the asthenosphere.

Scientists think that ridge push may help drive plate motions. However, most scientists agree that ridge push is not the main driving force of plate motion. So, scientists looked to convergent boundaries for other clues to the forces that drive plate motion.

Reading Check **How may density differences in the rock at a mid-ocean ridge help drive plate motion? (See the Appendix for answers to Reading Checks.)**

QuickLAB 30 min

Tectonic Plate Boundaries

Procedure

1. Using a **ruler,** draw two 7 cm × 12 cm rectangles on a **piece of paper.** Cut them out with **scissors.**
2. Use a rolling pin to flatten **two different colored pieces of clay** to about 1/2 cm thick.
3. Use a **plastic knife** to cut each piece of clay into a 7 cm × 12 cm rectangle. Place a paper rectangle on each piece of clay.
4. Place the two clay models side by side on a flat surface and paper side down.
5. Place one hand on each piece of clay, and slowly push the blocks together until the edges begin to buckle and rise off the surface of the table.
6. Turn the clay models around so that the unbuckled edges are touching each other.
7. Place one hand on each clay model. Apply slight pressure toward the plane where the two blocks meet. Slide one clay model forward 7 cm and the other model backward about 7 cm.

Analysis

1. What type of plate boundary are you modeling in step 5?
2. What type of plate boundary are you modeling in step 7?
3. How do you think the processes modeled in this activity might affect the appearance of Earth's surface?

QuickLAB

Skills Acquired

- Modeling
- Inferring

Teacher's Notes: You may want to prepare the models before class to save time. Students should line tables or desks with wax paper or newspaper before using clay.

Answers

1. convergent
2. transform
3. Convergent boundaries may cause mountains to form. Rocks at transform boundaries may get deformed by the plate motion.

Answer to Reading Check

When denser lithosphere sinks into the asthenosphere, the asthenosphere must move out of the way. As the asthenosphere moves, it drags or pushes on other parts of the lithosphere, which causes movement.

Close

Reteaching — BASIC

Tectonic Processes Reinforce the idea of tectonic processes by having students build a small model of two tectonic plates by using clay. Have students model divergent plates or convergent plates that result in either subduction or mountain building. Then, have students identify plates that are currently undergoing these processes. LS **Kinesthetic**

Quiz — GENERAL

1. How are tectonic plate boundaries related to earthquakes? (Most earthquakes happen as plates move at plate boundaries.)
2. What are the three types of plate boundaries? (convergent, divergent, and transform)
3. What is the main driving force of plate tectonics? (convection)

CHAPTER RESOURCES

Chapter Resource File

- Datasheet for Quick Lab GENERAL

Technology

Transparencies

- 52 Ridge Push and Slab Pull (with worksheet)

Close, continued

Alternative Assessment — Basic

Tectonic Art Have students draw examples of the three types of plate boundaries and label all of the components. Next to each drawing, students should list the geologic features that are common at each boundary. Encourage students to include these features as part of their drawings. **LS Visual**

Answers to Section Review

1. Earth's surface is covered by plates of solid lithosphere that move around on the plastic rock of the asthenosphere. The tectonic plates move as a result of convection in Earth's mantle.
2. Earthquakes occur when tectonic plates suddenly shift relative to each other. When plates interact, magma may form and rise to the surface to form volcanoes.
3. Convergent plate boundaries form where two plates collide as they move toward each other. Divergent plate boundaries form where two plates pull apart from each other. Transform boundaries form where two plates move past each other horizontally in opposite directions.
4. Convergent boundaries usually cause deep-ocean trenches, mountains, and volcanoes. Divergent boundaries are characterized by mid-ocean ridges.
5. As hot mantle material rises, cooler, denser mantle material flows away from the hot material and sinks to replace the rising material. As the mantle material moves, it drags the overlying tectonic plate along with it.
6. As cooling rock sinks, the asthenosphere below it "pushes" the plate away from the ridge. In slab pull, the leading edge of a subducting plate sinks and pulls the rest of the plate along behind it.
7. As one plate subducts, fluids from the plate cause partial melting of the mantle. The resulting magma rises to the surface and forms volcanoes.
8. As hot material moves toward Earth's surface through convection currents in the mantle, thermal energy is transferred toward the outer parts of Earth. This convection drives plate tectonics.
9. *Tectonic plates* meet at three types of *boundaries—convergent,* which form *subduction zones; divergent,* which form *mid-ocean ridges;* and *transform*—as they move over Earth's surface as a result of *convection,* which includes *slab pull* and *ridge push.*

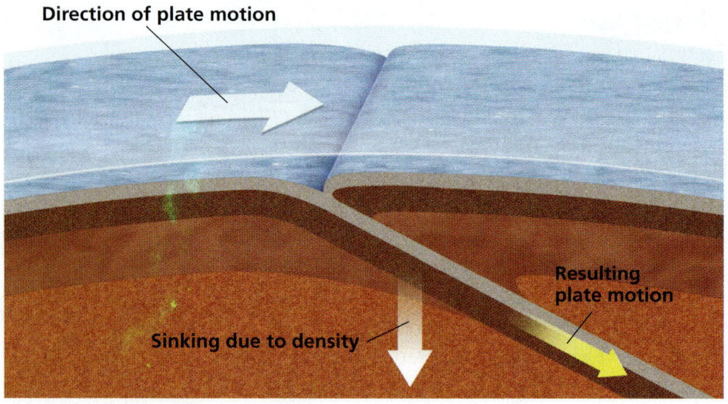

Figure 8 ▶ The leading edge of the subducting plate pulls the rest of the subducting plate into the asthenosphere in a process called *slab pull.*

SciLinks — NSTA — Developed and maintained by the National Science Teachers Association

For a variety of links related to this subject, go to www.scilinks.org

Topic: Plate Tectonics
SciLinks code: HQ61171

Slab Pull

Where plates pull away from each other at mid-ocean ridges, magma from the asthenosphere rises to the surface. The magma then cools to form new lithosphere. As the lithosphere moves away from the mid-ocean ridge, the lithosphere cools and becomes denser. Where the lithosphere is dense enough, it begins to subduct into the asthenosphere. As the leading edge of the plate sinks, it pulls the rest of the plate along behind it. The force exerted by the sinking plate is called *slab pull.* This process is shown in **Figure 8.** In general, plates that are subducting move faster than plates that are not subducting. This evidence indicates that the downward pull of the subducting lithosphere is a strong driving force for tectonic plate motion.

All three mechanisms of Earth's convecting system—drag on the bottoms of tectonic plates, ridge push, and slab pull—work together to drive plate motions. These mechanisms form a system that makes Earth's tectonic plates move constantly.

Section 2 Review

1. **Summarize** the theory of plate tectonics.
2. **Explain** why most earthquakes and volcanoes happen along plate boundaries.
3. **Identify and describe** the three major types of plate boundaries.
4. **Compare** the changes in Earth's surface that happen at a convergent boundary with those that happen at a divergent boundary.
5. **Describe** the role of convection currents in plate movement.
6. **Describe** how ridge push and slab pull contribute to the movement of tectonic plates.

CRITICAL THINKING

7. **Making Inferences** How do convergent boundaries add material to Earth's surface?
8. **Determining Cause and Effect** Explain how the outward transfer of heat energy from inside Earth drives the movement of tectonic plates.

CONCEPT MAPPING

9. Use the following terms to create a concept map: *tectonic plate, divergent, convergent, convection, transform, ridge push, slab pull, subduction zone,* and *mid-ocean ridge.*

CHAPTER RESOURCES

Chapter Resource File
- Section Quiz GENERAL

Workbooks
- Study Guide (also in Spanish)

Section 3 The Changing Continents

The continents did not always have the same shapes that they do today. And geologic evidence indicates that they will not stay the same shape forever. In fact, the continents are always changing. Slow movements of tectonic plates change the size and shape of the continents over millions of years.

OBJECTIVES

- **Identify** how movements of tectonic plates change Earth's surface.
- **Summarize** how movements of tectonic plates have influenced climates and life on Earth.
- **Describe** the supercontinent cycle.

KEY TERMS

rifting
terrane
supercontinent cycle
Pangaea
Panthalassa

Reshaping Earth's Crust

All of the continents that exist today contain large areas of stable rock, called *cratons,* that are older than 540 million years. Rocks within the cratons that have been exposed at Earth's surface are called *shields.* Cratons represent ancient cores around which the modern continents formed.

Rifting and Continental Reduction

One way that continents change shape is by breaking apart. **Rifting** is the process by which a continent breaks apart. New, smaller continents may form as a result of this process. The reason that continents rift is not entirely known. Because continental crust is thick and has a high silica content, continental crust acts as an insulator. This insulating property prevents heat in Earth's interior from escaping. Scientists think that as heat from the mantle builds up beneath the continent, continental lithosphere becomes thinner and begins to weaken. Eventually, a rift forms in this zone of weakness, and the continent begins to break apart, as shown in **Figure 1.**

rifting the process by which Earth's crust breaks apart; can occur within continental crust or oceanic crust

Figure 1 ▶ The East African Rift Valley formed as Africa began rifting about 30 million years ago.

Section 3

Focus

Overview

In this section, students learn about past and future movements of the continents. The section describes how continents change by rifting and accretion and how continental change affects climate and life on Earth.

Bellringer

Have students answer the following questions:

1. What might the United States be like if it were centered over the equator?
2. What might the United States be like if it were centered over the North Pole? LS Logical

Motivate

Activity — GENERAL

Modeling Rifting Have students flatten a piece of clay or plastic play putty until it is about 2 cm thick. Then, have students hold the edges of the clay or putty in their hands and pull apart slowly. The center of the clay or putty should thin and eventually pull apart. Explain to students that as continents rift, the crust in the rift zone thins and then eventually separates. LS Kinesthetic/Visual

CHAPTER RESOURCES

Chapter Resource File
- Directed Reading BASIC

Technology

Transparencies
- Bellringer

Student Edition on CD-ROM

One-Stop Planner CD-ROM
- Lesson Plan

Teach

Demonstration—GENERAL

Modeling Accretion Create a "continent" from one color of clay. Create several smaller "terranes" from other colors of clay. Place all of the terranes on a piece of wax paper at various intervals. Hold the continent just off the edge of a table, and pull the wax paper under the continent so that the terranes move toward the continent. Each terrane should collide with the continent to form several terranes of different colors at the edge of the continent. LS **Visual**

Using the Figure—GENERAL

Accretion Have students analyze the figure that shows how terranes accrete. Have students make a drawing that shows what the continent would look like when the remaining oceanic crust has subducted. (Drawings should show that Terrane B has also become part of the continent.) Answer to caption question: Most of the sediment on the subducting plate gets scraped off and added to the edge of a continent. Some of the sediment is subducted along with the plate. LS **Visual**

Answer to Reading Check

As a plate subducts beneath another plate, islands and other land features on the subducting plate are scraped off the subducting plate and become part of the overriding plate.

terrane a piece of lithosphere that has a unique geologic history and that may be part of a larger piece of lithosphere, such as a continent

Terranes and Continental Growth

Continents change not only by breaking apart but also by gaining material. Most continents consist of cratons surrounded by a patchwork of terranes. A **terrane** is a piece of lithosphere that has a unique geologic history that differs from the histories of surrounding lithosphere. A terrane can be identified by three characteristics. First, a terrane contains rock and fossils that differ from the rock and fossils of neighboring terranes. Second, there are major faults at the boundaries of a terrane. Third, the magnetic properties of a terrane generally do not match those of neighboring terranes.

Terranes become part of a continent at convergent boundaries. When a tectonic plate carrying a terrane subducts under a plate made of continental crust, the terrane is scraped off the subducting plate, as shown in **Figure 2.** The terrane then becomes part of the continent. Some terranes may form mountains, while other terranes simply add to the surface area of a continent. The process in which a terrane becomes part of a continent is called *accretion* (uh KREE shuhn).

A variety of materials can form terranes. Terranes may be small volcanic islands or underwater mountains called *seamounts*. Small coral islands, or *atolls*, can also form terranes. And large chunks of continental crust can be terranes. When large terranes and continents collide, major mountain chains often form. For example, the Himalaya Mountains formed when India began colliding with Asia about 50 million years ago (during the Cenozoic Era).

Reading Check **Describe the process of accretion. (See the Appendix for answers to Reading Checks.)**

Figure 2 ▶ As oceanic crust subducts, a terrane is scraped off the ocean floor and becomes part of the continental crust. ***What would you expect to happen to sediments on the sea floor when the plate they are on subducts?***

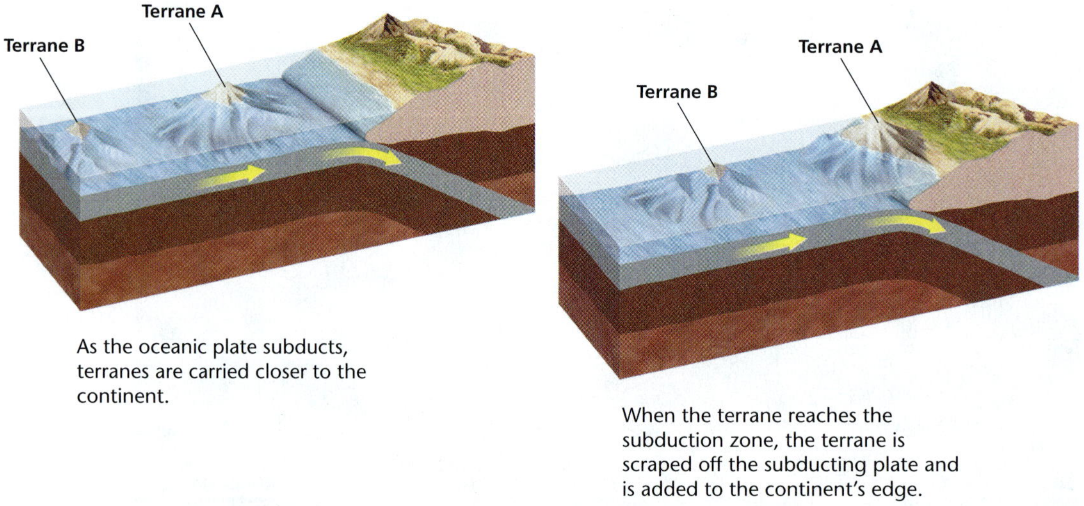

SKILL BUILDER—ADVANCED

Vocabulary The word *accrete* means "to grow by being added to." It comes from the Latin *accrescere,* which means "to increase or to grow." Ask students how knowing the etymology of *accrete* can help them understand the process of accretion. (Accretion is the process by which a continent grows when smaller pieces are added to it.) LS **Verbal** English Language Learners

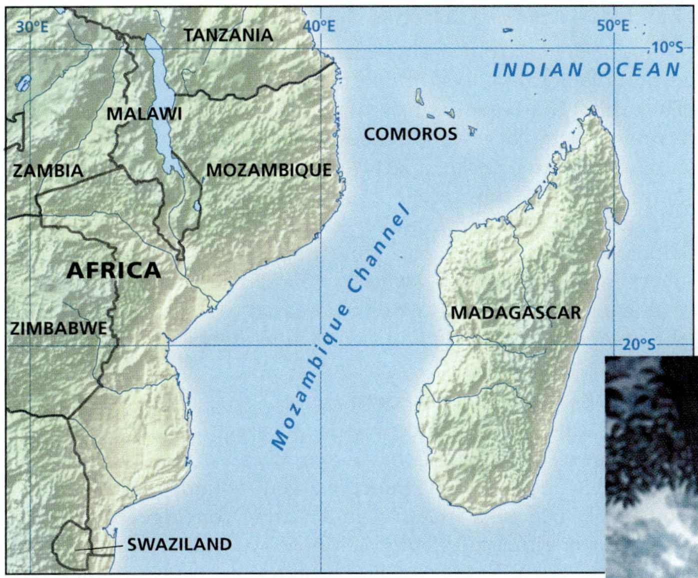

Figure 3 ▶ Madagascar separated from Africa about 165 million years ago and separated from India about 88 million years ago. This separation isolated the plants and animals on the island of Madagascar. As a result, unique species of plants and animals evolved on Madagascar. These species, such as the fossa (below), are found nowhere else on Earth.

Effects of Continental Change

Modern climates are a result of past movements of tectonic plates. A continent's location in relation to the equator and the poles affects the continent's overall climate. A continent's climate is also affected by the continent's location in relation to oceans and other continents. Mountain ranges affect air flow and wind patterns around the globe. Mountains also affect the amount of moisture that reaches certain parts of a continent. When continents move, the flow of air and moisture around the globe changes and causes climates to change.

Changes in Climate

Geologic evidence shows that ice once covered most of Earth's continental surfaces. Even the Sahara in Africa, one of the hottest places on Earth today, was once covered by a thick ice sheet. This ice sheet formed when all of the continents were close together and were located near the South Pole. As continents began to drift around the globe, however, global temperatures changed and much of the ice sheet melted.

Changes in Life

As continents rift or as mountains form, populations of organisms are separated. When populations are separated, new species may evolve from existing species. Sometimes, isolation protects organisms from competitors and predators and may allow the organismsm to evolve into unique organisms, as shown in **Figure 3.**

Group Activity — ADVANCED

Responses to Climate Change

Organize the class into four groups of students. Have all students stand together in the center of the room as if they were a population of one species that lives on one continent. Tell the students that parts of the continent are rifting apart, and tell them which groups of students should separate from the main group and move to another part of the room. Tell one group that the new climate is very cold. Tell another group that the climate is very dry, and tell a third group that the environment is now swampy. Have each group identify how its population may change to adapt to the new climates. (Cold-climate organisms may develop thick fur, small pine-like needles, or a light coloring. Dry-climate organisms may develop thick skins, nocturnal habits, or other means of preserving water. Swampy-environment organisms may adapt to a more aquatic lifestyle.) **LS Kinesthetic**

READING SKILL BUILDER — BASIC

Reading Organizer As students read this section, encourage them to take Power Notes, KWL Notes, or Two-Column Notes as described in the Skills Handbook section of the Appendix. Later, students can use these notes as a study guide for assessments. Students may want to use KWL notes as a place to start when doing independent research for class projects or reports. **LS Logical**

BIOLOGY CONNECTION

Extinction One of the assumptions of evolution by natural selection is that organisms are adapted to their environment. Because of this adaptation, changes in the environment affect the organisms. Only those organisms that can adapt to the environmental changes will survive. Organisms that cannot survive—in other words, those that are unfit to live and reproduce in the changing environment—become extinct.

ENVIRONMENTAL CONNECTION

Climate Changes Increased public concern about global warming has prompted much research into the environmental factors that may cause global warming. By studying climatic changes that occurred in Earth's geologic past, scientist hope to understand changes that are occurring in Earth's present climate.

Teach, *continued*

Identifying Preconceptions — Basic

Continental Outlines Students may think that the basic outlines of the continents have been the same throughout geologic time. Have students make sketches of what they think the continents looked like 300 million years ago. Explain to students that much of North America was underwater for most of the past 500 million years. Then, explain that much of the topography of western North America formed as a result of subduction, volcanism, and accretion in the past 100 million years. **LS Visual**

Teaching Tip — General

Connect to Prior Knowledge Complete the concept of the supercontinent cycle by asking students to draw on prior knowledge about continental drift. Ask students what evidence of continental drift can also be used to reconstruct the past positions of continents. (matching mountain ranges and rock types, matching fossils, and climatic evidence) Ask students how magnetic reversals and sea-floor spreading can be used to reconstruct the movements of continents over time. (Continents move away from each other during sea-floor spreading. Using the rates of sea-floor spreading based on the magnetic reversal patterns, scientists can show the relative positions of continents at specific points in the past.) **LS Logical**

The Supercontinent Cycle

Using evidence from many scientific fields, scientists can construct a general picture of continental change throughout time. They think that at several times in the past, the continents were arranged into large landmasses called *supercontinents*. These supercontinents broke apart to form smaller continents that moved around the globe. Eventually, the smaller continents joined again to form another supercontinent. When the last supercontinent broke apart, the modern continents formed. A new supercontinent is likely to form in the future. The process by which supercontinents form and break apart over time is called the **supercontinent cycle** and is shown in **Figure 4.**

supercontinent cycle the process by which supercontinents form and break apart over millions of years

Pangaea the supercontinent that formed 300 million years ago and that began to break up beginning 250 million years ago

Panthalassa the single, large ocean that covered Earth's surface during the time the supercontinent Pangaea existed

Why Supercontinents Form

The movement of plates toward convergent boundaries eventually causes continents to collide. Because continental lithosphere does not subduct, the convergent boundary between two continents becomes inactive, and a new convergent boundary forms. Over time, all of the continents collide to form a supercontinent. Then, heat from Earth's interior builds up under the supercontinent, and rifts form in the supercontinent. The supercontinent breaks apart, and plates carrying separate continents move around the globe.

Formation of Pangaea

The supercontinent **Pangaea** (pan JEE uh) formed about 300 million years ago (during the Paleozoic Era). As the continents collided to form Pangaea, mountains formed. The Appalachian Mountains of eastern North America and the Ural Mountains of Russia formed during these collisions. A body of water called the Tethys Sea cut into the eastern edge of Pangaea. The single, large ocean that surrounded Pangaea was called **Panthalassa.**

Figure 4 ▶ Over millions of years, supercontinents form and break apart in a cycle known as the *supercontinent cycle*.

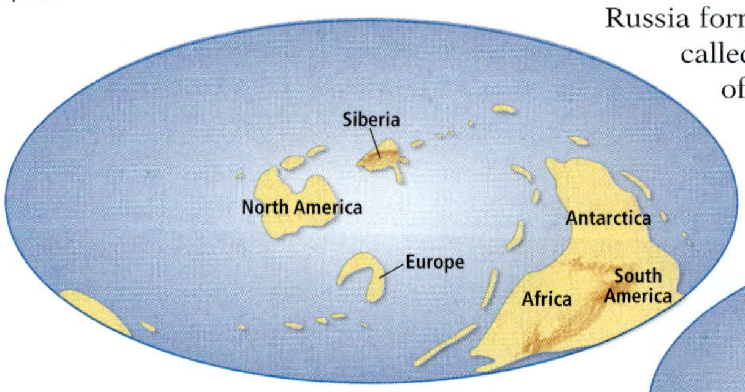

About 450 million years ago Earth's continents were separated, as they are today.

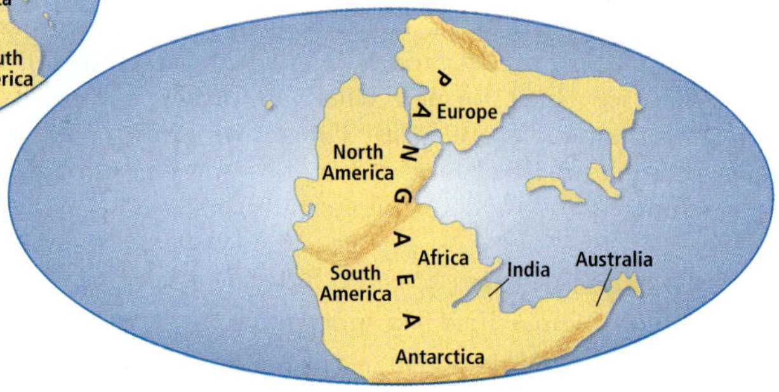

225 million to 200 million years ago Pangaea had formed and was beginning to break apart.

CHAPTER RESOURCES

Chapter Resource File

- **Internet Activity**
  The Paleomap Project **Basic**

Technology

Transparencies

- 53 The Supercontinent Cycle (with worksheet)

Internet Activity — Basic

The Paleomap Project The Paleomap Project Web site offers a variety of maps, graphs, and animations that illustrate the movement of continents throughout Earth's history and into Earth's projected future. The site also includes data about changes in Earth's climate. Ask students to browse the Paleomap Project site. Have them report on some aspect of the site that interests them. A worksheet designed to direct student research on this topic can be found in the **Chapter Resource File** booklet or by visiting go.hrw.com and entering the keyword **HQ6TECX.** **LS Visual**

Breakup of Pangaea

About 200 million years ago (during the Mesozoic Era), Pangaea began to break into two continents—*Laurasia* and *Gondwanaland*. A large rift split the supercontinent from east to west. Then, Laurasia began to drift northward and rotate slowly, and a new rift formed. This rift separated Laurasia into the continents of North America and Eurasia. The rift eventually formed the North Atlantic Ocean. The rotation of Laurasia also caused the Tethys Sea to close. The Tethys Sea eventually became the Mediterranean Sea.

As Laurasia began to break apart, Gondwanaland also broke into two continents. One continent broke apart to become the continents of South America and Africa. About 150 million years ago (during the Mesozoic Era), a rift between Africa and South America opened to form the South Atlantic Ocean. The other continent separated to form India, Australia, and Antarctica. As India broke away from Australia and Antarctica, it started moving northward, toward Eurasia. About 50 million years ago (during the Cenozoic Era), India collided with Eurasia, and the Himalaya Mountains began to form.

The Modern Continents

Slowly, the continents moved into their present positions. As the continents drifted, they collided with terranes and other continents. These collisions welded new crust onto the continents and uplifted the land. Mountain ranges, such as the Rocky Mountains, the Andes, and the Alps, formed. Tectonic plate motion also caused new oceans to open up and caused others to close.

Reading Check **What modern continents formed from Gondwanaland? (See the Appendix for answers to Reading Checks.)**

SCILINKS. NSTA Developed and maintained by the National Science Teachers Association

For a variety of links related to this subject, go to www.scilinks.org

Topic: Pangaea
SciLinks code: HQ61105

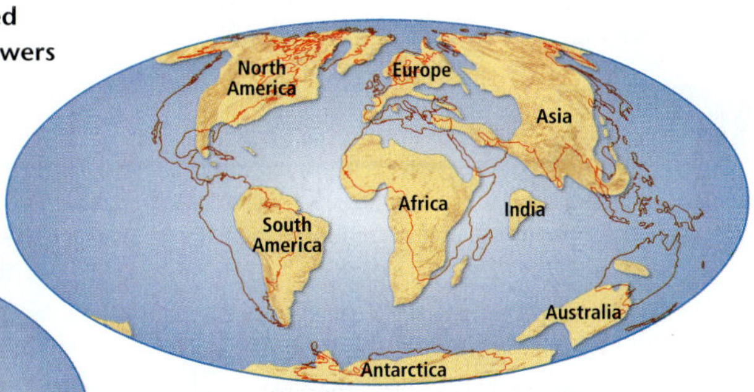

70 million to 50 million years ago
The continents were moving toward their current positions. The current positions of the continents are shown here in red.

160 million to 140 million years ago
Pangaea split into two continents—Laurasia to the north and Gondwanaland to the south.

Answer to Reading Check

The continents Africa, South America, Antarctica, and Australia formed from Gondwanaland. The subcontinent of India was also part of Gondwanaland.

Close

Reteaching — BASIC

Peer Reviewing Divide the class into pairs or small groups of students. Have each student write five questions that can be answered by reading the section. Have students use the questions to quiz each other on the information in the section. **LS Interpersonal**

Quiz — GENERAL

1. By what process do continents break apart? (rifting)
2. How does mountain formation cause changes in climate? (Mountain ranges affect air flow and wind patterns around the globe as well as the amount of moisture in an area.)
3. What was the name of the last supercontinent? When did it form, and when did it begin to break apart? (Pangaea; about 300 million years ago; about 200 million years ago)

MISCONCEPTION ALERT

Supercontinents Students may think that Pangaea was the only supercontinent in the past or that all supercontinents were called Pangaea. In fact, scientists think that at least five supercontinents have existed at different times since the first crust formed on Earth. The supercontinent Rodinia broke apart about 750 million years ago. It later reassembled to form the supercontinent Pannotia, which broke apart about 540 million years ago. Pangaea assembled from the scattered continental fragments a few hundred million years later. Scientists do not call the landmasses that became the modern continents by their modern names, either. Previous continents included Congo, Laurentia, Gondwana, Cimmeria, and Baltica.

Close, *continued*

Alternative Assessment — ADVANCED

Process Models Have students create models that illustrate the process of rifting or the process of accretion. The model should show steps in the process rather than just the final result of the process. Therefore, the models should be easy to reset to the original position or able to work in reverse. Students may wish to create computer models or animations for this purpose. **LS Kinesthetic**

Answers to Section Review

1. Continents get smaller as they break apart during rifting. Continents get larger as smaller pieces of crust collide during accretion.
2. A terrane is a piece of lithosphere that has a unique geologic history but has become part of a continent through accretion.
3. As continents rift apart or as mountain chains form, populations of plants and animals become separated and evolve in different ways in response to their changing environments.
4. Over time, movements of the tectonic plates cause all of the continents to collide to form a supercontinent and then break apart to form individual continents. This repeated pattern is called the *supercontinent cycle.*
5. Pangaea was a supercontinent that formed when all of the continents of Earth collided more than 300 million years ago. Continuing movements of the plates resulted in the breakup of Pangaea.
6. Pangaea was the supercontinent that formed about 300 million years ago. Gondwanaland was the large continent that broke away from Pangaea and eventually split to form Africa, South America, Australia, and Antarctica.
7. In about 150 million years, Africa will collide with Eurasia. Australia will also eventually collide with Eurasia. In 250 million years, a new supercontinent will form.
8. Because Pangaea was made of all of the continents combined, the area of the continent must have been much larger than any single continent today. Thus, the interior of the continent was larger, and the climate was drier.
9. Marine fossils formed on the ocean floor. Then, that part of the ocean became part of a convergent boundary, and the sea floor became part of the continent during accretion.
10. *Supercontinents,* such as *Pangaea,* form as *continents* collide with *terranes,* such as *seamounts* and *atolls,* in the process of *accretion,* and break apart as a result of *rifting.*

Geography of the Future

As tectonic plates continue to move, Earth's geography will change dramatically. If plate movements continue at current rates, in about 150 million years, Africa will collide with Eurasia, and the Mediterranean Sea will close. A new ocean will form as east Africa separates from the rest of Africa and moves eastward. New subduction zones will form off the east coast of North and South America. North and South America will then move east across the Atlantic Ocean. The Atlantic Ocean will close as North and South America collide with Africa.

In North America, Mexico's Baja Peninsula and the part of California that is west of the San Andreas Fault will move to where Alaska is today. If this plate movement occurs as predicted, Los Angeles will be located north of San Francisco's current location. Scientists predict that in 250 million years, the continents will come together again to form a new supercontinent, as shown in **Figure 5.**

Figure 5 ▶ Scientists predict that movements of tectonic plates will cause a supercontinent to form in the future.

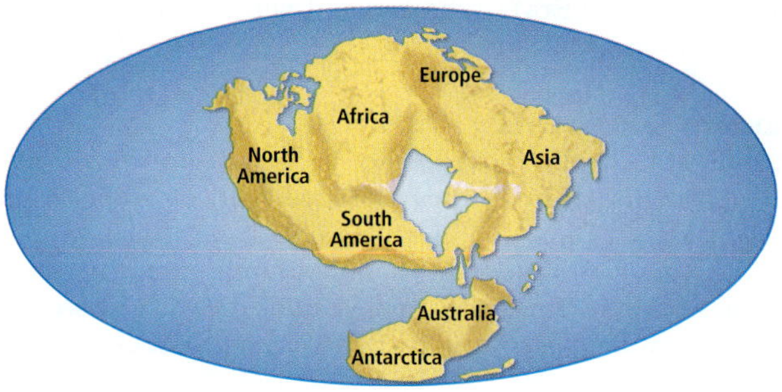

Section 3 Review

1. **Identify** how rifting and accretion change the shapes of continents.
2. **Describe** why a terrane has a different geologic history from that of the surrounding area.
3. **Summarize** how continental rifting may lead to changes in plants and animals.
4. **Describe** the supercontinent cycle.
5. **Explain** how the theory of plate tectonics relates to the formation and breakup of Pangaea.
6. **Compare** Pangaea and Gondwanaland.
7. **List** three changes in geography that are likely to happen in the future.

CRITICAL THINKING

8. **Identifying Relationships** The interior parts of continents generally have drier climates than coastal areas do. How does this fact explain the evidence that the climate on Pangaea was drier than many modern climates?
9. **Determining Cause and Effect** Explain how mountains on land can be composed of rocks that contain fossils of marine animals.

CONCEPT MAPPING

10. Use the following terms to create a concept map: *supercontinent, rifting, atoll, continent, terrane, seamount, accretion,* and *Pangaea.*

CHAPTER RESOURCES

Chapter Resource File

- Section Quiz GENERAL

Workbooks

- Study Guide (also in Spanish)

Chapter 10

Highlights

Sections

1 Continental Drift

Key Terms

continental drift, 239
mid-ocean ridge, 242
seafloor spreading, 243
paleomagnetism, 243

Key Concepts

- Fossil, rock, and climatic evidence supports Wegener's hypothesis of continental drift. However, Wegener could not explain the mechanism by which the continents move.
- New ocean floor is constantly being produced through sea-floor spreading, which creates mid-ocean ridges and changes the topography of the sea floor.
- Sea-floor spreading provides a mechanism for continental drift.

2 The Theory of Plate Tectonics

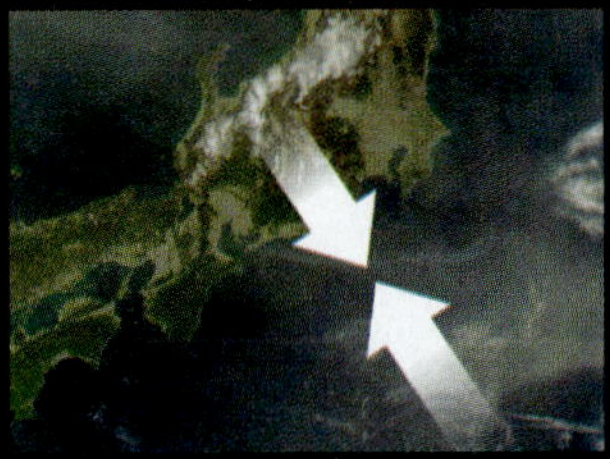

plate tectonics, 247
lithosphere, 247
asthenosphere, 247
divergent boundary, 249
convergent boundary, 250
transform boundary, 251

- The theory of plate tectonics proposes that changes in Earth's crust are caused by the very slow movement of large tectonic plates.
- Earthquakes, volcanoes, and young mountain ranges tend to be located in belts along the boundaries between tectonic plates.
- Tectonic plates meet at three types of boundaries—divergent, convergent, and transform. The geologic activity that occurs along the three types of plate boundaries differs according to the way plates move relative to each other.
- Tectonic plates may be part of a convecting system that is driven by differences in density and heat.

3 The Changing Continents

rifting, 255
terrane, 256
supercontinent cycle, 258
Pangaea, 258
Panthalassa, 258

- Continents grow through the accretion of terranes. Continents break apart through rifting.
- Movements of tectonic plates have altered climates on continents and have created conditions that lead to changes in plants and animals.
- Continents collide to form supercontinents and then break apart in a cycle called the *supercontinent cycle*.
- Earth's tectonic plates continue to move, and in the future, the continents will likely be in a different configuration.

Chapter Highlights

Alternative Assessment — ADVANCED

Comparing Scientific Ideas
Have students create a table that compares Wegener's hypothesis of continental drift with the modern theory of plate tectonics. Students should include as much detail in each column as possible. When students have completed their tables, have them write a paragraph that explains how the hypothesis of continental drift provided a base that scientists worked from to develop the theory of plate tectonics. LS Verbal

CHAPTER RESOURCES

Chapter Resource File
- **Concept Review** GENERAL
- **Critical Thinking** ADVANCED
- **Math Skills** GENERAL
- **Graphing Skills** GENERAL
- **Chapter Test A** GENERAL
- **Chapter Test B** ADVANCED

Workbooks
- **Study Guide** (also in Spanish)
- **Assessments** (Spanish)

Technology

Classroom Videos
- Brain Food Video Quiz

HRW Earth Science Video
- Segment 3: Plate Tectonics

Chapter Review

Assignment Guide	
Section	**Questions**
1	1, 9–12, 20–21, 24, 30, 32–35
2	2–3, 5, 7, 13–16, 19, 23, 25
3	4, 8, 18, 22, 26
1 and 2	6, 17, 31
2 and 3	28–29
1–3	27

Using Key Terms

1–8. Answers may vary but should show that students understand the definitions of and differences between key terms.

Understanding Key Concepts

9. a **10.** c
11. c **12.** c
13. b **14.** c
15. a **16.** a

Short Answer

17. As scientists developed technology, such as sonar and magnetic sensing devices, to study the ocean floor, they became able to identify features of the ocean floor that helped identify plate boundaries and explain how plates move.

18. Pangaea rifted apart to form several smaller continents. Over many millions of years, the continents moved as a result of interactions of the plates until the continents reached their present locations.

19. Most earthquakes are sudden movements of crust that occur as tectonic plates grind past each other. Most volcanoes form as a result of upwelling magma at divergent boundaries or the interaction of fluids and heat at convergent plate boundaries.

20. During sea-floor spreading, new sea floor forms. At convergent boundaries, sea floor is destroyed. Thus, sea-floor spreading continually makes new sea floor to replace the old sea floor.

21. As rocks cool, the magnetic fields of some minerals in the rocks align with Earth's magnetic field. When the rocks are solid, the minerals hold that magnetic orientation.

22. As continents separate, populations of organisms are separated. These populations then change in different ways to adapt to their new and changing environments.

Critical Thinking

23. Tectonic plates fit together to form a connected pattern on Earth's surface.

24. He may have suggested that the continents had not separated until after 10 million years ago—otherwise, the same fossils would not be in both places.

25. Sample answer: For the same surface area to be maintained, the same amount of material that is created must be destroyed. I would expect to find material being destroyed at other plate boundaries.

Chapter 10 Review

Using Key Terms

Use each of the following terms in a separate sentence.

1. *sea-floor spreading*
2. *convection*
3. *divergent boundary*
4. *terrane*

For each pair of terms, explain how the meanings of the terms differ.

5. *convergent boundary* and *subduction zone*
6. *continental drift* and *plate tectonics*
7. *ridge push* and *slab pull*
8. *Pangaea* and *Panthalasa*

Understanding Key Concepts

9. Support for Wegener's hypothesis of continental drift includes evidence of changes in
a. climatic patterns. **c.** terranes.
b. Panthalassa. **d.** subduction.

10. New ocean floor is constantly being produced through the process known as
a. subduction.
b. continental drift.
c. sea-floor spreading.
d. terranes.

11. An underwater mountain chain that formed by sea-floor spreading is called a
a. divergent boundary.
b. subduction zone.
c. mid-ocean ridge.
d. convergent boundary.

12. Scientists think that the upwelling of mantle material at mid-ocean ridges is caused by the motion of tectonic plates and comes from
a. the lithosphere. **c.** the asthenosphere.
b. terranes. **d.** rift valleys.

13. The layer of plastic rock that underlies the tectonic plates is the
a. lithosphere. **c.** oceanic crust.
b. asthenosphere. **d.** terrane.

14. The region along tectonic plate boundaries where one plate moves beneath another is called a
a. rift valley.
b. transform boundary.
c. subduction zone.
d. convergent boundary.

15. Two plates grind past each other at a
a. transform boundary.
b. convergent boundary.
c. subduction zone.
d. divergent boundary.

16. Convection occurs because heated material becomes
a. less dense and rises.
b. denser and rises.
c. denser and sinks.
d. less dense and sinks.

Short Answer

17. Explain the role of technology in the progression from the hypothesis of continental drift to the theory of plate tectonics.

18. Summarize how the continents moved from being part of Pangaea to their current locations.

19. Why do most earthquakes and volcanoes happen at or near plate boundaries?

20. Explain the following statement: "Because of sea-floor spreading, the ocean floor is constantly renewing itself."

21. Describe how rocks that form at a mid-ocean ridge become magnetized.

22. How may continental rifting influence the evolution of plants and animals?

Critical Thinking

23. Making Comparisons How are tectonic plates like the pieces of a jigsaw puzzle?

24. Making Inferences If Alfred Wegener had found identical fossil remains of plants and animals that had lived no more than 10 million years ago in both eastern Brazil and western Africa, what might he have concluded about the breakup of Pangaea?

25. Identifying Relationships Assume that the total surface area of Earth is not changing. If new material is being added to Earth's crust at one boundary, what would you expect to be happening at another boundary?

26. Making Predictions One hundred fifty million years from now, the continents will have drifted to new locations. How might these changes affect life on Earth?

Concept Mapping

27. Use the following terms to create a concept map: *asthenosphere, lithosphere, divergent boundary, convergent boundary, transform boundary, subduction zone, mid-ocean ridge, plates, convection, continental drift, theory of plate tectonics,* and *sea-floor spreading*.

Math Skills

28. Making Calculations The coasts of Africa and South America began rifting about 150 million years ago. Today, the coast of South America is about 6,660 km from the coast of Africa. Using the equation *velocity = distance ÷ time*, determine how fast the continents moved apart in millimeters per year.

29. Using Equations Assume that scientists know the rate at which the North American and Eurasian plates are moving away from each other. If t = time, d = distance, and v = velocity, what equation can they use to determine when North America separated from Eurasia during the breakup of Pangaea?

Writing Skills

30. Writing Persuasively Imagine that you are Alfred Wegener. Write a persuasive essay to explain your idea of continental drift. Use only evidence originally used by Wegener to support his hypothesis.

31. Communicating Main Ideas Explain how the research of Wegener, Hess, and others led to the theory of plate tectonics.

Interpreting Graphics

The graph below shows the relationship between the age of sea floor rocks and the depth of the sea floor beneath the ocean surface. Use the graph to answer the questions that follow.

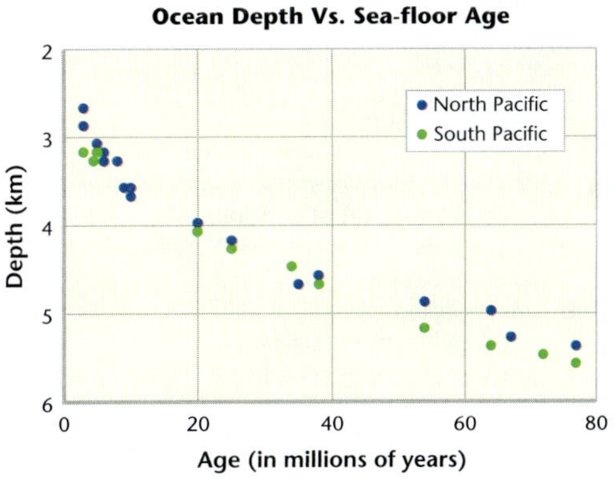

32. How old is the sea floor at a depth of 4 km?

33. Approximately how deep is the sea floor when it is 55 million years old?

34. What can you infer about the age of very deep sea floor from the data in the graph?

35. The ridge in the South Pacific Ocean is spreading faster than the ridge in the North Pacific Ocean. If this graph showed the depth of the sea floor in relation to the distance from the ridge, how would the graphs of the North Pacific and South Pacific ridges differ?

26. As the continents move, they affect climate. Thus, the climates of the various continents and of Earth as a whole would change. Also, rifting and collisions between continents would cause environmental changes that would cause species to change or become extinct.

Concept Mapping

27. Answers may vary but should include all of the terms listed. Sample answers appear at the end of this Teacher Edition.

Math Skills

28. 1,000 m/km × 100 cm/m × 10 mm/cm = 1,000,000 mm/km; 6,660 km × 1,000,000 mm/1 km = 6,660,000,000 mm; 6,660,000,000 mm ÷ 150,000,000 years = 44.4 mm/year

29. *time = distance ÷ velocity* ($t = d/v$)

Writing Skills

30. Answers may vary. Accept all reasonable answers.

31. Answers may vary. Accept all reasonable answers.

Interpreting Graphics

32. 20 million years

33. 5 km

34. that the sea floor is very old

35. The graph of the North Pacific would have a steeper slope than the graph of the South Pacific would, because the sea floor closer to the North Pacific ridge would be older and thus deeper than sea floor at the same distance from the South Pacific ridge.

Estimated Time

To give students practice under more realistic testing conditions, allow them 30 minutes to answer all of the questions in this practice test.

TEST DOCTOR

Question 1 Answer A is correct. Answer A is correct because the density of each colliding plate is more important in determining whether a plate will subduct or uplift than the size of the plate, answer B, the magnetic properties of the rock that makes up the plate, answer C, or the length of the boundary formed by the meeting of two plates, answer D.

Question 9 Answer H is correct. Answer F is incorrect because the nine mountains that rise above 8,000 m are not the nine tallest mountains on Earth. Answer G is incorrect because the longest mountain chain on Earth is the Mid-Atlantic Ridge, which is underwater. The longest above-water mountain chain is the Andes Mountains. Answer I is incorrect because the mountain chain is currently changing and growing.

Chapter 10 Standardized Test Prep

Understanding Concepts

Directions (1–4): **For** ***each*** **question, write on a separate sheet of paper the letter of the correct answer.**

1 Which of the following factors is most important when determining the type of collision that forms when two lithospheric plates collide?
A. the density of each plate
B. the size of each plate
C. the paleomagnetism of the rock
D. the length of the boundary

2 At locations where sea-floor spreading occurs, rock is moved away from a mid-ocean ridge. What replaces the rock as it moves away?
F. molten rock
G. older rock
H. continental crust
I. compacted sediment

3 Which of the following was a weakness of Wegener's proposal of continental drift when he first proposed the hypothesis?
A. an absence of fossil evidence
B. unsupported climatic evidence
C. unrelated continent features
D. a lack of proven mechanisms

4 Which of the following statements describes a specific type of continental growth?
F. Continents change not only by gaining material but also by losing material.
G. Terranes become part of a continent at convergent boundaries.
H. Ocean sediments move onto land because of sea-floor spreading.
I. Rifting adds new rock to a continent and causes the continent to become wider.

Directions (5–6): **For** ***each*** **question, write a short response.**

5 What is the name for the process by which the Earth's crust breaks apart?

6 What is the name for the layer of plastic rock directly below the lithosphere?

Reading Skills

Directions (7–9): **Read the passage below. Then, answer the questions.**

The Himalaya Mountains

The Himalaya Mountians are a range of mountains that is 2,400 km long and that arcs across Pakistan, India, Tibet, Nepal, Sikkim, and Bhutan. The Himalaya mountains are the highest Mountains on Earth. Nine mounains in the chain, including Mount Everest, the tallest above-water mountain on Earth, rise to heights of more than 8,000 m above sea-level. Mount Everest stands 8,850 m tall.

The formation of the Himalaya Mountains began about 50 million years ago. A tectonic plate carrying the Indian subcontinent collided with the Eurasian plate. The Indian plate was denser than the Eurasian plate. This difference in density caused the uplifting of the Eurasian plate and the subsequent formation of the Himalaya Mountains. This process continues today. The Indian plate continues to push under the Eurasian plate. New measurements show that Mount Everest is moving to the northeast by as much as 10 cm per year.

7 According to the passage, what geologic process formed the Himalaya Mountains?
A. divergence
B. continental rifting
C. strike-slip faulting
D. convergence

8 Which of the following statements is a fact according to the passage?
F. The nine tallest mountains on Earth are located in the Himalaya Mountains.
G. The Himalaya Mountains are the longest mountain chain on Earth.
H. The Himalaya Mountains are located within six countries.
I. The Himalaya Mountains had completely formed by 50 million years ago.

9 Which plate is being subducted along the fault that formed the Himalaya Mountains?
A. The Indian plate is being subducted.
B. The Eurasian plate is being subducted.
C. Both plates are being equally subducted.
D. Neither plate is being subducted.

Answers

Understanding Concepts

1. A
2. F
3. D
4. G
5. rifting
6. the asthenosphere

Reading Skills

7. D
8. H
9. D

Interpreting Graphics

10. B
11. Answers may vary. See Test Doctor for a detailed scoring rubric.
12. I
13. Answers may vary. See Test Doctor for a detailed scoring rubric.

Interpreting Graphics

Directions (10–13): **For *each* question below, record the correct answer on a separate sheet of paper.**

The map below shows the locations of the Earth's major tectonic plate boundaries. Use this map to answer questions 10 and 11.

Earth's Tectonic Plates

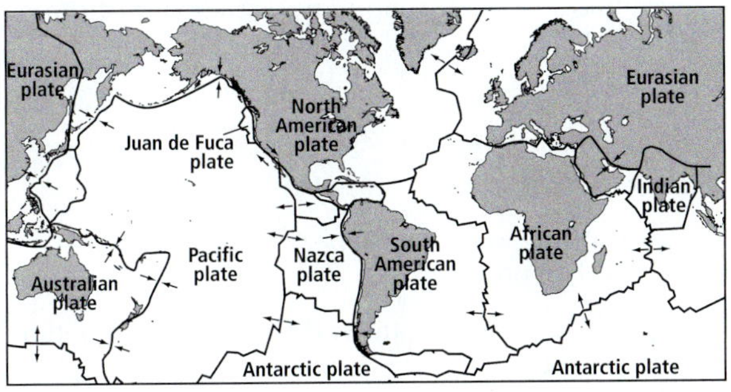

10 What type of boundary is found between the South American plate and the African plate?

A. convergent
B. divergent
C. transform
D. subduction

11 What type of boundary is found between the South American plate and the African plate? What surface features are most often found at boundaries of this type?

The graphic below shows a strike-slip fault along a transform boundary. Use the graphic to answer questions 12 and 13.

Plate Boundaries

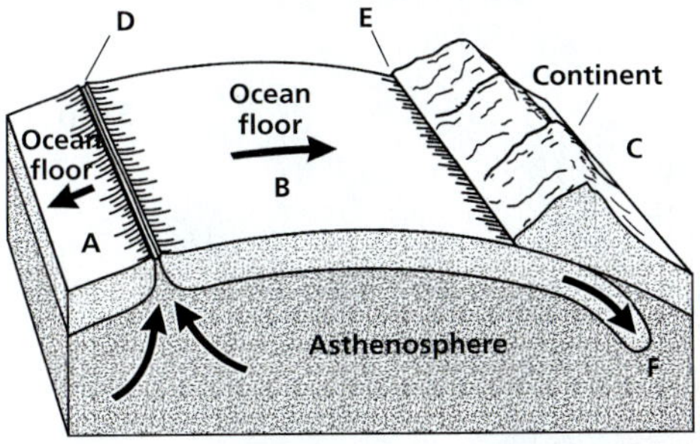

12 What type of crustal interaction is indicated by the letter E?

F. continental rifting
G. sea-floor spreading
H. divergence
I. subduction

13 Describe how a transform boundary differs from the boundaries shown by letters D and E in terms of plate movement and magmatic activity.

Test TIP

If you become short on time, quickly scan the unanswered questions to see which questions are easiest to answer.

Standardized Test Prep

TEST DOCTOR

Question 11 Full-credit answers should include the following points:

- the boundary between the South American plate and the African plate is a divergent boundary
- most divergent boundaries are located on the ocean floor and produce mid-ocean ridges and underwater mountain ranges

Question 13 Full-credit answers should include the following points:

- students should demonstrate a conceptual understanding that, unlike the plates at a convergent boundary, shown by letter E, or divergent boundary, shown by letter D, plates at a transform boundary move past one another, not into or away from one another
- transform boundaries produce a number of earthquakes, but they do not produce magma or cause mountain formation

Test Prep Correlations ***National Science Education Standards***

ES 1b: items 1, 2, 9, 10, 11, 12
ES 3c: items 4, 7, 8
HNS 3c: item 3
HNS 3d: item 3
UCP 1: item 6
UCP 2: items 5, 13

CHAPTER RESOURCES

State Resources

For specific resources for your state, visit go.hrw.com and type in the keyword **HSHSTR.**

Sea-Floor Spreading

Teacher's Notes

Time Required
two 45-minute class periods

Lab Ratings
EASY → HARD

Teacher Preparation (1)
Student Setup (1)
Concept Level (3)
Cleanup (1)

Skills Acquired
- Constructing Models
- Observing
- Identifying and Recognizing Patterns
- Interpreting Models

The Scientific Method
In this lab, students will
- Make Observations
- Analyze the Results
- Draw Conclusions

Materials
The materials listed on the page are enough for groups of two to four students. If you have your first-period students prepare the shoe boxes, use the same shoe boxes for the rest of your classes.

Tips and Tricks
Students may also wish to tape the trailing ends of the paper strips together to ensure that the strips are pulled equally. However, because rock at many divergent boundaries does not break down the center, students may try pulling one strip slightly faster than the other to make a more realistic model.

Objectives
- **Model** the formation of sea floor.
- **Identify** how magnetic patterns are caused by sea-floor spreading.

Materials
- marker
- paper, unlined
- ruler, metric
- scissors or utility knife
- shoebox

Safety

Sea-Floor Spreading

The places on Earth's surface where plates pull apart have many names. They are called divergent boundaries, mid-ocean ridges, and spreading centers. The term *spreading center* refers to the fact that sea-floor spreading happens at these locations. In this lab, you will model the formation of new sea floor at a divergent boundary. You will also model the formation of magnetic patterns on the sea floor.

PROCEDURE

1. Cut two identical strips of unlined paper, each 7 cm wide and 30 cm long.
2. Cut a slit 8 cm long in the center of the bottom of a shoebox.
3. Lay the strips of paper together on top of each other end to end so that the ends line up. Push one end of the strips through the slit in the shoe box, so that a few centimeters of both strips stick out of the slit.

Step 3

CHAPTER RESOURCES

Chapter Resource File
- Datasheet for Chapter Lab GENERAL
- Lab Notes and Answers

4. Place the shoe box flat on a table open side down, and make sure the ends of the paper strips are sticking up.
5. Separate the strips, and hold one strip in each hand. Pull the strips apart. Then, push the strips down against the shoe box.
6. Use a marker to mark across the paper strips where they exit the box. One swipe with the marker should mark both strips.
7. Pull the strips evenly until about 2 cm have been pulled through the slit.
8. Mark the strips with the marker again.
9. Repeat steps 7 and 8, but vary the length of paper that you pull from the slit. Continue this process until both strips are pulled out of the box.

Step 6

ANALYSIS AND CONCLUSION

1. **Evaluating Models** How does this activity model sea-floor spreading?
2. **Analyzing Models** What do the marker stripes in this model represent?
3. **Analyzing Methods** If each 2 cm marked on the paper is equal to 3 million years, how could you use your model to determine the age of certain points on the sea floor?
4. **Applying Conclusions** You are given only the paper strips with marks already drawn on them. How would you use the paper strips to reconstruct the way in which the sea-floor was formed?

Extension

1. **Making Models** Design a model that shows what happens at a convergent boundary and what happens at a transform boundary. Present these models to the class.

Making Models Lab

Answers to Analysis and Conclusion

1. The paper strips represent magma that rises to the surface between plates that are moving apart. The slit in the shoe box represents the mid-ocean ridge.
2. The marker stripes represent the magnetic orientation of the rock.
3. Sample answer: I could measure the length of the strip in centimeters from the "spreading center" to the point I want to date. Then, I could multiply the number of centimeters by 3 to get the age in millions of years.
4. Sample answer: I could use the symmetry of the marker stripes to match the magnetic patterns of the "sea floor." The stripes will tell me the order in which each part of the sea floor formed.

Answers to Extension

1. Answers may vary. Accept all reasonable answers. Students may use a variety of materials to build their models.

Randa Flinn
Northeast
High School
Ft. Lauderdale, FL

Locations of Earthquakes in South America, 2002–2003

Group Activity —— ADVANCED

3-D Model Have groups of students use the information in the map to make a three-dimensional model of the South American plate boundary. The model should have two tectonic plates, asthenosphere, and mesosphere. It should show the depth and thickness of each component to the same vertical and horizontal scale. Students may wish to research the thickness and geometry of the South American plate to make their model more realistic.
LS Kinesthetic/Visual

Answers to Map Skills Activity

1. about 11
2. shallow earthquakes
3. In the time period shown, no earthquakes with magnitude greater than 5.0 occurred on the east coast, but many occurred on the west coast.
4. along the western coast of South America
5. Most deep earthquakes happen toward the center of the continent. As you move east from the west coast, the earthquakes tend to get deeper.
6. The plate boundary on the west coast of South America is probably a convergent boundary (subduction zone). The subducting plate is moving eastward under the continent. As the subducting plate moves down under the continent, the earthquakes get deeper because the plate boundary is deeper.

CHAPTER RESOURCES

Technology

Transparencies
- 54 Locations of Earthquakes in South America, 2002–2003 (with worksheet)

MAPS in Action

Locations of Earthquakes in South America, 2002–2003

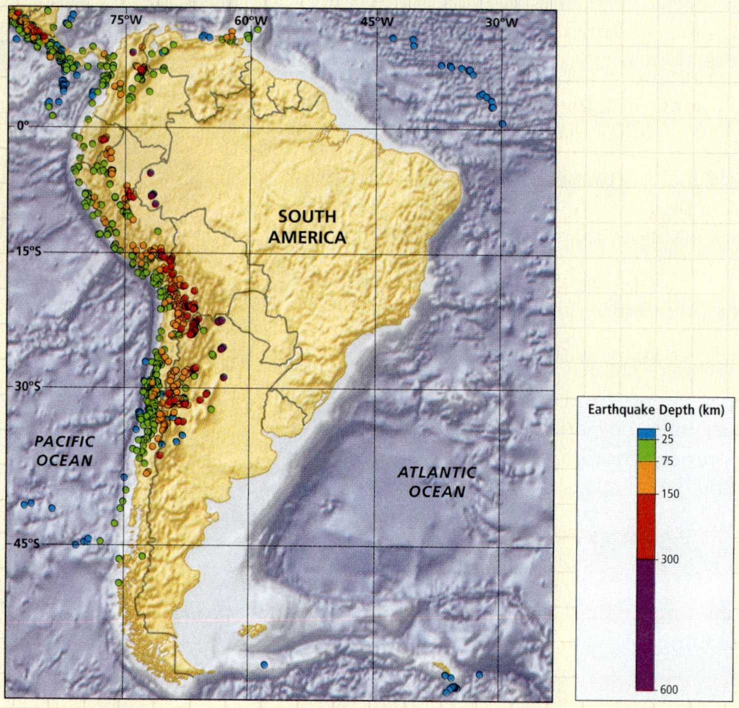

Map Skills Activity

This map shows the locations and depths of earthquakes that registered magnitudes greater than 5 and that happened in South America in 2002 and 2003. Use the map to answer the questions below.

1. **Using the Key** How many earthquakes happened at a depth greater than 300 km?
2. **Analyzing Data** Deep earthquakes are earthquakes that happen at a depth greater than 300 km. Which earthquakes happen more frequently: deep earthquakes or shallow earthquakes?
3. **Making Comparisons** How does the earthquake activity on the eastern edge of South America differ from the earthquake activity on the western edge?
4. **Inferring Relationships** The locations of earthquakes and plate boundaries are related. Where would you expect to find a major plate boundary?
5. **Identifying Trends** In what part of South America do most deep earthquakes happen? What relationships do you see between the locations of shallow and deep earthquakes in South America?
6. **Analyzing Relationships** Most deep earthquakes happen where subducting plates move deep into the mantle. What type of plate boundary is indicated by the earthquake activity in South America? Explain your answer.

IMPACT on Society

The Mid-Atlantic Ridge

Deep in the Atlantic Ocean lies a mountain range so vast that it dwarfs the Himalaya Mountains. This mountain range, called the *Mid-Atlantic Ridge,* is the mid-ocean ridge at the diverging boundary between the North American and Eurasian plates and also between the South American and African plates.

Sea-Floor Spreading on Land

Most of Earth's mid-ocean ridges are underwater, but part of the Mid-Atlantic Ridge rises above sea level just south of the Arctic Circle. The exposed section of the Mid-Atlantic Ridge forms the country of Iceland. Since Iceland was founded by Vikings more than 1,000 years ago, its inhabitants have contended with constant geologic activity associated with sea-floor spreading.

Lots and Lots of Lava

Separation of Earth's crust along the Mid-Atlantic Ridge affects Iceland's landscape in several ways. The movement of magma causes frequent earthquakes. Iceland is also one of the most volcanically active areas in the world. It contains about 200 volcanoes and averages one eruption every five years. Magma flowing up from the mantle creates numerous hot springs, geysers, and sulfurous gas vents. Scientists estimate that one-third of the total lava flow from Earth in the last 500 years has occurred on Iceland.

Despite Iceland's numerous volcanoes, much of its lava comes not from isolated eruptions but from cracks, or *fissures,* in the crust. In a recent rifting episode that lasted nearly 10 years, a series of fissures spit out enough molten basalt to cover 35 km^2 of land and individual fissures grew as much as 8 m in width. At present, sea-floor spreading adds an average of 2.5 cm of new material to Iceland each year. At this rate, Iceland will grow 25 km in width during the next million years.

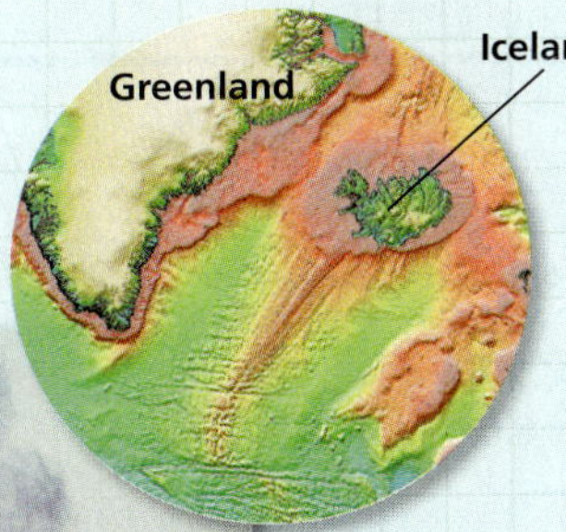

◀ The Helgafjell volcano erupted in a curtain of fire that rained black ash on the town of Reykjavik Iceland, in 1973.

Extension

1. **Making Inferences** If geologists want to locate the youngest rocks on Iceland, where should they look? Where should they look to find the oldest rocks? Explain your answers.

SCILINKS. NSTA Developed and maintained by the National Science Teachers Association

For a variety of links related to this chapter, go to www.scilinks.org

Topic: Mid-Atlantic Ridge
SciLinks code: HQ60960

Impact on Society

The Mid-Atlantic Ridge

Internet Activity — BASIC

The Heimaey Eruption Students may enjoy visiting sites that show photos and give historical accounts of the volcanic eruptions on Iceland. One particularly interesting event was the eruption that threatened the town of Vestmannaeyjar on the Icelandic island of Heimaey in 1973. Lava from the eruption threatened to destroy both the town and the harbor of the town. The townspeople used every resource available to them to save their town and their harbor. Have students research this eruption and the attempts to save the town. Then, have students submit an artistic piece that shows how the people dealt with the eruption. Possible projects include dioramas, paintings, drawings, dances, or songs. A worksheet designed to direct student research on this topic can be found in the **Chapter Resource File** booklet or by visiting go.hrw.com and entering the keyword **HQ6TECX.** LS **Auditory/Kinesthetic**

CHAPTER RESOURCES

Chapter Resource File

- **Internet Activity** The Heimaey Eruption BASIC

Answer to Extension

1. The youngest rocks on Iceland will be in the rift zone and near the volcanoes. These rocks form as the plates pull apart and as lava cools on the surface. The oldest rocks will probably be farthest from the ridge.

Chapter 11

Deformation of the Crust

Planning Guide

Compression Guide

To shorten instruction because of time limitations, omit Section 1.

OBJECTIVES | LABS, DEMONSTRATIONS, AND ACTIVITIES | TECHNOLOGY RESOURCES

PACING • 45 min pp. 270–278

Chapter Opener

Technology Resources
- OSP **Parent Letter** ■
- CD **Student Edition on CD-ROM**
- CD **Chapter Summaries Audio CD** ■
- VID **Brain Food Video Quiz**

Section 1 How Rock Deforms

- Summarize the principle of isostasy.
- Identify the three main types of stress.
- Compare folds and faults.

Labs, Demonstrations, and Activities
- TE **Demonstration** Deformation of Earth's Crust, p. 271 ◆ GENERAL
- SE **Quick Lab** Modeling Isostasy, p. 272 GENERAL
- CRF **Datasheet for Quick Lab*** GENERAL
- TE **Demonstration** Types of Stress, p. 273 ◆ BASIC
- SE **Quick Lab** Modeling Stress and Strain, p. 274 ◆ GENERAL
- CRF **Datasheet for Quick Lab*** GENERAL
- TE **Physics Connection** Elastic Modulus, p. 274 ADVANCED
- TE **Activity** Folded Art, p. 275 ADVANCED
- TE **Group Activity** It's Your Fault, p. 277 ADVANCED
- SE **Maps in Action** Shear Strain in New Zealand, p. 292 GENERAL
- TE **Activity** Geophysics of New Zealand, p. 292 GENERAL
- CRF **Inquiry Lab** Rock Deformation* GENERAL
- CRF **Skills Practice Lab** Hooke's Law* GENERAL

Technology Resources
- OSP **Lesson Plans** (also in print)
- TR **Bellringer***
- TR 55 Isostatic Adjustment*
- TR 56 Types of Stress*
- TR 57 Folds*
- TR 58 Faults*
- TR 61 Shear Strain in New Zealand*
- VID **HRW Earth Science Video** Deformation of the Crust

PACING • 90 min pp. 279–284

Section 2 How Mountains Form

- Identify the types of plate collisions that form mountains.
- Identify four types of mountains.
- Compare how folded and fault-block mountains form.

Labs, Demonstrations, and Activities
- TE **Discussion** Mountain Features, p. 279 ◆ GENERAL
- TE **Group Activity** Plates and Mountain Formation, p. 282 GENERAL
- SE **Making Models Lab** Continental Collisions, pp. 290–291 ◆ GENERAL
- CRF **Datasheet for Chapter Lab*** GENERAL
- TE **Activity** The Mediterranean of the Past, p. 293 ADVANCED

Technology Resources
- OSP **Lesson Plans** (also in print)
- TR **Bellringer***
- TR 59 How Mountains Form*
- TR 60 Types of Mountains in the United States*
- TE **Internet Activity** "Dead" Grabens, p. 282 ADVANCED
- CRF **Internet Activity** "Dead" Grabens* ADVANCED

PACING • 90 min

CHAPTER REVIEW, ASSESSMENT, AND STANDARDIZED TEST PREPARATION

- SE **Chapter Highlights**, p. 285
- SE **Chapter Review**, pp. 286–287
- SE **Standardized Test Prep**, pp. 288–289
- CRF **Concept Review*** ■ GENERAL
- CRF **Critical Thinking*** ADVANCED
- CRF **Math Skills*** GENERAL
- CRF **Graphing Skills*** GENERAL
- CRF **Chapter Test A*** ■ GENERAL
- CRF **Chapter Test B*** ADVANCED
- OSP **Lesson Plans** (also in print)
- OSP **Test Generator**
- OSP **Test Item Listing**

Online and Technology Resources

Visit go.hrw.com for access to Holt Online Learning, or enter the keyword **HQ6 Home** for a variety of free online resources.

One-Stop Planner® CD-ROM

This CD-ROM package includes
- Lab Materials QuickList Software
- Holt Calendar Planner
- Customizable Lesson Plans
- Printable Worksheets
- ExamView® Test Generator
- Interactive Teacher Edition
- Holt PuzzlePro®
- Holt PowerPoint® Resources

KEY

SE Student Edition	**OSP** One-Stop Planner	**VID** Classroom Video/DVD
TE Teacher Edition	**TR** Transparencies and Transparency Worksheets	* Also on One-Stop Planner
CRF Chapter Resource File	**CD** CD or CD-ROM	◆ Requires advance prep
LTP Long-Term Projects		■ Also available in Spanish

SKILLS DEVELOPMENT RESOURCES	REVIEW AND ASSESSMENT	CORRELATIONS
SE **Pre-Reading Activity,** p. 270 GENERAL **TE** **Using the Figure,** p. 270 GENERAL		National Science Education Standards
CRF **Directed Reading*** BASIC **TE** **Inclusion Strategies,** p. 272 **TE** **Skill Builder** Vocabulary, p. 274 ADVANCED **SE** **Math Practice,** p. 275 GENERAL **TE** **Using the Figure** Identifying Limbs and Hinges, p. 275 GENERAL **TE** **Reading Skill Builder** Reading Organizer, p. 276 BASIC	**SE** **Reading Checks,** pp. 273, 275, 277 GENERAL **SE** **Section Review,** p. 278 GENERAL **TE** **Homework,** p. 276 BASIC **TE** **Reteaching,** p. 277 BASIC **TE** **Quiz,** p. 277 GENERAL **TE** **Alternative Assessment,** p. 277 ADVANCED **CRF** **Section Quiz*** ■ GENERAL	HNS 2a, UCP 3, UCP 4
CRF **Directed Reading*** BASIC **SE** **Graphic Organizer** Cause-and-Effect Map, p. 280 GENERAL **TE** **Using the Figure** Lithospheric Collisions and Mountain Types, p. 280 BASIC **TE** **Skill Builder** Writing, p. 280 GENERAL **TE** **Reading Skill Builder** Discussion, p. 282 BASIC **TE** **Inclusion Strategies,** p. 283	**SE** **Reading Checks,** pp. 281, 283 GENERAL **SE** **Section Review,** p. 284 GENERAL **TE** **Homework,** p. 282 GENERAL **TE** **Reteaching,** p. 283 BASIC **TE** **Quiz,** p. 283 GENERAL **TE** **Alternative Assessment,** p. 283 ADVANCED **CRF** **Section Quiz*** ■ GENERAL	ES 1b, ES 3c, HNS 2a, UCP 3, UCP 4

Holt Lab Generator CD-ROM

Search for labs by topic, standard, difficulty level, and/or time frame. Edit those labs to fit your needs, or create your own labs. Use the Lab Materials QuickList to order just what you need.

Holt Earth Science Interactive Tutor CD-ROM

This CD-ROM consists of interactive activities that give students a fun way to extend their knowledge of Earth science concepts.

Chapter Summaries Audio CDs

These CDs include audio summaries of the key concepts presented in each chapter. (Audio summaries are also available in Spanish.)

SCI LINKS. NSTA

www.scilinks.org

Maintained by the **National Science Teachers Association.** See Chapter Enrichment pages that follow for a complete list of topics.

See Chapter Enrichment pages for Video Resources.

Refer to our ***Professional Reference for Teachers*** for additional teaching resources, including articles written by science education professionals about relevant and timely issues facing today's science teachers.

Chapter 11

Chapter Enrichment

This Chapter Enrichment provides relevant and interesting information to expand and enhance your classroom instruction of the chapter material.

Section 1 How Rock Deforms

The Geosyncline Model

Before the acceptance of the plate tectonic theory in the 1960s, the dominant model for the formation of mountains and the deformation of the rock that composed them was the model of geosynclines. According to this model, a mountain originated from a trough, called a geosyncline, in which sediments were gradually deposited. As this material accumulated, the trough sank, and as it did so, the sides of the geosyncline became more steeply folded. Eventually, the pressure from the surrounding rock would push inward from the sides of the geosyncline, causing the compressed sediments to rise and form mountains.

While the geosyncline model did provide a dynamic model for the vertical deformation of rock in mountains, other aspects of the theory were not so easily explained, such as the source of the horizontal stresses. These were accounted for by contraction of Earth's crust, a vaguely explained process that made the overall model more complicated. One of the great triumphs of plate tectonics is that its simple assumptions account for observations that even the complex features of the geosyncline model could not explain.

Gravity Measurements

The presence of different kinds of rocks beneath Earth's surface causes the density of the rock to vary, so that the mass of rock in a given volume varies. This variation in mass causes the gravitational force exerted by the rock to vary slightly.

Although the variations are extremely small, devices called *gravimeters* are sensitive enough to detect them. A gravimeter, in its simplest form, consists of a mass and a spring, so that changes in the local gravitational field cause a slight change in the position of the mass, which can be compared to a standard position. Thus, relative changes in gravity can be recorded. To assure accuracy, the elevation and latitude at which the measurement is made must be precisely known. Once all corrections are accounted for, a graph of the gravitational changes over the surveyed area can be made.

Since the 1970s, a type of gravimeter has been developed that uses the interaction of light reflected from a falling mass with light reflected from a fixed mass to measure the absolute gravitational acceleration at a particular location. These absolute-measurement gravimeters were initially large and difficult to set up, but recent design improvements have made them smaller and easier to transport, so that now they are suitable for field surveys.

Section 2 How Mountains Form

Mountains Under the Sea

If Earth's oceans were drained, so that the ocean floor were seen as the surface, Earth's mountain systems would look very different from those composed only of continental ranges. Many extensive ranges of mountains lie along the ocean floor. The Mid-Atlantic Ridge, for example, forms the longest single range of mountains on Earth. It stretches 16,000 km from the North Pole to the South Atlantic Ocean, more than twice the distance covered by the longest continental range, the Andes Mountains. Unlike continental mountains, which form from the collisions between plates, the mountains of the Mid-Atlantic Ridge are the result of diverging plates. As the plates pull apart, magma from the asthenosphere rises to fill the rift, then solidifies to form rock. Because the new rock is warmer and less dense than surrounding sea-floor rock,

◀ Study of the Appalachian Mountains in the nineteenth century provided a basis for the geosyncline model.

▲ Some volcanic mountains that form on the ocean floor break the ocean surface to form islands.

isostasy is reached when the new lithosphere is higher in the mantle than the surrounding lithosphere is. This relatively higher region of sea floor is the mid-ocean ridge, which is also a range of mountains. The tallest peaks of this range form the islands that appear in the Atlantic Ocean along the ridge.

The Highest Peak

The question "what is the highest peak on Earth" is tricky, as it depends on what one is measuring. If the measurement is from sea level to the peak, then Mt. Everest is the tallest, at 8850 m. If the measurement is from the mountain base to its peak, then the Hawaiian volcanoes Mauna Kea and Mauna Loa form two peaks of the tallest mountain, which reaches 9200 m from ocean floor to peak. The farthest point from Earth's center is the summit of Mt. Chimborazo in Ecuador. This mountain, believed to be the tallest mountain on Earth until the Himalayas were surveyed in the 1850s, lies close to the Equator. Because of Earth's rotation, the solid but ductile rock of Earth bulges outward at the Equator, where the rotational speed is greatest. Thus, sea level at the Equator is 21.4 km farther from the center of Earth's core than sea level is at the poles. This extra distance, added to Chimborazo's height of 6267 m, places its peak 27.7 km farther from Earth's center than sea level at the poles is.

Video Resources

Brain Food Video Quizzes

Brain Food Video Quizzes These videos contain game-show style quizzes that assess students' progress and motivate students to study the chapter material.

HRW Earth Science Video

This video introduces Earth science topics and includes a geology field trip. The video segment listed below complements this chapter.

Segment 4: Deformation of the Crust Plate movements create stress in the earth's crust, and the specific stresses that are produced most commonly at different plate boundaries are the topic of this segment. The folds (anticlines, monoclines, and synclines) and faults (normal, oblique, reverse, and strike-slip) that are related to a particular kind of stress are also discussed. (2 min)

NOVA Videos

To order **NOVA** videos related to this chapter, visit go.hrw.com and enter the keyword **HQ6DEFV.**

SciLinks

Developed and maintained by the National Science Teachers Association

SciLinks is maintained by the National Science Teachers Association to provide you and your students with interesting, up-to-date links that will enrich your classroom presentation of the chapter.

Visit www.scilinks.org and enter the SciLinks code for more information about the topic listed.

Topic: Folding and Faulting
SciLinks code: HQ60589

Topic: Types of Mountains
SciLinks code: HQ61568

Topic: Mountain Building
SciLinks code: HQ60999

Topic: Mountain Formation
SciLinks code: HQ61000

Chapter 11

Chapter Overview

The rock of Earth's crust deforms when the weight above the crust changes, as well as when tectonic plates collide. The results of the stress applied to rock are folds or faults. Deformation processes lead to the formation of mountains, and the type of mountain formed depends on the type of stress and the resulting strain.

Using the Figure — GENERAL

Point out to students that places where highways have been cut through rock often reveal the geological processes that have taken place there. Ask students what structures indicate the type of rock that is shown in the figure. (The layers in the rock indicate that the rock may be sedimentary.) LS **Visual**

PRE-READING ACTIVITY

FOLDNOTES

Have pairs of students use their FoldNotes to study key terms from the chapter. Instruct one student to use the FoldNote to provide the key term, and have the other student give the definition. Have the student who provides the key term correct the other student's definition.

Chapter 11 Deformation of the Crust

Sections

1 How Rock Deforms

2 How Mountains Form

What You'll Learn

- How Earth's crust responds to stress
- What forms deformed rock takes
- How forces in the crust cause mountains to form

Why It's Relevant

Many of the most dramatic features of Earth's surface are the result of deformation of the crust. Knowing how rock responds to stress provides a strong basis for understanding why and how Earth's surface changes.

PRE-READING ACTIVITY

FOLDNOTES

Key-Term Fold Before you read this chapter, create the FoldNote entitled "Key-Term Fold" described in the Skills Handbook section of the Appendix. Write a key term from the chapter on each tab of the key-term fold. Under each tab, write the definition of the key term.

▶ To build this portion of Highway 14 in California, construction crews had to cut through a hill. The exposed rock in the roadcut shows layers of sedimentary rock that have been folded by the stress caused by the nearby San Andreas fault.

Chapter Correlations — *National Science Education Standards*

ES 1b The outward transfer of earth's internal heat...propels the plates comprising earth's surface across the face of the globe. **(Sections 1 and 2)**

ES 3c Interactions among the solid earth...have resulted in the ongoing evolution of the earth system. ...Many processes, such as mountain building and plate movements, take place over hundreds of millions of years. **(Section 2)**

HNS 2a Science distinguishes itself from other...bodies of knowledge through...empirical standards, logical arguments, and skepticism, as scientists strive for the best possible explanations about the natural world. **(Sections 1 and 2)**

UCP 3 Although most things are in the process of becoming different—changing—some properties of objects and processes are characterized by constancy...Changes might occur, for example, in properties of materials, position of objects, motion, and form and function of systems. Interactions within and among systems result in change. Changes vary in rate, scale, and pattern, including trends and cycles. **(Sections 1 and 2)**

UCP 4 Evolution is a series of changes, some gradual and some sporadic, that accounts for the present form and function of...systems. The general idea of evolution is that the present arises from materials and forms of the past. **(Section 2)**

Section 1 How Rock Deforms

The Himalayas, the Rockies, and the Andes are some of Earth's most majestic mountain ranges. Mountain ranges are visible reminders that the shape of Earth's surface changes constantly. These changes result from **deformation**, or the bending, tilting, and breaking of Earth's crust.

OBJECTIVES

- **Summarize** the principle of isostasy.
- **Identify** the three main types of stress.
- **Compare** folds and faults.

KEY TERMS

deformation
isostasy
stress
strain
fold
fault

Isostasy

Deformation sometimes occurs because the weight of some part of Earth's crust changes. Earth's crust is part of the lithospheric plates that ride on top of the plastic part of the mantle called the *asthenosphere*. When parts of the lithosphere thicken and become heavier, they sink deeper into the asthenosphere. If parts of the lithosphere thin and become lighter, the lithosphere rises higher in the asthenosphere.

Vertical movement of the lithosphere depends on two opposing forces. One force is the force due to gravity, or weight, of the lithosphere pressing down on the asthenosphere. The other force is the buoyant force of the asthenosphere pressing up on the lithosphere. When these two forces are balanced, the lithosphere and asthenosphere are in a state called **isostasy.** However, when the weight of the lithosphere changes, the lithosphere sinks or rises until a balance of the forces is reached again. The movements of the lithosphere to reach isostasy are called *isostatic adjustments*. One type of isostatic adjustment is shown in **Figure 1.** As these isostatic adjustments occur, areas of the crust are bent up and down. This bending causes rock in that area to deform.

deformation the bending, tilting, and breaking of Earth's crust; the change in shape or volume of rock in response to stress

isostasy a condition of gravitational and buoyant equilibrium between Earth's lithosphere and asthenosphere

Figure 1 ▶ Isostatic Adjustments as a Result of Erosion

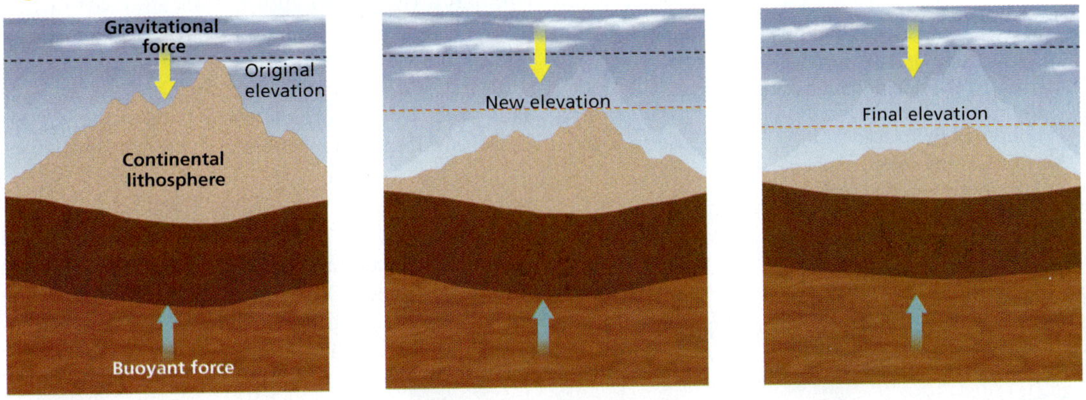

When the gravitational force equals the buoyant force, the lithosphere and asthenosphere are in isostasy.

As erosion wears away the crust, the lithosphere becomes lighter and is pushed up by the asthenosphere.

As erosion continues, the isostatic adjustment also continues.

Section 1

Focus

Overview

This section describes how changes in the weight of Earth's crust causes deformation. The section also explains how stress causes the bending and breaking of rock.

Bellringer

Have students imagine that they placed a bowling ball on a soft pillow. What would happen to the ball and to the pillow? (The ball would sink into the pillow, changing the pillow's shape.) **LS Verbal**

Motivate

Demonstration — GENERAL

Deformation of Earth's Crust

Fill a large, clear container halfway with a mud or plaster thick enough so that a brick placed in it will sink very slowly, or sink only partway into the mixture. Level the top of the mixture, and then place a brick on top of the mixture. Ask students to describe what happens. (The mixture beneath the brick is pressed down, while the mixture around the brick rises up.) Point out that isostatic adjustment involves a similar process. **LS Visual**

CHAPTER RESOURCES

Chapter Resource File

- Directed Reading BASIC

Technology

Transparencies

- Bellringer
- 55 Isostatic Adjustment (with worksheet)

Student Edition on CD-ROM

One-Stop Planner CD-ROM

- Lesson Plan

Teach

INCLUSION Strategies

- **Learning Disabled**
- **Developmentally Delayed**

Text with large words is often hard for students with reading delays to understand. The description of isostasy is an example. Help students understand these paragraphs by using this procedure: Read the first sentence aloud. Reword the sentence, replacing large words with definitions. Type or write the reworded sentence on a piece of paper. Repeat these steps with the remaining sentences. Make a copy of the reworded text for each student. LS **Verbal**

QuickLAB

Skills Acquired

- Observing
- Interpreting

Teacher's Notes Be sure that students do not break the glass beaker. Spills should be cleaned up quickly and completely. Students should wait for the water levels to become steady before making marks with the grease pencils.

Answers

1. The block sinks deeper into the water.
2. The model depicts the type of isostatic adjustment that occurs when lithosphere gets heavier, such as when mountains form or when glaciers or ice sheets form.

CHAPTER RESOURCES

Chapter Resource File

- Datasheet for Quick Lab GENERAL

Figure 2 ▶ Mt. Katahdin in Baxter State Park, Maine, has been worn down by weathering and erosion. As the mountain shrinks, the crust underneath it is uplifted.

QuickLAB 5 min

Modeling Isostasy

Procedure

1. Fill a **1 L beaker** with **500 mL of water.**
2. Place a **wooden block** in the water. Use a **grease pencil** to mark on the side of the beaker the levels of the top and the bottom of the block.
3. Place a **small mass,** of about 1 g, on the wooden block. Use a **second grease pencil** to mark the levels of the top and the bottom of the block.

Analysis

1. What happens to the block of wood when the weight is added?
2. What type of isostatic adjustment does this activity model?

Mountains and Isostasy

In mountainous regions, isostatic adjustments constantly occur. Over millions of years, the rock that forms mountains is worn away by the erosive actions of wind, water, and ice. This erosion can significantly reduce the height and weight of a mountain range, such as the one shown in **Figure 2.** As a mountain becomes smaller and lighter, the area may rise by isostatic adjustment in a process called *uplift*.

Deposition and Isostasy

Another type of isostatic adjustment occurs in areas where rivers carrying large amounts of mud, sand, and gravel flow into larger bodies of water. When a river flows into the ocean, most of the material that the river carries is deposited on the nearby ocean floor. The added weight of the deposited material causes the ocean floor to sink by isostatic adjustment in a process known as *subsidence*. This process is occurring in the Gulf of Mexico at the mouth of the Mississippi River, where a thick accumulation of deposited materials has formed.

Glaciers and Isostasy

Isostatic adjustments also occur as a result of the growth and retreat of glaciers and ice sheets. When a large amount of water is held in glaciers and ice sheets, the weight of the ice causes the lithosphere beneath the ice to sink. Simultaneously, the ocean floor rises because the weight of the overlying ocean water is less. When glaciers and ice sheets melt, the land that was covered with ice slowly rises as the weight of the crust decreases. As the water returns to the ocean, the ocean floor sinks.

PHYSICS CONNECTION

Archimedes' Principle An object floats on a fluid's surface when the buoyant force exerted upward on the object equals the downward force of the object's weight. To determine whether the buoyant force is great enough to balance an object's weight, one must apply Archimedes' principle, which states that the weight of a displaced fluid equals the buoyant force on the object displacing the fluid. If the object is less dense than the fluid, the displaced fluid will weigh more than the object. The buoyant force will be greater than the object's weight and will push the object upward until the two forces are balanced. Thus, when only a part of the object's volume is submerged in the fluid, the object will float. When the liquid is less dense than the object, the weight of the displaced fluid is less than the weight of the object, even when the maximum volume of fluid is displaced, and the object is completely submerged.

Stress

As Earth's lithosphere moves, the rock in the crust is squeezed, stretched, and twisted. These actions exert force on the rock. The amount of force that is exerted on each unit of area is called **stress.** For example, during isostatic adjustments, the lithosphere sinks and rises atop the asthenosphere. As the lithosphere sinks, the rock in the crust is squeezed and the direction of stress changes. As the lithosphere rises, the rock in the crust is stretched and the direction of stress changes again. Similarly, stress occurs in Earth's crust when tectonic plates collide, separate, or scrape past each other. **Figure 3** shows the three main types of stress.

stress the amount of force per unit area that acts on a rock

Compression

The type of stress that squeezes and shortens a body is called *compression.* Compression commonly reduces the amount of space that rock occupies. In addition to reducing the volume of rock, compression pushes rocks higher up or deeper down into the crust. Much of the stress that occurs at or near convergent boundaries, where tectonic plates collide, is compression.

Tension

Another type of stress is tension. *Tension* is stress that stretches and pulls a body apart. When rocks are pulled apart by tension, they tend to become thinner. Much of the stress that occurs at or near divergent boundaries, where tectonic plates pull apart, is tension.

Shear Stress

The third type of stress is shear stress. *Shear stress* distorts a body by pushing parts of the body in opposite directions. Sheared rocks bend, twist, or break apart as they slide past each other. Shear stress is common at transform boundaries, where tectonic plates slide horizontally past each other. However, each type of stress occurs at or near all types of plate boundaries and in various other regions of the crust, too.

Reading Check **Which two kinds of stress pull rock apart? (See the Appendix for answers to Reading Checks.)**

Figure 3 ▶ Types of Stress

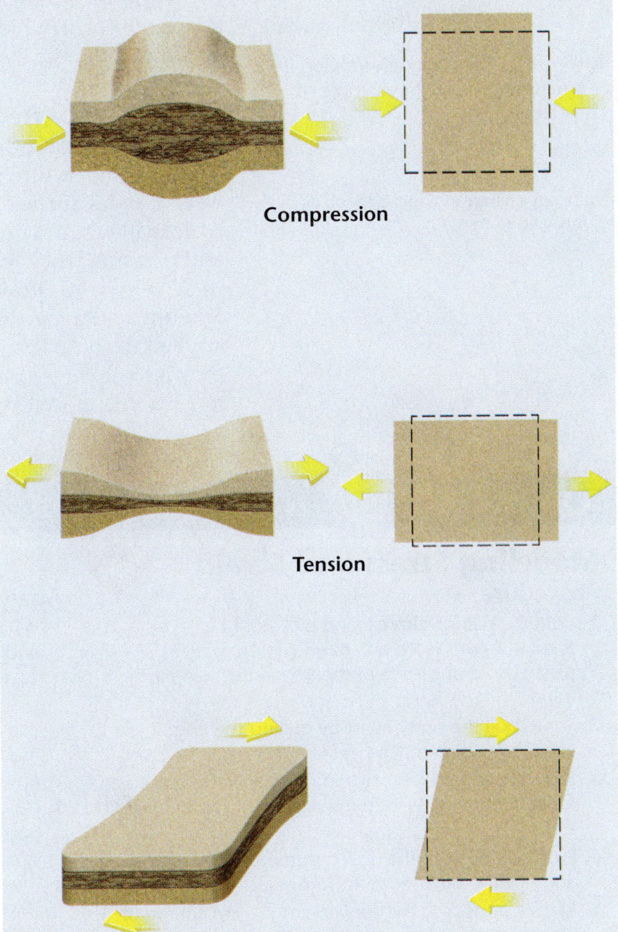

Demonstration — Basic

Types of Stress To help students visualize the different kinds of stress, take a block of clay about 5 cm long, 5 cm wide, and 3 cm thick. Demonstrate compression by pushing on the short sides of the block. Have students note that you pushed your hands toward each other and that the clay thickened in the middle. Reshape the clay block or use another, identical block to demonstrate tension by slowly pulling the sides of the block in opposite directions. Again, have students note that you moved your hands away from each other and that the clay thinned in the center. Return the clay to its original shape, or use a third identical block to demonstrate shear stress by placing your hands flat on opposite sides of the clay block, then moving your hands in opposite directions. Have students note that you moved your hands in opposite directions past each other and that the clay deformed by becoming offset at the edges. **LS Visual**

Answer to Reading Check

Tension and shear stress can both pull rock apart.

CHAPTER RESOURCES

Chapter Resource File

- **Inquiry Lab** Rock Deformation GENERAL

Technology

Transparencies
- 56 Types of Stress (with worksheet)

MISCONCEPTION ALERT

Ductility Students may be confused as to how a solid and seemingly rigid substance like rock can undergo deformation without breaking. Explain that all solid materials possess a property called *ductility,* which is a measure of how much the shape or volume of a solid can be changed without causing the substance to break. Ductility differs from elasticity, in that an elastic material will return to its original shape once the stress that deforms it has been removed. A ductile material does not break, but it will not return to its original shape when the stress is removed. An example of this difference is the difference between a rubber band and a piece of copper wire. When the rubber band is stretched and then released, it regains its original shape. When wire is bent or stretched and then released, the wire will stay bent or stretched. The ductility of rock varies with its composition and structure.

Figure 4 ▶ This rock deformation in Kingman, Arizona, is an example of brittle strain.

Strain

When stress is applied to rock, rock may deform. Any change in the shape or volume of rock that results from stress is called **strain.** When stress is applied slowly, the deformed rock may regain its original shape when the stress is removed. However, the amount of stress that rock can withstand without permanently changing shape is limited. This limit varies with the type of rock and the conditions under which the stress is applied. If a stress exceeds the rock's limit, the rock's shape permanently changes.

strain any change in a rock's shape or volume caused by stress

Types of Permanent Strain

Materials that respond to stress by breaking or fracturing are *brittle*. Brittle strain appears as cracks or fractures, as **Figure 4** shows. *Ductile* materials respond to stress by bending or deforming without breaking. Ductile strain is a change in the volume or shape of rock in which the rock does not crack or fracture. Brittle strain and ductile strain are types of permanent strain.

Factors That Affect Strain

The composition of rock determines whether rock is ductile or brittle. Temperature and pressure also affect how rock deforms. Near Earth's surface, where temperature and pressure are low, rock is likely to deform in a brittle way. At higher temperature and pressure, rock is more likely to deform in a ductile way.

The type of strain that stress causes is determined by the amount and type of stress and by the rate at which stress is applied to rock. The greater the stress on rock is, the more likely rock is to undergo brittle strain. The more quickly stress is applied to rock, the more likely rock is to respond in a brittle way.

Quick LAB — 15 min

Modeling Stress and Strain

Procedure

1. Put on a pair of **gloves,** and pick up a 5 cm × 5 cm square of **frozen plastic play putty.** Hold one edge of the frozen putty in each hand.
2. Try to pull the putty apart by pulling the edges away from each other.
3. Push the edges of the frozen putty toward each other. (You may have to reshape the putty between steps.)
4. Push one edge of the frozen putty away from you, and pull the other edge toward you.
5. Repeat steps 2–4, but use a 5 cm × 5 cm square of **warm plastic play putty.**

Analysis

1. What types of stress did you model in steps 2, 3, and 4?
2. Make a table that lists the characteristics of the two substances that you modeled, the stresses that you modeled, and the resulting strain on each model.
3. How does the frozen putty respond to the stress? How does the warm putty's response to the stress differ from the frozen putty's response?

Teach, continued

SKILL BUILDER — Advanced

Vocabulary The terms *stress* and *strain* have a specific meaning in physics, geology, and materials science. *Stress,* which is defined as the ratio of an applied force to the area over which it is applied, might seem similar to pressure. However, the applied force in stress can push, pull, or move along the side of the surface, whereas in pressure the force only pushes. *Strain* is defined as the ratio of the amount of change in a dimension to the total value of a dimension to which the change occurred. Thus, in the case of compression or tension, the strain would be measured as the amount the rock has compressed or stretched divided by the original length of the rock before the stress was applied. In the case of shear stress, the strain is calculated by using the degree to which structures that were previously perpendicular have been shifted from 90°. LS **Verbal**

Quick LAB

Skills Acquired

- Experimenting
- Observing
- Identifying/Recognizing Patterns
- Interpreting

Teacher's Notes Be sure that the frozen putty is not too cold, so that students are not accidentally "burned" by it. Students should make the observations of the frozen putty quickly, before it becomes too warm and pliable.

Answers

1. tension, compression, and shear stress, respectively
2. Answers may vary. Accept all reasonable answers.
3. Answers may vary depending on the stresses applied. In general, the frozen putty should respond in a brittle way and the warm putty should respond in a ductile way.

CHAPTER RESOURCES

Chapter Resource File

- **Datasheet for Quick Lab** General
- **Skills Practice Lab** Hooke's Law General

PHYSICS CONNECTION — Advanced

Elastic Modulus Different materials respond differently to a given amount of stress—some recover completely from the stress, others show permanent deformation, and still others break under the stress. A measure of a material's elasticity is given by the ratio of the stress on a material to the amount of strain it exhibits. In general, this ratio is called the *elastic modulus* of the substance. Have students research the kinds of elastic modulus —stretch modulus, shear modulus, and bulk modulus—and relate them to the kinds of stress that rock undergoes. LS **Verbal**

Folds

When rock responds to stress by deforming in a ductile way, folds commonly form. A **fold** is a bend in rock layers that results from stress. A fold is most easily observed where flat layers of rock were compressed or squeezed inward. As stress was applied, the rock layers bent and folded. Cracks sometimes appear in or near a fold, but most commonly the rock layers remain intact. Although a fold commonly results from compression, it can also form as a result of shear stress.

fold a form of ductile strain in which rock layers bend, usually as a result of compression

Anatomy of a Fold

Folds have features by which they can be identified. Scientists use these features to describe folds. The main features of a fold are shown by the illustration in **Figure 5.** The sloping sides of a fold are called *limbs*. The limbs meet at the bend in the rock layers, which is called the *hinge*. Some folds also contain an additional feature. If a fold's structure is such that a plane could slice the fold into two symmetrical halves, the fold is symmetrical. The plane is called the fold's *axial plane*. However, the two halves of a fold are rarely symmetrical.

Many folds bend vertically, but folds can have many other shapes, as shown by the photograph in **Figure 5.** Folds can be asymmetrical. Sometimes, one limb of a fold dips more steeply than the other limb does. If a fold is *overturned*, the fold appears to be lying on its side. Folds can have open shapes or be as tight as a hairpin. A fold's hinge can be a smooth bend or may come to a sharp point. Each fold is unique because the combination of stresses and conditions that caused the fold was unique.

Reading Check Name two features of a fold. (See the Appendix for answers to Reading Checks.)

MATHPRACTICE

Units of Stress
Two units are commonly used to describe stress or pressure. One unit is the pascal (Pa). A pascal is a measure of force (in newtons) divided by area (in square meters). The other unit of stress is the pound per square inch (psi). If the pressure in a region of Earth's crust is measured as 25 MPa (megapascals) and as 3,626 psi, how many pounds per square inch does 1 MPa equal? (Note: 1 MPa = 1,000,000 Pa)

Figure 5 ▶ Although not every fold is symmetrical, every fold has a hinge and limbs. ***Can you identify the limbs and hinge of each fold in the photo?***

MATHPRACTICE

Answer
$x = (3{,}626 \text{ psi} \div 25 \text{ MPa}) \times 1 \text{ MPa} = 145 \text{ psi}$

Answer to Reading Check
limbs and hinges

Using the Figure — GENERAL

Identifying Limbs and Hinges
Explain to students that limbs of two folds meet at a point called an *inflection point*. At this point, the limbs that slant away from one hinge change direction slightly as they slant toward a different hinge. The two limbs on either side of the inflection point are counted as separate limbs because they are part of separate folds. Ask students how many hinges and limbs they see in the diagram. (one hinge and two limbs) Ask students how many hinges and limbs extend from the folds in the upper left corner of the photograph. (two hinges and four limbs) Answer to caption question: The rock in the photo shows three clear hinges. The first is pointed toward the top of the photo at the lower left corner. The second is slightly to the right of the first hinge and points downward and slightly to the right. The third is the large hinge that points toward the upper left of the photo and is located at the center of the photo. The limbs are on either side of these hinges. **LS Visual**

Activity — ADVANCED

Folded Art Have interested students create drawings, paintings, or sculptures that represent symmetric and asymmetric folds. Students could also research and include art of isoclinal, overturned, recumbant, and chevron folds. **LS Visual/Kinesthetic**

CHAPTER RESOURCES

Technology

Transparencies
- 57 Folds (with worksheet)

Teach, continued

READING SKILL BUILDER — BASIC

Reading Organizer Before students read this page, have them make a table that has three columns and four rows. Have students label the columns with "Direction of curve" and "Location of oldest rock." Have students label the rows with "Anticline," "Syncline," and "Monocline." Then, have students read the page and insert into the table the correct information for each type of fold. Students should keep these tables as study guides for later assessments. English Language Learners
LS Verbal

Homework — BASIC

Locating Types of Folds Have students research areas of the United States in which each of the three types of folds are common. Students should note the geographical location of the folds, the general size of the folds, and whether they appear in ridges or valleys. Students should then compile their findings into a short, written and illustrated report. Encourage students to use creativity in the style of their writing. **LS Verbal**

Figure 6 ▶ The three major types of folds are anticlines (top), synclines (middle), and monoclines (bottom).

Types of Folds

To categorize a fold, scientists study the relative ages of the rocks in the fold. The rock layers of the fold are identified by age from youngest to oldest. An *anticline* is a fold in which the oldest layer is in the center of the fold. Anticlines are commonly arch shaped. A *syncline* is a fold in which the youngest layer is in the center of the fold. Synclines are commonly bowl shaped. A *monocline* is a fold in which both limbs are horizontal or almost horizontal. Monoclines form when one part of Earth's crust moves up or down relative to another part. The three major types of folds are shown in **Figure 6.**

Sizes of Folds

Folds, which appear as wavelike structures in rock layers, vary greatly in size. Some folds are small enough to be contained in a hand-held rock specimen. Other folds cover thousands of square kilometers and can be seen only from the air.

Sometimes, a large anticline forms a ridge. A *ridge* is a large, narrow strip of elevated land that can occur near mountains. Nearby, a large syncline may form a valley. The ridges and valleys of the Appalachian Mountains are examples of landforms that were formed by anticlines and synclines.

Connection to ENGINEERING

Oil Traps

Oil and natural gas form where the remains of organisms, especially marine plants, are buried in an environment that prevents the remains from rapidly decomposing. Over millions of years, chemical reactions slowly change the organic remains into oil and natural gas.

When prospecting for oil and natural gas, oil companies look for porous and permeable rock layers. Porous rock has spaces between rock particles. Permeable rock is rock in which the pore spaces are connected, so fluids can flow through the rock. When a rock layer is both porous and permeable and contains oil or gas, the layer is called a *reservoir*.

Because oil and natural gas are fluids that have low densities, they move upward through rock toward Earth's surface. Oil and gas move through rock layers until they meet an impermeable rock layer or structure, which then traps the oil and gas below it.

In addition to looking for porous and permeable rock layers, petroleum engineers look for rock layers that have been folded. Many folds are anticlines in which layers of impermeable rock overlay layers of permeable rock. Because the limbs of the fold slope upward, the oil and natural gas rise through the permeable layer to the crest of the anticline and are trapped there by the impermeable layer. Engineers can then drill through the impermeable layer to reach the oil or gas reservoir.

This oil pump brings oil and natural gas up to the surface.

Career

Geophysicist Geophysicists apply the principles of physics to study different properties of Earth. Depending on the area of specialization, a geophysicist may study electrical, magnetic, or gravitational fields; wave motion; the production and transfer of thermal energy; or radioactivity. The problems studied by geophysicists are just as wide-ranging, from investigating how and why earthquakes happen to studying the causes of climate change to analyzing the ways in which violent atmospheric storms develop.

Geophysicists are divided among theoreticians, who develop models to explain phenomena; experimentalists, who test those models; and applied geophysicists, who use the accumulated knowledge to solve practical problems. For instance, in oil and natural gas exploration, geophysicists measure changes in the gravitational forces exerted by rock under Earth's surface. They then analyze these data to determine whether oil or natural gas reservoirs exist in the area, thus increasing the chances of drilling a productive well. Geophysicists have college degrees, for which they study physics, geology, mathematics, and computer science.

Figure 7 ▶ Normal and Reverse Faults

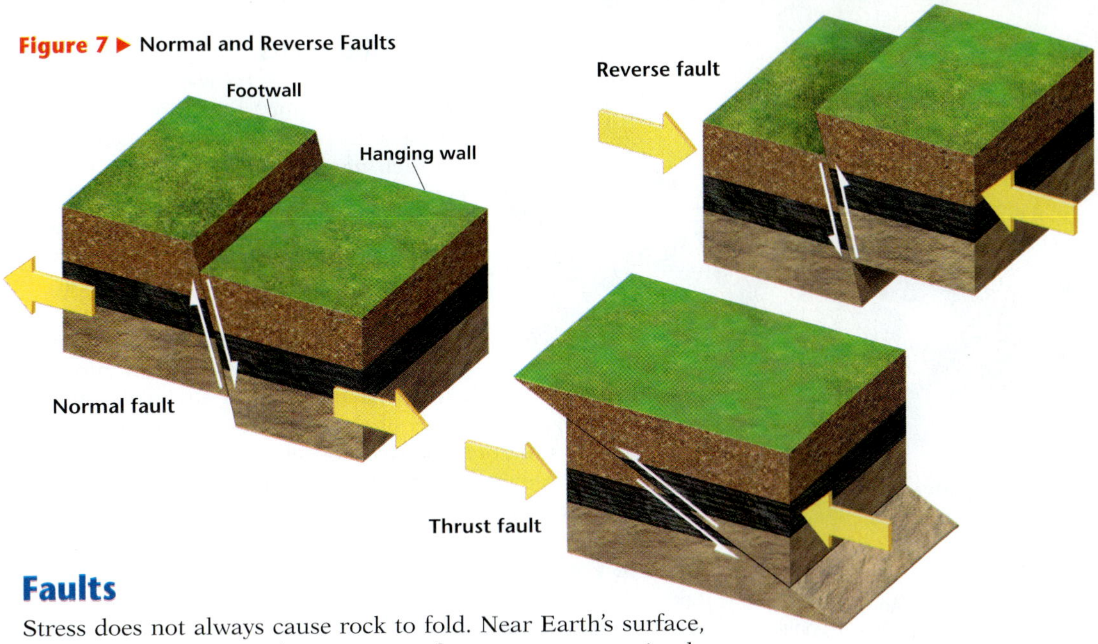

Faults

Stress does not always cause rock to fold. Near Earth's surface, where temperatures and pressure are low, stresses may simply cause rock to break. Breaks in rock are divided into two categories. A break along which there is no movement of the surrounding rock is called a *fracture*. A break along which the surrounding rock moves is called a **fault.** The surface or plane along which the motion occurs is called the *fault plane*. In a nonvertical fault, the *hanging wall* is the rock above the fault plane. The *footwall* is the rock below the fault plane.

fault a break in a body of rock along which one block slides relative to another; a form of brittle strain

Normal Faults

As shown in **Figure 7,** a *normal fault* is a fault in which the hanging wall moves downward relative to the footwall. Normal faults commonly form at divergent boundaries, where the crust is being pulled apart by tension. Normal faults may occur as a series of parallel fault lines, forming steep, steplike landforms. The Great Rift Valley of East Africa formed by large-scale normal faulting.

Reverse Faults

When compression causes the hanging wall to move upward relative to the footwall, also shown in **Figure 7,** a *reverse fault* forms. A *thrust fault* is a special type of reverse fault in which the fault plane is at a low angle or is nearly horizontal. Because of the low angle of the fault plane, the rock of the hanging wall is pushed up and over the rock of the footwall. Reverse faults and thrust faults are common in steep mountain ranges, such as the Rockies and the Alps.

Reading Check How does a thrust fault differ from a reverse fault? (See the Appendix for answers to Reading Checks.)

SCILINKS. NSTA Developed and maintained by the National Science Teachers Association

For a variety of links related to this subject, go to www.scilinks.org

Topic: Folding and Faulting
SciLinks code: HQ60589

Answer to Reading Check

A thrust fault is a type of reverse fault in which the fault plane is at a low angle relative to the surface.

Close

Reteaching — Basic

Deformation Flashcards Make a set of flashcards, and draw on each card a picture of either a type of stress (compression, tension, or shear stress) or a type of strain (different types of folds and faults). Show the cards rapidly to students, and have them identify whether the card illustrates stress or strain, what type of fold or fault is shown, and how that structure forms. **LS Visual**

Quiz — General

1. What happens to lithosphere when large amounts of mud, sand and gravel are deposited onto it? (It sinks into the asthenosphere.)
2. What type of fault commonly forms at transform boundaries? (strike-slip)
3. What type of stress commonly results in folds? (compression)

Alternative Assessment — Advanced

Modeling Isostasy Assign a particular isostatic adjustment to students, and have them create a series of models to demonstrate the changes that occur during the adjustment. The models can be physical models or computer models. The models should show at least four of the steps that take place during the isostatic adjustment. **LS Kinesthetic/Logical**

CHAPTER RESOURCES

Technology

Transparencies
- 58 Faults (with worksheet)

Group Activity — Advanced

It's Your Fault Separate the class into four or five groups of three or four students, and have each group research the major faults on one of the continents. Each student within the group should examine a region within the group's continent; locate any faults within the area; identify each fault as normal, reverse (or thrust), or strike-slip; and describe how the fault has contributed to the topography of the region. Have each group present their findings as a group to the rest of the class.
LS Verbal Co-op Learning

Figure 8 ▶ The San Andreas fault system stretches more than 1,200 km across California and is the result of two tectonic plates moving in different directions.

Strike-Slip Faults

In a *strike-slip fault*, the rock on either side of the fault plane slides horizontally in response to shear stress. Strike-slip faults got their name because they slide, or *slip*, parallel to the direction of the length, or *strike*, of the fault. Some strike-slip fault planes are vertical, but many are sloped.

Strike-slip faults commonly occur at transform boundaries, where tectonic plates grind past each other as they move in opposite directions. These motions cause shear stress on the rocks at the edges of the plates. Strike-slip faults also occur at fracture zones between offset segments of mid-ocean ridges. Commonly, strike-slip faults occur as groups of smaller faults in areas where large-scale deformation is happening.

Sizes of Faults

Like folds, faults vary greatly in size. Some faults are so small that they affect only a few layers of rock in a small region. Other faults are thousands of kilometers long and may extend several kilometers below Earth's surface. Generally, large faults that cover thousands of kilometers are composed of systems of many smaller, related faults, rather than of a single fault. The San Andreas fault in California, shown in **Figure 8,** is an example of a large fault system.

Section 1 Review

1. **Summarize** how isostatic adjustments affect isostasy.
2. **Identify and describe** three types of stress.
3. **Compare** stress and strain.
4. **Describe** one type of strain that results when rock responds to stress by permanently deforming without breaking.
5. **Identify** features that all types of folds share and features that only some types of folds have.
6. **Describe** four types of faults.
7. **Compare** folding and faulting as responses to stress.

CRITICAL THINKING

8. **Applying Ideas** Why is faulting most likely to occur near Earth's surface and not deep within Earth?
9. **Making Comparisons** How would the isostatic adjustment that results from the melting of glaciers differ from the isostatic adjustment that may occur when a large river empties into the ocean?
10. **Analyzing Relationships** You are examining a rock outcrop that shows a fold in which both limbs are horizontal but occur at different elevations. What type of fold does this outcrop show, and what can you say about the type of stress that the rock underwent?
11. **Predicting Consequences** You are watching a lab experiment in which a rock sample is being gently heated and slowly bent. Would you expect the rock to fold or to fracture? Explain your reasoning.

CONCEPT MAPPING

12. Use the following terms to create a concept map: *stress, compression, strain, tension, shear stress, folds,* and *faults.*

Close, *continued*

Answers to Section Review

1. Isostatic adjustments cause the lithosphere to rise or sink until the weight of the lithosphere equals the buoyant force from the asthenosphere and the two layers reach isostasy.
2. *Compression* is stress that squeezes and shortens a body. *Tension* is stress that stretches and thins a body. *Shear stress* is stress that acts by pushing different parts of a body in opposite directions.
3. *Stress* is the force applied per unit area on a rock or other material, whereas *strain* is the change in the shape or volume of rock that is affected by stress.
4. Folds are a type of ductile strain.
5. All folds have hinges and limbs. Folds that are symmetrical have axial planes.
6. A *normal fault* occurs when the hanging wall moves downward relative to the footwall. A *reverse fault* occurs when the hanging wall moves upward relative to the footwall. A *thrust fault* is a reverse fault that has a very low angle. A *strike-slip fault* occurs when the fault blocks slide horizontally past each other.
7. Folding is an example of ductile strain. Faulting is an example of brittle strain.
8. Rock temperatures and pressures are lower near Earth's surface, so the rock tends to be more brittle.
9. When glaciers melt, the continental crust becomes lighter and the lithosphere rises upward. Where material is deposited from a river into the ocean, the weight of the crust increases, and the lithosphere sinks.
10. The fold is a monocline, which formed when part of the rock had undergone vertical stress but the rest of the rock had not.
11. Because the rock is being heated, it is probably becoming more ductile. As a result, the rock is more likely to fold than to fracture.
12. *Stress,* such as *compression, tension,* and *shear stress,* may result in *strain,* such as *folds* and *faults.*

CHAPTER RESOURCES

Chapter Resource File

- **Section Quiz** GENERAL

Workbooks

- **Study Guide** (also in Spanish)

Section 2 How Mountains Form

A mountain is the most extreme type of deformation. Mount Everest, whose elevation is more than 8 km above sea level, is Earth's highest mountain. Forces inside Earth cause Mount Everest to grow taller every year. Mount St. Helens, a volcanic mountain, captured the world's attention in 1980 when its explosive eruption devastated the surrounding area.

OBJECTIVES

- **Identify** the types of plate collisions that form mountains.
- **Identify** four types of mountains.
- **Compare** how folded and fault-block mountains form.

KEY TERMS

mountain range
folded mountain
fault-block mountain
dome mountain

Mountain Ranges and Systems

A group of adjacent mountains that are related to each other in shape and structure is called a **mountain range.** Mount Everest is part of the Great Himalaya Range, and Mount St. Helens is part of the Cascade Range. A group of mountain ranges that are adjacent is called a *mountain system.* In the eastern United States, for example, the Great Smoky, Blue Ridge, Cumberland, and Green mountain ranges make up the Appalachian mountain system.

The largest mountain systems are part of two larger systems called *mountain belts.* Earth's two major mountain belts, the circum-Pacific belt and the Eurasian-Melanesian belt, are shown in **Figure 1.** The circum-Pacific belt forms a ring around the Pacific Ocean. The Eurasian-Melanesian belt runs from the Pacific islands through Asia and southern Europe and into northwestern Africa.

mountain range a series of mountains that are closely related in orientation, age, and mode of formation

Figure 1 ▶ Most mountain ranges lie along either the Eurasian-Melanesian mountain belt or the circum-Pacific mountain belt.

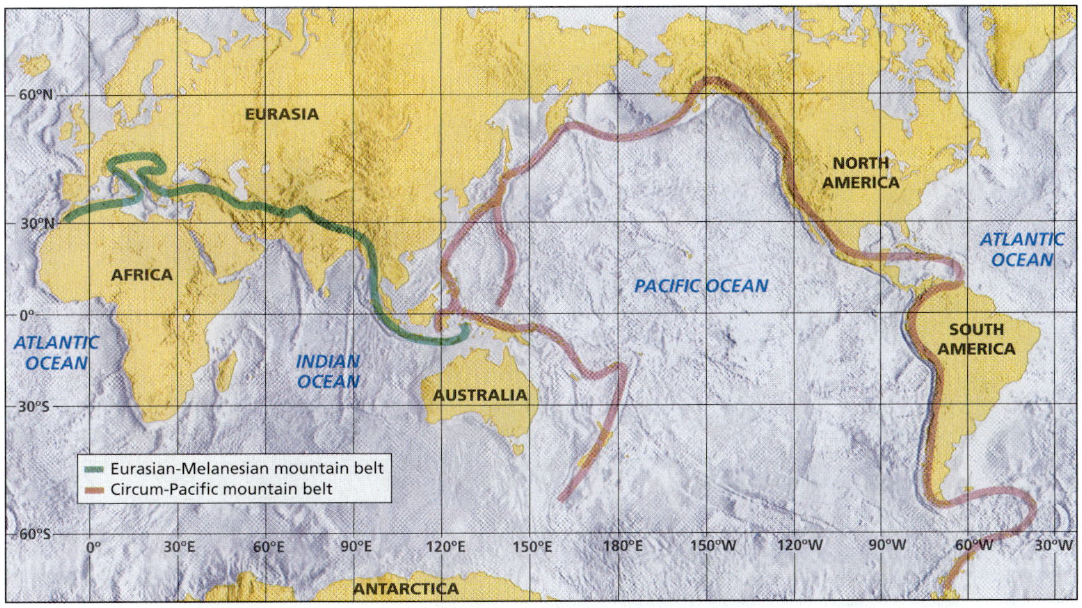

Section 2

Focus

Overview

This section describes the types of plate collisions that form mountains, and compares the ways that four different types of mountains form.

Bellringer

Have students list as many mountain ranges as they can and name the continents on which the ranges are located. (Answers may vary.) LS **Verbal**

Motivate

Discussion — GENERAL

Mountain Features Show students pictures of mountains or ask them to describe any mountains they may have visited. Discuss what features impressed students most. (Answers may vary. Students may note the height, steepness of the sides, their massiveness, how smooth and weathered the range was, or what types of vegetation grew there.) Point out that the features of each mountain range depend on a number of factors, including the material from which the mountains formed and how old the range is. LS **Verbal/Intrapersonal**

CHAPTER RESOURCES

Chapter Resource File

- Directed Reading BASIC

Technology

Transparencies
- Bellringer

Student Edition on CD-ROM

One-Stop Planner CD-ROM
- Lesson Plan

Teach

Graphic Organizer GENERAL

Cause-and-Effect Map

You may want to use this Graphic Organizer to assess students' understanding of mountain formation due to the interaction of tectonic plates. Collect the Graphic Organizers from the students, and correct any mistakes before returning the Graphic Organizers.

Using the Figure — BASIC

Lithospheric Collisions and Mountain Types Have students examine the figures of three different ways that mountain ranges form. Point out that although two plates collide in each case, different types of mountains form because the composition of the crust on each plate is different. Ask students to explain how the collision that formed the Himalayas differs from the collisions that formed the Andes Mountains and the Mariana Islands. (In the Himalayas, after the brief subduction of oceanic crust, there is little subduction at all, but rather the pushing upward of two continental crusts.) Ask students which surface features result from the subduction of oceanic crust. (volcanoes and trenches) LS **Visual/Logical**

CHAPTER RESOURCES

Technology

Transparencies
- 59 How Mountains Form (with worksheet)

Graphic Organizer **Cause-and-Effect Map**

Create the Graphic Organizer entitled "Cause-and-Effect Map" described in the Skills Handbook section of the Appendix. Label the effect with "Mountain formation." Then, fill in the map with causes of mountain formation and details about why mountains form.

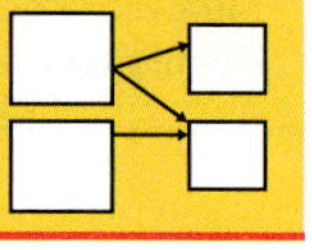

Plate Tectonics and Mountains

Both the circum-Pacific and the Eurasian-Melanesian mountain belts are located along convergent plate boundaries. Scientists think that the location of these two mountain belts provides evidence that most mountains form as a result of collisions between tectonic plates. Some mountains, such as the Appalachians, do not lie along active convergent plate boundaries. However, evidence indicates that the places at which these ranges formed were previously active plate boundaries.

Collisions Between Continental and Oceanic Crust

Some mountains form when oceanic lithosphere and continental lithosphere collide at convergent plate boundaries. When the moving plates collide, the oceanic lithosphere subducts beneath the continental lithosphere, as shown in **Figure 2.** This type of collision produces such large-scale deformation of rock that high mountains are uplifted. In addition, the subduction of the oceanic lithosphere causes partial melting of the overlying mantle and crust. This melting produces magma that may eventually erupt to form volcanic mountains on Earth's surface. The mountains of the Cascade Range in the northwest region of the United States formed in this way. The Andes mountains on the western coast of South America are another example of mountains that formed by this type of collision.

Some mountains at the boundary between continental lithosphere and oceanic lithosphere may form by a different process. As the oceanic lithosphere subducts, pieces of crust called *terranes* are scraped off. These terranes then become part of the continent and may form mountains.

Figure 2 ▶ The Andes, shown below, are being uplifted as the Pacific plate subducts beneath the South American plate.

SKILL BUILDER GENERAL

Writing Mountains have inspired many writers partly because of their grandeur and partly because of discoveries made there during periods of exploration. For example, Hiram Bingham's *Lost City of the Incas* told of the rediscovery after 400 years of Machu Picchu in the Andes Mountains. Fiction writers Thornton Wilder and H. G. Wells wrote stories set in the Andes. Have students read other mountain-inspired writings and write a book report or make an oral report. Encourage students to pay close attention to the author's knowledge of mountains, the accuracy of descriptions, and the role the mountainous setting had on plot, outcome, or character development. LS **Verbal**

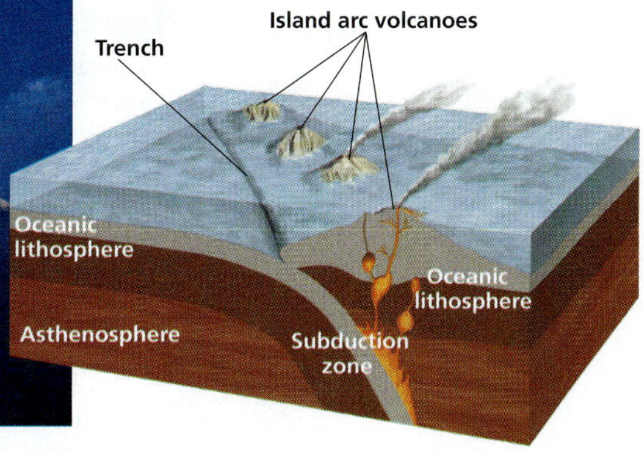

Figure 3 ▶ The Mariana Islands in the North Pacific Ocean are volcanic mountains that formed by the collision of two oceanic plates.

Collisions Between Oceanic Crust and Oceanic Crust

Volcanic mountains commonly form where two plates whose edges consist of oceanic lithosphere collide. In this collision, the denser oceanic plate subducts beneath the other oceanic plate, as shown in **Figure 3.** As the denser oceanic plate subducts, fluids from the subducting lithosphere cause partial melting of the overlying mantle and crust. The resulting magma rises and breaks through the oceanic lithosphere. These eruptions of magma form an arc of volcanic mountains on the ocean floor. The Mariana Islands are the peaks of volcanic mountains that rose above sea level.

Collisions Between Continents

Mountains can also form when two continents collide, as **Figure 4** shows. The Himalaya Mountains formed from such a collision. About 165 million years ago, India broke apart from Africa and Antarctica and became a separate continent. The Indian plate then began moving north toward Eurasia. The oceanic lithosphere of the Indian plate subducted beneath the Eurasian plate. This subduction continued until the continental lithosphere of India collided with the continental lithosphere of Eurasia. Because the two continents have equally dense lithosphere, subduction stopped, but the collision continued. The intense deformation that resulted from the collision uplifted the Himalayas. Because the plates are still colliding, the Himalayas are still growing taller.

Reading Check **Why are the Himalayas growing taller today? (See the Appendix for answers to Reading Checks.)**

Figure 4 ▶ The Himalayas formed when India collided with Eurasia.

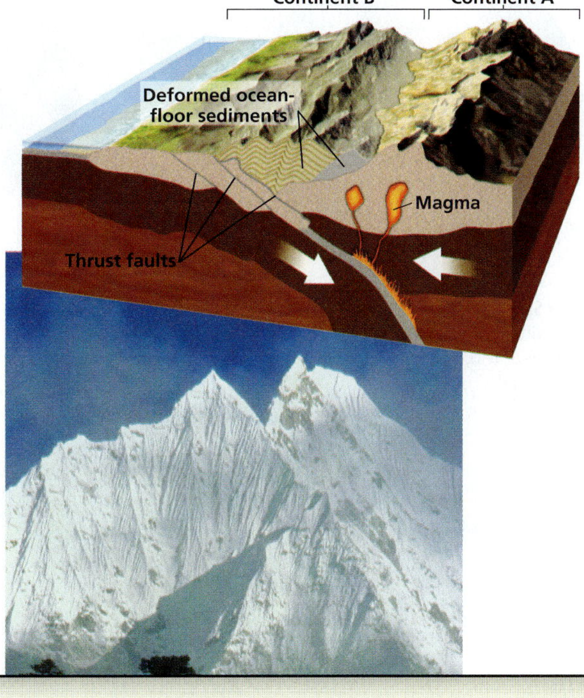

BRAIN FOOD

High Volcanoes While the Andes and Himalayan mountain ranges contain the highest mountains in the world, the two ranges are dramatically different in one respect: there are numerous volcanoes, both active and extinct, in the Andes, whereas there are very few in the Himalayas. This difference has resulted from the nature of the tectonic forces that formed the two ranges. The volcanoes of the Andes result from the subduction of dense oceanic crust. The Himalayan range formed when the continental crust of the Indian subcontinent collided with the continental crust of Asia. Once the two continental plates collided, their similar densities prevented either plate from subducting. Instead, the plates pushed each other upward. This process did not generate the melted rock that occurs in subduction, so no significant volcanic activity occurred.

Answer to Reading Check

The Himalayas are growing taller because the two plates are still colliding and causing further compression of the rock, which further uplifts the mountains.

MISCONCEPTION ALERT

Diverging Plates Students may think that all mountain ranges are produced by collisions between tectonic plates. Remind students that large amounts of lava erupt onto Earth's surface on the ocean floor where two plates diverge. From such lava, volcanic mountain ranges form. Point out that one of the most extensive mountain ranges on Earth, the Mid-Atlantic Ridge, lies mostly underwater, with only a few peaks extending above the surface as islands. Also note that mountain ranges can also form on dry land from the divergence of plates, as is the case in the East African Rift. A range of volcanoes, extending from Malawi in southern Africa to Ethiopia, has formed as a result of lava produced by the separation of the Somali and African plates. The East African Rift Valley is an example of a graben surrounded by fault-block mountains.

Teach, continued

Group Activity — GENERAL

Plates and Mountain Formation
Assign a small group of students to research the specific processes that formed each of the following mountain ranges: the Alps, the Himalayas, the Appalachians, and the Urals. Have students find out what plates interacted to form that particular range; when the range began forming; if and when formation ended; and, if applicable, what major processes have altered the mountains since the end of their growth. Students may present the results of their findings in a short written report, an oral presentation, or as a poster project. LS **Verbal**

Homework — GENERAL

Valleys, Plateaus, and Grabens
The formation of mountains causes other geologic features to form in the same area. Have students research the following features, noting what type of features they are, where they are located, when they formed, how they formed, and any mountains in the region that formed from similar tectonic forces: Massif Central, Great Rift Valley, Altiplano, and Rhine Valley. Students may present findings in a brief written report, an oral presentation, or as a poster project. LS **Logical**

CHAPTER RESOURCES

Chapter Resource File
- **Internet Activity**
  - "Dead" Grabens ADVANCED

Technology
- **Transparencies**
  - 60 Types of Mountains in the United States (with worksheet)

SCILINKS. NSTA. Developed and maintained by the National Science Teachers Association

For a variety of links related to this subject, go to www.scilinks.org

Topic: Types of Mountains
SciLinks code: HQ61568

Types of Mountains

Mountains are more than just elevated parts of Earth's crust. Mountains are complicated structures whose rock formations provide evidence of the stresses that created the mountains. Scientists classify mountains according to the way in which the crust was deformed and shaped by mountain-building stresses. Examples of several types of mountains are shown in **Figure 5.**

Folded Mountains and Plateaus

The highest mountain ranges in the world consist of folded mountains that form when continents collide. **Folded mountains** form when tectonic movements squeeze rock layers together into accordion-like folds. Parts of the Alps, the Himalayas, the Appalachians, and Russia's Ural Mountains consist of very large and complex folds.

folded mountain a mountain that forms when rock layers are squeezed together and uplifted

The same stresses that form folded mountains also uplift plateaus. *Plateaus* are large, flat areas of rock high above sea level. Most plateaus form when thick, horizontal layers of rock are slowly uplifted so that the layers remain flat instead of faulting and folding. Most plateaus are located near mountain ranges. For example, the Tibetan Plateau is next to the Himalaya Mountains, and the Colorado Plateau is next to the Rockies. Plateaus can also form when layers of molten rock harden and pile up on Earth's surface or when large areas of rock are eroded.

Figure 5 ▶ Mountains in the United States

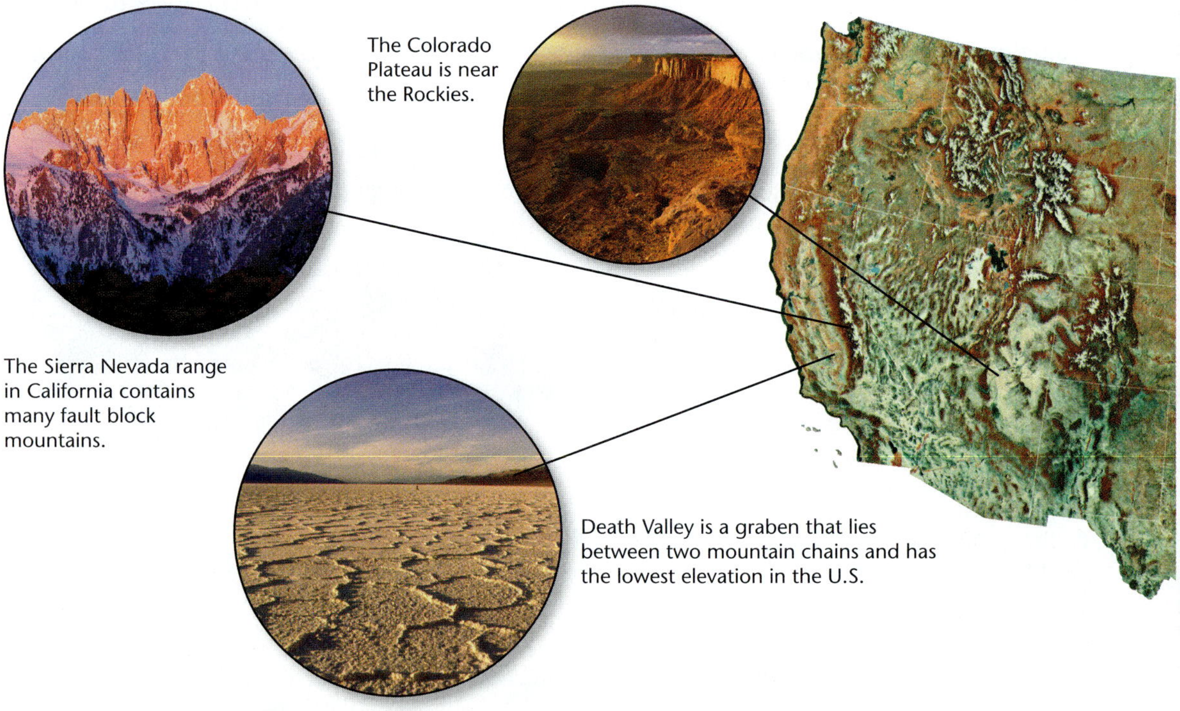

Internet Activity — ADVANCED

"Dead" Grabens The structural geology of Death Valley in California and of the Dead Sea in Israel is strikingly similar—both exist at low elevation and have an evaporating body of water. However, these two regions also display geologic differences. Have students research Death Valley and the Dead Sea and report on the similarities and differences between them. A worksheet designed to direct student research on this topic can be found in the **Chapter Resource File** booklet or by visiting **go.hrw.com** and entering the keyword **HQ6DEFX.** LS **Verbal**

READING SKILL BUILDER — BASIC

Discussion Geologic evidence indicates that the Appalachian Mountains formed between 390 and 210 million years ago, and that they are folded mountains, which tend to be very high. Ask students to explain why the Appalachians are not a higher range, like the Himalayas or Urals. (The Appalachians are relatively old mountains, and have had time for weathering and erosion to reduce their relief. Tell students that weathering is also indicated by sedimentary rock deposits east and west of the Appalachians.) LS **Logical**

Fault-Block Mountains and Grabens

Where parts of Earth's crust have been stretched and broken into large blocks, faulting may cause the blocks to tilt and drop relative to other blocks. The relatively higher blocks form **fault-block mountains.** The Sierra Nevada range of California consists of many fault-block mountains.

The same type of faulting that forms fault-block mountains also forms long, narrow valleys called *grabens*. Grabens develop when steep faults break the crust into blocks and one block slips downward relative to the surrounding blocks. Grabens and fault-block mountain ranges commonly occur together. For example, the Basin and Range Province of the western United States consists of grabens separated by fault-block mountain ranges.

fault-block mountain a mountain that forms where faults break Earth's crust into large blocks and some blocks drop down relative to other blocks

dome mountain a circular or elliptical, almost symmetrical elevation or structure in which the stratified rock slopes downward gently from the central point of folding

Dome Mountains

A rare type of mountain forms when magma rises through the crust and pushes up the rock layers above the magma. The result is a **dome mountain,** a circular structure made of rock layers that slope gently away from a central point. Dome mountains may also form when tectonic forces gently uplift rock layers. The Black Hills of South Dakota and the Adirondack Mountains of New York are examples of dome mountains.

Reading Check **Name three types of mountains found in the United States. (See the Appendix for answers to Reading Checks.)**

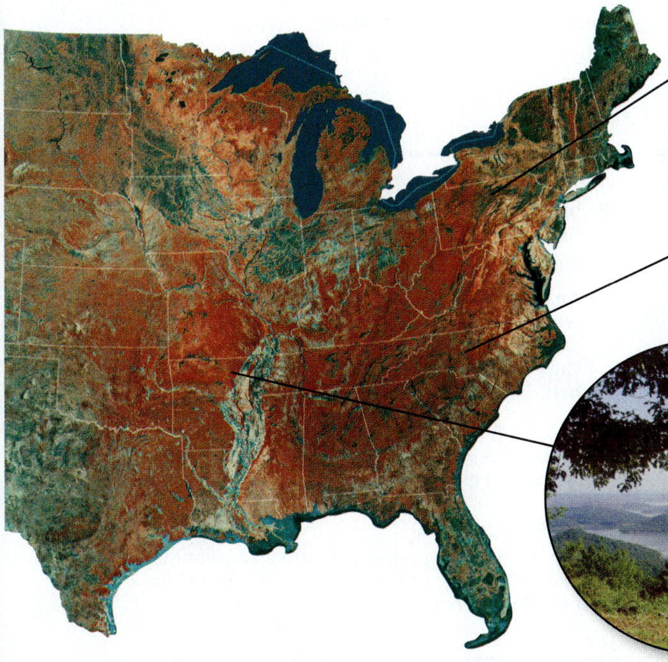

This dome mountain is part of the Adirondacks in New York.

The Appalachian Mountains stretch from Georgia to Canada and contain many older, more rounded mountains.

The Ouachita Plateau in Arkansas is much wetter than the Colorado Plateau is.

Answer to Reading Check

Answers may include three of the following: folded mountains, fault-block mountains, dome mountains, and volcanic mountains.

Close

Reteaching — BASIC

Mountain Formation Processes

On the board, list the four main types of mountains. In another column, write descriptions of how each type of mountain forms. Have students indicate which items in the first column are described by the items in the second column. LS **Logical**

Quiz — GENERAL

Determine whether each of the following statements is true or false.

1. The Eurasian-Melanesian belt is one of Earth's two major mountain belts. (true)
2. Volcanic mountains form when continental crust collides with continental crust. (false)
3. Terranes are the remains of fault-block mountain formation. (false)

Alternative Assessment — ADVANCED

Modeling Plate Interactions

Have students create models to show how interactions between different tectonic plates cause different types of mountains to form. The models can be physical or computer models. The models should show at least four steps during the plate collision. LS **Kinesthetic/Logical**

INCLUSION *Strategies*

- ***Hearing Impaired***
- ***Learning Disabled***
- ***Developmentally Delayed***

To show how converging plates form mountains, perform this small-group activity. Give each group two paper plates, two 8 1/2 × 11-inch pieces of paper, and tape. Have each group perform the following steps: "Tape the two pieces of paper together to form one 11 × 17-inch paper. Place the two paper plates side-by-side on the desk, and place the taped paper on top of the two paper plates. Hold the paper to the edges of the paper plates with your thumbs, and slide one plate over the top of the other." The taped paper should buckle upward, simulating mountain formation. LS **Visual**

Close, *continued*

Answers to Section Review

1. When continental lithosphere collides with oceanic lithosphere, the oceanic plate subducts beneath the continental plate, and magma that results forms volcanic mountains. When oceanic lithosphere collides with oceanic lithosphere, the denser plate subducts, and magma that results forms an arc of volcanic mountains. When continental lithosphere collides with continental lithosphere, neither plate subducts, and the crust is pushed upward to form mountains.
2. Two pieces of continental lithosphere collide, causing the rock to be pushed upward, producing folded mountains.
3. Plateaus have been uplifted by the same stresses that form folded mountains. However, plateaus are flat layers of slowly uplifted rock that are not folded.
4. Faulting in Earth's crust causes large blocks of the crust that have been broken to tilt and drop. The higher blocks make up fault-block mountains.
5. Magma beneath the crust pushes up the rock layers that lie over the magma.
6. Volcanic mountains form when melted rock within Earth's crust erupts onto Earth's surface. As more magma erupts, the volcano grows larger.
7. Volcanic mountains could become smaller during an explosive eruption by destroying part of the mountain, or, once extinct, could become smaller by erosion and weathering.
8. Grabens are the regions where blocks of Earth's crust have fallen between faults. The higher crustal blocks that remain above the graben form the fault-block mountains. The two features are part of the same process.
9. Plateaus often form by the same processes that form folded mountains, whereas grabens form by the same processes that form fault-block mountains. Because the nearby mountains have large folds, the flat area is likely a plateau.
10. volcanic mountains
11. A *mountain range,* which can include *folded mountains, fault-block mountains, dome mountains,* and *volcanic mountains,* may be part of a *mountain system,* which is part of a *mountain belt.*

Figure 6 ▶ Mount St. Helens (front) and Mount Rainier (back) in the Cascade Range of the western United States are volcanic mountains that formed along a convergent boundary.

Volcanic Mountains

Mountains that form when magma erupts onto Earth's surface are called *volcanic mountains.* Volcanic mountains commonly form along convergent plate boundaries. The Cascade Range of Washington, Oregon, and northern California is composed of this type of volcanic mountain, two of which are shown in **Figure 6.**

Some of the largest volcanic mountains are part of the mid-ocean ridges along divergent plate boundaries. Magma rising to Earth's surface at divergent boundaries makes mid-ocean ridges volcanically active areas. The peaks of these volcanic mountains sometimes rise above sea level to form volcanic islands, such as the Azores in the North Atlantic Ocean.

Other large volcanic mountains form on the ocean floor at hot spots. *Hot spots* are volcanically active areas that lie far from tectonic plate boundaries. These areas seem to correspond to places where hot material rises through Earth's interior and reaches the lithosphere. The Hawaiian Islands are an example of this type of volcanic mountain. The main island of Hawaii is a volcanic mountain that reaches almost 9 km above the ocean floor and has a base that is more than 160 km wide.

Section 2 Review

1. **Describe** three types of tectonic plate collisions that form mountains.
2. **Summarize** the process by which folded mountains form.
3. **Compare** how plateaus form with how folded mountains form.
4. **Describe** the formation of fault-block mountains.
5. **Explain** how dome mountains form.
6. **Explain** how volcanic mountains form.

CRITICAL THINKING

7. **Making Connections** Explain two ways in which volcanic mountains might get smaller.
8. **Making Connections** Explain why fault-block mountains and grabens are commonly found near each other.
9. **Analyzing Ideas** You are standing on a large, flat area of land and are examining the nearby mountains. You notice that many of the mountains have large folds. Are you standing on a plateau or a graben? Explain your answer.
10. **Making Predictions** Igneous rocks form from cooled magma. Near what types of mountains would you expect to find new igneous rocks?

CONCEPT MAPPING

11. Use the following terms to create a concept map: *mountain range, fault-block mountains, mountain belt, folded mountains, mountain system, dome mountains,* and *volcanic mountains.*

CHAPTER RESOURCES

Chapter Resource File

- Section Quiz GENERAL

Workbooks

- Study Guide (also in Spanish)

Chapter 11 Highlights

Sections

1 How Rock Deforms

Key Terms

deformation, 271
isostasy, 271
stress, 273
strain, 274
fold, 275
fault, 277

Key Concepts

- Tectonic plate movement and isostatic adjustments cause stress on the rock in Earth's crust.
- Stress can squeeze rock together, pull rock apart, and bend and twist rock.
- Stress on rock can cause strain, or the deformation of rock. Rock can deform by folding or by breaking to form fractures or faults.
- Three types of faults occur in rock: normal faults, reverse faults (including thrust faults), and strike-slip faults.

2 How Mountains Form

mountain range, 279
folded mountain, 282
fault-block mountain, 283
dome mountain, 283

- Mountains make up mountain ranges, which, in turn, make up mountain systems. The largest mountain systems form two major mountain belts.
- Four types of mountains are folded mountains, fault-block mountains, dome mountains, and volcanic mountains.
- Mountains commonly form as the result of the collision of tectonic plates.
- A mountain is classified according to the way in which the crust deforms when the mountain forms.

Chapter Highlights

Alternative Assessment — GENERAL

Poster Project Have students create a poster that shows the processes that rocks undergo that produce different types of mountains. Have students depict a given stress, show how the stress results from tectonic plate interaction, identify the resulting strain, and then show the final form that the mountain takes. Students should include labels to identify features and steps involved in each process. **LS Visual/Logical**

CHAPTER RESOURCES

Chapter Resource File
- Concept Review GENERAL
- Critical Thinking ADVANCED
- Math Skills GENERAL
- Graphing Skills GENERAL
- Chapter Test A GENERAL
- Chapter Test B ADVANCED

Workbooks
- **Study Guide** (also in Spanish)
- **Assessments** (Spanish)

Technology

Classroom Videos
- Brain Food Video Quiz

HRW Earth Science Video
- Segment 4: Deformation of the Crust

Chapter Review

Assignment Guide

Section	Questions
1	1–6, 9–15, 18–19, 22, 24–26, 28, 30–35
2	7–8, 16–17, 20–21, 23, 27, 29

Using Key Terms

1–8. Answers may vary but should show that students understand the definitions of and differences between key terms.

Understanding Key Concepts

9. a **10.** d
11. a **12.** d
13. a **14.** c
15. c **16.** c
17. d

Short Answer

18. Folds and faults are two types of deformation in Earth's crust. When rock is ductile, it is able to bend with applied stress, so it folds. When rock is brittle, it breaks under stress, so it forms faults or fractures.

19. In an anticline, the outermost layers of the fold have the youngest rock. An anticline fold may be arch shaped.

20. The mountain belts are the circum-Pacific belt and the Eurasian-Melanesian belt.

21. Mountains are categorized as folded mountains, which commonly form when two continental plates collide; fault-block mountains, which commonly form when broken blocks of crust fall away from other blocks that become the mountains; dome mountains, which form when magma pushes upward on overlying rock, forming dome-like structures; and volcanic mountains, which form when magma erupts onto Earth's surface and accumulates to form a mountain.

22. gravitational force and buoyant force

23. Dome mountains are round or elliptical and have gentle slopes that descend symmetrically from the mountain's central peak. Fault-block mountains form from the broken blocks that lie between faults and have not dropped down or tilted. These blocks are relatively higher than the surrounding blocks.

Chapter 11 Review

Using Key Terms

Use each of the following terms in a separate sentence.

1. *isostasy*

2. *compression*

3. *shear stress*

For each pair of terms, explain how the meanings of the terms differ.

4. *stress* and *strain*

5. *fold* and *fault*

6. *syncline* and *monocline*

7. *dome mountains* and *volcanic mountains*

8. *folded mountains* and *fault-block mountains*

Understanding Key Concepts

9. When the weight of an area of Earth's crust increases, the lithosphere
a. sinks. **c.** rises.
b. melts. **d.** collides.

10. The force per unit area that changes the shape and volume of rock is
a. footwall. **c.** rising.
b. isostasy. **d.** stress.

11. Shear stress
a. bends, twists, or breaks rock.
b. causes isostasy.
c. causes rock to melt.
d. causes rock to expand.

12. When stress is applied under conditions of high pressure and high temperature, rock is more likely to
a. fracture. **c.** fault.
b. sink. **d.** fold.

13. Folds in which both limbs remain horizontal are called
a. monoclines. **c.** synclines.
b. fractures. **d.** anticlines.

14. When a fault is not vertical, the rock above the fault plane makes up the
a. tension. **c.** hanging wall.
b. footwall. **d.** compression.

15. A fault in which the rock on either side of the fault plane moves horizontally in nearly opposite directions is called a
a. normal fault. **c.** strike-slip fault.
b. reverse fault. **d.** thrust fault.

16. The largest mountain systems are part of still larger systems called
a. continental margins.
b. ranges.
c. belts.
d. synclines.

17. Large areas of flat-topped rock high above the surrounding landscape are
a. grabens. **c.** hanging walls.
b. footwalls. **d.** plateaus.

Short Answer

18. Name two types of deformation in Earth's crust, and explain how each type occurs.

19. Explain how to identify an anticline.

20. Identify the two major mountain belts on Earth.

21. Describe how the various types of mountains are categorized.

22. Identify the two forces that are kept in balance by isostatic adjustments.

23. Compare the features of dome mountains with those of fault-block mountains.

Critical Thinking

24. The lithosphere under the continents would sink under the added weight of the continental ice sheets. The reduced amount of water in the oceans would reduce the weight over the lithosphere under the oceans, so that portion of the lithosphere would rise.

Critical Thinking

24. Evaluating Ideas If thick ice sheets covered large parts of Earth's continents again, how would you expect the lithosphere to respond to the added weight of the continental ice sheets? Explain your answer.

25. Analyzing Relationships When the Indian plate collided with the Eurasian plate and produced the Himalaya Mountains, which type of stress most likely occurred? Which type of stress is most likely occurring along the Mid-Atlantic Ridge? Which type of stress would you expect to find along the San Andreas fault? Explain your answers.

26. Making Predictions If the force that causes a rock to deform slightly begins to ease, what may happen to the rock? What might happen if the force causing the deformation became greater?

27. Analyzing Processes Why do you think that dome mountains do not always become volcanic mountains?

Concept Mapping

28. Use the following terms to create a concept map: *stress, strain, brittle, ductile, folds, fault, normal fault, reverse fault, thrust fault,* and *strike-slip fault*.

Math Skills

29. Making Calculations Scientists calculate that parts of the Himalayas are growing at a rate of 6.1 mm per year. At this rate, in how many years will the Himalayas have grown 1 m taller?

30. Analyzing Data Rock stress is measured as 48 MPa at point A below Earth's surface. At point B nearby, stress is measured as 12 MPa. What percentage of the stress at point A is the stress at point B equal to?

Writing Skills

31. Creative Writing Write a short story from the perspective of a rock that is being deformed. Describe the stresses that are affecting the rock and the final result of the stress.

32. Writing from Research Look for photos or illustrations of folding and faulting in a particular area. Then, research the geologic history of the area, and write a report based on your findings. Use any photos or drawings that you find to illustrate your report.

Interpreting Graphics

The diagram below shows a fault. Use the diagram below to answer the questions that follow.

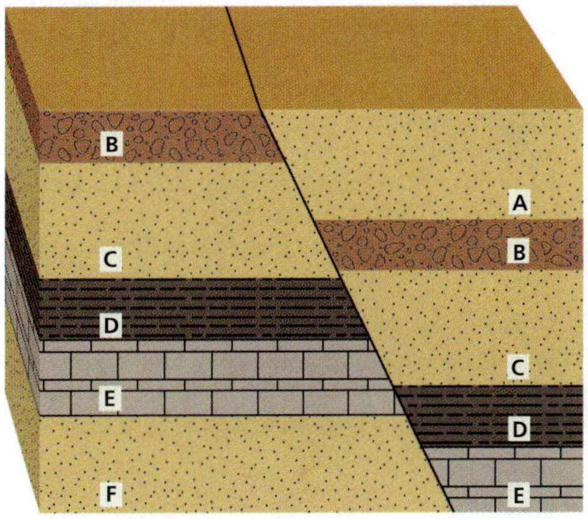

Block 1 **Block 2**

33. Is Block 2 a footwall or a hanging wall? Explain your answer.

34. What type of fault is illustrated? Explain your answer.

35. What type of stress generally causes this type of fault?

Chapter Review

25. The collision between the Indian and Eurasian plates mainly involved compression. The plates that meet at the Mid-Atlantic Ridge are separating and are affected mainly by tension. The blocks on either side of the San Andreas fault are moving in opposite directions, which causes shear stress.

26. If the force is reduced, the rock may return to its original condition. If the force is increased, the rock may permanently deform by folding or by faulting or fracturing.

27. The properties of the rock overlying the magma are such that the rock does not melt, and the magma is unable to seep upward through it. Thus, the magma is only able to lift the rock upward, forming the dome mountain.

Concept Mapping

28. Answers may vary but should include all of the terms listed. Sample answers appear at the end of this Teacher Edition.

Math Skills

29. number of years = increase in height ÷ amount of growth per year = [1 m ÷ (6.1 mm/year)] × 1000 mm/m = 164 years

30. percentage stress = (stress at B ÷ stress at A) × 100; percentage stress = (12 MPa/48 MPa) × 100 = 25%

Writing Skills

31. Answers may vary. Accept all reasonable answers.

32. Answers may vary. Accept all reasonable answers.

Interpreting Graphics

33. Block 2 is a hanging wall, because the rock lies above the fault plane.

34. A normal fault is shown, because the hanging wall has moved downward relative to the footwall.

35. tension

Standardized Test Prep

Estimated Time

To give students practice under more realistic testing conditions, allow them 30 minutes to answer all of the questions in this practice test.

TEST DOCTOR

Question 4 Answer H describes a type of isostatic adjustment that occurs when a river that is carrying mud, sand, and gravel flows into the ocean. The added weight of this material, when deposited on the ocean floor, causes the floor to sink—not form mountains. Answers F, G and I all describe situations in which mountain formation may occur.

Question 9 Answer G is correct. Answer F is incorrect because it describes flat, smooth surfaces, which are not produced by stress. Students who choose answer H may have a misunderstanding of the anology. Answer I is one example of stress. Students should assume by the wording in the passage that rocks undergo stress at other times.

Chapter 11 Standardized Test Prep

Understanding Concepts

Directions (1–4): **For *each* question, write on a separate sheet of paper the letter of the correct answer.**

1. Where are most plateaus located?
   A. near mountain ranges
   B. bordering ocean basins
   C. beneath grabens
   D. alongside diverging boundaries

2. Which of the following features form where parts of the crust have been broken by faults?
   F. monoclines
   G. plateaus
   H. synclines
   I. grabens

3. Which of the following statements describes the formation of rock along strike-slip faults?
   A. Rock on either side of the fault plane slides vertically.
   B. Rock on either side of the fault plane slides horizontally.
   C. Rock in the hanging wall is pushed up and over the rock of the footwall.
   D. Rock in the hanging wall moves down relative to the footwall.

4. Which does not result in mountain formation?
   F. collisions between continental and oceanic crust
   G. subduction of one oceanic plate beneath another oceanic plate
   H. deposition and isostasy
   I. deformation caused by collisions between two or more continents

Directions (5–7): **For *each* question, write a short response.**

5. What is the term for a condition of gravitational equilibrium in Earth's crust?

6. What is the term for a type of stress that squeezes and shortens a body?

7. As a volcanic mountain range is built, isostatic adjustment will cause the crust beneath the mountain range to do what?

Reading Skills

Directions (8–10): **Read the passage below. Then, answer the questions.**

Stress and Strain

Stress is defined as the amount of force per unit area on a rock. When enough stress is placed on a rock, the rock becomes strained. This strain causes the rock to deform, usually by bending and breaking. For example, if you put a small amount of pressure on the ends of a drinking straw, the straw may not bend—even though you have put stress on it. However, when you put enough pressure on it, the straw bends, or becomes strained.

One example of stress is when tectonic plates collide. When plates collide, a large amount of stress is placed on the rocks that make up the plate, especially the rocks at the edge of the plates involved in the collision. Because of the stress, these rocks become extremely strained. In fact, even the shapes of the tectonic plates can change as a result of these powerful collisions.

8. Based on the passage, which of the following statements is not true?
   A. Strain can cause a rock to deform by bending or breaking.
   B. Rocks, like drinking straws, will not bend when pressure is applied to them.
   C. Stress is defined as amount of force per unit area that is put on a rock.
   D. A large amount of stress is placed on the rocks involved in tectonic plate collisions.

9. Which of the following statements can be inferred from the information in the passage?
   F. The stress of tectonic plate collisions often creates large, smooth plains of rock.
   G. The stress of tectonic plate collisions often creates large, mountain chains.
   H. Bending a drinking straw requires the same amount of pressure that is needed to bend a rock.
   I. The only time a rock has stress is when the rock is involved in a tectonic collision.

10. What happens to rocks when plates collide?

Answers

Understanding Concepts

1. A
2. I
3. B
4. H
5. isostasy
6. compression
7. sink

Reading Skills

8. A
9. G
10. Rocks suffer severe stress in a tectonic collision, and they are usually deformed by the extremly high amounts of pressure.

Interpreting Graphics

11. D
12. Answers may vary. See Test Doctor for a detailed scoring rubric.
13. F
14. Answers may vary. See Test Doctor for a detailed scoring rubric.

Interpreting Graphics

Directions (11–14): **For *each* question below, record the correct answer on a separate sheet of paper.**

The diagrams below show a divergent and a convergent plate boundary. Use these diagrams to answer questions 11 and 12.

Divergent and Convergent Plate Boundaries

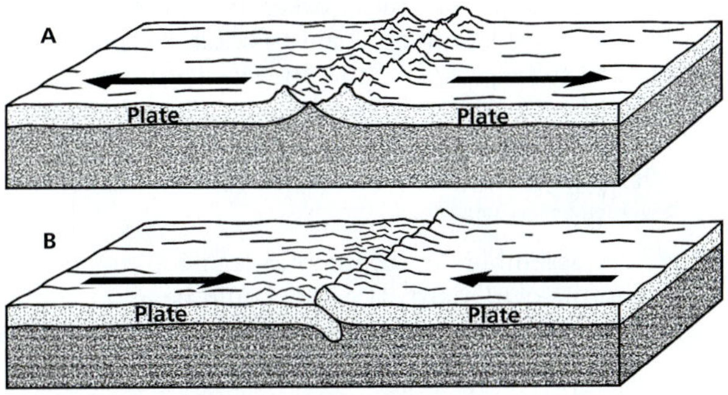

11 Which of the following is not likely to be found at or occur at the boundary found in diagram A?

A. volcanoes
B. lava flows
C. earthquakes
D. subduction

12 How does the subduction of the oceanic crust shown in diagram B produce volcanic mountains?

The diagram below shows two possible outcomes when pressure, which is represented by the large arrows, is applied to the rock on the left. Use this diagram to answer questions 13 and 14.

Rock Deformation

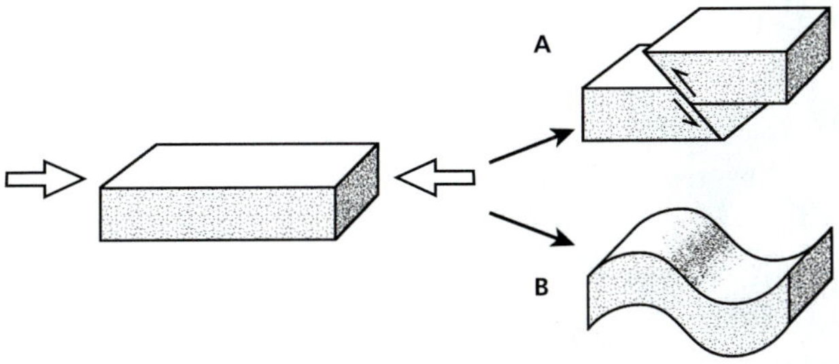

13 What type of deformation is seen in the rock labeled A?

F. brittle
G. ductile
H. folding
I. monocline

14 Describe the type of rock deformation shown in Figure B. Under what conditions is this type of deformation likely to occur?

Test TIP

Carefully study all of the details of a diagram before answering the question or questions that refer to it.

Standardized Test Prep

TEST DOCTOR

Question 12 Full-credit answers should include the following points:

- subduction causes partial melting of the overlying mantle, which produces magma that may erupt to form volcanic mountains
- as the oceanic plate is subducted by the denser continental plate, the oceanic plate heats up and releases water
- the water causes a partial melting of the mantle, which changes rock into magma
- this magma rises to the surface and forms volcanic mountains

Question 14 Full-credit answers should include the following points:

- when stress is applied, a rock can respond with brittle or ductile strain
- ductile strain is common in rock that is hot or has pressure evenly exerted
- ductile strain may be a change in volume or shape without breaking
- ductile strain is usually bending or folding

Test Prep Correlations *National Science Education Standards*

ES 1b: items 4, 5, 9, 10, 12
ES 3c: items 1, 2, 3, 4, 5, 6, 7, 10, 11, 12, 13, 14
UCP 2: items 4, 7, 10, 12, 14

CHAPTER RESOURCES

State Resources

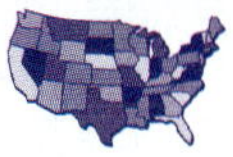

For specific resources for your state, visit go.hrw.com and type in the keyword HSHSTR.

Continental Collisions

Teacher's Notes

Time Required
two 45-minute class periods

Lab Ratings
EASY → HARD

TEACHER PREPARATION 2
STUDENT SETUP 3
CONCEPT LEVEL 2
CLEANUP 2

Skills Acquired
- Experimenting
- Constructing Models
- Predicting
- Interpreting
- Identifying and Recognizing Patterns

The Scientific Method
In this lab, students will
- Make Observations
- Test the Hypothesis
- Analyze the Results
- Draw Conclusions

Materials
The materials listed on the page are enough for groups of two students. Tissue paper can be used in place of napkins. Napkins or tissue paper can be of any color.

Tips and Tricks
Students should make sure that the strip of paper moves easily through the slit before proceeding to step 2. In step 6, students should be sure that the napkins are securely attached to the paper strip. In step 8, one student can hold the cardboard, while the other student pulls the paper strip.

Continental Collisions

Objectives

- **Model** collisions between continents.
- **Explain** how mountains form at convergent boundaries.

Materials

- blocks, wooden, 2.5 cm × 2.5 cm × 6 cm
- bobby pins, long (5)
- cardboard, thick, 15 cm × 30 cm
- napkins, paper, light- and dark-colored
- paper, adding-machine, 6 cm × 35 cm
- ruler, metric
- scissors
- tape, masking

Safety

When the subcontinent of India broke away from Africa and Antarctica and began to move northward toward Eurasia, the oceanic crust on the northern side of India began to subduct beneath the Eurasian plate. The deformation of the crust resulted in the formation of the Himalaya Mountains. Earthquakes in the Himalayan region suggest that India is still pushing against Eurasia. In this lab, you will create a model to help explain how the Himalaya Mountains formed as a result of the collision of the Indian and Eurasian tectonic plates.

PROCEDURE

1. To assemble the continental-collision model, cut a 7 cm slit in the cardboard. The slit should be about 6 cm from (and parallel to) one of the short edges of the cardboard. Cut the slit wide enough such that the adding-machine paper will feed through the slit without being loose.
2. Securely tape one wood block along the slit between the slit and the near edge of the cardboard. Tape the other block across the paper strip about 6 cm from one end of the paper. The blocks should be parallel to one another, as shown in the illustration on the next page.
3. Cut two strips of the light-colored paper napkin that are about 6 cm wide and 16 cm long. Cut two strips of the dark-colored paper napkin that are about 6 cm wide and 32 cm long. Fold all four strips in half along their width.

Step 8

CHAPTER RESOURCES

Chapter Resource File
- Datasheet for Chapter Lab GENERAL
- Lab Notes and Answers

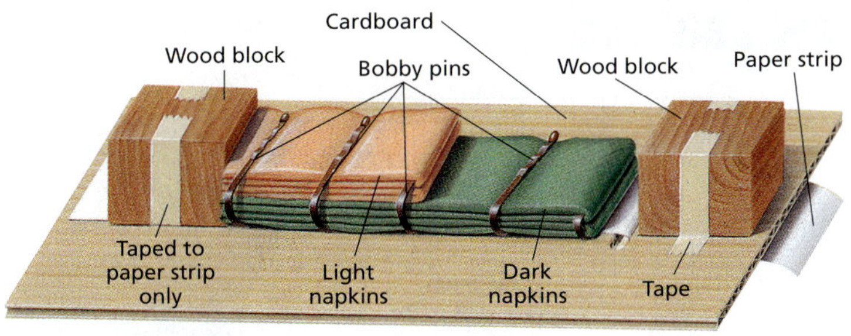

4. Stack the napkin strips on top of each other such that all of the folds are along the same side. Place the two dark-colored napkins on the bottom.
5. Place the napkin strips lengthwise on the paper strip. The nonfolded ends of the napkin strips should be butted up against the wood block that is taped to the paper strip.
6. Using the bobby pins, attach the napkins to the paper strip, as shown in the illustration above.
7. Push the long end of the paper strip through the slit in the cardboard until the first fold of the napkin rests against the fixed wood block.
8. Hold the cardboard at about eye level, and pull down gently on the paper strip. You may need a partner's help. Observe what happens as the dark-colored napkins contact the fixed wood block and as you continue to pull down on the paper strip. Stop pulling when you feel resistance from the strip.

ANALYSIS AND CONCLUSION

1. **Evaluating Methods** Explain what is represented by the dark napkins, the light napkins, and the wood blocks.
2. **Analyzing Processes** What plate-tectonics process is represented by the motion of the paper strip in the model? Explain your answer.
3. **Applying Ideas** What type of mountain would result from the kind of collision shown by the model?
4. **Evaluating Models** Explain how the process modeled here differs from the way the Himalaya Mountains formed.

Extension

1. **Analyzing Data** Obtain a world map of earthquake epicenters. Study the map. Describe the pattern of epicenters in the Himalayan region. Does the pattern suggest that the Himalaya Mountains are still growing?
2. **Writing from Research** Read about the breakup of Gondwanaland and the movement of India toward the Northern hemisphere. Write about stages in India's movement. List the time frame in which each important event occurred.

Answers to Analysis and Conclusion

1. The dark napkins represent oceanic crust, the light napkins represent continental crust, and the blocks represent the mass of the rock in the two colliding plates.
2. The motion of the paper strip represents the subduction of the oceanic lithosphere into the asthenosphere.
3. Folded mountains would result from this type of collision.
4. In the model, the light napkins represent the material in one plate that is compressed. In the actual formation of the Himalayas, two continental plates collided, and rock from both continents deformed.

Answers to Extension

1. Answers should indicate that the Himalayan Mountains are still growing.
2. Answers may vary. The essay should indicate that the Indian plate broke away from Africa about 165 million years ago and broke away from Madagascar about 88 million years ago. The Indian plate rotated as it traveled northward at a rate of about 15 cm/year. It collided with the Eurasian plate about 40 to 50 million years ago.

TEACHER TESTED & APPROVED

Alonda Droege
Highline High School
Burien, WA

Shear Strain in New Zealand

Activity — GENERAL

Geophysics of New Zealand
Have students research New Zealand's geology, notably the volcanic activity that exists on the North Island and the absence of such activity on the South Island. Have students predict what New Zealand may look like in 1 million years, given that its two islands are on separate plates. LS **Verbal/Visual**

Answers to Map Skills Activity

1. 0.4 ppm/yr
2. This strain lies between 169° and 174° east longitude, and between 42° and 44° south latitude.
3. The area of maximum shear strain is approximately 400 km long.
4. The plate boundary is a convergent boundary that is bordered by a trough, which forms when an ocean plate subducts beneath another plate.
5. Compressional stress should be located near the convergent boundaries.
6. The Alpine Fault is most likely a strike-slip fault. The area surrounding the Alpine Fault has very high rates of shear strain, which indicates that the lithosphere is moving in opposite directions in that region.
7. Answers may vary. Sample answer: The compression and shear stress in this region would most likely result in the formation of folded mountains.

CHAPTER RESOURCES

Technology

Transparencies
- 61 Shear Strain in New Zealand (with worksheet)

MAPS in Action

Shear Strain in New Zealand

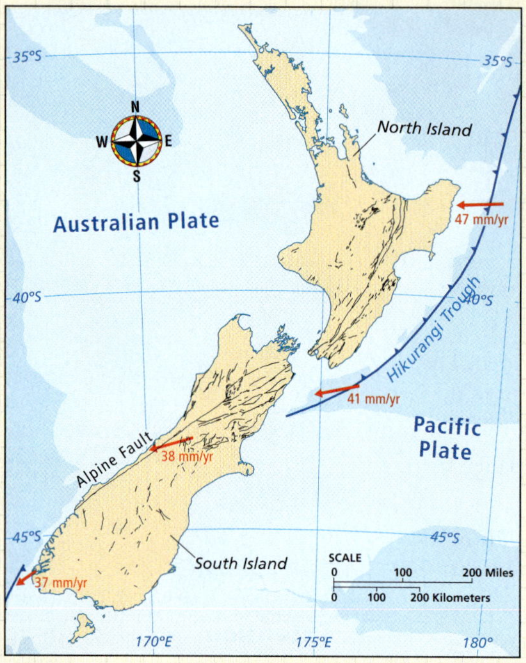

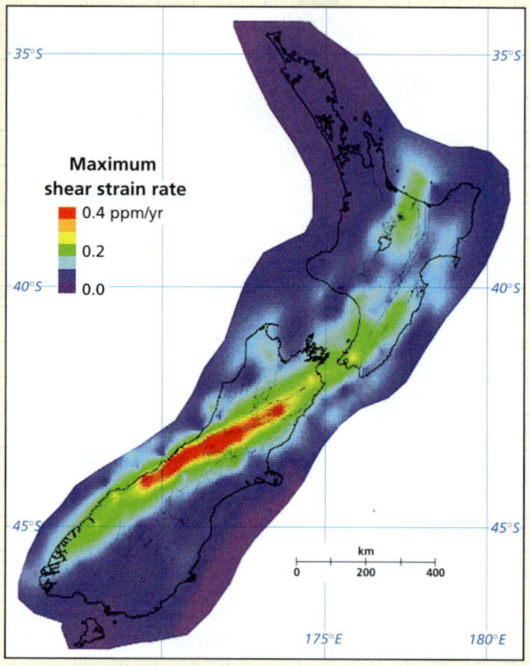

Map Skills Activity

The map above shows the plate boundary zone of New Zealand. In this region, the Australian plate is moving north, while the Pacific plate is moving west. These complex plate movements create areas of tension, compression, and shear stress, which result in strain. Strain is measured in parts per million (ppm) per year (yr). The map on the right shows strain in New Zealand. Use the two maps to answer the questions below.

1. **Using a Key** What is the highest amount of shear strain shown on the map?
2. **Identifying Locations** Using latitude and longitude, describe the location of the area that has the highest amount of shear strain.
3. **Using a Key** What is the approximate length of the area of maximum shear strain?
4. **Understanding Relationships** What type of plate boundary is located along the east coast of the North Island? Explain your answer.
5. **Comparing Areas** In what areas might you expect to find compression? Explain your answer.
6. **Making Inferences** What type of fault is the Alpine Fault? Explain your answer.
7. **Drawing Conclusions** A mountain range known as the *Southern Alps* runs through the center of the South Island. What type of mountains do you think the Southern Alps are? Explain your answer.

IMPACT on Society

The Disappearing Mediterranean

Two of the most breathtaking regions of the world are the Alps and the Mediterranean. The Alps, considered to be among Earth's most beautiful mountains, have become a vast natural playground for skiers, hikers, and climbers. The Mediterranean plays host to travelers from around the world who wish to sample the diverse cultures, balmy climate, and famous beach resorts that surround the Mediterranean Sea. Local residents depend on the sea for their economic well-being.

Push and Pull

The same natural forces that uplifted the Alps are slowly swallowing up the Mediterranean. The Alps were formed—and are still being shaped—by the collision of two tectonic plates. Italy, part of which rides on the African plate, collided with Eurasia sometime in the past. The collision formed the Alps, but it did not stop the movement of the African plate.

The History of Tomorrow

The northern oceanic crust of the African plate, which is the sea floor of the Mediterranean, is still subducting beneath the continental crust of Eurasia. As more oceanic crust subducts, the Mediterranean Sea will become smaller. Italy, which continues to be pushed into Eurasia, will eventually cease to exist as we know it. When the northern coast of the African continent finally collides with Eurasia, the Mediterranean Sea will disappear completely. Of course, this process will take millions of years, because tectonic plates move so slowly.

Extension

1. **Applying Ideas** What do you think will happen to the Alps as the African plate continues to push northward?

▼ The Alps formed when the African plate collided with Eurasia.

▼ The Aegean Sea is part of the larger Mediterranean Sea, which is slowly disappearing.

◀ Santorini, Greece, is an island in the Aegean Sea.

The Disappearing Mediterranean

Activity — ADVANCED

The Mediterranean of the Past Have students research the history of the Mediterranean Sea, from its formation to its current geography. Students should note in particular how the Mediterranean has interacted with the surrounding land and bodies of water, such as the Atlantic Ocean. In particular, students may be interested in researching the Miocene Messinian events and the formation of massive salt deposits in the Mediterranean basin. Once students have completed their research, have them report their findings in a written report, oral presentation, or poster presentation. LS **Verbal/Logical**

Answers to Extension

1. As the African plate continues to move northward, the Alps will continue to be raised, making them even higher.

Chapter 12 Earthquakes
Planning Guide

Compression Guide
To shorten instruction because of time limitations, omit the Chapter Lab.

OBJECTIVES	LABS, DEMONSTRATIONS, AND ACTIVITIES	TECHNOLOGY RESOURCES
PACING • 45 min pp. 294–300 **Chapter Opener**		**OSP Parent Letter** **CD Student Edition on CD-ROM** **CD Chapter Summaries Audio CD** **VID Brain Food Video Quiz**
Section 1 How and Where Earthquakes Happen • Describe elastic rebound. • Compare body waves and surface waves. • Explain how the structure of Earth's interior affects seismic waves. • Explain why earthquakes generally occur at plate boundaries.	**TE Demonstration** Elastic Rebound, p. 295 GENERAL **TE Group Activity** Modeling Locked Faults, p. 296 GENERAL **TE Technology Connection** Seismic Profiling, p. 297 GENERAL **TE Discussion** Surface Waves, p. 297 BASIC **TE History Connection** Moho Discontinuity, p. 298 GENERAL **SE Mapping Expeditions** A Case of the Tennessee Shakes, pp. 834–835 GENERAL **CRF Inquiry Lab** Simulating Earthquakes* GENERAL	**OSP Lesson Plans** (also in print) **TR Bellringer*** **TR** 62 Anatomy of an Earthquake* **TR** 63 Seismic Waves and Earth's Interior* **TR** 64 Earthquakes and Tectonic Plate Boundaries* **TE Internet Activity** Distribution Patterns, p. 296 GENERAL **CRF Internet Activity** Distribution Patterns* GENERAL **VID HRW Earth Science Video** Earthquakes **VID NOVA Video** Earthquake
PACING • 90 min pp. 301–304 **Section 2 Studying Earthquakes** • Describe the instrument used to measure and record earthquakes. • Summarize the method scientists use to locate an epicenter. • Describe the scales used to measure the magnitude and intensity of earthquakes.	**TE Group Activity** Model Seismograph, p. 301 GENERAL **SE Quick Lab** Seismographic Record, p. 302 GENERAL **CRF Datasheet for Quick Lab*** GENERAL **SE Skills Practice Lab** Finding an Epicenter, pp. 314–315 GENERAL **CRF Datasheet for Chapter Lab*** GENERAL	**OSP Lesson Plans** (also in print) **TR Bellringer*** **CD Interactive Tutor** Volcanoes and Earthquakes
PACING • 45 min pp. 305–308 **Section 3 Earthquakes and Society** • Discuss the relationship between tsunamis and earthquakes. • Describe two possible effects of a major earthquake on buildings. • List three safety techniques to prevent injury caused by earthquake activity. • Identify four methods scientists use to forecast earthquake risks.	**TE Demonstration** Giant Wave, p. 305 GENERAL **SE Quick Lab** Earthquake-Safe Buildings, p. 306 GENERAL **CRF Datasheet for Quick Lab*** GENERAL **TE Group Activity** Earthquake Safety, p. 306 GENERAL **TE Debate** Worth the Price?, p. 307 ADVANCED **SE Maps in Action** Earthquake Hazard Map, p. 316 GENERAL **CRF Skills Practice Lab** Earthquakes and Soil* GENERAL	**OSP Lesson Plans** (also in print) **TR Bellringer*** **TR** 65 Seismic Gaps* **TR** 66 Earthquake Hazard Map* **TE Internet Activity** Hazards in the Americas, p. 316 GENERAL **CRF Internet Activity** Hazards in the Americas* GENERAL

PACING • 90 min

CHAPTER REVIEW, ASSESSMENT, AND STANDARDIZED TEST PREPARATION

- **SE Chapter Highlights,** p. 309
- **SE Chapter Review,** pp. 310–311
- **SE Standardized Test Prep,** pp. 312–313
- **CRF Concept Review*** GENERAL
- **CRF Critical Thinking*** ADVANCED
- **CRF Math Skills*** GENERAL
- **CRF Graphing Skills*** GENERAL
- **CRF Chapter Test A*** GENERAL
- **CRF Chapter Test B*** ADVANCED
- **OSP Lesson Plans** (also in print)
- **OSP Test Generator**
- **OSP Test Item Listing**

Online and Technology Resources

go.hrw.com Holt Online Learning

Visit go.hrw.com for access to Holt Online Learning, or enter the keyword **HQ6 Home** for a variety of free online resources.

One-Stop Planner® CD-ROM

This CD-ROM package includes
- Lab Materials QuickList Software
- Holt Calendar Planner
- Customizable Lesson Plans
- Printable Worksheets
- ExamView® Test Generator
- Interactive Teacher Edition
- Holt PuzzlePro®
- Holt PowerPoint® Resources

KEY

SE Student Edition **TE** Teacher Edition **CRF** Chapter Resource File **LTP** Long-Term Projects	**OSP** One-Stop Planner **TR** Transparencies and Transparency Worksheets **CD** CD or CD-ROM	**VID** Classroom Video/DVD * Also on One-Stop Planner ◆ Requires advance prep ■ Also available in Spanish

SKILLS DEVELOPMENT RESOURCES	REVIEW AND ASSESSMENT	CORRELATIONS
SE Pre-Reading Activity, p. 294 GENERAL **TE Using the Figure** The Kobe Quake, p. 294 GENERAL		National Science Education Standards
CRF Directed Reading* BASIC **TE Skill Builder** Vocabulary, p. 296 GENERAL **TE Inclusion Strategies,** p. 298 **SE Graphic Organizer** Spider Map, p. 299 GENERAL	**SE Reading Checks,** pp. 297, 298 GENERAL **SE Section Review,** p. 300 GENERAL **TE Reteaching,** p. 299 BASIC **TE Quiz,** p. 299 GENERAL **TE Alternative Assessment,** p. 300 GENERAL **CRF Section Quiz*** ■ GENERAL	SAI 1a, SAI 1b, SAI 1d, SAI 1f, PS 6a, ES 1b, ES 3c, ST 1d, ST 1e, UCP 3, UCP 4
CRF Directed Reading* BASIC **TE Reading Skill Builder** Reading Organizer, p. 302 BASIC **TE Inclusion Strategies,** p. 302 **SE Math Practice,** p. 303 GENERAL **TE Skill Builder** Graphing, p. 303 ADVANCED **TE Using the Figure,** Discussion, p. 317 GENERAL	**SE Reading Check,** p. 303 GENERAL **SE Section Review,** p. 304 GENERAL **TE Reteaching,** p. 303 BASIC **TE Quiz,** p. 303 GENERAL **TE Alternative Assessment,** p. 304 ADVANCED **CRF Section Quiz*** ■ GENERAL	SAI 1b, SAI 1c, SAI 1d, SAI 1f, ST 1d, ST 1e
CRF Directed Reading* BASIC **TE Using the Figure,** p. 306 BASIC **TE Reading Skill Builder** Paired Summarizing, p. 306 BASIC	**SE Reading Check,** p. 307 GENERAL **SE Section Review,** p. 308 GENERAL **TE Reteaching,** p. 307 BASIC **TE Quiz,** p. 307 GENERAL **TE Alternative Assessment,** p. 308 GENERAL **CRF Section Quiz*** ■ GENERAL	SAI 1b, SAI 1d, SAI 1f, ST 1e, SPSP 4a, SPSP 5a

Holt Lab Generator CD-ROM

Search for labs by topic, standard, difficulty level, and/or time frame. Edit those labs to fit your needs, or create your own labs. Use the Lab Materials QuickList to order just what you need.

Holt Earth Science Interactive Tutor CD-ROM

This CD-ROM consists of interactive activities that give students a fun way to extend their knowledge of Earth science concepts.

Chapter Summaries Audio CDs

These CDs include audio summaries of the key concepts presented in each chapter. (Audio summaries are also available in Spanish.)

SCI LINKS. NSTA

www.scilinks.org

Maintained by the **National Science Teachers Association.** See Chapter Enrichment pages that follow for a complete list of topics.

See Chapter Enrichment pages for Video Resources.

Chapter 12

Chapter Enrichment

This Chapter Enrichment provides relevant and interesting information to expand and enhance your classroom instruction of the chapter material.

Section 1 How and Where Earthquakes Happen

New Madrid Earthquake

A series of extraordinary earthquakes struck in the region of New Madrid, Missouri, along the Mississippi River during the years 1811 and 1812. The damage covered about 130,000 km^2, including parts of Missouri, Tennessee, Arkansas, and Kentucky. The tremors were felt as far away as New York, Washington, D.C., and Charleston, South Carolina. The ground collapsed in some places, completely recarving parts of the landscape. The usually calm water of the Mississippi River overflowed its banks, forming new lakes and bayous, and reportedly even reversed direction. This collection of seismic events has come to be known as the New Madrid earthquake sequence. For a long time, geologists were puzzled by these tremors that occurred in the heartland of North America, so far from seismically active plate boundaries where earthquakes are normally expected. In the late 1970s, the technique of seismic profiling, or the creation of artificial earthquakes in order to probe underground geologic structures, was applied to the Mississippi Valley. Geologists found that layers of rock deep underground were riddled with cracks that formed part of a large fault system. This system is the remnant of an old geologic rift buried under hundreds of meters of sediment and is similar to a mid-oceanic ridge. This fault system may become reactivated because of changing tectonic forces and should be monitored carefully as a source of future earthquakes.

▲ The seismograph is a key device for studying earthquakes.

Section 2 Studying Earthquakes

Invention of the Seismograph

The first modern instruments for recording earthquakes came from Italy, where Nicholas Cirrillo used pendulums to detect ground motion in 1731. However, Cirrillo's device did not make a permanent record. The electromagnetic seismograph, developed by the Italian geologist Luigi Palmeri in 1856, recorded the local time of the earthquake. A major advance in seismography occurred in Japan in the 1890s when visiting British engineer John Milne and his colleagues, James Ewing and Thomas Gray, developed instruments that recorded earthquake ground shaking over time, producing the familiar record called a *seismogram.* The principle of modern seismography appeared when Milne incorporated the fact that the ground moves in three directions: vertically, horizontally east and west, and horizontally north and south. He constructed a device that combined records from three separate seismographs, one for each direction of ground motion. Within a few years of returning to England in 1895, Milne organized the first global network of seismic reporting stations and began the systematic analysis of earthquake patterns.

◀ Many major faults have characteristic surface features.

Section 3 Earthquakes and Society

Geologic Hazards and Damage Control

The geology of a local area can have a major impact on the amount of earthquake damage suffered. For example, during the Loma Prieta Earthquake of 1989, geologic conditions specific to the Bay Area greatly affected earthquake intensity. San Francisco's Marina district, located along the shores of a lagoon over 70 miles from the epicenter, experienced significant damage due to soil liquefaction. *Liquefaction* is the process by which loose, sandy soils with a high moisture content separate when shaken by an earthquake. These soils can flow much like quicksand, causing buildings to shift on their foundations or sink into the ground. Other areas along the Pacific Coast built on recently deposited soils produce so-called "sand volcanoes," where underlying sand is ejected onto the surface.

Generally speaking, it is best not to build on loose sands, unconsolidated soils, cliffs, permanent ice or snow, or on unstable slopes. These conditions can result in ground failure during an earthquake, leading to mass movement, landslides, or soil liquefaction, which are all common causes of structural failures. These effects underscore the importance of incorporating information from geologic surveys into land-use plans and building codes.

▲ Earthquake waves may cause more damage to some structures than to others.

Video Resources

Brain Food Video Quizzes

Brain Food Video Quizzes These videos contain game-show style quizzes that assess students' progress and motivate students to study the chapter material.

HRW Earth Science Video

This video introduces Earth science topics and includes a geology field trip. The video segment listed below complements this chapter.

Segment 5, Earthquakes This segment describes seismic waves and explains how seismologists calculate the location of an earthquake epicenter. (2 min)

NOVA Videos

The **NOVA** video below complements this chapter.

Earthquake This program shows how today's advanced technology helps geologists predict earthquakes. (60 min)

To order other **NOVA** videos related to this chapter, visit **go.hrw.com** and enter the keyword **HQ6EQKV.**

SciLinks

Developed and maintained by the National Science Teachers Association

SciLinks is maintained by the National Science Teachers Association to provide you and your students with interesting, up-to-date links that will enrich your classroom presentation of the chapter.

Visit www.scilinks.org and enter the SciLinks code for more information about the topic listed.

Topic: Earthquakes
SciLinks code: HQ60453

Topic: Seismic Waves
SciLinks code: HQ61371

Topic: Earthquake Measurement
SciLinks code: HQ60452

Topic: Tsunamis
SciLinks code: HQ61561

Topic: Earthquakes and Society
SciLinks code: HQ60455

Topic: Careers in Earth Science
SciLinks code: HQ60222

Chapter 12

Chapter Overview

This chapter describes the causes of earthquakes and the tectonic settings where earthquakes are most likely to happen. The chapter also examines how earthquakes are measured and how earthquakes affect humans.

Using the Figure — GENERAL

The Kobe Quake During the earthquake that caused the damage in the photo, over 5,000 people lost their lives. Part of the Nojima Fault, which ruptured during the Kobe earthquake, lies directly beneath the city. Stresses in the region come from the junction of the Pacific, Eurasian, and Philippine plates. Ask students to speculate what effect damage to power lines, water mains, rail lines, and highways would have during a disaster of this magnitude. (Accept all reasonable answers.) **LS Interpersonal**

PRE-READING ACTIVITY

FOLDNOTES

Have students exchange FoldNotes and check each other's notes on earthquakes. Encourage students to use a colored pen to add additional notes about earthquakes.

Chapter 12 Earthquakes

Sections

What You'll Learn

- What causes earthquakes
- How scientists measure earthquakes
- How earthquakes cause damage

Why It's Relevant

Understanding how, where, and why earthquakes happen can help scientists and engineers reduce earthquake damage and save lives. Studying earthquakes also helps scientists understand Earth's interior.

PRE-READING ACTIVITY

FOLDNOTES

Pyramid Before you read this chapter, create the FoldNote entitled "Pyramid" described in the Skills Handbook section of the Appendix. Label the sides of the pyramid with "How earthquakes happen," "How earthquakes are studied," and "How earthquakes affect society." As you read the chapter, write characteristics of each topic on the appropriate pyramid side.

▶ This expressway in Kobe, Japan, was toppled by the ground shaking of an earthquake that lasted 20 seconds and had a moment magnitude of 6.9.

Chapter Correlations — *National Science Education Standards*

PS 6 Interactions of energy and matter **(Sections 1–3)**

PS 6a Waves, including . . . seismic waves . . . have energy and can transfer energy when they interact with matter. **(Sections 1 and 2)**

ST 1 Abilities of Technological design **(Section 3)**

SPSP 5 Natural and human induced hazards **(Section 3)**

SPSP 5a Normal adjustments of earth may be hazardous for humans… As societies have grown, become stable … vulnerability to natural processes of change has increased. **(Section 3)**

SPSP 5c Some hazards, such as earthquakes, volcanic eruptions, and severe weather, are rapid and spectacular. **(Sections 1–3)**

UCP 3 Constancy, change, and measurement: Interactions within and among systems result in change. Changes vary in rate, scale, and pattern, including trends and cycles. **(Sections 2 and 3)**

Section 1 How and Where Earthquakes Happen

Earthquakes are one of the most destructive natural disasters. A single earthquake can kill thousands of people and cause millions of dollars in damage. **Earthquakes** are defined as movements of the ground that are caused by a sudden release of energy when rocks along a fault move. Earthquakes usually occur when rocks under stress suddenly shift along a fault. A *fault* is a break in a body of rock along which one block slides relative to another.

Why Earthquakes Happen

The rocks along both sides of a fault are commonly pressed together tightly. Although the rocks may be under stress, friction prevents them from moving past each other. In this immobile state, a fault is said to be *locked*. Parts of a fault remain locked until the stress becomes so great that the rocks suddenly grind past each other. This slippage causes the trembling and vibrations of an earthquake.

Elastic Rebound

Geologists think that earthquakes are a result of elastic rebound. **Elastic rebound** is the sudden return of elastically deformed rock to its undeformed shape. In this process, the rocks on each side of a fault are moving slowly. If the fault is locked, stress in the rocks increases. When the rocks are stressed past the point at which they can maintain their integrity, they fracture. The rocks then separate at their weakest point and *rebound*, or spring back to their original shape. This process is shown in **Figure 1.**

OBJECTIVES

- **Describe** elastic rebound.
- **Compare** body waves and surface waves.
- **Explain** how the structure of Earth's interior affects seismic waves.
- **Explain** why earthquakes generally occur at plate boundaries.

KEY TERMS

earthquake
elastic rebound
focus
epicenter
body wave
surface wave
P wave
S wave
shadow zone
fault zone

earthquake a movement or trembling of the ground that is caused by a sudden release of energy when rocks along a fault move

elastic rebound the sudden return of elastically deformed rock to its undeformed shape

Figure 1 ▶ Elastic Rebound

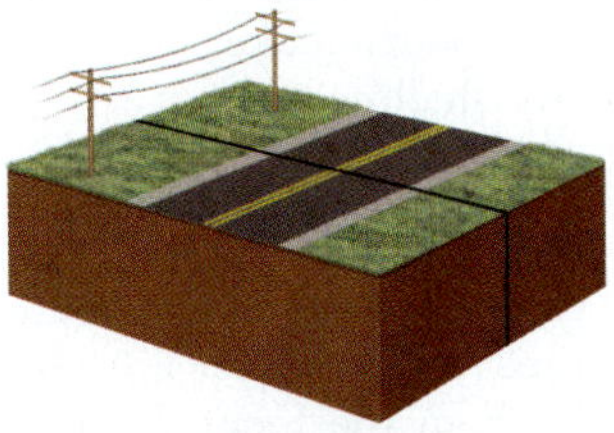

Two blocks of crust pressed against each other at a fault are under stress but do not move because friction holds them in place.

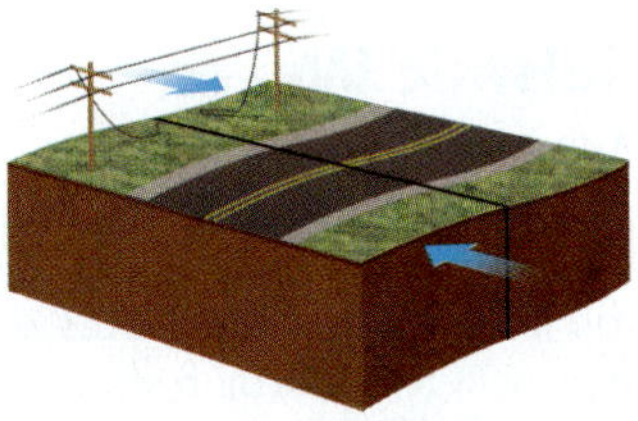

As stress builds up at the fault, the crust deforms.

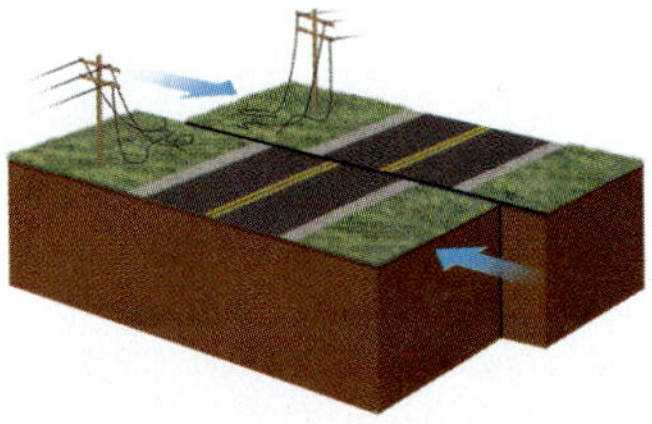

The rock fractures and then snaps back into its original shape, which causes an earthquake.

Section 1

Focus

Overview

This section explains the concept of elastic rebound and describes different types of seismic waves. It also explains how Earth's structure affects seismic waves and why earthquakes commonly occur at plate boundaries.

Bellringer

Have students describe what happens when a rock is thrown into a pond. Ask them to explain how this might be similar to an earthquake. (Sample answer: In each case, waves spread out from the source. However, the ripples you see on a pond's surface are two dimensional.) LS Visual

Motivate

Demonstration — GENERAL

Elastic Rebound Take two ends of a plastic ruler in each hand. Push down on the ends to form an arc. Then, stop pushing down and have students observe the rebounding motion of the ruler. Take a tongue depressor and bend it until it snaps. Explain that bending stores energy in the stick. That energy is released when the ruler rebounds or when the stick snaps. LS Visual

CHAPTER RESOURCES

Chapter Resource File

- **Directed Reading** BASIC
- **Inquiry Lab** Simulating Earthquakes GENERAL

Technology

Transparencies
- Bellringer

Student Edition on CD-ROM

One-Stop Planner CD-ROM
- Lesson Plan

Teach

Group Activity — GENERAL

Modeling Locked Faults Cover pairs of wood blocks with sandpaper. Use masking tape to hold the sandpaper in place. Have students hold one block in each hand. Tell them to push the sandpaper-covered edges of the blocks tightly together. While continuing to push the blocks together, have them try to slide the blocks in different directions. Ask someone to describe what happens. (The blocks lock and then move forward with a jerky motion.) Ask students how the model is like what happens during earthquakes. (Sample answer: The blocks are like rocks under stress. Friction prevents them from moving until suddenly they slip past each other and create the vibrations of an earthquake.)
LS **Kinesthetic**

SKILL BUILDER — GENERAL

Vocabulary Point out that *center* refers to the source of an event or phenomenon. Explain that the Greek prefix *epi-* is used to mean "above" or "on the surface." For example, the *epidermis* is the outer part of the skin. Similarly, the *epicenter* is the point located on the surface of Earth directly above the focus of the earthquake. The Greek prefix *hypo-* means "under." The focus of an earthquake is sometimes called the *hypocenter*. It is the point below the surface at which the earthquake originated.
LS **Verbal** English Language Learners

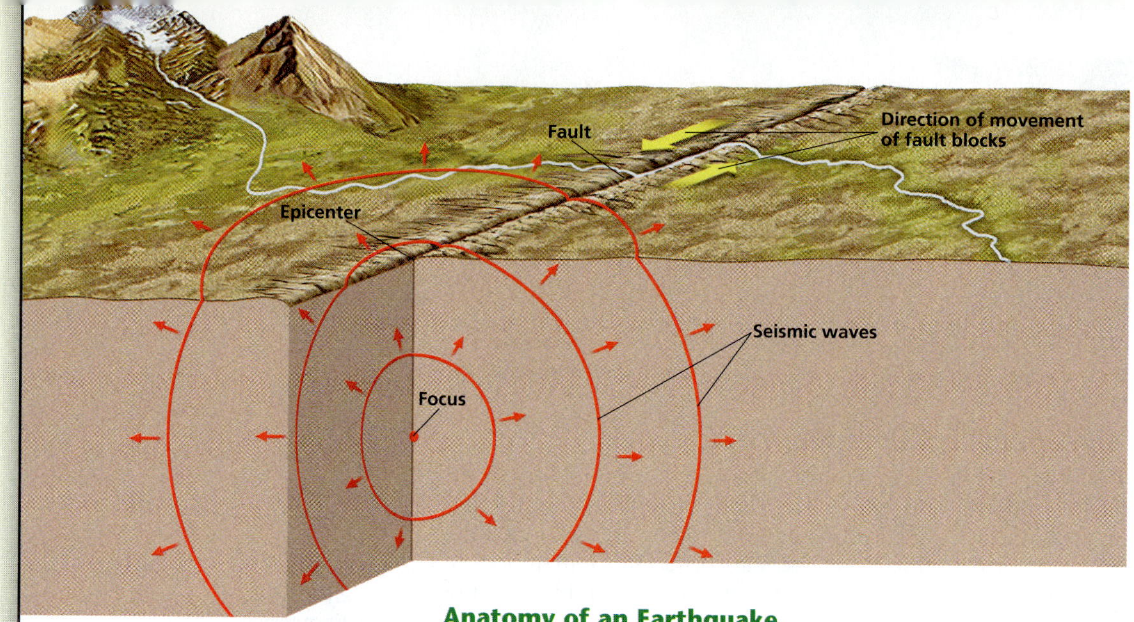

Figure 2 ▶ The epicenter of an earthquake is the point on the surface directly above the focus.

focus the location within Earth along a fault at which the first motion of an earthquake occurs

epicenter the point on Earth's surface directly above an earthquake's starting point, or focus

body wave in geology, a seismic wave that travels through the body of a medium

surface wave in geology, a seismic wave that travels along the surface of a medium and that has a stronger effect near the surface of the medium than it has in the interior

Anatomy of an Earthquake

The location within Earth along a fault at which the first motion of an earthquake occurs is called the **focus** (plural, *foci*). The point on Earth's surface directly above the focus is called the **epicenter** (EP i SENT uhr), as shown in **Figure 2.**

Although the focus depths of earthquakes vary, about 90% of continental earthquakes have a shallow focus. Earthquakes that have shallow foci take place within 70 km of Earth's surface. Earthquakes that have intermediate foci occur at depths between 70 km and 300 km. Earthquakes that have deep foci take place at depths between 300 km and 650 km. Earthquakes that have deep foci usually occur in subduction zones and occur farther from the plate boundary than shallower earthquakes do.

By the time the vibrations from an earthquake that has an intermediate or deep focus reach the surface, much of their energy has dissipated. For this reason, the earthquakes that usually cause the most damage usually have shallow foci.

Seismic Waves

As rocks along a fault slip into new positions, the rocks release energy in the form of vibrations called *seismic waves*. These waves travel outward in all directions from the focus through the surrounding rock. This wave action is similar to what happens when you drop a stone into a pool of still water and circular waves ripple outward from the center.

Earthquakes generally produce two main types of waves. **Body waves** are waves that travel through the body of a medium. **Surface waves** travel along the surface of a body rather than through the middle. Each type of wave travels at a different speed and causes different movements in Earth's crust.

CHAPTER RESOURCES

Chapter Resource File
- Internet Activity
  - Distribution Patterns GENERAL

Technology
- Transparencies
  - 62 Anatomy of an Earthquake (with worksheet)

Internet Activity — GENERAL

Distribution Patterns Ask students why shallow-foci earthquakes cause the greatest damage. (because their energy is released closer to Earth's surface) Invite interested students to research how earthquake foci are distributed relative to plate boundaries and to identify the most seismically active areas of North America. A worksheet designed to direct student research on this topic can be found in the **Chapter Resource File** booklet or by visiting **go.hrw.com** and entering the key word **HQ6EQKX.** LS **Visual**

Body Waves

Body waves can be placed into two main categories: P waves and S waves. **P waves,** also called *primary waves* or *compression waves*, are the fastest seismic waves and are always the first waves of an earthquake to be detected. P waves cause particles of rock to move in a back-and-forth direction that is parallel to the direction in which the waves are traveling, as shown in **Figure 3.** P waves can move through solids, liquids, and gases. The more rigid the material is, the faster the P waves travel through it.

S waves, also called *secondary waves* or *shear waves*, are the second-fastest seismic waves and arrive at detection sites after P waves. S waves cause particles of rock to move in a side-to-side direction that is perpendicular to the direction in which the waves are traveling. Unlike P waves, however, S waves can travel through only solid material.

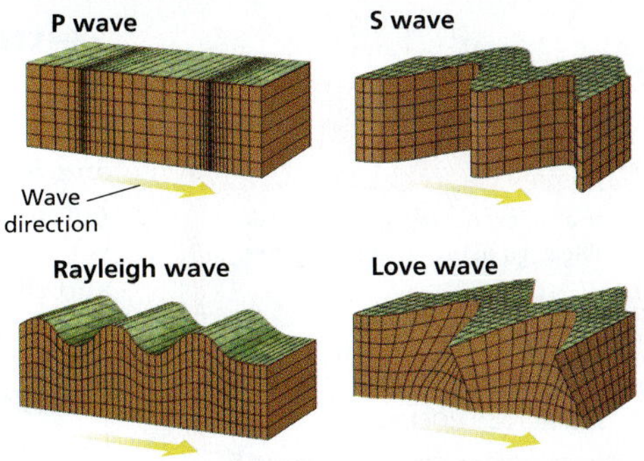

Figure 3 ▶ The different types of seismic waves cause different rock movements, which have different effects on Earth's crust.

Surface Waves

Surface waves form from motion along a shallow fault or from the conversion of energy when P waves and S waves reach Earth's surface. Although surface waves are the slowest-moving waves, they may cause the greatest damage during an earthquake. The two types of surface waves are Love waves and Rayleigh waves. *Love waves* cause rock to move side-to-side and perpendicular to the direction in which the waves are traveling. *Rayleigh waves* cause the ground to move with an elliptical, rolling motion.

Reading Check **Describe the two types of surface waves. (See the Appendix for answers to Reading Checks.)**

P wave a primary wave, or compression wave; a seismic wave that causes particles of rock to move in a back-and-forth direction parallel to the direction in which the wave is traveling; P waves are the fastest seismic waves and can travel through solids, liquids, and gases

S wave a secondary wave, or shear wave; a seismic wave that causes particles of rock to move in a side-to-side direction perpendicular to the direction in which the wave is traveling; S waves are the second-fastest seismic waves and can travel only through solids

Connection to ENGINEERING

Seismic Reflection Surveying

Many surveying companies have discovered the usefulness of seismic waves in mapping underground features. These features can be used to identify possible mineral deposits and oil or natural gas reservoirs.

In seismic surveying, a seismic shock is generated by using explosives, air guns, or a mechanical thumper. The seismic waves produced by the shock travel through the ground and reflect off bedding planes or other features below the surface. The reflected waves travel back to the surface, where they are recorded by an array of geophones. *Geophones* are instruments that convert the motion of a seismic wave into an electrical signal.

The geophones are set up in a straight line to collect data that are used to construct a two-dimensional profile of the underground layers. Scientists can also arrange the geophones in more-complex patterns to create three-dimensional images of the layers. Because each underground layer reflects waves at a different time, scientists can plot all of the data to construct an accurate picture of the underground layers.

Seismic reflection can be used to study phenomena at a variety of scales. It may be used to study just the top few tens of meters of soil and rock, or it may be used to study the structure of the deep crust. In particular, this technology has been adapted for use by oil and natural gas exploration companies to locate oil and gas reservoirs.

TECHNOLOGY CONNECTION — GENERAL

Seismic Profiling Oceanographers use a technique called *seismic profiling* to build a picture of the sea bottom that shows the thickness and geometry of layers of sediment. Oil companies looking for offshore oil pioneered the method, in which a compressed-air gun directs a sound pulse at the sea floor and an array of listening devices called *hydrophones* record the returning sound waves. These hydrophones are similar to the geophones used to map features of the underground surface. Invite interested students to create models or diagrams of how this technique works. **LS Kinesthetic/Visual**

Discussion — BASIC

Surface Waves Ask students to describe the effects of the two different types of surface waves on Earth's surface. (Sample answer: Rayleigh waves cause the ground to move with a rolling motion like ocean waves. Love waves cause the ground to move horizontally and perpendicular to the direction the wave is traveling.) Next, ask students why surface waves may cause the most extensive damage even though they travel much slower than body waves do. (because their energy is concentrated at the surface) **LS Verbal**

Answer to Reading Check

Rayleigh waves cause the ground to move in an elliptical, rolling motion. Love waves cause rock to move side-to-side and perpendicular to the direction the waves are traveling.

Teach, *continued*

HISTORY CONNECTION GENERAL

Moho Discontinuity The Croatian scientist Andrija Mohorovičić discovered the boundary layer between Earth's mantle and crust by analyzing the change in speed of seismic waves as they passed through different materials. The boundary layer was named the "Mohorovičić Discontinuity," or the "Moho," in his honor. He also contributed to our understanding of earthquakes' effects on buildings, location of earthquake epicenters, and Earth models. Invite students to investigate the work of this scientist, who also increased our knowledge of the energy of winds, and report back to the class. **LS Verbal**

Answers to Reading Check

The speed of seismic waves changes as they pass through different layers of Earth.

INCLUSION *Strategies*

- ***Learning Disabled***
- ***Developmentally Delayed***
- ***Hearing Impaired***

Use the following game to help students review material in this section: Divide the class into groups of three or four students. Assign each team one term or concept from the section. Have each team plan to act out the assigned term silently. When they are ready, have them write three terms on the board for the class to guess—only one of which is the correct answer. After a team acts out its term, have the class vote on which of the three terms the team was depicting. **LS Kinesthetic**

SCILINKS® NSTA Developed and maintained by the National Science Teachers Association

For a variety of links related to this subject, go to www.scilinks.org

Topic: Earthquakes
SciLinks code: HQ60453
Topic: Seismic Waves
SciLinks code: HQ61371

shadow zone an area on Earth's surface where no direct seismic waves from a particular earthquake can be detected

Seismic Waves and Earth's Interior

Seismic waves are useful to scientists in exploring Earth's interior. The composition of the material through which P waves and S waves travel affects the speed and direction of the waves. For example, P waves travel fastest through materials that are very rigid and are not easily compressed. By studying the speed and direction of seismic waves, scientists can learn more about the makeup and structure of Earth's interior.

Earth's Internal Layers

In 1909, Andrija Mohorovičić (MOH hoh ROH vuh CHICH), a Croatian scientist, discovered that the speed of seismic waves increases abruptly at about 30 km beneath the surface of continents. This increase in speed takes place because the mantle is denser than the crust. The location at which the speed of the waves increases marks the boundary between the crust and the mantle. The depth of this boundary varies from about 10 km below the oceans to about 30 km below continents. By studying the speed of seismic waves, scientists have been able to locate boundaries between other internal layers of Earth. The three main compositional layers of Earth are the *crust,* the *mantle,* and the *core*. Earth is also composed of five mechanical layers—the *lithosphere,* the *asthenosphere,* the *mesosphere,* the *outer core,* and the *inner core*.

Shadow Zones

Recordings of seismic waves around the world reveal shadow zones. **Shadow zones** are locations on Earth's surface where no body waves from a particular earthquake can be detected. Shadow zones exist because the materials that make up Earth's interior are not uniform in rigidity. When seismic waves travel through materials of differing rigidities, the speed of the waves changes. The waves will also bend and change direction as they pass through different materials.

As shown in **Figure 4,** a large S-wave shadow zone covers the side of Earth that is opposite an earthquake. S waves do not reach the S-wave shadow zone because they cannot pass through the liquid outer core. Although P waves can travel through all of the layers, the speed and direction of the waves change as the waves pass through each layer. The waves bend in such a way that a P-wave shadow zone forms.

Reading Check **What causes the speed of a seismic wave to change? (See the Appendix for answers to Reading Checks.)**

Figure 4 ▶ P waves and S waves behave differently as they pass through different structural layers of Earth.

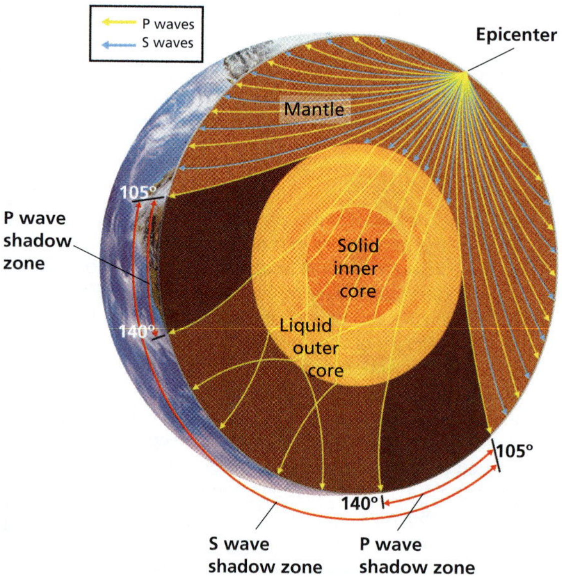

BRAIN FOOD

Probing the Depths If all of the layers of Earth's interior were the same, P and S waves would travel through Earth at constant speed and in a straight line. But the speed and direction of the waves varies depending on the composition of the layers waves pass through. Scientists measure the velocity at which seismic waves move through different Earth materials by using the technique of ultrasonic interferometry. Within an interferometer, ultrasonic waves of known frequency (*f*) are produced by a quartz plate and are reflected by a metallic plate. Standing waves have formed if the separation between these plates is exactly a whole multiple of the wavelength. By using the wavelength, the velocity can be obtained by using the following relationship: *velocity = wavelength × frequency.*

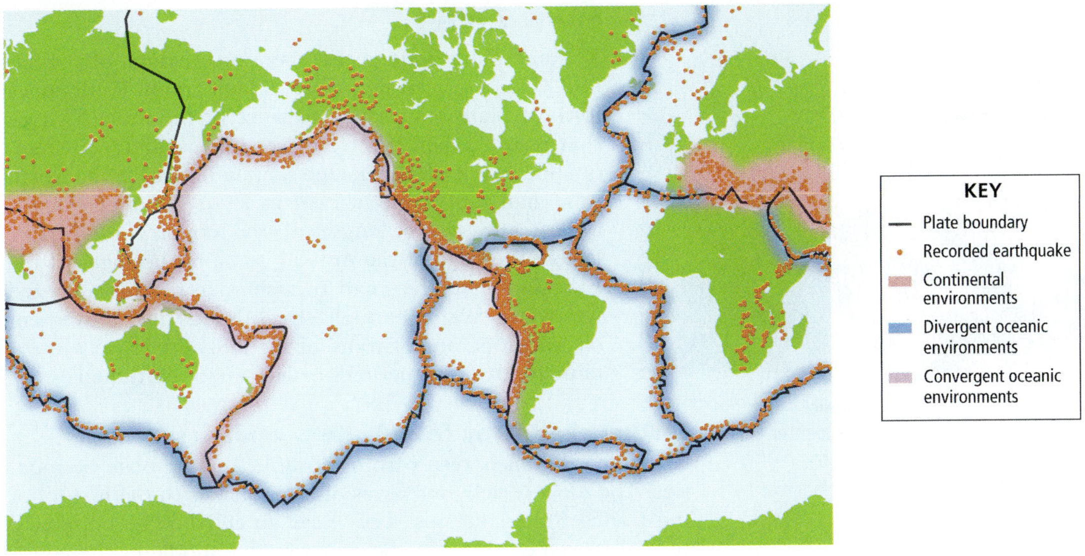

Figure 5 ▶ Earthquakes are the result of tectonic stresses in Earth's crust and occur in three main tectonic settings—mid-ocean ridges, subduction zones, and continental collisions.

Earthquakes and Plate Tectonics

Earthquakes are the result of stresses in Earth's lithosphere. Most earthquakes occur in three main tectonic environments, as shown in **Figure 5.** These settings are generally located at or near tectonic plate boundaries, where stress on the rock is greatest.

Convergent Oceanic Environments

At convergent plate boundaries, plates move toward each other and collide. The plate that is denser subducts, or sinks into the asthenosphere below the other plate. As the plates move, the overriding plate scrapes across the top of the subducting plate, and earthquakes occur. Convergent oceanic boundaries can occur between two oceanic plates or between one oceanic plate and one continental plate.

Divergent Oceanic Environments

At the divergent plate boundaries that make up the mid-ocean ridges, plates are moving away from each other. Earthquakes occur along mid-ocean ridges because oceanic lithosphere is pulling away from both sides of each ridge. This spreading motion causes earthquakes along the ocean ridges.

Continental Environments

Earthquakes also occur at locations where two continental plates converge, diverge, or move horizontally in opposite directions. As the continental plates interact, the rock surrounding the boundary experiences stress. The stress may cause mountains to form and also causes frequent earthquakes.

Graphic Organizer **Spider Map**

Create the Graphic Organizer entitled "Spider Map" described in the Skills Handbook section of the Appendix. Label the circle "Major earthquake zones." Create a leg for each type of earthquake zone. Then, fill in the map with details about each type of earthquake zone.

Close

Reteaching — BASIC

Earthquake Maps Have students create a flip-book that illustrates the process of elastic rebound. The book should show how the fault blocks change before the earthquake and how the blocks respond to the earthquake. Books should include at least 10 drawings. **LS Visual**

Quiz — GENERAL

1. Why do earthquakes occur along plate boundaries? (As plates separate, collide, or grind past each other, they jerk suddenly as they move.)
2. What causes seismic waves to change direction within Earth? (Differences in the composition of the layers cause the waves to bend and travel in a different direction or to be reflected.)

CHAPTER RESOURCES

Technology

Transparencies
- 63 Seismic Waves and Earth's Interior (with worksheet)
- 64 Earthquakes and Tectonic Plate Boundaries (with worksheet)

MISCONCEPTION ALERT

Earthquake Locations Students may think that earthquakes occur only at faults or at places where tectonic plates are colliding or subducting. However, volcanically active areas are also the sites of many shallow, small-magnitude earthquakes that are associated with volcanic eruptions and the movement of magma. As magma moves, surrounding rocks crack as a result of changes in temperature and pressure. This cracking is felt as earthquakes.

Graphic Organizer GENERAL

You may want to have students work in groups to create this Spider Map. Have one student draw the map and fill in information provided by other students from the group.

Close, *continued*

Alternative Assessment — GENERAL

Earthquake Patterns Give students a map of recent earthquake activity in the United States. Have students look for patterns in the data that indicate where the most earthquakes occurred or where the largest-magnitude earthquakes occurred. Have them explain the patterns they identify in terms of tectonic activity or the location of major fault zones. **LS Visual**

Answers to Section Review

1. the sudden return of deformed rock back to its original shape
2. The focus is located within Earth. The epicenter is a point on the surface above the focus.
3. Body waves travel through Earth's interior, while surface waves travel only along the outer surface.
4. Seismic waves behave differently as they pass through layers composed of different materials. These differences help create a picture of Earth's interior.
5. When seismic waves reach boundaries between rock layers, their speed changes. They are reflected or change direction as they pass through layers of different composition.
6. Earthquakes result from stress within the lithosphere. They commonly occur at plate boundaries because stress on rock is greatest near plate boundaries, which may be faulted, and thus weaker than plate interiors.
7. A fault zone is a region of numerous, closely spaced faults. When enough stress builds in the zone, movement occurs along one or more of the faults in the fault zone.
8. within about 70 km of Earth's surface
9. If the seismologic station records no S waves, you could conclude that the location of the epicenter was on the opposite side of Earth, because the S waves could not pass through the liquid outer core.
10. Because the center of Brazil is far away from any plate boundaries, you might conclude that a major fault zone lies deep below the region.
11. *Earthquakes* produce two types of *seismic waves*: *body waves*, which consist of *P waves* and *S waves*; and *surface waves*, which consist of *Rayleigh waves* and *Love waves*.

Figure 6 ▶ A series of parallel transform faults, seen in the center of this photo, make up part of the North Anatolian fault zone in Turkey.

fault zone a region of numerous, closely spaced faults

Fault Zones

At some plate boundaries, there are regions of numerous, closely spaced faults called **fault zones.** Fault zones form at plate boundaries because of the intense stress that results when the plates separate, collide, subduct, or slide past each other. One such fault zone is the North Anatolian fault zone, shown in **Figure 6,** that extends almost the entire length of the country of Turkey. Where the edge of the Arabian plate pushes against the Eurasian plate, the small Turkish microplate is squeezed westward. When enough stress builds up, movement occurs along one or more of the individual faults in the fault zone and sometimes causes major earthquakes.

Earthquakes Away from Plate Boundaries

Not all earthquakes result from movement along plate boundaries. The most widely felt series of earthquakes in the history of the United States did not occur near an active plate boundary. Instead, these earthquakes occurred in the middle of the continent, near New Madrid, Missouri, in 1811 and 1812. The vibrations from the earthquakes that rocked New Madrid were so strong that they caused damage as far away as South Carolina.

It was not until the late 1970s that studies of the Mississippi River region revealed an ancient fault zone deep within Earth's crust. This zone is thought to be part of a major fault zone in the North American plate. Scientists have determined that the fault formed at least 600 million years ago and that it was later buried under many layers of sediment and rock.

Section 1 Review

1. **Describe** elastic rebound.
2. **Explain** the difference between the epicenter and the focus of an earthquake.
3. **Compare** body waves and surface waves.
4. **Explain** how seismic waves help scientists learn about Earth's interior.
5. **Explain** how the structure of Earth's interior affects seismic wave speed and direction.
6. **Explain** why earthquakes generally take place at plate boundaries.
7. **Describe** a fault zone, and explain how earthquakes occur along fault zones.

CRITICAL THINKING

8. **Applying Ideas** In earthquakes that cause the most damage, at what depth would movement along a fault most likely occur?
9. **Identifying Patterns** If a seismologic station measures P waves but no S waves from an earthquake, what can you conclude about the earthquake's location?
10. **Making Inferences** If an earthquake occurs in the center of Brazil, what can you infer about the geology of that area?

CONCEPT MAPPING

11. Use the following terms to create a concept map: *earthquake, seismic wave, body wave, surface wave, P wave, S wave, Rayleigh wave,* and *Love wave.*

CHAPTER RESOURCES

Chapter Resource File
- Section Quiz GENERAL

Workbooks
- Study Guide (also in Spanish)

Section 2 Studying Earthquakes

The study of earthquakes and seismic waves is called *seismology*. Many scientists study earthquakes because earthquakes are the best tool Earth scientists have for investigating Earth's internal structure and dynamics. These scientists have developed special sensing equipment to record, locate, and measure earthquakes.

OBJECTIVES

- **Describe** the instrument used to measure and record earthquakes.
- **Summarize** the method scientists use to locate an epicenter.
- **Describe** the scales used to measure the magnitude and intensity of earthquakes.

KEY TERMS

seismograph
seismogram
magnitude
intensity

Recording Earthquakes

Vibrations in the ground can be detected and recorded by using an instrument called a **seismograph** (SIEZ MUH graf), such as the one shown in **Figure 1.** A modern three-component seismograph consists of three sensing devices. One device records the vertical motion of the ground. The other two devices record horizontal motion—one for east-west motion and the other for north-south motion. Seismographs record motion by tracing wave-shaped lines on paper or by translating the motion into electronic signals. The electronic signals can be recorded on magnetic tape or can be loaded directly into a computer that analyzes seismic waves. A tracing of earthquake motion that is recorded by a seismograph is called a **seismogram.**

Because they are the fastest-moving seismic waves, P waves are the first waves to be recorded by a seismograph. S waves travel much slower than P waves. Therefore, S waves are the second waves to be recorded by a seismograph. Surface waves, or Rayleigh and Love waves, are the slowest-moving waves. Thus, Rayleigh and Love waves are the last waves to be recorded by a seismograph.

seismograph an instrument that records vibrations in the ground

seismogram a tracing of earthquake motion that is recorded by a seismograph

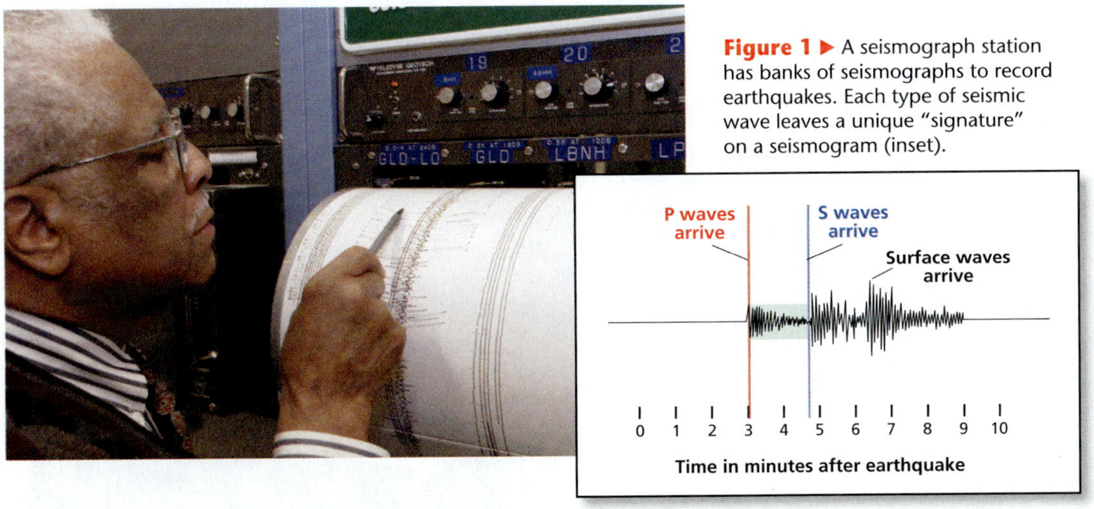

Figure 1 ▶ A seismograph station has banks of seismographs to record earthquakes. Each type of seismic wave leaves a unique "signature" on a seismogram (inset).

Section 2

Focus

Overview

This section describes the tools scientists use to measure earthquakes. The section explains how the source of an earthquake is identified and how its strength and destructiveness are measured.

Bellringer

Ask students to write down what information they would want to know if they were measuring or describing a major earthquake. (Accept all reasonable answers.) LS Verbal

Motivate

Group Activity — GENERAL

Model Seismograph Divide the class into groups of three. Tell each group to tape a sheet of paper around the cylinder of a rolling pin and secure it with a rubber band. Next, have them cut two notches in the sides of a shoebox to hold the ends of the roller and insert the drum into the notches. Tape the box securely to a table. Have one student hold a marking pen steady against the paper drum while the second turns the drum. The third student should shake the end of the table to simulate an earthquake. Have them compare the tracings from several tests. LS Kinesthetic

CHAPTER RESOURCES

Chapter Resource File
- Directed Reading BASIC

Technology

Transparencies
- Bellringer

Student Edition on CD-ROM

One-Stop Planner CD-ROM
- Lesson Plan

Teach

Quick LAB

Skills Acquired

- Making Models
- Observing
- Analyzing

Teacher's Notes You could substitute a brick or a heavy paperweight for the ball. Remind students to move the paper slowly at a steady rate of speed. You could have students compare the effects of waves traveling through other materials by filling the bag with gravel or water (make sure the bag is tightly sealed), or stuffing the box with aluminum foil or foam peanuts.

Answers

1. a graph of motion vs. time
2. different types of materials that make up Earth's crust
3. The sand vibrated less than the paper. The sand is more rigid than the crumpled paper and therefore vibrated less when the ball struck it.
4. they might deflect or change the speed of the waves
5. The farther the waves have to travel, the more energy may dissipate.

CHAPTER RESOURCES

Chapter Resource File

- Datasheet for Quick Lab GENERAL

SCILINKS. NSTA Developed and maintained by the National Science Teachers Association

For a variety of links related to this subject, go to www.scilinks.org

Topic: Earthquake Measurement
SciLinks code: HQ60452

Locating an Earthquake

To determine the distance to an epicenter, scientists analyze the arrival times of the P waves and the S waves. The longer the lag time between the arrival of the P waves and the arrival of the S waves is, the farther away the earthquake occurred. To determine how far an earthquake is from a given seismograph station, scientists consult a lag-time graph. This graph translates the difference in arrival times of the P waves and S waves into distance from the epicenter to each station. The start time of the earthquake can also be determined by using this graph.

To locate the epicenter of the earthquake, scientists use computers to perform complex triangulations based on information from several seismograph stations. Before computers were widely available, scientists performed this calculation in a simpler and more imprecise way. On a map, they drew circles around at least three seismograph stations that recorded vibrations from the earthquake. The radius of each circle was equal to the distance from that station to the earthquake's epicenter. The point at which all of the circles intersected indicated the location of the epicenter of the earthquake.

Quick LAB 20 min

Seismographic Record

Procedure

1. Line a **shoe box** with a **plastic bag**. Fill the box to the rim with **sand**. Put on the lid.
2. Mark an X near the center of the lid.
3. Fasten a **felt-tip pen** to the lid of the box with a tight **rubber band** so that the pen extends slightly beyond the edge of the box.
4. Have a partner hold a **pad of paper** so that the paper touches the pen.
5. Hold a **ball** over the X at a height of 30 cm. As your partner slowly moves the paper horizontally past the pen, drop the ball on the X.
6. Label the resulting line with the type of material in the box.
7. Replace about 2/3 of the sand with crumpled **newspaper**. Put on the lid, and fasten the pen to the lid with the rubber band.
8. Repeat steps 4–6.

Analysis

1. What do the lines on the paper represent?
2. What do the sand and newspaper represent?
3. Compare the lines made in steps 4–6 with those made in step 8. Which material vibrated more when the ball was dropped on it? Explain why one material might vibrate more than the other.
4. How might different types of crustal material affect seismic waves that pass through it?
5. How might the distance of the epicenter of an earthquake from a seismograph affect the reading of a seismograph?

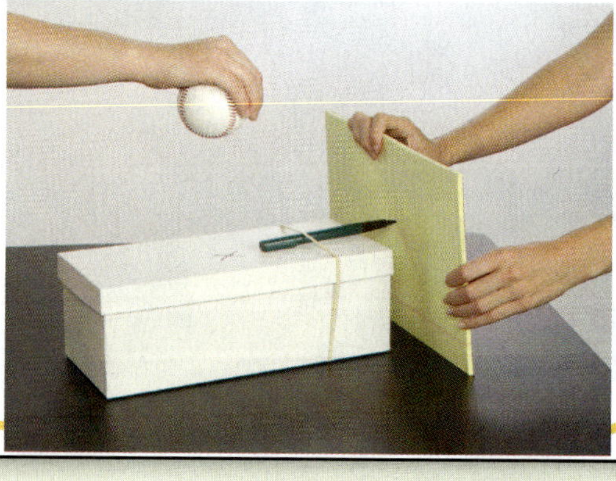

READING SKILL BUILDER BASIC

Reading Organizer As students read this section, encourage them to outline the way scientists record and identify the source and measure the strength and intensity of an earthquake. They can use the section subheads as major outline heads and record details under each head. Later, students can use these outlines as a study guide for assessments. LS **Verbal** English Language Learners

INCLUSION Strategies

- *Learning Disabled*
- *Behavior Control Issues*
- *Attention Deficit Disorder*

Ask volunteers to use text circles and lines to diagram groups of information on the board, such as:

1. Earthquake intensity: magnitude, extent of affected area from epicenter, local geology, and duration
2. Magnitude: Richter Scale, moment magnitude scale, used for most of last century, preferred by scientists now

Figure 2 ▶ The 2005 earthquake in northern Pakistan had a moment magnitude of 7.6, killed more than 75,000 people, and left about 3 million homeless.

Earthquake Measurement

Scientists who study earthquakes are interested in the amount of energy released by an earthquake. Scientists also study the amount of damage done by the earthquake. These properties are studied by measuring magnitude and intensity.

Magnitude

The measure of the strength of an earthquake is called **magnitude.** Magnitude is determined by measuring the amount of ground motion caused by an earthquake. Seismologists express magnitude by using a magnitude scale, such as the Richter scale or the moment magnitude scale.

The *Richter scale* measures the ground motion from an earthquake to find the earthquake's strength. While the Richter scale was widely used for most of the 20th century, scientists now prefer the moment magnitude scale. *Moment magnitude* is a measurement of earthquake strength based on the size of the area of the fault that moves, the average distance that the fault blocks move, and the rigidity of the rocks in the fault zone. Although the moment magnitude and the Richter scales provide similar values for small earthquakes, the moment magnitude scale is more accurate for large earthquakes.

The moment magnitude of an earthquake is expressed by a number. The larger the number, the stronger the earthquake. The largest earthquake that has been recorded (in Chile) registered a moment magnitude of 9.5. The earthquake in Pakistan in 2005 that caused the damage shown in **Figure 2** had a moment magnitude of 7.6. Earthquakes that have moment magnitudes of less than 2.5 usually are not felt by people.

Reading Check **What is the difference between the Richter scale and the moment magnitude scale? (See the Appendix for answers to Reading Checks.)**

magnitude a measure of the strength of an earthquake

MATHPRACTICE

Magnitudes On both the moment magnitude scale and the Richter scale, the energy of an earthquake increases by a factor of about 30 for each increment on the scale. Thus, a magnitude 4 earthquake releases 30 times as much energy as a magnitude 3 earthquake does. How much more energy does a magnitude 6 earthquake release than a magnitude 3 earthquake does?

Answer to Reading Check

Moment magnitude is more accurate for larger earthquakes than the Richter scale is. Moment magnitude is directly related to rock properties and so is more closely related to the cause of the earthquake than the Richter scale is.

MATHPRACTICE

Answer

$30 \times 30 \times 30 = 27,000$ times as much energy

Close

Reteaching — BASIC

Earthquake Posters Have students draw a series of pictures to illustrate the increasing levels of damage represented by earthquakes of different intensities. For example, they could show objects swaying, bookcases toppling, cracks appearing in structures, and so on. Tell them to label their earthquake scale with phrases that describe the effects at each level. **LS Visual**

Quiz — GENERAL

1. Describe the sensing equipment used to record earthquakes. (Seismographs use sensing devices to record the vibrations of the ground.)
2. Compare the Richter scale with the Mercalli scale in describing an earthquake. (The Richter scale gives a numerical value to the amount of ground motion. The Mercalli scale describes the earthquake's intensity, including its effects on an area.) **LS Verbal**

SKILL BUILDER — ADVANCED

Graphing Have students graph the seismic data for these earthquakes: 8.1: New Madrid, MO (1811); 2.2: Pyrenees (2004); 9.5: Chile (1960); 6.9: Loma Prieta, CA (1989); 4.5: Honshu, Japan (2004); 7.9: Gulf of Alaska (1987); 1.5: Straits of Gibraltar (2004); 8.6: Anchorage, AK (1960); 5.8: Jabalpur, India (1997). Place the magnitude on the horizontal axis and the approximate energy level represented by the magnitude of each event on the vertical axis. Have students compare the magnitude of New Madrid with one or more recent seismic events. **LS Logical/Visual**

BRAIN FOOD

Largest Earthquakes The Great Alaskan Earthquake of 1964 was the largest earthquake in U.S. history at estimated magnitude 9.2. The quake destroyed buildings and railroads and caused land to drop by more than 3 m. The largest earthquake of the last century was the Chilean earthquake of 1960 at estimated magnitude 9.5. It unleashed a tsunami that crossed the Pacific and devastated parts of Hawaii and Japan.

Close, *continued*

Alternative Assessment — ADVANCED

Sizing Up Quakes Cover a workspace with a drop cloth. Fill a box lid with popped corn or salt. Have pairs of students drop the lid from several heights and compare the amount of disturbance produced by the model earthquakes. Have students devise a scale to measure the magnitude of their quakes. Then, have them compare their scale with the existing measurement scales. **LS Kinesthetic/ Verbal**

Answers to Section Review

1. A seismograph consists of three sensing devices that record ground motion: one records vertical motion, one records horizontal east-west motion, and the third records horizontal north-south motion.
2. A seismograph is the sensing device. The seismogram is the tracing of earthquake motion made by the seismograph.
3. Scientists drew circles on a map around three seismograph stations. They used the lag time between the arrival of the P waves and the S waves to determine the distance between each site and the epicenter. The epicenter was located where the three circles intersected.
4. The Richter scale determines earthquake strength based on ground motion and adjusts for distance. Moment magnitude measures earthquake strength based on the size of the area of the fault that moves, the average distance the fault blocks move, and the rigidity of the rocks in the fault zone.
5. Magnitude is the measure of an earthquake's strength. Intensity a measure of the effects of an earthquake on a particular area.
6. because the two circles representing the distance of two locations from the epicenter would intersect at two points, while three circles would intersect at only one point
7. Intensity depends on the local geology, the distance from the epicenter, and the earthquake's duration, as well as magnitude.
8. *Seismographs* record *P waves* and *S waves* to make *seismograms,* which can be used to determine an earthquake's *magnitude, intensity,* and location of the *epicenter.*

Table 1 ▼

Modified Mercalli Intensity Scale

Intensity	Description
I	is not felt except by very few under especially favorable conditions
II	is felt by only few people at rest; delicately suspended items may swing
III	is felt by most people indoors; vibration is similar to the passing of a large truck
IV	is felt by many people; dishes and windows rattle; sensation is similar to a building being struck
V	is felt by nearly everyone; some objects are broken; and unstable objects are overturned
VI	is felt by all people; some heavy objects are moved; causes very slight damage to structures
VII	causes slight to moderate damage to ordinary buildings; some chimneys are broken
VIII	causes considerable damage (including partial collapse) to ordinary buildings
IX	causes considerable damage (including partial collapse) to earthquake-resistant buildings
X	destroys some to most structures, including foundations; rails are bent
XI	causes few structures, if any, to remain standing; bridges are destroyed and rails are bent
XII	causes total destruction; distorts lines of sight; objects are thrown into the air

Intensity

intensity in Earth science, the amount of damage caused by an earthquake

Before the development of magnitude scales, the size of an earthquake was determined based on the earthquake's effects. A measure of the effects of an earthquake is the earthquake's **intensity.** The modified *Mercalli scale,* shown in **Table 1,** expresses intensity in Roman numerals from I to XII and provides a description of the effects of each earthquake intensity. The highest-intensity earthquake is designated by Roman numeral XII and is described as total destruction. The intensity of an earthquake depends on the earthquake's magnitude, the distance between the epicenter and the affected area, the local geology, the earthquake's duration, and human infrastructure.

Section 2 Review

1. **Describe** the instrument that is used to record seismic waves.
2. **Compare** a seismograph and a seismogram.
3. **Summarize** the method that scientists used to identify the location of an earthquake before computers became widely used.
4. **Describe** the scales that scientists use to measure the magnitude of an earthquake.
5. **Explain** the difference between magnitude and intensity of an earthquake.

CRITICAL THINKING

6. **Analyzing Methods** Explain why it would be difficult for scientists to locate the epicenter of an earthquake if they have seismic wave information from only two locations.
7. **Evaluating Data** Explain why an earthquake with a moderate magnitude might have a high intensity?

CONCEPT MAPPING

8. Use the following terms to create a concept map: *seismograph, seismogram, epicenter, P wave, S wave, magnitude,* and *intensity.*

CHAPTER RESOURCES

Chapter Resource File
- Section Quiz GENERAL

Workbooks
- Study Guide (also in Spanish)

Section 3 Earthquakes and Society

Movement of the ground during an earthquake seldom directly causes many deaths or injuries. Instead, most injuries result from the collapse of buildings and other structures or from falling objects and flying glass. Other dangers include landslides, fires, explosions caused by broken electric and gas lines, and floodwaters released from collapsing dams.

OBJECTIVES

- **Discuss** the relationship between tsunamis and earthquakes.
- **Describe** two possible effects of a major earthquake on buildings.
- **List** three safety techniques to prevent injury caused by earthquake activity.
- **Identify** four methods scientists use to forecast earthquake risks.

KEY TERMS

tsunami
seismic gap

Tsunamis

An earthquake whose epicenter is on the ocean floor may cause a giant ocean wave called a **tsunami** (tsoo NAH mee), which may cause serious destruction if it crashes into land. A tsunami may begin to form when a sudden drop or rise in the ocean floor occurs because of faulting associated with undersea earthquakes. The drop or rise of the ocean floor causes a large mass of sea water to also drop or rise. This mass of water moves up and down as it adjusts to the change in sea level. This movement sets into motion a series of long, low waves that increase in height as they near the shore. These waves are tsunamis.

tsunami a giant ocean wave that forms after a volcanic eruption, submarine earthquake, or landslide

On December 26, 2004, a tsunami was triggered by an earthquake that struck beneath the Indian Ocean. The tsunami devastated coastal areas of Indonesia, India, Sri Lanka, and Thailand, leaving more than 200,000 people dead or missing.

Destruction to Buildings and Property

Most buildings are not designed to withstand the swaying motion caused by earthquakes. Buildings whose walls are weak may collapse completely. Very tall buildings may sway so violently that they tip over and fall onto lower neighboring structures, as shown in **Figure 1.**

The type of ground beneath a building can affect the way in which the building responds to seismic waves. A building constructed on loose soil and rock is much more likely to be damaged during an earthquake than a building constructed on solid ground is. During an earthquake, the loose soil and rock can vibrate like jelly. Buildings constructed on top of this kind of ground experience exaggerated motion and sway violently.

Figure 1 ▶ Rescue workers surround a building that collapsed in Taipei, Taiwan, during the earthquake of 1999.

Section 3

Focus

Overview

This section describes the damage caused by earthquakes. The section also explains earthquake safety procedures and how scientists are attempting to forecast future earthquake events.

Bellringer

Have students think about disaster movies, news footage, or documentaries that include scenes of earthquakes. Have them write a brief description of some earthquake effects. (Answers may vary but can be used to identify preconceptions about earthquake damage.) LS **Verbal**

Motivate

Demonstration — GENERAL

Giant Wave Model how a tsunami forms. Cut a round hole about 10 cm in diameter in the bottom of a foil cake pan. Cover the hole with plastic wrap and tape it so that it is watertight. Fill the pan with water. Tap gently in the bottom of the plastic wrap. Then, increase the strength of the tap. Ask students to describe what happens. (Sample answer: The mass of water rises and falls as it adjusts to the change in water level produced by the force from below.) LS **Visual**

CHAPTER RESOURCES

Chapter Resource File

- **Directed Reading** BASIC
- **Skills Practice Lab** Earthquakes and Soil GENERAL

Technology

Transparencies
- Bellringer

Student Edition on CD-ROM

One-Stop Planner CD-ROM
- Lesson Plan

Teach

Using the Figure — BASIC

Direct students' attention to the photo showing the disaster control center. Explain that the items shown are sold in Japan for earthquake preparedness. Ask students to identify the items in the figure and explain how they would be useful during or after an earthquake. (Sample answer: The padded headgear would be useful in avoiding bumps from falling debris; the fire extinguishers would be used to put out fires caused by broken gas mains or electric lines.) LS Verbal

QuickLAB

Skills Acquired

- Modeling
- Observing
- Analyzing

Teacher's Notes Using small plastic tables or empty cardboard boxes as building surfaces will allow several groups to work at the same location. Rubber hammers would provide a more steady, even source of vibrations.

Answers

1. The model building that was held together by rubber bands was more resistant to damage by the model earthquake.
2. Structures will be safer from earthquake damage if weak points are reinforced.

CHAPTER RESOURCES

Chapter Resource File

- Datasheet for Quick Lab GENERAL

Figure 2 ▶ In Tokyo, Japan—an area that has a high earthquake-hazard level—earthquake safety materials are available at disaster control centers.

QuickLAB 10 min

Earthquake-Safe Buildings

Procedure

1. On a tabletop, build one structure by stacking **building blocks** on top of each other.
2. Pound gently on the side of the table. Record what happens to the structure.
3. Using **rubber bands,** wrap sets of three blocks together. Build a second structure by using these blocks.
4. Repeat step 2.

Analysis

1. Which of your structures was more resistant to damage caused by the "earthquake"?
2. How could this model relate to building real structures, such as elevated highways?

Earthquake Safety

A destructive earthquake may take place in any region of the United States. However, destructive earthquakes are more likely to occur in certain geographic areas, such as California or Alaska. People who live near active faults should be ready to follow a few simple earthquake safety rules. These safety rules may help prevent death, injury, and property damage.

Before an Earthquake

Before an earthquake occurs, be prepared. Keep on hand a supply of canned food, bottled water, flashlights, batteries, and a portable radio. Some safety material is shown in **Figure 2.** Plan what you will do if an earthquake strikes while you are at home, in school, or in a car. Discuss these plans with your family. Learn how to turn off the gas, water, and electricity in your home.

During an Earthquake

When an earthquake occurs, stay calm. During the few seconds between tremors, you can move to a safer position. If you are indoors, protect yourself from falling debris by standing in a doorway or crouching under a desk or table. Stay away from windows, heavy furniture, and other objects that might topple over. If you are in school, follow the instructions given by your teacher or principal. If you are in a car, stop in a place that is away from tall buildings, tunnels, power lines, or bridges. Then, remain in the car until the tremors cease.

After an Earthquake

After an earthquake, be cautious. Check for fire and other hazards. Always wear shoes when walking near broken glass, and avoid downed power lines and objects touched by downed wires.

Group Activity — GENERAL

Earthquake Safety Have students work in groups to explore ways of staying safe during an earthquake. One group could interview local officials regarding disaster relief plans in your community. Another could do library or Internet research about earthquake hazards in the region. Still others could develop a poster that outlines earthquake safety guidelines. The entire class can put its knowledge to work by conducting a mock earthquake safety drill. LS Kinesthetic Co-op Learning

READING SKILL BUILDER — BASIC

Paired Summarizing Group students into pairs and have them read silently about earthquake warnings and forecasts. Then, have one student summarize the methods scientists have used to forecast earthquakes. The other student should listen to the retelling and point out any inaccuracies or ideas that were left out. Allow students to refer to the text as needed. LS Verbal/Auditory English Language Learners

Earthquake Warnings and Forecasts

Humans have long dreamed of being able to predict earthquakes. Accurate earthquake predictions could help prevent injuries and deaths that result from earthquakes.

Today, scientists study past earthquakes to predict where future earthquakes are most likely to occur. Using records of past earthquakes, scientists can make approximate forecasts of future earthquake risks. However, there is currently no reliable way to predict exactly when or where an earthquake will occur. Even the best forecasts may be off by several years.

To make forecasts that are more accurate, scientists are trying to detect changes in Earth's crust that can signal an earthquake. Faults near many population centers have been located and mapped. Instruments placed along these faults measure small changes in rock movement around the faults and can detect an increase in stress. Currently, however, these methods cannot provide reliable or accurate predictions of earthquakes.

SCILINKS. NSTA Developed and maintained by the National Science Teachers Association

For a variety of links related to this subject, go to www.scilinks.org

Topic: Earthquakes and Society
SciLinks code: HQ60455

Seismic Gaps

Scientists have identified zones of low earthquake activity, or seismic gaps, along some faults. A **seismic gap** is an area along a fault where relatively few earthquakes have occurred recently but where strong earthquakes occurred in the past. Some scientists think that seismic gaps are likely locations of future earthquakes. Several gaps that exist along the San Andreas Fault zone may be sites of major earthquakes in the future. One of these locations, Loma Prieta, California, is shown in **Figure 3.**

seismic gap an area along a fault where relatively few earthquakes have occurred recently but where strong earthquakes are known to have occurred in the past

Reading Check **Why do scientists think that seismic gaps are areas where future earthquakes are likely to occur? (See the Appendix for answers to Reading Checks.)**

Figure 3 ▶ Each red dot in the cross section of the San Andreas Fault represents an earthquake or aftershock before the 1989 Loma Prieta earthquake. Note how seismic gap 2 was filled by the 1989 earthquake and its aftershocks, which are represented by the blue dots.

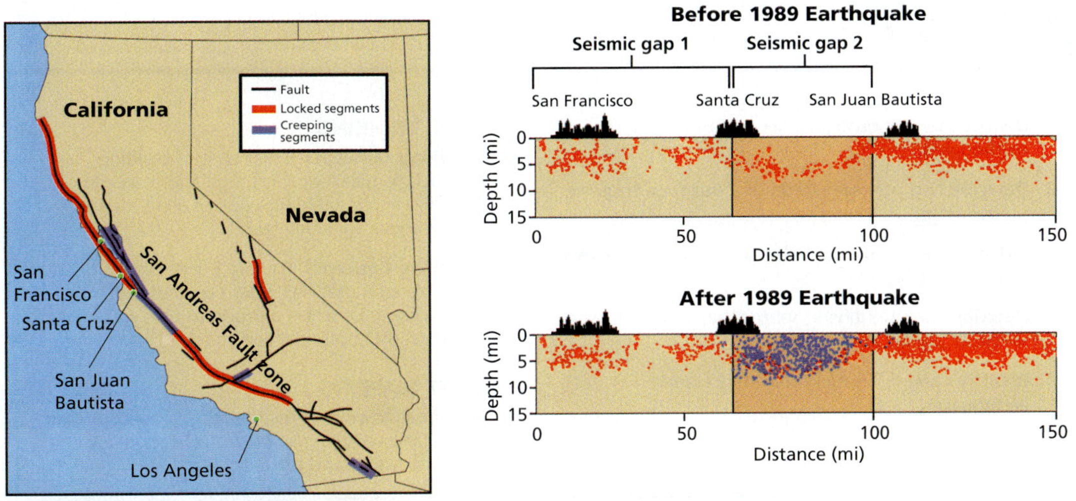

Close

Reteaching — BASIC

Questions Divide the class into small groups. Have students write questions on index cards about how earthquakes affect society. Invite a volunteer to act as host for a quiz game based on the questions and answers. **LS Verbal**

Quiz — GENERAL

1. Why might tsunamis accompany undersea earthquakes? (Faults along the sea bottom or undersea landslides cause the ocean floor to drop or change suddenly. As a result, the water creates giant waves that cross the ocean.)
2. What causes most of the injuries associated with earthquakes? (collapsing structures, falling objects, landslides, fires, and flooding)
3. How are geologists attempting to forecast future earthquakes? (They study past earthquakes, place instruments in fault zones to identify changes in rock motion, map seismic gaps, and look for signs of future seismic activity.)

CHAPTER RESOURCES

Technology

Transparencies
• 65 Seismic Gaps (with worksheet)

Answer to Reading check

Scientists think that stress on a fault builds up to a critical point and is then released as an earthquake. Seismic gaps are areas in which no earthquakes have happened in a long period of time and thus are likely to be under a high amount of stress.

Debate — ADVANCED

Worth the Price? Have interested students research and debate the following proposition: "What if a scientist developed a completely reliable prediction that a major earthquake would strike on a particular date in a specific location? Many businesses and individuals would relocate, and tourism would likely suffer. The economic loss from the release of the prediction could be greater than the economic losses suffered from the actual earthquake. Would you release the prediction?" Allow students time to prepare arguments for or against the release. **LS Intrapersonal**

Close, *continued*

Alternative Assessment — GENERAL

Making Models Give students shoeboxes, construction paper, scissors, tape, containers, rulers, small objects, and human figures. Have them construct dioramas and working models to demonstrate the effects of earthquakes on structures, how to remain safe during seismic events, or some of the methods used for earthquake forecasting, such as tiltmeters. They should supplement their models with explanatory labels and diagrams. LS **Visual**

Answers to Section Review

1. Undersea earthquakes cause sudden changes in the level of the ocean floor. This makes the water above drop or rise and sets in motion a series of long, low waves.
2. They may sway with the force of the earthquake so that they tip over onto nearby structures or they may collapse.
3. Remain calm and move to a safe position between tremors; stand in a doorway or crouch under a desk to avoid falling debris; stay away from windows or large objects that may fall; in a car, avoid buildings, power lines, or bridges.
4. Locations where relatively few earthquakes have occurred recently but have occurred in the past may be sites of future earthquakes if stress builds up there.
5. ground tilt, cracks in rocks caused by stress, changes in water content that could alter the rock's magnetic or electrical properties, or gas seepage
6. Sample answer: Buildings or dams should be constructed on solid rock. Plans should be approved by experts to ensure that buildings are constructed from materials that won't loosen and fall off and that they are designed to withstand the motion of the ground during an earthquake. Utility lines and cables should be protected to prevent fires and explosions.
7. Sample answer: I would map and locate all faults near population centers, place instruments along faults to measure changes in rock movements and increases in stress, identify seismic gaps, and monitor changes in nearby rock.
8. *Earthquake hazard levels* are assigned according to the number and size of *earthquakes,* the likelihood of *tsunamis,* and the type of *damage* that may occur, which can be reduced by following *safety* rules and setting up earthquake *prediction* programs.

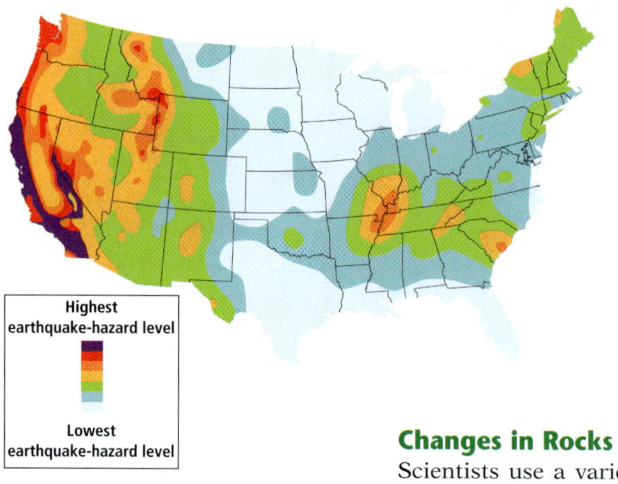

Figure 4 ▶ California, which has experienced severe earthquakes recently, has the highest earthquake-hazard level of the contiguous United States.

Foreshocks

Some earthquakes are preceded by little earthquakes called *foreshocks*. Foreshocks can precede an earthquake by a few seconds or a few weeks. In 1975, geophysicists in China recorded foreshocks near the city of Haicheng, which had a history of earthquakes. The city was evacuated the day before a major earthquake. The earthquake caused widespread destruction, but few lives were lost thanks to the warning. However, the Haicheng earthquake is the only example of a successful prediction made by using this method.

Changes in Rocks

Scientists use a variety of sensors to detect slight tilting of the ground and to identify the strain and cracks in rocks caused by the stress that builds up in fault zones. When these cracks in the rocks are filled with water, the magnetic and electrical properties of the rocks may change. Scientists also monitor natural gas seepage from rocks that are strained or fractured from seismic activity. Scientists hope that they will one day be able to use these signals to predict earthquakes.

Reliability of Earthquake Forecasts

Unfortunately, not all earthquakes have foreshocks or other precursors. Earthquake prediction is mostly unreliable. However, scientists have been able to determine areas that have a high earthquake-hazard level, as shown in **Figure 4.** Scientists continue to study seismic activity so that they may one day make accurate forecasts and save more lives.

Section 3 Review

1. **Discuss** the relationship between tsunamis and earthquakes.
2. **Describe** two possible effects of a major earthquake on buildings.
3. **List** three safety rules to follow when an earthquake strikes.
4. **Describe** how identifying seismic gaps may help scientists predict earthquakes.
5. **Identify** changes in rocks that may signal earthquakes.

CRITICAL THINKING

6. **Applying Concepts** What type of building construction and location regulations should be included in the building code of a city that is located near an active fault?
7. **Applying Concepts** You are a scientist assigned to study an area that has a high earthquake-hazard level. Describe a program that you could set up to predict potential earthquakes.

CONCEPT MAPPING

8. Use the following terms to create a concept map: *earthquake, earthquake-hazard level, damage, tsunami, safety,* and *prediction.*

CHAPTER RESOURCES

Chapter Resource File

- Section Quiz GENERAL

Workbooks

- Study Guide (also in Spanish)

Chapter 12 Highlights

Sections

1 How and Where Earthquakes Happen

Key Terms

- **earthquake,** 295
- **elastic rebound,** 295
- **focus,** 296
- **epicenter,** 296
- **body wave,** 296
- **surface wave,** 296
- **P wave,** 297
- **S wave,** 297
- **shadow zone,** 298
- **fault zone,** 300

Key Concepts

- In the process of elastic rebound, stress builds in rocks along a fault until they break and spring back to their original shape.
- There are two major types of seismic waves: body waves and surface waves.
- Different seismic waves act differently depending on the material of Earth's interior through which they pass.
- Most earthquakes occur near tectonic plate boundaries.

2 Studying Earthquakes

Key Terms

- **seismograph,** 301
- **seismogram,** 301
- **magnitude,** 303
- **intensity,** 304

Key Concepts

- Scientists use seismographs to record earthquake vibrations.
- The difference in the times that P waves and S waves take to arrive at a seismograph station helps scientists locate the epicenter of an earthquake.
- Earthquake magnitude scales describe the strength of an earthquake. Intensity is a measure of the effects of an earthquake.

3 Earthquakes and Society

Key Terms

- **tsunami,** 305
- **seismic gap,** 307

Key Concepts

- Most earthquake damage is caused by the collapse of buildings and other structures.
- Tsunamis often are caused by ocean-floor earthquakes.
- People who follow safety guidelines are less likely to be harmed by an earthquake.
- Seismic gaps, tilting ground, and variations in rock properties are some of the changes in Earth's crust that scientists use when trying to predict earthquakes.

Chapter Highlights

Alternative Assessment — GENERAL

News Broadcast Have students work in small groups to write a script and perform a mock radio or TV broadcast that describes the effects of an earthquake on a local community. They may wish to include an interview with a "geologist," who could describe the seismic event and its cause; an interview with an "engineer," who could detail the effects on structures; safety bulletins issued by local officials or Red Cross personnel; or first-person accounts by individuals who experienced the disaster. Students could create animations, plan sound effects, or compile video sequences to lend realism to their broadcast simulation. LS **Verbal/Auditory**

Co-op Learning

CHAPTER RESOURCES

Chapter Resource File

- Concept Review GENERAL
- Critical Thinking ADVANCED
- Math Skills GENERAL
- Graphing Skills GENERAL
- Chapter Test A GENERAL
- Chapter Test B ADVANCED

Workbooks

- **Study Guide** (also in Spanish)
- **Assessments** (Spanish)

Technology

Classroom Videos
- Brain Food Video Quiz

HRW Earth Science Video
- Segment 5: Earthquakes

Assignment Guide	
Section	**Questions**
1	1–2, 4–6, 9–14, 17–18, 24
2	7–8, 19–20, 28–31
3	3, 15–16, 21–23, 26
1 and 2	33–36
2 and 3	25, 32
1–3	27

Using Key Terms

1–8. Answers may vary but should show that students understand the definitions of and differences between key terms.

Understanding Key Concepts

9. b **10.** a
11. c **12.** c
13. c **14.** c
15. a **16.** b

Short Answer

17. Scientists can learn about the composition and structure of Earth's interior by studying changes in the speed and direction of seismic waves.

18. The S-wave shadow zone covers a larger area on the opposite side of Earth from the epicenter because S waves cannot travel through the liquid outer core as P waves do.

19. Scientists use computers to perform triangulations from several seismograph stations to locate the epicenters of earthquakes.

20. because the moment magnitude scale is more closely related to the cause of earthquakes and is more accurate for larger earthquakes

21. by swaying so violently that they topple and fall over onto neighboring structures

22. Park away from tall buildings, tunnels, power lines, or bridges and stay there until the tremors stop.

23. Answers may vary but should include three of the following changes: changes in the tilt of the ground, the appearance of cracks, changes in water content that can affect rock's magnetic or electrical properties, and natural gas leakage from fractures.

Chapter 12 Review

Using Key Terms

Use each of the following terms in a separate sentence.

1. *elastic rebound*

2. *fault zone*

3. *seismic gap*

For each pair of terms, explain how the meanings of the terms differ.

4. *focus* and *epicenter*

5. *body wave* and *surface wave*

6. *P wave* and *S wave*

7. *seismograph* and *seismogram*

8. *intensity* and *magnitude*

Understanding Key Concepts

9. Vibrations in Earth that are caused by the sudden movement of rock are called
- **a.** epicenters.
- **b.** earthquakes.
- **c.** faults.
- **d.** tsunamis.

10. In the process of elastic rebound, as a rock becomes stressed, it first
- **a.** deforms.
- **b.** melts.
- **c.** breaks.
- **d.** shrinks.

11. Earthquakes that cause severe damage are likely to have what characteristic?
- **a.** a deep focus
- **b.** an intermediate focus
- **c.** a shallow focus
- **d.** a deep epicenter

12. Most earthquakes occur
- **a.** in mountains.
- **b.** along major rivers.
- **c.** at plate boundaries.
- **d.** in the middle of tectonic plates.

13. P waves travel
- **a.** only through solids.
- **b.** only through liquids and gases.
- **c.** through solids, liquids, and gases.
- **d.** only through liquids.

14. S waves cannot pass through
- **a.** solids.
- **b.** the mantle.
- **c.** Earth's outer core.
- **d.** the asthenosphere.

15. Most injuries during earthquakes are caused by
- **a.** the collapse of buildings.
- **b.** cracks in Earth's surface.
- **c.** the vibration of S waves.
- **d.** the vibration of P waves.

16. Which of the following is *not* a method used to forecast earthquake risks?
- **a.** identifying seismic gaps
- **b.** determining moment magnitude
- **c.** recording foreshocks
- **d.** detecting changes in the rock

Short Answer

17. How do seismic waves help scientists understand Earth's interior?

18. Why is the S-wave shadow zone larger than the P-wave shadow zones are?

19. How do scientists determine the location of an earthquake's epicenter?

20. Why do scientists prefer the moment magnitude scale to the Richter scale?

21. How might tall buildings respond during a major earthquake?

22. What should you do if you are in a car when an earthquake happens?

23. List three changes in rock that may one day be used to help forecast earthquakes.

Critical Thinking

24. because they are most powerful at the surface of Earth where structures are plentiful, and their energy has not dissipated as a result of traveling through the layers of Earth

Critical Thinking

24. Understanding Relationships Why might surface waves cause the greatest damage during an earthquake?

25. Determining Cause and Effect Two cities are struck by the same earthquake. The cities are the same size, are built on the same type of ground, and have the same types of buildings. The city in which the earthquake produced a maximum intensity of VI on the Mercalli scale suffered $1 million in damage. The city in which the earthquake produced a maximum intensity of VIII on the Mercalli scale suffered $50 million in damage. What might account for this great difference in the costs of the damage?

26. Recognizing Relationships Would an earthquake in the Rocky Mountains in Colorado be likely to form a tsunami? Explain your answer.

Concept Mapping

27. Use the following terms to create a concept map: *earthquake, elastic rebound, surface wave, body wave, seismic wave, tsunami, seismograph, magnitude, intensity, moment magnitude scale,* and *Richter scale.*

Math Skills

28. Making Calculations If a P wave traveled 6.1 km/s, how long would the P wave take to travel 800 km?

29. Using Equations An earthquake with a magnitude of 3 releases 30 times more energy than does an earthquake with a magnitude of 2. How much more energy does an earthquake with a magnitude of 8 release than an earthquake with a magnitude of 6 does?

30. Making Calculations Of the approximately 420,000 earthquakes recorded each year, about 140 have a magnitude greater than 6. What percentage of all earthquakes have a magnitude greater than 6?

Writing Skills

31. Writing from Research Find out how and why the worldwide network of seismograph stations was formed. Also, find out how all the stations in the network work together. Prepare a report about your findings.

32. Communicating Main Ideas Find out which earthquake registered the highest intensity in history. Write a brief report that describes the effects of this earthquake.

Interpreting Graphics

The graph below shows three seismograms from a single earthquake. Use the graph to answer the questions that follow.

Seismic Wave Lapse Time

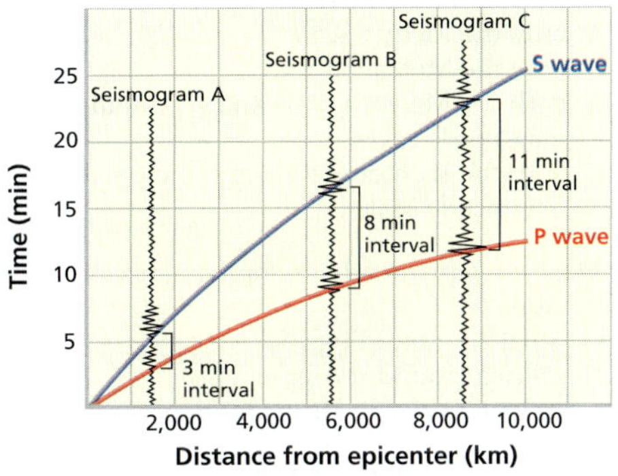

33. How far from the epicenter is seismograph B?

34. How far from the epicenter is seismograph C?

35. Which seismograph is farthest from the epicenter?

36. Why is there an 8 min interval between P waves and S waves in seismogram B but an 11 min interval between P waves and S waves in seismogram C?

Chapter Review

25. Sample answer: At VI on the intensity scale there is only slight damage to structures, while at VIII there is considerable damage, including the collapse of some buildings. Thus, the city that experienced greater maximum intensity had higher damage costs.

26. no; The Colorado Rockies are located within the North American continent, far from the ocean. Tsunamis result from underwater earthquakes and other disturbances on the ocean floor.

Concept Mapping

27. Answers may vary but should include all of the terms listed. Sample answers appear at the end of this Teacher Edition.

Math Skills

28. 800 km ÷ 6.1 km/s = 131 s

29. 30 × 30 = 900; The magnitude 8 earthquake releases 900 times as much energy as the magnitude 6 earthquake does.

30. 140 ÷ 420,000 = 0.00033 × 100 = 0.033%

Writing Skills

31. Answers may vary. Accept all reasonable answers. Funding for the worldwide network came about partly because of the cold war. Western powers wanted to be able to detect Soviet nuclear test sites. The scientific community supported the project because it furthered earthquake research.

32. Answers may vary. Accept all reasonable answers. Some remarkable effects were reported during the Assam India earthquake in 1897 and during the April 1868 Hawaii earthquake.

Interpreting Graphics

33. 5,500 km

34. 8,600 km

35. seismograph C

36. There is a larger gap between the P-wave and the S-wave arrival times at seismograph C because the waves had to travel farther from the epicenter.

Estimated Time

To give students practice under more realistic testing conditions, allow them 30 minutes to answer all of the questions in this practice test.

TEST DOCTOR

Question 3 Answer D is correct. The Mercalli scale measures the intensity of an earthquake rather than its magnitude, so answers B and C are incorrect. The Richter scale expresses the magnitude of an earthquake, but it is not the only scale to do so, so answer A is incorrect.

Question 5 Answer C is correct. Students can eliminate answers A, B, and D by carefully reading the question. People suffer more harm from the damage that is done to human constructions than from the movement of the ground.

Question 9 Answer D is correct. Students should use their knowledge of the different types of seismic waves when answering this question. P and S waves are specific types of body waves and cause little surface damage. So, answers A, B, and C are incorrect. The passage describes destruction on the surface, which is most likely caused by the powerful surface waves.

Chapter 12 Standardized Test Prep

Understanding Concepts

Directions (1–5): **For *each* question, write on a separate sheet of paper the letter of the correct answer.**

1. Energy waves that produce an earthquake begin at what location on or within Earth?
   A. the epicenter
   B. the seismic gap
   C. the focus
   D. the shadow zone

2. The fastest-moving seismic waves produced by an earthquake are called
   F. P waves
   G. S waves
   H. Raleigh waves
   I. surface waves

3. The magnitude of an earthquake can be expressed numerically by using
   A. only the Richter scale
   B. only the Mercalli scale
   C. both the Mercalli scale and the moment magnitude scale
   D. both the Richter scale and the moment magnitude scale

4. Most earthquake-related injuries are caused by
   F. tsunamis
   G. collapsing buildings
   H. rolling ground movements
   I. sudden cracks in the ground

5. Which of the following is least likely to cause deaths during an earthquake?
   A. floodwaters from collapsing dams
   B. falling objects and flying glass
   C. actual ground movement
   D. fires from broken electric and gas lines

Directions (6–8): **For *each* question, write a short response.**

6. What is the name of the instrument that is used to detect and record seismic waves?

7. What is the term for waves that move through a medium instead of along its surface?

8. Where is the Ring of Fire located?

Reading Skills

Directions (9–11): **Read the passage below. Then, answer the questions.**

The Loma Prieta Earthquake

At 5:04 P.M. on October 17, 1989, life in California's San Francisco Bay Area seemed relatively normal. While more than 62,000 excited fans filled Candlestick Park to watch the third game of baseball's World Series, other people were still rushing home from a long day's work or picking their children up from extracurricular activites. By 5:05 P.M., the situation had changed drastically. The area was rocked by the Loma Prieta earthquake, which was 6.9 on the moment magnitude scale. The earthquake lasted 20 seconds and caused 62 deaths, 3,757 injuries, and the destruction of more than 1,000 homes and businesses. By midnight, the city was fighting more than 20 large structural fires resulting from the earthquake. People suffered injuries from collapses in weakened structures for days following the initial earthquake. Considering that the earthquake was of such a high magnitude and that it happened during the busy rush hour, it is amazing that more people were not injured or killed.

9. What type of waves are the most likely to have caused the damage described during the Loma Prieta earthquake?
   A. P waves
   B. S waves
   C. body waves
   D. surface waves

10. Which of the following statements can be inferred from the information in the passage?
   F. *Loma Prieta* is the Spanish term for "deadly earthquake."
   G. The damage caused by the earthquake continued even after the waves had passed.
   H. There were fewer people injured in this earthquake than in most earthquakes.
   I. The Loma Prieta earthquake has the highest magnitude of any earthquake ever recorded.

11. The 6.9 rating of the Loma Prieta earthquake is a rating on what measurement scale?

Answers

Understanding Concepts

1. C
2. F
3. D
4. G
5. C
6. seismograph
7. body waves
8. The Ring of Fire surrounds the Pacific Ocean.

Reading Skills

9. D
10. G
11. the Richter scale

Interpreting Graphics

12. C
13. Answers may vary. See Test Doctor for a detailed scoring rubric.
14. Answers may vary. See Test Doctor for a detailed scoring rubric.

Interpreting Graphics

Directions (12–14): **For *each* question below, record the correct answer on a separate sheet of paper.**

The diagram shows a recording of data by a seismograph. Use this diagram to answer questions 13 and 14.

Reading a Seismogram

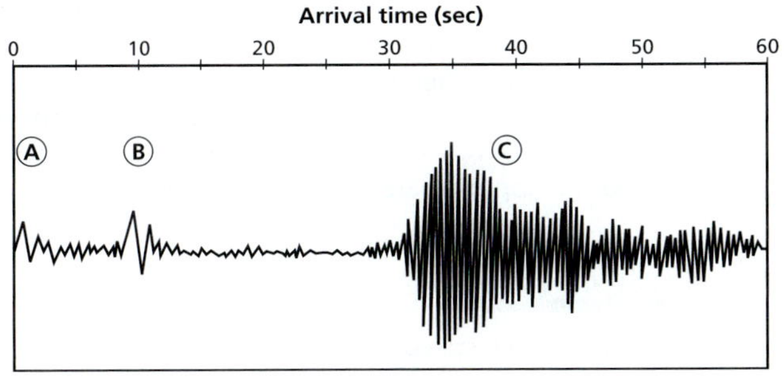

12 What type of seismic waves are indicated by the points on the seismogram marked by the letter A?

A. Love waves
B. Rayleigh waves
C. P waves
D. S waves

13 What type of seismic waves are indicated by the point on the seismogram marked by the letter C? How are these waves connected to the smaller waves that preceded them?

The illustration below shows the damage caused following an earthquake. Objects shown in this illustration are not drawn to scale. Use this illustration to answer question 15.

Earthquake Damage

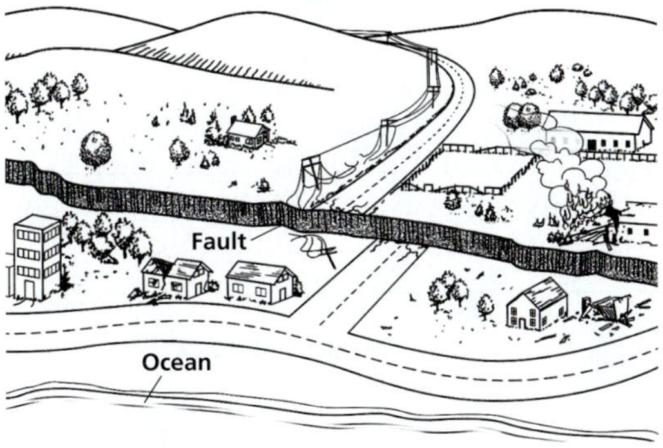

14 What safety hazards can you identify in this scene? What safety advice would you give to someone approaching the scene above? How should people prepare for dealing with such post-earthquake safety hazards?

Test TIP

Always read the full question to make sure you understand what is being asked before you look at the answer choices.

Standardized Test Prep

TEST DOCTOR

Question 13 Full-credit answers should include the following points:

- letter C shows the surface waves of an earthquake
- these waves are generated when the potential energy of P and S waves are converted into kinetic energy
- surface waves are the last waves to form. They are slower-moving than P or S waves and produce drastic vibrations
- surface waves cause the most damage to surface features and human constructions

Question 14 Full-credit answers should include the following points:

- students should carefully examine the illustration in order to determine the dangers present in the situation
- immediate hazards that non-rescue personnel should avoid include downed power lines, fires, and structural damage
- since this setting is near the ocean, hidden or delayed dangers may include flooding or tsunamis. People should move inland and find higher ground
- people in earthquake-prone areas should always have emergency plans in place. These plans may include pre-arranged meeting places for family members and friends, emergency supplies for dealing with power outages or injuries, and pre-determined evacuation routes

Test Prep Correlations ***National Science Education Standards***

ES 3c: items 1, 3, 8, 9, 11
SAI 2c: items 3, 6, 12, 13
SAI 2e: items 2, 7, 9, 12, 13
SPSP 5a: items 3, 4, 5, 9, 10, 11, 14
SPSP 5c: items 2, 10

CHAPTER RESOURCES

State Resources

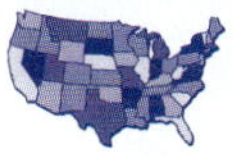

For specific resources for your state, visit go.hrw.com and type in the keyword HSHSTR.

Finding an Epicenter

Teaching Notes

Time Required
one 45-minute class period

Lab Ratings
EASY → HARD

TEACHER PREPARATION (1)
STUDENT SETUP (1)
CONCEPT LEVEL (2)
CLEANUP (1)

Skills Acquired
- Experimenting
- Constructing Models
- Organizing and Analyzing Data
- Measuring
- Interpreting
- Communicating

The Scientific Method
In this lab, students will
- Ask a Question
- Test the Hypothesis
- Analyze Results
- Communicate Results

Materials
Because the materials required for this investigation are relatively simple, students could work alone or in pairs.

Tips and Hints
If you choose to have students work in pairs, try to pair a student who is proficient in math with one who is not as adept.

Chapter 12 Skills Practice Lab

Finding an Epicenter

Objectives

- USING SCIENTIFIC METHODS **Analyze** P waves and S waves to determine the distance from a city to the epicenter of an earthquake.
- **Determine** the location of an earthquake epicenter by using the distance from three different cities to the epicenter of an earthquake.

Materials

calculator
drawing compass
ruler

An earthquake releases energy that travels through Earth in all directions. This energy is in the form of waves. Two kinds of seismic waves are P waves and S waves. P waves travel faster than S waves and are the first to be recorded at a seismograph station. The S waves arrive after the P waves. The time difference between the arrival of the P waves and the S waves increases as the waves travel farther from their origin. This difference in arrival time, called *lag time,* can be used to find the distance to the epicenter of the earthquake. Once the distance from three different locations is determined, scientists can find the approximate location of the epicenter.

PROCEDURE

1. The average speed of P waves is 6.1 km/s. The average speed of S waves is 4.1 km/s. Calculate the lag time between the arrival of P waves and S waves over a distance of 100 km.

2. The graph below shows seismic records made in three cities following an earthquake. These traces begin at the left. The arrows indicate the arrival of the P waves. The beginning of the next wave on each seismograph record indicates the arrival of the S wave. Use the time scale to find the lag time between the P wave and the S waves for each city. Draw a table similar to **Table 1.**

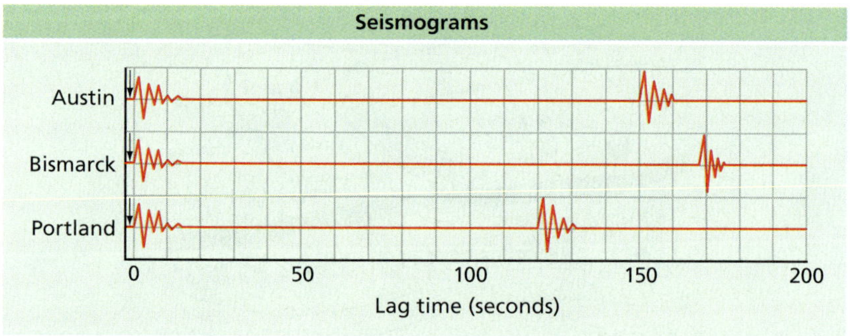

CHAPTER RESOURCES

Chapter Resource File

- Datasheet for Chapter Lab GENERAL
- Lab Notes and Answers

Table 1

City	Lag time (seconds)	Distance from city to epicenter
Austin		
Bismarck		
Portland		

DO NOT WRITE IN THIS BOOK

3. Record lag time for each city in the table.
4. Use the lag times found in step 2 and the lag time per 100 km found in step 1 to calculate the distance from each city to the epicenter of the earthquake by using the equation below.

$$distance = \frac{measured\ lag\ time\ (s) \times 100\ km}{lag\ time\ for\ 100\ km}$$

5. Record distances in the table.
6. Copy the map at right, which shows the location of the three cities. Using the map scale on your copy of the map, adjust the compass so that the radius of the circle with Austin at the center is equal to the calculation for Austin in step 2. Put the point of the compass on Austin. Draw a circle on your copy of the map.
7. Repeat step 6 for Bismarck and for Portland. The epicenter of the earthquake is located near the point at which the three circles intersect.

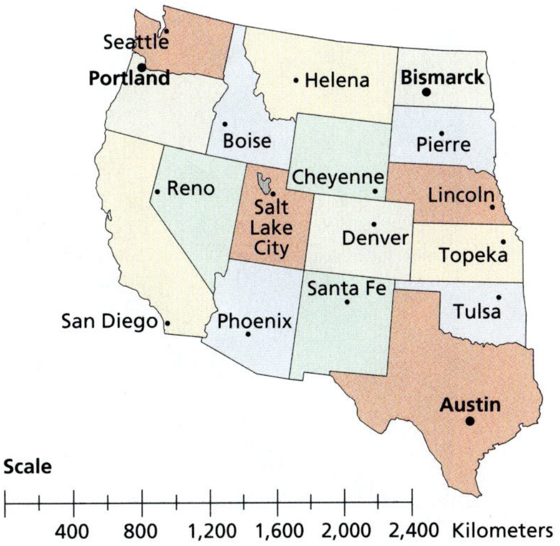

ANALYSIS AND CONCLUSION

1. **Evaluating Data** Describe the location of the earthquake's epicenter. To which city is the location of the earthquake's epicenter closest?
2. **Analyzing Processes** Why must measurements from three locations be used to find the epicenter of an earthquake?

Extension

1. **Evaluating Data** Research earthquakes in the United States. What is the probability of a major earthquake occurring in the area where you live? If an earthquake did occur in your area, what would most likely cause the earthquake?

Skills Practice Lab

Answers to Procedure

1. P wave: 100 km ÷ 6.1 km/s = 16.4 s; S wave: 100 km ÷ 4.1 km/s = 24.4 s; lag time = 24.4 s − 16.4 s = 8 s
2. The lag times for the cities would be: Austin 150 s; Bismarck 170 s; and Portland 120 s.
4. The computed distance for each station would be: For Austin: (150 s × 100 km) ÷ 8 s = 15,000 km ÷ 8 = 1,875 km; For Bismarck: (170 s × 100 km) ÷ 8 s = 17,000 km ÷ 8 = 2,125 km; For Portland: (120 s × 100 km) ÷ 8 s = 12,000 km ÷ 8 = 1,500 km.

Answers to Analysis and Conclusions

1. After drawing a circle around each seismograph station with the compass, the intersection of the three circles is the epicenter. It is located close to San Diego, California.
2. Three locations are necessary because sometimes the circles intersect in more than one place. The third circle intersects the other two in only one place. This ensures a more accurate result.

Answers to Extension

1. Answers may vary depending on location. Common causes would be living near a major fault or rift zone or in the case of the central United States earthquakes could result from an old buried fault being reactivated or the rebound of land from the retreat of a glacier.

TEACHER TESTED & APPROVED

Eric Cohen
Westhampton Beach High School
Westhampton Beach, NY

Maps in Action

Earthquake Hazard Map

Internet Activity — GENERAL

Hazards in the Americas Give students an outline map of North and South America. Have them go to the USGS Earthquake Hazards Program Web site for real-time earthquake information. Have students plot recent earthquakes on the regional map, using the map on this page as a model. Ask them to include earthquakes that are magnitude 5.0 or larger and color code the earthquakes by magnitude. Have them sketch the tectonic plate boundaries on their map. A worksheet designed to direct student research on this topic can be found in the **Chapter Resource File** booklet or by visiting **go.hrw.com** and entering keyword **HQ6EQKX.** LS **Visual**

Answers to Map Skills Activity

1. Eurasia north of the Indian and Saudi Arabian peninsulas, Japan, and Indonesia
2. the continent of Africa, northern Europe, and most of central Russia
3. southern Asia, just north of India, just northeast of Saudi Arabia, and off the west coast near Japan; and the Mediterranean region of Europe
4. Old rifts or other tectonic boundaries may exist in these regions.
5. The earthquakes happen on the deep-ocean floor and have little effect on people.
6. in East Africa, parallel to the eastern coast; The entire region represents a moderate hazard zone.

CHAPTER RESOURCES

Chapter Resource File

Internet Activity
- Hazards in the Americas GENERAL

Technology

Transparencies
- 66 Earthquake Hazard Map (with worksheet)

MAPS in Action

Earthquake Hazard Map

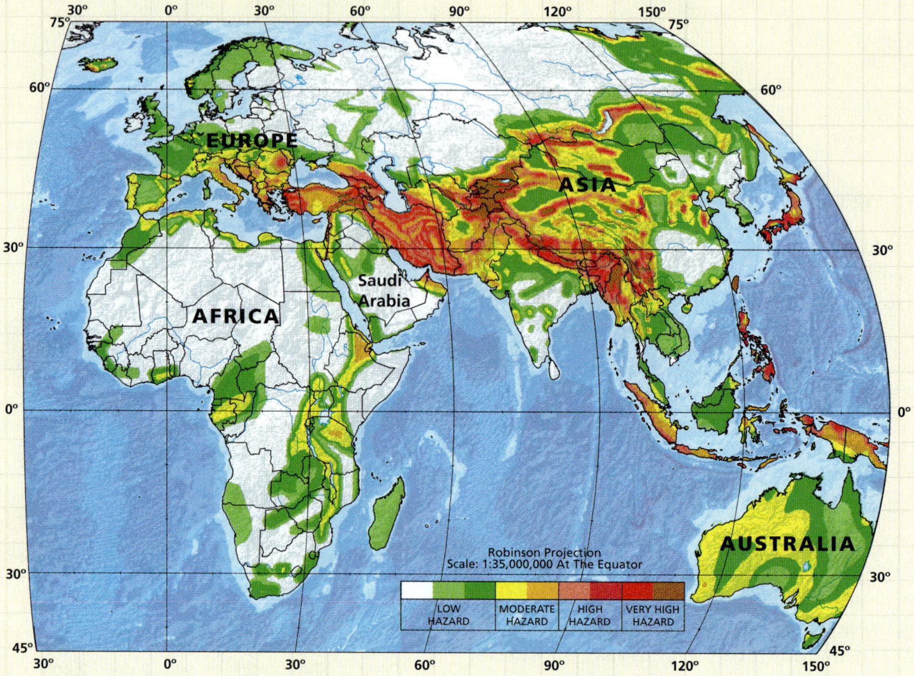

Map Skills Activity

This map shows the earthquake-hazard levels for Europe, Asia, Africa, and Australia. Use the map to answer the questions below.

1. **Using a Key** Which areas of the map have a very high earthquake-hazard level?
2. **Using a Key** Determine which areas of the map have very low earthquake-hazard levels.
3. **Inferring Relationships** Most earthquakes take place near tectonic plate boundaries. Based on the hazard levels, describe the areas of the map where you think tectonic plate boundaries are located.
4. **Analyzing Relationships** In Asia, just below 60° north latitude, there are areas that have high earthquake-hazard levels but no plate boundaries. Explain why these areas might experience earthquakes.
5. **Forming a Hypothesis** There is a tectonic plate boundary between Africa and Saudi Arabia. However, the earthquake-hazard level in that region is low. Explain the low earthquake-hazard level.
6. **Analyzing Relationships** A divergent plate boundary began to tear apart the continent of Africa about 30 million years ago. Where on the continent of Africa would you expect to find landforms created by this boundary? Explain your answer.

CAREER FOCUS

Geophysicist

Earth's surface may appear solid and unmoving. But Wayne Thatcher sees daily evidence that Earth's surface is always changing. As a research geophysicist for the U.S. Geological Survey, Thatcher studies the forces that shape Earth's crust, such as earthquakes.

Tools of the Trade

To measure ground deformation, Thatcher and his colleagues use global positioning system (GPS) and radar satellite images. GPS is a network of satellites that is commonly used for navigation of ships and aircraft. Thatcher uses GPS to measure changes in Earth's crust. By measuring ground elevation changes of a few millimeters, GPS enables discoveries that would not have been made many years ago.

Thatcher also uses Interferometric Synthetic Aperture Radar (InSAR), a remote-sensing technique that uses satellite radio waves to create three-dimensional images of Earth's surface. In this way, InSAR is able to map ground displacement from one satellite pass to the next, which helps scientists form mappings of deformation over months or years.

Predicting Future Earthquakes

For Thatcher and other geophysicists, the ability to forecast an earthquake remains an ongoing research aim. "Prediction is one of our ultimate goals," he notes. "We can estimate the level of risk based on GPS or InSAR readings. For example, we can track fault movements and note that 'something's got to give' and a quake is due."

For his research, Thatcher travels to fault zones throughout the western United States, where he sets up GPS benchmark sites. Trips to the sites in later years enable scientists to measure the rate and extent of movement of Earth's crust. Thatcher also journeys to other geologically active parts of the world, such as Japan, New Zealand, and Greece.

"Geophysics is a very international field of study. It's good to compare similarities and differences with other researchers who study earthquakes and volcanoes around the world."

—Wayne Thatcher, Ph.D.

◀ This colored satellite map shows the ground displacement after a 7.3 moment magnitude earthquake in Landers, California, on June 28, 1992.

SCILINKS. NSTA Developed and maintained by the National Science Teachers Association

For a variety of links related to this subject, go to www.scilinks.org

Topic: Careers in Earth Science
SciLinks code: HQ60222

Geophysicist

Using the Figure — GENERAL

Discussion Use the figure on this page to lead a discussion of how new satellite technologies offer new strategies for detecting the stresses that lead to earthquakes. Explain that the Interferometric Synthetic Aperture Radar (InSAR) works by combining two radar images so that scientists can observe changes in ground motion around a fault in greater detail and figure out where the stresses are building up. Point out that the black line in the colored satellite map is the fault line. The shift in colors represents a change in ground height. The closer together the banded colors are, the greater the deformation in that area is. Ask students how scientists might be able to use a network of such satellites to forecast earthquakes in the future. (Sample answers: The satellites could be deployed to monitor major fault zones around the world. The data could be used to anticipate where stresses are high enough that earthquakes may happen, as well as to show ground displacement afterward. The satellites could also be used to identify areas where ground-based equipment should be placed.) **LS Visual**

Extension — GENERAL

Research Have students identify four other careers that are related to seismology, geophysics, or plate tectonics. Have students write a brief synopsis of each career and present them to class. **LS Interpersonal**

Chapter 13 Volcanoes

Planning Guide

Compression Guide
To shorten instruction because of time limitations, omit the Chapter Lab.

OBJECTIVES	LABS, DEMONSTRATIONS, AND ACTIVITIES	TECHNOLOGY RESOURCES
PACING • 45 min pp. 318–324 **Chapter Opener**		**OSP** Parent Letter ■ **CD** Student Edition on CD-ROM **CD** Chapter Summaries Audio CD ■ **VID** Brain Food Video Quiz
Section 1 Volcanoes and Plate Tectonics • Describe the three conditions under which magma can form. • Explain what volcanism is. • Identify three tectonic settings where volcanoes form. • Describe how magma can form plutons.	**SE** **Quick Lab** Changing Melting Point, p. 321 ◆ GENERAL **CRF** **Datasheet for Quick Lab*** GENERAL **TE** **Social Studies Connection** Geothermal Energy, p. 322 BASIC **SE** **Maps in Action** The Hawaiian-Emperor Seamount Chain, p. 338 GENERAL **TE** **Group Activity** Charting Volcanic Activity, p. 338 ADVANCED **CRF** **Making Models Lab** Magma in Earth's Crust* GENERAL	**OSP** **Lesson Plans** (also in print) **TR** **Bellringer*** **TR** 67 Volcanoes and Tectonic Plate Boundaries* **TR** 68 Hot Spots and Mantle Plumes* **TR** 70 The Hawaiian-Emperor Seamount Chain* **VID** **HRW Earth Science Video** Volcanoes
PACING • 90 min pp. 325–330 **Section 2 Volcanic Eruptions** • Explain how the composition of magma affects volcanic eruptions and lava flow. • Describe the five major types of pyroclastic material. • Identify the three main types of volcanic cones. • Describe how a caldera forms. • List three events that may signal a volcanic eruption.	**TE** **Demonstration** Viscosity, p. 325 ◆ GENERAL **TE** **Activity** Poster Project, p. 327 ADVANCED **TE** **Discussion** Mount St. Helens, p. 327 GENERAL **TE** **Art Connection** The Scream, p. 327 GENERAL **TE** **Astronomy Connection** Extraterrestrial Volcanoes, p. 328 ADVANCED **TE** **Group Activity** Historic Eruptions, p. 328 GENERAL **SE** **Quick Lab** Volcanic Cones, p. 329 GENERAL **CRF** **Datasheet for Quick Lab*** GENERAL **SE** **Making Models Lab** Volcano Verdict, pp. 336–337 GENERAL **TE** **Interpreting Statistics** Temperature and Volcanic Dust, p. 339 ADVANCED **CRF** **Datasheet for Chapter Lab*** GENERAL **CRF** **Inquiry Lab** Lava Flows* GENERAL	**OSP** **Lesson Plans** (also in print) **TR** **Bellringer*** **TR** 69 Types of Volcanoes* **TE** **Internet Activity** The Carbon Cycle, p. 339 ADVANCED **CRF** **Internet Activity** The Carbon Cycle* ADVANCED **CD** **Interactive Tutor** Volcanoes and Earthquakes

PACING • 90 min

CHAPTER REVIEW, ASSESSMENT, AND STANDARDIZED TEST PREPARATION

- **SE** **Chapter Highlights**, p. 331
- **SE** **Chapter Review**, pp. 332–333
- **SE** **Standardized Test Prep**, pp. 334–335
- **CRF** **Concept Review*** ■ GENERAL
- **CRF** **Critical Thinking*** ADVANCED
- **CRF** **Math Skills*** GENERAL
- **CRF** **Graphing Skills*** GENERAL
- **CRF** **Chapter Test A*** ■ GENERAL
- **CRF** **Chapter Test B*** ADVANCED
- **OSP** **Lesson Plans** (also in print)
- **OSP** **Test Generator**
- **OSP** **Test Item Listing**

Online and Technology Resources

go.hrw.com Holt Online Learning

Visit **go.hrw.com** for access to Holt Online Learning, or enter the keyword **HQ6 Home** for a variety of free online resources.

One-Stop Planner® CD-ROM

This CD-ROM package includes
- Lab Materials QuickList Software
- Holt Calendar Planner
- Customizable Lesson Plans
- Printable Worksheets
- ExamView® Test Generator
- Interactive Teacher Edition
- Holt PuzzlePro®
- Holt PowerPoint® Resources

KEY

SE Student Edition	**OSP** One-Stop Planner	**VID** Classroom Video/DVD
TE Teacher Edition	**TR** Transparencies and Transparency Worksheets	* Also on One-Stop Planner
CRF Chapter Resource File	**CD** CD or CD-ROM	◆ Requires advance prep
LTP Long-Term Projects		■ Also available in Spanish

SKILLS DEVELOPMENT RESOURCES	REVIEW AND ASSESSMENT	CORRELATIONS
SE Pre-Reading Activity, p. 318 GENERAL **TE Using the Figure** Forming Islands, p. 318 GENERAL		National Science Education Standards
CRF Directed Reading* BASIC **TE Using the Figure** Pressure and Magma Formation, p. 319 GENERAL **TE Using the Figure** Volcano Locations, p. 320 GENERAL **TE Reading Skill Builder** Reading Organizer, p. 320 BASIC **TE Skill Builder** Vocabulary, p. 320 GENERAL **TE Inclusion Strategies,** p. 320 **TE Skill Builder** Graphing, p. 323 ADVANCED	**SE Reading Checks,** pp. 321, 323 GENERAL **SE Section Review,** p. 324 GENERAL **TE Homework,** p. 322 ADVANCED **TE Reteaching,** p. 323 BASIC **TE Quiz,** p. 323 GENERAL **TE Alternative Assessment,** p. 324 ADVANCED **CRF Section Quiz*** ■ GENERAL	ES 3c, UCP 4, ST 1e
CRF Directed Reading* BASIC **TE Using the Figure** Lava Flows, p. 326 BASIC **SE Math Practice,** p. 326 GENERAL **SE Graphic Organizer** Spider Map, p. 327 GENERAL **TE Using the Figure** Volcano Types, p. 328 GENERAL **TE Reading Skill Builder** Reading Hint, p. 328 BASIC **TE Inclusion Strategies,** p. 329	**SE Reading Checks,** pp. 326, 329 GENERAL **SE Section Review,** p. 330 GENERAL **TE Reteaching,** p. 329 BASIC **TE Quiz,** p. 329 GENERAL **TE Alternative Assessment,** p. 330 ADVANCED **CRF Section Quiz*** ■ GENERAL	ES 3c, SPSP 5a, UCP 2, ST 1e

Holt Lab Generator CD-ROM

Search for labs by topic, standard, difficulty level, and/or time frame. Edit those labs to fit your needs, or create your own labs. Use the Lab Materials QuickList to order just what you need.

Holt Earth Science Interactive Tutor CD-ROM

This CD-ROM consists of interactive activities that give students a fun way to extend their knowledge of Earth science concepts.

Chapter Summaries Audio CDs

These CDs include audio summaries of the key concepts presented in each chapter. (Audio summaries are also available in Spanish.)

SciLinks. NSTA

www.scilinks.org

Maintained by the **National Science Teachers Association.** See Chapter Enrichment pages that follow for a complete list of topics.

See Chapter Enrichment pages for Video Resources.

Refer to our ***Professional Reference for Teachers*** for additional teaching resources, including articles written by science education professionals about relevant and timely issues facing today's science teachers.

Chapter 13

Chapter Enrichment

This Chapter Enrichment provides relevant and interesting information to expand and enhance your classroom instruction of the chapter material.

Section 1 Volcanoes and Plate Tectonics

Heating Earth's Interior

Three processes contribute to heating Earth's interior. One process began when matter clumped together to form Earth and other planets. As matter built up, or accreted, the mass and gravitational field of Earth increased, further compressing the matter and causing the temperature to rise. In addition, bodies such as meteors were attracted to the more massive Earth, and collisions added more heat. Much of the heat from early planet formation and collisions remains in Earth's interior. A second important process that contributes to Earth's internal heat is the decay of radioactive isotopes. The third process, called tidal braking, produces internal heat as the gravitational pulls of Earth, the sun, and the moon slow Earth's rotation. Tidal braking is even more important in the volcanic activity of bodies that are experiencing stronger gravitational pull, such as Jupiter's close moon Io, which has many active volcanoes.

Volcanic Concentrations

Almost all of the world's volcanoes are caused by activity along subduction zones, mid-ocean ridges, or hot spots. But the number of active volcanoes in tectonically active regions varies greatly. The most active region is Indonesia, which has 130 active volcanoes. The Aleutian Islands arc of Alaska has more than 80 volcanoes, 41 of which are active, making this region the second most active on Earth. In the Pacific Ocean's "Ring of Fire," northern New Zealand has active volcanoes, but southern New Zealand does not. Central America and the northwest U.S. have numerous active volcanoes, while the southwestern coast of the U.S. and western Mexico have few. Hot spots, such as those of the Hawaiian Islands, tend to produce smaller volcano populations. Other hot spot volcanoes are the six that make up the Tibesti Complex in Chad, Africa, and the Galapagos Islands volcanoes off the coast of Ecuador.

Section 2 Volcanic Eruptions

Classifying Eruptions

Volcanic eruptions are generally classified by their descriptive attributes, so they often carry the names of volcanoes or volcanic locations that had similar previous eruptions. Lava eruptions are traditionally classified as Icelandic, Hawaiian, or Strombolian. An Icelandic eruption consists of mostly smooth lava flows from fissures, producing a lava field. In a Hawaiian eruption, the flows accumulate to form shield volcanoes. Both of these types of eruptions are generally "quiet." In contrast, Strombolian eruptions, named after the volcano in Stromboli, Italy, are more violent because the lava is more viscous. The more violent pyroclastic eruptions are classified into categories: Vulcanian, Vesuvian, Plinian, Pelean, and Krakatoan. Of these, only the Plinian eruption is not named for a volcano. Instead, it is named for the scholars Pliny the Elder and his nephew Pliny the Younger, who made one of the earliest attempts to study a volcanic eruption in the year 79 CE, when Vesuvius erupted. Pliny the Elder perished in the attempt, but his nephew recorded the eruption that had been observed. As a result, the eruption of Vesuvius is known as a Plinian eruption. These eruption categories are marked by increasing amounts

Volcanoes commonly form in chains along subduction zones at convergent plate boundaries.

▲ Smooth lava flows are typical of Hawaiian-type eruptions.

of pyroclastic material, higher ash columns, and more widespread dispersal of pyroclastic material. In the case of the Pelean-type eruption, the direction in which pyroclastic material is ejected is also important. The dense, hot cloud—called a *nuée ardente* (noo WAY ahr DAHNT, "glowing cloud")—of flowing material that swiftly rolled down the side of Mt. Pelée, on the Caribbean island of Martinique, during its 1902 eruption was particularly devastating, killing all but two residents of the nearby town.

Volcanic Soils

While the destructive forces of volcanic eruptions draw a great deal of attention, it is also important to note that many types of volcanic rock make very fertile soil. Soils from felsic magma are less consistently fertile, though many areas around volcanoes that have violent eruptions with a high output of pyroclastic material show rapid recovery of plant growth. In many cases, the time the cooled lava takes to erode and form soil is the primary factor that affects the recovery of vegetation after an eruption. Lava flows in 1960 near Hilo, Hawaii, showed plant growth in just six years. On the Caribbean island of St. Vincent, soil formed within two decades after the last eruption, and plants began growing shortly after the ash had cooled.

Video Resources

Brain Food Video Quizzes

Brain Food Video Quizzes These videos contain game-show style quizzes that assess students' progress and motivate students to study the chapter material.

HRW Earth Science Video

This video introduces Earth science topics and includes a geology field trip. The video segment listed below complements this chapter.

Segment 6: Volcanoes Most volcanoes are located along active plate boundaries. This segment examines volcanism at convergent and divergent plate boundaries. The formation of a volcanic island chain as a tectonic plate moves over a hot spot is also discussed. (2 min)

NOVA® Videos

To order **NOVA** videos related to this chapter, visit **go.hrw.com** and enter the keyword **HQ6VOLV.**

SciLinks®

Developed and maintained by the National Science Teachers Association

SciLinks is maintained by the National Science Teachers Association to provide you and your students with interesting, up-to-date links that will enrich your classroom presentation of the chapter.

Visit www.scilinks.org and enter the SciLinks code for more information about the topic listed.

Topic: Volcanic Zones
SciLinks code: HQ61618

Topic: Volcanoes
SciLinks code: HQ61619

Topic: Predicting Volcanic Eruptions
SciLinks code: HQ61209

Chapter 13

Chapter Overview

This chapter describes the ways volcanoes form, the different types of volcanoes, and what causes different kinds of volcanic eruptions.

Using the Figure — GENERAL

Forming Islands The photograph shows lava from Puu Oo (POO oo OH oh), a volcanic opening in Hawaii Volcanoes National Park. The lava flows to the southeastern coast of the island of Hawaii. Ask students what details of the photograph indicate that lava is hot, flowing rock. (The red glow of the lava in the foreground and the steam in the background indicate that the material is hot, while the smooth shape of the rock suggests that the lava flows like a thick liquid.) Ask how the lava becomes solid land. (The hot lava cools on contact with air or water. It solidifies and adds land to the island.) LS **Visual**

PRE-READING ACTIVITY

FOLDNOTES

You may want students to work in pairs to create this FoldNote. Have one student fill in the first two columns, and have the other student fill in the last column.

Chapter 13 Volcanoes

Sections

1 Volcanoes and Plate Tectonics

2 Volcanic Eruptions

What You'll Learn

- Where volcanoes are located
- What causes volcanic eruptions
- How lava affects the shape of volcanoes

Why It's Relevant

Volcanoes can form features such as mountains or islands. Volcanic eruptions can also endanger human life. Learning about volcanoes can help scientists better predict eruptions and can prepare people to evacuate dangerous areas.

PRE-READING ACTIVITY

FOLDNOTES

Table Fold Before you read this chapter, create the FoldNote entitled "Table Fold" described in the Skills Handbook section of the Appendix. Label the columns of the table fold with "Slope," "Lava," and "Eruption style." Label the rows with the three different types of volcanic cones. As you read the chapter, write examples of each topic under the appropriate column.

▶ Because volcanic eruptions add new material to Earth's surface, they have formed many islands, such as the Hawaiian Island chain. There, you can witness the volcanic processes that form islands, as shown in this photo.

Chapter Correlations — *National Science Education Standards*

ES 3c Interactions among the solid earth, the oceans, the atmosphere, and organisms have resulted in the ongoing evolution of the earth system. We can observe some changes such as...volcanic eruptions on a human time scale... **(Sections 1 and 2)**

SPSP 5a Normal adjustments of earth may be hazardous for humans.... As societies have grown...vulnerability to natural processes of change has increased. **(Section 2)**

UCP 2 Evidence consists of observations and data on which to base scientific explanations. Using evidence to understand interactions allows individuals to predict changes in natural and designed systems....Models help scientists and engineers understand how things work. Models take many forms, including physical objects, plans, mental constructs, mathematical equations, and computer simulations. **(Section 2)**

UCP 4 Evolution is a series of changes... that accounts for the present form and function of objects, organisms, and natural and designed systems. The general idea of evolution is that the present arises from materials and forms of the past... **(Section 1)**

ST 1e Communicate the problem, process, and solution. Students should present their results to students, teachers, and others in a variety of ways, such as orally, in writing, and in other forms—including models, diagrams, and demonstrations. **(Sections 1 and 2)**

Section 1 Volcanoes and Plate Tectonics

Volcanic eruptions can cause some of the most dramatic changes to Earth's surface. Some eruptions can be more powerful than the explosion of an atomic bomb. The cause of many of these eruptions is the movement of tectonic plates. The movement of tectonic plates is driven by Earth's internal heat.

By studying temperatures within Earth, scientists can learn more about volcanic eruptions. **Figure 1** shows estimates of Earth's inner temperatures and pressures. As the graph shows, the combined temperature and pressure in the lower part of the mantle keeps the rocks below their melting point.

OBJECTIVES

- **Describe** the three conditions under which magma can form.
- **Explain** what volcanism is.
- **Identify** three tectonic settings where volcanoes form.
- **Describe** how magma can form plutons.

KEY TERMS

magma
volcanism
lava
volcano
hot spot

Formation of Magma

Despite the high temperature in the mantle, most of this zone remains solid because of the large amount of pressure from the surrounding rock. Sometimes, however, solid mantle and crust melt to form **magma**, or liquid rock that forms under Earth's surface.

Magma can form under three conditions. First, if the temperature of rock rises above the melting point of the minerals the rock is composed of, the rock will melt. Second, if enough pressure is removed from the rock, the melting point will decrease and the rock will melt. Third, the addition of fluids, such as water, may decrease the melting point of some minerals in the rock and cause the rock to melt.

magma liquid rock produced under Earth's surface

Figure 1 ▶ Temperature and pressure increase as depth beneath Earth's surface increases. So, rock in the lower mantle stays below its melting point.

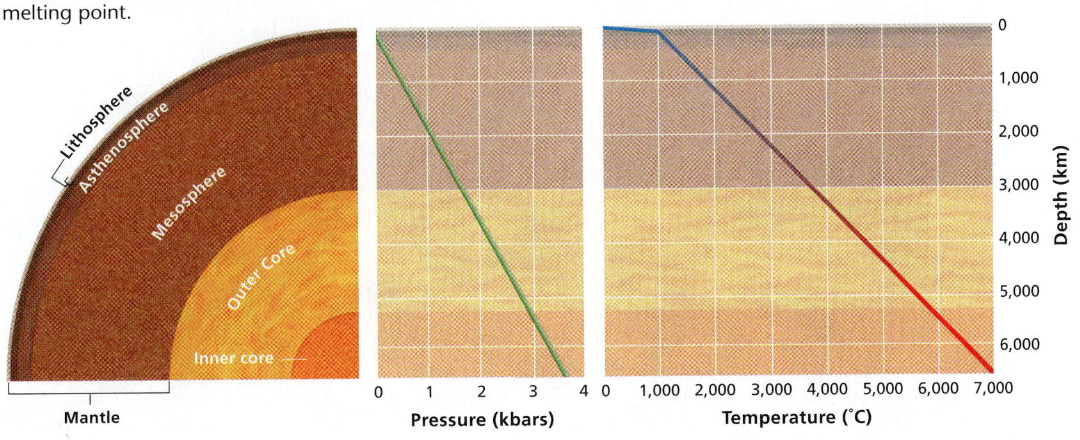

Section 1

Focus

Overview

This section describes how plate tectonics affects the formation of volcanoes. The section explains how different types of activity between and within lithospheric plates produce magma, and the ways in which magma forms different types of volcanoes and intrusions.

Bellringer

Have students describe what they think of when they hear the terms *volcano* and *volcanic eruption*. Ask students to explain how and where volcanoes form. (Answers may vary. You may want to use students' answers to address misconceptions about volcanoes.)
LS Intrapersonal

Motivate

Using the Figure — GENERAL

Pressure and Magma Formation
Explain that pressure keeps heated rock in a solid state. Reducing pressure causes the rock to melt without an increase in temperature. Ask students to look at the figure and explain why the rock in the mantle is solid. (Pressure from the overlying layers of rock keeps the mantle solid.) **LS Visual/Logical**

CHAPTER RESOURCES

Chapter Resource File
- Directed Reading BASIC

Technology
- Transparencies
  - Bellringer
- Student Edition on CD-ROM
- One-Stop Planner CD-ROM
  - Lesson Plan

Teach

Using the Figure — GENERAL

Volcano Locations Have students note where most of the active volcanoes are located on the map, and have them identify five regions that have high active volcano concentrations. (Lists may include the following areas: Indonesia, the western Pacific Rim, Alaska and the Aleutian Islands, Central America, the west coast of South America, Iceland, and eastern Africa.) Answer to caption question: Volcanoes commonly form along tectonic plate boundaries. **LS Visual**

READING SKILL BUILDER — BASIC

Reading Organizer Have students make a concept map that relates volcanism to the movement of magma, the increase in the amount of magma as it rises to Earth's surface, and the difference between magma and lava. Students can later use the concept map as a study guide. **LS Logical** English Language Learners

SKILL BUILDER — GENERAL

Vocabulary Many of the early terms that relate to volcanism are Latin in origin. The word *volcano* is derived from the name of the volcanic island *Vulcano,* which the Romans believed to be the home of the fire god Vulcan. The word *lava,* though it originates from the Latin *lavare* (to wash), meant "a stream caused suddenly by rain." Living near Vesuvius, which was frequently active, the Neapolitans applied the term to the sudden streams of molten rock that flowed down the volcano. **LS Verbal**

CHAPTER RESOURCES

Technology

Transparencies
- 67 Volcanoes and Tectonic Plate Boundaries (with worksheet)

volcanism any activity that includes the movement of magma toward or onto Earth's surface

lava magma that flows onto Earth's surface; the rock that forms when lava cools and solidifies

volcano a vent or fissure in Earth's surface through which magma and gases are expelled

Volcanism

Any activity that includes the movement of magma onto Earth's surface is called **volcanism.** Magma rises upward through the crust because the magma is less dense than the surrounding rock. As bodies of magma rise toward the surface, they can become larger in two ways. First, because they are so hot, they can melt some of the surrounding rock. Second, as the magma rises, it is forced into cracks in the surrounding rock. This process causes large blocks of overlying rock to break off and melt. Both of these processes add material to the magma body.

When magma erupts onto Earth's surface, the magma is called **lava.** As lava flows from an opening, or *vent,* the material may build up as a cone of material that may eventually form a mountain. The vent in Earth's surface through which magma and gases are expelled is called a **volcano.**

Major Volcanic Zones

If you were to plot the locations of the volcanoes that have erupted in the past 50 years, you would see that the locations form a pattern across Earth's surface. Like earthquakes, most active volcanoes occur in zones near both convergent and divergent boundaries of tectonic plates, as shown in **Figure 2.**

A major zone of active volcanoes encircles the Pacific Ocean. This zone, called the Pacific Ring of Fire, is formed by the subduction of plates along the Pacific coasts of North America, South America, Asia, and the islands of the western Pacific Ocean. The Pacific Ring of Fire is also one of Earth's major earthquake zones.

Figure 2 ▶ This map shows the locations of major tectonic plate boundaries and of active volcanoes. ***What is the relationship between the volcanoes and tectonic plate boundaries?***

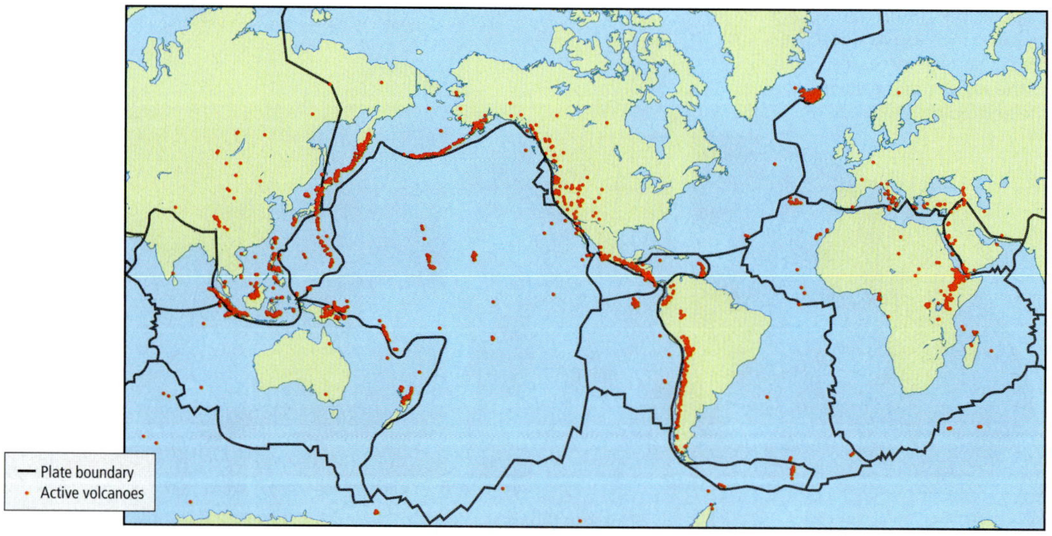

Cultural Awareness — BASIC

Geomyths Have interested students research the various "geomyths" that different cultures have created to explain volcanoes and volcanic eruptions. Have students explore the groups' connection to the role these natural events play in their culture. Native Americans of the Pacific Northwest, the Polynesians of Hawaii and the inhabitants of Iceland are examples of cultures that have developed volcano mythology. Students may work individually or in groups. They may present findings to the class as an oral report, a skit, or a song. **LS Kinesthetic/Auditory**

INCLUSION Strategies

- *Behavior Control Issues*
- *Attention Deficit Disorder*

Tell students to mark answers to these three directives by using sticky notes as they read the section:

1. Find the relationship between volcanoes and mountains.
2. Find out how deep trenches sometimes form on the ocean floor.
3. Decide if "Devils Tower" is a good name for a pluton. **LS Logical**

Subduction Zones

Many volcanoes are located along *subduction zones,* where one tectonic plate moves under another. When a plate that consists of oceanic lithosphere meets one that consists of continental lithosphere, the denser oceanic lithosphere moves beneath the continental lithosphere. A deep *trench* forms on the ocean floor along the edge of the continent where the plate is subducted. The plate that consists of continental lithosphere buckles and folds to form a line of mountains along the edge of the continent.

As the oceanic plate sinks into the asthenosphere, fluids such as water from the subducting plate combine with crust and mantle material. These fluids decrease the melting point of the rock and cause the rock to melt and form magma. When the magma rises through the lithosphere and erupts on Earth's surface, lines of volcanic mountains form along the edge of the tectonic plate.

If two plates that have oceanic lithosphere at their boundaries collide, one plate subducts, and a deep trench forms. As in the case of continental lithosphere colliding with oceanic lithosphere, magma also forms as fluids are introduced into the mantle during oceanic plate collisions. Some of the magma breaks through the overriding plate to Earth's surface. Over time, a string of volcanic islands, called an *island arc,* forms on the overriding plate, as shown in **Figure 3.** The early stages of this type of subduction produce an arc of small volcanic islands, such as the Aleutian Islands, which are in the North Pacific Ocean and between Alaska and Siberia. As more magma reaches the surface, the islands become larger and join to form one landmass, such as the volcanic islands that joined to form present-day Japan.

Reading Check **When a plate that consists of oceanic crust and one that consists of continental crust meet, which plate subducts beneath the other plate? (See the Appendix for answers to Reading Checks.)**

QuickLAB 5 min

Changing Melting Point

Procedure

1. Place a **piece of ice** on a small **paper plate.**
2. Wait 1 min, and observe how much ice has melted. Remove the meltwater from the plate.
3. Pour **1/4 teaspoon of salt** onto a **second piece of ice.**
4. Wait 1 min, and observe how much ice has melted.

Analysis

1. What happened to the rate of melting when you added salt to the ice?
2. In this model, what is represented by the ice? by the salt?

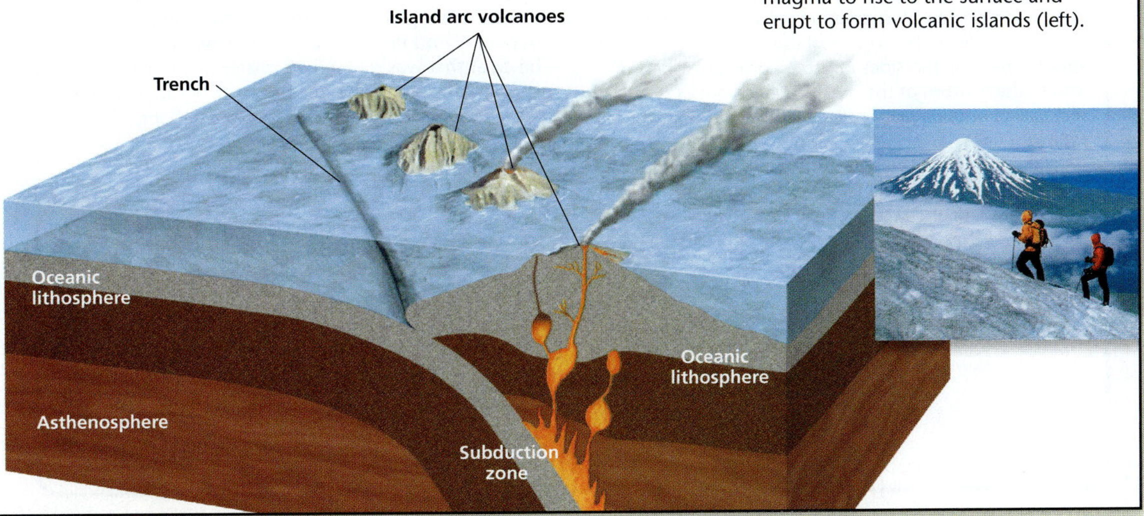

Figure 3 ▶ The Aleutian Islands (below) formed when oceanic lithosphere subducted beneath oceanic lithosphere and caused magma to rise to the surface and erupt to form volcanic islands (left).

QuickLAB

Skills Acquired

- Experimenting
- Observing
- Interpreting
- Inferring

Teacher's Notes Be sure that both pieces of ice are initially the same size and that both are equally unmelted at the beginning of each observation.

Answers

1. The rate of melting increased.
2. The ice represents solid rock. The salt represents fluids that were added to the rock.

Answer to Reading Check

The denser plate of oceanic lithosphere subducts beneath the less dense plate of continental lithosphere.

CHAPTER RESOURCES

Chapter Resource File

- **Datasheet for Quick Lab** GENERAL

Career

BASIC

Volcanologist Volcanologists study the geologic processes that cause volcanoes to form and erupt. They classify the types of rock and study rock's chemical composition. The information indicates how a volcano formed, the various stages in its life, the types of eruptions that occurred, and if still active, the types of eruptions that may happen. One goal of many volcanologists is to predict eruptions. These volcanologists work in and around volcanoes and make seismic and ground temperature measurements to determine how close the volcano is to erupting. Other volcanologists use a wide range of satellite data to remotely study the effects of volcanoes on Earth's atmosphere and climate as well as on heat output and ground deformation. Yet another group of volcanologists researches how past volcanic activity affected the development of Earth's early crust and atmosphere. Still others study volcanic processes in order to understand the composition and activities of Earth's interior. Volcanologists have college degrees, for which they study physics, chemistry, geology, and geography.

Teach, *continued*

MISCONCEPTION ALERT

Expanding Earth Students may think of mid-ocean ridges in isolation. If that were the case, the continuing growth of the ocean floor would cause Earth to expand. Ask students to consider what would happen if Earth were to grow continually. (Earth's diameter would get larger.) Then, point out that Earth has remained at a fairly constant size because as new lithosphere forms at a mid-ocean ridge, an equivalent amount of lithosphere is consumed by subduction, or is crumpled up during mountain formation. Because these processes are linked, the surface area of Earth remains constant.

Homework — ADVANCED

Touring the Mid-Atlantic Ridge Suggest that students plan a tour of the longest mountain range in the world—the Mid-Atlantic Ridge. Have students research the highest peaks of the range and identify the islands to which these peaks correspond. Have them indicate the distances between peaks, as well as which parts of the ridge are most volcanically active. Students should then compile their findings in a short, written report. Encourage students to use creativity in the style of their writing. LS **Verbal**

Figure 4 ▶ When water rapidly cools hot lava, a hard, pillow-shaped crust forms. As the crust cools, it contracts and cracks. Hot lava flows through the cracks in the crust and then cools quickly to form another pillow-shaped structure.

Mid-Ocean Ridges

The largest amount of magma comes to the surface where plates are moving apart at mid-ocean ridges. Thus, the interconnected mid-ocean ridges that circle Earth form a major zone of volcanic activity. As plates pull apart, magma flows upward along the rift zone. The upwelling magma adds material to the mid-ocean ridge and creates new lithosphere along the rift. This magma erupts to form underwater volcanoes. **Figure 4** shows pillow lava, an example of volcanic rock that forms underwater at a mid-ocean ridge. Pillow lava is named for its pillow shape, which is caused by the water that rapidly cools the outer surface of the lava.

Most volcanic eruptions that happen along mid-ocean ridges are unnoticed by humans because the eruptions take place deep in the ocean. An exception is found on Iceland. Iceland is one part of the Mid-Atlantic Ridge that is above sea level. One-half of Iceland is on the North American plate and is moving westward. The other half is on the Eurasian plate and is moving eastward. The middle of Iceland is cut by large *fissures,* which are cracks through which lava flows to Earth's surface.

Connection to GEOLOGY

Sea-floor Formation

At mid-ocean ridges, magma moves upward from the asthenosphere. When the magma reaches the surface, the magma cools to form new sea-floor rock. When new sea floor forms at a mid-ocean ridge, a special sequence of rocks forms. This sequence forms because of the way the magma rises and cools. The thickness of the layers in this sequence varies greatly worldwide.

At the base of the new lithosphere, where the plates pull apart, a magma chamber forms. This chamber is cooled by circulating sea water. As the magma slowly cools, large crystals form. The crystals stick to the roof and sides of the magma chamber or sink to the bottom of the chamber and form a type of rock called *gabbro*. Gabbro forms the base layer in the rock sequence.

In the rift where the plates pull apart, the magma repeatedly intrudes. This repeated intrusion forms a series of vertical dikes called *sheeted dikes*. The structure of these dikes appears to be similar to a deck of cards standing on end. The sheeted-dike complex forms the middle layer in the rock sequence.

At the top of the new lithosphere, where the magma comes into contact with the cold ocean water, the magma freezes rapidly. This rapid freezing causes pillow lava to form. Pillow lava forms the uppermost layer in the rock sequence.

Sea-floor rock that formed millions of years ago can be seen on land today. When one plate subducts under another plate, some of the subducting crust is scraped off and becomes part of the overriding plate. The crust that was scraped off is later uplifted and exposed on land, where geologists can study the rock to learn more about the formation of sea floor.

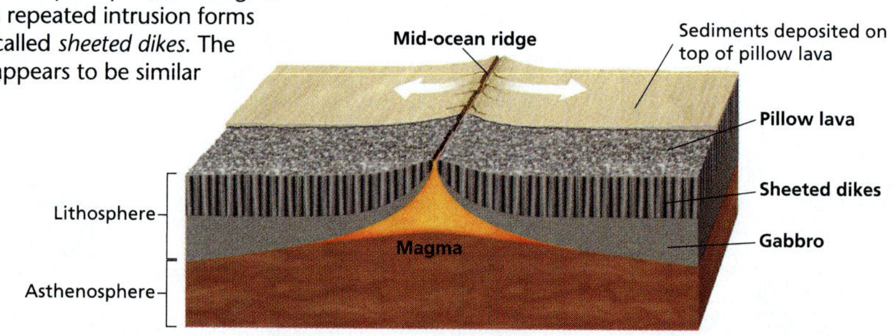

SOCIAL STUDIES CONNECTION — BASIC

Geothermal Energy Iceland is the largest volcanic island in the world, and one of the most volcanically active areas along the Mid-Atlantic Ridge. Bodies of groundwater indirectly heated by magma beneath the surface of Iceland have provided the island with hot water for heating since 1930. Steam from some of these hot springs is used to generate electric power, which has been done at the capital, Reykjavik, since 1964. Have students research and write a short report on other cases in Iceland where volcanic activity has been utilized for public benefit. LS **Verbal**

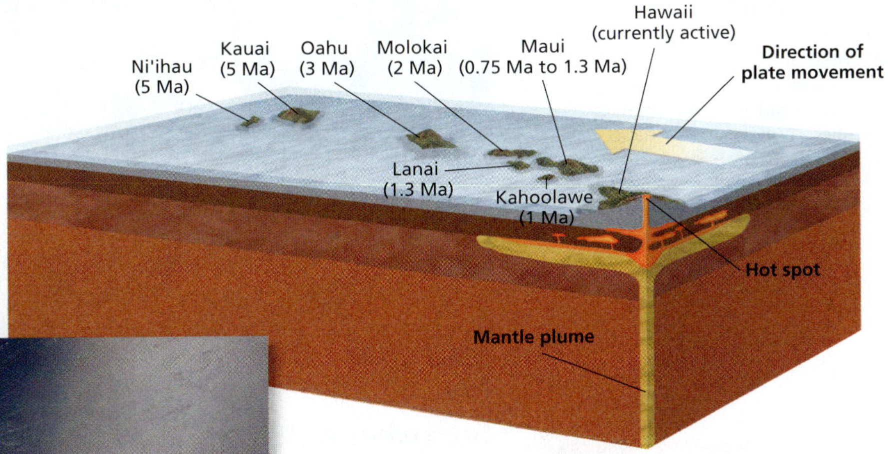

Figure 5 ▶ As a tectonic plate moves over a stationary mantle plume, volcanic mountains form (above). The Hawaiian Islands (left) formed in this way.

Hot Spots

Not all volcanoes develop along plate boundaries. Areas of volcanism within the interiors of lithospheric plates are called **hot spots.** Most hot spots form where columns of solid, hot material from the deep mantle, called *mantle plumes,* rise and reach the lithosphere. When a mantle plume reaches the lithosphere, the plume spreads out. As magma rises to the surface, it breaks through the overlying crust. Volcanoes can then form in the interior of a tectonic plate, as shown in **Figure 5.**

Mantle plumes appear to remain nearly stationary. However, the lithospheric plate above a mantle plume continues to drift slowly. So, the volcano on the surface is eventually carried away from the mantle plume. The activity of the volcano stops because a hot spot that contains magma no longer feeds the volcano. However, a new volcano forms where the lithosphere has moved over the mantle plume.

Other scientists think that hot spots are the result of cracks in Earth's crust. The theory argues that hot-spot volcanoes occur in long chains because they form along cracks in Earth's crust. Both theories may be correct.

Reading Check **Explain how one mantle plume can form several volcanic islands. (See the Appendix for answers to Reading Checks.)**

hot spot a volcanically active area of Earth's surface, commonly far from a tectonic plate boundary

Answer to Reading Check

As the lithosphere moves over the mantle plume, older volcanoes move away from the mantle plume. A new hot spot forms in the lithosphere above the mantle plume as a new volcano begins to form.

SKILL BUILDER — ADVANCED

Graphing Have students research hot spot activity in the Yellowstone National Park area. Have them learn the locations and dates of volcanic activity, and have them graph results to show the direction and speed the North American plate is moving. **LS Logical/Visual**

Close

Reteaching — BASIC

Volcanic Settings On the board, draw three diagrams to illustrate the separation of two lithospheric plates, the subduction of one plate below another, and a mantle plume rising to a hot spot. Then, write the words *hot spot, mid-ocean ridge,* and *subduction zone* in a column to the side. Have students indicate which words go with each image. **LS Visual**

Quiz — GENERAL

1. What happens to solid rock in the upper mantle and crust when pressure drops? (The rock melts, forming magma.)
2. Which tectonic setting that may result in volcanoes is not at a plate boundary? (hot spots)
3. What are structures that form when magma solidifies beneath Earth's surface called? (plutons)

CHAPTER RESOURCES

Chapter Resource File

- **Making Models Lab** Magma in Earth's Crust GENERAL

Technology

Transparencies

- 68 Hot Spots and Mantle Plumes (with worksheet)

MISCONCEPTION ALERT

Mantle Plumes and Hot Spots
Students may think that mantle plumes are melted rock. Stress that the hot rock in the mantle plume, like all the rock in the mantle, is solid. The plume spreads out as it rises to the lithosphere, but remains solid until pressure decreases and some of the rock melts to form magma. The resulting volcanoes on the surface, not the plume, are the hot spot.

Close, *continued*

Alternative Assessment — ADVANCED

Modeling Volcanic Processes
Have students create a computer model or a series of diagrams that shows the various stages in which a volcano forms by one of the three processes discussed in the section. The models should show the dynamic steps that lead to the formation of magma and its movement through the crust. **LS Kinesthetic/Visual**

Answers to Section Review

1. Magma forms when the temperature of the rock rises above the melting point of the minerals that compose it, when the pressure of the rock decreases faster than its temperature does, or when it is mixed with fluids that decrease its melting point.
2. Magma reaches Earth's surface by rising through less dense rock in the crust.
3. Magma is melted rock beneath Earth's surface; lava is magma that erupts onto Earth's surface.
4. The denser, subducted plate releases fluids to the surrounding crust and mantle materials, thus lowering their melting points and causing magma to form.
5. subduction zones, mid-ocean ridges, and hot spots
6. The hot spot lies above a column of solid, hot material from the mantle. When the mantle material spreads out at the top of the plume, its pressure drops and the rock melts. The resulting magma forms volcanoes at the hot spot.
7. Dikes are small plutons that are tabular in shape and may be only a few centimeters wide. Batholiths are much larger plutons that are exposed on Earth's surface over an area of at least 100 km^2.
8. Ocean water in crustal rock can lower the melting point of the rock and aid the formation of magma.
9. The volcanic activity in Yellowstone National Park would be due to mantle plume that caused a hot spot within the continental crust.
10. *Volcanism* is the movement of *magma*, which may reach Earth's surface through a *vent*, which is the center of a *volcano*, at a *subduction zone* or a *hot spot;* or may cool under Earth's surface to form a *pluton*, such as a *dike*.

Figure 6 ▶ Devils Tower in Wyoming is an example of a pluton, a formation caused by cooling of magma beneath Earth's surface. Erosion of surrounding rock has revealed this igneous formation.

SCILINKS. NSTA Developed and maintained by the National Science Teachers Association

For a variety of links related to this subject, go to www.scilinks.org

Topic: Volcanic Zones
SciLinks code: HQ61618

Intrusive Activity

Because magma is less dense than solid rock, magma rises through the crust toward the surface. As the magma moves upward, it comes into contact with, or *intrudes*, the overlying rock. Because of magma's high temperature, magma affects surrounding rock in a variety of ways. Magma may melt surrounding rock or may change the rock. Magma may also fracture surrounding rock and cause fissures to form or cause the surrounding rock to break apart and fall into the magma. Rock that falls into the magma may eventually melt, or the rock may combine with the new *igneous rock*, which is rock that forms when the magma cools.

When magma does not reach Earth's surface, the magma may cool and solidify inside the crust. This process results in large formations of igneous rock called *plutons*, as shown in **Figure 6.** Plutons can vary greatly in size and shape. Small plutons called *dikes* are tabular in shape and may be only a few centimeters wide. *Batholiths* are large plutons that cover an area of at least 100 km^2 when they are exposed on Earth's surface.

Section 1 Review

1. **Describe** three conditions that affect whether magma forms.
2. **Explain** how magma reaches Earth's surface.
3. **Compare** magma with lava.
4. **Describe** how subduction produces magma.
5. **Identify** three tectonic settings where volcanoes commonly occur.
6. **Summarize** the formation of hot spots.
7. **Describe** two igneous structures that form under Earth's surface.

CRITICAL THINKING

8. **Identifying Relationships** Describe how the presence of ocean water in crustal rocks might affect the formation of magma.
9. **Applying Ideas** Yellowstone National Park in Wyoming is far from any plate boundary. How would you explain the volcanic activity in the park?

CONCEPT MAPPING

10. Use the following terms to create a concept map: *magma, volcanism, vent, volcano, subduction zone, hot spot, dike,* and *pluton.*

CHAPTER RESOURCES

Chapter Resource File

- Section Quiz GENERAL

Workbooks

- Study Guide (also in Spanish)

Section 2 Volcanic Eruptions

Volcanoes can be thought of as windows into Earth's interior. Lava that erupts from them provides an opportunity for scientists to study the nature of Earth's crust and mantle. By analyzing the composition of volcanic rocks, geologists have concluded that there are two general types of magma. **Mafic** (MAF ik) describes magma or rock that is rich in magnesium and iron and is commonly dark in color. **Felsic** (FEL sik) describes magma or rock that is rich in light-colored silicate materials. Mafic rock commonly makes up the oceanic crust, whereas felsic and mafic rock commonly make up the continental crust.

OBJECTIVES

- **Explain** how the composition of magma affects volcanic eruptions and lava flow.
- **Describe** the five major types of pyroclastic material.
- **Identify** the three main types of volcanic cones.
- **Describe** how a caldera forms.
- **List** three events that may signal a volcanic eruption.

KEY TERMS

mafic
felsic
pyroclastic material
caldera

Types of Eruptions

The *viscosity,* or resistance to flow, of magma affects the force with which a particular volcano will erupt. The viscosity of magma is determined by the magma's composition. Because mafic magmas produce runny lava that has a low viscosity, they typically cause quiet eruptions. Because felsic magmas produce sticky lava that has a high viscosity, they typically cause explosive eruptions. Magma that contains large amounts of trapped, dissolved gases is more likely to produce explosive eruptions than is magma that contains small amounts of dissolved gases.

mafic describes magma or igneous rock that is rich in magnesium and iron and that is generally dark in color

felsic describes magma or igneous rock that is rich in feldspar and silica and that is generally light in color

Quiet Eruptions

Oceanic volcanoes commonly form from mafic magma. Because of mafic magma's low viscosity, gases can easily escape from mafic magma. Eruptions from oceanic volcanoes, such as those in Hawaii, shown in **Figure 1,** are usually quiet.

Figure 1 ▶ Lava flows from a quiet eruption like a red-hot river would flow. This lava flowed several miles from the Kilauea volcano to the sea.

Section 2

Focus

Overview

This section describes the effect of composition on the viscosity of magma, the types of pyroclastic materials, and the three main types of volcanoes. It also explains how calderas form, and how eruptions are predicted.

Bellringer

Have students draw a picture of a volcanic eruption. Use the pictures to start a discussion of the factors that determine the characteristics of an eruption. LS Visual

Motivate

Demonstration — GENERAL

Viscosity Wear a lab apron and goggles and cover a table with newpapers. Be sure students observe from more than three feet away. Obtain three 500 mL beakers. Place the beakers on the newspapers. Fill one beaker with 100 mL of water, the second with 100 mL of cooking oil, and the third with 100 mL of honey. Using a drinking straw, blow bubbles gently into each of the liquids. Have students note which liquid traps the most bubbles. (honey) Explain that more viscous fluids trap more gases. Increased pressure of trapped gases causes more violent eruptions. LS Visual

CHAPTER RESOURCES

Chapter Resource File

- Directed Reading BASIC

Technology

Transparencies
- Bellringer

Student Edition on CD-ROM

One-Stop Planner CD-ROM
- Lesson Plan

Teach

Using the Figure — BASIC

Lava Flows Emphasize to students that these flows apply specifically to mafic magma. Help students remember the names by noting the smooth, flowing sound of the word *pahoehoe* is like the smooth, flowing look of that rock. The word *aa* has a broken, abrupt sound, like the broken, sharp rock of that name. Point out that blocky lava is rougher and larger than the aa lava, as if made of aa blocks. Ask students why scientists use Hawaiian words to describe lava from mafic magma. (All of the lavas in Hawaii are from mafic magma.) **LS Visual**

MATHPRACTICE

Answer

number of years for lava flow = 2003 − 1986 = 17 years

$(2.5 \text{ km}^3/17 \text{ years}) \times (1{,}000{,}000{,}000 \text{ m}^3/\text{km}^3) = 2{,}500{,}000{,}000 \text{ m}^3/17 \text{ years} = 147 \text{ million m}^3/\text{year}$

Answer to Reading Check

The faster the rate of flow is and the higher the gas content is, the more broken up and rough the resulting cooled lava will be.

CHAPTER RESOURCES

Chapter Resource File

- **Inquiry Lab** Lava Flows GENERAL

Figure 2 ▶ Types of Mafic Lava Flow

Pahoehoe is the least viscous type of mafic lava. It forms wrinkly volcanic rock when it cools.

Aa lava is more viscous than pahoehoe lava and forms sharp volcanic rock when it cools.

Blocky lava is the most viscous type of mafic lava and forms chunky volcanic rock when it cools.

MATHPRACTICE

A Lot of Lava Since late 1986, Kilauea volcano in Hawaii has been erupting mafic lava. In 2003, the total volume of lava that had been produced by this eruption was 0.6 mi^3, or 2.5 km^3. Calculate the average amount of lava, in cubic meters, that erupts from Kilauea each year.

Lava Flows

When mafic lava cools rapidly, a crust forms on the surface of the flow. If the lava continues to flow after the crust forms, the crust wrinkles to form a volcanic rock called *pahoehoe* (pah HOH ee HOH ee), which is shown in **Figure 2.** Pahoehoe forms from hot, fluid lava. As it cools, it forms a smooth, ropy texture. Pahoehoe actually means "ropy" in Hawaiian.

If the crust deforms rapidly or grows too thick to form wrinkles, the surface breaks into jagged chunks to form *aa* (AH AH). Aa forms from lava that has the same composition as pahoehoe lava. Aa lava's texture results from differences in gas content and in the rate and slope of the lava flow.

Blocky lava has a higher silica content than aa lava does, which makes blocky lava more viscous than aa lava. The high viscosity causes the cooled lava at the surface to break into large chunks, while the hot lava underneath continues to flow. This process gives the lava flow a blocky appearance.

✓ Reading Check **How do flow rate and gas content affect the appearance of lavas? (See the Appendix for answers to Reading Checks.)**

Explosive Eruptions

Unlike the fluid lavas produced by oceanic volcanoes, the felsic lavas of continental volcanoes, such as Mount St. Helens, tend to be cooler and stickier. Felsic lavas also contain large amounts of trapped gases, such as water vapor and carbon dioxide. When a volcano erupts, the dissolved gases within the lava escape and send molten and solid particles shooting into the air. So, felsic lava tends to explode and throw pyroclastic material into the air. **Pyroclastic material** consists of fragments of rock that form during a volcanic eruption.

pyroclastic material fragments of rock that form during a volcanic eruption

Cultural Awareness — GENERAL

Vesuvius—Then and Now An eruption of Vesuvius, a volcano near what is now Naples, Italy, destroyed the thriving Roman towns of Pompeii and Herculaneum. In a short time, tons of erupted material buried people and animals alive. The eruption also engulfed—and preserved—public buildings, homes, and important art and engineering works. Scholars have excavated the sites and study the artifacts. Today, visitors can stroll through parts of preserved 2000-year-old homes, markets, and public baths. Colorful painted murals, mosaics, and statuary reflect how people lived. An archeological museum holds many additional artifacts. Have interested students research: what kind of eruption buried these towns; how volcanic ash preserves objects; how the sites were excavated; and what types of information scholars have learned. Students may work individually or in groups, and should prepare a written or oral report or create a diorama. **LS Verbal/Kinesthetic**

Types of Pyroclastic Material

Some pyroclastic materials form when magma breaks into fragments during an eruption because of the rapidly expanding gases in the magma. Other pyroclastic materials form when fragments of erupting lava cool and solidify as they fly through the air.

Scientists classify pyroclastic materials according to the sizes of the particles, as shown in **Figure 3.** Pyroclastic particles that are less than 2 mm in diameter are called *volcanic ash*. Volcanic ash that is less than 0.25 mm in diameter is called *volcanic dust*. Most volcanic dust and ash settles on the land that immediately surrounds the volcano. However, some of the smallest dust particles may travel around Earth in the upper atmosphere.

Large pyroclastic particles that are less than 64 mm in diameter, are called *lapilli* (luh PIL IE), which is from a Latin word that means "little stones." Lapilli generally fall near the vent.

Large clots of lava may be thrown out of an erupting volcano while they are red-hot. As they spin through the air, they cool and develop a round or spindle shape. These pyroclastic particles are called *volcanic bombs*. The largest pyroclastic materials, known as *volcanic blocks*, form from solid rock that is blasted from the vent. Some volcanic blocks are the size of a small house.

Graphic Organizer **Spider Map**

Create the **Graphic Organizer** entitled "Spider Map" described in the Skills Handbook section of the Appendix. Label the circle "Pyroclastic material." Create a leg for each type of pyroclastic material. Then, fill in the map with details about each type of pyroclastic material.

Figure 3 ▶ During an explosive eruption, like this one at Mount St. Helens, ash, blocks, and other pyroclastic materials are ejected violently from the volcano.

Graphic Organizer GENERAL

Spider Map

You may want to have students work in groups to create this spider map. Have one student draw the map and fill in information provided by other students from the group.

Activity — ADVANCED

Poster Project Have interested students research the classification system for volcanic explosivity—the Volcano Explosivity Index (VEI)—developed by geologists C. G. Newhall and Steven Self in 1982. Students should report the various eruption properties, such as amount of material ejected and height of eruptive column, that contribute to the number assigned a type of eruption. The roles of qualitative description and the Mercalli eruption classification system in the VEI should be included. Have students present their findings on a poster or make a report to the class. **LS Verbal/Visual**

Discussion — GENERAL

Mount St. Helens Have students research the 2004 eruption of Mt. St. Helens. Have students determine what tectonic setting led to the formation of this volcano. Have them determine whether the eruption was explosive or quiet, list the types of materials that were expelled from the volcano, and describe how the eruption affected people in the surrounding area. **LS Verbal/Interpersonal**

ART CONNECTION — GENERAL

The Scream Volcanoes and their eruptions have often influenced artists. One example is the work of the Norwegian painter Edvard Munch (ED vard MOONK), who in the 1880s painted a number of emotionally vivid works in a style called *expressionism*. The most famous of these is a work called "The Scream," which shows a screaming figure on a road against a reddish-orange sky. Recent research into Munch's paintings of this period indicate that the inspiration for "The Scream" and its companion works "Despair" and "Anxiety" occurred in the autumn or winter of 1883, shortly after the eruption of Krakatau. Have students research other works of art that may have been influenced by volcanoes or volcanic eruptions. Provide students with the names of artists, such as Katsushika Hokusai, Andy Warhol, and Jerry Garcia, as a starting point. Have students present their findings in a written report or as a poster or multimedia presentation. **LS Visual/Verbal**

Teach, *continued*

Using the Figure — GENERAL

Volcano Types Explain that the three cross-sectional models of volcanoes shown are not to the same scale. Because shield volcanoes are produced from lava that flows over a large area, the ratio of their height to the area of their base is small. The pyroclastic material that forms cinder cones is ejected into the air and falls close to the vent, so the ratio of the height to the base is large. Ask students what they would expect the ratio of height to base area to be for a composite volcano, compared to shield volcanoes and cinder cones. (The height to base area ratio for a composite volcano would be larger than for a shield volcano, but smaller than for a cinder cone.) Answer to caption question: A cinder cone forms from highly viscous lava. **LS Visual**

READING SKILL BUILDER — BASIC

Reading Hint Help students remember which types of lavas form which types of volcanoes. They can remember shield volcanoes by thinking of the flat sheets of rock that are layered, like the sheets of metal or leather on a shield. A cinder cone is made up of ash, which is another word for cinders. A *com*posite volcano is a *com*bination of both kinds of volcanic material. **LS Verbal/Visual** English Language Learners

CHAPTER RESOURCES

Technology

Transparencies
- 69 Types of Volcanoes (with worksheet)

Types of Volcanoes

Volcanic activity produces a variety of characteristic features that form during both quiet and explosive eruptions. The lava and pyroclastic material that are ejected during volcanic eruptions build up around the vent and form volcanic cones. Volcanic cones are classified as three main types, as described in **Table 1.**

The funnel-shaped pit at the top of a volcanic vent is known as a *crater*. The crater forms when material is blown out of the volcano by explosions. A crater usually becomes wider as weathering and erosion break down the walls of the crater and allow loose materials to collapse into the vent. Sometimes, a small cone forms within a crater. This formation occurs when subsequent eruptions cause material to build up around the vent.

Table 1 ▶ Volcanic cones are classified into three main categories. ***Which type of volcano would form from lava that is highly viscous?***

Types of Volcanoes

Shield Volcanoes Volcanic cones that are broad at the base and have gently sloping sides are called *shield volcanoes*. A shield volcano covers a wide area and generally forms from quiet eruptions. Layers of hot, mafic lava flow out around the vent, harden, and slowly build up to form the cone. The Hawaiian Islands form a chain of shield volcanoes that built up from the ocean floor at a hot spot.

Cinder Cones A type of volcano that has very steep slopes is a cinder cone. The slope angles of the cinder cones can be close to 40°, and the slopes are rarely more than a few hundred meters high. Cinder cones form from explosive eruptions and are made of pyroclastic material.

Composite Volcanoes Composite volcanoes are made of alternating layers of hardened lava flows and pyroclastic material. During a quiet eruption, lava flows cover the sides of the cone. Then, when an explosive eruption occurs, large amounts of pyroclastic material are deposited around the vent. The explosive eruption is followed again by quiet lava flows. Composite volcanoes, also known as *stratovolcanoes*, commonly develop to form large volcanic mountains.

ASTRONOMY CONNECTION — ADVANCED

Extraterrestrial Volcanoes Volcanic formations and ancient lava flows have been identified on the planets Mars and Venus. In fact, Mars boasts the largest volcano in the solar system: Olympus Mons. This shield volcano has a base of nearly 300,000 km^2 and a summit that rises 24 km above the surrounding surface. Jupiter's moon Io and Neptune's moon Triton also exhibit volcanic-like behaviors. Have interested students research volcanic activity on other planets and moons. Students should present their findings in a written or oral report. **LS Verbal**

Group Activity — GENERAL

Historic Eruptions Organize the class into small groups. Have each group research a volcanic eruption, such as Santorini, Vesuvius, Tambora, Krakatau, Pelée, and Mt. St. Helens. One or two students can research what made the event catastrophic, such as caldera collapse, heavy fall of ash, or pyroclastic flows. Another can report the effects on nearby populations. The remaining student can research the effects at greater distances, such as the effect on climate and on coastal areas affected by tsunamis. **LS Verbal** Co-op Learning

Figure 4 ▶ The Formation of a Caldera

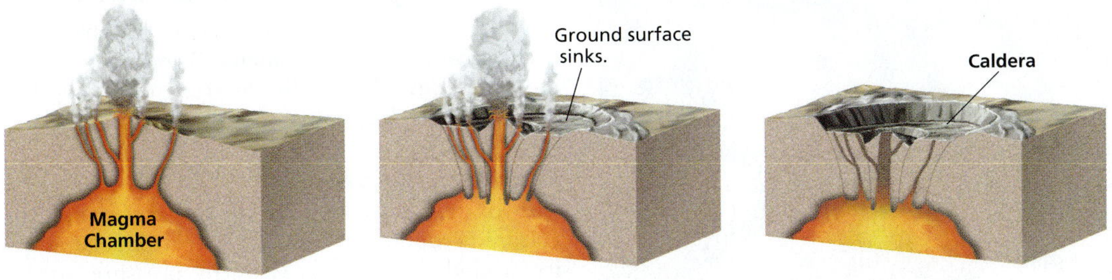

A cone forms from volcanic eruptions.

Volcanic eruptions partially empty the magma chamber.

The top of the cone collapses inward to form a caldera.

Calderas

When the magma chamber below a volcano empties, the volcanic cone may collapse and leave a large, basin-shaped depression called a **caldera** (kal DER uh). The process of caldera formation is shown in **Figure 4.**

Eruptions that discharge large amounts of magma can also cause a caldera to form. Krakatau, a volcanic island in Indonesia, is an example of this type of caldera. When the volcanic cone exploded in 1883, a caldera with a diameter of 6 km formed.

Calderas may later fill with water to form lakes. Thousands of years ago, the cone of Mount Mazama in Oregon collapsed and formed a caldera. The caldera eventually filled with water and is now called Crater Lake.

caldera a large, circular depression that forms when the magma chamber below a volcano partially empties and causes the ground above to sink

Reading Check Describe two ways that calderas form. (See the Appendix for answers to Reading Checks.)

QuickLAB 25 min

Volcanic Cones

Procedure

1. Pour **1/2 cup (about 4 oz) of dry plaster of Paris** into a **measuring cup.**
2. Use a **graduated cylinder** to measure **60 mL of water,** and add the water to the dry plaster in the measuring cup. Use a **mixing spoon** to blend the mixture until it is smooth.
3. Hold the measuring cup about 2 cm over a **paper plate.** Pour the contents slowly and steadily onto the center of the plate. Allow the plaster to dry.
4. On a clean paper plate, pour **dry oatmeal or potato flakes** slowly until the mound is approximately 5 cm high.
5. Without disturbing the mound, use a **protractor** to measure its slope.
6. When the plaster cone has hardened, remove it from the plate. Measure the average slope angle of the cone.

Analysis

1. Which cone represents a cinder cone? Which cone represents a shield volcano? Compare the slope angles formed by these cones.
2. How would the slope be affected if the oatmeal were rounder, and how would the slope be affected if the oatmeal were thicker?
3. How would you use the same supplies to model a composite volcano?

QuickLAB

Skills Acquired

- Observing
- Classifying
- Interpreting
- Constructing Models

Teacher's Notes Be sure that the plaster of Paris is not too thick, so that it pours easily.

Answers

1. The oatmeal represents a cinder cone. The plaster represents a shield volcano. The cinder cone has a steeper slope than the shield volcano does.
2. The oatmeal would spread out more if its pieces were more round. It would form a steeper cone if it were thicker.
3. The plaster and oatmeal would be applied in alternating layers to model a composite volcano.

INCLUSION Strategies

- *Learning Disabled*
- *Developmentally Delayed*

Ask each student to choose three (no duplicates) active or once-active volcanoes and create a poster that shows which of the following attributes apply to each volcano: Quiet eruptions, Pahoehoe, Aa, Mafic lava, Felsic lava, Caldera, Shield volcano, Cinder cone, Composite volcano, Explosive, eruptions, Ash, and Pyroclastic material. **LS Logical**

Answer to Reading Check

A caldera may form when a magma chamber empties or when large amounts of magma are discharged, causing the ground to collapse.

Close

Reteaching BASIC

Volcanoes, Magma, and Eruptions Make several cards with one of the following terms on each: *mafic magma, felsic magma, quiet eruption, explosive eruption, shield volcano, cinder cone.* Have students combine the cards to form two groups containing three cards that describe one type of volcano. **LS Verbal**

Quiz GENERAL

1. What can you tell about the viscosity of the magma that formed a steep volcano of ash and blocks? (The magma was very viscous.)
2. How would you describe the slope of a mountain that formed from smooth lava flows? (The slope would be gentle with a low incline.)

CHAPTER RESOURCES

Chapter Resource File

- Datasheet for Quick Lab GENERAL

Close, *continued*

Alternative Assessment — ADVANCED

Modeling Volcano Formation Have students create a model that shows the process of volcano formation or the process of caldera collapse. The model should show several stages and may be made of plaster, papier mâché, or clay, or it may be a computer model. **LS Kinesthetic**

Answers to Section Review

1. Mafic magma has more magnesium and iron, is darker, and is less viscous than felsic magma, which is rich in silica and feldspars.
2. More viscous magma traps gases more easily than less viscous magma, which may lead to more explosive eruptions.
3. Pahoehoe is smooth, ropy lava rock. When the same magma flows at a different rate or has a different gas content, it may break into jagged chunks called aa. If the magma is more viscous than aa lava, the resulting large chunks are called *blocky lava.*
4. Pyroclastic material consists of fragments of rock that form during an explosive eruption, such as volcanic dust, volcanic ash, lapilli, volcanic bombs, and volcanic blocks.
5. shield volcanoes, cinder cones, and composite volcanoes
6. A caldera forms when the magma chamber of a volcano empties, causing the volcanic cone to collapse in upon it, or when magma is ejected violently and the cone is destroyed.
7. Earthquake activity may increase, rock temperatures may increase, and the volcano's surface may begin to bulge.
8. Explosive eruptions are more likely to increase volcano height, because the pyroclastic materials rise upward and fall close to the volcanic vent.
9. Sudden earthquake activity could be caused by magma moving upward through the rock around the volcano.
10. Volcanoes form from *mafic lava,* which forms *shield volcanoes* from *pahoehoe* and *aa,* or from *felsic lava,* which produces *pyroclastic material* such as *volcanic dust, volcanic ash, lapilli, volcanic blocks,* and *volcanic bombs.*

Figure 5 ▶ These scientists are sampling gases emitted from the fumarole field on Vulcano Island in Italy.

Predicting Volcanic Eruptions

A volcanic eruption can be one of Earth's most destructive natural phenomena. Scientists, such as those in **Figure 5,** look for a variety of events that may signal the beginning of an eruption.

Earthquake Activity

One of the most important warning signals of volcanic eruptions is changes in earthquake activity around the volcano. Growing pressure on the surrounding rocks from magma that is moving upward causes small earthquakes. Temperature changes within the rock and fracturing of the rock around a volcano also cause small earthquakes. An increase in the strength and frequency of earthquakes may be a signal that an eruption is about to occur.

SCILINKS. NSTA Developed and maintained by the National Science Teachers Association

For a variety of links related to this subject, go to www.scilinks.org

Topic: Predicting Volcanic Eruptions
SciLinks code: HQ61209

Patterns in Activity

Before an eruption, the upward movement of magma beneath the surface may cause the surface of the volcano to bulge outward. Special instruments can measure small changes in the tilt of the ground surface on the volcano's slopes.

Predicting the eruption of a particular volcano also requires some knowledge of its previous eruptions. Scientists compare the volcano's past behavior with current daily measurements of earthquakes, surface bulges, and changes in the amount and composition of the gases that the volcano emits. Unfortunately, only a few of the active volcanoes in the world have been studied by scientists long enough to establish any activity patterns. Also, volcanoes that have been dormant for long periods of time may, with little warning, suddenly become active.

Section 2 Review

1. **Summarize** the difference between mafic and felsic magma.
2. **Explain** how the composition of magma affects the force of volcanic eruptions.
3. **Compare** three major types of lava flows.
4. **Define** *pyroclastic material,* and list three examples.
5. **Identify** the three main types of volcanic cones.
6. **Describe** how calderas form.
7. **List** three events that may precede a volcanic eruption.

CRITICAL THINKING

8. **Applying Ideas** Would quiet eruptions or explosive eruptions be more likely to increase the steepness of a volcanic cone? Explain your answer.
9. **Drawing Conclusions** Why would a sudden increase of earthquake activity around a volcano indicate a possible eruption?

CONCEPT MAPPING

10. Use the following terms to create a concept map: *mafic lava, felsic lava, pahoehoe, aa, shield volcano, pyroclastic material, lapilli, volcanic bomb, volcanic block, volcanic ash,* and *volcanic dust.*

CHAPTER RESOURCES

Chapter Resource File
- Section Quiz GENERAL

Workbooks
- Study Guide (also in Spanish)

Chapter 13 Highlights

Sections

1 Volcanoes and Plate Tectonics

Key Terms

magma, 319
volcanism, 320
lava, 320
volcano, 320
hot spot, 323

Key Concepts

- Magma can form when temperature or pressure changes in mantle rock. Magma also may form when water is added to hot rock.
- Volcanism is any activity that includes the movement of magma onto Earth's surface.
- Volcanism is common at convergent and divergent boundaries between tectonic plates.
- Hot spots are areas of volcanic activity that are located over rising mantle plumes that can exist far from tectonic plate boundaries.
- Magma that cools below Earth's surface forms intrusive igneous rock bodies called plutons.

2 Volcanic Eruptions

mafic, 325
felsic, 325
pyroclastic material, 326
caldera, 329

- Lava and magma can be described as mafic or felsic.
- Hot, less viscous, mafic lava commonly causes quiet eruptions. Cool, more viscous, felsic lava commonly causes explosive eruptions, especially if it contains trapped gases.
- Volcanic cones are classified into three categories based on composition and form.
- A caldera forms where a volcanic cone collapses and leaves a large, basin-shaped depression.
- Events that might signal a volcanic eruption include changes in earthquake activity, changes in the volcano's shape, changes in composition and amount of gases emitted, and changes in the patterns of the volcano's normal activity.

Chapter Highlights

Alternative Assessment — GENERAL

Chain-of-Events Charts Have students create three chain-of-events charts, one for each type of volcano: shield, cinder cone, and composite. Direct students to start each chart with a plate tectonic setting. Students should then introduce as side chains the magma type and viscosity, the resulting erupted materials (from smooth lavas to pyroclastic materials) and the type of eruption that will occur (explosive or quiet). Each chain will end with one particular type of volcano. LS **Logical/Visual**

CHAPTER RESOURCES

Chapter Resource File

- **Concept Review** GENERAL
- **Critical Thinking** ADVANCED
- **Math Skills** GENERAL
- **Graphing Skills** GENERAL
- **Chapter Test A** GENERAL
- **Chapter Test B** ADVANCED

Workbooks

- **Study Guide** (also in Spanish)
- **Assessments** (Spanish)

Technology

Classroom Videos
- Brain Food Video Quiz

HRW Earth Science Video
- Segment 6: Volcanoes

Assignment Guide	
Section	**Questions**
1	1, 2, 9–10, 16–19, 24–25, 33
2	3–8, 11–15, 20–23, 26, 28, 30–32, 34–36
1 and 2	27, 29

Using Key Terms

1–7. Answers may vary but should show that students understand the definitions of and differences between key terms.

Understanding Key Concepts

8. b **9.** d
10. a **11.** b
12. b **13.** b
14. b **15.** b

Short Answer

16. Magma becomes lava when it erupts onto Earth's surface.

17. Tectonic movement can increase the temperature of rock, reduce the pressure on rock, or add fluids to rock, causing the rock to melt. This molten rock, or magma, rises through the crust to form volcanoes.

18. volcanism

19. Magma that doesn't reach Earth's surface may solidify beneath the surface, resulting in a formation called a *pluton*.

20. Composition affects magma's viscosity. Lower viscosity magma yields quiet eruptions; more viscous magma yields explosive eruptions.

21. Small pyroclastic material (less than 2 mm in diameter) consists of volcanic ash and volcanic dust. Pyroclastic material less than 64 mm in diameter but greater than 2 mm in diameter is called *lapilli*. The largest pyroclastic materials are volcanic bombs and volcanic blocks.

22. Shield volcanoes are large, gently sloping mountains that form from layers of low-viscosity lava. Cinder cones are volcanoes that have steep sides made of accumulated pyroclastic material. Composite volcanoes consist of alternating layers of lava flows and pyroclastic material.

23. increasing earthquake activity, rising rock temperature, and bulging of the volcano's surface

Chapter 13 Review

Using Key Terms

Use each of the following terms in a separate sentence.

1. *volcanism*
2. *hot spot*
3. *pyroclastic material*

For each pair of terms, explain how the meanings of the terms differ.

4. *magma* and *lava*
5. *mafic* and *felsic*
6. *shield volcano* and *composite volcano*
7. *crater* and *caldera*

Understanding Key Concepts

8. A characteristic of lava that determines the force of a volcanic eruption is
a. color. **c.** density.
b. viscosity. **d.** age.

9. Island arcs form when oceanic lithosphere subducts under
a. continental lithosphere.
b. calderas.
c. volcanic bombs.
d. oceanic lithosphere.

10. Areas of volcanism within tectonic plates are called
a. hot spots. **c.** calderas.
b. cones. **d.** fissures.

11. Explosive volcanic eruptions commonly result from
a. mafic magma. **c.** aa lava.
b. felsic magma. **d.** pahoehoe lava.

12. Pyroclastic materials that form rounded or spindle shapes as they fly through the air are called
a. ash. **c.** lapilli.
b. volcanic bombs. **d.** volcanic blocks.

13. A cone formed by only solid fragments built up around a volcanic opening is a
a. shield volcano.
b. cinder cone.
c. composite volcano.
d. stratovolcano.

14. The depression that results when a volcanic cone collapses over an emptying magma chamber is a
a. crater. **c.** vent.
b. caldera. **d.** fissure.

15. Scientists have discovered that before an eruption, earthquakes commonly
a. stop.
b. increase in number.
c. have no relationship with volcanism.
d. decrease in number.

Short Answer

16. At what point does magma become lava?

17. Describe how tectonic movement can form volcanoes.

18. Name the process that includes the movement of magma onto Earth's surface.

19. What may happen to magma that does not reach Earth's surface?

20. How is the composition of magma related to the force of volcanic eruptions?

21. List and describe the major types of pyroclastic material.

22. Compare the three main types of volcanic cones.

23. What signs can scientists study to try to predict volcanic eruptions?

Critical Thinking

24. Most lava comes to Earth's surface where two oceanic plates diverge, deep under the ocean surface. The lava is therefore unobserved.

Critical Thinking

24. Analyzing Ideas Why is most lava that forms on Earth's surface unnoticed and unobserved?

25. Identifying Relationships The Pacific Ring of Fire is a zone of major volcanic activity because of tectonic plate boundaries. Identify another area of Earth where you might expect to find volcanic activity.

26. Analyzing Processes Why does felsic lava tend to form composite volcanoes and cinder cones rather than shield volcanoes?

27. Making Inferences How might geologists distinguish an impact crater on Earth, such as Meteor Crater in Arizona, from a volcanic crater?

28. Making Comparisons Sinkholes form when the roof of an underground cave is not supported by groundwater. Compare this process to the process by which calderas form.

Concept Mapping

29. Use the following terms to create a concept map: *magma, lava, volcano, pluton, mafic lava, felsic lava, pyroclastic material, volcanic ash, volcanic dust, lapilli, volcanic bomb, volcanic block, volcanic cone, shield volcano, cinder cone,* and *composite volcano*.

Math Skills

30. Making Calculations On day 1, a volcano expelled 5 metric tons of sulfur dioxide. On day 2, the same volcano expelled 12 metric tons of sulfur dioxide. What is the percentage increase of sulfur dioxide expelled from day 1 to day 2?

31. Interpreting Statistics A lava flow travels for 7.3 min before it flows into the ocean. The velocity of the lava is 3 m/s. How far did the lava flow travel?

Writing Skills

32. Outlining Topics Outline the essential steps in the process of caldera formation.

33. Communicating Main Ideas Write an essay describing the formation of a volcano.

Interpreting Graphics

The graphs below show data about earthquake activity; the slope angle, or *tilt*, of the ground; and the amount of gas emitted for a particular volcano over a period of 10 days. Use the graphs to answer the questions that follow.

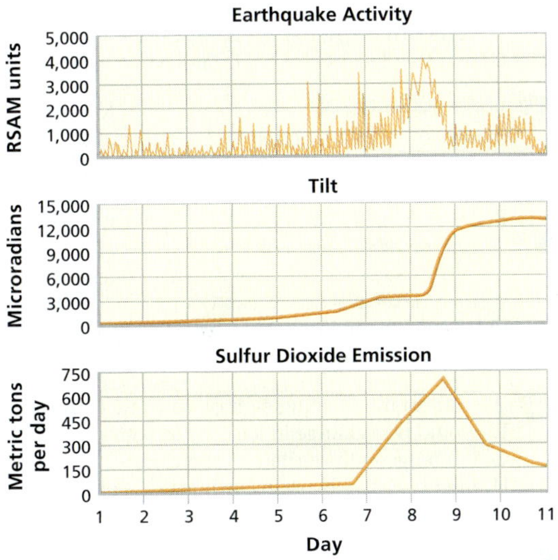

34. On what day did the volcano erupt? Explain your answer.

35. For how many days before the eruption did gas emission increase?

36. Why do you think the slope angle of the ground did not return to its original angle after the eruption?

Chapter Review

25. Answers may vary. Sample answer: The Mid-Atlantic Ridge, the Mediterranean fault zone, the rift zone along eastern Africa, and the hot spot near Yellowstone National Park are all locations of volcanic activity.

26. Lava from felsic magma is too viscous to flow smoothly and form shield volcanoes. Gas trapped in the felsic magmas contributes to explosive eruptions, which form the pyroclastic materials found in cinder cones and composite volcanoes.

27. An impact crater would not have the large amounts of lava flows or pyroclastic materials that would be found in a volcanic crater.

28. Both sinkholes and calderas form when there is no fluid to support the ground above the empty underground chamber.

Concept Mapping

29. Answers may vary, but should include all of the terms listed. Sample answers appear at the end of this Teacher Edition.

Math Skills

30. percent increase = [(new amount − old amount)/old amount] × 100; percent increase = [(12 metric tons − 5 metric tons)/5 metric tons] × 100 = 140%

31. distance = velocity × time; distance = 3 m/s × 7.3 min × 60 s/min = 1,314 m, or 1.314 km

Writing Skills

32. Answers may vary. Accept all reasonable answers.

33. Answers may vary. Accept all reasonable answers.

Interpreting Graphics

34. The volcano erupted in the middle of the eighth day, when earthquake activity, slope, and gas emission increased suddenly.

35. 2 days

36. The ground surrounding the volcano may have been permanently distorted or changed by the eruption, either by being covered with new lava or pyroclastic materials, or by being blasted away by the eruption.

Standardized Test Prep

Estimated Time

To give students practice under more realistic testing conditions, allow them 30 minutes to answer all of the questions in this practice test.

TEST DOCTOR

Question 1 Answer B is correct. Answers A and D are incorrect because basalt and mafic magma make up most of Earth's *oceanic* crust. Answer C is incorrect because although limestone is a common crustal rock, it is formed from sediments and is not volcanic in origin.

Question 10 Full-credit answers should include the following points:

- erupting volcanoes throw out dust, ash, fragments of rock, and lava, as well as dissolved CO_2 gas and sulphur compounds
- eruptions can send gases and volcanic dust high into the atmosphere, where they are able to travel all over the globe
- volcanic particles could contribute to global warming by providing surfaces for ozone reactions or by adding CO_2, a greenhouse gas, to the air
- multiple large-scale eruptions over a short period of time, would be required to produce longer-term effects to the climate
- if dust blocked sunlight for a long period of time, a reduction in plant growth could lead to ecological imbalances, possibly long-term

Chapter 13 Standardized Test Prep

Understanding Concepts

Directions (1–5): **For *each* question, write on a separate sheet of paper the letter of the correct answer.**

1 What type of volcanic rock commonly makes up much of the continental crust?
- **A.** basalt rock that is rich in olivines
- **B.** felsic rock that is rich in silicates
- **C.** limestone that is rich in calcium carbonate
- **D.** mafic rock that is rich in iron and magnesium

2 Which of the following formations results from magma that cools before it reaches Earth's surface?
- **F.** batholiths
- **G.** mantle plumes
- **H.** volcanic blocks
- **I.** aa lava

3 How does volcanic activity contribute to plate margins where new crust is being formed?
- **A.** Where plates collide at subduction zones, rocks melt and form pockets of magma.
- **B.** Between plate boundaries, hot spots may form a chain of volcanic islands.
- **C.** When plates pull apart at oceanic ridges, magma creates new ocean floor.
- **D.** At some boundaries, new crust is formed when one plate is forced on top of another.

4 An important warning sign of volcanic activity
- **F.** would be a change in local wind patterns
- **G.** is a bulge in the surface of the volcano
- **H.** might be a decrease in earthquake activity
- **I.** is a marked increase in local temperatures

5 Which aspect of mafic lava is important in the formation of smooth, ropy pahoehoe lava?
- **A.** a fairly high viscosity
- **B.** a fairly low viscosity
- **C.** rapidly deforming crust
- **D.** rapid underwater cooling

Directions (6–7): **For *each* question, write a short response.**

6 What is the name for rounded blobs of lava formed by the rapid, underwater cooling of lava?

7 Where is the Ring of Fire located?

Reading Skills

Directions (8–10): **Read the passage below. Then, answer the questions.**

Volcanoes That Changed the Weather

In 1815, Mt. Tambora in Indonesia erupted violently. Following this eruption, one of the largest recorded weather-related disruptions of the last 10,000 years occurred throughout North America and Western Europe. The year 1816 became known as "the year without a summer." Snowfalls and a killing frost occurred during the summer months of June, July, and August of that year. A similar, but less severe episode of cooling followed the 1991 eruption of Mt. Pinatubo. Eruptions such as these can send gases and volcanic dust high into the atmosphere. Once in the atmosphere the gas and dust travel great distances, block sunlight, and cause short-term cooling over large areas of the globe. Some scientists have even suggested a connection between volcanoes and the ice ages.

8 What can be inferred from the passage?
- **A.** Earthquakes can create the same atmospheric effects as volcanoes do.
- **B.** Volcanic eruptions can have effects far beyond their local lava flows.
- **C.** Major volcanic eruptions are common events.
- **D.** The year 1815 also had a number of earthquakes and other natural disasters.

9 According to the passage, which of the following statements is false?
- **F.** The year 1816 became known as "the year without a summer."
- **G.** The world experienced a period of unusually warm weather after Mt. Pinatubo erupted
- **H.** Mt. Pinatubo erupted in 1991.
- **I.** Eruptions send gas and dust into the atmosphere, where they travel around the globe.

10 The eruptions described in the passage changed the weather briefly. Some scientists believe that periods of severe volcanic activity can produce long-term changes to the climate. Suggest one specific way in which the materials sent into the atmosphere by volcanoes might cause long-term changes to global climate and temperature?

Answers

Understanding Concepts

1. B
2. F
3. C
4. G
5. B
6. pillow lava
7. The Ring of Fire surrounds the Pacific Ocean.

Reading Skills

8. B
9. G
10. Answers may vary. See Test Doctor for a detailed scoring rubric.

Interpreting Graphics

11. Answers may vary. See Test Doctor for a detailed scoring rubric.
12. D
13. Answers may vary. See Test Doctor for a detailed scoring rubric.

Interpreting Graphics

Directions (11–13): **For *each* question below, record the correct answer on a separate sheet of paper.**

Base your answers to question 11 on the cross-section below, which shows volcanic activity in the Cascade region of the Pacific West Coast.

Cross-Section of the Juan de Fuca Ridge

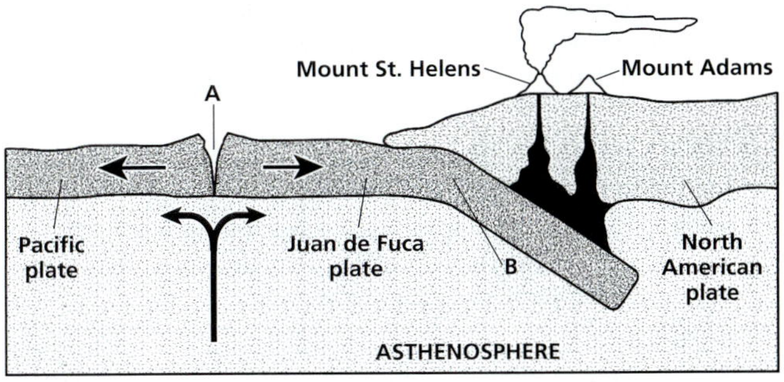

11 Explain how the tectonic activity near point B causes the volcanic activity at Mount St. Helens and Mount Adams in the Cascade Range?

Base your answers to questions 12 and 13 on the diagram of the interior of a volcano shown below.

Interior of a Volcano

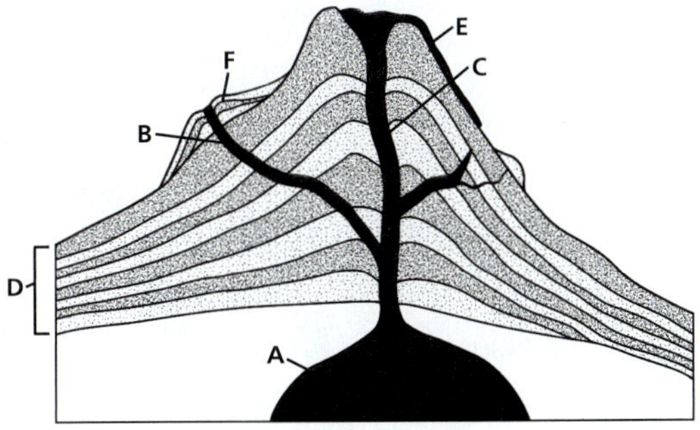

12 What is the term for the underground pool of molten rock, marked by the letter A, that feeds the volcano?
A. fissure
B. intrusion
C. lava pool
D. magma chamber

13 Letter D shows alternating layers in the volcanic cone. What are these layers made of, and what does this lead you to believe about the type of volcano that is represented in the diagram above?

Test TIP

When using a diagram to answer questions, carefully study each part of the figure as well as any lines or labels used to indicate parts of the diagram.

TEST DOCTOR

Question 11 Full-credit answers should include the following points:
- students should understand that where the oceanic crust of the Juan de Fuca plate meets the continental crust of the North American plate, a deep-ocean trench forms. Along this trench, the Juan de Fuca plate subducts beneath the North American plate
- the subducting oceanic crust and some continental material melt and supply mixed mafic and felsic magma
- the magma rises through the crust to form the volcanoes of the Cascade Range

Question 13 Full-credit answers should include the following points:
- students should show a conceptual understanding that volcanoes can form in several different ways and that the way in which a volcano forms determines the shape of the cone and type of volcanic structure
- the diagram shows alternating layers and a steep angle of the cone
- the shape and layered composition of the volcano indicate that this volcano formed over a long period of time as different eruptions caused cooled lava and other pyroclastic materials to build up
- composite volcanoes form in this manner

Test Prep Correlations *National Science Education Standards*

ES 1b: items 3, 7, 11
ES 1d: items 8, 9, 10
SAI 2a: items 3, 4, 8, 10, 11, 12, 13
SAI 2e: items 3, 5, 10, 11
SPSP 5a: items 4, 8, 9, 10
SPSP 5c: items 4, 8, 9, 10
ST 2c: item 10
UCP 1: items 3, 11

CHAPTER RESOURCES

State Resources

For specific resources for your state, visit go.hrw.com and type in the keyword **HSHSTR.**

Volcano Verdict

Teacher's Notes

Time Required
one 45-minute class period

Lab Ratings

EASY → HARD

Teacher Preparation ▲▲▲
Student Setup ▲▲▲
Concept Level ▲▲
Cleanup ▲▲▲

Skills Acquired
- Constructing Models
- Observing
- Designing Experiments
- Identifying and Recognizing Patterns
- Inferring

The Scientific Method
In this lab, students will
- Make Observations
- Analyze the Results
- Draw Conclusions

Materials
The materials listed on the page are enough for groups of two to four students. You may wish to have your first-period students prepare the clay plugs and straws, and then use these same plugs for all of the following classes.

The limewater mixture should be made by adding 1.40 g of $Ca(OH)_2$ to 1 L of water. Using bromothymol blue instead of limewater may provide a more easily-seen reaction.

Tips and Tricks
Students should be sure that the clay plug is tightly and completely sealed in steps 5, 9, and 10. This will assure that differences in results are due to different amounts of carbon dioxide being produced, and not to stray leaks. Students should also be sure that the end of the straw is submerged at the bottom of the glass of limewater.

Volcano Verdict

You will need to have a partner for this exploration. You and your partner will act as geologists who work in a city located near a volcano. City officials are counting on you to predict when the volcano will erupt next. You and your partner have decided to use limewater as a gas-emissions tester. You will use this tester to measure the levels of carbon dioxide emitted from a simulated volcano. The more active the volcano is, the more carbon dioxide it releases.

Objectives
- **Create** a working apparatus to test carbon dioxide levels.
- USING SCIENTIFIC METHODS **Analyze** the levels of carbon dioxide emitted from a model volcano.
- **Predict** the possibility of an eruption from a model volcano.

Materials
- baking soda, 15 cm^3
- drinking bottle, 16 oz
- box or stand for plastic cup
- clay, modeling
- coin
- cup, clear plastic, 9 oz
- graduated cylinder
- limewater, 1 L
- straw, drinking, flexible
- tissue, bathroom (2 sheets)
- vinegar, white, 140 mL
- water, 100 mL

Safety

PROCEDURE

1. Carefully pour limewater into the plastic cup until the cup is three-fourths full. Place the cup on a box or stand. This will be your gas-emissions tester.
2. Now, build a model volcano. Begin by pouring 50 mL of water and 70 mL of vinegar into the drink bottle.
3. Form a plug of clay around the short end of the straw. The clay plug must be large enough to cover the opening of the bottle. Be careful not to get the clay wet.

Step 5

CHAPTER RESOURCES

Chapter Resource File
- Datasheet for Chapter Lab GENERAL
- Lab Notes and Answers

4. Sprinkle 5 cm^3 of baking soda along the center of a single section of bathroom tissue. Then, roll the tissue, and twist the ends so that the baking soda can't fall out.

5. Drop the tissue into the drink bottle, and immediately put the short end of the straw inside the bottle to make a seal with the clay.

6. Put the other end of the straw into the limewater.

7. Record your observations. You have just taken your first measurement of gas levels from the volcano.

8. Imagine that it is several days later and that you need to test the volcano again to collect more data. Before you continue, toss a coin. If it lands heads up, go to step 9. If it lands tails up, go to step 10. Write down the step that you follow.

9. Repeat steps 1–7. But use 2 cm^3 of baking soda in the tissue in step 4 instead of 5 cm^3. (Note: You must use fresh water, vinegar, and limewater.) Record your observations.

10. Repeat steps 1–7. But use 8 cm^3 of baking soda in the tissue in step 4 instead of 5 cm^3. (Note: You must use fresh water, vinegar, and limewater.) Record your observations.

Mount Usu, 770 kilometers from Tokyo in Japan, erupted on March 31, 2000.

ANALYSIS AND CONCLUSION

1. **Explaining Events** How do you explain the difference in the appearance of the limewater from one trial to the next?

2. **Recognizing Patterns** What does the data that you collected tell you about the activity in the volcano?

3. **Evaluating Results** Based on your results in step 9 or 10, do you think it would be necessary to evacuate the city?

4. **Applying Conclusions** How would a geologist use a gas-emissions tester to predict volcanic eruptions?

Extension

1. **Evaluating Data** Scientists base their predictions of eruptions on a variety of evidence before recommending an evacuation. What other forms of evidence would a scientist need to know to predict an eruption?

Making Models Lab

Answers to Analysis and Conclusion

1. The difference in the limewater's appearance results from different amounts of carbon dioxide being added to the limewater.
2. The greater the activity of the volcano, the more carbon dioxide that is released by the volcano. A slight increase in baking soda does not greatly increase the amount of carbon dioxide produced, whereas the large increase in baking soda results in a greater release of carbon dioxide.
3. Results will vary with the coin toss. If step 9 is followed, evacuation is probably not necessary. If step 10 is followed, evacuation would be strongly advised.
4. Rapid increases in gas emissions would be an indicator of increased volcanic activity.

Answers to Extension

1. Ground temperature in the vicinity of the volcano, tilt of the ground indicating bulges due to rising magma, and increased earthquake activity would also be used to predict an eruption.

CLASSROOM TESTED & APPROVED

Tammie Niffenegger
Science Chair and Science Teacher
Port Washington High School
Port Washington, WI

The Hawaiian-Emperor Seamount Chain

Group Activity — ADVANCED

Charting Volcanic Activity Organize the class into groups of three or four students, and have each group study a geophysical map that shows long-term volcanic activity. The Yellowstone hot spot, the Mid-Atlantic Ridge, and the Aleutian Islands arc are three possible choices. Students should obtain information about the ages of the various volcanic features. From that information, they should determine the speed at which the tectonic plate is moving. LS Visual/Interpersonal

Answers to Map Skills Activity

1. Kilauea
2. Suiko 2
3. seamounts; They would be reduced in size by erosion.
4. Daikakuji; It is located at the end of the southbound Emperor chain and at the beginning of the southeast-bound Hawaiian chain.
5. northwest
6. about 43 million years ago
7. distance traveled by Emperor chain = 1,500 km; distance traveled by Hawaiian chain = 3,000 km; total distance = 1,500 km + 3,000 km = 4,500 km; average speed of plate = total distance/time; average speed of plate = 4500 km/65,000,000 y × 1,000 m/km × 100 cm/m = 6.15 cm/y
8. At 6.15 cm/y × 1,000,000 y, Kilauea would move 61.5 km to the northwest, and a new volcano would form at Kilauea's present location.

CHAPTER RESOURCES

Technology

Transparencies
- 70 The Hawaiian-Emperor Seamount Chain (with worksheet)

MAPS in Action

The Hawaiian-Emperor Seamount Chain

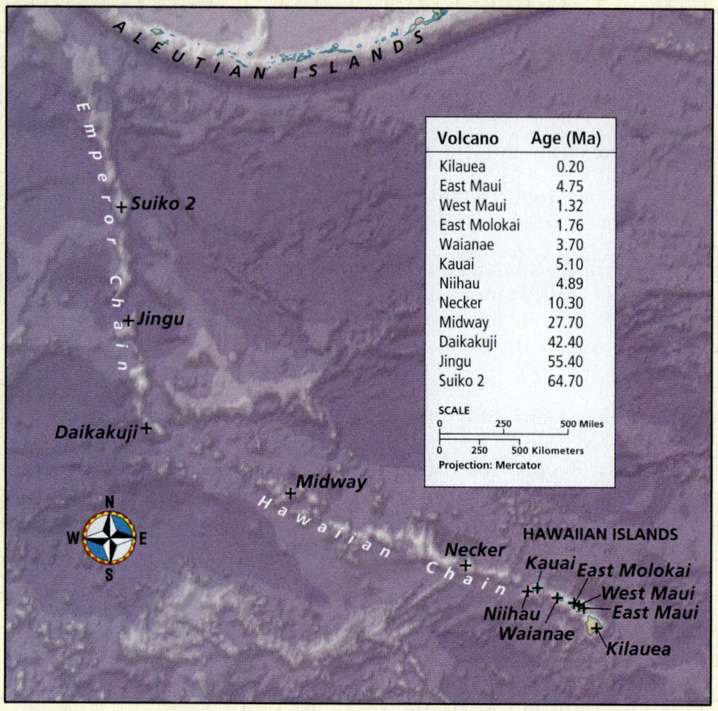

Volcano	Age (Ma)
Kilauea	0.20
East Maui	4.75
West Maui	1.32
East Molokai	1.76
Waianae	3.70
Kauai	5.10
Niihau	4.89
Necker	10.30
Midway	27.70
Daikakuji	42.40
Jingu	55.40
Suiko 2	64.70

Map Skills Activity

This map shows the locations and ages of islands and seamounts in the Hawaiian-Emperor seamount chain, which is located in the Pacific Ocean. Use the map to answer the questions below.

1. **Inferring Relationships** Under which volcano is the hot spot presently located?
2. **Using the Key** Which volcano is the oldest?
3. **Evaluating Data** A seamont is a submarine volcanic mountain. Would you expect older volcanoes to be seamounts or islands? Explain your answer.
4. **Analyzing Data** Which island signifies a change in direction of the movement of the Pacific plate? Explain your answer.
5. **Identifying Trends** In which direction has the Pacific plate been moving since the formation of the islands in the seamount chain changed direction?
6. **Analyzing Relationships** How many years ago did the Pacific plate change its direction?
7. **Analyzing Data** What is the average speed of the Pacific plate over the last 65 million years?
8. **Predicting Consequences** Where would you expect a new volcano to form 1 million years from now?

The Effects of Volcanoes on Climate

Some scientists think that average temperatures around the globe will rise by 2°C by the year 2050 because of the air pollution released by the use of fossil fuels by humans. To understand the possible effects of human activity on Earth's environment, scientists are studying another source of atmospheric pollution—volcanoes.

Volcanic Ash

The lava, gases, and ash that erupt into the atmosphere during a volcanic eruption can darken the skies for hundreds of kilometers. In the months after an eruption, some of this material settles to Earth, but much of it remains in the atmosphere, where wind currents disperse it around the world. While the dust and gases are not visible, they affect the climate of the entire planet.

Blocking the Sun's Energy

In June of 1991, Mount Pinatubo began a long series of eruptions. The resulting lava, ash, and mud flows devastated 20,000 km^2 of land. The eruption also released more than 18 million metric tons of sulfur dioxide into the atmosphere. Sulfur dioxide combines with water to become sulfuric acid, which reflects the sun's energy back into space. Scientists predicted that these large amounts of sulfur dioxide would have a cooling effect on Earth's surface. As predicted, average global surface temperatures dropped about 0.6°C by late 1992 and began to recover slowly after that.

By using data collected from several volcanic eruptions, scientists are developing computer models that may help them better understand and predict global climate changes.

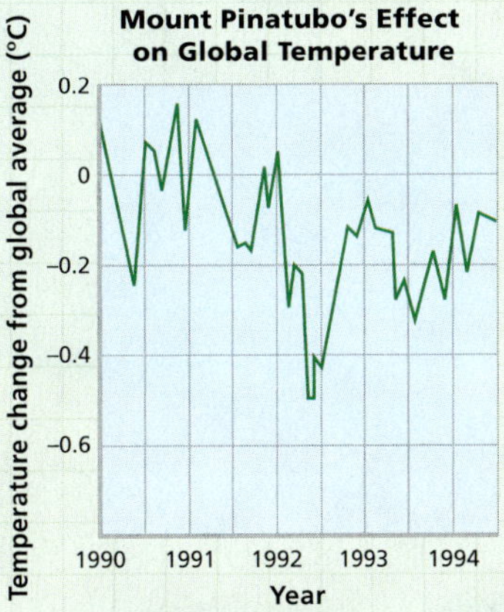

▲ The low average global temperatures in 1992 are most likely caused by Mount Pinatubo's eruption in 1991.

Learning how the atmosphere is affected by such natural events will also help scientists understand how pollution from human sources affects the atmosphere.

▼ During the 1991 eruption of Mount Pinatubo in the Philippines, ash fell like snow falls for several days.

Extension

1. **Applying Ideas** How would you determine if the 1991 eruption of Mount Pinatubo affected the average monthly temperatures in your community?
2. **Research** Find out how the 1980 eruption of Mount St. Helens changed nearby ecosystems, and identify which effects are still noticeable today.

Answer to Extension

1. Sample answer: By checking the local temperatures immediately before and after the eruption, the eruption's effect could be determined.
2. Answers may vary. Accept all reasonable answers. Students may report about different details of the eruption and its effects. For example, it had a small effect on global temperature and it had devastating effects on the immediate vicinity.

The Effects of Volcanoes on Climate

Internet Activity — ADVANCED

The Carbon Cycle Have students use the Internet to research the carbon cycle, which involves Earth's atmosphere and oceans. Be sure they identify the role that volcanoes play in generating carbon dioxide and how the greenhouse warming contributed by volcanic carbon dioxide offsets the cooling from volcanic dust in the atmosphere after an eruption. Have students present their findings in a written report or as an oral or poster presentation. A worksheet designed to direct student research on this topic can be found in the **Chapter Resource File** booklet or by visiting **go.hrw.com** and entering the keyword **HQ6VOLX.** **LS Verbal/Logical**

Interpreting Statistics — ADVANCED

Temperature and Volcanic Dust Have students research the major volcanic eruptions of the last 200 years. They should include, if available, the temperature information for different parts of the world immediately before and after the eruptions. Students should note if the eruptions were quiet or violent, and whether a large amount of material was ejected into the atmosphere. Have students compile their findings in a chart or table. **LS Logical**

CHAPTER RESOURCES

Chapter Resource File

Internet Activity
- The Carbon Cycle GENERAL

Unit 5 RESHAPING THE CRUST

Unit 5 Outline

CHAPTER 14
Weathering and Erosion

CHAPTER 15
River Systems

CHAPTER 16
Groundwater

CHAPTER 17
Glaciers

CHAPTER 18
Erosion by Wind and Waves

▶ Earth's surface is constantly shaped by the action of wind and water. Winds in the Sahara move enormous amounts of sand and form huge sand dunes, such as these near the town of Kerzaz in Algeria, Africa.

Chapter 14 Weathering and Erosion

Planning Guide

Compression Guide

To shorten instruction because of time limitations, omit the Chapter Lab.

OBJECTIVES	LABS, DEMONSTRATIONS, AND ACTIVITIES	TECHNOLOGY RESOURCES
PACING • 45 min pp. 342–348 **Chapter Opener**		OSP **Parent Letter** CD **Student Edition on CD-ROM** CD **Chapter Summaries Audio CD** VID **Brain Food Video Quiz**
Section 1 Weathering Processes • Identify three agents of mechanical weathering. • Compare mechanical and chemical weathering processes. • Describe four chemical reactions that decompose rock.	TE **Activity** Weathering Walk, p. 343 GENERAL TE **Activity** Ice Wedging, p. 344 GENERAL SE **Quick Lab** Mechanical Weathering, p. 345 GENERAL CRF **Datasheet for Quick Lab*** GENERAL TE **Group Activity** Finding Evidence, p. 345 GENERAL TE **Demonstration** Rusting, p. 346 GENERAL TE **Discussion** Weathering on the Moon, p. 346 GENERAL TE **Activity** Chalk Sculptures, p. 347 GENERAL	OSP **Lesson Plans** (also in print) TR **Bellringer*** TR 71 Chemical Weathering* VID **HRW Earth Science Video** Weathering and Erosion CD **Interactive Tutor** Physical and Chemical Weathering
PACING • 45 min pp. 349–352 **Section 2 Rates of Weathering** • Explain how rock composition affects the rate of weathering. • Discuss how surface area affects the rate at which rock weathers. • Describe the effects of climate and topography on the rate of weathering.	TE **Activity** Create a Cave, p. 350 GENERAL SE **Quick Lab** Surface Areas, p. 350 GENERAL CRF **Datasheet for Quick Lab*** GENERAL	OSP **Lesson Plans** (also in print) TR **Bellringer*** TR 72 Surface Area* TE **Internet Activity** National Parks, p. 349 GENERAL CRF **Internet Activity** National Parks* GENERAL
PACING • 90 min pp. 353–356 **Section 3 Soil** • Summarize how soils form. • Explain how the composition of parent rock affects soil composition. • Describe the characteristic layers of mature residual soils. • Predict the type of soil that will form in arctic and tropical climates.	TE **Group Activity** What Is Soil?, p. 353 GENERAL SE **Skills Practice Lab** Soil Chemistry, pp. 370–371 GENERAL CRF **Datasheet for Chapter Lab*** GENERAL SE **Maps in Action** Soil Map of North Carolina, p. 372 GENERAL TE **Discussion** Local Soils, p. 372 GENERAL TE **Activity** Guest Speaker, p. 373 GENERAL CRF **Inquiry Lab** Acid Rain and Soils* GENERAL CRF **Making Models Lab** Soil Profiles* GENERAL	OSP **Lesson Plans** (also in print) TE **Bellringer*** TR 73 Soil Horizons of Residual Soils* TR 75 Soil Map of North Carolina*
PACING • 45 min pp. 357–364 **Section 4 Erosion** • Define erosion, and list four agents of erosion. • Identify four farming methods that conserve soil. • Discuss two ways gravity contributes to erosion. • Describe the three major landforms shaped by weathering and erosion.	TE **Demonstration** Soil Conservation, p. 360 BASIC TE **Activity** Plants and Soil, p. 360 BASIC TE **Social Studies Connection** The Hills of Southern California, p. 361 GENERAL TE **Discussion** Protecting Against Mass Movements, p. 361 GENERAL TE **Demonstration** Solifluction, p. 362 BASIC	OSP **Lesson Plans** (also in print) TR **Bellringer*** TR 74 Soil Erosion Vulnerability Map* TE **Internet Activity** NRCS, p. 360 GENERAL CRF **Internet Activity** NRCS* GENERAL CD **Interactive Tutor** Soils

PACING • 90 min

CHAPTER REVIEW, ASSESSMENT, AND STANDARDIZED TEST PREPARATION

- SE **Chapter Highlights**, p. 365
- SE **Chapter Review**, pp. 366–367
- SE **Standardized Test Prep**, pp. 368–369
- CRF **Concept Review*** GENERAL
- CRF **Critical Thinking*** ADVANCED
- CRF **Math Skills*** GENERAL
- CRF **Graphing Skills*** GENERAL
- CRF **Chapter Test A*** GENERAL
- CRF **Chapter Test B*** ADVANCED
- OSP **Lesson Plans** (also in print)
- OSP **Test Generator**
- OSP **Test Item Listing**

Online and Technology Resources

Holt Online Learning

Visit **go.hrw.com** for access to Holt Online Learning, or enter the keyword **HQ6 Home** for a variety of free online resources.

One-Stop Planner® CD-ROM

This CD-ROM package includes

- Lab Materials QuickList Software
- Holt Calendar Planner
- Customizable Lesson Plans
- Printable Worksheets
- ExamView® Test Generator
- Interactive Teacher Edition
- Holt PuzzlePro®
- Holt PowerPoint® Resources

KEY

SE Student Edition
TE Teacher Edition
CRF Chapter Resource File
LTP Long-Term Projects
OSP One-Stop Planner
TR Transparencies and Transparency Worksheets
CD CD or CD-ROM
VID Classroom Video/DVD
* Also on One-Stop Planner
◆ Requires advance prep
■ Also available in Spanish

SKILLS DEVELOPMENT RESOURCES	REVIEW AND ASSESSMENT	CORRELATIONS
SE **Pre-Reading Activity,** p. 342 GENERAL TE **Using the Figure** Uluru, p. 342 GENERAL		National Science Education Standards
CRF **Directed Reading*** BASIC TE **Reading Skill Builder** Paired Summarizing, p. 344 BASIC SE **Math Practice,** p. 346 GENERAL	SE **Reading Checks,** p. 344, 346 GENERAL SE **Section Review,** p. 348 GENERAL TE **Reteaching,** p. 347 BASIC TE **Quiz,** p. 347 GENERAL TE **Alternative Assessment,** p. 347 GENERAL CRF **Section Quiz*** ■ GENERAL	SAI 1b, SAI 1c, SAI 1d, SAI 1f, ES 3c, ST 1e, SPSP 4a, SPSP 5c, UCP 1, UCP 2, UCP 3, UCP 4
CRF **Directed Reading*** BASIC TE **Inclusion Strategies,** p. 350	SE **Reading Check,** p. 350 GENERAL SE **Section Review,** p. 352 GENERAL TE **Homework,** p. 351 GENERAL TE **Reteaching,** p. 351 BASIC TE **Quiz,** p. 351 GENERAL TE **Alternative Assessment,** p. 351 GENERAL CRF **Section Quiz*** ■ GENERAL	SAI 1b, SAI 1d, SAI 1f, ES 3c, ST 1e, SPSP 4a, SPSP 5c, UCP 2, UCP 3, UCP 4
CRF **Directed Reading*** BASIC TE **Using the Figure** Soil Horizons, p. 354 GENERAL TE **Skill Builder** Math, p. 354 BASIC	SE **Reading Check,** p. 355 GENERAL SE **Section Review,** p. 356 GENERAL TE **Reteaching,** p. 355 BASIC TE **Quiz,** p. 355 GENERAL TE **Alternative Assessment,** p. 355 GENERAL CRF **Section Quiz*** ■ GENERAL	SAI 1a, SAI 1b, SAI 1c, SAI 1d, SAI 1f, ST 1d, ST 1e, SPSP 4a, UCP 2, UCP 3, UCP 4
CRF **Directed Reading*** BASIC TE **Using the Figure** Local Vulnerability, p. 357 GENERAL TE **Skill Builder** Math, p. 359 GENERAL SE **Graphic Organizer** Spider Map, p. 360 GENERAL TE **Skill Builder** Writing, p. 361 ADVANCED TE **Reading Skill Builder** Reading Organizer, p. 362 BASIC TE **Inclusion Strategies,** p. 362	SE **Reading Checks,** pp. 358, 361, 363 GENERAL SE **Section Review,** p. 364 GENERAL TE **Reteaching,** p. 363 BASIC TE **Quiz,** p. 363 GENERAL TE **Alternative Assessment,** p. 363 GENERAL CRF **Section Quiz*** ■ GENERAL	SAI 1b, SAI 1c, SAI 1f, ES 3c, ST 1e, SPSP 4a, SPSP 5c, UCP 2, UCP 3, UCP 4

Holt Lab Generator CD-ROM

Search for labs by topic, standard, difficulty level, and/or time frame. Edit those labs to fit your needs, or create your own labs. Use the Lab Materials QuickList to order just what you need.

Holt Earth Science Interactive Tutor CD-ROM

This CD-ROM consists of interactive activities that give students a fun way to extend their knowledge of Earth science concepts.

Chapter Summaries Audio CDs

These CDs include audio summaries of the key concepts presented in each chapter. (Audio summaries are also available in Spanish.)

SciLinks. NSTA

www.scilinks.org

Maintained by the **National Science Teachers Association.** See Chapter Enrichment pages that follow for a complete list of topics.

See Chapter Enrichment pages for Video Resources.

Chapter 14

Chapter Enrichment

This Chapter Enrichment provides relevant and interesting information to expand and enhance your classroom instruction of the chapter material.

Section 1 Weathering Processes

Telling Time with Lichens

By measuring the size of lichen colonies growing on the exposed surfaces of rocks, Earth scientists can estimate when landscape changes, caused by such events as earthquakes and glaciers, have occurred in the distant past. Lichens are often the first organisms to colonize bare rock surfaces. For example, when glaciers, fires, or earthquakes expose new rock surfaces, lichen colonies will start to grow on freshly exposed rocks and live there for centuries.

▲ Red soil indicates that the parent rock is rich in iron.

Scientists have determined that certain lichen species grow at a constant rate. By measuring the size of lichens, scientist can work backwards to calculate when an earthquake or other event occurred to within 20 to 40 years. Scientists hope that this technique will help them determine the pattern of earthquakes at major fault zones to better predict when the next earthquake will occur.

Section 2 Rates of Weathering

Karst and Caves

The name "karst" derives from a region of Slovenia where the landscape is made up of well-eroded hills, numerous sinkholes, disappearing streams, caves, tunnels, and natural bridges. Geologists now the use the term to describe any landscape that has these features. Karst landscapes may develop on any land with soluble bedrock, usually carbonate rocks such as limestone and dolomite. A karst landscape forms as rainwater picks up carbon dioxide from the air and plant acids and percolates through the ground. This slightly acidic water seeps through joints and fractures in the bedrock, dissolving some of the rock. Over time, as the bedrock dissolves, caves form. Sometimes cave roofs collapse, creating sinkholes.

▲ Karst topography is common in areas that are underlain by limestone.

Section 3 Soil

Soil Orders

Soils of the world have many different characteristics. They differ in color, composition, fertility, acidity, and numerous other properties. In the United States alone, scientists have identified some 17,000 different types of soil! To create some order, scientists have classified soils into 12 main groups, or orders. *Entisols* are soils with little or no development that closely resemble their parent rock. Soils with a little more subsurface horizon development are *inceptisols*. *Alfisols* develop in humid forests and have clay subsurface zones saturated with water to greater than 35%. *Utilisols* are similar but less saturated than alfisols. Soils on prairies with well-developed, rich topsoil are called *mollisols*. *Aridisols* contain little organic matter and develop in dry environments. *Vertisols* also develop in warm, arid regions, but are clay-like and develop deep cracks when dry. Highly weathered, reddish tropical soils are known as *oxisols*. *Spodosols* develop in cool, humid climates, are acidic, and usually have large amounts of aluminum, iron, and organic matter in the B-horizon. *Histosols* are highly organic soils that form in swamps and bogs. *Andisols* form in volcanic ash. *Gelisols* form in regions with permafrost within 2 m of the surface, such as arctic or mountain regions.

Section 4 Erosion

Lahars

Landslides, a form of erosion, can be triggered by volcanic eruptions. When ash and debris mix with water from rivers or melting ice and snow, a thick slurry of mud much like concrete, called *lahar,* forms. Lahars flow down river channels or valleys at extremely high speeds, up to 50 km per hour. Eventually they slow down and settle out in a sticky, muddy mess. A deadly lahar occurred in 1985 in the Andes Mountains in Columbia. The eruption of the volcano Nevado del Ruiz melted its thick snowcap and created a scalding lahar that swept into the sleeping town of Armero 60 km away, destroying 90% of the town's buildings and killing over 20,000 of its residents.

▼ The sharp peeks of these mountains will eventually be weathered down to rounded slopes.

Video Resources

Brain Food Video Quizzes

Brain Food Video Quizzes These videos contain game-show style quizzes that assess students' progress and motivate students to study the chapter material.

HRW Earth Science Video

This video introduces Earth science topics and includes a geology field trip. The video segment listed below complements this chapter.

Segment 9, Weathering and Erosion
Weathering and erosion, which continually reshape the landscape of Earth, are the focuses of this segment. The processes that bring about weathering and erosion are briefly examined, as well. Both naturally occurring and human-caused slope instability are also described. (1.5 min)

NOVA® Videos

To order **NOVA** videos related to this chapter, visit go.hrw.com and enter the keyword **HQ6WAEV.**

SciLinks®

NSTA Developed and maintained by the National Science Teachers Association

SciLinks is maintained by the National Science Teachers Association to provide you and your students with interesting, up-to-date links that will enrich your classroom presentation of the chapter.

Visit www.scilinks.org and enter the SciLinks code for more information about the topic listed.

Topic: Weathering
SciLinks code: HQ61648

Topic: Acid Precipitation
SciLinks code: HQ61690

Topic: Rates of Weathering
SciLinks code: HQ61269

Topic: Soil
SciLinks code: HQ61407

Topic: Soil Erosion
SciLinks code: HQ61410

Topic: Soil Conservation
SciLinks code: HQ61409

Topic: Careers in Earth Science
SciLinks code: HQ60222

Chapter 14

Chapter Overview

Weathering and erosion are major forces that change Earth's surface over time. Chemical and mechanical weathering break down rock and help form one of our most important natural resources—soil.

Using the Figure — GENERAL

Uluru This photograph shows Ayers Rock, also known as *Uluru* by aboriginal peoples, located in central Australia. Uluru towers 360 m above the desert plain and was originally part of a large mountain composed of beds of sandstone. The sandstone of Uluru resisted erosion over the years, while sandstone in the other beds did not. All that remains of the mountain are Uluru and the adjacent Olgas. Ask students if they have seen other landforms that illustrate differential weathering. (Answers may vary. Accept all reasonable answers.) **LS Logical**

PRE-READING ACTIVITY

FOLDNOTES

You may want to collect students' FoldNotes to determine their prior knowledge. You may also want to modify your lesson plan to answer questions that students list in the "Want" column of their table.

Chapter 14 Weathering and Erosion

Sections

What You'll Learn

- How rock breaks down
- How soil forms
- What agents erode rock and soil

Why It's Relevant

Weathering and erosion change the shape of Earth's surface and are essential to the formation of soil, one of our most important natural resources.

PRE-READING ACTIVITY

FOLDNOTES

TriFold
Before you read this chapter, create the FoldNote entitled "TriFold" described in the Skills Handbook Section of the Appendix. Write what you know about weathering and erosion in the column labeled "Know." Then, write what you want to know in the column labeled "Want." As you read the chapter, write what you learn about weathering and erosion in the column labeled "Learn."

▶ Ayers Rock in Australia, also called *Uluru* by the aboriginal peoples of Australia, is gray under its red surface. The surface appears red because iron in the rock oxidizes when it is exposed to air and water.

Chapter Correlations — *National Science Education Standards*

ES 3c Interactions among the solid earth, the oceans, the atmosphere, and organisms have resulted in the ongoing evolution of the earth system. We can observe some changes such as earthquakes and volcanic eruptions on a human time scale, but many processes such as mountain building and plate movements take place over hundreds of millions of years. **(Sections 1–4)**

SPSP 4a Natural ecosystems provide an array of basic processes that affect humans. Those processes include maintenance of the atmosphere, generation of soils, control of the hydrologic cycle, disposal of wastes, and recycling of nutrients. Humans may be changing many of these basic processes, and the changes may be detrimental to humans. **(Sections 2, 3, and 4)**

SPSP 5c Some hazards, such as earthquakes, volcanic eruptions, and severe weather are rapid and spectacular. But there are slow progressive changes that also result in problems for individuals and societies. For example, change in stream channel position, erosion of bridge foundations, sedimentation in lakes and harbors, coastal erosion, and continuing erosion and wasting of soil and landscapes can all negatively affect society. **(Section 4)**

UPC 3 Constancy, change and measurement: . . . Changes might occur . . . in properties of materials, positions of objects, motion, and form and function of systems. Interactions within and among systems result in change. Changes vary in rate, scale, and pattern, . . . **(Sections 1–4)**

UPC 4 Evolution and equilibrium: Evolution is a series of changes, some gradual and some sporadic, that accounts for the present form and function of objects . . . The general idea of evolution is that the present arises from materials and forms of the past. . . . **(Sections 1–4)**

Section 1 Weathering Processes

Most rocks deep within Earth's crust formed under conditions of high temperature and pressure. When these rocks are uplifted to the surface, they are exposed to much lower temperature and pressure. Uplifted rock is also exposed to the gases and water in Earth's atmosphere.

Because of these environmental factors, surface rocks undergo changes in their appearance and composition. The physical breakdown or chemical decomposition of rock materials exposed at Earth's surface is called **weathering.** There are two main types of weathering processes—mechanical weathering and chemical weathering. Each type of weathering has different effects on rock.

OBJECTIVES

- **Identify** three agents of mechanical weathering.
- **Compare** mechanical and chemical weathering processes.
- **Describe** four chemical reactions that decompose rock.

KEY TERMS

weathering
mechanical weathering
abrasion
chemical weathering
oxidation
hydrolysis
carbonation
acid precipitation

Mechanical Weathering

The process by which rock is broken down into smaller pieces by physical means is **mechanical weathering.** Mechanical weathering is strictly a physical process and does not change the composition of the rock. Common agents of mechanical weathering are ice, plants and animals, gravity, running water, and wind.

Physical changes within the rock also affect mechanical weathering. For example, when overlying rocks are eroded, granite that formed deep beneath Earth's surface can be exposed, decreasing the pressure on the granite. As a result of the decreasing pressure, the granite expands. Long, curved cracks, called *joints,* develop in the rock. When joints are parallel to the surface of the rock, the rock breaks into curved sheets that peel away from the underlying rock in a process called *exfoliation.* One example of granite exfoliation on a dome in Yosemite National Park is shown in **Figure 1.**

weathering the natural process by which atmospheric and environmental agents, such as wind, rain, and temperature changes, disintegrate and decompose rocks

mechanical weathering the process by which rocks break down into smaller pieces by physical means

Figure 1 ▶ This area of Yosemite National Park is part of a dome of granite that is shedding large sheets of rock through the process of exfoliation.

Section 1

Focus

Overview

This section explains the processes of mechanical and chemical weathering. Wind, water, ice, gravity, and living organisms are all weathering agents.

Bellringer

Show students a large rock. Ask them to list some natural agents that might be able to break the rock into smaller pieces. (Answers may vary. Possible answers include other rocks, changes in temperature, wind, water, and chemical reactions.) **LS Logical**

Motivate

Activity — GENERAL

Weathering Walk Have students walk around the school or their home neighborhood to look for examples of weathering. Students should make a list of what they find. As a class, have them discuss their findings, try to determine the weathering agent, and classify their examples as mechanical or chemical weathering. As an extension, students could return to their sites days or weeks later to observe how much has changed. **LS Visual/Kinesthetic**

CHAPTER RESOURCES

Chapter Resource File

- Directed Reading BASIC

Technology

- **Transparencies**
  - Bellringer
- **Student Edition on CD-ROM**
- **One-Stop Planner CD-ROM**
  - Lesson Plan

Teach

Teaching Tip — GENERAL

Connect to Familiar Processes Students may be familiar with potholes in streets. Street potholes form by thermal expansion and contraction of asphalt or concrete. Potholes may also form by the abrasive actions of pebbles and other materials that swirl around in natural depressions. These processes are similar to those that weather rock. Potholes may be especially bad in regions that have pronounced seasonal variation in temperature. **LS Logical/Visual**

Activity — GENERAL

Ice Wedging Divide the class into pairs. Have each pair moisten some clay with water and roll it into two balls. Wrap both in plastic wrap. Put one ball in the freezer and keep the other at room temperature. After 24 hours, unwrap both balls and compare them. Students should notice that the frozen clay ball has small cracks. The other ball should remain smooth and intact. Sometimes it takes two cycles of freezing to see changes.
LS Kinesthetic

Answer to Reading Check

Two types of mechanical weathering are ice wedging and abrasion. Ice wedging is caused by water that seeps into cracks in rock and freezes. When water freezes, it expands and creates pressure on the rock, which widens and deepens cracks. Abrasion is the grinding away of rock surfaces by other rocks or sand particles. Abrasive agents may be carried by gravity, water, and wind.

Figure 2 ▶ Water flows into a crack in a rock's surface. When the water freezes, it expands and causes the crack to widen. Ice wedging is responsible for most of the cracks shown in the photograph.

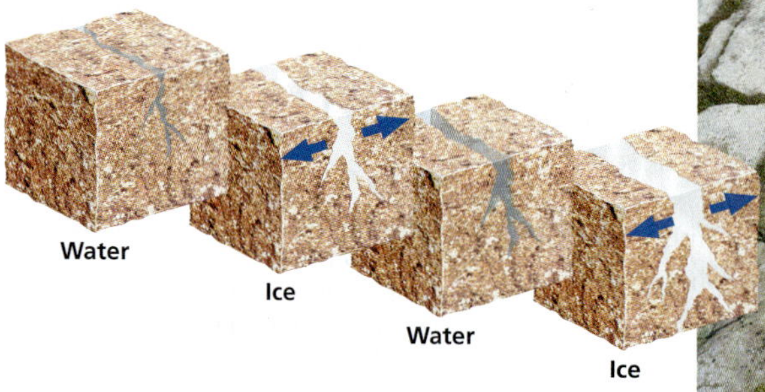

SCILINKS — NSTA — Developed and maintained by the National Science Teachers Association

For a variety of links related to this subject, go to www.scilinks.org

Topic: Weathering
SciLinksCode: HQ61648
Topic: Acid Precipitation
SciLinksCode: HQ61690

Ice Wedging

A type of mechanical weathering that occurs in cold climates is called *ice wedging*. Ice wedging occurs when water seeps into cracks in rock and then freezes. When the water freezes, its volume increases by about 10% and creates pressure on the surrounding rock. Every time the ice thaws and refreezes, cracks in the rock widen and deepen. This process eventually splits the rock apart, as shown in **Figure 2.** Ice wedging commonly occurs at high elevations and in cold climates. It also occurs in climates where the temperature regularly rises above and then falls below freezing, such as in the northern United States.

Abrasion

The collision of rocks that results in the breaking and wearing away of the rocks is a form of mechanical weathering called **abrasion.** Abrasion is caused by gravity, running water, and wind. Gravity causes loose soil and rocks to move down the slope of a hill or mountain. Rocks break into smaller pieces as they fall and collide. Running water can carry particles of sand or rock that scrape against each other and against stationary rocks. Thus, exposed surfaces are weathered by abrasion.

Wind is another agent of abrasion. When wind lifts and carries small particles, it can hurl them against surfaces, such as rocks. As the airborne particles strike the rock, they wear away the surface in the same way that a sandblaster would.

abrasion the grinding and wearing away of rock surfaces through the mechanical action of other rock or sand particles

Reading Check **Describe two types of mechanical weathering. (See the Appendix for answers to Reading Checks.)**

READING SKILL BUILDER — BASIC

Paired Summarizing Group students into pairs, and have them read silently about ice wedging and abrasion. Then, have one student summarize each weathering process and its effect on rocks. The other student should listen to the retelling and point out any inaccuracies or ideas that were left out. Allow students to refer to the text as needed. **LS Auditory/Verbal** English Language Learners

Figure 3 ▶ This gray wolf (left) is burrowing into soil to make a den. Prairie dogs (above) also dig into soil and rock to form extensive burrows, where an entire prairie dog community may live.

Organic Activity

Plants and animals are important agents of mechanical weathering. As plants grow, the roots grow and expand to create pressure that wedges rock apart. The roots of small plants cause small cracks to form in the rocks. Eventually, the roots of large plants and trees can fit in the cracks and make the cracks bigger.

The digging activities of burrowing animals, shown in **Figure 3,** also cause weathering. Common burrowing animals include ground squirrels, prairie dogs, ants, earthworms, coyotes, and rabbits. Earthworms and other animals that move soil expose new rock surfaces to both mechanical and chemical weathering. Animal activities and plants can weather rocks dramatically over a long period of time.

Quick LAB — 15 min

Mechanical Weathering

Procedure

1. Examine some **silicate rock chips** by using a **hand lens.** Observe the shape and surface texture.
2. Fill a **plastic container** that has a **tight-fitting lid** about half full of rock chips. Add **water** to just cover the chips.
3. Tighten the lid, and shake the container 100 times.
4. Hold a **strainer** over another container. Pour the water and rock chips into the strainer.
5. Move your finger around the inside of the empty container. Describe what you feel.
6. Use the hand lens to observe the rock chips.
7. Pour the water into a **glass jar,** and examine the water with the hand lens.
8. Put the rock chips and water back into the container that has the lid. Repeat steps 3–7.
9. Repeat step 8 two more times.

Analysis

1. Did the amount and particle size of the sediment that was left in the container change during your investigation? Explain your answer.
2. How did the appearance of the rock chips change? How did the appearance of the water change?
3. How does the transport of rock particles by water, such as in a river, affect the size and shape of the rock particles?

Quick LAB

Skills Acquired

- Experimenting
- Interpreting

Teacher's Notes Make sure students firmly secure lids before shaking the jar.

Answers

1. Because of the abrasive effect of the rocks on each other, there should be more and finer sediment in the container after each time it is shaken.
2. The rock chips should become smoother and smaller as they are worn down. The water will likely become more clouded with sediment particles.
3. Rocks will likely become smoother, more rounded and smaller as they are transported by water in a river.

Group Activity — GENERAL

Finding Evidence As a class, visit a local park and look for examples of weathering caused by plants and animals. Look for plants wedging rocks apart, animal burrows, anthills and other organic activities. Ask students to consider if human activities may also be contributing to weathering in the park. (Answers may vary. Students may notice that human activities, such as hiking, biking, and off-road-ing in the park may be contributing to weathering processes.) LS **Visual**

BRAIN FOOD

Earth Movers Earthworms are very important to weathering processes and the formation of soil. There are more than 15,000 different species of annelids (earthworms and their relatives), and they range in length from 1 mm to 3 m. Worms dig through soil looking for food. As they burrow, they bring rock and mineral particles to the surface for further weathering. Their tunnels aerate the soil and allow water to percolate to rocks below the soil surface. Earthworms have voracious appetites. They eat and digest about half their body weight each day.

CHAPTER RESOURCES

Chapter Resource File

- Datasheet for Quick Lab GENERAL

MATHPRACTICE

Answer

0.2 cm = 0.002 m

15 m ÷ 0.002 m/100 y

= 750,000 y

CHEMISTRY CONNECTION ADVANCED

Redox Reactions Chemically speaking, oxidation is not just the process of elements combining with oxygen. By definition, oxidation is the loss of electrons from a substance. Some substance must accept the electrons and in the process is reduced, which is why these reactions are often called *redox reactions*. In many cases, oxygen accepts the electrons lost by the substance, which is why loss of electrons is called *oxidation*. Iron is easily oxidized and forms several different oxides, depending on how many electrons it loses. These oxides include FeO, Fe_2O_3 and Fe_3O_4. **LS Logical**

Answer to Reading Check

Two effects of chemical weathering are changes in the chemical composition and changes in the physical appearance of a rock.

chemical weathering the process by which rocks break down as a result of chemical reactions

oxidation a reaction that removes one or more electrons from a substance such that the substance's valence or oxidation state increases; in geology, the process by which a metallic element combines with oxygen

MATHPRACTICE

Rates of Weathering Limestone is dissolved by chemical weathering at a rate of 0.2 cm every 100 years. At this rate, after how many years would a layer of limestone 15 m thick completely dissolve?

Chemical Weathering

The process by which rock is broken down because of chemical interactions with the environment is **chemical weathering.** Chemical weathering, or decomposition, occurs when chemical reactions act on the minerals in rock. Chemical reactions commonly occur between rock, water, carbon dioxide, oxygen, and acids. Acids are substances that form *hydronium ions,* or H_3O^+, in water. Hydronium ions are electrically charged and can pull apart the chemical bonds of the minerals in rock. Bases can also chemically weather rock. Bases are substances that form *hydroxide ions,* or OH^-, in water. Chemical reactions with either acids or bases can change the structure of minerals, which leads to the formation of new minerals. Chemical weathering changes both the chemical composition and physical appearance of the rock.

Oxidation

The process by which elements combine with oxygen is called **oxidation.** Oxidation commonly occurs in rock that has iron-bearing minerals, such as hematite and magnetite. Iron, Fe, in rocks and soil combines quickly with oxygen, O_2, that is dissolved in water to form rust, or iron oxide, Fe_2O_3:

$$4Fe + 3O_2 \rightarrow 2Fe_2O_3$$

The red color of much of the soil in the southeastern United States, as shown in **Figure 4,** is due to mainly the presence of iron oxide produced by oxidation. Similarly, the color of many red-colored rocks is caused by oxidized, iron-rich minerals.

Reading Check **Describe two effects of chemical weathering. (See the Appendix for answers to Reading Checks.)**

Figure 4 ▶ The red tint of the soil surrounding this farmhouse in Georgia is caused by the chemical interaction of iron-bearing minerals in the soil with oxygen in the atmosphere.

Demonstration GENERAL

Rusting Place a pad of steel wool moistened with water in a dish. Cover it with a glass and observe daily. The steel wool should turn reddish in a few days and crumble when touched as the iron oxidizes. Ask students if they can think of ways to prevent oxidation. (Answers may vary. Students may notice that the steel wool was not rusted to begin with. Iron in dry air is not easily oxidized. Painting protects steel from water and helps prevent rust and corrosion.) **LS Visual**

Discussion GENERAL

Weathering on the Moon Ask students to consider whether the surface of the moon has been weathered mechanically and/or chemically. (Mechanical weathering occurs when meteorites hit the moon surface, creating craters. But mechanical weathering by wind, water, and ice is absent because the moon has no water and no atmosphere. Similarly, chemical weathering is unlikely because the moon lacks water, an atmosphere, and life.) **LS Visual**

Rain, weak acids, and air chemically weather granite.

The bonds between mineral grains weaken as weathering proceeds.

Sediment forms from the weathered granite.

Figure 5 ▶ Thousands of years of chemical weathering processes, such as hydrolysis and carbonation, can turn even a hard rock such as granite into sediment.

Hydrolysis

Water plays a crucial role in chemical weathering, as shown in **Figure 5.** The change in the composition of minerals when they react chemically with water is called **hydrolysis.** For example, a type of feldspar combines with water and produces a common clay called *kaolin*. In this reaction, hydronium ions displace the potassium and calcium atoms in the feldspar crystals, which changes the feldspar into clay.

Minerals that are affected by hydrolysis often dissolve in water. Water can then carry the dissolved minerals to lower layers of rock in a process called *leaching*. Ore deposits, such as bauxite, the aluminum ore, sometimes form when leaching causes a mineral to concentrate in a thin layer beneath Earth's surface.

hydrolysis a chemical reaction between water and another substance to form two or more new substances

carbonation the conversion of a compound into a carbonate

Carbonation

When carbon dioxide, CO_2, from the air dissolves in water, H_2O, a weak acid called *carbonic acid*, H_2CO_3, forms:

$$H_2O + CO_2 \quad H_2CO_3$$

Carbonic acid has a higher concentration of hydronium ions than pure water does, which speeds up the process of hydrolysis. When certain minerals come in contact with carbonic acid, they combine with the acid to form minerals called carbonates. The conversion of minerals into a carbonate is called **carbonation.**

One example of carbonation occurs when carbonic acid reacts with calcite, a major component of limestone, and converts the calcite into calcium bicarbonate. Calcium bicarbonate dissolves easily in water, so the limestone eventually weathers away.

Organic Acids

Acids are produced naturally by certain living organisms. Lichens and mosses grow on rocks and produce weak acids that can weather the surface of the rock. The acids seep into the rock and produce cracks that eventually cause the rock to break apart.

Close

Reteaching — BASIC

Weathering Processes Have students write descriptions of various types of mechanical and chemical weathering processes on index cards. Have them work in pairs to use the cards to quiz each other about weathering processes. **LS Visual/Logical**

Quiz — GENERAL

1. What kinds of animals can cause weathering? (Burrowing animals such as coyotes, foxes, rabbits, squirrels, earthworms and ants may contribute to weathering.)
2. Why is acid precipitation decreasing in some areas? (because of pollution controls required by the Clean Air Act)

Alternative Assessment — GENERAL

Table Have students create a table that compares mechanical and chemical weathering. Students should include as much detail as possible in each column, including the agents of weathering and the various types of weathering processes. Students should summarize the table by writing a brief paragraph noting the similarities and differences between mechanical and chemical weathering. **LS Logical/Verbal**

CHAPTER RESOURCES

Technology

Transparencies
- 71 Chemical Weathering (with worksheet)

ENVIRONMENTAL CONNECTION

Acid From the Sky Despite the Acid Rain Control Program, acid precipitation still occurs in parts of the United States and around the world. Vehicles remain a major source of nitrogen oxides and sulfur dioxides. At times, precipitation as acidic as the acid in a car battery (pH 2) has been recorded. This can weather rocks and damage forests, bodies of water, and buildings. Have interested students investigate where acid rain remains a problem and what solutions have been proposed to remedy it. They can report their findings to the class. **LS Verbal/Logical**

Activity — GENERAL

Chalk Sculptures Have students make a statue by carving a piece of chalk. Then, use a small bit of clay to stand the statue upright in a dish. Have students put on goggles, spray their statues with white vinegar, and observe what happens. The statues should begin wearing away because chalk is calcium carbonate. Vinegar contains acetic acid, a weak acid that can eat away carbonates. Have students write a brief paragraph that describes how accurately this activity models the effects of acid rain. **LS Visual/Verbal**

Close, *continued*

Answers to Section Review

1. wind, water, and temperature change
2. Water that seeps into cracks freezes and expands, which widens and deepens cracks with each freeze/thaw cycle.
3. Plant roots grow and expand, physically wedging rocks apart. Animal dig and burrow, which exposes new rock to weathering processes.
4. Mechanical weathering is a purely physical process, breaking large rocks into smaller rocks of the same chemical composition. Chemical weathering involves chemical reactions that break down rocks.
5. In oxidation, iron-bearing minerals combine with oxygen to form red-colored iron oxide. In hydrolysis, water and other substances in rock react chemically to form two or more new substances. In carbonation, carbonic acid converts minerals into carbonates.
6. Oxidation, hydrolysis and carbonation are all chemical processes that weather rocks. Water plays a role in both hydrolysis and carbonation and it can speed up oxidation. Oxidation involves elements combining with oxygen. Hydrolysis changes the composition of minerals that react with water. Carbonation occurs when minerals react with carbonic acid.
7. Acid precipitation forms when nitrogen oxides and sulfur dioxide released during fossil fuel combustion combine with water in the atmosphere to produce nitric, nitrous, or sulfuric acid. When these acids fall back to Earth, they are called *acid precipitation.*
8. Answers may vary. Sample answer: Running water and vegetation are agents of weathering that would rarely occur in a desert. Rainfall is low in desert areas, so fewer streams and rivers exist to erode valley areas. Low precipitation also limits plant growth, so fewer plants mean fewer roots to break up the soil and less shelter for animals, which in greater numbers also expose soil to mechanical and chemical weathering.
9. Nitrogen oxides from exhaust could lead to the formation of acid precipitation and to an increase in chemical weathering.
10. *Weathering* includes *mechanical weathering,* such as *ice wedging* and *abrasion,* and *chemical weathering,* caused by *oxidation, hydrolysis, carbonation,* and *acid precipitation.*

Figure 6 ▶ This stone lion sits outside Leeds Town Hall in England. It was damaged by acid precipitation.

Acid Precipitation

Natural rainwater is slightly acidic because it combines with small amounts of carbon dioxide. But when fossil fuels, especially coal, are burned, nitrogen oxides and sulfur dioxides are released into the air. These compounds combine with water in the atmosphere to produce nitric acid, nitrous acid, or sulfuric acid. When these acids fall to Earth, they are called **acid precipitation.**

acid precipitation precipitation, such as rain, sleet, or snow, that contains a high concentration of acids, often because of the pollution of the atmosphere

Acid precipitation weathers some rock faster than ordinary precipitation does. In fact, many historical monuments and sculptures have been damaged by acid precipitation, as shown in **Figure 6.** Between 1940 and 1990, acid precipitation fell regularly in some cities in the United States. In 1990, the Acid Rain Control Program was added to the Clean Air Act of 1970. These regulations gave power plants 10 years to decrease sulfur dioxide emissions. The occurrence of acid precipitation has been greatly reduced since power plants have installed scrubbers that remove much of the sulfur dioxide before it can be released.

Section 1 Review

1. **Identify** three agents of mechanical weathering.
2. **Describe** how ice wedging weathers rock.
3. **Explain** how two activities of plants or animals help weather rocks or soil.
4. **Compare** mechanical and chemical weathering processes.
5. **Identify** and describe three chemical processes that weather rock.
6. **Compare** hydrolysis, carbonation, and oxidation.
7. **Summarize** how acid precipitation forms.

CRITICAL THINKING

8. **Making Connections** What two agents of weathering would be rare in a desert? Explain your reasoning.
9. **Understanding Relationships** Automobile exhaust contains nitrogen oxides. How might these pollutants affect chemical weathering processes?

CONCEPT MAPPING

10. Use the following terms to create a concept map: *weathering, oxidation, mechanical weathering, ice wedging, hydrolysis, abrasion, chemical weathering, carbonation,* and *acid precipitation.*

CHAPTER RESOURCES

Chapter Resource File
- Section Quiz GENERAL

Workbooks
- Study Guide (also in Spanish)

Section 2 Rates of Weathering

The processes of mechanical and chemical weathering generally work very slowly. For example, carbonation dissolves limestone at an average rate of only about one-twentieth of a centimeter (0.2 cm) every 100 years. At this rate, it could take up to 30 million years to dissolve a layer of limestone that is 150 m thick.

The pinnacles in Nambung National Park in Australia are shown in **Figure 1.** These large, jutting pieces of limestone are all that remains of a thick limestone formation that covered the area millions of years ago. Most of the limestone was weathered away by agents of both chemical and mechanical weathering until only the pinnacles remained. The rate at which rock weathers depends on a number of factors, including rock composition, climate, and topography.

OBJECTIVES

- **Explain** how rock composition affects the rate of weathering.
- **Discuss** how surface area affects the rate at which rock weathers.
- **Describe** the effects of climate and topography on the rate of weathering.

KEY TERM

differential weathering

Differential Weathering

The composition of rock greatly affects the rate at which rock weathers. The process by which softer, less weather-resistant rock wears away and leaves harder, more resistant rock behind is called **differential weathering.** When rocks that are rich in the mineral quartz are exposed on Earth's surface, they remain basically unchanged, even after all of the surrounding rock has weathered away. They remain unchanged because the chemical composition and crystal structure of quartz make quartz resistant to chemical weathering. These same characteristics make quartz a very hard mineral, so it also resists mechanical weathering.

differential weathering the process by which softer, less weather resistant rocks wear away at a faster rate than harder, more weather resistant rocks do

Rock Composition

Limestone and other sedimentary rocks that contain calcite are weathered most rapidly. They weather rapidly because they commonly undergo carbonation. Other sedimentary rocks are affected mainly by mechanical weathering processes. The rates at which these rocks weather depend mostly on the material that holds the sediment grains together. For example, shales and sandstones that are not firmly cemented together gradually break up to become clay and sand particles. However, conglomerates and sandstones that are strongly cemented by silicates resist weathering. Some of these strongly cemented sedimentary rocks can resist weathering longer than some igneous rocks do.

Figure 1 ▶ Differences in rock composition and structure are the reasons for the different rates of weathering that formed these limestone pinnacles at Nambung National Park in Australia.

Section 2

Focus

Overview

This section describes how composition; exposure; climate; topography; and plant, animal, and human activities affect rates of weathering.

Bellringer

Ask students, "Do all rocks weather at the same rate? What factors might affect rates of weathering?" (Answers may vary but students may answer that some rocks weather more easily than others. Students may indicate that rock composition, surface area, temperature, and precipitation affect rates of weathering.) LS **Verbal**

Motivate

Internet Activity — GENERAL

National Parks Have students visit the National Park Service Web site and choose one park in which weathering has played a role in the formation of the park's landforms. Students should present their findings to the class. A worksheet designed to direct student research on this topic can be found in the **Chapter Resource File** booklet or by visiting go.hrw.com and entering the keyword **HQ6WAEX.** LS **Verbal**

CHAPTER RESOURCES

Chapter Resource File
- **Directed Reading** BASIC
- **Internet Activity** National Parks GENERAL

Technology
- **Transparencies**
  - Bellringer
- **Student Edition on CD-ROM**
- **One-Stop Planner CD-ROM**
  - Lesson Plan

Teach

Activity — General

Create a Cave Have students pour about an inch of sand in the bottom of a tall, clear plastic cup and place a layer of sugar cubes on top of the sand. Then, ask them to cover the cubes with wood glue, making sure it seeps down into the cracks between the cubes. When the glue is dry, have students add a layer of sand and then poke a hole in the bottom of the cup with a nail. Have them place the cup in a dish and slowly pour warm water over the sand layer. As water seeps through the sand, it will dissolve the sugar but not the wood glue. "Caverns" will be left in the cup. LS **Kinesthetic**

Answer to Reading Check

Fractures and joints in a rock increase surface area and allow weathering to occur more rapidly.

INCLUSION *Strategies*

- *Hearing Impaired*
- *Learning Disabled*
- *Attention Deficit Disorder*

Being able to see the face of a speaker helps others to understand what that person is saying. This communication aid is especially helpful for students who have hearing impairments. Divide the class into groups of five to eight students. Have each group sit so each person can see everyone else. Ask each group to read the subheadings in this section and discuss the content under each subheading. LS **Interpersonal**

CHAPTER RESOURCES

Chapter Resource File

- Datasheet for Quick Lab GENERAL

Technology

Transparencies
- 72 Surface Area (with worksheet)

Figure 2 ▶ The Ratio of Total Surface Area to Volume

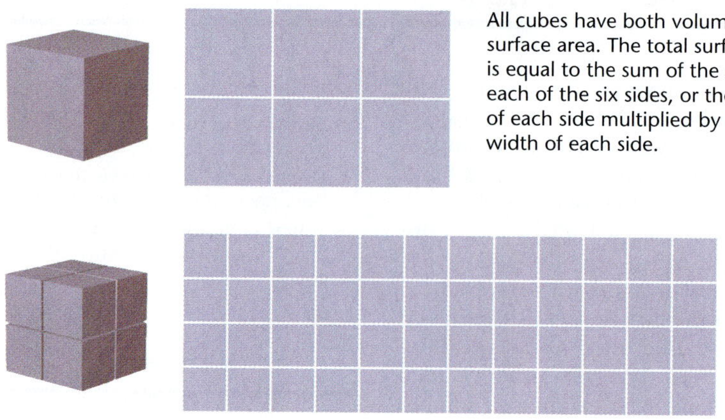

All cubes have both volume and surface area. The total surface area is equal to the sum of the areas of each of the six sides, or the length of each side multiplied by the width of each side.

If you split the first cube into eight smaller cubes, you have the same amount of material (volume), but the surface area doubles. If you split the eight small cubes in the same way, the original surface area is doubled again.

Amount of Exposure

The more exposure to weathering agents a rock receives, the faster the rock will weather. The amount of time the rock is exposed and the amount of the rock's surface area that is available for weathering are important factors in determining the rate of weathering.

Surface Area

Both chemical and mechanical weathering may split rock into a number of smaller rocks. The part of a rock that is exposed to air, water, and other agents of weathering is called the rock's *surface area*. As a rock breaks into smaller pieces, the surface area that is exposed increases. For example, imagine a block of rock as a cube that has six sides exposed. Splitting the block into eight smaller blocks, as shown in **Figure 2**, doubles the total surface area available for weathering.

Fractures and Joints

Most rocks on Earth's surface contain natural fractures and joints. These structures are natural zones of weakness within the rock. Fractures and joints increase the surface area of a rock and allow weathering to take place more rapidly. They also form natural channels through which water flows. Water may penetrate the rock through these channels and break the rock by ice wedging. As water moves through these channels, it chemically weathers the rock that is exposed in the fracture or joint. The chemical weathering removes rock material and makes the jointed or fractured area weaker.

Reading Check How do fractures and joints affect surface area? (See the Appendix for answers to Reading Checks.)

QuickLAB — 10 min

Surface Areas

Procedure

1. Fill **two small containers** about half full with **water.**
2. Add **one sugar cube** to one container.
3. Add **1 tsp of granulated sugar** to the other container.
4. Use **two different spoons** to stir the water and sugar in each container at the same rate.
5. Use a **stopwatch** to measure how long the sugar in each container takes to dissolve.

Analysis

1. Did the sugar dissolve at the same rate in both containers?
2. Which do you think would wear away faster—a large rock or a small rock? Explain your answer.

QuickLAB

Skills Acquired

- Experimenting
- Collecting data
- Measuring
- Interpreting

Teacher's Notes This activity will work well as a demo. Using warm water will speed up dissolution of the sugar.

Answers

1. The granulated sugar should dissolve more quickly than the cube sugar because it has a much greater surface area.
2. If both rocks have the same composition and are exposed to the same weathering processes, the smaller rock should wear away faster because it has a larger surface area to its volume.

Figure 3 ▶ The photo on the left shows Cleopatra's Needle before it was moved to New York City. The photograph on the right shows the 3,500 year old carving after only one century in New York City.

Climate

In general, climates that have alternating periods of hot and cold weather allow the fastest rates of weathering. Freezing and thawing can cause the mechanical breakdown of rock by ice wedging. Chemical weathering can then act quickly on the fractured rock. When temperatures rise, the rate at which chemical reactions occur also accelerates. In warm, humid climates, chemical weathering is also fairly rapid. The constant moisture is highly destructive to exposed surfaces.

The slowest rates of weathering occur in hot, dry climates. The lack of water limits many weathering processes, such as carbonation and ice wedging. Weathering is also slow in very cold climates.

The effects of climate on weathering rates can be seen on Cleopatra's Needle, which is shown in **Figure 3.** Cleopatra's needle is an obelisk that is made of granite. For 3,000 years, the obelisk stood in Egypt, where the hot, dry climate scarcely changed its surface. Then, in 1880, Cleopatra's Needle was moved to New York City. After the obelisk was exposed to more than 100 years of moisture, the pollution, ice wedging, and acid precipitation caused more weathering than was caused in the preceding 3,000 years in the Egyptian desert.

SCILINKS® NSTA Developed and maintained by the National Science Teachers Association

For a variety of links related to this subject, go to www.scilinks.org

Topic: Rates of Weathering
SciLinksCode: HQ61269

Topography and Elevation

Topography, or the elevation and slope of the land surface, also influences the rate of weathering. Because temperatures are generally cold at high elevations, ice wedging is more common at high elevations than at low elevations. On steep slopes, such as mountainsides, weathered rock fragments are pulled downhill by gravity and washed out by heavy rains. As the rocks slide down the mountain or are carried away by mountain streams, rocks smash against each other and break apart. As a result of the removal of these surface rocks, new surfaces of the mountain are continually exposed to weathering.

Homework — GENERAL

Research Paper Have interested students research efforts to preserve and restore famous monuments from the effects of weathering. Have them present their findings in a research paper or newspaper-style article highlighting preservation and restoration methods. **LS Logical/Verbal**

Close

Reteaching — BASIC

Peer Reviewing Have students work in pairs to write three to five questions about the material in this section. Then, have students quiz each other. **LS Verbal/Interpersonal** Co-op Learning

Quiz — GENERAL

1. What is the name of the process by which softer, less weather-resistant rock wears away, leaving harder, more resistant rock behind? (differential weathering)
2. Where do the slowest rates of weathering occur? (in hot, dry climates)

Alternative Assessment — GENERAL

Concept Collage Have students collect pictures or photographs that illustrate differential weathering and factors that affect rates of weathering. Have students use the pictures to create a visual concept map of the key points in this section. **LS Visual/Logical**

ENVIRONMENTAL CONNECTION

Off-Road Damage Riding off-road vehicles (ORVs), such as all-terrain vehicles and motorbikes, in backcountry areas is an increasingly popular recreational activity. However, it can dramatically increase weathering and threaten certain landscapes. When ORVs leave established roads and trails, they make ruts and expose rocks and soil. Advertising for these vehicles often encourages users to go off marked paths, through streams, up hillsides and ravines, and across deserts and dunes. Such areas are extremely sensitive to the effects of weathering and erosion. While the Forest Service and Bureau of Land Management are required by law to monitor ORV use and impact, they rarely are able to do so adequately. Conservationists around the country are pushing for stricter regulation of ORVs. Have interested students research current regulation of ORV use on public lands and report their findings to the class. **LS Verbal/Logical**

Close, *continued*

Answers to Section Review

1. Rocks that contain calcite or other soft minerals weather rapidly. Rocks that are rich in quartz or other hard minerals weather slowly.
2. Increased surface area can lead to increased weathering because more areas are exposed to agents of weathering.
3. Climates that have alternating periods of hot and cold increase weathering rates because they allow for ice wedging. Warm, humid climates increase rates of weathering because the constant moisture wears away rock surfaces.
4. Ice wedging is more common at higher elevation because the temperature is colder. With steep topography, weathered rock is more easily pulled down by gravity and washed away by water. As a result, new rocks are continually exposed to weathering on steep mountain slopes.
5. Mining exposes rock surfaces to agents of weathering and may add strong chemicals to increase the rate. Construction removes soil and exposes rock surfaces. Recreational activities such as hiking, biking, and using all-terrain vehicles remove soil and expose previously unexposed rock.
6. Burrowing animals can expose rock surfaces to weathering agents. Animal waste can accelerate chemical weathering and attract small animals to a site, further speeding up the weathering process.
7. Reducing the rate of mechanical weathering might involve restoring soil and/or vegetation cover to the area, keeping humans and animals away, and protecting the area from too much water.
8. Cleopatra's Needle would probably have suffered less damage in Siberia because in a cold, dry climate weathering is slow compared to the wet, temperate climate of New York.
9. Factors that can affect *weathering* include rock *composition; surface area,* which determines amount of *exposure; climate,* which includes *temperature* and *precipitation* patterns; *topography,* or *elevation;* and *human activities.*

Figure 4 ▶ All-terrain vehicles cause mechanical weathering on exposed surfaces. Because of this weathering, these vehicles are banned from some areas where erosion is a concern.

Human Activities

Rock can be chemically and mechanically broken down by the action of humans. Mining and construction often expose rock surfaces to agents of weathering. Mining also often exposes rock to strong acids and other chemical compounds that are used in mining processes. Construction often removes soil and exposes previously unexposed rock surfaces. Recreational activities such as hiking or riding all-terrain vehicles, as shown in **Figure 4,** can also speed up weathering by exposing new rock surfaces. Rock that is disturbed or broken by human activities weathers more rapidly than undisturbed rock does.

Plant and Animal Activities

Rock that is disturbed or broken by plants or animals also weathers more rapidly than undisturbed rock does. The roots of plants and trees often break apart rock. Burrowing animals dig holes into rock and soil. Some biological wastes of animals can cause chemical weathering. For example, caves that have large populations of bats also have large amounts of bat guano on the cave floors. Bat guano attracts small animals such as millipedes and other insects. The presence of these insects speeds up mechanical weathering, and the presence of the guano speeds up certain chemical weathering processes.

Section 2 Review

1. **Explain** how rock composition affects the rate of weathering.
2. **Discuss** how the surface area of a rock can affect the rock's weathering rate.
3. **Identify** two ways that climate can influence weathering rates.
4. **Describe** two ways the topography of a region affects weathering rates.
5. **Summarize** three ways human actions can affect the rate of weathering.
6. **Explain** two ways that animals can affect the rate of weathering.

CRITICAL THINKING

7. **Applying Concepts** Imagine that there is an area of land where mechanical weathering has caused damage. Describe two ways to reduce the rate of mechanical weathering.
8. **Identifying Relationships** How would Cleopatra's Needle probably have been affected if it had been in the cold, dry climate of Siberia for 100 years?

CONCEPT MAPPING

9. Use the following terms to create a concept map: *composition, exposure, precipitation, surface area, climate, temperature, topography, weathering, elevation,* and *human activities.*

CHAPTER RESOURCES

Chapter Resource File

- **Section Quiz** GENERAL

Workbooks

- **Study Guide** (also in Spanish)

Section 3 Soil

One result of weathering is the formation of *regolith*, the layer of weathered rock fragments that covers much of Earth's surface. *Bedrock* is the solid, unweathered rock that lies beneath the regolith. The lower regions of regolith are partly protected by those above and thus do not weather as rapidly as the upper regions do. The uppermost rock fragments weather to form a layer of very fine particles. This layer of small rock particles provides the basic components of soil. **Soil** is a complex mixture of minerals, water, gases, and the remains of dead organisms.

OBJECTIVES

- **Summarize** how soils form.
- **Explain** how the composition of parent rock affects soil composition.
- **Describe** the characteristic layers of mature residual soils.
- **Predict** the type of soil that will form in arctic and tropical climates.

KEY TERMS

soil
soil profile
horizon
humus

Characteristics of Soil

The characteristics of soil depend mainly on the rock from which the soil was weathered, which is called the soil's *parent rock*. Soil that forms and stays directly over its parent rock is called *residual soil*. However, the weathered mineral grains that form soil may be carried away from the location of the parent rock by water, wind, or glaciers. Soil that results from the deposition of this material is called *transported soil*, and this soil may have different characteristics than the bedrock on which it rests.

soil a loose mixture of rock fragments and organic material that can support the growth of vegetation

Soil Composition

Parent rock that is rich in feldspar or other minerals that contain aluminum weathers to form soils that contain large amounts of clay. Rocks that contain large amounts of quartz, such as granite, weather to form sandy soils. Soil composition refers to the materials of which it is made. The color of soil is related to the composition of the soil. Black soils are commonly rich in organic material, while red soils may form from iron-rich parent rocks. Soil moisture can also affect color, as shown in **Figure 1.**

Figure 1 ▶ These soil scientists are testing soil moisture to determine whether irrigation will be needed. Moister soils are generally darker than drier soils are.

Section 3

Focus

Overview

This section summarizes how soil forms and describes the characteristic layers of soils. It also explains how soil composition is affected by parent rock, climate, and topography.

Bellringer

Ask students to define *soil*. (Answers may vary.) LS Verbal

Motivate

Group Activity — GENERAL

What Is Soil? Have groups of students do the following: Dig up and weigh a sample of soil from the school grounds. Then, spread the soil out on newspaper and break up large clumps. Examine the soil with a magnifying glass. Pick out animals and plants (alive or dead) and set aside. Sift the remaining soil through a wide mesh sieve, collecting what goes through on another sheet of newspaper. Observe and weigh what remains in the sieve (coarse soil). Sift the soil on the newspaper through fine mesh, collecting what goes through (fine soil). The soil that remains in the sieve is medium. Observe and weigh both samples. Have students make a graph that shows the composition of the soil and discuss what soil is. LS Kinesthetic

CHAPTER RESOURCES

Chapter Resource File

- Directed Reading BASIC
- Inquiry Lab Acid Rain and Soils GENERAL

Technology

Transparencies
- Bellringer

Student Edition on CD-ROM

One-Stop Planner CD-ROM
- Lesson Plan

Teach

Using the Figure —GENERAL

Soil Horizons The descriptions on the left side of the figure provide the more thorough horizon classification used by soil scientists (OAEBCR). O = surface litter, or the organic layer; A = topsoil; E = zone of leaching; B = subsoil; C = rock particles; R = bedrock. Not all soils show all horizons, For example, agricultural soil often has little or no O horizon because plant remains are removed from fields. The E horizon is found only in heavily leached soils. Have interested students create a profile for soil on your school grounds by digging a hole and measuring the depths of the various layers. LS **Visual/ Logical**

SKILL BUILDER —BASIC

Math Forming soil takes an enormous amount of time. Estimates for forming 2.5 cm of topsoil range from 200 to 1000 years, depending on various climatic and geographic factors. Have students determine the range in rates of soil formation in mm/year.
1 cm = 10 mm; 2.5 cm = 25 mm
25 mm ÷ 200 y = 0.125 mm/y
25 mm ÷ 1000 y = 0.025 mm/y;
range in mm/y =
0.025 to 0.125 mm/y LS **Logical**

CHAPTER RESOURCES

Chapter Resource File

- **Making Models Lab** Soil Profiles GENERAL

Technology

Transparencies
- 73 Soil Horizons of Residual Soils (with worksheet)

Soil Texture

Rock material in soil consists of three main types: clay, silt, and sand. Clay particles have a diameter of less than 0.002 mm. Silt particles have diameters from 0.002 to 0.05 mm. Silt particles are too small to be seen easily, but they make soil feel gritty. Sand particles have diameters from 0.05 to 2 mm. The proportion of clay, silt, and sand in soil depends on the soil's parent rock.

Soil Profile

Transported soils are commonly deposited in unsorted masses by water or wind. However, residual soils commonly develop distinct layers over time. To determine a soil's composition, scientists study a soil profile. A **soil profile** is a cross section of the soil and its bedrock. The different layers of soil are called **horizons.**

Residual soils generally consist of three main horizons. The *A horizon,* or *topsoil,* is a mixture of organic materials and small rock particles. Almost all organisms that live in soil inhabit the A horizon. As organisms die, their remains decay and produce **humus,** a dark, organic material. The A horizon is also the zone from which surface water leaches minerals. The *B horizon* or *subsoil,* contains the minerals leached from the topsoil, clay, and, sometimes, humus. In dry climates, the B horizon also may contain minerals that accumulate as water in the soil evaporates. The *C horizon* consists of partially-weathered bedrock. The first stages of mechanical and chemical change happen in this bottom layer. **Figure 2** shows the relationships between the three soil horizons.

soil profile a vertical section of soil that shows the layers of horizons

horizon a horizontal layer of soil that can be distinguished from the layers above and below it; *also* a boundary between two rock layers that have different physical properties

humus dark, organic material formed in soil from the decayed remains of plants and animals

Figure 2 ▶ **Soil Horizons of Residual Soils**

BRAIN FOOD

Slash and Burn Although tropical soils are thick, they are not particularly fertile. Plant and animal remains quickly decay in the moist environment and are taken back up by plant roots. Most nutrients in a rainforest are locked up in the trees, and farming this soil is difficult. Using a practice called *shifting agriculture,* or "slash and burn," indigenous peoples have farmed the rainforests for thousands of years with little harm. They clear and burn small plots of forest and then farm for a few years, until nutrients are depleted. Then, they move on to a new patch, allowing forest to regrow on the old plot. On a small scale this worked well, but with growing populations, it contributes to deforestation. About half of tropical deforestation is related to this subsistence farming, and up to 12% of the plots are cleared not to raise food for locals, but to raise cattle for export to the developed world!

Soil and Climate

Climate is one of the most important factors that influences soil formation. Climate determines the weathering processes that occur in a region. These weathering processes, in turn, help determine the composition of soil.

Tropical Soils

In humid tropical climates, where much rain falls and where temperatures are high, chemical weathering causes thick soils to develop rapidly. These thick, tropical soils, called *laterites* (LAT uhr IETS), contain iron and aluminum minerals that do not dissolve easily in water. Leached minerals from the A horizon sometimes collect in the B horizon. Heavy rains, which are common in tropical climates, cause a lot of leaching of the topsoil, and thus keep the A horizon thin. But because of the dense vegetation in humid, warm climates, organic material is continuously added to the soil. As a result, a thin layer of humus usually covers the B horizon, as shown in **Figure 3.**

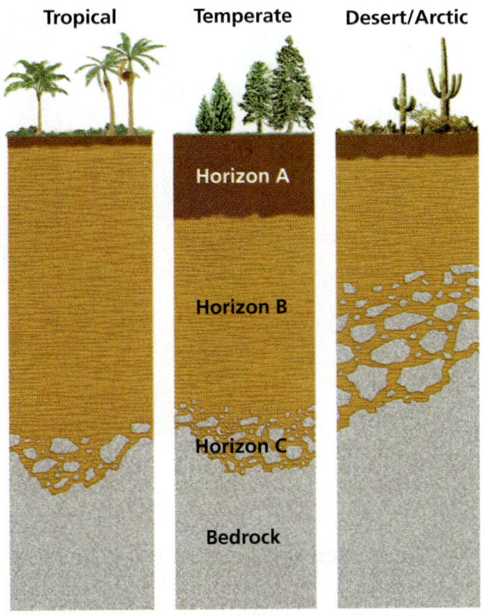

Figure 3 ▶ Tropical climates produce thick, infertile soils. Temperate climates produce thick, fertile soils. Desert and arctic climates produce thin soils.

Temperate Soils

In temperate climates, where temperatures range between cool and warm and where rainfall is not excessive, both mechanical and chemical weathering occur. All three soil horizons in temperate soils may reach a thickness of several meters, as shown in **Figure 3.**

Two main soil types form in temperate climates. In areas that receive more than 65 cm of rain per year, a type of soil called *pedalfer* (pi DAL fuhr) forms. Pedalfer soils contain clay, quartz, and iron compounds. The Gulf Coast states and states east of the Mississippi River have pedalfer soils. In areas that receive less than 65 cm of rain per year, a soil called *pedocal* (PED oh KAL) forms. Pedocal soils contain large amounts of calcium carbonate, which makes pedocal soil very fertile and less acidic than pedalfer soil. The southwestern states and most states west of the Mississippi River have pedocal soils.

Reading Check **Compare the formation of tropical soils and temperate soils. (See the Appendix for answers to Reading Checks.)**

Desert and Arctic Soils

In desert and arctic climates, rainfall is minimal and chemical weathering occurs slowly. As a result, the soil is thin and consists mostly of regolith—evidence that soil in these areas forms mainly by mechanical weathering. Desert and arctic climates are also often too warm or too cold to sustain life, so their soils have little humus.

Answer to Reading Check

Large amounts of rainfall and high temperatures cause thick soils to form in both tropical and temperate climates. Tropical soils have thin A horizons because of the continuous leaching of topsoil. Temperate soils have three thick layers, because leaching of the A horizon in temperate climates is much less than leaching of the A horizon in tropical climates.

Close

Reteaching — Basic

Flipbook Have students write the main soil types on index cards with a description of their soil profile and the factors that influence their formation. Students can also illustrate the cards. Have students bind the cards together with string or a binder ring and use them for review. **LS Visual**

Quiz — General

1. What type of parent rock forms sandy soil? (one with large amounts of quartz)
2. What is regolith? (the layer of weathered rock fragments that covers much of Earth's surface)

Alternative Assessment — General

Soil Profiles Have students create a model soil profile for the climate of their choice. They should use a glass jar, organic litter, soil, and rock samples. Have them write a brief paragraph that describes their soil and the factors that influence its formation. **LS Visual**

ENVIRONMENTAL CONNECTION

Acid Rain and Soil Not only does acid rain have an impact on weathering, it can affect fertility by altering soil chemistry. Excess acid washes some essential nutrients from soil, such as calcium and magnesium, but it can also make other minerals, such as manganese and aluminum, more available. Aluminum in high doses is toxic to plants. Vegetation shows more adverse effects from acid precipitation in some areas than in others. For example, forests in the northeastern United States seem more susceptible to the effects of acid rain than those in the western United States. Scientists believe this may be due to the different soil types found in the eastern and western parts of the United States. Have students hypothesize why this may be. (Pedocal soils in the west are alkaline and thus are more able to buffer the effects of acid precipitation. Pedalfer soils in the east are already acidic; extra acid is not buffered, and this upsets soil chemistry.) **LS Logical**

Close, *continued*

Answers to Section Review

1. Soils form by the mechanical and chemical weathering of rocks to form regolith. When the finer particles of regolith mix with organic matter and water, soil forms.
2. Soil composition, color, and texture are determined by parent rock composition.
3. The A horizon, or topsoil, is a mixture of small rock particles and organic matter. The B horizon, or subsoil, contains clay and minerals leached from the topsoil. The C horizon is partially weathered bedrock.
4. Thin soil, consisting mostly of regolith with little organic matter, will form in arctic climates. In the tropics, thick soils with thin A horizons will form.
5. A temperate climate would be ideal because it has well-developed soils with thick A horizons.
6. Sample answer: Crop growth would probably be more successful on level ground because the soil depth is thicker on level ground than it is on slopes.
7. Tropical soil has too thin an A horizon to be good for sustained farming. Crop plants rapidly use nutrients in the thin topsoil and good harvests cannot be sustained for long.
8. Arctic and desert soils are similar because of low rainfall, which reduces the rate of chemical weathering. In addition, both climates support fewer life forms, so humus is lacking in both soil types.
9. *Soil* is a mixture of *regolith* and *humus* and is classified as either *transported soil*, which is often different from the *bedrock* on which it rests, or *residual soil* which is greatly influenced by its *parent rock*, *climate*, and *topography* and develops a characteristic *soil profile* with distinct *horizons*.

Figure 4 ▶ Soil is thick at the top and bottom of a slope. Soil is thin along the slope.

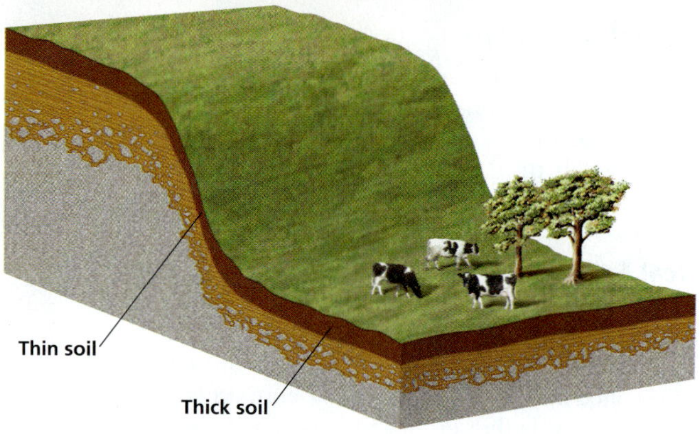

SCI LINKS. NSTA Developed and maintained by the National Science Teachers Association

For a variety of links related to this subject, go to www.scilinks.org

Topic: Soil
SciLinksCode: HQ61407

Soil and Topography

The shape of the land, or topography, also affects soil formation. Because rainwater runs down slopes, much of the topsoil of the slope washes away. Therefore, as shown in **Figure 4,** the soil at the top and bottom of a slope tends to be thicker than the soil on the slope.

One study of soil in Canada showed that A horizons on flat areas were more than twice as thick as those on 10° slopes. Topsoil that remains on a slope is often too thin to support dense plant growth. The lack of vegetation contributes to the development of a poor-quality soil that lacks humus. The soils on the sides of mountains are commonly thin and rocky, and the soils have few nutrients. Lowlands that retain water tend to have thick, wet soils and a high concentration of organic matter, which forms humus. A fairly flat area that has good drainage provides the best surface for formation of thick, fertile layers of residual soil.

Section 3 Review

1. **Summarize** how soils form.
2. **Explain** how the composition of parent rock affects soil composition.
3. **Describe** the three horizons of a residual soil.
4. **Predict** the type of soil that will form in arctic and tropical climates.

CRITICAL THINKING

5. **Applying Ideas** What combination of soil and climate would be ideal for growing deep-rooted crops? Explain your answer.
6. **Analyzing Relationships** Would you expect crop growth to be more successful on a farm that has an uneven topography or on a farm that has level land? Explain your answer.
7. **Analyzing Ideas** Why would tropical soil not be good for sustained farming?
8. **Making Comparisons** Although desert and arctic climates are extremely different, their soils may be somewhat similar. Explain why.

CONCEPT MAPPING

9. Use the following terms to create a concept map: *soil, bedrock, regolith, humus, parent rock, residual soil, transported soil, horizon, soil profile, climate,* and *topography.*

CHAPTER RESOURCES

Chapter Resource File
- Section Quiz GENERAL

Workbooks
- Study Guide (also in Spanish)

ENVIRONMENTAL CONNECTION

Section 4 Erosion

When rock weathers, the resulting rock particles do not always stay near the parent rock. Various forces may move weathered fragments of rock away from where the weathering occurred. The process by which the products of weathering are transported is called **erosion.** The most common agents of erosion are gravity, wind, glaciers, and water. Water can move weathered rock in several different ways including by ocean waves and currents, by streams and runoff, and by the movements of groundwater.

OBJECTIVES

- **Define** erosion, and list four agents of erosion.
- **Identify** four farming methods that conserve soil.
- **Discuss** two ways gravity contributes to erosion.
- **Describe** the three major landforms shaped by weathering and erosion.

KEY TERMS

erosion
sheet erosion
mass movement
solifluction
creep
landform

Soil Erosion

As rock weathers, it eventually becomes very fine particles that mix with water, air, and humus to form soil. The erosion of soil occurs worldwide and is normally a slow process. Ordinarily, new soil forms about as fast as existing soil erodes. However, some forms of land use and unusual climatic conditions can upset this natural balance. Once the balance is upset, soil erosion often accelerates.

Some farming and ranching practices increase soil erosion. For example, plants anchor soil with their roots and prevent wind and water from eroding the soil. Clearing plants or allowing animals to overgraze destroys this groundcover and increases erosion rates. Soil erosion is considered by some scientists to be the greatest environmental problem that faces the world today. As shown in **Figure 1,** vulnerability to erosion affects fertile topsoil around the world. This erosion prevents some countries from growing the crops needed to prevent widespread famine.

erosion a process in which the materials of Earth's surface are loosened, dissolved, or worn away and transported from one place to another by a natural agent, such as wind, water, ice, or gravity

Figure 1 ▶ This map shows the vulnerability of soils worldwide to erosion by water.

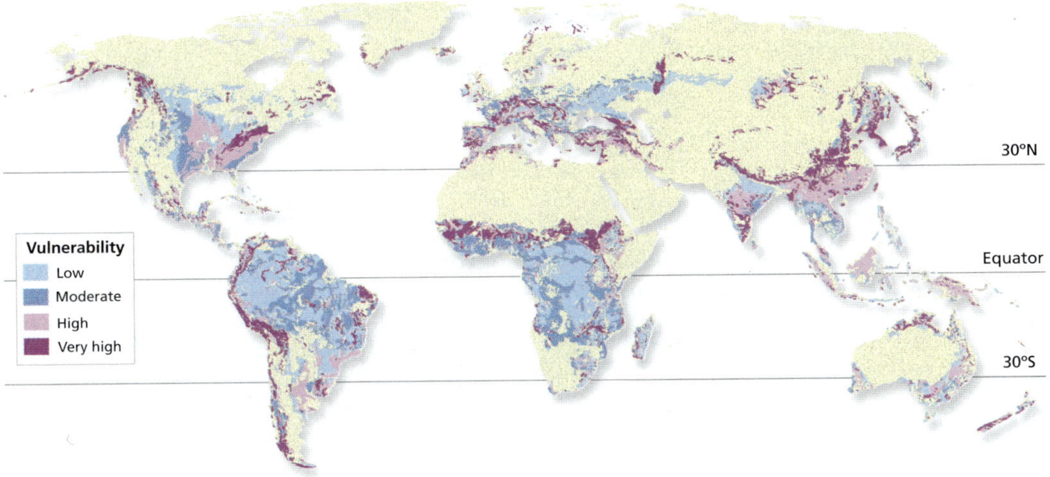

Section 4

Focus

Overview

This section defines erosion, introduces agents of erosion, describes the basic landforms that are shaped by weathering and erosion, and explains soil conservation practices to reduce erosion. The section also describes mass movements and the catastrophic nature of rapid mass movement.

Bellringer

Ask students, "Why do the Rocky Mountains have jagged, rugged peaks, while the Appalachian Mountains have more rounded, gentle slopes?" (The Rocky Mountains are much younger than the Appalachian Mountains and have not undergone as many years of weathering and erosion that wear down rugged peaks to gentle slopes.) LS **Logical**

Motivate

Using the Figure — GENERAL

Local Vulnerability Have students look at the world map that shows vulnerability of soils worldwide to erosion by water. Have them determine the vulnerability in your area. Discuss with students why your area may or may not be vulnerable to erosion by water and what conservation practices might be implemented to reduce or prevent erosion. LS **Visual**

CHAPTER RESOURCES

Chapter Resource File

- Directed Reading BASIC

Technology

Transparencies
- Bellringer
- 74 Soil Erosion Vulnerability Map (with worksheet)

Student Edition on CD-ROM

One-Stop Planner CD-ROM
- Lesson Plan

Teach

HISTORY CONNECTION GENERAL

The Dust Bowl From 1930 to 1937, drought spread across the Great Plains from Mexico into Canada. In the years before the drought, native prairie grasses had been plowed under and replaced with wheat. The hardy native grasses could have withstood the drought, but the wheat could not. Repeated crop failure left the soil completely exposed. Winds swept across barren soils, causing massive dust storms and eroding millions of acres of topsoil. More than 74 million acres of land were severely damaged during the Dust Bowl years. *The Grapes of Wrath* by John Steinbeck and *Out of the Dust* by Karen Hesse movingly capture the plight of families living through this disaster. Have students read excerpts from these books and/or visit the PBS Web site, *Surviving the Dust Bowl,* to help them understand the impact of soil erosion. **LS Interpersonal**

Answer to Reading Check

Dust storms may form during droughts when the soil is made dry and loose by lack of moisture and wind-caused sheet erosion carries it away in clouds of dust. If all the topsoil is removed, the remaining subsoil will not contain enough nutrients to raise crops.

Figure 2 ▶ This land in Madagascar can no longer be used for farming because of a form of rapid erosion called *gullying*.

SCILINKS. NSTA Developed and maintained by the National Science Teachers Association

For a variety of links related to this subject, go to www.scilinks.org

Topic: Soil Erosion
SciLinksCode: HQ61410
Topic: Soil Conservation
SciLinksCode: HQ61409

Gullying and Sheet Erosion

One farming technique that can accelerate soil erosion is the improper plowing of furrows, or long, narrow rows. Furrows that are plowed up and down slopes allow water to run swiftly over soil. As soil is washed away with each rainfall, a furrow becomes larger and forms a small gully. Eventually, land that is plowed in this way can become covered with deep gullies. This type of accelerated soil erosion is called *gullying*. The farmland shown in **Figure 2** has been ruined by gullying.

Another type of soil erosion strips away layers of topsoil. Eventually, erosion can expose the surface of the subsoil. This process is called **sheet erosion.** Sheet erosion may occur where continuous rainfall washes away layers of the topsoil. Wind also can cause sheet erosion during unusually dry periods. The soil, which is made dry and loose by a lack of moisture, is carried away by the wind as clouds of dust and drifting sand. These wind-borne particles may produce large dust storms.

sheet erosion the process by which water flows over a layer of soil and removes the topsoil

Reading Check **Describe one way a dust storm may form, and explain how a dust storm can affect the fertility of land. (See the Appendix for answers to Reading Checks.)**

Results of Soil Erosion

Constant erosion reduces the fertility of the soil by removing the A horizon, which contains the fertile humus. The B horizon, which does not contain much organic matter, is difficult to farm because it is much less fertile than the A horizon. Without plants, the B horizon has nothing to protect it from further erosion. So, within a few years, all of the soil layers could be removed by continuous erosion.

BRAIN FOOD

Worldwide Soil Losses Estimates indicate that we lose topsoil at a rate of 25 to 75 billion metric tons a year globally. This costs the global economy a staggering $400 billion dollars a year! Loss of topsoil reduces crop yield significantly because crops cannot be grown successfully if topsoil is too thin. In some areas, productivity has declined up to 50% due to erosion and desertification.

GEOGRAPHY CONNECTION

The Badlands In South Dakota, gully erosion helped create the rugged sandstone peaks of Badlands National Park. The Sioux tribes named the area *malosche shica*–"bad land." French trappers called the region *mauvaises terres a traverser*–"bad lands to cross." Sudden rainstorms that cut deep into sandstone and soil created these rugged peaks. After the rain, the soil dries out quickly only to undergo the same process with the next storm. Over time, this process has created the deep gullies separated by sharp peaks.

Soil Conservation

Erosion rates are affected not only by natural factors but also by human activities. Certain farming and grazing techniques and construction projects can also increase the rate of erosion. In developing urban areas, vegetation is removed to build houses and roads, such as those shown in **Figure 3.** This land clearing removes protective ground cover plants and accelerates topsoil erosion. In some areas, such as deserts or mountainous regions, it may take hundreds or thousands of years for the topsoil to be replenished.

But rapid, destructive soil erosion can be prevented by soil conservation methods. People, including city planners and some land developers, have begun to recognize the environmental impact of land development and are beginning to implement soil conservation measures. Some land development projects are leaving trees and vegetation in place whenever possible. Other projects are planting cover plants to hold the topsoil in place. Farmers are also looking for new ways to prevent soil erosion to preserve fertile topsoil.

Figure 3 ▶ This housing development in England is encroaching on cropland. The clearing of land for development can accelerate topsoil erosion.

SKILL BUILDER — GENERAL

Math A severe rainstorm may wash away 1 mm of soil. Over a 20 yr period, this loss can add up to 2.5 cm. If 1 mm of soil from 1 hectare of land weighs 13 metric tons, how much topsoil may be lost in 20 years?
2.5 cm = 25 mm; 25 mm × 13 metric tons/mm = 325 metric tons **LS Logical**

Teaching Tip — GENERAL

Make Concepts Relevant
Students may be unfamiliar with farming practices to reduce erosion, but they may have seen urban examples of erosion control. At construction sites, fabric fences, straw bales, slope regrading, groundcover planting on steep slopes, and detention ponds are just a few examples of methods used in urban settings to reduce erosion. Ask students to describe urban erosion control techniques they have seen. If possible, have them visit a local construction site to observe such techniques. English Language Learners
LS Logical/Visual

Connection to ENVIRONMENTAL SCIENCE

Land Degradation

Soil that can support the growth of healthy plants is called *fertile soil*. Land that can be used to grow crops is called *arable land*. A limited area of arable land is available on Earth. As humans use this land for agriculture, they plow, fertilize, irrigate, and otherwise change the natural processes of soil formation.

Human activity and natural processes may damage land to the point that it can no longer support the local ecosystem. This process is called *land degradation*. Several activities may lead to land degradation. Three common factors in land degradation are urbanization, overgrazing, and deforestation.

Urbanization is the movement of people from rural areas to cities. As cities expand rapidly, people begin to develop the surrounding area, in a process called *urban sprawl*. As a result, arable land is paved and demand for resources may overwhelm the water and land resources in the area.

Overgrazing occurs when more animals are allowed to graze an area than the plants in that area can support. When animals overgraze, too many plants are eaten or trampled and too few plants are left to protect the soil from eroding.

If this clearcut in Willamette National Forest in Oregon was not reclaimed, the land may have continued to undergo degradation.

Deforestation is the clearing of trees from an area without replacing them. This process destroys wildlife habitat and results in accelerated soil erosion.

Extreme land degradation may cause desertification. *Desertification* is the process by which land in dry areas becomes more desertlike because of human activity or climate change. Desertification can cause land to become useless for farming or human habitation.

Many government and independent organizations are working to develop laws and guidelines to protect both wilderness and agricultural land. These groups hope that by protecting these lands now, the resources will be available for future generations.

ENVIRONMENTAL CONNECTION

Land Degradation By the year 2020, global population is expected to be a third higher than in 1995, and we must find a way to feed, clothe, and shelter these future generations. At the same time, urbanization, deforestation, overgrazing, desertification, and other processes are degrading land around the world. These processes may severely lower productivity of agricultural lands at a time when we will most need to increase food and fiber production.

Land degradation is often most severe in regions experiencing the largest increase in population, such as Africa and Asia. Population growth may actually be causing some of the land degradation as burgeoning countries try to feed their people. The biggest challenge of the 21st century may be how to feed the world without destroying the land on which we depend. Have students look up the UN Global Assessment of Soil Degradation (GLASOD) on the Internet for further information on the problem of land degradation. **LS Logical/Verbal**

Teach, *continued*

Graphic Organizer — General

Spider Map
You may want to have students work in groups to create this Spider Map. Have one student draw the map and fill in information provided by other students from the group.

Demonstration — Basic

Soil Conservation Fill two pans with packed soil. Raise one end of each pan with books. Make furrows across the slope of one pan to model contour plowing. Sprinkle both pans with water from a sprinkling can and observe what happens. The water should erode soil and wash it away, but the pan with furrows across the slope should experience less erosion. Challenge students to use the setup to model gullying and terracing. LS **Visual/Kinesthetic**

CHAPTER RESOURCES

Chapter Resource File

- Internet Activity
  - NRCS General

Contour Plowing These fields are plowed in contours (or curves) that follow the shape of the land.

Strip-Cropping These fields were planted with alternating strips of different crops.

Terracing These terraced fields help slow runoff and prevent rapid soil erosion.

Figure 4 ▶ Soil Conservation Methods

Contour Plowing

Farmers in countries around the world use planting techniques to reduce soil erosion. In one method, called *contour plowing*, soil is plowed in curved bands that follow the contour, or shape, of the land. This method of planting, shown in **Figure 4,** prevents water from flowing directly down slopes, so the method prevents gullying.

Strip-Cropping

In *strip-cropping*, crops are planted in alternating bands, also shown in **Figure 4.** A crop planted in rows, such as corn, may be planted in one band, and another crop that fully covers the surface of the land, such as alfalfa, will be planted next to it. The *cover crop* protects the soil by slowing the runoff of rainwater. Strip-cropping is often combined with contour plowing. The combination of these two methods can reduce soil erosion by 75%.

Terracing

The construction of steplike ridges that follow the contours of a sloped field is called *terracing*, as shown in **Figure 4.** Terraces, especially those used for growing rice in Asia, prevent or slow the downslope movement of water and thus prevent rapid erosion.

Crop Rotation

In *crop rotation*, farmers plant one type of crop one year and a different type of crop the next. For example, crops that expose the soil to the full effects of erosion may be planted one year, and a cover crop will be planted the next year. Crop rotation stops erosion in its early stages, which allows small gullies that formed during one growing season to fill with soil during the next one.

Graphic Organizer

Spider Map
Create the **Graphic Organizer** entitled "Spider Map" described in the Skills Handbook section of the Appendix. Label the circle "Soil conservation methods." Then, fill in the map with at least four methods for soil conservation.

Internet Activity — General

NRCS The Natural Resources Conservation Service (NRCS) of the U.S. Department of Agriculture works with farmers and landowners to prevent loss of topsoil. Have students use the NRCS Web site to learn about soil conservation programs in your area and their effectiveness. A worksheet designed to direct student research on this topic can be found in the **Chapter Resource File** booklet or by visiting **go.hrw.com** and entering the keyword **HQ6WAEX.** LS **Verbal**

Activity — Basic

Plants and Soil To demonstrate the importance of plant cover in reducing soil erosion have students observe the root system of a potted plant. Run a knife around the edge of the pot to loosen the plant. Have students turn the pot upside down to remove the plant. Ask students to observe how the roots hold the soil together in a compact mass. If possible, have students observe the root system of grasses in a piece of sod. Have students write a brief paragraph that explains why strip-cropping and crop rotation are effective soil conservation methods. LS **Visual**

Gravity and Erosion

Gravity causes rock fragments to move down inclines. This movement of fragments down a slope is called **mass movement.** Some mass movements occur rapidly, and others occur very slowly.

mass movement the movement of a large mass of sediment or a section of land down a slope

Rockfalls and Landslides

The most dramatic and destructive mass movements occur rapidly. The fall of rock from a steep cliff is called a *rockfall*. A rockfall is the fastest kind of mass movement. Rocks in rockfalls often range in size from tiny fragments to giant boulders.

When masses of loose rock combined with soil suddenly fall down a slope, the event is called a *landslide*. Large landslides, in which loosened blocks of bedrock fall, generally occur on very steep slopes. You may have seen a small landslide on cliffs and steep hills overlooking highways, such as the one shown in **Figure 5.** Heavy rainfall, spring thaws, volcanic eruptions, and earthquakes can trigger landslides.

Reading Check **What is the difference between a rockfall and a landslide? (See the Appendix for answers to Reading Checks.)**

Mudflows and Slumps

The rapid movement of a large amount of mud creates a *mudflow*. Mudflows occur in dry, mountainous regions during sudden, heavy rainfall or as a result of volcanic eruptions. Mud churns and tumbles as it moves down slopes and through valleys, and it frequently spreads out in a large fan shape at the base of the slope. The mass movements that sometimes occur in hillside communities, such as the one shown in **Figure 5,** are often referred to as landslides, but they are actually mudflows.

Sometimes, a large block of soil and rock becomes unstable and moves downhill in one piece. The block of soil then slides along the curved slope of the surface. This type of movement is called a *slump*. Slumping occurs along very steep slopes. Saturation by water and loss of friction with underlying rock causes loose soil to slip downhill over the solid rock.

Figure 5 ▶ An earthquake in El Salvador caused this dramatic landslide (left). Heavy rains in the Philippines caused this destructive mudflow (right).

SKILL BUILDER - ADVANCED

Writing Have students research a major landslide, mudflow, slump or other mass movement event that has occurred recently. Have students write a newspaper style article that describes the event and the conditions that contributed to its occurrence. **LS Verbal**

Answer to Reading Check

Landslides are masses of loose rock combined with soil that suddenly fall down a slope. A rockfall consists of rock falling from a steep cliff.

Discussion — GENERAL

Protecting Against Mass Movements Although some mass movements are unpredictable and therefore unpreventable, others can be avoided. Geologists locate areas where mass movements have occurred, try to identify landforms and soils prone to mass movement, and look for signs of potential slump, creep, or other movement. Ask students to name landforms and soils prone to mass movement (clays, alpine and tundra soils, steep slopes, barren slopes, saturated soils) and signs of movement that geologists may look for. (tilted utility lines, cracks in building foundations, deep ridges and cracks in soil) Geologists use this information to create landslide potential maps that show regions susceptible to mass movement. If signs of movement are detected or land is prone to movement, common sense suggests not building there. If the area is already developed, steps can be taken to restabilize the slope. Ask students to discuss ways of doing this. (Answers may vary. Sample answers: replanting barren slopes, regrading slopes, terracing slopes, constructing retaining walls and other safety enclosures.) **LS Interpersonal**

SOCIAL STUDIES CONNECTION — GENERAL

The Hills of Southern California The hills of southern California are extremely prone to mass movements of a catastrophic nature for several reasons. The region is hot and dry in the summer and supports mainly scrub and desert vegetation. Wildfires occur frequently leaving steep hillsides barren. When occasional heavy rains come, water quickly sinks into the sparsely covered hills, adding weight to the slopes, and making them unstable. Southern California also lies along an active plate boundary, the San Andreas fault. Numerous small earthquakes rock the area and shake those unstable masses free. Many Californians pay huge sums to live in the hills along the Pacific coast, even with the ever-present risk of mass movements that may destroy their homes. Over the years, homeowners and insurance companies have lost millions of dollars just to have "a room with a view." Have students debate the value of building in scenic areas that are prone to mass movements. **LS Verbal/Logical**

Teach, *continued*

READING SKILL BUILDER — BASIC

Reading Organizer As students read this section, encourage them to take Power Notes, KWL notes, or Two-Column Notes as described in the Skills Handbook section of the Appendix. Later, students can use these notes as a study guide for assessments. LS Verbal English Language Learners

Demonstration — BASIC

Solifluction Build a slope out of modeling clay along one side of an aquarium or deep tray. Pour fine, dry sand over the slope, and then spray water on the sand until it becomes saturated. Have students describe what happens to the sand. (As the sand becomes saturated, it will slowly slip down the slope.) How does this model solifluction? (The clay subsoil is impermeable to water, so the sand above becomes saturated and slowly flows down the slope.) LS Visual

BRAIN FOOD

Creeping Trees While you may not notice soil creep, other objects do. As soil slowly moves down slopes, power poles and gravestones tilt, fences and walls bend and crack, building foundations sink, and, believe it or not, trees bend. Trees that continue to grow on creeping soil have a pronounced curve at their base, as if they are trying to remain upright in opposition to the slide downhill!

solifluction the slow, downslope flow of soil saturated with water in areas surrounding glaciers at high elevations

creep the slow downhill movement of weathered rock material

Solifluction

Although most slopes appear to be unchanging, some slow mass movement commonly occurs. Catastrophic landslides are the most hazardous mass movement. However, more rock material on the whole is moved by the greater number of slow mass movements than by catastrophic landslides.

One form of slow mass movement is called solifluction. **Solifluction** is the process by which water-saturated soil slips over hard or frozen layers. Solifluction occurs in arctic and mountainous climates where the subsoil is permanently frozen. In spring and summer, only the top layer of soil thaws. The moisture from this layer cannot penetrate the frozen layers beneath. So, the surface layer becomes muddy and slowly flows downslope, or downhill. Solifluction can also occur in warmer regions, where the subsoil consists of hard clay. The clay layer acts like the frozen subsoil in arctic climates by forming a waterproof barrier.

Creep

The extremely slow downhill movement of weathered rock material is known as **creep.** Soil creep moves the most soil of all types of mass movements. But creep may go unnoticed unless buildings, fences, or other surface objects move along with soil.

Many factors contribute to soil creep. Water separates rock particles, which allows them to move freely. Growing plants produce a wedgelike pressure that separates particles and loosens the soil. The burrowing of animals and repeated freezing and thawing loosen rock particles and allow gravity to slowly pull the particles downhill.

As rock fragments accumulate at the base of a slope, they form piles called *talus* (TAY luhs), as shown in **Figure 6.** Talus weathers into smaller fragments, which move farther down the slope. The fragments wash into gullies, are carried into successively larger waterways, and eventually flow into rivers.

Figure 6 ▶ The movement of rock fragments downslope formed these talus cones at the base of the Canadian Rockies.

INCLUSION Strategies

- Developmentally Delayed
- Hearing Impaired
- Attention Deficit Disorder

Use the following procedure to help students understand the idea of erosion:

1. Collect about 20 L of dirt, about 16 L of water, a 1 L plastic container, a sprinkling can, and an 8-1/2" × 11" piece of cardboard.
2. Divide the class into three teams. Assign a section of sidewalk to each team.
3. Have each team spread a thin layer of dirt about 1 m^3 in size.
4. Ask each team to simulate one of the following: blowing wind (fan with cardboard), gentle rain (water from the sprinkling can), or gushing rain (water from the plastic container).
5. Compare and contrast the effects of each of the three forms of erosion.
6. Clean up the sidewalk. LS Kinesthetic

Figure 7 ▶ The mountains in the Patagonian Andes, shown on the left, are still being uplifted and are more rugged than the more eroded Appalachian mountains on the right.

Erosion and Landforms

Through weathering and erosion, Earth's surface is shaped into different physical features, or **landforms.** There are three major landforms that are shaped by weathering and erosion—*mountains, plains,* and *plateaus*. Minor landforms include hills, valleys, and dunes. The shape of landforms is also influenced by rock composition.

All landforms are subject to two opposing processes. One process bends, breaks, and lifts Earth's crust and thus creates elevated, or uplifted, landforms. The other process is weathering and erosion, which wears down land surfaces.

landform a physical feature of Earth's surface

Erosion of Mountains

During the early stages in the history of a mountain, the mountain undergoes uplift. Generally, while tectonic forces are uplifting the mountain, it rises faster than it is eroded. Mountains that are being uplifted tend to be rugged and have sharp peaks and deep, narrow valleys. When forces stop uplifting the mountain, weathering and erosion wear down the rugged peaks to rounded peaks and gentle slopes. The formations in **Figure 7** show how the shapes of mountains are influenced by uplift and erosion.

Over millions of years, mountains that are not being uplifted become low, featureless surfaces. These areas are called *peneplains* (PEE nuh PLAYNZ), which means "almost flat." A peneplain commonly has low, rolling hills, as seen in New England.

Reading Check **Describe how a mountain changes after it is no longer uplifted. (See the Appendix for answers to Reading Checks.)**

Close

Reteaching — Basic

Section Highlights Have students work in small groups to summarize the main points under each heading. After the groups have finished their summaries, quiz students orally to ensure their understanding. **LS Verbal**

Quiz — General

1. Where is solifluction most likely to occur? (in arctic or mountain climates where the subsoil remains permanently frozen or in regions where the subsoil is hard clay)
2. What is talus? (piles of rock fragments that accumulate at the base of a slope)

Alternative Assessment — General

Conservation Plans Have students work in groups to develop a conservation plan to reduce weathering and erosion around their school or home. Have each group prepare a poster or pamphlet that illustrates their plan and the steps they would take to implement it. The Natural Resources Conservation Service Web site has information that may help students develop their plans. **LS Visual/Verbal**

GEOLOGY CONNECTION

What Goes Up There is a limit to how high mountains on Earth can grow for two basic reasons. First, as soon as mountains uplift, weathering and erosion start to attack and wear them down. Second, rocks are not infinitely strong. The weight of uplifted rock weighs down on rock buried deep below. The buried rocks gradually warm, soften, and flow. Mountains essentially begin to collapse under their own weight in a process geologists call *orogenic collapse*.

Answer to Reading Check

When a mountain is no longer being uplifted, weathering and erosion wear down its jagged peaks to low, featureless surfaces called *peneplains*.

Close, continued

Answers to Section Review

1. Erosion is the process by which materials of Earth's surface are loosened, dissolved, or worn away and transported from one place to another by natural agents.
2. wind, water, glaciers, and gravity
3. Answers may vary. Sample answer: Gullying occurs when furrows are plowed up and down a slope. Rainfall runs swiftly through the furrows widening and deepening them. Sheet erosion occurs when water or wind flows over soil and removes the top layers in entire sheets.
4. contour plowing, strip cropping, terracing, and crop rotation
5. Gravity causes water to move down slopes and carry away topsoil. It also causes mass movements of rocks and soil.
6. Rapid mass movements include rockfalls, landslides, mudflows, and slumping and can be catastrophic in nature. Creep and solifluction are slow mass movements, and while they usually are not catastrophic, over time they can move more material than rapid movement does.
7. Mountains are eroded from rugged peaks into rounded gentle slopes over millions of years. Plains are flat landforms near sea level and that are usually not subject to as much erosion as plateaus are. Plateaus, broad flat landforms with a high elevation, may erode into mesas and buttes over time.
8. Sample answer: Place a stick or pole in the ground on the slope and monitor its position over time.
9. contour planting with strip cropping between the rows of vines or terracing to reduce hillside erosion
10. The butte would probably have shallower slopes and a more rounded top.
11. Answers may vary. Possible answers include improving ground cover so soils are held by roots and do not become saturated with heavy rains, preventing or containing wildfires so slopes do not become empty of vegetation during the dry season, building retaining walls, regrading the hillside, or forbidding further building on the hillside.
12. *Landforms* such as *mountains, plains,* and *plateaus,* which erode into *mesas* and *buttes,* are shaped by *erosion* processes, such as *gullying, sheet erosion, landslides, mudflows, slump, solifluction,* and *creep,* which produces rock piles called *talus*.

Figure 8 ▶ Ancient rivers carved plateaus into mesas, which eventually eroded into the buttes of Monument Valley in Arizona.

Erosion of Plains and Plateaus

A *plain* is a relatively flat landform near sea level. A *plateau* is a broad, flat landform that has a high elevation. A plateau is subject to much more erosion than a plain. Young plateaus, such as the Colorado Plateau in the southwestern United States, commonly have deep stream valleys that separate broad, flat regions. Older plateaus, such as those in the Catskill region in New York State, have been eroded into rugged hills and valleys.

The effect of weathering and erosion on a plateau depends on the climate and the composition and structure of the rock. In dry climates, resistant rock produces plateaus that have flat tops. As a plateau ages, erosion may dissect the plateau into smaller, tablelike areas called *mesas* (MAY suhz). Mesas ultimately erode to small, narrow-topped formations called *buttes* (BYOOTS). In dry regions, such as in the area shown in **Figure 8**, mesas and buttes have steep walls and flat tops. In areas that have wet climates, humidity and precipitation weather landforms into round shapes.

Section 4 Review

1. **Define** erosion.
2. **List** four agents of erosion.
3. **Summarize** two processes of soil erosion.
4. **Identify** four farming methods that result in soil conservation.
5. **Discuss** two ways gravity contributes to erosion.
6. **Compare** rapid mass movements and slow mass movements.
7. **Describe** the erosion of the three major landforms.

CRITICAL THINKING

8. **Analyzing Relationships** Describe an experiment that could help you determine whether a nearby hill is undergoing creep.
9. **Applying Ideas** Suppose you wanted to grow grapevines on a hillside in Italy. What farming methods would you use? Explain your answer.
10. **Predicting Consequences** Describe two ways a small butte would change if it was in a wet climate, rather than a dry climate.
11. **Drawing Conclusions** A hillside community has asked you to help brainstorm ways to prevent future mudflows. Describe three of your ideas.

CONCEPT MAPPING

12. Use the following terms to create a concept map: *erosion, gullying, sheet erosion, landslide, mudflow, slump, solifluction, creep, talus, landform, mountain, plain, plateau, mesa,* and *butte.*

CHAPTER RESOURCES

Chapter Resource File
- Section Quiz GENERAL

Workbooks
- Study Guide (also in Spanish)

Chapter 14 Highlights

Sections

1 Weathering Processes

Key Terms

weathering, 343
mechanical weathering, 343
abrasion, 344
chemical weathering, 346
oxidation, 346
hydrolysis, 347
carbonation, 347
acid precipitation, 348

Key Concepts

- Agents of mechanical weathering break rock into smaller pieces but do not change its chemical composition.
- Chemical weathering changes the mineral composition of rock.
- Types of chemical weathering include hydrolysis, carbonation, oxidation, and acid precipitation.

2 Rates of Weathering

differential weathering, 349

- Rock weathers at different rates, which depend partly on its mineral composition.
- The greater the amount of exposure a rock has, the faster a rock weathers.
- Rock weathers more rapidly in regions where rainfall is abundant and where alternating freezes and thaws occur.

3 Soil

soil, 353
soil profile, 354
horizon, 354
humus, 354

- The parent rock from which soil forms is the major factor that determines the composition of the soil.
- Thick soils form in tropical climates and temperate climates. Thin soils form in arctic climates and desert climates, where rainfall is minimal.

4 Erosion

erosion, 357
sheet erosion, 358
mass movement, 361
solifluction, 362
creep, 362
landform, 363

- Natural agents often move weathered rock away from where the weathering occurred. This movement leads to erosion.
- Planting crops helps to conserve soil.
- Slow and rapid mass movements of rock, soil, and mud cause massive amounts of soil erosion.
- Erosion wears away landforms as it levels Earth's surface.

Chapter Highlights

Alternative Assessment — ADVANCED

Life as a Rock Have students write the life story of a rock. They should include details such as the landform where it originated; how it has been weathered, eroded, and transported; and whether it has become part of the soil. Students can illustrate their story if they wish. Some may want to write their stories as children's books. Encourage creativity, but make sure the stories are factually possible. LS Verbal

CHAPTER RESOURCES

Chapter Resource File

- **Concept Review** GENERAL
- **Critical Thinking** ADVANCED
- **Math Skills** GENERAL
- **Graphing Skills** GENERAL
- **Chapter Test A** GENERAL
- **Chapter Test B** ADVANCED

Workbooks

- **Study Guide** (also in Spanish)
- **Assessments** (Spanish)

Technology

Classroom Videos
- Brain Food Video Quiz

HRW Earth Science Video
- Segment 9: Weathering and Erosion

Chapter Review

Assignment Guide

Section	Questions
1	1, 5–6, 9–10, 17, 19
2	11–12, 18, 24, 27, 29
3	2, 7, 13–14, 36–37
4	3, 8, 15–16, 20–22, 25, 28, 32–34
2 and 4	4, 23, 30
1 and 2	26, 35
1–4	31

Using Key Terms

1-8. Answers may vary but should show that students understand the definitions of and differences between key terms.

Understanding Key Concepts

9. b **10.** b
11. d **12.** d
13. c **14.** a
15. d **16.** d
17. c **18.** c

Short Answer

19. Natural rain is slightly acidic due to the dissolution of carbon dioxide in water to create weak carbonic acid. Acid precipitation is considerably more acidic and contains nitric, nitrous, and sulfuric acids.

20. Answers may vary. Sample answer: Soil conservation is important to retain soil fertility mainly found in the A horizon and to avoid complete removal of all soil layers.

21. A mountain changes from a rugged mountain to gently rounded hills and finally to low, featureless surfaces called *peneplains* through processes of weathering and erosion over millions of years.

22. mountains, plains (flat landforms near sea level), and plateaus (broad, flat landforms with high elevation)

23. Weathering breaks down rocks into material that can be easily eroded, and both weathering and erosion are involved in shaping Earth's landforms.

Chapter 14 Review

Using Key Terms

Use each of the following terms in a separate sentence.

1. *abrasion*
2. *humus*
3. *landform*

For each pair of terms, explain how the meanings of the terms differ.

4. *weathering* and *erosion*
5. *mechanical weathering* and *chemical weathering*
6. *oxidation* and *carbonation*
7. *soil profile* and *horizon*
8. *solifluction* and *creep*

Understanding Key Concepts

9. A common kind of mechanical weathering is called
a. oxidation.
b. ice wedging.
c. carbonation.
d. leaching.

10. Oxides of sulfur and nitrogen that combine with water vapor cause
a. hydrolysis.
b. acid rain.
c. mechanical weathering.
d. carbonation.

11. The surface area of rocks exposed to weathering is increased by
a. burial.
b. accumulation.
c. leaching.
d. jointing.

12. Chemical weathering is most rapid in
a. hot, dry climates.
b. cold, dry climates.
c. cold, wet climates.
d. hot, wet climates.

13. The chemical composition of soil depends to a large extent on
a. topography.
b. the soil's A horizon.
c. the parent material.
d. soil's B horizon.

14. The soil in tropical climates is often
a. thick.
b. thin.
c. dry.
d. fertile.

15. All of the following farming methods prevent gullying, *except*
a. terracing.
b. strip-cropping.
c. contour plowing.
d. irrigation.

16. The type of mass movement that moves the most soil is
a. a landslide.
b. a mudflow.
c. a rockfall.
d. creep.

17. The grinding away of rock surfaces through the mechanical action of rock or sand particles is called
a. carbonation.
b. hydrolysis.
c. abrasion.
d. erosion.

18. The process by which softer rocks wear away and leave harder rocks behind is
a. chemical weathering.
b. mechanical weathering.
c. differential weathering.
d. erosion.

Short Answer

19. What is the difference between natural rain and acid precipitation?

20. Explain two reasons why soil conservation is important.

21. Describe how a mountain changes from a rugged mountain to a peneplain.

22. Describe three landforms that are shaped by weathering and erosion.

24. Climates with wide seasonal variation in temperature have higher rates of weathering because of numerous freeze/thaw cycles. Warm, humid climates increase rates of chemical weathering because of the constant exposure to moisture. Lack of water in deserts and cold, dry climates decreases rates of weathering.

25. plateaus, mesas, and buttes

Critical Thinking

26. A rock at the top of a mountain would be greatly affected by gravity and might be involved in rockslides. It would also be mechanically and chemically weathered because of exposure. Rock beneath the ground surface might be mechanically and chemically weathered by water seeping down to it through joints and fractures in the overlying ground, but lack of exposure to the atmosphere might prevent much weathering.

23. Explain two ways that weathering and erosion are related.

24. Identify three ways that climate affects the rate of weathering.

25. Name three landforms that you would expect to find in a desert.

Critical Thinking

26. Making Comparisons Compare the weathering processes that affect a rock on top of a mountain and those that affect a rock beneath the ground surface.

27. Understanding Relationships Which do you think would weather faster, a sculpted marble statue or a smooth marble column? Explain your answer.

28. Making Inferences Mudflows in the southern California hills are usually preceded by a dry summer and widespread fires, which are followed by torrential rainfall. Explain why these phenomena are followed by mudflows.

29. Evaluating Ideas How can differential weathering help you determine whether a rock is harder or softer than the rock that surrounds it?

30. Inferring Relationships Suppose that a mountain has been wearing down at the rate of about 2 cm per year for 10 years. After 10 years, scientists find that the mountain is no longer losing elevation. Why do you think the mountain is no longer losing elevation?

Concept Mapping

31. Use the following terms to create a concept map: *composition, mechanical weathering, chemical weathering, topography, erosion, conservation, exposure, weathering, surface area,* and *climate.*

Math Skills

32. Making Calculations A group of scientists calculates that an acre of land has crept 18 cm in 15 years. What is the average rate of creep in millimeters per year?

33. Making Calculations For a given area of land, the average rate of creep is 14 mm per year. How long will it take the area to move 1 m?

Writing Skills

34. Writing from Research Research a mudslide or landslide that occurred in the past. Describe the conditions that led to that mass movement and the impact it had.

35. Communicating Ideas You are in charge of preserving a precious marble statue. Write a paragraph that describes how you would protect the statue from weathering.

Interpreting Graphics

The graph below shows land use in the United States. Use the graph to answer the questions that follow.

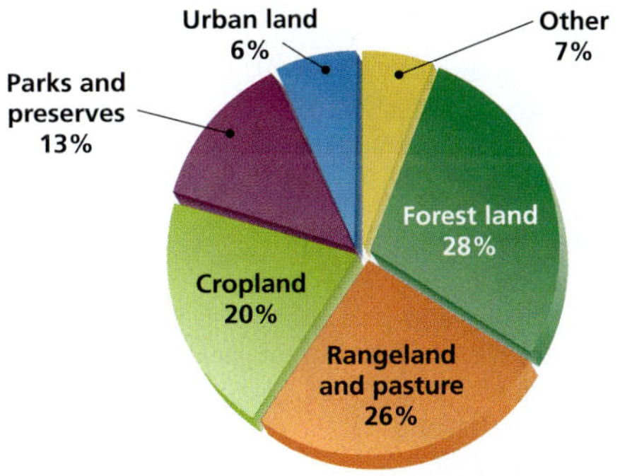

36. How much more land is rangeland and pasture than is urban land?

37. If cropland increased to 25% and all other categories remain the same except for forests, what would the percentage of forests be?

27. Assuming the two structures are located in the same climate, a sculpted marble statue would probably weather faster because it would have more surface area and cracks and crevices than a smooth marble column would.

28. Dry summers and wildfires strip hillsides of vegetation that could hold the soil in place. Then, heavy rains saturate the soils, creating heavy, runny mud that can suddenly slip down slopes.

29. Hard rock weathers less readily then softer rock does. So, if a rock layer sticks out above surrounding layers, it is harder than surrounding layers. A relatively soft rock layer will be surrounded by protruding layers.

30. Answers may vary. Accept all reasonable answers. Sample answers: The mountain is undergoing a new cycle of uplifting. Rocks now exposed to weathering are harder and less susceptible to the effects of weathering and erosion. Climate conditions may be changing.

Concept Mapping

31. Answers may vary but should include all of the terms listed. Sample answers appear at the end of this Teacher Edition.

Math Skills

32. 18 cm = 180 mm
180 mm/15 y = 12 mm/y

33. 1 m = 1000 mm,
1000 mm ÷ 14 mm/y = 71.4 y

Writing Skills

34. Answers may vary. Accept all reasonable answers.

35. Answers may vary. Accept all reasonable answers.

Interpreting Graphics

36. 26% − 6% = 20%; Twenty percent more of the total land is rangeland and pasture than is urban land.

37. 25% − 20% = 5%; 28% − 5% = 23%; Forests must lose 5%, so they would account for 23% of land use in the United States.

Estimated Time

To give students practice under more realistic testing conditions, allow them 30 minutes to answer all of the questions in this practice test.

TEST DOCTOR

Question 3 Answer A is correct. The reddish color is due to chemical processes. Mechanical weathering, answer B, abrasion, answer C, and erosion, answer D, are all physical processes and do not change the chemical composition of rocks and soil minerals.

Question 4 Answer H is correct. The chemical reactions involved in weathering are generally accelerated at higher temperatures, so answers F and G are incorrect. Many chemical and mechanical weathering processes are associated with the presence of water, so answer I is incorrect.

Question 10 Full-credit answers should include the following points:

- almost any activity initiated by a living organism that exposes rock and soil to wind or water would be an acceptable answer
- construction by humans breaks up rocks and creates quarries, which contribute to weathering
- tunneling animals, other than worms, churn and expose rocks in soil to the surface
- rotting vegetation and animal waste may release acids that can help destroy rocks when the acids are mixed with groundwater

Chapter 14 Standardized Test Prep

Understanding Concepts

Directions (1–5): **For *each* question, write on a separate sheet of paper the letter of the correct answer.**

1. The processes of physical weathering and erosion shape Earth's landforms by
   A. expanding the elevation of Earth's surface
   B. decreasing the elevation of Earth's surface
   C. changing the composition of Earth's surface
   D. bending rock layers near Earth's surface

2. Which of the following rocks is most likely to weather quickly?
   F. a buried rock in a mountain
   G. an exposed rock on a plain
   H. a buried rock in a desert
   I. an exposed rock on a slope

3. The red color of rocks and soil containing iron-rich minerals is caused by
   A. chemical weathering
   B. mechanical weathering
   C. abrasion
   D. erosion

4. In which of the following climates does chemical weathering generally occur most rapidly?
   F. cold, wet climates
   G. cold, dry climates
   H. warm, humid climates
   I. warm, dry climates

5. Which of the following has the greatest impact on soil composition?
   A. activity of plants and animals
   B. characteristics of the parent rock
   C. amount of precipitation
   D. shape of the land

Directions (6–7): **For *each* question, write a short response.**

6. In what type of decomposition reaction do hydrogen ions from water displace elements in a mineral?

7. Sand carried by wind is responsible for what type of mechanical weathering?

Reading Skills

Directions (8–10): **Read the passage below. Then, answer the questions.**

How Rock Becomes Soil

Earthworms are crucial for forming soil. As they search for food by digging tunnels, they expose rocks and minerals to the effects of weathering. Over time, this process creates new soil.

Worms are not the only living things that help create soil. Plants also play a part in the weathering process. As the roots of plants grow and seek out water and nutrients, they help break large rock fragments into smaller ones. Have you ever seen a plant growing in a sidewalk? As the plant grows, its roots spread into tiny cracks in the sidewalk. These roots apply pressure to the cracks, and over time, the cracks become larger. As the plants make the cracks larger, ice wedging can occur more readily. As the cracks expand, more water can flow into them. When the water freezes, it expands and presses against the walls of the crack, which makes the crack larger. Over time, the weathering caused by water, plants, and worms helps break down rock to form soil.

8. Which of the following statements can be inferred from the passage?
   A. Weathering can occur only when water freezes in cracks in rocks.
   B. Only large plants have roots that are powerful enough to increase the rate of weathering.
   C. Local biological activity may increase the rate of weathering in a given area.
   D. Plant roots often prevent weathering by filling cracks and keeping water out of cracks.

9. Ice wedging, as described in the passage, is an example of which of the following?
   F. oxidation
   G. mechanical weathering
   H. chemical weathering
   I. hydrolysis

10. What are some ways not mentioned in the passage in which the activity of biological organisms may increase weathering?

Answers

Understanding Concepts

1. B
2. I
3. A
4. H
5. B
6. hydrolysis
7. abrasion

Reading Skills

8. C
9. G
10. Answers may vary. See Test Doctor for a detailed scoring rubric.

Interpreting Graphics

11. A
12. H
13. Answers may vary. See Test Doctor for a detailed scoring rubric.

Interpreting Graphics

Directions (11–13): **For *each* question below, record the correct answer on a separate sheet of paper.**

The diagram shows the soil profile of a mature soil. Use this diagram to answer questions 11 and 12.

Mature Soil Profile

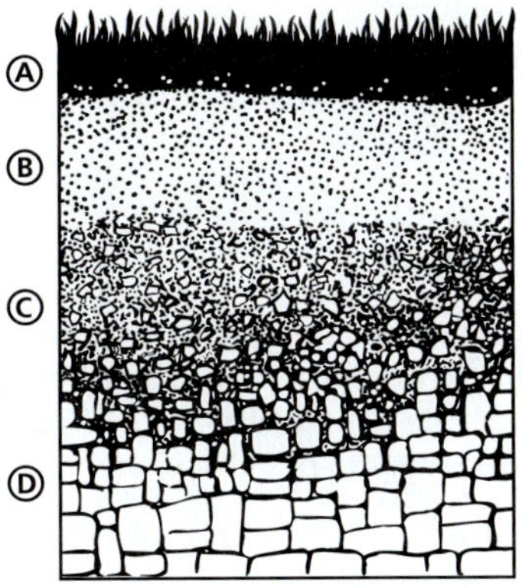

11 Which of the layers in the soil profile above contains the greatest number of soil organisms?

A. layer A
B. layer B
C. layer C
D. layer D

12 Which two layers in the soil profile above are least likely to contain the dark, organic material humus?

F. layers A and B
G. layers B and C
H. layers C and D
I. layers A and D

Use the diagram stone blocks below to answer question 13.

Blocks of Identical Volume

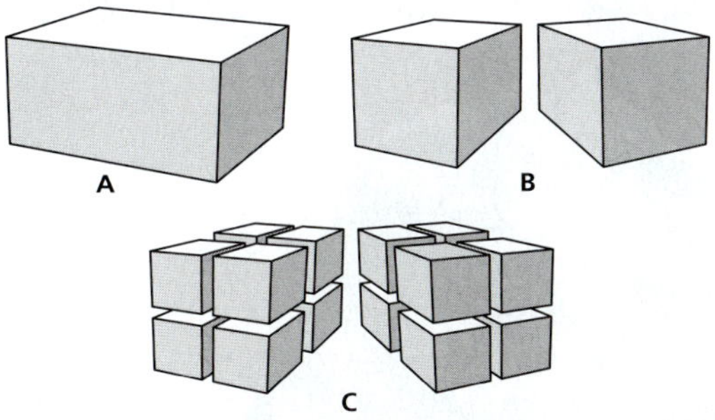

13 If the blocks shown in diagrams A, B, and C above contain identical volumes and are made of the same types of minerals, will they weather at the same rate? Explain your answer.

Test TIP

Whenever possible, highlight or underline important numbers or words that are critical to correctly answering a question.

TEST DOCTOR

Question 13 Full-credit answers should include the following points:

- the eight small blocks in part C will weather most rapidly
- students should relate the rate of weathering to the amount of exposed surface area available
- students should examine the diagram and consider that the total surface area of each rectangular solid, or cube, is equal to the sum of the areas of each of the cube's six sides. The block shown in figure A has far less exposed surface than the blocks shown in figures B or C. If all conditions are equal, the more surface area that is available to experience weathering, the faster weathering will occur
- even though the figures show the same volume and type of rock, the blocks of figure C would weather at the fastest rate due to their larger total surface area. The blocks of figure A would weather at the slowest rate

Test Prep Correlations *National Science Education Standards*

HNS 2a: items 1, 3, 4, 6, 7, 13
HNS 2b: items 2, 4, 8, 9, 13
SAI 2e: item 13
SPSP 4a: items 8, 9, 10, 11, 12
UCP 1: items 11, 12

CHAPTER RESOURCES

State Resources

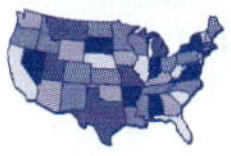

For specific resources for your state, visit go.hrw.com and type in the keyword HSHSTR.

Skills Practice Lab

Soil Chemistry

Teacher's Notes

Time Required
one 45-minute class period

Lab Ratings
EASY → HARD

Teacher Preparation 2
Student Setup 3
Concept Level 2
Cleanup 2

Skills Acquired
- Measuring
- Collecting data
- Predicting
- Organizing and Analyzing Data

The Scientific Method
In this lab, students will
- Make Observations
- Analyze Results
- Draw Conclusions
- Communicate Results

Materials
The materials listed on this page are enough for groups of two to four students.

Tips and Tricks
Make sure students are familiar with the safety precautions required when working with acids and bases. They should wear a lab apron, gloves, and goggles throughout the entire period. Make sure students understand pH and how to read pH paper. Remind students to review the differences between pedalfer and pedocal soils so that they understand why they are performing steps 6–11.

Chapter 14 Skills Practice Lab

Objectives
- **Test** the acidity of soil samples.
- **Identify** the composition of soil samples.

Materials
- ammonia solution
- stoppers, cork (9)
- hydrochloric acid, dilute
- medicine dropper
- pH paper
- subsoil sample (B and C horizons)
- test tubes, 9
- test-tube rack
- topsoil sample (A horizon)
- water

Safety

Soil Chemistry

Different soil types contain different kinds and amounts of minerals. To support plant life, soil must have a proper balance of minerals and nutrients. For plants to take in the minerals they need, the soil must also have the proper acidity.

Acidity is measured on a scale called the *pH scale.* The pH scale ranges from 0 (acidic) to 14 (alkaline). A pH of 7 is neutral (neither acidic nor alkaline). In this lab, you will test the acidity of soil samples.

PROCEDURE

1. pH paper changes color in the presence of an acidic or alkaline substance. Wet a strip of pH paper with tap water. Compare the color of the wet pH paper with the pH color scale. What is the pH of the tap water?
2. Fill a clean test tube to 1/8 full with a small amount of the topsoil. Add water to the test tube until it is 3/4 full. Place a cork stopper on the test tube, and shake the test tube.
3. Set the soil and water mixture aside in the test-tube rack to settle. When the water is fairly clear, test it with a piece of pH paper. What is the pH of the soil sample? Is the soil acidic or alkaline?
4. Repeat step 2 and step 3 with the subsoil sample.
5. Pedalfer soils tend to be acidic. Pedocal soils tend to be alkaline. Based on the pH results in steps 3 and 4, predict whether your soil is pedalfer or pedocal.

Step 3

CHAPTER RESOURCES

Chapter Resource File
- Datasheet for Chapter Lab GENERAL
- Lab Notes and Answers

6 To test your prediction, you will need to test the soil's composition. Take five rock particles from the subsoil sample. Place each particle in a separate test tube. Use the dropper to add two drops of hydrochloric acid, HCl, to the test tubes. **CAUTION** If you spill any acid on your skin or clothing, rinse immediately with cool water and alert your teacher.

7 HCl has little or no effect on silicates, but HCl decomposes calcium carbonate and causes CO_2 gas to bubble out of solution. How many of the rock particles were silicates? How many were calcium carbonate?

8 Fill a test tube 1/8 full with the subsoil. Slowly add HCl to the test tube until it is about 2/3 full. Cork the tube, and gently shake it. **CAUTION** Always shake the test tube by pointing it away from yourself and other students.

9 After shaking the test tube, remove the stopper and set the test tube in the rack. Record your observations. After the mixture has settled, draw the test tube and its contents. Label each layer. If iron is present, the solution may look brown. What color is the liquid above the soil sample?

10 Use a medicine dropper to place 10 drops of the liquid in a clean test tube. Carefully add 12 drops of ammonia to the test tube. Test the pH of the solution. If the pH is greater than 8, any iron should settle out as a reddish-brown residue. The remaining solution will be colorless.

11 If the pH is less than 8, add two more drops of ammonia and test the pH again. Continue adding ammonia until the pH reaches 8 or higher. Record your observations, and draw a diagram of the test tube. Label each layer of material in the test tube.

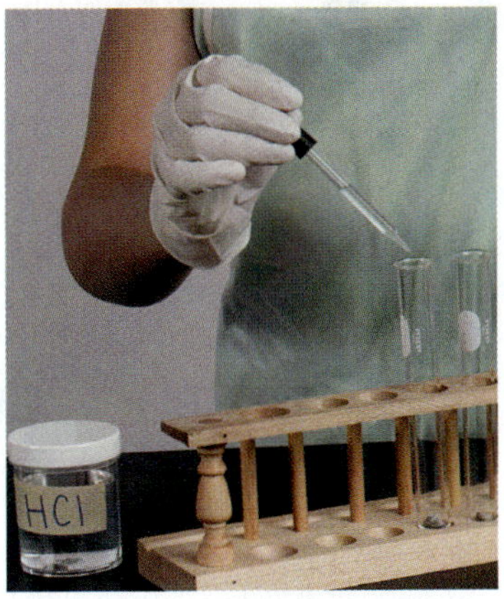

Step 6

ANALYSIS AND CONCLUSION

1 **Analyzing Results** Is your soil sample most likely pedalfer or pedocal? Explain your answer.

2 **Drawing Conclusions** What type of soil, pedalfer or pedocal, would you treat with acidic substances, such as phosphoric acid, sulfur, or ammonium sulfate, to help plant growth? Explain your answer.

3 **Recognizing Relationships** Explain why acidic substances are usually spread on the surface of the soil.

Extension

1 **Research** Use the library or the Internet to learn why the use of phosphate and nitrate detergents has been banned in some areas. Report your findings to the class.

Skills Practice Lab

Answers to Analysis and Conclusion

1. Answers may vary depending on the samples provided and the results of the tests. Pedalfer soil should be acidic and have silicates and iron. Pedocal should be alkaline and be composed mainly of carbonates.
2. Pedocal soils tend to be alkaline. They would be treated with acidic substances to reduce their pH, bringing it closer to neutral, which helps make certain minerals more available for plant uptake.
3. Acidic substances are spread on the surface of the soil because as water percolates through the soil, the acids will dissolve in the water and will aid in making certain minerals and nutrients found in the topsoil more available for plant uptake.

Answers to Extension

1. Answers may vary. Accept all reasonable answers. Students will likely find that phosphate detergents are banned in some areas because they run off into waterways and promote algae growth and cultural eutrophication of streams and lakes. Fertilizers contain nitrates and phosphates. When too much fertilizer is applied to soils, the excess may runoff in streams, rivers and lakes, promoting eutrophication. Phosphates and nitrates promote algae growth because they contain phosphorous and nitrogen, which (along with potassium) are the primary nutrients needed for plant growth.

TEACHER TESTED & APPROVED

Daniel Brownstein, MA/MAT
Geology
Hastings High School
Hastings-on-Hudson, NY

Soil Map of North Carolina

Discussion — GENERAL

Local Soils Obtain a map of your state that identifies the various soil types in your region. Discuss with students the various soil types and the conditions that led to their development. Discuss which, if any, regions in your state are most fertile, whether the area is currently used for food production, and whether it is threatened by land degradation. LS Visual

Answers to Map Skills Activity

1. 4
2. Large River Valleys and Flood Plain Systems; Outer Banks System; Brackish and Freshwater Marsh Systems; Organic Soil System; Lower Coastal Plain—Pamlico System; and Lower Coastal Plain—Wicomico and Talbot System
3. Answers may vary. Both Fayetteville and Wilmington are located within the "Large River Valleys and Flood Plain System" of soils, which indicates that they are located near a river.
4. mountains; Asheville is surrounded by the mountain soil region.
5. The Brackish and Freshwater Marsh Systems are located where major rivers empty into the ocean.
6. The average elevation would decrease from west to east, as you move from the mountains, through the piedmont (or foothills), and into the coastal plains. This trend is indicated both by the locations of the mountain, piedmont, and coastal plain soil regions and by the flow of the rivers eastward toward the ocean.

CHAPTER RESOURCES

Technology

Transparencies
- 75 Soil Map of North Carolina (with worksheet)

MAPS in Action

Soil Map of North Carolina

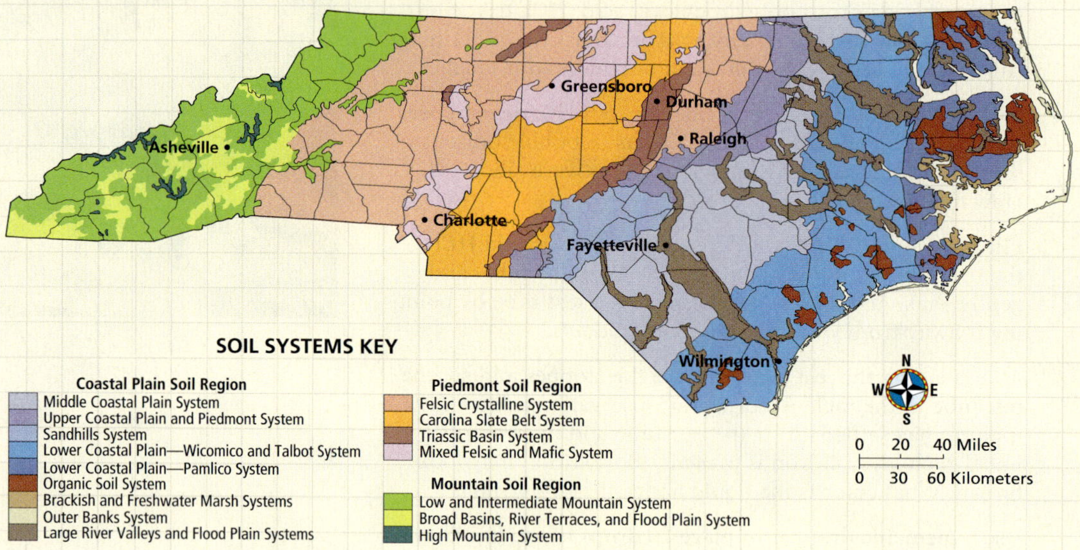

Map Skills Activity

This map shows soil systems in the state of North Carolina, including the Outer Banks barrier island system. Use the map to answer the questions below.

1. **Using a Key** How many colors on the map represent soil systems in the Piedmont Soil Region?
2. **Using a Key** What soil systems are present along the eastern shore of North Carolina?
3. **Analyzing Data** What city is located on the banks of a large river or in a river valley? Explain your answer.
4. **Inferring Relationships** What landforms would you expect to surround the town of Asheville? Explain your answer.
5. **Analyzing Relationships** Brackish water is water that is somewhat salty but is not as salty as sea water. How does this fact explain the location of the Brackish and Freswater Marsh soil systems?
6. **Identifying Trends** How would you describe the change in elevation of North Carolina from west to east based on the locations of soil systems? Explain your answer.

Career Focus

Soil Conservationist

One hundred and fifty years ago, Boone County, Illinois, consisted mainly of rolling prairies. Today, farms and homes have replaced much of the prairies. As a soil conservationist, Lewis Nichols helps farmers and homeowners to more effectively manage their land.

Down-to-Earth Solutions

"Erosion is our number one issue here," says Nichols. "Gully erosion, the kind of erosion that forms large trenches, is the most visible kind. Farmers also have problems with sheet erosion, which can remove topsoil from entire fields." Nichols helps farmers implement conservation practices to keep valuable soil in the fields and to keep the soil out of streams and rivers. As more rural areas give way to housing developments, Nichols also visits new neighborhoods. Though most of his work is done in the field, Nichols returns to his office to develop specific conservation practices for the problems he is studying.

For Nichols, the career path to soil conservation began in a high school greenhouse. Nichols turned his interests into a profession. He earned a bachelor's degree in agronomy, which is the study of crop production and soil management.

"Everything ties back to the soil."

—Lewis Nichols

Hamburger and the Farmer

Like other soil conservationists, Nichols reaches out to communities and schools to teach the value of natural resources and to encourage conservation efforts. "We go to schools to talk to kids about conservation. We want kids to know the connection between a hamburger and the farmer that produced it," says Nichols. "I find it very rewarding to know I'm able to help society by looking out for the future and providing a better world for our children—a world where the words *natural resources* still have meaning."

◀ Irrigation in this peanut field in Oklahoma has caused erosion. Soil conservationists work with farmers to reduce erosion when possible.

SCILINKS. NSTA Developed and maintained by the National Science Teachers Association

For a variety of links related to this subject, go to www.scilinks.org

Topic: Careers in Earth Science

SciLinksCode: HQ60222

Career Focus

Soil Conservationist

Activity — GENERAL

Guest Speaker Invite someone from your local soil conservation district to visit your class and discuss local soil erosion problems and practices that can be implemented to reduce erosion. Most counties have a soil conservation district with public outreach programs designed to educate people about conservation practices. Visit the Web site of the National Association of Conservation Districts to locate your local district. LS **Verbal**

Extension — GENERAL

Research Have students identify four other careers that are related to weathering and erosion. Have students write a brief synopsis of each career and present them to the class. LS **Interpersonal**

Chapter 15

River Systems

Planning Guide

Compression Guide

To shorten instruction because of time limitations, omit Section 1.

OBJECTIVES	LABS, DEMONSTRATIONS, AND ACTIVITIES	TECHNOLOGY RESOURCES
PACING • 45 min pp. 374–378 **Chapter Opener**	**LTP** **Long-Term Project** Water Clarity* ADVANCED **LTP** **Long-Term Project** Precipitation and the Water Table* ADVANCED	**OSP** **Parent Letter** **CD** **Student Edition on CD-ROM** **CD** **Chapter Summaries Audio CD** **VID** **Brain Food Video Quiz**
Section 1 The Water Cycle • Outline the stages of the water cycle. • Describe factors that affect a water budget. • List two approaches to water conservation.	**TE** **Group Activity** Modeling the Water Cycle, p. 375 GENERAL **SE** **Quick Lab** Modeling the Water Cycle, p. 377 GENERAL **CRF** **Datasheet for Quick Lab*** GENERAL **CRF** **Inquiry Lab** Eutrophication* GENERAL **CRF** **Skills Practice Lab** Stream Quality Monitoring* GENERAL	**OSP** **Lesson Plans** (also in print) **TR** **Bellringer*** **TR** 76 The Water Cycle* **CD** **Interactive Tutor** Ice, Water, and Vapor **CD** **Interactive Tutor** The Water Cycle
PACING • 90 min pp. 379–382 **Section 2 Stream Erosion** • Summarize how a river develops. • Describe the parts of a river system. • Explain factors that affect the erosive ability of a river. • Describe how erosive factors affect the evolution of a river channel.	**TE** **Group Activity** Field Trip, p. 380 GENERAL **TE** **Art Connection** The Big Muddy, p. 380 ADVANCED **SE** **Inquiry Lab** Sediments and Water, pp. 392–393 GENERAL **CRF** **Datasheet for Chapter Lab*** GENERAL **SE** **Maps in Action** World Watershed Sediment Yield, p. 394 GENERAL **TE** **Discussion** Sediment Load and Land Use, p. 394 GENERAL **SE** **Mapping Expeditions** What Comes Down Must Go. . . Where?, pp. 838–839 GENERAL	**OSP** **Lesson Plans** (also in print) **TR** **Bellringer*** **TR** 77 Stream Gradient and Channel Erosion* **TR** 78 World Watershed Sediment Yield* **TE** **Internet Activity** Watersheds, p. 379 GENERAL **CRF** **Internet Activity** Watersheds* GENERAL **VID** **HRW Earth Science Video** Water and Erosion (River Systems) **CD** **Interactive Tutor** Physical and Chemical Weathering
PACING • 45 min pp. 383–386 **Section 3 Stream Deposition** • Explain the two types of stream deposition. • Describe one advantage and one disadvantage of living in a floodplain. • Identify three methods of flood control. • Describe the life cycle of a lake.	**TE** **Demonstration** Changing Land Use, p. 383 GENERAL **TE** **Debate** To Stay or Not to Stay, p. 384 ADVANCED **SE** **Quick Lab** Soil Erosion, p. 385 GENERAL **CRF** **Datasheet for Quick Lab*** GENERAL	**OSP** **Lesson Plans** (also in print) **TR** **Bellringer*** **TE** **Internet Activity** Restoring Rivers, p. 395 GENERAL **CRF** **Internet Activity** Restoring Rivers* GENERAL

PACING • 90 min

CHAPTER REVIEW, ASSESSMENT, AND STANDARDIZED TEST PREPARATION

- **SE** **Chapter Highlights**, p. 387
- **SE** **Chapter Review**, pp. 388–389
- **SE** **Standardized Test Prep**, pp. 390–391
- **CRF** **Concept Review*** GENERAL
- **CRF** **Critical Thinking*** ADVANCED
- **CRF** **Math Skills*** GENERAL
- **CRF** **Graphing Skills*** GENERAL
- **CRF** **Chapter Test A*** GENERAL
- **CRF** **Chapter Test B*** ADVANCED
- **OSP** **Lesson Plans** (also in print)
- **OSP** **Test Generator**
- **OSP** **Test Item Listing**

Online and Technology Resources

go.hrw.com Holt Online Learning

Visit go.hrw.com for access to Holt Online Learning, or enter the keyword **HQ6 Home** for a variety of free online resources.

One-Stop Planner® CD-ROM

This CD-ROM package includes
- Lab Materials QuickList Software
- Holt Calendar Planner
- Customizable Lesson Plans
- Printable Worksheets
- ExamView® Test Generator
- Interactive Teacher Edition
- Holt PuzzlePro®
- Holt PowerPoint® Resources

KEY

- **SE** Student Edition
- **TE** Teacher Edition
- **CRF** Chapter Resource File
- **LTP** Long-Term Projects
- **OSP** One-Stop Planner
- **TR** Transparencies and Transparency Worksheets
- **CD** CD or CD-ROM
- **VID** Classroom Video/DVD
- * Also on One-Stop Planner
- ◆ Requires advance prep
- ■ Also available in Spanish

SKILLS DEVELOPMENT RESOURCES	REVIEW AND ASSESSMENT	CORRELATIONS
SE Pre-Reading Activity, p. 374 GENERAL **TE Using the Figure** Developing Wilderness, p. 374 GENERAL		National Science Education Standards
CRF Directed Reading* BASIC **TE Skill Builder** Math, p 376 GENERAL **TE Reading Skill Builder** Paired Summarizing, p. 376 BASIC	**SE Reading Check,** p. 376 GENERAL **SE Section Review,** p. 378 GENERAL **TE Reteaching,** p. 377 BASIC **TE Quiz,** p. 377 GENERAL **TE Alternative Assessment,** p. 378 GENERAL **CRF Section Quiz*** ■ GENERAL	SPSP 3c, UCP 4, ES 3c
CRF Directed Reading* BASIC **SE Math Practice,** p. 380 GENERAL **TE Skill Builder** Vocabulary, p 381 BASIC **SE Graphic Organizer** Cause-and-Effect Map, p. 382 GENERAL	**SE Reading Check,** p. 381 GENERAL **SE Section Review,** p. 382 GENERAL **TE Reteaching,** p. 381 BASIC **TE Quiz,** p. 381 GENERAL **TE Alternative Assessment,** p. 381 GENERAL **CRF Section Quiz*** ■ GENERAL	SPSP 5c, UCP 4
CRF Directed Reading* BASIC **TE Inclusion Strategies,** p. 384 **TE Skill Builder** Writing, p. 384 ADVANCED **TE Reading Skill Builder** Reading Organizer, p. 384 BASIC	**SE Reading Check,** p. 385 GENERAL **SE Section Review,** p. 386 GENERAL **TE Reteaching,** p. 385 BASIC **TE Quiz,** p. 385 GENERAL **TE Alternative Assessment,** p. 386 BASIC **CRF Section Quiz*** ■ GENERAL	SPSP 5c, UCP 4

Holt Lab Generator CD-ROM

Search for labs by topic, standard, difficulty level, and/or time frame. Edit those labs to fit your needs, or create your own labs. Use the Lab Materials QuickList to order just what you need.

Holt Earth Science Interactive Tutor CD-ROM

This CD-ROM consists of interactive activities that give students a fun way to extend their knowledge of Earth science concepts.

Chapter Summaries Audio CDs

These CDs include audio summaries of the key concepts presented in each chapter. (Audio summaries are also available in Spanish.)

SciLinks NSTA

www.scilinks.org

Maintained by the **National Science Teachers Association.** See Chapter Enrichment pages that follow for a complete list of topics.

See Chapter Enrichment pages for Video Resources.

Refer to our ***Professional Reference for Teachers*** for additional teaching resources, including articles written by science education professionals about relevant and timely issues facing today's science teachers.

Chapter 15 Chapter Enrichment

This Chapter Enrichment provides relevant and interesting information to expand and enhance your classroom instruction of the chapter material.

Section 1 The Water Cycle

The Chemistry of Life

When scientists look for life on other planets, one of the first things they look for is water. Why? Because life as we know it cannot exist without water. Between 70% and 90% of all living tissue is water. What is so special about water, and why is it essential to life?

Water is a simple molecule, composed of two hydrogen atoms attached to an oxygen atom. Water molecules have a special affinity for each other, called *hydrogen bonding*. Hydrogen bonding is a result of water's bent structure that gives it a slight negative charge at one end and a positive charge at the other. Because opposite charges attract, water molecules stick together like the opposite poles of magnets. This stickiness explains many of water's properties that make life possible: relatively high boiling point, expansion upon freezing, ability to dissolve many substances, and high capacity to hold heat.

Most molecules of water's size are gases at average surface temperatures, but water stays liquid until its temperature reaches 100°C. If water were mainly a solid or gas at average global temperatures, life on Earth might never have developed. Water can dissolve many substances, which is important because most of the chemical reactions that make life possible must take place in a liquid environment. Blood is mostly water, and supports metabolic processes by carrying dissolved nutrients and gases throughout the body to feed and energize body tissues, and to eliminate waste products.

Water also has a high specific heat. This means that heating up a given amount of water requires a relatively large amount of energy compared to molecules of other compounds of similar size. Thus, water can store and transport heat around Earth, buffering the environment against extreme temperature fluctuations, making coastal areas cooler in the summer and warmer in the winter.

Section 2 Stream Erosion

Divides and Drainage Basins

Watersheds, also known as *drainage basins,* are separated by divides that direct the flow of water. Divides affect local and regional water flow. On a grander scale, continental divides separate drainage basins that flow into different oceans. The Continental Divide of North America runs along the spine of the Rocky Mountains from Canada to the U.S. border with Mexico and separates watersheds that flow into the Pacific Ocean from those that flow into the Atlantic Ocean. Another important divide exists along the Appalachian Mountains and separates watersheds that flow directly into the Atlantic from those that flow into the Gulf of Mexico. Additionally, a divide just south of the U.S. border with Canada separates Gulf of Mexico drainage basins from Hudson Bay basins. In South America, the continental divide follows the crest of the Andes mountains. Because the Andes are so close to the Pacific Ocean, the Pacific drainage basin is relatively small, while the Amazon basin is immense. Some tributaries of the Amazon actually lie within 100 km of the Pacific coast!

◀ **Joshua trees grow in the Mojave Desert of California, where rainfall amounts are low.**

▲ People have adapted to living on this floodplain in Cambodia by building houses on stilts.

Section 3 Stream Deposition

Disappearing Delta

Not only is the health of the Mississippi River delta threatened by excessive pollution, the delta is actually disappearing into the Gulf of Mexico. Large portions of marshy wetlands have literally been washed away. The Mississippi River delta is one of the world's largest and richest estuaries. It provides breeding grounds and habitat for fish, oysters, crabs, and shrimp, and wintering grounds for many migratory birds. While the delta is always changing as it subsides and is eroded by the sea, recent losses are of great concern to local residents, the seafood industry, and many other groups. One of the main causes of the changes is the decreased sediment load carried by the lower Mississippi River since the 1950s.

Dams built along the Missouri and Mississippi Rivers trap much of the sediment that would normally reach the delta. Diversions and underwater channels built to control flooding bypass the delta and send sediments into the Gulf of Mexico. With less sediment coming in, natural subsidence outpaces the rebuilding process, and marshes are lost to the sea. In recent years, local, state and federal agencies, as well as citizens and environmental groups, have been working to protect and restore the delta. They have been filling in channels that divert sediments and developing methods to rebuild marshes and wetlands. Continued efforts may help reduce the threat to this important natural and cultural resource.

Video Resources

Brain Food Video Quizzes

Brain Food Video Quizzes These videos contain game-show style quizzes that assess students' progress and motivate students to study the chapter material.

HRW Earth Science Video

This video introduces Earth science topics and includes a geology field trip. The video segment listed below complements this chapter.

Segment 10: Water and Erosion (River Systems) In this segment, the subject of rivers as agents of erosion is examined. The transport of sediment is followed from the source of a river to the valley floor to deposition in the ocean, where with time and deep burial the sediment solidifies and becomes rock. Physical features of rivers, including cutbanks, meanders, oxbow lakes, and point bars, are discussed. (1.5 min)

NOVA® Videos

To order **NOVA** videos related to this chapter, visit **go.hrw.com** and enter the keyword **HQ6RVSV.**

SciLinks®

NSTA Developed and maintained by the National Science Teachers Association

SciLinks is maintained by the National Science Teachers Association to provide you and your students with interesting, up-to-date links that will enrich your classroom presentation of the chapter.

Visit www.scilinks.org and enter the SciLinks code for more information about the topic listed.

Topic: Water Cycle
SciLinks code: HQ61626

Topic: River Systems
SciLinks code: HQ61314

Topic: Flooding and Society
SciLinks code: HQ60585

Topic: Stream Deposition
SciLinks code: HQ61457

Chapter 15

Chapter Overview

This chapter describes how water moves continuously between land, oceans, and atmosphere through the water cycle. The chapter also explains how rivers shape the land by erosion and deposition.

Using the Figure — GENERAL

Developing Wilderness This photograph shows the Copper River delta in Alaska's Chugach National Forest, one of the most beautiful and untamed wetlands in the world. Some groups want more roads built to open the area for mining, logging, oil and gas exploration, and other activities. Lead a discussion exploring the impact of development on the river. (Possible adverse impacts include reducing wildlife populations, damaging scenic beauty, and increasing river pollution. Benefits may include better access to important natural resource deposits and improved local economy.) LS **Logical**

PRE-READING ACTIVITY

FOLDNOTES

Encourage students to use their FoldNote as a study guide to quiz themselves for a test on the chapter material. Students may want to create Layered Book FoldNotes for different topics within the chapter.

Chapter 15 River Systems

Sections

1 The Water Cycle
2 Stream Erosion
3 Stream Deposition

What You'll Learn

- How water moves between Earth's land, oceans, and atmosphere
- How rivers shape the land by erosion
- How rivers shape the land by deposition

Why It's Relevant

The continuous movement of water is necessary for the survival of humans and other lifeforms on Earth. Water movement also shapes the land by erosion and deposition.

PRE-READING ACTIVITY

FOLDNOTES

Layered Book Before you read this chapter, create the FoldNote entitled "Layered Book" described in the Skills Handbook section of the Appendix. Label the tabs of the layered book with "Evapotranspiration," "Condensation," "Precipitation," and "Conservation." As you read the chapter, write information you learn about each category under the appropriate tab.

▶ Rivers, such as the Copper River in Alaska shown here, are major forces in eroding sediment from one place and depositing sediment in another.

Chapter Correlations *National Science Education Standards*

SPSP 3c Humans use many natural systems as resources. Natural systems have the capacity to reuse waste, but that capacity is limited. Natural systems can change to an extent that exceeds the limits of organisms to adapt naturally or humans to adapt technologically. **(Section 1)**

SPSP 5c Some hazards, such as earthquakes, volcanic eruptions, and severe weather are rapid and spectacular. But there are slow progressive changes that also result in problems for individuals and society. **(Sections 1, 2 and 3)**

UCP 4 Evolution is a series of changes, some gradual and some sporadic, that accounts for the present form and function of objects, organisms, and natural and designed systems. The general idea of evolution is that the present arises from materials and forms of the past. Although evolution is most commonly associated with … the process of descent with modification of organisms from common ancestors, evolution also describes changes in the universe. **(Sections 2 and 3)**

ES 3c Interactions among the solid earth, the oceans, the atmosphere and organisms have resulted in the ongoing evolution of the earth system. We can observe some changes … on a human time scale, but many processes … take place over hundreds of millions of years. **(Sections 1, 2 and 3)**

Section 1 The Water Cycle

The origin of Earth's water supply has puzzled people for centuries. Aristotle and other ancient Greek philosophers believed that rivers such as the Nile and the Danube could be supplied by rain and snow alone. It was not until the middle of the 17th century that scientists could accurately measure the amount of water received on Earth and the amount flowing in rivers. These measurements showed that Earth's surface receives up to 5 times as much water as rivers carry off. So, a more puzzling question than "Where does Earth's water come from?" is "Where does the water go?"

OBJECTIVES

- **Outline** the stages of the water cycle.
- **Describe** factors that affect a water budget.
- **List** two approaches to water conservation.

KEY TERMS

water cycle
evapotranspiration
condensation
precipitation
desalination

Movement of Water on Earth

Water is essential for humans and all other organisms. Its availability in different forms is critical for the continuation of life on Earth. More than two-thirds of Earth's surface is covered with water. Water flows in streams and rivers. It is held in lakes, oceans, and icecaps at Earth's poles. It even flows through the rock below Earth's surface as groundwater. Water is found not only in these familiar bodies of water but also in the tissues of all living creatures. In the atmosphere, water occurs as an invisible gas. This gas is called *water vapor*. Liquid water also exists in the atmosphere as small particles in clouds and fog, as shown in **Figure 1.**

Earth's water is constantly changing from one form to another. Water vapor falls from the sky as rain. Glaciers melt to form streams. Rivers flow into oceans, where liquid water escapes into the atmosphere as water vapor. This continuous movement of water on Earth's surface from the atmosphere to the land and oceans and back to the atmosphere is called the **water cycle.**

SCILINKS. NSTA Developed and maintained by the National Science Teachers Association

For a variety of links related to this subject, go to www.scilinks.org

Topic: Water Cycle
SciLinks code: HQ61626

water cycle the continuous movement of water between the atmosphere, the land, and the oceans

Figure 1 ▶ The snow, the fog, and the river water in this photo are three of the forms that water takes on Earth. Invisible water vapor is also present in the air.

Section 1

Focus

Overview

This section describes the processes of the water cycle: evapotranspiration, condensation, and precipitation. It explains water budgets and local factors that affect them, and describes methods of water conservation.

Bellringer

Have students list as many properties and characteristics of water as they can think of. (Answers may vary. Accept all reasonable answers.) **LS Logical**

Motivate

Group Activity — GENERAL

Modeling the Water Cycle Have students work in small groups to build a terrarium in a jar to model the water cycle. Students should fill the bottom of their jar with soil and plants. A small plastic cup filled with water can model a lake. When their terrarium is complete, students should spray the soil with water and put a lid on the jar. Place the terrariums in a sunny window or near lights. Have each group observe what happens in their jar. Challenge students to keep their terrariums alive and active as long as possible. **LS Kinesthetic**

CHAPTER RESOURCES

Chapter Resource File
- Directed Reading BASIC

Technology

Transparencies
- Bellringer

Student Edition on CD-ROM

One-Stop Planner CD-ROM
- Lesson Plan

Teach

SKILL BUILDER — GENERAL

Math Have students determine the transpiration rate of a plant by calculating the amount of water that collects over a period of time on a specific number of leaves on the plant. First, have students count the number of leaves. Then, have them tightly secure a plastic bag around the plant and place it in a sunny window or near a light. Have students measure in mL the amount of water that collects in the bag over 24 hours. Have students calculate the transpiration rate using this formula: Transpiration rate = mL water ÷ number of leaves ÷ number of hours. (The unit of time could also be days if the transpiration rate is very slow.) LS **Logical**

READING SKILL BUILDER — BASIC

Paired Summarizing Group students into pairs and have them read silently about evapotranspiration, condensation, and precipitation. Then, have one student summarize the role of these processes in the water cycle. The other student should listen to the retelling and point out any inaccuracies or ideas that were left out. Allow students to refer to the text as needed. LS **Verbal** English Language Learners Co-op Learning

Answer to Reading Check

Precipitation is any form of water that falls to Earth from the clouds, including rain, snow, sleet, and hail.

Evapotranspiration

The process by which liquid water changes into water vapor is called *evaporation*. Each year, about 500,000 km^3 of water evaporates into the atmosphere. About 86% of this water evaporates from the ocean. The remaining water evaporates from lakes, streams, and the soil. Water vapor also enters the air by *transpiration*, the process by which plants release water vapor into the atmosphere. The total loss of water from an area, which equals the sum of the water lost by evaporation from the soil and other surfaces and the water lost by transpiration from organisms, is called **evapotranspiration.** Evapotranspiration is one part of the water cycle, which is shown in **Figure 2.**

evapotranspiration the total loss of water from an area, which equals the sum of the water lost by evaporation from the soil and other surfaces and the water lost by transpiration from organisms

condensation the change of state from a gas to a liquid

precipitation any form of water that falls to Earth's surface from the clouds; includes rain, snow, sleet, and hail

Condensation

Another process of the water cycle is condensation. **Condensation** is the change of state from a gas to a liquid. When water vapor rises in the atmosphere, it expands and cools. As the vapor becomes cooler, some of it condenses, or changes into tiny liquid water droplets, and forms clouds.

Precipitation

The third major process of the water cycle is *precipitation*, the process by which water falls from the clouds. **Precipitation** is any form of water that falls to Earth's surface from the clouds and includes rain, snow, sleet, and hail. More than 75% of all precipitation falls on Earth's oceans. The rest falls on land and becomes runoff or groundwater. Eventually, all of this water returns to the atmosphere by evapotranspiration, condenses, and falls back to Earth's surface to begin the cycle again.

Reading Check List the forms of precipitation. (See the Appendix for answers to Reading Checks.)

Figure 2 ▶ Evapotranspiration, condensation, and precipitation make up the continuous process called the *water cycle.*

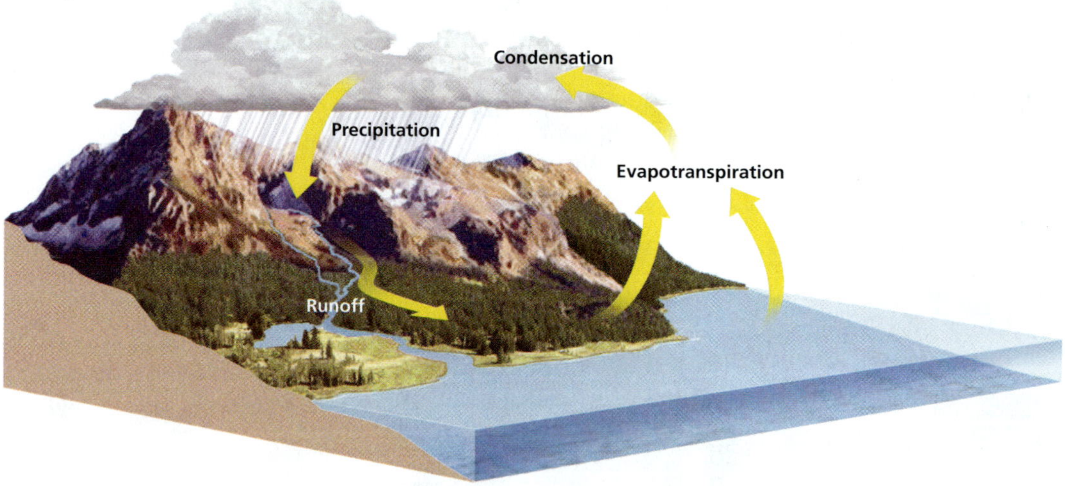

MISCONCEPTION ALERT

Water, Water Anywhere? Students may not realize how little water is available as freshwater for human consumption. Ninety-seven percent of Earth's water exists as undrinkable saltwater in the oceans. The remaining 3% is freshwater. A little more than 2% is in glaciers, and thus is fairly inaccessible. Less than one half of 1% of all the water on Earth is readily available for human use as surface water in streams, rivers, and lakes.

CHAPTER RESOURCES

Technology

Transparencies
- 76 The Water Cycle (with worksheet)

Water Budget

The continuous cycle of evapotranspiration, condensation, and precipitation establishes Earth's *water budget.* A financial budget is a statement of expected income—money coming in—and expenses—money going out. In Earth's water budget, precipitation is the income. Evapotranspiration and runoff are the expenses. The water budget of Earth as a whole is balanced because the amount of precipitation is equal to the amount of evapotranspiration and runoff. However, the water budget of a particular area, called the *local water budget,* usually is not balanced.

Factors That Affect the Water Budget

Factors that affect the local water budget include temperature, vegetation, wind, and the amount and duration of rainfall. When precipitation exceeds evapotranspiration and runoff in an area, the result is moist soil and possible flooding. When evapotranspiration exceeds precipitation, the soil becomes dry and irrigation may be necessary. Vegetation reduces runoff in an area but increases evapotranspiration. Wind increases the rate of evapotranspiration.

The factors that affect the local water budget vary geographically. For example, the Mojave Desert in California receives much less precipitation than do the tropical rain forests of Queensland, Australia, as **Figure 3** shows.

The local water budget also changes with the seasons in most areas of Earth. In general, cooler temperatures slow the rate of evapotranspiration. During the warmer months, evapotranspiration increases. As a result, streams generally transport more water in cooler months than they do in warmer months.

Figure 3 ▶ Tropical rain forests, such as the one in Queensland, Australia (top photo), require large amounts of rainfall annually. Deserts, such as the Mojave Desert in California (bottom photo), receive small amounts of rainfall each year.

ENVIRONMENTAL CONNECTION

Water Conservation The World Health Organization of the United Nations states that severe water shortages affecting 400 million people today will affect 2.7 billion people by 2025 and 4 billion by 2050 if the world continues to consume water at the current rate. States with little precipitation, such as Arizona in the southwestern U.S., will face severe freshwater shortages by 2025.

QuickLAB 35 min

Modeling the Water Cycle

Procedure

1. Place a **short glass** inside a **large plastic mixing bowl.** Add cold water to the mixing bowl until about three-fourths of the glass is covered with water. Make sure to keep the inside of the glass dry.
2. Add drops of **food coloring** (red, blue, or green) to the water in the bowl until the water has a strong color.
3. Now, add about 1 cup of dry **dirt** to the water, and stir gently until the water is muddy as well as colored.
4. Cover the bowl tightly with a **piece of plastic wrap** secured to the bowl with a **rubber band,** and place a **coin or stone** in the middle of the plastic wrap above the glass.
5. Set the bowl in the sun or under a **heat lamp** for 30 minutes to several hours. Then, observe the water that has collected in the glass.

Analysis

1. What are the processes that have taken place to allow water to collect in the glass?
2. Why is the water in the glass not muddy?
3. Is the water in the glass colored? What does this say about pollutants in water systems and the water cycle?

Close

Reteaching — Basic

Water Budget Diagram Provide students with photos or figures of a tropical rainforest, arctic tundra, deserts and/or other areas. Have students draw arrows on the photos that show each of the four processes in the water cycle. The size of the arrows in each figure should represent the relative contribution of each process to the region's overall water budget. Have students explain whether each regional water budget is balanced or not. LS **Visual/Logical**

Quiz — General

1. Why is desalination currently not a widely used method of obtaining freshwater? (At this time, it is too expensive on a large scale.)
2. In Earth's water budget, what processes are losses, or expenses? (evapotranspiration and runoff)

QuickLAB

Skills Acquired

- Observing
- Analyzing

Teacher's Notes Tell students that this apparatus is a solar still and can be used to obtain clean water from muddy or salty water in an emergency.

Answers

1. evaporation and condensation
2. As the sun heats the water, the water turns to vapor, and the mud is left behind. Clean water vapor then condenses on the plastic and drips into the glass.
3. Answers may vary depending on the type of food coloring used. If the water in the glass is not colored, the dye molecules do not evaporate with the water. Evaporation can purify water of some dissolved pollutants. However, some pollutants are transferred through evaporation.

CHAPTER RESOURCES

Chapter Resource File

- **Datasheet for Quick Lab** GENERAL
- **Inquiry Lab** Eutrophication GENERAL
- **Skills Practice Lab** Stream Quality Monitoring GENERAL

Close, continued

Alternative Assessment — GENERAL

Skit Have students work in small groups to act out key processes in the water cycle. Students should play the role of water in its various states and depict evapotranspiration, condensation, and precipitation. Different groups can model how water budgets differ in a tropical region versus a desert. LS Kinesthetic

Answers to Section Review

1. Water reaches the ocean from rivers and from precipitation.
2. Evapotranspiration is loss of water due to the combined effects of evaporation from surfaces and transpiration by plants and animals. Condensation occurs when water changes state from gaseous water vapor to liquid water. Precipitation occurs when water falls to Earth's surface.
3. Condensation occurs when water vapor changes to liquid water droplets as it cools. These droplets can form clouds and the water may fall back to Earth as precipitation.
4. Most local water budgets are not balanced because precipitation and evapotranspiration vary seasonally and geographically. Sometimes, precipitation will exceed evapotranspiration, leading to excess water and possible flooding. At other times and places, the opposite may occur.
5. Vegetation reduces runoff but may increase evapotranspiration. Precipitation increases available water and leads to excess moisture and to possible flooding, especially if vegetation is lacking.
6. Two ways to ensure continued water are to conserve water and to find alternative sources of freshwater.
7. Answers may vary but may include taking shorter showers, recycling water, turning off water when brushing teeth, using low water-use landscaping and low flow toilets and showers, and using front-loading washing machines.
8. because oceans cover more than 70% of Earth's surface and oceans have a large water surface that is exposed to the atmosphere
9. The local *water budget* is controlled by the *water cycle,* which includes *condensation, precipitation,* and *evapotranspiration,* which is the sum of *evaporation* and *transpiration.*

Figure 4 ▶ Waste from this paper mill has polluted the Qingai River in China.

Water Use

On average, each person in the United States uses about 95,000 L (20,890.5 gal) of water each year. Water is used for bathing, washing clothes and dishes, watering lawns, carrying away wastes, and drinking. Agriculture and industry also use large amounts of water. As the population of the United States increases, so does the demand for water.

About 90% of the water used by cities and industry is returned to rivers or to the oceans as wastewater. Some of this wastewater contains harmful materials, such as toxic chemicals and metals, as shown in **Figure 4.** These toxic materials can pollute rivers and can harm plants and animals in the water.

Conservation of Water

While Earth holds a lot of water, only a small percentage of that water is fresh water that can be used by humans. Scientists have identified two ways to ensure that enough fresh water is available today and in the future. One way is through conservation, or the wise use of water resources. Individuals can conserve water by limiting their water use as much as possible. Governments can help conserve water by enforcing conservation laws and antipollution laws that prohibit the dumping of waste into bodies of water.

A second way to protect the water supply is to find alternative methods of obtaining fresh water. One such method is called **desalination,** which is the process of removing salt from ocean water. However, this method is expensive and is impractical for supplying water to large populations. Currently, the best way of maintaining an adequate supply of fresh water is the wise use and conservation of the fresh water that is now available.

desalination a process of removing salt from ocean water

Section 1 Review

1. **List** two ways in which water reaches the oceans.
2. **Outline** the major stages of the water cycle.
3. **Explain** the difference between condensation and precipitation.
4. **Explain** why most local water budgets are not balanced.
5. **Describe** how vegetation and rainfall affect the local water budget.
6. **List** two ways to ensure the continued supply of fresh water.

CRITICAL THINKING

7. **Applying Concepts** Describe five ways that you can conserve water at home.
8. **Analyzing Processes** Why are the oceans the location of most evaporation and precipitation?

CONCEPT MAPPING

9. Use the following terms to create a concept map: *water cycle, evaporation, transpiration, evapotranspiration, condensation, precipitation,* and *water budget.*

CHAPTER RESOURCES

Chapter Resource File

- Section Quiz GENERAL

Workbooks

- Study Guide (also in Spanish)

Section 2 Stream Erosion

A river system begins to form when precipitation exceeds evapotranspiration in a given area. After the soil in the area soaks up as much water as the soil can hold, the excess water moves downslope as runoff. As runoff moves across the land surface, it erodes rock and soil and eventually may form a narrow ditch, called a *gully*. Eventually, the processes of precipitation and erosion form a fully developed valley with a permanent stream.

OBJECTIVES

- **Summarize** how a river develops.
- **Describe** the parts of a river system.
- **Explain** factors that affect the erosive ability of a river.
- **Describe** how erosive factors affect the evolution of a river channel.

KEY TERMS

tributary
watershed
stream load
discharge
gradient
meander
braided stream

Parts of a River System

A river system is made up of a main stream and **tributaries,** which are all of the feeder streams that flow into the main stream. The land from which water runs off into these streams is called a **watershed.** The ridges or elevated regions that separate watersheds are called *divides*. A river system is shown in **Figure 1.**

The relatively narrow depression that a stream follows as it flows downhill is called its *channel*. The edges of a stream channel that are above water level are called the stream's *banks*. The part of the stream channel that is below the water level is called the stream's *bed*. A stream channel gradually becomes wider and deeper as it erodes its banks and bed.

tributary a stream that flows into a lake or into a larger stream

watershed the area of land that is drained by a river system

Channel Erosion

River systems change continuously because of erosion. In the process of *headward erosion*, channels lengthen and branch out at their upper ends, where runoff enters the streams. Erosion of the slopes in a watershed can also extend a river system and can add to the area of the watershed. In the process known as *stream piracy*, a stream from one watershed is "captured" by a stream from another watershed that has a higher rate of erosion. The captured stream then drains into the river system that has done the capturing.

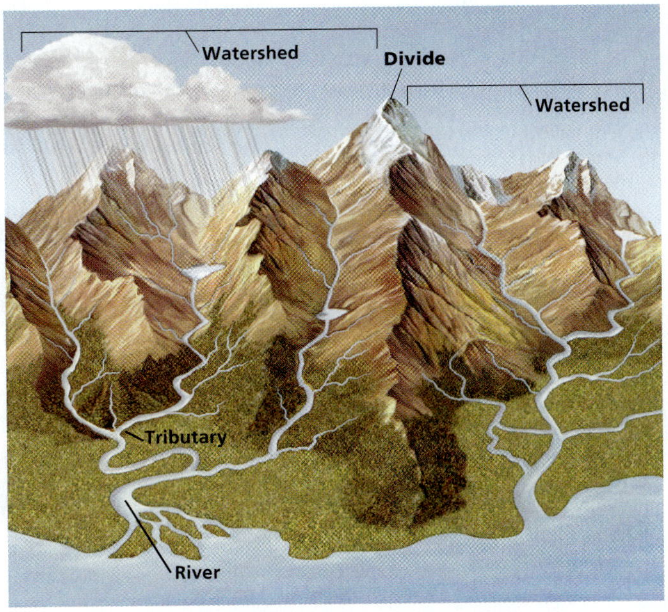

Figure 1 ▶ The tributaries that run into this river are fed by runoff from surrounding land. All of the land that drains into a single river makes up the watershed of the river.

Section 2

Focus

Overview

This section explains how a river develops and describes the parts of a river system. It explains the factors that affect the erosive capability of a river and how these factors influence the development of a river.

Bellringer

Ask students to make a list of words that describe rivers. The words can describe the shape of the river, how it affects the surrounding landscape, or the water in the river. (Answers may vary. Accept all reasonable answers.) LS Verbal

Motivate

Internet Activity — GENERAL

Watersheds Have students identify the watershed in which they live by using the Environmental Protection Agency's "Surf Your Watershed" Web site. Ask them to investigate ways to preserve and protect the watershed. A worksheet designed to direct student research on this topic can be found in the **Chapter Resource File** booklet or by visiting **go.hrw.com** and entering the keyword **HQ6RVSX.** LS Verbal

CHAPTER RESOURCES

Chapter Resource File
- Directed Reading BASIC
- Internet Activity
  - Watersheds GENERAL

Technology
- Transparencies
  - Bellringer
- Student Edition on CD-ROM
- One-Stop Planner CD-ROM
  - Lesson Plan

Teach

Group Activity — GENERAL

Field Trip Visit a local stream or river in your watershed. Have students walk along the channel and note stream gradient and flow, water quality, composition of stream bed, presence of meandering, bank erosion, sediment bars, flooding, evidence of human activities, human channel modifications, and any other features of interest. Have students make sketches, write down observations, and/or take photographs.

To measure stream velocity, have students mark off a section of known distance, drop a leaf or other floating object close to the center of the stream, and measure the time the object takes to travel the distance. For example, if the object travels 60 m in 20 s, the stream's surface velocity is 60 m/20 s, or 3 m/s. To calculate an average for the stream as a whole, multiply by 0.8, because water flows faster near the surface of the stream and slower near the bottom. If possible, visit the site again after a heavy rainfall, and have students note changes in the stream. Have small groups of students design posters that illustrate the key characteristics of that section of the watershed. **LS Visual**

MATHPRACTICE

Answer

Discharge = velocity × area
Discharge = 1.5 m/s × 520 m^2
= 780 m^3/s

CHAPTER RESOURCES

Technology

Transparencies
- 77 Stream Gradient and Channel Erosion (with worksheet)

stream load the materials other than the water that are carried by a stream

discharge the volume of water that flows within a given time

gradient the change in elevation over a given distance

MATHPRACTICE

Water Discharge of a River River channels can carry an enormous volume of water. The water that rivers discharge can be calculated by using the following equation:

discharge = *velocity of the water* × *cross-sectional area of the river channel*

In cubic meters per second (m^3/s), what is the discharge of water carried by a river that moves 1.5 m/s through a cross-sectional area of 520 m^2?

Stream Load

A stream transports soil, loose rock fragments, and dissolved minerals as it flows downhill. The materials carried by a stream are called the **stream load.** Stream load takes three forms: suspended load, bed load, and dissolved load. The *suspended load* consists of particles of fine sand and silt. The velocity, or rate of downstream travel, of the water keeps these particles suspended, so they do not sink to the stream bed. The *bed load* is made up of larger, coarser materials, such as coarse sand, gravel, and pebbles. This material moves by sliding and jumping along the bed. The *dissolved load* is mineral matter transported in liquid solution.

Stream Discharge

The volume of water moved by a stream in a given time period is the stream's **discharge.** The faster a stream flows, the higher its discharge and the greater the load that the stream can carry. Thus, a swift stream carries more sediment and larger particles than a slow stream does. A stream's velocity also affects how the stream cuts down and widens its channel. Swift streams erode their channels more quickly than slow-moving streams do.

Stream Gradient

The velocity of a stream depends mainly on gradient. **Gradient** is the change in elevation of a stream over a given horizontal distance. In other words, gradient is the steepness of the stream's slope. Near the *headwaters,* or the beginning of a stream, the gradient generally is steep. This area of the stream has a high velocity, which causes rapid channel erosion. As the stream nears its *mouth,* where the stream enters a larger body of water, its gradient often becomes flatter. As a result, the river's velocity and erosive power decrease. The stream channel eventually is eroded to a nearly flat gradient by the time the stream channel reaches the sea. Streams with different gradients are shown in **Figure 2.**

Figure 2 ▶ Streams that have steep gradients, such as the stream on the left, have a higher velocity than streams that have low gradients, such as the stream on the right, do.

ART CONNECTION — ADVANCED

The Big Muddy The Mississippi River has had a profound effect on American culture, inspiring classic works of literature and art. The celebrated American writer Mark Twain chronicled river life and culture in *The Adventures of Huckleberry Finn, The Adventures of Tom Sawyer,* and *Life on the Mississippi.* The Mississippi River delta is also the birthplace of a uniquely American form of music—the blues. The main action of *Showboat,* America's first important musical drama, which is an original American art form, took place along the Mississippi. Have students read passages of Mark Twain's works and listen to blues music and songs from *Showboat,* including "Ol' Man River." The PBS Web site "River of Song" is a good source for information about blues music. Discuss with students why the Mississippi River plays such an important role in American life and culture. **LS Auditory/Verbal**

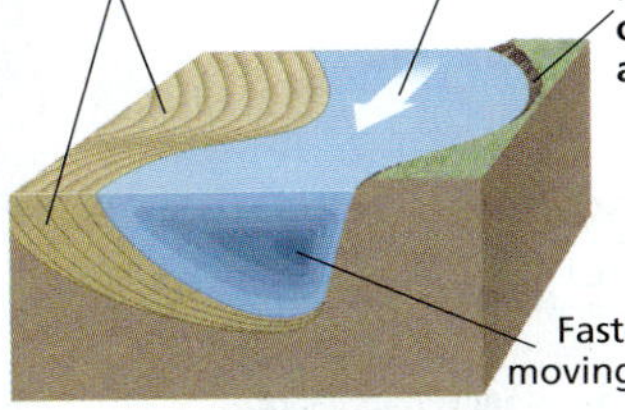

Figure 3 ▶ Decreased velocity on the inside of a river's curve leads to the deposition of sediment, as this photo of a river in the Banff National Park in Alberta, Canada, shows.

Evolution of River Channels

As the stream's load, discharge, and gradient decrease, the erosive power of the stream decreases, which influences the evolution of the stream's channel. Over time, as the channel erodes, it becomes wider and deeper. When the stream becomes longer and wider, it is called a *river*.

Meandering Channels

As a river evolves, it may develop curves and bends. A river that has a low gradient tends to have more bends than a river that has a steep gradient does. A winding pattern of wide curves, called **meanders,** develops because as the gradient decreases, the velocity of the water decreases. When the velocity of the water decreases, the river is less able to erode down into its bed. As the water flows through the channel, more energy is directed against the banks, which causes erosion of the banks.

When a river rounds a bend, the velocity of the water on the outside of the curve increases. The fast-moving water on the outside of a river bend erodes the outer bank of that bend. However, on the inside of the curve, the velocity of the water decreases. This decrease in velocity leads to the formation of a *bar* of deposited sediment, such as sand or gravel, as shown in **Figure 3.**

As this process continues, the curve enlarges while further sediment deposition takes place on the opposite bank, where the water is moving more slowly. Meanders can become so curved that they almost form a loop, separated by only a narrow neck of land. When the river cuts across this neck, the meander can become isolated from the river, and an *oxbow lake* forms.

Reading Check **How would you describe the gradient of a river that has meanders? (See the Appendix for answers to Reading Checks.)**

meander one of the bends, twists, or curves in a low-gradient stream or river

SCILINKS. NSTA Developed and maintained by the National Science Teachers Association

For a variety of links related to this subject, go to www.scilinks.org

Topic: River Systems
SciLinks code: HQ61314

Answer to Reading Check
A river that has meanders probably has a low gradient.

Close

Reteaching — BASIC

Flowchart Have students construct a flowchart of the stages in the evolution of a stream. Have them identify factors that play a role in the features of the stream. **LS Visual**

Quiz — GENERAL

1. What separates one watershed from another? (ridges or elevated regions called *divides*)
2. What is a bar? (an area in a river channel where sediments, such as sand and gravel, have been deposited)
3. What is a watershed? (the area of land that is drained by a water system)

Alternative Assessment — GENERAL

Build a Watershed Have students work in small groups to build a model of a watershed out of sand and clay in a plastic container. Have students use clay to create the underlying structure, then layer sand over the clay. The model should show the slope of the watershed, the main channel, and several tributaries. Have students demonstrate the flow of runoff in their watershed by simulating rain with a spray bottle or by using a watering can. **LS Kinesthetic**

SKILL BUILDER — BASIC

Vocabulary The word *tributary* comes from the Latin *tribuere,* which means "to pay or contribute." Feeder streams, or tributaries, contribute their flow to the main river. The word *meander* comes from the Latin *maeander,* which means "circuitous winding." The Maeander River in Phyrgia, an ancient country in Asia Minor, was famous for its winding course. Have students think of other uses for the base root meanings of tributary and meander. (Answers may vary. Volunteer contributions help pay for charitable projects. People meander aimlessly in the park.) **LS Verbal** English Language Learners

Teaching Tip — BASIC

Make Concepts Relevant Students may understand the erosive power of river systems if you show them examples of landforms carved out of Earth by rivers. Show students a picture of the Grand Canyon with the Colorado River flowing through it. Ask them to think about how the action of river water could form something as vast and deep as the Grand Canyon. (Over millions of years, the relentless action of the river eroded the land, gradually cutting deeper into the rock to create steep canyons.) **LS Visual**

Close, *continued*

Graphic Organizer GENERAL

You may want to use this Graphic Organizer to assess students' understanding of the topic. Collect the Graphic Organizers from the students and correct any mistakes before returning the Graphic Organizers.

Answers to Section Review

1. When precipitation exceeds evapotranspiration, water that does not soak into the soil runs off, eroding rock and soil and forming a narrow gully. Gradually, the gully enlarges to form a channel that has a permanent stream.
2. A river system consists of the main stream, all of its tributaries, and their watersheds.
3. Headward erosion occurs when a stream lengthens and branches out at its upper end. If headward erosion cuts through a divide between watersheds, piracy, in which a stream captures water from another watershed, can occur.
4. suspended load, bed load, and dissolved load
5. A stream with a high discharge can carry a higher load and thus erodes its channel more quickly than slow-moving streams do. The steeper the gradient is, the faster a stream flows and the more it erodes its channel.
6. Braided rivers occur when streams have large sediment loads, especially coarse sand and gravel. The river cannot move all of the load and deposits it as bars along the bottom, dividing the river into multiple channels. A low gradient tends to cause meanders because, as velocity slows, water is less able to erode the riverbed but can still erode the banks.
7. Meanders can become so curved that the banks of the river are separated only by a narrow neck of land. If the river cuts across the neck, the loop may become isolated from the main channel, forming an oxbow lake.
8. If uplift occurred, the gradient of the stream would increase, causing an increase in stream velocity and channel erosion.
9. The higher the velocity is, the more sediment and larger particles a stream can carry.
10. At the *headwaters*, the *stream gradient* is steep, so *stream load*, composed of *bed load*, *suspended load*, and *dissolved load*, is high, but *meanders* and *braided channels* develop as gradient and load change.

Figure 4 ▶ Braided streams, such as the Chisana River in Alaska, divide into multiple channels.

Graphic Organizer — Cause-and-Effect Map

Create the Graphic Organizer entitled "Cause-and-Effect Map" described in the Skills Handbook section of the Appendix. Label the effects with "Meandering streams" and "Braided streams." Then, fill in the map with causes of meandering streams and braided streams and details about the causes and effects.

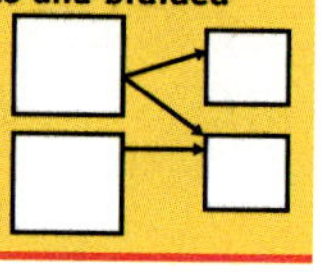

braided stream a stream or river that is composed of multiple channels that divide and rejoin around sediment bars

Braided Streams

Most rivers are single channels. However, under certain conditions, the presence of sediment bars between a river's banks can divide the flow of the river into multiple channels. A stream or river that is composed of multiple channels that divide and rejoin around sediment bars is called a **braided stream.** Braided streams are a direct result of a large sediment load, particularly when a high percentage of the load is composed of coarse sand and gravel. The bars form on the channel floor when the river is unable to move all of the available load.

Although braided streams, such as those in **Figure 4,** look very different from meandering channels, they can cause just as much erosion. The channel location shifts constantly such that bars between channels erode and new bars form. Sometimes, a single river can change from a braided stream to a meandering stream as the gradient and discharge change.

Section 2 Review

1. **Summarize** how a river develops.
2. **Describe** the parts of a river system.
3. **Explain** the processes of headward erosion and stream piracy.
4. **List** the three types of stream load.
5. **Explain** how stream discharge and gradient affect the erosive ability of a river.
6. **Describe** the factors that control whether a river is braided or meandering.
7. **Summarize** the process that forms an oxbow lake.

CRITICAL THINKING

8. **Predicting Consequences** If geologic forces were to cause an uplift of the land surface, what would the effect on stream channel erosion be?
9. **Analyzing Processes** Explain how the velocity of a stream affects the suspended load.

CONCEPT MAPPING

10. Use the following terms to create a concept map: *braided channels, stream load, suspended load, dissolved load, bed load, meanders, stream gradient,* and *headwaters.*

CHAPTER RESOURCES

Chapter Resource File

- **Section Quiz** GENERAL

Workbooks

- **Study Guide** (also in Spanish)

Section 3 Stream Deposition

The total load that a stream can carry is greatest when a large volume of water is flowing swiftly. When the velocity of the water decreases, the ability of the stream to carry its load decreases. As a result, part of the stream load is deposited as sediment.

Deltas and Alluvial Fans

A stream may deposit sediment on land or in water. For example, the load carried by a stream can be deposited when the stream reaches an ocean or a lake. As a stream empties into a large body of water, the velocity of the stream decreases sharply. The load is usually deposited at the mouth of the stream in a triangular shape. A triangular-shaped deposit that forms where the mouth of a stream enters a larger body of water is called a **delta.** The exact shape and size of a delta are determined by waves, tides, offshore depths, and the sediment load of the stream.

When a stream descends a steep slope and reaches a flat plain, the speed of the stream suddenly decreases. As a result, the stream deposits some of its load on the level plain at the base of the slope. A fan-shaped deposit called an **alluvial fan** forms on land, and its tip points upstream. In arid and semi-arid regions, temporary streams commonly form alluvial fans. Alluvial fans differ from deltas in that alluvial fans form on land instead of being deposited in water. This difference is shown in **Figure 1.**

OBJECTIVES

- **Explain** the two types of stream deposition.
- **Describe** one advantage and one disadvantage of living in a floodplain.
- **Identify** three methods of flood control.
- **Describe** the life cycle of a lake.

KEY TERMS

delta
alluvial fan
floodplain

delta a fan-shaped mass of rock material deposited at the mouth of a stream; for example, deltas form where streams flow into the ocean at the edge of a continent

alluvial fan a fan-shaped mass of rock material deposited by a stream when the slope of the land decreases sharply; for example, alluvial fans form when streams flow from mountains to flat land

Figure 1 ▶ A delta, such as this one in Alaska's Prince William Sound (above), forms when a stream deposits sediment into another body of water. An alluvial fan, such as this one in California's Death Valley (right), forms when a stream deposits sediment on land.

Section 3

Focus

Overview

This section explains sediment deposition in deltas and alluvial fans. The section also describes floodplains and flood control and the life cycle of lakes.

Bellringer

Ask students to write a short paragraph that describes what they think happens to all of the sediment eroded and carried by a river. (Sediments are deposited when the water slows down or when the river reaches the ocean.) LS **Verbal**

Motivate

Demonstration — GENERAL

Changing Land Use Have students use a stream table or model watershed to investigate how land development may affect runoff. First, have students cover the banks of their stream with leaves to simulate vegetation. Then, have students make rain by using a watering can. Next, have students remove the vegetation and repeat the rain. Finally, have students cover the banks with plastic wrap or aluminum foil to simulate paving and repeat the rain. Discuss why flooding may become more frequent as a watershed is developed. LS **Kinesthetic**

CHAPTER RESOURCES

Chapter Resource File
- Directed Reading BASIC

Technology

Transparencies
- Bellringer

Student Edition on CD-ROM

One-Stop Planner CD-ROM
- Lesson Plan

Teach

Debate — Advanced

To Stay or Not to Stay Have students debate this issue: Should communities be allowed to rebuild in an area prone to repeated and severe flooding? Have students investigate the pros and cons of living in a floodplain. Have students consider the costs of protecting, insuring, and rebuilding homes and businesses versus the costs of relocating the community. Students may wish to address who should pay for those costs. Have students present their views to the class, allowing time for questions, answers, and rebuttals. **LS Interpersonal/Logical**

SKILL BUILDER — Advanced

Writing Have interested students research an historic U.S. flood, such as the 1889 Johnstown flood, the 1927 or the 1993 Mississippi River flood, or the 1976 Big Thompson Canyon flood. Have students write a newspaper style article that reports on the flood, its causes and consequences, and its aftermath. **LS Verbal**

READING SKILL BUILDER — Basic

Reading Organizer As students read this section, encourage them to take Power Notes, KWL notes, or Two-Column Notes as described in the Skills Handbook section of the Appendix. Later, students can use these notes as a study guide for assessments. **LS Visual** English Language Learners

Figure 2 ▶ People who live in the Tonle Sap Floodplain in Cambodia have adapted to frequent flooding by building houses that are raised above the water level on stilts.

floodplain an area along a river that forms from sediments deposited when the river overflows its banks

Floodplains

The volume of water in nearly all streams varies depending on the amount of rainfall and snowmelt in the watershed. A dramatic increase in volume can cause a stream to overflow its banks and to wash over the valley floor. The part of the valley floor that may be covered with water during a flood is called a **floodplain.**

Natural Levees

When a stream overflows its banks and spreads out over the floodplain, the stream loses velocity and deposits its coarser sediment load along the banks of the channel. The accumulation of these deposits along the banks eventually produces raised banks, called *natural levees*.

Finer Flood Sediments

Not all of the load deposited by a stream in a flood will form levees. Finer sediments are carried farther out into the floodplain by the flood waters and are deposited there. A series of floods produces a thick layer of fine sediment, which becomes a source of rich floodplain soils. Swampy areas are common on floodplains because drainage is usually poor in the area between the levees and the outer walls of the valley. Despite the hazards of periodic flooding, people choose to live on floodplains, as shown in **Figure 2.** Floodplains provide convenient access to the river for shipping, fishing, and transportation. The rich soils, which are good for farming, also draw people to live on floodplains.

Connection to ENVIRONMENTAL SCIENCE

The Dead Zone

Oceanographers have discovered that water in the Gulf of Mexico west of the mouth of the Mississippi River delta has a low level of oxygen during the summer. This condition, known as *hypoxia,* can suffocate crabs, shrimp, and other fish that live on the sea floor. For this reason, the area has been dubbed the Dead Zone. Hypoxia has severely affected the marine food chain in this area as well as the fishing industry.

What is the origin of the Dead Zone? Studies indicate that water flowing from the Mississippi River to the Gulf of Mexico has a high level of dissolved nitrogen and phosphorus from fertilizers that are used to grow crops in the watershed. These substances increase the growth of phytoplankton—small floating marine plants—in shallow gulf waters where the river water ends up. When these plants die and sink to the sea floor, they are broken down by bacteria that use oxygen in the process. As a result, the ocean water becomes depleted of oxygen. This process usually

Mississippi River watershed
Area of hypoxia in Gulf of Mexico

occurs in the summer when river flow is high, more sunlight is available for plant growth, and shallow gulf waters are poorly mixed with offshore water by winds.

To reduce the effects of hypoxia in the Gulf of Mexico, government officials are considering regulating the amount and type of fertilizers that can be used in the Mississippi watershed.

INCLUSION Strategies

- *Behavior Control Issues*
- *Developmentally Delayed*
- *Attention Deficit Disorder*

Have students join with a partner. Ask the partners to take turns reading alternate paragraphs quietly aloud. Then, have the students discuss what the paragraph means and draw a simple sketch that goes with the paragraph. When pairs are finished, ask partners to share their favorite sketches with the class. **LS Auditory/Verbal**

QuickLAB 40 min

Soil Erosion

Procedure

1. Fill a **23 cm × 33 cm pan** about half full with **moist, fine sand.**
2. Place the pan in a **sink** so that one end of the pan is resting on a **brick** and is under the water faucet.
3. Position an **additional pan or container** such that it catches any sand and water that flow out of the first pan.
4. Slowly open the faucet until a gentle trickle of water falls onto the sand in the raised end of the pan. Let the water run for 15 to 20 s.
5. Turn off the water, and draw the pattern of water flow over the sand.
6. Press the sand back into place, and carefully smooth the surface by using a **ruler.** Repeat steps 4 and 5 three more times. Each time, increase the rate of water flow slightly without splashing the sand.

Analysis

1. Describe how the rate of water flow affects erosion.
2. How does the rate of water flow affect gullies?
3. How could erosion on a real hillside be reduced without changing the rate of water flow?
4. How does the shape of a river bend change as water flows?

Human Impacts on Flooding

Human activity can contribute to the size and number of floods in many areas. Vegetation, such as trees and grass, protects the ground surface from erosion by taking in much of the water that would otherwise run off. Where this natural ground cover is removed, water can flow more freely across the surface. As a result, the likelihood of flooding increases. Logging and the clearing of land for agriculture or housing development can increase the volume and speed of runoff, which leads to more frequent flooding. Natural events, such as forest fires, can also increase the likelihood of flooding.

Flood Control

Indirect methods of flood control include forest and soil conservation measures that prevent excess runoff during periods of heavy rainfall. More-direct methods include the building of artificial structures that redirect the flow of water.

The most common method of direct flood control is the building of *dams*. The artificial lakes that form behind dams act as reservoirs for excess runoff. The stored water can be used to generate electricity, supply fresh water, and irrigate farmland. Another direct method of flood control is the building of *artificial levees*. However, artificial levees must be protected against erosion by the river. As **Figure 3** shows, when artificial levees break, flooding and property damage can result. Permanent overflow channels, or floodways, can also help prevent flooding. When the volume of water in a river increases, floodways carry away excess water and keep the river from overflowing.

Reading Check Describe two ways that floods can be controlled. (See the Appendix for answers to Reading Checks.)

Figure 3 ▶ Week-long storms in Modesto, California, in 1997 broke this levee and caused flooding.

ENVIRONMENTAL CONNECTION

Choking Lake Okeechobee Have interested students report on how building dikes and straightening the 103-mile long Kissimmee River into a canal to control flooding impacts the ecology of Lake Okeechobee and the Everglades in Florida.

Answer to Reading Check

Floods can be controlled indirectly through forest and soil conservation measures that reduce or prevent runoff, or directly by building artificial structures, such as dams, levees, and floodways, to redirect water flow.

Close

Reteaching — Basic

Peer Reviews Have students work in pairs to write three to five questions about the material in this section. Then, have students use their questions to quiz each other. LS Interpersonal Co-op Learning

Quiz — General

1. What is a floodway? (a permanent overflow channel that diverts excess water at times of high river volume)
2. Define *natural levee.* (a raised bank of coarse sediments that are deposited when water velocity decreases as a stream overflows its banks)

CHAPTER RESOURCES

Chapter Resource File

- Datasheet for Quick Lab GENERAL

QuickLAB

Skills Acquired

- Observing
- Identifying/Recognizing Patterns

Teacher's Notes Stream tables can be used to investigate many variables that affect stream-channel evolution, erosion, deposition, and flooding.

Answers

1. As the rate of flow increases, erosion should increase.
2. The faster water flow increases the size of gullies and the rate of gully formation.
3. Erosion on real hillsides could be reduced by planting natural ground cover, such as trees and grass, or by constructing barriers to water flow.
4. Answers may vary. As the channel gets deeper, it may also get wider or begin to meander a little.

Close, *continued*

Alternative Assessment — BASIC

Sediment Stories Have students create a collage of photos of river deposits, floods, floodplains and lakes to illustrate the places where and ways in which rivers deposit their sediment loads. LS Visual

Answers to Section Review

1. A delta forms where a stream or river enters a larger body of water, such as a lake or ocean. An alluvial fan forms on land where a steep stream reaches a flat plain.
2. Deltas and alluvial fans are triangle shaped because the flow is no longer confined by a channel. When rivers overflow their banks, coarse sediments are deposited along the banks as natural levees, but finer sediments are carried farther out onto the floodplain and form rich, thick soils.
3. The disadvantage of living on a floodplain is that frequent flooding damages homes and businesses. But living on a floodplain provides convenient access to rivers for shipping, fishing, and transportation. The fertile soils of most floodplains are also very good for farming.
4. When humans remove natural ground cover for agricultural or urban development or pave areas, runoff increases, which can lead to more frequent and intense flooding. Humans can restore ground cover to reduce runoff and the potential for flooding.
5. Methods of flood control include forest and soil conservation and building dams, artificial levees, and floodways.
6. Lakes are usually short lived because water drains away or evaporates faster than it is replenished. They also fill in with sediments carried by streams, rivers, and overland runoff.
7. Answers may vary. When snow melts in the spring, a large amount of water flows through the area. The soil cannot soak up the water, so it runs off into stream channels that cannot hold all of the excess water.
8. Answers may vary. Artificial levees should be constructed of materials that are hard to erode, otherwise they will quickly be breached by floodwaters.
9. *Stream deposition* can fill in a *lake,* or may form *deltas, alluvial fans,* or *natural levees* in a *floodplain,* which can be protected from floods by *artificial levees* and *dams.*

Precipitation collects in a depression and forms a lake.

A lake loses its water as the water drains away or evaporates.

As water is lost, the lake basin may eventually become dry land.

Figure 4 ▶ Compared to rivers, lakes are short lived, and some lakes may eventually dry up.

SCILINKS. NSTA Developed and maintained by the National Science Teachers Association

For a variety of links related to this subject, go to www.scilinks.org

Topic: Flooding and Society
SciLinks code: HQ60585
Topic: Stream Deposition
SciLinks code: HQ61457

The Life Cycle of Lakes

Not all streams flow from the land to the ocean. Sometimes, water from streams collects in a depression in the land and forms a lake. Most lakes are located at high latitudes and in mountainous areas. Most of the water in lakes comes from precipitation and the melting of ice and snow. Springs, rivers, and runoff coming directly from the land are also sources of lake water.

Most lakes are relatively short lived in geologic terms. Many lakes eventually disappear because too much of their water drains away or evaporates, as shown in **Figure 4.** A common cause of excess drainage is an outflowing stream that erodes its bed below the level of a lake basin. Lakes may also lose water if the climate becomes drier and evaporation exceeds precipitation.

Lake basins may also disappear if they fill with sediments. Streams that feed a lake deposit sediments in the lake. Sediments also are carried into the lake by water that runs off the land but does not enter a stream. Most of these sediments are deposited near the shore. These sediments build up over time, which creates new shorelines and gradually fills in the lake. Organic deposits from vegetation also may accumulate in the bottom of a shallow lake. As these deposits grow denser, a bog or swamp may form. The lake basin may eventually become dry land.

Section 3 Review

1. **Identify** the differences between a delta and an alluvial fan.
2. **Explain** the differences between the deposition of sediment in deltas and alluvial fans with the deposition of sediment on a floodplain.
3. **Describe** the advantages and disadvantages of living in a floodplain.
4. **Summarize** how human activities can affect the size and number of floods.
5. **Identify** three methods of flood control.
6. **Explain** why lakes are usually short lived.

CRITICAL THINKING

7. **Analyzing Ideas** Why are spring floods common in rivers where the headwaters are in an area of cold, snowy winters?
8. **Making Inferences** If you were picking a material to make an artificial levee, what major characteristic would you look for? Explain your answer.

CONCEPT MAPPING

9. Use the following terms to create a concept map: *stream deposition, delta, alluvial fan, floodplain, natural levee, dam, artificial levee,* and *lake.*

CHAPTER RESOURCES

Chapter Resource File
- Section Quiz GENERAL

Workbooks
- Study Guide (also in Spanish)

Chapter 15 Highlights

Sections

1 The Water Cycle

Key Terms

water cycle, 375
evapotranspiration, 376
condensation, 376
precipitation, 376
desalination, 378

Key Concepts

- The continuous movement of water from the atmosphere to the land and oceans and back to the atmosphere is called the *water cycle*.
- A region's water budget is affected by temperature, vegetation, wind, and the amount and duration of rainfall.
- The availability of fresh water can be enhanced through conservation efforts.

2 Stream Erosion

tributary, 379
watershed, 379
stream load, 380
discharge, 380
gradient, 380
meander, 381
braided stream, 382

- A river system is made of a main stream and feeder streams, called *tributaries*.
- A watershed is the area of land that is drained by a river system.
- The erosive ability of a river is affected by stream load, stream discharge, and stream gradient.
- A bend in a low-gradient stream or river is called a *meander*.
- A river that is composed of multiple channels that divide and rejoin around sediment bars is called a *braided stream*.

3 Stream Deposition

delta, 383
alluvial fan, 383
floodplain, 384

- Where a stream slows significantly, it can deposit its stream load to form deltas and alluvial fans.
- In floodplains, flooding commonly brings in new, rich soil for farming but can cause property damage.
- Floods can be controlled through forest and soil conservation and by building structures such as levees and dams.

Chapter Highlights

Alternative Assessment — ADVANCED

Biography of a River Have students research the history and development of one river and write its story. The biography should include an explanation of when and where the river formed, what its original course was, and how it has changed over time; a map of the river that notes major cities and land uses along its route; a list of historical events or people of importance to the river; and a description of past and current uses of the river and any environmental challenges it is facing. **LS Verbal/Visual**

CHAPTER RESOURCES

Chapter Resource File

- **Concept Review** GENERAL
- **Critical Thinking** ADVANCED
- **Math Skills** GENERAL
- **Graphing Skills** GENERAL
- **Chapter Test A** GENERAL
- **Chapter Test B** ADVANCED

Workbooks

- **Study Guide** (also in Spanish)
- **Assessments** (Spanish)

Technology

Classroom Videos
- Brain Food Video Quiz

HRW Earth Science Video
- Segment 10: Water and Erosion (River Systems)

Assignment Guide

Section	Questions
1	1, 3, 5, 9–10, 17–18, 23–24, 29–30
2	2, 6–7, 11–13, 19, 25, 28, 32–35
3	4, 8, 14–16, 21–22, 31
2 and 3	20
1–3	26–27

Using Key Terms

1-8. Answers may vary but should show that students understand the definitions of and differences between key terms.

Understanding Key Concepts

9. d **10.** a
11. b **12.** a
13. d **14.** d
15. b **16.** a

Short Answer

17. The water budget of Earth as a whole is balanced, but local water budgets commonly are not balanced.

18. Conservation of water means wise use of water. Polluting surface and ground water makes it unfit for many uses. Preventing pollution will keep existing water supplies fit for most uses, so we do not have to find other sources of freshwater.

19. As a river slows down it cannot erode its bed as effectively, but it still can erode its banks. At curves, along the outside bank, velocity is highest, causing increased erosion in that area. Along the inside of the curve, the velocity is slower, and sediment loads are deposited as bars of sand and gravel. Over time, the process continues and curves enlarge and progress along the length of the stream.

20. Sediment load is a function of the volume of water a stream carries and the stream's velocity. Thus, to carry a large suspended and dissolved sediment load, a river needs a high volume of water or a high velocity.

21. Streams that feed lakes deposit their sediment load into the lake. Water that runs off land but does not enter streams also can deposit sediments into lakes. Over time, the lake is gradually filled in and may become a swamp, bog, or even dry land.

22. Indirect methods of flood control try to prevent excess runoff during periods of high rainfall to avoid flooding. Direct methods try to redirect the flow during a flood to reduce damage from floodwaters.

Critical Thinking

23. If the sun's rays were blocked by dust or other contaminants, the amount of evapotranspiration on Earth would be reduced. Less water in the atmosphere would mean less precipitation. But overall, Earth's water budget would remain balanced.

Chapter 15 Review

Using Key Terms

Use each of the following terms in a separate sentence.

1. *water cycle*

2. *gradient*

3. *evapotranspiration*

4. *floodplain*

For each pair of terms, explain how the meanings of the terms differ.

5. *condensation* and *precipitation*

6. *watershed* and *tributary*

7. *stream load* and *discharge*

8. *delta* and *alluvial fan*

Understanding Key Concepts

9. The change of water vapor into liquid water is called
a. runoff. **c.** evaporation.
b. desalination. **d.** condensation.

10. In a water budget, the income is precipitation and the expense is
a. evapotranspiration and runoff.
b. condensation and saltation.
c. erosion and conservation.
d. conservation and sedimentation.

11. The land area from which water runs off into a stream is called a
a. tributary. **c.** divide.
b. watershed. **d.** gully.

12. Tributaries branch out and lengthen as a river system develops by
a. headward erosion.
b. condensation.
c. saltation.
d. runoff.

13. The stream load that includes gravel and large rocks is the
a. suspended load. **c.** dissolved load.
b. runoff load. **d.** bed load.

14. A fan-shaped formation that develops when a stream deposits its sediment at the base of a steep slope is called a(n)
a. delta.
b. meander.
c. oxbow lake.
d. alluvial fan.

15. The part of a valley floor that may be covered during a flood is the
a. floodway.
b. floodplain.
c. meander.
d. artificial levee.

16. One way to control floods indirectly is through
a. soil conservation.
b. dams.
c. floodways.
d. artificial levees.

Short Answer

17. How does a local water budget differ from the water budget of the whole Earth?

18. How is reducing the pollution in streams and groundwater linked to water conservation?

19. Describe how bank erosion can cause a river to meander.

20. Why do most rivers that have a large sediment load also have high water velocity?

21. Describe how lakes fill with sediment.

22. What is the difference between direct and indirect methods of flood control?

Critical Thinking

23. Evaluating Ideas How would Earth's water cycle be affected if a significant percentage of the sun's rays were blocked by dust or other contaminants in the atmosphere?

24. Making Comparisons Use an atlas to determine the geographic location of Calcutta, India, and Stockholm, Sweden. How might the local water budgets of these two cities differ? Explain your answer.

25. Making Inferences The Colorado River is usually grayish brown as it flows through the Grand Canyon. What causes this color?

26. Making Predictions What do you think would happen to cities in the southwestern U.S. if rivers in that area could not be dammed?

Concept Mapping

27. Use the following terms to create a concept map: *water vapor, condensation, precipitation, channel, stream load, bar, alluvial fan, delta, divides, watersheds, tributaries, floodplains, dams,* and *artificial levees*.

Math Skills

28. Making Calculations If a river is 3,705 km long from its headwaters to its delta and the average downstream velocity of its water is 200 cm/s, use the equation *time = distance ÷ velocity* to determine how many days a water molecule takes to make the trip.

29. Using Equations You wish to examine the annual water budget for the state of Colorado. If p = total precipitation, e = total evapotranspiration, r = total stream runoff, and g = total water soaking into the ground, what equation will allow you to determine whether Colorado experiences a net loss or net gain of water over the course of a year?

Writing Skills

30. Writing Persuasively Write a persuasive essay of at least 300 words that suggests ways in which your community can conserve water and reduce water pollution.

31. Communicating Main Ideas Discuss the dangers and advantages of living in a river floodplain. Outline the options for adapting to living in a river floodplain.

Interpreting Graphics

The graph below shows the gradients of several rivers of the United States. Use the graph to answer the questions that follow.

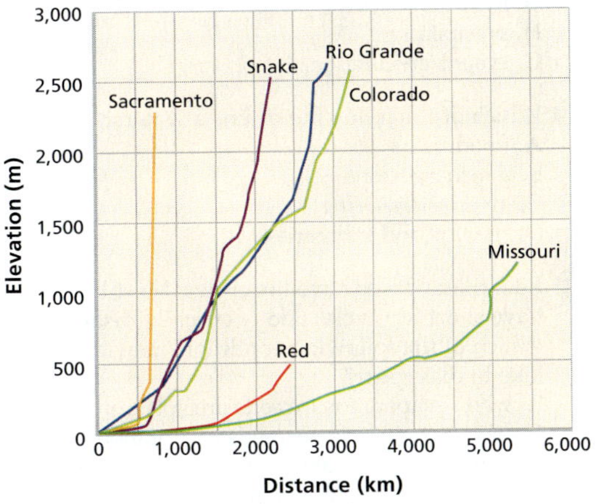

32. Which river has the shallowest average gradient over its entire course?

33. Which river has the steepest average gradient over its entire course?

34. Based only on gradient, how would the velocity of the Snake River compare with the velocity of the Missouri River?

35. Which end of each line on the graph represents the headwaters of the river system? Explain your answer.

Chapter Review

24. Calcutta, India, is much more tropical than Stockholm, Sweden. Evapotranspiration would likely be higher in Calcutta, while precipitation may be less frequent in Stockholm.

25. Sediments from upstream give the river its grayish brown color.

26. If rivers in desert areas of the U.S. could not be dammed, much of the western U.S. would have significantly lower population densities and reduced agriculture. Cities such as Phoenix, Las Vegas, and others would be much smaller, and farming would be very difficult without irrigation.

Concept Mapping

27. Answers may vary but should include all of the terms listed. Sample answers appear at the end of this Teacher Edition.

Math Skills

28. 100,000 cm = 1 km,
so 200 cm/s = 0.002 km/s
time = distance/velocity
= 3,705 km ÷ 0.002 km/s
= 18,525,000 s;
1 day = 24 h × 60 min × 60 s
= 86,400 s, so

$$\frac{18{,}525{,}000 \text{ s}}{86{,}400 \text{ s/day}} = 214.4 \text{ days}$$

29. net budget = (p + g) − (e + r)

Writing Skills

30. Answers may vary. Accept all reasonable answers.

31. Answers may vary. Accept all reasonable answers.

Interpreting Graphics

32. The Red River has the shallowest average gradient (0.2 m/km) over its course.

33. The Sacramento River has the steepest gradient (3.1 m/km) over its course.

34. Based only on gradient, the Snake River would have a higher velocity than the Missouri River does.

35. The headwaters of each river are represented by the point on each line that is at the highest elevation and that is farthest from the origin of the graph.

Standardized Test Prep

Estimated Time

To give students practice under more realistic testing conditions, allow them 30 minutes to answer all of the questions in this practice test.

TEST DOCTOR

Question 4 Answer F is correct. Answer G is incorrect because excessive precipitation would keep a lake full and may even cause the lake to grow in size. Answer H is incorrect because sediments may actually be harmful to a lake if they build up. Removing sediments will often extend the lifespan of a lake. Answer I is incorrect because a balanced water budget would keep the lake's size stable.

Question 9 Answer C is correct. Students who do not thoroughly read the passage may choose answer A. They may think that half of the river, not the wetlands, has been lost.

Question 11 Full-credit answers should include the following points:

- the river was altered to accommodate human society and human inventions, such as boats
- students should realize that the river was being altered to benefit humans and that any attempts to alter nature in order to benefit human society may have unexpected—and unwanted—consequences

Chapter 15 Standardized Test Prep

Understanding Concepts

Directions (1–4): **For *each* question, write on a separate sheet of paper the letter of the correct answer.**

1. Condensation is often triggered as water vapor rising in the atmosphere
   A. cools
   B. warms
   C. contracts
   D. breaks apart

2. The continuous movement of water from the ocean, to the atmosphere, to the land, and back to the ocean is
   F. condensation.
   G. the water cycle.
   H. precipitation.
   I. evapotranspiration.

3. Which of the following drains a watershed?
   A. floodplains
   B. a recharge zone
   C. an artesian spring
   D. streams and tributaries

4. Like rivers, lakes have life cycles. Most lakes have short life cycles and eventually disappear. Which of the following conditions may cause a lake to disappear?
   F. when evaporation exceeds precipitation
   G. when precipitation exceeds evaporation
   H. when sediments are removed from the lake
   I. when a local water budget is balanced

Directions (5–8): **For *each* question, write a short response.**

5. What is the term for a volume of water that is moved by a stream during a given amount of time?

6. The gradient of a river is defined as a change in what over a given distance?

7. Streams are said to have varying loads. What makes up a stream's load?

8. Desalination removes what naturally occurring compound from ocean water?

Reading Skills

Directions (9–11): **Read the passage below. Then, answer the questions.**

The Mississippi Delta

In the Mississippi River Delta, long-legged birds step lightly through the marsh and hunt fish or frogs for breakfast. Hundreds of species of plants and animals start another day in this fragile ecosystem. This delta ecosystem, like many other ecosystems, is in danger of being destroyed.

The threat to the Mississippi River Delta ecosystem comes from efforts to make the river more useful. Large parts of the river bottom have been dredged to deepen the river for ship traffic. Underwater channels were built to control flooding. What no one realized was that the sediments that once formed new land now pass through the channels and flow out into the ocean. Those river sediments had once replaced the land that was lost every year to erosion. Without them, the river could no longer replace land lost to erosion. So, the Mississippi River Delta began shrinking. By 1995, more than half of the wetlands were already gone—swept out to sea by waves along the Lousiana coast.

9. Based on the passage, which of the following statements about the Mississippi River is true?
   A. The Mississippi River never floods.
   B. The Mississippi River is not wide enough for ships to travel on it.
   C. The Mississippi River's delicate ecosystem is in danger of being lost.
   D. The Mississippi River is disappearing.

10. Based on the passage, which of the following statements is true?
    F. By 1995, more than half of the Mississippi River was gone.
    G. Underwater channels may control flooding.
    H. Channels help form new land.
    I. Sediment cannot replace lost land.

11. The passage mentions that damage to the ecosystem came from efforts to make the river more useful. For who or what was the river being made more useful?

Answers

Understanding Concepts

1. A
2. G
3. D
4. F
5. discharge
6. elevation
7. small particles and dissolved minerals
8. salt

Reading Skills

9. C
10. G
11. Answers may vary. See Test Doctor for a detailed scoring rubric.

Interpreting Graphics

12. D
13. Answers may vary. See Test Doctor for a detailed scoring rubric.
14. G
15. Answers may vary. See Test Doctor for a detailed scoring rubric.

Interpreting Graphics

Directions (12–15): **For *each* question below, record the correct answer on a separate sheet of paper.**

The diagram below shows how a hydropower plant works. Use this diagram to answer questions 12 and 13.

Hydroelectric Dam

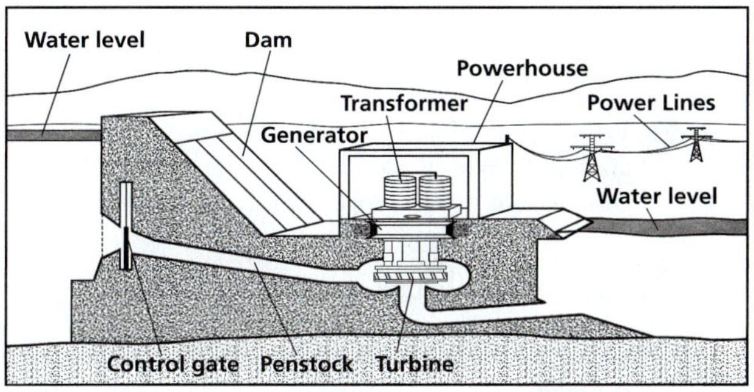

12 Hydroelectric dams are used to generate electricity for human use. As water rushes past the machinery inside, an electric current is generated. What does water rush past to turn the generator, which produces the current?

A. a transformer
B. the control gate
C. an intake
D. a turbine

13 Look at the diagram above. What direction does the water flow? What makes the water flow in this direction?

The graphic below shows the formation of an oxbow lake. Use this graphic to answer questions 14 and 15.

Formation of an Oxbow Lake

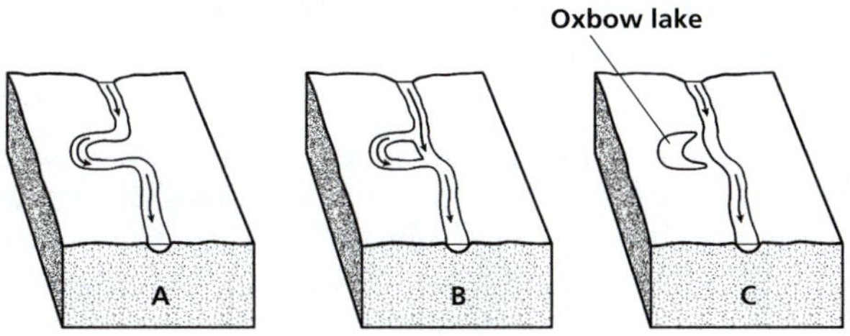

14 What is the term for the wide curves whose development causes the formation of oxbow lakes?

F. wonders
G. meanders
H. bows
I. loops

15 How does the speed at which the water flows contribute to the process of forming an oxbow lake?

Test TIP

If you are permitted to, draw a line through each incorrect answer choice as you eliminate it.

Standardized Test Prep

TEST DOCTOR

Question 13 Full-credit answers should include the following points:

- water flows from the left to the right in the diagram. This flow can be deduced from the difference in water levels
- students should understand that water naturally seeks to equalize the levels of the two pools and that, in situations such as those shown in the graphic, the water in the deeper pool will move into the shallower pool, if possible
- water is propelled from the deep reservoir on the left through the penstock by gravity and into the more shallow reservoir on the right

Question 15 Full-credit answers should include the following points:

- water on the outside edges of the river bend flows faster, which erodes the banks, and makes the meander wider
- students should know that meanders form when fast-moving water that is opposite to a bar deposition erodes the adjacent bank
- when meanders become so curved that they form a loop, the river may reconnect to itself and the meander may become isolated from the river, which forms an oxbow lake
- the faster the flow of water is, the faster this process of erosion and meander growth occurs

Test Prep Correlations ***National Science Education Standards***

ES 1c: items 1, 2
ES 2a: item 2
ES 3c: items 3, 4, 5, 7, 14, 15
LS 4e: items 9, 11
LS 5e: items 9, 11
PS 4b: item 13
SAI 2e: items 13, 15
SPSP 2c: item 11
SPSP 3a: item 12
SPSP 3c: items 8, 9
SPSP 5b: items 9, 11
ST 2c: item 12
ST 2d: items 10, 11
UCP 1: items 1, 2, 4, 15

CHAPTER RESOURCES

State Resources

For specific resources for your state, visit go.hrw.com and type in the keyword HSHSTR.

Inquiry Lab

Sediments and Water

Teacher's Notes

Time Required
one 45-minute class period

Lab Ratings

EASY → HARD

Teacher Preparation: 1
Student Setup: 2
Concept Level: 2
Cleanup: 2

Skills Acquired
- Observing
- Measuring
- Experimenting
- Predicting
- Inferring
- Collecting Data
- Organizing and Analyzing Data

The Scientific Method
In this lab, students will
- Make Observations
- Ask Questions
- Test a Hypothesis
- Analyze the Results
- Draw Conclusions
- Communicate Results

Materials
The materials listed on this page are enough for groups of two to four students. You may want to ask students to bring in juice containers from home.

Tips and Tricks
Have students observe the sediments with a magnifying glass or dissecting microscope. Students can also use the sediments in a stream table setup to see how resistant each type is to erosion under different conditions.

Chapter 15 Inquiry Lab

Using Scientific Methods

Sediments and Water

Objectives
- **Measure** the amount of water that sediment can hold.
- **Identify** the properties that affect how sediment interacts with water.

Materials
- graduated cylinder, 100 mL
- grease pencil
- juice containers, 12 oz (2)
- metric ruler
- nail, large
- pan, 23 cm × 33 cm × 5 cm or larger
- sand, dry and coarse
- silt, dry
- stopwatch or clock with second hand
- water

Safety

Running water erodes some types of soil more easily than it erodes others. How rapidly a soil erodes depends on how well the soil holds water. In this lab, you will determine the erosive effect of water on various types of sediment.

ASK A QUESTION

1. Which type of soil would hold more water: sandy soil or silty soil? Which soil would water flow through faster and thus would erode more rapidly: sandy soil or silty soil?

FORM A HYPOTHESIS

2. Write a hypothesis that is a possible answer to the questions above.

TEST THE HYPOTHESIS

3. Use a graduated cylinder to pour 300 mL of water into each of two juice containers.
4. Place the containers on a flat surface. Using a grease pencil, draw a line around the inside of the containers to mark the height of the water. Label one container "A" and the other "B." Empty and dry the containers.
5. Using silt, fill container A up to the line drawn inside the container. Tap the container gently to even out the surface of the sediment. Repeat this step for container B, but use sand.
6. Fill the graduated cylinder with 100 mL of water. Slowly pour the water into container A. Stop every few seconds to allow the soil to absorb the water. Continue pouring until a thin film of water forms on the surface of the sediment. If more than 100 mL of water is needed, refill the graduated cylinder and continue this step.

Step 5

CHAPTER RESOURCES

Chapter Resource File
- Datasheet for Chapter Lab GENERAL
- Lab Notes and Answers

Workbooks

Long-Term Projects
- Water Clarity ADVANCED
- Precipitation and the Water Table ADVANCED

7. Record the volume of water poured into the container.

8. Using container B, repeat steps 6 and 7. Record your observations.

9. Use a metric ruler to measure 1 cm above the surface of the sediment in each container. Using the grease pencil, draw a line to mark this height on the inside of each container. Pour water from the graduated cylinder into containers A and B until the water reaches the 1 cm mark.

10. Poke a nail through the very bottom of the side of container A. Place the container inside the pan. At the same time, start recording the time by using a stopwatch and pull the nail out of the container.

11. Observe the water level, and record the amount of time that the water takes to drop to the sediment surface.

12. Using container B, repeat steps 10 and 11. Record your observations.

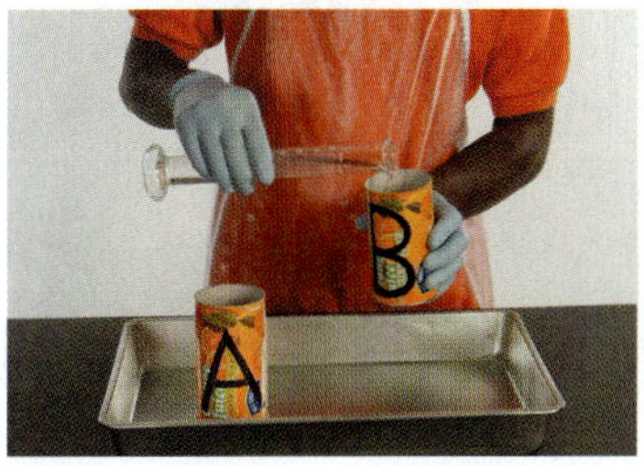

Step 8

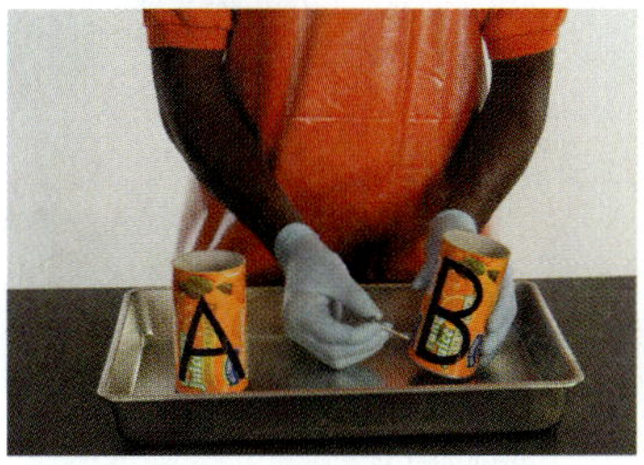

Step 12

ANALYZE THE RESULTS

1. **Analyzing Results** In step 8, which type of sediment held more water?

2. **Analyzing Results** Which type of sediment was the water able to flow through faster?

3. **Summarizing Results** What properties of the sediment do you think affected how the water flowed through the sediment? Explain your answer.

DRAW CONCLUSIONS

4. **Analyzing Results** On the basis of your answers to the questions above, which would water erode more quickly: an area of silt or an area of sand? Explain your answer.

5. **Drawing Conclusions** In which sediment do you think a deep stream channel is most likely to form? In which sediment is a meandering stream likely to form? Explain your answers.

Extension

1. **Applying Conclusions** Describe three ways to make slopes covered with soil more resistant to erosion.

Inquiry Lab

Answers to Analyze the Results

1. Silt should hold more water than sand.
2. Water should flow faster through the sand than through the silt.
3. The size of the particles and the spaces between particles influence how fast water flows through the sediments. Sand is coarser than silt and will have much larger pore spaces for water to flow through than silt does.

Answers to Draw Conclusions

4. Answers may vary. Because water can soak into and flow more quickly through the sand, there is less runoff to cause erosion. Because silt is finer than sand and because water cannot flow through silt as easily, the water would erode and carry away the silt.
5. Deep stream channels would most likely form in silty soils because water can erode the stream bed easily. Meandering streams are more likely to form in sandy sediments because the stream bed is less easily eroded.

Answers to Extension

1. Planting ground cover, building barriers, reducing the steepness of the slope, terracing, and contour planting can all slow runoff and make an area more resistant to soil erosion.

Scott Robertson
North Warren Central School
Chestertown, NY

World Watershed Sediment Yield

Discussion — GENERAL

Sediment Load and Land Use Draw students' attention to the relatively high sediment yields in Southeast Asia around Indonesia, Thailand, Myanmar, Bangladesh, and Malaysia. Have students propose a hypothesis to explain the high loads that come from this area. (Answers may vary; possibilities include high precipitation rates, high topographic relief, regular flooding during the monsoon season, and extensive loss of rainforests leading to increased runoff.) LS **Visual/Logical**

Answers to Map Skills Activity

1. Southeast Asia, around Indonesia (3,000 million tons)
2. less than 10 tons to 500 tons per square kilometer
3. 1,311 + 150 + 18 + 154 + 28 = 1,661 million tons.
4. Southeast Asia appears to have higher relief than Africa. Sediment yields in Southeast Asia tend to be more than 1,000 tons per square kilometer compared to less than 100 tons per square kilometer in Africa.
5. The total area of land in the Amazon basin is much larger (3 to 4 times greater) than the area of land in the Indian basin, which would explain why the total yield per year is so much larger in the Amazon basin. In addition, the Amazon basin receives a large part of its sediment load from the Andes Mountains, while most of the high relief areas in northern India drain east.

CHAPTER RESOURCES

Technology

Transparencies
- 78 World Watershed Sediment Yield (with worksheet)

MAPS *in Action*

World Watershed Sediment Yield

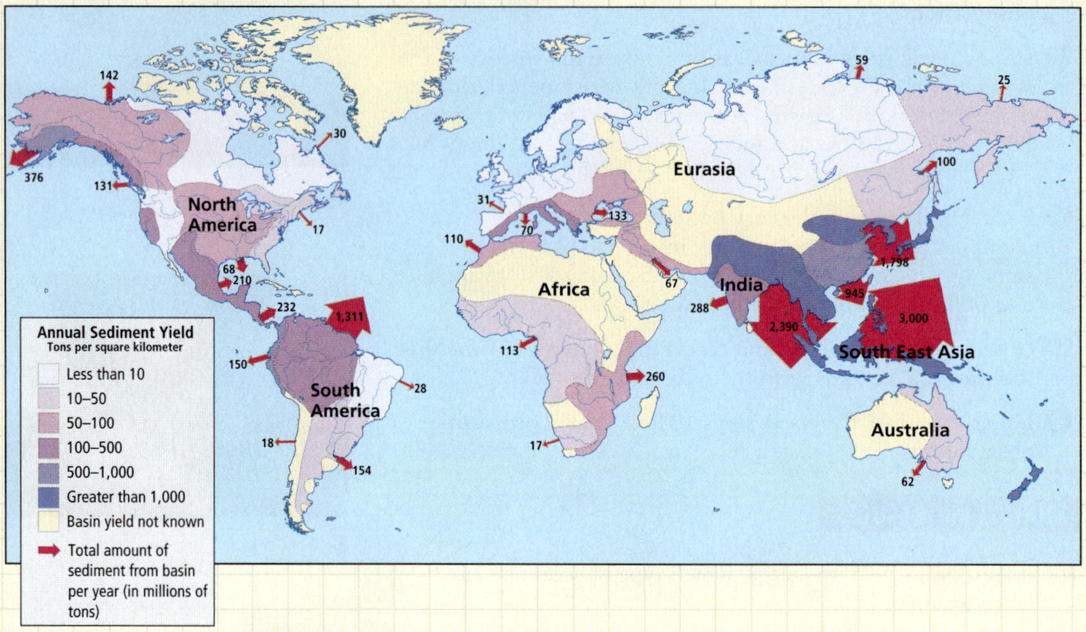

Map Skills Activity

This map shows the world's watersheds and identifies the sediment yield of each watershed basin in tons per square kilometer and the total amount of sediment that each basin dumps into the ocean in millions of tons per year. Use the map to answer the questions below.

1. **Using a Key** What area has the highest total annual sediment yield from one basin?
2. **Using a Key** What is the range of the annual sediment yield in tons per square kilometer for the United States, excluding Alaska?
3. **Analyzing Data** What is the total amount of sediment that basins in South America yield per year?
4. **Analyzing Relationships** Areas that have high relief, where the range of elevations is great, tend to have higher sediment yields than areas that have low relief, where the topography is flatter, do. Which area would you conclude has higher relief: Africa or South East Asia? Explain your answer.
5. **Making Comparisons** Both the Amazon basin, which is in northern South America, and the India basin have an annual sediment yield range of 100 to 500 tons per square kilometer. However, the total amount of sediment per year from the Amazon basin is 1,311 million tons, while the total amount of sediment per year from the India basin is 288 million tons. Explain why these two basins differ so significantly in their total sediment yield per year.

The Three Gorges Dam

▲ A resident on the Yangtze River carries his belongings to higher ground after his home was destroyed when the dam caused the water level to rise.

China's Yangtze River is the third-longest river in the world. The Yangtze River flows through the Three Gorges region of central China, which is famous for its natural beauty and historical sites. This region is also where the Three Gorges Dam—the largest hydroelectric dam project in the world—is currently being built. When the dam is complete, the Yangtze River will rise to form a reservoir that is 595 km long—as long as Lake Superior. In other words, the reservoir will be about as long as the distance between Los Angeles and San Francisco in California!

Benefits of the Dam

The dam has several purposes, one of which is to control the water level of the Yangtze River to prevent flooding. About 1 million people died in the last century from flooding along the river. The other purpose is to provide millions of people with hydroelectric power. China now burns air-polluting coal to meet 75% of the country's energy needs. Engineers project that when the dam is completed, its turbines will provide enough electrical energy to power a city that is 10 times the size of Los Angeles. When its flow is controlled, the Yangtze River will be deep enough for ships to navigate on it, so the dam will also increase trade in a relatively poor region of China.

Disadvantages of the Dam

The project has several drawbacks, however. The reservoir behind the dam will flood an enormous area. About 1.3 million people living in the affected areas are being relocated—there are 13 cities and hundreds of villages in the area of the reservoir. As the reservoir's waters rise, fragile ecosystems and valuable archeological sites will be destroyed and scenic recreation areas will be submerged.

Opponents of the project also claim that the dam will increase pollution levels in the Yangtze River. Most of the cities and factories along the river dump untreated wastes directly into the water. Some people think the reservoir will become the world's largest sewer when 1 billion tons of sewage flows into the reservoir every year.

Because the dam lies over a fault, scientists question whether the dam will be able to withstand earthquakes. If the dam were to burst, towns and cities downstream would be destroyed by the ensuing flood.

◀ An engineer on the Three Gorges Dam project celebrates the successful test of sluicing, which is the rushing of water through channels in the dam.

Extension

1. **Research** What do you think of the Three Gorges dam? Research other dam projects. After analyzing the benefits and risks of dams, write an essay describing your opinion about dam projects.

Answer to Extension

1. Answers may vary. Accept all well-reasoned answers.

Eye on the Environment

The Three Gorges Dam

Internet Activity — GENERAL

Restoring Rivers Large dams continue to be built in many developing countries around the world to generate power, control flooding, provide clean water, and increase trade and navigation. However, in the U.S., the era of big dam building is over and a new movement is afoot—restoring rivers by removing dams. Aging, unsafe dams have been removed in many states, including Maine, North Carolina, Wisconsin, Oregon, and Vermont. While most of the dams removed in the last twenty years were small, privately owned dams, some river activists would like to see even large dams, such as the Glen Canyon Dam on the Colorado River, removed to restore rivers to a more natural state. Have students investigate the issue on the Internet. Have students research the following questions: "Which dams have been removed recently? Which dams are likely to be removed or are currently being debated for removal? Why are these dams slated for removal? What are the effects of dam removal on a river and its watershed? What are the effects on the communities upstream and downstream from the dam?" A worksheet designed to direct student research on this topic can be found in the **Chapter Resource File** booklet or by visiting go.hrw.com and entering the keyword **HQ6RVSX.** LS Verbal

CHAPTER RESOURCES

Chapter Resource File

- Internet Activity
  - Restoring Rivers GENERAL

Chapter 16 Groundwater

Planning Guide

Compression Guide

To shorten instruction because of time limitations, omit Section 2.

OBJECTIVES	LABS, DEMONSTRATIONS, AND ACTIVITIES	TECHNOLOGY RESOURCES
PACING • 90 min pp. 396–404 **Chapter Opener**	**LTP Long-Term Project** Precipitation and the Water Table* ADVANCED	**OSP Parent Letter** □ **CD Student Edition on CD-ROM** **CD Chapter Summaries Audio CD** □ **VID Brain Food Video Quiz**
Section 1 Water Beneath the Surface • Identify properties of aquifers that affect the flow of groundwater. • Describe the water table and its relationship to the land surface. • Compare wells, springs, and artesian formations. • Describe two land features formed by hot groundwater.	**SE Quick Lab** Permeability, p. 398 □ GENERAL **CRF Datasheet for Quick Lab*** GENERAL **TE Demonstration** Comprehension Check, p. 398 GENERAL **TE Group Activity** Model an Aquifer, p. 399 GENERAL **SE Skills Practice Lab** Porosity, pp. 414–415 □□ GENERAL **CRF Datasheet for Chapter Lab*** GENERAL **SE Maps in Action** Water Level in the Southern Ogallala, p. 416 GENERAL **TE Debate** Limiting Development?, p. 417 GENERAL **SE Mapping Expeditions** What Goes Up Must Come Down . . . Where?, pp. 838–839 GENERAL **CRF Making Models Lab** Making a Classroom Geyser* GENERAL	**OSP Lesson Plans** (also in print) **TR Bellringer*** **TR** 79 Porosity and Permeability* **TR** 80 Zones of Aquifers* **TR** 81 Topography and the Water Table* **TR** 82 Water Level in the Southern Ogallala* **TE Internet Activity** Your Watershed, p. 400 ADVANCED **CRF Internet Activity** Your Watershed* ADVANCED **TE Internet Activity** Agriculture and the Aquifer, p. 416 GENERAL **CRF Internet Activity** Agriculture and the Aquifer* GENERAL **TE Internet Activity** Sinkholes, p. 417 GENERAL **CRF Internet Activity** Sinkholes* GENERAL
PACING • 45 min pp. 405–408 **Section 2 Groundwater and Chemical Weathering** • Describe how water chemically weathers rock. • Explain how caverns and sinkholes form. • Identify two features of karst topography.	**SE Quick Lab** Chemical Weathering, p. 405 GENERAL **CRF Datasheet for Quick Lab*** GENERAL **CRF Inquiry Lab** Cave Formations and Ecology* GENERAL	**OSP Lesson Plans** (also in print) **TR Bellringer*** **CD Interactive Tutor** Physical and Chemical Weathering

PACING • 90 min

CHAPTER REVIEW, ASSESSMENT, AND STANDARDIZED TEST PREPARATION

- **SE Chapter Highlights,** p. 409
- **SE Chapter Review,** pp. 410–411
- **SE Standardized Test Prep,** pp. 412–413
- **CRF Concept Review*** □ GENERAL
- **CRF Critical Thinking*** ADVANCED
- **CRF Math Skills*** GENERAL
- **CRF Graphing Skills*** GENERAL
- **CRF Chapter Test A*** □ GENERAL
- **CRF Chapter Test B*** ADVANCED
- **OSP Lesson Plans** (also in print)
- **OSP Test Generator**
- **OSP Test Item Listing**

Online and Technology Resources

go.hrw.com — Holt Online Learning

Visit **go.hrw.com** for access to Holt Online Learning, or enter the keyword **HQ6 Home** for a variety of free online resources.

One-Stop Planner® CD-ROM

This CD-ROM package includes

- Lab Materials QuickList Software
- Holt Calendar Planner
- Customizable Lesson Plans
- Printable Worksheets
- ExamView® Test Generator
- Interactive Teacher Edition
- Holt PuzzlePro®
- Holt PowerPoint® Resources

KEY

- **SE** Student Edition
- **TE** Teacher Edition
- **CRF** Chapter Resource File
- **LTP** Long-Term Projects
- **OSP** One-Stop Planner
- **TR** Transparencies and Transparency Worksheets
- **CD** CD or CD-ROM
- **VID** Classroom Video/DVD
- * Also on One-Stop Planner
- ◆ Requires advance prep
- ■ Also available in Spanish

SKILLS DEVELOPMENT RESOURCES	REVIEW AND ASSESSMENT	CORRELATIONS
SE **Pre-Reading Activity,** p. 396 GENERAL TE **Using the Figure** Warmth in a Cold Climate, p. 396 GENERAL		National Science Education Standards
CRF **Directed Reading*** BASIC TE **Using the Figure** Compare, p. 397 GENERAL TE **Using the Figure** Replenishing Groundwater, p. 398 GENERAL TE **Using the Figure** Extreme Conditions, p. 399 GENERAL TE **Reading Skill Builder** Paired Summarizing, p. 399 BASIC SE **Math Practice,** p. 400 GENERAL TE **Inclusion Strategies,** p. 400 TE **Using the Figure** Going Down the Drain, p. 401 BASIC SE **Graphic Organizer** Venn Diagram, p. 402 GENERAL TE **Using the Figure** Slope, p. 403 ADVANCED TE **Skill Builder** Vocabulary, p. 403 GENERAL	SE **Reading Checks,** pp. 399, 400, 403 GENERAL TE **Homework,** p. 402 GENERAL SE **Section Review,** p. 404 GENERAL TE **Reteaching,** p. 403 BASIC TE **Quiz,** p. 403 GENERAL TE **Alternative Assessment,** p. 404 GENERAL CRF **Section Quiz*** ■ GENERAL	LS 4e, SPSP 6b, UCP 1
CRF **Directed Reading*** BASIC TE **Using the Figure** Cave Formations, p. 406 GENERAL TE **Reading Skill Builder** Reading Hint, p. 406 BASIC TE **Biology Connection** Cave Vision, p. 406 ADVANCED TE **Skill Builder** Writing, p. 407 ADVANCED TE **Inclusion Strategies,** p. 407	SE **Reading Check,** p. 407 GENERAL SE **Section Review,** p. 408 GENERAL TE **Reteaching,** p. 407 BASIC TE **Quiz,** p. 407 GENERAL TE **Alternative Assessment,** p. 408 ADVANCED CRF **Section Quiz*** ■ GENERAL	PS 3c, UCP 3, UCP 4

Holt Lab Generator CD-ROM

Search for labs by topic, standard, difficulty level, and/or time frame. Edit those labs to fit your needs, or create your own labs. Use the Lab Materials QuickList to order just what you need.

Holt Earth Science Interactive Tutor CD-ROM

This CD-ROM consists of interactive activities that give students a fun way to extend their knowledge of Earth science concepts.

Chapter Summaries Audio CDs

These CDs include audio summaries of the key concepts presented in each chapter. (Audio summaries are also available in Spanish.)

SCI LINKS. NSTA

www.scilinks.org

Maintained by the **National Science Teachers Association.** See Chapter Enrichment pages that follow for a complete list of topics.

See Chapter Enrichment pages for Video Resources.

Refer to our ***Professional Reference for Teachers*** for additional teaching resources, including articles written by science education professionals about relevant and timely issues facing today's science teachers.

Chapter 16 Chapter Enrichment

This Chapter Enrichment provides relevant and interesting information to expand and enhance your classroom instruction of the chapter material.

Section 1 Water Beneath the Surface

Groundwater Contamination

About 51% of the U.S. population gets drinking water from groundwater, including 95% of rural Americans. A 1999 report by the U.S. Geological Survey (USGS) revealed that 47% of urban wells and 14% of rural wells were contaminated with volatile organic compounds (VOCs), such as benzene or vinyl chloride that are often byproducts of human activities. Wells in agricultural areas are far more likely to be contaminated with pesticides, herbicides, and nitrates. USGS research also shows that the gasoline additive MTBE (methyl tertiary butyl ether) is the second most common VOC in groundwater. MTBE, which improves gasoline combustion, can leak into groundwater supplies from improperly maintained underground storage tanks. Other VOCs that have entered U.S. groundwater resources include solvents, such as trichloromethane and trichloroethene, and hydrocarbons, such as benzene and naphthalene.

Water, Water Everywhere....

There is far more to groundwater than what humans tap with wells. Some of the oldest groundwater in the world lies too deep beneath Earth's surface to be easily reached. About 417,000 km^3 of groundwater lie within 0.8 km of the surface, and thus are usable. Yet an equal amount lies far below this layer and so is not economical to exploit. Some of this is *connate water,* or water that was deposited at the same time as the ancient rocks in which it occurs. Because these rocks are so deeply buried, nearly all connate water has been trapped beneath the surface for millions of years, exhibiting little or no flow. Many of these rocks formed from marine sediments, so much connate groundwater is salty. Of course, humans cannot drink salty water, so even if this water could be easily accessed, it would not be usable for drinking or agriculture.

Section 2 Groundwater and Chemical Weathering

Cave Formations

There are many types of magnificent and beautiful cave formations besides stalactites and stalagmites:

Soda straws: quarter-inch diameter hollow tubes, up to 6 feet long, that grow as water flows through them and deposits rings of calcite on their tips;

Columns: huge, floor-to-ceiling formations that occur when stalactites and stalagmites meet and grow together;

Flowstone: wave-like sheets of calcium carbonate that form when films of water flow over the walls and floor of a cave;

Draperies: large, downward flowing "curtains" that form when beads of water trickle down the underside of an inclined surface, depositing thin, often translucent, sheets of calcium carbonate;

◀ **These men are testing water from a well for contamination. Such testing is essential for maintaining a supply of safe drinking water.**

▲ The eerie beauty of the Carlsbad Caverns is created by a colorful and varied array of cave formations.

Bacon formation: draperies that have alternating light and dark bands that result from variations in the mineral content of the flowing water;

Cave Grapes: irregular clusters of calcium carbonate nodules that build up on the walls and floors in flooded cave chambers;

Gypsum Flowers: wavy, petal-like formations that form on the walls of drier caves and grow from their bases when flowing water deposits gypsum;

Dogtooth Spar: six inch long, pyramid-shaped calcium carbonate crystals that form underwater in the flooded chambers of caves.

Life in a Cave

Though many people might think that caves are inhospitable places, many caves support their own ecosystems. For example, Carlsbad Caverns is home to about half a million roosting bats. The bat droppings provide sustenance to a wide variety of spiders and insects, some of which are blind or even eyeless. In caves that have standing water, cave lakes are home to a complex web of animals, from microorganisms to predatory fish. These and other moist areas of a cave may also support amphibians, some of which are translucent. Though plants cannot live without light, fungi can. Many caves harbor bizarre fungi, such as living mold drapery, that depend on decaying organic material for their survival.

Video Resources

Brain Food Video Quizzes

Brain Food Video Quizzes These videos contain game-show style quizzes that assess students' progress and motivate students to study the chapter material.

NOVA® Videos

To order **NOVA** videos related to this chapter, visit **go.hrw.com** and enter the keyword **HQ6GWEV.**

SCILINKS®

NSTA Developed and maintained by the National Science Teachers Association

SciLinks is maintained by the National Science Teachers Association to provide you and your students with interesting, up-to-date links that will enrich your classroom presentation of the chapter.

Visit www.scilinks.org and enter the SciLinks code for more information about the topic listed.

Topic: Groundwater
SciLinks code: HQ60699

Topic: Aquifers
SciLinks code: HQ60089

Topic: Caverns (and Karst Topography)
SciLinks code: HQ60234

Chapter 16

Chapter Overview

Groundwater is water that seeps through soil and flows beneath Earth's surface. Water reaches the surface through wells and springs. Hot springs and geysers occur when hot, subsurface rock heats groundwater. As groundwater seeps through the ground, the water may react chemically with rock to form caverns and other geological formations.

Using the Figure — GENERAL

Warmth in a Cold Climate Ask students to look at the background in the photograph and use it to describe the winter climate in this part of Japan. Ask students to explain how a hot spring can occur in such a cold climate. (A hot spring can occur in a cold climate because the water is heated by hot rocks or magma beneath Earth's surface.) LS **Logical**

PRE-READING ACTIVITY

FOLDNOTES Have pairs of students use their FoldNotes to study key terms from the chapter. Instruct one student to use the FoldNote to provide the key term and have the other student give the definition. Have the student who provides the key term correct the other student's definition.

Chapter 16 Groundwater

Sections

1 Water Beneath the Surface

2 Groundwater and Chemical Weathering

What You'll Learn

- How water moves under Earth's surface
- How humans use groundwater
- How groundwater shapes Earth's surface

Why It's Relevant

Groundwater is a major source of drinking water for humans and a source of fresh water for agriculture and industry. Groundwater is also a major force in shaping Earth's landforms.

PRE-READING ACTIVITY

FOLDNOTES **Key-Term Fold** Before you read this chapter, create the FoldNote entitled "Key-Term Fold" described in the Skills Handbook section of the Appendix. Write a key term from the chapter on each tab of the key-term fold. Under each tab, write the definition of the key term.

▶ These macaque monkeys are bathing in hot springs in Jigokudani National Park in Japan. Hot springs form where molten rock in Earth's crust heats surrounding rock and the groundwater in the rock. The water then rises to the surface to form pools and springs.

Chapter Correlations — *National Science Education Standards*

LS 4e Human beings live within the world's ecosystems. Increasingly, humans modify ecosystems as a result of population growth, technology, and consumption.... **(Section 1)**

SPSP 6b Understanding basic concepts and principles of science and technology should precede active debate about the economics, policies, politics, and ethics of various science—and technology—related challenges. However, understanding science alone will not resolve local, national, or global challenges. **(Section 1)**

UCP 1 The natural and designed world is complex; it is too large and complicated to investigate and comprehend all at once. Scientists and students learn to define small portions for the convenience of investigation. The units of investigation can be referred to as "systems." A system is an organized group of related objects or components that form a whole. **(Section 1)**

PS 3c A large number of important reactions involve the transfer of either electrons (oxidation/reduction reactions) or hydrogen ions (acid/base reactions) between reacting ions, molecules, or atoms. **(Section 2)**

UCP 3 Interactions within and among systems result in change. Changes vary in rate, scale, and pattern, including trends and cycles. **(Section 2)**

UCP 4 Evolution is a series of changes, some gradual and some sporadic.... Although evolution is most commonly associated with the biological theory..., evolution also describes changes in the universe. **(Section 2)**

Section 1 Water Beneath the Surface

Surface water that does not run off into streams and rivers may seep down through the soil into the upper layers of Earth's crust. There, the water fills spaces, or *pores,* between rock particles. Water may also fill fractures or cavities in rock that were caused by erosion. Water that fills and moves through these spaces in rock and sediment is called **groundwater.** Groundwater is an important source of fresh water in the United States.

Properties of Aquifers

A body of rock or sediment in which large amounts of water can flow and be stored is called an **aquifer.** For water to flow freely through an aquifer, the pores or fractures in the aquifer must be connected. The ease with which water flows through an aquifer is affected by many factors, including porosity and permeability.

Porosity

In a set volume of rock or sediment, the percentage of the rock or sediment that consists of open spaces is **porosity.** One factor that affects porosity is sorting. *Sorting* is the amount of uniformity in the size of the rock or sediment particles, as **Figure 1** shows. Most particles in a well-sorted sediment are about the same size, and a few smaller particles fill the spaces between them. Poorly sorted sediment contains particles of many sizes. Small particles fill the spaces between large particles, which makes the rock less porous. Particle packing also affects porosity. Loosely packed particles leave many open spaces that can store water, so the rock has high porosity. Rock that has tightly packed particles contains few open spaces and thus has low porosity. Grain shape also affects porosity. In general, the more irregular the grain shape is, the more porous the rock or sediment is.

OBJECTIVES

- **Identify** properties of aquifers that affect the flow of groundwater.
- **Describe** the water table and its relationship to the land surface.
- **Compare** wells, springs, and artesian formations.
- **Describe** two land features formed by hot groundwater.

KEY TERMS

groundwater
aquifer
porosity
permeability
water table
artesian formation

groundwater the water that is beneath Earth's surface

aquifer a body of rock or sediment that stores groundwater and allows the flow of groundwater

porosity the percentage of the total volume of a rock or sediment that consists of open spaces

Figure 1 ▶ Differences in Porosity

Well-sorted, coarse-grained sediment has high porosity.

Well-sorted, fine-grained sediment has high porosity equal to the porosity of coarse-grained sediment.

Poorly sorted sediment that contains grains of many sizes has low porosity.

Section 1

Focus

Overview

This section describes aquifers, how water enters aquifers, aquifer porosity, permeability, and zonation. The section describes groundwater movement, water table levels, wells and springs, as well as the formation of hot springs and geysers.

Bellringer

Ask students where the water that flows from the faucets in their home comes from. (In most locations, the water that is used by humans comes from groundwater. In other locations, it may be purified surface water.) LS **Logical**

Motivate

Using the Figure — GENERAL

Compare Have students describe the three types of sediment shown in the figure at the bottom of the page. Have them think about what may happen when grains of many sizes occur together. Ask them to hypothesize why poorly sorted sediment that has grains of different sizes has low porosity. (When grains of many sizes occur together, small grains may become trapped or compacted between large grains, thereby reducing porosity.) LS **Visual**

CHAPTER RESOURCES

Chapter Resource File

- Directed Reading BASIC

Technology

Transparencies
- Bellringer
- 79 Porosity and Permeability (with worksheet)

Student Edition on CD-ROM

One-Stop Planner CD-ROM
- Lesson Plan

Teach

Using the Figure — GENERAL

Replenishing Groundwater Have students describe the differences they see between the porous rock and the permeable rock shown in the figure at the top of this page. Ask students to speculate on how water is replenished in porous rock (by precipitation only) as compared with permeable rock. (Water may flow from one water-filled space to another.) LS Visual

QuickLAB

Skills Acquired

- Measuring
- Calculating
- Observing
- Analyzing

Teacher's Notes You may wish to try this lab ahead of time with the materials students will be using. Most soils contain clay, which may make the soil less permeable than the sand. Use alternative materials if necessary to best illustrate permeability. Ask the students before the lab begins to predict which material will be the most permeable. After the lab is completed, ask the students if their prediction was correct.

Answers

1. The gravel sample had the highest drainage rate.
2. Answers may vary. Gravel should be the most permeable; the sand or soil should be the least permeable.

Rock is considered porous if it has many empty spaces that can fill with water.

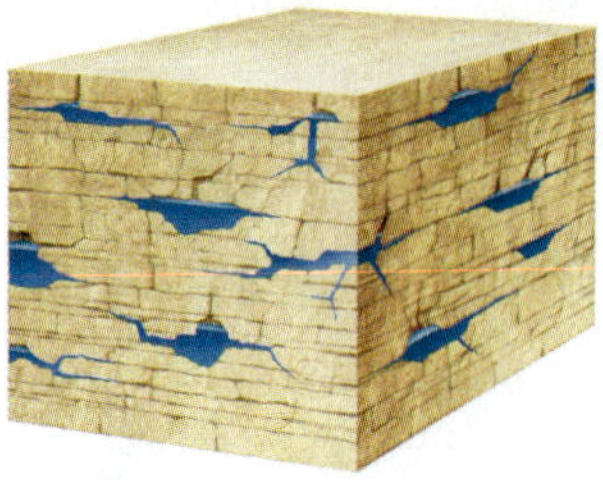

Rock is considered permeable if its empty spaces are connected so that water may flow from one space to the next.

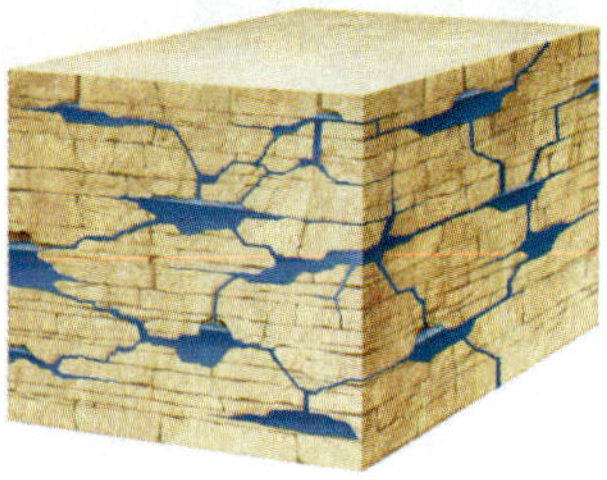

Figure 2 ▶ Porous rocks do not make good aquifers unless water can move freely through the rocks.

permeability the ability of a rock or sediment to let fluids pass through its open spaces, or pores

Permeability

The ease with which water passes through a porous material is called **permeability.** For a rock to be permeable, the open spaces must be connected, as shown in **Figure 2.** A rock that has high porosity is not permeable if the pores or fractures are not connected. Permeability is also affected by the size and sorting of the particles that make up a rock or sediment. The larger and better sorted the particles are, the more permeable the rock or sediment tends to be. The most permeable rocks, such as sandstone, are composed of coarse particles. Other rocks, such as limestone, may be permeable if they have interconnected cracks. Clay is a sediment composed of flat, very fine-grained particles. Because of this characteristic composition, clay is essentially *impermeable,* which means that water cannot flow through it.

QuickLAB 20 min

Permeability

Procedure

1. With a sharpened pencil, make seven tiny holes in the bottom of each of three paper or plastic cups. Stretch cheesecloth tightly over the bottom of each cup. Secure the cloth with a rubber band.
2. Mark a line 2 cm from the top of one cup. Stand the cup on three thread spools in a saucer or pie pan, and fill the cup to the line with sand.
3. Pour 120 mL of water into the cup. Use a stopwatch to time how long the water takes to drain.
4. Pour the water from the saucer into a measuring cup. Record the amount of water.
5. Repeat steps 2–4 with the two other cups, but fill one cup with soil and one with gravel, not sand.
6. Calculate the rates of drainage for each cup by dividing the amount of water that drained by the time the water took to drain.
7. For each cup, calculate the percentage of water retained by subtracting the amount of water drained from 120 mL. Divide this volume by 120.

Analysis

1. Which cup had the highest drainage rate?
2. A sample with a high drainage rate and a low percentage of water retained is highly permeable. Which sample was the most permeable? Which sample was the least permeable?

CHAPTER RESOURCES

Chapter Resource File

- Datasheet for Quick Lab GENERAL

Demonstration — GENERAL

Comprehension Check Demonstrate the concepts of permeability and impermeability. Get two shallow bowls or cups. In one bowl, press potters' clay around the bottom and up the sides to a height of 2.5 cm; put 2.5 cm of loose potting soil in the other bowl. Have students watch what happens as you pour about 60 mL of water into each bowl. The water should sit atop the clay because the clay is impermeable. The water should enter the soil because the soil is permeable. LS Visual English Language Learners

MISCONCEPTION ALERT

Underwater Lakes and Rivers Many students think of groundwater as an underground lake or as a river of water that flows through large open channels in rock. Explain to students that while groundwater does form underground lakes or rivers in some places, in most cases, groundwater flows slowly through tiny spaces between rock particles. The rate at which water flows depends on the permeability of the rock or sediment.

Zones of Aquifers

The attraction of water molecules to the soil holds water in the soil. When the soil becomes full, or saturated, gravity overcomes this attraction and pulls water down through the rock layers until it reaches impermeable rock. As more water soaks into the ground, the water level rises underground and forms two distinct zones of groundwater, as shown in **Figure 3.**

Zone of Saturation

The layer of an aquifer in which the pore space is completely filled with water is the *zone of saturation*. The term *saturated* means "filled to capacity." The zone of saturation is the lower of the two zones of groundwater. The upper surface of the zone of saturation is called the **water table.**

water table the upper surface of underground water; the upper boundary of the zone of saturation

Zone of Aeration

The zone that lies between the water table and Earth's surface is called the *zone of aeration*. The zone of aeration is composed of three regions. The uppermost region of the zone of aeration holds soil moisture—water that forms a film around grains of topsoil. The bottom region, just above the water table, is the capillary fringe. Water is drawn up from the zone of saturation into the capillary fringe by capillary action. *Capillary action* is caused by the attraction of water molecules to other materials, such as soil. For example, when a paper towel soaks up a spill, capillary action draws moisture into the towel. Between the soil moisture region and the capillary fringe is a region that is dry—except during periods of rain—and thus contains air in its pores.

Reading Check What are the two zones of groundwater? (See the Appendix for answers to Reading Checks.)

SCILINKS. NSTA Developed and maintained by the National Science Teachers Association

For a variety of links related to this chapter, go to www.scilinks.org

Topic: Groundwater
SciLinks code: HQ60699

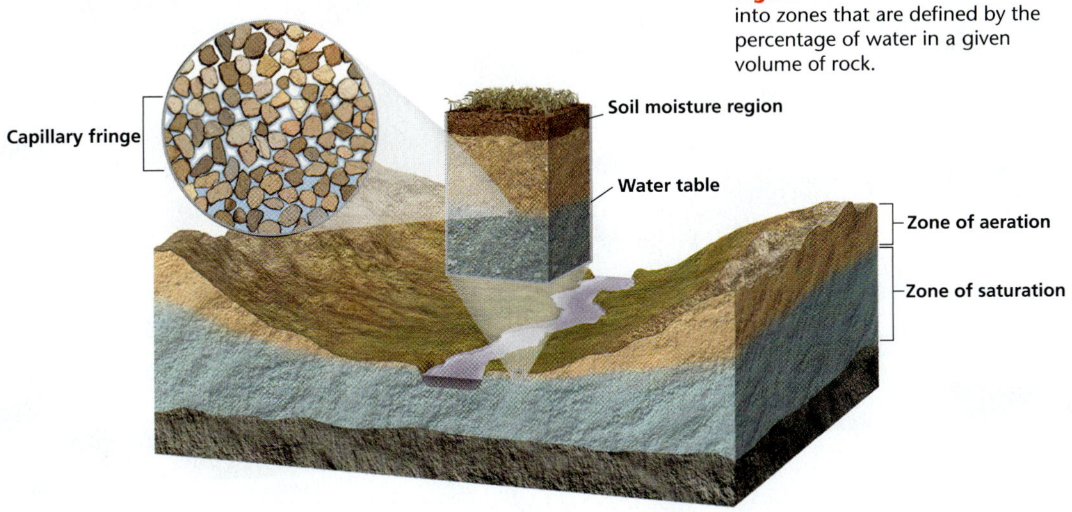

Figure 3 ▶ Groundwater is divided into zones that are defined by the percentage of water in a given volume of rock.

Using the Figure — GENERAL

Extreme Conditions Have students look at the illustration of the zones of groundwater. Have students identify and describe each layer shown. Ask students to identify the lowest layer, which is not labeled in the diagram. (the impermeable layer) Then, have students draw cross sections of an aquifer under two extreme conditions—during a drought and during an extended, heavy rain. Ask them to refer to their drawing to explain how each condition affects the zones of the aquifer. (During a drought, the water table is lower and the dry region expands; during a period of heavy rain, the water table rises and the dry region and the soil moisture layer may become saturated.) **LS Visual**

Answer to Reading Check

The two zones of groundwater are the zone of saturation and the zone of aeration.

Group Activity — GENERAL

Model an Aquifer Organize the class into groups of 5 or 6 students. Provide each group with a clear glass bowl or shallow jar, clay, small pebbles, soil, and a small amount of water. Have students create a model of an aquifer. Students should observe the water and then describe how their model is like an aquifer. (In the models, the water should collect on top of the impermeable clay and be stored in the empty spaces in the layer of pebbles.) **LS Kinesthetic**

READING SKILL BUILDER — BASIC

Paired Summarizing Group students into pairs and have them read silently about the zones of aquifers. Then, have one student summarize the ideas of the zone of saturation and the zone of aeration. The other student should listen to the retelling and should point out any inaccuracies or ideas that were left out. Allow students to refer to the text as needed.
LS Verbal English Language Learners Co-op Learning

CHAPTER RESOURCES

Technology

Transparencies
- 80 Zones of Aquifers (with worksheet)

Teach, continued

MATHPRACTICE

Answer

1,500 million m³/y − 575 million m³/y = 925 million m³/y; The yearly loss of groundwater is 925 million m³. 6,475 million m³ ÷ 925 million m³/y = 7 y; The aquifer should be depleted in 7 years.

Answer to Reading Check

The depth of a water table depends on topography, aquifer permeability, the amount of rainfall, and the rate at which humans use the groundwater.

Internet Activity — ADVANCED

Your Watershed Have students visit the Web site of the U.S. Environmental Protection Agency and go to their watershed page. Here, students can locate and learn what watershed your area is part of and find out about both surface water and groundwater resources. By analyzing the areas overlying the watershed, students can speculate about what types of contaminants enter surface water and groundwater and how these pollutants might affect the overall water supply. Have a group of students report to the class about what they learned. They may make a poster that shows the boundaries of your watershed, and they may discuss what may be entering your groundwater. A worksheet designed to direct student research on this topic can be found in the **Chapter Resource File** booklet or by visiting **go.hrw.com** and entering the keyword **HQ6GWEX.** LS **Logical**

CHAPTER RESOURCES

Chapter Resource File

- **Internet Activity**
  - Your Watershed ADVANCED

Technology

- **Transparencies**
  - 81 Topography and the Water Table (with worksheet)

MATHPRACTICE

Rate of Groundwater Depletion In some areas, more groundwater is removed than is naturally replaced. In one area, for example, 575 million cubic meters of water enters the rock every year, while 1,500 million cubic meters of water is removed each year. What is the rate of groundwater depletion in that area? The total amount of groundwater available is 6,475 million cubic meters. If groundwater use continues at the current rate, in how many years will the water be completely depleted?

Movement of Groundwater

Like water on Earth's surface, groundwater flows downward in response to gravity. Water passes quickly through highly permeable rock and slowly through rock that is less permeable. The rate at which groundwater flows horizontally depends on both the permeability of the aquifer and the gradient of the water table. *Gradient* is the steepness of a slope. The velocity of groundwater increases as the water table's gradient increases.

Topography and the Water Table

The depth of the water table below the ground surface depends on surface topography, the permeability of the aquifer, the amount of rainfall, and the rate at which humans use the water. Generally, shallow water tables match the contours of the surface, as shown in **Figure 4.** During periods of prolonged rainfall, the water table rises. During periods of drought, the water table falls and flattens because water that leaves the aquifer is not replaced.

Only one water table exists in most areas. In some areas, however, a layer of impermeable rock lies above the main water table. This rock layer prevents water from reaching the main zone of saturation. Water collects on top of this upper layer and creates a second water table, which is called a *perched water table*.

Reading Check **What four factors affect the depth of a water table?** **(See the Appendix for answers to Reading Checks.)**

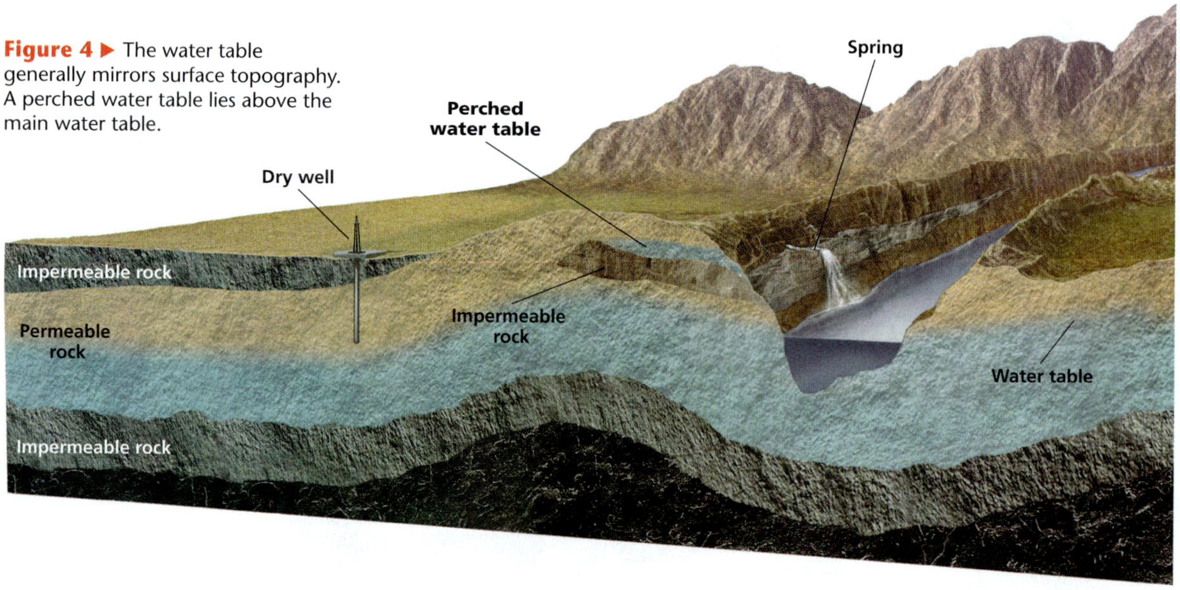

Figure 4 ▶ The water table generally mirrors surface topography. A perched water table lies above the main water table.

INCLUSION Strategies

- *Gifted and Talented*
- *Attention Deficit Disorder*
- *Behavior Control Issues*

Many students, especially those who are gifted and talented, find textbook material more interesting and meaningful if they extrapolate through research. Have students use the library, interviews, or the Internet to find the following details: 1–The names of the aquifers that supply water to the state you live in, 2–Other states that receive water from the same aquifer system, 3–The total number of aquifer systems that supply water to the United States.

Conserving Groundwater

In many communities, groundwater is the only source of fresh water. Although groundwater is renewable, its long renewal time limits its supply. Groundwater collects and moves slowly, and the water taken from aquifers may not be replenished for hundreds or thousands of years. Communities often regulate the use of groundwater to help conserve this valuable resource. They can monitor the level of the local water table and discourage excess pumping. Some communities recycle used water. This water is purified and may be used to replenish the groundwater supply.

Surface water enters an aquifer through an area called a recharge zone. A *recharge zone* is anywhere that water from the surface can travel through permeable rock to reach an aquifer, as shown in **Figure 4.** Recharge zones are environmentally sensitive areas because pollution in the recharge zone can enter the aquifer. Therefore, recharge zones are often labeled by signs like the one shown in **Figure 5.** Pollution can enter an aquifer from waste dumps and underground storage tanks for toxic chemicals, from fertilizers and pesticides used in agriculture and on lawns, or from leaking sewage systems. If too much groundwater is pumped from an aquifer that is near the ocean, salt water from the ocean can then flow into the aquifer and contaminate the groundwater supply.

Figure 5 ▶ Water that enters this drain runs off into the Charles River and surrounding aquifers in Massachusetts.

Identifying Preconceptions — GENERAL

What Gets into Groundwater Perform this exercise before students begin to read this page. Supply students with a list of substances and ask them to determine whether or not each one enters the groundwater. These items may include lawn weed killer, dog droppings, oil leaks from cars, discarded medications, and household waste from landfills. After students have discussed the items on the list, invite them to consider how all of these items, and countless others, do end up in groundwater as they wash off streets or leach through landfills. LS **Verbal/Logical**

ENVIRONMENTAL CONNECTION

Munching Microbes Once pollutants get into groundwater, removing the pollutants is extremely difficult because groundwater is so inaccessible. In addition, the aquifer material may get coated with pollutants such as oil, which remain for a long time and recontaminate the water. Biologists are continuing to search for and find bacteria that "eat" groundwater pollutants. The bacteria they have in mind would ingest a contaminant and render it harmless. Experiments so far are promising, but scientists must make sure the bacteria are not harmful. Have students do research to find out more details about such experiments, including what kinds of pollution the bacteria ingest and how they make the pollutants harmless. Have them write a report presenting their findings. LS **Verbal**

Using the Figure — BASIC

Going Down the Drain Invite a volunteer to read the sign painted on the sidewalk in the photo at the top of the page. Ask students to identify the location of this drain. (It is on the curb next to the gutter and a storm drain.) Invite students to discuss why the sign just above the drain is probably insufficient to keep all contaminants out of the drain and thus out of groundwater. (When it rains, everything on the street will wash into the drain, even if nothing is deliberately dumped into it.) LS **Visual**

Teach, *continued*

Graphic Organizer GENERAL

Venn Diagram
You may want to have students work in groups to create this Venn Diagram. Have one student draw the diagram and fill in information provided by other students in the group. LS Visual/Logical

Teaching Tip — ADVANCED

Make Concepts Relevant Tell students that when a new well is dug for a home, the well driller is required by law to test to see how much water the well can supply per minute. Ask students to discuss why this information is one of many factors that is important to the homeowner and why it might also be important to neighbors. (The homeowner must understand how much water can be used without depleting the well. Neighbors may wish to know because one neighbor could lower the overall water table and cause surrounding wells to go dry.) LS Interpersonal

Graphic Organizer **Venn Diagram**
Create the **Graphic Organizer** entitled "Venn Diagram" described in the Skills Handbook section of the Appendix. Label the circles "Ordinary wells" and "Artesian wells." Then, fill in the diagram with characteristics that are common to both types of wells.

Wells and Springs

Groundwater reaches Earth's surface through wells and springs. A *well* is a hole that is dug to below the level of the water table and through which groundwater is brought to Earth's surface. A *spring* is a natural flow of groundwater to Earth's surface in places where the ground surface dips below the water table. Wells and springs are classified into two groups—ordinary and artesian.

Ordinary Wells and Springs

Ordinary wells work only if they penetrate highly permeable sediment or rock below the water table. If the rock is not permeable enough, groundwater cannot flow into the well quickly enough to replace the water that is withdrawn.

Pumping water from a well lowers the water table around the well and forms a *cone of depression*, as shown in **Figure 6.** If too much water is taken from a well, the cone of depression may drop to the bottom of the well and the well will go dry. The lowered water table may extend several kilometers around the well and may cause surrounding wells to become dry.

Ordinary springs are usually found in rugged terrain where the ground surface drops below the water table. These springs may not flow continuously if the water table in the area has an irregular depth as a result of variable rainfall. Springs that form from perched water tables that intersect the ground surface are very sensitive to the amount of local precipitation. Thus, these springs may go dry during dry seasons or severe droughts.

Figure 6 ▶ A cone of depression develops in the water table around a pumping well.

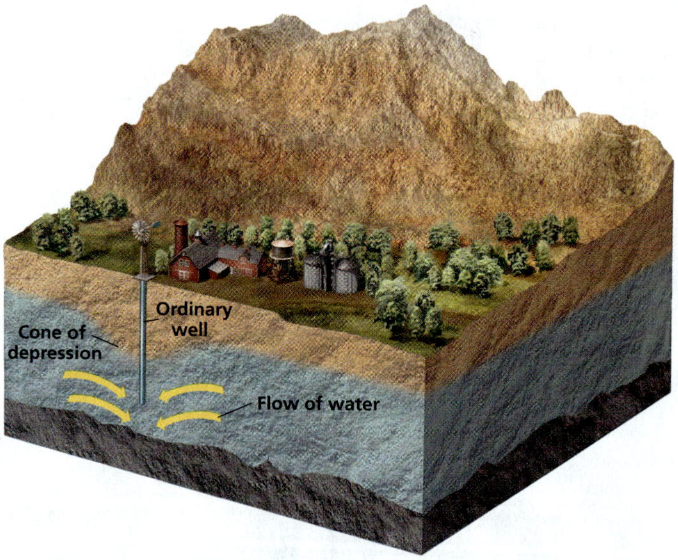

Homework — GENERAL

Home Water Use Tell students that no matter the immediate source of their own water supply, conserving water is crucial. Because fresh water supplies replenish slowly, if present usage patterns continue, the water supply will be lower in the future than it is now. Have students create a checklist of fixtures and appliances in their homes that use water. Have them check which ones are designed to use "less" water and which are not. Students may also make a checklist of water wasting at home, such as taking long showers, running half-loads of laundry or of dishes in the dishwasher, and leaving the water running when they brush their teeth. Then, have students create a plan for conserving water at home. LS Logical

Figure 7 ▶ The aquifer in an artesian formation dips under the impermeable caprock. When a well is drilled into an artesian aquifer, pressure is released and the water rushes upward. These men are testing the quality of water from an artesian well in Pakistan.

Artesian Wells and Springs

The groundwater that supplies many wells comes from local precipitation. However, the water in some wells may come from as far away as hundreds of kilometers. Water may travel through an aquifer to a distant location. Because the aquifer is so extensive, it may become part of an artesian formation, an arrangement of permeable and impermeable rock.

An **artesian formation** is a sloping layer of permeable rock that is sandwiched between two layers of impermeable rock, as shown in **Figure 7.** The permeable rock is the aquifer, and the top layer of impermeable rock is called the *caprock*. Water enters the aquifer at a recharge zone and flows downhill through the aquifer. As the water flows downward, the weight of the overlying water causes pressure in the aquifer to increase. Because the water is under pressure, when a well is drilled through the caprock, the water quickly flows up through the well and may even spout from the surface. An *artesian well* is a well through which water flows freely without being pumped.

artesian formation a sloping layer of permeable rock sandwiched between two layers of impermeable rock and exposed at the surface

Artesian formations are also the source of water for some springs. When cracks occur naturally in the caprock, water from the aquifer flows through the cracks. This flow forms *artesian springs*.

Reading Check What is the difference between ordinary springs and artesian springs? (See the Appendix for answers to Reading Checks.)

Using the Figure — Advanced

Slope Have students study the diagram of an artesian formation. Ask them to speculate about why the slope enhances the formation. Have them describe how the angle of slope would affect the rate of release of the water at the surface. (The slope causes water to flow downward more quickly, which adds greater pressure to the water accumulated at the base. The greater the slope, the more pressure builds up at the base. Therefore, water emerges faster from an artesian well or spring at the base of a steep slope.) LS **Visual/Logical**

Close

Reteaching — Basic

Diagrams Have students make sketches or diagrams of the different formations that hold groundwater. Have them label all of the layers, zones, and surface features. LS **Visual**

Quiz — General

1. What type of sediment has the lowest porosity and therefore is able to hold the least amount of water? (poorly sorted sediment)
2. How does a water table's gradient affect the flow of groundwater? (the steeper the gradient, the faster the movement of groundwater)
3. What is travertine, and how does it form? (Travertine is the form of calcite that is deposited when hot spring waters cool as they reach the surface.)

Answer to Reading Check

Ordinary springs occur where the ground surface drops below the water table. An artesian spring occurs where groundwater flows to the surface through natural cracks in the overlying caprock.

CHAPTER RESOURCES

Chapter Resource File

Making Models Lab
- Making a Classroom Geyser GENERAL

SKILL BUILDER — General

Vocabulary Point out to students that the word *artesian* comes from the French term *puits artésien,* which means "wells of Artois." Artois is a small city in northern France. Anthropologists think that the first artesian wells were dug there in the eighteenth century. The name *Artois* is derived from the *Atrebates,* an ancient tribe that lived in the area. You may want to have students find Artois on a map of France or Europe. Interested students may also be interested in learning about the types of rock found in the Artois area. LS **Visual**

Close, *continued*

Alternative Assessment — GENERAL

Assessing Groundwater Have students imagine they are experts charged with assessing groundwater resources in one region of your state. Have them make a list of criteria that they would study in preparing a report about groundwater resources. Students may create an illustration or diagram to accompany their checklist. **LS Logical/Visual**

Answers to Section Review

1. Porosity is the percentage of open space in rock or sediment. Permeability refers to the ease with which water passes through a porous material. The more permeable the rock is, the more easily groundwater flows through the rock.
2. The zone of saturation is completely filled with water. The zone of aeration lies above the zone of saturation and is composed of the soil water region, a dry region, and the capillary fringe.
3. The contours of a shallow water table generally match the land's topography.
4. because they depend on the amount of local precipitation
5. A cone of depression is an area where the water table is lowered as a result of the withdrawal of water from a well.
6. In an ordinary aquifer the rock layers may be horizontal and lack an impermeable cap. In an artesian formation the permeable rock layer slopes and is covered by an impermeable layer called the *caprock*.
7. In an ordinary well, the water is mechanically pumped to the surface; in an artesian well, natural pressure pushes the water to the surface.
8. Answers may vary. An artesian well may provide a more constant water source, because it is not subject to local weather conditions.
9. Pollutants produced in a recharge zone will likely find their way into the aquifer below.
10. Shallow pools do not erupt because the water is not under pressure, so it boils before it can reach the superheated state that causes a geyser to erupt.
11. The *water table* separates the *zone of aeration* from the *zone of saturation*, which stores *groundwater.*

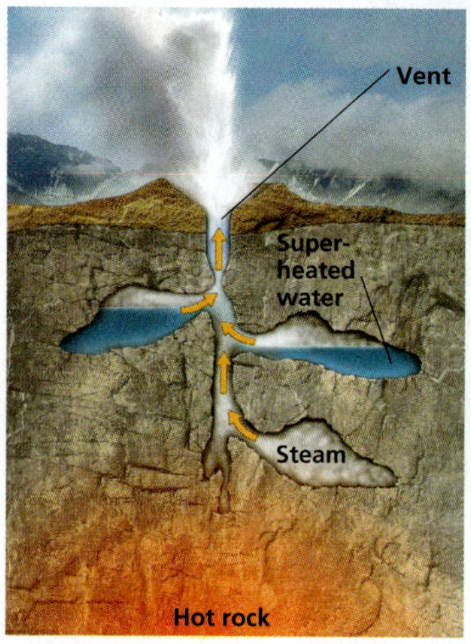

Figure 8 ▶ The vent and underground chambers of a geyser enable water to become superheated and to eventually erupt to the surface.

Hot Springs

Groundwater is heated when it passes through rock that has been heated by magma. Hot groundwater that is at least 37°C and that rises to the surface before cooling produces a *hot spring*. When water in a hot spring cools, the water deposits minerals around the spring's edges. The deposits form steplike terraces of calcite called *travertine*. *Mud pots* form when chemically weathered rock mixes with hot water to form a sticky, liquid clay that bubbles at the surface. Mud pots are called *paint pots* when the clay is brightly colored by minerals or organic materials.

Geysers

Hot springs that periodically erupt from surface pools or through small vents are called *geysers*. A geyser consists of a narrow vent that connects one or more underground chambers with the surface. The hot rocks that make up the chamber walls superheat the groundwater. The water in the vent exerts pressure on the water in the chambers, which keeps the water in the chambers from boiling for a time. When the water in the vent finally begins to boil, the boiling water produces steam that pushes the water above it to the surface. Release of the water near the top of the vent relieves the pressure on the superheated water farther down. With the sudden release of pressure, the superheated water changes into steam and explodes toward the surface, as shown in **Figure 8.** The eruption continues until most of the water and steam are emptied from the vent and chambers. After the eruption, groundwater begins to collect again and the process is repeated, often at regular intervals.

Section 1 Review

1. **Identify** the difference between porosity and permeability, and explain how permeability affects the flow of groundwater.
2. **Name and describe** the two zones of groundwater.
3. **Describe** how the contour of a shallow water table compares with the local topography.
4. **Explain** why ordinary springs often flow intermittently.
5. **Define** the term *cone of depression*.
6. **Compare** the rock layers in an artesian formation with those in an ordinary aquifer.
7. **Compare** artesian wells and ordinary wells.

CRITICAL THINKING

8. **Making Inferences** Which type of well would provide a community with a more constant source of water: an ordinary well or an artesian well? Explain your answer.
9. **Identifying Relationships** Why is protecting the environment from pollution important for communities in recharge zones?
10. **Analyzing Ideas** Why don't shallow pools of hot water erupt the way that geysers erupt?

CONCEPT MAPPING

11. Use the following terms to create a concept map: *groundwater, water table, zone of saturation,* and *zone of aeration.*

CHAPTER RESOURCES

Chapter Resource File
- Section Quiz GENERAL

Workbooks
- Study Guide (also in Spanish)

Section 2 Groundwater and Chemical Weathering

As groundwater passes through permeable rock, minerals in the rock dissolve. The warmer the rock is and the longer it is in contact with water, the greater the amount of dissolved minerals in the water. Water that contains relatively high concentrations of dissolved minerals, especially minerals rich in calcium, magnesium, and iron, is called *hard water*. Water that contains relatively low concentrations of dissolved minerals is called *soft water*.

Many people think that using hard water is unappealing. For example, more soap is needed to produce suds in hard water than in soft water. Also, many people prefer not to drink hard water because of its metallic taste. Some household appliances or fixtures may be damaged by the buildup of mineral deposits from hard water. **Figure 1** shows some results of the long-term presence of hard water.

OBJECTIVES

- **Describe** how water chemically weathers rock.
- **Explain** how caverns and sinkholes form.
- **Identify** two features of karst topography.

KEY TERMS

cavern
sinkhole
karst topography

Results of Weathering by Groundwater

One way that minerals become dissolved in groundwater is through chemical weathering. As water moves through soil and other organic materials, the water combines with carbon dioxide to form carbonic acid. This weak acid chemically weathers the rock that the acid passes through by breaking down and dissolving the minerals in the rock.

Figure 1 ▶ Soap scum forms when soap reacts with calcium carbonate in hard water (inset). During high-water stages, hard water deposited a residue of calcium carbonate on the canyon walls that border this creek.

QuickLAB 25 min

Chemical Weathering

Procedure

1. Place **limestone, granite, pyrite,** and **chalk chips** into separate **small beakers.**
2. Cover the rocks in **1% HCl solution.**
3. After 20 min, observe the rocks.

Analysis

1. How have the rocks changed?
2. How is this process of change like the process of chemical weathering by groundwater?

Section 2

Focus

Overview

This section explains the weathering of rock by groundwater, the formation and characteristics of caves, the formation of sinkholes and natural bridges, and the formation of karst topography.

Bellringer

Have students look closely at the two photographs on this page. Ask them to describe how minerals in water can affect objects that are are in extensive contact with the water.(Students may think of mineral buildup in pipes or rusty stains under water faucets.) **LS Visual**

Motivate

QuickLAB

Skills Acquired

- Observing
- Analyzing

Teacher's Notes Instruct students on the correct handling and disposal of the acid. You can use lemon juice instead of 1% HCl solution.

1. The limestone and chalk became pitted and/or reduced in size.
2. The HCl is like the carbonic acid in groundwater that chemically weathers rock.

CHAPTER RESOURCES

Chapter Resource File

- Directed Reading BASIC
- Datasheet for Quick Lab GENERAL

Technology

Transparencies
- Bellringer

Student Edition on CD-ROM

One-Stop Planner CD-ROM
- Lesson Plan

Teach

Using the Figure — GENERAL

Cave Formations Use the photo of a cavern at the top of this page to lead a discussion of the sometimes bizarre formations seen in caves. Answer to caption question: Stalagmites are the upward-pointing cones on the floor of a cave. LS **Visual**

CHEMISTRY CONNECTION — ADVANCED

Rock and Soil After discussing the chemistry feature with students, invite interested students to research the connection between bedrock and the type of soil that forms in your area. Students may use soil pH tests to determine if the soil is acidic or basic. Then, they may research this trait in relation to the type of rock in and beneath the soil. Have students present an oral report, perhaps with a diagram, to the class. LS **Logical/Visual**

READING SKILL BUILDER — BASIC

Reading Hint Tell students an easy mnemonic they can use to remember the words for cave features. Tell them that the word stalagmite has a **g** and grows upward from the **g**round. The word *stalactite* has a **c** and **c**lings "tite" to the **c**eiling of a cave. LS **Verbal** English Language Learners

CHAPTER RESOURCES

Chapter Resource File

- **Inquiry Lab** Cave Formations and Ecology GENERAL

Figure 2 ▶ The formations in Carlsbad Caverns in New Mexico are made of calcite. ***Which formations in this photo are stalagmites?***

cavern a natural cavity that forms in rock as a result of the dissolution of minerals; also a large cave that commonly contains many smaller, connecting chambers

Caverns

Rocks that are rich in the mineral calcite, such as limestone, are especially vulnerable to chemical weathering. Although limestone is not porous, vertical and horizontal cracks commonly cut through limestone layers. As groundwater flows through these cracks, carbonic acid slowly dissolves the limestone and enlarges the cracks. Eventually, a cavern may form. A **cavern** is a large cave that may consist of many smaller connecting chambers. Carlsbad Caverns in New Mexico is a good example of a large limestone cavern, as shown in **Figure 2.**

Stalactites and Stalagmites

Although a cavern that lies above the water table does not fill with water, water still passes through the rock surrounding the cavern. When water containing dissolved calcite drips from the ceiling of a limestone cavern, some of the calcite is deposited on the ceiling. As this calcite builds up, it forms a suspended, cone-shaped deposit called a *stalactite* (stuh LAK TIET). When drops of water fall on the cavern floor, calcite builds up to form an upward-pointing cone called a *stalagmite* (stuh LAG MIET). Often, a stalactite and a stalagmite will grow until they meet and form a calcite deposit called a *column*.

Connection to CHEMISTRY

How Water Dissolves Limestone

Water that falls as precipitation contains dissolved carbon dioxide, CO_2. This dissolved CO_2 causes the water to be slightly acidic. The formula for the dissolution of CO_2 is shown below.

Carbon dioxide reacts with **water** to form **carbonic acid:** $H_2O + CO_2 = H_2CO_3$.

Water that reaches the ground seeps into the soil. As the water passes through the soil, more CO_2 dissolves and the water becomes more acidic. The slightly acidic water enters fractures in limestone and dissolves the rock as the water moves through. The formula for the dissolution of limestone is shown below.

Carbonic acid dissolves **limestone** to form **calcium bicarbonate:** $CaCO_3 + H_2CO_3 = Ca(HCO_3)_2$.

Fractures widen over time, and caves develop underground. The calcium bicarbonate that forms when the limestone dissolves is deposited as the water drips and flows through the caves. This deposited calcium bicarbonate forms structures such as stalagmites and stalactites.

Cultural Awareness — GENERAL

Prehistoric Cave Art Display pictures of prehistoric cave art, such as the cave art at Lascaux, France. Discuss with students how early humans painted the walls of caves with remarkable pictures of animals and other objects. Invite interested students to read about prehistoric cave art and to report to the class on how cave painting was done and what its significance may have been to prehistoric people. LS **Visual/Interpersonal**

BIOLOGY CONNECTION — ADVANCED

Cave Vision Tell students that many animals that live in caves are blind. Ask them to discuss why this is the case. Invite interested students to research and present an illustrated report on some of the creatures (blind or with other adaptations) that spend their lives in caves. LS **Verbal/Visual**

Sinkholes

A circular depression that forms at the surface when rock dissolves, when sediment is removed, or when caves or mines collapse is a **sinkhole.** Most sinkholes form by dissolution, in which the limestone or other rock dissolves where weak areas in the rock, such as fractures, previously existed. The dissolved material is carried away from the surface, and a small depression forms. *Subsidence sinkholes* form by a similar process except that as rock dissolves, overlying sediments settle into cracks in the rock and a depression forms.

Collapse sinkholes may form when sediment below the surface is removed and an empty space forms within the sediment layer. Eventually, the overlying sediments collapse into the empty space below. Collapse sinkholes may also form during dry periods, when the water table is low and caverns are not completely filled with water. Because water no longer supports the roof of the cavern, the roof may collapse. Collapse sinkholes may develop abruptly and cause extensive damage. A collapse sinkhole is shown in **Figure 3.**

Figure 3 ▶ When land overlying a cavern collapses to form a sinkhole, human-made structures, such as this highway, can be damaged.

sinkhole a circular depression that forms when rock dissolves, when overlying sediment fills an existing cavity, or when the roof of an underground cavern or mine collapses

Natural Bridges

When the roof of a cavern collapses in several places, a relatively straight line of sinkholes forms. The uncollapsed rock between each pair of sinkholes forms an arch of rock called a *natural bridge*, such as the one shown in **Figure 4.** When a natural bridge first forms, it is thick, but erosion causes the bridge to become thinner. Eventually, the natural bridge may collapse.

Reading Check **How are sinkholes related to natural bridges? (See the Appendix for answers to Reading Checks.)**

Figure 4 ▶ This natural bridge near San Marcos, Texas, formed when the roofs of two adjoining caverns collapsed.

Answer to Reading Check

A natural bridge may form when two sinkholes form close to each other. The bridge is the uncollapsed rock between the sinkholes.

SKILL BUILDER — ADVANCED

Writing Ask students to imagine they live on a street where a sinkhole "swallowed" several cars and damaged some homes. Have students write a letter to the editor of the local newspaper to explain how they feel about the situation and what, if anything, they think should be done to prevent future occurrences. **LS Intrapersonal**

Close

Reteaching — BASIC

Illustrating Cave Formation
Have students draw illustrations showing how water flows through rock to create the caves and cave formations, as well as the sinkholes and natural bridges, covered in the section. Student pictures should show direction of water flow and label the formation of features (e.g., stalactites and stalagmites) within caves. Students may create their pictures as they read through the section. **LS Visual**

Quiz — GENERAL

1. What is the difference between hard water and soft water? (Hard water has more minerals dissolved in it than soft water does.)
2. How does a subsidence sinkhole form? (Rock dissolves and overlying sediments settle into cracks in the rock. Eventually, a depression forms at the surface.)

INCLUSION Strategies

- *Learning Disabled*
- *Attention Deficit Disorder*

Many students learn better if they have a chance to move around. Give these students a chance to get out of their seats by having them fill in an empty chart you have drawn on the board. Use these column headings: Caverns, Stalactites and Stalagmites, Sinkholes, and Natural Bridges. Use these questions as row labels: Caused by rock dissolving? Caused by rock building up? Usually seen above ground? Takes years to form? (Caverns—yes, no, no, yes; Stalactites and Stalagmites—yes, yes, no, yes; Sinkholes: yes, no, yes, no; Natural Bridges: yes, no, yes, no) **LS Logical/Visual**

Close, *continued*

Alternative Assessment — Advanced

Guided Tour Have students pretend that they are guides who lead visitors through a cavern. Have them write a brochure, which they may illustrate, that describes the cavern's history and all of the features that visitors will see when they tour the cavern. LS **Visual/Verbal**

Answers to Section Review

1. Water that moves through soil combines with CO_2 to form carbonic acid. This acid breaks down and dissolves minerals in the rock.
2. Caverns form when groundwater flows through cracks in limestone. Over time, the limestone dissolves and large, open spaces form.
3. Stalactites form as calcite precipitates from water that drips down from the cavern ceiling. Stalagmites build up as calcite precipitates from water that drips onto the floor of a cavern.
4. Three common features of karst topography are many closely spaced sinkholes, many caverns, and disappearing and emerging streams.
5. A natural bridge may form when the roof of a cavern collapses in several places. Natural bridges may also form from uncollapsed rock between two sinkholes.
6. Caverns are empty spaces within rock that form as a result of chemical weathering. Sinkholes are circular depressions that form when rock dissolves or when the roofs of caverns collapse.
7. Answers may vary but may include that karst topography may form in a dry climate where the water table has fallen and caused sinkholes to form.
8. Answers may vary. Sample answer: Limestone layers are commonly aquifers. Anywhere the water table reaches the surface, you would expect to find a spring.
9. *Groundwater* erodes limestone to form *caverns* that have *stalactites* on the ceilings and *stalagmites* on the ground, and dissolves rock to form *sinkholes,* which may leave a *natural bridge.*

Figure 5 ▶ The Stone Forest in Yunnan, China, is a dramatic example of karst topography.

Karst Topography

Irregular topography caused by the chemical weathering of limestone or other soluble rock by groundwater is called **karst topography.** Common features of karst topography include many closely spaced sinkholes and caverns. In karst regions, streams often disappear into cracks in the rock and then emerge in caves or through other cracks many kilometers away. In the United States, there is karst topography in Kentucky, Tennessee, southern Indiana, northern Florida, and Puerto Rico.

Generally, karst topography forms in regions where the climate is humid and where limestone formations exist at or near the surface. The plentiful precipitation in these regions commonly becomes groundwater. The groundwater flows through the limestone and reacts chemically with the calcite in the limestone. As the groundwater dissolves the limestone, cracks in the rock enlarge to form cave systems. Features of karst topography can form in relatively dry regions, too. In these areas, sinkholes may form very close together and leave dramatic arches and spires, as shown in **Figure 5.** Karst topography in these regions may indicate that the climate is becoming drier.

karst topography a type of irregular topography that is characterized by caverns, sinkholes, and underground drainage and that forms on limestone or other soluble rock

Section 2 Review

1. **Describe** how water chemically weathers rock.
2. **Explain** how caverns form.
3. **Explain** the difference between stalactites and stalagmites.
4. **Identify** three common features of karst topography.
5. **Describe** two ways in which a natural bridge might form.
6. **Compare** sinkholes and caverns.

CRITICAL THINKING

7. **Making Inferences** If an area has a dry climate, how can the area have karst topography?
8. **Identifying Relationships** Why might you expect to find springs in regions that have karst topography?

CONCEPT MAPPING

9. Use the following terms to create a concept map: *groundwater, stalagmite, stalactite, natural bridge, cavern,* and *sinkhole.*

CHAPTER RESOURCES

Chapter Resource File
- Section Quiz GENERAL

Workbooks
- **Study Guide** (also in Spanish)

Chapter 16 Highlights

Sections

1 Water Beneath the Surface

Key Terms

groundwater, 397
aquifer, 397
porosity, 397
permeability, 398
water table, 399
artesian formation, 403

Key Concepts

- Porosity and permeability determine how water moves through rock or sediment.
- Aquifers have two main zones—the zone of aeration and the zone of saturation. The upper surface of the zone of saturation is called the *water table*.
- The depth of the water table depends on the topography of the land, the permeability of the rock, the amount of rainfall, and the rate at which groundwater is used by humans.
- Groundwater can be polluted by wastes, agricultural and lawn fertilizers, agricultural and lawn pesticides, and sea water.
- Wells and springs may be ordinary or artesian. Artesian formations are the source of artesian wells and springs.
- Hot springs and geysers form when hot rock beneath Earth's surface heats groundwater.

2 Groundwater and Chemical Weathering

Key Terms

cavern, 406
sinkhole, 407
karst topography, 408

Key Concepts

- Caverns form as a result of the chemical weathering of limestone.
- Stalagmites are calcite formations that form on the floor of a cavern. Stalactites are calcite formations that form on the roof of a cavern.
- Sinkholes form when rock dissolves, when sediment is removed, or when the roof of a cavern or mine collapses.
- Karst topography features closely spaced sinkholes, caverns, and streams that disappear into cracks in the rock and then emerge several kilometers away.

Chapter Highlights

Alternative Assessment — GENERAL

The Water Beneath Your Feet

Have students work in groups to research the groundwater resources and characteristics in your region or your state. Have one group of students identify and map aquifers in your area. A second group should research and map wells and springs. A third group may find out how to test local water for pH and hardness and conduct the tests. A fourth group should research and map or illustrate any caves or caverns in your region or state. Students may also research groundwater's effects on limestone in your area. If caves do occur, have students in this group find out how they formed and what physical characteristics they have. Students may put all of their research and materials together to give a class demonstration and "teach-in" on groundwater in your area. LS **Visual/Kinesthetic**
Co-op Learning

CHAPTER RESOURCES

Chapter Resource File

- **Concept Review** GENERAL
- **Critical Thinking** ADVANCED
- **Math Skills** GENERAL
- **Graphing Skills** GENERAL
- **Chapter Test A** GENERAL
- **Chapter Test B** ADVANCED

Workbooks

- **Study Guide** (also in Spanish)
- **Assessments** (Spanish)

Technology

Classroom Videos
- Brain Food Video Quiz

Chapter Review

Assignment Guide

Section	Questions
1	1–2, 4–7, 9–14, 18–19, 21–23, 25–29, 31–38
2	3, 8, 15–17, 20, 24, 30

Using Key Terms

1–8. Answers may vary but should show that students understand the definitions of and differences between key terms.

Understanding Key Concepts

9. b **10.** d
11. a **12.** a
13. a **14.** a
15. c **16.** b
17. a **18.** d

Short Answer

19. lakes, rivers, and swamps

20. As groundwater drips from the ceiling to the floor of a cavern, calcite is left behind to form stalactites and stalagmites. Icicles form in a similar process.

21. A mud pot forms when chemically weathered rock mixes with hot water to form a sticky liquid clay that bubbles at the surface.

22. In an aquifer, the zone in which the pore spaces are completely filled with water is the zone of saturation. The zone of aeration is the zone that contains dry soil and the capillary fringe and that lies between the ground surface and the zone of saturation.

23. Groundwater may reach the surface through wells or through springs.

Chapter 16 Review

Using Key Terms

Use each of the following terms in a separate sentence.

1. *groundwater*

2. *water table*

3. *karst topography*

For each pair of terms, explain how the meanings of the terms differ.

4. *geyser* and *hot spring*

5. *porosity* and *permeability*

6. *well* and *spring*

7. *ordinary well* and *artesian well*

8. *stalactite* and *stalagmite*

Understanding Key Concepts

9. Any body of rock or sediment in which water can flow and be stored is called a(n)
a. well.
b. aquifer.
c. sinkhole.
d. artesian formation.

10. The percentage of open space in a given volume of rock is the rock's
a. viscosity.
b. capillary fringe.
c. permeability.
d. porosity.

11. The ease with which water can pass through a rock or sediment is called
a. permeability.
b. carbonation.
c. porosity.
d. velocity.

12. The slope of a water table is called the
a. gradient.
b. porosity.
c. permeability.
d. aquifer.

13. A natural flow of groundwater that has reached the surface is a(n)
a. spring.
b. well.
c. aquifer.
d. travertine.

14. Pumping water from a well causes a local lowering of the water table known as a
a. cone of depression.
b. horizontal fissure.
c. hot spring.
d. sinkhole.

15. Calcite formations that hang from the ceiling of a cavern are called
a. stalagmites.
b. sinks.
c. stalactites.
d. aquifers.

16. Regions where the results of weathering by groundwater are clearly visible have
a. sink topography.
b. karst topography.
c. limestone topography.
d. artesian formations.

17. When the roofs of several caverns collapse, the uncollapsed rock between sinkholes can form
a. natural bridges.
b. stalactites.
c. limestone topography.
d. artesian formations.

18. A layer of permeable rock that is sandwiched between layers of impermeable rock is called
a. a natural bridge.
b. karst topography.
c. limestone topography.
d. an artesian formation.

Short Answer

19. In regions where the water table is at the surface of the land, what type of terrain would you expect to find?

20. Explain the process that forms stalactites and stalagmites. Name another process in nature that produces shapes similar to the shapes of stalactites.

21. How does a mud pot form?

22. Describe the zones of an aquifer.

23. What are two ways that groundwater reaches Earth's surface?

24. How are caverns and sinkholes related?

25. What causes a geyser to erupt?

26. Why does it take a long time to replenish a depleted aquifer?

Critical Thinking

27. Making Inferences In what type of location might pumping too much water from an aquifer lead to contamination of the groundwater supply? Explain how the water becomes contaminated.

28. Analyzing Relationships Describe an artesian formation, and explain how the water in an artesian well may have entered the ground many hundreds of kilometers away.

29. Analyzing Ideas Explain how a rock can be both porous and impermeable.

30. Identifying Relationships Do you think that an area that has karst topography would have many surface streams or few surface streams? Explain your answer.

Concept Mapping

31. Use the following terms to create a concept map: *porosity, sorting, permeability, ordinary well, artesian formation, highly permeable rock,* and *impermeable rock*.

Math Skills

32. Evaluating Data People in Oklahoma use 11 billion gallons of water every day. The renewable water supply in Oklahoma is 68.7 billion gallons per day. What percentage of the renewable water supply do Oklahomans use every day?

33. Making Conversions In an average aquifer, groundwater moves about 50 m per year. At this rate, how long would the groundwater take to flow 1 km?

Writing Skills

34. Writing Persuasively Write a persuasive essay about the importance of conserving groundwater.

35. Communicating Main Ideas Explain how overpumping at one well can affect groundwater availability in surrounding areas.

Interpreting Graphics

The graph below shows the average annual decline in water level for the Ogallala Aquifer over 30 years. Use the graph below to answer the questions that follow.

Average Annual Rate of Decline by Decade for the Ogallala Aquifer, 1969–1999

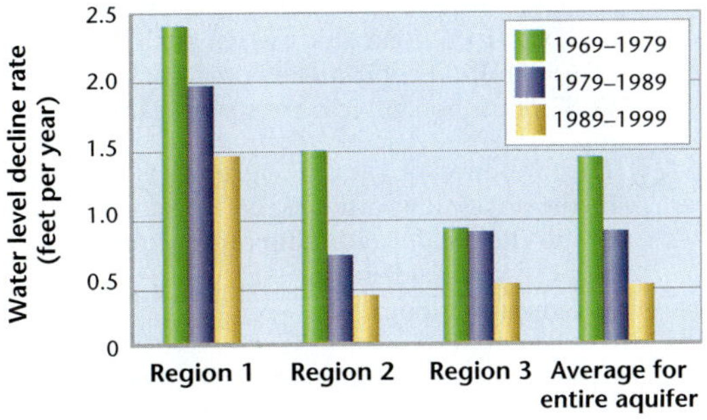

36. Which region had the highest rate of decline from 1969 to 1999?

37. Over which decade did the aquifer have the lowest rate of decline?

38. Which years would you expect to have a higher rate of decline: the years 1999 to 2009 or the years 1989 to 1999? Explain your answer.

Chapter Review

24. Both caverns and sinkholes result from the weathering of rock by groundwater or surface water.

25. A geyser erupts when magma or hot rock superheats water and causes steam to push the water to the surface.

26. Groundwater moves slowly through rock and is replenished very slowly by rain.

Critical Thinking

27. In areas near the ocean, pumping too much groundwater may cause an influx of saltwater from the sea.

28. The water in an artesian formation may have entered the aquifer far away and then traveled a long distance down the slope of the aquifer. The water in an artesian formation can emerge only when the overlying impermeable rock layer is cracked or drilled.

29. If the open spaces in a porous rock are not connected, the rock will not be very permeable.

30. Karst areas have few surface streams because there are so many cracks into which streams may disappear.

Concept Mapping

31. Answers may vary but should include all of the terms listed. Sample answers appear at the end of this Teacher Edition.

Math Skills

32. 11 billion gal ÷ 68.7 billion gal = 0.16 × 100 = 16%

33. 50 m = 0.050 km/y; 1 km ÷ 0.050 km/y = 20 y

Writing Skills

34. Answers may vary. Accept all reasonable answers.

35. Answers may vary. Accept all reasonable answers.

Interpreting Graphics

36. Region 1

37. 1989–1999

38. Answers may vary. Sample answer: I would expect 1989–1999 to have a greater rate of decline because the overall trend is decreasing.

Estimated Time

To give students practice under more realistic testing conditions, allow them 30 minutes to answer all of the questions in this practice test.

TEST DOCTOR

Question 1 Answer C is false. Gravity pulls water down through soil and rock layers, which causes the underground water level to rise and form *two* distinct zones of groundwater: the zone of saturation and the zone of aeration.

Question 3 Answer D is correct. Simple subtraction can be used to determine that groundwater was depleted at a rate of 82.28 million cubic meters in 2002.

Question 4 Answer G is correct. In a region where there is plentiful precipitation, groundwater flows through limestone and dissolves it. This chemical reaction results in the enlargement of cracks in the rock to form cave systems. Answers F and H therefore describe the process of karst topography formation in *humid* regions. Answer I is not applicable to karst topography, which is due to chemical weathering, not physical weathering.

Chapter 16 Standardized Test Prep

Understanding Concepts

Directions **(1–5): For** ***each*** **question, write on a separate sheet of paper the letter of the correct answer.**

1. Which of the following statements is false?
   A. Permeability affects flow through an aquifer.
   B. Groundwater can be stored in an aquifer.
   C. Aquifers are always a single rock layer.
   D. Well-sorted sediment holds the most water.

2. The amount of surface water that seeps into the pores between rock particles is influenced by which of the following factors?
   F. rock type, land slope, and climate
   G. rock type, land slope, and capillary fringe
   H. rock type, land slope, and sea level
   I. rock type, land slope, and recharging

3. Shanghai removed 96.03 million cubic meters of groundwater in 2002 but replaced only 13.75 million cubic meters. What was the rate of groundwater depletion in Shanghai that year?
   A. 109.78 million cubic meters per year
   B. 1,320.41 million cubic meters per year
   C. 6.98 million cubic meters per year
   D. 82.28 million cubic meters per year

4. The formation of karst topography is caused by
   F. the physical weathering of limestone.
   G. the chemical weathering of limestone.
   H. closely-spaced sinkholes.
   I. irregular topography.

5. What quality distinguishes an ordinary well from an artesian well?
   A. Water flows freely from an ordinary well.
   B. Water is pressurized in an ordinary well.
   C. Water must be pumped from an ordinary well.
   D. Water comes from rainfall in an ordinary well.

Directions **(6–7): For** ***each*** **question, write a short response.**

6. What is a watershed?

7. What is the term for a local lowering of the water table caused by the pumping water from a well?

Reading Skills

Directions **(8–10): Read the passage below. Then answer the questions.**

Land Subsidence

Land subsidence is the settling or sinking of earth in response to the movement of materials under its surface. The greatest contributor to land subsidence is aquifer depletion. As groundwater is removed, the surface above may sink. Rocks may settle and pores may close, which leaves less area for water to be stored. In areas where aquifers are replenished, the surface of Earth may subside and then return almost to its previous level. However, in areas where water is not pumped back into aquifers, subsidence is substantial and whole regions may sink. Human activites can contribute to land subsidence. These activities include the pumping of water, gas, and oil from underground reservoirs and the collapse of mine tunnels.

8. According to the passage, which of the following statements is not true?
   A. Land subsidence is the settling or sinking of earth.
   B. As groundwater is removed, the earth above may sink.
   C. The greatest contributor to land subsidence is aquifer depletion.
   D. Rocks settle and pores close, which leaves more area for water to be stored.

9. Which of the following statements can be inferred from the information in the passage?
   F. Subsidence sinkholes occur most often in rural areas.
   G. The majority of all subsidence sinkholes are formed through natural processes
   H. Subsidence sinkholes form both naturally and because of the activities of humans.
   I. Older sinkholes are easily recovered by refilling the area with water.

10. Subsidence due to groundwater depletion may occur slowly or very abruptly. Which type of subsidence presents a greater chance for recovery? Why?

Answers

Understanding Concepts

1. C
2. F
3. D
4. G
5. C
6. the area of land that is drained by a river
7. a cone of depression

Reading Skills

8. D
9. H
10. Abrupt subsidence may involve greater compression of pore space and leave less chance for recovery.

Interpreting Graphics

11. A
12. Answers may vary. See Test Doctor for a detailed scoring rubric.
13. Answers may vary. See Test Doctor for a detailed scoring rubric.

Interpreting Graphics

Directions (11–13): **For *each* question below, record the correct answer on a separate sheet of paper.**

This graphic shows an example of the water cycle. Use this graphic to answer question 11.

The Water Cycle

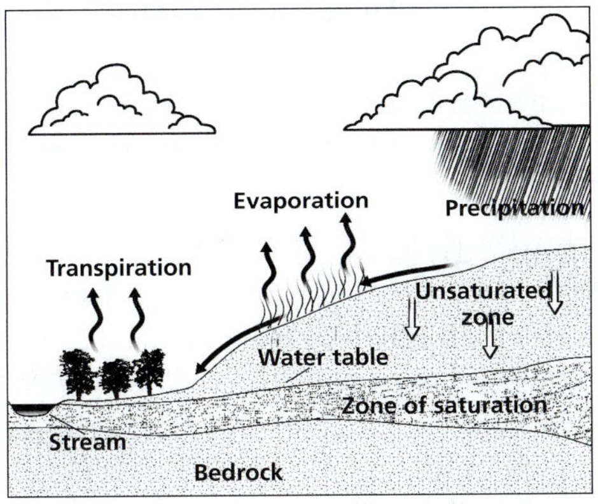

11 Which process occurs where the water table intersects the surface?
- **A.** stream formation
- **B.** runoff
- **C.** groundwater movement
- **D.** saturation

The graph below shows indoor water use for a typical family in the United States. Use this graph to answer questions 12 and 13.

Water Use for a Family of Four

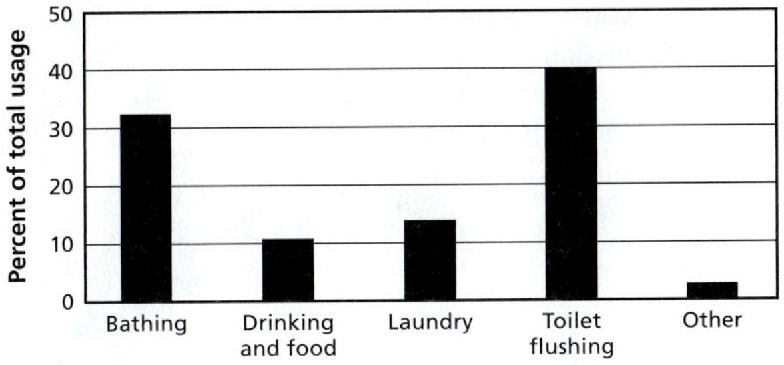

12 According to the graph, what is the largest use of indoor water for a family in the United States? Name some ways that people can reduce the amount of water consumed by this task.

13 What total percentage of household indoor water is consumed by the two largest uses of indoor water? Round your answer to the nearest 10. How could this knowledge be used to help people reduce water usage?

Test TIP

Remember that if you can eliminate two of four answer choices, your chances of choosing the correct answer will double.

Standardized Test Prep

TEST DOCTOR

Question 12 Students should bring to this test item the ability to propose creative and viable solutions for implementation.

Full-credit answers should include the following points:
- toilet flushing is the largest use of indoor water
- an understanding that toilets are a necessity, but may be made more water-efficient.
- suggested solutions may include using low-flow toilets, not flushing items such as tissue unnecessarily, and keeping toilets in good working order to help reduce the need for water

Question 13 Full-credit answers should include the following points:
- students should estimate the percentage of water used for the two largest uses of indoor water, toilet flushing (40%) and bathing (32%). Rounded to the nearest 10, these uses account for 70% of water usage
- proposed suggestions for conservation that target the largest uses of water first
- students may connect ecological benefits of water conservation to the economic benefits of water conservation
- students may suggest that people may be more easily convinced to reduce water usage if they are shown how much that usage costs them

Test Prep Correlations — *National Science Education Standards*

ES 3c: items 4, 10, 11
LS 4e: items 12, 13
SAI 2d: items 3, 13
SPSP 3a: items 7, 12, 13
SPSP 3b: items 9, 12, 13
SPSP 4a: items 3, 7, 9
SPSP 4b: items 3, 9, 12, 13
SPSP 5b: items 3, 9, 12, 13
SPSP 5d: items 12, 13
ST 1e: items 12, 13
ST 2c: items 12, 13
UCP 1: items 6, 11

CHAPTER RESOURCES

State Resources

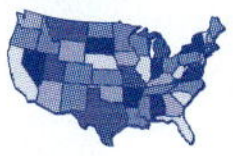

For specific resources for your state, visit go.hrw.com and type in the keyword HSHSTR.

Skills Practice Lab

Porosity

Teacher's Notes

Time Required

one 45-minute class period

Lab Ratings

EASY → HARD

Teacher Preparation

Student Setup

Concept Level

Cleanup

Skills Acquired

- Constructing Models
- Measuring
- Observing
- Interpreting Data

The Scientific Method

In this lab, students will

- Construct Models
- Make Observations
- Interpret Data
- Analyze Data
- Draw Conclusions

Materials

The materials listed on the page are enough for groups of two to four students. This activity could also work with different sized marbles or with well-sorted pebbles.

Tips and Tricks

You may wish to have students take turns, with one actively doing one step of the lab while the other records the data. Then, have the students exchange roles.

Chapter 16

Skills Practice Lab

Objectives

- USING SCIENTIFIC METHODS **Measure** the porosity of a given volume of beads for each of three samples: large beads, small beads, and a mix of large and small beads.
- **Describe** how particle size and sorting of a material affect porosity.

Materials

beads, plastic, 4 mm (400)

beads, plastic, 8 mm (200)

beaker, 100 mL

graduated cylinder, 100 mL

Safety

Porosity

Whether soil is composed of coarse pieces of rock or very fine particles, some pore space remains between the pieces of solid material. Porosity is calculated by dividing the volume of the pore space by the total volume of the soil sample. Thus, if 50 cm^3 of soil contains 5.0 cm^3 of pore space, the porosity of the soil sample is

$$5.0\ cm^3/50\ cm^3 = 0.10 \times 100 = 10\%.$$

The result is generally written as a percentage. In this lab, you will measure and compare the porosity of three samples that represent rock particles.

PROCEDURE

1. Fill a beaker to the top with water. Pour the water into a graduated cylinder and record the volume of water.
2. Dry the beaker, and fill it to the top with large (8 mm) plastic beads. Gently tap the beaker to settle and compact the beads. Add more beads to fill the beaker until the beads are level with the top. Record the total volume of the beads, which includes the pore space volume.
3. Fill the graduated cylinder with water to the top mark, and record the volume of water. Carefully pour the water from the cylinder into the beaker filled with the large beads until the water level just reaches the top of the beads.

Step 3

CHAPTER RESOURCES

Chapter Resource File

- Datasheet for Chapter Lab GENERAL
- Lab Notes and Answers

Workbooks

Long-Term Projects

- Precipitation and the Water Table ADVANCED

4. To determine the amount of water that you added to the beaker, subtract the volume of water in the graduated cylinder from the volume that you recorded in step 3. This difference is the volume of the pore space between the beads. Record the volume of the pore space.
5. Calculate the porosity of the beads. Record the porosity as a decimal and as a percentage.
6. Repeat steps 2–5 using small (4 mm) beads.
7. Drain and dry both sets of beads. Mix together equal volumes of the small and large beads. Using the mixed-size beads, repeat steps 2–5.

Step 4

ANALYSIS AND CONCLUSION

1. **Analyzing Methods** Do the 8 mm beads in step 2 represent well-sorted large rock particles, well-sorted small rock particles, or unsorted rock particles?
2. **Analyzing Methods** Do the 4 mm beads in step 6 represent well-sorted large rock particles, well-sorted small rock particles, or unsorted rock particles?
3. **Analyzing Methods** Do the mixed beads in step 7 represent well-sorted or unsorted rock particles?
4. **Making Graphs** Compare the porosity of the large beads with the porosity of the small beads. Make a graph that shows bead size on the *x*-axis and porosity on the *y*-axis.
5. **Drawing Conclusions** In well-sorted sediment, does porosity depend on particle size? Explain your answer.
6. **Determining Cause and Effect** How did mixing the bead sizes affect the porosity? Explain the effect.

Extension

1. **Designing Experiments** How would mixing coarse gravel with fine sand affect the porosity of the gravel? Conduct an experiment to find out if your answer is correct.

Skills Practice Lab

Answers to Analysis and Conclusion

1. The 8 mm beads represent well-sorted large rock particles.
2. The 4 mm beads represent well-sorted small rock particles.
3. The mixed beads represent unsorted rock particles.
4. Graphs may vary slightly but should show that 4 mm and 8 mm beads have basically the same porosity.
5. If you have well-sorted sediment, both coarse- and fine-grained sediment will have the same porosity.
6. Mixing the two bead sizes together lowers the porosity because the smaller beads fill in some of the spaces between the bigger beads.

Answers to Extension

1. Sand would reduce the porosity of the gravel because some of the sand grains fill in the spaces between the gravel.

CLASSROOM TESTED & APPROVED

Mrs. Teresa Tucker
Northwest High School
Jackson, MI

Water Level in the Southern Ogallala

Internet Activity — GENERAL

Agriculture and the Aquifer
Most of the Ogallala aquifer (or High Plains aquifer) underlies agricultural land. Divide the class into two groups to do Internet or library research about this aquifer. One group should research the effects of center-pivot irrigation on withdrawal rates from the aquifer and on conservation of the Ogallala's water resources. The other group should research the amounts of pesticides and other agricultural chemicals applied to land overlying the aquifer and their effects on groundwater and drinking water in the region. Have students prepare a presentation, with illustrations, for the class when their research is complete. A worksheet designed to direct student research on this topic can be found in the **Chapter Resource File** booklet or by visiting **go.hrw.com** and entering the keyword **HQ6GWEX**. **LS Verbal**

Answers to Map Skills Activity

1. About six areas had a decline in water level of 50 to 100 feet up to 1980.
2. More than twelve areas had a decline of more than 60 feet between 1980 and 1999.
3. It is larger.
4. The decline has increased since 1980.
5. Answers may vary. The trend may be related to increases in population in the region, to increased water use on agricultural land, or to changes in rainfall.

MAPS in Action

Water Level in the Southern Ogallala

KEY

WATER–LEVEL CHANGE, IN FEET, 1980 TO 1999

Declines
- More than 60
- 40 to 60
- 20 to 40
- 10 to 20
- 5 to 10

No significant change
- −5 to 5

Rises
- 5 to 10
- 10 to 20
- 20 to 40
- More than 40

- Area of little or no saturated thickness
- Area with 50 to 100 feet of water–level decline, predevelopment to 1980 (Luckey and others, 1981)
- Area with 100 to 175 feet of water–level decline, predevelopment to 1980 (Luckey and others, 1981)

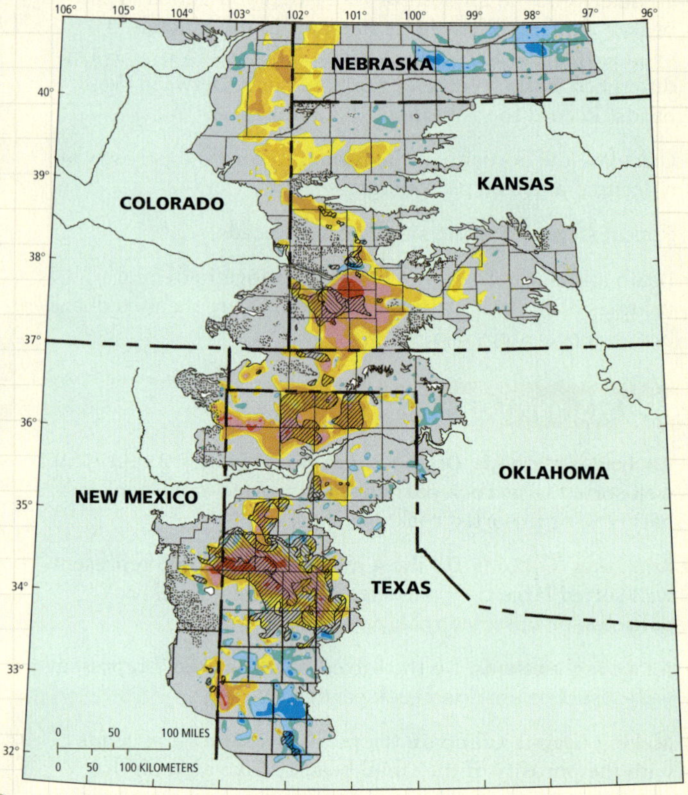

Map Skills Activity

This map shows water-level change in regions of the Ogallala Aquifer, which supplies much of the drinking water in the midwestern United States. Use the map to answer the questions below.

1. **Using a Key** From predevelopment to 1980, how many areas had a decline in water level of 50 to 100 ft?
2. **Using a Key** From 1980 to 1999, how many areas had a decline in water level of more than 60 ft?
3. **Making Comparisons** Is the area in which the water level declined 50 to 100 ft before 1980 larger than or smaller than the area in which the water level declined more than 60 ft from 1980 to 1999?
4. **Identifying Trends** Has the overall amount of decline in the water-table level increased or decreased since 1980?
5. **Analyzing Relationships** What may have caused the trend that you identified in question 4? What may have changed between 1980 and 1999 that could account for the trend?

CHAPTER RESOURCES

Chapter Resource File

Internet Activity
- Agriculture and the Aquifer GENERAL

Technology

Transparencies
- 82 Water Level in the Southern Ogallala (with worksheet)

IMPACT on Society

Disappearing Land

On May 8, 1981, in Winter Park, Florida, a resident saw a sycamore tree disappear. A sinkhole had swallowed the tree, and that disappearance was only the beginning.

When it first appeared, at about 8:00 P.M., the cone-shaped sinkhole was 13 m wide and 7 m deep. Overnight, the hole expanded to a diameter of 27 m. Before noon the next day, the hole expanded to a size of 300 m in diameter—the size of a football field—and 37 m deep. In the process, it consumed about 160,000 m^3 of ground.

The Winter Park Sinkhole

This destructive sinkhole resulted from a combination of natural processes and human activities. Collapse sinkholes, such as the one in Winter Park, occur when the sediment overlying a cavern collapses into a cavity formed by groundwater. However, the rapid expansion of this sinkhole may have been the result of the collapse of the cavern roof. Such a collapse may have resulted from the removal of too much water from an aquifer that was already depleted after a two-year drought.

The Money Pit

Eventually, the Winter Park sinkhole swallowed up several houses, part of a four-lane highway, a swimming pool, a parking lot, five cars, and a truck. The cost of the damage reached about $2 million. When the sinkhole finally stabilized, the city of Winter Park turned the hole into a municipal lake.

Extension

1. **Applying Ideas** What might Winter Park residents do to prevent the formation of more sinkholes?

▲ Florida's landscape is dotted with small sinkhole lakes.

Answer to Extension

1. Answers may vary. Sample answer: Winter Park residents might assess the condition of groundwater resources on a more regular and rigorous basis. Under drought conditions, public authorities could limit water withdrawals to essential purposes only. A special tax could be added to water bills during those times. On a longer-term basis, residents should be concerned about maintaining a balance between water resources and water usage.

Impact on Society

Disappearing Land

Debate — GENERAL

Limiting Development? Tell students that throughout the United States, the opportunity to develop land almost anywhere is seen by many people as a constitutional right. However, there are many places where an ever-increasing human population threatens to overwhelm limited water resources. Have students form two debate teams. Some members of each team should research data on the pros and cons of limiting development rights to reduce the impact of development on water resources. Students should consider legal, economic, and social issues. Other students of each team should prepare debate speeches. Set aside time for students to have a formal debate on the question: "Should there be limits placed on development where water resources are limited or negatively impacted by the development?" After the debate, have the whole class discuss what they learned from the debate. LS Verbal Co-op Learning

Internet Activity — GENERAL

Sinkholes Have students research occurrences of sinkholes that have caused property damage, as in Florida in recent years. They may search area newspapers to find out what caused the sinkholes and what is being done to prevent others. A worksheet designed to direct student research on this topic can be found in the **Chapter Resource File** booklet or by visiting go.hrw.com and entering the keyword HQ6GWEX. LS Verbal

CHAPTER RESOURCES

Chapter Resource File

- **Internet Activity**
  - Sinkholes GENERAL

Chapter 17 Glaciers
Planning Guide

Compression Guide

To shorten instruction because of time limitations, omit the Chapter Lab.

OBJECTIVES	LABS, DEMONSTRATIONS, AND ACTIVITIES	TECHNOLOGY RESOURCES
PACING • 45 min pp. 418–422 **Chapter Opener**		OSP **Parent Letter** ■ CD **Student Edition on CD-ROM** CD **Chapter Summaries Audio CD** ■ VID **Brain Food Video Quiz**
Section 1 Glaciers: Moving Ice • Describe how glaciers form. • Compare two main kinds of glaciers. • Explain two processes by which glaciers move. • Describe three features of glaciers.	TE **Activity** Modeling Glacial Ice Formation, p. 419 GENERAL SE **Quick Lab** Slipping Ice, p. 421 GENERAL CRF **Datasheet for Quick Lab*** GENERAL SE **Maps in Action** Gulkana Glacier, p. 442 GENERAL	OSP **Lesson Plans** (also in print) TR **Bellringer*** TR 83 Internal Plastic Flow* TR 88 Gulkana Glacier* TE **Internet Activity** Glacial Sounds, p. 420 GENERAL CRF **Internet Activity** Glacial Sounds* GENERAL TE **Internet Activity** Photographic Evidence, p. 442 GENERAL CRF **Internet Activity** Photographic Evidence* GENERAL
PACING • 90 min pp. 423–430 **Section 2 Glacial Erosion and Deposition** • Describe the landscape features that are produced by glacial erosion. • Name and describe five features formed by glacial deposition. • Explain how glacial lakes form.	TE **Demonstration** Glacial Erosion and Deposition Simulation, p. 423 GENERAL SE **Quick Lab** Glacial Erosion, p. 425 ◆ GENERAL CRF **Datasheet for Quick Lab*** GENERAL TE **Activity** Glacial Erosion Theories, p. 425 ADVANCED TE **Discussion** Comprehension Check, p. 426 GENERAL TE **Activity** Effects of Glacial Erosion and Deposition, p. 428 BASIC TE **Group Activity** Types of Glacial Lakes, p. 428 GENERAL CRF **Inquiry Lab** Glacial Deposition* GENERAL CRF **Making Models Lab** Melted Glacier Formations* GENERAL	OSP **Lesson Plans** (also in print) TR **Bellringer*** TR 84 Landforms Created by Glacial Erosion* TR 85 Features of Glacial Deposition* TE **Internet Activity** More Glacial Landforms and Features, p. 428 GENERAL CRF **Internet Activity** More Glacial Landforms and Features* GENERAL CD **Interactive Tutor** Physical and Chemical Weathering
PACING • 90 min pp. 431–434 **Section 3 Ice Ages** • Describe glacial and interglacial periods within an ice age. • Summarize the theory that best accounts for the ice ages.	TE **Discussion** Ice Age Mammals, p. 431 GENERAL TE **Debate** Global Warming, p. 432 ADVANCED SE **Making Models Lab** Glaciers and Sea Level Change, pp. 440–441 GENERAL CRF **Datasheet for Chapter Lab*** GENERAL	OSP **Lesson Plans** (also in print) TR **Bellringer*** TR 86 Ice Ages of Earth's History* TR 87 The Milankovitch Theory* TE **Internet Activity** Vanishing Ice Caps, p. 432 GENERAL CRF **Internet Activity** Vanishing Ice Caps* GENERAL CD **Interactive Tutor** Glaciers and Glaciation CD **Interactive Tutor** Past Climate

PACING • 90 min

CHAPTER REVIEW, ASSESSMENT, AND STANDARDIZED TEST PREPARATION

- SE **Chapter Highlights,** p. 435
- SE **Chapter Review,** pp. 436–437
- SE **Standardized Test Prep,** pp. 438–439
- CRF **Concept Review*** ■ GENERAL
- CRF **Critical Thinking*** ADVANCED
- CRF **Math Skills*** GENERAL
- CRF **Graphing Skills*** GENERAL
- CRF **Chapter Test A*** ■ GENERAL
- CRF **Chapter Test B*** ADVANCED
- OSP **Lesson Plans** (also in print)
- OSP **Test Generator**
- OSP **Test Item Listing***

Online and Technology Resources

go.hrw.com Holt Online Learning

Visit go.hrw.com for access to Holt Online Learning, or enter the keyword **HQ6 Home** for a variety of free online resources.

One-Stop Planner® CD-ROM

This CD-ROM package includes

- Lab Materials QuickList Software
- Holt Calendar Planner
- Customizable Lesson Plans
- Printable Worksheets
- ExamView® Test Generator
- Interactive Teacher Edition
- Holt PuzzlePro®
- Holt PowerPoint® Resources

KEY

SE	Student Edition	OSP	One-Stop Planner	VID	Classroom Video/DVD
TE	Teacher Edition	TR	Transparencies and Transparency Worksheets	*	Also on One-Stop Planner
CRF	Chapter Resource File	CD	CD or CD-ROM	◆	Requires advance prep
LTP	Long-Term Projects			■	Also available in Spanish

SKILLS DEVELOPMENT RESOURCES	REVIEW AND ASSESSMENT	CORRELATIONS
SE **Pre-Reading Activity**, p. 418 GENERAL TE **Using the Figure** Blue and White Ice, p. 418 ADVANCED		National Science Education Standards
CRF **Directed Reading*** BASIC	SE **Reading Check**, p. 420 GENERAL SE **Section Review**, p. 422 GENERAL TE **Reteaching**, p. 421 BASIC TE **Quiz**, p. 421 GENERAL TE **Alternative Assessment**, p. 421 GENERAL CRF **Section Quiz*** ■ GENERAL	UCP 4, SPSP 5c
CRF **Directed Reading*** BASIC TE **Skill Builder** Vocabulary, p. 424 GENERAL TE **Using the Figure** Comprehension Check, p. 424 GENERAL SE **Graphic Organizer** Spider Map, p. 427 GENERAL TE **Using the Figure** Describing Glacial Landforms, p. 427 GENERAL TE **Inclusion Strategies**, p. 427 TE **Reading Skill Builder** Reading Organizer, p. 429 BASIC	SE **Reading Checks**, pp. 424, 427, 428 GENERAL TE **Homework**, p. 424 GENERAL TE **Homework**, p. 425 GENERAL SE **Section Review**, p. 430 GENERAL TE **Reteaching**, p. 429 BASIC TE **Quiz**, p. 429 GENERAL TE **Alternative Assessment**, p. 429 ADVANCED CRF **Section Quiz*** ■ GENERAL	UCP 4, SPSP 5c
CRF **Directed Reading*** BASIC TE **Inclusion Strategies**, p. 432 SE **Math Practice**, p. 433 GENERAL	SE **Reading Check**, p. 432 GENERAL SE **Section Review**, p. 434 GENERAL TE **Reteaching**, p. 433 BASIC TE **Quiz**, p. 433 GENERAL TE **Alternative Assessment**, p. 433 GENERAL CRF **Section Quiz*** ■ GENERAL	UCP 4, SPSP 5c, ES 3c

Holt Lab Generator CD-ROM

Search for labs by topic, standard, difficulty level, and/or time frame. Edit those labs to fit your needs, or create your own labs. Use the Lab Materials QuickList to order just what you need.

Holt Earth Science Interactive Tutor CD-ROM

This CD-ROM consists of interactive activities that give students a fun way to extend their knowledge of Earth science concepts.

Chapter Summaries Audio CDs

These CDs include audio summaries of the key concepts presented in each chapter. (Audio summaries are also available in Spanish.)

SCILINKS. NSTA

www.scilinks.org

Maintained by the **National Science Teachers Association.** See Chapter Enrichment pages that follow for a complete list of topics.

See Chapter Enrichment pages for Video Resources.

Chapter 17

Chapter Enrichment

This Chapter Enrichment provides relevant and interesting information to expand and enhance your classroom instruction of the chapter material.

Section 1 Glaciers: Moving Ice

Surging Glaciers

Despite the massive size of glaciers, most glacial movement is as slow as a snail. The average rate of speed of a glacier is usually less than 20 cm per day, with valley glaciers moving only a few centimeters per day. The exception is *surging* glaciers. During a surge, a glacier's velocity can reach 100 times its usual rate of speed. Surging glaciers may have a regular, periodic surge or may flow at a permanently surging velocity. Short-term advances can begin and end suddenly and can last for a few months or several years.

Mass Budget and Balance

Glaciers accumulate mass through several processes, including through precipitation, avalanches, and the cycles of melting and freezing that create firn ice. *Ablation* is the way glaciers lose mass. Ablation processes include melting, sublimation of ice, and calving (the breaking off of a part of the glacier). The difference between a glacier's accumulation and ablation during a single year is its *mass budget*. A glacier's *mass balance* is the difference between the total accumulation and ablation over a period of years.

Most glaciers have a high-elevation area, or *accumulation zone,* where their mass balance is positive, or where accumulation is greater than ablation. They also have a low-elevation area, or *ablation zone,* where their mass balance is negative, or where ablation is greater than accumulation. The dividing line between these two areas is the *equilibrium line altitude* (ELA), the point where the amount of accumulation and ablation are equal. The ELA is an indicator of whether a glacier is advancing or retreating. If approximately 70% of a glacier is above the ELA, the glacier is in a steady state. When glaciers have a positive mass balance, they advance. When their mass balance is negative, they retreat. During the last century, most glaciers have retreated as their equilibrium line altitude has risen due to increasing temperatures, less accumulation, or a combination of these factors.

Section 2 Glacial Erosion and Deposition

Glacial Erosion Model

Scientists have compared an important component of glacial erosion to the process that causes a very cold can of carbonated liquid to form ice crystals when it is opened. Glaciohydraulic supercooling allows glacial meltwater, although at subfreezing temperature, to remain liquid because it is under pressure from the overlying ice. If the slope of the glacier is not too steep, friction produces enough heat not only to keep the meltwater liquid but also to widen the flow path of the meltwater, which is carrying sediment that erodes the flow bed. However, when the slope becomes steep enough, the pressure drops, fine particles of ice form, the meltwater flow slows, and sediment no longer is carried away. This halts the erosion process.

Outburst Floods

When the huge amount of water trapped behind or underneath a glacier is suddenly released, a *jökulhlaup* (YO kool LOWP), or outburst flood, occurs. In 1996, volcanic activity beneath the Vatnajökull (VAHT nah YUH kool) ice cap in Iceland triggered one of the largest floods on record. The consequences were dramatic. The landscape changed dramatically—a plume of sediment formed in the ocean, and a 6-km-long canyon averaging 100 meters deep formed. A jökulhlaup ended the surge of the Bering glacier in 1994, causing large iceberg calving. The Great Missoula floods are thought to have been jökulhlaups.

▼ **Kettle lakes form as a result of glacial deposition.**

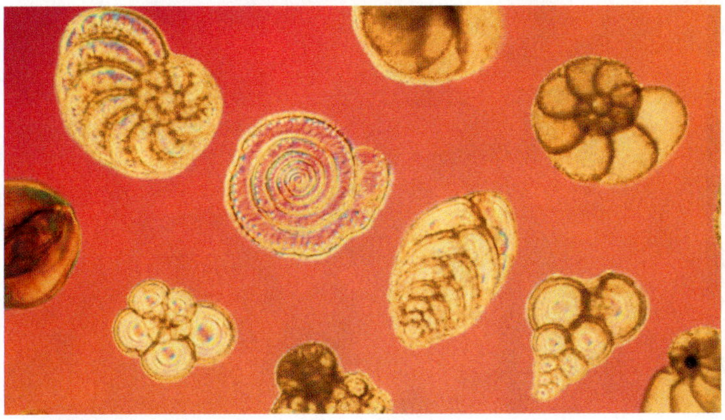

▲ Fossils of past organisms provide clues about Earth's climate history.

Section 3 Ice Ages

The Nature of Ice Sheets

The Antarctic Ice Sheet comprises two distinctively different parts—the East Antarctic Ice Sheet and the West Antarctic Ice Sheet. The East Antarctic Ice Sheet is land-based. The bottom of a land-based ice sheet lies mostly above sea level and has little or no meltwater or moving sediment base to facilitate basal slip. It is slow moving. These characteristics suggest that this type of ice sheet will respond slowly and steadily to environmental changes. The West Antarctic Ice Sheet is a marine-based ice sheet. Most of its bottom lies significantly below sea level. A marine-based ice sheet flows rapidly.

Glacial Clues to the Past

As Earth's glaciers retreat, fossils and artifacts preserved in the ice are becoming accessible. Finds in the Yukon Territory of Canada include preserved birds, small mammals, and caribou droppings; 90-year-old horseshoes; and 7,000-year-old hunting darts. In Siberia, the preserved body of a woolly mammoth and a 4,000-year-old basket were uncovered. Fossils of two previously unknown dinosaur species were found in Antarctica. Through such finds, scientists learn new information about ancient and more recent climate changes, plant and animal life, and human behaviors.

Video Resources

Brain Food Video Quizzes

Brain Food Video Quizzes These videos contain game-show style quizzes that assess students' progress and motivate students to study the chapter material.

HRW Earth Science Video

This video introduces Earth science topics and includes a geology field trip. The video segment listed below complements this chapter.

Segment 11: Glaciers and Erosion
Glaciers scour and break apart solid Bedrock and transport anormous amounts of sand, silt, and rock. In this segment, two types of glaciers—continental glaciers and valley glaciers—are described. In addition, the segment discusses the great Pleistocene ice sheets that covered northern Europe, Canada, and parts of the United States and discusses some of their effects on terrain. (1.5 min)

NOVA Videos

To order **NOVA** videos related to this chapter, visit **go.hrw.com** and enter the keyword **HQ6GLAV.**

SciLinks

Developed and maintained by the National Science Teachers Association

SciLinks is maintained by the National Science Teachers Association to provide you and your students with interesting, up-to-date links that will enrich your classroom presentation of the chapter.

Visit www.scilinks.org and enter the SciLinks code for more information about the topic listed.

Topic: Glaciers
SciLinks code: HQ60675

Topic: Glaciers and Landforms
SciLinks code: HQ60676

Topic: Ice Ages
SciLinks code: HQ60781

Chapter 17

Chapter Overview

This chapter discusses the nature of glaciers and their effect on Earth's surface. The chapter explains how glaciers act as indicators of climate change and concludes with an explanation of Earth's glacial cycles.

Using the Figure —ADVANCED

Blue and White Ice Over time, compression forces the air from between the ice grains in glaciers. This loss of air makes the ice very dense. When light travels into the ice, the ice grains scatter the blue light and absorb the red light. As a result, glaciers appear blue. Unlike glaciers, the ice in icebergs contains many different-sized air bubbles that are very close together in the ice. This causes all of the colors of light to scatter, making icebergs appear white. Ask students why the ice on a pond or skating rink does not appear blue. (The ice is not thick and heavy enough to produce the density of grains necessary to scatter the blue light.) **LS Visual**

PRE-READING ACTIVITY

FOLDNOTES

Encourage students to use the FoldNotes as a study guide to quiz themselves for a test on the chapter material. Students may want to create Double Door FoldNotes for different topics within the chapter.

Chapter 17 Glaciers

Sections

1 Glaciers: Moving Ice

2 Glacial Erosion and Deposition

3 Ice Ages

What You'll Learn

- How glaciers form and move
- What landforms glaciers create
- What factors drive glacial cycles

Why It's Relevant

Earth's surface was reshaped by glaciers during the last glacial period. Glaciers provide information about past and present climates and are key indicators of current climatic change.

PRE-READING ACTIVITY

FOLDNOTES

Double Door Before you read the chapter, create the FoldNote entitled "Double Door" described in the Skills Handbook section of the Appendix. Write "Alpine glaciers" on one flap of the double door and "Continental glaciers" on the other flap. As you read the chapter, write the characteristics of each type of glacier under the appropriate flap.

▶ The brilliant blue color of Alaska's Mendenhall Glacier is characteristic of glacial ice. Ice crystals in the glacier scatter more blue light than any other color, which makes the ice look blue.

Chapter Correlations *National Science Education Standards*

UCP 4 Evolution is a series of changes, some gradual and some sporadic, that accounts for the present form and function of objects, organisms, and natural and designed systems.... Although evolution is most commonly associated with the biological theory explaining the process of descent with modification of organisms from common ancestors, evolution also describes changes in the universe. **(Sections 1, 2, and 3)**

SPSP 5c Some hazards, such as earthquakes, volcanic eruptions, and severe weather, are rapid and spectacular. But there are slow and progressive changes that also result in problems for individuals and societies. For example, change in stream channel position, erosion of bridge foundations, sedimentation in lakes and harbors, coastal erosions, and continuing erosion and wasting of soil and landscapes can all negatively affect society. **(Sections 1, 2, and 3)**

ES 3c Interactions among the solid earth, the oceans, the atmosphere, and organisms have resulted in the ongoing evolution of the earth system. We can observe some changes such as earthquakes and volcanic eruptions on a human time scale, but many processes such as mountain building and plate movements take place over hundreds of millions of years. **(Section 3)**

ENVIRONMENTAL CONNECTION

Section 1 Glaciers: Moving Ice

A single snowflake is lighter than a feather. However, if you squeeze a handful of snow, you make a firm snowball. In a process similar to making a snowball, natural forces compact snow to make a large mass of moving ice called a **glacier.**

Formation of Glaciers

At high elevations and in polar regions, snow may remain on the ground all year and form an almost motionless mass of permanent snow and ice called a *snowfield*. Snowfields form as ice and snow accumulate above the snowline. The *snowline* is the elevation above which ice and snow remain throughout the year, as shown in **Figure 1.**

Average temperatures at high elevations and in polar regions are always near or below the freezing point of water. So, snow that falls there accumulates year after year. Cycles of partial melting and refreezing change the snow into grainy ice called *firn*.

In deep layers of snow and firn, the pressure of the overlying layers flattens the ice grains and squeezes the air from between the grains. The continued buildup of snow and firn forms a glacier that moves downslope or outward under its own weight.

The size of a glacier depends on the amount of snowfall received and the amount of ice lost. When new snow is added faster than ice and snow melt, the glacier gets bigger. When the ice melts faster than snow is added, the glacier gets smaller. Small differences in average yearly temperatures and snowfall may upset the balance between snowfall and ice loss. Thus, changes in the size of a glacier may indicate climatic change.

OBJECTIVES

- **Describe** how glaciers form.
- **Compare** two main kinds of glaciers.
- **Explain** two processes by which glaciers move.
- **Describe** three features of glaciers.

KEY TERMS

glacier
alpine glacier
continental glacier
basal slip
internal plastic flow
crevasse

glacier a large mass of moving ice

Figure 1 ▶ The snowline on the Grand Teton Mountains at Grand Teton National Park, Wyoming, is more than 3000 m above sea level.

Section 1

Focus

Overview

This section explains the nature of glaciers, the types of glaciers, and the features of glaciers. It also explains, how glaciers form and how they move.

Bellringer

Ask students what they know about ice and glaciers. (Answers may vary. Students may mention that glaciers consist of heavy, thick areas of ice and snow, are found at the poles, and move.)

Motivate

Activity — GENERAL

Modeling Glacial Ice Formation

Give each student a square-shaped piece of white paper and scissors. Tell them to make a paper snowflake by following this process: "Fold the paper in half diagonally and in half again. Next, fold the paper in thirds. Then, cut several small areas from each of the three edges and unfold." Tell students that a snowflake contains 90–95% air; firn contains 20–30% air, and glacial ice contains less than 10% air. Have students model glacial ice formation by crushing the paper snowflake in their hands first into golf ball size to represent firn and then into as small and compact a ball as possible to represent glacial ice. **LS Visual/Kinesthetic**

CHAPTER RESOURCES

Chapter Resource File

- Directed Reading BASIC

Technology

Transparencies
- Bellringer

Student Edition on CD-ROM

One-Stop Planner CD-ROM
- Lesson Plan

Teach

ENVIRONMENTAL CONNECTION

Harvesting Glacial Ice Alaska has more than 100,000 glaciers, covering 75,000 km². Seventy-five percent of its water is stored in these glaciers. Alaska issues permits to companies, allowing them to take more than 18 metric tons of glacier ice. Without a permit, individuals can take up to 20 cubic meters of ice per year. Have students research glacier harvesting and write an essay that discusses its potential positive and negative environmental impacts.

Internet Activity — GENERAL

Glacial Sounds The Web site of public radio station WBUR has audio files of glacial sounds. Ask students what sounds they think glaciers make and what might cause them. Then, have students listen to the sounds of glacier ice popping; glaciers calving, or losing material; and icebergs cracking. Discuss how the actual sounds compared to the students' predictions. A worksheet designed to direct student research on this topic can be found in the **Chapter Resource File** booklet or by visiting **go.hrw.com** and entering the keyword **HQ6GLAX.** LS **Auditory**

Answer to Reading Check

Continental glaciers exist only in Greenland and Antarctica.

CHAPTER RESOURCES

Chapter Resource File

- **Internet Activity**
  - Glacial Sounds GENERAL

Figure 2 ▶ An alpine glacier (above) descends through Thompson Pass in Alaska. A continental glacier (right) covers much of the land surface in Greenland.

alpine glacier a narrow, wedge-shaped mass of ice that forms in a mountainous region and that is confined to a small area by surrounding topography; examples include valley glaciers, cirque glaciers, and piedmont glaciers

continental glacier a massive sheet of ice that may cover millions of square kilometers, that may be thousands of meters thick, and that is not confined by surrounding topography

SCILINKS. NSTA Developed and maintained by the National Science Teachers Association

For a variety of links related to this subject, go to www.scilinks.org

Topic: Glaciers
SciLinks code: HQ60675

Types of Glaciers

The two main categories used to classify glaciers are alpine and continental. An **alpine glacier** is a narrow, wedge-shaped mass of ice that forms in a mountainous region and that is confined to a small area by surrounding topography, as shown in **Figure 2.** Alpine glaciers are located in Alaska, the Himalaya Mountains, the Andes, the Alps, and New Zealand.

Continental glaciers are massive sheets of ice that may cover millions of square kilometers, that may be thousands of meters thick, and that are not confined by surrounding topography, as shown in **Figure 2.** Today, continental glaciers, also called *ice sheets*, exist only in Greenland and Antarctica. The Antarctic ice sheet covers an area of more than 13 million km² and is more than 4,000 m thick in some places. The Greenland ice sheet covers 1.7 million km² of land, and its maximum thickness is more than 3,000 m. If these ice sheets melted, the water they contain would raise the worldwide sea level by more than 80 m.

Reading Check **Where can you find continental glaciers today? (See the Appendix for answers to Reading Checks.)**

Movement of Glaciers

Glaciers are sometimes called "rivers of ice." Gravity causes both glaciers and rivers to flow downward. However, glaciers and rivers move in different ways. Unlike water in a river, glacial ice cannot move rapidly or flow easily around barriers. In a year, some glaciers may travel only a few centimeters, while others may move a kilometer or more. Glaciers move by two basic processes—basal slip and internal plastic flow.

BRAIN FOOD

Glacier Classification Glaciers are sometimes classified by their environments. There are three glacial environments: polar, subpolar, and temperate.

Polar environments have cold temperatures, little glacial melting, and a permanent ice pack. Summer temperatures are below 0°C. Antarctica and Northern Greenland are examples of polar environments.

In subpolar environments, the temperatures in summer are less than 10°C, but are warm enough for some vegetation to grow. Moderate summer rain causes some glacial surface melting. Southern Greenland and Iceland have subpolar environments.

In temperate environments, summer temperatures are higher than 10°C, and considerable melting and precipitation occur. The glaciers in Patagonia, New Zealand, and Alaska are in temperate environments.

Basal Slip

One way that glaciers move is by slipping over a thin layer of water and sediment that lies between the ice and the ground. The weight of the ice in a glacier exerts pressure that lowers the melting point of ice. As a result, the ice melts where the glacier touches the ground. The water mixes with sediment at the base of the glacier. This mixture acts as a lubricant between the ice and the underlying surface. The process that lubricates a glacier's base and causes the glacier to slide forward is called **basal slip.**

Basal slip also allows a glacier to work its way over small barriers in its path by melting and then refreezing. For example, if the ice pushes against a rock barrier, the pressure causes some of the ice to melt. The water from the melted ice travels around the barrier and freezes again as the pressure is removed.

basal slip the process that causes the ice at the base of a glacier to melt and the glacier to slide

internal plastic flow the process by which glaciers flow slowly as grains of ice deform under pressure and slide over each other

Internal Plastic Flow

Glaciers also move by a process called **internal plastic flow.** In this process, pressure deforms grains of ice under a glacier. As the grains deform, they slide over each other and cause the glacier to flow slowly. However, the rate of internal plastic flow varies for different parts of a glacier, as shown in **Figure 3.** The slope of the ground and the thickness and temperature of the ice determine the rate at which ice flows at a given point. The edges of a glacier move more slowly than the center because of friction with underlying rock. For this same reason, a glacier moves more quickly near its surface than near its base.

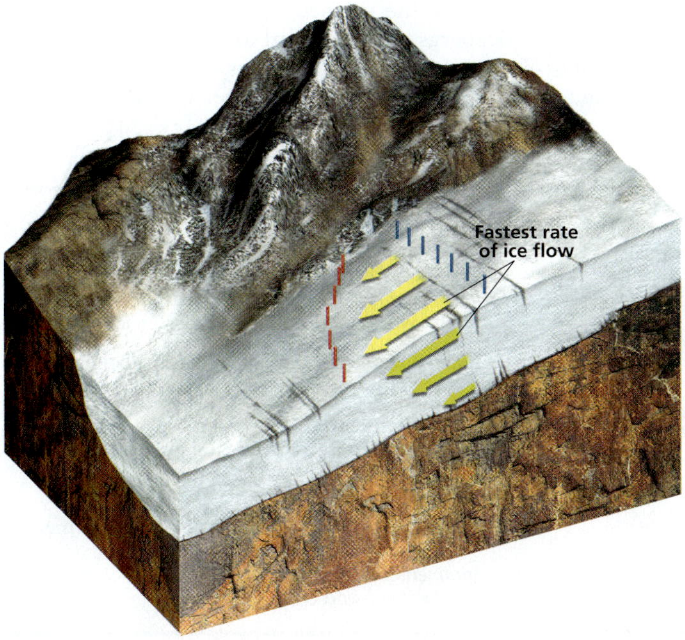

Figure 3 ▶ A line of blue stakes driven into a alpine glacier moves to the position of the red stakes as the glacier flows. This measurement shows that the central part of the glacier moves faster than its edges.

QuickLAB 15 min

Slipping Ice

Procedure

1. Squeeze flat a handful of **snow** or **shaved ice** until you notice a change in the particles.
2. Place the squeezed ice on a **tray.** Place a **heavy weight** on the squeezed ice.
3. Lift one end of the tray to a 30° incline, and record your observations.

Analysis

1. What does squeezing do to the particles of snow or ice?
2. How is squeezing the snow or ice similar to the formation of a glacier?
3. What did you observe when you placed the squeezed ice on the tray? What glacial movement process is modeled here?

Close

Reteaching — BASIC

Glacial Movement Have students work in pairs to review the steps of both types of glacial movement. Have one student write each step of basal slip on an index card and have the other student write each step of internal plastic flow. Next, have students shuffle their sets of cards and exchange them. Then have students put the sets of cards in the correct order. English Language Learners
LS Verbal/Logical
Co-op Learning

Quiz — GENERAL

1. What factors affect the size of a glacier? (the amounts of snowfall received and of ice lost due to raised temperature)
2. What is another name for an ice sheet? (continental glacier)
3. What is firn and how does it form? (Firn is grainy ice that forms by the melting and refreezing of snow.)

Alternative Assessment — GENERAL

Glacial Variety and Features Have students make a slide presentation that shows images of different glacial types and features. Tell students to identify the type of glacier or feature and the location of the glacier. The Web site of the United States Geological Survey has many images of glaciers. LS Visual/Verbal

CHAPTER RESOURCES

Chapter Resource File

- Datasheet for Quick Lab GENERAL

Technology

Transparencies

- 83 Internal Plastic Flow (with worksheet)

QuickLAB

Skills Acquired

- Observing
- Identifying/Recognizing Patterns

Teacher's Notes Caution students to squeeze the ice strongly but briefly and into a flat shape that can accommodate a weight.

Answers

1. Squeezing the ice removes air from between the particles and makes the ice more compact.
2. Glaciers form when pressure flattens ice grains and squeezes air out.
3. At first, the ice does not move. A little while after the weight was placed on the ice, the ice started to slide. Basal slip is being modeled.

Close, *continued*

Answers to Section Review

1. Snow accumulates year after year at high elevations and in polar regions.
2. The pressure from overlying snow flattens the snow into grains of ice and squeezes out the air. As snow and firn build up, a glacier forms.
3. An alpine glacier is narrow and wedge-shaped. It forms in mountainous areas and is confined by the surrounding topography. A continental glacier is a massive sheet of ice that is not confined by surrounding topography and may cover large expanses of land.
4. In internal plastic flow, the glacier's interior moves as grains of ice deform under pressure and slip over each other. In basal slip, the entire glacier moves by slipping over a thin layer of water and sediment at the glacier's base.
5. Answers may vary. Sample answer: Crevasses are large cracks on the surface of a glacier. Ice shelves are part of a continental glacier that has moved out over the ocean.
6. A snowfield is a nearly motionless mass of permanent snow and ice that forms as ice and snow accumulate above the snowline. A glacier is a moving mass of snow and firn whose ice crystals have been flattened and compacted by the pressure of overlying layers of firn and ice.
7. Tension and compression under the surface of a glacier caused by uneven flow make the glacier's brittle surface crack, forming crevasses.
8. Answers may vary. Sample answer: All glaciers would move more slowly because moving by internal flow is more complex and gradual than moving by basal slip.
9. Small differences in average yearly temperatures and snowfall amounts affect the size of the glacier. If average temperatures go up, a glacier may get smaller. If average temperatures drop, a glacier may get larger.
10. Most of the ice in an iceberg is underwater and cannot be easily seen.
11. *Glaciers* form from *firn* in *snowfields* above the *snowline;* form features such as *ice shelves, icebergs,* and *crevasses;* may be *alpine glaciers* or *continental glaciers;* and move by *internal plastic flow* and *basal slip*.

Figure 4 ▶ Crevasses (right) are large cracks in a glacier. The composite photograph below shows what an iceberg might look like if you could see the entire iceberg.

crevasse in a glacier, a large crack or fissure that results from ice movement

Features of Glaciers

While the interior of a glacier moves by internal plastic flow and the entire glacier moves by basal slip, the low pressure on the surface ice causes the surface ice to remain brittle. The glacier flows unevenly beneath the surface, and regions of tension and compression build under the brittle surface. As a result, large cracks, called **crevasses** (kruh VAS uhz), form on the surface, as shown in **Figure 4.** Some crevasses may be as deep as 50 m!

Continental glaciers move outward in all directions from their centers toward the edges of their landmasses. Some parts of the ice sheets may move out over the ocean and form *ice shelves*. When the tides rise and fall, large blocks of ice, called *icebergs,* may break from the ice shelves and drift into the ocean. Because most of an iceberg is below the surface of the water, as shown in **Figure 4,** icebergs pose a hazard to ships. The area above water of one of the largest icebergs ever observed in the Antarctic was twice the size of Connecticut!

Section 1 Review

1. **Identify** two regions in which snow accumulates year after year.
2. **Describe** the process by which glaciers form.
3. **Compare** an alpine glacier and a continental glacier.
4. **Explain** how internal plastic flow and basal slip move glaciers.
5. **Describe** two features of glaciers.
6. **Compare** a glacier and a snowfield.
7. **Explain** how a crevasse forms.

CRITICAL THINKING

8. **Making Inferences** If glaciers could move only by internal plastic flow, what might happen to the rate at which glaciers move? Explain your answer.
9. **Identifying Relationships** How can changes in the size of a glacier indicate climate change?
10. **Analyzing Ideas** If icebergs are visible at sea level, why do they pose a hazard to ships?

CONCEPT MAPPING

11. Use the following terms to create a concept map: *glacier, firn, snowline, snowfield, alpine glacier, continental glacier, basal slip, internal plastic flow, ice shelf, iceberg,* and *crevasse.*

CHAPTER RESOURCES

Chapter Resource File
- **Section Quiz** GENERAL

Workbooks
- **Study Guide** (also in Spanish)

Section 2 Glacial Erosion and Deposition

Many of the landforms in Canada and in the northern United States were created by glaciers. Large lakes, solitary boulders on flat plains, and jagged ridges are just a few examples of landforms created by glaciers. Glaciers created these landforms through the processes of erosion and deposition.

OBJECTIVES

- **Describe** the landscape features that are produced by glacial erosion.
- **Name** and describe five features formed by glacial deposition.
- **Explain** how glacial lakes form.

KEY TERMS

cirque
arête
horn
erratic
glacial drift
till
moraine
kettle
esker

Glacial Erosion

Like rivers, glaciers are agents of erosion. Both a river and a glacier can pick up and carry rock and sediment. However, because of the size and density of glaciers, landforms that result from glacial action are very different from those that rivers form. For example, deep depressions in rock form when a moving glacier loosens and dislodges, or plucks, a rock from the bedrock at the base or side of the glacier. The rock plucked by the glacier is then dragged across the bedrock and causes abrasions. As shown in **Figure 1,** long parallel grooves in the bedrock are left behind and show the direction of the glacier's movement.

Figure 1 ▶ As a glacier moves, it picks up and carries rocks from the bedrock. These grooves at Kelly's Island in Ohio were carved by a glacier 35,000 years ago.

Section 2

Focus

Overview

This section describes the different features that are produced by glacial erosion and deposition and how glacial lakes form.

Bellringer

Ask students to describe the actions of a steam shovel, a chisel, sandpaper, and a conveyor belt. Then, ask them to predict how glaciers behave like each of these objects. (Answers may vary. Accept all reasonable answers.) LS **Logical**

Motivate

Demonstration — GENERAL

Glacial Erosion and Deposition Simulation Before class, prepare about 50 mL of a mixture of table salt, coarse salt, and a small amount of rock salt and spread it evenly over a sheet of paper. Cover a 3×5 index card with double-sided tape; and, with the sticky side facing down, move the index card across the length of the paper exerting even pressure on the card. Ask students to describe how the salt mixture and the card look after the card has moved across the paper. Ask students what effect the movement of glaciers has on the land. (Glaciers remove material by picking it up, they leave grooves on the land, and they push material to another place.) LS **Visual/ Verbal**

CHAPTER RESOURCES

Chapter Resource File

- Directed Reading BASIC

Technology

Transparencies
- Bellringer

Student Edition on CD-ROM

One-Stop Planner CD-ROM
- Lesson Plan

Teach

SKILL BUILDER — GENERAL

Vocabulary The word *arête* comes from the Old French *areste,* which means "fishbone spine." Ask students to explain how this information may help them remember the meaning of the term *arête.* English Language Learners **LS Verbal**

Using the Figure — GENERAL

Comprehension Check Tell students that scientists study landforms created by glaciers to learn more about the path and action of glaciers. Ask students why one side of a roche moutonnée is steep and jagged and what roches moutonnées reveal about the movement of a glacier. (One side is steep and jagged because rock was pulled away as the glacier passed. The smooth, sloping side faces the direction from which the glacier came.) Answer to caption question: The jagged ridge that separates cirques are called *arêtes.* **LS Logical**

Teaching Tip — BASIC

Connect to Prior Knowledge Obtain a photograph of the Matterhorn. Show students the photograph. Suggest that they may be familiar with this well-known mountain peak because "horn" is part of its name. Ask students to name the mountain peak. (the Matterhorn) **LS Visual**

Answer to Reading Check

A moving glacier forms a cirque by pulling blocks of rock from the floor and walls of a valley and leaving a bowl-shaped depression.

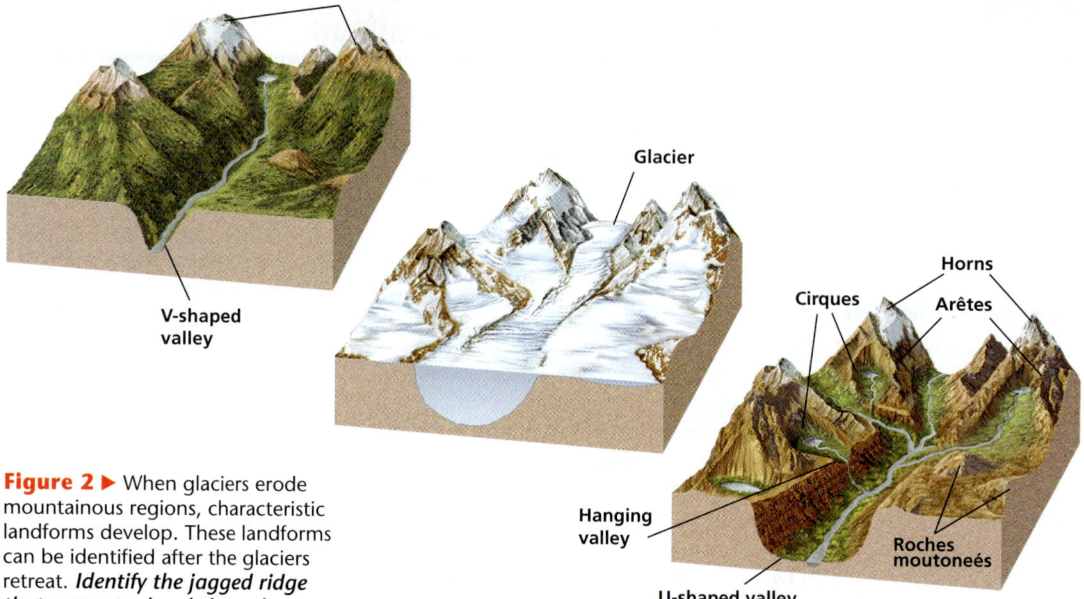

Figure 2 ▶ When glaciers erode mountainous regions, characteristic landforms develop. These landforms can be identified after the glaciers retreat. ***Identify the jagged ridge that separates bowl-shaped depressions.***

cirque a deep and steep bowl-like depression produced by glacial erosion

arête a sharp, jagged ridge that forms between cirques

horn a sharp, pyramid-like peak that forms because of the erosion of cirques

Landforms Created by Glacial Erosion

Glaciers have shaped many mountain ranges and have created unique landforms by erosive processes. The glacial processes that change the shape of mountains begin in the upper end of the valley where an alpine glacier forms. As a glacier moves through a narrow, V-shaped river valley, rock from the valley walls breaks off and the walls become steeper. The moving glacier also pulls blocks of rock from the floor of the valley. These actions create a bowl-shaped depression called a **cirque** (SUHRK). A sharp, jagged ridge called an **arête** (uh RAYT) forms between cirques. When several arêtes join, they form a sharp, pyramid-like peak called a **horn,** as shown in **Figure 2.**

As the glacier flows down through an existing valley, the glacier picks up large amounts of rock. These rock fragments, which range in size from microscopic particles to large boulders, become embedded in the ice.

Rock particles embedded in the ice may polish solid rock as the ice moves over the rock. Large rocks carried by the ice may gouge deep grooves in the bedrock. Glaciers may also round large rock projections. These rounded projections usually have a smooth, gently sloping side facing the direction from which the glacier came. The other side is steep and jagged because rock is pulled away as the ice passes. The resulting rounded knobs of rock are called *roches moutonnées* (ROHSH MOO tuh NAY), which means "sheep rocks" in French.

Reading Check How does a glacier form a cirque? (See the Appendix for answers to Reading Checks.)

Homework — GENERAL

Glacial Deposition and Erosion Gallery
As students read this section, direct them to make a scrapbook of photos or drawings, from the Internet or magazines, that show the different effects of glacial erosion. Tell students to include labels with every photo that identify the landform in the photo, as well as the location of the landform and any other information they can find about the location. **LS Visual/Verbal**

CHAPTER RESOURCES

Technology

Transparencies
- 84 Landforms Created by Glacial Erosion (with worksheet)

U-Shaped Valleys

A stream forms the V shape of a valley. As a glacier scrapes away a valley's walls and floor, this original V shape becomes a U shape, as shown in **Figure 3.** Because glacial erosion is the only way by which U-shaped valleys form, scientists can use this feature to determine whether a valley has been glaciated in the past.

Small tributary glaciers in adjacent valleys may flow into a main alpine glacier. Because a small tributary glacier has less ice and less cutting power than the main alpine glacier does, the small glacier's U-shaped valley is not cut as deeply into the mountains. When the ice melts, the tributary valley is suspended high above the main valley floor and is called a *hanging valley*. When a stream flows from a hanging valley, a waterfall forms.

Figure 3 ▶ Jollie Valley, a U-shaped glaciated valley, is located in the Southern Alps of New Zealand.

Erosion by Continental Glaciers

The landscape eroded by continental glaciers differs from the sharp, rugged features eroded by alpine glaciers. Continental glaciers erode by leveling landforms to produce a smooth, rounded landscape. Continental glaciers smooth and round exposed rock surfaces in a way similar to the way that bulldozers flatten landscapes. Rock surfaces are also scratched and grooved by rocks carried at the base of the ice sheet. These scratches and grooves are parallel to the direction of glacial movement.

QuickLAB 25 min

Glacial Erosion

Procedure

1. Put a **mixture of sand, gravel,** and **rock** in the bottom of a **15 cm × 10 cm × 5 cm plastic container.** Fill the container with **water** to a depth of about 4 cm. Freeze the container until the water is solid. Remove the **ice block** from the container.
2. Use a **rolling pin** or **large dowel** to flatten some **modeling clay** into a rectangle about 20 cm × 10 cm × 1 cm.
3. Grasp the ice block firmly with a **hand towel.** Place the block with the gravel-and-rock side down at one end of the clay. Press down on the ice block, and push it along the length of the flat clay surface.
4. Sketch the pattern made in the clay by the ice block.
5. Next, press a 2 cm layer of **damp sand** into the bottom of a **shallow, rectangular box.** As in step 3, push the ice block along the surface of the sand, but press down lightly.
6. Repeat steps 3 and 4, but use a **soft, wooden board** in place of the clay.

Analysis

1. Describe the effects of the ice block on the clay, on the sand, and on the wood.
2. Did any clay, sand, or wood become mixed with material from the ice block? Did the ice deposit material on any surface?
3. What glacial land features are represented by the features of the clay model? the sand model? the wood model?

QuickLAB

Skills Acquired

- Constructing Models
- Observing
- Analyzing Data

Teacher's Notes Prepare the sand, gravel, rock, and ice and freeze them before class. If possible, use an ice block fresh from the freezer for each part of the lab. You may wish to conduct this lab as a demonstration and have students collect and analyze the data.

Answers

1. The ice block made a wide depression in the clay; it made a wide depression with grooves in the sand, and grooves in the wood.
2. None of the clay or wood mixed with the ice block, but some of the sand did. The ice deposited very, very little material on the clay and only some material on the wood. The ice deposited the most material on the sand.
3. The clay represents a smooth, rounded landscape. The sand and the wood represent a rock surface with scratches and grooves.

CHAPTER RESOURCES

Chapter Resource File

- Datasheet for Quick Lab GENERAL

Homework GENERAL

Fiord Formation Tell students that a fiord forms when a narrow U-shaped valley that extends to a coastline becomes partly flooded by sea water after the glacier melts. Have students research the Harriman Fiord in Alaska, which was discovered in 1899. Ask students to imagine that they were participants in the Harriman expedition. Have each student write a series of journal entries that describe the days leading up to the discovery and the day of the discovery itself.
LS Verbal/Intrapersonal

Activity ADVANCED

Glacial Erosion Theories Direct interested students to research the historical development of glacial erosion hypotheses up through the most recent. Have students use their research results to write an essay in which they explain each hypothesis and describe how past hypotheses helped lead to modern understanding of glaciers. LS Logical

Teach, continued

Discussion — GENERAL

Comprehension Check Tell students to imagine that they have discovered a large boulder in a local park. It is approximately 3 m high, has scratches on one side, and is uncharacteristic of the rocks found in the area. Ask students to identify the glacial deposit that they have discovered and to explain what the rock tells them about the area. (The rock is an erratic. It tells me that a glacier once covered the area but dropped the rock as it melted.)
LS **Visual/Kinesthetic**

MISCONCEPTION ALERT

Different Drifts Students may confuse *glacial drift* and *snowdrift* because they sound similar and both involve cold conditions. Explain that each term involves a different substance and is a result of a different process. A snowdrift is the result of wind blowing snow into a mound. Glacial drift is sediment, such as sand, rocks, or gravel, that has been carried and deposited by glaciers.

Glacial Deposition

Glaciers are also agents of deposition. Deposition occurs when a glacier melts. A glacier will melt if it reaches low, warm elevations or if the climate becomes warmer. As the glacier melts, it deposits all of the material that it has accumulated, which may range in size from fine sediment to large rocks.

Large rocks that a glacier transports from a distant source are called **erratics.** Because a glacier carries an erratic a long distance, the composition of an erratic usually differs from that of the bedrock over which the erratic lies.

Various other landforms develop as glaciers melt and deposit sediment, as shown in **Figure 4.** The general term for all sediments deposited by a glacier is **glacial drift.** Unsorted glacial drift that is deposited directly from a melting glacier is called the **till.** Till is composed of sediments from the base of the glacier and is commonly left behind when glacial ice melts. Another type of glacial drift is stratified drift. *Stratified drift* is material that has been sorted and deposited in layers by streams flowing from the melted ice, or *meltwater*.

erratic a large rock transported from a distant source by a glacier

glacial drift rock material carried and deposited by glaciers

till unsorted rock material that is deposited directly by a melting glacier

Figure 4 ▶ Moraines, glacial lakes, drumlins, meltwater streams, and outwash plains are some examples of landforms created by glacial deposition.

CHAPTER RESOURCES

Chapter Resource File

- **Inquiry Lab**
  Glacial Deposition GENERAL
- **Making Models Lab**
  Melted Glacier Formations GENERAL

Technology

Transparencies
- 85 Features of Glacial Deposition (with worksheet)

BRAIN FOOD

Glacier Flour and Milk Glacier flour forms when a glacier's drift grinds against Earth's rocky surface. Glacier flour, which consists of very finely ground rock, silt, and clay, is washed out of glaciers in meltwater. The sediment-laden meltwater is called *glacier milk* and often is white in color.

Till Deposits

Landforms that result when a glacier deposits till are called *moraines*. **Moraines** are ridges of unsorted sediment on the ground or on the glacier itself. There are several types of moraines, as shown in **Figure 4.** A *lateral moraine* is a moraine that is deposited along the sides of an alpine glacier, usually as a long ridge. When two or more alpine glaciers join, their adjacent lateral moraines combine to form a *medial moraine*.

The unsorted material left beneath the glacier when the ice melts is the *ground moraine*. The soil of a ground moraine is commonly very rocky. An ice sheet may mold ground moraine into clusters of drumlins. *Drumlins* are long, low, tear-shaped mounds of till. The long axes of the drumlins are parallel to the direction of glacial movement.

Terminal moraines are small ridges of till that are deposited at the leading edge of a melting glacier. These moraines have many depressions that may contain lakes or ponds. Large terminal moraines, some of which are more than 100 km long, can be seen across the Midwest, especially south of the Great Lakes.

Reading Check **Which glacial deposit is a tear-shaped mound of sediment? (See the Appendix for answers to Reading Checks.)**

moraine a landform that is made from unsorted sediments deposited by a glacier

Graphic Organizer **Spider Map**

Create the Graphic Organizer entitled "Spider Map" described in the Skills Handbook section of the Appendix. Label the circle "Moraines." Create a leg for each type of moraine. Then, fill in the map with details about each type of moraine.

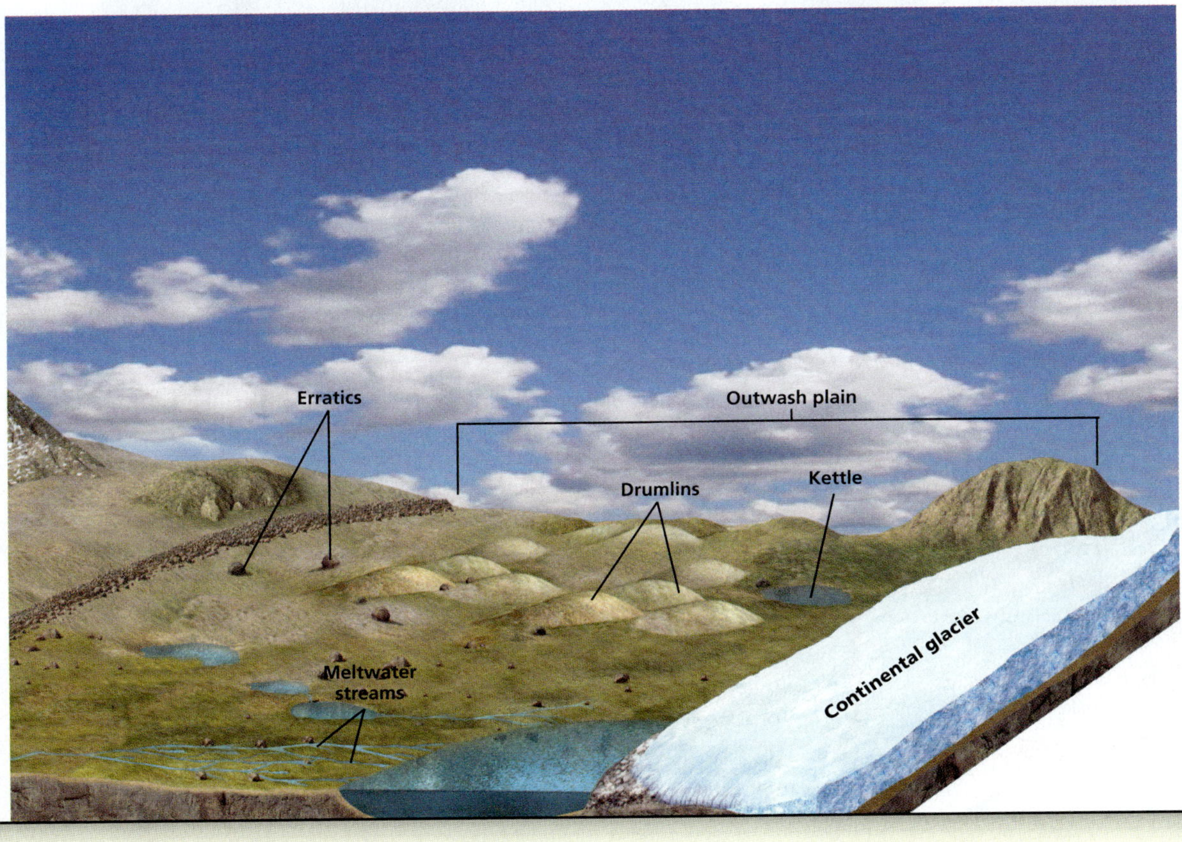

Graphic Organizer GENERAL

Spider Map

You may want to have students work in groups to create this Spider Map. Have one student draw the map and fill in the information provided by other students from the group.

Answer to Reading Check

A drumlin is a long, low, tear-shaped mound of till.

Using the Figure — GENERAL

Describing Glacial Landforms

Direct students to list each glacial landform that is labeled in the figure on the bottom of these pages. Then, have students write a description of the feature and an explanation of how that feature formed. (Medial moraines form when two adjacent lateral moraines combine. Lateral moraines are unsorted sediment that is deposited along the sides of a glacier. Terminal moraines are small ridges of till that are deposited at the leading edge of a melting glacier. Erratics are large rocks that were transported from a distant source by a glacier and that were deposited when the glacier melted. Meltwater streams are streams that flow from the melted ice of a glacier. Outwash plains are areas of stratified drift deposited by meltwater in front of terminal moraines. Kettles are depressions in glacial drift that form when chunks of ice buried in the drift melt. Drumlins are long, low, tear-shaped mounds of till.) **LS Visual/Verbal**

INCLUSION *Strategies*

- ***Developmentally Delayed***
- ***Hearing Impaired***
- ***Learning Disabled***

Divide the class into small groups. Have each group choose one or more paragraphs from the text. Help them to identify in these paragraphs words (other than ones that are defined and explained) that they don't understand. Have each group rewrite their paragraphs using definitions or synonyms in place of the identified words. Ask groups to read their new paragraphs to the class. **LS Verbal**

Teach, continued

Activity — Basic

Effects of Glacial Erosion and Deposition Have students make a table that has two columns. Have them label one column "Erosion" and the other column "Deposition." Have them label the rows "Continental glacier" and "Alpine glacier." Tell students to list the landforms that result from each process in the appropriate row. LS Verbal

Internet Activity — General

More Glacial Landforms and Features Have interested students use the Internet to research topographical features formed by glacial deposition, such as chatter scratches, kames, scours, whale-backs, and braided streams. Direct students to write descriptions of the appearance and formation of each feature. A worksheet designed to direct student research on this topic can be found in the **Chapter Resource File** booklet or by visiting go.hrw.com and entering the keyword **HQ6GLAX.** LS Verbal

Answer to Reading Check

Eskers form when meltwater from receding continental glaciers flows through ice tunnels and deposits long, winding ridges of gravel and sand.

CHAPTER RESOURCES

Chapter Resource File

- Internet Activity
  - More Glacial Landforms and Features GENERAL

Figure 5 ▶ This kettle lake in Saskatchewan, Canada, formed as a result of glacial deposition.

kettle a bowl-like depression in a glacial drift deposit

esker a long, winding ridge of gravel and coarse sand deposited by glacial meltwater streams

SCILINKS. NSTA Developed and maintained by the National Science Teachers Association

For a variety of links related to this subject, go to www.scilinks.org

Topic: Glaciers and Landforms
SciLinks code: HQ60676

Outwash Plains

When a glacier melts, streams of meltwater flow from the edges, the surface, and beneath the glacier. Glacial meltwater may have beautiful colors, such as milky white, emerald green, or turquoise blue, because it carries very fine sediment. The meltwater carries drift as well as rock particles and deposits them in front of the glacier as a large outwash plain. An *outwash plain* is a deposit of stratified drift that lies in front of a terminal moraine and is crossed by many meltwater streams.

Kettles

Most outwash plains are pitted with depressions called **kettles.** A kettle forms when a chunk of glacial ice is buried in drift. As the ice melts, a cavity forms in the drift. The drift collapses into the cavity and produces a depression. Kettles commonly fill with water to form kettle lakes, such as the one shown in **Figure 5.**

Eskers

When continental glaciers recede, **eskers** (ES kuhrz)—long, winding ridges of gravel and sand—may be left behind. These ridges consist of stratified drift deposited by streams of meltwater that flow through ice tunnels within the glaciers. Eskers may extend for tens of kilometers, like raised, winding roadways.

✔ **Reading Check** How do eskers form? (See the Appendix for answers to Reading Checks.)

Geography Connection — Advanced

Local Glacial Formations Tell students that glaciers significantly changed the surface of large parts of Europe, Asia, and North America. Have students work in pairs to research the glacial history of the continents. Tell students to make a map of one continent that shows the location of different glacial formations. Students should use a different symbol to indicate each type of landform or deposit. LS Visual

Group Activity — General

Types of Glacial Lakes Tell students that there are three types of glacial lakes: kettle lakes, cirque lakes, and moraine-formed lakes. Organize the class into three groups. Have each group research one type of glacial lake and create a poster or multimedia presentation that describes how that type of lake forms and what its characteristics are. LS Verbal

Glacial Lakes

Lake basins commonly form where glaciers erode surfaces and leave depressions in the bedrock. Thousands of lake basins in Canada and the northern United States were gouged from solid rock by a continental glacier. Thousands of other glacial lakes form as a result of deposition rather than as a result of erosion. Many lakes form in the uneven surface of ground moraine deposited by glaciers. These lakes exist in many areas of North America and Europe.

Long, narrow *finger lakes,* such as those in western New York, form where terminal and lateral moraines block existing streams. The area south of the Great Lakes, from Minnesota to Ohio, has belts of moraines and lakes. Minnesota, also called the "Land of 10,000 Lakes," was completely glaciated and has evidence of all types of glacial lakes.

Figure 6 ▶ Many streams and rivers carry dissolved minerals to the Great Salt Lake in Utah. However, because there is no outlet, the lake becomes concentrated with these minerals as continual evaporation removes water.

Formation of Salt Lakes

Many lakes existed during the last glacial advance. But because of topographic and climatic changes, outlet streams no longer leave these lakes. Water leaves the lakes only by evaporation. When the water evaporates, salt that was dissolved in the water is left behind, which makes the water increasingly salty. Salt lakes, such as the one shown in **Figure 6,** commonly form in dry climates, where evaporation is rapid and precipitation is low.

Connection to CHEMISTRY

Precipitation of Minerals

Salt lakes form when a high rate of evaporation removes only water from the lake and leaves the dissolved minerals, such as salt, in the lake. Water can dissolve a limited amount of minerals. Therefore, when the concentration of minerals in the lake water becomes too high, the minerals crystallize in a process called *precipitation*. The crystallized minerals then settle to the bottom of the lake.

The minerals that precipitate from evaporating water are called *evaporites*. Common evaporite minerals include halite, or salt (NaCl); gypsum ($CaSO_4 \cdot 2H_2O$); calcite ($CaCO_3$); and borax ($Na_2B_4O_7 \cdot 10H_2O$).

When the minerals settle out of the water, the concentration of chemicals in the water changes. This change in concentration causes different minerals to precipitate at different times. The first minerals to precipitate as water evaporates are carbonates, such as calcite. Continued evaporation leads to the formation of gypsum. Finally, salts such as halite begin to form. This process leads to the characteristic sequences of mineral deposits found in natural salt formations.

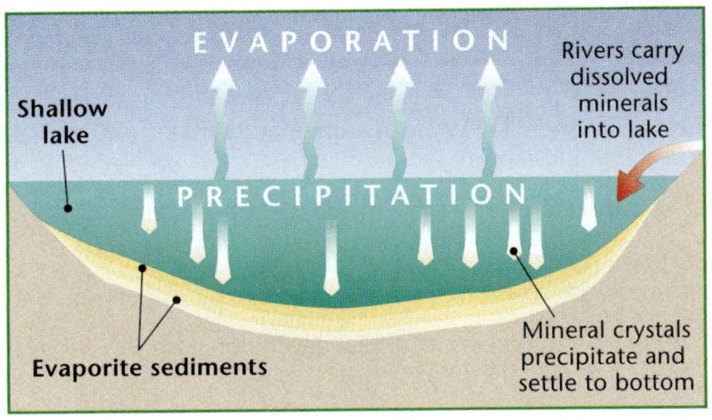

Close

Reteaching — BASIC

Chart Glacial Landforms Make a class chart of glacial landforms. Divide the class into three groups. Ask students in one group to provide the names of different types of glacial landforms. Make a list of them on the board. Have the second group come up with a description for each landform. Add the descriptions to the board. Ask the third group to indicate whether each landform results from glacial erosion or deposition. Record their answers on the chart. Students may make copies of the finished chart to help them study glacial landforms.
LS **Verbal** Co-op Learning

Quiz — GENERAL

1. How does a hanging valley form and what happens when a stream flows from it? (A hanging valley forms when a tributary glacier cuts a shallower valley than the main valley glacier. The tributary valley is suspended above the floor of the main valley. A waterfall forms when a stream flows from a hanging valley.)
2. What is meltwater? (Meltwater is the stream of melted ice that flows from a receding glacier.)
3. Why do salt lakes most commonly form in dry climates? (Evaporation is rapid and precipitation is low in dry climates. Dissolved salt remains behind as water evaporates from lakes.)

Alternative Assessment — ADVANCED

Modeling Glacial Landforms Direct students to make a three-dimensional model, using clay, papier-mâché, or plaster of Paris, that shows different landforms created by glaciation. Tell them to label each landform.
LS **Kinesthetic/Visual**

READING SKILL BUILDER — BASIC

Reading Organizer As students read this section, encourage them to take Power Notes, KWL Notes, or Two-Column Notes as described in the Skills Handbook section of the Appendix. Later, students can use these notes as a study guide for assessments. LS **Verbal** Co-op Learning English Language Learners

Close, *continued*

Answers to Section Review

1. A cirque is a deep, steep bowl-shaped depression produced by glacial erosion. An arête is a sharp, jagged ridge that forms between cirques. A horn is a sharp, jagged ridge that forms when several arêtes are joined.
2. Features formed by glacial deposition include erratics, drumlins, moraines, outwash plains, kettles, eskers, glacial lakes, and meltwater streams.
3. Long narrow lakes form when terminal and lateral moraines block existing streams.
4. Glacial deposition occurs as a glacier melts and deposits material it has collected. Glacial erosion occurs when a glacier carries material away, leaving depressions in the bedrock, and when this material is dragged, causing abrasions in the rock.
5. A kettle forms when a chunk of glacial ice that has been buried in drift melts. A cavity forms in the drift, which collapses into the cavity to produce a depression.
6. An alpine glacier erodes mountainous regions, creating landforms such as cirques, arêtes, and hanging valleys.
7. Both rivers and glaciers pick up and carry rocks and other sediment. However, while rivers create V-shaped valleys and floodplains, glaciers, because of their size and density, form unique landforms, such as U-shaped valleys, horns, and cirques.
8. Ground moraine is the rocky, unsorted material left beneath the glacier when the ice melts. An esker is long winding ridge of gravel and sand that is deposited by meltwater streams.
9. Answers may vary. Sample answer: The hypothesis is that a glacier once moved across the area. I would try to find erratics, eskers, kettle lakes, and drumlins.

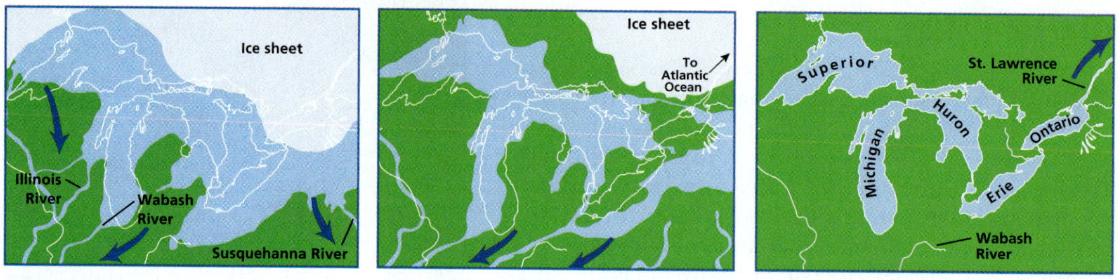

Early Ice Retreat The ice sheet that covered the northern part of North America formed enormous lakes.

Late Ice Retreat As the ice sheet retreated, the lakes became smaller and the drainage pattern changed.

Today's Great Lakes Uplifting of the land reduced the Great Lakes to their present sizes.

Figure 7 ▶ The Great Lakes in the northern United States were formed by a massive continental glacier.

History of the Great Lakes

The Great Lakes of North America formed as a result of erosion and deposition by a continental glacier, as shown in **Figure 7.** Glacial erosion widened and deepened existing river valleys. Moraines to the south blocked off the ends of these valleys. As the ice sheet melted, the meltwater was trapped in the valleys by the moraines and lakes formed.

In their early stages, the lakes emptied to the south into the Wabash and Illinois Rivers, which flowed into the Mississippi River. Later, the lakes grew larger and also drained into the Atlantic Ocean through the Susquehanna, Mohawk, and Hudson River valleys.

After the glacial period, the crust rose as the weight of the ice was removed. The lake beds uplifted and shrank. The uplift of the land caused the lakes to drain to the northeast through the St. Lawrence River. As a result of this northeasterly flow, Niagara Falls formed between Lake Erie and Lake Ontario.

Section 2 Review

1. **Describe** the following landscape features: a cirque, an arête, and a horn.
2. **List** five features that form by glacial deposition.
3. **Explain** how terminal and lateral moraines can form glacial lakes.
4. **Compare** the process of glacial deposition with the process of glacial erosion.
5. **Describe** how a kettle forms.
6. **Explain** how an alpine glacier can change the topography of a mountainous area.
7. **Compare** the process of erosion by glaciers with the process of erosion by rivers.

CRITICAL THINKING

8. **Making Comparisons** Compare the processes that form ground moraines with the processes that form eskers.
9. **Analyzing Predictions** On a field trip, you find rock that has long, parallel grooves. Form a hypothesis that explains this feature. What other landforms would you try to find to test this hypothesis?
10. **Making Comparisons** Compare glacial sediment deposited directly by glacial ice with the sediment deposited by glacial meltwater.

CONCEPT MAPPING

11. Use the following terms to create a concept map: *cirque, stratified drift, roches moutonnées, moraine, till, glacial drift, kettle, outwash plains,* and *glacier.*

10. The glacial sediment deposited by glacial ice is till, which is unsorted sand and rock. The glacial sediment deposited by glacial meltwater is stratified drift, which is sorted and deposited in layers.
11. A *glacier* erodes land, forming *cirques* and *roches moutonnées,* and deposits *glacial drift,* such as *till, moraines,* and *stratified drift,* on *outwash plains* that are commonly pitted with *kettles.*

CHAPTER RESOURCES

Chapter Resource File

- Section Quiz GENERAL

Workbooks

- Study Guide (also in Spanish)

ENVIRONMENTAL CONNECTION

Section 3 Ice Ages

Today, continental glaciers are located mainly in latitudes near the North and South Poles. However, thousands of years ago, ice sheets covered much more of Earth's surface. An **ice age** is a long period of climatic cooling during which the continents are glaciated repeatedly. Several major ice ages have occurred during Earth's geologic history, as shown in **Figure 1.** The earliest known ice age began about 800 million years ago. The most recent ice age began about 4 million years ago. The last advance of this ice age's massive ice sheets reached its peak about 18,000 years ago. Ice ages probably begin with a long, slow decrease in Earth's average temperatures. A drop in average global temperature of only about 5°C may be enough to start an ice age.

OBJECTIVES

- **Describe** glacial and interglacial periods within an ice age.
- **Summarize** the theory that best accounts for the ice ages.

KEY TERMS

ice age
Milankovitch theory

ice age a long period of climatic cooling during which the continents are glaciated repeatedly

Glacial and Interglacial Periods

Continental glaciers advance and retreat several times during an ice age. The ice sheets advance during colder periods and retreat during warmer periods. A period of cooler climate that is characterized by the advancement of glaciers is called a *glacial period*. A period of warmer climate that is characterized by the retreat of glaciers is called an *interglacial period*. Currently, Earth is in an interglacial period of the most recent ice age.

Figure 1 ▶ Glacial and Interglacial Periods

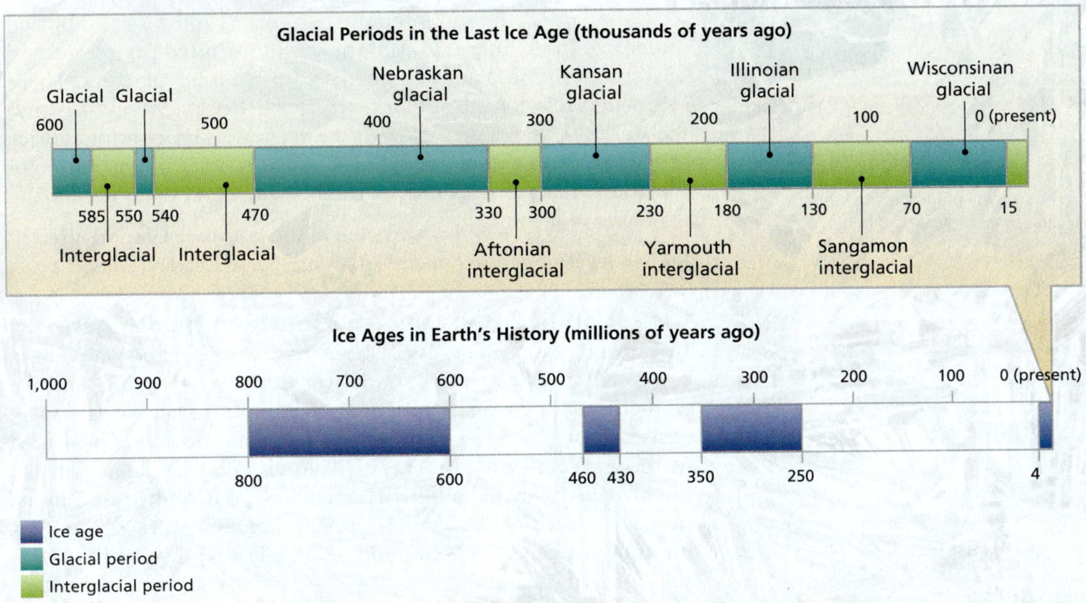

Section 3

Focus

Overview

This section describes the history and effect of Earth's ice ages, the difference between glacial and interglacial periods, and hypotheses about the causes of ice ages.

Bellringer

Ask students to describe what they think Earth would be like during an ice age. (Answers may vary. Students may describe a snow-covered landscape populated by woolly mammoths and other animals from the last ice age.) **LS Verbal**

Motivate

Discussion — GENERAL

Ice-Age Mammals Show images of ice-age animals such as saber-toothed cats, woolly rhinos, mammoths, and giant ground sloths. Images can be found at the Web site of the National Museum of Natural History in Washington, D.C. Explain that about 80 species of mammals became extinct at the end of the last ice age and that there are several hypotheses about the cause. Have students suggest reasons why these mammals may have become extinct. (Answers may vary. Current hypotheses include climate change, human hunting, disease, and a combination of these factors.) **LS Logical**

CHAPTER RESOURCES

Chapter Resource File
- Directed Reading BASIC

Technology

Transparencies
- Bellringer
- 86 Ice Ages of Earth's History (with worksheet)

Student Edition on CD-ROM

One-Stop Planner CD-ROM
- Lesson Plan

Teach

ENVIRONMENTAL CONNECTION

Earth is currently in the middle of an interglacial period. The last advance of this age's massive ice sheets started to retreat about 15,000 years ago. Since that time, climate has slowly warmed, but shorter and less-extreme cooling periods have occurred. Some scientists think that current trends in global warming are part of the natural glacial-interglacial cycle.

Debate — ADVANCED

Global Warming Have interested students research and debate whether global warming might stall the next glacial period. You may wish to assign students to prepare arguments for and against this hypothesis.
LS Verbal/Logical

Internet Activity — GENERAL

Vanishing Ice Caps Tell students that recent research has found that Greenland's ice cap is thinning. Direct students to view NASA Goddard Space Flight Center's Scientific Visualization Studio Web site. Have students present a summary of the significance and implications of Greenland's vanishing ice. A worksheet designed to direct student research on this topic can be found in the **Chapter Resource File** booklet or by visiting **go.hrw.com** and entering the keyword **HQ6GLAX.** **LS Verbal**

Answer to Reading Check

The sea level was up to 140 m lower than it is now.

CHAPTER RESOURCES

Chapter Resource File

Internet Activity
- Vanishing Ice Caps GENERAL

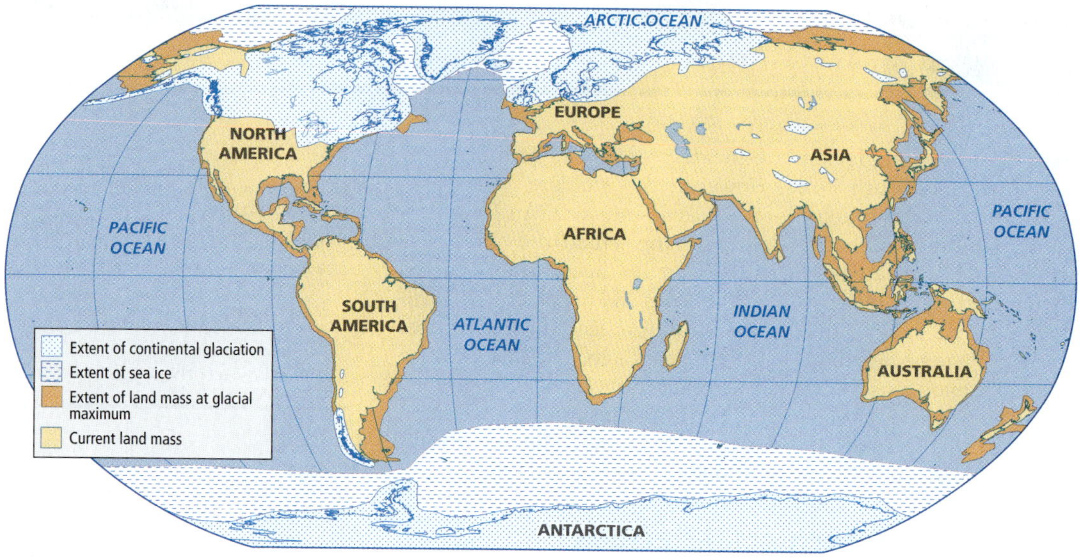

Figure 2 ▶ During the last glacial period, about 30% of Earth's surface was covered with ice.

SCILINKS. NSTA Developed and maintained by the National Science Teachers Association

For a variety of links related to this subject, go to www.scilinks.org

Topic: Ice Ages
SciLinks code: HQ60781

Glaciation in North America

Glaciers covered about one-third of Earth's surface during the last glacial period. Most glaciation took place in North America and Eurasia. In some parts of North America, the ice was several kilometers thick. So much water was locked in ice during the last glacial period that sea level was as much as 140 m lower than it is today. As a result, the coastlines of the continents extended farther than they do today, as shown in **Figure 2.**

Canada and the mountainous regions of Alaska were buried under ice. In the mountains of the western United States, numerous small alpine glaciers joined to form larger glaciers. These large glaciers flowed outward from the Rocky Mountains and the Cascade and Sierra Nevada Ranges. A great continental ice sheet that was centered on what is now the Hudson Bay region of Canada spread as far south as the Missouri and Ohio Rivers.

Reading Check **How did glaciation in the last glacial period affect the sea level? (See the Appendix for answers to Reading Checks.)**

Glaciation in Eurasia and the Southern Hemisphere

In Europe, a continental ice sheet that was centered on what is now the Baltic Sea spread south over Germany, Belgium, and the Netherlands and west over Great Britain and Ireland. It flowed eastward over Poland and Russia. Long alpine glaciers formed in the Alps and the Himalayas. A continental ice sheet formed in Siberia. In the Southern Hemisphere, the Andes Mountains in South America and much of New Zealand were covered by mountainous ice fields and alpine glaciers. Many land features that formed during the last glacial period are still recognizable.

INCLUSION *Strategies*

- *Learning Disabled*
- *Developmentally Delayed*

To help students understand the idea of a geologic timeline, ask them to make a timeline of their own lives. Ask students to draw a one-pointed arrow that starts at the left edge of the paper and goes to the right edge of the paper. Then have them divide the arrow into equal parts by twenty short lines. Under the short lines, have the students write numbers from one to twenty. Tell them to plot the following important life events under the correct years: I was born, I learned to walk, I started school, and Today. Then, ask them to add three important events that they choose themselves. **LS Logical**

Causes of Glaciation

The repeated glaciation of Earth's surface is caused by a variety of complex factors. These factors range from very rapid events to very slow processes. For example, the movement of tectonic plates happens very slowly but can cause changes in global circulation patterns of air and ocean water. These changes may lead to ice ages. Scientists have also proposed a number of ideas to explain glacial and interglacial periods. The leading explanation is the Milankovitch theory.

The Milankovitch Theory

A Serbian scientist named Milutin Milankovitch proposed a theory to explain the cause of glacial periods. Milankovitch noticed that climate change occurs in cycles. He thought that these cycles could be linked to cycles in Earth's movement relative to the sun. The **Milankovitch theory** is the theory that cyclical changes in Earth's orbit and in the tilt of Earth's axis occur over thousands of years and cause climatic changes.

Three periodic changes occur in the way that Earth moves around the sun, as **Figure 3** shows. First, the shape of Earth's orbit, or *eccentricity,* changes from nearly circular to elongated and back to nearly circular every 100,000 years. The second change occurs in the tilt of Earth's axis. Every 41,000 years, the tilt of Earth's axis varies between about 22.2° and 24.5°. A third periodic change is caused by the circular motion, or *precession,* of Earth's axis. Precession causes the axis to change its position, which is often described as a wobble. The axis of Earth traces a complete circle every 25,700 years.

Milankovitch calculated how these three factors may affect the distribution of solar energy that reaches Earth's surface. Changes in the distribution of solar energy affects global temperatures, which may cause periods of glacial advance and retreat.

MATHPRACTICE

Earth's Tilt
The current angle of Earth's tilt is 23.5°. That angle is decreasing. If the tilt moves from 24.5° to 22.2° over 41,000 years, estimate how many years will pass before Earth's tilt reaches 22.2°.

Milankovitch theory the theory that cyclical changes in Earth's orbit and in the tilt of Earth's axis occur over thousands of years and cause climatic changes

Figure 3 ▶ According to the Milankovitch theory, the distribution of solar radiation that Earth receives varies because of three kinds of changes in Earth's position relative to the sun.

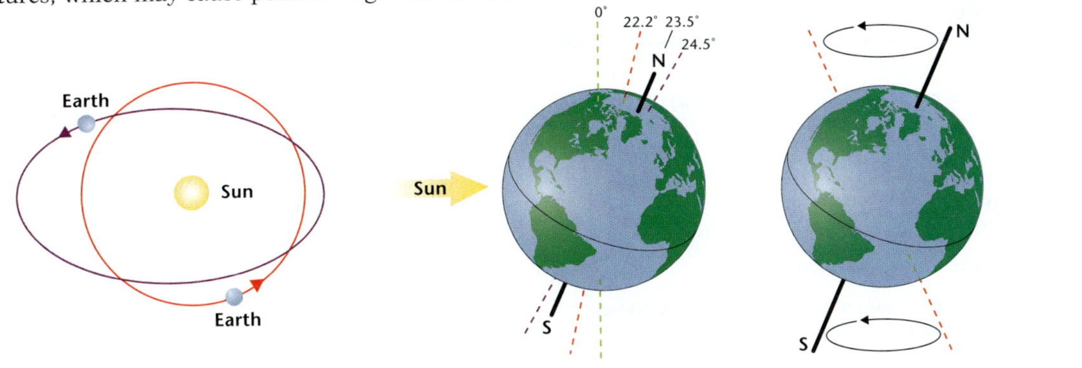

Eccentricity Changes in orbital eccentricity cause an increase in seasonality in one hemisphere and reduce seasonality in the other hemisphere.

Tilt Over a period of 41,000 years, the tilt of Earth's axis varies between 22.2° and 24.5°. The poles receive more solar energy when the tilt angle is greater.

Precession The wobble of Earth's axis affects the amount of solar radiation that reaches different parts of Earth's surface at different times of the year.

MATHPRACTICE

Answer
24.5° − 22.2° = 2.3°
23.5° − 22.2° = 1.3°
(41,000 years/2.3°) ×
1.3° = 23,174 years

Close

Reteaching — BASIC

Reading Organizer Have students outline the chapter material using the section headings and subheadings. Then, have students fill in the outline by providing additional detail for each heading. **LS Verbal**

Quiz — GENERAL

1. What is an ice age? (An ice age is a long period of climatic cooling during which the continents are glaciated repeatedly.)
2. How much of Earth's surface was covered by ice during the last glacial period? (one-third)
3. Explain the differences among Earth's eccentricity, tilt, and precession. (Eccentricity is the shape of Earth's orbit. Tilt is the angle of Earth's axis. Precession is the circular motion of Earth's axis.)

Alternative Assessment — GENERAL

Poster Project Direct students to make a three-panel poster that shows Earth's ice ages and glacial and interglacial periods, the areas of Earth covered by ice during the last glacial period, and the various explanations about what causes the start of an ice age.
LS Verbal/Visual

Career

Glaciologist Glaciologists are scientists who study the physical properties of snow and ice. You might call them "ice experts." These scientists design experiments to learn more about global sea level and climate changes. Glaciologists also monitor ice movement and changes in elevation; they measure ice markers, take ice cores, and analyze trapped atmospheric gases from ancient times to gain information. Glaciologists have a college degree in geology.

MISCONCEPTION ALERT

What Is an Ice Age? The term *ice age* is commonly used interchangeably with *glacial period.* Make sure students know that ice ages are long intervals that may contain several glacial and interglacial periods.

CHAPTER RESOURCES

Technology

Transparencies
• 87 The Milankovitch Theory (with worksheet)

Close, *continued*

Answers to Section Review

1. A glacial period is a period in which the climate is cooler and glaciers advance. An interglacial period is a period in which the climate is warmer and glaciers retreat.
2. The sea level decreases during a glacial period because more water is stored in glacial ice.
3. In North America: Canada, the mountainous region of Alaska, the Rocky Mountains, the Cascade and Sierra Nevada ranges, and from the Canadian border to the Missouri and Ohio rivers in the United States; in Europe and Asia: Belgium, the Netherlands, Great Britain, Poland, Russia, Siberia, Tibetan Plateau, the mountains of Asia, and the Baltic Sea over Germany; and parts of South America, Africa, and New Zealand
4. The Milankovitch theory states that Earth's eccentricity, tilt, and precession affect the distribution of solar energy to Earth's surface. When the intensity of solar energy that reaches Earth is less, global temperature decreases, which can cause a glacial period.
5. Because the shells of Foraminifera curl to the right instead of to the left when they live in water warmer than 8°C, scientists can determine the temperature of the water when the sediment layer containing Foraminifera fossils formed.
6. Glacial periods may start when dust from volcanic eruptions blocks the sun's rays. Glacial periods may be caused by a change the amount of solar energy that the sun produces.
7. Volcanic eruptions have a short-term effect on Earth's atmosphere, so they are not likely to cause long ice ages.
8. Answers may vary. Sample answer: If Earth's orbit were circular, Earth would be more likely to have ice ages because when Earth's orbit is more circular, Earth receives less energy from the sun.
9. An *ice age* has *glacial periods* and *interglacial periods* and may be caused by *volcanic eruptions* or by changes in Earth's *eccentricity, precession,* and *tilt,* which are described in the *Milankovitch theory*.

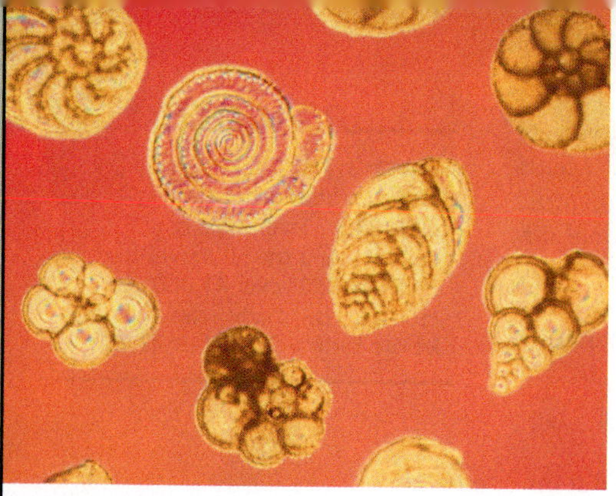

Figure 4 ▶ The microscopic shells of organisms from the order Foraminifera give clues to climate change.

Biological Evidence of Glaciation

Evidence for past glaciation has been discovered in the shells of dead marine animals found on the ocean floor. The formation of the shells of organisms from the order Foraminifera, shown in **Figure 4,** is affected by the temperature of the ocean water. Temperature of the ocean water affects the amount of oxygen that the water dissolves. The amount of oxygen in turn affects how these organisms form their shells. In addition, organisms that lived in ocean waters that were warmer than 8°C coiled their shells to the right. Organisms that lived in ocean waters that were much cooler coiled their shells to the left.

By studying Foraminifera shells in the layers of sediment on the ocean floor, scientists have discovered evidence of different glacial periods. Scientists have found that the record of changes in marine sediments closely follows the cycle of cooling and warming predicted by the Milankovitch theory.

Other Explanations for Glaciation

Other explanations for the causes of glacial periods have been suggested. Most of these explanations indicate that the glacial periods were caused by changes in the amount of solar energy that reached Earth's surface that were not caused by changes in Earth's position relative to the sun. Some scientists propose that changes in solar energy are caused by varying amounts of energy produced by the sun. Other scientists suggest that glacial periods start when volcanic dust blocks the sun's rays. Volcanic dust in Earth's atmosphere prevents solar radiation from reaching Earth's surface. As a result, Earth's average global temperature drops.

Section 3 Review

1. **Describe** glacial periods and interglacial periods.
2. **Explain** what happens to global sea level during a glacial period.
3. **Identify** the areas of Earth's surface that were covered by ice during the last glacial period.
4. **Summarize** the Milankovitch theory.
5. **Explain** how fossils of marine animals provide evidence of past glaciation.
6. **Describe** two explanations for glacial periods other than the Milankovitch theory.

CRITICAL THINKING

7. **Evaluating Hypotheses** Use the information that you learned about glacial periods within an ice age to explain why volcanic eruptions may not be the cause of ice ages.
8. **Predicting Consequences** If Earth's orbit were always circular, would Earth be more likely to experience ice ages or less likely to experience ice ages? Explain your answer.

CONCEPT MAPPING

9. Use the following terms to create a concept map: *ice age, glacial period, interglacial period, Milankovitch theory, eccentricity, tilt, volcanic eruption,* and *precession.*

CHAPTER RESOURCES

Chapter Resource File

- **Section Quiz** GENERAL

Workbooks

- **Study Guide** (also in Spanish)

Chapter 17 Highlights

Sections

1 Glaciers: Moving Ice

Key Terms

glacier, 419
alpine glacier, 420
continental glacier, 420
basal slip, 421
internal plastic flow, 421
crevasse, 422

Key Concepts

- A glacier is a large mass of moving ice formed by the compaction of snow.
- The two main types of glaciers are alpine glaciers and continental glaciers.
- Glaciers move by basal slip and by internal plastic flow.
- Three features of glaciers are crevasses, ice shelves, and icebergs.

2 Glacial Erosion and Deposition

cirque, 424
arête, 424
horn, 424
erratic, 426
glacial drift, 426
till, 426
moraine, 427
kettle, 428
esker, 428

- Glaciers create land features by eroding the land and by depositing rock and sediment.
- Glaciers erode the valleys through which they flow and produce characteristic landforms such as cirques, arêtes, horns, hanging valleys, and roches moutonneés.
- When glaciers melt, they deposit sediments called *glacial drift*.
- Glacial deposits may form erratics, kettles, eskers, drumlins, and moraines.
- Glaciers may form lake basins by eroding the land or by depositing sediments.

3 Ice Ages

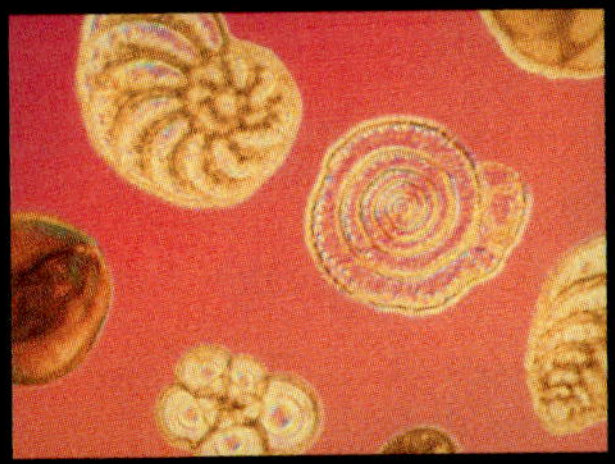

ice age, 431
Milankovitch theory, 433

- An ice age occurs when a long period of climatic cooling causes continental glaciers to cover large areas of Earth's surface.
- During an ice age, cooler glacial periods alternate with warmer interglacial periods.
- The Milankovitch theory suggests that glacial periods are caused by changes in the distribution of solar energy Earth receives. These changes are caused by regular changes in the eccentricity of Earth's orbit, the tilt of Earth's axis, and precession.
- Variations in solar activity, volcanic activity, and plate tectonics also affect climate.

Chapter Highlights

Alternative Assessment — GENERAL

Glacier Brochure Have students make a six-panel brochure that explains the following: how a glacier forms, how it moves, what it looks like, and how it affects both the surface and the climate of Earth. Display the brochures and allow students to review other students' brochures. LS **Verbal/Visual**

CHAPTER RESOURCES

Chapter Resource File

- **Concept Review** GENERAL
- **Critical Thinking** ADVANCED
- **Math Skills** GENERAL
- **Graphing Skills** GENERAL
- **Chapter Test A** GENERAL
- **Chapter Test B** ADVANCED

Workbooks

- **Study Guide** (also in Spanish)
- **Assessments** (Spanish)

Technology

Classroom Videos
- Brain Food Video Quiz

HRW Earth Science Video
- Segment 11: Glaciers and Erosion

Chapter Review

Assignment Guide

Section	Questions
1	1–2, 4, 6, 9–12, 21, 23, 26
2	5, 7, 13–14, 18–20
3	3, 8, 15–17, 24, 30–33
1 and 2	25
1–3	22, 27–29

Using Key Terms

1–8. Answers may vary but should show that students understand the definitions of and differences between key terms.

Understanding Key Concepts

9. b **10.** a
11. a **12.** b
13. a **14.** b
15. a **16.** c

Short Answer

17. The climate cools during an ice age.

18. A *lateral moraine* is sediment deposited along the sides of a valley glacier. A *medial moraine* is a combination of two adjacent lateral moraines that occurs when two or more valley glaciers join. A *ground moraine* is the unsorted material left when the ice melts. A *terminal moraine* is a small ridge of till deposited at the leading edge of a melting glacier.

19. Three types of landforms created by alpine glaciers are cirques, or bowl-shaped depressions; arêtes, or sharp, jagged ridges between cirques; horns, or sharp, pyramid-like peaks that form when several arêtes are joined; and hanging valleys, tributary valleys suspended on mountains high above the main valley floor.

20. Some glacial lakes form when glaciers erode surfaces and leave depressions. Others form in the uneven surfaces of ground moraine deposits.

21. Basal slip is the process by which an entire glacier moves by slipping over a thin layer of water and sediment that is the result of the melting of the ice that is in contact with the ground. Internal plastic flow is the process by which the interior of a glacier moves as grains of ice deform under pressure and slip over each other.

Chapter 17 Review

Using Key Terms

Use each of the following terms in a separate sentence.

1. *glacier*
2. *crevasse*
3. *ice age*

For each pair of terms, explain how the meanings of the terms differ.

4. *alpine glacier* and *continental glacier*
5. *till* and *moraine*
6. *basal slip* and *internal plastic flow*
7. *cirque* and *arête*
8. *glacial period* and *interglacial period*

Understanding Key Concepts

9. Glaciers that form in mountainous areas are called
 a. continental glaciers.
 b. alpine glaciers.
 c. icebergs.
 d. ice shelves.

10. A glacier will move by sliding when the base of the ice and the underlying rock are separated by a thin layer of
 a. water and sediment.
 b. snow.
 c. pebbles.
 d. drift.

11. What part of a glacier moves fastest when the glacier moves by internal plastic flow?
 a. The center of the glacier moves fastest.
 b. The bottom of the glacier moves fastest.
 c. The edges of the glacier move fastest.
 d. The whole ice mass moves at the same speed.

12. Icebergs form when ice breaks off of a(n)
 a. crevasse. **c.** alpine glacier.
 b. ice shelf. **d.** esker.

13. As a glacier moves through a valley, it carves out a(n)
 a. U shape. **c.** V shape.
 b. esker. **d.** moraine.

14. A deposit of stratified drift is called a(n)
 a. drumlin.
 b. outwash plain.
 c. ground moraine.
 d. roche moutonnée.

15. One component of the Milankovitch theory is
 a. the circular motion of Earth's axis.
 b. continental drift.
 c. volcanic activity.
 d. landslide activity.

16. Which of the following is *not* a theory for the cause of ice ages?
 a. volcanic eruptions
 b. variations in Earth's orbit
 c. Foraminifera shell coils
 d. changes in tectonic plate position

Short Answer

17. How does climate change during an ice age?

18. What are the four types of moraines, and how are they different from each other?

19. Identify three types of landforms created by alpine glaciers.

20. In what two ways do glacial lakes form?

21. How do the processes of basal slip and internal plastic flow differ?

Critical Thinking

22. Identifying Relationships Why is it important for scientists to monitor and study the continental ice sheets that cover Greenland and Antarctica?

23. Applying Concepts Antarctic explorers need special training to travel safely over the ice sheet. Besides the cold, what structural aspects of the glaciers may be dangerous?

24. Evaluating Data What phenomenon other than decreased temperature and increased snowfall might signal the beginning of a glacial period?

Concept Mapping

25. Use the following terms to create a concept map: *snowfield, erosion, deposition, glacier, horn, arête, kettle, moraine, basal slip,* and *internal plastic flow.*

Math Skills

26. Using Equations The area of Earth's surface that is covered with water is 361,000,000 km^2. The volume of water locked in ice in the Antarctic ice sheet is about 26,384,368 km^3. Use the following equation to find the average worldwide rise in sea level, in meters, that would occur if the Antarctic ice sheet melted.

$$\text{rise in water level} = \frac{\text{volume of water in ice sheet}}{\text{area of Earth covered by water}}$$

27. Evaluating Data New York City has an elevation of 27 m above sea level. Although highly unlikely, what would happen to the city if the Antarctic ice sheet suddenly melted and raised the worldwide sea level by 50 m?

Writing Skills

28. Writing Persuasively Write a research proposal to the National Science Foundation that details a plan of action and reasons for studying the ice sheet in Greenland.

29. Writing from Research Research how ocean currents affect the polar icecaps. Write a short essay that describes global ocean currents and explains how they affect the formation and advancement of polar icecaps.

Interpreting Graphics

The graph below shows the relationship between cycles of eccentricity, tilt, and precession. Use the graph to answer the questions that follow.

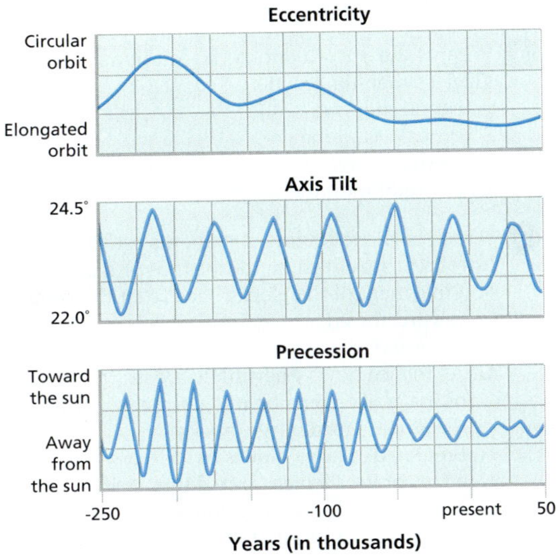

30. What was the angle of Earth's tilt 50,000 years ago?

31. Describe the shape of Earth's orbit 200,000 years ago.

32. How would the seasons 50,000 years from now be different from the seasons 50,000 years ago?

Chapter Review

Critical Thinking

22. If these ice sheets melted, the worldwide sea level could rise and could flood many coastal cities.

23. The surface of a glacier is brittle, which causes crevasses to form. Large pieces of an ice shelf may break off to form an iceberg.

24. The advancement of ice sheets and a drop in sea level may be indications of the beginning of a glacial period.

Concept Mapping

25. Answers may vary but should include all of the terms listed. Sample answers appear at the end of this Teacher Edition.

Math Skills

26. 26,384,368 km^3 ÷ 361,000,000 km^2 = 0.073 km = 73 m

27. Because the average water level would rise 50 m, New York would be 23 m under water (50 m − 27 m = 23 m).

Writing Skills

28. Answers may vary. Accept all reasonable answers.

29. Answers may vary. Accept all reasonable answers.

Interpreting Graphics

30. approximately 24.3°

31. more circular than now

32. There will be less seasonality 50,000 years from now than there was 50,000 years ago. There will be less snow melt in summer and more snowfall in winter. Although Earth's orbit and precession will be about the same as they were then, the tilt of the axis will be much less.

Estimated Time

To give students practice under more realistic testing conditions, allow them 30 minutes to answer all of the questions in this practice test.

TEST DOCTOR

Question 1 Answer C is correct. Though glaciers are sometimes called *rivers of ice*, frozen water moves differently than water in its liquid state. So answer D is incorrect. Glacial ice cannot flow rapidly or move around obstacles easily, so answers A and B are incorrect. Both ice and water, however, do flow downward in response to gravity.

Question 4 Answer G is correct. Outward movement of ice sheets forms *ice shelves*, so answer answer F is incorrect. The breakage of large blocks of ice from the edges of ice shelves forms *icebergs*, so answer H is incorrect. Narrow, wedge-shaped masses of ice confined to small areas by the surrounding topography are *alpine glaciers*, so answer I is incorrect.

Question 8 From the information in the passage, the best choice is answer I. The text does not discuss how often ice ages occur, so students can rule out answer F. Answers G and H use the word *always*, which should serve as a warning flag to students. The text describes one instance but does not discuss what always happens.

Chapter 17 Standardized Test Prep

Understanding Concepts

Directions (1–4): **For *each* question, write on a separate sheet of paper the letter of the correct answer.**

1 Which statement *best* compares the movement of glacial ice to the movement of river water?
- **A** Glacial ice moves more rapidly than water.
- **B.** Glacial ice cannot easily flow around barriers.
- **C.** Glacial ice moves in response to gravity.
- **D.** Glacial ice moves in the same way as water.

2 What landform created by glaciers has a bowl-like shape?
- **F.** cirque
- **G.** arête
- **H.** horn
- **I.** roches moutonnée

3 What is the unsorted material left beneath a glacier when the ice melts?
- **A.** lateral moraine
- **B.** ground moraine
- **C.** medial moraine
- **D.** terminal moraine

4 Which of the following statements *best* describes how crevasses form on the surface of a glacier?
- **F.** Movement of the glacier's ice from the center toward the edges forms large cracks on the surface of the glacier.
- **G.** As the ice flows unevenly beneath the surface of the glacier, tension and compression on the surface form large cracks.
- **H.** Breakage of large blocks of ice from the edges of ice shelves forms large cracks.
- **I.** Narrow, wedge-shaped masses of ice confined to a small area form large cracks.

Directions (5–6): **For *each* question, write a short response.**

5 What is the term for all types of sediments deposited by a glacier?

6 What is the name of a jagged ridge that is formed between two or more cirques that cut into the same mountain?

Reading Skills

Directions (7–9): **Read the passage below. Then, answer the questions.**

Glacial and Interglacial Periods

Ice ages are periods during which ice collects in high latitudes and moves toward lower latitudes. During an ice age, there are periods of cold and of warmth. These periods are called *glacial and interglacial periods*. During glacial periods, enormous sheets of ice advance, grow bigger, and cover a large area. Because a large amount of sea water is frozen during glacial periods, the sea level around the world drops.

Warmer time periods that occur between glacial periods are known as interglacial periods. During an interglacial period, the large ice sheets begin to melt and the sea levels begin to rise again. Scientists believe that the last interglacial period began approximately 10,000 years ago and is still happening. For nearly 200 years, scientists have been debating what the current interglacial period might mean for humans and the possibility of a future glacial period.

7 According to the passage, which of the following statements is true?
- **A.** The last interglacial period began approximately 1,000 years ago.
- **B.** Scientists have been thinking about the next glacial period for two centuries.
- **C.** Ice ages are periods during which ice collects in the lower latitudes and moves toward higher latitudes.
- **D.** During glacial periods, enormous sheets of ice tend to melt, so they become smaller and cover less area.

8 Which of the following statements can be inferred from the information in the passage?
- **F.** On average, ice ages occur every 50,000 years and always start with a glacial period.
- **G.** Interglacial periods always last 10,000 years.
- **H.** Glacial periods always last 10,000 years.
- **I.** The current interglacial period will likely be followed by a glacial period.

9 If a new glacial period began tomorrow, what might happen to coastal cities

Answers

Understanding Concepts

1. C
2. F
3. B
4. G
5. glacial drift
6. arêtes

Reading Skills

7. B
8. I
9. As sea levels decreased, some coastal cities might become landlocked. Due to the withdrawal of ocean water, they may also experience changes in local wind and weather patterns.

Interpreting Graphics

10. A
11. Answers may vary. See Test Doctor for a detailed scoring rubric.
12. Only Kiev would remain above sea level. The other cities would be below sea level by as few as 23 m (New York) and as much as 48 m (Amsterdam).
13. F

Interpreting Graphics

Directions (10–13): **For *each* question below, record the correct answer on a separate sheet of paper.**

Base your answers to questions 10 through 11 on the diagram below.

Basal Slip

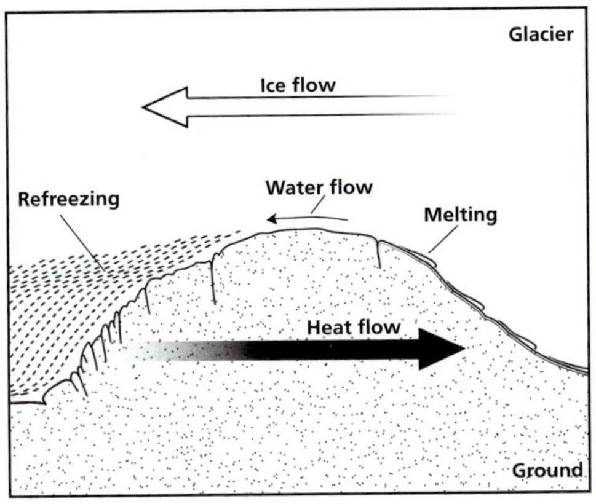

10 What causes the ice to melt in the diagram above?

A. Pressure decreases the melting point of the ice.
B. Pressure increases the melting point of the ice.
C. The ground heats the ice until it melts.
D. Ice at the base of a glacier does not melt.

11 How does meltwater influence basal slip?

Base your answers to question 12 on the table below.

World Cities and Their Elevations

City	Elevation (m)
New York City	27
Kiev, Ukraine	168
Buenos Aries, Argentina	25
Amsterdam, Netherlands	2

12 Although unlikely, what would happen to each of the cities listed in the table if the Antarctic ice sheet were to melt and raise the sea level by 50 m?

13 The movement of a glacier was recorded over a period of 180 days. During that time, the glacier moved a total of 36 m. What was the average speed of the glacier each day?

F. 0.20 m/day
G. 0.50 m/day
H. 2.00 m/day
I. 5.00 m/day

Test TIP

For a group of questions that refer to a diagram, graph, or table, read all of the questions quickly to determine what information you will need to glean from the graphic.

Standardized Test Prep

TEST DOCTOR

Question 11 Full-credit answers should include the following points:

- as pressure builds up, heat energy is created which causes the ice to melt
- meltwater acts as a lubricant, decreasing friction between the glacier and the underlying obstacle, and allows the glacier to slide over obstacles
- once past the obstacle, the pressure decreases and the meltwater refreezes

Question 13 Answer F is the correct choice. Speed is the ratio between the distance moved and the time taken to move that distance. Students can use the following equation to find the answer:

x 36 m 180 days

Students should arrive at an answer of 0.20 meters per day.

Test Prep Correlations ***National Science Education Standards***

ES 3c: items 1, 2, 3, 4, 5, 6, 8, 13
HNS 2b: items 1, 4, 10
SAI 2d: item 12
SAI 2e: items 9, 10, 11, 13
SPSP 4a: items 12, 13
SPSP 5a: items 8, 9, 12, 13
SPSP 5c: items 7, 8, 13
SPSP 5d: item 13
SPSP 6d: items 9, 13
UCP 1: items 7, 8, 9
UCP 3: items 5, 7, 8, 9, 10, 13

CHAPTER RESOURCES

State Resources

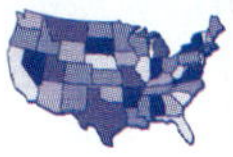

For specific resources for your state, visit go.hrw.com and type in the keyword HSHSTR.

Glaciers and Sea Level Change

Teacher's Notes

Time Required

two 45-minute class periods

Lab Ratings

EASY → HARD

Teacher Preparation ▲▲▲
Student Setup ▲
Concept Level ▲▲▲
Cleanup ▲▲▲▲

Skills Acquired

- Constructing Models
- Observing
- Measuring
- Collecting Data
- Analyzing Models
- Interpreting Models

The Scientific Method

In this lab, students will

- Make Observations
- Analyze the Results
- Draw Conclusions

Materials

The materials listed on the page are enough for groups of two to four students. Blocks of ice must be prepared prior to the day the lab is conducted to allow sufficient time for the water to freeze.

Tips and Tricks

To speed up the melting of the ice block, you may wish to use a hair drier or other safe heat source. This lab also can be done as a classroom demonstration. Tell the students to record the measurements as you make them.

Objectives

- **Model** the melting of an ice sheet.
- USING SCIENTIFIC METHODS **Analyze** the effects of melting ice on sea level.

Materials

- block, ice, 5 cm × 5 cm × 5 cm
- block, wooden, 5 cm × 5 cm × 5 cm
- pan, 30 cm × 40 cm × 10 cm
- pebbles (1 kg)
- ruler, metric
- sand (1 kg)
- water

Safety

Glaciers and Sea Level Change

Today, glaciers hold only about 2.2% of Earth's water. But if the polar ice sheets melted, the coastal areas of many countries would flood. Many major cities, such as New York, New Orleans, Houston, and Los Angeles, would flood if the sea level rose only a few meters. In this lab, you will construct a model to simulate what would happen if the Antarctic ice sheet melted.

PROCEDURE

1. Calculate and record the approximate surface area of the bottom of the pan. Area (A) is equal to length (l) times width (w), or $A = l \times w$, and is expressed in square units.
2. Calculate and record the overall volume of the ice block and the area of one side of the ice block. Volume (V) is equal to length (l) times width (w) times height (h), or $V = l \times w \times h$, and is expressed in cubic units.
3. Add the sand and small pebbles to one end of the pan so that they cover about half of the area of the pan and slope toward the middle of the pan. Use the wooden block to elevate the end of the pan containing the sand and pebbles.
4. Slowly add water to the opposite end of the pan. Be sure that the water does not cover the sand and pebbles and touches only the edge of the sand.

Step 3

CHAPTER RESOURCES

Chapter Resource File

- Datasheet for Chapter Lab GENERAL
- Lab Notes and Answers

5 Measure and record the depth of the water at the deepest point.

6 Measure and record the distance from the end of the pan covered with sand to the point where the sand touches the water.

7 Place the block of ice in the pan on top of the sand. Calculate and record the percentage of the total area of thc pan that is covered by ice.

8 As the ice begins to melt, pick up the ice block. Note the appearance of the bottom of the ice block and the appearance of the sand under the ice block. Record what is happening to the ice block and what is happening to the sand under the ice block. Place the ice block back on the sand.

9 While the ice is melting, calculate the expected rise in the pan's water level by using the following formula:

$$\text{rise in water level} = \frac{\text{volume of water in ice block}}{\text{area of pan covered by water}}$$

10 When the ice is completely melted, measure and record the depth of water at the deepest point.

11 Measure and record the distance from the end of the pan covered with sand to the point where the sand touches the water.

Step 4

ANALYSIS AND CONCLUSION

1 **Making Comparisons** Compare the depth of water at the beginning of the lab with the depth at the end of the lab. Explain any differences.

2 **Analyzing Results** How did the distance from the end of the pan covered with sand to the point where the sand touches the water change? Explain your answer.

3 **Compare and Contrast** How does the ice-block model differ from a glacier on Earth?

4 **Drawing Conclusions** How does the ice-block model represent what would happen on Earth if the Antarctic polar ice sheet melted?

Extension

1 **Evaluating Models** In this lab, you used a physical model to simulate an occurrence in nature. In what other ways do scientists use models? What kinds of errors can occur when models are used?

Answers to Analysis and Conclusion

1. The depth of the water is greater at the end of the exercise than at the beginning because the ice melted.
2. The distance decreased because the surface of the liquid water increased.
3. Answers may vary. Sample answer: The ice-block model does not have all of the physical features of a glacier on Earth. For example, the ice in the model has not undergone the compression of a real glacier. Therefore, the volume of water in the ice block is less than the volume of water in actual glacier ice.
4. The water on Earth would enter the ocean, which would cause sea level to rise and to cover land just as the water covered some of the sand in the model.

Answers to Extension

1. Answers may vary. Sample answer: Scientists study the universe by making models of things that are too large or too small to be easily observed in nature. Models can be simpler than an actual item or situation or they can be out of proportion. These factors can cause errors.

Randa Flinn
Northeast High School
Fort Lauderdale, FL

MAPS in Action

Gulkana Glacier

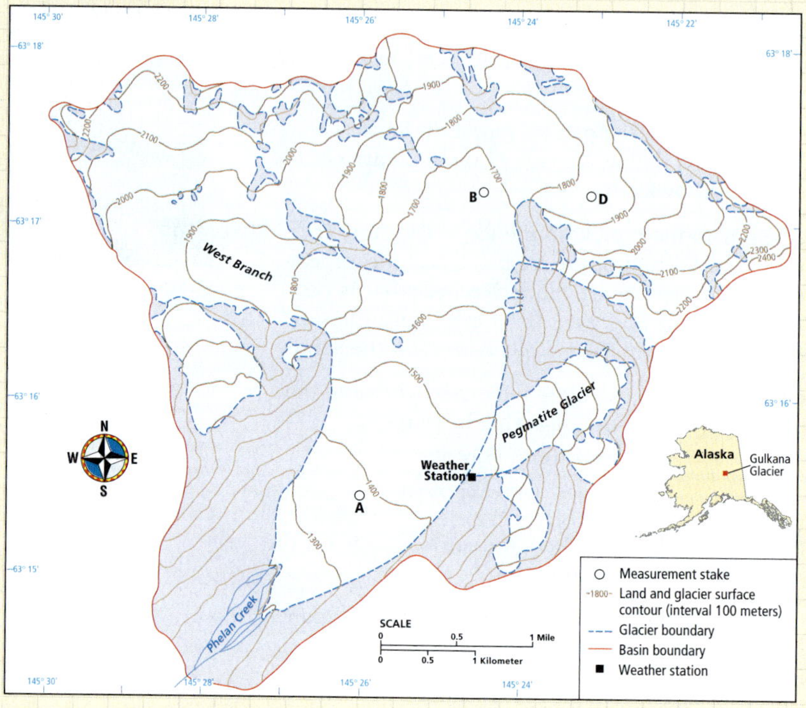

Gulkana Glacier

Internet Activity — GENERAL

Photographic Evidence Have students compare photographs of the Gulkana Glacier with the topographic map. The Gulkana Glacier home page of the United States Geological Survey Web site contains aerial photographs and other images of the glacier. Have students write a short essay that describes the Gulkana landscape. A worksheet designed to direct student research on this topic can be found in the **Chapter Resource File** booklet or by visiting **go.hrw.com** and entering the keyword **HQ6GLAX.** **LS Visual**

Answers to Map Skills Activity

1. about 6.25 km
2. 63° 16.9'N latitude 145° 21'W longitude
3. northwest
4. southwest
5. Answers may vary. Sample answer: The movement of the glacier is toward the lower elevation. For measurement stake A, the elevation becomes lower toward the southwest; for measurement stake B, the elevation becomes lower toward the south. The direction of flow is constrained by surrounding topography.
6. Answers may vary: Sample answer: Scientists are probably measuring the rate at which the glacier is flowing.

CHAPTER RESOURCES

Chapter Resource File

- Internet Activity
  Photographic Evidence GENERAL

Technology

Transparencies
- 88 Gulkana Glacier (with worksheet)

Map Skills Activity

This contour map shows the elevation and boundaries of the Gulkana Glacier located in Alaska. Use the map to answer the questions below.

1. **Using the Key** Estimate the length of the Gulkana Glacier from its northernmost point to its southernmost point.
2. **Analyzing Data** Estimate the latitude and longitude of the glacier's highest point.
3. **Identifying Trends** In which direction do you think Gulkana Glacier is moving at measurement stake D?
4. **Identifying Trends** In which direction do you think Gulkana Glacier is moving at measurement stake A?
5. **Predicting Consequences** Why might you think that the glacier moves in different directions at measurement stake A and measurement stake B?
6. **Analyzing Data** What do you think the scientists are measuring at the measurement stakes?

EYE on the Environment

The Missoula Floods

In an area called the Channeled Scablands in Washington State, scientists discovered numerous gravel bars and ridges that are roughly parallel to each other. The ridges are 10 m high, extend about 115 m from crest to crest, and can be more than 3 km long. The scientists were puzzled and set out to determine how the ridges formed.

Mystery Ripples

After extensive research and some aerial photography, scientists realized that the gravel ridges are giant ripple marks that formed as a huge amount of water poured across the land. Scientists had observed similar ripples created by the movement of flowing water in rivers and on beaches, but no one had ever seen ripples of this size!

Catastrophic Floods

What caused such enormous ripples? During the last glacial period, a series of ice dams formed across the Clark Fork River, and a giant glacial lake—Lake Missoula—periodically formed. This lake was as big as Lake Erie and Lake Ontario combined! About 14,000 years ago, the last ice dam broke, and a wave 650 m tall raced across eastern Washington at a speed of about 90 km/h. The wave left huge piles of gravel and ripple marks in its wake. Scientists estimate that the water rushed through the region as fast as 400 million cubic feet per second and emptied the entire lake in only two days.

Extension

1. **Research and Communications** Research the type of sediment that makes up the gravel ridges in the Channeled Scablands. Write a brief essay that explains whether the type of sediment in the gravel ridges supports the hypothesis of a giant flood.

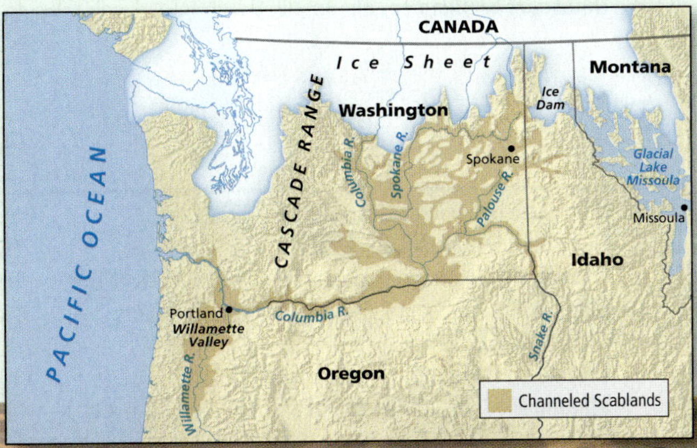

▼ The Channeled Scablands formed when an ice dam broke 14,000 years ago.

Eye on the Environment

The Missoula Floods

Group Activity — ADVANCED

Skit Tell students that the first evidence of the catastrophic floods that created the Channeled Scablands was presented by J. Harlen Bretz in the 1920s. His findings were so controversial that he was publicly challenged by six other geologists in a debate in Washington, D.C. Have students first research Bretz's evidence for the formation of the Channeled Scablands and the scientific community's objections to it. Then, have them write a skit about the debate and present the skit to the class. LS Kinesthetic/Interpersonal

Answer to Extension

1. Students should conclude from their research that the gravel ridges are made up of basalt. Essays will vary but should support the hypothesis of a giant flood or floods.

Chapter 18

Erosion by Wind and Waves

Planning Guide

Compression Guide
To shorten instruction because of time limitations, omit Section 3.

OBJECTIVES	LABS, DEMONSTRATIONS, AND ACTIVITIES	TECHNOLOGY RESOURCES
PACING • 45 min pp. 444–450 **Chapter Opener**		**OSP** Parent Letter **CD** Student Edition on CD-ROM **CD** Chapter Summaries Audio CD **VID** Brain Food Video Quiz
Section 1 Wind Erosion • Describe two ways that wind erodes land. • Compare the two types of wind deposits.	**TE Demonstration** Watching Sand Jump, p. 445 GENERAL **TE Activity** Shaping, p. 447 GENERAL **TE History Connection** Dust Bowl, p. 447 ADVANCED **SE Quick Lab*** Modeling Desert Winds, p. 449 GENERAL **CRF Datasheet for Quick Lab*** GENERAL **CRF Inquiry Lab** Soil Erosion* GENERAL	**OSP Lesson Plans** (also in print) **TR Bellringer*** **TR** 89 Types of Dunes* **TE Internet Activity** Great Lakes Dunes, p. 448 ADVANCED **CRF Internet Activity** Great Lakes Dunes* ADVANCED **VID HRW Earth Science Video** Weathering and Erosion
PACING • 90 min pp. 451–454 **Section 2 Wave Erosion** • Compare the formation of six features produced by wave erosion. • Explain how beaches form. • Describe the features produced by the movement of sand along a shore.	**TE Teaching Tip** Connect to Prior Knowledge, p. 452 GENERAL **SE Inquiry Lab** Beaches, p. 464–465 GENERAL **CRF Datasheet for Chapter Lab*** GENERAL	**OSP Lesson Plans** (also in print) **TR Bellringer*** **TR** 90 Wave Erosion and Landforms* **TE Internet Activity** Landform Photographs, p. 452 GENERAL **CRF Internet Activity** Landform Photographs* GENERAL **CD Interactive Tutor** Physical and Chemical Weathering
PACING • 45 min pp. 455–458 **Section 3 Coastal Erosion and Deposition** • Explain how changes in sea level affect coastlines. • Describe the features of a barrier island. • Analyze the effect of human activity on coastal land.	**SE Quick Lab** Graphing Tides, p. 456 GENERAL **CRF Datasheet for Quick Lab*** GENERAL **TE Geography Connection** Coastal Features, p. 457 GENERAL **SE Maps in Action** Coastal Erosion Near the Beaufort Sea, p. 466 GENERAL **TE Group Activity** Debate Project, p. 467 ADVANCED **CRF Making Models Lab** Erosion of a Submerging Coastal Profile* GENERAL	**OSP Lesson Plans** (also in print) **TR Bellringer*** **TR** 91 Submergent Coastlines* **TR** 92 Coastal Erosion Near the Beaufort Sea* **VID Nova Video** Hurricane! **CD Interactive Tutor** Coastal Zone

PACING • 90 min

CHAPTER REVIEW, ASSESSMENT, AND STANDARDIZED TEST PREPARATION

- **SE Chapter Highlights**, p. 459
- **SE Chapter Review**, pp. 460–461
- **SE Standardized Test Prep**, pp. 462–463
- **CRF Concept Review*** GENERAL
- **CRF Critical Thinking*** ADVANCED
- **CRF Math Skills*** GENERAL
- **CRF Graphing Skills*** GENERAL
- **CRF Chapter Test A*** GENERAL
- **CRF Chapter Test B*** ADVANCED
- **OSP Lesson Plans** (also in print)
- **OSP Test Generator**
- **OSP Test Item Listing**

Online and Technology Resources

Holt Online Learning

Visit **go.hrw.com** for access to Holt Online Learning, or enter the keyword **HQ6 Home** for a variety of free online resources.

One-Stop Planner® CD-ROM

This CD-ROM package includes
- Lab Materials QuickList Software
- Holt Calendar Planner
- Customizable Lesson Plans
- Printable Worksheets
- ExamView® Test Generator
- Interactive Teacher Edition
- Holt PuzzlePro®
- Holt PowerPoint® Resources

KEY

SE Student Edition	**OSP** One-Stop Planner	**VID** Classroom Video/DVD
TE Teacher Edition	**TR** Transparencies and Transparency Worksheets	* Also on One-Stop Planner
CRF Chapter Resource File	**CD** CD or CD-ROM	◆ Requires advance prep
LTP Long-Term Projects		■ Also available in Spanish

SKILLS DEVELOPMENT RESOURCES	REVIEW AND ASSESSMENT	CORRELATIONS
SE **Pre-Reading Activity,** p. 444 GENERAL TE **Using the Figure** Rock Formation, p. 444 GENERAL		National Science Education Standards
CRF **Directed Reading*** BASIC TE **Inclusion Strategies,** p. 446 TE **Reading Skill Builder** Paired Summarizing, p. 446 BASIC TE **Using the Figure** Dune Formation, p. 447 BASIC SE **Graphic Organizer** Comparison Table, p. 448 GENERAL TE **Using the Figure** Dune Migration, p. 449 BASIC	SE **Reading Checks,** pp. 446, 448 GENERAL SE **Section Review,** p. 450 GENERAL TE **Homework,** p. 446 BASIC TE **Reteaching,** p. 449 BASIC TE **Quiz,** p. 449 GENERAL TE **Alternative Assessment,** p. 450 GENERAL CRF **Section Quiz*** ■ GENERAL	ES 3c, UCP 3, UCP 4, SPSP5c
CRF **Directed Reading*** BASIC TE **Using the Figure** Breaking Waves, p. 451 GENERAL TE **Reading Skill Builder** Reading Organizer, p. 452 BASIC TE **Inclusion Strategies,** p. 452 SE **Math Practice,** p. 453 GENERAL TE **Skill Builder** Writing, p. 453 BASIC	SE **Reading Check,** p. 452 GENERAL SE **Section Review,** p. 454 GENERAL TE **Homework,** p. 453 ADVANCED TE **Reteaching,** p. 453 BASIC TE **Quiz,** p. 453 GENERAL TE **Alternative Assessment,** p. 454 ◆ GENERAL CRF **Section Quiz*** ■ GENERAL	ES 3c, UCP 3, UCP 4
CRF **Directed Reading*** BASIC TE **Using the Figure** Sea-Level Changes, p. 455 GENERAL TE **Skill Builder** Writing, p. 457 BASIC	SE **Reading Check,** p. 457 GENERAL SE **Section Review,** p. 458 GENERAL TE **Reteaching,** p. 457 BASIC TE **Quiz,** p. 457 GENERAL TE **Alternative Assessment,** p. 458 GENERAL TE **Homework,** p. 466 ADVANCED CRF **Section Quiz** * ■ GENERAL	ES 3c, SPSP5c, SAI 1a, SAI 1f, ST 1e, UCP 2, UCP 3, UCP 4

Holt Lab Generator CD-ROM

Search for labs by topic, standard, difficulty level, and/or time frame. Edit those labs to fit your needs, or create your own labs. Use the Lab Materials QuickList to order just what you need.

Holt Earth Science Interactive Tutor CD-ROM

This CD-ROM consists of interactive activities that give students a fun way to extend their knowledge of Earth science concepts.

Chapter Summaries Audio CDs

These CDs include audio summaries of the key concepts presented in each chapter. (Audio summaries are also available in Spanish.)

SCiLINKS. NSTA

www.scilinks.org

Maintained by the **National Science Teachers Association.** See Chapter Enrichment pages that follow for a complete list of topics.

See Chapter Enrichment pages for Video Resources.

Refer to our ***Professional Reference for Teachers*** for additional teaching resources, including articles written by science education professionals about relevant and timely issues facing today's science teachers.

Chapter 18 Chapter Enrichment

This Chapter Enrichment provides relevant and interesting information to expand and enhance your classroom instruction of the chapter material.

Section 1 Wind Erosion

The Dust Bowl of the 1930s

At the beginning of the 20th century, the Great Plains of the United States consisted of open grassland. As settlers moved west in the 1920s, many laid claim to the vast land, and millions of acres of grassland were plowed. In the early 1930s, the farmers of southeast Colorado, southwest Kansas, the panhandles of Texas and Oklahoma, and northeast New Mexico suffered the effects of a severe drought. Because the native grasses, which served to anchor the topsoil, had been plowed under, the winds swept the dry soil away in huge black clouds. These dust storms were massive in size and intensity. Swirling clouds of dust seemed to change day into night and left people gasping for air. Breathing was so difficult that most outdoor activities were impossible. The dust that was left behind seeped into even the most tightly closed buildings.

The dust remained as a thin, gritty film, in layers several inches thick, or even as drifts, similar to snow. The winds carried the dust unbelievable distances—at times more than 900 miles from its starting point. The amount of dust carried by the winds was also remarkable—one ton per square mile was common.

The dust storms fostered a whole new era of soil conservation. In 1937 the Shelterbelt Project paid farmers to plant and cultivate native trees such as red cedar and green ash along fence lines that separated properties. In 1938 farmers began the practice of plowing their land into furrows. Together, these conservation practices resulted in a 65 percent reduction of soil erosion from blowing winds. The drought finally ended in 1939, and the barren plains soon returned golden with wheat.

▼ Dust storms are common where wind erosion is greatest.

▲ The energy of crashing waves causes erosion.

Section 2 Wave Erosion

Water Currents and Swimming Safety

Many lives are tragically lost each year from drowning accidents along the shores of lakes and oceans. Often, individuals were not in deep water or far from shore. They were caught in longshore and rip currents and were not able to fight the strength of the pull. Longshore currents flow parallel to the shore. They can form circulatory patterns and leave the shore in narrow rip currents. The longshore current is strongest when the waves are high and it is approaching at a greater angle from perpendicular to shore. Longshore currents have their highest velocities near the water's surface. Swimmers can escape longshore currents by swimming or wading toward the shore.

Rip currents occur when gravity pulls water from a wave back in a narrow, river-like current moving away from shore. Rip currents vary in size and intensity. They can be from 15 to 45 m wide. They may flow just past the surf line or hundreds of meters offshore. Rip currents can appear suddenly or intensify after a set of waves, or they may pull continuously. The United States Lifeguard Association offers guidelines for swimmers caught in a rip current. First, try to relax and do not try to swim against the current. Tread water, float, and wave for assistance, or try to swim parallel to the shore until the end of the current is reached, then swim toward shore.

Section 3 Coastal Erosion and Deposition

Methods to Shield the Shoreline

Sea walls are constructed parallel to the shore to help reduce erosion from pounding waves. Jetties that run perpendicular to the shore are designed to block currents that carry away sand. However, these barriers don't last forever. Water action eventually erodes the barrier foundations. In addition, barriers can solve and create problems at the same time. For example, sand may collect on one side of the jetty, but swirling currents may scour the shore on the opposite side of the jetty.

The Dutch have fought the sea for centuries, reclaiming over 7,000 km^2 of land for farming and housing. They created a network of protective dikes, water pumping stations, ditches, and canals. However, in 1953 the North Sea overflowed the sea walls and killed nearly 2,000 people. In some cases, human attempts to slow down erosion have instead sped up the process. In Redondo, California, a misplaced jetty resulted in the loss of an entire city block before the jetty was stabilized. Santa Barbara, California, has also suffered losses of kilometers of beach due to misplaced jetties.

▼ **Shoreline preservation requires constant vigilance.**

Video Resources

Brain Food Video Quizzes

Brain Food Video Quizzes These videos contain game-show style quizzes that assess students' progress and motivate students to study the chapter material.

HRW Earth Science Video

This video introduces Earth science topics and includes a geology field trip. The video segment listed below complements this chapter.

Segment 9: Weathering and Erosion
Weathering and erosion, which continually reshape the landscape of Earth, are the focuses of this segment. The processes that bring about weathering, as well as are the principal agents of erosion are briefly examined. Both naturally occurring and human-caused slope instability are also described. (1.5 min)

NOVA® Videos

The **NOVA** video below complements this chapter.

Hurricane! Watch scientists fly into the center of hurricanes so that they can learn more about the hurricanes, and listen to firsthand accounts of people affected by Hurricane Camille. (60 min)

To order other **NOVA** videos related to this chapter, visit **go.hrw.com** and enter the keyword **HQ6EWWV**.

SciLinks®

NSTA Developed and maintained by the National Science Teachers Association

SciLinks is maintained by the National Science Teachers Association to provide you and your students with interesting, up-to-date links that will enrich your classroom presentation of the chapter.

Visit www.scilinks.org and enter the SciLinks code for more information about the topic listed.

Topic: Wind Erosion
SciLinks code: HQ61669

Topic: Waves
SciLinks code: HQ61641

Topic: Wave Erosion
SciLinks code: HQ61638

Topic: Coastal Changes
SciLinks code: HQ60307

Chapter 18

Chapter Overview

This chapter describes the powerful influence of wind and waves on Earth's surface. Agents of erosion constantly modify the appearance of coastlines and challenge humans to find creative solutions to protect coastal lands.

Using the Figure — GENERAL

Rock Formation Ask students if they have seen rock formations similar to the one in the photo. Have students describe where they saw these formations and what the rock looked like. Provide pictures of rock formations, coastline erosion, dunes, and oceans to demonstrate the power of erosion by wind and water. LS Visual

PRE-READING ACTIVITY

FOLDNOTES

Have students exchange FoldNotes and check each other's notes on wind and wave erosion. Encourage students to use a colored pen to add additional notes about deposition by wind and waves.

Chapter 18 Erosion by Wind and Waves

Sections

1 Wind Erosion
2 Wave Erosion
3 Coastal Erosion and Deposition

What You'll Learn

- How wind erodes the land
- How waves erode shorelines
- How coastlines change

Why It's Relevant

Wind and wave erosion change features of Earth's surface. Soil that is needed for farming is carried away by wind. Wave action affects beaches and coasts, which have high economic and recreational value.

PRE-READING ACTIVITY

FOLDNOTES

Pyramid Before you read the chapter, create the FoldNote entitled "Pyramid" described in the Skills Handbook section of the Appendix. Label the sides of the pyramid with "Wind erosion," "Wave erosion," and "Coastal erosion and deposition." As you read the chapter, define each type of erosion, and write characteristics of each type on the appropriate pyramid side.

▶ This unusual rock sits on Kangaroo Island, which is located off the southern shore of Australia. This rock, like many other rocks, has been shaped by the forces of the wind and the waves.

Chapter Correlations

National Science Education Standards

ES 3c Interactions among the solid earth, the oceans, the atmosphere, and organisms have resulted in the ongoing evolution of the earth system.... **(Sections 1–3)**

SPSP 5c ...But there are slow and progressive changes that also result in problems for individuals and societies. For example, ...erosion of bridge foundations, sedimentation in lakes and harbors, coastal erosions, and continuing erosion and wasting of soil and landscapes can all negatively affect society. **(Sections 1–3)**

SAI 1a ...Formulate a testable hypothesis and demonstrate the logical connections between the scientific concepts ...and the design of an experiment... **(Section 3)**

SAI 1f Students...should develop the abilities associated with accurate and effective communication....expressing concepts, reviewing information, summarizing data, using language appropriately, developing diagrams and charts... **(Section 3)**

ST 1e Students should present their results to students, teachers, and others in a variety of ways, such as orally, in writing, and in other forms—including models, diagrams, and demonstrations. **(Section 3)**

UCP 2 ... Using evidence to understand interactions allows individuals to predict changes in natural and designed systems...Models take many forms, including physical objects, plans, mental constructs.... **(Section 3)**

UCP 3 ...Interactions within and among systems result in change. Changes vary in rate, scale, and pattern, including trends and cycles. **(Sections 1–3)**

UCP 4 Evolution is a series of changes, some gradual and some sporadic, that accounts for the present form and function of objects, organisms, and natural and designed systems ....evolution also describes changes in the universe. **(Sections 1–3)**

Section 1

Wind Erosion

Wind contains energy. Some of this energy can move a sailboat or turn a wind turbine, but this energy can also erode the land. As wind passes over the land, the wind can carry sand or dust. Sand is loose fragments of weathered rocks and minerals. Most grains of sand are made of quartz. Other common minerals that make up sand are mica, feldspar, and magnetite.

Dust consists of particles that are smaller than the smallest sand grain. Most dust particles are microscopic fragments of rocks and minerals that come from the soil or volcanic eruptions. Other sources of dust are plants, animals, bacteria, pollution from the burning of fuels, and certain manufacturing processes.

OBJECTIVES

- **Describe** two ways that wind erodes land.
- **Compare** the two types of wind deposits.

KEY TERMS

saltation
deflation
ventifact
dune
loess

How Wind Moves Sand and Dust

Wind cannot keep sand aloft. Instead, sand grains are rolled by the wind along the ground (surface creep) or are moved by a series of jumps and bounces called **saltation.** Saltation occurs when rolling sand grains collide and some bounce up, as shown in **Figure 1.** Once in the air, a sand grain moves ahead a short distance and then falls. As a sand grain falls, it strikes other sand grains. Saltating sand grains move in the same direction that the wind blows. However, the grains rarely rise more than 1 m above the ground, even in very strong winds.

saltation the movement of sand or other sediments by short jumps and bounces that is caused by wind or water

Because dust particles are very small and light, even gentle air currents can keep dust particles suspended in the air. Dust from volcanic eruptions can remain in the atmosphere for several years. Strong winds may lift large amounts of dust and create dust storms, such as the one shown in **Figure 1.** Some dust storms cover hundreds of square kilometers and darken the sky for several days.

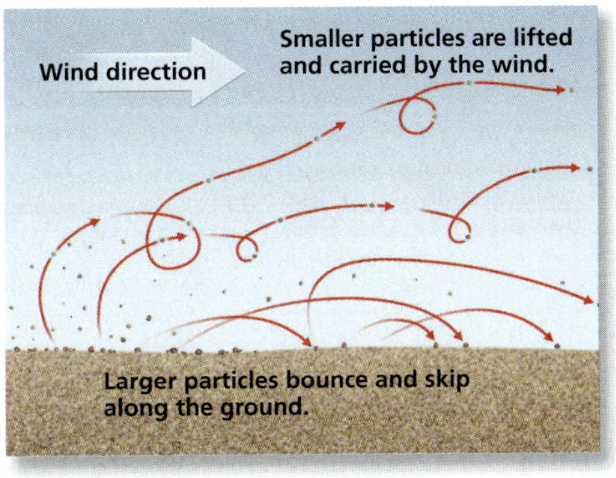

Figure 1 ▶ Heavy sand gains move by rolling or by making low, arcing jumps when blown by the wind (above). Dust is light enough to be suspended for days, as show in this dust storm in Phoenix, Arizona (left).

Section 1

Focus

Overview

This section explains how wind moves sand and dust. It also describes the effects of wind erosion and wind deposition, as well as the creation of dunes, dune migration, and the deposition of loess.

Bellringer

Ask students how something as tiny as sand or dust can erode hard surfaces. (Over long periods of time wind-borne particles of sand or dust become abrasive against solid objects by breaking off tiny pieces.) **LS Logical**

Motivate

Demonstration — GENERAL

Watching Sand Jump Place a wood block at the end of an ice cube tray. Spoon a mixture of fine and coarse sand onto the block in the shape of a dune and blow the sand into the tray. The finer sand should travel farthest. Have students observe the remains of the sand dune and the sand particles that traveled into the tray. The heaviest particles will remain closest to the dune and the lightest particles will travel the farthest distance. **LS Visual**

CHAPTER RESOURCES

Chapter Resource File

- **Directed Reading** BASIC
- **Inquiry Lab** Soil Erosion GENERAL

Technology

Transparencies
- Bellringer

Student Edition on CD-ROM

One-Stop Planner CD-ROM
- Lesson Plan

Teach

READING SKILL BUILDER — BASIC

Paired Summarizing Group students into pairs, and have them read silently about the effects of wind erosion. Then, have one student summarize the ideas of desert pavement and deflation hollows. The other student should listen to the retelling and should point out any inaccuracies or ideas that were left out. Allow students to refer to the text as needed. English Language Learners
LS **Verbal**

Answer to Reading Check

Moisture makes soil heavier, so the soil sticks and is more difficult to move. Therefore, erosion happens faster in dry climates.

Homework — BASIC

Soil Conservation The process of producing rich fertile soil takes hundreds or thousands of years, but soil can be depleted of its nutrients or washed away in just a few years. Farming practices that have not taken this problem into consideration have caused losses of millions of acres of soil. Ask students to research some of the unsound agricultural methods that have caused such erosion. (over-tilling, plowing without regard to contours, cutting down trees, overgrazing herds of cattle) Have students research ways that farmers prevent soil erosion. (wind breaks, ground cover, contour plowing, and terracing) Have them make a poster that shows what the technique looks like, describes what the technique does, and identifies why and where the technique is used. LS **Logical/Visual**

Figure 2 ▶ Desert pavement, such as the example above from Calico Hills, California, prevents erosion of the material beneath it.

SCILINKS. NSTA Developed and maintained by the National Science Teachers Association

For a variety of links related to this subject, go to www.scilinks.org

Topic: Wind Erosion
SciLinks code: HQ61669

Effects of Wind Erosion

While wind erosion happens everywhere there is wind, the landscapes that are most dramatically shaped by wind erosion are deserts and coastlines. In these areas, fewer plant roots anchor soil and sand in place to reduce the amount of wind erosion. Also, in the desert, where there is little moisture, soil layers are thin and are likely to be swept away by the wind. Moisture makes soil heavy and causes some soil and rock particles to stick together, which makes them difficult to move.

Reading Check **Why does wind erosion happen faster in dry climates than in moist climates? (See the Appendix for answers to Reading Checks.)**

Desert Pavement

deflation a form of wind erosion in which fine, dry soil particles are blown away

One common form of wind erosion is deflation. **Deflation** is the process by which wind removes the top layer of fine, very dry soil or rock particles and leaves behind large rock particles. These remaining rock particles often form a surface of closely packed small rocks called *desert pavement*, or *stone pavement*, as shown in **Figure 2.** Desert pavement protects the underlying land from erosion by forming a protective barrier over underlying soil.

Deflation Hollows

Deflation is a serious problem for farmers because it blows away the best soil for growing crops. Deflation may form shallow depressions in areas where the natural plant cover has been removed. As the wind strips off the topsoil, a shallow depression called a *deflation hollow* forms. A deflation hollow may expand to a width of several kilometers and to a depth of 5 to 20 m.

INCLUSION Strategies

- *Learning Disabled*
- *Hearing Impaired*
- *Developmentally Delayed*

Many students with special needs have trouble organizing and summarizing information while they read. Ask students to physically organize and summarize the section by using the following procedure: Start with a blank piece of paper. Write "Wind Erosion" (the red section heading) at the top left. Skip three lines. Indent one inch and write "How Wind Moves Sand and Dust" (first blue subheading). Skip three lines. In line with the first blue heading, write "Effects of Wind Erosion" (second blue subheading). Skip three lines. Indent two inches and write "Desert Pavement" (first green subheading). Skip three lines. Continue in this fashion until all remaining subheadings are included and are arranged correctly. Then, under each heading, write one sentence that summarizes the information in that part of the text. LS **Logical**

Ventifacts

When pebbles and small stones in deserts and on beaches are exposed to wind abrasion, the surfaces of the rocks become flattened and polished on two or three sides. Rocks that have been pitted or smoothed by wind abrasion are called **ventifacts.** The word *ventifact* comes from the Latin word *ventus,* which means "wind." The direction of the prevailing wind in an area can be determined by the appearance of ventifacts.

Scientists once thought that large rock structures, such as desert basins, natural bridges, rock pinnacles, and rocks perched on pedestals, were formed by wind erosion. However, scientists now think that it is more likely that such large features were produced by erosion due to surface water and weathering. Erosion of large masses of rock by wind-blown sand happens very slowly and happens only close to the ground, where saltation occurs.

ventifact any rock that is pitted, grooved, or polished by wind abrasion

dune a mound of wind-deposited sand that moves as a result of the action of wind

Wind Deposition

The wind drops particles when it slows down and can no longer carry them. These deposited particles are continually covered by additional deposits. Eventually, cementation and pressure from overlying layers bind the fragments together. This process is one way that sedimentary rocks form.

Dunes

The best-known wind deposits are **dunes,** which are mounds of wind-deposited sand. Dunes form where the soil is dry and unprotected and where the wind is strong, such as in deserts and along the shores of oceans and large lakes. A dune begins to form when a barrier slows the speed of the wind. When wind speed slows, sand accumulates on both sides of the barrier, as shown in **Figure 3.** As more sand is deposited, the dune itself acts as a barrier, grows, and buries the original barrier.

Figure 3 ▶ Dune Formation

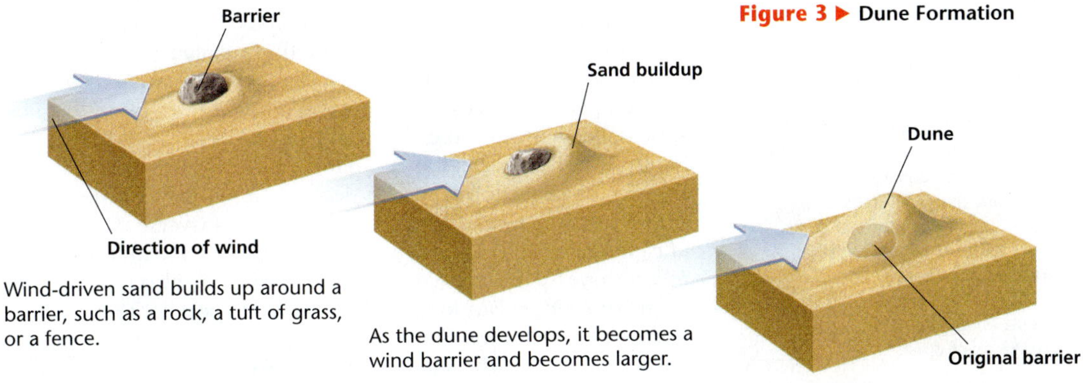

Wind-driven sand builds up around a barrier, such as a rock, a tuft of grass, or a fence.

As the dune develops, it becomes a wind barrier and becomes larger.

The fully formed dune covers the original barrier.

Activity — GENERAL

Shaping Take an emery board and rub it back and forth across the ridges of a six-sided pencil. What happens to the pencil? (The ridges are worn down.) As the file moves back and forth across the pencil, its rough grainy surface cuts tiny pieces from the pencil. Ask students: How is this process like abrasion by blowing sand? (Grains of sand act like a file as they cut away surfaces of rocks, which pits and polishes the rock.) **LS Visual**

HISTORY CONNECTION — ADVANCED

Dust Bowl The Great Plains of the U.S. were covered with grassland before World War I. Over the following decade, grasslands were plowed to make room for wheat crops. A severe drought in the early 1930s caused topsoil to be blown into huge dust storms. Have students research the Dust Bowl and make a presentation to the class. Students' reports should include answers to the following questions: Why did the dust storms occur? What was done to stop them? What damage did they cause? Are similar dust storms a threat today? **LS Verbal**

Using the Figure — BASIC

Dune Formation Use the figure on this page to lead a discussion of how dunes form. Ask students to explain the steps in the process and to identify the most important factor in dune formation. (a barrier that slows the speed of the wind and lets sand accumulate around it) **LS Visual**

Career

Geomorphologist Geomorphologists describe and classify the surface features of Earth and try to explain how these features formed as a result of events like earthquakes or erosion. They also study how human changes to the land affect erosion. Identify the education and training needed to be a geomorphologist and the types of tools these scientists commonly use and explain that this field is influenced by advances in other fields of Earth science.

BRAIN FOOD

Sand Blasting Sand blasting, or sand abrasion, is a type of wind erosion in which blown sand grains work like tiny chisels, eventually grinding down even the hardest materials. Glass bottles have been worn down to look dull and frosted. Cars caught in sandstorms for less than half an hour have had all of the paint worn off and the windshields permanently frosted. Even wooden telephone poles can be worn away by sand blasting.

Teach, *continued*

Identifying Preconceptions — BASIC

Living and Dead Dunes To assess students' ideas regarding the movement of dunes, ask them: What do you think is the difference between a living dune and a "dead" dune? Can a dead dune come back to life?

The four types of dunes discussed in the Student Edition are called *living dunes,* which means that they continue to move under the action of the wind. Once a dune is invaded by vegetation, it can become dead, or stabilized. When a dune is dead, the wind does not affect it. However, if the vegetation is destroyed, erosion begins again and the dune is revitalized. LS Verbal

Graphic Organizer — GENERAL

Comparison Table

You may want to use this Graphic Organizer in a game to review material before the test. Divide the class into two teams. Ask students questions about material from the Comparison Table. Give points to each team that provides correct answers.

Answer to Reading Check

Barchan dunes are crescent shaped; transverse dunes form linear ridges.

CHAPTER RESOURCES

Chapter Resource File

- Internet Activity
  Great Lakes Dunes ADVANCED

Technology

Transparencies

- 89 Types of Dunes (with worksheet)

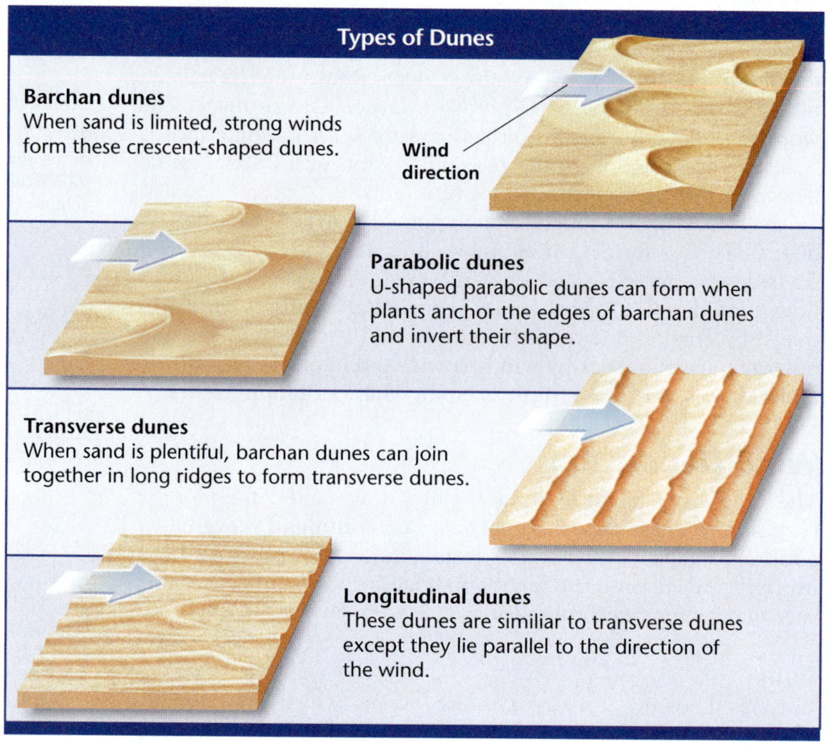

Types of Dunes

The force and direction of the wind shapes sand dunes. Commonly, the gentlest slope of a dune is the side that faces the wind. Sand that is blown over the crest of the dune tumbles down the opposite side, which is called the *slipface*. The slipface has a steeper slope than the windward side does. Two long, pointed extensions may form as wind sweeps around the ends of the dune and gives the dune a crescent shape. Crescent-shaped dunes that have an open side facing away from the wind are called *barchan dunes* (BAHR kahn doonz). A *parabolic dune* is a crescent-shaped dune whose open side faces into the wind. These dunes often form as sand collects around the rim of a deflation hollow.

In desert or coastal areas that have a large amount of sand, a series of ridges of sand may form in long, wavelike patterns. These ridges are called *transverse dunes*. Transverse dunes form at right angles to the wind direction. A type of dune that is similar to a transverse dune, called a *longitudinal dune,* also forms in the shape of a ridge. But longitudinal dunes lie parallel to the direction the wind blows. **Table 1** illustrates several types of dunes and explains how the shape of dunes relates to wind direction.

Graphic Organizer — Comparison Table

Create the **Graphic Organizer** entitled "Comparison Table" described in the Skills Handbook section of the Appendix. Label the columns with "Barchan dunes," "Parabolic dunes," "Transverse dunes," and "Longitudinal dunes." Label the rows with "Shape" and "Formation." Then, fill in the table with details about the four kinds of dunes.

Reading Check How do barchan dunes differ from transverse dunes? (See the Appendix for answers to Reading Checks.)

BRAIN FOOD

Dune Movement All living dunes move in some manner. How fast and how far they move depends on wind strength, wind direction, and the size of the dune. Most dunes travel from about 1 to 20 m per year. Generally, the larger the dune is, the more slowly it moves, but large barchan dunes are among the fastest moving dunes, traveling up to 30 m per year.

Internet Activity — ADVANCED

Great Lakes Dunes Great Lakes Shoreline Geography is a Web site that offers a wealth of information about how the sand dunes formed around the North American Great Lakes. Ask students to research the history of the dunes. Have students identify some of the problems people in that area have with dune migration and the solutions they have used to combat this problem. A worksheet designed to direct student research on this topic can be found in the **Chapter Resource File** booklet or by visiting **go.hrw.com** and entering the keyword **HQ6EWWX.** LS Verbal

Figure 4 ▶ Wind erodes sand from the windward side of the dune and deposits it on the slipface. ***Which direction is this sand dune migrating?***

Dune Migration

The movement of dunes is called *dune migration*. If the wind usually blows from the same direction, dunes will move downwind. Dune migration occurs as sand is blown over the crest from the windward side and builds up on the slipface, as shown in **Figure 4.** In mostly level areas, dunes migrate until they reach a barrier. To prevent dunes from drifting over highways and farmland, people often build fences or plant grasses, trees, and shrubs.

QuickLAB — 20 min

Modeling Desert Winds

Procedure

1. Spread a mixture of **dust, sand,** and **gravel** on a table placed outdoors.
2. Place an **electric fan** at one end of the table.
3. Put on **safety goggles** and a **filter mask.** Aim the fan across the sediment you have laid out. Start the fan on the lowest speed. Record any observations.
4. Turn the fan to a medium speed and record any observations. Then, turn the fan to its highest speed to imitate a desert windstorm. Record any observations.

Analysis

1. In what direction did the sediment move?
2. What was the relationship between the wind speed and the sediment size that was moved?
3. How does the remaining sediment compare to desert pavement?
4. How did the sand move in this activity? How do your observations relate to dune migration?
5. Using the same materials, how would you model dune migration? How would you model the formation of the different kinds of dunes?

Using the Figure — Basic

Dune Migration Have students study the photograph at the top of the page. Ask them to identify which direction the wind is blowing. (The wind is blowing to the right.) Answer to caption question: The sand dune is moving from left to right in the picture. **LS Visual**

Close

Reteaching — Basic

Crossword Puzzle Have students create a crossword puzzle that uses all of the vocabulary words from this section. Provide graph paper so that students can place words in a grid format and then write the clues to match. Students should work in pairs and then switch papers with another pair to fill in the puzzle and to check for any inaccuracies. **LS Logical**

Quiz — General

1. What is deflation? (a form of wind erosion in which fine, dry soil particles are blown away)
2. What is the composition of loess? (very fertile, fine sediments of quartz, feldspar, hornblende, and mica)

CHAPTER RESOURCES

Chapter Resource File

- **Datasheet for Quick Lab** General

QuickLAB

Skills Acquired

- Experimenting
- Constructing Models
- Organizing and Analyzing Data

Teacher's Notes It is recommended that you perform this lab outside if possible. This lab will require an outdoor electrical outlet or extension cord. Also, make sure students stand away from blowing sand.

Answers

1. away from the fan
2. The higher the wind speed, the larger the sediment that was moved.
3. It is similar to desert pavement. Particles that remain are too large for the wind to move.
4. The sand moves with a skipping motion. The sand would form a dune when the sand is stopped by a barrier.
5. Answers may vary. Accept all resonable answers.

Close, *continued*

Alternative Assessment — GENERAL

Landform Description Have students pretend that they are flying over a desert to get to the coast. Ask them to describe the landforms they would see that were created by wind erosion and deposition. They must describe at least two features of each landform and explain how each feature formed. LS **Verbal**

Answers to Section Review

1. Sand grains move by jumps and bounces as they collide with other grains.
2. Deflation is a form of wind erosion in which dry soil particles are blown away, leaving a shallow depression called a *deflation hollow*.
3. Desert pavement is the surface of closely packed small rocks left behind after the top layer of very fine, dry soil is removed by the wind.
4. Dunes form where sediment is dry and unprotected and where the wind is strong. A barrier slows the speed of the wind, and the sand accumulates on both sides of the barrier, eventually burying it. Dune migration occurs as the sand is blown over the crest of the dune and builds up on the slipface.
5. Wind carries loess by suspending it in the air.
6. There are few plants and roots in the desert to anchor the soil and sand, and the dry, thin soil is easily swept away by the wind.
7. Barchan and parabolic dunes are crescent shaped. Barchan dunes have an open side that faces away from the wind, and parabolic dunes have an open side that faces the wind. Transverse and longitudinal dunes are ridge-shaped. Transverse dunes form at right angles to the wind direction, and longitudinal dunes form parallel to the wind direction.
8. Wind cannot carry large rock particles, which are all that exist in desert pavement. Also, desert pavement commonly occurs in dry areas that have little water to cause erosion.
9. They provide a barrier to stop the wind from carrying dune material onto and across a road.
10. Dunes move continuously.
11. *Wind* can move sediment by *deflation*, which lifts dust from the *top soil layer* and forms *desert pavement* and *deflation hollows;* and by *saltation*, which moves *sand particles* and may form *dunes* that move by *migration*.

Figure 5 ▶ These loess deposits are located in Vicksburg, Mississippi. Much of the surrounding land is fertile farmland.

Loess

loess fine-grained sediments of quartz, feldspar, hornblende, mica, and clay deposited by the wind

The wind carries dust higher and much farther than it carries sand. Fine dust may be deposited in such thin layers that it is not noticed. However, thick deposits of yellowish, fine-grained sediment, called **loess** (LOH ES), can form by the accumulation of windblown dust. Although loess is soft and easily eroded, it sometimes forms steep bluffs, such as those shown in **Figure 5.**

A large area in northern China is covered in a deep layer of loess. The material in this deposit came from the Gobi Desert, in Mongolia. Deposits of loess are also located in central Europe. In North America, loess is located in the midwestern states, along the eastern border of the Mississippi River valley, and in eastern Oregon and Washington State. These deposits probably formed as dust from dried beds of glacial lakes and from outwash plains blew across the region. Loess deposits are extremely fertile and provide excellent soil for grain-growing regions.

Section 1 Review

1. **Describe** how wind transports sediment by saltation.
2. **Define** *deflation,* and explain how *deflation hollows* form.
3. **Describe** how desert pavement forms.
4. **Describe** how sand dunes form and how dunes migrate.
5. **Explain** how the wind moves loess.
6. **Identify** two reasons why wind erosion has a major affect on deserts.
7. **Compare** the four main shapes of dunes.

CRITICAL THINKING

8. **Analyzing Processes** Explain how scientists know that desert pavement forms by wind erosion.
9. **Determining Cause and Effect** Why does planting grass, trees, or shrubs help prevent dunes from covering roads?
10. **Analyzing Processes** Explain why the position of dunes would not be helpful for navigation in the desert.

CONCEPT MAPPING

11. Use the following terms to create a concept map: *saltation, deflation, dune, deflation hollow, desert pavement, wind, sand particle, top soil layer,* and *migration.*

CHAPTER RESOURCES

Chapter Resource File

- **Section Quiz** GENERAL

Workbooks

- **Study Guide** (also in Spanish)

Section 2 Wave Erosion

As wind moves over the ocean, the wind produces waves and currents that erode the coastline. Wave erosion changes the shape of shorelines, the places where the ocean and the land meet.

OBJECTIVES

- **Compare** the formation of six features produced by wave erosion.
- **Explain** how beaches form.
- **Describe** the features produced by the movement of sand along a shore.

KEY TERMS

headland
beach
longshore current

Shoreline Erosion

The power of waves striking rock along a shoreline can sometimes shake the ground as much as a small earthquake would. The great force of waves may break off pieces of rock and throw the pieces back against the shore. These sediments grind together in the tumbling water. This abrasive action, which is known as *mechanical weathering,* eventually reduces most of the rock fragments to small pebbles and sand grains.

Much of the erosion along a shoreline takes place during storms, which cause large waves that release tremendous amounts of energy, as shown in **Figure 1.** A severe storm can noticeably change the appearance of a shoreline in a single day.

Chemical weathering also affects the rock along a shoreline. The waves force salt water and air into small cracks in the rock. Chemicals in the air and water react with the rock and enlarge the cracks. Enlarged cracks expose more of the rock to mechanical and chemical weathering.

Figure 1 ▶ Large waves break apart rock on shorelines and change the shoreline's appearance. ***Where is erosion occurring in the photo shown here?***

Section 2

Focus

Overview

This section describes and compares six coastal features produced by wave erosion and explains how these features form. It also explains how beaches form and what features form when longshore currents transport sand along the shoreline.

Bellringer

Show students a picture of a coastline with steep cliffs. Have students draw pictures of what they think the shoreline will look like in 50 years and in 500 years. Students can share their drawings with each other. (Answers may vary.) **LS Visual**

Motivate

Using the Figure — GENERAL

Breaking Waves Have students study the photograph at the bottom of this page. Ask students to imagine the force with which the waves are striking the rock in the foreground. Answer to caption question: Erosion is occurring in the cracks and on the surface of the rock in the foreground where the water strikes the rock. **LS Visual**

CHAPTER RESOURCES

Chapter Resource File

- Directed Reading BASIC

Technology

Transparencies
- Bellringer

Student Edition on CD-ROM

One-Stop Planner CD-ROM
- Lesson Plan

Teach

Internet Activity — GENERAL

Landform Photographs Have students search the Internet for examples, descriptions, and photos of the shoreline feature formations caused by wave erosion. Have them search by using either the name of the formation or the broad heading of wave erosion. Some sites have short movies that show how wave erosion has changed coastlines over time. A worksheet designed to direct student research on this topic can be found in the **Chapter Resource File** booklet or by visiting go.hrw.com and entering the keyword **HQ6EWWX.** LS Visual

READING SKILL BUILDER — BASIC

Reading Organizer As students read this section, encourage them to take Power Notes, KWL Notes, or Two-Column Notes as described in the Skills Handbook section of the Appendix. Later, students can use these notes as a study guide for assessments.
LS Visual

Answer to Reading Check

Answers should include three of the following: sea cliffs, sea caves, sea arches, sea stacks, wave-cut terraces, and wave-built terraces.

CHAPTER RESOURCES

Chapter Resource File

- **Internet Activity**
  Landform Photographs GENERAL

Technology

Transparencies
- 90 Wave Erosion and Landforms (with worksheet)

Sea Cliffs

In places where waves strike directly against rock, the waves slowly erode the base of the rock. The waves cut under the overhanging rock, until the rock eventually collapses to form a steep *sea cliff*. The rate at which sea cliffs erode depends on the amount of wave energy and on the resistance of the rock along the shoreline. Soft rock, such as limestone, erodes very rapidly. Harder rock, such as granite, shows little change over hundreds of years. Resistant rock formations that project out from shore are called **headlands.** Areas that have less resistant rock form *bays*. **Figure 2** shows bays and several other coastal landforms produced by wave erosion.

headland a high and steep formation of rock that extends out from shore into the water

Sea Caves, Arches, and Stacks

Waves often cut deep into fractured and weak rock along the base of a cliff to form a large hole, or a *sea cave*. When waves cut completely through a headland, a *sea arch* forms. Offshore columns of rock that once were connected to a sea cliff or headland, are called *sea stacks*.

Terraces

As a sea cliff is worn, a nearly level platform, called a *wave-cut terrace,* usually remains beneath the water at the base of the cliff. Eroded material may be deposited offshore to create an extension to the wave-built terrace called a *wave-built terrace*.

Reading Check List three features that are caused by shoreline erosion. (See the Appendix for answers to Reading Checks.)

Figure 2 ▶ Wave erosion of sea cliffs causes cliff retreat and forms isolated sea stacks. Sea cliffs develop where waves strike directly against rock that is along a shoreline.

INCLUSION Strategies

- *Visually Imparied*
- *Learning Disabled*
- *Developmentally Delayed*

Students can work in small groups or pairs to demonstrate different shore formations by using their hands. For example, a sideways-cupped hand with a hand over the top represents a cave. Or, have students sculpt formations with modeling clay, based on verbal descriptions or from feeling the hands or models of others.
LS Kinesthetic

Teaching Tip — GENERAL

Connect to Prior Knowledge Ask students to recall visits to lake and/or ocean beaches. Have them describe landscapes and landforms around the beaches they have seen. After they listen to other students' descriptions, have them discuss the similarities and differences between ocean shoreline formations and features found around lakes.
LS Verbal

Beaches

Waves create features by eroding the land and depositing sediment. A deposit of sediment along an ocean or lake shore is called a **beach.** Beaches form where more sediment is deposited than is removed. After a beach forms, the rate at which sediment is deposited and the rate at which sediment is removed may vary.

beach an area of the shoreline that is made up of deposited sediment

Composition of Beaches

The sizes and kinds of materials that make up beaches vary. In general, the smaller the particle is, the farther it traveled before it was deposited. The composition of beach materials depends on the minerals in the source rock. Some beaches may consist of fragments of shells and coral that are washed ashore. In other locations, sand beaches form from sediment deposited by rivers or glaciers. Other beaches are composed of large pebbles.

The Berm

Each wave that reaches the shore moves sand slightly. The sand piles up to produce a sloping surface. During high tides or large storms, sand is deposited at the back of this slope. So, most beaches have a raised section called the *berm,* as shown in **Figure 2.** The berm is high and steep during the winter because large storms remove sand from the beach on the seaward side of the berm. The sand that is removed may be deposited offshore to form a long underwater ridge called a *sand bar*. In the summer, waves may move the sand back to the shore to widen the beach.

MATHPRACTICE

Wave Depth A wave will break when the depth of the wave is equal to 3/2 the height of the wave. This statement is represented by the formula $D = 3/2\ H$. If the tallest wave in a specific area is 6 m, what is the maximum depth at which wave erosion would occur in that area?

MATHPRACTICE

Answer

$$D = \frac{3}{2}H$$

If the tallest wave is 6 m, then the maximum depth at which wave erosion would occur is

$$\frac{3}{2} \times 6\ m = 9\ m$$

Close

Reteaching — Basic

Vocabulary Flash Cards Have students make vocabulary and picture flashcards of the 12 different landforms described in the section. Have students write the term on one side of the card and place a description and picture of the landform on the other side. Students can use the cards to quiz each other on identifying the formations. LS **Visual**

Quiz — General

1. How is mechanical weathering different from chemical weathering? (Mechanical weathering is the abrasive action that occurs when waves break off pieces of rock and they tumble against each other. Chemical weathering occurs when chemicals in air and water react with rock to break the rock down.)
2. Name two sand deposits formed by longshore currents. (spits and tombolos)
3. Describe how a sea cliff forms. (When waves strike directly against rock, the base of the rock erodes, or wears away. Eventually the overlying rock collapses and a steep sea cliff remains.)

Homework — Advanced

Beach Composition Have students research the composition of sand beaches in different locations around the world. Students should report what minerals are found in each beach and the color of each beach. Students should learn that the color of sand on a beach depends on the rock that was eroded to produce the sand. For example, white sand commonly forms from coral and seashells, pale yellow sand commonly forms from quartz, black sand commonly forms from basaltic lava, and green sand commonly forms from the mineral olivine. LS **Verbal**

SKILL BUILDER — Basic

Writing After students read the paragraphs that explain beaches, ask them to write a short story about a grain of sand or a pebble that becomes part of a beach. The story should be an adventure story that describes where the particle came from and how the particle reaches the beach. Students may choose to write the story in the first person or in the third person. LS **Verbal**

Close, *continued*

Alternative Assessment — GENERAL

Landform Identification Show students photographs of coastal formations. Have students identify the formations and describe how they formed. LS **Visual**

Answers to Section Review

1. Sea cliffs form when water erodes the base of a rock and the rock collapses. Sea caves form when waves cut a large hole into the base of a cliff. Sea arches form when waves cut through a headland. Sea stacks form when part of a sea cliff becomes isolated by water. A wave-cut terrace is a flat platform that forms when a sea cliff is worn down. A wave-built terrace is an extension to a wave-cut terrace that is made of deposited material.
2. the composition of the source rock and the distance that the particles travel
3. Beaches form where more sediment is deposited than is removed by wave action.
4. A sea cave is a cave at the base of a headland. When waves cut a hole completely through the headland, an arch is made. When the arch collapses to leave a column of rock surrounded by water, a sea stack has formed.
5. A beach is an area of the shoreline made of deposited sediment. A spit is a long, narrow deposit of sand connected at one end to the shore. Tombolos are ridges of sand that connect an island to the mainland.
6. During winter, large storms remove sand from the beach on the seaward side of the berm. In summer, waves move the sand toward the land, widening the beach.
7. Only very rapidly moving water can carry large particles. When the speed of the water decreases, the water is able to move only smaller particles, so it deposits the larger particles. Very fine particles are deposited only in very still water.
8. Wave action causes *shoreline erosion,* which can result in a *sea cliff,* a *sea arch,* a *sea cave,* or a *terrace,* which could be a *wave-cut terrace* or a *wave-built terrace.*

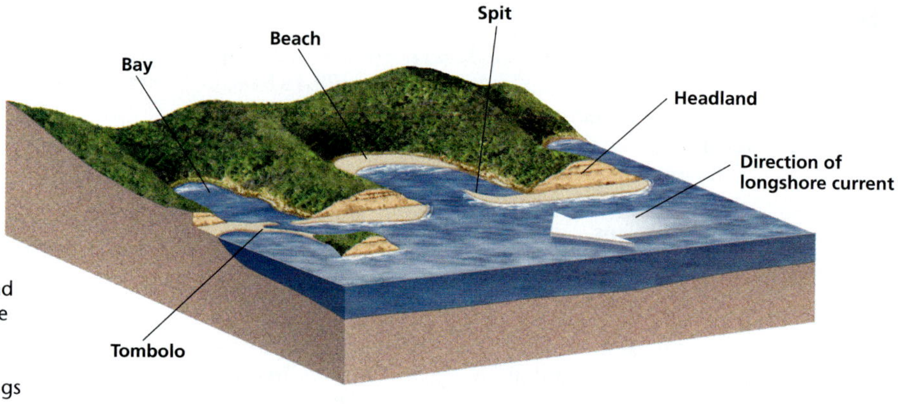

Figure 3 ▶ Spits and tombolos form where longshore currents deposit sand at headlands, at openings to a bay, or on offshore islands.

Longshore-Current Deposits

The direction in which a wave approaches the shore determines how the wave will move sediment. Most waves approach the beach at an angle and retreat in a direction that is more perpendicular to the shore. So, waves move individual sand grains in a zig-zag motion. The general movement of sand along the beach is in the direction in which the waves strike the shore.

Waves moving at an angle to the shoreline often create longshore currents. A **longshore current** is a movement of water parallel to and near the shoreline. Longshore currents transport sand parallel to the shoreline, as shown in **Figure 3.**

longshore current a water current that travels near and parallel to the shoreline

Along a relatively straight coastline, sand keeps moving until the shoreline changes direction at bays and headlands. The longshore current slows, and sand is deposited at the far end of the headland. A long, narrow deposit of sand connected at one end to the shore is called a *spit*. Currents and waves may curve the end of a spit into a hook shape. Beach deposits may also connect an offshore island to the mainland. Such connecting ridges of sand are called *tombolos*.

Section 2 Review

1. **Compare** the formation of six features that are produced by shoreline erosion.
2. **Identify** two factors that determine the composition of beach materials.
3. **Explain** how beaches form.
4. **Compare** sea arches, sea caves, and sea stacks.
5. **Describe** three features produced by the movement of sand along a shore.

CRITICAL THINKING

6. **Making Inferences** How do seasonal changes affect beaches?
7. **Identifying Relationships** How does the speed at which water moves affect the deposition of materials of differing sizes?

CONCEPT MAPPING

8. Use the following terms to create a concept map: *shoreline erosion, sea arch, sea cave, sea cliff, wave-cut terrace, wave-built terrace,* and *terrace.*

CHAPTER RESOURCES

Chapter Resource File
- Section Quiz GENERAL

Workbooks
- Study Guide (also in Spanish)

ENVIRONMENTAL CONNECTION

Section 3 Coastal Erosion and Deposition

The boundaries between land and the ocean are among the most rapidly changing parts of Earth's surface. Coastal areas extend from relatively shallow water to several kilometers inland. Coastlines are affected by the long-term rise and fall of sea level and by the long-term uplifting or sinking of the land that borders the water. These and other more rapid processes, such as wave erosion and deposition, constantly change the appearance of coastlines.

OBJECTIVES

- **Explain** how changes in sea level affect coastlines.
- **Describe** the features of a barrier island.
- **Analyze** the effect of human activity on coastal land.

KEY TERMS

estuary
barrier island
lagoon

Absolute Sea-Level Changes

A change in the amount of ocean water causes sea level to rise or fall, so coastlines are covered or exposed. During the last glacial period, which ended about 15,000 years ago, some of the water that is now in the ocean existed as continental ice sheets. Scientists estimate that the ice sheets held about 70 million cubic kilometers of ice. Now, the ice sheets in Antarctica and Greenland hold only about 25 million cubic kilometers of ice.

During the last glacial period, the water that made up the additional 45 million cubic kilometers of ice is thought to have come from the oceans. As a result, sea level was as much as 140 m lower during the last glacial period than it is today. Since the last glacial period, the ice sheets have been melting and sea level has been rising at a rate of about 1 mm per year, as shown in **Figure 1.** If, in the distance future, the polar icecaps were to melt completely, the oceans would rise about 60 m and submerge low-lying coastal regions. The locations of many large cities, such as New York, Los Angeles, Miami, and Houston would be submerged.

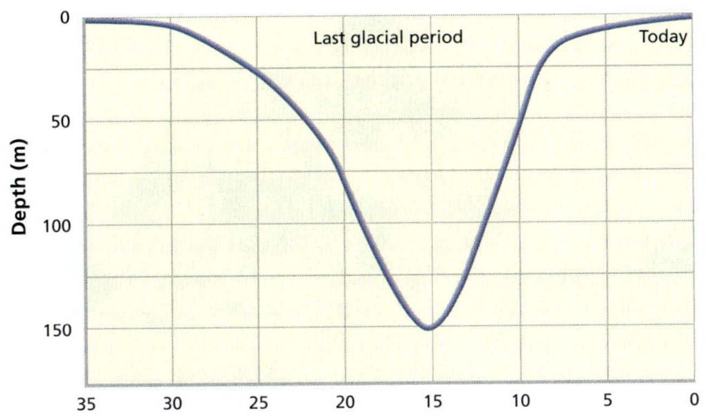

Figure 1 ▶ This graph shows how sea level has changed during the past 35,000 years.

Section 3

Focus

Overview

This section distinguishes between submergent and emergent coastlines and describes the formation of barrier islands and lagoons and the effects of human activities on coastlines.

Bellringer

Ask students to imagine how life on Earth would be different if the sea level was 140 m (420 feet) lower than it is today. Ask them: What would the land look like? How would plant and animal life change? (Answers may vary.) **LS Logical**

Motivate

Using the Figure — GENERAL

Sea-Level Changes Use the figure on this page to lead a discussion about sea-level changes. Global warming increases the melting of the polar ice caps and glaciers, but thermal expansion of the water may be a greater factor affecting sea-level changes. As the water in Earth's oceans warms, it becomes less dense and expands. In addition, ice sheets melt and break off faster than new ice can form. Ask students what they know about global warming and how it would affect sea-level changes and consequently coastal erosion. **LS Verbal/Logical**

CHAPTER RESOURCES

Chapter Resource File

- Directed Reading BASIC

Technology

Transparencies
- Bellringer

Student Edition on CD-ROM

One-Stop Planner CD-ROM
- Lesson Plan

QuickLAB

Skills Acquired

- Collecting Data
- Identifying Patterns
- Organizing and Analyzing Data

Teacher's Notes Tidal data may be found on the Internet. If data are already graphed, have students use the graphs. Students should plot the daily range between high and low tides on a graph and compare that information with the phases of the moon.

Answers

1. Answers may vary but should accurately reflect the data.
2. Tides represent relative sea-level changes.
3. At first and last quarter moons, the sun, the moon, and Earth are at right angles and represent neap tides, when the tidal range is the smallest. At new and full moons, the sun, the moon, and Earth are in alignment and represent spring tides, when the tidal range is greatest.

CHAPTER RESOURCES

Chapter Resource File

- **Datasheet for Quick Lab** GENERAL
- **Making Models Lab** Erosion of a Submerging Coastal Profile GENERAL

Technology

Transparencies

- 91 Submergent Coastlines (with worksheet)

QuickLAB 35 min

Graphing Tides

Procedure

1. Research the daily tidal data for a certain area for a calendar month.
2. Graph the tide measurements on a line graph.

Analysis

1. When was high tide? When was low tide?
2. What kind of sea-level changes do tides represent?
3. Research the full and new moon dates for the time period you graphed. How does the moon correspond with your high- and low-tides?

Relative Sea-Level Changes

Absolute sea-level changes occur when the amount of water in the ocean changes. Relative sea level changes when the land or features near the coast change. These changes can be caused by large-scale geologic processes or by localized coastal changes. For example, movements of Earth's crust can cause coastlines to sink or to rise. Coastlines near a tectonic plate boundary may change as tectonic plates move. When coastlines change, the relative sea-level of that area also changes.

Submergent Coastlines

When sea level rises or when land sinks, a *submergent coastline* forms. Divides between neighboring valleys become headlands separated by bays and inlets, and submerged peaks may form offshore islands, as shown in **Figure 2.** Beaches are generally short, narrow, and rocky. When U-shaped glacial valleys become flooded with ocean water as sea level rises, spectacular narrow, deep bays that have steep walls, called *fiords* (FYAWRDZ), form.

The mouth of a river valley that is submerged by ocean water may become a wide, shallow bay that extends far inland. This type of bay, where salt water and fresh water mix, is called an **estuary** (ES tyoo er ee).

estuary an area where fresh water from rivers mixes with salt water from the ocean; the part of a river where the tides meet the river current

Emergent Coastlines

When the land rises or when sea level falls, an *emergent coastline* forms. If an emergent coastline has a steep slope and is exposed rapidly, the coastline will erode to form sea cliffs, narrow inlets, and bays. A series of wave-cut terraces may be exposed as well.

A gentle slope forms when part of the continental shelf is slowly lifted and exposed. The gentle slope forms a smooth coastal plain that has few bays or headlands and that has many long, wide beaches.

Figure 2 ▶ The features of a submergent coastline erode over time as sea level rises.

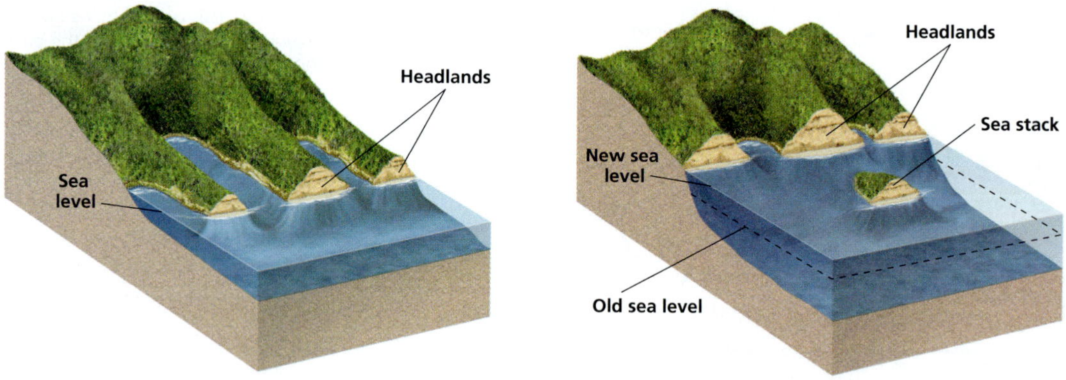

Cultural Awareness GENERAL

Tides Before students perform the Quick Lab, ask them what they know about tides. Then, tell them that ancient people came up with their own ideas and explanations for tides. Chinese people believed that a huge Earth monster was breathing in and out, making the tides rise and fall. In Scandinavian legends, Thor, the god of aerial forces, was believed to make the tides as he blew on the waters of the sea and then let them recede.

ENVIRONMENTAL CONNECTION

Coastal Erosion Virtual field trips available on the internet offer opportunities for students to observe coastal erosion in different locations around the globe. They provide students the chance to examine natural as well as human-induced causes of coastal erosion, ways to control erosion, and ways to deal with the effects of erosion. Many sites are created by experts in the field who provide photos and text that richly supplement this section's content. LS Visual

Barrier Islands

As sea level rises over a flat coastal plain, the shoreline moves inland and isolates dunes from the old shoreline to form barrier islands, such as the one shown in **Figure 3.** **Barrier islands** are long, narrow ridges of sand that lie nearly parallel to the shoreline. Barrier islands can be 3 to 30 km offshore and can be more than 100 km long. Between a barrier island and the shoreline is a narrow region of shallow water called a **lagoon.**

Barrier islands also form when sand spits are separated from the land by storms or when waves pile up ridges of sand that were scraped from the shallow, offshore sea bottom. These deposits are then moved toward the shore by waves, currents, and winds. This motion causes most barrier islands to migrate toward the shoreline. Winds blowing toward the land often create a line of dunes that are 3 to 6 m high on the side of the island that faces the shore.

Large waves from storms, especially waves from hurricanes, may severely erode barrier islands. During a storm, sand washes from the ocean side toward the inland side of the island. Some barrier islands are eroding at a rate of about 20 m per year.

Reading Check How do barrier islands form? (See the Appendix for answers to Reading Checks.)

Figure 3 ▶ Santa Rosa Island is a long, narrow barrier island that is located off the coast of Florida.

barrier island a long ridge of sand or narrow island that lies parallel to the shore

lagoon a small body of water separated from the sea by a low, narrow strip of land

Connection to BIOLOGY

Coral Reefs

A common coastal feature called a *reef* forms when small marine animals called *corals,* which live in warm, shallow sea water, grow. Corals extract calcium carbonate from ocean water and use it to build a hard outer skeleton. Corals attach to each other to form a large colony made of millions of coral skeletons.

Fringing reefs form when a coral colony grows in the shallow water around a tropical volcanic island. As the sea floor bends under the weight of the volcano, both the volcano and the reef sink. The coral builds higher to form a *barrier reef* around the remnant of the volcanic island. When the island is completely submerged, a nearly circular coral reef, called an *atoll,* remains around a shallow lagoon.

A coral reef completely surrounds the island of Bora Bora in French Polynesia.

Answer to Reading Check

As sea levels rise over a flat coastal plain, the shoreline moves inland and isolates dunes from the old shoreline. These dunes become barrier islands.

Close

Reteaching — BASIC

Organizing Ideas Ask students to re-read the text under the heading "Relative Sea-Level Changes" and create a chart with the following headings: "Sea-Level Change," "Resulting Coastline," and "Formations." Under the first heading should be two entries—"Increasing" and "Decreasing." Have students summarize what happens in each situation. (Increasing–submergent–fiords, estuaries, barrier islands, and lagoons. Decreasing–emergent–if steep, sea cliffs, narrow inlets, and bays; if gently sloping, few bays or headlands and long, wide beaches.) **LS Logical**

Quiz — GENERAL

1. What effect would the melting of polar ice caps have on coastlines around the world? (Absolute sea level would rise and submergent coastlines would form.)

2. What landforms result from the formation of an emergent coastline? (If steep slopes are present and the coastline is exposed rapidly, then sea cliffs, narrow bays, and inlets will form. If an emergent coastline forms slowly, then a gently-sloping coastal plain that has few headlands and bays and many wide beaches forms.)

SKILL BUILDER — BASIC

Writing Ask students to read the text under the heading "Preserving the Coastline." Then, have them create bumper stickers that promote coastal preservation. Students should use information from the text or from the Internet pages introduced by the SciLinks topic on Coastal Changes. **LS Verbal/Visual**

GEOGRAPHY CONNECTION — GENERAL

Coastal Features Have students research fiords, estuaries, and lagoons and create posters that show examples of each feature from different parts of the world. Ask students to identify whether these three features are related geographically in any way and how each feature forms. (Estuaries form anywhere that a fresh-water body meets the ocean. Fiords form where ocean water fills in glacial valleys. Lagoons form where shallow ocean is protected by a sandbar or barrier island.) **LS Visual**

Close, *continued*

Alternative Assessment — GENERAL

Landform Description Ask students to read the imaginary headlines below and write the news story that would accompany each. Students should include a description of the landscapes formed and how humans, animals, and plants would be affected.

1. Polar Ice Caps Melt!—Ocean Levels Rise 60 Meters! (submergent coastline)
2. Lake Michigan Rises Unexpectedly and a 3 km Expanse of Sand Dunes is Surrounded by Water! (barrier island and lagoon) LS Verbal

Answers to Section Review

1. Changes in sea level affect how waves and currents erode the land and deposit sediment. These changes alter the appearance of the shoreline.
2. A submergent coastline forms when the sea level rises or land sinks. An emergent coastline forms when sea level falls or land rises.
3. A barrier island is a long narrow ridge of sand that lies parallel to the shore and is separated from the shore by a narrow region of shallow water.
4. Barrier islands are sensitive to erosion because they are made of sand and they are exposed to the full force of the ocean.
5. Oil spills and garbage and sewage from towns and industries affect coastlines.
6. A season of heavy storms would severely erode or possibly destroy a barrier island by removing much of the sand.
7. Sea levels would drop and emergent coastlines would form.
8. A *coastline* is a boundary between land and the ocean, and may be a *submergent coastline,* or an *emergent coastline,* or may include a *barrier island* that is isolated from the main shoreline by a *lagoon.*

Figure 4 ▶ This beach area near Biloxi, Mississippi was devastated by Hurricane Katrina.

SCILINKS. NSTA Developed and maintained by the National Science Teachers Association

For a variety of links related to this subject, go to www.scilinks.org

Topic: Coastal Changes
SciLinks code: HQ60307

Preserving the Coastline

Coastal lands are used for commercial fishing, shipping, industrial and residential development, and recreation. While development of coastal areas is economically important, it can also damage coastal areas in several ways. Pollution is a serious threat to coastal resources. Oil spills are a threat because tankers travel near shorelines and because oil wells are drilled offshore. Garbage, pollution from industry, and sewage from towns on the coast can pollute the coastline. This pollution can damage habitats and kill marine birds and other animals.

To preserve the coastal zone, private owners and government agencies often work together to set guidelines for coastal protection. Beaches damaged by hurricanes such as Katrina in 2005 will require years of rebuilding, as shown in **Figure 4.** Coastal development in some environmentally sensitive areas, such as the North Carolina coast has been slowed or stopped completely in an attempt to protect these important areas.

Section 3 Review

1. **Explain** how changes in sea level affect coastlines.
2. **Explain** how the formation of a submergent coastline differs from the formation of an emergent coastline.
3. **Describe** two features of a barrier island.
4. **Explain** why barrier islands are particularly sensitive to erosion.
5. **Describe** two ways in which human activity affects coastlines.

CRITICAL THINKING

6. **Making Predictions** Predict the effect that a season of heavy storms would have on a barrier island.
7. **Identifying Relationships** If Earth were to enter a new glacial period, how might coastlines around the world change?

CONCEPT MAPPING

8. Use the following terms to create a concept map: *coastline, emergent coastline, submergent coastline, barrier island,* and *lagoon.*

CHAPTER RESOURCES

Chapter Resource File
- Section Quiz GENERAL

Workbooks
- Study Guide (also in Spanish)

Chapter 18 Highlights

Sections

1 Wind Erosion

Key Terms

saltation, 445
deflation, 446
ventifact, 447
dune, 447
loess, 450

Key Concepts

- Wind transports sediment by saltation and by deflation. Saltation is the movement of particles by a series of jumps or bounces. Deflation is the process of carrying small particles that are suspended in air currents.
- Wind erosion forms desert pavement, deflation hollows, and ventifacts.
- The two types of wind deposits are dunes, which are generally made of sand, and loess, which is made of dust particles.

2 Wave Erosion

headland, 452
beach, 453
longshore current, 454

- Waves weather and erode the shoreline and produce characteristic features, such as sea cliffs, sea caves, sea arches, sea stacks, terraces, and beaches.
- Beaches form from the deposition of sediments by waves.
- Longshore currents move sediments parallel to a shoreline.

3 Coastal Erosion and Deposition

estuary, 456
barrier island, 457
lagoon, 457

- Coastlines are exposed or submerged as sea level changes.
- Barrier islands are long, narrow offshore ridges of sand.
- Human activities, including development and pollution, affect land along the coasts.

Chapter Highlights

Alternative Assessment — General

Quiz Show To help students review for a final test or to see how they have grasped the ideas from the chapter, set up a quiz-style game. Categories can include:

- Name that Formation (show a picture of a landform and have students name it)
- Match the Process (describe an erosional process and have students name it)
- Wind or Wave? (list a process or formation and have students identify whether it forms as a result of wind or wave action)
- Spelling Bee (have students spell vocabulary terms)
- Dunes Day (have students identify characteristics of the different dunes)

One variation of the game may be to divide the class into teams and have each student take a turn choosing a question. Students can use their notes and teammates to find answers. LS Verbal/Auditory

CHAPTER RESOURCES

Chapter Resource File

- Concept Review General
- Critical Thinking Advanced
- Math Skills General
- Graphing Skills General
- Chapter Test A General
- Chapter Test B Advanced

Workbooks

- Study Guide (also in Spanish)
- Assessments (Spanish)

Technology

Classroom Videos
- Brain Food Video Quiz

HRW Earth Science Video
- Segment 9: Weathering and Erosion

Chapter Review

Assignment Guide	
Section	**Questions**
1	1, 3, 4, 7–10, 19, 21–22, 28–32
2	2, 5, 11–13, 18, 23, 24, 26
3	6, 14–17, 20, 25
1–3	27

Using Key Terms

1–6. Answers may vary but should show that students understand the definitions of and differences between key terms.

Understanding Key Concepts

7. c **8.** a
9. b **10.** b
11. a **12.** a
13. b **14.** d
15. d

Short Answer

16. An emergent coastline forms when the sea level falls or land rises. A submergent coastline occurs when the sea level rises or when land sinks.

17. If shoreline resources are not protected, then habitats can be damaged and shoreline birds and other animals killed.

18. The factors that determine the composition of a beach are the composition of the source rock and the distance the particles travel before being deposited.

19. Barchan and parabolic dunes are both crescent shaped. However, the center of a Barchan dune faces away from the wind, and the center of a parabolic dune faces into the wind. Transverse and longitudinal dunes take the shape of ridges in long wavelike patterns. Transverse dunes form perpendicular to the direction of the wind, while longitudinal dunes form parallel to the direction of the wind.

20. Absolute sea level changes when the amount of water in the ocean changes. Relative sea level changes when the land or features near the land change.

Critical Thinking

21. Sample answer: The tall, sculpted rock formations were most likely produced by water erosion because most wind erosion occurs very close to the ground.

Chapter 18 Review

Using Key Terms

Use each of the following terms in a separate sentence.

1. *ventifact*

2. *longshore current*

3. *loess*

For each pair of terms, explain how the meanings of the terms differ.

4. *saltation* and *deflation*

5. *headland* and *beach*

6. *barrier island* and *lagoon*

Understanding Key Concepts

7. Wind forms desert pavement by removing fine sediment and by leaving large rocks behind in a process called
a. saltation. **c.** deflation.
b. abrasion. **d.** ventifact.

8. Wind moves sand by
a. saltation. **c.** abrasion.
b. emergence. **d.** depression.

9. Dunes move primarily by the process called
a. abrasion. **c.** deflation.
b. migration. **d.** submergence.

10. Thinly layered, yellowish, fine-grained deposits are called
a. beaches.
b. loess.
c. dunes.
d. desert pavement.

11. The most important erosion agent along shorelines is
a. wave action. **c.** wind.
b. weathering. **d.** the tide.

12. Which of the following shoreline features is *not* produced by wave erosion of sea cliffs?
a. spits **c.** wave-cut terraces
b. sea stacks **d.** sea arches

13. Longshore-current deposition of sand at the end of a headland produces a
a. sand bar. **c.** dune.
b. spit. **d.** sea cliff.

14. Sea level is now
a. stationary.
b. falling about 1 mm per year.
c. rising about 1 cm per year.
d. rising about 1 mm per year.

15. Barrier islands tend to migrate
a. away from the shore.
b. along the shore.
c. in the summer.
d. toward the shore.

Short Answer

16. What is the difference between an emergent coastline and a submergent coastline?

17. Explain what may happen if shoreline resources are not protected.

18. What factors affect the composition of a beach?

19. Explain the difference between the four main types of dunes.

20. Explain the difference between absolute sea-level change and relative sea-level change.

Critical Thinking

21. Identifying Relationships The deserts of the southwestern United States contain many tall, sculpted rock formations. Was wind or water erosion the most likely agent responsible for these formations? Explain.

22. **Making Inferences** Suppose that one time each month for a year a satellite orbiting Earth takes a photograph of the same sandy, 1 km^2 area of the Sahara. Would the surface features shown in these 12 photographs remain essentially the same, or would they vary? Explain your answer.

23. **Inferring Relationships** Wave energy decreases when waves travel through shallow water. Based on this information, what effect do you think development of a wave-built terrace has on erosion of the shoreline? Explain your answer.

Concept Mapping

24. Use the following terms to create a concept map: *wave erosion, beaches, spit, berm,* and *tombolos*.

Math Skills

25. **Making Calculations** Suppose that sea level continuously rises at the rate of 1 mm per year and that other factors affecting the coastlines do not change. How many kilometers will sea level rise in 1 million years?

26. **Applying Quantities** Every year, 25 km^3 of sand is deposited on a beach by a nearby river, and 28 km^3 of sand is removed by wave action. Is the size of the beach increasing or decreasing? Explain.

Writing Skills

27. **Creative Writing** Imagine that you are a newspaper reporter who has traveled to another planet. Scientists know that, at one time, both wind and water eroded the surface of the planet. Prepare a news release that describes the landscape you see and explains the processes that produced it.

28. **Writing Persuasively** You have learned that beautifully colored sunsets and sunrises are the result of dust in the atmosphere. Write a letter or essay to your doubting friend to convince him or her that sunsets and sunrises are caused by dust. Use the evidence that remarkable sunsets were visible around the world for two years after the 1883 eruption of Krakatau, a volcanic island in Indonesia, in your essay.

Interpreting Graphics

The graph below shows soil erosion in the United States by wind and water from 1982 to 1997. Use this graph to answer the questions that follow.

Soil Erosion in the United States, 1982–1997

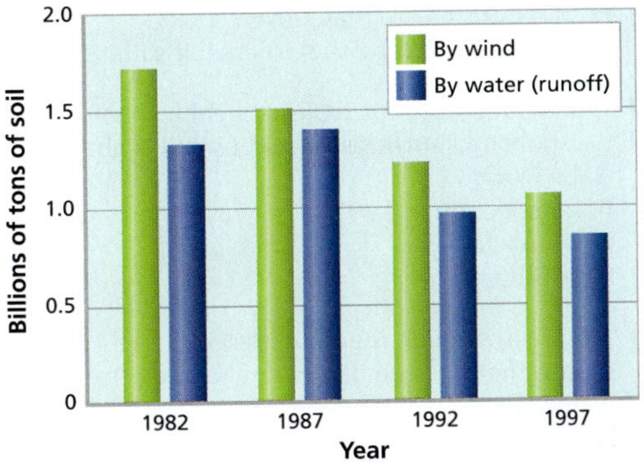

29. Which year had the most combined soil erosion?

30. Which year had the most soil erosion due to water?

31. Do you predict more or less soil erosion by wind in future years? Explain your answer.

32. Would you expect that the amount of soil erosion in the United States would ever be zero? Explain your answer.

Chapter Review

22. Sample answer: Photos taken of the same area of the Sahara Desert over a year's time would vary because the sand dunes would migrate and change shape slightly over that time.

23. The wave-built terrace would make the depth of water decrease closer to the shore. If wave energy decreases through shallow water, then the development of wave-built terraces would slow the effects of erosion on the coastline.

Concept Mapping

24. Answers may vary but should include all of the terms listed. Sample answers appear at the end of this Teacher Edition.

Math Skills

25. 1 mm/y × 1,000,000 y = 1,000,000 mm; 1,000,000 mm × 1 m/1,000 mm = 1,000 m; 1,000 m × 1 km/1,000 m = 1 km

26. The beach is gradually getting smaller by 3 km^3 each year; 25 km^3 − 28 km^3 = −3 km^3.

Writing Skills

27. Answers may vary. Accept all reasonable answers.

28. Answers may vary. Accept all reasonable answers.

Interpreting Graphics

29. 1982

30. 1987

31. Sample answer: less; The downward trend indicates that efforts are being made to control erosion, and the trend should continue.

32. Sample answer: no; Soil erosion could never be completely stopped, but the amount of soil eroded could remain relatively stable if conservation methods are employed and no extreme weather conditions occur.

Estimated Time

To give students practice under more realistic testing conditions, allow them 30 minutes to answer all of the questions in this practice test.

TEST DOCTOR

Question 2 Answer G is correct. Dust particles are much smaller than sand grains. Answer F is incorrect because most sand particles are not lifted high but move by short jumps along the ground. Answer H is incorrect because both dust and sand are composed of rock fragments. Answer I is incorrect because sand, not dust, is moved by saltation.

Question 10 Full-credit answers should include the following points:

- the dust came from topsoil that was loosened by overworking
- clearing land for planting removes plant roots that hold the soil in place. The over plowed, overgrazed, and overworked lands were devastated by erosion
- without rain, soil drys out, is lifted up by high winds, is suspended in the air, and forms dark dust clouds
- when the dust fell, it suffocated crops that had survived the drought

Understanding Concepts

Directions (1–5): **For *each* question, write on a separate sheet of paper the letter of the correct answer.**

1. Which of the following factors most affects the rate at which waves erode land features along the shore?
   - **A** temperature of the waves
   - **B.** direction in which waves approach shores
   - **C.** shape of the rock formation
   - **D.** compostion of the rock formation

2. Why are dust particles more likely to remain in the atmosphere longer and travel farther than sand particles?
   - **F.** Sand grains are carried higher and fall.
   - **G.** Dust particles are smaller and lighter.
   - **H.** Sand grains are made from rocks.
   - **I.** Dust is moved by the process of saltation.

3. What is the term for rocks or pebbles that have flat, polished surfaces caused by wind abrasion?
   - **A.** bedrock
   - **B.** compaction
   - **C.** pinnacles
   - **D.** ventifacts

4. What is the name for a submerged river valley mouth that forms a bay where salt and fresh water mix?
   - **F.** estuary
   - **G.** fiord
   - **H.** atoll
   - **I.** lagoon

5. Why is erosion by wind more common in arid climates than in other regions of the world?
   - **A.** arid climates have much thicker soil layers
   - **B.** arid climates have less frequent dust storms
   - **C.** arid climates have less plant cover to anchor soil
   - **D.** arid climates have more moisture to hold soil

Directions (6–7): **For *each* question, write a short response.**

6. What is the term for dune movement?

7. What factor is most important in determining the composition of beach materials?

Reading Skills

Directions (8–10): **Read the passage below. Then, answer the questions.**

Black Blizzards

The area that covers parts of Colorado, Kansas, New Mexico, Oklahoma, and Texas had been converted from natural grassland to farmland in the early 1900s. Many of the plants brought in to replace the natural prairie grasses had shallow root systems that could not hold soil in place. Much of the rest of the grassland was turned over to grazing land for hungry livestock.

During the 1930s, a long period of drought set in and the already dry soil turned to dust. In the spring of 1934, high winds blew black dust clouds across the dry wheat fields of these states. Some of the dust settled only when it reached Boston and New York City. The sky turned black at mid-day. And when the dust fell, houses were coated with thick layers of dust. Roads and fences were covered by dust. The remaining crops that had survived the droughts suffocated on the ground as the dust blocked the sunlight and other nutrients. Millions of people left their farms in search of a better life.

8. According to the passage, how far did some of the dust travel during the dust storms of 1934?
   - **A.** all the way to the Pacific Ocean
   - **B.** all the way to the Atlantic Ocean
   - **C.** only as far as the Rocky Mountains
   - **D.** only as far as the Great Smokey Mountains

9. Which of the following statements can be inferred from the information in the passage?
   - **A.** The dust storms from the 1930s continued well into the 1940s.
   - **B.** Black blizzards are common occurances in the states of Texas and Colorado.
   - **C.** One of the main causes of the dust storms of the 1930s was misuse of land by humans.
   - **D.** Damage from the dust storms of the 1930s can still be seen today in states such as Texas and Oklahoma.

10. Briefly describe how high winds and an extensive drought could combine to produce the terrible conditions seen during the 1930s.

Answers

Understanding Concepts

1. D
2. G
3. D
4. F
5. C
6. migration
7. the source rock

Reading Skills

8. B
9. C
10. Answers may vary. See Test Doctor for a detailed scoring rubric.

Interpreting Graphics

11. D
12. Answers may vary. See Test Doctor for a detailed scoring rubric.

Interpreting Graphics

Directions (11–12): **For *each* question below, record the correct answer on a separate sheet of paper.**

Base your answers to question 11 on the image of an eroding sea cliff . The projecting headland of the cliff is composed of granite, and the hillside below is made of limestone.

Erosion of a Sea Cliff

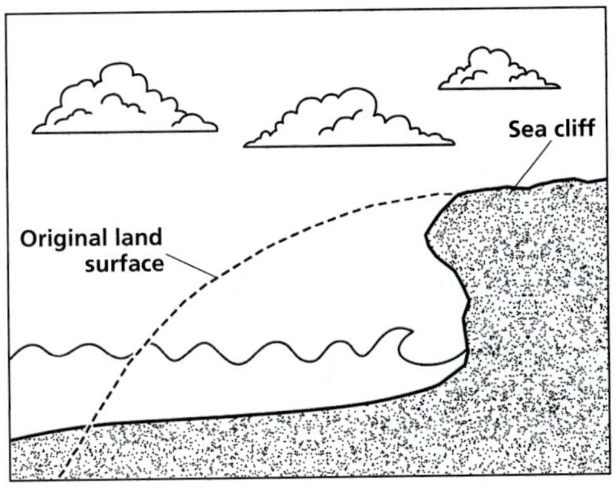

11 The coastal landforms shown were most likely formed as a result of the action of
A. glacial movements
B. seismic activity
C. sea level changes
D. waves and weathering

Base your answers to question 12 on the map below, which shows shoreline changes across the United States.

Map of Coastal Erosion Patterns

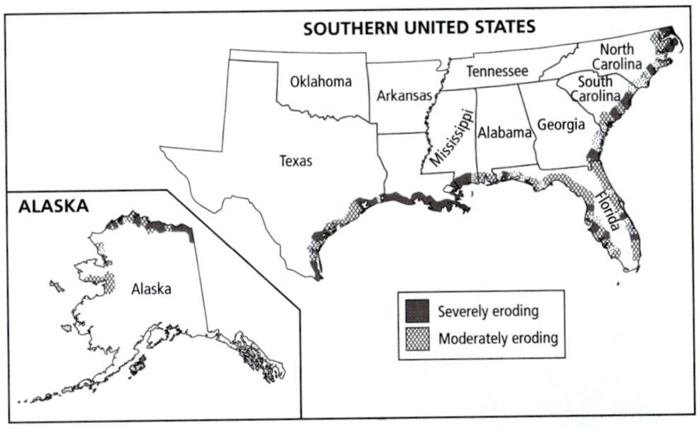

12 How might human activities along the rivers that empty into the Gulf of Mexico and the coastal shoreline play a part in the increasing rate of erosion that is affecting the region?

Test TIP

Do not spend a long period of time on any single question. Mark a question that you cannot answer quickly, and come back to it.

TEST DOCTOR

Question 12 Full-credit answers should include the following points:
- students must use critical-thinking skills and their knowledge of deposition to answer this question
- an understanding that under natural conditions, rivers dump tons of sediment in the areas where they meet the ocean
- under natural circumstances, the rate of this deposition may effectively balance the rate of erosion in the area. Currents that run parallel to the shore carry deposited sediments that replace sand lost to erosion
- as humans build dams that slow the flow of water and filter sediments, less deposition takes place in the mouth of the stream. The amount of sediment that reaches the Gulf of Mexico is decreased significantly
- erosion happens faster than deposition under these circumstances
- other human activities, such as construction and the pumping of oil and groundwater, also contribute to the problem by weakening the shoreline's physical integrity

Test Prep Correlations *National Science Education Standards*

ES 3c: items 1, 3, 4, 5, 6, 7, 10, 11, 12
LS 4e: items 8, 9, 12
ST 2d: item 12
SPSP 2c: items 9, 12
SPSP 3a: items 9, 12
SPSP 3c: items 9, 10
SPSP 4a: items 9, 12
SPSP 4c: items 9, 10
SPSP 5b: items 9, 12
SPSP 5c: items 9, 10, 12
SPSP 4c: item 9
UCP 1: items 1, 2, 5, 6, 7, 11
UCP 3: items 1, 3, 6, 11

CHAPTER RESOURCES

State Resources

For specific resources for your state, visit go.hrw.com and type in the keyword **HSHSTR.**

Beaches

Teacher's Notes

Time Required
one 45-minute class period

Lab Ratings
EASY → HARD

Teacher Preparation: 1
Student Setup: 1
Concept Level: 1
Cleanup: 2

Skills Acquired
- Constructing Models
- Observing
- Predicting
- Collecting Data
- Experimenting
- Interpreting
- Organizing and Analyzing Data
- Inferring
- Communicating

The Scientific Method
In this lab, students will
- Make Observations
- Ask Questions
- Test Hypotheses
- Analyze the Results
- Communicate

Materials
Have extra containers available for students to store the sandy water after the activity. Use plastic toy shovels or large serving spoons to help scoop the sand out of the containers.

Tips and Tricks
Organizing students into groups of three would allow each student to contribute one breakwater design to test in step 7. Make a mark on the large plastic containers to indicate water level. Encourage students to create a data table that describes the shore before waves and after each type of breakwater is constructed.

Have students communicate the results of their designs with the class. Each student or group can share their drawings with the class to see other students' designs. Show actual pictures of breakwater designs to compare with those the students designed.

Chapter 18 **Inquiry Lab** Using Scientific Methods

Beaches

Objectives
- **Model** the effects of wave action and longshore currents on a beach.
- **Identify** ways to decrease the effects of wave action on beach sand.

Materials
- block, plaster (2)
- block, wooden, large
- container, plastic, large
- pebbles
- ruler, metric
- sand, 5 to 10 lb
- water

Safety

Coastal management is a growing concern because beaches are increasingly used for resources and recreation. The supply of sand for many beaches has been cut off by dams built on rivers and streams that are used to carry sand to the sea. Waves generated by storms also continuously wear away beaches. In some places, breakwaters have been built offshore to protect beaches from washing away. In this lab, you will examine how wave action may change the shape of beaches and how these changes can be reduced.

Ask a Question

1. How does wave action affect the amount of sand on a beach? How can these effects be reduced?

Form a Hypothesis

2. Form a hypothesis that answers your question. Explain your reasoning.

Test the Hypothesis

3. Make a beach in a large, shallow container by placing a mixture of sand and small pebbles at one end of the container. The beach should occupy about one-fourth of the length of the container.
4. In front of the sand, add water to a depth of 2 to 3 cm. Record what happens.
5. Use the large wooden block to generate several waves by moving the block up and down in the water at the end of the container opposite the beach. Continue this wave action until about half the beach has moved. Describe the beach after this wave action has taken place.
6. Remove the sand, and rebuild the beach.

Step 5

CHAPTER RESOURCES

Chapter Resource File
- Datasheet for Chapter Lab GENERAL
- Lab Notes and Answers

7. Design three breakwaters that change the flow of water along the beach. Draw your designs on a piece of paper. The two top photos at right are samples of some breakwater arrangements.

8. Have your teacher approve your designs before you build them into your model beach.

9. Use the two plaster blocks to model the first breakwater that you designed. Use a wooden block to generate waves as in step 5. Record your observations.

10. Use the wooden block to generate waves that move parallel to the beach. Record your observations.

11. Repeat steps 9 and 10 for each of your other two designs. Record your observations.

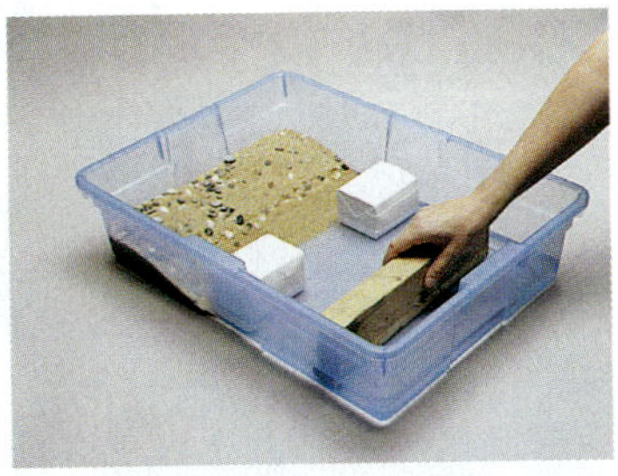

Step 7

Step 7

Step 10

ANALYZE THE RESULTS

1. **Making Comparisons** How does wave action build up a beach? How does wave action wear away a beach?

2. **Explaining Events** Describe what happened to the shape of the waves along the beach in step 10.

3. **Analyzing Results** How do breakwaters modify the effect that longshore currents have on the shape of a beach?

DRAW CONCLUSIONS

4. **Making Predictions** Predict what will happen to a beach that is affected by wave action if it had no source of additional sand.

5. **Drawing Conclusions** What effect would a series of jetties have on a beach?

Extension

1. **Research and Communications** Research what can be done to preserve a recreational beach from erosion that is caused by excessive use by people. Write a letter to a local authority outlining a plan of action to protect that beach.

Inquiry Lab

Answers to Analyze the Results

1. Beaches form where more sediment is deposited than is removed. Wave action during storms or high tides generally carries sediment onto the beach.
2. Answers may vary. Waves should transport sand parallel to the "shore."
3. Answers may vary. Breakwaters should reduce the amount of sand that longshore currents remove from the beach.

Answers to Conclusions

4. Answers may vary. The beach, over time, would lose sediment and would disappear or would become covered with water.
5. Sample answer: A series of jetties would impede movement of sand along the beach, which would cause sand to accumulate on the side of the jetty that faces the current and to be carried away from the back side of the jetty.

Answers to Extension

1. Answers may vary. Accept all reasonable answers.

Tammie Niffenegger
Science Chair and Science Teacher
Port Washington High School
Port Washington, WI

Coastal Erosion Near the Beaufort Sea

Homework — Advanced

Coastal Landforms Coastal landforms can be classified as erosional or depositional. Many coasts are a combination of both, to varying degrees. Each type of coastal landform has defining characteristics. Have students research the characteristics of both erosional and depositional coastal landforms. They should identify the rock formations and beach composition as well as places in the United States where each type is located. After students present their research, ask them to use the clues in the map to identify the type of coastal region depicted in the map.
LS **Visual**

Answers to Map Skills Activity

1. 11
2. Most areas show no detectable erosion.
3. The estimated overall shoreline change shows a negative value, indicating erosion.
4. The coastal area toward the south shows more erosion.
5. The area appears to be more affected by waves along the shoreline than by depositional actions of rivers entering the ocean. Many areas show moderate to rapid erosion. The areas of deposition are fewer, smaller, and located only in sheltered areas.
6. Answers may vary. Sample answer: The islands would be eroded more drastically over time and may eventually disappear.

CHAPTER RESOURCES

Technology

Transparencies
- 92 Coastal Erosion Near the Beaufort Sea (with worksheet)

MAPS *in Action*

Coastal Erosion Near the Beaufort Sea

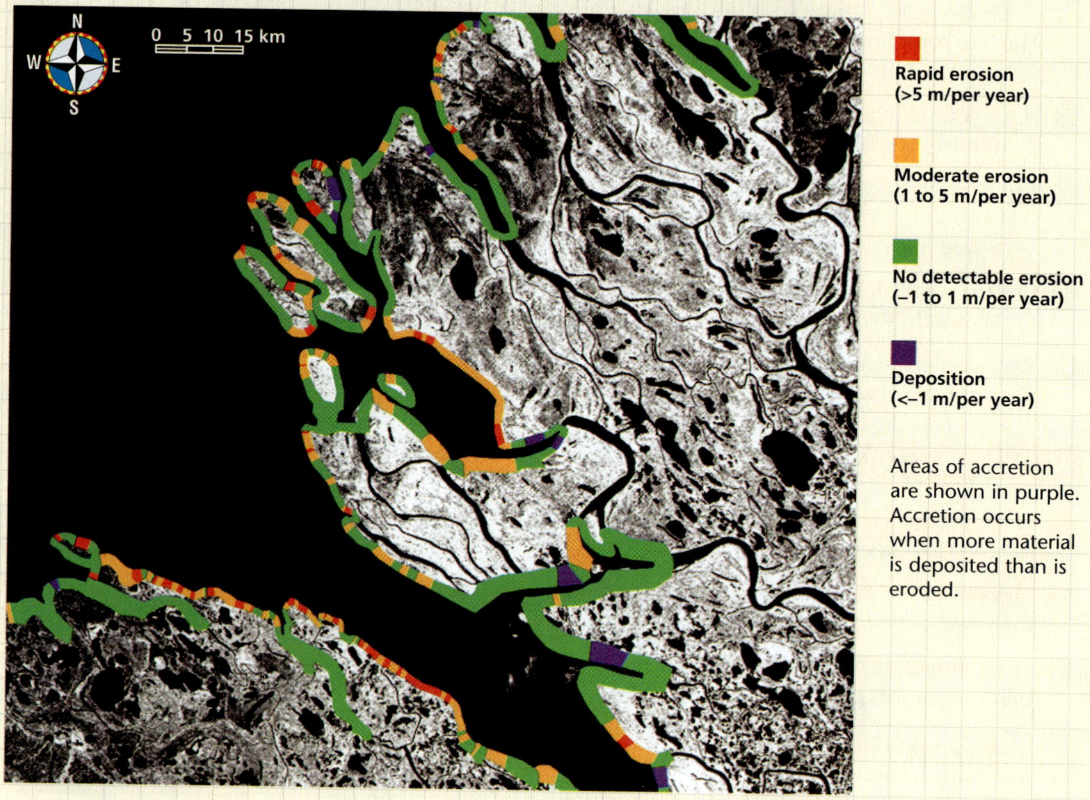

Map Skills Activity

This map shows the coastline of the Beaufort Sea in Canada along with computed amounts of shoreline erosion and accretion. Use the map to answer the questions below.

1. **Using the Key** How many areas of accretion are shown on the map?
2. **Using the Key** What is the level of erosion that is present in most areas shown?
3. **Analyzing Data** Is the estimated overall shoreline change for the entire area shown a positive value (accretion) or a negative value (erosion)?
4. **Making Comparisons** Is there more erosion on the coastal area toward the north or on the area toward the south?
5. **Inferring Relationships** Is the area shown in the map more significantly affected by the effects of waves along the shoreline or by the depositional actions of rivers that enter the ocean? Explain your answer.
6. **Identifying Trends** If present conditions remain the same, what would you expect to happen to the five small islands that are located on the northwest area of the map.

The Flooding of Venice

The city of Venice, Italy, is famous for its canals and priceless art treasures. Venice is built on about 120 islands in the Adriatic Sea.

◀ Flooding in St. Mark's square invites residents to be creative about their transportation.

A City Underwater

The waters of the Adriatic Sea constantly threaten Venice. Every year, winter storms cause sea water to flood the city's public squares, walkways, and buildings. Venice's famous St. Mark's Square may flood more than 100 times per year. Overexposure to floodwaters is slowly eroding the foundations of the city's Byzantine and Renaissance architecture. The city's sidewalks are buckled because the water has eroded the ground underneath them.

These floods are the result of three factors: erosion, rising sea level, and sinking land. Erosion of barrier beaches and sandbars allows high tides to reach farther into Venice. Also, the Adriatic Sea is rising. Its average sea level has risen 11 cm over the last 100 years, and Venice has sunk 13 cm.

The weight of added sediments is causing the city to sink about 1 mm per year. Removal of too much groundwater and the weight of the buildings within the city have caused additional sinkage.

An Uncertain Future

To prevent future and permanent damage to Venice, the city restricts the pumping of groundwater. Pavement is often raised to help keep the city above water and to minimize erosion caused by floodwaters. Massive floodgates have been built to block high tides that threaten the city. In addition, city planners are constructing a system to pump fluids into the aquifer below the lagoon to raise the level of water in the aquifer.

◀ Flooding causes erosion in wood, brick, and stone. Many building foundations are slowly dissolving into the waters of the Adriatic Sea.

Extension

1. **Researching Trends** Research another area of the world that is grappling with regular flooding. How is that city or town dealing with the erosion that is caused by flooding?

Impact on Society

The Flooding of Venice

Group Activity — Advanced

Debate Project A project called MOSE (Modulo Sperimentale Elettromeccanico, or Experimental Electromechanical Module) is under construction in Venice as a means of controlling flooding. It is a system of huge gates designed to keep the high tides of the Adriatic Sea, which cause the flooding, away from the city. While the majority of the population of Venice endorses the plan, there are opponents, including environmentalists, architects and engineers, and other Venetians who believe there are better solutions.

Divide the class into two groups: one to take a position in favor of Project MOSE and the other opposing it. Ask students to research the issue and accumulate facts to support their cases. Tell students that the subject often depends on one's historical perspective on the causes of the flooding. This is a topic that lends itself well to debate. Or, have students write letters to the editor on the issue. LS Logical/Verbal

Answers to Extension

1. Answers may vary. Students may want to research one of the following areas: Denmark, the Netherlands, New Zealand, Bangladesh, or London, England.

Unit 6 OCEANS

Unit 6 Outline

CHAPTER 19
The Ocean Basins

CHAPTER 20
Ocean Water

CHAPTER 21
Movements of the Ocean

▶ Oceans affect coastlines, climates, and aquatic life around the world. As shown in this satellite image of the Bahamas in the Caribbean Sea, ocean currents and tides have shaped the sand and seaweed beds so that from space the beds appear to be blue flames of a fire.

Chapter 19 The Ocean Basins

Planning Guide

Compression Guide

To shorten instruction because of time limitations, omit Section 3.

OBJECTIVES	LABS, DEMONSTRATIONS, AND ACTIVITIES	TECHNOLOGY RESOURCES
PACING • 45 min pp. 470–474 **Chapter Opener**		**OSP** **Parent Letter** **CD** **Student Edition on CD-ROM** **CD** **Chapter Summaries Audio CDs** **VID** **Brain Food Video Quiz**
Section 1 The Water Planet • Name the major divisions of the global ocean. • Describe how oceanographers study the ocean. • Explain how sonar works.	**TE** **Language Arts Connection** 20,000 Leagues Under the Sea Today, p. 472 ADVANCED **SE** **Quick Lab** Sonar, p. 473 GENERAL **CRF** **Datasheet for Quick Lab*** GENERAL **TE** **Biology Connection** Seeing with Sound, p. 473 GENERAL	**OSP** **Lesson Plans** (also in print) **TR** **Bellringer*** **TR** 93 The Global Ocean*
PACING • 45 min pp. 475–478 **Section 2 Features of the Ocean Floor** • Describe the main features of the continental margins. • Describe the main features of the deep-ocean basin.	**CRF** **Making Models Lab** Island to Guyot* GENERAL	**OSP** **Lesson Plans** (also in print) **TR** **Bellringer*** **TR** 94 Features of the Ocean Floor* **VID** **HRW Earth Science Video** Features of the Ocean Floor
PACING • 90 min pp. 479–482 **Section 3 Ocean-Floor Sediments** • Describe the formation of ocean-floor sediments. • Explain how ocean-floor sediments are classified by their physical composition.	**TE** **Demonstration** Sediments Sinking, p. 479 GENERAL **TE** **Demonstration** Turbidity Currents and Submarine Canyons, p. 480 GENERAL **SE** **Quick Lab** Diatoms, p. 481 GENERAL **CRF** **Datasheet for Quick Lab*** GENERAL **SE** **Skills Practice Lab** Ocean-Floor Sediments, p. 488–489 GENERAL **CRF** **Datasheet for Chapter Lab*** GENERAL **SE** **Maps in Action** Total Sediment Thickness of Earth's Oceans, p. 490 GENERAL **CRF** **Inquiry Lab** Sediment Loading* GENERAL	**OSP** **Lesson Plans** (also in print) **TR** **Bellringer*** **TR** 95 Total Sediment Thickness of Earth's Oceans* **TE** **Internet Activity** Deep-Sea Cores, p. 490 ADVANCED **CRF** **Internet Activity** Deep-Sea Cores* ADVANCED **CD** **Interactive Tutor** Ocean Basins **CD** **Interactive Tutor** Ocean Depths **CD** **Interactive Tutor** Coastal Zone

PACING • 90 min

CHAPTER REVIEW, ASSESSMENT, AND STANDARDIZED TEST PREPARATION

- **SE** **Chapter Highlights**, p. 483
- **SE** **Chapter Review**, pp. 484–485
- **SE** **Standardized Test Prep**, pp. 486–487
- **CRF** **Concept Review*** GENERAL
- **CRF** **Critical Thinking*** ADVANCED
- **CRF** **Math Skills*** GENERAL
- **CRF** **Graphing Skills*** GENERAL
- **CRF** **Chapter Test A*** GENERAL
- **CRF** **Chapter Test B*** ADVANCED
- **OSP** **Lesson Plans** (also in print)
- **OSP** **Test Generator**
- **OSP** **Test Item Listing**

Online and Technology Resources

go.hrw.com Holt Online Learning

Visit go.hrw.com for access to Holt Online Learning, or enter the keyword **HQ6 Home** for a variety of free online resources.

One-Stop Planner® CD-ROM

This CD-ROM package includes
- Lab Materials QuickList Software
- Holt Calendar Planner
- Customizable Lesson Plans
- Printable Worksheets
- ExamView® Test Generator
- Interactive Teacher Edition
- Holt PuzzlePro®
- Holt PowerPoint® Resources

KEY

SE Student Edition	OSP One-Stop Planner	VID Classroom Video/DVD
TE Teacher Edition	TR Transparencies and Transparency Worksheets	* Also on One-Stop Planner
CRF Chapter Resource File	CD CD or CD-ROM	◆ Requires advance prep
LTP Long-Term Projects		■ Also available in Spanish

SKILLS DEVELOPMENT RESOURCES	REVIEW AND ASSESSMENT	CORRELATIONS
SE **Pre-Reading Activity,** p. 470 GENERAL TE **Using the Figure** The Ora Verde, p. 470 GENERAL		National Science Education Standards
CRF **Directed Reading*** BASIC TE **Using the Figure** Global Oceans, p. 471 GENERAL TE **Reading Skill Builder** Paired Summarizing, p. 472 BASIC TE **Skill Builder** Writing, p. 472 ADVANCED TE **Skill Builder** Math, p. 473 BASIC	SE **Reading Check,** p. 472 GENERAL SE **Section Review,** p. 474 GENERAL TE **Reteaching,** p. 473 BASIC TE **Quiz,** p. 474 GENERAL TE **Alternative Assessment,** p. 474 GENERAL CRF **Section Quiz*** ■ GENERAL	SAI 2c
CRF **Directed Reading*** BASIC TE **Using the Figure** Modeling the Ocean Floor, p. 475 GENERAL SE **Graphic Organizer** Spider Map, p. 476 GENERAL TE **Inclusion Strategies,** p. 476 ◆	SE **Reading Check,** p. 476 GENERAL SE **Section Review,** p. 478 GENERAL TE **Reteaching,** p. 477 BASIC TE **Quiz,** p. 477 GENERAL TE **Alternative Assessment,** p. 477 ADVANCED CRF **Section Quiz*** ■ GENERAL	
CRF **Directed Reading*** BASIC TE **Inclusion Strategies,** p. 480 TE **Reading Skill Builder** Reading Organizer, p. 480 BASIC SE **Math Practice,** p. 481 GENERAL	SE **Reading Check,** p. 481 GENERAL SE **Section Review,** p. 482 GENERAL TE **Reteaching,** p.481 BASIC TE **Quiz,** p. 481 GENERAL TE **Alternative Assessment,** p. 482 BASIC CRF **Section Quiz*** ■ GENERAL	SAI 2c, UCP 4

Holt Lab Generator CD-ROM

Search for labs by topic, standard, difficulty level, and/or time frame. Edit those labs to fit your needs, or create your own labs. Use the Lab Materials QuickList to order just what you need.

Holt Earth Science Interactive Tutor CD-ROM

This CD-ROM consists of interactive activities that give students a fun way to extend their knowledge of Earth science concepts.

Chapter Summaries Audio CDs

These CDs include audio summaries of the key concepts presented in each chapter. (Audio summaries are also available in Spanish.)

SCILINKS. NSTA

www.scilinks.org

Maintained by the **National Science Teachers Association.** See Chapter Enrichment pages that follow for a complete list of topics.

See Chapter Enrichment pages for Video Resources.

Refer to our ***Professional Reference for Teachers*** for additional teaching resources, including articles written by science education professionals about relevant and timely issues facing today's science teachers.

Chapter 19

Chapter Enrichment

This Chapter Enrichment provides relevant and interesting information to expand and enhance your classroom instruction of the chapter material.

Section 1 The Water Planet

Challenges of Ocean Exploration

The oceans have fascinated humans for ages, but only within in the last 150 years have humans really begun to uncover some of the mysteries of the deep. Why have the oceans remained "the last frontier" for so long? The great physical challenges posed by the ocean provide part of the answer. Working on the ocean surface, scientists have to deal with constant rolling waves, cold winds, deadly storms, corrosive seawater, slippery decks, and cramped ship's quarters for months at a time. In the depths of the ocean, humans simply cannot survive the extreme cold temperatures and pressures that can crush all but the strongest reinforced vehicles. Humans can visit the abyss only in sophisticated, specially designed submersible undersea vehicles.

Only a few submersibles can actually reach the deepest parts of the ocean. France's *Nautile,* Russia's *Mir II,* and the United States' *Sea Cliff* can dive to 6 km, which is the depth of 97 percent of the ocean. Only two submersibles, *Trieste I* and *Shinkai 6500,* have reached the very deepest sites.

▲ Scientists aboard the research ship *JOIDES Resolution* performed scientific studies of the ocean.

▲ Unique species such as this angler fish live in the farthest depths of the ocean.

Section 2 Features of the Ocean Floor

Black Smokers and Creatures of the Abyss

Hydrothermal vents dot mid-ocean ridges and volcanic hot spots along the ocean basins. In these spots, water seeps into cracks in the oceanic crust and is superheated by hot rocks. The superheated water (up to 360°C) spews back into the icy ocean in a plume, along with toxic gases and metals that solidifies into tube-like structures, named *black smokers.* Amazingly, vibrant but bizarre communities of giant clams, mussels, blind shrimp, and giant tubeworms surround these sites. In all, some 300 new species have been identified in these vent communities. The base of the food chain in these communities is an amazing group of bacteria that convert sulfur compounds in the toxic vent gases into food and energy in a process known as *chemosynthesis.* Many scientists now think that life on Earth may have begun in places such as the hydrothermal vents.

Section 3 Ocean Floor Sediments

Sediment Stories

Scientists can uncover clues about Earth's past climates by studying sediment cores. Cores recovered from the ocean can represent millions of years of Earth's history. Because the number and type of plankton in the sea changes as climate changes, changes in the remains of plankton that exist in sediment layers give clues to how Earth's climate has varied in the past. For instance, sea-floor sediments have aided scientists in trying to understand climate conditions during the last Ice Age. Understanding past climates can also help scientists predict how Earth's systems will behave in the future. It could help us predict the effects of global warming, which results from increasing atmospheric carbon dioxide concentrations.

Mining the Sea

Ocean sediments commonly contain nodules, or lumps of minerals that are rich in oxides of manganese, iron, and nickel. Copper, zinc, and cobalt are also part of nodules in lower concentrations. Nodules could be a source of concentrated ores if they could be economically retrieved. Unfortunately, the largest accumulations of nodules occur in the Pacific basin at depths of more than 4 km. Deposits around hydrothermal vents are another potentially rich source of minerals. Copper, lead, zinc, gold, silver, and mercury have been found in hydrothermal vent deposits. In 1998 an Australian mining company was granted rights to explore hydrothermal vents off the coast of Papua New Guinea.

◀ **The remains of tiny diatoms are important components of biogenic sediments on the ocean floor.**

Video Resources

Brain Food Video Quizzes

Brain Food Video Quizzes These videos contain game-show style quizzes that assess students' progress and motivate students to study the chapter material.

HRW Earth Science Video

This video introduces Earth science topics and includes a geology field trip. The video segment listed below complements this chapter.

Segment 14: Features of the Ocean Floor The oceans comprise three-quarters of Earth's surface and contain some of the planet's largest topographic structures. One such structure is the island of Hawaii, the subject of this segment, which measures 9,000 m from its base on the bottom of the Pacific Ocean to its summit. (1 min)

NOVA Videos

To order **NOVA** videos related to this chapter, visit **go.hrw.com** and enter the keyword **HQ6OBAV.**

SciLinks

NSTA Developed and maintained by the National Science Teachers Association

SciLinks is maintained by the National Science Teachers Association to provide you and your students with interesting, up-to-date links that will enrich your classroom presentation of the chapter.

Visit www.scilinks.org and enter the SciLinks code for more information about the topic listed.

Topic: The Oceans
SciLinks code: HQ61069

Topic: Ocean-Floor Features
SciLinks code: HQ61067

Topic: Ocean-Floor Sediments
SciLinks code: HQ61068

Topic: Careers in Earth Science
SciLinks code: HQ60222

Chapter 19

Chapter Overview

This chapter describes how scientists study the deep trenches, huge mountain ranges, submarine canyons, and the abyssal plains of the ocean basins. This chapter also describes ocean-floor sediments.

Using the Figure — GENERAL

The Ora Verde This photograph shows the Ora Verde, a ship lost off the Grand Cayman Islands. With advances in undersea exploration, scientists and adventurers are uncovering more and more shipwrecks. Have students discuss some difficult issues that may arise from exploring shipwrecks. (Answers may vary but may include: showing respect for the dead; deciding who owns the site and who should profit from salvaged "treasures"; and undertaking dangerous expeditions solely for profit.) **LS Interpersonal**

PRE-READING ACTIVITY

FOLDNOTES

You may want to assign this FoldNote activity as homework. Collect the FoldNotes to check students' understanding of the material.

Chapter 19 The Ocean Basins

Sections

1 The Water Planet
2 Features of the Ocean Floor
3 Ocean-Floor Sediments

What You'll Learn

- Why scientists study the ocean
- What the features of the ocean basins are
- How ocean-floor sediments are classified

Why It's Relevant

The oceans interact with the atmosphere and the land to affect weather, climate, and the shape of continents. Exploration of the oceans explains how these interactions shape Earth's surface.

PRE-READING ACTIVITY

FOLDNOTES

Booklet Before you read the chapter, create the FoldNote entitled "Booklet" described in the Skills Handbook section of the Appendix. Label each page of the booklet with a main idea from the chapter. As you read the chapter, write what you learn about each main idea on the appropriate page of the booklet.

▶ The ocean floors contain evidence of changes in Earth over time and of human endeavors. This is the wreckage of the *Ora Verde,* a ship that was lost near the Grand Cayman Islands.

Chapter Correlations *National Science Education Standards*

SAI 2c Scientists rely on technology to enhance the gathering and manipulation of data. New techniques and tools provide new evidence to guide inquiry and new methods to gather data, thereby contributing to the advance of science. The accuracy and precision of the data, and therefore the quality of the exploration, depends on the technology used. **(Section 1 and Section 3)**

UCP 4 Evolution is a series of changes, some gradual and some sporadic, that accounts for the present form and function of objects, organisms, and natural and designed systems. The general idea of evolution is that the present arises from materials and forms of the past. Although evolution is most commonly associated with the biological theory explaining the process of descent with modification of organisms from common ancestors, evolution also describes changes in the universe. **(Section 3)**

Section 1

The Water Planet

Nearly three-quarters of Earth's surface lies beneath a body of salt water called the **global ocean.** No other known planet has a similar covering of liquid water. Only Earth can be called the *water planet*.

The global ocean contains more than 97% of all of the water on Earth. Although the ocean is the most prominent feature of Earth's surface, the ocean is only about 1/4,000 of Earth's total mass and only 1/800 of Earth's total volume.

OBJECTIVES

- **Name** the major divisions of the global ocean.
- **Describe** how oceanographers study the ocean.
- **Explain** how sonar works.

KEY TERMS

global ocean
sea
oceanography
sonar

Divisions of the Global Ocean

As shown in **Figure 1,** the global ocean is divided into five major oceans. These major oceans are the Atlantic, Pacific, Indian, Arctic, and Southern Oceans. Each ocean has special characteristics. The Pacific Ocean is the largest ocean on Earth's surface. It contains more than one-half of the ocean water on Earth. With an average depth of 4.3 km, the Pacific Ocean is also the deepest ocean. The next largest ocean is the Atlantic Ocean. The Atlantic Ocean has an average depth of 3.9 km. The Indian Ocean is the third-largest ocean and has an average depth of 3.9 km. The Southern Ocean extends from the coast of Antarctica to 60°S latitude. The Arctic Ocean is the smallest ocean, and it surrounds the North Pole.

A **sea** is a body of water that is smaller than an ocean and that may be partially surrounded by land. Examples of major seas include the Mediterranean, Caribbean, and South China Seas.

global ocean the body of salt water that covers nearly three-fourths of Earth's surface

sea a large, commonly saline body of water that is smaller than an ocean and that may be partially or completely surrounded by land

Figure 1 ▶ The global ocean is divided into oceans and seas. ***How many oceans are on Earth?***

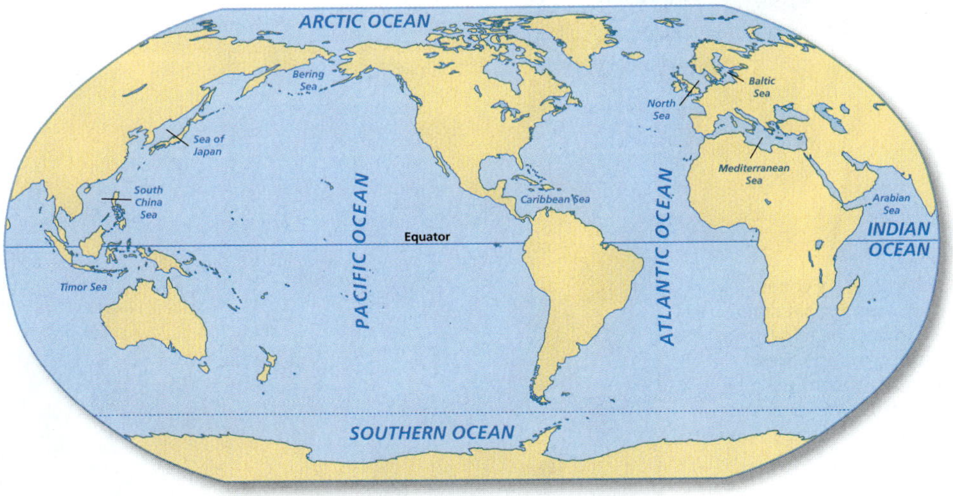

Section 1

Focus

Overview

This section reviews the major divisions of the global oceans, differentiates between oceans and seas, and describes how oceanographers study the ocean by using sonar and submersibles for underwater research.

Bellringer

Ask students why they think the famous science fiction novelist Arthur C. Clarke once remarked: "How inappropriate to call this planet Earth, when clearly it is Ocean." (More than 70% of Earth's surface is covered with water. It is known as the global ocean.) **LS Logical**

Motivate

Using the Figure — GENERAL

Global Oceans The map at left shows the divisions of the global ocean. Have students use this map to determine the area of Earth covered by the various oceans. Have students trace the map onto graph paper and count the number of squares covered by all of the oceans compared to the total number of squares covered by the map. (The area of the oceans should equal about 70% of the total area of the map.) Answer to caption question: five **LS Visual**

CHAPTER RESOURCES

Chapter Resource File

- Directed Reading BASIC

Technology

Transparencies
- Bellringer
- 93 The Global Ocean (with worksheet)

Student Edition on CD-ROM

One-Stop Planner CD-ROM
- Lesson Plan

Teach

Answer to Reading Check

Oceanographers study the physical characteristics, chemical composition, and life forms of the ocean.

SKILL BUILDER — Advanced

Writing Have students visit the Web site of the Woods Hole Oceanographic Institution, the home of *Alvin*. Have them find information about one of *Alvin's* expeditions and write a short newspaper-style article about the expedition's findings. LS **Verbal**

Teaching Tip — Basic

Make Concepts Relevant Invite students to discuss why studying the ocean is important. (Answers will vary but could include: the importance of understanding the influence of oceans on weather and climate change; the importance of understanding life in the oceans; and the importance of exploring for natural resources.) LS **Logical**

READING SKILL BUILDER — Basic

Paired Summarizing Group students into pairs and have them read silently about the tools oceanographers use to study the oceans. Then, have one student summarize the ideas of sonar and submersibles. The other student should listen to the retelling and should point out any inaccuracies or ideas that were left out. Allow students to refer to the text as needed. English Language Learners

LS **Verbal** Co-op Learning

oceanography the scientific study of the ocean, including the properties and movement of ocean water, the characteristics of the ocean floor, and the organisms that live in the ocean

SCILINKS. NSTA Developed and maintained by the National Science Teachers Association

For a variety of links related to this subject, go to www.scilinks.org

Topic: The Oceans

SciLinks code: HQ61069

Exploration of the Ocean

The study of the physical characteristics, chemical composition, and life-forms of the ocean is called **oceanography.** Although some ancient civilizations studied the ocean, modern oceanography did not begin until the 1850s.

The Birth of Oceanography

An American naval officer named Matthew F. Maury used records from navy ships to learn about ocean currents, winds, depths, and weather conditions. In 1855, he published these observations as one of the first textbooks about the oceans. Then, from 1872 to 1876, a team of scientists aboard the British Navy ship HMS *Challenger* crossed the Atlantic, Indian, and Pacific Oceans. The scientists measured water temperatures at great depths and collected samples of ocean water, sediments, and thousands of marine organisms. The voyages of the HMS *Challenger* laid the foundation for the modern science of oceanography.

Today, many ships perform oceanographic research. In the 1990s and in the beginning of the 21st century, the research ship *JOIDES Resolution* was the world's largest and most sophisticated scientific drilling ship. Samples drilled by *JOIDES Resolution*, shown in **Figure 2,** provide scientists with valuable information about plate tectonics and the ocean floor. The Japanese ship *CHIKYU*, which is operated by the Integrated Ocean Drilling Program, is one of the most advanced drilling ship now in use.

Reading Check **List three characteristics of the ocean that oceanographers study. (See the Appendix for answers to Reading Checks.)**

Figure 2 ▶ Reentry cones (above) are used so that core samples can later be taken from the same place on the ocean floor. Scientists aboard the research ship *JOIDES Resolution* (right) perform scientific studies of the ocean floor.

LANGUAGE ARTS CONNECTION — Advanced

20,000 Leagues Under the Sea Today The sea has fascinated humans for centuries. Many famous novels, short stories, and poems have been written about the sea. In 1870, Jules Verne published *20,000 Leagues Under the Sea*, a science fiction account of the adventures of Captain Nemo as he piloted the deep-sea vehicle *Nautilus*. In many ways, this story was visionary. Have students read excerpts from the book. Then, have students write an updated version of an important scene, using information about modern methods of undersea exploration. LS **Verbal**

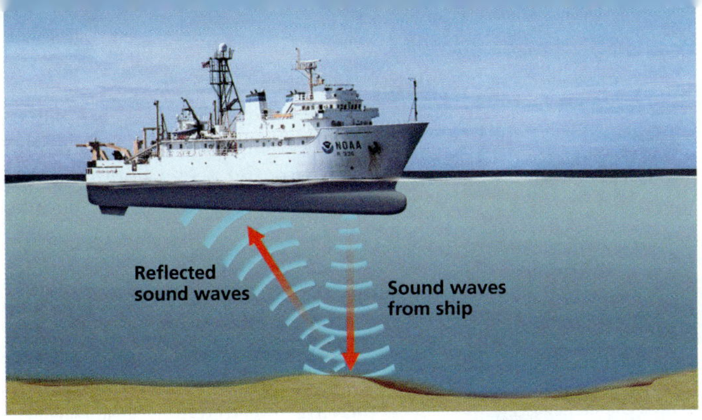

Figure 3 ▶ Active sonar sends out a pulse of sound. The pulse, called a *ping* because of the way it sounds, reflects when it strikes a solid object.

Sonar

Oceanographic research ships are often equipped with sonar. **Sonar** is a system that uses acoustic signals and returned echoes to determine the location of objects or to communicate. Sonar is an acronym for ***so**und **n**avigation **a**nd **r**anging*. A sonar transmitter sends out a continuous series of sound waves from a ship to the ocean floor, as shown in **Figure 3.** The sound waves travel about 1,500 m/s through sea water and bounce off the solid ocean floor. The waves reflect back to a receiver. Scientists measure the time that the sound waves take to travel from the transmitter, to the ocean floor, and to the receiver in order to calculate the depth of the ocean floor. Scientists then use this information to make maps and profiles of the ocean floor.

sonar *sound navigation and ranging,* a system that uses acoustic signals and returned echoes to determine the location of objects or to communicate

QuickLAB 30 min

Sonar

Procedure

1. Use **heavy string** to tie one end of a **spring** securely to a **doorknob.** Pull the spring taut and parallel to the floor. You will need to keep the tension of the spring constant throughout the lab.
2. Use **masking tape** to mark the floor directly beneath the hand that is holding the spring taut. Use a **meterstick** to measure and record the distance from that hand to the doorknob.
3. Note the time on a **stopwatch or clock with a second hand.** Hold the spring taut, and hit the spring horizontally to create a compression wave.
4. Check the time again to see how long the pulse takes to travel to the doorknob and back to your hand. Record the time.
5. Repeat steps 2–4 three times. Each time, hold the spring 60 cm closer to the doorknob. Keep tension constant by gathering coils as necessary.
6. Calculate the rate of travel for each trial by multiplying the distance between your hand and the doorknob by 2. Then, divide by the number of seconds the pulse took to travel to the doorknob and back.

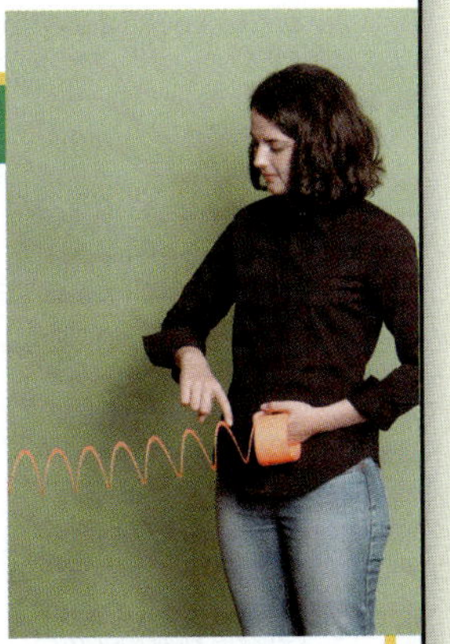

Analysis

1. Did the rate the pulse traveled change during the course of the investigation?
2. If a pulse took 3 s to travel to the doorknob and back to your hand, what is the distance from the doorknob to your hand?
3. How is the apparatus you used similar to sonar? How is the apparatus different than sonar? Explain.

SKILL BUILDER — BASIC

Math Sound waves travel 1,500 m/s in sea water. Calculate the depth of the ocean floor if a sonar ping takes 10 s to return to a research vessel.

$$\text{depth} = \frac{\text{rate} \times \text{time}}{2 \text{ (to account for time to the ocean floor and back)}}$$

$$\text{depth} = \frac{1{,}500 \text{ m/s} \times 10 \text{ s}}{2}$$

depth = 7,500 m

LS **Logical**

BIOLOGY CONNECTION — GENERAL

Seeing with Sound Sonar is like "seeing" with sound. Explain to students that some animals use sonar, or *echolocation,* to navigate and to find prey. Whales can locate objects, such as krill schools or members of the pod, by echolocation. Using low and high frequency clicks, whales can precisely map size, shape, speed, distance, and density of objects. Bats are well known for their echolocation abilities. They produce frequency-modulated ultrasound to find and track insect prey and to navigate at night.

Close

Reteaching — BASIC

Outline Have students make a brief timeline of advances in oceanographic research techniques, identifying important vessels and methods. LS **Visual**

CHAPTER RESOURCES

Chapter Resource File

- Datasheet for Quick Lab GENERAL

QuickLAB

Skills Acquired

- Measuring
- Calculating

Teacher's Notes Make sure that students strike the spring in a horizontal direction to form a compression wave and that they keep the tension on the spring constant by pulling the coil taut as they shorten the distance to the doorknob.

Answers

1. If the spring tension is kept constant, the rate should not change.
2. Answer will depend on the rate calculated. Sample answer: If the rate is 100 cm/s, then 100 cm/s × 3 s = 300 cm to doorknob and back, distance to doorknob is 150 cm (300 ÷ 2).
3. They both use waves to measure distances. Sonar uses sound waves, while the spring model uses a compression wave.

Close, *continued*

Quiz — GENERAL

1. How is a bathyscaph different from a bathysphere? (A bathyscaph is self-propelled and free moving. A bathysphere is tethered to a research ship for communications and life support.)
2. Where have strange life forms, such as giant clams and tubeworms been discovered? (near hydrothermal vents along mid-ocean ridges)
3. When did modern oceanography start? (in the 1850s)

Alternative Assessment — GENERAL

Oceanographic Interview Have students form pairs and pretend that one of them is a journalist and the other is an oceanographer who is being interviewed for a magazine article. The interviewer should ask, "What is the most important advance in oceanographic research in the last 100 years? What is the most important future oceanographic question to address?" When one student has answered, have students switch roles. LS **Verbal/ Interpersonal**

Answers to Section Review

1. The five major divisions of the global ocean are the Atlantic, the Pacific, the Indian, the Arctic, and the Southern oceans.
2. A sea is a body of water that is smaller than the ocean, and that may be partially surrounded by land.
3. Oceanography is the study of the physical, chemical, and biological characteristics of the ocean.
4. Oceanographers use sonar and submersibles to study the ocean.
5. Sonar uses acoustic signals and returned echoes to determine the position of objects or the depth of the ocean floor.
6. Submersibles can collect samples from the ocean floor and take photographs of the ocean floor.
7. The bathyscaph and the bathysphere are both piloted by humans. Submarine robots are remotely operated.
8. Answers may vary but should explain how sonar waves bounce off objects in a vessel's path, which allows the vessel to move around any obstacles.
9. Answers may vary. Sample answer: Submarine robots can stay in the ocean depths for much longer periods than human-piloted submersibles can.
10. The science of *oceanography* uses *sonar* and *submersibles* such as *bathyspheres, bathyscaphs,* and *robot submersibles.*

Figure 4 ▶The submersible *Nautile* (top) carries enough oxygen to keep a three-person crew underwater for more than five hours. Deep-sea submersibles have discovered many strange organisms in the deep ocean, such as this angler fish (bottom).

Submersibles

Underwater research vessels, called *submersibles,* also enable oceanographers to study the ocean depths. Some submersibles are piloted by people. One such submersible is the *bathysphere,* a spherical diving vessel that remains connected to the research ship for communications and life support. Another type of piloted submersible, called a *bathyscaph,* is a self-propelled, free-moving submarine. One of the most well-known bathyscaphs is the *Alvin.* Another modern submersible, called *Nautile* (NOH teel), is shown in **Figure 4.**

Other modern submersibles are submarine robots. They can take photographs, collect mineral samples from the ocean floor, and perform many other tasks. These robot submersibles are remotely piloted and allow oceanographers to study the ocean depths for long periods of time.

Underwater Research

Submersibles have helped scientists make exciting discoveries about the deep ocean. During one dive in a submersible, startled oceanographers saw communities of unusual marine life living at depths and temperatures where scientists thought that almost no life could exist. Giant clams, blind white crabs, and giant tube worms were some of the strange life-forms that were discovered. Many of these life-forms have unusual adaptations that allow them to live in hostile environments. The angler fish, shown in **Figure 4,** can produce its own light, which attracts prey.

Section 1 Review

1. **Name** the five major divisions of the global ocean.
2. **Explain** the difference between an ocean and a sea.
3. **Define** *oceanography.*
4. **Describe** two ways that oceanographers study the ocean.
5. **Explain** how sonar works.
6. **Describe** two aspects of the ocean that submersibles are used to study.
7. **List** three types of submersibles.

CRITICAL THINKING

8. **Evaluating Ideas** Most submarines use sonar as a navigation aid. How would sonar enable an underwater vessel to move through the ocean depths?
9. **Analyzing Methods** Why are submarine robots more practical for deep-ocean research than submersibles designed to carry people are?

CONCEPT MAPPING

10. Use the following terms to create a concept map: *oceanography, submersible, bathysphere, bathyscaph, robot submersible,* and *sonar.*

CHAPTER RESOURCES

Chapter Resource File

- Section Quiz GENERAL

Workbooks

- **Study Guide** (also in Spanish)

Section 2 Features of the Ocean Floor

The ocean floor can be divided into two major areas, as shown in **Figure 1.** The **continental margins** are shallow parts of the ocean floor that are made of continental crust and a thick wedge of sediment. The other major area is the **deep-ocean basin,** which is made of oceanic crust and a thin sediment layer, is the deep part of the ocean beyond the continental margin.

OBJECTIVES

- **Describe** the main features of the continental margins.
- **Describe** the main features of the deep-ocean basin.

KEY TERMS

continental margin
deep-ocean basin
trench
abyssal plain

Continental Margins

The line that divides the continental crust from the oceanic crust is not abrupt or distinct. Shorelines are not the true boundaries between the oceanic crust and the continental crust. The boundaries are actually some distance offshore and beneath the ocean and the thick sediments of the continental margin.

Continental Shelf

Continents are outlined in most places by a zone of shallow water where the ocean covers the edge of the continent. The part of the continent that is covered by water is called a *continental shelf*. The shelf usually slopes gently from the shoreline and drops about 0.12 m every 100 m. The average depth of the water covering a continental shelf is about 60 m. Although it is underwater, a continental shelf is part of the continental margin, not the deep-ocean basin.

Changes in sea level affect the continental shelves. During glacial periods, continental ice sheets hold large amounts of water. So, sea level falls and exposes more of the continental shelf to weathering and erosion. But if ice sheets melt adding water to the oceans, sea level rises and covers the continental shelf.

continental margin the shallow sea floor that is located between the shoreline and the deep-ocean bottom

deep-ocean basin the part of the ocean floor that is under deep water beyond the continental margin and that is composed of oceanic crust and a thin layer of sediment

SCILINKS. NSTA Developed and maintained by the National Science Teachers Association

For a variety of links related to this subject, go to www.scilinks.org

Topic: Ocean-Floor Features
SciLinks code: HQ61067

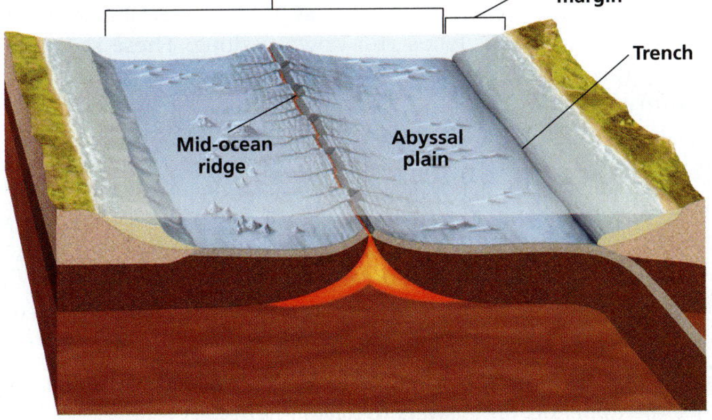

Figure 1 ▶ The ocean floor includes the continental margins and the deep-ocean basin. ***What are three other major features of the ocean floor?***

Section 2

Focus

Overview

This section reviews the features of the ocean floor. It describes the continental margin, including the continental shelf, slope, and rise. It explains features of the deep ocean basins, such as trenches, abyssal plains, mid-ocean ridges, and seamounts.

Bellringer

Have students make a quick sketch of what they think the ocean floor looks like from the Mid-Atlantic United States to Spain. (Sketches may vary but can be used to start a discussion of the ocean floor.) **LS Visual**

Motivate

Using the Figure — GENERAL

Modeling the Ocean Floor The diagram at left shows the features of the ocean floor. Have students work in groups to build a model ocean floor out of modeling clay in a plastic tub. Have students cover the top and walls of the tub with black paper so that the ocean floor cannot be seen. Have groups exchange models and try to map the model ocean floor by using a dowel or craft stick to measure depths at various locations. Answer to caption question: mid-ocean ridges, trenches, and abyssal plains **LS Kinesthetic/Visual**

CHAPTER RESOURCES

Chapter Resource File

- Directed Reading BASIC

Technology

Transparencies
- Bellringer

Student Edition on CD-ROM

One-Stop Planner CD-ROM
- Lesson Plan

Teach

Identifying Preconceptions — GENERAL

Where Do the Continents End? Perform this exercise before students begin to read this section. On a map of the eastern continental margin of North America, have students mark where they think the continental crust ends and the oceanic crust begins. Students may place the boundary right at the shoreline. While presenting this section, emphasize that the boundary is some distance offshore and is not necessarily abrupt or distinct. **LS Verbal/Visual**

Graphic Organizer — GENERAL

You may want to have students work in groups to create this Spider Map. Have one student draw the map and fill in information provided by other students from the group.

Answer to Reading Check

Trenches; broad, flat plains; mountain ranges; and submerged volcanoes are part of the deep-ocean basins.

CHAPTER RESOURCES

Chapter Resource File

- **Making Models Lab** Island to Guyot GENERAL

Technology

Transparencies
- 94 Features of the Ocean Floor (with worksheet)

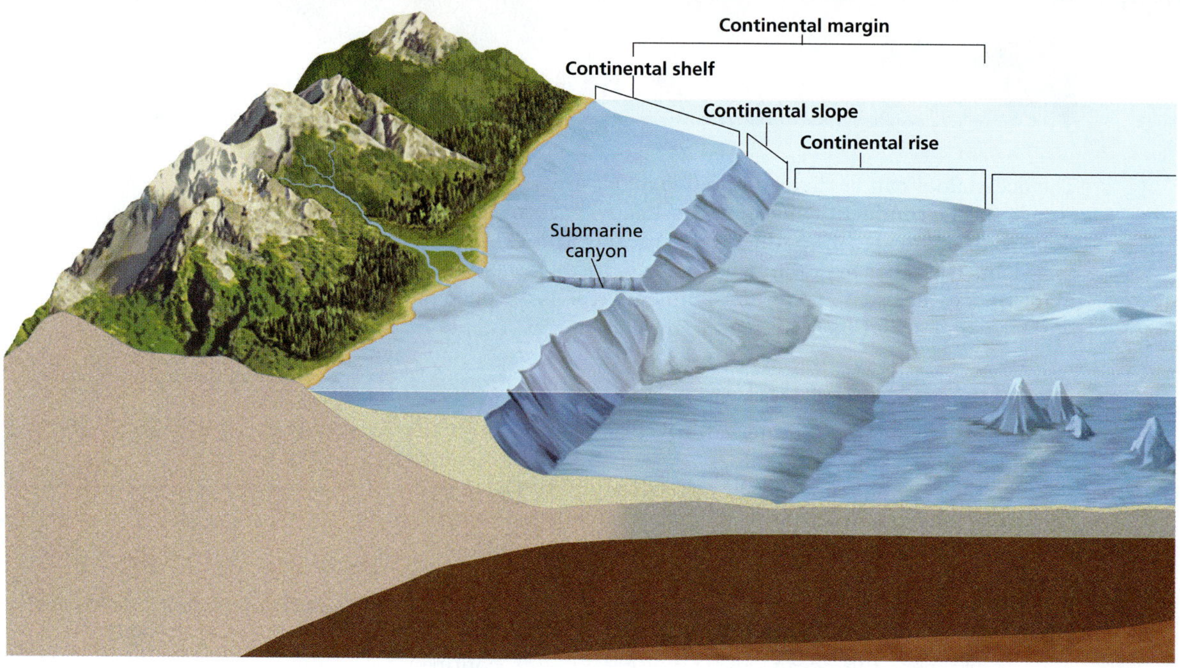

Figure 2 ▶ The ocean floor is made of distinct areas and features.

Graphic Organizer — Spider Map

Create the Graphic Organizer entitled "Spider Map" described in the Skills Handbook section of the Appendix. Label the circle "Ocean Basin Features." Create a leg for each ocean basin feature. Then, fill in the map with details about each ocean basin feature.

Continental Slope and Continental Rise

At the seaward edge of a continental shelf is a steep slope called a *continental slope*. The boundary between the continental crust and the oceanic crust is located at the base of the continental slope. Along the continental slope, the ocean depth increases by several thousand meters within a distance of a few kilometers, as shown in **Figure 2.** The continental shelf and continental slope may be cut by deep V-shaped valleys. These deep valleys are called *submarine canyons*. These deep canyons are often found near the mouths of major rivers. Other canyons may form over time as very dense currents called *turbidity currents* carry large amounts of sediment down the continental slopes. Turbidity currents form when earthquakes cause underwater landslides or when large sediment loads run down a slope. These sediments form a raised wedge at the base of the continental slope called a *continental rise*.

Deep-Ocean Basins

Deep-ocean basins also have distinct features, as shown in **Figure 2.** These features include broad, flat plains; submerged volcanoes; gigantic mountain ranges; and deep trenches. In the deep-ocean basins, the mountains are higher and the plains are flatter than any features found on the continents are.

✓ Reading Check **What features are located in the deep-ocean basins? (See the Appendix for answers to Reading Checks.)**

INCLUSION Strategies

- *Visually Impaired*
- *Learning Disabled*
- *Attention Deficit Disorder*

Work with students to build a replica of the different levels of the ocean floor that are pictured in the figures on the first three pages of this section. Shape modeling clay to show elevation changes, ridges, seamounts, trenches, and abyssal plains that they can touch with their hands. Let the clay dry overnight. **LS Kinesthetic**

BRAIN FOOD

Ridges and Rises At the slow-moving Mid-Atlantic Ridge, only approximately 3 cm of new sea floor forms each year. The topography is rough because the magma supply does not keep up with spreading, creating rift valleys and faults. In contrast, at the East Pacific Rise as much as 8 cm of new sea floor forms each year. Here the topography is smoother because magma supply keeps up with extension, so less faulting occurs.

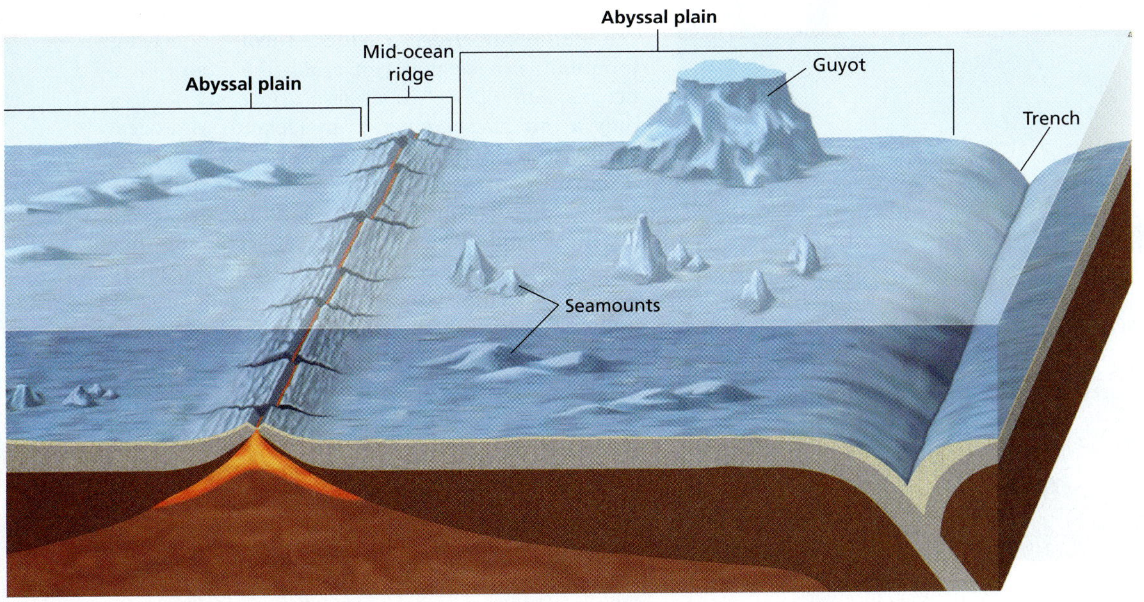

Trenches

Long, narrow depressions located in the deep-ocean basins are called **trenches.** At more than 11,000 m deep, the Mariana Trench, in the western Pacific Ocean, is the deepest place in Earth's crust. Trenches form where one tectonic plate subducts below another plate. Earthquakes occur near trenches. Volcanic mountain ranges and volcanic island arcs also form near trenches.

trench a long, narrow, and steep depression that forms on the ocean floor as a result of subduction of a tectonic plate, that runs parallel to the trend of a chain of volcanic islands or the coastline of a continent, and that may be as deep as 11 km below sea level; also called an *ocean trench* or a *deep-ocean trench*

abyssal plain a large, flat, almost level area of the deep-ocean basin

Abyssal Plains

The vast, flat areas of the deep-ocean basins where the ocean is more than 4 km deep are called **abyssal plains** (uh BIS uhl PLAYNZ). Abyssal plains cover about half of the deep-ocean basins and are the flattest regions on Earth. In some places, the ocean depth changes less than 3 m over more than 1,300 km.

Layers of fine sediment cover the abyssal plains. Ocean currents and wind carry some sediments from the continental margins. Other sediment is made by organisms that live in the ocean and settle to the ocean floor when they die.

The thickness of sediments on the abyssal plains is determined by three factors. The age of the oceanic crust is one factor. Older crust is generally covered with thicker sediments than younger crust is. The distance from the continental margin to the abyssal plain also determines how much sediment reaches the plain from the continent. Third, the sediment cover on abyssal plains that are bordered by trenches is generally thinner than the sediment cover on abyssal plains that are not bordered by trenches.

MISCONCEPTION ALERT

Lifeforms on Abyssal Plains Although the abyssal plains may be some of the flattest places on Earth, they are neither featureless nor lifeless. Using seismic profilers, oceanographers have determined that the floor of the abyssal plain is actually quite rugged, like the rest of the ocean. However, the abyssal plain has been covered by deep layers of sediments, such as "ocean snow," the continual fall of waste and debris from organisms at the ocean's surface. Many creatures, such as sea urchins, sea cucumbers, clams, and mussels, live in and feed on the sediments of the abyssal plains. In the 1980s, a team of scientists studying deep-ocean sediments off the eastern United States found 90,677 individual organisms from 798 species in an area of only 54 km². Sixty percent of these organisms were new to humans. Based on their results, the scientists estimated that a new species might be found in every square kilometer of deep-sea floor—100 million more species to be discovered!

Teaching Tip — GENERAL

Connect to Prior Knowledge Remind students that plate tectonics is a theory that explains how continents and ocean basins form and change over time. Trenches result when one plate slides beneath another. As a result, earthquakes and volcanic activity are frequent in these areas. The Peru-Chile trench, resulting from the Nazca plate's subduction beneath the South American plate, is an example of a trench that formed along a continental margin. LS Logical

Close

Reteaching — BASIC

Diagram Have students label an ocean floor diagram with the various features discussed in the section. LS Visual

Quiz — GENERAL

1. Is the shoreline generally the boundary between the continental crust and the oceanic crust? (No, the boundary is generally some distance offshore at the base of the continental slope.)
2. Where are submarine canyons commonly located? (near the mouths of major rivers)
3. What regions of the ocean basin are the flattest places on Earth? (the abyssal plains)
4. Where do seamounts form? (near hot spots, or regions of the oceanic crust that have increased volcanic activity)

Alternative Assessment — ADVANCED

Deep-Sea Tour Have students design a "travel brochure" for an ocean floor vacation tour. Their brochure should include descriptions of key sites they will visit (e.g., the continental margin, submarine canyons, the abyssal plains, mid-ocean ridges, trenches, hydrothermal vents and seamounts). Students may also wish to illustrate their brochures. LS Verbal/Visual

Close, *continued*

Answers to Section Review

1. The continental shelf is the part of the continent submerged underwater. The continental slope is the steep, seaward edge of the continental shelf. The continental rise is a raised wedge of sediment at the base of the slope.
2. The boundary between the continental crust and the oceanic crust is generally offshore, at the base of the continental slope.
3. Submarine canyons commonly form where large loads of sediment run down a slope as part of turbidity currents.
4. Trenches are very deep and form where one tectonic plate subducts under another. Abyssal plains are vast flat areas of the ocean more than 4 km deep that are covered by layers of fine sediment. Mid-ocean ridges form where plates are moving away from each other and have a narrow rift at the center. Seamounts are submerged volcanoes that may rise above the ocean surface to form islands.
5. Guyots and atolls form from islands. When an island sinks and becomes eroded by waves, a flat-topped guyot forms. An atoll is an intermediate stage before the eroding island is completely submerged.
6. The continental shelf, the continental slope, and the continental rise are parts of the continental margin.
7. The Atlantic Ocean probably has thicker sediments because its crust is, in general, older, so sediment has had more time to accumulate. It is not surrounded by trenches, which collect sediment that washes down the continental margins. Also, it is smaller than the Pacific Ocean, so more sediment would reach its abyssal plains.
8. The continental shelves would be subject to much more erosion by wind, water, and ice if sea levels were to fall significantly.
9. The ocean floor has distinct features such as the *continental margin*, which consists of the *continental shelf*, *continental slope*, and *continental rise;* and the *deep-ocean basin*, which may include *trenches*, *abyssal plains*, and *mid-ocean ridges*.

Figure 3 ▶ The white ridges in this photo are coral reefs of an atoll that formed in the shallow waters around a volcanic island. Erosion is changing the island into a guyot.

Mid-Ocean Ridges

The most prominent features of ocean basins are the *mid-ocean ridges,* which form underwater mountain ranges that run along the floors of all oceans. Mid-ocean ridges rise above sea level in only a few places, such as in Iceland. Mid-ocean ridges form where plates pull away from each other. A narrow depression, or rift, runs along the center of the ridge. Through this rift, magma reaches the sea floor and forms new lithosphere. This new lithosphere is less dense than the old lithosphere. As the new lithosphere cools, it becomes denser and begins to sink as it moves away from the rift. Fault-bounded blocks of crust that form parallel to the ridges as the lithosphere cools and contracts are called *abyssal hills*.

As ridges adjust to changes in the direction of plate motions, they break into segments that are bounded by faults. These faults create areas of rough topography called *fracture zones*, which run perpendicular across the ridge.

Seamounts

Submerged volcanic mountains that are taller than 1 km are called *seamounts*. Seamounts form in areas of increased volcanic activity called *hot spots*. Seamounts that rise above the ocean surface form oceanic islands. As tectonic plate movements carry islands away from a hot spot, the islands sink and are eroded by waves to form flat-topped, submerged seamounts called *guyots* (GEE OHZ) or *tablemounts*. An intermediate stage in this process, called an *atoll*, is shown in **Figure 3.**

Section 2 Review

1. **Describe** the three main sections of the continental margins.
2. **Describe** where the boundary between the oceanic crust and the continental crust is located.
3. **Explain** how turbidity currents are related to submarine canyons.
4. **List** four main features of the deep-ocean basins, and describe one characteristic of each feature.
5. **Compare** seamounts, guyots, and atolls.
6. **Explain** the difference between the meanings of the terms *continental margin, continental shelf, continental slope,* and *continental rise.*

CRITICAL THINKING

7. **Making Inferences** The Pacific Ocean is surrounded by trenches, but the Atlantic Ocean is not. In addition, the Pacific Ocean is wider than the Atlantic Ocean, and much of the crust under the Pacific Ocean is very young. Which ocean's abyssal plain has thicker sediments? Explain your answer.
8. **Determining Cause and Effect** If sea level were to fall significantly, what would happen to the continental shelves?

CONCEPT MAPPING

9. Use the following terms to create a concept map: *continental margin, deep-ocean basin, continental shelf, continental slope, continental rise, trench, abyssal plain,* and *mid-ocean ridge.*

CHAPTER RESOURCES

Chapter Resource File
- Section Quiz GENERAL

Workbooks
- Study Guide (also in Spanish)

Section 3 Ocean-Floor Sediments

Continental shelves and slopes are covered with sediments. Sediments are carried into the ocean by rivers, are washed away from the shoreline by wave erosion, or settle to the ocean bottom when the organisms that created them die. The composition of ocean sediments varies and depends on which part of the ocean floor the sediments form in. The sediments are fairly well sorted by size. Coarse gravel and sand are usually found close to shore because these heavier sediments do not move easily offshore. Lighter particles are suspended in ocean water and are usually deposited at a great distance from shore.

OBJECTIVES

- **Describe** the formation of ocean-floor sediments.
- **Explain** how ocean-floor sediments are classified by their physical composition.

KEY TERMS

core sample
nodule

Sources of Deep Ocean–Basin Sediments

Sediments found in the deep-ocean basin, which is beyond the continental margin, are generally finer than those found in shallow water. Samples of the sediments in the deep-ocean basins can be gathered by scooping up sediments or by taking core samples. **Core samples** are cylinders of sediment that are collected by drilling into sediment layers on the ocean floor. **Figure 1** shows a core sample being studied aboard the research vessel *JOIDES Resolution*.

core sample a cylindrical piece of sediment, rock, soil, snow, or ice that is collected by drilling

The study of sediment samples shows that most of the sediments in the deep-ocean basins are made of materials that settle slowly from the ocean water above. These materials may come from organic or inorganic sources.

Figure 1 ▶ A scientist studies a core sample that was brought up from the drill aboard the research ship *JOIDES Resolution*.

Section 3

Focus

Overview

This section explains the formation and composition of ocean-floor sediments and how they are studied and classified. The section explains that ocean sediments vary depending on their source and where they are deposited.

Bellringer

Ask students the following question: Where does sediment at the bottom of the ocean come from and what is it like? (Answers may vary but may include rock and mud from rivers and glaciers, and the remains of dead sea creatures. The size of the pieces vary, and they may be either organic or inorganic in origin.) LS Verbal

Motivate

Demonstration — GENERAL

Sediments Sinking Show students sediments of various sizes and different shapes. Ask students to predict how the sediments will settle out if suspended in water. Put some sediment of each size and shape in a jar with water and shake the jar. Compare students' predictions with the results. LS Visual/Kinesthetic

CHAPTER RESOURCES

Chapter Resource File

- Directed Reading BASIC

Technology

Transparencies
- Bellringer

Student Edition on CD-ROM

One-Stop Planner CD-ROM
- Lesson Plan

Teach

INCLUSION Strategies

- *Behavior Control Issues*
- *Attention Deficit Disorder*
- *Developmentally Delayed*

Have students join with a partner. Ask the partners to take turns reading alternate paragraphs quietly aloud. Then, have the students discuss what the paragraph means and draw a simple sketch that goes with the paragraph. When finished, ask partners to share their favorite sketches with the class. **LS Verbal**

Demonstration — GENERAL

Turbidity Currents and Submarine Canyons Use this demonstration to show students what turbidity currents look like and how they may form submarine canyons. Pour a slurry of mixed sediments (or a mixture of sand, cleaning powder, and water colored with food coloring) through a funnel onto an underwater slope you have constructed in a plastic tub or tank. Then, allow sediments to settle and view the graded bed formed at the base of the slope, which is a model of the continental rise. **LS Visual/ Kinesthetic**

Figure 2 ▶ This picture of sediment emptying out of the Mahakam River in Indonesia was taken by astronauts aboard the space shuttle *Columbia.*

Inorganic Sediments

Some ocean-basin sediments are rock particles that were carried from land by rivers. When a river empties into the ocean, the river deposits its sediment load, as shown in **Figure 2.** Most of these sediments are deposited along the shore and on the continental shelf. However, large quantities of these sediments occasionally slide down continental slopes to the ocean floor below. The force of the slide creates powerful turbidity currents that spread the sediments over the deep-ocean basins. Other deep ocean–basin sediments consist of fine particles of rock, including volcanic dust, that have been blown great distances out to sea by the wind. These particles land on the surface of the water, sink, and gradually settle to the bottom of the ocean.

Icebergs also provide sediments that can end up on the ocean basins. As a glacier moves across the land, the glacier picks up rock. The rock becomes embedded in the ice and moves with the glacier. When an iceberg breaks from the glacier, drifts out to sea, and melts, the rock material sinks to the ocean floor.

Even meteorites contribute to deep ocean-basin sediments. Much of a meteorite vaporizes as it enters Earth's atmosphere. The remaining cosmic dust falls to Earth's surface. Because most of Earth's surface is ocean, most meteorite fragments fall into the ocean and become part of the sediments on the ocean floor.

Connection to PHYSICS

Turbidity Currents

Underwater landslides can be caused by earthquakes or can happen when the sediment-water mixture becomes denser than the surrounding water. These landslides form currents called *turbidity currents.*

Gravity powers turbidity currents. As the dense sediment mixture moves downhill, it picks up sediment. This added sediment increases the current's density, which increases the current's speed. Turbidity currents can travel at speeds of more than 100 km/h.

The speed and composition of these currents make them powerful agents of erosion. One turbidity current may move billions of kilograms of mud, rock, and sand down a slope. Sediments are deposited as the speed of the current decreases. A turbidity current may gain enough momentum that it does not stop at the continental rise. Thus, it may spread sediment hundreds of kilometers onto the abyssal plain.

This simulated turbidity current was created in a laboratory. The density of the sediment-filled water causes it to move downhill like a landslide does.

MISCONCEPTION ALERT

Sediment vs. Sand Students may think most ocean sediments are like the sand found on ocean beaches. Sand is sometimes found on the deep sea floor because of turbidity currents that are caused by earthquakes or underwater landslides. However, most sediments on the deep ocean floor are very fine inorganic and organic particles that settle to the ocean floor from the water above.

READING SKILL BUILDER — BASIC

Reading Organizer As students read this section, encourage them to take Power Notes, KWL notes, or Two-Column Notes as described in the Skills Handbook section of the Appendix. Later, students can use these notes as a study guide for assessments. English Language Learners

Figure 3 ▶ Nodules, such as these mined from the East Pacific Rise, are rich in a variety of minerals.

Biogenic Sediments

In many places on the ocean floor, almost all of the sediments are *biogenic,* meaning that the sediments were originally produced by living organisms. Biogenic sediments are the remains of marine plants and animals. The two most common compounds that make up organic sediments are silica, SiO_2, and calcium carbonate, $CaCO_3$. Silica comes primarily from microscopic organisms called *diatoms* and *radiolarians*. Calcium carbonate comes mostly from the skeletons of tiny organisms called *foraminiferans*.

Chemical Deposits

When substances that are dissolved in ocean water crystallize, these materials can form mineral deposits on the ocean floor. Some of these mineral deposis are potato-shaped lumps called **nodules.** Nodules, such as the ones shown in **Figure 3,** are commonly located on the abyssal plains. Nodules are composed mainly of the oxides of manganese, nickel, copper, and iron. Other minerals, such as phosphorite, are also carried in the ocean water before they crystallize and form mineral deposits on the ocean floor.

Reading Check **How do nodules form? (See the Appendix for answers to Reading Checks.)**

MATHPRACTICE

Ocean-floor Sediments Ocean-floor sediments are composed of an average of 54% biogenic sediments, 45% Earth rocks and dust, less than 1% precipitation of dissolved materials (nodules and phosphorite), and less than 1% of rocks and dust from space. If you collected 10,000 kg of ocean-floor sediment, how many kilograms of each type of ocean-floor sediment would you expect to find?

QuickLAB 20 min

Diatoms

Procedure

1. Observe **diatoms** under a **microscope.**
2. Sketch what you see. Make sure to note the magnification.

Analysis

1. What characteristics of the diatoms did you observe?
2. Propose one possible function for each of the structures you observed.

nodule a lump of minerals that is made of oxides of manganese, iron, copper, or nickel and that is found in scattered groups on the ocean floor

MATHPRACTICE

Answer

Amount per source = total amount of sediment × percentage of given source

Biogenic = (10,000 kg × 0.54) = 5,400 kg

Earth rocks and dust = (10,000 kg × 0.45) = 4,500 kg

Dissolved materials = (10,000 kg × 0.01) = <100 kg

Space rocks and dust = (10,000 kg × 0.01) = <100 kg

Answer to Reading Check

When chemical reactions take place in the ocean, dissolved substances can crystallize to form nodules that settle to the ocean floor.

Close

Reteaching — BASIC

Peer Reviewing Have students work in pairs to write three to five questions about the material in this section. Then, have students use their questions to quiz each other. **LS Interpersonal**

Co-op Learning

Quiz — GENERAL

1. What are core samples? (cylinders of sediment collected by drilling into the ocean floor)
2. Are sediments in the deep-ocean basins generally coarse or fine particles? (fine)
3. Where are nodules most commonly located? (on the abyssal plains)

CHAPTER RESOURCES

Chapter Resource File

- Datasheet for Quick Lab GENERAL
- Inquiry Lab Sediment Loading GENERAL

QuickLAB

Skills Acquired

- Observing
- Identifying/Recognizing Patterns
- Analyzing

Teacher's Notes Tell students that diatoms are photosynthetic algae. Diatomaceous earth is the crumbly remains of diatom shells. Diatomaceous earth is used for applications such as paints, polishes, and filters. Show students a sample of diatomaceous earth.

Answers

1. Answers may vary but should be based on students' sketches. Students may note box-like structures (some diatoms look like a box and its tightly fitted lid), two-part shells, a golden brown coloration, or that diatoms may be grouped singly or in chains or colonies.
2. Answers may vary. Accept all reasonable answers.

Close, *continued*

Alternative Assessment — BASIC

Comparing and Contrasting Sediments Have students create a table that compares and contrasts the three types of sediment found on the sea floor: inorganic, biogenic, and chemical. **LS Visual**

Answers to Section Review

1. Inorganic sediments are rock particles that are carried to the sea by streams, rivers, glaciers, and wind. Biogenic sediments are the remains of marine plants and animals.
2. Icebergs form from glaciers that scour rocks, embedding particles in the ice. Icebergs break off glaciers and float out to sea. When the iceberg melts, any rock materials in the iceberg will sink to the ocean floor.
3. Substances that are dissolved in ocean water can undergo chemical reactions that cause them to crystallize into nodules, which form on or settle to the ocean floor.
4. Sediments are classified as muds or oozes. Muds are fine particles of rock. Oozes are muds mixed with at least 30% biogenic material.
5. A mud is sediment composed of fine silt or clay-sized particles.
6. Calcareous ooze is mostly calcium carbonate, whereas siliceous ooze is mostly silicon dioxide.
7. Answers may vary. The presence of volcanic dust indicates that a volcano erupted somewhere on Earth because volcanic dust can be blown great distances.
8. Answers may vary. Possible factors include the expected price of the ore extracted from the nodules; the distribution and density of nodules in any given area; the costs of collecting nodules from the ocean floor; and the cost of extracting ore from the nodules once they are collected.
9. *Ocean floor sediments* consist of *inorganic sediment* made from rocks and volcanic dust; *biogenic sediment,* which may include the remains of *diatoms*; and *chemical deposits,* such as *nodules.*

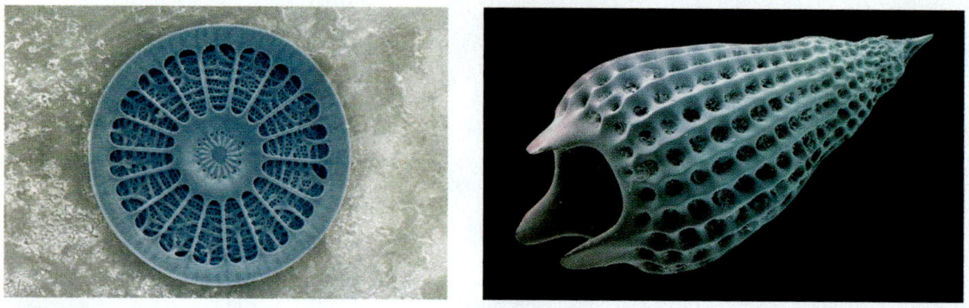

Figure 4 ▶ The remains of diatoms (left) and radiolarians (right), both magnified hundreds of times in these photos, are important components of biogenic sediments on the ocean floor.

SCILINKS. NSTA Developed and maintained by the National Science Teachers Association

For a variety of links related to this topic, go to www.scilinks.org

Topic: Ocean-Floor Sediments
SciLinks code: HQ61068

Physical Classification of Sediments

Deep ocean-floor sediments can be classified into two basic types. *Muds* are very fine silt- and clay-sized particles of rock. One common type of mud on the abyssal plains is red clay. Red clay is made of at least 40% clay particles and is mixed with silt, sand, and biogenic material. This clay can vary in color from red to gray, blue, green, or yellow-brown. About 40% of the ocean floor is covered with soft, fine sediment called *ooze*. At least 30% of the ooze is biogenic materials, such as the remains of microscopic sea organisms. The remaining material is fine mud.

Ooze can be classified into two types. *Calcareous ooze* is ooze that is made mostly of calcium carbonate. Calcareous ooze is never found below a depth of 5 km, because at depths between 3 km and 5 km, calcium carbonate dissolves in the deep, cold ocean water. *Siliceous ooze*, which can be found at any depth, is made of mostly silicon dioxide, which comes from the shells of radiolarians and diatoms. Examples of remains of these organisms are shown in **Figure 4.** Most siliceous ooze is found in the cool, nutrient-rich ocean waters around Antarctica because of the abundance of diatoms and radiolarians in that location.

Section 3 Review

1. **Describe** the formation of two different types of ocean-floor sediments.
2. **Summarize** how icebergs contribute to deep ocean-basin sediments.
3. **Explain** how substances that are dissolved in ocean water travel to the ocean floor.
4. **Explain** how ocean-floor sediments are classified by physical composition.
5. **Describe** how scientists define the word *mud.*
6. **Compare** the compositions of calcareous ooze and siliceous ooze.

CRITICAL THINKING

7. **Making Inferences** What could you infer from a core sample of a layer of sediment that contains volcanic ash and dust?
8. **Applying Ideas** Some businesses have tried to develop methods of extracting nodules from the ocean. Name two factors that businesses should consider when they are determining whether extracting nodules is profitable.

CONCEPT MAPPING

9. Use the following terms to create a concept map: *nodule, inorganic sediment, biogenic sediment, diatom, chemical deposit,* and *ocean-floor sediment.*

CHAPTER RESOURCES

Chapter Resource File

- Section Quiz GENERAL

Workbooks

- Study Guide (also in Spanish)

Chapter 19 Highlights

Sections

1 The Water Planet

Key Terms

global ocean, 471
sea, 471
oceanography, 472
sonar, 473

Key Concepts

- The global ocean can be divided into five major oceans—the Pacific, Atlantic, Indian, Arctic, and Southern Oceans—and many smaller seas.
- Oceanography is the study of the oceans and the seas. Oceanographers study the ocean using research ships, sonar, and submersibles.
- Sonar is a system that uses acoustic signals and echo returns to determine the location of objects or to communicate.

2 Features of the Ocean Floor

continental margin, 475
deep-ocean basin, 475
trench, 477
abyssal plain, 477

- The areas around continents that are shallow parts of the ocean floor are called *continental margins*.
- Continental margins include the continental shelf, the continental slope, and the continental rise.
- Features of deep-ocean basins include trenches, abyssal plains, mid-ocean ridges, and seamounts.

3 Ocean-Floor Sediments

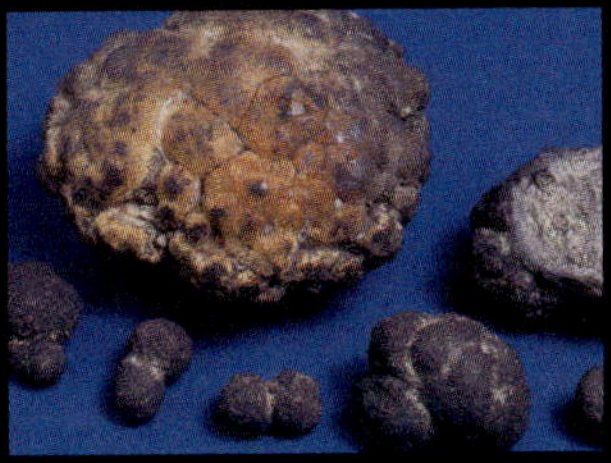

core sample, 479
nodule, 481

- Core samples are taken by drilling into sediment layers. Scientists study core samples to learn about the composition and characteristics of ocean-floor sediments.
- Ocean-floor sediments form from inorganic and biogenic materials as well as from chemical deposits.
- Based on physical characteristics, deep ocean–floor sediments are classified as mud or as ooze.

Chapter Highlights

Alternative Assessment — ADVANCED

Debate Many oceanographers argue that we know more about the surface of the moon or even Mars than we do about the oceans on Earth. Divide students into groups and have them debate the validity of this statement. Give students time to prepare their arguments using class material as well as supplementary research. Have the opposing groups present their cases to each other and allow for rebuttals. LS **Verbal/Interpersonal**

CHAPTER RESOURCES

Chapter Resource File

- Concept Review GENERAL
- Critical Thinking ADVANCED
- Math Skills GENERAL
- Graphing Skills GENERAL
- Chapter Test A GENERAL
- Chapter Test B ADVANCED

Workbooks

- Study Guide (also in Spanish)
- Assessments (Spanish)

Technology

Classroom Videos
- Brain Food Video Quiz

HRW Earth Science Video
- Segment 14: Features of the Ocean Floor

Chapter Review

Assignment Guide	
Section	**Questions**
1	1–2, 4, 7, 9–10, 18, 23, 28, 31
2	6, 11–14, 17, 22, 24, 30
3	3, 8, 15–16, 22, 25–26
1 and 2	29
2 and 3	20, 27
1–3	5, 19

Using Key Terms

1–8. Answers may vary but should show that students understand the definitions of and differences between key terms.

Understanding Key Concepts

9. c **10.** d
11. a **12.** d
13. a **14.** a
15. b **16.** a

Short Answer

17. A guyot forms when an island sinks, becoming eroded by waves. A seamount is an underwater mountain that never reaches the ocean surface.

18. Sonar uses sound signals and returned echoes to determine the depth of the ocean floor. Depth information from various locations can be compiled to make maps and profiles of the ocean floor.

19. sonar, submersibles, satellites, and core samples

20. Turbidity currents can carry sediments hundreds of kilometers out to the deep ocean basins. Volcanic dust can be carried by winds far out to sea and deposited in the deep ocean basins. Fine sediments are also derived from biogenic materials.

21. Abyssal plains surrounded by deep ocean trenches generally have thinner sediment cover.

22. The three main types of sediments are inorganic sediments, biogenic sediments, and chemical deposits. Inorganic sediments are derived from rocks. They can be deposited by rivers that flow into the ocean, by volcanoes that spew dust and ash into the air and out to sea, by icebergs that carry rock particles out to sea, and by meteorites. Biogenic sediments are the deposited remains and waste products of ocean plants and animals. Chemical deposits form when reactions take place in the ocean, causing some substances to crystallize out of the water at the ocean floor.

Critical Thinking

23. Both the ocean and space are very harsh environments where humans cannot survive without special equipment because of the extreme temperatures and pressures and the lack of oxygen and light. Thus, exploration requires special methods and/or vehicles. Exploration is often done by robots which are remotely controlled by humans.

Chapter 19 Review

Using Key Terms

Use each of the following terms in a separate sentence.

1. *oceanography*
2. *sonar*
3. *core sample*

For each pair of terms, explain how the meanings of the terms differ.

4. *ocean* and *sea*
5. *submersible* and *nodule*
6. *continental margin* and *deep-ocean basin*
7. *global ocean* and *sea*
8. *mud* and *ooze*

Understanding Key Concepts

9. A self-propelled, free-moving submarine that is equipped for ocean research is a
a. turbidity. **c.** bathyscaph.
b. bathysphere. **d.** guyot.

10. A system that is used for determining the depth of the ocean floor is
a. a guyot. **c.** a bathysphere.
b. radiolarians. **d.** sonar.

11. The parts of the ocean floor that are made up of continental crust are called
a. continental margins.
b. abyssal plains.
c. mid-ocean ridges.
d. trenches.

12. The accumulation of sediments at the base of the continental slope is called the
a. trench.
b. turbidity current.
c. continental margin.
d. continental rise.

13. The deepest parts of the ocean are called
a. trenches.
b. submarine canyons.
c. abyssal plains.
d. continental rises.

14. Large quantities of the inorganic sediment that makes up the continental rise come from
a. turbidity currents.
b. earthquakes.
c. diatoms.
d. nodules.

15. Potato-shaped lumps of minerals on the ocean floor are called
a. guyots. **c.** foraminiferans.
b. nodules. **d.** diatoms.

16. Very fine particles of silt and clay that have settled to the ocean floor are called
a. muds. **c.** guyots.
b. seamounts. **d.** nodules.

Short Answer

17. What is the differences between a seamount and a guyot?

18. Explain how sonar is used to study the oceans.

19. List four ways that scientists can learn about the deep ocean.

20. How do fine sediments reach the deep-ocean bottom?

21. What effects do deep-ocean trenches have on the sediment thickness of the abyssal plain?

22. List the three main types of ocean-floor sediments, and describe how they are deposited.

Critical Thinking

23. Making Comparisons The exploration of the ocean depths has been compared with the exploration of space. What similarities exist between these two environments and the attempts by people to explore them?

24. Making Predictions What may be the eventual fate of seamounts as they are carried along the spreading oceanic crust?

25. Analyzing Ideas A type of fish is known to exist only in one river in the central United States. Explain how the fossilized remains of this fish might become part of the sediments on the ocean floor.

26. Analyzing Relationships Explain how it is possible that scientists have found some red clays on the ocean floor that contain material from outer space.

Concept Mapping

27. Use the following terms to create a concept map: *deep-ocean basin, continental shelf, mud, ooze, calcareous ooze, siliceous ooze,* and *sediment*.

Math Skills

28. Making Calculations The total area of Earth is approximately 511,000,000 km^2. About 71% of Earth's surface is covered with water. Calculate the area of Earth, in square kilometers, that is covered with water.

Writing Skills

29. Writing from Research Prepare a brief report on the different types of submersibles. Your report should explain the special features of each type of submersible as well as how each type has contributed to oceanographers' knowledge of the oceans.

30. Creative Writing Create an imaginary walking tour of the ocean basins. Your tour should begin at the edge of a continent—perhaps at a beach on the east coast of Florida. Explain exactly what tourists should look for along the continental margin and the ocean floor on their way to the western coast of Africa.

Interpreting Graphics

The graph below compares elevations of land and depths of oceans on Earth's surface. Use the graph to answer the questions that follow.

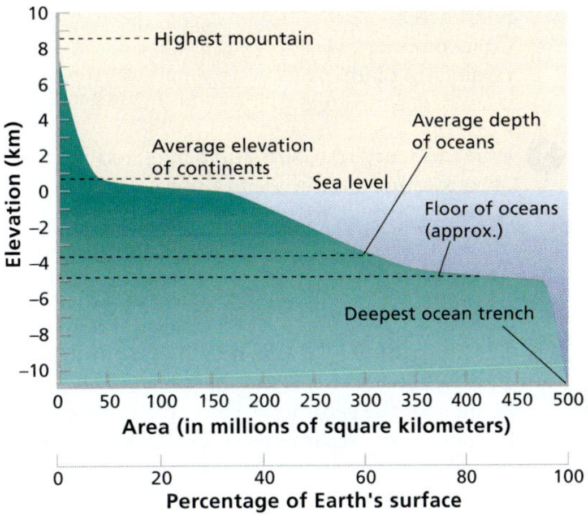

31. What percentage of Earth's surface is covered by land?

32. Which is greater: the elevation of the highest mountain above sea level or the depth of the deepest ocean trench below sea level?

33. Relative to sea level, how many times greater is the average depth of the ocean than the average elevation of land?

34. According to this diagram, how many millions of square kilometers of crust is under ocean water?

24. Seamounts tall enough to emerge as islands and that are carried along by the spreading of Earth's crust may become atolls and eventually guyots as the crust beneath them subsides. They may eventually become part of a continent at subduction zones where one tectonic plate collides with another.

25. As the fossil-containing rocks erode and are washed into rivers, the sediments eventually may be carried into the ocean where they may be deposited on the ocean floor.

26. Because most of Earth's surface is water, meteorite fragments often fall into the oceans and become part of the sediments of the ocean floor.

Concept Mapping

27. Answers may vary but should include all of the terms listed. Sample answers appear at the end of this Teacher Edition.

Math Skills

28. area of Earth covered in water = 511,000,000 $km^2 \times 0.71$ = 362,810,000 km^2

Writing Skills

29. Answers may vary. Accept all reasonable answers.

30. Answers may vary. Accept all reasonable answers.

Interpreting Graphics

31. About 34 percent of Earth's surface is covered by land.

32. The depth of the deepest trench below sea level is about 11 km.

33. The average depth of the ocean is about 5 times the average elevation of land. (avg. elevation of land = 0.75 km, avg. depth of ocean = 3.75 km; 3.75 km ÷ 0.75 km = 5)

34. About 340 million km^2 of Earth's surface is underwater.

Estimated Time

To give students practice under more realistic testing conditions, allow them 30 minutes to answer all of the questions in this practice test.

TEST DOCTOR

Question 2 Answer H is correct. Though the features of deep-ocean topography have a greater magnitude than those found on the continents—with the mountains being higher and the plains being flatter in the deep ocean—students can use what they know about continental landforms to arrive at the correct answer. In this case, *plains* are large, flat areas.

Question 6 The correct answer is 178,850,000 km^2. Students should multiply 511,000,000 km^2 by 0.70 to determine the area of Earth that is covered by water. Dividing the result by 2 yields the correct answer.

Question 9 Answer G is the best choice because the passage clearly states that coral needs sunlight to grow. Answer H is incorrect because, while coral does thrive in warm waters, this connection is not clearly stated in the passage. Answers F and I are not mentioned in the passage.

Chapter 19 Standardized Test Prep

Understanding Concepts

Directions (1–5): **For** ***each*** **question, write on a separate sheet of paper the letter of the correct answer.**

1. The global ocean is divided into which of the following oceans, in order of decreasing size?
   A. Atlantic, Pacific, Arctic, Indian
   B. Arctic, Indian, Atlantic, Pacific
   C. Pacific, Arctic, Indian, Atlantic
   D. Pacific, Atlantic, Indian, Arctic

2. What in the name for a vast, flat area of a deep-ocean basin?
   F. trench
   G. seamount
   H. abyssal plain
   I. mid-ocean ridge

3. What are very fine, silt- and clay-sized particles of rock found on the ocean floor called?
   A. muds
   B. calcareous ooze
   C. siliceous ooze
   D. sand

4. The study of deep-ocean sediment samples shows that
   F. most of the sediments came from the crust.
   G. most of the sediments settled from above.
   H. sediments cannot be organic.
   I. sediments cannot be inorganic.

5. Which of the following affects the ocean's salinity?
   A. number of fish
   B. wave size
   C. evaporation
   D. wave speed

Directions (6–7): **For** ***each*** **question, write a short response.**

6. The surface area of Earth is about 511,000,000 km^2. About 70% of the Earth's surface is covered by water and the Pacific Ocean makes up 50% of this amount. Calculate the surface area of Earth that is covered by the Pacific Ocean.

7. What is the name of the process used to remove salt from seawater?

Reading Skills

Directions (8–10): **Read the passage below. Then, answer the questions.**

Life on a Continental Shelf

While fish, mammals, and other forms of life can be found throughout these ocean waters, most life in the ocean is concentrated near the continental shores. The shallow waters of the continental shelf, which make up less than 10% of the ocean's total surface area, are home to an amazing array of plants, animals, and microscopic organisms.

Organisms such as coral and seaweed can grow on the ocean floor and still receive much needed sunlight that cannot penetrate deeper waters. The sunlight also makes the shallow waters much warmer than deeper abyssal waters. Algae flourishes in these warm, nutrient-rich waters and serves as food for many small ocean organisms. These organisms are in turn eaten by larger organisms. Even humans have become part of the food chain on the shelf. The vast majority of fish caught for human consumption are caught in waters above a continental shelf.

8. Which of the following statements about why humans catch so many fish in the waters over a continental shelf can be inferred from the information in the passage?
   A. There are no fish in deeper waters.
   B. Fish from deeper waters are inedible.
   C. Humans do not have the technological ability to catch fish in deeper ocean waters.
   D. There are larger and more varied fish populations over a continental shelf.

9. Coral reefs stop actively growing at depths of aboout 70 m. According to the passage, why might this be true?
   F. Coral feed on algae in shallow waters.
   G. Coral need sunlight to live, and sunlight can penetrate water only to a certain depth.
   H. Coral need warmth, and the deeper ocean waters are too cold for them to survive.
   I. Coral at greater depths are eaten by fish.

10. Why might the waters of a continental shelf have more nutrients than abyssal waters?

Answers

Understanding Concepts

1. D
2. H
3. A
4. G
5. C
6. 178,850,000 km^2
7. desalination

Reading Skills

8. D
9. G
10. Continental shelf waters have many benefits over deeper ocean waters. The sun makes waters of the shelf rich in plant life. The proximity to land constantly cycles new nutrients into the system by land and river runoff. Upwelling also brings nutrients up from the deeper waters.

Interpreting Graphics

11. A
12. Answers may vary. See Test Doctor for a detailed scoring rubric.
13. Chemical sediment precipitated from the sea water; rocks and dust formed on land and were carried to the ocean by erosion.

Interpreting Graphics

Directions (11–13): **For *each* question below, record the correct answer on a separate sheet of paper.**

Base your answer to question 11 on this image which shows how sonar equipment works.

Studying the Ocean Floor with Sonar

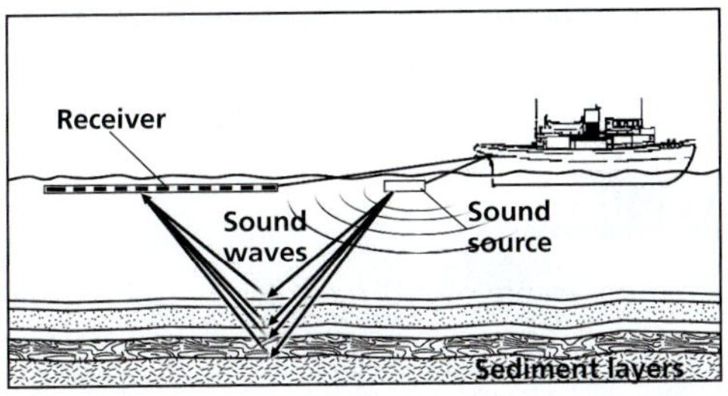

11 Which of the following best summarizes how sound waves are used?
A. A sound source dragged behind the boat emits waves that penetrate the different layers of the sea floor and bounce back to the receiver.
B. A sound source in front of the boat emits waves that penetrate the different layers of the sea floor and then bounce back to the receiver.
C. A receiver dragged behind the boat emits waves that penetrate the different layers of the sea floor and then bounce back to the receiver.
D. A receiver in front of the boat emits waves that penetrate the different layers of the sea floor and then bounce back to the receiver.

Base your answers to questions 12 and 13 on the pie graph below, which shows the composition of ocean-floor sediments.

Composition of Ocean-Floor Sediments

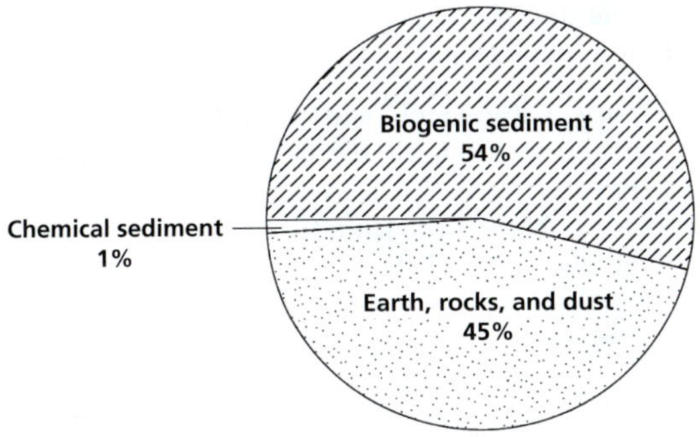

12 Why is there such a large difference between the percentage of biogenic sediment and the percentage of chemical sediment?

13 How did the inorganic materials in the two kinds of inorganic sediment shown on the pie graph above form and become part of the ocean floor?

Test TIP

Before choosing an answer to a question, try to answer the question without looking at the answer choices on the test.

TEST DOCTOR

Question 11 Answer A is the best choice. Students need to know how sonar is used to correctly answer this question. A sound source dragged behind the boat emits sound waves. These waves travel to the ocean floor. When they strike the ocean floor, they bounce off of it. Some of the waves are lost as they bounce in various directions, though others return to the receiver. A computer can then calculate the time the sound wave took to make its complete journey and plot the depth of the location. When numerous readings are taken, scientists are able to create three dimensional models of the ocean's floor.

Question 12 Full-credit answers should include the following points:
- biologic productivity in the oceans far outstrips the inorganic chemical reactions taking place in the oceans
- biologic organisms are also able to reproduce and thus provide a continuous source of sediment as the organisms die
- biogenic sediment is found mostly in the form of animal remains

Test Prep Correlations *National Science Education Standards*

ES 1d: item 9
ES 3c: items 1, 2, 4, 5, 12, 13
LS 4d: items 8, 9, 10
LS 4e: item 8
LS 5b: item 9
LS 5e: items 8, 9, 10
SPSP 3a: items 7, 8
SPSP 4a: item 8
ST 2b: item 11
UCP 1: items 2, 4, 5, 10, 12, 13
UCP 2: item 11

CHAPTER RESOURCES

State Resources

For specific resources for your state, visit go.hrw.com and type in the keyword HSHSTR.

Ocean-Floor Sediments

Teacher's Notes

Time Required

one or two 45-minute class periods

Lab Ratings

EASY → HARD

TEACHER PREPARATION (2 of 4)
STUDENT SETUP (3 of 4)
CONCEPT LEVEL (2 of 4)
CLEANUP (2 of 4)

Skills Acquired

- Observing
- Measuring
- Experimenting
- Predicting
- Inferring
- Collecting Data
- Organizing and Analyzing Data

The Scientific Method

In this lab, students will

- Make Observations
- Ask Questions
- Analyze the Results
- Draw Conclusions
- Communicate Results

Materials

The materials listed on this page are enough for groups of two to four students. If soil products with labeled sizes are not available in your area, you can use sieves to separate soil into different sizes. Generally, coarse soil has a diameter of 2.0 to 1.0 mm; medium, 1.0 to 0.5 mm; medium-fine, 0.5 to 0.25 mm; and fine, 0.25 to 0.10 mm, but any divisions of size will work. You may also want to have students measure the diameters of particles in the given samples.

Objectives

- USING SCIENTIFIC METHODS **Observe and record** the settling rates of four different sediments.
- **Draw conclusions** about how particle size affects settling rate.
- **Identify** factors that affect settling rate of sediments besides particle size.

Materials

balance, metric
column, clear plastic, 80 cm × 4 cm
cup, paper
pencil, grease
ring stand with clamp
ruler, metric
sieve, 4 mm, 2 mm, 0.5 mm
soil, coarse, medium, medium-fine, and fine grain
stopper, rubber
stopwatch
tape, adhesive
teaspoon
towels, paper
water

Safety

Ocean-Floor Sediments

Most of the ocean floor is covered with a layer of sediment that varies in thickness from 0.3 km to more than 1 km. Much of this sediment is thought to have originated on land through the process of weathering. Through erosion, the sediment has made its way to the deep-ocean basins. In this lab, you will use sediment samples of four particle sizes to determine the relationship between the size of particles and the settling rate of the particles in water.

PROCEDURE

1. Take one sample of sediment from each of the following size ranges: coarse, medium, medium-fine, and fine.
2. Plug one end of the plastic column with a rubber stopper, and secure the stopper to the column with tape. Place the column in a vertical position using the ring stand and clamp. Carefully fill the column with water to a level about 5 cm from the top, and allow the water to stand until all large air bubbles have escaped.
3. Use the grease pencil to mark the water level on the column. This will be the starting line.
4. Next, draw a line about 5 cm from the bottom of the column. This will be the finish line.
5. Have a member of your lab group put 1 tsp of the coarse sample into the water column. The other group member should record three time measurements as follows:

   a. Using a stopwatch, start timing when the first particles hit the start line on the column, and stop timing when they reach the finish line. Perform this procedure three times. Record the time for each trial in a table similar to the one shown below.

Soil samples	*Trial 1*	*Trial 2*	*Trial 3*	*Average*
Coarse	First time measurment:			
	Second time measurement:			
Medium	First time measurement:			
	Secont time measurement:			

DO NOT WRITE IN THIS BOOK

CHAPTER RESOURCES

Chapter Resource File

- Datasheet for Chapter Lab GENERAL
- Lab Notes and Answers

b. Next, use the stopwatch to determine how long it takes the last particle in the sample to travel from the start line to the finish line. Perform this procedure three times. Record the time for each trial in your table.

6. Determine the average time of the three trials for the first measurement. Do the same for the second measurement. Record the averages.

7. Pour the soil and water from the column into the container provided by your teacher. (Note: Do not pour the soil into the sink.)

8. Refill the plastic column with water up to the original level marked with the grease pencil.

9. Repeat steps 5, 6, and 7 for the remaining sediment sizes. Record the measurements and the averages in your table.

10. With the plastic column filled with water, pour 20 g of unsieved soil into the column, and allow the soil to settle for 5 minutes. After 5 minutes, look at the column and record your observations of both the settled sediment and the water. Repeat step 7, and then answer the questions below.

Step 5

ANALYSIS AND CONCLUSION

1. **Organizing Data** Which particles settled fastest? Which particles settled slowest?

2. **Making Comparisons** Compare the settling time of the medium particles with the settling time of the medium-fine particles. Do similar-sized particles fall at the same rate?

3. **Making Inferences** In step 10, why did the water remain slightly cloudy even after most of the particles had settled?

4. **Evaluating Methods** How do the results in step 10 help to explain why the deep-ocean basins are covered with a very fine layer of sediment while areas near the shore are covered with coarse sediment?

5. **Making Predictions** Other than size, what factors do you think would influence the speed at which particles fall in water? Explain your answer.

Extension

1. **Analyzing Predictions** Obtain particles of different shapes, such as long, cylindrical grains; flat, disk-shaped grains; round grains; and angular grains. Test the settling times of these grains, and write a brief paragraph that explains how grain shape affects the settling rate of particles in water.

Tips and Tricks

Have students observe the sediments with a magnifying glass or dissecting microscope. You might have students weigh out a given mass of each sediment (or have students weigh their sample) so they can see how the same mass of a given sediment may not have the same density. If possible, leave a column from step 10 set up until the next period. Have students predict whether the water will still be cloudy or if it will have cleared.

Answers to Analysis and Conclusion

1. Students should find that the coarse particles settled fastest. The finest particles settled slowest.
2. Medium-fine particles should settle more slowly than medium particles because of their size.
3. Some very fine particles are still suspended in the water because it takes them more than five minutes to settle to the bottom of the column.
4. Because fine particles remain suspended in the water for a longer time, they can be carried out farther into the deep-ocean regions than coarse particles can. Coarse particles usually settle close to shore.
5. The shape and density of particles also influences their settling rate because these characteristics affect the drag of the particle through the water.

Answer to Extension

1. Answers may vary but should be based on student data. Flat and disk-shaped particles will take longer to settle than more rounded particles will.

Alexander Dvorak
Heritage School
New York City, NY

Total Sediment Thickness of Earth's Oceans

Internet Activity—ADVANCED

Deep-Sea Cores Have students visit the Lamont-Doherty Earth Observatory Deep-Sea Sample Repository website to learn about questions scientists have answered or hope to answer by studying deep sea sediment core samples. Have students present their findings to the class. A worksheet designed to direct student research on this topic can be found in the **Chapter Resource File** booklet or by visiting **go.hrw.com** and entering the keyword **HQ6OBAX.** LS **Logical/Verbal**

Answers to Map Skills Activity

1. The approximate thickness is about 2,000 m.
2. Two areas with the thickest sediments are 25°N, 95°W (in the Gulf of Mexico), and 20°N, 90°E (off the coast of India and Bangladesh).
3. The amount of sediment in the middle of the ocean is less than the amount of sediment around the continental margins. Near the middle of the oceans, sediment thickness is 500 m or less; near the continental margins, sediment thickness is 1,000 m or more.
4. Major rivers empty into the Gulf of Mexico (the Mississippi) and into the Bay of Bengal (the Ganges).
5. The west coast is bordered by a trench. The sediment layer on the west coast is relatively thin all the way up to the coastline.
6. Areas denoted with white spots probably have no valid data on sediment thickness. These areas are predominantly located in the Arctic Circle where the sea is covered with ice. Sampling in those areas may be difficult, if not impossible.

MAPS in Action

Total Sediment Thickness of Earth's Oceans

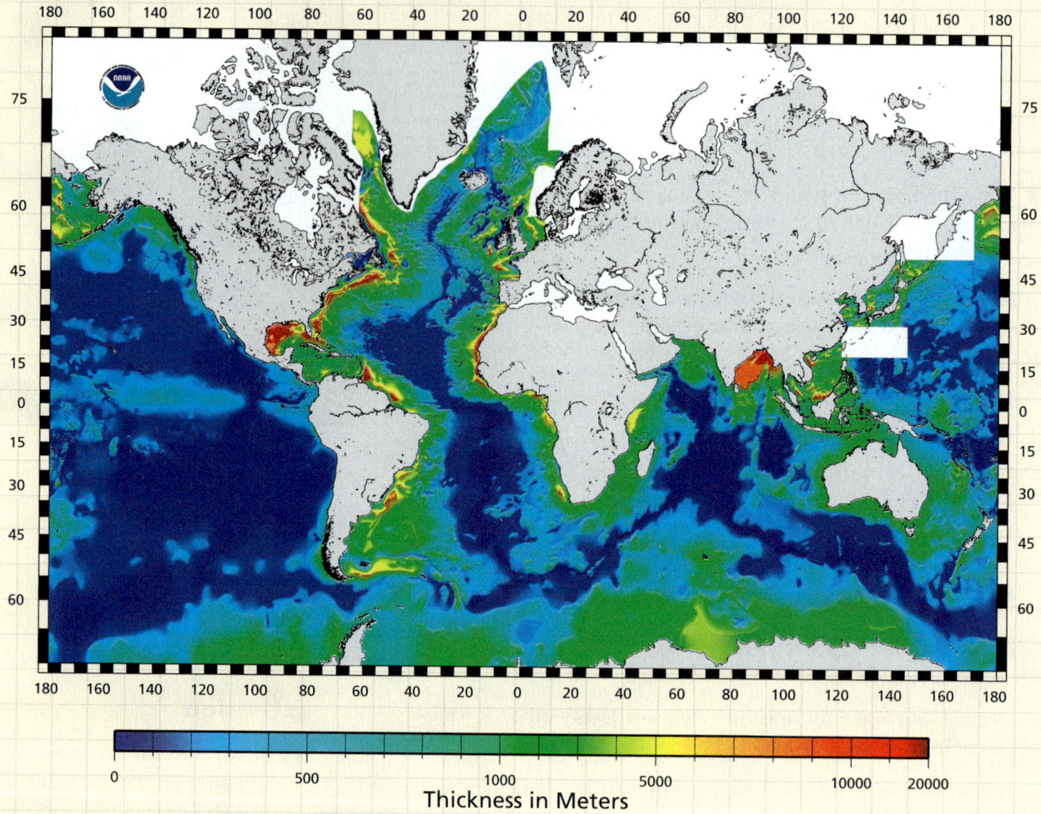

Map Skills Activity

This map shows the total thicknesses of sediments on Earth's ocean floors. Use the map to answer the questions below.

1. **Using the Key** What is the approximate thickness of the sediment located at 45°S and 45°W?
2. **Analyzing Data** Use latitude and longitude to identify two areas that have the thickest sediments.
3. **Comparing Areas** Compare the amount of sediment near the middle of the oceans with the amount of sediment on the continental margins.
4. **Identifying Trends** Rivers deposit massive amounts of sediment when they reach the ocean. Based on this map, at what locations would you expect to find mouths of major rivers?
5. **Inferring Relationships** Which coast of South America—east or west—is most likely bordered by a trench?
6. **Analyzing Relationships** Why does this map contain white spaces even though the key lists no thickness that corresponds with the color white?

CHAPTER RESOURCES

Chapter Resource File

Internet Activity
- Deep-Sea Cores ADVANCED

Technology

Transparencies
- 95 Total Sediment Thickness of Earth's Oceans (with worksheet)

Career Focus

Oceanographer

Lynne Talley is a physical oceanographer. Physical oceanographers study waves, tides, and currents in the ocean and the interactions between the ocean and the atmosphere. Talley's research focuses on large-scale patterns of ocean circulation that govern the worldwide movement of ocean waters. Her oceanography career began with a love for physics. Scientists also enter the field of oceanography from the fields of mathematics, geology, engineering, chemistry, biology, or ecology.

Going to Sea

Unlike the many oceanographers who use remote-sensing aircraft and Earth-orbiting satellites to collect data, Talley is a seagoing oceanographer. She uses direct measurement techniques to collect data. Every two years, she goes to sea on a month-long research cruise. Scientists on the cruise measure the temperature, salinity, and dissolved oxygen of ocean waters at various depths. Water samples are analyzed for chemical "fingerprints," such as the isotope helium-3. The presence of helium-3 may indicate that the sampled waters were near the surface during thermonuclear weapons testing that began in the 1950s.

After each cruise, Talley uses computer programs to analyze the data collected on the cruise. "Because oceans cover almost three-fourths of the world's surface, they have a huge impact on climate," says Talley. By studying the global movements of ocean waters, Talley can trace the global movement of heat. She uses her data as a basis for building numerical models that help scientists predict future oceanic and atmospheric conditions.

▼ Lynne Talley uses this rosette sampler to take samples of ocean water at various depths.

"I chose oceanography because it's so environmental and large scale. You can see it and touch it."

—Lynne Talley Ph.D.

Predicting Future Climate

Discovering how heat moves around Earth gives scientists a better understanding of global warming. Nations around the world are developing regulations to control global warming. Talley says that it is critical that such regulations be based on accurate data. "You must understand the actual processes occurring right now in order to predict change. You can't make up a model unless you understand the system."

SCILINKS. NSTA Developed and maintained by the National Science Teachers Association

For a variety of links related to this chapter, go to www.scilinks.org

Topic: Careers in Earth Science
SciLinks code: HQ60222

Career Focus

Oceanographer

Activity — GENERAL

Job Ad Have students write a classified advertisement for an ocean-related career aboard the research vessel JOIDES Resolution. The advertisement should include a position title, how the position relates to oceanography, required qualifications, and a salary range. For an additional activity, have students post their advertisments. Then, have students reply to each other's advertisments with a letter of application and a resume. LS Verbal

Extension — GENERAL

Research Have students identify four other careers that are related to oceans. Have students write a brief synopsis of each career and present them to the class. LS Verbal

Chapter 20 Ocean Water

Planning Guide

Compression Guide

To shorten instruction because of time limitations, omit Section 2.

OBJECTIVES	LABS, DEMONSTRATIONS, AND ACTIVITIES	TECHNOLOGY RESOURCES
PACING • 90 min pp. 492–500 **Chapter Opener**	**LTP** **Long-Term Project** Water Clarity* ADVANCED	**OSP** **Parent Letter** **CD** **Student Edition on CD-ROM** **CD** **Chapter Summaries Audio CD** **VID** **Brain Food Video Quiz**
Section 1 Properties of Ocean Water • Describe the chemical composition of ocean water. • Describe the salinity, temperature, density, and color of ocean water.	**TE** **Activity** Losing Fizz, p. 493 GENERAL **SE** **Quick Lab** Dissolving Solids, p. 495 GENERAL **CRF** **Datasheet for Quick Lab*** GENERAL **TE** **Activity** Ocean Surface Temperatures, p. 497 GENERAL **SE** **Quick Lab** Density Factors, p. 499 GENERAL **CRF** **Datasheet for Quick Lab*** GENERAL **SE** **Skills Practice Lab** Ocean Water Density, pp. 514–515 GENERAL **CRF** **Datasheet for Chapter Lab*** GENERAL **SE** **Maps in Action** Sea Surface Temperatures in August, p. 516 GENERAL **CRF** **Skills Practice Lab** The Blue-Green Ocean* GENERAL	**OSP** **Lesson Plans** (also in print) **TR** **Bellringer*** **TR** 96 Dissolved Gases in the Ocean* **TR** 97 Average Surface Salinity of the Global Ocean* **TR** 100 Sea Surface Temperatures in August* **CD** **Interactive Tutor** Seawater **CD** **Interactive Tutor** Ocean Color
PACING • 45 min pp. 501–504 **Section 2 Life in the Oceans** • Explain how marine organisms alter the chemistry of ocean water. • Explain why plankton can be called *the foundation of life in the ocean.* • Describe the major zones of life in the ocean.	**TE** **Group Activity** Poster Project, p. 501 GENERAL **TE** **Group Activity** Brine Shrimp, p. 509 GENERAL	**OSP** **Lesson Plans** (also in print) **TR** **Bellringer*** **TR** 98 Upwelling* **TR** 99 Marine Environments* **VID** **NOVA Video** Adrift on the Gulf Stream **CD** **Interactive Tutor** Ocean Depths
PACING • 45 min pp. 505–508 **Section 3 Ocean Resources** • Describe three important resources of the ocean. • Explain the threat water pollution poses to marine organisms.	**TE** **Discussion** Fresh Water from Salt Water, p. 505 GENERAL **TE** **Activity** Where's the Oil?, p. 506 BASIC **TE** **Group Activity** Curing Pollution in Boston Harbor, p. 517 ADVANCED **CRF** **Inquiry Lab** Oil Spill!* GENERAL	**OSP** **Lesson Plans** (also in print) **TR** **Bellringer*** **TE** **Internet Activity** The Romance of Aquaculture, p. 507 ADVANCED **CRF** **Internet Activity** The Romance of Aquaculture* ADVANCED

PACING • 90 min

CHAPTER REVIEW, ASSESSMENT, AND STANDARDIZED TEST PREPARATION

- **SE** **Chapter Highlights,** p. 509
- **SE** **Chapter Review,** pp. 510–511
- **SE** **Standardized Test Prep,** pp. 512–513
- **CRF** **Concept Review*** GENERAL
- **CRF** **Critical Thinking*** ADVANCED
- **CRF** **Math Skills*** GENERAL
- **CRF** **Graphing Skills*** GENERAL
- **CRF** **Chapter Test A*** GENERAL
- **CRF** **Chapter Test B*** ADVANCED
- **OSP** **Lesson Plans** (also in print)
- **OSP** **Test Generator**
- **OSP** **Test Item Listing**

Online and Technology Resources

go.hrw.com Holt Online Learning

Visit **go.hrw.com** for access to Holt Online Learning, or enter the keyword **HQ6 Home** for a variety of free online resources.

One-Stop Planner® CD-ROM

This CD-ROM package includes
- Lab Materials QuickList Software
- Holt Calendar Planner
- Customizable Lesson Plans
- Printable Worksheets
- ExamView® Test Generator
- Interactive Teacher Edition
- Holt PuzzlePro®
- Holt PowerPoint® Resources

KEY

SE	Student Edition	OSP	One-Stop Planner	VID	Classroom Video/DVD
TE	Teacher Edition	TR	Transparencies and Transparency Worksheets	*	Also on One-Stop Planner
CRF	Chapter Resource File	CD	CD or CD-ROM	◆	Requires advance prep
LTP	Long-Term Projects			■	Also available in Spanish

SKILLS DEVELOPMENT RESOURCES	REVIEW AND ASSESSMENT	CORRELATIONS
SE **Pre-Reading Activity**, p. 492 GENERAL TE **Using the Figure** Water of Life, p. 492 GENERAL		National Science Education Standards
CRF **Directed Reading*** BASIC TE **Skill Builder** Math, p. 495 GENERAL TE **Using the Figure** Salinity Around the World, p. 496 GENERAL TE **Reading Skill Builder** Paired Summarizing, p. 498 BASIC TE **Using the Figure** Thermoclines, p. 498 GENERAL TE **Using the Figure** Liquid Density, p. 499 BASIC	SE **Reading Checks**, p. 495, 497, 499 GENERAL SE **Section Review**, p. 500 GENERAL TE **Homework**, p. 497 BASIC TE **Reteaching**, p. 499 BASIC TE **Quiz**, p. 499 GENERAL TE **Alternative Assessment**, p. 500 ADVANCED CRF **Section Quiz*** ■ GENERAL	SPSP 4a
CRF **Directed Reading*** BASIC TE **Reading Skill Builder** Anticipation Guide, p. 502 BASIC TE **Using the Figure** Upwelling, p. 502 ADVANCED TE **Inclusion Strategies**, p. 502 SE **Graphic Organizer** Comparison Table, p. 503 GENERAL	SE **Reading Check**, p. 503 GENERAL SE **Section Review**, p. 504 GENERAL TE **Reteaching**, p. 503 BASIC TE **Quiz**, p. 503 GENERAL TE **Alternative Assessment**, p. 504 GENERAL CRF **Section Quiz*** ■ GENERAL	LS 4c
CRF **Directed Reading*** BASIC SE **Math Practice**, p. 506 GENERAL TE **Skill Builder** Vocabulary, p. 506 GENERAL TE **Inclusion Strategies**, p. 506	SE **Reading Check**, p. 507 GENERAL SE **Section Review**, p. 508 GENERAL TE **Reteaching**, p. 507 BASIC TE **Quiz**, p. 507 GENERAL TE **Alternative Assessment**, p. 508 ADVANCED CRF **Section Quiz*** ■ GENERAL	LS 4e, SPSP 3a

Holt Lab Generator CD-ROM

Search for labs by topic, standard, difficulty level, and/or time frame. Edit those labs to fit your needs, or create your own labs. Use the Lab Materials QuickList to order just what you need.

Interactive Tutor — Holt Earth Science Interactive Tutor CD-ROM

This CD-ROM consists of interactive activities that give students a fun way to extend their knowledge of Earth science concepts.

Chapter Summaries Audio CDs

These CDs include audio summaries of the key concepts presented in each chapter. (Audio summaries are also available in Spanish.)

SciLinks. NSTA

www.scilinks.org

Maintained by the **National Science Teachers Association.** See Chapter Enrichment pages that follow for a complete list of topics.

DVD VIDEO See Chapter Enrichment pages for Video Resources.

Refer to our ***Professional Reference for Teachers*** for additional teaching resources, including articles written by science education professionals about relevant and timely issues facing today's science teachers.

Chapter 20 Chapter Enrichment

This Chapter Enrichment provides relevant and interesting information to expand and enhance your classroom instruction of the chapter material.

Section 1 Properties of Ocean Water

Enhanced Global Warming and Ocean Waters

Some scientists think that Earth may be in the early stages of the most important climate change since the "Little Ice Age" that occurred between about 1400 and 1800 CE across large portions of North America and Europe. This climate change may be happening because in recent decades human activity has produced enormous amounts of "greenhouse gases" such as carbon dioxide. These excess amounts of greenhouse gases may cause *enhanced global warming* that affects the complex interactions between ocean waters and Earth's atmosphere. Additional stresses such as these actually may *cool* select regions. Here is what many scientists think happens: Earth's atmosphere acts like the glass in a greenhouse by absorbing heat from Earth's surface and warming the planet. Burning fossil fuels such as oil and coal in power plants, the release of emissions from automobiles, and deforestation by burning, among other human activities, release millions of tons of carbon dioxide and other gases into Earth's atmosphere. These gases absorb more solar energy so that the average global temperature increases. This process affects ocean currents by changing the temperature and salinity, and thus the density, of ocean water.

Some scientists think that enhanced global warming increases the amount of warmer, less salty water at the poles, by increasing the input of fresh water into the ocean as fresh water ice melts. The resulting decrease in water density slows the rate at which dense water sinks to form deep ocean currents, and in turn the rate at which the thermal energy in those currents is distributed. The warm Gulf Stream Current, for example, flows from the U.S. eastern seaboard to Europe, where it moderates coastal climates on both sides of the Atlantic Ocean. Any change in current patterns could seriously disrupt weather patterns and affect human and animal and plant populations. Some experts theorize that the recent series of very cold winters across the eastern U.S. may be related to changes occurring in the Atlantic Ocean. Other scientists think that recent warming is part of natural climatic variability.

◀ Icebergs form when ice from ice shelves breaks off in the ocean.

Section 2 Life in the Oceans

Marine Deserts

Marine life is not evenly distributed across the globe. In fact, some areas could be called "marine deserts." The concentration of phytoplankton, or surface-dwelling microscopic plant organisms, varies from place to place and from season to season. Phytoplankton are part of the base of the food chain that supports marine life and are most abundant near coastlines and throughout the North Atlantic Ocean during spring and summer. Fish and marine mammals, which are more mobile, exist throughout Earth's oceans.

Interestingly, many tropical waters have very little phytoplankton. Oceanographer John H. Martin of California's Moss Landing Marine Laboratories conducted experiments in the 1990s to determine why. His theory states that phytoplankton are absent from tropical waters because of the absence of upwelling currents that transport iron, a necessary micronutrient for phytoplankton, to the surface. As a result, he proposed a potential solution to global warming. He suggested dumping fine particles of iron into ocean water to stimulate the growth of phytoplankton. In turn, this would stimulate the uptake of atmospheric carbon dioxide, potentially enough to reverse the effects of global warming. Other scientists later verified Martin's theory. However, despite the success of his experiment, the oceans have not been seeded for ethical reasons and because of concerns about unintended consequences.

▲ Fresh water is an ocean resource that is exploited through desalination.

Section 3 Ocean Resources

Potential Opportunities

Oceans are a significant source of food, minerals, transportation, and even, via desalination, fresh water, for Earth's human, plant, and animal life. The waters of Earth's ocean are renewable—up to a point. Environmental stresses offer important opportunities for forward-looking people to develop ways of maximizing the ocean's enormous life-sustaining potential. Human activity and population growth, especially in the last century, have dangerously polluted and depleted fresh water reservoirs and ground water supplies. The cost of reversing these effects has risen. But improved desalination techniques have reduced costs and boosted production of fresh water. New techniques in heating water in distillation plants, such as capturing waste heat from other processes, also may lower costs. Opportunities abound to continue ocean research and applied scientific discoveries for the purpose of feeding Earth's inhabitants—without destroying ocean resources. Improving fishing practices and techniques, solving problems due to over-harvesting, and developing new techniques in aquaculture represent just a few of the ways responsible humans can ensure that valuable ocean resources remain available.

Video Resources

Brain Food Video Quizzes

Brain Food Video Quizzes These videos contain game-show style quizzes that assess students' progress and motivate students to study the chapter material.

NOVA Videos

The NOVA video below complements this chapter.

Adrift on the Gulf Stream NOVA explores the Gulf Stream's importance to ocean life and climate from a variety of vantage points. (60 min)

To order other **NOVA** videos related to this chapter, visit **go.hrw.com** and enter the keyword **HQ6OWAV.**

SciLinks

NSTA Developed and maintained by the National Science Teachers Association

SciLinks is maintained by the National Science Teachers Association to provide you and your students with interesting, up-to-date links that will enrich your classroom presentation of the chapter.

Visit www.scilinks.org and enter the SciLinks code for more information about the topic listed.

Topic: Properties of Ocean Water
SciLinks code: HQ61232

Topic: Marine Life
SciLinks code: HQ60912

Topic: Ocean Resources
SciLinks code: HQ61065

Topic: Ocean Pollution
SciLinks code: HQ61063

Chapter 20

Chapter Overview

Ocean water is a complex mixture of dissolved gases and solids from different sources. Its composition is affected by a number of factors, such as temperature and evaporation rates. The ocean's dissolved gases and solids are essential to marine life, which, in turn, maintains the chemical balance of ocean water and is supported by nutrients in the water. The oceans are a source of many resources, which are threatened by human pollution.

Using the Figure — GENERAL

Water of Life Ask students to write a paragraph that describes what they see in the image. Have them observe the behavior of the fish in the photo. Ask them to discuss what conditions might be required to support a school of fish both directly and indirectly. (Students may suggest such conditions as the salinity of the water, the temperature, sunlight, and the availability of food.) **LS Visual/Logical**

PRE-READING ACTIVITY

FOLDNOTES

Encourage students to use their FoldNote as a study guide to quiz themselves for a test on the chapter material. Students may want to create Four-Corner FoldNotes for different topics within the chapter.

Chapter 20 Ocean Water

Sections

1 Properties of Ocean Water

2 Life in the Oceans

3 Ocean Resources

What You'll Learn

- What the properties of ocean water are
- How life survives in the ocean
- Why ocean resources are important

Why It's Relevant

Earth's oceans play a vital role in Earth's ecology. Resources from the ocean provide humans with food, fuel, and fresh water.

PRE-READING ACTIVITY

FOLDNOTES

Four-Corner Fold

Before you read the chapter, create the FoldNote entitled "Four-Corner Fold" described in the Skills Handbook section of the Appendix. Label each flap of the four-corner fold with a topic. Write what you know about each topic under the appropriate flap. As you read the chapter, add other information that you learn.

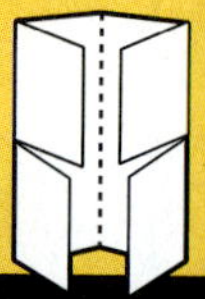

▶ To avoid predators, chevron barracuda school in a spiraling tornado near the ocean surface in Kimbe Bay, Papua New Guinea. These fish get all of the nutrients they need for life from the ocean water in which they live.

Chapter Correlations — *National Science Education Standards*

SPSP 4a Natural ecosystems provide an array of basic processes that affect humans. Those processes include maintenance of the quality of the atmosphere, generation of soils, control of the hydrologic cycle, disposal of wastes, and recycling of nutrients. Humans are changing many of these basic processes, and the changes may be detrimental to humans. **(Sections 1–3)**

LS 4c Organisms both cooperate and compete in ecosystems. The interrelationships and interdependencies of these organisms may generate ecosystems that are stable for hundreds or thousands of years. **(Section 2)**

LS 4e Human beings live within the world's ecosystems. Increasingly, humans modify ecosystems as a result of population growth, technology, and consumption. Human destruction of habitats through direct harvesting, pollution, atmospheric changes, and other factors is threatening current global stability, and if not addressed, ecosystems will be irreversibly affected. **(Section 3)**

SPSP 3a Human populations use resources in the environment in order to maintain and improve their existence. Natural resources have been and will continue to be used to maintain human populations. **(Section 3)**

Section 1 Properties of Ocean Water

Pure liquid water is tasteless, odorless, and colorless. However, the water in the ocean is not pure. Many solids and gases are dissolved in the ocean. In addition to dissolved substances, small particles of matter and tiny organisms are also suspended in ocean water. Ocean water is a complex mixture of chemicals that sustains a variety of plant and animal life.

Scientists describe ocean water by using a variety of properties, such as the presence of dissolved gases and the presence of dissolved solids, salinity, temperature, density, and color. Scientists study all of these properties to understand the complex interactions between the oceans, the atmosphere, and the land.

OBJECTIVES

- **Describe** the chemical composition of ocean water.
- **Describe** the salinity, temperature, density, and color of ocean water.

KEY TERMS

salinity
pack ice
thermocline
density

Dissolved Gases

The two principal gases in the atmosphere are nitrogen, N_2, and oxygen, O_2. These two gases are also the main gases dissolved in ocean water. While carbon dioxide, CO_2, is not a major component of the atmosphere, a large amount of this gas is dissolved in ocean water. Other atmospheric gases are also present in the ocean in small amounts.

Ocean water dissolves gases from a variety of sources, as shown in **Figure 1.** Gases may enter ocean water from water in streams and rivers. Some of the gases in ocean water come from volcanic eruptions beneath the ocean. Gases are also released directly into ocean water by organisms that live in the ocean. For example, many plants in the ocean make oxygen as a product of photosynthesis. However, most oxygen in the ocean enters at the surface of the ocean from the atmosphere.

Figure 1 ▶ Gases can enter the ocean from streams, volcanoes, organisms, and the atmosphere.

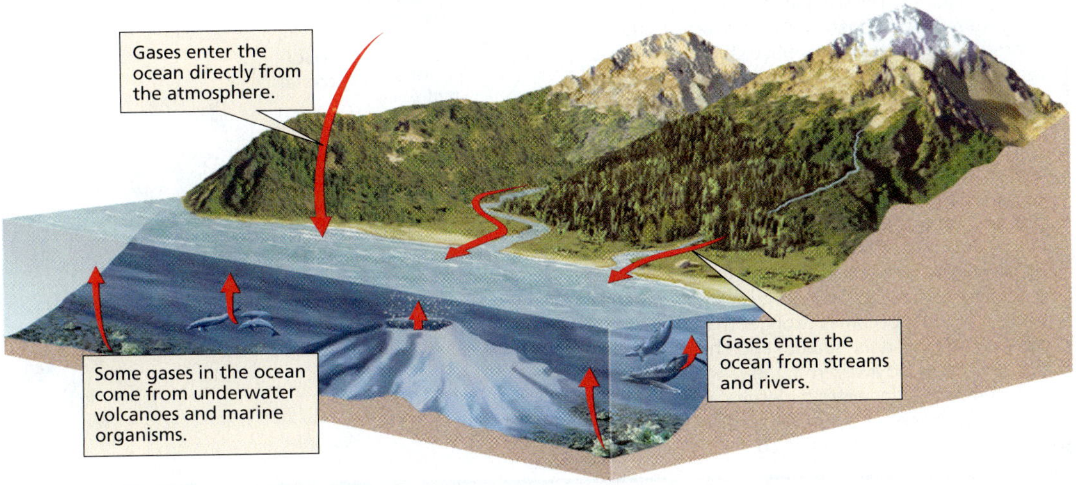

Section 1

Focus

Overview

This section introduces students to the composition of ocean water and describes the sources of the dissolved substances. The section also covers ocean composition, salinity, temperature, density, and color.

Bellringer

Have students list everything they think is found in ocean water. (The list may include such things as dissolved salt, various marine animals, plants, rocks, sand, and dissolved gases.) **LS Logical**

Motivate

Activity — GENERAL

Losing Fizz Provide students with two samples, about 250 mL each, of fresh club soda in beakers. One sample should be chilled, the other should be at room temperature. Direct students to observe each sample for 5 minutes and to record their observations. Then, ask students what occurred in each sample. (Gas was bubbling from the liquid.) What was the variable in this experiment? (temperature) What difference was observed in the two samples? (The gas bubbled more rapidly in the room temperature sample.) **LS Kinesthetic/Visual**

CHAPTER RESOURCES

Chapter Resource File

- Directed Reading BASIC

Technology

Transparencies
- Bellringer
- 96 Dissolved Gases in the Ocean (with worksheet)

Student Edition on CD-ROM

One-Stop Planner CD-ROM
- Lesson Plan

Teach

MISCONCEPTION ALERT

Solubility Trends Because students have learned that gases are more soluble in cooler liquids, they may assume that the same is true of solids. Explain that the solubility trends for solids and gases are generally opposite. Most, though not all, solids become more soluble as water temperature rises, while gases become more soluble as water cools.

ENVIRONMENTAL CONNECTION

Carbon Dioxide and Sinks The oceans are an important sink for atmospheric carbon dioxide. As a result, the oceans moderate changes in the concentration of atmospheric carbon dioxide. Ask students what they think might happen if the surface waters of the oceans become warmer. (Dissolved CO_2 would be released into the atmosphere. The atmosphere would grow warmer because increased CO_2 would trap more thermal energy from the sun. This would probably have a pronounced effect on the global climate system.) **LS Logical/Verbal**

SCILINKS. NSTA Developed and maintained by the National Science Teachers Association

For a variety of links related to this subject, go to www.scilinks.org

Topic: Properties of Ocean Water
SciLinks code: HQ61232

Temperature and Dissolved Gases

The temperature of water affects the amount of gas that dissolves in water. Gases dissolve more readily in cold water than in warm water. You may have noticed this phenomenon when your glass of soda quickly goes "flat" on a warm day. The soda goes flat quickly because the CO_2 that makes the soda bubbly escapes into the air. But if the soda is kept in the refrigerator, the soda will retain its fizz longer. Because cold water dissolves gases more readily, water at the surface of the ocean in cold regions dissolves larger amounts of gases than water in warm tropical regions does.

Gases also can return to the atmosphere from the ocean. If the water temperature rises, less gas will remain dissolved, and the excess gas will be released into the atmosphere. For example, warm equatorial ocean waters tend to release CO_2 into the atmosphere, but ocean waters at cooler, higher latitudes take up large amounts of CO_2. Therefore, the ocean and the atmosphere are continuously exchanging gases as water temperatures change.

The Oceans as a Carbon Sink

Oceans contain more than 60 times as much carbon as the atmosphere does. Dissolved CO_2 may be trapped in the oceans for hundreds to thousands of years. Because of this ability to dissolve and contain a large amount of CO_2, the oceans are commonly referred to as a *carbon sink*. Because gaseous CO_2 affects the atmosphere's ability to trap thermal energy from the sun, the oceans are important in the regulation of climate.

Connection to CHEMISTRY

How Substances Dissolve

A water molecule is one oxygen atom bonded to two hydrogen atoms. The oxygen atom pulls electrons away from the hydrogen atoms, which gives the hydrogen atoms a partial positive charge and gives the oxygen atom a partial negative charge. This uneven distribution of charges allows the water molecule to attract both positive and negative ions.

The figure at right shows how a sodium chloride, NaCl, crystal dissolves in water. The partially negative oxygen atoms in water molecules attract the positively charged sodium ions of the salt. The partially positive hydrogen atoms in the water molecules attract the negatively charged chloride ions of the salt. When the force of attraction between the ions and the water molecules becomes stronger than the force of attraction between the sodium and chloride atoms, the ions are pulled away from the crystal. The ions are then surrounded by water molecules. Eventually, all of the ions in the crystal are pulled into solution, and the substance is completely dissolved.

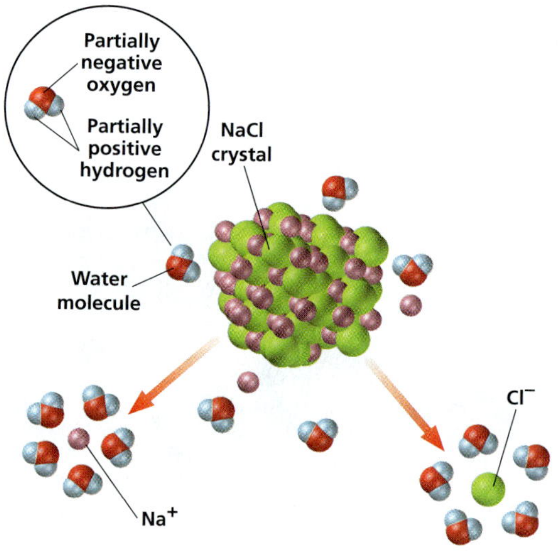

METEOROLOGY CONNECTION

Positive Feedback Explain to students that warming ocean waters may result in *positive feedback* between the oceans, the atmosphere, and Earth's climate system. *Positive feedback* is a cycle in which one event leads to other events, which in turn increase the effects of the first event. For example, when increased concentrations of greenhouse gases, such as CO_2, cause atmospheric warming, this warming leads to increased ocean warming. This, in turn, results in less absorption of CO_2 by these waters. The lower absorption rate results in an even greater concentration of CO_2 in the atmosphere, and even greater atmospheric warming. Have students draw a simple diagram or concept map that shows this process. You may want to tell students that positive feedback is rare in nature. Most feedbacks are negative; that is, after one event begins, another event occurs to shut it down. For example, strong daytime heating can produce thunderstorms, which produce rain. The rain cools Earth's surface and helps weaken the thunderstorm.

Figure 2 ▶ Dissolved solids make up 3.5% of the mass of ocean water. More than 85% of these dissolved solids are sodium and chlorine.

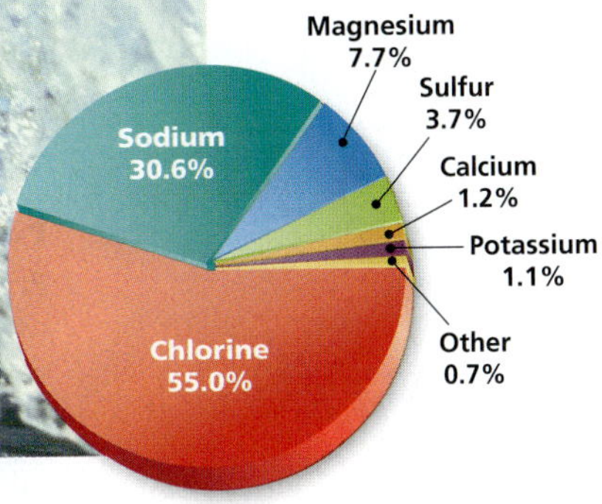

Dissolved Solids

Ocean water is 96.5% pure water, or H_2O. Dissolved solids make up about 3.5% of the mass of ocean water. These dissolved solids, commonly called *sea salts*, give the ocean its salty taste.

Most Abundant Elements

Solids dissolved in ocean water are composed of about 75 chemical elements. The six most abundant elements in ocean water are chlorine, sodium, magnesium, sulfur, calcium, and potassium. The salt halite, which is made of sodium and chloride ions, makes up more than 85% of the ocean's dissolved solids. The remaining dissolved solids consist of various other salts and minerals, as shown in **Figure 2.** *Trace elements* are elements that exist in very small amounts. Gold, zinc, and phosphorus are some of the trace elements that are found in the ocean.

Sources of Dissolved Solids

Most of the elements that form sea salts come from three main sources—volcanic eruptions, chemical weathering of rock on land, and chemical reactions between sea water and newly formed sea-floor rocks. Each year, rivers carry about 400 billion kilograms of dissolved solids into the ocean. Most of these dissolved solids are salts. As water evaporates from the ocean, salts and other minerals remain in the ocean. Only a small fraction of these salts and minerals are returned to the land in the water that falls as rain and snow during the water cycle.

Reading Check **How do dissolved solids enter the ocean? (See the Appendix for answers to Reading Checks.)**

QuickLAB 10 min

Dissolving Solids

Procedure

1. Heat **200 ml of water** in a **beaker** over a **hot plate** until the water is about 60°C.
2. Dissolve **table salt** in the water 1 tsp at a time until no more salt will dissolve. Record the total amount of salt that dissolves.
3. Dissolve table salt 1 tsp at a time into **200 ml of water** that has been chilled in the refrigerator to about 5°C. Record the total amount of salt that dissolves.

Analysis

1. Which water sample dissolved the most salt?
2. Describe what would happen to the dissolved salt in the hot water if the hot water was chilled to 10°C.

SKILL BUILDER—GENERAL

Math Tell students that 500 g of ocean salts were recovered from a sample of ocean water. Ask them to use the pie chart in the figure on this page to determine the mass of each element in the 500 g sample.

Chlorine: 55.0% = 0.55; 500 g × 0.55 = 275 g
Sodium: 30.6% = 0.306; 500 g × 0.306 = 153 g
Magnesium: 7.7% = 0.077; 500 g × 0.077 = 38.5 g
Sulfur: 3.7% = 0.037; 500 g × 0.037 = 18.5 g
Calcium: 1.2% = 0.012; 500 g × 0.012 = 6 g
Potassium: 1.1% = 0.011; 500 g × 0.011 = 5.5 g
All others: 0.7% = 0.007; 500 g × 0.007 = 3.5 g

LS Visual/Logical

QuickLAB

Skills Acquired

- Observing
- Measuring
- Analyzing

Teacher's Notes Make sure students stir well after each teaspoon of salt. You may want to do this as a demonstration and have students record the amount of salt that dissolved in each solution.

Answers

1. The warmer water will dissolve more salt than the cold water will.
2. Some salt would precipitate, or settle, out of the solution.

CHAPTER RESOURCES

Chapter Resource File

- Datasheet for Quick Lab GENERAL

Answer to Reading Check

Dissolved solids enter the oceans from the chemical weathering of rock on land, from volcanic eruptions, and from chemical reactions between sea water and newly formed sea-floor rocks.

Teach, *continued*

MISCONCEPTION ALERT

Salinity of Polar Waters Students may think that only the evaporation of warm ocean water will result in an increase in salinity. However, the freezing of cold ocean water also increases salinity. This increase in salinity occurs because only the water freezes. The dissolved salts are left behind and increase the salinity of the water that doesn't freeze. In addition, the higher salinity prevents the ocean water from freezing until it is a few degrees below 0°C.

The map on this page shows large areas of highly saline water in the "sun belt," which includes regions of global high pressure centered about 30° north and south of the Equator. The areas of highly saline waters in polar regions are much smaller—too small to be represented on such a small map.

Using the Figure — GENERAL

Salinity Around the World Use the figure on this page to lead a discussion of ocean salinity. Ask students to generalize about the location of areas of high salinity and suggest what conditions produce them. (The areas of high salinity on the map are located between 30° north and 30° south of the equator, where temperatures are warm and evaporation rates are high.) Answer to caption question: River mouths tend to lower the salinity of nearby ocean waters because river water is fresh water, which is much less saline than the ocean water. LS **Visual/Logical**

salinity a measure of the amount of dissolved salts in a given amount of liquid

Salinity of Ocean Water

One of the biggest differences between ocean water and fresh water is the high concentration of salts in ocean water. **Salinity** is a measure of the amount of dissolved salts and other solids in a given liquid. Salinity is measured by the number of grams of dissolved solids in 1,000 g of ocean water. For example, if 1,000 g of ocean water contained 35 g of solids, the salinity of the sample would be about 35 parts salt per 1,000 parts ocean water. This measurement is written as *salinity* = 35 parts per thousand, or 35‰. Thus, the ocean is about 3.5% salts. However, fresh water is less than 0.1% salt or has a salinity of 1‰.

Factors That Change Salinity

Precipitation such as rain and snow is composed only of freash water. When ocean surface water evaporates or freezes, only water molecules are removed from the ocean; dissolved salts and other solids remain. Where the rate of evaporation is higher than the rate of precipitation, the salinity of surfaces water increases. Therefore, in equatorial waters, where the rate of precipitation is highest, the salinity is lower than it is in subtropical waters, where rates of evaporation are highest.

Over most of the surface of the ocean, salinity ranges from 33‰ to 36‰. The global ocean has an average salinity of 34.7‰. However, salinity at particular locations can vary greatly, as shown in **Figure 3.** The salinity of the Red Sea, for example, is more than 40‰. The high salinity is due to the hot, dry climate around the Red Sea, which causes high levels of evaporation.

Figure 3 ▶ The average surface salinity of the global ocean varies from one location to another. ***What effect do river mouths tend to have on the salinity of the surrounding ocean water?***

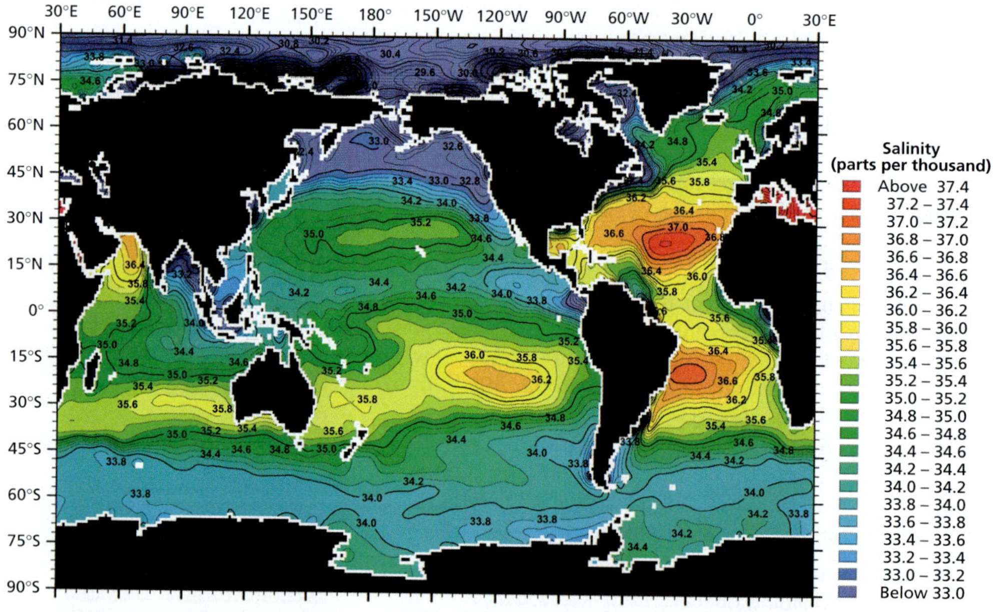

BRAIN FOOD

Salt Deposits During the Miocene Epoch, the Straits of Gibraltar were closed off because of moving tectonic plates. Thus, the Mediterranean Sea was cut off from the Atlantic Ocean and completely surrounded by land. Over time, the entire Mediterranean Sea evaporated. This event produced a layer of halite, or rock salt, along the sea floor of this region and caused temperatures to soar to over 50°C in neighboring landmasses.

CHAPTER RESOURCES

Technology

Transparencies
- 97 Average Surface Salinity of the Global Ocean (with worksheet)

Temperature of Ocean Water

Like ocean salinity, ocean temperature varies depending on depth and location on the surface of the oceans. The range of ocean temperatures is affected by the amount of solar energy an area receives and by the movement of water in the ocean.

Surface Water

The mixing of the ocean's surface water distributes heat downward to a depth of 100 to 300 m. Thus, the temperature of this zone of surface water is relatively constant and decreases only slightly as depth increases. However, the temperature of surface water does decrease as latitude increases. Therefore, polar surface waters are much cooler than the surface waters in the Tropics, as explained in **Figure 4.**

The total amount of solar energy that reaches the surface of the ocean is much greater at the equator than in areas near the North and South Poles. In tropical waters, ocean surface temperatures of about 30°C are common. Surface temperatures in polar oceans, however, often drop as low as –1.9°C. Because ocean water freezes at about –1.9°C, vast areas of sea ice exist in polar oceans. A floating layer of sea ice that completely covers an area of the ocean surface is called **pack ice.** Usually, pack ice is no more than 5 m thick because the ice insulates the water below and prevents it from freezing. In the middle latitudes, the ocean surface temperature varies depending on the seasons. In some areas, the ocean surface temperature may vary by as much as 10°C to 20°C between summer and winter.

Reading Check **What factors affect the surface temperature of the ocean? (See the Appendix for answers to Reading Checks.)**

Figure 4 ▶ The surface temperature of tropical ocean water (right) can be as high as 30°C. However, the surface temperature of polar ocean water (left) is below the freezing point of fresh water.

pack ice a floating layer of sea ice that completely covers an area of the ocean surface

Activity — GENERAL

Ocean Surface Temperatures Divide the class into groups. Have each group fill a deep plastic tank with room-temperature water. Have students tape a thermometer to the tank so that the bulb is about 5 mm below the surface of the water. Have students aim a heat lamp at the water's surface so that the lamp's light strikes the water at a 90° angle. Leave the lamps on over the water for about 5 minutes. Ask students to check and record the water temperature each minute. Then, have groups repeat the activity using fresh water and placing the lamp so that the light strikes the water surface at a 30° angle. They should keep all other factors constant, including the lamp's distance from the water's surface. Have them compare data collected in the two trials. Have students graph their data and discuss the differences. (The water should have warmed more significantly in the first trial.) Ask students how this relates to the world's oceans. (Tropical latitudes receive more direct sunlight year round than polar latitudes do.) You might want to do this activity as a demonstration. **LS Kinesthetic/Visual**

Answer to Reading Check

Ocean surface temperatures are affected by the amount of solar energy an area receives and by the movement of water in the ocean.

Homework — BASIC

Freezing Fresh and Salt Water Ask students to place 150mL of water in each of two cups. Have them place 3 teaspoons of table salt in one cup; they should add nothing to the other cup. Have students place the cups in their freezers and check each cup every 20 minutes. Have them record how long the liquid in each cup took to freeze. Also ask a number of students who have access to a freezer thermometer to report the temperature of their freezer at home. Ask all students to make graphs recording the information they collect. In class, start a discussion by asking them which liquid froze first. (The fresh water will freeze before the salt solution does.) **LS Kinesthetic/Visual** English Language Learners

Teach, *continued*

READING SKILL BUILDER — BASIC

Paired Summarizing Group students into pairs, and have them read silently about the thermocline. Then, have one student summarize the description of the thermocline. The other student should listen to the retelling and should point out any inaccuracies or ideas that were left out. Allow students to refer to the text as needed. English Language Learners
LS Verbal/Auditory

Using the Figure — GENERAL

Thermoclines Use the graph of the thermocline on this page to review with students the axes of graphs. Remind them that the vertical axis on this particular graph represents the depth in meters of the ocean, not the decrease in temperature. The water temperature is represented by the horizontal axis, beginning with the lowest temperature to the left. Ask students what the depth range of the surface layer is on the graph. (from 0 m to about 280 m) Ask them the extent of the thermocline layer. (from about 280 m to about 500 m) Have them use the graph to determine the drop in temperature in degrees that occurs in the thermocline layer. (The temperature decreases from about 22.5°C to about 5°C, for a decline of 17.5°C.) LS **Visual/Logical**

The Thermocline

thermocline a layer in a body of water in which water temperature drops with increased depth faster than it does in other layers

Because the sun cannot directly heat ocean water below the surface layer, the temperature of the water decreases sharply as depth increases. In most places in the ocean, this sudden decrease in temperature begins close to the surface. The layer in a body of water in which water temperature drops with increased depth faster than it does in other layers is called the **thermocline.**

The thermocline exists because the water near the surface becomes less dense as energy from the sun warms the water. This warm water cannot mix easily with the cold, dense water below. Thus, a thermocline marks the distinct separation between the warm surface water and the cold deep water. Below the thermocline, the temperature of the water continues to decrease, but it decreases very slowly, as shown in **Figure 5.** Changing temperature or shifting currents may alter the depth of the thermocline or cause the thermocline to disappear. Nevertheless, a thermocline is usually present beneath much of the ocean surface.

Deep Water

In the deep zones of the ocean, the temperature of the water is usually about 2°C. The colder the water is, the denser it is. The density of cold, deep water controls the slow movement of deep ocean currents. This movement occurs when the cold, dense water at the poles sinks and flows beneath warm water toward the equator. Cold, deep ocean water also holds more dissolved gases than warm, shallow ocean water does.

Figure 5 ▶ The temperature of ocean water decreases as depth increases. Just below the surface is a thermocline, an area where the water temperature decreases sharply.

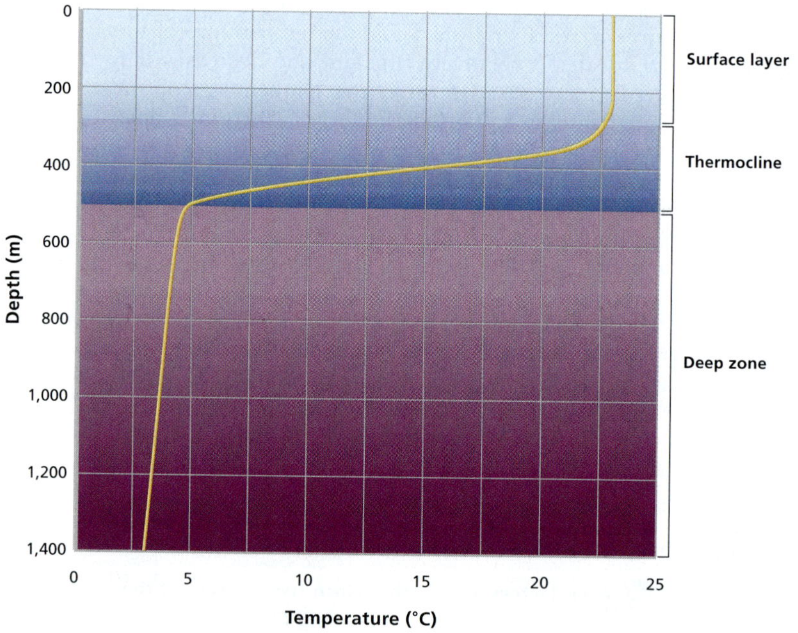

ENVIRONMENTAL SCIENCE CONNECTION — GENERAL

Surface-Water Temperatures While surface-water temperatures are greatly affected by solar elevation or by the intensity of incoming solar radiation, other factors have an impact on the sea surface also. Surface ocean currents and cold water moving upward play a significant role, for example, in keeping Pacific surface waters along the west coast of the United States quite cool all summer in comparison to Atlantic waters at similar latitudes along the east coast of the United States. Invite interested students to do research on surface ocean currents along the coasts of the United States and present their findings in illustrated oral reports to the class. LS **Visual/Verbal**

Density of Ocean Water

The mass of a substance per unit volume is that substance's **density.** For example, 1 cm^3 of pure water has a mass of 1 g. So, the density of pure water is 1 g/cm^3. Different liquids have different densities, as shown in **Figure 6.** Two factors affect the density of ocean water: salinity and the temperature of the water. Dissolved solids, which are mainly salts, add mass to the water. The large amount of dissolved solids in ocean water makes it denser than pure fresh water. Ocean water has a density between 1.026 g/cm^3 and 1.028 g/cm^3.

Ocean water becomes denser as it becomes colder and less dense as it becomes warmer. Water temperature usually affects the density of ocean water more than salinity does. Therefore, the densest ocean water is found in the polar regions, where the ocean surface is coldest. This cold, dense water sinks and moves through the ocean basins near the ocean floor.

Figure 6 ▶ This graduated cylinder contains six liquids that have different densities. From top to bottom they are corn oil, water, shampoo, dish detergent, anti-freeze, and maple syrup. ***Which liquid has the lowest density?***

density the ratio of the mass of a substance to the volume of the substance; commonly expressed as grams per cubic centimeter for solids and liquids and as grams per liter for gases

✓ Reading Check **Explain why ocean water is denser than fresh water.** **(See the Appendix for answers to Reading Checks.)**

QuickLAB 20 min

Density Factors

Procedure

1. Fill a deep, clear plastic container half full with room temperature water.
2. In a 1 L beaker, mix 1/8 cup of table salt, a few drops of red food coloring, and 1 L of room temperature water. Stir the mixture until the salt is dissolved.
3. Add the red saltwater mixture to the water in the clear plastic container. Record your observations.
4. In the 1 L beaker, mix a few drops of blue food coloring with water that is 8°C.
5. Slowly add the cold, blue water to the clear plastic container in step 3. Record your observations.

Analysis

1. Describe what happened when you added the red salt water to the fresh water. Which is denser: fresh water or salt water?
2. What happened when you added the cold water to the room temperature water? Which is denser: cold water or room temperature water?
3. What would you expect to happen if the blue water was heated, instead of cooled?
4. Based on your observations, where would you expect the water in the ocean to be the least dense? the most dense?
5. Describe water layering where a river empties into the ocean.

QuickLAB

Skills Acquired

- Observing
- Interpreting
- Analyzing

Teacher's Notes You may want to do this experiment as a demonstration.

Answers

1. The salt water sank below the fresh water. Salt water is denser.
2. The cold water sank below the room temperature water. Cold water is denser.
3. The warm water would float on the room temperature water, because warm water is less dense.
4. It would be least dense where water is warm and less saline (near the equator), and most dense where water is cold and highly saline (near the poles).
5. Fresh water from the river would "float" on the denser ocean water.

Using the Figure — BASIC

Liquid Density You may want to create a cylinder with six layers as shown in the figure on this page. Ask students whether a layer of water and dissolved salts would come above or below the layer of fresh water. Ask them to explain why. (The salt water would come below the fresh water because the dissolved salts would add mass to the solution, making it more dense.) Answer to caption question: The least dense liquid is the corn oil at the top of the column. LS **Visual/Logical**

Answer to Reading Check

Ocean water contains dissolved solids (mostly salts) that add mass to a given volume of water. The large amount of dissolved solids in ocean water makes ocean water denser than fresh water.

Close

Reteaching — BASIC

Basic Diagrams Have students create diagrams that illustrate the substances found in the oceans and their sources. LS **Visual**

Quiz — GENERAL

1. If polar waters warm, what will happen to their ability to absorb atmospheric carbon dioxide? (It will decrease; less CO_2 will dissolve in the oceans.)
2. In what part of the ocean are the densest waters located? (near the poles and at the bottom of the sea, where the water is very cold)

CHAPTER RESOURCES

Chapter Resource File

- **Datasheet for Quick Lab** GENERAL
- **Skills Practice Lab** The Blue-Green Ocean GENERAL

Close, *continued*

Alternative Assessment — ADVANCED

Essay Ask students to write an essay that ties dissolved carbon dioxide in the oceans to atmospheric carbon dioxide, ocean temperature, and climate change. Have them draw diagrams to illustrate the processes described in their essay. LS Verbal/Visual

Answers to Section Review

1. Warmer water warms is less able to dissolve gases than cooler water is.
2. Freezing and evaporation can increase salinity. As water evaporates or freezes, it leaves behind dissolved solids that increase the salinity of the remaining water.
3. Ocean water is a mixture of dissolved solids, such as chlorine and sodium, and gases, such as oxygen, nitrogen, and carbon dioxide.
4. The thermocline is a layer in a body of water in which the temperature drops faster with increased depth than it does in other layers.
5. Ocean water becomes denser as temperature decreases and as salinity increases.
6. Cold, saline, dense water at the poles sinks and flows beneath warmer water toward the equator as deep currents.
7. When sunlight penetrates the surface of the ocean, the ocean water reflects the blue wavelengths of light. The other colors are absorbed.
8. because the amount of incoming solar radiation varies greatly from season to season in these latitudes
9. Surface waters in the North Sea are colder. Because cold waters can hold more dissolved gases than warm waters can, more disolved gases would be found in the cold North Sea than in the warm Caribbean waters.
10. If global temperatures increase, the oceans may warm. Warmer waters can't hold as much CO_2, thus more CO_2 may accumulate in the atmosphere.
11. Because phytoplankton are at the bottom of the food chain, a decrease in phytoplankton will have a negative effect on all ocean animals, and those organisms may decrease in number.
12. The properties of *ocean water* include the amounts of *dissolved gases* and of *dissolved solids,* which is called *salinity,* and which together with *temperature* affects *density*.

Figure 7 ▶ Ocean water appears blue as far as 100 m below the surface.

Color of Ocean Water

Have you ever wondered why the ocean appears blue, as shown in **Figure 7?** The color of ocean water is determined by the way it absorbs or reflects sunlight. White light from the sun contains light from all the visible wavelengths of the electromagnetic spectrum. Much of the sunlight penetrates the surface of the ocean and is absorbed by the water. Water absorbs most of the wavelengths, or colors, of visible light. Only the blue wavelengths tend to be reflected. The reflection of this blue light makes ocean water appear blue.

Why Is Ocean Color Important?

Substances or organisms in ocean water, such as phytoplankton, can affect the color of the water. *Phytoplankton* are microscopic plants in the ocean that provide food to many of the ocean's organisms. Phytoplankton absorb red and blue light, but reflect green light. Therefore, the presence and amount of phytoplankton can affect the shade of blue of the ocean.

By studying variations in the color of the ocean, scientists can determine the presence of phytoplankton in the ocean. Because phytoplankton require nutrients, the presence or absence of phytoplankton can indicate the health of the ocean. If the color of an area of the ocean indicates that no phytoplankton is present, pollution may have prevented phytoplankton growth.

Section 1 Review

1. **Describe** how water temperature affects the ability of the ocean water to dissolve gases.
2. **Summarize** how freezing and evaporation affect salinity.
3. **Describe** the composition of ocean water.
4. **Define** *thermocline.*
5. **Describe** how temperature and salinity affect the density of ocean water.
6. **Explain** how the density of ocean water drives the movement of deep ocean currents.
7. **Explain** why shallow ocean water appears to be blue in color.

CRITICAL THINKING

8. **Making Inferences** Why does the surface temperature of ocean water in middle latitudes vary during the year?
9. **Understanding Relationships** Why would surface water in the North Sea be more likely to contain a high percentage of dissolved gases than the surface water in the Caribbean Sea would?
10. **Predicting Consequences** If global temperatures increase, how would this change affect the ability of the oceans to absorb CO_2?
11. **Identifying Relationships** If an area of the ocean has a large decrease in phytoplankton, how would this change affect other ocean organisms? Explain your answer.

CONCEPT MAPPING

12. Use the following terms to create a concept map: *ocean water, salinity, temperature, density, dissolved solids,* and *dissolved gas.*

CHAPTER RESOURCES

Chapter Resource File
- Section Quiz GENERAL

Workbooks
- Study Guide (also in Spanish)

Section 2 Life in the Oceans

Most marine organisms depend on two major factors for their survival—the essential nutrients available in ocean water and sunlight. Variations in either of these factors affect the ability of aquatic organisms to survive and flourish.

OBJECTIVES

- **Explain** how marine organisms alter the chemistry of ocean water.
- **Explain** why plankton can be called *the foundation of life in the ocean.*
- **Describe** the major zones of life in the ocean.

KEY TERMS

upwelling
plankton
nekton
benthos
benthic zone
pelagic zone

Ocean Chemistry and Marine Life

The chemistry of the ocean is a balance of dissolved gases and solids that are essential to marine life. Marine organisms help maintain the chemical balance of ocean water. They do this by removing nutrients and gases from the ocean while returning other nutrients and gases to the ocean. For example, marine plants absorb large amounts of carbon, hydrogen, oxygen, and sulfur. They also absorb other elements such as nitrogen, phosphorus, and silicon. Marine organisms also return nutrients and gases to the ocean. For example, photosynthetic marine plants remove carbon dioxide from ocean water to produce oxygen.

Marine organisms, such as the sea horse shown in **Figure 1,** also help recycle nutrients in the ocean. During a marine organism's lifetime, the organism absorbs and stores nutrients from the ocean. These nutrients are eventually returned to the water when the organism dies. For example, bacteria in the water digest the remains of the dead organisms. The bacteria then release the essential nutrients from the dead organisms into the ocean.

Figure 1 ▶ Like many marine organisms, this sea horse gets many of the nutrients it needs from the ocean water.

Section 2

Focus

Overview

This section describes life in the oceans and the various environmental zones within the oceans and along the ocean floor. This section discusses how marine life changes the chemistry of sea water and discusses the importance of plankton.

Bellringer

Ask students to make a list of marine organisms that they are familiar with, and have them identify whether each organism lives on the ocean floor or swims freely through ocean water. (Sample answers: sea-floor dwellers: sea stars, octopuses, sponges; free swimmers: fish, dolphins) **LS Logical**

Motivate

Group Activity — GENERAL

Poster Project Have students read the two paragraphs under the heading "Ocean Chemistry and Marine Life." Then, have students work in groups to create posters with flowcharts that illustrate how marine organisms help maintain the chemical balance of ocean water. Have students prepare prior to the lesson to save time. **LS Visual/Logical**

CHAPTER RESOURCES

Chapter Resource File

- Directed Reading BASIC

Technology

Transparencies
- Bellringer

Student Edition on CD-ROM

One-Stop Planner CD-ROM
- Lesson Plan

Teach

READING SKILL BUILDER — BASIC

Anticipation Guide Before students read this page, ask them to predict the relationship between ocean chemistry and marine life. Then, ask students to read the page to find out if their predictions were accurate. English Language Learners
LS Verbal/Logical

Using the Figure — ADVANCED

Upwelling The figure at the top of this page illustrates *prevailing,* or consistent, surface winds that blow parallel to the coastline. This action results in the upwelling of coastal waters. Ask students to consider what might happen if the prevailing winds were to shift and blow more perpendicular to the coastline. (Upwelling would cease.) Ask students how this change would affect marine life in the area. (If upwelling ceases, marine life along the coast will decrease. This decrease would affect industries such as fishing and the food supply for humans and larger marine organisms.) Answer to caption question: Storms may change the velocity or direction of the prevailing wind, or cause deeper waves that mix the water. Thus, storms may intensify upwelling or temporarily cut it off. LS Logical

CHAPTER RESOURCES

Technology

Transparencies
- 98 Upwelling (with worksheet)
- 99 Marine Environments (with worksheet)

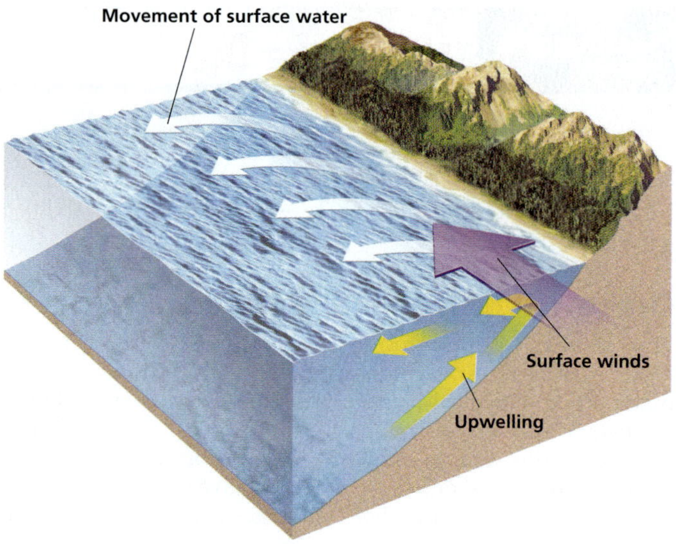

Figure 2 ▶ Upwelling is caused by offshore movement of surface water. ***How might stormy weather affect the process of upwelling?***

upwelling the movement of deep, cold, and nutrient-rich water to the surface

plankton the mass of mostly microscopic organisms that float or drift freely in the waters of aquatic (freshwater and marine) environments

nekton all organisms that swim actively in open water, independent of currents

benthos organisms that live at the bottom of oceans or bodies of fresh water

Upwelling

The distribution of life in the ocean depends on the way life-supporting nutrients cycle in the ocean water. In general, all of the elements necessary for life are consumed by organisms near the surface. Elements are then released back into the ocean water when organisms die, sink to lower depths, and decay. Thus, deep water is a storage area for the nutrients needed for life. These nutrients must, however, return to the surface before most organisms in the ocean can use them.

One way that nutrients return to the surface is through a process called upwelling. **Upwelling** is the movement of deep, cold, and nutrient-rich water to the surface, as shown in **Figure 2.** When wind blows steadily parallel to a coastline, surface water moves farther offshore. The deep, cold water then rises to replace the surface water that has moved away from the shore.

Figure 3 ▶ Plankton are so tiny that you need a microscope to see them.

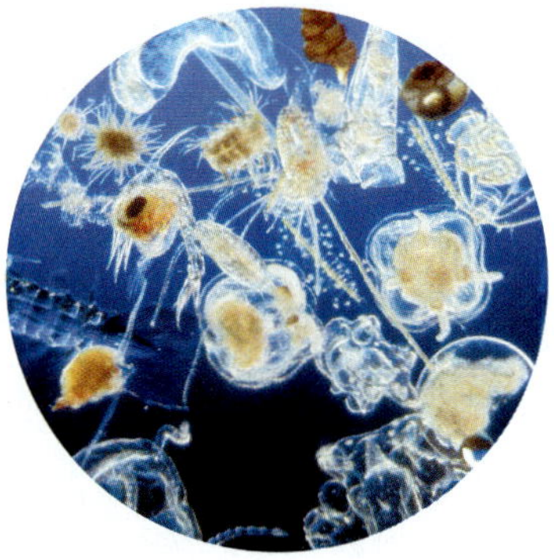

Marine Food Webs

Because most marine organisms need sunlight as well as nutrients, most marine organisms live in the upper 100 m of water. Free-floating, microscopic plants and animals called **plankton** live within the sunlit zone. Plankton, shown in **Figure 3,** form the base of the complex food webs in the ocean. The plankton are consumed primarily by small marine organisms, which, in turn, become food for larger marine animals. These larger animals fall into two groups. All organisms that swim actively in open water, such as fish, dolphins, and squid, are called **nekton.** The organisms that live on the ocean floor are called **benthos.** Benthos include marine plants and animals, such as oysters, sea stars, and crabs, that live in sunlit, shallow waters.

INCLUSION Strategies

• *Hearing Impaired*

Students who have hearing impairments can participate more easily in class discussions if they can see everyone's face. Before beginning a class discussion, ask students to arrange their chairs in a circle and to talk with their faces unobstructed. Tell the group that, when talking with people who have hearing impairments, they can help the hearing-impaired person understand by always maintaining eye contact. Explain that people who have hearing impairments benefit both from the opportunity to read lips and from meaning they can glean from facial expressions. Furthermore, point out that people who have normal hearing can also better understand others when they have eye contact and can see facial expressions. LS Interpersonal

Ocean Environments

The ocean can be divided into two basic environments, as shown in **Figure 4.** These zones are the bottom region, or **benthic zone,** and the upper region, or **pelagic zone.** The amount of sunlight, the water temperature, and the water pressure determine the distribution of marine life within these zones.

benthic zone the bottom region of oceans and bodies of fresh water

pelagic zone the region of an ocean or body of fresh water above the benthic zone

Benthic Zones

The shallowest benthic zone lies between the low-tide and high-tide lines and is called the *intertidal zone*. Shifting tides and breaking waves make this zone a continually changing environment for the marine organisms that flourish there.

Most of the organisms that live in the benthic zone live in the shallow *sublittoral zone*. This continuously submerged zone is located on the continental shelves and is populated by organisms such as sea stars, brittle stars, and sea lilies.

The *bathyal zone* begins at the continental slope and extends to a depth of 4,000 m. Because little or no sunlight reaches this zone, plant life is scarce. Examples of animals that live in the bathyal zone are octopuses, sea stars, and brachiopods.

The *abyssal zone* has no sunlight because it begins at a depth of 4,000 m and extends to a depth of 6,000 m. Organisms that live in the abyssal darkness include sponges and worms.

The *hadal zone* is confined to the ocean trenches, which are deeper than 6,000 m below the surface of the water. This zone is virtually unexplored, and scientists think that life in the hadal zone is sparse.

Reading Check **Which benthic zone has the most marine life? Why? (See the Appendix for answers to Reading Checks.)**

Graphic Organizer — Comparison Table

Create the Graphic Organizer entitled "Comparison Table" described in the Skills Handbook section of the Appendix. Label the columns with the different benthic zones. Label the rows with "Depth," "Marine life," and "Sunlight." Then, fill in the table with details about the different benthic zones.

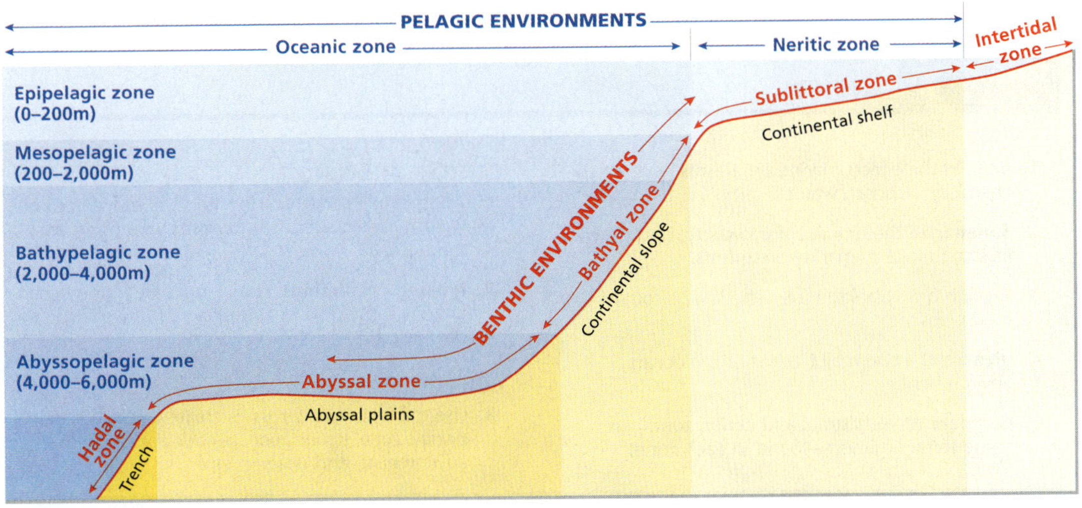

Figure 4 ▶ This diagram shows the classification and location of marine environments.

Graphic Organizer GENERAL

Comparison Table

You may want to use this activity in a game to review material before a test. Divide the class into two teams. Ask students questions about material from the Comparison Table. Give points to each team that provides correct answers.

Answer to Reading Check

Most marine life is found in the sublittoral zone. Life in this zone is continuously submerged, but waters are still shallow enough to allow sunlight to penetrate.

Close

Reteaching — BASIC

Peer Reviewing Divide the class into pairs or small groups. Ask each student to write five questions that can be answered after reading the section. Have students use the questions to quiz each other on the information in the section. **LS Interpersonal**

Quiz — GENERAL

1. In what direction must surface winds blow relative to a coastline for upwelling currents to develop? (parallel to the coastline)
2. Why are the sublittoral and neritic oceanic zones important? (Most marine organisms live within these zones.)
3. Which oceanic zone is the least explored? (the hadal)

ENVIRONMENTAL CONNECTION

Photosynthesis Most organisms derive their energy either directly from photosynthesis or by consuming organisms that engage in photosynthesis. Light is required in order for this process to take place. At best, sunlight can penetrate to a depth of 200 m below the ocean's surface. Thus, most marine organisms can't live too far beneath the ocean surface. Ask students to research some marine organisms that live too far beneath the surface to depend on photosynthesis. Have students present their findings in an illustrated oral report to the class.

MISCONCEPTION ALERT

The Benthic Zone Students may think that the benthic zone consists only of the deepest part of the ocean, because it runs along the bottom. Explain to students that although the benthic zone does include areas of extreme depth, it also includes very shallow regions, such as the intertidal and sublittoral zones. The important concept when thinking about the benthic zone is *bottom-dwelling*. The depth of sea water above the sea floor is not important.

Close, *continued*

Alternative Assessment — GENERAL

Modeling the Ocean Have groups of students use deep, clear plastic containers to create 3-dimensional models of the benthic and pelagic zones. Have them use clay and sand as needed to develop the model. Models of various marine organisms should be placed in appropriate regions, with toothpicks to place organisms within the pelagic zone, well above the sea floor. **LS Kinesthetic/Visual**

Answers to Section Review

1. Marine organisms help maintain the chemical balance of ocean water. They remove some nutrients and gases from ocean water while returning others to the ocean.
2. Upwelling is a process that brings deep, cold, nutrient-rich waters to the surface. Winds that blow parallel to the coastline cause surface waters to move offshore. Deep water then moves upward to replace the coastal surface waters. The nutrients carried by these deep waters are vital for the survival of organisms that live near the ocean surface.
3. Some types of plankton use sunlight to make their own food. Other organisms that cannot produce their own food eat these organisms.
4. the benthic zone and the pelagic zone
5. The sublittoral zone is the benthic environment that lies along the continental shelf. It is populated by sea stars, brittle stars, and sea lilies. The neritic zone is the pelagic zone that is located above the sublittoral zone. This zone has abundant sunlight, moderate temperatures and is populated by fish and crustaceans.
6. Answers may vary. Sample answer: Wind currents can cause upwelling, which brings deep water nutrients to the surface, cools the surface waters and enables more gases to dissolve in the ocean waters.
7. If upwelling decreased, fewer nutrients would reach the surface. Thus, the amount of life would decrease, because organisms depend on deep-water nutrients to live.
8. The *ocean environment* is subdivided into the *benthic zone* and the *pelagic zone*, which contains the *neritic zone* and the *oceanic zone.*

Figure 5 ▶ Fish and marine mammals are examples of organisms that live in the pelagic zone.

SCILINKS. NSTA Developed and maintained by the National Science Teachers Association

For a variety of links related to this chapter, go to www.scilinks.org

Topic: Marine Life
SciLinks code: HQ60912

Pelagic Zones

The region of the ocean above the benthic zone is the pelagic zone. The area of the pelagic zone above the continental shelves is called the *neritic zone*. The neritic zone has abundant sunlight, moderate temperatures, and relatively low water pressure, which are ideal factors for marine life. Nekton fill the zone's waters and are the source of much of the fish and seafood that humans eat.

The *oceanic zone* extends into the deep waters beyond the continental shelf. It is divided into four zones, based on depth. The epipelagic zone is the uppermost area of the oceanic zone. It is sunlit and populated by marine life, such as the dolphins shown in **Figure 5.** The mesopelagic, bathypelagic, and abyssopelagic zones occur at increasingly greater depths. The amount of marine life in the pelagic zone decreases as depth increases.

Section 2 Review

1. **Explain** the effects marine organisms have on the chemistry of ocean water.
2. **Summarize** the process of upwelling, and describe its importance to marine organisms.
3. **Explain** how plankton form the base of ocean food webs.
4. **Identify** the two major zones of the ocean environment.
5. **Compare** the sublittoral and neritic zones, and name some organisms found in each zone.

CRITICAL THINKING

6. **Analyzing Processes** How can the movement of wind currents alter the chemistry of a given area of the ocean?
7. **Making Predictions** How would life in the ocean change if the area of all regions of upwelling decreased?

CONCEPT MAPPING

8. Use the following terms to create a concept map: *pelagic zone, neritic zone, benthic zone, ocean environment,* and *oceanic zone.*

CHAPTER RESOURCES

Chapter Resource File
- Section Quiz GENERAL

Workbooks
- Study Guide (also in Spanish)

ENVIRONMENTAL CONNECTION

Section 3 Ocean Resources

The ocean supplies humans with a number of natural resources. It is a major source of food and minerals, and it provides a means of transportation. Furthermore, the growth of Earth's population has created new interest in the sea as a source of fresh water.

OBJECTIVES

- **Describe** three important resources of the ocean.
- **Explain** the threat water pollution poses to marine organisms.

KEY TERMS

desalination
aquaculture

Fresh Water from the Ocean

The increasing demand for fresh water for things such as drinking water, industry, and irrigation can be met by converting ocean water to fresh water. One way of increasing the freshwater supply is through desalination, as shown in **Figure 1.** **Desalination** is the extraction of fresh water from salt water. Although desalination may provide needed fresh water, the process is generally costly.

Methods of Desalination

One method of desalination is distillation. During *distillation,* ocean water is heated to remove salt. Heat causes liquid water to evaporate and leaves dissolved salts behind. When the water vapor condenses, the result is pure fresh water. However, the process of evaporating liquid water often requires a large amount of costly heat energy.

Another method of desalination is *freezing*. When water freezes, the first ice crystals that form do not contain salt. The ice can be removed and melted to obtain fresh water. This process requires about one-sixth the energy needed for distillation.

Reverse osmosis desalination is a popular method for desalinating ocean water. It includes the use of special membranes that allow water under high pressure to pass through and that block the dissolved salts.

desalination a process of removing salt from ocean water

Figure 1 ▶ After salt water from the Persian Gulf has undergone desalination, much of the resulting fresh water is stored in these towers in Kuwait.

Section 3

Focus

Overview

This section describes ocean resources and their importance to humans. The section identifies ocean resources, including fresh water, mineral and energy resources, and food. Finally, pollution is discussed as a threat to some of these resources.

Bellringer

Ask students to list as many resources as they can think of that may come from the ocean. Ask students to identify whether the resources they list are renewable or nonrenewable. (Answers may vary. Accept all reasonable responses.) **LS Logical**

Motivate

Discussion — GENERAL

Fresh Water from Salt Water

Have small groups of students research desalination and outline key points for discussion. Students should present information about: areas of the world where desalination is used successfully to convert salt water to freshwater; what areas may be likely candidates for desalination; different types of methods used; costs involved in desalination compared to other methods; how human, plant, and animal life would be affected; any barriers other than cost. **LS Verbal**

CHAPTER RESOURCES

Chapter Resource File

- Directed Reading BASIC

Technology

Transparencies
- Bellringer

Student Edition on CD-ROM

One-Stop Planner CD-ROM
- Lesson Plan

Teach

Activity — Basic

Where's the Oil? Have students use the U.S. Department of Energy Web site or other sources to learn where the world's oil reserves are and how much oil they produce. Ask students to select one region of the world and to mark its oil reserves on a map of the world. Instruct students to create a color key that uses several colors, in which each color represents an amount of oil (in thousands of barrels) produced per day. Students should color each oil reserve area on their map according to the data they collect. Have students compare their regional maps and discuss the world's oil reserves. LS Visual/Logical

MATH PRACTICE

Answer

$\$1 \times 4{,}000{,}000\ L/\$0.04 = 100{,}000{,}000\ L$; or $\$1/\$0.04 = 25$; $25 \times 4{,}000{,}000\ L = 100{,}000{,}000\ L$

SKILL BUILDER — General

Vocabulary Students may have heard the word *petrol* used to mean "oil." However, this usage is misleading because the Greek root *petra,* means "stone" or "rock." The same root is found in *petrify,* meaning "to turn to stone" or "to frighten." It is the Latin root *oleum,* that means "oil." So *petroleum,* literally and appropriately means, "rock oil." LS Verbal

English Language Learners

Figure 2 ▶ Offshore oil rigs, such as this one in the Gulf of Mexico, produce about one-fourth of the world's oil.

MATH PRACTICE

Ocean's Gold One cubic kilometer of ocean water contains about 6 kg of gold. If you must process 4 million liters of ocean water to get an amount of gold worth 4¢, how many liters of water would have to be processed to get gold worth $1?

Mineral and Energy Resources

Salt is one mineral resource that can be obtained from the ocean. Other minerals and energy resources can also be extracted from the oceans. While some valuable minerals are easily extracted from the oceans, others are costly or difficult to extract.

Petroleum

The most valuable resource in the ocean is the petroleum found beneath the sea floor. Offshore oil and natural gas deposits exist along continental margins around the world. About one-fourth of the world's oil is now obtained from offshore wells, such as the one shown in **Figure 2.** As a result of new drilling techniques, oil and gas can be extracted far offshore and from great depths.

Nodules

Potato-shaped lumps of minerals, called *nodules,* are found on the abyssal floor of the ocean. Nodules are a valuable source of manganese, iron, copper, nickel, cobalt, and phosphates. However, the recovery of nodules is expensive and difficult because they are located in very deep water. Because country borders are observed only close to land, the question of who has the right to mine minerals from the ocean floor has not been answered.

Trace Minerals

The ocean is also the main source of magnesium and bromine. However, the concentration of most other useful chemicals that are dissolved in the oceans is very small. The extraction of minerals found only in trace amounts is too costly to be practical.

INCLUSION Strategies

• *Gifted and Talented*

Give gifted and talented students a chance to expand their knowledge and to share their information. Ask them to explore common foods that are surprisingly connected to the world's oceans. For example, seaweed is added to ice cream to prevent ice crystals from forming. Have them present the results of their research as an oral report to the class. LS Verbal/Auditory

ENVIRONMENTAL CONNECTION

Fresh Water Sources Present students with a scenario in which the fresh water resources of a coastal community are so polluted that they will be unusuable for the next 50 years. Have students discuss alternative sources of water. (Students may suggest bringing in water by boat or truck, producing water by chemical processes, or removing the salt from ocean water.) Use this activity to begin a discussion of conservation and stewardship.

Food from the Ocean

Of all of the resources that the ocean supplies, the one in greatest demand is food. Seafood, which is an important source of protein, can be harvested through fishing or through aquaculture.

Fishing

Because fish are a significant food source for people around the world, fishing has become an important industry. But when the ocean is overfished, or overharvested, over a long period of time, fish populations can collapse. A collapse may damage the ecosystem and threaten the fishing industry. To prevent overharvesting, many governments have passed laws to manage fishing.

Aquaculture

Another way to deal with the high demand for seafood is by farming aquatic life. **Aquaculture** is the raising of aquatic plants and animals for human use or consumption. Catfish, salmon, oysters, and shrimp are already grown on large aquatic farms. Similar methods may be used to breed fish and seaweed in ocean farms, such as the one shown in **Figure 3.** A major problem for aquaculturalists is that the ocean farms are susceptible to pollution and that the farms may be a local source of pollution.

Under the best conditions, an ocean farm could produce more food than an agricultural farm of the same size does. For example, in agriculture, only the top layers of soil can be used. In contrast, ocean farms may use a wide range of depths to produce food. Someday, the nutrient-rich bottom water may be pumped to the surface as a way of fertilizing aquatic farms.

Reading Check **List the benefits and problems of aquaculture. (See the Appendix for answers to Reading Checks.)**

SCILINKS. NSTA Developed and maintained by the National Science Teachers Association

For a variety of links related to this chapter, go to www.scilinks.org

Topic: Ocean Resources
SciLinks code: HQ61065

aquaculture the raising of aquatic plants and animals for human use or consumption

Figure 3 ▶ Aquaculture establishments, such as this seaweed farm in Madagascar, provide a reliable, economical source of food.

Internet Activity — ADVANCED

The Romance of Aquaculture
Several organizations, such as the World Aquaculture Society and Aquaculture for Youth and Youth Educators, have Web sites where students can explore aquaculture in greater depth. Invite students to learn about the latest techniques, issues related to food production, and aquaculture related projects. Have them report on some aspect of aquaculture that interests them or carry out one of the suggested projects. A worksheet designed to direct student research on this topic can be found in the **Chapter Resource File** booklet or by visiting **go.hrw.com** and entering the keyword **HQ6OWAX.** LS **Interpersonal**

Close

Reteaching — BASIC

Summarizing Ask students to list each of the ways humans make use of the oceans and to write a brief summary of the benefits derived from each use. (Answers may vary but should include desalination to provide fresh water, petroleum to provide energy, minerals for various purposes, and fishing and aquaculture to provide food.) LS **Verbal**

Quiz — GENERAL

1. Why is desalination not widely used as a method of providing fresh water? (because it is expensive)
2. What is the connection between oceans and petroleum? (About 1/4 of the world's oil is obtained from deposits beneath the sea floor.)
3. Why is aquaculture potentially important to society? (It is an additional food source for the growing human population.)

CHAPTER RESOURCES

Chapter Resource File

- **Inquiry Lab** Oil Spill! GENERAL

Internet Activity

- The Romance of Aquaculture ADVANCED

Answer to the Reading Check

Aquaculture provides a reliable, economical source of food. However, aquatic farms are susceptible to pollution and they may become local sources of pollution.

Close, *continued*

Alternative Assessment — ADVANCED

Debate Have students research and engage in a debate on ocean oil drilling. Have them answer the following questions: What are ocean drilling's effects on the oceans, their marine life, and other resources? What, if any, are the risks to fishing and aquaculture? Can humans continue to drill for off-shore oil and expand their efforts to fish and farm the oceans? You may wish to assign students to prepare arguments for and against off-shore oil drilling. **LS Logical**

Answers to Section Review

1. Distillation involves heating water to cause evaporation and then condensing the vapor to yield pure water. Freezing causes crystals of pure water to form. This ice is collected and melted to yield pure water. In reverse osmosis, sea water is forced through special membranes that let pure water through but block the salts.
2. Distillation can be expensive because it requires costly energy to heat the water until it evaporates.
3. Answers may vary but should include two of the following: salt, manganese, iron, copper, nickel, cobalt, phosphates, magnesium, and bromine.
4. petroleum
5. Aquaculture is the raising of aquatic plants and animals for human use and consumption. It is important because it is reliable, economical source of food.
6. because they are closest to the likely sources of pollution
7. Nodules are normally found in deep waters outside of national boundaries, so there may be problems determining who has the rights to them.
8. Pollution can destroy fish populations, thus destroying the fishing industry.
9. Mercury in the microscopic organisms becomes concentrated in the bodies of marine organisms that eat the microscopic organisms. The concentrated mercury will end up in human bodies when humans eat these larger marine animals.
10. *Ocean resources* include *petroleum;* food, produced by *fishing* and *aquaculture,* which are threatened by *pollution;* and fresh water, which is produced by *desalination,* in which *salt* is removed from sea water by *distillation, freezing,* and *reverse osmosis.*

Figure 4 ▶ Pollution can damage the ocean's ecosystem and make seafood unsafe to eat.

Ocean-Water Pollution

The oceans have been used as a dumping ground for many kinds of wastes including garbage, sewage, and nuclear waste. Until recently, most wastes were diluted or destroyed as they spread throughout the ocean. But the growth of the world population and the increased use of more-toxic substances have reduced the ocean's ability to absorb wastes and renew itself.

Productive coastal areas and beaches are in the greatest danger of being polluted because they are closest to sources of pollution, as shown in **Figure 4.** Pollution has destroyed clam and oyster beds, sea birds have become tangled in plastic products, and beaches have been closed because of sewage and oil spills.

Besides being found in coastal waters, pollutants can be found in most other areas of the oceans. Traces of mercury, of the insecticide DDT, and of lead from gasoline have been detected in the ocean. In some areas of the world, concentrations of pollutants are so high that the fish have become unsafe for humans to eat. Recognizing the affects of dumping waste in the ocean, scientists and governments have been working to reduce pollution. For example, the use of DDT has been banned in the United States and the use of leaded gasoline has been reduced.

Section 3 Review

1. **Describe** three methods of desalinating ocean water.
2. **Explain** why distillation can be an expensive method of desalination.
3. **List** two important mineral resources in the ocean.
4. **Identify** the most valuable resource that can be obtained from the ocean.
5. **Define** the term *aquaculture,* and explain why aquaculture is important.
6. **Explain** why beaches are especially vulnerable to ocean pollution.

CRITICAL THINKING

7. **Making Inferences** Describe how the mining of nodules may create problems between countries.
8. **Predicting Consequences** How would pollution of oceans affect the fishing industry?
9. **Analyzing Relationships** How could humans be affected if microscopic marine organisms absorb small amounts of mercury?

CONCEPT MAPPING

10. Use the following terms to create a concept map: *desalination, distillation, freezing, reverse osmosis, petroleum, aquaculture, fishing, salt, ocean resource,* and *pollution.*

CHAPTER RESOURCES

Chapter Resource File
- Section Quiz GENERAL

Workbooks
- Study Guide (also in Spanish)

Chapter 20 Highlights

Sections

1 Properties of Ocean Water

Key Terms

salinity, 496
pack ice, 497
thermocline, 498
density, 499

Key Concepts

- Cold ocean water dissolves gases more readily than warm ocean water does.
- The ocean is a carbon sink that dissolves CO_2 from the atmosphere.
- Dissolved solids make up 3.5% of the mass of ocean water.
- Salinity is a measure of the amount of dissolved salts in ocean water.
- Temperature of ocean water is dependent on depth and latitude.
- Density of ocean water is dependent on temperature and salinity.
- The color of ocean water is affected by the presence of phytoplankton.

2 Life in the Oceans

upwelling, 502
plankton, 502
nekton, 502
benthos, 502
benthic zone, 503
pelagic zone, 503

- Marine organisms help maintain the chemical balance of ocean water by using nutrients for life processes and by returning the nutrients to the water after death.
- Plankton form the base of complex ocean food webs by acting as food for other marine organisms.
- There are two major zones of life in the ocean: benthic and pelagic. Each zone supports different types of organisms.

3 Ocean Resources

desalination, 505
aquaculture, 507

- The ocean is valuable as a source of fresh water, minerals, and food.
- Fresh water can be obtained from the ocean by the methods of desalination, freezing, and reverse osmosis desalination.
- Ocean-water pollution threatens both marine organisms and humans by damaging food resources in the ocean.

Alternative Assessment — GENERAL

Ocean Video Have students work in groups of four or five to produce plans for an animated video or film about the concepts covered in this chapter. Have students prepare a detailed presentation of their plans that they might give to a video producer. The proposal should show how they will cover the concepts using techniques they have seen in films and videos. For example, invite them to create animated characters to narrate the video and to plan illustrations, graphs, and charts to present the material. Then, have them "pitch" their video to the class. If possible, they should show sketches and samples of visual materials they plan to use, including storyboards. They may want to suggest and play appropriate music and sound effects for various parts of the video. LS Visual/Auditory

CHAPTER RESOURCES

Chapter Resource File

- Math Skills GENERAL
- Concept Review GENERAL
- Critical Thinking ADVANCED
- Chapter Test A GENERAL
- Chapter Test B ADVANCED

Workbooks

- Study Guide (also in Spanish)
- Assessments (Spanish)

Technology

Classroom Videos
- Brain Food Video Quiz

Group Activity — GENERAL

Brine Shrimp Have groups of students gather information from the library or a local aquarium store on raising brine shrimp. Have them culture and raise a batch of brine shrimp by using what they have learned. Ask students to examine the shrimp under a microscope at various developmental stages and to draw what they see. Then, have students produce an exhibit that includes their brine shrimp habitat, drawings, and the information they have gathered. LS Kinesthetic/Visual

Chapter Review

Assignment Guide

Section	Questions
1	1–2, 4–7, 9–14, 18–19, 21–23, 25–29, 31–38
2	3, 8, 15–17, 20, 24, 30

Using Key Terms

1–7. Answers may vary but should show that students understand the definitions of and differences between key terms.

Understanding Key Concepts

8. a **9.** b
10. d **11.** a
12. c **13.** a
14. d **15.** b
16. c

Short Answer

17. Upwelling is a process that brings deep, cold, nutrient-rich waters to the surface. Winds parallel to the coastline cause surface waters to move offshore. Deep water then moves upward to replace the coastal surface waters. The nutrients carried by these deep waters are vital for the survival of organisms living near the ocean surface.

18. chlorine, sodium, magnesium, sulfur, calcium, and potassium

19. Density increases as ocean water temperature decreases and/or as salinity increases.

20. Oil is a serious pollutant. Fish become covered in oil and then die. Thus, an oil spill would be harmful to the fishing industry.

21. Answers may vary. Sample answer: Petroleum is obtained by drilling into oil deposits beneath the sea floor. Fish and other foods are acquired by fishing and aquaculture. Salt is obtained by evaporating ocean water.

22. Humans are at the top of the food chain. When humans eat marine organisms that have taken in pollutants, such as mercury, these pollutants enter our bodies and can make us sick.

Chapter 20 Review

Using Key Terms

Use each of the following terms in a separate sentence.

1. *thermocline*

2. *upwelling*

3. *desalination*

For each pair of terms, explain how the meanings of the terms differ.

4. *salinity* and *density*

5. *plankton* and *nekton*

6. *benthic zone* and *pelagic zone*

7. *upwelling* and *aquaculture*

Understanding Key Concepts

8. The amount of dissolved salts in ocean water is called the water's
a. salinity. **c.** plankton.
b. nekton. **d.** density.

9. When liquid water is warmed, its density
a. increases.
b. decreases.
c. remains the same.
d. doubles.

10. Although most of the various wavelengths of visible light are absorbed by ocean water, the one wavelength that is most often reflected is the color
a. violet. **c.** yellow.
b. green. **d.** blue.

11. Drifting marine plants and animals are known as
a. plankton. **c.** nekton.
b. benthos. **d.** sea stars.

12. Marine animals that can swim to search for food and avoid predators are called
a. phytoplankton. **c.** nekton.
b. zooplankton. **d.** benthos.

13. Which of the following ocean environments experiences the most change?
a. intertidal zone **c.** bathyal zone
b. abyssal zone **d.** neritic zone

14. Which of the following methods is *not* used for producing fresh water by desalinating ocean water?
a. distillation **c.** reverse osmosis
b. evaporation **d.** aquaculture

15. Lumps of minerals on the ocean floor are called
a. nekton. **c.** benthos.
b. nodules. **d.** plankton.

16. Aquaculture is another name for
a. desalination.
b. distillation.
c. ocean farming.
d. rapid temperature changes.

Short Answer

17. Describe the process of upwelling, and explain its effects on marine life.

18. What are the six most abundant elements dissolved in ocean water?

19. How are temperature, salinity, and density related?

20. Describe how an oil spill would affect a fishing industry.

21. List three important resources from the ocean, and describe how they are obtained.

22. What effects does ocean pollution have on humans?

Chapter Review

Critical Thinking

23. Predicting Consequences If climatic conditions over Earth's oceans caused upwelling and wave action to stop, what would happen to marine life? Explain your answer.

24. Identifying Relationships How would a significant and global decrease in sunlight affect plankton and other marine organisms?

25. Applying Concepts If you were to start an aquatic farm, in which of the zones of marine life would you locate your farm? Explain your answer.

26. Making Inferences When oceanographers first explored the deep-ocean basin along mid-ocean ridges, they discovered a variety of marine life, including sightless crabs. Explain why sightlessness is not a disadvantage to these crabs.

Concept Mapping

27. Use the following terms to create a concept map: *fishing, marine life, plankton, fish, color, dissolved gas, dissolved solid, desalination, salt, ocean water characteristics,* and *aquaculture*.

Math Skills

28. Using Equations Using the equation *density = mass ÷ volume*, determine the mass of a 3 cm^3 sample of ocean water if the water's density is 1.027 g/cm^3.

29. Making Calculations What percentage of dissolved salts would be present in water that has a salinity of 40‰?

30. Making Calculations A 1,000 g sample of ocean water contains 35 g of dissolved solids. Magnesium makes up 7.7% of the 35 g of dissolved solids. How many grams of magnesium are in the 1,000 g sample of ocean water?

Writing Skills

31. Creative Writing Write a descriptive essay about the deep-ocean waters of the oceanic zone. The essay should include a description of the marine organisms in this zone as well as a description of what life is like for the marine organisms in this zone.

32. Writing from Research Research the new foods that are being produced through aquaculture and the nations that are investing in this method of farming. Write a short essay that describes these foods, where they are grown, and their nutritional values.

Interpreting Graphics

The graph shows the depths at which different wavelengths of light penetrate ocean water. Use this graph to answer the questions that follow.

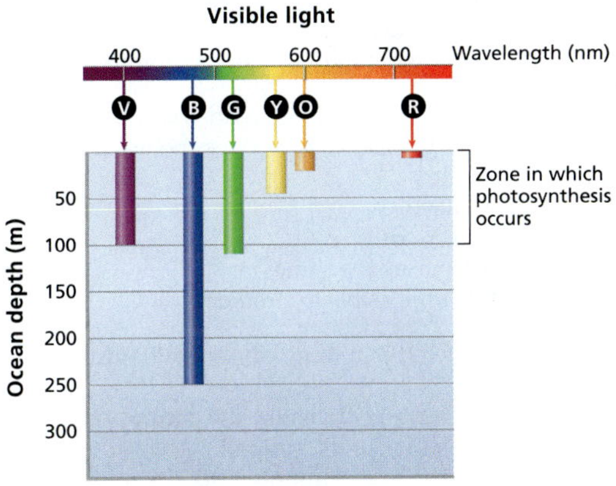

33. Estimate the depth at which yellow light can no longer penetrate ocean water.

34. Which colors can penetrate the ocean at a depth of 50 m?

35. Would an object that is painted red appear red at a depth of 50 m? Explain your answer.

Critical Thinking

23. Many forms of marine life would die because they would lack the nutrients that are stored in deep water and are supplied by upwelling. This would adversely affect all organisms in the food chain, including humans.

24. Without sunlight, plankton that form the base of food webs in the ocean would die, and so would the small marine life that feeds on them. The larger marine animals would, in turn lack nourishment and die.

25. In the sublittoral and neritic zones. Organ-isms that live in these zones are continu-ously submerged, yet waters are shallow. Sunlight is abundant and temperatures are moderate. These are the most hospitable zones to marine life.

26. Because no sunlight reaches the deep-ocean basin, vision would not be useful to these crabs.

Concept Mapping

27. Answers may vary but should include all of the terms listed. Sample answers appear at the end of this Teacher Edition.

Math Skills

28. D = m/v can be rearranged to m = Dv. So, m = 1.027 g/cm^3 × 3 cm^3 = 3.081 g

29. 40‰ = 40 parts/1000 = 4parts/100 = .04 = 4%

30. 7.7% = 0.077; 35 g × 0.077 = 2.695 g magnesium

Writing Skills

31. Answers may vary. Accept all reasonable answers.

32. Answers may vary. Accept all reasonable answers.

Interpreting Graphics

33. about 50 meters

34. green, blue, and violet

35. No; red wavelengths do not penetrate that deeply, so the object would not look red.

Estimated Time

To give students practice under more realistic testing conditions, allow them 30 minutes to answer all of the questions in this practice test.

TEST DOCTOR

Question 2 Answer I is correct. Adding fresh water would not remove salt, but it would lower the salt concentration. During *distillation*, answer F, water is heated to remove salt. During *freezing*, answer G, the first ice crystals that form do not contain salt and are removed and melted to obtain fresh water. During *reverse osmosis*, answer H, water is forced, under pressure, through a membrane that allows water, but not salt, to pass through.

Question 9 Answer G is correct. The passage indicates that, normally, weather patterns in the Pacific move from the east to the west. During an El Niño event, this pattern changes. Warm waters move from the west to the east, and bring storms with them. Using their knowledge of geography, students should realize that this pattern pushes these storms toward the United States. Answers F, H, and I contain information that cannot be inferred from the passage.

Chapter 20 Standardized Test Prep

Understanding Concepts

Directions (1–5): **For *each* question, write on a separate sheet of paper the letter of the correct answer.**

1. Organisms that live on the ocean floor are called
   A. benthos
   B. nekton
   C. plankton
   D. phytoplankton

2. Which of the following cannot be used to remove salt from sea water to make the water safe for drinking?
   F. distillation
   G. freezing
   H. reverse osmosis
   I. adding fresh water

3. The temperature of ocean water is dependent on all of the following except
   A. depth
   B. the amount of solar energy it receives
   C. water movement
   D. the number of organisms living in it

4. As the temperature of ocean water increases from 10°C to 30°C, how does the water's density change?
   F. It increases.
   G. It decreases.
   H. It remains the same.
   I. It is impossible to predict.

5. Barriers to the mining of mineral nodules include
   A. that mining rights for the ocean floor have not yet been determined.
   B. that they contain only traces of minerals and therefore are not worth the effort to gather.
   C. that they are readily accessible and therefore not valuable.
   D. that they primarily contain elements that are dangerous to humans.

Directions (6–7): **For *each* question, write a short response.**

6. What is the cause of deep ocean currents?

7. What is the name for the top layer of ocean water that extends to 300 m below sea level?

Reading Skills

Directions (8–10): **Read the passage below. Then answer the questions.**

The Effects of El Niño

The interaction between the ocean and the atmosphere can profoundly affect weather conditions. Occurring, on average, every four years and lasting about 18 months, El Niño is one event that triggers global weather changes. El Niño is characterized by changes in wind patterns that allow warmer water from the western Pacific Ocean to surge eastward. Normally, east-to-west winds cause warm water to accumulate in the western Pacific Ocean. During El Niño, the trade winds shift warm water east. Sea surface temperatures from the coast of Peru to the equatorial central Pacific rise. The warm waters cause the thermocline to sink and contribute to the formation of convective clouds, which cause heavy rains that shift eastward at the same rate as the waters. In areas on the western coast of the Pacific Ocean, droughts become common.

8. According to the passage, which of the following statements is not true?
   A. During El Niño, the trade winds shift warm water eastward.
   B. El Niño is one event that triggers global weather changes.
   C. An El Niño weather event lasts about two years on average.
   D. An El Niño event lead to the formation of convective clouds that shift eastward.

9. Which of the following statements can be inferred from the reading passage?
   F. An El Niño is usually followed by a weather event that moves cold water westward.
   G. The changes caused by El Niño directly affect the weather in the United States.
   H. El Niño causes severe disruptions to international trade and travel.
   I. El Niño weather cycles are a relatively recent phenomenon.

10. During an El Niño weather event, what happens to the thermocline and what effect might this have on upwelling?

Answers

Understanding Concepts

1. A
2. I
3. D
4. G
5. A
6. the flow of cold, dense polar water beneath warm water toward the equator
7. the surface zone

Reading Skills

8. C
9. G
10. The thermocline sinks. A deeper thermocline limits the nutrient-rich cold water that is returned by upwelling.

Interpreting Graphics

11. G
12. All parts of the oceans are connected, which causes a continuous mixing of waters and dissolved solids.
13. Answers may vary. See Test Doctor for a detailed scoring rubric.

Interpreting Graphics

Directions (11–13): **For each question below, record the correct answer on a separate sheet of paper.**

Base your answers to questions 11 and 12 on the pie graph below.

Solids Dissolved in Ocean Water

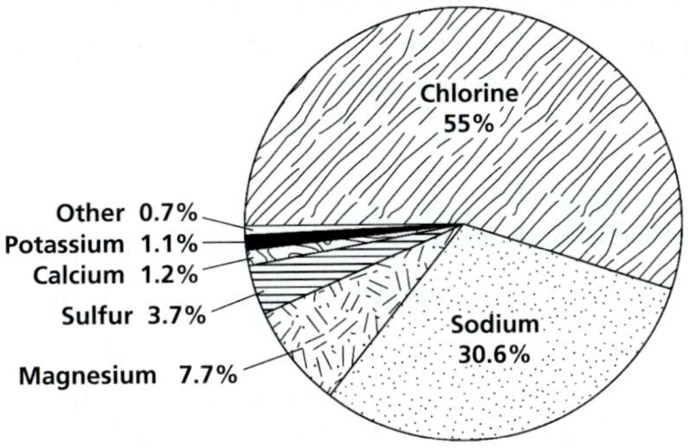

11 The two elements that make up the largest percentage of the dissolved solids combine to make what common solid found in ocean water?

F. sand
G. salt
H. siliceous ooze
I. calcareous ooze

12 While the salinity of ocean waters varies from one area to another, the proportions among the different solids dissolved in ocean water do not change. What is the reason for this equilibrium?

Base your answer to question 13 on the diagram below, which shows the basic mechanics of upwelling.

Diagram of Upwelling

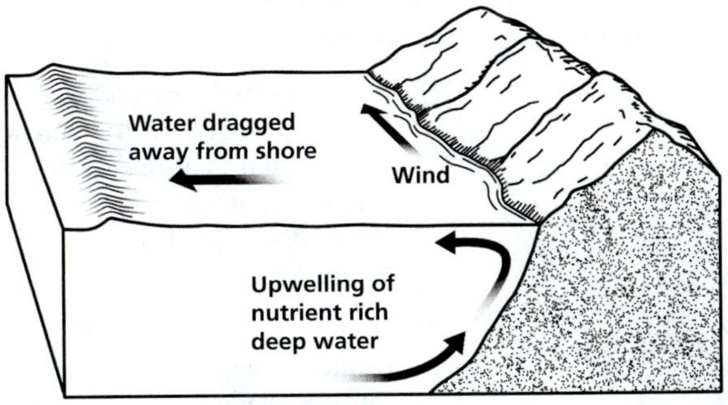

13 During summer months, beaches may sometimes close because of the appearance of large phytoplankton blooms. How might an upwelling contribute to these beach closings during warmer months?

Test TIP

Scan the answer set for words such as "never" and "always." Such words are often used in statements that are incorrect because they are too general.

Standardized Test Prep

TEST DOCTOR

Question 11 Answer G is correct. Salt, which is made up of sodium and chloride ions, is the most abundant of dissolved solids in ocean water. The pie graph breaks down solids into their most basic elements, but salt is the most abundant dissolved compound. The scientific name for salt is sodium chloride, which alludes to the elements that compose the substance.

Question 13 Full-credit answers should include the following points:

- nutrients are returned to the surface during upwelling
- an understanding that these nutrients support life near the surface of the ocean
- this may lead to an overgrowth of phytoplankton, which contributes to the "dirtiness"
- a conceptual understanding of how living and natural systems are connected and may affect one another

Test Prep Correlations ***National Science Education Standards***

ES 1c: item 6
ES 2a: items 11, 12
LS 4c: item 1
LS 4d: item 13
LS 5e: items 10, 13
PS 2c: item 11
SAI 2e: items 4, 6, 10, 12, 13
SPSP 3a: item 2
SPSP 4a: items 8, 9, 13
SPSP 5a: items 8, 9
SPSP 6b: item 5
SPSP 6c: item 5
ST 2b: item 2
ST 2d: item 5
ST 2b: item 2
UCP 1: items 8, 9, 10, 12

CHAPTER RESOURCES

State Resources

For specific resources for your state, visit go.hrw.com and type in the keyword **HSHSTR.**

Ocean Water Density

Teacher's Notes

Time Required
two 45-minute class periods

Lab Ratings
EASY → HARD

Teacher Preparation: 1

Student Setup: 2

Concept Level: 3

Cleanup: 1

Skills Acquired
- Observing
- Measuring
- Collecting Data
- Analyzing
- Inferring

The Scientific Method
In this lab, students will
- Make Observations
- Analyze the Results
- Draw Conclusions

Materials
The materials listed on the page are enough for groups of two to four students.

Tips and Tricks
You may want to tell students that the instrument they are making from a straw and clay is called a *hydrometer*. Marking the straw with the grease pencil may be easiest if one student grasps the straw gently just above the water line while another student marks the straw just below the first student's fingers.

Objectives
- **Measure** the temperature and density of water.
- USING SCIENTIFIC METHODS **Analyze** the effects of temperature and salinity on the density of water.

Materials
- beaker, 250 mL
- clay, modeling
- freezer (optional)
- gloves, heat-resistant
- graduated cylinder, 100 mL
- hot plate
- pencil, grease, red
- pencil, grease, yellow
- ruler, metric
- scissors
- straw, plastic
- table salt
- teaspoon
- thermometer
- water, distilled

Safety

Step 8

Ocean Water Density

The density of ocean water varies. Density is affected by the salinity of the water and the temperature of the water. Furthermore, the salinity of an area of the ocean is affected by the rate of evaporation or freezing and by the addition of fresh water and salts. The temperature of the ocean is determined by the amount of solar radiation that reaches Earth's surface. In this lab, you will observe the effects of temperature and salinity on the density of salt water.

PROCEDURE

1. Make a hydrometer by filling 5 cm of one end of the straw with modeling clay.
2. Pour 100 mL of distilled water at room temperature into a glass jar or beaker. Float the straw upright in the jar. If the straw does not float upright, cut off the open end at 1 cm intervals until it floats upright.
3. Use a red grease pencil to mark the water level on the straw. Remove the straw from the water, and draw a continuous line around the straw at the mark.
4. Use a yellow grease pencil to draw lines around the straw at 1 cm intervals above and below the red line. The red line will be used as a reference point.
5. Add 2 tsp of salt to the water, and stir until all of the salt has dissolved. Draw a table similar to **Table 1**. Measure and record the water temperature in **Table 1.**
6. Place the hydrometer in the salt water. In **Table 1,** record the density by counting the marks above or below the red line to the water's surface. The higher the hydrometer floats out of the water, the more dense the water.
7. Turn the hot plate on low. Place the beaker of salt water on the hot plate. **CAUTION** Wear heat-resistant gloves.
8. Hold a thermometer in the water. Do not let the thermometer touch the bottom of the beaker.

CHAPTER RESOURCES

Chapter Resource File
- Datasheet for Chapter Lab GENERAL
- Lab Notes and Answers

Workbooks

Long-Term Projects
- Water Clarity ADVANCED

9. When the water's temperature reaches 25°C, turn off the hot plate. Immediately place the hydrometer in the water. Record the relative density in **Table 1.**

10. Repeat steps 7–9, and heat the water until it is 30°C.

11. Turn the hot plate on high. Heat the salt water until it begins to boil. Boil the water for 5 min. Turn off the hot plate.

12. Place the hydrometer in the water. Draw a table similar to **Table 2**. Measure and record the water's density in **Table 2.**

13. Boil the water for another 5 min. Measure and record the water's density.

14. Repeat step 13. Once the water is cool, measure the amount of water that remains in the beaker.

Temperature (°C)	*Density (cm above or below the red line)*
25	
30	

Table 1

Minutes of boiling	*Density (cm above or below the red line)*
5	
10	
15	

Table 2

ANALYSIS AND CONCLUSION

1. **Analyzing Data** In which trial was the water the most dense? the least dense? Explain your answers.

2. **Identifying Trends** As the temperature of the water increases, does the water's density increase or decrease?

3. **Making Inferences** Based on your observations, infer the density of polar ocean waters, and compare it with the density of equally saline water near the equator. Explain your answer.

4. **Analyzing Processes** Why did the amount of water in the beaker change? Explain why boiling the water affected its density.

5. **Forming a Hypothesis** How would you expect the density of the water to change if the water was frozen instead of boiled? Explain your answer.

Extension

1. **Evaluating Hypotheses** Place a beaker of salt water in a freezer until a crust of ice forms. Break up and remove the ice from the water, and record the density of the remaining water. Is the water denser or less dense than before it was frozen? Explain.

Skills Practice Lab

Answers to Analysis and Conclusion

1. The final trial, after step 14, yielded the densest water because the most water had evaporated, leaving a higher percentage of salt in the remaining solution. The least dense water is that at 30°C, because it is not hot enough to evaporate but is warmer, and so less dense, than the 25°C water.
2. Density decreases as temperature increases, until the water starts to evaporate as it boils.
3. Polar waters will be denser than equally saline water near the equator because they are colder.
4. The amount of water changed because some water evaporated as it boiled. The remaining water became more saline because the salt remained behind.
5. If some of the water is converted into ice, the remaining water will become more saline and thus, denser, like the water did when it was boiled. However, the freezing water will be denser than the hot water because colder water is more dense than warm water.

Answers to Extension

1. denser; The ice contains no salt, so the remaining water contains more salt than before some water froze. In addition, it is cooler than it was before it was frozen.

TEACHER TESTED & APPROVED

Scott Robertson
North Warren Central School
Chestertown, NY

Sea Surface Temperatures in August

Discussion — ADVANCED

Thermal Energy Show students a global solar energy chart (available on the Internet on NASA's Web site and others) and prompt them to identify areas where discrepancies exist between the solar energy map and the sea surface temperature map on this page. Ask them to hypothesize reasons for any discrepancies they find. (Ocean currents may distribute thermal energy in a pattern that differs from the patterns of insolation intensity.) **LS Visual**

Answers to Map Skills Activity

1. 20–25°C
2. Answers may vary, but should include regions where sea surface temperatures are warmest. These are along and up to 20°S of the equator in the Eastern Pacific Ocean and Eastern Atlantic Ocean.
3. Answers may vary. Pack ice is most likely found where sea surface temperatures are coldest—near the North Pole and along the coast of Antarctica.
4. North Africa
5. Answers may vary. Sample answer: I would expect density to increase as surface temperature decreases, except in areas where evaporation is high.
6. Answers may vary. Sample answer: I would expect the water density to increase in December as the water got colder, provided that the evaporation rate did not change drastically.

CHAPTER RESOURCES

Technology

Transparencies
- 100 Sea Surface Temperatures in August (with worksheet)

MAPS in Action

Sea Surface Temperatures in August

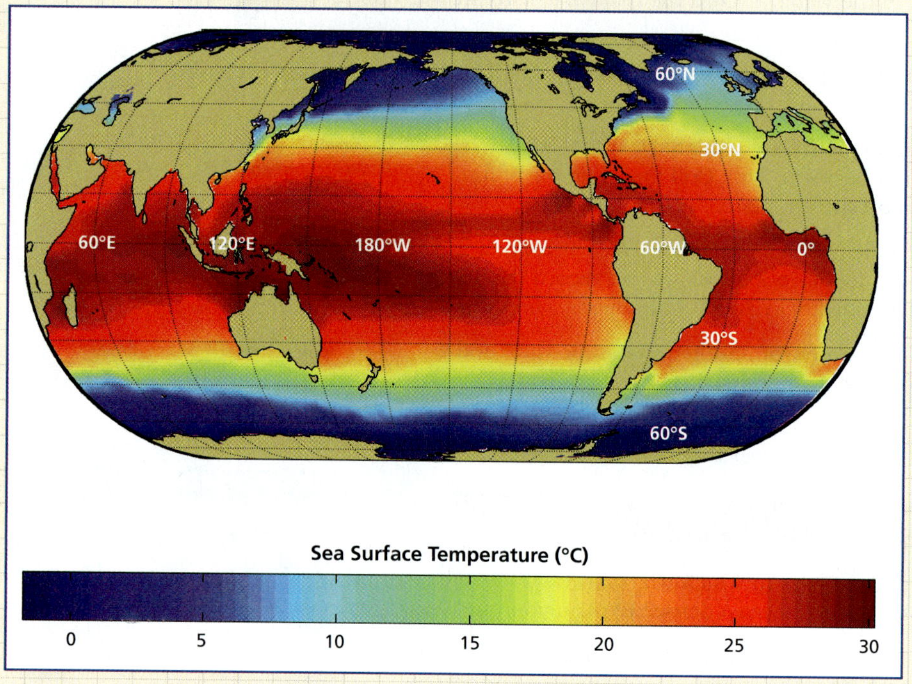

Map Skills Activity

This map shows global sea surface temperatures in the month of August. Latitude and longitude are shown in 10° intervals. The latitude of New York City is 41°N and 74°W. Use the map to answer the questions below.

1. **Analyzing Data** Estimate the sea surface temperatures off the east coast of Florida.
2. **Identifying Relationships** Identify areas of the globe that receive the most solar energy.
3. **Inferring Relationships** At which locations would you most likely find pack ice?
4. **Making Comparisons** Where would you expect to find higher surface salinity values, off the coast of North Africa or off the coast of the southern tip of South America?
5. **Analyzing Relationships** As latitude increases, surface temperatures decrease. How would you expect the density of surface water to change as latitude increases? Explain your answer.
6. **Making Predictions** Off the coast of New York City in December, would you expect the density of the ocean waters to increase or decrease compared to the density of the waters in August? Explain.

EYE on the Environment

A Harbor Makes a Comeback

Not long ago, the harbor in Boston, Massachusetts, was one of the most polluted harbors in the United States. In recent years, however, Boston Harbor has made one of the most impressive comebacks in history.

Polluted Since Colonial Times

During America's colonial period, residents of the city of Boston dealt with their sewage by dumping it directly into Boston Harbor. They hoped that the tides would carry the sewage out to sea. However, much of the raw sewage remained in the harbor. For hundreds of years, waste and sewage washed up along the shore and sometimes caused outbreaks of disease. Over time, most of the species of marine life disappeared from the harbor.

In the late 1960s, a sewage-treatment plant was built to treat the city's sewage before it was dumped into the harbor. Unfortunately, the plant could not meet the demands of the fast-growing city. Even as late as the 1980s, millions of tons of sewage flowed directly into the harbor every year. Beaches were closed, and the shell-fishing industry in Boston Harbor was shut down. Divers reported that a "gray, mayonnaise-like substance" covered the harbor's floor.

Taking Action

In the mid 1980s, a new sewage-treatment complex was built at Deer Island, which lies in the outskirts of the harbor. By 1995, the beaches were reopened and were significantly safer. Plants, fish, and other organisms returned to the harbor, and the ocean floor improved.

The task of protecting the harbor is far from over, however. Officials continue to monitor the sewage-treatment process. In addition, officials are working to expand the cleanup project to include surrounding watersheds.

▲ For many years, Boston Harbor was known as the most polluted harbor in the world.

◀ The Deer Island sewage-treatment facility is one of the largest facilities of its kind in the world. The large, egg-shaped tanks are used in the primary treatment of sewage.

Extension

1. **Determining Cause and Effect** Why would the health of Boston Harbor depend on the health of the surrounding watersheds?
2. **Research** Write a brief summary of how sewage from your area is treated and disposed and how long the current sewage-treatment system has been used.

Eye on the Environment

A Harbor Makes a Comeback

Group Activity — ADVANCED

Curing Pollution in Boston Harbor Divide the class into three groups. Ask the first group to research the pre-1960 history of the pollution in Boston Harbor, focusing on the impact pollution had on human health and on the fishing industry. Ask the second group to identify and research the various programs, completed since 1960, that were designed to alleviate the harbor pollution. Ask the third group to research the current state of the harbor and to identify how long the harbor may take to return to good health. In addition to using the Internet, students should be encouraged to consult local sources such as historical societies, city officials, pollution control agencies, recreational interests, and fishing interests. Then, have each group prepare a presentation for the class. LS Interpersonal/Verbal Co-op Learning

Answers to Extension

1. Sample answer: Surrounding watersheds contain rivers that flow into Boston Harbor. If the waters in these watersheds are polluted, those pollutants will eventually reach Boston Harbor and compromise efforts to clean up the harbor.
2. Answers may vary. Accept all reasonable answers.

Chapter 21

Movements of the Ocean

Planning Guide

Compression Guide

To shorten instruction because of time limitations, omit Section 2.

OBJECTIVES	LABS, DEMONSTRATIONS, AND ACTIVITIES	TECHNOLOGY RESOURCES
PACING • 45 min pp. 518–524 **Chapter Opener**	LTP **Long-Term Project** Tides at the Shoreline* ADVANCED	OSP **Parent Letter** CD **Student Edition on CD-ROM** CD **Chapter Summaries Audio CD** VID **Brain Food Video Quiz**
Section 1 Ocean Currents • Describe how wind patterns, the rotation of Earth, and continental barriers affect surface currents in the ocean. • Identify the major factor that determines the direction in which a surface current circulates. • Explain how differences in the density of ocean water affect the flow of deep currents.	TE **Discussion** Beachcombing, p. 519 GENERAL TE **Demonstration** Modeling Ocean Currents, p. 520 GENERAL TE **Geography Connection** Research, p. 521 ADVANCED SE **Quick Lab** Ocean Currents, p. 522 GENERAL CRF **Datasheet for Quick Lab*** GENERAL TE **History Connection** Research, p. 522 GENERAL TE **Biology Connection** Vertical Ocean Movements, p. 523 GENERAL SE **Maps in Action** Roaming Rubber Duckies, p. 542 GENERAL TE **Activity** Exploring Water Power, p. 543 ADVANCED TE **Group Activity** Power Plants, p. 543 ADVANCED CRF **Skills Practice Lab** Ocean Currents and Water Temperature* GENERAL	OSP **Lesson Plans** (also in print) TR **Bellringer*** TR 101 Major Surface Currents of Earth's Oceans* TR 105 Roaming Rubber Duckies* TE **Internet Activity** Beachcomber Tales, p. 542 ADVANCED CRF **Internet Activity** Beachcomber Tales* ADVANCED VID **NOVA Video** Adrift on the Gulf Stream CD **Interactive Tutor** Heating the Ocean CD **Interactive Tutor** Ocean Circulation
PACING • 90 min pp. 525–530 **Section 2 Ocean Waves** • Describe the formation of waves and the factors that affect wave size. • Explain how waves interact with the coastline. • Identify the cause of destructive ocean waves.	TE **Demonstration** Student Wave, p. 525 GENERAL SE **Quick Lab** Waves, p. 527 GENERAL CRF **Datasheet for Quick Lab*** GENERAL TE **Activity** Wave Refraction in a Tank, p. 528 GENERAL TE **Discussion** Water Safety, p. 528 BASIC SE **Making Models Lab** Wave Motion, pp. 540–541 GENERAL CRF **Datasheet for Chapter Lab*** GENERAL CRF **Inquiry Lab** Tsunami!* GENERAL	OSP **Lesson Plans** (also in print) TR **Bellringer*** TR 102 Wave Motion and Wave Energy* TR 103 The Formation of Breakers*
PACING • 45 min pp. 531–534 **Section 3 Tides** • Describe how the gravitational pull of the moon causes tides. • Compare spring tides and neap tides. • Describe how tidal oscillations affect tidal patterns. • Explain how the coastline affects tidal currents.	TE **Astronomy Connection** Same Face, p. 533 GENERAL TE **Discussion** Bay of Fundy, p. 533 GENERAL	OSP **Lesson Plans** (also in print) TR **Bellringer*** TR 104 Spring Tides and Neap Tides* VID **HRW Earth Science Video** Tides CD **Interactive Tutor** Waves and Tides

PACING • 90 min

CHAPTER REVIEW, ASSESSMENT, AND STANDARDIZED TEST PREPARATION

- SE **Chapter Highlights**, p. 535
- SE **Chapter Review**, pp. 536–537
- SE **Standardized Test Prep**, pp. 538–539
- CRF **Concept Review*** GENERAL
- CRF **Critical Thinking*** ADVANCED
- CRF **Math Skills*** GENERAL
- CRF **Graphing Skills*** GENERAL
- CRF **Chapter Test A*** GENERAL
- CRF **Chapter Test B*** ADVANCED
- OSP **Lesson Plans** (also in print)
- OSP **Test Generator**
- OSP **Test Item Listing**

Online and Technology Resources

go.hrw.com Holt Online Learning

Visit go.hrw.com for access to Holt Online Learning, or enter the keyword **HQ6 Home** for a variety of free online resources.

One-Stop Planner® CD-ROM

This CD-ROM package includes
- Lab Materials QuickList Software
- Holt Calendar Planner
- Customizable Lesson Plans
- Printable Worksheets
- ExamView® Test Generator
- Interactive Teacher Edition
- Holt PuzzlePro®
- Holt PowerPoint® Resources

KEY

SE Student Edition	**OSP** One-Stop Planner	**VID** Classroom Video/DVD
TE Teacher Edition	**TR** Transparencies and Transparency Worksheets	* Also on One-Stop Planner
CRF Chapter Resource File	**CD** CD or CD-ROM	◆ Requires advance prep
LTP Long-Term Projects		■ Also available in Spanish

SKILLS DEVELOPMENT RESOURCES	REVIEW AND ASSESSMENT	CORRELATIONS
SE **Pre-Reading Activity,** p. 518 GENERAL **TE** **Using the Figure** Sunset Tide, p. 518 GENERAL		National Science Education Standards
CRF **Directed Reading*** BASIC **TE** **Inclusion Strategies,** p. 520 **TE** **Reading Skill Builder** Paired Summarizing, p. 521 BASIC **TE** **Using the Figure** The Ocean Conveyor Belt, p. 521 GENERAL **TE** **Skill Builder** Vocabulary, p. 522 GENERAL **SE** **Graphic Organizer** Comparison Table, p. 523 GENERAL	**SE** **Reading Checks,** p. 521, 523 GENERAL **SE** **Section Review,** p. 524 GENERAL **TE** **Reteaching,** p. 523 BASIC **TE** **Quiz,** p. 523 GENERAL **TE** **Alternative Assessment,** p. 524 ADVANCED **CRF** **Section Quiz*** ■ GENERAL	ES 1c
CRF **Directed Reading*** BASIC **TE** **Inclusion Strategies,** p. 526 **TE** **Skill Builder** Math, p. 526 GENERAL **TE** **Reading Skill Builder** Anticipation Guide, p. 527 BASIC	**SE** **Reading Checks,** p. 526, 528 GENERAL **SE** **Section Review,** p. 530 GENERAL **TE** **Homework,** p. 529 ADVANCED **TE** **Reteaching,** p. 529 BASIC **TE** **Quiz,** p. 529 GENERAL **TE** **Alternative Assessment,** p. 530 GENERAL **CRF** **Section Quiz*** ■ GENERAL	PS 6a
CRF **Directed Reading*** BASIC **TE** **Using the Figure** Tidal Bulges, p. 531 GENERAL **SE** **Math Practice,** p. 532 GENERAL **TE** **Using the Figure** Lunar Motions, p. 532 GENERAL **TE** **Skill Builder** Graphing, p. 532 ◆ ADVANCED	**SE** **Reading Check,** p. 532 GENERAL **SE** **Section Review,** p. 534 GENERAL **TE** **Reteaching,** p. 533 BASIC **TE** **Quiz,** p. 533 GENERAL **TE** **Alternative Assessment,** p. 534 ADVANCED **CRF** **Section Quiz*** ■ GENERAL	

Holt Lab Generator CD-ROM

Search for labs by topic, standard, difficulty level, and/or time frame. Edit those labs to fit your needs, or create your own labs. Use the Lab Materials QuickList to order just what you need.

Holt Earth Science Interactive Tutor CD-ROM

This CD-ROM consists of interactive activities that give students a fun way to extend their knowledge of Earth science concepts.

Chapter Summaries Audio CDs

These CDs include audio summaries of the key concepts presented in each chapter. (Audio summaries are also available in Spanish.)

SCILINKS. NSTA

www.scilinks.org

Maintained by the **National Science Teachers Association.** See Chapter Enrichment pages that follow for a complete list of topics.

See Chapter Enrichment pages for Video Resources.

Chapter 21

Chapter Enrichment

This Chapter Enrichment provides relevant and interesting information to expand and enhance your classroom instruction of the chapter material.

Section 1 Ocean Currents

The Sargasso Sea

The Sargasso Sea is an unusually calm, warm body of water located in the middle of the Atlantic Ocean. It lies in the center of a gyre, or a huge circle of moving water, formed by the Gulf Stream and the North Equatorial Currents, which cause the water to move with a slow, clockwise drift. Portuguese sailors named these waters *sargaço,* from their word for "grape," because of the brown seaweed that floats in this sea. Although the Sargasso Sea lacks the nutrients to attract larger fish, many small marine animals, including tiny crabs, shrimp, and octopus, live among the *sargassum* seaweed, forming a unique ecosystem. Eels migrate to the Sargasso Sea from thousands of miles away in North America and Europe to mate and lay their eggs. The Sargasso Sea entered the realm of fiction through the writings of Jules Verne, who described the voyage of the submarine *Nautilus* through this fascinating body of water in his novel *Twenty Thousand Leagues Under the Sea.*

Ocean Currents and Climates

A complex system of ocean currents moves heat around Earth, significantly affecting climates. More direct radiation from the sun reaches Earth's equator than reaches the poles, so equatorial regions have more thermal energy and are warmer than the poles. This uneven heating creates winds that help drive surface currents. Warm surface currents moderate adjacent land climates. For example, the British Isles are warmer than portions of Canada at the same latitude directly across the frigid North Atlantic Ocean, due to the warm Gulf Stream Current. Water retains heat longer and cools more slowly than land does, so land areas near water have less temperature variation than inland areas. For example, St. Louis, Missouri, located in the middle of the U.S., has hot, sunny summers and snowy winters. San Francisco, California, located on the same latitude but on the coast of the Pacific Ocean, typically has cloud cover and rain all year.

◀ This anemone and fish live in the Sargasso Sea.

Section 2 Ocean Waves

The Power of Tsunamis

While tsunamis can occur in almost any ocean, they are most common throughout the Ring of Fire, the belt of volcanoes that encircles the Pacific Ocean. Interestingly, tsunamis are only a few feet high in deep water, which makes them difficult to track by air. But as tsunamis approach shore, they increase in energy and height. If the trough of the wave hits land first, water drains from harbors and beaches, stranding fish and boats. This may tempt bystanders to explore the exposed ocean floor—the most dangerous thing they could do. The wave crest, a wall of water that may be more than 30 m high, follows soon after.

The energy of these huge waves can smash buildings, snap trees, flood harbors, capsize boats, and carry objects great distances. Tsunamis do the most damage to low-lying coastal areas near their point of origin. An effective early-warning system is crucial to minimizing destruction. An international tsunami early-warning system for the entire Pacific Rim was set up in Hawaii in 1948. It collects seismic data from member countries via satellite. A second center is located in Alaska to collect seismic data on tsunamis likely to threaten the Pacific coast of North America. Following earthquake activity, instruments detect whether a tsunami was generated. A warning is then sent out to evacuate people from vulnerable areas.

Section 3 Tides

The Earth-Moon System

Earth and its large natural satellite, the moon, form a double-planet system that is held together by the force of gravity. They revolve around a common center point located about 1,610 km deep within Earth. The moon's gravitational pull causes oceans to shift periodically—the rise and fall of the tides. Scientists speculate that the moon was once much closer to Earth and that the whole system once spun much faster. Because the force of gravity is indirectly proportional to distance, tides would have been much larger millions of years ago, when the moon was closer to Earth. The forces that affect tides also affect Earth's crust. Friction occurs as ocean waters are dragged over the ocean floor. Over time, this friction has slowed Earth down, resulting in the current 24-hour day. The system has stayed in equilibrium as the moon has spun farther away from Earth. The weaker gravitational pull produces a smaller range of tidal activity.

The Bay of Fundy's Giant Tides

As the waters of the Atlantic Ocean rise in response to the gravitational pull of the moon, a huge surge of water pours into the mouth of the Bay of Fundy in southeastern Canada. Because the shape of the bay is wider at the mouth and narrows at the back, the waters pile up, causing unusually high tides. Also present is a natural rocking motion called a *seiche,* or tidal oscillation. This is a standing wave set up in an enclosed or semi-enclosed body of water caused by strong winds or changes in atmospheric pressure.

▼ **The tidal range in the Bay of Fundy is one of the largest on Earth.**

Video Resources

Brain Food Video Quizzes

Brain Food Video Quizzes These videos contain game-show style quizzes that assess students' progress and motivate students to study the chapter material.

HRW Earth Science Video

This video introduces Earth science topics and includes a geology field trip. The video segment listed below complements this chapter.

Segment 15: Tides The combination of Earth's rotation and the gravitational forces of the sun and moon causes daily high and low tides. This segment examines how the Earth-moon system creates the daily rise and fall of sea level and discusses how the alignment of the sun and the moon with Earth causes the highest tides. The regular and predictable intervals of the tides are also described. (3 min)

NOVA® Videos

The NOVA video below complements this chapter.

Adrift on the Gulf Stream NOVA explores the Gulf Stream's importance to ocean life and climate from a variety of vantage points. (60 min)

To order other **NOVA** videos related to this chapter, visit **go.hrw.com** and enter the keyword **HQ6MOVV.**

SciLinks®

NSTA Developed and maintained by the National Science Teachers Association

SciLinks is maintained by the National Science Teachers Association to provide you and your students with interesting, up-to-date links that will enrich your classroom presentation of the chapter.

Visit www.scilinks.org and enter the SciLinks code for more information about the topic listed.

Topic: Ocean Currents
SciLinks code: HQ61061

Topic: Tides
SciLinks code: HQ61525

Topic: Tsunamis
SciLinks code: HQ61561

Topic: Ocean Waves
SciLinks code: HQ61066

Chapter 21

Chapter Overview

This chapter describes the forces that affect the motion of ocean waters, including winds, Earth's spinning motion, land barriers that affect surface currents, differences in water density, and the gravitational pull of the moon.

Using the Figure — GENERAL

Sunset Tide The photo shows a tide coming in at sunset. Tell students that in some places, high tide is 15 m higher than low tide. Ask students to predict what would happen to low-lying features such as harbors, beaches, and rocks if that amount of water came into a bay regularly. (Sample answer: Many features would be covered with water. The energy of so much water could be very destructive.) LS **Visual**

PRE-READING ACTIVITY

FOLDNOTES

Encourage students to use their FoldNotes as a study guide to quiz themselves for a test on the chapter material. Students may want to create Layered Book FoldNotes for different topics within the chapter.

Chapter 21 Movements of the Ocean

Sections

1 Ocean Currents

2 Ocean Waves

3 Tides

What You'll Learn

- What factors cause surface and deep currents
- How waves form and move
- How the moon affects tides

Why It's Relevant

Ocean water moves in currents. Ocean currents are a major method of heat transport that affects climate around the globe.

PRE-READING ACTIVITY

FOLDNOTES

Layered Book Before you read this chapter, create the FoldNote entitled "Layered Book" described in the Skills Handbook section of the Appendix. Label the tabs of the layered book with "Surface currents," "Deep currents," "Waves," and "Tides." As you read the chapter, write information you learn about each category under the appropriate tab.

▶ Tides are caused by the moon's gravitational pull on the oceans. In some places, the moon can cause tides that are as high as 15 m above the low tide.

Chapter Correlations *National Science Education Standards*

PS 6a Waves, including sound and seismic waves, waves on water, and light waves, have energy and can transfer energy when they interact with matter. **(Section 2)**

ES 1c Heating of Earth's surface and atmosphere by the sun drives convection within the atmosphere and oceans, producing winds and ocean currents. **(Section 1)**

Section 1

Ocean Currents

The water in the ocean moves in giant streams called **currents.** Many ocean currents are complex and difficult to trace. Oceanographers identify ocean currents by studying the physical and chemical characteristics of the ocean water. They also identify currents by mapping the paths of debris that is dumped or washed overboard from ships, as shown in **Figure 1.** From these data, scientists have mapped a detailed pattern of ocean currents around the world. Scientists place ocean currents into two major categories: surface currents and deep currents.

OBJECTIVES

- **Describe** how wind patterns, the rotation of Earth, and continental barriers affect surface currents in the ocean.
- **Identify** the major factor that determines the direction in which a surface current circulates.
- **Explain** how differences in the density of ocean water affect the flow of deep currents.

KEY TERMS

current
surface current
Coriolis effect
gyre
Gulf Stream
deep current

Factors That Affect Surface Currents

Currents that move on or near the surface of the ocean and are driven by winds are called **surface currents.** Surface currents are controlled by three factors: air currents, Earth's rotation, and the location of the continents.

All surface currents are affected by winds. Winds are caused by the uneven heating of the atmosphere. Variations in air temperature lead to variations in air density and pressure. Colder, denser air sinks and forms areas of high pressure. Air moves away from high-pressure areas to lower pressure areas. This movement gives rise to wind.

Because *wind* is moving air, wind has kinetic energy. The wind passes this energy to the ocean as the air moves across the ocean surface. As energy is transferred from the air to the ocean, the water at the ocean's surface begins to move.

current in geology, a horizontal movement of water in a well-defined pattern, such as a river or stream

surface current a horizontal movement of ocean water that is caused by wind and that occurs at or near the ocean's surface

Figure 1 ▶ Glass floats, which are used to hold up Japanese fishing nets, have been carried by surface currents from off the coasts of Japan to this beach in northwest Hawaii.

Section 1

Focus

Overview

This section identifies factors that affect global surface currents and describes the motions of major surface currents. The section also examines how variations in water density affect the formation of deep-ocean currents.

Bellringer

Ask students to think about movies or pictures they have seen that show an ocean or a beach. Have students list words or phrases that describe the motion of ocean water. (Answers may vary. Students may describe waves, ripples, swells, whitecaps, or tides.) LS **Verbal**

Motivate

Discussion — GENERAL

Beachcombing Use the figure on the bottom of this page as a starting point for discussion. Ask students to list the kinds of objects they might expect to find on a beach. (Sample answers: shells, driftwood, bottles, seeds, toys, paper cups, rocks, boxes, and seaweed) Ask them where they think these objects came from. (Sample answers: from things living in the sea, from nearby coasts, from boats, or from people who live on the coast) Explain that the motion of ocean currents often deposits floating objects on the shore. LS **Verbal**

CHAPTER RESOURCES

Chapter Resource File

- Directed Reading BASIC

Technology

Transparencies
- Bellringer

Student Edition on CD-ROM

One-Stop Planner CD-ROM
- Lesson Plan

Teach

Demonstration — GENERAL

Modeling Ocean Currents Set up an overhead projector. Fill a wide, flat, clear plastic container with water to simulate an ocean. Place the container on the projector and focus as needed. Put a drop of food color at one end of the container. Using a straw, blow gently across the water's surface. Have students describe what happens. Students also may draw what the see, both at the surface and below. (Air blown through the straw caused streams of water, just as the wind creates currents in the ocean.) Put a small cup upside down in the center of the model ocean to represent land. Add a drop of food color in front of the barrier. Gently blow through the straw and have students describe and draw what happens. Ask what effect a land mass has on the currents. (The currents divide and flow around the barrier.) Repeat with a jar lid placed under the surface. Ask whether the currents always flow in the direction of the wind. (No, sometimes the currents flow in circles around barriers.) LS **Visual**

INCLUSION Strategies

- ***Visually Impaired***

Help students understand the Coriolis effect by using a 3-D model of Earth in the form of a globe. Pair a visually impaired student with a sighted partner. Have them place a lump of clay on the equator and another closer to one of the poles. Together they can spin the globe to show how Earth rotates and demonstrate how objects at different points on the globe move at different speeds. LS **Kinesthetic**

Global Wind Belts

Global wind belts, such as the trade winds and westerlies shown in **Figure 2,** are a major factor affecting the flow of ocean surface water. The *trade winds* are located just north and south of the equator. In the Northern Hemisphere, trade winds blow from the northeast. In the Southern Hemisphere, they blow from the southeast. In both hemispheres, trade-wind belts push currents westward across the tropical latitudes of all three major oceans.

The *westerlies* are located in the middle latitudes. In the Northern Hemisphere, westerlies blow from the southwest. In the Southern Hemisphere, they blow from the northwest. Westerlies push ocean currents eastward in the higher latitudes of the Northern and Southern Hemispheres.

Coriolis effect the apparent curving of the path of a moving object from an otherwise straight path due to Earth's rotation

gyre a huge circle of moving ocean water found above and below the equator

Continental Barriers

The continents are another major influence on surface currents. The continents act as barriers to surface currents. When a surface current flows against a continent, the current is deflected and divided.

The Coriolis Effect

Global wind belts and ocean currents do not flow in straight lines. Wind belts and ocean currents follow a curved or circular pattern that is caused by Earth's rotation. As Earth spins on its axis, ocean currents and wind belts curve. The curving of the path of oceans and winds due to Earth's rotation is called the **Coriolis effect**. The wind belts and the Coriolis effect cause huge circles of moving water, called **gyres,** to form. In the Northern Hemisphere, water flow in gyres is to the right, or clockwise. In the Southern Hemisphere, the flow is to the left, or counterclockwise.

Figure 2 ▶ Global winds and the Coriolis effect together drive the surface currents of the oceans in great circular patterns. The photo at right shows a gyre in the Pacific Ocean off the east coast of Japan.

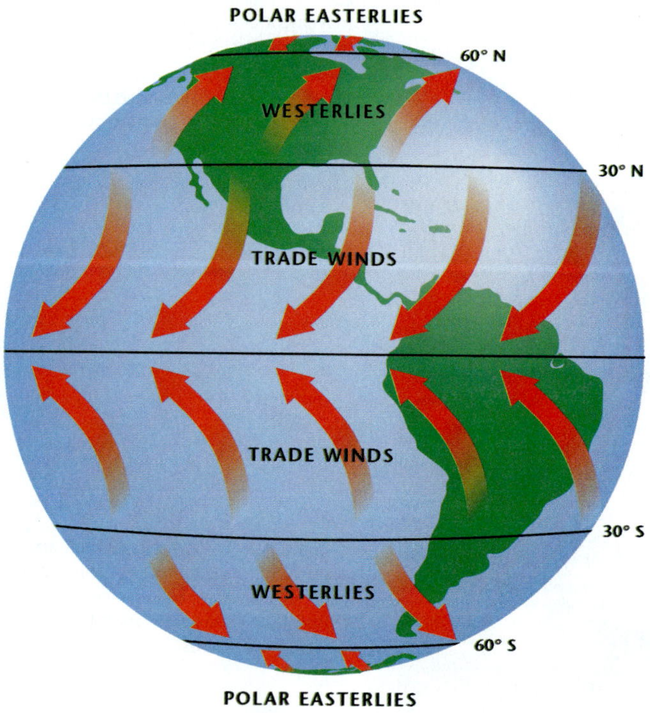

MISCONCEPTION ALERT

Plumbing and the Coriolis Effect
Students may think that the way water drains in a circular motion down sinks or in a flushing toilet is related to the Coriolis effect. The rotation of Earth affects large-scale phenomena, such as prevailing winds and ocean currents. However, the deflection is so small that it plays no part in how water moves through household plumbing. Explain that toilets and sinks drain in the directions they do because of the way water is directed into them or pulled away. If water enters in a swirling motion, it will exit with the same swirling pattern.

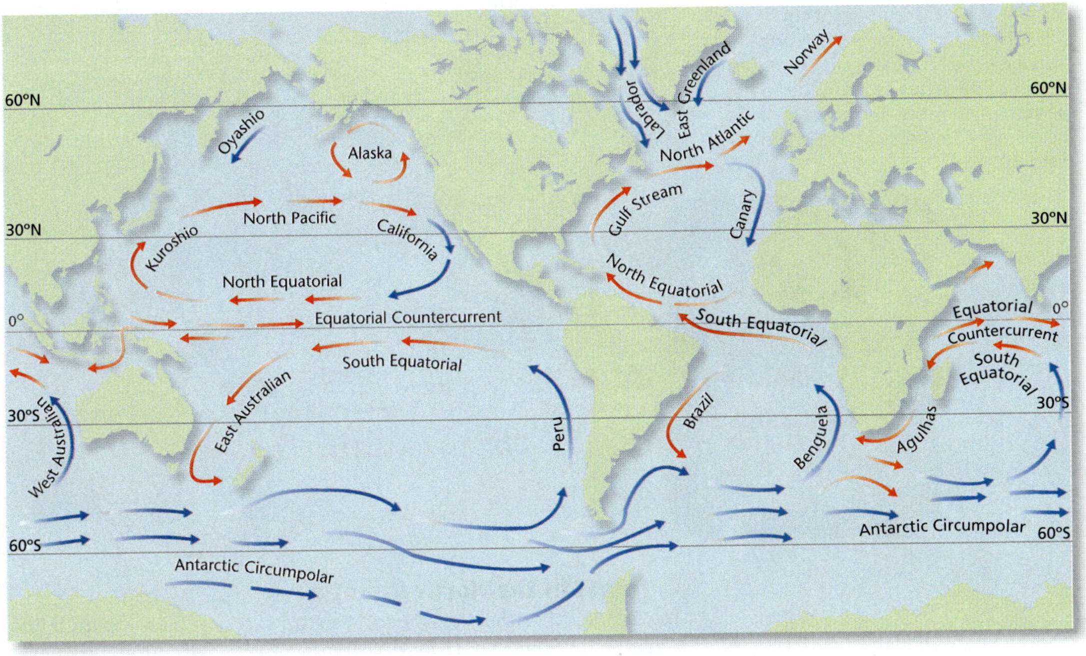

Figure 3 ▶ This map shows the major surface currents of the oceans of the world. Warm-water currents are shown in red; cold-water currents are shown in blue.

Major Surface Currents

The major surface currents of the world are shown in **Figure 3.** The major currents in the equatorial region and in the Northern and Southern Hemispheres are described below.

Equatorial Currents

Warm equatorial currents are located in the Atlantic, Pacific, and Indian Oceans. Each of these oceans has two warm-water equatorial currents that move in a westward direction. Between these westward-flowing currents lies a weaker, eastward-flowing current called the *Equatorial Countercurrent*.

Currents in the Southern Hemisphere

In the Southern Hemisphere, the currents in gyres move counterclockwise. In the most southerly regions of the oceans, constant westward winds produce the world's largest current, the *Antarctic Circumpolar Current,* also known as *West Wind Drift*. No continents interrupt the movement of this current that completely circles Antartica and crosses all three major oceans.

The Indian Ocean surface currents follow two patterns. Currents in the southern Indian Ocean follow a circular, counterclockwise gyre. Currents in the northern Indian Ocean are governed by *monsoons,* winds whose directions change seasonally.

Reading Check **What is the world's largest ocean current? (See the Appendix for answers to Reading Checks.)**

SCILINKS. Developed and maintained by the National Science Teachers Association

For a variety of links related to this subject, go to www.scilinks.org

Topic: Ocean Currents
SciLinks code: HQ61061

Using the Figure — GENERAL

The Ocean Conveyor Belt

Explain that sunlight that shines on the surface of the ocean adds thermal energy to the ocean's surface. Currents help transfer this energy to other regions of the globe. Ask students how the temperature of each water current is indicated on the world map. Note that *warm* means "warmer than surrounding ocean waters." (Warm water currents are shown in red; cold-water currents are shown in blue.) Ask where most of the warm water currents originate. (near the equator) Then, ask students to identify the temperature of currents that flow from the poles. (cold) **LS Visual**

Answer to Reading Check

Because no continents interrupt the flow of the Antarctic Circumpolar Current, also called the *West Wind Drift*, it completely encircles Antarctica and crosses three major oceans. All other surface currents are deflected and divided when they meet a continental barrier.

CHAPTER RESOURCES

Chapter Resource File

- **Skills Practice Lab** Ocean Currents and Water Temperature GENERAL

Technology

Transparencies

- 101 Major Surface Currents of Earth's Oceans (with worksheet)

GEOGRAPHY CONNECTION — ADVANCED

Research Have students research the effect of the Gulf Stream on the climate of the British Isles. Have them compare weather and precipitation with that of a North American location in the same latitude. Ask them to hypothesize what would happen to the climates of these regions if the Gulf Stream passed closer to North America and farther from the British Isles. (The climate of the British Isles would probably be colder with more snow; the North American location would probably be warmer, with more rain.) **LS Verbal**

READING SKILL BUILDER — BASIC

Paired Summarizing Group students into pairs and have them read silently about major surface currents. Then, have one student summarize the paths of major surface currents. The other student should listen to the retelling and should point out any inaccuracies or ideas that were left out. Allow students to refer to the text as needed. **LS Verbal** English Language Learners

Teach, *continued*

QuickLAB

Skills Acquired

- Constructing models
- Observing
- Analyzing
- Communicating

Teacher's Notes You could substitute pepper grains, talcum powder, or other things that float for the confetti. An aluminum pie pan works well to replace the plate. Students could model the effects of land barriers by using small lumps of clay to model landmasses that deflect currents.

Answers

1. Student diagrams may vary but should accurately show the positions of the straw. Arrows should indicate that the water flows in the same direction as the moving air.
2. Surface currents in the ocean are produced by energy transferred to water by the winds, just as the water in the plate received energy from the air that was blown through the straw.

CHAPTER RESOURCES

Chapter Resource File

- **Datasheet for Quick Lab** GENERAL

Figure 4 ▶ Surface currents in the Atlantic Ocean form the North Atlantic Gyre. The Sargasso Sea in the center of the gyre results from this pattern of currents. The organisms below are commonly found in the Sargasso Sea.

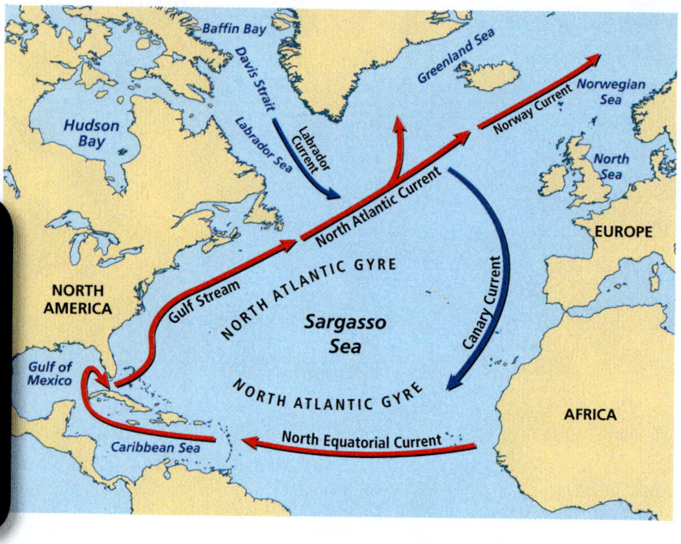

Gulf Stream the swift, deep, and warm Atlantic current that flows along the eastern coast of the United States toward the north

QuickLAB 10 min

Ocean Currents

Procedure

1. Fill a **shallow pan** with **water.**
2. Sprinkle **paper confetti** on the surface of the water.
3. Blow across the surface of the water through a **drinking straw** to produce a clockwise current.
4. Blow through the straw to make a counterclockwise current. Try to make two currents.

Analysis

1. Draw a diagram of the currents. Draw the straw's positions, and use arrows to show air and water direction.
2. How does this activity relate to what happens in ocean currents?

Currents in the North Atlantic

In the North Atlantic Ocean, warm water moves through the Caribbean Sea and Gulf of Mexico and north along the east coast of North America in a swift, warm current called the **Gulf Stream.** Farther north, the cold-water Labrador Current, which flows south, joins the Gulf Stream. South of Greenland, the Gulf Stream widens and slows until it becomes a vast, slow-moving warm current known as the *North Atlantic Current.* Near western Europe, the North Atlantic Current splits. One part becomes the Norway Current, which flows northward along the coast of Norway and keeps that coast ice-free all year. The other part is deflected southward and becomes the cool Canary Current, which eventually warms and rejoins the North Equatorial Current.

As **Figure 4** shows, the Gulf Stream, the North Atlantic Current, the Canary Current, and the North Equatorial Current form the North Atlantic Gyre. At the center of this gyre lies a vast area of calm, warm water called the *Sargasso Sea.* The Sargasso Sea is named after *sargassum,* the brown seaweed that floats on the sea's surface in this area. The pattern of winds and currents around the Sargasso Sea concentrates all kinds of floating debris, such as orange peels and plastic cups, in this area.

Currents in the North Pacific

The pattern of currents in the North Pacific is similar to that in the North Atlantic. The warm Kuroshio Current, the Pacific equivalent of the Gulf Stream, flows northward along the east coast of Asia. This current then flows toward North America as the North Pacific Drift. It eventually flows southward along the California coast as the cool California Current.

SKILL BUILDER GENERAL

Vocabulary Here are three words with Latin roots that students will encounter in ocean study. The term *salinity* comes from *sal,* which means "salt." The term *density* comes from *densus,* which means "thick." The term *turbidity* comes from *turbidus,* which means "disturbed." Ask students to identify how knowing these roots can help them remember the meanings of these terms. **LS Verbal**

HISTORY CONNECTION GENERAL

Research When Benjamin Franklin was deputy postmaster general of the American colonies from 1753 through 1774, mail ships had been making the journey from America to England weeks faster than in the reverse direction. Franklin set out to determine why. He was able to document the motion of the Gulf Stream current. His findings became one of the earliest published charts of North Atlantic ocean currents. Have students find out more about Franklin's maritime observations and report back to the class. **LS Verbal**

Deep Currents

In addition to having wind-driven surface currents, the ocean has **deep currents,** cold, dense currents far below the surface. Deep currents move much more slowly than surface currents do. Deep currents form as cold, dense water of the polar regions sinks and flows beneath warmer ocean water.

deep current a streamlike movement of ocean water far below the surface

The movement of polar waters is a result of differences in density. When water cools, it contracts and the water molecules move closer together. This contraction makes the water denser, and the water sinks. When water warms, it expands and the water molecules move farther apart. The warm water is less dense, so it rises above the cold water. Temperature determines density.

Salinity, too, determines the density of water. The water in polar regions has high salinity because of the large amount of water frozen in icebergs and sea ice. When water freezes, the salt in the water does not freeze but stays in the unfrozen water. So, unfrozen polar water has a high salt concentration and is denser than water that has a lower salinity. This dense polar water sinks and forms a deep current that flows beneath less dense surface currents, as shown in **Figure 5.**

Antarctic Bottom Water

The temperature of the water near Antarctica is very cold, –2°C. The water's salinity is high. These two factors make the water off the coast of Antarctica the densest and coldest ocean water in the world. This dense, cold water sinks to the ocean bottom and forms a deep current called the *Antarctic Bottom Water.* The Antarctic Bottom Water moves slowly northward along the ocean bottom for thousands of kilometers to a latitude of about 40°N. It takes hundreds of years for the current to make the trip.

Reading Check **Why is Antarctic Bottom Water the densest in the world? (See the Appendix for answers to Reading Checks.)**

Graphic Organizer **Comparison Table**

Create the Graphic Organizer entitled "Comparison Table" described in the Skills Handbook section of the Appendix. Label the columns with "Surface currents" and "Deep currents." Label the rows with "Salinity," "Temperature," and "Density." Then, fill in the table with details about the salinity, temperature, and density of each current.

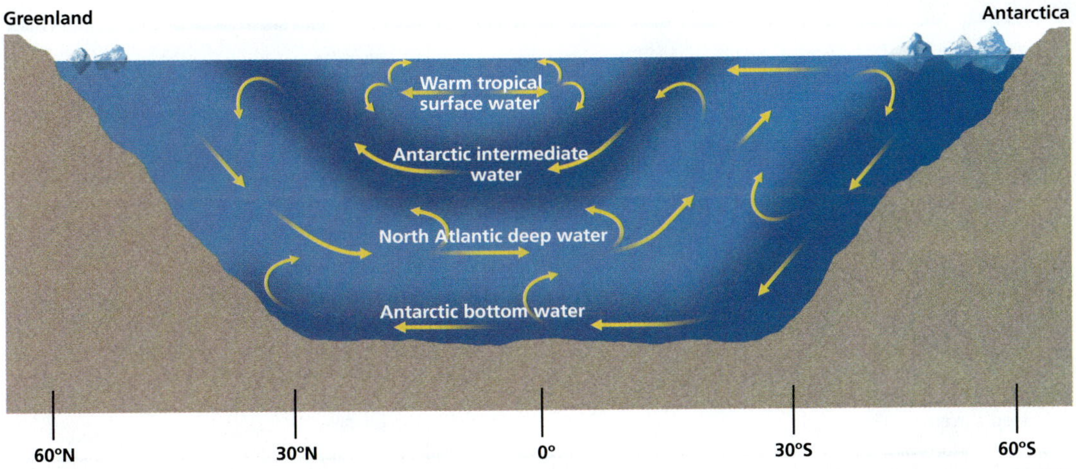

Figure 5 ▶ The very dense and highly saline Antarctic Bottom Water travels beneath less dense North Atlantic Deep Water.

BIOLOGY CONNECTION GENERAL

Vertical Ocean Movements Deepwater ocean currents driven by differences in water density distribute thermal energy and nutrients in a continuous system that sustains ocean life. When deeper water moves upward (upwelling), colder, nutrient-rich waters come to the surface. Ask students why deeper waters have more nutrients. (Dying marine organisms fall to the bottom. Decomposition releases nutrients.) Nutrients moving to the surface nourish marine plants called *plankton,* which in turn nourish other marine organisms. Surface water moving downward is known as *downwelling*. It is caused in part by evaporation. Ask students to explain how evaporation affects the density and movement of ocean water. (Evaporation makes ocean water saltier, colder, and more dense, causing it to sink.) Explain that these vertical upward and downward motions of ocean water occur as a constantly moving global system. **LS Verbal/Logical**

Graphic Organizer GENERAL

Comparison Table
You may want to use this graphic organizer in a game to review material before the test. Divide the class into two teams. Ask students questions about material from the Comparison Table. Give points to each team that provides correct answers.

Answer to Reading Check

Antarctic Bottom Water is very cold. It also has a high salinity. The extreme cold and high salinity combine to make the water extremely dense.

Close

Reteaching — BASIC

Venn Diagrams Have students work in pairs to construct a Venn diagram that describes the similarities and differences between surface currents and deep currents. Then, have each student write a short summary of what the diagram shows. **LS Visual**

Quiz — GENERAL

1. What three factors control the motion of surface currents? Describe their effects.
(1. Prevailing winds transfer energy to the surface to produce surface currents. 2. Earth's rotation causes water currents to curve. 3. Land barriers cause currents to be deflected or divided.)
2. Explain how water density produces deep-ocean currents.
(Differences in water temperature and salinity produce differences in density. The denser water sinks and flows beneath surface currents. Sediment from landslides or seismic activity also makes water denser so it flows down slopes on the ocean floor.)

Close, continued

Alternative Assessment — ADVANCED

Adventures of a Drop Have students write a story about the adventures of a drop of water as it moves through the ocean. They should incorporate what they have learned about the motions of surface and deep-water currents and the factors that affect them. Students may include drawings to illustrate their narratives. LS Verbal/Visual

Answers to Section Review

1. Uneven heating of Earth's atmosphere produces moving air, or winds. As winds move across the ocean surface, they transfer kinetic energy to the water. This energy drives the horizontal motion of water at the surface.
2. The trade wind belts, located north and south of the equator, push surface currents west across the tropical latitudes of all three major oceans.
3. Westerlies are the wind belt system located in the middle latitudes. They push the currents in an easterly direction.
4. Denser water sinks below less-dense water. This motion of the water is the basis of deep-ocean currents.
5. Temperature, salinity, and turbidity affect the density of water.
6. surface currents: Gulf Stream, Kuroshio Current, Antarctic Circumpolar Current; deep currents: North Atlantic Deep Water, Antarctic Bottom Water
7. Earth's rotation causes the path of ocean currents to curve. If Earth did not rotate, waters would flow in straight paths until they were interrupted by land.
8. Sunlight that shines on the surface of Earth adds thermal energy to the air. Differences in the amount of solar radiation leads to differential heating of Earth's surface. This heating pattern creates winds, which drive surface currents.
9. *Ocean currents* can be *deep currents,* such as the *Antarctic Bottom Water,* or *surface currents,* such as the *Gulf Stream* and the *North Atlantic Current,* which are part of a *gyre* that is caused by the *Coriolis effect.*

Figure 6 ▶ The dense, highly saline water of the Mediterranean Sea forms a deep current as it flows through the strait of Gibraltar and into the less dense Atlantic Ocean.

North Atlantic Deep Water

In the North Atlantic, south of Greenland, the water is very cold and has a high salinity. This cold, salty water forms a deep current that moves southward under the northward-flowing Gulf Stream. Near the equator, this deep current divides. One part begins to rise, reverse direction, and flow northward again. The rest of the current continues southward toward Antarctica and flows over the colder, denser Antarctic Bottom Water.

Deep Atlantic currents exist near the Mediterranean Sea, too. Every summer in the region, evaporation increases and precipitation decreases. These changes increase the salinity, and thus the density, of the water of the Mediterranean. This denser water sinks and flows through the strait of Gibraltar into the Atlantic Ocean. In turn, surface water from the Atlantic, which is less saline and less dense than deep current water is, flows into the Mediterranean Sea, as shown in **Figure 6.**

Turbidity Currents

A turbidity current is a strong current caused by an underwater landslide. A turbidity current occurs when large masses of sediment that have accumulated along a continental shelf or continental slope suddenly break loose and slide downhill. The landslide mixes the nearby water with sediment. The sediment causes the water to become cloudy, or turbid, and denser than the surrounding water. The dense water mass of the turbidity current moves beneath the less dense, clear water.

Section 1 Review

1. **Describe** the force that drives most surface currents.
2. **Identify** the winds that affect the surface currents on either side of the equator.
3. **Identify** the winds that affect the surface currents in the middle latitudes.
4. **Describe** how density affects the flow of deep currents.
5. **List** the factors that affect the density of ocean water.
6. **List** three major surface currents and two major deep currents.

CRITICAL THINKING

7. **Predicting Consequences** Describe how surface currents would be affected if Earth did not rotate.
8. **Identifying Relationships** Explain how the distribution of solar energy around Earth affects ocean surface currents.

CONCEPT MAPPING

9. Use the following terms to create a concept map: *ocean currents, surface currents, deep currents, Gulf Stream, North Atlantic Current, Antarctic Bottom Water, gyres,* and *Coriolis effect.*

CHAPTER RESOURCES

Chapter Resource File
- Section Quiz GENERAL

Workbooks
- Study Guide (also in Spanish)

Section 2 Ocean Waves

A **wave** is a periodic disturbance in a solid, liquid, or gas as energy is transmitted through the medium. One kind of wave is described as the periodic up-and-down movement of water. Such a wave has two basic parts—a *crest* and a *trough*, as shown in **Figure 1.** The crest is the highest point of a wave. The trough is the lowest point between two crests. The *wave height* is the vertical distance between the crest and the trough of a wave. The *wavelength* is the horizontal distance between two consecutive crests or between two consecutive troughs. The **wave period** is the time required for two consecutive wave crests to pass a given point. The speed at which a wave moves is calculated by dividing the wave's wavelength by its period.

$$\textit{wave speed} = \frac{\textit{wavelength}}{\textit{wave period}}$$

OBJECTIVES

- **Describe** the formation of waves and the factors that affect wave size.
- **Explain** how waves interact with the coastline.
- **Identify** the cause of destructive ocean waves.

KEY TERMS

wave
wave period
fetch
refraction

Wave Energy

The uneven heating of Earth's atmosphere causes pressure differences that make air move. This moving air is called *wind*. Wind then transfers the energy received from the sun to the ocean and forms waves. Small waves, or ripples, form as a result of friction between the moving air and water. As a ripple receives more energy from wind, the ripple grows into a larger wave. The longer that wind blows from a given direction, the more energy is transferred from wind to water and the larger the wave becomes.

The smoothness of the ocean's surface is generally disrupted by many small waves moving in different directions. Because of their large surface area, larger waves receive more energy from the wind than smaller waves do. Thus, larger waves grow larger, and smaller waves die out.

wave a periodic disturbance in a solid, liquid, or gas as energy is transmitted through a medium

wave period the time required for two consecutive wave crests to pass a given point

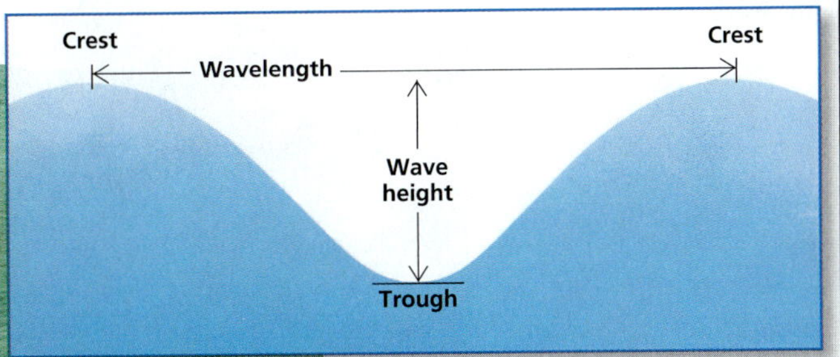

Figure 1 ▶ The vertical distance between the crest and the trough of a wave is the wave height. In a series of waves, the wave height may be similar for all of the waves, as shown in the photo at left.

Section 2

Focus

Overview

This section introduces wave characteristics, explains how waves form, and identifies different types of ocean waves.

Bellringer

Have students list as many words as they can think of that describe the motion of ocean waves. (Answers may vary. Sample answers may include adjectives such as rolling, rocking, bobbing, crashing, breaking, rising and falling, and heaving.) **LS Verbal**

Motivate

Demonstration — GENERAL

Student Wave Have 10 to 12 students stand side-by-side at the front of the room. Have them face the same direction and link arms. Tell them that they are going to mimic the motion of an ocean wave. Stand at one end and gently pull the closest student forward and then back. Continue until all the students are in motion. Have the seated students observe. Ask students where the wave energy came from and how far it traveled. (Sample answer: The teacher supplied the energy that pulled the first student. The energy traveled all the way down the line.) Emphasize that waves transfer energy over great distances. **LS Kinesthetic/Visual**

CHAPTER RESOURCES

Chapter Resource File
- Directed Reading BASIC

Technology

Transparencies
- Bellringer

Student Edition on CD-ROM

One-Stop Planner CD-ROM
- Lesson Plan

Teach

MISCONCEPTION ALERT

Movement of Water Molecules Students may think that the waves of water breaking on the beach have traveled for many miles before reaching shore. Emphasize that it is only the wave energy or the waveform that moves. Water molecules within a wave actually circulate more or less in place, with little or no forward motion.

INCLUSION Strategies

• Learning Disabled

Have students create a chart in which they list key words from the text, such as *crest*, *trough*, *wave period*, and *wavelength*, and give their meanings. The process of organizing the information into a graphic format will help students process the information visually. Ask students to summarize the main points. **LS Verbal**

Answer to Reading Check

Because waves receive energy from wind that pushes against the surface of the water, the amount of energy decreases as the depth of water increases. As a result, the diameter of the water molecules' circular path also decreases.

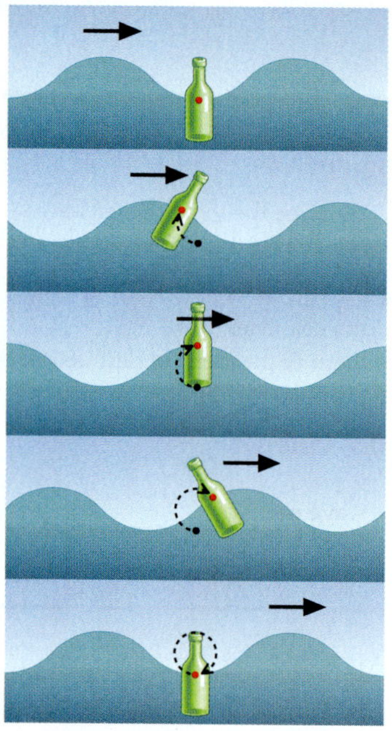

Figure 2 ▶ Like the bottle in this diagram, water molecules do not travel horizontally through the water with the wave.

Water Movement in a Wave

Although the energy of a wave moves from water molecule to water molecule in the direction of the wave, the water itself moves very little. This fact can be demonstrated by observing the movement of a bottle floating on the water as a wave passes. The bottle appears to move up and down, but it moves in a circular path, as shown in **Figure 2.** As the wave passes, the bottle returns to where it started.

As a wave moves across the surface of the ocean, only the energy of the wave, not the water, moves in the direction of the wave. The water molecules within the wave move in a circular motion. During a single wave period, each water particle moves in one complete circle. At the end of the wave period, a circling water particle ends up almost exactly where it started.

As a wave passes a given point, the circle traced by a water particle on the ocean surface has a diameter that is equal to the height of the wave. Because waves receive their energy from wind pushing against the surface of the ocean, the energy received decreases as the depth of the water increases. As a result, water at various depths receives varying amounts of energy. Thus, the diameter of a water molecule's circular path decreases as water depth increases, as shown in **Figure 3.** Below a depth of about one-half the wavelength, there is almost no circular motion of water molecules.

Reading Check **Why does the diameter of a water molecule's circular path in a wave decrease as depth increases? (See the Appendix for answers to Reading Checks.)**

Figure 3 ▶ Wave energy decreases as depth increases. As a result, the diameter of a water molecule's circular path in the wave also decreases.

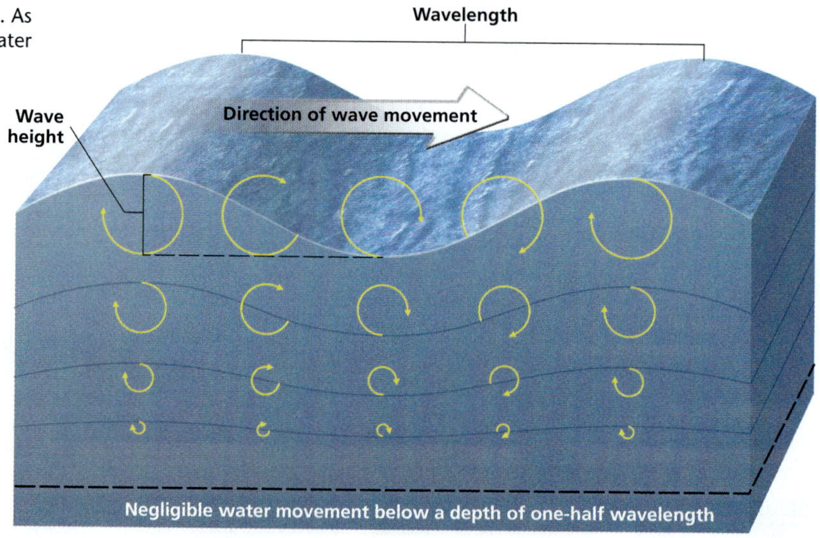

CHAPTER RESOURCES

Technology

Transparencies
- 102 Wave Motion and Wave Energy (with worksheet)

SKILL BUILDER GENERAL

Math Have students calculate the wave speed of a wave that has a wavelength of 1530 m and a period of 30 s.

(wavespeed $= \dfrac{\text{wavelength}}{\text{wave period}}$

wavespeed $= \dfrac{1530 \text{ m}}{30 \text{ s}} = 51$ m/s) **LS Logical**

Wave Size

Figure 4 ▶ Whitecaps, such as the ones shown here off the coast of North Carolina, may form during storms.

Three factors determine the size of a wave. These factors are the speed of the wind, the length of time the wind blows, and fetch. **Fetch** is the distance that the wind can blow across open water. Very large waves are produced by strong, steady winds blowing across a long fetch.

During a storm, steady high winds can cause some waves to gather enough energy to reach great size. Strong, gusty winds, on the other hand, produce choppy water that has waves of various heights and lengths and which may come from various directions. Nevertheless, the size of a wave will increase to only a certain height-to-length ratio before the wave collapses.

On calm days, small, smooth waves move steadily across the ocean's surface. One of a group of long, rolling waves that are of similar size is called a *swell.* Swells move in groups in which one wave follows another. Swells that reach the shore may have formed thousands of kilometers out in the ocean.

fetch the distance that wind blows across an area of the sea to generate waves

Whitecaps

When winds blow the crest of a wave off, *whitecaps* form, as shown in **Figure 4.** Because whitecaps reflect solar radiation, they allow less radiation to reach the ocean. Scientists have been studying how this characteristic may affect climate.

QuickLAB 15 min

Waves

Procedure

1. Fill a **rectangular pan** (40 cm × 30 cm × 10 cm) with **water** to a depth of 7 cm.
2. Float a **cork** near the center of the pan. On each side of the pan, mark the location of the cork with a small piece of **tape.**
3. Hold a **spoon** in the water at one end of the pan. Carefully move the spoon up and down in the water to make a slow, regular pattern of waves.
4. Observe the movement of the cork for 1 minute. Sketch how the cork moves in relation to the waves.
5. Remove the cork from the pan.
6. Use the spoon to make a strong, steady series of waves.
7. Remove the spoon from the pan. Observe what happens when the waves reach the edges of the pan. Write down or sketch what you observe.

Analysis

1. Describe the motion of the cork when a wave passes.
2. How does the cork move relative to the tape on the sides of the pan? Explain your answer.
3. When a wave breaks on the shore, the water is carried in the direction of the wave. Based on your observations in step 4, does this statement contradict your model? Explain your answer.

QuickLAB

Skills Acquired

- Constructing Models
- Experimenting
- Recognizing Patterns
- Interpreting

Teacher's Notes Another way to make waves is by using a fan or hairdryer set at different speeds. You could also substitute wood blocks for the spoon. If you have a larger tank available, such as an aquarium, you can use fishing sinkers to float several corks at different depths below the water's surface so that students can observe the effect of increasing wave size on the movement of water at different depths.

Answers

1. Answers may vary. The cork appears to move slightly forward, then slightly backward about the same distance.
2. The cork stays almost even with the tape, bobbing up and down and moving slightly forward and back. It does not show much movement in the direction of the wave.
3. No, the waves created in the pan resemble waves in deep water instead of waves breaking on a beach. Because the pan is the same depth throughout, the water moves back and forth instead of being carried forward.

CHAPTER RESOURCES

Chapter Resource File

- Datasheet for Quick Lab GENERAL

READING SKILL BUILDER BASIC

Anticipation Guide Before students read this page, ask them to predict what factors determine the size of a wave. Then, ask students to read the page and find out if their predictions were accurate. LS Logical English Language Learners

ENVIRONMENTAL CONNECTION

Whitecaps and Climate A study by Scripps Institution of Oceanography shows that whitecaps, because they are white, impact climate by reflecting sunlight back into space, thus reducing the amount of warming in ocean water. This effect is especially significant in regions that have high winds and cloudless skies, such as the Arabian Sea and the Indian Ocean. Ask students to find out more about what factors scientists include in their computerized models and report back to the class on some aspect of the study that interests them. LS Verbal

Teach, *continued*

Activity — GENERAL

Wave Refraction in a Tank Have individual groups of students investigate the effects of waves in shallow water. Divide the class into small groups. Give each group a clear plastic container or aquarium tank and a bowl or other solid object. Tell students to follow these instructions:

1. Fill the container or tank with a small amount of water.
2. Place the solid object on the bottom at one end. (Make sure students understand that the object represents a piece of land).
3. Direct waves toward the object using a plastic lid or wood block.

Ask students to describe what happens to the wave when it reaches the object. (The wave splits when it touches the object. Part of the wave lags behind and develops a curved front, or bend.) Ask students to name this bending process that occurs in shallow coastal waters. (refraction) **LS Visual/Kinesthetic**

Answer to Reading Check

Contact with the ocean floor slows down the bottom of the wave but not the top of the wave. Because of the difference in speed between the top and bottom of the wave, the top gets farther ahead of the bottom until the wave becomes unstable and falls over.

Waves and the Coastline

In shallow water near the coastline, the bottom of a wave touches the ocean floor. A wave touches the ocean bottom where the depth of the water is about half the wavelength. Contact with the ocean floor causes the wave to slow and eventually break, as shown in **Figure 5.**

Breakers

The height of a wave changes as the wave approaches the coastline. The water involved in the motion of a wave extends to a depth of one-half wavelength. As the wave moves into shallow water, the bottom of the wave is slowed. The top of the wave, however, continues to move at its original speed. The top of the wave gets farther and farther ahead of the bottom of the wave. Finally, the top of the wave topples over and forms a *breaker*, a foamy mass of water that washes onto the coastline. The height of the wave when the wave topples over is 1 to 2 times the height of the original wave.

Breaking waves scrape sediments off the ocean floor and move the sediments along the coastline. The waves also erode rocky coastlines. The size and force of breakers are determined by the original wave height, wavelength, and the steepness of the ocean floor close to the coastline. If the slope of the ocean floor is steep, the height of the wave increases rapidly and the wave breaks with great force. If the coastline slopes gently, the wave rises slowly. The wave spills forward with a rolling motion that continues as the wave advances up the coastline.

Reading Check **As a wave moves into shallow water, what causes the top of the wave to break and topple over? (See the Appendix for answers to Reading Checks.)**

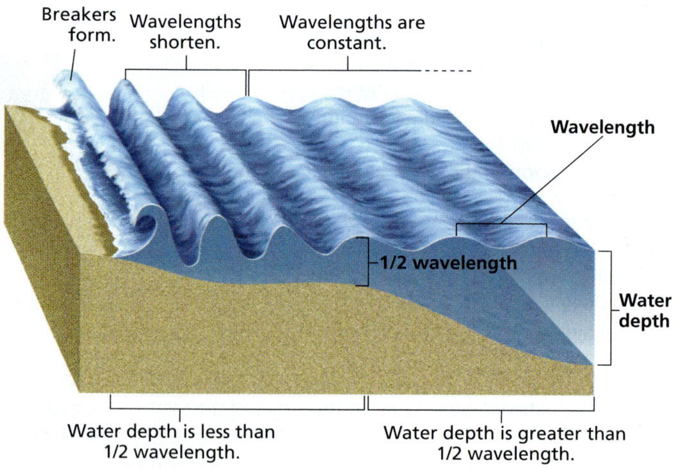

Figure 5 ▶ Breakers begin to form as the wave approaches the coastline. As a wave nears the coastline, wave height increases and wavelength decreases.

CHAPTER RESOURCES

Chapter Resource File

- **Inquiry Lab** Tsunami! GENERAL

Technology

Transparencies

- 103 The Formation of Breakers (with worksheet)

Discussion — BASIC

Water Safety Ask students to explain what a rip current is and how to identify one. (a strong seaward current created by channels underwater along sandbars or piers; muddy water caused by sand being stirred up, or a gap in a line of breakers) Encourage students to identify ways to protect themselves from these dangerous currents. (Learn to swim. Do not swim alone. If caught in a rip current, swim parallel to the beach until free.) **LS Logical**

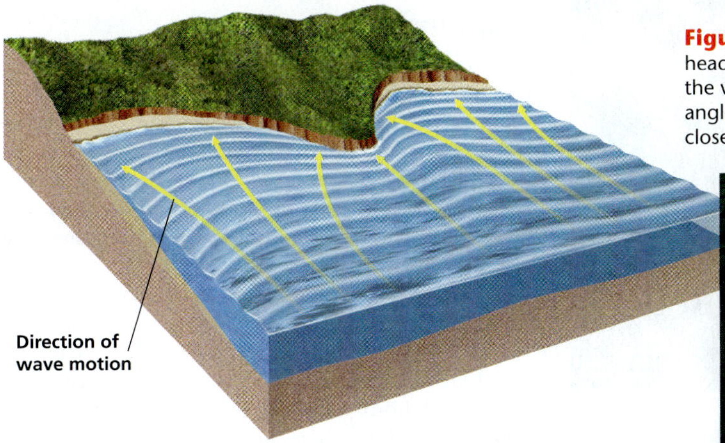

Figure 6 ▶ Waves strike the shore head-on as a result of refraction. Notice the waves approaching the shore at an angle. These waves bend as they draw closer to the shore.

Refraction

Most waves approach the coastline at an angle. When a wave reaches shallow water, however, the wave bends. This bending is called refraction. **Refraction** is the process by which ocean waves bend toward the coastline as they approach shallow water. As a wave approaches the coastline, the part of the wave that is in shallower water slows, and the part of the wave that is in deeper water maintains its speed. The wave gradually bends toward the beach and strikes the shore head-on, as shown in **Figure 6.**

refraction the process by which ocean waves bend directly toward the coastline as they approach shallow water, the part of the wave that is traveling in shallow water travels more slowly than the part of the wave that is still advancing in deeper water

Undertows and Rip Currents

Water carried onto a beach by breaking waves is pulled back into deeper water by gravity. This motion forms an irregular current called an *undertow.* An undertow is seldom strong, and only along shorelines that have steep drop-offs do undertows create problems for swimmers.

The generally weak undertow is often confused with the more dangerous *rip current.* Rip currents form when water from large breakers returns to the ocean through channels that cut through underwater sandbars that are parallel to the beach. Rip currents flow perpendicularly to shore through those channels and may be strong enough to quickly carry a swimmer away from shore. The presence of rip currents can usually be detected by a gap in a line of breakers or by turbid water, water in which sand has been stirred up by the current.

Longshore Currents

Longshore currents form when waves approach the beach at an angle. Longshore currents flow parallel to the shore. Great quantities of sand are carried by longshore currents. If there is a bay or inlet along the shoreline where waves refract, sand will be deposited as the energy of the waves decreases. These sand deposits form low ridges of sand called *sandbars*.

Close

Reteaching — BASIC

Learn by Teaching Divide the class into pairs or small groups. Have each pair or group create a presentation for younger students that uses diagrams or demonstrations to explain how wind causes the formation of ocean waves and how waves affect coastlines and erode beaches. **LS Interpersonal**

Quiz — GENERAL

Ask students to write the letter of the best answer to each of the following questions.

1. What characteristic describes the distance between two consecutive wave crests?
   a. wave period
   b. wave height
   c. wavelength
   d. wave speed
2. What causes ocean waves to bend toward the shore as they approach shallow water?
   a. fetch
   b. reflection
   c. rip currents
   d. refraction
3. Which of these factors causes sand to be deposited in ridges called sandbars?
   a. decrease in wave energy
   b. increase in wave height
   c. decrease in wavelength
   d. slope of the ocean floor
4. What type of wave or wave action appears to affect both ocean temperature and global climate?
   a. breakers
   b. swells
   c. tsunamis
   d. whitecaps

Cultural Awareness — BASIC

Tsunamis The literal translation of tsunami is "harbor wave." This name refers to the fact that these gigantic waves can fill an entire harbor and bury low-lying coastal areas. People all over the world call these waves by their Japanese name. Show the famous Japanese print, Hokusai's "The Great Wave Off Kanagawa," which hangs in the Metropolitan Museum of Modern Art in New York. The print depicts a tsunami off the Japanese coast. Point out that this work has influenced the work of Western impressionist artists such as Monet and Degas.

Homework — ADVANCED

Longshore Currents Have students research and write a report about longshore currents and their effects on the environment. Students may work individually or as a group. Students should consider how wave action shapes beaches and affects other coastal landforms, what human efforts are being made to control the movements of sand and sediment, and why these efforts are being made. The report should show examples of localities that are replenishing beach sand and building sea walls, beach dunes, and jetties. **LS Logical/Verbal**

Close, *continued*

Alternative Assessment — GENERAL

Wave Haikus The Japanese poetry form *haiku* is a 3-line poem that has five syllables in the first line, seven syllables in the second line, and five syllables in the last line. Have students write a haiku about waves. Tell them to incorporate what they have learned about how waves form and move. Encourage students to illustrate their work with pen and ink drawings. **LS Verbal**

Answers to Section Review

1. Wave speed equals wavelength divided by wave period.
2. Uneven heating of the atmosphere causes pressure differences that produce moving air, or winds. Waves form when friction between moving air and water causes ripples.
3. Three factors that determine wave size are length of time the wind blows in one direction; the speed of the wind; and the fetch, or distance the wind can blow across open water.
4. Most waves approach the coastline at an angle. As they get closer, the part in shallower water slows, while the part in deeper water keeps its speed. This causes the wave to bend gradually toward the shoreline.
5. Tsunamis are giant waves that are caused by underwater earthquakes, volcanic eruptions, and landslides.
6. Waves touch bottom at about one-half their wavelength. Contact with the ocean floor causes the wave to slow down.
7. No; where a wave breaks depends on original wave height and wavelength, which would vary over time.
8. Whitecaps reflect solar radiation, which allows less radiation to reach the ocean. If less thermal energy is transferred to the water, then there is less heat available to moderate climates along the path of the currents.
9. Two types of *waves* are *tsunamis* and *swells,* which have a constant *wave height,* which is the distance between a *crest* and a *trough,* and form when wind blows across a *fetch,* which may also form *whitecaps.*

Figure 7 ▶ The tsunami of 2004 left coastal communities of Sri Lanka in ruins. The tsunami was triggered by an earthquake that had a moment magnitude of 9.0.

Tsunamis

The most destructive waves in the ocean are not powered by the wind. *Tsunamis* are giant seismic ocean waves. Most tsunamis are caused by earthquakes on the ocean floor, but some can be caused by volcanic eruptions and underwater landslides. Tsunamis are commonly called *tidal waves,* which is misleading because tsunamis are not caused by tides.

Tsunamis have a long wavelength. In deep water, the wave height of a tsunami is usually less than 1 m, but the wavelength may be as long as 500 km. A tsunami commonly has a wave period of about 1 hour and may travel at speeds of up to 890 km/h (as fast as a jet airplane). Because the wave height of a tsunami is so low in the open ocean, the tsunami cannot be felt by people aboard ships.

Tsunami as a Destructive Force

A tsunami has a tremendous amount of energy. Because its wavelength is so great, the entire depth of the water is involved in the wave motion of a tsunami. All of the energy of this mass of water is released against the shore and causes a great deal of destruction, as shown in **Figure 7.** Near the shore, the height of the tsunami greatly increases as the tsunami's speed decreases. As it approaches the shore, the wave may reach a height of 30 to 40 m. The arrival of a tsunami may be signaled by the sudden pulling back of the water along the shore. This pulling back occurs when the trough of the tsunami arrives before the crest. If the crest arrives first, a sudden, rapid rise in the water level occurs.

Tsunamis generated by the earthquakes in Chile in 1960 and in the Indian Ocean in 2004 both caused wide-spread destruction. The Chilean tsunami struck the coast of South America and then Hawaii and crossed 17,000 km of ocean to strike Japan.

SCILINKS® NSTA Developed and maintained by the National Science Teachers Association

For a variety of links related to this subject, go to www.scilinks.org

Topic: Tsunamis
SciLinks code: HQ61561

Section 2 Review

1. **Explain** how wavelength and wave period can be used to calculate wave speed.
2. **Describe** the formation of waves.
3. **List** three factors that determine the size of a wave.
4. **Explain** why incoming waves refract toward the beach until they strike the shore head-on.
5. **Describe** what factors cause tsunamis.
6. **Explain** why waves slow down in shallow water.

CRITICAL THINKING

7. **Analyzing Processes** Would the breakers at a specific beach always form at the same distance from shore? Explain your answer.
8. **Predicting Consequences** Explain how whitecaps could affect climate.

CONCEPT MAPPING

9. Use the following terms to create a concept map: *wave, wave height, whitecap, trough, crest, fetch, swell,* and *tsunami.*

CHAPTER RESOURCES

Chapter Resource File
- Section Quiz GENERAL

Workbooks
- Study Guide (also in Spanish)

Section 3 Tides

The periodic rise and fall of the water level in the oceans is called the **tide.** *High tide* is when the water level is highest. *Low tide* is when the water level is lowest. The tide change is most noticeable on the coastline. If you stand on a beach long enough, you can see how the ocean retreats and returns with the tides.

OBJECTIVES

- **Describe** how the gravitational pull of the moon causes tides.
- **Compare** spring tides and neap tides.
- **Describe** how tidal oscillations affect tidal patterns.
- **Explain** how the coastline affects tidal currents.

KEY TERMS

tide
tidal range
tidal oscillation
tidal current

The Causes of Tides

In the late 1600s, Isaac Newton identified the force that causes the rise and fall of tides along coastlines. According to Newton's law of gravitation, the gravitational pull of the moon on Earth and Earth's waters is the major cause of tides. The sun also causes tides, but they are smaller because the sun is so much farther from Earth than the moon is.

As the moon revolves around Earth, the moon exerts a gravitational pull on the entire Earth. However, because the force of the moon's gravity decreases with distance from the moon, the gravitational pull of the moon is strongest on the side of Earth that is nearest to the moon. As a result, the ocean on Earth's near side bulges slightly, which causes a high tide within the area of the bulge.

tide the periodic rise and fall of the water level in the oceans and other large bodies of water

At the same time, another tidal bulge forms on the opposite side of Earth. This tidal bulge forms because the solid Earth, which acts as though all of its mass were at Earth's center, is pulled more strongly toward the moon than the ocean water on Earth's far side is. The result is a smaller tidal bulge on Earth's far side. **Figure 1** shows the Earth-moon system and the position of the moon in relation to the tidal bulges.

Low tides form halfway between the two high tides. Low tides form because as ocean water flows toward the areas of high tide, the water level in other areas of the oceans drops.

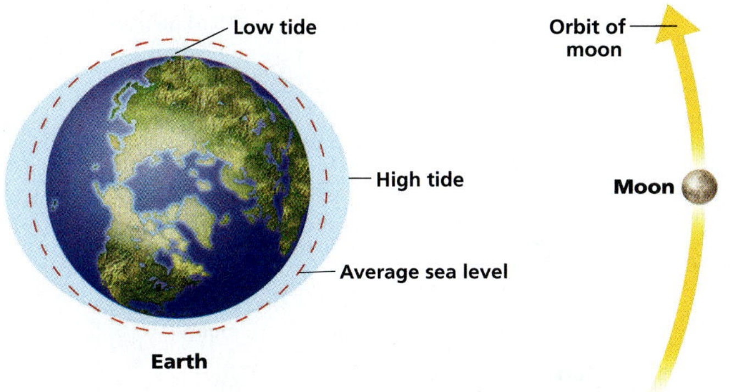

Figure 1 ▶ Tides form because the gravitational pull of the moon decreases with distance from the moon. Because of Earth's rotation, most locations in the ocean have two high tides and two low tides daily.

Section 3

Focus

Overview

This section explains how the moon's gravitational pull affects the waters of Earth. The section describes difference between spring and neap tides and identifies how variations in ocean depth, the shape of ocean basins, and coastline features affect tidal patterns.

Bellringer

Have students answer the following questions: What does the phrase "high and dry" mean? Where do you think this phrase comes from? (Boats anchored near shore during a high tide are left high and dry when the tide goes out.) LS **Intrapersonal**

Motivate

Using the Figure — GENERAL

Tidal Bulges The "circle of water" around the globe in the diagram uses artistic license to illustrate the concept of tidal bulges. The ocean is not really that deep. If it were, it would cover the land masses. At the scale in the figure, however, the tidal bulge would be too small to see if it were not exaggerated. LS **Visual**

CHAPTER RESOURCES

Chapter Resource File

- Directed Reading BASIC

Technology

Transparencies
- Bellringer

Student Edition on CD-ROM

One-Stop Planner CD-ROM
- Lesson Plan

Teach

MATHPRACTICE

Answer

$10.8 \text{ min} \times 60 \text{ s/min} = 648 \text{ s}$
$6.5 \times 10^7 \text{ years} \div 6.48 \times 10^2 \text{ s} =$
$1.0030864 \times 10^{7-2} =$
1.0030864×10^5 years;
Earth's rotation slows by 1 s in 100,308.6 years.

Using the Figure — GENERAL

Lunar Motions Ask students to analyze the illustration of spring tides and neap tides. Ask students to identify the phases of the moon that would coincide with the highest tides. (full moon and new moon) Answer to caption question: Spring tides occur twice a month. **LS Visual**

Answer to Reading Check

When the tidal range is small, the sun and the moon are at right angles to each other relative to Earth's orbit.

CHAPTER RESOURCES

Technology

Transparencies
- 104 Spring Tides and Neap Tides (with worksheet)

MISCONCEPTION ALERT

Spring Tides Students may think that spring tides have something to do with the season of spring. Emphasize that spring tides occur at all seasons of the year around the time of new and full moons.

MATHPRACTICE

Tidal Friction As the tidal bulges move around Earth, friction between the water and the ocean floor slows Earth's rotation slightly. Scientists estimate that the average length of a day has increased by 10.8 min in the last 65 million years. How many years does it take for Earth's rotation to slow by 1 s?

Behavior of Tides

Earth rotates on its axis once every 24 h. In that 24 h, the moon moves through about 1/29 of its orbit. Because the moon orbits Earth in the same direction that Earth rotates, all areas of the ocean pass under the moon every 24 h 50 min. As seen from above the North Pole, Earth rotates counterclockwise and the tidal bulges appear to move westward around Earth.

Because there are two tidal bulges, most locations in the ocean have two high tides and two low tides daily. The difference in levels of ocean water at high tide and low tide is called the **tidal range.** The tidal range can vary widely from place to place. Because the moon rises about 50 minutes later each day, the times of high and low tides are about 50 minutes later each day.

tidal range the difference in levels of ocean water at high tide and low tide

Spring Tides

The sun's gravitational pull can add to or subtract from the moon's influence on the tides. During the new moon and the full moon, Earth, the sun, and the moon are aligned, as shown in **Figure 2.** The combined gravitational pull of the sun and the moon results in higher high tides and lower low tides. So, the daily tidal range is greatest during the new moon and the full moon. During these two monthly periods, tides are called *spring tides.*

Neap Tides

During the first- and third-quarter phases of the moon, the moon and the sun are at right angles to each other in relation to Earth, also shown in **Figure 2.** The gravitational forces of the sun and moon work against each other. As a result, the daily tidal range is small. Tides that occur during this time are called *neap tides.*

Reading Check **Describe the location of the sun and moon in relation to Earth when the tidal range is small. (See the Appendix for answers to Reading Checks.)**

Figure 2 ▶ The alignment of the sun, moon, and Earth during spring tides differs from their alignment during neap tides. ***How often do spring tides occur?***

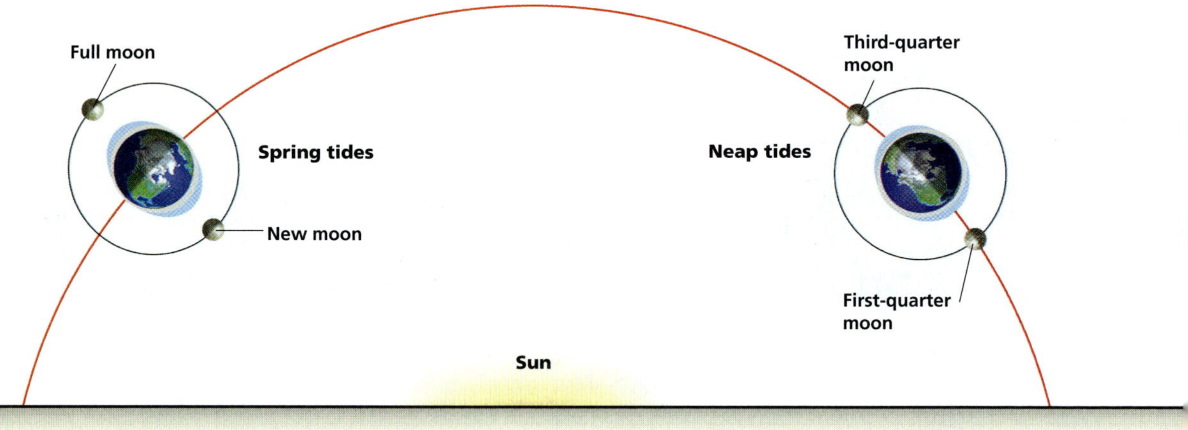

SKILL BUILDER — ADVANCED

Graphing Use the Internet or local newspapers to obtain a listing of the times of high and low tides for a month. Be aware that some sources may use military time, which students will have to convert to standard time. Give students a calendar that has the phases of the moon marked. Have students create a graph that lists the dates and times for one month on the horizontal axis. The vertical axis should be the water height above or below sea level in meters. Tell students to note the phase of the moon beside the appropriate dates by using the following symbols: an empty circle (new moon), two half-filled circles (first quarter and last quarter moons), and a fully darkened circle (full moon). Have students plot the tidal data from the Internet or the newspaper on this graph. Have them analyze the data and record how many high and low tides occurred during a 24-hour period. Encourage them to note how the times for the tides vary each day and that the moon's phases correlate with the tidal cycle. **LS Logical**

Tidal Variations

Although the global ocean is one body of water, continents and irregularities in the ocean floor divide the ocean into several basins. Tidal patterns are greatly influenced by the size, shape, depth, and location of the ocean basin in which the tides occur.

Along the Atlantic Coast of the United States, two high tides and two low tides occur each day and have a fairly regular tidal range. Along the shore of the Gulf of Mexico, however, only one high tide and one low tide occur each day. Along the Pacific Coast, the tides follow a mixed pattern of tidal ranges. Pacific Coast tides commonly have a very high tide followed by a very low tide and then a lower high tide, followed by a higher low tide.

Tidal Oscillations

Tidal patterns are also affected by tidal oscillations. **Tidal oscillations** (TIE duhl AHS uh LAY shunz) are slow, rocking motions of ocean water that occur as the tidal bulges move around the ocean basins. Along straight coastlines and in the open ocean, the effects of tidal oscillations are not very obvious. In some enclosed seas, such as the Baltic and Mediterranean Seas, tidal oscillations reduce the effects of the tidal bulges. As a result, these seas have a very small tidal range. However, in small basins and narrow bays located off major ocean basins, tidal oscillations may amplify the effects of the tidal bulges. An example of the effects of tidal oscillations is shown in **Figure 3.**

Figure 3 ▶ A great tidal range of as much as 15 m in the V-shaped Bay of Fundy in Canada is caused by tidal oscillations.

tidal oscillation the slow, rocking motion of ocean water that occurs as the tidal bulges move around the ocean basins

Connection to ASTRONOMY

The Earth-Moon System

The force of attraction that exists between all matter in the universe is called *gravity.* The force of attraction between any two objects depends on their masses and the distance between them. Earth's mass is estimated to be 6.0×10^{24} kg, and the moon's mass is estimated to be 7.3×10^{22} kg. The average distance between Earth and the moon is 384,000 km. These masses and distance make the attractive force of gravity strong between Earth and the moon. Because the gravitational forces between these two bodies is strong, the moon causes tides on Earth's surface. At the same time, Earth's gravity affects the rate at which the moon rotates.

Earth rotates on its axis once every 24 hours. While Earth completes one rotation, the moon travels only 1/29 of the way around Earth relative to the sun. As a result, the moon rises and sets about 50 minutes later each day. However, relative to the stars, the moon revolves around Earth once every 27.3 days.

In addition to orbiting Earth and revolving around the sun, the moon spins very slowly on its axis. It rotates on its axis once every 27.3 days, or about once every lunar cycle. Because the rotation and revolution of the moon take the same amount of time, observers on Earth always see the same side of the moon.

Discussion — GENERAL

Bay of Fundy Use the two photographs on this page to initiate a discussion about why tides in the Bay of Fundy are among the highest in the world. Explain that the period of the tidal oscillation depends on the basin's length and depth. The rocking movement in the Bay of Fundy coincides with the tides, which drives the water farther into the bay, reinforcing the lunar tide. **LS Logical**

Close

Reteaching — BASIC

Tidal Games Divide the class into pairs. Have students work together to prepare questions, with answers, about causes of tides and tidal patterns. Have them use the questions in a game show style review to study for the section assessment. **LS Auditory/ Interpersonal**

Quiz — GENERAL

Determine whether each of the following statements is true or false. If false, correct the statement to make the statement true.

1. The major cause of tides is the gravitational pull of the sun. (false, moon)
2. When the sun and moon are in line with each other, neap tides occur. (false, spring tides)
3. Movement of tidal waters toward the coast is known as ebb tide. (false, flood tide)
4. A tidal bore may occur when ocean water enters a river through a narrow bay. (true)

ASTRONOMY CONNECTION — GENERAL

Same Face Explain that Earth's gravity pulling on the moon has slowed the moon's rate of rotation. As a result, the moon always presents the same face (side) to people on Earth. Use this individual activity to show that the moon's orbital period exactly equals its rotational period. Have students draw a large circle with a compass on a sheet of paper to represent the moon's orbit. Then, have them make a smaller circle in the center and label it Earth. Next, have them place a round sticky label on one side of a pencil and move the pencil around the large circle once, keeping the label facing the smaller circle. Ask students: How many times during one orbit does the moon rotate on its axis around Earth? (once) This is why the moon always keeps the same face pointing toward Earth. It also means that the lunar year (orbital period) is the same length as its day (rotational period). Ask them: About how long are both lunar periods? (27.3 days) **LS Visual**

Close, *continued*

Alternative Assessment — ADVANCED

Making Mobiles Have students create mobiles of the sun-Earth-moon system to illustrate how the relationships among these three bodies cause tides to form. Remind them that Earth and the moon spin around a common center of gravity. Have students use these models to explain tides. LS **Kinesthetic**

Answers to Section Review

1. As the moon revolves around Earth, its gravity exerts a pull on Earth's waters. Two high tidal bulges are created on opposite sides of the planet. Low tides occur in between.
2. The sun contributes to the size of tides by pulling on Earth's waters in the same direction as the moon or in the opposite direction.
3. When the sun and moon pull together in a straight line, high spring tides occur. When the sun and moon are arranged at right angles to each other, their pulls oppose, causing lower neap tides.
4. Variations in the size, shape, depth, and location of the ocean basins with respect to landmasses affect the number of tides, as well as tidal range.
5. In enclosed seas, tidal oscillations reduce the effects of the tidal bulges, producing small tidal range.
6. Tidal currents in open oceans are much smaller than those at the coasts. In narrow bays, tidal oscillations magnify the effects of tidal bulges, producing larger tidal ranges than in the open ocean.
7. Where a river enters an ocean through a long bay, a surge of tidal water may flow upstream in a wave called a *tidal bore*.
8. Where adjacent coastal regions have differences in heights of the tides, ships approaching these coasts must watch out for strong tidal currents. Also, in bays or areas that have narrow coastlines, ships must deal with rapid tidal currents.
9. Answers may vary. Accept all reasonable answers.
10. The moon's gravity produces *tides*, which result in *tidal ranges* that are affected by *spring tides, neap tides, tidal oscillations,* and *tidal currents,* which include *ebb tides* and *flood tides.*

Figure 4 ▶ The photo to the right shows a tidal bore in early spring at Turnagain Arm of Cook Inlet, Alaska.

tidal current the movement of water toward and away from the coast as a result of the rise and fall of the tides

SCILINKS. NSTA Developed and maintained by the National Science Teachers Association

For a variety of links related to this subject, go to www.scilinks.org

Topic: Tides

SciLinks code: HQ61525

Tidal Currents

As ocean water rises and falls with the tides, it flows toward and away from the coast. This movement of the water is called a **tidal current.** When the tidal current flows toward the coast, it is called *flood tide.* When the tidal current flows toward the ocean, it is called *ebb tide.* When there are no tidal currents, the time period between flood tide and ebb tide is called *slack water.*

Tidal currents in the open ocean are much smaller than those at the coastline. Tidal currents are strongest between two adjacent coastal regions that have large differences in the height of the tides. In bays and along other narrow coastlines, tides may create rapid currents. Some tidal currents may reach speeds of 20 km/h.

Where a river enters the ocean through a long bay, the tide may enter the river mouth and create a *tidal bore,* a surge of water that rushes upstream, such as the one shown in **Figure 4.** In some cases, the tidal bore rushes upstream in the form of a large wave up to 5 m high that eventually loses energy. The tidal bores in the River Severn in England travel almost 20 km/h and reach as far as 33 km inland.

Section 3 Review

1. **Describe** how the moon causes tides.
2. **Explain** how the sun can influence the moon's effect on tides.
3. **Compare** spring tides and neap tides.
4. **Describe** how ocean basins affect tidal patterns.
5. **Explain** how tidal oscillations in an enclosed sea would affect tidal patterns in that sea.
6. **Compare** the movement of ocean water in the open ocean with the movement of ocean water in narrow bays.
7. **Describe** how a tidal bore forms.

CRITICAL THINKING

8. **Predicting Consequences** Predict where tidal currents may be of concern to ships that are approaching the land.
9. **Identifying Relationships** Describe ways in which tides could be affected if Earth had two moons.

CONCEPT MAPPING

10. Use the following terms to create a concept map: *tide, tidal range, spring tide, neap tide, tidal oscillation, tidal current, flood tide,* and *ebb tide.*

CHAPTER RESOURCES

Chapter Resource File

- **Section Quiz** GENERAL

Workbooks

- **Study Guide** (also in Spanish)

Chapter 21 Highlights

Sections

1 Ocean Currents

Key Terms

- **current,** 519
- **surface current,** 519
- **Coriolis effect,** 520
- **gyre,** 520
- **Gulf Stream,** 522
- **deep current,** 523

Key Concepts

- ▶ Surface currents of the ocean are the result of global wind belts, the Coriolis effect, and continental land barriers.
- ▶ Major ocean currents, such as the Gulf Stream and the Canary Current, help create gyres.
- ▶ Deep currents are produced as dense water near the North and South Poles sinks and moves toward the equator beneath less-dense water.
- ▶ The Antarctic Bottom Water current is a deep-water current that moves slowly north along the ocean bottom.

2 Ocean Waves

Key Terms

- **wave,** 525
- **wave period,** 525
- **fetch,** 527
- **refraction,** 529

Key Concepts

- ▶ The speed of a wave can be calculated by dividing the wavelength by the wave period.
- ▶ Wind is the primary source of wave energy.
- ▶ Wave size is determined by wind speed, by the length of time wind blows, and fetch.
- ▶ As a wave comes into contact with the ocean floor, the wave may undergo refraction or form breakers.
- ▶ Waves near the shoreline can cause currents such as an undertow and rip current.
- ▶ Tsunamis are giant, destructive waves.

3 Tides

Key Terms

- **tide,** 531
- **tidal range,** 532
- **tidal oscillation,** 533
- **tidal current,** 534

Key Concepts

- ▶ The gravitational effects of the moon and, to a lesser extent, the sun cause tides.
- ▶ Tidal ranges are greatest during spring tides and smallest during neap tides.
- ▶ Variations in the tides are influenced by the size, shape, depth, and location of the ocean basins or seas in which the tides occur.
- ▶ Tidal currents are generally small in the open ocean but may create rapid currents in narrow bays along the coastline.

Chapter Highlights

Alternative Assessment — GENERAL

Wave Dioramas Have groups of students create dioramas or models representing the movement of waves, tides, and currents and their effects on coastal areas. Students may have seen representations like these in museums. They can use empty shoeboxes to construct their displays. Those groups that are illustrating tides may create astronomical models. Encourage students to draw a sketch or plan before they start building their model. They can remove the box top to create a large viewing window, and use marking pens, paints, and construction paper to decorate the back and the sides of the shoebox. Students can use pictures cut from magazines or draw typical seashore scenery inside. The setting might also include foreground sand and small rocks to represent a beach. After students finish the background, they can use modeling clay or other art materials to create models of the typical seaside land features. They can also add charts, diagrams, or captions to help viewers interpret their work. LS Visual

CHAPTER RESOURCES

Chapter Resource File

- Concept Review GENERAL
- Critical Thinking ADVANCED
- Math Skills GENERAL
- Graphing Skills GENERAL
- Chapter Test A GENERAL
- Chapter Test B ADVANCED

Workbooks

- **Study Guide** (also in Spanish)
- **Assessments** (Spanish)

Technology

Classroom Videos
- Brain Food Video Quiz

HRW Earth Science Video
- Segment 15: Tides

Assignment Guide

Section	Questions
1	1–2, 5–7, 9–12, 15–16, 19, 21, 24, 27, 29
2	4, 13–14, 17, 20, 22, 25, 30
3	3, 8, 18, 23, 26, 31–37
1–3	28

Chapter 21 Review

Using Key Terms

Use each of the following terms in a separate sentence.

1. *current*
2. *gyre*
3. *tide*
4. *wave*

For each pair of terms, explain how the meanings of the terms differ.

5. *surface current* and *deep current*
6. *Coriolis effect* and *gyre*
7. *fetch* and *refraction*
8. *tidal range* and *tidal oscillation*

Understanding Key Concepts

9. The water in the ocean moves in giant streams called
   - **a.** currents.
   - **b.** westerlies.
   - **c.** waves.
   - **d.** tides.

10. The effect of Earth's rotation on winds and ocean currents is called the
    - **a.** neap-tide effect.
    - **b.** refraction effect.
    - **c.** Coriolis effect.
    - **d.** tsunami effect.

11. Which of the following currents is the westward warm-water current in the North Atlantic Gyre?
    - **a.** Canary Current
    - **b.** North Atlantic Current
    - **c.** North Equatorial Current
    - **d.** Gulf Stream

12. Deep currents are the result of
    - **a.** the Coriolis effect.
    - **b.** changes in the density of ocean water.
    - **c.** the trade winds.
    - **d.** neap tides.

13. The periodic disturbance in water as energy is transmitted through the water is a
    - **a.** current.
    - **b.** breaker.
    - **c.** fetch.
    - **d.** wave.

14. The highest point of a wave is the
    - **a.** trough.
    - **b.** period.
    - **c.** crest.
    - **d.** length.

15. The distance that wind blows across an area of the sea to generate waves is the
    - **a.** trough.
    - **b.** sargassum.
    - **c.** fetch.
    - **d.** wave period.

16. The movement of water toward and away from the coasts due to tidal forces is called a
    - **a.** tidal bore.
    - **b.** tidal current.
    - **c.** tidal range.
    - **d.** tidal oscillation.

Short Answer

17. Describe how a breaker forms.
18. Define *tide,* and explain why tides form.
19. What factors control most ocean surface currents?
20. How does the depth of the ocean floor affect the shape and speed of a wave?
21. How do deep currents form?
22. Explain how wind is the primary source for wave energy.
23. What is a tidal bore?

Using Key Terms

1–8. Answers may vary but should show that students understand the definitions of and differences between key terms.

Understanding Key Concepts

9. a **10.** c
11. c **12.** b
13. d **14.** c
15. c **16.** b

Short Answer

17. Breakers form as waves approach the coast. As a wave moves into shallow water, the bottom of the wave is slowed. The top of the wave continues at its original speed. The top gets farther ahead of the bottom, until the wave becomes unstable and topples over.

18. A tide is the regular rise and fall of the level of the ocean in response to the gravitational pull of the moon on Earth's waters. This produces two tidal bulges on opposite sides of Earth. Low tides form as ocean waters flow toward the areas of high tide.

19. Ocean surface currents are controlled by global wind patterns; by Earth's spinning motion, which cause the currents to flow in huge circles called *gyres*; and by the location of continents, which act as barriers to the current flow.

20. As waves approach a coast, the part in shallow water slows down. The part that is in deeper water maintains its speed, causing the waves to bend, or refract.

21. Deep currents form as a result of water density differences. The denser water sinks and flows below the less-dense water. Factors that produce these density differences include salt concentration, temperature differences, and turbidity.

22. Uneven heating of the atmosphere produces pressure differences that cause winds. Winds transfer their energy to the ocean water through friction between moving air and the water surface.

23. A tidal bore is a surge of ocean water that rushes upstream through a narrow bay or river mouth.

Critical Thinking

24. The direction of monsoon winds will directly affect the direction in which surface currents flow in the northern part of the Indian Ocean. The surface currents will change direction when winds change direction with the seasons.

Critical Thinking

24. Determining Cause and Effect During winter in the northern Indian Ocean, winds called *monsoons* blow in a direction opposite to the direction that they blow during summer. What effect do these winds have on surface currents?

25. Analyzing Processes Suppose that a retaining wall is built along a shoreline. What will happen to waves as they pass over the retaining wall?

26. Making Inferences Imagine that you are fishing from a small boat anchored off the shore of the Gulf of Mexico. You are lulled to sleep by the gently rocking boat but wake up to find your boat on wet sand. What happened?

27. Making Predictions What effect would Earth's rotating in the direction opposite that in which it now rotates have on the movement of ocean currents?

Concept Mapping

28. Use the following terms to create a concept map: *currents, surface currents, trade winds, deep currents, Coriolis effect, wave, breaker, rip current, tide, Antarctic Bottom Water* and *tidal current*.

Math Skills

29. Applying Quantities The Gulf Stream can move 100 million cubic meters of water per second. The Mississippi River moves 15,400 m^3 of water per second. How many times more water per second does the Gulf Stream move than the Mississippi River does?

30. Making Calculations If a wave has a wavelength of 216 m and a period of 12 s, what is the wave's speed?

Writing Skills

31. Writing from Research Write a report that describes the La Rance, France, tidal power plant project, the amount of electricity provided by the project, and the impact of the project on the environment of the area.

32. Outlining Topics Create an outline that shows the steps of tide formation. Provide diagrams as needed to illustrate the steps.

Interpreting Graphics

The graph below shows the measurements of tides in one location on the Atlantic coast of North America. Use the graph to answer the questions that follow.

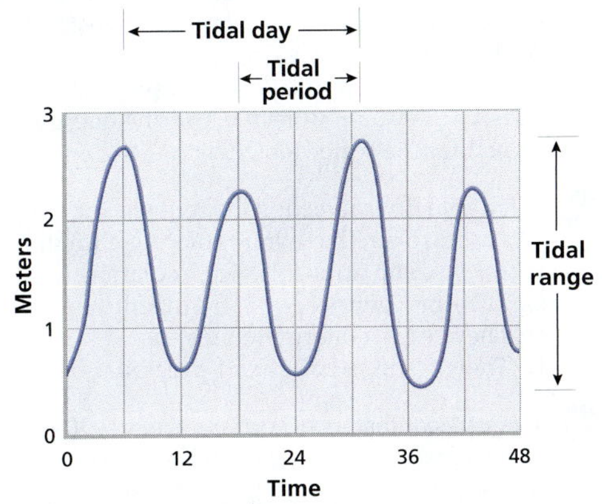

33. How many high tides occur every day in this location?

34. How many hours is the tidal period?

35. How many hours apart are the low tides?

36. What is the average tidal range in meters?

37. What is the difference in height (in meters) between the first high tide and the second high tide?

Chapter Review

25. Answers may vary. Sample answer: Retaining walls represent a barrier between waves and the shore. The waves are deflected back from the walls onto the adjoining beaches.

26. There is only one high tide and one low tide in the Gulf of Mexico. The boat was anchored offshore during high tide, and the ebb tide swept the waters away from the coast, leaving the boat aground on the sand.

27. If Earth rotated in the opposite direction, both the global wind patterns and the surface currents would be affected. Gyres would flow counterclockwise in the Northern Hemisphere and clockwise in the Southern Hemisphere.

Concept Mapping

28. Answers may vary but should include all of the terms listed. Sample answers appear at the end of this Teacher Edition.

Math Skills

29. 100,000,000 $m^3/s \div$ 15,400 m^3/s = 6494 times as much water

30. $$\textit{wave speed} = \frac{\text{wavelength } (\lambda)}{\text{wave period}}$$

$$\textit{wave speed} = 216\text{ m} \div 12\text{ s} = 18\text{ m/s}$$

Writing Skills

31. Answers may vary. Accept all reasonable answers based upon student research.

32. Answers may vary. Accept all reasonable answers. Outlines should include key steps in the process of tide formation and illustrative diagrams as needed.

Interpreting Graphics

33. 2 high tides daily

34. 12 hours

35. 12 hours

36. 2 meters

37. 0.3 meters

Estimated Time

To give students practice under more realistic testing conditions, allow them 30 minutes to answer all of the questions in this practice test.

TEST DOCTOR

Question 2 Answer C is correct. The time required for two consecutive wave crests to pass a given point is the wave period. Wave speed is calculated by dividing the wavelength by the wave period. Answers must be given in units of length divided by time, such as meters per second.

Question 3 Answer A is correct. Energy from wind moves the wave across the ocean surface and passes from molecule to molecule as the wave moves horizontally. A water molecule moves in a circle and ends up where it started, but the molecule moves very little, if at all, in the direction of the wave.

Question 4 Answer F is true because when the sun heats Earth's surface, the surface heats the atmosphere above it. This causes winds that circulate air. Answer G is incorrect because convection currents actually help drive deep-water currents. Answer H is incorrect because they redistribute energy globally. Answer I is incorrect because precipitation is not balanced across the Earth's surface.

Chapter 21 Standardized Test Prep

Understanding Concepts

Directions (1–5): **For** *each* **question, write on a separate sheet of paper the letter of the correct answer.**

1. Which of the following factors is a cause of surface currents?
   A. Earth's rotation on its axis
   B. water salinity
   C. human activity
   D. sea-floor spreading

2. What is the speed of an ocean wave that has 12 s between crests and a wavelength of 216 m?
   F. 6 m/s
   G. 3 km
   H. 18 m/s
   I. 12 m

3. When an ocean wave travels 100 m west, which of the following also travels 100 m west?
   A. the energy in the wave
   B. the water molecules in the wave
   C. both the water molecule and the energy
   D. neither the water molecule nor the energy

4. What role do convection currents in the ocean and atmosphere have in regulating climate?
   F. They set up atmospheric circulation.
   G. They prevent deep-water currents.
   H. They restrict energy to local use.
   I. They ensure a balance of precipitation.

5. The vertical distance from the trough of a wave to the crest of a wave is called the
   A. wave height
   B. wave length
   C. wave speed
   D. wave distance

Directions (6–8): **For** *each* **question, write a short response.**

6. What is a main factor that causes the movements of deep-water currents?

7. What happens to a wave's height as the wave approaches the shore?

8. Most waves are generated by energy transferred to water from what?

Reading Skills

Directions (9–11): **Read the passage below. Then, answer the questions.**

Tsunamis

Tsunamis are the most destructive waves in the ocean. Most tsunamis are caused by earthquakes on the ocean floor, but some can be caused by volcanic eruptions and underwater landslides. Tsunamis are sometimes called *tidal waves*, which is misleading because tsunamis have no connections to tides.

Tsunamis commonly have a wave period of about 1 hour and a wave speed of about 890 km/h, which is about as fast as a commercial airplane. By the time the tsunami reaches the shore, the tsunami's height may be 40 m.

Tsunamis can travel thousands of kilometers. One tsunami was triggered by an earthquake off the coast of South America in 1960. The tsunami was so powerful that it crossed the Pacific Ocean and hit the city of Hilo, on the coast of Hawaii, approximately 10,000 km away. The same tsunami then continued and struck Japan.

9. Why is the word *misleading* used to describe the use of the term *tidal waves* in the reading passage?
   A. Tsunamis are really large tides.
   B. Tsunamis can cause extensive damage to coastal areas.
   C. Tsunamis are related to earthquakes.
   D. Tsunamis are not related to tides.

10. Which of the following statements is a fact from the passage?
   F. All tsunamis are caused by earthquakes.
   G. A tsunami can travel as fast as an airplane.
   H. The tsunami of 1960 only struck Japan.
   I. Tsunamis are caused by surface currents.

11. Once triggered, how far can a tsunami travel?
   A. Tsunamis are short-lived and usually dissipate within just a few kilometers.
   B. Tsunamis travel about 100 km before dissipating in the ocean.
   C. Tsunamis travel about 1,000 km before dissipating in the ocean.
   D. Tsunamis can travel thousands of kilometers before dissipating or striking land.

Answers

Understanding Concepts

1. A
2. H
3. A
4. F
5. A
6. water density differences
7. Its height increases.
8. wind

Reading Skills

9. D
10. G
11. D

Interpreting Graphics

12. F
13. Answers may vary. See Test Doctor for a detailed scoring rubric.
14. A
15. Answers may vary. See Test Doctor for a detailed scoring rubric.

Interpreting Graphics

Directions (12–15): **For *each* question below, record the correct answer on a separate sheet of paper.**

Base your answers to questions 12 and 13 on the diagrams of the Earth, moon, and sun system below.

Effect of Sun and Moon on Earth's Tides

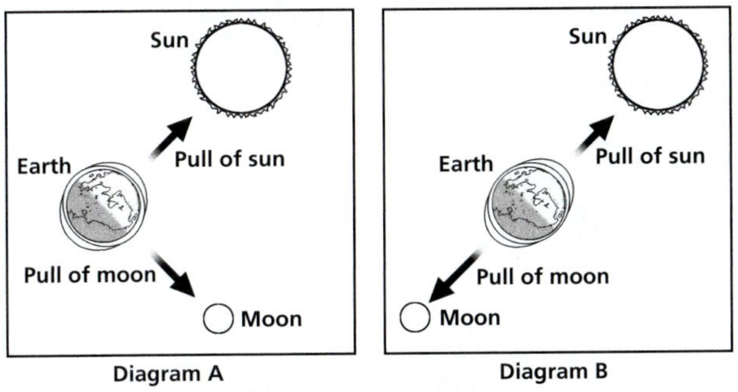

12 What type of tide is produced by the arrangement in Diagram B?
F. spring tide
G. neap tide
H. winter tide
I. weak tide

13 Using the diagrams above, explain in general terms how the gravitational effects of astronomical bodies cause tides on Earth.

Base your answers to questions 14 and 15 on the climate graphs shown below, which combine temperature and precipitation data for San Francisco, California and Wichita, Kansas.

Average Yearly Weather Data for San Francisco and Wichita

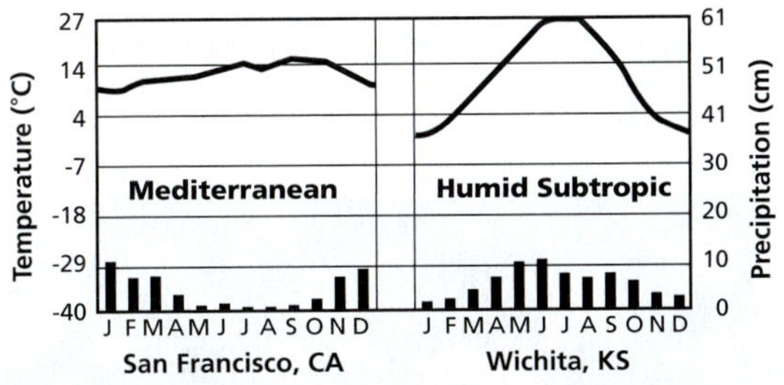

14 Which location shows the most extreme climate variation?
A. Wichita, Kansas shows the most extreme climate variation.
B. San Francisco, California shows the most extreme climate variation.
C. Both climates are equally mild.
D. Both climates are equally variable.

15 How do the location of these cities and the nearby currents help explain the differences in their climates?

Test TIP

Allow a few minutes at the end of the test-taking period to check for careless mistakes such as marking two answers for a single question.

Standardized Test Prep

TEST DOCTOR

Question 13 Full-credit answers should include the following points:
- students should show an understanding that the moon's gravity, and to a lesser extent the sun's gravity, tugs on the surface of Earth and its waters
- as the moon revolves around Earth, the moon exerts a gravitational pull on Earth's surface and its ocean waters
- two high tidal bulges are created on opposite sides of the planet as the moon's gravity affects Earth. Low tides occur in between these bulges
- because the waters flow more easily they are more affected than the solid Earth is

Question 15 Full-credit answers should include the following points:
- students should use concepts of the differential energy absorption patterns of water and land to explain the observed differences in climate
- an understanding that water heats and cools more slowly than does land
- cities that are located near large bodies of water have fewer temperature variations because of the mediating effect of the water

Test Prep Correlations *National Science Education Standards*

ES 1c: items 4, 6, 15
ES 1d: items 14, 15
ES 3c: items 1, 4, 7, 8, 10, 11, 12, 13, 14, 15
PS 6a: items 3, 8
SPSP 4a: items 1, 4, 10, 11, 12, 13, 14, 15
SPSP 5a: items 10, 11
SPSP 5c: items 10, 11
SPSP 5d: items 10, 11
UCP 1: items 1, 4, 6, 12, 13, 14, 15
UCP 2: items 12, 13, 14, 15

CHAPTER RESOURCES

State Resources

For specific resources for your state, visit go.hrw.com and type in the keyword HSHSTR.

Wave Motion

Teacher's Notes

Time Required

one 45-minute lab period

Lab Ratings

EASY → HARD

TEACHER PREPARATION 1

STUDENT SETUP 1

CONCEPT LEVEL 2

CLEANUP 1

Skills Acquired

- Experimenting
- Collecting Data
- Organizing and Analyzing Data
- Analyzing Relationships
- Interpreting
- Communicating

The Scientific Method

In this lab, students will

- Make Observations
- Analyze the Results
- Draw Conclusions

Materials

The materials listed are enough for groups of two to four students. You could substitute small wooden beads with a diameter large enough to slip over rope for ties.

Tips and Tricks

You may wish to review graphing techniques with students before they begin this lab.

Objectives

- **Model** the movement of waves.
- **Compare** the characteristics of waves when wave speed changes.

Materials

cloth ties, about 50 cm in length (2)
marker
meterstick
paper, 2 m × 1 m
paper, graph
pen or pencil, colored (3)
rope, thin, 2.5 m in length

Wave Motion

The source of wave movement in water is energy, which is generated primarily from wind. Waves of water appear to move horizontally. However, only the energy of the waves moves horizontally; the water moves horizontally very little. In this lab, you will work with two partners to simulate wave motion and to observe how energy generates wave motion in water. You will also observe the properties of waves.

PROCEDURE

1. Tie one end of the rope securely to the leg of a chair or table.
2. On the large sheet of paper, use the meterstick to draw a grid like the one shown in the illustration on the next page. Draw and label the grid using the measurements shown.
3. Place the sheet of paper on the floor, and line up the rope along the 2 m line of the grid.
4. To make waves, move the free end of the rope from side to side. (Note: Be sure to maintain a constant motion with the rope.)
5. While one person moves the rope, another person marks the paper where a crest of a wave hits. The third group member marks the paper where a trough of a wave hits.
6. On the graph paper, make a graph that has wavelength (in meters) as the x-axis and wave height (in meters) as the y-axis.

Step 5

CHAPTER RESOURCES

Chapter Resource File

- Datasheet for Chapter Lab GENERAL
- Lab Notes and Answers

Workbooks

Long-Term Projects

- Tides at the Shoreline ADVANCED

7. Plot a wave that represents the wave that you observed in step 5. Plot the wave height and the wavelength. Indicate the direction of the wave's motion.

8. Move the rope at a fast speed. Do not change the side-to-side distance that you move the free end of the rope.

9. As soon as a constant motion has been established, repeat steps 5 and 6. Plot a wave that represents the wave that you observed when repeating step 5. Use a pen or pencil whose color differs from the color of the first wave plot.

10. Next, generate very small waves. Repeat steps 5 and 6 using a third color of pen or pencil.

11. On your graph, label a crest and a trough on each of the waves that you plotted. Measure the wave height and wavelengths of each wave that you plotted.

12. Use the following formula to calculate the wave speeds of the three waves represented on the graph if each wave period is 6 s:

$$\textit{wave speed} = \frac{\textit{wavelength}}{\textit{wave period}}$$

13. Tie the two pieces of cloth around the middle of the rope about 15 cm apart.

14. Make waves by moving the end of the rope from side to side. Observe and record the motion of the cloth ties relative to the motion of the waves.

ANALYSIS AND CONCLUSION

1. **Examining Data** How do the wave motions of the waves shown on your graph differ from each other? If these waves were real water waves, what might be the cause(s) of the differences in motion?

2. **Recognizing Relationships** How is the movement of the rope similar to wave movement in water?

3. **Analyzing Relationships** How do the motions of the cloth ties differ from the wave's motion?

4. **Drawing Conclusions** What do the motions of the cloth ties tell you about wave movement in water?

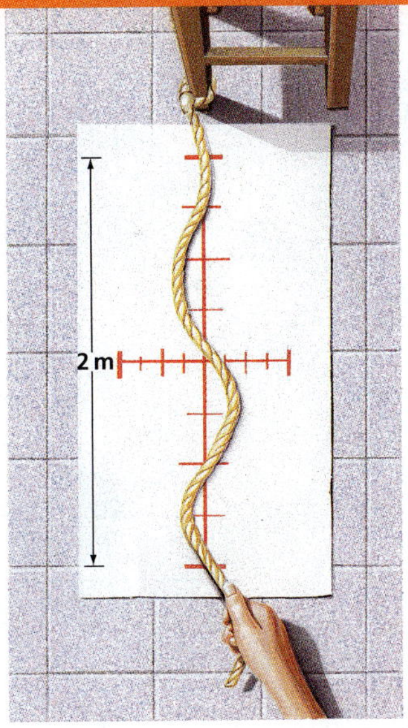

Extension

1. **Making Comparisons** Use a 4 m rope to repeat the investigation. Construct a graph similar to your first graph, but extend the *x*-axis to provide room to plot a 4 m length. Observe and plot five waves of varying speeds and heights. Compare waves generated on a 2 m rope with those generated on a 4 m rope. Describe your results. Using 6 s as the wave period for each wave that you plot, calculate wave speeds.

Making Models Lab

Answers to Analysis and Conclusion

1. Answers may vary. Students should note that wave motion in the water varies according to certain wind factors such as strength, fetch, and the length of time the wind blows.
2. The waves created on the rope moved toward the secured end, just as waves move toward the shore. Also, the energy of hand movements determines the speed of the rope motion, just as energy from the wind determines the speed of waves.
3. The ties move from side to side, while the waves move along the rope toward the chair or table leg.
4. As waves pass, the ties move side to side with little forward movement. These motions show that water molecules actually move forward very little as a wave passes.

Answers to Extension

1. Answers may vary. Students should note that the extra length of the rope enables them to create more varied waveforms in terms of speed and height.

CLASSROOM TESTED & APPROVED

Randa Flinn
Northeast High School
Fort Lauderdale, FL

Maps in Action

Roaming Rubber Duckies

Internet Activity —ADVANCED

Beachcomber Tales Ask students to research how other types of floating objects—from seeds and bottles to toys, hockey equipment, and candy—have helped add to the understanding of ocean currents. Tell them to collect any unusual stories in booklet form and to add their own drawings to illustrate the tales. A worksheet designed to direct student research on this topic can be found in the **Chapter Resource File** booklet or by visiting go.hrw.com and entering the keyword **HQ6MOVX.** LS **Verbal/Visual**

Answers to Map Skills Activity

1. A container ship spilled the bathtub toys in the middle of the North Pacific Ocean on January 10, 1992.
2. 9 months
3. North Equatorial Current
4. The California Current is a cold water current because it is shown in blue on the ocean current map, and because it flows away from the pole but has not yet reached the waters around the Equator.
5. Some of the toys might have floated north past Japan and back toward where they were first dropped.
6. Oyashio Current
7. The toys traveled from January 10, 1992 to July 26, 2003—about 11 years and 6 months, or 4,215 days.

CHAPTER RESOURCES

Chapter Resource File

- **Internet Activity** Beachcomber Tales ADVANCED

Technology

Transparencies
- 105 Roaming Rubber Duckies (with worksheet)

MAPS *in Action*

Roaming Rubber Duckies

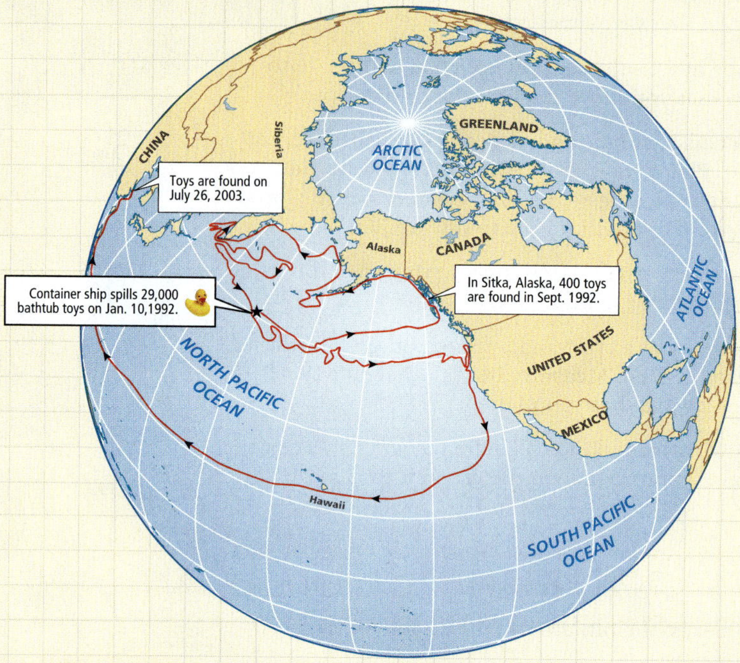

Map Skills Activity

This map shows the estimated route taken by bathtub toys spilled from a cargo ship in the North Pacific Ocean. Use the map to answer the questions below:

1. **Analyzing Data** Describe where the toys started their journey.
2. **Evaluating Data** How long did it take the toys to travel to Sitka, Alaska, by the most direct route?
3. **Identifying Relationships** Compare the map above with the map of the major surface currents in the section entitled *Ocean Currents*. Then, name the current that carried the toys past Hawaii.
4. **Evaluating Sources** Is the current that carries the toys along the coast of the western United States cold or warm? Explain your answer.
5. **Predicting Consequences** Predict where the toys might have been located in December 2003 if tracking data were plotted on the map.
6. **Identifying Relationships** What is the name of the current that carried the toys south along the coast of Siberia?
7. **Evaluating Data** How long did it take the toys to travel from the location where they were dumped to their location on the coast of China on July 26, 2003?

Energy from the Ocean

The search for renewable energy sources has led to some ingenious ways of tapping the ocean for energy. Scientists have found ways to generate electricity from three ocean features: waves, tides, and heat.

Wave Energy

Most wave-power systems use buoys and the up-and-down motion of waves to run pumps that force water or air through turbine generators. A newer design uses piezoelectric plastic (pie EE zoh ee LEK trik PLAS tik), a material that produces electricity when stretched.

Tidal Energy

Tides can generate electricity in a way similar in principle to the way that hydroelectric plants on rivers generate electricity. At high tide, a dam built across a bay or inlet traps water. As the tide ebbs, this water is released through the dam's turbines to drive electric generators.

Thermal Energy

The process of producing electricity from the heat energy contained in ocean water is called *ocean thermal energy conversion* (OTEC). OTEC plants rely on steam to turn their turbine generators. Warm surface water is pumped through a vacuum chamber, which turns the water to steam. The steam then turns special low-pressure turbines.

Ocean power is not as cost effective as electricity from fuel-burning or nuclear power plants is. However, as ocean power technology becomes more reliable and more efficient, the ocean's renewable energy may become more affordable.

Extension

1. **Determining Cause and Effect** What do all three types of ocean power have in common?
2. **Research** Write a short report that explains how using energy from waves, tides, or heat in the ocean could help reduce pollution of the air and ocean.

▼ Tidal power plants, such as this one in France, rely on the flow of tides to turn turbines that generate electricity.

◀ This wave-power energy station in Scotland generates 500 kilowatts of electricity, enough to power 300 homes.

Science and Technology

Energy from the Ocean

Activity — Advanced

Exploring Water Power Modern water wheels use the power of flowing water to generate electricity. Have students follow these instructions to make a simple model: Remove the top from a 2-liter plastic bottle. Make two V-shaped notches in the rim. Mold a ball of clay. Push a straw through the middle of the clay ball. Cut four paddles from the top of the bottle. Insert the paddles at even distances into the ball of clay. Mount the paddle wheel into the V-shaped notches so that the straw turns freely. Then, have students test the model water wheel in a sink under a running faucet. Challenge students to find a way to make the device do some work such as lifting a small weight made from paper clips tied to a piece of string. LS Kinesthetic

Group Activity — Advanced

Power Plants Invite students to do further research and to draw diagrams to explain how the mechanical motion of turbines generates electricity. The diagrams should show how the electricity is moved from power stations to places where it is used. Place these diagrams around the room or on a bulletin board as an aid to discussion. LS Visual

Answers to Extension

1. All three types of ocean power convert the energy stored in ocean waves, tides, and temperature into electrical energy using turbine generators.
2. Answers may vary. Accept all reasonable responses. Reports should focus on ocean power as a clean source of energy generation.

Unit 7 ATMOSPHERIC FORCES

Unit 7 Outline

CHAPTER 22
The Atmosphere

CHAPTER 23
Water in the Atmosphere

CHAPTER 24
Weather

CHAPTER 25
Climate

▶ Wall clouds, such as this one near Adrian, Texas, are small clouds that form underneath large storm clouds. The beginning of a tornado can be seen in the circulation of dust on the ground under this wall cloud. Tornadoes tend to form under wall clouds.

Chapter 22 The Atmosphere

Planning Guide

Compression Guide
To shorten instruction because of time limitations, omit the Chapter Lab.

OBJECTIVES	LABS, DEMONSTRATIONS, AND ACTIVITIES	TECHNOLOGY RESOURCES
PACING • 45 min pp. 546–554 **Chapter Opener**	SE **Long-Term Project** Air Pollution Watch, pp. 852–855 ADVANCED LTP **Long-Term Project** Air Pollution Watch* ADVANCED	OSP **Parent Letter** ■ CD **Student Edition on CD-ROM** CD **Chapter Summaries Audio CD** ■ VID **Brain Food Video Quiz**
Section 1 Characteristics of the Atmosphere • Describe the composition of Earth's atmosphere. • Explain how two types of barometers work. • Identify the layers of the atmosphere. • Identify two effects of air pollution.	TE **Demonstration** Air Force, p. 547 ◆ GENERAL TE **Demonstration** Observing Photosynthesis, p. 548 ◆ GENERAL TE **Activity** Dust Collectors, p. 549 BASIC TE **Activity** Create a Vacuum, p. 550 BASIC TE **Activity** Weather Maps, p. 551 GENERAL SE **Quick Lab** Barometric Pressure, p. 551 GENERAL CRF **Datasheet for Quick Lab*** GENERAL	OSP **Lesson Plans** (also in print) TR **Bellringer*** TR 106 Layers of the Atmosphere* CD **Interactive Tutor** Earth's Atmosphere CD **Interactive Tutor** Troposphere Thickness CD **Interactive Tutor** Hail, Northern Lights, and Rainbows
PACING • 90 min pp. 555–560 **Section 2 Solar Energy and the Atmosphere** • Explain how radiant energy reaches Earth. • Describe how visible light and infrared energy warm Earth. • Summarize the processes of radiation, conduction, and convection.	TE **Activity** Magic with Beads, p. 555 ◆ GENERAL TE **Activity** Blue and Red Skies, p. 556 GENERAL TE **Group Activity** Comparing Albedos, p. 557 ◆ GENERAL SE **Quick Lab** Light and Latitude, p. 559 GENERAL CRF **Datasheet for Quick Lab*** GENERAL SE **Inquiry Lab** Energy Absorption and Reflection, pp. 570–571 ◆ GENERAL CRF **Datasheet for Chapter Lab*** GENERAL SE **Maps in Action** Absorbed Solar Radiation, p. 572 GENERAL TE **Discussion** The Tropics, p. 572 GENERAL CRF **Inquiry Lab** Ultraviolet Protection* GENERAL	OSP **Lesson Plans** (also in print) TR **Bellringer*** TR 107 The Electromagnetic Spectrum* TR 108 The Greenhouse Effect and Latitude and Season* TR 111 Absorbed Solar Radiation* TE **Internet Activity** Global Warming, p. 558 GENERAL CRF **Internet Activity** Global Warming* GENERAL CD **Interactive Tutor** Greenhouse Effect CD **Interactive Tutor** Energy CD **Interactive Tutor** Light and Electromagnetic Energy
PACING • 45 min pp. 561–564 **Section 3 Atmospheric Circulation** • Explain the Coriolis effect. • Describe the global patterns of air circulation, and name three global wind belts. • Identify two factors that form local wind patterns.	TE **Demonstration** Modeling the Coriolis Effect, p. 561 ◆ GENERAL TE **Group Activity** It's a Breeze, p. 562 ◆ GENERAL CRF **Making Models Lab** Global Air Movement* GENERAL	OSP **Lesson Plans** (also in print) TR **Bellringer*** TR 109 The Coriolis Effect* TR 110 Global Wind Belts* TE **Internet Activity** Harnessing Wind, p. 573 GENERAL CRF **Internet Activity** Harnessing Wind* GENERAL VID **HRW Earth Science Video** Global Winds VID **HRW Earth Science Video** Local Winds CD **Interactive Tutor** Global Wind Patterns

PACING • 90 min

CHAPTER REVIEW, ASSESSMENT, AND STANDARDIZED TEST PREPARATION

- SE **Chapter Highlights**, p. 565
- SE **Chapter Review**, pp. 566–567
- SE **Standardized Test Prep**, pp. 568–569
- CRF **Concept Review*** ■ GENERAL
- CRF **Critical Thinking*** ADVANCED
- CRF **Math Skills*** GENERAL
- CRF **Graphing Skills*** GENERAL
- CRF **Chapter Test A*** ■ GENERAL
- CRF **Chapter Test B*** ADVANCED
- OSP **Lesson Plans** (also in print)
- OSP **Test Generator**
- OSP **Test Item Listing**

Online and Technology Resources

go.hrw.com Holt Online Learning

Visit go.hrw.com for access to Holt Online Learning, or enter the keyword **HQ6 Home** for a variety of free online resources.

One-Stop Planner® CD-ROM

This CD-ROM package includes
- Lab Materials QuickList Software
- Holt Calendar Planner
- Customizable Lesson Plans
- Printable Worksheets
- ExamView® Test Generator
- Interactive Teacher Edition
- Holt PuzzlePro®
- Holt PowerPoint® Resources

KEY

SE	Student Edition	OSP	One-Stop Planner	VID	Classroom Video/DVD
TE	Teacher Edition	TR	Transparencies and Transparency Worksheets	*	Also on One-Stop Planner
CRF	Chapter Resource File	CD	CD or CD-ROM	◆	Requires advance prep
LTP	Long-Term Projects			■	Also available in Spanish

SKILLS DEVELOPMENT RESOURCES	REVIEW AND ASSESSMENT	CORRELATIONS
SE **Pre-Reading Activity,** p. 546 GENERAL TE **Using the Figure** Storm Chasing, p. 546 GENERAL		National Science Education Standards
CRF **Directed Reading*** BASIC TE **Reading Skill Builder** Paired Summarizing, p. 548 BASIC SE **Math Practice,** p. 550 GENERAL TE **Using the Figure** Mercurial Barometers, p. 551 BASIC TE **Inclusion Strategies,** p. 552	SE **Reading Checks,** pp. 548, 551, 553 GENERAL SE **Section Review,** p. 554 GENERAL TE **Reteaching,** p. 553 BASIC TE **Quiz,** p. 553 GENERAL TE **Alternative Assessment,** p. 553 GENERAL CRF **Section Quiz*** ■ GENERAL	PS 3c
CRF **Directed Reading*** BASIC TE **Skill Builder** Writing, p. 556 ADVANCED TE **Using the Figure** Phantom Images, p. 557 GENERAL TE **Skill Builder** Vocabulary, p. 557 BASIC	SE **Reading Checks,** p. 559 GENERAL SE **Section Review,** p. 560 GENERAL TE **Reteaching,** p. 559 BASIC TE **Quiz,** p. 559 GENERAL TE **Alternative Assessment,** p. 559 GENERAL CRF **Section Quiz*** ■ GENERAL	PS 3c, PS 6a, PS 6b, ES 1c, ES 1d
CRF **Directed Reading*** BASIC TE **Using the Figure** Coriolis Effect, p. 562 ADVANCED TE **Inclusion Strategies,** p. 562 TE **Reading Skill Builder** Reading Organizer, p. 562 BASIC SE **Graphic Organizer** Comparison Table, p. 563 GENERAL	SE **Reading Check,** p. 562 GENERAL SE **Section Review,** p. 564 GENERAL TE **Reteaching,** p. 563 BASIC TE **Quiz,** p. 563 GENERAL TE **Alternative Assessment,** p. 563 ADVANCED CRF **Section Quiz*** ■ GENERAL	ES 1c, ES 1d

Holt Lab Generator CD-ROM

Search for labs by topic, standard, difficulty level, and/or time frame. Edit those labs to fit your needs, or create your own labs. Use the Lab Materials QuickList to order just what you need.

Holt Earth Science Interactive Tutor CD-ROM

This CD-ROM consists of interactive activities that give students a fun way to extend their knowledge of Earth science concepts.

Chapter Summaries Audio CDs

These CDs include audio summaries of the key concepts presented in each chapter. (Audio summaries are also available in Spanish.)

SCILINKS. NSTA

www.scilinks.org

Maintained by the **National Science Teachers Association.** See Chapter Enrichment pages that follow for a complete list of topics.

See Chapter Enrichment pages for Video Resources.

Chapter 22 Chapter Enrichment

This Chapter Enrichment provides relevant and interesting information to expand and enhance your classroom instruction of the chapter material.

Section 1 Characteristics of the Atmosphere

Earth's Early Atmosphere

Although Earth's atmosphere today consists of a relatively stable mixture of life-sustaining gases, it was not always so. The early atmosphere was radically different from today's atmosphere. Geologic events and early life-forms helped change the atmosphere into the complex mixture of gases that has sustained life on Earth for the last half-billion years.

Scientists think that the early atmosphere was a "reducing" atmosphere, or one that lacked free oxygen. As Earth cooled, volcanic activity formed an atmosphere that was rich in carbon dioxide, nitrogen, and water vapor, with trace amounts of methane, ammonia, and sulfur dioxide. This mixture of gases would be toxic to most modern life-forms. As life evolved in this noxious brew, it began to modify the atmosphere. Sometime around 3.5 billion years ago, the first photosynthetic organisms, cyanobacteria, appeared and began to use the carbon dioxide in the atmosphere while producing oxygen. Over millions of years, the atmosphere changed from a reducing atmosphere to an oxidizing one as oxygen began to accumulate. After sufficient oxygen had accumulated in the atmosphere, the ozone layer formed. Once a layer of ozone was in place to shield organisms from some of the sun's harmful radiation, organisms began to leave the protective blanket of the seas. Terrestrial plants and animals began to evolve and to modify the atmosphere in different ways. Human activities now may be one of the biggest contributors to the continuing development of the atmosphere. Air pollution from human activities modifies the composition of the atmosphere and may alter the global climate.

Section 2 Solar Energy and the Atmosphere

The Solar-Hydrogen Economy

Increasing carbon dioxide emissions from humans' enormous dependence on fossil fuels, such as coal, oil, and natural gas, to power everything we do at home, work, and play, may be affecting Earth's atmosphere. Many scientists agree that alternatives to energy derived from fossil fuels must be found. Harnessing energy from the sun is one possible alternative. The sun regularly provides more energy than we could ever use; humans use energy at a rate of about 13 terawatts (13,000 billion joules per second), while the light and heat of the sun deliver energy to Earth at a rate of 80,000 terawatts (8×10^{16} joules per second). The problem with using the sun's energy is that it may be dilute, diffuse, and intermittent—not what is needed from a reliable energy source.

However, solar energy technologies are rapidly improving. One such technology is photovoltaic (PV) cells that convert sunlight directly into electricity. PV cells are thin films of silicon treated so that they generate electricity when exposed to the sun. They are found in calculators, watches, and many other applications, even roof shingles. Electricity produced by PV cells could also be used to make hydrogen from water. Hydrogen is a clean, transportable fuel that could be used to heat buildings, run appliances, and power vehicles, much like natural gas does. Hydrogen can produce electricity when burned in fuel cells. With continued research in this area, the future of energy may be bright.

▶ **As altitude increases, the characteristics of the atmosphere change.**

Section 3 Atmospheric Circulation

Halley, Hadley, and Ferrell

The trade winds and westerlies were used by sailors for hundreds of years before scientists could really explain why they blew so regularly. And it wasn't even a scientist who came up with the best explanation!

The first explanation came from Sir Edmund Halley (of Halley's comet fame) in 1686. He proposed correctly that heat from the sun at the equator makes air rise and that the trade winds result as surface air is drawn toward the equator from the north and south. This explained what drove the winds, but not their direction nor the other wind belts.

In 1735, George Hadley, a lawyer by profession, refined Halley's model by suggesting that the winds curve because of Earth's rotation. The westerlies result as warm air rises at the equator and moves toward the poles. There, the air sinks and is drawn back toward the equator in one giant convection cell.

Hadley's insights were useful but not complete; a single convection cell cannot explain all of the global wind belts. It was not until 1856 that the currently accepted three-cell model was proposed by William Ferrell, an American schoolteacher who was dedicated to the study of wind.

Measuring Winds

One way to estimate wind strength without actually measuring wind speed is to use the Beaufort wind scale, devised in 1806 by Admiral Sir Francis Beaufort. Originally designed for sailing ships, this scale divides wind speed into 13 forces by observing the effects of winds of different speeds on the sea or on objects on land such as flags, trees, and structures. At force 0 on the scale, the sea is smooth like a mirror. Force 12 means a hurricane is in progress: the air is filled with foam and the sea is white with driving spray. The scale is still used today, especially when instruments such as anemometers are not available.

Video Resources

Brain Food Video Quizzes

Brain Food Video Quizzes These videos contain game-show style quizzes that assess students' progress and motivate students to study the chapter material.

HRW Earth Science Video

This video introduces Earth science topics and includes a geology field trip. The video segment listed below complements this chapter.

HRW Earth Science Video

Segment 16: Global Winds This segment focuses on winds that blow steadily over large areas of the planet's surface. The uneven heating of Earth that produces pressure belts at different latitudes is examined, and the convection cells that are created by these differences in pressure are discussed. The Coriolis effect—the deflection of surface winds by Earth's rotation about its axis—is described, and global wind patterns are shown. (2.5 min)

Segment 17: Local Winds Local winds are created by temperature differences brought on by the uneven heating and cooling of land and water. In this segment, the relative rates of warming and cooling over land and water and the ascending and descending air masses that result in land and sea breezes are discussed. Also explored are local temperature differences in mountainous areas that result in mountain and valley breezes. (1.5 min)

NOVA Videos

To order **NOVA** videos related to this chapter, visit **go.hrw.com** and enter the keyword **HQ6ATMV.**

SciLinks

Developed and maintained by the National Science Teachers Association

SciLinks is maintained by the National Science Teachers Association to provide you and your students with interesting, up-to-date links that will enrich your classroom presentation of the chapter.

Visit www.scilinks.org and enter the SciLinks code for more information about the topic listed.

Topic: The Atmosphere
SciLinks code: HQ60112

Topic: Winds
SciLinks code: HQ61670

Topic: Coriolis Effect
SciLinks code: HQ60357

Topic: Greenhouse Effect
SciLinks code: HQ60694

Chapter 22

Chapter Overview

The atmosphere is the protective blanket of gases that surrounds Earth. The atmosphere not only contains the oxygen we breathe, but makes life on Earth possible by protecting us from harmful solar radiation and by moderating global temperatures.

Using the Figure — General

Storm Watch Point out how the clouds in the photo seem to top out at a certain level—the tropopause. This is the boundary between the troposphere, the atmospheric layer closest to Earth, and the stratosphere. Most clouds form in the troposphere. However, the storm at the center of the photo rises through the tropopause. Ask students why many pilots prefer to fly at or above the tropopause. (fewer clouds, better visibility, less likely to encounter storms and other adverse weather conditions) **LS Logical**

PRE-READING ACTIVITY

FOLDNOTES You may want to assign this FoldNote activity as homework. Collect the FoldNotes to check students' understanding of the material.

Chapter 22 The Atmosphere

Sections

1 Characteristics of the Atmosphere
2 Solar Energy and the Atmosphere
3 Atmospheric Circulation

What You'll Learn

- Which gases make up the atmosphere
- How solar energy interacts with the atmosphere
- What causes wind

Why It's Relevant

The atmosphere affects all living things on Earth. It provides the air that we breathe, protection from solar radiation, and the insulation that maintains the global temperature of Earth.

PRE-READING ACTIVITY

FOLDNOTES **Booklet** Before you read this chapter, create the FoldNote entitled "Booklet" described in the Skills Handbook section of the Appendix. Label each page of the booklet with a main idea from the chapter. As you read the chapter, write what you learn about each main idea on the appropriate page of the booklet.

▶ This image was captured by a satellite that was monitoring a large storm over the Atlantic Ocean. Storms are only one small feature of Earth's dynamic atmosphere.

Chapter Correlations — *National Science Education Standards*

PS 3c ... In other reactions, chemical bonds are broken by heat or light to form very reactive radicals with electrons ready to form new bonds. Radical reactions control many processes such as the presence of ozone and greenhouse gases in the atmosphere... **(Sections 1 and 2)**

PS 6a Waves, including sound and seismic waves, waves on water, and light waves have energy and can transfer energy when they interact with matter. **(Section 2)**

PS 6b Electromagnetic waves result when a charged object is accelerated or decelerated. Electromagnetic waves include radio waves (the longest wavelength), microwaves, infrared radiation, visible light, ultraviolet radiation, x-rays and gamma rays. **(Section 2)**

ES 1c Heating of the earth's surface and atmosphere by the sun drives convection within the atmosphere and oceans, producing winds and ocean currents. **(Sections 2 and 3)**

ES 1d Global climate is determined by energy transfer from the sun at and near the earth's surface. This energy transfer is influenced by dynamic processes such as cloud cover, the earth's rotation and static conditions such as the position of mountain ranges and oceans. **(Sections 2 and 3)**

Section 1 Characteristics of the Atmosphere

The layer of gases that surrounds Earth is called the **atmosphere**. The atmosphere is made up of a mixture of chemical elements and compounds that is commonly called *air.* The atmosphere protects Earth's surface from the sun's radiation and helps regulate the temperature of Earth's surface.

Composition of the Atmosphere

As the graph in **Figure 1** shows, the most abundant elements in air are the gases nitrogen, oxygen, and argon. The composition of dry air is nearly the same everywhere on Earth's surface and up to an altitude of about 80 km. The two most abundant compounds in air are the gases carbon dioxide, CO_2, and water vapor, H_2O. In addition to containing gaseous elements and compounds, the atmosphere commonly carries various kinds of tiny solid particles, such as dust and pollen.

Nitrogen in the Atmosphere

Nitrogen makes up about 78% of Earth's atmosphere. Nitrogen in the atmosphere is maintained through a process called the *nitrogen cycle*. During the nitrogen cycle, nitrogen moves from air to the soil and then to plants and animals and eventually returns to the air, as shown in **Figure 1.**

Nitrogen is removed from the air mainly by the action of nitrogen-fixing bacteria. These microscopic organisms live in the soil and on the roots of certain plants. The bacteria chemically change nitrogen from the air into nitrogen compounds that are vital to the growth of all plants. When animals eat plants, nitrogen compounds enter the animals' bodies. Nitrogen compounds are then returned to the soil through animal wastes or by the decay of dead organisms. Decay releases nitrogen back into the atmosphere. A similar nitrogen cycle takes place between marine organisms and ocean water.

OBJECTIVES

- **Describe** the composition of Earth's atmosphere.
- **Explain** how two types of barometers work.
- **Identify** the layers of the atmosphere.
- **Identify** two effects of air pollution.

KEY TERMS

atmosphere
ozone
atmospheric pressure
troposphere
stratosphere
mesosphere
thermosphere

atmosphere a mixture of gases that surrounds a planet, such as Earth

Figure 1 ▶ The pie graph shows the composition of dry air by volume at sea level. Nitrogen in the atmosphere is kept at a relatively constant level by the nitrogen cycle.

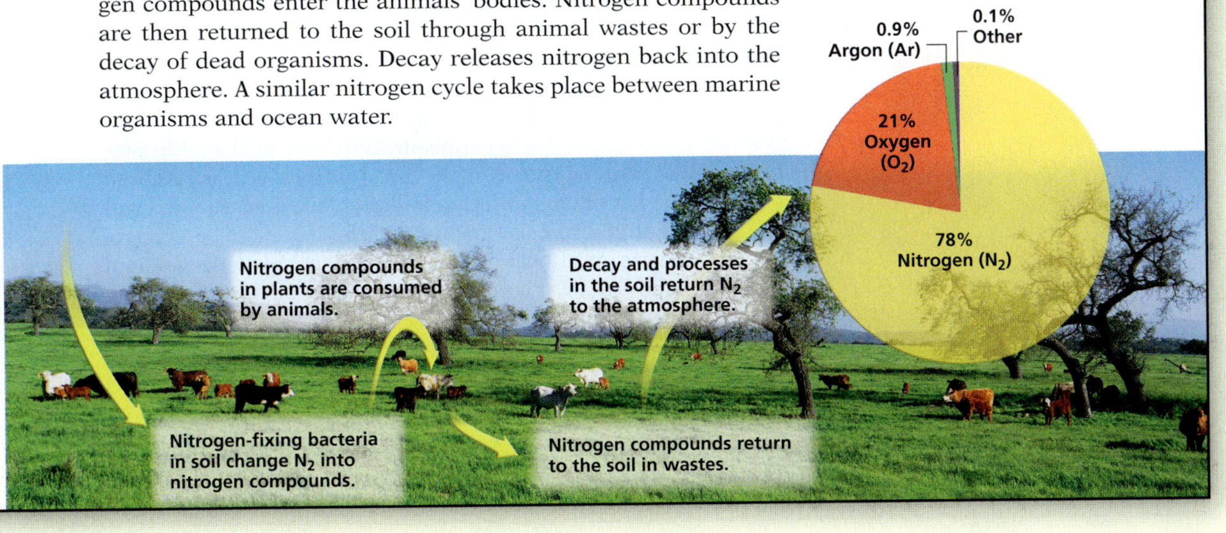

Section 1

Focus

Overview

This section explains the major components and the four layers of Earth's atmosphere. It describes ways to measure atmospheric pressure and describes air pollution and its effects on people, animals, plants, and property.

Bellringer

Have students predict what will happen to a lighted candle that is covered by a glass jar. (Combustion will use up the oxygen in the jar, and the candle will go out.)

Motivate

Demonstration — GENERAL

Air Force Demonstrate air pressure using a 600 mL glass milk bottle or similar glass bottle with an opening of about 4 cm and a peeled hard-boiled egg. Make sure the egg can sit on the opening of the bottle without falling in. Light three matches at once and drop them quickly into the bottle. Then, immediately set the egg on the opening. The egg will be forced into the bottle. Ask students why they think this happened. (The burning matches caused a partial vacuum inside the bottle, which decreased the air pressure in the bottle. The greater pressure of the outside air pushed the egg into the bottle.) LS **Visual**

CHAPTER RESOURCES

Chapter Resource File
- Directed Reading BASIC

Technology

Transparencies
- Bellringer

Student Edition on CD-ROM

One-Stop Planner CD-ROM
- Lesson Plan

Teach

Teaching Tip — Basic

Connect to Prior Knowledge Ask students to think about a puddle that forms after a rain shower in the summer. Ask them where the water in the puddle came from. (Some water came from the sky as rain. Some water ran into the puddle from the surrounding ground.) Ask them what happens to the puddle after a few hot, dry days. (The puddle shrinks and the water disappears. This happens as the water evaporates.) Explain to students that water cycles between the land, ocean, and atmosphere through the water cycle. **LS Logical**

Demonstration — General

Observing Photosynthesis Use aquatic plants to show students the production of oxygen. Place some pondweed or *Elodea* in a jar with water. Place a clear funnel over the plant and a test tube over the spout. Submerge the apparatus in water to fill up the test tube. Place the apparatus under a lamp or in a sunny window. Students should carefully observe the tips of the leaves. Ask students: "What is in the bubbles that form on the leaves and also fill up the test tube?" (oxygen) Ask students to propose a way to verify that the bubbles they are collecting are oxygen. (Sample answer: Students could light a wooden splint, blow it out, and quickly put the glowing end into the test tube. The splint would burst into flame because of the oxygen filling the tube.) **LS Visual**

Answer to Reading Check

Transpiration increases the amount of water vapor in the atmosphere.

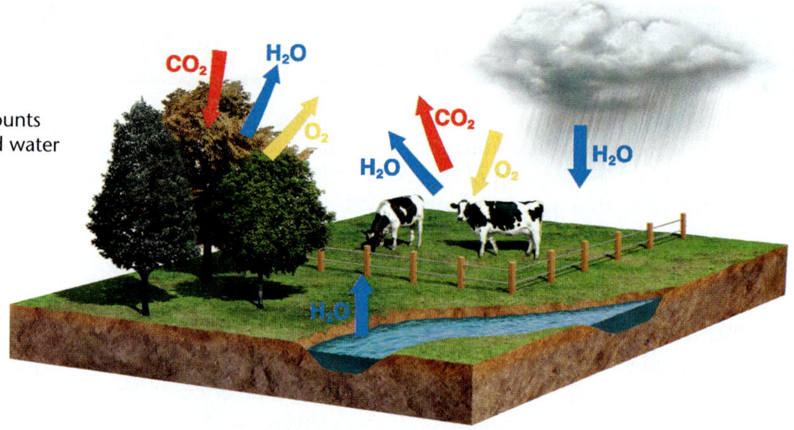

Figure 2 ▶ Several processes interact to maintain stable amounts of oxygen, carbon dioxide, and water in the atmosphere.

SCI LINKS. NSTA Developed and maintained by the National Science Teachers Association

For a variety of links related to this subject, go to www.scilinks.org

Topic: The Atmosphere
SciLinks code: HQ60112

Oxygen in the Atmosphere

Oxygen makes up about 21% of Earth's atmosphere. As shown in **Figure 2,** natural processes maintain the chemical balance of oxygen in the atmosphere. Animals, bacteria, and plants remove oxygen from the air as part of their life processes. Forest fires, the burning of fuels, and the weathering of some rocks also remove oxygen from air. These processes would quickly use up most atmospheric oxygen if various processes that add oxygen to air did not take place.

Land and ocean plants produce large quantities of oxygen in a process called *photosynthesis*. During photosynthesis, plants use sunlight, water, and carbon dioxide to produce their food, and they release oxygen as a byproduct. The amount of oxygen produced by plants each year equals the amount consumed by all animal life processes. Thus, the oxygen content of the air remains at about 21% of Earth's atmosphere.

Water Vapor in the Atmosphere

As water evaporates from oceans, lakes, streams, and soil, it enters air as the invisible gas *water vapor*. Plants and animals give off water vapor during transpiration, one of their life processes. But as water vapor enters the atmosphere, it is removed by the processes of condensation and precipitation. The percentage of water vapor in the atmosphere varies depending on factors such as time of day, location, and season. Because the amount of water vapor in air varies, the composition of the atmosphere is usually given as that of dry air. Dry air has less than 1% water vapor. Moist air may contain as much as 4% water vapor.

Reading Check **Does transpiration increase the amount of water vapor in the atmosphere or decrease the amount of water vapor in the atmosphere? (See the Appendix for answers to Reading Checks.)**

Reading Skill Builder — Basic

Paired Summarizing Group students into pairs, and have them read silently about oxygen and water vapor in the atmosphere. Then, have one student summarize the processes that maintain stable amounts of these gases in the atmosphere. The other student should listen to the retelling and point out any inaccuracies or ideas that were left out. Allow students to refer to the text as needed. **LS Verbal** English Language Learners

Biology Connection — General

The Breath of Life Respiration is the process by which our bodies use oxygen to convert the energy stored in the chemical bonds of carbohydrates, fats, and proteins into ATP, a compound the cells can use for energy. Without oxygen, the cells would become starved for the energy needed to carry out all of our bodies' functions, and we would quickly die. Invite interested students to research this process and make diagrams that illustrate it. **LS Logical**

Ozone in the Atmosphere

Although it is present only in small amounts, a form of oxygen called **ozone** is an important component of the atmosphere. The oxygen that we breathe, O_2, has two atoms per molecule, but ozone, O_3, has three atoms. Ozone in the upper atmosphere forms the *ozone layer,* which absorbs harmful ultraviolet radiation from the sun. Without the ozone layer, living organisms would be severely damaged by the sun's ultraviolet rays. Unfortunately, a number of human activities damage the ozone layer. Compounds known as *chlorofluorocarbons,* or CFCs, which were previously used in refrigerators and air conditioners, and exhaust compounds, such as nitrogen oxide, break down ozone and have caused parts of the ozone layer to weaken, as **Figure 3** shows.

ozone a gas molecule that is made up of three oxygen atoms

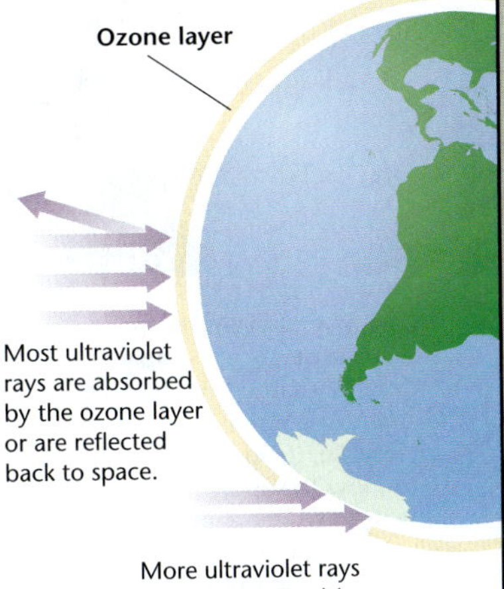

Figure 3 ▶ Harmful ultraviolet radiation can reach Earth's surface through the weakened ozone layer over Antarctica.

Particulates in the Atmosphere

In addition to containing gases, the atmosphere contains various tiny solid particles, called *particulates*. Particulates can be volcanic dust, ash from fires, microscopic organisms, or mineral particles lifted from soil by winds. Pollen from plants and particles from meteors that have vaporized are also particulates. When tiny drops of ocean water are tossed into the air as sea spray, the drops evaporate. Left behind in the air are tiny crystals of salt, another type of particulate. Four common sources of particulates are shown in **Figure 4.** Large, heavy particles remain in the atmosphere only briefly, but tiny particles can remain suspended in the atmosphere for months or years.

Figure 4 ▶ Sources of Particulates

Volcanic ash and dust can remain in the atmosphere for years.

The wind carries pollen from plant to plant.

Tornadoes and windstorms carry dirt and dust high into the atmosphere.

As seaspray evaporates, salt particles are left in the atmosphere.

ENVIRONMENTAL CONNECTION

Ozone, Good or Bad? While stratospheric ozone is essential to life on Earth, in the troposphere ozone is a pollutant. Have students research the effects of ground-level ozone. Have them prepare a brochure that explains how ground-level ozone forms; how it affects humans, animals, and plants; and what actions can be taken to reduce it. LS **Verbal/Visual**

Activity — BASIC

Dust Collectors Have students coat one side of several glass slides with petroleum jelly. Have them place the slides outdoors for 24 hours in various locations. Then, have students view the slides under a microscope and describe what they see to the class. Discuss which locations are "dirtiest" and the possible sources of airborne particulates. LS **Visual/Kinesthetic**

Teaching Tip — GENERAL

Make Concepts Relevant For every 1% loss of ozone, the amount of UV radiation that reaches the ground increases by 2%. The increase in exposure to UV radiation is linked to greater incidence of skin cancers and immune system suppression. Have interested students research either the latest findings about the causes of skin cancer, or local, regional, and global efforts to protect the ozone layer. Have them make an oral report to the class. LS **Logical**

MISCONCEPTION ALERT

The Ozone Hole Most of the ozone molecules in the atmosphere are located in a layer between 10 and 40 km above Earth's surface. During the Antarctic winter, a vortex of atmospheric circulation results in many ozone molecules being destroyed. So, the "ozone hole" is, in actuality, an area that has fewer ozone molecules in that layer. Because Antarctica is largely uninhabited by humans, students might assume that this phenomenon could not affect them. But at times the thinning has extended over parts of South America, Australia, and New Zealand. Furthermore, recent studies show that over much of the United States and most of the middle latitudes around the world, ozone levels have declined 5% in the summer and up to 10% in the winter, compared to pre-1980 levels. There is good news: the level of CFCs in the stratosphere seems to have peaked, so some scientists think that the ozone layer will recover within 40 to 50 years.

Teach, continued

MATHPRACTICE

Answer
If the force exerted on Earth's surface by a column of air that has a base of 1 m^2 = 101,325 N, then the force exerted by a column of air that has a base of 3 m^2 = 101,325 N/m^2 × 3 m^2 = 303,975 N

Activity — BASIC

Create a Vacuum Have students fill a plastic soda bottle half full with hot water and screw the top on. Then, have them place the bottle in a pan. Have them cover the bottle with ice and cold water and observe what happens. (The hot water vapor cools and condenses, forming a partial vacuum in the bottle. The bottle is crushed because the air pressure is greater outside than inside the bottle.) LS **Visual/Kinesthetic**

HISTORY CONNECTION

Incontestable Experiment In 1643, Galileo asked his assistant, Evangelista Torricelli, to find out why water could never be pumped higher than 33 ft (10 m). Torricelli conducted experiments using a tube filled not with water, but with mercury, the heaviest liquid. As a result of his experiments, Torricelli developed a precursor to the mercurial barometer, which is used to measure differences in pressure due to changes in weather and altitude. He also established that "We live submerged at the bottom of an ocean of elementary air, which is known by incontestable experiments to have weight."

Figure 5 ▶ At high altitudes, climbers must carry a supply of oxygen to breathe from because the density of the atmosphere there is very low. So, oxygen levels in the air are too low to supply the climber. In 2001, Eric Weihenmeyer (shown above) was the first blind person to reach the summit of Mount Everest.

Atmospheric Pressure

Gravity holds the gases of the atmosphere near Earth's surface. As a result, the air molecules are compressed together and exert force on Earth's surface. The pressure exerted on a surface by the atmosphere is called **atmospheric pressure.** Atmospheric pressure is exerted equally in all directions—up, down, and sideways.

Earth's gravity keeps 99% of the total mass of the atmosphere within 32 km of Earth's surface. The remaining 1% extends upward for hundreds of kilometers but gets increasingly thinner at high altitudes, as shown in **Figure 5.** Because the pull of gravity is not as strong at higher altitudes, the air molecules are farther apart and exert less pressure on each other at higher altitudes. Thus, atmospheric pressure decreases as altitude increases.

Atmospheric pressure also changes as a result of differences in temperature and in the amount of water vapor in the air. In general, as temperature increases, atmospheric pressure at sea level decreases. The reason is that molecules move farther apart when the air is heated. So, fewer particles exert pressure on a given area, and the pressure decreases. Similarly, air that contains a lot of water vapor is less dense than drier air because water vapor molecules have less mass than nitrogen or oxygen molecules do. The lighter water vapor molecules replace an equal number of heavier oxygen and nitrogen molecules, which makes the volume of air less dense.

atmospheric pressure the force per unit area that is exerted on a surface by the weight of the atmosphere

MATHPRACTICE

Force of the Air
On average, a column of air 1 m^2 at its base that reaches upward from sea level has a mass of 10,300 kg and exerts a force of 101,325 N (newtons) on the ground. So, at sea level, on every square meter of Earth's surface, the atmosphere presses down with an average force of 101,325 N. What would the average force of a column of air that has a 3 m^2 base be?

Measuring Atmospheric Pressure

Meteorologists use three units for atmospheric pressure: atmospheres (atm), millimeters or inches of mercury, and millibars (mb). *Standard atmospheric pressure,* or 1 atmosphere, is equal to 760 mm of mercury, or 1000 millibars. The average atmospheric pressure at sea level is 1 atm. Meteorologists measure atmospheric pressure by using an instrument called a *barometer.*

BRAIN FOOD

Cooking at High Altitude Cooking at altitudes above 1.5 km can challenge even the best chef. As altitude increases, pressure decreases, so cakes rise faster. To offset lower atmospheric pressure, the proportions of ingredients in recipes are usually modified. For instance, the amounts of leavening agents such as baking soda and baking powder are usually decreased. Another difference is that water boils at a lower temperature—as low as 80°C compared to 100°C at sea level. Thus, it would take longer for an egg to become hard boiled at higher altitude.

Mercurial Barometers

Meteorologists use two main types of barometers. One type is the *mercurial barometer,* a model of which is shown in **Figure 6.** Atmospheric pressure presses on the liquid mercury in a well at the base of the barometer. The pressure holds the mercury up to a certain height inside a tube. The height of the mercury inside the tube varies with the atmospheric pressure. The greater the atmospheric pressure is, the higher the mercury rises.

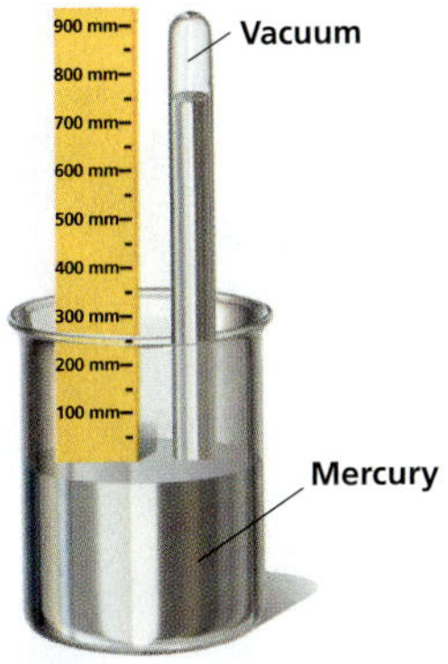

Figure 6 ▶ The height of the mercury in this mercurial barometer indicates barometric pressure. ***What is the barometric pressure shown?***

Aneroid Barometers

The type of barometer most commonly used today is called an *aneroid barometer.* Inside an aneroid barometer is a sealed metal container from which most of the air has been removed to form a partial vacuum. Changes in atmospheric pressure cause the sides of the container to bend inward or bulge out. These changes move a pointer on a scale. Aneroid barometers can be constructed to keep a continuous record of atmospheric pressure.

An aneroid barometer can also measure altitude above sea level. When used for this purpose, an aneroid barometer is called an *altimeter*. The scale on an altimeter registers altitude instead of pressure. At high altitudes, the atmosphere is less dense and exerts less pressure than at low altitudes. So, a lowered pressure reading can be interpreted as an increased altitude reading.

✓ **Reading Check** **What is inside an aneroid barometer? (See the Appendix for answers to Reading Checks.)**

QuickLAB 25 min (over 5 days)

Barometric Pressure

Procedure

1. Use a **rubber band** to secure **plastic wrap** tightly over the open end of a **coffee can**.
2. Use **tape** to secure one end of a **10 cm drinking straw** onto the plastic wrap near the center of the can.
3. Use **scissors** and a **metric ruler** to cut a piece of **cardboard** 10 cm wide. The cardboard should also be at least 13 cm taller than the can.
4. Fold the cardboard so that it stands upright and extends at least 3 cm above the top of the straw.
5. Place the cardboard near the can so that the free end of the straw just touches the front of the cardboard. Mark an *X* where the straw touches.
6. Draw three horizontal lines on the cardboard: one that is level with the *X*, one that is 2 cm above the *X*, and one that is 2 cm below the *X*.
7. Position the cardboard so that the straw touches the *X*. Tape the base of the cardboard in place.
8. Observe the level of the straw at least once per day over a 5-day period. Record any changes that you see.

Analysis

1. What factors affect how your model works? Explain.
2. What does an upward movement of the straw indicate? What does a downward movement indicate?
3. Compare your results with the barometric pressures listed in your local newspaper. What may have caused your results to differ from the newspaper's?

QuickLAB

Skills Acquired

- Observing
- Analyzing

Teacher's Notes Tell students that this apparatus is a "Cape Cod Barometer," which measures air density. A balloon can also be used in place of the plastic wrap. Make sure that the temperature remains constant. Students also may find weather data on the Internet.

Answers

1. The can must be well sealed and remain at about the same temperature.
2. upward: increasing pressure is pushing down on the plastic; downward: decreasing pressure causes the plastic to bulge up
3. Answers may vary. Possible reasons for differences: different instruments are used; the model uses a relative scale; and instruments are in different locations.

Using the Figure — BASIC

Mercurial Barometers Have students study the diagram of a mercurial barometer. Ask students: "If a low pressure system moved into the area, what would happen to the height of the mercury column in this figure?" (The height of the mercury in the vacuum would drop to lower than 750 mm of Hg.) Answer to caption question: The barometric pressure shown is 750 mm of Hg. **LS Logical**

Answer to Reading Check

An aneroid barometer contains a sealed metal container that has a partial vacuum.

Activity — GENERAL

Weather Maps Have students bring in current weather maps from the DataStreme Atmosphere Web site, posted by the American Meteorological Society, or from other sources. Discuss with students how to read these maps. In a high-pressure area, the number increases as you move toward the center of the circles; lows are the opposite. The spacing of the isobars indicates the pressure gradient, or change in pressure over a given distance. The closer the spacing is, the steeper the pressure gradient is and the stronger the winds associated with a weather system are. Concentric rings of isobars indicate high and low pressure systems. **LS Visual/Logical**

CHAPTER RESOURCES

Chapter Resource File

- **Datasheet for Quick Lab** GENERAL

Teach, *continued*

Teaching Tip — GENERAL

Connect to Prior Knowledge
Most students have probably seen a thunderstorm developing or pictures of thunderstorm clouds. Tell students that the anvil-shaped, flat top of storm clouds is due to the tropopause. As the cumulonimbus thunderclouds build up in the atmosphere, fed by warm, moist air, they rise and eventually reach the tropopause. Because the atmosphere in the tropopause is extremely stable and does not tend to move up or down, the cloud spreads out horizontally, which accounts for its flat top. Occasionally a very strong thunderstorm can break through the tropopause, and a bulge of clouds may protrude above the flat top. Bring in photos of thunderclouds or search the Internet for photos of thunderstorms. **LS Visual**

BRAIN FOOD

Gas Mass and Volume While the gases that make up the atmosphere are relatively light, they do have mass. The troposphere has about 80% of the gases of the atmosphere by mass. The stratosphere contains about 19%, and the mesosphere and thermosphere each have less than 1% of the atmosphere's gases by mass. However, on a volume basis, most of the atmosphere, about 87.3%, is located in the thermosphere because of this layer's thickness. The mesosphere holds about 4.5% of the atmosphere's volume, the stratosphere holds about 5.8%, and the troposphere contains only about 1.9%.

CHAPTER RESOURCES

Technology

Transparencies
- 106 Layers of the Atmosphere (with worksheet)

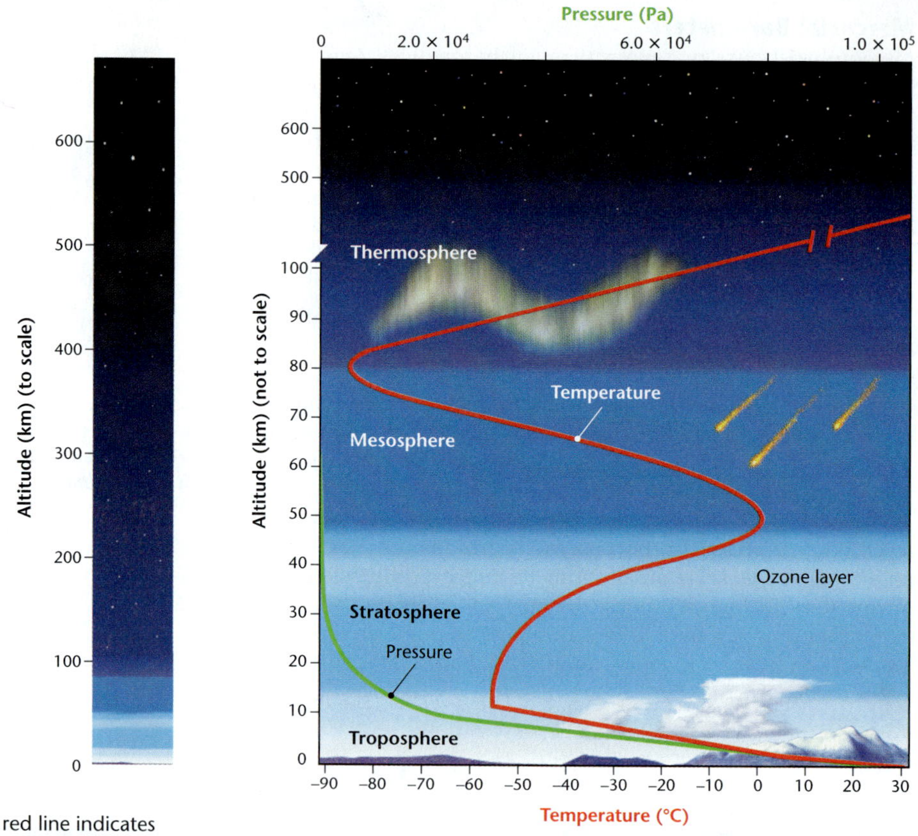

Figure 7 ▶ The red line indicates the temperature at various altitudes in the atmosphere. The green line indicates atmospheric pressure at various altitudes.

Layers of the Atmosphere

Earth's atmosphere has a distinctive pattern of temperature changes with increasing altitude, as shown in **Figure 7.** The temperature differences mainly result from how solar energy is absorbed as it moves through the atmosphere. Scientists identify four main layers of the atmosphere based on these differences.

The Troposphere

troposphere the lowest layer of the atmosphere, in which temperature drops at a constant rate as altitude increases; the part of the atmosphere where weather conditions exist

The atmospheric layer that is closest to Earth's surface and in which nearly all weather occurs is called the **troposphere.** Almost all of the water vapor and carbon dioxide in the atmosphere is found in this layer. Temperature within the troposphere decreases as altitude increases because air in this layer is heated from below by thermal energy that radiates from Earth's surface. The temperature within the troposphere decreases at the average rate of 6.5°C per kilometer as the distance from Earth's surface increases. However, at an average altitude of 12 km, the temperature stops decreasing. This zone is called the *tropopause* and represents the upper boundary of the troposphere. The altitude of this boundary varies with latitude and season.

INCLUSION Strategies

- **Visually Impaired**
- **Developmentally Delayed**

Use tactile skills to help students better understand the layers of the atmosphere. Gather a supply of fabrics and papers that have different textures. Divide the class into groups of three to five students. Have each group use the different textured materials to make a poster that shows the different layers of the atmosphere. Students who have visual impairments should feel the poster elements to understand the order and altitudes of the layers.

The Stratosphere

The layer of the atmosphere called the **stratosphere** extends from the tropopause to an altitude of nearly 50 km. Almost all of the ozone in the atmosphere is concentrated in this layer. In the lower stratosphere, the temperature is almost –60°C. In the upper stratosphere, the temperature increases as altitude increases because air in the stratosphere is heated from above by absorption of solar radiation by ozone. The temperature of the air in this layer rises steadily to a temperature of about 0°C at an altitude of about 50 km above Earth's surface. This zone, called the *stratopause,* marks the upper boundary of the stratosphere.

The Mesosphere

Located above the stratopause and extending to an altitude of about 80 km is the **mesosphere.** In this layer, temperature decreases as altitude increases. The upper boundary of the mesosphere, called the *mesopause*, has an average temperature of nearly –90°C, which is the coldest temperature in the atmosphere. Above this boundary temperatures again begin to increase.

The Thermosphere

The atmospheric layer above the mesopause is called the **thermosphere.** In the thermosphere, temperature increases steadily as altitude increases because nitrogen and oxygen atoms absorb solar radiation. Because air particles in the thermosphere are very far apart, they do not strike a thermometer often enough to produce an accurate temperature reading. Therefore, special instruments are needed. These instruments have recorded temperatures of more than 1,000°C in the thermosphere.

The lower region of the thermosphere, at an altitude of 80 to 400 km, is commonly called the *ionosphere*. In the ionosphere, solar radiation that is absorbed by atmospheric gases causes the atoms of gas molecules to lose electrons and to produce ions and free electrons. Interactions between solar radiation and the ionosphere cause the phenomena known as *auroras,* which are shown in **Figure 8.**

There are not enough data about temperature changes in the thermosphere to determine its upper boundary. However, above the ionosphere is the region where Earth's atmosphere blends into the almost complete vacuum of space. This zone of indefinite altitude, called the *exosphere,* extends for thousands of kilometers above the ionosphere.

Reading Check **What is the lower region of the thermosphere called? (See the Appendix for answers to Reading Checks.)**

stratosphere the layer of the atmosphere that lies between the troposphere and the mesosphere and in which temperature increases as altitude increases; contains the ozone layer

mesosphere the coldest layer of the atmosphere, between the stratosphere and the thermosphere, in which temperature decreases as altitude increases

thermosphere the uppermost layer of the atmosphere, in which temperature increases as altitude increases; includes the ionosphere

Figure 8 ▶ Auroras can be seen from space as well as from the ground.

Answer to Reading Check

The lower region of the thermosphere is called the *ionosphere.*

Close

Reteaching — BASIC

Name that Sphere Have students write characteristics of the atmospheric layers on index cards. Have them work in pairs, using the cards to quiz each other about the atmosphere. **LS Visual/Logical**

Quiz — GENERAL

1. What are auroras? (Interactions between solar radiation and the ionosphere create displays of light in the night sky.)
2. What layer of Earth's atmosphere blends into space? (the exosphere)
3. What units are used to measure atmospheric pressure? (atmospheres, millimeters or inches of mercury, and millibars)

Alternative Assessment — GENERAL

Modeling Have students work in small groups to design a model of Earth's atmosphere. The model should demonstrate the layers of Earth's atmosphere and how temperature, pressure, and chemical composition vary in the different layers. **LS Kinesthetic**

BRAIN FOOD

Radio Waves The ionosphere makes worldwide radio communication possible. Tiny, charged particles act as transmitters, bouncing radio waves back to Earth. On the ground, the signals are received and can be transmitted back up to the ionosphere. Thus, radio signals can be sent around Earth without the satellites or cable required for higher frequency television transmissions.

ENVIRONMENTAL CONNECTION

Deadly Blankets of Smog One fall evening in 1948, dirty fog crept into the small industrial valley town of Donora, Pennsylvania. Due to thermal inversion, the smog stayed for several days. More than 7,000 people were hospitalized with breathing difficulties. In London, about 4,000 people died in one week due to "the Great Smog" in 1952. Instances such as these led to the passage of clean air laws in the United States, culminating in the federal Clean Air Act of 1970.

Close, *continued*

Answers to Section Review

1. 78% nitrogen (N_2), 21% oxygen (O_2), 0.9% argon (Ar), and less than 0.1% other gases such as carbon dioxide
2. nitrogen, oxygen, water vapor, ozone, and particulates
3. Atmospheric pressure is caused by the weight of atmospheric gases as a result of gravity pulling the gas molecules toward Earth.
4. Atmospheric pressure pushes on liquid mercury in a well at the base of a mercurial barometer. This holds the mercury in a tube at a height that varies with pressure. Changes in air pressure cause the sides of a sealed metal container that holds a partial vacuum to bend inward or bulge outward, moving the pointer on a scale of an aneroid barometer.
5. the troposphere
6. Answers may vary but should reflect the information about composition, altitude, temperature, and pressure that is presented in this section.
7. troposphere and thermosphere
8. because molecules move farther apart when they are heated, so there are fewer particles in a given area to exert pressure
9. 9 km × 6.5°C/km = 58.5°C; Temperature on Mount Everest is about 58.5°C colder than on the coast.
10. A city on the Great Plains would have fewer problems with temperature inversions because the topography is flatter, allowing air to circulate more readily and not be trapped in valleys or against the mountains.
11. Answers may vary. Sample answer: First, temperatures at that altitude are relatively cold and air pressure relatively low, which could make breathing difficult. Second, he would have to find a way to land his lawn chair safely.
12. The *atmosphere* is a mixture of gases commonly referred to as *air* that includes *nitrogen, oxygen,* and *ozone;* compounds such as *water vapor;* and tiny particles called *particulates*.

Figure 9 ▶ During a temperature inversion, polluted cool air becomes trapped beneath a warm-air layer.

Temperature Inversions

Any substance that is in the atmosphere and that is harmful to people, animals, plants, or property is called an *air pollutant*. Today, the main source of air pollution is the burning of fossil fuels, such as coal and petroleum. As these fuels burn, they may release harmful chemical substances, such as sulfur dioxide gas, hydrocarbons, nitrogen oxides, carbon monoxide, and lead, into the air.

Certain weather conditions can make air pollution worse. One such condition is a *temperature inversion*, the layering of warm air on top of cool air. Warm air, which is less dense than cool air is, can trap cool, polluted air beneath it. In some areas, topography may make air pollution even worse by keeping the polluted inversion layer from dispersing, as **Figure 9** shows. Under conditions in which air cannot circulate up and away from an area, trapped automobile exhaust can produce *smog*, a general term for air pollution that indicates a combination of smoke and fog.

Air pollution can be controlled only by preventing the release of pollutants into the atmosphere. International, federal, and local laws have been passed to reduce the amount of air pollutants produced by automobiles and industry.

Section 1 Review

1. **Describe** the composition of dry air at sea level.
2. **Identify** five main components of the atmosphere.
3. **Explain** the cause of atmospheric pressure.
4. **Explain** how the two types of barometers measure atmospheric pressure.
5. **Identify** the layer of the atmosphere in which weather occurs.
6. **Compare** the four main layers of the atmosphere.
7. **Identify** the two atmospheric layers that contain air as warm as 25°C.

CRITICAL THINKING

8. **Drawing Conclusions** Why is atmospheric pressure generally lower beneath a mass of warm air than beneath a mass of cold air?
9. **Making Calculations** Calculate how much colder air is at the top of Mount Everest, which is almost 9 km above sea level, than air is at the Indian coastline. (Hint: On average, the temperature in the troposphere decreases by 6.5°C per kilometer of altitude.)
10. **Applying Ideas** Which industrial city would have fewer air-pollution incidents related to temperature inversions: one on the Great Plains or one near the Rocky Mountains? Explain your answer.
11. **Applying Concepts** In 1982, Larry Walters rose to an altitude of approximately 4,900 m on a lawn chair attached to 45 helium-filled weather balloons. Give two reasons why Walters' trip was dangerous.

CONCEPT MAPPING

12. Use the following terms to create a concept map: *oxygen, atmosphere, air, nitrogen, water vapor, ozone,* and *particulates.*

CHAPTER RESOURCES

Chapter Resource File

- Section Quiz GENERAL

Workbooks

- Study Guide (also in Spanish)

Section 2

Solar Energy and the Atmosphere

Earth's atmosphere is heated by the transfer of energy from the sun. Some of the heat in the atmosphere comes from the absorption of the sun's rays by gases in the atmosphere. Some heat enters the atmosphere indirectly as ocean and land surfaces absorb solar energy and then give off that energy as heat.

OBJECTIVES

- **Explain** how radiant energy reaches Earth.
- **Describe** how visible light and infrared energy warm Earth.
- **Summarize** the processes of radiation, conduction, and convection.

KEY TERMS

electromagnetic spectrum
albedo
greenhouse effect
conduction
convection

Radiation

All of the energy that Earth receives from the sun travels through space between Earth and the sun as radiation. *Radiation* includes all forms of energy that travel through space as waves. Visible light is the form of radiation that human eyes can detect. However, there are many other forms of radiation that humans cannot see, such as ultraviolet light, X rays, and radio waves.

Radiation travels through space in the form of waves at a very high speed—approximately 300,000 km/s. The distance from any point on a wave to the identical point on the next wave, for example from crest to crest, is called the *wavelength* of a wave. The various types of radiation differ in the length of their waves. Visible light, for example, consists of waves that have various wavelengths that are seen as different colors. The wavelengths of ultraviolet rays, X rays, and gamma rays are shorter than those of visible light. Infrared waves and radio waves have relatively long wavelengths. The waves that make up all forms of radiation are called *electromagnetic waves*. Almost all of the energy that reaches Earth from the sun is in the form of electromagnetic waves. The **electromagnetic spectrum,** shown at the bottom of **Figure 1,** consists of the complete range of wavelengths of electromagnetic waves.

electromagnetic spectrum all of the frequencies or wavelengths of electromagnetic radiation

Figure 1 ▶ The sun emits radiation whose wavelengths range throughout the electromagnetic spectrum. Five images of the sun are shown above. Each image shows radiation emitted at different wavelengths.

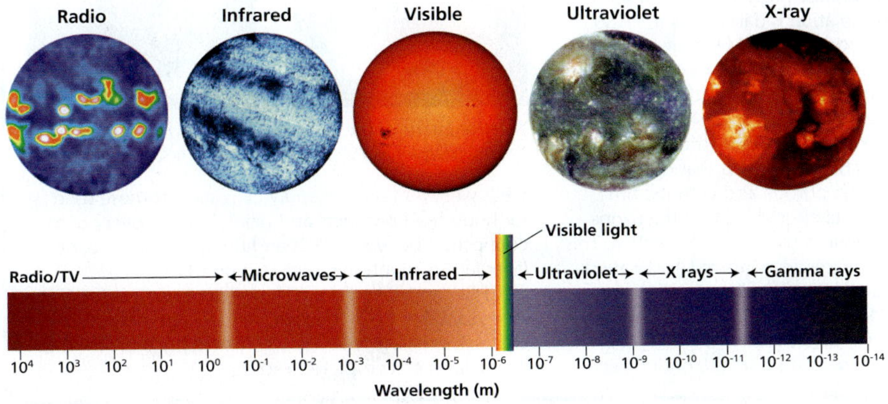

Section 2

Focus

Overview

This section explains how radiant energy from the sun interacts with Earth and its atmosphere. The section reviews the processes of conduction and convection and explains the greenhouse effect.

Bellringer

Ask students to describe what effect the sun has on Earth. (Answers may vary. The sun warms Earth's surface and provides light and energy.)

Motivate

Activity — GENERAL

Magic with Beads Use ultraviolet sensing beads (small beads that change color when exposed to UV light) to demonstrate how sunlight differs from indoor light. These beads are available through science supply catalogs. Have students expose the beads to light from an incandescent light bulb and then to sunlight and observe what happens. (The beads should change color only in the sun because white light does not have UV.) Then, have students use a prism and sunlight to create a spectrum. Place beads just outside the red and blue ends. In what regions do the beads change color? (only in the region just beyond the violet end of the spectrum, the UV region) LS **Visual**

CHAPTER RESOURCES

Chapter Resource File

- **Directed Reading** BASIC
- **Inquiry Lab** Ultraviolet Protection GENERAL

Technology

Transparencies
- Bellringer
- 107 The Electromagnetic Spectrum (with worksheet)

Student Edition on CD-ROM

One-Stop Planner CD-ROM
- Lesson Plan

Teach

Activity — GENERAL

Blue and Red Skies Skies appear blue because the gases and dust in the atmosphere scatter blue light most. Sunsets are commonly yellow or red because the sun is low on the horizon and the light has to travel through so much of the atmosphere that only red and yellow light make it through without being scattered. To model this effect, have students shine a flashlight at different angles through a glass of water that has about half a teaspoon of milk in it. The milk acts like the dust and gases in the atmosphere that scatter light. The color changes are subtle and best viewed in a darkened room. Have students add more milk, a teaspoon at a time, and see how this affects color. LS **Visual/Logical**

SKILL BUILDER — ADVANCED

Writing CFCs were used extensively for many purposes, including as propellants, refrigerants, fire retardants, and solvents, because they were extremely stable and nontoxic, at least in the troposphere. Now that they have been banned internationally, consumers want substitutes that have similar properties. Have students research and write a paper that discusses the chemicals being used in place of CFCs. The paper should compare the substitute products to CFCs in terms of use, cost, and possible adverse environmental effects. LS **Verbal/Logical**

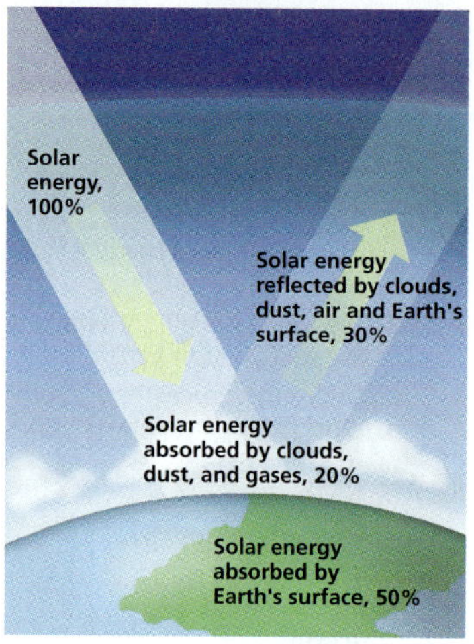

Figure 2 ▶ About 70% of the solar energy that reaches Earth is absorbed by Earth's land and ocean surfaces and by the atmosphere. The remainder is reflected back into space.

The Atmosphere and Solar Radiation

As solar radiation passes through Earth's atmosphere, the atmosphere affects the radiation in several ways. The upper atmosphere absorbs almost all radiation that has a wavelength shorter than the wavelengths of visible light. Molecules of nitrogen and oxygen in the thermosphere and mesosphere absorb the X rays, gamma rays, and ultraviolet rays. In the stratosphere, ultraviolet rays are absorbed and act upon oxygen molecules to form ozone.

Most of the solar rays that reach the lower atmosphere, such as visible and infrared waves, have longer wavelengths. Most incoming infrared radiation is absorbed by carbon dioxide, water vapor, and other complex molecules in the troposphere. As visible light waves pass through the atmosphere, only a small amount of this radiation is absorbed. **Figure 2** shows the percentage of solar energy that is reflected and absorbed by the atmosphere.

Scattering

Clouds, dust, water droplets, and gas molecules in the atmosphere disrupt the paths of radiation from the sun and cause scattering. Scattering occurs when particles and gas molecules in the atmosphere reflect and bend the solar rays. This deflection causes the rays to travel out in all directions without changing their wavelengths. Scattering sends some of the radiation back into space. The remaining radiation continues toward Earth's surface. As a result of scattering, sunlight that reaches Earth's surface comes from all directions. In addition, scattering makes the sky appear blue and makes the sun appear red at sunrise and sunset.

Connection to PHYSICS

The Ozone "Hole"

Ozone, O_3, is a naturally occurring gas that is present primarily in the stratosphere. The thin layer of ozone that surrounds Earth prevents most of the sun's ultraviolet (UV) radiation from reaching Earth's surface. Overexposure to UV radiation is dangerous to living things because it damages DNA. DNA is the genetic material that carries the information for inherited characteristics. UV radiation also makes the body more susceptible to skin cancer.

The protective ozone layer is not distributed around Earth evenly. Scientists have observed that ozone concentrations vary with latitude and with the time of year. In 1985, scientists discovered that the ozone layer was unusually thin in regions over Antarctica. This "ozone hole" allows greater amounts of UV radiation to reach Earth's surface. Scientists discovered that chemicals called chlorofluorocarbons (CFCs) were causing the ozone layer to break down. CFCs were used as coolants in refrigerators and air conditioners.

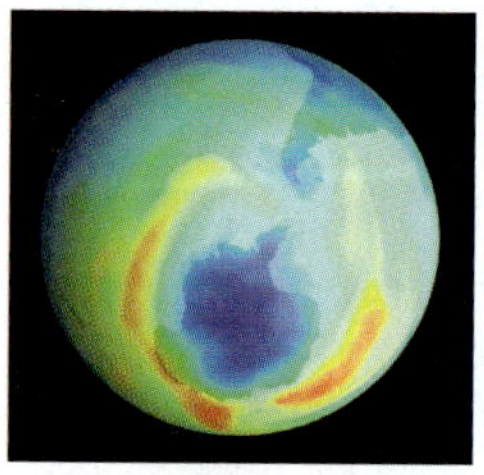

Satellite image of the ozone "hole" (purple) in 1980

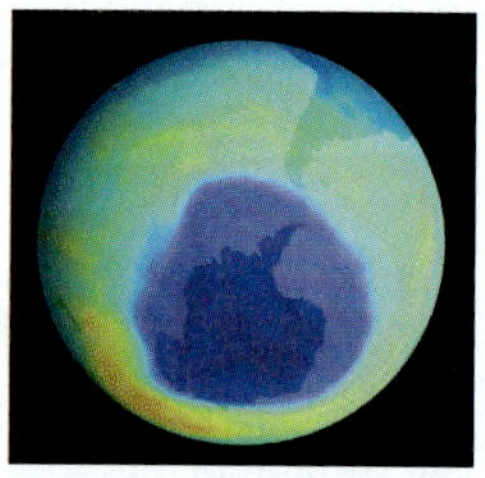

Satellite image of the same ozone "hole" in 2000

CFCs were also used in spray cans such as those used for household products and paint. The discovery of a connection between CFCs and the weakening ozone layer led to an international ban on CFCs.

CFCs can act to destroy ozone continuously for 60 to 120 years. So, CFCs released 30 years ago may still be destroying ozone today. It will take many years for the ozone layer to completely recover.

CHEMISTRY CONNECTION — ADVANCED

Catalytic Conversions CFCs are extremely damaging to the ozone layer because one CFC molecule can destroy thousands of ozone molecules by a catalytic reaction. Strong sunlight breaks down CFCs that drift into the stratosphere, releasing chlorine, the catalyst of ozone breakdown. The chlorine destroys ozone by reacting with the ozone to form unstable chlorine monoxide (ClO) and oxygen (O_2). The chlorine monoxide readily reacts with free oxygen atoms in the atmosphere, to form more O_2, and releasing the destructive chlorine. The same chlorine that destroyed the first ozone molecule reacts with and destroys another molecule of ozone. This process can occur over and over again, and thus one chlorine molecule can destroy thousands of ozone molecules before it is washed out of the atmosphere by precipitation. This process occurs much more readily on the surfaces of ice crystals, which explains why the ozone thinning is most prominent over Antarctica.

Table 1 ▼

Table 1 ▶ Reflection and Absorption Rates of Various Materials

Percentage of Solar Radiation		
Surface	**Reflected**	**Absorbed**
Soils (dark colored)	5–10	90–95
Desert	20–40	60–80
Grass	5–25	75–95
Forest	5–10	90–95
Snow	50–90	10–50
Water (high sun angle)	5–10	90–95
Water (low sun angle)	50–80	20–50

Reflection

When solar energy reaches Earth's surface, the surface either absorbs or reflects the energy. The amount of energy that is absorbed or reflected depends on characteristics such as the color, texture, composition, volume, mass, transparency, state of matter, and specific heat of the material on which the solar radiation falls. The intensity and amount of time that a surface material receives radiation also affects how much energy is reflected or absorbed.

albedo the fraction of solar radiation that is reflected off the surface of an object

The fraction of solar radiation that is reflected by a particular surface is called the **albedo.** Because 30% of the solar energy that reaches Earth's atmosphere is either reflected or scattered, Earth is said to have an albedo of 0.3. **Table 1** shows the amount of incoming solar radiation that is absorbed and reflected by various surfaces.

Absorption and Infrared Energy

The sun constantly emits radiation. Solar radiation that is not reflected is absorbed by rocks, soil, water, and other surface materials. When Earth's surface absorbs solar radiation, the radiation's short-wavelength infrared rays and visible light heat the surface materials. Then, the heated materials convert the energy into infrared rays of longer-wavelengths and reemit it as those waves. Gas molecules, such as water vapor and carbon dioxide, in the atmosphere absorb these infrared rays. The absorption of thermal energy from the ground heats the lower atmosphere and keeps Earth's surface much warmer than it would be if there were no atmosphere. Sometimes, warm air near Earth's surface bends light rays to produce an effect called a *mirage,* as **Figure 3** shows.

Figure 3 ▶ Hot air near the surface of this road bends light rays. ***What objects in this photo appear to be reflected?***

Group Activity — GENERAL

Comparing Albedos Have students work in small groups to fill two jars with dry sand and to cover the top of one jar with white paper and the other jar with black paper. Have students seal both jars by using tape. Have students place the jars side by side in the sun. After about a half an hour, have students measure the temperature just above the surface of the paper and just below the surface of the sand. Have each group explain their observations based on what they know about albedo. LS **Kinesthetic**

Using the Figure — GENERAL

Phantom Images Mirages form when light passes through layers of air that have different densities. The light is refracted, and forms an image above or below an object's true location. The mirage in this figure is an inferior mirage because it forms below the object's true location. Inferior mirages are most likely to form above hot land surfaces such as deserts or highways. Superior mirages form above an object's true location, most often above large bodies of water or fields of ice and snow. While mirages may seem to be delusions of the mind, they are real and can be photographed, as this figure shows. Answer to caption question: The sky and the poles appear to be reflected in this photo. LS **Visual/Logical**

SKILL BUILDER — BASIC

Vocabulary The term *albedo* comes from the Latin, *albedo*, which means "whiteness." Objects that are white tend to reflect light, so a surface that has high albedo reflects much of the sunlight it receives. Snow and ice, for example, have a very high albedo. LS **Verbal** English Language Learners

Teach, *continued*

Internet Activity — GENERAL

Global Warming Many scientists are concerned that the global climate is warming due to a human-enhanced greenhouse effect. Have students investigate the causes and potential impact of an enhanced greenhouse effect on Earth, as well as the controversies that surround predictions of global warming. A worksheet designed to direct student research on this topic can be found in the **Chapter Resource File** booklet or by visiting **go.hrw.com** and entering the keyword **HQ6ATMX.** **LS Verbal**

ENVIRONMENTAL CONNECTION

CO_2 Concentrations Recent increases in carbon dioxide in the atmosphere concern scientists who study global warming. The CO_2 concentration has risen from 0.028% to over 0.036% since 1860, which is commonly considered to be the beginning of the Industrial Age. Scientists expect CO_2 to double pre-industrial levels by the middle of this century. Burning fossil fuels in vehicles and power plants contributes significantly to greenhouse gas emissions.

CHAPTER RESOURCES

Chapter Resource File
- Internet Activity Global Warming GENERAL

Technology

Transparencies
- 108 The Greenhouse Effect and Latitude and Season (with worksheet)

Figure 4 ▶ One process that helps heat Earth's atmosphere is similar to the process that heats a greenhouse.

The Greenhouse Effect

One of the ways in which the gases of the atmosphere absorb and reradiate infrared rays, shown in **Figure 4,** can be compared to the process that keeps a greenhouse warm. The glass of a greenhouse allows visible light and infrared rays from the sun to pass through and warm the surfaces inside of the greenhouse. But the glass prevents the infrared rays that are emitted by the warmed surfaces within the greenhouse from escaping. Similarly, Earth's atmosphere slows the escape of energy that radiates from Earth's surface. Because this process is similar to the process that heats a greenhouse, it is called the **greenhouse effect.**

greenhouse effect the warming of the surface and lower atmosphere of Earth that occurs when carbon dioxide, water vapor, and other gases in the air absorb and reradiate infrared radiation

Human Impact on the Greenhouse Effect

Generally, the amount of solar energy that enters Earth's atmosphere is about equal to the amount that escapes into space. However, human activities, in addition to natural causes, are changing this balance and are causing the average temperature of the atmosphere to increase. For example, measurements indicate that the amount of carbon dioxide in the atmosphere has been increasing in recent years. These increases have been attributed to the burning of more fossil fuels. These increases seem likely to continue in the future. Increases in the amount of carbon dioxide may intensify the greenhouse effect and may cause Earth to become warmer in some areas and cooler in others.

SCILINKS. NSTA Developed and maintained by the National Science Teachers Association

For a variety of links related to this subject, go to www.scilinks.org

Topic: Greenhouse Effect
SciLinks code: HQ60694

Variations in Temperature

Radiation from the sun does not heat Earth equally at all places at all times. In addition, a slight delay occurs between the absorption of energy and an increase in temperature. Earth's surface must absorb energy for a time before enough heat has been absorbed and reradiated from the ground to change the temperature of the atmosphere. For a similar reason, the warmest hours of the day are usually mid- to late afternoon even though solar radiation is most intense at noon. The temperature of the atmosphere in any region on Earth's surface depends on several factors, including latitude, surface features, and the time of year and day.

MISCONCEPTION ALERT

Greenhouse Goofs A common misconception about the greenhouse effect is the belief that greenhouse gases reflect heat back to Earth. They do not. Greenhouse gases, such as carbon dioxide, methane, water vapor, and even CFCs, absorb or retain heat that radiates from Earth, then reradiate it back to Earth. Another common misconception is that the greenhouse effect has only negative effects to life on Earth. The greenhouse effect is actually essential to life on Earth. Without the greenhouse effect, scientists estimate that Earth would be 15.5°C (60°F) cooler, and life on Earth would be very different. On the other hand, too much of a greenhouse effect would make Earth much like Venus—too hot to support life as we know it.

Latitude and Season

Latitude is the primary factor that affects the amount of solar energy that reaches any point on Earth's surface. Because Earth is a sphere, the sun's rays do not strike all areas at the same angle, as shown in **Figure 5.** The rays of the sun strike the ground near the equator at an angle near 90°. At the poles, the sunlight strikes the ground at a much smaller angle. When sunlight hits Earth's surface at an angle smaller than 90°, the energy is spread out over a larger area and is less intense. Thus, the energy that reaches the equator is more intense than the energy that strikes the poles, so average temperatures are higher near the equator than near the poles.

Temperature varies seasonally because of the tilt of Earth's axis. As Earth revolves around the sun once each year, the portion of Earth's surface that receives the most intense sunlight changes. For part of the year, the Northern Hemisphere is tilted toward the sun and receives more direct sunlight. During this time of year, temperatures are at their highest. For the other part of the year, the Southern Hemisphere is tilted toward the sun. During this time, the Northern Hemisphere receives less direct sunlight, and the temperatures there are at their lowest.

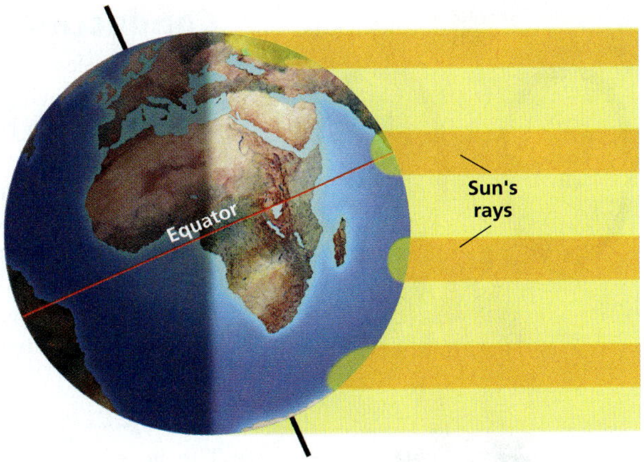

Figure 5 ▶ Temperatures are higher at the equator because solar energy is concentrated in a small area. Farther north and south, the same amount of solar energy is spread out over a larger area.

Water in the Air and on the Surface

Because water vapor stores heat, the amount of water in the air affects the temperature of a region. The thinner air at high elevations contains less water vapor and carbon dioxide to absorb the heat. As a result, those areas become warm during the day but cool very quickly at night. Similarly, desert temperatures may vary widely between day and night because little water vapor is present to hold the heat of the day.

Land areas close to large bodies of water generally have more moderate temperatures. In other words, these areas will be cooler during the day and warmer at night than inland regions that have the same general weather conditions are. The reason for these moderate temperatures is that water heats up and cools down faster than air does, so the temperature of water changes less than the temperature of land does.

The wind patterns in an area also affect temperature. A region that receives winds off the ocean waters has more moderate temperatures than does a similar region in which the winds blow from the land.

✓ Reading Check **Why are deserts generally colder at night than other areas are? (See the Appendix for answers to Reading Checks.)**

QuickLAB — 5 min

Light and Latitude

Procedure

1. Hold a **flashlight** so that the beam shines directly down on a **white piece of paper.** Use a **pencil** to trace the outline of the beam of light.
2. Move the flashlight so that the light shines on the paper at an angle. Trace the outline of the beam of light.

Analysis

1. How does the area of the direct beam differ from the area of the angled beam?
2. How does this exercise illustrate how latitude affects incoming solar radiation?

QuickLAB

Skills Acquired

- Observing
- Analyzing

Teacher's Notes Make sure the height of the flashlight does not change when students change the angle of the flashlight. You may want to perform this experiment as a demonstration.

Answers

1. The area of the direct beam is smaller than the area of the angled beam.
2. At the equator, energy from the sun hits Earth at a 90° angle, so the energy is concentrated on a small area. At latitudes far north and south of the equator, sunlight hits Earth at angles less than 90°. The same amount of energy is spread out over a greater area. Therefore, each square unit of surface receives less energy.

Answer to Reading Check

Deserts are colder at night than other areas are because the air in deserts contains little water vapor that can absorb heat during the day and release heat slowly at night.

Close

Reteaching — BASIC

Peer Reviewing Have students work in pairs to write three to five questions about the material in this section. Then, have students use their questions to quiz each other. **LS Interpersonal** Co-op Learning

Quiz — GENERAL

1. What property makes a substance a good conductor? (molecules that are close together, such as in solid metals)
2. What may happen to Earth's climate if the amount of carbon dioxide in the air continues to increase? (The greenhouse effect may intensify and the average annual temperature of most areas will be warmer.)
3. What is the main cause of increasing carbon dioxide concentrations in the atmosphere? (The burning of fossil fuels.)

Alternative Assessment — GENERAL

The Spectrum of Life Have students keep an "electromagnetic journal" for one week. They should record each time they observe or encounter electromagnetic radiation. The record should include the types of radiation, explanations of their encounters, and what type of radiation they encounter most. **LS Verbal**

CHAPTER RESOURCES

Chapter Resource File

- **Datasheet for Quick Lab** GENERAL

Close, *continued*

Answers to Section Review

1. Radiation from the sun travels through space in the form of electromagnetic waves.
2. long wavelength waves: radio/TV, microwaves, infrared, and visible light; short wavelength waves: UV rays, X rays, and gamma rays
3. Nitrogen, oxygen, and ozone in the upper atmosphere absorb most radiation that has a wavelength shorter than that of visible light. In the lower atmosphere, carbon dioxide and water vapor absorb infrared radiation. About 50% of solar energy reaches Earth's surface. Clouds, particulates, and gases in the lower atmosphere scatter and reflect much of the light.
4. Infrared and visible light that reach Earth's surface are absorbed. The heated surfaces radiate heat back into the air, where greenhouse gases absorb some of the heat.
5. If the sunlight is concentrated over a small area, as happens near the equator, the sun's energy is more intense and the surface becomes hotter. When sunlight is spread over a larger area, as happens at higher latitudes, the sun's energy and the resulting heat are less intense.
6. Conduction is the transfer of energy from one substance to another by direct contact. Convection is the transfer of energy through movement of matter due to differences in density as a result of temperature differences.
7. Scientists study the entire electromagnetic spectrum because Earth is affected by radiation of all of the wavelengths in the spectrum.
8. In convection ovens, the fan mixes the air, so that hot or cold spots are less likely to form, resulting in more even cooking.
9. The hottest hours of the day will probably be in the late afternoon when Earth's surface has had time to absorb and reradiate energy and when convection is at its peak.
10. The atmosphere interacts with *electromagnetic waves* from the sun, such as *ultraviolet waves, visible light* and *infrared waves,* by *scattering* or *absorption,* which is part of the *greenhouse effect.*

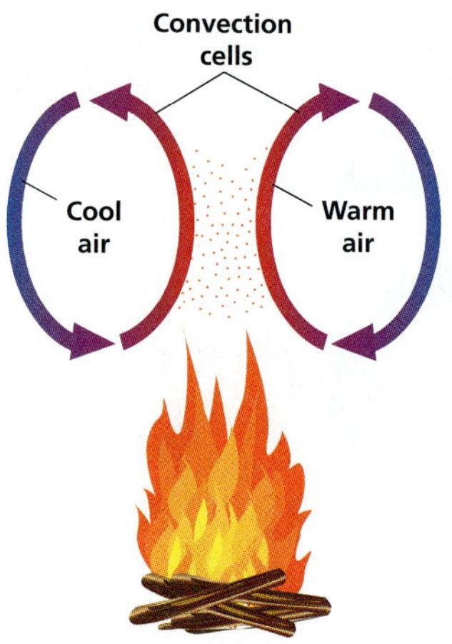

Figure 6 ▶ During convection, energy is carried away by heated air as it rises above cooler, denser air.

Conduction

The molecules in a substance move faster as they become heated. These fast-moving molecules cause other molecules to move faster. Collisions between the particles result in the transfer of energy, which warms the substance. The transfer of energy as heat from one substance to another by direct contact is called **conduction**. Solid substances, in which the molecules are close together, make relatively good conductors. Because the molecules of air are far apart, air is a poor conductor. Thus, conduction heats only the lowest few centimeters of the atmosphere, where air comes into direct contact with the warmed surface of Earth.

Convection

The heating of the lower atmosphere is primarily the result of the distribution of heat through the troposphere by convection. **Convection** is the process by which air, or other matter, rises or sinks because of differences in temperature. Convection occurs when gases or liquids are heated unevenly. As air is heated by radiation or conduction, it becomes less dense and is pushed up by nearby cooler air. In turn, this cooler air becomes warmer, and the cycle repeats, as shown in **Figure 6.**

The continuous cycle in which cold air sinks and warm air rises warms Earth's atmosphere evenly. Because warm air is less dense than cool air is, warm air exerts less pressure than the same volume of cooler air does. So, the atmospheric pressure is lower beneath a mass of warm air. As dense, cool air moves into a low-pressure region, the less dense, warmer air is pushed upward. These pressure differences, which are the result of the unequal heating that causes convection, create winds.

conduction the transfer of energy as heat through a material

convection the movement of matter due to differences in density that are caused by temperature variations; can result in the transfer of energy as heat

Section 2 Review

1. **Explain** how radiant energy reaches Earth.
2. **List and describe** the types of electromagnetic waves.
3. **Describe** how gases and particles in the atmosphere interact with light rays.
4. **Describe** how visible light and infrared energy warm Earth.
5. **Explain** how variations in the intensity of sunlight can cause temperature differences on Earth's surface.
6. **Summarize** the processes of conduction and convection.

CRITICAL THINKING

7. **Making Inferences** Why do scientists study all wavelengths of the electromagnetic spectrum?
8. **Applying Concepts** Explain how fans in convection ovens help cook food more evenly.
9. **Applying Conclusions** You decide not to be outside during the hottest hours of a summer day. When will the hottest hours probably be? How do you know?

CONCEPT MAPPING

10. Use the following terms to create a concept map: *electromagnetic waves, infrared waves, greenhouse effect, ultraviolet waves, visible light, scattering,* and *absorption.*

CHAPTER RESOURCES

Chapter Resource File
- **Section Quiz** GENERAL

Workbooks
- **Study Guide** (also in Spanish)

Section 3 Atmospheric Circulation

Pressure differences in the atmosphere cause the movement of air worldwide. The air near Earth's surface generally flows from the poles toward the equator. The reason for this flow is that air moves from high-pressure regions to low-pressure regions. High-pressure regions form where cold air sinks toward Earth's surface. Low-pressure regions form where warm air rises away from Earth's surface.

OBJECTIVES

- **Explain** the Coriolis effect.
- **Describe** the global patterns of air circulation, and name three global wind belts.
- **Identify** two factors that form local wind patterns.

KEY TERMS

Coriolis effect
trade winds
westerlies
polar easterlies
jet stream

The Coriolis Effect

The circulation of the atmosphere and of the oceans is affected by the rotation of Earth on its axis. Earth's rotation causes its diameter to be greatest through the equator and smallest through the poles. Because each point on Earth makes one complete rotation every day, points near the equator travel farther and faster in a day than points closer to the poles do. When air moves toward the poles, it travels east faster than the land beneath it does. As a result, the air follows a curved path. The tendency of a moving object to follow a curved path rather than a straight path because of the rotation of Earth is called the **Coriolis effect,** which is shown in **Figure 1.**

Coriolis effect the curving of the path of a moving object from an otherwise straight path due to Earth's rotation

Winds that blow from high-pressure areas to lower-pressure areas curve as a result of the Coriolis effect. The Coriolis effect deflects moving objects along a path that depends on the speed, latitude, and direction of the object. Objects are deflected to the right in the Northern Hemisphere and are deflected to the left in the Southern Hemisphere.

The faster an object travels, the greater the Coriolis effect on that object is. The Coriolis effect also noticeably changes the paths of large masses that travel long distances, such as air or ocean currents. In general, the Coriolis effect is detectable only on objects that move very fast or that travel over long distances.

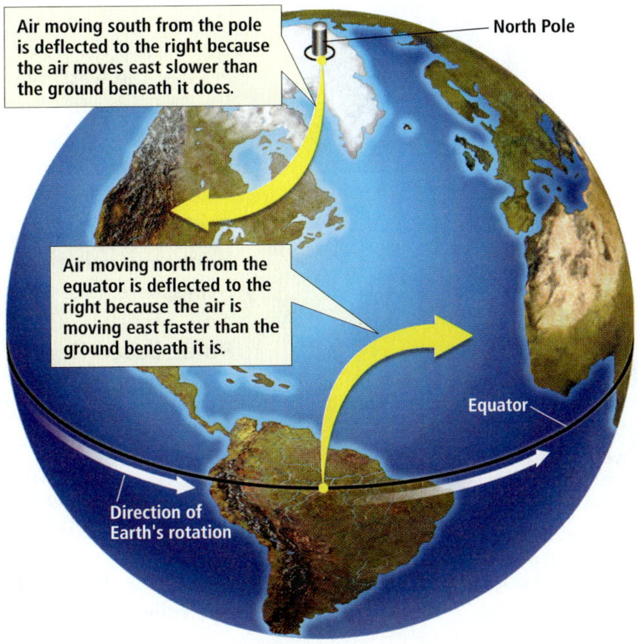

Figure 1 ▶ Because of Earth's rotation, an object that travels north from the equator will curve to the east. This curving is called the *Coriolis effect.*

Section 3

Focus

Overview

This section explains the Coriolis effect and global patterns of air circulation that result from pressure differences in Earth's atmosphere. It also describes the local winds that result from land features such as mountains and bodies of water.

Bellringer

Ask students to describe what causes air to move. (Answers may vary. Use student answers to begin a discussion of the causes of winds.) **LS Logical**

Motivate

Demonstration — GENERAL

Modeling the Coriolis Effect

Center a paper plate on a turntable or "lazy susan." Draw a straight line from the center of the plate to the outside edge without spinning the turntable. Then, spin the turntable counterclockwise (the direction that Earth rotates), and draw another line from the center to the edge of the plate. Because the paper is spinning, the second line will spiral around the plate to the right. Ask students what the spinning plate and curved line model. (The spinning plate models the rotation of Earth on its axis; the curved line models the Coriolis effect on air in the Northern Hemisphere.) **LS Visual**

CHAPTER RESOURCES

Chapter Resource File
- Directed Reading BASIC

Technology

Transparencies
- Bellringer
- 109 The Coriolis Effect (with worksheet)

Student Edition on CD-ROM

One-Stop Planner CD-ROM
- Lesson Plan

Teach

Using the Figure — ADVANCED

Coriolis Effect Explain to students that the convection cells from the equator to the subtropics are known as *Hadley cells*, after George Hadley, a lawyer, who in 1735 proposed a partially correct explanation for the direction of the trade winds. The convection cells from the subtropics to the subpolar regions are known as *Ferrell cells*, after William Ferrell, an American schoolteacher, who in 1855 correctly proposed the three-cell model that accounts for the global wind belts. Answer to caption question: Winds curve clockwise in the Northern Hemisphere. LS **Visual**

Group Activity — GENERAL

It's a Breeze Have students work in small groups to fill one baking dish with sand and warm it up in an oven on low heat. Have them fill another baking pan with ice and place the pans side by side. Have students make a screen 20–25 cm high out of cardboard and use it to surround the pans on three sides. Then, have them light an incense stick and hold it between the two pans. Ask students to describe and explain what happens. (The sand warms the air above it, making the air rise. The air above the ice is cold and dense and flows in to take the place of the warm air rising above the sand, creating a breeze.) LS **Verbal/ Visual**

Answer to Reading Check

They flow in opposite directions from each other, and they occur at different latitudes.

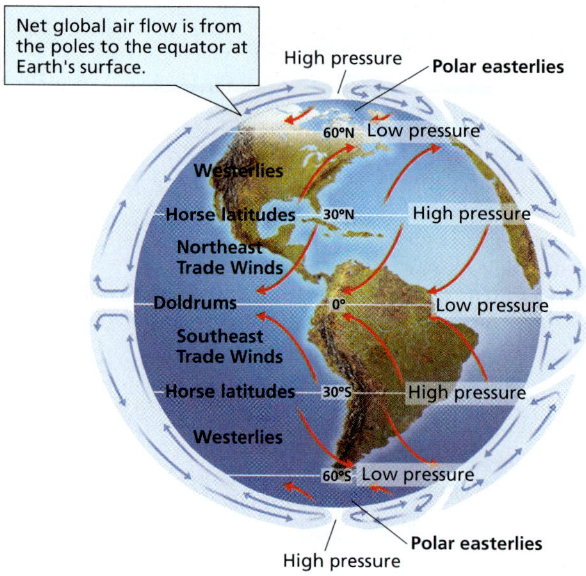

Figure 2 ▶ Each hemisphere has three wind belts. Wind belts are the result of pressure differences at the equator, the subtropics, the subpolar regions, and the poles. Winds in the belts curve because of the Coriolis effect. ***Do winds in the Northern Hemisphere curve clockwise or counterclockwise?***

Global Winds

The air that flows from the poles toward the equator does not flow in a single, straight line. Each hemisphere contains three looping patterns of flow called *convection cells*. Each of these convection cells correlates to an area of Earth's surface, called a *wind belt*, that is characterized by winds that flow in one main direction. These winds are called *prevailing winds*. All six wind belts are shown in **Figure 2.**

Trade Winds

In both hemispheres, the winds that flow toward the equator between 30° and 0° latitude are called **trade winds.** Like all winds, the trade winds are named according to the direction from which they flow. In the Northern Hemisphere, the trade winds flow from the northeast and are called the *northeast trade winds*. In the Southern Hemisphere, the trade winds are called the *southeast trade winds*. These wind belts are called *trade winds* because many trading ships sailed on these winds from Europe in the 18th and 19th centuries.

Westerlies

Between 30° and 60° latitude, some of the descending air moving toward the poles is deflected by the Coriolis effect. This flow creates the **westerlies,** which exist in another wind belt in each hemisphere. In the Northern Hemisphere, the westerlies are southwest winds. In the Southern Hemisphere, they are northwest winds. The westerlies blow throughout the contiguous United States.

Reading Check **Name two ways in which the trade winds of the Northern Hemisphere differ from the westerlies of the Northern Hemisphere. (See the Appendix for answers to Reading Checks.)**

trade winds prevailing winds that blow from east to west from 30° latitude to the equator in both hemispheres

westerlies prevailing winds that blow from west to east between 30° and 60° latitude in both hemispheres

polar easterlies prevailing winds that blow from east to west between 60° and 90° latitude in both hemispheres

Polar Easterlies

Toward the poles, or poleward, of the westerlies—at about 60° latitude—is a zone of low pressure. This zone of low pressure separates the westerlies from a third wind belt in each hemisphere. Over the polar regions themselves, descending cold air creates areas of high pressure. Surface winds created by the polar high pressure are deflected by the Coriolis effect and become the **polar easterlies.** The polar easterlies are strongest where they flow off Antarctica. Where the polar easterlies meet warm air from the westerlies, a stormy region known as a *front* forms.

CHAPTER RESOURCES

Chapter Resource File

- **Making Models Lab** Global Air Movement GENERAL

Technology

Transparencies

- 110 Global Wind Belts (with worksheet)

INCLUSION *Strategies*

- ***Visually Impaired***
- ***Developmentally Delayed***
- ***Attention Deficit Disorder***

Students with disabilities may learn the terminology of science lessons more easily by using this method: Divide the class into three teams. Assign one of the three sections to each team. Ask each team to divide up the list of terms and create drawings that depict the terms. Have teams display their drawings. LS **Visual/ Verbal**

READING SKILL BUILDER — BASIC

Reading Organizer As students read this section, encourage them to take Power Notes, KWL notes, or Two-Column Notes as described in the Skills Handbook section of the Appendix. Later, students can use these notes as a study guide for assessments. LS **Verbal** English Language Learners

The Doldrums and Horse Latitudes

As **Figure 2** shows, the trade wind systems of the Northern Hemisphere and Southern Hemisphere meet at the equator in a narrow zone called the *doldrums*. In this warm zone, most air movement is upward and surface winds are weak and variable. As the air approaches 30° latitude, it descends and a high-pressure zone forms. These subtropical high-pressure zones are called the *horse latitudes*. Here, too, surface winds are weak and variable.

Wind and Pressure Shifts

As the sun's rays shift northward and southward during the changing seasons of the year, the positions of the pressure belts and wind belts shift. Although the area that receives direct sunlight can shift by up to 47° north and south of the equator, the average shift for the pressure belts and wind belts is only about 10° of latitude. However, even this small change causes some areas of Earth's surface to be in different wind belts during different times of the year. In southern Florida, for example, westerlies prevail in the winter, but trade winds dominate in the summer.

Jet Streams

Narrow bands of high-speed winds that blow in the upper troposphere and lower stratosphere are **jet streams.** These winds exist in the Northern Hemisphere and Southern Hemisphere.

One type of jet stream is a polar jet stream. Polar jet streams form as a result of density differences between cold polar air and the warmer air of the middle latitudes. These bands of winds, which are about 100 km wide and 2 to 3 km thick, are located at altitudes of 10 to 15 km. Polar jet streams can reach speeds of 500 km/h and can affect airline routes and the paths of storms.

Another type of jet stream is a subtropical jet stream. In the subtropical regions, very warm equatorial air meets the cooler air of the middle latitudes, and the *subtropical jet streams* form. Unlike the polar jet streams, the subtropical jet streams do not change much in speed or position. A subtropical jet stream is shown in **Figure 3.**

Figure 3 ▶ Clouds in this jet stream are traveling high over Egypt. This remarkable photograph was taken by *Gemini 12* astronauts.

Graphic Organizer **Comparison Table**

Create the Graphic Organizer entitled "Comparison Table" described in the Skills Handbook section of the Appendix. Label the columns with "Trade winds," "Westerlies," "Polar easterlies," and "Jet streams." Label the rows with "Latitude" and "Direction.". Then, fill in the table with details about each type of wind.

jet stream a narrow band of strong winds that blow in the upper troposphere

Graphic Organizer GENERAL

Comparison Table
You may want to use this Graphic Organizer in a game to review material before the test. Divide the class into two teams. Ask students questions about material from the Comparison Table. Give points to each team that provides correct answers.

Close

Reteaching — BASIC

Labeling Have students use a laminated map or a globe to identify the location and prevailing direction of the major wind belts. LS Visual

Quiz — GENERAL

1. What are the doldrums? (a region of low pressure near the equator where the trade wind systems meet and surface winds are very weak and variable)
2. What is a convection cell? (a vertical looping pattern of air flow that corresponds to the global wind belts)
3. What is a breeze? (a gentle wind that extends over a distance of less than 100 km)

Alternative Assessment — ADVANCED

Ask students to write a short essay that explains why it is correct to say that wind energy is related to solar energy. LS Verbal/Logical

BRAIN FOOD

The Horse Latitudes The region where trade winds and the westerlies diverge, at about 30° North latitude, got its name for a macabre reason. Here, winds are light or even calm over vast areas of the ocean. Ships sailing to the New World often were becalmed here for weeks at a time. When food and water ran low, horses carried as cargo were tossed overboard to conserve remaining supplies. Their carcasses floated around these waters, a grim reminder for the next passing ship.

Close, *continued*

Answers to Section Review

1. Air moves from regions of high pressure toward regions of low pressure.
2. The Coriolis effect causes winds to curve because the rate at which the air travels differs from the rate of travel of the ground beneath the air. In the Northern Hemisphere, winds curve to the right; in the Southern Hemisphere, they curve to the left.
3. Polar easterlies are prevailing winds that blow from east to west between 60° and 90° latitude in both hemispheres. The westerlies are winds that blow from the southwest in the Northern Hemisphere and from the northwest in the Southern Hemisphere in the belts between 30° and 60° latitude. The trade winds are prevailing winds that blow from the northeast from 30° N to the equator and from the southeast from 30° S to the equator.
4. Jet streams are narrow bands of high-speed winds that blow in the upper troposphere and lower stratosphere. They are important because they can affect the paths of storms and airline routes.
5. Temperature differences between land and sea and between mountains and valleys influence local wind patterns.
6. Wind moving southward from the equator will curve to the east because of the Coriolis effect.
7. The air in my lungs has lower pressure. Because air moves from regions of higher pressure to areas of lower pressure, the air pressure in my lungs must be lower than the pressure outside my body.
8. Because sea breezes blowing from the water to land generally form in the afternoon, I would walk into the wind to reach the ocean.
9. *Winds* may be *global winds,* such as *polar easterlies, westerlies,* and *trade winds,* or *local winds,* such as *sea breezes, land breezes, mountain breezes,* and *valley breezes.*

Figure 4 ▶ Sea breezes keep these kites aloft during the afternoon. Overnight, land breezes will blow the flags toward the ocean.

Local Winds

Winds also exist on a scale that is much smaller than a global scale. Movements of air are influenced by local conditions, and local temperature variations commonly cause local winds. Local winds are not part of the global wind belts. Gentle winds that extend over distances of less than 100 km are called *breezes*.

Land and Sea Breezes

Equal areas of land and water may receive the same amount of energy from the sun. However, land surfaces heat up faster than water surfaces do. Therefore, during daylight hours, a sharp temperature difference develops between a body of water and the land along the water's edge. This temperature difference is apparent in the air above the land and water. The warm air above the land rises as the cool air from above the water moves in to replace the warm air. A cool wind moving from water to land, called a *sea breeze,* generally forms in the afternoon, as shown in **Figure 4.** Overnight, the land cools more rapidly than the water does, and the sea breeze is replaced by a *land breeze*. A land breeze flows from the cool land toward the warmer water.

Mountain and Valley Breezes

During the daylight hours in mountainous regions, a gentle valley breeze blows upslope. This *valley breeze* forms when warm air from the valleys moves upslope. At night, the mountains cool more quickly than the valleys do. At that time, cool air descends from the mountain peaks to create a *mountain breeze*. Areas near mountains may experience a warm afternoon that turns to a cold evening soon after sunset. This evening cooling happens because cold air flows down mountain slopes and settles in valleys.

Section 3 Review

1. **Describe** the pattern of air circulation between an area of low pressure and an area of high pressure.
2. **Explain** how the Coriolis effect affects wind flow.
3. **Name and describe** Earth's three global wind belts.
4. **Summarize** the importance of the jet streams.
5. **Identify** two factors that create local wind patterns.

CRITICAL THINKING

6. **Applying Concepts** Determine whether wind moving south from the equator will curve eastward or westward because of the Coriolis effect.
7. **Inferring Relationships** Which has a lower pressure: the air in your lungs as you inhale or the air outside your body? Explain.
8. **Applying Ideas** While visiting the Oregon coast, you decide to hike toward the ocean, but you are not sure of the direction. The time is 4:00 P.M. How might the breeze help you find your way?

CONCEPT MAPPING

9. Use the following terms to create a concept map: *wind, sea breeze, global winds, trade winds, westerlies, local winds, polar easterlies, land breeze, mountain breeze,* and *valley breeze.*

CHAPTER RESOURCES

Chapter Resource File
- Section Quiz GENERAL

Workbooks
- Study Guide (also in Spanish)

Chapter 22 Highlights

Sections

1 Characteristics of the Atmosphere

Key Terms

atmosphere, 547
ozone, 549
atmospheric pressure, 550
troposphere, 552
stratosphere, 553
mesosphere, 553
thermosphere, 553

Key Concepts

- Earth's atmosphere is the mixture of gases, called *air*, that surrounds Earth. Mixed with the gases that make up air are solid particles called *particulates*.
- Atmospheric pressure is the force exerted on Earth's surface by the weight of the atmosphere. It is measured by using a barometer.
- The atmosphere is divided into four major layers whose temperature and pressure vary.
- Air pollution can be harmful to people, animals, plants, and property.

2 Solar Energy and the Atmosphere

electromagnetic spectrum, 555
albedo, 557
greenhouse effect, 558
conduction, 560
convection, 560

- Most of the energy that reaches Earth from the sun is in the form of electromagnetic radiation.
- Visible light and infrared rays from the sun penetrate Earth's atmosphere and heat materials on the surface.
- The upper atmosphere is heated by absorption of radiation from the sun. The lower atmosphere is heated by conduction from Earth's surface and by convection of air.

3 Atmospheric Circulation

Coriolis effect, 561
trade winds, 562
westerlies, 562
polar easterlies, 562
jet stream, 563

- Air-pressure differences due to the unequal heating of Earth in combination with Earth's rotation cause the global wind belts.
- A surface feature, such as a body of water, a mountain, or a valley, can influence local wind patterns.

Chapter Highlights

Alternative Assessment — ADVANCED

Life's a Gas Have students imagine that they are one of the main gases in the atmosphere (N_2, O_2, Ar, H_2O, ozone, or CO_2) and have them write a story that describes their life by using words and illustrations. They should identify what gas they are, what layers of the atmosphere they have resided in, and what has happened to them during their life in the atmosphere. Students may want to describe whether they have cycled through living or nonliving things on Earth, interacted with solar radiation, or reacted with pollutants in the air. Encourage creativity and accept all reasonable answers but make sure their stories and illustrations are factually possible. **LS Verbal/Visual**

CHAPTER RESOURCES

Chapter Resource File

- Concept Review GENERAL
- Critical Thinking ADVANCED
- Math Skills GENERAL
- Graphing Skills GENERAL
- Chapter Test A GENERAL
- Chapter Test B ADVANCED

Workbooks

- **Study Guide** (also in Spanish)
- **Assessments** (Spanish)

Technology

Classroom Videos
- Brain Food Video Quiz

HRW Earth Science Video
- Segment 16: Global Winds
- Segment 17: Local Winds

Chapter Review

Assignment Guide

Section	Questions
1	1, 4–5, 9–12, 17–21, 33
2	2, 6, 13–14, 22–25, 32, 34, 36
3	3, 7–8, 15–16, 26–30, 35, 37–39
1 and 2	31

Using Key Terms

1–8. Answers may vary but should show that students understand the definitions of and differences between key terms.

Understanding Key Concepts

9. a **10.** b
11. a **12.** c
13. a **14.** d
15. b **16.** c
17. a **18.** d

Short Answer

19. nitrogen, oxygen, and argon

20. Atmospheric pressure is the force per unit area exerted on a surface by the weight of the atmosphere. It is measured by using a barometer.

21. The *troposphere* is the lowest layer of the atmosphere and is the place where most weather occurs. Temperature decreases with altitude in the troposphere, until the tropopause, or boundary between the troposphere and the *stratosphere*, is reached. In the stratosphere, temperature increases with increasing altitude as solar radiation is absorbed by ozone. Almost all of the ozone in the atmosphere is concentrated in this layer, which extends to an altitude of about 50 km. The *mesosphere*, extending to an altitude of about 80 km, is the coldest layer of the atmosphere. Temperature decreases with increasing altitude. In the *thermosphere*, temperature increases with increasing altitude.

22. Visible light that enters the atmosphere is reflected, absorbed, or scattered by the atmosphere and by Earth's surface.

23. by radiation, conduction, and convection

24. Because Earth is a sphere, sunlight does not hit the surface at the same angle all over the globe. When sunlight hits Earth's surface at angles less than 90°, as at high latitudes, its energy is spread out over a larger area and thus is less intense and temperatures are lower.

25. The greenhouse effect warms Earth's lower atmosphere as carbon dioxide, water vapor and other greenhouse gases absorb and reradiate infrared radiation, which slows the escape of heat from Earth.

26. the rotation of Earth

27. The trade winds are prevailing winds that blow from the northeast from 30° North latitude to the equator and southeast from 30° South latitude to the equator. The westerlies are winds that blow from west to east in the belt between 30° and 60° North and South latitude. Polar easterlies are prevailing winds that blow from east to west between 60° and 90° North and South latitudes.

28. Surface features, such as bodies of water, mountains, and valleys, create local winds as a result of uneven heating and cooling of the ground surface.

Chapter 22 Review

Using Key Terms

Use each of the following terms in a separate sentence.

1. *atmosphere*

2. *electromagnetic spectrum*

3. *Coriolis effect*

For each pair of terms, explain how the meanings of the terms differ.

4. *troposphere* and *stratosphere*

5. *mesosphere* and *thermosphere*

6. *conduction* and *convection*

7. *trade winds* and *westerlies*

8. *polar easterlies* and *westerlies*

Understanding Key Concepts

9. During one part of the nitrogen cycle, nitrogen is removed from the air mainly by nitrogen-fixing
- **a.** bacteria.
- **b.** waves.
- **c.** minerals.
- **d.** crystals.

10. The atmosphere contains tiny solid particles called
- **a.** gases.
- **b.** particulates.
- **c.** meteors.
- **d.** nitrogen.

11. A barometer measures
- **a.** atmospheric pressure.
- **b.** wind speed.
- **c.** ozone concentration.
- **d.** wavelengths.

12. Almost all of the water and carbon dioxide in the atmosphere is in the
- **a.** exosphere.
- **b.** ionosphere.
- **c.** troposphere.
- **d.** stratosphere.

13. The process by which the atmosphere slows Earth's loss of heat to space is called the
- **a.** greenhouse effect.
- **b.** Coriolis effect.
- **c.** doldrums.
- **d.** convection cell.

14. Energy as heat can be transferred within the atmosphere in three ways—radiation, conduction, and
- **a.** transpiration.
- **b.** temperature inversion.
- **c.** weathering.
- **d.** convection.

15. A vertical looping pattern of airflow is known as
- **a.** the Coriolis effect.
- **b.** a convection cell.
- **c.** a trade wind.
- **d.** a westerly.

16. A gentle wind that covers less than 100 km is called
- **a.** a jet stream.
- **b.** the doldrums.
- **c.** a breeze.
- **d.** a trade wind.

17. Which of the following layers of the atmosphere is closest to the ground?
- **a.** troposphere
- **b.** thermosphere
- **c.** mesosphere
- **d.** exosphere

18. Which of the following layers of the atmosphere is closest to space?
- **a.** troposphere
- **b.** ionosphere
- **c.** mesosphere
- **d.** exosphere

Short Answer

19. List the three main elemental gases that compose the atmosphere.

20. What is atmospheric pressure, and how is it measured?

21. List and describe the four main layers of the atmosphere.

22. What happens to visible light that enters Earth's atmosphere?

23. How is heat energy transferred by Earth's atmosphere?

24. How does latitude affect the temperature of a region?

25. Explain how the greenhouse effect helps warm the atmosphere.

26. What causes the Coriolis effect?

27. Name and describe the three main wind belts in both hemispheres.

28. How do surface features influence local wind patterns?

Critical Thinking

29. **Making Inferences** If a breeze is blowing from the ocean to the land on the coast of Maine, about what time of day is it? Explain your answer.

30. **Evaluating Ideas** What effect might jet streams have on airplane travel?

31. **Inferring Relationships** Most aerosol sprays that contain CFCs have been banned in the United States. Which of the four layers of the atmosphere does this ban help protect? Explain your answer.

32. **Evaluating Information** You hear a report about Earth's weather. The reporter says that visible light rays coming from Earth's surface heat the atmosphere in a way similar to the way a greenhouse is heated. Explain why the reporter's statement is incorrect.

Concept Mapping

33. Use the following terms to create a concept map: *atmosphere, troposphere, stratosphere, temperature, mesosphere, thermosphere, atmospheric pressure, altitude,* and *exosphere*.

Math Skills

34. **Applying Quantities** The albedo of the moon is 0.07. What percent of the total solar radiation that reaches the moon is reflected?

35. **Making Calculations** Maximum local wind speeds for each of the last seven days were 12 km/h, 20 km/h, 11 km/h, 6 km/h, 8 km/h, 19 km/h, and 17 km/h. What was the average maximum wind speed?

Writing Skills

36. **Writing from Research** Research the debate about global warming. Write one paragraph that includes evidence that supports global warming and one paragraph that includes evidence that does not support global warming.

Interpreting Graphics

The graph below shows how the Coriolis effect changes as latitude and wind speed change. Use the graph to answer the questions that follow.

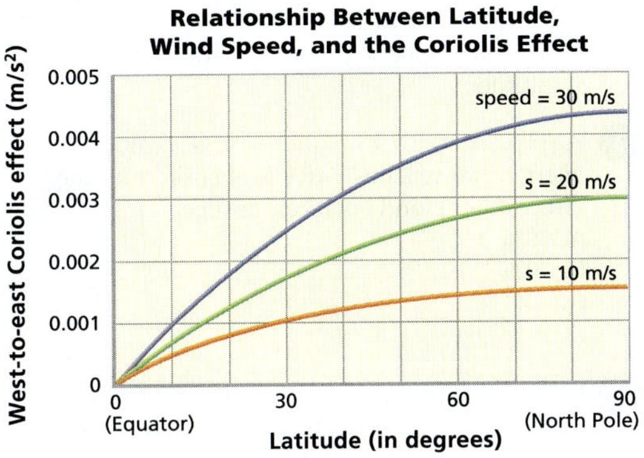

37. At what wind speed is the Coriolis effect the greatest?

38. At what latitude is the Coriolis effect the smallest?

39. At 90° latitude, is there a direct relationship between the Coriolis effect and wind speed?

Critical Thinking

29. It is most likely afternoon, because warm air above the land rises and is replaced by cool air from the ocean.

30. Because jet streams influence storm paths, airplanes might tend to avoid the jet streams when major storms are brewing. On the other hand, pilots may "ride" jet streams in fair weather to take advantage of tailwinds by cutting fuel costs and flight times.

31. The ban on CFCs helps protect the stratosphere, the location of most of the ozone in the atmosphere, because CFCs destroy ozone.

32. Infrared rays, not visible light rays, radiate from Earth's surface and are trapped by gases in the atmosphere much as heat is trapped in a greenhouse.

Concept Mapping

33. Answers may vary but should include all of the terms listed. Sample answers appear at the end of this Teacher Edition.

Math Skills

34. Albedo is the fraction of solar radiation that is reflected off the surface of an object, and the moon's albedo is 0.07, 7% of solar radiation that reaches the moon is reflected. $(0.07 \times 100 = 7\%)$

35. Average maximum wind speed = sum of maximum wind speed each day ÷ number of days = $(12 + 20 + 8 + 11 + 17 + 19 + 6) \div 7 = 13.3$ km/h

Writing Skills

36. Answers may vary. Accept all reasonable answers.

Interpreting Graphics

37. The Coriolis effect is greatest at the highest wind speed, 30 m/s.

38. The Coriolis effect is smallest at 0°, the equator.

39. yes; For every 10 m/s rise in wind speed, the Coriolis effect increases by 0.0015 m/s^2.

Estimated Time

To give students practice under more realistic testing conditions, allow them 30 minutes to answer all of the questions in this practice test.

TEST DOCTOR

Question 1 Answer D is correct. Students should understand that oxidation, respiration, and combustion are processes that consume, not produce, oxygen. So, answers A, B, and C are incorrect.

Question 9 Full-credit answers should include the following points:

- students should demonstrate an understanding that chinook winds affect human beings and agriculture in a number of ways
- while the winds have a definite benefit in reducing exposure to the cold, some of their other effects may be less beneficial
- chinook winds are hot and arid, which lowers local humidity levels and removes moisture from the soil
- in the summer, the hot winds are especially destructive to small and non-native crops
- the sudden changes in temperature may also make animals more susceptible to disease
- in the winter months, chinook winds remove snow cover and make grazing easier for animals. But in the summer the winds may harm suitable grazing lands

Chapter 22 Standardized Test Prep

Understanding Concepts

Directions (1–4): **For *each* question, write on a separate sheet of paper the letter of the correct answer.**

1. Which of the following processes is the source of the oxygen gas found in Earth's atmosphere?
   A. oxidation
   B. combustion
   C. respiration
   D. photosynthesis

2. Which of the following statements best describes the relationship of atmospheric pressure to altitude?
   F. The atmospheric pressure increases as the altitude increases.
   G. The atmospheric pressure increases as the altitude decreases.
   H. The atmospheric pressure varies unpredictably at different altitudes.
   I. The atmospheric pressure is constant at all altitudes.

3. Approximately how much of the solar energy that reaches Earth is absorbed by the atmosphere, land surfaces, and ocean?
   A. 30%
   B. 50%
   C. 70%
   D. 100%

4. In the Northern Hemisphere, the Coriolis effect causes winds moving toward the North Pole to be deflected in which of the following ways?
   F. Winds are deflected to the right.
   G. Winds are deflected to the left.
   H. Winds are deflected in unpredictable patterns.
   I. Winds are not deflected by the Corilois effect.

Directions (5–6): **For *each* question, write a short response.**

5. What is the most abundant gas in Earth's atmosphere?

6. In which atmospheric layer do interactions between gas molecules and solar radiation produce the aurora borealis phenomenon?

Reading Skills

Directions (7–9): **Read the passage below. Then, answer the questions.**

The Snow Eater

The chinook, or "snow eater," is a dry wind that blows down the eastern side of the Rocky Mountains from New Mexico to Canada. Arapaho gave the chinook its name because of its ability to melt large amounts of snow very quickly. Chinooks form when moist air is forced over a mountain range. The air cools as it rises. As the air cools, it releases moisture in the form of rain or snow, which nourishes the local flora. As the dry air flows over the mountaintop, it descends, compressing and heating the air below. The warm, dry wind that results can melt half of a meter of snow in just a few hours.

The temperature change caused when a chinook rushes down a mountainside can be dramatic. In 1943, in Spearfish, South Dakota, the temperature at 7:30 A.M. was –4°F. But only two minutes later, a chinook caused the temperature to soar to 45°F.

7. Why are the chinook winds of the Rocky Mountains called "snow eaters?"
   A. Chinook winds pick up snow and carry it to new locations.
   B. Chinook winds drop all of their snow on the western side of the mountains.
   C. Chinook winds cause the temperature to decrease, which causes snow to accumulate.
   D. Chinook winds cause the temperature to increase rapidly, which causes snow to melt.

8. Which of the following statements can be inferred from the information in the passage?
   F. Chinook winds are a relatively new phenomenon related to global warming.
   G. The Rocky Mountains are more arid on their eastern side than on their western side.
   H. The only type of wind that blows down from mountaintops are chinook winds.
   I. When they blow up the western side of the Rocky Mountains, chinook winds are very hot.

9. How might chinook winds affect agriculture on the eastern side of the Rocky Mountains?

Answers

Understanding Concepts

1. D
2. G
3. C
4. F
5. nitrogen
6. ionosphere

Reading Skills

7. D
8. G
9. Answers may vary. See Test Doctor for a detailed scoring rubric.

Interpreting Graphics

10. A
11. Location C has northeast winds. Location D has southeast winds. Cool air flows toward low pressure caused by rising warm air near the equator. Both winds are somewhat easterly because of Earth's rotation.
12. Answers may vary. See Test Doctor for a detailed scoring rubric.

Interpreting Graphics

Directions (10–12): **For *each* question below, record the correct answer on a separate sheet of paper.**

The diagram below shows global wind belts and convection cells at different latitudes. Use this diagram to answer questions 10 and 11.

Global Wind Belts

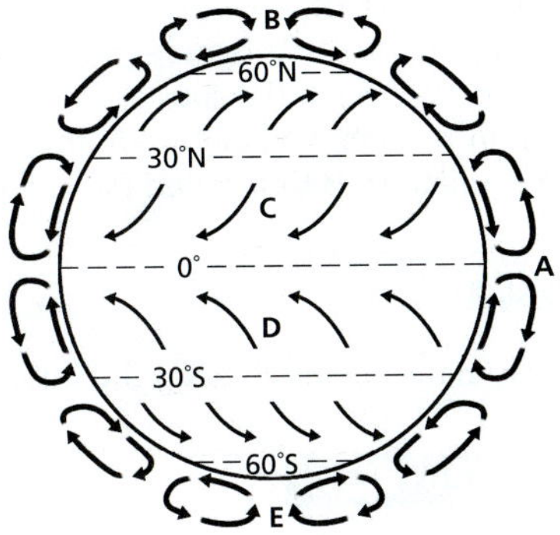

10 What happens to the air around location A as the air warms and decreases in density?

A. It rises.
B. It sinks.
C. It stagnates.
D. It contracts.

11 Compare and contrast the wind patterns in the global wind belts labeled C and D. Why do the winds move in the directions shown?

The graphic below shows a typical coastal area. Use this graphic to answer question 12.

Coastal Land Area on a Summer Day

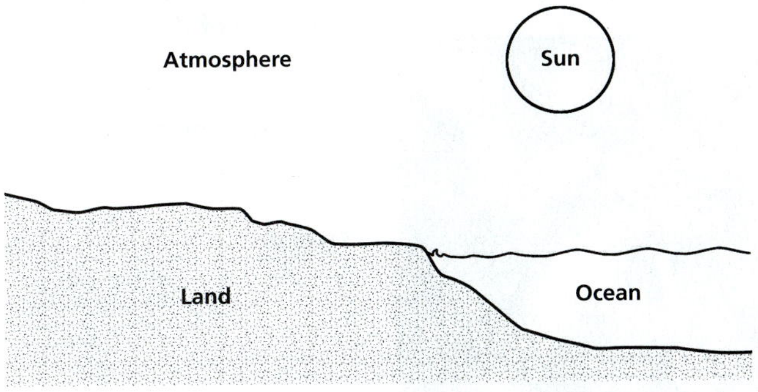

12 Would the direction of local winds be the same during the night as they would during the day in this location? Explain your answer in terms of the wind's direction and cause during the day and during the night.

Test TIP

Do not be fooled by answers that may seem correct to you just because they contain unfamiliar words.

TEST DOCTOR

Question 10 Answer A is correct. Students should know that as air becomes warmer and less dense, it rises. Some students may mistakenly select answer B by following the arrows in a circle until the air sinks back toward Earth. However, the air sinks only after cooling again.

Question 12 Full-credit answers should include the following points:

- students should demonstrate a conceptual understanding that the directions of local winds would be different at this location during the night from those during the day
- students should also demonstrate an understanding that the sun will heat land and water at different rates. This concept should lead students to the idea that temperature and density differences will exist in the air masses above the land and the water
- during the day, solar radiation will heat the land faster than it will heat the ocean. This will produce a wind that moves from the ocean to the land as warm land air rises
- just as land warms quicker during the day than water, land cools more quickly at night than ocean water does
- at night, cool air from the land moves seaward as warmer ocean air rises

Test Prep Correlations ***National Science Education Standards***

ES 1a: items 3, 6
ES 1d: item 12
ES 2a: item 5
HNS 1c: items 7
LS 1e: item 1
SPSP 4a: item 9
UCP 2: items 2, 4, 8, 10, 11

CHAPTER RESOURCES

State Resources

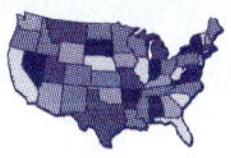

For specific resources for your state, visit **go.hrw.com** and type in the keyword **HSHSTR.**

Inquiry Lab

Air Density and Temperature

Teacher's Notes

Time Required
one 45-minute class period

Lab Ratings
EASY → HARD

Teacher Preparation: 1
Student Setup: 2
Concept Level: 3
Cleanup: 2

Skills Acquired
- Predicting
- Experimenting
- Observing
- Measuring
- Collecting Data
- Organizing and Analyzing Data
- Inferring

The Scientific Method
In this lab, students will
- Ask Questions
- Test a Hypothesis
- Make Observations
- Analyze the Results
- Draw Conclusions
- Communicate Results

Materials
The materials listed on this page are enough for groups of two to four students.

Tips and Tricks
You may want to limit the number of materials that students can use for this activitiy. Have all materials cut to the appropriate size before students begin to use them. To save time, you may want to paint materials before students arrive.

Chapter 22

Inquiry Lab

Using Scientific Methods

Energy Absorption and Reflection

When solar energy reaches Earth's surface, the energy is either reflected or absorbed by the material that the surface is made of. Whether the material absorbs or reflects energy, and the amount of energy that is reflected or absorbed, depends on several characteristics of the material. These characteristics include the material's composition, its color and texture, how transparent the material is, the mass and volume of the substance, and the specific heat of the substance. In this lab, you will study these characteristics to determine which material is best suited for use as roofing material.

Objectives
- **Determine** which material would keep the inside of a house coolest.
- **Explain** which properties of that material determine whether it is a conductor or an insulator.

Materials
- cardboard, 4 cm × 4 cm × 1 cm (4 pieces)
- paint, black, white, and light blue tempera
- metal, 4 cm × 4 cm × 1 cm
- rubber, beige or tan, 4 cm × 4 cm × 1 cm
- sandpaper, 4 cm × 4 cm × 1 cm
- thermometers, Celsius (4)
- watch, or clock
- wood, beige or tan, 4 cm × 4 cm × 1 cm

Safety

ASK A QUESTION

1. Which material would keep the interior of a house coolest?

FORM A HYPOTHESIS

2. Identify the material that you think will keep the inside of a house coolest. List the characteristics of that material that caused you to choose that material.

TEST THE HYPOTHESIS

3. Brainstorm with a partner or with a small group of classmates to design a procedure that will help you determine which materials absorb the most energy and which materials keep the surface below them coolest. You do not have to test all of the materials, if you can explain why you think those materials would not be coolest. Write down your experimental procedure.

Step 3

CHAPTER RESOURCES

Chapter Resource File
- Datasheet for Chapter Lab GENERAL
- Lab Notes and Answers

Workbooks

Long-Term Projects
- Air Pollution Watch ADVANCED

4. Have your teacher approve your experimental design.

5. Create a table like the one shown at right. Use this table to record the data you collect as you perform your experiment.

6. Following your design, measure the temperatures that the materials and the surfaces beneath them reach.

Material	Color	Surface temperature (°C)	Temperature below material (°C)
Cardboard	white		
Rubber	beige		
Sandpaper	beige		

DO NOT WRITE IN THIS BOOK

ANALYZE THE RESULTS

1. **Graphing Data** Use the data you collected to create a bar graph whose x-axis you label with the materials you tested and y-axis you label with a range of temperatures.

2. **Analyzing Data** Which material reached the highest temperature on its surface? Which material caused the surface below it to reach the highest temperature?

3. **Analyzing Data** Which material stayed the lowest temperature at its surface? Which material kept the temperature of the surface beneath it lowest?

4. **Evaluating Results** Did the color of the materials affect whether they absorbed or reflected solar energy? Explain your answer.

DRAW CONCLUSIONS

5. **Drawing Conclusions** Based on your results, which material would you use for the roof of a house? Did your experimental results support your original hypothesis?

6. **Inferring Relationships** What properties of the material you identified in question 5 do you think make it best for this purpose? Explain your answer.

7. **Making Predictions** Do you think the material you chose would keep the inside of a house warm in colder weather? Explain your answer.

8. **Analyzing Methods** Name two changes in your experimental design that you would make if you were going to repeat the experiment. Explain why you would make each change.

Extension

1. **Applying Ideas** Use the materials you tested in this lab to create a model of Earth's surface that represents how different parts of Earth's surface absorb or reflect solar energy. Which areas of Earth's surface absorb the least energy?

Answers to Analyze the Results

1. Graphs may vary.
2. Answers may vary but should accurately reflect student data.
3. Answers may vary but should accurately reflect student data.
4. Answers may vary. In general, darker colors will absorb more energy than lighter colors will.

Answers to Draw Conclusions

5. Answers may vary. Accept all reasonable answers.
6. Answers may vary. Accept all reasonable answers.
7. Answers may vary. Accept all reasonable answers.
8. Answers may vary. Accept all reasonable answers.

Answers to Extension

1. Models may vary. The areas of Earth's surface that absorb the least amount of energy are the oceans and icecaps.

Classroom Tested & Approved

Alonda Droege
Highline High School
Burien, WA

Absorbed Solar Radiation

Discussion — GENERAL

The Tropics Draw students' attention to the relatively low amounts of absorbed radiation in both January and July at both poles. Why would the poles tend not to absorb radiation very well? (They have high albedo because of ice and snow cover throughout the year, especially in the Antarctic.) Discuss with students the effects global warming might have on the albedo in these regions and how those effects might affect the regional climate. (Global warming could cause the snow and ice in these areas to melt and be replaced with a surface that has lower albedo. This would result in higher absorption and higher temperatures.) **LS Visual/Logical**

Answers to Map Skills Activity

1. In January, the highest amount of absorbed solar radiation is at about 30°S in bands on the Pacific, Atlantic, and Indian Oceans.
2. In July, the highest amount of absorbed solar radiation is at about 30°N, 20°E, in the Mediterranean Sea just north of Libya and Egypt.
3. Southern Australia
4. The amount of radiation absorbed is least changed from January to July at the equator, 0° latitude.
5. In the Northern Hemisphere in January, when absorbed solar radiation is very low, temperatures are low, and winter is occurring. In the Northern Hemisphere in July, absorbed radiation is much higher, so temperatures are higher, and summer occurs.

CHAPTER RESOURCES

Technology

Transparencies
- 111 Absorbed Solar Radiation (with worksheet)

MAPS in Action

Absorbed Solar Radiation

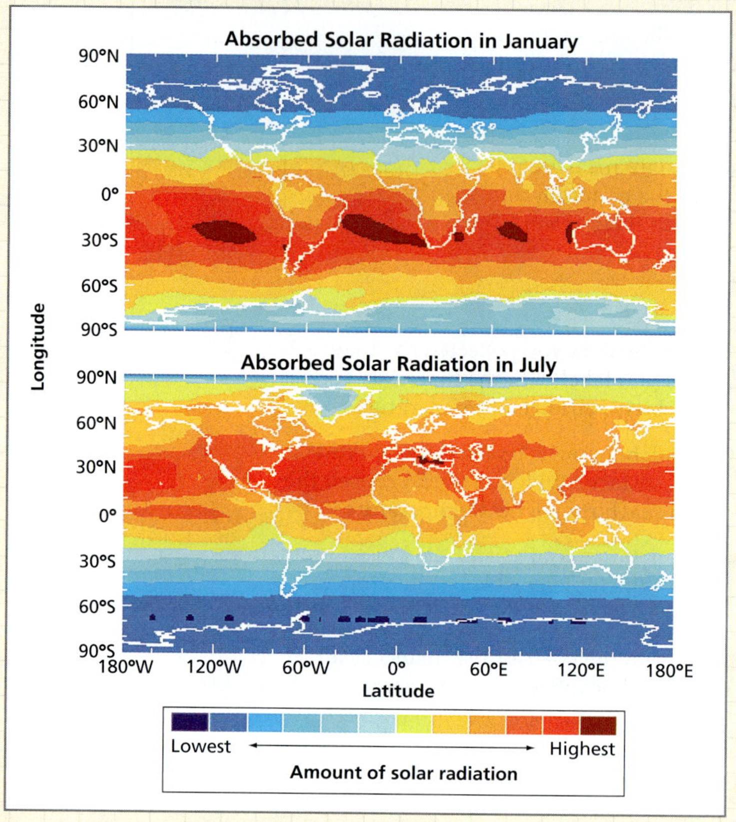

Map Skills Activity

These maps show the total amount of solar radiation that is absorbed by Earth in January and July. Use the maps to answer the questions below.

1. **Using a Key** In January, which region has the highest amount of absorbed solar radiation?
2. **Using a Key** In July, which region has the highest amount of absorbed solar radiation?
3. **Comparing Areas** Which area has the greatest difference between the amount of absorbed solar radiation in January and the amount in July: southern Australia or northeastern South America?
4. **Analyzing Data** At what latitudes are January's and July's amounts of absorbed solar radiation similar?
5. **Inferring Relationships** How do these maps explain the differences between the Northern Hemisphere's weather in January and its weather in July?

SCIENCE AND TECHNOLOGY

Energy from the Wind

For more than a thousand years, people have used windmills for tasks such as grinding grain or pumping water. Today, more and more people are rediscovering the benefits of wind power, and the number of wind turbines used to generate electricity continues to grow.

A New Take on an Old Idea

One result of the renewed interest in wind power is the reintroduction of the Darrieus turbine. Invented in 1920, the Darrieus turbine resembles an upside-down eggbeater. The Darrieus turbine is designed to catch winds blowing from all directions, and most of its moving parts are located on the ground. The bend in the blades can be changed to maximize efficiency, and the turbine's shape—taller than it is wide—minimizes land use.

Other, more recent technological advances in wind turbines include the use of lighter and less costly materials and the use of faster airfoil blades. Instead of being flat like earlier blades, an airfoil blade is shaped like an airplane wing. This shape allows air to flow smoothly over the blades and enables the airfoil to use the lifting force of the wind more effectively.

Energy of the Future

Using the wind to generate electricity consumes no fuels and produces no wastes. Also, wind itself costs nothing and is available in certain areas in a limitless supply. However, wind power has some disadvantages. One disadvantage is the unpredictability of wind speed and wind direction. Wind farms can also be noisy, unattractive, and deadly to migrating birds.

Thousands of wind turbines have been installed in California, Texas, and other states. Because of the steady increase in wind power use in European countries such as the Netherlands, Denmark, Germany, and England, wind power is quickly becoming the sustainable energy source of choice for utility companies worldwide.

Extension

1. **Making Predictions** What problems might arise in wind turbines placed between 20° and 40° latitude?

▲ A modern Darrieus turbine

▼ A modern wind farm

Energy from the Wind

Internet Activity — GENERAL

Harnessing Wind Have students use the Internet to research the benefits and drawbacks of wind energy. They should determine what percentage of total energy is currently produced by wind power, both in the United States and worldwide. They also might investigate how wind energy compares to other forms of energy production in term of efficiency, costs, technical limitations, and environmental impact. Students should use this information to prepare a persuasive essay that argues for or against the expansion of wind power generation capacity. A worksheet designed to direct student research on this topic can be found in the **Chapter Resource File** booklet or by visiting go.hrw.com and entering the keyword **HQ6ATMX.** LS Verbal

CHAPTER RESOURCES

Chapter Resource File

- **Internet Activity** Harnessing Wind GENERAL

Answer to Extension

1. Turbines placed between 20° and 40°N or S latitude would be close to the horse latitudes. Surface winds in that region tend to be weak and variable, so power output would be low and unreliable.

Chapter 23

Water in the Atmosphere

Planning Guide

Compression Guide

To shorten instruction because of time limitations, omit the Chapter Lab.

OBJECTIVES	LABS, DEMONSTRATIONS, AND ACTIVITIES	TECHNOLOGY RESOURCES
PACING • 90 min pp. 574–580 **Chapter Opener**		OSP **Parent Letter** ▫▫ CD **Student Edition on CD-ROM** CD **Chapter Summaries Audio CD** ▫▫ VID **Brain Food Video Quiz**
Section 1 Atmospheric Moisture • Explain how heat energy affects the changing phases of water. • Explain what absolute humidity and relative humidity are, and describe how they are measured. • Describe what happens when the temperature of air decreases to the dew point or below the dew point.	TE **Activity** Cooling Effect, p. 575 ▫ GENERAL TE **Physics Connection** Atmospheric Effects of Light, p. 576 ADVANCED TE **Biology Connection** Desert Survival, p. 578 ADVANCED TE **Activity** Scientific Instruments, p. 578 ▫▫ GENERAL SE **Quick Lab** Dew Point, p. 579 GENERAL CRF **Datasheet for Quick Lab*** GENERAL SE **Skills Practice Lab** Relative Humidity, pp. 596–597 GENERAL CRF **Datasheet for Chapter Lab*** GENERAL SE **Mapping Expeditions** What Comes Down Must Go… Where?, pp. 838–839 GENERAL	OSP **Lesson Plans** (also in print) TR **Bellringer*** TR 112 Phases of Water* TR 113 Vapor Pressure* CD **Interactive Tutor** Ice, Water, and Vapor CD **Interactive Tutor** The Water Cycle
PACING • 45 min pp. 581–586 **Section 2 Clouds and Fog** • Describe the conditions that are necessary for clouds to form. • Explain the four processes of cooling that can lead to the formation of clouds. • Identify the three types of clouds. • Describe four ways in which fog can form.	TE **Demonstration** Condensation Nuclei, p. 581 ▫▫ GENERAL TE **Geography Connection** Rain Shadow, p. 583 GENERAL SE **Quick Lab** Cloud Formation, p. 584 GENERAL CRF **Datasheet for Quick Lab*** GENERAL	OSP **Lesson Plans** (also in print) TR **Bellringer*** TR 114 Formation of a Water Droplet* TR 115 Classification of Clouds* TE **Internet Activity** Latent Heat and Thunderstorms, p. 582 GENERAL CRF **Internet Activity** Latent Heat and Thunderstorms* GENERAL CD **Interactive Tutor** Where Do Clouds Form?
PACING • 45 min pp. 587–590 **Section 3 Precipitation** • Identify the four forms of precipitation. • Compare the two processes that cause precipitation. • Describe two ways that precipitation is measured. • Explain how rain can be produced artificially.	TE **Demonstration** Indoor Rain, p. 587 ▫▫▫ GENERAL TE **Group Activity** Supercooled: Test Tube Hail, p. 588 ▫▫▫ GENERAL SE **Maps in Action** Annual Precipitation in the United States, p. 598 GENERAL TE **Activity** Research, p. 599 GENERAL CRF **Inquiry Lab** How Big Is a Raindrop?* GENERAL CRF **Making Models Lab** What Is the Shape of a Raindrop?* GENERAL	OSP **Lesson Plans** (also in print) TR **Bellringer*** TR 116 Annual Precipitation in the United States* TE **Internet Activity** Climate and Precipitation, p. 598 GENERAL CRF **Internet Activity** Climate and Precipitation* GENERAL CD **Interactive Tutor** Hail, Northern Lights, and Rainbows VID **NOVA Video** Hurricane!

PACING • 90 min

CHAPTER REVIEW, ASSESSMENT, AND STANDARDIZED TEST PREPARATION

SE **Chapter Highlights,** p. 591
SE **Chapter Review,** pp. 592–593
SE **Standardized Test Prep,** pp. 594–595
CRF **Concept Review*** ▫▫ GENERAL
CRF **Critical Thinking*** ADVANCED
CRF **Math Skills*** GENERAL
CRF **Graphing Skills*** GENERAL
CRF **Chapter Test A*** ▫▫ GENERAL
CRF **Chapter Test B*** ADVANCED
OSP **Lesson Plans** (also in print)
OSP **Test Generator**
OSP **Test Item Listing**

Online and Technology Resources

go.hrw.com — Holt Online Learning

Visit go.hrw.com for access to Holt Online Learning, or enter the keyword **HQ6 Home** for a variety of free online resources.

One-Stop Planner® CD-ROM

This CD-ROM package includes
- Lab Materials QuickList Software
- Holt Calendar Planner
- Customizable Lesson Plans
- Printable Worksheets
- ExamView® Test Generator
- Interactive Teacher Edition
- Holt PuzzlePro®
- Holt PowerPoint® Resources

KEY

SE	Student Edition	OSP	One-Stop Planner	VID	Classroom Video/DVD
TE	Teacher Edition	TR	Transparencies and Transparency Worksheets	*	Also on One-Stop Planner
CRF	Chapter Resource File	CD	CD or CD-ROM	◆	Requires advance prep
LTP	Long-Term Projects			■	Also available in Spanish

SKILLS DEVELOPMENT RESOURCES	REVIEW AND ASSESSMENT	CORRELATIONS
SE **Pre-Reading Activity**, p. 574 GENERAL TE **Using the Figure** Clouds and Weather, p. 574 GENERAL		National Science Education Standards
CRF **Directed Reading*** BASIC TE **Reading Skill Builder** Reading Hint, p. 577 BASIC SE **Math Practice**, p. 578 GENERAL	SE **Reading Checks**, pp. 576, 578 GENERAL SE **Section Review**, p. 580 GENERAL TE **Reteaching**, p. 579 BASIC TE **Quiz**, p. 579 GENERAL TE **Alternative Assessment**, p. 580 ADVANCED CRF **Section Quiz*** ■ GENERAL	ES 2b, ES 1c, ES 1d
CRF **Directed Reading*** BASIC TE **Using the Figure** Discussion, p. 582 BASIC TE **Skill Builder** Math, p. 582 GENERAL TE **Reading Skill Builder** Paired Summarizing, p. 583 BASIC TE **Inclusion Strategies**, p. 583 TE **Using the Figure** Cloud Bulletin Board, p. 584 GENERAL TE **Skill Builder** Vocabulary, p. 584 GENERAL SE **Graphic Organizer** Comparison Table, p. 585 GENERAL	SE **Reading Checks**, pp. 582, 585 GENERAL SE **Section Review**, p. 586 GENERAL TE **Reteaching**, p. 585 BASIC TE **Quiz**, p. 585 GENERAL TE **Alternative Assessment**, p. 585 ADVANCED CRF **Section Quiz*** ■ GENERAL	ES 2b, ES 1c, ES 1d, PS 2d
CRF **Directed Reading*** BASIC TE **Inclusion Strategies**, p. 589	SE **Reading Check**, p. 589 GENERAL SE **Section Review**, p. 590 GENERAL TE **Homework**, p. 589 GENERAL TE **Reteaching**, p. 589 BASIC TE **Quiz**, p. 589 GENERAL TE **Alternative Assessment**, p. 590 GENERAL CRF **Section Quiz*** ■ GENERAL	ES 2b, ES 1c, LS 4e

Holt Lab Generator CD-ROM

Search for labs by topic, standard, difficulty level, and/or time frame. Edit those labs to fit your needs, or create your own labs. Use the Lab Materials QuickList to order just what you need.

Holt Earth Science Interactive Tutor CD-ROM

This CD-ROM consists of interactive activities that give students a fun way to extend their knowledge of Earth science concepts.

Chapter Summaries Audio CDs

These CDs include audio summaries of the key concepts presented in each chapter. (Audio summaries are also available in Spanish.)

SciLinks. NSTA

www.scilinks.org

Maintained by the **National Science Teachers Association.** See Chapter Enrichment pages that follow for a complete list of topics.

See Chapter Enrichment pages for Video Resources.

Chapter 23

Chapter Enrichment

This Chapter Enrichment provides relevant and interesting information to expand and enhance your classroom instruction of the chapter material.

Section 1 Atmospheric Moisture

Ice Crystals and Optics

Light that shines through ice crystals in cirrus clouds is refracted, which causes interesting optical effects similar to those produced by sunlight shining through rain droplets. Different types of ice crystals cause different optical effects. One strange atmospheric phenomenon, called a *mock sun* or *sun dog*, occurs when sunlight refracts, or bends, as it passes through six-sided, pencil-shaped ice crystals called *hexagonal columns*. Plate-shaped crystals refract light in a way that causes halos, or rings, around the sun and moon. Sometimes these halos are pale, while at other times they may be brightly colored. A brilliant feather of light called a *solar pillar* can be seen above the rising or setting sun when light is reflected from capped-columns and plates. A winter fog made from ice crystals may cause pillars to form above street lamps.

Special optical effects occur when light bends as it passes through water or ice in the atmosphere.

Drinking Bird Toy

A novelty toy called the *Drinking Bird* provides a fun way to introduce the physical principles that govern evaporation. A glass tube connects two hollow glass bulbs at the bird's head and bottom. The bulbs are filled with a liquid such as freon that has a low evaporation temperature. The bird's body rotates on a pivot joint above the feet. Once its head is dipped in a glass of water, the bird endlessly bobs its head into the water and stands back up. How does the toy work? The parts of the bird that are not occupied by liquid contain vapor from the liquid. Wetting the head starts the process of evaporation. As the water on the bird's head evaporates, it draws heat from inside the tube, causing the vapor inside to contract and condense. This results in an imbalance in the vapor pressure between the bottom and top bulbs, with lower pressure in the top bulb. In order to equalize the pressure in the two bulbs, gas in the bottom bulb expands and forces liquid up the tube. Liquid rising in the tube causes the toy to become top heavy and tip forward into the water, wetting the head again. When bubbles of gas rise in the tube, it equalizes the pressure difference between the two bulbs. The liquid drains out of the head, and the bird stands upright again.

Section 2 Clouds and Fog

Unusual Clouds

Airplanes flying at very high altitudes produce condensation trails, or contrails. In the cold air at those altitudes, water vapor from the engine exhaust changes into ice crystals before it has a chance to disperse, forming cirrus-like clouds.

Another unusual phenomenon occurs when air crossing over a mountain range is first lifted and then sinks to its former level, producing lenticular clouds. Sometimes lenticular clouds look like a stack of dinner plates, a formation of flying saucers, or a tall lid hovering over the mountain peak. The hovering effect results from the fact that the cloud is continuously being renewed as the winds flow over the mountain.

Smog looks like a fog or mist that hangs over an urban area. A common source of smog is pollution from factories, smoke, and auto exhaust that combines

Lenticular clouds form as air rises over barriers, such as mountains.

with water droplets. Smog can be particularly dangerous to human health when accompanied by a temperature inversion. People who have respiratory diseases such as asthma usually suffer the most on smoggy days.

▶ Ice crystals form as water vapor condenses on freezing nuclei.

Section 3 Precipitation

Snowflakes

Snow forms in super-cooled clouds. Each snowflake is composed of many tiny ice crystals joined together. The water molecules hook up in a hexagonal crystal lattice, like tiny building blocks. In addition to the lacy ice sculptures that are commonly seen, snow can also be shaped like needles, hollow columns, capped columns, and irregular crystals. They have been classified into ten broad categories by using an international classification system based on the structure of the crystals. These categories include hexagonal plates, stellar crystals, spatial dendrites, sleet, graupel (snow pellets), and hail. Cloud height, air temperature, and humidity determine the shape snowflakes take. Capped crystals, which look somewhat like tiny spools, form in the highest clouds and at the coldest temperatures. Hexagonal plates form at mid-level altitudes where the temperature is slightly warmer. Simpler shapes tend to form at lower humidity, while needles and plates form at higher humidity levels. Snowflakes appear white because light that shines on them is reflected and scattered.

Video Resources

Brain Food Video Quizzes

Brain Food Video Quizzes These videos contain game-show style quizzes that assess students' progress and motivate students to study the chapter material.

NOVA Videos

The **NOVA** video below complements this chapter.

Hurricane! Watch scientists fly into the center of hurricanes so that they can learn more about the hurricanes, and listen to firsthand accounts of people affected by Hurricane Camille. (60 min)

To order other **NOVA** videos related to this chapter, visit **go.hrw.com** and enter the keyword **HQ6WIAV.**

SciLinks

NSTA Developed and maintained by the National Science Teachers Association

SciLinks is maintained by the National Science Teachers Association to provide you and your students with interesting, up-to-date links that will enrich your classroom presentation of the chapter.

Visit www.scilinks.org and enter the SciLinks code for more information about the topic listed.

Topic: Atmospheric Moisture
SciLinks code: HQ60113

Topic: Clouds and Fog
SciLinks code: HQ60304

Topic: Precipitation
SciLinks code: HQ61202

Topic: Water Cycle
SciLinks code: HQ61626

Chapter 23

Chapter Overview

This chapter describes how moisture in the atmosphere is described and measured. The chapter also explains how clouds form and what causes precipitation.

Using the Figure — GENERAL

Clouds and Weather Unlike other aspects of weather, clouds can easily be seen and studied. Different types of clouds can indicate approaching weather systems. The leading edge of the clouds in the photo is a squall line that indicates approaching storms. Ask students what types of weather they associate with particular clouds. (Sample answers: Dark clouds bring lightning, thunder, and heavy rain. High, layered clouds may mean rain is coming.) **LS Visual**

PRE-READING ACTIVITY

FOLDNOTES

You may want to use this Two-Panel Flip Chart in a classroom discussion to review material from the chapter. On the board, write each category from the chart. Then, ask students to provide information for each category. Under the appropriate category on the board, write the information students provide.

Chapter 23 Water in the Atmosphere

Sections

1 Atmospheric Moisture
2 Clouds and Fog
3 Precipitation

What You'll Learn

- How water enters the atmosphere
- How clouds form
- How precipitation forms

Why It's Relevant

Severe weather conditions can put lives in danger. Understanding how water moves through the atmosphere provides a basis for understanding weather.

PRE-READING ACTIVITY

FOLDNOTES

Two-Panel Flip Chart Before you read this chapter, create the FoldNote entitled "Two-Panel Flip Chart" described in the Skills Handbook section of the Appendix. Label the flaps of the two-panel flip chart with "Clouds" and "Precipitation." As you read the chapter, write information you learn about each category under the appropriate flap.

▶ Dark clouds, such as the clouds near Rosston, Oklahoma, shown here, usually signal the coming of a thunderstorm. The clouds look dark because they are so thick that little solar radiation passes through them.

Chapter Correlations — *National Science Education Standards*

ES 2b Movement of matter between reservoirs is driven by the earth's internal and external sources of energy.... **(Sections 1, 2, and 3)**

PS 2d The physical properties of compounds reflect the nature of the interactions among its molecules.... determined by the structure of the molecule, including...atoms and the distances and angles between them. **(Section 2)**

LS 4e Human beings live within the world's ecosystems. Increasingly, humans modify ecosystems as a result of population growth, technology, and consumption.... **(Section 3)**

ES 1c Heating of earth's surface and atmosphere by the sun drives convection within the atmosphere and oceans, producing winds and ocean currents. **(Sections 1, 2, and 3)**

ES 1d Global climate is determined by energy transfer from the sun at and near the earth's surface. This energy transfer is influenced by dynamic processes such as cloud cover... **(Sections 1 and 2)**

Section 1 Atmospheric Moisture

Water in the atmosphere exists in three states, or *phases*. One phase is known as a gas called *water vapor.* The other two phases of water are the solid phase known as *ice* and the liquid phase known as *water*.

OBJECTIVES

- **Explain** how heat energy affects the changing phases of water.
- **Explain** what absolute humidity and relative humidity are, and describe how they are measured.
- **Describe** what happens when the temperature of air decreases to the dew point or below the dew point.

KEY TERMS

latent heat
sublimation
dew point
absolute humidity
relative humidity

Changing Forms of Water

Water changes from one phase to another when heat energy is absorbed or released, as shown in **Figure 1.** Molecules of ice are held almost stationary in a definite crystalline arrangement. However, when energy is absorbed by the ice, the molecules move more rapidly. They break from their fixed positions and slide past each other in the fluid form of a liquid.

When more energy is absorbed by liquid water, the water changes from a liquid to a gas. Because the additional energy causes the movement of molecules in liquid water to speed up, the molecules collide more frequently with each other. Such collisions can cause the molecules to move so rapidly that the fastest-moving molecules escape from the liquid to form invisible water vapor in a process called *evaporation*.

Latent Heat

The heat energy that is absorbed or released by a substance during a phase change is called **latent heat.** When liquid water evaporates, the water absorbs energy from the environment. This energy becomes potential energy between the molecules. When water vapor changes back into a liquid through the process of *condensation*, energy is released to the surrounding air and the molecules move closer together. Likewise, latent heat is absorbed when ice thaws, and latent heat is released when water freezes.

latent heat the heat energy that is absorbed or released by a substance during a phase change

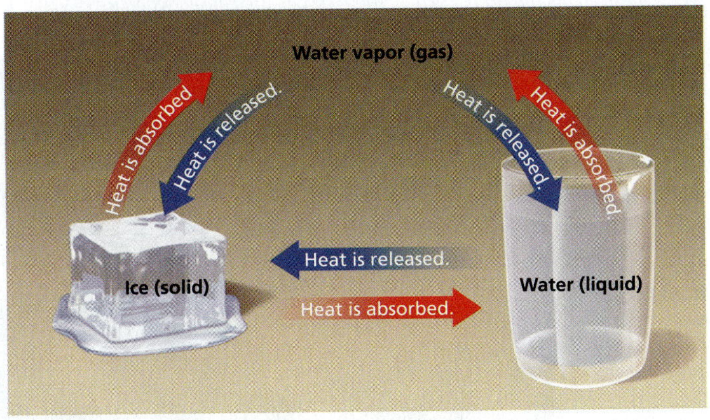

Figure 1 ▶ Water exists in three states, called *phases*. As it changes from one phase to another, water either absorbs or releases heat energy.

Section 1

Focus

Overview

This section explains the phases of water, the concepts of absolute and relative humidity and how they are measured, and the effects of temperature changes on moisture in the air.

Bellringer

Ask students to describe what must happen to ice to cause it to melt and what must happen to water to cause it to boil. (Temperature must increase, or more thermal energy must be added to the system.) LS Visual

Motivate

Activity — GENERAL

Cooling Effect Divide the class into pairs and distribute isopropyl alcohol swabs to each team. Explain that alcohol evaporates quickly. Have students rub the alcohol swab on a patch of skin. (Caution students not to participate if they have skin allergies.) Ask students to describe how the swabbed skin feels, and why. (cool; Because energy is required to change the liquid to a gas, heat leaves the skin during evaporation. This process leaves the skin feeling cooler.) LS Kinesthetic

CHAPTER RESOURCES

Chapter Resource File

- Directed Reading BASIC

Technology

Transparencies
- Bellringer
- 112 Phases of Water (with worksheet)

Student Edition on CD-ROM

One-Stop Planner CD-ROM
- Lesson Plan

Teach

BRAIN FOOD

Hidden Energy Water's unusual properties and behavior, including its high latent heat value, are largely due to its chemical structure. Water is a V-shaped polar molecule that consists of two atoms of hydrogen and one atom of oxygen, arranged so the oxygen side has a slight negative charge and the hydrogen side has a slight positive charge. This arrangement allows each hydrogen atom to share electrons with the oxygen atom through a stable covalent bond. In addition, a force called *hydrogen bonding* exists between water molecules. Water's latent heat—the energy that is absorbed or released during phase changes—results from hydrogen bonding. The angle between the atoms of hydrogen in the water molecule varies between 105° in the liquid phase and 109° in the solid phase. In the solid state, water molecules are bound together in a crystal lattice. In liquid water, they are arranged more irregularly but are able to glide past each other. In water vapor, the hydrogen bonds between the molecules are broken, and water exists as separate molecules. The energy used to create and maintain the separation between the water molecules is latent heat.

Evaporation

Most water enters the atmosphere through the process of evaporation. Because the largest amounts of solar energy reach Earth near the equator, most evaporation takes place over the oceans of the equatorial region. However, water vapor also enters the atmosphere by evaporation from lakes, ponds, streams, and soil. Plants release water into the atmosphere in a process called transpiration. Volcanoes and burning fuels also release small amounts of water vapor into the atmosphere.

Sublimation

sublimation the process in which a solid changes directly into a gas (the term is sometimes also used for the reverse process)

Ice commonly changes into a liquid before changing into a gas. However, in some cases, ice can change directly into water vapor without becoming a liquid. The process by which a solid changes directly into a gas is called **sublimation.** When the air is dry and the temperature is below freezing, ice and snow may sublimate into water vapor. Water vapor can also turn directly into ice without becoming a liquid.

Reading Check **Summarize the conditions under which sublimation commonly occurs. (See the Appendix for answers to Reading Checks.)**

Connection to PHYSICS

Light and Water in the Atmosphere

The interaction of water and light in the atmosphere can create a number of visual effects. When visible light passes through raindrops, the raindrops *refract*, or bend the light rays, which separates the white light into the colors that make up the entire visible spectrum. You can see these colors in rainbows, glories, and coronas. *Coronas* are small, multicolored circles that surround the sun or moon. *Glories* are small, multicolored circles that surround the shadow of an object on which light is shining.

Sun dogs, sun pillars, halos, and rings are caused by the interaction of light rays and ice crystals in the atmosphere. In general, the ice crystals do not separate the light into different colors, so these phenomena involve white light. *Sun dogs* are bright spots that appear in the sky, often on either side of the sun. *Sun pillars* are bright shafts of light that extend up and down from the sun or moon. *Halos* and *rings* are fuzzy or bright rings that are commonly seen around the sun and the moon.

Crepuscular rays are slanting sunbeams that form when sunlight is intermittently interrupted by clouds or fog. The clouds cast shadows that appear to break up the sunlight to form smaller, individual rays.

A *mirage* is one of two types of optical illusions that form when light rays are refracted as they pass through a boundary between hot air and cool air. The first type, called an *inferior mirage*, occurs when light passes from a layer of cool air to a layer of hot air close to the ground. When the light rays enter the hot air, they bend upward and cause the sky to appear as a pool of water on the ground. The other type, a *superior mirage*, is an image of an object that seems to be suspended in the sky. This type occurs when light rays pass through a layer of warm air into a layer of cool air below.

Sun dogs, such as these below, are caused when ice crystals in the atmosphere interact with light rays.

Answer to Reading Check

When the air is very dry and the temperature is below freezing, ice and snow change directly into water vapor by sublimation.

PHYSICS CONNECTION — ADVANCED

Atmospheric Effects of Light Most students will be familiar with rainbows and how raindrops bend light rays to produce a continuous spectrum of colors from white light. They will probably be less familiar with unusual visual effects, such as sun dogs, mirages, and glories. Invite interested students do further research into these phenomena. Encourage them to create visual presentations to share with the class on how unusual temperature conditions and the interaction of light with water droplets and ice crystals produce these interesting effects. **LS Visual**

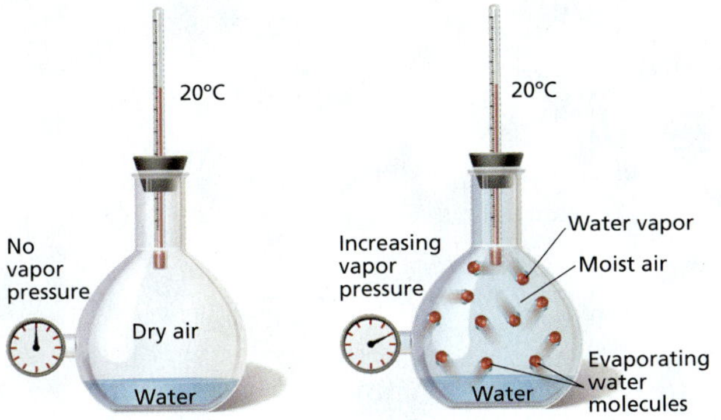

Figure 2 ▶ When water comes into contact with dry air, some of the water molecules evaporate into the dry air. The addition of the water molecules to the air causes the air pressure to increase. This increase in pressure is due to vapor pressure.

Humidity

Water vapor in the atmosphere is known as *humidity*. Humidity is controlled by rates of condensation and evaporation. The rate of evaporation is determined by the temperature of the air. The higher the temperature is, the higher the rate of evaporation is. The rate of condensation is determined by vapor pressure. Vapor pressure is the part of the total atmospheric pressure that is caused by water vapor, as shown in **Figure 2.** When vapor pressure is high, the condensation rate is high.

When the rate of evaporation and the rate of condensation are in equilibrium, the air is said to be "saturated." The temperature at which the condensation rate equals the evaporation rate is called the **dew point.** At temperatures below the dew point, net condensation occurs, and liquid water droplets form.

dew point at constant pressure and water vapor content, the temperature at which the rate of condensation equals the rate of evaporation

absolute humidity the mass of water vapor per unit volume of air that contains the water vapor, usually expressed as grams of water vapor per cubic meter of air

Absolute Humidity

One way to express the amount of moisture in air is by absolute humidity. **Absolute humidity** is the mass of water vapor contained in a given volume of air. In other words, absolute humidity is a measure of the actual amount of water vapor in the air. Absolute humidity is calculated by using the following equation:

$$\textit{absolute humidity} = \frac{\textit{mass of water vapor (grams)}}{\textit{volume of air (cubic meters)}}$$

However, as air moves, its volume changes as a result of temperature and pressure changes. Therefore, meteorologists prefer to describe humidity by using the mixing ratio of air. The *mixing ratio* is the mass of water vapor in a unit of air relative to the mass of the dry air. For example, the very moist air in tropical regions might have 18 g of water vapor in 1 kg of air, or a mixing ratio of 18 g/kg. On the other hand, the cold, dry air in polar regions commonly has a mixing ratio of less than 1 g/kg. Because this measurement uses only units of mass, it is not affected by changes in temperature or pressure.

SCILINKS® NSTA Developed and maintained by the National Science Teachers Association

For a variety of links related to this subject, go to www.scilinks.org

Topic: Atmospheric Moisture
SciLinks code: HQ60113

MISCONCEPTION ALERT

Holding Capacity of Air
Students may think that air has an inherent capacity for holding water that is dependent on temperature. They also may think that condensation occurs (dew point is reached) when air holds as much water as possible. However, there is no inherent limit to how much water air can hold. Water molecules are constantly moving back and forth between solid, liquid, and gas phases. If more molecules leave the water's surface than enter the water, the evaporation rate exceeds the condensation rate. Warm air tends to "hold" more water than cooler air does because the evaporation rate is usually greater than the condensation rate in warm air. LS **Verbal**

Teaching Tip — GENERAL

Make Concepts Relevant
Discuss some of the effects of humidity on humans. For example, when humidity is high, sweat does not evaporate easily, and people have difficulty cooling their bodies. Invite students to describe some of the effects of low humidity. (Sample answers: Skin becomes drier. Drier nasal passages make it easier for germs to enter the body. Forest fires are more common when dry conditions persist for a long time.) LS **Verbal**

READING SKILL BUILDER — BASIC

Reading Hint Before students read this page and the next, ask them to think about occasions when they have heard relative humidity discussed in weather reports. Ask them how humidity level relates to physical comfort. (Sample answers: When the humidity is high, especially during summer, people may feel hot and "sticky." Sometimes in the winter when humidity is low, and the air is very dry, some people may have trouble breathing.) Explain that this section describes what determines the moisture content of the air. LS **Verbal** English Language Learners

CHAPTER RESOURCES

Technology

Transparencies
- 113 Vapor Pressure (with worksheet)

Teach, *continued*

BIOLOGY CONNECTION — ADVANCED

Desert Survival Southwest Africa's Namib Desert is one of the hottest and driest places on Earth. This region is home to a dazzling variety of creatures with unusual adaptations to help them survive in this harsh environment. Like other places around the globe, dawn in the Namib is a time when moisture condenses and collects on desert vegetation. There is no rain, but warm air from the Atlantic Ocean regularly sweeps over the cold waters of the Benguela current, creating thick coastal fogs. Creatures such as the long-legged beetle, dune ants, desert crickets, and the side-winder snake all have special ways to get water from condensation. Ask students to research these creatures' ingenious survival methods and report their findings to the class. **LS Logical**

MATH PRACTICE

Answer

10 g/kg ÷ 14 g/kg = 0.71 × 100 = 71%

Figure 3 ▶ Dew forms on surfaces such as grass and spider webs when the temperature of air reaches the dew point.

relative humidity the ratio of the amount of water vapor in the air to the amount of water vapor needed to reach saturation at a given temperature

Relative Humidity

A more common way to express the amount of water vapor in the atmosphere is by *relative humidity.* **Relative humidity** is a ratio of the actual water vapor content of the air to the amount of water vapor needed to reach saturation. In other words, relative humidity is a measure of how close the air is to reaching the dew point. For example, at 25°C, air is saturated when it contains 20 g of water vapor per 1 kg of air. If air that is 25°C contains 5 g of water vapor, the relative humidity is expressed as 5/20, or 25%.

If the temperature does not change, the relative humidity will increase if moisture enters the air. Relative humidity can also increase if the moisture in the air remains constant but the temperature decreases. If the temperature increases as the moisture in the air remains constant, the relative humidity will decrease.

MATH PRACTICE

Relative Humidity
Relative humidity can be calculated by using the following equation:

$$\text{relative humidity} = \left[\frac{\text{amount of water vapor in air}}{\text{amount of water vapor needed to reach saturation}}\right] \times 100$$

Air at 20°C is saturated when it contains 14 g/kg of water vapor. What is the relative humidity of a volume of air that is 20°C and that contains 10 g/kg of water vapor?

Reaching the Dew Point

When the air is nearly saturated with a relative humidity of almost 100%, only a small temperature drop is needed for air to reach its dew point. Air may cool to its dew point by conduction when the air is in contact with a cold surface. During the night, grass, leaves, and other objects near the ground lose heat. Their surface temperatures often drop to the dew point of the surrounding air. Air, which normally remains warmer than surfaces near the ground do, cools to the dew point when it comes into contact with cooler objects, such as grass. The resulting form of condensation, shown in **Figure 3,** is called *dew.* Dew is most likely to form on cool, clear nights when there is little wind.

If the dew point falls below the freezing temperature of water, water vapor may change directly into solid ice crystals, or *frost.* Because frost forms when water vapor turns directly into ice, frost is not frozen dew. Frozen dew is relatively uncommon. Unlike frost, frozen dew forms as clear beads of ice.

✓ Reading Check **How does dew differ from frost? (See the Appendix for answers to Reading Checks.)**

Answer to Reading Check

Dew is liquid moisture that condenses from air on cool objects when the air is nearly saturated and the temperature drops. Frost is water vapor that condenses as ice crystals onto a cool surface directly from the air when the dew point is below freezing.

Activity — GENERAL

Scientific Instruments Bring in real meteorological instruments for measuring relative humidity, such as dew cells, psychrometers, or hair and electric hygrometers, so students can see how the tools work. Set up the instruments in workstations and allow students to rotate though the stations. Have volunteers demonstrate and explain the use of each tool. **LS Kinesthetic**

Measuring Humidity

Meteorologists are interested in measuring humidity so that they can better predict weather conditions. Relative humidity can be measured by using a variety of instruments, such as a thin polymer film, a psychrometer, a dew cell, and a hair hygrometer.

Using Thin Polymer Film to Measure Humidity

Humidity is commonly measured by a humidity sensor that uses a *thin polymer film*. The relative humidity of the surrounding air affects the ability of the thin polymer film to absorb or release water vapor. The amount of water vapor the thin polymer film contains changes the film's ability to conduct electricity. The polymer film's ability to conduct electricity is affected by the relative humidity of the surrounding air. Thus, by measuring the polymer film's ability to store electricity, relative humidity can be determined.

Using Psychrometers to Measure Humidity

A *psychrometer*, shown in **Figure 4,** is another instrument that is used to measure relative humidity. It consists of two identical thermometers. The bulb of one thermometer is covered with a damp wick, while the bulb of the other thermometer remains dry. When the psychrometer is held by a handle and whirled through the air, the air circulates around both thermometers. As a result, the water in the wick of the wet-bulb thermometer evaporates. Evaporation requires heat, so heat escapes from the thermometer. Consequently, the temperature of the wet-bulb thermometer is lower than that of the dry-bulb thermometer. The difference between the dry-bulb temperature and the wet-bulb temperature is used to calculate relative humidity. If there is no difference between the wet-bulb temperature and dry-bulb temperature, no water evaporated from the wet-bulb thermometer. Thus, the air is saturated and the relative humidity is 100%.

QuickLAB 10 min

Dew Point

Procedure

1. Pour **room-temperature water** into a **glass container,** such as a drinking glass, until the water level is near the top of the cup.
2. Observe the outside of the glass container, and record your observations.
3. Add **one or two ice cubes** to the container of water.
4. Watch the outside of the container for 5 min for any changes.

Analysis

1. What happened to the outside of the container?
2. What is the liquid on the container?
3. Where did the liquid come from? Explain your answer.

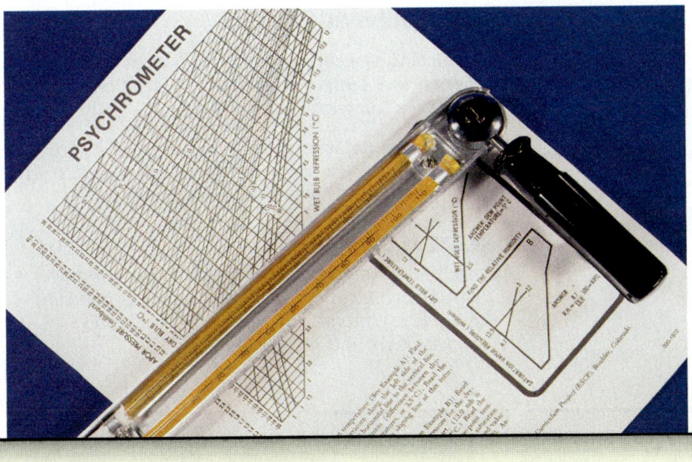

Figure 4 ▶ A psychrometer is commonly used with a table that lists relative humidity based on differences between wet-bulb and dry-bulb readings.

Close

Reteaching — Basic

Making Diagrams Have pairs of students create diagrams with labels and captions to illustrate chapter concepts such as latent heat in phase changes, dew point and vapor pressure, and humidity. After students have completed their diagrams, invite them to take turns using the diagrams to explain the illustrated concepts to each other. **LS Visual/Auditory**

Quiz — General

1. How is thermal energy involved in the evaporation of water? (Water absorbs heat energy from the environment, energizing the molecules to break the bonds holding them together.)
2. What happens to water in the air when the rate of condensation exceeds the rate of evaporation? (Net condensation results in visible moisture condensing out of the air.)
3. How is thermal energy involved in the condensation of water? (The heat energy is released into the surrounding air as molecules come closer together.)
4. What forms when water vapor changes directly into solid form? (frost)

CHAPTER RESOURCES

Chapter Resource File

- **Datasheet for Quick Lab** General

QuickLAB

Skills Acquired

- Experimenting
- Observing
- Analyzing
- Communicating

Teacher's Notes Provide lab thermometers so students can determine the current dew point of the air. Tell them to record the starting air temperature next to the glass and the temperature when a thin layer of moisture (dew) forms on the outside of the container.

Answers

1. Moisture appears outside the container.
2. water
3. from the surrounding air; The ice lowered the temperature of the air to below the dew point, which caused the water vapor in the air to condense on the cool surface of the container.

Close, *continued*

Alternative Assessment — ADVANCED

Song Lyrics Have students compose a song about energy and phase changes. The completed songs should include at least two verses and a chorus and should incorporate what students have learned about such processes as evaporation and condensation. Remind students that song lyrics are similar to a poem, except that they are set to music. One way to get started is to use a children's tune (such as Pop Goes the Weasel or Three Blind Mice), keep the music, and change the words. A rhyming dictionary can be helpful in composing the lyrics. Students may want to perform their songs for the class and make a video or an audio recording. **LS Verbal/Auditory**

Answers to Section Review

1. When water absorbs heat energy, the energy causes the molecules to move so rapidly that they escape the liquid and enter the air.
2. the oceans in the equatorial region
3. sublimation
4. Humidity is water vapor in the atmosphere.
5. Absolute humidity is the mass water vapor per unit volume of air and represents a measure of the actual amount of water in the air. Relative humidity is the ratio of the amount of water vapor in the air compared with the amount of water vapor needed to reach saturation at a given temperature.
6. As air temperature drops, the rate of condensation increases, and the rate of evaporation decreases. At the dew point, the rates of condensation and evaporation are equal. Below the dew point, net condensation occurs, and visible water droplets form on surfaces.
7. dew cells, psychrometers, hair hygrometers, and polymer films
8. The rate of condensation would increase.
9. Because the solar energy that reaches Earth's surface at the equator is more intense than the energy that reaches the poles, the equatorial region would probably have a higher overall absolute humidity than polar regions, because evaporation is higher where temperature is higher.
10. *Water vapor* in the air is known as *humidity*, which can be expressed as *absolute humidity* or *relative humidity*, which can be measured by using a *dew cell*, a *hygrometer*, or a *psychrometer;* and which is controlled by rates of *condensation* and *evaporation*, which determine the *dew point*.

Figure 5 ▶ Scientists use weather balloons, such as this one in Antarctica, to send electric hygrometers into the high altitudes of the atmosphere.

Other Methods for Measure Humidity

Another instrument that has been used to measure relative humidity is the *dew cell.* Dew cells consist of a ceramic cylinder with electrodes attached to it and treated with lithium chloride, LiCl. When LiCl absorbs water from the air, the dew cells ability to conduct electricity increases. By detecting the electrical resistance of LiCl as it is heated and cooled, the dew cell can determine the dew point.

The *hair hygrometer* determines relative humidity based on the principle that hair becomes longer as relative humidity increases. As relative humidity decreases, hair becomes shorter.

Measuring Humidity at High Altitudes

To measure humidity at high altitudes, scientists use an electric hygrometer. The hygrometer may be carried up into the atmosphere in an instrument package known as a *radiosonde.* The radiosonde is attached to a weather balloon, such as the one shown in **Figure 5.** The electric hygrometer is triggered by passing an electric current through a moisture-attracting chemical substance. The amount of moisture changes the electrical conductivity of the chemical substance. The change can then be expressed as the relative humidity of the surrounding air.

Section 1 Review

1. **Explain** how most water vapor enters the air.
2. **Identify** the principal source from which most water vapor enters the atmosphere.
3. **Identify** the process by which ice changes directly into a gas.
4. **Define** *humidity*.
5. **Compare** relative humidity with absolute humidity.
6. **Describe** what happens when the temperature of air decreases to the dew point or below the dew point.
7. **Identify** four instruments that are used to measure relative humidity.

CRITICAL THINKING

8. **Predicting Consequences** Explain what would happen to a sample of air whose relative humidity is 100% if the temperature decreased.
9. **Identifying Relationships** Which region of Earth would you expect to have a higher absolute humidity: the equatorial region or the polar regions?

CONCEPT MAPPING

10. Use the following terms to create a concept map: *humidity, water vapor, dew point, absolute humidity, dew cell, psychrometer, hygrometer, evaporation, condensation,* and *relative humidity*.

CHAPTER RESOURCES

Chapter Resource File
- Section Quiz GENERAL

Workbooks
- Study Guide (also in Spanish)

Section 2 Clouds and Fog

Clouds are collections of small water droplets or ice crystals that fall slowly through the air. Ice crystals and water droplets form when condensation or crystallization occurs more rapidly than evaporation or sublimation does. People commonly think that clouds are high in the sky and fog is close to the ground. However, clouds are not limited to high altitudes. Fog is actually a cloud that forms near or on Earth's surface.

Cloud Formation

For water vapor to condense and form a cloud, a solid surface on which condensation can take place must be available. Although the lowest layer of the atmosphere, the *troposphere,* does not contain any large solid surfaces, it contains millions of suspended particles of ice, salt, dust, and other materials. Because the particles are so small—less than 0.001 mm in diameter—they remain suspended in the atmosphere for a long time. The suspended particles that provide the surfaces necessary for water vapor to condense are called **condensation nuclei.** As water molecules collect on the nuclei, water droplets form, as **Figure 1** shows.

In addition, for clouds to form, the rate of evaporation must initially be in equilibrium with the rate of condensation. When this condition occurs, the air is said to be "saturated" with water vapor. When the temperature of the saturated air drops, condensation occurs more rapidly than evaporation does. As a result of this net condensation, clouds begin to form. Because the rate of evaporation decreases as temperature decreases, cooling of air may lead to net condensation. Four major processes can cause the cooling that is necessary for clouds to form.

Figure 1 ▶ Formation of a Water Droplet

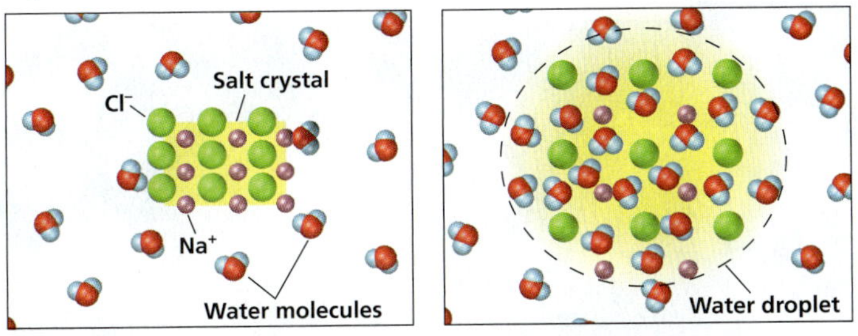

Water molecules are electrically attracted to the sodium ions, Na^+, and the chlorine ions, Cl^-, in a salt crystal.

The water molecules and ions form a solution. Additional water molecules are attracted to the solution, and the droplet gets bigger.

OBJECTIVES

- **Describe** the conditions that are necessary for clouds to form.
- **Explain** the four processes of cooling that can lead to the formation of clouds.
- **Identify** the three types of clouds.
- **Describe** four ways in which fog can form.

KEY TERMS

cloud
condensation nucleus
adiabatic cooling
advective cooling
stratus cloud
cumulus cloud
cirrus cloud
fog

cloud a collection of small water droplets or ice crystals suspended in the air, which forms when the air is cooled and condensation occurs

condensation nucleus a solid particle in the atmosphere that provides the surface on which water vapor condenses

Section 2

Focus

Overview

This section describes the conditions necessary for cloud formation and the processes that cause clouds and fog to form. The section also explains how clouds are classified.

Bellringer

Ask students to think about kinds of clouds they have seen. Then, have them write brief descriptions or draw sketches of the different cloud formations. (Answers may vary. Accept all reasonable answers.) LS **Visual**

Motivate

Demonstration — GENERAL

Condensation Nuclei Put a few grains of salt on a glass microscope slide. Place the slide in a petri dish and use several bent paper clips to elevate the slide above the bottom of the dish. Add just enough water to wet the bottom of the dish without wetting the slide. Ask someone to describe the shape of the salt grains. (cubes) Cover the petri dish. Point out that the closed container has a very high relative humidity. After 15 to 20 minutes, invite students to describe what happened. (The water evaporates from the bottom of the dish, dissolves the grains of salt, and forms droplets of water on the slide.) LS **Visual**

CHAPTER RESOURCES

Chapter Resource File
- Directed Reading BASIC

Technology

Transparencies
- Bellringer
- 114 Formation of a Water Droplet (with worksheet)

Student Edition on CD-ROM

One-Stop Planner CD-ROM
- Lesson Plan

Teach

Internet Activity — GENERAL

Latent Heat and Thunderstorms Latent heat plays a critical role in many weather processes, including the development of thunderstorms and tornadoes. As a result of rising and expanding, lifting air begins to cool. At a certain elevation, the dew point is reached, resulting in condensation and cloud formation. This condensation releases huge quantities of latent heat into the air, and produces further uplift, which helps fuel the developing thunderstorm. Severe weather associated with some these clouds includes hail, strong winds, thunder, lightning, intense rain, and tornadoes. Invite interested students to do further Internet research into the role of latent heat in these exceptional weather patterns. Have them create diagrams that help explain the processes. A worksheet designed to direct student research on this topic can be found in the **Chapter Resource File** booklet or by visiting go.hrw.com and entering the keyword **HQ6WIAX**. LS **Logical**

Using the Figure — BASIC

Discussion Direct students' attention to the diagram of dew-point temperature versus cloud height on this page. Point out that the flat bottom of the clouds indicates the altitude of the condensation level, where net condensation begins. Ask students at what temperature the dew point and the atmospheric temperature are equal. (at 10°C) LS **Logical**

CHAPTER RESOURCES

Chapter Resource File

- **Internet Activity** Latent Heat and Thunderstorms GENERAL

Adiabatic Cooling

As a mass of air rises, the surrounding atmospheric pressure decreases. Because of the lower pressure, the molecules in the rising air move farther apart. Thus, fewer collisions between the molecules happen. The resulting decrease in the amount of energy that transfers between molecules decreases the temperature of the air. The process by which the temperature of a mass of air decreases as the air rises and expands is called **adiabatic cooling** (AD ee uh BAT ik KOOL ing).

adiabatic cooling the process by which the temperature of an air mass decreases as the air mass rises and expands

Adiabatic Lapse Rate

The rate at which the temperature of a parcel of air changes as the air rises or sinks is called the *adiabatic lapse rate*. The adiabatic lapse rate of clear air is about –1°C for every 100 m that the air rises. Air that is below the dew point—and thus is cloudy—cools more slowly, however. The average adiabatic lapse rate for cloudy air varies between –0.5°C and –0.9°C per 100 m that the air rises. The slower rate of cooling of moist air results from the release of latent heat as the water condenses.

Condensation Level

The process through which clouds form by adiabatic cooling is shown in **Figure 2**. Earth's surface absorbs energy from the sun and then reradiates that energy as heat. The air close to Earth's surface absorbs the heat. As the air warms, it rises, expands, and then cools. When the air cools to a temperature that is below the dew point, net condensation causes clouds to form. The altitude at which this net condensation begins is called the *condensation level*. The condensation level is marked by the base of the clouds. Further condensation allows clouds to rise and expand above the condensation level.

Reading Check **What is the source of heat that warms the air and leads to cloud formation? (See the Appendix for answers to Reading Checks.)**

Figure 2 ▶ Notice in this illustration that temperature and dew point are the same at an altitude of 1,000 m. Above that altitude, condensation begins and clouds, such as the clouds in the image on the right, form.

SKILL BUILDER — GENERAL

Math Tell students that the adiabatic lapse rate for the dew point in a cloud is −0.2°C per 100 m. Have them calculate how much the dew point would drop for elevation heights of 300 m, 500 m and, 1500 m. (For 300 m: −0.2°C/100 m × 300 m = −0.2°C × 3 = −0.6°C. For 500 m: −0.2°C/100 m × 500 m = −0.2°C × 5 = −1°C. For 1500 m: −0.2°C/100 m × 1500 m = −0.2°C × 15 = −3°C.) LS **Logical**

Answer to Reading Check

The source of heat that warms the air and leads to cloud formation is solar energy that is reradiated as heat by Earth's surface. As the process continues, latent heat released by the condensation may allow the clouds to expand beyond the condensation level.

Figure 3 ▶ Clouds can form as air is pushed up along a mountain slope and is cooled to below the dew point.

Mixing

Some clouds form when one body of moist air mixes with another body of moist air that has a different temperature. The combination of the two bodies of air causes the temperature of the air to change. This temperature change may cool the combined air to below its dew point, which results in cloud formation.

Lifting

The forced upward movement of air commonly results in the cooling of air and in cloud formation. Air can be forced upward when a moving mass of air meets sloping terrain, such as a mountain range. As the rising air expands and cools, clouds form. As **Figure 3** shows, entire mountaintops can be covered with clouds that formed in this way.

The large cloud formations associated with storm systems also form by lifting. These clouds form when a mass of cold, dense air enters an area and pushes a less dense mass of warmer air upward.

SCILINKS. NSTA Developed and maintained by the National Science Teachers Association

For a variety of links related to this subject, go to www.scilinks.org

Topic: Clouds and Fog
SciLinks code: HQ60304

Advective Cooling

Another cooling process that is associated with cloud formation is advective cooling. **Advective cooling** is the process by which the temperature of an air mass decreases as the air mass moves over a cold surface, such as a cold ocean or land surface. As air moves over a surface that is colder than the air is, the cold surface absorbs heat from the air and the air cools. If the air cools to below its dew point, clouds form.

advective cooling the process by which the temperature of an air mass decreases as the air mass moves over a cold surface

READING SKILL BUILDER — BASIC

Paired Summarizing Group students into pairs and have them read silently about the four processes of cooling that lead to cloud formation. Then, have one student summarize the concept of cloud formation. The other student should listen to the retelling and point out any inaccuracies or ideas that were left out. Allow students to refer to the text as needed. English Language Learners
LS Auditory

GEOGRAPHY CONNECTION — GENERAL

Rain Shadow Deserts form on the leeward side of mountain ranges due to the rain shadow effect. A rain shadow is one result of the effect topography has on weather. The phenomenon occurs when moisture-rich air moves inland from the ocean, is lifted over a coastal mountain range, and cools. Clouds build up along the windward side of the mountain range, and when the atmosphere becomes saturated, rain occurs. Once the air reaches the other side of the mountain, the cloud no longer contains enough moisture to produce rain. One example of this effect in the United States is the Sierra Nevada range in California and the deserts of Nevada and Arizona. The Andes Mountains of Chile also produce a rain shadow that affects the Atacama Desert. Have students research the rain shadow effect and find other examples. Have them draw diagrams that explain the process. LS Visual

Teaching Tip — GENERAL

Connect to Familiar Processes Help students understand the process of cloud formation better by comparing cloud formation to the cloud that forms when you boil a kettle of water on the stove. The mist that comes out of the spout is popularly called "steam," but it is really a mixed cloud composed of water droplets, not water vapor. Another example of cloud formation is the cloud that forms when you exhale on a cold day. LS Verbal/Visual

INCLUSION Strategies

• *Visually Impaired*

Help students with visual impairments understand the appearance of stratus, cumulus, and cirrus clouds by using cotton balls, glue, and a piece of corrugated cardboard (10" × 10") for each type of cloud. On the cardboard, shape the cotton to resemble the appropriate cloud. Glue the cloud shapes in place. Ask students to feel and experience the different clouds.

Teach, *continued*

QuickLAB

Skills Acquired

- Experimenting
- Observing
- Analyzing

Teacher's Notes Substitute a metal dish for the jar lid, and use additional ice cubes to chill the air above the water's surface more rapidly. To introduce the concept of condensation nuclei, use matches to add smoke particles to the jar, and then seal the jar quickly. Have students compare the size and speed of cloud formation.

Answers

1. Evaporation takes place near the water's surface, where latent heat is absorbed. Condensation in the form of a cloud forms at the top of the jar, where latent heat is released.
2. The conversion of liquid water to a gas requires energy to break the attractive forces between water molecules. When the process is reversed, the energy reenters the air.

CHAPTER RESOURCES

Chapter Resource File

- Datasheet for Quick Lab GENERAL

Technology

Transparencies

- 115 Classification of Clouds (with worksheet)

stratus cloud a gray cloud that has a flat, uniform base and that commonly forms at very low altitudes

QuickLAB 15 min

Cloud Formation

Procedure

1. Use a **bottle opener** to puncture one or two holes into the metal lid of a **glass jar.**
2. Pour **1 mL of hot water** into the jar.
3. Place an **ice cube** over the holes in the lid of the jar. Make sure the holes are completely covered.
4. Observe the changes that occur within the jar.

Analysis

1. Draw a diagram of the jar. Label the areas of the diagram where evaporation and condensation take place. Also, label areas where latent heat is released and absorbed.
2. Explain why latent heat was released and absorbed in the areas that you labeled on the diagram.

Classification of Clouds

Clouds are classified by their shape and their altitude. The three basic cloud forms are stratus clouds, cumulus clouds, and cirrus clouds. There are also three altitude groups: low clouds (0 to 2,000 m), middle clouds (2,000 to 6,000 m), and high clouds (above 6,000 m). This classification system is shown in **Figure 4.**

Stratus Clouds

Clouds that have a flat, uniform base and that begin to form at very low altitudes are called **stratus clouds.** *Stratus* means "sheetlike" or "layered." The base of stratus clouds is low and may almost touch Earth's surface. Stratus clouds form where a layer of warm, moist air lies above a layer of cool air. When the overlying warm air cools below its dew point, wide clouds appear. Stratus clouds cover large areas of sky and often block out the sun. Usually, very little precipitation falls from stratus clouds.

Two variations of stratus clouds are known as *nimbostratus* and *altostratus*. The prefix *nimbo-* and the suffix *-nimbus* mean "rain." Unlike other stratus clouds, the dark nimbostratus clouds can cause heavy precipitation. Altostratus clouds form at the middle altitudes. They are generally thinner than the low stratus clouds and usually produce very little precipitation.

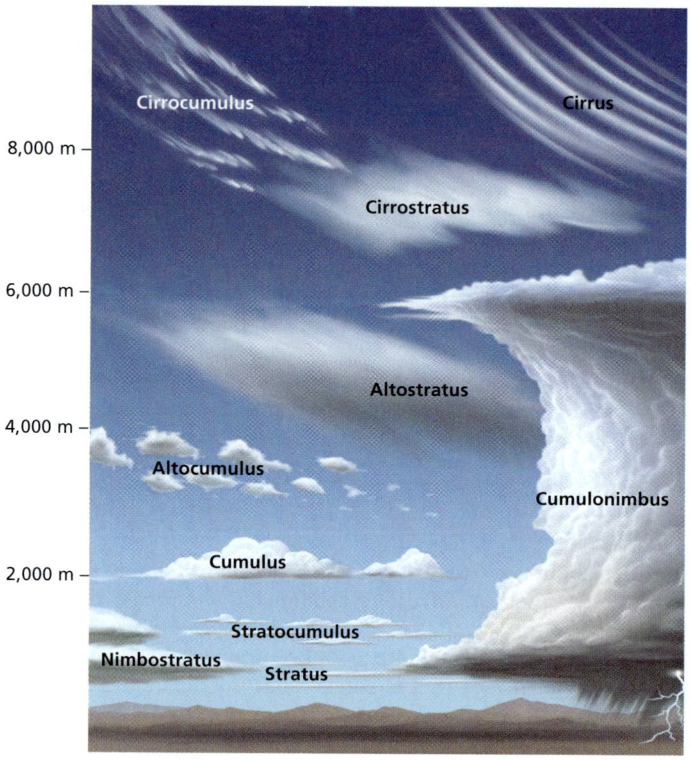

Figure 4 ▶ A variety of cloud types can be identified by their altitude and shape. ***What cloud types form at or above 6,000 m?***

Using the Figure GENERAL

Cloud Bulletin Board The figure summarizes the basic cloud types and the altitudes at which they form. Invite interested students to use this diagram to make a bulletin board that shows different cloud types. Students should use a background of blue construction paper and pictures of clouds cut from magazines or photos of clouds. Answer to caption question: Above 6000 m, cirrus, cirrostratus, and cirrocumulus clouds form. LS **Visual** Co-op Learning

SKILL BUILDER GENERAL

Vocabulary The names assigned to the three basic cloud shapes have Latin roots. The name *cumulus* comes from *cumulo-*, which means "piled" or "heaped." The term *stratus* comes from *stratum,* which means "layered," and the word *cirrus* comes from *cirro-*, which means "hair" or "curl." The word *nimbus,* meaning "rain," is used to describe the dark clouds that bring heavy rains. The name *alto,* from *altus,* meaning "high," describes the height of the cloud base. Various combinations of these prefixes and suffixes give the ten basic cloud groupings. LS **Verbal** English Language Learners

Figure 5 ▶ Cumulus clouds (left) are puffy, vertically growing clouds, while cirrus clouds (right) are wispy.

Cumulus Clouds

Low-altitude, billowy clouds that commonly have a top that resembles cotton balls and a dark bottom are called **cumulus clouds.** *Cumulus* means "piled" or "heaped." Cumulus clouds usually look fluffy, as shown in **Figure 5.** These clouds form when warm, moist air rises and cools. As the cooling air reaches its dew point, the clouds form. The flat base that is characteristic of most cumulus clouds represents the condensation level.

The height of a cumulus cloud depends on the stability of the troposphere, which is the layer of the atmosphere that touches Earth's surface, and on the amount of moisture in the air. On hot, humid days, cumulus clouds reach their greatest heights. High, dark storm clouds known as *cumulonimbus clouds,* or thunderheads, are often accompanied by rain, lightning, and thunder. If the base of cumulus clouds begins at middle altitudes, the clouds are called *altocumulus clouds*. Low clouds that are a combination of stratus and cumulus clouds are called *stratocumulus clouds*.

cumulus cloud a low-level, billowy cloud that commonly has a top that resembles cotton balls and a dark bottom

cirrus cloud a feathery cloud that is composed of ice crystals and that has the highest altitude of any cloud in the sky

Cirrus Clouds

Feathery clouds that are composed of ice crystals and that have the highest altitude of any cloud in the sky are **cirrus clouds.** Cirrus clouds are also shown in **Figure 5.** *Cirro-* and *cirrus* mean "curly." Cirrus clouds form at altitudes above 6,000 m. These clouds are made of ice crystals because the temperatures are low at such high altitudes. Because these clouds are thin, light can easily pass through them.

Cirrocumulus clouds are rare, high-altitude, billowy clouds composed entirely of ice crystals. Cirrocumulus clouds commonly appear just before a snowfall or a rainfall. Long, thin clouds called *cirrostratus clouds* form a high, transparent veil across the sky. A halo may appear around the sun or moon when either is viewed through a cirrostratus cloud. This halo effect is caused by the bending of light rays as they pass through the ice crystals.

Reading Check Why are cirrus clouds commonly composed of ice crystals? (See the Appendix for answers to Reading Checks.)

Graphic Organizer Comparison Table

Create the **Graphic Organizer** entitled "Comparison Table" described in the Skills Handbook section of the Appendix. Label the columns with "Stratus clouds," "Cumulus clouds," and "Cirrus clouds." Label the rows with "Altitude" and "Shape." Then, fill in the table with details about the altitude and shape of each cloud.

Close

Reteaching — BASIC

Cloud Lingo Have students work in pairs to write on index cards key terms and italicized words from the section. They can write the word on one side of the card and write a definition or draw an explanatory diagram on the back. Then, bring the groups together to play a game that uses the cards to review the vocabulary words. **LS Auditory**

Quiz — GENERAL

1. Why are condensation nuclei necessary for cloud formation? (The suspended particles provide a surface on which water droplets can form.)
2. How is advective cooling involved in cloud and fog formation? (As the air moves over the cooler surface, the surface absorbs heat from the air, and cools the air to below the dew point.)

Alternative Assessment — ADVANCED

Cloud Atlas Have students create a Cloud Atlas based on the modern cloud classification system. Students can illustrate the atlas and guide to cloud types by using photos or drawings of the common types of clouds and fog, giving the cloud names, and showing the altitude and general appearance of each cloud. They should include details about how the clouds form and what weather conditions are related to the clouds. **LS Visual**

Answer to Reading Check

because cirrus clouds form at very high altitudes where air temperature is low

Graphic Organizer GENERAL

Comparison Table

You may want to use this Graphic Organizer in a game to review material before the test. Divide the class into two teams. Ask students questions about material from the Comparison Table. Give points to each team that provides correct answers.

Close, *continued*

Answers to Section Review

1. There must be suspended particles around which water droplets can form, and the temperature of the saturated air must be below the dew point so that the rate of evaporation is slower than the rate of condensation.
2. rising and expansion of an air mass (adiabatic); mixing of moist air with another air mass at a different temperature; forced upward movement of air as it moves over sloped terrain; and an air mass moving over a cooler surface (advective)
3. cirrus and cirrocumulus
4. Stratus clouds are layered and form when warm moist air lies above a layer of cooler air. Cumulus clouds are puffy, vertically growing clouds. Cirrus clouds are wispy, high-altitude clouds composed of ice crystals.
5. cumulonimbus
6. Both form as a result of condensation of water vapor in the air. Fog occurs near the surface of Earth, while clouds occur at higher altitudes.
7. Radiation fog occurs as a result of the nightly cooling of Earth when the layer of air in contact with the ground is chilled. Advection fog occurs when warm moist air moves across a cold surface, as when air moves from water to flow over cooler land. Upland fog forms when moist air rises over a slope. Steam fog forms when cool air moves over a warm body of water such as a river.
8. When air rises, it enters an area of lower pressure. The lower pressure allows the air molecules to move farther apart.
9. Sample answer: The pollutants would provide additional surfaces on which water droplets can condense.
10. Stratus clouds condense at the lowest altitudes. Cirrus clouds condense at the highest altitudes, above 6,000 m.
11. *Fog* and *clouds*, such as *stratus*, *cirrus*, and *cumulus*, can form through *adiabatic cooling* or *advective cooling* at different *condensation levels*.

Figure 6 ▶ Steam fog covers the Yakima River in Washington.

Fog

Like clouds, **fog** is the result of the condensation of water vapor in the air. The obvious difference between fog and clouds is that fog is very near the surface of Earth. However, fog also differs from clouds because of how fog forms.

Radiation Fog

One type of fog forms from the nightly cooling of Earth. The layer of air in contact with the ground becomes chilled to below the dew point, and the water vapor in that layer condenses into droplets. This type of fog is called *radiation fog* because it results from the loss of heat by radiation. Radiation fog is thickest in valleys and low places because dense, cold air sinks to low elevations. Radiation fog is often quite thick around cities, where smoke and dust particles act as condensation nuclei.

fog water vapor that has condensed very near the surface of Earth because air close to the ground has cooled

Other Types of Fog

Another type of fog, *advection fog*, forms when warm, moist air moves across a cold surface. Advection fog is common along coasts, where warm, moist air from above the water moves in over a cooler land surface. Advection fog forms over the ocean when warm, moist air is carried over cold ocean currents.

An *upslope fog* forms by the lifting and cooling of air as air rises along land slopes. *Steam fog* is a shallow layer of fog that forms when cool air moves over an inland warm body of water, such as a river, as shown in **Figure 6**.

Section 2 Review

1. **Describe** the conditions that are necessary for clouds to form.
2. **Explain** the four processes of cooling that can lead to cloud formation.
3. **Identify** the cloud types that form at 8,000 m.
4. **Compare** cirrus, cumulus, and stratus clouds.
5. **Identify** the type of cloud that is known for causing thunderstorms.
6. **Compare** clouds with fog.
7. **Describe** four ways in which fog can form.

CRITICAL THINKING

8. **Applying Ideas** Explain why air expands when it rises.
9. **Making Predictions** How might an increase in pollution affect cloud formation?
10. **Making Comparisons** Which type of cloud has the lowest condensation level? Which type has the highest condensation level?

CONCEPT MAPPING

11. Use the following terms to create a concept map: *cloud, cirrus, condensation level, advective cooling, adiabatic cooling, stratus, cumulus,* and *fog.*

CHAPTER RESOURCES

Chapter Resource File
- Section Quiz GENERAL

Workbooks
- Study Guide (also in Spanish)

ENVIRONMENTAL CONNECTION

Section 3 Precipitation

Any moisture that falls from the air to Earth's surface is called **precipitation.** The four major types of precipitation are rain, snow, sleet, and hail.

Forms of Precipitation

Rain is liquid precipitation. Normal raindrops are between 0.5 and 5 mm in diameter. They may vary from a fine mist to large drops in a torrential rainstorm. If the raindrops are smaller than 0.5 mm in diameter, the rain is called *drizzle*. Drizzle results in only a small amount of total precipitation.

The most common form of solid precipitation is *snow,* which consists of ice particles. These particles may fall as small pellets, as individual crystals, or as crystals that combine to form snowflakes. Snowflakes tend to be large at temperatures near 0°C and become smaller at lower temperatures.

When rain falls through a layer of freezing air near the ground, clear ice pellets, called *sleet,* can form. In some cases, the rain does not freeze until it strikes a surface near the ground. There, it forms a thick layer of ice called *glaze ice,* as shown in **Figure 1.** The condition in which glaze ice is produced is commonly referred to as an *ice storm.*

Hail is solid precipitation in the form of lumps of ice. The lumps can be either spherical or irregularly shaped. Hail usually forms in cumulonimbus clouds. Convection currents within the clouds carry raindrops to high levels, where the drops freeze before they fall. If the frozen raindrops are carried upward again, they can accumulate additional layers of ice until they are too heavy for the convection currents to carry them. They then fall to the ground. Large hailstones can damage crops and property.

Figure 1 ▶ Glaze ice forms as rain freezes on surfaces near the ground, such as on these flowers.

OBJECTIVES

- **Identify** the four forms of precipitation.
- **Compare** the two processes that cause precipitation.
- **Describe** two ways that precipitation is measured.
- **Explain** how rain can be produced artificially.

KEY TERMS

precipitation
coalescence
supercooling
cloud seeding

precipitation any form of water that falls to Earth's surface from the clouds; includes rain, snow, sleet, and hail

Section 3

Focus

Overview

This section describes precipitation and how it forms. The section also explains how rain and snow are measured and describes cloud seeding.

Bellringer

Ask students why they think the water droplets that form clouds often fall to Earth as rain or other forms of precipitation.

Motivate

Demonstration — GENERAL

Indoor Rain You will need a hot plate, a wide-mouth jar, ice cubes, a zip-top plastic bag, and aluminum foil. Heat some water to the boiling point on a hot plate. Pour about a cup of the hot water into the jar and cover it tightly with aluminum foil. Put ice cubes into the zip-top bag and place the bag on top of the foil. Have students observe for a few minutes. Invite a volunteer to describe what happens. (Sample answer: Water evaporates from the hot water, and condenses on the cool surface of the aluminum foil. Water droplets fall back down into the jar, simulating rain.) LS Visual

CHAPTER RESOURCES

Chapter Resource File

- **Directed Reading** BASIC
- **Inquiry Lab** How Big Is a Raindrop? GENERAL
- **Making Models Lab** What Is the Shape of a Raindrop? GENERAL

Technology

Transparencies
- Bellringer

Student Edition on CD-ROM

One-Stop Planner CD-ROM
- Lesson Plan

Teach

Group Activity — GENERAL

Supercooled: Test-Tube Hail You will need beakers, clean test tubes, thermometers, crushed ice, and salt. Divide the class into small groups. Have students fill beakers about three-quarters full with ice and cold water and add salt to the mixture until no more salt will dissolve. Ask students to fill the test tube with cold water so that when it is placed in the beaker the water levels of the beaker and the tube will be the same. Have students place the test tubes and thermometers into the beakers and take the starting temperature, then watch for ten minutes. Then, have them gently remove the test tube and immediately drop a piece of ice into it. Have someone describe what happens when the ice is dropped into the test tube. (Ice immediately begins to form.) Take the temperature of the water in the beaker again. Ask students why it was important that the test tube be clean. (If there were particles in the test tube, ice would form before the ice crystal was added and before the water became supercooled.) LS **Visual/Kinesthetic**

MISCONCEPTION ALERT

Shape of Raindrops
Although in popular culture raindrops are often visualized as tear-shaped, small raindrops (less than 1 mm) are actually spherical. As they get larger they take on a flattened shape, something like a hamburger bun. When they become larger than 4.5 mm, raindrops assume a bag-like shape and then split into smaller drops. This transformation is the result of the interaction of two forces—surface tension between the water molecules and air pressure pushing upward against the bottom of the drop.

Figure 2 ▶ During coalescence, as cloud droplets fall, they collide and combine with small droplets. The resulting larger droplets fall as rain.

Causes of Precipitation

Most cloud droplets have a diameter of about 20 micrometers, which is smaller than the period at the end of this sentence. Droplets of this size fall very slowly through the air. A droplet must increase in diameter by about 100 times to fall as precipitation. Two natural processes cause cloud droplets to grow large enough to fall as precipitation: coalescence and supercooling.

Coalescence

The formation of a large droplet by the combination of smaller droplets is called **coalescence** (KOH uh LES uhnts) and is shown in **Figure 2.** Large droplets fall much faster through the air than small ones do. As these larger droplets drift downward, they collide and combine with smaller droplets. Each large droplet continues to coalesce until it contains a million times as much water as it did originally.

Supercooling

Precipitation also forms by the process of supercooling. **Supercooling** is a condition in which a substance is cooled to below its freezing point, condensation point, or sublimation point without changing state. Supercooled water droplets may have a temperature as low as –50°C. Yet even at this low temperature, the water droplets do not freeze. They cannot freeze because too few *freezing nuclei* on which ice can form are available. Freezing nuclei are solid particles that are suspended in the air and that have structures similar to the crystal structure of ice. Most water from the supercooled water droplets evaporates. The water vapor then condenses on the ice crystals that have formed on the freezing nuclei. The ice crystals rapidly increase in size until they gain enough mass to fall as snow, as shown in **Figure 3.** If ice crystals melt and turn into rain as they pass through air whose temperature is above freezing, they form the big raindrops that are common in summer thunderstorms.

coalescence the formation of a large droplet by the combination of smaller droplets

supercooling a condition in which a substance is cooled below its freezing point, condensation point, or sublimation point without going through a change of state

Figure 3 ▶ Most of the rain and snow in the middle and high latitudes of Earth are the result of the formation of ice crystals in supercooled clouds.

ENVIRONMENTAL CONNECTION

Climate and Precipitation The global water cycle lies at the heart of Earth's climate system. A recent study by scientists at NASA's Goddard Flight Center suggests that a 20-year warming trend may accelerate patterns of evaporation and precipitation and lead to increased warm rain over tropical oceans. The study found that warm rains account for over 30 percent of global rainfall and play a critical role in the overall water cycle. Scientists think that warm tropical rains deplete clouds of rain with more efficiency than cold rains do. The information on liquid water and precipitation rates was gathered by a variety of space satellites used to collect data on cloud heights and temperatures and to monitor tropical rainfall patterns. The NASA scientists think that the warmer climate may be associated with a more vigorous water cycle and more extreme weather patterns. Invite interested students to learn more about precipitation rates and climate and about the satellite imaging programs (GPM and TRMM) designed to collect data on rainfall over oceans.

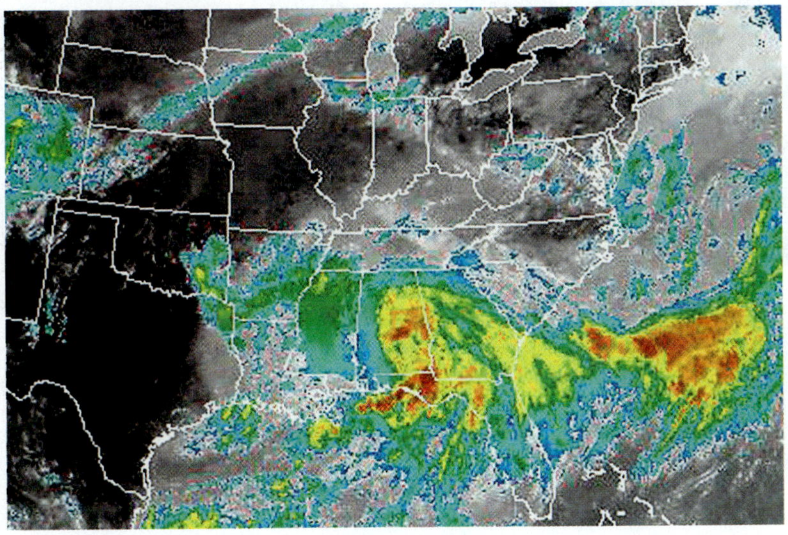

Figure 4 ▶ Doppler radar helps meteorologists track storms, such as Tropical Storm Barry over the southeastern United States. The colors represent the intensity of rainfall. Reds and yellows indicate areas of heaviest rainfall, while blues and greens denote areas of lighter rainfall.

Measuring Precipitation

Meteorologists use a variety of instruments to measure precipitation. For example, a *rain gauge* may be used to measure rainfall.

Amount of Precipitation

In one type of rain gauge, rainwater passes through a funnel into a calibrated container, where the amount of rainfall can then be measured. In another type of rain gauge, rain caught in a funnel fills a bucket. Each time the bucket fills with a given amount of rainwater, the bucket tips and sets off an electrical device that records the amount. As the bucket tips, it activates a switch that releases the water from the bucket.

Snow depth is simply determined with a measuring stick. The water content of the snow is determined by melting a measured volume of snow and by measuring the amount of water that results. On average, 10 cm of snow will melt to produce about 1 cm of water.

Doppler Radar

The intensity of precipitation can be measured using Doppler radar. Doppler radar images, such as the one in **Figure 4,** are commonly used by meteorologists for communicating weather forecasts. Doppler radar works by bouncing radio waves off rain or snow. By timing how long the wave takes to return, meteorologists can detect the location, direction of movement, and intensity of precipitation. This information is very valuable in saving lives because people can be warned of an approaching storm.

✓ Reading Check What aspects of precipitation can Doppler radar measure? (See the Appendix for answers to Reading Checks.)

SCILINKS. NSTA Developed and maintained by the National Science Teachers Association

For a variety of links related to this subject, go to www.scilinks.org

Topic: Precipitation
SciLinks code: HQ61202

Answer to the Reading Check

Doppler radar measures the location, direction of movement, and intensity of precipitation.

Close

Reteaching — BASIC

Comparison Table Divide students into pairs. Have students create a comparison table to compare the different forms of precipitation according to their mode of formation, their state of matter, their shape or structure, and the means of measurement. **LS Logical**

Quiz — GENERAL

1. Compare how sleet forms with how hail forms. (Sleet forms from raindrops falling through a layer of freezing air near the ground. Hail forms when raindrops are carried by convection currents in clouds to high levels where they freeze before they fall.)
2. How is precipitation by supercooling similar to the formation of clouds? (Both processes require solid nuclei on which water vapor can condense.)
3. Explain why large raindrops are often the result of freezing. (Large ice crystals form as a result of supercooling. If the crystals melt as they pass through warmer air, they will fall as large rain drops.)

Homework — GENERAL

Measuring Rainfall Have students make a simple rain gauge from a clean, empty 2-liter plastic bottle. Have students cut off the top of the bottle to use as a funnel. Have them make a scale on the outside by taping a thin plastic ruler to the bottle. They should put marbles inside the bottle to add stability and insert the funnel upside-down into the mouth of the rain gauge. Tell students to find a secure place in the open for the instrument. Ask them to record the daily rainfall amounts for at least a week and graph their results. **LS Kinesthetic**

INCLUSION *Strategies*

• ***Behavior Control Issues*** • ***Learning Disabled***

Students who have behavior control issues and learning disabilities will better understand how Doppler radar works if they watch actual screens in motion. Ask students to use a VCR at home to tape Doppler radar reports from news shows. Have them bring their tapes to school and show them to the class. Have the class watch the reports and discuss the information given by each. **LS Visual**

Close, *continued*

Alternative Assessment — GENERAL

Instruction Manuals Have students write an instruction manual on how to form rain, snow, sleet, and hail in the atmosphere. Students should include information about how to measure these types of precipitation. LS Logical

Answers to Section Review

1. rain, snow, sleet, and hail
2. Coalescence is the combination of many smaller droplets into a larger drop. Supercooling is the cooling of water to below its freezing point without a change of state until the water encounters freezing nuclei and forms ice crystals.
3. a rain gauge
4. Snow depth is determined by using a measuring stick. Melting a measured volume of snow, and then measuring the amount of water that forms from the melted snow determines the water content of the snow.
5. Doppler radar works by bouncing radio waves off rain or snow. By measuring the waves that return, the intensity of the precipitation can be determined.
6. Precipitation can be stimulated artificially by introducing condensation nuclei in the form of dry ice or other chemical crystals into a cloud.
7. Because there are generally too few freezing nuclei on which snow crystals can form naturally, if water could not be supercooled below its freezing point, much of the precipitation in colder climates would be in the form of sleet or hail.
8. It might result in severe storms, or it might actually decrease precipitation.
9. *Precipitation,* such as *rain, sleet, snow, hail, drizzle,* and *glaze ice,* results from the processes of *coalescence* and *super-cooling,* which involves *freezing nuclei.*

Figure 5 ▶ Special equipment attached to the wings of cloud-seeding planes releases freezing nuclei into clouds. Meteorologists hope that cloud seeding will induce rain to fall on drought-stricken areas.

cloud seeding the process of introducing freezing nuclei or condensation nuclei into a cloud in order to cause rain to fall

Weather Modification

In areas suffering from drought, scientists may attempt to induce precipitation through cloud seeding, as shown in **Figure 5.** **Cloud seeding** is the process of introducing freezing nuclei or condensation nuclei into a cloud to cause rain to fall.

Methods of Cloud Seeding

One method of cloud seeding uses silver iodide crystals, which resemble ice crystals, as freezing nuclei. The silver iodide is released from burners on the ground or from flares dropped from aircraft. Another method of cloud seeding uses powdered dry ice, which is dropped from aircraft to cool cloud droplets and to cause ice crystals to form. As the ice crystals fall, they may melt to form raindrops.

Improving Cloud Seeding

In some cases, seeded clouds produce more precipitation than unseeded clouds do. In other experiments, cloud seeding does not cause a significant increase in precipitation. In some instances, cloud seeding appears to cause less precipitation. Thus, meteorologists have concluded that cloud seeding may increase precipitation under some conditions but decrease it under others. Research is underway to identify the conditions that cause increased precipitation. Eventually, cloud seeding may become a way to overcome many drought-related problems. Cloud seeding could also help control severe storms by releasing precipitation from clouds before a storm can become too large.

Section 3 Review

1. **Identify** four forms of precipitation.
2. **Compare** coalescence and supercooling.
3. **Identify** the instrument that measures amounts of rainfall.
4. **Describe** how the amount of snowfall can be measured.
5. **Explain** how Doppler radar can be used to measure the intensity of precipitation.
6. **Describe** how precipitation can be induced or increased artificially.

CRITICAL THINKING

7. **Predicting Consequences** If water could not remain liquid during supercooling, how would the potential for precipitation in colder climates be affected?
8. **Making Inferences** Explain how cloud seeding could be dangerous if it is not done properly.

CONCEPT MAPPING

9. Use the following terms to create a concept map: *precipitation, rain, snow, glaze ice, hail, coalescence, supercooling, freezing nucleus, sleet,* and *drizzle.*

CHAPTER RESOURCES

Chapter Resource File
- Section Quiz GENERAL

Workbooks
- Study Guide (also in Spanish)

Chapter 23 Highlights

Sections

1 Atmospheric Moisture

Key Terms

latent heat, 575
sublimation, 576
dew point, 577
absolute humidity, 577
relative humidity, 578

Key Concepts

- Latent heat is released or absorbed when water changes from one state to another.
- Relative humidity is a ratio of the actual amount of water vapor in the air to the amount of water vapor needed to reach saturation.
- When air reaches the dew point, the rate of condensation equals the rate of evaporation. Below the dew point, net condensation causes dew to form.

2 Clouds and Fog

cloud, 581
condensation nucleus, 581
adiabatic cooling, 582
advective cooling, 583
stratus cloud, 584
cumulus cloud, 585
cirrus cloud, 585
fog, 586

- Clouds form when water vapor cools and condenses on condensation nuclei.
- Water vapor can cool and condense by adiabatic cooling, by the mixing of two bodies of moist air that have different temperatures, by lifting of air, and by advective cooling.
- The three major forms of clouds are stratus clouds, cumulus clouds, and cirrus clouds.
- The four types of fog are radiation fog, advection fog, upslope fog, and steam fog.

3 Precipitation

precipitation, 587
coalescence, 588
supercooling, 588
cloud seeding, 590

- The major forms of precipitation are rain, snow, sleet, and hail.
- Coalescence and supercooling are two processes by which cloud droplets become large enough to fall as precipitation.
- A rain gauge is used to measure liquid precipitation. Snow is measured by its depth and water content.
- Cloud seeding is one way in which meteorologists try to induce precipitation.

Chapter Highlights

Alternative Assessment — GENERAL

Water Cycle Skit Invite students to write and perform a skit in which they explain how water moves around Earth through the water cycle. They can describe the processes of evaporation and condensation, the roles of latent heat, vapor pressure, dew point, and relative humidity; how water droplets gather together to form clouds; and how water returns to Earth's surface as precipitation. Students can take on the roles of water in the ocean, and in different types of clouds, and the various forms of precipitation—rain, sleet, dew, snowflakes, frost, and hail. They can make props, scenery, and include diagrams to make their explanations clearer or more convincing. LS Kinethetic

CHAPTER RESOURCES

Chapter Resource File

- Concept Review GENERAL
- Critical Thinking ADVANCED
- Math Skills GENERAL
- Graphing Skills GENERAL
- Chapter Test A GENERAL
- Chapter Test B ADVANCED

Workbooks

- **Study Guide** (also in Spanish)
- **Assessments** (Spanish)

Technology

Classroom Videos
- Brain Food Video Quiz

Chapter Review

Assignment Guide

Section	Questions
1	1, 7, 9, 18–21, 25–26, 32, 35–39
2	2, 5–6, 8, 10–14, 24, 27, 29, 33
3	3–4, 15–17, 22–23, 28, 31
2 and 3	34
1–3	30

Using Key Terms

1–8. Answers may vary but should show that students understand the definitions of and differences between key terms.

Understanding Key Concepts

9. d **10.** c
11. a **12.** b
13. c **14.** c
15. b **16.** d
17. a **18.** d

Short Answer

19. Water changes from one phase to another when energy is transferred. When energy is absorbed by liquid water, the water changes to a gas. When water vapor condenses back into a liquid form, energy is released back into the surrounding air.

20. A pychrometer consists of two identical thermometers. The bulb of one thermometer is covered with a damp cloth. They are whirled through the air, which causes the water surrounding the wet bulb to evaporate rapidly. Heat leaves as water evaporates from the wet bulb. The difference in the temperature between the two thermometer readings is used to calculate relative humidity.

21. Frost forms when the dew point temperature is below freezing and water vapor changes directly to solid ice crystals on a cold surface.

22. Precipitation is measured by using a calibrated container in which the amount of rainfall can be directly measured or by using an electrical device that records the amount of precipitation that tips into a divided bucket. The water content of snow is measured by melting, or is directly measured by using a meter stick to determine the snowfall depth.

23. Cloud seeding increases precipitation by artificially introducing condensation nuclei or freezing nuclei on which the water droplets may form.

24. Clouds are categorized by both their appearance and their height above Earth's surface.

Critical Thinking

25. Air would be likely to contain more water vapor over Panama, because it is located near the equator where more solar energy reaches Earth's surface to cause evaporation of the ocean water.

26. The air that has a relative humidity of 97% is closer to its dew point because it is nearly saturated. Only a small increase in moisture or a small drop in temperature is needed for the air to reach its dew point.

Chapter 23 Review

Using Key Terms

Use each of the following terms in a separate sentence.

1. *latent heat*

2. *condensation nucleus*

3. *precipitation*

For each pair of terms, explain how the meanings of the terms differ.

4. *coalescence* and *supercooling*

5. *stratus cloud* and *cumulus cloud*

6. *adiabatic cooling* and *advective cooling*

7. *relative humidity* and *absolute humidity*

8. *cloud* and *fog*

Understanding Key Concepts

9. When the temperature of the air decreases, the rate of evaporation
a. increases. **c.** stays the same.
b. varies. **d.** decreases.

10. The type of fog that results when moist air moves across a cold surface is
a. radiation fog. **c.** advection fog.
b. ground fog. **d.** steam fog.

11. Changes in temperature that result from the cooling of rising air or the warming of sinking air are
a. adiabatic. **c.** advective.
b. relative. **d.** latent.

12. Clouds form when the water vapor in air condenses as
a. the air is heated.
b. the air is cooled.
c. snow falls.
d. the air is superheated.

13. The prefix *nimbo-* and suffix *-nimbus* mean
a. high. **c.** rain.
b. billowy. **d.** layered.

14. The fog that results from the nightly cooling of Earth is called
a. steam fog. **c.** radiation fog.
b. upslope fog. **d.** advection fog.

15. Rain that freezes when it strikes a surface produces
a. sleet. **c.** hail.
b. glaze ice. **d.** frost.

16. Clouds in which the water droplets remain liquid below 0°C are said to be
a. saturated. **c.** superheated.
b. supersaturated. **d.** supercooled.

17. In one method of cloud seeding, silver iodide crystals are used as
a. freezing nuclei. **c.** dry ice.
b. cloud droplets. **d.** latent heat.

18. An instrument that uses the electrical conductance of the chemical lithium chloride to measure relative humidity is the
a. hygrometer. **c.** psychrometer.
b. rain gauge. **d.** dew cell.

Short Answer

19. Explain how the transfer of energy affects the changing forms of water.

20. Explain how a psychrometer measures humidity.

21. Describe how frost forms.

22. Describe how precipitation is measured.

23. Describe how cloud seeding may increase precipitation.

24. Explain how clouds are classified.

Critical Thinking

25. **Making Inferences** Where would air contain more water vapor—over Panama or over Antarctica? Explain your answer.

26. **Identifying Relationships** One body of air has a relative humidity of 97%. Another has a relative humidity of 44%. At the same temperature, which body of air is closer to its dew point? Explain your answer.

27. **Applying Ideas** Why would polluted air be more likely to form fog than clean air would?

28. **Analyzing Relationships** In tropical regions, surface temperatures are very high. However, some precipitation in these regions forms by supercooling. Why might this be true?

29. **Predicting Consequences** How would a significant decrease in condensation nuclei in the world's atmosphere affect cloud formation and climate?

Concept Mapping

30. Use the following terms to create a concept map: *hygrometer, condensation nucleus, stratus, cirrus, cloud, cumulus, precipitation, relative humidity, saturated, rain, supercooling, snow, sleet, coalescence, dew cell,* and *psychrometer.*

Math Skills

31. **Applying Quantities** One day in January, 6 cm of snow falls on your area. If all of this snow melts quickly, how deep will the water from the melted snow be? Explain your answer.

32. **Making Calculations** At 15°C, air reaches saturation when it contains 10 g of water vapor per 1 kg of air. What is the relative humidity of air at 15°C that contains 7 g of water vapor per 1 kg of air?

Writing Skills

33. **Writing from Research** Write a report that describes weather conditions necessary to form each type of cloud. Propose regions and describe climates where each cloud type is most likely to be found.

34. **Outlining Topics** Create an outline of how clouds form and a separate outline of how precipitation forms. Then, explain how the two differ.

Interpreting Graphics

The graph below shows variations of temperature and humidity in a 24 h period. Use this graph to answer the questions that follow.

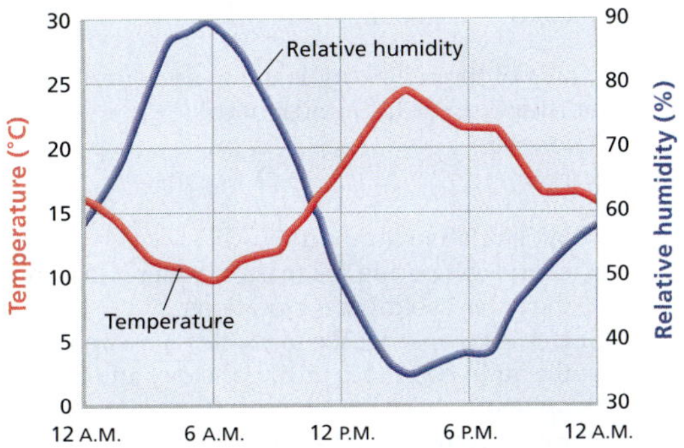

35. Estimate the relative humidity at 6:00 A.M.

36. Estimate the temperature at 6:00 A.M.

37. Explain why relative humidity might be highest at 6:00 A.M.

38. When is relative humidity lowest?

39. How does humidity vary relative to temperature?

Chapter Review

27. Polluted air has smoke, dust, and chemicals that can serve as condensation nuclei for fog formation.

28. Supercooling happens at high altitudes where air temperatures are low. If the air rises quickly and there are not enough freezing nuclei, supercooling will occur.

29. Answers may vary. Sample answer: Cloud formation would decrease, and possibly worldwide precipitation would also change. This might sharply increase drought conditions temporarily or contribute to desertification.

Concept Mapping

30. Answers may vary but should include all of the terms listed. Sample answers appear at the end of this Teacher Edition.

Math Skills

31. On average, 10 cm of snow melts to produce 1 cm of water. 6 cm/10 cm = 0.6 × 1 cm = 0.6 cm of water

32. relative humidity = (amount of H_2O in air ÷ amount needed to reach saturation) × 100; relative humidity = 7 g/kg ÷ 10 g/kg = 7/10 or 0.7 x 100 = 70%

Writing Skills

33. Answers may vary. Accept all reasonable answers.

34. Answers may vary. Accept all reasonable answers.

Interpreting Graphics

35. 89%

36. 10°C

37. Relative humidity is highest in the early morning when temperature is lowest because the air is close to the dew point. Evaporation is slowest compared to condensation.

38. at about 3 P.M.

39. Relative humidity seems to have an inverse relationship to temperature. Humidity tends to decrease during the day when temperatures are higher and to increase at night when the air temperature is cooler.

Standardized Test Prep

Estimated Time

To give students practice under more realistic testing conditions, allow them 30 minutes to answer all of the questions in this practice test.

TEST DOCTOR

Question 2 Answer G is correct. A greater area of Earth's surface is covered by water than by ice, so answer F is incorrect. Advective cooling, answer H, and convective cooling, answer I, do not produce water vapor.

Question 4 Answer H is correct. Relative humidity is a measurement of the moisture in the atmosphere at a given time in one location. Answers F, G, and I do not accurately describe this measurement.

Question 9 Answer F is correct. Answer G is incorrect because, while some species may adapt to low acidic levels, acid rain is destructive to most plant and animal species. Answer H is incorrect because the acid in precipitation raises the acidic levels of lakes and streams, it is not neutralized. Answer I is incorrect because the text does not describe the number of points on the pH scale.

Chapter 23 Standardized Test Prep

Understanding Concepts

Directions (1–4): **For *each* question, write on a separate sheet of paper the letter of the correct answer.**

1. What type of fog is formed when cool air moves across a warm river or lake?
   A. radiation fog
   B. advection fog
   C. upslope fog
   D. steam fog

2. Which of the following processes produces most of the water vapor in the atmosphere?
   F. sublimation
   G. evaporation
   H. advective cooling
   I. convective cooling

3. Which of the following is the main source of moisture in Earth's atmosphere?
   A. lakes
   B. rivers
   C. polar icecaps
   D. oceans

4. What is relative humidity?
   F. a ratio comparing the mass of water vapor in the air at two different locations
   G. a ratio comparing the mass of water vapor in the air at two times during the day and in the same location
   H. a ratio comparing the actual amount of water vapor in the air with the capacity of the air to hold moisture at a given temperature
   I. a ratio comparing the mass of water vapor that air can hold at two different altitudes at noon and at midnight

Directions (5–7): **For *each* question, write a short response.**

5. What instrument is used to measure atmospheric pressure?

6. Particles called condensation nuclei, which are suspended in the atmosphere, are necessary in allowing what process to take place?

7. Water vapor will turn into what when the dew point falls below the freezing point of water?

Reading Skills

Directions (8–9): **Read the passage below. Then, answer the questions.**

Acid Precipitation

Thousands of lakes thoughout the world are affected by acid precipitation, often known simply as acid rain. Acid precipitation is precipitation, such as rain, sleet, or snow, that contains high concentrations of acids. When fossil fuels are burned, they release oxides of sulfur and nitrogen. When the oxides combine with water in the atmosphere, they form sulfuric acid and nitric acid, which fall as precipitation. This acidic water flows over and through the ground, and then flows into lakes, rivers, and streams. Acid precipitation can kill living things and can result in the decline or loss of some local animal and plant populations.

A pH (power of hydrogen) number is a measure of how acidic or basic a substance is. The lower the number on the pH scale is, the more acidic a substance is; the highter a pH number is, the more basic a substance is. Each whole number on the pH scale indicates a tenfold change in acidity.

8. According to the passage, which of the following statements is true?
   A. Acid precipitation always falls as rain.
   B. Acid precipitation seeps into local water supplies and may pose a danger to living things in the area.
   C. Sulfur and nitrogen mix with oxygen in the atmosphere and become acids.
   D. The amount of acidic precipitation is balanced in nature by an equal amount of basic precipitation.

9. Which of the following statements can be inferred from the information in the passage?
   F. A reduction in the usage of fossil fuels may help alleviate the problem of acid rain.
   G. Local animal and plant species will most likely adapt to acid rain.
   H. The acid in precipitation is effectively neutralized once it is in a lake or stream.
   I. The amount of acid in a substance can be measured by using a 10-point scale.

Answers

Understanding Concepts

1. D
2. G
3. D
4. H
5. barometer
6. cloud formation
7. frost

Reading Skills

8. B
9. F

Interpreting Graphics

10. B
11. F
12. cloud formation
13. C
14. Answers may vary. See Test Doctor for a detailed scoring rubric.

Interpreting Graphics

Directions (10–14): **For *each* question below, record the correct answer on a separate sheet of paper.**

The diagram below shows shows the direction of air movement over a mountain. Use this diagram to answer questions 10 through 12.

Movement of Air Over a Mountain

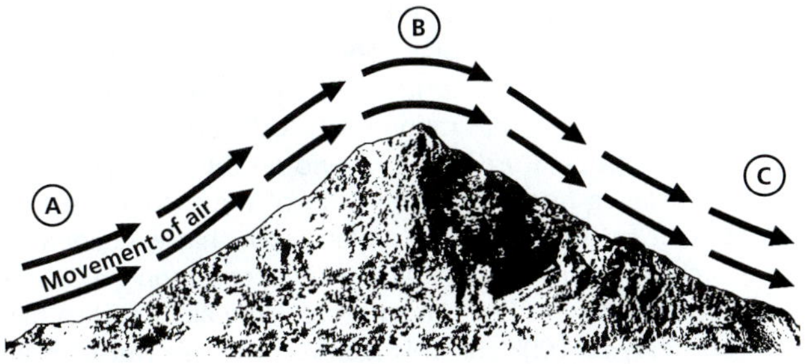

10 As air moves from point A to point B, the air temperature
A. increases
B. decreases
C. stays the same
D. impossible to predict

11 Air moving from point B to point C will become compressed and gain energy as it moves down the mountain, which will cause the air to undergo
F. adiabatic warming
G. adiabatic cooling
H. condensation
I. sublimation

12 If moist air moves up the mountain from point A, what process is likely to occur when the moist air moves near point B?

The diagram below shows the parts of a psychrometer. Use this diagram to answer questions 13 and 14.

Parts of a Psychrometer

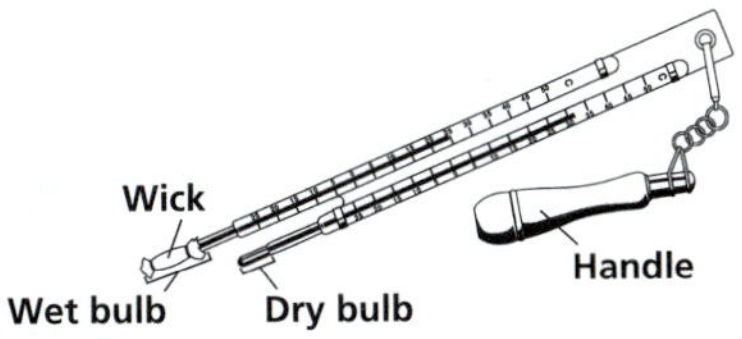

13 How does a meteorologist use a psychrometer, such as the one shown in the diagram above?
A. It is placed in a pan of water and exposed to the air for one hour.
B. The handle is used to dip it into a body of water such as a lake.
C. It is held by the handle and twirled in the air.
D. It is held until the readings on both thermometers are equal.

14 Why is it necessary to obtain two readings? What measurements of atmospheric moisture can be determined from these readings?

Test TIP

If you are not sure of an answer, go to the next question, but remember to skip that number on your answer sheet.

TEST DOCTOR

Questions 10 and 11 For question 10, answer B is correct. For question 11, answer F is correct. Students can use the diagram to determine that air is forced upward or downward at the specified locations. Students should demonstrate an understanding of the adiabatic temperature changes that result from expansion or compression of air. As the air rises between points A and B, it loses energy and undergoes adiabatic cooling. As the air falls from point B toward point C, it gains energy and undergoes adiabatic warming.

Question 14 Full-credit answers should include the following points:
- dew point temperature and relative humidity are calculated through dry-bulb and wet-bulb temperature readings
- dry-bulb temperature readings indicate air temperature; wet-bulb temperature readings indicate the cooling effect of evaporation
- these readings indicate the evaporation rate at the current temperature

Test Prep Correlations *National Science Education Standards*

ES 1d: items 10, 11, 12
HNS 2a: items 1, 4
HNS 2b: items 2, 3, 6, 7
SAI 2c: items 5, 13, 14
SAI 2d: item 4
SPSP 4b: items 8, 9
SPSP 5b: items 8, 9

CHAPTER RESOURCES

State Resources

For specific resources for your state, visit go.hrw.com and type in the keyword **HSHSTR.**

Skills Practice Lab

Relative Humidity

Teacher's Notes

Time Required

one 45-minute class period

Lab Ratings

EASY → HARD

TEACHER PREPARATION (1)
STUDENT SETUP (2)
CONCEPT LEVEL (2)
CLEANUP (1)

Skills Acquired

- Predicting
- Experimenting
- Collecting Data
- Interpreting Results

The Scientific Method

In this lab, students will

- Make Observations
- Analyze Results
- Communicate Results

Materials

The materials listed are enough for groups of 2 to 4 students. A rubber band or piece of string will keep the cloth securely fastened around the wet-bulb thermometer. If ring stands are unavailable, thermometers can also be securely mounted on a piece of stiff poster board by using tape. This also makes transporting and using the psychrometer later in an outside weather shelter easier.

Tips and Hints

The wet bulb is chilled because it releases heat to the evaporating water. The drier the air is, the faster the water will evaporate, and the more the bulb will be chilled.

Students will need to take readings from both bulbs to obtain the relative humidity from the Relative Humidity Table in the Appendix.

Chapter 23

Skills Practice Lab

Objectives

- **Measure** humidity in the classroom.
- **Calculate** relative humidity.

Materials

- cloth, cotton, at least 8 cm × 8 cm
- container, plastic
- piece of paper
- ring stand with ring
- rubber band
- string
- thermometer, Celsius (2)
- water

Safety

Relative Humidity

Earth's atmosphere acts as a reservoir for water that evaporates from Earth's surface. However, the amount of water vapor in the atmosphere depends on the relative rates of condensation and evaporation. When the rates of condensation and evaporation are equal, the air is said to be "saturated." However, when the rate of condensation exceeds the rate of evaporation, water droplets begin to form in the air or on nearby surfaces. The point at which the condensation rate equals the evaporation rate is called the *dew point* and depends on the temperature of the air and on the atmospheric pressure.

Relative humidity is the ratio of the amount of water vapor in the air to the amount of water vapor that is needed for the air to become saturated. This ratio is most commonly expressed as a percentage. When the air is saturated, the air is said to have a relative humidity of 100%. In this lab, you will use wet-bulb and dry-bulb thermometer readings to determine the relative humidity of the air in your classroom.

PROCEDURE

1. Hang two thermometers from a ring stand, as shown in the illustration below.
2. Using a rubber band, fasten a piece of cotton cloth around the bulb of one thermometer. Adjust the length of the string so that only the cloth, not the thermometer bulb, is immersed in the water. By using this setup, you can measure both the air temperature and the cooling effect of evaporation.

Step 2

CHAPTER RESOURCES

Chapter Resource File

- Datasheet for Chapter Lab GENERAL
- Lab Notes and Answers

3 Predict whether the two thermometers will have the same reading or which thermometer will have the lower reading.

4 Using a piece of paper, fan both thermometers rapidly until the reading on the wet-bulb thermometer stops changing. Read the temperature on each thermometer.
a. What is the temperature on the dry-bulb thermometer?
b. What is the temperature on the wet-bulb thermometer?
c. What is the difference in the two temperature readings?

5 Use the table entitled "Relative Humidity" in the Reference Tables section of the Appendix to find the relative humidity based on your temperature readings in **Step 4.** Look at the left-hand column labeled "Dry-Bulb Temperature." First, find the temperature that you recorded in **Step 4a.** Follow along to the right in the table until you come to the number that is directly below the column entitled "Difference in Temperature" (top row of the table) and that you recorded in **Step 4c.** This number, expressed as a percentage, is the relative humidity. What is the relative humidity of the air in your classroom?

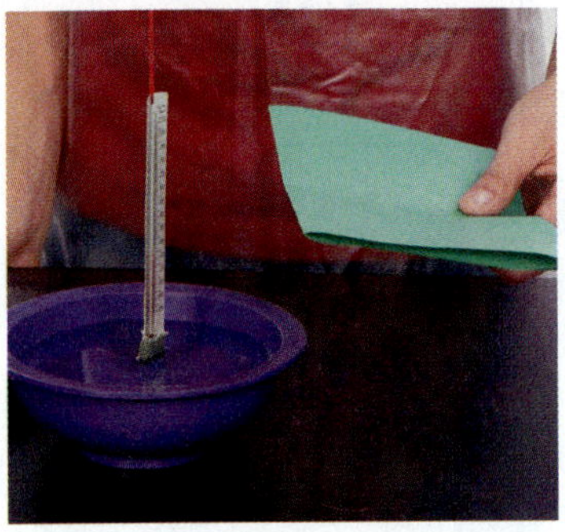

Step 4

ANALYSIS AND CONCLUSION

1 **Drawing Conclusions** On the basis of the relative humidity you calculated, is the air in your classroom close to or far from the dew point? Explain your answer.

2 **Applying Conclusions** If you wet the back of your hand, would the water evaporate and cool your skin?

Extension

1 **Making Inferences** Suppose that you exercise in a room in which the relative humidity is 100%.
a. Would the moisture on your skin from perspiration evaporate easily?
b. Would you be able to cool off readily? Explain your answer.

2 **Applying Ideas** Suppose that you have just stepped out of a swimming pool. The relative humidity is low, about 30%. Would you feel warm or cool? Explain your answer.

Skills Practice Lab

Answers to Procedure

3. Answers may vary.
4a–c. Answers may vary.
5. Answers may vary.

Answers to Analysis and Conclusion

1. Answers may vary depending on how close the relative humidity is to 100%.
2. Unless the relative humidity is 100%, at which point the net condensation equals evaporation, yes.

Answers to Extension

1a. no
b. no; Because the air is saturated, no more water will evaporate from the skin and, thus, there would be little cooling effect.
2. cool; Because the relative humidity is low, rapid evaporation would cause the skin to cool.

Randa Flinn
Northeast High School
Fort Lauderdale, FL

Annual Precipitation in the United States

Internet Activity — GENERAL

Climate and Precipitation Have students research the annual rates of two forms of precipitation for one region of the United States. Have them identify what factors, such as presence of large bodies of water, elevations, or air temperature, have led to this pattern of precipitation. A worksheet designed to direct student research on this topic can be found in the **Chapter Resource File** booklet or by visiting **go.hrw.com** and entering the keyword **HQ6WIAX.** **LS Visual**

Answers to Map Skills Activity

1. Answers may vary.
2. the Pacific Northwest
3. More moisture is in the air in coastal areas that are bordered by mountains.
4. Answers may vary. Sample answers: rain: Oregon; snow or freezing rain: Northeast or Midwest
5. southwestern states such as Nevada, Utah, Arizona, and New Mexico
6. southern and southwestern states
7. the Pacific Northwest and the Gulf coast states

CHAPTER RESOURCES

Chapter Resource File

- **Internet Activity**
  Climate and Precipitation
  GENERAL

Technology

Transparencies
- 116 Annual Precipitation in the United States (with worksheet)

MAPS in Action

Annual Precipitation in the United States

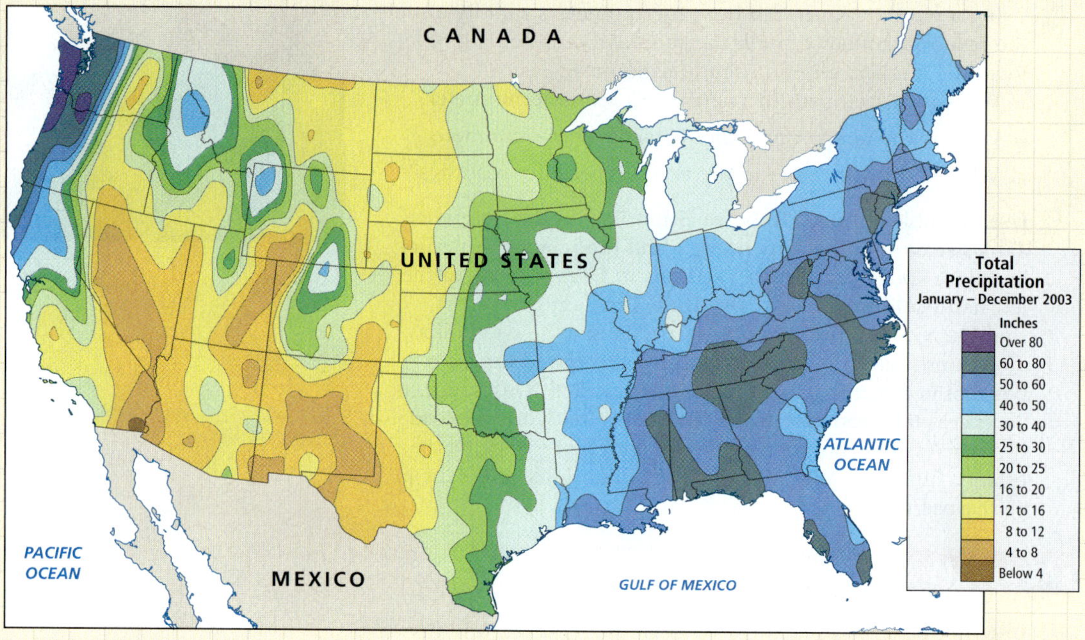

Map Skills Activity

This map shows the total precipitation for the continental United States for the year 2003. Use the map to answer the questions below.

1. **Using a Key** What is the highest total amount of precipitation for any area in your state?
2. **Making Comparisons** Which area of the United States has the highest total annual precipitation?
3. **Analyzing Methods** Using what you have learned about the formation of precipitation, explain why one area of the United States might have a higher total annual precipitation than another area has.
4. **Making Inferences** List the forms of precipitation that occur in the United States. Identify areas of the United States where you would likely encounter each form.
5. **Evaluating Data** Describe the location in the United States that might be classified as desert.
6. **Making Comparisons** Describe the location in the United States that might have the highest rate of evaporation.
7. **Making Comparisons** Describe the location in the United States that you think might have the highest relative humidity.

IMPACT on Society

Hail

Hail is precipitation in the form of a lump of ice that starts as a raindrop falling from clouds. Updraft convection currents carry the raindrop back into freezing cloud regions, where layer after layer of supercooled water is added to the frozen drops. The added water also freezes and increases the sizes of the hailstones. Some hailstones have grown from this process to the size of grapefruits. The largest hailstone ever recorded had a circumference of more than 47 cm.

Hailstones often consist of alternating layers of clear ice and cloudy ice. The clear ice forms as a hailstone passes through a layer of very moist air. The cloudy layer forms when water droplets freeze to the surface of the hailstone and trap air bubbles between them.

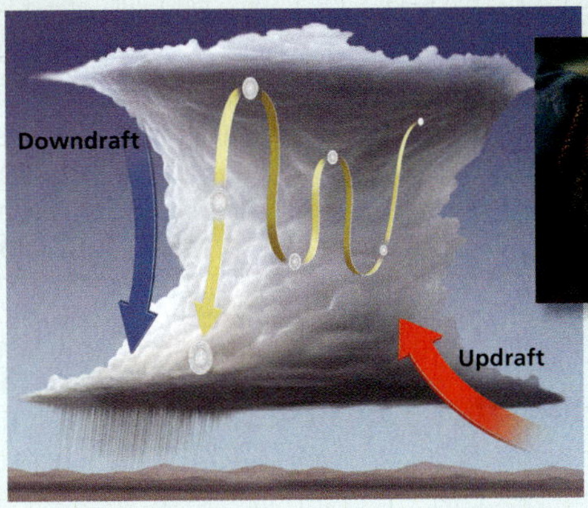

▲ Hailstones commonly consist of alternating layers of clear ice and cloudy ice.

Hail's Destructive Power

A hailstorm that lasts only a few minutes can cause tremendous damage. For example, hail has been responsible for some airplane crashes. It has torn holes in the roofs of houses and automobiles and in the cabins of aircraft. Hail has killed people, livestock, and wildlife, and it has stripped plants of their leaves.

In July 1984, a hailstorm in Germany caused $1 billion worth of damage to crops, trees, buildings, and vehicles. Trees were stripped of their bark. Crops were destroyed, and 400 people were injured.

The Cost of Hail

Hail destroys more than $200 million worth of wheat, corn, soybeans, and other crops in the United States each year. Hailstorms can have a devastating effect on individual farmers. Unlike drought, hail does not cause gradual damage. Instead, a family's entire crop, which may represent work over an entire year, can be wiped out within minutes.

◀ A strawberry farmer in Atwater, California, investigates damage to his crop. Seventy-five to eighty percent of his crop was destroyed by a hailstorm.

Extension

1. **Analyzing Processes** If you cut a hailstone in half, you would see layering. Explain why.

Impact on Society

Hail

Activity — GENERAL

Research Divide the class into groups. Have one group find out where hail occurs most commonly and where it does the most damage. Other students can investigate record setting hailstones and how the storms that produce large hail are different from other storms. Some groups could research cloud-seeding programs that use silver iodide (AgI) crystals to reduce damage by hailstones. Other groups can investigate the different methods that have been tried to prevent hail and how successful these methods have been. Have the groups report back to the class on the results of their research using visual aids such as photos or diagrams, first person accounts of the damage, models, or demonstrations.

LS Visual Co-op Learning

Answer to Extension

1. Hail starts out as falling rain, but an updraft sends the droplets upward instead of letting them fall. Some of the raindrops freeze in the colder temperatures and act as freezing nuclei. The pellets of ice fall and are lifted again and again; each time another layer of ice is added. This cycle is responsible for the layers seen in cross-section.

Chapter 24 Weather

Planning Guide

Compression Guide

To shorten instruction because of time limitations, omit Sections 1 and 3.

OBJECTIVES	LABS, DEMONSTRATIONS, AND ACTIVITIES	TECHNOLOGY RESOURCES
PACING • 45 min pp. 600–604 **Chapter Opener**	**SE** **Long-Term Project** Correlating Weather Variables, pp. 856–857 Advanced **SE** **Long-Term Project** Weather Forecasting, pp. 858–861 Advanced	**OSP** **Parent Letter** ■ **CD** **Student Edition on CD-ROM** **CD** **Chapter Summaries Audio CD** ■ **VID** **Brain Food Video Quiz**
Section 1 Air Masses • Explain how an air mass forms. • List the four main types of air masses. • Describe how air masses affect the weather of North America.	**TE** **Demonstration** Coriolis Effect, p. 601 ◆ General **TE** **Demonstration** Comparing Air Masses, p. 602 ◆ General **TE** **Activity** Air On the Move, p. 601 ◆ General **TE** **Activity** Bulletin Board Project, p. 603 General	**OSP** **Lesson Plans** (also in print) **TR** **Bellringer*** **TR** 117 Air Masses* **CD** **Interactive Tutor** North American Weather
PACING • 45 min pp. 605–610 **Section 2 Fronts** • Compare the characteristic weather patterns of cold fronts with those of warm fronts. • Describe how a midlatitude cyclone forms. • Describe the development of hurricanes, thunderstorms, and tornadoes.	**TE** **Demonstration** Make a Cold Front, p. 605 ◆ General **TE** **Group Activity** Weather Front Pop-Ups, p. 606 ◆ General **TE** **Demonstration** Spiraling Winds, p. 607 ◆ General **TE** **Demonstration** Updrafts, p. 608 ◆ General **TE** **Math Connection** Coming or Going?, p. 608 General **SE** **Maps in Action** Weather-Related Disasters, 1980–2003, p. 628 General **TE** **Activity** Hurricane Hunters, p. 629 General	**OSP** **Lesson Plans** (also in print) **TR** **Bellringer*** **TR** 118 Types of Fronts* **TR** 119 Stages of a Midlatitude Cyclone* **TR** 120 Anatomy of a Hurricane* **TR** 123 Weather-Related Disasters, 1980–2005* **VID** **NOVA Video** Hurricane!
PACING • 45 min pp. 611–614 **Section 3 Weather Instruments** • Identify four instruments that measure lower-atmospheric weather conditions. • Describe how scientists measure conditions in the upper atmosphere. • Explain how computers help scientists understand weather.	**TE** **Group Activity** Simple Thermometer, p. 611 ◆ General **SE** **Quick Lab** Wind Chill, p. 612 General **CRF** **Datasheet for Quick Lab*** General **SE** **Mapping Expeditions** Snapshots of the Weather, pp. 842–843 General **CRF** **Inquiry Lab** Making a Weather Station* General	**OSP** **Lesson Plans** (also in print) **TR** **Bellringer*** **TE** **Internet Activity** Weather Images, p. 613 General **CRF** **Internet Activity** Weather Images* General **CD** **Interactive Tutor** Tornadoes and Hurricanes
PACING • 90 min pp. 615–620 **Section 4 Forecasting the Weather** • Explain how weather stations communicate weather data. • Explain how a weather map is created. • Explain how computer models help meteorologists forecast weather. • List three types of weather that meteorologists have attempted to control.	**TE** **Group Activity** Cloud Cover, p. 616 ◆ General **TE** **Activity** Local Weather Maps, p. 617 ◆ General **SE** **Quick Lab** Gathering Weather Data, p. 618 General **CRF** **Datasheet for Quick Lab*** General **TE** **Group Activity** Weather Safety Project, p. 619 General **TE** **Debate** Weather Modification, p. 619 Advanced **SE** **Skills Practice Lab** Weather Map Interpretation, pp. 626–627 General **CRF** **Datasheet for Chapter Lab*** General **CRF** **Making Models Lab** Blowing in the Wind* General	**OSP** **Lesson Plans** (also in print) **TR** **Bellringer*** **TR** 121 Weather Symbols* **TR** 122 Weather Map of the United States* **CD** **Interactive Tutor** Old-Time Forecasting **CD** **Interactive Tutor** Modern Forecasting **CD** **Interactive Tutor** Lightning **CD** **Interactive Tutor** Other Storms and El Niño

PACING • 90 min

CHAPTER REVIEW, ASSESSMENT, AND STANDARDIZED TEST PREPARATION

- **SE** **Chapter Highlights**, p. 621
- **SE** **Chapter Review**, pp. 622–623
- **SE** **Standardized Test Prep**, pp. 624–625
- **CRF** **Concept Review*** ■ General
- **CRF** **Critical Thinking*** Advanced
- **CRF** **Math Skills*** General
- **CRF** **Graphing Skills*** General
- **CRF** **Chapter Test A*** ■ General
- **CRF** **Chapter Test B*** Advanced
- **OSP** **Lesson Plans** (also in print)
- **OSP** **Test Generator**
- **OSP** **Test Item Listing**

Online and Technology Resources

go.hrw.com Holt Online Learning

Visit go.hrw.com for access to Holt Online Learning, or enter the keyword **HQ6 Home** for a variety of free online resources.

One-Stop Planner® CD-ROM

This CD-ROM package includes

- Lab Materials QuickList Software
- Holt Calendar Planner
- Customizable Lesson Plans
- Printable Worksheets
- ExamView® Test Generator
- Interactive Teacher Edition
- Holt PuzzlePro®
- Holt PowerPoint® Resources

KEY

SE Student Edition	**OSP** One-Stop Planner	**VID** Classroom Video/DVD
TE Teacher Edition	**TR** Transparencies and Transparency Worksheets	* Also on One-Stop Planner
CRF Chapter Resource File	**CD** CD or CD-ROM	◆ Requires advance prep
LTP Long-Term Projects		■ Also available in Spanish

SKILLS DEVELOPMENT RESOURCES	REVIEW AND ASSESSMENT	CORRELATIONS
SE **Pre-Reading Activity,** p. 600 GENERAL **TE** **Using the Figure** Light Show, p. 600 GENERAL		National Science Education Standards
CRF **Directed Reading*** BASIC **TE** **Reading Skill Builder** Reading Hint, p. 602 BASIC **TE** **Inclusion Strategies,** p. 602	**SE** **Reading Check,** p. 603 GENERAL **SE** **Section Review,** p. 604 GENERAL **TE** **Reteaching,** p. 603 BASIC **TE** **Quiz,** p. 603 GENERAL **TE** **Alternative Assessment,** p. 604 ADVANCED **CRF** **Section Quiz*** ■ GENERAL	SAI 1b, SAI 1d, SAI 1f, ST 1e, HNS 2a, UCP 1, UCP 2, UCP 3
CRF **Directed Reading*** BASIC **TE** **Reading Skill Builder** Paired Summarizing, p. 606 BASIC **TE** **Skill Builder** Vocabulary, p. 607 BASIC **SE** **Math Practice,** p. 608 GENERAL **SE** **Graphic Organizer** Venn Diagram, p. 609 GENERAL	**SE** **Reading Checks,** pp. 607, 609 GENERAL **SE** **Section Review,** p. 610 GENERAL **TE** **Reteaching,** p. 609 BASIC **TE** **Quiz,** p. 609 GENERAL **TE** **Alternative Assessment,** p. 609 GENERAL **CRF** **Section Quiz*** ■ GENERAL	SAI 1b, SAI 1c, SAI 1d, SAI 1f, SAI 2a, ST 1e, SPSP 1a, UCP 1, UCP 2, UCP 3
CRF **Directed Reading*** BASIC **TE** **Skill Builder** Graphing, p. 612 GENERAL **TE** **Inclusion Strategies,** p. 612	**SE** **Reading Check,** p. 612 GENERAL **SE** **Section Review,** p. 614 GENERAL **TE** **Reteaching,** p. 613 BASIC **TE** **Quiz,** p. 613 GENERAL **TE** **Alternative Assessment,** p. 614 GENERAL **CRF** **Section Quiz*** ■ GENERAL	SAI 1b, SAI 1c, SAI 1d, SAI 1f, SAI 2a, ST 1e, HNS 2a, UCP 2, UCP 3
CRF **Directed Reading*** BASIC **TE** **Using the Figure** Classroom Station Model, p. 616 GENERAL **TE** **Using the Figure** Isobars, p. 617 GENERAL	**SE** **Reading Checks,** pp. 617, 618 GENERAL **SE** **Section Review,** p. 620 GENERAL **TE** **Homework,** p. 618 ADVANCED **TE** **Reteaching,** p. 619 BASIC **TE** **Quiz,** p. 619 GENERAL **TE** **Alternative Assessment,** p. 620 GENERAL **CRF** **Section Quiz*** ■ GENERAL	SAI 1a, SAI 1b, SAI 1c, SAI 1d, SAI 1f, SAI 2a, ST 1d, ST 1e, SPSP 1a, HNS 2a, UCP 1, UCP 2, UCP 3

Holt Lab Generator CD-ROM

Search for labs by topic, standard, difficulty level, and/or time frame. Edit those labs to fit your needs, or create your own labs. Use the Lab Materials QuickList to order just what you need.

INTERACTIVE TUTOR

Holt Earth Science Interactive Tutor CD-ROM

This CD-ROM consists of interactive activities that give students a fun way to extend their knowledge of Earth science concepts.

Chapter Summaries Audio CDs

These CDs include audio summaries of the key concepts presented in each chapter. (Audio summaries are also available in Spanish.)

SCILINKS. NSTA

www.scilinks.org

Maintained by the **National Science Teachers Association.** See Chapter Enrichment pages that follow for a complete list of topics.

DVD VIDEO

See Chapter Enrichment pages for Video Resources.

Chapter 24 Chapter Enrichment

This Chapter Enrichment provides relevant and interesting information to expand and enhance your classroom instruction of the chapter material.

Section 1 Air Masses

Jet Streams

Jet streams are narrow rivers of fast-moving air thousands of kilometers long found in the mid-latitudes. They generally flow from west to east in the upper atmosphere and form over the boundaries between air masses of different temperatures. These winds cause air to rise from and sink toward Earth's surface and generate regions of high and low pressure associated with precipitation and storm patterns. The bomber pilots who flew missions across the Pacific Ocean during World War II originally discovered jet streams. Flying at altitudes of above 7 km, the pilots reported wind speeds of over 170 knots (about 315 km/h).

▲ Weather on Earth, such as this hurricane, is controlled by the movement of air masses.

Section 2 Fronts

The Fujita Scale

Like hurricanes, tornadoes are rated on an intensity scale, based on the damage produced. The scale was developed by meteorologist and tornado researcher Ted Fujita (1920–1998) of the University of Chicago. The ratings are: F-0 Gale force, light damage, branches broken off trees (72 mph winds); F-1 moderate damage, mobile homes pushed off foundations (73–112 mph winds); F-2 significant damage, large trees snapped or uprooted, mobile homes demolished (113–157 mph winds); F-3 severe damage, heavy cars lifted (158–206 mph winds); F-4 devastating damage, houses leveled, cars and large missiles thrown (207–260 mph winds); and F-5 incredible damage, houses lifted off foundations (winds above 261 mph).

▲ Lightning is a dangerous aspect of severe weather.

Section 3 Weather Instruments

Go Fly a Kite!

Kite flying has played an important, and largely unsung role in the history of meteorological science. Most people have heard the story of the American scientist, inventor, and founding father Benjamin Franklin's famous weather experiment of 1752—flying a kite attached to a metal key in a thunderstorm—to prove that lightning is an electrical discharge. But kites were first used for meteorological purposes in 1749, when a Scottish professor named Alexander Wilson used them to send thermometers aloft to obtain rough atmospheric temperature readings. Around 1840, American meteorologist James Pillard Espy used kites to determine cloud heights and relate this data to surface temperature and dew point. Later, kites were used as a tool for measuring changes in wind velocity. During the 19th and early 20th centuries, kites became a standard meteorological tool for probing the upper atmosphere and measuring temperature, pressure, and humidity. With the introduction of weather balloons and aircraft equipped with meteorological instruments, interest in kites as a scientific tool declined. Recently, University of Colorado meteorologist Ben Balsley has reawakened interest in the meteorological uses of kites. He says kites have advantages over satellite technology, airplanes, and weather balloons. Kites are inexpensive, provide a stable platform, can be kept aloft for days to provide a continuous profile, and can carry scientific payloads as high as several kilometers over remote regions.

Section 4 Weather Forecasting

Weather and Human Health

In 1877, a Philadelphia doctor named S. Weir Mitchell described the relationship between human pain and the onset of approaching storms. Generally, the pain continues until the air pressure rises and the storm system moves on. Conditions such as arthritis and old injuries are common causes for such sensitivities. Since that time, the effect has been confirmed by controlled studies. This research has given rise to a new science known as *biometeorology,* which focuses on the effects of weather on living things. Other observed weather-health connections include the effect of cold weather and pollution on respiratory and heart-related ailments and on people with allergies, the combined effects of extreme heat and humidity in producing heat stroke, UV rays and skin cancer, and the beneficial effects of sunshine on human moods.

▼ **Meteorologists study weather patterns to better forecast the weather.**

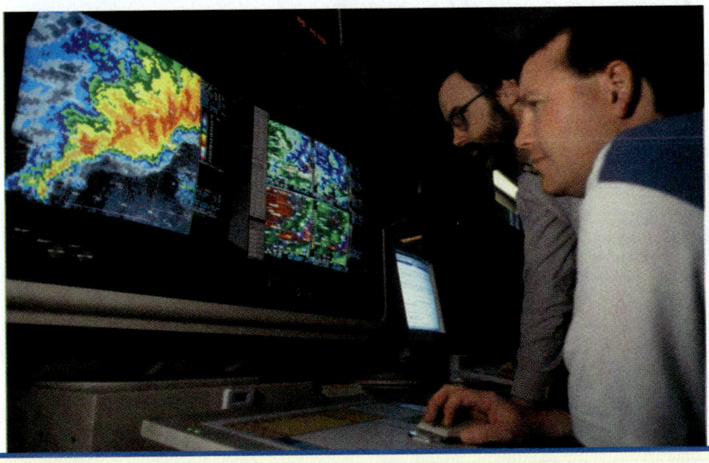

Video Resources

Brain Food Video Quizzes

Brain Food Video Quizzes These videos contain game-show style quizzes that assess students' progress and motivate students to study the chapter material.

NOVA Videos

The NOVA video below complements this chapter.

Hurricane! Watch scientists fly into the center of hurricanes so that they can learn more about the hurricanes, and listen to firsthand accounts of people affected by Hurricane Camille. (60 min)

To order other **NOVA** videos related to this chapter, visit **go.hrw.com** and enter the keyword **HQ6WTHV.**

SCILINKS® — NSTA — Developed and maintained by the National Science Teachers Association

SciLinks is maintained by the National Science Teachers Association to provide you and your students with interesting, up-to-date links that will enrich your classroom presentation of the chapter.

Visit www.scilinks.org and enter the SciLinks code for more information about the topic listed.

Topic: Air Masses
SciLinks code: HQ60031

Topic: Fronts and Severe Weather
SciLinks code: HQ60624

Topic: Weather Instruments
SciLinks code: HQ61646

Topic: Weather Maps
SciLinks code: HQ61647

Topic: Weather Forecasting
SciLinks code: HQ61645

Topic: Careers in Earth Science
SciLinks code: HQ60222

Chapter 24

Chapter Overview

This chapter describes how air masses affect weather and how fronts produce severe weather. The chapter also explains how scientists measure atmospheric conditions and forecast weather.

Using the Figure — GENERAL

Light Show Cumulonimbus clouds discharge lightning to balance electrical charges. Ice crystals at the top of the clouds are positively charged; rain near the bottom becomes negatively charged. As a result of this charge separation, the ground below also becomes positively charged. When the difference in charge is great enough, an electric spark flows between the cloud and the ground. Thunder is the sound of rapidly expanding gases from the channel of the lightning. Ask students why thunder always follows lightning. (Sound and light travel at different speeds.) LS **Visual**

PRE-READING ACTIVITY

FOLDNOTES

Have pairs of students use their FoldNotes to study key terms from the chapter. Instruct one student to use the FoldNote to provide the key term, and have the other student give the definition. Have the student who provides the key term correct the other student's definitions.

Chapter 24

Weather

Sections

1 Air Masses
2 Fronts
3 Weather Instruments
4 Forecasting the Weather

What You'll Learn

- How air masses affect weather
- How fronts lead to severe weather
- How scientists forecast weather

Why It's Relevant

Weather affects many different aspects of our lives every day. Studying the weather helps scientists make forecasts and warn people of dangerous weather.

PRE-READING ACTIVITY

FOLDNOTES

Key Term Fold Before you read this chapter, create the FoldNote entitled "Key-Term Fold" described in the Skills Handbook section of the Appendix. Write a key term from the chapter on each tab of the key-term fold. Under each tab, write the definition of the key term.

▶ During a thunderstorm, clouds may discharge a spark of electricity called lightning. This bolt of lightning struck near Sugar Loaf Mountain in Rio De Janeiro, Brazil.

Chapter Correlations

National Science Education Standards

SAI 2a Scientists usually inquire about how physical, living, or designed systems function. **(Sections 1–4)**

SPSP 1a Hazards and the potential for accidents exist. Regardless of the environment, the possibility of injury …or death may be present. Humans have a variety of mechanisms… social, and technological—that can reduce and modify hazards. **(Section 4)**

HNS 2a Science distinguishes itself from other ways of knowing and from other bodies of knowledge through the use of empirical standards, logical arguments, and skepticism…**(Sections 1–4)**

UCP 1 Systems, order, and organization: The natural and designed world is complex; it is too large and complicated to investigate and comprehend all at once. **(Sections 3 and 4)**

UCP 2 Evidence, models, and explanation:…Using evidence to understand interactions allows individuals to predict changes in natural…systems. …Models are tentative schemes or structures that correspond to real objects, events, or classes of events, and that have explanatory power. Models help scientists and engineers understand how things work… Models take many forms, including… computer simulations. **(Sections 1–4)**

UCP 3 Constancy, change, and measurement: Although most things are in the process of becoming different—changing—some properties of objects and processes are characterized by constancy… interactions within and among systems result in change. **(Sections 1–4)**

Section 1 Air Masses

Differences in air pressure are caused by unequal heating of Earth's surface. The region along the equator receives more solar energy than the regions at the poles do. The heated equatorial air rises and creates a low-pressure belt. Conversely, cold air near the poles sinks and creates high-pressure centers. Differences in air pressure at different locations on Earth create wind patterns.

OBJECTIVES

- **Explain** how an air mass forms.
- **List** the four main types of air masses.
- **Describe** how air masses affect the weather of North America.

KEY TERM

air mass

How Air Moves

Air moves from areas of high pressure to areas of low pressure. Therefore, there is a general, worldwide movement of surface air from the poles toward the equator. At high altitudes, the warmed air flows from the equator toward the poles. Temperature and pressure differences on Earth's surface create three wind belts in the Northern Hemisphere and three wind belts in the Southern Hemisphere. The *Coriolis effect,* which occurs when winds are deflected by Earth's rotation, also influences wind patterns. The processes that affect air movement also influence storms, such as the one shown in **Figure 1.**

air mass a large body of air throughout which temperature and moisture content are similar

Formation of Air Masses

When air pressure differences are small, air remains relatively stationary. If the air remains stationary or moves slowly over a uniform region, the air takes on the characteristic temperature and humidity of that region. A large body of air throughout which temperature and moisture are similar is called an **air mass.** Air masses that form over frozen polar regions are very cold and dry. Air masses that form over tropical oceans are warm and moist.

Figure 1 ▶ The motion of Earth's atmosphere can lead to the formation of powerful storms such as Hurricane Katrina, photographed by satellite over the Gulf of Mexico in 2005.

Section 1

Focus

Overview

This section tells how air masses form, introduces the four main types of air masses, and explains how they affect the weather patterns of North America.

Bellringer

Ask students to compare what they know about characteristics of land air and ocean air, and characteristics of polar air and tropical air. (Land air is dry and ocean air is humid; polar air is cold and tropical air is warm.) LS **Verbal**

Motivate

Demonstration — GENERAL

Coriolis Effect Gather the class around a globe. Ask a student to draw a straight chalk line from the North Pole to the equator. Ask the class to imagine what would happen to the line if the globe were spinning. Then, have the volunteer repeat the experiment while you spin the globe slowly from west to east. What happens to the chalk line? (The line curves to the right) Earth's rotation causes winds to move in curved paths around the surface of Earth. These curving wind patterns create massive, twisting storms such as the one shown on this page. LS **Visual**

CHAPTER RESOURCES

Chapter Resource File

- Directed Reading BASIC

Technology

Transparencies
- Bellringer

Student Edition on CD-ROM

Student Edition on CD-ROM
- Lesson Plan

Teach

Demonstration — GENERAL

Comparing Air Masses To model the formation of air masses on a small scale, fill two dishpans with about 2.5 cm of water to represent oceans. Place wet-dry bulb hygrometers on trays to keep them dry. Put them inside the pans so the wet and dry temperatures (or humidity of the air) can be compared. Cover the pans with plastic wrap and label them as indicated. Put one in the sun (mT) and the other in the shade (mP). After 1/2 hour, measure the temperature and humidity of each air mass. Dry out the pans to represent continents, and repeat the set-up. Put one in the sun (cT) and the other in the shade (cP). Repeat the measurements. Ask students to explain how stable air masses form, and have them compare the characteristics of the different air masses. (Sample answer: When air remains over an area, it takes on the temperature and humidity of the region; mP is cool and moist; mT is warm and moist; cT is dry and warm and cP is dry and cool.) LS **Kinesthetic**

Activity — GENERAL

Air On the Move Have students identify the sources of air masses by performing the following steps: "On an outline map of North America, use a blue pencil to draw the outline of an air mass moving south over the northern U.S. from Canada. Label it *Air mass 1*. Use a red pencil to draw the outline of a second air mass moving northeast from the south Pacific. Label it *Air mass 2*. Draw arrows to show the direction of motion." Ask students to describe the characteristics of each air mass shown. (Sample answer: Air mass 1 is continental polar, dry and cold; Air mass 2 is maritime tropical, warm and moist). LS **Visual**

Table 1 ▼

Air Masses

Source region	Type of air	Symbol
Continental	dry	c
Maritime	moist	m
Tropical	warm	T
Polar	cold	P

Types of Air Masses

Air masses are classified according to their source regions. The source regions also determine the temperature and the humidity of the air mass. The source regions for cold air masses are polar areas. The source regions for warm air masses are tropical areas. Air masses that form over the ocean are called *maritime*. Air masses that form over land are called *continental*. Maritime air masses are moist, and continental air masses are dry. Air masses and the symbols used to designate them are listed in **Table 1.** The combination of tropical or polar air and continental or maritime air results in air masses that have distinct characteristics.

Continental Air Masses

Continental air masses form over large landmasses, such as northern Canada, northern Asia, or the southwestern United States. Because these air masses form over land, the level of humidity is very low. An air mass may remain over its source region for days or weeks. However, the air mass will eventually move into other regions because of global wind patterns. In general, continental air masses bring dry weather conditions when they move into another region. There are two types of continental air masses: *continental polar* (cP) and *continental tropical* (cT). Continental polar air masses are cold and dry. Continental tropical air masses are warm and dry.

Maritime Air Masses

Maritime air masses form over oceans or other large bodies of water. These air masses take on the characteristics of the water over which they form. The humidity in these air masses tends to be higher than that of the continental air masses. When these very moist masses of air travel to a new location, they commonly bring more precipitation and fog, as shown in **Figure 2.**

The two different maritime air masses are *maritime polar* (mP) and *maritime tropical* (mT). Maritime polar air masses are moist and cold. Maritime tropical air masses are moist and warm.

Figure 2 ▶ A maritime air mass brings fog that rolls in off the coast of California.

READING SKILL BUILDER — BASIC

Reading Hint The adjectives *polar,* "related to the poles" and *continental,* "related to continents," can be used to clearly indicate the source region of an air mass. The adjective *maritime* provides a similar hint. Maritime comes from the Latin word *mare* meaning "a sea or ocean," plus the adjective suffix form *–timus*. Thus, maritime air masses form over large bodies of water. The word *mare* gives us the word *marine,* which means, "of or pertaining to the sea, navigation, or sailing." LS **Verbal** English Language Learners

INCLUSION Strategies

- **Behavior Control Issues**
- **Hearing Impaired**
- **Attention Deficit Disorder**

Use the following activity to encourage some student movement. On the board, write numbered sentences with fill-in blanks related to section content. Assign numbers to specific students. Ask the students to read to find the answers for their assigned sentences, and then to go to the board and fill in the blanks. LS **Kinesthetic/ Verbal**

North American Air Masses

The four types of air masses that affect the weather of North America come from six regions. These air masses, their source locations, their movements, and the weather they bring are summarized in **Table 2.** The general directions of the air masses' movements are shown in **Figure 3.** An air mass usually brings the weather of its source region, but an air mass may change as it moves away from its source region. For example, cold, dry air may become warmer and more moist as it moves from land to the warm ocean. As the lower layers of air are warmed, the air rises. This warmed air may then create clouds and precipitation.

Table 2 ▼

Air Masses of North America

Air mass	Source location	Movement	Weather
cP	polar regions in Canada	south-southeast	cold and dry
mP	polar Pacific; polar Atlantic	southeast; southwest-south	cold and moist
cT	U.S. southwest	north-northeast	warm and dry
mT	tropical Pacific; tropical Atlantic	northeast; north-northwest	warm and moist

Tropical Air Masses

Continental tropical air masses form over the deserts of the southwestern United States. These air masses bring dry, hot weather in the summer. They do not form in the winter. Maritime tropical air masses form over the warm water of the tropical Atlantic Ocean. They bring mild, often cloudy weather to the eastern United States in the winter. In the summer, they bring hot, humid weather and thunderstorms. Maritime tropical air masses also form over warm areas of the Pacific Ocean. But these air masses do not usually reach the Pacific coast. In the winter, maritime tropical air masses bring moderate precipitation to the coast and the southwestern deserts.

Reading Check Which air mass brings dry, hot weather in the summer? (See the Appendix for answers to Reading Checks.)

SCILINKS. NSTA Developed and maintained by the National Science Teachers Association

For a variety of links related to this subject, go to www.scilinks.org

Topic: Air Masses

SciLinks code: HQ60031

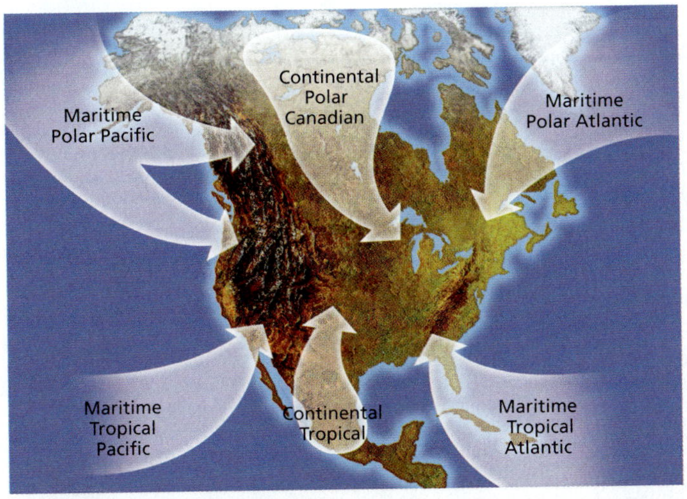

Figure 3 ▶ The four types of air masses that influence the weather in North America come from six regions and are named according to their source regions.

Answer to Reading Check

a continental tropical air mass

Close

Reteaching — BASIC

Spider Maps Have students create a spider map around the concept of air masses. Have them create a leg for each type of air mass and fill in the branches with details about its source and direction of movement and about the type of weather associated with each air mass. Students can use these graphic organizers to study for assessments. LS **Visual**

Quiz — GENERAL

1. What determines the distinct characteristics of an air mass? (the region over which it forms)
2. What type of air mass forms over the Gulf of Mexico and the warm waters of the Atlantic Ocean in summer? What kind of weather does it bring when it travels northward? (maritime tropical; hot, humid weather and thunderstorms)
3. What are the properties of a continental polar air mass? Identify a place where this air mass is likely to form. (cold and dry; over northern Canada)
4. What type of air mass develops over the North Pacific Ocean? How might the character of this air mass change as it moves inland over the central United States? (maritime polar; It might lose some of its moisture and warm slightly.)

Activity — GENERAL

Bulletin Board Project Have interested students create a weather bulletin board using the information on North American air masses and their associated weather phenomena. They can place a large map in the center and use colored yarn to connect the source region of each air mass to poems or drawings that describe the weather it brings. As students learn more, they can add additional information about weather fronts and storm systems, weather instruments, and weather forecasting to the bulletin board. LS **Visual**

CHAPTER RESOURCES

Technology

Transparencies
- 117 Air Masses (with worksheet)

Figure 4 ▶ Maritime polar Atlantic air masses can bring heavy snowfall, such as in this snowstorm that hit New York City in 2003.

Polar Air Masses

Polar air masses from three regions—northern Canada and the northern Pacific and Atlantic Oceans—influence weather in North America. Continental polar air masses form over ice- and snow-covered land. These air masses move into the northern United States and can occasionally reach as far south as the Gulf Coast of the United States. In summer, the air masses usually bring cool, dry weather. In winter, they bring very cold weather to the northern United States.

Maritime polar air masses form over the North Pacific Ocean and are very moist, but they are not as cold as continental polar Canadian air masses. In winter, these maritime polar Pacific air masses bring rain and snow to the Pacific Coast. In summer, they bring cool, often foggy weather. As they move inland and eastward over the Cascades, the Sierra Nevada, and the Rocky Mountains, these cold air masses lose much of their moisture and warm slightly. Thus, they may bring cool and dry weather by the time they reach the central United States.

Maritime polar Atlantic air masses move generally eastward toward Europe. But they sometimes move westward over New England and eastern Canada. In winter, they can bring cold, cloudy weather and snow, as shown in **Figure 4.** In summer, these air masses can produce cool weather, low clouds, and fog.

Section 1 Review

1. **Define** *air mass.*
2. **Explain** how an air mass forms.
3. **Identify** the location where a cold, dry air mass would form.
4. **List** the four main types of air masses.
5. **Describe** how the four main types of air masses affect the weather of North America.
6. **Describe** the air mass that forms over the warm waters of the Atlantic Ocean. What letters designate the source region of this air mass?

CRITICAL THINKING

7. **Making Predictions** How would temperature and humidity at a given location change when a maritime tropical air mass is replaced by a continental polar air mass?
8. **Recognizing Relationships** In which direction would you expect a tropical air mass near the coast of Europe to travel? Explain your answer.

CONCEPT MAPPING

9. Use the following terms to create a concept map: *maritime polar Pacific, maritime polar, continental polar Canadian, air mass, continental polar,* and *maritime polar Atlantic.*

Close, *continued*

Alternative Assessment — ADVANCED

Electric Quiz Game Invite groups of students to create a homemade, electric match game about the characteristics of air masses. Students may use a string of tree lights, tag board, 2 C batteries in a battery holder, paper clips, a hole punch, strips of aluminum foil, and electrical tape. Students can either enlarge the map of North America or use a question and answer format so that a bulb will light up when a paper clip switch touches items that match. Tell them to mix up the questions and answers so that they are not lined up. By fitting a new panel of tag board with new questions over the circuit, the game can be reused with other topics. **LS Kinesthetic**

Answers to Section Review

1. a large body of air that has the same temperature and moisture content throughout
2. When differences in air pressure are small, air remains stationary or moves very slowly and takes on the characteristics of the region.
3. over land in a polar region
4. continental polar, continental tropical, maritime polar, and maritime tropical
5. continental polar: cool, dry weather in summer and cold weather in the north in winter; maritime polar: rain and snow along the Pacific coast in winter and cool, foggy weather in summer; continental tropical: hot, dry weather; Atlantic maritime tropical: hot, humid weather or thunderstorms in summer and mild, cloudy weather in winter; Pacific maritime tropical: moderate precipitation
6. The maritime tropical air mass that forms over the Atlantic is warm and moist. The symbol is mT.
7. Warm, moist air of a maritime tropical air mass would be replaced with cool, dry air.
8. Answers may vary. Sample answer: A tropical air mass near the coast of Europe would have originated in the South Atlantic Ocean and moved north and east. I would expect the air mass to continue in the same direction, moving across continental Europe. The tendency of colder air from the North Pole to move into the low-pressure area and push the warm air south would be offset by the Coriolis effect.
9. *Air masses* may be polar which includes *maritime polar,* such as *maritime polar Pacific* and *maritime polar Atlantic,* or *continental polar,* which includes *continental polar Canadian.*

CHAPTER RESOURCES

Chapter Resource File
- Section Quiz GENERAL

Workbooks
- Study Guide (also in Spanish)

Section 2 Fronts

When two unlike air masses meet, density differences usually keep the air masses separate. A cool air mass is dense and does not mix with the less-dense air of a warm air mass. Thus, a boundary, called a *front,* forms between air masses. A typical front is several hundred kilometers long. However, some fronts may be several thousand kilometers long. Changes in middle-latitude weather usually take place along the various types of fronts. Fronts do not exist in the Tropics because no air masses that have significant temperature differences exist there.

OBJECTIVES

- **Compare** the characteristic weather patterns of cold fronts with those of warm fronts.
- **Describe** how a midlatitude cyclone forms.
- **Describe** the development of hurricanes, thunderstorms, and tornadoes.

KEY TERMS

- cold front
- warm front
- stationary front
- occluded front
- midlatitude cyclone
- thunderstorm
- hurricane
- tornado

Types of Fronts

For a front to form, one air mass must collide with another air mass. The kind of front that forms is determined by how the air masses move in relationship to each other.

Cold Fronts

When a cold air mass overtakes a warm air mass, a **cold front** forms. The moving cold air lifts the warm air. If the warm air is moist, clouds will form. Large cumulus and cumulonimbus clouds typically form along fast-moving cold fronts. Storms that form along a cold front are usually short-lived and are sometimes violent. A long line of heavy thunderstorms, called a *squall line,* shown in **Figure 1,** may occur in the warm, moist air just ahead of a fast-moving cold front. A slow-moving cold front lifts the warm air ahead of it more slowly than a fast-moving front does. A slow-moving cold front typically produces weaker storms and lighter precipitation than a fast-moving cold front does.

cold front the front edge of a moving mass of cold air that pushes beneath a warmer air mass like a wedge

Figure 1 ▶ As a cold air mass overtakes a warm air mass, a line of thunderstorms called a *squall line* forms. The photo shows a squall line over the North Atlantic Ocean.

Section 2

Focus

Overview

This section explains what happens when air masses collide and describes different kinds of fronts. The section also describes cyclones and anticyclones and violent weather systems.

Bellringer

Provide the following list: blizzard, thunderstorm, drought, hurricanes, thunderstorm, and tornado. Have students select a weather condition from the list and write a paragraph that describes how it affects people. LS **Verbal/Interpersonal**

Motivate

Demonstration — GENERAL

Make a Cold Front You can demonstrate the effects of meeting air masses by using liquids. Fill an empty jar two-thirds full with warm water. Then, pour cold milk gently down the side of the jar. Ask students to describe what happens. (The cold milk stays at the bottom and the warm water remains on top. It takes some time before the liquids start to mix.) Have them repeat the experiment using cold milk and cold tap water. Ask what happens now. (The liquids should mix fairly quickly.) LS **Visual**

CHAPTER RESOURCES

Chapter Resource File
- Directed Reading BASIC

Technology

Transparencies
- Bellringer
- 118 Types of Fronts (with worksheet)

Student Edition on CD-ROM

One-Stop Planner CD-ROM
- Lesson Plan

Teach

Group Activity — GENERAL

Weather Front Pop-Ups Have students work in small groups to create booklets or cards using pop-up folds and 3-D effects to show what happens when cold, warm, stationary, or occluded fronts form. Provide stiff cards, scissors, glue, and reference materials with basic instructions for paper-engineering projects. You may want to show students samples of pop-up books or cards. Tell students to write captions to explain what is happening in their paper models. After they construct their models, students can display them in the classroom. Discuss the weather patterns associated with each front. LS Kinesthetic

READING SKILL BUILDER — BASIC

Paired Summarizing Group students into pairs, and have them read silently about cold, warm, stationary, and occluded fronts. Then, have one student summarize the definitions and effects of these fronts, while the other student listens to the retelling and points out any inaccuracies or ideas that were left out. Allow students to refer to the text as needed.
LS Verbal/Auditory English Language Learners
Co-op Learning

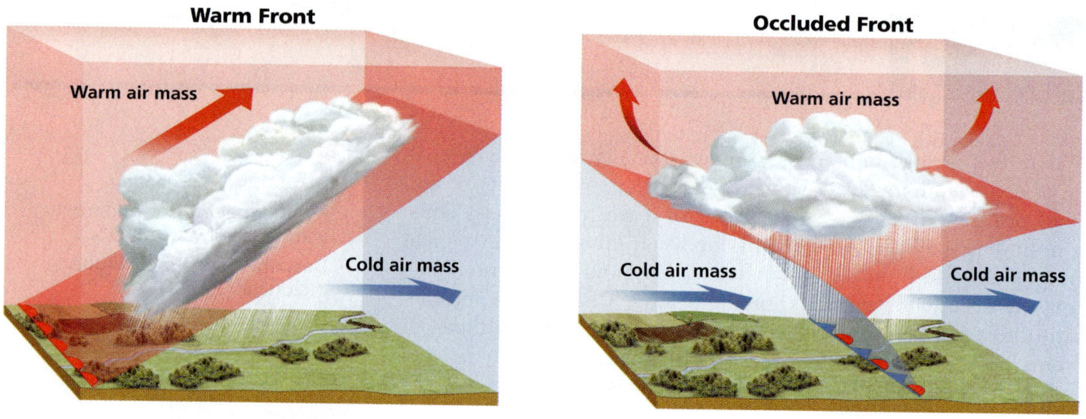

Figure 2 ▶ As a warm air mass rises over a cold air mass (left), a warm front forms at the boundary of the two air masses. An occluded front (right) forms when a cold air mass lifts a warm air mass off the ground.

warm front the front edge of advancing warm air mass that replaces colder air with warmer air

stationary front a front of air masses that moves either very slowly or not at all

occluded front a front that forms when a cold air mass overtakes a warm air mass and lifts the warm air mass off the ground and over another air mass

midlatitude cyclone an area of low pressure that is characterized by rotating wind that moves toward the rising air of the central low-pressure region

Warm Fronts

When a cold air mass retreats from an area, a **warm front** forms. The less dense warm air rises over the cooler air. The slope of a warm front is gradual, as shown in **Figure 2.** Because of this gentle slope, clouds may extend far ahead of the surface location, or *base,* of the front. A warm front generally produces precipitation over a large area and may cause violent weather.

Stationary and Occluded Fronts

Sometimes, when two air masses meet, the cold air moves parallel to the front, and neither air mass is displaced. A front at which air masses move either very slowly or not at all is called a **stationary front.** The weather around a stationary front is similar to that produced by a warm front. An **occluded front** usually forms when a fast-moving cold front overtakes a warm front and lifts the warm air off the ground completely, as shown in **Figure 2.**

Polar Fronts and Midlatitudes Cyclones

Over each of Earth's polar regions is a dome of cold air that may extend as far as 60° latitude. The boundary where this cold polar air meets the tropical air mass of the middle latitudes, especially over the ocean, is called the *polar front.* Waves commonly develop along the polar front. A *wave* is a bend that forms in a cold front or a stationary front. This wave is similar to the waves that moving air produces when it passes over a body of water. However, waves that form in a cold front or stationary front are much larger. They are the beginnings of low-pressure storm centers called midlatitude cyclones or *wave cyclones.* **Midlatitude cyclones** are areas of low pressure that are characterized by rotating wind that moves toward the rising air of the central, low-pressure region. These cyclones strongly influence weather patterns in the middle latitudes.

HISTORY CONNECTION — ADVANCED

Pioneer Meteorologist The Norwegian scientist Vilhelm Bjerknes was one of the founders of the field of meteorology and weather forecasting. A professor at Stockholm University, he studied the circulation of the atmosphere and the oceans. Together with his son Jacob, Bjerknes developed the theory of air masses and fronts. Encourage students to learn more about the contributions of this family of meteorologists to the science of predicting the weather. Invite them to share their discoveries through oral and written reports. LS Verbal

Stages of a Midlatitude Cyclone

A midlatitude cyclone usually lasts several days. The stages of formation and dissipation of a midlatitude cyclone are shown in **Figure 3.** In North America, midlatitude cyclones generally travel about 45 km/h in an easterly direction as they spin counterclockwise. They follow several storm tracks, or routes, as they move from the Pacific coast to the Atlantic coast. As they pass over the western mountains, they may lose their moisture and energy.

Anticyclones

Unlike the air in a midlatitude cyclone, the air of an *anticyclone* sinks and flows outward from a center of high pressure. Because of the Coriolis effect, the circulation of air around an anticyclone is clockwise in the Northern Hemisphere. Anticyclones bring dry weather, because their sinking air does not promote cloud formation. If an anticyclone stagnates over a region for a few days, the anticyclone may cause air pollution problems. After being stationary for a few weeks, anticyclones may cause droughts.

Reading Check **How is the air of an anticyclone different from that of a midlatitude cyclone? (See the Appendix for answers to Reading Checks.)**

SCILINKS. NSTA Developed and maintained by the National Science Teachers Association

For a variety of links related to this subject, go to www.scilinks.org

Topic: Fronts and Severe Weather
SciLinks code: HQ60624

Figure 3 ▶ **Stages of a Midlatitude Cyclone**

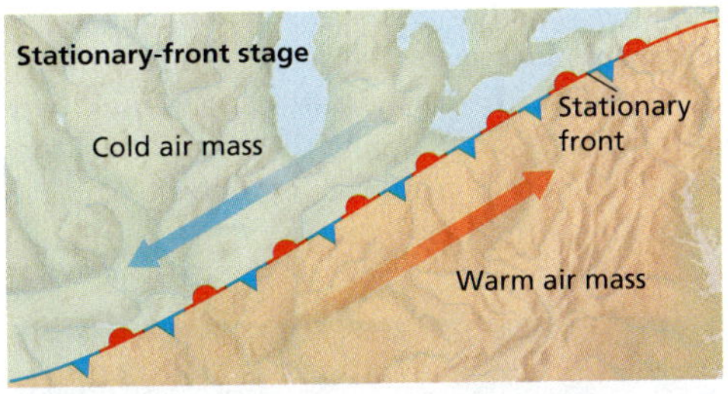

1 Midlatitude cyclones occur along a cold or a stationary front. Winds move parallel to the front but in opposite directions on each side of the front.

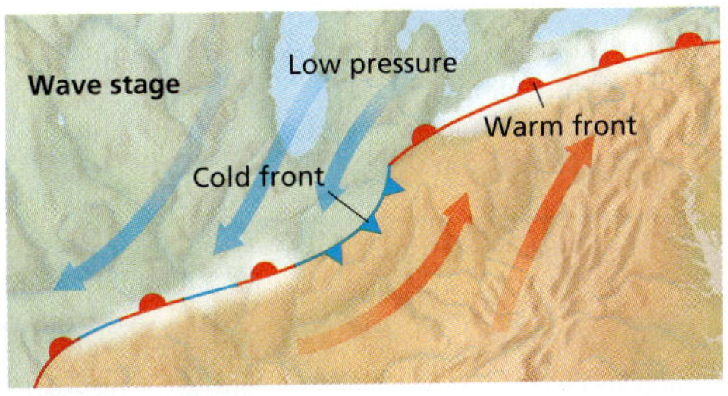

2 A wave forms when a bulge of cold air develops and advances slightly ahead of the rest of the front.

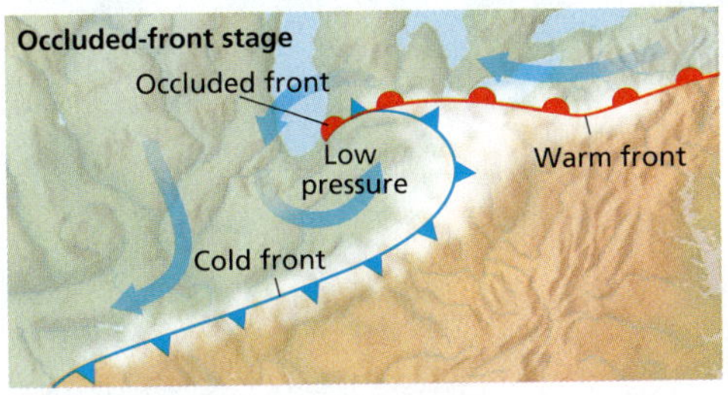

3 As the fast-moving part of the cold front overtakes the warm front, an occluded front forms and the storm reaches its highest intensity.

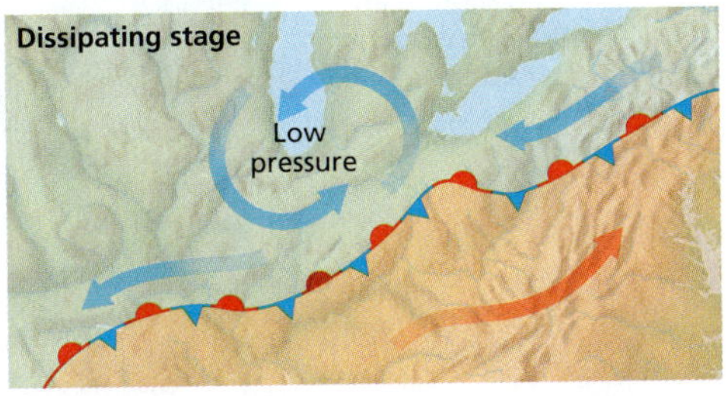

4 Eventually, the system generally loses all of its energy and the midlatitude cyclone dissipates.

Demonstration — GENERAL

Spiraling Winds Set a pan of tap water over a hot plate. Heat the water until it steams gently. While the water heats, draw a circle 7.5 cm in diameter on a sheet of printer paper. Cut out the circle to form a spiral shape. Tape a 30 cm length of thread to the center of the paper spiral. Hold the free end of the thread so that the paper spiral hangs about 5 cm above the pan of hot water. Invite students to describe what happens to the paper spiral. Ask them what causes the paper to move and where it gets its energy. (The paper model swirls around with a circular motion. Rising air currents strike the paper model, causing it to turn. The energy comes from the heat of the water.) **LS Visual**

Discussion — GENERAL

Highs and Lows Remind students that cold air masses are normally associated with high pressure because cold air is denser. Thus, the air pressure is higher. In contrast, warmer air masses are associated with low-pressure areas. Point out that cyclones form around central low-pressure areas, whereas anticyclones sink around high-pressure areas. Ask students how these facts can help them remember the characteristics of cyclones and anticyclones. (Sample answers: With cyclones, air flows counterclockwise into the low-pressure zone. As this air rises, it cools. The moisture in the air condenses to form clouds and often violent storms. In an anticyclone, the air flows in the opposite direction, clockwise (in the Northern Hemisphere). As the air sinks around the high-pressure area, it warms, which inhibits the formation of clouds. This is why highs bring fair weather and lows tend to bring storms.) **LS Verbal/Logical**

SKILL BUILDER — BASIC

Vocabulary Explain to students that the word *cyclone* comes from the Greek word *kyklos,* which means "a circle." Winds form a circle around a low-pressure area. This root word was combined with the prefix *anti-* meaning "the opposite of," to describe the wind pattern that forms around a high-pressure area. Invite students to name other science words that use this prefix. (Sample answers: antibiotic, antibody, antidote, antigravity, and antigen) English Language Learners **LS Verbal**

Answer to Reading Check

The air of an anticyclone sinks and flows outward from a center of high pressure. The air of a midlatitude cyclone rotates toward the rising air of a central low-pressure region.

CHAPTER RESOURCES

Technology

Transparencies
- 119 Stages of a Midlatitude Cyclone (with worksheet)

Teach, *continued*

Demonstration — GENERAL

Updrafts To help students visualize the lifting power of winds within severe storms, invite a student to balance a table-tennis ball in the airstream produced by an up-turned hair dryer. With a little practice, the ball will be supported. Now, have the volunteer try the experiment with a tennis ball. Compare the tennis ball to a growing raindrop or hailstone. **LS Visual**

MATH CONNECTION — GENERAL

Coming or Going? Explain to students that they can use the difference in speed between thunder and lightning to determine not only approximately how far the thunderstorm is from them, but also whether it is moving toward them or away. Suggest that they try this activity the next time they are indoors during a storm. Have them estimate the storm's distance for several successive lightning flashes and thunder crashes by dividing the lapse time (in seconds) by 3 s/km to obtain the distance in kilometers. If the distance increases, the storm is moving away. If the distance decreases, the storm is approaching their location. **LS Logical**

MATHPRACTICE

Answer

thunderstorm distance = 27 s ÷ 3 s/km = 9 km

HISTORY CONNECTION

A Record-Setting Season The hurricane season of 2005 was the worst on record for the United States. More hurricanes formed than in any other year (15). Four, Emily, Katrina, Rita, and Wilma, were Category 5 storms with winds in excess of 155 mph. Katrina, the costliest hurricane ever to hit the United States, did close to $100 billion in damage and killed about 1,800 people.

Severe Weather

Severe weather is weather that may cause property damage or loss of life. Severe weather may include large quantities of rain, lightning, hail, strong winds, or tornadoes. This type of weather causes billions of dollars in damage each year.

thunderstorm a usually brief, heavy storm that consists of rain, strong winds, lightning, and thunder

Thunderstorms

A heavy storm that is accompanied by rain, thunder, lightning, and strong winds is called a **thunderstorm.** Thunderstorms develop in three distinct stages. In the first stage, or *cumulus stage,* warm, moist air rises, and the water vapor within the air condenses to form a cumulus cloud. In the next stage, called the *mature stage,* condensation continues as the cloud rises and becomes a dark cumulonimbus cloud. Heavy, torrential rain and hailstones may fall from the cloud. While strong updrafts continue to rise, downdrafts form as air is dragged downward by the falling precipitation. During the final stage, or *dissipating stage,* the strong downdrafts stop air currents from rising. The thunderstorm dissipates as the supply of water vapor decreases.

MATHPRACTICE

Thunderstorm Distance The time between when a person sees a lightning strike and when he or she hears thunder indicates how far away the lightning bolt was from that person. Sound travels approximately 1 km in 3 s. The lapse time in seconds divided by 3 is roughly the number of kilometers between the viewer and the lightning. If 27 seconds pass between a flash of lightning and the sound of thunder, how far away was the lightning strike from the viewer?

Lightning

During a thunderstorm, clouds discharge electricity in the form of *lightning.* The released electricity heats the air, and the air expands rapidly and produces the loud noise known as *thunder.* For lightning to occur, the clouds must have areas that carry distinct electrical charges. The upper part of the cloud usually carries a positive charge, while the lower part carries mainly a negative charge. Lightning is a huge spark that travels within the cloud or between the cloud and ground to equalize electrical charges. **Figure 4** shows an example of lightning.

Figure 4 ▶ The average lightning flash lasts only about a quarter of a second, but lightning causes more than $330,000,000 in damage per year in the United States.

PHYSICS CONNECTION

Lightning Discharge Air currents cause friction between the air, water molecules, and ice particles within clouds. This creates static electricity. Positively charged particles build up at the top of cloud layers, and heavier, negatively charged particles collect at the bottom. A charge separation occurs on Earth as well. When the difference between the charges becomes strong enough to overcome the insulating properties of the air, electrons jump the gap. A giant spark is produced as this difference in charge is transformed into electrical energy in the cloud and then into light, sound, and heat energy. Lightning can flash in three ways: within clouds, between clouds, and between a cloud and the ground. Only about 20% of lightning strikes are cloud to ground. Lightning rods are tall pointed conductors attached to the tops of buildings and grounded to Earth. Because they are made of materials that conduct electricity, lightning rods provide a safe, direct path for the static electricity to travel harmlessly to Earth.

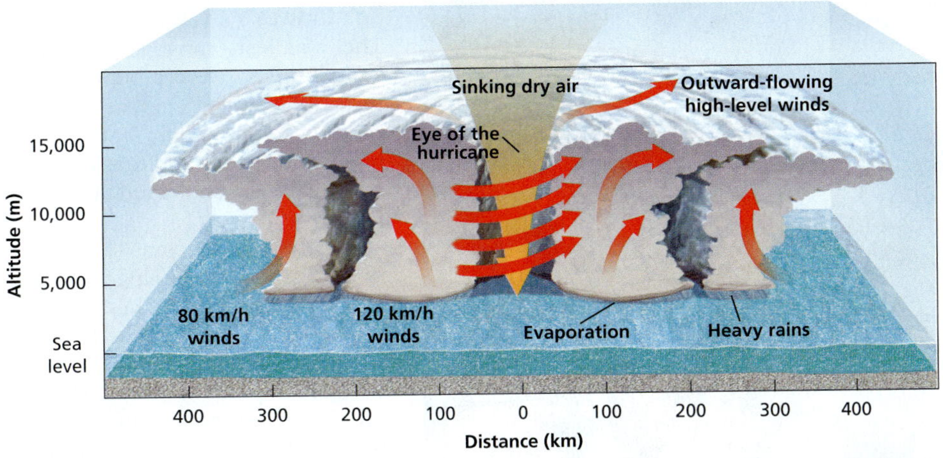

Figure 5 ▶ Although hurricanes are the most destructive storms, the eye at the center of the hurricane is relatively calm.

Hurricanes

Tropical storms differ from midlatitude cyclones in several ways. Tropical storms are concentrated over a small area. They lack warm and cold fronts. Also, they are usually much more violent and destructive than midlatitude cyclones. A tropical storm that has strong wind speeds of more than 120 km/h that spiral in toward its intense low pressure center is called a **hurricane.**

Hurricanes develop over warm, tropical oceans. A hurricane begins when warm, moist air over the ocean rises rapidly. When moisture in the rising warm air condenses, a large amount of energy in the form of latent heat is released. *Latent heat* is heat energy that is absorbed or released during a phase change. This heat increases the force of the rising air.

A fully developed hurricane consists of a series of thick cumulonimbus cloud bands that spiral upward around the center of the storm, as shown in **Figure 5.** Winds increase toward the center, or eye, of the storm and reach speeds of up to 275 km/h along the eyewall. The eye itself, however, is a region of calm, clear, sinking air.

At about 700 km in diameter, hurricanes are the most destructive storms that occur on Earth. The most dangerous aspect of a hurricane is a rising sea level and large waves, called a *storm surge*. A storm surge can submerge vast low-lying coastal areas. This flooding is the reason why most deaths during hurricanes are caused by drowning.

Every hurricane is categorized on the *Saffir-Simpson scale* by using several factors. These factors include central pressure, wind speed, and storm surge. The Saffir-Simpson scale has five categories. Category 1 storms cause the least damage. Category 5 storms can result in catastrophic damage.

hurricane a severe storm that develops over tropical oceans and whose strong winds of more then 120 km/h spiral in toward the intensely low-pressure storm center

Graphic Organizer **Venn Diagram**

Create the **Graphic Organizer** entitled "Venn Diagram" described in the Skills Handbook section of the Appendix. Label the circles with "Hurricanes," "Cyclones," and "Anticyclones." Then, fill in the diagram with characteristics that each weather event shares with the other weather events.

Reading Check Where do hurricanes develop? (See the Appendix for answers to Reading Checks.)

Answer to Reading Check

over warm tropical seas

CHAPTER RESOURCES

Technology

Transparencies
- 120 Anatomy of a Hurricane (with worksheet)

Graphic Organizer GENERAL

Venn Diagram

You may want to have students work in groups to create this Venn diagram. Have one student draw the map and fill in information provided by other students from the group.

Close

Reteaching — BASIC

Game Show Have groups of students brainstorm a series of 4 or 5 clues or phrases that describe each of the weather patterns associated with warm or cold fronts, cyclones or anticyclones, and severe weather systems such as thunderstorms, hurricanes, or tornados. Have them write their clues on cards—one card for each weather pattern. Shuffle the cards and have teams try to guess the weather pattern from as few clues as possible. LS **Verbal**

Quiz — GENERAL

1. What is a front? (the boundary between two colliding air masses)
2. What is the Saffir-Simpson scale? (a scale used to classify the destructiveness of hurricanes based on pressure, wind speed, and storm surge)
3. Compare the characteristics of a midlatitude cyclone with those of an anticyclone. (cyclone: Air moves counterclockwise toward the low-pressure center, often bringing stormy weather. anticyclone: Air moves outward from the high-pressure center, often bringing fair weather.)

Alternative Assessment — GENERAL

Radio Play Divide the class into small groups. Have groups develop a short radio script that describes the development of a complex weather system such as a midlatitude cyclone, a thunderstorm, a hurricane, or a tornado. Tell them to include a description of each stage in the process. When they have completed writing the scripts, have each group perform its play for the class or create an audio tape recording of the performance. They can use simple props to create sound effects. LS **Auditory**

Tornadoes

The smallest, most violent, and shortest-lived severe storm is a tornado. A **tornado** is a destructive, rotating column of air that has very high wind speeds and that is visible as a funnel-shaped cloud, as shown in **Figure 6.**

A tornado forms when a thunderstorm meets high-altitude, horizontal winds. These winds cause the rising air in the thunderstorm to rotate. A storm cloud may develop a narrow, funnel-shaped, rapidly spinning extension that reaches downward and may or may not touch the ground. If the funnel does touch the ground, it generally moves in a wandering, haphazard path. Frequently, the funnel rises and touches down again a short distance away. Tornadoes generally cover paths not more than 100 m wide. Usually, everything in that path is destroyed. Tornadoes occur in many locations, but they are most common in *Tornado Alley* in the late spring or early summer. Tornado Alley stretches from Texas up through the midwestern United States.

The destructive power of a tornado is due to mainly the speed of the winds in the funnel. These winds may reach speeds of more than 400 km/h. Most injuries and deaths caused by tornadoes occur when people are trapped in collapsing buildings or are struck by objects blown by the wind.

Figure 6 ▶ A powerful tornado in Texas embedded this bucket in a wooden door (inset).

tornado a destructive, rotating column of air that has very high wind speeds and that maybe visible as a funnel-shaped cloud

Section 2 Review

1. **Describe** the four main types of fronts.
2. **Compare** the characteristic weather patterns of cold fronts with those of warm fronts.
3. **Identify** the type of front that may form a squall line.
4. **Summarize** how a midlatitude cyclone forms.
5. **Describe** the stages in the development of thunderstorms.
6. **Describe** the stages in the development of hurricanes.
7. **Explain** why tornadoes are destructive.

CRITICAL THINKING

8. **Evaluating Methods** What areas of Earth should meteorologists monitor to detect developing hurricanes? Explain your answer.
9. **Making Comparisons** Compare the destructive power of midlatitude cyclones, hurricanes, and tornadoes in terms of size, wind speed, and duration.

CONCEPT MAPPING

10. Use the following terms to create a concept map: *tornado, hurricane, warm front, squall line, cold front, severe weather, stationary front, front, midlatitude cyclone,* and *occluded front.*

Close, *continued*

Answers to Section Review

1. *Cold fronts* form when a cold air mass pushes beneath a warm air mass. A *warm front* is the leading edge of an advancing warm air mass that replaces colder air. *Stationary fronts* move slowly or not at all. An *occluded front* occurs when cold air overtakes warm air, lifting the warm air over another cold air mass.
2. Storms along cold fronts are short-lived. A warm front generally produces precipitation over a wider area. Both may produce violent weather.
3. a fast-moving cold front
4. Midlatitude cyclones form along cold or stationary fronts. Winds move parallel to the front, but in opposite directions on each side. A wave of rotating air forms around the central low-pressure region as cold air advances ahead of the rest of the front. An occluded front forms as the cold front overtakes the warm front.
5. Warm, moist air rises and condenses, forming clouds. Condensation continues, and the clouds rise, changing into cumulonimbus clouds. Strong updrafts continue to rise and downdrafts form. Heavy rain and hailstones fall. Strong downdrafts stop air currents from rising, and so the storm's energy dissipates.
6. Warm, moist air rises rapidly over the oceans. Latent heat is released as moisture in the warm air condenses. This feeds development of cumulonimbus clouds and winds spiraling around a calm region.
7. because the funnel may touch the ground with winds that may reach speeds of more than 400 km/h
8. Scientists should monitor storms developing over warm tropical seas, where hurricanes start. The tropical air would flow north toward the continent from the ocean, and the Coriolis effect would deflect the air to the right, or east.
9. Midlatitude cyclones produce storms with winds speeds of 45 km/h. Hurricanes and tropical storms produce wind speeds of 120–275 km/h. They can be as large as 700 km in diameter. Tornados have narrow paths generally no more than 100 m wide, but wind speeds as high as 400 km/h. Increasing storm diameter means a greater area may be damaged. Higher winds increase the potential for damage. Rain may lead to flooding.
10. Weather, which includes *severe weather* such as *tornadoes, hurricanes,* and *midlatitude cyclones,* is caused by the movement of *fronts,* such as *warm fronts, stationary fronts, occluded fronts, and cold fronts* that may be preceded by *squall lines*.

CHAPTER RESOURCES

Chapter Resource File

- **Section Quiz** GENERAL

Workbooks

- **Study Guide** (also in Spanish)

Section 3 Weather Instruments

Weather observations are based on a variety of measurements, including atmospheric pressure, humidity, temperature, wind speed, and precipitation. These measurements are made with special instruments. Meteorologists then use the measurements to forecast weather patterns.

OBJECTIVES

- **Identify** four instruments that measure lower-atmospheric weather conditions.
- **Describe** how scientists measure conditions in the upper atmosphere.
- **Explain** how computers help scientists understand weather.

KEY TERMS

thermometer
barometer
anemometer
wind vane
radiosonde
radar

Measuring Lower-Atmospheric Conditions

During the course of a day, the lower-atmospheric conditions at a given location can change drastically. Meteorologists use the magnitude and speed of these changes to make predictions of future weather events. To obtain accurate data from the lower atmosphere, scientists use instruments such as those shown in **Figure 1.**

Air Temperature

An instrument that measures and indicates temperature is called a **thermometer.** A common type of thermometer uses a liquid—usually mercury or alcohol—sealed in a glass tube to indicate temperature. A rise in temperature causes the liquid to expand and fill more of the tube. A drop in temperature causes the liquid to contract and fill less of the tube. A scale marked on the glass tube indicates the temperature.

Another type of thermometer is an *electrical thermometer.* As the temperature rises, the electric current that flows through the material of the electrical thermometer increases and is translated into temperature readings. A *thermistor,* or thermal resistor, is a type of electrical thermometer that responds very quickly to temperature changes. For this reason, thermistors are extremely useful where temperature change occurs rapidly.

thermometer an instrument that measures and indicates temperature

SCILINKS® NSTA Developed and maintained by the National Science Teachers Association

For a variety of links related to this subject, go to www.scilinks.org

Topic: Weather Instruments
SciLinks code: HQ61646

Figure 1 ▶ Weather instruments, such as these at Elk Mountain weather research facility in Wyoming, indicate wind speed and direction.

Section 3

Focus

Overview

This section explains how meteorologists measure weather conditions.

Bellringer

Have students brainstorm a list of weather conditions that scientists monitor. (Sample answers: temperature, wind direction, and precipitation) **LS Visual**

Motivate

Group Activity — GENERAL

Simple Thermometer Have students work in pairs. Provide empty water bottles, see-through straws, rubbing alcohol, clay, and food color. Have students fill the bottles one quarter full with equal parts rubbing alcohol and water and then add a few drops of food color. Have them put the straw in the bottle and seal the neck with the clay so the straw stays in place. Have them cut two slits in a piece of paper and slip the paper over the straw and tape it in place. Have students mark the present level of the liquid in the straw, and then observe what happens to the level of the liquid. Ask them to explain why, as air temperature changes, the level rises or falls. (Sample answer: Heat makes the liquid expand and rise. A decrease in temperature makes the liquid contract and fall.) **LS Kinesthetic**

CHAPTER RESOURCES

Chapter Resource File

- **Directed Reading** BASIC
- **Inquiry Lab** Making a Weather Station GENERAL

Technology

Transparencies
- Bellringer

Student Edition on CD-ROM

One-Stop Planner CD-ROM
- Lesson Plan

Teach

SKILL BUILDER — GENERAL

Graphing Have students use a mercury or alcohol thermometer to take hourly air temperature readings from 9 A.M. to 6 P.M. You may want to assign this project as weekend homework. Have students make a simple line graph of their temperature data and identify temperature patterns. When is air temperature hottest? When is it coolest? (Temperatures generally peak in mid-afternoon and get lower toward evening.) LS **Logical**

Answer to Reading Check

A barometer is used to measure atmospheric pressure.

INCLUSION Strategies

- Gifted and Talented
- Behavior Control Issues

Encourage students who can benefit from additional challenges to track daily local atmospheric pressure, humidity, temperature, wind speed, and precipitation. If possible, record the measurements during class, or get readings from news media or the Internet. Also record the weather for each day—cloudy or sunny, rain or snow, or no precipitation. Then, have students create graphs that show the relationships of the measurements to the weather. LS **Logical**

CHAPTER RESOURCES

Chapter Resource File

- Datasheet for Quick Lab GENERAL

Figure 2 ▶ A meteorologist uses an anemometer during Hurricane Luis to measure wind speed.

barometer an instrument that measures atmospheric pressure

anemometer an instrument used to measure wind speed

wind vane an instrument used to determine direction of the wind

Air Pressure

Changes in air pressure affect air masses. The approach of a front is usually indicated by a drop in air pressure. Scientists use instruments called **barometers** to measure atmospheric pressure.

Wind Speed

An instrument called an **anemometer** (AN uh MAHM uht uhr) measures wind speed. A typical anemometer consists of small cups that are attached by spokes to a shaft that rotates freely. The wind pushes against the cups and causes them to rotate, as shown in **Figure 2.** This rotation triggers an electrical signal that registers the wind speed in meters per second or in miles per hour.

Wind Direction

The direction of the wind is determined by using an instrument called a **wind vane.** The wind vane is commonly an arrow-shaped device that turns freely on a pole as the tail catches the wind. Wind direction may be described by using one of 16 compass directions, such as north-northeast. Wind direction also may be recorded in degrees by moving clockwise and beginning with 0° at the north. Thus, east is 90°, south is 180°, and west is 270°.

Reading Check **Which instrument is used to measure air pressure?** **(See the Appendix for answers to Reading Checks.)**

QuickLAB 15 min

Wind Chill

Procedure

1. Place a **23 cm × 33 cm pan** on a level table. Fill the pan to a depth of 1 cm with **room temperature water.**
2. Lay a **thermometer** in the center of the pan with the bulb submerged. After 5 minutes, record the water temperature. Do not touch the thermometer.
3. Place an **electric fan** facing the pan and a few centimeters from the pan. Turn on the fan at a low speed. **CAUTION** Do not get the fan or cord wet.
4. Record the water temperature every minute until the temperature remains constant.

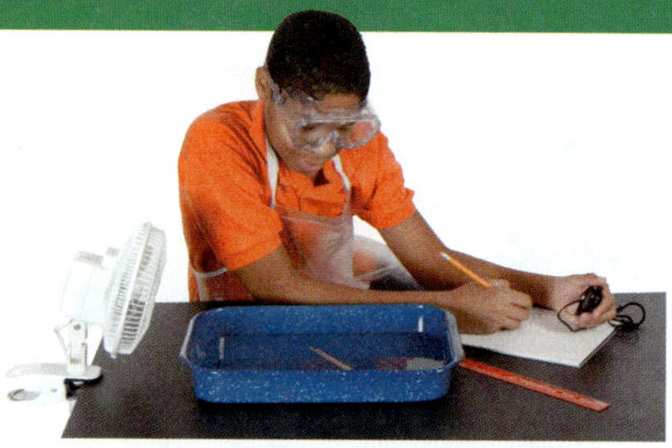

Analysis

1. How does the moving air affect the temperature of the water?
2. If the moving air is the same temperature as the still air in the room, what causes the water temperature to change?
3. How would you dress on a cool, windy day to stay comfortable? Explain your answer.

QuickLAB

Skills Acquired

- Experimenting
- Observing
- Analyzing

Teacher's Notes Small portable desk fans will work for this activity. If you are not able to obtain enough fans for a small group activity, present this lab as a demonstration. Assign student assistants to perform the demonstration for the class.

Answers

1. The temperature of the water decreases.
2. The wind from the fan increases the rate of evaporation of the water. Evaporation takes heat from the water, decreasing the temperature.
3. Answers may vary. Sample answer: minimize the effects of wind chill by exposing as little skin as possible.

Connection to TECHNOLOGY

Doppler Radar

Conventional radar helps scientists estimate the distance to a storm and the intensity of the storm by measuring how many radio waves return. To learn more about storms, meteorologists developed a different form of radar to study weather. This form of radar, called *Doppler radar,* can measure not only the distance to a storm and the storm's overall direction and speed but also the direction that rain droplets or ice particles inside the storm are moving. This information allows scientists to identify conditions inside the storm.

Doppler radar uses the Doppler effect to read the apparent shift in wavelength of reflected radio waves as the particles that the waves reflect from move. You may have experienced the Doppler effect when an ambulance sped by you. The sound of the siren appeared to change from high-pitched as the ambulance approached to low-pitched as it moved farther away. Because the wind causes the rain or ice particles to move around, the frequency of the radar signals they reflect also appear to shift. The Doppler radar measures that apparent frequency shift. Computer models then convert this information into an overall picture of the movement of the particles.

The use of Doppler radar allows meteorologists to identify dangerous conditions within a storm and to track the movement of these features. This ability allows meteorologists to warn communities of weather threats in time to save lives.

Lightning strikes near this Doppler radar facility in Oklahoma.

Measuring Upper-Atmospheric Conditions

Conditions of the atmosphere near Earth's surface are only a part of the complete weather picture. Scientists use several instruments to measure conditions in the upper atmosphere to obtain a better understanding of local and global weather patterns.

Radiosonde

An instrument package that is carried high into the atmosphere by a helium-filled weather balloon to measure relative humidity, air pressure, and air temperature is called a **radiosonde.** The radiosonde sends measurements as radio waves to a receiver that records the information. The path of the balloon is tracked to determine the direction and speed of high-altitude winds. When the balloon reaches a very high altitude, the balloon expands and bursts, and the radiosonde parachutes back to Earth.

radiosonde a package of instruments that is carried aloft by balloons to measure upper atmospheric conditions, including temperature, dew point, and wind velocity

radar **ra**dio **d**etection **a**nd **r**anging, a system that uses reflected radio waves to determine the velocity and location of objects

Radar

Another instrument for determining weather conditions in the atmosphere is radar. **Radar,** which stands for **ra**dio **d**etection **a**nd **r**anging, is a system that uses reflected radio waves to determine the velocity and location of objects. For example, large particles of water in the atmosphere reflect radar pulses. Thus, precipitation and storms, such as thunderstorms, tornadoes, and hurricanes, are visible on a radar screen. The newest Doppler radar can indicate the precise location, movement, and extent of a storm. It can also indicate the intensity of precipitation and wind patterns within a storm.

Close

Reteaching — BASIC

Tools of the Trade Have students work in pairs to create a poster that illustrates what instruments scientists use to measure weather variables and how each tool works. The poster should be divided into surface and upper atmosphere conditions. It should include a picture or sketch of each instrument and a diagram or verbal description of how each instrument is used. **LS Visual**

Quiz — GENERAL

1. How may temperature changes be used to predict weather? (Sample answer: A sharp drop in temperature may signal the arrival of a cold front, which may produce precipitation.)
2. What devices do scientists use to measure wind direction and speed? (Scientists use wind vanes to determine wind direction and anemometers to measure wind speed.)
3. What kinds of information do weather balloons help scientists collect? (air pressure, temperature, humidity at different altitudes, and the direction and speed of high-altitude winds)

CHAPTER RESOURCES

Chapter Resource File

Internet Activity
- Weather Images GENERAL

Internet Activity — GENERAL

Weather Images Two kinds of satellites monitor weather conditions from orbit. Geostationary satellites orbit the equator at speeds that match Earth's rotation. Polar orbiting satellites orbit the North and South Poles. Students may visit the National Weather Service Web site and report on an aspect that interests them. A worksheet designed to direct student research on this topic can be found in the **Chapter Resource File** booklet or by visiting **go.hrw.com** and entering the keyword **HQ6WTHX.** **LS Visual**

Close, *continued*

Alternative Assessment — GENERAL

Weather Riddles Have students create a booklet from construction paper in which they write and illustrate a set of riddles, each of which describes a different weather instrument. Have them exchange riddle books and try to solve each others' word puzzles. LS **Verbal/Visual**

Answers to Section Review

1. thermometer, barometer, anemometer, and weather vane
2. Upper-level atmospheric conditions affect local and global weather patterns.
3. Answers may vary but should accurately describe radiosondes, radar, and weather satellites.
4. Meteorologists use helium-filled balloons to carry instrument packages into the upper atmosphere. Their paths can be tracked using radio signals. At high altitudes, the balloon bursts and the package parachutes back to Earth.
5. Satellites provide images that cannot be obtained from the ground, such as wind speed and direction within clouds, temperatures at the tops of clouds, and observations of ocean currents and wave heights.
6. Computers can be used to model the behavior of weather conditions that require complex equations. They can also store weather records for rapid retrieval and help improve weather forecasts.
7. That is the direction from which developing weather systems will come.
8. Sample answers: Air pressure would go down at the higher elevation. Air temperature would be lower, and wind speed readings would increase. In the valley, the instruments were protected from upper-level wind patterns.
9. *Weather instruments* that measure conditions in the *lower atmosphere,* such as *thermometers, barometers,* and *anemometers;* and in the *upper atmosphere,* such as *radar, satellites,* and *radiosondes,* provide weather data to scientists.

Figure 3 ▶ This satellite image captured Hurricane Andrew in 1992 as it approached Louisiana.

Weather Satellites

Instruments carried by weather satellites also collect important information about the atmosphere. Satellite images, such as the one shown in **Figure 3,** provide weather information for regions where observations cannot be made from the ground.

The direction and speed of the wind at the level of the clouds can also be measured by examining a continuous sequence of cloud images. For night monitoring, satellite images made by using infrared energy reveal temperatures at the tops of clouds, at the surface of the land, and at the ocean surface. Satellite instruments can also measure marine conditions. For example, the instruments can measure the temperature and flow of ocean currents and the height of ocean waves.

Computers

Meteorologists also use supercomputers to understand the weather. Before computers were available, solving the mathematical equations that describe the behavior of the atmosphere was very difficult, and sometimes impossible. In addition to solving many of these equations, computers can store weather data from around the world. These data can provide information that is useful in forecasting weather changes. Computers can also store weather records for quick retrieval. In the future, powerful computers may greatly improve weather forecasts and provide a much better understanding of the atmosphere.

Section 3 Review

1. **Identify** four instruments scientists use to measure lower-atmospheric conditions.
2. **Explain** why scientists are interested in weather conditions in the upper atmosphere.
3. **Describe** the instruments used to measure conditions in the upper atmosphere.
4. **Explain** how meteorologists send weather instruments into the upper atmosphere.
5. **Summarize** how satellites help meteorologists study weather.
6. **Summarize** how computers help scientists study weather.

CRITICAL THINKING

7. **Recognizing Relationships** Wind is named according to the direction from which it blows. Why would a meteorologist need to know the direction wind is blowing from?
8. **Making Inferences** If weather instruments were moved from a location in a valley to the top of a hill, what changes would you expect in the data? Explain your answer.

CONCEPT MAPPING

9. Use the following terms to create a concept map: *thermometer, barometer, anemometer, radar, radiosonde, satellite, upper atmosphere, lower atmosphere,* and *weather instruments.*

CHAPTER RESOURCES

Chapter Resource File

- Section Quiz GENERAL

Workbooks

- Study Guide (also in Spanish)

Section 4 Forecasting the Weather

Predicting the weather has challenged people for thousands of years. People in many early civilizations attributed control of weather conditions, such as wind, rain, and thunder, to gods. Some people attempted to forecast the weather by using the position of the moon and stars as the basis for their predictions.

Scientific weather forecasting began with the invention of basic weather instruments, such as the thermometer and the barometer. The invention of the telegraph in 1844 enabled meteorologists to share information about weather conditions quickly and led to the creation of national weather services. For example, the United States formed a weather-forecasting agency called the Weather Bureau. In 1970, it was renamed the *National Weather Service*. Because weather events in the United States commonly originate beyond U.S. borders, the National Weather Service exchanges weather data with other nations around the world.

OBJECTIVES

- **Explain** how weather stations communicate weather data.
- **Explain** how a weather map is created.
- **Explain** how computer models help meteorologists forecast weather.
- **List** three types of weather that meteorologists have attempted to control.

KEY TERM

station model

Global Weather Monitoring

Weather observers at stations around the world report weather conditions frequently, sometimes hourly. They record the barometric pressure and how it has changed as well as the speed and direction of surface wind. They measure precipitation, temperature, and humidity. They note the type, amount, and height of cloud cover. Observers also record visibility and general weather conditions. Similar data are gathered continuously by automated observing systems. Each station in the system sends its data to a collection center. Weather centers around the world exchange the weather information they have collected.

The World Meteorological Organization (WMO) sponsors a program called *World Weather Watch* to promote the rapid exchange of weather information. The organization helps developing countries establish or improve their meteorological services, as shown in **Figure 1.** It also offers advice on the effect of weather on natural resources and on human activities, such as farming and transportation. WMO was founded in 1873 and is now part of the United Nations.

Figure 1 ▶ One major role of the World Meteorological Organization is to train professionals to use weather instruments, such as this Bobson spectrophotometer installed at Maun, Botswana.

Section 4

Focus

Overview

This section explains how global weather is monitored, how maps display data, and how computers are used. The section also provides weather safety tips.

Bellringer

Have students think about weather reports they have seen or heard. Have them write down the kinds of information they expect a complete weather forecast to include. (Sample answers: cloud coverage, temperature data, expected precipitation, wind speed, highs and lows, wind chill factor, and humidity levels) LS **Verbal**

Motivate

Discussion — GENERAL

Weather Lore Discuss with students stories and sayings from other cultures about the weather. For example, some sayings claim that the color of the evening and morning sky can be used to predict weather. Tell students that many myths and folk tales attempt to explain powerful natural forces. LS **Interpersonal**

CHAPTER RESOURCES

Chapter Resource File

- Directed Reading BASIC

Technology

Transparencies
- Bellringer

Student Edition on CD-ROM

One-Stop Planner CD-ROM
- Lesson Plan

Teach

Using the Figure — GENERAL

Classroom Station Model Invite interested students to use the chart of meteorological symbols to create a large Classroom Station Model as part of a Weather Bulletin Board. Have students make the model with symbols for cloud coverage, wind speed and direction, and weather conditions that they can move around on the board. Assign students to change the classroom station model as outside weather conditions change. LS **Kinesthetic**

Discussion — BASIC

Comprehension Check Direct students' attention to the temperature data contained in the station model on this page. Ask students how they could use the current temperature and the dew point to determine the likelihood of precipitation. (Sample answer: The closer the dew-point temperature is to the air temperature, the greater the likelihood of fog, rain, or snow.) LS **Verbal**

CHAPTER RESOURCES

Chapter Resource File

- **Making Models Lab** Blowing in the Wind GENERAL

Technology

Transparencies

- 121 Weather Symbols (with worksheet)

Weather Maps

The data that weather stations collect are transferred onto weather maps. Weather maps allow meteorologists to understand the current weather and to predict future weather events. To communicate weather data on a weather map, meteorologists use symbols and colors. These symbols and colors are understood and used by meteorologists around the world.

Weather Symbols

station model a pattern of meteorological symbols that represents the weather at a particular observing station and that is recorded on a weather map

On some weather maps, clusters of meteorological symbols show weather conditions at the locations of weather stations. Such a cluster of symbols is called a **station model.** Common weather symbols describe cloud cover, wind speed, wind direction, and weather conditions, such as type of precipitation and storm activity. These symbols and a station model are shown in **Figure 2.** Notice that the symbols for cloud cover, wind speed, and wind direction are combined in one symbol in the station model.

Other information included in the station model are the air temperature and the dew point. The *dew point* is the temperature to which the air must cool in order for more water to condense than to evaporate in a given amount of time. The dew point indicates how high the humidity of the air is, or how much water is in the air.

The station model also indicates the atmospheric pressure by using a three-digit number in the upper right hand corner. If this number starts with 0, then the pressure is higher than 1,000 millibars. The position of a straight line under this figure—horizontal or angled up or down—shows whether the atmospheric pressure is steady or is rising or falling.

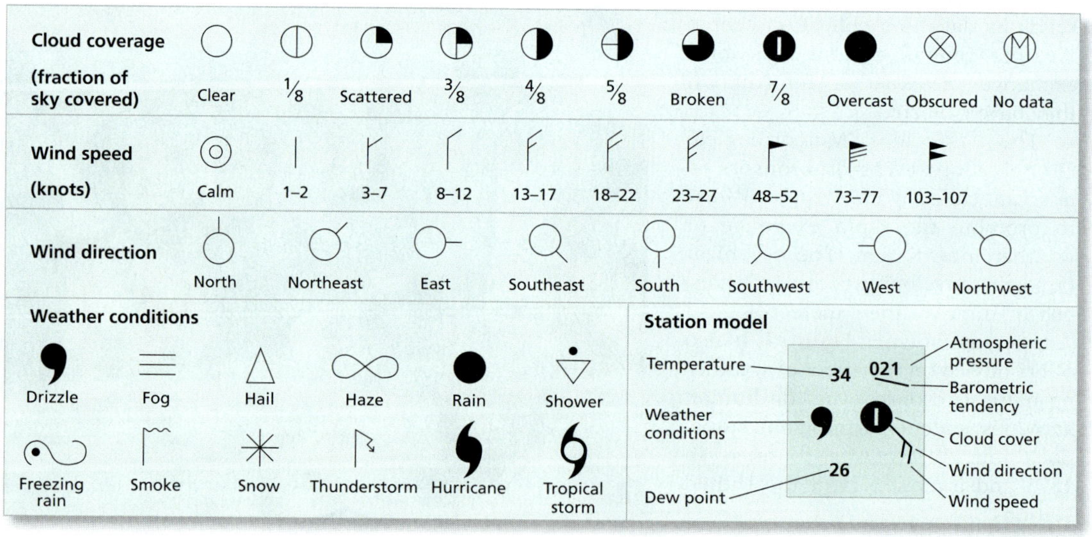

Figure 2 ▶ Meteorologists use symbols to indicate weather conditions. The station model (lower right) shows an example of conditions around a weather station.

Group Activity — GENERAL

Cloud Cover Divide a mirror into a grid by using a ruler and a wax crayon. Mark the cardinal points of the compass on the outside edges of the mirror. When the sky is partly covered with clouds, put the mirror on the ground outside, and position it with N facing north. Have groups of students count the number of squares covered by clouds and divide by the total number of squares to determine the percentage of the sky that is covered by clouds. Use the motion of the clouds across the mirror to determine upper atmospheric wind direction. LS **Visual**

Discussion — GENERAL

Barometric Pressure A change in the barometric tendency—the line angled up or down under the number that indicates pressure—generally predicts a change in the weather pattern. Ask students what a falling barometric reading would indicate. (Because warm air accompanies low-pressure systems, a drop in barometric pressure indicates that a warm front may be approaching, which may lead to precipitation.) Ask students what rising barometric pressure indicates. (High pressure is associated with colder air, which may be associated with clear weather.) LS **Verbal**

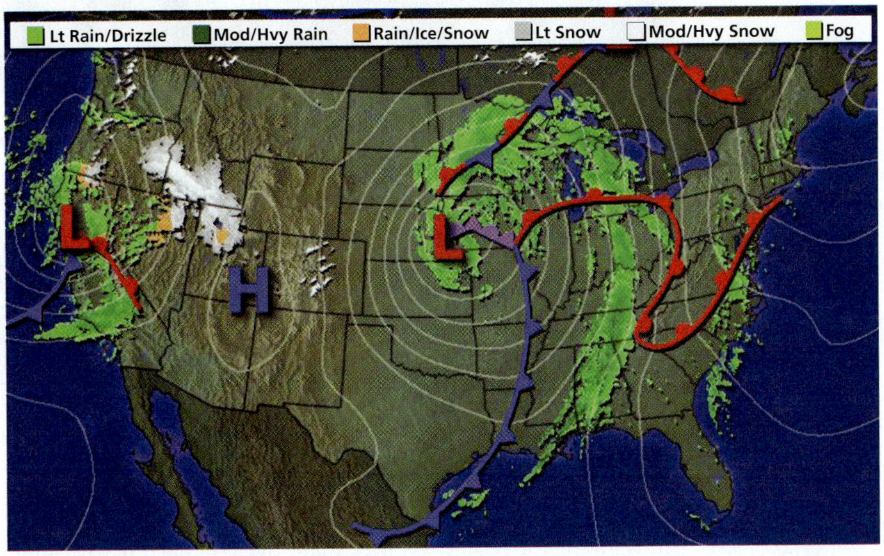

Figure 3 ▶ A typical weather map shows isobars, highs and lows, fronts, and precipitation. ***In what parts of the United States are low-pressure areas located?***

Plotting Temperature and Pressure

Scientists use lines on weather maps to connect points of equal measurement. Lines that connect points of equal temperature are called *isotherms*. Lines that connect points of equal atmospheric pressure are called *isobars*. The spacing and shape of the isobars help meteorologists interpret their observations about the speed and direction of the wind. Closely spaced isobars indicate a rapid change in pressure and high wind speeds. Widely spaced isobars generally indicate a gradual change in pressure and low wind speeds. Isobars that form circles indicate centers of high or low air pressure. Such centers that are marked with an *H* represent high pressure, as you can see in **Figure 3.** Centers that are marked with an *L* represent low pressure.

Plotting Fronts and Precipitation

Most weather maps mark the locations of fronts and areas of precipitation. The weather map in **Figure 3** shows examples of a warm front, a cold front, an occluded front, and a stationary front. Fronts are identified by sharp changes in wind speed and direction, temperature, or humidity.

Areas of precipitation are commonly marked by using colors or symbols. Different forms of precipitation are represented by different colors or symbols. For example, the weather map in **Figure 3** indicates light rain by using light green, while snow is represented by gray and white. Some weather maps use colors to represent different amounts of precipitation so that the amount of precipitation that falls in different areas can be compared.

Reading Check How do meteorologists mark precipitation on a weather map? (See the Appendix for answers to Reading Checks.)

SCILINKS. NSTA Developed and maintained by the National Science Teachers Association

For a variety of links related to this subject, go to www.scilinks.org

Topic: Weather Maps
SciLinks code: HQ61647
Topic: Weather Forecasting
SciLinks code: HQ61645

Using the Figure — GENERAL

Isobars Explain that weather maps with isobars allow us to identify low- and high-pressure areas. Isobars are typically spaced at intervals of 4 millibars (mb) apart. Ask how this pressure data helps forecast the weather. (Sample answer: High-pressure regions are commonly associated with dry weather, and low-pressure centers bring precipitation.) Ask students to identify the precipitation pattern associated with the low in the midwestern part of the United States. (light rain and drizzle) Answer to caption question: There are low-pressure areas centered over eastern Nebraska and northern California. **LS Visual**

Discussion — BASIC

Isotherms Isotherms connect areas of equal temperature and are typically spaced at intervals of 2°C apart. Ask students what key weather characteristic maps with isotherms would help us locate. (Sample answer: where large air masses with similar weather conditions are, and how quickly temperature changes can be expected) **LS Verbal**

Answer to Reading Check

Areas of precipitation are marked by using colors or symbols.

CHAPTER RESOURCES

Technology

Transparencies
- 122 Weather Map of the United States (with worksheet)

Activity — GENERAL

Local Weather Maps Have students bring in weather maps from local newspapers for two or more days. Tell students to make a list of the symbols they find. Explain that meteorologists around the globe use the same symbols. Ask students to identify symbols for cold and warm fronts. (cold fronts: triangles connected by a line representing the front boundary; warm fronts: half circles connected by a boundary line) Have students compare the previous day's weather map with the current one to see if the forecast was accurate. **LS Verbal**

Career

Meteorologist Meteorology is the science of the atmosphere. Meteorologists work in fields such as atmospheric research, weather forecasting, and classroom teaching. They study pollution, the behavior of tornadoes and hurricanes, and global climate change, among many other weather phenomena. Their tools range from basic weather instruments to research aircraft and satellites with remote-sensing capabilities. A career in meteorology requires a bachelor's degree or higher, including courses in chemistry, physics, mathematics, and other fields.

Teach, *continued*

LITERATURE CONNECTION GENERAL

The Perfect Storm The storm described in Sebastian Junger's best-selling book *The Perfect Storm* resulted from a collision between a cold high-pressure system from Canada; a low-pressure system moving east; and the warm, moist air from a dying hurricane. Supercomputers, Doppler radar, and satellites have greatly improved forecasting technology since that time, so today's maritime vessels can usually avoid these kinds of dangerous storms. Share excerpts from this exciting narrative with the class, or invite interested students to read the book and make a written or oral presentation. **LS Verbal**

QuickLAB

Skills Acquired

- Experimenting
- Measuring
- Observing
- Analyzing

Teacher's Notes You may have students monitor other conditions such as barometric pressure. They can use instruments they have already made or commercial instruments you have access to. You can have them use the Beaufort wind scale to estimate wind speed.

Answers to Analysis

1. Answers will vary depending on conditions.
2. Answers may vary but should use recognized weather symbols.

Figure 4 ▶ With the help of Doppler radar, meteorologists can track severe storms from radar stations, such as this one in Kansas.

Weather Forecasts

To forecast the weather, meteorologists regularly plot the intensity and path of weather systems on maps. Meteorologists then study the most recent weather map and compare it with maps from previous hours. This comparison allows them to follow the progress of large weather systems. By following the progress of weather systems, meteorologist can forecast the weather.

QuickLAB 10 min

Gathering Weather Data

Procedure

1. Select an area outside your school building that is in the shade away from buildings and pavement.
2. Use a **thermometer** to measure the air temperature.
3. Estimate the percentage of cloud cover.
4. Estimate wind speed. Use a **magnetic compass** to estimate the wind direction.

Analysis

1. What are the current weather conditions outside your school?
2. Using the data you collected, create a station model that describes the weather at your school.

Weather Data

Doppler radar, shown in **Figure 4,** and satellite images supply important information, such as intensity of precipitation. Meteorologists input these data into computers to create weather models. Computer models can show the possible weather conditions for several days. However, meteorologists must carefully interpret these models because computer predictions are based on generalized descriptions.

Some computer models may be better at predicting precipitation for a particular area, while other computer models may be better at predicting temperature and pressure. Comparing models helps meteorologists better predict weather. If weather information on two or more models is similar, a meteorologist will be more confident about the weather prediction. By using all of the weather data available, meteorologists can issue an accurate forecast of the weather.

Temperature, wind direction, wind speed, cloudiness, and precipitation can usually be forecasted accurately. But it is often difficult to predict precisely when precipitation will occur or the exact amount. By using computers, scientists can manipulate data on temperature and pressure to simulate errors in measuring these data. Forecasts are then compared to see if slight data changes cause substantial differences in forecasts. From what they learn, meteorologists can make more accurate forecasts.

Reading Check **Why do meteorologists compare models? (See the Appendix for answers to Reading Checks.)**

Answer to Reading Check

Meteorologists compare computer models because different models are better at predicting different weather variables. If information from two or more models matches, scientists can be more confident of their predictions.

CHAPTER RESOURCES

Chapter Resource File

- **Datasheet for Quick Lab** GENERAL

Homework ADVANCED

Weather Station Invite groups of interested students to build a class weather station that consists of homemade weather instruments. Store equipment out of direct sunlight. Encourage students to make simple observational equipment such as a rain gauge or a weather vane. Assign students to keep a weekly chart of weather observations, and have them graph the data collected. After reviewing their data, have them forecast the weather and then compare their predictions with the actual weather. **LS Kinesthetic/Logical**

Types of Forecasts

Meteorologists make four types of forecasts. *Nowcasts* mainly use radar and enable forecasters to focus on timing precipitation and tracking severe weather. *Daily forecasts* predict weather conditions for a 48-hour period. *Extended forecasts* look ahead 3 to 7 days. *Medium range forecasts* look ahead 8 to 14 days. *Long-range forecasts* cover monthly and seasonal periods.

Accurate weather forecasts can be made for 0 to 7 days. However, accuracy decreases with each day. Extended forecasts of 8 to 14 days are made by computer analysis of slowly changing large-scale movements of air. These changes help meteorologists predict the general weather pattern. For example, the changes indicate if temperature will be warmer or cooler than normal or if conditions will be dry or wet.

Severe Weather Watches and Warnings

One main goal of meteorology is to reduce the amount of destruction caused by severe weather by forecasting severe weather early. When meteorologists forecast severe weather, they issue warnings and watches. A *watch* is issued when the conditions are ideal for severe weather. A *warning* is given when severe weather has been spotted or is expected within 24 hours. Meteorologists use these alerts to provide people in areas facing severe weather with instructions on how to be safer during the event. **Table 1** lists some safety tips to follow for different types of severe weather.

Table 1 ▼

Severe Weather Safety Tips

Type of weather	How to prepare	Safety during the event
Thunderstorm	Have a storm preparedness kit that includes a portable radio, fresh batteries, flashlights, rain gear, blankets, bottled water, canned food, and medicines.	Listen to weather updates. Stay or go indoors. Avoid electrical appliances, running water, metal pipes, and phone lines. If outside, avoid tall objects, stay away from bodies of water, and get into a car, if possible.
Tornado	Have a storm preparedness kit as described above. Plan and practice a safety route.	Listen to weather updates. Stay or go indoors. Go to a basement, storm cellar, or small, inner room, closet, or hallway that has no windows. Stay away from areas that are likely to have flying debris or other dangers. If outside, lie in a low-lying area. Protect your head and neck.
Hurricane	Have a storm preparedness kit as described above. Secure loose objects, doors, and windows. Plan and practice an evacuation route.	Listen to weather updates. Be prepared to follow instructions and planned evacuation routes. Stay indoors and away from areas that are likely to have flying debris or other dangers.
Blizzard	Have a storm preparedness kit as described above. Make sure you have a way to safely make heat in the event of power outages.	Listen to weather updates. Stay or go indoors. Dress warmly. Avoid walking or driving in icy conditions.

Close

Reteaching — Basic

Matching Game Have pairs of students make drawings of two meteorological symbols on index cards with an identifying label. The cards are shuffled and placed face down. Students take turns turning over cards and trying to match the weather symbols. The object of the game is to find all the matching pairs of cards. As they play, students will familiarize themselves with the weather symbols. **LS Kinesthetic**

Quiz — General

1. Which organization helps developing countries improve their weather data collection? (the World Meteorological Organization)
2. How do meteorologists present a cluster of weather information from one location on weather maps? (They use a station model to record weather data.)
3. What are the circles on weather maps formed by isobars? (low- or high-pressure areas)
4. What safety precautions should you take during thunderstorms? (Avoid tall objects such as trees and electrical wiring or metals; remain indoors; and listen for updates.)
5. How reliable are extended weather forecasts? (Weather forecasting requires the correlation of so many variables that accuracy declines with each added day.)

Group Activity — General

Weather Safety Project Divide the class into groups, and assign each group a different type of weather. Have students explore ways of remaining safe during weather emergencies. Some students could do library or Internet research to identify the hazards associated with the assigned weather condition. Another student could interview local officials regarding your community's disaster plan. Other students could develop posters to heighten awareness of safety guidelines. Groups could plan and conduct safety drills. **LS Interpersonal** Co-op Learning

Debate — Advanced

Weather Modification Have interested students research and debate the ethical and environmental issues surrounding weather modification experiments. For example, could cloud seeding potentially cause harm by changing the direction of hurricane wind patterns, cause unexpected flooding, or even aggravate drought conditions if not done properly? Also, what are the potential long-term effects on the environment? You may wish to assign students to prepare arguments for or against the use of weather modification practices. **LS Verbal/Logical**

Close, *continued*

Alternative Assessment — GENERAL

Station Models Obtain a local weather report from the National Weather Service Interactive Weather Information Network or from a local newspaper. Remove the weather map or graphics. Have students create a station model to record the weather data. Have them include sky conditions, wind direction and speed, temperature, dew point, barometric pressure and the pressure tendency, precipitation, and any other recorded weather conditions. LS Visual

Answers to Section Review

1. Weather stations record precipitation, temperature, humidity, changes in air pressure, wind speed and direction, cloud cover, and visibility. Some functions are automated. Weather data collected is exchanged all over the world.
2. The data collected at local weather stations are transferred to weather maps in the form of meteorological symbols. Clusters are plotted around a station model.
3. Closely spaced isobars would indicate strong winds.
4. cloud cover, wind speed, wind direction, temperature, barometric pressure, and dew point
5. Comparison of new weather data with old weather maps allows meteorologists to follow changing weather systems and forecast future conditions.
6. Meteorologists input data into computers to create weather models and accurately predict temperature, wind direction, wind speed, and precipitation for a particular area. Computers allow them to manipulate the data, but they must make use of experience to interpret the results and make their forecasts.
7. precipitation levels, storm intensity or wind strength, and lightning
8. Cloud seeding may reduce the amount of moisture in air to prevent hailstones from forming.
9. Answers may vary, but should indicate well-reasoned solutions based upon risks, benefits, and costs to individuals, government, and businesses.
10. Meteorologists create *weather maps* containing *isobars, isotherms,* and *station models,* which are clusters of *meteorological symbols,* in order to *forecast* severe weather and issue weather *warnings* and *watches*.

Figure 5 ▶ An outdoor ultrahigh-voltage laboratory generates artificial lightning to test its effects on electrical utility equipment. Research has led to the development of equipment that suffers less damage from lightning.

Controlling the Weather

Some meteorologists are investigating methods of controlling rain, hail, and lightning. Currently, the most researched method for producing rain has been *cloud seeding*. In this process, particles are added to clouds to cause the clouds to precipitate. Cloud seeding can also be used to prevent more-severe precipitation. Scientists in Russia have used cloud seeding with some success on potential hail clouds by causing rain, rather than hail, to fall.

Hurricane Control

Hurricanes have also been seeded with freezing nuclei in an effort to reduce the intensity of the storm. During Project Stormfury, which took place from 1962 to 1983, four hurricanes were seeded, and the project had mixed results. Scientists have, for the most part, abandoned storm and hurricane control because it is not an attainable goal with existing technology. They do, however, continue to seed clouds to cause precipitation.

Lightning Control

Attempts have also been made to control lightning. Seeding of potential lightning storms with silver-iodide nuclei has seemed to modify the occurrence of lightning. However, no conclusive results have been obtained. Researchers have also generated artificial lightning at research facilities to learn more about lightning and how it affects objects it strikes. An example of one of these facilities is shown in **Figure 5.**

Section 4 Review

1. **Summarize** how global weather is monitored.
2. **Explain** how a weather map is made.
3. **Explain** which would show stronger winds—widely spaced isobars or closely spaced isobars.
4. **List** six different pieces of information that you can obtain from a station model.
5. **Explain** why meteorologists compare new weather maps and weather maps that are 24 hours old.
6. **Describe** how computer models help meteorologists forecast weather.
7. **List** three types of weather that meteorologists have tried to control.

CRITICAL THINKING

8. **Making Inferences** Why might cloud seeding reduce the amount of hail from a storm?
9. **Making Reasoned Judgment** Seeding hurricanes may or may not yield positive results. Each attempt costs a lot of money. If you were in charge of deciding whether to seed a potentially dangerous hurricane, what factors would you consider when deciding what to do? Explain your answer.

CONCEPT MAPPING

10. Use the following terms to create a concept map: *isobar, isotherm, weather map, forecast, watch, warning, station model,* and *meteorological symbol.*

CHAPTER RESOURCES

Chapter Resource File
- Section Quiz GENERAL

Workbooks
- Study Guide (also in Spanish)

Chapter 24 Highlights

Sections

1 Air Masses

Key Terms

air mass, 601

Key Concepts

- An air mass is a large body of air that has uniform temperature and humidity.
- Air masses can be described as polar, tropical, continental, and maritime. Their characteristics, which affect how they influence weather in North America, depend on their source region.

2 Fronts

cold front, 605
warm front, 606
stationary front, 606
occluded front, 606
midlatitude cyclone, 606
thunderstorm, 608
hurricane, 609
tornado, 610

- Cold and warm fronts are associated with characteristic weather conditions.
- A midlatitude cyclone is a storm that has a low-pressure center, rotating winds, and high-speed winds.
- Hurricanes, thunderstorms, and tornadoes are destructive storms. Thunderstorms and tornadoes are caused by the interaction of air masses with different properties.

3 Weather Instruments

thermometer, 611
barometer, 612
anemometer, 612
wind vane, 612
radiosonde, 613
radar, 613

- Thermometers, barometers, anemometers, and wind vanes measure lower-atmospheric weather conditions.
- Radiosondes, radar, satellite equipment, and computers are used to measure upper-atmospheric weather conditions.
- Computers are used to solve complicated mathematical equations that describe weather.

4 Forecasting the Weather

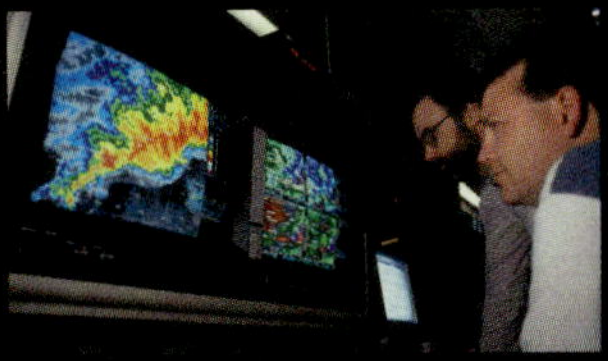

station model, 616

- Meteorologists prepare weather maps that are based on information from weather stations around the world.
- Meteorologists use different instruments to make daily and long-term forecasts of the weather.
- Meteorologists have attempted to control rain, hurricanes, and lightning with only limited success.

Chapter Highlights

Alternative Assessment — GENERAL

Weatherwise Museum Divide the class into working groups of 3 or 4 students. Have each group create models, diagrams, demonstrations, and hands-on activities that will help visitors to a weather museum understand the following ideas: (1) how air masses form and which air masses affect weather in North America, (2) how interactions between weather fronts cause cyclones, storms, and other severe weather systems, and (3) how scientists forecast the weather and warn people about hazardous atmospheric conditions. Students can use any working models that they have made of weather instruments. Suggest that students label the parts of the exhibits and displays they create, and provide simple instructions for any activities that they set up. Remind them to think about safety precautions.

LS Kinesthetic Co-op Learning

CHAPTER RESOURCES

Chapter Resource File

- Concept Review GENERAL
- Critical Thinking ADVANCED
- Math Skills GENERAL
- Graphing Skills GENERAL
- Chapter Test A GENERAL
- Chapter Test B ADVANCED

Workbooks

- Study Guide (also in Spanish)
- Assessments (Spanish)

Technology

Classroom Videos
- Brain Food Video Quiz

Chapter Review

Assignment Guide	
Section	**Questions**
1	1, 10, 23, 32
2	2, 4, 7, 11–15, 18–20, 24, 30–31, 33–35
3	5–6, 8, 16, 22, 27, 29
4	3, 9, 17, 21, 26
2 and 4	25
1–3	28

Using Key Terms

1–8. Answers may vary but should show that students understand the definitions of and differences between the key terms.

Understanding Key Concepts

9. c
10. a
11. a
12. b
13. c
14. b
15. c
16. d
17. a

Short Answer

18. Before the occluded front forms, a fast moving cold front overtakes a warm front. These conditions often produce cumulonimbus clouds and a line of heavy thunderstorms called a *squall line*. After the occluded front forms, the storm reaches its highest intensity. Then, the system loses its energy and dissipates within 24 hours.
19. Lightning is caused by electrical differences within clouds or between the ground and a cloud.
20. Hurricanes develop over warm tropical oceans away from the equator during intense tropical storms. Latent heat released during the formation of storm clouds increases the force of the rising air, generating powerful rotating winds.
21. The barometric tendency indicates whether pressure is rising or falling. Rising pressure may indicate the approach of a cold front. The station model could also be compared to previous ones to identify patterns in pressure, temperature, wind direction, and wind speed.
22. northwest; The wind vane would point to the northwest.
23. continental tropical air mass (cT)

Critical Thinking

24. Sample answer: Maritime polar air masses form over the North Pacific Ocean. They move southeasterly and do bring rain and snow to the Pacific coast. In all probability, if a midlatitude cyclone forms over the Gulf of Alaska and travels at about 45 km/h, much of the storm's energy will probably dissipate before the storm reaches Vancouver Island (over 1800 km away).
25. Sample answer: The center of a hurricane is a region of calm air. The eye of the storm might be passing directly overhead. The hurricane could be up to 700 km in diameter and its violent winds could easily pick up again. The person should remain indoors, away from flying debris.

Chapter 24 Review

Using Key Terms

Use each of the following terms in a separate sentence.

1. *air mass*
2. *stationary front*
3. *station model*

For each pair of terms, explain how the meanings of the terms differ.

4. *midlatitude cyclone* and *hurricane*
5. *wind vane* and *anemometer*
6. *radiosonde* and *radar*
7. *cold front* and *warm front*
8. *thermometer* and *barometer*

Understanding Key Concepts

9. Which of the following is information you would not find from a station model?
   a. precipitation
   b. cloud cover
   c. front
   d. wind speed

10. Continental polar Canadian air masses generally move
    a. southeasterly.
    b. northerly.
    c. northeasterly.
    d. westerly.

11. The type of front that forms when two air masses move parallel to the front between them is called
    a. stationary.
    b. occluded.
    c. polar.
    d. warm.

12. The type of front that is completely lifted off the ground by cold air is called
    a. cold.
    b. occluded.
    c. polar.
    d. warm.

13. The eye of a hurricane is a region of
    a. hailstorms.
    b. torrential rainfall.
    c. calm, clear air.
    d. strong winds.

14. The winds of a midlatitude cyclone blow in circular paths around a
    a. front.
    b. low-pressure center.
    c. high-pressure center.
    d. jet stream.

15. In the mature stage of a thunderstorm, a cumulus cloud grows until it becomes a
    a. stratocumulus cloud.
    b. altocumulus cloud.
    c. cumulonimbus cloud.
    d. cirrocumulus cloud.

16. An instrument package attached to a weather balloon is
    a. an anemometer.
    b. a wind vane.
    c. a thermograph
    d. a radiosonde.

17. The lines that connect points of equal atmospheric pressure on a weather map are called
    a. isobars.
    b. isotherms.
    c. highs.
    d. lows.

Short Answer

18. Describe the weather before and after an occluded front.
19. What causes lightning?
20. What is the most likely location for hurricane development? Explain your answer.
21. How could a meteorologist use a station model to determine whether a cold front is approaching?
22. Identify the wind direction of wind given as 315°. What direction would a wind vane point in that case?
23. Identify the type of air mass that would most likely be responsible if the air in your region is warm and dry. What letters designate this air mass?

Critical Thinking

24. Making Predictions Suppose people on Vancouver Island, off the west coast of Canada, hear reports of a midlatitude cyclone in the Gulf of Alaska. Is it likely that the midlatitude cyclone will reach their area? Explain why.

25. Making Inferences Suppose a hurricane is passing over a Caribbean island. Suddenly, the rain and winds stop and the air becomes calm and clear. Can a person safely go outside? Explain your answer.

26. Applying Ideas Is it safe to be in an automobile during a tornado? Explain your answer.

27. Making Inferences An air traffic controller is monitoring nearby airplanes by radar. The controller warns an incoming pilot of a storm a few miles away. How did radar help the controller detect the storm?

Concept Mapping

28. Use the following terms to create a concept map: *air mass, front, warm front, cold front, cyclones, thunderstorm, thermometer, hurricane, barometer,* and *anemometer.*

Math Skills

29. Making Calculations The temperature at a station is given as 47°F . Using the equation, °C = 5/9 × (°F − 32), find the temperature in degrees Celsius.

30. Making Calculations An average of 124 tornadoes occur each year in Texas. If that is equivalent to 4.7 tornadoes per 10,000 mi^2, what is the area of Texas in square miles?

Writing Skills

31. Creative Writing Imagine that you are traveling with friends through the desert in the southwestern United States and a thunderstorm occurs. You then tell them about the type of air mass that may have brought the storm. Describe what the stages might look like by types of clouds formed, types of precipitation, and sky color.

32. Communicating Main Ideas Explain how cP and mT air masses travel across the United States, and explain why this information helps meteorologists make forecasts.

Interpreting Graphics

The graph below shows the number of tornadoes that happened in Kansas at different times of day between January 1980 and July 2003. Use the graph to answer the questions that follow.

Number of Tornadoes by Time of Day in Kansas (1980–2003)

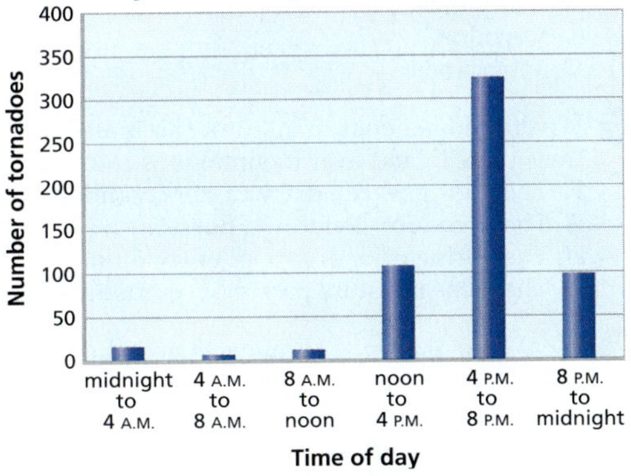

33. During which time of day did the least tornadoes occur?

34. Would Kansas students be more likely to experience a tornado while they are at school or while they are at home?

35. During which time of day did most tornadoes occur? Why would most tornadoes happen at this time of day?

Chapter Review

26. Sample answer: no; An automobile would not provide adequate protection from the violent winds. It is best to remain indoors in a basement or underground shelter away from flying debris or from windows that could break.

27. The air traffic controller could see the approaching thunderstorms on the radar screen and estimate how far away they were from the airplanes he was monitoring.

Concept Mapping

28. Answers may vary but should include all of the terms listed. Sample answers appear at the end of this Teacher Edition.

Math Skills

29. °C = 5/9 × (°F − 32) = 5/9 × (47°F − 32) = 5/9 × 15° = 75/9 = 8.3°C

30. area of Texas = 124 ÷ 4.7 = 26.23 × 10,000 mi^2 = 262,300 mi^2

Writing Skills

31. Answers may vary. Accept all reasonable answers. Answers should include the fact that maritime tropical air masses bring moderate precipitation to the southwestern desert.

32. Answers may vary. Accept all reasonable answers. Answers should include the facts that air masses bring the weather of the source region and that the general direction of movement of the air masses is known.

Interpreting Graphics

33. between 4 A.M. and 8 A.M.

34. at home; More tornadoes occur during the period from 4 P.M. to 8 A.M. than between 8 A.M. and 4 P.M.

35. Most tornadoes occur between 4 P.M. and 8 P.M. because daily temperatures tend to peak during mid-afternoon hours, which is necessary for the development of the strong updrafts that produce destructive wind patterns.

Estimated Time

To give students practice under more realistic testing conditions, allow them 30 minutes to answer all of the questions in this practice test.

TEST DOCTOR

Question 2 Answer I is correct. Answer F is incorrect because in an occluded front, warm air is lifted up and cut off by cold air. Answer G is incorrect because polar fronts form at the boundary between a dome of cold air and warm air. Answer H is incorrect because in a warm front, warm air rises over cool air.

Question 4 Answer I is correct. Air masses that rise over mountains often lose moisture through precipitation. Answers F and H are incorrect because most air masses bring weather from their source region. Maritime air masses are most likely to bring wet conditions because they form over water. Answer G is incorrect. Though maritime air masses may spawn hurricanes, the formation of these storms is not triggered by the air mass moving over mountainous terrain.

Chapter 24 Standardized Test Prep

Understanding Concepts

Directions (1–5): **For *each* question, write on a separate sheet of paper the letter of the correct answer.**

1. What tool do meteorologists use to analyze particle movements within storms?
   A. an anemometer
   B. a radiosonde balloon
   C. doppler radar
   D. satellite imaging

2. What kind of front forms when two air masses move parallel to the boundary located between them?
   F. an occluded front
   G. a polar front
   H. a warm front
   I. a stationary front

3. Which of the following weather systems commonly forms over warm tropical oceans?
   A. thunderstorms
   B. hurricanes
   C. tornadoes
   D. anticyclones

4. What often happens to maritime air masses as they move inland over mountainous country?
   F. They bring warm, dry weather conditions.
   G. They produce clouds and hurricanes.
   H. They bring cold, dry weather conditions.
   I. They lose moisture passing over mountains.

5. What type of air mass originates over the southwestern desert of the United States in summer?
   A. continental polar air mass
   B. continental tropical air mass
   C. maritime polar air mass
   D. maritime tropical air mass

Directions (6–7): **For *each* question, write a short response.**

6. What type of front is formed when a warm air mass is overtaken by a cold air mass, which causes the warm air to lift above the cold air?

7. What do closely spaced isobars indicate about the wind on a weather map?

Reading Skills

Directions (8–10): **Read the passage below. Then, answer the questions.**

Tornado Alley

Though tornadoes are not unique to the area, the violent, rotating, funnel-shaped clouds and their trails of destruction are so common in the central United States that the area is called Tornado Alley. These severe thunderstorms and the super-cell tornadoes that they spawn are formed when warm, moist air from the Gulf of Mexico becomes trapped beneath hot, dry air from the southwest desert region. Above that hot, dry air, cold, dry air sweeps in from the Rocky Mountains. The interaction between high-altitude winds and thunderstorms creates the funnel-shaped vortex of high-speed winds known as a tornado.

The largest outbreak of tornadoes in this region occurred in April of 1974. Before the storms ended, 148 separate tornadoes roared through 13 different states. More than 300 people lost their lives, and another 5,000 people were injured. More than 1,300 buildings were destroyed.

8. Why is the central part of the United States also known as Tornado Alley?
   A. Tornadoes in the area move in straight lines known as alleys.
   B. The destruction left by tornadoes made the area look like an unkempt alley.
   C. Areas between buildings are the safest places to be during of a tornado.
   D. Tornadoes are common occurances in this particular part of the country.

9. Which of the following statements can be inferred from the information in the passage?
   F. In the United States, tornadoes are more common in some areas than in other areas.
   G. Tornadoes can form only in the area near the Rocky Mountains.
   H. All tornadoes cause injuries to humans.
   I. Multiple tornadoes are a rare occurance.

10. What makes tornadoes so much more difficult to predict than other severe weather systems?

Answers

Understanding Concepts

1. C
2. I
3. B
4. I
5. B
6. cold front
7. high-speed winds

Reading Skills

8. D
9. F
10. Tornadoes are difficult to predict because they form suddenly from unstable conditions.

Interpreting Graphics

11. D
12. E is wind direction, and F is wind speed; Currently the station model is showing a 30-knot wind that is blowing in from south.
13. B
14. Answers may vary. See Test Doctor for a detailed scoring rubric.

Interpreting Graphics

Directions (11–14): **For *each* question below, record the correct answer on a separate sheet of paper.**

The diagram below shows a station model. Use this diagram to answer questions 11 and 12.

Interpreting a Station Model

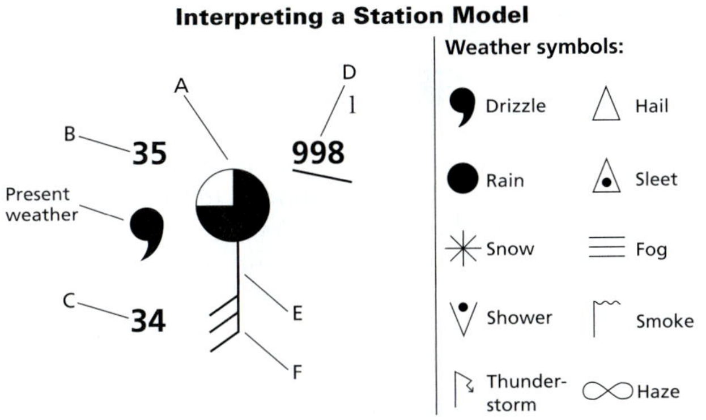

11 What letter represents the current barometric reading shown in the model?

A. letter A
B. letter B
C. letter C
D. letter D

12 What weather information do the symbols indicated by the letters E and F provide? Interpret this part of the station model.

The diagram below shows a home weather station. Use this diagram to answer questions 13 and 14.

Weather Instruments

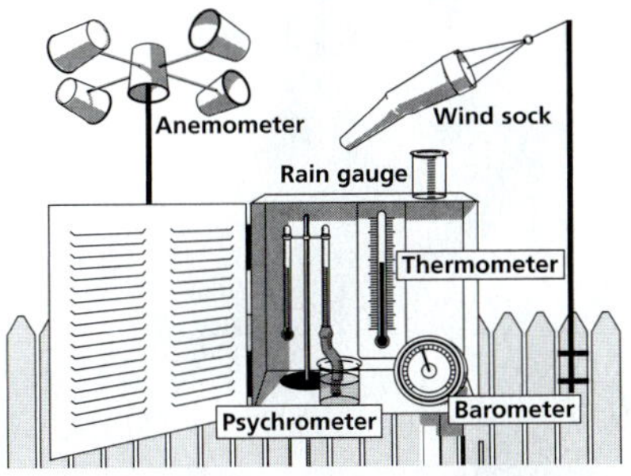

13 Which of the following weather instruments shown uses the cooling effect of evaporation to take measurements?

A. a rain gauge
B. a psychrometer
C. a wind sock
D. a thermometer

14 Describe how an anemometer is used to calculate wind speed.

Test TIP

Sometimes, only one part of a diagram, graph, or table is needed to answer a question. In such cases, focus on only that information to answer the question.

Standardized Test Prep

TEST DOCTOR

Question 11 Answer D is correct. The symbols for cloud cover, wind direction, and wind speed are combined in one symbol. The cloud cover circle on the station model, which is indicated by the letter A, is mostly darkened to show an overcast sky. This symbol shows that 70% to 80% of the sky is covered. The temperature, shown by letter B, is 25°C; the dew point, shown by letter C, is 20°C. Letter D shows the barometric reading, which is 25 mb. The wind barb, which is indicated by the letter E, points in the direction the wind is coming from. In this case, the wind is coming from the south.

Question 14 Full-credit answers should include the following points:

- as the cups on the anemometer catch the wind, the device begins to rotate
- the speed of this rotation, usually given in revolutions per minute, and the circumference of the circle made by the cups are used to calculate wind speed
- rudimentary devices rely on the user to count the revolutions per minute of the device and to perform the necessary math to determine the wind speed
- in modern computerized devices, a number of factors may be considered to obtain the most accurate measurement possible. These factors include the circumference of the device, friction of the air, and drag

Test Prep Correlations *National Science Education Standards*

SAI 2c: items 1, 13, 14
HNS 2b: items 2, 3, 4, 5, 6, 8, 9, 10
UCP 2: items 7, 11, 12

CHAPTER RESOURCES

State Resources

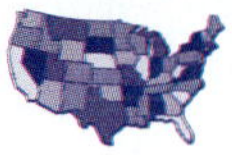

For specific resources for your state, visit go.hrw.com and type in the keyword HSHSTR.

Skills Practice Lab

Weather Map Interpretation

Teacher's Notes

Time Required
one 45-minute class period

Lab Ratings
EASY → HARD

TEACHER PREPARATION (1 of 4)
STUDENT SETUP (1 of 4)
CONCEPT LEVEL (3 of 4)
CLEANUP (1 of 4)

Skills Acquired
- Collecting Data
- Organizing and Analyzing Data
- Identifying and Recognizing Patterns
- Interpreting
- Communicating

The Scientific Method
In this lab, students will
- Make Observations
- Analyze Results
- Draw Conclusions

Materials
Students can either do this investigation alone or with a partner. Students should make an enlarged photocopy of the weather map from the Reference Tables section of the Appendix.

Tips and Tricks
To read this weather map, students must be familiar with the meanings of the map symbols.

Chapter 24

Skills Practice Lab

Objectives
- **Construct** a pressure and temperature map.
- **Interpret** a weather map.
- **Explain** how weather patterns are related to pressure systems.

Materials
paper
pencil
pencils, colored, red, blue

Weather Map Interpretation

Weather maps use various map symbols and lines to illustrate the weather conditions in an area at a given time. In this lab, you will study the symbols used on a weather map to gain an understanding of the relationships between temperature, pressure, and winds.

PROCEDURE

1. Make a copy of the weather map on the following page. This map can also be found in the Reference Tables section of the Appendix. You will use the map symbols on the same page of the Appendix to interpret the weather map. The number on the right of each station on the map represents atmospheric pressure. The number on the left represents temperature.
2. On your copy of the weather map, find stations that have a temperature of 10.0°C. Use a red pencil to draw a light line through these stations. If two adjacent stations have temperatures above and below 10°C, there is an estimated point between them that is 10.0°C. Draw a line through these estimated points to connect the stations that have temperatures of 10.0°C with a 10.0°C isotherm.

Step 2

CHAPTER RESOURCES

Chapter Resource File
- Datasheet for Chapter Lab GENERAL
- Lab Notes and Answers

Workbooks

Long-Term Projects
- Correlating Weather Variables ADVANCED
- Weather Forecasting ADVANCED

3. Using the same method as in step 2, draw isotherms for every two degrees of temperature. Examples are isotherms of 12.0°C, 14.0°C, and 16.0°C. Label each isotherm with the temperature it represents.

4. Find a station that has a barometric pressure of 1,004 millibars. Use a blue pencil, and follow the same method that you used in step 2 to create a 1,004 millibars isobar.

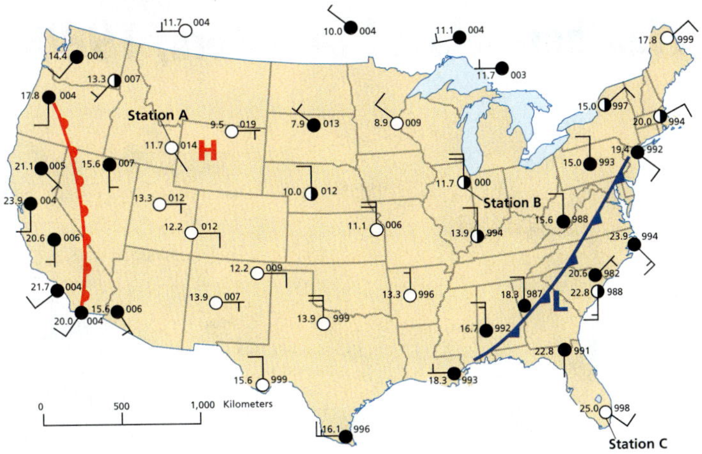

Step 1

6. Using the same method as in step 3, lightly draw isobars for every 4 millibars of pressure. Examples are isobars of 1,000 mb, 1,008 mb, and 1,012 mb. Label each isobar with the pressure it represents.

ANALYSIS AND CONCLUSION

1. **Identifying Trends** What is the lowest temperature for which you have drawn an isotherm? What is the highest temperature for which you have drawn an isotherm? Is either isotherm a closed loop? If so, which one?
2. **Making Inferences** Is the air mass that is identified by the closed isotherms a cold air mass or a warm air mass? Explain your answer.
3. **Analyzing Data** Is there a shift in wind direction associated with either front shown on your map? Describe the shift.
4. **Identifying Trends** What is the value of the lowest-pressure isobar that was drawn? What is the value of the highest-pressure isobar that was drawn? Is either isobar a closed loop? If so, which one?
5. **Drawing Conclusions** At the time that the map represents, were there any areas of low pressure? of high pressure? Identify these areas. What weather conditions would you expect to find in those areas?

Extension

1. **Making Predictions** Predict the weather conditions at Station A 24 hours after the observations for your map were made. Record your predictions in a table with columns for pressure, wind direction, wind speed, temperature, and sky condition. Also, make and record predictions for Station B and Station C.

Answers to Analysis and Conclusion

1. The lowest temperature isotherm is 8°C and the highest temperature isotherm is 24°C. The lowest isotherm was a closed loop.
2. cold; Low temperatures are associated with cold air masses.
3. yes; The winds ahead of the cold front shift from a south-south-easterly direction to a northerly direction as the front passes, but the winds ahead of the warm front show little or no shift in direction.
4. The lowest pressure isobar is 984 mb. The highest pressure isobar is 1016 mb. Both form closed loops.
5. There is a high-pressure area over Wyoming and a low-pressure area over South Carolina and Georgia. The high-pressure area is associated with clear weather. Rain, clouds, rapid temperature changes, and wind shifts occur in low-pressure areas.

Answers to Extension

1. Answers may vary. Accept all reasonable predictions based on the weather data.

TEACHER TESTED & APPROVED

Daniel Brownstein, MA/ MAT Geology
Hastings High School
Hastings-on-Hudson, NY

Maps in Action

Weather-Related Disasters 1980–2005

Answers to the Map Skills Activity

1. eight
2. floods
3. western U.S.: primarily fires and floods; eastern U.S.: mixture of tropical storms, hurricanes, ice and snow; The convergence of 2 polar and 2 maritime air masses over the eastern U.S. and the general west to east movement of weather systems may help account for the difference.
4. Icy conditions are not common in southern states so they are not prepared with the proper equipment to deal with them.
5. The Atlantic Ocean and Gulf of Mexico are the sources of maritime tropical air masses that form hurricanes.
6. Snow and ice that fall in the winter melt to form water that may cause flooding. If more ice and snow fell than normal, the spring meltwaters would be higher than normal as well.
7. Lightning strikes from electrical storms may cause brush fires that destroy acres of forested land.

CHAPTER RESOURCES

Workbooks

Transparencies
- 123 Weather-Related Disasters, 1980–2005 (with worksheet)

MAPS *in Action*

Weather-Related Disasters, 1980–2005

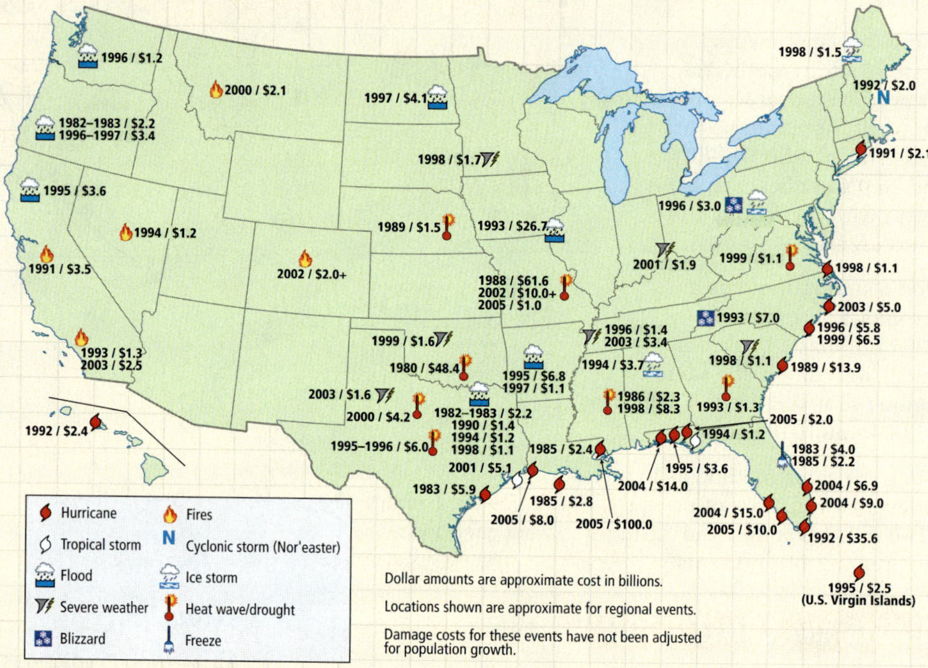

Map Skills Activity

This map shows the types and locations of weather disasters in the United States that caused at least $1 billion in damage. Use the map to answer the questions below.

1. **Using the Key** How many severe weather events caused more than $10 billion in damage between 1980 and 2005?
2. **Analyzing Data** Which type of weather disaster is more common—floods or fires?
3. **Making Comparisons** How do the types of disasters that happen in the western United States differ from the types of disasters that happen in the eastern United States? Explain why this difference exists.
4. **Inferring Relationships** Why might an ice storm in Alabama cause more damage than an ice storm in Maine?
5. **Identifying Trends** Almost all hurricane damage during this period happened along the coasts of the Atlantic Ocean and the Gulf of Mexico. Explain why.
6. **Analyzing Relationships** In 1996, a blizzard and floods caused $3.0 billion in damage in Ohio, Pennsylvania, and West Virginia. How might these events be related? Explain your answer.
7. **Analyzing Processes** Explain why fires are included in this map of weather-related disasters.

CAREER FOCUS

Meteorologist

When a hurricane threatens, most people flee, but not Shirley Murillo. Murillo boards a research aircraft and flies into the hurricane! As the plane flies through the hurricane, instruments record wind speed, wind direction, temperature, and air pressure.

Through the Eye of the Storm

Most of the flight is remarkably smooth, until the aircraft reaches the eyewall. The eyewall is a donutlike ring of turbulent thunderstorms that surround the calm eye of the storm. Once past the eyewall, the plane enters calm air. "We fly for hours going in and out of the eye, dropping instruments … into the eye," says Murillo. Some of these instruments measure air pressure. These data help hurricane forecasters decide if the storm is becoming weaker or stronger.

Improving Hurricane Forecasting

When she is not flying into hurricanes, Murillo is in her office at the Hurricane Research Division of the Atlantic Oceanographic and Meteorological Laboratory, an NOAA facility in Miami, Florida. Murillo studies how hurricane wind speeds change at landfall.

Murillo's research helps hurricane forecasters at the National Hurricane Center create their forecasts and advisories. Emergency managers also use it to determine areas along the coast that need to be evacuated. Scientists also use Murillo's work to predict storm surge, the onshore rush of seawater caused by a hurricane's high winds and low pressure.

"Flying into a hurricane is a thrill. The data that are collected … help save lives and property."

—Shirley Murillo

Rewards and Benefits

There is no doubt that Murillo's job can be exciting—and gratifying. "The most rewarding part of my job is the way in which our understanding of hurricanes and our improvement in forecasting benefit the scientific community and the public."

◀ The eye of the storm can be seen just below the plane *NOAA P-3* as it flies through Hurricane Caroline.

SCILINKS. NSTA Developed and maintained by the National Science Teachers Association

For a variety of links related to this subject, go to www.scilinks.org

Topic: Careers in Earth Science
SciLinks code: HQ60222

Meteorologist

Activity — GENERAL

Hurricane Hunters Encourage students to learn more about how aircraft are used to study hurricanes. For example, how does this research help forecasters better understand the storm processes generally, and during hurricanes, make more accurate predictions of their strength and point of landfall. Have students imagine what it would be like to fly one of these research craft directly into the path of a hurricane and write a first-person account. They might incorporate factual details about the sophisticated weather equipment used on these flying research vessels. Other students might look into the historical record of the Air Forces Weather Reconnaissance Squadron and imagine they were flying during that exciting period. Still others might investigate some of the other jobs that support the role of the National Hurricane Center in tracking hurricanes and protecting the public, incorporating details about these career paths into their narratives. LS **Verbal**

Extension — GENERAL

Research Have students identify four other careers that are related to meteorology. Have students write a brief synopsis of each career and present them to the class. LS **Interpersonal**

Chapter 25 Climate

Planning Guide

Compression Guide

To shorten instruction because of time limitations, omit the Chapter Lab.

OBJECTIVES	LABS, DEMONSTRATIONS, AND ACTIVITIES	TECHNOLOGY RESOURCES
PACING • 90 min pp. 630–636 **Chapter Opener**	SE **Long-Term Project** Comparing Climate Features, pp. 862–865 ADVANCED LTP **Long-Term Project** Comparing Climate Features* ADVANCED	OSP **Parent Letter** CD **Student Edition on CD-ROM** CD **Chapter Summaries Audio CD** VID **Brain Food Video Quiz**
Section 1 Factors That Affect Climate • Identify two major factors used to describe climate. • Explain how latitude determines the amount of solar energy received on Earth. • Describe how the different rates at which land and water are heated affect climate. • Explain the effects of topography on climate.	TE **Demonstration** Latitude and Temperature, p. 632 GENERAL SE **Quick Lab** Evaporation, p. 634 GENERAL CRF **Datasheet for Quick Lab*** GENERAL TE **Physics Connection** Latent Heat, p. 634 GENERAL TE **Activity** El Niño, p. 635 GENERAL SE **Inquiry Lab** Factors That Affect Climate, pp. 652–653 GENERAL CRF **Datasheet for Chapter Lab*** GENERAL SE **Maps in Action** Climates of the World, p. 654 GENERAL TE **Discussion** Climate Classification, p. 654 GENERAL SE **Mapping Expeditions** Where the Hippos Roam, pp. 840–841 GENERAL	OSP **Lesson Plans** (also in print) TR **Bellringer*** TR 124 Average Sea-Level Temperatures During Winter in the Northern Hemisphere* TR 129 Climates of the World* TE **Internet Activity** Gaia Hypothesis, p. 655 GENERAL CRF **Internet Activity** Gaia Hypothesis* GENERAL CD **Interactive Tutor** Other Storms and El Niño CD **Interactive Tutor** Greenhouse Effect
PACING • 45 min pp. 637–640 **Section 2 Climate Zones** • Describe the three types of tropical climates. • Describe the five types of middle-latitude climates. • Describe the three types of polar climates. • Explain why city climates may differ from rural climates.	TE **Discussion** Name That Climate, p. 637 GENERAL TE **Activity** Adaptations, p. 638 GENERAL CRF **Skills Practice Lab** Microclimates* GENERAL	OSP **Lesson Plans** (also in print) TR **Bellringer*** TR 125 Tropical Climates* TR 126 Middle-Latitude Climates* TR 127 Polar Climates* CD **Interactive Tutor** Climate Zones
PACING • 45 min pp. 641–646 **Section 3 Climate Change** • Compare four methods used to study climate change. • Describe four factors that may cause climate change. • Identify potential impacts of climate change. • Identify ways that humans can minimize their effect on climate change.	TE **Activity** Tree Rings, p. 641 GENERAL TE **Debate** Global Warming, p. 642 ADVANCED TE **Chemistry Connection** Oxygen Isotopes, p. 642 ADVANCED SE **Quick Lab** Hot Stuff, p. 644 GENERAL CRF **Datasheet for Quick Lab*** GENERAL TE **Demonstration** Icecaps Melting, p. 644 BASIC TE **Law Connection** Kyoto Protocol, p. 645 ADVANCED CRF **Inquiry Lab** Particulates in the Atmosphere* GENERAL	OSP **Lesson Plans** (also in print) TR **Bellringer*** TR 128 Orbital Changes and Climate* TE **Internet Activity** Climate Models, p. 642 GENERAL CRF **Internet Activity** Climate Models* GENERAL CD **Interactive Tutor** Past Climate CD **Interactive Tutor** Recent Climate

PACING • 90 min

CHAPTER REVIEW, ASSESSMENT, AND STANDARDIZED TEST PREPARATION

- SE **Chapter Highlights,** p. 647
- SE **Chapter Review,** pp. 648–649
- SE **Standardized Test Prep,** pp. 650–651
- CRF **Concept Review*** GENERAL
- CRF **Critical Thinking*** ADVANCED
- CRF **Math Skills*** GENERAL
- CRF **Graphing Skills*** GENERAL
- CRF **Chapter Test A*** GENERAL
- CRF **Chapter Test B*** ADVANCED
- OSP **Lesson Plans** (also in print)
- OSP **Test Generator**
- OSP **Test Item Listing**

Online and Technology Resources

Holt Online Learning

Visit go.hrw.com for access to Holt Online Learning, or enter the keyword **HQ6 Home** for a variety of free online resources.

One-Stop Planner® CD-ROM

This CD-ROM package includes

- Lab Materials QuickList Software
- Holt Calendar Planner
- Customizable Lesson Plans
- Printable Worksheets
- ExamView® Test Generator
- Interactive Teacher Edition
- Holt PuzzlePro®
- Holt PowerPoint® Resources

KEY

SE	Student Edition	OSP	One-Stop Planner	VID	Classroom Video/DVD
TE	Teacher Edition	TR	Transparencies and Transparency Worksheets	*	Also on One-Stop Planner
CRF	Chapter Resource File	CD	CD or CD-ROM	◆	Requires advance prep
LTP	Long-Term Projects			■	Also available in Spanish

SKILLS DEVELOPMENT RESOURCES	REVIEW AND ASSESSMENT	CORRELATIONS
SE **Pre-Reading Activity**, p. 630 GENERAL TE **Using the Figure** Life Adapts, p. 630 GENERAL		National Science Education Standards
CRF **Directed Reading*** BASIC TE **Using the Figure** Temperature Ranges, p. 631 GENERAL TE **Reading Skill Builder** Paired Summarizing, p. 632 BASIC TE **Using the Figure** Shifting Winds, p. 633 ADVANCED SE **Math Practice**, p. 634 GENERAL	SE **Reading Checks**, pp. 633, 634 GENERAL SE **Section Review**, p. 636 GENERAL TE **Reteaching**, p. 635 BASIC TE **Quiz**, p. 635 GENERAL TE **Alternative Assessment**, p. 635 GENERAL CRF **Section Quiz*** ■ GENERAL	ES 1d
CRF **Directed Reading*** BASIC TE **Skill Builder** Math, p. 638 GENERAL TE **Inclusion Strategies**, p. 639	SE **Reading Checks**, p. 638 GENERAL SE **Section Review**, p. 640 GENERAL TE **Reteaching**, p. 639 BASIC TE **Quiz**, p. 639 GENERAL TE **Alternative Assessment**, p. 639 GENERAL CRF **Section Quiz*** ■ GENERAL	ES 1d, SPSP 5b
CRF **Directed Reading*** BASIC SE **Graphic Organizer** Cause-and-Effect Map, p. 643 GENERAL TE **Inclusion Strategies**, p. 643 TE **Using the Figure** Rising Seas, p. 645 GENERAL TE **Reading Skill Builder** Reading Organizer, p. 645 BASIC	SE **Reading Checks**, pp. 642, 644 GENERAL SE **Section Review**, p. 646 GENERAL TE **Homework**, p. 644 ADVANCED TE **Reteaching**, p. 645 BASIC TE **Quiz**, p. 645 GENERAL TE **Alternative Assessment**, p. 645 GENERAL CRF **Section Quiz*** ■ GENERAL	ES 1d, SPSP 4a, SPSP 4b, SPSP 4c, SPSP 5a, SPSP 5b, SPSP 5d, UCP 2

Holt Lab Generator CD-ROM

Search for labs by topic, standard, difficulty level, and/or time frame. Edit those labs to fit your needs, or create your own labs. Use the Lab Materials QuickList to order just what you need.

Holt Earth Science Interactive Tutor CD-ROM

This CD-ROM consists of interactive activities that give students a fun way to extend their knowledge of Earth science concepts.

Chapter Summaries Audio CDs

These CDs include audio summaries of the key concepts presented in each chapter. (Audio summaries are also available in Spanish.)

SCILINKS. NSTA

www.scilinks.org

Maintained by the **National Science Teachers Association.** See Chapter Enrichment pages that follow for a complete list of topics.

See Chapter Enrichment pages for Video Resources.

Refer to our ***Professional Reference for Teachers*** for additional teaching resources, including articles written by science education professionals about relevant and timely issues facing today's science teachers.

Chapter 25

Chapter Enrichment

This Chapter Enrichment provides relevant and interesting information to expand and enhance your classroom instruction of the chapter material.

Section 1 Factors That Affect Climate

The Little Boy

South American fishermen have known for centuries about El Niño, which often begins around Christmas. El Niño is Spanish for "little boy," a name for the infant Jesus. For the fishermen, El Niño meant drastically reduced fish catches. Warm water off the western shore of South America kept the cold, nutrient-rich waters that sustained their fish stocks trapped in the ocean depths. Scientists now understand more about the causes and consequences of El Niño and know that it affects weather around the world.

▲ **Monsoons bring regular flooding into parts of south Asia.**

Normally, strong trade winds push warm surface water westward, away from South America toward Southeast Asia. The warm waters in the western Pacific fuel the monsoon rains of India and Asia. But in some years, for reasons that are not entirely clear, trade winds weaken, allowing warm water to build up along the western coast of South America. This results in heavy rains and flooding. El Niño does not just affect the climate in South America. Its impact is felt around the globe. During an El Niño event, India and Southeast Asia experience disastrous droughts as monsoon rains fail to come. In the United States, the Southwest experiences heavy rains, while the Northern states have mild winters. El Niño has even been implicated in the 1993 hantavirus outbreak in the Four Corners region of the United States. Scientists believe heavy rains increased food sources for deer mice, which are carriers of the virus. With more food, the deer mouse populations exploded, setting the stage for greater contact with humans and outbreaks in Arizona, Colorado, and other regions.

An El Niño event lasts one to two years and seems to occur in cycles of three to ten years. By monitoring the ocean and atmosphere using satellites and weather buoys, scientists are able to follow the evolution of an El Niño event better. But they still cannot always predict when one will occur or all the climatic disturbances it may cause.

Section 2 Climate Zones

Great Lakes Snowfall

Some of the snowiest cities in the United States are located along the eastern and southern shores of the Great Lakes. Buffalo and Rochester, New York, on the shores of Lake Erie and Lake Ontario, receive an average of 3 m of snow each winter. Watertown, New York, has received 3 m of snow in a single storm! These cities receive so much snow because of the lake-effect "snow machine," which is a consequence of the differential heating of land and water.

During the summer, the Great Lakes slowly absorb huge amounts of heat. The surrounding land also heats up under the summer sun, but it cannot store heat as effectively as the water of the lakes can. As autumn and winter approach, the land rapidly cools while the Great Lakes retain much of their heat. From late November into January, the water is significantly warmer than the surrounding land, and the lake-effect snow machine revs up.

When cold, dry air from Canada flows over the warm water, it absorbs water vapor and heat from the lakes. When the air mass reaches land, the air mass rises and quickly cools. The water vapor condenses as snow that falls along the leeward shores of the Great Lakes. Lake-effect snow belts occur in Michigan, Ohio, Pennsylvania, and New York and usually extend no more than 40 km from the shores of the lakes.

Video Resources

Brain Food Video Quizzes

Brain Food Video Quizzes These videos contain game-show style quizzes that assess students' progress and motivate students to study the chapter material.

NOVA® Videos

To order **NOVA** videos related to this chapter, visit **go.hrw.com** and enter the keyword **HQ6CLIV.**

Section 3 Climate Change

Trapping Carbon Dioxide

Most climatologists think that increasing atmospheric levels of carbon dioxide contribute to global warming. Some scientists are concerned that future climatic change may take place more abruptly than previously thought, and more quickly than Earth's inhabitants can adapt. In efforts to slow down climate change, scientists are beginning to develop technologies to capture, store, and/or recycle carbon-dioxide emissions.

The United States Department of Energy (DOE) has several projects that test trapping carbon-dioxide emissions from fossil fuel fired power plants and then injecting them into geologic formations such as oil and gas reservoirs, coal seams, and deep saline formations. Another study involves injecting trapped carbon dioxide into the deep ocean for long-term storage.

Pilot studies by scientists from New Zealand, Germany, and the United States have shown that fertilizing the ocean with micro- and macronutrients increases carbon dioxide uptake in the short term by the vast sea of phytoplankton. The environmental consequences of projects like these over the long-term and on a large-scale are unknown and hotly debated. The Research Institute of Innovative Technology for the Earth (RITE) in Japan and the DOE are also investigating ways to use photosynthetic microorganisms to remove carbon dioxide from waste gases and recycle it into useful fuels and chemicals.

Because it is likely that fossil fuels will remain a major source of energy in many parts of the world for some time, it is imperative to develop ways to reduce their global warming potential as we transition to cleaner forms of energy.

▼ **Human activities, such as logging, affect local and global climates.**

SciLinks®

Developed and maintained by the National Science Teachers Association

SciLinks is maintained by the National Science Teachers Association to provide you and your students with interesting, up-to-date links that will enrich your classroom presentation of the chapter.

Visit www.scilinks.org and enter the SciLinks code for more information about the topic listed.

Topic: What Affects Climate?
SciLinks code: HQ61652

Topic: Polar Climates
SciLinks code: HQ61175

Topic: Climate Zones
SciLinks code: HQ60301

Topic: Changes in Climate
SciLinks code: HQ60252

Topic: Global Warming
SciLinks code: HQ60681

Chapter 25

Chapter Overview

Many factors help shape climates, and latitude plays a key role in controlling climate. Climates change over time for various reasons. Scientists study past climate changes to better understand the effects climate change may have on life on Earth.

Using the Figure — GENERAL

Life Adapts In all environments, plants and animals adapt to the climate. For example, polar bears can live in the icy Arctic realm because they have traits that allow them to thrive in the cold. Discuss with students how plants and animals adapt to their surroundings. Ask students whether humans adapt in the same ways. (Answers may vary. Accept all reasonable answers.) LS **Logical**

PRE-READING ACTIVITY

FOLDNOTES

You may want to collect students' FoldNotes to determine their prior knowledge. You may also want to modify your lesson plan to answer questions that students list in the "Want" column of their table.

Chapter 25 Climate

Sections

1 Factors That Affect Climate
2 Climate Zones
3 Climate Change

What You'll Learn

- What factors affect climate
- How regional climates vary
- How scientists study past climate changes and predict future climate changes

Why It's Relevant

Earth sustains life because the temperature and moisture conditions are right. By learning about climate, we can understand factors that might affect life on Earth.

PRE-READING ACTIVITY

FOLDNOTES

Tri-Fold Before you read this chapter, create the FoldNote entitled "TriFold" described in the Skills Handbook section of the Appendix. Write what you know about climate in the column labeled "Know." Then, write what you want to know in the column labeled "Want." As you read the chapter, write what you learn about climate in the column labeled "Learn."

▶ Thick fur and other adaptations allow polar bears to survive and thrive in the freezing temperatures of the tundra climate.

Chapter Correlations — *National Science Education Standards*

ES 1d Global climate is determined by energy transfer from the sun at and near the earth's surface. This energy transfer is influenced by dynamic processes such as cloud cover and the earth's rotation, and static conditions such as the position of mountain ranges and oceans. **(Sections 1, 2, and 3)**

SPSP 5b Human activities can enhance the potential for hazards. **(Sections 2 and 3)**

SPSP 4a Natural ecosystems provide an array of basic processes that affect humans. Those processes include maintenance of the quality of the atmosphere....Humans are changing many of these basic processes, and the changes may be detrimental to humans. **(Section 3)**

SPSP 4b Materials from human societies affect both physical and chemical cycles of the earth. **(Section 3)**

SPSP 4c Many factors influence environmental quality. Factors that students might investigate include population growth, resource use, population distribution, overconsumption, the capacity of technology to solve problems, poverty, the role of economic, political, and religious views, and different ways humans view the earth. **(Section 3)**

SPSP 5d Natural and human-induced hazards present the need for humans to assess potential danger and risk.... **(Section 3)**

SPSP 5a Normal adjustments of earth may be hazardous for humans. Humans live at the interface between the atmosphere driven by solar energy and the upper mantle where convection creates changes in the earth's solid crust. As societies have grown, become stable, and come to value aspects of the environment, vulnerability to natural processes of change has increased. **(Section 3)**

UCP 2 Models are tentative schemes or structures that correspond to real objects, events, or classes of events, and that have explanatory power. Models help scientists and engineers understand how things work. Models take many forms, including...computer simulations. **(Section 3)**

ENVIRONMENTAL CONNECTION

Section 1 Factors That Affect Climate

The average weather conditions for an area over a long period of time are referred to as **climate.** Climate is different from weather in that weather is the condition of the atmosphere at a particular time. Weather conditions, such as temperature, humidity, wind, and precipitation, vary from day to day. To understand climate, scientists study the features that define different climates.

OBJECTIVES

- **Identify** two major factors used to describe climate.
- **Explain** how latitude determines the amount of solar energy received on Earth.
- **Describe** how the different rates at which land and water are heated affect climate.
- **Explain** the effects of topography on climate.

KEY TERMS

climate
specific heat
El Niño
monsoon

Temperature and Precipitation

Climates are chiefly described by using average temperature and precipitation. To estimate the average daily temperature, add the high and low temperatures of the day and divide by two. The monthly average is the average of all of the daily averages for a given month. The yearly average temperature can be found by averaging the 12 monthly averages. However, using only average temperatures to describe climate can be misleading. As you can see in **Figure 1,** areas that have similar average temperatures may have very different temperature ranges. Another way scientists describe climate is by using the *yearly temperature range,* or the difference between the highest and lowest monthly averages.

Another major factor that affects climate is precipitation. It is also described by using monthly and yearly averages as well as ranges. As with temperature, average yearly precipitation alone cannot describe a climate. The months that have the largest amount of precipitation are important for determining climate. When describing climates, extremes of temperature and precipitation as well as averages have to be considered. The factors that have the greatest influence on both temperature and precipitation are latitude, heat absorption and release, and topography.

climate the average weather conditions in an area over a long period of time

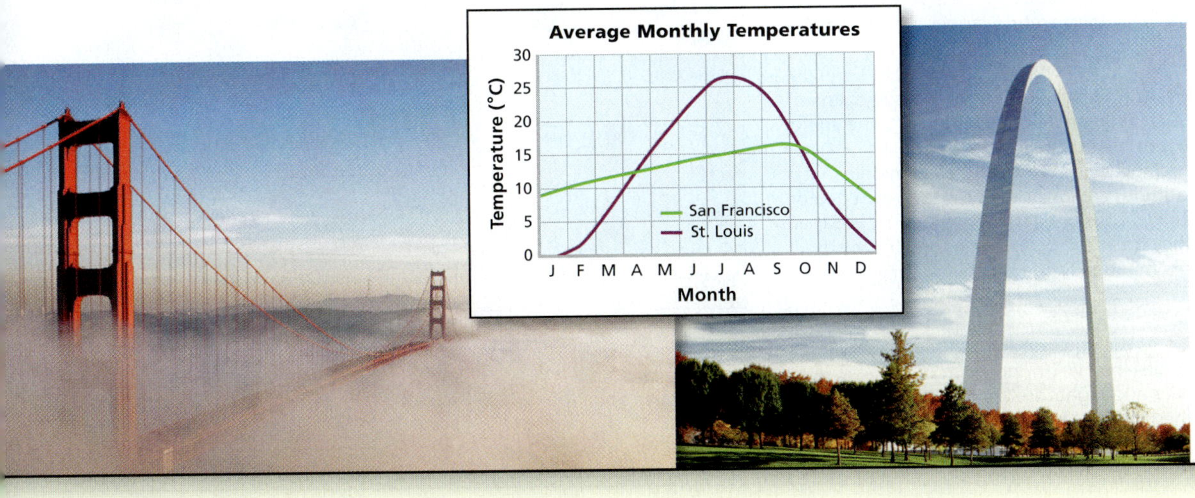

Figure 1 ▶ Both St. Louis and San Francisco have the same average yearly temperature. However, St. Louis (right) has a climate of cold winters and hot summers, while San Francisco (left) has a generally mild climate all year.

Section 1

Focus

Overview

This section presents temperature and precipitation as the two major factors used to describe climate. The section also explains how latitude, proximity to large bodies of water, and topography affect climate.

Bellringer

Ask students, "What is the difference between weather and climate?" (Weather is the day-to-day variations in temperature, humidity, wind, and precipitation. Climate is the average weather conditions for an area over a long period.) LS **Logical**

Motivate

Using the Figure — GENERAL

Temperature Ranges Have students use the figure that shows the average monthly temperatures of St. Louis and San Francisco to determine and compare the average yearly temperatures of the two cities. Then, have them determine the yearly temperature range in the two cities. Students will find that, while the average yearly temperature is similar, (about 13°C) the temperature range for St. Louis is much greater than that for San Francisco. (about 28°C for St. Louis and about 8°C for San Francisco) LS **Logical/Visual**

CHAPTER RESOURCES

Chapter Resource File

- Directed Reading BASIC

Technology

Transparencies
- Bellringer

Student Edition on CD-ROM

One-Stop Planner CD-ROM
- Lesson Plan

Teach

Teaching Tip — GENERAL

Make Concepts Relevant Remind students that the tilt of Earth's axis determines the seasons. In December, the northern end of the axis is tilted away from the sun, so the Northern Hemisphere receives less solar energy than at other times of the year, and the Northern Hemisphere experiences winter. At the same time, the Southern Hemisphere receives more solar energy than at other times of the year, and the Southern Hemisphere experiences summer. In June the opposite occurs: the Northern Hemisphere begins to experience summer and the Southern Hemisphere begins its winter. LS **Logical/Visual**

ENVIRONMENTAL CONNECTION

As Earth Tilts While most people understand that seasons are due to the tilt of Earth's axis, it is less well-known that the degree of tilt changes from 22.2° to 24.5° every 41,000 years. As the angle changes, the exposure to incoming solar energy (insolation) and the seasonal distribution change at high latitudes. For example, as the angle decreases, temperature differences between the seasons at high latitudes reduces (because insolation does not vary much with the seasons).

CHAPTER RESOURCES

Technology

Transparencies

- 124 Average Sea-Level Temperatures During Winter in the Northern Hemisphere (with worksheet)

Latitude

One of the most important factors that determines a region's climate is latitude. Different latitudes on Earth's surface receive different amounts of solar energy. Solar energy determines the temperature and wind patterns of an area, which influence the average annual temperature and precipitation.

SCILINKS. NSTA Developed and maintained by the National Science Teachers Association

For a variety of links related to this subject, go to www.scilinks.org

Topic: What Affects Climate?
SciLinks code: HQ61652

Solar Energy

The higher the latitude of an area is, the smaller the angle at which the sun's rays hit Earth is and the smaller the amount of solar energy received by the area is. At the equator, or 0° latitude, the sun's rays hit Earth at a 90° angle. So, temperatures at the equator are high. At the poles, or 90° latitudes, the sun's rays hit Earth at a smaller angle, and solar energy is spread over a large area. So, temperatures at the poles are low.

Because Earth's axis is tilted, the angle at which the sun's rays hit an area changes as Earth orbits the sun. During winter in the Northern Hemisphere, the northern half of Earth is tilted away from the sun. Thus, light that reaches the Northern Hemisphere hits Earth's surface at a smaller angle than it does in summer, when the axis is tilted toward the sun. Because of the tilt of Earth's axis during winter in the Northern Hemisphere, areas of Earth at higher northern latitudes directly face the sun for less time than during the summer. As a result, the days are shorter and the temperatures are lower during the winter months than during the summer months. **Figure 2** describes these effects.

Figure 2 ▶ Average Sea-Level Temperatures During Winter in the Northern Hemisphere

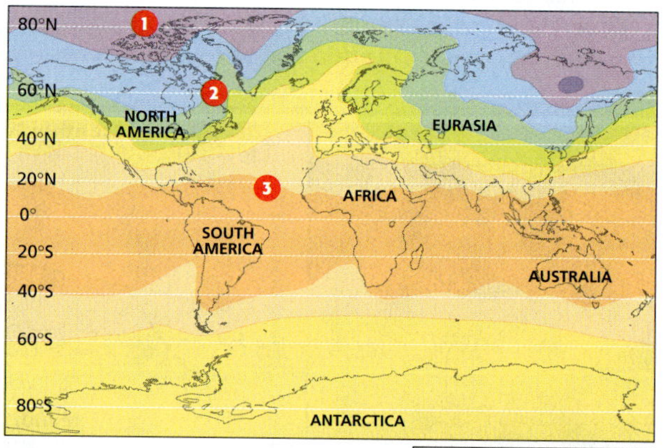

1. In polar regions, the amount of daylight varies from 24 hours of daylight in the summer to 0 hours in the winter. Thus, the annual temperature range is very large, but the daily temperature ranges are very small.

2. At middle latitudes, the sun's rays strike Earth at an angle of less than 90°. The energy of the rays is spread over a large area. Thus, average yearly temperatures at middle latitudes are lower than those at the equator. The lengths of days and nights vary less than they do at the poles. Therefore, the yearly temperature range is large.

3. At the equator, the sun's rays always strike Earth at a very large angle—nearly 90° for much of the year. In equatorial regions, both days and nights are about 12 hours long throughout the year. So, these regions have steady, high temperatures year-round.

Temperatures (°C)	
21 to 32	−23 to −12
10 to 21	−34 to −23
−1 to 10	−46 to −34
−12 to −1	Below −46

READING SKILL BUILDER — BASIC

Paired Summarizing Have students join with a partner and read silently about how the tilt of Earth's axis affects the amount of solar energy Earth receives. Then, have one student summarize the effect of the axial tilt on temperatures in the Northern Hemisphere over the course of the year. The other student should listen to the retelling and point out any inaccuracies or ideas that were left out. Allow students to refer to the text as needed.
LS **Verbal** Co-op Learning English Language Learners

Demonstration — GENERAL

Latitude and Temperature Place three thermometers at angles of 0°, 45°, and 90° at the same distance from a light source. Record the temperature of each thermometer. Have students predict which will show the greatest rise in temperature. At the end of class, record the temperatures. Discuss the results and how they relate to the effect of latitude on Earth's temperatures. (The one at 0° represents the poles and should show the least change; the one at 90° represents the equator and should show the greatest increase.) LS **Visual**

Global Wind Patterns

Because Earth receives different amounts of solar energy at different latitudes, belts of cool, dense air form at latitudes near the poles, while belts of warm, less dense air form near the equator. Because cool air is dense, it forms regions of high pressure, while warm air forms regions of low pressure. Differences in air pressure create wind. Because air pressure is affected by latitude, the atmosphere is made up of global wind belts that run parallel to lines of latitude. Winds affect many weather conditions, such as precipitation, temperature, and cloud cover. Thus, regions that have different global wind belts often have different climates.

In the equatorial belt of low pressure, called the *doldrums*, the air rises and cools, and water vapor condenses. Thus, this region generally has large amounts of precipitation. The amount of precipitation generally decreases as latitude increases. In the regions between about 20° and 30° latitude in both hemispheres, or the *subtropical highs*, the air sinks, warms, and dries. Thus, little precipitation occurs in these regions. Most of the world's deserts are located in these regions. In the middle latitudes, at about 45° to 60° latitude in both hemispheres, warm tropical air meets cold polar air, which leads to belts of greater precipitation. In the high-pressure areas, above 60° latitude, the air masses are cold and dry, and average precipitation is low.

As seasons change, global wind belts shift in a north or south direction, as shown in **Figure 3.** As the wind and pressure belts shift, the belts of precipitation associated with them also shift.

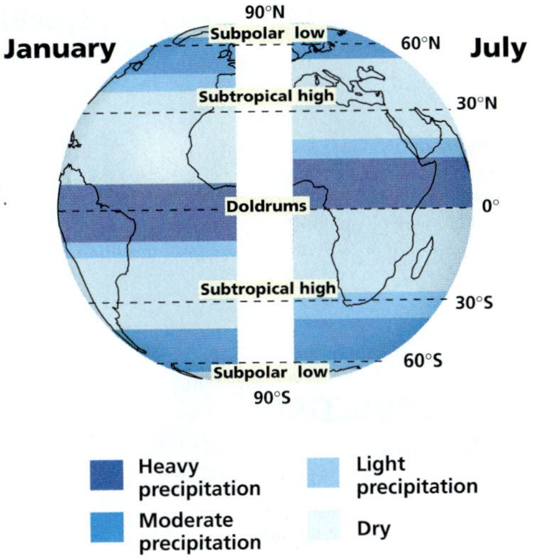

Figure 3 ▶ During winter in the Northern Hemisphere, global wind and precipitation belts shift to the south.

Heat Absorption and Release

Latitude and cloud cover affect the amount of solar energy that an area receives. However, different areas absorb and release heat differently. Land heats faster than water and thus can reach higher temperatures in the same amount of time. One reason for this difference is that the land surface is solid and unmoving. Surface ocean water, on the other hand, is liquid and moves continuously. Waves, currents, and other movements continuously replace warm surface water with cooler water from the ocean depths. This action prevents the surface temperature of the water from increasing rapidly. However, the surface temperature of the land can continue to increase as more solar energy is received. In turn, the temperature of the land or ocean influences the amount of heat that the air above the land or ocean absorbs or releases. The temperature of the air then affects the climate of the area.

Reading Check **How do wind and ocean currents affect the surface temperature of oceans? (See the Appendix for answers to Reading Checks.)**

Using the Figure — ADVANCED

Shifting Winds The figure at the top of this page shows how global wind and precipitation belts shift south during winter in the Northern Hemisphere. The shift occurs because in the winter, a very strong high-pressure system develops over Asia and a weaker high-pressure system develops over North America. These systems push global wind belts and the doldrums, or the Intertropical Convergence Zone, south. As the continents warm in the summer, low-pressure systems develop, especially in India and to a lesser extent in the southwestern United States, pushing the wind belts back north. Ask students to explain why climates near the equator may not have temperature-based seasons, but commonly do have wet and dry seasons. (The intensity of radiation from the sun does not vary much throughout the year at the equator. Therefore, temperature does not change significantly. However, as the wind belts shift north and south over the course of the year, some regions will experience changes in precipitation, such as in the central part of Africa [about 15°N] and in South America [about 15°S]) **LS Visual**

Answer to Reading Check

Waves, currents, and other water motions continually replace warm surface waters with cooler water from the ocean depths, which keeps the surface temperature of the water from increasing rapidly.

BRAIN FOOD

Deep-Ocean Currents and Climate Water circulates deep within the ocean because of differences in density, which in turn are the result of differences in temperature and salinity. Cold, salty water tends to sink because it is denser than warm, less salty water. Water near the poles is cold; it is also very saline because as water freezes salts remain behind in the water. Cold, dense water in the Arctic and Antarctic regions sinks and travels toward the equator along the ocean floor. Near the equator, these waters rise and join surface currents in a global conveyor belt that circulates water through the ocean. Water may take several hundred to thousands of years to complete a circuit. This global conveyer belt influences climate because the deep ocean currents act as a reservoir for carbon dioxide and for thermal energy.

Teach, *continued*

MATHPRACTICE

Answer

6°C = 279 K
(4,186 J/kg*K) × 200 kg × 279 K = 23,357,888,000 J

PHYSICS CONNECTION GENERAL

Latent Heat The evaporation of water requires a large amount of thermal energy. Much of the thermal energy absorbed by water is used not to increase temperature but to increase the motion of the water molecules, allowing them to escape the water surface and to become vapor. When the higher-temperature molecules escape as water vapor, they leave cooler molecules (molecules that are vibrating less rapidly) behind, which is why evaporation has a cooling effect. Because this thermal energy is not used for temperature change, it is referred to as latent (hidden) heat. Latent heat is another reason why the temperature of land and water at the same latitude differ: water surfaces are much more influenced by the cooling effect of evaporation than land is.

Answer to Reading Check

The temperature of land increases faster than that of water does because the specific heat of land is lower than that of water, and thus the land requires less energy to heat up than the water does.

CHAPTER RESOURCES

Chapter Resource File

- Datasheet for Quick Lab GENERAL

specific heat the quantity of heat required to raise a unit mass of homogeneous material 1 K or 1°C in a specified way given constant pressure and volume

MATHPRACTICE

Specific Heat
Use the following equation to calculate the amount of energy needed to heat 200 kg of water 6°C given that the specific heat of water is 4,186 J/kg•K.

energy = specific heat × mass × temperature change

Specific Heat and Evaporation

Even if not in motion, water warms more slowly than land does. Water also releases heat energy more slowly than land does. This is because the specific heat of water is higher than that of land. **Specific heat** is the amount of energy needed to change the temperature of 1 g of a substance by 1°C. A given mass of water requires more energy than land of the same mass does to experience an increase in temperature of the same number of degrees.

The average temperatures of land and water at the same latitude also vary because of differences in the loss of heat through evaporation. Evaporation affects water surfaces much more than it affects land surfaces.

Ocean Currents

The temperature of ocean currents that come in contact with the air influences the amount of heat absorbed or released by the air. If winds consistently blow toward shore, ocean currents have a strong effect on air masses over land. For example, the combination of a warm Atlantic current and steady westerly winds gives northwestern Europe a high average temperature for its latitude. In contrast, the warm Gulf Stream has little effect on the eastern coast of the United States. This is because westerly winds usually blow the Gulf Stream and its warm tropical air away from the coast.

Reading Check **Why does land heat faster than water does? (See the Appendix for answers to Reading Checks.)**

QuickLAB 15 min

Evaporation

Procedure

1. On a **piece of paper**, make a data table similar to the one shown here.
2. Assemble a **ring stand** on a **table**. Use a **meterstick** to place the support rings at heights of 20 cm and 40 cm above the base. Position a **portable clamp lamp that has an incandescent bulb** directly over the rings and at a height of 60 cm.
3. Place **three Petri dishes** or **watch glasses** as follows: one on the base of the stand and one on each of the two rings.
4. Take **three thermometers,** and lay one across each dish. Turn on the lamp. Use a **stopwatch** to record the temperature every 3 min for 9 min.
5. Remove the thermometers, and add **30 mL of water** to each of the three dishes.
6. Keep the lamp on and over the dishes for 24 h.

Dish	Temperature	Amount of water evaporated
1		
2		
3		

DO NOT WRITE IN THIS BOOK

7. Turn off the lamp. Carefully pour the water from the first dish into a **graduated cylinder,** and record any change in volume. Repeat this process for the other two dishes.

Analysis

1. At what distance from the lamp did the most water evaporate? the least water evaporate?
2. Explain the relationship between temperature and the rate of evaporation.
3. Explain why puddles of water dry out much more quickly in summer than they do in fall or winter.

QuickLAB

Skills Acquired

- Measuring
- Analyzing

Teacher's Notes Position the lamp on a separate support. Lamps may have to be turned off overnight for safety or insurance reasons. If so, similar results can be achieved by using the lamp for two school days. If the lamp is turned off, cover dishes to prevent evaporation overnight.

Answers

1. The most water evaporated from the dish closest to the lamp. The least evaporated from the dish at the base of the stand.
2. As temperature increases, rate of evaporation increases.
3. Water evaporates more quickly from puddles in summer because the air and the water are warmer in summer than in fall or winter.

El Niño–Southern Oscillation

The *El Niño–Southern Oscillation*, or *ENSO,* is a cycle of changing wind and water-current patterns in the Pacific Ocean. Every 3 to 10 years, **El Niño,** which is the warm-water phase of the ENSO, causes surface-water temperatures along the west coast of South America to rise. The event changes the interaction of the ocean and the atmosphere, which can change global weather patterns. During El Niño, typhoons, cyclones, and floods may occur in the Pacific Ocean region and southeastern United States. Droughts may strike other areas around the world, such as Indonesia and Australia. The ENSO also has a cool-water phase called *La Niña*. La Niña also affects weather patterns.

El Niño the warm-water phase of the El Niño–Southern Oscillation; a periodic occurrence in the eastern Pacific Ocean in which the surface-water temperature becomes unusually warm

monsoon a seasonal wind that blows toward the land in the summer, bringing heavy rains, and that blows away from the land in the winter, bringing dry weather

Seasonal Winds

Temperature differences between the land and the oceans sometimes cause winds to shift seasonally in some regions. During the summer, the land warms more quickly than the ocean. The warm air rises and is replaced by cool air from the ocean. Thus, the wind moves toward the land. During the winter, the land loses heat more quickly than the ocean does, and the cool air flows away from the land. Thus, the wind moves seaward. Such seasonal winds are called **monsoons.**

Monsoon climates, such as that in southern Asia, are caused by heating and cooling of the northern Indian peninsula. In the winter, continental winds bring dry weather and sometimes drought. In the summer, winds carry moisture to the land from the ocean and cause heavy rainfall and flooding, as shown in **Figure 4.** Monsoon conditions also occur in eastern Asia and affect the tropical regions of Australia and East Africa.

Figure 4 ▶ Effects of Monsoon Climates

Because monsoon rains cause regular flooding, such as this flood in eastern India, people who live in monsoon regions have adapted to living in flood conditions.

People who live in monsoon climates must adjust to periodic droughts, such as the drought that affected this cropland in southern India.

Close

Reteaching — BASIC

Climate Factors Have students write a brief description on index cards of major factors that affect climate. Have them work in pairs using the cards to quiz one another. LS Visual/Logical

Quiz — GENERAL

1. Why does northwestern Europe have a milder climate than its latitude would indicate? (A warm Atlantic current [the Gulf Stream] and steady westerly winds moderate its climate.)
2. Where are most of the world's deserts located? (in the region of the subtropical highs, 20° to 30° N and S latitude)

Alternative Assessment — GENERAL

Graphing Have students graph the monthly average precipitation (noting if and when it is snow) and the high and low temperatures for your city over the last year. Then, have them write a short description of your climate and the factors that influence it. LS Logical

Activity — GENERAL

El Niño Ask interested students to research recent El Niño events, how they are predicted, and their effects on wildlife and on weather patterns around the world. Some students may also want to research La Niña. Have students prepare a poster that illustrates their findings to display for the class. LS Visual/Verbal

Cultural Awareness — ADVANCED

Climates and Culture Climate has a profound effect on people's clothing, housing, food, and customs. Have students work in small groups to study a specific culture. They may choose a traditional or a modern culture anywhere in the world. Have them research how climate has shaped the culture they have chosen. Groups should prepare a creative presentation that illustrates the culture and the effects of climate on the culture of the people they studied. LS Visual/Verbal

Close, *continued*

Answers to Section Review

1. temperature and precipitation
2. The higher the latitude, the smaller the angle at which the sun's rays hit Earth and the smaller the amount of solar energy an area receives.
3. Because Earth receives different amounts of solar energy at different latitudes, belts of air of different densities form, with cool, dense air at the poles and warm, less dense air at the equator. These differences in air pressure create winds. Global wind belts run roughly parallel to lines of latitude.
4. Land heats and cools faster than water does. Because of this difference, large bodies of water can affect the climate of land areas near them.
5. El Niño is the warm-water phase of the ENSO that periodically warms surface water along the west coast of South America. This warming changes atmospheric and oceanic interactions, which can affect global weather patterns.
6. In summer, winds blow from the ocean, as air warmed by the land is replaced by air cooled by the ocean. These winds carry moisture to the land and bring heavy rain. In the winter, winds blow from the cooler land, bringing dry weather.
7. Temperature decreases with increasing altitude, so even mountain peaks at the equator can be covered with snow.
8. When an air mass reaches a mountain, it rises and cools. The water vapor it contains condenses and falls as precipitation. Air flowing down the other side of the slope, in the rain shadow, is dry and warms up as it sinks.
9. Climate would be less varied. There would be fewer differences in air pressure and thus fewer winds. The moderating effect of bodies of water on climate would be lost.
10. I would expect more vegetation on the side facing the prevailing winds because that side receives the most precipitation.
11. Snow-capped mountains occur in Hawaii because temperature decreases with altitude.
12. *Climate* is described by using *temperature range* and precipitation, and is affected by *topography* and *winds*, which can be global, which cause the *doldrums* and *subtropical highs;* and seasonal, such as *monsoons* and *El Niño*.

Figure 5 ▶ Mountains cause air to rise, cool, and lose moisture as the air passes over the mountains. This process affects the climate on both sides of the mountain.

Topography

The surface features of the land, or *topography,* also influence climate. Topographical features, such as mountains, can control the flow of air through a region.

Elevation

The elevation, or height of landforms above sea level, produces distinct temperature changes. Temperature generally decreases as elevation increases. For example, for every 100 m increase in elevation, the average temperature decreases by 0.7°C. Even along the equator, the peaks of high mountains can be cold enough to be covered with snow.

Rain Shadows

When a moving air mass encounters a mountain range, the air mass rises, cools, and loses most of its moisture through precipitation, as shown in **Figure 5.** As a result, the air that flows down the other side of the range is usually warm and dry. This effect is called a *rain shadow*. One type of warm, dry wind that forms in this way is the *foehn* (FAYN), a dry wind that flows down the slopes of the Alps. Similar dry, warm winds that flow down the eastern slopes of the Rocky Mountains are called *chinooks*.

Section 1 Review

1. **Identify** two factors that are used to describe climate.
2. **Explain** how latitude determines the amount of solar energy received on Earth.
3. **Describe** how latitude determines wind patterns.
4. **Describe** how the different rates at which land and water are heated affect climate.
5. **Explain** the El Niño–Southern Oscillation cycle.
6. **Summarize** the conditions that cause monsoons.
7. **Explain** how elevation affects climate.
8. **Describe** a rain shadow and the resulting local winds.

CRITICAL THINKING

9. **Making Inferences** If land and water had the same specific heat, how might climate be different around the world?
10. **Analyzing Processes** On a mountain, are you likely to find more vegetation on the side facing prevailing winds or on the side facing away from them?
11. **Recognizing Relationships** Why might you find snow-capped mountains in Hawaii even though Hawaii is closer to the equator than Florida is?

CONCEPT MAPPING

12. Use the following terms to create a concept map: *climate, temperature range, wind, doldrums, subtropical high, monsoon, El Niño,* and *topography.*

CHAPTER RESOURCES

Chapter Resource File
- Section Quiz GENERAL

Workbooks
- Study Guide (also in Spanish)

ENVIRONMENTAL CONNECTION

Section 2 Climate Zones

Earth has three major types of climate zones: tropical, middle-latitude, and polar. Each zone has distinct temperature characteristics, including a specific range of temperatures. Each of these zones has several types of climates because the amount of precipitation within each zone varies.

OBJECTIVES

- **Describe** the three types of tropical climates.
- **Describe** the five types of middle-latitude climates.
- **Describe** the three types of polar climates.
- **Explain** why city climates may differ from rural climates.

KEY TERMS

tropical climate
middle-latitude climate
polar climate
microclimate

Tropical Climates

Climates that are characterized by high temperatures and are located in the equatorial region are referred to as **tropical climates.** These climates have an average monthly temperature of at least 18°C, even during the coldest month of the year. Within the tropical zone, there are three types of tropical climates: tropical rain forest, tropical desert, and savanna. These climates are described in **Table 1.**

tropical climate a climate characterized by high temperatures and heavy precipitation during at least part of the year; typical of equatorial regions

Tropical rain-forest climates are humid and warm and support a diverse variety of life including dense, rain-forest vegetation. The warm, moist, rising air produces an annual rainfall of 200 cm. Central Africa, the Amazon River basin of South America, Central America, and Southeast Asia have areas that have tropical rain-forest climates.

Tropical desert climates receive less than 25 cm of precipitation every year and have little or no vegetation. The largest belt of tropical deserts extends across north Africa and southwestern Asia.

Savanna climates support open grasslands that have drought-resistant trees and shrubs. Savannas have very wet summers and very dry winters. Savanna climates are located in South America, Africa, Southeast Asia, and northern Australia.

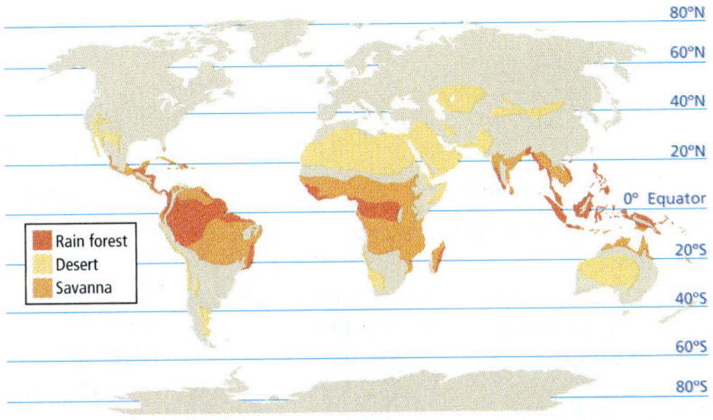

Table 1 ▼

Tropical Climates

Climate	Temperature and precipitation	Description
Rain forest	small temperature range; annual rainfall of 200 cm	characterized by dense, lush vegetation; broadleaf plants; and high biodiversity
Desert	large temperature range, hot days, and cold nights; annual rainfall of less than 25 cm	characterized by little to no vegetation and organisms adapted to dry conditions
Savanna	small temperature range; annual rainfall of 50 cm; alternating wet and dry periods	characterized by open grasslands that have clumps of drought-resistant shrubs

Section 2

Focus

Overview

This section describes the three major climate zones and the subclimate zones within them. The section also explains factors that may alter local climates.

Bellringer

Ask students to describe the climate and physical features of the region where they live and to identify any other regions of the world that have similar climates. (Answers may vary. Accept all reasonable answers.) **LS Verbal**

Motivate

Discussion — GENERAL

Name That Climate Ask students to name and briefly describe as many climates as they can think of that exist on Earth. Use student responses to address misconceptions about climate zones and introduce the major climates and their subclimates. **LS Verbal/Logical**

CHAPTER RESOURCES

Chapter Resource File
- Directed Reading BASIC

Technology

Transparencies
- Bellringer
- 125 Tropical Climates (with worksheet)

Student Edition on CD-ROM

One-Stop Planner CD-ROM
- Lesson Plan

Teach

Activity — GENERAL

Adaptations Collect photos of a variety of leaves and/or plants from various climates around the world, such as cacti, jade plants, grasses, bromeliads, orchids, sedges, mosses, lichens, ferns, conifers, and deciduous trees. (If it is possible, collect actual plants.) Have students observe the plants and try to determine under what conditions the plant is most likely to grow. Then, lead a discussion about the adaptations of the various plants and how the traits enable the plants to thrive in their home climate. LS **Visual**

SKILL BUILDER — GENERAL

Math Have students use the table and text that describe middle latitude climates to convert temperature and precipitation values from SI units to English units for one of the subclimates. Have students use the following conversion factors: °F = (°C × 1.8) + 32, and 1 in. = 2.54 cm.
Sample answer—marine west coast: summer temp = (20°C × 1.8) + 32 = 68°F
winter temp = (7°C × 1.8) + 32 = 45°F
precipitation range = (60 cm ÷ 2.54 cm/in.) to (150 cm ÷ 2.54 cm/in.) = 24 to 59 in.
LS **Logical**

Answer to Reading Check

marine west coast, humid continental, and humid subtropical

CHAPTER RESOURCES

Chapter Resource File

- **Skills Practice Lab** Microclimates GENERAL

Technology

Transparencies

- 126 Middle-Latitude Climates (with worksheet)
- 127 Polar Climates (with worksheet)

Table 2 ▼

Middle-Latitude Climates		
Climate	**Temperature and precipitation**	**Description**
Marine west coast	low annual temperature range; frequent rainfall throughout the year	characterized by deciduous trees and dense forests; mild winters and summers
Steppe	large annual temperature range; annual precipitation of less than 40 cm	characterized by drought-resistant vegetation; cold, dry winters and warm, wet summers
Humid continental	large annual temperature range; annual precipitation of greater than 75 cm	characterized by a wide variety of vegetation and evergreen trees; variable weather
Humid subtropical	large annual temperature range; annual precipitation of 75 to 165 cm	characterized by broadleaf and evergreen trees; high humidity
Mediterranean	low annual temperature range; average annual precipitation of about 40 cm	characterized by broadleaf and evergreen trees; long, dry summers and mild, wet winters

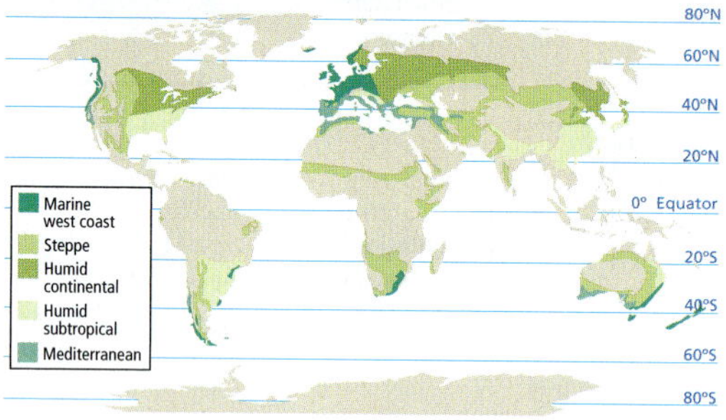

Middle-Latitude Climates

Climates that have an average maximum temperature of 8°C in the coldest month and an average minimum temperature of 10°C in the warmest month are referred to as **middle-latitude climates.** There are five middle-latitude climates, which are described in **Table 2.**

Marine west coast climates receive about 60 to 150 cm of precipitation annually. The average temperature is 20°C in the summer and 7°C in the winter. The Pacific Northwest of the United States has a marine west coast climate.

Steppe climates are dry climates that receive less than 40 cm of precipitation per year. The average summer temperature is about 23°C. The winters are very cold and have an average temperature of −1°C. The Great Plains of the United States has a steppe climate.

The *humid continental climate* and *humid subtropical climate* have a high annual precipitation. However, the humid continental has a much greater temperature range between the summers and winters than the humid subtropical. In the United States, the humid subtropical climate is in the southeast and the humid continental climate is located in the northeast.

The *mediterranean climate* is a mild climate that has a small temperature range between summer and winter. This climate is named after the sea between Africa and Europe, where this climate is located. However, this climate is also located along the coast of central and southern California.

middle-latitude climate a climate that has an average maximum temperature of 8°C in the coldest month and an average minimum temperature of 10°C in the warmest month

Reading Check **Which subclimates have a high annual precipitation? (See the Appendix for answers to Reading Checks.)**

ENVIRONMENTAL CONNECTION

The North American Rain Forest The term *rain forest* may conjure up images of a steamy tropical jungle. But within the marine west coast environment that extends from northern California into southern Alaska lies a temperate rain forest—the Olympic National Park in Washington state. In this long stretch of coast, the prevailing winds blow humid air onshore from the Pacific Ocean. The winds moderate the climate, giving the region mild winters and cool summers. The winds also bring precipitation—lots of it. As winds hit the coastal mountains, the air rises and dumps large amounts of rain and snow throughout the year. Thick forests of majestic trees, such as Sitka spruce, Douglas fir, Western cedar, and giant redwoods grow in this lush, mild region. The climate does not extend inland because of the coastal mountains. Have students investigate where else in the world the marine west coast environment exists and whether rain forests occur in these places. (southern Australia, much of western Europe, tip of South America; Unfortunately, rain forests that once existed in these locations have largely disappeared as a result of development and logging.)

Figure 1 ▶ Subarctic climates, as shown here at Tombstone Valley in Yukon, Canada, support sparse tree growth.

SCILINKS. NSTA Developed and maintained by the National Science Teachers Association

For a variety of links related to this subject, go to www.scilinks.org

Topic: Polar Climates
SciLinks code: HQ61175

Polar Climates

The climates of the polar regions are referred to as the **polar climates.** There are three types of polar climates: the subarctic climate, shown in **Figure 1,** the tundra climate, and the polar icecap climate. The *subarctic climate* has the largest annual temperature range of all climates. The difference between summer and winter temperatures in the subarctic climate has been as much as 63°C. The *tundra climate* has a smaller annual temperature range than the subarctic climate does. But the average temperature of the tundra is colder than that of the subarctic climate. In the *polar icecap climate,* most of the land surface and much of the ocean are covered in thick sheets of ice year-round. The average temperature in these regions never rises above freezing. The polar climates are described in **Table 3.**

polar climate a climate that is characterized by average temperatures that are near or below freezing; typical of polar regions

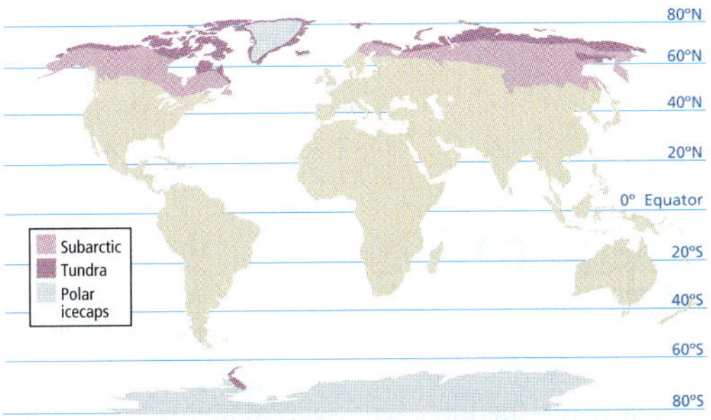

Table 3 ▼

Polar Climates		
Climate	**Temperature and precipitation**	**Description**
Subarctic	largest annual temperature range (63°C); annual precipitation of 25 to 50 cm	characterized by evergreen trees; brief, cool summers and long, cold winters
Tundra	average temperature below 4°C; annual precipitation of 25 cm	characterized by treeless plains; nine months of temperatures below freezing
Polar icecaps	average temperature below 0°C; low annual precipitation	characterized by little or no life; temperatures below freezing year-round and high winds

Close

Reteaching — BASIC

Flipbook Have students write all the subclimates on index cards with a description of temperature, precipitation, and other characteristics. Have them organize the subclimates into major climate zones and connect the cards together by using a binder ring. Students can illustrate the cards and use them for review. **LS Visual**

Quiz — GENERAL

1. What is a highland climate? (A highland climate is located in the mountains and has large variations in temperature and precipitation over short distances because of changes in elevation.)
2. What is a heat island? (an area that shows a higher temperature than surrounding areas; The rise in temperature commonly occurs over cities, where buildings and pavement absorb and reradiate a lot of heat.)

Alternative Assessment — GENERAL

Travel Agent Have students pretend they are travel agents. They should choose a travel destination and prepare a brochure that describes the destination, its climate, what types of wildlife lives there, and what travelers should pack for their trip. Encourage students to be creative and to use photos or illustrations in their brochure. They might also describe side trips to destinations that have different climates. **LS Verbal/Visual**

BRAIN FOOD

Mountain Climates Based on the type of vegetation that grows in different regions of mountains, mountain climates are divided into the montane, subalpine, and alpine zones. The montane zone is very similar to a coniferous forest. In the subalpine, no trees grow, only scrubby plants similar to those found on the tundra grow. Farthest up is the alpine, a harsh, variable climate that is populated by grasses, lichens, and mosses.

INCLUSION Strategies

- *Learning Disabled*
- *Behavior Control Issues*

Help students relate personally to different climate zones by making a chart called "Climates We Have Known." Label columns with the names of the eleven subclimates. Ask students to list their names on the left side of the chart and to write places where they have visited or lived under the correct headings to show which subclimates they have "known." **LS Intrapersonal**

Close, *continued*

Answers to Section Review

1. tropical, middle-latitude, and polar
2. rain forest: small temperature range, annual precipitation of 200 cm, lush, dense vegetation and high biodiversity; desert: large temperature range, precipitation less than 25 cm, little vegetation; savanna: small temperature range, alternating dry and wet periods with precipitation of 50 cm, open grasslands
3. marine west coast: small temperature range, high annual rainfall, mild winters and summers, dense forests; steppe: large temperature range, low precipitation, drought-resistant vegetation, cold, dry winters, and warm, wet summers; humid continental: large temperature range, precipitation greater than 75 cm, diverse vegetation, variable weather; humid subtropical: large temperature range, high precipitation, broadleaf and evergreen trees, high humidity; mediterranean: small temperature range, precipitation about 40 cm, broadleaf and evergreens, long, dry summers and mild, wet winters
4. subarctic: largest temperature range, precipitation of 25 to 50 cm, evergreen trees, cool summers and long, cold winters; tundra: average temp below 4°C, 9 months below freezing, precipitation of 25 cm, treeless plains; polar icecaps: average temp below 0°C, low annual precipitation, high winds, little or no life
5. vegetation, elevation, and proximity to bodies of water
6. City climates have surfaces, such as pavement and buildings, that absorb and reradiate large amounts of heat, so temperatures may be several degrees warmer than in surrounding rural areas.
7. Temperatures would be higher because pavement absorbs and reradiates more heat than vegetation does.
8. Climate zones do not always line up with latitude because of the effect of prevailing winds, elevation, and proximity to oceans and other large bodies of water.
9. Climate zones include *tropical climates*, such as *rain forests, savannas*, and *deserts; middle-latitude* climates, such as *mediterranean* climates and *steppes;* and *polar climates*, such as the *subarctic, tundra*, and *polar icecap* climates.

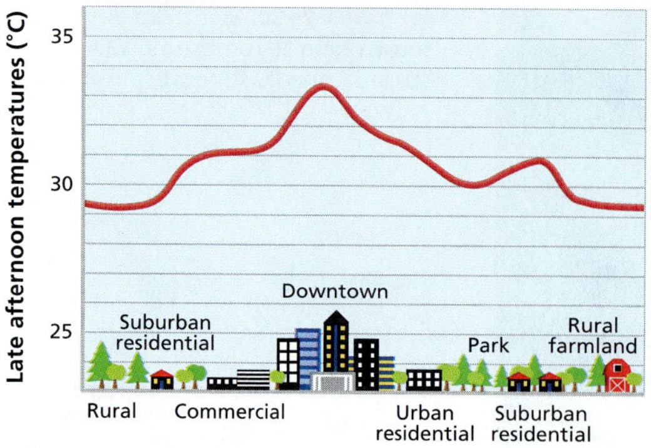

Figure 2 ▶ The less vegetation and more pavement and buildings an area has, the higher the temperatures in the area tend to be.

microclimate the climate of a small area

Local Climates

The climate of a small area is called a **microclimate.** Microclimates are influenced by density of vegetation, by elevation, and by proximity to large bodies of water. For example, in a city, pavement and buildings absorb and reradiate a lot of solar energy, which raises the temperature of the air above and creates a "heat island," as shown in **Figure 2.** As a result, the average temperature may be a few degrees higher in the city than it is in surrounding rural areas. In contrast, vegetation in rural areas does not reradiate as much energy, so temperatures in those areas are lower.

Effects of Elevation

Elevation also may affect local climates. As elevation increases, temperature decreases and the climate changes. For example, the *highland climate* is characterized by large variation in temperatures and precipitation over short distances because of changes in elevation. Highland climates are commonly located in mountainous regions—even in tropical areas.

Effects of Large Bodies of Water

Large bodies of water, such as lakes, influence local climates. The water absorbs and releases heat slower than land does. Thus, the water moderates the temperature of the nearby land. Large bodies of water can also increase precipitation. Therefore, microclimates near large bodies of water have a smaller range of temperatures and higher annual precipitation than other locations at the same latitude do.

Section 2 Review

1. **Identify** the three types of climate zones.
2. **Describe** the three types of tropical climates.
3. **Describe** the five types of middle-latitude climates.
4. **Describe** the three types of polar climates.
5. **Identify** three factors that influence microclimates.
6. **Explain** why city climates may differ from rural climates.

CRITICAL THINKING

7. **Making Inferences** What would happen to the temperature of a rural location if the vegetation were replaced with a parking lot?
8. **Compare and Contrast** Compare latitude lines with the boundaries of major climate zones. Why do they align in some regions but not in others?

CONCEPT MAPPING

9. Use the following terms to create a concept map: *tropical climate, subarctic, tundra, steppe, polar icecap, mediterranean, middle-latitude climate, rain forest, savanna, desert,* and *polar climate.*

CHAPTER RESOURCES

Chapter Resource File
- Section Quiz GENERAL

Workbooks
- Study Guide (also in Spanish)

ENVIRONMENTAL CONNECTION

Section 3 Climate Change

When discussing changes in climate, scientists must answer two questions. First, is the climate really changing, or are year-to-year changes just natural variation? Second, if the climate is changing, what is the cause? **Climatologists** are scientists who try to answer these questions by gathering data to study and compare past and present climates.

OBJECTIVES

- **Compare** four methods used to study climate change.
- **Describe** four factors that may cause climate change.
- **Identify** potential impacts of climate change.
- **Identify** ways that humans can minimize their effect on climate change.

KEY TERMS

climatologist
global warming

Studying Climate Change

Climatologists look to past climates to find patterns in the changes that occur. Identifying those patterns allows the scientists to make predictions about future climates. Climatologists use a variety of techniques to reconstruct changes in climate.

Collecting Climate Data

Today, scientists use thousands of weather stations around the world to measure recent precipitation and temperature changes. However, when trying to learn about factors that influence climate change, scientists need to study the evidence left by past climates. This evidence can be left in the remains of plants and animals from earlier time periods. For example, *fossils* of a plant or animal may show adaptations to a particular environment that can reveal clues about the environment's climate. Even polar icecaps contain evidence of past climates. By studying the concentration of gases in *ice cores,* scientists can learn about the gas composition of the atmosphere thousands of years ago. **Table 1** describes methods used to study past climates.

climatologist a scientist who gathers data to study and compare past and present climates and to predict future climate change

Table 1 ▼

Methods of Studying Past Climates

Method	What is measured	What is indicated	Length of time measured
Ice cores	concentrations of gases in ice and meltwater	High levels of CO_2 indicate warmer climate; ice ages follow decreases in CO_2.	hundreds of thousands of years
Sea-floor sediment	concentration of ^{18}O in shells of microorganisms	High ^{18}O levels indicate cool water; lower ^{18}O levels indicate warm water.	hundreds of thousands of years
Fossils	pollen types, leaf shapes, and animal body adaptations	Flower pollens and broad leaves indicate warm climates; evergreen pollens and small, waxy leaves indicate cool climates. Animal fossils show adaptations to climate changes.	millions of years
Tree rings	ring width	Thin rings indicate cool weather and less precipitation.	hundreds to thousands of years

Section 3

Focus

Overview

This section explains how scientists study past climate change and model future climate change. It describes natural factors and human activities that may alter climate and ways to minimize human effects on climate.

Bellringer

Have students imagine that the climate of the area in which they live has changed so that it is now colder than it used to be. Have students write down five different ways that they think the area would be affected by cooler temperatures. LS **Logical**

Motivate

Activity — GENERAL

Tree Rings Obtain a cross-section of a tree trunk. Have students look closely at the rings. Ask, "Are all of the rings the same thickness?" (no) Have students discuss under what conditions growth rings might be thicker or thinner. (Thicker rings grow during years that have good growing conditions, thinner rings form during years that have poor growing conditions. Tree rings show more about short-term climate changes than long-term climate changes. Tree rings also reveal the history of forest fires, disease, and weather.) LS **Visual**

CHAPTER RESOURCES

Chapter Resource File
- Directed Reading BASIC

Technology

Transparencies
- Bellringer

Student Edition on CD-ROM

One-Stop Planner CD-ROM
- Lesson Plan

Teach

Internet Activity — GENERAL

Climate Models Have interested students visit Web sites of government agencies, such as NASA's Goddard Institute for Space Studies and the University Corporation for Atmospheric Research, that use general circulation models (GCMs). Have them investigate the history of GCMs, how they are set up, what variables they incorporate, what predictions they make, and what limitations they have. Have students present their findings in a short oral report. A worksheet designed to direct student research on this topic can be found in the **Chapter Resource File** booklet or by visiting go.hrw.com and entering the keyword **HQ6CLIX**. LS **Verbal**

Debate — ADVANCED

Global Warming While many climatologists think that human activities are inducing global warming, some remain skeptical. Given the variety of other factors known to influence climate and the evidence that climates have changed in the past without human influence, these scientists do not think that increasing greenhouse gases are solely responsible for the temperature increases during the last century. Have interested students research different sides of this issue and present a debate to the class. LS **Verbal/Logical**

Answer to Reading Check

Scientists use computer models to incorporate as much data as possible to sort out the complex variables that influence climate and to make predictions about climate.

CHAPTER RESOURCES

Chapter Resource File

- Internet Activity
  - Climate Models GENERAL

Figure 1 ▶ Scientists need to use powerful computers to process the amount of data required to study climates.

Modeling Climates

Because so many factors influence climate, studying climate change is a complicated process. Currently, scientists use computers to create models to study climate, as shown in **Figure 1.** The models incorporate millions of pieces of data and help sort the complex sets of variables that influence climate. These models are called *general circulation models,* or GCMs. GCMs simulate changes in one variable when other variables are unchanged. For example, if the sulfur dioxide level is raised in a particular model, the model indicates a decrease in incoming solar radiation because sulfur dioxide reflects sunlight.

Climate models simulate many factors of climate, including temperature, precipitation, wind patterns, and sea-level changes. These computer models are complex because they model interactions between oceans, wind, land, clouds, and vegetation. As computers become more powerful, computer-generated climate models will provide greater detail about the global climate system and will help scientists better understand climate change.

Reading Check **Why do scientists use computers to model climate? (See the Appendix for answers to Reading Checks.)**

Connection to CHEMISTRY

Oxygen Isotopes

By studying the shells of certain marine organisms, scientists have found evidence that the oceans were much warmer during the Jurassic period than they are today. Marine shells form differently depending on the temperature of the ocean, because the concentration of oxygen isotopes in the marine shells depends on the temperature of the water.

Isotopes are atoms of the same element that have a different number of neutrons. All elements are made up of atoms that contain protons, electrons, and neutrons. Isotopes of the same element have the same number of protons and electrons, but have a different number of neutrons.

Because isotopes have a different number of neutrons, they have different mass numbers. For example, oxygen's most common isotope has a mass number of 16 and is written as ^{16}O. A less common isotope of oxygen is ^{18}O.

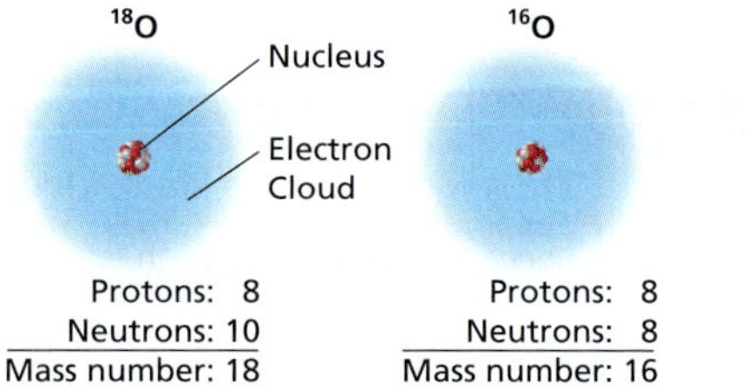

Marine shells contain both isotopes ^{16}O and ^{18}O. When the temperature of the sea water increases, the concentration of ^{18}O decreases and the concentration of ^{16}O increases. Scientists study the fossils of marine shells from different time periods to determine the concentration of each oxygen isotope in the marine shells. Upon determining the concentrations, scientists can predict the temperature of the oceans during the time period in which the shells formed.

CHEMISTRY CONNECTION — ADVANCED

Oxygen Isotopes The isotopes of oxygen differ in several physical and chemical ways. Most importantly, ^{18}O is heavier than ^{16}O. Because it is lighter, ^{16}O tends to evaporate more readily than ^{18}O does, so precipitation is rich in ^{16}O. Ocean water, on the other hand, is rich in ^{18}O. When glaciers are extensive and ocean is water cold, ^{18}O in seawater increases and the ratio of ^{18}O to ^{16}O in marine shells is higher than during warm periods. Scientists use a mass spectrometer, an instrument that can measure the molecular weight of substances, to determine the ratio of oxygen isotopes. And from this they can make inferences about the history of Earth's climate conditions. LS **Logical**

Potential Causes of Climate Change

By studying computer-generated climate models, scientists have determined several potential causes of climate change. Factors that might cause climate change include the movement of tectonic plates, changes in the Earth's orbit, human activity, and atmospheric changes.

Plate Tectonics

The movement of continents over millions of years caused by tectonic plate motion may affect climate changes. The changing position of the continents changes wind flow and ocean currents around the globe. These changes affect the temperature and precipitation patterns of the continents and oceans. Thus, the climate of any particular continent is not the same as it was millions of years ago.

Orbital Changes

Changes in the shape of Earth's orbit, changes in Earth's tilt, and the wobble of Earth on its axis can lead to climate changes, as shown in **Figure 2.** The combination of these factors is described by the *Milankovitch theory*. Each change of motion has a different effect on climate. Variation in the shape of Earth's orbit, from elliptical to circular, affects Earth's distance from the sun. Earth's distance from the sun affects the temperature of Earth and therefore affects the climate. Decreasing tilt decreases temperature differences between seasons. The wobble of Earth on its axis changes the direction of Earth's tilt and can reverse the seasons. These changes occur in cycles of 21,000 to 100,000 years.

Figure 2 ▶ Earth's Orbital Changes

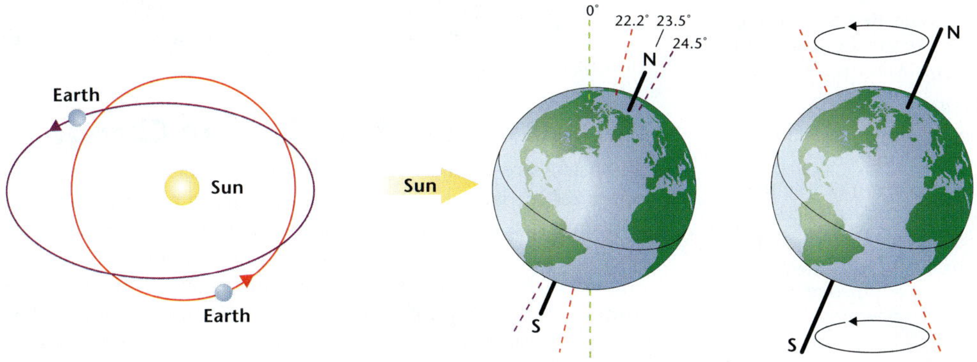

Eccentricity Earth encounters more variation in the energy that it receives from the sun when Earth's orbit is elongated than it does when Earth's orbit is circular.

Tilt The tilt of Earth's axis varies between 22.2° and 24.5°. The greater the tilt angle is, the more solar energy the poles receive.

Precession The wobble of Earth's axis affects the amount of solar radiation that reaches different parts of Earth's surface at different times of the year.

Graphic Organizer **Cause-and-Effect Map**

Create the **Graphic Organizer** entitled "Cause-and-Effect Map" described in the Skills Handbook section of the Appendix. Label the effect with "Climate change." Then, fill in the map with causes of climate change and details about the causes and effects.

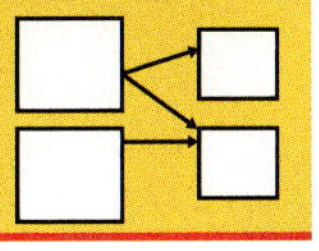

Graphic Organizer GENERAL

Cause-and-Effect Map
You may want to use this Graphic Organizer to assess students' understanding of the causes of climate change. Collect the Graphic Organizers from the students and correct any mistakes before returning the Graphic Organizers.

BRAIN FOOD

The Little Ice Age While the last glacial period ended more than 15,000 years ago, many parts of the world experienced what is known as the "Little Ice Age" between the 15th and the mid-19th centuries. In Europe, winters were harsh, springs and summers were cold and wet, crops failed, growing seasons were shorter, and mountain glaciers advanced. North America and China experienced similar trends. Around 1900, temperatures started rising again and continued to rise through the 20th century.

CHAPTER RESOURCES

Technology

Transparencies
- 128 Orbital Changes and Climate (with worksheet)

ASTRONOMY CONNECTION

A Variable Star The sun is a variable star; its energy output varies over time. During the last 200 to 300 years, the sun's brightness has increased about 0.4 percent. Some GCMs indicate that this increase may be responsible for 40 percent of the observed global warming since 1900. Changes in the number and duration of sunspots, the dark, cooler regions on the surface of the sun, also correlate with changes in Earth's temperature. Current climate changes may be the result of both natural events and human activities.

INCLUSION Strategies

- *Learning Disabled*
- *Behavior Control Issues*
- *Developmentally Delayed*

To help students stay involved with the text, ask them to think of and write possible test questions while they are reading the text. Then, have students review by asking other students their questions. After the question-and-answer session, collect all of the questions. Tell students you are going to use some of their questions for a test or quiz—and do so. **LS Verbal**

Teach, continued

Homework — ADVANCED

Volcanoes and Climate Volcanic eruptions can have dramatic effects on climates around the world. In June 1991, Mount Pinatubo erupted, spewing millions of tons of soot and ash into the atmosphere. Within a year, average global temperatures dropped about 0.5°C and many regions experienced a harsher winter and cooler summer. Other eruptions have had similar effects, perhaps most notably, the "year without a summer" that followed the 1815 eruption of Mount Tambora in Indonesia. Have interested students investigate major volcanic eruptions throughout history and prepare a timeline documenting the effects on climate. LS **Verbal**

Answers to Reading Check

Climate change influences humans, plants, and animals. It also affects nearby climates, sea level, and precipitation rates.

QuickLAB

Skills Acquired

- Measuring
- Analyzing

Teacher's Notes Make sure the thermometer is not directly in the sun and that students measure temperature at the same height over the grass and the pavement.

Answers

1. The temperature should be hotter over the pavement.
2. Pavement absorbs and reradiates more heat than a grassy field does, heating the air above it to a higher temperature.
3. Make sure to have some trees and other landscaping around the new store and not just a massive expanse of parking lot.

Figure 3 ▶ Most deforestation in Brazil is caused by farmers who clear the land for planting crops.

QuickLAB — 15 min

Hot Stuff

Procedure

1. In mid-afternoon, use a **thermometer** to measure the air temperature over a grassy field or other vegetated area. Make sure to shield the thermometer from direct sunlight.
2. Measure the air temperature over a parking lot. Take the measurement at the same height above the surface as the height of the measurement of the vegetated area. Again, make sure the thermometer is not directly in the sunlight.

Analysis

1. How did the results differ for each location?
2. How would you explain the difference in the results?
3. What suggestion for how to keep cooling costs low would you give to someone who is building a new store?

Human Activity

Human activity affects climate through emissions and land use. Pollution from transportation and industry releases carbon dioxide, CO_2, into the atmosphere. Increases in CO_2 concentrations may lead to global warming, an increase in temperatures around Earth. CO_2 is also released into the atmosphere when trees are burned to provide land for agriculture and urban development. Because vegetation uses CO_2 to make food, deforestation, as shown in **Figure 3,** also affects one of the natural ways of removing CO_2 from the atmosphere. As scientists continue to study climate, they will learn more about how human activity affects climate and about how changes in climate may affect us.

Volcanic Activity

Large volcanic eruptions can influence climates around the world. Sulfur and ash from eruptions can decrease temperatures by reflecting sunlight back into space. These changes last from a few weeks to several years and depend on the strength and duration of the eruption.

Potential Impacts of Climate Change

Scientists are concerned about climate changes because of the potential impacts of these changes. Earth's atmosphere, oceans, and land are all connected, and each influences both local and global climates. Changes in the climate of one area can affect climates around the world. Climate change affects not only humans but also plants and animals. Even short-term changes in the climate may lead to long-lasting effects that may make the survival of life on Earth more difficult for both humans and other species. Some of these potential climate changes include global warming, sea-level changes, and changes in precipitation.

✓ Reading Check **What things are influenced by climate change?** (See the Appendix for answers to Reading Checks.)

CHAPTER RESOURCES

Chapter Resource File

- **Datasheet for Quick Lab** GENERAL
- **Inquiry Lab** Particulates in the Atmosphere GENERAL

Demonstration — BASIC

Icecaps Melting Sea levels may rise if icecaps melt but only if those icecaps cover land, as the Antarctic ice sheet does. Fill a bowl with water, and add a few ice cubes to simulate floating icecaps. Mark the water level, and then let the ice melt, observing any changes as the ice melts. (The water level should stay the same.) Repeat with the same level of water, but with an island made of clay in the bowl. Place the ice cubes on the clay, and mark the water level. Let the ice melt, and observe how the water level rises. LS **Visual**

Global Warming

Global temperatures have increased approximately 1°C over the last 100 years. Researchers are trying to determine if this increase is a natural variation or the result of human activities, such as deforestation and pollution. A gradual increase in average global temperatures is called **global warming.** This process may result from an increase in the concentration of greenhouse gases, such as CO_2, in the atmosphere.

An increase in global temperature can lead to an increase in evaporation. Increased evaporation could cause some areas to become drier than they are now. Some plants and animals would not be able to live in these drier conditions. An increase in evaporation in other areas could cause crops to suffer damage. However, an increase in temperatures due to global warming might improve conditions for crops in colder, northern regions.

An increase in global temperatures could also cause ice at the poles to melt. If a significant amount of ice melts, sea levels around the world could rise. This rise in sea levels would cause flooding around coastlines, where many cities are located.

global warming a gradual increase in the average global temperature that is due to a higher concentration of gases such as carbon dioxide in the atmosphere

SCILINKS. NSTA Developed and maintained by the National Science Teachers Association

For a variety of links related to this subject, go to www.scilinks.org

Topic: Global Warming
SciLinks code: HQ60681

Sea-Level Changes

Using computer models, some scientists have predicted an increase in global temperature of 2°C to 4°C during this century. An increase of only a few degrees worldwide could melt the polar icecaps and raise sea level by adding water to the oceans. On a shoreline that has a gentle slope, the shoreline could shift inland many miles, as shown in **Figure 4.** Many coastal inhabitants would be displaced, and freshwater and agricultural land resources would be diminished. Because approximately 50% of the world population lives near coastlines, this sea-level rise would have devastating effects.

Figure 4 ▶ As sea level rises, shorelines could shift inland many miles. ***Which two states would lose the most area if sea level were to rise by 3 m?***

READING SKILL BUILDER — BASIC

Reading Organizer As students read this section, encourage them to take Power Notes, KWL notes, or Two-Column Notes as described in the Skills Handbook section of the Appendix. Later, students can use these notes as a study guide for assessments. LS **Verbal** English Language Learners

Close

Reteaching — BASIC

Peer Reviewing Have students work in pairs to write three to five questions about the material in this section. Then, have students quiz each other by using the questions they wrote. LS **Interpersonal** Co-op Learning

Quiz — GENERAL

1. What does a climatologist do? (studies and compares past and present climates to help predict future climate changes)
2. What is a hybrid car and how can its use reduce pollution? (A hybrid car uses both gasoline and electricity. It releases less CO_2 into the atmosphere.)

Alternative Assessment — GENERAL

Global Warming in the News Have students write a newspaper style article about global warming. The article should describe how scientists study and model past and present climate change and what their models predict for the future. LS **Verbal**

LAW CONNECTION — ADVANCED

Kyoto Protocol In 1997, 160 nations met in Japan to negotiate a treaty limiting greenhouse gas emissions in an effort to forestall climate change. The nations agreed to reduce emissions from 1990 levels by an average of 5.2%. Today, 104 countries, accounting for about 44% of emissions, have ratified the treaty. Nineteen countries, including the United States, have signed but not ratified it. Have interested students investigate the status of the treaty and the position of the United States. Have them present findings in a poster or oral report. LS **Logical/Verbal**

Using the Figure — GENERAL

Rising Seas If global warming continues, rising sea levels pose real threats to more than half of the world's people, those who live near coastlines. Recent estimates indicate that sea levels could rise almost a meter by 2100. Cities along coasts would be prone to flooding. Some islands might disappear. Bangladesh could lose up to 18% of its land, displacing millions of people. Have students investigate the causes and consequences of rising sea levels. Answer to caption question: Louisiana and Florida LS **Interpersonal**

Close, *continued*

Answers to Section Review

1. ice cores: measure concentrations of CO_2 in ice and meltwater; sea-floor sediments: measure concentrations of oxygen isotopes in microorganism shells; fossils: measure changes in pollen types, leaf shapes, and animal adaptations; tree rings: measure ring widths; Ice cores, sea-floor sediments and tree rings help assess climate change over hundreds of thousands of years, while fossils help reconstruct changes over millions of years.
2. movement of continents due to plate tectonics; changes in Earth's orbit and tilt; volcanic activity; human activity
3. When Earth's orbit is elliptical, Earth encounters more variation in energy from the sun. The tilt of the Earth's axis also changes. When tilt decreases, temperature differences between seasons decreases. Wobble changes the direction of tilt and can reverse the seasons.
4. Carbon dioxide is a greenhouse gas that absorbs heat radiated by Earth's surface. Increasing carbon dioxide could increase the greenhouse effect and lead to global warming.
5. Answers may vary. Sample answer: One potential negative impact of global warming would be the flooding of coastal communities caused by rising sea levels as polar icecaps melt.
6. Treaties and laws aimed at reducing pollution can help countries reduce the impacts of global warming. Monitoring of industrial practices can also play a role in reducing those impacts.
7. By turning off lights, reducing heating and cooling, recycling, and using mass transportation, individuals can reduce the impacts of global warming.
8. The melting of small icebergs would not affect sea level because they are floating in the ocean water. They displace the same amount of water as their volume.
9. Short-term climate changes could not be explained by the Milankovitch theory because these orbital changes occur on cycles of 21,000 to 100,000 years.
10. A *climatologist* may study *fossils, tree rings, ice cores,* oxygen *isotopes* in sea-floor sediments, and *general circulation models* to study climate change and *global warming.*

Figure 5 ▶ People from the Wangari Maathai Green Belt Movement in Kenya, Africa, prepare seedlings for planting.

What Humans Can Do

Many countries are working together to reduce the potential effects of global warming. Treaties and laws have been passed to reduce pollution. Industrial practices are being monitored and changed. Even community projects to reforest areas, such as the one shown in **Figure 5,** have been developed on a local level.

Individual Efforts

Each individual person can also help reduce pollution that is caused by the burning of fossil fuels, such as running automobiles and using electricity. These activities increase CO_2 concentrations in the atmosphere. Therefore, humans can have a significant effect on pollution rates by turning lights off when they are not in use, by turning down the heat in winter, and by reducing air conditioner use in the summer. Recycling is also helpful because less energy is needed to recycle some products than to create them.

Transportation Solutions

Using public transportation and driving fuel-efficient vehicles also help release less CO_2 into the atmosphere. All vehicles burn fuel more efficiently when they are properly tuned and the tires are properly inflated. Driving at a consistent speed also allows a vehicle to burn fuel efficiently. Car manufacturers have been developing cars that are more fuel efficient. For example, *hybrid cars* use both gasoline and electricity. These cars release less CO_2 into the atmosphere from burning fuel than other cars do.

Section 3 Review

1. **Compare** four methods that climatologists use to study climate.
2. **Identify** four factors that may cause climate change.
3. **Describe** how orbital changes may affect climate.
4. **Explain** how changes in CO_2 concentrations affect global temperatures.
5. **Explain** one potential negative impact of global warming.
6. **Identify** two ways that countries can work together to reduce the potential effects of global warming.
7. **Identify** four ways that an individual can reduce the potential effects of global warming.

CRITICAL THINKING

8. **Making Predictions** How would the melting of small icebergs affect sea level? Explain your answer.
9. **Evaluating Models** Can short-term climate changes be explained by using the cycles described by the Milankovitch theory? Explain your answer.

CONCEPT MAPPING

10. Use the following terms to create a concept map: *climatologist, general circulation models, global warming, ice cores, tree rings, fossils,* and *isotopes.*

CHAPTER RESOURCES

Chapter Resource File

- **Section Quiz** GENERAL

Workbooks

- **Study Guide** (also in Spanish)

Chapter 25 Highlights

Sections

1 Factors That Affect Climate

Key Terms

climate, 631
specific heat, 634
El Niño, 635
monsoon, 635

Key Concepts

- The climate of a region is described by the region's temperature and precipitation.
- Latitude affects climate by determining the intensity of solar energy received by an area.
- The rates at which land and water are heated affect climate.
- Topography affects climate by causing temperature variations due to elevation and by creating rain shadows.

2 Climate Zones

tropical climate, 637
middle-latitude climate, 638
polar climate, 639
microclimate, 640

- Tropical subclimates are located near the equator and include tropical rain-forest, tropical desert, and savanna climates.
- Middle-latitude climates include marine west coast, steppe, humid continental, humid subtropical, and mediterranean climates.
- Polar climates include subarctic, tundra, and polar icecap climates.
- Microclimates are influenced by large bodies of water, by elevation, and by vegetation and urban development.

3 Climate Change

climatologist, 641
global warming, 645

- By using ice cores, sea-floor sediment, fossils, and tree rings, scientists have been able to study past climates.
- Computer-generated climate models help scientists predict possible consequences of changing certain variables in the climate system.
- Natural processes and human activities may be causing changes in Earth's climate, including global warming.

Chapter Highlights

Alternative Assessment — ADVANCED

Subclimate Story Have students choose a subclimate other than their own and write a story about a family that lives in that subclimate. The family should include at least one teenager. The story plot may be in the form of "a day in the life" of the family or something more complex. The story should include references to the factors that have shaped the climate, adaptations people make to the climate in their daily lives, and how climate change might alter this climate and its inhabitants. Allow students time for additional research. Have volunteers read their stories aloud to the class. **LS Verbal/Interpersonal**

CHAPTER RESOURCES

Chapter Resource File

- **Concept Review** GENERAL
- **Critical Thinking** ADVANCED
- **Math Skills** GENERAL
- **Graphing Skills** GENERAL
- **Chapter Test A** GENERAL
- **Chapter Test B** ADVANCED

Workbooks

- **Study Guide** (also in Spanish)
- **Assessments** (Spanish)

Technology

Classroom Videos
- Brain Food Video Quiz

Chapter Review

Assignment Guide

Section	Questions
1	1, 5, 7, 9–12, 22, 24
2	2, 6, 13–15, 18, 20, 27
3	3, 8, 16–17, 25–26, 28–30
1 and 2	4, 21
1–3	19, 23

Using Key Terms

1-6. Answers may vary but should show that students understand the definitions of and differences between key terms.

Understanding Key Concepts

7. b **8.** b
9. a **10.** b
11. c **12.** c
13. c **14.** a
15. b

Short Answer

16. The Milankovitch theory proposes that cyclic changes in Earth's orbit from circular to elliptical and changes in the tilt and wobble of Earth's axis can lead to climate change by altering the amount of energy Earth receives from the sun. Changes to incoming solar energy would affect Earth's temperature and thus could cause some climate changes.

17. Answers may vary. Accept all reasonable answers. The possible effects of global warming include changes in global precipitation patterns; melting of polar icecaps and rising sea levels, leading to coastal flooding; and changes in agricultural production.

18. The marine west coast climate has less seasonal temperature fluctuation than the humid continental climate does. Both receive significant amounts of precipitation throughout the year. Marine west coast climates are characterized by mild winters and summers, while humid continental climates have variable weather.

Critical Thinking

19. If all of the trees in California were cut down, the climate might be significantly warmer during the day because vegetation absorbs solar energy but does not reradiate as much energy as bare ground does. Vegetation also moderates night temperatures, so nights might be colder.

20. Vegetation in the tundra is sparse because the temperature is so cold (<4°C) almost all of the time. Some layers of the soil are permanently frozen, and few plants can survive these cold conditions.

Chapter 25 Review

Using Key Terms

Use each of the following terms in a separate sentence.

1. *specific heat*

2. *microclimate*

3. *climatologist*

For each pair of terms, explain how the meanings of the terms differ.

4. *climate* and *microclimate*

5. *El Niño* and *monsoon*

6. *tropical climate* and *polar climate*

Understanding Key Concepts

7. At the equator, the sun's rays always strike Earth
- **a.** at a low angle.
- **b.** at nearly a 90° angle.
- **c.** 18 hours each day.
- **d.** no more than 8 hours each day.

8. Which of the following is *not* used as evidence of past climates?
- **a.** ice cores
- **b.** general circulation models
- **c.** tree rings
- **d.** fossils

9. Water cools
- **a.** more slowly than land does.
- **b.** more quickly than land does.
- **c.** only during evaporation.
- **d.** during global warming.

10. Ocean currents influence temperature by
- **a.** eroding shorelines.
- **b.** heating or cooling the air.
- **c.** washing warm, dry sediments out to sea.
- **d.** dispersing the rays of the sun.

11. Winds that blow in opposite directions in different seasons because of the differential heating of the land and the oceans are called
- **a.** chinooks.
- **b.** foehn.
- **c.** monsoons.
- **d.** El Niño.

12. When a moving air mass encounters a mountain range, the air mass
- **a.** stops moving.
- **b.** slows and sinks.
- **c.** rises and cools.
- **d.** reverses its direction.

13. In regions that have a mediterranean climate, almost all of the yearly precipitation falls
- **a.** during monsoons.
- **b.** in the summer.
- **c.** in the winter.
- **d.** during hurricanes.

14. The climate that has the largest annual temperature range is the
- **a.** subarctic climate.
- **b.** middle-latitude desert climate.
- **c.** mediterranean climate.
- **d.** humid continental climate.

15. The pavement and buildings in cities affect the local climate by
- **a.** decreasing the temperature.
- **b.** increasing the temperature.
- **c.** increasing the precipitation.
- **d.** decreasing the precipitation.

Short Answer

16. Describe the Milankovitch theory, including how it may explain some climate changes.

17. What are the possible effects of global warming?

18. Compare marine west coast and humid continental climates.

Critical Thinking

19. Making Predictions Describe how the climate in California might be affected if all of the trees in California were cut down.

20. Making Inferences Explain why the vegetation in areas that have a tundra climate is sparse even though these areas receive enough precipitation to support plant life.

21. Analyzing Ideas Explain why climates cannot be classified only by latitude.

22. Predicting Consequences How would global climate be affected if Earth were not tilted on its axis? Explain your reasoning.

Concept Mapping

23. Use the following terms to create a concept map: *fossil, ice cores, climate, polar climate, climatologist, steppe, temperature range, tropical climate, middle-latitude climate,* and *savanna.*

Math Skills

24. Using Equations Temperature generally decreases about 6.5°C for every kilometer above sea level. If T_N = temperature at new altitude, a = altitude in kilometers, and T_I = the initial temperature at sea level, what equation can be used to find the temperature at a given altitude?

25. Making Calculations From 1970 to 1997, nitrogen oxides, NO_X, emissions increased from about 18.7 million to about 20.8 million tons per year. By what percentage did NO_X emissions increase from 1970 to 1997?

Writing Skills

26. Researching Topics Research greenhouse gases to determine how they are produced. Then, write a brief essay that outlines how they can be reduced.

27. Communicating Main Ideas Imagine that you are going to build a vacation house. Research three locations where you would like to build your house, and outline the climate features that would make each location ideal for your vacation home.

Interpreting Graphics

The pie graphs below show world emissions of carbon dioxide, CO_2, in 1995 and predict emissions in 2035. Use these graphs to answer the questions that follow.

Total World Emissions of Carbon Dioxide

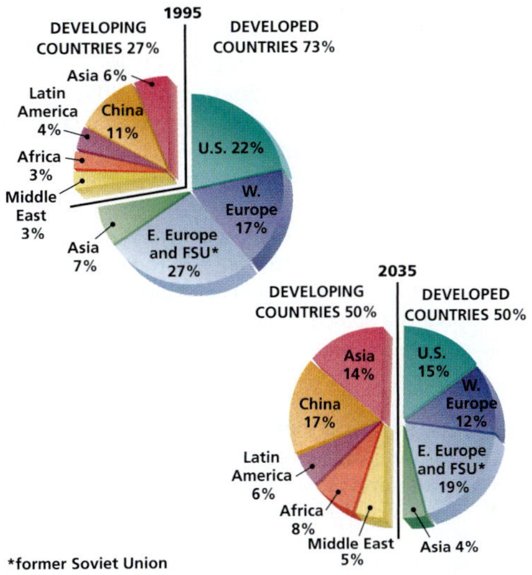

*former Soviet Union

28. In 1995, which country or region emitted the most CO_2? Which emitted the least CO_2?

29. What percentage of the total CO_2 was emitted by developing countries in 1995?

30. Why do you think researchers predict that CO_2 emissions by developing countries will equal that of developed countries by 2035?

Chapter Review

21. Climates cannot be classified by latitude alone because land features such as mountains, dense vegetation, proximity to large bodies of water, and ocean currents can greatly affect climate.

22. If Earth were not tilted on its axis, there would be no seasons in the temperate zones of the Northern and Southern hemispheres. Seasons result from the tilt of Earth's axis. The tilt causes changes in the angle at which the sun's rays hit Earth in the temperate zones as the planet orbits the sun. In winter, the northern part of the axis tilts away from the sun, the angle at which sunlight strikes the Northern Hemisphere is smaller, and the sun's energy is weaker at higher latitudes of the Northern Hemisphere. In summer, the opposite occurs.

Concept Mapping

23. Answers may vary but should include all of the terms listed. Sample answers appear at the end of this Teacher Edition.

Math Skills

24. $T_N = T_I - (6.5°C \times a)$

25. Percentage increase = $[(20.8 \text{ million} - 18.7 \text{ million}) \div 18.7 \text{ million}] \times 100 = 11\ \%$

Writing Skills

26. Answers may vary. Accept all reasonable answers.

27. Answers may vary. Accept all reasonable answers.

Interpreting Graphics

28. Eastern Europe and the former Soviet Union emitted the most CO_2 in 1995 (27%), though the United States emitted the most of any single country (22%). Africa and the Middle East emitted the least (3% each).

29. Developing countries emitted 27% of total CO_2 emissions in 1995.

30. Answers may vary. Possible answers include rapid population growth and/or rapid economic development in developing countries, both of which will require higher energy consumption. Also, many developed countries are attempting to reduce their CO_2 emissions.

Estimated Time

To give students practice under more realistic testing conditions, allow them 30 minutes to answer all of the questions in this practice test.

TEST DOCTOR

Question 1 Answer B is correct. Water has a higher specific heat than land, which means that if both are exposed to the same amount of solar energy, the temperature of the water will increase less. Therefore, answers A, C, and D are incorrect.

Question 2 Answer G is correct. The El Niño–Southern Oscillation is a complex phenomenon that involves *both* winds and ocean currents, so answers F and I are incorrect. As wind patterns change at the equator, the movement of surface water warmed, not cooled, by the sun changes, so answer H is incorrect.

Question 8 Answer I is correct. Answer F is incorrect because, while the greenhouse effect has been amplified by the use of fossil fuels by humans, it did not necessarily cause humans to increase the use of such fuels. Answer G is incorrect because humans did not create the greenhouse effect. Answer H is incorrect because humans are not the only producers of greenhouse gases.

Chapter 25 Standardized Test Prep

Understanding Concepts

Directions (1–4): **For** *each* **question, write on a separate sheet of paper the letter of the correct answer.**

1 Which statement best compares how land and water are heated by solar energy?
- **A.** Water heats up faster and to a higher temperature than land does.
- **B.** Land heats up faster and to a higher temperature than water does.
- **C.** Water heats up more slowly but reaches a higher temperature than land does.
- **D.** Land heats up more slowly and reaches a lower temperature than water does.

2 Which of the following statements best describes the El Niño–Southern Oscillation?
- **F.** a change in global wind patterns that occurs in the Southern Hemisphere
- **G.** a warming of surface waters in the eastern Pacific due to the effects of changing wind patterns on ocean currents near the equator
- **H.** a cooling of surface waters in the eastern Pacific due to the effects of changing wind patterns on ocean currents near the equator
- **I.** a global wind and precipitation belt between 20°N and 30°N latitude

3 A seasonal wind that blows toward the land in the summer and brings heavy rains is called a
- **A.** trade wind
- **B.** jet stream
- **C.** doldrum
- **D.** monsoon

4 In samples of atmospheric gases taken from ice cores, high levels of carbon dioxide indicate that the sample is from a time period that had
- **F.** a warm climate
- **G.** a cool climate
- **H.** high amounts of precipitation
- **I.** low amounts of precipitation

Directions (5–6): **For** *each* **question, write a short response.**

5 What is the term for the area around a mountain that receives warm, dry winds?

6 What is the term for the average weather in an area over a long period of time?

Reading Skills

Directions (7–9): **Read the passage below. Then, answer the questions.**

The Greenhouse Effect

The greenhouse effect is Earth's natural heating process, in which gases in the atmosphere trap thermal energy. Earth's atmosphere acts like the glass windows of a car. Imagine that it is a hot day and that you are about to get inside a car. You immediately notice that it feels hotter inside the car than it does outside the car.

Many scientists hypothesize that the rise in global temperatures is due to an increase carbon dioxide that is produced as a result of human activity. Most evidence indicates that the increase in carbon dioxide is caused by the burning of fossil fuels that release carbon dioxide into the atmosphere. Fossil fuels are organic compounds that are formed from the buried remains of ancient plants and animals. These fuels are used by humans for many things such as heating homes and providing fuel for automobiles.

7 Based on the passage, which of the following statements is not true?
- **A.** The atmosphere of Earth traps thermal heat in a similar manner to the way a car window traps heat.
- **B.** The greenhouse effect is a natural heating process for Earth.
- **C.** Earth absorbs sunlight and reradiates it as carbon dioxide.
- **D.** Human activity is the one producer of the greenhouse gas carbon dioxide.

8 Which of the following statements can be inferred from the information in the passage?
- **F.** The greenhouse effect is responsible for an increase in the use of fossil fuels by humans.
- **G.** Humans created the greenhouse effect by burning coal for industrial uses.
- **H.** Human activity is the only producer of gases that create the greenhouse effect.
- **I.** Human activity may play a role in amplifying the natural process of the greenhouse effect.

9 Name some fossil fuels that are contributors to the production of carbon dioxide?

Answers

Understanding Concepts

1. B
2. G
3. D
4. F
5. rain shadow
6. climate

Reading Skills

7. C
8. I
9. coal, natural gas, and oil

Interpreting Graphics

10. A
11. city A; City A is located in the rain shadow of a mountain range.
12. F
13. Answers may vary. See Test Doctor for a detailed scoring rubric.

Interpreting Graphics

Directions (10–13): **For** *each* **question below, record the correct answer on a separate sheet of paper.**

The diagram shows the locations of two cities at the same latitude. Use this map to answer questions 10 and 11.

Two Cities Separated by Coastal Mountains

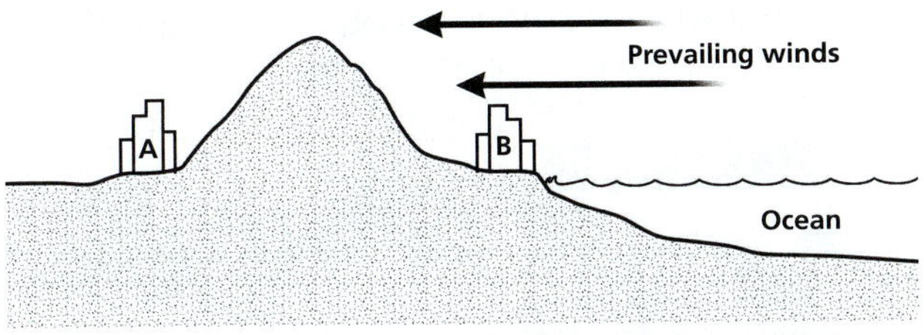

10 Which city is most likely to have the largest yearly temperature range?
A. City A would likely have the largest yearly temperature range.
B. City B would likely have the largest yearly temperature range.
C. Both cities would likely have the same temperature range.
D. There is not enough information to answer the question.

11 Which city is most likely to have a dry climate? Explain what would cause this city's climate to be drier than the other city's climate.

The climatograms below summarize average monthly precipitation and temperature data measured in two locations over a period of one year. Use these climatograms to answer questions 12 and 13.

Climatograms for Two Cities

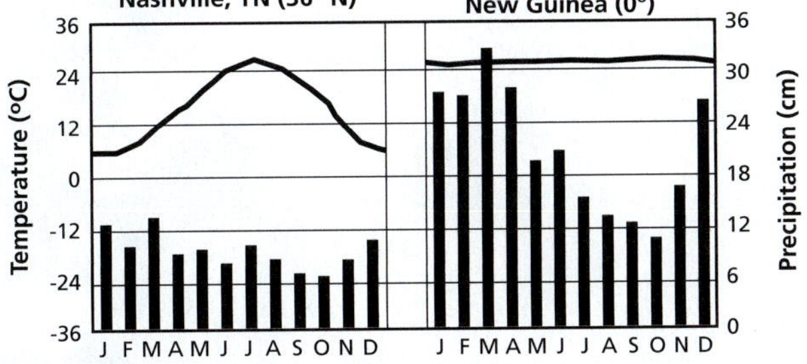

12 Which month shows the most rainfall for both climates in the climatograms?
F. March
G. June
H. September
I. December

13 Based on the data in the climatograms, write a description of the climate found in each location and the type of vegetation that is likely to occur as a result of the climate in each location.

Test TIP

Read all of the information, including the heads, in a table or chart before answering the questions that refer to it.

TEST DOCTOR

Question 10 Answer A is correct. Students should understand that city A is located inland, far from the ocean. Students that miss this question may not understand that land absorbs and releases heat more rapidly than water does. This produces a larger temperature range around the location city A than the temperature range at the location of city B.

Question 13 Full-credit answers should include the following points:
- students can locate summer and winter months on the x-axis of each climatogram and use the temperature and precipitation data for these months to start their analyses
- locations that have large amounts of consistent rainfall are likely to be lush year-round. Cities that have more-distinct seasons are more likely to have deciduous forest vegetation
- climatogram A shows a climate that has warm, humid summers and cold winters
- the area near climatogram A likely has deciduous forest vegetation
- climatogram B shows a climate that is hot and rainy all year
- the area near climatogram B likely has tropical rain-forest vegetation

Test Prep Correlations — *National Science Education Standards*

ES 1a: items 1, 7, 10
ES 1d: items 2, 3, 5, 6, 11, 12
ES 3c: item 4
LS 4d: item 13
SPSP 4b: items 8, 9

CHAPTER RESOURCES

State Resources

For specific resources for your state, visit go.hrw.com and type in the keyword HSHSTR.

Inquiry Lab

Factors That Affect Climate

Teacher's Notes

Time Required
one 45-minute class period

Lab Ratings
EASY → HARD

Teacher Preparation (1)
Student Setup (2)
Concept Level (2)
Cleanup (2)

Skills Acquired
- Observing
- Measuring
- Experimenting
- Predicting
- Collecting Data
- Organizing and Analyzing Data

The Scientific Method
In this lab, students will
- Ask Questions
- Test a Hypothesis
- Make Observations
- Analyze the Results
- Draw Conclusions

Materials
The materials listed on this page are enough for groups of two to four students.

Tips and Tricks
Make sure that the two containers are the same size and shape and that the lamp is positioned so that it shines evenly on both containers. The thermometers should be positioned the same distance below the surface of the soil and the water, and the same volume of soil and water should be used in the containers. Also, make sure that the soil is dry. Have students predict the results of this activity if they used different soil types or sand or gravel.

Chapter 25

Inquiry Lab

Using Scientific Methods

Objectives
- **Determine** whether land or water absorbs heat faster.
- **Explain** how the properties of land and water affect climate.

Materials
- container (2)
- heat lamp
- meterstick
- soil
- thermometer, Celsius (2)
- water

Safety

Factors That Affect Climate

Many factors affect climate. One of the most significant factors that influence climate is the distribution of land and water. Because land and water absorb and release heat energy differently, they affect the atmosphere differently. In turn, the differences between land and water affect the climate. In this lab, you will explore how the properties of land and water affect climate.

Ask a Question

1. How do the properties of land and water affect climate?

Form a Hypothesis

2. On a separate piece of paper, write a hypothesis that is a possible answer to the question above.

Test the Hypothesis

3. Fill one container with soil and the other with water. Place both containers on a flat surface next to each other.
4. Place the thermometer in the soil, as shown below, and record the temperature.

Step 5

CHAPTER RESOURCES

Chapter Resource File
- Datasheet for Chapter Lab GENERAL
- Lab Notes and Answers

Workbooks

Long-Term Projects
- Comparing Climate Features ADVANCED

5 Place the second thermometer in the container of water, as shown in the illustration on the previous page. The bulbs of both thermometers should be placed so that they are covered by no more than 0.5 cm of water or soil.

6 Place the heat lamp 25 cm above both containers. Turn on the heat lamp.

7 Create a data table like the one shown to the right. In the table, record the temperature of each sample at 1, 3, 5, and 10 min intervals.

Data Table

Time (min)	Temperature of soil (°C)	Temperature of water (°C)
1		
3		
5	DO NOT WRITE IN THIS BOOK	
10		
5 (after light off)		

8 Disconnect the lamp, and record the temperature of the soil and water after 5 min. **CAUTION** Be sure to let the heat lamp cool before storing it.

ANALYZE THE RESULTS

1 **Analyzing Data** Which substance absorbed more heat energy: water or soil?

2 **Analyzing Results** Which substance lost heat energy faster when the heat source was turned off: water or soil?

DRAW CONCLUSIONS

3 **Evaluating Conclusions** What conclusion can you draw about how land and water on Earth are heated by the sun?

4 **Analyzing Methods** Does this experiment describe how proximity to a body of water affects the temperature of a region? If so, explain your answer. If not, how could you test that variable?

Extension

1 **Applying Ideas** Repeat this experiment, but modify the angle at which the light strikes the surface of the soil and the water. How do your results differ from the results of the original experiment? How does the angle of the light affect temperature change in water and soil?

Inquiry Lab

Answers to Analyze the Results

1. The soil absorbed more heat than the water did.
2. The soil lost heat faster than the water did.

Answers to Draw Conclusions

3. The sun heats the land faster than it does the water. The water, on the other hand, will retain heat from the sun longer than the land does and the water will cool down more slowly.
4. While this experiment provides evidence that proximity to a body of water might have an effect on the temperature of a region, it does not really test that variable specifically. One way to test that variable would be to take measurements of soil temperatures in areas near a large body of water and away from the water at various times on a sunny day and again at different times throughout the following night.

Answers to Extension

1. At angles of less than 90°, the soil and water should not heat up as much or as quickly as they did in the original experiment. The smaller the angle at which the light strikes the soil and water, the smaller the temperature increase should be.

CLASSROOM TESTED & APPROVED

Alonda Droege
Highline
High School
Burien, WA

Maps in Action

Climates of the World

Discussion — GENERAL

Climate Classification Discuss with students the fact that classification schemes are the product of human ideas and are not natural phenomena. The value of a system is determined by how it is used. A classification system may be useful for one purpose and of no use for others. Have interested students investigate other climate classification systems and their uses or devise their own system. LS Logical

Answers to Map Skills Activity

1. about 15°N to 30°N
2. The eastern coast has a different climate than the western coast because a warm current flows northward along the Atlantic coast, while a cold current flows southward along the Pacific coast.
3. If the current on the west coast of Australia were a warm current, the climate of western Australia would probably be wetter and more like the climate on the eastern coast of Australia.
4. Monsoons are located from about 0° (the equator) to 40°N, near India and Southeast Asia.
5. The western coast of South America is desert because the cold ocean current that flows along the coast chills and stabilizes air masses, preventing rainfall. Prevailing winds also blow away from the continent. The inland part of the continent at the same latitude is humid because it is located in the region of the trade winds. The trade winds blow across warm Atlantic currents, bringing the region large amounts of precipitation.

CHAPTER RESOURCES

Technology

Transparencies
- 129 Climates of the World (with worksheet)

MAPS in Action

Climates of the World

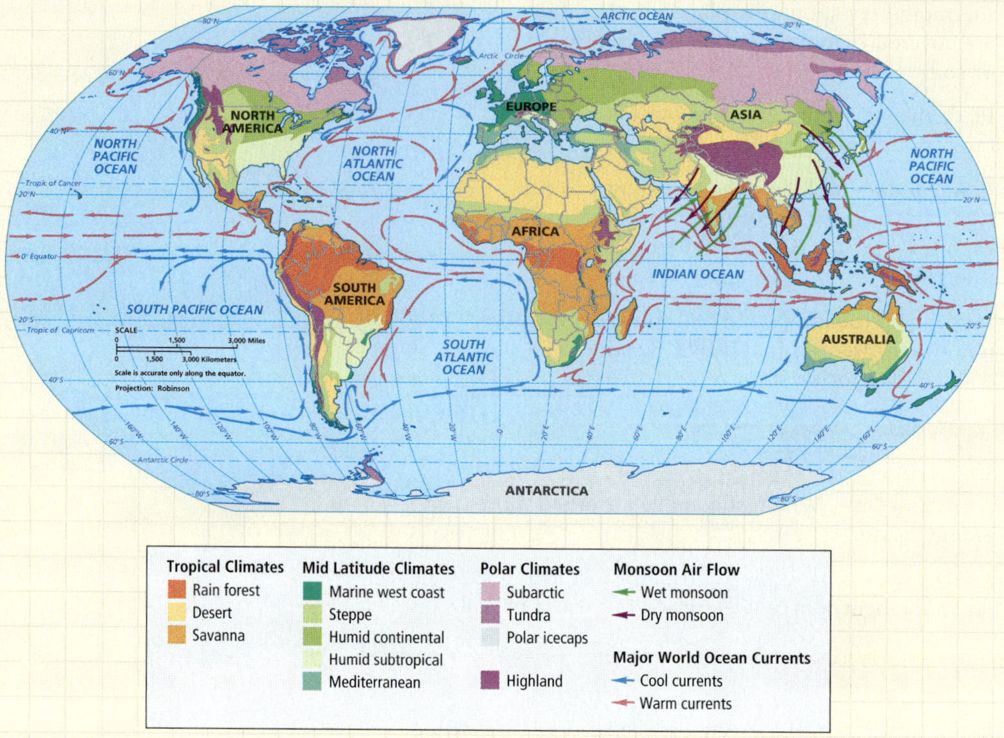

Map Skills Activity

This map shows the climate regions of Earth and the locations of warm and cold ocean currents. Use the map to answer the questions below.

1. **Analyzing Data** Estimate the latitude range for the desert climate of northern Africa.
2. **Making Comparisons** Why does the eastern coast of the United States have a different climate than the western coast does even though the coasts are at similar latitudes?
3. **Analyzing Ideas** If the ocean current that flows off the western coast of Australia were a warm current, how would this type of current affect the climate of western Australia?
4. **Using a Key** Identify the latitudes where monsoons are located.
5. **Evaluating Data** Explain why the western coast of South America is desert while the inland part of the continent at the same latitude is humid.

EYE on the Environment

Keeping Cool with Algae

Earth is constantly receiving energy from the sun. At the same time, Earth emits energy into space. By balancing these processes, Earth maintains its temperature. What keeps Earth from getting too hot or too cold? In seeking to answer this question, scientists look to microorganisms called *coccolithophores*.

Disappearing Carbon Dioxide

Each year, humans release more than 6 billion tons of carbon dioxide, CO_2, but only about half of that amount can be detected in the atmosphere. Where does the rest of the CO_2 go? Much of it is absorbed by coccolithophores in the oceans. These tiny algae are the primary users of CO_2. They use CO_2 to build chalky disks made of calcium carbonate. By absorbing the CO_2, these algae have a significant impact on the greenhouse effect.

DMS and Sulfur Cycles

Coccolithophores also may affect the number of clouds that form. These algae absorb sulfur compounds from ocean water to produce dimethyl sulfide, or DMS, which is then released into the air. DMS causes water vapor to condense and form clouds. As more clouds form, sunlight is blocked and photosynthesis is reduced. The number of algae begins to decline. Less DMS is produced, and fewer clouds form. When more sunlight hits the ocean, more coccolithophores can grow, and the cycle continues.

▲ This tiny microorganism may help regulate the temperature of our planet.

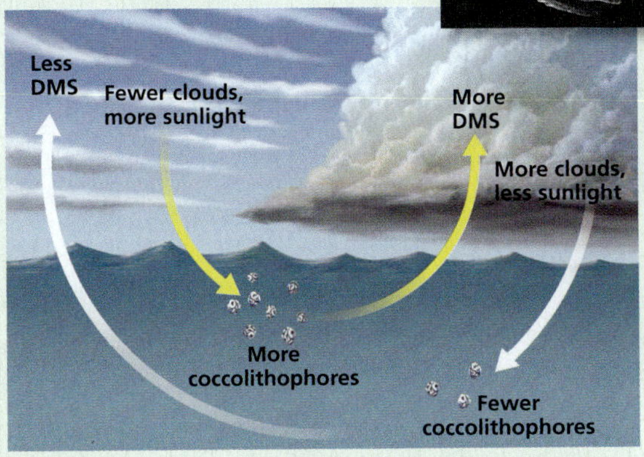

Scientists think that this process may help serve as a natural thermostat for the entire planet.

Tough Puzzles to Solve

Scientists now have a better understanding of how coccolithophores affect the atmosphere. However, scientists do not know if this knowledge can be used to help regulate climates. If scientists could promote coccolithophore growth in the ocean, atmospheric CO_2 could be reduced and more clouds could be created. As a result, the planet would cool down. However, because scientists do not know how marine ecosystems would be affected, this plan would be risky. As research continues, scientists will probably learn more about how to apply this research.

Extension

1. **Making Inferences** Why would predicting the effects of artificially promoting coccolithophore growth be difficult?
2. **Understanding Relationships** Find out about other marine microorganisms. How are they important to other organisms in marine ecosystems?

Eye on the Environment

Keeping Cool with Algae

Internet Activity — GENERAL

Gaia Hypothesis In 1972, Dr. James Lovelock proposed the idea that life on Earth and the environment are an interconnected system that work together to maintain optimal conditions for life to exist. He went so far as to propose that Earth was a kind of living organism with systems that help regulate the environment to keep Earth livable. His idea came to be known as the *Gaia hypothesis*, after the 'mother earth' goddess of the ancient Greeks. Have students work in small groups to investigate the Gaia hypothesis on the Internet. They should find out about its history, whether other scientists accept it, and if experiments are being conducted to test the hypothesis. Have groups present their findings to the class and discuss whether the Gaia hypothesis is helpful in understanding how Earth's climates work. A worksheet designed to direct student research on this topic can be found in the **Chapter Resource File** booklet or by visiting **go.hrw.com** and entering the keyword **HQ6CLIX.** LS Verbal

CHAPTER RESOURCES

Chapter Resource File

Internet Activity
- Gaia Hyptohesis GENERAL

Answers to Extension

1. Scientists may not know exactly how much CO_2 coccolithophores absorb, how much DMS they produce, nor how fast they grow in response to some promoting factor. Without this knowledge, an overgrowth of coccolithophores could occur. This might result in the absorption of too much CO_2 and/or the release of too much DMS, which could cool the planet too much.
2. Answers may vary. Accept all reasonable answers. Marine microorganisms known as *phytoplankton* form the base of almost all marine food chains.

Unit 8 SPACE

Unit 8 Outline

CHAPTER 26
Studying Space

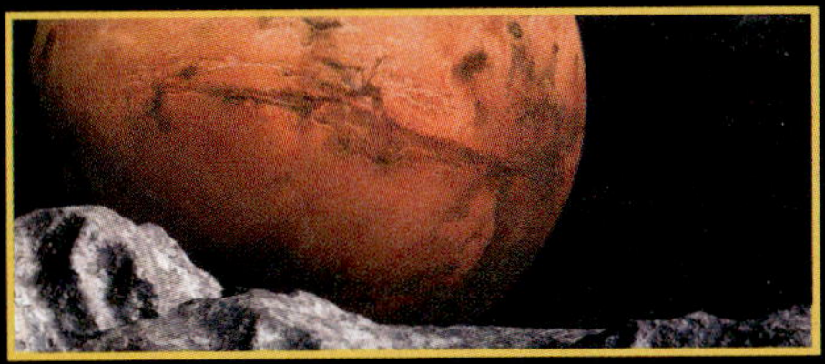

CHAPTER 27
Planets of the Solar System

CHAPTER 28
Minor Bodies of the Solar System

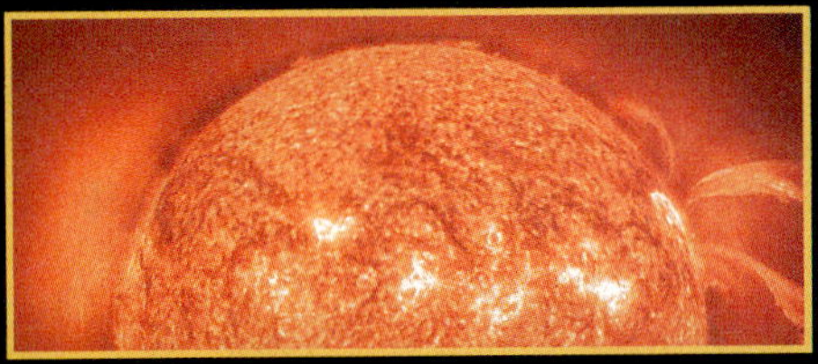

CHAPTER 29
The Sun

CHAPTER 30
Stars, Galaxies, and the Universe

▶ The Starfire telescope in New Mexico provides clear images by using a laser to compensate for atmospheric turbulance.

Chapter 26 Studying Space

Planning Guide

Compression Guide

To shorten instruction because of time limitations, omit the Chapter Lab.

PACING • 90 min — pp. 658–666

OBJECTIVES

Chapter Opener

LABS, DEMONSTRATIONS, AND ACTIVITIES

- SE **Long-Term Project** Positions of Sunrise and Sunset, pp. 848–851 ADVANCED
- LTP **Long-Term Project** Positions of Sunrise and Sunset* ADVANCED
- LTP **Long-Term Project** Apparent Motions of the Moon* ADVANCED

TECHNOLOGY RESOURCES

- OSP **Parent Letter** ■
- CD **Student Edition on CD-ROM**
- CD **Chapter Summaries Audio CD** ■
- VID **Brain Food Video Quiz**

OBJECTIVES

Section 1 Viewing the Universe

- Describe characteristics of the universe in terms of time, distance, and organization.
- Identify the visible and nonvisible parts of the electromagnetic spectrum.
- Compare refracting telescopes and reflecting telescopes.
- Explain how telescopes for nonvisible electromagnetic radiation differ from light telescopes.

LABS, DEMONSTRATIONS, AND ACTIVITIES

- TE **Activity** Descriptive Writing, p. 659 GENERAL
- TE **Group Activity** Cosmic Timeline, p. 660 ◆ GENERAL
- TE **Discussion** Galactic Address, p. 660 BASIC
- TE **Demonstration** Colors of Light, p. 661 ◆ BASIC
- TE **Group Activity** Splitting Light, p. 661 ◆ GENERAL
- TE **Demonstration** Heating with Invisible Light, p. 662 ◆ GENERAL
- TE **History Connection** Seeing the Light, p. 662 ADVANCED
- TE **Activity** Telescope Optics, p. 662 ◆ BASIC
- TE **Debate** Space Exploration, p. 665 GENERAL
- SE **Maps in Action** Light Sources, p. 682 GENERAL
- CRF **Inquiry Lab** Comet Meets Jupiter* GENERAL
- CRF **Making Models Lab** Telescopes* GENERAL

TECHNOLOGY RESOURCES

- OSP **Lesson Plans** (also in print)
- TR **Bellringer***
- TR 130 Reflecting and Refracting Telescopes*
- TE **Internet Activity** Space Spinoffs, p. 665 GENERAL
- CRF **Internet Activity** Space Spinoffs* GENERAL
- VID **NOVA Video** Runaway Universe
- CD **Interactive Tutor** Eyes and Mind
- CD **Interactive Tutor** Telescopes and Spectroscopes
- CD **Interactive Tutor** Satellite Explorations
- CD **Interactive Tutor** Light and Electromagnetic Energy

PACING • 135 min — pp. 667–674

OBJECTIVES

Section 2 Movements of Earth

- Describe two lines of evidence for Earth's rotation.
- Explain how the change in apparent positions of constellations provides evidence of Earth's rotation and revolution around the sun.
- Summarize how Earth's rotation and revolution provide a basis for measuring time.
- Explain how the tilt of Earth's axis and Earth's movement cause seasons.

LABS, DEMONSTRATIONS, AND ACTIVITIES

- TE **Activity** Role-Playing, p. 667 ◆ GENERAL
- SE **Quick Lab** Modeling a Pendulum, p. 668 GENERAL
- CRF **Datasheet for Quick Lab*** GENERAL
- TE **Demonstration** Coriolis Effect p. 668 ◆ BASIC
- TE **Discussion** Comprehension Check, p. 669 BASIC
- TE **Activity** Star Viewers, p. 669 ◆ GENERAL
- TE **History Connection** It's About Time, p. 670 GENERAL
- TE **Math Connection** What Time Is It?, p. 671 GENERAL
- TE **Debate** Daylight Savings Time, p. 672 GENERAL
- SE **Quick Lab** The Angle of the Sun's Rays, p. 673 GENERAL
- CRF **Datasheet for Quick Lab*** GENERAL
- SE **Inquiry Lab** Earth-Sun Motion, pp. 680–681 ◆ GENERAL
- CRF **Datasheet for Chapter Lab*** GENERAL
- TE **Group Activity** Community Survey, p. 682 GENERAL

TECHNOLOGY RESOURCES

- OSP **Lesson Plans** (also in print)
- TR **Bellringer***
- TR 131 Earth's Orbit*
- TR 132 The Apparent Motion of Constellations*
- TR 133 Time Zones*
- TR 134 How the Tilt of Earth's Axis Affects Seasons*
- TR 135 Light Sources*
- TE **Internet Activity** International Zones, p. 671 GENERAL
- CRF **Internet Activity** International Zones* GENERAL
- TE **Internet Activity** Landsat Images, p. 683 GENERAL
- CRF **Internet Activity** Landsat Images* GENERAL
- CD **Interactive Tutor** Gravity, Orbits, Motion, and Forces

---

PACING • 90 min

CHAPTER REVIEW, ASSESSMENT, AND STANDARDIZED TEST PREPARATION

- SE **Chapter Highlights**, p. 675
- SE **Chapter Review**, pp. 676–677
- SE **Standardized Test Prep**, pp. 678–679
- CRF **Concept Review*** ■ GENERAL
- CRF **Critical Thinking*** ADVANCED
- CRF **Math Skills*** GENERAL
- CRF **Graphing Skills*** GENERAL
- CRF **Chapter Test A*** ■ GENERAL
- CRF **Chapter Test B*** ADVANCED
- OSP **Lesson Plans** (also in print)
- OSP **Test Generator**
- OSP **Test Item Listing**

Online and Technology Resources

Holt Online Learning

Visit go.hrw.com for access to Holt Online Learning, or enter the keyword **HQ6 Home** for a variety of free online resources.

One-Stop Planner® CD-ROM

This CD-ROM package includes

- Lab Materials QuickList Software
- Holt Calendar Planner
- Customizable Lesson Plans
- Printable Worksheets
- ExamView® Test Generator
- Interactive Teacher Edition
- Holt PuzzlePro®
- Holt PowerPoint® Resources

KEY

SE	Student Edition	OSP	One-Stop Planner	VID	Classroom Video/DVD
TE	Teacher Edition	TR	Transparencies and Transparency Worksheets	*	Also on One-Stop Planner
CRF	Chapter Resource File	CD	CD or CD-ROM	◆	Requires advance prep
LTP	Long-Term Projects			■	Also available in Spanish

SKILLS DEVELOPMENT RESOURCES	REVIEW AND ASSESSMENT	CORRELATIONS
SE **Pre-Reading Activity,** p. 658 GENERAL TE **Using the Figure** Man in Space, p. 658 GENERAL		National Science Education Standards
CRF **Directed Reading*** BASIC SE **Math Practice,** p. 660 GENERAL TE **Skill Builder** Math, p. 660 GENERAL TE **Reading Skill Builder** Paired Summarizing, p. 661 BASIC TE **Inclusion Strategies,** p. 661 TE **Using the Figure** Comparing Telescopes, p. 663 BASIC	SE **Reading Checks,** pp. 661, 663, 664 GENERAL SE **Section Review,** p. 666 GENERAL TE **Homework,** p. 664 ADVANCED TE **Reteaching,** p. 665 BASIC TE **Quiz,** p. 665 GENERAL TE **Alternative Assessment,** p. 665 ADVANCED CRF **Section Quiz*** ■ GENERAL	SAI 2c, SAI 2d, PS 5a, ES 4a, ES 4b, HNS 3c, SPSP 6b, SPSP 6c, SPSP 6d, UCP 2
CRF **Directed Reading*** BASIC TE **Skill Builder** Vocabulary, p. 668 GENERAL SE **Graphic Organizer** Venn Diagram, p. 670 GENERAL TE **Reading Skill Builder** Reading Hint, p. 670 BASIC TE **Inclusion Strategies,** p. 670 TE **Using the Figure** Time Zone Zigzags, p. 671 GENERAL	SE **Reading Checks,** pp. 669, 671, 672 GENERAL SE **Section Review,** p. 674 GENERAL TE **Reteaching,** p. 673 BASIC TE **Quiz,** p. 673 GENERAL TE **Alternative Assessment,** p. 674 ADVANCED CRF **Section Quiz*** ■ GENERAL	SAI 2c, SAI 2d, UCP 2

Holt Lab Generator CD-ROM

Search for labs by topic, standard, difficulty level, and/or time frame. Edit those labs to fit your needs, or create your own labs. Use the Lab Materials QuickList to order just what you need.

Holt Earth Science Interactive Tutor CD-ROM

This CD-ROM consists of interactive activities that give students a fun way to extend their knowledge of Earth science concepts.

Chapter Summaries Audio CDs

These CDs include audio summaries of the key concepts presented in each chapter. (Audio summaries are also available in Spanish.)

SCILINKS. NSTA

www.scilinks.org

Maintained by the **National Science Teachers Association.** See Chapter Enrichment pages that follow for a complete list of topics.

See Chapter Enrichment pages for Video Resources.

Chapter Enrichment

This Chapter Enrichment provides relevant and interesting information to expand and enhance your classroom instruction of the chapter material.

Section 1 Viewing the Universe

Competition and the Private Space Industry

Inspired by early aviation prizes—including the $25,000 Ortieg Prize, which stimulated the transatlantic flight of Charles Lindbergh in 1927 and helped build the air transportation industry—the X Prize Foundation, which is based in St. Louis, Missouri, offered the $10 million Ansari X Prize in 1996. The prize would go to the first privately funded space venture to launch a reusable manned space vehicle to an altitude of 100 km twice in a two-week period. The purpose was to stimulate the private sector to develop space technology and to pave the way for making commercial spaceports around the globe available to nongovernment users. On October 4, 2004, the American Mojave Aerospace Ventures team won the X prize with the successful flight of SpaceShipOne.

◀ Space telescopes help scientists study the near and far reaches of space.

Humans and Long-Term Space Travel

With prospects of return flights to the moon and future missions to the planet Mars, scientists are investigating health issues associated with space travel. Among their focuses are the effects of microgravity, radiation, and other space-related factors on living systems. The human body undergoes many changes as a result of space flight. Among the biggest dangers to space travelers are the effects of microgravity (very low gravity). Terrestrial life has developed in a gravity field, and protected by Earth's atmosphere and magnetic field. Microgravity leads to the weakening of muscles and to thinning bone tissue, which is due to calcium loss. On long missions, it is possible for astronauts to lose as much as 20 to 30 percent of their bone mass. Worse still, the calcium could be deposited as kidney stones, a dangerous prospect on a three-year mission. In addition, microgravity can cause swelling of body tissues and motion sickness. Another potential health risk is associated with radiation. Astronauts are exposed to greater doses of radiation in space than they receive on Earth. Scientists at NASA's Space Radiation Laboratory are investigating the possibility that radiation could damage key blood-making stem cells. Antioxidants (special substances to halt dangerous bone loss and stimulate the immune system's natural repair mechanisms), specially shielded areas within spacecraft, and "artificial gravity" are among the solutions proposed to counteract these health threats. What about the psychological affects of long space voyages? Boredom, stress, and lengthy isolation from family could all take a toll. This, in turn, could affect the immune system's ability to fight infectious disease. All these issues deserve careful study to protect space travelers during long-term missions.

Section 2 Movements of Earth

The Constellations

The ancients divided the night sky into distinctive patterns of stars called *constellations*. Many cultures related these star patterns to their traditional myths and stories about gods and heroes. Some of the constellations we know today go back to ancient Babylonia and Egypt. Many constellations were probably invented by sailors for navigation, or by farmers to know when to plant or harvest their crops. Modern star charts still use constellations as a way of identifying individual stars, because each constellation shows the relative positions of stars as viewed from Earth. Lines are used on these maps to help observers find stars in the sky. The early star catalogue of Ptolemy in the 2nd century CE grouped the stars into 48 constellations. Many of the constellations visible in the Southern Hemisphere were named by German astronomer Johann Bayer or by French astronomer La Caille. The official modern list of 88 constellations was adopted by the International Astronomical Union in 1930. It is important to remember that the patterns we see in the night sky are due to chance. The stars that make them up are not necessarily located near each other in space or at equal distances. Some groups of stars within constellations have their own names, called *asterisms*. Among the best known is the Big Dipper.

Calendar Systems

Throughout history, the motions of astronomical bodies—the sun, moon, planets, and stars—have been used to measure the passage of time. As early as 15,000 years ago, ice-age hunters painted dots and squares on cave walls of Lascaux in France to represent the 29-day cycle of the moon. The ancient civilizations that followed needed reliable calendars to divide up the seasons for planting and to record historic events and mark religious observances. Five thousand years ago, the Sumerians used a calendar that divided the year into 30-day months, and the day into 12 periods roughly equivalent to two hours each. The Egyptian calendar was first based on the lunar cycle. Their observation of the star called *Sirius* later led them to adopt the first solar calendar of 365 days (12 months × 30 days + 5 days), which they used to predict the annual flooding of the Nile. The Babylonians used a calendar divided into 12 months with alternating lengths of 29 and 30 days, giving them a 354-day lunar year. The Mayan calendar used three different dating systems based on the motions of the sun, the moon, and Venus. The Mayans had a religious calendar of 260 days and a civil calendar based on the solar year of 365 days, which was adopted by many Mesoamerican nations, including the Aztecs and the Toltecs. The Islamic calendar is a purely lunar calendar that contains 12 months (12 months × 29.53 = 354.36 days), so it shifts with respect to the Gregorian calendar. The Chinese and the Hebrew calendars are combined solar/lunar calendars. Both have leap years in which an extra month is inserted periodically.

◀ **The Aztec culture used calendars to track agricultural seasons and holidays.**

Video Resources

Brain Food Video Quizzes

Brain Food Video Quizzes These videos contain game-show style quizzes that assess students' progress and motivate students to study the chapter material.

NOVA® Videos

The **NOVA** video below complements this chapter.

Runaway Universe Join two competing teams in their quest to discover the secrets to the stars and the ultimate fate of the universe. (60 min)

To order other **NOVA** videos related to this chapter, visit **go.hrw.com** and enter the keyword **HQ6SSPV.**

SciLinks®

NSTA Developed and maintained by the National Science Teachers Association

SciLinks is maintained by the National Science Teachers Association to provide you and your students with interesting, up-to-date links that will enrich your classroom presentation of the chapter.

Visit www.scilinks.org and enter the SciLinks code for more information about the topic listed.

Topic: Telescopes
SciLinks code: HQ61500

Topic: Space Exploration
SciLinks code: HQ61429

Topic: Space Probes
SciLinks code: HQ61432

Topic: Seasons
SciLinks code: HQ61363

Chapter 26

Chapter Overview

This chapter describes how scientists study the universe and how telescopes work. It also explains how Earth moves and how the movements are related to timekeeping, calendars, and seasons.

Using the Figure — GENERAL

Man in Space This photo shows an astronaut wearing a spacesuit specially designed for protection and maneuverability while floating in space. The suit enables astronauts to perform experiments and make repairs outside the space shuttle. Ask students what scientists or engineers might study that they could learn about better in space. (Sample answer: the effects of lower gravity, Earth's weather, pollution, medical research, and whether we could manufacture goods in space) LS **Logical**

PRE-READING ACTIVITY

FOLDNOTES You may want to use this FoldNote in a classroom discussion to review material from the chapter. On the board, write each category from the Two-Panel Flip Chart. Then, ask students to provide information for each category. Under the appropriate category on the board, write the information the students provide.

Chapter 26 Studying Space

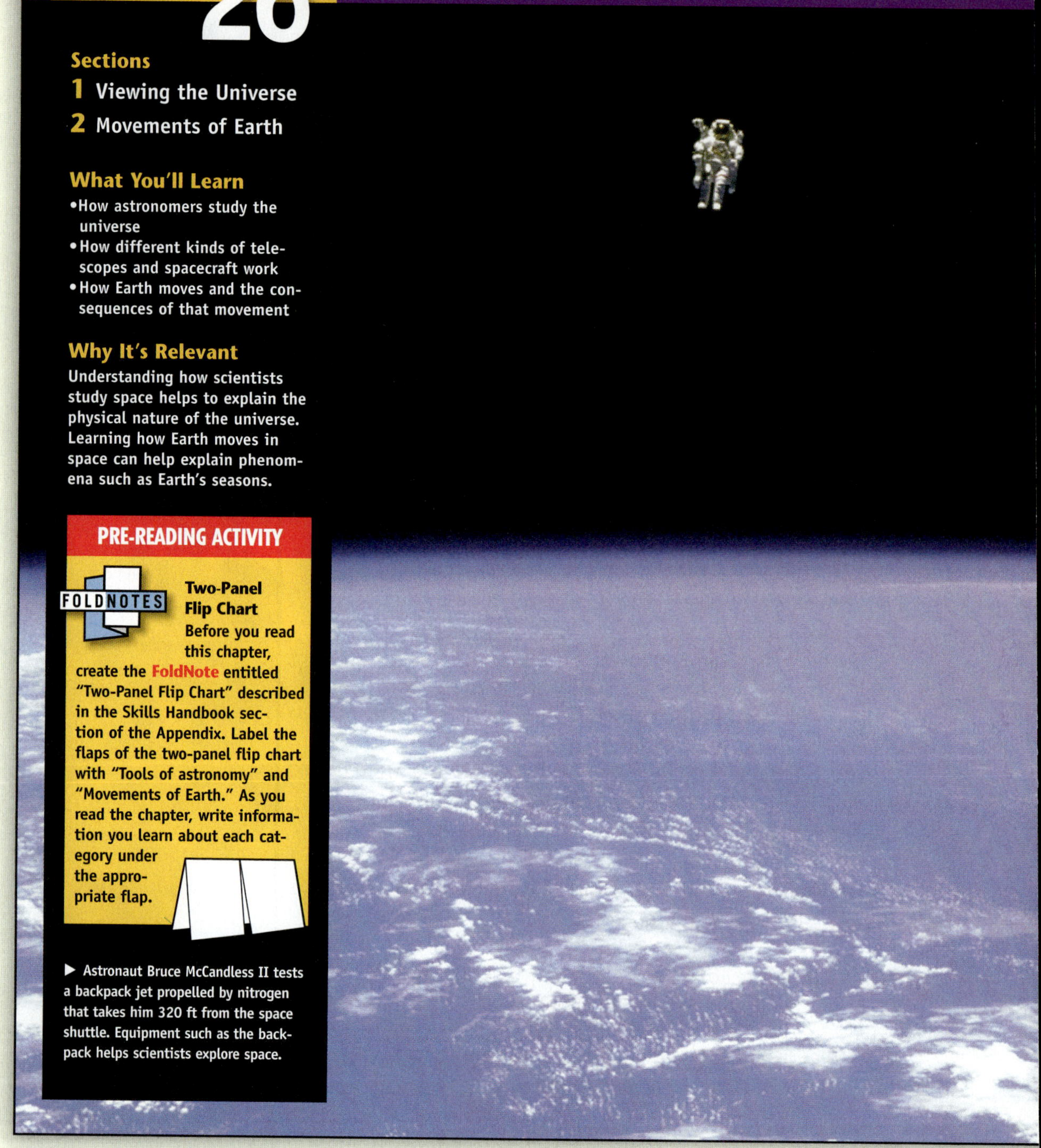

Sections

1 Viewing the Universe

2 Movements of Earth

What You'll Learn

- How astronomers study the universe
- How different kinds of telescopes and spacecraft work
- How Earth moves and the consequences of that movement

Why It's Relevant

Understanding how scientists study space helps to explain the physical nature of the universe. Learning how Earth moves in space can help explain phenomena such as Earth's seasons.

PRE-READING ACTIVITY

FOLDNOTES **Two-Panel Flip Chart** Before you read this chapter, create the FoldNote entitled "Two-Panel Flip Chart" described in the Skills Handbook section of the Appendix. Label the flaps of the two-panel flip chart with "Tools of astronomy" and "Movements of Earth." As you read the chapter, write information you learn about each category under the appropriate flap.

▶ Astronaut Bruce McCandless II tests a backpack jet propelled by nitrogen that takes him 320 ft from the space shuttle. Equipment such as the backpack helps scientists explore space.

Chapter Correlations — *National Science Education Standards*

SAI 2c Scientists rely on technology to enhance the gathering and manipulation of data. New techniques and tools provide new evidence to guide inquiry and new methods to gather data .... **(Sections 1 and 2)**

SAI 2d Mathematics is essential in scientific inquiry. **(Sections 1 and 2)**

PS 5a The total energy of the universe is constant. Energy can be transferred by...light waves and other radiations. ... **(Section 1)**

ES 4a The origin of the universe remains one of the greatest questions of science. The big bang theory places the origin of the universe between 10 and 20 billion years ago...; according to this theory the universe has been expanding ever since. **(Section 1)**

ES 4b Early in the history of the universe, matter ... clumped together ... to form ... trillions of stars. Billions of galaxies ... now form most of the visible mass in the universe. **(Section 1)**

HNS 3c Occasionally, there are advances in science and technology that have important and long-lasting effects on science and society. Examples of such advances include ... Galactic universe;... **(Section 1)**

SPSP 6b Understanding basic concepts and principles of science and technology should precede active debate about economics, policies, and ethics of various science and technology related challenges... **(Section 1)**

SPSP 6c Progress in science and technology can be affected by social issues and challenges. Funding priorities ...serve as examples of ways that social issues influence science and technology. **(Section 1)**

SPSP 6d Individuals and society must decide on proposals involving research and the introduction of new technologies into society. Decisions involve the assessment of alternatives, risks, costs, and benefits, and consideration of who benefits and who suffers, who pays, and who gains, and what the risks are and who bears them. **(Section 1)**

UCP Evidence consists of observations and data on which to base scientific explanations. **(Sections 1 and 2)**

Section 1 Viewing the Universe

People studied the sky long before the telescope was invented. For example, farmers observed changes in daylight throughout the year to track seasons and to predict floods and droughts. Sailors focused on the stars to navigate through unknown territory. Today, most interest in studying the sky comes from a curiosity to discover what lies within the universe. This scientific study of the universe is called **astronomy.** Scientists who study the universe are called *astronomers.*

OBJECTIVES

- **Describe** characteristics of the universe in terms of time, distance, and organization.
- **Identify** the visible and nonvisible parts of the electromagnetic spectrum.
- **Compare** refracting telescopes and reflecting telescopes.
- **Explain** how telescopes for nonvisible electromagnetic radiation differ from light telescopes.

KEY TERMS

astronomy
galaxy
astronomical unit
electromagnetic spectrum
telescope
refracting telescope
reflecting telescope

astronomy the scientific study of the universe

The Value of Astronomy

In the process of observing the universe, astronomers have made exciting discoveries, such as new planets, stars, black holes, and nebulas, such as the one shown in **Figure 1.** By studying these objects, astronomers have been able to learn more about the origin of Earth and the processes involved in the formation of our solar system.

Studying the universe is also important for the potential benefits to humans. For example, studies of how stars shine may one day lead to improved or new energy sources on Earth. Astronomers may also learn how to protect us from potential catastrophes, such as collisions between asteroids and Earth. Because of these and other contributions, astronomical research is supported by federal agencies, such as the National Science Foundation and NASA. Private foundations and industry also fund research in astronomy.

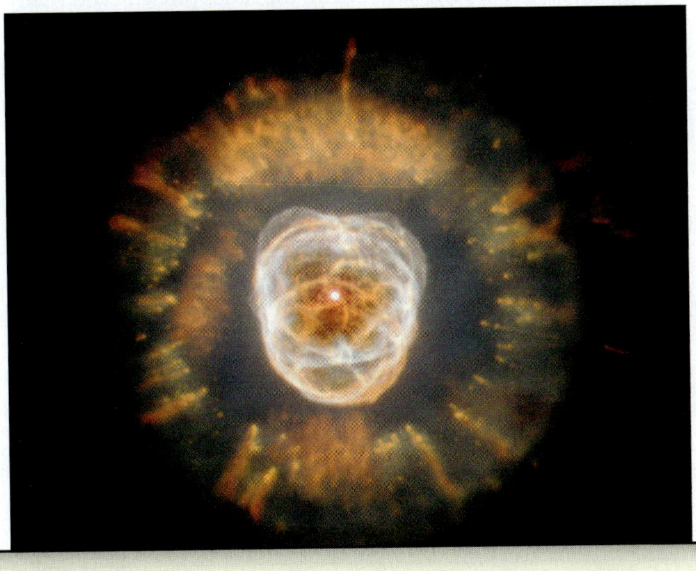

Figure 1 ▶ A nebula is a large cloud of gas and dust in space. This nebula is called the Eskimo nebula. It formed as a result of the outer layers of the central star drifting away. By studying nebulas like this one, scientists may learn if our sun will ever reach this state.

Section 1

Focus

Overview

This section summarizes the nature of the universe in terms of its size, age, and arrangement; describes the electromagnetic spectrum; and explains various tools astronomers use to study space.

Bellringer

Ask students what comes to mind when they hear about new astronomical discoveries. Have them think about what an astronomer does and then draw a picture, with caption, of an astronomer at work. (Answers may vary.) LS Visual

Motivate

Activity — GENERAL

Descriptive Writing Have students write a brief essay that compares what they know about ancient astronomy with what they know about present-day space science. Guide students to explain why the sky interested ancient peoples and how celestial bodies and their cycles affected daily lives. Ask students to consider what observations people could make with only their eyes, how modern tools extend knowledge of space, and what benefits space science has for modern societies. LS Verbal

CHAPTER RESOURCES

Chapter Resource File

- Directed Reading BASIC

Technology

Transparencies
- Bellringer

Student Edition on CD-ROM

One-Stop Planner CD-ROM
- Lesson Plan

Teach

Group Activity — GENERAL

Cosmic Timeline Help students visualize the time span since the beginning of the universe by making a timeline that corresponds to a 12-month calendar. Hang 12 pieces of poster board or part of a role of shelf paper labeled with the calendar months in order along one wall. Have students brainstorm a list of major events that occurred between the big bang and the present. Here are some ideas, followed by the cosmic timeframe and corresponding calendar date: the big bang, 14 billion years ago (Jan. 1); the birth of the solar system, 4.5 billion years ago (Sept. 5); earliest on Earth, 3.8 billion years ago (Sept. 22); extinction of the dinosaurs, 65.5 million years ago (Dec. 29); appearance of *homo sapiens*, 600,000 years ago (Dec. 31, 11:30 PM), advent of agriculture, 11,500 years ago (Dec. 31, 11:59 PM). Other major events may include the origin of life, the oldest fossils, and the space age. Have students research and determine when each event occurred. Then, have students draw a picture of each event on note cards and place the cards at the appropriate place on the calendar, making a connecting timeline with dates marked. **LS Logical/ Kinesthetic** Co-op Learning

Discussion — BASIC

Galactic Address Propose this scenario: "Suppose you want to send a message to an unknown civilization in a different part of the universe. You would probably include your own location in space. Describe your galactic address, starting with your seat in the classroom and expanding outward." (Sample answers: Classroom seat and row, the school name, address, city, state, country, Earth [third planet from the sun], solar system [Sol], Milky Way galaxy, the universe) Invite students to create maps that illustrate their descriptions. **LS Verbal**

Figure 2 ▶ The Whirlpool galaxy, M51 (above), is 28 million light-years from the Milky Way. Abell 1689 (right) is one of the most massive galaxy clusters known. The galaxies' mass has distorted even farther objects into the giant arcs shown here.

Characteristics of the Universe

The study of the origin, properties, processes, and evolution of the universe is called *cosmology*. Most astronomers agree that the universe began about 14 billion years ago in one giant explosion, called the *big bang*. Since that time, the universe has continued to expand. The universe is very large, and the objects within it are extremely far apart. Telescopes are used to study some distant objects. However, astronomers also commonly use computer and mathematical models to study the universe.

MATHPRACTICE

Astronomical Unit
An astronomical unit is the average distance between the sun and Earth, or about 150 million km. Venus orbits the sun at a distance of 0.7 AU. Venus is how many kilometers from the sun?

galaxy a collection of stars, dust, and gas bound together by gravity

astronomical unit the average distance between the Earth and the sun; approximately 150 million kilometers (symbol, AU)

Organization of the Universe

The nearest part of the universe to Earth is our solar system. The solar system includes the sun, Earth, the other planets, and many smaller objects such as asteroids and comets. The solar system is part of a **galaxy**, which is a large collection of stars, dust, and gas bound together by gravity. The galaxy in which the solar system resides is called the *Milky Way galaxy*. Beyond the Milky Way galaxy, there are billions of other galaxies, a few of which are shown in **Figure 2.**

Measuring Distances in the Universe

Because the universe is so large, the units of measurement used on Earth are too small to represent the distance between objects in space. To measure distances in the solar system, astronomers often use astronomical units. An **astronomical unit** (symbol, AU) approximates the average distance between Earth and the sun, which is 149,597,870.691 km or about 150 million km.

Astronomers also use the speed of light to measure distance. Light travels at 300,000 km/s. In one year, light travels 9.46×10^{12} km. This distance is known as a *light-year*. Aside from the sun, the closest star to Earth is 4.22 light-years away.

MATHPRACTICE

Answer
150,000,000 km × 0.7 AU = 105,000,000 km, or 105 million km

SKILL BUILDER — GENERAL

Math Distances in astronomy are difficult for students to comprehend because the distances are so huge. Have students calculate how far (in km) light travels in one year based on the speed of light (300,000 km/s). Allow them to round their figures. (Answer: 60 s/min × 60 min/h = 3600 s/h × 24 h/d = 86,400 s/d × 365 d/y = 31,536,000 s/y × 300,000 km/s = 9.4608×10^{12} km or 9,461 trillion (9,461,000,000,000) km) Tell them the star Sirius is located 84,321 trillion km from Earth. Have them calculate this distance in light-years. (84,321 trillion km ÷ 9,461 trillion km/ly = 8.91 ly) **LS Logical**

Observing Space

Light enables us to see the world around us and to make observations. When people look at the night sky, they see stars and other objects in space because of the light these objects emit. This visible light is only a small amount of the energy that comes from these objects. By studying other forms of energy, astronomers are able to learn more about the universe. Recall that planets do not emit light. They reflect the light from stars.

Electromagnetic Spectrum

Visible light is a form of energy that is part of the electromagnetic spectrum. The **electromagnetic spectrum** is all of the wavelengths of electromagnetic radiation. Light, radio waves, and X rays are all examples of electromagnetic radiation. The radiation is composed of traveling waves of electric and magnetic fields that oscillate at fixed frequencies and wavelengths.

electromagnetic spectrum all of the frequencies or wavelengths of electromagnetic radiation

Visible Electromagnetic Radiation

The human eye can see only radiation of wavelengths in the visible light range of the spectrum. When white light passes through a prism, the light is broken into a continuous set of colors, as shown in **Figure 3.** Every rainbow formed in the sky and any color spectrum formed by a prism will always have the same colors in the same order. The different colors result because each color of light has a characteristic wavelength. Though all light travels at the same speed, different colors of light have different wavelengths. For example, the shortest wavelengths of visible light are blue and violet, while the longest wavelengths of light are orange and red.

Electromagnetic radiation that has wavelengths that are shorter than the wavelengths of violet light or longer than the wavelengths of red light cannot be seen by humans. But these wavelengths can be detected by instruments that are designed to detect electromagnetic radiation that cannot be seen by human eyes. These invisible wavelengths include infrared waves, microwaves, radio waves, ultraviolet rays, X rays, and gamma rays.

Reading Check Which type of electromagnetic radiation can be seen by humans? (See the Appendix for answers to Reading Checks.)

Figure 3 ▶ Visible light is broken into different colors because each color has a different wavelength. These colors can be seen when visible light is passed through a prism.

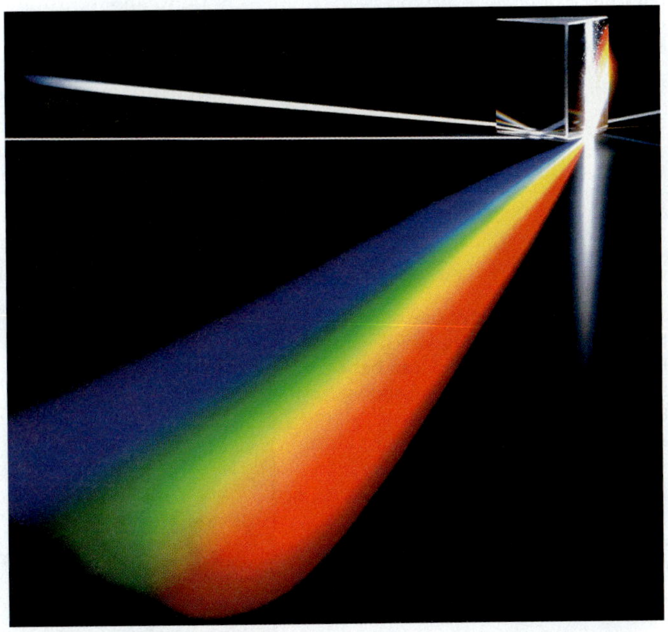

Demonstration — Basic

Colors of Light Shine a flashlight beam onto a white sheet of paper. Put a prism into the path of the beam of light. Turn the prism slightly until the spectrum appears on the paper. Ask students to describe what they see. Ask them: Which color light seems to bend the most and which color bends the least? (Students should see the colors of the rainbow. The prism bends violet light the most and red light the least.) **LS Visual**

Group Activity — General

Splitting Light Explain that astronomers use an instrument called a *spectrograph* to analyze the light from stars. Spectrographs use prisms or diffraction gratings to break up the light. Students can see the spectrum directly through a diffraction grating cut into 3 cm squares. (You can obtain diffraction grating in sheets or rolls from a science supply house.) Set up a table lamp with the shade removed. Distribute diffraction gratings to students. Darken the room and turn on the lamp. Direct students to look at the light source through the diffraction grating. Tell them to handle the grating by the edges, as oil from their fingers will clog the openings. Have students use crayons or colored pencils to draw the spectra they see. **LS Visual**

Answer to Reading Check

The only kind of electromagnetic radiation the human eye can detect is visible light.

Reading Skill Builder — Basic

Paired Summarizing Group students into pairs and have them read silently about the electromagnetic spectrum. Then, have one student summarize what types of radiation make up the visible and invisible parts of the electromagnetic spectrum and how astronomers use these forms of energy. Direct the other student to listen to the retelling and point out any inaccuracies or ideas that were left out. Allow students to refer to the text as needed. English Language Learners **LS Auditory**

Inclusion Strategies

- *Attention Deficit Disorder*
- *Hearing Impaired*
- *Behavior Control Issues*

Have students draw and color a rainbow in the center of a piece of paper. Then, have them add, in order and in black and white, the wavelengths that are too short or too long to see. Infrared and radio wavelengths should be beyond the red end. Ultraviolet, X-ray, and gamma-ray wavelengths should appear beyond the violet end. **LS Visual/Kinesthetic**

Teach, *continued*

Demonstration — GENERAL

Heating with Invisible Light Demonstrate the thermal properties of infrared radiation by repeating William Herschel's historic experiment. Set up a prism in front of a window so that it breaks white sunlight into a color spectrum. Place three thermometers at different points in the spectrum, one at the violet end, one in the center, and one just beyond the red end. Ask a volunteer to take starting temperatures. Wait at least 5 minutes and take second readings. Ask students what they observe about the temperature data. (There are readings in all parts of the spectrum even in the invisible part beyond the red.) **LS Logical**

HISTORY CONNECTION — ADVANCED

Seeing the Light Isaac Newton once declared, "If I have seen further, it is by standing on the shoulders of giants." His laws of gravity and motion paved the way to carrying modern spacecraft into orbit. Newton also contributed to understanding the nature of light. His experiments with prisms revealed the spectrum contained in white light, which led to his improvements of the telescope and the particle theory of light. Encourage students to research the contributions of Newton and other "giants," such as James Clerk Maxwell and William Herschel, to our understanding of electromagnetic radiation. Have them make a brief oral report to the class. **LS Verbal**

Invisible Electromagnetic Radiation

If you place a thermometer in any wavelength of the visible spectrum, the temperature reading on the thermometer will increase. In 1800, the scientist Sir Frederick William Herschel moved a thermometer beyond the red end of the visible spectrum. Even though he could not see any light on the thermometer, the temperature reading on the thermometer increased. He had discovered *infrared,* which means "below the red." Infrared is electromagnetic radiation that has waves that are longer than waves of visible light. Later, other scientists discovered radio waves, which have even longer wavelengths than infrared waves do.

The ulraviolet wavelengths, which are invisible to humans, are shorter than the wavelengths of violet light. *Ultraviolet* means "beyond the violet." The X-ray wavelengths are shorter than the ultraviolet wavelengths are. The shortest wavelengths are the gamma-ray wavelengths.

Telescopes

Our eyes can see detail, but some things are too small or too far away to see. Our ability to see the detail of distant objects in the sky began with the Italian scientist Galileo. In 1609, he heard of a device that used two lenses to make distant objects appear closer. He built one of the devices and turned it toward the sky. For the first time, he could see that there are craters on the moon and that the Milky Way is made of stars. An example of one of these early devices is shown in **Figure 4.**

telescope an instrument that collects electromagnetic radiation from the sky and concentrates it for better observation

A **telescope** is an instrument that collects electromagnetic radiation from the sky and concentrates it for better observation. While modern telescopes are able to collect and use invisible electromagnetic radiation, the first telescopes that were de veloped collected only visible light. Telescopes that collect only visible light are called *optical telescopes.* The two types of optical telescopes are refracting telescopes and reflecting telescopes.

Figure 4 ▶ A model of one of the first reflecting telescopes, which was invented by Isaac Newton, can be seen at the Royal Society in London, England.

Activity — BASIC

Telescope Optics To show how a refracting telescope produces images, give students two magnifying glasses and some printed material. Preferably, one lens should be larger than the other. Have them hold one magnifying glass between them and the paper. Tell them to put the second magnifier between one eye and the first magnifying glass. Have them move the first magnifier backward and forward until the image comes into sharp focus. Ask students what they notice about the image. (Answers may vary. Students should note that the print is magnified and is also upside down). **LS Visual**

BRAIN FOOD

Inventing the Telescope Dutch lens-grinder Hans Lippershey is credited with inventing the telescope, which was first used for military purposes. Galileo directed the new instrument, using a combination of concave and convex lenses, at the night sky. Galileo observed the moon and resolved the light of the Milky Way into stars. In 1611, Johannes Kepler built a telescope that combined two convex lenses.

Figure 5 ▶ Reflecting and Refracting Telescopes

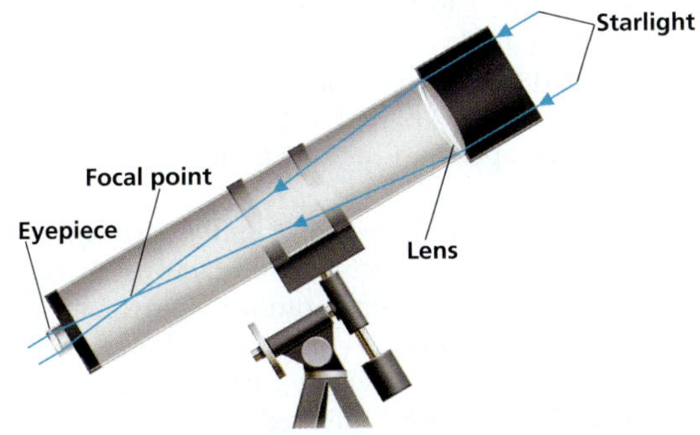

Refracting telescopes use lenses to gather and focus light from distant objects.

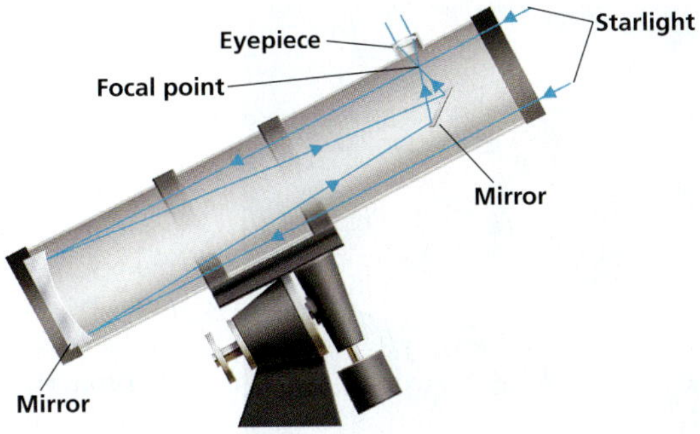

Reflecting telescopes use mirrors to gather and focus light from distant objects.

Refracting Telescopes

Lenses are clear objects shaped to bend light in special ways. The bending of light by lenses is called *refraction*. Telescopes that use a set of lenses to gather and focus light from distant objects are called **refracting telescopes.** Refracting telescopes have an objective lens that bends light that passes through the lens and focuses the light to be magnified by an eyepiece, as shown in **Figure 5.**

One problem with refracting telescopes is that the lens focuses different colors of light at different distances. For example, if an object is in focus in red light, the object will appear out of focus in blue light. Another problem with refracting telescopes is that it is difficult to make very large lenses of the required strength and clarity. The amount of light collected from distant objects is limited by the size of the objective lens.

refracting telescope a telescope that uses a set of lenses to gather and focus light from distant objects

reflecting telescope a telescope that uses a curved mirror to gather and focus light from distant objects

Reflecting Telescopes

In the mid-1600s, Isaac Newton solved the problem of color separation that resulted from the use of lenses. He invented the **reflecting telescope,** as shown in **Figure 5,** which used a curved mirror to gather and focus light from distant objects. When light enters a reflecting telescope, the light is reflected by a large curved mirror to a second mirror. The second mirror reflects the light to the eyepiece, which is a lens that magnifies and focuses the image.

Unlike objective lenses in refracting telescopes, mirrors in reflecting telescopes can be made very large without affecting the quality of the image. Thus, reflecting telescopes can be much larger and can gather more light than refracting telescopes can. The largest reflecting telescopes are a pair called the Keck Telescopes in Hawaii. Each telescope is 10 m in diameter. Astronomers are tentatively planning to build an OWL, Overwhelmingly Large Telescope, that would be 100 m in diameter.

Reading Check What are the problems with refracting telescopes? (See the Appendix for the answers to Reading Checks.)

SCILINKS. NSTA Developed and maintained by the National Science Teachers Association

For a variety of links related to this subject, go to www.scilinks.org

Topic: Telescopes
SciLinks code: HQ61500

Using the Figure — BASIC

Comparing Telescopes Use the pictures of optical telescopes on this page to compare and contrast the features of reflecting and refracting telescopes. Ask students what part of each telescope gathers light. (In refracting telescopes, lenses gather the incoming light. In the reflecting telescopes, mirrors gather the light.) Ask students what is similar about how images are focused. (Both telescopes use an eyepiece or lens to focus the image. In the refracting telescope, the first lens bends light toward eyepiece. In the reflecting telescope a flat mirror directs the light to the eyepiece.) **LS Visual**

Answer to Reading Check

Images produced by refracting telescopes are subject to distortion because of the way different colors of visible light are focused at different distances from the lens and because of weight limitations on the size of the objective lens.

BRAIN FOOD

Large Telescopes The primary mirrors of the Keck Telescopes are constructed of 36 lightweight hexagonal segments. A computer controls the cooperative work of the segments, providing even sharper images than those from the *Hubble Space Telescope*. The OWL project would use a similar mirror design in a telescope about the size of a football field. The likely site for the OWL telescope is the Atacama Desert of Argentina and Chile.

MISCONCEPTION ALERT

Students may think the purpose of a telescope is to magnify distant objects. For astronomers, a telescope's main job is to gather light from distant objects so that they are bright enough to study. As magnifying power doubles, the field of view and surface brightness decrease to 1/4 of the original values. Usually, the Earth's atmosphere instead of the size of the telescope or the eyepiece's magnification limits the amount of detail that is visible. Thus, the ability to detect faint objects is a better indication of usefulness.

CHAPTER RESOURCES

Chapter Resource File

- **Inquiry Lab** Comet Meets Jupiter GENERAL
- **Making Models Lab** Telescopes GENERAL

Technology

Transparencies

- 130 Reflecting and Refracting Telescopes (with worksheet)

Teach, *continued*

ENVIRONMENTAL CONNECTION

Light Pollution Sky glow is the brightening of the night sky that occurs when water droplets and dust scatter the large amounts of artificial light produced in urban areas, which profoundly reduces the visibility of stars. Astronomers were first to raise the alarm about this phenomenon. Light pollution not only wastes energy, but also concerns naturalists because it may impact wildlife. Light pollution may produce false signals that disrupt biological rhythms and interfere with the normal behavior patterns of nocturnal animals. For example, migrating bird populations often use the moon and stars to navigate. Artificial light confuses the birds, causing them to fly into buildings and towers. Many animal behaviors in response to light also change their hormonal systems, which can become toxic if overstimulated. Too much exposure to light can even affect human physiology. Invite students to learn more about the consequences of light pollution and ways to reduce this waste by turning down the lights. Have them write letters to community leaders about their concerns. **LS Verbal**

Answer to Reading Check

Scientists launch spacecraft into orbit to detect radiation screened out by Earth's atmosphere and to avoid light pollution and other atmospheric distortions.

Figure 6 ▶ Radio telescopes, such as this one at the National Radio Astronomy Observatory in New Mexico, provide scientists with information about objects in space.

Telescopes for Invisible Electromagnetic Radiation

Each type of electromagnetic radiation provides scientists with information about objects in space. Scientists have developed telescopes that detect invisible radiation. For example, a radio telescope, such as the one shown in **Figure 6,** detects radio waves. There are also telescopes that detect gamma rays, X rays, and infrared rays.

One problem with using telescopes to detect invisible electromagnetic radiation is that Earth's atmosphere acts as a shield against many forms of electromagnetic radiation. Atoms and molecules in the atmosphere prevent short wavelengths like gamma rays, X rays, and most ultraviolet rays from reaching Earth's surface. Water vapor blocks infrared rays, so ground-based telescopes that are used to study infrared work best at high elevations, where the air is thin and dry. But the only way to study many forms of radiation is from space.

Space-Based Astronomy

While ground-based telescopes have been critical in helping astronomers learn about the universe, valuable information has also come from spacecraft. Spacecraft that contain telescopes and other instruments have been launched to investigate planets, stars, and other distant objects. In space, Earth's atmosphere cannot interfere with the detection of electromagnetic radiation.

Reading Check **Why do scientists launch spacecraft beyond Earth's atmosphere? (See the Appendix for answers to Reading Checks.)**

Connection to ENVIRONMENTAL SCIENCE

Light Pollution

About 6,000 stars shine brightly enough to be seen with the unaided eye. Of those, about half are above the horizon at a time. But we can see those 3,000 stars only if the sky is very dark. Most people, however, live in or close to cities, where street lighting and other lighting casts a glow into the sky. This glow masks the fainter stars. In big cities, you may see only a handful of stars, instead of thousands!

The glow in the sky that obstructs our view of the stars is known as *light pollution*. Light pollution not only affects humans but also affects animals. Near the ocean, for example, street lighting can affect where and whether turtles come ashore to lay their eggs. On the beaches of Florida's developed coastline, bright lights may discourage female leatherback and green turtles from coming ashore to nest.

In recent years, efforts have been made to reduce light pollution. Reflectors on tops of streetlights would keep the light aimed at the ground instead of into the sky. Reflectors would also reduce wasted electricity. Outdoor house lights and floodlights in parking lots can be directed downward to decrease light pollution. Turning off outdoor lighting when it is not needed, such as when businesses are closed, also contributes to limiting light pollution. Taking steps such as these can improve the view of the night sky without compromising safety in our neighborhoods.

Homework — ADVANCED

Space Telescopes and Probes Invite interested students to choose two of the space telescopes or probes discussed in the text and learn more about them. Students may work individually or in pairs and may use library or Internet sources. Students should compare project goals, types of remote-sensing technologies, and the science benefits expected. If either or both of the projects have been completed, students should identify successes and problems. Have them write a report on their findings or make an oral presentation to the class. **LS Verbal** Co-op Learning

Figure 7 ▶ The *Hubble Space Telescope* is in orbit around Earth, where the telescope can detect visible and nonvisible electromagnetic radiation without the obstruction of Earth's atmosphere.

Space Telescopes

The *Hubble Space Telescope,* shown in **Figure 7,** is an example of a telescope that has been launched into space to collect electromagnetic radiation from objects in space. Another example, the *Chandra X-ray Observatory* makes remarkably clear images using X rays from objects in space, such as the remnants of exploded stars. The *Compton Gamma Ray Observatory* is no longer in space, but it detected gamma rays from objects, such as black holes. The *Spitzer Space Telescope* was launched in 2003 to detect infrared radiation. The *James Webb Space Telescope* is scheduled to be launched in 2011. When deployed in space, this telescope will detect infrared radiation from objects in space.

Other Spacecraft

Since the early 1960s, spacecraft have been sent out of Earth's orbit to study other planets. Launched in 1977, the *Voyager 1* and *Voyager 2* spacecraft investigated Jupiter, Saturn, Uranus, and Neptune. These two spacecraft collected images of these planets and their moons. The *Galileo* spacecraft was in orbit around Jupiter and its moons from 1995 to 2003. This spacecraft gathered information about the composition of Jupiter's atmosphere and storm systems, which are several times larger than Earth's storm systems. The *Cassini-Huygens* spacecraft was launched in 1997 and began orbiting Saturn in 2004. In December 2004, the *Huygens* probe detached from the *Cassini* orbiter to study the atmosphere and surface of Titan, Saturn's largest moon. The twin Mars rovers *Spirit,* shown in **Figure 8,** and *Opportunity* landed on Mars in January 2004. Early in their mission, they confirmed that water had once been present in at least one area of Mars.

Figure 8 ▶ By April 2006, Mars rover *Spirit* and its twin, *Opportunity,* had transmitted over 150,000 images of the planet to Earth. The mission was planned to last only three months, but the rovers continued to operate for several years.

Close

Reteaching — BASIC

Two-column Notes Organize students into pairs. Have them work together to create a chart that outlines the idea of observing the universe. Have them write main ideas or section heads in the left column and details, examples, or explanations in the right column. Have them use the guide to quiz each other and to study for assessments. **LS Auditory**

Quiz — GENERAL

1. Name the two types of optical telescopes and explain how they differ. (Reflecting telescopes use mirrors while refracting telescopes use lenses to gather light.)
2. How are distances between the stars measured? (in light-years)
3. List the forms of electromagnetic radiation other than visible light. (gamma rays, infrared, radio waves, X rays, microwaves, and ultraviolet radiation) **LS Verbal**

Alternative Assessment — ADVANCED

Skit Have students present a skit about the history of astronomy. Students should consider how long light takes to travel through space. They should cover naked-eye observations, the invention of telescopes, how electromagnetic radiation brings details into focus, and how space-based astronomy provides a more accurate picture of the universe than was previously obtained. **LS Verbal/Kinesthetic**

Debate — GENERAL

Space Exploration Some people think that the costs and risks associated with human exploration of space are so great that space missions should be limited to robotic technologies or space probes. Others believe that space is such an important frontier that it would be a mistake to curtail humanity's reach toward the moon, Mars, and the space beyond. Assign students to prepare arguments for and against this public policy issue and stage a classroom debate. **LS Verbal** Co-op Learning

Internet Activity — GENERAL

Space Spinoffs Divide students into small groups and have them identify products of space technology that now have "everyday" applications on Earth. Each group should select products in a specific field, such as weather prediction, medicine, industry, computers, or communications. Have each group create a multimedia presentation to share with the class. A worksheet designed to direct student research on this topic can be found in the **Chapter Resource File** booklet or by visiting **go.hrw.com** and entering the keyword **HQ6SSPX.** **LS Verbal/Visual** Co-op Learning

CHAPTER RESOURCES

Chapter Resource File

- Internet Activity
  - Space Spinoffs GENERAL

Close, *continued*

Answers to Section Review

1. The universe began 14 billion years ago, and is organized into solar systems, nebulas of dust and gas, star islands called *galaxies,* and galaxy clusters. Vast distances are measured in astronomical units or light years.
2. visible: continuous color spectrum from violet to red; invisible: radio waves, infrared, microwaves, ultraviolet rays, X rays, and gamma rays
3. Optical telescopes gather visible light. Special telescopes such as radio telescopes detect invisible radiation. Spacecraft and probes study space without interference from Earth's atmosphere.
4. Refracting telescopes use lenses to gather and focus light, while reflecting telescopes use curved mirrors.
5. Optical telescopes make use of lenses or mirrors to gather visible light. Radio telescopes detect radio wave emissions.
6. space telescopes: *Hubble Space Telescope*, *James Webb Space Telescope*, *Chandra X-ray Observatory*; space probes: *Voyager 1* and *2*, *Galileo*, *Cassini-Huygens*
7. The lenses of refracting telescope focus each color differently. Newton's experiments with lenses provided insight into color separation. Concluding that lenses would never give sharp color images, he designed a reflecting telescope.
8. Answers may vary but should show that students have evaluated the pros and cons of human space flight and should take a stand supported by factual information.
9. *Astronomy* is the science of studying the *universe* using *telescopes,* which include *reflecting telescopes, refracting telescopes,* and special telescopes that detect other forms of *electromagnetic radiation;* and *probes* such as *Cassini* and *Voyager.*

Figure 9 ▶ Astronaut Jerry L. Ross conducts space assembly experiments while anchored to the foot restraint on the remote manipulator system on the space shuttle.

Human Space Exploration

Spacecraft that carry only instruments and computers are described as *robotic*. These spacecraft can explore space and travel beyond the solar system. Crewed spacecraft, or those that carry humans, have never gone beyond Earth's moon.

The first humans went into space in the 1960s. Between 1969 and 1972, NASA landed 12 people on the moon. Now, crewed spaceflights only orbit Earth. Flights, such as those aboard the space shuttles, allow people to release or repair satellites and to perform scientific experiments, as shown in **Figure 9.**

Eventually, NASA would like to send people to explore Mars. However, such a voyage would be expensive, difficult, and dangerous. The loss of two space shuttles and their crews, the *Challenger* in 1986 and the *Columbia* in 2003, have focused public attention on the risks of human space exploration.

Spinoffs of the Space Program

Space programs have brought benefits to areas outside of the field of astronomy. Satellites in orbit provide information about weather all over Earth. This information helps scientists make accurate weather predictions days in advance. Other satellites broadcast television signals from around the world or allow people to navigate cars and airplanes. Inventing ways to make objects smaller and lighter so that they can go into space has also led to improved electronics. These technological developments have been applied to radios, televisions, and other equipment. Even medical equipment has benefited from space programs. For example, heart pumps have been improved based on NASA's research on the flow of fluids through rockets.

Section 1 Review

1. **Describe** characteristics of the universe in terms of time, distance, and organization.
2. **Identify** the parts of the electromagnetic spectrum, both visible and invisible.
3. **Explain** how astronomers use electromagnetic radiation to study space.
4. **Compare** reflecting telescopes and refracting telescopes.
5. **Explain** how a radio telescope differs from an optical telescope.
6. **Identify** two examples of space telescopes and two examples of probes.

CRITICAL THINKING

7. **Identifying Relationships** Using the development of reflecting telescopes as an example, explain how scientific inquiry leads to advances in technology.
8. **Analyzing Processes** Human space exploration is expensive and dangerous. Should NASA continue human spaceflight?

CONCEPT MAPPING

9. Use the following terms to create a concept map: *electromagnetic radiation, reflecting telescope, refracting telescope, telescope, probe, astronomy, universe,* Voyager, and Cassini.

CHAPTER RESOURCES

Chapter Resource File

- Section Quiz GENERAL

Workbooks

- Study Guide (also in Spanish)

Section 2 Movements of Earth

Understanding the basic motions of Earth helps scientists understand the motions of other bodies in the solar system and the universe. These movements of Earth are also responsible for the seasons and the changes in weather.

OBJECTIVES

- **Describe** two lines of evidence for Earth's rotation.
- **Explain** how the change in apparent positions of constellations provides evidence of Earth's rotation and revolution around the sun.
- **Summarize** how Earth's rotation and revolution provide a basis for measuring time.
- **Explain** how the tilt of Earth's axis and Earth's movement cause seasons.

KEY TERMS

rotation
revolution
perihelion
aphelion
equinox
solstice

The Rotating Earth

The spinning of Earth on its axis is called **rotation.** Each complete rotation takes about one day. The most observable effects of Earth's rotation on its axis are day and night. As Earth rotates from west to east, the sun appears to rise in the east in the morning. The sun then appears to cross the sky and set in the west. At any given moment, the hemisphere of Earth that faces the sun experiences daylight. At the same time, the hemisphere of Earth that faces away from the sun experiences nighttime.

rotation the spin of a body on its axis

The Foucault Pendulum

In the 19th century, the scientist Jean-Bernard-Leon Foucault, provided evidence of Earth's rotation by using a pendulum. He created a long, heavy pendulum that rocks back and forth by attaching a wire to the ceiling and then attaching a weight to the wire. Throughout the day, the bob would swing back and forth. The path of the pendulum appeared to change over time. However, it was the floor that was moving while the pendulum's path stayed constant. Because the floor was attached to Earth, one can conclude that Earth rotates. A Foucault pendulum is shown in **Figure 1.**

Figure 1 ▶ The 12 ft arc of this Foucault pendulum in Spokane, Washington, appears to change throughout the day. However, the path of the pendulum does not actually change. Instead, Earth moves the floor as Earth rotates on its axis.

Section 2

Focus

Overview

This section explains Earth's rotation on its axis and its revolution around the sun. The section also relates Earth's motions to the measurement of time and to the passage of seasons.

Bellringer

Ask students to answer these questions: In what direction did the sun rise this morning? Where is the sun in the sky at noon? In which direction will the sun set tonight? (The sun rose in the east. It is directly overhead at noon. It will set in the west.) LS **Verbal**

Motivate

Activity — GENERAL

Role-Playing Dim the room lights and have one student be the *sun*, standing still and holding an illuminated flashlight. Have another student be *Earth*. Have *Earth* walk around the *sun* while turning counterclockwise. Tell *sun* to keep the flashlight beam pointed at *Earth* as that student moves. Have *Earth* announce "day" when he or she sees the light, and "night" when his or her back is turned to the *sun*. LS **Kinesthetic**

CHAPTER RESOURCES

Chapter Resource File
- Directed Reading BASIC

Technology

Transparencies
- Bellringer

Student Edition on CD-ROM

One-Stop Planner CD-ROM
- Lesson Plan

Teach

Demonstration — Basic

Coriolis Effect To show how objects on a rotating surface are affected by the spinning motion, use a record turntable or a lazy Susan. Gather the class around. Keeping the turntable or tray motionless, roll a marble from the center to the edge. Ask students to describe the marble's path (The marble moves in a straight line.) Start the turntable moving and release another marble. Ask students how the path changes. (The path of the marble seems to curve as the marble rolls because the turntable is moving.) LS **Visual**

QuickLAB

Skills Acquired

- Experimenting
- Observing
- Analyzing

Teacher's Notes Students may work in pairs. Have one student get the pendulum moving at a steady rate. The other student can introduce a twisting motion to the string. Remind them to twist the string in only one direction.

Answers

1. Yes, the pendulum was deflected to the side in the direction in which the string was twisted.
2. No, the pendulum keeps moving back and forth, but its path is altered slightly and it tends to slow down.
3. There is no string controlling the spinning motion or rotation of Earth. The twisting motion is introduced to the string rather than being caused by Earth rotating beneath the pendulum.

QuickLAB 10 min

Modeling a Pendulum

Procedure

1. Use a **yo-yo,** or tie a **small object** at one end of a **length of string for a pendulum.**
2. Hold the end of the string in your fingers, and swing the pendulum.
3. Twist the string in your fingers as you allow the pendulum to continue swinging.

Analysis

1. Does the direction in which the yo-yo swings change?
2. Does the yo-yo twist around with the string?
3. How does this experiment differ from the Foucault pendulum?

revolution the motion of a body that travels around another body in space; one complete trip along an orbit

perihelion the point in the orbit of a planet at which the planet is closest to the sun

aphelion the point in the orbit of a planet at which the planet is farthest from the sun

The Coriolis Effect

Evidence of the rotation of Earth can also be seen in the movment of ocean surface currents and wind belts. Ocean currents and wind belts do not move in a straight path. The rotation of Earth causes ocean currents and wind belts to be deflected to the right in the Northern Hemisphere. In the Southern Hemisphere, ocean currents and wind belts deflect to the left. This curving of the path of wind belts and ocean currents is caused by Earth's rotation and is called the *Coriolis effect*.

The Revolving Earth

As Earth spins on its axis, Earth also revolves around the sun. Even though you cannot feel Earth moving, it is traveling around the sun at an average speed of 29.8 km/s. The motion of a body that travels around another body in space is called **revolution.** Each complete revolution of Earth around the sun takes 365 1/4 days, or about one year.

Earth's Orbit

The path that a body follows as it travels around another body in space is called an *orbit*. Earth's orbit around the sun is not quite a circle. Earth's orbit is an ellipse. An *ellipse* is a closed curve whose shape is determined by two points, or foci, within the ellipse. In planetary orbits, one focus is usually located within the sun. No object may be located at the other focus.

Because its orbit is an ellipse, Earth is not always the same distance from the sun. The point in the orbit of a planet at which the planet is closest to the sun is the **perihelion.** The point in the orbit of a planet at which the planet is farthest from the sun is the **aphelion** (uh FEE lee uhn). As shown in **Figure 2,** Earth's aphelion distance is 152 million km. Its perihelion distance is 147 million km.

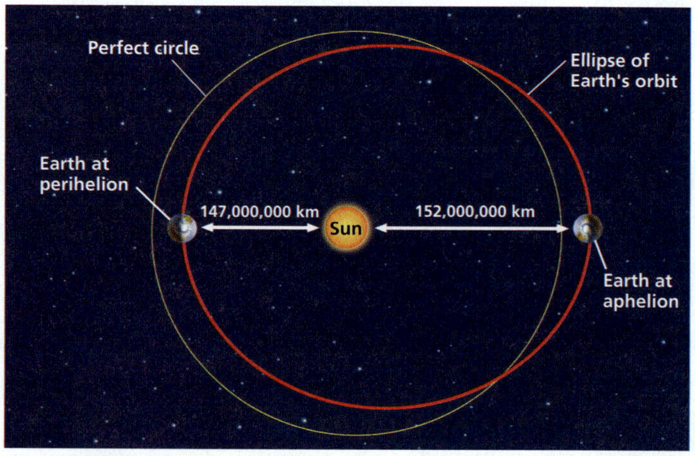

Figure 2 ▶ As Earth revolves around its elliptical orbit, the planet is farthest from the sun in July and closest to the sun in January. The elliptical orbit in this illustration has been exaggerated for emphasis.

CHAPTER RESOURCES

Chapter Resource File

- Datasheet for Quick Lab GENERAL

Technology

Transparencies

- 131 Earth's Orbit (with worksheet)
- 132 The Apparent Motion of Constellations (with worksheet)

SKILL BUILDER — General

Vocabulary The words *aphelion* and *perihelion* come from the Greek root *-helios*, which means "sun." *Ap-* is a variant of *apo*, which means "away or apart." Thus, the aphelion is the point in a planet's orbit at which the planet is farthest from the sun. *Peri-*, in contrast, means "near." The perihelion is the point in a planet's orbit at which the planet is closest to the sun. LS **Verbal** English Language Learners

Constellations and Earth's Motion

Evidence of Earth's revolution and rotation around the sun can be seen in the motion of constellations. A *constellation* is a group of stars that are organized in a recognizable pattern. In 1930, the International Astronomical Union divided the sky into 88 constellations. Many of the names given to these constellations came from the ancient Greeks more than 2,000 years ago. Taurus, the bull, and Orion, the hunter, are some examples of names from Greek mythology that have been given to constellations.

Evidence of Earth's Rotation

If you gaze up at a constellation in the evening sky over a period of several hours, you may notice that the constellation appears to have changed its position in the sky. However, the constellation's movement has not caused the constellation's change in position. The rotation of Earth on its axis causes the change in position. Thus, Earth is moving, and the constellation is not moving.

Evidence of Earth's Revolution

A constellation's position in the evening sky will change not only because of Earth's rotation but also because of Earth's revolution around the sun. Over a period of several weeks, at the same time of the evening, a constellation's position will appear to change, as shown in **Figure 3.** But Earth's revolution around the sun causes the constellation to have different positions in the evening sky over a period of several weeks. As Earth revolves around the sun, the night side of Earth faces a different direction of the universe. Thus, different constellations will appear in the night sky as the seasons change.

Reading Check **How does the movement of constellations provide evidence of Earth's rotation and revolution? (See the Appendix for answers to Reading Checks.)**

Figure 3 ▶ In one month's period, the position of the constellations in the sky seen from Denver, Colorado, at 10 PM change because of the revolution of Earth.

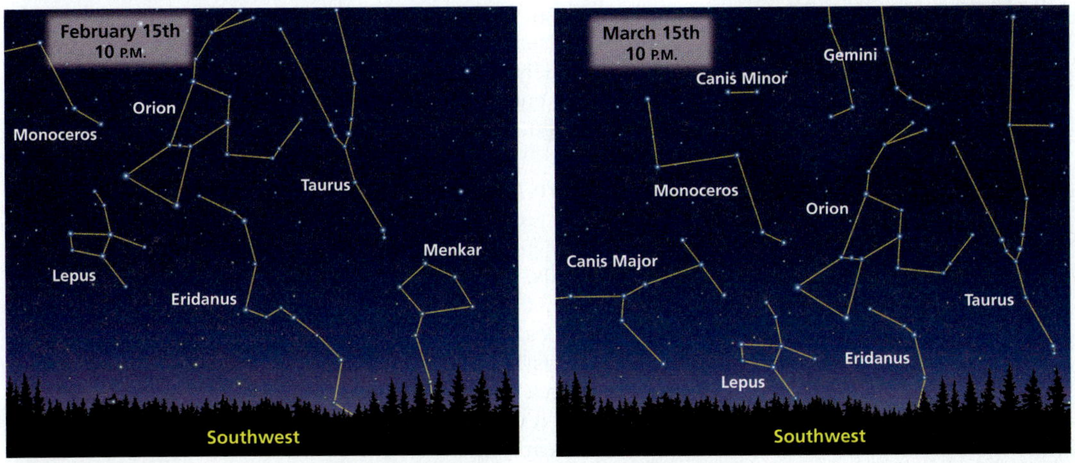

Discussion — BASIC

Comprehension Check Explain that the stars that make up constellations are not necessarily located near each other, but are simply in the same part of the sky. The patterns were devised as a memory aid to help people identify the stars. Ask students why people in ancient cultures wanted to remember the stars. Tell them to think about how constellations change throughout the year. (Answers may vary. Sample answers: Because constellations are visible at different but predictable times of the year, people could anticipate and plan for planting and harvesting seasons and for winter.) **LS Verbal**

Activity — GENERAL

Star Viewers To help students become familiar with star patterns, have them make a star viewer. They will need a cardboard tube, black construction paper, masking tape, and pins of various sizes. Have them select a constellation and find out the names of its main stars. Tell them to put the end of the tube on top of the construction paper and trace a circle the same size as the opening. Have them copy the star pattern on the circle, and use pins to poke holes in the paper for each star. Have them cut out the circle and attach it to the front of the tube with the front of the paper facing the tube. They can hold the tube up to a bright light to view the constellation. **LS Visual**

Cultural Awareness — BASIC

Constellation Stories Have students identify constellations that people from different cultures saw in the night sky. Then, have students work individually, in pairs, or in small groups to select one constellation to research. Provide star guides and mythology books for reference. Students should identify when and where the constellation appears. Ask them to retell the myth or legend in their own words and to illustrate or act out the story. Have a class storytelling session in which students share their stories. **LS Verbal/Visual/Auditory**

Answer to Reading Check

Constellations provide two kinds of evidence of Earth's motion. As Earth rotates, the stars appear to change position during the night. As Earth revolves around the sun, Earth's night sky faces a different part of the universe. As a result, different constellations appear in the night sky as the seasons change.

HISTORY CONNECTION GENERAL

It's About Time Calendars are typically based on natural astronomical cycles. Divide the class into small groups and have each group choose a calendar system to research, such as the Mayan, Islamic, Jewish, Gregorian, Aztec, Egyptian, Babylonian, or Chinese systems. Three basic types of calendars have been created: solar calendars designed to match the year, lunar calendars based on the lunar phase cycle, and lunisolar calendars that have months based on the lunar phase cycle with an extra month periodically intercalculated. Have student groups find out the scientific basis of the calendar they chose, interesting facts about its evolution, and its usefulness in helping a society organize its activities. Have them share what they discover with the class. LS Verbal

Graphic Organizer GENERAL

Venn Diagram

You may want to have students work in groups to create this Venn Diagram. Have one student draw the map and fill in information provided by other students from the group.

Figure 4 ▶ This is a reconstruction of a stone calendar that the Aztecs created to help determine when to plant crops.

Measuring Time

Earth's motion provides the basis for measuring time. For example, the day and year are based on periods of Earth's motion. The day is determined by Earth's rotation on its axis. Each complete rotation of Earth on its axis takes one day, which is then broken into 24 hours.

The year is determined by Earth's revolution around the sun. Each complete revolution of Earth around the sun takes 365 1/4 days, or one year.

A month is based on the moon's motion around Earth. A month was originally determined by the period between successive full moons, which is 29.5 days. The word *month* actually comes from the word *moon*. However, the number of full moons in a year is not a whole number. Therefore, a month is now determined as roughly one-twelfth of a year.

Graphic Organizer **Venn Diagram**

Create the Graphic Organizer entitled "Venn Diagram" described in the Skills Handbook section of the Appendix. Label the circles with "Day," "Month," and "Year." Then, fill in the diagram with characteristics that each period of time shares with the other periods of time.

Formation of the Calendar

A *calendar* is a system created for measuring long intervals of time by dividing time into periods of days, weeks, months, and years. Many ancient civilizations created versions of calendars based on astronomical cycles. The ancient Egyptians were the first to use a calendar based on a solar year. The Babylonians used a 12 month lunar year. The Aztecs, who lived in what is now Mexico, also created a calendar, which is shown in **Figure 4.**

Because the year is 365 1/4 days long, the extra 1/4 day is usually ignored to make the number of days on a calendar a whole number. To keep the calendars on the same schedule as Earth's movements, we must account for the extra time. So, every four years, one day is added to the month of February. Any year that contains an extra day is called a *leap year*.

More than 2,000 years ago, Julius Caesar, of the Roman Empire, revised the calendar so that an extra day every four years was added. His successor, Augustus Caesar, made the extra day come at the end of the shortest month, February. He also made July and August long months with 31 days each.

The Modern Calendar

Because the year is not exactly 365 1/4 days long, over centuries, the calendar gradually became misaligned with the seasons. In the late 1500s, Pope Gregory XIII formed a committee to create a calendar that would keep the calendar aligned with the seasons. We use this calendar today. In this Gregorian calendar, century years, such as 1800 and 1900, are not leap years unless the century years are exactly divisible by 400. Thus, 2000 was a leap year even though it was a century year. However, 2100, 2200, and 2300 will not be leap years.

READING SKILL BUILDER BASIC

Reading Hint Pair students to practice the "skimming and scanning" technique on the passages about measuring time and calendar formation. Have students brainstorm ways to get the most out of this reading technique, such as looking for italicized words, heads, dates, and captions. First, they can read quickly to identify the main topic, then scan for specific information. They can use the related section review questions as their focus or work together to create an outline to use as a study guide. LS Verbal English Language Learners

INCLUSION Strategies

- Hearing Impaired
- Learning Disabled
- Developmentally Delayed

Gather several calendars, one for every two students, from a variety of years, including some leap years. Have students sit with a partner, and give each pair a calendar. Read through this page as a group. Have students explore each calendar reference on their individual calendars. LS Verbal

Time Zones

Using the sun as the basis for measuring time, we define noon as the time when the sun is highest in the sky. Because of Earth's rotation, the sun is highest above different locations on Earth at different times of day. Earth's surface has been divided into 24 standard time zones, as shown in **Figure 5,** to avoid problems created by different local times. In each zone, noon is set as the time when the sun is highest over the center of that zone. Because Earth is nearly spherical, its circumference equals 360°. If you divide 360° by the 24 hours needed for one rotation, you find that Earth rotates at a rate of 15° per hour. Therefore, each of Earth's 24 standard time zones covers about 15°. The time in each zone is one hour earlier than the time in the zone to the east of each zone.

International Date Line

There are 24 standard time zones and 24 hours in a day. But there must be some point on Earth's surface where the date changes. The *International Date Line* was established to prevent confusion. The International Date Line is a line that runs from north to south through the Pacific Ocean. When it is Friday west of the International Date Line, it is Thursday east of the line. The line is drawn so that it does not cut through islands or continents. Thus, everyone living within one country has the same date. Note where the line is drawn between Alaska and Siberia in **Figure 5.**

Reading Check What is the purpose of the International Date Line? (See the Appendix for answers to Reading Checks.)

Figure 5 ▶ Earth has been divided into 24 standard time zones. Irregular time zones may differ between 15 to 45 minutes compared to regular time zones. ***At 6 P.M. on the east coast of South America what time is it on the west coast of South America?***

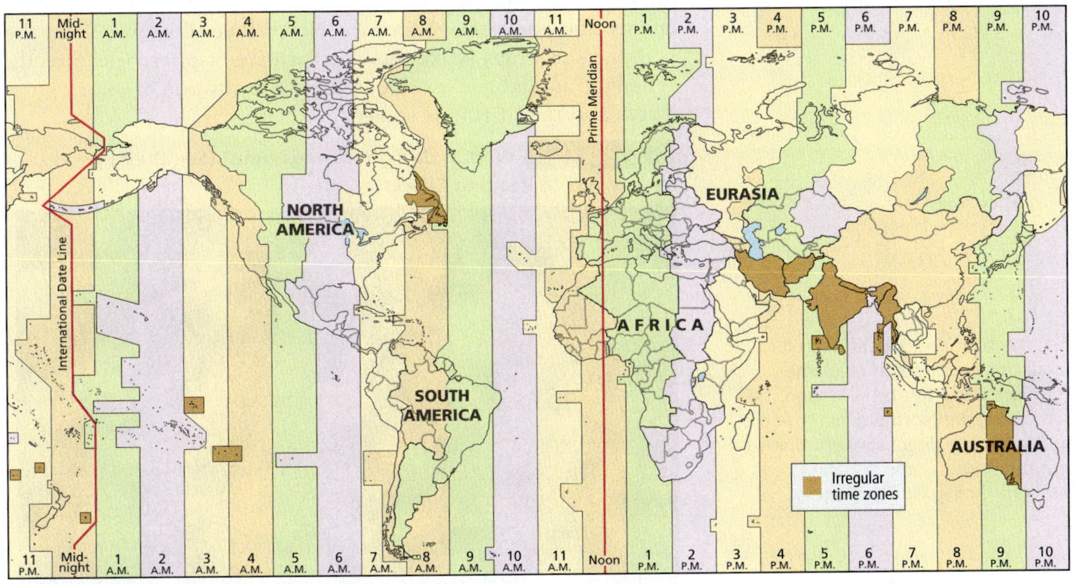

Using the Figure — GENERAL

Time Zone Zigzags Explain to students that political borders and other considerations sometimes override the drawing of time zone boundaries strictly on the basis of longitude. Ask students to imagine what it would be like if a time zone line went directly through the middle of your town. Ask students what happens as you move west from one time zone to the next. (As you move west, clock time is one hour earlier.) Answer to caption question: When it is 6 P.M. on the east coast of South America, it would be 4 P.M. in most of the northern half of the west coast of South America, but 5 P.M. in Chile because of the staggered time zone. **LS Visual**

MATH CONNECTION — GENERAL

What Time Is It? Ask students to use the map of time zones to figure out what time it would be in different parts of the world. If it is 6 A.M. on Tuesday in Beijing, China, what day and time is it in San Francisco, California? (Going east, you move through 8 time zones: 6 A.M. + 8 h = 2 P.M.; When you cross the International Dateline, you lose a day, so it is 2 P.M. on Monday in San Francisco.) If it is 4 P.M. on Saturday in Greenland, what day and time is it in Siberia? (Going west, you move through 9 zones: 4 P.M. − 9 h = 7 A.M.; By crossing the International Dateline, you gain a day, so it is 7 A.M. on Sunday in Siberia.) **LS Logical**

Internet Activity — GENERAL

International Zones Invite students to research the evolution of time zones. Students may also investigate what local variations of the time-zone system are used in China, the Middle East, and South Asia. Have students report back to the class on an aspect of this topic that interests them. A worksheet designed to direct student research on this topic can be found in the **Chapter Resource File** booklet or by visiting **go.hrw.com** and entering the keyword **HQ6SSPX.** **LS Verbal**

Answer to Reading Check

Because time zones are based on Earth's rotation, as you travel west you eventually come to a location where, on one side of the time zone border, the calendar moves ahead one day. The purpose of the International Dateline is to locate the border so that the transition would affect the least number of people. So that it will affect the least number of people, the International Dateline is in the middle of the Pacific Ocean, instead of on a continent.

CHAPTER RESOURCES

Chapter Resource File

- **Internet Activity**
  - International Zones GENERAL

Technology

- **Transparencies**
  - 133 Time Zones (with worksheet)

Teach, continued

Debate — GENERAL

Daylight Savings Time For thousands of years, many societies based their clocks on local solar time. During World War I, daylight savings time was adopted in an effort to save fuel. Today, the system is followed in more than 70 countries. Some groups object to the practice—for example, farmers must then do chores in the morning darkness. Invite students to offer additional examples and to debate the merits of daylight savings time as a public policy. Assign students to prepare arguments for and against the practice and stage a debate. **LS Verbal/Logical**

Answer to Reading Check

Daylight savings time is an adjustment that is made to standard time by setting clocks ahead one hour to take advantage of longer hours of daylight in the summer months and to save energy.

BRAIN FOOD

Planetary Seasons
Variations in the seasons of other planets are due to axial tilt and how elliptical a planet's orbit is. Venus and Jupiter have axial tilts of about 3° compared with Earth's 23.5°, so their seasonal variations are smaller. Mercury's pattern of rotating three times for every two orbits of the sun produces two week-long seasons. Mars has a more eccentric orbit and a slightly larger axial tilt, so its seasons are different lengths. Uranus's extreme axial tilt of 98° produces seasons that last about 21 years. The coming of spring warmth triggers huge storms in Uranus' atmosphere.

CHAPTER RESOURCES

Technology

Transparencies
- 134 How the Tilt of Earth's Axis Affects Seasons (with worksheet)

SCILINKS. NSTA Developed and maintained by the National Science Teachers Association

For a variety of links related to this subject, go to www.scilinks.org

Topic: Seasons
SciLinks code: HQ61363

Daylight Savings Time

Because of the tilt of Earth's axis, daylight time is shorter in the winter months than in the summer months. During the summer months, days are longer so that the sun rises earlier in the morning when many people are still sleeping. To take advantage of that daylight time, the United States uses *daylight savings time*. Under this system, clocks are set one hour ahead of standard time in April, which provides an additional hour of daylight during the evening. The additional hour also saves energy because the use of electricity decreases. In October, clocks are set back one hour to return to standard time. Countries in the equatorial region do not observe daylight savings time because there are not significant changes in the amount of daylight time in the equatorial region. Daylight is about 12 hours every day of the year.

The Seasons

Earth's axis is tilted at 23.5°. As Earth revolves around the sun, Earth's axis always points toward the North Star. Thus, during each revolution, the North Pole sometimes tilts toward the sun and sometimes tilts away from the sun, as shown in **Figure 6.** When the North Pole tilts toward the sun, the Northern Hemisphere has longer periods of daylight than the Southern Hemisphere does. When the North Pole tilts away from the sun, the Southern Hemisphere has longer periods of daylight.

The angle at which the sun's rays strike each part of Earth's surface changes as Earth moves through its orbit. When the North Pole tilts toward the sun, the sun's rays strike the Northern Hemisphere more directly. When the sun's rays strike Earth directly, that region receives a higher concentration of solar energy and is warmer. When the North Pole tilts away from the sun, the sun's rays strike the Northern Hemisphere less directly. When the sunlight is less direct, the concentration of solar energy is less and that region is cooler.

Reading Check **What is daylight savings time? (See the Appendix for answers to Reading Checks.)**

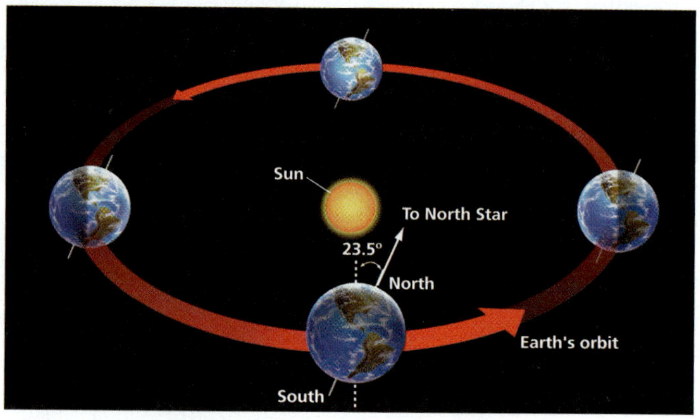

Figure 6 ▶ The direction of tilt of Earth's axis remains the same throughout Earth's orbit around the sun. Thus, the Northern Hemisphere receives more direct sunlight during summer months and less direct sunlight during winter months.

MISCONCEPTION ALERT

Seasons Students may believe that seasons are the result of the changing distance between Earth and the sun, with hotter weather occurring in both hemispheres when Earth is closer to the sun. Because Earth's orbit is slightly elliptical in shape, Earth's distance from the sun does vary. But Earth is actually closer to the sun in January and farther from the sun in July, the opposite of what students might expect. Emphasize that the actual cause of seasons is Earth's tilt on its axis. The axial tilt changes the angle at which the sun's rays strike Earth, and also affects the number of daylight hours, resulting in differences in energy and temperatures. In January, when the North Pole is tilted away from the sun, the sun's rays strike the Northern Hemisphere at a low angle, producing colder weather. At the same time, the sun's rays strike the Southern Hemisphere at a higher angle, producing warmer weather. In effect, the seasons are reversed north and south of the equator. **LS Verbal**

Seasonal Weather

Changes in the angle at which the sun's rays strike Earth's surface cause the seasons. When the North Pole tilts away from the sun, the angle of the sun's rays falling on the Northern Hemisphere is low. As a result, the sun's rays spread solar energy over a large area, which leads to lower temperatures. The tilt of the North Pole away from the sun also causes the Northern Hemisphere to experience fewer daylight hours. Fewer daylight hours also mean less energy and lower temperatures. Lower temperatures cause the winter seasons. At the time, the Northern Hemisphere tilts away from the sun, the Southern Hemisphere tilts toward the sun. The sun's rays strike the Southern Hemisphere at a greater angle than they do in the Northern Hemisphere, and there are more daylight hours in the Southern Hemisphere. Therefore, the Southern Hemisphere experiences the warm summer season.

Equinoxes

The seasons fall and spring begin on days called equinoxes. An **equinox** is the moment when the sun appears to cross the celestial equator. The *celestial equator* is a line drawn on the sky directly overhead from the equator on Earth. At an equinox, the sun's rays strike Earth at a 90° angle along the equator. The hours of daylight and darkness are approximately equal everywhere on Earth on that day. The *autumnal equinox* occurs on September 22 or 23 of each year and marks the beginning of fall in the Northern Hemisphere. The *vernal equinox* occurs on March 21 or 22 of each year and marks the beginning of spring in the Northern Hemisphere.

equinox the moment when the sun appears to cross the celestial equator

QuickLAB 10 min

The Angle of the Sun's Rays

Procedure

1. Turn the lights down low or off in the classroom.
2. Place a **piece of paper** on the floor. Hold a **flashlight** 1 m above the paper, and shine the light of the flashlight straight down on the piece of paper.
3. Have a partner outline the perimeter of the circle of light cast by the flashlight on the paper. Label the circle "90° angle." Place a clean piece of paper on the floor.
4. At a height of 1/2 m from the floor, shine the light of the flashlight on the paper at an angle. Make sure the distance between the flashlight and the paper is 1 m.
5. Have a partner outline the perimeter of the circle of light cast by the flashlight on the paper. Label the circle "low angle."

Analysis

1. Compare the two circles drawn in steps 3 and 5. Which circle concentrates the light in a smaller area?
2. Which circle would most likely model the sun's rays striking Earth during the summer season?

Close

Reteaching — Basic

Evidence Have students develop a cause-and-effect organizer to help them understand the evidence for Earth's motions. Guide them to put Earth's rotation and revolution in the cause boxes and list the different lines of evidence in corresponding effects boxes. Students may work in pairs. LS Logical Co-op Learning

Quiz — General

1. How does Earth's axial tilt cause seasonal temperature changes? (At a lower angle, the sun's rays spread over a larger area, reducing their heating effect. At a higher angle, the solar energy is more concentrated.)
2. How are time zones related to Earth's rotation? (Earth's surface was divided along longitude lines into 24 time zones of 15° each. Earth rotates through the longitudes, represented by each time zone, one for each hour, in 24 hours.)
3. What would happen if we did not add an extra day to the solar calendar for leap year? (Because of the extra quarter day in the solar year, over time the calendar would drift out of synch with Earth's motions and be misaligned with the seasons.) LS Verbal

CHAPTER RESOURCES

Chapter Resource File

- Datasheet for Quick Lab General

QuickLAB

Skills Acquired

- Experimenting
- Observing
- Analyzing

Teacher's Notes Although the distance between the flashlight and the floor in step 4 is 1 m, the actual distance between the light beam and the circle remains 2 m, as in step 2. Students must lower their hands and turn the flashlight at a low angle. Students may need to move the piece of paper to the side to draw the circle. Students can place thermometers in the circles of light to compare temperatures.

Answers

1. the circle in step 3
2. the circle in step 3

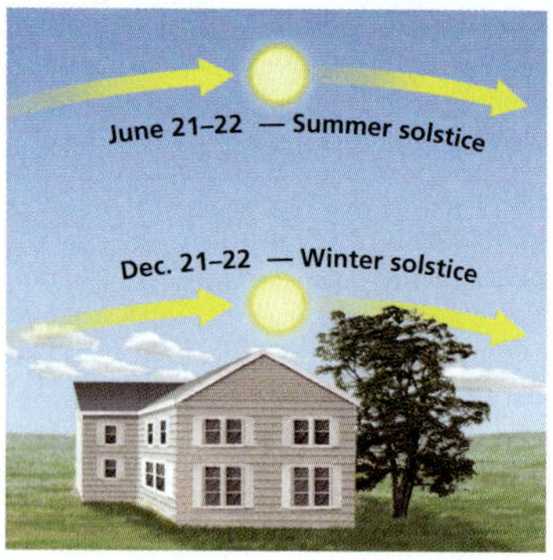

Figure 7 ▶ In the Northern Hemisphere, the sun appears to follow its highest path across the sky on the summer solstice and its lowest path across the sky on the winter solstice.

solstice the point at which the sun is as far north or as far south of the equator as possible

Summer Solstices

The seasons of summer and winter begin on days called **solstices.** Each year on June 21 or 22, the North Pole's tilt toward the sun is greatest. On this day, the sun's rays strike Earth at a 90° angle along the Tropic of Cancer, which is located at 23.5° north latitude. This day is called the *summer solstice* and marks the beginning of summer in the Northern Hemisphere. *Solstice* means "sun stop" and refers to the fact that in the Northern Hemisphere, the sun follows its highest path across the sky on that day, as shown in **Figure 7.**

The Northern Hemisphere has the most hours of daylight at the summer solstice. The farther north of the equator you are, the longer the period of daylight you have. North of the Arctic Circle, which is located at 66.5° north latitude, there are 24 hours of daylight at the summer solstice. At the other extreme, south of the Antarctic Circle, there are 24 hours of darkness at that time.

Winter Solstices

By December, the North Pole is tilted to the farthest point away from the sun. On December 21 or 22, the sun's rays strike Earth at a 90° angle along the Tropic of Capricorn, which is located at 23.5° south latitude. This day is called the *winter solstice.* It marks the beginning of winter in the Northern Hemisphere. At the winter solstice, the Northern Hemisphere has the fewest daylight hours. The sun follows its lowest path across the sky. Places that are north of the Arctic Circle then have 24 hours of darkness. However, places that are south of the Antarctic Circle have 24 hours of daylight at that time.

Section 2 Review

1. **Explain** how the apparent change of position of constellations over time provides evidence of Earth's revolution around the sun.
2. **Describe** two lines of evidence that indicate that Earth is rotating.
3. **Summarize** how movements of Earth provide a basis for measuring time.
4. **Explain** why today's calendars have leap years.
5. **Identify** two advantages in using daylight savings time.
6. **Explain** how the tilt of Earth's axis and Earth's movements cause seasons.
7. **Identify** the position of Earth in relation to the sun that causes winter in the Northern Hemisphere.
8. **Describe** the position of Earth in relation to the sun during the Northern Hemisphere's summer solstice.

CRITICAL THINKING

9. **Understanding Relationships** How can it be that Earth is at perihelion during wintertime in the Northern Hemisphere?
10. **Predicting Consequences** Explain how measurements of time might differ if Earth rotated on its axis only once per year.

CONCEPT MAPPING

11. Use the following terms to create a concept map: *revolution, perihelion, aphelion, rotation, ellipse, orbit, rotation, Foucault pendulum, Coriolis effect, Earth,* and *constellation.*

Close, *continued*

Alternative Assessment — ADVANCED

Science Fiction Stories Have students write stories in which the evidence for Earth's motions plays a key role in solving a mystery. The key might focus on time zones, seasons, the Coriolis effect, or the apparent movement of constellations. The story must include correct explanations. **LS Verbal**

Answers to Section Review

1. During different seasons, different constellations appear in the night sky because we see the stars from a different position in Earth's orbit.
2. A Foucault pendulum traces out a changing path because Earth is rotating underneath. Earth's rotation causes wind belts and currents to deflect.
3. Earth's revolution around the sun takes about 365 days. The Earth also spins on its axis every 24 hours. The side that faces the sun has day and the other side has night. The whole lunar cycle takes about four weeks, the basis of our month.
4. The year is actually 365 1/4 days long. In the 1500s the calendar did not match seasons, so leap year was introduced, with an extra day every four years.
5. an additional hour of daylight; saving energy
6. During summer, half of Earth is tilted toward the sun and during winter that half is tilted away. Regions pointing toward the sun are warmer because daylight lasts longer and the sun's rays are more vertical.
7. Earth is closer to the sun in winter, but the North Pole is tilted away from the sun. The sun's rays strike at a low angle, producing lower winter temperatures in the Northern Hemisphere.
8. During the summer solstice, Earth is located near the aphelion of its orbit, or the farthest distance from the sun.
9. It is hours of daylight and the angle of the sun's rays that cause the seasons, not distance from the sun.
10. If Earth rotated only once a year, one side of the planet would face the sun at all times and the other half of Earth would be in darkness. Because Earth rotates at a rate of 15° of its circumference each hour, we would also not have time measurements dividing the day into 24-hour periods. Earth's rotation on its axis is related to the division of the year into four seasons. If Earth's axis were not tilted at an angle, these seasonal variations would not exist.
11. The movements of *Earth* include its *revolution* around the sun in an *orbit* shaped like an *ellipse* from *perihelion* to *aphelion;* and *rotation* on its axis, which is evidenced by the *Foucault pendulum,* the *Coriolis effect,* and the apparent movement of *constellations.*

CHAPTER RESOURCES

Chapter Resource File

- Section Quiz GENERAL

Workbooks

- Study Guide (also in Spanish)

Chapter 26 Highlights

Sections

1 Viewing the Universe

Key Terms

astronomy, 659
galaxy, 660
astronomical unit, 660
electromagnetic spectrum, 661
telescope, 662
refracting telescope, 663
reflecting telescope, 663

Key Concepts

- The universe formed about 14 billion years ago in a giant explosion and has been expanding since that time.
- The electromagnetic spectrum contains all of the frequencies or wavelengths of electromagnetic radiation. Scientists use this radiation to study the universe.
- Refracting telescopes use lenses to gather and focus light, while reflecting telescopes use curved mirrors to gather and focus light.
- Space programs have led to improved technology in areas such as airplane navigation, weather forecasting, and medical equipment.

2 Movements of Earth

rotation, 667
revolution, 668
perihelion, 668
aphelion, 668
equinox, 673
solstice, 674

- Earth's rotation is evidenced by the change from day to night, the apparent motion of constellations, the Foucault pendulum, and the Coriolis effect.
- Earth's revolution around the sun is evidenced by the apparent motion of constellations.
- Movements of Earth provide a basis for measuring time. One revolution of Earth around the sun is equal to one year. One rotation of Earth on its axis is equal to one day.
- The angle of the sun's rays changes throughout the year and leads to seasonal change on Earth's surface.

Chapter Highlights

Alternative Assessment — GENERAL

Astronomy Picture Books Have students work in small groups to create picture books that will explain to younger children one of the following topics: (1) how scientists study the universe using ground-based observations and telescopes, space telescopes, probes and manned space missions; or (2) how Earth motions are related to time measurement, the calendar, and seasonal change. Provide good models in the form of children's nonfiction books for students to refer to. For students who want to make illustrations using three-dimensional effects, you may want to provide reference materials on paper engineering. Suggest that students plan the contents by making an outline and thumbnail before starting to create the books. Students can draw diagrams to explain concepts and illustrate the books using photos from library sources, from the Internet, or pictures that they have taken themselves of the night sky. Good sources include NASA Web sites. Display the completed picture books in the classroom. LS Visual/Verbal Co-op Learning

CHAPTER RESOURCES

Chapter Resource File

- Concept Review GENERAL
- Critical Thinking ADVANCED
- Math Skills GENERAL
- Graphing Skills GENERAL
- Chapter Test A GENERAL
- Chapter Test B ADVANCED

Workbooks

- **Study Guide** (also in Spanish)
- **Assessments** (Spanish)

Technology

Classroom Videos
- Brain Food Video Quiz

Chapter Review

Assignment Guide	
Section	**Questions**
1	1–2, 4, 7, 10, 12–13, 16–17, 22, 25, 27, 29
2	3, 5–6, 8–9, 11, 14–15, 18–21, 23–24, 28, 30–34
1 and 2	26

Using Key Terms

1–6. Answers may vary but should show that students understand the definitions of and differences between key terms.

Understanding Key Concepts

7. c
8. d
9. b
10. b
11. d
12. a
13. c
14. a
15. d

Short Answer

16. X rays and gamma rays
17. Orbiting telescopes are above the distorting effects of Earth's atmosphere and light pollution. They can also study the electromagnetic radiation that does not penetrate our atmosphere.
18. We use Earth's regular rotation around the sun to keep time. Establishing noon as the time when the sun is highest, and dividing Earth into time zones establishes standards that everyone on Earth can use, no matter where they are located.
19. A leap year is any calendar year to which an extra day has been added. The day is added in order to keep the calendar aligned with the seasons over a period of centuries.
20. the International Dateline, drawn through the Pacific Ocean
21. The side of Earth that is tilted toward the sun has longer periods of daylight, receives more concentrated solar radiation at a more vertical angle, and has warmer temperatures. The side that is tilted away from the sun has fewer hours of daylight, receives less concentrated solar radiation at a less direct angle, and has cooler temperatures. Because of Earth's tilt, seasons are reversed north and south of the equator.

Chapter 26 Review

Using Key Terms

Use each of the following terms in a separate sentence.

1. *electromagnetic spectrum*
2. *galaxy*
3. *perihelion*

For each pair of terms, explain how the meanings of the terms differ.

4. *reflecting telescope* and *refracting telescope*
5. *solstice* and *equinox*
6. *rotation* and *revolution*

Understanding Key Concepts

7. Stars organized into a pattern are
   a. perihelions.
   b. satellites.
   c. constellations.
   d. telescopes.

8. Days are caused by Earth's
   a. perihelion.
   b. aphelion.
   c. revolution.
   d. rotation.

9. The seasons are caused by
   a. Earth's distance from the sun.
   b. the tilt of Earth's axis.
   c. the sun's temperature.
   d. the calendar.

10. Which of the following is a tool that is used by astronomers to study radiation?
    a. a computer model
    b. a ground-based telescope
    c. a Foucault pendulum
    d. a calendar

11. Which of the following is evidence of Earth's revolution?
    a. the Foucault pendulum
    b. the Coriolis effect
    c. night and day
    d. constellation movement

12. Which of the following forms of radiation can be shielded by Earth's atmosphere?
    a. gamma rays
    b. radio waves
    c. visible light
    d. All of the above

13. Which of the following is not a space telescope?
    a. *Hubble Space Telescope*
    b. *Chandra X-ray Observatory*
    c. *Challenger*
    d. *Spitzer Space Telescope*

14. Which of the following marks the beginning of spring in the Northern Hemisphere?
    a. vernal equinox
    b. autumnal equinox
    c. summer solstice
    d. winter solstice

15. Which of the following is evidence of Earth's rotation?
    a. the Foucault pendulum
    b. day and night
    c. the Coriolis effect
    d. All of the above

Short Answer

16. Which two forms of electromagnetic radiation have the shortest wavelengths?
17. What is an advantage of using orbiting telescopes rather than ground-based telescopes?
18. Why does the rotation of Earth require people to establish time zones?
19. What is a leap year, and what purpose does it serve?
20. What line on Earth's surface marks where the date changes?
21. How does the tilt of Earth's axis cause the seasons?

Critical Thinking

22. Evaluating Data If telescopes had not been developed, how would our knowledge of the universe be different?

23. Analyzing Ideas In each time zone, it gets dark earlier on the eastern side of the zone than on the western side. Explain why.

24. Applying Ideas How would seasons be different if Earth was not tilted on its axis?

25. Making Inferences What limitation of a refracting telescope could be overcome by placing the telescope in space? Explain your answer.

Concept Mapping

26. Use the following terms to create a concept map: *Galileo, spacecraft, telescope, constellation, rotation, revolution, Foucault pendulum, Coriolis effect, equinox, solstice,* and *astronomy.*

Math Skills

27. Making Calculations A certain star is 1.135×10^{14} km away from Earth. If light travels at 9.4607×10^{12} km per year, how long will it take for light from the star to reach Earth?

28. Applying Quantities At aphelion, Earth is 152,000,000 km from the sun. At perihelion, the two bodies are 147,000,000 km apart. What is the difference in kilometers between Earth's farthest point from the sun and Earth's closest point to the sun?

Writing Skills

29. Creative Writing Imagine that you are the head of a space program that has created the first orbiting telescope. Write a press release that explains to the public why your space agency has spent billions of dollars to build and launch a space telescope.

30. Communicating Main Ideas Explain how the Foucault pendulum and the Coriolis effect provide evidence of Earth's rotation.

Interpreting Graphics

The diagram below shows the different time zones of the world by looking down at the North Pole. Use the diagram to answer the questions that follow.

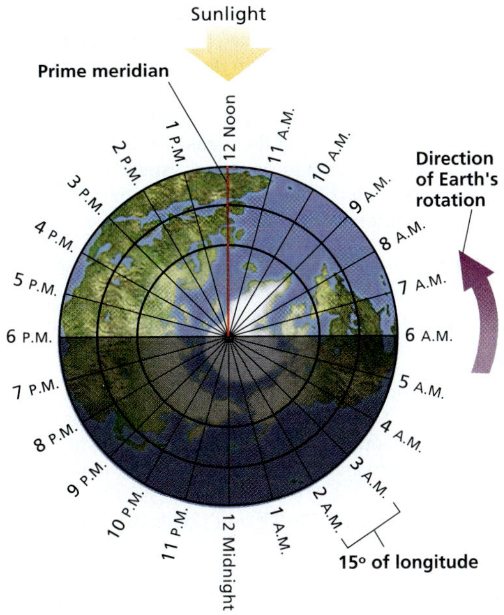

31. If it is 6 P.M. at the prime meridian, what is the time on the opposite side of the world?

32. On the diagram, it is 9 P.M. in Japan and 2 A.M. in Alaska. How many degrees apart are Alaska and Japan?

33. If it is 6 A.M. in Alaska, what time is it in Japan?

34. How many hours are in 120°?

Chapter Review

Critical Thinking

22. Answers may vary. Sample answer: We would be limited to naked-eye observations of the sky and to the interpretations we could make of Earth's motions.

23. Earth rotates in a counterclockwise direction from west to east, exposing the eastern part of each time zone to the sun's light first. The eastern half of the time zone likewise rotates into darkness before the western side does.

24. The sun's rays would strike the northern and southern hemispheres at the same angle throughout the year. Seasonal temperature changes would be minimal and would be related more to distance from the sun.

25. Answers may vary: Sample answers: If the objective lens of a refracting telescope is too large, the lens will sag under its own weight and distort the images. In orbit, the effects of Earth's gravity are less. Larger lenses can be used to view more distant objects without the distortion caused by excess weight.

Concept Mapping

26. Answers may vary but should include all of the terms listed. Sample answers appear at the end of this Teacher Edition.

Math Skills

27. 1.135×10^{14} km $\div 9.4607 \times 10^{12}$ km/y $= 1.135 \times 10^{(14-12)} \div 9.461$ y $= 0.11996 \times 10^{2}$ y $=$ 11.996 y, or about 12 y

28. 152,000,000 km − 147,000,000 km = 5,000,000 km

Writing Skills

29. Answers may vary. Accept all reasonable answers.

30. Answers may vary. Accept all reasonable answers.

Interpreting Graphics

31. 6 A.M.

32. 9 P.M. to 2 A.M. = 5 h
$15° \times 5 = 75°$

33. 1 A.M.

34. 120°/15° per hour = 8 h

Estimated Time

To give students practice under more realistic testing conditions, allow them 30 minutes to answer all of the questions in this practice test.

TEST DOCTOR

Question 3 Answer D is correct. Earth makes one complete revolution around the sun every 365.25 days. The other answers do not deal with time markers that involve the sun. Earth rotates on its axis once every 24 hours and rotates 15° every hour, so answers A and B are incorrect. Answer C is incorrect because the moon revolves around Earth about once every month.

Question 10 Answer I is correct. Scientists need to account for the Chandler wobble when using telescopes and when determining the orbit of artificial satellites. Answer F is incorrect because Earth's axis moves constantly, though it takes about 14 months to complete one wobble. Answers G and H are incorrect because these things may cause the wobble to occur, they do not result from the wobble.

Chapter 26 Standardized Test Prep

Understanding Concepts

Directions (1–5): **For *each* question, write on a sheet of paper the letter of the correct answer.**

1. Earth is closest to the sun at which of the following points in its orbit?
   A. aphelion
   B. perihelion
   C. an equinox
   D. a solstice

2. What object is located at one of the focus points for the orbit of each planet in the solar system?
   F. Earth is located at one of the focus points in the orbit of each planet in the solar system.
   G. A moon of each planet is located at one of the focus points in that planet's orbit.
   H. The sun is one of the focus points in the orbit of each planet in the solar system.
   I. The orbits of the planets do not share any common focus points.

3. Earth revolves around the sun about once every
   A. 1 hour
   B. 24 hours
   C. 1 month
   D. 365 days

4. Which of the following statements describes the position of Earth during the equinoxes?
   F. The North Pole tilts 23.5° toward the sun.
   G. The South Pole tilts 23.5° toward the sun.
   H. Rays from the sun strike the equator at a 90° angle.
   I. Earth's axis tilts 90° and points directly at the sun.

5. Which of the following statements about the electromagnetic spectrum is true?
   A. It moves slower than the speed of light.
   B. It consists of waves of varying lengths.
   C. The shortest wavelengths are orange and red.
   D. Scientists can only detect waves of visible light.

Directions (6–8): **For *each* question, write a short response.**

6. In what year did NASA first land astronauts on the moon?

7. What is the term that describes a spacecraft sent from Earth to another planet?

8. How does the wavelength of gamma rays compare to the wavelength of visible light?

Reading Skills

Directions (9–10): **Read the passage below. Then, answer the questions.**

The Chandler Wobble

In 1891, an American astronomer named Seth Carlo Chandler, Jr., discovered that Earth "wobbles" as it spins on its axis. This change in the spin of Earth's axis, known as the Chandler wobble, can be visualized if you imagine that Earth is penetrated by an enormous pen at the South Pole. This pen emerges at the North Pole and draws the pattern of rotation of Earth on its axis on a gigantic paper placed directly at the tip of the pen. If Earth did not have a wobble, you would expect the pen to draw a dot as Earth rotated on its axis. Because of the wobble, however, the pen draws a small circle. Over the course of 14 months, the pen will draw a spiral.

While the exact cause of the Chandler wobble is not known, scientists believe that it is related to the movement of the liquid center of Earth or to fluctuating pressure at the bottom of the ocean. This wobble affects celestial navigation. Because of the wobble, navigators' star charts must reflect new reference points for the North Pole and South Pole every 14 months.

9. Because of the Chandler wobble, celestial navigators must chart new reference points for the poles every 14 months. Changes in determining the location of the North Pole by using a compass are not required. Why?
   A. Compasses point to Earth's magnetic north pole, not Earth's geographic North Pole.
   B. Compasses automatically adapt and move with the wobble.
   C. The wobble is related to stellar movements.
   D. The wobble improves compass accuracy.

10. Which of the following statements can be inferred from the information in the passage?
    F. Earth's axis moves once every 14 months.
    G. The Chandler wobble prevents the liquid center of Earth from solidifying.
    H. The Chandler wobble causes the oceans to move and fluctuate in pressure.
    I. To locate a star, scientists must account for the wobble when using telescopes.

Answers

Understanding Concepts

1. B
2. H
3. D
4. H
5. B
6. 1969
7. Answers may vary. Correct answers include probes, missions, explorers.
8. Gamma rays have shorter wavelengths.

Reading Skills

9. A
10. I

Interpreting Graphics

11. C
12. G
13. Answers may vary. See Test Doctor for a detailed scoring rubric.
14. On both the vernal and autumnal equinoxes, the tilt of Earth's axis causes sunlight to strike the equator at a 90° angle.

Interpreting Graphics

Directions (11–14): **For *each* question below, record the correct answer on a separate sheet of paper.**

The diagram below shows the position of Earth during the four seasons. Use this diagram to answer questions 11 and 12.

Seasons and Tilt

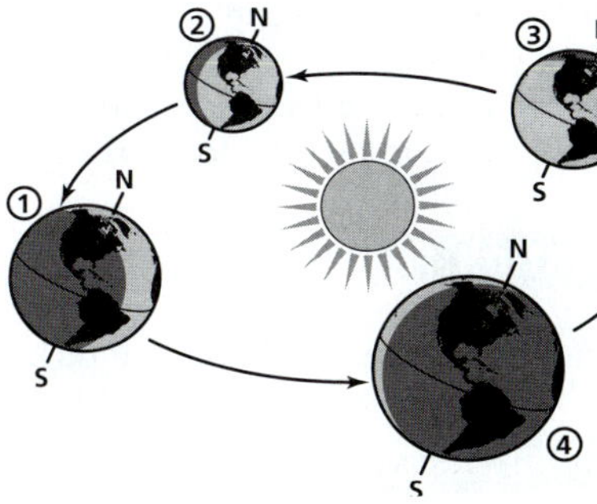

11 The Northern Hemisphere tilts toward the sun during which season?
A. winter
B. spring
C. summer
D. fall

12 The Northern Hemisphere experiences a vernal equinox when it is at which of the following positions on the diagram?
F. position 1
G. position 2
H. position 3
I. position 4

The diagram below shows the dates of specific events in Earth's orbit around the sun. Use this diagram to answer questions 13 and 14.

Orbit of Earth Around the Sun

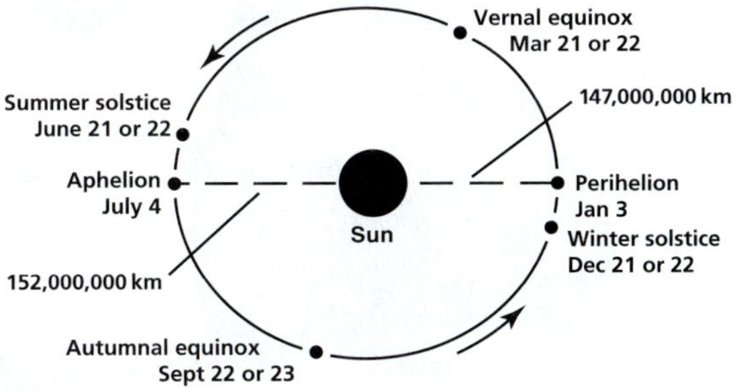

13 Use the diagram to describe the shape of Earth's orbit around the sun, and explain how the solstices differ from the aphelion and perihelion.

14 What is the relationship between Earth and the sun on March 21 or 22? Compare this relationship with the relationship between Earth and the sun on September 22 or 23?

Test TIP

Keep an eye on your time limit. If you begin to run short on time, quickly read the remaining questions to which questions might be the easiest for you to answer.

TEST DOCTOR

Question 13 Full-credit answers should include the following points:
- students should demonstrate a conceptual understanding that Earth's orbit is elliptical, and the sun is located at one of the foci; the two focal points of the ellipse of Earth's orbit are very close together
- the perihelion marks the closest Earth comes to the sun; the aphelion marks the farthest point Earth moves from the sun
- the solstices mark points at which the tilt of Earth's axis is away from the sun at its maximum angle in the Northern Hemisphere in the winter and is toward the sun at its maximum angle in the Northern Hemisphere in the summer
- during the winter solstice, the sun's rays strike the Tropic of Capricorn in the Southern Hemisphere at a 90° angle; during the summer solstice, the sun's rays strike the Tropic of Cancer in the Northern Hemisphere at a 90° angle
- the perihelion and winter solstice occur close together but are two different events; the same is true of the aphelion and summer solstice

Test Prep Correlations *National Science Education Standards*

HNS 1a: items 6, 7
HNS 2a: items 13, 14
PS 6b: items 5, 8
UCP 1: items 1, 2, 3, 4, 5, 11, 12, 13, 14
UCP 2: items 9, 10

CHAPTER RESOURCES

State Resources

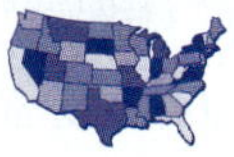

For specific resources for your state, visit go.hrw.com and type in the keyword HSHSTR.

Inquiry Lab

Earth-Sun Motion

Teacher's Notes

Time Required
one 45-minute class period

Lab Ratings
EASY → HARD

TEACHER PREPARATION ▲▲
STUDENT SETUP ▲
CONCEPT LEVEL ▲▲▲
CLEANUP ▲

Skills Acquired
- Designing Experiments
- Measuring
- Identifying and Recognizing Patterns
- Interpreting
- Organizing and Analyzing Data
- Communicating

The Scientific Method
In this lab, students will
- Test the Hypothesis
- Make Observations
- Analyze Results
- Draw Conclusions

Materials
Materials listed are enough for 2 to 3 students to construct the sundial apparatus and measure the movement of Earth. You may wish to make predrilled holes in the wooden board to hold the dowel rod upright, or provide modeling clay so students can change its position. Tell students that another term for a shadow stick is a *gnomon*.

Tips and Tricks
The experiment is best performed on a sunny day. In order to measure the movement of the sun's shadow, students will need to orient the apparatus so that the side with the upright gnomon is toward the south. Once they find north with the compass, students should label the sheet of notebook paper with the proper directions.

Have students present their plan for approval before they begin. One way to proceed would be to use the ruler to record the direction and length of the shadow at 5-minute intervals. If students plan to perform the extension, have them locate a familiar landmark so they can place the apparatus in the same spot each time they use it. Encourage students to graph their data to show how the shadow length changes during the day.

Chapter 26

Inquiry Lab

Using Scientific Methods

Earth-Sun Motion

Objectives
- **Design** an experiment to measure the movement of Earth.
- **Analyze** the effectiveness of your experimental design.
- **Demonstrate** how shadows can be used to measure time.

Materials
- board, wooden, 20 cm × 30 cm
- clock or watch
- compass, magnetic
- dowel, 30 cm long, 1/4 in. diameter
- paper, lined
- pencil
- ruler, metric
- tape, masking

During the course of a day, the sun moves across the sky. This motion is due to Earth's rotation. In ancient times, one of the earliest devices used by people to study the sun's motion was the shadow stick. The shadow stick is a primitive form of a sundial. Before clocks were invented, sundials were one of the only means of telling time.

In this lab, you will use a shadow stick to identify how changes in a shadow are related to Earth's rotation. You will also determine how a shadow stick can be used to measure time.

ASK A QUESTION

1. How can I measure the movement of Earth?

FORM A HYPOTHESIS

2. With a partner, build a shadow stick apparatus that is similar to the one in the illustration on the following page. Brainstorm with your partner a way in which you can use the apparatus in an experiment to measure the movement of Earth for 30 minutes. Write a few sentences that describe your design and your hypothesis about how this experiment will measure Earth's motion.

Step 2

CHAPTER RESOURCES

Chapter Resource File
- Datasheet for Chapter Lab GENERAL
- Lab Notes and Answers

Workbooks

Long-Term Projects
- Positions of Sunrise and Sunset ADVANCED
- Apparent Motions of the Moon ADVANCED

TEST THE HYPOTHESIS

3. When you complete your experimental design, have your teacher approve your design before you begin. **CAUTION** Never look directly at the sun.
4. Follow your design to set up and to complete your experiment.
5. Take measurements every 5 min, and record this information in a data table.

ANALYZE THE RESULTS

1. **Analyzing Data** In what direction did the sun appear to move in the 30 min period?
2. **Evaluating Methods** If you made your shadow stick half as long, would its shadow move the same distance in 30 min? Explain your answer.
3. **Evaluating Methods** Would you make any changes to your experimental design? Explain your answer.

DRAW CONCLUSIONS

4. **Drawing Conclusions** In what direction does Earth rotate?
5. **Applying Conclusions** How might a shadow stick be used to tell time?

Extension

1. **Evaluating Methods** Repeat this lab at different hours of the day. Perform the lab early in the morning, early in the afternoon, and early in the evening. Record the results and any differences that you observe. Explain how shadow sticks can be used to tell direction.

Inquiry Lab

Answers to Analyze the Results

1. The sun appears to move across the sky from east to west while its shadow moves clockwise from west to east.
2. The distance traveled would be unaffected by the length of the stick. This factor is controlled by the rotation of the Earth and the path of the sun across the sky.
3. Answers may vary. Students should evaluate their procedures and suggest appropriate changes.

Answers to Draw Conclusions

4. Earth rotates west to east.
5. Set up a shadow stick to track the movement of the shadow throughout the day. Mark the tip of the shadow line at hourly intervals using a different letter or number for each mark. The shortest shadow represents local noon, when the sunlight comes directly from the south. Once you have a set of measurements for the whole day, mark them permanently so you can use movements of the stick's shadow as a clock to measure time.

Answer to Extension

1. Use a length of string to make a circle around the shadow stick the length of the shadow in the morning. Place a marker at the point where the shadow reaches. The shadow will get shorter, but eventually it will again touch the tip of the circle. Mark that point. Measure the distance between the two markers to find the halfway point. This is the north-south line. Make a straight line where the arc crosses the circle on each side. This line will be at right angles to the north-south line and will show east-west.

Erich Landstrom
Boynton Beach Community High School
NASA Education Ambassador
Boynton Beach, FL

Light Sources

Group Activity — GENERAL

Community Survey Conduct a discussion of the environmental effects of light pollution on other organisms as well as the costs in wasted energy. Invite interested students to prepare a survey of the usage patterns of outdoor lights in your community. Identify sources to look for, such as unshielded floodlights and streetlights, lights along bridges that shine into the water, lights along highways, at malls, in housing developments, and in rural areas. Have the survey group document their findings in the form of a map and summary statement of their conclusions. Other students may produce posters to increase public awareness. LS **Visual/Verbal**

Answers to Map Skills Activity

1. Antarctica, Greenland and the Arctic Circle, the Amazon rain forest, northern and central Africa, the mountainous region north of India, and Central Australia
2. India, Northeastern China, Japan, United States, Western Europe, Mexico and Central America, and parts of Southwest Asia
3. Rio de Janeiro, Brazil; Santiago Chile; Lima, Peru; New York; Canberra and Sidney, Australia; Rabat, Algiers, Tripoli, and Cairo in North Africa; Pretoria in South Africa; Reykjavik in Iceland
4. You can easily identify the international borders between the United States and Canada and between India and China.

CHAPTER RESOURCES

Technology

Transparencies
- 135 Light Sources (with worksheet)

MAPS in Action

Light Sources

Map Skills Activity

This image of Earth as seen from space at night shows light sources that are almost all created by humans. The image is a composite image made from hundreds of nighttime images taken by orbiting satellites. Use the map to answer the questions below.

1. **Comparing Areas** Some climatic conditions on Earth, such as extreme cold, heat, wetness, or a thin atmosphere, make parts of our planet less habitable than other parts. Examples of areas on our planet that do not support large populations include deserts, high mountains, polar regions, and tropical rain forests. Using the image, identify regions of Earth where climatic conditions may not be able to support large human populations.
2. **Inferring Relationships** Using a map of the world and the brightness of the light sources on the image as a key, identify the locations of some of the most densely populated areas on Earth.
3. **Finding Locations** Many large cities are seaports on the coastlines of the world's oceans. By using the image, can you look along coastlines and locate light sources that might indicate the sites of large ports? Using a map of the world, name some of these cities.
4. **Inferring Relationships** By looking at the differences in the density of the light sources on the map, can you locate any borders between countries? Identify the countries on both sides of these borders.

SCIENCE AND TECHNOLOGY

Landsat Maps of Earth

Landsat satellites have been recording images of Earth for more than three decades. In that time, these Earth-scanning satellites have logged more than a million images. As Landsat satellites periodically rescan regions, the satellites create a visual history of Earth's changing landscapes.

Say Cheese!

Landsat images resemble aerial photographs. Each image records about 30,000 km^2 of Earth's surface. Landsat images are not ordinary photographs, however. Each satellite uses a scanning sensor system called a *thematic mapper,* or *TM,* to create images. The TM sensors detect visible light, which is the light recorded by an ordinary camera. The TM sensors also detect other parts of the electromagnetic spectrum that humans cannot see, such as infrared light. This capability gives Landsat images much more detail than conventional photographs have.

Once a satellite sends an image back to Earth, cartographers can use computers to create a thematic map. A thematic map is a map that illustrates a particular subject or feature. Selecting different combinations of data allows cartographers to highlight features such as river deltas, geologic faults, and mineral deposits.

▲ *Landsat 7* is the latest mission in the Landsat series of Earth-observation satellites.

Earth's Changing Surface

Landsat images appeal to Earth scientists in a variety of fields. Landsat images have enabled cartographers to map remote areas of the world. The images have allowed hydrologists to find uncharted lakes. Landsat images have also helped geologists discover oil in the Sudan, tin in Brazil, and copper in Mexico.

Recently, ecologists have begun to use Landsat images to observe changes in Earth's environment. With a series of images of the same region, taken over years, ecologists can monitor the effects of processes such as urbanization, deforestation, and soil erosion. New, useful applications of Landsat satellites may arise as the satellites continue to record Earth's changing surface.

◀ The Ganges River empties into the Bay of Bengal and forms a delta. The delta is covered by a swamp forest known as the Sunderbans.

Extension

1. **Extension** How might cartographers use Landsat images to check the accuracy of maps?

Science and Technology

Landsat Maps of Earth

Internet Activity — GENERAL

Landsat Images For over thirty years, NASA has photographed Earth in visible light and in other parts of the electromagnetic spectrum using the Landsat satellite's remote sensing equipment to study changes on Earth. Divide the class into groups and have them search the Internet to find various applications of this satellite technology. Have some students learn more about how the polar synchronous orbital satellite works. Others can investigate how Landsat images are interpreted. Still other students can look at its applications to fields such as cartography, environmental science, land management, and geology. Have student groups create visual presentations to share what they learn with the class. Students can create photomontages, diagrams, posters, and videotape recordings. A worksheet designed to direct student research on this topic can be found in the **Chapter Resource File** booklet or by visiting **go.hrw.com** and entering the keyword **HQ6SSPX.** LS **Visual** Co-op Learning

CHAPTER RESOURCES

Chapter Resource File

Internet Activity
- Landsat Images GENERAL

Answer to Extension

1. Because Landsat images record data using different wavelengths of light, including ultraviolet, microwave, and infrared, as well as different forms of visible light, cartographers can compare different aspects of a region to get a more complete picture. They can identify the effects of human activities on the landscape and use repeated images of the same region to measure environmental changes.

Chapter 27 Planets of the Solar System

Planning Guide

Compression Guide

To shorten instruction because of time limitations, omit the Chapter Lab.

OBJECTIVES	LABS, DEMONSTRATIONS, AND ACTIVITIES	TECHNOLOGY RESOURCES
PACING • 45 min pp. 684–690 **Chapter Opener**	**SE** **Long-Term Project** Positions of Sunrise and Sunset, pp. 848–851 ADVANCED **LTP** **Long-Term Project** Positions of Sunrise and Sunset* ADVANCED **SE** **Long-Term Project** Planetary Motions, pp. 866–869 ADVANCED **LTP** **Long-Term Project** Planetary Motions* ADVANCED	**OSP** **Parent Letter** ■ **CD** **Student Edition on CD-ROM** **CD** **Chapter Summaries Audio CD** ■ **VID** **Brain Food Video Quiz**
Section 1 Formation of the Solar System • Explain the nebular hypothesis of the origin of the solar system. • Describe how the planets formed. • Describe the formation of the land, the atmosphere, and the oceans of Earth.	**TE** **Group Activity** Spinning Nebula, p. 685 GENERAL **TE** **Group Activity** Skit, p. 686 GENERAL **SE** **Quick Lab** Water Planetesimals, p. 687 GENERAL **CRF** **Datasheet for Quick Lab*** GENERAL	**OSP** **Lesson Plans** (also in print) **TR** **Bellringer*** **TR** 136 The Nebular Model of the Formation of the Solar System* **TR** 137 Differentiation of Earth and Formation of Earth's Atmosphere* **CD** **Interactive Tutor** Our Sun and Origin of the Solar System
PACING • 45 min pp. 691–694 **Section 2 Models of the Solar System** • Compare the models of the universe developed by Ptolemy and Copernicus. • Summarize Kepler's three laws of planetary motion. • Describe how Newton explained Kepler's laws of motion.	**TE** **Activity** Role Play, p. 691 GENERAL **SE** **Quick Lab** Ellipses, p. 692 GENERAL **CRF** **Datasheet for Quick Lab*** GENERAL **TE** **Group Activity** Interview an Astronomer, p. 717 GENERAL	**OSP** **Lesson Plans** (also in print) **TR** **Bellringer*** **TR** 138 Kepler's Law of Equal Area* **CD** **Interactive Tutor** Gravity, Orbits, Motion, and Forces
PACING • 90 min pp. 695–700 **Section 3 The Inner Planets** • Identify the basic characteristics of the inner planets. • Compare the characteristics of the inner planets. • Summarize the features that allow Earth to sustain life.	**TE** **Demonstration** Global Greenhouse, p. 696 GENERAL **TE** **Group Activity** Remote Sensing, p. 697 GENERAL **SE** **Making Models Lab** Crater Analysis, pp. 714–715 ◆ GENERAL **CRF** **Datasheet for Chapter Lab*** GENERAL **SE** **Maps in Action** MOLA Map of Mars, p. 716 GENERAL **CRF** **Inquiry Lab** Probing for Information* GENERAL	**OSP** **Lesson Plans** (also in print) **TR** **Bellringer*** **TR** 139 MOLA Map of Mars* **TE** **Internet Activity** Life on Mars?, p. 699 ADVANCED **CRF** **Internet Activity** Life on Mars?* ADVANCED **CD** **Interactive Tutor** The Inner Planets
PACING • 45 min pp. 701–708 **Section 4 The Outer Planets** • Identify the basic characteristics that make the outer planets different from terrestrial planets. • Compare the characteristics of the outer planets. • Explain why Pluto is now considered a dwarf planet.	**TE** **Demonstration** Play Ball, p. 701 GENERAL **TE** **Demonstration** Gas Storms, p. 703 GENERAL **TE** **Demonstration** Radically Tilted Planet, p. 705 GENERAL **TE** **Discussion** Neptune's Weather, p. 706 GENERAL **TE** **Group Activity** To Be or Not to Be a Planet, p. 706 ADVANCED **CRF** **Making Models Lab** It's a Long Way to Pluto* GENERAL	**OSP** **Lesson Plans** (also in print) **TR** **Bellringer*** **TE** **Internet Activity** Gravitational Microlensing, p. 707 ADVANCED **CRF** **Internet Activity** Gravitational Microlensing* ADVANCED **CD** **Interactive Tutor** The Outer Planets

PACING • 90 min

CHAPTER REVIEW, ASSESSMENT, AND STANDARDIZED TEST PREPARATION

SE **Chapter Highlights,** p. 709
SE **Chapter Review,** pp. 710–711
SE **Standardized Test Prep,** pp. 712–713
CRF **Concept Review*** ■ GENERAL
CRF **Critical Thinking*** ADVANCED
CRF **Math Skills*** GENERAL
CRF **Graphing Skills*** GENERAL
CRF **Chapter Test A*** ■ GENERAL
CRF **Chapter Test B*** ADVANCED
OSP **Lesson Plans** (also in print)
OSP **Test Generator**
OSP **Test Item Listing**

Online and Technology Resources

go.hrw.com Holt Online Learning

Visit **go.hrw.com** for access to Holt Online Learning, or enter the keyword **HQ6 Home** for a variety of free online resources.

One-Stop Planner® CD-ROM

This CD-ROM package includes
- Lab Materials QuickList Software
- Holt Calendar Planner
- Customizable Lesson Plans
- Printable Worksheets
- ExamView® Test Generator
- Interactive Teacher Edition
- Holt PuzzlePro®
- Holt PowerPoint® Resources

KEY

SE	Student Edition	OSP	One-Stop Planner	VID	Classroom Video/DVD
TE	Teacher Edition	TR	Transparencies and Transparency Worksheets	*	Also on One-Stop Planner
CRF	Chapter Resource File	CD	CD or CD-ROM	◆	Requires advance prep
LTP	Long-Term Projects			■	Also available in Spanish

SKILLS DEVELOPMENT RESOURCES	REVIEW AND ASSESSMENT	CORRELATIONS
SE **Pre-Reading Activity**, p. 684 GENERAL TE **Using the Figure** Martian Canyons, p. 684 GENERAL		National Science Education Standards
CRF **Directed Reading*** BASIC TE **Using the Figure** Discussion, p. 686 GENERAL TE **Reading Skill Builder** Reading Organizer, p. 686 BASIC SE **Graphic Organizer** Chain-of-Events Chart, p. 688 GENERAL	SE **Reading Checks**, pp. 687, 689 GENERAL SE **Section Review**, p. 690 GENERAL TE **Homework**, p. 689 GENERAL TE **Reteaching**, p. 689 BASIC TE **Quiz**, p. 689 GENERAL TE **Alternative Assessment**, p. 689 ADVANCED CRF **Section Quiz*** ■ GENERAL	ES 3a, UCP 1
CRF **Directed Reading*** BASIC TE **Skill Builder** Vocabulary, p. 692 GENERAL SE **Math Practice**, p. 693 GENERAL TE **Inclusion Strategies**, p. 693	SE **Reading Check**, p. 692 GENERAL SE **Section Review**, p. 694 GENERAL TE **Reteaching**, p. 693 BASIC TE **Quiz**, p. 693 GENERAL TE **Alternative Assessment**, p. 693 ADVANCED CRF **Section Quiz*** ■ GENERAL	ST 1e, UCP 2
CRF **Directed Reading*** BASIC SE **Math Practice**, p. 697 GENERAL TE **Using the Figure** Computer Generated Image, p. 697 BASIC TE **Reading Skill Builder** Paired Summarizing, p. 698 BASIC TE **Skill Builder** Writing, p. 698 GENERAL	SE **Reading Checks**, pp. 697, 699 GENERAL SE **Section Review**, p. 700 GENERAL TE **Reteaching**, p. 699 BASIC TE **Quiz**, p. 699 GENERAL TE **Alternative Assessment**, p. 700 ADVANCED CRF **Section Quiz*** ■ GENERAL	SAI 1a, ST 1d, ST 1e
CRF **Directed Reading*** BASIC TE **Using the Figure** Discussion, p. 702 GENERAL SE **Graphic Organizer** Comparison Table, p. 704 GENERAL TE **Inclusion Strategies**, p. 704	SE **Reading Checks**, pp. 702, 704, 707 GENERAL SE **Section Review**, p. 708 GENERAL TE **Reteaching**, p. 707 BASIC TE **Quiz**, p. 707 GENERAL TE **Alternative Assessment**, p. 708 GENERAL CRF **Section Quiz*** ■ GENERAL	ST 1e

Holt Lab Generator CD-ROM

Search for labs by topic, standard, difficulty level, and/or time frame. Edit those labs to fit your needs, or create your own labs. Use the Lab Materials QuickList to order just what you need.

Holt Earth Science Interactive Tutor CD-ROM

This CD-ROM consists of interactive activities that give students a fun way to extend their knowledge of Earth science concepts.

Chapter Summaries Audio CDs

These CDs include audio summaries of the key concepts presented in each chapter. (Audio summaries are also available in Spanish.)

SciLinks NSTA

www.scilinks.org

Maintained by the **National Science Teachers Association.** See Chapter Enrichment pages that follow for a complete list of topics.

See Chapter Enrichment pages for Video Resources.

Chapter 27 Chapter Enrichment

This Chapter Enrichment provides relevant and interesting information to expand and enhance your classroom instruction of the chapter material.

Section 1 Formation of the Solar System

Theories of Origin

Modern scientific calculations support the nebular hypothesis developed in the late 18th century to explain the formation of the solar system. The hypothesis states that the solar system formed from a huge rotating cloud of interstellar gas and dust that contracted under the force of its own gravity. Other theories also contributed to scientific understanding. In 1905, University of Chicago scientists proposed that the planets formed from material pulled from the sun by the gravitational attraction of a passing star. They introduced the idea of planetesimals that grew larger by attracting more material. Other early ideas suggested that tidal forces between colliding stars formed the planets. The most modern and widely accepted theory—called the *protoplanet theory*—is based on the nebular hypothesis. In 1945, Carl Weizsacker and Gerard Kuiper proposed independently that a shock wave from a supernova caused the nebular disk to contract. The sun and planets formed at roughly the same time from the contracting and flattening disk. Other scientists extended and modified the hypothesis.

▲ Stars form from the matter in nebulas.

Section 2 Models of the Solar System

Galileo's Telescopes

Galileo built some of the finest early telescopes in 1609, modeled after those invented in Holland, and he was the first not only to point them skyward but also to interpret the results. Galileo discovered mountains and craters on the moon and four satellites of Jupiter, which are still called the *Galilean moons*. He published these discoveries in *Sidereus Nuncius* (The Starry Messenger). Galileo also detected sunspots and measured their motion. In addition, Galileo observed the phases of Venus and the rings of Saturn, though he could not identify separate rings. Galileo's discovery of Jupiter's moons showed that not all celestial bodies circle the sun, helping to pave the way for the eventual acceptance of Copernicus's heliocentric model of the solar system. These views brought Galileo into conflict with theologians who supported Ptolemy's geocentric model of the solar system.

Section 3 Models of the Solar System

Impact Craters

The pockmarked land surface that Galileo observed on the moon is caused by impact craters, which are a common feature of most of the rocky bodies of the solar system, including all of the inner planets. Craters form when a space rock strikes the surface of a planet or moon, sending out shock waves that blast surface material away from the impact site. Most objects like these that reach Earth burn up in Earth's atmosphere, but some get through. About 150 impact craters have been identified on the surface of Earth. Among the most famous are Barringer Crater located in the Arizona desert (also called

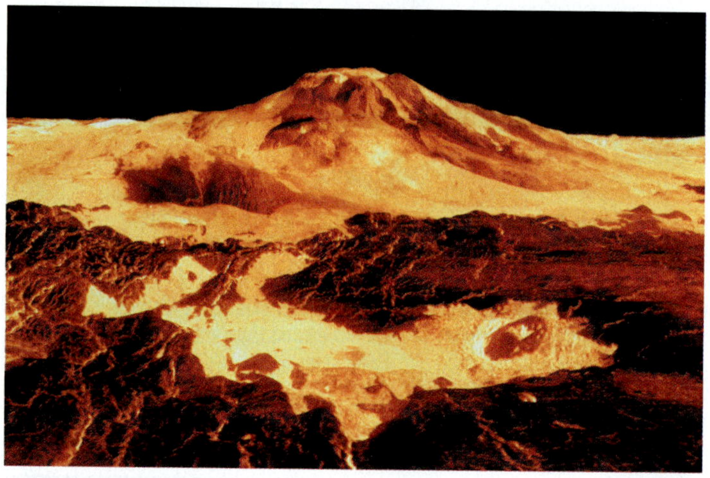

▲ Volcanism on Venus has covered many of the planet's impact craters.

"Meteor Crater"), and Chicxulub Crater (pronounced CHEEK shoo loob) in the ocean off the Yucatan peninsula, Mexico, which scientists speculate contributed to the extinction of the dinosaurs.

Impacts were more common in the early period of the solar system. On the moon, Mercury, and Mars, which have thin or no atmosphere to buffer against bombardment and no flowing surface water to cause erosion, craters tend to remain as they originally formed. However, on geologically active planets that have thicker atmospheres and weather systems, such as Earth, the evidence of impacts diminishes over time.

◀ Scientists think that exoplanets are common throughout the universe.

Section 4 The Outer Planets

Exploring the Kuiper Belt

Astronomers call the group of objects made of rock and ice that orbit the sun in an icy ring beyond Neptune the *Kuiper Belt*. The objects seem to be the matter left behind from the formation of the solar system. This matter is similar in some ways to the rocky debris field known as the *asteroid belt*, which is located between Mars and Jupiter. However, the Kuiper Belt consists of much more material than all of the asteroids put together. In January 2006, NASA launched an interplanetary probe to explore Pluto and the Kuiper Belt. The robot explorer will swing by Pluto and then enter the Kuiper Belt. Studying these icy bodies could provide important information about the early history of the solar system.

Video Resources

Brain Food Video Quizzes

Brain Food Video Quizzes These videos contain game-show style quizzes that assess students' progress and motivate students to study the chapter material.

HRW Earth Science Video

This video introduces Earth science topics and includes a geology field trip. The video segment listed below complements this chapter.

Segment 18: The Solar System In this segment, the birth of the solar system is examined. The formation of planets from planetesimals and the differences in composition and size between the inner four planets and the outer five are explained. The composition, orbit, origin, and physical properties of asteroids and comets are described, and information about meteors and meteorites is given. (4 min)

NOVA Videos

To order **NOVA** videos related to this chapter, visit **go.hrw.com** and enter the keyword **HQ6PSSV.**

SciLinks

NSTA Developed and maintained by the National Science Teachers Association

SciLinks is maintained by the National Science Teachers Association to provide you and your students with interesting, up-to-date links that will enrich your classroom presentation of the chapter.

Visit www.scilinks.org and enter the SciLinks code for more information about the topic listed.

Topic: Origins of the Solar System
SciLinks code: HQ61087

Topic: Early Astronomers
SciLinks code: HQ60441

Topic: Inner Planets
SciLinks code: HQ60798

Topic: Outer Planets
SciLinks code: HQ61091

Topic: Galileo
SciLinks code: HQ60633

Topic: Careers in Earth Science
SciLinks code: HQ60222

Chapter 27

Chapter Overview

This chapter describes how Earth's solar system formed, the laws that govern the movements of the planets, and how the rocky inner planets differ from the gas giants.

Using the Figure —GENERAL

Martian Canyons This photo shows the Valles Marineris, which is a system of canyons located south of the Martian equator. Explain to students that this system of canyons is over 4,000 km long—the coast-to-coast width of the United States on Earth. Point out that in places the canyon is many times deeper than the Grand Canyon. LS Visual

PRE-READING ACTIVITY

FOLDNOTES

You may want to use this FoldNote in a classroom discussion to review material from the chapter. On the board, write each category from the Three-Panel Flip Chart. Then, ask students to provide information for each category. Under the appropriate category on the board, write the information that students provide.

Chapter 27 Planets of the Solar System

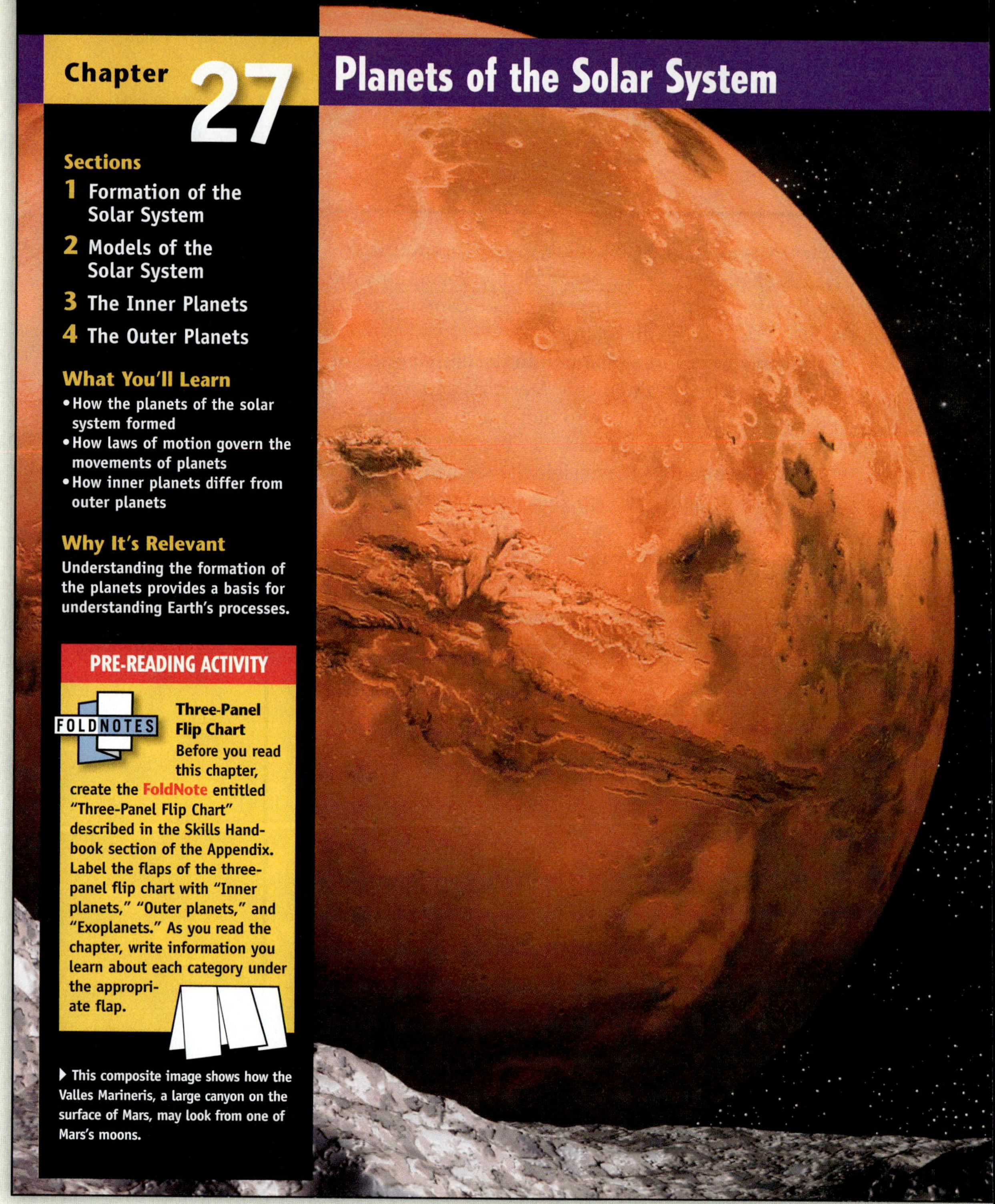

Sections

1 Formation of the Solar System
2 Models of the Solar System
3 The Inner Planets
4 The Outer Planets

What You'll Learn

- How the planets of the solar system formed
- How laws of motion govern the movements of planets
- How inner planets differ from outer planets

Why It's Relevant

Understanding the formation of the planets provides a basis for understanding Earth's processes.

PRE-READING ACTIVITY

FOLDNOTES

Three-Panel Flip Chart Before you read this chapter, create the FoldNote entitled "Three-Panel Flip Chart" described in the Skills Handbook section of the Appendix. Label the flaps of the three-panel flip chart with "Inner planets," "Outer planets," and "Exoplanets." As you read the chapter, write information you learn about each category under the appropriate flap.

▸ This composite image shows how the Valles Marineris, a large canyon on the surface of Mars, may look from one of Mars's moons.

Chapter Correlations *National Science Education Standards*

PS 4b Gravitation is a universal force that each mass exerts on any other mass. The strength of the gravitational attractive force between two masses is proportional to the masses and inversely proportional to the square of the distance between them. **(Sections 1 and 2)**

ES 3a The sun, the earth, and the rest of the solar system formed from a nebular cloud of dust and gas 4.6 billion years ago. The early earth was very different from the planet we live on today. **(Section 1)**

HNS 2a Science distinguishes itself from other ways of knowing and from other bodies of knowledge through the use of empirical standards, logical arguments, and skepticism. . . . **(Sections 1 and 2)**

HNS 2b Scientific explanations must meet certain criteria. First and foremost, they must be consistent with experimental and observational evidence about nature, and must make accurate predictions. . . . **(Sections 1–4)**

HNS 2c Because all scientific ideas depend on experimental and observational confirmation, all scientific knowledge is, in principle, subject to change as new evidence becomes available. **(Section 2)**

HNS 3c Occasionally, there are advances in science and technology that have important and long-lasting effects on science and society. Examples of such advances include the following: Copernican revolution; Newtonian mechanics;. . . . **(Section 2)**

UCP 2 Evidence consists of observations and data on which to base scientific explanations. **(Section 2)**

UCP 3 Although most things are in the process of becoming different—changing—some properties of objects and processes are characterized by constancy. . . **(Sections 1–4)**

Section 1 Formation of the Solar System

The **solar system** consists of the sun, the planets, and all of the other bodies that revolve around the sun. **Planets** are the primary bodies that orbit the sun. Scientists have long debated the origins of the solar system. In the 1600s and 1700s, many scientists thought that the sun formed first and threw off the materials that later formed the planets. But in 1796, the French mathematician Pierre-Simon, marquis de Laplace, advanced a hypothesis that is now known as the *nebular hypothesis*.

OBJECTIVES

- **Explain** the nebular hypothesis of the origin of the solar system.
- **Describe** how the planets formed.
- **Describe** the formation of the land, the atmosphere, and the oceans of Earth.

KEY TERMS

solar system
planet
solar nebula
planetesimal

The Nebular Hypothesis

Laplace's hypothesis states that the sun and the planets condensed at about the same time out of a rotating cloud of gas and dust called a *nebula*. Modern scientific calculations support Laplace's hypothesis and help explain how the sun and the planets formed from an original nebula of gas and dust.

Matter is spread throughout the universe. Some of this matter gathers into clouds of dust and gas, such as the one shown in **Figure 1.** Almost 5 billion years ago, the amount of gravity near one of these clouds increased as a result of a nearby supernova or other forces. The rotating cloud of dust and gas from which the sun and planets formed is called the **solar nebula.** Energy from collisions and pressure from gravity caused the center of the solar nebula to become hotter and denser. When the temperature at the center became high enough—about 10,000,000°C—hydrogen fusion began. A star, which is now called the sun, or *Sol*, formed. The sun is composed of about 99% of all of the matter that was contained in the solar nebula.

solar system the sun and all of the planets and other bodies that travel around it

planet a celestial body that orbits the sun, is round because of its own gravity, and has cleared the neighborhood around its orbital path

solar nebula a rotating cloud of gas and dust from which the sun and planets formed; *also* any nebula from which stars and exoplanets may form

Figure 1 ▶ The Orion nebula is about 1,500 light-years from Earth. Scientists study the nebula to learn about the processes that give birth to stars.

Section 1

Focus

Overview

This section describes the formation of the sun and planets of Earth's solar system. The section also describes the origin of the land, atmosphere, and oceans of Earth.

Bellringer

Have students write a list of all of the objects they think make up Earth's solar system. (Student may suggest such objects as planets, moons, the asteroid belt, comets, meteoroids, and the sun.) LS **Verbal**

Motivate

Group Activity — GENERAL

Spinning Nebula Divide the class into groups, and give each group a small plastic container three-quarters full of water, perlite, and a stirring rod. Have each group add perlite to the water, and stir the mixture vigorously. Once the mixture is rotating, have them remove the rod. Have students describe their model solar nebulae. (Answers may vary. Students should notice that most of the material tends to concentrate at the center of the rotating disk. They may notice that the inner parts of the disk move more rapidly than the outer parts do.) LS **Kinesthetic/Visual**

CHAPTER RESOURCES

Chapter Resource File
- Directed Reading BASIC

Technology

Transparencies
- Bellringer

Student Edition on CD-RO M

One-Stop Planner CD-ROM
- Lesson Plan

Teach

Using the Figure — GENERAL

Discussion Explain that the explosion of a nearby supernova could have produced a shock wave that led to the collapse of the solar nebula. Ask students "What would cause the center of the solar nebula to become hotter?" (Sample answer: energy from collisions between the particles and the energy released by gravitational contraction) Then, have students summarize the steps that led to the formation of planets. (Sample answers: The accumulation of dust particles formed small planetesimals. Collisions between the planetesimals led to the formation of protoplanets. The protoplanets eventually developed into planets and moons.) LS Verbal/Visual

READING SKILL BUILDER — BASIC

Reading Organizer As students read this section, encourage them to take Power Notes, KWL Notes, or Two Column Notes about the nebular hypothesis. These organizers are described in the Skills Handbook section of the Appendix. Later, students can use their notes as a study guide for assessments. English Language Learners LS Logical

CHAPTER RESOURCES

Technology

Transparencies
- 136 The Nebular Model of the Formation of the Solar System (with worksheet)

planetesimal a small body from which a planet originated in the early stages of development of the solar system

SCILINKS. NSTA Developed and maintained by the National Science Teachers Association

For a variety of links related to this subject, go to www.scilinks.org
Topic: Origins of the Solar System
SciLinks code: HQ61087

Formation of the Planets

While the sun was forming in the center of the solar nebula, planets were forming in the outer regions, as shown in **Figure 2.** Small bodies from which a planet originated in the early stages of formation of the solar system are called **planetesimals.** Some planetesimals joined together through collisions and through the force of gravity to form larger bodies called *protoplanets*. The protoplanets' gravity attracted other planetesimals in the solar nebula. These planetesimals collided with the protoplanets and added their masses to the protoplanets.

Eventually, the protoplanets became very large and condensed to form planets and moons. *Moons* are the smaller bodies that orbit the planets. Planets and moons are smaller and denser than the protoplanets.

Formation of the Inner Planets

The features of a newly formed planet depended on the distance between the protoplanet and the developing sun. The four protoplanets that became Mercury, Venus, Earth, and Mars were close to the sun. They contained large percentages of heavy elements, such as iron and nickel. These planets lost their less dense gases because at the temperature of the gases, gravity was not strong enough to hold the gases. Other lighter elements may have been blown or boiled away by radiation from the sun. As the denser material sank to the centers of the planets, layers formed. The less dense material was on the outer part of the planet, and the denser material was at the center. Today, the inner planets have solid surfaces that are similar to Earth's surface. The inner planets are smaller, rockier, and denser than the outer planets.

Figure 2 ▶ The Nebular Model of the Formation of the Solar System

The young solar nebula begins to collapse because of gravity.

As the solar nebula rotates, it flattens and becomes warmer near its center.

Planetesimals begin to form within the swirling disk.

Teaching Tip — GENERAL

Connect to Familiar Processes Explain that the solar nebula in which the solar system formed would have spun faster and faster as it contracted. Students may better understand why this happened if you compare it to the rotation of a spinning skater. A skater spins faster with arms folded close to the body (contracted) and slower with arms spread apart. LS Visual

Group Activity — GENERAL

Skit Invite interested students to develop a script about the formation of the solar system and perform it for the rest of the class. Ask students to focus on the differences between the planets that formed close to the developing sun and those that formed in the outer regions of the solar nebula. If there are enough performers, they can assume the roles of the sun, the inner planets, the outer planets, the dwarf planets, and even the debris, such as comets and asteroids, that remained after the formation of the new solar system. LS Auditory/Kinesthetic

Formation of the Outer Planets

Four other protoplanets became Jupiter, Saturn, Uranus, and Neptune. As a group, these outer planets are very different from the small, rocky inner planets. These outer planets formed in the colder regions of the solar nebula. They were far from the sun and therefore were cold. Thus, they did not lose their lighter elements, such as helium and hydrogen, or their ices, such as water ice, methane ice, and ammonia ice.

At first, thick layers of ice surrounded small cores of heavy elements. However, because of the intense heat and pressure in the planets' interiors, the ices melted to form layers of liquids and gases. Today, these planets are referred to as *gas giants* because they are composed mostly of gases, have low density, and are huge planets. Jupiter, for example, has a density of only 24% of Earth's density but a diameter that is 11 times Earth's diameter.

Pluto—The First Dwarf Planet

From its discovery in 1930, Pluto was known as the ninth planet. However, it is quite unlike the other outer planets, which are gas giants. In fact, Pluto is smaller than Earth's moon, and it can be best described as an ice ball that is made of frozen gases and rock. Because of its characteristics, many astronomers disagreed with Pluto's classification as a planet.

In 2006, astronomers from around the world revised the definition of *planet.* The new definition includes the first eight familiar planets, but it excludes Pluto. Pluto is part of a new category of solar system bodies called *dwarf planets.*

Reading Check **How is Pluto different from the outer planets?** **(See the Appendix for answers to Reading Checks.)**

QuickLAB **5 min**

Water Planetesimals

Procedure

1. Use a **medicine dropper** to place two drops of **water** about 3 cm apart on a **piece of wax paper.**
2. Lift one edge of the wax paper so that one drop of water moves toward the other drop until the drops collide.
3. Add a third drop of water to the wax paper. Then, repeat step 2.

Analysis

1. What happened when the water droplets collided?
2. How does this activity model the formation of protoplanets?

As planetesimals grow, their gravitational pull increases. The largest planetesimals begin to collect more of the gas and dust of the nebula.

Small planetesimals collide with larger ones, and the planets begin to grow.

The excess dust and gas is gradually removed from the solar nebula, which leaves planets around the sun and thus creates a new solar system.

QuickLAB

Skills Acquired

- Constructing Models
- Observing

Teacher's Notes Students could also use a straightened paper clip to draw the water droplets together. You might point out that the force of attraction in this model is cohesion, or the attractive force between particles of the same substance, while in the solar nebula the attractive force is gravity.

Answers

1. They merged to form larger droplets.
2. Answers may vary. Sample answer: Planetesimals grow by clumping together and attracting more matter. Collisions between planetesimals form larger bodies, just as the water droplets do.

Answer to Reading Check

Unlike the outer planets, Pluto is very small and is composed of rock and frozen gas, instead of thick layers of gases.

CHAPTER RESOURCES

Chapter Resource File

- **Datasheet for Quick Lab** GENERAL

MISCONCEPTION ALERT

What Happened to Pluto? In the summer of 2006, astronomers from around the world met in Prague (Czech Republic) for the 26th General Assembly of the International Astronomical Union (IAU). At that meeting, the IAU voted on a new definition of *planet,* which includes the first eight planets, but excludes Pluto. Pluto was reclassified as a *dwarf planet.* While students may be familiar with nine planets, which had been the number of planets since soon after the discovery of Pluto in 1930, the decision made by the IAU in 2006 has affected the presentation in this textbook. Pluto, another trans-Neptunian object called Eris, and the asteroid Ceres are the first objects to be named dwarf planets.

Teach, *continued*

MISCONCEPTION ALERT

Gravitational Separation Point out that the layered structure of Earth resulted because accumulated heat caused the early planetary body to become completely molten, or liquid. Only then could gravitational differentiation of the formerly uniform mixture of materials cause them to separate into distinct layers. A solid body cannot be differentiated in this manner. Only later did most of the layers of the planet cool and solidify.

Graphic Organizer GENERAL

Chain-of-Events Chart
You may want to use this Graphic Organizer to assess students' prior knowledge before beginning a discussion of the steps in the formation of the solid Earth. You may also choose to use a similar activity as a quiz to assess students' knowledge after students have read the description of the formation of Earth.

CHAPTER RESOURCES

Technology

Transparencies
- 137 Differentiation of Earth and Formation of Earth's Atmosphere (with worksheet)

Graphic Organizer **Chain-of-Events Chart**

Create the **Graphic Organizer** entitled "Chain-of-Events Chart" described in the Skills Handbook section of the Appendix. Then, fill in the chart with details about each step of the formation of Earth.

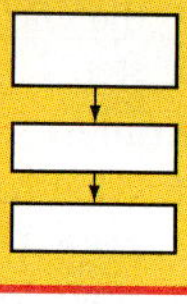

Formation of Solid Earth

When Earth first formed, it was very hot. Three sources of energy contributed to the high temperature on the new planet. First, much of the energy was produced when the planetesimals that formed the planet collided with each other. Second, the increasing weight of Earth's outer layers compressed the inner layers, which generated more energy. Third, radioactive materials that emit high-energy particles were very abundant when Earth formed. When surrounding rocks absorbed these particles, the energy of the particles' motion led to higher temperatures.

Early Solid Earth

Young Earth was hot enough to melt iron, the most common of the existing heavy elements. As Earth developed, denser materials, such as molten iron, sank to its center, and less dense materials were forced to the outer layers. This process is called *differentiation*. Differentiation caused Earth to form three distinct layers, as shown in **Figure 3**. At the center is a dense *core* that is composed mostly of iron and nickel. Around the core is the very thick layer of iron- and magnesium-rich rock called the *mantle*. The outermost layer of Earth is a thin *crust* of less dense, silica-rich rock. Today, processes that shape Earth, such as plate tectonics, are driven by heat transfer and differences in density.

Present Solid Earth

Eventually, Earth's surface cooled enough for solid rock to form. The solid rock at Earth's surface formed from less dense elements that were pushed toward the surface during differentiation. Earth's surface continued to change as a result of the heat in Earth's interior as well as through impacts and through interactions with the newly forming atmosphere.

Figure 3 ▶ Differentiation of Earth

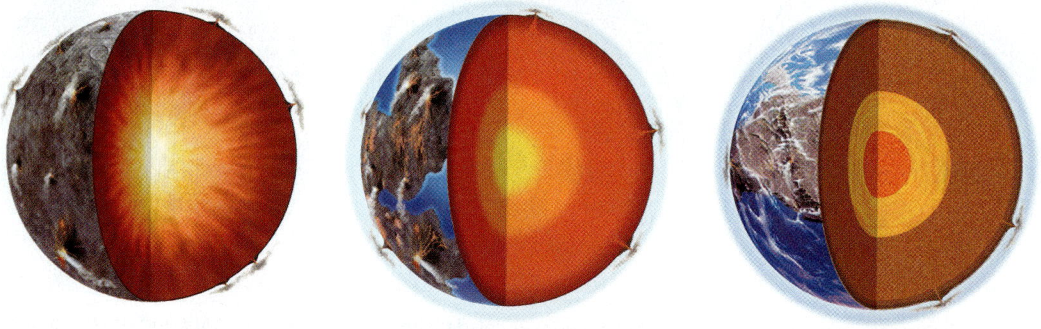

During its early history, Earth cooled to form three distinct layers.

An atmosphere began to form from the water vapor and carbon dioxide released by volcanic eruptions.

Organisms produced oxygen from photosynthesis to create an oxygenated atmosphere.

ENVIRONMENTAL CONNECTION

Earth's Primitive Atmosphere Direct students' attention to the illustration of volcanoes on the opposite page. Ask students to consider what volcanic activity has to do with Earth's atmosphere. (Sample answer: After the light gases acquired from the solar nebula escaped Earth's gravity, a new atmosphere composed of carbon dioxide, water vapor, and other gases developed from the eruption of volcanoes on Earth's surface.) You might point out that during the formation of Earth's atmosphere, comets composed of ice and rock debris also plunged into Earth, releasing quantities of water vapor and other gases. Ask students where they think the oxygen in Earth's atmosphere came from. (Sample answer: Initially, it came from the breakdown of water and carbon dioxide by ultraviolet radiation. Then, after life developed, photosynthesizing bacteria and plants produced oxygen as a byproduct of food production.) LS **Verbal**

Formation of Earth's Atmosphere

Like solid Earth, the atmosphere formed because of differentiation. During the original differentiation of Earth, less dense gas molecules, such as hydrogen and helium, rose to the surface. Thus, the original atmosphere of Earth consisted primarily of hydrogen and helium.

Earth's Early Atmosphere

The high concentrations of hydrogen and helium did not stay with Earth's atmosphere. Earth's gravity is too weak to hold these gases. The sun heated the gases enough so that they escaped Earth's gravity. These gases were probably blown away by the solar wind, which might have been stronger at that time than it is today. Also, Earth's magnetic field, which protects the atmosphere from the solar wind, might not have been fully developed.

Outgassing

As Earth's surface continued to form, volcanic eruptions were much more frequent than they are today. The volcanic eruptions released large amounts of gases, mainly water vapor, carbon dioxide, nitrogen, methane, sulfur dioxide, and ammonia, as shown in **Figure 4.** This process, known as *outgassing,* formed a new atmosphere.

The gases released during outgassing interacted with radiation from the sun. The solar radiation caused the ammonia and some of the water vapor in the atmosphere to break down. Most of the hydrogen that was released during this breakdown escaped into space. Some of the remaining oxygen formed *ozone,* a molecule that contains three oxygen atoms. The ozone collected in a high atmospheric layer around Earth and shielded Earth's surface from the harmful ultraviolet radiation of the sun.

Earth's Present Atmosphere

Organisms that could survive in Earth's early atmosphere developed. Some of these organisms, such as cyanobacteria and early green plants, used carbon dioxide during photosynthesis. Oxygen, a byproduct of photosynthesis, was released. So, the amount of oxygen in the atmosphere slowly increased. About 2 billion years ago, the percentage of oxygen in the atmosphere increased rapidly. Since that time, the chemical composition of the atmosphere has been similar to the present composition of the atmosphere, as shown in **Figure 5.**

Reading Check **How did green plants contribute to Earth's present-day atmosphere? (See the Appendix for answers to Reading Checks.)**

Figure 4 ▶ Earth's early atmosphere formed as volcanic eruptions released nitrogen, N_2; water vapor, H_2O; ammonia, NH_3; methane, CH_4; argon, Ar; sulfur dioxide, SO_2; and carbon dioxide, CO_2.

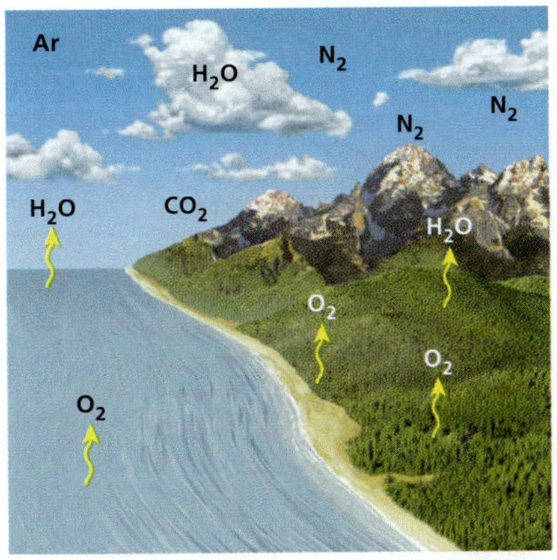

Figure 5 ▶ As Earth's surface changed, the gases in the atmosphere changed. Today, the atmosphere is 78% nitrogen, N_2; 21% oxygen, O_2; and 1% other gases.

Answer to Reading Check

Green plants release free oxygen as part of photosynthesis, which caused the concentration of oxygen gas in the atmosphere to gradually increase.

Close

Reteaching — Basic

Flashcards Have students work in pairs to create flashcards that have drawings and descriptions of the stages in the formation of Earth. Then, students can shuffle the cards and take turns organizing them into chronological order. LS **Logical**

Quiz — General

1. What process led to the formation of layers in Earth? (differentiation)
2. What process helped form Earth's atmosphere through the release of gases during volcanic eruptions? (outgassing)
3. What is the rotating cloud of dust and gas that formed the solar system called? (solar nebula)

Alternative Assessment — Advanced

Cosmic Cartoons Have students create a series of cartoon strips that illustrates the steps in the formation of the sun and the inner and outer planets from the solar nebula. Have students use captions to explain what happened at each stage in the process. LS **Visual**

CHEMISTRY CONNECTION

Ocean-Atmosphere Interface The formation of the oceans and atmosphere were closely interrelated. UV light disassociated water from the oceans, producing gases that allowed the atmosphere to develop further. Changes in the atmosphere changed ocean chemistry, allowing life to develop. That life produced gases such as free oxygen that altered the atmospheric composition. The increasing concentration of oxygen led to the formation of the protective ozone shield. Thus, a complex feedback system governed changes in both the oceans and atmosphere.

Homework — General

Earth Spheres Posters Have students work in small groups to design a poster around the theme of one of the following Earth spheres: the lithosphere (rock), the atmosphere (air), the hydrosphere (water), and the biosphere (life). Students can use library and Internet resources to learn more about the origin of the Earth sphere that they have chosen. Posters should trace the history of the sphere, identify the forces that have shaped it, and explain its present form. LS **Visual**

Close, *continued*

Answers to Section Review

1. The sun, planets, and other bodies of the solar system formed out of a spinning cloud of gas and dust as a result of collisions and gravitational contraction.
2. Planetesimals were small bodies formed when particles in the nebula stuck together. Protoplanets were larger bodies formed from the planetesimals.
3. through collisions and gravitational attraction of planetesimals and protoplanets
4. Because the outer planets formed far from the hot center of the nebula, they could retain lighter gases that would have escaped at higher temperatures.
5. because the collisions between planetesimals that formed Earth produced heat; because the outer layers of the planet compressed the inner layers, creating heat; and because radioactive elements common in the material of early Earth emitted high-energy particles that heated Earth's rocks
6. As Earth developed, denser materials sank to its center and less dense materials were forced outward. This gave the planet its layered structure. An early atmosphere that consisted of light gases formed, but Earth was too small and too close to the sun to hold them. Large amounts of other gases and water vapor produced by volcanic eruptions and introduced by comets formed a new atmosphere. As Earth cooled further, water vapor condensed and fell as rain, forming oceans. Life began in the oceans. Green plants contributed oxygen to the atmosphere as a byproduct of photosynthesis.
7. The amount of gas in an outer planet is much greater than that in an inner planet because outer planets were too far from the sun to lose their gaseous elements through its radiation. Also, they are more massive, and thus able to hold more gas.
8. because it has abundant liquid water, because it has a moderate temperature, and because it has an ozone layer that shields the surface from harmful ultraviolet radiation
9. The *solar system* formed from the *solar nebula,* within which *planetesimals* joined together to form *protoplanets,* which became *planets,* including the terrestrial planets and the *gas giants*.

CHAPTER RESOURCES

Chapter Resource File

- **Section Quiz** GENERAL

Workbooks

- **Study Guide** (also in Spanish)

Figure 6 ▶ Salt from the ocean can be harvested from salt flats, such as this one in Habantota, Sri Lanka.

Formation of Earth's Oceans

Some scientists think that part of Earth's water may have come from space. Icy bodies, such as comets, collided with Earth. Water from these bodies then became part of Earth's atmosphere. As Earth cooled, water vapor condensed to form rain. This liquid water collected on the surface to form the first oceans.

The first ocean was probably made of fresh water. Over millions of years, rainwater fell to Earth and ran over the land, through rivers, and into the ocean. The rainwater dissolved some of the rocks on land and carried those dissolved solids into the oceans. As more dissolved solids were carried to the oceans, the concentration of certain chemicals in the oceans increased. As the water cycled back into the atmosphere through evaporation, some of these chemicals combined to form salts. Over millions of years, water has cycled between the oceans and the atmosphere. Through this process, the oceans have become increasingly salty. Where shallow ocean water has evaporated completely, the salt precipitates and is left behind. This salt may be harvested for human use, as shown in **Figure 6.**

The Ocean's Effects on the Atmosphere

The oceans affect global temperatures in a variety of ways. One way the oceans affect temperature is by dissolving carbon dioxide from the atmosphere. Scientists think that early oceans also affected Earth's early climate by dissolving carbon dioxide. However, Earth's early atmosphere contained less carbon dioxide than the Earth's atmosphere does today. Thus, Earth's early climate was probably cooler than the global climate is today.

Section 1 Review

1. **Describe** the nebular hypothesis.
2. **Explain** how planetesimals differ from protoplanets.
3. **Describe** how planets developed.
4. **Explain** why the outer planets are more gaseous than the inner planets.
5. **List** three reasons that Earth was hot when it formed.
6. **Summarize** the process by which the land, atmosphere, and oceans of Earth formed.

CRITICAL THINKING

7. **Identifying Relationships** How does the amount of gas in an outer planet differ from the amount of gas in an inner planet? Explain your answer.
8. **Analyzing Ideas** Explain why Earth is capable of supporting life.

CONCEPT MAPPING

9. Use the following terms to make a concept map: *solar system, solar nebula, protoplanet, planetesimal, planet,* and *gas giant.*

Section 2 Models of the Solar System

The first astronomers who studied the sky thought that the stars, planets, and sun revolved around Earth. This idea led to the first model of the solar system. However, the model changed as scientists learned more about how the solar system works.

OBJECTIVES

- **Compare** the models of the universe developed by Ptolemy and Copernicus.
- **Summarize** Kepler's three laws of planetary motion.
- **Describe** how Newton explained Kepler's laws of motion.

KEY TERMS

eccentricity
orbital period
inertia

Early Models of the Solar System

More than 2,000 years ago, the Greek philosopher Aristotle suggested an Earth-centered, or *geocentric,* model of the solar system. In this model, the sun, the stars, and the planets revolved around Earth. However, this model did not explain why some planets sometimes appeared to move backward in the sky relative to the stars—a pattern called *retrograde motion*.

Around 130 CE, the Greek astronomer Claudius Ptolemy (TAHL uh mee) proposed changes to this model. Ptolemy thought that planets moved in small circles, called *epicycles,* as they revolved in larger circles around Earth. These epicycles seemed to explain why planets sometimes appeared to move backward.

In 1543 CE, a Polish astronomer named Nicolaus Copernicus proposed a sun-centered, or *heliocentric,* model of the solar system. In this model, the planets revolved around the sun in the same direction but at different speeds and distances from the sun. Fast-moving planets passed slow-moving planets. Therefore, planets that were slower than Earth appeared to move backward. **Figure 1** compares Ptolemy's and Copernicus's models. Later, the Italian scientist Galileo Galilei observed that four moons traveled around Jupiter. This observation showed him that objects can revolve around objects other than Earth.

Figure 1 ▶ Early Solar System Models Ptolemy's solar system model (left) is Earth centered and has the planets moving in epicycles around Earth. Copernicus's solar model (right) is heliocentric and has the planets moving at different speeds around the sun.

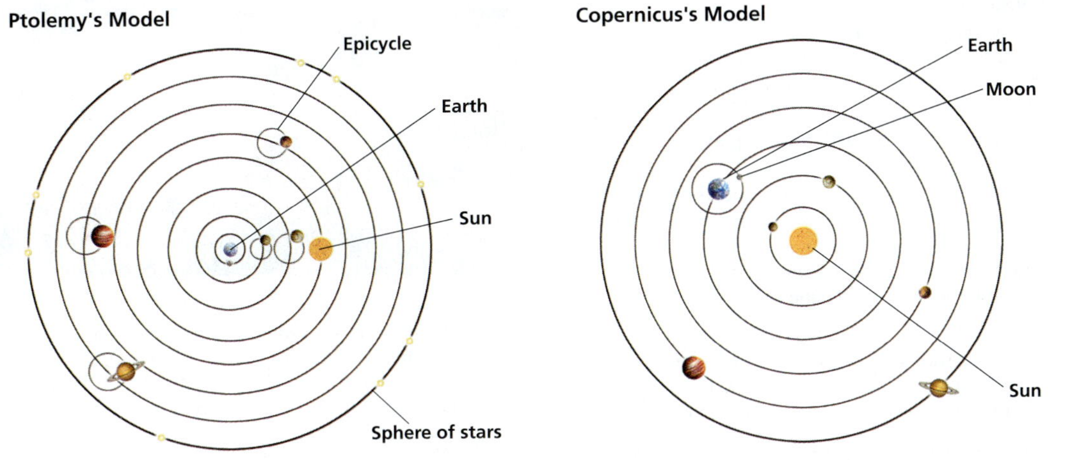

Section 2

Focus

Overview

This section compares models of the solar system and summarizes the laws of planetary motion.

Bellringer

Ask students to draw a diagram of the solar system and to label all of the parts of the diagram. **LS Verbal**

Motivate

Activity GENERAL

Role Play To help students understand how the planets might appear to move backward, try this activity. Group students into pairs. Have one student in each pair represent Earth, the other Mars. Have them stand next to each other. Ask "Mars" to begin moving forward slowly, while "Earth" observes background objects as "Mars" passes them. Have "Earth" begin moving faster while still looking at the same objects in relation to "Mars." Invite students to report what happens when "Earth" passes "Mars." (Sample answer: "Earth" has to look backward in order to view the objects in relation to its partner.) **LS Kinesthetic/Visual**

CHAPTER RESOURCES

Chapter Resource File

- Directed Reading BASIC

Technology

Transparencies
- Bellringer

Student Edition on CD-ROM

One-Stop Planner CD-ROM
- Lesson Plan

Teach

QuickLAB

Skills Acquired

- Constructing Models
- Measuring
- Identifying/Recognizing patterns

Teacher's Notes Using graph paper that has 1 cm squares will make measuring easier. Masking tape will hold paper securely to the corkboard and keep it steady while students draw. As an extension, you might give students a table of planetary distances in AUs and have them make a scale model of the solar system.

Answers

1. The ellipse made using only one pin (step 5) has an eccentricity close to 0 and is most like a circle. The ellipse drawn in step 4 has an eccentricity closest to 1 and is least like a circle.
2. As I increased the distance between foci, the ellipses became more elongated. If I increased string length without changing the foci, the ellipse would become larger and more circular.

CHAPTER RESOURCES

Chapter Resource File

- **Datasheet for Quick Lab** GENERAL

SCILINKS. NSTA Developed and maintained by the National Science Teachers Association

For a variety of links related to this subject, go to www.scilinks.org

Topic: Early Astronomers
SciLinks code: HQ60441

Kepler's Laws

Twenty years before Galileo used a telescope, a Danish astronomer named Tycho Brahe made detailed observations of the solar system. After Tycho's death, one of his assistants, Johannes Kepler, discovered patterns in Tycho's observations. These patterns led Kepler to develop three laws that explained planetary motion.

Law of Ellipses

Kepler's first law, the *law of ellipses,* states that each planet orbits the sun in a path called an ellipse, not in a circle. An *ellipse* is a closed curve whose shape is determined by two points, or *foci,* within the ellipse. In planetary orbits, one focus is located within the sun. No object is located at the other focus. The combined length of two lines, one from each focus to any one point on the ellipse, would always be the same as the length of two lines from each focus to any other one point on the same ellipse.

Elliptical orbits can vary in shape. Some orbits are elongated ellipses. Other orbit shapes are almost perfect circles. The shape of an orbit can be describe in a numerical form called eccentricity. **Eccentricity** is the degree of elongation of an elliptical orbit (symbol, e). Eccentricity is determined by dividing the distance between the foci of the ellipse by the length of the major axis. Therefore, the eccentricity of a circular orbit is $e = 0$. The eccentricity of an extreme elongated orbit, or parabolic orbit, is $e = 1$.

eccentricity the degree of elongation of an elliptical orbit (symbol, e)

Reading Check **Define and describe an ellipse. (See the Appendix for answers to Reading Checks.)**

QuickLAB 15 min

Ellipses

Procedure

1. Cover a **cork board** with a **piece of paper.** Put **two push pins** into the cork board 5 cm apart.
2. Tie together the ends of a **string** that is 25 cm long. Loop the string around the pins.
3. Hold the string taut with a **pencil,** and move the pencil to outline an ellipse.
4. Replace the paper with another **piece of paper.** Place the pins 10 cm apart. Outline another ellipse by using the string.
5. Use another **piece of paper.** Loop the string around one pin, and outline another ellipse.

Analysis

1. Which ellipse has an eccentricity closest to 0? Which has an eccentricity closest to 1? Describe the shape of your ellipse in terms of eccentricity.
2. Describe the ellipse as you increased the distance between the foci. Describe what would happen to the ellipse if you increased the length of the string without changing the distance between the foci.

Answer to Reading Check

An ellipse is a closed curve whose shape is defined by two points inside the curve. An ellipse looks like an oval.

SKILL BUILDER GENERAL

Vocabulary The word *ellipse* is from the Greek word *elleipsis*. The word *ellipsoid* also derives from this Greek word. The word *eccentricity* is from the Greek *ekkentros,* which is a combination of *ex-* meaning "out of" and *kentron* meaning "center." Thus, *eccentricity* is the extent to which a planet's orbit is elongated, or pulled away from the center. A related word, *eccentric,* is often used to describe people who live their lives outside of the social norms. **LS Verbal** English Language Learners

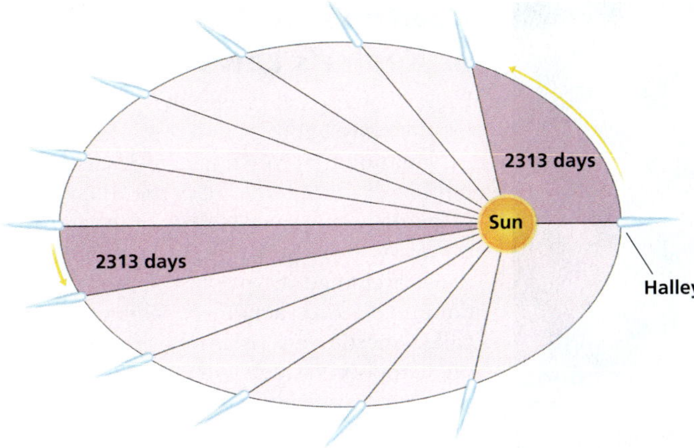

Figure 2 ▶ Kepler's second law states that the areas through which an object sweeps in a given period of time are equal.

Law of Equal Areas

Kepler's second law, the *law of equal areas,* describes the speed at which objects travel at differnet points in their orbits. Kepler discovered Mars moves fastest in its elliptical orbit when it is closest to the sun. He calculated that a line from the center of the sun to the center of an object sweeps through equal areas in equal periods of time. This principle is illustrated in **Figure 2.**

Imagine a line that connects the center of the sun to the center of an object in orbit around the sun. When the object is near the sun, the imaginary line is short. The object moves relatively rapidly, and the line sweeps through a short, wide, pie-shaped sector. When the object is far from the sun, the line is long. However, the object moves relatively slowly when it is far from the sun, and the imaginary line sweeps through a long, thin, pie-shaped sector in the same period. Kepler's second law states that equal areas are covered in equal amounts of time as an object orbits the sun.

MATHPRACTICE

Law of Periods
Scientists discover an asteroid that is 0.38 AU from the sun. If 1 AU = 150 million km, how long would the asteroid's orbital period be in Earth years?

Laws of Periods

Kepler's third law, the *law of periods,* describes the relationship between the average distance of a planet from the sun and the orbital period of the planet. The **orbital period** is the time required for a body to complete a single orbit. According to Kepler's third law, the cube of the average distance (a) of a planet from the sun is always proportional to the square of the period (p). The mathematical formula that describes this relationship is $K \times a^3 = p^2$, where K is a constant. When distance is measured in astronomical units (AU) and the period is measured in Earth years, $K = 1$ and $a^3 = p^2$.

Scientists can find out how far away the planets are from the sun by using this law, because they can measure the orbital periods by observing the planets. Jupiter's orbital period is 11.9 Earth years. The square of 11.9 is 142. The cubed number that is equal to 142 is 5.2, so Jupiter is 5.2 AU from the sun.

orbital period the time required for a body to complete a single orbit

MATHPRACTICE

Answer
$a^3 = p^2$
$a^3 = 0.38 \text{ AU} \times 0.38 \text{ AU} \times 0.38 \text{ AU} = 0.055 \text{ AU}^3$
$p^2 = 0.055; p = \sqrt{0.055} = 0.23 \text{ y}$

Close

Reteaching — BASIC

Have students create a Venn diagram to compare the geocentric and heliocentric models of the solar system. Students can use the diagram to study for future assessments. LS **Visual**

Quiz — GENERAL

Determine whether each of the following statements is true or false.

1. A planet moves relatively slower when it is farther from the sun than it does when it is closer to the sun. (true)
2. Kepler's first law states that each planet orbits the sun, not in a circle, but in an ellipse. (true)
3. Kepler's third law states that the square of the average distance of a planet from the sun is proportional to the cube of the orbital period. (false)

Alternative Assessment — ADVANCED

Ask students to create an advertising "jingle" for each of Kepler's laws of motion. Each jingle should state the law and describe how to use it. Students may wish to perform their jingles for the class. LS **Auditory**

CHAPTER RESOURCES

Technology

Transparencies
- 138 Kepler's Law of Equal Area (with worksheet)

INCLUSION Strategies

• Learning Disabled • Developmentally Delayed

Mnemonics they create themselves can help students remember scientists' names and accomplishments. Divide the class into groups, and have each team create a mnemonic to help them remember the following people and their accomplishments: Ptolemy, Copernicus, Galileo, Tycho, Kepler, and Newton. Explain that mnemonics can be any word or sentence that is easy to remember. LS **Verbal/Auditory**

Close, *continued*

Answers to Section Review

1. Ptolemy's model was geocentric, with the planets moving in small circles as they revolved in larger circles around Earth. Copernicus's model was heliocentric, showing planets moving around the sun at different speeds and distances.
2. Galileo's observations with the telescope provided support for the heliocentric model because he found smaller objects (Jupiter's moons) that revolved around celestial bodies other than Earth.
3. The orbits are ellipses, with one of two foci located within the sun.
4. The law of equal areas describes the speed at which objects travel at different points in their orbits. Objects move faster when they are closest to the sun. An imaginary line between the two bodies sweeps equal areas in equal amounts of time.
5. The law of periods describes the mathematical relationship between the average distance of a planet from the sun and its orbital period. The cube of the average distance in astronomical units is equal to the square of the planet's orbital period in Earth years, or $a^3 = p^2$.
6. Newton's model showed that the gravitational force that pulls a planet toward the sun combines with the straight-line motion that results from the planet's inertia to cause the planet to move in an elliptical orbit.
7. When it is located in the inner region of the solar system, the comet is closer to the sun and moves relatively faster than when it is farther from the sun. Also, a much larger fraction of its orbit is in the outer solar system. Thus, it spends a proportionally shorter time in the inner solar system.
8. Kepler explained his observations of the motions of planets in terms of the elliptical shape of their orbits and the position of the foci. He also described the effects of the sun on the planets' speed at different points in their orbits. Newton supplied the physical causes for what Kepler observed and described: the inertia of planets and gravitational pull of the sun on the planets.
9. The *geocentric* model of the solar system was replaced by the *heliocentric* model that accounts for apparent *retrograde motion*, and that describes how planets travel in orbits that are governed by the laws of *gravity* and *inertia* and that have the shape of *ellipses* that have two *foci*.

Figure 3 ▶ In 1997, Comet Hale-Bopp was visible in the sky above the observatory on Mauna Kea in Hawaii.

Newton's Explanation of Kepler's Laws

Isaac Newton asked why the planets move in the ways that Kepler observed. The explanation that Newton gave described the motion of objects on Earth and the motion of planets in space. He hypothesized that a moving body will remain in motion and resist a change in speed or direction until an outside force acts on it. This concept is called **inertia.** For example, a ball rolling on a smooth surface will continue to move unless a force causes it to stop or change direction.

inertia the tendency of an object to resist being moved or, if the object is moving, to resist a change in speed or direction until an outside force acts on the object

Newton's Model of Orbits

Because a planet does not follow a straight path, an outside force must cause the orbit to curve. Newton gave this force the name *gravity,* and he realized that this attractive force exists between any two objects in the universe. The gravitational pull of the sun keeps objects, such as the comet shown in **Figure 3,** in orbit around the sun. While gravity pulls an object toward the sun, inertia keeps the object moving forward in a straight line. The sum of these two motions forms the ellipse of a stable orbit.

The farther from the sun a planet is, the weaker the sun's gravitational pull on the planet is. So, the outer planets are not pulled toward the sun as strongly as the inner ones are. As a result, the orbits of the outer planets are larger and are curved more gently, and the outer planets have longer periods of revolution than the inner planets do.

Section 2 Review

1. **Compare** Ptolemy's and Copernicus's models of the universe.
2. **Identify** the role that Galileo played in developing the heliocentric theory.
3. **Describe** the shape of planetary orbits.
4. **Explain** the law of equal areas.
5. **Summarize** Kepler's third law of planetary orbits.
6. **Describe** how Newton explained Kepler's laws by combining the effects of two forces.

CRITICAL THINKING

7. **Applying Ideas** A comet's orbit is a highly elongated ellipse. So, why does a comet spend so little time in the inner solar system?
8. **Making Comparisons** How did Kepler's explanation of the orbits of planets differ from Newton's explanation?

CONCEPT MAPPING

9. Use the following terms to create a concept map: *retrograde motion, geocentric, heliocentric, ellipse, foci, gravity,* and *inertia.*

CHAPTER RESOURCES

Chapter Resource File
- Section Quiz GENERAL

Workbooks
- Study Guide (also in Spanish)

Section 3 The Inner Planets

The planets closest to the sun are called the *inner planets*. These planets are Mercury, Venus, Earth, and Mars. The inner planets are also called **terrestrial planets,** because they are similar to Earth. These planets consist mostly of solid rock and have metallic cores. The number of moons per planet varies from zero to two. The surfaces of inner planets have bowl-shaped depressions, called *impact craters,* that were caused by collisions of the planets with other objects in space.

OBJECTIVES

- **Identify** the basic characteristics of the inner planets.
- **Compare** the characteristics of the inner planets.
- **Summarize** the features that allow Earth to sustain life.

KEY TERM

terrestrial planet

Mercury

Mercury, the planet closest to the sun, circles the sun every 88 days. The ancient Romans named the planet after the messenger of the gods, who moved quickly. Mercury rotates on its axis once every 59 days.

Images of Mercury reveal a surface that is heavily cratered, as shown in **Figure 1.** The images also show a line of cliffs hundreds of kilometers long. These cliffs may be wrinkles that developed in the crust when the molten core cooled and shrank.

The absence of a significant atmosphere and the planet's slow rotation contributes to the large daily temperature range on Mercury. During the day, the temperature may reach as high as 427°C. At night, the temperature may plunge to –173°C.

terrestrial planet one of the highly dense planets nearest to the sun; Mercury, Venus, Mars, and Earth

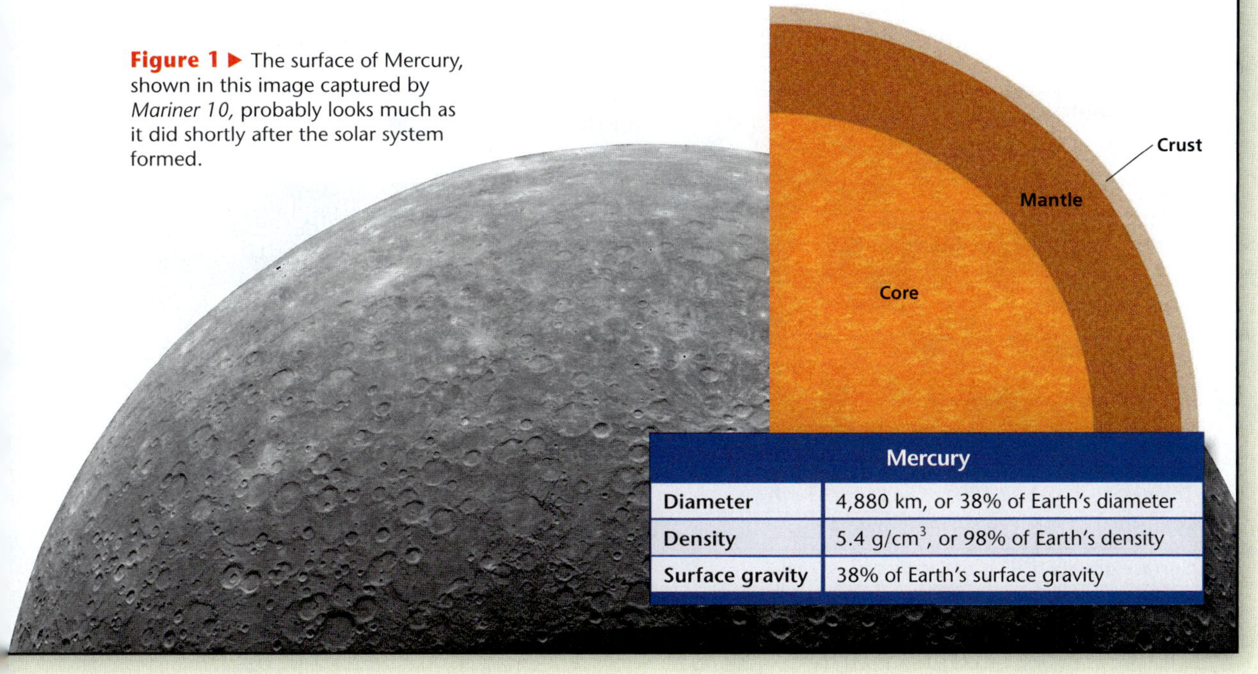

Figure 1 ▶ The surface of Mercury, shown in this image captured by *Mariner 10,* probably looks much as it did shortly after the solar system formed.

Mercury	
Diameter	4,880 km, or 38% of Earth's diameter
Density	5.4 g/cm^3, or 98% of Earth's density
Surface gravity	38% of Earth's surface gravity

Section 3

Focus

Overview

This section identifies the common and unique characteristics of the inner (terrestrial) planets and summarizes the life-sustaining features of Earth.

Bellringer

Have students write a brief description of the characteristics of Earth. (Answers may vary but should lead to a discussion of characteristics of the planet that are similar to the characteristics of other terrestrial planets.) **LS Verbal/Visual**

Motivate

Discussion — GENERAL

Terrestrial Planets Ask students to describe how they think Earth would be different if it were closer to or farther from the sun. (Answers may vary. Students should comment that Earth would be hotter if it were closer to the sun and may be colder if it were farther from the sun.) Use this discussion as a launching point for discussing why the terrestrial planets have unique characteristics. **LS Verbal**

CHAPTER RESOURCES

Chapter Resource File

- Directed Reading BASIC

Technology

Transparencies
- Bellringer

Student Edition on CD-ROM

One-Stop Planner CD-ROM
- Lesson Plan

Teach

Demonstration — GENERAL

Global Greenhouse

1. Label two tall plastic tubes or jars "Earth" and "Venus." Use double-sided tape to attach a thermometer to the inside of each container near the top, facing outward. Leave empty space at the base.
2. Place a desk lamp above the model planetary atmospheres. Take a starting temperature reading for each jar.
3. Put a medicine bottle full of vinegar inside the Venus model and drop a twist of paper full of baking soda into the vinegar before closing the jar again. Ask students what gas the reaction between the vinegar and baking soda releases into the jar. (carbon dioxide, CO_2)
4. Turn on the light. Have a student record the temperatures of the containers at 5-minute intervals for 30 minutes.
5. Remove the heat source and continue to take readings for another 10 minutes.
6. Examine the data. Have students identify which model was hotter after 30 minutes. (The Venus model will be hotter.) Ask students to compare the temperature changes that occurred in each model. (The model representing Earth's atmosphere cools off faster. Venus, the runaway greenhouse model with additional CO_2, retains the heat much longer.) LS Logical

Venus

Venus is the second planet from the sun and has an orbital period of 225 days. However, Venus rotates very slowly, only once every 243 days. In some ways, Venus is Earth's twin. The two planets are of almost the same size, mass, and density. However, Venus and Earth differ greatly in other areas.

Venus's Atmosphere

The biggest difference between Earth and Venus is Venus's atmosphere. Venus's atmospheric pressure is about 90 times the pressure on Earth. The high concentration of carbon dioxide in Venus's atmosphere and Venus's relative closeness to the sun have the strongest influences on surface temperatures. Venus's atmosphere is about 96% carbon dioxide.

Solar energy that penetrates the atmosphere heats the planet's surface. The high concentration of carbon dioxide in the atmosphere blocks most of the infrared radiation from escaping. This type of heating is called a *greenhouse effect*. On Earth, the greenhouse effect warms Earth enough to allow organisms to live on the planet. But the greenhouse effect on Venus makes the average surface temperature 464°C! This phenomenon is commonly referred to as a *runaway greenhouse effect* and makes Venus's surface temperature the highest known in the solar system.

Venus also has sulfur dioxide droplets in its upper atmosphere. These droplets form a cloud layer that reflects sunlight. The cloud layer reflects the sunlight so strongly that from Earth, Venus appears to be the brightest object in the night sky, aside from Earth's moon and the sun. Because Venus appears near the sun, Venus is usually visible from Earth only in the early morning or evening. Therefore, Venus is commonly called the *evening star* or the *morning star*.

Figure 2 ▶ Venus's surface is composed of basalt and granite rocks. However, Venus's dense atmosphere is composed primarily of carbon dioxide.

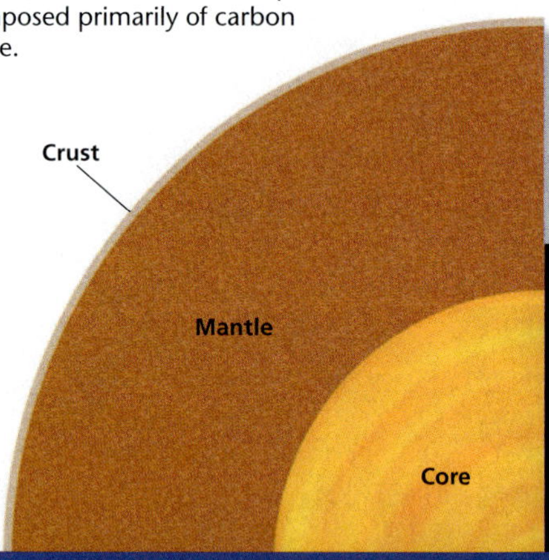

Venus	
Diameter	12,100 km, or 95% of Earth's diameter
Density	5.2 g/cm^3, or 95% of Earth's density
Surface gravity	91% of Earth's surface gravity

ENVIRONMENTAL CONNECTION

Venus: Earth's Twin Astronomers have long referred to Venus as "Earth's twin" because the masses, densities, and compositions of the two planets are similar. However, Venus is actually very different from Earth in some ways. Venus's massive atmosphere has caused a runaway greenhouse effect. Venus's atmosphere prevents heat from solar radiation from radiating back into space. As a result, Venus's surface temperature is hot enough to melt lead. Water vapor boils away, so Venus has no oceans. Sulfuric acid makes Venus's upper atmosphere poisonous. The milder greenhouse effect on Earth, however, keeps Earth's temperatures moderate.

One of the most common greenhouse gases on Venus is carbon dioxide (CO_2). On Earth, this compound is present in the atmosphere, but most CO_2 is held in non-atmospheric reservoirs such as the ocean and in rocks. Scientists fear that burning fossil fuels will release too much CO_2 into the atmosphere, causing Earth's surface temperature to rise, which would drastically alter Earth's environment.

Figure 3 ▶ The scale of this computer-generated image of the Maat Mons volcano on Venus has been stretched vertically to make the topography look more dramatic. The image has also been color enhanced.

Missions to Venus

In the 1970s, the Soviet Union sent six probes to explore the surface of Venus. The probes survived in the atmosphere long enough to transmit surface images of a rocky landscape. The images showed a smooth plain and some rocks. Other instruments carried by the probes indicated that the surface of Venus is composed of basalt and granite. These two types of rock are also common on Earth.

The United States's *Magellan* satellite orbited Venus for four years in the 1990s before the satellite was steered into the planet to collect atmospheric data. *Magellan* also bounced radio waves off Venus to produce radar images of Venus's surface.

Surface Features of Venus

From the radar mapping produced by *Magellan,* scientists discovered landforms such as mountains, volcanoes, lava plains, and sand dunes. Volcanoes and lava plains are the most common features on Venus. At an elevation of 8 km, the volcano Maat Mons, which is shown in **Figure 3,** is Venus's highest volcano.

The surface of Venus is also somewhat cratered. All of the craters are of about the same age, and they are surprisingly young. This evidence and the abundance of volcanic features on Venus's surface have led some scientists to speculate that Venus undergoes a periodic resurfacing as a result of massive volcanic activity. Heat inside the planet builds up over time, which causes the volcanoes to erupt and cover the planet's surface with lava. However, scientists think that another 100 million years may pass before volcanic activity again covers Venus's surface with lava. Venus's surface is very different from Earth's surface, which is constantly changing because of the motion of tectonic plates.

Reading Check **How is Venus different from Earth? (See the Appendix for answers to Reading Checks.)**

MATHPRACTICE

Distance from the Sun Earth is about 150 million kilometers from the sun. Venus is 108.2 million kilometers from the sun. How much closer to the sun is Venus than Earth? Express your answer as a percentage.

Using the Figure — BASIC

Computer Generated Image Point out that the vertical scale of this photo of Maat Mons has been exaggerated about 20 times to show clearly the structures and topography of the region around the volcano. The volcanic slope is actually very gentle. Color was added to match the reflectivity of the surface to radar energy. **LS Visual**

Discussion — GENERAL

Radar Revelations Invite volunteers to describe the surface features of Venus that have been revealed by radar. (volcanoes, lava flows, mountainous landscape, impact craters, and sand dunes) Ask students to explain why the craters on Venus appear to be in their original condition and almost unaffected by erosion. (The craters are relatively young. There is no surface water to erode them. Also, Venus lacks moving tectonic plates.) **LS Verbal**

MATHPRACTICE

Answer

150 million km − 108.2 million km = 41.8 million km ÷ 150 million km = 0.2787. Venus is 27.9% closer to the sun than Earth is.

Answer to Reading Check

Answers may vary but should address differences in distance from the sun, density, atmospheric pressure and density, and tectonics.

Group Activity — GENERAL

Remote Sensing Explain that Venus's dense atmosphere makes viewing Venus's surface features difficult. Have small groups of students model how radar-imaging probes map a hidden surface.

1. Build a clay terrain model in a box. Do not show the surface to other groups.
2. Label a piece of graph paper with letters in alphabetical order along the top and numbers in sequence along the left side. Tape the grid on the box.
3. Mark a dowel rod with different colors at 1 cm intervals to use as a probe. Use a pencil to poke a hole in each grid square.
4. Make a similar grid on a fresh sheet of graph paper to use as a map.
5. Gently insert the probe into the first hole labeled A1 until it hits the surface. The deeper the probe drops, the lower the elevation of the surface feature is.
6. On the probe, note the color that is even with the grid. Color the A1 square on the map a matching color.
7. Repeat steps 5 and 6 until the entire surface is mapped. **LS Kinesthetic**

READING SKILL BUILDER — BASIC

Paired Summarizing Group students into pairs and have them read silently the introductory description of Earth as it might be viewed from elsewhere in our solar system. Then, have one student summarize Earth's characteristics. The other student can listen to the description and make corrections or identify any elements that were left out. Allow students to refer to the text as needed.
LS Auditory English Language Learners
Co-op Learning

ENVIRONMENTAL CONNECTION

Suitable for Life Earth is the only planet that has both oceans of liquid water and areas of frozen ice on its surface. The idea that Earth is at the perfect distance from the sun is known as the "Goldilocks" principle. But the composition of Earth's atmosphere may be the key factor. If Earth had a dense atmosphere, Earth could be as hot as Venus; If Earth's atmosphere were thin, like the thin Martian atmosphere, Earth would likely be a rocky ball of ice. Instead, Earth has an oxygen-rich atmosphere that keeps the planet at a comfortable temperature and that protects the surface from harmful radiation. Earth's oceans also prevent the greenhouse effect that warms our planet from getting out of control by dissolving excess CO_2 gas. These environmental conditions make Earth ideally suited to support a complex biosphere.

Earth

The third planet from the sun is Earth. The orbital period of Earth is 365 1/4 days, and Earth completes one rotation on its axis every day. Earth has one large moon.

Earth has had an extremely active geologic history. Geologic records indicate that over the last 250 million years, Earth's continents separated from a single landmass and drifted to their present positions. Weathering and erosion have changed and continue to change the surface of Earth.

Water on Earth

Earth's unique atmosphere and distance from the sun allow water to exist in a liquid state. Mercury and Venus are so close to the sun that any liquid water on those planets would boil. The outer planets are so far from the sun that water freezes. Earth is the only planet known to have oceans of liquid water, as shown in **Figure 4.** However, scientists think that Jupiter's moon Europa may have an ocean under its icy crust.

Life on Earth

Scientists theorize that as oceans formed on Earth, liquid water dissolved carbon dioxide from the atmosphere. Because of this process, carbon dioxide did not build up in the atmosphere and solar heat was able to escape. Thus, Earth maintained the moderate temperatures needed to support life. Plants and cyanobacteria contributed free oxygen to the atmosphere. Earth is the only known planet that has the proper combination of water, temperature, and oxygen to support life.

SCILINKS. NSTA Developed and maintained by the National Science Teachers Association

For a variety of links related to this subject, go to www.scilinks.org

Topic: Inner Planets
SciLinks code: HQ60798

Figure 4 ▶ Oceans of water and an atmosphere that can support life make Earth a unique planet.

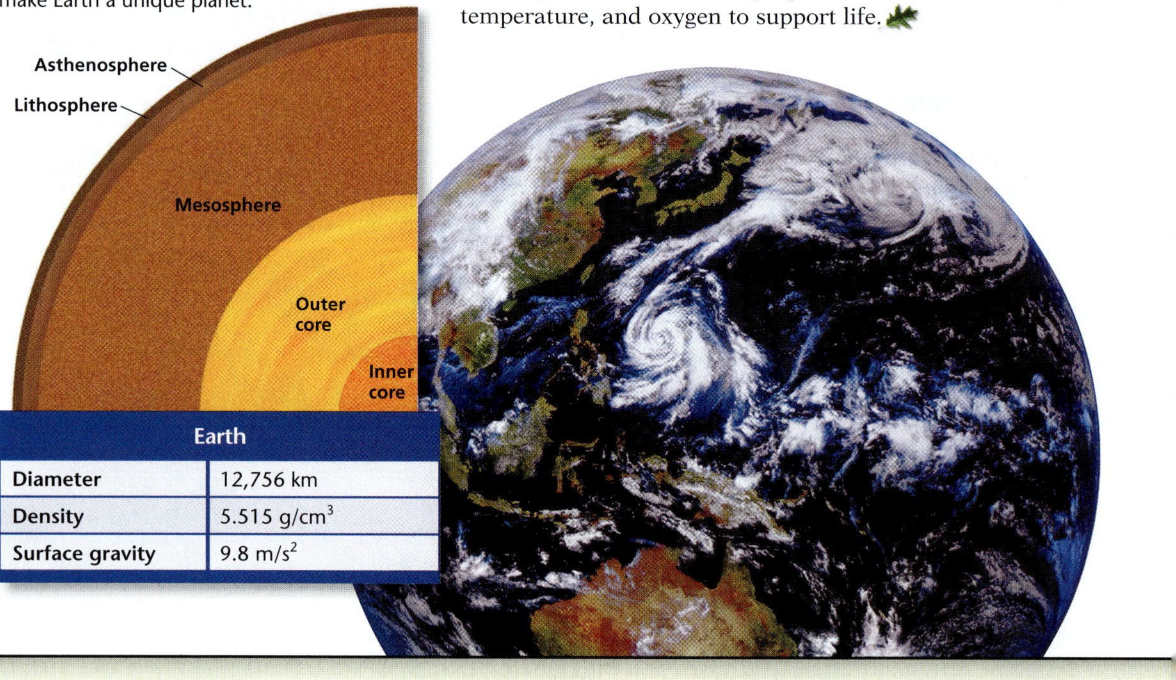

Earth	
Diameter	12,756 km
Density	5.515 g/cm^3
Surface gravity	9.8 m/s^2

SKILL BUILDER — GENERAL

Writing The terrestrial planets have inspired numerous science fiction tales. For example, Mars was home to Ray Bradbury's *Martian Chronicles*. Venus is often pictured as a steamy jungle. Isaac Asimov wrote a "locked room" mystery located on Mercury. Have interested students share these and other stories with the class. Or, have interested students write a short story or poem set on an inner planet other than Earth. Encourage students to make the planet's physical properties important elements in the story. Students' descriptions of the planets should be consistent with what they have learned. LS Verbal

Cultural Awareness — BASIC

Classical Heritage When early scientists discovered the planets of the solar system, they gave most of them names associated with ancient Western mythology. For example, Mars was the Roman god of war. Have interested students research some ancient myths and legends and explain how the names of the inner and outer planets and some of their moons reflect their characteristics. Students may present their results as an oral or written report. LS Verbal

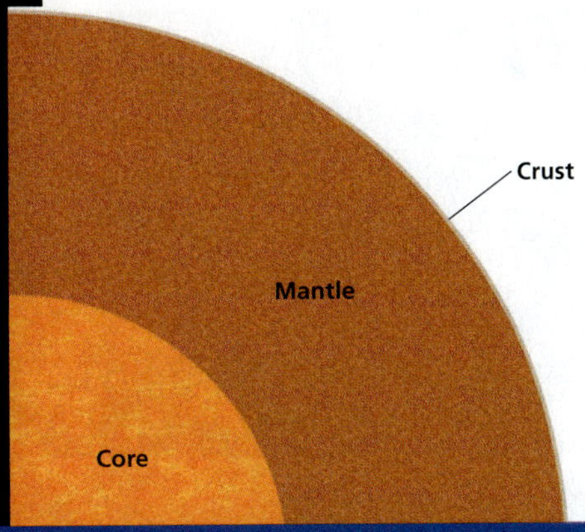

Mars	
Diameter	6,800 km, or 53% of Earth's diameter
Density	3.9 g/cm^3, or 71% of Earth's density
Surface gravity	38% of Earth's surface gravity

Figure 5 ▶ Mars is called the *Red Planet* because the oxidized rocks on the planet's surface give the planet a red color.

Mars

Mars, shown in **Figure 5,** is the fourth planet from the sun. At an average distance of about 228 million kilometers from the sun, Mars is about 50% farther from the sun than Earth is. Its orbital period is 687 days, and it rotates on its axis every 24 h and 37 min. Because its axis tilts at nearly the same angle that Earth's does, Mars's seasons are much like Earth's seasons.

Mars has been geologically active in its past, which is shown in part by the presence of massive volcanoes. A system of deep canyons also covers part of the surface. Valles Marineris is a series of canyons that is as long as the United States is wide—4,000 km. The canyon is thought to be a crack that formed in the crust as the planet cooled. It was later eroded by water.

Martian Volcanoes

Tharsis Montes is one of several volcanic regions on Mars. Volcanoes in this region are 100 times as large as Earth's largest volcano. The largest volcano on Mars is Olympus Mons, which is nearly 24 km tall. It is three times as tall as Mount Everest. At 550 km across, the base of Olympus Mons is about the size of Nebraska. Scientists think that the volcano has grown so large because Mars has no moving tectonic plates. So, Olympus Mons may have had a magma source for millions of years.

Whether Martian volcanoes are still active is a question scientists have yet to answer. A *Viking* landing craft detected two geological events that produced seismic waves. These events, called *marsquakes,* may indicate that volcanoes on Mars are active.

Reading Check Why are Martian volcanoes larger than Earth's volcanoes? (See the Appendix for answers to Reading Checks.)

Answer to Reading Check

Martian volcanoes are larger than volcanoes on Earth because Mars has no moving tectonic plates. Magma sources remain in the same spot for millions of years and produce volcanic material that builds the volcanic cone higher and higher.

Close

Reteaching — Basic

Same and Different Organize students into small groups. Have students write questions about the four inner planets based on the section content. Tell students to focus on ways in which the planets are the same and ways they are different. Then, have students join their assigned group and quiz each other using their questions.
LS Auditory

Quiz — General

Determine whether each of the following statements is true or false. If false, provide the correct word(s) to make the statement true.

1. The biggest differences between Earth and Venus involve mass and density. (false; atmospheric pressure and composition)
2. The surface of Mercury has many craters. (true)
3. Because of the massive geysers on its surface, scientists know that Mars has been geologically active in the past. (false; volcanoes) **LS Verbal**

CHAPTER RESOURCES

Chapter Resource File

- **Inquiry Lab** Probing for Information GENERAL

Internet Activity

- Life on Mars? ADVANCED

Internet Activity — Advanced

Life on Mars? The possibility of liquid water on Mars is important because water is considered essential for life. Some scientists think life may exist in protected niches on Mars. Fossil-like structures and organic chemicals have been found in ancient Martian meteorites. Invite students to research more about the evidence for life on Mars. A worksheet designed to direct student research on this topic can be found in the **Chapter Resource File** booklet or by visiting go.hrw.com and entering the keyword **HQ6PSSX.** **LS Verbal/Logical**

Close, *continued*

Alternative Assessment — Advanced

Planetary Base Have students work in small groups to plan a permanent colony or base on an inner planet other than Earth. Students may create a model, diagram, or written description. They should include a mission plan that describes the purpose of the development. They should also detail how to protect inhabitants from extremes of temperature, radiation, or other unusual atmospheric conditions and how to provide for personal needs. **LS Interpersonal**

Answers to Section Review

1. Mercury lacks a dense atmosphere to hold heat and has a very slow rotational rate.
2. Venus has nearly the same diameter, density, and surface gravity as Earth. Rocks are composed of similar materials. Both are geologically active. Venus has a much denser atmosphere with higher pressure and a much higher surface temperature than Earth's.
3. Earth's distance from the sun, the presence of surface liquid water, moderate surface temperature, and free oxygen in its atmosphere
4. Mars lacks moving tectonic plates, so volcanoes remain above the magma source for a very long time.
5. Atmospheric pressure and temperature are too low for liquid water to exist on the surface.
6. Answers may vary. Accept all reasonable answers.
7. On Earth, the greenhouse effect warms the planet's surface by absorbing heat radiated by Earth's surface. Because of its large concentration of carbon dioxide (96%), Venus's atmosphere absorbs massive amounts of heat from the surface. This heat makes the average surface temperature of Venus about 464°C.
8. As the rock on Mars's surface cooled, it shrank and cracked, causing the canyon to form.
9. The *terrestrial planets* include *Mercury; Venus,* whose highest volcano is *Maat Mons; Earth,* whose surface has *liquid water;* and *Mars,* whose highest volcano is *Olympus Mons.*

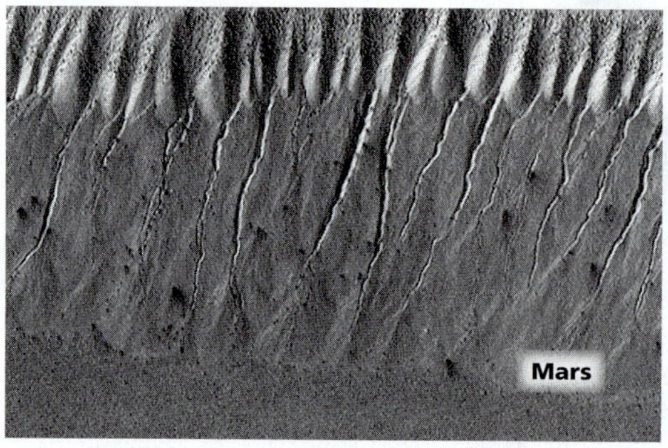

Figure 6 ▶ The images above compare the formation of gullies by possible liquid water runoff on Mars (left) with the formation of similar gullies at Mungo National Park in Australia on Earth (right). The Mungo National Park was the site of a large lake that dried up more than 10,000 years ago.

Water on Mars

The pressure and temperature of Mars's atmosphere are too low for water to exist as a liquid on Mars's surface. However, several NASA spacecraft—such as the Mars rovers, *Spirit* and *Opportunity,* which landed on Mars in 2004—have found evidence that liquid water did exist on Mars's surface in the past. Mars has many surface features that are characteristic of erosion by water, such as branching paths that look like gullies, as shown in **Figure 6.** Scientists think that other features on Mars might be evidence of vast flood plains produced by a volume of water equal to that of all five of Earth's Great Lakes.

The surface temperature on Mars ranges from 20°C near the equator during the summer to as low as –130°C near the poles during the winter. Although most of the water on Mars is trapped in polar icecaps, data from the *Mars Global Surveyor* suggest that water may also exist as permanent frost or as a liquid just below the surface. If liquid water does exist below Mars's surface, the odds of life existing on Mars will dramatically increase. However, no solid evidence of life on Mars has been found.

Section 3 Review

1. **Explain** why Mercury has such drastically different temperatures during its day and during its night.
2. **Describe** the main ways in which Venus is similar to and different from Earth.
3. **Identify** the aspects that make Earth hospitable for life.
4. **Explain** why Mars's volcanoes became so tall.
5. **Explain** why Mars does not have liquid water on its surface.
6. **Compare** the characteristics of the inner planets.

CRITICAL THINKING

7. **Making Comparisons** Describe the difference between the greenhouse effect on Venus and the greenhouse effect on Earth.
8. **Understanding Relationships** As rock cools, it contracts. How could this fact explain the presence of Valles Marineris on Mars?

CONCEPT MAPPING

9. Use the following terms to create a concept map: *Mercury, Venus, Earth, Mars, terrestrial planet, Olympus Mons, liquid water,* and *Maat Mons.*

CHAPTER RESOURCES

Chapter Resource File
- Section Quiz GENERAL

Workbooks
- Study Guide (also in Spanish)

Section 4 The Outer Planets

The four planets farthest from the sun are called the *outer planets*. They are separated from the inner planets by a ring of debris called the *asteroid belt*. Jupiter, Saturn, Uranus, and Neptune, which are shown in **Figure 1,** are also called **gas giants** because they are large planets that have deep, massive atmospheres made mostly of gas. Usually found past the orbit of Neptune is Pluto. Prior to 2006, Pluto was considered to be the most distant outer planet. However, in August of 2006, Pluto was reclassified as a *dwarf planet*.

OBJECTIVES

- **Identify** the basic characteristics that make the outer planets different from terrestrial planets.
- **Compare** the characteristics of the outer planets.
- **Explain** why Pluto is now considered a dwarf planet.

KEY TERMS

gas giant
Kuiper Belt

Gas Giants

Although the gas giants are much larger and more massive than the terrestrial planets, the gas giants are much less dense than the terrestrial planets. Unlike the terrestrial planets, the gas giants did not lose their original gases during their formation. Their large masses give them a huge amount of gravity, which helps them retain the gases. Each of the gas giants has a thick atmosphere that is made mostly of hydrogen and helium gases. A cloud layer prevents scientists from directly observing more than the topmost part of the atmosphere of the gas giants. But each planet probably has a core made of rock and metals.

Although Saturn's rings may be the most impressive, all four gas giants have ring systems that are made of dust and icy debris that orbit the planets. Most of the debris probably came from comets or other bodies.

gas giant a planet that has a deep massive atmosphere, such as Jupiter, Saturn, Uranus, or Neptune

Figure 1 ▶ The four gas giants are much larger than Earth, which is the terrestrial planet shown here at the lower left.

Section 4

Focus

Overview

This section compares characteristics of the outer planets and explains how the outer planets differ from the inner planets. It also explains that Pluto, once considered a planet, is now called a dwarf planet.

Bellringer

Ask students to list what they know about the properties of gases. (Sample answers: no definite shape or volume; atoms have more energy than atoms in solids and liquids; low density; exert pressure and can be compressed) **LS Verbal**

Motivate

Demonstration — GENERAL

Play Ball Compare the relative sizes of all of the planets in the solar system by using a marble for Mercury, two tennis balls for Earth and Venus, a ping pong ball for Mars, a basketball for Jupiter, a soccer ball for Saturn, and two softballs for Uranus and Neptune. Use a beach ball for the sun. Tell students that sizes are only approximate and not to scale. Ask students which planets are about the same size. (Earth and Venus, Uranus and Neptune) Ask which is the largest and which is the smallest. (largest, Jupiter; smallest, Mercury) **LS Visual/Logical**

CHAPTER RESOURCES

Chapter Resource File

- Directed Reading BASIC

Technology

Transparencies
- Bellringer

Student Edition on CD-ROM

One-Stop Planner CD-ROM
- Lesson Plan

Teach

METEOROLOGY CONNECTION GENERAL

Lightning Discharge Model During storms on Earth, electric charges build up in the clouds. Lightning flashes between areas of the clouds that have opposite charges. A similar process occurs on Jupiter. To model this effect, use an aluminum pie pan, two large rubber bands, a plastic tumbler, wool fabric, and a foam dinner plate, and follow these instructions:

1. Make a handle by putting a tumbler upside-down in the center of the pie pan. Stretch two rubber bands over the pan and the tumbler to form an "X" to hold the tumbler firmly in place.
2. Place the foam plate on a table and rub it with the fabric vigorously for 2 to 3 minutes.
3. Use the handle to place the aluminum pan onto the foam plate. Lift the pan and plate by the handle. Dim the lights. Slowly touch your finger to the edge of the pie pan.

You may need a few tries to build up the appropriate charge. Ask students to describe what happened. (Sample answers: A spark jumped from the pie pan to the finger. Some students may hear a crackling sound accompanying the spark.) LS **Kinesthetic**

Answer to Reading Check

When Jupiter formed, it did not have enough mass for nuclear fusion to begin.

SCILINKS. NSTA Developed and maintained by the National Science Teachers Association

For a variety of links related to this subject, go to www.scilinks.org

Topic: Outer Planets
SciLinks code: HQ61091
Topic: Galileo
SciLinks code: HQ60633

Jupiter

Jupiter, shown in **Figure 2**, is the fifth planet from the sun and is by far the largest planet in the solar system. Its mass is more than 300 times that of Earth and is twice that of all the other planets combined. Jupiter's orbital period is almost 12 years. Jupiter rotates on its axis faster than any other planet rotates—once every 9 h and 50 min. Jupiter has at least 60 moons, 4 of which are the size of small planets. It also has several thin rings that are made up of millions of particles.

Jupiter's Atmosphere

Hydrogen and helium make up 92% of Jupiter, so Jupiter's composition is much like the sun's. However, when Jupiter formed about 4.6 billion years ago, it did not have enough mass to allow nuclear fusion to begin. So, Jupiter never became a star.

The alternating light and dark bands on its surface make Jupiter unique in our solar system. Orange, gray, blue, and white bands spread out parallel to the equator. The colors suggest the presence of organic molecules mixed with ammonia, methane, and water vapor. Jupiter's rapid rotation causes these gases to swirl around the planet and form the bands. The average temperature of Jupiter's outer atmospheric layers is –160°C. Jupiter also has lightning storms and thunderstorms that are much larger than those on Earth.

Reading Check **Why didn't Jupiter become a star? (See the Appendix for answers to Reading Checks.)**

Figure 2 ▶ Jupiter is easily identified by its large size and alternating light and dark bands. One of Jupiter's larger moons can be seen in front of the planet.

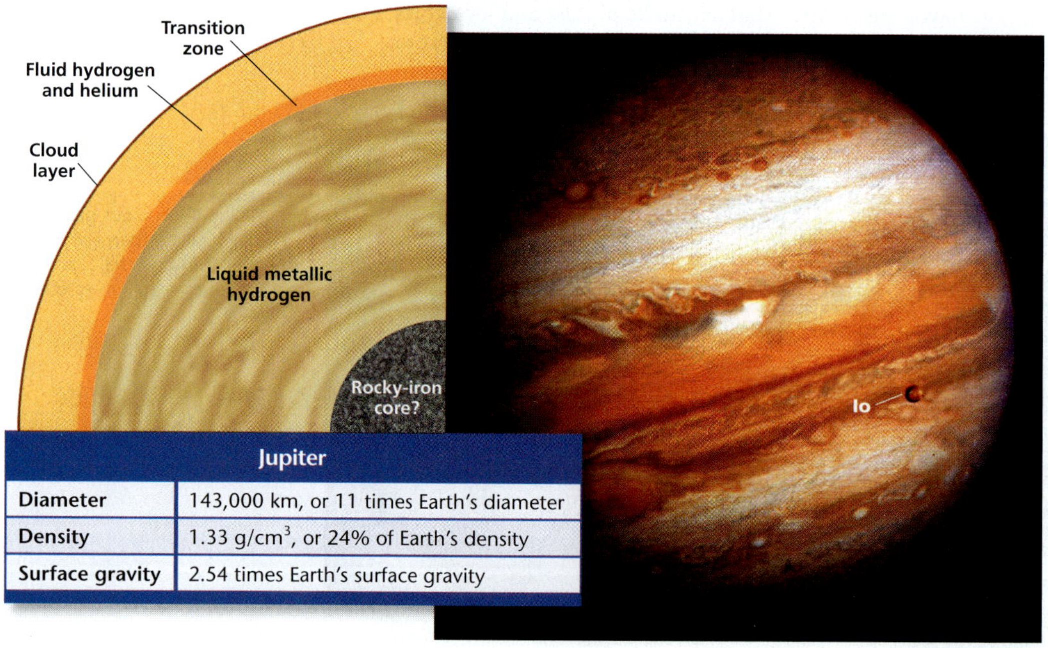

Jupiter	
Diameter	143,000 km, or 11 times Earth's diameter
Density	1.33 g/cm^3, or 24% of Earth's density
Surface gravity	2.54 times Earth's surface gravity

BRAIN FOOD

Asteroid Belt The asteroid belt is a ring of space debris that separates the inner planets from the gas giants. Because there are five to ten distinct asteroid families and mineral types for meteorites, astronomers speculate that a handful of large asteroids may have been shattered by impacts to form the belts we see today.

Using the Figure GENERAL

Discussion Direct students' attention to the diagram that shows Jupiter's internal structure. Ask the class to describe Jupiter's internal structure, and have a student write key points on the board. (Sample answer: The outer cloud layer is composed of gases including ammonia, methane, and water vapor; the next deeper layer is a layer of fluid hydrogen and helium; the next deeper is a layer of liquid metallic hydrogen; the center is a rocky-iron core.) LS **Visual**

Weather and Storms on Jupiter

Figure 3 ▶ Jupiter's Great Red Spot is an ongoing, massive, hurricane-like storm that is about twice the diameter of Earth.

Jupiter's most distinctive feature is its *Great Red Spot,* shown in **Figure 3.** The Great Red Spot is a giant rotating storm, similar to a hurricane on Earth, that has been raging for at least several hundred years. Several other oval spots, or storms, can be seen on Jupiter, although they are usually white. Sometimes, the smaller storms are swallowed up by the larger ones. While storms are common on Jupiter's surface, only a few of the largest storms persist for a long time.

A probe dropped by the *Galileo* spacecraft measured wind speeds on Jupiter of up to 540 km/h. Because winds are caused by temperature differences, scientists have concluded that Jupiter's internal heat affects the planet's weather more than heat from the sun does. From Earth, even by using a small telescope, you can see bands of clouds on Jupiter. These bands, which vary depending on latitude, show regions of different wind speeds.

Jupiter's Interior

Jupiter's large mass causes the temperature and pressure in Jupiter's interior to be much greater than they are inside Earth. The intense pressure and temperatures as high as 30,000°C have changed Jupiter's interior into a sea of liquid, metallic hydrogen. Electric currents in this hot liquid may be the source of Jupiter's enormous magnetic field. Scientists think that Jupiter has a solid, rocky, iron core at its center.

Connection to TECHNOLOGY

Galileo Probes Jupiter

In 1995, the spacecraft *Galileo* arrived at Jupiter and began monitoring an atmospheric probe that Galileo had launched five months earlier. The 336 kg probe transmitted data about the composition and meteorology of Jupiter's atmosphere.

The data surprised mission scientists in many ways. The wind speeds that were measured were much higher than the expected wind speeds. But the most surprising discovery was that although the levels of carbon and nitrogen that were measured were consistent with what was anticipated, only one-fifth of the expected amount of water was found.

The *Galileo* spacecraft studied Jupiter and its moons through 2003. It was then crashed into Jupiter to avoid a collision with one of Jupiter's moons, Europa. No spacecraft is now in orbit around Jupiter. NASA is considering launching a mission to study Europa. *Galileo* discovered evidence of a possible subsurface ocean on Europa. So, scientists are excited about the possibility of life on this moon of Jupiter.

Demonstration — GENERAL

Gas Storms Explain to students that Jupiter's storm systems are similar to hurricanes on Earth. They are produced by vortex-like winds that spin around an area of relative calm. You will need a large glass jar, such as a pasta-sauce jar or a mayonnaise jar. Remove the label by soaking the jar in water. Fill the jar about three-quarters full with water. Add 2 tsp (10 mL) of white vinegar, 1 tsp (5 mL) of dish liquid, and some glitter or food coloring. Screw the cap firmly onto the jar, and hold the jar securely around the middle. Begin to move the jar by using a circular motion to get the contents swirling. Hold the jar so students can observe the motion of the liquids from all sides. Ask students to explain how the pattern that the different liquids make relates to storms on Jupiter. (Sample answers: The swirling water is similar to the violent winds caused by streams of gases, which are represented by the different liquids and the glitter, that move at different speeds and form colored bands.) **LS Visual**

Discussion — GENERAL

What If...? Ask the class to speculate how the solar system might be different if Jupiter had become a star with a larger mass than it now has. (Sample answers: In a binary star system, with two stars that orbit each other, planetary orbits might be unstable; seasonal variations and patterns of night and day would be more complex; atmospheric composition and weather patterns would be more complex.) **LS Verbal/Logical**

BRAIN FOOD

Extreme Magnetism, Brilliant Auroras
Jupiter's magnetic field may be nearly 20,000 times stronger than Earth's. Scientists think the strong magnetic field is created by the flow of electricity through the planet's layer of liquid hydrogen. Moving electric charges are the source of all magnetism. One effect of this strong magnetism is the occurrence of auroras, or brilliant curtains of light, in Jupiter's upper atmosphere. While Earth's auroras are caused by interactions between Earth's magnetosphere and particles from solar wind, at least some of Jupiter's auroras appear to be caused by particles that are spewed out of volcanoes on Io, one of Jupiter's moons. These charged particles become trapped by Jupiter's gigantic magnetic field and produce a current that flows along the planet's magnetic field lines to its north and south magnetic poles.

Teach, *continued*

Graphic Organizer — GENERAL

Comparison Table
You may want to use this Graphic Organizer in a game to review material before the test. Divide the class into two teams. Ask students questions about material from the Comparison Table. Give points to each team that provides correct answers.

Discussion

Saturn's Rings Ask students what feature Saturn is especially noted for. (the extensive ring system around its equator) Though they look solid, Saturn's rings are made up of chunks of ice, rock, and dust particles. Ask students where most of the material that makes up this ring system came from. (debris from comets or other rocky bodies) LS **Verbal**

Answer to Reading Check

Saturn and Jupiter are made almost entirely of hydrogen and helium and have rocky-iron cores, ring systems, many satellites, rapid rotational periods, and bands of colored clouds.

Graphic Organizer — Comparison Table

Create the **Graphic Organizer** entitled "Comparison Table" described in the Skills Handbook section of the Appendix. Label the columns with "Jupiter," "Saturn," Uranus," and "Neptune." Label the rows with "Diameter," "Density," "Orbital period," and "Composition." Then, fill in the table with details about the gas giants.

Saturn

Saturn, shown in **Figure 4**, is the sixth planet from the sun and has an orbital period of 29.5 years. Because it is so far from the sun, Saturn is very cold and has an average cloud-top temperature of –176°C. Saturn has at least 47 moons, and additional small moons continue to be discovered. Its largest moon, Titan, which has a diameter of 5,150 km, is half the size of Earth.

Saturn, like Jupiter, is made almost entirely of hydrogen and helium and has a rocky, iron core at its center. However, Saturn is much less dense than Jupiter. In fact, Saturn is the least dense planet in the solar system.

Saturn's Bands and Rings

Saturn is known for its rings, which are 2 times the planet's diameter. While the other gas giants also have rings, Saturn has the most complex and extensive system of rings. The rings are made of billions of dust and ice particles. Most of the ring debris probably came from comets or other bodies.

Like Jupiter, Saturn also has bands of colored clouds that run parallel to its equator. These bands are caused by Saturn's rapid rotation. Saturn rotates on its axis every 10 h and 30 min. This rapid rotation, paired with Saturn's low density, causes Saturn to bulge at its equator and to flatten at its poles.

Scientists are learning more about Saturn and its moons from NASA's *Cassini* spacecraft, which reached Saturn on July1, 2004. Cassini also carried the European Space Agency's *Huygens* probe, which landed on Titan, Saturn's largest moon.

Reading Check How is Saturn similar to Jupiter? (See the Appendix for answers to Reading Checks.)

Figure 4 ▶ This composite image of Saturn taken from the *Cassini* spacecraft clearly shows the planet's rings.

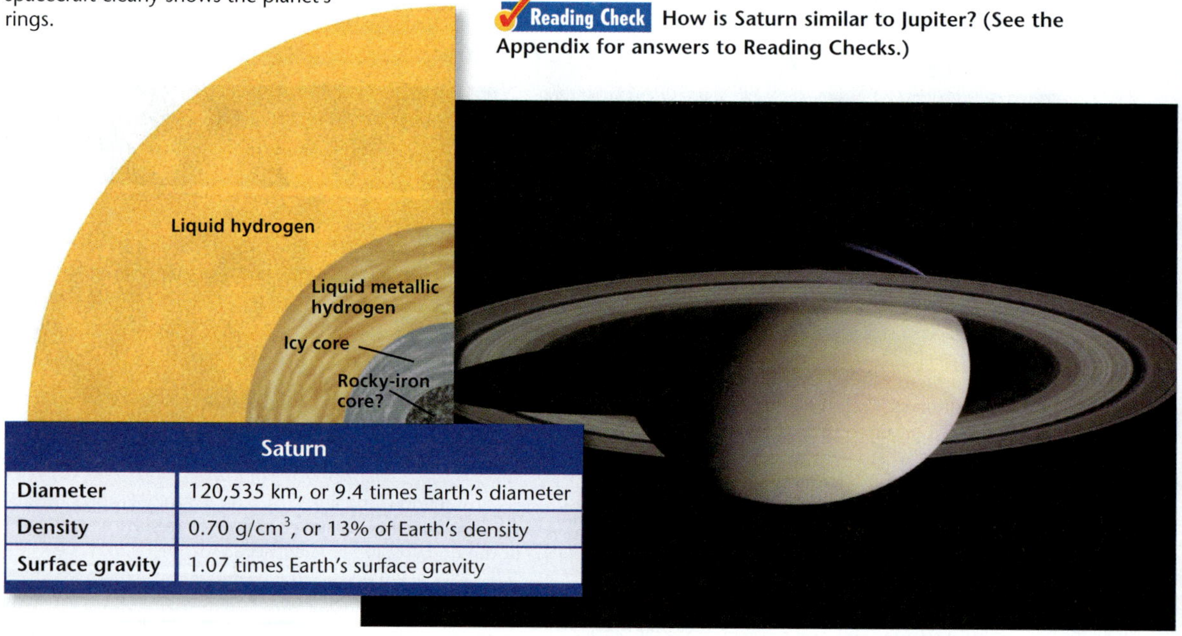

Saturn	
Diameter	120,535 km, or 9.4 times Earth's diameter
Density	0.70 g/cm^3, or 13% of Earth's density
Surface gravity	1.07 times Earth's surface gravity

INCLUSION *Strategies*

- *Behavior Control Issues*
- *Hearing Impaired*
- *Developmentally Delayed*

Help students clarify the similarities and differences between the planets by creating a chart on the board and filling it in as a class. Use these headings for the chart: Number; Order from Sun; Rock or Gas?; Large or Small?; Length of One Day; Length of One Year; Hot or Cold?; Average Daytime Temperature; and Average Nighttime Temperature. LS **Logical**

TECHNOLOGY CONNECTION — ADVANCED

***Cassini* Mission** The *Cassini* mission to Saturn is a joint project of NASA, the European Space Agency (ESA) and the Italian space agency. The *Cassini* orbiter conducts in-depth studies of the planet, its moons, its rings, and its magnetic environment. Have interested students research how the *Huygens* scientific probe and other instruments on board perform their jobs and what information scientists want to collect. Students may present their findings in an oral or written report. LS **Verbal**

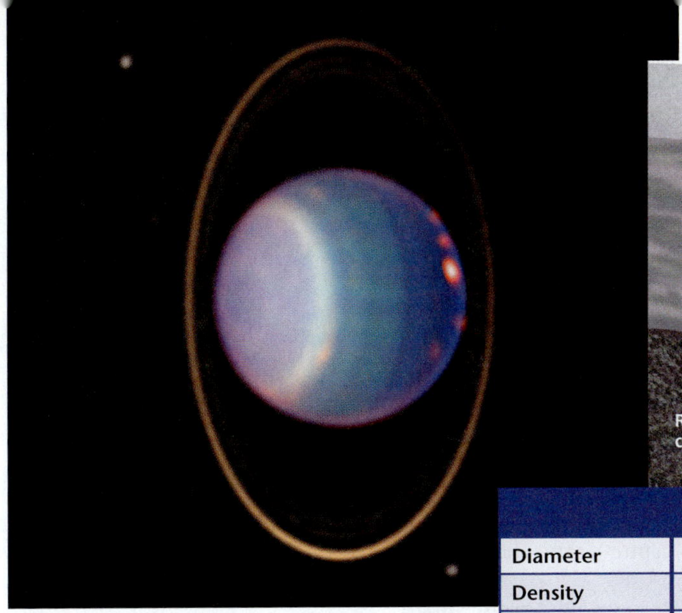

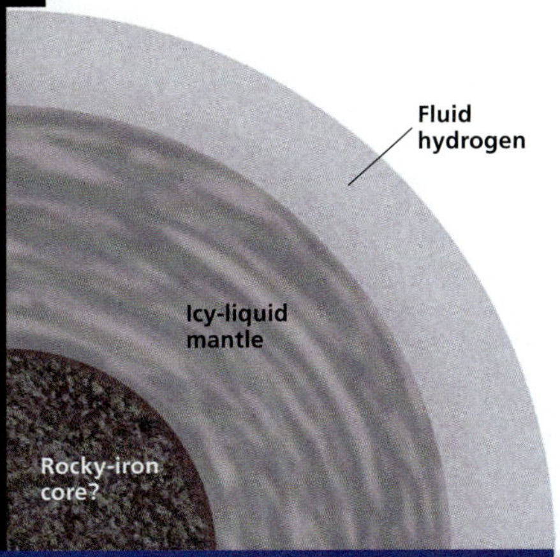

Uranus	
Diameter	51,120 km, or 4 times Earth's diameter
Density	1.30 g/cm^3, or 24% of Earth's density
Surface gravity	91% of Earth's surface gravity

Figure 5 ▶ This exaggerated-color image from the *Hubble Space Telescope* shows Uranus, two of its moons, and some of its rings.

Uranus

Uranus, shown in **Figure 5,** is the seventh planet from the sun and the third-largest planet in the solar system. Sir William Herschel discovered Uranus in 1781. Because Uranus is nearly 3 billion kilometers from the sun, Uranus is a difficult planet to study. But the *Hubble Space Telescope* has taken images that show changes in Uranus's atmosphere. Uranus has at least 24 moons and at least 11 thin rings. Its orbital period is almost 84 years.

Uranus's Rotation

The most distinctive feature of Uranus is its unusual orientation. Most planets, including Earth, rotate with their axes perpendicular to their orbital planes as they revolve around the sun. However, Uranus's axis is almost parallel to the plane of its orbit. The rotation rate of Uranus was not discovered until 1986, when *Voyager 2* passed by Uranus. Astronomers were then able to determine that Uranus rotated once about every 17 h.

Uranus's Atmosphere

Like the other gas giants, Uranus has an atmosphere that contains mainly hydrogen and helium. The blue-green color of Uranus indicates that the atmosphere also contains significant amounts of methane. The average cloud-top temperature of Uranus is –214°C. However, astronomers believe that the planet's temperature is much higher below the clouds. There may be a mixture of liquid water and methane beneath the atmosphere. Scientists also think that the center of Uranus, which has a temperature of about 7,000°C, is a core of rock and melted elements.

Demonstration — GENERAL

Radically Tilted Planet Direct students' attention to the image of Uranus. Ask students to describe the planet's most unusual feature. (The planet is tilted so that it appears to lie on its side.) Tell students that, even though Uranus' axis is almost parallel to the plane of its orbit, its poles do not always point toward the sun. Demonstrate by putting an axis through a sphere of some kind, tilt it as Uranus is tilted, and move the planet around a "sun." Tell students that, because of the tilt, each pole is totally dark for about 21 years out of an orbit of 84 years. Invite students to speculate as to what may have caused this axial tilt. (Perhaps during the period when much of the solar system was bombarded with debris, a large planet-sized object collided with Uranus, knocking the planet on its side.) **LS Visual**

Discussion — BASIC

Composition and Structure Explain that in addition to helium and hydrogen, Uranus' atmosphere apparently also contains organic compounds such as hydrocarbons. Ask students what gives Uranus a blue-green color. (Methane in the atmosphere. The compound absorbs red wavelengths of light, giving the planet a bluish tint.) Invite students to describe what scientists think about the planet's structure. (The planet consists of a cloud layer of fluid hydrogen over a mantle of liquid water, methane, and ammonia, which surrounds a core of rock and metals.) **LS Verbal**

HISTORY CONNECTION — ADVANCED

Caroline Herschel Only six planets, including Earth, were known until the eighteenth century. In 1781, amateur astronomer William Herschel made the next discovery: Uranus. His sister Caroline assisted her brother in his hobby of making telescopes and shared his passion for astronomy. Caroline also discovered eight comets, published several astronomical catalogues, and received a gold medal from the Royal Astronomical Society for her research—in an era when women stayed mostly in the home. Invite interested students to learn more about the achievements of this remarkable astronomer, and have them make an oral or written report on their findings. **LS Verbal**

Teach, *continued*

Discussion — GENERAL

Neptune's Weather Point out that Neptune is a very dynamic planet that has giant, hurricane-like storms similar to Jupiter's Great Red Spot. Ask students to name the weather system that *Voyager 2* identified. (the Great Dark spot) Explain that the Great Dark Spot had disappeared when the *Hubble Space Telescope* was directed at Neptune, but another appeared in the planet's northern hemisphere. Ask students which planet in the solar system is the windiest and what wind speeds have been recorded. (Neptune; 2,000 km/h) Ask students what causes winds on Earth and on Neptune. (Differences in atmospheric temperatures cause differences in pressure. The atmospheric gases move from areas of high pressure to areas of lower pressure.) LS **Verbal**

Group Activity — ADVANCED

To Be or Not to Be a Planet
Many small objects have been discovered in the Kuiper Belt beyond Neptune's orbit. These objects have orbits and other properties that are similar to those of Pluto. In 2006, the International Astronomical Union (IAU) categorized Pluto and another Kuiper-Belt object, named Eris, as dwarf planets. Have interested students research the various IAU resolutions leading up to this reclassification of Pluto, the protests by some astronomers that followed, and the reaction of the general public. Have students prepare short presentations of their findings for the class. LS **Verbal/Logical** Co-op Learning

Neptune

Neptune, shown in **Figure 6,** is the eighth planet from the sun and is similar to Uranus in size and mass. Neptune's orbital period is nearly 164 years, and the planet rotates about every 16 h. Neptune has at least 13 moons and six rings.

The Discovery of Neptune

Neptune's existence was predicted before Neptune was actually discovered. After Uranus was discovered, astronomers noted variations from its calculated orbit. They suspected that the gravity of an unknown planet was responsible for the variation. In the mid-1800s, John Couch Adams, an English mathematician, and Urbain Leverrier, a French astronomer, independently calculated the position of the unknown planet. A German astronomer, Johann Galle, discovered a bluish-green disk where Leverrier had predicted the planet would be. Astronomers named the planet Neptune after the Roman god of the sea.

Neptune's Atmosphere

Data from the *Voyager 2* spacecraft indicate that Neptune's atmosphere is made up mostly of hydrogen, helium, and methane. Neptune's upper atmosphere contains some white clouds of frozen methane. These clouds appear as continually changing bands between the equator and the poles of Neptune.

Images taken by *Voyager 2* and the *Hubble Space Telescope* indicate that Neptune has an active weather system. Neptune has the solar system's strongest winds, which exceed 1,000 km/h. A storm that is the size of Earth and that is known as the Great Dark Spot appeared and disappeared on Neptune's surface. Neptune's average cloud-top temperature is about –225°C.

Figure 6 ▶ This Great Dark Spot on Neptune was a giant storm that was similar to the Great Red Spot on Jupiter.

Neptune	
Diameter	49,530 km, or 3.9 times Earth's diameter
Density	1.64 g/cm^3, or 30% of Earth's density
Surface gravity	1.2 times Earth's surface gravity

MATH CONNECTION

Discovery of Neptune and Pluto The discoveries of Neptune and Pluto were the result of mathematical analysis of irregularities in the motions of Uranus and Neptune, based upon Newton's law of gravitation. In the case of Uranus, English mathematician John Couch Adams and French astronomer Urbain LeVerrier independently calculated not only the orbit, but also the predicted mass, of an unknown object whose gravitational pull was affecting Uranus's orbit. These calculations were the first time Newton's theory of gravitation was used to predict orbital position by observing the effects of a planet's gravity. Later, astronomers searching for an unknown planet that disturbed Uranus's orbit observed Neptune. When Uranus still did not exactly follow its predicted orbit, astronomers began to look for a planet beyond Neptune to explain the data. They found Pluto after a careful search, although it had been predicted erroneously.

Figure 7 ▶ Pluto, the prototype for dwarf planets in our solar system, is even smaller than Earth's moon. Quaoar and Sedna, as shown by these artist renditions, have a smaller diameter than Pluto.

Objects Beyond Neptune

Pluto was discovered in 1930. Until 2006, it was known as the ninth planet. Now, it is defined as a dwarf planet. Pluto orbits the sun in an unusual elongated and tilted ellipse. Pluto is usually far beyond Neptune but is closer to the sun than Neptune is for about 20 years out of its 248-year orbital period. With a diameter of 2,302 km, Pluto is smaller than Earth's moon. Astronomers think that Pluto is made up of frozen methane, rock, and ice and has an average temperature of –235°C. Infrared images show that Pluto has extensive methane icecaps and a very thin nitrogen atmosphere.

Kuiper-Belt Objects

In recent years, scientists have discovered hundreds of objects in our solar system beyond Neptune's orbit. Some of these objects are similar to Pluto in size and composition, but most are simply small chunks of ice. This region beyond Neptune's orbit is called the **Kuiper Belt** (KIE puhr BELT). Quaoar, shown in **Figure 7,** is more than half of Pluto's size. Another object, called Eris, is larger than Pluto. Like Pluto, Eris is considered to be a dwarf planet. A *dwarf planet* is any object that orbits the sun, is round because of its own gravity, but has not cleared its orbital path.

One of the most distant objects in the solar system was found beyond the Kuiper Belt in March 2004. This object, named Sedna, is about three-fourths the size of Pluto. It is currently more than twice as far from Earth as Pluto is.

Kuiper Belt a region of the solar system that starts just beyond the orbit of Neptune and that contains dwarf planets and other small bodies made mostly of ice

Reading Check Where is the Kuiper Belt located? (See the Appendix for answers to Reading Checks.)

MISCONCEPTION ALERT

What Happened to Pluto? In the summer of 2006, astronomers from around the world met in Prague (Czech Republic) for the 26th General Assembly of the International Astronomical Union (IAU). At that meeting, the IAU voted on a new definition of *planet,* which includes the first eight planets, but excludes Pluto. Pluto was reclassified as a *dwarf planet.* While students may be familiar with nine planets, which had been the number of planets since soon after the discovery of Pluto in 1930, the decision made by the IAU in 2006 has affected the presentation in this textbook. Pluto, another trans-Neptunian object called Eris, and the asteroid Ceres are the first objects to be named dwarf planets.

Close

Reteaching — BASIC

Concept Webs Organize students into pairs. Have each pair create a concept web about the gas giants and identify the similarities and differences between the planets. LS **Visual**

Quiz — GENERAL

1. Explain how Pluto is similar to the terrestrial planets. (It is smaller and denser than the outer planets, and it is composed of ice and rock.)
2. What do scientists expect to learn by studying exoplanets? (how solar systems form; the general arrangement of solar systems; and possibility of life in other solar systems)

Answer to Reading Check

The Kuiper Belt is located beyond the orbit of Neptune.

CHAPTER RESOURCES

Chapter Resource File

- **Making Models Lab** It's a Long Way to Pluto GENERAL

Internet Activity

- Gravitational Microlensing ADVANCED

Internet Activity — ADVANCED

Gravitational Microlensing In 2004, astronomers used a technique first proposed by Albert Einstein to identify exoplanets that orbit distant stars. Gravitational microlensing uses a star or exoplanet as a cosmic lens to magnify and brighten a more distant star lined up behind it. Scientists may now be able to find small exoplanets in other solar systems. Have interested students report on this technique. A worksheet designed to direct student research on this topic can be found in the **Chapter Resource File** booklet or by visiting **go.hrw.com** and entering the keyword **HQ6PSSX.** LS **Verbal**

Close, *continued*

Alternative Assessment — GENERAL

Outer Planets Museum Create a mini-space museum around the topic of the outer planets. Have students design displays that feature each of the gas giants, Pluto and the other dwarf planets, the asteroid belt, and the Kuiper Belt. Exhibits should be based on facts about each object and should look as realistic as possible and include explanatory captions. LS **Visual/Kinesthetic**

Answers to Section Review

1. Jupiter is extremely large and has numerous satellites and huge electrical fields. It consists mainly of helium and hydrogen gases; it has very high internal temperatures.
2. All four are extremely large and have massive atmospheres, mainly composed of gases. They are far from the sun and have huge gravity fields, many satellites, and ring systems.
3. Jupiter's Great Red Spot is a giant rotating storm similar to a hurricane on Earth. However, the Great Red Spot is twice the diameter of Earth and has wind speeds of up to 540 km/h.
4. Most planets rotate with their axis nearly perpendicular to the orbital plane. But Uranus's axis is tilted so that its axis is nearly parallel to its orbital plane.
5. Pluto is a dwarf planet made of water and methane ice and rock. The outer planets are very large and are made mostly of hydrogen and helium gases.
6. Located just beyond Neptune's orbit, these small bodies are composed mainly of ice; some are similar in mass to Pluto.
7. Most were detected because of gravitational effects on the stars they orbit, or shifts produced in light coming from those stars; most orbit stars similar to Earth's sun and are larger than Uranus.
8. The compositions of the sun and of the outer planets are similar because all are made mostly of hydrogen and helium.
9. Pluto's elongated orbit takes it into the Kuiper Belt, which Pluto shares with thousands of other objects. Planets do not share their orbits with other objects.
10. Answers may vary. Accept all reasonable answers.
11. The *outer planets* consist of the *gas giants*, which include *Jupiter*, which has a *Great Red Spot*; *Saturn*; *Uranus*; and *Neptune*. *Pluto* is a dwarf planet located in the *Kuiper belt*.

Figure 8 ▶ This illustration shows an artist's idea of a Jupiter-sized exoplanet in the foreground that was recently discovered. This exoplanet orbits a sun-like star called HD209458, which is located 150 light years from Earth.

Exoplanets

Until the 1990s, all of the planets that astronomers had discovered were in Earth's solar system. Since then, however, more than 200 planetlike objects have been attributed to stars other than Earth's sun. Because these objects circle stars other than Earth's sun, they are called *exoplanets*. The prefix *exo-* means "outside." **Figure 8** shows an artist's rendition of an exoplanet. Most known exoplanets orbit stars that are similar to Earth's sun. Therefore, the existence of exoplanets leads some scientists to wonder if life could exist in another solar system.

Exoplanets cannot be directly observed with telescopes or satellites. Most exoplanets can be detected only because their gravity tugs on stars that they orbit. When scientists study some distant stars, they notice that the light coming from the stars shifts in wavelength. This shifting could be explained by the stars' movement slightly toward and then away from Earth. Scientists know that the gravity of an object that cannot be seen can affect a star's movement. In these cases, that object is most likely an exoplanet that orbits the star.

Most of the exoplanets that have been identified are larger than Uranus, but recently two exoplanets have been discovered that are closer to Earth in mass. Many of these exoplanets, though more massive than Jupiter, are closer to their stars than Mercury is to Earth's sun. From studying these many solar systems, scientists hope to learn more about the formation and basic arrangement of solar systems.

Section 4 Review

1. **Explain** what makes Jupiter similar to the sun.
2. **Compare** the characteristics of Jupiter, Saturn, Uranus, and Neptune.
3. **Compare** Jupiter's Great Red Spot with weather on Earth.
4. **Describe** the way in which the tilt of the axis of Uranus's rotation is unusual.
5. **Explain** how Pluto differs from the outer planets of the solar system.
6. **Summarize** the features of objects in the Kuiper Belt.
7. **Describe** what scientists know about exoplanets, planetlike objects outside the solar system.

CRITICAL THINKING

8. **Making Comparisons** How are the compositions of the gas giants similar to the composition of the sun?
9. **Making Inferences** Why is Pluto not considered a planet even though Neptune is sometimes farther from the sun than Pluto is?
10. **Evaluating Conclusions** Should scientists reconsider Pluto to be a planet? Explain your answer.

CONCEPT MAPPING

11. Use the following terms to create a concept map: *outer planet, Jupiter, Saturn, Uranus, Neptune, Pluto, gas giant, Kuiper belt,* and *Great Red Spot.*

CHAPTER RESOURCES

Chapter Resource File
- Section Quiz GENERAL

Workbooks
- Study Guide (also in Spanish)

Chapter 27 Highlights

Sections

1 Formation of the Solar System

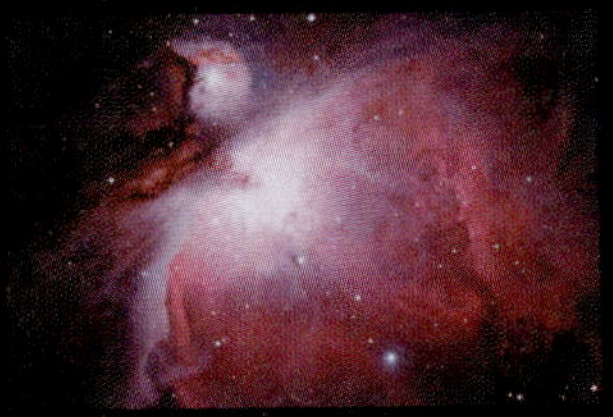

Key Terms

- solar system, 685
- planet, 685
- solar nebula, 685
- planetesimal, 686

Key Concepts

- The solar system formed from a rotating and contracting region of gas and dust about 5 billion years ago.
- The planets formed from collisions of smaller bodies called *planetesimals*.
- As Earth cooled, differentiation caused three distinct compositional layers—the crust, the mantle, and the core—to form.

2 Models of the Solar System

Key Terms

- eccentricity, 692
- orbital period, 693
- inertia, 694

Key Concepts

- Geocentric models of the solar system, such as those developed by Aristotle and Ptolemy, were replaced by the heliocentric model proposed by Copernicus.
- Kepler's three laws describe the motion of the planets in their orbits around the sun.
- The planets travel in elliptical orbits around the sun. Planets nearer to the sun travel faster than those farther from the sun do.

3 The Inner Planets

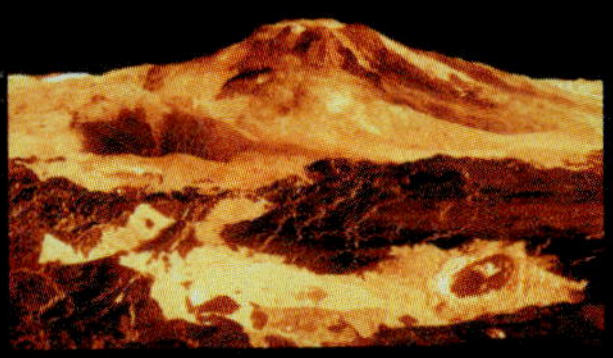

Key Terms

- terrestrial planet, 695

Key Concepts

- The four inner planets share similar characteristics and are called the *terrestrial planets*. Earth, Venus, and Mars have a history of geologic activity.
- The terrestrial planets are denser and smaller than the gas giants.

4 The Outer Planets

Key Terms

- gas giant, 701
- Kuiper Belt, 707

Key Concepts

- The outer planets consist of the four *gas giants*—Jupiter, Saturn, Uranus, and Neptune. Pluto is a dwarf planet usually located in the Kuiper Belt.
- The Kuiper Belt is a region of the solar system that is beyond Neptune's orbit and that contains small bodies made mostly of ice.
- Exoplanets orbit stars other than the sun.

BRAIN FOOD

What's in a Name? At age fifteen, the Danish astronomer Tyge Brahe adopted the Latinized version of his name—Tycho. As is the case with Galileo, Tycho is commonly referred to by his first name only today. Tycho reported his observations of a *supernova*, or explosion of a large star, in 1572. He also compiled the most precise planetary observations of his time by devising the best instruments available before the telescope. Kepler used Tycho's planetary data to develop his Laws of Orbital Motion.

Chapter Highlights

Alternative Assessment — GENERAL

Solar System Highlights Have students work in small groups to design fantasy Space Explorer Guides for the solar system. Provide sample travel brochures that students can use as models. Students can either pick a single planet, other than Earth, as the destination for their tour, or plan a highlights tour of the entire system. Students might include descriptions of interesting places such as canyons, volcanoes, polar icecaps, or craters, and side trips to nearby moons or planetary rings. Students should explain how travelers will reach their destinations, what protective clothing they should wear, and how to deal with dangers such as storms, extreme temperatures, or unusual geologic activity. Students may include drawings or photos from the Internet to illustrate their guides. They should try to use language that will make the trip seem inviting and choose lettering and design elements that will help sell their tour package. LS **Visual/Verbal**

CHAPTER RESOURCES

Chapter Resource File

- **Concept Review** GENERAL
- **Critical Thinking** ADVANCED
- **Math Skills** GENERAL
- **Graphing Skills** GENERAL
- **Chapter Test A** GENERAL
- **Chapter Test B** ADVANCED

Workbooks

- **Study Guide** (also in Spanish)
- **Assessments** (Spanish)

Technology

Classroom Videos
- Brain Food Video Quiz

HRW Earth Science Video
- Segment 18: The Solar System

Chapter Review

Assignment Guide

Section	Questions
1	5, 7, 16–19, 23–25, 28
2	1, 8–9, 14, 20–22, 27, 30–31
3	2, 12, 30, 32–33
4	3, 10–11, 13, 15, 26, 34–37
2 and 4	4
3 and 4	6
1–4	29

Using Key Terms

1–7. Answers may vary but should show that students understand the definitions of and differences between the key terms.

Understanding Key Concepts

8. d **9.** a **10.** a
11. c **12.** c **13.** b
14. c **15.** d **16.** a
17. c **18.** b

Short Answer

19. Earth's early atmosphere contained mainly water vapor, carbon dioxide, nitrogen, methane, sulfur dioxide, and ammonia from outgassing. Today, Earth's atmosphere contains mainly nitrogen, oxygen, and argon, with some water vapor and carbon dioxide.

20. elliptical

21. Kepler's first law states that planets orbit the sun in curved paths called ellipses, whose shape is determined by two points called foci, one of which is located within the sun.

22. Kepler described the elliptical shape of planetary orbits and the planets' speeds at different points along their orbits. Newton gave the causes of these same orbital motions: the curved elliptical orbital motion is the sum of the straight-line motion of a planet that results from inertia and the attractive force of gravity provided by the sun.

23. the presence of liquid water on the surface, a moderate surface temperature, and free oxygen in the atmosphere

24. Dust particles in a solar nebula stick together to form larger clumps of matter called *planetesimals*. Through collisions caused by motion of the particles and the gravity between them, the planetesimals form larger bodies called *protoplanets*. When protoplanets grow very large, they form bodies called *planets*.

25. When the early Earth was molten, denser materials such as iron sank to the center of the planet, and less dense materials were forced to the outer layers through differentiation. This process gave the planet its layered structure of an iron and nickel core surrounded by a thick layer of iron and magnesium-rich rock called the *mantle*, and a less dense, silica-rich crust.

Critical Thinking

26. Answers may vary. Sample answer: Astronomers can hypothesize that the solar system formed in a similar way to how Earth's solar system formed.

27. Sample answer: If you know the distance of a planet from the sun, you can determine the planet's orbital period. This is done by applying Kepler's law of periods. After converting the distance into astronomical units, cube that number (a^3), and then find the square root of a^3 to obtain the orbital period (p), because $a^3 = p^2$.

Chapter 27 Review

Using Key Terms

Use each of the following terms in a separate sentence.

1. *inertia*

2. *terrestrial planet*

3. *Kuiper Belt*

For each pair of terms, explain how the meanings of the terms differ.

4. *Kuiper Belt* and *orbital period*

5. *planet* and *planetesimal*

6. *terrestrial planet* and *gas giant*

7. *solar nebula* and *solar system*

Understanding Key Concepts

8. Copernicus's model of the solar system is
- **a.** geocentric.
- **b.** lunocentric.
- **c.** ethnocentric.
- **d.** heliocentric.

9. Kepler's first law states that each planet orbits the sun in a path called a(n)
- **a.** ellipse.
- **b.** circle.
- **c.** epicycle.
- **d.** period.

10. The most distinctive feature of Jupiter is its
- **a.** Great Red Spot.
- **b.** Great Dark Spot.
- **c.** ring.
- **d.** elongated orbit.

11. The planet that has an axis of rotation that is almost parallel to the plane of its orbit is
- **a.** Venus.
- **b.** Jupiter.
- **c.** Uranus.
- **d.** Neptune.

12. The tilt of the axis of Mars is nearly the same as that of
- **a.** Mercury.
- **b.** Venus.
- **c.** Earth.
- **d.** Jupiter.

13. The planet that rotates faster than any other planet in the solar system is
- **a.** Earth.
- **b.** Jupiter.
- **c.** Uranus.
- **d.** Pluto.

14. Kepler's law that describes how fast planets travel at different points in their orbits is called the law of
- **a.** ellipses.
- **b.** equal speeds.
- **c.** equal areas.
- **d.** periods.

15. All of the outer planets rotate with their axes perpendicular to their orbital planes *except*
- **a.** Saturn.
- **b.** Jupiter.
- **c.** Nepture.
- **d.** Uranus.

16. The first atmosphere of Earth contained a large amount of
- **a.** helium.
- **b.** oxygen.
- **c.** carbon dioxide.
- **d.** methane.

17. The hypothesis that states that the sun and the planets developed out of the same cloud of gas and dust is called the
- **a.** Copernicus hypothesis.
- **b.** solar hypothesis.
- **c.** nebular hypothesis.
- **d.** Galileo hypothesis.

18. In the process of photosynthesis, green plants give off
- **a.** hydrogen.
- **b.** oxygen.
- **c.** carbon dioxide.
- **d.** helium.

Short Answer

19. Explain how Earth's early atmosphere differs from Earth's atmosphere today.

20. What is the shape of the planets' orbits?

21. What is Kepler's first law?

22. How did Newton's ideas about the orbits of the planets differ from Kepler's ideas?

23. List three features of Earth that allow it to sustain life.

24. Describe how a planet might form.

25. How did differentiation help to form solid Earth?

Critical Thinking

26. Applying Ideas Suppose astronomers discover that exoplanets orbiting stars that are similar to Earth's sun have similar compositions to the planets in Earth's solar system. What can the astronomers hypothesize about the formation of those solar systems?

27. Identifying Trends If you know the distance from the sun to a planet, what other information can you determine about the orbit of the planet? Explain your answer.

28. Making Inferences How would the layers of Earth be different if the planet had never been hotter than it is today?

Concept Mapping

29. Use the following terms to create a concept map: *solar system, planet, protoplanet, planetesimal, differentiation, core, mantle, crust, geocentric, heliocentric, Aristotle, Ptolemy, Copernicus, ellipse, Earth, terrestrial planet, outer planet, Jupiter, Saturn, gas giant, Kuiper Belt, solar nebula* and *inner planet*.

Math Skills

30. Making Calculations Mercury has a period of rotation equal to 58.67 Earth days. Mercury's period of revolution is equal to 88 Earth days. How many times does Mercury rotate during one revolution around the sun?

31. Applying Quantities Uranus's orbital period is 84 years. What is its distance from the sun in astronomical units?

32. Making Calculations Venus's orbital period is 225 days. Calculate your age in Venus years.

Writing Skills

33. Creative Writing Imagine that you are the first astronaut to land on Mars. In a short essay, describe what you hope and expect to find.

34. Communicating Main Ideas Create your own definition for *planet*. Then, write an explanation for why Pluto is or is not a planet.

Interpreting Graphics

The graph below shows density in relation to mass for Earth, Uranus, and Neptune. Mass is given in Earth masses. The mass of Earth is equal to 1. Use the graph to answer the questions that follow:

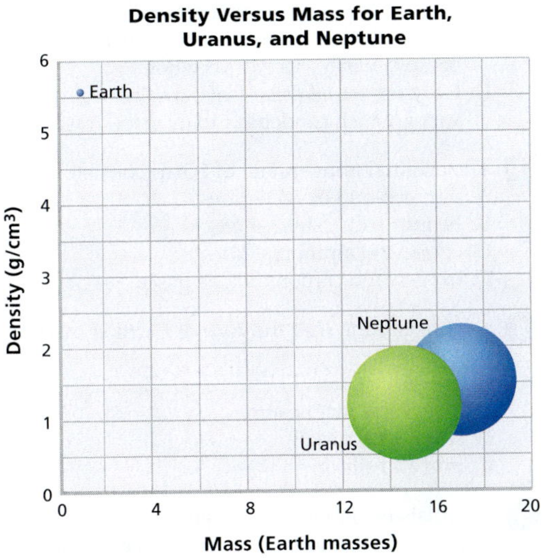

35. Which planet is denser, Uranus or Neptune? Explain your answer.

36. Which planet has the smallest mass?

37. How can Earth be the densest of the three planets even though Uranus and Neptune have so much more mass than Earth does?

28. Answers may vary. Sample answer: If Earth had never been hotter than it is today, the process of differentiation would not have occurred. If Earth had not been in a molten state, substances of different densities would not have separated out into layers. Earth would be more uniform in composition.

Concept Mapping

29. Answers may vary but should include all of the terms listed. Sample answers appear at the end of this Teacher Edition.

Math Skills

30. 88 days ÷ 58.67 days = 1.5 times

31. $p^2 = a^3$; $p^2 = 84 \times 84 = 7056$; so $a^3 = 7056$. The cube root of 7056 = 19.18. The average distance of Uranus from the sun is 19.18 AU.

32. Answers may vary depending on the students' ages. Sample answer: I am exactly 15 years old, so 15 years × 365 days/year = 5475 days ÷ 225 days/Venusian year = 24.3 years. I am 24.3 years old in Venus years.

Writing Skills

33. Answers may vary. Accept all reasonable answers.

34. Answers may vary. Accept all reasonable answers.

Interpreting Graphics

35. Neptune: its density is 1.76 g/cm^3, while Uranus's density is 1.30 g/cm^3.

36. Earth, with a mass of 1

37. Density is mass per unit volume. Both Neptune and Uranus contain more matter than Earth does, so they have more total mass. But that matter occupies a larger volume than the matter of Earth does, so Earth's density is higher.

Standardized Test Prep

Estimated Time

To give students practice under more realistic testing conditions, allow them 30 minutes to answer all of the questions in this practice test.

TEST DOCTOR

Question 1 Answer C is correct. Protoplanets form before planets do, so answer A is incorrect. A solar nebula is the precursor to a star such as our sun, so answer B is incorrect. Gas giants are the outer planets of the solar system, so answer D is incorrect.

Question 11 Full-credit answers should include the following points:

- the ancient Greeks realized that the stars seem to move as a whole
- the Greeks observed that stars move slowly across the sky each night, and each year, in predictible and regular patterns
- the Greeks believed in a geocentric model of the solar system, in which the planets and sun were thought to revolve around Earth

Chapter 27 Standardized Test Prep

Understanding Concepts

Directions (1–5): **For** *each* **question, write on a separate sheet of paper the letter of the correct answer.**

1. Small bodies that join to form protoplanets in the early stages of the development of the solar system are
   - **A.** planets
   - **B.** solar nebulas
   - **C.** plantesimals
   - **D.** gas giants

2. Scientists hypothesize that Earth's first oceans were made of fresh water. How did oceans obtain fresh water?
   - **F.** Water vapor in the early atmosphere cooled and fell to Earth as rain.
   - **G.** Frozen comets that fell to Earth melted as they traveled through the atmosphere.
   - **H.** As soon as icecaps formed, they melted because Earth was still very hot.
   - **I.** Early terrestrial organisms exhaled water vapor, which condensed to form fresh water.

3. The original atmosphere of Earth consisted of
   - **A.** nitrogen and oxygen gases
   - **B.** helium and hydrogen gases
   - **C.** ozone and ammonia gases
   - **D.** oxygen and carbon dioxide gases

4. Scientists think that the core of Earth is made of molten
   - **F.** iron and nickel
   - **G.** nickel and magnesium
   - **H.** silicon and nickel
   - **I.** iron and silicon

5. Scientists estimate that the sun originated as a solar nebula and began to produce its own energy through nuclear fusion approximately how many years ago?
   - **A.** 50 million years
   - **B.** 500 million years
   - **C.** 1 billion years
   - **D.** 5 billion years

Directions (6–7): **For** *each* **question, write a short response.**

6. What four planets make up the group known as the inner planets?

7. The Great Red Spot is found on what planet?

Reading Skills

Directions (8–10): **Read the passage below. Then, answer the questions.**

Movement of the Planets

Imagine that it is the year 200 BCE and that you are an apprentice to a famous Greek astronomer. After many years of observing the sky, the astronomer knows all of the constellations as well as he knows the back of his hand. He shows you how all the stars move together—how the whole sky spins slowly as the night goes on. He also shows you that among the thousands of stars in the sky, some of the brighter ones slowly change their position in relation to the other stars. The astronomer names these stars plantetai, the Greek word that means "wanderers."

Building on the observations of the ancient Greeks, we now know that the planetai are actually planets, not wandering stars. Because of their proximity to Earth and their orbits around the sun, the planets appear to move relative to the stars.

8. According to the passage, which of the following statements is not true?
   - **A.** It is possible to determine planets in the night sky by the way they move relative to the other stars.
   - **B.** The word planetai means "wanderers" in the Greek language.
   - **C.** Some of the earliest astronomers to detect the presence of planets were Roman.
   - **D.** Ancient Greeks were studying astronomy more than 2,200 years ago.

9. What can you infer from the passage about the ancient Greek astronomers?
   - **F.** They were patient and observant.
   - **G.** They knew much more about astronomy than we do today.
   - **H.** They spent all of their time counting the number of stars in the sky.
   - **I.** They invented astronomy and were the first people to observe the skies.

10. What did the Greek astronomers note about the movement of stars and constellations?

Answers

Understanding Concepts

1. C
2. F
3. B
4. F
5. D
6. Mercury, Venus, Earth, and Mars
7. Jupiter

Reading Skills

8. C
9. F
10. Answers may vary. See Test Doctor for a detailed scoring rubric.

Interpreting Graphics

11. D
12. Answers may vary. See Test Doctor for a detailed scoring rubric.
13. G
14. 1.5 rotations

Interpreting Graphics

Directions (11–14): **For** *each* **question below, record the correct answer on a separate sheet of paper.**

The pie graphs below show the percentages of different gases in the atmospheres of three planets. Use these graphs to answer questions 11 and 12.

Atmospheres of Venus, Earth, and Mars

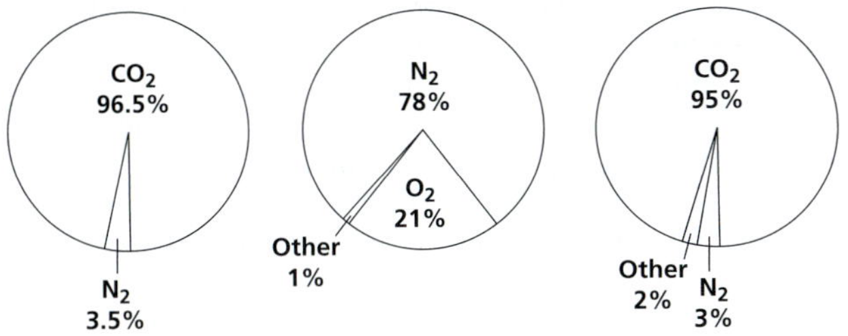

11 What is the percentage of carbon dioxide in the atmosphere of Venus?
A. 3.5%
B. 21%
C. 95%
D. 96.5%

12 Today, Earth's atmosphere includes a large amount of oxygen. Describe how the oxygen in Earth's atmosphere formed, and using this information, predict the likelihood that Mars will someday have oxygen in its atmosphere.

The table below shows the orbital and rotational periods of the planets in the solar system. Use this table to answer questions 13 and 14.

Planets of the Solar System

Planet	Orbital period	Rotational period
Mercury	88 days	59 days
Venus	225 days	243 days
Earth	365.25 days	23 hours 56 minutes
Mars	687 days	24 hours 37 minutes
Jupiter	12 years	9 hours 50 minutes
Saturn	29.5 years	10 hours 30 minutes
Uranus	84 years	17 hours
Neptune	164 years	16 hours

13 Which planet's day length is nearly the same as Earth's?
F. Mercury
G. Mars
H. Saturn
I. Neptune

14 How many rotations does Neptune complete in one Earth day?

Test TIP

Even if you are sure of the answer to a test question, read all of the answer choices before selecting your response.

Standardized Test Prep

TEST DOCTOR

Question 12 Full-credit answers should include the following points:
- the graphs show that Earth's atmosphere consists of approximately 21% oxygen, whereas Mars has no oxygen in its atmosphere
- Earth obtained oxygen from organisms that used liquid water and carbon dioxide in photosynthesis and released oxygen in the process
- without liquid water, it is unlikely that Mars could currently support photosynthetic organisms
- without photosynthetic organisms, no oxygen could be released into Mars's atmosphere
- some students may mention that it is possible that Mars may have had such organisms in the past

Question 14 The correct answer is 1.5 rotations. Neptune completes its rotation in 16 hours. One day on Earth is 24 hours. Students should determine how many times 16 goes into 24 by dividing 24 by 16. A common mistake that some students may make is to divide 16 by 24 and reach an answer of 2/3 rotations.

Test Prep Correlations *National Science Education Standards*

ES 3a: items 2, 3, 5, 7, 12
ES 4b: items 1,
HNS 1a: items 8, 9, 10
HNS 1c: item 10
HNS 2b: item 4
LS 1e: item 12
SAI 2e: item 8, 9, 10
UCP 1: items 6, 11, 13, 14

CHAPTER RESOURCES

State Resources

For specific resources for your state, visit go.hrw.com and type in the keyword **HSHSTR.**

Crater Analysis

Teacher's Notes

Time Required

one 45-minute class period

Lab Ratings

EASY → HARD

Teacher Preparation: 1

Student Setup: 2

Concept Level: 2

Cleanup: 3

Skills Acquired

- Experimenting
- Collecting Data
- Interpreting Results

The Scientific Method

In this lab, students will

- Make Observations
- Analyze Results
- Communicate Results

Materials

The materials listed are enough for groups of 2 to 4 students. Have students spread newspapers over the work area for easier disposal after completing the lab. Direct students to wrap up the used plaster of Paris mixture in newspaper and toss it into the trash. Tell students not to dispose of plaster in sinks because plaster will clog drains and sinks if it hardens in plumbing fixtures.

Tips and Tricks

Tell students that hyper-velocity impacts (those at or above 20 km/s) tend to form round craters, despite the projectile's angle when it approaches, as long as the angle is not much below 5°. That is why most craters on planets and moons appear round rather than elliptical.

The book *Craters* produced by the National Science Teachers Association contains a CD disk that has hundreds of images of craters on planets and moons.

The following list describes the craters and the surrounding areas produced in this lab.

Crater A large, deep crater that has very high walls; A great amount of material has splashed out around the crater and may be far from the center of the crater.

Crater B circular crater that has moderately high walls; Some material has splashed out.

Crater C shallow circular crater that has low walls; Little or no visible material was ejected.

Crater D deep, circular crater that has high walls; Much material has splashed out.

CHAPTER RESOURCES

Chapter Resource File

- Datasheet for Chapter Lab GENERAL
- Lab Notes and Answers

Workbooks

Long-Term Projects

- Positions of Sunrise and Sunset ADVANCED
- Planetary Motions ADVANCED

Chapter 27

Making Models Lab

Crater Analysis

Objectives

- **Create** a model that demonstrates the formation of impact craters.
- USING SCIENTIFIC METHODS **Analyze** how an object's speed and projectile angle affect the impact crater that the object forms on planets and moons.

Materials

- marble, large (1)
- marbles, small (5)
- marker
- meter stick
- plaster of Paris
- protractor
- scissors
- shoe box
- tape, masking
- toothpicks (6)
- tweezers

Safety

All of the inner planets—Mercury, Venus, Earth, and Mars—have many features in common. They are made of mostly solid rock and have metallic cores. They have no rings and have from zero to two moons each. And they have bowl-shaped depressions called *impact craters*. Impact craters are caused by collisions between the planets and rocky objects that travel through space. Most of these collisions took place during the formation of the solar system.

Mercury's entire surface is covered with these craters, while very few craters are still evident on the surface of Earth. Many of the moons of the inner and outer planets are also heavily cratered. In this lab, you will experiment with making craters to discover the effect of speed and projectile angle on the way craters form.

PROCEDURE

1. Place the top of a toothpick in the center of a piece of masking tape that is 6 cm long. Fold the tape in half around the toothpick to form a "small flag" and "flagpole." On the flag, write the letter *A*. Repeat this step for the other toothpicks, and label them with the letters *B* through *F*.

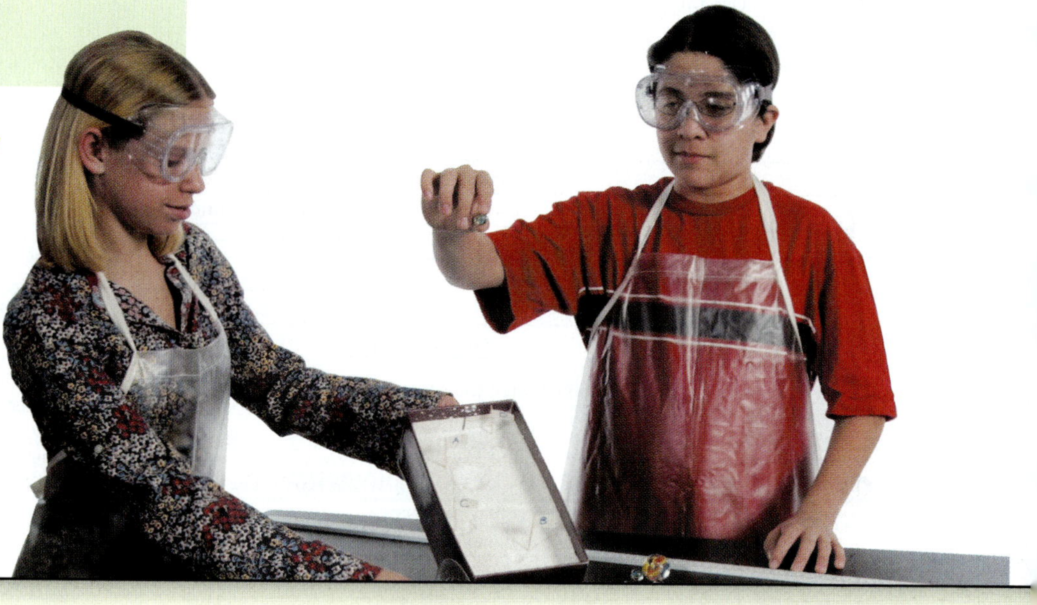

Step 7

2 Mix plaster of Paris with water, according to instructions for making plaster of Paris. Spread your mixture in the bottom of the shoe box. Make your plaster layer about 4 cm thick. The surface should be as smooth as possible.

3 Allow the plaster to dry until it is no longer soupy, but not yet rigid.

4 Drop a large marble onto the plaster from a height of 50 cm above the surface. Quickly remove the marble with tweezers, but do not damage the crater that formed. Place flag A next to the crater to label the crater.

5 Repeat **step 4** by using a small marble dropped from a height of 50 cm and another small marble dropped from a height of 25 cm. Use the flags to label craters *B* (50 cm drop) and *C* (25 cm drop).

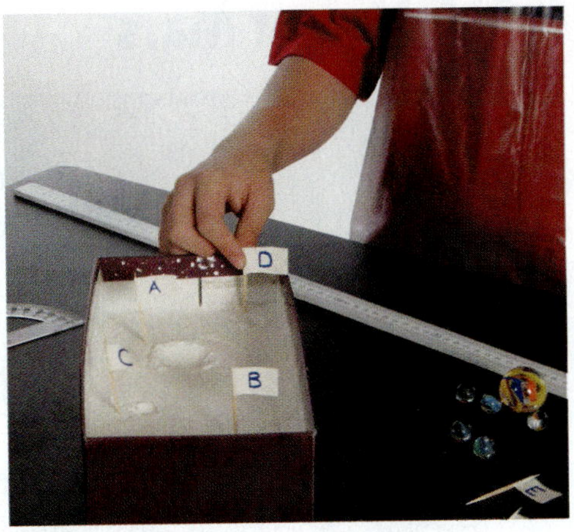

Step 6

6 Repeat **step 4** by using a small marble dropped from a height of 1 m. Label the crater *D*.

7 Using a protractor as a guide, have your partner tilt the box at a 30° angle to the table. Be sure your partner holds the box steady. Then, drop a small marble vertically from a height of 50 cm. Label the crater *E*.

8 Repeat **step 7** by using an angle of 45°. Label the crater *F*.

9 Allow the plaster of Paris to harden. Write a description of each crater and the surrounding area.

ANALYSIS AND CONCLUSION

1 **Examining Data** Which crater was formed by the marble that had the highest velocity? What is the effect of velocity on the characteristics of the crater formed?

2 **Explaining Events** Study the shapes of craters B, E, and F. How did the angle of the plaster of Paris affect the shape of the craters that formed?

3 **Making Comparisons** Compare craters A, B, and D. How do they differ from each other? What caused this difference? Is the difference in the masses of the objects a factor? Explain your answer.

Extension

1 **Applying Conclusions** Find a map of the surface craters on one of the terrestrial planets. Identify craters that were made by different angles of impact. Label the craters on the diagram, and present your findings to the class.

Crater E circular crater of moderate depth; One wall is slightly higher than the other. Material that has splashed out of the crater is located more to one side of the lower wall.

Crater F very similar to Crater E

Answers to Analysis and Conclusion

1. Crater D; a higher velocity impact produces a larger crater that has more material ejected around the crater
2. All three craters are circular. However, the walls of craters E and F are higher on the side of the crater toward which the marble was traveling. No noticeable difference occurred between the craters formed by marbles dropped from different angles.
3. Craters A and D are deeper and have larger diameters. More material was ejected from craters A and D. Also, the material is ejected farther from the craters than the material ejected from crater B was; crater A was formed by a larger object. The difference in mass is a factor because the energy of two objects moving at the same velocity is greater for a more massive object; crater D formed a larger crater due to its greater speed.

Answers to Extension

1. Answers may vary depending on the maps and the terrestrial planet chosen to investigate. Some students may decide to investigate impacts on Earth. In general, the ejected materials would be thrown preferentially downrange of the impacting objects as was observed in the lab

TEACHER TESTED & APPROVED

Mrs. Teresa Tucker
Northwest High School
Jackson, MI

MOLA Map of Mars

Answers to Map Skills Activity

1. Elysium Mons is approximately 22° N latitude,140° E longitude.
2. Answers may vary but should include three of the following for below 0: Utopia Planitia, Hellas Planitia, Argyre Planitia, and Valles Marineris; and three of the following for above 0: Tharsis Monte, Olympus Mons, Alba Patera, and Elysium Mons.
3. the south pole
4. Agyre Planitia
5. Olympus Mons is more likely to be a volcano than Hellas Planitia because Olympus Mons is shaped like a mountain and Hellas Planitia is a crater.
6. The distance between Olympus Mons and Ellysium Mons is approximately 280° (or 80°) of longitude.
7. Answers may vary. Sample answers: *Mons* refers to a large isolated mountain, perhaps of volcanic origin; and *planitia* is a plain located at a lower elevation than the surrounding terrain.

CHAPTER RESOURCES

Technology

Transparencies
- 139 MOLA Map of Mars (with worksheet)

MAPS in Action

MOLA Map of Mars

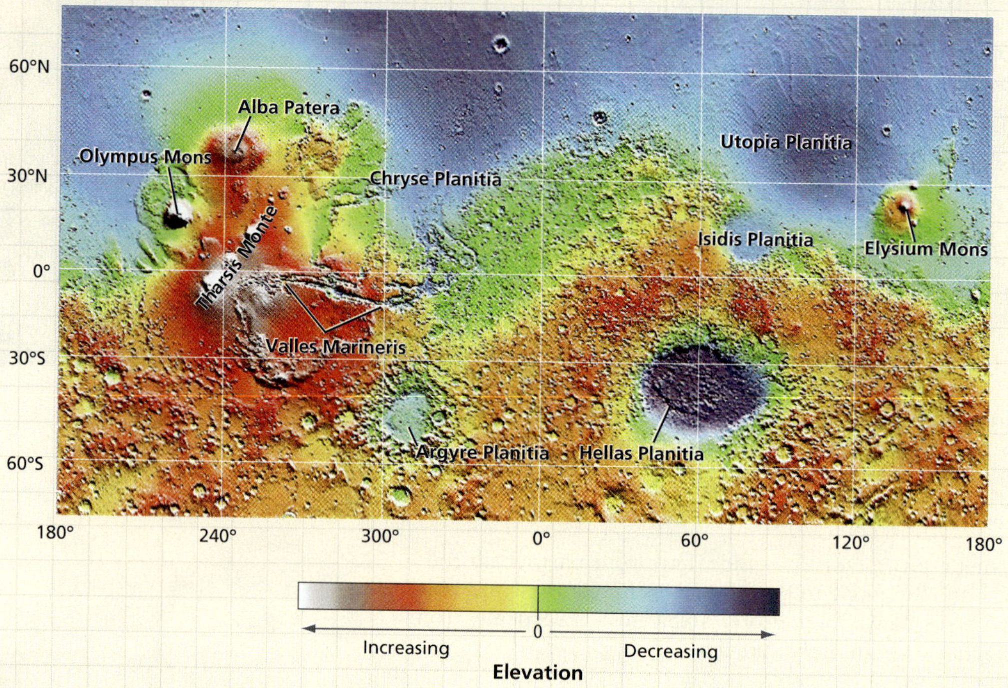

Map Skills Activity

The above map shows the relative elevation of surface features on Mars's surface. The number 0 on the elevation scale marks the average elevation at the equator. This map was created from data that was collected by the Mars Orbiter Laser Altimeter (MOLA) on the *NASA Mars Global Surveyor*. Use the map to answer the questions below.

1. **Analyzing Data** Estimate the longitude and latitude of Elysium Mons.
2. **Using the Key** Identify three features that have elevations below 0. Identify three features that have elevations above 0.
3. **Comparing Areas** In general, which pole on Mars, the north pole or south pole, has higher elevations?
4. **Using the Key** Which feature, Isidis Planitia or Argyre Planitia, has a higher elevation?
5. **Making Comparisons** Which feature, Hellas Planitia or Olympus Mons, would most likely be a volcano?
6. **Using the Key** Estimate the distance between Olympus Mons and Elysium Mons in degrees of longitude.
7. **Inferring Relationships** Based on what you have learned from the map, what type of features do you think the words *planitia,* and *mons* refer to?

CAREER FOCUS

Astronomer

"As a five-year-old, I had binoculars—I'd lie on the lawn at night and use them to look at the sky," says Sandra Faber, professor of astronomy at the University of Santa Cruz, and staff member at the Lick Observatory in Santa Cruz, California.

Like a Detective Story

Faber's research focuses on the formation and evolution of galaxies and the evolution of structures in the universe. To gather data, Faber uses various kinds of telescopes, including ground-based telescopes and the *Hubble Space Telescope*. "Astronomy is like a big detective story," she says. "Information is gathered and shared among scientists, who then draw scientific conclusions about how the universe came to be." Faber is currently a core member of the Deep Extragalactic Evolutionary Probe (DEEP) project. DEEP uses the Keck II telescope in Hawaii and the *Hubble Space Telescope* to survey faint, faraway galaxies.

A Look Back in Time

To see these galaxies, the DEEP project uses a spectrographic instrument called DEIMOS, which stands for *deep imaging multi-object spectrograph*. DEIMOS allows Faber and her colleagues to collect and analyze light that has traveled for billions of years from its original source. Astronomers call such research *look back studies*. By collecting data, Faber can look back billions of years to find answers to scientific questions about the origin of the universe.

Faber's research has given her appreciation for Earth's uniqueness. She says, "Earth is far more varied and beautiful than any other planet. It's all we've got." She also has a special perspective on the Earth's vulnerability: "When seen from space, the Earth is small. It floats in a hostile void. Its atmosphere is thin, like the skin on an apple. Even from space, one can see visual signs of pollutants. It provides a motivation for humans to save this planet and take advantage of its enormous possibilities." In many ways, notes Faber, "Earth is like a spaceship. It's up to the crew members to maintain control."

"Astronomy offers profound messages for our future. It provides a motivation for humans to save this planet and take advantage of its enormous possibilities."

—Sandra Faber

◀ Faber studies distant galaxies from the W. M. Keck Observatory on top of the Hawaiian volcano Mauna Kea.

SCILINKS. NSTA Developed and maintained by the National Science Teachers Association

For a variety of links related to this subject, go to www.scilinks.org

Topic: Careers in Earth Science
SciLinks code: HQ60222

Career Focus

Astronomer

Group Activity — GENERAL

Interview an Astronomer Invite an astronomer from a planetarium or space museum, an observatory, or a local university to talk to the class about astronomy careers. Or, have a group of students interview the person by phone. Have the class prepare a list of questions. Students should record the interviewee's answers on standard notebook paper, or they can write questions on one side of a set of note cards and record answers on the reverse side. (Sample questions: What education do you need to become a professional astronomer? What is the job outlook in astronomy? How were you first introduced to astronomy? What area of astronomy do you specialize in? What kind of equipment do you use in your research? What is a typical day like? How do astronomers get jobs? What inspired you to become an astronomer? What advice do you have for someone who is interested in astronomy?) Have students take notes and summarize the responses. If the interview is by phone, have students report back to the class. **LS Auditory/Verbal**

Extension — GENERAL

Research Have students identify four other careers that are related to astronomy. Have students write a brief synopsis of each career and present them to the class. **LS Interpersonal**

Chapter 28 Minor Bodies of the Solar System

Planning Guide

Compression Guide

To shorten instruction because of time limitations, omit Sections 2 and 3.

OBJECTIVES	LABS, DEMONSTRATIONS, AND ACTIVITIES	TECHNOLOGY RESOURCES
PACING • 45 min pp. 718–724 **Chapter Opener**	LTP **Long-Term Project** Tides at the Shoreline* ADVANCED LTP **Long-Term Project** Apparent Motions of the Moon* ADVANCED	OSP **Parent Letter** ■ CD **Student Edition on CD-ROM** CD **Chapter Summaries Audio CD** ■ VID **Brain Food Video Quiz**
Section 1 Earth's Moon • List four kinds of lunar surface features. • Describe the three layers of the moon. • Summarize the three stages by which the moon formed.	TE **Group Activity** Moon Rescue, p. 719 GENERAL SE **Quick Lab** Liquid and Solid Cores, p. 722 ◆ GENERAL CRF **Datasheet for Quick Lab*** GENERAL TE **Debate** Lunar Origins, p. 723 ADVANCED SE **Maps in Action** Lunar Landing Sites, p. 752 GENERAL TE **Demonstration** Electrolysis of Water, p. 753 ◆ GENERAL CRF **Making Models Lab** Crater Eraser* ◆ GENERAL	OSP **Lesson Plans** (also in print) TR **Bellringer*** TR 140 Formation of the Moon* TR 145 Lunar Landing Sites* TE **Internet Activity** Lunar Timeline, p. 752 BASIC CRF **Internet Activity** Lunar Timeline* BASIC
PACING • 90 min pp. 725–732 **Section 2 Movements of the Moon** • Describe the shape of the moon's orbit around Earth. • Explain why eclipses occur. • Describe the appearance of four phases of the moon. • Explain how the movements of the moon affect tides on Earth.	TE **Group Activity** Barycenter, p. 725 ◆ GENERAL TE **Demonstration** Synchronous Rotation, p. 726 ◆ GENERAL TE **Activity** Eclipsed, p. 727 ◆ GENERAL SE **Quick Lab** Eclipses, p. 728 GENERAL CRF **Datasheet for Quick Lab*** GENERAL TE **Group Activity** Many Moons, p. 730 ◆ GENERAL CRF **Inquiry Lab** Inconstant Moon* GENERAL	OSP **Lesson Plans** (also in print) TR **Bellringer*** TR 141 The Earth-Moon System* TR 142 Solar and Lunar Eclipses* TR 143 Phases of the Moon* TR 144 Causes of Tides* CD **Interactive Tutor** Waves and Tides
PACING • 90 min pp. 733–738 **Section 3 Satellites of Other Planets** • Compare the characteristics of the two moons of Mars. • Describe how volcanoes were discovered on Io. • Name one distinguishing characteristic of each of the Galilean moons. • Compare the characteristics of the rings of Saturn with the rings of the other outer planets.	TE **Discussion** Mythology Connection, p. 733 GENERAL TE **Demonstration** Shepherd Moons, p. 736 ◆ GENERAL TE **Meteorology Connection** Titan's Atmosphere, p. 736 GENERAL TE **Activity** Planetary Rings, p. 737 ◆ GENERAL SE **Skills Practice Lab** Galilean Moons of Jupiter, pp. 750–751 GENERAL CRF **Datasheet for Chapter Lab*** GENERAL	OSP **Lesson Plans** (also in print) TR **Bellringer*** TE **Internet Activity** Martian Moons, p. 734 GENERAL CRF **Internet Activity** Martian Moons* GENERAL
PACING • 45 min pp. 739–744 **Section 4 Asteroids, Comets, and Meteoroids** • Describe the physical characteristics of asteroids and comets. • Describe where the Kuiper Belt is located. • Compare meteoroids, meteorites, and meteors. • Explain the relationship between the Oort cloud and comets.	TE **Discussion** Asteroid Motions, p. 739 GENERAL TE **Physics Connection** Unusual Orbits, p. 740 ADVANCED	OSP **Lesson Plans** (also in print) TR **Bellringer*** TE **Internet Activity** Kuiper Belt Objects, p. 742 GENERAL CRF **Internet Activity** Kuiper Belt Objects* GENERAL VID **HRW Earth Science Video** The Solar System CD **Interactive Tutor** Asteroids, Comets, and Meteors

PACING • 90 min

CHAPTER REVIEW, ASSESSMENT, AND STANDARDIZED TEST PREPARATION

- SE **Chapter Highlights,** p. 745
- SE **Chapter Review,** pp. 746–747
- SE **Standardized Test Prep,** pp. 748–749
- CRF **Concept Review*** ■ GENERAL
- CRF **Critical Thinking*** ADVANCED
- CRF **Math Skills*** GENERAL
- CRF **Graphing Skills*** GENERAL
- CRF **Chapter Test A*** ■ GENERAL
- CRF **Chapter Test B*** ADVANCED
- OSP **Lesson Plans** (also in print)
- OSP **Test Generator**
- OSP **Test Item Listing**

Online and Technology Resources

Holt Online Learning

Visit go.hrw.com for access to Holt Online Learning, or enter the keyword **HQ6 Home** for a variety of free online resources.

One-Stop Planner® CD-ROM

This CD-ROM package includes
- Lab Materials QuickList Software
- Holt Calendar Planner
- Customizable Lesson Plans
- Printable Worksheets
- ExamView® Test Generator
- Interactive Teacher Edition
- Holt PuzzlePro®
- Holt PowerPoint® Resources

KEY

SE Student Edition	OSP One-Stop Planner	VID Classroom Video/DVD
TE Teacher Edition	TR Transparencies and Transparency Worksheets	* Also on One-Stop Planner
CRF Chapter Resource File	CD CD or CD-ROM	◆ Requires advance prep
LTP Long-Term Projects		■ Also available in Spanish

SKILLS DEVELOPMENT RESOURCES	REVIEW AND ASSESSMENT	CORRELATIONS
SE **Pre-Reading Activity,** p. 718 GENERAL TE **Using the Figure** Asteroids, p. 718 GENERAL		National Science Education Standards
CRF **Directed Reading*** BASIC TE **Skill Builder** Math, p. 720 GENERAL TE **Using the Figure** Research, p. 720 GENERAL TE **Using the Figure** Footprints on the Moon, p. 721 BASIC TE **Reading Skill Builder** Paired Summarizing, p. 723 BASIC	SE **Reading Checks,** pp. 720, 722 GENERAL SE **Section Review,** p. 724 GENERAL TE **Reteaching,** p. 723 BASIC TE **Quiz,** p. 723 GENERAL TE **Alternative Assessment,** p. 724 GENERAL CRF **Section Quiz*** ■ GENERAL	SAI 2c, SAI 2d, SAI 2f, PS 4b, ES 3a, ST 2c, HNS 2c, HNS 3c, UCP 1, UCP 3
CRF **Directed Reading*** BASIC TE **Inclusion Strategies,** p. 726 TE **Using the Figure** Solar Eclipse, p. 727 BASIC TE **Skill Builder** Vocabulary, p. 731 GENERAL	SE **Reading Checks,** pp. 726, 728, 731 GENERAL SE **Section Review,** p. 732 GENERAL TE **Homework,** p. 730 GENERAL TE **Reteaching,** p. 731 BASIC TE **Quiz,** p. 731 GENERAL TE **Alternative Assessment,** p. 731 ADVANCED CRF **Section Quiz*** ■ GENERAL	SAI 2c, SAI 2d, SAI 2f, PS 4b, ST 2c, HNS 2c, HNS 3c, UCP 1, UCP 3
CRF **Directed Reading*** BASIC TE **Reading Skill Builder** Reading Organizer, p. 734 BASIC TE **Using the Figure** Geologic Pasts, p. 735 GENERAL TE **Inclusion Strategies,** p. 735 SE **Graphic Organizer** Comparison Table, p. 736 GENERAL TE **Using the Figure** Ringmaster, p. 736 BASIC	SE **Reading Checks,** pp. 735, 737 GENERAL SE **Section Review,** p. 738 GENERAL TE **Reteaching,** p. 737 BASIC TE **Quiz,** p. 737 GENERAL TE **Alternative Assessment,** p. 737 GENERAL CRF **Section Quiz*** ■ GENERAL	SAI 2c, SAI 2d, SAI 2f, PS 4b, ST 2c, HNS 2c, HNS 3c, UCP 1, UCP 3
CRF **Directed Reading*** BASIC TE **Using the Figure** Comparing Craters, p. 740 GENERAL TE **Using the Figure** Discussion, p. 741 GENERAL TE **Using the Figure** Oort Cloud Distances, p. 742 GENERAL SE **Math Practice,** p. 743 GENERAL TE **Using the Figure** Leonid Meteor Shower, p. 743 GENERAL	SE **Reading Checks,** pp. 740, 743 GENERAL SE **Section Review,** p. 744 GENERAL TE **Reteaching,** p. 743 BASIC TE **Quiz,** p. 743 GENERAL TE **Alternative Assessment,** p. 743 GENERAL CRF **Section Quiz*** ■ GENERAL	SAI 2c, SAI 2d, SAI 2f, PS 4b, ST 2c, HNS 2c, HNS 3c, UCP 1, UCP 3

Holt Lab Generator CD-ROM

Search for labs by topic, standard, difficulty level, and/or time frame. Edit those labs to fit your needs, or create your own labs. Use the Lab Materials QuickList to order just what you need.

Holt Earth Science Interactive Tutor CD-ROM

This CD-ROM consists of interactive activities that give students a fun way to extend their knowledge of Earth science concepts.

Chapter Summaries Audio CDs

These CDs include audio summaries of the key concepts presented in each chapter. (Audio summaries are also available in Spanish.)

SCILINKS. NSTA

www.scilinks.org

Maintained by the **National Science Teachers Association.** See Chapter Enrichment pages that follow for a complete list of topics.

See Chapter Enrichment pages for Video Resources.

Chapter 28 Chapter Enrichment

This Chapter Enrichment provides relevant and interesting information to expand and enhance your classroom instruction of the chapter material.

Section 1 Earth's Moon

Moon Rocks

The Apollo missions brought back more than 380 kg of lunar samples from six different locations on the moon. The samples were gathered using tongs to pick up rocks, scoops to collect soil samples, and rakes to collect smaller pebbles. Hammers were used to break samples from large boulders. Core tubes were used to obtain soil samples from below the surface, and ejected materials were gathered from around crater sites. Samples were photographed before collection and sealed in environmentally controlled containers for the return trip. Robot probes from Soviet spacecraft also returned with samples from three additional lunar sites. How do the moon rocks differ from Earth rocks? They contain no water in their crystal structure, particles of clay formed through sedimentary processes are absent, and the rocks are riddled with tiny craters produced by micrometeorites. The study of these samples has revealed important information about the history of the moon, Earth, and conditions in the inner solar system during its early history.

▲ This lunar rock is between 4.3 billion and 4.5 billion years old.

Section 2 Movements of the Moon

Moon's Orbital Tilt

Some planetary scientists think that the moon's mysterious orbital tilt may be the natural result of the impact event that created the moon. Most other satellites have an orbital tilt of only 1° or 2°, but Earth's moon has an inclination of 5°. Drs. William Ward and Robin Canup used a computer to model the impact event. They suggest that the collision would have actually placed two large masses of material into Earth orbit. The material at the outer edge of the debris field would have rapidly formed the moon. The material in the inner region would have been prevented from clumping together by Earth's gravity. The moon could have continued to coexist with the debris left over from the impact for as long as a century. Ward and Canup think that the gravitational interaction between the moon and the debris from the inner region produced resonance waves that affected the moon's motion, and gradually increased the moon's orbital tilt.

Section 3 Satellites of Other Planets

Ring Systems

All four gas giants—Jupiter, Saturn, Uranus, and Neptune—have planetary ring systems composed of tiny bits of ice and rock that circle the planets like miniature moons. Saturn's rings actually consist of thousands of separate rings that blend together when observed from Earth. The gravity of small moons called "shepherds" located on the edges of ring systems or in gaps between the rings causes rings to form distinct bands. The origin of ring systems has been a mystery since they were

▲ Saturn has the most extensive system of rings in the solar system.

discovered. One hypothesis links the formation of rings to the fact that large moons never orbit inside the ring system. There is a zone called the Roche limit close to the parent planet within which tidal forces will rip apart a rocky moon or prevent smaller bodies from joining together. It is in this zone that ring systems form. If this idea is correct, planetary ring systems would be unstable, lasting only about 100 million years, and would be relatively young phenomena. Though short-lived, ring systems would form and reform as orbits of moons decay and the moons are destroyed as they pass the Roche limit.

▲ **Barringer Meteoroid Crater, in Arizona, formed when a meteoroid collided with Earth.**

Section 4 Asteroids, Comets, and Meteoroids

Formation of Impact Craters

When two bodies collide at high velocity, most of the projectile vaporizes on impact. The exploding projectile's energy of motion is transformed into heat and pressure, which produce two shock waves. The first wave melts the surface rock, and the second wave creates the circular cavity. The explosion blasts out surface material, hurling it outward, and creating a raised rim around the cavity. The typical impacting body creates a crater roughly 25 to 40 times its diameter, depending on its velocity, composition, and the existing geological structures at the target. Sometimes, meteorites are buried at the bottom of smaller craters. Ejecta rays, or streaks of ejected target material, such as those surrounding the Tycho crater on the moon, nearly always accompany relatively young craters.

Video Resources

Brain Food Video Quizzes

Brain Food Video Quizzes These videos contain game-show style quizzes that assess students' progress and motivate students to study the chapter material.

HRW Earth Science Video

This video introduces Earth science topics and includes a geology field trip. The video segment listed below complements this chapter.

Segment 18: The Solar System This segment examines the formation of planets from planetesimals and the differences between the inner four planets and the outer five. The composition, orbit, origin, and physical properties of asteroids, comets, meteors, and meteorites are described. (4 min)

Segment 19: The Earth's Moon This segment explores the various alignments of Earth, the moon, and the sun that give rise to the phases of the moon. This segment also examines the alignments that create lunar and solar eclipses. (3 min)

NOVA Videos

To order **NOVA** videos related to this chapter, visit **go.hrw.com** and enter the keyword **HQ6MBSV.**

SciLinks

Developed and maintained by the National Science Teachers Association

SciLinks is maintained by the National Science Teachers Association to provide you and your students with interesting, up-to-date links that will enrich your classroom presentation of the chapter.

Visit www.scilinks.org and enter the SciLinks code for more information about the topic listed.

Topic: Earth's Moon
SciLinks code: HQ60449

Topic: Lunar Cycle
SciLinks code: HQ60887

Topic: Moons of Other Planets
SciLinks code: HQ60993

Topic: Comets, Asteroids, and Meteoroids
SciLinks code: HQ60317

Chapter 28

Chapter Overview

This chapter describes the structure, composition, and movements of Earth's moon. It explains how the moon's orbital motions affect eclipses and tides. The chapter also describes other natural satellites and small bodies in the solar system.

Using the Figure — GENERAL

Asteroids Explain that the majority of asteroids orbit in a belt of debris between the orbits of Mars and Jupiter. Ask students where they think these bodies came from and what clues they might offer about the early solar system. (Sample answer: Asteroids may be the remains of planetesimals that were unable to form a planet due to the effects of Jupiter's gravity. Their composition may be similar to the materials from which the inner planets and many moons formed.) **LS Visual**

PRE-READING ACTIVITY

FOLDNOTES

You may want to assign this FoldNote activity as homework. Collect the FoldNotes to check students' understanding of the material.

Chapter 28 Minor Bodies of the Solar System

Sections

What You'll Learn

- What Earth's moon is made of and how it moves
- How satellites of other planets differ from Earth's moon
- What comets, asteroids, and meteoroids are

Why It's Relevant

Small orbiting objects can provide information about the conditions that existed at the beginnings of our solar system.

PRE-READING ACTIVITY

FOLDNOTES

Booklet Before you read this chapter, create the FoldNote entitled "Booklet" described in the Skills Handbook section of the Appendix. Label each page of the booklet with a main idea from the chapter. As you read the chapter, write what you learn about each main idea on the appropriate page of the booklet.

▶ This is one idea of how an asteroid might look as it moves toward Earth. Asteroids are one type of small body that travels through our solar system.

Chapter Correlations — *National Science Education Standards*

SAI 2c Scientists rely on technology to enhance the gathering and manipulation of data. **(Sections 1–4)**

SAI 2d Mathematics is essential in scientific inquiry. **(Sections 1–4)**

SAI 2f Results of scientific inquiry—new knowledge and methods—emerge from different types of investigations and public communication among scientists. **(Sections 1–4)**

PS 4b Gravitation is a universal force that each mass exerts on any other mass. . . . **(Sections 1–4)**

ES 3a The sun, the earth, and the rest of the solar system formed from a nebular cloud of dust and gas 4.6 billion years ago. **(Section 1)**

ST 2c Creativity, imagination, and a good knowledge base are all required in the work of science and engineering. **(Sections 1–4)**

HNS 2c Because all scientific ideas depend on experimental and observational confirmation, all scientific knowledge is, in principle, subject to change as new evidence becomes available. **(Sections 1–4)**

HNS 3c Occasionally, there are advances in science and technology that have important and long-lasting effects on science and society...: Copernican revolution... **(Sections 1–4)**

UCP 1 Systems, order, and organization: The natural and designed world is complex; it is too large and complicated to investigate and comprehend all at once...A system is an organized group of related objects or components that form a whole. **(Sections 1–4)**

UCP 3 Constancy, change, and measurement: Although most things are in the process of becoming different—changing—some properties of objects and processes are characterized by constancy. **(Sections 1–4)**

Section 1 Earth's Moon

A body that orbits a larger body is called a **satellite.** Six of the planets in our solar system have smaller bodies that orbit around them. These natural satellites are also called **moons.** Our moon is Earth's natural satellite.

In 1957, the Soviet Union launched *Sputnik,* which was the first *artificial satellite* launched into space. In 1958, the United States launched its first artificial satellite, which was named *Explorer 1.* Thousands of artificial satellites are now in orbit around Earth, including weather satellites and space telescopes, such as the *Hubble Space Telescope.*

Exploring the Moon

Between 1969 and 1972, the United States sent six spacecraft to the moon as part of the Apollo space program. Apollo astronauts found that the moon's weak gravity affected the way they moved. They discovered that bouncing was more efficient than walking. Apollo astronauts also explored the moon's surface in a variety of specially-designed vehicles, such as the one shown in **Figure 1.**

The moon has much less mass than Earth does, so the gravity on the moon's surface is about one-sixth of the gravity on Earth. As a result, a person who has a mass of 61.2 kg and who exerts about 600 newtons (N) of force on Earth would exert only about 100 N on the moon. The gravity at the moon's surface is not strong enough to hold gases and therefore has no significant atmosphere. Because the moon has no atmosphere to absorb and transport heat, the moon's surface temperature varies greatly, from 134°C during the day to –170°C at night.

Figure 1 ▶ *Apollo 17* astronaut Eugene Cernan explores the lunar surface in a Lunar Roving Vehicle.

OBJECTIVES

- ▶ **List** four kinds of lunar surface features.
- ▶ **Describe** the three layers of the moon.
- ▶ **Summarize** the three stages by which the moon formed.

KEY TERMS

satellite
moon
mare
crater

satellite a natural or artificial body that revolves around a larger celestial body

moon a celestial body that revolves around a body that is larger in mass; a natural satellite

SCILINKS. NSTA Developed and maintained by the National Science Teachers Association

For a variety of links related to this subject, go to www.scilinks.org

Topic: Earth's Moon
SciLinks code: HQ60449

Section 1

Focus

Overview

This section describes the moon's surface features and its layers. It also explains the giant impact hypothesis and summarizes how the moon formed.

Bellringer

Ask students to write a paragraph that describes what they know about the moon. (Answers may vary.) LS Verbal

Motivate

Group Activity — GENERAL

Moon Rescue Have small groups of students imagine that they are setting up a moon base. Ask groups to make a list of the supplies and materials they would need to build a base for 100 people. Students may wish to organize their group in a way to allow one person to be in charge of identifying resources needed to supply each major need. (Answers may vary but students should provide for the need for temperature control, oxygen, food, water, electricity, exercise, and living quarters.) LS Logical/Interpersonal Co-op Learning

CHAPTER RESOURCES

Chapter Resource File
- **Directed Reading** BASIC
- **Making Models Lab** Crater Eraser GENERAL

Technology

Transparencies
- Bellringer

Student Edition on CD-ROM

One-Stop Planner CD-ROM
- Lesson Plan

Teach

SKILL BUILDER — GENERAL

Math Use this exercise to help students gain perspective on the enormous distances involved as you teach about the Earth-moon system. Construct a paper model of the Earth-moon system. Write these measurements on the board: Earth's diameter = 12,800 km; Moon's diameter = 3,500 km; Mean distance from Earth to moon = 385,000 km. Select a scale to use. For example, with a 10 cm diameter circle representing Earth, how large must the moon be? (Calculate the conversion factor for the diameters of the two bodies: 12,800 km (Earth) ÷ 3,500 km (moon) = 3.7; rounding up, for any scale model, Earth's diameter must be approximately 4 times larger than the moon's. Divide the size of the Earth model by the scale factor: 10 cm ÷ 4 = 2.5 cm.) Use construction paper to make two paper circles to represent Earth (10 cm) and the moon (2.5 cm). Next, ask how far apart the models must be placed. (If 10 cm = 12,800 km, then 1 cm = 1,280 km; 385,000 km ÷ 1,280 km/cm = 301 cm) Measure out a string 301 cm long to represent this distance. Tape the two circles on a wall with the string between the models. The system model is now to scale for both size and distance. **LS Logical**

Answer to Reading Check

Answers should include two of the following features: maria, highlands, craters, ridges, and rilles.

mare a large, dark area of basalt on the moon (plural, *maria*)

crater a bowl-shaped depression that forms on the surface of an object when a falling body strikes the object's surface or when an explosion occurs

The Lunar Surface

Because *luna* is the Latin word for "moon," any feature of the moon is referred to as *lunar*. Light and dark patches on the moon can be seen with the unaided eye. The lighter areas are rough highlands that are composed of rocks called *anorthosites*. The darker areas are smooth, reflect less light, and are called *maria* (MAHR ee uh). Each dark area is a **mare** (MAHR AY). *Mare* is Latin for "sea." Galileo named these dark areas *maria* because he thought that they looked like Earth's seas. Today, astronomers know that maria are plains of dark, solidified lava. These lava plains formed more than 3 billion years ago when lava slowly filled basins that were created by impacts of massive asteroids.

Craters, Rilles, and Ridges

The surface of the moon, shown in **Figure 2,** is covered with numerous bowl-shaped depressions, called **craters.** Most of the moon's craters formed when debris left over from the formation of the solar system struck the moon about 4 billion years ago. Younger craters are characterized by bright streaks, called *rays*, that extend outward from the impact site. Even these younger craters, however, are billions of years old.

Long, deep channels called *rilles* run through the maria in some places. The moon's rilles are thought to be leftover lava channels from the formation of the maria. Some rilles are as long as 240 km. Another surface feature of the moon is ridges. Ridges are long, narrow elevations of rock that rise out of the surface and criss-cross the maria.

Reading Check **Name two features of the moon. (See the Appendix for answers to Reading Checks.)**

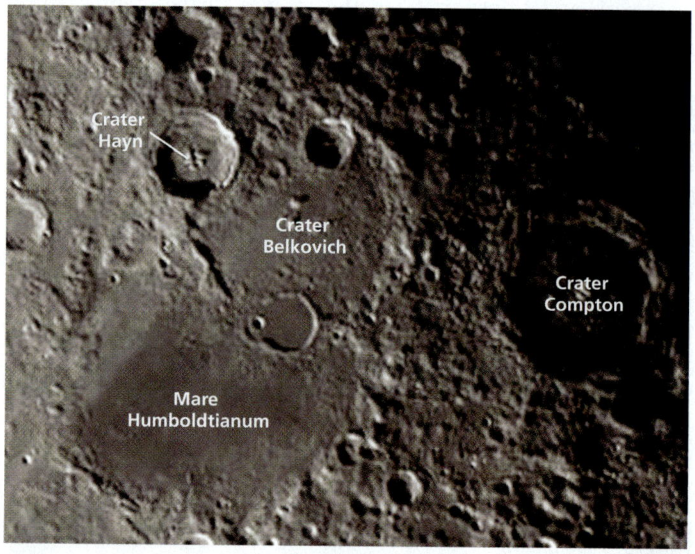

Figure 2 ▶ The largest lunar craters are named for famous scholars and scientists. The lunar surface also has millions of small, overlapping craters. ***How does the shape of Mare Humboldtianum indicate that this feature once was an impact crater?***

Using the Figure — GENERAL

Research Invite interested students to learn about the contributions of the scientists and other famous people whose names are used to identify lunar features. For example, Mare Humboldtianum (Humboldt's Sea) was named for Alexander von Humboldt, a famous German naturalist and explorer. The USGS Gazetteer of Planetary Nomenclature on the Internet lists the origin of many other lunar names. Answer to caption question: The mare's bowl shape is common to impact craters. The circular depression was later filled in with darker lava. **LS Logical**

Figure 3 ▶ This footprint is one of the first marks left on the moon by humans. The footprint is visible because of the fine layer of rock and dust, called *regolith,* that covers the moon's surface.

Regolith

More meteorites have reached the surface of the moon than have reached Earth's surface because the moon has no atmosphere for protection. Over billions of years, these meteorites crushed much of the rock on the lunar surface into dust and small fragments. Today, almost all of the lunar surface is covered by a layer of dust and rock, called *regolith*. Regolith is shown in **Figure 3.** The depth of the regolith layer varies from 1 m to 6 m.

Lunar Rocks

Many lunar rocks are very similar to rocks on Earth. Lunar rocks, including the one shown in **Figure 4,** contain many of the same elements as Earth's rocks do, but lunar rocks contain different proportions of those elements. Lunar rocks are igneous, and most rocks near the surface are composed mainly of oxygen and silicon. These surface rocks are similar to the rocks in Earth's crust. Rocks from the lunar highlands are light-colored, coarse-grained anorthosites. Highland rocks are rich in calcium and aluminum. Rocks from the maria are fine-grained basalts and contain large amounts of titanium, magnesium, and iron.

Nevertheless, lunar surface rocks have only small amounts of some elements that are common on Earth. Many of these elements have low melting points and may have boiled off early in the moon's history when the moon was still molten. Also, the minerals in lunar rocks do not contain water.

One type of rock that occurs in both maria and the highlands is *breccia*. Lunar breccia contains fragments of other rocks that have been fused together. These breccias formed when meteorites struck the moon. The force of these impacts broke up rocks, and the heat from the impacts partially melted the fragments.

Figure 4 ▶ This rock is 4.3 billion to 4.5 billion years old; it is the oldest rock discovered on the moon. The rock's texture indicates that the rock has a complicated history.

Discussion — GENERAL

Moon Rocks Ask students to compare the lunar rocks with rocks found on Earth. For example, are the same kinds of rocks found on both bodies? (Answers may vary. Sample answer: Lunar rocks are almost all igneous, while Earth has sedimentary and metamorphic rocks as well.) How are the composition and structure of moon rocks similar or different from those of Earth's rocks? (Sample answers: Moon rocks are made of the same elements but in different proportions. The lunar samples do not have water in their crystal structure.) **LS Verbal**

Using the Figure — BASIC

Footprints on the Moon Ask students to identify the feature of the lunar environment in which Neil Armstrong's footprints were preserved when he walked on the moon. (the regolith) Have them explain why those first marks left by humans visiting the moon will remain visible for an extremely long time. (The moon lacks water and an atmosphere, which would cause erosion. It also lacks tectonic processes, which also would erase traces of human activity.) **LS Logical**

LIFE SCIENCE CONNECTION — GENERAL

Moon Trees When *Apollo 14* was launched in January 1971, hundreds of tree seeds (of five different long-lived varieties of trees) orbited in the personal kit of Air Force Colonel Stuart Roosa in the command module *Kitty Hawk*. Roosa was a former Forest Service smoke jumper, and the seeds were part of a biological experiment to see whether the trip to the moon would affect the ability of seeds to germinate. Upon their return to Earth, the seeds were planted by the Forest Service throughout the United States as part of the country's bicentennial celebration. Despite the fact that they were exposed to vacuum conditions during decontamination, nearly all of the seeds germinated. The "Moon Trees" remain a living legacy of the space program. No systematic records were kept of all of the plantings, so the Forest Service and NASA are trying to track them down. A NASA-sponsored Web site lists the known Moon Trees. Use the Web site to locate any known Moon Trees in your home state, and plan a class visit. Perhaps your students can even help NASA find some as yet unidentified Moon Trees. **LS Logical**

Teach, *continued*

Quick LAB

Skills Acquired

- Experimenting
- Observing
- Analyzing

Teacher's Notes

Use solid foods that are uniform in content, and about the same mass as the soup can, such as a jar of peanut butter or a can of refried beans.

Answers

1. The uncooked egg stops first. The can of soup stops first.
2. The object with a solid core would rotate with a more uniform motion. Because of inertia, the liquid would cause the object to wobble slightly.

CHAPTER RESOURCES

Chapter Resource File

- Datasheet for Quick Lab GENERAL

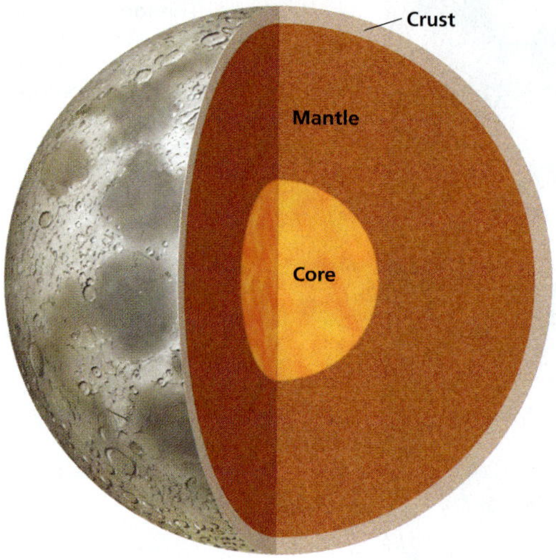

Figure 5 ▶ The moon, like Earth, has three compositional layers: the crust, the mantle, and the core.

The Interior of the Moon

Rocks of the lunar surface are about as dense as those on Earth's surface. However, the overall density of the moon is only three-fifths the density of Earth. The difference in overall density indicates that the interior of the moon is less dense than the interior of Earth.

Most of the information about the interior of the moon comes from seismographs that were placed on the moon by the Apollo astronauts. Seismographs have recorded numerous weak moonquakes, which are similar to earthquakes. More than 10,000 moonquakes have been detected. Most moonquakes occur in the mantle at a depth that is 10 times deeper than the depth at which most earthquakes occur on Earth. From these moonquakes, scientists learned that the moon's interior is layered, as shown in **Figure 5.**

The Moon's Crust

One side of the moon always faces Earth. That side is therefore called the *near side*. The other side always faces away from Earth and is called the *far side*. The pull of Earth's gravity during the moon's formation caused the crust on the far side of the moon to become thicker than the crust on the near side. On the near side, the lunar crust is about 60 km thick. On the far side, the lunar crust is up to 100 km thick. Images of the far side show that the far side's surface is mountainous and has only a few small maria. The crust of the far side appears to consist of materials that are similar to those of the rocks in the highlands on the near side.

Reading Check **Name two features of the far side of the moon. (See the Appendix for answers to Reading Checks.)**

The Moon's Mantle and Core

Beneath the crust is the moon's mantle. The mantle is thought to be made of rock that is rich in silica, magnesium, and iron. Of the moon's 1,738 km radius, the mantle makes up more than half of that distance and reaches 1,000 km below the crust.

Scientists think that the moon has a small iron core that has a radius of less than 700 km. When laser beams were bounced off of small mirrors placed on the moon, scientists discovered that the moon's rotation is not uniform. This nonuniform rotation indicates that the core is neither completely solid nor completely liquid. This characteristic may explain why the moon has almost no overall magnetic field. There are, however, small areas on the moon that exhibit local magnetism.

Quick LAB — 5 min

Liquid and Solid Cores

Procedure

1. Take **one uncooked egg** and **one hardboiled egg**. With your thumb and forefinger, spin both eggs.
2. Record the amount of time each egg spins.
3. Lay **one can of solid food** and **one can of soup** on their sides, and spin both cans.
4. Record the amount of time each can spins.

Analysis

1. Which egg stopped spinning first? Which can of food stopped spinning first?
2. Which rotates more steadily: an object that has a solid core or an object that has a liquid core? Explain your answer.

Discussion — GENERAL

Comprehension Check Pose this scenario to students. Suppose someone said that the interior of the moon is very similar to the interior of Earth. Would you agree or disagree with this assessment? (Answers may vary. Sample answer: Like Earth, the moon consists of a thin crust, a mantle and an iron core. However, the moon's overall density is substantially less than that of Earth. No convection currents have been detected in the lunar mantle. The lunar core is also considerably smaller.) **LS Verbal**

Answer to Reading Check

The crust of the far side of the moon is thicker than the crust of the near side is. The crust of the far side also consists mainly of mountainous terrain and has only a few small maria.

The Formation of the Moon

Rocks taken from the moon by Apollo astronauts provided evidence to help astronomers understand the moon's history. Most scientists generally agree that the moon formed in three stages.

The Giant Impact Hypothesis

Most scientists think that the moon's development began when a large object collided with Earth more than 4 billion years ago. This *giant impact hypothesis* states that a Mars-sized body struck Earth early in the history of the solar system. Before the impact, Earth was molten, or heated to an almost liquid state. The collision ejected chunks of Earth's mantle into orbit around Earth. The debris eventually clumped together to form the moon, as shown in **Figure 6.**

Most of the ejected materials came from Earth's silica-rich mantle rather than from Earth's dense, metallic core. This hypothesis explains why moon rocks share many of the chemical characteristics of Earth's mantle. As the material clumped together, it continued to revolve around Earth because of Earth's gravitational pull.

Differentiation of the Lunar Interior

Early in its history, the lunar surface was covered by an ocean of molten rock. Over time, the densest materials moved toward the center of the moon and formed a small core. The least dense materials formed an outer crust. The other materials settled between the core and the outer layer to form the moon's mantle.

Meteorite Bombardment

The outer surface of the moon eventually cooled to form a thick, solid crust over the molten interior. At the same time, debris left over from the formation of the solar system struck the solid surface and produced craters and regolith.

About 3 billion years ago, the number of small objects in the solar system decreased. Less material struck the lunar surface, and few new craters formed. Craters that have rays formed during the most recent meteor impacts. During this stage of lunar development, virtually all geologic activity stopped. Because the moon cooled more than 3 billion years ago, it looks today almost exactly as it did 3 billion years ago. Therefore, the moon is a valuable source of information about the conditions that existed in the solar system long ago.

Figure 6 ▶ The First Stage of Moon Formation

Scientists think that a Mars-sized object collided with Earth and blasted part of Earth's mantle into space.

The resulting debris then began to revolve around Earth.

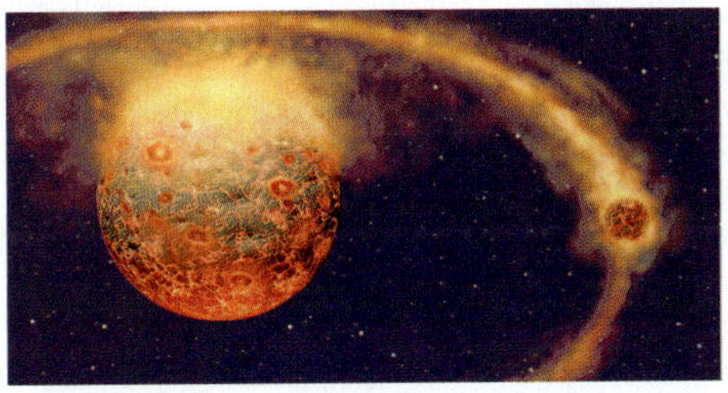

The material eventually joined to form Earth's moon.

Close

Reteaching — Basic

Storyboards Ask students to use a storyboard organizer to sort out the sequence of events in the formation of the moon. Tell them to make sketches that illustrate the steps in the formation process and include an explanatory note with each storyboard. LS **Visual**

Quiz — General

1. Describe how the surface features of the far side of the moon compare with those on the near side. (The near side has several large basins filled with lava (maria). The far side has fewer maria and more mountains and craters.)
2. How do scientists know the moon has a layered structure? (Astronauts placed seismographs that have detected weak moonquakes and helped reveal the moon's structure. Observations of the moon's rotation suggest that its core is not completely solid.)
3. How have meteorite impacts affected the surface of the moon? (Impacts have produced numerous craters and pulverized the crust, which covered the surface with a fine layer of dust and rock, and melted together fragments of rock to produce breccias.) LS **Verbal**

CHAPTER RESOURCES

Technology

Transparencies
- 140 Formation of the Moon (with worksheet)

Reading Skill Builder — Basic

Paired Summarizing Group students into pairs, and have them read silently about the giant impact hypothesis of the origin of the moon and the events that followed the impact. Then, have one student summarize the stages in the formation of the moon. The other student should listen to the retelling and should point out any inaccuracies or details that were left out. Allow students to refer to the text as needed.

LS **Auditory** Co-op Learning English Language Learners

Debate — Advanced

Lunar Origins To gain acceptance, any explanation of lunar origins had to explain these facts: (1) why Earth has such a large moon for its size; (2) why the moon has such a small iron core relative to Earth's; (3) the similarity in oxygen isotopes and trace elements in the two bodies; and (4) the existence of a lava "ocean" on the moon. Invite students to research and debate the merits of the giant impact hypothesis and competing hypotheses on these points. LS **Logical**

Close, *continued*

Alternative Assessment — GENERAL

Lunar Picture Book Have students create a picture book for younger children that explains the topics in this section, including lunar exploration, the moon's structure, and the origin of the moon. Provide fiction and nonfiction examples for students to use as models. LS **Verbal/Visual**

Answers to Section Review

1. Galileo called the smooth dark patches on the lunar surface maria, because he thought they looked like seas.
2. The crust on the far side of the moon is thicker than the crust on the near side is.
3. The maria formed more than 3 billion years ago, when lava filled craters that formed during previous impacts.
4. Many more craters would have been produced and the layer of debris would be much thicker. Some maria may have formed, but they would have been covered with overlapping craters, breccias, and regolith.
5. Breccias are made up of fragments of rocks from the highlands and maria that are cemented by material that was melted and cooled. Breccias formed when the force of impacts broke up existing rocks and heat from the impacts melted some of the fragmented material.
6. Scientists think that the moon formed when a Mars-sized body struck Earth, throwing part of Earth's mantle into orbit. The debris eventually clumped together to form the moon.
7. The pull of Earth's gravity during the moon's formation caused the crust on the far side of the moon to become thicker than the crust on the near side.
8. Sample answer: Impact events formed craters, breccias, and the lunar regolith. Lava flows formed the smooth maria, which fill in many of the largest impact craters.
9. The surface of the *moon* has light areas called *highlands* and dark areas called *maria,* which formed when *craters* formed by *meteorite* impacts filled with *basalt,* which also formed *rilles.*

Figure 7 ▶ The near side of the moon (top) has fewer visible craters than the far side (bottom) does in part because lava flows on the near side covered many of the impact sites with maria.

Lava Flows on the Moon

After impacts on the moon's surface formed deep basins, lava flowed out of cracks, or *fissures,* in the lunar crust. This lava flooded the crater basins to form maria. The presence of the maria suggests that fissure eruptions once characterized the moon, even though there is no evidence that large active volcanoes have ever been present on the moon.

Because the moon's crust is thinner on the near side than on the far side, much more lava flowed onto the surface on the near side than onto the surface of the far side of the moon. The near side of the moon has several smooth maria, but the far side has few maria and many more craters, as shown in **Figure 7.**

Scientists do not yet know how magma formed in the lunar interior or how the magma reached the surface. There is no evidence of plate tectonics or convection currents in the moon's mantle, so the magma must have formed in some other way. A large amount of energy would have been needed to produce the magma in the upper layers of the moon. Some scientists think this energy may have come from a long period of intense meteorite bombardment. Other scientists think that radioactive decay of materials may have also heated the moon's interior enough to cause magma to form. Scientists agree that the lava flows ended about 3.1 billion years ago, when the interior cooled completely.

Section 1 Review

1. **Describe** what maria on the surface of the moon look like and how they came to be known as maria.
2. **Compare** the thickness of the moon's crust on the near side with the thickness of the crust on the far side.
3. **Summarize** how and when the maria formed.
4. **Describe** how the surface of the moon would be different today if meteorites had continued to hit it at the same rate as they did 3 billion years ago.
5. **Describe** breccias and how they formed on the moon.
6. **Summarize** how scientists think the moon formed.

CRITICAL THINKING

7. **Analyzing Ideas** Explain how Earth's gravity affected the moon's near side and far side differently.
8. **Making Comparisons** Compare the features of the lunar surface created by lava flows and the features created by impacts.

CONCEPT MAPPING

9. Use the following terms to create a concept map: *moon, meteorite, crater, rille, maria, highlands,* and *basalt.*

CHAPTER RESOURCES

Chapter Resource File

- Section Quiz GENERAL

Workbooks

- Study Guide (also in Spanish)

Section 2

Movements of the Moon

If you looked down on the moon from above its north pole, you would see the moon rotate once on its axis every 27.3 days. However, if you stood on the moon's surface and measured the lunar day by the amount of time between sunrises, you would find that a lunar day is 29.5 Earth days long. This discrepancy is due to the fact that, while the moon is revolving around Earth, Earth and the moon are also revolving around the sun.

OBJECTIVES

- **Describe** the shape of the moon's orbit around Earth.
- **Explain** why eclipses occur.
- **Describe** the appearance of four phases of the moon.
- **Explain** how the movements of the moon affect tides on Earth.

KEY TERMS

apogee
perigee
eclipse
solar eclipse
lunar eclipse
phase

The Earth-Moon System

To observers on Earth, the moon appears to orbit Earth. However, if you could observe Earth and the moon from space, you would see that Earth and the moon revolve around each other. Together, they form a single system that orbits the sun.

The mass of the moon is only 1/80 that of Earth. So, the balance point of the Earth-moon system is not halfway between the centers of the two bodies. The balance point is located within Earth's interior because Earth's mass is greater than the moon's mass. This balance point is called the *barycenter*. The barycenter follows a smooth orbit around the sun, as shown in **Figure 1.**

The Moon's Elliptical Orbit

The orbit of the moon around Earth forms an ellipse that is about 5% more elongated than a circle is. Therefore, the distance between Earth and its moon varies over a month's time. When the moon is farthest from Earth, the moon is at **apogee.** When the moon is closest to Earth, the moon is at **perigee.** The average distance of the moon from Earth is 384,000 km.

apogee in the orbit of a satellite, the point at which the satellite is farthest from Earth

perigee in the orbit of a satellite, the point at which the satellite is closest to Earth

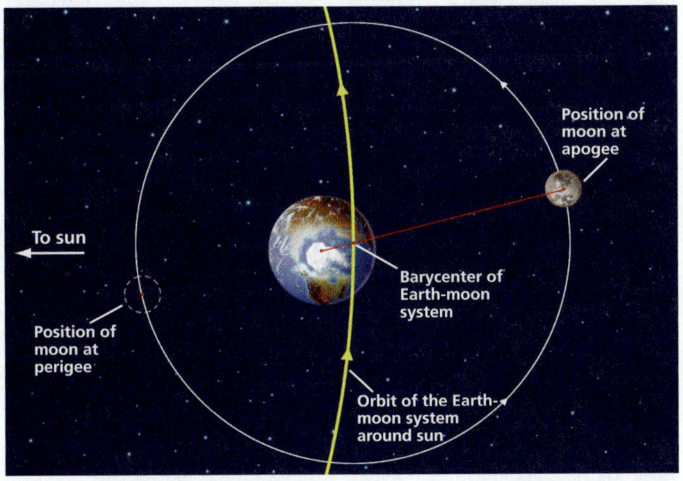

Figure 1 ▶ The barycenter of the Earth-moon system orbits the sun in a smooth ellipse. The orbits of both Earth and the moon "wobble" as the bodies move around the sun. This diagram is not to scale.

Section 2

Focus

Overview

This section describes the Earth-moon system, the shape of the moon's orbit, and how the moon's gravity causes tides. The section also explains why eclipses and the phases of the moon occur.

Bellringer

Ask students to create a series of drawings that shows how the moon's appearance changes over several weeks. **LS Visual**

Motivate

Group Activity — GENERAL

Barycenter Have students make a simple model that shows how the Earth-moon system rotates around a common center point. Students should follow these steps: Tie a 30 cm length of string about 3 cm from the end of a pencil. Stick an apple-sized clay ball on the end of the pencil. Mold the clay ball around the string so that the string lies just inside the edge of the ball. Add a grape-sized ball of clay to the other end of the pencil. Holding the model by the end of the string, add small bits of clay to the smaller ball until the whole system is in balance. Ask what the parts of the model represent. (The large ball is Earth, the small ball stands for the moon. The string is located at the balance point, or the barycenter.) **LS Kinesthetic**

CHAPTER RESOURCES

Chapter Resource File
- Directed Reading BASIC

Technology

Transparencies
- Bellringer
- 141 The Earth-Moon System (with worksheet)

Student Edition on CD-ROM

One-Stop Planner CD-ROM
- Lesson Plan

Teach

Discussion — BASIC

Lunar Orbit The elliptical shape of the moon's orbit means that the distance from Earth to the moon varies as the moon travels around Earth. Ask students what observable effects this variation in distance produces. (Sample answers: The moon moves faster when it is closest to Earth, and it moves slower when it is farther away. This difference in speed affects the times of moonrise and moonset and the distance affects the apparent size and brightness of the moon. The moon looks larger and brighter when it is closer, and it looks smaller and dimmer when it is farther away.) **LS Verbal**

Demonstration — GENERAL

Synchronous Rotation Use a foam ball to represent the moon. Place a dowel rod in the center to form a handle. Mark a series of letters from A to D with a marker or pen at 90° intervals around the face of the ball. Invite a student to sit in a swiveling desk chair a few feet from a desk lamp, which represents the sun. Attach the ball to the armrest of the chair. Slowly turn the chair (Earth), while the seated student observes the view of the moon model. Ask the seated volunteer to report what he or she observes. (The visible part of the ball always remains the same.) Ask the rest of the class what they see and how this differs from what the seated student sees. (Each letter marked on the ball appears in turn as the chair rotates.) Thus, the moon turns once on its axis in the time it takes to complete one orbit, so the same half of the moon always faces Earth. **LS Kinesthetic**

Answer to Reading Check

The far side of the moon is never visible from Earth, because the moon's rotation on its axis and the moon's revolution around Earth take the same amount of time.

Moonrise and Moonset

The moon appears to rise and set at Earth's horizon because of Earth's rotation on its axis. If you were to watch the moon rise or set on successive nights, however, you would notice that it rises or sets approximately 50 minutes later each night. This happens because of both Earth's rotation and the moon's revolution. While Earth completes one rotation each day, the moon also moves in its orbit around Earth. It takes an extra 1/29 of Earth's rotation, or 50 minutes, for the horizon to catch up to the moon.

Lunar Rotation

In addition to orbiting Earth and revolving around the sun, the moon also spins on its axis. The moon rotated rapidly when it formed, but the pull of Earth's gravity has slowed the moon's rate of rotation. The moon now spins very slowly and completes a rotation only once during each orbit around Earth. The moon revolves only once around Earth in about 27.3 days relative to the stars. Because the rotation and the revolution of the moon take the same amount of time, observers on Earth always see the same side of the moon. Therefore, images of the far side of the moon must be taken by spacecraft orbiting the moon.

As the moon orbits Earth, the part of the moon's surface that is illuminated by sunlight changes. The sun's light always illuminates half of the moon and, as shown in **Figure 2**, half the Earth. The near side of the moon is sometimes fully illuminated by the sun. At other times, depending on where the moon is in its orbit, the near side is partly or completely darkened.

Reading Check **Why are we unable to photograph the far side of the moon from Earth? (See the Appendix for answers to Reading Checks.)**

Figure 2 ▶ The lunar landing module, which was also called *The Eagle,* flew to meet the command module at the end of the *Apollo 11* mission. Seen from this viewpoint, Earth is illuminated from above.

BRAIN FOOD

Rotating Together Both Martian moons, Phobos and Deimos, are locked in synchronous orbits with Mars. Pluto and Charon are locked in orbits in which both bodies keep the same faces pointing toward each other as they rotate. Artificial satellites such as weather satellites are commonly placed in synchronous orbits to make continuous observations of one location on the globe. **LS Verbal**

INCLUSION Strategies

- *Attention Deficit Disorder*
- *Visually Impaired*
- *Developmentally Delayed*

To help students understand positions and orbits in the Earth-moon system, have one student pose as the sun, one as Earth, and one as the moon. Ask "Earth" and "moon" to revolve around each other as both revolve around "sun." Have them maintain an orbit with the sun off-center. **LS Kinesthetic**

Figure 3 ▶ During a solar eclipse, the shadow of the moon falls on Earth. The distance between Earth and the moon in this diagram is not to scale.

Eclipses

Bodies orbiting the sun, including Earth and its moon, cast long shadows into space. An **eclipse** occurs when one celestial body passes through the shadow of another. Shadows cast by Earth and the moon have two parts. In the inner, cone-shaped part of the shadow, the *umbra*, sunlight is completely blocked. In the outer part of the shadow, the *penumbra*, sunlight is only partially blocked, as shown in **Figure 3.**

Solar Eclipses

When the moon is directly between the sun and part of Earth, the shadow of the moon falls on Earth and causes a **solar eclipse.** During a *total solar eclipse*, the sun's light is completely blocked by the moon. The umbra falls on the area of Earth that lies directly in line with the moon and the sun. Outside the umbra, but within the penumbra, people see a *partial solar eclipse*. The penumbra falls on the area that immediately surrounds the umbra.

The umbra of the moon is too small to make a large shadow on Earth's surface. The part of the umbra that hits Earth during an eclipse, as shown in **Figure 4,** is never more than a few hundred kilometers across. So, a total eclipse of the sun covers only a small part of Earth and is seen only by people in particular parts of Earth along a narrow path. A total solar eclipse also never lasts more than about seven minutes at any one location. A total eclipse will not be visible in the United States until 2017, even though there is a total eclipse somewhere on Earth about every 18 months.

eclipse an event in which the shadow of one celestial body falls on another

solar eclipse the passing of the moon between Earth and the sun; during a solar eclipse, the shadow of the moon falls on Earth

Figure 4 ▶ The dark area (right) is the shadow of the moon cast on cloudtops over Europe during a total solar eclipse in 1999. The photo was taken from the orbiting space station *Mir*. The composite (left) shows the partial and total phases of a solar eclipse over several hours.

Using the Figure — Basic

Solar Eclipse Direct students' attention to the diagram that shows a solar eclipse. Point out that the dark inner shadow or umbra of the moon covers only a very small portion of Earth, sometimes less than 160 km wide. Ask students where people on Earth must be in order to observe a total solar eclipse. (Sample answer: They must be within the area covered by the umbra.) Ask them what observers located within the penumbra will see. (a partial eclipse of the sun) Make sure students understand that the distance between Earth and the moon in this diagram is not to scale. The distance between Earth and its satellite is actually about 60 times the radius of Earth. **LS Visual**

Activity — General

Eclipsed Demonstrate how a smaller body like the moon can completely block out the light of a larger body like the sun to produce a total eclipse. Provide a coin or small marble for each student. Have them close one eye and look at a distant large object such as a tree by using their open eye. Tell them to hold the coin or marble in front of the open eye and slowly bring it closer until it is directly in front of the eye. As the object nears their faces, the larger object gradually disappears from view. **LS Kinesthetic/Visual**

CHAPTER RESOURCES

Technology

Transparencies
- 142 Solar and Lunar Eclipses (with worksheet)

BRAIN FOOD

Size is Relative Earth is unusual in the inner solar system in that it has an extremely large companion body. The diameter of the sun is about 400 times greater than the diameter of the moon, but the sun is also roughly 400 times farther away from Earth. The combination of the moon's size and distance from Earth causes the moon to appear roughly the same size as the sun does, and helps explain why eclipses can occur. **LS Verbal**

Teach, *continued*

QuickLAB

Skills Acquired
- Making models
- Observing
- Analyzing

Teacher's Notes

Point out that with a penlight as the light source, the light spreads out to form a cone-shape, whereas the sun's rays are actually parallel. You may want to use a larger light, such as a wide flashlight or overhead projector, and place waxed paper over the lens's surface to create a disk of uniform brightness that is larger than the two balls are.

Answers

1. The penlight represents the sun. The larger ball represents Earth and the smaller ball represents the moon.
2. As viewed from Earth, step 3 represents a lunar eclipse. From the moon, it would represent a solar eclipse because the sun's light is blocked by Earth.
3. As viewed from Earth, step 4 represents a solar eclipse in which part of Earth is in the moon's shadow. As viewed from the moon, it would represent an eclipse of Earth in which part of Earth passes through the moon's shadow.
4. Answers may vary. Sample answers: If the clay balls that represent the moon and Earth were placed on dowel rods, they could be positioned more accurately in space. It would also be more realistic if the models were moving around each other instead of remaining stationary.

Figure 5 ▶ The diamond-ring effect produced by a solar eclipse can be stunning for observers on the part of Earth that falls under the moon's shadow.

Effects of Solar Eclipses

During a total solar eclipse, people on the ground are in the moon's umbra. In the areas on Earth's surface under the umbra, the sky becomes as dark as it does at twilight. During this period of darkness, the sunlight that is not eclipsed by the moon shows the normally invisible outer layers of the sun's atmosphere. The last bits of normal sunlight before darkness often glisten like the diamond on a ring and cause what is known as the *diamond-ring effect*. The diamond-ring effect is shown in **Figure 5.** Therefore, many people think that total solar eclipses are very beautiful.

If the moon is at or near apogee when it comes directly between Earth and the sun, the moon's umbra does not reach Earth. If the umbra fails to reach Earth, a ring-shaped eclipse occurs. This type of eclipse is called an *annular eclipse,* because *annulus* is the Latin word for "ring." During an annular eclipse, the sun is never completely blocked out. Instead, a thin ring of sunlight is visible around the outer edge of the moon. The brightness of this thin ring of ordinary sunlight prevents observers from seeing the outer layers of the sun's atmosphere that are visible during a total solar eclipse.

Reading Check **What is one difference between a total solar eclipse and an annular eclipse? (See the Appendix for answers to Reading Checks.)**

QuickLAB 15 min

Eclipses

Procedure

1. Make two balls from **modeling clay,** one about 4 cm in diameter and one about 1 cm in diameter.
2. Using a **metric ruler,** position the balls about 15 cm apart on a **sheet of paper,** as shown in the photo at right.
3. Turn off any nearby lights. Place a **penlight** approximately 15 cm in front of and almost level with the larger ball. Shine the light on the larger ball. Sketch your model, and note the effect of the beam of light.
4. Repeat step 3, but reverse the positions of the two balls. You may need to raise the smaller ball slightly to center its shadow on the larger ball. Sketch your model, and again note the effect of the light beam.

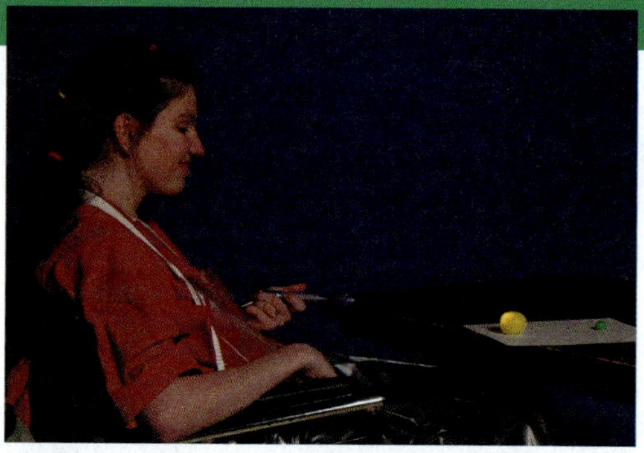

Analysis

1. Which planetary bodies do the larger clay ball, the smaller clay ball, and the penlight represent?
2. As viewed from Earth, what event did your model in step 3 represent? As viewed from the moon, what would your model represent?
3. As viewed from Earth, what event did your model in step 4 represent? As viewed from the moon, what would your model represent?
4. In what ways could you modify this activity to more closely model how eclipses occur?

CHAPTER RESOURCES

Chapter Resource File
- Datasheet for Quick Lab GENERAL

Answer to Reading Check

During a total eclipse, the entire disk of the sun is blocked, and the outer layers of the sun become visible. During an annular eclipse, the disk of the sun is never completely blocked out, so the sun is too bright for observers on Earth to see the outer layers of the sun's atmosphere.

Figure 6 ▶ During a lunar eclipse, the shadow of Earth falls on the moon. The distance between Earth and the moon in this diagram is not to scale.

Lunar Eclipses

A **lunar eclipse** occurs when Earth is positioned between the moon and the sun and when Earth's shadow crosses the lighted half of the moon. For a total lunar eclipse to occur, the entire moon must pass into Earth's umbra, as shown in **Figure 6.** When only part of the moon passes into Earth's umbra, a *partial lunar eclipse* occurs. The remainder of the moon passes through Earth's penumbra. When the entire moon passes through Earth's penumbra, a *penumbral eclipse* occurs. During a penumbral eclipse, the moon darkens so little that the eclipse is barely noticeable.

A lunar eclipse may last for more than an hour. Even during a total lunar eclipse, sunlight is bent around Earth through our atmosphere. Mainly red light reaches the moon, so the totally eclipsed moon appears to have a reddish color, as shown in the middle portion of the composite image in **Figure 7.**

lunar eclipse the passing of the moon through Earth's shadow at full moon

Frequency of Solar and Lunar Eclipses

As many as seven eclipses may occur during a calendar year. Four may be lunar, and three may be solar or vice versa. However, total eclipses of the sun and the moon occur infrequently. Solar and lunar eclipses do not occur during every lunar orbit. This is because the orbit of the moon is not in the same plane as the orbit of Earth around the sun. The moon crosses the plane of Earth's orbit only twice in each revolution around Earth. A solar eclipse will occur only if this crossing occurs when the moon is between Earth and the sun. If this crossing occurs when Earth is between the moon and the sun, a lunar eclipse will occur.

Lunar eclipses are visible everywhere on the dark side of Earth. A total solar eclipse, however, can be seen only by observers in the small path of the moon's shadow as it moves across Earth's lighted surface. A partial solar eclipse can be seen for thousands of kilometers on either side of the path of the umbra.

Figure 7 ▶ This composite image shows a total lunar eclipse as seen from Earth over several hours.

Teaching Tip — GENERAL

Connect to Familiar Processes Students may better understand why the moon sometimes has a copper color during a lunar eclipse if you compare the color of the moon with the color of the sky at sunset. Sunsets are red because all of the other colors in sunlight are scattered by dust and molecules in Earth's atmosphere. Only red and orange wavelengths pass through the atmosphere to shed a rosy glow over the early evening sky. The same effect that creates the spectacular shades of red during a sunset also paints the moon a coppery red during a total lunar eclipse. **LS Verbal**

MISCONCEPTION ALERT

Eclipse Frequency Students may think that both a lunar and solar eclipse should happen every month as the moon completes its orbit of Earth. However, the moon is rarely in direct alignment with Earth's orbital plane. Because the moon's orbit is tilted about 5° from the plane of Earth's orbit around the sun, which is called the *ecliptic plane*, the shadows do not fall on Earth. An eclipse can occur only when the moon crosses the ecliptic while it is in the new-moon phase to create a solar eclipse, or in full-moon phase to create a lunar eclipse. This happens only about seven times a year.

LIFE SCIENCE CONNECTION — GENERAL

Biological Rhythms Biologists and zoologists sometimes use eclipses to learn more about the rhythms of living creatures. Totality resembles the arrival of night, and animals and plants respond accordingly. From monitoring the activities of a colony of bats or of fish in coral reefs, researchers study how animals and plants react to the brief moments of darkness.

For example, a team of Polish zoologists observed mammal, bird, and insect behavior from 1954 to 1975 during seven eclipses with different levels of totality. They discovered that insects and birds were influenced the most. Many bird species seemed anxious, stopped singing, and exhibited roosting behaviors. Bees returned to their hives, nocturnal insects such as mosquitoes and moths appeared, and butterflies settled for the night. Invite interested students to learn more about the effects of such events on circadian rhythms of living things and to present their findings in an oral report. **LS Verbal**

Teach, *continued*

BRAIN FOOD

Phases of Earth From the moon, Earth also appears to have phases. However, the phases of Earth would be exactly opposite those of the moon. For example, when the moon is in the new-moon phase, an observer on the near side of the moon would see a "full Earth." As the moon waxes, Earth would appear to wane. LS Verbal

Group Activity — GENERAL

Many Moons Collect enough foam balls for each student to have one. Put a lamp without its shade on a stand in the front of the classroom. Distribute the balls and have each student place the point of a pencil into their ball to form a handle. Have students stand in a semicircle facing the light. Tell students to imagine that the ball is the moon. Their bodies represent the spinning Earth. Darken the room lights. Have students turn around counter-clockwise slowly, holding the ball level with their heads at arm's length so that the light reflects off the model moon. Tell them to stop every quarter turn to observe the dark portion of the ball. Ask students to compare the positions of the moon model, the light (sun), and their heads (Earth) with the diagram of the moon phases, and explain what causes the different parts of the model moon to light up. LS Visual/Kinesthetic

CHAPTER RESOURCES

Chapter Resource File

- **Inquiry Lab** Inconstant Moon GENERAL

Technology

Transparencies
- 143 Phases of the Moon (with worksheet)
- 144 Causes of Tides (with worksheet)

Phases of the Moon

On some nights, the moon shines brightly enough for you to read a book by its light. But moonlight is not produced by the moon. The moon merely reflects light from the sun. Because the moon is spherical, half of it is always lit by sunlight. As the moon revolves around Earth, however, different amounts of the near side of the moon, which faces Earth, are lighted. Therefore, the apparent shape of the visible part of the moon varies. These varying shapes, lighted by reflected sunlight, are called **phases** of the moon and are shown in **Figure 8.**

When the moon is directly between the sun and Earth, the sun's rays strike only the far side of the moon. As a result, the entire near side of the moon is dark. When the near side is dark, the moon is said to be in the *new-moon* phase. During this phase, no lighted area of the moon is visible from Earth.

phase in astronomy, the change in the illuminated area of one celestial body as seen from another celestial body; phases of the moon are caused by the changing positions of Earth, the sun, and the moon

Figure 8 ▶ Phases of the Moon

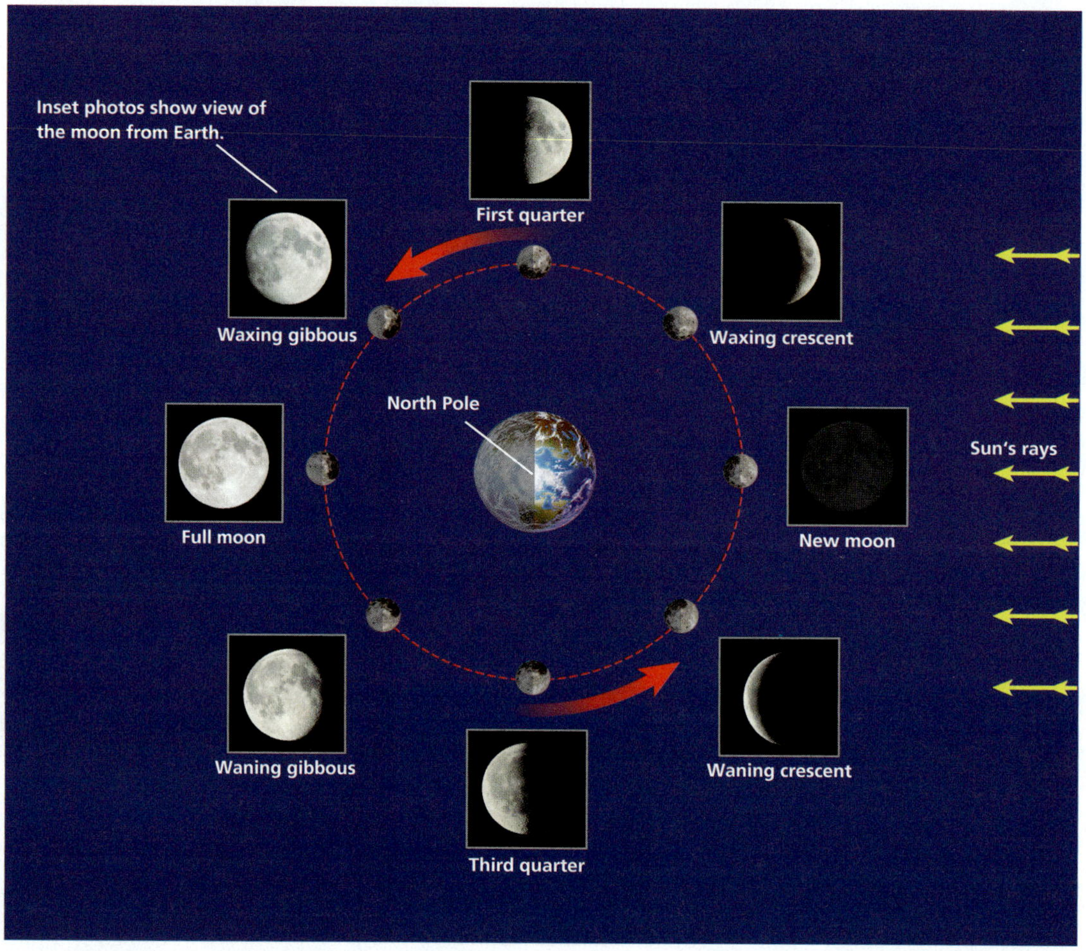

Cultural Awareness — GENERAL

Lunar Calendar The period from one new moon to the next is one lunar month. Islamic, Jewish, and other cultures base their calendars on the lunar month. The problem with these calendars is that the lunar year does not contain a whole number of lunar months. It is 11 days short. Different cultures had to find ways of reconciling this difference. Invite interested students to find out more about these calendars and how events are timed according to the lunar phases, or how the calendar has been reconciled. LS Logical

Homework — GENERAL

Moon Watching Invite students to keep a daily moon observation journal for two weeks or more. Students can make naked eye observations or they may use binoculars. Have them record the date and time of their observations, the phase of the moon, and where it appears in the sky. Depending on the phase, students may be able to make observations during daylight hours as well as at night. Have them note the phase, make a picture of what the moon looks like, and list its rising and setting times. Guide students to look for any patterns they observe. LS Visual

Waxing Phases of the Moon

As the moon continues to move in its orbit around Earth, part of the near side becomes illuminated. When the size of the lighted part of the moon is increasing, the moon is said to be *waxing*. When a sliver of the moon's near side is illuminated, the moon enters its *waxing-crescent* phase.

When the moon has moved through one-quarter of its orbit after the new moon phase, the moon appears to be a semicircle. Half of the near side of the moon is lighted. When a waxing moon becomes a semicircle, the moon enters its *first-quarter* phase. When the lighted part of the moon's near side is larger than a semicircle and still increasing in size, the moon is in its *waxing-gibbous* phase. The moon continues to wax until it appears as a full circle. At *full moon*, Earth is between the sun and the moon. Consequently, the entire near side of the moon is illuminated by the light of the sun.

Waning Phases of the Moon

After the full moon phase, when the lighted part of the near side of the moon appears to decrease in size, the moon is *waning*. When it is waning but the lighted part is still larger than a semicircle, the moon is in the *waning-gibbous* phase. When the lighted part of the near side becomes a semicircle, the moon enters the *last-quarter* phase. When only a sliver of the near side is visible, the moon enters the *waning-crescent* phase. After the waning-crescent phase, the moon again moves between Earth and the sun. The moon once more becomes a new moon, and the cycle of phases begins again.

Before and after a new moon, only a small part of the moon shines brightly. However, the rest of the moon is not completely dark. It shines dimly from sunlight that reflects first off Earth's clouds and oceans and then reflects off the moon. Sunlight that is reflected off Earth is called *earthshine*. The darker part of the moon shown in **Figure 9** is lit by earthshine.

Time from New Moon to New Moon

Although the moon revolves around Earth in 27.3 days, a longer period of time is needed for the moon to go through a complete cycle of phases. The period from one new moon to the next one is 29.5 days. This difference of 2.2 days is due to the orbiting of the Earth-moon system around the sun. In the 27.3 days in which the moon orbits Earth, the two bodies move slightly farther along their orbit around the sun. Therefore, the moon must go a little farther to be directly between Earth and the sun. About 2.2 days are needed for the moon to travel this extra distance. The position directly between Earth and the sun is the position of the moon in each new moon phase.

Reading Check **Describe two phases of the waning moon. (See the Appendix for answers to Reading Checks.)**

SCILINKS. NSTA Developed and maintained by the National Science Teachers Association

For a variety of links related to this subject, go to www.scilinks.org

Topic: Lunar Cycle
SciLinks code: HQ60887

Figure 9 ▶ The darker portion of this crescent moon is not completely dark because some sunlight is reflected from Earth, to the moon, and back to Earth.

Close

Reteaching — BASIC

Lunar Flipbook Have students create a flipbook about the order of the moon's phases. Students can make drawings based on the text on index cards and label the phases. Stack the cards in order with the new moon on the bottom and attach them with a binder clip so they can flip the cards to view the phases. **LS Visual**

Quiz — GENERAL

1. Explain why the barycenter of the Earth-moon system is not located halfway between the two bodies. (Because Earth is so much more massive than the moon, the balance point is located within Earth's interior.)
2. How does the shape of the lunar orbit cause an annular eclipse? (Because the moon's orbit is elliptical, if the moon is at or near apogee when it passes between Earth and the sun, the moon is too far from Earth for its umbra to reach Earth.)

Alternative Assessment — ADVANCED

System Models Give groups of students foam balls, modeling clay, dowel rods, pencils, string, paper clips, tape, and poster board. Have them make a model to explain to younger students the phases of the moon, how eclipses occur, or what causes tides. Remind them that Earth and the moon share a common center of gravity. Have each student use the model to explain the process involved. **LS Kinesthetic**

Answer to Reading Check

When the lighted part of the moon is larger than a semicircle but the visible part of the moon is shrinking, the phase is called *waning gibbous*. When only a sliver of the near side is visible, the phase is a waning crescent.

SKILL BUILDER — GENERAL

Vocabulary The term *crescent* comes from the Latin word *crescere*, which means "to grow." A crescent moon is growing from the dark new-moon phase. The term *gibbous* is derived from the Latin *gibbosus*, which means "humpbacked." A gibbous moon is larger than a quarter moon but smaller than a full moon. The words that describe whether the moon is getting larger or smaller have Germanic roots. *Waxing* comes from the German *wachsen*, which means "to grow." *Waning* comes from the Middle English word *wanen*, which means "to decrease in size." **LS Verbal** English Language Learners

Close, *continued*

Answers to Section Review

1. As Earth completes one rotation each day, the moon also moves in its orbit around Earth. It takes an extra 1/29 of Earth's rotation (or 50 minutes) for Earth's horizon to catch up with the moon.
2. The moon is directly between the sun and Earth, and the cone-shaped inner shadow or umbra of the moon falls on part of Earth.
3. The moon's orbit arond Earth is not in the same plane as Earth's orbit around the sun. A total lunar or solar eclipse occurs only when the moon and Earth are near one of the two crossings of the orbits at the same time.
4. Solar eclipses occur when the moon passes between the sun and Earth. Lunar eclipses occur when Earth is between the sun and the moon and the moon passes through Earth's shadow.
5. The moon is between Earth and the sun.
6. The lighted portion appears to grow larger.
7. Although the moon revolves around Earth in 27.3 days, a longer period of time is needed for the moon to go through a complete cycle of phases. The period from one new moon to the next one is 29.5 days. This difference of 2.2 days is due to the orbiting of the Earth-moon system around the sun.
8. On the near side of Earth, the gravitational force of the moon pulls ocean water toward the moon. The mass of the solid Earth is subject to less gravitational force than the ocean water is. On the far side of Earth, the water is subject to less gravitational force than the solid Earth is. These gravitational differences cause the water on the near side and on the far side of Earth to form two tidal bulges.
9. Observers on Earth always see the same side of the moon because the moon revolves around Earth at the same rate that it rotates on its axis.
10. Lunar eclipses are visible everywhere on the dark side of Earth. Only observers in the narrow path of the moon's shadow see a total solar eclipse as it crosses Earth's lighted surface.

Figure 10 ▶ The moon's pull on Earth is greatest at point A, on Earth's near side, and weakest at point C, on Earth's far side. Point B represents Earth's center of mass. Earth's rotation causes two low tides and two high tides each day on most shorelines.

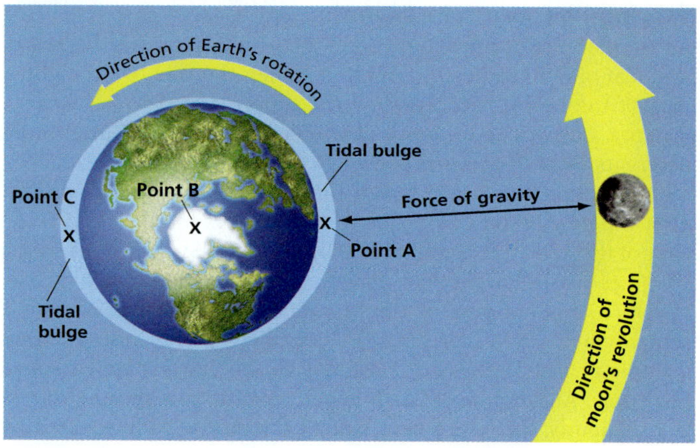

Tides on Earth

Bulges in Earth's oceans, called *tidal bulges,* form because the moon's gravitational pull on Earth decreases with distance from the moon. As a result, the ocean on Earth's near side is pulled toward the moon with the greatest force. The solid Earth, which acts as though all of its mass were at Earth's center, experiences a lesser force. The ocean on Earth's far side is subject to less force than the solid Earth is. As shown in **Figure 10,** these differences cause Earth's tidal bulges. Because Earth rotates, tides occur in a regular rhythm at any given point on Earth's surface each day. The sun also causes tides, but they are smaller because the sun is so much farther from Earth than the moon is.

Section 2 Review

1. **Explain** why the moon rises and sets about 50 minutes later each successive night.
2. **Describe** the conditions that cause a total solar eclipse to occur.
3. **Explain** why a lunar eclipse does not occur every time the moon revolves around Earth in its monthly orbit.
4. **Summarize** how a solar eclipse differs from a lunar eclipse.
5. **Describe** the relative locations of the sun, Earth, and the moon during a new moon phase.
6. **Describe** how the appearance of the moon changes when it is waxing.
7. **Explain** why the moon repeats phases every 29.5 days even though it orbits Earth every 27.3 days.
8. **Explain** how the moon causes tidal bulges on Earth.

CRITICAL THINKING

9. **Analyzing Ideas** Explain why observers on Earth always see the same side of the moon.
10. **Making Comparisons** Explain why more people see each total lunar eclipse than each total solar eclipse.
11. **Analyzing Relationships** The sun's gravity also affects tides on Earth. Why does the moon have a larger effect on tides than the sun does?

CONCEPT MAPPING

12. Use the following terms to create a concept map: *Earth, moon, sun, eclipse, solar eclipse, lunar eclipse, umbra,* and *penumbra.*

CHAPTER RESOURCES

Chapter Resource File

- Section Quiz GENERAL

Workbooks

- Study Guide (also in Spanish)

Section 3 Satellites of Other Planets

Until the 1600s, astronomers thought that Earth was the only planet that had a moon. In 1610, Galileo discovered four moons orbiting Jupiter. He also observed what later were identified as the rings of Saturn. Since the time of Galileo, astronomers have discovered that all of the planets in our solar system except Mercury and Venus have moons. In addition, the gas giants Saturn, Jupiter, Uranus, and Neptune all have rings.

OBJECTIVES

- **Compare** the characteristics of the two moons of Mars.
- **Describe** how volcanoes were discovered on Io.
- **Name** one distinguishing characteristic of each of the Galilean moons.
- **Compare** the characteristics of the rings of Saturn with the rings of the other outer planets.

KEY TERM

Galilean moon

Moons of Mars

Mars has two tiny moons, named Phobos and Deimos. They revolve around Mars relatively quickly. Phobos and Deimos are irregularly shaped chunks of rock and are thought to be captured asteroids. Phobos is 27 km across at its longest, and Deimos is about 15 km across at its longest.

The surfaces of Phobos and Deimos are dark like maria on Earth's moon. Both moons have many craters. The large number of craters shows that the moons have been hit by many asteroids and comets, and suggests that the moons are fairly old.

Moons of Jupiter

Galileo observed four large moons revolving around Jupiter. Since that discovery was made, scientists have observed dozens of smaller moons around Jupiter. Smaller moons continue to be discovered today. Most of Jupiter's moons have diameters of less than 200 km, but of the largest four, known as the **Galilean moons,** three are bigger than Earth's moon. Until spacecraft flew near the moons, scientists knew little about them. Now, scientists have identified many unique characteristics of the Galilean moons. One of the four Galilean moons is shown in **Figure 1.**

Galilean moon any one of the four largest satellites of Jupiter—Io, Europa, Ganymede, and Callisto—that were discovered by Galileo in 1610

Figure 1 ▶ The stormy surface of Jupiter is visible in the background. In the foreground is Io, which orbits Jupiter once every 42 hours.

Section 3

Focus

Overview

This section identifies the distinguishing features of the moons and ring systems of planets in our solar system and describes how space missions have helped us learn more about these objects.

Bellringer

Ask students to write a definition of a moon that identifies the essential characteristics of a moon. (Answers may vary.) Use student responses to assess prior knowledge and encourage deeper thinking. **LS Verbal**

Motivate

Discussion — GENERAL

Mythology Connection Have students share what they know about historical origins for names of the planets. Invite interested students to learn more about the names of planetary moons. Students may recognize some lunar names that originated in classical stories. Guide students to explain why a particular moon may be aptly named and why the same name may fit well in a related modern setting. Have students share their findings with the class. **LS Verbal**

CHAPTER RESOURCES

Chapter Resource File

- Directed Reading BASIC

Technology

- **Transparencies**
  - Bellringer
- **Student Edition on CD-ROM**
- **One-Stop Planner CD-ROM**
  - Lesson Plan

Teach

Internet Activity — GENERAL

Martian Moons Invite interested students to research the moons of Mars. While the most commonly accepted theory is that they are captured asteroids, some scientists have suggested that the Martian moons could be fragments that formed during the break up of a much larger body that once circled the Red Planet. The nearer moon, Phobos, may eventually spiral in and crash on the planet's surface or be torn apart by gravity. A worksheet designed to direct student research on this topic can be found in the **Chapter Resource File** booklet or by visiting **go.hrw.com** and entering the keyword **HQ6MBSX.** **LS Visual/Auditory**

READING SKILL BUILDER — BASIC

Reading Organizer As students read the section about the moons of Jupiter, encourage them to take Power Notes, KWL Notes, or Two-Column Notes as described in the Skills Handbook section of the Appendix to identify the chief characteristics of the Galilean moons. Later, students can use their notes as a study guide to prepare for assessments. **LS Verbal** English Language Learners

CHAPTER RESOURCES

Chapter Resource File

Internet Activity
- Martian Moons GENERAL

Figure 2 ▶ In this image taken by the *Galileo* spacecraft you can see a volcanic eruption on the left side of Io against the background of space.

Io

Io is the innermost of Jupiter's four Galilean moons. An engineer examining images from the *Voyager* spacecraft discovered volcanoes on Io. Io is the first extraterrestrial body on which active volcanoes have been seen. Since the discovery of Io's volcanoes, scientists realize that volcanism is more widespread in the solar system than they had thought. Volcanoes on Io eject thousands of metric tons of material each second. The lava that erupts on Io is much hotter than the lava that erupts on Earth. The temperature of the lava on Io is higher because the lava has more magnesium and iron than lava on Earth does. Plumes of volcanic material on Io reach heights of hundreds of kilometers, as shown in **Figure 2.** Because parts of Io's surface are yellow-red, scientists think that the volcanic material is mostly sulfur and sulfur dioxide.

Io moves inward and outward in its orbit around Jupiter because of the gravitational pull of the other moons of Jupiter. These *tidal forces* are caused by the difference between the force on one side of Io and that on the other side. These forces are similar to tides on Earth caused by the pull of the moon. As Io is pulled back and forth, its surface also moves in and out. Calculations show that tidal forces make Io's surface move in and out by 100 m. Heat from the friction caused by this surface flexing results in the melting of the interior of Io and leads to volcanism. Data from the *Galileo* spacecraft show that Io has a giant iron core and may possess a magnetic field. Much of what we know about Jupiter's moons came from information gathered by the *Galileo* spacecraft, which orbited Jupiter from 1995 to 2003.

Connection to GEOLOGY

Extraterrestrial Volcanism

Volcanoes on Earth can be compared to volcanoes that exist elsewhere in the universe. Recent discoveries have revealed that volcanism was very common in our solar system's past. Mars and Venus each have giant volcanoes that are now extinct. Some of these volcanoes are much bigger than the largest volcanoes on Earth. Mars's Olympus Mons is 600 km across and 25 km tall. The largest volcano on Earth, Hawaii's Mauna Loa, is only 17 km tall. Radar images show lava flows on Venus that extend for hundreds of kilometers. Venus and Mars also have many smaller volcanic landforms.

It was a surprise when the *Voyager I* spacecraft, flying by in 1979, revealed interesting and complex structures on the Galilean moons. Because of its small size, Io was expected to be a geologically dead world, like Earth's moon. Some long-exposure images were taken of Io to show the stars behind it and to help track the spacecraft's position. Linda Morabito, an engineer working at the California Institute of Technology's Jet Propulsion Laboratory, examined the images and noticed something extending over Io's edge. It could not be a cloud, because Io had no atmosphere. A volcano was the only possibility. Since that time, scientists have realized that Io is the most volcanically active place in the solar system.

The only other place in the solar system where active volcanism is known is on Neptune's moon Triton. The *Voyager 2* spacecraft observed dark streaks on Triton's surface that show where nitrogen that erupted from below Triton's surface was carried downwind.

The black plumes at the bottom left of the photo are from an erupting volcano on Io.

BRAIN FOOD

Volcanoes and Auroras Volcanism on Io affects Jupiter's entire system because Io's volcanoes are the primary source of matter in Jupiter's magnetic field. The moon acts as an electrical generator as it moves through Jupiter's magnetic field, generating an electric current that flows along the magnetic field lines to the planet's ionosphere. The current's power output is greater than all of the power stations in the United States. Rotating with the planet, the magnetosphere sweeps past Io and strips away particles erupted by Io's volcanoes and accelerates the particles to high speeds. The material forms a doughnut-shaped cloud of ions that completely encircles Jupiter. The pressure produced by these ions inflates the magnetic field to more than twice the size scientists predicted it would be. Many of the more-energetic ions are pulled along the magnetic field lines, creating auroras in Jupiter's upper atmosphere.

Europa

Europa is the second closest Galilean moon to Jupiter. Europa is about the size of Earth's moon, but Europa is much less dense than Earth's moon. Astronomers think that Europa has a rock core that is covered with a crust of ice that is about 100 km thick. Images of Europa, such as the one shown in **Figure 3,** show cracks in this enormous ice sheet. Scientists have concluded from observations made from spacecraft that an ocean of liquid water may exist under this blanket of ice. If liquid water exists, simple forms of life could also exist there. Astronomers have no evidence of life on Europa, but many think Europa would be a good place to investigate the possibility of extraterrestrial life.

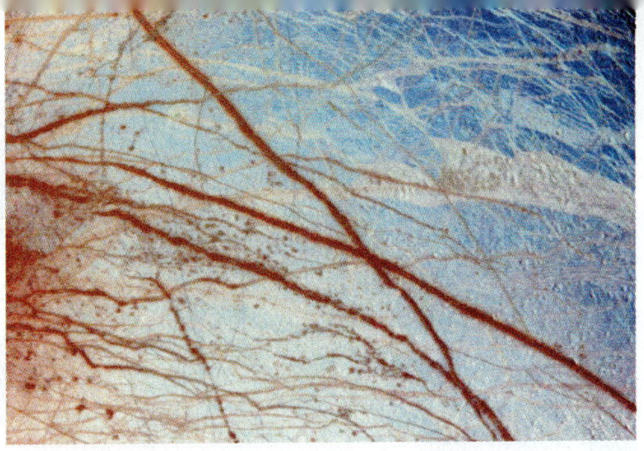

Figure 3 ▶ This false-color image of Europa shows immense cracks across its ice sheets.

Ganymede

Ganymede is the third Galilean moon from Jupiter. Ganymede is also the largest moon in the solar system, even larger than the planet Mercury. However, Ganymede has a relatively small mass because it is probably composed mostly of ice mixed with rock.

Images of Ganymede, such as **Figure 4,** show dark, crater-filled areas. Other light areas show marks that are thought to be long ridges and valleys. The Galileo spacecraft provided evidence to support the existence of a magnetic field around Ganymede. Ganymede and Io are the only Galilean moons that have strong magnetic fields. Both moons' magnetic fields are completely surrounded by Jupiter's much more powerful magnetic field.

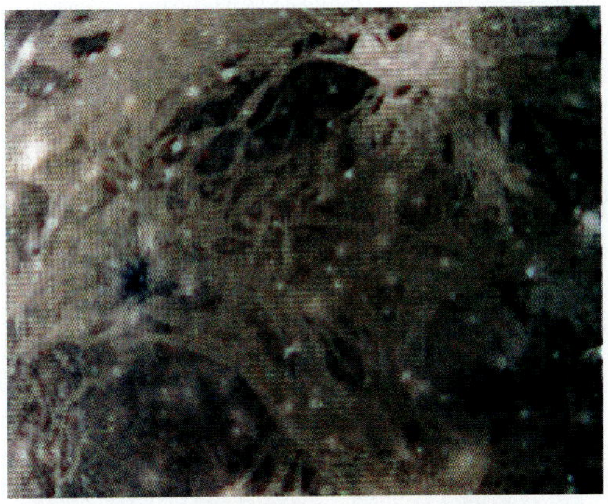

Figure 4 ▶ Much of Ganymede's surface is covered with ridges and valleys.

Callisto

Of the four Galilean moons, Callisto is the farthest from Jupiter. Callisto is similar to Ganymede in size, density, and composition. However, Callisto has a much rougher surface than Ganymede does. In fact, Callisto may be one of the most densely cratered moons in our solar system.

Like craters on Earth's moon and other bodies in our solar system, craters on Callisto are the result of collisions that occurred early in the history of the solar system. **Figure 5** shows a giant impact basin that is 600 km across and a set of concentric rings that extend about 1,500 km outward in all directions from the crater.

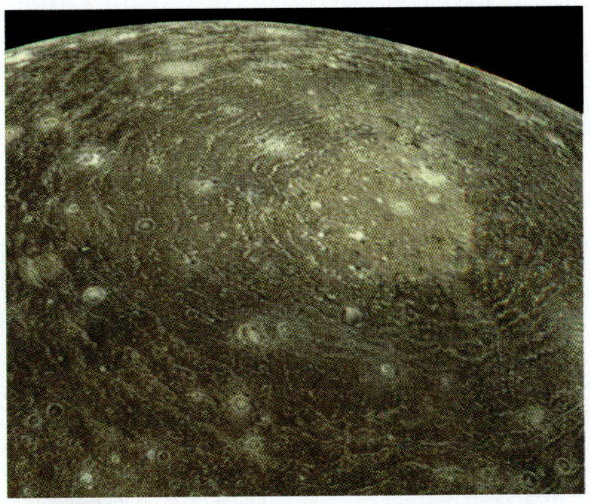

Figure 5 ▶ "Ripples" of ice and rock radiate out from this impact crater, called *Valhalla*, on Callisto.

Reading Check **Name one feature of each of the Galilean moons. (See the Appendix for answers to Reading Checks.)**

Using the Figure — GENERAL

Geologic Pasts Use the photos of the surfaces of Io, Europa, Ganymede, and Callisto to stimulate a discussion of the diverse geology of Jupiter's moons. Ask students which moon is the most geologically active and why they know that. (Massive volcanoes make Io the most geologically active. Io is continuously being resurfaced as lava flows erase any craters.) Point out that the ice-covered surface of Europa is also remarkably smooth and uncratered. Its craters have been erased, perhaps by liquid water flowing through its cracked surface. Ask students to describe the surface of Ganymede. (It has mountains, ridges, valleys, craters, and dark lava flows.) Explain that some of the grooved features on Ganymede's surface appear to be the result of tectonic processes. Ask students which moon's surface exhibits the most ancient surface features. (Callisto, because its surface is covered by large impact craters from the early days of the solar system.) **LS Visual**

Answer to Reading Check

Io's surface is covered with many active volcanoes. Europa's surface is covered by an enormous ice sheet. Ganymede is the largest moon in the solar system and has a strong magnetic field. Callisto's surface is heavily cratered.

ENVIRONMENTAL CONNECTION

Hidden Ocean, Hidden Life? Scientists who are looking for extraterrestrial life are focusing on Jupiter's moon Europa. Oceanographer David Karl has been studying microscopic life in Antarctica, an environment similar to Europa's. Karl hopes to discover ways to sample liquids under frozen ice without contaminating the liquids. This technology could be used to search for signs of life on Europa. Invite students to research the evidence for life on Europa and the biology of microbes in extreme environments on Earth. **LS Auditory/Verbal**

INCLUSION Strategies

- *Learning Disabled*
- *Visually Impaired*
- *Developmentally Delayed*

Using mnemonic techniques may help students remember the names for the main moons of each planet. Encourage students to invent mnemonic devices for each planet and its main moons. Example 1: Mark phoned Desi (Mars—Phobos, Deimos). Example 2: Judy ignored Eddie Gand's calls (Jupiter—Io, Europa, Ganymede, Callisto). **LS Verbal**

Teach, *continued*

Using the Figure — BASIC

Ringmaster Direct students' attention to the picture of Saturn's family. Have them describe the bodies that make up the Saturn system. (Saturn, the second largest planet in the solar system, is encircled by hundreds of icy rings, one large moon called Titan, and at least 46 other moons of various sizes.) Ask students to estimate the relative size of Saturn based on the scale of the image. (Answers may vary but students should realize that Saturn would be extremely large. Students may wish to use butcher paper to try to draw a scale version of the planet.) **LS Visual**

Demonstration — GENERAL

Shepherd Moons You will need a rotating stand or turntable. Scatter talcum powder or fine sand on the rotating surface. Explain that the powder represents the chunks of ice that make up planetary rings. Tape two pencils together with their points even. The points represent the inner moons of the Saturn system. Spin the rotating disk. Lower the pencil points into the powder while spinning the turntable. Ask students to describe what happens. (Sample answer: The pencils push the powder aside to clear two paths.) Invite a volunteer to compare the model to how inner moons affect planetary rings. (Sample answer: Moons moving within the rings cause the materials that make up the rings to form separate bands.) **LS Visual**

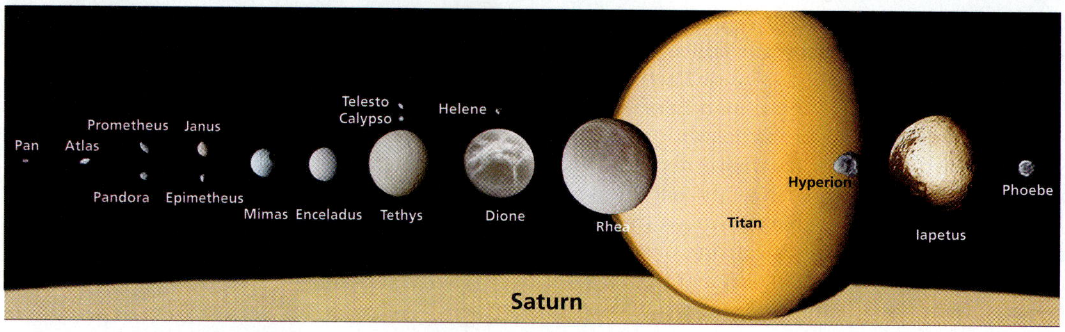

Figure 6 ▶ This composite image shows Saturn and many of Saturn's largest moons. The distances of the moons from Saturn and from each other are not to scale.

Moons of Saturn

Saturn has at least 47 moons. Most of them are small, icy bodies that have many craters. However, five of Saturn's moons are fairly large. These five moons and many of Saturn's other moons are shown in **Figure 6.**

Graphic Organizer — **Comparison Table**

Create the Graphic Organizer entitled "Comparison Table" described in the Skills Handbook section of the Appendix. Label the columns with "Moons of Mars," "Moons of Jupiter," "Moons of Saturn, and "Moons of Uranus and Neptune." Fill in the table with details about each moon as you read this section.

Titan

Saturn's largest moon, called Titan, has a diameter of more than 5,000 km. Only Jupiter's moon Ganymede is larger. Unlike any of the other moons in our solar system, Titan has a thick atmosphere that is composed mainly of nitrogen. Titan's atmosphere is so thick that hydrocarbon smog conceals the surface.

In 2005, the *Huygens* (HIE guhnz) probe, part of the Cassini mission, gathered data about Titan's atmosphere. The information it gathered is giving scientists clues about how Titan and its atmosphere formed. When the probe got below much of the atmosphere's haze, it sent back clear images of the surface, which showed signs of flowing liquid, probably methane. *Cassini* also sent back many images of Titan, an example of which is shown in **Figure 7.**

Saturn's Other Moons

Saturn's icy moons resemble Jupiter's icy Galilean moons. Saturn's other smaller moons have irregular shapes. Scientists think that many of the smallest moons, such as Janus, were captured by Saturn's gravity.

Figure 7 ▶ This false-color image of Titan was taken by the Cassini mission. The image shows Titan in the ultraviolet and infrared wavelenghts.

Graphic Organizer — GENERAL

Comparison Table

You may want to use this Graphic Organizer in a game to review material before the test. Divide the class into two teams. Ask students questions about material from the Comparison Table. Give points to each team that provides correct answers.

METEOROLOGY CONNECTION — GENERAL

Titan's Atmosphere The chemistry of Titan's atmosphere appears to be similar to that of Earth's nitrogen atmosphere before living things introduced oxygen and modified it. However, the surface temperature is much colder. On Titan, methane may play the same role as water does on Earth: forming clouds, condensing as rain, and forming oceans on the surface. Studying Titan's atmosphere and surface may give scientists important clues to how life got started on Earth. Invite students to learn more about the Cassini/Huygens mission to Saturn and Titan. **LS Verbal/Auditory**

Moons of Uranus and Neptune

Uranus's four largest moons, Oberon, Titania, Umbriel, and Ariel, were known by the mid-1800s. A fifth, Miranda, was discovered in 1948 and is shown in **Figure 8.** Other much smaller moons have been discovered recently by using spacecraft and orbiting observatories such as the *Hubble Space Telescope*. Astronomers know that Uranus has at least two dozen small moons.

Neptune has at least 13 moons. Triton, an icy moon, is unusual because it revolves around Neptune in a backward, or *retrograde*, orbit. Some astronomers think that Triton has an unusual orbit because the moon was probably captured by the gravity of Neptune after forming elsewhere in the solar system and then coming too close to the planet. Triton's diameter is 2,705 km, and the moon has a thin atmosphere.

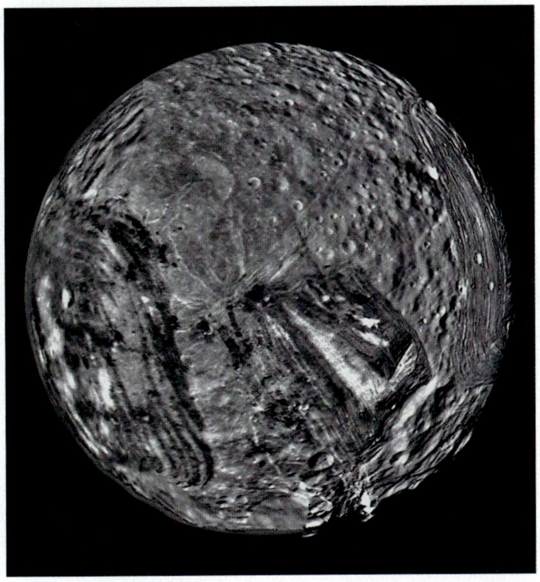

Figure 8 ▶ Uranus's moon called Miranda shows intriguing evidence of past geologic activity.

Pluto's Moons

Although Pluto is no longer officially considered a planet, it does have at least three moons. Pluto's largest moon, Charon (KER uhn), is almost half the size of Pluto. In fact, because Pluto and Charon are so similar in mass, they both orbit a common balance point, or *barycenter,* that is located between them. Charon orbits Pluto in 6.4 days, the same length of time as a day on Pluto. Because of these equal lengths, Charon stays in the same place in Pluto's sky. In the same way that one side of Earth's moon always faces Earth, one side of Pluto always faces Charon.

Pluto's other two moons, Hydra and Nix, are much smaller. These moons were discovered in 2005 by astronomers using the *Hubble Space Telescope*. These moons also orbit the barycenter between Pluto and Charon.

Reading Check **Identify two ways Charon is different from other moons. (See the Appendix for answers to Reading Checks.)**

SCiLINKS. NSTA Developed and maintained by the National Science Teachers Association

For a variety of links related to this subject, go to www.scilinks.org

Topic: Moons of Other Planets
SciLinks code: HQ60993

Figure 9 ▶ Charon has a diameter almost half as large as Pluto's. Pluto was discovered in 1930, Charon was discovered in 1978, and two smaller moons, Nix and Hydra, were discovered in 2005.

Close

Reteaching — BASIC

Peer Reviewing Divide the class into pairs or small groups. Have each student write five questions that can be answered by using information from the section. Have students use the questions to quiz each other on the section material. LS **Interpersonal**

Quiz — GENERAL

1. What are the names of the moons of Mars? (Phobos and Deimos)
2. What was the role of the Voyager mission in uncovering the secrets of Io's geology? (An engineer scanning images taken of Io discovered erupting volcanoes. Scientists later determined the role of gravitation, including tidal forces.)
3. How is Pluto's largest moon unique? (Charon is unique because it is so close in size to the body that it orbits.)

Alternative Assessment — GENERAL

Satellite Bingo Have students work in groups to create a bingo game that focuses on the characteristics of the moons of the solar system. Head the columns with the names of moons and identify four distinct features under each name. Use pictures as well as words and phrases to describe the features. Each card should have a different combination of moons and features. Students can play the game using pennies as markers. Four matching features in a row (up, down, or across) wins. LS **Verbal/Kinesthetic**

Answer to Reading Check

Charon is almost half the size of the body it orbits. Charon's orbital period is the same length as Pluto's day, so one side of Pluto always faces its largest moon.

Activity — GENERAL

Planetary Rings Draw a Venn diagram on the board. Label one circle with "Saturn's Rings" and the other with "Rings of Other Gas Giants." Ask students to write words and phrases that describe the differences between the two topics in the outer parts of the two circles. Then, have students list similarities in the area where the two circles overlap. Ask students why the outer planets have ring systems. (Tidal forces close to the gas giants cause planetary rings.) LS **Visual/Verbal**

Close, *continued*

Answers to Section Review

1. Both are small, irregularly shaped bodies that have cratered surfaces and dark regions; Phobos is larger than Deimos.
2. Answers may vary. Sample answer: Io—active volcanoes; Europa—liquid water beneath its icy surface; Ganymede—larger than Mercury; Callisto—heavily cratered
3. by researchers studying the images taken by spacecraft
4. because heat is created in Io's crust by friction caused by the gravitational pull of Jupiter; the heat melts the rocks in the moon's interior.
5. by observatories such as the *Hubble Space Telescope*
6. Triton revolves around Neptune in a backward orbit, which suggests that Triton was "captured" when it got too close to Neptune.
7. Saturn has the most extensive ring system. The other gas giants have much thinner ring systems. Saturn's may be the remains of a comet ripped apart by tidal forces, while Jupiter's may result from particles given off by Io, or debris from meteor collisions. Uranus has a dozen thin rings. Neptune's rings are clumpy and not uniform.
8. Although Ganymede is larger than Mercury, its density is relatively low, which suggests that it may contain mostly ice in its interior rather than rock or iron.
9. Sample answer: Many moons may have formed from materials remaining after the planet they orbit formed, or they may have resulted from collisions. Other moons were captured later. Ring systems appear to result from the break up of comets and moons or from debris left over from collisions.
10. Triton probably retains a thin atmosphere because it is much larger and cooler than Phobos and has sufficient mass to allow its gravity to hold onto the volatile gases.
11. Most planets have *natural satellites,* called *moons,* such as *Mars's* moons *Phobos* and *Deimos, Jupiter's Galilean moons, Saturn's* moon *Titan, Uranus's* moon *Titania. Pluto's* largest moon is named *Charon.* Many planets also have *rings* such as those around *Saturn.*

CHAPTER RESOURCES

Chapter Resource File

- **Section Quiz** GENERAL

Workbooks

- **Study Guide** (also in Spanish)

Figure 10 ▶ Saturn has the most extensive system of rings in the solar system. The angle at which its rings are visible changes as Saturn orbits the sun.

Rings of the Gas Giants

Saturn's spectacular set of rings, shown in **Figure 10,** was discovered more than 300 years ago. Each of the rings circling Saturn is divided into hundreds of small ringlets. The ringlets are composed of billions of pieces of rock and ice. These pieces range in size from particles the size of dust to chunks the size of a house. Each piece follows its own orbit around Saturn. The ring system of Saturn is very thin.

Originally, astronomers thought that the rings formed from material that was unable to clump together to form moons while Saturn was forming. However, evidence indicates that the rings are much younger than originally thought. Now, most scientists think that the rings are the remains of a large cometlike body that entered Saturn's system and was ripped apart by tidal forces. Particles from the rings continue to spiral into Saturn, but the rings are replenished by particles given off by Saturn's moons.

The other gas giants have rings as well. These rings are relatively narrow. Jupiter's were not discovered until the *Voyager 1* spacecraft flew by Jupiter in 1979. Jupiter has a single, thin ring made of microscopic particles that may have been given off by Io or one of Jupiter's other moons. The particles may also be debris from collisions of comets or meteorites with Jupiter's moons. Uranus also has a dozen thin rings. Neptune's relatively small number of rings are clumpy rather than thin and uniform.

Section 3 Review

1. **Compare** the characteristics of Phobos and Deimos.
2. **List** the four moons of Jupiter that were discovered by Galileo, and identify one distinguishing characteristic of each.
3. **Describe** how volcanoes were discovered on Io.
4. **Explain** why Io remains volcanically active.
5. **Describe** how the smaller moons of Uranus were discovered.
6. **Explain** why Triton has an unusual orbit.
7. **Compare** the characteristics of Saturn's rings with the rings of the other outer planets.

CRITICAL THINKING

8. **Analyzing Relationships** Explain why scientists think that Ganymede's interior includes ice.
9. **Inferring Relationships** Compare and contrast the way in which moons and ring systems form.
10. **Making Comparisons** Explain why Triton retains an atmosphere while Phobos does not.

CONCEPT MAPPING

11. Use the following terms to create a concept map: *moon, ring, Mars, Uranus, Jupiter, Saturn, Phobos, Deimos, Pluto, Galilean moon, natural satellite, Charon, Titania,* and *Titan.*

Section 4 Asteroids, Comets, and Meteoroids

In addition to the sun, the planets, and the planets' moons, our solar system includes millions of smaller bodies. Some of these small bodies are tiny bits of dust or ice that orbit the sun. Other bodies are as big as small moons. Astronomers theorize that these smaller bodies are leftover debris from the formation of the solar system.

OBJECTIVES

- **Describe** the physical characteristics of asteroids and comets.
- **Describe** where the Kuiper Belt is located.
- **Compare** meteoroids, meteorites, and meteors.
- **Explain** the relationship between the Oort cloud and comets.

KEY TERMS

asteroid
comet
Oort cloud
Kuiper Belt
meteoroid
meteor

Asteroids

The largest of the smaller bodies in the solar system are called asteroids. **Asteroids** are fragments of rock that orbit the sun. Astronomers have discovered more than 50,000 asteroids. Millions of asteroids may exist in the solar system. The orbits of asteroids, like those of the planets, are ellipses. The largest known asteroid, Ceres, has a diameter of about 1,000 km. Because it is large enough that gravity has caused it to become round, Ceres is also considered to be a dwarf planet. Two other asteroids are shown in **Figure 1.**

Most asteroids are located in a region between the orbits of Mars and Jupiter known as the *asteroid belt*. This main belt extends from about 299 million to about 598 million kilometers from the sun. However, not all asteroids are located in the main asteroid belt. The closest asteroids to the sun are inside the orbit of Mars, about 224 million kilometers from the sun. The *Trojan asteroids* are concentrated in groups just ahead of and just behind Jupiter as it orbits the sun. In fact, the Trojan asteroids almost share Jupiter's orbit. These asteroids are name for the Trojan and Greek warriors of the famous Trojan War of Greek mythology. Asteroids also exist beyond Jupiter's orbit.

asteroid a small, rocky object that orbits the sun; most asteroids are located in a band between the orbits of Mars and Jupiter

Figure 1 ▶ This image of the asteroids Ida (left) and Dactyl (right) were taken by the spacecraft *Galileo* as it passed through the asteroid belt on its way to Jupiter. Ida is 56 km long, and Dactyl is 1.5 km across.

Section 4

Focus

Overview

This section describes characteristics of asteroids, various kinds of meteors, and comets and explains what these small bodies can teach us about the early history of Earth's solar system.

Bellringer

Ask students to write brief descriptions of meteorites, comets, and asteroids or to draw pictures of what these bodies are like. (Answers may vary.) LS **Visual**

Motivate

Discussion — GENERAL

Asteroid Motions Explain that asteroids are essentially large rocks and that they range in size from the size of boulders to the size of a small moon. Ask students to describe where these small objects are located and how they move. (Sample answers: Most orbit the sun in elliptical orbits like the planets in a ring or belt between the orbits of Mars and Jupiter.) LS **Verbal**

CHAPTER RESOURCES

Chapter Resource File
- Directed Reading BASIC

Technology

Transparencies
- Bellringer

Student Edition on CD-ROM

One-Stop Planner CD-ROM
- Lesson Plan

Teach

PHYSICS CONNECTION ADVANCED

Unusual Orbits Explain that the Trojan asteroids travel in two clouds approximately 60° in front of and behind Jupiter, where the gravitational forces of Jupiter and the sun are equal. Held captive by these combined gravities, the Trojans oscillate back and forth along Jupiter's orbit. Three groups of asteroids called the *Atens*, the *Apollos*, and the *Amors* move in eccentric orbits near Earth's orbit. Invite interested students to research asteroids that have unusual orbits, and have them create orbital diagrams to share with the class. LS **Visual**

Using the Figure—GENERAL

Comparing Craters Explain that even if a surface was solid before an impact, when a body is large and fast enough, the ground partially liquefies during impact, forming more complex crater features such as a central peak. Provide photos of craters on the moon or other planets for students to compare with Barringer Meteor Crater. Good sources for images are NASA Web sites or the National Science Teachers Association book about craters. LS **Visual**

Answer to Reading Check

The most common type is made mostly of carbon materials. Other asteroids are made mostly of silicate minerals. The third type is composed mostly of iron and nickel.

SCILINKS. NSTA Developed and maintained by the National Science Teachers Association

For a variety of links related to this subject, go to www.scilinks.org

Topic: Comets, Asteroids, and Meteoroids

SciLinks code: HQ60317

Composition of Asteroids

The composition of asteroids is similar to that of the inner planets. Asteroids are classified according to their composition into three main categories. The most common of the three types of asteroids is made mostly of carbon materials, which give this type of asteroid a dark color. The second type of asteroid is made of mostly silicate minerals. These asteroids look like Earth rocks. The third, and rarest, type of asteroid is composed of mostly iron and nickel. These asteroids have a shiny, metallic appearance, especially on fresh surfaces.

Many astronomers think that asteroids in the asteroid belt are made of material that was not able to form a planet because of the strong gravitational force of Jupiter. Scientists estimate that the total mass of all asteroids is less than the mass of Earth's moon.

Near-Earth Asteroids

More than a thousand asteroids have orbits that sometimes bring them very close to Earth. These asteroids have wide, elliptical orbits that bring them near Earth's orbit. Thus they are called *near-Earth asteroids*. Near-Earth asteroids make up only a small percentage of the total number of asteroids in the solar system.

Interest in near-Earth asteroids has increased in recent years with the realization that these asteroids could inflict great damage on Earth if they were to strike the planet. Meteor Crater, in Arizona, which is shown in **Figure 2**, formed when a small asteroid that had a diameter of less than 50 m struck Earth about 40,000 years ago. Several recently established asteroid detection programs have begun to track all asteroids whose orbits may approach Earth. By identifying and monitoring these asteroids, scientists hope to predict and possibly avoid future collisions.

Reading Check **What are the three types of asteroids by composition? (See the Appendix for answers to Reading Checks.)**

Figure 2 ▶ Barringer Meteorite Crater, also known simply as Meteor Crater, in Arizona, has a diameter of more than 1 km. Dozens of such craters have resulted from past impacts on Earth, but most craters have eroded or have been covered by sediment.

ENVIRONMENTAL CONNECTION

Target Earth Several asteroid groups follow orbits that bring them very close to Earth. Research suggests that terrestrial impacts may seriously disrupt the environment and influence life on Earth. For example, an asteroid that struck 65 million years ago may have played a part in the extinction of the dinosaurs. The rock vaporized when it struck, releasing kinetic energy that blasted a hole in the ocean floor near what is now the Yucatan Peninsula, generated enormous tsunamis, and produced heat waves that created global forest fires. The impact sent a cloud of dust and debris into the atmosphere that remained for months, blocking sunlight and interfering with photosynthesis. This event probably temporarily cooled the climate, caused acid rain, and introduced greenhouse gases. It left behind a chemical signature and impact-melted glass that attest to the projectile's extraterrestrial origin. More than 150 impact craters of various ages have been identified on Earth. Invite interested students to investigate the potential effects of impacts and share their findings in oral reports and presentations. LS **Verbal**

Figure 3 ▶ A comet, such as Comet Hale-Bopp, consists of a nucleus, a coma, and two tails. The blue streak is the *ion tail,* and the white streak is the *dust tail.*

Comets

Every few years, an object that looks like a star that has a tail is visible in the evening sky. This object is a comet. **Comets** are small bodies of ice, rock, and cosmic dust that follow highly elliptical orbits around the sun. The most famous is Halley's Comet, which passes by Earth every 76 years. It last passed Earth in 1986 and will return in 2061. Every 5 to 10 years, another very bright comet will be visible from Earth. Comet Hale-Bopp, shown in **Figure 3,** was particularly spectacular as it passed by Earth in 1997.

comet a small body of rock, ice, and cosmic dust that follows an elliptical orbit around the sun and that gives off gas and dust in the form of a tail as it passes close to the sun

Composition of Comets

A comet has several parts. The core, or nucleus, of a comet is made of rock, metals, and ice. Cores of comets are commonly between 1 km and 100 km in diameter. A spherical cloud of gas and dust, called the *coma,* surrounds the nucleus. The coma can extend as far as 1 million kilometers from the nucleus. A comet's bright appearance largely results from sunlight reflected by the comet's coma. The nucleus and the coma form the head of the comet. In 2004, the spacecraft *Stardust* flew by a comet named Wild 2. It returned samples from the comet's coma to Earth in 2006. In 2005, the spacecraft, *Deep Impact,* slammed a probe into the nucleus of comet Temple 1. The probe released a plume of dust, giving scientists a look at what a comet's nucleus is made of.

The most spectacular parts of a comet are its tails. Tails form when sunlight causes the comet's ice to change to gas. The gas, or ion, tail of a comet streams from the comet's head. The solar wind—electrically charged particles expanding away from the sun—pushes the gas away from the comet's head. Thus, regardless of the direction the comet travels, its ion tail points away from the sun. The comet's second tail is made of dust and curves backward along the comet's orbit. Some comets have tails that are more than 80 million kilometers long.

MISCONCEPTION ALERT

Comet Tails Students may think that comet tails always extend backward in the direction the comet is coming from. The nucleus of a comet is composed of ices, along with bits of dust and rock. The sun's heat makes these ices change directly into gases and produces jets of glowing charged ions—the ion tail. The solar wind pushes these charged particles outward, so that the ion tail always points away from the sun, even as the comet swings around the sun to leave the inner solar system. Some comets also have a second curved tail composed of dust that does point back along the path that the comet has traveled.

Using the Figure—GENERAL

Discussion Direct students' attention to the picture of comet Hale-Bopp, one of the most spectacular comets of the last century. Explain that Hale-Bopp is a long-period comet that is believed to have last visited the inner solar system more that 4,000 years ago. No two comets ever look exactly alike. Ask students to describe the basic features of this comet that are visible in the picture. (Sample answer: The comet has a glowing head, or coma, and has both a blue ion tail and a white dust tail.) LS **Visual**

Cultural Awareness — GENERAL

Legendary Comets In ancient times, the appearance of comets was thought to foretell wars, death, and disaster. The word *comet* comes from the Greek words *aster kometes,* which means "hairy star," because its long tail resembled flowing hair. Comets were associated with death because hair is a symbol of mourning and with war because some people thought the comet's tail looked like a sword. A comet appeared in 44 BCE, around the time of Julius Caesar's assassination. The arrival of a comet, later called *Halley's Comet,* was associated with the victory of the Norman king William the Conqueror over the Anglo-Saxons at the Battle of Hastings in 1066. This comet was also seen in 1456 when the Turks conquered Constantinople. As recently as 1910, people panicked when Earth passed through a comet's tail, because they incorrectly feared that the gases, which would be poisonous if inhaled, would reach them. Comets were demystified when British astronomer Edmond Halley showed that the comet that now bears his name travels in an elongated path around the sun and returns regularly every 76 years. Invite students to research comets of legend and history and have them make a visual presentation to the class. LS **Verbal/Visual**

Using the Figure—GENERAL

Oort Cloud Distances Help students to visualize the vast distances involved with a region of space as far from Earth as the Oort cloud by comparing the distance to a familiar landmark. Tell students that at the scale in which the drawing of the Oort cloud was made, the outer edge of the sphere would be 0.4 km in diameter. Have students identify a landmark or city that is about that distance (about 1/4 mile) from the school. Explain that the diagram would have to be stretched from their school to that landmark for the image to be to scale. LS **Logical**

Internet Activity—GENERAL

Kuiper Belt Objects Pluto shares a similar orbit and composition with thousands of other small bodies of rock and ice that travel around the sun in a region called the *Kuiper Belt*. These objects may be among the most primitive objects in the solar system. Invite interested students to research icy Kuiper-Belt objects such as Quaoar/Varuna, Sedna, and the dwarf planet Eris, and report their findings to the class. A worksheet designed to direct student research on this topic can be found in the **Chapter Resource File** booklet or by visiting **go.hrw.com** and entering the keyword **HQ6MBSX.** LS **Verbal**

CHAPTER RESOURCES

Chapter Resource File

Internet Activity
- Kuiper Belt Objects GENERAL

Figure 4 ▶ Most comets come from the Oort cloud, a region in the outer solar system that is far beyond the orbit of Neptune.

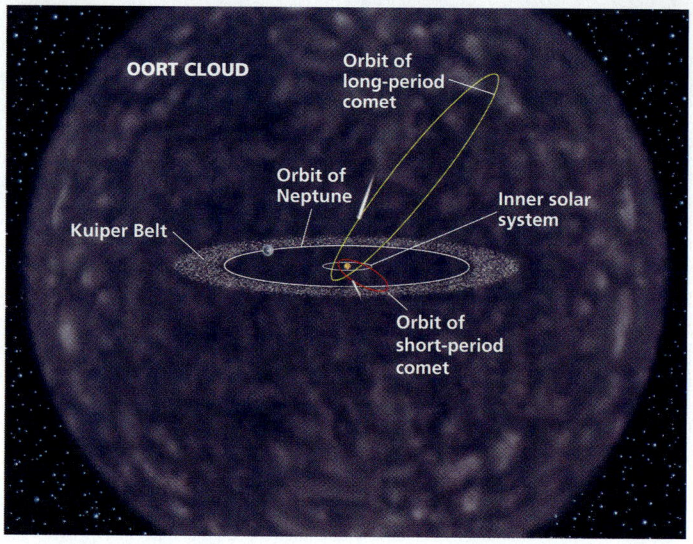

Oort cloud a spherical region that surrounds the solar system, that extends from the Kuiper Belt to almost halfway to the nearest star, and that contains billions of comets

Kuiper Belt a region of the solar system that starts just beyond the orbit of Neptune and that contains dwarf planets and other small bodies made mostly of ice

The Oort Cloud

Astronomers think that most comets originate in the Oort cloud, which is illustrated in **Figure 4.** The **Oort cloud** is a spherical cloud of dust and ice that lies far beyond Neptune's orbit and that contains the nuclei of billions of comets. The total mass of the Oort cloud is estimated to be between 10 and 40 Earth masses.

The Oort cloud surrounds the solar system and may reach as far as halfway to the nearest star. Scientists think that the matter in the Oort cloud was left over from the formation of the solar system. Studying this distant matter helps scientists understand the early history of the solar system.

Bodies within the Oort cloud circle the sun so slowly that they take a few million years to complete one orbit. But the gravity of a star that passes near the solar system may cause a comet to fall into a more elliptical orbit around the sun. The orbits of comets that pass by Jupiter may also be changed by Jupiter's gravitational force. If a comet takes more than 200 years to complete one orbit of the sun, the comet is called a *long-period comet*.

The Kuiper Belt

Recent advances in technology have allowed scientists to observe thousands of small objects beyond the orbit of Neptune. Most of these objects, including some comets, are from a flat ring called the **Kuiper Belt** (KIE puhr BELT), which is located just beyond Neptune's orbit. The Kuiper Belt is also illustrated in **Figure 4.** The dwarf planets Pluto and Eris are located in the Kuiper Belt during much of their orbits. Several other objects that are candidates for dwarf-planet status also reside in the Kuiper Belt.

BRAIN FOOD

Comet Decay Comets are very fragile bodies. The gas and dust blown off most comets is lost into space. In a natural aging process, some comets simply turn into dark, rocky bodies that resemble asteroids. Many short-lived comets break apart as a result of tidal forces caused by the gravity of large planets or of the sun. The debris trail from a comet's break-up may rain down on the inner planets as meteor showers. If comets are pulled completely from their orbits, they may collide with the sun or with one of the other planets. Comet Shoemaker-Levy 9 broke apart in 1994 and crashed into Jupiter's atmosphere.

Short-Period Comets

Comets called *short-period comets* take less than 200 years to complete one orbit around the sun. In recent years, astronomers have discovered that most short-period comets come from the Kuiper Belt. Some of the comets that originate in the Kuiper Belt have been forced outward into the Oort cloud by Jupiter's gravity. Many comets in the Kuiper Belt are the result of collisions between larger Kuiper-Belt objects there. Halley's comet, which has a period of 76 years, is a short-period comet.

MATHPRACTICE

Matter From Space Astronomers estimate that about 1 million kg of matter from meteoroids falls to Earth each day. Based on this estimate, how many kilograms of matter from meteoroids would fall on Earth in three weeks?

Meteoroids

In addition to relatively large asteroids and comets, very small bits of rock or metal move throughout the solar system. These small, rocky bodies are called **meteoroids.** Most meteoroids have a diameter of less than 1 mm. Scientists think that most meteoroids are pieces of matter that become detached from passing comets. Large meteoroids—more than 1 cm in diameter—are probably the result of collisions between asteroids.

meteoroid a relatively small, rocky body that travels through space

meteor a bright streak of light that results when a meteoroid burns up in Earth's atmosphere

Meteors

Meteoroids that travel through space on an orbit that takes them directly into Earth's path may enter Earth's atmosphere. When a meteoroid enters Earth's atmosphere, friction between the object and the air molecules heats the meteoroid's surface. As a result of this friction and heat, most meteoroids burn up in the atmosphere. As a meteoroid burns up in Earth's atmosphere, the meteoroid produces a bright streak of light called a **meteor.** Meteors are commonly called *shooting stars*. Meteoroids sometimes also vaporize very quickly in a brilliant flash of light called a *fireball*. Observers on Earth may hear a loud noise as a fireball disintegrates.

When a large number of small meteoroids enter Earth's atmosphere in a short period of time, a *meteor shower* occurs. During the most spectacular of these showers, several meteors are visible every minute. A composite photo of a meteor shower is shown in **Figure 5.** Meteor showers occur at the same time each year. This happens because Earth intersects the orbits of comets that have left behind a trail of dust. As these particles burn up in Earth's atmosphere, they appear as meteors streaking across the sky.

Reading Check **What is the difference between a meteor and a meteoroid? (See the Appendix for answers to Reading Checks.)**

Figure 5 ▶ The straight lines in this composite photo are meteors burning up as they move through Earth's atmosphere.

MATHPRACTICE

Answer

1 million kg/d × 7 d/wk × 3 wk = 21,000,000 kg of matter

Close

Reteaching — BASIC

Crater Features Spread newspapers beneath a box lid. Fill the lid about half full with flour. Sprinkle a layer of tempera paint over the flour and smooth the surface. Have students drop a pebble from various heights and then draw the features of the craters that form. **LS Kinesthetic**

Quiz — GENERAL

1. How are asteroids and meteoroids related? (Scientists think most large meteoroids are fragments that break off when asteroids collide.)
2. How are scientists attempting to meet the threat represented by near-Earth asteroids? (They have assembled a computer database and are attempting to monitor and track the positions of these asteroids to predict future collisions.)

Alternative Assessment — GENERAL

Science Nonfiction Have students write a description of a trip through the solar system on one of the small solar system bodies—for example, the return of a long-lived comet from the Oort cloud to the inner solar system. Students should research their topic to make their stories accurate and entertaining. Have students outline the main points of the story before they begin writing. **LS Verbal**

Using the Figure — GENERAL

Leonid Meteor Shower The photo of the Leonid meteor shower is a composite of 30 separate 1-minute exposures taken over Spain in 2003. The short exposures explain the lack of star trails. The Leonid meteor shower is associated with comet Tempel-Tuttle. The best time to spot meteors is during an annual meteor shower, when Earth passes through a concentration of meteoroids. Meteor showers are named for the constellation in which they appear. Ask students what constellation the Leonid shower was named for. (Leo, the lion.) **LS Visual**

Answer to Reading Check

A meteoroid is a rocky body that travels through space. When a meteoroid enters Earth's atmosphere and begins to burn up, the meteoroid produces a meteor.

Close, *continued*

Answers to Section Review

1. between the orbits of Mars and Jupiter
2. Asteroids are fragments of rock, metals, and carbon that travel in elliptical orbits; they are the largest of the small bodies that orbit the sun.
3. nucleus—a core of rock, metals, and ice; coma—a cloud of gas and dust that surrounds the nucleus; ion tail; and dust tail
4. ion tail—electrically charged particles that point away from the sun; dust tail—debris that curves back along the comet's path
5. Most comets are believed to originate in the Oort cloud.
6. beyond Neptune's orbit
7. A meteoroid is a small body made of rock or metal that travels through space. The streak of light that results when a meteoroid enters an atmosphere is a meteor. A meteoroid that strikes the planetary surface without burning up is a meteorite.
8. The solar wind pushes the gases and particles of the ion tail away from the comet's head so that the gases and particles stream away from the sun.
9. Sample Answers: Pluto and Ceres are both dwarf planets because they orbit the sun, are round because of their own gravity, and neither has cleared its orbital path. Pluto is in the Kuiper Belt and Ceres is in the Asteroid belt.
10. Answers may vary. Sample answer: Most meteorites are stony, but metallic ones are easier to identify, so it is likely a stony or iron meteorite. Step one: A collision between two asteroids sends a meteoroid hurtling toward the inner solar system. Step two: The meteoroid crosses Earth's orbit, streaks through the atmosphere, and strikes Earth's surface.
11. Minor bodies include *asteroids; comets,* such as *short-period comets* that originate in the *Kuiper Belt* and *long-period comets* that originate in the *Oort cloud;* and *meteoroids.*

Figure 6 ▶ Types of Meteorites

Meteorites

Millions of meteoroids enter Earth's atmosphere each day. A few of these meteoroids do not burn up entirely in the atmosphere because they are relatively large. These meteoroids fall to Earth's surface. A meteoroid or any part of a meteoroid that is left when a meteoroid hits Earth is called a *meteorite*. Most meteorites are small and have a mass of less than 1 kg. However, large meteorites occasionally strike Earth's surface with the force of a large bomb. These impacts leave large impact craters.

Meteorites can be classified into three basic types: stony, iron, and stony-iron. These three types of meteorites are shown in **Figure 6.** *Stony meteorites* are similar in composition to rocks on Earth. Some stony meteorites contain carbon-bearing compounds that are similar to the carbon compounds in living organisms. Although most meteorites are stony, *iron meteorites* are easier to find. Iron meteorites are easier to find because they have a distinctive metallic appearance. This distinctive appearance makes iron meteorites easy to distinguish from common Earth rocks. The third type of meteorites, called *stony-iron meteorites,* contain iron and stone. Stony-iron meteorites are rare.

Astronomers think that almost all meteorites come from collisions between asteroids. The oldest meteoroids may be 100 million years older than Earth and its moon. Therefore, meteorites may provide information about how the early solar system formed.

Some rare meteorites originated on the moon or Mars. Computer simulations have shown that meteorites that hit the moon or Mars can eject rocks that then fall to Earth. Many of these rare meteorites have been found in Antarctica. Finding meteorites in Antarctica is relatively easy because they stand out against the background of snow and ice.

Section 4 Review

1. **Identify** where the asteroid belt is located in the solar system.
2. **Describe** the physical characteristics of asteroids.
3. **List** the four main parts of a comet, and identify their physical characteristics.
4. **Compare** the ion and dust tails of a comet.
5. **Explain** the relationship between the Oort cloud and comets.
6. **Describe** the location of the Kuiper Belt.
7. **Distinguish** between a meteor, a meteoroid, and a meteorite.

CRITICAL THINKING

8. **Analyzing Relationships** Explain why a comet's ion tail always points away from the sun.
9. **Making Comparisons** Explain how both Pluto (a Kuiper-Belt object) and Ceres (an asteroid) can both be considered dwarf planets.
10. **Making Comparisons** You find a meteorite on the ground. What kind of meteorite did you most likely find? Describe two steps of its journey from space.

CONCEPT MAPPING

11. Use the following terms to create a concept map: *comet, asteroid, Kuiper Belt, Oort cloud, long-period comet, short-period comet,* and *meteoroid.*

CHAPTER RESOURCES

Chapter Resource File

- **Section Quiz** GENERAL

Workbooks

- **Study Guide** (also in Spanish)

Chapter 28 Highlights

Sections

1 Earth's Moon

Key Terms

satellite, 719
moon, 719
mare, 720
crater, 720

Key Concepts

- The surface of the moon is covered by craters, rilles, and ridges.
- The structure of the moon's interior was determined by using seismographs.
- The moon may have formed when an object hit Earth more than 4 billion years ago. Lunar surface features have changed little since they formed 3 billion years ago.

2 Movements of the Moon

Key Terms

apogee, 725
perigee, 725
eclipse, 727
solar eclipse, 727
lunar eclipse, 729
phase, 730

Key Concepts

- Eclipses occur when one planetary body passes through the shadow of another.
- The moon spins on its axis once during each orbit of Earth. As the moon orbits, the amount of the near side that is lighted increases and decreases, which causes phases.
- Tides on Earth are the result of the gravitational pull of the sun and the moon.

3 Satellites of Other Planets

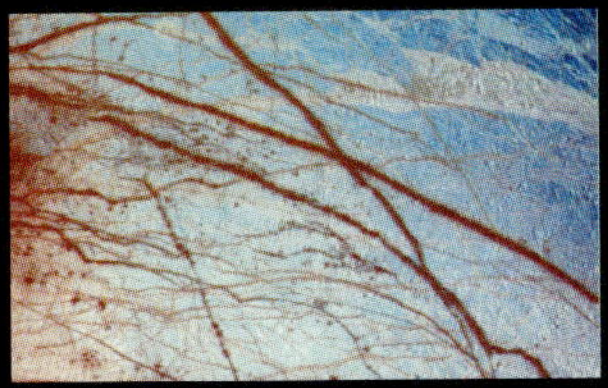

Key Terms

Galilean moon, 733

Key Concepts

- Phobos and Deimos, the two tiny moons of Mars, are much smaller than Earth's moon.
- The Galilean moons are the four largest satellites of Jupiter. Jupiter, Saturn, Uranus, and Neptune have rings.
- Uranus and Neptune each have several moons. Pluto has at least three moons.

4 Asteroids, Comets, and Meteoroids

Key Terms

asteroid, 739
comet, 741
Oort cloud, 742
Kuiper Belt, 742
meteoroid, 743
meteor, 743

Key Concepts

- Most asteroids are located in an asteroid belt that is located between the orbits of Mars and Jupiter.
- Comets come from the Oort cloud, which is far beyond Neptune's orbit, and from the Kuiper Belt, which starts just beyond Neptune's orbit.
- Meteoroids are small rocks in space that either burn up in Earth's atmosphere (as meteors) or stay intact to hit Earth (as meteorites).

Chapter Highlights

Alternative Assessment — GENERAL

Poster Project Have groups of students create posters about how studying objects such as Earth's moon illuminates the early history of the solar system. Posters could illustrate typical features of the lunar landscape and explain how moon rocks provided insight into lunar origins, or compare the moon's cratered surface with the surfaces of other natural satellites. Other posters might depict what we have learned about the origin of life on Earth from space missions to other moons or how craters throughout the solar system provided clues to mass extinctions, or they may summarize the nature and composition of asteroids, meteorites, and comets or the ancient features preserved in the objects of the Kuiper Belt and Oort cloud. LS Visual

CHAPTER RESOURCES

Chapter Resource File

- Concept Review GENERAL
- Critical Thinking ADVANCED
- Math Skills GENERAL
- Graphing Skills GENERAL
- Chapter Test A GENERAL
- Chapter Test B ADVANCED

Workbooks

- **Study Guide** (also in Spanish)
- **Assessments** (Spanish)

Technology

Classroom Videos
- Brain Food Video Quiz

HRW Earth Science Video
- Segment 18: The Solar System
- Segment 19: The Earth's Moon

Chapter Review

Assignment Guide

Section	Questions
1	1–2 9–10, 17–18, 23, 25–26, 32
2	4, 6, 11–13, 19, 24, 27, 29–30, 33
3	3, 14, 20–21, 31
4	5, 7–8, 15–16, 22, 28, 34–37

Using Key Terms

1–8. Answers may vary but should show that students understand the definitions of and differences between key terms.

Understanding Key Concepts

9. c **10.** d
11. a **12.** b
13. d **14.** a
15. d **16.** a

Short Answer

17. Maria formed when lava from the moon's interior slowly filled the basins left behind by prior meteorite impacts.

18. Craters on the moon were caused by impacts with other bodies. The circular shape, presence of ejected materials, regolith layer, and texture of lunar rocks provides evidence.

19. no; During the full moon phase, Earth is located between the moon and the sun. Solar eclipses occur only during the new moon when the moon is located between Earth and the sun, and the moon's shadow falls on part of Earth.

20. yes; Both Titan and Ganymede are larger than Mercury.

21. Saturn has the most extensive ring system. Jupiter, Uranus, and Neptune also have rings.

22. Oort cloud and Kuiper Belt

23. A natural satellite is smaller body such as a moon that orbits around a larger body. An artificial satellite is an object launched by humans, such as weather satellites or space observatories.

Chapter 28 Review

Using Key Terms

Use each of the following terms in a separate sentence.

1. *crater*
2. *mare*
3. *Galilean moon*

For each pair of terms, explain how the meanings of the terms differ.

4. *perigee* and *apogee*
5. *Oort cloud* and *Kuiper Belt*
6. *solar eclipse* and *lunar eclipse*
7. *comet* and *asteroid*
8. *meteoroid* and *meteorite*

Understanding Key Concepts

9. Dark areas on the moon that are smooth and that reflect little light are called
a. rilles. **c.** maria.
b. rays. **d.** breccia.

10. What happened in the most recent stage in the development of the moon?
a. The densest material sank to the core.
b. The crust began to break.
c. Earth's gravity captured the moon.
d. The number of meteorites hitting the moon decreased.

11. During each orbit around Earth, the moon spins on its axis
a. 1 time. **c.** about 27 times.
b. about 29 times. **d.** 365 times.

12. In a lunar eclipse, the moon
a. casts a shadow on Earth.
b. is in Earth's shadow.
c. is between Earth and the sun.
d. blocks part of the sun from view.

13. When the size of the lighted part of the moon's near side is decreasing, the moon is
a. full. **c.** annular.
b. waxing. **d.** waning.

14. Compared with the other moons of Jupiter, the four Galilean moons are
a. larger.
b. farther from Jupiter.
c. lighter.
d. younger.

15. The main asteroid belt exists in a region between the orbits of
a. Mercury and Venus. **c.** Venus and Earth.
b. Earth and Mars. **d.** Mars and Jupiter.

16. Meteorites can provide information about
a. the composition of the solar system before the planets formed.
b. the size of Earth.
c. the destiny of the solar system.
d. the size of the universe.

Short Answer

17. Describe how maria formed on the moon.

18. Are craters on the moon caused by volcanism or by impacts with other bodies? Explain your answer.

19. Do total eclipses of the sun occur only at full moons? Explain your answer.

20. Are any moons in the solar system bigger than planets? Explain.

21. Which planets have rings?

22. Which two places in the solar system do comets come from?

23. What is the difference between natural and artificial satellites?

Critical Thinking

24. Answers may vary. The location and size of the tidal bulges would be different and may not exist at all.

25. If the moon had a dense atmosphere with water, fewer meteorites would have penetrated to reach the surface and many of the craters that formed during the early bombardment period would have been erased by erosion.

26. The moon's surface would be smooth and would not be heavily cratered.

Critical Thinking

24. Analyzing Relationships If Earth had two moons that traveled on the same orbit and were the same distance from Earth, but formed a 90° angle with Earth, how would Earth's tides be different?

25. Making Inferences How would the craters on the moon be different today if the moon had developed a dense atmosphere that moved as wind and that contained water?

26. Determining Cause and Effect If meteorites had stopped hitting the moon before the outer surface of the moon cooled, how would the moon's surface be different than it is today?

27. Evaluating Information Suppose that the moon spun twice on its axis during each orbit around Earth. How would the study of the moon from Earth be easier than it is currently?

28. Applying Ideas The surfaces of some asteroids reflect only small amounts of light. Other asteroids reflect up to 40% of the light that falls on them. Of what kind of materials would each type of asteroid probably be composed?

Concept Mapping

29. Use the following terms to create a concept map: *moon, Earth, apogee, perigee, new moon, full moon, waxing, waning, solar eclipse, lunar eclipse, umbra, penumbra,* and *phase*.

Math Skills

30. Making Calculations There are 60 s in 1 min, 60 min in 1 h, 24 h in 1 day, and 365 1/4 days in a year. How many seconds are in a year?

31. Making Calculations The radius of Earth's moon is 1,738 km. The diameter of Neptune's moon Triton is 2,705 km. What percentage of Earth's moon's size is Triton?

Writing Skills

32. Creative Writing Imagine that you want to live on the moon. Describe how you would get your water and how you would acquire food and other supplies.

33. Communicating Ideas Summarize how the moon's gravity and the rotation of Earth cause tides.

Interpreting Graphics

The graph below shows the number of near-Earth asteroids discovered each year. Use the graph below to answer the questions that follow.

Near-Earth Asteroid Discoveries

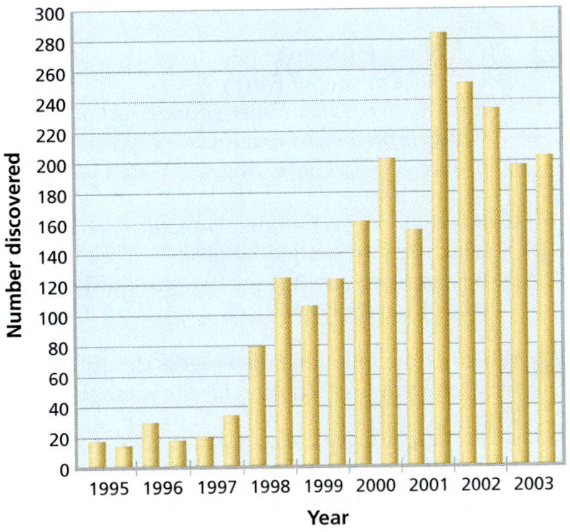

34. Was the rate of discovery of near-Earth asteroids in 2003 higher or lower than the rate in 2000?

35. How many near-Earth asteroids were discovered in the half-year in which the most discoveries were made?

36. Which calendar year had the highest total number of near-Earth asteroid discoveries?

37. What is the total number of near-Earth asteroids discovered in the last three years shown on the graph?

27. If the moon spun on its axis twice in each orbit around Earth, it would present both sides to observers on Earth. We would be able to map what we now think of as the far side of the moon using ordinary telescopes from Earth.

28. The highly reflective asteroids are probably composed mainly of metals such as iron and nickel. Those that reflect only a small amount of light are probably made of rocky materials or carbon.

Concept Mapping

29. Answers may vary but should include all of the terms listed. Sample answers appear at the end of this Teacher Edition.

Math Skills

30. 60 s/min × 60min/h = 3600 s/h × 24 h/d = 86,400 s/d × 365.25 d/year = 31,557,600 s/year

31. 1,738 km × 2 = 3476 km; 2705 km ÷ 3476 km = 0.7781 × 100 = 77.81%

Writing Skills

32. Answers may vary. Accept all reasonable answers.

33. Answers may vary. Sample answer: On the near side of Earth, the gravitational force of the moon pulls ocean water toward the moon. The mass of the solid Earth is subject to less gravitational force than the ocean water is. On the far side of Earth, the water is subject to less gravitational force than the solid Earth is. These gravitational differences cause the water on the near side and on the far side of Earth to form two tidal bulges. Rotation causes tides to occur all over Earth.

Interpreting Graphics

34. The overall rate of discovery was higher in 2003.

35. Approximately 285 near-Earth asteroids were discovered in the second half of 2001.

36. 2002

37. Answers may vary. Accept all reasonable answers. The total number of near-Earth asteroids discovered was about 1325.

Estimated Time

To give students practice under more realistic testing conditions, allow them 30 minutes to answer all of the questions in this practice test.

TEST DOCTOR

Question 1 Answer B is correct. Because the moon has much less mass than Earth does, the moon's surface gravity is about one-sixth the surface gravity of Earth. As a result, a person who weighs 600 newtons (600 N) on Earth would weigh about 100 N on the moon. A person who weighs 360 N on Earth would weigh about 60 N on the moon. Answer A is incorrect because the answer uses one-tenth the gravity of Earth in its math. Answer C is incorrect because the answer uses one-half the gravity of Earth. Answer D is incorrect because the answer uses one-fourth the gravity of Earth.

Question 12 Full-credit answers should include the following points:

- Kuiper Belt objects are usually very small
- the Kuiper Belt was not discovered until the middle of the 20th century because previous technology had not been powerful enough to detect the small objects that populate it
- even with today's advanced technology, scientists speculate that hundreds of Kuiper Belt objects remain undiscovered

Chapter 28 Standardized Test Prep

Understanding Concepts

Directions (1–4): **For *each* question, write on a separate sheet of paper the letter of the correct answer.**

1. Because of differences in surface gravity, how much does a person who weighs 360 newtons (360 N) on Earth weigh on the moon?
   A. 36 N **B.** 60 N **C.** 180 N **D.** 90 N

2. The point in the orbit of a satellite at which the satellite is farthest from Earth is the satellite's
   F. apogee
   G. perigee
   H. barycenter
   I. phase

3. Which of the following statements accurately describes each ring of Saturn?
   A. It is divided into smaller ringlets, all of which orbit Saturn together.
   B. It consists of a single ring composed of rock and ice pieces.
   C. It is divided into smaller ringlets, each of which has an individual orbit.
   D. It is part of a set of rings that are unlike those found anywhere else.

4. Which of the following statements describes why temperature variation on the moon is so large?
   F. The moon has no atmosphere to provide insulation.
   G. The atmosphere of the moon is made up of cold gases.
   H. Gases are dense and close to the surface.
   I. Dark, smooth rocks absorb the sun's heat.

Directions (5–7): **For *each* question, write a short response.**

5. Approximately how long does it take the moon to make one orbit around Earth?

6. What are the names of the four moons of Jupiter known as the Galilean moons?

7. When the moon is at its apogee, what part of its shadow cannot reach Earth during an eclipse?

Reading Skills

Directions (9–12): **Read the passage below. Then, answer the questions.**

Kuiper Belt Objects

To explain the source of short-period comets, or comets that have a relatively short orbit around the sun, the Dutch-American astronomer Gerard Kuiper proposed in 1949 that a belt of icy bodies must lie beyond the orbits of Neptune and Pluto. Kuiper argued that comets were icy planetesimals that formed from the condensation that happened during the formation of our galaxy.

Because the icy bodies are so far from any large planet's gravitational field (30 to 100 AU), they are able to remain on the fringe of the solar system. Some theorists speculate that the large moons Triton and Charon were once independent members of the Kuiper Belt before they were captured by Neptune and Pluto, respectively. These moons and short-period comets have similar physical and chemical properties. Scientists now believe that the Kuiper Belt may be home to thousands of objects that have diameters of more than 100 km.

9. According to the information in the passage, which of the following did Gerard Kuiper think were actually icy planetesimals?
   A. outer planets
   B. comets
   C. moons of every planet
   D. inner planets

10. What two bodies do some scientists believe were once Kuiper-Belt objects?
   F. Neptune and Charon
   G. Neptune and Pluto
   H. Triton and Neptune
   I. Triton and Charon

11. What did the moon Triton orbit before it was captured by the gravity of Neptune?
   A. the sun **B.** Pluto **C.** the solar system **D.** Charon

12. Why did it take until the middle of the 20th century for astronomers to discover the presence of the Kuiper Belt?

Answers

Understanding Concepts

1. B
2. F
3. C
4. F
5. about 27.3 days (or one month)
6. Io, Europa, Ganymede, and Callisto
7. the umbra

Reading Skills

9. B
10. I
11. A
12. Answers may vary. See Test Doctor for a detailed scoring rubric.

Interpreting Graphics

13. C
14. F
15. Answers may vary. See Test Doctor for a detailed scoring rubric.
16. D

Interpreting Graphics

Directions (13–16): **For** *each* **question below, record the correct answer on a separate sheet of paper.**

The diagram below shows the waxing and waning of the moon. Use this diagram to answer questions 13 and 14.

Phases of the Moon in the Northern Hemisphere

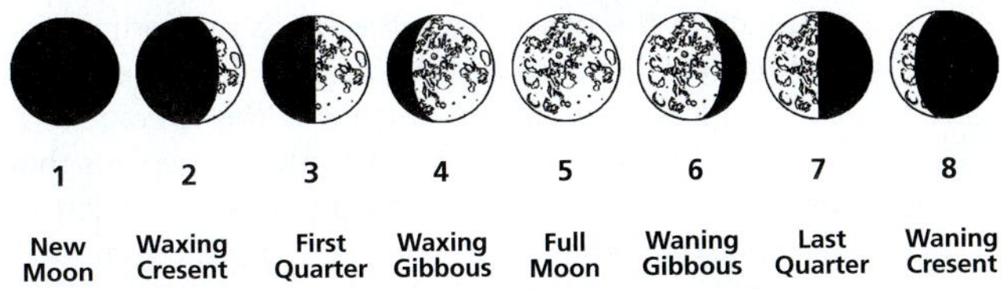

13 How would the appearance of the moon in the Southern Hemisphere be different from its appearance in the Northern Hemisphere?
A. The phases of the moon would appear exactly the same.
B. The Southern Hemisphere would see a full moon when the Northern Hemisphere sees a new moon.
C. The moon would wax from left to right instead of from right to left.
D. The Southern Hemisphere would see a waxing moon when the Northern Hemisphere sees a waning moon.

14 What part of the moon is facing Earth during the new moon in stage 1?
F. the near side
G. the far side
H. the north pole
I. the south pole

15 The word *wax* means "to grow larger," while *wane* means "to grow smaller." If the lighted portion of a waxing crescent is the same size as that of a waning crescent, why do you think these terms are used?

The diagram below shows data about the interior structure of the moon. Use this diagram to answer question 16.

Structure of the Moon

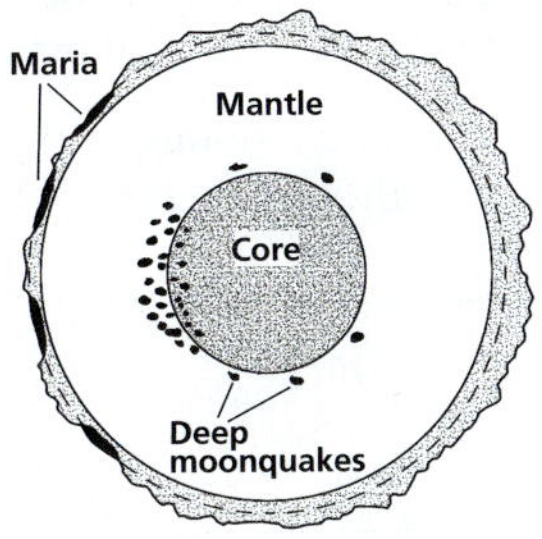

16 Where is the crust of the moon the thickest?
A. at the poles
B. at the equator
C. on the near side
D. on the far side

Test TIP

Test questions are not necessarily arranged in order of difficulty. If you are unable to answer a question, mark it and move on to other questions.

Standardized Test Prep

TEST DOCTOR

Question 15 Full-credit answers should include the following points:

- students should demonstrate a conceptual understanding that waxing and waning describe the only the lighted portion of the moon
- during a waxing moon, the lighted portion appears to enlarge each subsequent night. The moon begins to wax after the new moon, when no light is visible
- during a waning moon, the lighted portion appears to shrink each subsequent night. The moon begins to wane after the full moon, when the most light is visible

Question 16 Answer D is correct. Students may use the information in the diagram, such as the location of the maria, to determine that the left side of the diagram represents the near side of the moon and that the right side represents the far side of the moon. Answer A is incorrect because the crust at the poles is not as thick as the crust on the far side of the moon. Answer B is incorrect because the crust at the equator, while thicker on the far side, is thinner on the near side. Answer C is incorrect because the crust on the near side is very thin.

Test Prep Correlations ***National Science Education Standards***

HNS 2a: items 9, 10, 11, 16
HNS 2b: items 1, 4, 13, 15
ST 2b: item 12
UCP 1: items 2, 3, 5, 6, 7, 11, 14

CHAPTER RESOURCES

State Resources

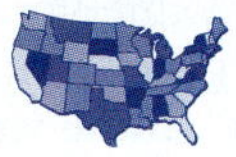

For specific resources for your state, visit **go.hrw.com** and type in the keyword **HSHSTR.**

Skills Practice Lab

Galilean Moons of Jupiter

Objectives

- **Calculate** the value of a constant, K.
- **Explain** how Kepler's law of periods explains orbits of moons of Jupiter.

Materials

calculator

metric ruler

Kepler's third law of motion—the law of periods—explains the relationship between a planet's distance from the sun and the planet's period (the time required to make one revolution around the sun.) According to the law of periods, the cube of the average distance of the planet from the sun is proportional to the square of the planet's period. Kepler's third law can be expressed mathematically as $K \times a^3 = p^2$, in which a is the average distance from the sun, p is the period, and K is a constant. Kepler's third law also may be applied to moons orbiting a planet, in which a is the average distance of a moon to the planet and p is the moon's period. In this activity, you will verify that the orbital motions of Jupiter's moons obey Kepler's third law.

PROCEDURE

1. Two telescope eyepiece views at the left show how Jupiter and its four largest, or Galilean, moons appear through a telescope on Earth at midnight on the 9th and 19th day of a month. Compare these illustrations with the chart below, which shows the path of each moon as it orbits Jupiter during the same month.
   a. List the days when each of Jupiter's moons crosses in front of the planet.
   b. List the days when each of the moons is behind Jupiter.
2. Use the data in the table on the next page to test Kepler's third law. Calculate p^2 and a^3 for each of the planets. Record your results in a table of your own. Then, calculate K for each planet by using Kepler's third law, $K = p^2/a^3$. Record your results in a similar table.
3. Draw Jupiter and its moons as they would appear from Earth at midnight on the 2nd and 26th of the month.

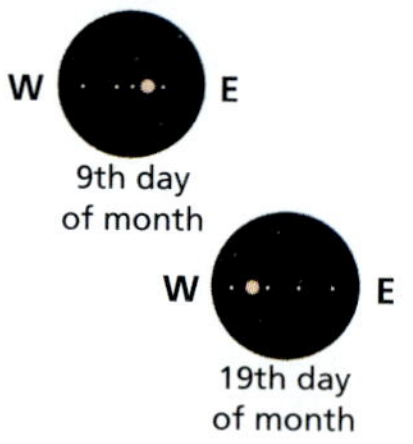

Step 1 The central horizontal band on the chart below represents Jupiter. When a moon's path crosses in front of this band, the moon is in front of the planet. When a moon's path crosses behind this band, the moon is behind Jupiter.

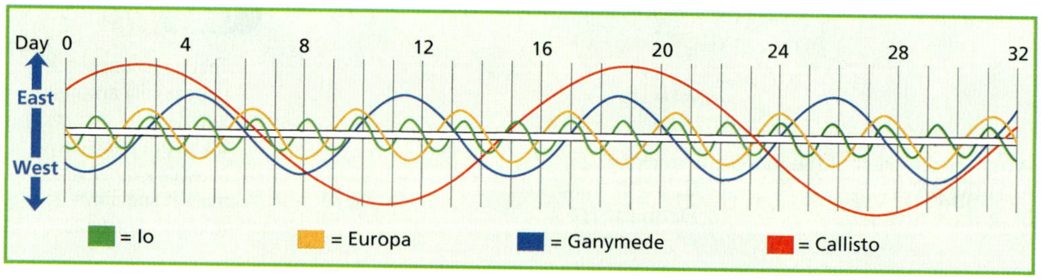

Skills Practice Lab

Galilean Moons of Jupiter

Teacher's Notes

Time Required

two 45-minute class periods

Lab Ratings

EASY → HARD

TEACHER PREPARATION: 1

STUDENT SETUP: 1

CONCEPT LEVEL: 3

CLEANUP: 1

Skills Acquired

- Organizing and Analyzing Data
- Identifying and Recognizing Patterns

The Scientific Method

In this lab, students will

- Make Observations
- Test the Hypothesis
- Analyze the Results
- Draw Conclusions

Materials

Students can do this investigation without a partner. However, you may want to try to pair a student who is proficient in math with one who is not as adept.

Tips and Hints

Have students identify the variables and the constant in the equation. You may want to review the use of exponents in recording large numbers. Check student results in step 2 of the procedure, before letting them continue. If they do not calculate p^2 and a^3 carefully, they will arrive at the wrong value for K. Although students may get various values for K, when rounded, these values should be approximately equal to 300.

Answers to Procedure

Note: Some student's answers may deviate by one day when the satellite crosses in the middle of a day.

1a. Io: 1, 3, 5, 6, 8, 10, 12, 13, 15, 17, 19, 20, 22, 24, 26, 28, 29, 31; Europa: 0, 3, 7, 10, 14, 17, 21, 24, 28, 31; Ganymede: 6, 13, 20, 27; Callisto: 6, 23

1b. Io: 0, 2, 4, 5, 7, 9, 11, 12, 14, 16, 18, 20, 21, 23, 25, 27, 28, 30; Europa: 1, 5, 8, 12, 16, 19, 23, 26, 30; Ganymede: 2, 9, 16, 23, 31; Callisto: 14, 31

2. Mercury: $a^3 = 0.000195$; $p^2 = 0.058$; $K = 297$. Venus: $a^3 = 0.00126$; $p^2 = 0.38$; $K = 302$. Earth: $a^3 = 0.00338$; $p^2 = 1$; $K = 296$. Mars: $a^3 = 0.01185$; $p^2 = 3.53$; $K = 298$. Jupiter: $a^3 = 0.47091$; $p^2 = 140.7$; $K = 299$. Saturn: $a^3 = 2.9058$; $p^2 = 867.9$; $K = 299$. Uranus: $a^3 = 23.665$; $p^2 = 7{,}022$; $K = 297$. Neptune: $a^3 = 90.943$; $p^2 = 26{,}798$; $K = 295$.

CHAPTER RESOURCES

Chapter Resource File

- Datasheet for Chapter Lab GENERAL
- Lab Notes and Answers

Workbooks

Long-Term Projects

- Tides at the Shoreline ADVANCED
- Apparent Motions of the Moon ADVANCED

4. Draw Jupiter's moons on the first day of the month that all four moons are on the same side of the planet. Identify the date.

5. Give a date when only two moons will be visible. Name the two visible moons.

6. Follow each moon's motion on the chart. Find the length of time, in Earth days, required for each moon to orbit Jupiter. To do this, measure the time between two points when the moon is in exactly the same position on the same side of Jupiter. Record your answers in a table listing the moons, p (in Earth days), a (in mm), p^2, a^3, and K.

Kepler's Third Law					
Planet	p (in Earth years)	a (in billions of km)	a^3	p^2	K
Mercury	0.24	0.058			
Venus	0.62	0.108			
Earth	1	0.150			
Mars	1.88	0.228			
Jupiter	11.86	0.778			
Saturn	29.46	1.427			
Uranus	83.8	2.871			
Neptune	163.7	4.497			

DO NOT WRITE IN THIS BOOK

7. Measure the scale distance between the maximum outward swing of each moon and the center of Jupiter in millimeters. Record your answers in your table.

8. Square each period measurement, and record the answer in your table. Cube each distance measurement, and record the answer.

9. Use your results to test Kepler's third law. Because $K = p^2/a^3$, divide p^2 by a^3 for each moon to find K. Record your results in your table.

ANALYSIS AND CONCLUSION

1. **Analyzing Events** Will you see all four of Jupiter's largest moons each time you look at Jupiter through a telescope or binoculars? Explain your answer.

2. **Making Inferences** If you look at Jupiter's moons through a telescope, they look like dots. If you had no charts, how could you identify each moon?

3. **Drawing Conclusions** After you solve for K for each moon, study your results. Is K a constant for the moons of Jupiter? Explain your answer.

Extension

1. **Making Calculations** Recalculate the values of K for the planets by using astronomical units instead of kilometers. How does this affect the amount of variation in the value of the constant?

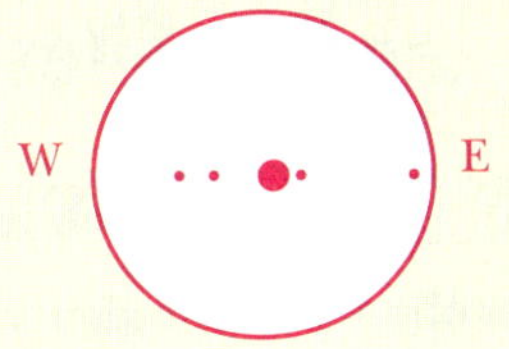

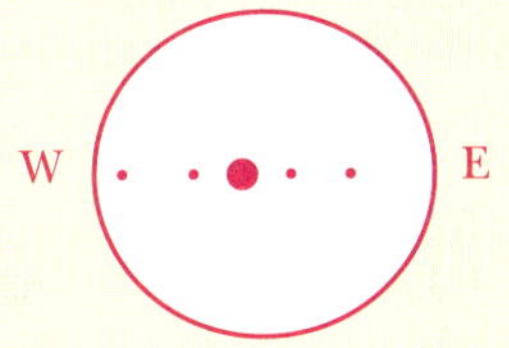

4. the 2nd

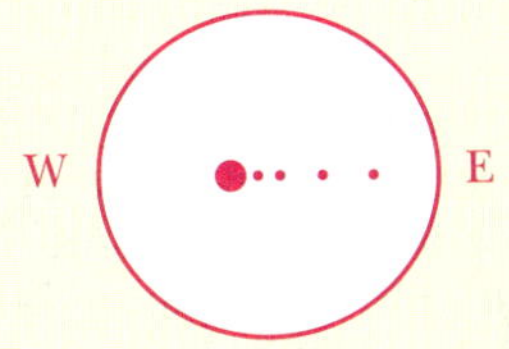

5. 3, 10: Ganymede and Callisto; 14: Europa and Ganymede; 23, 30: Ganymede and Callisto; 31: Io and Europa

6–9. Io: p = 1.8 days; a = 2.5 mm; p^2 = 3.24; a^3 = 16; K = 0.2025
Europa: 3.6 days; 4 mm; 12.96; 64; 0.2025
Ganymede: 7.4 days; 6.5 mm; 54.76; 27.5; 0.1991
Callisto: 16.5 days; 11 mm; 272.35; 1331; 0.2045

Answers to Analysis and Conclusion

1. No, sometimes a moon will be behind or in front of Jupiter.
2. by observing their maximum outward swing and by determining the periods
3. K is a constant because all the numbers are about equal; average value of K = 0.20.

Answers to Extension

1. The value of K would remain a constant, but when distance is measured in astronomical units that constant would be equal to 1, and $a^3 = p^2$.

TEACHER TESTED & APPROVED
Alexander Dvorak
Heritage School
New York City, NY

Maps in Action

Lunar Landing Sites

Internet Activity — **BASIC**

Lunar Timeline Have students research the lunar missions shown on the map. They can investigate dates, mission objectives, sites, discoveries, or mission successes. They could also investigate the Lunar Prospector and Selene Lunar Lander missions. Have students create a timeline for these missions. Have them use photos or drawings to illustrate each event on the timeline. A worksheet designed to direct student research on this topic can be found in the **Chapter Resource File** booklet or by visiting **go.hrw.com** and entering the keyword **HQ6MBSX.** **LS Visual**

Answers to Map Skills Activity

1. 8
2. 3
3. three; *Apollo 11* landed near *Surveyor 5*, and *Apollo 12* and *14* landed near *Surveyor 3*.
4. There are no landing sites north of 40°N latitude, south of 40°S latitude, or between 20°S and 30°S latitude.
5. between 5°N and 5°S of the lunar equator
6. maria and craters
7. Sample answer: none, because the radio waves used for communication would be blocked by the moon

CHAPTER RESOURCES

Chapter Resource File

Internet Activity
- Lunar Timeline **BASIC**

Technology

Transparencies
- 145 Lunar Landing Sites (with worksheet)

MAPS *in Action*

Lunar Landing Sites

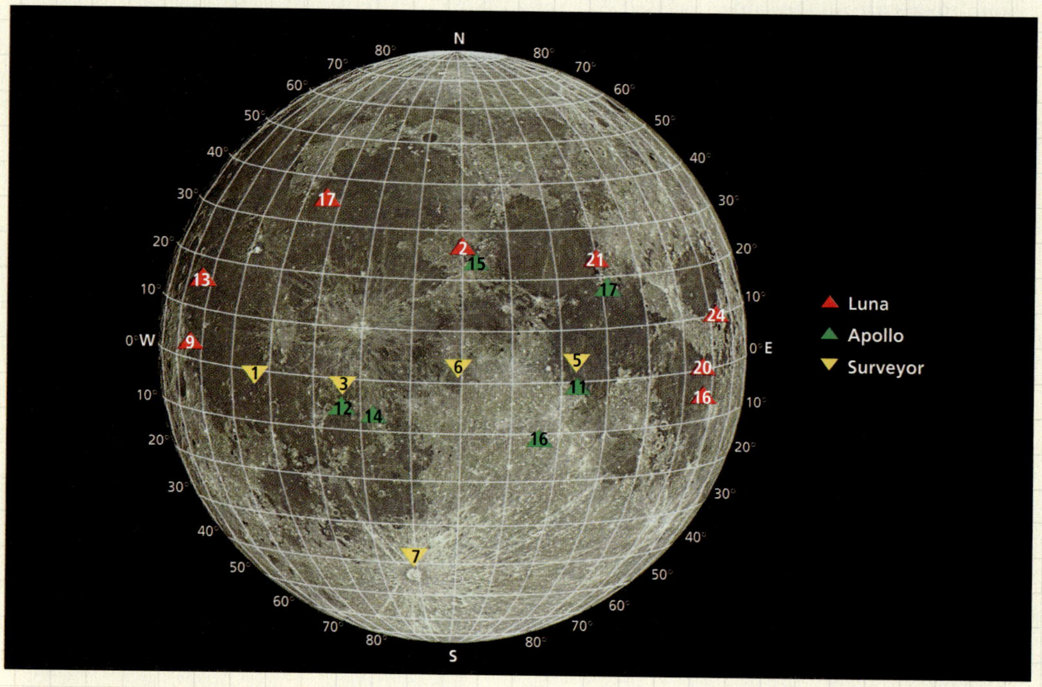

Map Skills Activity

This map shows the surface of the near side of the moon and the landing sites of both lunar missions that had crews and lunar missions that did not have crews. Most Surveyor missions took place before the Apollo missions. The Luna missions were launched by the former Soviet Union. Use the map to answer the questions below.

1. **Using the Key** How many Luna missions landed on the moon?
2. **Using the Key** How many Surveyor missions landed on the moon's southern hemisphere?
3. **Analyzing Data** How many Apollo missions landed close to Surveyor mission landing sites?
4. **Making Comparisons** At which areas of latitude are no landing sites located?
5. **Making Comparisons** In what 10° range of latitude are most landing sites located?
6. **Inferring Relationships** Based on the locations of most landing sites, what surface features do you think interested scientists?
7. **Identifying Trends** Many of the missions to the moon used radio communications. Radio communications require a clear path between the transmitter and receiver. Taking these facts into consideration, how many landing sites would you expect to find on the far side of the moon? Explain your answer.

SCIENCE AND TECHNOLOGY

Mining on the Moon

Since astronauts landed on the moon in 1969, scientists have been working toward permanent lunar bases. The biggest obstacle is the cost of transporting construction materials and life-support systems. So, scientists are developing ways to produce the materials by mining the moon's own natural resources.

Moon Mining Methods

One of the promising mining methods uses electrolysis, a process in which an electrical current is passed through a substance to change the chemical composition of the substance. Scientists hope to use electrolysis to extract iron and oxygen from lunar rocks. To test the procedure, scientists created a silicate rock like the rocks that are common on the moon. After melting the rock, they passed an electrical current through the molten rock, which caused iron and oxygen to separate from the other elements. On the moon, the extracted iron could be used to manufacture steel, and the oxygen could support human life.

Researchers also imagine using a solar-powered satellite to provide energy for everything from lunar mining to life-support systems. The satellite might use a solar concentrator and lasers to beam energy to the moon's surface. There, the laser light would be converted into electricity.

Ice Mines

Another important substance that might be mined on the moon is water in the form of ice. Many scientists think that the amount of ice that has been discovered in the craters near the poles could be of practical use. This ice was discovered where the crater rims have shielded the ice from the sun's rays and prevented it from melting.

▲ Lunar mining operations might look something like this.

Mining the moon may not be possible for many years to come. But scientists continue to develop the technology to make it possible. In addition to developing the necessary technology, many more geological studies of the moon are necessary before undertaking such a project. If mining the moon were profitable, it would help secure funding for future space exploration by opening the door to private commercial ventures.

◄ Apollo astronauts discovered valuable materials on the moon.

Extension

1. **Making Inferences** Why would solar energy be important for a lunar mining operation?

Science and Technology

Mining on the Moon

Demonstration — GENERAL

Electrolysis of Water Demonstrate the chemical process of electrolysis to help students understand this moon-mining method. Explain that the process uses an electrical current to separate or extract materials by means of a chemical change. Fill a drinking glass with distilled water to about 2.5 cm (1 in.) from the top of the glass. Add about 1/4 teaspoon of table salt to the water and stir to dissolve the salt. Straighten out one end of two paper clips, leaving a hook on the other end of each paper clip. Insert the paper clips through a square of stiff card about 2.5 cm (1 in.) apart. Place the bare ends of the paper clips in the water and rest the card on top of the glass. Cut two strips of aluminum foil about 10 cm (4 in.) wide and 30 cm (12 in.) long. Wrap the foil around the hooked ends of the paper clips. Have students observe as you hold the free ends of the foil strips around the two ends of a D-cell battery. Ask someone to describe what happens at each of the bare ends of the paper clips. (Bubbles of gas appear at each paper clip.) Explain that the current caused the water to change into its component gases, oxygen and hydrogen. Because each element is located at a different terminal of the simple battery, the elements are easy to extract. LS **Visual**

Answer to Extension

1. Using solar-powered equipment would reduce the need to transport energy sources from Earth to the moon, which would be prohibitively expensive because of the weight involved. The sun could provide an almost unlimited source of energy, because the sun illuminates half of the moon's surface at any given time.

Chapter 29

The Sun

Planning Guide

Compression Guide

To shorten instruction because of time limitations, omit Section 2.

OBJECTIVES	LABS, DEMONSTRATIONS, AND ACTIVITIES	TECHNOLOGY RESOURCES
PACING • 90 min pp. 754–760 **Chapter Opener**	**SE Long-Term Project** Positions of Sunrise and Sunset, pp. 848–851 ADVANCED **LTP Long-Term Project** Positions of Sunrise and Sunset* ADVANCED	**OSP Parent Letter** **CD Student Edition on CD-ROM** **CD Chapter Summaries Audio CD** **VID Brain Food Video Quiz**
Section 1 Structure of the Sun • Explain how the sun converts matter into energy in its core. • Compare the radiative and convective zones of the sun. • Describe the three layers of the sun's atmosphere.	**SE Quick Lab** Modeling Fusion, p. 757 GENERAL **CRF Datasheet for Quick Lab*** GENERAL **TE History Connection** Reinventing Time and Space, p. 757 ADVANCED **SE Quick Lab** The Size of Our Sun, p. 758 GENERAL **CRF Datasheet for Quick Lab*** GENERAL **SE Skills Practice Lab** Energy of the Sun, pp. 770–771 GENERAL **CRF Datasheet for Chapter Lab*** GENERAL **SE Maps in Action** SXT Composite Image of the Sun, p. 772 GENERAL **CRF Inquiry Lab** Solar Cooker* GENERAL **CRF Making Models Lab** Light Fingerprints* GENERAL	**OSP Lesson Plans** (also in print) **TR Bellringer*** TR 146 Nuclear Fusion* TR 147 The Sun's Interior* TR 148 SXT Composite Image of the Sun* **TE Internet Activity** SOHO Images, p. 772 GENERAL **CRF Internet Activity** SOHO Images* GENERAL **CD Interactive Tutor** What's the Matter? (Atomic Theory and Nuclear Energy) **CD Interactive Tutor** Energy **CD Interactive Tutor** Light and Electromagnetic Energy
PACING • 45 min pp. 761–764 **Section 2 Solar Activity** • Explain how sunspots are related to powerful magnetic fields on the sun. • Compare prominences, solar flares, and coronal mass ejections. • Describe how the solar wind can cause auroras on Earth.	**TE Discussion** Sunspots, p. 761 GENERAL **TE Group Activity** Geomagnetic Storms, p. 763 GENERAL **TE Discussion** Collecting Solar Samples, p. 773 BASIC	**OSP Lesson Plans** (also in print) **TR Bellringer*** **TE Internet Activity** Solar Activity and Climate, p. 762 ADVANCED **CRF Internet Activity** Solar Activity and Climate* ADVANCED **CD Interactive Tutor** Hail, Northern Lights, and Rainbows **CD Interactive Tutor** Our Sun and Origin of the Solar System

PACING • 90 min

CHAPTER REVIEW, ASSESSMENT, AND STANDARDIZED TEST PREPARATION

- **SE Chapter Highlights,** p. 765
- **SE Chapter Review,** pp. 766–767
- **SE Standardized Test Prep,** pp. 768–769
- **CRF Concept Review*** GENERAL
- **CRF Critical Thinking*** ADVANCED
- **CRF Math Skills*** GENERAL
- **CRF Graphing Skills*** GENERAL
- **CRF Chapter Test A*** GENERAL
- **CRF Chapter Test B*** ADVANCED
- **OSP Lesson Plans** (also in print)
- **OSP Test Generator**
- **OSP Test Item Listing***

Online and Technology Resources

Holt Online Learning

Visit **go.hrw.com** for access to Holt Online Learning, or enter the keyword **HQ6 Home** for a variety of free online resources.

One-Stop Planner® CD-ROM

This CD-ROM package includes

- Lab Materials QuickList Software
- Holt Calendar Planner
- Customizable Lesson Plans
- Printable Worksheets
- ExamView® Test Generator
- Interactive Teacher Edition
- Holt PuzzlePro®
- Holt PowerPoint® Resources

KEY

- **SE** Student Edition
- **TE** Teacher Edition
- **CRF** Chapter Resource File
- **LTP** Long-Term Projects
- **OSP** One-Stop Planner
- **TR** Transparencies and Transparency Worksheets
- **CD** CD or CD-ROM
- **VID** Classroom Video/DVD
- * Also on One-Stop Planner
- ◆ Requires advance prep
- ■ Also available in Spanish

SKILLS DEVELOPMENT RESOURCES	REVIEW AND ASSESSMENT	CORRELATIONS
SE **Pre-Reading Activity,** p. 754 GENERAL TE **Using the Figure** The Explosive Sun, p. 754 GENERAL		National Science Education Standards
CRF **Directed Reading*** BASIC TE **Using the Figure** Spectra, p. 755 GENERAL SE **Graphic Organizer** Chain-of-Events Chart, p. 756 GENERAL TE **Skill Builder** Math, p. 756 ADVANCED TE **Using the Figure** Discussion, p. 757 GENERAL TE **Reading Skill Builder** Reading Organizer, p. 758 BASIC TE **Inclusion Strategies,** p. 758 TE **Skill Builder** Vocabulary, p. 759 GENERAL	SE **Reading Checks,** pp. 757, 759 GENERAL SE **Section Review,** p. 760 GENERAL TE **Reteaching,** p. 759 BASIC TE **Quiz,** p. 759 GENERAL TE **Alternative Assessment,** p. 759 ADVANCED CRF **Section Quiz*** ■ GENERAL	PS 5c, PS 5d, ES 4b, ES 4c
CRF **Directed Reading*** BASIC TE **Using the Figure** Sunspot Cycles, p. 762 GENERAL TE **Reading Skill Builder** Paired Summarizing, p. 762 BASIC TE **Inclusion Strategies,** p. 762 TE **Skill Builder** Writing, p. 763 GENERAL SE **Math Practice,** p. 763 GENERAL	SE **Reading Check,** p. 763 GENERAL SE **Section Review,** p. 764 GENERAL TE **Reteaching,** p. 763 BASIC TE **Quiz,** p. 763 GENERAL TE **Alternative Assessment,** p. 764 ADVANCED CRF **Section Quiz*** ■ GENERAL	PS 4e

Holt Lab Generator CD-ROM

Search for labs by topic, standard, difficulty level, and/or time frame. Edit those labs to fit your needs, or create your own labs. Use the Lab Materials QuickList to order just what you need.

Holt Earth Science Interactive Tutor CD-ROM

This CD-ROM consists of interactive activities that give students a fun way to extend their knowledge of Earth science concepts.

Chapter Summaries Audio CDs

These CDs include audio summaries of the key concepts presented in each chapter. (Audio summaries are also available in Spanish.)

SciLinks NSTA

www.scilinks.org

Maintained by the **National Science Teachers Association.** See Chapter Enrichment pages that follow for a complete list of topics.

See Chapter Enrichment pages for Video Resources.

Refer to our ***Professional Reference for Teachers*** for additional teaching resources, including articles written by science education professionals about relevant and timely issues facing today's science teachers.

Chapter 29

Chapter Enrichment

This Chapter Enrichment provides relevant and interesting information to expand and enhance your classroom instruction of the chapter material.

Section 1 Structure of the Sun

Observing the Sun

Scientists use a wide variety of instruments to study different aspects of the solar disk. Telescopes used for viewing the sun are commonly equipped with special filters or are designed to project an image onto a screen to overcome the inherent risk of blindness associated with direct solar observation. Telescopes must be fitted with devices called *coronagraphs* to photograph the solar corona.

Seeing the Light

Spectroscopy is the technique of using spectral lines to identify the composition, temperature, or density of distant astronomical objects. Each element emits light waves at a distinctive combination of wavelengths, and so the emission spectrum provides a unique "fingerprint" for each element. This spectrum can be detected by using prisms to refract or bend light or by using diffraction gratings, which are plates of plastic, glass, or reflective metal that have finely spaced lines that diffract light. In addition to data obtained from the visible spectrum, there is much to be learned by observing the sun by using special instruments that detect ultraviolet, infrared, X-ray, and radio emissions. Instruments mounted on spacecraft and satellites allow scientists to study aspects of the sun without interference by Earth's atmosphere.

▲ **The sun's corona is visible to observers on Earth only when the sun is blocked by an object such as the moon.**

Energy Transfers Within the Layered Sun

The sun is a hot sphere of gases that is fueled by fusion in its core. A series of reactions converts hydrogen nuclei into helium nuclei. The energy that is released is emitted as high-energy photons. Radiation is the principle means of energy transport deep within the sun's interior. The radiative zone transfers energy from the core toward the surface as high-energy photons make countless collisions with atomic nuclei. Convection is the principle means of energy transport in the sun's topmost layers. The mottled surface of the sun, called *granulation,* is the visible manifestation of the convection process. The seething bubbles of gas called *granules* form and dissolve within minutes. The photosphere is the zone in which the sun becomes transparent to visible light, the form in which much of this energy leaves the sun. As gases churn in the photosphere, they produce shock waves that form the chromosphere as spikes of hot gas called *spicules*. The hot corona is a strong source of X rays and the charged particles of the solar wind.

Section 2 Solar Activity

Sunspots

The dark, cool areas of the photosphere called *sunspots* usually appear in pairs. Their intense magnetic fields leave the surface through one sunspot and re-enter through the second spot. The end of one eleven-year sunspot cycle and the beginning of the next one is accompanied by a reversal in the direction of the sun's overall magnetic field. Thus, the complete magnetic cycle of the sun actually takes about 22 years.

Earth's Protective Magnetosphere

Earth has a strong internal magnetic field, probably generated by electric currents that result from the movement of the metallic elements in Earth's interior core. This magnetic field protects Earth from most of the sun's emissions. Charged particles become trapped within Earth's magnetic field lines, forming the magnetosphere. The

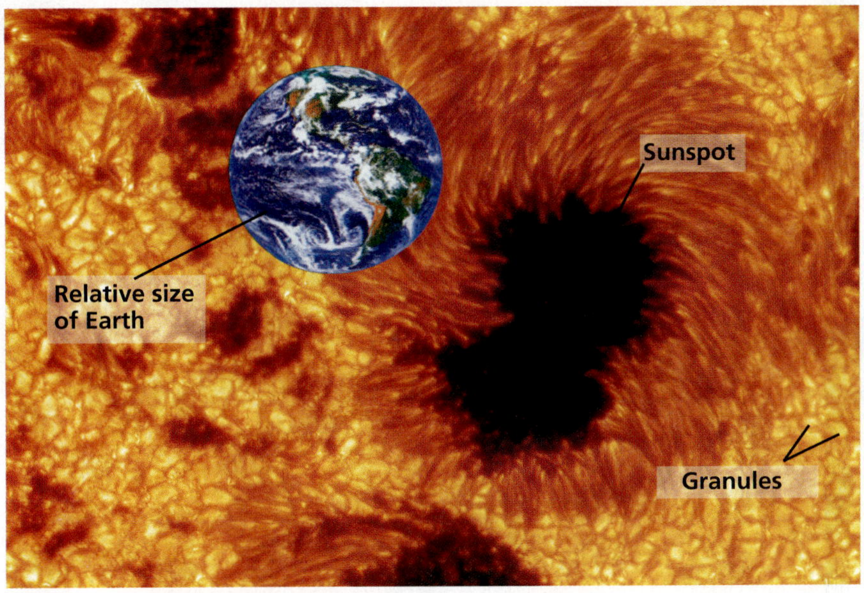

▲ An image of Earth was superimposed on this photo to show the scale of a sunspot.

magnetosphere changes its shape constantly in response to the pressure of the solar wind. The solar wind pushes the magnetosphere in on the side nearest the sun and stretches it into a long tail on the opposite side. Two large donut-shaped regions of the magnetosphere, called the *Van Allen Radiation Belts,* contain high-energy particles that drift around the planet. Particles that enter the magnetosphere spiral around toward the poles, where they produce shimmering curtains of light called *auroras*.

The flow of plasma inside the sun twists and snarls its magnetic field into loops that poke out through the corona and cause wild eruptions. When conditions are just right, the sudden release of this magnetic energy, as manifested in solar flares and coronal mass ejections, can merge with Earth's magnetic field and pump tremendous amounts of energy into the magnetosphere, causing magnetic storms. These storms can disrupt communications equipment and satellites and cause power blackouts. They affect radio communications by ionizing parts of Earth's atmosphere, which can distort radio waves.

Video Resources

Brain Food Video Quizzes

Brain Food Video Quizzes These videos contain game-show style quizzes that assess students' progress and motivate students to study the chapter material.

NOVA Videos

To order **NOVA** videos related to this chapter, visit **go.hrw.com** and enter the keyword **HQ6SUNV.**

SciLinks

Developed and maintained by the National Science Teachers Association

SciLinks is maintained by the National Science Teachers Association to provide you and your students with interesting, up-to-date links that will enrich your classroom presentation of the chapter.

Visit www.scilinks.org and enter the SciLinks code for more information about the topic listed.

Topic: The Sun
SciLinks code: HQ61477

Topic: Electromagnetic Spectrum
SciLinks code: HQ60482

Topic: Nuclear Fusion in the Sun
SciLinks code: HQ61051

Topic: Solar Activity
SciLinks code: HQ61413

Chapter 29

Chapter Overview

This chapter describes the structure of the sun, how the sun produces energy, and how its activity cycle affects Earth.

Using the Figure — GENERAL

The Explosive Sun The outer layers of the sun's atmosphere are extremely volatile and subject to explosive disturbances. Ask students to describe some of the characteristic solar features seen in the photo. (Sample answer: Huge loops and arcs of hot gas are erupting from the surface. Hotter and cooler areas appear on the surface.) Many of these features are associated with the sun's magnetic field. The smallest visible solar surface features—small bubbles of gas called granules—are not visible in this photo. These granules may be as big as the state of Texas and carry energy from below the photosphere. LS **Visual**

PRE-READING ACTIVITY

FOLDNOTES

You may want students to work in pairs to create this FoldNote. Have one student fill in the first two columns, and have the other student fill in the last two columns.

Chapter 29 The Sun

Sections

1 Structure of the Sun

2 Solar Activity

What You'll Learn

- How the sun's interior differs from its surface layers
- How sunspots and other solar activity vary during an 11-year cycle

Why It's Relevant

The sun is the energy source that fuels most life on Earth. Understanding how the sun's energy affects Earth helps people understand how Earth interacts with the sun.

PRE-READING ACTIVITY

FOLDNOTES

Table Fold Before you read this chapter, create the FoldNote entitled "Table Fold" described in the Skills Handbook section of the Appendix. Label the columns of the table fold with "Characteristics of the sun," "Energy of the sun," "Composition of the sun," and "Layers of the sun." As you read the chapter, write examples of each topic under the appropriate column.

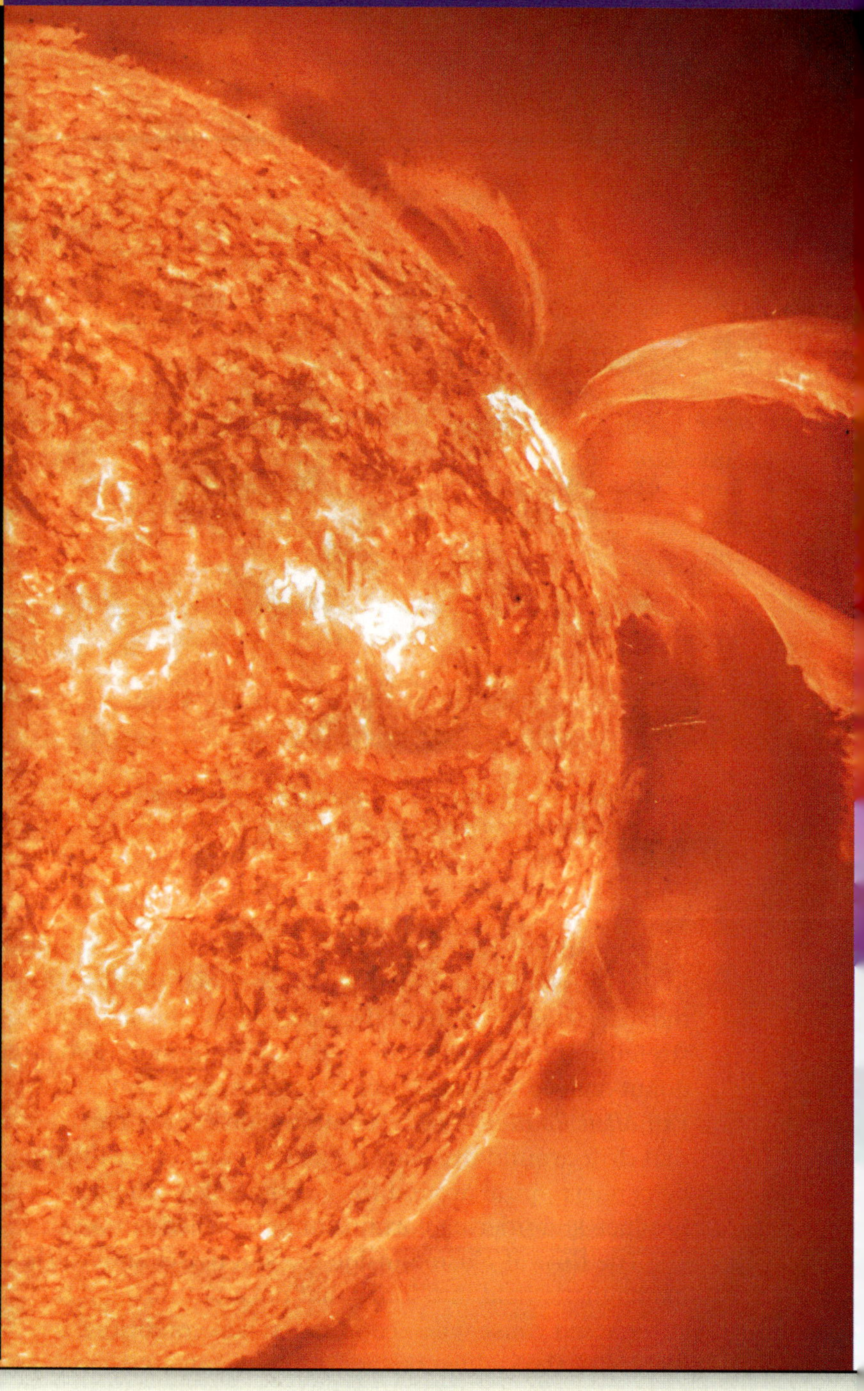

▶ Huge plumes of hot gas that are many times the size of Earth, called *prominences*, are occasionally visible in the sun's atmosphere. The hottest parts of the sun shown in this false-color image are white, and the coolest areas are dark red.

Chapter Correlations — *National Science Education Standards*

PS 4e Electricity and magnetism are two aspects of a single electromagnetic force. Moving electric charges produce magnetic forces, and moving magnets produce electric forces… **(Section 2)**

PS 5c Heat consists of random motion and the vibrations of atoms, molecules, and ions. The higher the temperature, the greater the atomic or molecular motion. **(Section 1)**

PS 5d Everything tends to become less organized and less orderly over time. Thus, in all energy transfers, the overall effect is that the energy is spread out uniformly. Examples are the transfer of energy from hotter to cooler objects by conduction, radiation, or convection.... **(Section 1)**

ES 4b Early in the history of the universe, matter, primarily the light atoms hydrogen and helium, clumped together by gravitational attraction to form countless trillions of stars.... **(Section 1)**

ES 4c Stars produce energy from nuclear reactions, primarily the fusion of hydrogen to form helium. These and other processes in stars have led to the formation of all the other elements. **(Section 1)**

Section 1 Structure of the Sun

Throughout much of human history, people thought that the sun's energy came from fire. People knew that burning a piece of coal or wood produced heat and light. They assumed that the sun, too, burned some type of fuel to produce its energy. But less than 100 years ago, scientists discovered that the source of the sun's energy is quite different from fire.

OBJECTIVES

- **Explain** how the sun converts matter into energy in its core.
- **Compare** the radiative and convective zones of the sun.
- **Describe** the three layers of the sun's atmosphere.

KEY TERMS

nuclear fusion
radiative zone
convective zone
photosphere
chromosphere
corona

The Sun's Energy

The sun appears to the unaided eye as a dazzling, brilliant ball that has no distinct features. Because the sun's brightness can damage your eyes if you look directly at the sun, astronomers look at the sun only through special filters. Astronomers often use other specialized scientific instruments to study the sun.

Composition of the Sun

Scientists break up the sun's light into a spectrum (plural, *spectra*) by using a device called a *spectrograph*. Dark lines form in the spectra of stars when gases in the stars' outer layers absorb specific wavelengths of the light that passes through the layers. The temperature of these outer layers determines which gases produce visible spectral lines. By studying the spectrum of a star, scientists can determine the amounts of elements that are present in a star's atmosphere. They can also deduce the temperature, density, and pressure of the gas. Because each element produces a unique pattern of spectral lines, astronomers can match the spectral lines of starlight to those of Earth's elements, as shown in **Figure 1,** and identify the elements in the star's atmosphere.

Both hydrogen and helium occur in the sun. About 75% of the sun's mass is hydrogen, and hydrogen and helium together make up about 99% of the sun's mass. The sun's spectrum reveals that the sun contains traces of almost all other chemical elements.

SCILINKS. NSTA Developed and maintained by the National Science Teachers Association

For a variety of links related to this subject, go to www.scilinks.org

Topic: The Sun
SciLinks code: HQ61477

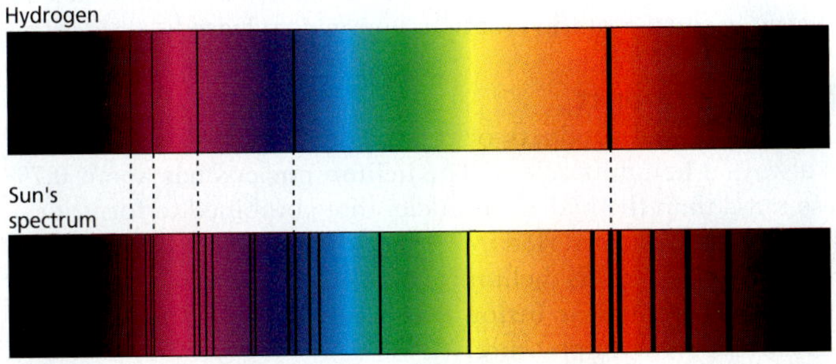

Figure 1 ▶ When light passes through hydrogen gas and then through a slit in a prism, dark lines appear in the spectrum. Hydrogen and lines from other elements in the solar spectrum are shown in the bottom spectrograph. ***How many lines are not accounted for by the presence of hydrogen in the sun's atmosphere?***

Section 1

Focus

Overview

This section describes the composition of the sun and how the sun produces energy. The section also details the layered structure of the sun and how scientists study the sun.

Bellringer

Have students answer the following questions:

1. What is the sun made of?
2. What does the sun use as a fuel to produce energy?

(Answers may vary but may provide insights into students' misconceptions or may act as a starting point for discussions.)

Motivate

Using the Figure — GENERAL

Spectra Point out the dark lines that cross the sun's spectrum in the bottom spectrograph. Explain that astronomers use the pattern of spectral lines present in the light from stars, such as Earth's sun, like a fingerprint to identify what elements are present. Answer to caption question: Nineteen of the twenty-four lines are not accounted for by hydrogen. LS **Logical/Visual**

CHAPTER RESOURCES

Chapter Resource File

- **Directed Reading** BASIC
- **Inquiry Lab** Solar Cooker GENERAL
- **Making Models Lab** Light Fingerprints GENERAL

Technology

Transparencies
- Bellringer

Student Edition on CD-ROM

One-Stop Planner CD-ROM
- Lesson Plan

Teach

SKILL BUILDER—ADVANCED

Math During the last step of fusion, 26 MeV (megaelectron Volts) of energy are released. 1 MeV equals approximately 1.6×10^{-13} J. Have students calculate how many fusion reactions it would take to release 1 J of energy.
(1 J = 26 $(1.6 \times 10^{-13}$ J$)$ × *N* fusion reactions;
N fusion reactions = 1 J ÷ $(4.16 \times 10^{-12}$ J$)$ =
$(10 \div 4.16) \times 10^{11} = 2.4 \times 10^{11}$, or 240,000,000,000 fusion reactions)

Graphic Organizer GENERAL

Chain-of-Events Chart

You may want to use this Graphic Organizer to assess students' prior knowledge before beginning a discussion of the process of nuclear fusion. You may also choose to use a similar activity as a quiz to assess students' knowledge of how the sun produces energy or to assess students' knowledge after students have read the description of nuclear fusion.

CHAPTER RESOURCES

Technology

Transparencies
- 146 Nuclear Fusion (with worksheet)

Figure 2 ▶ In the core of the sun, the nuclei of hydrogen atoms fuse to form helium. The fusion process converts mass into energy.

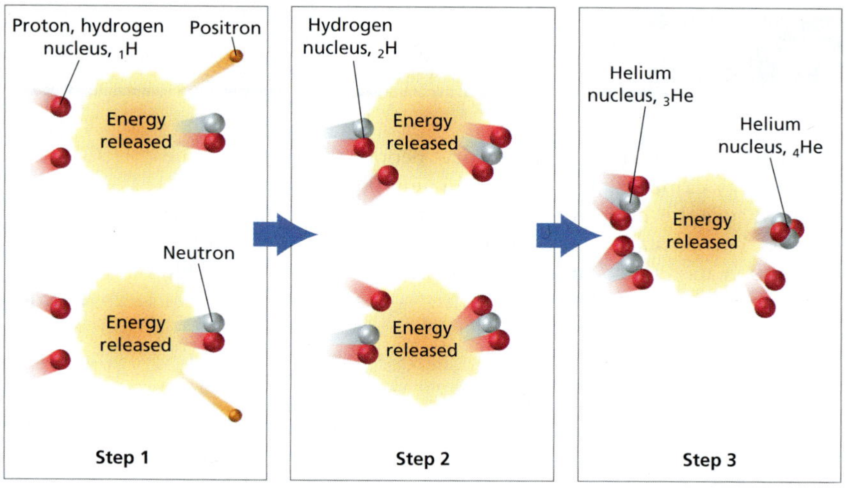

nuclear fusion the process by which nuclei of small atoms combine to form a new, more massive nucleus; the process releases energy

Nuclear Fusion

A powerful atomic process known as nuclear fusion occurs inside the sun. **Nuclear fusion** is the process of combining nuclei of small atoms to form more-massive nuclei. Fusion releases huge amounts of energy. Nuclei of hydrogen atoms are the primary fuel for the sun's fusion. A hydrogen atom, the simplest of all atoms, commonly consists of only one electron and one proton. Inside the sun, however, electrons are stripped from the protons by the sun's intense heat.

Nuclear fusion produces most of the sun's energy and consists of three steps, as shown in **Figure 2.** In the first step, two hydrogen nuclei, or *protons,* collide and fuse. In this step, the positive charge of one of the protons is neutralized as that proton emits a particle called a *positron*. As a result, the proton becomes a neutron and changes the original two protons into a proton-neutron pair. In the second step, another proton combines with this proton-neutron pair to produce a nucleus made up of two protons and one neutron. In the third step, two nuclei made up of two protons and one neutron collide and fuse. As this fusion happens, two protons are released. The remaining two protons and two neutrons are fused together and form a helium nucleus. During each step of the reaction, energy is released.

Graphic Organizer **Chain-of-Events Chart**

Create the **Graphic Organizer** entitled "Chain-of-Events Chart" described in the Skills Handbook section of the Appendix. Then, fill in the chart with details about each step of the process of nuclear fusion.

The Final Product

One of the final products of the fusion of hydrogen in the sun is always a helium nucleus. The helium nucleus has about 0.7% less mass than the hydrogen nuclei that combined to form it do. The lost mass is converted into energy during the series of fusion reactions that forms helium. The energy released during the three steps of nuclear fusion causes the sun to shine and gives the sun its high temperature.

BRAIN FOOD

Solar Life Cycle Earth's sun formed about 5 billion years ago when a cloud of interstellar gas began to collapse as a result of its gravity. The gas in the center got hotter and hotter as it absorbed thermal energy that resulted from the kinetic energy of subatomic particles. Eventually, electrons were stripped away from their atoms, and protons (hydrogen nuclei) moved so fast that they collided violently and began the fusion reactions that fuel the sun to this day.

In another 5 billion years, the sun's nuclear furnace will be forced into hotter reactions that fuse helium nuclei to form elements such as carbon. When this happens, the sun will swell in size as its outer shell cools. It will become a red giant that will engulf Mercury, Venus, and perhaps even Earth. When a star uses up the nuclear fuel in its center, it collapses. In this final phase, the sun will shrink in size and become a white dwarf.

Mass Changing into Energy

The sun's energy comes from fusion, and the mass that is lost during fusion becomes energy. In 1905, the physicist Albert Einstein, then an unknown patent-office worker, proposed that a small amount of matter yields a large amount of energy. At the time, the existence of nuclear fusion was unknown. In fact, scientists had not yet discovered the nucleus of the atom. Einstein's proposal was part of his special theory of relativity. This theory included the equation $E = mc^2$. In this equation, E represents energy produced; m represents the mass, or the amount of matter, that is changed; and c represents the speed of light, which is about 300,000 km/s. Einstein's equation can be used to calculate the amount of energy produced from a given amount of matter.

By using Einstein's equation, astronomers were able to explain the huge quantities of energy produced by the sun. The sun changes more than 600 million tons of hydrogen into helium every second. Yet this amount of hydrogen is small compared with the total mass of hydrogen in the sun. During fusion, a type of subatomic particle, called a *neutrino,* is given off. Neutrinos escape the sun and reach Earth in about eight minutes. Studies of these particles indicate that the sun is fueled by the fusion of hydrogen into helium. One apparatus that collects these particles is shown in **Figure 3.** Elements other than hydrogen can fuse, too. In stars that are hotter than the sun, energy is produced by fusion reactions of the nuclei of carbon, nitrogen, and oxygen.

Figure 3 ▶ In Japan, this giant tank of pure water, which was only partly filled when the photo was taken, captures subatomic particles that fly out of the sun during nuclear fusion.

Reading Check **How did the equation $E = mc^2$ help scientists understand the energy of the sun? (See the Appendix for answers to Reading Checks.)**

QuickLAB — 5 min

Modeling Fusion

Procedure

1. Mark **six coins** by using a **marker** or **wax pencil.** Put a *P* for "proton" on the head side of each coin and an *N* for "neutron" on the tail side of the coins.
2. Place two coins P-side up. These two protons each represent hydrogen's simplest isotope, H. Model the fusion of these two H nuclei by placing them such that their edges touch. When they touch, flip one of them to be N-side up. This flip represents a proton becoming a neutron during fusion. The resulting nucleus, which consists of one proton and one neutron, represents the isotope hydrogen-2, ^{2}H.
3. To model the next step of nuclear fusion, place a third coin, P-side up, against the ^{2}H nucleus from step 2. This forms the isotope helium-3, or ^{3}He.
4. Repeat steps 2 and 3 to form a second ^{3}He nucleus.
5. Next, model the fusion of two ^{3}He nuclei. Move the two ^{3}He nuclei formed in step 3 so that their edges touch. When the two ^{3}He nuclei touch, move two of the protons in the two ^{3}He nuclei away from the other four particles. These four particles form a new nucleus: helium-4, or ^{4}He.

Analysis

1. Large amounts of energy are released when nuclei combine. How many energy-producing reactions did you model?
2. Create a diagram that shows the formation of ^{4}He.

QuickLAB

Skills Acquired

- Constructing Models
- Recognizing Patterns

Teacher's Notes You may wish to have students provide their own coins. This lab can also be performed by using paper circles instead of coins.

Answers

1. Five reactions were modeled: the formation of two hydrogen-2 atoms; the formation of two helium-3 atoms; and the formation of one helium-4 atom.
2. Student diagrams may vary but should accurately depict the stages of nuclear fusion.

Using the Figure — GENERAL

Discussion The light-sensitive detectors in the tank illustrated in the figure on this page were designed to detect subatomic particles called *neutrinos,* that are released during fusion reactions in the sun. Neutrinos were first proposed to help explain how mass is conserved when particles such as protons or neutrons change into other particles during nuclear decay reactions. Review the steps in the fusion process. Ask students: "At which step in this process would these subatomic particles most likely be released?" (During the first step, when a proton spontaneously changes into a neutron.) **LS Logical/Visual**

Answer to Reading Check

Einstein's equation helped scientists understand the source of the sun's energy. The equation explained how the sun could produce huge amounts of energy without burning up.

HISTORY CONNECTION — ADVANCED

Reinventing Time and Space
Invite interested students to research the life and work of Albert Einstein, whose analysis of space, time, gravity, electromagnetic waves, and matter paved the way for a revolution in physics. Einstein also had a major impact on atomic research as a result of his formula E=mc^2. Students may also focus on Einstein's humanitarian and social achievements. Encourage students to share the results of their research with classmates as oral reports, demonstrations, or visual presentations. **LS Verbal/Visual**

CHAPTER RESOURCES

Chapter Resource File

- **Datasheet for Quick Lab** GENERAL

Teach, *continued*

QuickLAB

Skills Acquired

- Constructing Models
- Measuring
- Analyzing Data
- Communicating

Teacher's Notes In this model, Jupiter will be 1/10 the size of the sun, and Earth will be on the order of 1/110 the size of the sun.

The planets could also be modeled from clay once the paper sun has been drawn to scale.

Sample calculation for step 3: If the diameter of the sun is 20 cm, then the diameter of Earth can be found by solving the equation:
$(1.4 \times 10^9)/(1.3 \times 10^7) = 1077$
$20 \text{ cm}/1077 = 0.2 \text{ cm}$

Answers

1. The diameter of the sun's core of 1.75×10^8 km is slightly larger than the diameter of Jupiter (1.4×10^8 km) and over 10 times the diameter of Earth (1.3×10^7 km).

READING SKILL BUILDER — BASIC

Reading Organizer As students read this section, encourage them to take KWL notes or Two Column notes, as described in the Skills Handbook section of the Appendix, about the sun's structure. Later, students can use their notes as a study guide for the assessments.
LS **Logical** English Language Learners

CHAPTER RESOURCES

Chapter Resource File

- **Datasheet for Quick Lab** GENERAL

Technology

Transparencies
- 147 The Sun's Interior (with worksheet)

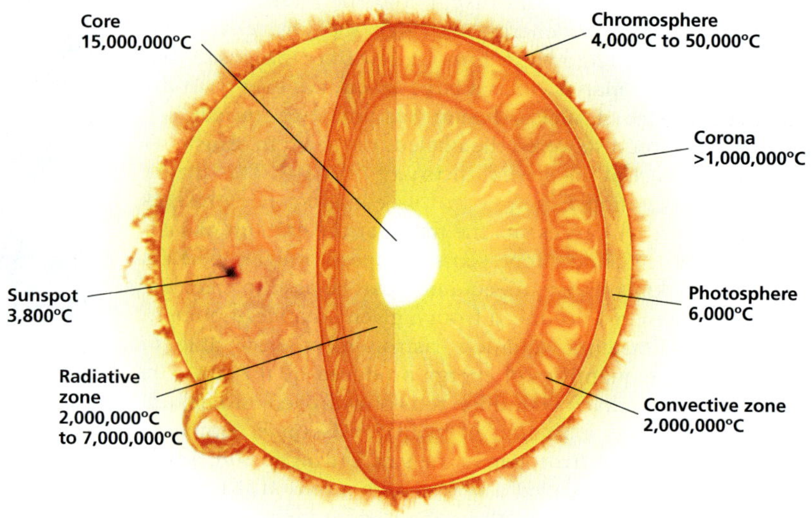

Figure 4 ▶ Energy released by fusion reactions in the core slowly works its way through the layers of the sun by the processes of radiation and convection.

QuickLAB — 20 min

The Size of Our Sun

Procedure

1. Using a **compass**, draw a large circle near the edge of a piece of **butcher paper** to represent the sun.
2. Measure the diameter (D) of your "sun."
3. Calculate the size of Earth and Jupiter, and compare the sizes with the size of the sun in step 1 by using the following values:
   $D(\text{sun}) = 1.4 \times 10^9$ m
   $D(\text{Jupiter}) = 1.4 \times 10^8$ m
   $D(\text{Earth}) = 1.3 \times 10^7$ m
4. Now, draw Earth and Jupiter to scale on your model.

Analysis

1. The diameter of the sun's core is about 175,000,000 m. How does size of the core compare with that of Earth and Jupiter?

The Sun's Interior

Scientists can't see inside the sun. But computer models have revealed what the invisible layers may be like. In recent years, careful studies of motions on the sun's surface have supplied more detail about what is happening inside the sun. The parts of the sun are shown in **Figure 4.**

The Core

At the center of the sun is the core. The core makes up 25% of the sun's total diameter of 1,390,000 km. The temperature of the sun's core is about 15,000,000°C. No liquid or solid can exist at such a high temperature. The core, like the rest of the sun, is made up entirely of ionized gas. The mass of the sun is 300,000 times the mass of Earth. Because of the sun's large mass, the pressure from the sun's material is so great that the center of the sun is more than 10 times as dense as iron.

The enormous pressure and high temperature of the sun's core cause the atoms to separate into nuclei and electrons. On Earth, atoms generally consist of a nucleus surrounded by one or more electrons. Within the core of the sun, however, the energy and pressure strip electrons away from the atomic nuclei. The nuclei have positive charges, so they tend to push away from each other. But the high temperature and pressure force the nuclei close enough to fuse. The most common nuclear reaction that occurs inside the sun is the fusion of hydrogen into helium.

INCLUSION Strategies

- ***Hearing Impaired***
- ***Learning Disabled***
- ***Developmentally Delayed***

Have students complete the following sentences to help clarify some of the cause and effect relationships that relate to the structure of the sun:
1—We don't really know what the inside of the sun looks like because (scientists cannot see inside the sun).
2—The layers of the sun are all made of gas because (the sun is too hot for liquids or solids to exist).
3—The center of the sun is more dense than iron because (due to its large mass, the pressure is great).
4—Atoms in the sun's core are different from atoms on Earth because (the energy and pressure at the core strip the electrons away from the atomic nuclei).

The Radiative Zone

Before reaching the sun's atmosphere, the energy produced in the core moves through two zones of the sun's interior. The zone surrounding the core is called the **radiative zone.** The temperature in this zone ranges from about 2,000,000°C to 7,000,000°C. In the radiative zone, energy moves outward in the form of electromagnetic waves, or radiation.

The Convective Zone

Surrounding the radiative zone is the **convective zone,** where temperatures are about 2,000,000°C. Energy produced in the core moves through this zone by convection. *Convection* is the transfer of energy by moving matter. On Earth, boiling water carries energy upward by convection. In the sun's convective zone, hot gases carry energy to the sun's surface. As the hot gases move outward and expand, they lose energy. The cooling gases become denser than the other gases and sink to the bottom of the convective zone. There, the cooled gases are heated by the energy from the radiative zone and rise again. Thus, energy is transferred to the sun's surface as the gases rise and sink.

radiative zone the zone of the sun's interior that is between the core and the convective zone and in which energy moves by radiation

convective zone the region of the sun's interior that is between the radiative zone and the photosphere and in which energy is carried upward by convection

photosphere the visible surface of the sun

The Sun's Atmosphere

Surrounding the convective zone is the sun's atmosphere. Although the sun is made of gases, the term *atmosphere* refers to the uppermost region of solar gases. This region has three layers—the photosphere, the chromosphere, and the corona.

The Photosphere

The innermost layer of the solar atmosphere is the **photosphere.** *Photosphere* means "sphere of light." The photosphere is made of gases that have risen from the convective zone. The temperature in the photosphere is about 6,000°C. Much of the energy given off from the photosphere is in the form of visible light. The layers above the photosphere are transparent, so the visible light is the light that is seen from Earth. A photo of the sun's photosphere is shown in **Figure 5.** The dark spots are cool areas of about 3,800°C and are called *sunspots*.

Reading Check What layers make up the sun's atmosphere? (See the Appendix for answers to Reading Checks.)

Figure 5 ▶ The photosphere is referred to as the sun's surface because this layer is the visible surface of the sun.

Close

Reteaching — BASIC

Diagramming the Sun Have students make a diagram that has six concentric circles. Ask them to label the sun's layers from the core outward to the corona. Underneath the diagram have them summarize the characteristics of each layer. Pairs of students can use their diagrams to quiz each other. **LS Visual/Verbal**

Quiz — GENERAL

1. What is the solar wind and in which solar layer does it originate? (The solar wind is a stream of electrically charged particles that flows outward from the corona.)
2. How have scientists learned about the sun's composition and internal structure? (by using devices such as spectrographs, computer models, and spacecraft)
3. What is the source of the sun's energy? (fusion reactions that convert hydrogen nuclei into helium and give off energy in the process)

Alternative Assessment — ADVANCED

Solar Tales Have students do a creative writing exercise in which they tell the life story of an atom of hydrogen inside the sun. Student stories should incorporate details about nuclear reactions, the temperatures and forces matter is subjected to in the different layers, and how energy travels within the sun. **LS Verbal**

Answer to Reading Check

The sun's atmosphere consists of the photosphere, the chromosphere, and the corona.

SKILL BUILDER — GENERAL

Vocabulary The names for the layers of the sun's atmosphere come from the following Greek roots: *chroma* means "color" and *photo* means "pertaining to light." Related scientific terms that students may encounter include *photon, photosynthesis, chromatid,* and *chromatography*. The combining form *-sphere,* from the Greek *sphaira,* meaning "ball," is commonly used to refer to a layer of gas surrounding a planet or star, as in the words *ionosphere* and *atmosphere*. **LS Verbal** English Language Learners

Close, *continued*

Answers to Section Review

1. The dark lines in spectra show what elements are present in the gases of the star.
2. hydrogen and helium
3. helium nuclei and energy
4. Nuclear fusion consists of a series of reactions in which hydrogen nuclei collide and fuse to form helium nuclei and release other particles. During each step in the process, mass is lost and energy is released.
5. The radiative zone is the region around the sun's core where energy moves in the form of electromagnetic waves. Temperatures range from 2,000,000°C to 7,000,000°C. The convective zone is located between the radiative zone and the sun's atmosphere and is where energy moves by convection currents. Temperatures are about 2,000,000°C.
6. The photosphere is the innermost layer of the sun's atmosphere. Its energy is made of visible light. The outer two layers are thin and transparent. The chromosphere glows with the reddish color of hydrogen. The corona is the extremely hot outermost layer.
7. The corona can be seen during an eclipse because the moon blocks the photosphere's light. As the sky darkens, the corona becomes visible.
8. Answers may vary. Sample answer: Because hydrogen nuclei combine to form helium nuclei during fusion, the amount of hydrogen will decrease over time.
9. Nuclear fusion takes enormous heat and pressure. The core of the sun is the only layer that has the necessary conditions for fusion.
10. Answers may vary. Sample answer: The corona helps prevent most subatomic particles from escaping. If the sun's corona were lost, the solar wind would be much stronger and would bombard Earth and the inner planets with charged particles.
11. The *sun* is composed of several layers, including the *core,* in which *nuclear fusion* of *hydrogen* nuclei forms *helium;* the *radiative zone;* and the *convective zone*.

Figure 6 ▶ The corona of the sun becomes visible during a total solar eclipse.

chromosphere the thin layer of the sun that is just above the photosphere and that glows a reddish color during eclipses

corona the outermost layer of the sun's atmosphere

The Chromosphere

Above the photosphere lies the **chromosphere,** or color sphere. This is a thin layer of gases that glows with reddish light that is typical of the color given off by hydrogen. The chromosphere's temperature ranges from 4,000°C to 50,000°C. The gases of the chromosphere move away from the photosphere. In an upward movement, gas regularly forms narrow jets of hot gas that shoot outward to form the chromosphere and then fade away within a few minutes. Some of these jets reach heights of 16,000 km.

Spacecraft study the sun from above Earth's atmosphere. These spacecraft can detect small details on the sun because they measure wavelengths of light that are blocked by Earth's atmosphere. Movies made from these spacecraft images show how features on the sun rise, change, and sometimes twist.

The Sun's Outer Parts

Just above the chromosphere is a thin zone called the transition region where the temperature rises dramatically. The outermost layer of the sun's atmosphere is the **corona** (kuh ROH nuh), or crown. The corona is a huge region of gas that has a temperature above 1,000,000°C. The corona is not very dense, but its magnetic field can stop most subatomic particles from escaping into space. However, electrons and electrically charged particles called *ions* do stream out into space as the corona expands. These particles make up the *solar wind,* which flows outward from the sun to the rest of the solar system.

The chromosphere and the corona are normally not seen from Earth because the sky during the day is too bright. Occasionally, however, the moon moves between Earth and the sun and blocks out the light of the photosphere. The sky darkens, and the corona becomes visible, as shown in **Figure 6.**

Section 1 Review

1. **Describe** how scientists use spectra to determine the composition of stars.
2. **Identify** the two elements that make up most of the sun.
3. **Identify** the end products of the nuclear fusion process that occurs in the sun.
4. **Explain** how the sun converts matter into energy in its core.
5. **Compare** the radiative and convective zones of the sun.
6. **Describe** the three layers of the sun's atmosphere.
7. **Explain** why the sun's corona can be seen during an eclipse but not at other times.

CRITICAL THINKING

8. **Making Inferences** Describe whether the amount of hydrogen in the sun will increase or decrease over the next few million years. Explain your reasoning.
9. **Analyzing Ideas** Why does fusion occur in the sun's core but not in other layers?
10. **Predicting Consequences** What might happen to the solar wind if the sun lost its corona?

CONCEPT MAPPING

11. Use the following terms to create a concept map: *sun, hydrogen, helium, nuclear fusion, core, radiative zone,* and *convective zone.*

CHAPTER RESOURCES

Chapter Resource File

• **Section Quiz** GENERAL

Workbooks

• **Study Guide** (also in Spanish)

Section 2 Solar Activity

The gases that make up the sun's interior and atmosphere are in constant motion. The energy produced in the sun's core and the force of gravity combine to cause the continuous rising and sinking of gases. The gases also move because the sun rotates on its axis. Because the sun is a ball of hot gases rather than a solid sphere, not all locations on the sun rotate at the same speed. Places close to the equator on the surface of the sun take 25.3 Earth days to rotate once. Points near the poles take 33 days to rotate once. On average, the sun rotates once every 27 days.

OBJECTIVES

- **Explain** how sunspots are related to powerful magnetic fields on the sun.
- **Compare** prominences, solar flares, and coronal mass ejections.
- **Describe** how the solar wind can cause auroras on Earth.

KEY TERMS

sunspot
prominence
solar flare
coronal mass ejection
aurora

Sunspots

The movement of gases within the sun's convective zone and the movements caused by the sun's rotation produce magnetic fields. These magnetic fields cause convection to slow in parts of the convective zone. Slower convection causes a decrease in the amount of gas that is transferring energy from the core of the sun to these regions of the photosphere. In some places, the magnetic field is thousands of times stronger than it is in other places. Because less energy is being transferred, these regions of the photosphere are up to 3,000°C cooler than surrounding regions.

Although they still shine brightly, these cooler areas of the sun appear darker than the areas that surround them do. These cool, dark areas of gas within the photosphere are called **sunspots.** The photosphere has a grainy appearance called *granulation.* The area around the sunspots shown in **Figure 1** has visible granulation. A large sunspot can have a diameter of more than 100,000 km, which is several times the diameter of Earth.

sunspot a dark area of the photosphere of the sun that is cooler than the surrounding areas and that has a strong magnetic field

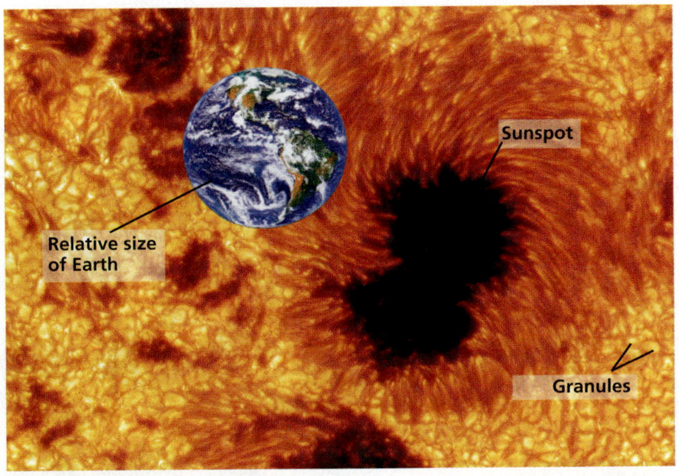

Figure 1 ▶ The diameter of this large sunspot is bigger than Earth's diameter. This image shows the granulation on the sun's surface.

Section 2

Focus

Overview

The section examines the relationship between the sun's magnetic field and sunspots, prominences, coronal mass ejections, and flares, and explains how these solar storms affect the environment around Earth.

Bellringer

Have students answer the following question: "What do you think the 'weather' in the sun's atmosphere is like?" (Answers may vary. Accept all reasonable answers.) LS Logical

Motivate

Discussion — GENERAL

Sunspots In 1610, Galileo was among the first to study sunspots in detail when he turned the newly invented telescope toward the sky. Invite students to discuss why these regions appear dark. (They are cooler than hotter, brighter surrounding regions.) Have students use a sheet of paper or a ruler to compare the diameter of Earth shown in the figure to the large sunspot. (At its widest point the sunspot is about 1/3 larger than Earth's diameter.) LS Verbal/Visual

CHAPTER RESOURCES

Chapter Resource File
- Directed Reading BASIC

Technology

Transparencies
- Bellringer

Student Edition on CD-ROM

One-Stop Planner CD-ROM
- Lesson Plan

Teach

Using the Figure — GENERAL

Sunspot Cycles Direct student attention to the graph of the sunspot cycle. Ask them how many sunspot cycles the graph shows. (three) Invite students to create a graph that estimates the sunspot cycle for the next 30 to 35 years. They can use this graph to answer the caption question.
Answer to caption question: High points will occur in 2011, 2022, and 2033. LS Visual/Logical

READING SKILL BUILDER — BASIC

Paired Summarizing Group students into pairs, and have them read silently about the events associated with the solar activity cycle, including sunspots, solar flares, prominences, and coronal mass ejections. Then, have one student summarize the relationship between these events and solar magnetic fields. The other student should listen to the retelling, and point out any inaccuracies or ideas that were left out. English Language Learners
LS Verbal

CHAPTER RESOURCES

Chapter Resource File

Internet Activity
- Solar Activity and Climate BASIC

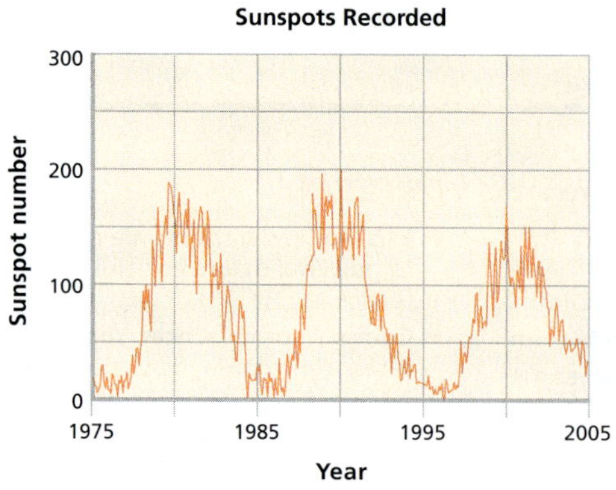

Figure 2 ▶ The sunspot cycle lasts an average of 11 years. ***When will the next high point in the cycle occur?***

The Sunspot Cycle

Astronomers have carefully observed sunspots for hundreds of years. Observations of sunspots showed astronomers that the sun rotates. Later, astronomers observed that the numbers and positions of sunspots vary in a cycle that lasts about 11 years.

A sunspot cycle begins when the number of sunspots is very low but begins to increase. Sunspots initially appear in groups about midway between the sun's equator and poles. The number of sunspots increases over the next few years until it reaches a peak of 100 or more sunspots. Then, sunspots at higher latitudes slowly disappear, and new ones appear closer to the sun's equator. **Figure 2** shows that after the peak, the number of sunspots begins to decrease until it reaches a minimum. Another 11-year cycle begins when the number of sunspots begins to increase again.

Solar Ejections

Many other solar activities are affected by the sunspot cycle. The *solar-activity cycle* is caused by the changing solar magnetic field. This cycle is characterized by increases and decreases in various types of solar activity, including solar ejections. Solar ejections are events in which the sun emits atomic particles. These events include prominences, solar flares, and coronal mass ejections.

Connection to ASTRONOMY

Total Solar Irradiance

Astronomers once thought that the sun's energy output remained steady. They called the amount of solar energy that a square centimeter of the top of Earth's atmosphere receives each second the *solar constant.* Earth's atmosphere causes an amount of energy that is less than the solar constant to reach Earth's surface.

Careful spacecraft observations made outside Earth's atmosphere have shown, however, that the solar constant isn't constant. Observations made by the Solar Maximum Mission first showed the effect. The solar constant is now being followed with special equipment on the Solar and Heliospheric Observatory (SOHO), the Active Cavity Radiometer Irradiance Monitor satellite (ACRIMSAT), and the Solar Radiation and Climate Experiment (SORCE) satellite. The ACRIMSAT and the SORCE satellite measure the entire range of solar radiation. Because this value changes, it is now officially called the *total solar irradiance* instead of the *solar constant. Irradiance* is simply the amount of energy that falls on a square centimeter of Earth each second. *Total* means that the whole spectrum of sunlight is included, not just the visible part.

The total solar irradiance varies by about 0.2% over the 11-year sunspot cycle. It is highest when the sunspot cycle is at its maximum. Individual sunspots, however, can temporarily block a small amount of the energy that comes from the sun and thus lower the total solar irradiance.

The SORCE satellite is being used to follow total solar irradiance.

Internet Activity — BASIC

Solar Activity and Climate Invite interested students to use the Internet to research the relationship between solar activity levels and climate on Earth. They may investigate the debate over the connection between sunspot activity and climate variations known as the *Medieval Maximum* and the *Little Ice Age,* or the variations in Earth's orbit relative to the sun and changes in solar intensity. A worksheet designed to direct student research on this topic can be found in the **Chapter Resource File** booklet or by visiting go.hrw.com and entering the keyword **HQ6SUNX.** LS Visual/Verbal/Logical

INCLUSION Strategies

- ***Gifted and Talented***

Advanced students may want to learn more about material referred to in the text. Challenge students to explore the frequency, appearances, and other qualities of solar ejections including prominences, solar flares, and coronal mass ejections. Have them create posters or multimedia presentations that show what they find.
LS Visual

Prominences

The magnetic fields that cause sunspots also create other disturbances in the sun's atmosphere. Great clouds of glowing gases, called **prominences**, form huge arches that reach high above the sun's surface. Each solar prominence follows curved lines of magnetic force from a region of one magnetic polarity to a region of the opposite magnetic polarity. Some prominences may last for several weeks, while others may erupt and disappear in hours. The gas in prominences is very hot and is commonly associated with the chromosphere.

Figure 3 ▶ A loop of coronal gas can arch half a million kilometers or more above the sun's surface.

Solar Flares

The most violent of all solar disturbances is a **solar flare**, a sudden outward eruption of electrically charged particles, such as electrons and protons. The trigger for these eruptions is unknown. However, scientists know that solar flares release the energy stored in the strong magnetic fields of sunspots. This release of energy can lead to the formation of coronal loops, such as the ones shown in **Figure 3.**

Solar flares may travel upward thousands of kilometers within minutes, but few eruptions last more than an hour. During a peak in the sunspot cycle, 5 to 10 solar flares may occur each day. The temperature of the gas in solar flares may reach 20,000,000°C. Some of the particles from a solar flare escape into space. These particles increase the strength of the solar wind.

Coronal Mass Ejections

Particles also escape into space as **coronal mass ejections,** or parts of the corona that are thrown off the sun. As the gusts of particles strike Earth's *magnetosphere*, or the space around Earth that contains a magnetic field, the particles can generate a sudden disturbance in Earth's magnetic field. These disturbances are called *geomagnetic storms*. Although several small geomagnetic storms may occur each month, the average number of severe storms is less than one per year.

Geomagnetic storms have been known to interfere with radio communications on Earth. The high-energy particles that circulate in Earth's outer atmosphere during geomagnetic storms can also damage satellites. They can also lead to blackouts when power lines become overloaded. Not all solar activity is so dramatic, but the activity of the sun affects Earth every day.

Reading Check **How do coronal mass ejections affect communications on Earth? (See the Appendix for answers to Reading Checks.)**

prominence a loop of relatively cool, incandescent gas that extends above the photosphere

solar flare an explosive release of energy that comes from the sun and that is associated with magnetic disturbances on the sun's surface

coronal mass ejection a part of coronal gas that is thrown into space from the sun

MATHPRACTICE

Magnetic Fields
Magnetic fields on the sun differ greatly from each other. Field densities at the poles are 0.001 teslas (T), while field densities near sunspots are up to 0.3 T. How many times the densities of fields at the sun's poles are densities of fields near sunspots?

SKILL BUILDER — GENERAL

Writing Have students write a descriptive passage about the spectacular light displays provided by auroras, and include a scientific explanation of how the interaction of the solar wind with Earth's magnetic field and atmospheric gases is responsible for this phenomenon. LS Verbal

MATHPRACTICE

Answer
(0.3 T) ÷ (0.001 T) = 300; Fields are 300 times more dense near the sunspots.

Close

Reteaching — BASIC

Solar Lingo Have students work in groups to create cards for each key term and italicized word in the section by writing the terms on one side of the card and definitions on the other. Then, have students shuffle the cards and play a game of twenty questions in which teams try to identify the terms on the cards by asking yes-and-no questions. LS Verbal

Quiz — GENERAL

1. What is the relationship between solar flares and sunspots? (Solar flares release energy stored in the strong magnetic fields of sunspots.)
2. Describe the role of Earth's magnetic field in causing auroras. (High-energy particles from the solar wind interact with Earth's magnetic field and are trapped by the magnetosphere. These particles strike atoms and gas molecules in the upper atmosphere and cause the molecules to emit light.)

Group Activity — GENERAL

Geomagnetic Storms Have groups of students explore methods of dealing with the effects of geomagnetic storms. Some students could interview experts about how these storms can be predicted. Others could do Internet or library research on the effects of these phenomena on pilots, astronauts, emergency personnel, and others. Other students could prepare posters with safety instructions for the general public. Still others could work on a presentation of their group findings.
LS Interpersonal Co-op Learning

Answer to Reading Check
Coronal mass ejections generate sudden disturbances in Earth's magnetic field. The high-energy particles that circulate during these storms can damage satellites, cause power blackouts, and interfere with radio communications.

Close, continued

Alternative Assessment — ADVANCED

The Active Sun Ask students to create a comic strip or poster that explains how electromagnetic radiation and particles from the active sun affect life on Earth. Before they begin the assessment, have them create a concept map or other graphic organizer that includes information about sunspots, the solar activity cycle, auroras, and geomagnetic storms. LS **Logical/Visual**

Answers to Section Review

1. The strong magnetic fields of sunspots tend to slow down movement of gases. Slower movement means less energy is transferred between particles, which makes the matter cooler.
2. At the peak of the cycle about 100 or more sunspots occur.
3. During the sunspot cycle, sunspots initially appear in groups about midway between the poles and the equator. After the peak, sunspots at higher latitudes begin to disappear, and new ones appear near middle latitudes.
4. Prominences are larger masses of glowing gases that form arches above the sun's surface. They follow lines of magnetic force and may last several weeks. A solar flare is a violent eruption of electrically charged particles that releases energy stored in the magnetic fields of sunspots. Flares may form coronal loops and travel thousands of kilometers. Eruptions seldom last longer than an hour.
5. Auroras are sheets of light produced by high-energy particles released by the solar wind. These particles excite gases in Earth's upper atmosphere.
6. Sunspots appear darker in contrast with the surrounding regions of the photosphere that are much brighter and hotter.
7. because the sun is made up of gases that do not move as a single unit as solid materials do
8. Patterns of *sunspot* activity are related to the *solar-activity cycle,* which affects solar ejections such as *solar flares, prominences,* and *coronal mass ejections.*

Figure 4 ▶ Auroras, such as these over Canada, can fill the entire sky. When high-energy particles strike oxygen atoms in the upper atmosphere, green curtains form. When low-energy particles strike oxygen atoms, red curtains form.

aurora colored light produced by charged particles from the solar wind and from the magnetosphere that react with and excite the oxygen and nitrogen of Earth's upper atmosphere; usually seen in the sky near Earth's magnetic poles

SCILINKS. NSTA Developed and maintained by the National Science Teachers Association

For a variety of links related to this subject, go to www.scilinks.org

Topic: Solar Activity
SciLinks code: HQ61413

Auroras

On Earth, a spectacular effect of the interaction between the solar wind and Earth's magnetosphere is the appearance in the sky of bands of light called **auroras** (aw RAWR uhz). **Figure 4** shows an example of an aurora. Auroras are usually seen close to Earth's magnetic poles because electrically charged particles are guided toward Earth's magnetic poles by Earth's magnetosphere. The electrically charged particles strike the atoms and gas molecules in the upper atmosphere and produce colorful sheets of light. Depending on which pole they are near, auroras are called *northern lights*—or *aurora borealis* (aw RAWR uh BAWR ee AL is)—or *southern lights*—or *aurora australis*.

Auroras normally occur between 100 and 1,000 km above Earth's surface. They are most frequent just after a peak in the sunspot cycle, especially after solar flares occur. Across the northern contiguous United States, auroras are visible about five times per year. In Alaska, however, people can see auroras almost every clear, dark night. Astronauts in orbit can also look down on Earth and see auroras. But Earth is not the only planet that has auroras. Spacecraft have recorded auroras on Jupiter and Saturn.

Section 2 Review

1. **Explain** why sunspots are cooler than surrounding areas on the sun's surface.
2. **Identify** the number of sunspots that are on the sun during the peak of the sunspot cycle.
3. **Summarize** how the latitude of sunspots varies during the sunspot cycle.
4. **Explain** how prominences are different from solar flares.
5. **Summarize** the cause of auroras on Earth.

CRITICAL THINKING

6. **Identifying Relationships** How can a sunspot be bright but look dark?
7. **Analyzing Ideas** Why doesn't the whole sun rotate at the same rate?

CONCEPT MAPPING

8. Use the following terms to create a concept map: *sunspot, prominence, solar flare, solar-activity cycle,* and *coronal mass ejection.*

CHAPTER RESOURCES

Chapter Resource File
- Section Quiz GENERAL

Workbooks
- Study Guide (also in Spanish)

Chapter 29 Highlights

Sections

1 Structure of the Sun

Key Terms

nuclear fusion, 756
radiative zone, 759
convective zone, 759
photosphere, 759
chromosphere, 760
corona, 760

Key Concepts

- The enormous pressure and heat in the sun's core converts matter into energy through the process of nuclear fusion.
- Nuclear fusion combines four hydrogen nuclei to form one helium nucleus.
- Energy from the sun's core moves through the radiative zone and the convective zone before it enters the sun's atmosphere.
- The sun's atmosphere is composed of the photosphere, the chromosphere, and the corona.
- Solar wind forms when electrons and electrically charged particles called *ions* from the corona travel into space.

2 Solar Activity

Key Terms

sunspot, 761
prominence, 763
solar flare, 763
coronal mass ejection, 763
aurora, 764

Key Concepts

- Sunspots are caused by powerful magnetic fields in the sun.
- The sunspot cycle is a periodic variation in the number of sunspots and occurs about every 11 years. The sunspot cycle is closely related to the cycles of other solar activity.
- Prominences, solar flares, and coronal mass ejections are examples of solar activity that are caused by changes in the sun's magnetic field.
- Auroras in Earth's polar regions occur when charged particles from the interaction between the solar wind and Earth's magnetosphere collide with atoms and molecules in Earth's atmosphere.

Chapter Highlights

Alternative Assessment — GENERAL

Our Daytime Star Divide the class into small groups of students. Have each group create models, diagrams, demonstrations, and hands-on activities that will enable them to explain the following concepts to a group of younger children: (1) the nuclear reactions that fuel the sun's energy source, (2) the sun's layered structure, and (3) how the active regions of the sun affect the atmospheric conditions on Earth that make life possible. Caution them to explain all the terms they use, and to include simple directions for doing the activities. If students are interested, have them present the materials they have created to a local school or community group. LS **Kinesthetic/Interpersonal**

CHAPTER RESOURCES

Chapter Resource File

- **Concept Review** GENERAL
- **Critical Thinking** ADVANCED
- **Math Skills** GENERAL
- **Graphing Skills** GENERAL
- **Chapter Test A** GENERAL
- **Chapter Test B** ADVANCED

Workbooks

- **Study Guide** (also in Spanish)
- **Assessments** (Spanish)

Technology

Classroom Videos
- Brain Food Video Quiz

Cultural Awareness — GENERAL

Auroras For thousands of years, people have observed auroras with wonder. Eskimos described them as the dancing souls of animals. The Maori of New Zealand believed that they were travelers that had become trapped by ice and cold. Auroras reminded the Finns of magical "fire foxes." The Lakota Sioux associated them with ghost dancers. Encourage students to research a folk belief about auroras and then create a children's book based on their choice. Have students work in pairs or groups, with some students doing the writing and others doing the illustrations. LS **Verbal/Visual** Co-op Learning

Chapter Review

Assignment Guide	
SECTION	QUESTIONS
1	1, 3, 5–6, 9–13, 19, 22, 24–27, 32, 35
2	2, 4, 7, 14–18, 20–21, 23, 28–31, 33–34, 36–39
1–2	8

Using Key Terms

1–8. Answers may vary but should show that students understand the definitions of and differences between key terms.

Understanding Key Concepts

9. c
10. b
11. a
12. b
13. b
14. a
15. c
16. d
17. b
18. b

Short Answer

19. the corona
20. The solar activity cycle is closely linked to changes in the sun's magnetic field. Because sunspots are regions with strong magnetic fields, variations in sunspots affect other disturbances in the sun's atmosphere.
21. Sunspots have magnetic fields that are thousands of times stronger than those in other parts of the photosphere.
22. The sun produces energy through nuclear fusion. In the first step, two hydrogen nuclei (protons) fuse, and one proton changes to a neutron. In step two, another proton joins the nucleus, forming a nucleus made up of one neutron and two protons. In step three, two nuclei from step two collide and fuse. Two protons are released, and a helium nucleus (two protons and two neutrons) is left. Each stage of the process releases energy.
23. Answers may vary. Sample answer: Prominences are huge arches of glowing gases that extend above the sun's surface. Solar flares are sudden eruptions of charged particles that travel upward thousands of kilometers.
24. The corona is the extremely hot outermost layer of the sun's atmosphere. This low-density layer helps prevent atomic particles from escaping into space. The corona becomes visible during a total solar eclipse when the moon blocks the photosphere, and the sky darkens enough for the corona to be seen.
25. In the sun's radiative zone, energy radiates in the form of electromagnetic waves. In the convective zone, hot moving gases carry energy toward the sun's surface. As heated gases move upward and expand, they cool and sink back toward the radiative zone.

Critical Thinking

26. The transfer of energy in each case involves the movement of matter. In both cases, the matter rises as it is heated from below. As it moves away from the heat source, the matter cools and sinks down toward the heat source again.
27. In both the radiative zone of the sun and in the region of space between the sun and Earth, energy transfers take place by means of electromagnetic waves, or radiation. These waves of energy can travel through both empty space and matter.
28. If parts of the sun's magnetic field suddenly increased in strength, the number of sunspots in those locations would likewise increase.
29. Auroras would be seen more frequently near the equator than near the poles.

Chapter 29 Review

Using Key Terms

Use each of the following terms in a separate sentence.

1. *photosphere*
2. *solar flare*
3. *solar wind*
4. *sunspot cycle*

For each pair of terms, explain how the meanings of the terms differ.

5. *chromosphere* and *corona*
6. *photosphere* and *core*
7. *solar flare* and *prominence*
8. *aurora* and *solar wind*

Understanding Key Concepts

9. According to Einstein's theory of relativity, in the formula $E = mc^2$, the c stands for
   a. corona.
   b. core.
   c. the speed of light.
   d. the length of time.

10. A nuclear reaction in which atomic nuclei combine is called
   a. fission.
   b. fusion.
   c. magnetism.
   d. granulation.

11. The part of the sun in which energy moves from atom to atom in the form of electromagnetic waves is called the
   a. radiative zone.
   b. convective zone.
   c. solar wind.
   d. chromosphere.

12. The number of hydrogen atoms that fuse to form a helium atom is
   a. two.
   b. four.
   c. six.
   d. eight.

13. The part of the sun that is normally visible from Earth is the
   a. core.
   b. photosphere.
   c. corona.
   d. solar nebula.

14. Sunspots are regions of
   a. intense magnetism.
   b. the core.
   c. high temperature.
   d. lighter color.

15. The sunspot cycle repeats about every
   a. month.
   b. 5 years.
   c. 11 years.
   d. 19 years.

16. Sudden outward eruptions of electrically charged particles from the sun are called
   a. prominences.
   b. coronas.
   c. sunspots.
   d. solar flares.

17. Gusts of solar wind can cause
   a. rotation.
   b. magnetic storms.
   c. nuclear fission.
   d. nuclear fusion.

18. *Northern lights* and *southern lights* are other names for
   a. prominences.
   b. auroras.
   c. granulations.
   d. total solar irradiance.

Short Answer

19. What is the outermost layer of the sun?
20. How is the solar activity cycle related to the sunspot cycle?
21. What is unusual about the magnetic field in a sunspot?
22. From what process does the sun gets its energy? What steps does this process follow?
23. Compare two types of solar activity.
24. Describe the corona, and identify when it is visible from Earth.
25. How does the transfer of energy in the radiative zone differ from the transfer of energy in the convective zone?

Critical Thinking

26. Making Comparisons How is the transfer of energy in a pan of hot water similar to the transfer of energy in the sun's convective zone?

27. Making Comparisons Explain how the radiative zone in the sun is similar to the region between the sun and Earth.

28. Making Predictions Predict what would happen to the number of sunspots if parts of the sun's magnetic field suddenly increased in strength.

29. Drawing Conclusions If Earth's magnetosphere shifted so that solar wind was not deflected toward the poles but was deflected toward the equator, what would happen to the area where auroras are most often visible?

30. Analyzing Relationships Magnetic fields create electric currents that can damage electric power grids and interrupt the flow of electricity. How does this information help explain why strong magnetic storms can knock out power in cities?

31. Predicting Consequences How do scientists predict magnetic storms? List two ways that scientists on Earth could help people prepare for a very large magnetic storm.

Concept Mapping

32. Use the following terms to create a concept map: *sun, nuclear fusion, core, radiative zone, convective zone, photosphere,* and *corona.*

Math Skills

33. Making Calculations On average, Earth is 150 × 10^6 km from the sun. A coronal mass ejection, or CME, can have a speed of 7 × 10^6 km/h. At this speed, how long would a CME take to reach Earth?

34. Applying Information A peak of the sunspot cycle occurred in the year 2000. In what years will the next two peaks occur?

Writing Skills

35. Creative Writing Write a short story that describes an imaginary trip to the center of the sun. Describe each layer and zone through which you would pass.

36. Writing from Research Research the northern lights. Write a short travel brochure that describes when and where to go to see the most spectacular and frequent displays of the auroras.

Interpreting Graphics

The graph below shows how the latitudes of sunspots vary over time. Use the graph to answer the questions that follow.

Average Daily Sunspot Area

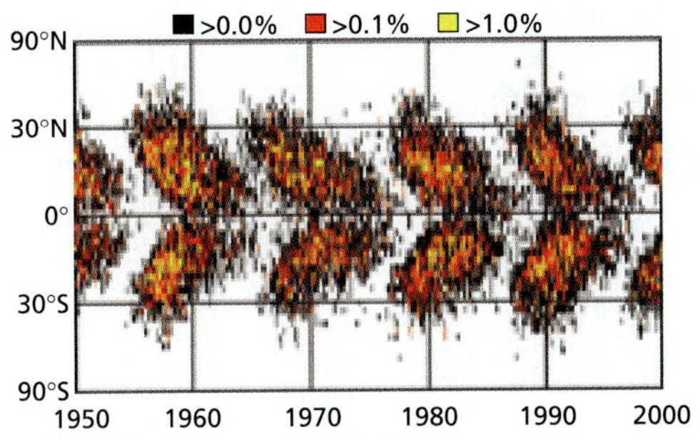

37. How many complete sunspot cycles are illustrated by the graph?

38. How does the range of latitudes of sunspots change over time? How is this change related to the sunspot cycle?

39. According to the graph, how many sunspots were located at the sun's north pole?

Chapter Review

30. Strong magnetic storms produce magnetic fields, which in turn produce the electrical currents that can overload power grids and knock out power in cities.

31. Scientists predict the likelihood of magnetic storms by watching for increases in the number of sunspots and related magnetic disturbances such as solar flares. Scientists could encourage individuals to prepare alternate light sources, such as candles and flashlights, that do not depend on the power grids and to expect communications equipment failures. Utilities and power stations should prepare for power surges.

Concept Mapping

32. Answers may vary but should include all of the terms listed. Sample answers appear at the end of this Teacher Edition.

Math Skills

33. time = distance ÷ speed
time = (150 × 10^6 km) ÷ (7 × 10^6 km/h) = 150 ÷ 7 = 21.4 h

34. Peaks in the sunspot cycle occur approximately every 11 years, so 2000 + 11 years = 2011 + 11 = 2022. The next two peaks could be expected in 2011 and 2022.

Writing Skills

35. Answers may vary. Accept all reasonable answers.

36. Answers may vary. Accept all reasonable answers.

Interpreting Graphics

37. Four complete cycles are shown.

38. Sunspots initially appear in groups about midway between the sun's equator and poles. Those spots are gradually replaced by spots closer to the equator. After about 11 years, a new sunspot cycle begins and sunspots begin to appear again midway between the equator and the poles.

39. No sunspots occurred at 90° north latitude.

Standardized Test Prep

Estimated Time

To give students practice under more realistic testing conditions, allow them 30 minutes to answer all of the questions in this practice test.

TEST DOCTOR

Question 3 Answer D is correct. Students should have a conceptual understanding that dense materials are often found in the cores of stellar objects. This is true of Earth and the sun. One reason that the material is so dense at the sun's core is pressure. The enormous pressure exerted on the materials creates a core that scientists believe has a density that is 10 times denser than iron.

Question 6 The convective zone is the best answer. Students may notice a clue in that energy is carried in this part of the sun by the motion of matter. Students should use their understanding of the sun to determine that, unlike other layers of the sun, which move energy by radiation or electromagnetic waves, the convective zone has convection currents, which transfer energy by the rise and fall of moving matter.

Chapter 29 Standardized Test Prep

Understanding Concepts

Directions (1–5): **For** ***each*** **question, write on a separate sheet of paper the letter of the correct answer.**

1. What is the source of the sun's energy?
   A. nuclear fission reactions that break down massive nuclei to form lighter atoms
   B. nuclear fusion reactions that combine smaller nuclei to form more massive ones
   C. reactions that strip away electrons to form lighter atoms
   D. reactions that strip away electrons to form more massive ones

2. What do electrically charged particles from the sun strike in Earth's magnetosphere to produce sheets of light known as auroras?
   F. gas molecules
   G. dust particles
   H. water vapor
   I. ice crystals

3. Which layer of the sun has the densest material?
   A. the corona
   B. the convection zone
   C. the radiative zone
   D. the core

4. Based on the amount of fuel the sun possessed at its formation, the life span of the sun is thought by scientists to be how long?
   F. 1 billion years
   G. 5 billion years
   H. 10 billion years
   I. 50 billion years

5. The solar activity cycle occurs regularly every
   A. 5 years
   B. 11 years
   C. 16 years
   D. 22 years

Directions (6–7): **For** ***each*** **question, write a short response.**

6. In which part of the sun's interior is energy carried to the sun's surface by moving matter?

7. What is the term for the innermost layer of the sun's atmosphere?

Reading Skills

Directions (8–10): **Read the passage below. Then, answer the questions.**

Studying the Sun

Sunlight that has been focused, especially through a magnifying glass, can produce a great amount of thermal energy—enough to start a fire. Imagine focusing the sun's rays by using a magnifying glass that has a diameter of 1.6 m. The resulting heat could easily melt metal. If a conventional telescope were pointed directly at the sun, its parts could melt and become useless.

To avoid a meltdown, the McMath-Pierce telescope uses a special mirror that produces an image of the sun. This mirror directs the sun's rays down a long, diagonal shaft to another mirror, which is located 50 m underground. This second mirror is adjustable, which allows it to focus the sunlight. The sunlight is then directed to a third mirror, which in turn directs the light to an observing room and instrument shaft. This system, while complex, not only protects the sensitive and expensive telescopic equipment but also protects the scientists that use it as well.

8. According to the information in the passage, which of the following statements about solar telescopes is true?
   A. Solar telescopes allow scientists to safely observe the sun.
   B. Solar telescopes do not need mirrors to focus the sun's rays.
   C. All solar telescopes are built 50 m underground.
   D. All solar telescopes are built with a diameter of 1.6 m.

9. Which of the following statements can be inferred from the information in the passage?
   F. Focusing sunlight can help avoid a meltdown.
   G. Unfocused sunlight produces little energy.
   H. A magnifying glass can focus sunlight to produce a great amount of thermal energy.
   I. Mirrors greatly increase the intensity and danger of studying sunlight.

10. Why do scientists have to use specialized equipment to study the sun?

Answers

Understanding Graphics

1. B
2. F
3. D
4. H
5. B
6. the convective zone
7. the photosphere

Reading Skills

8. A
9. H
10. Conventional equipment would be at risk of melting under the intensity of the sun's rays. Scientists' eyesight could also be put at risk by looking directly at the sun.

Interpretting Graphics

11. A
12. sunspots
13. Answers may vary. See Test Doctor for a detailed scoring rubric.

Interpreting Graphics

Directions (11–13): **For *each* question below, record the correct answer on a separate sheet of paper.**

The graphic below shows the structure of the sun. Use this diagram to answer questions 11 and 12.

Structure of the Sun

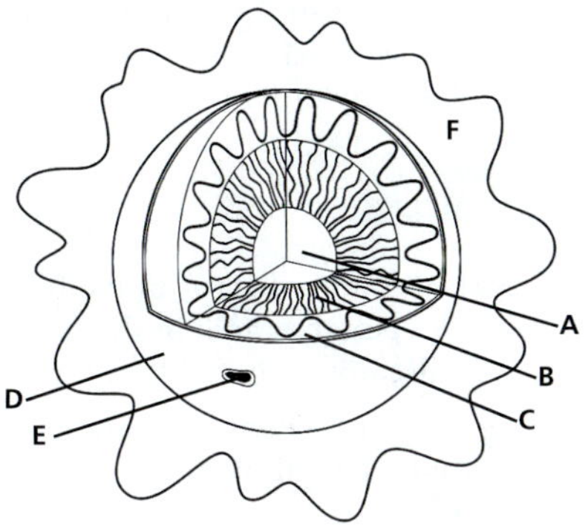

11 Fusion reactions provide power for the stars, such as the sun. In which part of the sun do these fusion reactions take place?
A. layer A
B. layer B
C. layer C
D. layer D

12 What is the term for the dark, cool regions of the sun, which are represented by the letter E on the diagram?

The diagram below shows what happens when Earth's magnetic field interacts with the solar wind. Use this graphic to answer question 13.

Earth's Magnetosphere

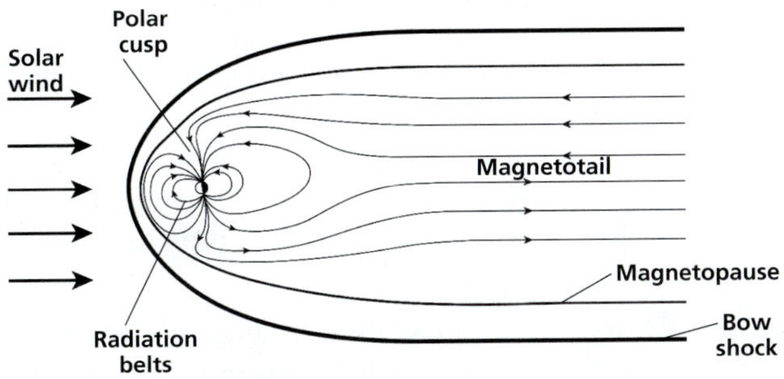

13 How does the solar wind affect humans and other living things on Earth, despite the protection provided by the magnetosphere? Use examples to explain your answer.

Test TIP

Choose the best possible answer for each question, even if you think there is another possible answer that is not given.

TEST DOCTOR

Question 12 The correct answer is sunspots. Students should first note the location of the spots. If students notice that these dark spots occur on the sun's surface, they will narrow the range of potential answers to include only surface features. Students should then remember that other surface features, such as solar flares, do not match the description in the question.

Question 13 Full-credit answers should include the following points:
- students should explain that the magnetosphere extends far into space and protects the planet from many of the effects of the solar wind, a stream of ionized particles from the sun's corona
- an example of weakness that students may mention is the fact that there are small gaps in the shield at the poles through which some particles can pass. These particles are responsible for colorful auroras
- students may also mention that radiation from solar flares can generate magnetic storms and may disrupt radio communications, or interfere with power grids

Test Prep Correlations ***National Science Education Standards***

ES 1a: items 2, 13
ES 4c: items 1, 4, 11
PS 1c: items 1, 6, 11
SAI 2a: items 8, 9, 10
SAI 2c: items 8, 9, 10
ST 2b: items 8, 9, 10
ST 2c: items 8, 9, 10
SPSP 5a: item 13
UCP 1: items 1, 3, 5, 6, 7, 11, 12, 13

CHAPTER RESOURCES

State Resources

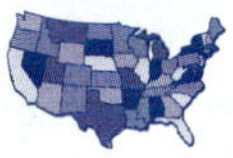

For specific resources for your state, visit go.hrw.com and type in the keyword **HSHSTR.**

Energy of the Sun

Teacher's Notes

Time Required

two 45-minute class periods

Lab Ratings

EASY ——→ HARD

TEACHER PREPARATION 3
STUDENT SETUP 2
CONCEPT LEVEL 2
CLEANUP 1

Skills Acquired

- Experimenting
- Measuring
- Collecting Data
- Calculating
- Interpreting Results

The Scientific Method

In this lab, students will

- Analyze Results
- Evaluate Models
- Communicate Results

Materials

The materials listed are enough for groups of 2 or 3 students. You should collect enough similar glass jars for as many groups as plan to do the lab. Pasta sauce jars work well for this purpose. An Erlenmeyer flask with a one-hole stopper can be used instead of a jar if available.

You may want to punch holes in the metal jar lids in advance and paint the metal sheets with black flat finish paint to save class time. Solar collector jars may be reused with later classes.

Tips and Tricks

The wings on the solar collector can be made bigger to exaggerate the temperature differences. A heat lamp bulb can also be used instead of a light bulb. Multiple solar collectors can be built and used to test and compare solar intensities in different conditions.

Make sure that the lab thermometers you are using will fit through the holes in the jar lids and that the holes can easily be plugged with the modeling clay. Another lump of clay or a bookend could be used to prop up the solar collector so that it remains tilted to catch the direct rays of the sun. If it is windy or cool outside, have students set up their experiments indoors on a sunny windowsill. Remind students to allow the thermometer to cool down to room temperature before repeating the experiment with the lamp. (The distance from the lamp to the thermometer has to be about 8 cm for the math to work correctly.)

CHAPTER RESOURCES

Chapter Resource File

- Datasheet for Chapter Lab GENERAL
- Lab Notes and Answers

Workbooks

Long-Term Projects

- Positions of Sunrise and Sunset ADVANCED

Chapter 29

Skills Practice Lab

Energy of the Sun

Objectives

- **Estimate** the sun's energy output.
- USING SCIENTIFIC METHODS **Evaluate** the differences between known values and experimental values.

Materials

- clay, modeling
- desk lamp with 100 W bulb
- jar, glass, with lid
- metal, sheet, very thin, at least 2 cm × 8 cm
- paint, black, flat finish
- pencil
- ruler, metric
- tape, masking
- thermometer, Celsius

Safety

The sun is, on average, 150 million kilometers away from Earth. Scientists use complicated astronomical instruments to measure the size and energy output of the sun. However, it is possible to estimate the sun's energy by using simple instruments and the knowledge of the relationship between the sun's size and the sun's distance from Earth. In this lab, you will collect energy from sunlight and estimate the amount of energy produced by the sun.

PROCEDURE

1. Construct a solar collector in the following way.
   a. Carefully punch a hole in the jar lid, or use a lid that is already prepared by your teacher.
   b. Shape the piece of sheet metal by gently bending it around a pencil. Bend the edges out so that they form "wings," as shown in the photo. Then, carefully place the metal piece around the thermometer bulb so that it fits snugly. Be careful not to press too hard.
   CAUTION Thermometers are fragile. Do not squeeze the bulb of the thermometer or let the thermometer strike any solid object. Bend the remaining metal outward to collect as much sunlight as possible.
   c. If the sheet metal is not already painted, paint the sheet metal black.
   d. Slip the top of the thermometer through the hole in the jar's lid. On the top and bottom of the lid, mold the clay around the thermometer to hold the thermometer steady. Place the lid on the jar. Adjust the thermometer so that the metal wings are centered in the jar. Then, secure the thermometer and clay to the lid with masking tape.
2. Place the solar collector in sunlight. Tilt the jar so that the sun shines directly on the metal wings. Carefully hold the jar in place. You may want to prop the jar up carefully with books.
3. Watch the temperature reading on the thermometer until it reaches a maximum value or until 5 min has elapsed. Record this value. Allow the collector to cool for 2 min.

4 Place the lamp or heat lamp at the end of a table. Remove any reflector or shade from the lamp.

5 Place the collector about 30 cm from the lamp, and turn the collector toward the lamp.

6 Turn on the lamp, and wait 1 min. Then, gradually move the collector toward the lamp in 2 cm increments. Watch the temperature carefully. At each position, let the collector sit until the temperature reading stabilizes. Stop moving the collector when the temperature reaches the maximum temperature that was achieved in sunlight.

7 Once the temperature has stabilized at the same level reached in sunlight, record the distance between the center of the lamp and the thermometer bulb.

Step 6

ANALYSIS AND CONCLUSION

1 **Analyzing Results** Because the collector reached the same temperature in both trials, the collector absorbed as much energy from the sun at a distance of 150 million km as it did from the light bulb at the distance that you measured. Using 1.5×10^{13} cm as the distance to the sun, calculate the power of the sun in watts by using the equation that follows. The power of the lamp is equal to the wattage of the light bulb.

$$\frac{power_{sun}}{(distance_{sun})^2} = \frac{power_{lamp}}{(distance_{lamp})^2}$$

2 **Evaluating Models** The sun's power is generally given as 3.7×10^{26} W. Calculate your experimental percentage error by first subtracting your experimental value from the accepted value. Divide this difference by the accepted value, and multiply by 100. Describe two possible sources for your calculated error.

Extension

1 **Evaluating Models** How would using a fluorescent bulb instead of an incandescent bulb in the experiment affect the results of the experiment? Explain your answer.

Skills Practice Lab

Answers to Analysis and Conclusion

1. Values of student temperature readings and the recorded distances between the center of the lamp and the thermometer bulb may vary. Though answers may vary, calculated values for solar power should be on the order of 3.7×10^{26} watts.
2. Calculated error may vary but should show correct calculations. Possible sources of error in the measurements may include some heat exchange between the jar and the room if the clay plug was affected by heat over time; heat from room lighting; variations in cloud cover affecting solar readings; and variations between the readings obtained for the temperature of the sun and for the incandescent bulb.

Answers to Extension

1. Answers may vary. The experiment would probably not have worked as well with a fluorescent bulb. Fluorescent bulbs produce light in a different way than incandescent bulbs do, and they do not produce as much thermal energy.

TEACHER TESTED & APPROVED

Scott Robertson
North Warren Central School
Chestertown, NY

SXT Composite Image of the Sun

Internet Activity — GENERAL

SOHO Images The light that humans can see represents only a small fraction of the total electromagnetic spectrum. To study all of the wavelengths of light that are emitted by the sun, scientists need to place observational equipment above Earth's atmosphere. The Solar and Heliospheric Observatory (SOHO), a project of the European Space Agency and NASA, was designed to provide a years-long uninterrupted view of the sun. Have students visit the SOHO Web site and look through the vast archive of solar images. Have them each select an image and use it to construct a map of some aspect of solar activity that interests them. A worksheet designed to direct student research on this topic can be found in the **Chapter Resource File** booklet or by visiting **go.hrw.com** and entering the keyword **HQ6SUNX**. **LS Visual**

Answers to Map Skills Activity

1. the active region numbered 7995
2. 30° north latitude and –15° west longitude
3. The lines of longitude appear to be closer together at the right and left sides of the map because of distortion produced by projecting a curved surface onto a flat map.
4. 0° to about –25° longitude
5. at locations that have the color yellow

CHAPTER RESOURCES

Chapter Resource File

Internet Activity
- SOHO Images GENERAL

Technology

Transparencies
- 148 SXT Composite Image of the Sun (with worksheet)

MAPS in Action

SXT Composite Image of the Sun

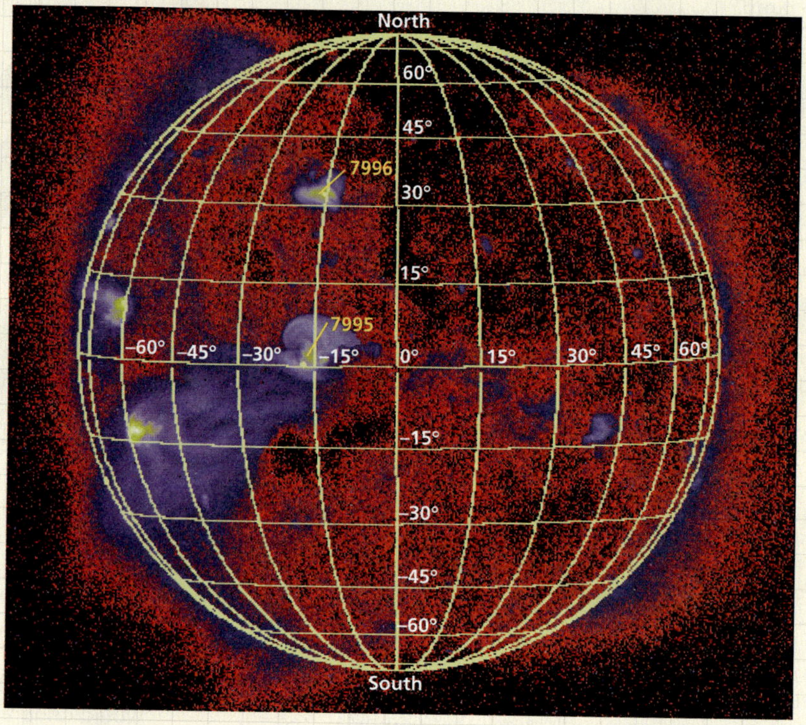

Map Skills Activity

This map is a soft X-ray telescope (SXT) image of the sun that includes latitude and longitude lines. Yellow represents the strongest X-ray radiation. Blue represents moderate X-ray radiation. Red represents the weakest X-ray radiation. The numbered areas are known active regions. Use the map to answer the questions below.

1. **Analyzing Data** Which numbered active region is near the sun's equator?
2. **Analyzing Data** What is the latitude and longitude of the active region numbered 7996?
3. **Inferring Relationships** Why do the lines of solar longitude appear to be close to each other near the left side and the right side of the map?
4. **Interpreting Data** Coronal holes produce almost no X-ray radiation, so they appear as black regions on the map. What is the range of longitudes covered by the coronal hole in the southwestern quadrant of the map?
5. **Analyzing Relationships** Sunspots emit large amounts of X-ray radiation. Where on this map would you expect to find sunspots?

IMPACT on Society

The Genesis Mission

What makes up the sun? Are the materials that make up Earth the same as those that make up other planets? These are only a few of the questions that NASA hoped an ambitious project called *The Genesis Mission* would answer.

Collecting Solar Wind

The Genesis mission began on August 8, 2001, when NASA launched its *Genesis* spacecraft. The spacecraft traveled 1.5 million kilometers toward the sun to an area called *Lagrange Point 1* (L1). L1 is a point in space outside Earth's magnetic field where the gravitational pulls of Earth and the sun are balanced.

Once at L1, the *Genesis* spacecraft collected electrically charged particles called *ions*, which make up solar wind. The spacecraft's solar collector contained pure materials that trapped the solar ions that collided with the collector. Another device, called the *concentrator*, consolidated the ions. The spacecraft's solar wind monitors collected data and helped coordinate the motions of the solar collectors and the concentrator. On April 1, 2004, after two years of collecting particles, the collectors were shut down in preparation for their scheduled September 8, 2004, return to Earth.

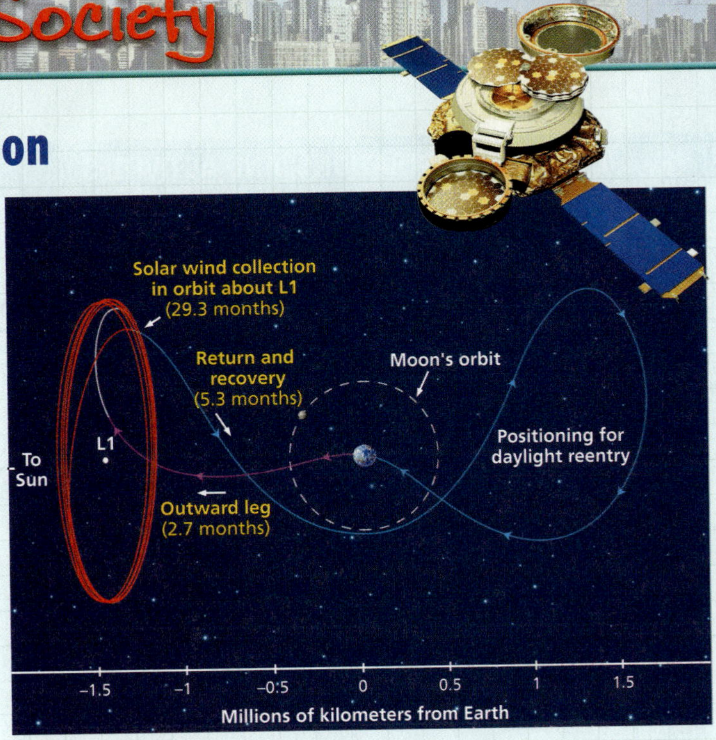

▶ The *Genesis* spacecraft collected solar wind particles.

Bringing the Sun to Earth

During the return to Earth, Genesis's parachute failed to open and the space capsule crashed into the Utah desert at nearly 200 mi/h. Scientists transported the damaged canister containing the collectors to a specially built clean room. Clean rooms are rooms in which specialized air filters and flooring reduce the number of airborne particles that may contaminate the samples.

Despite the crash, scientists remain hopeful that the samples recovered from the canister will provide enough data to help them measure the composition of the solar wind. The composition of the solar wind can be used to indirectly measure the abundance of isotopes and elements in the sun. Scientists hope to compare isotope data with data collected from comets, asteroids, and other bodies that formed at the beginning of our solar system.

Extension

1. **Evaluating Conclusions** Research the results of the scientists' investigation of Genesis's canister.

Impact on Society

The Genesis Mission

Discussion — BASIC

Collecting Solar Samples Have students summarize the mission objectives of the *Genesis* project. (Sample answer: to collect electrically charged particles, or ions, that make up the solar wind and bring samples back to Earth to analyze their composition) Then, have students compare the procedures of the Genesis Mission with those used to collect, recover, and study lunar rock samples recovered during the Apollo missions. Explain that studying the lunar rocks has helped scientists to better understand the origin and history of the moon just as the Genesis Mission may help them to better understand the origins of this solar system and its local star. **LS Verbal**

Answer to Extension

1. Answers may vary. Accept all reasonable answers.

Chapter 30 Stars, Galaxies, and the Universe

Planning Guide

Compression Guide
To shorten instruction because of time limitations, omit Section 1.

OBJECTIVES	LABS, DEMONSTRATIONS, AND ACTIVITIES	TECHNOLOGY RESOURCES
PACING • 90 min pp. 774–780 **Chapter Opener**		**OSP** Parent Letter **CD** Student Edition on CD-ROM **CD** Chapter Summaries Audio CD **VID** Brain Food Video Quiz
Section 1 Characteristics of Stars • Describe how astronomers determine the composition and temperature of stars. • Explain why stars appear to move in the sky. • Describe one way astronomers measure the distances to stars. • Explain the difference between absolute magnitude and apparent magnitude.	**TE Demonstration** Spectra, p. 775 GENERAL **TE Activity** Observing Stars, p. 778 ADVANCED **SE Quick Lab** Parallax, p. 779 GENERAL **CRF Datasheet for Quick Lab*** GENERAL **SE Making Models Lab** Star Magnitudes, pp. 802–803 GENERAL **CRF Datasheet for Chapter Lab*** GENERAL **TE Activity** Space Telescopes, p. 805 GENERAL **CRF Inquiry Lab** Curving Space-Time* GENERAL **CRF Skills Practice Lab** Blackbody Radiation* GENERAL	**OSP Lesson Plans** (also in print) **TR Bellringer*** **TR** 149 The Doppler Effect* **TR** 150 Apparent Magnitude* **TE Internet Activity** Proper Motion, p. 777 ADVANCED **CRF Internet Activity** Proper Motion* ADVANCED **CD Interactive Tutor** Distances
PACING • 90 min pp. 781–788 **Section 2 Stellar Evolution** • Describe how a protostar becomes a star. • Explain how a main-sequence star generates energy. • Describe the evolution of a star after its main-sequence stage.	**TE Group Activity** Main-Sequence Stars, p. 783 ADVANCED **TE Physics Connection** Star Size and Spectra, p. 784 GENERAL **TE Debate** SETI, p. 785 ADVANCED **TE Physics Connection** Supernova Physics, p. 787 ADVANCED	**OSP Lesson Plans** (also in print) **TR Bellringer*** **TR** 151 The Hertzsprung-Russell Diagram* **TR** 152 The Life Cycle of Stars* **CD Interactive Tutor** Stars
PACING • 45 min pp. 789–792 **Section 3 Star Groups** • Describe the characteristics that identify a constellation. • Describe the three main types of galaxies. • Explain how a quasar differs from a typical galaxy.	**TE Activity** Constellations, p. 789 GENERAL **TE Activity** Multimedia Project, p. 791 GENERAL **TE Astronomy Connection** Identifying a Galaxy from Within, p. 791 ADVANCED **SE Maps in Action** The Milky Way, p. 804 GENERAL **SE Mapping Expeditions** Stars in Your Eyes, pp. 844–845 GENERAL **TE Group Activity** Charting the Galaxy, p. 804 GENERAL	**OSP Lesson Plans** (also in print) **TR Bellringer*** **TR** 153 The Constellation Orion* **TR** 155 The Milky Way*
PACING • 45 min pp. 793–796 **Section 4 The Big Bang Theory** • Explain how Hubble's discoveries lead to an understanding that the universe is expanding. • Summarize the big bang theory. • List evidence for the big bang theory.	**TE Discussion** The Expanding Universe, p. 793 GENERAL **TE Physics Connection** New Physics, Old Universe, p. 794 ADVANCED **SE Quick Lab** The Expanding Universe, p. 795 GENERAL **CRF Datasheet for Quick Lab*** GENERAL	**OSP Lesson Plans** (also in print) **TR Bellringer*** **TR** 154 Timeline of the Big Bang* **VID NOVA Video** Runaway Universe **CD Interactive Tutor** Galaxies and the Universe

PACING • 90 min

CHAPTER REVIEW, ASSESSMENT, AND STANDARDIZED TEST PREPARATION

- **SE Chapter Highlights,** p. 797
- **SE Chapter Review,** pp. 798–799
- **SE Standardized Test Prep,** pp. 800–801
- **CRF Concept Review*** GENERAL
- **CRF Critical Thinking*** ADVANCED
- **CRF Math Skills*** GENERAL
- **CRF Graphing Skills*** GENERAL
- **CRF Chapter Test A*** GENERAL
- **CRF Chapter Test B*** ADVANCED
- **OSP Lesson Plans** (also in print)
- **OSP Test Generator**
- **OSP Test Item Listing**

Online and Technology Resources

Holt Online Learning

Visit **go.hrw.com** for access to Holt Online Learning, or enter the keyword **HQ6 Home** for a variety of free online resources.

One-Stop Planner® CD-ROM

This CD-ROM package includes
- Lab Materials QuickList Software
- Holt Calendar Planner
- Customizable Lesson Plans
- Printable Worksheets
- ExamView® Test Generator
- Interactive Teacher Edition
- Holt PuzzlePro®
- Holt PowerPoint® Resources

KEY

SE Student Edition
TE Teacher Edition
CRF Chapter Resource File
LTP Long-Term Projects
OSP One-Stop Planner
TR Transparencies and Transparency Worksheets
CD CD or CD-ROM
VID Classroom Video/DVD
* Also on One-Stop Planner
◆ Requires advance prep
■ Also available in Spanish

SKILLS DEVELOPMENT RESOURCES	REVIEW AND ASSESSMENT	CORRELATIONS
SE Pre-Reading Activity, p. 774 GENERAL **TE Using the Figure** Tarantula Nebula, p. 774 GENERAL		National Science Education Standards
CRF Directed Reading* BASIC	**SE Reading Checks,** pp. 777, 778 GENERAL **SE Section Review,** p. 780 GENERAL **TE Reteaching,** p. 779 BASIC **TE Quiz,** p. 779 GENERAL **TE Alternative Assessment,** p. 779 ADVANCED **CRF Section Quiz*** ■ GENERAL	SAI 2e
CRF Directed Reading* BASIC **TE Using the Figure** Star Mass and Classification, p. 781 GENERAL **SE Math Practice,** p. 782 GENERAL **TE Using the Figure** Masses of Stars, p. 783 GENERAL **TE Inclusion Strategies,** p. 783 **TE Using the Figure** Identifying Stars, p. 784 BASIC **TE Reading Skill Builder** Paired Summarizing, p. 785 BASIC **SE Graphic Organizer** Chain-of-Events Chart, p. 786 GENERAL **TE Using the Figure** Identifying Stars, p. 786 BASIC **TE Skill Builder** Vocabulary, p. 786 GENERAL	**SE Reading Checks,** pp. 783, 784, 787 GENERAL **SE Section Review,** p. 788 GENERAL **TE Homework,** p. 786 GENERAL **TE Reteaching,** p. 787 BASIC **TE Quiz,** p. 787 GENERAL **TE Alternative Assessment,** p. 787 ADVANCED **CRF Section Quiz*** ■ GENERAL	ES 4b, ES 4c, UCP 3, SAI 2e
CRF Directed Reading* BASIC **TE Skill Builder** Writing, p. 790 ADVANCED **TE Reading Skill Builder** Anticipation Guide, p. 791 BASIC	**SE Reading Check,** p. 790 GENERAL **SE Section Review,** p. 792 GENERAL **TE Reteaching,** p. 791 BASIC **TE Quiz,** p. 792 GENERAL **TE Alternative Assessment,** p. 792 ADVANCED **CRF Section Quiz*** ■ GENERAL	UCP 1, SAI 2e
CRF Directed Reading* BASIC **TE Inclusion Strategies,** p. 794	**SE Reading Check,** p. 794 GENERAL **SE Section Review,** p. 796 GENERAL **TE Reteaching,** p. 795 BASIC **TE Quiz,** p. 795 GENERAL **TE Alternative Assessment,** p. 796 ADVANCED **CRF Section Quiz*** ■ GENERAL	ES 4a, SAI 2e

Holt Lab Generator CD-ROM

Search for labs by topic, standard, difficulty level, and/or time frame. Edit those labs to fit your needs, or create your own labs. Use the Lab Materials QuickList to order just what you need.

Interactive Tutor — Holt Earth Science Interactive Tutor CD-ROM

This CD-ROM consists of interactive activities that give students a fun way to extend their knowledge of Earth science concepts.

Chapter Summaries Audio CDs

These CDs include audio summaries of the key concepts presented in each chapter. (Audio summaries are also available in Spanish.)

SCILINKS. NSTA

www.scilinks.org

Maintained by the **National Science Teachers Association.** See Chapter Enrichment pages that follow for a complete list of topics.

DVD VIDEO **See Chapter Enrichment pages for Video Resources.**

Chapter 30 Chapter Enrichment

This Chapter Enrichment provides relevant and interesting information to expand and enhance your classroom instruction of the chapter material.

Section 1 Characteristics of Stars

Spectra and Star Temperatures

The absorption spectrum for a given star provides more information than just the chemical composition of the star's surface. Because elements and compounds have different electron transition energies, a star's spectrum reveals whether atoms or ions of a given substance are present in the star, and therefore, what the star's surface temperature is. In the coolest (red) stars, molecules such as titanium oxide (TiO), carbon monoxide (CO), and cyanogen (CN) remain intact, so these stars have complex spectra. In warmer orange and yellow stars, most molecules dissociate, but some smaller molecules, such as OH and CH, persist and show up in the stars' spectra. At the higher temperatures of yellow-white and white stars, almost all metal atoms are ionized, and the spectral lines indicate the presence of the various ionized species.

For the hottest stars, the surface temperatures are so high that only the elements that have the greatest ionization energies can still absorb light and therefore produce absorption lines.

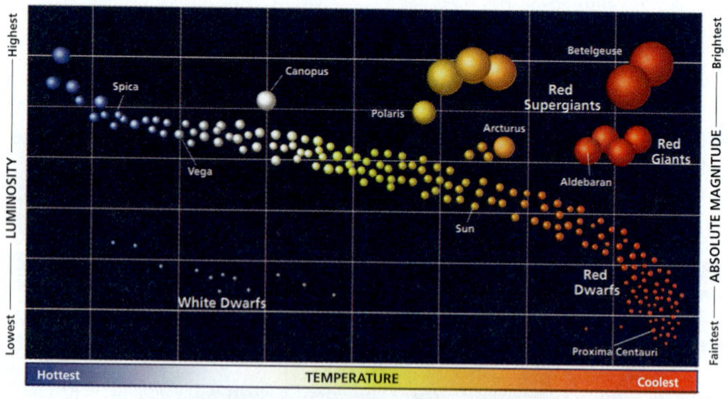

▲ **Cepheid variables evolve from main-sequence stars in the upper left corner of the HR diagram.**

Section 2 Stellar Evolution

Creating the Elements

While many elements such as carbon, oxygen, and neon are produced from the long-term fusion reactions within massive stars, the process that creates most elements with large atomic masses and distributes them into space is sudden and violent. During the last stages of a massive star's existence, its iron core collapses as energy-absorbing fusion begins. The material in the iron core is converted into neutrons, producing a flood of neutrinos. These particles do not interact strongly with matter, but so many are released within a fraction of a second that they push outward on the collapsing star. At the same time, layers of overlying matter come crashing down on the neutron core surface, which causes this material to rebound in a shock wave. If conditions are right, the interaction of the rapidly rising matter, outward-flowing neutrinos, and inward-falling matter compress nuclei to form a variety of elements. These particles also cause the star's outer layers to fly outward into space. This material becomes part of nebulae from which new stars and planets form.

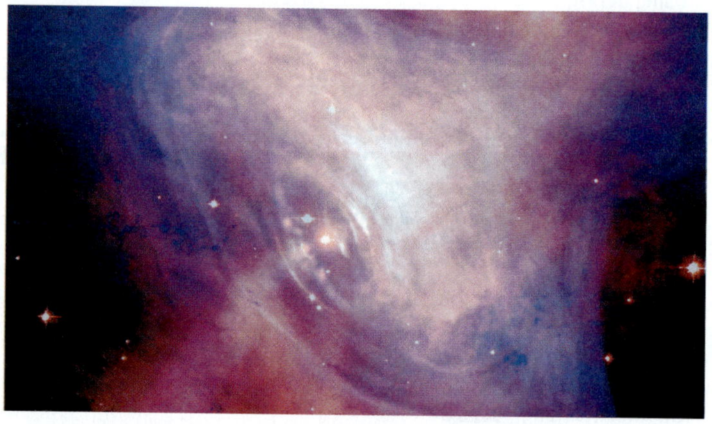

▲ **Supernovas produce various elements and distribute them into space.**

Section 3 Star Groups

Cepheid Variables and Stellar Distances

Of the indirect methods of measurement, the most powerful and versatile has involved Cepheid variable stars. Cepheid variables, so-called because Delta Cephei was the first variable star of this type to be discovered, are giant yellow-white stars that have reached a stage in their evolution at which their brightness varies uniformly over a regular period of time. In 1912, Henrietta Leavitt discovered the period-luminosity correlation for Cepheid variables, noting that the longer the period over which the star's brightness varied, the greater the star's absolute brightness is. By observing the time a Cepheid variable takes to slightly brighten, dim, and brighten again, the period is measured and, thus, the star's absolute brightness is determined. By then measuring the star's apparent brightness and comparing it to absolute brightness, the distance to the star—and the distance to nearby galaxies—can be calculated. However, using Cepheid variables to measure distance is complex: they are not the only uniformly variable stars. Edwin Hubble didn't realize that there were two types of Cepheid variables, and thus initially underestimated the distance to the galaxy M31 by a factor of seven.

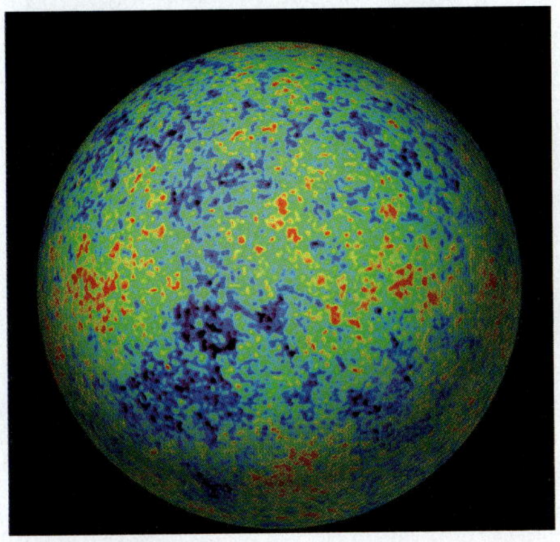

◀ **Temperature "ripples," shown as different colors in the sky as viewed from Earth, may indicate the first stages of galaxy development.**

Section 4 The Big Bang Theory

Galaxies and Missing Mass

One of the great problems of cosmology is also a problem in galactic astronomy: that of missing mass. In the 1960s, Vera Rubin noticed that the rotational speeds of spiral galaxies did not conform to classical orbital mechanics. The outer parts of the galactic disk moved at the same speed as the inner parts. This result could be explained only if a large amount of matter—90% of the galaxy's total mass—was distributed throughout the galaxy. Yet, this extra mass could not be visibly detected. It was to this problem, rather than the problem of cosmology and the missing mass of the universe, that the idea of dark matter was first applied. The solution to the problem remains open. Thus far, traditional kinds of matter (atoms) in the form of dim stars, large planet-like objects (brown dwarfs), and globular clusters have failed to account for the missing mass. Still more exotic, an unknown factor known as *dark energy* is making the universe expand at an increasing rate.

Video Resources

Brain Food Video Quizzes

Brain Food Video Quizzes These videos contain game-show style quizzes that assess students' progress and motivate students to study the chapter material.

NOVA Videos

The NOVA video below complements this chapter.

Runaway Universe Join two competing teams in their quest to discover the secrets to the stars and the ultimate fate of the universe. (60 min)

To order other **NOVA** videos related to this chapter, visit **go.hrw.com** and enter the keyword **HQ6STGV.**

SciLinks

Developed and maintained by the National Science Teachers Association

SciLinks is maintained by the National Science Teachers Association to provide you and your students with interesting, up-to-date links that will enrich your classroom presentation of the chapter.

Visit www.scilinks.org and enter the SciLinks code for more information about the topic listed.

Topic: Stars
SciLinks code: HQ61448

Topic: How Stars Evolve
SciLinks code: HQ60764

Topic: Big Bang
SciLinks code: HQ60146

Topic: Galaxies
SciLinks code: HQ60632

Topic: Milky Way Galaxy
SciLinks code: HQ60964

Chapter 30

Chapter Overview

This chapter describes what stars are and how they evolve. The chapter discusses how to determine a star's temperature, brightness, composition, size, motion, and distance. The chapter also explains star groups and galaxies and describes how each of the structures formed. Finally, the chapter explains the current model for the structure, formation, and evolution of the universe.

Using the Figure — GENERAL

Tarantula Nebula The Tarantula Nebula is a gaseous region in the Large Magellanic Cloud, a small galaxy that orbits the Milky Way galaxy. Explain that this particular nebula is called an *emission nebula* because the hot, glowing gases emit light. Ask students what provides the energy for the nebula's gases to emit light. (The high-energy light from bright stars within the nebula causes the gases to emit light.)

PRE-READING ACTIVITY

FOLDNOTES Encourage students to use their FoldNote as a study guide to quiz themselves for a test on the chapter material. Students may want to create Layered Book FoldNotes for different topics within the chapter.

Chapter 30 Stars, Galaxies, and the Universe

Sections

1 Characteristics of Stars
2 Stellar Evolution
3 Star Groups
4 The Big Bang Theory

What You'll Learn

- How stars are organized by their characteristics
- How stars form and die
- How stars are grouped into clusters and galaxies
- What the universe is like and how it formed

Why It's Relevant

Our solar system is only one small part of the universe. By studying stars and galaxies, we can learn more about the formation and evolution of our universe.

PRE-READING ACTIVITY

FOLDNOTES **Layered Book** Before you read this chapter, create the FoldNote entitled "Layered Book" described in the Skills Handbook section of the Appendix. Label the tabs of the layered book with "Star types," "Stellar evolution," "Star groups," and "Origin of the universe." As you read the chapter, write information you learn about each category under the appropriate tab.

▶ New stars are currently forming in the Tarantula nebula, an enormous region of dust and ionized gas.

Chapter Correlations *National Science Education Standards*

ES 4a The "big bang" theory places the origin of the universe between 10 and 20 billion years ago, when the universe began in a hot dense state; according to this theory, the universe has been expanding ever since. **(Section 4)**

ES 4b Early in the history of the universe, matter...clumped together by gravitational attraction to form...stars. Billions of galaxies, each...a gravitationally bound cluster of billions of stars,...form most of the visible mass in the universe. **(Section 2)**

ES 4c Stars produce energy from nuclear reactions, primarily the fusion of hydrogen to form helium. These and other processes in stars have led to the formation of the other elements. **(Section 2)**

UCP 1 Scientists...define...units of investigation...as "systems." A system is an organized group of related objects... **(Section 3)**

UCP 3 Although...some properties of objects and processes are characterized by constancy, including the speed of light...and the total mass plus energy of the universe, changes might occur. **(Section 2)**

SAI 2e ...Scientific explanations must adhere to criteria such as: a proposed explanation must be logically consistent; it must abide by the rules of evidence; it must be open to questions and possible modification; and it must be based on historical and current scientific knowledge. **(Sections 1–4)**

Section 1 Characteristics of Stars

A **star** is a ball of gases that gives off a tremendous amount of electromagnetic energy. This energy comes from nuclear fusion within the star. *Nuclear fusion* is the combination of light atomic nuclei to form heavier atomic nuclei.

As seen from Earth, most stars in the night sky appear to be tiny specks of white light. However, if you look closely at the stars, you will notice that they vary in color. For example, the star Antares shines with a slightly reddish color, the star Rigel shines blue-white, and the star Arcturus shines with an orange tint. Our own star, Sol, is a yellow star.

OBJECTIVES

- **Describe** how astronomers determine the composition and temperature of stars.
- **Explain** why stars appear to move in the sky.
- **Describe** one way astronomers measure the distances to stars.
- **Explain** the difference between absolute magnitude and apparent magnitude.

KEY TERMS

star
Doppler effect
light-year
parallax
apparent magnitude
absolute magnitude

Analyzing Starlight

Astronomers learn about stars primarily by analyzing the light that the stars emit. Astronomers direct starlight through *spectrographs,* which are devices that separate light into different colors, or wavelengths. Starlight passing through a spectrograph produces a display of colors and lines called a *spectrum*. There are three types of spectra: *emission,* or bright-line; *absorption,* or dark-line; and *continuous*.

All stars have *dark-line spectra*—bands of color crossed by dark lines where the color is diminished, as shown in **Figure 1.** A star's dark-line spectrum reveals the star's composition and temperature.

star a large celestial body that is composed of gas and that emits light

Stars are made up of different elements in the form of gases. While the inner layers of a star are very hot, the outer layers are somewhat cooler. Elements in the outer layers absorb some of the light radiating from within the star. Because different elements absorb different wavelengths of light, scientists can determine the elements that make up a star by studying its spectrum.

Figure 1 ▶ The spectrum of the sun consists of bands of color crossed by dark absorption lines. This spectrum appears as though it has been cut into strips that have been arranged vertically.

Section 1

Focus

Overview

This section explains how light is used to determine the chemical composition and temperature of a star's surface, the apparent motion of stars due to Earth's motion, and the parallax method of measuring a star's distance.

Bellringer

Have students write down the names of stars that they are familiar with. (Sample answers: North Star [Polaris], Rigel, Betelgeuse, Antares, Sirius, Vega, Alpha Centauri) LS **Verbal**

Motivate

Demonstration — GENERAL

Spectra Darken the room and place a hydrogen discharge tube into a power source. Turn on the power source, have students look at the light through diffraction gratings. Turn off the power, use heat-resistant gloves to remove the hydrogen tube, and insert a neon spectrum tube. Repeat the demonstration. Ask students how the lines from the two gases differ. (Hydrogen produces violet, dark blue, blue-green, and red lines; neon produces green, yellow, and orange lines.) Explain that the emission lines are produced from hot gases and that each element produces a unique combination of colors. LS **Visual**

CHAPTER RESOURCES

Chapter Resource File
- **Directed Reading** BASIC
- **Skills Practice Lab** Blackbody Radiation GENERAL

Technology
- **Transparencies**
  - Bellringer
- **Student Edition on CD-ROM**
- **One-Stop Planner CD-ROM**
  - Lesson Plan

Teach

MISCONCEPTION ALERT

Stellar Composition The composition of a star is based not on materials deep within the star but on the spectra of substances in the star's photosphere. The photosphere, the region of the star that emits the light we see, is considered the star's surface. The term "surface," when applied to a star, does not mean a solid region. Like all parts of a star, the surface is made up of gas or plasma. Because the material in the photosphere is cooler than the material underneath, the atoms of the photosphere will absorb certain colors of light from the continuous spectrum of light that comes through the photosphere, producing the characteristic absorption spectrum.

BRAIN FOOD

The Size of the Sun Help students to understand how large a distance 1,390,000 km is by noting that Earth's diameter is about 12,700 km. Thus, if nearly 110 Earths were placed end to end they would extend from one side of the sun to the other. More than 1 million Earths would be needed to fill a volume equal to that of the sun. The sun's mass is nearly a thousand times greater than that of the next most massive object in the solar system, Jupiter. In fact, more than 99% of the entire solar system's mass is contained within the sun.

Classification of Stars

Color	Surface temperature (°C)	Examples
Blue	above 30,000	10 Lacertae
Blue-white	10,000–30,000	Rigel, Spica
White	7,500–10,000	Vega, Sirius
Yellow-white	6,000–7,500	Canopus, Procyon
Yellow	5,000–6,000	sun, Capella
Orange	3,500–5,000	Arcturus, Aldebaran
Red	less than 3,500	Betelgeuse, Antares

Figure 2 ▶ Stars in the sky show tinges of different colors, which reveal the stars' temperatures. Blue stars shine with the hottest temperatures, and red stars shine with the coolest.

The Compositions of Stars

Every chemical element has a characteristic spectrum in a given range of temperatures. The colors and lines in the spectrum of a star indicate the elements that make up the star. Through spectrum analysis, scientists have learned that stars are made up of the same elements that compose Earth. But while the most common element on Earth is oxygen, the most common element in stars is hydrogen. Helium is the second most common element in stars. Elements such as carbon, oxygen, and nitrogen, usually in small quantities, make up most of the remaining mass of stars.

The Temperatures of Stars

The surface temperature of a star is indicated by the star's color, as shown in **Figure 2.** The temperature of most stars ranges from 2,800°C to 24,000°C, although a few stars are hotter. Generally, a star that shines with blue light has an average surface temperature of 35,000°C. However, the surface temperatures of some blue stars are as high as 50,000°C. Red stars are the coolest stars and have average surface temperatures of 3,000°C. Yellow stars, such as the sun, have surface temperatures of about 5,500°C.

The Sizes and Masses of Stars

Stars also vary in size and mass. Some dwarf stars are about the same size as Earth. The sun, a medium-sized star, has a diameter of about 1,390,000 km. Some giant stars have diameters that are 1,000 times the sun's diameter. Most stars visible from Earth are medium-sized stars that are similar to our sun.

Many stars also have about the same mass as the sun, though some stars may be significantly more or less massive. Stars that are very dense may have more mass than the sun and still be much smaller than the sun. Less-dense stars may have a larger diameter than the sun has but still have less mass than the sun.

PHYSICS CONNECTION

Blackbody Radiation The temperature of stars is based on the assumption that they behave like *blackbodies,* which are perfect absorbers and radiators of energy. The intensity of the electromagnetic radiation emitted by a blackbody increases with temperature. Also, the color of a blackbody ranges from a dull red at lower temperatures through a brighter yellow to a brilliant blue-white at higher temperatures. This property of blackbodies allows their temperatures to be determined from the colors of the light they emit. By applying this method to starlight, the star's surface temperature can be determined. For blackbodies, there is a peak wavelength at which the body radiates light most intensely. Because other wavelengths are also emitted, this color is not the same as the observed color. For instance, the sun emits most strongly in the green portion of the spectrum, yet its overall color is yellow. A curve of stellar intensities usually deviates from an ideal blackbody curve because of absorption by elements at the star's surface.

Stellar Motion

Two kinds of motion are associated with stars—actual motion and apparent motion. Because stars are so far from Earth, their actual motion can be measured only with high-powered telescopes and other specialized instruments. Apparent motion, on the other hand, is much more noticeable.

Apparent Motion of Stars

The *apparent motion* of stars, or motion visible to the unaided eye, is caused by the movement of Earth. By aiming a camera at the sky and leaving the shutter open for a few hours, you can photograph the apparent motion of the stars. The curves of light in **Figure 3** record the apparent motion of stars in the northern sky. The circular trails make it seem as though the stars are moving counter-clockwise around a central star called Polaris, or the North Star. The circular pattern is caused by the rotation of Earth on its axis. Polaris is almost directly above the North Pole, and thus the star does not appear to move much.

Earth's revolution around the sun causes the stars to appear to move in a second way. Stars located on the side of the sun opposite Earth are obscured by the sun. As Earth orbits the sun, however, different stars become visible during different seasons. The visible stars appear to shift slightly to the west every night. Each night, most stars appear a small distance farther across the sky than they were the night before. After many months, they may finally disappear below the western horizon.

Reading Check **Why does Polaris appear to remain stationary in the night sky? (See the Appendix for answers to Reading Checks.)**

SCILINKS® — NSTA — Developed and maintained by the National Science Teachers Association

For a variety of links related to this subject, go to www.scilinks.org

Topic: Stars
SciLinksCode: HQ61448

Figure 3 ▶ Stars appear as curved trails in this long-exposure photograph. These trails result from the rotation of Earth on its axis.

Internet Activity — ADVANCED

Proper Motion Explain to students that a star's actual motion is divided into two components: its radial motion, which is the amount a star moves toward or away from Earth along the observer's line of sight; and its proper motion, which is the amount a star moves in the plane of the sky. Proper motion varies with how close the star is to Earth and how fast it is actually moving. Have students use the Internet to research the topic of proper motion for stars. Have them note the speeds at which some stars actually move, how far those stars are from Earth, and how much they move relative to the background sky as a result of their speed and distance. Have students present their findings in a written report or oral presentation. A worksheet designed to direct student research on this topic can be found in the **Chapter Resource File** booklet or by visiting **go.hrw.com** and entering the keyword **HQ6STGX.** LS **Verbal**

Answer to Reading Check

Polaris is almost exactly above the pole of Earth's rotational axis, so Polaris moves only slightly around the pole during one rotation of Earth.

CHAPTER RESOURCES

Chapter Resource File

Internet Activity
• Proper Motion GENERAL

Cultural Awareness — BASIC

Earth's Rotation The idea that the apparent motion of stars results from Earth's rotation was suggested as long ago as the fourth century BCE, by a Greek scholar named Heraclides (hair RAH kli deez). However, Aristotle's belief that Earth could not move, and therefore neither rotated nor traveled through the heavens, prevailed over Western classical and medieval thinking. The concept of Earth's rotation was renewed, along with the notion that Earth revolved around the sun, in the early sixteenth century CE by Nicolaus Copernicus.

BRAIN FOOD

Precession Earth wobbles around its rotational axis much as a top does. This process, called *precession,* causes the North Pole to point at different parts of the northern sky, so that Polaris is not always the "pole star." Other stars that have been "the north star" include Vega, Thuban, and Deneb. During most of the 26,000-year precession cycle, no bright stars are situated above the North Pole.

Teach, *continued*

Activity — ADVANCED

Observing Stars Explain that the location of circumpolar stars in the sky depends on the observer's location on Earth. When standing at the North Pole, all of the stars in the sky are circumpolar, but as a person moves south, more and more of the stars seen from the northern hemisphere disappear for a time beneath the horizon. At the equator, all stars rise above the eastern horizon and set in the west. Polaris makes a small circular path above and below the northern horizon. Students can gain a better sense of which stars appear at different latitudes by performing the following exercise. Have students draw sketches of the northern half of Earth and the position of Polaris with respect to the North Pole. Have students draw an observer at different latitudes on Earth, and then have them show where Polaris will appear above the observer's horizon. Students should then show that the angle between Polaris and the northern horizon is equal to the angle of northern latitude of the observer. LS **Logical**

Answer to Reading Check

Starlight is shifted toward the red end of the spectrum when the star is moving away from the observer.

CHAPTER RESOURCES

Technology

Transparencies
- 149 The Doppler Effect (with worksheet)

Circumpolar Stars

Some stars are always visible in the night sky. These stars never pass below the horizon in either their nightly or annual movements. In the Northern Hemisphere, the movement of these stars makes them appear to circle Polaris, the North Star. These circling stars are called *circumpolar stars*. The stars of the Little Dipper are circumpolar for most observers in the Northern Hemisphere. At the North Pole, all visible stars are circumpolar. The farther the observer moves from the North Pole toward the equator, the fewer circumpolar stars the observer will be able to see.

Actual Motion of Stars

Most stars have several types of *actual motion*. First, they rotate on an axis. Second, they may revolve around another star. Third, they either move away from or toward our solar system.

From a star's spectrum, astronomers can learn more about how that star is moving in space. The spectrum of a star that is moving toward or away from Earth appears to shift, as shown in **Figure 4.** The apparent shift in the wavelength of light emitted by a light source moving toward or away from an observer is called the **Doppler effect.** The colors in the spectrum of a star moving toward Earth are shifted slightly toward blue. This shift, called *blue shift*, occurs because the light waves from a star appear to have shorter wavelengths as the star moves toward Earth.

A star moving away from Earth has a spectrum that is shifted slightly toward red. This shift, called *red shift*, occurs because the wavelengths of light appear to be longer. Distant galaxies have red-shifted spectra, which indicates that these galaxies are moving away from Earth.

Doppler effect an observed change in the frequency of a wave when the source or observer is moving

Reading Check What causes starlight to shift toward the red end of the spectrum? (See the Appendix for answers to Reading Checks.)

Figure 4 ▶ The light from stars is shifted based on the star's movement in relationship to Earth. For example, light from stars that are moving away from Earth is shifted slightly toward the red end of the spectrum.

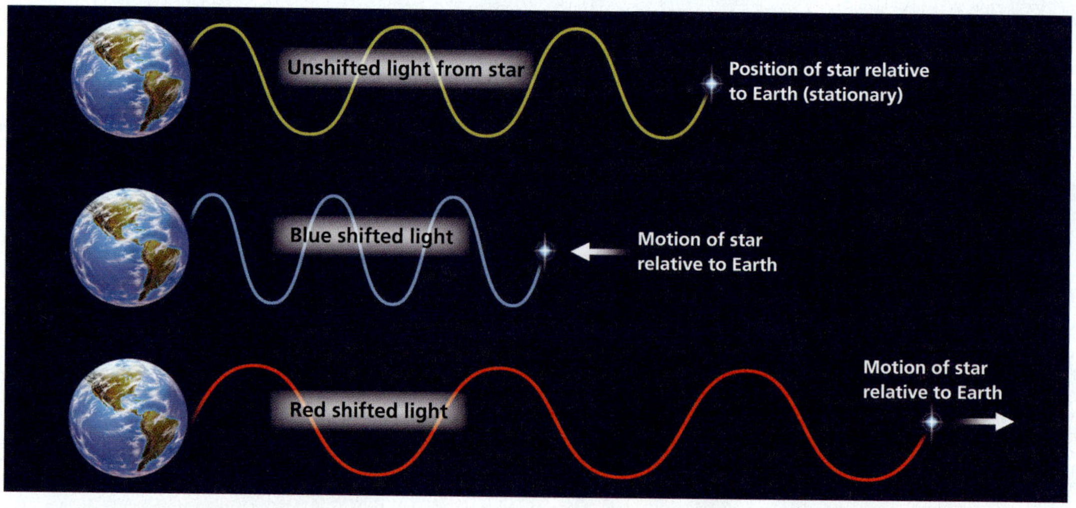

BRAIN FOOD

Causes of Red Shifts While the term *red shift* always indicates that an object's light has been shifted toward the red end of the spectrum, it does not always happen for the same reason. A *gravitational red shift* occurs when light is affected by strong gravitational forces, like those at the surface of a star or in the vicinity of a black hole. The unusual conditions that give rise to this type of red shift indicate whether strong gravitational forces are at work or not. A *cosmological red shift* only begins to affect the light from galaxies at great distances from Earth. This type of red shift is the result of the overall expansion of the universe, which causes more distant objects to move away at greater speeds. The amount by which light is cosmologically red-shifted is an indication of how far these objects are from Earth.

Distances to Stars

Because space is so vast, distances between the stars and Earth are measured in light-years. A **light-year** is the distance that light travels in one year. Because the speed of light is 300,000 km/s, light travels about 9.46 trillion km in one year. The light you see when you look at a star left that star sometime in the past. Light from the sun, for example, takes about 8 minutes to reach Earth. The sun is therefore 8 light-minutes from Earth. When we witness an event on the sun, such as a solar flare, the event actually took place about 8 minutes before we saw it.

Apart from the sun, the star nearest Earth is Proxima Centauri. This star is 4.2 light-years from Earth, nearly 300,000 times the distance from Earth to the sun. Polaris is 700 light-years from Earth. When you look at Polaris, you see the star the way it was 700 years ago.

For relatively close stars, scientists can determine a star's distance by measuring **parallax,** the apparent shift in a star's position when viewed from different locations. As Earth orbits the sun, observers can study the stars from different perspectives, as shown in **Figure 5.** As Earth moves halfway around its orbit, a nearby star will appear to shift slightly relative to stars that are farther from Earth. The closer the star is to Earth, the larger the shift will be. Using this method, astronomers can calculate the distance to any star within 1,000 light-years of Earth.

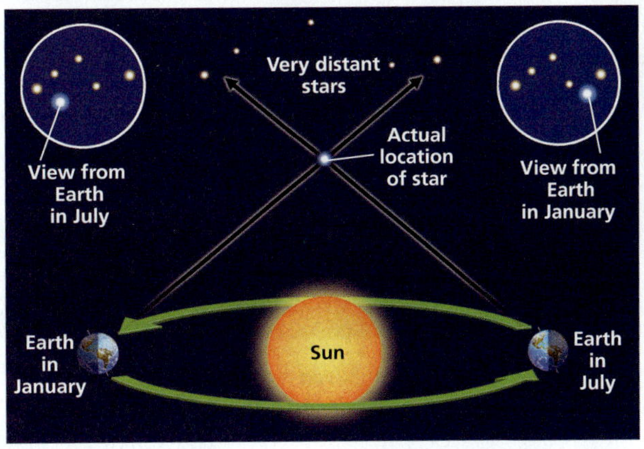

Figure 5 ▶ Observers on Earth see nearby stars against those in the distant background. The movement of Earth causes nearby stars to appear to move back and forth each year.

light-year the distance that light travels in one year

parallax an apparent shift in the position of an object when viewed from different locations

QuickLAB 15 min

Parallax

Procedure

1. Use a **metric ruler** and **scissors** to cut **five 1 m lengths of thread.** Use **masking tape** to tape one end of each piece of thread to the edge of a **paper plate.** Each plate should have the same diameter. One plate should be red, and four should be blue.
2. Stand on a **ladder,** and tape the free end of each piece of thread to the ceiling at various heights. Place the thread 30 cm apart in a staggered pattern. Hang the plates in a location that allows the widest field of view and movement.
3. Stand directly in front of and facing the red plate at a distance of several meters.
4. Close one eye, and sketch the position of the red plate in relation to the blue plates.
5. Take several steps back and to the right. Repeat step 4.
6. Take several more steps and make another sketch.
7. Repeat step 6.

Analysis

1. Compare your drawings. Did the red plate change position as you viewed it from different locations? Explain your answer.
2. What results would you expect if you continued to repeat step 6? Explain your answer.
3. If you noted the positions of several stars by using a powerful telescope, what would you expect to observe about their positions if you saw the same stars six months later? Explain.

Close

Reteaching — Basic

Star Temperatures and Colors Create cards that have circles colored red, orange, yellow, yellow-white (pale yellow), white, blue-white (pale blue), and blue. Mix these cards up and have students place these "stars" in the correct order from coolest to hottest. **LS Visual/Logical**

Quiz — General

1. What color are the hottest stars? (blue)
2. What causes a star's spectrum to be blue-shifted? (the Doppler effect acting on light from a star that is moving toward the observer)
3. What are stars that never set below the horizon called? (circumpolar stars)

Alternative Assessment — Advanced

Moving Stars Have students create a series of posters or a computer model that shows how the stars in a chosen constellation will move over time. The models should show how the constellation looks after several 100,000-year periods have passed. **LS Kinesthetic/Logical**

CHAPTER RESOURCES

Chapter Resource File

- Datasheet for Quick Lab GENERAL
- Inquiry Lab
  Curving Space-Time GENERAL

Technology

Transparencies
- 150 Apparent Magnitude (with worksheet)

QuickLAB

Skills Acquired

- Experimenting
- Observing
- Interpreting

Teacher's Notes You may want to hang the plates from the ceiling to save time, to improve accuracy, and to avoid safety issues related to the use of ladders.

Answers

1. yes; The red plate appeared to shift to the side relative to the blue plates as the observer moved to the sides. The shift was less extreme when the observer was farther away.
2. Eventually the red plate would not appear to shift relative to the blue plates.
3. Closer stars would have shifted the most relative to the background, while more distant stars would have shifted less.

Close, *continued*

Answers to Section Review

1. Dark lines in a star's spectrum tell which substances are present at the star's surface. The color of the star's light indicates its surface temperature.
2. The sun is an average star, so most stars share its general size and mass characteristics. However, very dense stars may have more mass than the sun but be much smaller. Less-dense stars may have larger diameters but have less mass.
3. Earth rotates toward the east, so that objects in the sky appear to move in the opposite direction (to the west).
4. Distances are described in terms of how far light will travel in a given time. The speed of light is 300,000 km/s. A light-minute is how far light travels in a minute. A light-year is how far light travels in a year.
5. Parallax is used to measure distances to stars that are close to Earth. When Earth has moved to the opposite side of the sun in six months, the stars will appear to have shifted slightly. The greater the shift is, the closer the star is.
6. Apparent magnitude is how bright a star appears from Earth. Absolute magnitude is the star's actual brightness if the star were a standard distance from Earth.
7. Stars appear to move in circles around points in the sky above Earth's poles, in a direction opposite Earth's rotational direction.
8. In one year, Earth will have returned to roughly the same position it had when the first measurement was made, so no parallax shift will be observed.
9. The closer star would appear to be brighter.
10. A *star* is described in terms of distance from Earth, which is measured in *light-years;* brightness, which is measured in terms of *apparent magnitude,* which is how bright it appears to be, and *absolute magnitude,* which is how bright it would be at a standard distance; and motion, in which the *Doppler effect* causes a shift in the wavelength of light, either away from Earth, which is called a *red shift,* or toward Earth, which is called a *blue shift*.

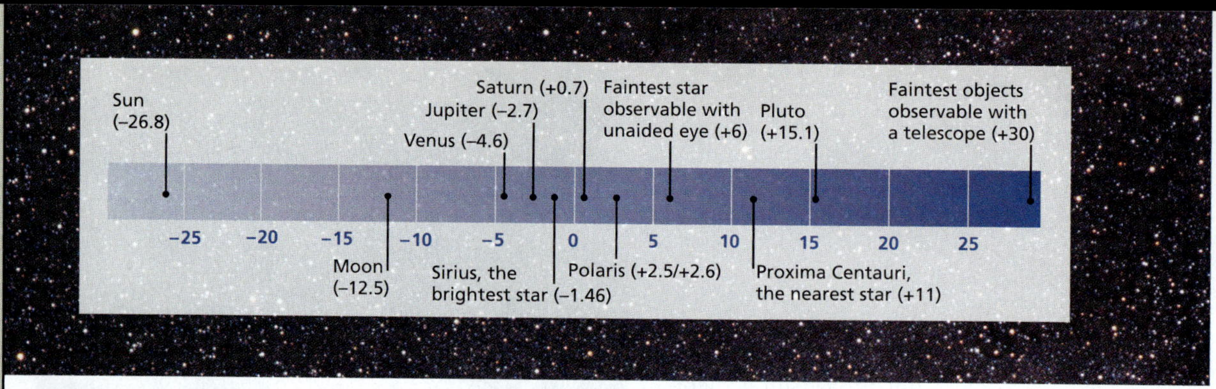

Figure 6 ▶ The sun, which has an apparent magnitude of –26.8, is the brightest object in our sky. All other objects appear dimmer in the sky, so their apparent magnitudes are higher on the scale.

Stellar Brightness

More than 3 billion stars can be seen through telescopes on Earth. Of these, only about 6,000 are visible without a telescope. Billions more stars can be observed from Earth-orbiting telescopes, such as the *Hubble Space Telescope*. The visibility of a star depends on its brightness and its distance from Earth. Astronomers use two scales to describe the brightness of a star.

The brightness of a star as it appears to us on Earth is called the star's **apparent magnitude.** The apparent magnitude of a star depends on both how much light the star emits and how far the star is from Earth. The lower the number of the star on the scale shown in **Figure 6,** the brighter the star appears to observers on Earth. The true brightness, or **absolute magnitude,** of a star is how bright the star would appear if all the stars were at a standard, uniform distance from Earth. The brighter a star actually is, the lower the number of its absolute magnitude.

apparent magnitude the brightness of a star as seen from the Earth

absolute magnitude the brightness that a star would have at a distance of 32.6 light-years from Earth

Section 1 Review

1. **Describe** what astronomers analyze to determine the composition and surface temperature of a star.
2. **Compare** the mass of the sun with the masses of most other stars in the universe.
3. **Explain** why, as you observe the night sky over time, stars appear to move westward across the sky.
4. **Describe** the units used to measure the distance to stars in terms of whether their starlight takes minutes or years to reach Earth.
5. **Describe** the method astronomers use to measure the distance to stars that are less than 1,000 light-years from Earth.
6. **Explain** the difference between apparent magnitude and absolute magnitude.

CRITICAL THINKING

7. **Identifying Relationships** How does the movement of Earth affect the apparent movement of stars in the sky?
8. **Analyzing Ideas** Why is it better for astronomers to measure parallax by observing every six months instead of observing every year?
9. **Understanding Relationships** If two stars have the same absolute magnitude, but one of the stars is farther from Earth than the other one, which star would appear brighter in the night sky?

CONCEPT MAPPING

10. Use the following terms to create a concept map: *star, apparent magnitude, red shift, Doppler effect, light-year, absolute magnitude,* and *blue shift.*

CHAPTER RESOURCES

Chapter Resource File
- Section Quiz GENERAL

Workbooks
- Study Guide (also in Spanish)

Section 2

Stellar Evolution

Because a typical star exists for billions of years, astronomers will never be able to observe one star throughout its entire lifetime. Instead, they have developed theories about the evolution of stars by studying stars in different stages of development.

OBJECTIVES

- **Describe** how a protostar becomes a star.
- **Explain** how a main-sequence star generates energy.
- **Describe** the evolution of a star after its main-sequence stage.

KEY TERMS

main sequence
nebula
giant
white dwarf
nova
neutron star
pulsar
black hole

Classifying Stars

Plotting the surface temperatures of stars against their *luminosity*, or the total amount of energy they give off each second, reveals a consistent pattern. The graph that illustrates this pattern is the *Hertzsprung-Russell diagram,* or *H-R diagram,* a simplified version of which is shown in **Figure 1.** The graph is named for Ejnar Hertzsprung and Henry Norris Russell, the astronomers who discovered the pattern nearly 100 years ago. The temperature of a star's surface is plotted on the horizontal axis. The luminosity of a star is plotted on the vertical axis.

Astronomers use the H-R diagram to describe the life cycles of stars. Astronomers always plot the highest temperatures on the left and the highest luminosities at the top. The temperature and luminosity for most stars falls within a band that runs diagonally through the middle of the H-R diagram. This band, which extends from cool, dim, red stars at the lower right to hot, bright, blue stars at the upper left, is known as the **main sequence.** Stars within this band are called *main-sequence stars.* The sun is one example of a main-sequence star.

main sequence the location on the H-R diagram where most stars lie; it has a diagonal pattern from the lower right to the upper left

Figure 1 ▶ The Hertzsprung-Russell Diagram

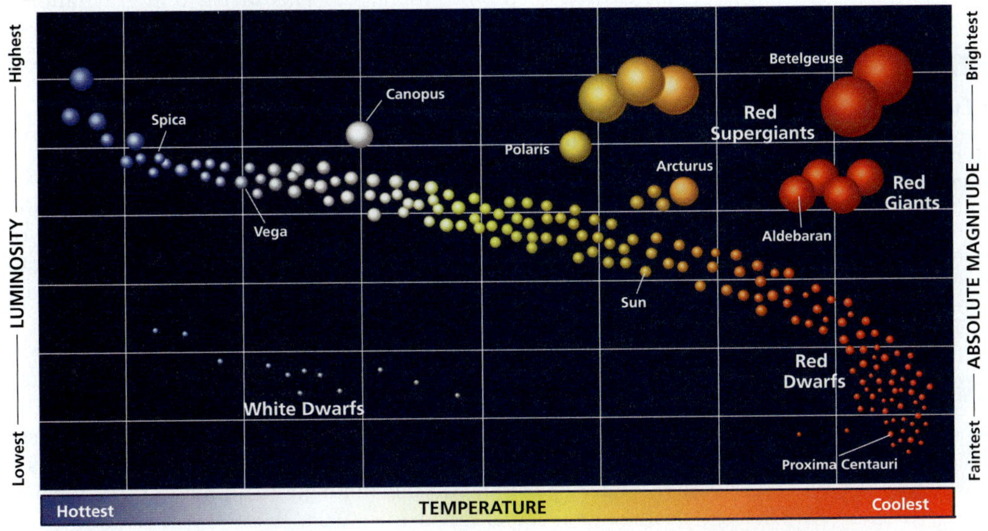

Section 2

Focus

Overview

This section explains how stars are classified and how they form and evolve. The section also describes how stars generate energy and describes the final stages of massive stars.

Bellringer

Have students write down two or three characteristics of stars that they might expect to change as the star ages. (Sample answers: the size of the star, surface color and temperature, brightness, and interior composition) LS **Logical**

Motivate

Using the Figure — GENERAL

Star Mass and Classification

Have students examine the Hertzsprung-Russell diagram and note what properties vary along the horizontal and vertical axes. Ask students what stars are hottest and brightest. (blue stars) Point out that the more massive stars at the upper left end of the H-R diagram convert fuel faster than the low-mass stars in the lower right do, so the more massive stars go through their life cycle more rapidly. LS **Visual/ Logical**

CHAPTER RESOURCES

Chapter Resource File

- Directed Reading BASIC

Technology

Transparencies
- Bellringer
- 151 The Hertzsprung-Russell Diagram (with worksheet)

Student Edition on CD-ROM

One-Stop Planner CD-ROM
- Lesson Plan

Teach

MISCONCEPTION ALERT

False Color Photographs Many photographs of distant astronomical objects, notably nebulae and galaxies, are reproduced in false colors. This is most commonly done when the object has been observed in a part of the electromagnetic spectrum other than visible light. In order to reproduce the image as a visible photograph with the proper contrast, areas of different intensities are assigned colors that visually convey the information effectively.

MATHPRACTICE

Answer
percent of mass radiated = (3.6 million metric tons/s ÷ 545 million metric tons/s) × 100 = (3.6/545) × 100 = 0.66%; 60 s/min × 60 min/h × 24 h/day × 365 days/y = 3.2×10^7 s/y; 3.6×10^6 metric tons/s × 3.2×10^7 s/y = 1.2×10^{14} metric tons/y

BRAIN FOOD

The Messier Catalogue The term "nebula" originally referred to any object in the night sky that was vaguely defined and cloudlike. The first serious investigation of these objects was conducted by French astronomer Charles Messier. While searching for comets in the mid-eighteenth century, Messier noticed objects that appeared hazy, like comets, but which did not change position. By 1784, Messier had published a description of 103 "nebulae."

Figure 2 ▶ The Eagle Nebula is a region in which star formation is currently taking place. This false-color image was captured by the *Hubble Space Telescope*.

Star Formation

nebula a large cloud of gas and dust in interstellar space; a region in space where stars are born

A star begins in a **nebula** (NEB yu luh), a cloud of gas and dust, such as the one shown in **Figure 2.** A nebula commonly consists of about 70% hydrogen, 28% helium, and 2% heavier elements. When an outside force, such as the explosion of a nearby star compresses the cloud, some of the particles move close to each other and are pulled together by gravity.

According to Newton's *law of universal gravitation,* all objects in the universe attract each other through gravitational force. This gravitational force increases as the mass of an object increases or as the distance between two objects decreases. Therefore, as gravity pulls particles closer together, the gravitational pull of the particles on each other increases. This increase in gravitational force causes more nearby particles to be pulled toward the area of increasing mass. As more particles come together, regions of dense matter begin to build up within the cloud.

MATHPRACTICE

Nuclear Fusion
The sun converts nearly 545 million metric tons of hydrogen to helium every second. In the process, approximately 3.6 million metric tons of that hydrogen mass is changed into energy and radiated into space. What percentage of the converted hydrogen is changed into radiated energy? If the sun loses 3.6 million metric tons of mass per second, how many metric tons of mass will it lose in one year?

Protostars

As gravity makes these dense regions more compact, any spin the region has is greatly amplified. The shrinking, spinning region begins to flatten into a disk that has a central concentration of matter called a *protostar*. Gravitational energy is converted into heat energy as more matter is pulled into the protostar. This heat energy causes the temperature of the protostar to increase.

The protostar continues to contract and increase in temperature for several million years. Eventually, the gas becomes so hot that its electrons are stripped from their parent atoms. The nuclei and free electrons move independently, and the gas is then considered a separate state of matter called plasma. *Plasma* is a hot, ionized gas that consists of an equal number of free-moving positive ions and electrons.

Teaching Tip GENERAL

Connect to Prior Knowledge To help students understand why a nebula spins faster to form a protostar, have them visualize an ice skater spinning. At first, the skater has his or her arms extended outward while spinning slowly. Then, by pulling the arms inward, the skater begins to spin faster. As gravity pulls particles toward the nebula's center (like a skater pulling in his or her arms), the circular motion around the center increases. The nebula flattens like pizza dough around the axis of rotation. Because most of the nebular matter is concentrated in the flattened disk, the matter that does not form the central star clumps together to form planets or, if there is enough matter in the nebula, companion stars. LS Visual

The Birth of a Star

Temperature continues to increase in a protostar to about 10,000,000°C. At this temperature, nuclear fusion begins. *Nuclear fusion* is a process that occurs when extremely high temperature and pressure cause less-massive atomic nuclei to combine to form more-massive nuclei and, in the process, release enormous amounts of energy. The onset of fusion marks the birth of a star. Once nuclear fusion begins in a star, the process can continue for billions of years.

A Delicate Balancing Act

As gravity increases the pressure on the matter within the star, the rate of fusion increases. In turn, the energy radiated from fusion reactions heats the gas inside the star. The outward pressures of the radiation and the hot gas resist the inward pull of gravity. The stabilizing effect of these forces is shown in **Figure 3.** This equilibrium makes the star stable in size. A main-sequence star maintains a stable size as long as the star has an ample supply of hydrogen to fuse into helium.

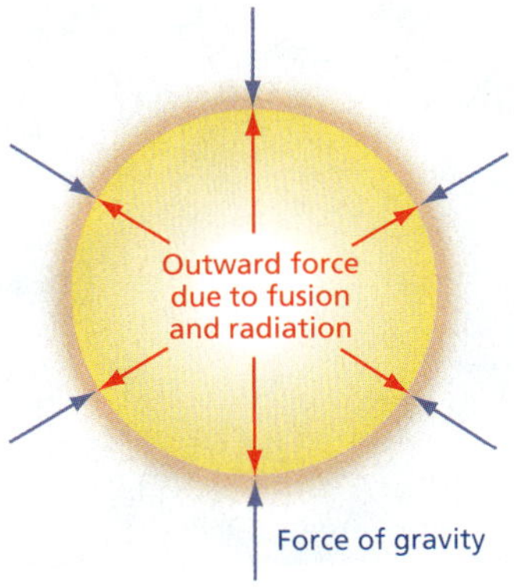

Figure 3 ▶ Stellar equilibrium is achieved when the inward force of gravity is balanced by the outward pressure from fusion and radiation inside the star.

Reading Check **How does the pressure from fusion and hot gas interact with the force of gravity to maintain a star's stability? (See the Appendix for answers to Reading Checks.)**

The Main-Sequence Stage

The second and longest stage in the life of a star is the main-sequence stage. During this stage, energy continues to be generated in the core of the star as hydrogen fuses into helium. Fusion releases enormous amounts of energy. For example, when only 1 g of hydrogen is converted into helium, the energy released could keep a 100 W light bulb burning for more than 200 years.

A star that has a mass about the same as the sun's mass stays on the main sequence for about 10 billion years. More-massive stars, on the other hand, fuse hydrogen so rapidly that they may stay on the main sequence for only 10 million years. Because the universe is about 14 billion years old, massive stars that formed long ago have long since left the main sequence. Less massive stars, which are at the bottom right of the main sequence on the H-R diagram, are thought to be able to exist for hundreds of billions of years.

The stages in the life of a star cover an enormous period of time. Scientists estimate that over a period of almost 5 billion years, the sun, shown in **Figure 4,** has converted only 5% of its original hydrogen nuclei into helium nuclei. After another 5 billion years, though, with 10% of the sun's original hydrogen converted, fusion will stop in the core. When fusion stops, the sun's temperature and luminosity will change and the sun will move off the main sequence.

Figure 4 ▶ Our sun is a yellow star. It is located in the diagonal band of main-sequence stars on the H-R diagram.

Using the Figure — GENERAL

Masses of Stars The mass of the material in a protostar determines what class of star will ultimately form. Have students look at the figure at the top of the page. Ask what would happen if a large amount of matter were in the protostar. (The gravitational force will be greater.) Ask how equilibrium is maintained in a more massive star. (More energy is produced within the star, so a greater outward pressure from radiation balances the greater inward pull of gravity.) Ask students what would happen if the mass of the protostar were smaller. (The gravitational force will be less. Less radiant energy will be needed and produced to balance the force of gravity.) Point out that nuclear fusion can occur only at pressures and temperatures equal to or greater than a certain minimum. This is why certain massive bodies, like the planet Jupiter, are not stars. **LS Visual**

Answer to Reading Check

The forces balance each other and keep the star in equilibrium. As gravity increases the pressure on the matter within a star, the rate of fusion increases. This increase in fusion causes a rise in gas pressure. As a result, the energy from the increased fusion and gas pressure generates outward pressure that balances the force of gravity.

Group Activity — ADVANCED

Main-Sequence Stars Separate students into small groups. Have each group research the properties and lifetimes of each class of main-sequence star. One student should research the range of masses for each class of star. Another should research the spectral classification system based on stellar colors and temperatures. A third student should learn how long the stars remain on the main sequence before entering more evolved stages. Each member should present his or her findings in a short written or oral report or as a poster. **LS Verbal/Logical**

INCLUSION Strategies

• Behavior Control Issues • Attention Deficit Disorder

Divide the class into small groups and assign one student who has behavior issues to be the "writer" for each group. Have each group read the section and summarize the most important idea for each of the section's subheadings. The writers should record the chosen ideas and each group should share its findings with the class.

Teach, *continued*

MISCONCEPTION ALERT

Shell and Core Fusion Students may think that the red-giant phase of stellar evolution occurs immediately as helium atoms in the core fuse into heavier carbon atoms. Suggest that students think of a star as having a dynamic, layered core, like an onion, whose composition evolves over time. Evolution between main-sequence stage and red-giant stage is a process that starts with the fusion of hydrogen into helium, a heavier element that "sinks" to the center of the star because of gravity. As helium, at much higher temperatures, fuses into carbon, the heavier carbon in turn sinks inward, to be surrounded by a shell of helium, which is surrounded by a shell of hydrogen.

Answer to Reading Check

Giants and supergiants appear in the upper-right part of the H-R diagram.

Using the Figure — BASIC

Identifying Stars Have students study the image of the constellation Orion. Ask why the blue-white stars in Orion are bright. (Blue-white stars are very hot and may be very luminous.) Ask why the red star Betelgeuse at the upper left of Orion is bright. (Betelgeuse is a red supergiant, so its size makes it very luminous.) Point out that the red object near the bottom of the photograph is the large Orion nebula, an emission nebula where stars are forming. Answer to caption question: The temperature of Betelgeuse, an orange-red star, is lower than that of the sun, a yellow star. **LS Visual**

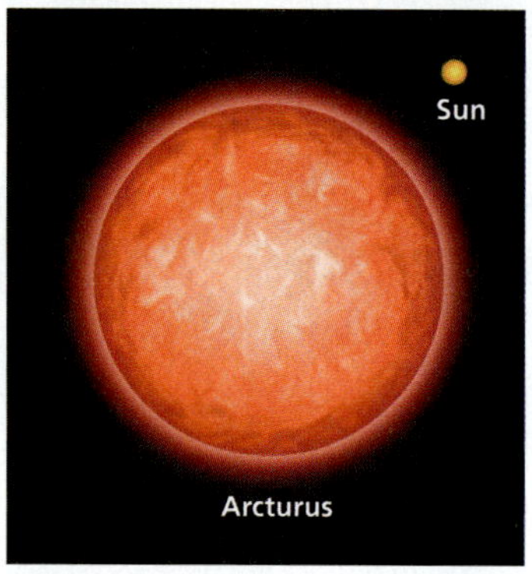

Figure 5 ▶ Arcturus is an orange giant that is about 23 times larger than the sun. Despite being about 1,000°C cooler than the sun, Arcturus gives off more than 100 times as much light as the sun does.

giant a very large and bright star whose hot core has used most of its hydrogen

Leaving the Main Sequence

A star enters its third stage when almost all of the hydrogen atoms within its core have fused into helium atoms. Without hydrogen for fuel, the core of a star contracts under the force of its own gravity. This contraction increases the temperature in the core. As the helium core becomes hotter, it transfers energy into a thin shell of hydrogen surrounding the core. This energy causes hydrogen fusion to continue in the shell of gas. The on-going fusion of hydrogen radiates energy outward, which causes the outer shell of the star to expand greatly.

Giant Stars

A star's shell of gases grows cooler as it expands. As the gases in the outer shell become cooler, they begin to glow with a reddish color. These large, red stars are known as **giants.**

Because of their large surface areas, giant stars are bright. Giants, such as the star Arcturus shown in **Figure 5,** are 10 or more times larger than the sun. Stars that contain about as much mass as the sun will become giants. As they become larger, more luminous, and cooler, they move off the main sequence. Giant stars are above the main sequence on the H-R diagram.

Supergiants

Main-sequence stars that are more massive than the sun will become larger than giants in their third stage. These highly luminous stars are called *supergiants*. These stars appear along the top of the H-R diagram. Supergiants are often at least 100 times larger than the sun. Betelgeuse, the large, orange-red star shown in **Figure 6,** is one example of a supergiant. Located in the constellation Orion, Betelgeuse is 1,000 times larger than the sun.

Though such supergiant stars make up only a small fraction of all the stars in the sky, their high luminosity makes the stars easy to find in a visual scan of the night sky. However, despite the high luminosity of supergiants, their surfaces are relatively cool.

Reading Check Where are giants and supergiants found on the H-R diagram? (See the Appendix for answers to Reading Checks.)

Figure 6 ▶ If the sun were replaced by the red supergiant Betelgeuse, the surface of this star would be farther out than Jupiter's orbit. ***How does the temperature of Betelgeuse compare with that of the sun?***

PHYSICS CONNECTION — GENERAL

Star Size and Spectra The absorption spectrum of a star is a ready indicator of whether the star is a main-sequence dwarf, a giant, or a supergiant. The atoms in the denser dwarf stars collide more often, causing a broadening of absorption lines. Collisions are less frequent in giant and supergiant stars, so the lines are thinner. Thus, the thickness of the absorption lines indicates the evolutionary stage of the star. Have students research how spectra are used to determine the size and temperature of stars. Have them present findings as a written or oral report or as a poster project. **LS Verbal**

The Final Stages of a Sunlike Star

In the evolution of a medium-sized star, fusion in the core will stop after the helium atoms have fused into carbon and oxygen. With energy no longer available from fusion, the star enters its final stages.

Planetary Nebulas

As the star's outer gases drift away, the remaining core heats these expanding gases. The gases appear as a *planetary nebula,* a cloud of gas that forms around a sunlike star that is dying. Some of these clouds may form a simple sphere or ring around the star. However, many planetary nebulas form more-complex shapes. For example, the Ant nebula has a double-lobed shape, as shown in **Figure 7.**

Figure 7 ▶ The Ant nebula is a planetary nebula that is located more than 3,000 light-years from Earth in the southern constellation Norma.

White Dwarfs

As a planetary nebula disperses, gravity causes the remaining matter in the star to collapse inward. The matter collapses until it cannot be pressed further together. A hot, extremely dense core of matter—a **white dwarf**—is left. White dwarfs shine for billions of years before they cool completely.

White dwarfs are in the lower left of the H-R diagram. They are hot but dim. These stars are very small, about the size of Earth. As white dwarfs cool, they become fainter. This is the final stage in the life cycle of many stars.

When a white dwarf no longer gives off light, the star will become a *black dwarf.* However, this process is long, and many astronomers do not believe that any black dwarf stars yet exist.

white dwarf a small, hot, dim star that is the leftover center of an old star

Connection to TECHNOLOGY

Searching for Extraterrestrial Life

In 1967, scientists discovered strange, regular pulses of radio waves coming from a specific point in space. Some scientists briefly thought these pulses might be coming from an intelligent source and called them LGMs, for *Little Green Men*. Further research showed that the source of these waves was a natural phenomenon, but the idea that there might be life in the universe continued to spur scientific interest.

In 1984, the SETI Institute was founded. The name SETI stands for the Search for Extraterrestrial Intelligence. SETI is dedicated to searching for evidence of extraterrestrial life and signs of alien intelligence.

The SETI program uses telescopes all over Earth to gather data. Many of these telescopes, such as the Arecibo Observatory in Puerto Rico, gather radio data, but optical searches are also performed.

Combing through all the data that the telescopes collect is no small job. In fact, the telescopes often collect more data than SETI's computers can process. In 1998, the SETI@home project was launched by scientists at the University of California at Berkeley. This program allows anyone with a computer and an internet connection to help process the data collected.

MISCONCEPTION ALERT

Planetary Nebulae While a planetary nebula is a true nebula (that is, a region of gas or dust, and not a distant grouping of stars), it has nothing to do with planets. The name has been handed down from the time these objects were first discovered. When observed through a small telescope, many planetary nebulae looked like dim, greenish, fuzzy disks, which is how distant large planets like Uranus and Neptune appeared through the same telescopes. The similarity in appearance to planets gave rise to the name "planetary" nebula.

READING SKILL BUILDER — BASIC

Paired Summarizing Group students into pairs, and have them read silently about white dwarfs. Then, have one student summarize how white dwarfs form, what properties they have, and how they ultimately evolve. The other student should listen to the retelling and should point out any inaccuracies or ideas that were left out. Allow students to refer to the text as needed. **LS Verbal** English Language Learners

Debate — ADVANCED

SETI Have interested students research and debate the advantages and disadvantages of the SETI program. Arguments in favor are the potential for increased knowledge and information exchange that could arise from the discovery of life on other worlds. Arguments against SETI are siphoning of resources away from more pressing research, the long time periods required for signals to travel through space, and the possibility that information may be missed. **LS Verbal/Logical**

BRAIN FOOD

White Dwarfs The mass of a white dwarf is typically about one solar mass. However, because the material in the star's core is so tightly compressed, the size of a white dwarf is roughly equal to that of Earth. The high density of a white dwarf is believed to make it an excellent energy insulator, so that the cooling from white dwarf to black dwarf stage is extremely slow, taking tens or even hundreds of billions of years.

Teach, *continued*

Graphic Organizer GENERAL

Chain-of-Events Chart
You may want to use this Graphic Organizer to assess students' prior knowledge before beginning a discussion of the evolution of a main-sequence star. You may also choose to use a similar activity as a quiz to assess students' understanding of how stars evolve after students have read the description of stellar evolution.

Using the Figure — BASIC

Identifying Stars Have students study the figure and then ask what causes the difference between the upper and lower paths. (The mass of the original nebula determines how massive the stars that form from it will be, and thus which evolutionary path the stars will undergo.) Point out that massive stars also go through several supergiant stages as fusion takes place in the carbon core and, in the more massive stars, in the various shells of heavier material. In the largest stars, the shells are like the layers of an onion, with the lighter outer shell producing material to fuel the shell below it. LS **Visual/Logical**

CHAPTER RESOURCES

Technology

Transparencies
- 152 The Life Cycle of Stars (with worksheet)

nova a star that suddenly becomes brighter

Graphic Organizer **Chain-of-Events Chart**
Create the **Graphic Organizer** entitled "Chain-of-Events Chart" described in the Skills Handbook section of the Appendix. Then, fill in the chart with details about each stage in the life cycle of a main-sequence star.

Novas and Supernovas

Some white dwarf stars are part of a binary star system. If a white dwarf revolves around a red giant, the gravity of the very dense white dwarf may capture gases from the red giant. As these gases accumulate on the surface of the white dwarf, pressure begins to build up. This pressure may cause large explosions, which release energy and stellar material into space, to occur. Such an explosion is called a **nova.**

A nova may cause a star to become many thousands of times brighter than it normally is. However, within days, the nova begins to fade to its normal brightness. Because these explosions rarely disrupt the stability of the binary system, the process may start again and a white dwarf may become a nova several times.

A white dwarf star in a binary system may also become a *supernova,* a star that has such a tremendous explosion that it blows itself apart. Unlike an ordinary nova, a white dwarf can sometimes accumulate so much mass on its surface that gravity overwhelms the outward pressure. The star collapses and becomes so dense that the outer layers rebound and explode outward. Supernovas are thousands of times more violent than novas. The explosions of supernovas completely destroy the white dwarf star and may destroy much of the red giant.

The Final Stages of Massive Stars

Stars that have masses of more than 8 times the mass of the sun may produce supernovas without needing a secondary star to fuel them. In 1054, Chinese astronomers saw a supernova so bright that it was seen during the day for more than three weeks. At its peak, the supernova radiated an amount of energy that was equal to the output of about 400 million suns.

Life Cycle of Stars

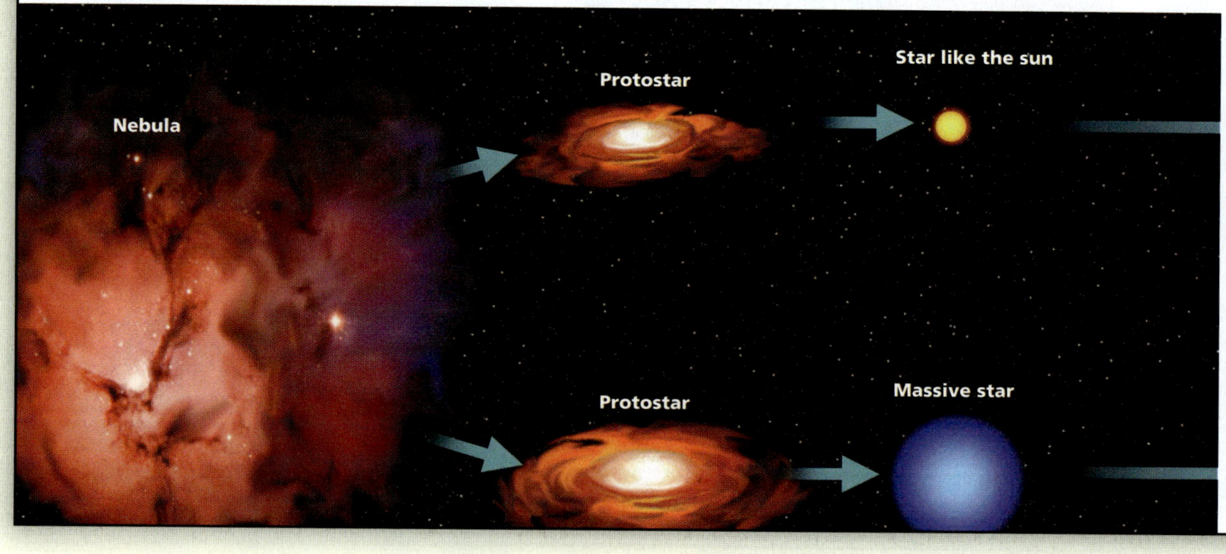

SKILL BUILDER GENERAL

Vocabulary The terms *nova, supernova,* and *nebula* are among the few pure Latin terms commonly used in astronomy. *Nova* is Latin for "new" and was applied to the bright stars that would suddenly appear in the night sky and fade after a few days. Because the process that causes novas to brighten was not understood, they were believed to be new stars. The prefix *super-* means "above" or "greater than," suggesting that these bright appearances were new stars with even greater brilliance. *Nebula* is the Latin word for "cloud." LS **Verbal**

Homework GENERAL

Star Remnants Have students research when the first white dwarf and neutron star were discovered, who made the discoveries, and by what means the discoveries were made. Students may present their findings in a short written report. LS **Verbal**

Supernovas in Massive Stars

While only a small percentage of white dwarfs become supernovas, massive stars become supernovas as part of their life cycle, which is shown in **Figure 8.** After the supergiant stage, these stars contract with a gravitational force that is much greater than that of small-mass stars. The collapse produces such high pressures and temperatures that nuclear fusion begins again. This time, carbon atoms in the core of the star fuse into heavier elements such as oxygen, magnesium, or silicon.

Fusion continues until the core is almost entirely made of iron. Because iron has a very stable nuclear structure, fusion of iron into heavier elements takes energy from the star rather than giving off energy. Having used up its supply of fuel, the core begins to collapse under its own gravity. Energy released as the core collapses is transferred to the outer layers of the star, which explode outward with tremendous force. Within a few minutes, the energy released by the supernova may surpass the amount of energy radiated by a sunlike star over its entire lifetime.

Reading Check **What causes a supergiant star to explode as a supernova? (See the Appendix for answers to Reading Checks.)**

SCILINKS® Developed and maintained by the National Science Teachers Association

For a variety of links related to this subject, go to www.scilinks.org

Topic: How Stars Evolve
SciLinksCode: HQ60764

Neutron Stars

Stars that contain about 8 or more times the mass of the sun do not become white dwarfs. After a star explodes as a supernova, the core may contract into a very small but incredibly dense ball of neutrons, called a **neutron star.** A single teaspoon of matter from a neutron star would weigh 100 million metric tons on Earth. A neutron star that has more mass than the sun may have a diameter of only about 20 km but may emit the same amount of energy as 100,000 suns. Neutron stars rotate very rapidly.

neutron star a star that has collapsed under gravity to the point that the electrons and protons have smashed together to form neutrons

Figure 8 ▶ A star the mass of the sun becomes a white dwarf near the end of its life cycle. A more massive star may become a neutron star.

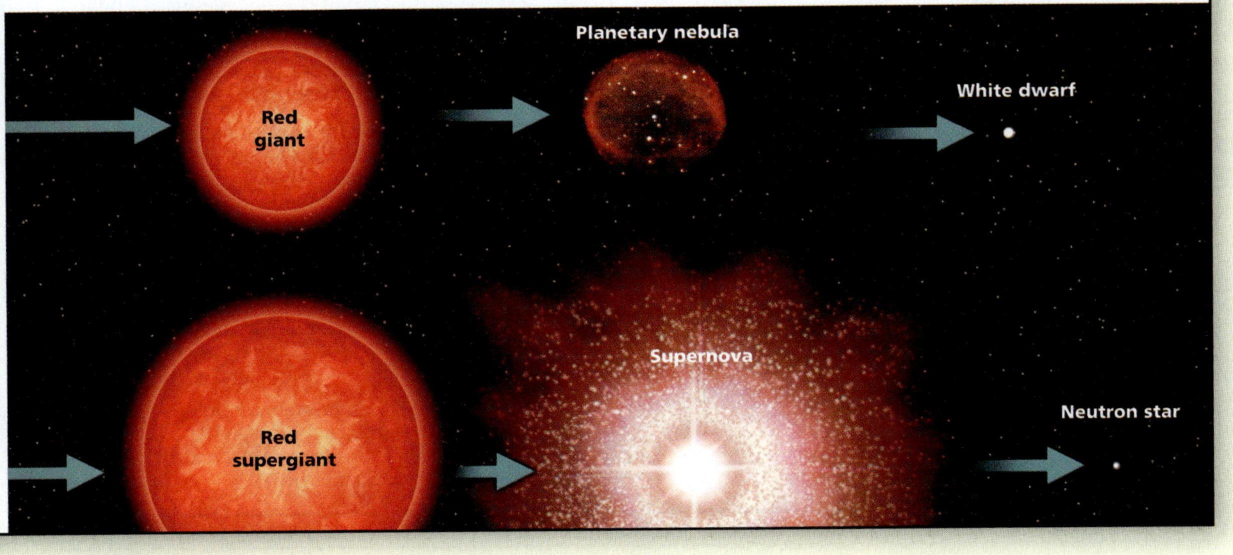

Answer to Reading Check

As supergiants collapse because of gravitational forces, fusion begins and continues until the supply of fuel is used up. The core begins to collapse under its own gravity and causes energy to transfer to the outer layers of the star. The transfer of energy to the outer layers causes the explosion.

Close

Reteaching — BASIC

Stellar Evolution On the board, write the words *red giant*, *main sequence*, *nebula*, *white dwarf*, *protostar*, and *planetary nebula*. Have students write these words on a sheet of paper and place each term in the correct order from earliest event to latest event. (From earliest to latest: nebula, protostar, main sequence, red giant, planetary nebula, white dwarf) **LS Logical**

Quiz — GENERAL

Determine whether each of the following statements is true or false.

1. Red stars are hotter and more luminous than blue stars. (false)
2. The force of gravity pulling a star's gases inward is balanced by the outward forces of gas pressure and radiation produced from fusion. (true)

Alternative Assessment — ADVANCED

Odd Stars Have students research the behavior of Wolf-Rayet stars and blue stragglers to learn how their evolution differs from the pattern of stellar evolution for a low-mass star. Students should note how each star's initial mass affects the unusual ways in which it evolves. Have students present their findings as a written report or oral presentation. **LS Logical**

PHYSICS CONNECTION — ADVANCED

Supernova Physics Supernovas are difficult to observe because they involve massive stars or massive close binary-star systems, both of which are relatively rare. Also, the supernova stage of a star represents a relatively short part of a star's life. The best-studied supernova is the event 1987A, a Type II (single massive star) supernova that occurred in the Large Magellanic Cloud, a small galaxy that orbits the Milky Way galaxy. Because the exploding star was relatively close and technology was available to study it, scientists obtained information about the physics of a supernova. One observation involved measuring the number of neutrinos—highly energetic and nearly massless particles that barely interact with matter—while the supernova was brightening. The two dozen neutrinos that were detected confirmed theoretical predictions about the nuclear processes that take place during a supernova. Have interested students research the various types of supernovas and the physical properties of each type. Students may present their findings in a written report or oral presentation. **LS Verbal**

Close, *continued*

Answers to Section Review

1. Gravity causes the gas and dust of a nebula to concentrate, flatten, and spin. The gas and dust become hotter as gravitational energy converts to thermal energy. The hydrogen nuclei at the center of the nebula undergo fusion, and the compact center of the nebula becomes a star.
2. During the fusion process, the tremendous pressure at the center of the star causes hydrogen nuclei to fuse and release energy as electromagnetic radiation.
3. The equilibrium between the inward force of gravity and the outward pressure from fusion and radiation makes the star stable in size.
4. Nuclear fusion in a main-sequence star takes place entirely within the star's core. In a giant star, fusion takes place in a shell of matter surrounding the core.
5. A star leaves the main sequence as hydrogen core fusion gives way to hydrogen shell fusion around the core. This causes the outer part of the star to expand and cool, so that the star becomes a giant.
6. When a white dwarf draws matter from its companion star, the pressure of the accumulated matter increases and may cause an explosion.
7. Only very massive stars have enough mass in their dense cores to prevent light or matter from escaping when the star dies.
8. A supernova may occur when a white dwarf accumulates matter from its larger companion star, and the star collapses and pushes matter outward in an explosive supernova. A supernova also may occur when a massive star whose core consists of iron that cannot fuse and produce energy to counter the weight of overlying layers collapses inward, and matter is pushed outward in a powerful explosion.
9. Because the supergiant star is much brighter, it must have a larger surface area.
10. Hydrogen fusion produces helium; so older stars will have fused more of their initial hydrogen supply into helium.
11. More mass increases the gravitational force exerted on gases, increasing pressure and temperature.
12. Matter near a black hole emits X rays when it is heated as it is pulled into the black hole.
13. Gases in a *nebula* contract to form a hot *protostar*, which may become a *main-sequence star*, which may evolve into a *giant* star, which may form a *planetary nebula*, until all that is left is a *white dwarf*, or into a *supergiant* star, which may explode in a *supernova* and leave behind a *black hole* or a *neutron star*, which may become a *pulsar*.

Figure 9 ▶ This pulsar, located in the heart of the Crab nebula, is still surrounded by the remains of a supernova explosion that took place less than 1,000 years ago.

Pulsars

Some neutron stars emit a beam of radio waves that sweeps across space like a lighthouse light beam sweeps across water. Because we detect pulses of radio waves every time the beam sweeps by Earth, these stars are called **pulsars.** For each pulse we detect, we know that the star has rotated within that period. Newly formed pulsars, such as the one shown in **Figure 9**, are commonly surrounded by the remnants of a supernova. But most known pulsars are so old that these remnants have long since dispersed and have left behind only the spinning star.

pulsar a rapidly spinning neutron star that emits pulses of radio and optical energy

black hole an object so massive and dense that even light cannot escape its gravity

Black Holes

Some massive stars produce leftovers too massive to become stable neutron stars. If the remaining core of a star contains more than 3 times the mass of the sun, the star may contract further under its greater gravity. The force of the contraction crushes the dense core of the star and leaves a **black hole.** The gravity of a black hole is so great that nothing, not even light, can escape it.

Because black holes do not give off light, locating them is difficult. But a black hole can be observed by its effect on a companion star. Matter from the companion star is pulled into the black hole. Just before the matter is absorbed, it swirls around the black hole. The gas becomes so hot that X rays are released. Astronomers locate black holes by detecting these X rays. Scientists then try to find the mass of the object that is affecting the companion star. Astronomers conclude that a black hole exists only if the companion star's motion shows that a massive, invisible object is present nearby.

Section 2 Review

1. **Explain** the steps that the gas in a nebula goes through as it becomes a star.
2. **Describe** the process that generates energy in the core of a main-sequence star.
3. **Explain** how a main-sequence star like the sun is able to maintain a stable size.
4. **Describe** how nuclear fusion in a main-sequence star is different from nuclear fusion in a giant star.
5. **Describe** how a star similar to the sun changes after it leaves the main-sequence stage of its life cycle.
6. **Describe** what causes a nova explosion.
7. **Explain** why only very massive stars can form black holes.
8. **Describe** the two types of supernovas.

CRITICAL THINKING

9. **Identifying Relationships** How do astronomers conclude that a supergiant star is larger than a main-sequence star of the same temperature?
10. **Analyzing Ideas** Why would an older main-sequence star be composed of a higher percentage of helium than a young main-sequence star?
11. **Compare and Contrast** Why does temperature increase more rapidly in a more massive protostar than in a less massive protostar?
12. **Analyzing Ideas** How can astronomers detect a black hole if it is invisible to a normal telescope?

CONCEPT MAPPING

13. Use the following terms to create a concept map: *main-sequence star, nebula, supergiant, white dwarf, planetary nebula, black hole, supernova, protostar, giant, pulsar,* and *neutron star.*

CHAPTER RESOURCES

Chapter Resource File

- Section Quiz GENERAL

Workbooks

- Study Guide (also in Spanish)

Section 3

Star Groups

When you look into the sky on a clear night, you see what appear to be individual stars. These visible stars are only some of the trillions of stars that make up the universe. Most of the ones we see are within 100 light-years of Earth. However, in the constellation Andromeda, there is a hazy region that is actually a huge collection of stars, gas, and dust. This region is more than two million light-years from Earth. It is the farthest one can see with the unaided eye.

OBJECTIVES

- **Describe** the characteristics that identify a constellation.
- **Describe** the three main types of galaxies.
- **Explain** how a quasar differs from a typical galaxy.

KEY TERMS

constellation
galaxy
quasar

Constellations

By using a star chart and observing carefully, you can identify many star groups that form star patterns or regions. Although the stars that make up a pattern appear to be close together, they are not all the same distance from Earth. In fact, they may be very distant from one another, as shown in **Figure 1.**

If you look at the same region of the sky for several nights, the positions of the stars in relation to one another do not appear to change. Because of the tremendous distance from which the stars are viewed, they appear fixed in their patterns. For more than 3,000 years, people have observed and recorded these patterns. These patterns of stars and the region of space around them are called **constellations.**

constellation one of 88 regions into which the sky has been divided in order to describe the locations of celestial objects; a group of stars organized in a recognizable pattern

Dividing Up the Sky

In 1930, astronomers around the world agreed upon a standard set of 88 constellations. The stars of these constellations and the regions around them divide the sky into sectors. Just as you can use a road map to locate a particular town, you can use a map of the constellations to locate a particular star. Star charts can be found in the Reference Map section of the Appendix.

Naming Constellations

Many of the modern names we use for the constellations come from Latin. Some constellations are named for real or imaginary animals, such as Ursa Major, which means "the great bear," and Draco, which means "the dragon." Other constellations are named for ancient gods or legendary heroes, such as Hercules and Orion. In some cases, smaller parts of constellations are well known and have their own names. The Big Dipper, for example, is a part of the constellation Ursa Major.

Figure 1 The stars that make up the constellation Orion appear close together when viewed from Earth. However, these stars are located at various distances from Earth and from each other.

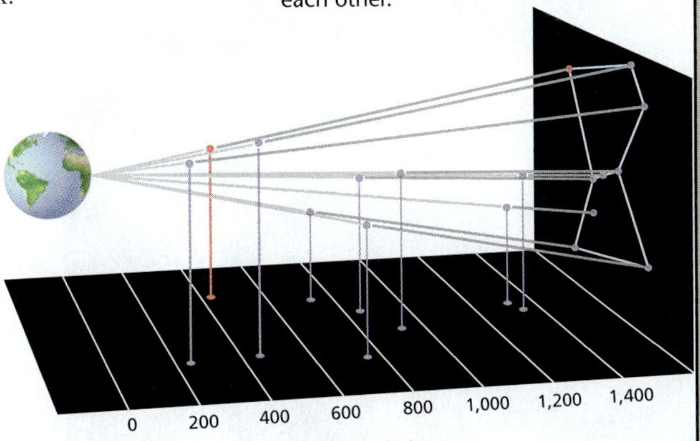

Section 3

Focus

Overview

This section describes groupings of stars, including constellations, star clusters, and galaxies. Types of galaxies and quasars are also discussed.

Bellringer

Ask students what they think a constellation is. (Sample answer: a group of stars that form a pattern) Have students write down the names of five constellations. (Sample answers: Ursa Major, Canis Major, Scorpius, Orion, Taurus, Pegasus, Gemini, Leo, Virgo, and Sagittarius) LS Verbal

Motivate

Activity — GENERAL

Constellations Have each student choose a constellation and research its name; its history, paying particular attention to the role of the constellation in agriculture, navigation, and cultural rituals; and its significance for cultures other than Greek or Roman. Students may present their findings as a written or oral report, a poster project, a skit, a poem, or a song. LS Verbal/Visual

CHAPTER RESOURCES

Chapter Resource File

- Directed Reading BASIC

Technology

Transparencies
- Bellringer
- 153 The Constellation Orion (with worksheet)

Student Edition on CD-ROM

One-Stop Planner CD-ROM
- Lesson Plan

Teach

SKILL BUILDER — ADVANCED

Writing Have students research the differences between the kinds of stars that make up open clusters and those that make up globular clusters. Also have them research where these two kinds of clusters exist relative to our galaxy, and how their locations relate to the process of galaxy formation. Students should then report their findings by writing a two- or three-page report. LS **Verbal**

Answer to Reading Check

More than 50% of all stars are in multiple-star systems.

MISCONCEPTION ALERT

Changing Constellations
Remind students that, because stars have their own motion, they will move toward or away from each other over thousands of years. Many constellations we see today will not be recognizable in 30,000 years.

BRAIN FOOD

Connect to Prior Knowledge
Galaxies range in size from a few thousand light-years across for some irregular galaxies to several hundred thousand light years across for giant elliptical galaxies. To help students appreciate the scale of these objects, explain to them that the Milky Way, our large spiral galaxy, is 100,000 light years across. The solar system, by contrast is 12 billion km (only a few light-hours) across if measured from one end of Pluto's orbit to the other. Usually, the larger diameter of 56 billion km is used, based on the point where the solar wind stops. Even using this larger distance, the Milky Way galaxy has a diameter equal to nearly 20 million solar systems.

Multiple-Star Systems

Stars are not always solitary objects isolated in space. When two or more stars are closely associated, they form multiple-star systems. *Binary stars* are pairs of stars that revolve around each other and are held together by gravity. In systems where the two stars have similar masses, the center of mass, or *barycenter,* will be somewhere between the stars. If one star is more massive than the other, the barycenter will be closer to the more massive star.

Multiple-star systems sometimes have more than two stars. In such a star system, two stars may revolve rapidly around a common barycenter, while a third star revolves more slowly at a greater distance from the pair. Astronomers estimate that more than half of all observed stars are part of multiple-star systems.

Reading Check What percentage of stars are in multiple-star systems? (See the Appendix for answers to Reading Checks.)

Star Clusters

Sometimes, nebulas collapse to form groups of hundreds or thousands of stars, called clusters. *Globular clusters* have a spherical shape and can contain up to 100,000 stars. An *open cluster,* such as the one shown in **Figure 2,** is loosely shaped and rarely contains more than a few hundred stars.

galaxy a collection of stars, dust, and gas bound together by gravity

Figure 2 ▶ The open cluster NGC 2516 is made up of about 100 stars. Located about 1,300 light-years from Earth, many of the stars in the cluster appear blue in color.

Galaxies

A large-scale group of stars, gas, and dust that is bound together by gravity is called a **galaxy.** Galaxies are the major building blocks of the universe. A typical galaxy, such as the Milky Way galaxy in which we live, has a diameter of about 100,000 light-years and may contain more than 200 billion stars. Astronomers estimate that the universe contains hundreds of billions of galaxies.

Distances to Galaxies

Some stars allow astronomers to find distances to the galaxies that contain the stars. For example, giant stars called *Cepheid* (SEF ee id) *variables* brighten and fade in a regular pattern. Most Cepheids have regular cycles that range from 1 to 100 days. The longer a Cepheid's cycle is, the brighter the star's visual absolute magnitude is. By comparing the Cepheid's absolute magnitude and the Cepheid's apparent magnitude, astronomers calculate the distance to the Cepheid variable. This distance, in turn, tells them the distance to the galaxy in which the Cepheid is located.

Cultural Awareness

The Pleiades The open cluster known as the *Pleiades*, which can be seen in the constellation Taurus during the Northern Hemisphere winter months, has long been one of the most recognizable star clusters in the sky. It is so close that its brightest stars can be seen individually, though overall it gives the impression of being a nebulous object. It is even classified in Charles Messier's catalogue of nebulae as *M45.*

Nearly every culture that has looked at the night sky has made some reference to the Pleiades. To the Greeks, they were the seven daughters of Atlas and Pleione. Early Hindus believed they were the Krittikas, the six nurses of the god of war. Different tribes of Aborigines in Australia assign different meanings to the stars, so that the cluster represents ancestral women, kangaroos, gum trees, and the realm of the dead. The Navajo called them the *Flint Boys*, who were raised to the heavens by the sky god. For the Inuit, the stars are called *Aggiattaat* and represent dogs that chased away a polar bear that threatened humanity. When the Aztecs viewed the Pleiades, the busy cluster reminded them of the market place.

Figure 3 ▶ The three main types of galaxies are spiral (left), elliptical (center), and irregular (right).

Types of Galaxies

In studying galaxies, astronomers found that galaxies could be classified by shape into the three main types shown in **Figure 3.** The most common type of large galaxy, called a *spiral galaxy,* has a nucleus of bright stars and flattened arms that spiral around the nucleus. The spiral arms consist of billions of young stars, gas, and dust. Some have a straight bar of stars that runs through the center. These galaxies are called *barred spiral galaxies*.

Galaxies of the second type vary in shape from nearly spherical to very elongated, like a stretched-out football. These galaxies are called *elliptical galaxies.* They are extremely bright in the center and do not have spiral arms. Elliptical galaxies have few young stars and contain little dust and gas.

The third type of galaxy, called an *irregular galaxy,* has no particular shape. These galaxies usually have low total masses and are fairly rich in dust and gas. Irregular galaxies make up only a small percentage of the total number of observed galaxies.

The Milky Way

If you look into the night sky, you may see what appears to be a cloudlike band that stretches across the sky. Because of its milky appearance, this part of the sky is called the Milky Way.

The *Milky Way galaxy* is a spiral galaxy in which the sun is one of hundreds of billions of stars. Each star orbits around the center of the Milky Way galaxy. It takes the sun about 225 million years to complete one orbit around the galaxy.

Two irregular galaxies, the Large Magellanic Cloud and Small Magellanic Cloud, are our closest neighbors. Even so, these galaxies are each more than 170,000 light-years away from Earth. Within 5 million light-years of the Milky Way are about 30 other galaxies. These galaxies and the Milky Way galaxy are collectively called the *Local Group*.

SCILINKS® NSTA Developed and maintained by the National Science Teachers Association

For a variety of links related to this subject, go to www.scilinks.org

Topic: Galaxies
SciLinksCode: HQ60632
Topic: Milky Way Galaxy
SciLinks code: HQ60964

READING SKILL BUILDER — BASIC

Anticipation Guide Before students read about the types of galaxies, ask them to predict the difference between spiral, barred spiral, elliptical, and irregular galaxies. Then, ask students to read silently the first half of the page to find out if their predictions were accurate. LS Verbal English Language Learners

Activity — GENERAL

Multimedia Project Have students research the three basic classes of galaxies. Students may wish to show the differences between the classes of galaxies by using drawings, photographs, or video footage. Have students display their projects in the classroom. LS Visual

Close

Reteaching — BASIC

Galaxy Types Draw examples or paste photographic reproductions of various galaxies onto flashcards. Show the cards in quick succession to students, and have students indicate whether the galaxy on the card is spiral, barred spiral, elliptical, or irregular. You may choose to separate the class into teams and let them compete to see who can answer most accurately and quickly. LS Logical

ASTRONOMY CONNECTION — ADVANCED

Identifying a Galaxy from Within Because Earth is part of the Milky Way galaxy, it is impossible for us to say for certain what the galaxy looks like from an outside point of view. Dust in the center of the galaxy obscures the view of the spiral arms opposite our solar system. This area is gradually being explored by monitoring forms of nonvisible electromagnetic energy. Recent data suggest that the Milky Way, long thought to be a "twin" of the spiral galaxy M31, may be a barred spiral galaxy.

Close, *continued*

Quiz — GENERAL

1. How does a constellation differ from a star cluster? (A constellation is an arrangement of stars into a pattern as seen from Earth at a particular time. A star cluster is a group of stars located near each other that formed when a nebula collapsed.)
2. What type of galaxy has flattened arms that extend outward from a central nucleus? (a spiral galaxy)
3. What type of galaxy has no particular shape and is rich in dust and gas from which new stars form? (an irregular galaxy)

Alternative Assessment — ADVANCED

Modeling Constellations Have students create a clay or plaster model that shows the actual relative positions of the stars in a chosen constellation. Students will need to scale the model so that the great distances between stars are manageable.
LS Kinesthetic/Logical

Answers to Section Review

1. A constellation is a group of stars organized into a recognizable pattern that occupies one of 88 regions into which the sky has been divided. Although the stars may appear close together, they are different distances from Earth.
2. spiral, elliptical, and irregular
3. It is a spiral galaxy, because it has flattened arms that spiral around a bright nucleus of stars.
4. A quasar is a very bright feature in some galaxies that produces intense radio emissions and has the appearance of a bright star. Typical galaxies do not emit such intense radio signals.
5. The pattern of stars in a constellation is based on the way they appear against the sky. To a viewer, the stars all appear to be in a plane the same distance from Earth along the line of sight, even though they are different distances from Earth.
6. $(4.6 \times 10^9$ y$)/(2.25 \times 10^8$ y/revolution$)$ = 20.4 revolutions
7. Earth is on the opposite side of the sun during winter, so Earth faces a different part of the universe during night in the winter.
8. A *galaxy* may be an *elliptical galaxy;* an *irregular galaxy;* a *spiral galaxy,* such as the *Milky Way galaxy,* which may be or a *barred spiral galaxy.*

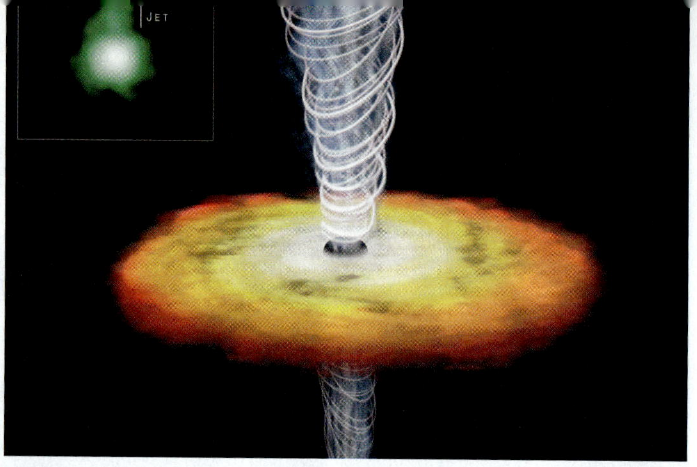

Figure 4 ▶ The jets of gas projected from a quasar can extend for more than 100,000 light-years. Quasars are too distant to be clearly photographed. The image shown is an artist's rendition of a quasar. The smaller inset is an actual image taken by the Chandra X-Ray Observatory.

Quasars

quasar quasi-stellar radio source; a very luminous object that produces energy at a high rate

First discovered in 1963, quasars used to be the most puzzling objects in the sky. Viewed through an optical telescope, a quasar appears as a point of light, almost in the same way that a small, faint star would appear. The word **quasar** is a shortened term for *quasi-stellar radio source*. The prefix *quasi-* means "similar to," and the word *stellar* means "star." Quasars are not related to stars, but quasars are related to galaxies. Some quasars project a jet of gas, as shown in **Figure 4.**

Astronomers have discovered that quasars are located in the centers of galaxies that are distant from Earth. Galaxies that have quasars in them differ from other galaxies in that the quasars in their centers are very bright. The large amount of energy emitted from such a small volume could be explained by the presence of a giant black hole. The mass of such black holes is estimated to be billions of times the mass of our sun. Quasars are among the most distant objects that have been observed from Earth.

Section 3 Review

1. **Identify** the characteristics of a constellation.
2. **List** the three basic types of galaxies.
3. **Describe** the Milky Way galaxy in terms of galaxy types.
4. **Describe** the difference between an typical galaxy and a quasar.

CRITICAL THINKING

5. **Identifying Relationships** Explain how stars can form a constellation when seen from Earth but can still be very far from each other.
6. **Making Calculations** The sun orbits the center of the Milky Way galaxy every 225 million years. How many revolutions has the sun made since the formation of Earth 4.6 billion years ago?
7. **Analyzing Ideas** Why are the constellations that are seen in the winter sky different from those seen in the summer sky?

CONCEPT MAPPING

8. Use the following terms to create a concept map: *galaxy, elliptical galaxy, Milky Way galaxy, irregular galaxy, barred spiral galaxy,* and *spiral galaxy.*

CHAPTER RESOURCES

Chapter Resource File
- Section Quiz GENERAL

Workbooks
- Study Guide (also in Spanish)

Section 4 The Big Bang Theory

The study of the origin, structure, and future of the universe is called **cosmology.** Cosmologists, or people who study cosmology, are concerned with processes that affect the universe as a whole. Like the parts found within it, the universe is always changing. While some astronomers study how planets, stars, or galaxies form and evolve, a cosmologist studies how the entire universe formed and tries to predict how it will change in the future.

Like all scientific theories, theories about the origin and evolution of the universe must constantly be tested against new observations and experiments. Many current theories of the universe began with observations made less than 100 years ago.

OBJECTIVES

- **Explain** how Hubble's discoveries lead to an understanding that the universe is expanding.
- **Summarize** the big bang theory.
- **List** evidence for the big bang theory.

KEY TERMS

cosmology
big bang theory
cosmic background radiation

Hubble's Observations

Just as the light from a single star can be used to make a stellar spectrum, scientists can also use the light given off by an entire galaxy to create the spectrum for that galaxy. In the early 1900s, finding the spectrum of a galaxy could take the whole night, or even several nights. Although collecting new spectra was very time consuming, the astronomer Edwin Hubble used these galactic spectra to uncover new information about our universe.

cosmology the study of the origin, properties, processes, and evolution of the universe

Measuring Red Shifts

Near the end of the 1920s, Hubble found that the spectra of galaxies, except for the few closest to Earth, were shifted toward the red end of the spectrum. By examining the amount of red shift, he determined the speed at which the galaxies were moving away from Earth. Hubble found that the most distant galaxies showed the greatest red shift and thus were moving away from Earth the fastest.

Many distant galaxies are shown in **Figure 1.** Modern telescopes that have electronic cameras can take images of hundreds of spectra per hour. To date, these spectra all confirm Hubble's original findings.

Figure 1 ▶ This image from the *Hubble Space Telescope* shows hundreds of galaxies. These galaxies all have large red shifts, so they are moving away from Earth very fast.

Section 4

Focus

Overview

This section describes how cosmologists study processes that affect the universe as a whole and explains current theories about the origin of the universe.

Bellringer

Ask students how they think the universe formed. (Answers may vary.) LS Logical/Verbal

Motivate

Discussion — GENERAL

The Expanding Universe Ask students to recall how the Doppler effect relates to light from a moving object. Ask how they could tell if all galaxies were moving toward the observer. (The light from all galaxies would be shifted toward the blue end of the spectrum.) Ask how they could tell if all galaxies were moving away from the observer. (The light from all galaxies would be shifted toward the red end of the spectrum.) Ask students to explain which condition corresponds to an expanding universe, and why. (a red shift; All galaxies moving away from the observer and each other would indicate that the space between them was growing larger, and therefore the universe was expanding.) LS Visual

CHAPTER RESOURCES

Chapter Resource File
- Directed Reading BASIC

Technology

Transparencies
- Bellringer

Student Edition on CD-ROM

One-Stop Planner CD-ROM
- Lesson Plan

Teach

MISCONCEPTION ALERT

Expanding Space-Time Students may think the expansion of the universe is simply a matter of the volume of space increasing. Point out that the situation is more complicated. According to the theory of general relativity, which provides the modern description for gravitation, the three spatial dimensions and time are treated as related and inseparable dimensions. To help simplify the process, it is easier for students to think of space in terms of just two spatial dimensions, like a sheet of paper, and then use these in conjunction with time. The expanding universe for this model does resemble a ball or balloon that is getting larger in time, which corresponds to the radius of the ball. The change in space occurs only on the ball's surface.

Answer to Reading Check

All matter and energy in the early universe was compressed into a small volume at an extremely high temperature until the temperature cooled and all of the matter and energy were forced outward in all directions.

Figure 2 ▶ The farther away one raisin is from another raisin, the faster they move away from each other. Similarly, the farther galaxies are from each other, the faster they move away from each other.

The Expanding Universe

Imagine a raisin cake rising in a kitchen oven. If you were able to sit on one raisin, you would see all the other raisins moving away from you, as shown in **Figure 2.** Raisins that are farther away in the dough when it begins rising move away faster because there is more cake between you and them and because the whole cake is expanding. The situation is similar with galaxies and the universe. By using Hubble's observations, astronomers were able to determine that the universe was expanding.

The Big Bang Theory Emerges

Although cosmologists have proposed several different theories to explain the expansion of the universe, the current and most widely accepted is the big bang theory. The **big bang theory** states that billions of years ago, all the matter and energy in the universe was compressed into an extremely small volume. If you trace the expanding universe back in time, all matter would have been close together at one point in time. About 14 billion years ago, a sudden event called the *big bang* sent all of the matter and energy outward in all directions.

As the universe expanded, some of the matter gathered into clumps that evolved into galaxies. Today, the universe is still expanding, and the galaxies continue to move apart from one another. This expansion of space explains the red shift that we detect in the spectra of galaxies. **Figure 3** shows a timeline of events following the big bang.

By the mid-20th century, almost all astronomers accepted the big bang theory. An important discovery in the 1960s finally convinced most of the remaining scientists that a sudden event, the big bang, had taken place.

Reading Check **What does the big bang theory tell us about the early universe? (See the Appendix for answers to Reading Checks.)**

big bang theory the theory that all matter and energy in the universe was compressed into an extremely small volume that 13 to 15 billion years ago exploded and began expanding in all directions

Figure 3 ▶ Following the big bang, matter and energy began to take shape, but the universe as we know it did not begin forming until the temperature cooled by many billions of degrees.

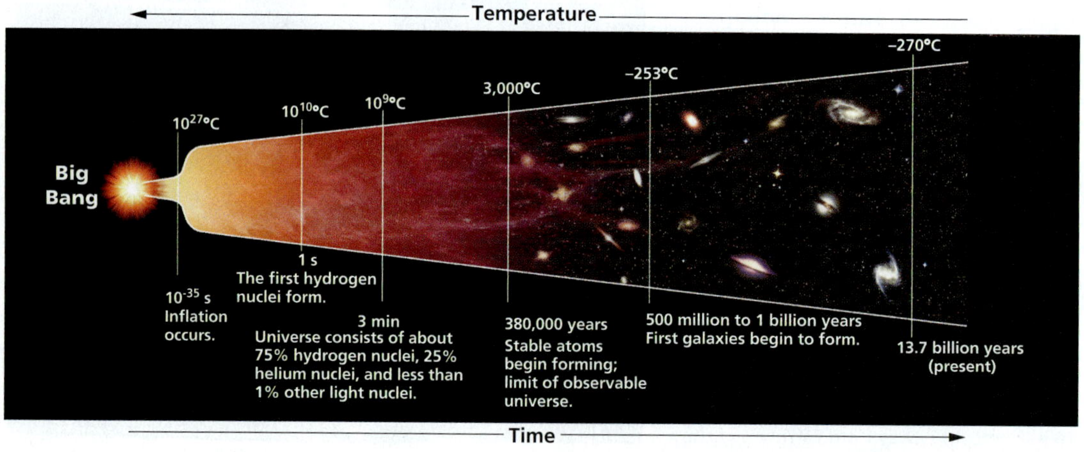

PHYSICS CONNECTION ADVANCED

New Physics, Old Universe Cosmology is an evolving and theoretical science. Many discoveries were made in just the last century. These discoveries include Einstein's work on general relativity and the quantum mechanical model for atomic and subatomic matter. More radical physics such as string theory, "brane" (membrane) theory, and multiple universes are part of cosmology. Have students research previous scientific models of the universe leading to the current model. Have them present their findings in a written report or oral presentation. LS **Verbal**

INCLUSION Strategies

- *Learning Disabled*
- *Hearing Impaired*
- *Developmentally Delayed*

To help students identify important information, use this summarizing technique. As a group, read a paragraph. Ask a volunteer to summarize the paragraph in one sentence. Adjust the sentence so it is a good summary sentence. Have students write the sentence. Continue until all paragraphs have been read and summarized, and guide students to use the sentences when they study. LS **Verbal**

QuickLAB 15 minutes

The Expanding Universe

Procedure

1. Use a **marker** to make 3 dots in a row on an noninflated **balloon.** Label them "A," "B," and "C." Dot B should be closer to A than dot C is to B.
2. Blow the balloon up just until it is taut. Pinch the balloon to keep it inflated, but do not tie the neck.
3. Use **string** and a **ruler** to measure the distances between A and B, B and C, and A and C.
4. With the balloon still inflated, blow into the balloon until its diameter is twice as large.
5. Measure the distances between A and B, B and C, and A and C. For each set of dots, subtract the original distances measured in step 3 from the new distances. Then, divide by 2, because the balloon is about twice as large. This calculation will give you the rate of change for each pair of dots.
6. Repeat steps 4 and 5.

Analysis

1. Did the distance between A and B, between B and C, or between A and C show the greatest rate of change?
2. Suppose dot A represents Earth and that dots B and C represent galaxies. How does the rate at which galaxies are moving away from us relate to how far they are from Earth?

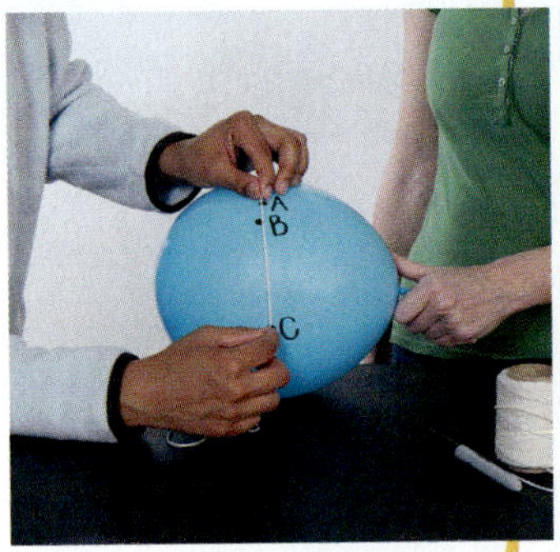

Cosmic Background Radiation

In 1965, researchers using radio telescopes detected **cosmic background radiation,** or low levels of energy evenly distributed throughout the universe. Astronomers think that this background radiation formed shortly after the big bang.

The universe soon after the big bang would have been very hot and would have cooled to a great extent by now. The energy of the background radiation has a temperature of only about 3°C above *absolute zero,* the coldest temperature possible. Because absolute zero is about –273°C, the cosmic background radiation's temperature is about 270°C below zero.

Like any theory, the big bang theory must continue to be tested against each new discovery about the universe. As new information emerges, the big bang theory may be revised, or a new theory may become more widely accepted.

cosmic background radiation radiation uniformly detected from every direction in space; considered a remnant of the big bang

Ripples in Space

Maps of cosmic background radiation over the whole sky look very smooth. But on a map that shows where temperatures differ from the average background temperature, "ripples" become apparent, as shown in **Figure 4.** These ripples are irregularities in the cosmic background radiation, which were caused by small fluctuations in the distribution of matter in the early universe. The ripples may indicate the first stages in the formation of the universe's first galaxies.

Figure 4 ▶ This display is shown on half a globe that represents the sky as seen from Earth. The temperature difference between the red spots and the blue spots is only 2/10,000°C.

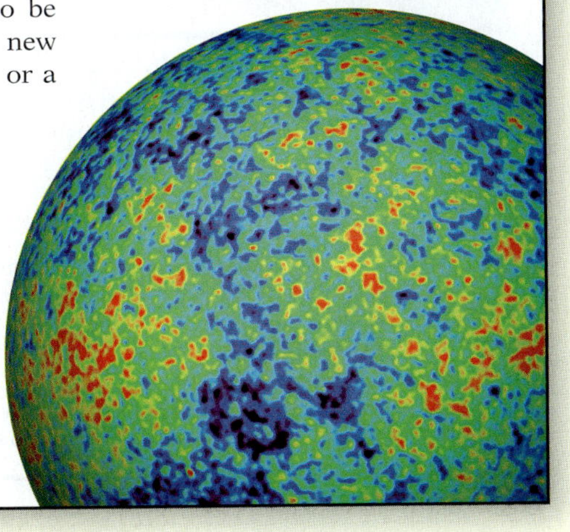

Close

Reteaching — BASIC

Timeline Have students create a timeline that indicates the major events in the development of the big bang model of the universe. Allow students to refer to the text as needed. LS **Logical**

Quiz — GENERAL

1. What indicates that the universe is expanding? (Almost all galaxies have spectra that are shifted toward the red end of the spectrum and thus are moving away from Earth and each other at a rate that increases with distance.)
2. Why is cosmic background radiation important to cosmological research? (Cosmic background radiation indicates that the universe has cooled over the last 14 billion years, and thus provides support for an initially hot, compact, outwardly-expanding universe.)
3. Why is dark matter an important concept? (Most matter that holds the universe together by gravity is invisible. Dark matter helps account for the estimated mass of the universe.)

CHAPTER RESOURCES

Chapter Resource File

- Datasheet for Quick Lab GENERAL

QuickLAB

Skills Acquired

- Experimenting
- Observing
- Interpreting

Teacher's Notes Be sure that students do not overinflate the balloon. Adhesive paper dots may be used instead of ink dots. To extend the lab, repeat steps 4 and 5 and chart or graph the different rates of change and physical changes in distance.

Answers

1. The distance between A and C showed the greatest change.
2. The faster a galaxy moves away from Earth, the farther the galaxy is from Earth.

Close, *continued*

Alternative Assessment — ADVANCED

Rate of Universal Expansion Have students research the Hubble constant, which is a measure of both the speed at which the universe expands and the age of the universe. Students should note why the value of this "constant" has changed, how stellar evolution limits the value that it can have, and how the Hubble constant relates to the big bang theory and the current "dark-energy" model of the universe. Students should present their findings in a written or oral report or poster project. **LS Verbal/Logical**

Answers to Section Review

1. The spectra of light from most galaxies were Doppler-shifted toward the red end, indicating that these galaxies were moving away from our galaxy.
2. All of the matter and energy in the universe was compressed into an extremely small volume that, about 13 to 15 billion years ago, exploded and began expanding in all directions.
3. cosmological red shifts, cosmic background radiation, and "ripples" in space
4. Only about 4% of matter and energy is visible.
5. The uneven distribution of matter in the early universe has caused subtle temperature ripples to form in the cosmic background radiation.
6. Because the universe is expanding outward, galaxies move away from each other on the large scale, even if galaxies within a group locally move toward each other. As a result, the spectra of all galaxies will show a cosmological red shift, not a blue shift. This effect is greater for more distant galaxies.
7. The observations of distant galaxies indicate that those galaxies are accelerating outward. The concept of dark energy is needed to help account for this acceleration.
8. Evidence for the *big bang theory* is found in the presence of *cosmic background radiation*, by the *red shift* in the spectra of *galaxies*, and the concepts of *dark energy* and *dark matter*.

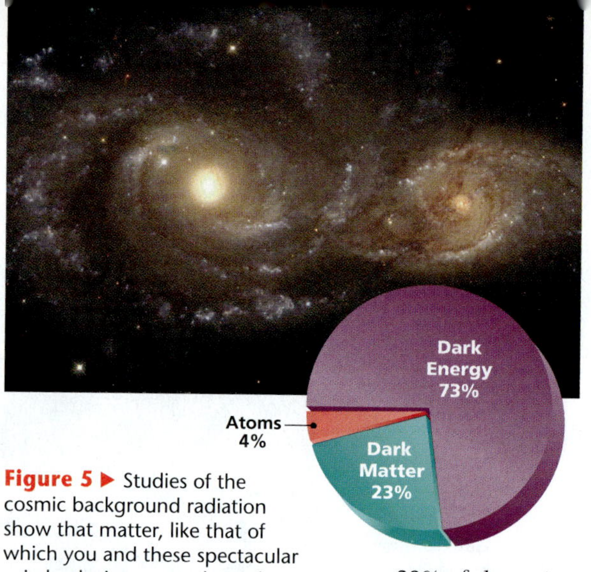

Figure 5 ▶ Studies of the cosmic background radiation show that matter, like that of which you and these spectacular spiral galaxies are made, makes up only 4% of the universe.

A Universe of Surprises

Recent data based on the ripples in the cosmic background radiation and studies of the distance to supernovas found in ancient galaxies have forced astronomers to rethink some of the theories about what makes up the universe. Astronomers now think that the universe is made up of more mass and energy than they can currently detect.

Dark Matter

Surprisingly, analyzing the ripples in the cosmic background radiation tells us that the kinds of matter that humans, the planets, the stars, and the matter between the stars are made of makes up only 4% of the universe, as shown in **Figure 5.** Another 23% of the universe is made up of a type of matter that does not give off light but that has gravity that we can detect. Because this type of matter does not give off light, it is called *dark matter*.

Dark Energy

Another surprise is that most of the universe is composed of something that we know almost nothing about. The unknown material is called *dark energy*, and scientists think that it acts as a force that opposes gravity. Recent evidence suggests that distant galaxies are farther from Earth than current theory would indicate. So, many scientists conclude that some form of undetectable dark energy is pushing galaxies apart. Because of dark energy, the universe is not only expanding, but the rate of expansion also seems to be accelerating.

SCILINKS. NSTA Developed and maintained by the National Science Teachers Association

For a variety of links related to this subject, go to www.scilinks.org

Topic: Big Bang
SciLinks code: HQ60146

Section 4 Review

1. **Describe** how red shifts were used by cosmologists to determine that the universe is expanding.
2. **Summarize** the big bang theory.
3. **List** evidence that supports the big bang theory.
4. **Compare** the amount of visible matter in the universe with the total amount of matter and energy.

CRITICAL THINKING

5. **Inferring Relationships** How did the distribution of matter in the early universe affect how we are able to detect cosmic background radiation today?
6. **Evaluating Theories** Use the big bang theory to explain why scientists do not expect to find galaxies that have large blue shifts.
7. **Identifying Relationships** Why do observations made of distant galaxies indicate that dark energy exists?

CONCEPT MAPPING

8. Use the following terms to create a concept map: *cosmic background radiation, dark energy, red shift, dark matter, big bang theory,* and *galaxies*.

CHAPTER RESOURCES

Chapter Resource File
- Section Quiz GENERAL

Workbooks
- **Study Guide** (also in Spanish)

Chapter 30 Highlights

Sections

1 Characteristics of Stars

Key Terms

- **star,** 775
- **Doppler effect,** 778
- **light-year,** 779
- **parallax,** 779
- **apparent magnitude,** 780
- **absolute magnitude,** 780

Key Concepts

- To determine the composition and surface temperature of a star, astronomers study the spectrum of the star.
- Stars appear to circle Polaris each night and appear to move westward across the sky on successive nights.
- To measure the distance to a star, astronomers use direct and indirect methods.

2 Stellar Evolution

Key Terms

- **main sequence,** 781
- **nebula,** 782
- **giant,** 784
- **white dwarf,** 785
- **nova,** 786
- **neutron star,** 787
- **pulsar,** 788
- **black hole,** 788

Key Concepts

- Plotting stars by temperature and luminosity on the H-R diagram groups stars by their current stage in their life cycles.
- A protostar becomes a star when hydrogen begins to fuse into helium.
- The main-sequence stage is the longest and most stable stage for most stars.
- A red giant is a large, relatively cool star that has a core in which helium fusion is occurring.

3 Star Groups

Key Terms

- **constellation,** 789
- **galaxy,** 790
- **quasar,** 792

Key Concepts

- A constellation is a region of the sky that contains a recognizable star pattern and is used to locate celestial objects.
- Astronomers have identified three main types of galaxies: spiral, elliptical, irregular.
- Quasars are very bright, distant galaxies that are thought to have enormous black holes in their centers.

4 The Big Bang Theory

Key Terms

- **cosmology,** 793
- **big bang theory,** 794
- **cosmic background radiation,** 795

Key Concepts

- The red shifts of distant galaxies show that the universe is expanding.
- Tracing the expansion backward indicates that everything in the universe was close together about 14 billion years ago.
- Tiny ripples in the temperature of the cosmic background radiation hint that ordinary matter makes up only a small percentage of the universe.

Chapter Highlights

Alternative Assessment — GENERAL

Poster Project Have students create a poster that shows the different classes of stars and their properties, as well as the types of star clusters or regions in galaxies in which different stars are most likely to be located. Posters should show the changing composition and physical evolution of the star. LS Logical

CHAPTER RESOURCES

Chapter Resource File

- Concept Review GENERAL
- Critical Thinking ADVANCED
- Math Skills GENERAL
- Graphing Skills GENERAL
- Chapter Test A GENERAL
- Chapter Test B ADVANCED

Workbooks

- Study Guide (also in Spanish)
- Assessments (Spanish)

Technology

Classroom Videos
- Brain Food Video Quiz

Chapter Review

Assignment Guide

Section	Questions
1	1, 8, 10, 12, 14, 16, 22–24, 30
2	11, 13, 15, 17–18, 25–26, 33–35
3	4–6, 19, 29, 32
4	2–3, 7, 9, 20–21, 27
1–3	28
2 and 3	31

Using Key Terms

1–7. Answers may vary but should show that students understand the definitions of and differences between key terms.

Understanding Key Concepts

8. b **9.** d
10. a **11.** c
12. d **13.** c
14. d **15.** a
16. a

Short Answer

17. The sun will enter a red giant stage, becoming larger and cooler.

18. The X rays emitted by material being pulled into the black hole provide a clue as to the black hole's whereabouts. The source of the material is usually a companion star, so by studying the motion of that star, the mass of the invisible object can be determined and compared to the mass necessary for a star to become a black hole.

19. Galaxies that have quasars at their centers are very bright because of the large amount of energy emitted by the quasar.

20. The red shift of the spectra of all galaxies indicates that the universe is expanding. All galaxies are moving away from the Milky Way and from each other at a rate that increases with distance.

21. The cosmic background radiation indicates that the universe was once much more compact and much hotter, and that it expanded, cooling in the process. This is the behavior predicted by the big bang theory.

Chapter 30 Review

Using Key Terms

Use each of the following terms in a separate sentence.

1. *light-year*
2. *cosmology*
3. *big bang theory*

For each pair of terms, explain how the meanings of the terms differ.

4. *constellation* and *cluster*
5. *spiral galaxy* and *elliptical galaxy*
6. *galaxy* and *quasar*
7. *cosmic background radiation* and *red shift*

Understanding Key Concepts

8. The most common element in most stars is
- **a.** oxygen.
- **b.** hydrogen.
- **c.** helium.
- **d.** sodium.

9. Cosmic background radiation
- **a.** is very hot.
- **b.** is blue-green.
- **c.** comes from supernovas.
- **d.** comes equally from all directions.

10. Stars appear to move in circular paths through the sky because
- **a.** Earth rotates on its axis.
- **b.** Earth orbits the sun.
- **c.** the stars orbit Polaris.
- **d.** the Milky Way is a spiral galaxy.

11. A nebula begins the process of becoming a protostar when the nebula
- **a.** develops a red shift.
- **b.** changes color from red to blue.
- **c.** begins to shrink and increases its spin.
- **d.** explodes as a nova.

12. The brightest star in the night sky is
- **a.** Polaris.
- **b.** Mars.
- **c.** Arcturus.
- **d.** Sirius.

13. A main-sequence star generates energy by fusing
- **a.** nitrogen into iron.
- **b.** helium into carbon.
- **c.** hydrogen into helium.
- **d.** nitrogen into carbon.

14. Which of the following choices lists the colors of stars from hottest to coolest?
- **a.** red, yellow, orange, white, blue
- **b.** orange, red, white, blue, yellow
- **c.** yellow, orange, red, blue, white
- **d.** blue, white, yellow, orange, red

15. The heaviest element formed in the core of a star is
- **a.** iron.
- **b.** carbon.
- **c.** helium.
- **d.** nitrogen.

16. The change in position of a nearby star as seen from different points on Earth's orbit compared with the position of a faraway star is called
- **a.** parallax.
- **b.** blue shift.
- **c.** red shift.
- **d.** a Cepheid variable.

Short Answer

17. Describe what scientists think will happen to the sun in the next 5 billion years.

18. How can a black hole be detected if it is invisible?

19. How does a galaxy that contains a quasar differ from an ordinary galaxy?

20. What evidence indicates that the universe is expanding?

21. How does the presence of cosmic background radiation support the big-bang theory?

Critical Thinking

22. The approximate surface temperature is less than 3,500°C.

23. During each season, Earth is at a different position relative to the sun. Thus, in each season, the night sky faces a different direction from the sun, and therefore different constellations are viewed.

24. Polaris is almost directly over Earth's North Pole and does not appear to move around the North Pole, as do the other stars. Polaris is bright enough that it can be used to find true north.

Critical Thinking

22. **Inferring Relationships** If the spectrum of a star indicates that the star shines with red light, what is the approximate surface temperature of the star?

23. **Analyzing Ideas** Why are different constellations visible during different seasons?

24. **Analyzing Ideas** Explain why Polaris is considered to be a very significant star even though it is not the brightest star in Earth's sky.

25. **Making Comparisons** Why does energy build up more rapidly in a massive protostar than in a less massive one?

26. **Analyzing Ideas** Explain why an old main-sequence star is made of a higher percentage of helium than a young main-sequence star is.

27 **Analyzing Ideas** If all galaxies began to show blue shifts, what would this change indicate about the fate of the universe?

Concept Mapping

28. Use the following terms to create a concept map: *galaxy, star, black hole, white dwarf, neutron star, giant, spiral galaxy, supergiant, elliptical galaxy, planetary nebula, main sequence, irregular galaxy,* and *protostar*.

Math Skills

29. **Making Calculations** The Milky Way galaxy has about 200 billion stars. If only 10% of an estimated 125 billion galaxies thought to exist in the universe were as large as the Milky Way, how many total stars would be in those galaxies?

30. **Making Calculations** Given that the nearest star is about 4 light-years from Earth and a light-year is about 10,000,000,000,000 km, how many years would it take to travel to the nearest star if your spaceship goes 100 times faster than a car traveling 100 km/h?

Writing Skills

31. **Creative Writing** Imagine that you are navigating through the galaxy and seeing many kinds of objects. Write a brief tour article for a magazine that describes your trip.

32. **Writing from Research** Use the Internet and library resources to research the function of constellations in ancient cultures. Write a short essay describing three different ways that ancient cultures used constellations.

Interpreting Graphics

The graph below shows the relationship between a star's age and mass. Use the graph to answer the questions that follow.

Relationship Between Age and Mass of a Star

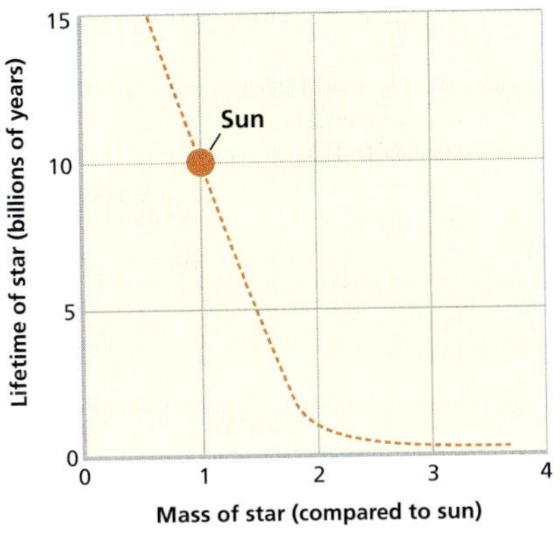

33. Which star would live longer, a star that has half the mass of the sun or a star that has 2 times the mass of the sun?

34. Approximately how long would a main-sequence star that has a mass about 1.5 times that of our sun live?

35. If the mass of the sun was reduced by one-half, approximately how much longer would the sun live than it would with its current mass?

Chapter Review

25. More mass equals more potential gravitational energy. As the mass clumps together, the gravitational energy increases, which causes even more mass to accumulate.

26. A young main-sequence star has just begun the core fusion process by which it converts hydrogen to helium, whereas an old main-sequence star has been converting hydrogen to helium for a long time. Therefore, the older star contains more helium.

27. A universal blue shift in the spectra of galaxies would indicate that all galaxies were moving toward each other, and therefore the universe would be contracting.

Concept Mapping

28. Answers may vary but should include all of the terms listed. Sample answers appear at the end of this Teacher Edition.

Math Skills

29. total number of stars = number of stars in Milky Way × percentage of Milky-Way sized galaxies × number of galaxies;
total number of stars = $(2.00 \times 10^{11}$ stars/galaxy$) \times (0.10) \times (1.25 \times 10^{11}$ galaxies);
total number of stars = 2.5×10^{21} stars, or 2,500,000,000,000,000,000,000 stars

30. time of travel = (number of light years × length of light-year)/speed of vehicle;
time of travel = $[(4 \text{ ly} \times 10^{13} \text{ km/ly})/(100 \times 100 \text{ km/h})] \times (1 \text{ d}/24 \text{ h}) \times (1 \text{ y}/365.25 \text{ d}) = [(4 \times 10^{13} \text{ km})/10^4 \text{ km/h}] \times (1 \text{ y}/8766 \text{ h}) = 4.6 \times 10^5$ y, or 460,000 y

Writing Skills

31. Answers may vary. Accept all reasonable answers.

32. Answers may vary. Accept all reasonable answers.

Interpreting Graphics

33. a star with half the mass of the sun

34. 5 billion years

35. more than 5 billion years longer

Standardized Test Prep

Estimated Time

To give students practice under more realistic testing conditions, allow them 30 minutes to answer all of the questions in this practice test.

TEST DOCTOR

Question 2 Answer I is correct. Answer F is incorrect because the composition of stars does not indicate their motion. Answer G is incorrect because these circular trails are evidence of Earth's motion, not the motion of the stars. Color separation alone does not reveal motion, so answer H is incorrect.

Question 4 Answer H is correct, because typical yellow stars are the most likely to be found on the main sequence. Answer F is incorrect. White dwarf stars evolve after the main sequence stage from a star with a sunlike mass. Answer G is incorrect because red supergiants evolve after the main sequence stage. Answer I is incorrect because neutron stars are the result of a red supergiant star that has experienced a supernova.

Question 10 Answer H is correct. Students should multiply the distance light travels in one second (300,000 km/s) by 60. Then, they should multiply the total by 8 minutes. The true average distance between Earth and sun is approximately 149,500,000 km.

Chapter 30 Standardized Test Prep

Understanding Concepts

Directions (1–5): **For *each* question, write on a separate sheet of paper the letter of the correct answer.**

1. What accounts for different stars being seen in the sky during different seasons of the year?
   - **A.** stellar motion around Polaris
   - **B.** Earth's rotation on its axis
   - **C.** Earth's revolution around the sun
   - **D.** position north or south of the equator

2. How do stellar spectra provide evidence that stars are actually moving?
   - **F.** Dark line spectra reveal a star's composition.
   - **G.** Long exposure photos show curved trails.
   - **H.** Light separates into different wavelengths.
   - **I.** Doppler shifts occur in the star's spectrum.

3. What happens to main sequence stars like the sun when energy from fusion is no longer available?
   - **A.** They expand and become supergiants.
   - **B.** They collapse and become white dwarfs.
   - **C.** They switch to fission reactions.
   - **D.** They contract and turn into neutron stars.

4. Which type of star is most likely to be found on the main sequence?
   - **F.** a white dwarf
   - **G.** a red supergiant
   - **H.** a yellow star
   - **I.** a neutron star

5. Evidence for the big-bang theory is provided by
   - **A.** cosmic background radiation
   - **B.** apparent parallax shifts
   - **C.** differences in stellar luminosity
   - **D.** star patterns called constellations

Directions (6–8): **For *each* question, write a short response.**

6. What type of galaxy has no identifiable shape?

7. What is the collective name for the Milky Way galaxy and a cluster of approximately 30 other galaxies located nearby?

8. What is the name for stars that seem to circle around Polaris and never dip below the horizon?

Reading Skills

Directions (9–11): **Read the passage below. Then, answer the questions.**

GEOMAGNETIC POLES

Today, we know that Copernicus was right—the stars are very far from Earth. In fact, stars are so distant that a new unit of length—the light-year—was created to measure their distance. A light-year is a unit of length equal to the distance that light travels through space in 1 year. Because the speed of light through space is about 300,000 km/s, light travels approximately 9.46 trillion kilometers in one year.

Even after astronomers figured out that stars were far from Earth, the nature of the universe was hard to understand. Some astronomers thought that our galaxy, the Milky Way, included every object in space. In the early 1920's, Edwin Hubble made one of the most important discoveries in astronomy. He discovered that the Andromeda galaxy, which is the closest major galaxy to our own, was past the edge of the Milky Way. This fact confirmed the belief of many astronomers that the universe is larger than our galaxy.

9. Why was Edwin Hubble's discovery important?
   - **A.** Hubble's discovery showed scientists that the universe was smaller than previously thought.
   - **B.** Hubble showed that the Andromeda galaxy was larger than the Milky Way galaxy.
   - **C.** Hubble's discovery showed scientists that the universe was larger than our own galaxy.
   - **D.** Hubble showed that all of the stars exist in two galaxies, the Andromeda and the Milky Way.

10. Because the sun and Earth are close together, the distance between the sun and Earth is measured in light-minutes. A light-minute is the distance light travels in 1 minute. The sun is about 8 light-minutes from Earth. What is the approximate distance between the sun and Earth?
    - **F.** 2,400,000 km
    - **G.** 18,000,000 km
    - **H.** 144,000,000 km
    - **I.** 1,000,000,000 km

11. Why might scientists use light-years as a measurement of distance between stars?

Answers

Understanding Concepts

1. C
2. I
3. B
4. H
5. A
6. irregular galaxy
7. the local group
8. circumpolar stars

Reading Skills

9. C
10. H
11. Light-years can express vast distances in compact form. When expressing distance between stars, using light-years is easier and more efficient than using kilometers.

Interpreting Graphics

12. Answers may vary. See Test Doctor for a detailed scoring rubric.
13. D
14. G
15. Answers may vary. See Test Doctor for a detailed scoring rubric.

Interpreting Graphics

Directions (12–15): For *each* question below, record the correct answer on a separate sheet of paper.

The diagram shows a group of stars called the Big Dipper moving over a period of 200,000 years. Use this map to answer question 12.

Changing Shape of the Big Dipper over Time

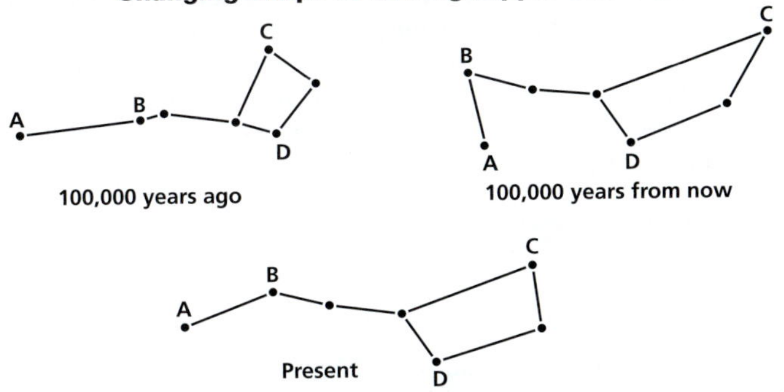

12 What does this series of drawings demonstrate about the individual stars in such a star group?

The table below shows data about several well-known stars. Distance is given in light-years. Use this table to answer questions 13 through 15.

Stellar Characteristics

Name	Color	Magnitude	Distance
Arcturus	orange	0.0	36.8 ly
Betelguese	red	0.5	400 ly
Canopus	yellow-white	-0.6	310 ly
Capella	yellow	0.1	42.2 ly
Mintaka	blue-violet	2.2	915 ly
Rigel	blue-white	0.2	800 ly
Sirius	white	-1.4	8.6 ly
Vega	white	0.0	25.3 ly

13 Which star has the brighest apparent magnitude as seen from Earth?
- **A.** Rigel
- **B.** Betalgeuse
- **C.** Mintaka
- **D.** Sirius

14 Which of these stars is the coolest?
- **F.** Arcturus
- **G.** Betalgeuse
- **H.** Mintaka
- **I.** Vega

15 Which star most likely has a temperature that is similar to the temperature of our sun? Explain how you are able to determine this information.

Test TIP

If you are unsure of an answer, eliminate the answers that you know are wrong before choosing your answer.

Standardized Test Prep

TEST DOCTOR

Question 12 Full-credit answers should include the following points:
- the diagrams show that the individual stars move at different rates and in different directions from one another
- constellations are arbitrary human distinctions. The stars within constellations move along individual paths, not as a group as many ancient civilizations thought they did
- the familiar patterns that stars form in Earth's sky change slowly over time as the stars that comprise the patterns move relative to each other
- star movement may take thousands of years to become apparent

Question 15 Full-credit answers should include the following points:
- the star in the table with the closest temperature to the sun in most likely Capella
- a star's temperature can be determined by its color. Stars that have similar colors share a common temperature range
- Capella is a yellow star like the sun and thus it is the most likely to have a temperature similar to that of the sun

Test Prep Correlations *National Science Education Standards*

ES 4a: item 5
ES 4b: items 6, 7, 9
HNS 1a: item 9
HNS 2a: items 1, 4, 6, 7, 8, 10, 13
HNS 2b: items 1, 2, 3, 5, 12, 14, 15
PS 6b: item 2
UCP 1: item 12

CHAPTER RESOURCES

State Resources

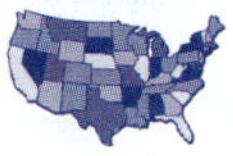

For specific resources for your state, visit go.hrw.com and type in the keyword **HSHSTR.**

Star Magnitudes

Teacher's Notes

Time Required
one 45-minute class period

Lab Ratings

EASY → HARD

TEACHER PREPARATION ⚗⚗
STUDENT SETUP ⚗
CONCEPT LEVEL ⚗⚗
CLEANUP ⚗

Skills Acquired
- Experimenting
- Collecting Data
- Organizing and Analyzing Data
- Interpreting
- Identifying and Recognizing Patterns

The Scientific Method
In this lab, students will
- Make Observations
- Analyze the Results
- Draw Conclusions

Materials
The materials listed on the page are enough for groups of two students. Be sure the paraffin bricks are new and smooth on all sides. If the light from the bulbs appears to be weak, replace the batteries.

Tips and Tricks
Students should be sure that the foil blocks off any light leaking between the paraffin bricks. This lab can be performed effectively on an overcast day, though the outside light will have a noticeably different color from that of unfiltered sunlight on a clear day. Be sure students are comparing light from only two sources, and that no other light is interfering with the observations.

Chapter 30

Making Models Lab

Objectives
- **Construct** a model photometer and two model stars.
- **Demonstrate** how distance affects brightness of stars.
- **Explain** how color is related to the temperature of stars.

Materials
- aluminum foil, 12 cm × 12 cm
- batteries, AA (3)
- desk lamp with incandescent bulb
- flashlight bulbs, 3-volt (2)
- paraffin, 12 cm × 6 cm bricks (2)
- rubber band, large
- ruler, metric
- tape, electrical
- wire, plastic-coated with stripped ends, 15 cm
- wire, plastic-coated with stripped ends, 20 cm

Safety

Star Magnitudes

Astronomers study the brightness, or magnitude, of stars. Except for the sun, stars are very faint and visible only at night. Thus, their brightness must be measured with a device called a *photometer*. An astronomical photometer consists of a surface that is sensitive to light and a device that measures the amount of light that reaches the surface. Photometers can also be used to compare the colors of different light sources. In this lab, you will determine the effect of distance on brightness and the relationship between temperature and color.

PROCEDURE

1. Construct two flashlights:
   a. Arrange the bulbs and batteries as shown in the figure below. Using electrical tape, attach the wires to the batteries and bulbs. The bulb should be on. If it is not on, study the illustration again and make adjustments.
   b. Tape the flashlight arrangement together so that it can be moved. Be sure to leave the wire loose at the negative end of the battery so that you can turn your flashlight on and off.
2. Construct a photometer by folding the aluminum foil in half with the shiny side facing out, and place it between two paraffin bricks. Hold the pieces together by using a rubber band.
3. Place the two flashlights about 2 m apart on a table. Place the photometer between them with the largest sides of the bricks facing each flashlight bulb, as shown in the figure below.

Step 1

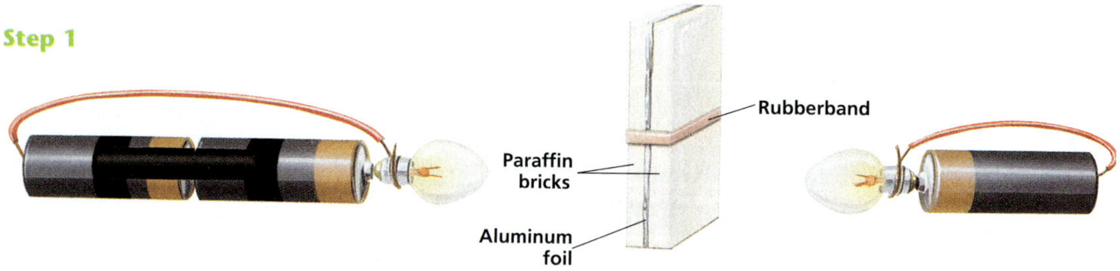

CHAPTER RESOURCES

Chapter Resource File
- Datasheet for Chapter Lab GENERAL
- Lab Notes and Answers

4. Turn on both flashlights, and turn off all room lights.

5. Move the photometer until both sides are equally bright. Measure the distance, in centimeters, from each flashlight bulb to the center of the photometer. Record these measurements.

6. Square the distances you recorded in step 5. Record these values.

7. Incandescent light bulbs have filaments that emit light at a temperature that is much cooler than the sun's surface. Place the photometer between the desk lamp and a window on a bright day. Sunlight coming through a window will be the same color as the sunlight outdoors. Turn off any fluorescent ceiling lighting, and turn on the desk lamp.

8. Compare the color differences between the paraffin sides of your photometer.

9. Darken the room once again, and compare the colors of the bulb powered by one battery with the colors of the bulb powered by two batteries.

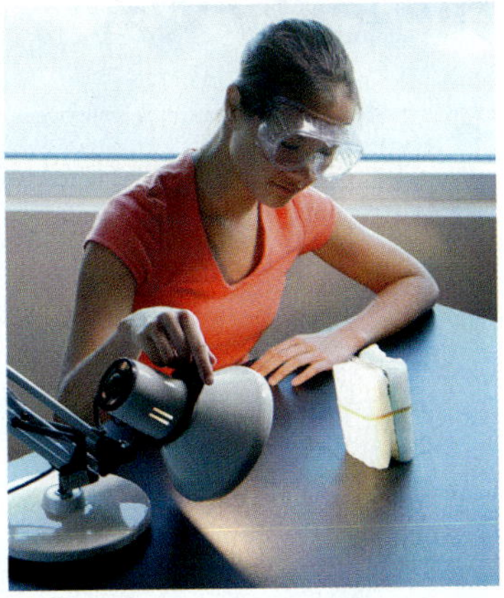

Step 8

ANALYSIS AND CONCLUSION

1. **Analyzing Data** The ratio of the square of the distances you calculated in step 6 is equal to the ratio of the brightnesses of the bulbs. What is the ratio of the square of the distance of the two-battery flashlight to that of the one-battery flashlight? What does this information tell you about the relationship between the brightness of the two flashlights?

2. **Drawing Conclusions** Based on the results of the investigation, would you expect a white star to be hotter or cooler than a yellow star?

3. **Applying Conclusions** Using your knowledge of the spectrum, would you expect a white star to be hotter or cooler than an orange star? Predict whether a blue star is hotter or cooler than a white star. Also, predict whether a red star is hotter or cooler than an orange star.

Extension

1. **Explaining Observations** Find an incandescent bulb controlled by a dimmer. Watch the color of the light as it fades. Does it become more yellow or more white? Explain why.

Making Models Lab

Answers to Analysis and Conclusion

1. Answers should be close to 2:1. The brightness of the flashlight with two batteries is twice as great as the brightness of the flashlight with one battery.
2. The white star is brighter and uses more energy in a given time. This means that a white star is hotter than a blue star is.
3. A white star is hotter than an orange star. A blue star is hotter than a white star. A red star is cooler than an orange star. (Temperature increases as light progresses from red to blue.)

Answers to Extension

1. As the light fades, it becomes more yellow. This is because the filament of the bulb is cooler when less current passes through it, and the light associated with this decreased temperature is redder.

Erich Landstrom
Boynton Beach Community High School
NASA Education Ambassador
Boynton Beach, FL

The Milky Way

Group Activity — GENERAL

Charting the Galaxy Divide the class into groups of two or three students and have them research those portions of the Milky Way's arms whose contents are well-identified, such as a segment of the Perseus arm that contains a number of prominent stars. Each group should create a map of their section, labeling the largest and/or brightest stars, star clusters, and nebulae. Have each group present its chart and explain the details to the class. LS **Visual**

Answers to Map Skills Activity

1. 90,000 to 100,000 ly
2. 15,000 to 20,000 ly
3. The Perseus and Cygnus arms branch off from the Norma arm.
4. They spread out and become less dense than the center. Single arms tend to branch into multiple arms.
5. The clusters on the map are not close to the point of view of the observer but are above or below the plane of the galaxy disc. This distant perspective and the fact that globular clusters are much smaller than the Milky Way galaxy account for why they appear to be small on the map.

CHAPTER RESOURCES

Technology

Transparencies
- 155 The Milky Way (with worksheet)

MAPS in Action

The Milky Way

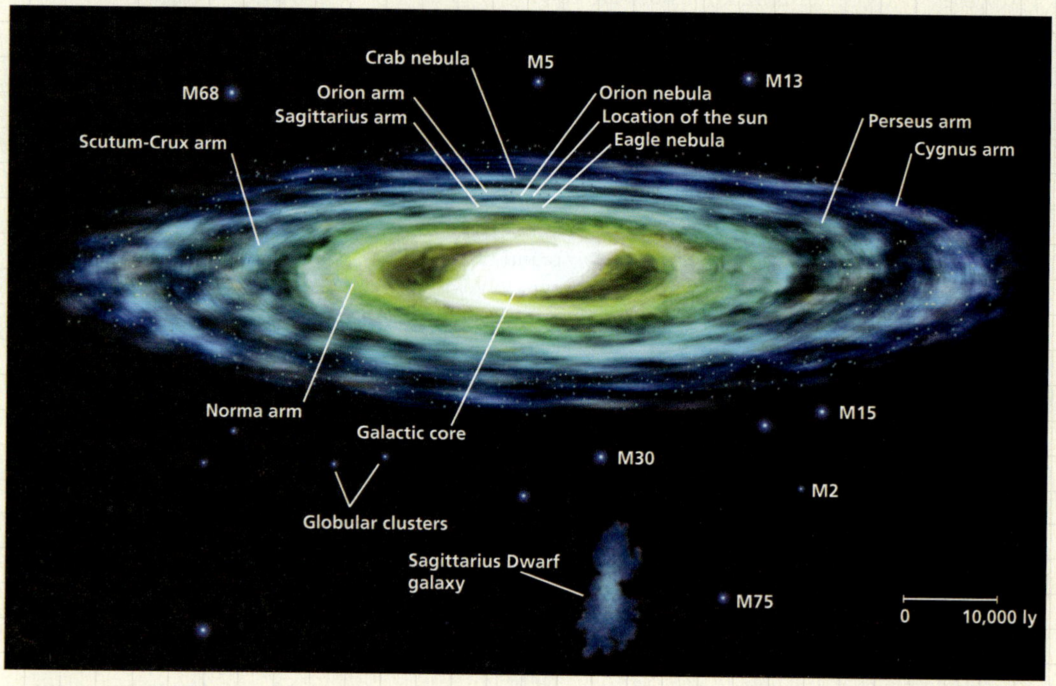

Map Skills Activity

The map above shows what astronomers think the Milky Way galaxy looks like. Because of Earth's position within the galaxy, scientists must hypothesize what our galaxy looks like from a perspective outside of the galaxy. They must also form a hypothesis about the shape and location of those spiral arms that are obscured by either the galactic core or other spiral arms that are closer to Earth. Use the map to answer the questions below.

1. **Using the Key** What is the approximate distance from one edge of the Milky Way to the other edge?
2. **Using the Key** What is the approximate width of the galactic core?
3. **Analyzing Relationships** How are the Perseus arm and the Cygnus arm related to the Norma arm?
4. **Identifying Trends** What happens to the arms of the Milky Way as they radiate outward from the center of the galaxy?
5. **Analyzing Relationships** The locations on the map marked with an M and a number are known as Messier objects. These objects are named for their discoverer, Charles Messier, who first catalogued them in the 1700s. The Messier objects shown on this map are all globular clusters. If each of these clusters contain hundreds, or even thousands, of stars, why might the clusters appear so small on this map compared to the size of the Milky Way?

SCIENCE AND TECHNOLOGY

Studying Stars in Formation

When stars are forming, they have a dim, reddish glow. As their temperature increases, they eventually become much hotter and brighter. But when they are dim and reddish, they give off most of their light in the infrared part of the spectrum. Most of this infrared light does not penetrate Earth's atmosphere, where water vapor and carbon dioxide block the light. So, astronomers have launched telescopes into space that allow them to observe stars and galaxies from above the layers of Earth's atmosphere.

The *Hubble Space Telescope*

The *Hubble Space Telescope* observes ultraviolet, visible, and infrared spectra. The telescope did not have infrared capability at first. But a camera called NICMOS, which stands for *Near Infrared Camera and Multi-Object Spectrometer,* was launched on an updating mission in which astronauts brought new equipment to the *Hubble Space Telescope*. Infrared cameras such as NICMOS have to be kept very cool, and the liquid nitrogen that originally cooled NICMOS was depleted after about two years. The next mission to the *Hubble Space Telescope* brought a new type of refrigerator to keep the camera cool.

The *Spitzer Space Telescope*

On August 25, 2003, NASA launched a new and more sensitive infrared telescope. Originally called *Space Infrared Telescope Facility,* it was renamed the *Spitzer Space Telescope*. This telescope is not orbiting Earth in the way that the *Hubble Space Telescope* does. Instead, it is far from Earth and trails Earth in its orbit around the sun. This keeps the light reflected by Earth from heating the camera.

▲ Astronauts conduct delicate repairs to the *Hubble Space Telescope* while in orbit.

The *Spitzer Space Telescope* takes images of infrared light, which has wavelengths as much as 100 times longer than the wavelengths in images taken by the *Hubble Space Telescope*. The *Spitzer Space Telescope's* high-quality images show glowing dust, cool stars, and the most distant galaxies especially well.

◀ This false-color image from the *Spitzer Space Telescope* shows the intensity of infrared emissions of the galaxy M81.

Extension

1. **Making Comparisons** What types of information can scientists determine from infrared images that they cannot determine from images taken in the visual spectrum?

Science and Technology

Studying Stars in Formation

Activity — GENERAL

Space Telescopes The *Spitzer Space Telescope*, originally called the *Space Infrared Telescope Facility*, was officially named only after it proved functional. The Spitzer telescope was renamed by a winning entry in a contest and is named for Lyman Spitzer, Jr., Ph.D., who first suggested placing telescopes in space.

By placing telescopes in space, the problems with seeing that are caused by the atmosphere are greatly reduced. This is especially true of telescopes designed to observe in the infrared, ultraviolet, and X-ray portions of the spectrum. Orbiting space telescopes designed to observe in particular portions of the spectrum have been in use for more than forty years.

Have students research several of the space telescopes, including some sent into orbit over the past forty years and those that are being planned for the next two decades. From this research, have students propose a project for a space telescope. The project should be based on some aspect of space that has sparked their imaginations, and should describe how a given space telescope could be used to explore or learn about this area of interest. Have students present their proposals in a written report, oral presentation, or poster project. LS **Verbal**

Answer to Extension

1. Infrared radiation passes through dust that dims or extinguishes visible light. Infrared imaging can therefore allow astronomers to peer into dusty nebulae or the galactic core and see details that are not detectable in the visible part of the spectrum.

APPENDIX CONTENTS

* Items marked with an oak leaf contain information directly related to environmental issues.

Analyzing Science Terms

You can unlock the meaning of an unfamiliar science term by analyzing its word parts. Many parts of scientific words carry a meaning that derives from Latin or Greek. The parts of words listed below provide clues to the meanings of many science terms.

Word part or root	Meaning	Application
a-	not, without	abiotic
astr-, aster-	star	astronomy
bar-, baro-	weight, pressure	barometer
batho-, bathy-	depth	batholith, bathysphere
circum-	around	circum-Pacific, circumpolar
-cline	lean, slope	anticline, syncline
-duct-	to lead, draw	conduction
eco-	environment	ecology, ecosystem
epi-	on	epicenter
ex-, exo-	out, outside of	exosphere, exfoliation, extrusion
geo-	earth	geode, geology, geomagnetic
-graph	write, writing	seismograph
hydro-	water	hydrosphere
hypo-	under	hypothesis
iso-	equal	isoscope, isostasy, isotope
-lith, -lithic	stone	Neolithic, regolith
-log-	study	ecology, geology, meteorology
magn-	great, large	magnitude
mar-	sea	marine
meta-	among, change	metamorphic, metamorphism
-meter	to measure	thermometer, spectrometer
micro-	small	microquake
-morph, -morphic	form, shape	metamorphic
nebula-	mist, cloud	nebula
neo-	new	Neolithic
paleo-	old	paleontology, Paleozoic
ped-, pedo-	ground, soil	pediment
per-	through	permeable
peri-	around	perigee, perihelion
seism-, seismo-	shake, earthquake	seismic, seismograph
sol-	sun	solar, solstice
spectro-	look at, examine	spectroscope, spectrum
-sphere	ball, globe	geosphere, lithosphere
strati-, strato-	spread, layer	stratification, stratovolcano
terra-	earth, land	terracing, terrane
thermo-	heat	thermosphere, thermometer
top-, topo-	place	topographic
trop-, tropo-	turn, respond to	tropopause, troposphere

How to Make Power Notes

Power notes help you organize the Earth science concepts you are studying by distinguishing main ideas from details. Similar to outlines, power notes are linear in form and provide you with a framework of important concepts. To make power notes, you assign a *power* of 1 to each main idea and a 2, 3, or 4 to each detail. You can use power notes to organize ideas while reading your text or to restructure your class notes for studying purposes. Practice first by using simple concepts. For example, start with a few headers or bold-faced vocabulary terms from this book. Later, you can strengthen your notes by expanding these simple words into more-detailed phrases and sentences. Use the following general format.

Power 1: Main idea
- Power 2: Detail or support for Power 1 idea
  - Power 3: Detail or support for Power 2 concept
    - Power 4: Detail or support for Power 3 concept

1 Pick a Power 1 word or phrase from the text.

The text you choose does not have to come from your Earth science textbook. You may make power notes from your lecture notes or from another source. We'll use the term *environmental problems* as an example of a main idea.

Power 1: environmental problems

2 Using the text, select some Power 2 words to support your Power 1 word.

We'll use the terms *resource depletion, pollution,* and *extinction,* which are the three main types of environmental problems.

Power 1: environmental problems
- Power 2: resource depletion
- Power 2: pollution
- Power 2: extinction

3 Select some Power 3 words to support your Power 2 words.

We'll use the terms *renewable resources* and *nonrenewable resources*. These two terms are related to *resource depletion,* which is one of the Power 2 concepts.

Power 1: environmental problems
- Power 2: resource depletion
  - Power 3: renewable resources
  - Power 3: nonrenewable resources
- Power 2: pollution
- Power 2: extinction

4 Continue to add powers to support and detail the main idea as necessary.

There are no restrictions on how many power numbers you can add to help you extend and organize your ideas. Words that have the same power number should have a similar relationship to the previous power but do not have to be related to each other.

Power 1: environmental problems
- Power 2: resource depletion
  - Power 3: renewable resources
  - Power 3: nonrenewable resources
- Power 2: pollution
  - Power 3: degradable pollutants
  - Power 3: nondegradable pollutants
- Power 2: extinction
  - Power 3: pollution
  - Power 3: habitat loss

Practice

1. Use this book's lesson on scientific methods and power notes structure to organize the following terms: *observing, hypothesizing and predicting, experimenting, organizing and analyzing data, drawing conclusions, repeating experiments, communicating results, observation, hypothesis, prediction, experiment, variable, experimental group, control group,* and *data.*

(See the last page of the Skills Handbook for the answers to practice problems.)

How to Make KWL Notes

KWL stands for *what I **K**now, what I **W**ant to know,* and *what I **L**earned*. The KWL strategy is somewhat different from other learning strategies because it prompts you to brainstorm about the subject matter before reading the assigned material. Relating new ideas and concepts with those that you have learned will help you to understand and apply the knowledge you obtain in this course. The section objectives throughout your text are ideal for using the KWL strategy. Read the objectives before reading each section, and follow the instructions in the example below.

1 Read the section objectives.

You may also want to scan headings, boldfaced terms, and illustrations in the section. We'll use a few sample objectives as examples.

- List and describe the steps of the scientific method.
- Describe why a good hypothesis is not simply a guess.
- Describe the two essential parts of a good experiment.

2 Divide a sheet of paper into three columns. Label the columns "What I know," "What I want to know," and "What I learned."

Here is an example table:

What I know	What I want to know	What I learned

3 Brainstorm about what you know about the information in the objectives, and write these ideas in the first column.

Because this table is designed to help you blend your knowledge with new information, you do not have to write complete sentences.

4 Think about what you want to know about the information in the objectives. Write these ideas in the second column.

You'll want to know the information you will be tested on, so include information from both the objectives and any other topics your teacher has given to you.

5 Use the third column to write down the information you learned. Do this while you read the text or just after reading the text.

While you read, pay close attention to any information about the topics you wrote in the column entitled "What I want to know." If you do not find all of the answers you are looking for, you may need to reread the text or reference a second source. Be sure to ask your teacher for help if you cannot find the information after reading the text a second time.

When you have completed reading the text, review the ideas you brainstormed. Compare your ideas in the first column with the information you wrote down in the third column. If you find that some of the ideas are incorrect, cross them out. Before you begin studying for your test, identify and correct any misconceptions you had prior to reading.

Here is an example of what your notes might look like after using the KWL strategy:

What I know	What I want to know	What I learned
• The steps of the experimental method are predict, test, and conclude.	• What are the steps of the experimental method?	• The steps of the experimental method are observing, hypothesizing, experimenting, organizing and analyzing data, drawing conclusions, communicating results, and repeating experiments.
• A hypothesis is similar to a guess, but when you form a hypothesis, you have an idea of what might happen.	• Why is a hypothesis not a guess?	• A hypothesis is more than a guess. You have to base a hypothesis on observations and really think about what you are trying to learn. You should also design an experiment that can test if your hypothesis is wrong, but an experiment cannot prove that your hypothesis is correct.
• A good experiment includes a hypothesis and a lot of equipment.	• What are the two important parts of a good experiment?	• The two important parts of a good experiment are a single variable and a control group.

Practice

1. Use the third column from the table above to identify and correct any misconceptions in the following list of ideas.

a. The first step of the experimental method is to predict.

b. A hypothesis is similar to a guess.

c. A good experiment includes a lot of equipment.

(See the last page of the Skills Handbook for the answers to practice problems.)

SKILLS HANDBOOK

How to Make Two-Column Notes

Two-column notes can be used to learn and review definitions of vocabulary terms, examples of multiple-step processes, or details of specific concepts. The two-column note strategy is simple: write the term, main idea, step-by-step process, or concept in the left-hand column, and write the definition, example, or detail on the right.

One strategy for using two-column notes is to organize main ideas and their details. You will write the main ideas from your reading in the left-hand column of your paper. You can write these ideas as questions, key words, or a combination of both. Then, write details that describe these main ideas in the right-hand column of your paper.

1 Identify the main ideas.

The main ideas for a chapter are listed in the section objectives. However, you decide which ideas to include in your notes. The example below shows some main ideas from possible objectives in a section of this book.

- Define Earth science, and compare Earth science with geology.
- List the four major fields of study that contribute to Earth science.

2 Divide a blank sheet of paper into two columns, and write the main ideas in the left-hand column.

Remind yourself that your two-column notes are precisely that—notes. Do not copy whole phrases out of the book or waste your time writing ideas in complete sentences. Summarize your ideas by using short phrases that are easy for you to understand and remember. Decide how many details you need for each main idea, and write that number in parentheses under the main idea.

Main idea	Detail notes
Earth science (two definitions)	
Goals of Earth science (one main goal)	
What is studied (two main areas)	
Related fields of study (four major fields)	

3 Write the detail notes in the right-hand column.

List as many details as you designated in the main-idea column.

Main idea	Detail notes
Earth science (two definitions)	Earth science is the study of Earth and the universe around it. Scientific methods are the organized, logical approaches to scientific research.
Goals of Earth science (one main goal)	to understand Earth and the universe around it
What is studied (two main areas)	Earth and the universe around it • the origin, history, and structure of the solid Earth and the processes that shape it • how Earth's atmosphere, oceans, and land interact
Related fields of study (four major fields)	geology • the study of the origin, history, and structure of Earth and the processes that shape it oceanography • the study of Earth's oceans, including waves, tides, ocean currents, the ocean floor, and life in the oceans meteorology • the study of Earth's atmosphere, including weather and climate astronomy • the study of the universe beyond Earth, including planets, stars, and galaxies

You can use two-column notes to study for a short quiz or for a test on the material in an entire chapter. Cover the information in the right-hand column with a sheet of paper. Recite what you know, and then uncover the notes to check your answers. Then, ask yourself what else you know about that topic. Linking ideas in this way will help you gain a more complete picture of Earth science.

SKILLS HANDBOOK FoldNote Instructions

Have you ever tried to study for a test or quiz but didn't know where to start? Or have you read a chapter and found that you can remember only a few ideas? Well, FoldNotes are a fun and exciting way to help you learn and remember the ideas you encounter as you learn science!

FoldNotes are tools that you can use to organize concepts. One FoldNote focuses on a few main concepts. FoldNotes help you learn and remember how the concepts fit together. FoldNotes can help you see the "big picture." Below, you will find instructions for building 10 different FoldNotes.

Pyramid

1. Place a **sheet of paper** in front of you. Fold the lower left-hand corner of the paper diagonally to the opposite edge of the paper.
2. Cut off the tab of paper created by the fold (at the top).
3. Open the paper so that it is a square. Fold the lower right-hand corner of the paper diagonally to the opposite corner to form a triangle.
4. Open the paper. The creases of the two folds will have created an X.
5. Using **scissors,** cut along one of the creases. Start from any corner, and stop at the center point to create two flaps. Use **tape** or **glue** to attach one of the flaps on top of the other flap.

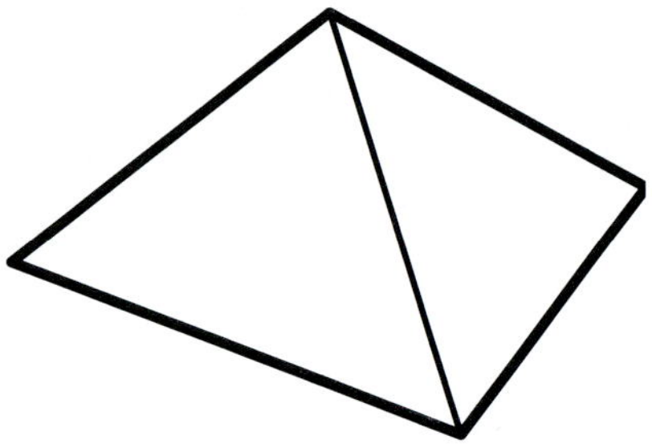

Double-Door Fold

1. Fold a **sheet of paper** in half from the top to the bottom. Then, unfold the paper.
2. Fold the top and bottom edges of the paper to the center crease.

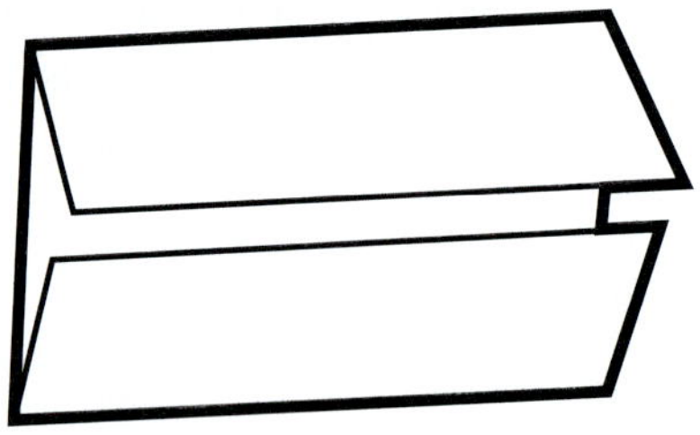

Table Fold

1. Fold a **piece of paper** in half from the top to the bottom. Then, fold the paper in half again.
2. Fold the paper in thirds from side to side.
3. Unfold the paper completely. Carefully trace the fold lines by using a pen or pencil.

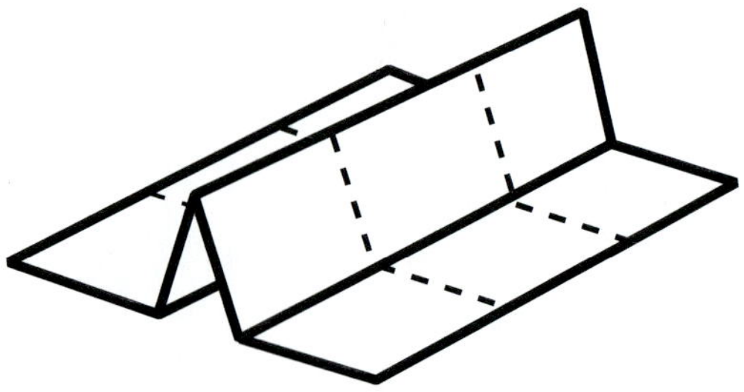

Booklet

1. Fold a **sheet of paper** in half from left to right. Then, unfold the paper.
2. Fold the sheet of paper in half again from the top to the bottom. Then, unfold the paper.
3. Refold the sheet of paper in half from left to right.
4. Fold the top and bottom edges to the center crease.
5. Completely unfold the paper.
6. Refold the paper from top to bottom.
7. Using **scissors,** cut a slit along the center crease of the sheet from the folded edge to the creases made in step 4. Do not cut the entire sheet in half.
8. Fold the sheet of paper in half from left to right. While holding the bottom and top edges of the paper, push the bottom and top edges together so that the center collapses at the center slit. Fold the four flaps to form a four-page book.

Layered Book

1. Lay **one sheet of paper** on top of **another sheet.** Slide the top sheet up so that 2 cm of the bottom sheet is showing.
2. Holding the two sheets together, fold down the top of the two sheets so that you see four 2 cm tabs along the bottom.
3. Using a **stapler,** staple the top of the FoldNote.

Two-Panel Flip Chart

1. Fold a **piece of paper** in half from the top to the bottom.
2. Fold the paper in half from side to side. Then, unfold the paper so that you can see the two sections.
3. From the top of the paper, cut along the vertical fold line to the fold in the middle of the paper. You will now have two flaps.

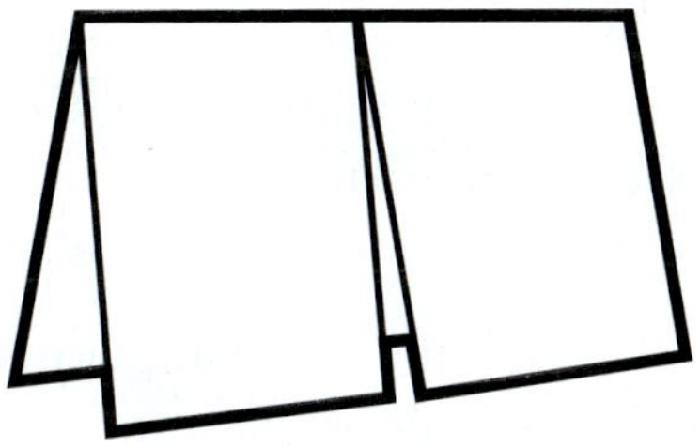

SKILLS HANDBOOK

Key-Term Fold

1. Fold a **sheet of lined notebook paper** in half from left to right.
2. Using **scissors,** cut along every third line from the right edge of the paper to the center fold to make tabs.

Four-Corner Fold

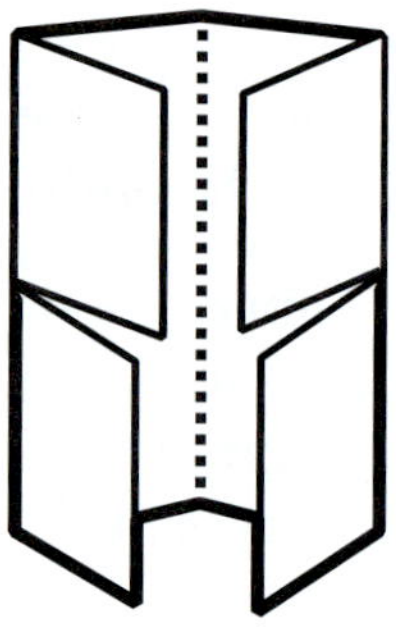

1. Fold a **sheet of paper** in half from left to right. Then, unfold the paper.
2. Fold each side of the paper to the crease in the center of the paper.
3. Fold the paper in half from the top to the bottom. Then, unfold the paper.
4. Using **scissors,** cut the top flap creases made in step 3 to form four flaps.

Three-Panel Flip Chart

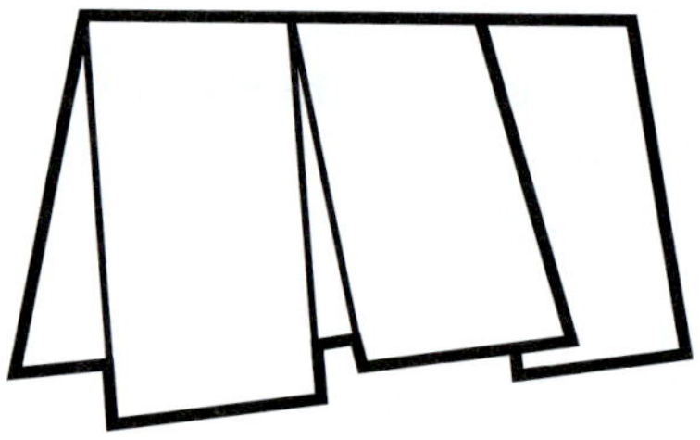

1. Fold a **piece of paper** in half from the top to the bottom.
2. Fold the paper in thirds from side to side. Then, unfold the paper so that you can see the three sections.
3. From the top of the paper, cut along each of the vertical fold lines to the fold in the middle of the paper. You will now have three flaps.

Tri-Fold

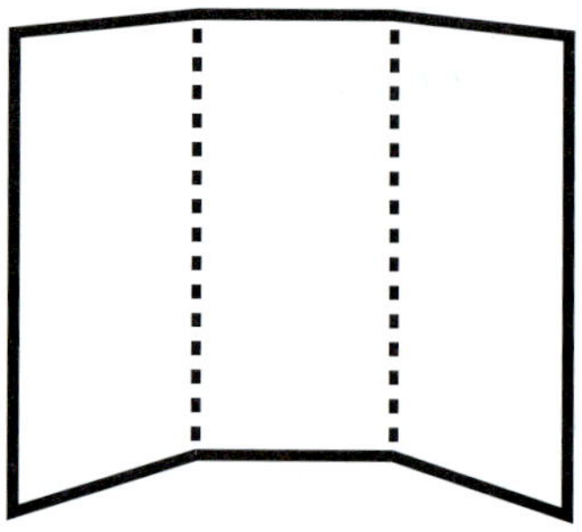

1. Fold a piece a paper in thirds from the top to the bottom.
2. Unfold the paper so that you can see the three sections. Then, turn the paper sideways so that the three sections form vertical columns.
3. Trace the fold lines by using a **pen** or **pencil.** Label the columns “Know,” “Want,” and “Learn.”

SKILLS HANDBOOK Graphic Organizer Instructions

Have you ever wished that you could draw the many concepts you learn in your science class? Sometimes, being able to see how concepts are related helps you remember what you've learned. Graphic Organizers help you see the concepts! They are a way to draw or map out concepts.

You need only a piece of paper and a pencil to make a Graphic Organizer. Below, you will find instructions for six different Graphic Organizers that are designed to help you organize the concepts you'll learn in this book.

Spider Map

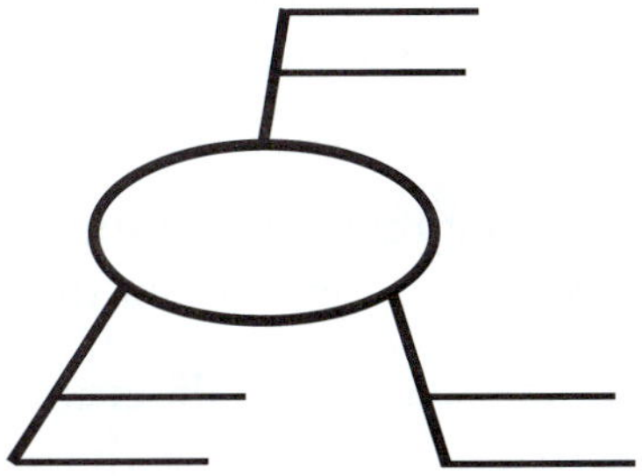

1. Draw a diagram like the one shown. In the circle, write the main topic.
2. From the circle, draw legs to represent different categories of the main topic. You can have as many categories as you want.
3. From the category legs, draw horizontal lines. As you read the chapter, write details about each category on the horizontal lines.

Comparison Table

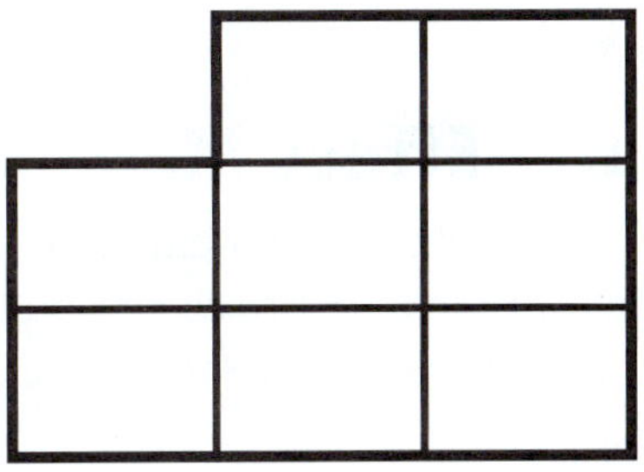

1. Draw a chart like the one shown. Your chart can have as many columns and rows as you want.
2. In the top row, write the topics that you want to compare.
3. In the left column, write characteristics of the topics that you want to compare. As you read the chapter, fill in the characteristics for each topic in the appropriate boxes.

Chain-of-Events-Chart

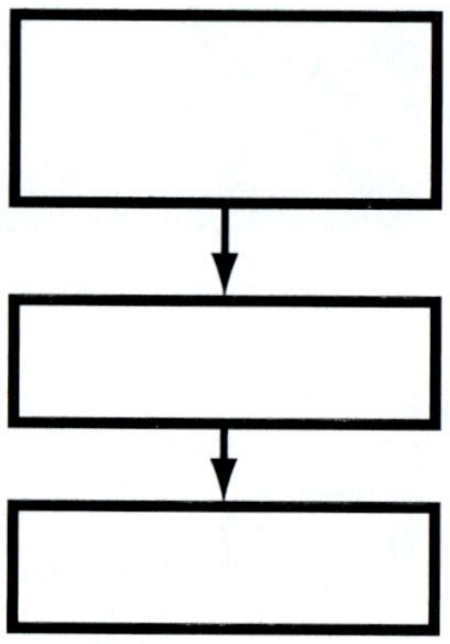

1. Draw a box. In the box, write the first step of a process or the first event of a timeline.
2. Under the box, draw another box, and use an arrow to connect the two boxes. In the second box, write the next step of the process or the next event in the timeline.
3. Continue adding boxes until the process or timeline is finished.

Venn Diagram

1. Draw a diagram like the one shown. You may have two or three circles depending on the number of topics. Make sure the circles overlap with each other.
2. In each circle, write a topic that you want to compare with a topic in another circle.
3. In the areas of the diagram where circles overlap, fill in characteristics that the topics in the overlapping circles share.
4. In the areas of the diagram where circles do not overlap, fill in characteristics that are unique to the topic of the particular circle.

Cause-and-Effect Map

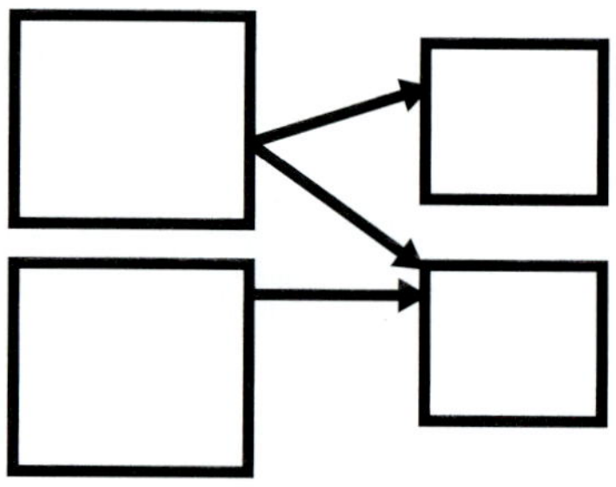

1. Draw a box, and write a cause in the box. You can have as many cause boxes as you want. The diagram shown here is one example of a cause-and-effect map.
2. Draw another box to represent an effect of the cause. You can have as many effect boxes as you want. Draw a line from each cause to the effect(s).
3. In the cause boxes, write a description, explanation, or details about the cause. In the effect boxes, explain the effects that result from the process or factor identified in the cause box.

Concept Map

A concept map is a simple drawing that shows how concepts are connected to each other. Concept maps may be a good tool for visual learners to use when studying and to use to test their understanding of information in the text. Concept maps may be based on key vocabulary terms from the text. These terms are usually nouns, which make good labels for major concepts. Linking words may be used to explain relationships. A group of connected words and lines show a proposition. A proposition is another way of stating a main idea or explaining a concept.

1. Identify main ideas from the text, and write those ideas as short phrases or single words. Concepts may be vocabulary terms, important phrases, or descriptions of processes.
2. List all of the important concepts. Select a main concept for the map, and place this concept at the top or center of a piece of paper.
3. Build the map by placing the other concepts under or around the main concept, according to their importance or their relationship to the main concept.
4. Draw lines between the concepts to show relationships between ideas. Add linking words to give meaning to the arrangement of concepts. To distinguish concepts from links, place concepts in circles, ovals, or rectangles.

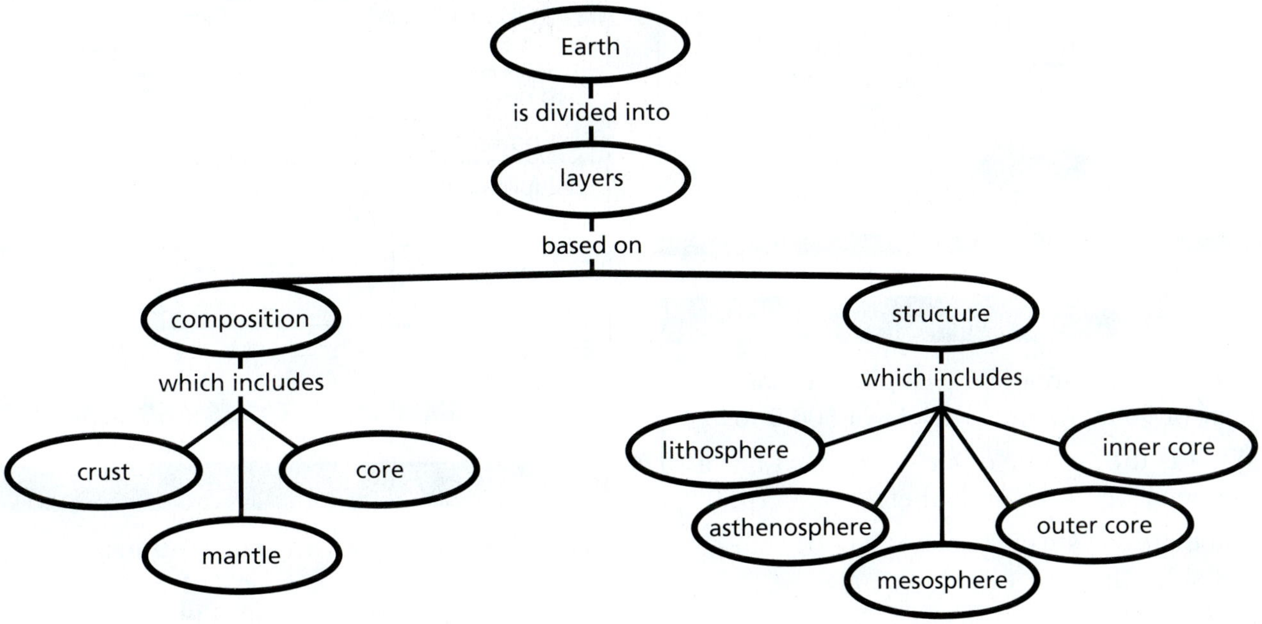

SKILLS HANDBOOK Math Skills Refresher

Geometry

A useful way to model the objects and substances studied in science is to consider them in terms of their shapes. For example, many of the properties of a wheel can be understood by pretending that the wheel is a perfect circle.

When you use shapes as models, the ability to calculate the area or volume of shapes is a useful skill. The table below provides equations for the area and volume of several geometric shapes.

Geometric Areas and Volumes

Geometric shape		Equations for shape
Rectangle	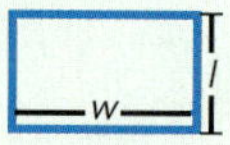	*area* = lw
Circle		*area* = πr^2 *circumference* = $2\pi r$
Triangle	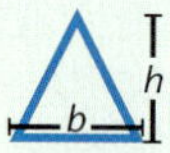	*area* = $\frac{1}{2}bh$
Sphere		*surface area* = $4\pi r^2$ *volume* = $\frac{4}{3}\pi r^3$
Cylinder		*surface area* = $2\pi r^2 + 2\pi rh$ *volume* = $\pi r^2 h$
Rectangular box	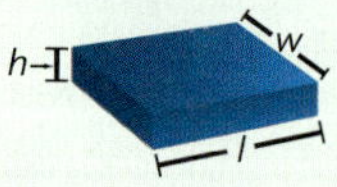	*surface area* = $2(lh + lw + hw)$ *volume* = lwh

Practice

1. Calculate the area of a triangle that has a base of 900.0 m and a height of 500.0 m.
2. What is the volume of a cylinder that has a diameter of 14 cm and a height of 8 cm?
3. Calculate the surface area of a cube that has sides that are 4 cm long.

(See the last page of the Skills Handbook for the answers to practice problems.)

Exponents

An exponent is a number that is written as superscript to the right of another number. The best way to explain how an exponent works is by using an example. In the value 5^4, the 4 is the exponent of the 5. The number and its exponent means that 5 is multiplied by itself 4 times as shown below:

$$5^4 = 5 \times 5 \times 5 \times 5 = 625$$

Exponents are also referred to as *powers*. Using this terminology, the above equation could be read "five to the fourth power equals 625," or "five to the power of four equals 625." Keep in mind that any number raised to the power of 0 is equal to 1: $5^0 = 1$. Also, any number raised to the power of 1 is equal to itself: $5^1 = 5$.

A scientific calculator is very helpful for solving most problems involving exponents. Many calculators have keys for squares and square roots, but scientific calculators usually have a special caret key, ^, for entering exponents. If you type in "5^4" and then press the equals sign or Enter, the calculator will determine that $5^4 = 625$ and display the answer 625.

Exponents

	Rule	Example
Zero power	$x^0 = 1$	$7^0 = 1$
First power	$x^1 = x$	$6^1 = 6$
Multiplication	$(x^n)(x^m) = (x^{n+m})$	$(x^2)(x^4) = x^{(2+4)} = x^6$
Division	$\frac{x^n}{x^m} = x^{(n-m)}$	$\frac{x^8}{x^2} = x^{(8-2)} = x^6$
Exponents raised to a power	$(x^n)^m = x^{nm}$	$(5^2)^3 = 5^6 = 15{,}625$

Practice

1. Perform the following calculations:
   a. $9^1 =$
   b. $(3^3)^5 =$
   c. $(14^2)(14^3) =$
   d. $11^0 =$

(See the last page of the Skills Handbook for the answers to practice problems.)

Order of Operations

Use this phrase to remember the correct order for long mathematical problems: "Please Excuse My Dear Aunt Sally." Some people just remember the acronym "PEMDAS". This acronym stands for *parentheses, exponents, multiplication, division, addition,* and *subtraction*. This is the correct order in which to complete mathematical operations. These rules are summarized in the table below.

Order of Operations
1. Simplify groups inside parentheses. Start with the innermost group and work out.
2. Simplify all exponents.
3. Perform multiplication and division in order from left to right.
4. Perform addition and subtraction in order from left to right.

Look at the following example.

$$4^3 + 2 \times [8 - (3 - 1)] = ?$$

First, simplify the operations inside parentheses. Begin with the innermost parentheses:

$$(3 - 1) = 2$$
$$4^3 + 2 \times [8 - 2] = ?$$

Then, move on to the next-outer parentheses:

$$[8 - 2] = 6$$
$$4^3 + 2 \times 6 = ?$$

Now, simplify all exponents:

$$4^3 = 64$$
$$64 + 2 \times 6 = ?$$

Next, perform the remaining multiplication:

$$2 \times 6 = 12$$
$$64 + 12 = ?$$

Finally, perform the addition:

$$64 + 12 = 76$$

Practice

1. $2^3 \div 2 + 4 \times (9 - 2^2) =$

2. $\dfrac{2 \times (6-3) + 8}{4 \times 2 - 6} =$

(See the last page of the Skills Handbook for the answers to practice problems.)

Algebraic Rearrangements

Algebraic equations contain *constants* and *variables*. Constants are simply numbers, such as 2, 5, and 7. Variables are represented by letters such as x, y, z, a, b, and c. Variables are unspecified quantities and are also called the *unknowns*. Often, you will need to determine the value of a variable in an equation that contains algebraic expressions.

An algebraic expression contains one or more of the four basic mathematical operations: addition, subtraction, multiplication, and division. Constants, variables, or terms made up of both constants and variables can be involved in the basic operations.

The key to finding the value of a variable in an algebraic equation is that the total quantity on one side of the equals sign is equal to the quantity on the other side. If you perform the same operation on either side of the equation, the results will still be equal. To determine the value of a variable in an algebraic expression, you try to reduce the equation into a simple value that tells you exactly what x (or some other variable) equals.

Look at the simple problem below:

$$8x = 32$$

If you wish to solve for x, you can multiply or divide each side of the equation by the same factor. You can perform any operation on one side of an equation as long as you do the same thing to the other side of the equation. In this example, if you divide both sides of the equation by 8, you have the following:

$$\frac{8x}{8} = \frac{32}{8}$$

The two 8s on the left side of the equation cancel each other out, and the fraction $\frac{32}{8}$ can be reduced to give the whole number *4*. Therefore, $x = 4$.

Next, consider the following equation:

$$2x + 4 = 16$$

If you divide each side by 2, you are left with $x + 2$ on the left and 8 on the right:

$$x + 2 = 8$$

SKILLS HANDBOOK

Now, you can subtract 2 from each side of the equation to find that $x = 6$. In all cases, the operation that is performed on the left side of the equals sign must also be performed on the right side.

Practice

1. Rearrange each of the following equations to give the value of the variable indicated with a letter.

a. $8x - 32 = 128$

b. $6 - 5(4a + 3) = 26$

c. $-3(y - 2) + 4 = 29$

d. $-2(3m + 5) = 14$

e. $\left[8\frac{(8+2z)}{32}\right] + 2 = 5$

f. $\frac{(6b + 3)}{3} - 9 = 2$

(See the last page of the Skills Handbook for the answers to practice problems.)

Scientific Notation

Many quantities that scientists deal with are very large or very small values. For example, light travels at about 300,000,000 m/s, and an electron has a mass of about 0.000 000 000 000 000 000 000 000 0009 g. Obviously, it is difficult to read, write, and keep track of numbers such as these. We avoid this problem by using a method that deals with powers of the number 10.

Study the positive powers of 10 shown in the following table. You should be able to check these numbers by using what you know about exponents. The number of zeros in the equivalent number corresponds to the exponent of the 10, or the power to which the 10 is raised. The equivalent of 10^4 is 10,000, so the number has four zeros.

But how can we use the powers of 10 to simplify large numbers such as the speed of light? The speed of light is equal to 3 × 100,000,000 m/s. The factor of 10 in this number has 8 zeros, so the number can be rewritten as 10^8. So, 300,000,000 can be expressed as 3×10^8.

Powers of 10

Power of 10	Decimal equivalent
10^4	10,000
10^3	1,000
10^2	100
10^1	10
10^0	1
10^{-1}	0.1
10^{-2}	0.01
10^{-3}	0.001

Negative exponents can be used to simplify numbers that are less than 1. Study the negative powers of 10 in the table above. In these cases, the exponent of 10 equals the number of decimal places you must move the decimal point to the right so that there is one digit just to the left of the decimal point. In the case of the mass of an electron, the decimal point has to be moved 28 decimal places to the right for the numeral 9 to be just to the left of the decimal point. The mass of the electron, about 0.000 000 000 000 000 000 000 000 0009 g, can be rewritten as about 9×10^{-28} g.

Scientific notation is a way to express numbers as a power of 10 multiplied by another number that has only one digit to the left of the decimal point. For example, 5,943,000,000 is written as 5.943×10^9 when expressed in scientific notation. The number 0.000 0832 is written as 8.32×10^{-5} when expressed in scientific notation.

Practice

1. Rewrite the following values using scientific notation.

a. 12,300,000 m/s

b. 0.000 000 000 0045 kg

c. 0.000 0653 m

d. 55,432,000,000,000 s

e. 273.15 K

f. 0.000 627 14 kg

(See the last page of the Skills Handbook for the answers to practice problems.)

Significant Digits

The following list can be used to review how to determine the number of *significant digits* (also called *significant figures*) in a given value or measurement. Significant digits are shown in red below.

Rules for Significant Digits:

1. All nonzero digits are significant. For example, 1,246 has four significant digits.
2. Any zeros between significant digits are also significant. For example, 1,206 has four significant digits.
3. Zeros at the end of a number but to the left of a decimal are significant if they have been measured or are the first estimated digit; otherwise, they are not significant. In this book, they will be treated as not significant. For example, 1,000 may contain from one to four significant digits, but in this book it will be assumed to have one significant digit.
4. If a value has no significant digits to the left of a decimal point, any zeros to the right of the decimal point and also to the left of a significant digit are not significant. For example, 0.0012 has only two significant digits.
5. If a value ends with zeros to the right of a decimal point, those zeros are significant. For example, 0.1200 has four significant digits.

After you have reviewed the rules, use the following table to check your understanding of the rules. Cover up the second column of the table, and try to determine how many significant digits each number in the first column has. If you get confused, refer to the rule given.

Significant Digits

Measurement	Number of significant digits	Rule
12,345	5	1
2,400 cm	2	3
305 kg	3	2
235.0 cm	4	1 and 5
234.005 K	6	2
12.340	5	5
0.001	1	4
0.002 450	4	4 and 5

Rounding and Significant Digits

When performing mathematical operations with measurements, you must remember to keep track of significant digits. If you are adding or subtracting two measurements, your answer can have only as many decimal positions as the value that has the fewest number of decimal places. When you multiply or divide measurements, your answer can have only as many significant digits as the value that has the fewest number of significant digits.

Practice

1. Determine the number of significant digits in each of the following measurements:
   - **a.** 65.04 mL
   - **b.** 564.00 m
   - **c.** 0.007 504 kg
   - **d.** 1,210 K
2. Perform each of the following calculations, and report your answer with the correct number of significant digits and units:
   - **a.** 0.004 dm + 0.12508 dm =
   - **b.** 340 m ÷ 0.1257 s =
   - **c.** 40.1 m × 0.2453 m =
   - **d.** 1.03 g − 0.0456 g =

(See the last page of the Skills Handbook for the answers to practice problems.)

SKILLS HANDBOOK Graphing Skills Refresher

Line Graphs

In laboratory experiments, you will usually control one variable and see how it affects another variable. Line graphs can show these relationships clearly. For example, you might perform an experiment in which you measure the volume of a gas at different temperatures to determine how volume is related to temperature. In this experiment, you are controlling the temperature intervals at which the gas's volume is measured. Therefore, temperature is the independent variable. The volume of the gas is the dependent variable. The table below gives some sample data for an experiment that measures the volume of gas.

The independent variable is plotted on the *x*-axis. This axis is labeled "Temperature (K)" and has a range from 0 to 400 K. Be sure to properly label each axis, including the units.

The dependent variable is plotted on the *y*-axis. This axis is labeled "Volume (L)" and has a range of 0.0 to 1.4 L.

Experimental Data for Gas Volume Versus Temperature

Temperature (K)	Gas volume (L)
0	0.0
100	0.35
200	0.70
300	1.05
400	1.4

Think of your graph as a grid that has lines running horizontally from the *y*-axis and vertically from the *x*-axis. To plot a point, find the *x* value on the *x*-axis. For the example above, plot each value for time on the *x*-axis. Follow the vertical line from the *x*-axis until it intersects the horizontal line from the *y*-axis at the corresponding *y* value. For the example, each temperature value has a corresponding volume value. Place your point at the intersection of these two lines.

The line graph below shows how the data in the table may be graphed.

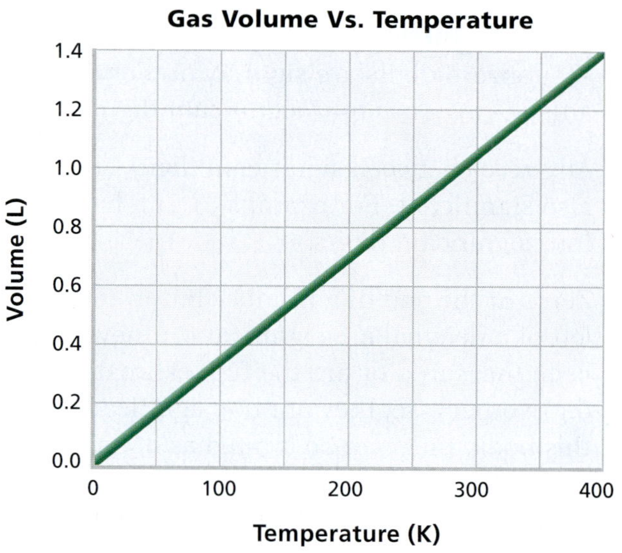

Bar Graphs

Bar graphs are useful for comparing data values. If you wanted to compare the area or depth of the major oceans, you might use a bar graph. The table below gives the data for each of these quantities.

Depth of the Major Oceans

Ocean	Depth (m)
Pacific Ocean	4,028
Atlantic Ocean	3,926
Indian Ocean	3,963
Arctic Ocean	1,205

To create a bar graph from the data in the table, begin on the *x*-axis by labeling four bar positions with the names of the four oceans. Label the *y*-axis "Depth (m)." Be sure the range on your *y*-axis includes 1,205 m and 4,028 m. Then, draw the bars to represent the area of each ocean.

Make sure the bar height on the *y*-axis matches each ocean's area value, as shown in the bar graph below.

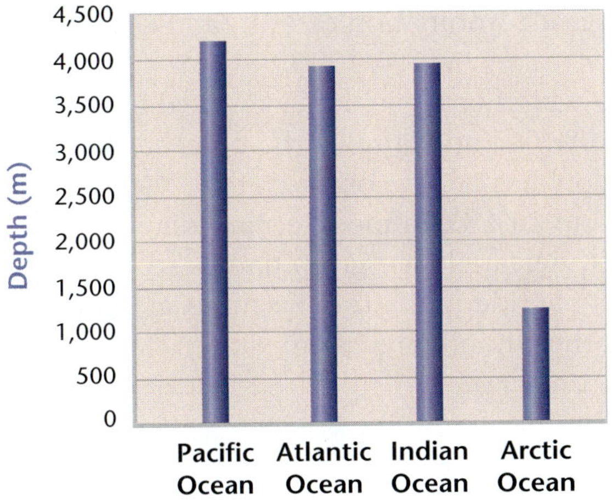

Pie Graphs

Pie graphs can help you visualize how many parts make up a whole. Frequently, pie graphs are made from percentage data. For example, you could create a pie graph that shows the percentage of different materials that make up the waste generated in cities of the United States. Study the example data in the table below.

United States Municipal Solid Waste	
Material	Percentage of total waste
Paper	38.1%
Yard waste	12.1%
Food waste	10.9%
Plastics	10.5%
Metals	7.8%
Rubber, leather, and textiles	6.6%
Glass	5.5%
Wood	5.3%
Other	3.2%

To create a pie graph from the data in the table, begin by drawing a circle to represent the whole, or total. Because all circles are 360°, 1% of a circle is equal to 3.6°. From this point, the pie graph can be constructed in two ways.

First, a protractor can be used to measure the number of degrees that are represented by a percentage of the circle. For example, if paper represents 38% of the municipal solid waste in the United States, that percentage would be equal to 38 × 3.6°, or 138.6°.

Second, the circle can be divided into 100 equal sections of 3.6° each. Then, you can shade in 38 consecutive sections and label that area "Paper." Continue to shade sections with other colors until the entire pie graph has been filled in and until each type of waste has a corresponding area in the circle, as shown in the pie graph below.

United States Municipal Solid Waste (Percentage by Weight)

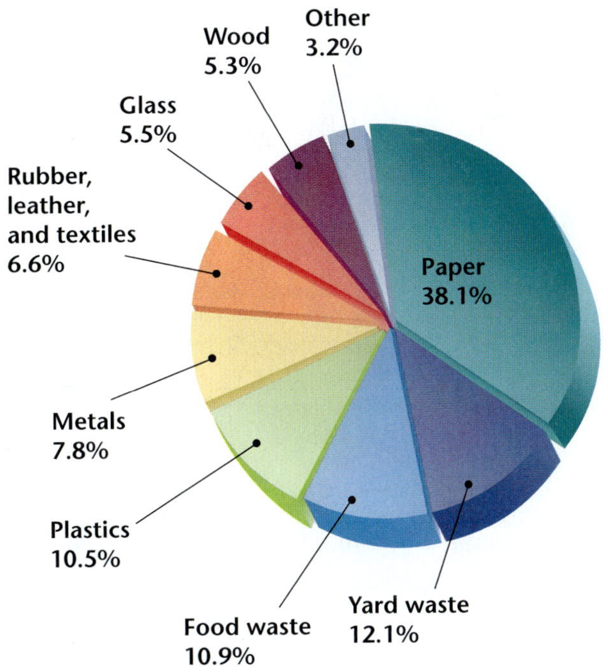

Ternary Diagrams

Ternary graphs, or ternary diagrams, show three variables on the same plot. Earth scientists use ternary diagrams to show composition of rocks and minerals and the physical states of rock material. The most common use of ternary diagrams is to represent the relative percentage of three components, such as three minerals or three elements.

The composition of any point on a ternary diagram can be described by first determining the percentage of each of the three components, as shown in the diagram below. In a ternary diagram, any point represents the relative percentage of three components: A, B, and C. The three components must always add up to 100%. In other words, the total composition of the mineral or rock represented by a given point on a ternary diagram is a combination of A, B, and C, so that x% A + y% B + z% C = 100%.

Readings of composition are stated as % A, % B, and % C. For example, the point in the diagram below has a composition of 40% A, 50% B, and 10% C. In most ternary diagrams, areas of the triangle are given names so that scientists can identify a rock or mineral by its name, rather than by its composition.

How to Read a Ternary Diagram

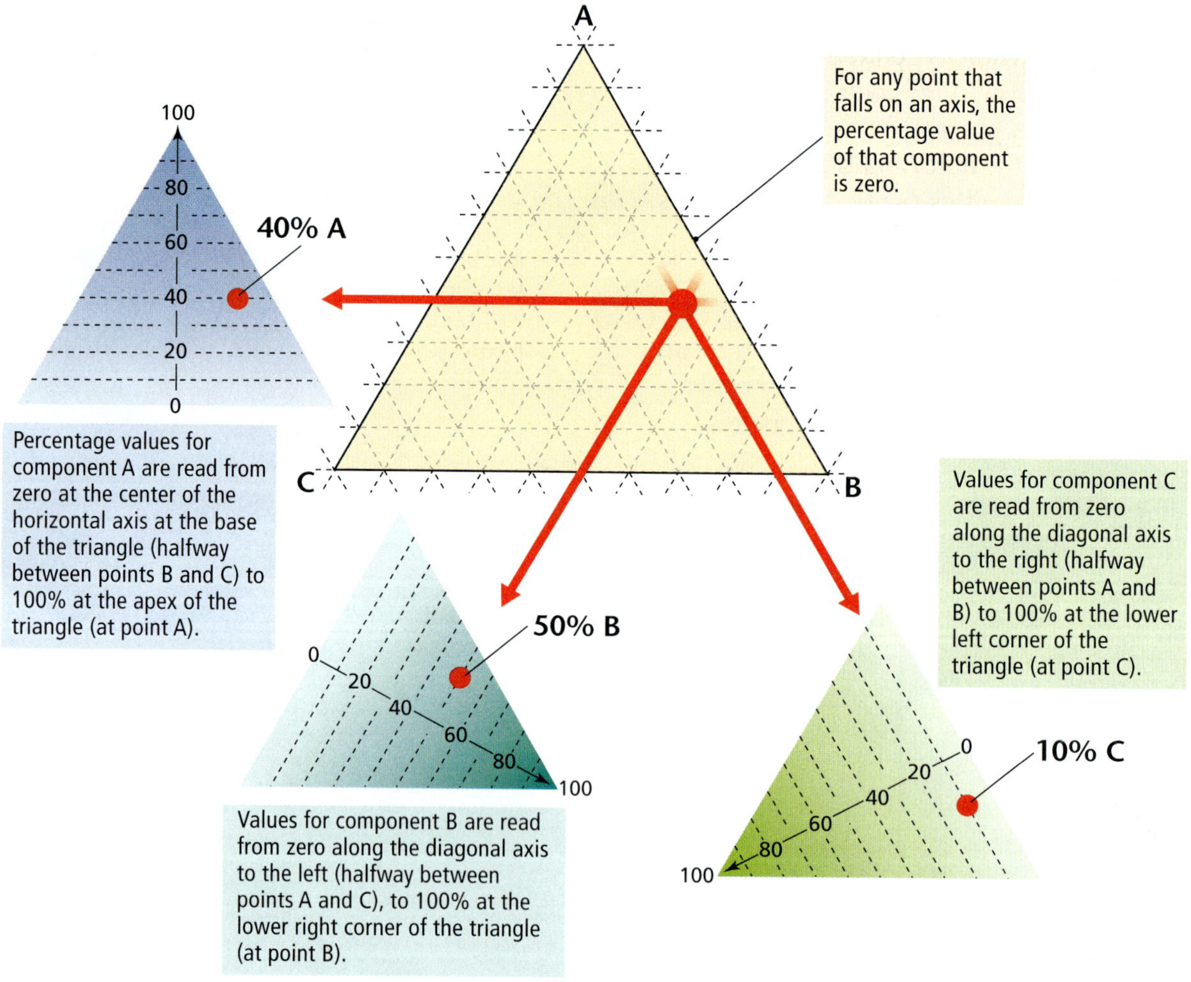

SKILLS HANDBOOK Chemistry Skills Refresher

Atoms and Elements

Every object in the universe is made up of particles of matter. Matter is anything that has mass and takes up space. An element is a substance that cannot be separated into simpler substances by chemical means. Elements cannot be separated in this way because each element consists of only one kind of atom. An atom is the smallest unit of an element that maintains the properties of that element.

Atomic Structure Atoms are made up of small particles called *subatomic particles*. The three major types of subatomic particles are **electrons, protons,** and **neutrons.** Electrons have a negative electrical charge, protons have a positive charge, and neutrons have no electrical charge. The protons and neutrons are packed close to one another and form the **nucleus.** The protons give the nucleus a positive charge. The electrons of an atom are located in a region around the nucleus known as an **electron cloud.** The negatively charged electrons are attracted to the positively charged nucleus.

Atomic Number To help in the identification of elements, scientists have assigned an **atomic number** to each kind of atom. The atomic number is equal to the number of protons in the atom. Atoms that have the same number of protons are all of the same element. An uncharged, or electrically neutral, atom has an equal number of protons and electrons. Therefore, the atomic number is also equal to the number of electrons in an uncharged atom. The number of neutrons, however, can vary for a given element. Atoms that have different numbers of neutrons but are of the same element are called **isotopes.**

Periodic Table of the Elements In a periodic table, the elements are arranged in order of increasing atomic number. Each element in the table is found in a separate box. In each horizontal row of the table, each element has one more electron and one more proton than the element to its left. Each row of the table is called a **period.** Changes in chemical properties across a period correspond to changes in the elements' electron arrangements. Each vertical column of the table, known as a **group,** contains elements that have similar properties. The elements in a group have similar chemical properties because they have the same number of electrons in their outer energy level. For example, the elements helium, neon, argon, krypton, xenon, and radon all have similar properties and are known as the *noble gases.*

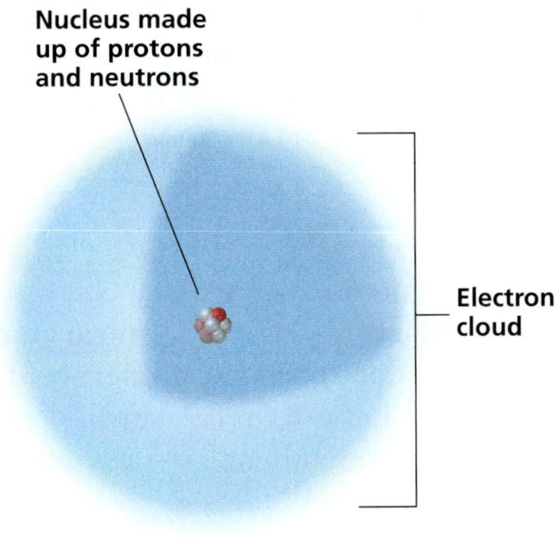

▶ The nucleus of the atom contains the protons and neutrons. The protons give the nucleus a positive charge. The negatively charged electrons are in the electron cloud surrounding the nucleus.

Molecules and Compounds

When the atoms of two or more elements are joined chemically, the resulting substance is called a **compound.** A compound is a new substance that has properties different from those of the elements that compose it. For example, water, H_2O, is a compound formed when atoms of hydrogen, H, and oxygen, O, combine. The smallest complete unit of a compound that has all of the properties of that compound is called a **molecule.**

Chemical Formulas

A chemical formula indicates the elements that make up a compound. The chemical formula also indicates the relative number of atoms of each element present. For example, the chemical formula for water is H_2O, which indicates that each water molecule consists of two atoms of hydrogen and one atom of oxygen.

Chemical Equations

A chemical reaction occurs when a chemical change takes place. (During a chemical change, new substances that have new properties form.) A chemical equation is a useful way of describing a chemical reaction by means of chemical formulas. The equation indicates the substances that react and the products. For example, when carbon and oxygen combine, they can form carbon dioxide. The equation for this reaction is as follows:

$$C + O_2 \rightarrow CO_2$$

Acids, Bases, and pH

An ion is an atom or group of atoms that has an electrical charge because it has lost or gained one or more electrons. When an acid, such as hydrochloric acid, HCl, is mixed with water, the acid separates into ions. An acid is a compound that produces hydrogen ions, H^+, in water. The hydrogen ions then combine with a water molecule to form a hydronium ion, H_3O^+. A solution that contains hydronium ions is an acidic solution. A base, on the other hand, is a substance that produces hydroxide ions, OH^-, in water.

To determine whether a solution is acidic or basic, scientists measure pH. **pH** is a measure of how many hydronium ions are in solution. The pH scale ranges from 0 to 14. The middle point, pH = 7, is neutral, neither acidic nor basic. Acids have a pH of less than 7; bases have a pH of more than 7. The lower the number is, the stronger the acid is. The higher the number is, the stronger the base is. A pH scale is shown below.

pH Measurements of Some Common Substances

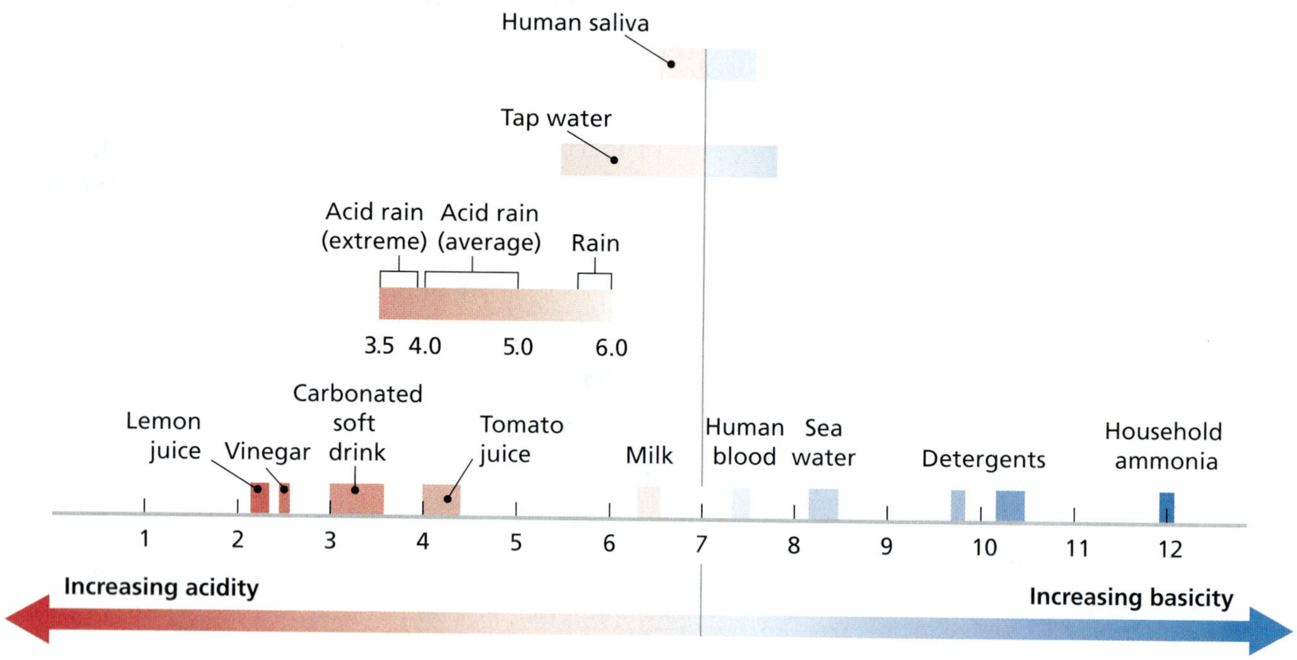

Physics Skills Refresher

Mass

All matter has mass. Mass is the amount of matter that makes up an object. For example, Earth is made of a very large amount of matter and therefore has a large mass. An object's mass can be changed only by changing the amount of matter in the object.

Weight

Weight is different from mass. Weight is a measure of the gravitational force that is exerted on an object. Objects that have large mass are heavier than objects that have a small mass, even if the objects are the same size.

Density

The mass per unit of volume of a substance is density. Thus, a material's density is the amount of matter it has in a given space. To find density, both mass and volume must be measured. Density is calculated by using the following equation:

$$density = \frac{mass}{volume}$$

Density is expressed in units of mass over units of volume. Most commonly, density is expressed as grams per cubic centimeter (g/cm^3) or as kilograms per cubic meter (kg/m^3).

The density of a particular substance is always the same at a given temperature and pressure. The density of one substance is usually different from the density of other substances. Therefore, density is a useful property for identifying substances.

Concentration

A measure of the amount of one substance that is dissolved in another substance is concentration. The substance that is dissolved is the solute. The substance that dissolves another substance is the solvent. Concentration is calculated by using the following equation:

$$concentration = \frac{mass\ of\ solute}{volume\ of\ solvent}$$

Concentration is expressed as mass of solute divided by volume of solvent. Most commonly, concentration is expressed as grams per milliliter (g/mL) or as kilograms per liter (kg/L).

Forces

In science, a force is simply a push or a pull. All forces have both magnitude and direction. Force is expressed using a unit called a newton (N). All forces are exerted by one object on another object.

More than one force can be exerted on an object at the same time. The net force is the force that results from combining all the forces exerted on an object. When forces are in the same direction, net force is calculated by using the following equation:

$$net\ force = force\ A + force\ B$$

When forces are in the opposite direction, net force is calculated by using the following equation:

$$net\ force = force\ A - force\ B$$

Pressure

The force exerted over a given area is pressure. Pressure can be calculated by using the following equation:

$$pressure = \frac{force}{area}$$

The SI unit for pressure is the pascal (Pa). Other common units of pressure include bars and atmospheres.

Speed

The rate at which an object moves is its speed. Speed depends on the distance traveled and the time taken to travel that distance. Speed is calculated by using the following equation:

$$speed = \frac{distance}{time}$$

The SI unit for speed is meters per second (m/s). Other units commonly used to express speed are kilometers per hour, feet per second, and miles per hour.

Velocity

The speed of an object in a particular direction is velocity. Speed and velocity are not the same, even though they are calculated using the same equation. Velocity must include a direction, so velocity is described as speed in a certain direction. For example, the speed of a plane that is traveling south at 600 km/h is 600 km/h. The velocity of a plane that is traveling south at 600 km/h is 600 km/h south.

Velocity can also be thought of as the rate of change of an object's position. An object's velocity remains constant only if its speed and direction don't change. Therefore, constant velocity occurs only along a straight line.

Acceleration

The rate at which velocity changes is called *acceleration*. Acceleration can be calculated by using the following equation:

$$acceleration = \frac{final\ velocity - starting\ velocity}{time\ it\ takes\ to\ change\ velocity}$$

Velocity is expressed in meters per second (m/s), and time is expressed in seconds (s). Therefore, acceleration is expressed in meters per second per second (m/s/s), or meters per second squared (m/s^2).

Inertia

The tendency of an object to resist any change in motion is called *inertia*. Because of inertia, an object at rest will remain at rest until something causes it to move. A moving object continues to move at the same speed and in the same direction unless something acts on it to change its speed or direction.

Momentum

The property of a moving object that is equal to the product of the object's mass and velocity is momentum. Momentum is calculated by using the following equation:

$$momentum = mass \times velocity, \text{ or } p = mv$$

The SI unit for momentum is kilograms multiplied by meters per second (kg•m/s)

When a moving object hits another object, some or all of the momentum of the first object is transferred to the other object. If only some of the momentum is transferred, the rest of the momentum stays with the first object.

Thermodynamics

The study of the behavior of the flow of energy in natural systems is thermodynamics. The laws of thermodynamics describe some of the basic truths of how energy behaves in the universe. Many Earth processes involve the flow of energy through the Earth system.

The First Law of Thermodynamics This law is often called the Law of Conservation of Energy. Simply stated, this law states that energy can be changed from one form to another but that it cannot be created or destroyed. Energy constantly changes from one form to another, but the total amount of energy available in the universe is constant.

The Second Law of Thermodynamics This law states that in all energy exchanges, if no energy enters or leaves the system, the potential energy of the new state will always be less than that of the initial state. In other words, no form of energy converts entirely to another form of energy without losing some energy as heat. So, the entropy of an isolated system always increases as time increases. Entropy is a measure of disorder, or randomness, of energy and matter.

The Third Law of Thermodynamics This law states that if all of the thermal motion of molecules, or kinetic energy, were removed from a system, a temperature called *absolute zero* would be reached. Absolute zero is in a temperature of 0 Kelvin or –273.15 degrees Celsius.

$$\text{absolute zero} = 0\text{K} = -273.15°\text{C}$$

ANSWERS Answers to Practice Problems

Reading and Study Skills

How to Make Power Notes

1. Sample answer:

The Experimental Method

Power 1: observing
- Power 2: observation

Power 1: hypothesizing and predicting
- Power 2: hypothesis
- Power 2: prediction

Power 1: experimenting
- Power 2: experiment
  - Power 3: variable
  - Power 3: experimental group
  - Power 3: control group

Power 1: organizing and analyzing data
- Power 2: data

Power 1: drawing conclusions

Power 1: repeating experiments

Power 1: communicating results

How to Make KWL Notes

1. a. The first step is observing.

b. A hypothesis is more than a guess. It must be based on observations and be testable by experiment.

c. A good experiment has a single variable and a control group.

Math Skills Refresher

Geometry

1. 225,000 m^2

2. 1,230 cm^3 (rounded to three significant figures)

3. 96 cm^2

Exponents

1. a. 9

b. 14,348,907

c. 537,824

d. 1

Order of Operations

1. 24

2. 7

Algebraic Rearrangements

1. a. $x = 20$

b. $a = -1.75$

c. $y = -6.3$

d. $m = -4$

e. $z = 2$

f. $b = 5$

Scientific Notation

1. a. 1.23×10^7 m/s

b. 4.5×10^{-12} kg

c. 6.53×10^{-5} m

d. 5.5432×10^{13} s

e. 2.7315×10^2 K

f. 6.2714×10^{-4} kg

Significant Digits

1. a. 4

b. 5

c. 4

d. 3

2. a. 0.129 dm

b. 2700 m/s

c. 9.84 m^2

d. 0.98 g

Journey to Red River

Teacher's Notes

Skills Acquired

- Classifying
- Communicating
- Inferring
- Interpreting
- Measuring

Chapter and Section Correlation

Students can find additional information to aid them in this expedition by reading the section entitled "Types of Maps" in the chapter entitled "Models of the Earth."

Materials

One set of materials is enough for a group of six students.

Tips and Tricks

1. Before beginning the exercise, review with students all of the features shown on the map, referring to the legend and to the table of topographic map symbols in the Reference Tables section of the Appendix. Have students study and describe the map's contour intervals.
2. Have groups of students plan hikes through different parts of the mapped region. For each hike, have a student use a pencil to draw the path on the map. Co-op Learning
3. Have each student write a brief description of the trail's difficulty, based on the terrain, and what features they would see on the hike. Co-op Learning

MAPPING EXPEDITIONS

Journey to Red River

Materials

- compass, magnetic, with degree markings (optional)
- ruler, metric

How do you get from one place to another when you don't know the route? Whether you are planning a trip on foot or by car or boat, a map can be very handy. The ability to read a map can help you reach your destination quickly and safely and can help you avoid becoming disoriented and lost.

The topographic map on the facing page shows the area around Red River, New Mexico. Imagine that you are traveling to Red River to do some camping, hiking, and sightseeing. Study the map for a few moments, and note the locations of roads, creeks, hills, and other features. Then, answer the questions below.

1. Red River lies in northeastern New Mexico near the Colorado border. The magnetic declination in Red River is about 13°E. Draw a diagram that shows how you would adjust a magnetic compass to determine true north in Red River. Why is distinguishing true north from geomagnetic north important?
2. You set up a tent at Mallette Campground. If you walk in a straight line from your campsite to the cemetery at the base of Graveyard Canyon, how far will you walk? Show your work.
3. You decide to hike from St. Edwin Chapel to location A. What is your elevation at location A? How much higher than your starting point is your destination? (Elevations on the map are given in feet.)
4. Notice that the road in the lower-right corner of the map winds back and forth to make a series of hairpin turns. Why did the road designers build the road this way?
5. Most United States Geological Survey maps, including this one, were created in the early 1960s. Like most towns, Red River has changed in the last few decades. Which features of the map might not reflect how Red River looks today? Which features are probably still accurate?

HEALTH CONNECTION BASIC

The Physically Challenged Have students imagine that they plan to visit this area with a physically challenged friend or relative who has trouble walking uphill or climbing. Have students study the map and choose a hiking trail that they can follow with their friend or relative. Then, have them mark the path by using a pencil. Have students explain how they used the contour intervals on the map to choose the least strenuous hike. LS Interpersonal

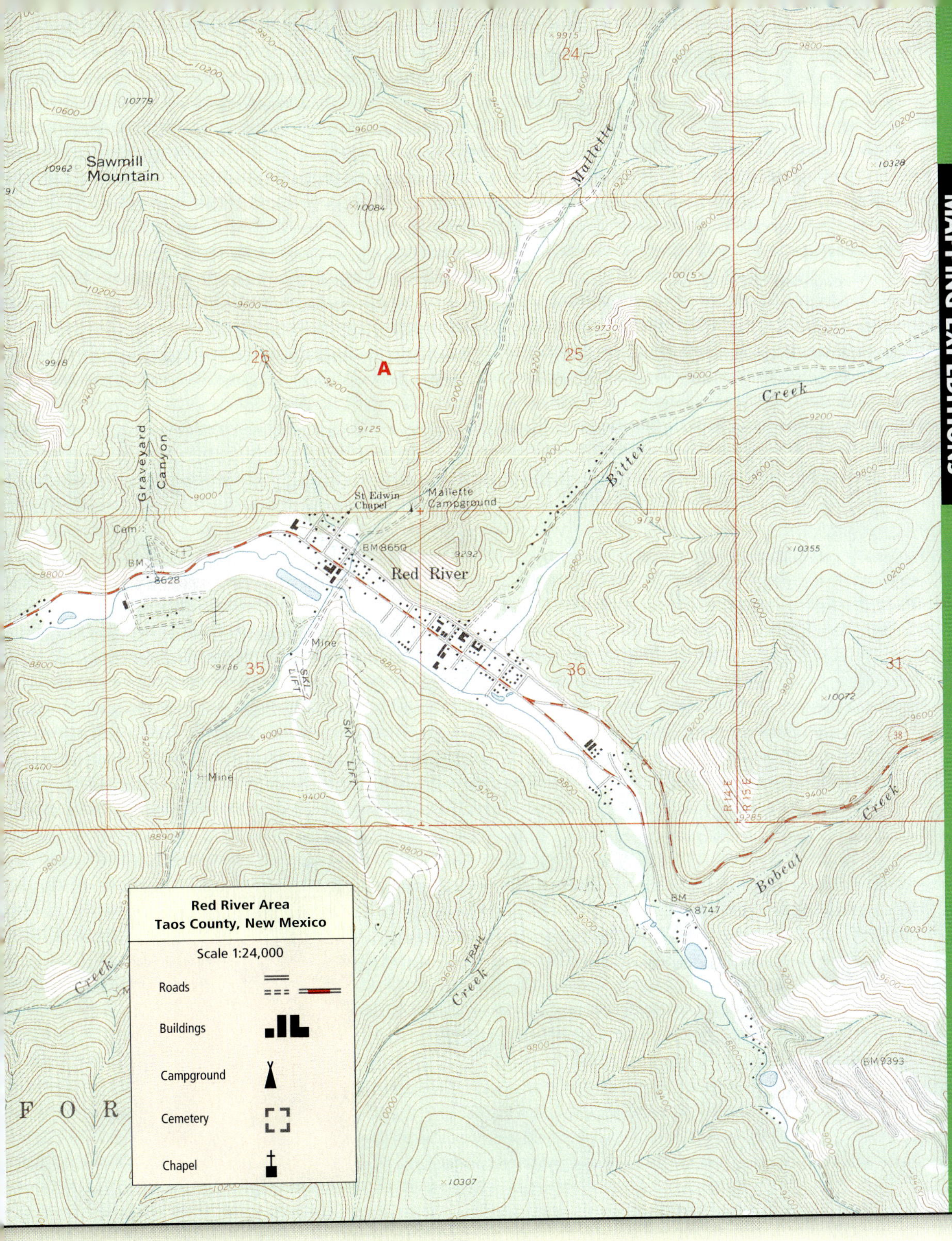

Mapping Expeditions

Answers

1. Drawings may vary but should show that a compass must be deflected by 13° clockwise to point to true north. True north and geomagnetic north are located in different places. Map directions are based on true north, but compasses point to geomagnetic north.
2. 6 cm × 24,000 = 144,000 cm ÷ 100,000 cm/km = 1.4 km
3. The elevation of location A is about 9,330 ft. The difference in elevation is 9,330 ft − 8,720 ft = 610 ft.
4. If the road led straight up the hill, it would be very steep. By winding back and forth, the road climbs the same hill without being as steep. The slope, or rise-to-run ratio, of the road is less when it winds back and forth.
5. Answers may vary. Accept all reasonable answers. Students may point out that the number and distribution of the buildings may have changed, while many of the natural features, such as hills and streambeds, are probably about the same as they were in the 1960s.

A Case of the Tennessee Shakes

Teacher's Notes

Skills Acquired

- Communicating
- Identifying and Recognizing Patterns
- Inferring
- Interpreting
- Predicting

Chapter and Section Correlation

Students can find additional information to aid them in this expedition by reading the section entitled "The Theory of Plate Tectonics" in the chapter entitled "Plate Tectonics," and the section entitled "Earthquakes and Society" in the chapter entitled "Earthquakes."

Materials

One road map of Tennessee is sufficient for a group of six students.

Tips and Tricks

1. Remind students that most earthquakes occur along plate boundaries, but that earthquakes may occur far from plate boundaries as a result of ancient fault zones or distant tectonic stresses.
2. Have groups of students create an evacuation plan for the towns and cities in the ETSZ in the event of a nuclear accident caused by an earthquake. Each student should create a plan for one town, and then coordinate their plan with the plans organized by other group members to ensure the most efficient, quickest, and least congested evacuation of the population. Co-op Learning

MAPPING EXPEDITIONS

MAPPING EXPEDITIONS

A Case of the Tennessee Shakes

Materials

- road map of Tennessee

Did you know that almost 10,000 earthquakes occur every day? In fact, an earthquake likely is occurring right now somewhere in the world. Fortunately, less than 2% of the earthquakes that seismographs record are strong enough to do serious damage.

You might think that scientists are most interested in strong earthquakes. But weak earthquakes can tell a seismologist (a scientist who studies earthquakes) as much as strong ones can.

Earthquake Frequency (based on observations since 1900)

Descriptor	Magnitude	Average occurring annually
Great	8.0 and higher	1
Major	7.0 to 7.9	18
Strong	6.0 to 6.9	120
Moderate	5.0 to 5.9	800
Light	4.0 to 4.9	about 6,200
Minor	3.0 to 3.9	about 49,000
Very minor	2.0 to 2.9 1.0 to 1.9	about 365,000 about 29,200,000

Part 1

1. Examine the map below. Which tectonic plates are involved in most earthquakes that occur in North America?
2. At tectonic plate boundaries, most earthquake epicenters are densely distributed, or closely packed. Why do most earthquakes occur along tectonic plate boundaries?

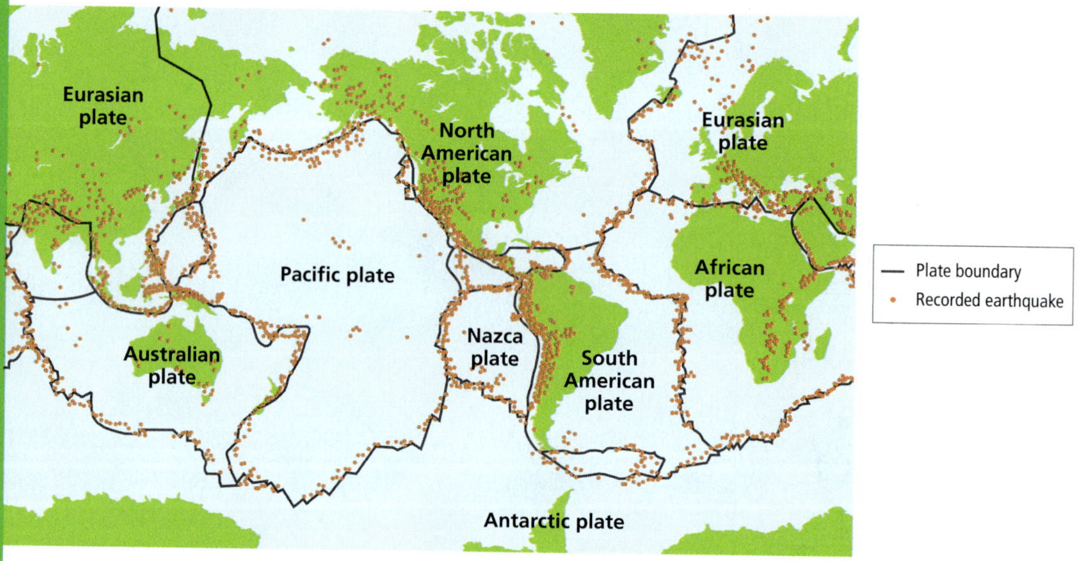

HISTORY CONNECTION

Biggest U.S. Quake The most damaging earthquake in the United States did not occur in California but in the interior of the continent along the New Madrid fault. On three separate days between December 16, 1811, and February 7, 1812, three intense earthquakes shook New Madrid, Missouri. Tremors were felt as far away as New York. No instruments were then available to measure the earthquakes' magnitudes, but scientists estimate that the largest of these tremors had a greater magnitude than the 1906 San Francisco earthquake.

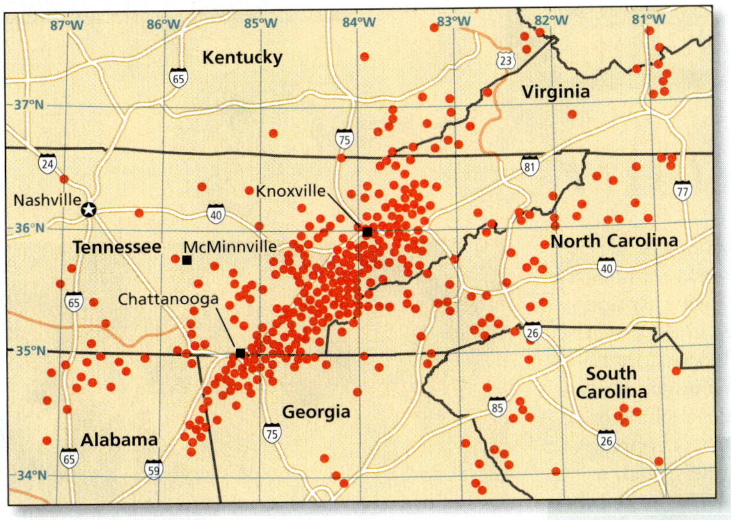

▶ Each dot on this map represents the epicenter of an earthquake. Most of these earthquakes, which occurred over a 20-year period, were too weak to be felt by people.

3 Some earthquakes, however, occur in the interior of the United States, which is far from any plate boundary. Propose a hypothesis that explains these earthquakes.

Part 2

The map above shows the epicenters of earthquakes in eastern Tennessee. However, Tennessee is far from any plate boundary. Some scientists think that the earthquakes in this region are the result of an ancient fault that has been reactivated. Other scientists think that a new fault zone is forming in eastern Tennessee. If they are correct, eastern Tennessee may experience a major earthquake in this zone.

1 Use the map to describe the location of the eastern Tennessee seismic zone (ETSZ) in terms of longitude and latitude.

2 Name at least two major cities that are located in the ETSZ. How could a major earthquake affect these cities?

3 Two nuclear power plants are located in the ETSZ. Imagine that a company has plans to build a plant near McMinnville, Tennessee. The United States Geological Survey has hired you to advise this company about the risk of a major earthquake. Briefly describe what you would say in a letter to the company. Explain your reasoning as clearly as possible.

▶ By using trench excavations, seismologist Karl Mueller can study sediments across the New Madrid fault in Tennessee to estimate the dates and magnitudes of past earthquakes.

Mapping Expeditions

Answers
Part 1

1. The North American and Pacific plates are involved in most earthquakes that occur in North America.
2. At plate boundaries, the lithosphere experiences stress caused by the movement of the plates against each other. Earthquakes occur when the stress becomes too great, and the rocks fracture and shift along faults.
3. Answers may vary. Accept all reasonable answers. Students may propose that ancient faults exist in those areas.

Part 2

1. The ETSZ runs diagonally from the northeast at about 37°N, 83°W, to the southwest at about 34°N, 86°W.
2. Chattanooga and Knoxville; An earthquake could damage roads, buildings, and other structures. It could disrupt communications and endanger people's lives.
3. Answers may vary but should indicate that McMinnville is located on the edge of the ETSZ. Therefore, there is some risk of an earthquake in McMinnville, but the risk is much less than if the plant were located in the center of the ETSZ.

Buried Treasure

Teacher's Notes

Skills Acquired

- Identifying and Recognizing Patterns
- Inferring
- Interpreting

Chapter and Section Correlation

Students can find additional information to aid them in this expedition by reading the section entitled "Combinations of Atoms" in the chapter entitled "Earth Chemistry," the section entitled "Identifying Minerals" in the chapter entitled "Minerals of Earth's Crust," and the section entitled "Mineral Resources" in the chapter entitled "Resources and Energy."

Materials

Detailed resource maps and population density maps for individual states may be available on the Internet.

Tips and Tricks

1. To help students answer question 6, you may want to display a physical map of the country that shows rivers and other natural features that would attract large populations.
2. Have each student in a group note the location of one type of resource and count or tally the concentration of this resource in each region. Students should pool their data to obtain information for all resources. Co-op Learning

ENVIRONMENTAL CONNECTION

MAPPING EXPEDITIONS

Buried Treasure

By many standards, the United States is one of the wealthiest countries in the world. Although this wealth is largely due to the ingenuity and hard work of the people who live in the country, it is also due to good fortune. The crust that lies beneath the United States holds a huge supply of natural resources. These resources include ores that contain precious metals, such as copper, silver, and gold, as well as fossil fuels, such as petroleum, coal, and natural gas. The availability and distribution of these resources have been important in shaping U.S. history.

▶ This worker in California cuts through steel, an iron alloy, at 2,000°F!

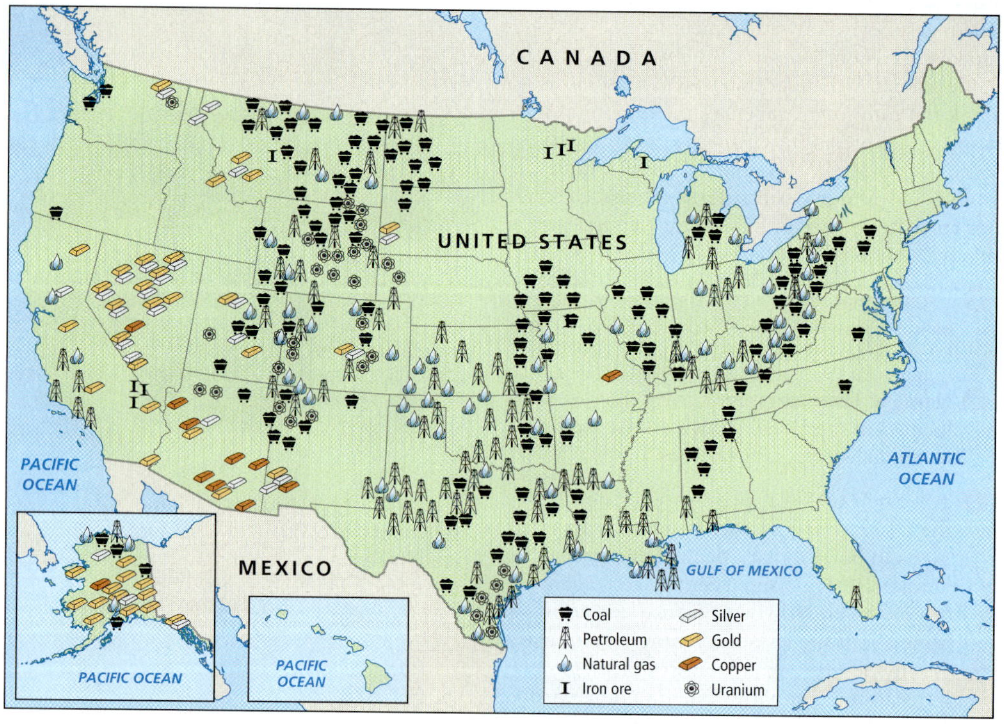

ENVIRONMENTAL CONNECTION

Limited Resources America's natural resources grew industry and created millions of jobs. Yet many of these resources are nonrenewable. Invite students to discuss the economic impact of resource depletion on regions where resources once were abundant. Interested students may find out if these areas are suitable for development of renewable energy resources that could provide jobs. LS Interpersonal

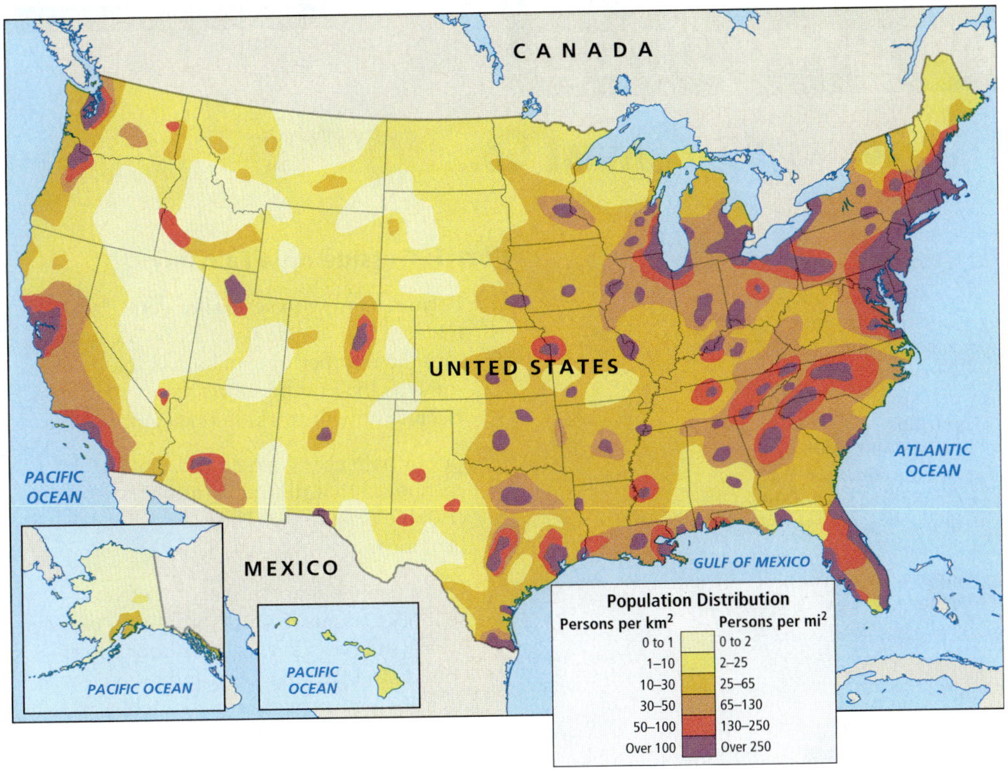

1. Find your state on the map of natural resources on the previous page. What resources are produced in your state?

2. Look at the locations of the various resources. Which resources are commonly located near each other? Why do certain resources commonly occur together?

3. When tectonic plates collide, pockets of hot magma may come into contact with cooler, solid rock. Using what you know about plate tectonics and the ways in which minerals form, describe why Washington has more iron-ore deposits than Nebraska does.

4. Describe the conditions that existed in the United States millions of years ago and that resulted in the formation of the modern petroleum and natural-gas deposits.

5. Compare the map on this page with the map on the previous page. What areas have both high concentrations of people and a large reserve of natural resources? What areas have many people but few resources? How could resources be transported to areas in which they are needed?

6. Using these two maps, would you say that most cities have grown up in places in or near which there are natural resources? Why or why not? What other factors could have influenced the location of cities?

7. Imagine that you work for a company that builds electrical equipment made primarily of copper. Why might southern Arizona be a good place to locate a new plant? What might be a disadvantage of locating your plant there?

Mapping Expeditions

Answers

1. Answers may vary but should reflect the natural resources in your state.
2. The fossil fuels (coal, petroleum, and natural gas) have roughly similar distributions, and the metals (silver, gold, copper, iron, and uranium) have roughly similar distributions. These patterns of distribution are probably due to similarities in the ways the deposits formed.
3. Washington lies in a fault zone, where tectonic activity is frequent and pockets of magma are likely to shift. Because ore deposits may form when magma cools or comes into contact with existing rocks, ore deposits are more likely to form in Washington than in Nebraska, which does not lie in a major fault zone.
4. Answers may vary but should indicate that fossil fuels commonly form from sediments that accumulated at the bottom of swamps, shallow lakes, and oceans, thus, swamps or an inland body of water may have covered most of the area that is now the central U.S.
5. Answers may vary. Accept all reasonable responses. The Great Lakes area, Texas, and California have large amounts of natural resources and many people. Parts of Florida have many people but few resources. Resources could be shipped by using water routes, trains, airplanes, or trucks.
6. In general, the locations of natural resources do not coincide with the metropolitan centers. Many cities are located along major transportation routes such as rivers and near coastlines.
7. Answers may vary. Sample answer: There are plentiful copper deposits in Arizona. However, few metropolitan centers are located nearby, which might make finding employees difficult.

What Comes Down Must Go...Where?

Teacher's Notes

Skills Acquired

- Collecting Data
- Measuring
- Predicting
- Identifying and Recognizing Patterns
- Organizing and Analyzing Data
- Interpreting
- Communicating

Chapter and Section Correlation

Students can find additional information to aid them in this expedition by reading the section entitled "Stream Erosion" in the chapter entitled "River Systems," the section entitled "Water Beneath the Surface" in the chapter entitled "Groundwater," and the section entitled "Precipitation" in the chapter entitled "Water in the Atmosphere."

Materials

Materials listed are enough for individual students or for small groups of two or three students.

Tips and Tricks

1. Have student teams prepare maps in advance, so they are ready to use the maps when rain occurs.
2. If you are in a region where rain does not occur often, you may want to use a garden hose to simulate rain on different parts of the school grounds.
3. Have groups of students divide the observation tasks so that one student observes pollution, one observes erosion, and one observes runoff. Then, have students pool their observations to make a final map. Co-op Learning

ENVIRONMENTAL CONNECTION

MAPPING EXPEDITIONS

What Comes Down Must Go...WHERE?

Materials

- cardboard, about 23 cm × 33 cm
- paper, about 23 cm × 33 cm
- pencils, red, blue, and purple
- permanent marker, fine-tipped
- plastic bag, reclosable, about 23 cm × 33 cm
- scissors
- umbrella, raincoat, or other rain gear

Imagine looking out your classroom window during a downpour. Billions of tiny raindrops splatter off of everything in sight. Streams of water fall from the roof and form dozens of puddles and streams on the ground. These miniature lakes and rivers swirl together, and tiny torrents carry away leaves, bits of trash, and other debris. A day or two later, the ground outside looks completely dry. Where did all of the water go?

In this activity, you will create a map of your school. After observing the type of ground cover and the slope of the terrain at various locations, you will predict whether rainwater will collect or run off at those locations. You will also look for possible sources of pollution and places where erosion might occur. Later, you will go outside in the rain and find out whether your predictions are correct.

Part 1: Outside on a Fair Day

1. Form a team with several of your classmates. Then, divide your school's campus into the number of equal areas that is the same as the number of teams in your class. Each team will work on one campus area.
2. Cut out a piece of paper and a piece of cardboard that fit exactly into a large reclosable plastic bag. The piece of paper will be your map.
3. On the paper, map one section of your school's campus. Include buildings, paved areas (such as sidewalks, outdoor sports courts, and parking lots), and vegetated areas (such as lawns, athletic fields, and wooded areas). The map on the next page is an example of the type of map that you will make.
4. **a.** Use a red pencil to mark the areas on your map. Draw arrows to indicate a downhill slope. Use a narrow arrow to indicate a steep slope and a wider arrow to indicate a gradual slope. Use circles to indicate flat areas.

ENVIRONMENTAL CONNECTION

Down the Drain Have interested students look for sewer drains located along the curbs around the school. Have them record the locations on their maps. Invite students to discuss what effect the sewer drains have on dispersing the pollutants or sediments they noted on their maps. Discuss how conditions at a single location (the school) often end up contaminating larger segments of the environment, such as the river the sewers might empty into. LS Logical

b. Use a blue pencil to draw arrows and circles that indicate where you think surface water may collect or flow during a steady rain. These areas may include low-lying areas, the roofs of buildings, gutters, and drainage ditches.

c. Use a purple pencil to mark the locations that you think might contribute pollution to the runoff. (These areas may include parking lots that contain oil stains or places where trash is usually found on the ground, such as near a dumpster.)

5 Seal your map and the piece of cardboard in the reclosable bag. When you are outside in the rain, use a permanent marker to write your observations on the outside of the bag.

Part 2: Outside During a Steady Rain

6 Dress appropriately, and go outside. Using the marker, write on your plastic-covered map the places where water collects and runs. In places where water moves along the ground, use arrows to show the water's direction. Use the letter *P* to mark the locations of pollutants that you observe in the water. Use the letter *E* to mark the locations where erosion seems to be occurring. (Look for soil or natural debris that is being washed along by moving water.)

Part 3: Back in the Classroom

7 Discuss your predictions for some of the locations. Were your predictions correct? How do you explain differences between your predictions and your observations?

8 Was the pollution that you observed suspended load, bed load, or dissolved load? Explain how there may have been pollution that you could not observe.

9 Explain how erosion on your school's campus could affect the erosion and deposition that occurs downstream from the campus.

10 Assemble the maps from your class into a single map of your school's campus. In your opinion, does most of the rainfall at your school become groundwater or runoff? Where does runoff go when it leaves your school's campus? Your school is probably part of a larger, local watershed. Find out what stream or other body of water the surface runoff in your area empties into.

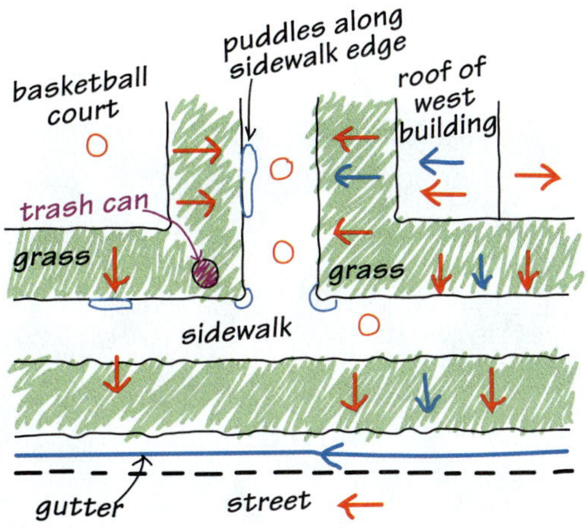

MAPPING EXPEDITIONS

Mapping Expeditions

Answers
Part 3

7. Answers may vary. Accept all reasonable answers. Not all of the subtle slopes of a terrain are visible on a dry day. Also, the movement of water on the ground depends on the ground's ability to absorb water.
8. Answers may vary depending on the pollution observed. Accept all reasonable answers. Students should indicate that dissolved pollutants may not be visible in water.
9. Erosion on campus could increase the erosion and deposition that occurs downstream. As materials are carried downstream, they may increase erosion by making the moving water more abrasive. Because the materials are carried downstream, deposition will also increase in areas downstream.
10. Answers may vary. Accept all reasonable answers.

Mapping Expeditions

Where the Hippos Roam

Teacher's Notes

Skills Acquired

- Communicating
- Identifying and Recognizing Patterns
- Inferring
- Interpreting
- Predicting

Chapter and Section Correlation

Students can find additional information to aid them in this expedition by reading the section entitled "The Fossil Record" in the chapter entitled "The Rock Record," the section entitled "The Mesozoic and Cenozoic Eras" in the chapter entitled "A View of Earth's Past," and the section entitled "Climate Change" in the chapter entitled "Climate."

Materials

Each student should use three different colored pencils. Students can share rulers.

Tips and Tricks

1. You may want to create a grid on the board so that students can check the correct placement of fossils in the grid.
2. Organize students into groups and have each student record the locations of one fossil type. Co-op Learning
3. After the grid is completed, have the group work together to answer the questions. Co-op Learning

MAPPING EXPEDITIONS

ENVIRONMENTAL CONNECTION

MAPPING EXPEDITIONS

Where the hippos Roam

Materials

- pencils, assorted colors
- ruler, metric

Millions of years ago, ancestors of modern crocodiles lurked in the shallow waters of lakes and other bodies of water. Like their current descendents, they hunted fish and other animals. If you could travel back in time to visit one of those lakes, you might see the ancestors of today's hippopotamuses there, too. Antelopes might browse along the edges of the lake, and rodents of various sizes might scurry back and forth.

When paleontologists examine the fossil of a prehistoric organism, they may discover clues about the organism's life. They may also answer questions about the organism's environment: Was the area hot or cold? Was it humid or dry? Then, by putting all of these clues together, the paleontologists may be able to learn a little more about how organisms and environments change over time.

Unfortunately, studying a fossil site is no easy task! Discoveries of complete organisms are rare. More often, a paleontologist may find a few teeth scattered over a very large area. In such cases, keeping track of where the fossils were found is very important. In this activity, you will use the data from a fossil site to create a map of fossil locations at that site. Then, you will draw some conclusions about the past environment, or *paleoenvironment,* at that location.

▶ The animals that lived near lakes millions of years ago probably had lives similar to the lives of animals that live near lakes today.

Map Answer Key

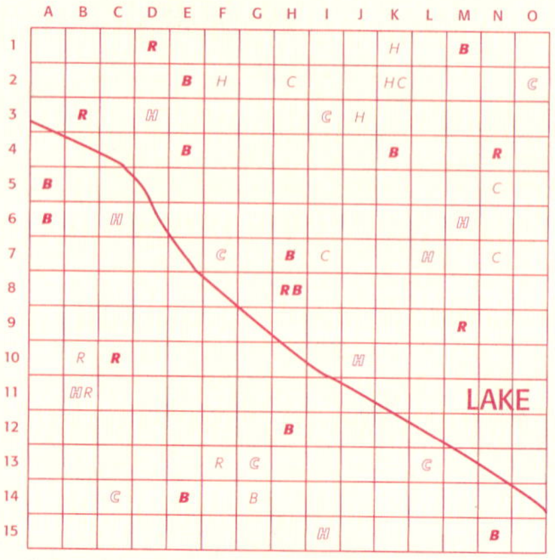

Key: Layer A, B, C

Location of Fossil Teeth				
Layer	Hippos	Rodents	Crocodiles	Bovids*
A	B11, C6, D3, I15, J10, L7, M6		C14, F7, G13, I3, L13, O2	
B	F2, J3, K1, K2	B10, B11, F13	H2, I7, K2, N5, N7	G14
C		B3, C10, D1, H8, M9, N4		A5, A6, E2, E4, E14, H7, H8, H12, K4, M1, N15

*Bovids are antelopes and other such animals.

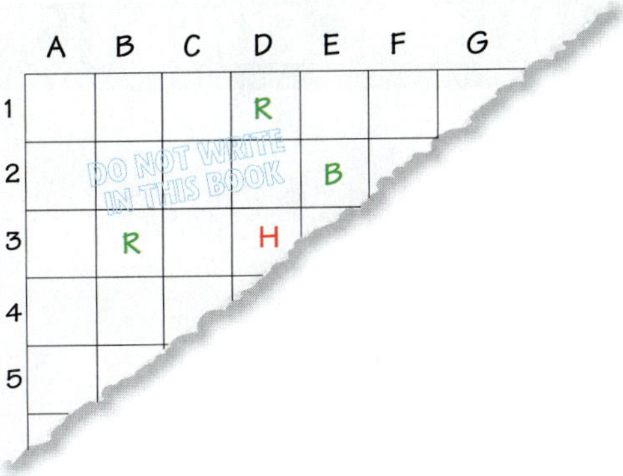

The table above shows the locations of fossils that were found spread out over 22,500 m^2. A team of paleontologists decided that this site, which measured 150 m × 150 m, was too large to work on all at once. So, the paleontologists decided to create a grid of 10 m squares. Starting in the northwest corner of the site, they labeled the squares from west to east with the letters *A–O*. Then, they numbered the squares 1–15 from north to south. Thus, each fossil could be labeled with a letter and a number that would identify where the fossil was found. For example, A1 would signify the 10 m × 10 m square in the northwest corner of the site, and O15 would signify the square in the southeast corner.

1. Create a map of the fossil site by drawing a grid similar to the one described above. The scale should be 1 cm = 10 m. Use letters to label across the top edge of the grid, and use numbers to label down the left edge of the grid. For each fossil, place a letter (*H* for a hippo fossil, *R* for a rodent fossil, *C* for a crocodile fossil, and *B* for a bovid fossil) in the square that corresponds to where the fossil was found. Use pencil color to represent the different layers of sediment, and make a key that shows which layer each color represents.
2. From the distribution of fossils in the layer of sediment just below the surface layer, what part of this site might have been underwater? Explain your answer, and devise a way to show that area on your map.
3. Describe how the environment at this site changed over time.
4. One team member wished to search this site for fossils of dry-climate plants. Which layer or layers would most likely yield fossils of such plants? Explain your answer.
5. One paleontologist suggested that tectonic uplift had raised the area's elevation over time and thus caused the climate to change. A second paleontologist thought that the area had probably lost elevation over time. With which scientist do you agree? Explain your answer.

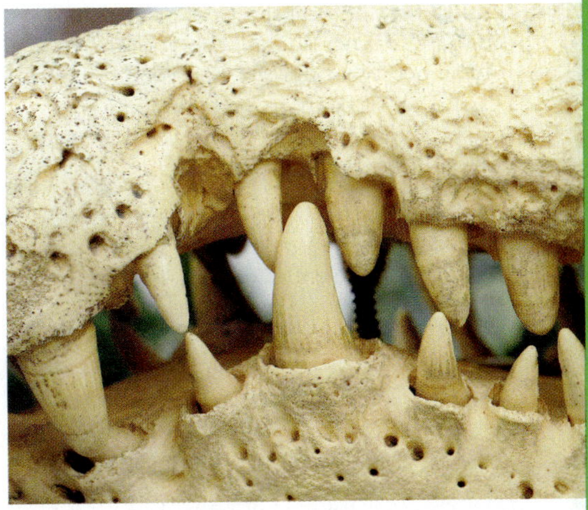

▶ Fossils, such as these crocodile teeth, help scientists learn what an area was like millions of years ago.

Mapping Expeditions

ENVIRONMENTAL CONNECTION

Hippos and Whales Hippos are members of the artiodactyl group of animals, which are ungulates that have an even number of toes. Peccaries, pigs, camels, and giraffes are artiodactyls. Some scientists think that hippos have a common ancestor with whales. Evidence indicates that ancient whales branched off from artiodactyls about 54 million years ago when some of these mammals took to the oceans. Hippos evolved from the main artiodactyl line beginning about 23.7 million years ago.

Answers

1. See the sample map on the bottom of the previous page.
2. the upper right-hand corner; Students should draw in the lake and label it.
3. The environment became wetter, changing from a dry area to an area covered by water as a lake or river advanced from the upper right corner.
4. Because the climate was driest during the earliest period of time, the lowest layer (layer C) would be the best place to find dry-climate plants. The lower lefthand corner of layer B would probably also yield dry-climate plants.
5. Answers may vary. Accept all reasonable answers. Sample answer: I think that the area lost elevation over time, because a lake eventually formed in the area, and water flows from higher elevations toward lower elevations.

Snapshots of the Weather

Teacher's Notes

Skills Acquired

- Communicating
- Collecting Data
- Constructing Models
- Identifying and Recognizing Patterns
- Inferring
- Interpreting
- Predicting

Chapter and Section Correlation

Students can find additional information to aid them in this expedition by reading the section entitled "Forecasting the Weather" in the chapter entitled "Weather."

Materials

You may wish to provide blank maps of the United States in step 5 rather than having students trace the maps.

Tips and Tricks

1. Before or after having students perform this activity, you may wish to have a discussion about the causes of weather. For item 3, if your area is not included on this map, you may wish to assign an area for students to focus on.
2. Place students in groups and have each student predict one aspect of the weather following the 4-day period shown in the maps. Have groups compare their completed weather maps and their predictions. Co-op Learning

MAPPING EXPEDITIONS

Snapshots of the Weather

Materials

- paper, tracing
- pencil

From looking at a weather map, you might get the impression that the clouds, fronts, and other features shown are standing still. However, weather patterns change constantly, and a weather map can show only what is happening at one particular instant. For this reason, meteorologists rely on a sequence of maps to make predictions about local weather.

In this activity, you will analyze a sequence of weather maps. The maps were taken from a daily newspaper and show weather patterns that occurred in the United States during a 4-day period. You will note what information the maps show and do not show, and you will make a few predictions based on your observations.

1. Look carefully at the maps on the next page. What weather information do they show? Now, look at the weather symbols in the Reference Tables section of the Appendix. What information is not included on these maps? Why might a newspaper exclude certain types of information on daily weather maps?
2. Why would a newspaper that serves only a specific geographic region publish the weather for the entire continental United States?
3. Describe how the weather patterns in your location changed during the 4-day period shown.
4. During what season do you think this 4-day period occurred? Explain your answer.
5. Trace the outline of one weather map on a separate piece of paper, but do not include any information on the map. Predict the locations of the fronts on the day following this 4-day period. Note the locations of the fronts on your new map.
6. Predict the temperature and precipitation patterns that occur in your location on the day following this 4-day period.

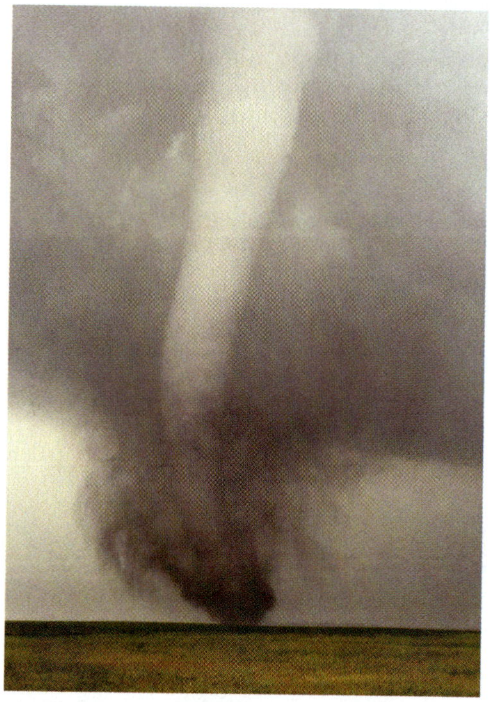

▶ This tornado twisted through Manchester, South Dakota, on June 24, 2003. On the same day, South Dakota had its largest recorded outbreak of twisters ever!

BRAIN FOOD

Awesome Power A medium-strength tornado that has wind speeds of about 200 mph (320 km/h) releases one billion watts of energy. Yet this energy pales in comparison with the energy in the supercell thunderstorm that generates the tornado. A supercell thunderstorm is 40,000 times more powerful than a tornado, generating 40 trillion watts of energy—enough energy to power the entire world for three years.

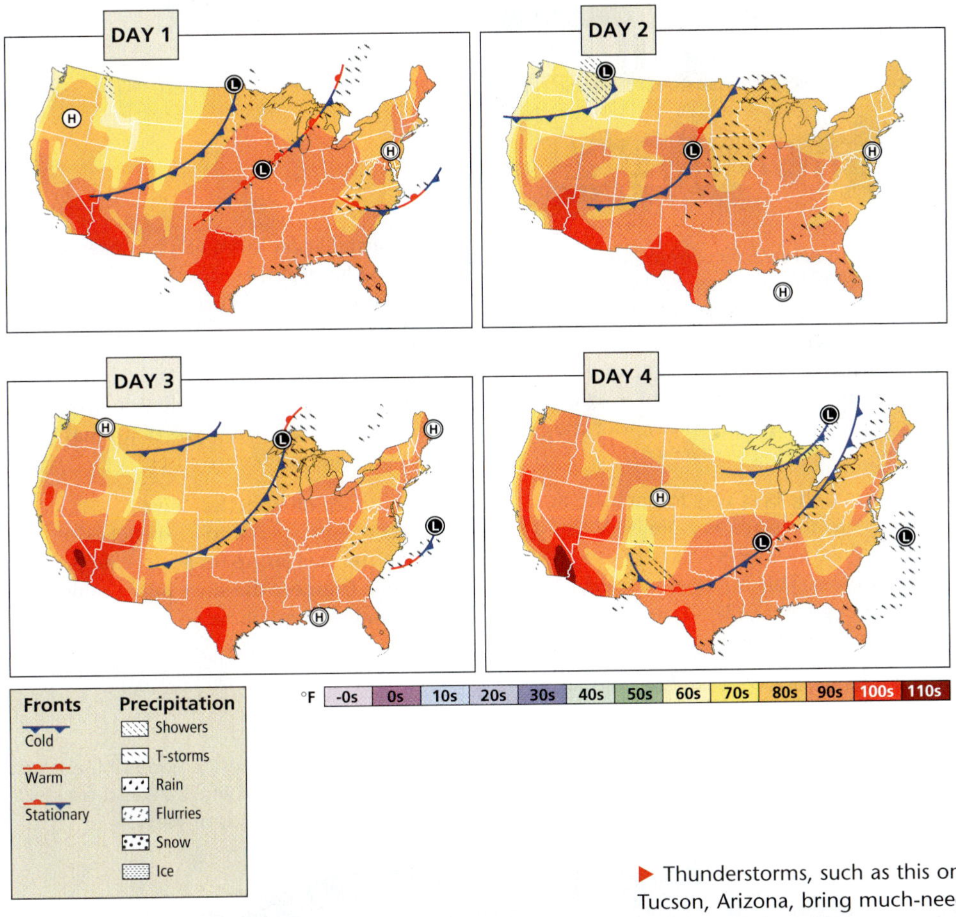

▶ Thunderstorms, such as this one in Tucson, Arizona, bring much-needed moisture to dry regions.

Mapping Expeditions

Answers

1. The maps show high temperatures and precipitation patterns. They also show the location of fronts and high- and low-pressure centers. The maps do not show wind speed or direction, nor do they show cloud cover, dew point, or specific atmospheric pressure readings. A daily newspaper might not include certain types of information because they are less relevant to a typical reader's life.
2. Answers may vary. Accept all reasonable responses. Travelers may need to know the weather in other parts of the country. In addition, knowing the weather conditions in other parts of the country can help readers understand weather in their own part of the country.
3. Answers may vary depending on location. Students should include both temperature and precipitation in their answers.
4. The temperatures shown are generally high. Therefore, this period probably occurred in summer.
5. Student maps should show two cold fronts–one moving across the East Coast and another stretching from the Midwest through upstate New York. The maps should show a stationary front across the western Gulf states. Student maps may also show a new warm or cold front entering the United States from the northwest.
6. Answers may vary depending on location. Accept all reasonable answers.

Stars in Your Eyes

Teacher's Notes

Skills Acquired

- Communicating
- Identifying and Recognizing Patterns
- Inferring
- Interpreting

Chapter and Section Correlation

Students can find additional information to aid them in this expedition by reading the section entitled "Star Groups" in the chapter entitled "Stars, Galaxies, and the Universe."

Materials

In addition to the star charts in the textbook, encourage students to find and study star charts for different locations on Earth and for different seasons.

Tips and Tricks

1. Bring in a book that illustrates the constellations, and the stars that make them, so students can easily see the shapes of constellations and identify how the stars are "connected."
2. Organize students into groups and have each student in a group trace one constellation from the map on the opposite page. The student should add the names, magnitudes, and temperatures of the stars that make up the constellation. Co-op Learning
3. Have all of the students in each group assemble their constellations and compare the different constellations in terms of which stars comprise them. Co-op Learning

MAPPING EXPEDITIONS

Stars in Your Eyes

A constellation is an arbitrary grouping of stars. The map to the right shows constellations that can be seen from the Northern Hemisphere of Earth. A Southern Hemisphere sky map, which is not provided, would show stars that can be seen from the Southern Hemisphere of Earth. Refer to the map to the right as you answer the questions below.

1. For many years, the best way to navigate at night was to use the stars as a guide. Many people used Polaris—the North Star—to orient themselves. This approach would not have worked for people all over the world, however. Why not?
2. What is the name of the constellation that contains Polaris?
3. What is the temperature of the star in the Bootes constellation, with magnitude 0?
4. The constellation Virgo can be observed from the Northern Hemisphere during the summer but not during the winter. Explain why.
5. Compare the constellation Draco on the map below with the constellation on the map to the right. What type of animal was this constellation named after?
6. Pick a group of stars that have not been connected into a constellation. Sketch the star group on a separate piece of paper. Connect the stars into a new constellation, and name your constellation.

▶ People, such as Sumerians, Greeks, Chinese, and Egyptians, have been grouping stars into constellations like these for thousands of years.

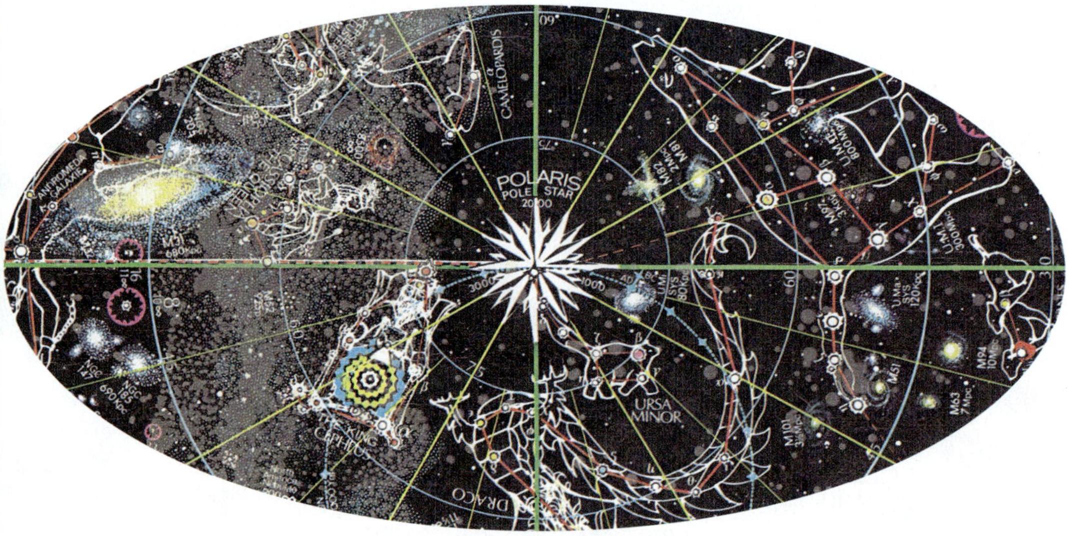

Activity — GENERAL

Astral Orientation Tell students that, in the Northern Hemisphere, the star that makes the outer, top edge of the cup in the Big Dipper constellation (part of Ursa Major) always points to Polaris, the North Star. Explain to students that they can use this relationship to orient themselves outside at night. Show students a picture that illustrates this concept. Have students look at the night sky to locate the Big Dipper and Polaris. Have them draw a picture of what they see. LS Visual

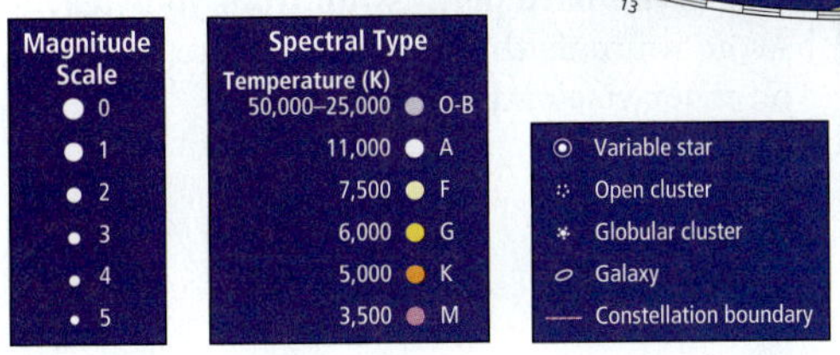

▶ There are 88 constellations recognized by modern astronomers. The Northern Hemisphere sky map shows those that can be seen from the Northern Hemisphere of Earth.

Mapping Expeditions

Group Activity —— ADVANCED

Hertzsprung-Russell Diagrams

Early in the twentieth century, two scientists developed a way to group stars based on their luminosity relative to our Sun. The Hertzsprung-Russell diagram is a graph that shows a star's luminosity on the *y*-axis and its temperature on the *x*-axis. You may provide an example of an H-R graph for students to examine. Have groups of students graph the luminosity of stars in one constellation, using the temperature data given and finding the luminosity of each star in a reference book. When their graphs are complete, have students compare the average luminosity of various constellations. **LS Logical**

Answers

1. Polaris is visible only in the Northern Hemisphere.
2. Ursa Minor, or Little Dipper
3. 5,000 K
4. The visible night sky changes as Earth circles the sun.
5. a dragon
6. Answers may vary. Accept all reasonable answers.

About Long-Term Projects

The six Long-Term Projects located in the Student Edition, and the additional four projects located in the Long-Term Projects workbook involve students working in astronomy, hydrology, oceanography, meteorology, and climatology. Ideally, the Long-Term Projects should demonstrate the relevance of Earth science to the student, the school, and the community.

Long-Term Projects provide students with the opportunity to go beyond library and Internet resources to explore scientific topics. You may wish to use the Long-Term Projects as

- suggestions for science fair projects
- field trip suggestions for individuals or the entire class
- science explorations carried out in a fun, nontraditional manner
- creative, hands-on projects for curious students in search of alternate learning methods
- ways for students to use multiple media to support their scientific findings
- ideas and inspirations for topics of further investigation

LONG-TERM PROJECTS

Introducing Long-Term Projects

Scientific investigations that lead to important discoveries are almost never short term. Usually, these investigations last months, years, and even decades before results are considered complete and dependable. Investigations in Earth science are no exception. The long-term projects included in this section will give you practical experience in investigating Earth science the way that Earth scientists do—over extended periods of time. You will observe changes over time, keep detailed records of your observations, and draw conclusions from your data. By following these steps, you will learn firsthand what it is like to be an Earth scientist.

Safety First!

Many of the long-term projects require you to make field trips to an observation site or to conduct your activities outdoors. Advance planning is essential. You should plan carefully for these investigations and should be certain that you are aware of the safety guidelines that must be followed. The following are general guidelines for fieldwork and lab work.

Conducting Fieldwork

Find out about on-site hazards before setting out. Determine whether there are poisonous plants or dangerous animals where you are going, and know how to identify them. Also, find out about other hazards, such as steep or slippery terrain.

Wear protective clothing. Dress in a manner that will keep you warm, comfortable, and dry. Wear sunglasses, a hat, gloves, rain gear, or other gear to suit local weather conditions. Wear waterproof shoes if you will be near water or mud.

Do not approach or touch wild animals unless you have permission from your teacher. Avoid animals that may sting, bite, scratch, or otherwise cause injury.

Do not touch wild plants or pick wildflowers without permission from your teacher. Many wild plants can cause irritation or can be toxic, and many are protected by law. Never taste a wild plant.

Do not wander away from the group. Do not go beyond where you can be seen or heard. Travel with a partner at all times.

Report any hazards or accidents to your teacher immediately. Even if an incident seems unimportant, tell your teacher about it.

Consider the safety of the ecosystem that you will be visiting as well as your own safety. Do not remove anything from a field site without your teacher's permission. Stay on trails when possible to avoid trampling delicate vegetation. Never leave garbage behind at a field site. Strive to leave natural areas just as you find them.

Conducting Lab Work

Be aware of safety hazards. Any field or lab exercises in which there are known safety hazards will include safety cautions and icons to identify specific hazards. By being aware of safety concerns, you may avoid accidents. Know where safety equipment and emergency exits are located so that you are prepared in the event of an emergency.

Do not engage in inappropriate behavior. Most laboratory accidents are caused by carelessness, lack of attention, or inappropriate behavior. Always be aware of your surroundings, and pay attention to safety cautions.

Be neat. Keep your work area free of unnecessary clutter. Tie back loose hair and loose articles of clothing. Do not wear dangling jewelry or open-toed shoes in the lab. Never eat or drink in the laboratory.

Clean your lab station when your lab time is over. Before leaving the lab, clean up your work area. Put away all equipment and supplies, and dispose of chemicals and other materials as directed by your teacher. Turn off water, gas, and burners, and unplug electrical equipment. Wash your hands with soap and water after working on any lab.

For additional information about safety in the lab and in the field, refer to the Lab and Field Safety section in the front of this book. Don't take any chances with safety!

Long-Term Projects

Using the Long-Term Projects

Long-Term Projects can be assigned prior to, during, or after teaching a related chapter. The observation schedules for the investigations range from about one week to several months. Where the time frame is longer, the intervals between observations generally are longer. Therefore, student effort is roughly equal for each investigation.

Several of the Long-Term Projects—Positions of Sunrise and Sunset, Comparing Climate Features, and Planetary Motions—should be assigned at the beginning of the academic year. These three investigations have extended observation periods. As the topics of these investigations become pertinent to classroom lessons, the selected students could report on their observations. The entire class could contribute to the analyses and conclusions for the investigation. Throughout the year, you may want to schedule time for periodic progress reports from the students who have been assigned these Long-Term Projects.

Each Long-Term Project has associated teacher's notes and answers. The Long-Term Projects are not arranged to follow the sequence of chapters in the Student Edition. However, chapter correlations have been included in the teacher's notes for each investigation. Datasheets for all ten Long-Term Projects are available in the workbook entitled *Long-Term Projects* and on the One-Stop Planner CD-ROM.

Positions of Sunrise and Sunset

Teacher's Notes

Purpose
To investigate the sun's change of position along the horizon at sunrise and sunset

Introduction
Students are aware that the sun rises in the east and sets in the west, but many may not be aware that the sun's position on the horizon changes throughout the year as the part of Earth that is tilted toward the sun changes. This project lets students observe those changes.

Duration
Approximately eight or nine months (depending on the school year)

Frequency
Two observations per month

Suggested Schedule
Students will be making two observations on or near the 21st of each month for approximately eight or nine months (depending on the school year).

Skills Acquired
- Collecting Data
- Organizing and Analyzing Data
- Measuring
- Inferring
- Constructing models

The Scientific Method
In this lab, students will:
- Make Observations
- Ask Questions
- Test the Hypothesis
- Analyze the Results
- Draw Conclusions
- Communicate Results

LONG-TERM PROJECT 1

Duration
8 or 9 months

Objectives
- USING SCIENTIFIC METHODS **Observe and record** the positions of sunrise and sunset once per month.
- USING SCIENTIFIC METHODS **Graph and analyze** collected data that describe the positions of sunrise and sunset.
- **Predict** the positions of sunrise and sunset for 3 or 4 months.

Materials
- compass, magnetic
- glue
- paper
- paper, graph
- pen
- pencil
- poster board
- scissors
- twist tie (or pipe cleaner)

Safety

Positions of Sunrise and Sunset

You are probably aware that the sun rises in the east and sets in the west each day. What may not be obvious to you is that the positions of sunrise and sunset along the horizon differ from day to day in a specific pattern. As the positions of sunrise and sunset change, the amount of sunlight that an area receives also changes. In this investigation, you will observe the changes in the sun's position along the horizon at sunrise and at sunset. You will be making two observations on or near the 21st of each month for approximately 8 or 9 months (depending on the schedule of your school year).

PROCEDURE

1. Construct the bearing chart before taking any measurements. Copy the bearing chart from the next page onto a piece of paper.
2. Glue your copy of the chart to a piece of poster board. When the glue is dry, trim away the excess poster board.
3. Wrap the center of a twist tie once around the center of a pencil. Poke the ends of the twist tie through the center of the bearing chart to make a pointer for your chart.

The sun appears to rise and set at different positions relative to landmarks, such as these skyscrapers in Los Angeles, California.

CHAPTER RESOURCES

Workbooks

Long-Term Projects
- Positions of Sunrise and Sunset ADVANCED

4. On a cloudless morning just before sunrise, place the bearing chart on a level spot where no buildings or trees block your view of sunrise and sunset.
CAUTION Although sunlight is less intense at sunrise and sunset, you should never stare at the sun for extended periods of time.

5. Use the magnetic compass to determine the direction of north. Set the bearing chart so that 0° is pointing north and 180° is pointing south. (Note: Have the chart face the same direction for every observation, even months from now.) Try to align an edge of the chart with some permanent object near your observation point.

Step 1 Copy the chart below on a separate piece of paper, and use it to construct your bearing chart.

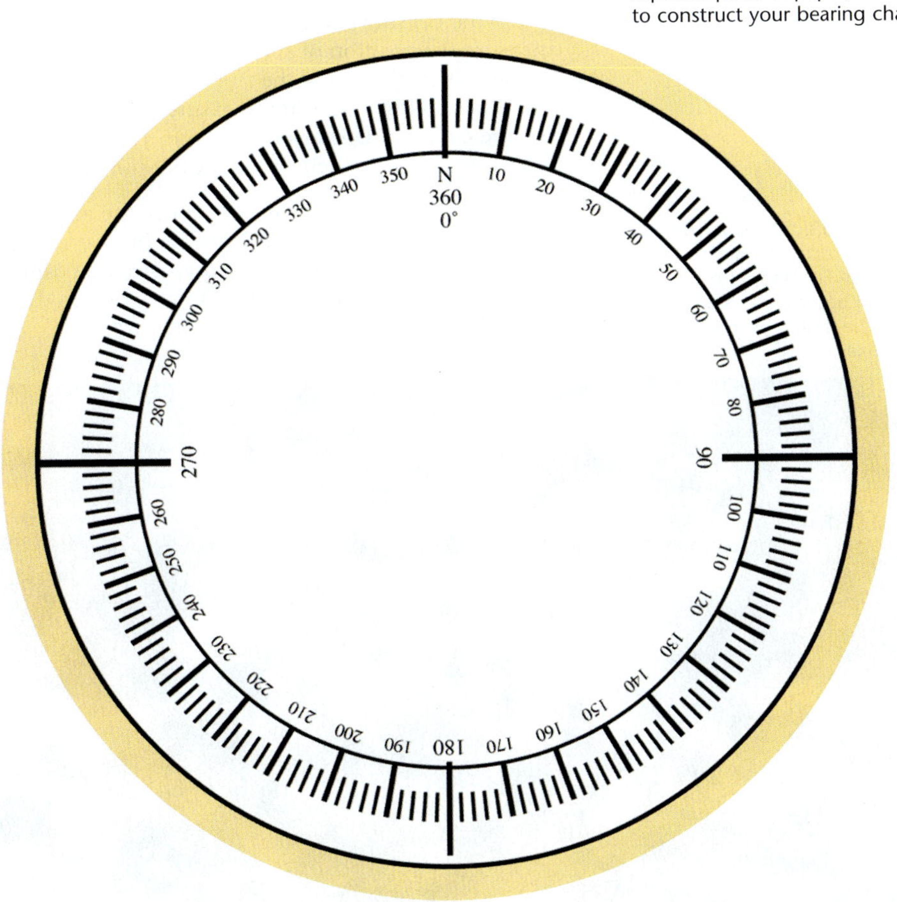

LONG-TERM PROJECTS

Long-Term Projects

Chapter Correlation

Students can find additional information to aid them in this investigation by reading the chapters entitled "Models of the Earth," "Studying Space," "Planets of the Solar System," and "The Sun."

Materials

The materials listed on this page are enough for one student.

Tips and Tricks

Advance Preparation

1. Prepare a list of open areas or high areas in your community, where the view of sunrise and sunset would not be blocked.
2. You may wish to review graphing skills and extrapolation with your students before they begin this activity. The Skills Handbook section of the Appendix contains a graphing skills refresher.
3. Students may need practice finding directions with a magnetic compass before they begin their observations.

Possible Problems

1. Some students may not be willing to get up before dawn to prepare for observations. Have students share the responsibilities for observing sunrise and sunset directions. Early risers could be matched with students who find it difficult to get up before dawn, or students could alternate so that one student does not have to make all of the observations. Co-op Learning
2. Consistent alignment of the bearing chart is critical. Students who are careless in the setup will find no pattern in the graphs.

Tips and Tricks, *continued*

3. The direction of magnetic north usually differs from true north by a few degrees. See the chapter entitled "Models of the Earth" for a map of magnetic declination throughout the United States. Have students adjust the alignment of the bearing chart to point true north so their graphs are accurate.
4. If the sky is cloudy or the sunrise or sunset is otherwise obscured, instruct students to perform the observation as close to the suggested date as possible.

Teaching Suggestions

1. A large magnetic compass marked from 0° to 360° can be used instead of the bearing chart. A small compass is too inaccurate for the range of directions in this investigation. The compass can be used like the bearing chart.
2. Have students correct for the local declination. Magnetic declination is discussed in the chapter entitled "Models of the Earth."
3. Emphasize to students that proper alignment of the bearing chart is essential.
4. Assigning students to staggered dates within each month will provide for nearly continuous observations. Co-op Learning

Step 7

Date	Position of sunrise (in degrees)	Position of sunset (in degrees)

DO NOT WRITE IN THIS BOOK

6. When the bearing chart is in place and properly aligned, measure the position of sunrise. (Note: Sunrise occurs at the instant the top of the sun appears on the horizon.) Without looking at the sun, measure its position by pointing the chart's pencil toward the sun. The shadow of the eraser end of the pencil will mark the sun's position.
7. In your lab notebook, draw a table like the one shown above. Record the position of sunrise (in degrees) in your table. That evening, measure and record the position of sunset (in degrees). (Note: Sunset occurs when the top edge of the sun drops below the horizon.)

Many scientists think that ancient people used structures, such as Stonehenge in England, to predict astronomical occurrences and to determine the timing of solstices and equinoxes.

8 Repeat steps 6 and 7 on the same date each month. (Note: On the equinoxes, the sun will rise due east and set due west. On the solstices, the sunrise and sunset will be shifted from these directions. The amount of the maximum shift will depend on the observer's latitude.)

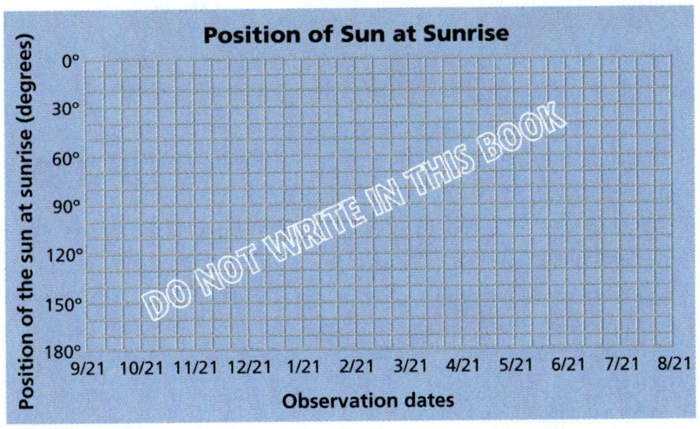

Step 9

9 On a sheet of graph paper, prepare a graph similar to the one shown at right. Label the graph's *y*-axis, which represents the position (in degrees) of the sun at sunrise, from 0° to 180°. The position of 90° will be in the center of the *y*-axis scale. The *x*-axis represents observation dates. Next, make a graph on which to plot the position of the sun at sunset. Label the *y*-axis of this graph from 180° to 360°.

10 On each graph, connect the points by drawing a smooth line. Estimate the positions of points between the plotted points.

ANALYSIS AND CONCLUSION

1. **Analyzing Data** On which date does the sun appear to follow the highest path across the sky? What happens to the length of daylight during this season?

2. **Analyzing Data** On which date does the sun appear to follow the lowest path across the sky? What happens to the length of daylight during this season?

3. **Analyzing Results** On which dates are the positions of the sun at sunrise and again at sunset about 180° apart? Those dates mark the beginning of which seasons?

4. **Describing Events** What general statement can you make about the pattern of sunrise and sunset according to the graphs?

5. **Forming a Hypothesis** Expand your graph to include the predicted position of the sun in June and July. Describe the pattern that you predicted.

Extension

1. **Evaluating Hypotheses** Use the process described in the investigation to chart the positions of sunrise and sunset for the months of June and July. How do your observations compare with your prediction?

2. **Making Comparisons** Use the same process to chart the positions of the moon at moonrise and moonset. Refer to an almanac when you choose the times at which you will make measurements. Do the positions of sunrise and sunset correlate to those of moonrise and moonset? What can you conclude from your observations?

Answers to Analysis and Conclusion

1. Sample answer: May 21; the length of daylight increases during this season.
2. Sample answer: December 21; the length of daylight increases during this season.
3. Sample answer: March 21 and September 22, respectively; these dates mark the beginning of spring and fall.
4. The graphs follow a curve with the highest recorded angle around June 21 and the lowest angle around December 21. There is a gradual curve upward from December 21 to June 21 and a gradual curve downward from June 21 to December 21.
5. Sample answer: I predict that the sun's angle will increase until June 21 and then begin to decrease.

Answers to Extension

1. Answers may vary but should indicate that the sun appeared successively higher in the sky until June 21 and then began to appear successively lower in the sky.
2. Answers may vary. The plane of the moon's orbit is about 5° off from the plane of Earth's orbit around the sun. The moon also moves around Earth at a different rate than Earth moves around the sun. Therefore, students should not see the same patterns in moonrise and moonset that they did in sunrise and sunset.

Air-Pollution Watch

Teacher's Notes

Purpose

To examine how the number of particles in the air may vary according to wind direction

Introduction

Any material that is foreign to the environment is called a *pollutant.* Some air pollutants in high concentrations may be dangerous to the general health of those in the area. The types of air pollutants vary significantly. Examples include dust and smoke particles, pollen, mold spores, and waste gases. If the tiny particles of pollutants remain suspended in the air for long periods of time, they are called *particulates.*

Wind direction has an effect on the number of particulates in the air. A microscope slide with a thin coating of petroleum jelly can be used to collect particulates. A microscope can then be used to view and count them. In this investigation, students collect and view a few types of particulates using basic laboratory equipment.

Duration

Two weeks

Frequency

Daily observations

Suggested Schedule

Students will need two weeks to complete this investigation. Have students make their observations in early fall or late spring. If you wish students to share their data and conclusions with the class, have them complete this investigation before you reach the chapter entitled "The Atmosphere."

LONG-TERM PROJECT 2

ENVIRONMENTAL CONNECTION

Duration

2 weeks

Objectives

- USING SCIENTIFIC METHODS **Count and record** the number of particulates in the air over a 2-week period.
- **Analyze** how wind direction and particulate source are related.

Materials

- compass, magnetic
- microscope
- microscope slides (8 or more)
- paper, graph
- pencil, grease
- petroleum jelly
- slide box
- tape, masking or packaging, or rubber bands

Safety

Air-Pollution Watch

When certain types of pollutants are present in high concentrations, they threaten the general health and well-being of humans. Some substances that can be air pollutants include dust and smoke particles, pollen, mold spores, and waste gases. If these tiny particles remain suspended in the air for long periods of time, they are called *particulates.*

Wind direction affects the number of particulates in the air. If there is a source of particulates in an area, there will be a large number of particulates in the air when the wind blows from the direction of that source. There will be fewer particulates in the air when the wind blows from the direction opposite the source.

In this investigation, you will collect and view a few types of particulates. You will collect particulates from an outdoor site every day for 2 weeks. Then, you will examine those particulates.

PROCEDURE

1. Select a collection site in an open area, such as a large field or pasture, where the wind can blow past the site from every direction.
2. Locate a four-sided post, such as a 4 in. × 4 in. fence post, that is firmly driven into the ground. Try to choose a post whose top is at least 1 m above ground level. If there are no fences in your area, look for another four-sided structure that you could use.

Haze and smog are common in large cities, such as Los Angeles, California.

CHAPTER RESOURCES

Workbooks

Long-Term Projects
- Air-Pollution Watch ADVANCED

Day	Date	Wind direction	Slide direction	Number of particulates 1	2	3	4	5	Total
1			N						
			S						
			E						
			W						

DO NOT WRITE IN THIS BOOK

Step 3

3. Use a compass to establish north, south, east, and west directions from the post's location. Determine the direction from which the wind is blowing by watching objects, such as a flag, move in the wind. Record the wind direction and the date in a table like the one shown above.

4. Look for any phenomena that may affect air quality, such as smokestacks or heavy traffic. Record these observations and the day's weather conditions.

5. Use a grease pencil to mark on the back of each microscope slide the direction that the slide will face when placed on the post. Place each slide on the appropriate side of the post.

6. Use tape or rubber bands to attach the slides to the post, as shown on the lower-right side of this page. Use your finger to spread a thin, even film of petroleum jelly on one side of each slide.

7. Return to the site the following day. Remove the slides. Place each slide carefully in the slide box. (Note: Do not touch the greased surface.)

Step 6

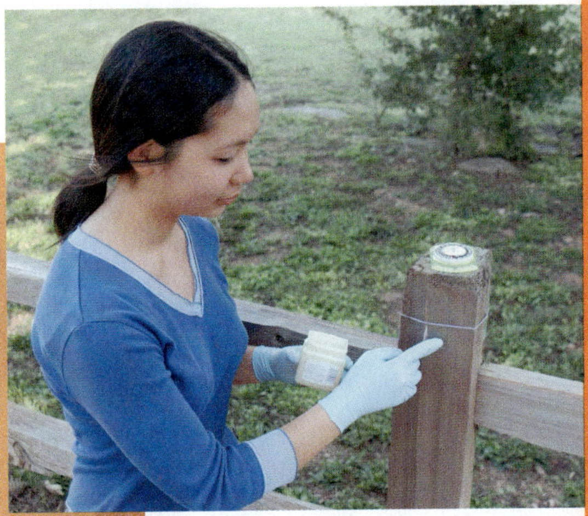

Long-Term Projects

Skills Acquired
- Measuring
- Organizing and Analyzing Data
- Inferring
- Constructing Models
- Identifying/Recognizing Patterns

The Scientific Method
In this lab, students will
- Make Observations
- Ask Questions
- Test the Hypothesis
- Analyze the Results
- Draw Conclusions
- Communicate the Results

Chapter Correlation
Students can find additional information to aid them in this investigation by reading the chapters entitled "Resources and Energy," and "The Atmosphere."

Materials
The materials listed for this project are enough for one student or one group collaborating on the project.

Tips and Tricks
Advance Preparation

1. Compile a list of collecting sites for students. These should be open areas where the wind can blow past the site from every direction. Check each site to be sure all are safe for students.
2. Make arrangements with the biology teacher to use microscopes for counting particulates.
3. You may want to review with students how to make a bar graph or histogram.
4. You may want to have students work with microscopes before they begin this investigation so that they can establish good microscope techniques.

Tips and Tricks, *continued*

Possible Problems

1. Students may need help in making accurate judgments of wind direction. You may wish to demonstrate this a few times before they begin the investigation.
2. Students may need help in using the microscope and deciding which objects observed on the slide are actually particulates and which are possibly blemishes in the grease film.

Teaching Suggestions

1. Class time will need to be set aside for students involved in this investigation because most students will have access to a microscope only at school.
2. If your school is in a suburban area, you may want to have students set up their observation posts in their backyards.
3. If your school is in an urban area, you may want to have students set up their observations on a rooftop, fire escape, or balcony. CAUTION: Emphasize to students that rooftops and fire escapes can be dangerous. Remind them not to set up their investigation near the edge of the roof or fire escape.
4. Have students place several slides on a windowsill before actually beginning the investigation. Students can make counts from these slides to become familiar with the techniques of greasing and counting the particulates on the slides.
5. If a certain wind direction seems to correlate with the highest particulate counts, investigation stations might be set up at varying intervals along this direction. The purpose would be to locate the source of the pollutants.
6. Different students could establish collecting sites at scattered locations or at varied altitudes. Such data might show that pollution is greater in different areas. Hypotheses could be established and tested.

Co-op Learning

8. Place new slides on the post, and record the wind direction and weather conditions.
9. Examine each slide under the microscope at 100×. Focus on one section that you have chosen at random. Count the number of particulates that you observe in the section. Record this number in column 1 of your table.
10. Move the slide, and examine another section that you have chosen at random. Count the number of particulates in this section. Record this number in column 2 of your table.
11. Repeat step 10 three more times so that you have a total of five observations per slide. Record the total number of particulates counted in the five sections.
12. Repeat steps 7 through 11 each weekday for 2 weeks.
13. When you have finished examining the slides and recording your results, total the number of particulates counted for each of the wind directions.
14. Using the data in your table, construct a bar graph. On the y-axis, plot the total number of particulates obtained in the past 5 days. On the x-axis, plot the day and wind direction. A sample graph is shown on the next page.

Step 9

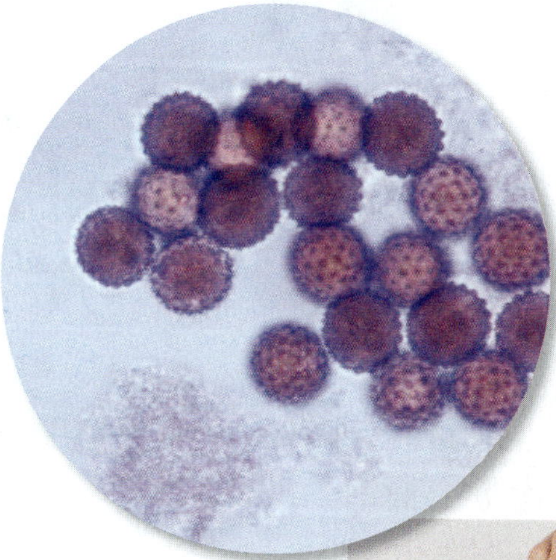

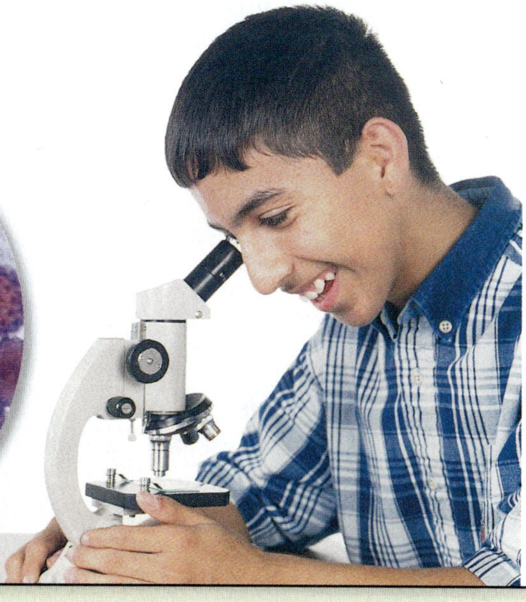

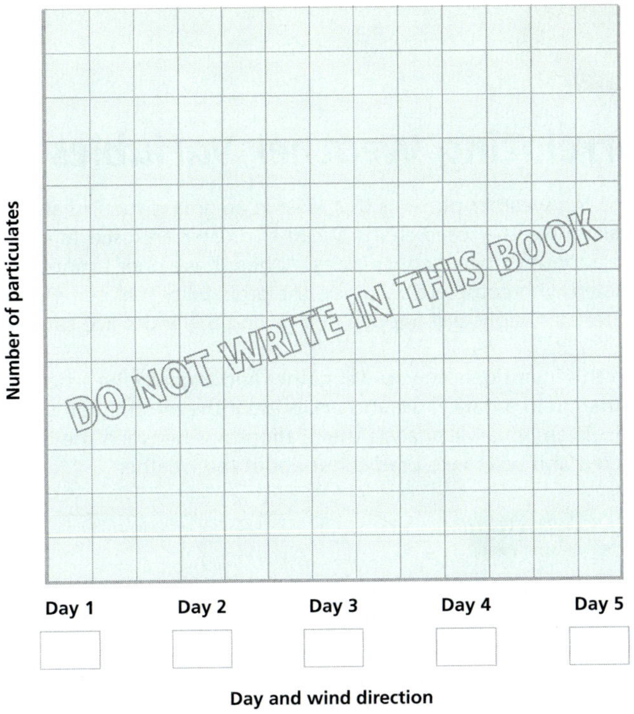

Step 14

ANALYSIS AND CONCLUSION

1. **Analyzing Results** For your location, did any one wind direction or group of wind directions result in more particulates than any other direction or directions did?
2. **Interpreting Information** What are possible sources of these particulates?
3. **Applying Conclusions** What would your results likely be if you set up this investigation near a populated urban area?
4. **Evaluating Methods** Did weather conditions have any effect on your results? Explain your answer.
5. **Graphing Data** Construct a second graph that uses data from all 10 days on which you collected samples. How does this graph differ from your previous graph?
6. **Identifying Patterns** Did you see a different pattern in the data when you plotted data from more than 5 days? Explain your answer.

Extension

1. **Research** Use the library or the Internet to research common particulates. Use your research to identify common particulates on several of the slides from this investigation. Which particulates are most common in your area? Explain why these particulates are most common.

LONG-TERM PROJECTS

Long-Term Projects

Answers to Analysis and Conclusion

1. Answers may vary depending upon local conditions.
2. Answers may vary depending upon local conditions such as proximity to industrial centers or highways.
3. Answers may vary. Numbers of particulates would most likely be higher due to heavy traffic and soot from buildings and factories. Wind direction might be affected by tall buildings.
4. Answers may vary. Humidity, rain, and wind speed are factors that could affect outcomes.
5. Answers may vary.
6. Answers may vary.

Answers to Extension

1. Answers may vary.

Correlating Weather Variables

Teacher's Notes

Purpose

To examine how the relationships between certain weather variables help meteorologists predict the weather

Introduction

To make a weather prediction, you need to identify weather patterns that exist in your area. Weather records can be used to identify relationships among variables. In this investigation, students record measurements of weather data. Then, they organize the information in a data table and a graph. Finally, students analyze the relationships among the data collected and make predictions about the weather.

Duration

Approximately one month

Frequency

Twice daily at about the same times each day

Suggested Schedule

Try to plan this activity so that it will terminate when you are about to start your weather unit.

Skills Acquired

- Collecting Data
- Identifying/Recognizing Patterns
- Predicting

The Scientific Method

In this lab, students will
- Make Observations
- Analyze the Results
- Draw Conclusions

Chapter Correlation

Students can find additional information to aid them in this investigation by reading the chapters entitled "Water in the Atmosphere" and "Weather."

LONG-TERM PROJECTS

LONG-TERM PROJECT 3

Duration

1 month

Objectives

- USING SCIENTIFIC METHODS **Measure and record** weather variables twice every day.
- **Predict** weather conditions based on data that you collected.

Materials

- aneroid barometer or barograph
- compass, magnetic
- paper, graph
- thermometer, Celsius

Safety

Correlating Weather Variables

Identifying weather patterns that exist in an area is the first step in making weather predictions. Weather records are used to identify relationships between variables. Although weather cannot be predicted with complete accuracy, the probability that certain weather conditions will happen at a given time and place can be established.

In this investigation, you will gather and organize weather information in a data table and will present the information as a graph. Then, you will analyze the relationships between the data collected and will make predictions about the weather.

PROCEDURE

1. Use graph paper to make two charts like the one shown at lower left. One chart will hold data from the first 15 days of the project. The other will hold data from the second 15 days of the project. At the top of each column, write the date on which the observation was made.
2. Twice daily for a month, at about the same times every day (about 10 hours apart), measure and record the following weather variables: temperature (°C), barometric pressure (mb), wind direction, cloud cover, and weather conditions. The chart on the following page explains how to measure and record these variables.
3. Use the month's data to make two graphs. Do so by connecting the points for temperature with one smooth line and the points for pressure with a second smooth line.

Step 1

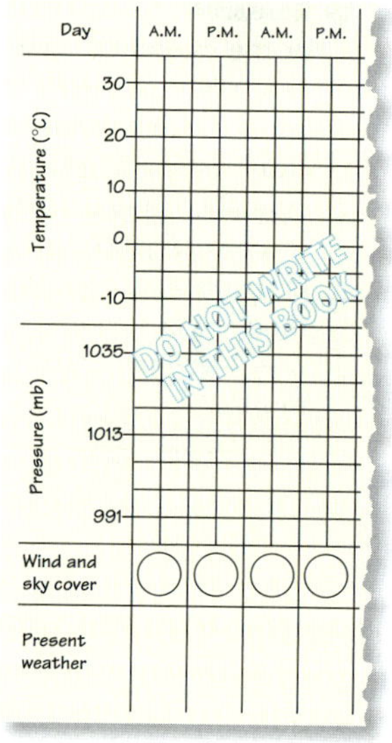

Materials

The materials listed for this project are enough for one student or one group collaborating on the project.

Tips and Tricks

Advance Preparation

1. If possible, secure the weather instruments ahead of time and set up a formal weather station.

CHAPTER RESOURCES

Workbooks

Long-Term Projects
- Correlating Weather Variables ADVANCED

How to Measure Weather Variables	
Weather variable	**How to measure and record the variable**
Temperature	Place a thermometer where it is in the shade and is not exposed to precipitation. Wait at least 3 min, and then read the temperature. Plot a point for that temperature on your chart.
Barometric pressure	Using a barometer, record the barometric pressure to the nearest tenth of a millibar. If your barometer is calibrated in "inches of mercury," change inches to millibars by using the Barometric Conversion Scale at right. Plot a point for the barometric pressure.
Wind direction	Determine the wind direction by using a weather vane or by observing objects moved by the wind. Wind direction is named according to the direction from which the wind blows. Use a compass to help determine direction. Using the symbols shown in the Table of Weather Symbols in the Reference Tables section of the Appendix of this book, record wind direction on the circles at the bottom of your chart.
Cloud cover	Estimate the amount of sky that is covered by clouds. Using the symbols shown in the Table of Weather Symbols in the Reference Tables section of the Appendix of this book, shade the circles at the bottom of your chart.
Weather conditions	Observe the present weather conditions. Using the Table of Weather Symbols in the Reference Tables section of the Appendix of this book, draw in your chart the symbol that most accurately indicates the weather conditions.

Step 2

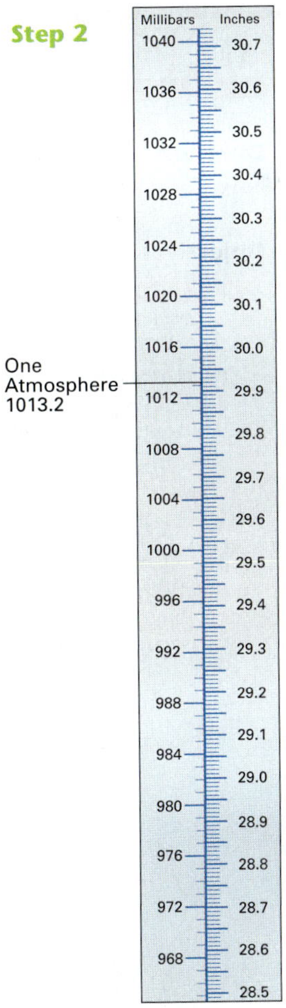

ANALYSIS AND CONCLUSION

1. **Evaluating Data** According to your graph, on how many days was the temperature falling? On how many days was the barometric pressure falling?
2. **Identifying Patterns** Of the days that had falling temperature, how many had rising barometric pressure?
3. **Inferring Relationships** In general, what is the relationship between temperature and pressure?
4. **Analyzing Results** What sky cover and wind direction are generally associated with falling barometric pressure?
5. **Interpreting Results** What weather conditions are generally associated with high barometric pressure? with low barometric pressure?
6. **Drawing Conclusions** How do the relationships between certain weather variables help you predict the weather?
7. **Evaluating Methods** Which weather variables are most useful in predicting precipitation?

Extension

1. **Research** Find out what the normal temperature, pressure, and precipitation are for your area during the time period in which you recorded your data. Do your observations match the normal conditions for that time period? How can you explain differences between your observations and the normal conditions?

Long-Term Projects

Teaching Suggestions

1. The relationships between pairs of weather data should be explained as a follow-up to the investigation.
2. Teams of students can share the responsibilities for measuring different variables at the same time or the same variables at different times. Co-op Learning

Answers to Analysis and Conclusion

1. Answers may vary.
2. Answers may vary but it is likely to be more than 50%.
3. Usually, as temperature increases, pressure decreases.
4. Answers may vary, but cloud cover of more than 50% and easterly winds typically accompany falling pressure.
5. A typical response for high pressure is cool temperatures, clear skies, and westerly winds. A typical response for low pressure is warm temperatures, cloudy skies, and easterly winds. Also, storms generally occur when barometric pressure is low.
6. If the wind direction changes to easterly, or the pressure is falling, stormy weather is likely to come soon. Conversely, fair weather can be predicted when the winds shift toward the west and the barometric pressure rises.
7. barometric pressure, temperature, cloud cover, and wind direction

Answers to Extension

1. Answers may vary.

Possible Problems

1. Interpretations of the amount of cloud cover may vary.
2. Many science textbooks use only SI units. The National Weather Service reports temperatures in degrees Fahrenheit, precipitation amounts in inches, and wind speeds in miles per hour or knots. Barometric pressure is plotted on maps using millibars, while it is reported in the news media in inches of mercury. Pressure in this investigation is indicated in millibars.

Weather Forecasting

Teacher's Notes

Purpose

To use a series of daily weather maps to track the movements of weather systems and then make a prediction of future weather conditions for a given location

Introduction

Every three hours, the National Weather Service produces a weather map. These maps are compiled from data received from about 800 weather stations located across the globe, even from ships and buoys at sea. Daily newspapers often summarize the most recent weather data in the form of a national weather map. The data usually include temperature, precipitation, cloud cover, and barometric pressure. Points of equal barometric pressure are connected by lines called *isobars*. The pattern of the isobars provides meteorologists with a picture of where high-pressure and low-pressure systems are located. Weather fronts can be identified, providing even more information about weather conditions in specific regions.

Duration

One week in winter and one week in spring

Frequency

Daily

Suggested Schedule

Students will need two weeks (one in winter and one in spring) to complete this investigation. If you plan to have students share their data and conclusions with the class, this investigation should be completed before you begin the chapter entitled "Weather."

LONG-TERM PROJECTS

LONG-TERM PROJECT 4

Duration

1 week

Objectives

- USING SCIENTIFIC METHODS **Observe and record** locations of weather fronts.
- **Predict** weather conditions based on data you collected.
- **Compare** the movement of weather fronts in different seasons.

Materials

- pencils, colored
- daily weather maps for consecutive days (5)

Weather Forecasting

Every three hours, the National Weather Service collects data from about 800 weather stations located around the world. Daily newspapers summarize this weather data in the form of national weather maps. The data include temperature, precipitation, cloud cover, and barometric pressure. The patterns produced by the data allow meteorologists to identify weather fronts and to provide information about weather conditions around the globe.

In this investigation, you will use a series of daily weather maps to track the movements of weather systems in the winter months. Then, you will use these data to predict weather conditions.

PROCEDURE

1. Find a local or national newspaper that prints a daily weather map from the National Weather Service. You may also use the Internet to find daily weather maps.
2. Cut out or print out the map, and write on it the date that it represents.
3. Make a data table similar to the sample table shown at the bottom of this page.
4. Fill in your table with the information from the weather map.
5. Make at least one copy of the blank weather map on the third page of this exercise.

	Week 1 (winter)					Week 2 (spring)		
	1	2	3	4	5	1	2	3
Temperature								
Barometric pressure								
Barometric trend (R = rising; F = falling; S = steady)								
Wind direction								
Cloud cover (C=clear; CL=cloudy; PC=partly cloudy; O=overcast)								
Present weather (rain, sleet, snow, etc.)								
Prediction								

DO NOT WRITE IN THIS BOOK

Step 3

CHAPTER RESOURCES

Workbooks

Long-Term Projects
- Weather Forecasting ADVANCED

A meteorologist from the National Weather Service tracks the path of a hurricane in the Gulf of Mexico.

6. On your copy of the map, put an *L* at any locations where low-pressure centers are shown on the daily weather map that you collected in step 1. Circle the *L*s with a colored pencil, and label each circle with the date.

7. Put an *H* on your map at the locations of any high-pressure centers. Circle the *H*s with a second colored pencil, and label each circle with the date.

8. Repeat steps 1–7, but use four consecutive daily weather maps that follow the first day's map, and reuse your data table and weather map. For each symbol, use the same colors that you used for the first day's map.

9. Draw arrows to connect the daily positions of each high-pressure center and of each low-pressure center.

10. Use the formula below to calculate the average velocity (in kilometers per day) of each high-pressure center and each low-pressure center. The average velocity equals the total distance traveled divided by the number of days traveled, or

$$\textit{average velocity} = \frac{\textit{total distance traveled}}{\textit{number of days}}$$

Long-Term Projects

Skills Acquired
- Collecting Data
- Organizing and Analyzing Data
- Inferring
- Interpreting
- Predicting

The Scientific Method
In this lab, students will
- Make Observations
- Ask Questions
- Test the Hypothesis
- Analyze the Results
- Draw Conclusions
- Communicate the Results

Chapter Correlation
Students can find additional information to aid them in this investigation by reading the chapter entitled "Weather."

Materials
The materials listed for this project are enough for one student.

Tips and Tricks
Advance Preparation
1. Make a list of local newspapers that contain a useful daily weather map. National newspapers contain weather maps that can also be used by students.
2. There are several other sources for daily weather maps, including:
   - Local airports have weather-service "facsimile" maps, which are often available to teachers after they have been used.
   - Large daily maps can also be obtained for a fee from: Superintendent of Documents Government Printing Office Washington, DC 20402
   - Weather maps are available at a variety of Web sites on the Internet.

Tips and Tricks, *continued*

Possible Problems

1. Students may have difficulty locating pressure centers on the weather map. Encourage students to use state borders as guides in locating these centers.
2. Some students may encounter problems with the computations of daily rates of movement of pressure centers.
3. Students will be concerned about being "wrong" with their forecasts. Emphasize that there is always a certain percentage of error involved with forecasting weather, especially because interpreting daily maps uses averages. Comparison with the Long-Term Project entitled "Correlating Weather Variables" may be helpful.

Teaching Suggestions

1. In order to make a long-term weather prediction, it is necessary to use student information on the rate of motion and the track of their pressure system. By daily observation of the student's maps, you can help the students determine both the speed and direction of one or more weather systems as they move across the United States.
2. Explain that a low-pressure system usually moves across North America with the prevailing westerly winds. The direction of movement of the low-pressure system tends to be across isotherms and parallel to isobars. Strong winds will tend to slow the motion of a low-pressure center. Emphasize that there can be significant variations from the average velocity when these strong winds exist. This information may help students explain why the movements of the pressure systems they are tracking are not always consistent.

Step 5

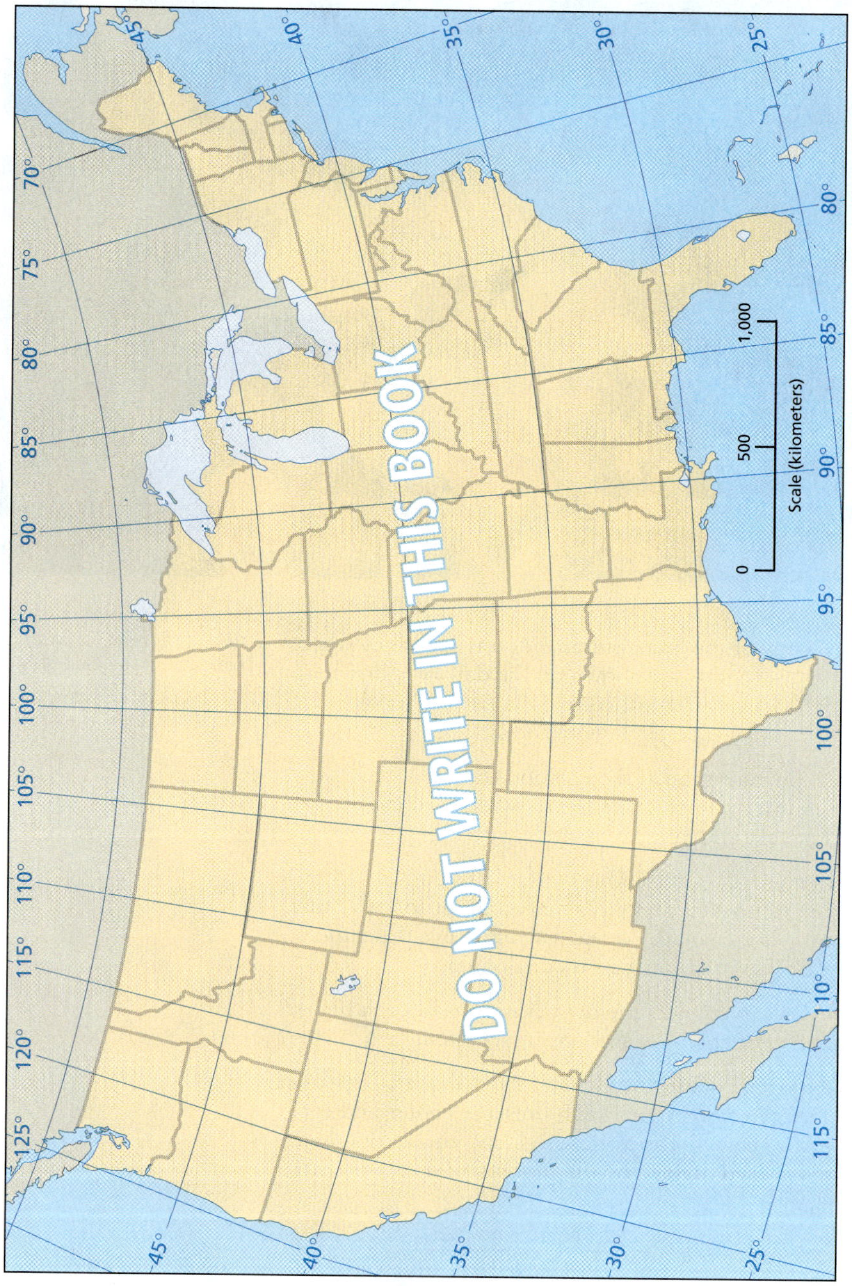

Many hurricane-prone areas have established evacuation routes to help people reach safety before a hurricane hits. This man is boarding up windows in his house to prepare for an imminent hurricane.

ANALYSIS AND CONCLUSION

1. **Analyzing Data** Generally, in which direction do the pressure centers over the United States move?
2. **Analyzing Results** From your calculations, what is the average rate of movement (in kilometers per day) of low- and high-pressure centers in winter?
3. **Making Predictions** Predict where the low- and high-pressure centers will be located on the day following the date of the last map in your series.
4. **Forming a Hypothesis** Refer to your series of daily weather maps to predict the weather for your hometown on the sixth day of the series. Write a forecast, and fill in the data table with your predictions of weather conditions.
5. **Evaluating Predictions** Was your prediction about the locations of the low- and high-pressure centers from question 3 accurate? Explain why or why not.
6. **Evaluating Hypotheses** Compare your weather prediction with the daily weather map for the appropriate day. Check the accuracy of your prediction. What factors could have caused errors in your prediction?
7. **Explaining Events** Describe the general weather conditions associated with regions of low and high atmospheric pressure.

Extension

1. **Designing Experiments** In the spring, repeat the entire investigation. What is the average rate of movement (in kilometers per day) of low- and high-pressure centers in the spring? Compare the rate of movement of pressure systems during spring and winter.

LONG-TERM PROJECTS

Long-Term Projects

3. Explain that high-pressure systems tend to move toward the area where the greatest increase in barometric pressure is occurring. A well-developed high-pressure system that is east of a low will tend to slow the low's movement or deflect it to the left or right. Also, two low-pressure centers tend to merge.
4. Groups of students can be assigned different weeks so that a continuous weather record is maintained. Co-op Learning

Answers to Analysis and Conclusion

1. The general direction is west to east.
2. Answers will depend on students' maps.
3. Locations will depend on the map sequence for the period of the investigation.
4. Predictions will depend upon weather patterns during the course of the investigation.
5. Answers may vary.
6. Factors that can result in errors in predictions are changes in the rate of movement of the highs and lows from the calculated average, a shift in the path of the pressure system, and a shift in the position of the jet stream.
7. Low-pressure regions have relatively warm temperatures, increased cloud cover, high humidity, surface winds flowing in a counterclockwise pattern, and high probability of precipitation. High-pressure regions have relatively cool temperatures, clear skies, low humidity, surface winds flowing in a clockwise pattern, and low probability of precipitation.

Answers to Extension

1. Answers may vary.

Comparing Climate Features

Teacher's Notes

Purpose
To have students use climate data to compare their local climate with those of other regions of the United States

Introduction
When the monthly temperatures and amounts of precipitation for an area are graphed for an entire year, the graph will reflect conditions that are similar to one of seven world climates. In this investigation, students use climate data to compare their local climate with climates of other regions of the United States. Then, they make comparisons of their graphed data with the different world climates and develop a conclusion about the type of climate in their location.

Duration
Six months, from October through March

Frequency
Daily

Suggested Schedule
This investigation will take six months to complete, from October through March. When you reach the chapter entitled "Climate," you might have students relate what their observations have indicated thus far.

Skills Acquired
- Collecting Data
- Measuring
- Organizing and Analyzing Data
- Classifying
- Inferring

LONG-TERM PROJECT 5

ENVIRONMENTAL CONNECTION

Duration
6 months (October 1 to April 1)

Objectives
- **Record** temperature and precipitation data for eight regions.
- USING SCIENTIFIC METHODS **Graph and analyze** climate features for eight regions.
- **Classify** regions by using two climate classification systems.

Materials
- almanac
- atlas
- paper, graph
- rain gauge (optional)
- thermometer (optional)
- weather reports, daily

Safety

Comparing Climate Features

A graph of the monthly temperatures and amounts of precipitation for a region is called a *climatograph*. Climatographs can be used to compare the climates of different areas or to classify an area's climate.

In this investigation, you will use climate data to compare your local climate with the climates of other regions of the United States. You will keep a daily temperature and precipitation log. You will record data every day from the first day of October until the first day of April. You will then compare your graphed data with information about the world's climates. You will use this comparison to develop a conclusion about the type of climate that your location has.

PROCEDURE

1. Listen to or watch a daily weather report for your area, or find this information in a daily newspaper or on the Internet. You may also keep your own records by using a thermometer and a rain gauge.
2. Beginning on the first day of October, keep a daily record of the high and low temperatures and of the amount of precipitation that occurs. During winter, snow should be melted before determining the amount of precipitation (in centimeters).

Rain gauges collect precipitation for scientists to measure.

CHAPTER RESOURCES

Workbooks

Long-Term Projects
- Comparing Climate Features ADVANCED

3. Calculate the average temperature for each day by dividing the sum of the day's high and low temperatures by 2. Record this information.

4. At the end of each month, record the average monthly temperature given by the weather report, or calculate the average monthly temperature by dividing the sum of the daily averages by the number of days in the month. Also, record the total monthly precipitation given by the weather report.

5. Use the Internet or an almanac to look up climate data for your town or city and for seven other cities in the United States. Select one city from each of the following regions: New England, the Gulf Coast, the Midwest, the Southwest, the Pacific Northwest, the interior of Alaska, and the Hawaiian Islands.

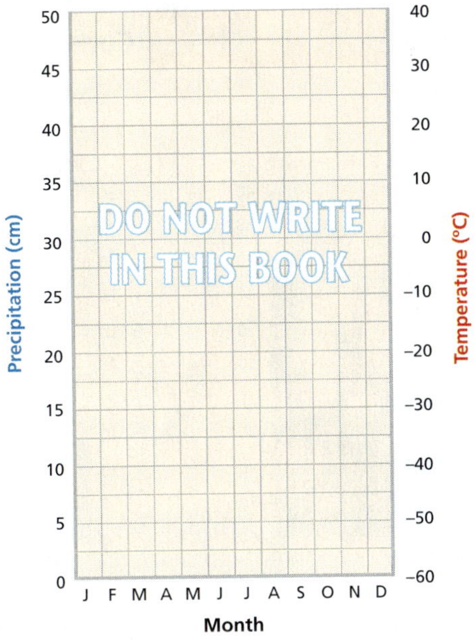

Step 7

6. Look up the average monthly temperatures and precipitation for your town or city and for each city that you chose. Record these data in a table.

7. On your graph paper, make eight copies of the blank climatograph shown on the upper-right side of this page.

8. Label each blank climatograph with the name of one of the seven cities. Label the eighth climatograph with the name of your town or city.

9. If you recorded temperature and precipitation in English (American) units, such as degrees Fahrenheit or inches, convert your measurements to SI units, such as degrees Celsius or centimeters. Use the SI Conversions table in the Appendix of this book to convert English units to SI units.

10. For each of the eight locations, plot the average temperature for January by placing a dot in the center of the square that is located in the column representing January and in the row representing that average temperature.

11. Using the same method, plot the average precipitation for January.

12. Repeat steps 10 and 11 for each month's data for each location. Then, connect the temperature points and connect the precipitation points in order of consecutive months.

Long-Term Projects

The Scientific Method

In this lab, students will

- Make Observations
- Ask Questions
- Analyze the Results
- Draw Conclusions
- Communicate the Results

Chapter Correlation

Students can find additional information to aid them in this investigation by reading the chapter entitled "Climate."

Materials

The materials listed for this project are enough for one student or one group collaborating on the project.

Tips and Tricks

Advance Preparation

1. Each student will need access to an almanac or an atlas. These can be made available in the classroom, obtained from a library, or located on the Internet.
2. If applicable, check to see if the students know how to use and read thermometers and rain gauges. Have several students make the same observation with a thermometer and a rain gauge, and look for differences in the readings.

Possible Problems

1. Students may be confused about estimating the amount of daily precipitation when rain continues for more than one day. Suggest that they divide the entire amount of precipitation into portions that are proportional to the hours of rainfall each day.
2. Because such a large file of daily data is collected, it is helpful to check student logs at the end of each month. The calculation of monthly data provides an opportunity for feedback. Periodic reminders are necessary to help students remember to make their observations.

Tips and Tricks, *continued*

Teaching Suggestions

1. If direct measurements will be taken, secure a location on the school grounds for students to observe temperature and precipitation. The instruments should be kept safe from theft or vandalism. An ideal arrangement would be to use a permanent weather station.
2. Maximum-minimum thermometers are best for recording temperatures in this project. If maximum-minimum thermometers are available, students should be instructed in the proper use of the thermometer.
3. Arrangements should be made that permit students to record data at appropriate times during the school day.
4. On days when direct measurements cannot be made, students should record data taken from newspaper, television, or radio reports.
5. A classroom calculator or computer may ease the task of monthly averaging. A computer spreadsheet or graphing program could be used to track, average, and graph data.
6. Data could be collected by several students and averaged. This would tend to produce more reliable observations and account for any missing data. Co-op Learning
7. A group of students could prepare climatographs for a large number of cities. Each member of the group should prepare climatographs for a few cities and share this information with other members of the group or class. Co-op Learning
8. Different groups of students could be assigned different days to monitor the local and national weather conditions. The groups could then share the data so that daily observations by each group would not be necessary. Co-op Learning

LONG-TERM PROJECT 5 *Comparing Climate Features, continued*

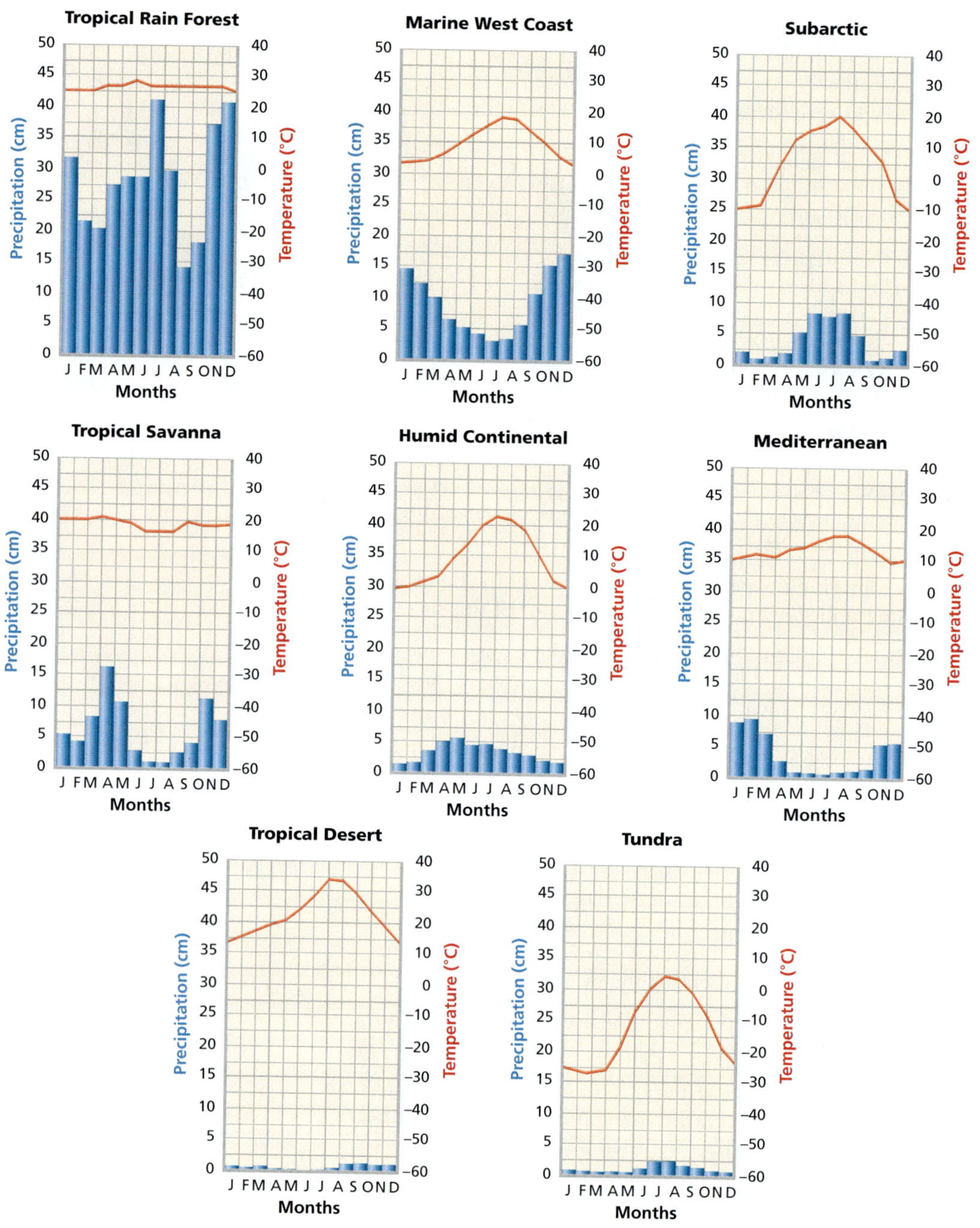

Although rain forests and deserts may have similar latitudes, the amount of precipitation that rain forests receive differs greatly from the amount that deserts receive.

ANALYSIS AND CONCLUSION

1. **Making Comparisons** Compare each of the climatographs of your seven chosen U.S. cities with the sample climatographs on the previous page. Identify the climate type or types for each location that you selected. What features of each climatograph helped you classify each region?

2. **Analyzing Results** Use the climatograph for your area to classify your regional climate. What features of your climatograph helped you identify the climate type?

3. **Examining Data** Compare the average temperatures and precipitation amounts for your location that you collected to the values that you obtained from the Internet or an almanac. Do you think that this year's climate data are typical for your region? Explain your answer.

4. **Classifying Information** How would each of the climatographs, including the one for your region, fit into the climate classification system outlined in the chapter entitled "Climate"?

5. **Evaluating Methods** In this investigation, you compared climates by looking at average precipitation and temperature. What other factors might affect the climate of an area? Give examples of each factor.

Extension

1. **Making Comparisons** Bermuda is a small island in the Atlantic Ocean. Bermuda is at about the same latitude as St. Louis, Missouri, which lies in the middle of a continent. In which of the two locations does the temperature vary least from month to month? Explain the cause of the temperature pattern in the location that has the more moderate pattern.

LONG-TERM PROJECTS

Long-Term Projects

Answers to Analysis and Conclusion

1. The New England region has a humid continental climate. The Gulf Coast region has a subtropical climate. The Midwest region has either a steppe or humid continental climate depending on the city selected. The Southwest has either a desert or steppe climate, depending on proximity to the west coast mountains. The Pacific west coast has either a Mediterranean or marine west coast climate. Interior Alaska has a subarctic climate. The Hawaiian Islands have both tropical rain forest and savanna climates. Temperature range and precipitation patterns were used to classify the climates.
2. Answers may vary depending upon local conditions.
3. Answers may vary.
4. Answers may vary. Students should attempt to correlate rainfall and temperature with each type of climate mentioned.
5. Factors include altitude, proximity to large bodies of water, patterns of winds, vegetation cover, and pressure systems. Examples may vary.

Answers to Extension

1. Bermuda; Water changes temperature much less than land that has been similarly heated or cooled does. Therefore, the surrounding water moderates island temperature.

Planetary Motions

Teacher's Notes

Purpose

To observe the motion of the planet Mars against background stars over a period of several months

Introduction

As viewed from Earth, planets show a relatively strange pattern in their motion at various times of the year. In this investigation, students observe the planet Mars over a period of nine months and use their observations to draw conclusions about planetary motion.

Duration

Approximately eight or nine months (depending on the school year)

Frequency

Twice a month on the 1st and 15th

Suggested Schedule

Students will need nine months to complete this project; therefore, students should start it at the beginning of the school year. If you plan to have students share their work, data, and conclusions with the class, be sure the investigation is complete before you begin the chapter entitled "Planets of the Solar System."

Skills Acquired

- Collecting Data
- Organizing and Analyzing Data
- Inferring
- Interpreting
- Identifying and Recognizing Patterns
- Predicting

LONG-TERM PROJECTS

LONG-TERM PROJECT 6

Duration
8 months

Objectives

- **Observe** the position of Mars in the night sky for 8 months.
- USING SCIENTIFIC METHODS **Graph** the apparent movement of Mars through the night sky.
- **Identify** changes in the relative positions of Earth and Mars.

Materials

- celestial sphere model (optional)
- compass, magnetic
- constellation charts
- flashlight
- metric ruler

Planetary Motions

While observing the evening sky over a period of time, you might have noticed that some objects look like stars but do not maintain a fixed position relative to the celestial sphere. These objects are the planets. As viewed from Earth at various times during the year, the patterns in the planets' motion differ from the patterns that you might expect.

In this investigation, you will observe the planet Mars in the night sky on the 1st and 15th of each month over a period of 8 months. Then, you will use your observations to draw conclusions and make predictions about planetary motion.

PROCEDURE

1. Obtain data on the positions of Mars in the night sky throughout the last year from an astronomical yearbook or from the Internet. Astronomical yearbooks can be found at most libraries.
2. Check the Internet, an almanac, or the weather section of a newspaper to find the time of night that Mars will be visible in your area.
3. Copy the star chart on the third and fourth pages of this lab onto a separate piece of paper.
4. Practice measuring angular distance by using the method shown at the bottom of this page. Always use the same hand when measuring angular distance. To have confidence in the accuracy of your measurements, you may need to practice measuring angular distance for a few days or weeks before beginning this lab.

Step 4

Estimating Angular Distance in Degrees
Hold your hand at arm's length.

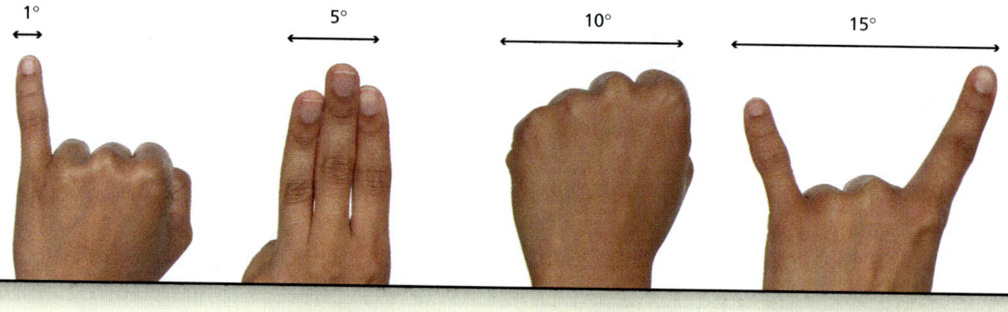

CHAPTER RESOURCES

Workbooks

Long-Term Projects
- Planetary Motions ADVANCED

In 2003, Mars was closer to Earth than it had been for about 60,000 years.

5. On the 1st and 15th of each month, go at night to an area that gives you a clear view of the eastern, southern, and western skies. (Note: If the skies are not clear on the 1st and 15th, make observations as close to these dates as possible.)
6. At the same time each night, locate Mars in the night sky. Mars will have a dull red appearance.
7. Use your magnetic compass to position yourself facing south, and observe the position of Mars relative to the background stars.
8. Choose a constellation. Use the method illustrated at the bottom of the previous page to estimate Mars's angular distance (in degrees) from the constellation.
9. On your star chart, locate the constellation from which you measured Mars's angular distance. Draw Mars in the appropriate position relative to the constellation, and label the planet's position with the date.
10. Compare the apparent brightness of Mars with that of the background stars. Record your observation on a separate sheet of paper.
11. Repeat steps 6–10 on the 1st and 15th of each month for the next 8 months.
12. After each observation, draw an arrow from Mars's previous position to the position that you just observed. The progression of arrows will show Mars's apparent path.

Long-Term Projects

The Scientific Method

In this lab, students will

- Make Observations
- Ask Questions
- Test the Hypothesis
- Analyze the Results
- Draw Conclusions
- Communicate the Results

Chapter Correlation

Students can find additional information to aid them in this investigation by reading the chapter entitled "Planets of the Solar System."

Materials

The materials listed for this project are enough for one student or one group collaborating on the project.

Tips and Tricks

Advance Preparation

1. Make copies of the constellation chart for students to use.
2. Explain to students how to locate constellations before the investigation starts. It is best for you to become familiar with the constellations and where to find them in the night sky so that you can assist students with this investigation.
3. Prepare a list of periodicals and newspapers that give accurate information on the time and location Mars will be visible in the sky. Distribute this list to the students.
4. Supply students with information about Mars's position in the sky for the last year so they can compare that position with its present position. This information can be obtained from a planetarium, a college astronomy department, or a local astronomer.

Tips and Tricks, *continued*

Possible Problems

1. Poor weather may obscure the sky. If observations cannot be made on the suggested day, have students make observations on the nearest possible date.
2. Students may forget to make their observations on the correct days of the month. They should be reminded to keep the schedule. Students should not skip observations. If they miss one, have them make observations as soon as possible after the missed date.
3. Students may confuse Mars with other planets in the night sky.

Teaching Suggestions

1. Explain that Mars has a longer period of revolution around the sun than Earth does. Periodically, Earth overtakes Mars and passes it in an easterly direction. As a result, for a period of months Mars appears to move backwards in a westerly direction against the stellar background. This is referred to as *retrograde motion.*
2. Explain to your students that the synodic period of a planet's revolution is the time required for the planet to return to the same position in the sky as seen from Earth. The synodic period of Mars is two years and forty-nine days.
3. Organize a field trip to a planetarium so students can observe the constellations and be better able to find them in the nighttime sky.
4. Students could make observations of Jupiter or Saturn. At the completion of the investigation, students should share their observations with the class and make comparisons with Mars.
5. Students could be assigned staggered schedules to produce a more detailed graph. Co-op Learning

LONG-TERM PROJECT 6 Planetary Motions, continued

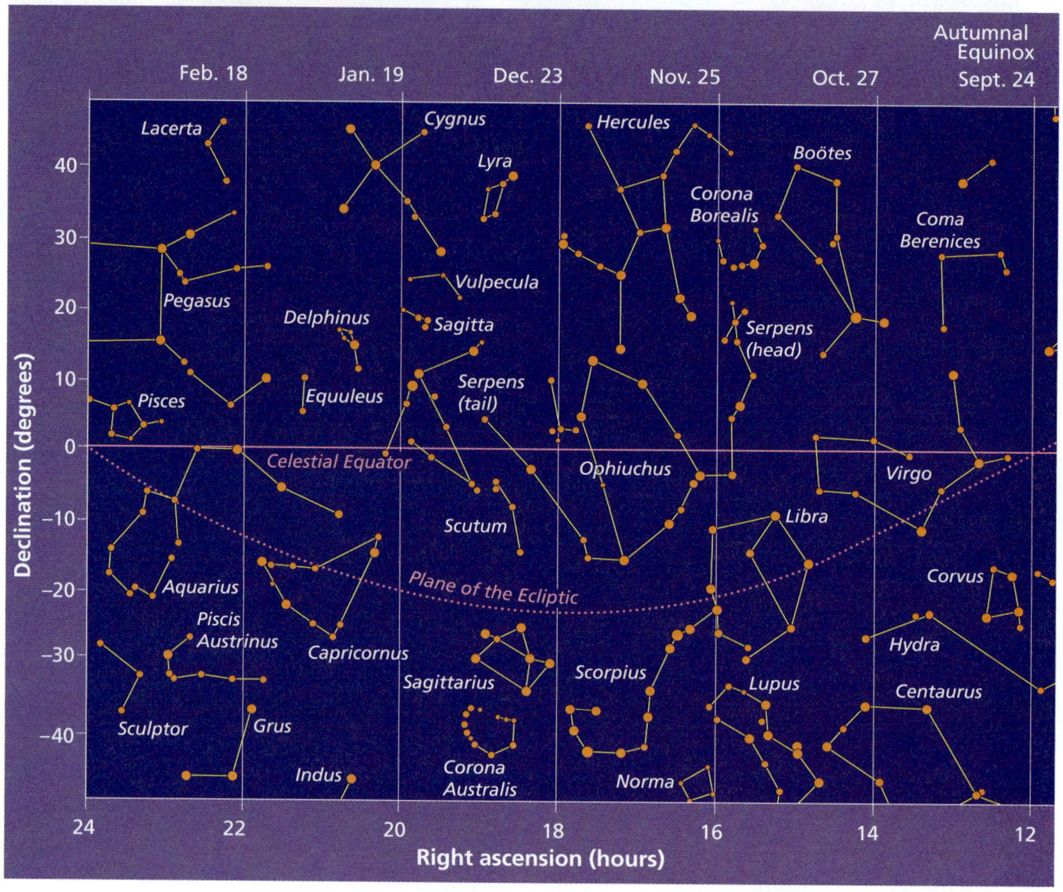

Step 3

ANALYSIS AND CONCLUSION

1. **Examining Data** In which months does Mars appear highest in the night sky? lowest in the night sky?
2. **Evaluating Results** In which direction does Mars appear to move across the sky? At any point during the year, does Mars appear to deviate from this apparent path?
3. **Making Predictions** In one year from today, will Mars be in the same position that it is in today? Explain.
4. **Drawing Conclusions** From your observations of the apparent brightness of Mars at different times of the year, what can you infer about the distance between Mars and Earth? Explain your answer.

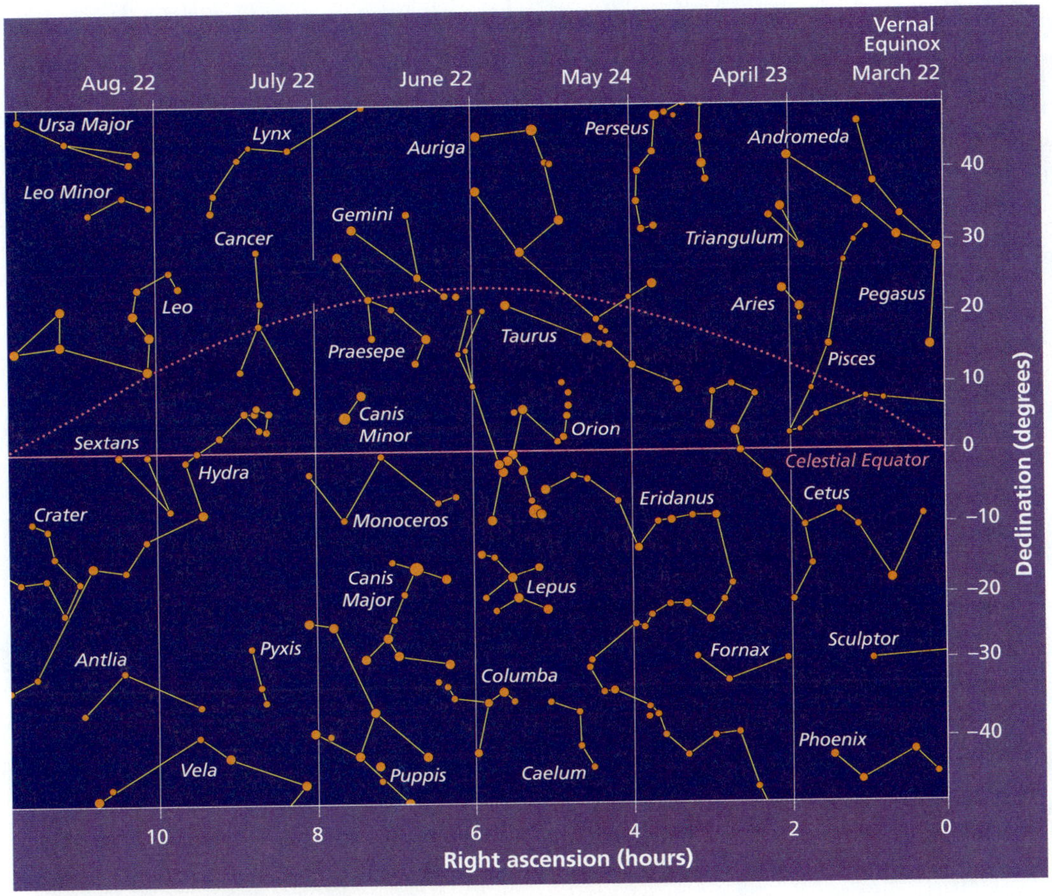

Extension

1 **Evaluating Predictions** In the Analysis and Conclusion section of this investigation, you predicted the position of Mars in one year. Use the Internet or library to research where astronomers predict Mars will be in one year. Was your prediction accurate? Explain why or why not.

2 **Evaluating Hypotheses** Repeat this investigation, but measure the positions of another planet, such as Venus, for the course of 8 months. Write a brief essay that explains how the paths of Mars and Venus differ.

Long-Term Projects

Answers to Analysis and Conclusion

1. Mars appears highest in September and April and lowest in March and June.
2. Mars appears to move from east to west, but for a few months it appears to be moving backwards.
3. no; Because Mars takes two years to orbit the sun, the differences in the orbital periods will not make Mars appear to be in the same position one year from today.
4. The distance between Mars and Earth varies because the brightness of Mars changes during the year.

Answers to Extension

1. Answers may vary.
2. Answers may vary. The five planets observable with the naked eye all experience retrograde motion. Planetary motion across Earth's sky varies, but Mercury is the fastest, followed by Venus, Mars, Jupiter, and then Saturn, the slowest. Because the orbits of Venus and Mercury are closer to the sun, we can observe those planets only near sunrise or sunset.

REFERENCE TABLES SI Conversions

The metric system is used for making measurements in science. The official name of this system is the Système Internationale d'Unités, or International System of Measurements (SI).

SI Units	From SI to English	From English to SI
Length		
kilometer (km) = 1,000 m	1 km = 0.62 mile	1 mile = 1.609 km
meter (m) = 100 cm	1 m = 3.28 feet	1 foot = 0.305 m
centimeter (cm) = 0.01 m	1 cm = 0.394 inch	1 inch = 2.54 cm
millimeter (mm) = 0.001 m	1 mm = 0.039 inch	
micrometer (µm) = 0.000 001 m		
nanometer (nm) = 0.000 000 001 m		
Area		
square kilometer (km^2) = 100 hectares	1 km^2 = 0.386 square mile	1 square mile = 2.590 km^2
hectare (ha) = 10,000 m^2	1 ha = 2.471 acres	1 acre = 0.405 ha
square meter (m^2) = 10,000 cm^2	1 m^2 = 10.765 square feet	1 square foot = 0.093 m^2
square centimeter (cm^2) = 100 mm^2	1 cm^2 = 0.155 square inch	1 square inch = 6.452 cm^2
Volume		
liter (L) = 1,000 mL = 1 dm^3	1 L = 1.06 fluid quarts	1 fluid quart = 0.946 L
milliliter (mL) = 0.001 L = 1 cm^3	1 mL = 0.034 fluid ounce	1 fluid ounce = 29.577 mL
microliter (µL) = 0.000 001 L		
Mass		* Equivalent weight at Earth's surface
kilogram (kg) = 1,000 g	1 kg = 2.205 pounds*	1 pound* = 0.454 kg
gram (g) = 1,000 mg	1 g = 0.035 ounce*	1 ounce* = 28.35 g
milligram (mg) = 0.001 g		
microgram (µg) = 0.000 001 g		
Energy		
British Thermal Units (BTU)	1 BTU = 1,055.056 joules	1 joule = 0.00095 BTU

Temperature

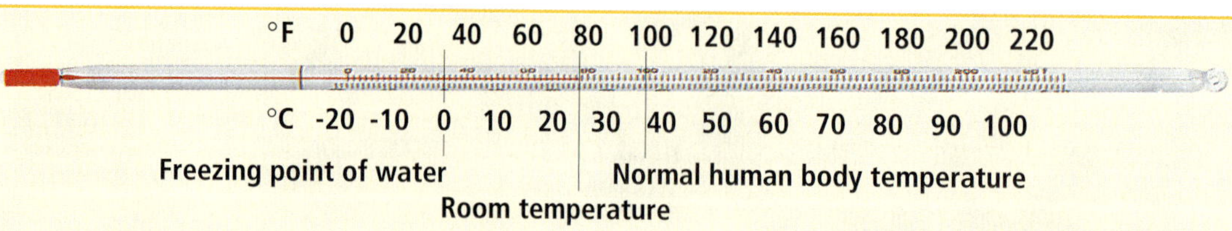

Conversion of Fahrenheit to Celsius:
$°C = \frac{5}{9}(°F - 32)$

Conversion of Celsius to Fahrenheit:
$°F = \frac{9}{5}(°C) + 32$

Mineral Uses

Metallic Minerals		
Mineral and chemical formula	**Location of economically important deposits**	**Important uses**
Chalcopyrite, $CuFeS_2$	Chile, U.S., and Indonesia	electrical and electronic products, wiring, telecommunications equipment, industrial machinery and equipment
Chromite, $FeCr_2O_4$	South Africa, Kazahkstan, and India	production of stainless steel, alloys, and metal plating
Galena, PbS	Australia, China, and U.S.	batteries, ammunition, glass and ceramics, and X-ray shielding
Gold, Au	South Africa, U.S., and Australia	computers, communications equipment, spacecraft, jet engines, dentistry, jewelry, and coins
Ilmenite, $FeTiO_3$	Australia, South Africa, and Canada	jet engines; missile components; and white pigment in paints, toothpaste, and candy
Magnetite, Fe_3O_4	China, Brazil, and Australia	steelmaking
Uraninite, UO_2	Canada and Australia	fuel in nuclear reactors and manufacture of radioisotopes

Nonmetallic Minerals		
Mineral and chemical formula	**Location of economically important deposits**	**Important uses**
Barite, $BaSO_4$	China, India, and U.S.	weighting agent in oil well drilling fluids, automobile paint primer, and X-ray diagnostic work
Borax, $Na_2B_4O_7 \bullet 10H_2O$	Turkey, U.S., and Russia	glass, soaps and detergents, agriculture, fire retardants, and plastics and polymer additives
Calcite, $CaCO_3$	China, U.S., and Russia	cement, lime production, crushed stone, glassmaking, chemicals, and optics
Diamond, C	Australia, Democratic Republic of the Congo, and Russia	jewelry, cutting tools, drill bits, and manufacture of computer chips
Fluorite, CaF_2	China, Mexico, and South Africa	hydrofluoric acid, steelmaking, water fluoridation, solvents, manufacture of glass, and enamels
Gypsum, $CaSO_4 \bullet 2H_2O$	U.S., Iran, and Canada	wallboard, building plasters, and manufacture of cement
Halite, NaCl	U.S., China, and Germany	chemical production, human and animal nutrition, highway deicer, and water softener
Sulfur, S	Canada, U.S., and Russia	sulfuric acid, fertilizers, gunpowder, and tires
Kaolinite, $Al_2Si_2O_5(OH)_4$	U.S., Uzbehkistan, and Czech Republic	glossy paper and whitener and abrasive in toothpaste
Orthoclase, $KAlSi_3O_8$	Italy, Turkey, and U.S.	glass, ceramics, and soaps
Quartz, SiO_2	U.S., Germany, and France	glass, computer chips, ceramics, abrasives, and water filtration
Talc, $Mg_3Si_4O_{10}(OH)_2$	China, U.S., and Republic of Korea	ceramics, plastics, paint, paper, rubber, and cosmetics

Guide to Common Minerals

This table is used in the chapter lab for the chapter entitled "Minerals of Earth's Crust."

	Luster		Hardness	Cleavage	Fracture	Color/opacity	
Nonmetallic; light color	glassy to pearly	Scratches glass	6	two cleavage planes at nearly right angles		various colors but often white or pink; opaque	
	glassy		6	two cleavage planes at 86° and 94°		colorless, white, pink, or various colors; translucent to opaque	
	glassy and waxy		7	no cleavage	conchoidal fracture	various colors; transparent to opaque	
	glassy		6.5–7	no cleavage	conchoidal to irregular fracture	olive green; transparent to translucent	
	glassy	Does not scratch glass	2.5–3	three cleavage planes at right angles		colorless to gray; transparent to opaque	
	glassy		3	three cleavage planes at 75° and 105°		colorless or white and may be tinted; transparent to opaque	
	glassy, pearly, or silky		1–2.5	one perfect cleavage plane	conchoidal and fibrous fracture	white, pink, or gray to colorless; transparent to opaque	
	pearly to waxy		1	one cleavage plane		white to green; opaque	
	glassy or pearly		2–2.5	one cleavage plane		colorless to light gray or brown; translucent to opaque	
	glassy		4	eight cleavage planes (octahedral)		green, yellow, purple, and other colors; transparent to translucent	
	glassy		4.5–5	no cleavage	conchoidal to irregular fracture	green, blue, violet, brown, or colorless; translucent to opaque	
	silky		3.5–4	no cleavage	irregular, splintery fracture	green; translucent to opaque	
Nonmetallic; dark color	glassy and silky	Scratches glass	5–6	two cleavage planes at 56° and 124°		dark green, brown, or black; translucent to opaque	
	resinous and glassy		6.5–7.5	no cleavage	irregular fracture	dark red or green; transparent to opaque	
	pearly and glassy	Does not scratch glass	2.5–3	one cleavage plane		black to dark brown; translucent to opaque	
	metallic to earthy		5.5–6.5	no cleavage	irregular fracture	reddish brown to black; opaque	
Metallic	metallic or earthy	Does not scratch glass	1–2	one cleavage plane		black to gray; opaque	
	metallic		2.5	three cleavage planes at right angles		lead gray; opaque	
	metallic	Scratches glass	5–6	two cleavage planes at 56° and 124°		iron black; opaque	
	metallic		6–6.5	no cleavage	conchoidal to irregular fracture	brass yellow; opaque	

Streak	Specific gravity	Other properties	Mineral name and chemical formula
white	2.6	prismatic, columnar, or tabular crystals	orthoclase, $KAlSi_3O_8$
blue-gray to white	2.6 to 2.7	striations	plagioclase, $(Na, Cl)(Al, Si)_4O_8$
white	2.65	six-sided crystals	quartz, SiO_2
white to pale green	3.2 to 3.3	stubby, prismatic crystals	olivine, $(Mg, Fe)_2SiO_4$
white	2.2	cubic crystals and salty taste	halite, NaCl
white	2.7	may produce double image when you look through it	calcite, $CaCO_3$
white	2.2 to 2.4	thin layers and flexible	gypsum, $CaSO_4 \bullet 2H_2O$
white	2.7 to 2.8	soapy feel and thin scales	talc, $Mg_3Si_4O_{10}(OH)_2$
white	2.7 to 3	thin sheets	muscovite, $KAl_2Si_3O_{10}(OH)_2$
white	3.2	fluorescent under UV light; cubic and six-sided crystals	fluorite, CaF_2
white or pale red-brown	3.1	six-sided crystals	apatite, $Ca_5(OH, F, Cl)(PO_4)_3$
emerald green	4	fibrous, radiating aggregates or circular, banded structure	malachite, $CuCO_3 \bullet Cu(OH)_2$
pale green or white	3.2	six-sided crystals	hornblende, $(Ca, Na)_{2\text{-}3}(Mg, Fe, Al)_5 Si_6(Si, Al)_2O_{22}(OH)_2$
white	4.2	12- or 24-sided crystals	garnet, $Fe_3Al_2(SiO_4)_3$
white to gray	2.7 to 3.2	thin, flexible sheets	biotite, $K(Mg, Fe)_3AlSi_3O_{10}(OH)_2$
red to red-brown	5.25	granular masses	hematite, Fe_2O_3
black to dark green	2.3	greasy feel, soft, and flaky	graphite, C
lead gray to black	7.4 to 7.6	very heavy	galena, PbS
black to dark green	5.2	8- or 12-sided crystals; may be magnetic	magnetite, Fe_3O_4
greenish black	5	cubic crystals	pyrite, FeS_2

Guide to Common Rocks

This table is used in the chapter lab for the chapter entitled "Rocks."

Rock class	Grain size	Description	Rock class	Rock name
Made of crystals	Coarse grained	mostly light in color; shades of pink, gray, and white are common	igneous	granite
		dark in color; commonly black and white; heavy heft	igneous	gabbro
		foliated; layers of different minerals give a banded appearance	metamorphic	gneiss
		foliated; contains abundant amount of quartz, and may contain garnet; flaky minerals	metamorphic	schist
		nonfoliated; reacts with acid	metamorphic	marble
	Fine grained	usually light in color; many holes and spongy appearance; may float in water	igneous	pumice
		light to dark in color; glassy luster; conchoidal fracture	igneous	obsidian
		dark in color; may ring like a bell when struck with a hammer	igneous	basalt
		fine grained; foliated; cleaves into thin, flat plates	metamorphic	slate
Made of rock particles	Coarse grained	coarse-grained particles, more than 2 mm; rounded pebbles; some sorting; clay and sand are visible	sedimentary	conglomerate
		well-preserved fossils are common; can be scratched with a knife; many colors but usually white-gray; reacts with acid	sedimentary	limestone
		cube-shaped crystals; commonly colorless; does not react with acid	sedimentary	halite
	Medium grained	1/16 to 2 mm grains; mostly quartz fragments; surface feels sandy	sedimentary	sandstone
	Fine grained	soft and porous; commonly white or buff color	sedimentary	chalk
		microscopic grains; clay composition; smooth surface; hardened mud appearance	sedimentary	shale

Radiogenic Isotopes and Half-Life

Unstable isotopes, called *radiogenic isotopes* or *radioactive isotopes,* decay to form different isotopes called *daughter isotopes.* Each radiogenic isotope breaks down at a predictable rate, called its *half-life,* into a daughter isotope. Because of this predictable decay pattern, radiogenic isotopes are used to determine numeric dates for rocks. The table below describes several common radiometric dating methods.

Radiometric dating method	How it works	Parent isotope	Daughter isotope	Half-life	Effective dating range
Argon-argon dating ($^{39}Ar/^{40}Ar$)	Comparison made between ^{39}Ar and ^{40}Ar in a sample specially irradiated to form ^{39}Ar; ^{39}Ar is equivalent to ^{40}K in potassium-argon dating.	potassium-40 (^{40}K) irradiated to form argon-39 (^{39}Ar)	argon-40, ^{40}Ar	1.25 billion years	10,000 to 4.6 billion years
Fission track dating	Tracks of damage created by charged particles from radioactive decay that pass through a mineral's crystal lattice are counted under an electron microscope.	uranium, U	ultimately, lead, Pb, but also several other daughter isotopes	not applicable	500 years to 1 billion years
Potassium-argon dating ($^{40}K/^{40}Ar$)	Comparison is made between the amount of ^{40}K and amount of ^{40}Ar; over time, ^{40}K decreases and ^{40}Ar increases.	potassium-40, ^{40}K	argon-40, ^{40}Ar	1.25 billion years	50,000 to 4.6 billion years
Radiocarbon dating ($^{14}C/^{12}C$)	Comparison is made between the amount of ^{14}C in organic matter and the amount of ^{12}C; ^{12}C remains constant over time, and ^{14}C breaks down.	carbon-14, ^{14}C	nitrogen-14, ^{14}N	5,730 years	<80,000 years
Rubidium-strontium dating ($^{87}Rb/^{87}Sr$)	Comparison made between the ratio of $^{87}Sr/^{86}Sr$ and the ratio of $^{87}Rb/^{86}Sr$ to find the amount of ^{87}Sr formed by radioactive decay.	rubidium-87, ^{87}Rb	strontium-87, ^{87}Sr	48.8 billion years	10 million to 4.6 billion years
Thorium-lead dating	Comparison made between amount of ^{232}Th and the ratio of $^{208}Pb/^{204}Pb$; ^{232}Th breaks into ^{208}Pb, and ^{204}Pb remains constant.	thorium-232, ^{232}Th	lead-208, ^{208}Pb	14.0 billion years	>200 million years
Uranium-lead dating ($^{235}U/^{207}Pb$)	Comparison made between amount of ^{235}U and the ratio of $^{207}Pb/^{204}Pb$; ^{235}U breaks into ^{207}Pb, and ^{204}Pb remains constant.	uranium-235, ^{235}U	lead-207, ^{207}Pb	704 million years	10 million to 4.6 billion years
Uranium-lead dating ($^{238}U/^{206}Pb$)	Comparison made between amount of ^{238}U and the ratio of $^{206}Pb/^{204}Pb$; ^{238}U breaks into ^{206}Pb, and ^{204}Pb remains constant.	uranium-238, ^{238}U	lead-206, ^{206}Pb	4.5 billion years	10 million to 4.6 billion years

Topographic and Geologic Map Symbols

Topographic Map Symbols

Elevation markers

Contour lines	
Index contour lines	100
Depression contour lines	
Water elevation	9600
Spot elevation	x 9136

Boundaries

National	
State	
County, parish, municipal	
Township, precinct, town	
Incorporated city, village, or town	
National or state reservation	
Small park, cemetery, airport, etc.	
Land grant	

Buildings and Structures

Buildings	
School	
Church	
Cemetery	cem
Barn and warehouse	
Wells (non-water)	○ oil ○ gas
Open-pit mine, quarry, or prospect	
Tunnel	
Benchmark	⊗BM △8025
National Park	
Campsite	
Bridge	

Roads and Railroads

Divided highway	
Road	
Trail	
Railroad	

Geologic Map Symbols

Sedimentary Rocks

Breccia	
Conglomerate	
Dolomite	
Limestone	
Mudstone	
Sandstone	
Siltstone	
Shale	

Igneous and Metamorphic Rocks

Rock	Symbol
Extrusive	
Intrusive	
Metamorphic	

Features

River	
Water well	water
Spring	
Lake	
Glacier	

Contour Map

This map is used in the chapter lab for the chapter entitled "Models of the Earth."

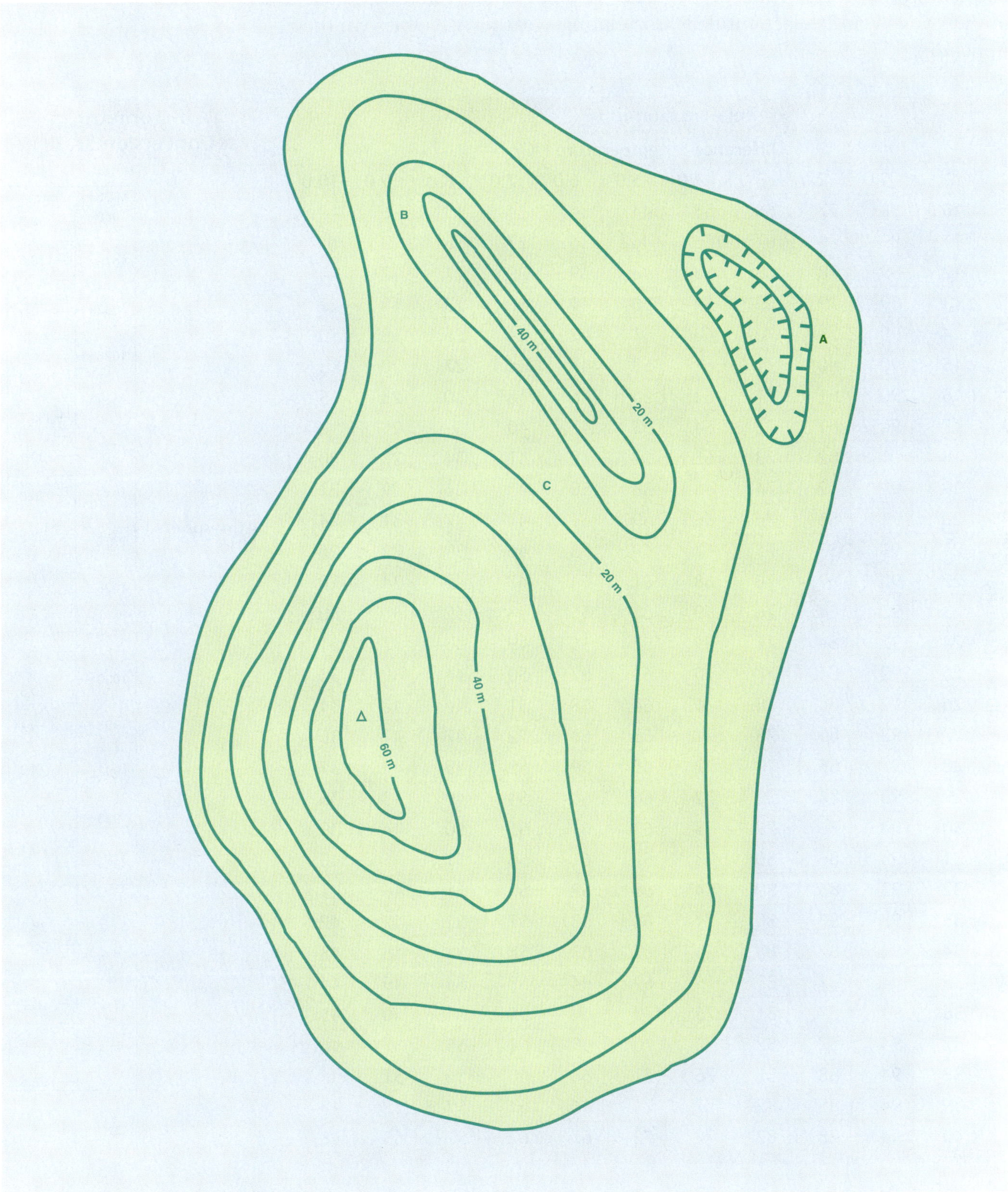

Humidity and Air Pressure

The Relative Humidity table below is used in the chapter lab for the chapter entitled "Water in the Atmosphere." The Barometric Conversion Scale is used in the Long-Term Project entitled "Correlating Weather Variables" in the Appendix.

Relative Humidity (%)

Dry-bulb temperature (°C)	Difference in temperature (°C)									
	1.0	2.0	3.0	4.0	5.0	6.0	7.0	8.0	9.0	10.0
10	88	77	66	55	44	34	24	15	6	—
11	89	78	67	56	46	36	27	18	9	—
12	89	78	68	58	48	39	29	21	12	—
13	89	79	69	59	50	41	32	23	15	7
14	90	79	70	60	51	42	34	26	18	10
15	90	80	71	61	53	44	36	27	20	13
16	90	81	71	63	54	46	38	30	23	15
17	90	81	72	64	55	47	40	32	25	18
18	91	82	73	65	57	49	41	34	27	20
19	91	82	74	65	58	50	43	36	29	22
20	91	83	74	66	59	51	44	37	31	24
21	91	83	75	67	60	53	46	39	32	26
22	92	83	76	68	61	54	47	40	34	28
23	92	84	76	69	62	55	48	42	36	30
24	92	84	77	69	62	56	49	43	37	31
25	92	84	77	70	63	57	50	44	39	33
26	92	85	78	71	64	58	51	46	40	34
27	92	85	78	71	65	58	52	47	41	36
28	93	85	78	72	65	59	53	48	42	37
29	93	86	79	72	66	60	54	49	43	38
30	93	86	79	73	67	61	55	50	44	39
31	93	86	80	73	67	61	56	51	45	40
32	93	86	80	74	68	62	57	51	46	41
33	93	87	80	74	68	63	57	52	47	42
34	93	87	81	75	69	63	58	53	48	43
35	94	87	81	75	69	64	59	54	49	44
36	94	87	81	75	70	64	59	54	50	45
37	94	87	82	76	70	65	60	55	51	46
38	94	88	82	76	71	66	60	56	51	47
39	94	88	82	77	71	66	61	57	52	48
40	94	88	82	77	72	67	62	57	53	48

Barometric Conversion Scale

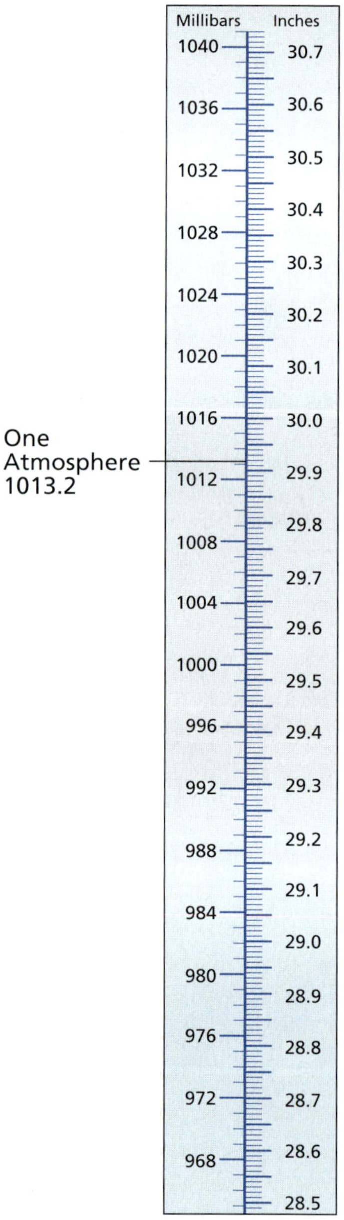

Weather Map of the United States

This map is used in the chapter lab for the chapter entitled "Weather."

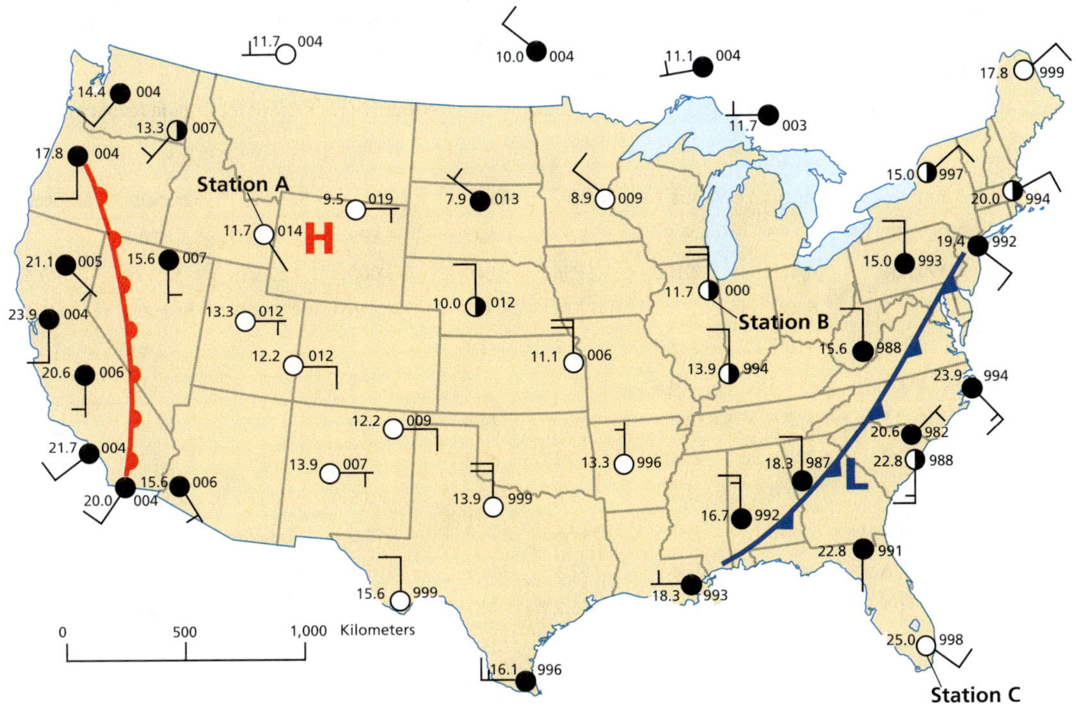

Weather Map Symbols

This chart is used in the Long-Term Project entitled "Weather Forecasting."

Cloud cover (fraction of sky covered)	Clear	1/8	Scattered	3/8	4/8
	5/8	Broken	7/8	Overcast	Obscured / No data
Wind speed (knots)	Calm	1 to 2	3 to 7	8 to 12	13 to 17
	18 to 22	23 to 27	48 to 52	73 to 77	103 to 107
Wind direction	North	Northeast	East	Southeast	
	South	Southwest	West	Northwest	
Weather conditions	Drizzle	Fog	Hail	Haze	Rain / Shower
	Freezing rain	Smoke	Snow	Thunderstorm	Hurricane / Tropical storm

Solar System Data

Notes

The **semimajor axis** is the average distance between an object and its primary (the body the object revolves around).

Surface gravity indicated for the gas giants is calculated for the altitude at which the atmospheric pressure equals 1 bar.

Rotation period and **orbital period** are sidereal measurements (relative to the stars, not the sun).

* This value indicates distance from the sun in AU.

† This value represents the rate of rotation at the sun's equator. The sun displays differential rotation; in other words, it rotates faster at its equator than at its poles.

R This value indicates retrograde rotation or retrograde revolution.

	Sun	Mercury	Venus	Earth	Mars	Jupiter	Saturn	Uranus	Neptune
Mass (10^{24} kg)	1,989,100	0.33	4.87	5.97	0.642	1,899	568	86.8	102
Diameter (km)	1,390,000	4,880	12,100	12,756	6,800	143,000	120,535	51,120	49,530
Density (kg/m^3)	1,408	5,427	5,243	5,515	3,933	1,326	687	1,270	1,638
Surface gravity (m/s^2)	274	3.7	8.9	9.8	3.7	23.1	9	8.7	11
Escape velocity (km/s)	617.7	4.3	10.4	11.2	5	59.5	35.5	21.3	23.5
Rotation period (h)	609.12	1,407.6	5,832.5 R	23.9	24.6	9.9	10.7	17.2 R	16.1
Length of day (hours)	609.6†	4,222.6	2,802	24	24.7	9.9	10.7	17.2	16.1
Semimajor axis (10^6 km)	N/A	57.9	108.2	149.6	227.9	778.6	1,433.5	2,872.5	4,495.1
Perihelion (10^6 km)	N/A	46	107.5	147.1	206.6	740.5	1,352.6	2,741.3	4,444.5
Aphelion (10^6 km)	N/A	69.8	108.9	152.1	249.2	816.6	1,514.5	3,003.6	4,545.7
Orbital period (days)	N/A	88	224.7	365.2	687	4,331	10,747	30,589	59,800
Orbital velocity (km/s)	N/A	47.9	35	29.8	24.1	13.1	9.7	6.8	5.4
Orbital inclination (degrees)	N/A	7	3.4	0	1.9	1.3	2.5	0.8	1.8
Orbital eccentricity	N/A	0.205	0.007	0.017	0.094	0.049	0.057	0.046	0.011
Axial tilt (degrees)	7.25	0.01	2.6	23.5	25.2	3.1	26.7	82.2	28.3
Mean surface temperature (°C)	6,073	167	464	15	–65	–110	–140	–195	–200
Global magnetic field?	yes	yes	no	yes	no	yes	yes	yes	yes

	Earth's moon	Major moons of Jupiter				Major moons of Saturn			
		Io	Europa	Ganymede	Callisto	Dione	Rhea	Titan	Iapetus
Mass (10^{20} kg)	0.073	893.2	480.0	1,481.9	1,075.9	0.375	11.0	1,345.5	15.9
Diameter (km)	3,475	3,643.2	3,121.6	5,262.4	4,820.6	1,120	1,528	5,150	1,436
Density (kg/m^3)	3,340	3,530	3,010	1,940	1,830	1,500	1,240	1,881	1,020
Rotation period (days)	655.7	1.77	3.55	7.15	16.69	2.74	4.52	15.95	79.33
Semimajor axis (10^3 km)	0.384*	421.6	670.9	1,070.4	1,882.7	377.40	527.04	1,221.83	3,561.3
Orbital period (days)	27.32	1.77	3.55	7.15	16.69	2.74	4.52	15.95	79.33

	Major moons of Uranus			Major moons of Neptune		Pluto's moon	Selected asteroids		Selected comets	
	Umbriel	Titania	Oberon	Triton	Nereid	Charon	Vesta	Ceres	Chiron	Hale-Bopp
Mass (10^{20} kg)	11.7	35.2	30.1	214	0.2	19	3	8.7	—	—
Diameter (km)	1,169	1,578	1,523	2,707	340	1,186	530	960 × 932	—	—
Density (kg/m^3)	1,400	1,710	1,630	2,050	1,000	2,000	—	—	—	—
Rotation period	4.14 days	8.71 days	13.46 days	5.87 days R	unknown	6.39 days	5.342 h	9.075 h	—	—
Semimajor axis (10^3 km)	266.30	435.91	583.52	354.76	5,513.4	19,600	2.362 *	2.767 *	13.7 *	250 *
Orbital period	4.14 days	8.71 days	13.46 days	5.87 days R	360.14 days	6.39 days	3.63 y	4.60 y	50.7 y	4,000 y

REFERENCE MAPS

Topographic Provinces of North America

REFERENCE MAPS

Geologic Map of North America

Mineral and Energy Resources of North America

Fossil Fuel Deposits of North America

Topographic Maps of the Moon

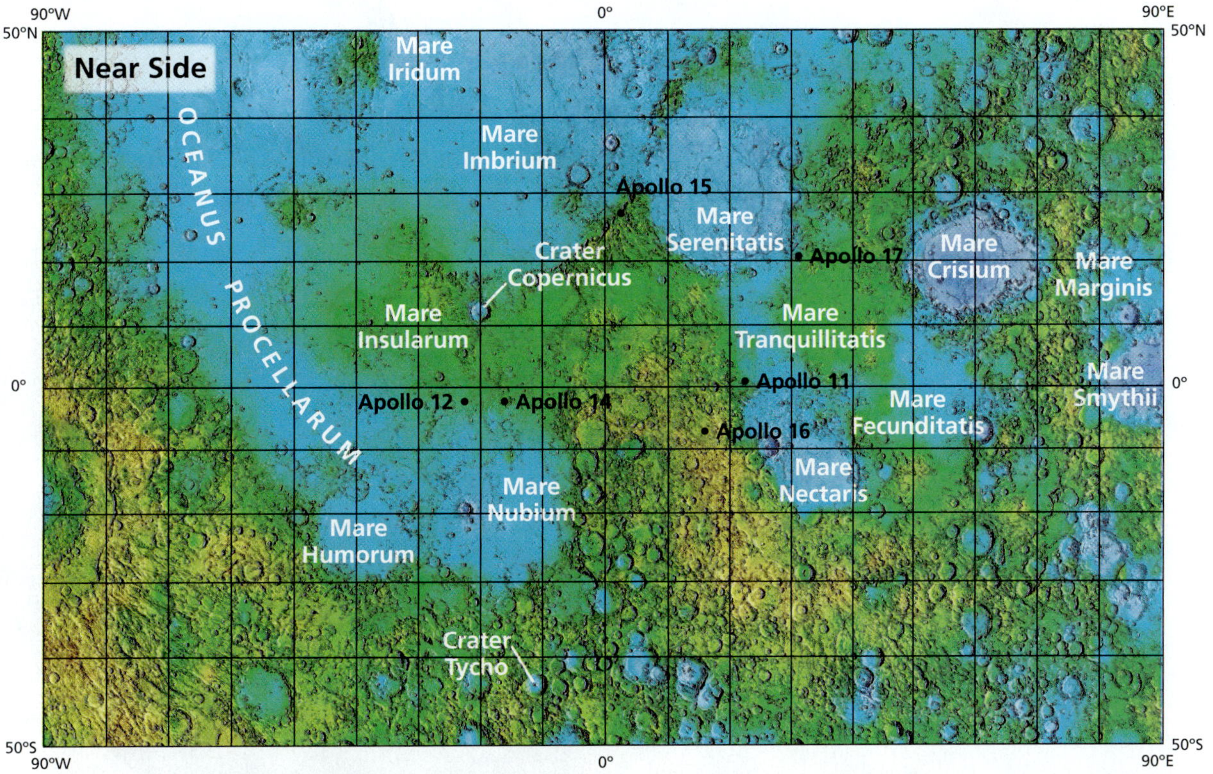

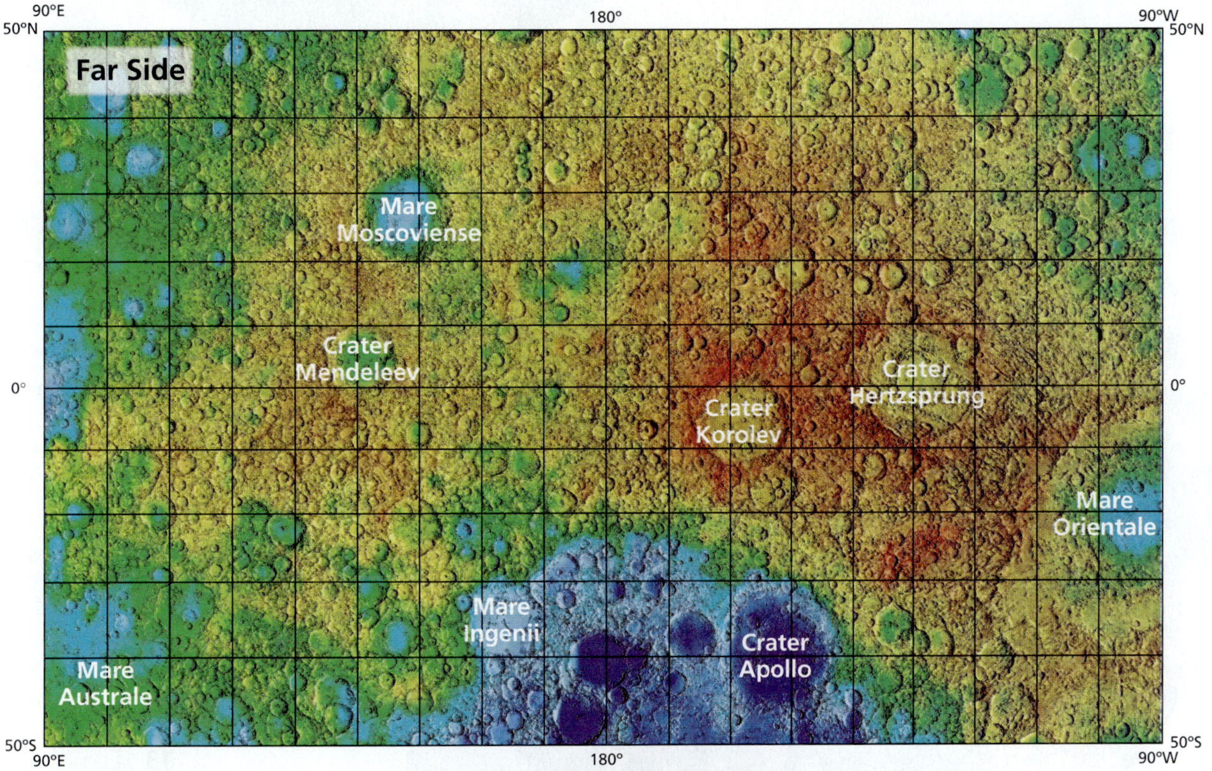

● Apollo landing sites

Color-coded topography key

Star Charts for the Northern Hemisphere

Spring

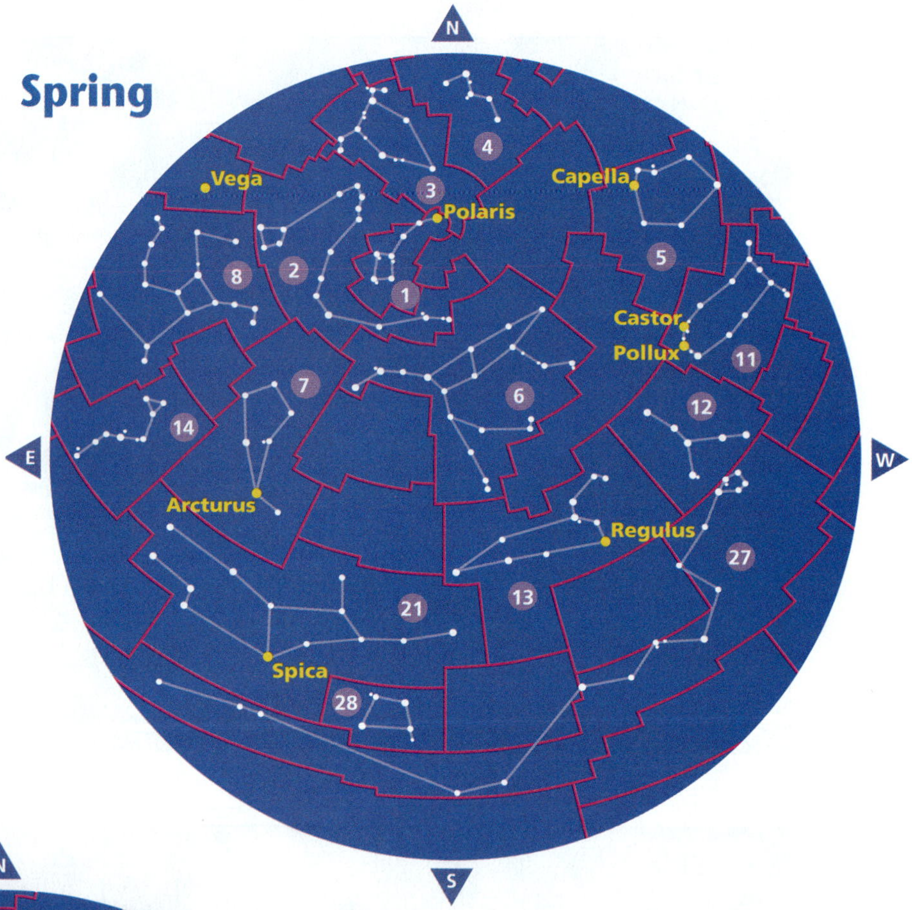

Summer

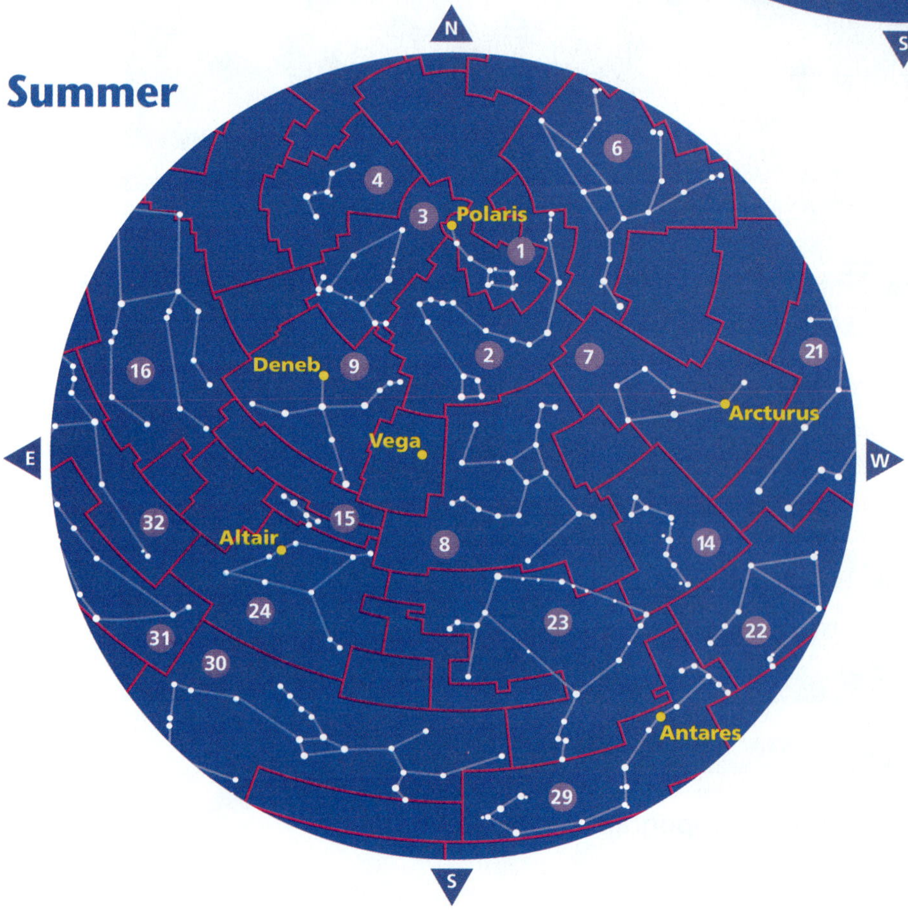

Constellations	
1	Ursa Minor
2	Draco
3	Cepheus
4	Cassiopeia
5	Auriga
6	Ursa Major
7	Boötes
8	Hercules
9	Cygnus
10	Perseus
11	Gemini
12	Cancer
13	Leo
14	Serpens
15	Sagitta
16	Pegasus
17	Pisces

Autumn

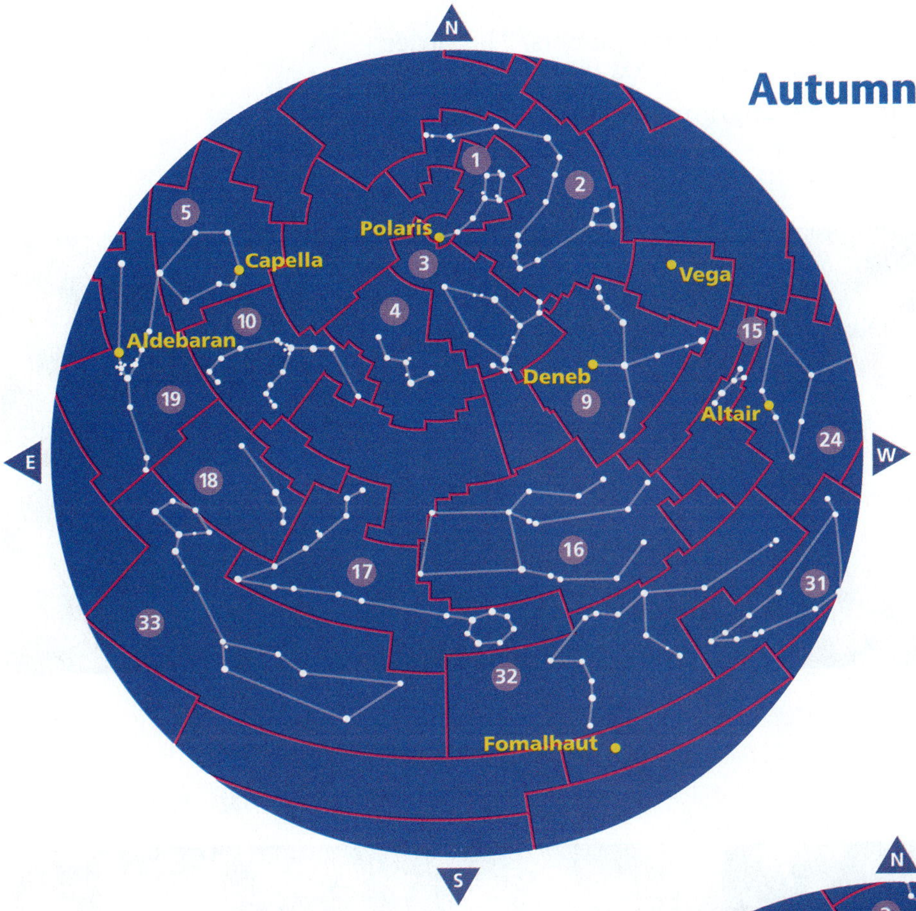

Winter

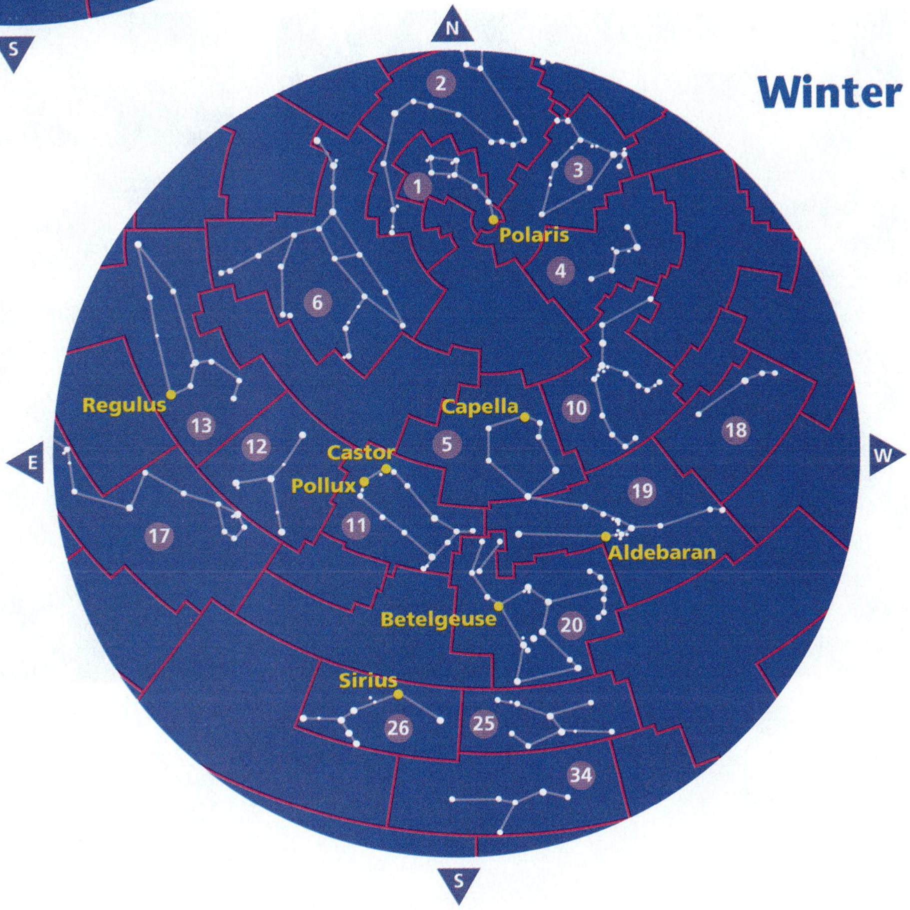

Constellations	
18	Aries
19	Taurus
20	Orion
21	Virgo
22	Libra
23	Ophiuchus
24	Aquila
25	Lepus
26	Canis Major
27	Hydra
28	Corvus
29	Scorpius
30	Sagittarius
31	Capricornus
32	Aquarius
33	Cetus
34	Columba

REFERENCE MAPS

Maps of the Solar System

The diagram at top shows the relative sizes of the nine planets. The order of the planets from the sun is the following: Mercury, Venus, Earth, Mars, Jupiter, Saturn, Uranus, and Neptune. The diagrams at bottom show the orbits of the planets around the sun.

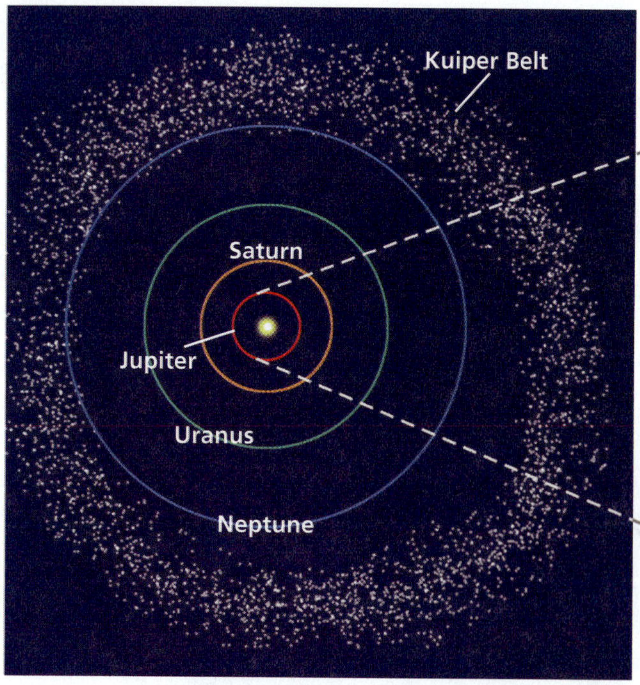

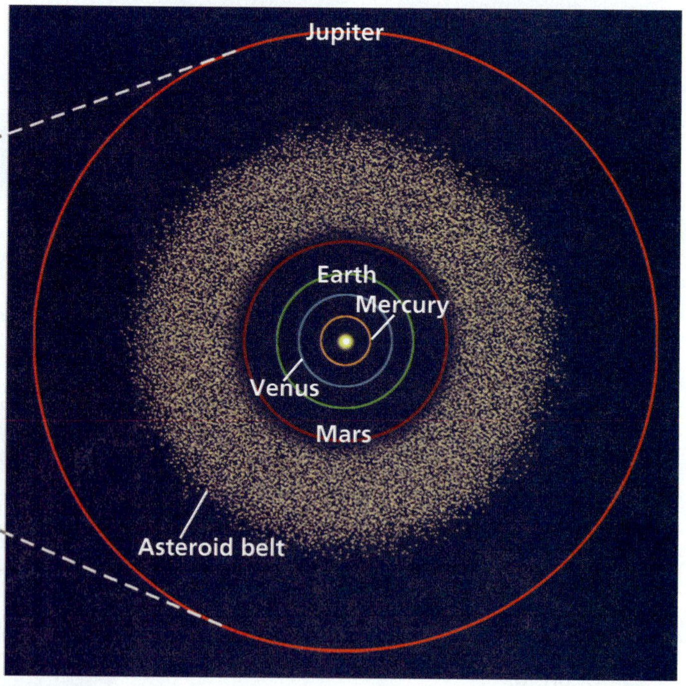

READING CHECK ANSWERS

Chapter 1 Introduction to Earth Science

Section 1

Page 7: The development of telescopes, satellites, and space probes has greatly expanded astronomers' understanding of the universe.

Section 2

Page 10: Observations may lead to interesting scientific questions and may help scientists formulate reasonable and testable hypotheses.

Page 13: Answers may vary but should include three of the following types of models: physical models, graphic models, conceptual models, computer models, and mathematical models.

Page 14: Scientists present the results of their work at professional meetings and in scientific journals.

Chapter 2 Earth as a System

Section 1

Page 28: Indirect observations are the only means available for exploring Earth's interior at depths too great to be reached by drilling.

Section 2

Page 32: Dust and rock come to Earth from space, while hydrogen atoms from the atmosphere enter space from Earth.

Page 34: An energy budget is the total distribution of energy to, from, and between Earth's various spheres.

Page 36: soil and plants

Section 3

Page 40: The amount of matter and energy in an ecosystem can supply a population of a given size, and no larger. This maximum population is the carrying capacity of the ecosystem.

Chapter 3 Models of the Earth

Section 1

Page 54: because the equator is the only parallel that divides Earth into halves

Section 2

Page 58: Because both the parallels and the meridians are equally spaced straight lines on a cylindrical projection, the parallels and meridians form a grid.

Page 61: by using a graphic scale, or a printed line divided into proportional parts that represent units of measure; a fractional scale, in which a ratio shows how distance on Earth relates to distance on a map; or a verbal scale, which expresses scale in sentence form

Section 3

Page 65: Water moves from areas of higher elevation to areas of lower elevation. Because the V shape points toward higher elevation, it points upstream.

Page 67: Scientists create soil maps to classify, map, and describe soils.

Chapter 4 Earth Chemistry

Section 1

Page 83: The atomic number is the number of protons in an atom's nucleus. The mass number is the sum of the number of protons and the number of neutrons in an atom. The atomic mass unit is used to express the mass of subatomic particles or atoms.

Section 2

Page 89: Atoms form chemical bonds by transferring electrons or by sharing electrons.

Page 91: The oxygen atom has a larger and more positively charged nucleus than the hydrogen atoms do. As a result, the oxygen nucleus pulls the electrons from the hydrogen atoms closer to it than the hydrogen nuclei pull the shared electrons from the oxygen. This unequal attraction forms a polar-covalent bond.

Chapter 5 Minerals of Earth's Crust

Section 1

Page 105: Nonsilicate minerals never contain compounds of silicon bonded to oxygen.

Page 106: The building block of the silicate crystalline structure is a four-sided structure known as the *silicon-oxygen tetrahedron*, which is one silicon atom surrounded by four oxygen atoms.

Section 2

Page 111: The strength and geometric arrangement of the bonds between the atoms that make up a mineral's internal structure determine the hardness of a mineral.

Page 113: Chatoyancy is the silky appearance of some minerals in reflected light. Asterism is the appearance of a six-sided star when a mineral reflects light.

Chapter 6 Rocks

Section 1

Page 127: As magma cools and solidifies, minerals crystallize out of the magma in a specific order that depends on their melting points.

Section 2

Page 131: Fine-grained igneous rock forms mainly from magma that cools rapidly; coarse-grained igneous rock forms mainly from magma that cools more slowly.

Page 133: A batholith is an intrusive structure that covers an area of at least 100 km^2. A stock covers an area of less than 100 km^2.

Section 3

Page 137: Three groups of clastic sedimentary rock are conglomerates and breccias, sandstones, and shales.

Page 139: Graded bedding is a type of stratification in which different sizes and types of sediments settle to different levels.

Section 4

Page 142: The high pressures and temperatures that result from the movements of tectonic plates may cause chemical changes in the minerals.

Chapter 7 Resources and Energy

Section 1

Page 156: Water creates ore deposits by eroding rock and releasing minerals and by carrying the mineral fragments and depositing them in streambeds.

Section 2

Page 161: Cap rock is a layer of impermeable rock at the top of an oil- or natural gas-bearing formation through which fluids cannot flow.

Page 162: As neutrons strike neighboring nuclei, the nuclei split and release additional neutrons that strike other nuclei and cause the chain to continue.

Section 3

Page 167: Answers may vary but should include three of the following: geothermal, solar, hydroelectric, and biomass.

Section 4

Page 170: The use of fossil fuels affects the environment when coal is mined from the surface, which destroys the land. When fossil fuels are burned, they affect the environment by creating air pollution.

Chapter 8 The Rock Record

Section 1

Page 186: Hutton reasoned that the extremely slow-working forces that changed the land on his farm had also slowly changed the rocks that make up Earth's crust. He concluded that large changes must happen over a period of millions of years.

Page 188: Because ripple marks form at the top of a rock layer, scientists can use the orientation of the ripple marks to determine which direction was "up" when the rock layers formed.

Section 2

Page 192: Varves are like tree rings in that varves are laid down each year. Thus, counting varves can reveal the age of sedimentary deposits.

Page 195: An isotope that has an extremely long half-life will not show significant or measurable changes in a young rock. In a very old rock, an isotope that has a short half-life may have decayed to the point at which too little of the isotope is left to give an accurate age measurement. So, the estimated age of the rock must be correlated to the dating method used.

Section 3

Page 199: A trace fossil is fossilized evidence of past animal movement, such as tracks, footprints, borings, or burrows, that can provide information about prehistoric life.

Chapter 9 A View of Earth's Past

Section 1

Page 211: You would find fossils of extinct animals in older layers of a geologic column.

Section 2

Page 216: Earth is approximately 4.6 billion years old.

Page 218: Answers may vary but should include three of the following: trilobites, brachiopods, jellyfish, worms, snails, and sponges.

Section 3

Page 222: Answers may vary but could include *Archaeopteryx*, pterosaurs, *Apatosaurus*, and *Stegosaurus*.

Page 225: During ice ages, water from the ocean was frozen as ice, so the amount of liquid water in the seas decreased and sea level fell.

Chapter 10 Plate Tectonics

Section 1

Page 241: Many scientists rejected Wegener's hypothesis because the mechanism that Wegener suggested was easily disproved by geologic evidence.

Page 243: New sea floor forms as magma rises to fill the rift that forms when two plates pull apart at a divergent boundary.

Page 245: The symmetrical magnetic patterns in sea-floor rocks show that rock formed at one place (at a ridge) and then broke apart and moved away from the center in opposite directions.

Section 2

Page 248: Scientists use the locations of earthquakes, volcanoes, trenches, and mid-ocean ridges to outline tectonic plates.

Page 250: Collisions at convergent boundaries can happen between two oceanic plates, between two continental plates, or between one oceanic plate and one continental plate.

Page 253: When denser lithosphere sinks into the asthenosphere, the asthenosphere must move out of the way. As the asthenosphere moves, it drags or pushes on other parts of the lithosphere, which causes movement.

Section 3

Page 256: As a plate subducts beneath another plate, islands and other land features on the subducting plate are scraped off the subducting plate and become part of the overriding plate.

Page 259: The continents Africa, South America, Antarctica, and Australia formed from Gondwanaland. The subcontinent of India was also part of Gondwanaland.

Chapter 11 Deformation of the Crust

Section 1

Page 273: Tension and shear stress can both pull rock apart.

Page 275: limbs and hinges

Page 277: A thrust fault is a type of reverse fault in which the fault plane is at a low angle relative to the surface.

Section 2

Page 281: The Himalayas are growing taller because the two plates are still colliding and causing further compression of the rock, which further uplifts the mountains.

Page 283: Answers may include three of the following: folded mountains, fault-block mountains, dome mountains, and volcanic mountains.

Chapter 12 Earthquakes

Section 1

Page 297: Rayleigh waves cause the ground to move in an elliptical, rolling motion. Love waves cause rock to move side-to-side and perpendicular to the direction the waves are traveling.

Page 298: The speed of seismic waves changes as they pass through different layers of Earth.

Section 2

Page 303: Moment magnitude is more accurate for larger earthquakes than the Richter scale is. Moment magnitude is directly related to rock properties and so is more closely related to the cause of the earthquake than the Richter scale is.

Section 3

Page 307: Scientists think that stress on a fault builds up to a critical point and is then released as an earthquake. Seismic gaps are areas in which no earthquakes have happened in a long period of time and thus are likely to be under a high amount of stress.

Chapter 13 Volcanoes

Section 1

Page 321: The denser plate of oceanic lithosphere subducts beneath the less dense plate of continental lithosphere.

Page 323: As the lithosphere moves over the mantle plume, older volcanoes move away from the mantle plume. A new hot spot forms in the lithosphere above the mantle plume as a new volcano begins to form.

Section 2

Page 326: The faster the rate of flow is and the higher the gas content is, the more broken up and rough the resulting cooled lava will be.

Page 329: A caldera may form when a magma chamber empties or when large amounts of magma are discharged, causing the ground to collapse.

Chapter 14 Weathering and Erosion

Section 1

Page 344: Two types of mechanical weathering are ice wedging and abrasion. Ice wedging is caused by water that seeps into cracks in rock and freezes. When water freezes, it expands and creates pressure on the rock, which widens and deepens cracks. Abrasion is the grinding away of rock surfaces by other rocks or sand particles. Abrasive agents may be carried by gravity, water, and wind.

Page 346: Two effects of chemical weathering are changes in the chemical composition and changes in the physical appearance of a rock.

Section 2

Page 350: Fractures and joints in a rock increase surface area and allow weathering to occur more rapidly.

Section 3

Page 355: Large amounts of rainfall and high temperatures cause thick soils to form in both tropical and temperate climates. Tropical soils have thin A horizons because of the continuous leaching of topsoil. Temperate soils have three thick layers, because leaching of the A horizon in temperate climates is much less than leaching of the A horizon in tropical climates.

Section 4

Page 358: Dust storms may form during droughts when the soil is made dry and loose by lack of moisture and wind-caused sheet erosion carries it away in clouds of dust. If all of the topsoil is removed, the remaining subsoil will not contain enough nutrients to raise crops.

Page 361: Landslides are masses of loose rock combined with soil that suddenly fall down a slope. A rockfall consists of rock falling from a steep cliff.

Page 363: When a mountain is no longer being uplifted, weathering and erosion wear down its jagged peaks to low, featureless surfaces called *peneplains*.

Chapter 15 River Systems

Section 1

Page 376: Precipitation is any form of water that falls to Earth from the clouds, including rain, snow, sleet, and hail.

Section 2

Page 381: A river that has meanders probably has a low gradient.

Section 3

Page 385: Floods can be controlled indirectly through forest and soil conservation measures that reduce or prevent runoff, or directly by building artificial structures, such as dams, levees, and floodways, to redirect water flow.

Chapter 16 Groundwater

Section 1

Page 399: The two zones of groundwater are the zone of saturation and the zone of aeration.

Page 400: The depth of a water table depends on topography, aquifer permeability, the amount of rainfall, and the rate at which humans use the groundwater.

Page 403: Ordinary springs occur where the ground surface drops below the water table. An artesian spring occurs where groundwater flows to the surface through natural cracks in the overlying cap rock.

Section 2

Page 407: A natural bridge may form when two sinkholes form close to each other. The bridge is the uncollapsed rock between the sinkholes.

Chapter 17 Glaciers

Section 1

Page 420: Continental glaciers exist only in Greenland and Antarctica.

Page 424: A moving glacier forms a cirque by pulling blocks of rock from the floor and walls of a valley and leaving a bowl-shaped depression.

Section 2

Page 427: A drumlin is a long, low, tear-shaped mound of till.

Page 428: Eskers form when meltwater from receding continental glaciers flows through ice tunnels and deposits long, winding ridges of gravel and sand.

Section 3

Page 432: The sea level was up to 140 m lower than it is now.

READING CHECK ANSWERS

Chapter 18 Erosion by Wind and Waves

Section 1

Page 446: Moisture makes soil heavier, so the soil sticks and is more difficult to move. Therefore, erosion happens faster in dry climates.

Page 448: Barchan dunes are crescent shaped; transverse dunes form linear ridges.

Section 2

Page 452: Answers should include three of the following: sea cliffs, sea caves, sea arches, sea stacks, wave-cut terraces, and wave-built terraces.

Section 3

Page 457: As sea levels rise over a flat coastal plain, the shoreline moves inland and isolates dunes from the old shoreline. These dunes become barrier islands.

Chapter 19 Ocean Basins

Section 1

Page 472: Oceanographers study the physical characteristics, chemical composition, and life-forms of the ocean.

Section 2

Page 476: Trenches; broad, flat plains; mountain ranges; and submerged volcanoes are part of the deep-ocean basins.

Section 3

Page 481: When chemical reactions take place in the ocean, dissolved substances can crystallize to form nodules that settle to the ocean floor.

Chapter 20 Ocean Water

Section 1

Page 495: Dissolved solids enter the oceans from the chemical weathering of rock on land, from volcanic eruptions, and from chemical reactions between sea water and newly formed sea-floor rocks.

Page 497: Ocean surface temperatures are affected by the amount of solar energy an area receives and by the movement of water in the ocean.

Page 499: Ocean water contains dissolved solids (mostly salts) that add mass to a given volume of water. The large amount of dissolved solids in ocean water makes ocean water denser than fresh water.

Section 2

Page 503: Most marine life is found in the sublittoral zone. Life in this zone is continuously submerged, but waters are still shallow enough to allow sunlight to penetrate.

Section 3

Page 507: Aquaculture provides a reliable, economical source of food. However, aquatic farms are susceptible to pollution and they may become local sources of pollution.

Chapter 21 Movements of the Ocean

Section 1

Page 521: Because no continents interrupt the flow of the Antarctic Circumpolar Current, also called the *West Wind Drift*, it completely encircles Antarctica and crosses three major oceans. All other surface currents are deflected and divided when they meet a continental barrier.

Page 523: Antarctic Bottom Water is very cold. It also has a high salinity. The extreme cold and high salinity combine to make the water extremely dense.

Section 2

Page 526: Because waves receive energy from wind that pushes against the surface of the water, the amount of energy decreases as the depth of water increases. As a result, the diameter of the water molecules' circular path also decreases.

Page 528: Contact with the ocean floor slows down the bottom of the wave but not the top of the wave. Because of the difference in speed between the top and bottom of the wave, the top gets farther ahead of the bottom until the wave becomes unstable and falls over.

Section 3

Page 532: When the tidal range is small, the sun and the moon are at right angles to each other relative to Earth's orbit.

Chapter 22 The Atmosphere

Section 1

Page 548: Transpiration increases the amount of water vapor in the atmosphere.

Page 551: An aneroid barometer contains a sealed metal container that has a partial vacuum.

Page 553: The lower region of the thermosphere is called the *ionosphere*.

Section 2

Page 559: Deserts are colder at night than other areas are because the air in deserts contains little water vapor that can absorb heat during the day and release heat slowly at night.

Section 3

Page 562: They flow in opposite directions from each other, and they occur at different latitudes.

Chapter 23 Water in the Atmosphere

Section 1

Page 576: When the air is very dry and the temperature is below freezing, ice and snow change directly into water vapor by sublimation.

Page 578: Dew is liquid moisture that condenses from air on cool objects when the air is nearly saturated and the temperature drops. Frost is water vapor that condenses as ice crystals onto a cool surface directly from the air when the dew point is below freezing.

Section 2

Page 582: The source of heat that warms the air and leads to cloud formation is solar energy that is reradiated as heat by Earth's surface. As the process continues, latent heat released by the condensation may allow the clouds to expand beyond the condensation level.

Page 585: because cirrus clouds form at very high altitudes where air temperature is low

Section 3

Page 589: Doppler radar measures the location, direction of movement, and intensity of precipitation.

Chapter 24 Weather

Section 1

Page 603: a continental tropical air mass

Section 2

Page 607: The air of an anticyclone sinks and flows outward from a center of high pressure. The air of a midlatitude cyclone rotates toward the rising air of a central, low-pressure region.

Page 609: over warm tropical seas

Section 3

Page 612: A barometer is used to measure atmospheric pressure.

Section 4

Page 617: Areas of precipitation are marked by using colors or symbols.

Page 618: Meteorologists compare computer models because different models are better at predicting different weather variables. If information from two or more models matches, scientists can be more confident of their predictions.

Chapter 25 Climate

Section 1

Page 633: Waves, currents, and other water motions continually replace warm surface waters with cooler water from the ocean depths, which keeps the surface temperature of the water from increasing rapidly.

Page 634: The temperature of land increases faster than that of water does because the specific heat of land is lower than that of water, and thus the land requires less energy to heat up than the water does.

Section 2

Page 638: marine west coast, humid continental, and humid subtropical

Section 3

Page 642: Scientists use computer models to incorporate as much data as possible to sort out the complex variables that influence climate and to make predictions about climate.

Page 644: Climate change influences humans, plants, and animals. It also affects nearby climates, sea level, and precipitation rates.

Chapter 26 Studying Space

Section 1

Page 661: The only kind of electromagnetic radiation the human eye can detect is visible light.

Page 663: Images produced by refracting telescopes are subject to distortion because of the way different colors of visible light are focused at different distances from the lens and because of weight limitations on the objective lens.

Page 664: Scientists launch spacecraft into orbit to detect radiation screened out by Earth's atmosphere and to avoid light pollution and other atmospheric distortions.

Section 2

Page 669: Constellations provide two kinds of evidence of Earth's motion. As Earth rotates, the stars appear to change position during the night. As Earth revolves around the sun, Earth's night sky faces a different part of the universe. As a result, different constellations appear in the night sky as the seasons change.

Page 671: Because time zones are based on Earth's rotation, as you travel west, you eventually come to a location where, on one side of the time zone border, the calendar moves ahead one day. The purpose of the International Dateline is to locate the border so that the transition would affect the least number of people. So that it will affect the least number of people, the International Dateline is in the middle of the Pacific Ocean, instead of on a continent.

Page 672: Daylight savings time is an adjustment that is made to standard time by setting clocks ahead one hour to take advantage of longer hours of daylight in the summer months and to save energy.

Chapter 27 Planets of the Solar System

Section 1

Page 687: Unlike the outer planets, Pluto is very small and is composed of rock and frozen gas, instead of thick layers of gases.

Page 689: Green plants release free oxygen as part of photosynthesis, which caused the concentration of oxygen gas in the atmosphere to gradually increase.

Section 2

Page 692: An ellipse is a closed curve whose shape is defined by two points inside the curve. An ellipse looks like an oval.

Section 3

Page 697: Answers may vary but should address differences in distance from the sun, density, atmospheric pressure and density, and tectonics.

Page 699: Martian volcanoes are larger than volcanoes on Earth because Mars has no moving tectonic plates. Magma sources remain in the same spot for millions of years and produce volcanic material that builds the volcanic cone higher and higher.

Section 4

Page 702: When Jupiter formed, it did not have enough mass for nuclear fusion to begin.

Page 704: Saturn and Jupiter are made almost entirely of hydrogen and helium and have rocky-iron cores, ring systems, many satellites, rapid rotational periods, and bands of colored clouds.

Page 707: The Kuiper Belt is located beyond the orbit of Neptune.

Chapter 28 Minor Bodies of the Solar System

Section 1

Page 720: Answers should include two of the following features: maria, highlands, craters, ridges, and rilles.

Page 722: The crust of the far side of the moon is thicker than the crust of the near side is. The crust of the far side also consists mainly of mountainous terrain and has only a few small maria.

Section 2

Page 726: The far side of the moon is never visible from Earth, because the moon's rotation and the moon's revolution around Earth take the same amount of time.

Page 728: During a total eclipse, the entire disk of the sun is blocked, and the outer layers of the sun become visible. During an annular eclipse, the disk of the sun is never completely blocked out, so the sun is too bright for observers on Earth to see the outer layers of the sun's atmosphere.

Page 731: When the lighted part of the moon is larger than a semicircle but the visible part of the moon is shrinking, the phase is called *waning gibbous*. When only a sliver of the near side is visible, the phase is a waning crescent.

Section 3

Page 735: Io's surface is covered with many active volcanoes. Europa's surface is covered by an enormous ice sheet. Ganymede is the largest moon in the solar system and has a strong magnetic field. Callisto's surface is heavily cratered.

Page 737: Charon is almost half the size of the body it orbits. Charon's orbital period is the same length as Pluto's day, so only one side of Pluto always faces the moon.

Section 4

Page 740: The most common type is made mostly of carbon materials. Other asteroids are made mostly of silicate minerals. The third type is composed mostly of iron and nickel.

Page 743: A meteoroid is a rocky body that travels through space. When a meteoroid enters Earth's atmosphere and begins to burn up, the meteoroid produces a meteor.

Chapter 29 The Sun

Section 1

Page 757: Einstein's equation helped scientists understand the source of the sun's energy. The equation explained how the sun could produce huge amounts of energy without burning up.

Page 759: The sun's atmosphere consists of the photosphere, the chromosphere, and the corona.

Section 2

Page 763: Coronal mass ejections generate sudden disturbances in Earth's magnetic field. The high-energy particles that circulate during these storms can damage satellites, cause power blackouts, and interfere with radio communications.

Chapter 30 Stars, Galaxies, and the Universe

Section 1

Page 777: Polaris is almost exactly above the pole of Earth's rotational axis, so Polaris moves only slightly around the pole during one rotation of Earth.

Page 778: Starlight is shifted toward the red end of the spectrum when the star is moving away from the observer.

Section 2

Page 783: The forces balance each other and keep the star in equilibrium. As gravity increases the pressure on the matter within a star, the rate of fusion increases. This increase in fusion causes a rise in gas pressure. As a result, the energy from the increased fusion and gas pressure generates outward pressure that balances the force of gravity.

Page 784: Giants and supergiants appear in the upper-right part of the H-R diagram.

Page 787: As supergiants collapse because of gravitational forces, fusion begins and continues until the supply of fuel is used up. The core begins to collapse under its own gravity and causes energy to transfer to the outer layers of the star. The transfer of energy to the outer layers causes the explosion.

Section 3

Page 790: More than 50% of all stars are in multiple-star systems.

Section 4

Page 794: All matter and energy in the early universe were compressed into a small volume at an extremely high temperature until the temperature cooled and all of the matter and energy were forced outward in all directions.

GLOSSARY/GLOSARIO

Terms and their definitions are listed in English in alphabetical order in the first column. The second column lists the equivalent term in Spanish.

A

abrasion the grinding and wearing away of rock surfaces through the mechanical action of other rock or sand particles (344)	**abrasion/abrasión** proceso por el cual las super-ficies de las rocas se muelen o desgastan por medio de la acción mecánica de otras rocas y partículas de arena (344)
absolute age the numeric age of an object or event, often stated in years before the present, as established by an absolute-dating process, such as radiometric dating (191)	**absolute age/edad absoluta** la edad numérica de un objeto o suceso, que suele expresarse en cantidad de años antes del presente, determinada por un proceso de datación absoluta, tal como la datación radiométrica (191)
absolute humidity the mass of water vapor per unit volume of air that contains the water vapor; usually expressed as grams of water vapor per cubic meter of air (577)	**absolute humidity/humedad absoluta** la masa de vapor de agua por unidad de volumen de aire que contiene al vapor de agua; normalmente se expresa por metro cúbico de aire (577)
absolute magnitude the brightness that a star would have at a distance of 32.6 light-years from Earth (780)	**absolute magnitude/magnitud absoluta** el brillo que una estrella tendría a una distancia de 32.6 años luz de la Tierra (780)
abyssal plain a large, flat, almost level area of the deep-ocean basin (477)	**abyssal plain/llanura abisal** un área amplia, llana y casi plana de la cuenca oceánica profunda (477)
adiabatic cooling the process by which the temperature of an air mass decreases as the air mass rises and expands (582)	**adiabatic cooling/enfriamiento adiabático** el proceso por medio del cual la temperatura de una masa de aire disminuye a medida que ésta se eleva y se expande (582)
advective cooling the process by which the temperature of an air mass decreases as the air mass moves over a cold surface (583)	**advective cooling/enfriamiento advectivo** el proceso por medio del cual la temperatura de una masa de aire disminuye a medida que ésta se mueve sobre una superficie fría (583)
air mass a large body of air throughout which temperature and moisture content are similar (601)	**air mass/masa de aire** un gran volumen de aire, cuya temperatura y cuyo contenido de humedad son similares en toda su extensión (601)
albedo the fraction of solar radiation that is reflected off the surface of an object (557)	**albedo/albedo** porcentaje de la radiación solar que la superficie de un objeto refleja (557)
alluvial fan a fan-shaped mass of rock material deposited by a stream when the slope of the land decreases sharply; for example, alluvial fans form when streams flow from mountains to flat land (383)	**alluvial fan/abanico aluvial** masa de materiales rocosos en forma de abanico, depositados por un arroyo cuando la pendiente del terreno disminuye bruscamente; por ejemplo, los abanicos aluviales se forman cuando los arroyos fluyen de una montaña a un terreno llano (383)
alpine glacier a narrow, wedge-shaped mass of ice that forms in a mountainous region and that is confined to a small area by surrounding topography; examples include valley glaciers, cirque glaciers, and piedmont glaciers (420)	**alpine glacier/glaciar alpino** una masa de hielo angosta, parecida a una cuña, que se forma en una región montañosa y que está confinada a un área pequeña por la topografía que la rodea; los glaciares de valle, los circos glaciares y los glaciares de pie de monte son algunos ejemplos de esto (420)

anemometer an instrument used to measure wind speed (612)

aphelion the point in the orbit of a planet at which the planet is farthest from the sun (668)

apogee in the orbit of a satellite, the point at which the satellite is farthest from Earth (725)

apparent magnitude the brightness of a star as seen from the Earth (780)

aquaculture the raising of aquatic plants and animals for human use or consumption (507)

aquifer a body of rock or sediment that stores groundwater and allows the flow of groundwater (397)

arête a sharp, jagged ridge that forms between cirques (424)

artesian formation a sloping layer of permeable rock sandwiched between two layers of impermeable rock and exposed at the surface (403)

asteroid a small, rocky object that orbits the sun; most asteroids are located in a band between the orbits of Mars and Jupiter (739)

asthenosphere the solid, plastic layer of the mantle beneath the lithosphere; made of mantle rock that flows very slowly, which allows tectonic plates to move on top of it (29, 247)

astronomical unit the average distance between the Earth and the sun; approximately 150 million kilometers (symbol, AU) (660)

astronomy the scientific study of the universe (7, 659)

atmosphere a mixture of gases that surrounds a planet or moon (33, 547)

atmospheric pressure the force per unit area that is exerted on a surface by the weight of the atmosphere (550)

atom the smallest unit of an element that maintains the chemical properties of that element (82)

anemometer/anemómetro un instrumento que se usa para medir la rapidez del viento (612)

aphelion/afelio el punto en la órbita de un planeta en que el planeta está más lejos del Sol (668)

apogee/apogeo en la órbita de un satélite, el punto en el que el satélite está más alejado de la Tierra (725)

apparent magnitude/magnitud aparente el brillo de una estrella como se percibe desde la Tierra (780)

aquaculture/acuacultura el cultivo de plantas y animales acuáticos para uso o consumo humano (507)

aquifer/acuífero un cuerpo rocoso o sedimento que almacena agua subterránea y permite que fluya (397)

arête/cresta una cumbre puntiaguda e irregular que se forma entre circos glaciares (424)

artesian formation/formación artesiana capa inclinada de rocas permeables que está en medio de dos capas de rocas impermeables y expuesta en la superficie (403)

asteroid/asteroide un objeto pequeño y rocoso que se encuentra en órbita alrededor del Sol; la mayoría de los asteroides se ubican en una banda entre las órbitas de Marte y Júpiter (739)

asthenosphere/astenosfera la capa sólida y plástica del manto, que se encuentra debajo de la litosfera; está formada por roca del manto que fluye muy lentamente, lo cual permite que las placas tectónicas se muevan en su superficie (29, 247)

astronomical unit/unidad astronómica la distancia promedio entre la Tierra y el Sol; aproximadamente 150 millones de kilómetros (símbolo: UA) (660)

astronomy/astronomía el estudio científico del universo (7, 659)

atmosphere/atmósfera una mezcla de gases que rodea un planeta o una luna (33, 547)

atmospheric pressure/presión atmosférica la fuerza por unidad de área que el peso de la atmósfera ejerce sobre una superficie (550)

atom/átomo la unidad más pequeña de un elemento que conserva las propiedades químicas de ese elemento (82)

aurora colored light produced by charged particles from the solar wind and from the magnetosphere that react with and excite the oxygen and nitrogen of Earth's upper atmosphere; usually seen in the sky near Earth's magnetic poles (764)

aurora/aurora luz de colores producida por partículas con carga del viento solar y de la magnetosfera, que reaccionan con los átomos de oxígeno y nitrógeno de la parte superior de la atmósfera de la Tierra y los excitan; normalmente se ve en el cielo cerca de los polos magnéticos de la Tierra (764)

B

barometer an instrument that measures atmospheric pressure (612)

barometer/barómetro un instrumento que mide la presión atmosférica (612)

barrier island a long ridge of sand or narrow island that lies parallel to the shore (457)

barrier island/isla barrera un largo arrecife de arena o una isla angosta ubicada paralela a la costa (457)

basal slip the process that causes the ice at the base of a glacier to melt and the glacier to slide (421)

basal slip/deslizamiento basal el proceso que hace que el hielo de la base de un glaciar se derrita y que éste se deslice (421)

beach an area of the shoreline that is made up of deposited sediment (453)

beach/playa un área de la costa que está formada por sedimento depositado (453)

benthic zone the bottom region of oceans and bodies of fresh water (503)

benthic zone/zona bentónica la región del fondo de los océanos y de las masas de agua dulce (503)

benthos organisms that live at the bottom of oceans or bodies of fresh water (502)

benthos/benthos organismos que viven en el fondo de los océanos o de las masas de agua dulce (502)

big bang theory the theory that all matter and energy in the universe was compressed into an extremely small volume that 13 billion to 15 billion years ago exploded and began expanding in all directions (794)

big bang theory/teoría del Big Bang la teoría que establece que toda la materia y la energía del universo estaban comprimidas en un volumen extremadamente pequeño que explotó hace aproximadamente 13 a 15 mil millones de años y empezó a expandirse en todas direcciones (794)

biomass plant material, manure, or any other or-ganic matter that is used as an energy source (167)

biomass/biomasa materia vegetal, estiércol o cualquier otra materia orgánica que se usa como fuente de energía (167)

biosphere the part of Earth where life exists; includes all of the living organisms on Earth (33)

biosphere/biosferavla parte de la Tierra donde existe la vida; abarca a todos los organismos vivos de la Tierra (33)

black hole an object so massive and dense that even light cannot escape its gravity (788)

black hole/hoyo negro un objeto tan masivo y denso que ni siquiera la luz puede salir de su campo gravitacional (788)

body wave in geology, a seismic wave that travels through the body of a medium (296)

body wave/onda interna en geología, una onda sísmica que se desplaza a través del cuerpo de un medio (296)

Bowen's reaction series the simplified pattern that illustrates the order in which minerals crystallize from cooling magma according to their chemical composition and melting point (127)

Bowen's reaction series/serie de reacción de Bowen el patrón simplificado que ilustra el orden en que los minerales se cristalizan a partir del magma que se enfría, de acuerdo con su composición química y punto de fusión (127)

braided stream a stream or river that is composed of multiple channels that divide and rejoin around sediment bars (382)

braided stream/corriente anastomosada una corriente o río compuesto por varios canales que se dividen y se vuelven a encontrar alrededor de barreras de sedimento (382)

C

caldera a large, circular depression that forms when the magma chamber below a volcano partially empties and causes the ground above to sink (329)

caldera/caldera una depresión grande y circular que se forma cuando se vacía parcialmente la cámara de magma que hay debajo de un volcán, lo cual hace que el suelo se hunda (329)

carbonation the conversion of a compound into a carbonate (347)

carbonation/carbonación la transformación de un compuesto a un carbonato (347)

carrying capacity the largest population that an environment can support at any given time (40)

carrying capacity/capacidad de carga la población más grande que un ambiente puede sostener en cualquier momento dado (40)

cavern a natural cavity that forms in rock as a result of the dissolution of minerals; *also* a large cave that commonly contains many smaller, connecting chambers (406)

cavern/caverna una cavidad natural que se forma en la roca como resultado de la disolución de minerales; *también,* una gran cueva que generalmente contiene muchas cámaras más pequeñas comunicadas entre sí (406)

cementation the process in which minerals precipitate into pore spaces between sediment grains and bind sediments together to form rock (135)

cementation/cementación el proceso en el cual los minerales se precipitan entre los poros de granos de sedimento y unen los sedimentos para formar rocas (135)

Cenozoic Era the current geologic era, which began 65.5 million years ago; also called the *Age of Mammals* (224)

Cenozoic Era/Era Cenozoica la era geológica actual, que comenzó hace 65.5 millones de años; también llamada *Edad de los Mamíferos* (224)

chemical sedimentary rock sedimentary rock that forms when minerals precipitate from a solution or settle from a suspension (136)

chemical sedimentary rock/roca sedimentaria química roca sedimentaria que se forma cuando los minerales precipitan a partir de una solución o se depositan a partir de una suspensión (136)

chemical weathering the process by which rocks break down as a result of chemical reactions (346)

chemical weathering/desgaste químico el proceso por medio del cual las rocas se fragmentan como resultado de reacciones químicas (346)

chromosphere the thin layer of the sun that is just above the photosphere and that glows a reddish color during eclipses (760)

chromosphere/cromosfera la delgada capa del Sol que se encuentra justo encima de la fotosfera y que resplandece con un color rojizo durante los eclipses (760)

cirque a deep and steep bowl-like depression produced by glacier erosion (424)

cirque/circo una depresión profunda y empinada, con forma de tazón, producida por erosión glaciar (424)

cirrus cloud a feathery cloud that is composed of ice crystals and that has the highest altitude of any cloud in the sky (585)

cirrus cloud/nube cirro una nube liviana formada por cristales de hielo, la cual tiene la mayor altitud de todas las nubes en el cielo (585)

clastic sedimentary rock sedimentary rock that forms when fragments of preexisting rocks are compacted and cemented together (137)

clastic sedimentary rock/roca sedimentaria clástica roca sedimentaria que se forma cuando los fragmentos de rocas preexistentes se unen por compactación y cementación (137)

cleavage in geology, the tendency of a mineral to split along specific planes of weakness to form smooth, flat surfaces (110)

cleavage/exfoliación en geología, la tendencia de un mineral a agrietarse a lo largo de planos débiles específicos y formar superficies lisas y planas (110)

climate the average weather conditions in an area over a long period of time (631)

climate/clima las condiciones promedio del tiempo en un área durante un largo período de tiempo (631)

climatologist a scientist who gathers data to study and compare past and present climates and to predict future climate change (641)

climatologist/climatólogo un científico que recopila datos para estudiar y comparar los climas del pasado y del presente y para predecir cambios climáticos en el futuro (641)

cloud a collection of small water droplets or ice crystals suspended in the air, which forms when the air is cooled and condensation occurs (581)

cloud/nube un conjunto de pequeñas gotitas de agua o cristales de hielo suspendidos en el aire, que se forma cuando el aire se enfría y ocurre condensación (581)

cloud seeding the process of introducing freezing nuclei or condensation nuclei into a cloud in order to cause rain to fall (590)

cloud seeding/sembrado de nubes el proceso de introducir núcleos congelados o núcleos de condensación en una nube para producir lluvia (590)

coalescence the formation of a larger droplet by the combination of smaller droplets (588)

coalescence/coalescencia la formación de una gota más grande al combinarse gotas más pequeñas (588)

cold front the front edge of a moving mass of cold air that pushes beneath a warmer air mass like a wedge (605)

cold front/frente frío el borde del frente de una masa de aire frío en movimiento que empuja por debajo de una masa de aire más caliente como una cuña (605)

comet a small body of ice, rock, and cosmic dust that follows an elliptical orbit around the sun and that gives off gas and dust in the form of a tail as it passes close to the sun (741)

comet/cometa un cuerpo pequeño formado por hielo, roca y polvo cósmico que sigue una órbita elíptica alrededor del Sol y que libera gas y polvo, los cuales forman una cola al pasar cerca del Sol (741)

compaction the process in which the volume and porosity of a sediment is decreased by the weight of overlying sediments as a result of burial beneath other sediments (135)

compaction/compactación el proceso en el que el volumen y la porosidad de un sedimento disminuyen por efecto del peso al quedar el sedimento enterrado debajo de otros sedimentos superpuestos (135)

compound a substance made up of atoms of two or more different elements joined by chemical bonds (87)

compound/compuesto una substancia formada por átomos de dos o más elementos diferentes unidos por enlaces químicos (87)

condensation the change of state from a gas to a liquid (376)

condensation/condensación el cambio de estado de gas a líquido (376)

condensation nucleus a solid particle in the atmos-phere that provides the surface on which water vapor condenses (581)

condensation nucleus/núcleo de condensación una partícula sólida en la atmósfera que proporciona la superficie en la que el vapor de agua se condensa (581)

conduction the transfer of energy as heat through a material (560)

conduction/conducción transferencia de energía en forma de calor a través de un material (560)

conservation the preservation and wise use of natural resources (171)

conservation/conservación la preservación y el uso inteligente de los recursos naturales (171)

constellation one of 88 regions into which the sky has been divided in order to describe the locations of celestial objects; a group of stars organized in a recognizable pattern (789)

constellation/constelación una de las 88 regiones en las que se ha dividido el cielo con el fin de describir la ubicación de los objetos celestes; un grupo de estrellas organizadas en un patrón reconocible (789)

contact metamorphism a change in the texture, structure, or chemical composition of a rock due to contact with magma (142)

contact metamorphism/metamorfismo de contacto un cambio en la textura, estructura o composición química de una roca debido al contacto con el magma (142)

continental glacier a massive sheet of ice that may cover millions of square kilometers, that may be thousands of meters thick, and that is not confined by surrounding topography (420)

continental glacier/glaciar continental una enorme capa de hielo que puede cubrir millones de kilómetros cuadrados, tener un espesor de miles de metros y que no está confinada por la topografía que la rodea (420)

continental drift the hypothesis that states that the continents once formed a single landmass, broke up, and drifted to their present locations (239)

continental drift/deriva continental la hipótesis que establece que alguna vez los continentes formaron una sola masa de tierra, se dividieron y se fueron a la deriva hasta terminar en sus ubicaciones actuales (239)

continental margin the shallow sea floor that is located between the shoreline and the deep-ocean bottom (475)

continental margin/margen continental el suelo marino poco profundo que se ubica entre la costa y el fondo profundo del océano (475)

contour line a line that connects points of equal elevation on a map (64)

contour line/curva de nivel una línea en un mapa que une puntos que tienen la misma elevación (64)

convection the movement of matter due to differences in density that are caused by temperature variations; can result in the transfer of energy as heat (560)

convection/convección el movimiento de la materia debido a diferencias en la densidad que se producen por variaciones en la temperatura; puede resultar en la transferencia de energía en forma de calor (560)

convective zone the region of the sun's interior that is between the radiative zone and the photosphere and in which energy is carried upward by convection (759)

convective zone/zona convectiva la región del interior del Sol que se encuentra entre la zona radiactiva y la fotosfera y en la cual la energía se desplaza hacia arriba por convección (759)

convergent boundary the boundary between tectonic plates that are colliding (250)

convergent boundary/límite convergente el límite entre placas tectónicas que chocan (250)

core the central part of the Earth below the mantle; *also* the center of the sun (28)

core/núcleo la parte central de la Tierra, debajo del manto; *también,* el centro del Sol (28)

core sample a cylindrical piece of sediment, rock, soil, snow, or ice that is collected by drilling (479)

core sample/muestra de sondeo un fragmento de sedimento, roca, suelo, nieve o hielo que se obtiene taladrando (479)

Coriolis effect the curving of the path of a moving object from an otherwise straight path due to the Earth's rotation (520, 561)

Coriolis effect/efecto de Coriolis la desviación de la trayectoria recta que experimentan los objetos en movimiento debido a la rotación de la Tierra (520, 561)

GLOSSARY/GLOSARIO

corona the outermost layer of the sun's atmosphere (760)

corona/corona la capa externa de la atmósfera del Sol (760)

coronal mass ejection a part of coronal gas that is thrown into space from the sun (763)

coronal mass ejection/eyección de masa coronal una parte de gas coronal que el Sol expulsa al espacio (763)

cosmic background radiation radiation uniformly detected from every direction in space; considered a remnant of the big bang (795)

cosmic background radiation/radiación cósmica de fondo radiación que se detecta de manera uniforme desde todas las direcciones en el espacio; se considera un resto del Big Bang (795)

cosmology the study of the origin, properties, processes, and evolution of the universe (793)

cosmology/cosmología el estudio del origen, propiedades, procesos y evolución del universo (793)

covalent bond a bond formed when atoms share one or more pairs of electrons (91)

covalent bond/enlace covalente un enlace formado cuando los átomos comparten uno más pares de electrones (91)

crater a bowl-shaped depression that forms on the surface of an object when a falling body strikes the object's surface or when an explosion occurs; a similar depression around the central vent of a volcano or geyser (720)

crater/cráter una depresión con forma de tazón, que se forma sobre la superficie de un objeto cuando un cuerpo en caída impacta sobre ésta o cuando se produce una explosión; una depresión similar alrededor de la chimenea de un volcán o géiser (720)

creep the slow downhill movement of weathered rock material (362)

creep/arrastre el movimiento lento y descendente de materiales rocosos desgastados (362)

crevasse in a glacier, a large crack or fissure that results from ice movement (422)

crevasse/grieta en un glaciar, una fractura o fisura grande debida al movimiento del hielo (422)

crust the thin and solid outermost layer of the Earth above the mantle (28)

crust/corteza la capa externa, delgada y sólida de la Tierra, que se encuentra sobre el manto (28)

crystal a solid whose atoms, ions, or molecules are arranged in a regular, repeating pattern (106)

crystal/cristal un sólido cuyos átomos, iones o moléculas están ordenados en un patrón regular y repetitivo (106)

cumulus cloud a low-level, billowy cloud that commonly has a top that resembles cotton balls and a dark bottom (585)

cumulus cloud/nube cúmulo una nube esponjada ubicada en un nivel bajo, cuya parte superior normalmente parece una bola de algodón y es obscura en la parte inferior (585)

current in geology, a horizontal movement of water in a well-defined pattern, such as a river or stream; the movement of air in a certain direction (519)

current/corriente en geología, un movimiento horizontal de agua en un patrón bien definido, como por ejemplo, un río o arroyo; el movimiento del aire en una cierta dirección (519)

D

deep current a streamlike movement of ocean water far below the surface (523)

deep current/corriente profunda un movimiento del agua del océano que es similar a una corriente y ocurre debajo de la superficie (523)

deep-ocean basin the part of the ocean floor that is under deep water beyond the continental margin and that is composed of oceanic crust and a thin layer of sediment (475)

deep-ocean basin/cuenca oceánica profunda la parte del fondo del océano que está bajo aguas profundas más allá del margen continental y que se compone de corteza oceánica y una delgada capa de sedimento (475)

GLOSSARY/GLOSARIO

deflation a form of wind erosion in which fine, dry soil particles are blown away (446)

deflation/deflación una forma de erosión del viento en la que se mueven partículas de suelo finas y secas (446)

deformation the bending, tilting, and breaking of Earth's crust; the change in the shape of rock in response to stress (271)

deformation/deformación el proceso de doblar, inclinar y romper la corteza de la Tierra; el cambio en la forma de una roca en respuesta a la tensión (271)

delta a fan-shaped mass of rock material deposited at the mouth of a stream; for example, deltas form where streams flow into the ocean at the edge of a continent (383)

delta/delta un depósito de materiales rocosos en forma de abanico ubicado en la desembocadura de un río; por ejemplo, los deltas se forman en el lugar donde las corrientes fluyen al océano en el borde de un continente (383)

density the ratio of the mass of a substance to the volume of the substance; commonly expressed as grams per cubic centimeter for solids and liquids and as grams per liter for gases (112, 499)

density/densidad la relación entre la masa de una substancia y su volumen; comúnmente se expresa en gramos por centímetro cúbico para los sólidos y líquidos, y como gramos por litro para los gases (112, 499)

dependent variable in an experiment, the factor that changes as a result of manipulation of one or more other factors (the independent variables) (11)

dependent variable/variable dependiente en un experimento, el factor que cambia como resultado de la manipulación de uno o más factores (las variables independientes) (11)

desalination a process of removing salt from ocean water (378, 505)

desalination/desalación (o desalinización) un proceso de remoción de sal del agua del océano (378, 505)

dew point at constant pressure and water vapor content, the temperature at which the rate of condensation equals the rate of evaporation (577)

dew point/punto de rocío a presión y contenido de vapor constantes, la temperatura a la cual la tasa de condensación iguala la tasa de evaporación (577)

differential weathering the process by which softer, less weather resistant rocks wear away at a faster rate than harder, more weather resistant rocks do (349)

differential weathering/desgaste diferencial el proceso por medio cual las rocas más blandas y menos resistentes al clima se desgastan a una tasa más rápida que las rocas más duras y resistentes al clima (349)

discharge the volume of water that flows within a given time (380)

discharge/descarga el volumen de agua que fluye en un tiempo determinado (380)

divergent boundary the boundary between two tectonic plates that are moving away from each other (249)

divergent boundary/límite divergente el límite entre dos placas tectónicas que se están separando una de la otra (249)

dome mountain a circular or elliptical, almost symmetrical elevation or structure in which the stratified rock slopes downward gently from the central point of folding (283)

dome mountain/domo una elevación o estructura circular o elíptica, casi simétrica, en la cual la roca estratificada se encuentra en una ligera pendiente hacia abajo a partir del punto central de plegamiento (283)

Doppler effect an observed change in the frequency of a wave when the source or observer is moving (778)

Doppler effect/efecto Doppler un cambio que se observa en la frecuencia de una onda cuando la fuente o el observador está en movimiento (778)

dune a mound of wind-deposited sand that moves as a result of the action of wind (447)

dune/duna un montículo de arena depositada por el viento que se mueve como resultado de la acción de éste (447)

E

earthquake a movement or trembling of the ground that is caused by a sudden release of energy when rocks along a fault move (295)

earthquake/terremoto un movimiento o temblor del suelo causado por una liberación súbita de energía que se produce cuando las rocas ubicadas a lo largo de una falla se mueven (295)

Earth science the scientific study of Earth and the universe around it (5)

Earth science/ciencias de la Tierra el estudio científico de la Tierra y del universo que la rodea (5)

eccentricity the degree of elongation of an elliptical orbit (symbol, *e*) (692)

eccentricity/excentricidad el grado de alargamiento de una orbita eliptica (símbolo: *e*) (692)

eclipse an event in which the shadow of one celestial body falls on another (727)

eclipse/eclipse un suceso en el que la sombra de un cuerpo celeste cubre otro cuerpo celeste (727)

ecosystem a community of organisms and their abiotic environment (39)

ecosystem/ecosistema una comunidad de organismos y su ambiente abiótico (39)

elastic rebound the sudden return of elastically deformed rock to its undeformed shape (295)

elastic rebound/rebote elástico ocurre cuando una roca deformada elásticamente vuelve súbitamente a su forma no deformada (295)

electromagnetic spectrum all of the frequencies or wavelengths of electromagnetic radiation (555, 661)

electromagnetic spectrum/espectro electromagnético todas las frecuencias o longitudes de onda de la radiación electromagnética (555, 661)

electron a subatomic particle that has a negative charge (82)

electron/electrón una partícula subatómica que tiene carga negativa (82)

element a substance that cannot be separated or broken down into simpler substances by chemical means; all atoms of an element have the same atomic number (81)

element/elemento una substancia que no se puede separar o descomponer en substancias más simples por medio de métodos químicos; todos los átomos de un elemento tienen el mismo número atómico (81)

elevation the height of an object above sea level (63)

elevation/elevación la altura de un objeto sobre el nivel del mar (63)

El Niño the warm-water phase of the El Niño–Southern Oscillation; a periodic occurrence in the eastern Pacific Ocean in which the surface-water temperature becomes unusually warm (635)

El Niño/El Niño la fase caliente de la Oscilación Sureña "El Niño"; un fenómeno periódico que ocurre en el océano Pacífico oriental en el que la temperatura del agua superficial se vuelve más caliente que de costumbre (635)

epicenter the point on Earth's surface directly above an earthquake's starting point, or focus (296)

epicenter/epicentro el punto de la superficie de la Tierra que queda justo arriba del punto de inicio, o foco, de un terremoto (296)

epoch a subdivision of geologic time that is longer than an age but shorter than a period (214)

epoch/época una subdivisión del tiempo geológico que es más larga que una edad pero más corta que un período (214)

equinox the moment when the sun appears to cross the celestial equator (673)

equinox/equinoccio el momento en que el Sol parece cruzar el ecuador celeste (673)

era a unit of geologic time that includes two or more periods (214)

era/era una unidad de tiempo geológico que incluye dos o más períodos (214)

erosion a process in which the materials of Earth's surface are loosened, dissolved, or worn away and transported from one place to another by a natural agent, such as wind, water, ice, or gravity (357)

erosion/erosión un proceso por medio del cual los materiales de la superficie de la Tierra se aflojan, disuelven o desgastan y son transportados de un lugar a otro por un agente natural, como el viento, el agua, el hielo o la gravedad (357)

erratic a large rock transported from a distant source by a glacier (426)

erratic/errática una piedra grande transportada de una fuente lejana por un glacial (426)

esker a long, winding ridge of gravel and coarse sand deposited by glacial meltwater streams (428)

esker/esker una cumbre larga y con curvas, compuesta por grava y arena gruesa depositada por corrientes de aguas glaciares (428)

estuary an area where fresh water from rivers mixes with salt water from the ocean; the part of a river where the tides meet the river current (456)

estuary/estuario un área donde el agua dulce de los ríos se mezcla con el agua salada del océano; la parte de un río donde las mareas se encuentran con la corriente del río (456)

evapotranspiration the total loss of water from an area, which equals the sum of the water lost by evaporation from the soil and other surfaces and the water lost by transpiration from organisms (376)

evapotranspiration/evapotranspiración la pérdida total de agua de un área, igual a la suma del agua perdida por evaporación del suelo y otras superficies, y el agua perdida debido a la transpiración de los organismos (376)

evolution a heritable change in the characteristics within a population from one generation to the next; the development of new types of organisms from preexisting types of organisms over time (215)

evolution/evolución un cambio hereditario en las características de una población que se produce de una generación a la siguiente; el desarrollo de nuevos tipos de organismos a partir de organismos preexistentes a lo largo del tiempo (215)

extrusive igneous rock rock that forms from the cooling and solidification of lava at Earth's surface (131)

extrusive igneous rock/roca ígnea extrusiva roca que se forma a partir del enfriamiento y la solidificación de la lava en la superficie de la Tierra (131)

F

fault a break in a body of rock along which one block slides relative to another; a form of brittle strain (277)

fault/falla una grieta en un cuerpo rocoso a lo largo de la cual un bloque se desliza respecto a otro; una forma de tensión quebradiza (277)

fault-block mountain a mountain that forms where faulting breaks Earth's crust into large blocks, which causes some blocks to drop down relative to other blocks (283)

fault-block mountain/montaña de bloque de falla una montaña que se forma cuando una falla rompe la corteza de la Tierra en grandes bloques, lo cual hace que algunos bloques se hundan respecto a otros bloques (283)

fault zone a region of numerous, closely spaced faults (300)

fault zone/zona de fallas una región donde hay muchas fallas, las cuales están cerca unas de otras (300)

felsic describes magma or igneous rock that is rich in feldspars and silica and that is generally light in color (132, 325)

felsic/félsica término que describe el magma o la roca ígnea que es rica en feldespato y sílice y que en general es de color claro (132, 325)

fetch the distance that wind blows across an area of the sea to generate waves (527)

fetch/alcance la distancia que el viento sopla en un área del mar para generar olas (527)

floodplain an area along a river that forms from sediments deposited when the river overflows its banks (384)

floodplain/llanura de inundación un área a lo largo de un río formada por sedimentos que se depositan cuando el río se desborda (384)

focus the location within Earth along a fault at which the first motion of an earthquake occurs (296)

focus/foco el lugar dentro de la Tierra a lo largo de una falla donde ocurre el primer movimiento de un terremoto (296)

fog water vapor that has condensed very near the surface of Earth because air close to the ground has cooled (586)

fog/niebla vapor de agua que se ha condensado muy cerca de la superficie de la Tierra debido al enfriamiento del aire próximo al suelo (586)

fold a form of ductile strain in which rock layers bend, usually as a result of compression (275)

fold/pliegue una forma de tensión dúctil en la cual las capas de roca se curvan, normalmente como resultado de la compresión (275)

folded mountain a mountain that forms when rock layers are squeezed together and uplifted (282)

folded mountain/montaña de plegamiento una montaña que se forma cuando las capas de roca se comprimen y se elevan (282)

foliation the metamorphic rock texture in which mineral grains are arranged in planes or bands (143)

foliation/foliación la textura de una roca metamórfica en la que los granos de mineral están ordenados en planos o bandas (143)

food web a diagram that shows the feeding relationships among organisms in an ecosystem (41)

food web/red alimenticia un diagrama que muestra las relaciones de alimentación entre los organismos de un ecosistema (41)

fossil the trace or remains of an organism that lived long ago, most commonly preserved in sedimentary rock (197)

fossil/fósil los indicios o los restos de un organismo que vivió hace mucho tiempo, comúnmente preservados en las rocas sedimentarias (197)

fossil fuel a nonrenewable energy resource formed from the remains of organisms that lived long ago; examples include oil, coal, and natural gas (159)

fossil fuel/combustible fósil un recurso energético no renovable formado a partir de los restos de organismos que vivieron hace mucho tiempo; algunos ejemplos incluyen el petróleo, el carbón y el gas natural (159)

fracture in geology, a break in a rock, which results from stress, with or without displacement, including cracks, joints, and faults; *also* the manner in which a mineral breaks along either curved or irregular surfaces (110)

fracture/fractura en geología, un rompimiento en una roca, que resulta de la tensión, con o sin desplazamiento, incluyendo grietas, fisuras y fallas; *también,* la forma en la que se rompe un mineral a lo largo de superficies curvas o irregulares (110)

G

galaxy a collection of stars, dust, and gas bound together by gravity (660, 790)

galaxy/galaxia un conjunto de estrellas, polvo y gas unidos por la gravedad (660, 790)

Galilean moon any one of the four largest satellites of Jupiter—Io, Europa, Ganymede, and Callisto—that were discovered by Galileo in 1610 (733)

Galilean moon/satélite galileano cualquiera de los cuatro satélites más grandes de Júpiter (Io, Europa, Ganímedes y Calisto) que fueron descubiertos por Galileo en 1610 (733)

gas giant a planet that has a deep, massive atmosphere, such as Jupiter, Saturn, Uranus, or Neptune (701)

gas giant/gigante gaseoso un planeta con una atmósfera masiva y profunda, como por ejemplo, Júpiter, Saturno, Urano o Neptuno (701)

gemstone a mineral, rock, or organic material that can be used as jewelry or an ornament when it is cut and polished (157)

gemstone/piedra preciosa un mineral, roca o material orgánico que se puede usar como joya u ornamento cuando se corta y se pule (157)

geologic column an ordered arrangement of rock layers that is based on the relative ages of the rocks and in which the oldest rocks are at the bottom (211)

geologic column/columna geológica un arreglo ordenado de capas de rocas que se basa en la edad relativa de las rocas y en el cual las rocas más antiguas están al fondo (211)

geology the scientific study of the origin, history, and structure of Earth and the processes that shape Earth (6)

geology/geología el estudio científico del origen, la historia y la estructura del planeta Tierra y los procesos que le dan forma (6)

geosphere the mostly solid, rocky part of the Earth; extends from the center of the core to the surface of the crust (33)

geosphere/geosfera la parte principalmente sólida y rocosa de la Tierra; se extiende del centro del núcleo a la superficie de la corteza (33)

geothermal energy the energy produced by heat within Earth (165)

geothermal energy/energía geotérmica la energía producida por el calor del interior de la Tierra (165)

giant a very large and bright star whose hot core has used most of its hydrogen (784)

giant/gigante una estrella muy grande y brillante que tiene un núcleo caliente que ha usado la mayor parte de su hidrógeno (784)

glacial drift the rock material carried and deposited by glaciers (426)

glacial drift/deriva glacial el material rocoso que es transportado y depositado por los glaciares (426)

glacier a large mass of moving ice (419)

glacier/glaciar una masa grande de hielo en movimiento (419)

global ocean the body of salt water that covers nearly three-fourths of Earth's surface (471)

global ocean/océano global la masa de agua salada que cubre cerca de tres cuartas partes de la superficie de la Tierra (471)

global warming a gradual increase in the average global temperature that is due to a higher concentration of gases such as carbon dioxide in the atmosphere (645)

global warming/calentamiento global un aumento gradual de la temperatura global promedio debido a una concentración más alta de gases (tales como dióxido de carbono) en la atmósfera (645)

gradient the change in elevation over a given distance (380)

gradient/gradiente el cambio en la elevación a lo largo de una distancia determinada (380)

greenhouse effect the warming of the surface and lower atmosphere of Earth that occurs when carbon dioxide, water vapor, and other gases in the air absorb and reradiate infrared radiation (558)

greenhouse effect/efecto de invernadero el calentamiento de la superficie terrestre y de la parte más baja de la atmósfera, el cual se produce cuando el dióxido de carbono, el vapor de agua y otros gases del aire absorben radiación infrarroja y la vuelven a irradiar (558)

groundwater the water that is beneath the Earth's surface (397)

groundwater/agua subterránea el agua que está debajo de la superficie de la Tierra (397)

Gulf Stream the swift, deep, and warm Atlantic current that flows along the eastern coast of the United States toward the northeast (522)

Gulf Stream/corriente del Golfo la corriente rápida, profunda y cálida del océano Atlántico que fluye por la costa este de los Estados Unidos hacia el noreste (522)

gyre a huge circle of moving ocean water found above and below the equator (520)

gyre/giro un círculo enorme de agua oceánica en movimiento que se encuentra debajo del ecuador (520)

H

half-life the time required for half of a sample of a radioactive isotope to break down by radioactive decay to form a daughter isotope (194)	**half-life/vida media** el tiempo que se requiere para que la mitad de una muestra de un isótopo radiactivo se descomponga por desintegración radiactiva y forme un isótopo hijo (194)
headland a high and steep formation of rock that extends out from shore into the water (452)	**headland/promontorio** una formación rocosa alta y empinada que se extiende de la costa hacia el agua (452)
horizon the line where the sky and the Earth appear to meet; *also* a horizontal layer of soil that can be distinguished from the layers above and below it; *also* a boundary between two rock layers that have different physical properties (354)	**horizon/horizonte** la línea donde parece que el cielo y la Tierra se unen; *también,* una capa horizontal de suelo que puede distinguirse de las capas que están por encima y por debajo de ella; *también,* un límite entre dos capas de roca que tienen propiedades físicas distintas (354)
horn a sharp, pyramid-like peak that forms because of the erosion of cirques (424)	**horn/cuerno** un pico puntiagudo en forma de pirámide que se forma debido a la erosión de los circos (424)
hot spot a volcanically active area of Earth's surface, commonly far from a tectonic plate boundary (323)	**hot spot/mancha caliente** un área volcánicamente activa de la superficie de la Tierra que comúnmente se encuentra lejos de un límite entre placas tectónicas (323)
humus dark, organic material formed in soil from the decayed remains of plants and animals (354)	**humus/humus** material orgánico obscuro que se forma en la tierra a partir de restos de plantas y animales en descomposición (354)
hurricane a severe storm that develops over tropical oceans and whose strong winds of more than 120 km/h spiral in toward the intensely low-pressure storm center (609)	**hurricane/huracán** tormenta severa que se desarrolla sobre océanos tropicales, con vientos fuertes que soplan a más de 120 km/h y que se mueven en espiral hacia el centro de presión extremadamente baja de la tormenta (609)
hydroelectric energy electrical energy produced by the flow of water (167)	**hydroelectric energy/energía hidroeléctrica** energía eléctrica producida por el flujo del agua (167)
hydrolysis a chemical reaction between water and another substance to form two or more new substances; a reaction between water and a salt to create an acid or a base (347)	**hydrolysis/hidrólisis** una reacción química entre el agua y otras substancias para formar dos o más substancias nuevas; una reacción entre el agua y una sal para crear un ácido o una base (347)
hydrosphere the portion of the Earth that is water (33)	**hydrosphere/hidrosfera** la porción de la Tierra que es agua (33)
hypothesis an idea or explanation that is based on observations and that can be tested (10)	**hypothesis/hipótesis** una idea o explicación que se basa en observaciones y que se puede probar (10)

I

ice age a long period of climatic cooling during which the continents are glaciated repeatedly (431)	**ice age/edad de hielo** un largo período de enfriamiento del clima, durante el cual los continentes se ven repetidamente sometidos a la glaciación (431)
igneous rock rock that forms when magma cools and solidifies (129)	**igneous rock/roca ígnea** una roca que se forma cuando el magma se enfría y se solidifica (129)

independent variable in an experiment, the factor that is deliberately manipulated (11)	**independent variable/variable independiente** en un experimento, el factor que se manipula deliberadamente (11)
index fossil a fossil that is used to establish the age of a rock layer because the fossil is distinct, abundant, and widespread and existed for only a short span of geologic time (200)	**index fossil/fósil guía** un fósil que se usa para establecer la edad de una capa de roca debido a que puede diferenciarse bien de otros y es abundante; está extendido y existió sólo por un corto período de tiempo geológico (200)
inertia the tendency of an object to resist being moved or, if the object is moving, to resist a change in speed or direction until an outside force acts on the object (694)	**inertia/inercia** la tendencia de un objeto a no moverse o, si el objeto se está moviendo, la tendencia a resistir un cambio en su rapidez o dirección hasta que una fuerza externa actúe en el objeto (694)
intensity in Earth science, the amount of damage caused by an earthquake (304)	**intensity/intensidad** en las ciencias de la Tierra, la cantidad de daño causado por un terremoto (304)
internal plastic flow the process by which glaciers flow slowly as grains of ice deform under pressure and slide over each other (421)	**internal plastic flow/flujo plástico interno** el proceso por medio del cual los glaciares fluyen lentamente a medida que los granos de hielo se deforman por efecto de la presión y se deslizan unos sobre otros (421)
intrusive igneous rock rock formed from the cooling and solidification of magma beneath the Earth's surface (131)	**intrusive igneous rock/roca ígnea intrusiva** una roca formada a partir del enfriamiento y solidificación del magma debajo de la superficie terrestre (131)
ion an atom, radical, or molecule that has gained or lost one or more electrons and has a negative or positive charge (90)	**ion/ion** un átomo, radical o molécula que ha ganado o perdido uno o más electrones y que tiene una carga negativa o positiva (90)
ionic bond the attractive force between oppositely charged ions, which form when electrons are transferred from one atom to another (90)	**ionic bond/enlace iónico** la fuerza de atracción entre iones con cargas opuestas, que se forman cuando se transfieren electrones de un átomo a otro (90)
isogram a line on a map that represents a constant or equal value of a given quantity (62)	**isogram/isograma** una línea en un mapa que representa un valor constante o igual de una cantidad dada (62)
isostasy a condition of gravitational and buoyant equilibrium between Earth's lithosphere and asthenosphere (271)	**isostasy/isostasia** una condición de equilibrio gravitacional y flotante entre la litosfera y la astenosfera de la Tierra (271)
isotope an atom that has the same number of protons (or the same atomic number) as other atoms of the same element do but that has a different number of neutrons (and thus a different atomic mass) (83)	**isotope/isótopo** un átomo que tiene el mismo número de protones (o el mismo número atómico) que otros átomos del mismo elemento, pero que tiene un número diferente de neutrones (y, por lo tanto, otra masa atómica) (83)

J

jet stream a narrow band of strong winds that blow in the upper troposphere (563)	**jet stream/corriente en chorro** un cinturón delgado de vientos fuertes que soplan en la parte superior de la troposfera (563)

K

karst topography a type of irregular topography that is characterized by caverns, sinkholes, and underground drainage and that forms on limestone or other soluble rock (408)

karst topography/topografía de karst una tipo de topografía irregular que se caracteriza por cavernas, depresiones y drenaje subterráneo y que se forma en piedra caliza o algún otro tipo de roca soluble (408)

kettle a bowl-shaped depression in a glacial drift deposit (428)

kettle/marmita una depresión con forma de tazón en un depósito de deriva glaciar (428)

Kuiper Belt a region of the solar system that starts just beyond the orbit of Neptune and that contains dwarf planets and other small bodies made mostly of ice (707, 742)

Kuiper Belt/cinturón de Kuiper una región del Sistema Solar que comienza justo después de la órbita de Neptuno y que contiene planetas enanos y otros cuerpos pequeños formados principalmente de hielo (707, 742)

L

lagoon a small body of water separated from the sea by a low, narrow strip of land (457)

lagoon/laguna una masa pequeña de agua separada del mar por una tira de tierra baja y angosta (457)

landform a physical feature of Earth's surface (363)

landform/accidente geográfico una característica física de la superficie terrestre (363)

latent heat the heat energy that is absorbed or released by a substance during a phase change (575)

latent heat/calor latente la energía calorífica que es absorbida o liberada por una substancia durante un cambio de fase (575)

latitude the distance north or south from the equator; expressed in degrees (53)

latitude/latitud la distancia hacia el norte o hacia el sur del ecuador; se expresa en grados (53)

lava magma that flows onto Earth's surface; the rock that forms when lava cools and solidifies (320)

lava/lava magma que fluye a la superficie terrestre; la roca que se forma cuando la lava se enfría y se solidifica (320)

law of crosscutting relationships the principle that a fault or body of rock is younger than any other body of rock that it cuts through (190)

law of crosscutting relationships/ley de las relaciones entrecortadas el principio de que una falla o un cuerpo rocoso siempre es más joven que cualquier otro cuerpo rocoso que atraviese (190)

law of superposition the principle that a sedimentary rock layer is older than the layers above it and younger than the layers below it if the layers are not disturbed (187)

law of superposition/ley de la sobreposición el principio de que una capa de roca sedimentaria es más vieja que las capas que se encuentran arriba de ella y más joven que las capas que se encuentran debajo de ella si las capas no han sido alteradas (187)

legend a list of map symbols and their meanings (61)

legend/leyenda una lista de símbolos de un mapas y sus significados (61)

light-year the distance that light travels in one year; about 9.46 trillion kilometers (779)

light-year/año luz la distancia que viaja la luz en un año; aproximadamente 9.46 trillones de kilómetros (779)

lithosphere the solid, outer layer of Earth that consists of the crust and the rigid upper part of the mantle (29, 247)

lithosphere/litosfera la capa externa y sólida de la Tierra que está formada por la corteza y la parte superior y rígida del manto (29, 247)

lode a mineral deposit within a rock formation (156)

lode/veta un depósito mineral que se encuentra dentro de una formación rocosa (156)

loess fine-grained sediments of quartz, feldspar, hornblende, mica, and clay deposited by the wind (450)

loess/loess sedimentos de grano fino de cuarzo, feldespato, horneblenda, mica y arcilla depositados por el viento (450)

longitude the angular distance east or west from the prime meridian; expressed in degrees (54)

longitude/longitud la distancia angular hacia el este o hacia el oeste del primer meridiano; se expresa en grados (54)

longshore current a water current that travels near and parallel to the shoreline (454)

longshore current/corriente de ribera una corriente de agua que se desplaza cerca de la costa y paralela a ella (454)

lunar eclipse the passing of the moon through the Earth's shadow at full moon (729)

lunar eclipse/eclipse lunar el paso de la Luna frente a la sombra de la Tierra cuando hay luna llena (729)

luster the way in which a mineral reflects light (110)

luster/brillo la forma en que un mineral refleja la luz (110)

M

mafic describes magma or igneous rock that is rich in magnesium and iron and that is generally dark in color (132, 325)

mafic/máfica término que describe el magma o la roca ígnea que es rica en magnesio y hierro y que en general es de color obscuro (132, 325)

magma liquid rock produced under the Earth's surface; igneous rocks are made of magma (319)

magma/magma roca líquida producida debajo de la superficie terrestre; las rocas ígneas están hechas de magma (319)

magnitude a measure of the strength of an earthquake (303)

magnitude/magnitud una medida de la fuerza de un terremoto (303)

main sequence the location on the H-R diagram where most stars lie; it has a diagonal pattern from the lower right (low temperature and luminosity) to the upper left (high temperature and luminosity) (781)

main sequence/secuencia principal la ubicación en el diagrama H-R donde se encuentran la mayoría de las estrellas; tiene un patrón diagonal de la parte inferior derecha (baja temperatura y luminosidad) a la parte superior izquierda (alta temperatura y luminosidad) (781)

mantle in Earth science, the layer of rock between Earth's crust and core (28)

mantle/manto en las ciencias de la Tierra, la capa de roca que se encuentra entre la corteza terrestre y el núcleo (28)

map projection a flat map that represents a spherical surface (58)

map projection/proyección cartográfica un mapa plano que representa una superficie esférica (58)

mare a large, dark area of basalt on the moon (plural, *maria*) (720)

mare/mar lunar una gran área oscura de basalto en la Luna (720)

mass extinction an episode during which large numbers of species become extinct (221)

mass extinction/extinción masiva un episodio durante el cual grandes cantidades de especies se extinguen (221)

mass movement the movement of a large mass of sediment or a section of land down a slope (361)

mass movement/movimiento masivo el movimiento hacia abajo por una pendiente de una gran masa de sedimento o una sección de terreno (361)

matter anything that has mass and takes up space (81)

matter/materia cualquier cosa que tiene masa y ocupa un lugar en el espacio (81)

meander one of the bends, twists, or curves in a low-gradient stream or river (381)

meander/meandro una de las vueltas, giros o curvas de un arroyo o río de bajo gradiente (381)

mechanical weathering the process by which rocks break down into smaller pieces by physical means (343)

mechanical weathering/desgaste mecánico el proceso por medio del cual las rocas se rompen en pedazos más pequeños mediante medios físicos (343)

meridian any semicircle that runs north and south around Earth from the geographic North Pole to the geographic South Pole; a line of longitude (54)

meridian/meridiano cualquier semicírculo que va de norte a sur alrededor de la Tierra, del Polo Norte geográfico al Polo Sur geográfico; una línea de longitud (54)

mesosphere literally, the "middle sphere"; the strong, lower part of the mantle between the asthenosphere and the outer core; *also* the coldest layer of the atmosphere, between the stratosphere and the thermosphere, in which temperature decreases as altitude increases (29, 553)

mesosphere/mesosfera literalmente, la "esfera media"; la parte fuerte e inferior del manto que se encuentra entre la astenosfera y el núcleo externo; *también,* la capa más fría de la atmósfera que se encuentra entre la estratosfera y la termosfera, en la cual la temperatura disminuye al aumentar la altitud (29, 553)

Mesozoic Era the geologic era that lasted from 251 million to 65.5 million years ago; also called the *Age of Reptiles* (221)

Mesozoic Era/Era Mesozoica la era geológica que comenzó hace 251 millones de años y terminó hace 65.5 millones de años; también llamada *Edad de los Reptiles* (221)

metamorphism the process in which one type of rock changes into metamorphic rock because of chemical processes or changes in temperature and pressure (141)

metamorphism/metamorfismo el proceso en el que un tipo de roca cambia a roca metamórfica debido a procesos químicos o cambios en la temperatura y la presión (141)

meteor a bright streak of light that results when a meteoroid burns up in Earth's atmosphere (743)

meteor/meteoro un rayo de luz brillante que se produce cuando un meteoroide se quema en la atmósfera de la Tierra (743)

meteoroid a relatively small, rocky body that travels through space (743)

meteoroid/meteoroide un cuerpo rocoso relativamente pequeño que viaja en el espacio (743)

meteorology the scientific study of Earth's atmosphere, especially in relation to weather and climate (7)

meteorology/meteorología el estudio científico de la atmósfera de la Tierra, sobre todo en lo que se relaciona al tiempo y al clima (7)

microclimate the climate of a small area (640)

microclimate/microclima el clima de un área pequeña (640)

middle-latitude climate a climate that has a maximum average temperature of 8°C in the coldest month and a minimum average temperature of 10°C in the warmest month (638)

middle-latitude climate/clima de latitud media un clima que tiene una temperatura máxima promedio de 8°C en el mes más frío y una temperatura mínima promedio de 10°C en el mes más caliente (638)

mid-latitude cyclone an area of low pressure that is characterized by rotating wind that moves toward the rising air of the central low-pressure region; the motion is counterclockwise in the Northern Hemisphere (606)

mid-latitude cyclone/ciclón de latitud media un área de baja presión caracterizada por la presencia de viento en rotación que se desplaza hacia el aire ascendente de la región central de baja presión; en el hemisferio norte, el movimiento se produce en sentido contrario al de las manecillas del reloj (606)

mid-ocean ridge a long, undersea mountain chain that has a steep, narrow valley at its center, that forms as magma rises from the asthenosphere, and that creates new oceanic lithosphere (sea floor) as tectonic plates move apart (242)	**mid-ocean ridge/dorsal oceánica** una larga cadena submarina de montañas que tiene un valle empinado y angosto en el centro, se forma a medida que el magma se eleva a partir de la astenosfera y produce una nueva litosfera oceánica (suelo marino) a medida que las placas tectónicas se separan (242)
Milankovitch theory the theory that cyclical changes in Earth's orbit and in the tilt of Earth's axis occur over thousands of years and cause climatic changes (433)	**Milankovitch theory/teoría de Milankovitch** la teoría que establece que los cambios cíclicos en la órbita de la Tierra y en la inclinación de su eje se producen a lo largo de miles de años y provocan cambios climáticos (433)
mineral a natural, usually inorganic solid that has a characteristic chemical composition, an orderly internal structure, and a characteristic set of physical properties (103)	**mineral/mineral** un sólido natural, normalmente inorgánico, que tiene una composición química característica, una estructura interna ordenada y propiedades físicas y químicas características (103)
mineralogist a person who examines, analyzes, and classifies minerals (109)	**mineralogist/minerólogo** una persona que examina, analiza y clasifica los minerales (109)
mixture a combination of two or more substances that are not chemically combined (92)	**mixture/mezcla** una combinación de dos o más substancias que no están combinadas químicamente (92)
Mohs hardness scale the standard scale against which the hardness of minerals is rated (111)	**Mohs hardness scale/escala de dureza de Mohs** la escala estándar que se usa para clasificar la dureza de un mineral (111)
molecule a group of atoms that are held together by chemical forces; a molecule is the smallest unit of matter that can exist by itself and retain all of a substance's chemical properties (87)	**molecule/molécula** un conjunto de átomos que se mantienen unidos por acción de las fuerzas químicas; una molécula es la unidad más pequeña de la materia capaz de existir en forma independiente y conservar todas las propiedades químicas de una substancia (87)
monsoon a seasonal wind that blows toward the land in the summer, bringing heavy rains, and that blows away from the land in the winter, bringing dry weather (635)	**monsoon/monzón** viento estacional que sopla hacia la tierra en el verano, ocasionando fuertes lluvias, y que se aleja de la tierra en el invierno, ocasionando tiempo seco (635)
moon a celestial body that revolves around a body that is larger in mass; a natural satellite (719)	**moon/luna** un cuerpo celeste que gira alrededor de un cuerpo que tiene mayor masa; un satélite natural (719)
moraine a landform that is made from unsorted sediments deposited by a glacier (427)	**moraine/morrena** un accidente geográfico que se forma a partir de varios tipos de sedimentos depositados por un glaciar (427)
mountain range a series of mountains that are closely related in orientation, age, and mode of formation (279)	**mountain range/cinturón de montañas** una serie de montañas que están íntimamente relacionadas en orientación, edad y modo de formación (279)

N

nebula a large cloud of gas and dust in interstellar space; a region in space where stars are born (782)

nebula/nebulosa una nube grande de gas y polvo en el espacio interestelar; una región en el espacio donde las estrellas nacen (782)

nekton all organisms that swim actively in open water, independent of currents (502)

nekton/necton todos los organismos que nadan activamente en las aguas abiertas, de manera independiente de las corrientes (502)

neutron a subatomic particle that has no charge and that is located in the nucleus of an atom (82)

neutron/neutrón una partícula subatómica que no tiene carga y que está ubicada en el núcleo de un átomo (82)

neutron star a star that has collapsed under gravity to the point that the electrons and protons have smashed together to form neutrons (787)

neutron star/estrella de neutrones una estrella que se ha colapsado debido a la gravedad hasta el punto en que los electrones y protones han chocado unos contra otros para formar neutrones (787)

nodule a lump of minerals whose composition differs from the composition of the surrounding sediment or rock; *also* a lump of minerals that is made of oxides of manganese, iron, copper, or nickel and that is found in scattered groups on the ocean floor (481)

nodule/nódulo un bulto de minerales que tienen una composición diferente a la de los sedimentos o rocas de los alrededores; *también,* un bulto de minerales compuesto por óxidos de manganeso, hierro, cobre o níquel y que se encuentra en grupos esparcidos en el fondo del océano (481)

nonfoliated the metamorphic rock texture in which mineral grains are not arranged in planes or bands (144)

nonfoliated/no foliada la textura de una roca metamórfica en la que los granos de mineral no están ordenados en planos ni bandas (144)

nonrenewable resource a resource that forms at a rate that is much slower than the rate at which it is consumed (159)

nonrenewable resource/recurso no renovable un recurso que se forma a una tasa que es mucho más lenta que la tasa a la que se consume (159)

nonsilicate mineral a mineral that does not contain compounds of silicon and oxygen (105)

nonsilicate mineral/mineral no-silicato un mineral que no contiene compuestos de sílice y oxígeno (105)

nova a star that suddenly becomes brighter (786)

nova/nova una estrella que súbitamente se vuelve más brillante (786)

nuclear fission the process by which the nucleus of a heavy atom splits into two or more fragments; the process releases neutrons and energy (162)

nuclear fission/fisión nuclear el proceso por medio del cual el núcleo de un átomo pesado se divide en dos o más fragmentos; el proceso libera neutrones y energía (162)

nuclear fusion the process by which nuclei of small atoms combine to form new, more massive nuclei; the process releases energy (164, 756)

nuclear fusion/fusión nuclear el proceso por medio del cual los núcleos de átomos pequeños se combinan y forman núcleos nuevos con mayor masa; el proceso libera energía (164, 756)

O

observation the process of obtaining information by using the senses; the information obtained by using the senses (10)

observation/observación el proceso de obtener información por medio de los sentidos; la información que se obtiene al usar los sentidos (10)

occluded front a front that forms when a cold air mass overtakes a warm air mass and lifts the warm air mass off the ground and over another air mass (606)

occluded front/frente ocluido un frente que se forma cuando una masa de aire frío supera a una masa de aire caliente y la levanta del suelo por encima de otra masa de aire (606)

oceanography the scientific study of the ocean, including the properties and movements of ocean water, the characteristics of the ocean floor, and the organisms that live in the ocean (6, 472)

oceanography/oceanografía el estudio científico del océano, incluyendo las propiedades y los movimientos del agua, las características del fondo y los organismos que viven en él (6, 472)

Oort cloud a spherical region that surrounds the solar system, that extends from the Kuiper Belt to almost halfway to the nearest star, and that contains billions of comets (742)

Oort cloud/nube de Oort una región esférica que rodea al Sistema Solar, que se extiende desde el cinturón de Kuiper hasta la mitad del camino hacia la estrella más cercana y contiene miles de millones de cometas (742)

orbital period the time required for a body to complete a single orbit (693)

orbital period/período de órbita el tiempo que se requiere para que un cuerpo complete una órbita (693)

ore a natural material whose concentration of economically valuable minerals is high enough for the material to be mined profitably (155)

ore/mena un material natural cuya concentración de minerales con valor económico es suficientemente alta como para que el material pueda ser explotado de manera rentable (155)

organic sedimentary rock sedimentary rock that forms from the remains of plants or animals (136)

organic sedimentary rock/roca sedimentaria orgánica roca sedimentaria que se forma a partir de los restos de plantas o animales (136)

oxidation a reaction that removes one or more electrons from a substance such that the substance's valence or oxidation state increases; in geology, the process by which a metallic element combines with oxygen (346)

oxidation/oxidación una reacción en la que uno o más electrones son removidos de una substancia, aumentado su valencia o estado de oxidación; en geología, el proceso por medio del cual un elemento metálico se combina con oxígeno (346)

ozone a gas molecule that is made up of three oxygen atoms (549)

ozone/ozono una molécula de gas que está formada por tres átomos de oxígeno (549)

P

pack ice a floating layer of sea ice that completely covers an area of the ocean surface (497)

pack ice/manto de hielo marino una capa flotante de hielo marino que cubre completamente un área de la superficie del océano (497)

paleomagnetism the study of the alignment of magnetic minerals in rock, specifically as it relates to the reversal of Earth's magnetic poles; *also* the magnetic properties that rock acquires during formation (243)

paleomagnetism/paleomagnetismo el estudio de la alineación de los minerales magnéticos en la roca, específicamente en lo que se relaciona con la inversión de los polos magnéticos de la Tierra; *también*, las propiedades magnéticas que la roca adquiere durante su formación (243)

paleontology the scientific study of fossils (197)

paleontology/paleontología el estudio científico de los fósiles (197)

Paleozoic Era the geologic era that followed Precambrian time and that lasted from 542 million to 251 million years ago (218)

Paleozoic Era/Era Paleozoica la era geológica que vino después del período Precámbrico; comenzó hace 542 millones de años y terminó hace 251 millones de años (218)

Pangaea the supercontinent that formed 300 million years ago and that began to break up 250 million years ago (258)

Pangaea/Pangea el supercontinente que se formó hace 300 millones de años y que comenzó a separarse hace 250 millones de años (258)

Panthalassa the single, large ocean that covered Earth's surface during the time the supercontinent Pangaea existed (258)

Panthalassa/Panthalassa el único gran océano que cubría la superficie de la Tierra cuando existía el supercontinente Pangea (258)

parallax an apparent shift in the position of an object when viewed from different locations (779)

parallax/paralaje un cambio aparente en la posición de un objeto cuando se ve desde lugares distintos (779)

parallel any circle that runs east and west around Earth and that is parallel to the equator; a line of latitude (53)

parallel/paralelo cualquier círculo que va hacia el Este o hacia el Oeste alrededor de la Tierra y que es paralelo al ecuador; una línea de latitud (53)

peer review the process in which experts in a given field examine the results and conclusions of a scientist's study before that study is accepted for publication (14)

peer review/evaluación de pares el proceso en el cual los expertos en un campo dado examinan los resultados y las conclusiones de un estudio científico antes de aceptar su publicación (14)

pelagic zone the region of an ocean or body of fresh water above the benthic zone (503)

pelagic zone/zona pelágica la región de un océano o una masa de agua dulce sobre la zona bentónica (503)

perigee in the orbit of a satellite, the point at which the satellite is closest to Earth (725)

perigee/perigeo en la órbita de un satélite, el punto en el que el satélite está más cerca de la Tierra (725)

perihelion the point in the orbit of a planet at which the planet is closest to the sun (668)

perihelion/perihelio el punto en la órbita de un planeta en el que el planeta está más cerca del Sol (668)

period a unit of geologic time that is longer than an epoch but shorter than an era (214)

period/período una unidad de tiempo geológico que es más larga que una época pero más corta que una era (214)

permeability the ability of a rock or sediment to let fluids pass through its open spaces, or pores (398)

permeability/permeabilidad la capacidad de una roca o sedimento de permitir que los fluidos pasen a través de sus espacios abiertos o poros (398)

phase in astronomy, the change in the illuminated area of one celestial body as seen from another celestial body; phases of the moon are caused by the changing positions of the Earth, the sun, and the moon (730)

phase/fase en astronomía, el cambio en el área iluminada de un cuerpo celeste según se ve desde otro cuerpo celeste; las fases de la Luna se producen como resultado de los cambios en la posición de la Tierra, el Sol y la Luna (730)

photosphere the visible surface of the sun (759)

photosphere/fotosfera la superficie visible del Sol (759)

placer deposit a deposit that contains a valuable mineral that has been concentrated by mechanical action (156)

placer deposit/yacimiento de aluvión un yacimiento que contiene un mineral valioso que se ha concentrado debido a la acción mecánica (156)

planet a celestial body that orbits the sun, is round because of its own gravity, and has cleared the neighborhood around its orbital path (685)

planet/planeta un cuerpo celeste que orbita alrededor del Sol, es redondo debido a su propia fuerza de gravedad y ha despejado los alrededores de su trayectoria orbital (685)

planetesimal a small body from which a planet originated in the early stages of development of the solar system (686)

planetesimal/planetesimal un cuerpo pequeño a partir del cual se originó un planeta en las primeras etapas de desarrollo del Sistema Solar (686)

GLOSSARY/GLOSARIO

plankton the mass of mostly microscopic organisms that float or drift freely in the waters of aquatic (freshwater and marine) environments (502)

plankton/plancton la masa de organismos casi microscópicos que flotan o se encuentran a la deriva en aguas (dulces y marinas) de ambientes acuáticos (502)

plate tectonics the theory that explains how large pieces of the lithosphere, called plates, move and change shape (247)

plate tectonics/tectónica de placas la teoría que explica cómo las grandes partes de litosfera, denominadas placas, se mueven y cambian de forma (247)

polar climate a climate that is characterized by average temperatures that are near or below freezing; typical of polar regions (639)

polar climate/clima polar un clima caracterizado por temperaturas cercanas o inferiores al punto de congelación; típico de las regiones polares (639)

polar easterlies prevailing winds that blow from east to west between 60° and 90° latitude in both hemispheres (562)

polar easterlies/vientos polares del este vientos preponderantes que soplan de este a oeste entre los 60° y los 90° de latitud en ambos hemisferios (562)

porosity the percentage of the total volume of a rock or sediment that consists of open spaces (397)

porosity/porosidad el porcentaje del volumen total de una roca o sedimento que está formado por espacios abiertos (397)

Precambrian time the interval of time in the geologic time scale from Earth's formation to the beginning of the Paleozoic Era, from 4.6 billion to 542 million years ago (216)

Precambrian time/período Precámbrico el intervalo en la escala de tiempo geológico que abarca desde la formación de la Tierra hasta el comienzo de la Era Paleozoica; comenzó hace 4,600 millones de años y terminó hace 542 millones de años (216)

precipitation any form of water that falls to Earth's surface from the clouds; includes rain, snow, sleet, and hail (376, 587)

precipitation/precipitación cualquier forma de agua que cae de las nubes a la superficie de la Tierra; incluye a la lluvia, nieve, aguanieve y granizo (376, 587)

prominence a loop of relatively cool, incandescent gas that extends above the photosphere and above the sun's edge as seen from Earth (763)

prominence/protuberancia una espiral de gas incandescente y relativamente frío que, vista desde la Tierra, se extiende por encima de la fotosfera y la superficie del Sol (763)

proton a subatomic particle that has a positive charge and that is located in the nucleus of an atom; the number of protons of the nucleus is the atomic number, which determines the identity of an element (82)

proton/protón una partícula subatómica que tiene una carga positiva y que está ubicada en el núcleo de un átomo; el número de protones que hay en el núcleo es el número atómico, y éste determina la identidad del elemento (82)

pulsar a rapidly spinning neutron star that emits pulses of radio and optical energy (788)

pulsar/pulsar una estrella de neutrones que gira rápidamente y emite pulsaciones de energía radioeléctrica y óptica (788)

P wave a primary wave, or compression wave; a seismic wave that causes particles of rock to move in a back-and-forth direction parallel to the direction in which the wave is traveling; P waves are the fastest seismic waves and can travel through solids, liquids, and gases (297)

P wave/onda P una onda primaria u onda de compresión; una onda sísmica que hace que las partículas de roca se muevan en una dirección de atrás hacia delante en forma paralela a la dirección en que viaja la onda; las ondas P son las ondas sísmicas más rápidas y pueden viajar a través de sólidos, líquidos y gases (297)

pyroclastic material fragments of rock that form during a volcanic eruption (326)

pyroclastic material/material piroclástico fragmentos de roca que se forman durante una erupción volcánica (326)

Q

quasar quasi-stellar radio source; a very luminous object that produces energy at a high rate; quasars are thought to be the most distant objects in the universe (792)

quasar/cuasar fuente de radio cuasi-estelar; un objeto muy luminoso que produce energía a una gran velocidad; se piensa que los cuasares son los objetos más distantes del universo (792)

R

radar radio detection and ranging, a system that uses reflected radio waves to determine the velocity and location of objects (613)

radar/radar detección y exploración a gran distancia por medio de ondas de radio; un sistema que usa ondas de radio reflejadas para determinar la velocidad y ubicación de los objetos (613)

radiative zone the zone of the sun's interior that is between the core and the convection zone and in which energy moves by radiation (759)

radiative zone/zona radiactiva la zona del interior del Sol que se encuentra entre el núcleo y la zona de convección y en la cual la energía se mueve por radiación (759)

radiometric dating a method of determining the absolute age of an object by comparing the relative percentages of a radioactive (parent) isotope and a stable (daughter) isotope (193)

radiometric dating/datación radiométrica un método para determinar la edad absoluta de un objeto comparando los porcentajes relativos de un isótopo radiactivo (precursor) y un isótopo estable (hijo) (193)

radiosonde a package of instruments that is carried aloft by balloons to measure upper atmosphere conditions, including temperature, dew point, and wind velocity (613)

radiosonde/radiosonda un conjunto de instrumentos que llevan los globos para medir condiciones de la atmósfera superior, como la temperatura, el punto de rocío y la velocidad del viento (613)

recycling the process of recovering valuable or useful materials from waste or scrap; the process of reusing some items (171)

recycling/reciclar el proceso de recuperar materiales valiosos o útiles de los desechos o de la basura; el proceso de reutilizar algunas cosas (171)

reflecting telescope a telescope that uses a curved mirror to gather and focus light from distant objects (663)

reflecting telescope/telescopio reflector un telescopio que utiliza un espejo curvo para captar y enfocar la luz de objetos lejanos (663)

refracting telescope a telescope that uses a set of lenses to gather and focus light from distant objects (663)

refracting telescope/telescopio refractante un telescopio que utiliza un conjunto de lentes para captar y enfocar la luz de objetos lejanos (663)

refraction the bending of a wavefront as the wavefront passes between two substances in which the speed of the wave differs; *also* the process by which ocean waves bend directly toward the coastline as they approach shallow water (529)

refraction/refracción el curvamiento de un frente de ondas a medida que el frente pasa entre dos substancias en las que la velocidad de las ondas difiere; *también,* el proceso por medio del cual las olas oceánicas se curvan directamente hacia la costa a medida que se acercan a agua poco profunda (529)

regional metamorphism a change in the texture, structure, or chemical composition of a rock due to changes in temperature and pressure over a large area, generally as a result of tectonic forces (142)

regional metamorphism/metamorfismo regional un cambio en la textura, estructura o composición química de una roca debido a cambios en la temperatura y presión en un área extensa, generalmente como resultado de la acción de fuerzas tectónicas (142)

relative age the age of an object in relation to the ages of other objects (186)

relative age/edad relativa la edad de un objeto en relación con la edad de otros objetos (186)

relative humidity the ratio of the amount of water vapor in the air to the amount of water vapor needed to reach saturation at a given temperature (578)

relative humidity/humedad relativa la proporción de la cantidad de vapor de agua que hay en el aire respecto a la cantidad de vapor de agua necesaria para alcanzar la saturación a una temperatura dada (578)

relief the difference between the highest and lowest elevations in a given area; the variations in elevation of a land surface (64)

relief/relieve la diferencia entre las elevaciones más altas y las más bajas en un área dada; las variaciones en elevación de una superficie de terreno (64)

remote sensing the process of gathering and analyzing information about an object without physically being in touch with the object (57)

remote sensing/teledetección el proceso de recopilar y analizar información acerca de un objeto sin estar en contacto físico con el objeto (57)

renewable resource a natural resource that can be replaced at the same rate at which the resource is consumed (165)

renewable resource/recurso renovable un recurso natural que puede reemplazarse a la misma tasa a la que se consume (165)

revolution the motion of a body that travels around another body in space; one complete trip along an orbit (668)

revolution/revolución el movimiento de un cuerpo que viaja alrededor de otro cuerpo en el espacio; un viaje completo a lo largo de una órbita (668)

rifting the process by which Earth's crust breaks apart; can occur within continental crust or oceanic crust (255)

rifting/fracturación el proceso por medio del cual la corteza de la Tierra se fractura; puede producirse dentro de la corteza continental u oceánica (255)

rock cycle the series of processes in which rock forms, changes from one type to another, is destroyed, and forms again by geologic processes (126)

rock cycle/ciclo de las rocas la serie de procesos por medio de los cuales una roca se forma, cambia de un tipo a otro, se destruye y se forma nuevamente por procesos geológicos (126)

rotation the spin of a body on its axis (667)

rotation/rotación el giro de un cuerpo alrededor de su eje (667)

S

salinity a measure of the amount of dissolved salts in a given amount of liquid (496)

salinity/salinidad una medida de la cantidad de sales disueltas en una cantidad determinada de líquido (496)

saltation the movement of sand or other sediments by short jumps and bounces that is caused by wind or water (445)

saltation/saltación el movimiento de la arena u otros sedimentos por medio de saltos pequeños y rebotes debido al viento o al agua (445)

satellite a natural or artificial body that revolves around a larger celestial body (719)

satellite/satélite un cuerpo natural o artificial que gira alrededor de un cuerpo celeste más grande (719)

scale the relationship between the distance shown on a map and the actual distance (61)

scale/escala la relación entre la distancia que se muestra en un mapa y la distancia real (61)

sea a large, commonly saline body of water that is smaller than an ocean and that may be partially or completely surrounded by land; *also* a subdivision of an ocean (471)

sea/mar una gran masa de agua, generalmente salada, que es más pequeña que un océano y que puede estar parcial o totalmente rodeada de tierra; *también,* una subdivisión de un océano (471)

sea-floor spreading the process by which new oceanic lithosphere (sea floor) forms as magma rises to Earth's surface and solidifies at a mid-ocean ridge (243)

sea-floor spreading/expansión del suelo marino el proceso por medio del cual se forma nueva litosfera oceánica (suelo marino) a medida que el magma se eleva a la superficie de la Tierra y se solidifica en una dorsal oceánica (243)

seismic gap an area along a fault where relatively few earthquakes have occurred recently but where strong earthquakes are known to have occurred in the past (307)

seismic gap/brecha sísmica un área a lo largo de una falla donde han ocurrido relativamente pocos terremotos recientemente, pero donde se sabe que han ocurrido terremotos fuertes en el pasado (307)

seismogram a tracing of earthquake motion that is recorded by a seismograph (301)

seismogram/sismograma una traza del movimiento de un terremoto registrada por un sismógrafo (301)

seismograph an instrument that records vibrations in the ground (301)

seismograph/sismógrafo un instrumento que registra las vibraciones en el suelo (301)

shadow zone an area on Earth's surface where no direct seismic waves from a particular earthquake can be detected (298)

shadow zone/zona de sombra un área de la superficie de la Tierra donde no se detectan ondas sísmicas directas de un determinado terremoto (298)

sheet erosion the process by which water flows over a layer of soil and removes the topsoil (358)

sheet erosion/erosión laminar el proceso por medio del cual el agua fluye sobre el suelo y remueve la capa superior de éste (358)

silicate mineral a mineral that contains a combination of silicon and oxygen and that may also contain one or more metals (104)

silicate mineral/mineral silicato un mineral que contiene una combinación de silicio y oxígeno y que también puede contener uno o más metales (104)

silicon-oxygen tetrahedron the basic unit of the structure of silicate minerals; a silicon ion chemically bonded to and surrounded by four oxygen ions (106)

silicon-oxygen tetrahedron/tetraedro de sílice-oxígeno la unidad fundamental de la estructura de los minerales silicatos: un ion de silicio unido químicamente a cuatro iones de oxígeno, los cuales lo rodean (106)

sinkhole a circular depression that forms when rock dissolves, when overlying sediment fills an existing cavity, or when the roof of an underground cavern or mine collapses (407)

sinkhole/depresión una depresión circular que se forma cuando la roca se funde, cuando el sedimento suprayacente llena una cavidad existente, o al colapsarse el techo de una caverna o mina subterránea (407)

soil a loose mixture of rock fragments and organic material that can support the growth of vegetation (353)

soil/suelo una mezcla suelta de fragmentos de roca y material orgánico en la que puede crecer vegetación (353)

soil profile a vertical section of soil that shows the layers of horizons (354)

soil profile/perfil del suelo una sección vertical de suelo que muestra las capas u horizontes (354)

solar eclipse the passing of the moon between Earth and the sun; during a solar eclipse, the shadow of the moon falls on Earth (727)

solar eclipse/eclipse solar el paso de la Luna entre la Tierra y el Sol; durante un eclipse solar, la sombra de la Luna cae sobre la Tierra (727)

solar energy the energy received by Earth from the sun in the form of radiation (166)

solar energy/energía solar la energía que la Tierra recibe del Sol en forma de radiación (166)

solar flare an explosive release of energy that comes from the sun and that is associated with magnetic disturbances on the sun's surface (763)

solar flare/erupción solar una liberación explosiva de energía que proviene del Sol y que se asocia con disturbios magnéticos en la superficie solar (763)

solar nebula a rotating cloud of gas and dust from which the sun and planets formed; *also* any nebula from which stars and exoplanets may form (685)

solar nebula/nebulosa solar una nube de gas y polvo en rotación a partir de la cual se formaron el Sol y los planetas; *también,* cualquier nebulosa a partir de la cual se pueden formar estrellas y exoplanetas (685)

solar system the sun and all of the planets and other bodies that travel around it (685)

solar system/Sistema Solar el Sol y todos los planetas y otros cuerpos que se desplazan alrededor de él (685)

solifluction the slow, downslope flow of soil saturated with water in areas surrounding glaciers at high elevations (362)

solifluction/soliflucción el flujo lento y descendente de suelo saturado con agua en áreas que rodean glaciares a altas elevaciones (362)

solstice the point at which the sun is as far north or as far south of the equator as possible (674)

solstice/solsticio el punto en el que el Sol está tan lejos del ecuador como es posible, ya sea hacia el norte o hacia el sur (674)

solution a homogeneous mixture throughout which two or more substances are uniformly dispersed (92)

solution/solución una mezcla homogénea en la cual dos o más sustancias se dispersan de manera uniforme (92)

sonar sound navigation and ranging, a system that uses acoustic signals and returned echoes to determine the location of objects or to communicate (473)

sonar/sonar navegación y exploración por medio del sonido; un sistema que usa señales acústicas y ondas de eco que regresan para determinar la ubicación de los objetos o para comunicarse (473)

specific heat the quantity of heat required to raise a unit mass of homogeneous material 1 K or $1^{\circ}C$ in a specified way given constant pressure and volume (634)

specific heat/calor específico la cantidad de calor que se requiere para aumentar una unidad de masa de un material homogéneo 1 K ó $1^{\circ}C$ de una manera especificada, dados un volumen y una presión constantes (634)

star a large celestial body that is composed of gas and that emits light; the sun is a typical star (775)

star/estrella un cuerpo celeste grande que está compuesto de gas y emite luz; el Sol es una estrella típica (775)

stationary front a front of air masses that moves either very slowly or not at all (606)

stationary front/frente estacionario un frente de masas de aire que se mueve muy lentamente o que no se mueve (606)

station model a pattern of meteorological symbols that represents the weather at a particular observing station and that is recorded on a weather map (616)

station model/estación modelo el modelo de símbolos meteorológicos que representan el tiempo en una estación de observación determinada y que se registra en un mapa meteorológico (616)

strain any change in a rock's shape or volume caused by stress; deformation (274)

strain/tensión cualquier cambio en la forma o volumen de una roca causado por el estrés; deformación (274)

stratosphere the layer of the atmosphere that lies between the troposphere and the mesosphere and in which temperature increases as altitude increases; contains the ozone layer (553)

stratosphere/estratosfera la capa de la atmósfera que se encuentra entre la troposfera y la mesosfera y en la cual la temperatura aumenta al aumentar la altitud; contiene la capa de ozono (553)

stratus cloud a gray cloud that has a flat, uniform base and that commonly forms at very low altitudes (584)

stratus cloud/nube estrato una nube gris que tiene una base plana y uniforme y que comúnmente se forma a altitudes muy bajas (584)

streak the color of a mineral in powdered form (110)

streak/veta el color de un mineral en forma de polvo (110)

stream load the materials other than the water that are carried by a stream (380)

stream load/carga de un arroyo los materiales que lleva un arroyo, además del agua (380)

stress the amount of force per unit area that acts on a rock (273)

stress/estrés la cantidad de fuerza por unidad de área que se ejerce sobre una roca (273)

sublimation the process in which a solid changes directly into a gas (the term is sometimes also used for the reverse process) (576)

sublimation/sublimación el proceso por medio del cual un sólido se transforma directamente en un gas (en ocasiones, este término también se usa para describir el proceso inverso) (576)

sunspot a dark area of the photosphere of the sun that is cooler than the surrounding areas and that has a strong magnetic field (761)

sunspot/mancha solar un área oscura en la fotosfera del Sol que es más fría que las áreas que la rodean y que tiene un campo magnético fuerte (761)

supercontinent cycle the process by which supercontinents form and break apart over millions of years (258)

supercontinent cycle/ciclo de los supercontinentes el proceso por medio del cual los supercontinentes se forman y se separan a lo largo de millones de años (258)

supercooling a condition in which a substance is cooled below its freezing point, condensation point, or sublimation point without going through a change of state (588)

supercooling/superfrío una condición en la que una sustancia se enfría por debajo de su punto de congelación, punto de condensación o punto de sublimación sin pasar por un cambio de estado (588)

surface current a horizontal movement of ocean water that is caused by wind and that occurs at or near the ocean's surface (519)

surface current/corriente superficial un movimiento horizontal del agua del océano que es producido por el viento y que ocurre en la superficie del océano o cerca de ella (519)

surface wave in geology, a seismic wave that travels along the surface of a medium and that has a stronger effect near the surface of the medium than it has in the interior (296)

surface wave/onda superficial en geología, una onda sísmica que se desplaza a lo largo de la superficie de un medio, cuyo efecto es más fuerte cerca de la superficie del medio que en el interior de éste (296)

S wave a secondary wave, or shear wave; a seismic wave that causes particles of rock to move in a side-to-side direction perpendicular to the direction in which the wave is traveling; S waves are the second-fastest seismic waves and can travel only through solids (297)

S wave/onda S una onda secundaria u onda rotacional; una onda sísmica que hace que las partículas de roca se muevan en una dirección de lado a lado, en forma perpendicular a la dirección en la que viaja la onda; las ondas S son las segundas ondas sísmicas en cuanto a velocidad y únicamente pueden viajar a través de sólidos (297)

system a set of particles or interacting components considered to be a distinct physical entity for the purpose of study (31)

system/sistema un conjunto de partículas o componentes que interactúan unos con otros, el cual se considera una entidad física independiente para fines de estudio (31)

T

telescope an instrument that collects electromagnetic radiation from the sky and concentrates it for better observation (662)

telescope/telescopio un instrumento que capta la radiación electromagnética del cielo y la concentra para mejorar la observación (662)

terrane a piece of lithosphere that has a unique geologic history and that may be part of a larger piece of lithosphere, such as a continent (256)

terrane/macizo autóctono un fragmento de litosfera que tiene una historia geológica única y que puede formar parte de un fragmento de litosfera mayor, como por ejemplo, un continente (256)

GLOSSARY/GLOSARIO

terrestrial planet one of the highly dense planets nearest to the sun; Mercury, Venus, Mars, and Earth (695)

terrestrial planet/planeta terrestre uno de los planetas muy densos que se encuentran más cerca del Sol; Mercurio, Venus, Marte y la Tierra (695)

theory an explanation for some phenomenon that is based on observation, experimentation, and reasoning; that is supported by a large quantity of evidence; and that does not conflict with any existing experimental results or observations (15)

theory/teoría una explicación sobre algún fenómeno que está basada en la observación, experimentación y razonamiento; que está respaldada por una gran cantidad de pruebas; y que no contradice ningún resultado experimental ni observación existente (15)

thermocline a layer in a body of water in which water temperature drops with increased depth faster than it does in other layers (498)

thermocline/termoclinal una capa en una masa de agua en la que, al aumentar la profundidad, la temperatura del agua disminuye más rápido de lo que lo hace en otras capas (498)

thermometer an instrument that measures and indicates temperature (611)

thermometer/termómetro un instrumento que mide e indica la temperatura (611)

thermosphere the uppermost layer of the atmosphere, in which temperature increases as altitude increases; includes the ionosphere (553)

thermosphere/termosfera la capa más alta de la atmósfera, en la cual la temperatura aumenta a medida que la altitud aumenta; incluye la ionosfera (553)

thunderstorm a usually brief, heavy storm that consists of rain, strong winds, lightning, and thunder (608)

thunderstorm/tormenta eléctrica una tormenta fuerte y normalmente breve que consiste en lluvia, vientos fuertes, relámpagos y truenos (608)

tidal current the movement of water toward and away from the coast as a result of the rise and fall of the tides (534)

tidal current/corriente de marea el movimiento del agua hacia la costa y de la costa hacia el mar, como resultado del ascenso y descenso de las mareas (534)

tidal oscillation the slow, rocking motion of ocean water that occurs as the tidal bulges move around the ocean basins (533)

tidal oscillation/oscilación de las mareas el movimiento lento y mecedor del agua del océano que se produce cuando los abultamientos de marea se mueven alrededor de las cuencas oceánicas (533)

tidal range the difference in levels of ocean water at high tide and low tide (532)

tidal range/rango de marea la diferencia en los niveles del agua del océano entre la marea alta y la marea baja (532)

tide the periodic rise and fall of the water level in the oceans and other large bodies of water (531)

tide/marea el ascenso y descenso periódico del nivel del agua en los océanos y otras masas grandes de agua (531)

till unsorted rock material that is deposited directly by a melting glacier (426)

till/arcilla glaciárica material rocoso desordenado que deposita directamente un glaciar que se está derritiendo (426)

topography the size and shape of the land surface features of a region, including its relief (63)

topography/topografía el tamaño y la forma de las características de una superficie de terreno, incluyendo su relieve (63)

tornado a destructive, rotating column of air that has very high wind speeds and that may be visible as a funnel-shaped cloud (610)

tornado/tornado una columna destructiva de aire en rotación cuyos vientos se mueven a velocidades muy altas y que puede verse como una nube con forma de embudo (610)

trace fossil a fossilized mark that formed in sedimentary rock by the movement of an animal on or within soft sediment (199)

trace fossil/fósil traza una marca fosilizada que se formó en una roca sedimentaria por el movimiento de un animal sobre sedimento blando o dentro de éste (199)

trade winds prevailing winds that blow from east to west from 30° latitude to the equator in both hemispheres (562)

trade winds/vientos alisios vientos prevalecientes que soplan de este a oeste desde los 30° de latitud hacia el ecuador en ambos hemisferios (562)

transform boundary the boundary between tectonic plates that are sliding past each other horizontally (251)

transform boundary/límite de transformación el límite entre placas tectónicas que se están deslizando horizontalmente una sobre otra (251)

trench a long, narrow, and steep depression that forms on the ocean floor as a result of subduction of a tectonic plate, that runs parallel to the trend of a chain of volcanic islands or the coastline of a continent, and that may be as deep as 11 km below sea level; also called an *ocean trench* or a *deep-ocean trench* (477)

trench/fosa submarina una depresión larga, angosta y empinada que se forma en el fondo del océano debido a la subducción de una placa tectónica; corre paralela al curso de una cadena de islas montañosas o a la costa de un continente; y puede tener una profundidad de hasta 11 km bajo el nivel del mar; también denominada *fosa oceánica* o *fosa oceánica profunda* (477)

tributary a stream that flows into a lake or into a larger stream (379)

tributary/afluente un arroyo que fluye a un lago o a otro arroyo más grande (379)

tropical climate a climate characterized by high temperatures and heavy precipitation during at least part of the year; typical of equatorial regions (637)

tropical climate/clima tropical un clima caracterizado por temperaturas altas y precipitación fuerte durante al menos una parte del año; típico de las regiones ecuatoriales (637)

troposphere the lowest layer of the atmosphere, in which temperature drops at a constant rate as altitude increases; the part of the atmosphere where weather conditions exist (552)

troposphere/troposfera la capa inferior de la atmósfera, en la que la temperatura disminuye a una tasa constante a medida que la altitud aumenta; la parte de la atmósfera donde se dan las condiciones del tiempo (552)

tsunami a giant ocean wave that forms after a volcanic eruption, submarine earthquake, or landslide (305)

tsunami/tsunami una ola gigante del océano que se forma después de una erupción volcánica, terremoto submarino o desprendimiento de tierras (305)

U

unconformity a break in the geologic record created when rock layers are eroded or when sediment is not deposited for a long period of time (189)

unconformity/disconformidad una ruptura en el registro geológico, creada cuando las capas de roca se erosionan o cuando el sedimento no se deposita durante un largo período de tiempo (189)

uniformitarianism a principle that geologic processes that occurred in the past can be explained by current geologic processes (185)

uniformitarianism/uniformitarianismo un principio que establece que es posible explicar los procesos geológicos que ocurrieron en el pasado en función de los procesos geológicos actuales (185)

upwelling the movement of deep, cold, and nutrient-rich water to the surface (502)

upwelling/surgencia el movimiento de las aguas profundas, frías y ricas en nutrientes hacia la superficie (502)

GLOSSARY/GLOSARIO

V

varve a banded layer of sand and silt that is de-posited annually in a lake, especially near ice sheets or glaciers, and that can be used to determine absolute age (192)

varve/sedimentos cíclicos estacionales una capa de arena y limo dispuestos en bandas, que se deposita en un lago durante un año, especialmente cerca de las capas de hielo o los glaciares, y que puede usarse para determinar la edad absoluta (192)

ventifact any rock that is pitted, grooved, or polished by wind abrasion (447)

ventifact/ventifacto cualquier roca que es marcada, estriada o pulida por la abrasión del viento (447)

volcanism any activity that includes the movement of magma toward or onto the Earth's surface (320)

volcanism/volcanismo cualquier actividad que incluye el movimiento de magma hacia la superficie de la Tierra o sobre ella (320)

volcano a vent or fissure in the Earth's surface through which magma and gases are expelled (320)

volcano/volcán una chimenea o fisura en la superficie de la Tierra a través de la cual se expulsan magma y gases (320)

W

warm front the front edge of an advancing warm air mass that replaces colder air with warmer air (606)

warm front/frente cálido el borde del frente de una masa de aire caliente en movimiento que reemplaza al aire más frío (606)

water cycle the continuous movement of water between the atmosphere, the land, and the oceans (375)

water cycle/ciclo del agua el movimiento continuo del agua entre la atmósfera, la tierra y los océanos (375)

watershed the area of land that is drained by a river system (379)

watershed/cuenca hidrográfica el área del terreno que es drenada por un sistema de ríos (379)

water table the upper surface of underground water; the upper boundary of the zone of saturation (399)

water table/capa freática el nivel más alto del agua subterránea; el límite superior de la zona de saturación (399)

wave a periodic disturbance in a solid, liquid, or gas as energy is transmitted through a medium (525)

wave/onda una perturbación periódica en un sólido, líquido o gas que se transmite a través de un medio en forma de energía (525)

wave period the time required for identical points on consecutive waves to pass a given point (525)

wave period/período de onda el tiempo que se requiere para que puntos idénticos de ondas consecutivas pasen por un punto dado (525)

weathering the natural process by which atmospheric and environmental agents, such as wind, rain, and temperature changes, disintegrate and decompose rocks (343)

weathering/meteorización el proceso natural por medio del cual los agentes atmosféricos o ambientales, como el viento, la lluvia y los cambios de temperatura, desintegran y descomponen las rocas (343)

westerlies prevailing winds that blow from west to east between 30° and 60° latitude in both hemispheres (562)

westerlies/vientos del oeste vientos preponderantes que soplan de oeste a este entre 30° y 60° de latitud en ambos hemisferios (562)

white dwarf a small, hot, dim star that is the left-over center of an old star (785)

white dwarf/enana blanca una estrella pequeña, caliente y tenue que es el centro sobrante de una estrella vieja (785)

wind vane an instrument used to determine the direction of the wind (612)

wind vane/veleta un instrumento que se usa para determinar la dirección del viento (612)

INDEX

Note: Page references followed by *f* refer to figures. Page references followed by *t* refer to tables.

A

B

INDEX

C

INDEX

INDEX

INDEX

F

INDEX

G

H

INDEX

INDEX

M

N

INDEX

O

INDEX

Q

R

S

INDEX

INDEX

T

INDEX

U

V

W

X

Y

Z

Acknowledgments, continued

Lou Goldstein
Staff Development Specialist for Middle Level Science (retired)
New York City Department of Education
New York, New York

Sheila Lightbourne
Curriculum Specialist
Okaloosa School District
Fort Walton Beach, Florida

Craig Seibert, M. Ed.
Science Coordinator of Collier County Public Schools
Collier County Public Schools
Naples, Florida

Staff Credits

The people who contributed to **Holt Earth Science** are listed below. They represent editorial, design, production, e-media, marketing, and permissions.

Mary Anderson, Melanie Baccus, Wesley Bain, Juan Baquera, Adriana Bardin-Prestwood, Kimberly Barr, Kate Bennett, Ed Blake, Bruce Bond, Sara Buller, Soojinn Choi, Andy Christiansen, Eddie Dawson, Juliet Dervin, Lydia Doty, Paul Draper, Marsh Flournoy, Katrina Gnader, Armin Gutzmer, Angela Hemmeter, Tim Hovde, Shari Husain, Wilonda Ieans, Jevara Jackson, Lana Kaupp, Simon Key, Jane A. Kirschman, Suzanne Krejci, Cathy Kuhles, Ivania Quant Lee, Laura Likon, Denise Mahoney, Carol Martin, Kristen McCardel, Richard Metzger, Cathy Murphy, Janice Noske, Shannon Oehler, Cathy Paré, Nanda Patel, Jenny Patton, Kristen Quiring, Sara Rider, Curtis Riker, Michael Rinella, Kelly Rizk, Jeff Robinson, Tara F. Ross, Marjorie Roueché, Margaret Sanchez, Kay Selke, Chris Smith, Dawn Marie Spinozza, Debbie Starr, JoAnn Stringer, Elaine Tate, Jeannie Taylor, Bob Tucek, Clay Walton, Kira J. Watkins, Holly Whittaker, Nadyne Wood, Patty Zepeda

Photo Credits

COVER PHOTO: John & Eliza Forder/Getty Images.

FRONTMATTER: vi (t), Terry Donnelly/Getty Images/The Image Bank; vi (c), Steve Bloom Images; vi (b), National Geophysical Data Center/National Oceanic and Atmospheric Administration/Department of Commerce; vii (Ch. 4), Steve Allen/PictureQuest; vii (Ch. 5), Richard Price/Getty Images; vii (Ch. 6), John Cleare/Worldwide Picture Library/Alamy Photos; vii (b), Richard Sisk/Panoramic Images/NGSimages.com; viii (Ch. 7), Lester Lefkowitz/CORBIS; viii (Ch. 8), Mark E. Gibson/CORBIS; viii (Ch. 9), John Gurche; viii (br), Andrew Leitch/(c)1992 the Walt Disney Co. Reprinted with permission of Discover Magazine; viii (bc), Bill Steele/A+C Anthology; ix (Ch. 10), Mats Wibe Lund; ix (Ch. 11), Roger Ressmeyer/CORBIS; ix (Ch. 12), Tom Wagner/CORBIS SABA; ix (Ch. 13), Gavriel Jecan/CORBIS; x (Ch. 14), Mark Laricchia/CORBIS ; x (Ch. 15), Jim Wark; x (Ch. 16), Joseph Van Os/Getty Images/The Image Bank; x (br), Harald Sund/Getty Images; xi (t), Liz Hymans/CORBIS; xi (Ch. 17), Andrew Wenzel/Masterfile; xi (Ch. 18), Nicole Duplaix/National Geographic Image Collection; xi (Ch. 19), Jeffrey L. Rotman/CORBIS; xii (tr), Sean Davey/AllSport/Getty Images; xii (Ch. 20), David Hall; xii (Ch. 21), eStock Photo/PictureQuest; xii (Ch. 22), Kevin Kelly/Getty Images/The Image Bank; xiii (Ch. 23), Charles Doswell III/Getty Images; xiii (Ch. 24), Getty Images/Taxi; xiii (Ch. 25), Theo Allofs/CORBIS; xiii (br), Rick Doyle/CORBIS; xiv (Ch. 26), NASA; xiv (Ch. 27), Denis Scott/CORBIS; xiv (Ch. 28), Denis Scott/CORBIS; xiv (br), Dr. Fred Espenak/Photo Researchers, Inc.; xv (t), Grant Faint/Getty Images; xv (Ch. 29), SOHO/ESA/NASA; xv (Ch. 30), Stocktrek/CORBIS; xv (inset), Barry Runk/Stan/Grant Heilman Photography.

UNIT ONE: 3 (tr), Terry Donnelly/Getty Images/The Image Bank; 3 (cr), Steve Bloom Images; 3 (br), National Geophysical Data Center/National Oceanic and Atmospheric Administration/Department of Commerce; 4, Terry Donnelly/Getty Images/The Image Bank; 5, Danny Lehman/CORBIS; 6 (bl), Roger Ressmeyer/CORBIS; 6 (c), Roger Ressmeyer/CORBIS; 6 (r), Bryan and Cherry Alexander Photography; 7, Bettmann/CORBIS; 8, Dr. Howard B. Bluestein; 9 (bl), (c)Michael Sewell/Peter Arnold, Inc.; 9 (br), CSIRO/Simon Fraser/SPL/Photo Researchers, Inc.; 11, NASA; 11, NASA; 12 (tl), Sam Dudgeon/HRW; 12 (tc), Sam Dudgeon/HRW; 12 (tr), Sam Dudgeon/HRW; 12 (b), Andy Christiansen/HRW; 13 (l), Kuni/AP/Wide World Photos; 13 (r), NASA/JPL/NIMA; 14, Ric Francis/AP/Wide World Photos; 15 (bl), Aventurier Patrick/Gamma; 15 (cl), V. L. Sharpton/LPI; 15 (cr), Dr. David A. Kring; 15 (br), Courtesy of Los Alamos National Laboratories; 16, Patrick J. Endres/Alaskaphotographics.com; 17 (t), Roger Ressmeyer/CORBIS; 22, HRW; 2-3, Tom Bean/CORBIS ; 25 (t), Courtesy Ken Pierce; 25 (b), Corbis Images; 26, Steve Bloom Images; 27 (bl), Getty Images/PhotoDisc; 31 (bkgd), Douglas Faulkner/CORBIS; 31 (inset), Stuart Westmorland/CORBIS; 31 (inset), Stuart Westmorland/CORBIS; 32 (l), Andy Christiansen/HRW; 32 (r), Andy Christiansen/HRW; 33, Darrell Gulin/CORBIS; 35, Bernhard Lang/The Image Bank/Getty Images; 38, Pam Ostrow/Index Stock Imagery, Inc.; 39, Martin Harvey/Gallo Images/CORBIS; 40 (t), Raymond Gehman/CORBIS ; 40 (b), Roine Magnusson/The Image Bank/Getty Images; 42, Bob Krist; 43 (t), Getty Images/PhotoDisc; 43 (c), Bernhard Lang/The Image Bank/Getty Images; 43 (b), Bob Krist; 49, Andy Christiansen/HRW; 49, Peter Van Steen/HRW; 50, Provided by the SeaWiFS Project, NASA/Goddard Space Flight Center and ORBIMAGE/NASA/Seawifs; 51 (b), Carlos Barria/REUTERS/NewsCom; 51 (t), Martin Harvey/CORBIS; 52 (all), National Geophysical Data Center/National Oceanic and Atmospheric Administration/Department of Commerce; 55 (tl), Layne Kennedy/CORBIS; 55 (br), National Maritime Museum, London; 57 (bl), Christopher Cormack/CORBIS; 60 (bl), Courtesy U.S. Geological Survey; 60 (inset), ; 61 (tl), Simon & Schuster, Inc.; 62 (tl), Bill Frymire/Masterfile; 64 (tl), Nathan Wier/Getty Images; 64 (br), Victoria Smith/HRW; 66 (t), Courtesy U.S. Geological Survey; 67 (bl), National Conservation Resource Service; 67 (br), National Conservation Resource Service; 68 (tl), Image courtesy NSSTC Lightning Team; 69 (tl), Layne Kennedy/CORBIS; 69 (cl), Christopher Cormack/CORBIS; 69 (bl), Nathan Wier/Getty Images; 74 (b), Victoria Smith/HRW; 74 (cr), Victoria Smith/HRW; 75 (tr), Victoria Smith/HRW; 76 (t), Courtesy U.S. Geological Survey; 77 (tr), Robert Goldstrom/The Newborn Group; 77 (bl), Image Makers /Getty Images; 79 (Ch. 4), Steve Allen/PictureQuest; 79 (Ch. 5), Richard Price/Getty Images; 79 (Ch. 6), John Cleare/Worldwide Picture Library/Alamy Photos.

UNIT TWO: 78-79, Jack Dykinga/Stone/Getty Images; 79 (Ch. 7), Lester Lefkowitz/CORBIS; 80, Steve Allen/PictureQuest; 87 (bl), Andrew Lambert Photography/Photo Researchers, Inc.; 87 (bc), Andrew Lambert Photography/Photo Researchers, Inc.; 87 (br), Tony Freeman/PhotoEdit; 89 (tr), Steve Chen/CORBIS; 90 (inset), Bruce Dale/National Geographic Image Collection; 91 (tr), Charlie Winters/HRW; 92 (tl), Craig Aurness/CORBIS; 93 (bl), Steve Chen/CORBIS; 99 (cr), Victoria Smith/HRW; 101 (b), Fermilab; 101 (cr), FERMILAB/SPL/Photo Researchers, Inc.; 102 (all), Richard Price/Getty Images; 103 (inset), Barry Runk/Grant Heilman Photography, Inc.; 103 (inset), Richard Cummins/CORBIS; 103 (inset), Doug Sokell/Visuals Unlimited; 103 (inset), Dr. E. R. Degginger/Color-Pic, Inc.; 103 (inset), Barry Runk/Stan/Grant Heilman Photography; 104 (tl), Martin Miller/Visuals Unlimited; 104 (tr), Dr. E. R. Degginger/Color-Pic, Inc.; 104 (cl), Charles D. Winters/Photo Researchers, Inc.; 105 (inset), Cabisco/Visuals Unlimited; 105 (inset), Mark Schneider/Visuals Unlimited; 105 (inset), Stonetrust, Inc.; 105 (inset), Paul Silverman/Fundamental Photographs; 105 (inset), Dr. E. R. Degginger/Color-Pic, Inc.; 105 (inset), Dr. E. R. Degginger/Color-Pic, Inc.; 105 (inset), Barry Runk/Stan/Grant Heilman Photography, Inc.; 105 (inset), Science VU/Visuals Unlimited; 105 (inset), Grace Davies Photography; 105 (inset), Geoff Tompkinson/SPL/Photo Researchers, Inc.; 105 (inset), Ken Lucas/Visuals Unlimited; 105 (inset), Ken Lucas/Visuals Unlimited; 108 (tl), Ken Lucas/Visuals Unlimited; 108 (tc), Jose Manuel Sanchez Calvate/CORBIS; 108 (tr), Breck P. Kent; 109 (bl), G. Tompkinson/Photo Researchers, Inc.; 109 (br), Mark Schneider/Visuals Unlimited; 110 (tl), E. R. Degginger/Dembinsky Photo Associates; 110 (cl), Mark A. Schneider/Visuals Unlimited; 110 (bc), Barry Runk/Stan/Grant Heilman Photography; 110 (bl), Tom Pantages Photography; 113 (tr), Dr. E. R. Degginger/Color-Pic, Inc.; 113 (cr), Dr. E. R. Degginger/Color-Pic, Inc.; 113 (inset), Victoria Smith/HRW; 114 (tl), Larry Stepanowicz / Visuals Unlimited; 115 (tl), Mark Schneider/Visuals Unlimited; 115 (bl), Mark Schneider/Visuals Unlimited; 120 (b), Victoria Smith/HRW; 123 (bl), Craig Aurness/CORBIS; 123 (tr), Courtesy of Jami Gerard-Dwyer; 124, John Cleare/Worldwide Picture Library/Alamy Photos; 125 (l), Breck P. Kent; 125 (c), Astrid & Hans-Frieder Michler/SPL/Photo Researchers, Inc.; 125 (r), Wally Eberhart/Visuals Unlimited ; 126 (br), Breck P. Kent; 126 (t), Joyce Photographics/Photo Researchers, Inc.; 126 (bl), Breck P. Kent; 128, Brand X Photos/Alamy Photos; 130 (b), Andy Christiansen/HRW; 131 (coarse), Image Copyright (c) Jerome Wyckoff; Image courtesy Earth Science World ImageBank; 131 (rhyolite), Breck P. Kent; 131 (porphyritic), Breck P. Kent; 131 (obsidian), G. R. Roberts/Natural Sciences Image Library (NSIL) of New Zealand; 131 (pumice), Wally Eberhart/Visuals Unlimited; 132 (tl), Martin Miller/Visuals Unlimited; 132 (tl, inset), Breck P. Kent; 132 (tr), Paul Harris/Getty Images/Stone; 132 (tr, inset), Victoria Smith/HRW; 134, Richard Sisk/Panoramic Images/NGSimages.com; 137 (tr, bl), Breck P. Kent; 137 (tr), Copyright Dorling Kindersley; 137 (br), Dr. E. R. Degginger/Color-Pic, Inc.; 139 (l), Russell Wood; 139 (r), Bernhard Edmaier/Photo Researchers, Inc.; 140, Tom Wagner/CORBIS SABA; 141 (br), Joyce Photographics/Photo Researchers, Inc.; 141 (cr), Wally Eberhart/Visuals Unlimited; 142, Martin Miller, University of Oregon; 143 (bl), Breck P. Kent; 143 (br), Galen Rowell/CORBIS; 144 (t), Adam Crowley/Photodisc Green/gettyimages; 144 (inset), Andy Christiansen/HRW; 145 (Sec. 1), Brand X Photos/Alamy Photos; 145 (Sec. 2, t), Image Copyright (c) Jerome Wyckoff; Image courtesy Earth Science World ImageBank; 145 (Sec. 2, b), Breck P. Kent; 145 (Sec. 3), Russell Wood; 145 (b), Adam Crowley/Photodisc Green/gettyimages; 150, Victoria Smith/HRW; 152, Ohio Department of Natural Resources, Division of Geological Survey; 153 (tr), NASA; 153 (c), NASA; 153 (bl), Roger Ressmeyer/CORBIS ; 154, Lester Lefkowitz/CORBIS; 156 (br), Neal Mishler/Getty Images/Photographer's Choice; 158, H. David Seawell/CORBIS; 159, Macduff Everton/CORBIS; 160 (all), Barry Runk/Stan/Grant Heilman Photography, Inc.; 164, Roger Ressmeyer/CORBIS; 165, Bob Krist; 166, Andy Christiansen/HRW; 168, Stefan Schott/Panoramic Images; 169, Peter Essick/Aurora; 170, John Lovretta, The Hawk Eye/AP/Wide World Photos; 171 (t), Jose Fuste Raga/CORBIS; 171 (b), Gary Klinkhammer/College of Oceanic and Atmospheric Sciences, Oregon State University; 172 (t), Photodisc/gettyimages; 173 (1), H. David Seawell/CORBIS; 173 (2), Macduff Everton/CORBIS; 173 (3), Bob Krist; 173 (4), Jose Fuste Raga/CORBIS; 181 (t), Applied Ecological Services; 181 (inset), Barry/Runk/Stan/Grant Heilman Photography, Inc.; 181 (b), Applied Ecological Services.

UNIT THREE: 182-183, Discovery Images/PictureQuest; 183 (Ch. 8), Mark E. Gibson/CORBIS; 183 (Ch. 7), John Gurche; 184, Mark E. Gibson/CORBIS; 185 (l), CORBIS; 185 (r), Tom Bean/CORBIS; 186, Deborah Long/Visuals Unlimited; 188 (tr), Tom Bean/CORBIS; 188 (tl), Christian Hass; 188 (tc), Gerald and Buff Corsi/Visuals Unlimited ; 191, Alan Smith/Getty Images/Stone; 192 (r), Pat O'Hara/CORBIS; 192 (l), Bruce Molnia/Terra Photo Graphics; 194 (t), Sam Dudgeon/HRW; 194 (t), Sam Dudgeon/HRW; 194 (t), Sam Dudgeon/HRW; 194 (t), Sam Dudgeon/HRW; 194 (t), Sam Dudgeon/HRW; 194, Victoria Smith/HRW; 196, James King-Holmes/SPL/Photo Researchers, Inc.; 197, Annie Griffiths Belt/CORBIS; 198 (mummy), Landmann Patrick/CORBIS SYGMA; 198 (amber), Layne Kennedy/CORBIS; 198 (tar), Nick Ut/AP/Wide World Photos; 198 (mammoth), Bettmann/CORBIS; 198 (log), Bernhard Edmaier/Photo Researchers, Inc.; 199 (leaf), Layne Kennedy/CORBIS ; 199 (mold), G. R. Roberts/Natural Sciences Image Library (NSIL) of New Zealand; 199 (coprolite), Sinclair Stammers/SPL/Photo Researchers, Inc.; 199 (gastrolith), Francios Gohier/Photo Researchers, Inc.; 200, James L. Amos/Photo Researchers, Inc.; 201 (b), James L. Amos/Photo Researchers, Inc.; 201

(c), Pat O'Hara/CORBIS; 201 (t), Tom Bean/CORBIS; 206, Victoria Smith/HRW; 207 (all), Victoria Smith/HRW; 208, Courtesy, Ohio Department of Natural Resources, Division of Geological Survey; 209 (t), Steve Bloom Images; 209 (b), Pablo Corral Vega/CORBIS; 210, John Gurche; 212, Jonathan Blair/CORBIS; 214, Reuters New Media Inc./CORBIS; 216 (l), Jim Brandenburg/Minden Pictures; 216 (r), Manfred Danegger/Peter Arnold; 217, John Reader/SPL/Photo Researchers, Inc.; 218, James L. Amos/CORBIS; 219, Kaj R. Svensson/SPL/Photo Researchers, Inc.; 220, Ken Lucas/Visuals Unlimited; 221 (bl), Chris Butler/SPL/Photo Researchers, Inc.; 221 (bc), Joe Tucciarone/SPL/Photo Researchers, Inc.; 222 (t), Doug Henderson/Reuters/NewsCom; 222 (b), James L. Amos/CORBIS; 223, Sue Ogrocki/Reuters/NewsCom; 224, Stuart Westmorland/CORBIS; 225 (bl), Wardene Weiser/Bruce Coleman, Inc.; 225, Jeff Gage/Florida Museum of Natural History; 226, Bettmann/CORBIS; 227 (t), Jonathan Blair/CORBIS; 227 (c), John Reader/SPL/Photo Researchers, Inc.; 227, Stuart Westmorland/CORBIS; 233, Jonathan Blair/CORBIS; 235 (tr), Andrew Leitch/(c)1992 the Walt Disney Co. Reprinted with permission of Discover Magazine; 235 (tc), Bill Steele/A+C Anthology; 235 (bl), Ira Block.

UNIT FOUR: 236-237, NASA; 237 (Ch. 10), Mats Wibe Lund; 237 (Ch. 11), Roger Ressmeyer/CORBIS; 237 (Ch. 12), Tom Wagner/CORBIS SABA; 237 (Ch. 13), Gavriel Jecan/CORBIS; 238 (all), Mats Wibe Lund; 239 (bl), The Granger Collection, New York.; 240 (inset), Paleontological Museum/University of Oslo, Norway; 241 (b), Galen Rowell/CORBIS; 241 (inset), British Antarctic Survey/SPL/Photo Researchers, Inc.; 242 (tl), P. Hickey/Woods Hole Oceanographic Institute; 246 (t), Courtesy of Dr. Donald Prothero; 249 (tl), NASA; 250 (br), Jacques Descloitres. MODIS Land Rapid Response Team, NASA/GSFC; 251 (tr), Tom Bean/CORBIS; 253 (br), Victoria Smith/HRW; 255 (b), Y. Arthus-B./Peter Arnold, Inc.; 255 (b), Y. Arthus-B./Peter Arnold, Inc.; 257 (cr), Michael Dick/Animals Animals/Earth Scenes; 261 (tl), British Antarctic Survey/SPL/Photo Researchers, Inc.; 261 (cr), Jacques Descloitres. MODIS Land Rapid Response Team, NASA/GSFC; 266 (b), Victoria Smith/HRW; 267 (tr), Victoria Smith/HRW; 269 (bl), Bettmann/CORBIS; 269 (inset), Walter H. F. Smith and David T. Sandwell; 270, Roger Ressmeyer/CORBIS; 272, Jeremy Woodhouse/Getty Images; 272, Jeremy Woodhouse/Getty Images; 274 (t), Tom Bean; 274 (b), Andy Christiansen/HRW; 275, Bill Bachman; 276, Photodisc Blue/Getty Images; 278, Lloyd Cluff/CORBIS; 280, Jeremy Woodhouse/Pixelchrome; 281 (t), Kim Westerskov/Getty Images/Stone; 281 (b), Alexander Stewart/Getty Images/The Image Bank; 282-283, Photodisc Green/gettyimages; 282 (sierra), Russ Bishop; 282 (colorado), George H. H. Huey/CORBIS; 282 (valley), Bob Krist/CORBIS; 283 (dome), Alan Schein Photography/CORBIS ; 283 (appalachian), James P. Blair/National Geographic Image Collection; 283 (arkansas), Zephyr Picture/Index Stock Imagery, Inc.; 284, Harvey Lloyd/Getty Images/Taxi; 285 (t), Kim Westerskov/Getty Images/Stone; 290, Victoria Smith/HRW; 292, Institute of Geological and Nuclear Sciences; 292, Institute of Geological and Nuclear Sciences; 293 (bl), Bill Bachmann/Rainbow; 293 (cr), Gavin Hellier/Nature Picture Library; 294 (all), Tom Wagner/CORBIS SABA; 300 (tl), Yann Arthus Bertrand/CORBIS; 301 (bl), Reuters/CORBIS; 302 (br), Michael S. Yamashita/CORBIS; 302 (br), Sam Dudgeon/HRW; 303, Paula Bronstein/Getty Images; 305 (br), Reuters/CORBIS; 306 (tr), Samuel Zuder/laif/Aurora; 309 (tl), Yann Arthus Bertrand/CORBIS; 309 (cl), Michael S. Yamashita/CORBIS; 309 (bl), Reuters/CORBIS; 316, Global Seismic Hazard Assessment Program; 317 (bl), Courtesy of Wayne Thatcher; 317 (bl), Massonet, CNES/Photo Researchers, Inc.; 317 (bkgd), Dewitt Jones/CORBIS; 318 (all), Gavriel Jecan/CORBIS; 321 (br), Barry Tessman/National Geographic Image Collection; 322 (tl), James Wall/Animals Animals/Earth Scenes; 323 (cl), Jacques Descloitres, MODIS Rapid Response Team, NASA/GSFC; 324 (t), Bill Ross/CORBIS; 325 (bl), Stuart Westmoreland/Getty Images; 326 (tl), Gary Braasch/CORBIS; 326 (tc), J. D. Griggs/CORBIS; 326 (tr), David Muench/CORBIS; 327 (b), Gary Braasch/CORBIS; 327 (cr), Robert Patrick/Corbis Sygma; 327 (cr), Juerg Alean, Switzerland/www.stromboli.net; 327 (bl), Michael Yamashita/CORBIS; 328 (cr), Mike Zens/CORBIS; 328 (cl), Yann Arthrus-Bertrand/CORBIS; 328 (br), Japack Company/CORBIS; 330 (tl), Roger Ressmeyer/CORBIS; 331 (tl), Jacques Descloitres, MODIS Rapid Response Team, NASA/GSFC; 331 (bl), Gary Braasch/CORBIS; 337 (tr), Koji Sasahara/AP/Wide World Photos; 339 (bl), InterNetwork Media/Getty Images; 339 (inset).

UNIT FIVE: 340-341, Frans Lemmens/Image Bank/Getty Images; Roger Ressmeyer/CORBIS; 341 (Ch. 14), Mark Laricchia/CORBIS ; 341 (Ch. 15), Jim Wark; 341 (Ch. 16), Joseph Van Os/Getty Images/The Image Bank; 341 (Ch. 17), Andrew Wenzel/Masterfile; 341 (Ch. 18), Nicole Duplaix/National Geographic Image Collection; 342 (all), Mark Laricchia/CORBIS ; 344 (bl), Dr. E. R. Degginger/Color-Pic, Inc.; 344 (tr), SuperStock; 345 (tl), Layne Kennedy/CORBIS; 345 (tr), W. Perry Conway/CORBIS; 345 (br), Victoria Smith/HRW; 346 (br), Kevin Fleming/CORBIS; 348 (tr), Adam Hart-Davis/SPL/Photo Researchers, Inc.; 349 (br), Tom Till/Getty Images; 351 (tl), Bettman/CORBIS; 351 (tc), Joen Iaconetti/Bruce Coleman, Inc.; 352 (tl), Richard Hamilton Smith/CORBIS; 353 (bl), Jeff Vanuga/USDA/NRCS; 358 (tr), Yann Arthus-Bertrand/CORBIS; 359 (tr), Jason Hawkes/CORBIS; 359 (cr), Daniel Dancer/Peter Arnold, Inc.; 360 (tl), Jim Richardson/CORBIS; 360 (tr), Keren Su/CORBIS; 360 (tc), Grant Heilman Photography, Inc.; 361 (br), AFP/CORBIS; 361 (bl), Handout/Malacanang/Reuters/CORBIS; 362 (br), CORBIS; 363 (tl), Galen Rowell/CORBIS; 363 (tr), Steve Terrill/CORBIS; 364 (t), (c) Robert Frerck/Odyssey/Chicago; 365 (tl), Adam Hart-Davis/SPL/Photo Researchers, Inc.; 365 (cl), Richard Hamilton Smith/CORBIS; 365 (cl), Jeff Vanuga/USDA/NRCS; 365 (bl), Galen Rowell/CORBIS; 370 (b), Victoria Smith/HRW; 371 (tr), Victoria Smith/HRW; 373 (tr), Courtesy of Lewis Nichols; 373 (bl), Grant Heilman/Grant Heilman Photography, Inc.; 374 (all), Jim Wark; 375 (bl), Annie Reynolds/PhotoLink/gettyimages; 377 (tr), Mark Taylor/Warren Photographic/Bruce Coleman, Inc.; 377 (cr), Brad Wrobleski/Masterfile; 378 (tl), Peter Turnley/CORBIS; 380 (bl), Nancy Simmerman/Getty Images; 380 (br), Rich Reid/National Geographic Image Collection; 381 (tl), Harald Sund/Getty Images; 382 (tr), Jim Wark/Airphoto; 383 (bl), Jim Wark/Airphoto; 383 (br), Martin Miller/Visuals Unlimited; 384 (tl), Kevin R. Morris/CORBIS; 385 (br), Modesto Bee/AP/Wide World Photos; 387 (tl), Annie Reynolds/PhotoLink/gettyimages; 387 (cl), Harald Sund/Getty Images; 387 (bl), Martin Miller/Visuals Unlimited; 392 (bl), Victoria Smith/HRW; 393 (tr), Victoria Smith/HRW; 393 (cr), Victoria Smith/HRW; 395 (bl), EPA/AP Wide World Photos; 395 (tr), Reuters New Media, Inc./CORBIS; 396, Joseph Van Os/Getty Images/The Image Bank; 398, Victoria Smith/HRW; 401, Charles River Watershed Association; 402, Tom Bean/CORBIS; 403, Pagasus/Visuals Unlimited; 405 (inset), Martyn F. Chillmaid/SPL/Photo Researchers, Inc.; 405 (bl), Peter Essick/Aurora; 406, Adam Woolfitt/CORBIS; 407 (t), Bettmann/CORBIS; 407, Natural Bridge Caverns; 408, Keren Su/CORBIS; 409 (t), Pagasus/Visuals Unlimited; 409, Peter Essick/Aurora; 414, Victoria Smith/HRW; 415, Victoria Smith/HRW; 416, McGuire, V.L., 2001, Water-level changes in the High Plains Aquifer, 1980 to 1999: U.S. Geological Survey Fact Sheet, FS-029-01, 2 p.; 417 (br), Leif Skoogfers/Woodfin Camp & Associates; 417 (bl), Joseph Melanson/Aero Photo-Aerials Only Gallery; 418, Andrew Wenzel/Masterfile; 419 (b), Kevin R. Morris/CORBIS; 420 (tl), Jim Wark/Airphoto; 420 (tr), Hanne & Jens Eriksen/Nature Picture Library; 422 (l), Ralph A. Clevenger/CORBIS; 422 (tr), Jim Brandenburg/Minden Pictures; 423 (bl), Ronald Gorbutt/Visuals Unlimited; 425 (tr), G. R. Roberts/Natural Sciences Image Library (NSIL) of New Zealand; 425 (br), Victoria Smith/HRW; 428 (t), Galen Rowell/CORBIS; 429 (tr), Scott T. Smith/CORBIS; 431 (bkgd), George D. Lepp/CORBIS; 434 (tl), SPL/Photo Researchers, Inc.; 435 (tl), Jim Brandenburg/Minden Pictures; 435 (cl), Scott T. Smith/CORBIS; 435 (bl), SPL/Photo Researchers, Inc.; 440 (br), Victoria Smith/HRW; 441 (tr), Victoria Smith/HRW; 443 (b), Craig Tuttle/CORBIS; 444, Nicole Duplaix/National Geographic Image Collection; 445, Mark J. Terrill/AP/Wide World Photos; 446, Jonathan Blair/CORBIS; 449 (t), John Warden/Getty Images/Stone; 449, Andy Christiansen/HRW; 450, Walter H. Hodge/Peter Arnold, Inc.; 451, David Welling; 452 (bc), Jeff Foott/DRK Photo; 452 (tl), Jeff Foott/Tom Stack & Associates; 453 (bl), G. R. Roberts/Natural Sciences Image Library (NSIL) of New Zealand; 453 (tc), Breck P. Kent; 453 (br), John S. Shelton ; 457 (t), Aerial by Caudell; 457 (bl), Frans Lanting/Minden Pictures; 457 (br), Jean-Pierre Pieuchot/Getty Images/The Image Bank; 458, Robert Sullivan/AFP/Getty Images; 459, Mark J. Terrill/AP/Wide World Photos; 459, David Welling; 459, Robert Sullivan/AFP/Getty Images; 464, HRW; 465 (t), Sam Dudgeon/HRW; 465 (c), Sam Dudgeon/HRW; 465 (b), Sam Dudgeon/HRW; 466, Joost Van der Sanden, Canada Centre for Remote Sensing/Natural Resources Canada, RADARSAT image: (c)1999 Canadian Space Agency; 467 (t), Jonathan Blair/CORBIS; 467 (b), Adam Woolfitt/CORBIS.

UNIT SIX: 468-469, USGS/NASA; 469 (Ch. 19), Jeffrey L. Rotman/CORBIS; 469 (Ch. 20), David Hall; 469 (Ch. 21), eStock Photo/PictureQuest; 470, Jeffrey L. Rotman/CORBIS; 472 (br), Ocean Drilling Program - Texas A&M University; 472 (l), Ocean Drilling Program - Texas A&M University; 473 (ship), National Oceanic and Atmospheric Administration/Department of Commerce; 473 (b), Victoria Smith/HRW; 474 (tl), Alexis Rosenfeld/SPL/Photo Researchers, Inc.; 474 (inset), David Shale/naturepl.com; 478, B. Tanaka/Getty Images/Taxi; 479, Ocean Drilling Program - Texas A&M University; 480 (t), Digital image (c) 1996 CORBIS; Original image courtesy of NASA/CORBIS; 480 (b), Courtesy Stephen Morris, Experimental Nonlinear Physics Group, University of Toronto; 481 (nodules), Tom McHugh/Photo Researchers, Inc.; 481 (bkgd), Darryl Torckler/Getty Images/Stone; 481 (nodules), Tom McHugh/Photo Researchers, Inc.; 482 (l), Jim Zuckerman/CORBIS; 482 (r), Andrew Syred/SPL/Photo Researchers, Inc.; 483 (t), Ocean Drilling Program - Texas A&M University; 483 (c), B. Tanaka/Getty Images/Taxi; 489, Victoria Smith/HRW; 490, NGDC/NOAA; 491 (t), Scripps Institution of Oceanography; 491 (b), Scripps Institution of Oceanography; 491 (bkgd), Royalty-Free/Corbis; 492 (all), David Hall; 495 (tl), Sean Davey/AllSport/Getty Images; 497 (tl), Tom Stewart/CORBIS; 497 (tr), CORBIS; 499 (tc), Sam Dudgeon/HRW; 499 (br), Victoria Smith/HRW; 500 (tl), Lawson Wood/CORBIS; 501 (bl), David Hall; 502 (bl), IQ3d/Bruce Coleman, Inc.; 504 (tr), Craig Tuttle/CORBIS; 505 (bl), Steve Raymer/National Geographic Image Collection; 506 (tr), Bohemian Nomad Picturemakers/CORBIS; 507 (b), Chris Hellier/CORBIS; 508 (tl), Simon Fraser/SPL/Photo Researchers, Inc.; 509 (cl), David Hall; 509 (bl), Chris Hellier/CORBIS; 509 (tl), Sean Davey/AllSport/Getty Images; 514 (bl), Victoria Smith/HRW; 516, Courtesy of N.R. Nalli, NOAA/NESDIS, Washington, D.C.; 517 (bl), Karen J. Dodge/MWRA/RVA; 517 (tr), Digital Vision; 518 (all), eStock Photo/PictureQuest; 519 (bl), Jonathan Blair/CORBIS; 520 (br), Provided by the SeaWiFS Project, NASA/Goddard Space Flight Center, and ORBIMAGE; 522 (inset), Peter David/Nature Picture Library; 524 (inset), CORBIS; 525 (bl), Jack Fields/CORBIS; 527 (tr), Darrell Wong/Getty Images; 527 (br), Victoria Smith/HRW; 529 (tr), G. R. Roberts/Natural Sciences Image Library (NSIL) of New Zealand; 530 (tl), John Stanmeyer/VII/AP/Wide World Photos; 533 (tl), CORBIS; 533 (cl), CORBIS; 533 (br), NASA; 534 (tl), Photo Researchers, Inc.; 535 (tl), CORBIS; 535 (cl), Darrell Wong/Getty Images; 535 (bl), NASA; 540 (b), Victoria Smith/HRW; 543 (bl), SPL/Photo Researchers, Inc.; 543 (br), R. Toms/OSF/Animals Animals/Earth Scenes.

UNIT SEVEN: 544-545, (c) Warren Faidley/Weatherstock; 545 (Ch. 22), Kevin

Kelly/Getty Images/The Image Bank; 545 (Ch. 23), Charles Doswell III/Getty Images; 545 (Ch. 24), Taxi/Getty; 545 (Ch. 25), Theo Allofs/CORBIS; 546, Kevin Kelly/Getty Images/The Image Bank; 547 (b), Joseph Sohm; Visions of America/CORBIS; 549 (bl), Carol Hughes; Gallo Images/CORBIS; 549 (tr), Wolfgang Kaehler/CORBIS; 549 (tl), Roger Ressmeyer/CORBIS ; 549 (tr), Royalty Free/CORBIS; 550 (tr), Chris Noble/Stone/Getty Images; 550 (tl), Didrik Johnck/CORBIS; 551, Andy Christiansen/HRW; 553, NASA/CORBIS; 554, Martin Thomas/Reuters NewMediInc./CORBIS; 555 (radio), NRAO/AUI/NSF; 555 (x-ray), NASA/JIAS/SPL/Photo Researchers, Inc.; 555 (visible), NASA/JIAS/SPL/Photo Researchers, Inc.; 555 (ultraviolet), SOHO/ESA/NASA; 555 (infrared), NOAO/SPL/Photo Researchers, Inc.; 556 (br), NASA; 556 (bl), NASA; 557, Jeremy Woodhouse/Photodisc/Getty; 563, NASA/Photo Researchers, Inc.; 564, Charles Benes/Index Stock Imagery, Inc.; 565 (t), Wolfgang Kaehler/CORBIS; 565 (c), Jeremy Woodhouse/Photodisc/Getty; 565 (b), NASA/Photo Researchers, Inc.; 572, NASA; 573 (bl), ML Sinibaldi/CORBIS; 573 (br), Roger Ressmeyer/CORBIS; 574, Charles Doswell III/Getty Images; 576 (br), Robert Wright/Ecoscence/CORBIS; 578 (tr), (c)National Geographic Image Collection; 579 (bl), Barry Runk/Stan/Grant Heilman Photography, Inc.; 580 (tl), Graham Neden/Ecoscene/CORBIS; 582 (br), Michael S. Yamashita/CORBIS; 583 (t), Chris Anderson/Aurora; 585 (tl), Royalty Free/CORBIS; 585 (tr), Gary Braasch/CORBIS; 586 (tl), Charles O'Rear/CORBIS; 587 (bl), Darrell Gulin/CORBIS; 587 (bc), Gary W. Carter/CORBIS; 588 (br), Barry Runk/Stan/Grant Heilman Photography, Inc.; 589 (tl), AFP/CORBIS; 590 (tl), Jim Brandenburg/Minden Pictures; 591 (tl), Robert Wright/Ecoscence/CORBIS; 591 (cl), Michael S. Yamashita/CORBIS; 591 (bl), Gary W. Carter/CORBIS; 596 (br), Victoria Smith/HRW; 597 (tr), Victoria Smith/HRW; 599 (tr), National Center for Atmospheric Research/University Corporation for Atmospheric Research; 599 (bl), Marci Stenberg/Merced Sun-Star/AP/Wide World Photos; 600, Getty Images/Taxi; 601, NOAA; 602, Rick Doyle/CORBIS; 604, Benjamin Lowy/Corbis; 605, CORBIS; 608, A & J Verkaik/CORBIS; 610 (inset), Jim Reed/CORBIS; 610 (tl), Howard B. Bluestein/Photo Researchers, Inc.; 611, Jonathan Blair/CORBIS; 612 (t), Philippe Giraud/CORBIS SYGMA; 612 (b), Victoria Smith/HRW; 613, Gene Rhoden/Visuals Unlimited; 614, NASA; 615, Courtesy World Meteorological Organization; 617 (t), Courtesy, The Weather Channel; 618, Jim Reed/CORBIS; 620, H. David Seawell/CORBIS; 621 (1), Benjamin Lowy/CORBIS; 621 (2), A & J Verkaik/CORBIS; 621 (3), NASA; 621 (4), Jim Reed/CORBIS; 626, HRW; 628, Data courtesy NOAA; 629 (t), Courtesy Shirley Murillo; 629 (b), NOAA; 629 (bkgd), Don Farrall/Getty Images/PhotoDisc; 630 (all), Theo Allofs/CORBIS; 631 (bl), Wolfgang Kaehler/CORBIS; 631 (br), Andrea Pistolesi/Getty Images; 635 (br), Indranil Mukherjee/AFP/Getty Images/Getty Images; 635 (bl), Reuters NewMedia, Inc./CORBIS; 639 (tl), (c) Bryan and Cherry Alexander Photography; 642 (tl), Bojan Brecelj/CORBIS; 644 (tr), Yann Arthus-Bertrand/CORBIS; 646 (tl), Wendy Stone/CORBIS; 646 (bl), Wendy Stone/CORBIS; 647 (tl), Wolfgang Kaehler/CORBIS; 647 (cl), (c) Bryan and Cherry Alexander Photography; 652 (b), Victoria Smith/HRW; 655 (c), Dee Breger/Lamont-Doherty Earth Observatory.

UNIT EIGHT: 656-657, Roger Ressmeyer/CORBIS; 657 (Ch. 28), Denis Scott/CORBIS; 657 (Ch. 26), NASA; 657 (Ch. 26), Denis Scott/CORBIS; 657 (Ch. 29), SOHO/ESA/NASA; 657 (Ch. 30), Stocktrek/CORBIS; 658 (all), NASA; 659 (bl), NASA; 660 (tr), NASA; 660 (tl), Jean-Charles Cuillandre/Canada-france-Hawaii Telescope/SPL/Photo Researchers, Inc.; 661 (br), Alfred Pasieka/SPL/Photo Researchers, Inc.; 662 (br), Michael Freeman/CORBIS; 664 (br), Bohemia Nomad Picturemakers/CORBIS; 664 (tl), Glen Alison/PhotoDisc Green/gettyimages; 665 (tl), NASA; 665 (br), NASA; 666 (tl), NASA; 667 (bl), Inga Spence/Visuals Unlimited; 667 (br), Inga Spence/Visuals Unlimited; 670 (tl), Gianni Dagli Orti/CORBIS; 673 (br), Andy Christiansen/HRW; 675 (bl), Inga Spence/Visuals Unlimited; 675 (tl), Jean-Charles Cuillandre/Canada-france-Hawaii Telescope/SPL/Photo Researchers, Inc.; 680 (br), Peter van Steen/HRW; 682 (t), C. Mayhew & R. Simmon/NASA/GSFO; 683 (tr), Lockheed Martin Corporation; 683 (bl), Image courtesy of NASA Landsat Project Science Office and USGS EROS Data Center; 684 (all), Denis Scott/CORBIS; 685 (bl), Royal Observatory, Edinburgh/SPL/Photo Researchers, Inc.; 690 (tl), Lindsay Hebberd/CORBIS; 692 (br), Victoria Smith/HRW; 694 (tl), David Nunuk/SPL/Photo Researchers, Inc.; 695 (b), JPL/NASA; 696 (br), NASA/SPL/Photo Researchers, Inc.; 697 (tl), NASA/SPL/Photo Researchers, Inc.; 698 (br), ESA/PLI/CORBIS; 699 (tl), World Perspectives/Getty Images; 700 (tl), AFP/CORBIS; 700 (tr), Gordon Garradd/Photo Researchers, Inc.; 701 (bl), Kevin Kelley/Getty Images; 702 (br), NASA; 703 (tr), CORBIS; 703 (br), NASA; 704 (br), NASA/JPL/Space Science Institute; 705 (tl), NASA; 706 (br), JPL/NASA; 707 (tl), NASA/JPL-Caltech; 708 (tl), Lynette R. Cook; 709 (t), Royal Observatory, Edinburgh/SPL/Photo Researchers, Inc.; 709 (cl), David Nunuk/SPL/Photo Researchers, Inc.; 709 (bl), NASA/SPL/Photo Researchers, Inc.; 709 (b), CORBIS; 714 (b), Victoria Smith/HRW; 715 (tr), Victoria Smith/HRW; 716 (tc), Photo Researchers, Inc.; 717 (tr), George Sakkestad; 717 (br), Roger Ressmeyer/CORBIS; 717 (bkgd), NASA; 719, NASA; 720, Galileo Project/JPL/NASA; 721 (t), NASA; 721 (b), NASA; 724 (t), USGS; 724 (b), USGS; 726, NASA; 727 (bl), George Post/SPL/Photo Researchers; 727 (br), CNES/GAMMA; 728 (b), Victoria Smith/HRW; 728 (t), Royalty Free/CORBIS; 729 (b), Dr. Fred Espenak/Photo Researchers, Inc.; 730, John Bova/Photo Researchers, Inc.; 730, John Bova/Photo Researchers, Inc.; 730, John Bova/Photo Researchers, Inc.; 730, John Bova/Photo Researchers, Inc.; 730, John Bova/Photo Researchers, Inc.; 730, John Bova/Photo Researchers, Inc.; 730, John Bova/Photo Researchers, Inc.; 730, NASA; 731, John Sanford/Photo Researchers, Inc.; 733, NASA/JPL/University of Arizona; 734 (t), Galileo Project/JPL/NASA; 734 (b), Galileo Project/JPL/NASA; 735 (t), Galileo Project/JPL/NASA; 735 (c), NASA; 735 (b), NASA; 736 (t), NASA; 736 (b), NASA/JPL/Space Science Institute; 737 (t), NASA; 737 (b), NASA, ESA, H. Weaver (JHU/APL), A. Stern (SWRI), and the HST Pluto Companion Search Team; 738, NASA; 739, NASA; 740, Getty Images/PhotoDisc; 741, Wally Pacholka/AstroPics.com; 743, Juan Carlos Casado; 744 (t), E.R. Degginger/Bruce Coleman, Inc.; 744 (c), Breck P. Kent/Animals Animals/Earth Scenes; 744 (b), Ken Nichols/Institute of Meteorites; 745 (1), NASA; 745 (2), CNES/GAMMA; 745 (3), Galileo Project/JPL/NASA; 745 (4), Wally Pacholka/AstroPics.com; 752, NASA; 753 (b), NASA; 753 (t), NASA; 754, SOHO/ESA/NASA; 757 (t), Kamioka Observatory, ICRR (Institute for Cosmic Ray Research), The University of Tokyo; 757 (b), Andy Christiansen/HRW; 759, NOAO/AURA/NSF; 760, Fred Espenek; 761 (bl), The Institute for Solar Physics, The Royal Swedish Academy of Sciences; 761(inset), Getty Images/Photodisc; 762, Laboratory for Atmospheric and Space Physics, University of Colorado-Boulder; 763, M. Aschwanden et al. (LMSAL), TRACE, NASA ; 764, Michael DeYoung/AlaskaStock Images; 765 (t), Fred Espenek; 765 (b), Michael DeYoung/AlaskaStock Images; 767, NASA; 770, HRW; 771, HRW; 772, Yohkoh Solar Observatory; 773 (tr), JPL/NASA; 774 (all), Stocktrek/CORBIS; 775 (b), N.A.Sharp, NOAO/NSO/Kitt Peak FTS/AURA/NSF; 776 (t), NASA/Hubble Heritage Team; 777 (b), Grant Faint/Getty Images; 779 (inset), Victoria Smith/HRW; 780 (bkgd), Roger Ressmeyer/CORBIS; 782 (tr), NASA; 783 (br), Tony Craddock/SPL/Photo Researchers, Inc.; 784 (br), Matthew Spinelli; 785 (tr), NASA The Heritage Hubble Team; 785 (inset), Photo courtesy of the NAIC/Aricebo Observatory/NSF; 788 (tl), NASA; 790 (bl), Celestial Image Co./SPL/Photo Researchers, Inc.; 791 (tl), NASA/Gemini Observatory, GMOS Team; 791 (tr), Royal Observatory, Edinburgh/AATB/SPL/Photo Researchers, Inc.; 791 (tc), NOAO/Photo Researchers, Inc.; 792 (tr), NASA/CXC/A.Siemiginowska et al.; Illustration: CXC/M.Weiss ; 793 (br), NASA; 795 (tr), Victoria Smith/HRW; 795 (br), NASA/WMAP Science Team; 796 (tl), NASA and Hubble Heritage Team; 797 (tl), Grant Faint/Getty Images; 797 (cl), NASA The Heritage Hubble Team; 797 (cl), Celestial Image Co./SPL/Photo Researchers, Inc.; 797 (bl), NASA; 803 (tr), Victoria Smith/HRW; 805 (bl), NASA/JPL-Caltech/S. Willner (Harvard-Smithsonian Center for Astrophysics); 805 (tr), Roger Ressmeyer/CORBIS.

APPENDIX: 806 (bl), R. L. Christiansen/CORBIS; 806 (bc), Stuart Westmorland/CORBIS /CORBIS; 806 (br), Jean Miele/CORBIS ; 832 (bkgd), David Muench/CORBIS; 832 (b), Victoria Smith/HRW; 833, USGS; 835, Karl Mueller/University of Colorado/AP/Wide World Photos; 836, David McNew/Getty Images/NewsCom; 838-839, Victoria Smith/HRW; 840, Chris Johns/Getty Images/National Geographic; 841, Greg Dimijian/Photo Researchers, Inc.; 842, A.T. Willett/Alamy Photos; 843, A.T. Willett/Alamy Photos; 844, Courtesy Earth and Sky; 846, HRW; 847 (tr), David Young-Wolff/PhotoEdit; 847 (c), Photodisc Green/Getty Images; 848, Digital Image copyright (c) 2006 Larry Brownstein/PhotoDisc; 850, David Gallant/CORBIS; 852-853, Robert Landau/CORBIS ; 853, Victoria Smith/HRW; 854 (r), Daniel Schaefer/HRW; 854 (l), Michael Abbey/Visuals Unlimited; 856, A & J Verkaik/CORBIS; 859, Hillery Smith/AP/Wide World Photos; 861, Michael Wirtz/Philadelphia Inquirer/NewsCom; 862, Tony Freeman/PhotoEdit; 865 (l), Galen Rowell/CORBIS; 865 (r), Paul Wakefield/Getty Images/Stone; 866 (all), HRW; 867 (r), WallyPacholka/AstroPics.com; 867 (l), Stocktrek/CORBIS.

TEACHER EDITION PHOTO CREDITS: 3C, Danny Lehman/CORBIS; 3D,©Michael Sewell/Peter Arnold, Inc.; 25D, Martin Harvey/Gallo Images/CORBIS; 51C (t), Christopher Cormack/CORBIS 51C (b), Layne Kennedy/CORBIS; 51D, Nathan Wier/Getty Images; 79D, Steve Chen/CORBIS; 101C, Jose Manuel Sanchez Calvate/CORBIS; 101D, Dr. E. R. Degginger/Color-Pic, Inc.; 123C, Brand X Photos/Alamy Photos; 153C (t), Roger Ressmeyer/CORBIS; 153C (b), H. David Seawell/CORBIS; 153D, Jose Fuste Raga/CORBIS; 183C (l), Tom Bean/CORBIS; 183C (r), Pat O'Hara/CORBIS; 183D, James L. Amos/Photo Researchers, Inc.; 209C, Jonathan Blair/CORBIS; 209D, Sue Ogrocki/Reuters/NewsCom; 237C (b), Y. Arthus-B./Peter Arnold, Inc.; 269C, James P. Blair/National Geographic Image Collection; 293C (t), Reuters/CORBIS; 293C (b), Yann Arthus Bertrand/CORBIS; 293D, Michael S. Yamashita/CORBIS; 317C, Barry Tessman/National Geographic Image Collection; 317D, Stuart Westmoreland/Getty Images; 373C, Brad Wrobleski/Masterfile; 373D, Kevin R. Morris/CORBIS; 341C (l), Kevin Fleming/CORBIS; 341C (r), Tom Till/Getty Images; 341D, Galen Rowell/CORBIS; 395C, Pagasus/Visuals Unlimited; 395D, Adam Woolfitt/CORBIS; 417C, Galen Rowell/CORBIS; 417D, Astrid & Hanns-Frieder Michler/SPL/Photo Researchers, Inc.; 443C (t), David Welling; 443C (b), Mark J. Terrill/AP/Wide World Photos; 443D, Tami Chappell/Rueters/NewsCom 469C (tr), David Shale/naturepl.com; 469C (bl), Ocean Drilling Program - Texas A&M University; 469D, Jim Zuckerman/CORBIS; 491C, Tom Stewart/CORBIS; 491D, Steve Raymer/National Geographic Image Collection; 517C, Peter David/Nature Picture Library; 517D, Paul A. Souders/CORBIS; 545C, Chris Noble/Stone/Getty Images; 573C (l), Robert Wright/Ecoscence/CORBIS; 573C (r), Chris Anderson/Aurora; 573D, Barry Runk/Stan/Grant Heilman Photography, Inc.; 599C (l), NASA/CORBIS; 599C (r), A & J Verkaik/CORBIS; 599D, Jim Reed/CORBIS; 629C, Reuters NewMedia, Inc./CORBIS; 629D, Yann Arthus-Bertrand/CORBIS; 657C , NASA; 657D, Gianni Dagli Orti/CORBIS; 683C (l), Royal Observatory, Edinburgh/SPL/Photo Researchers, Inc.; 683C (r), NASA/SPL/Photo Researchers, Inc.; 683D, Lynette R. Cook 717C, NASA; 717D, Getty Images/PhotoDisc; 753C, Fred Espenek; 753D, The Institute for Solar Physics, The Royal Swedish Academy of Sciences; 773C (r), NASA; 773D, NASA/WMAP Science Team.

Answers to Concept Mapping Questions

The following pages contain sample answers to all of the concept mapping questions that appear in the Chapter Reviews. Because there is more than one way to do a concept map, students' answers may vary.

Chapter 1 Introduction to Earth Science

30.

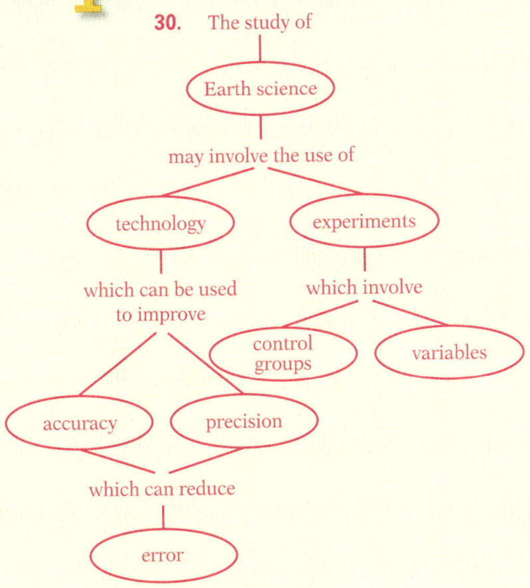

Chapter 2 Earth as a System

30.

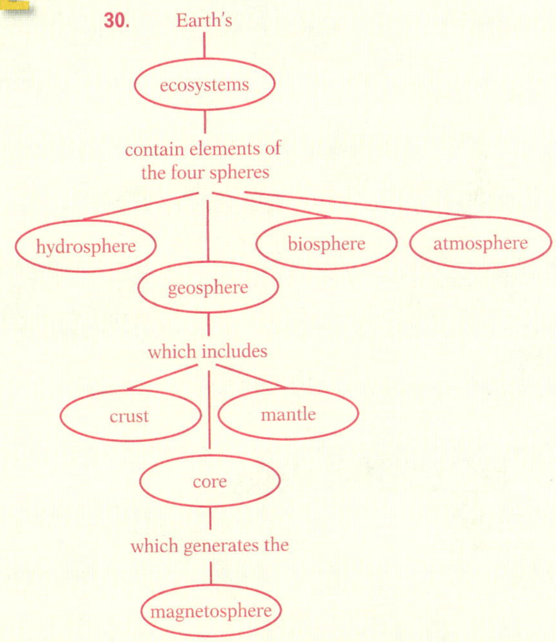

Chapter 3 Models of the Earth

28.

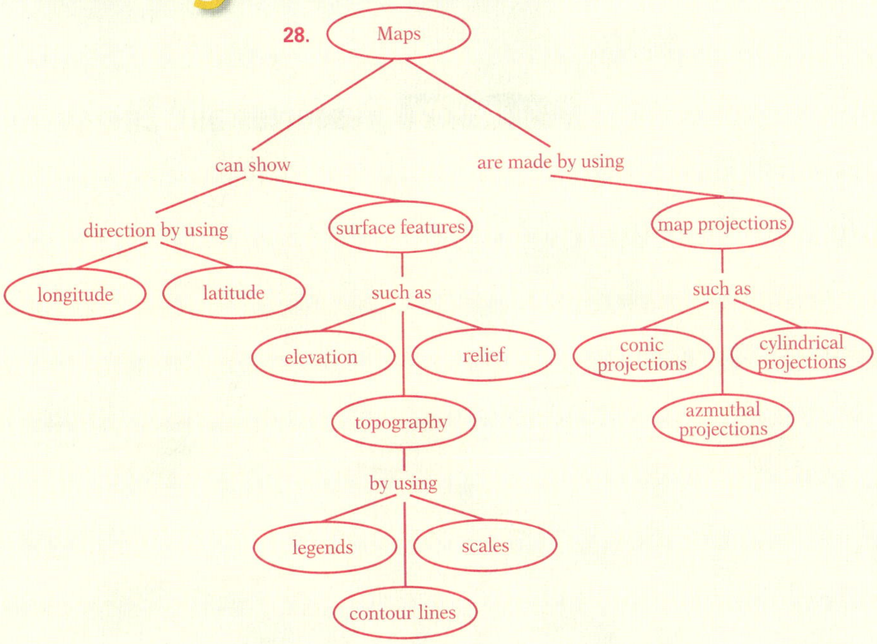

Chapter 4 Earth Chemistry

28.

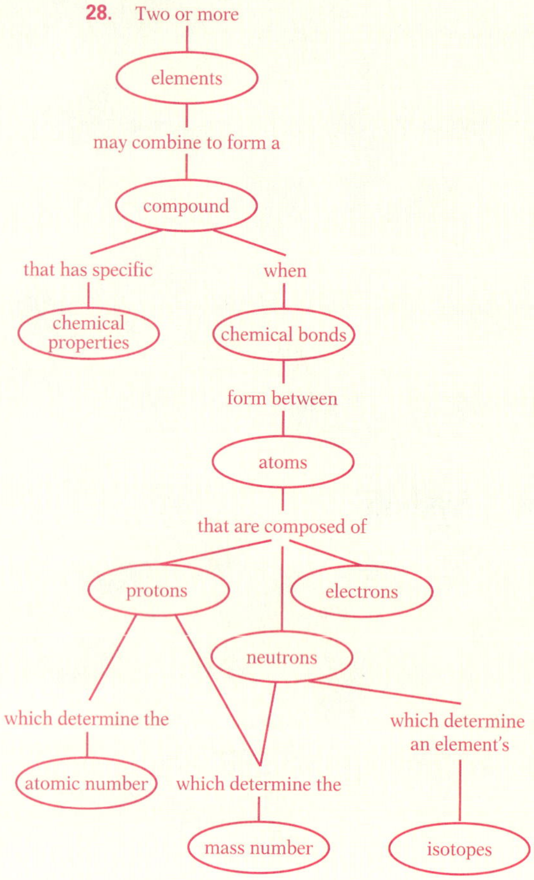

Chapter 5 Minerals of Earth's Crust

29.

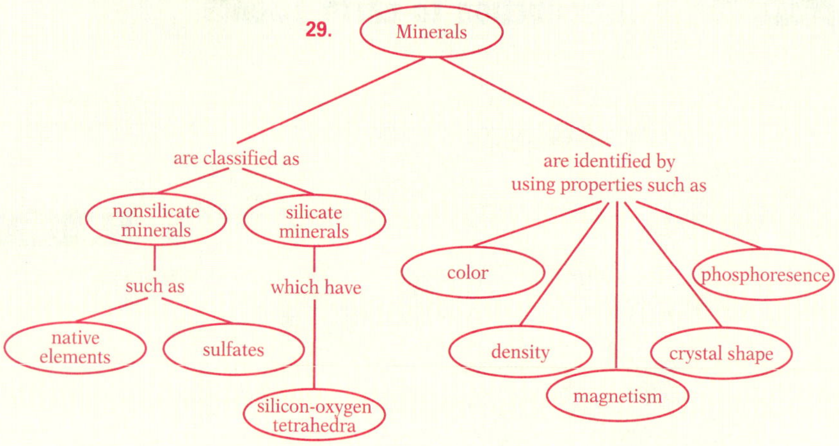

Chapter 6 Rocks

28.

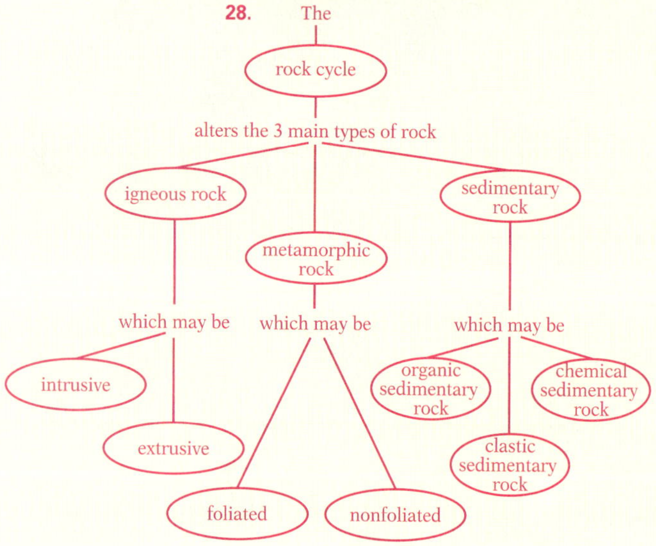

Chapter 7 Resources and Energy

28.

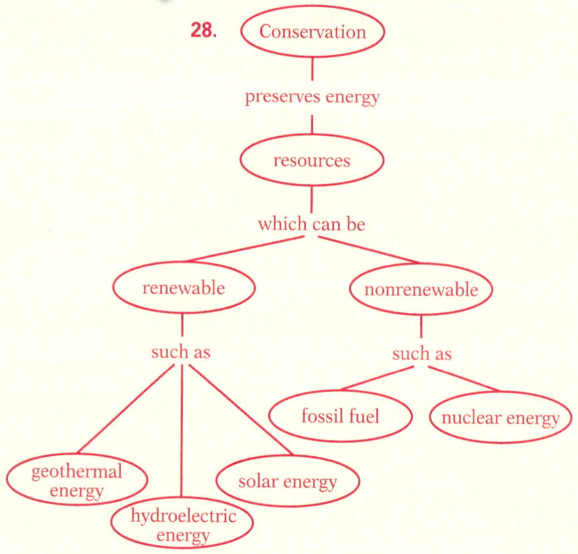

Chapter 8 The Rock Record

27.

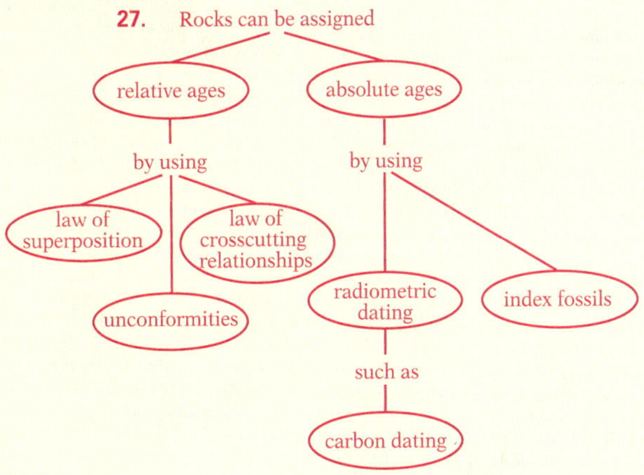

Chapter 9 A View of Earth's Past

26.

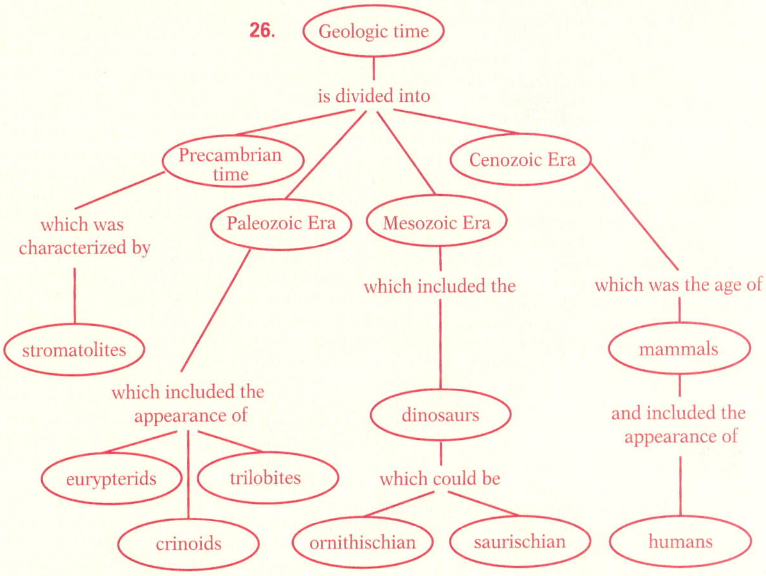

Chapter 10 Plate Tectonics

27.

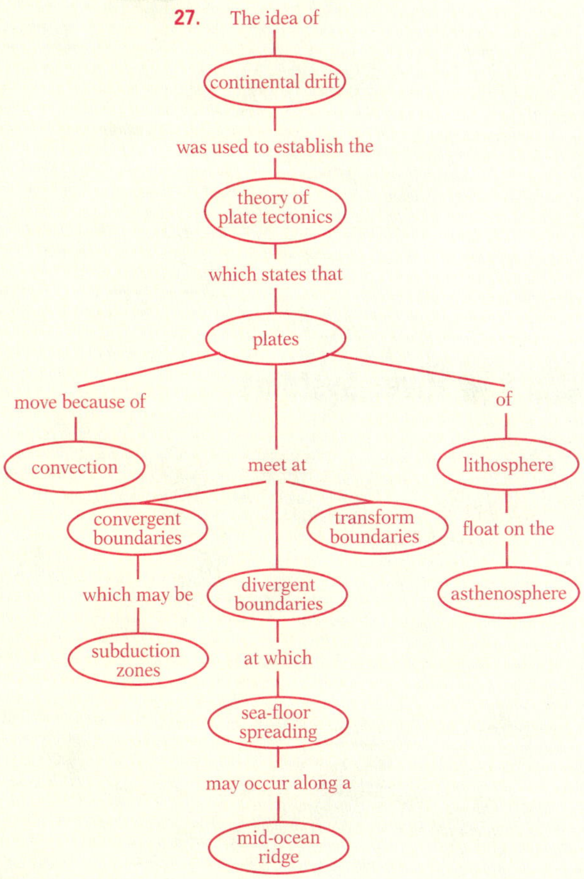

Chapter 11 Deformation of the Crust

28.

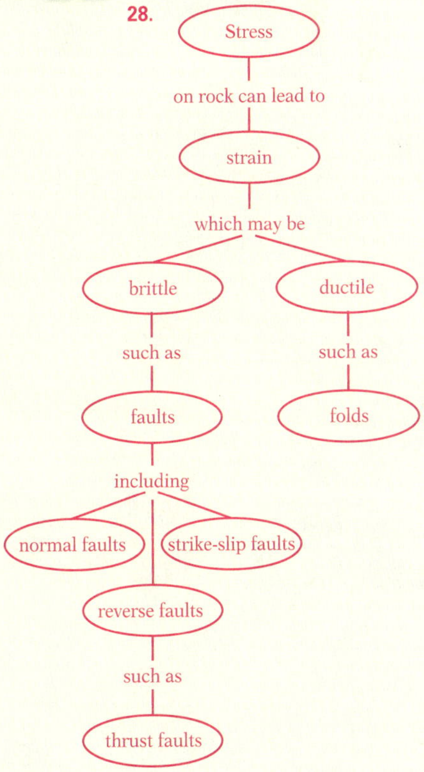

Chapter 12 Earthquakes

27.

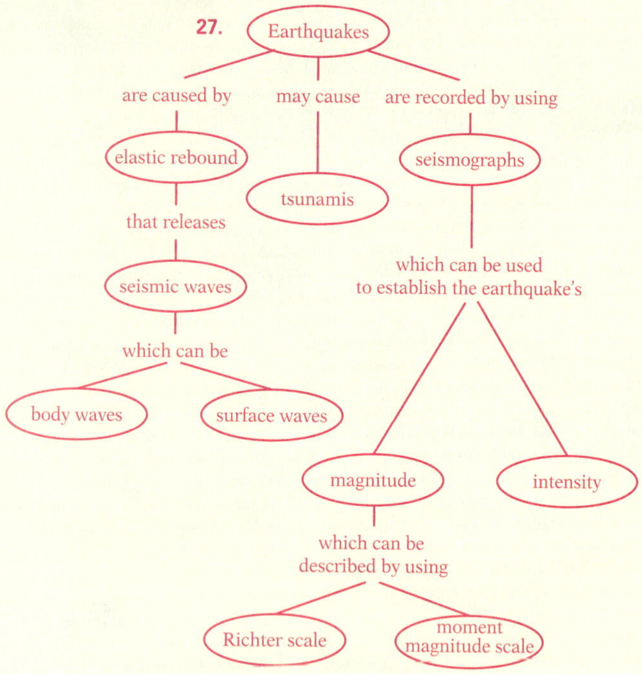

Chapter 13 Volcanoes

29.

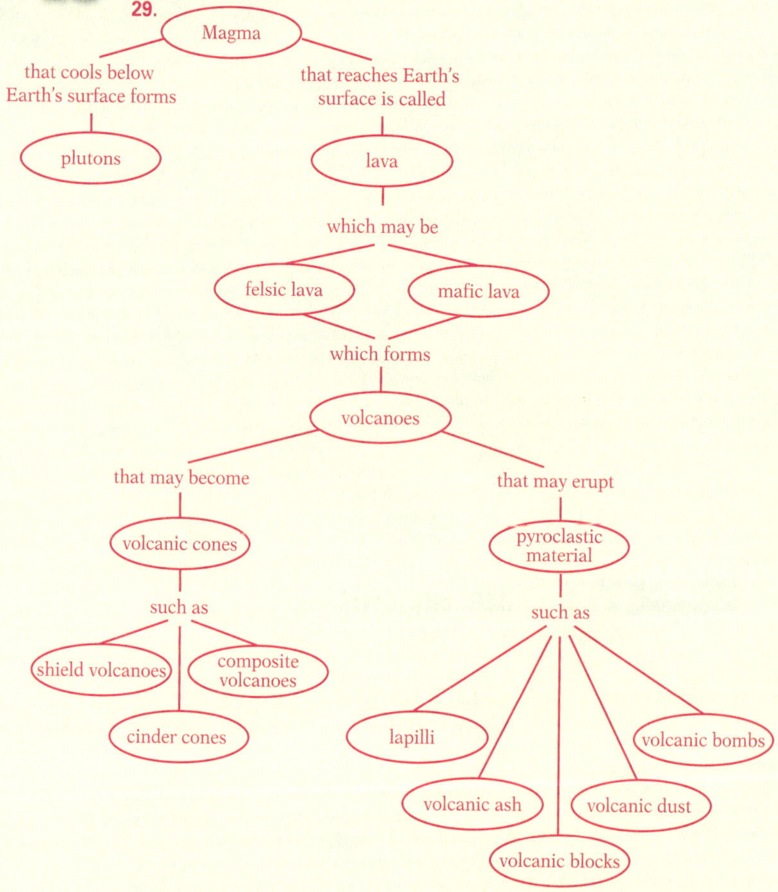

Chapter 14 Weathering and Erosion

31.

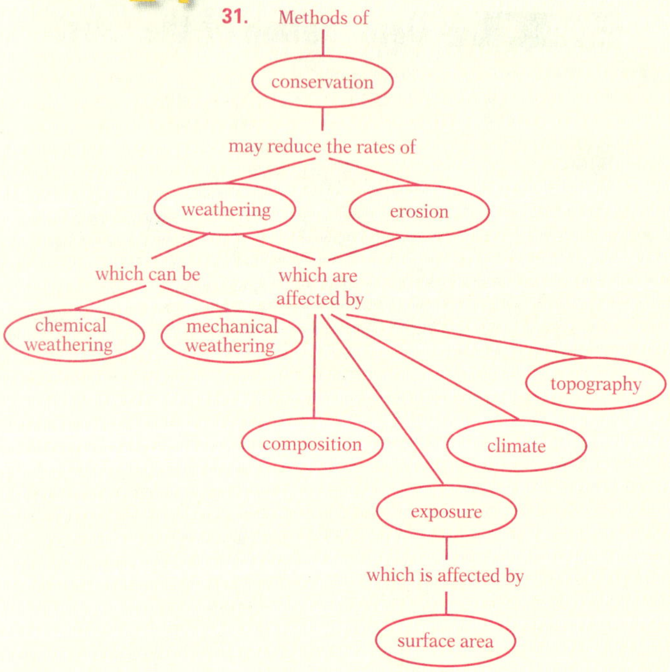

Chapter 15 River Systems

27.

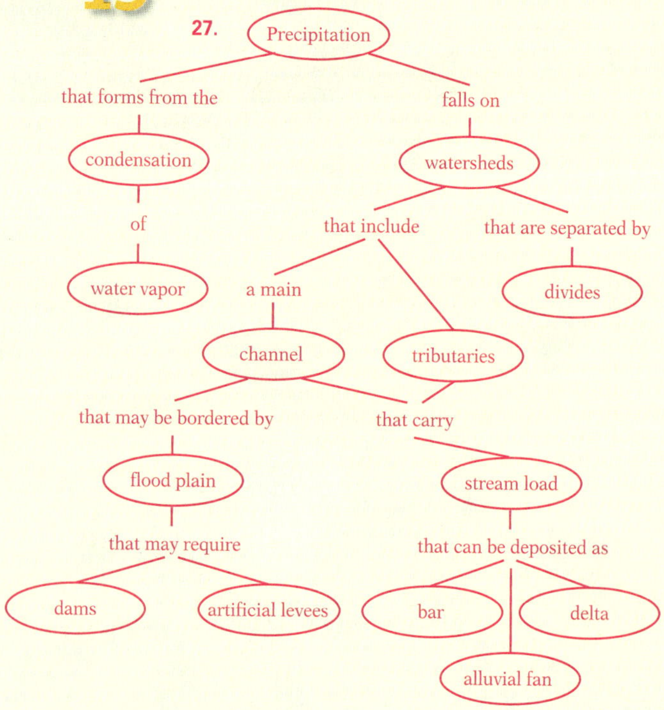

Chapter 16 Groundwater

31.

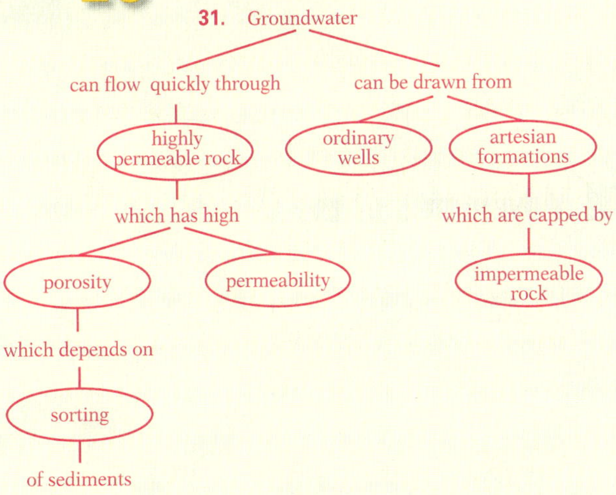

Chapter 17 Glaciers

25.

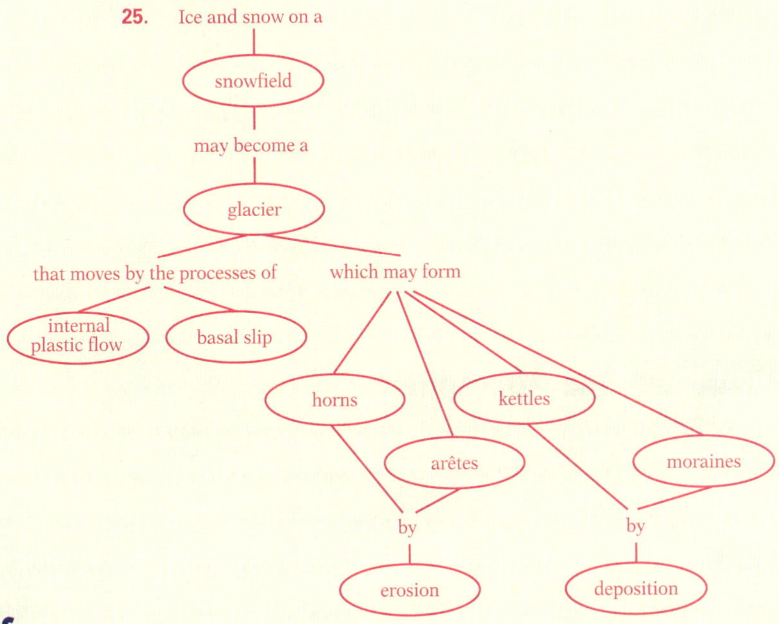

Chapter 18 Erosion by Wind and Waves

24.

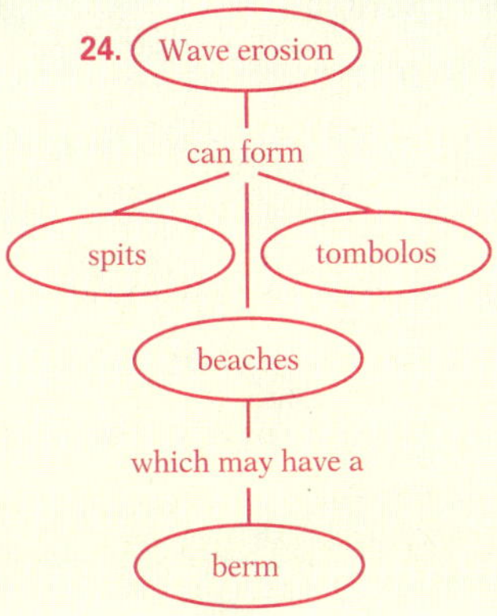

Chapter 19 The Ocean Basins

27.

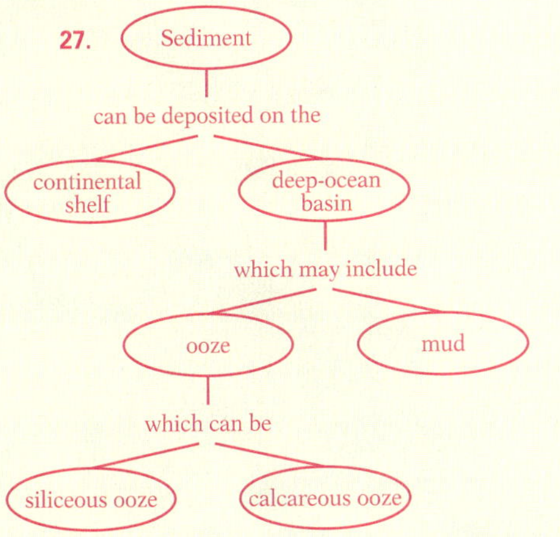

Chapter 20 Ocean Water

27.

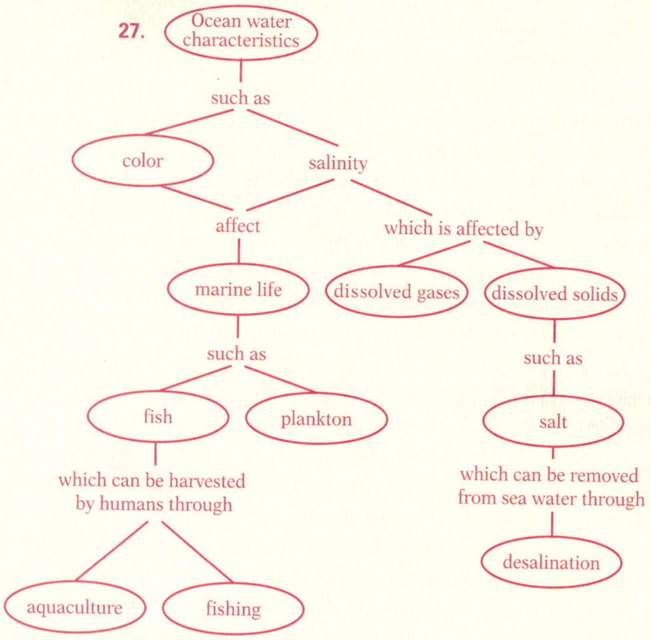

Chapter 21 Movements of the Ocean

28.

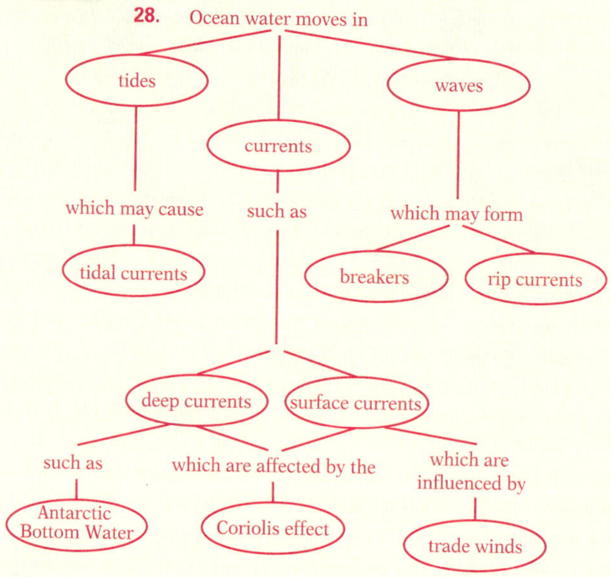

Chapter 22 The Atmosphere

33.

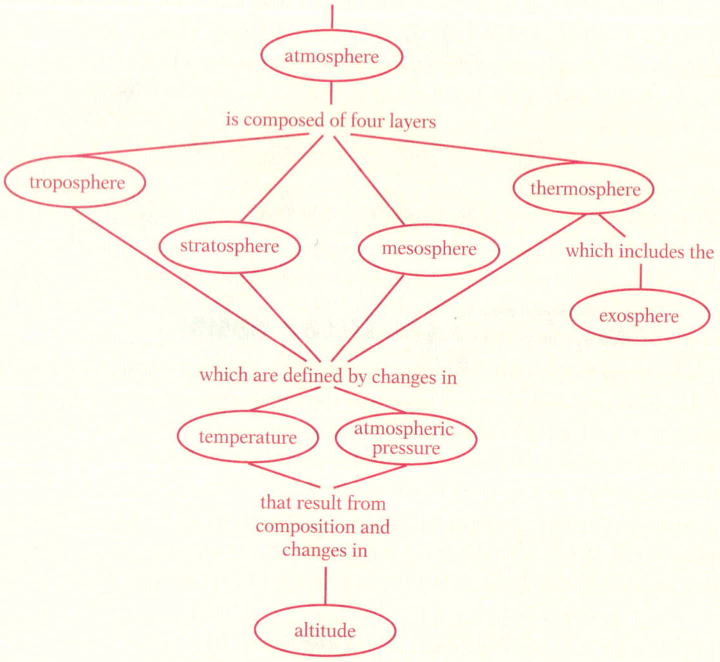

Chapter 23 Water in the Atmosphere

23.

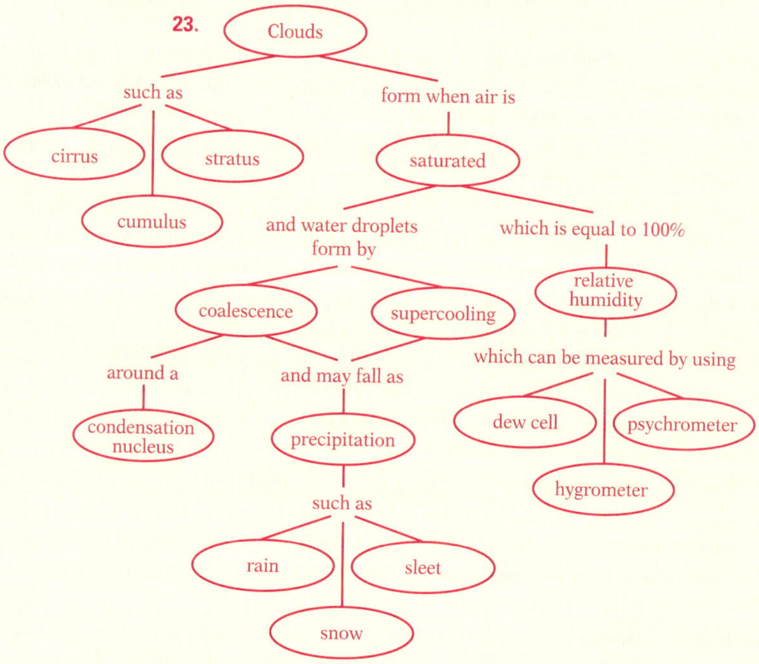

Chapter 24 Weather

28.

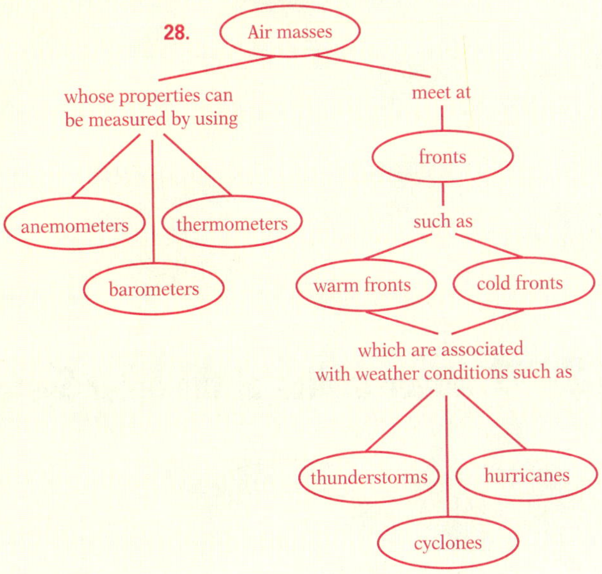

Chapter 25 Climate

23.

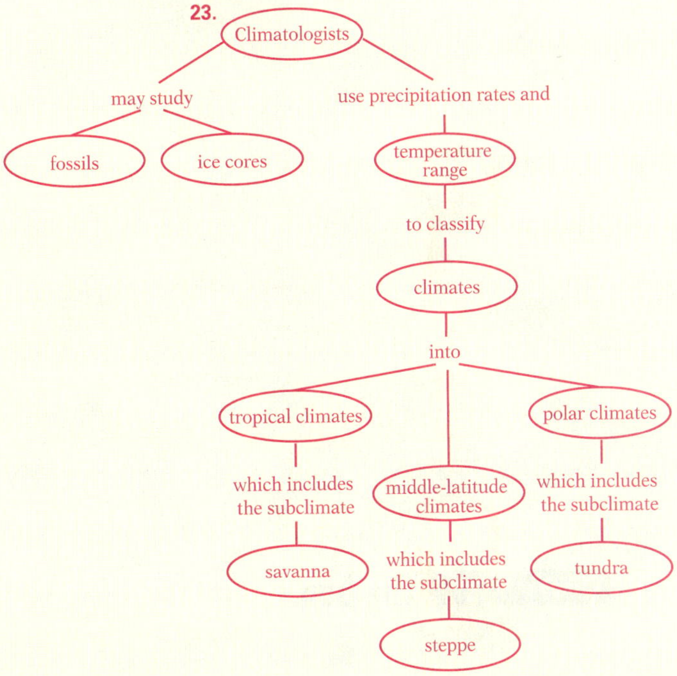

Chapter 26 Studying Space

26.

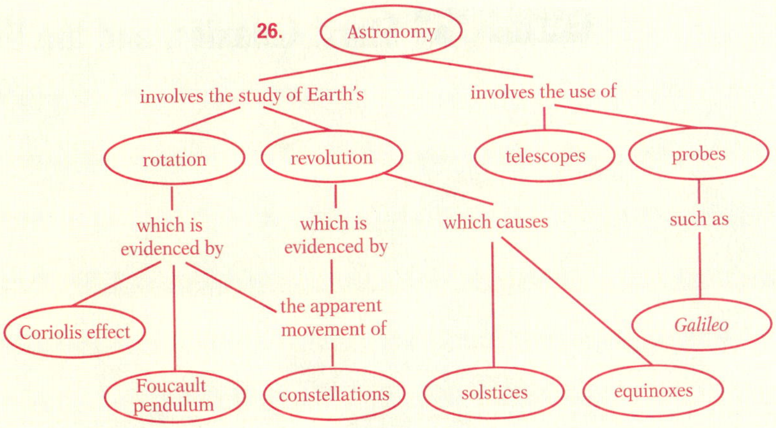

Chapter 27 Planets of the Solar System

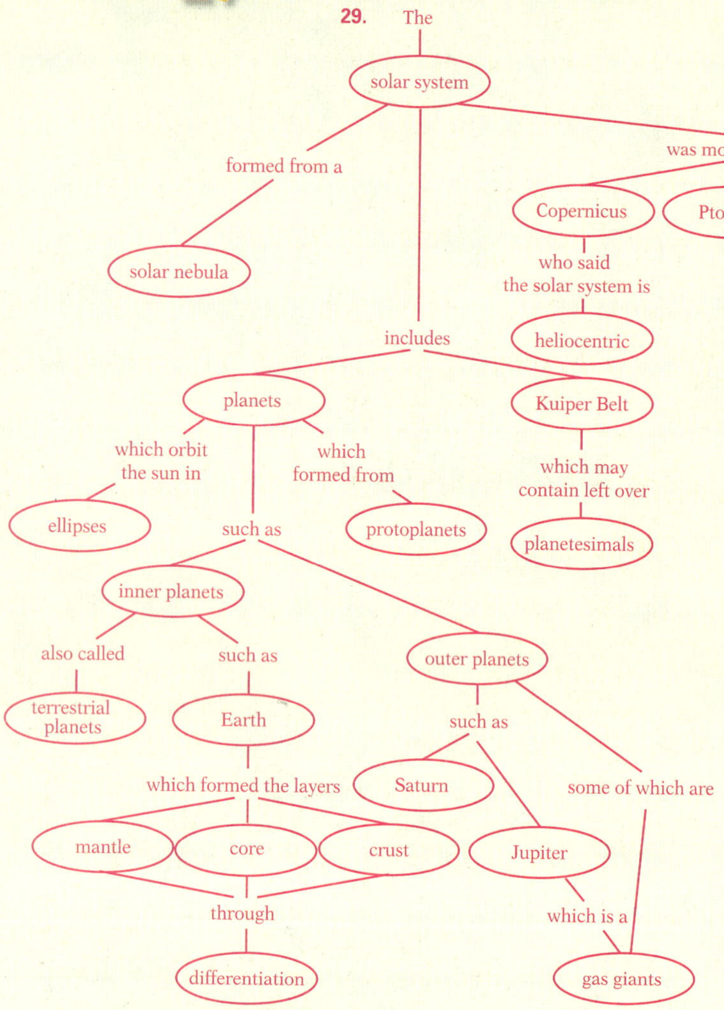

Chapter 28 Minor Bodies of the Solar System

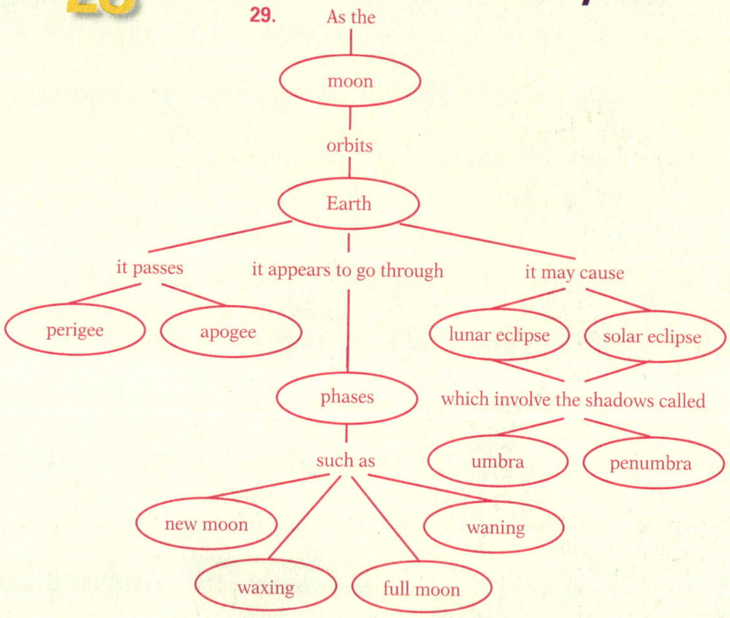

Chapter 29 The Sun

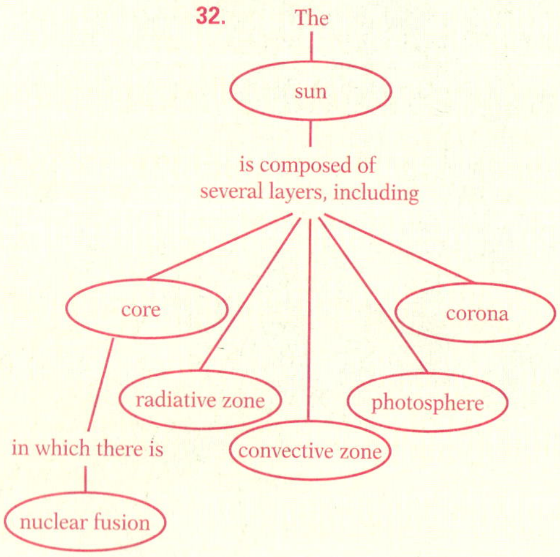

Chapter 30 Stars, Galaxies, and the Universe

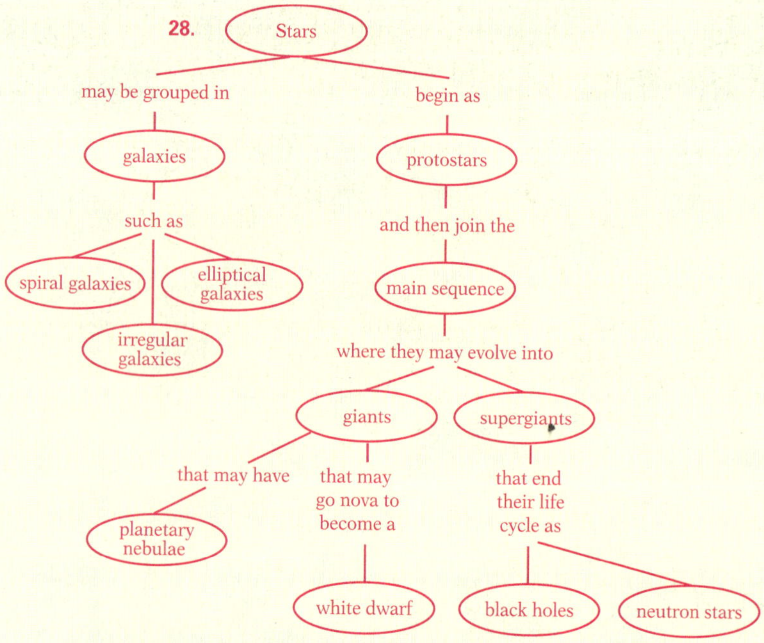

The Periodic Table of Elements

Period	Group 1	Group 2	Group 3	Group 4	Group 5	Group 6	Group 7	Group 8	Group 9
1	1 **H** Hydrogen 1.007 94								
2	3 **Li** Lithium 6.941	4 **Be** Beryllium 9.012 182							
3	11 **Na** Sodium 22.989 769 28	12 **Mg** Magnesium 24.3050							
4	19 **K** Potassium 39.0983	20 **Ca** Calcium 40.078	21 **Sc** Scandium 44.955 912	22 **Ti** Titanium 47.867	23 **V** Vanadium 50.9415	24 **Cr** Chromium 51.9961	25 **Mn** Manganese 54.938 045	26 **Fe** Iron 55.845	27 **Co** Cobalt 58.933 195
5	37 **Rb** Rubidium 85.4678	38 **Sr** Strontium 87.62	39 **Y** Yttrium 88.905 85	40 **Zr** Zirconium 91.224	41 **Nb** Niobium 92.906 38	42 **Mo** Molybdenum 95.94	43 **Tc** Technetium (98)	44 **Ru** Ruthenium 101.07	45 **Rh** Rhodium 102.905 50
6	55 **Cs** Cesium 132.905 4519	56 **Ba** Barium 137.327	57 **La** Lanthanum 138.905 47	72 **Hf** Hafnium 178.49	73 **Ta** Tantalum 180.947 88	74 **W** Tungsten 183.84	75 **Re** Rhenium 186.207	76 **Os** Osmium 190.23	77 **Ir** Iridium 192.217
7	87 **Fr** Francium (223)	88 **Ra** Radium (226)	89 **Ac** Actinium (227)	104 **Rf** Rutherfordium (261)	105 **Db** Dubnium (262)	106 **Sg** Seaborgium (266)	107 **Bh** Bohrium (264)	108 **Hs** Hassium (277)	109 **Mt** Meitnerium (268)

Key:

Atomic number	6
Symbol	**C**
Name	Carbon
Average atomic mass	12.0107

58 **Ce** Cerium 140.116	59 **Pr** Praseodymium 140.907 65	60 **Nd** Neodymium 144.242	61 **Pm** Promethium (145)	62 **Sm** Samarium 150.36
90 **Th** Thorium 232.038 06	91 **Pa** Protactinium 231.035 88	92 **U** Uranium 238.028 91	93 **Np** Neptunium (237)	94 **Pu** Plutonium (244)

* The systematic names and symbols for elements greater than 111 will be used until the approval of trivial names by IUPAC.

internet connect

go.hrw.com

Topic: Periodic Table
Go To: go.hrw.com
Keyword: Holt Periodic

Visit the HRW Web site for updates on the periodic table.